Magnetism and Magnetic Materials—1974

(20th Annual Conference—San Francisco)

AIP Conference Proceedings
Series Editor: Hugh C. Wolfe
Number 24

Magnetism and Magnetic Materials—1974

(20th Annual Conference—San Francisco)

Editors

C.D. Graham, Jr.
University of Pennsylvania

G.H. Lander
Argonne National Laboratory

and

J.J. Rhyne
National Bureau of Standards

American Institute of Physics
New York 1975

CONFERENCE ORGANIZATION

General Chairman

J. M. Lommel

Steering Committee

J. M. Lommel, Chairman
M. K. Wilkinson, Secretary
D. I. Gordon, Treasurer
R. L. White, Chairman-Elect
E. W. Pugh
I. S. Jacobs
R. C. Byloff (IEEE)
H. C. Wolfe (AIP)

Program Committee

E. M. Gyorgy) Co-Chairmen
R. E. Watson)
A. Arrott
R. B. Clover
R. E. Dietz
M. A. Gilleo
C. D. Graham, Jr.
F. B. Humphrey
T. A. Kitchens
G. H. Lander
F. E. Luborsky
A. H. Luther
J. E. Mee
R. Orbach
C. A. Patton
J. J. Rhyne
J. C. Slonczewski
J. B. Torrance, Jr.
E. P. Valstyn
J. E. Wernick
R. M. White
D. K. Wohlleben

Local Committee

K. Lee, Chairman
E. Catalano
R. B. Clover
M. Comstock
R. L. Comstock
D. E. Kaplan
A. J. Kurtzig
G. B. Street
J. C. Suits
R. L. White
R. M. White

Publications Committee

C. D. Graham, Jr. G. H. Lander J. J. Rhyne

Advisory Committee

Chairman

F. B. Hagedorn

Term Expires 1974

G. Bate
R. F. Elfant
G. Fedde
A. Goldman
B. Hershenov
G. H. Lander
K. Lee
N. Menyuk
G. T. Rado
M. K. Wilkinson

Term Expires 1975

W. A. Baker
W. D. Doyle
D. I. Gordon
E. M. Gyorgy
V. Jaccarino
S. Kern
C. J. Kriessman
J. M. Lommel
R. L. White
W. P. Wolf

Term Expires 1976

S. H. Charap
P. L. Donoho
F. B. Hagedorn
I. S. Jacobs
A. V. Pohm
E. W. Pugh
M. P. Sarachik
B. F. Stein
E. J. Torok
F. E. Werner

Sponsoring Society Representatives

R. C. Byloff (IEEE) H. C. Wolfe (AIP)

Co-operating Society Representatives

J. W. Shilling (Met.Soc.AIME)
J. O. Dimmock (ONR)
D. H. Jones (ASTM)

L.C. Catalog Card No. 75–2647
ISBN 0–88318–123–1
AEC CONF–741202

American Institute of Physics
335 East 45th Street
New York, N.Y. 10017

Printed in the United States of America

TWENTIETH ANNUAL CONFERENCE

on

MAGNETISM AND MAGNETIC MATERIALS

December 3-6, 1974
San Francisco, California

Sponsored by

The American Institute of Physics
The Magnetics Society
of the
Institute of Electrical and Electronics Engineers

In Co-operation with

The Office of Naval Research
The American Physical Society
The Metallurgical Society
of the
American Institute of Mining, Metallurgical, and Petroleum Engineers
The American Society for Testing and Materials

The Conference is especially grateful to

THE OFFICE OF NAVAL RESEARCH

for its support of the expenses of foreign and interdisciplinary speakers
under Contract NONR (G)-00004-74.

Contributions to the Conference from the following firms
are gratefully acknowledged:

Ampex Corporation
Applied Magnetics Corporation
The Arnold Engineering Company
Audio Magnetics Corporation
Automatic Electric, Inc.
Bell Telephone Laboratories
Ceramic Magnetics, Inc.
Colt Industries
Digital Equipment Corporation
E.I. du Pont de Nemours and Company
Eastman Kodak Company
Electronic Memories and Magnetics Corporation
Eriez Magnetics
Ford Motor Company
General Electric Company
General Magnetic Company
Hewlett Packard
International Business Machines Corporation
International Telephone and Telegraph Corporation
Magnetic Metals Company
National Micronetics, Inc.
The Permanent Magnet Company, Inc.
Pfizer, Inc., Minerals, Pigments, and Metals Div.
Raytheon Company
RCA Laboratories
Rockwell International, Electronics Research Div.
Siemens Corporation, Electronics Systems Div.
Texas Instruments, Inc.
United Aircraft Corporation
Walker Scientific, Inc.
Westinghouse Electric Corp., Research & Development Ctr.
Xerox Corporation

All papers in this volume, and in previous Proceedings of the Conference on Magnetism and Magnetic Materials published in this series, have been reviewed for technical content. The selection of referees, review guidelines, and all other editorial procedures are in accordance with standards prescribed by the American Institute of Physics.

PREFACE

The success of a technical conference cannot be measured fully when it is concluded. Only references and citations to work presented at the meeting will demonstrate whether the technical information exchanged has had lasting importance. A short-term assessment of the 20th Annual Conference on Magnetism and Magnetic Materials can be made by looking at its registration lists and by comments from participants. By and large, participants felt the meeting was a good one, in the well established tradition of previous Magnetism and Magnetic Materials Conferences.

After some initial concern as to whether a West Coast site would bring together the number of participants needed to make a conference worthwhile, the attendance by 650 scientists and engineers clearly demonstrated the drawing power of San Francisco and the strong local interest in magnetism. Early indications of interest were received when over 500 abstracts were submitted to the program committee, a number significantly higher than has been received in the past. The resulting program, based on a critical selection of abstracts, represented only a portion of the magnetism r&d being conducted around the world. The 30 invited papers and 360 contributed technical papers covered the widespread interests of the magnetism community.

Two special sessions challenged participants to be concerned with societal issues. Emilio Q. Daddario, Director of the Office of Technology Assessment of the United States Congress, pointed out that "government must never lose sight of the fact that the fundamental principles and truths revealed by basic research make possible the practical applications which can so dramatically improve people's lives" and that "development of knowledge alone is no guarantee that it will be wisely or even helpfully put to use". The theme of his address was the taming of technology so that it becomes a servant of man. These ideas were particularly relevant to the conference since its goal is to bring together the developers of new fundamental principles and the practitioners of these results.

Becoming increasingly aware of the role that women and minorities must play in science and technology, the Steering Committee heartily endorsed a workshop on "Women in Science" organized by Dr. Jill C. Bonner. The purpose of the workshop was to bring to the attention of conference attendees the attitudes, concerns and legal bases for assuring that women receive a larger role in science. That a panel representing many points of view could be assembled from conference participants demonstrates that women have been making contributions to the field of magnetism; the workshop pointed out that a much greater role can be expected in the future.

For a number of years, Dr. Hugh C. Wolfe has shown his interest and given of his time in the work of the Magnetism and Magnetic Materials Conference. As representative to the Steering Committee from the American Institute of Physics, he helped to shape our constitution, to guide our publications and to aid in the administration of the conference. In recognition of this valued work, the Steering Committee expressed its appreciation with the presentation of a plaque at an evening plenary session.

When human activities reach decades in their life span, special recognition seems to be required. The occasion of the 20th annual conference should be therefore suitably noted. The conference continues to carry out its interdisciplinary mission of exchanging information between basic and applied practitioners of magnetism. The many new ideas and results presented in these proceedings are clear evidence that the conference is successfully fulfilling the role it has established for itself. Because it exists, is no reason for its continuing, however. The role of the program chairmen will continue to be the key one in determining how well the conference will serve the needs of its constituency in the future. They must continually forecast those areas of new results which will be of use to the field. They must be able to identify speakers who can clearly point out the direction that r&d should take for fruitful results. This year's chairmen have ably met that challenge. Future conferences must continue to have that as a critical challenge. For the work of the program committee and all others associated with the conference, I express my appreciation and gratitude, not only personally but from all conference attendees as well.

James M. Lommel
General Chairman
General Electric Research & Development Center
Schenectady, New York 12301

TABLE OF CONTENTS

Ferromagnetism in Amorphous Solids......138
G.S. Cargill, III
Spin Waves and Magnetic Interactions in the Ferromagnetic Heusler Alloys......145
Y. Ishikawa and Y. Noda
Magnetic Excitations in Rare Earth Al_2 Compounds......152
P. Bak

Section 9. SPIN WAVES

Temperature Dependence of Magnetic Excitations in Pd_3Fe......159
T.M. Holden
Single-Ion Resonant States in Two Spin Wave Spectra......160
S.T. Chiu-Tsao, C. Paulson and P.M. Levy
Magnetic Excitations in Mn-Doped CoF_2......161
E.C. Svensson, S.M. Kim, W.J.L. Buyers, S. Rolandson, R.A. Cowley and D.A. Jones
Statistical Theory of Magnetic Excitations in Heisenberg Paramagnets......163
S.H. Liu and P.A. Swanson
Spin Waves in the Laves Compound $Ho_{.88}Tb_{.12}Fe_2$......165
R.M. Nicklow, N.C. Koon, C.M. Williams and J.B. Milstein
Theoretical Magnon Dispersion Curves for Gd......165
B.N. Harmon, P.-A. Lindgård, A.J. Freeman and J. Rath
Antiferromagnetic Resonance Studies of $NaMnF_3$......166
G.O. White and R.L. White
AFMR Versus Orientation in Weakly Ferromagnetic $BaMnF_4$......168
E.L. Venturini and F.R. Morgenthaler
New Resonance Effects in Antiferromagnetic MnF_2......170
W.E. Tennant, R.B. Bailey and P.L. Richards
Theory of Multimagnon Relaxation in Antiferromagnets......172
R.M. White, S.M. Rezende and L.C.M. Miranda
Thermal Transport by Magnetoelastic Modes in $CoCl_2 \cdot 6H_2O$......174
J.E. Rives and S.N. Bhatia
Magnon Heat Conduction and Magnon Lifetimes in the Metallic Ferromagnet $Fe_{68}Co_{32}$ at Low Temperatures......176
Y. Hsu and L. Berger
Effect of Spin Wave Amplification on the Magnetoresistance of Ferromagnetic Semiconductors......178
M.D. Coutinho, F°, L.C.M. Miranda and S.M. Rezende
Effect of Single-Ion Anisotropy on Two-Magnon Raman Spectra of Heisenberg Ferromagnets......180
B.J. Choudhury and P.D. Loly

Section 10. MAGNETO OPTICS

Magnetic Circular Dichroism in Si and Sn Doped Garnets......182
B. Antonini, A. Paoletti, P. Paroli, A. Tucciarone, J.F. Dillon, Jr., E.M. Gyorgy and J.P. Remeika
Observation of the Twin-Walls Movement in $KNiF_3$ by Dichroism......184
R.H. Petit, J. Ferré and J. Badoz
A New Class of Room-Temperature Magneto-Optic Insulators: The Cobalt Ferrites......186
R.K. Ahrenkiel, T.J. Coburn, D. Pearlman, E. Carnall, Jr., T.W. Martin and S.L. Lyu
Magnetooptical and Magnetic Properties of Terbium Iron Garnet at Low Temperatures......188
T.K. Vien, H. Le Gall, A. Lepaillier-Malécot, D. Minella and M. Guillot
Magneto-Optical Studies of Ferromagnetic Metals......190
J.L. Erskine
Magnetic Tricritical Behavior of $FeCl_2$......195
J.A. Griffin and S.E. Schnatterly
Magnetooptical Studies of Metamagnets......200
J.F. Dillon, Jr., E. Y. Chen and H.J. Guggenheim

Section 11. EXCHANGE, COVALENCY AND HYPERFINE FIELDS

Multiplet Splitting of X-Ray Photoemission Spectra Core Levels in Magnetic Metals......207
S.P. Kowalczyk, F.R. McFeely, L. Ley and D.A. Shirley
Covalency Effects on Tetrahedral and Octahedral Fe^{3+} Sites in YIG: Charge and Spin Densities and Neutron Form Factors......209
E. Byrom, A.J. Freeman and D.E. Ellis
Covalency of Hydrated Complexes of Co^{2+} from ^{17}O ENDOR......210
J.W. Culvahouse, H. Glotfelty and W.P. Unruh
A Systematic Study of the Orbitally Dependent Rare Earth - Gd^{3+} Exchange Parameters in $GdCl_3$......211
R.S. Meltzer, D.K. Braswell and R.L. Cone
Exchange Interactions in Isolated Trinuclear Clusters of Fe^{3+}: Magnetic and Mössbauer Studies......213
L.N. Mulay and G.H. Ziegenfuss
Magnetic Hyperfine Structure and Relaxation Effects in Bromobis(Morpholyldithiocarbamato)Iron(III).215
J.M. Grow and H.H. Wickman
Temperature Independent ^{55}Mn Magnetic Hyperfine Fields in $ErMn_2$ and $TmMn_2$......217
R.G. Barnes and B.K. Lunde
On Superexchange Theory......218
R.S. Tu and T.A. Kaplan
Mössbauer Effect Study of $Co^{57}Cl_2 \cdot 2H_2O$ and $Co^{57}Cl_2 \cdot 6H_2O$......219
B.A. Alavi, D.F. Goble and W. Leiper

Section 16. CRITICAL PHENOMENA II

Section 17. THEORY

Section 18. ACTINIDES AND CHAINS

Section 19. MATERIALS SYNTHESIS AND NEW MATERIALS

Section 20. SURFACES AND SMALL PARTICLES

Section 21. METALS AND ALLOYS I

Section 22. METALS AND ALLOYS II

Section 23. METALS AND DILUTE ALLOYS

Section 24. DILUTE ALLOYS

Section 25. MICROWAVE MATERIALS AND DEVICES

Section 26. MICROWAVE RESONANCE AND RELAXATION

Section 27. NON-VOLATILE MAGNETIC STORAGE

Section 31. BUBBLE PROPAGATION AND RELATED TOPICS

Section 32. MAGNETO-ELASTIC AND ANISOTROPY EFFECTS

Section 37. LUNAR AND OTHER MATTER: HARD MAGNETIC MATERIALS

Section 38. GENERAL INTEREST

INTRODUCTION

In the past, little technical information has passed between people with applied and those with basic research interests although the intent of the M^3 conferences has been quite the reverse. This year, three speakers were asked to give tutorials for the nonspecialist which would be accessible to the majority of conference attendees. One subject was applied, two were basic. In the Program Committee's view, the tutors successfully fulfilled their mandate.

MAGNETIC BUBBLES--FOR THE NON-SPECIALIST

F. B. Hagedorn
Bell Laboratories
Murray Hill, New Jersey 07974

SUMMARY

A tutorial talk on magnetic bubble domains and their applications was presented at the 20th Annual Conference on Magnetism and Magnetic Materials. This talk is briefly summarized below, and certain illustrative and representative references are given. However, these references are neither complete nor exhaustive. Instead, this summary is presented partly to help in providing a relatively complete written record of the 20th CM[3] and partly to provide a starting point for nonspecialists who want to learn about magnetic bubbles.

Bubble domains have been known to exist in thin slabs of uniaxial magnetic materials for many years: Sherwood et al.[1] published photographs of bubbles in 1959; Kooy and Enz[2] presented an approximate theory of the bubble collapse field in 1960; Bobeck[3] gave the first account of bubble domain devices in 1967, and Thiele[4] developed a complete theory of bubble stability in 1969.

The theory[2,4] of bubble stability takes into account three major contributions to the magnetic free energy of the magnetic system: the interaction energy between the externally applied stabilizing field and the magnetic slab, the energy of the domain wall which defines the bubble, and the change in demagnetizing energy which is introduced by the presence of the bubble. Detailed analysis[4] shows bubbles to be stable over only a very limited range of values for the external magnetic field. For larger values, the bubbles collapse and disappear; for smaller values, the bubbles expand out into strip domains. A simple and frequently useful approximation of Thiele's complete theory[4] has been given by Callen and Josephs[5].

Stability is thus an intrinsic property of bubble domains and does not require that the magnetic slab possess imperfections so as to introduce coercivity. A second useful intrinsic property of magnetic bubbles is that they are mobile. Motion of bubbles within the plane of the magnetic slab requires a driving force, and a generalized treatment of bubble forces has been given by Thiele et al.[6] The most frequently used force arises from a gradient in the externally applied magnetic field, and Thiele[7] shows that one expects the steady state velocity of a bubble domain to be proportional to the magnitude of the field gradient. The constant of proportionality is simply related to the domain wall mobility of the material, and the material parameters which affect the domain wall mobility were reviewed recently by the present author.[8]

The linear relationship between the bubble's velocity and driving force pertains until a critical velocity is reached. Two basically different kinds of mechanisms are known to create critical velocities: Walker breakdown (see Ref. 8 for a review and earlier references) and Slonczewski breakdown[9]. Both mechanisms have been analyzed most completely for the case of a planar wall. Walker breakdown takes place simultaneously over the entire wall (assumed to be infinite in extent) while Slonczewski breakdown arises from an interplay between the motion of the wall and the nonuniformities of the wall structure near the surface of the magnetic slab. Because these critical velocities depend differently on the material parameters, it is possible for either mechanism to be the limiting one; however, Slonczewski breakdown generally occurs at lower velocities in practical cases and is the mechanism which presently seems to be the main limitation on the high frequency operation of bubble devices.

Applying[10] the Slonczewski breakdown mechanism to the cylindrical structure of the bubble domain leads to interesting results: it shows how vertical Bloch lines may be dynamically developed in bubble domains, thereby accounting at least qualitatively for the experimentally observed phenomenon of dynamic conversion[11]. It also enables one better to understand the formation of hard bubbles, the interesting static[12,13] and dynamic[14,15,16,17] properties of which have been reported previously. For most of the present generation of bubbles devices, it is important to avoid hard bubbles, and several methods of hard bubble suppression[18,19,20] have been developed.

The intrinsic stability and mobility of bubble domains make them potentially useful for device applications. In order to realize this potential, it is necessary to have available slabs of magnetic material with the following properties: (1) the uniaxial anisotropy must be large enough so that the bubble will be stable throughout the range of conditions that exist during device operation, (2) the bubble size must be controlled so as to fit the application, (3) the bubbles must be able to move easily and rapidly under the application of modest driving forces, (4) the above-mentioned three properties should persist over a range of temperature of the order of 100°C, with room temperature being typically near the center of this range, (5) slabs of such material with uniform properties over areas of at least several cm^2 should be available at modest cost. The relationship between many of these requirements and the magnetic material parameters has been emphasized by Gianola et al.[21]

Only a flexible and versatile material system will be able to meet all of these requirements simultaneously. At the present time, magnetic rare earth garnets[22] grown by liquid phase epitaxy from supercooled melts[23] seem to represent the best solution to this material problem. Wafers of gadolinium gallium garnet, typically 1" to 2" in diameter and 0.01" to 0.03" in thickness, are used as substrates for the epitaxial growth process, and considerable progress[24] has been reported in developing processes for supplying substantial quantities of epitaxial garnet material.

An alternative class of materials for magnetic bubbles may be found in the amorphous magnetic metal films. Bubble domains have been shown to be stable in several different compositions,[25] and this work is presently being pursued. It is not so advanced as the liquid phase epitaxial garnet work, however, and it remains to be seen whether or not the amorphous materials will develop into a serious competitor for the garnets.

To make a bubble domain device from an appropriate slab of magnetic material, one must fabricate a structure to create, manipulate, and detect bubble domains within the magnetic slab. Bobeck and Scovil[26] have recently reviewed the methods for accomplishing these bubble device functions. A broadly-used concept in these devices is that of a re-entrant loop created from a periodic array of magnetic potential wells. The separation between adjacent wells is several bubble diameters, and the presence or absence of a bubble trapped in each of the potential wells represents a string of binary data. Several ingenious methods have been devised to cause the array of magnetic potential wells to move like a travelling wave. In this way, a data string stored in a loop can be made to circulate around that loop. As a consequence, a single bubble detector can be located on each bubble device chip, and interrogation of all data stored in the chip may be accomplished by propagating the data pattern past the detector.

Bubble devices are presently attractive because of their potential to provide relatively inexpensive, compact, and nonvolatile memory for a variety of digital electronic applications. Costs in the range of tens of millicents per bit for bubble circuit densities in the range of millions of bits per square inch are foreseen, using bubble domains with diameters in the range of several microns. The basic simplicity of the bubble device is the reason that such objectives appear to be possible in the near future. Starting with the slab of magnetic material which has been treated so as to suppress hard bubbles, a typical procedure for complete fabrication of the device is as follows: (1) Deposit on one entire surface of the magnetic slab a layer of conducting metal about 1/2 micron in thickness; then create by conventional photolithographic techniques the patterns of conducting loops required to perform the various control functions such as creation, annihilation, transfer, and replication of bubbles. (2) Deposit a continuous dielectric layer about 1 micron in thickness; this layer covers the conducting pattern wherever it exists but is directly on the magnetic slab where the conducting layer was removed during the creation of the conductor pattern. (3) Deposit a continuous layer of permalloy about 1/2 micron in thickness on top of the dielectric layer; then pattern it using conventional photolithographic techniques so as to create both the propagation pattern and the magnetoresistive detector pattern. (4) Deposit a final protective layer of the dielectric film of about 1 micron in thickness over the entire surface; follow this step by a photolithographically defined plasma-etch removal of the dielectric material so as to expose the bonding pad areas of the conductor and detector patterns.

Bubble memory chips are presently being made with capacities in the range of 10^4 to 10^5 bits, using the fabrication approach outlined above. It is anticipated that electron-beam lithography, rather than photolithography, may extend this capacity range by an order of magnitude in the foreseeable future. Even, with existing chips, however, memory modules with capacities of the order of 10^6 bits have been operated with latency times of the order of milliseconds and data rates of the order of megabits/sec. Assembling many such modules into a single system appears to offer an attractive alternative for the memory function presently performed by fixed-head disk files. Custom memories, where the total requirement is in the range from 10^3 to 10^5 bits, is another immediate area of possible application. Examples of this type of application might be repertory dialer telephones or pocket calculators. Finally, applications wherein bubble logic and memory are intermixed are foreseen for the somewhat more distant future. An example of this kind of application could be a digital switching network.

REFERENCES

1. R. C. Sherwood, J. P. Remeika, and H. J. Williams, J. Appl. Phys. 30, 217-225 (1959).
2. C. Kooy and U. Enz, Philips Res. Reports 15, 7-29 (1960).
3. A. H. Bobeck, Bell Syst. Tech. J. 46, 1901-1925 (1967).
4. A. A. Thiele, Bell Syst. Tech. J. 48, 3287-3335 (1969).
5. H. Callen and R. M. Josephs, J. Appl. Phys. 42, 1977-1982 (1971).
6. A. A. Thiele, A. H. Bobeck, E. Della Torre, and U. F. Gianola, Bell Syst. Tech. J. 50, 711-724 (1971).
7. A. A. Thiele, Bell Syst. Tech. J. 50, 725-773 (1971).
8. F. B. Hagedorn, AIP Conf. Proc. Series 5, 72-90 (1972).
9. J. C. Slonczewski, J. Appl. Phys. 44, 1759-1770 (1973).
10. F. B. Hagedorn, J. Appl. Phys. 45, 3129-3140 (1974).
11. G. P. Vella-Coleiro, F. B. Hagedorn, Y. S. Chen, and S. L. Blank, Appl. Phys. Lett. 22, 324-325 (1973).
12. A. Rosencwaig, W. J. Tabor, and T. J. Nelson, Phys. Rev. Lett. 29, 946-948 (1972).
13. A. P. Malozemoff, Appl. Phys. Lett. 21, 149-151 (1972).
14. G. P. Vella-Coleiro, A. Rosencwaig, and W. J. Tabor, Phys. Rev. Lett. 29, 949-952 (1972).
15. A. P. Malozemoff and J. C. Slonczewski, Phys. Rev. Lett. 29, 952-955 (1972).
16. J. C. Slonczewski, Phys. Rev. Lett. 29, 1679-1682 (1972).
17. A. A. Thiele, F. B. Hagedorn, and G. P. Vella-Coleiro, Phys. Rev. B8, 241-245 (1973).
18. A. H. Bobeck, S. L. Blank, and H. J. Levinstein, Bell Syst. Tech. J. 51, 1431-1435 (1972).
19. R. Wolfe and J. C. North, Bell Syst. Tech. J. 51, 1436-1440 (1972).
20. Y. S. Lin and G. E. Keefe, Appl. Phys. Lett. 22, 603-604 (1973).
21. U. F. Gianola, D. H. Smith, A. A. Thiele, and L. G. Van Uitert, IEEE Trans. on Magnetics MAG-5, 558-561 (1969).
22. L. J. Varnerin, IEEE Trans. on Magnetics MAG-7, 404-409 (1971).
23. H. J. Levinstein, S. J. Licht, R. W. Landorf, and S. L. Blank, Appl. Phys. Lett. 19, 486-488 (1972).
24. B. S. Hewitt, R. D. Pierce, S. L. Blank, and S. Knight, IEEE Trans. on Magnetics MAG-9, 366-372 (1973).
25. P. Chaudhari, J. J. Cuomo, R. J. Gambino, IBM J. Res. & Dev. 17, 66-68 (1973).
26. A. H. Bobeck and H. E. D. Scovil, Scientific American, 78-90, (June 1971).

IMPURITIES AND THEIR MAGNETIC MOMENTS, OR "...INTO THE PROMISED LAND."

R. Orbach*
Physics Department, University of California,
Los Angeles, California 90024

ABSTRACT

The basic ideas behind magnetic impurities in non-magnetic metallic hosts are reviewed. The atomic, or insulator, picture is used as a beginning. The alterations in wave function and energy level width upon solution in a metal are explored. The resulting covalent exchange is shown to lead to Kondo ℓnT behavior in the resistivity and susceptibility. The Kondo temperature is defined. Discrepancies between perturbation theoretic results and experiment are exhibited below this temperature. The Anderson-Yuval-Hamann concept of scaling is presented, and Wilson's low temperature results are discussed using Noziere's perturbation model. A pedestrian version of scaling is used to find the stable exchange fixed point. The implications of scaling are discussed.

Recently, P. W. Anderson wrote,[1]

Kondo Effect IV: Out of the Wilderness

I have delayed over two years since the last of these articles because I had hoped that Comment IV in this series on Kondo effect could represent the solution, in full detail. Unfortunately, while a quantum jump, in sophistication and power of the methodology and in our qualitative understanding of the Kondo problem, has taken place, the last few steps in achieving and experimentally confirming an actual numerical solution, rather than a solution in principle, have yet to be completed; so what I have to give you here is yet another progress report.

Fig. 1. Quote from P.W. Anderson (Ref. 1).

After Anderson's article appeared, remarkable quantitative progress has been made, justifying the sub-title of this tutorial.

The purpose of this presentation is to provide a physical picture for a magnetic moment dissolved substitutionally in a non-magnetic host metal. We begin with the simplest limit - a free magnetic ion (e.g., Fe^{2+} or Mn^{2+}), and consider the consequences of metallic solution. Pictorially, the free ion possesses outer orbitals of the form

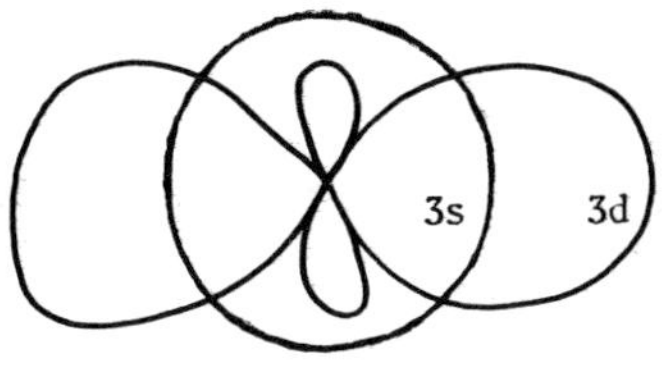

Fig. 2. Schematic of outer shell transition metal orbitals.

where the 3s are double occupied, and only a single one of the occupied 3d orbitals has been exhibited. The Hamiltonian governing the electronic motion can be written as

$$H_e = \sum_i p_i^2/2m - \sum_i Ze^2/|\vec{R}-\vec{r}_i| + \sum_{i<j} e^2/|\vec{r}_i-\vec{r}_j| \qquad (1)$$

where the first two (one electron) terms are not difficult to treat. The last, the electron-electron Coulombic interaction, is another matter. In the one electron Hartree-Fock approximation, (1) can be solved and yields the well known Hund's rule ground state:

(1) Maximum S
(2) Maximum L consistent with maximum S.

Thus, the ground determinants have the form

$$(\overset{+}{2}\overset{+}{1}\overset{+}{0}\text{-}\overset{+}{1}\text{-}\overset{+}{2}) \quad S = 5/2,\ L = 0 \quad \text{for } Mn^{2+}\ (3d^5)$$

$$(\overset{+}{2}\overset{+}{1}\overset{+}{0}\text{-}\overset{+}{1}\text{-}\overset{+}{2}\text{-}\overset{-}{2}) \quad S = 2,\ L = 2 \quad \text{for } Fe^{2+}\ (3d^6) \qquad (2)$$

where the symbol over the orbital state labels the z axis spin projection (+ is "up," - is "down").

From these requirements, we can predict the magnetic moments using the definition

$$\vec{\mu} = \mu_\beta(\vec{L}+2\vec{S}) = \mu_\beta\Lambda_J\vec{J}\ , \text{ where } \vec{J} = \vec{L}+\vec{S} \qquad (3)$$

where Λ_J is the Landé g factor. Thus, for Mn^{2+} we expect five Bohr magnetons. For Fe^{2+} in a cubic (octahedral) environment, the spin-orbit coupling and crystalline field combine to yield[2] an effective angular momentum of one, with a magnetic moment of four Bohr magnetons. In insulators,[2] covalency can reduce this slightly (a common value[3] is 3.4 μ_B). But, in Al, Fe has no moment, and in Pd, each Fe impurity contributes 13 Bohr magnetons to the magnetic moment. Even worse, Clogston et al.[4] found for Fe impurities in 4d metals and alloys

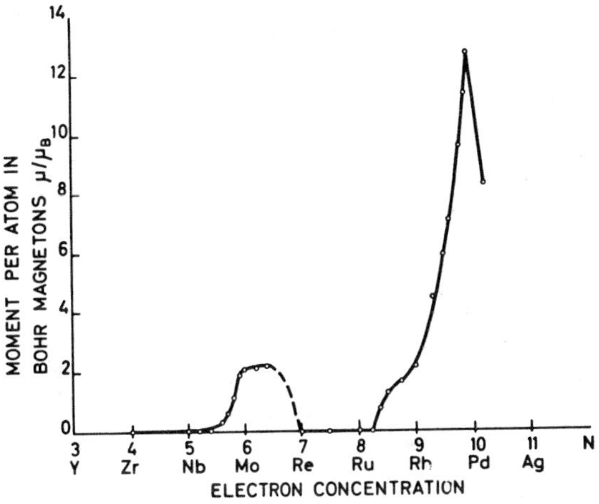

Fig. 3. Magnetic moment per Fe impurity as a function of the number of electrons per atom (Ref. 4).

Heeger[5] summarizes when the 3d transition metals have moments in metallic hosts (+), and when they do not (-):

	Au	Cu	Ag	Al
Sc				
Ti	—			—
V	?			—
Cr	+	+	+	—
Mn	+	+	+	?
Fe	+	+		—
Co	?	?		—
Ni	—	—		—

Fig. 4. Characterization of the magnetic state of transition metal impurities in simple metal hosts. + is magnetic, - is not.

Clearly, something is altering the ionic picture when a magnetic moment is dissolved in a metal.

To understand the reasons for this behavior, imagine superposing the energy levels of the free atom (or ion) upon those of the conduction electrons in a metal.

Here, the horizontal lines represent the energy of the magnetic impurity for different occupancies, and we have not allowed the impurity to interact with the conduction electrons (yet). The energy required to add or

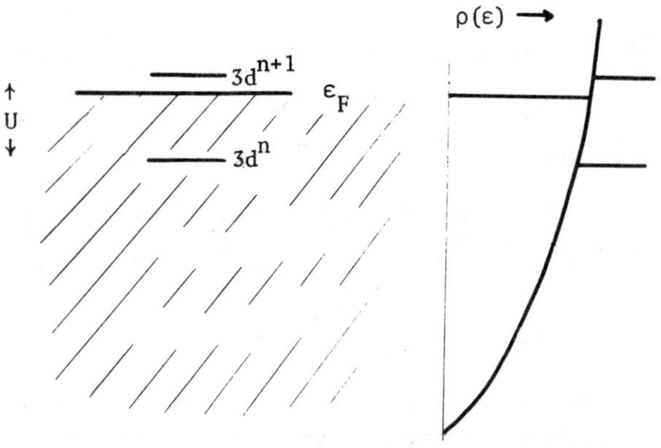

Fig. 5. Schematic of the energy density of states for a transition metal impurity in a metal, in the absence of the mixing interaction.

remove an electron from the impurity is denoted by U, representing primarily the last term in (1). This simplistic view is not so far from reality (Charlotte E. Moore[6] shows that the difference in the successive d electron ionization potentials is rather constant in the 3d series. For Cu, this energy change is of the order of 14 eV, though this is likely to be substantially smaller in the metal because of screening). It enables us to write for the localized orbital energy

$$H_{loc} = \varepsilon_d \sum_\sigma n_\sigma + U n_\sigma n_{-\sigma} \quad \text{where} \quad \sigma = \pm 1 \quad , \tag{4}$$

and n_σ is the number operator for electrons on the localized site. Hence, if we have one localized electron with spin up ($\sigma = +1$), it costs us an energy U to add to it in the same orbital another localized electron with spin down ($\sigma = -1$). In this model, $\langle n_\sigma \rangle \leq 1$ because we have not allowed for degeneracy (the 5 orbital states on the d site). This can be easily grafted onto our description, but only serves to complicate the algebra. On the right of Fig. 5, we plot the density of states of all the electrons. Because the localized levels are not allowed to interact, their density of states are delta functions of energy, while the conduction electrons exhibit the usual smoothly varying parabolic density of free electron states.

We now "turn-on" the interaction between the localized and conduction electrons. Physically, this is caused by the fact that the 3d atomic functions are not eigenfunctions of the full crystal Hamiltonian. In the Hartree-Fock sense, there exists a one electron "mixing" Hamiltonian which "connects" the localized electrons with the conduction electrons, even if the two sets of electrons are orthogonal. This "Anderson-Clogston[7]-Kondo[8]" mixing Hamiltonian has the form

$$H_{mix} = V_{kd} \sum_{\vec{k},\sigma} (c^*_{\vec{k}\sigma} a_\sigma + a^*_\sigma c_{\vec{k}\sigma}) \tag{5}$$

where $c^*_{\vec{k}\sigma} a_\sigma$ destroys a localized electron with spin σ and creates a conduction electron with wave vector $\vec{k}$ and spin σ. Thus, H_{mix} allows "tunnelling" between localized and conduction electron states. Simple perturbation theory tells us that this will give rise to a finite lifetime for occupancy of the localized state, and hence an energy width appropriate to the occupancy of that state. This width equals

$$\Delta = \pi |V_{kd}|^2 \rho(\varepsilon_d) \tag{6}$$

where $\rho(\varepsilon_d)$ is the conduction electron density of states at the energy of the localized d level. Pictorially, one has (See Fig. 6)

where we have displayed separately the density of states for the two different spin directions. The sharp levels of Fig. 5 have now been broadened out to half-width Δ, and have equal up-down occupancy. This

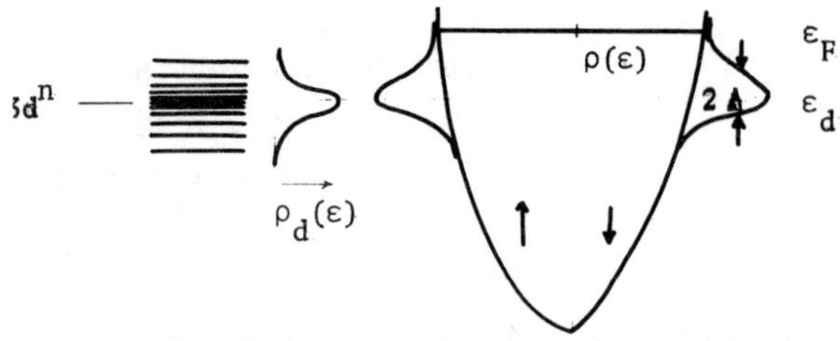

Fig. 6. Schematic of the energy density of states for a transition metal impurity in a metal, in the presence of the mixing interaction.

is the Friedel[9]-Anderson[10] picture for the non-magnetic d level impurity state in a metal. The magnitude of the localized moment is given by the difference in occupancy of the up and down spin populations. They are equal in Fig. 6, and no moment is therefore present on the localized site.

Anderson[10] was first to make quantitative the transition to the magnetic state. Magnetization results from the Coulomb interaction [second term of (4)] dominating the effects of mixing [Eq. (5)]. For U sufficiently large, it is energetically favorable to occupy the localized site with different up or down populations. Increasing U beyond $\pi\Delta$ causes the down spin localized level to "empty," leaving a net up spin density on the impurity site (a magnetic state, in the Hartree-Fock sense). The condition for magnetization is

$$\pi\Delta/U < 1 \tag{7}$$

and the magnitude of the moment is then

$$\vec{\mu} = \mu_\beta (n_{+1} - n_{-1}) \tag{8}$$

Self-consistent values, as functions of Δ/U, are plotted by Anderson.[10]

Pictorially, one has

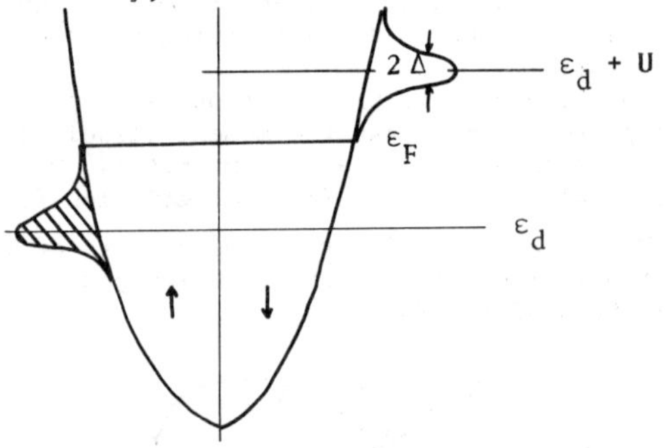

Fig. 7. Schematic of the energy density of states for a magnetic transition metal impurity in a metal.

where the shaded regions represent occupied states.

The weakness of this picture lies in our use of the Hartree-Fock one electron approximation. To obtain the requirement (7), we have had to resort to a mean field (albeit self consistent) result. That is, we replace

$$U n_{+1} n_{-1} \quad \text{by} \quad U n_{+1} \langle n_{-1} \rangle \quad . \tag{9}$$

This causes one to over-estimate the energy of the double occupancy (non-magnetic) state. Said another way, the Hartree-Fock approximation leads too easily to the magnetic state. Breaking the problem into uncorrelated local up-down occupancies does not describe the lowest energy of the double occupancy state[11] very well. Unfortunately, however, no one has yet done better than the Hartree-Fock result of Anderson. We must use the one electron ideas more as an intuitional guide than as a quantitative tool.[12]

With this warning in mind, it is instructive to inspect the consequences of our construction of the magnetic state. Imagine that a moment exists, so that the conditions of Fig. 7 are met. Now, what are the consequences, not of the Coulomb interaction U, but rather of the mixing potential V_{sd}? In addition to broadening the one electron occupancy levels, (5) also allows electron transfer from (to) the impurity levels to (from) the conduction electron Fermi surface states. This is pictured below.

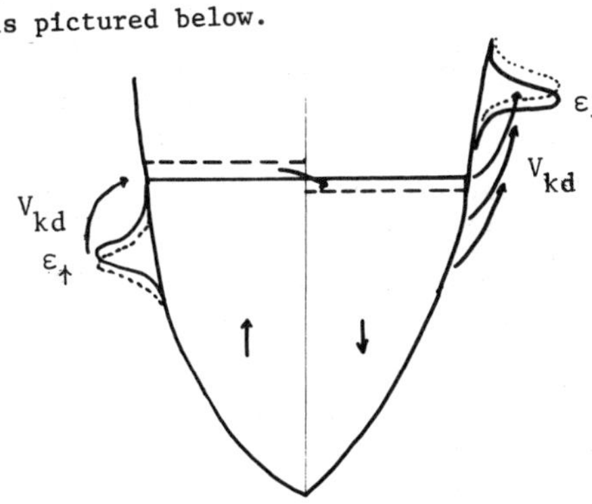

Fig. 8. Shift of the energy levels in the magnetic states caused by the mixing interaction between localized and Fermi surface electrons.

The solid lines represent the energy levels before Fermi surface electron admixture. The matrix element of a perturbing Hamiltonian between two levels results in a repulsion of the two levels from one another. The level position after repulsion is represented by the dashed lines in Fig. 8. The difficulty is that the Fermi levels now appear to be different for up and down spin. To correct this non-equilibrium situation, up-spin conduction electrons "flow" into down-spin states. This means that there are more down-spin conduction electrons than up-spin conduction electrons, i.e., a polarization has been created. This polarization is opposite to the impurity moment. It is as though an antiferromagnetic coupling existed between the localized moment and the conduction electron spins. Indeed, one can show[13] that this coupling has the form

$$H = -J\,\vec{S}\cdot\vec{s} \qquad \text{where}$$

$$J = |V_{kd}|^2\{(\varepsilon_d-\varepsilon_F)^{-1} + (\varepsilon_F-\varepsilon_d-U)^{-1}\} \qquad (10)$$

for electrons at the Fermi surface. Both terms are negative in (10) (as can be seen from the position of the energy levels in Fig. 8), so that $J < 0$. In addition to the energy shift of the conduction electrons, leading to the negative spin polarization, there is also an admixture arising from matrix elements of (5) between the impurity state and the unoccupied conduction electron states above the Fermi level. This also results in a conduction electron "moment." In reality, it is only a syphoning of moment off the localized impurity into conduction electron states locked to the impurity moment. For a constant density of conduction electron states, this (positive spin) admixture exactly cancels the (negative spin) polarization contribution to the conduction electron spin density (the "compensation theorem"[7,14]). There remains, of course, a diminution of the local moment - that part which was syphoned off. For transport properties or dynamics near the Fermi surface, only the polarization contribution is significant. We therefore focus on (10) in what follows.

There is substantial experimental evidence for (10). Probably the most direct is the recent magnetic resonance determination[15] that the g shift is negative for Ag:Mn dilute alloys. The idea is as follows. A magnetic field polarizes the conduction electron spin

$$\langle\vec{s}\rangle = \chi_p\vec{H} \qquad (11)$$

where χ_p is the Pauli conduction electron susceptibility. Equation (10) can be rewritten as

$$H = -J\,\vec{S}\cdot\vec{s} = -J\vec{S}\cdot\langle\vec{s}\rangle = -J\vec{S}\cdot\chi_p\vec{H} \equiv \mu_\beta g\vec{S}\cdot\Delta\vec{H} \qquad (12)$$

defining the additional field ΔH experienced by the localized moment (in addition to the bare applied field). Noting the Zeeman interaction between the impurity moment and applied magnetic field is written as

$$H_Z = -\vec{\mu}\cdot\vec{H} = -g\mu_\beta\vec{S}\cdot\vec{H} \qquad (13)$$

(12) can be interpreted as a shift in g value:

$$\frac{\Delta g}{g} = -\frac{\Delta H}{H} \quad \text{or, from (12),} \quad \Delta g = J\chi_p \qquad (14)$$

Thus, measurement of the g shift leads directly to the determination of the sign of the exchange coupling J. A phenomenon called the magnetic resonance bottleneck[16] heretofore obscured the observation of Δg for transition metal impurities. Very recently, the bottleneck was broken in dilute Ag:Mn alloys[15] (by adding Sb), and a negative g shift ($g < 2$) observed.

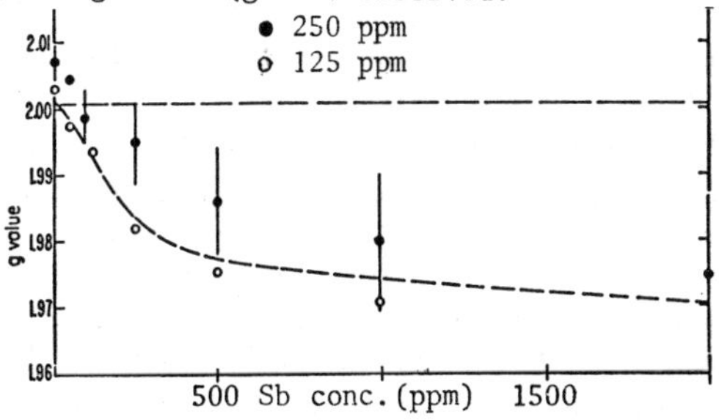

Fig. 9. The g value for dilute Ag:Mn alloys as a function of Sb concentration.

From (14), this confirms the exchange as antiferromagnetic, and gives unequivocal proof for the presence of the polarization process. For Ag:Mn

$$\Delta g = -0.035 \pm 0.01 \quad \text{yielding} \quad J = -0.25 \pm 0.1\ \text{eV}. \qquad (15)$$

The consequences of antiferromagnetic character for the impurity spin-conduction electron spin exchange coupling are extraordinary. The unusual high temperature (to be defined later) transport properties of dilute magnetic alloys (electrical conductivity, thermopower, magnetic susceptibility) were first calculated by Kondo[8] (the last by Scalapino[17]), and hence bear the name "Kondo effect." Using conventional perturbation theory in the parameter of smallness $J\rho(\varepsilon_f)$, Kondo found to third order that the electrical conductivity equaled

$$(1/3)\int\{(J^3/2N^2)S(S+1)cNg(\varepsilon)\}^{-1}(df^o/d\varepsilon)\rho(\varepsilon)d\varepsilon$$

where the term

$$g(\varepsilon) = (1/N)\Sigma_{\vec{k}}f_{\vec{k}}^{o}/(\varepsilon_{\vec{k}}-\varepsilon) \qquad (16)$$

Here, S is the impurity moment spin, c the fractional impurity concentration, and f^o the Fermi function. The important aspect of (16) is $g(\varepsilon)$. It contains an integrand which varies as the inverse of the excitation energy. As the temperature is lowered, this energy (characteristically of order kT) goes to zero, and $g(\varepsilon)$ diverges. Physically, (16) originates with sum of the perturbation processes,

(See Fig. 10)

Though these look imposing, they are straightforward. In the top line, a conduction electron in the state $\vec{k}$ with spin "up" ($\uparrow$) interacts with the magnetic impurity, itself in a state characterized by the z component of spin, M_s, via the exchange interaction (10).

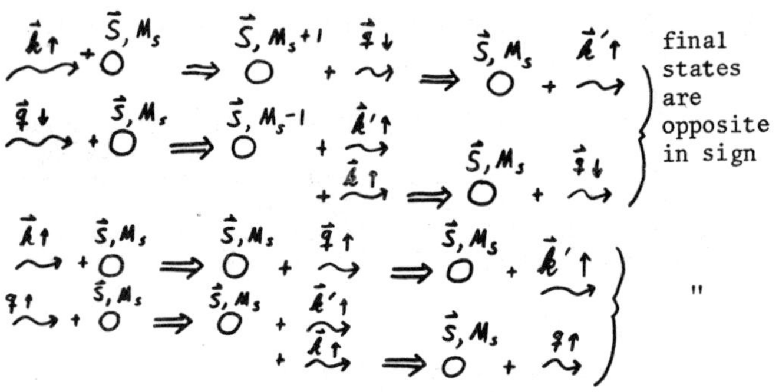

Fig. 10. Schematic of the pertubation processes included in the Kondo's calculation of the resistivity.

The local moment experiences S^+ and so "flips" up to a state M_s+1, while the conduction electron scatters to $\vec{q}$ and simultaneously flips to a down spin state (↓). Then, the conduction electron, now in the state $\vec{q},\downarrow$, interacts with the same local moment again. This time, however, H (10) acts with S^- and therefore lowers the local moment spin back to M_s, and scatters the electron into the state $\vec{k}\uparrow$. The net result of all this is to leave the local moment as it was initially, and to scatter the conduction electron from $\vec{k}\uparrow$ to $\vec{k}'\uparrow$. Involved, however, were the local spin operators S^-S^+, in that order, and an initial-intermediate energy difference of $\varepsilon_{\vec{k}\uparrow}-\varepsilon_{\vec{q}\downarrow}$. Now, the same final state could have been achieved by reversing the order of the scattering. As shown in the second line in Fig. 10, we could first scatter $\vec{q}\downarrow$ into the final state $\vec{k}\uparrow$, and thereby acting with S^+ on the localized spin, and then scatter $\vec{k}\uparrow$ to $\vec{q}\downarrow$, acting with S^- on the localized spin. The final result is the same: no electron in $\vec{k}\uparrow$, one in $\vec{k}'\uparrow$, and none in $\vec{q}\downarrow$. However, the local spin operators were S^+S^- (in that order), and the initial-intermediate energy difference was $\varepsilon_{\vec{q}\downarrow}-\varepsilon_{\vec{k}}$. In addition, the order of set-ups was opposite to that of the first line. Fermi statistics require that this changes the sign of the final states. This means the two processes subtract. The transition probability for the scattering process is proportional to the difference $S^-S^+ - S^-S^+$. This difference is not zero because the spin operators do not commute. The lack of commutivity is representative of internal dynamics of the local moment. If only potential scattering (no spin flip) was involved, the cancellation would have been complete and no contribution to the scattering $\vec{k}\uparrow - \vec{k}'\uparrow$ would result from terms of this sort (as in the last two lines of Fig. 10).

The reason the terms derived from the first two processes of Fig. 10 are so important is that, when they exist, the energy denominators $\varepsilon_{\vec{q}} - \varepsilon_{\vec{k}}$ can get very small (scattering on, or close to, the Fermi surface). Indeed, it is precisely these initial-intermediate energy differences which lead to the denominator in $g(\varepsilon)$, displayed in (16). Though terms with small denominators are always present in higher order perturbation theory, they exactly cancel for potential scattering when all processes are calculated and summed. The remarkable contribution of Kondo[8] was his observation that they did not cancel for exchange scattering (10) because of the lack of commutivity of the local spin operators.

Evaluation of $g(\varepsilon)$ in (16) leads to

$$g(\varepsilon_{\vec{k}}) = (3Z/2\varepsilon_F)\{1 + (k/2k_F)\ln\left|\frac{k-k_F}{k+k_F}\right|\}$$

which, in turn, leads to the celebrated expression for the resistivity

$$\rho_K = \{3\pi mJ^2S(S+1)/8e^2\hbar\varepsilon_F\}c\{1+2J\rho(\varepsilon_F)\ln(k_BT/D)\}$$
$$\equiv \rho_M\{1+2J\rho(\varepsilon_F)\ln(k_BT/D)\} \qquad (17)$$

defining ρ_M, the magnetic scattering to second order $J\rho$. Z in (17) represents the number of conduction electrons/atom, and D the conduction electron bandwidth [more physically, the high energy cut-off for the exchange (10), typically of order 10,000K]. It is somewhat surprising to see the cut-off explicitly enter an expression for ρ_K. There should be no reason why such high energy excitations can affect the electrical resistivity, the result of low energy scattering. We shall return to this point later.

According to (17), as T decreases, $\ln(k_BT/D)$ becomes more negative. If J is itself negative, then ρ_K increases as T diminishes. But the other contributions to ρ (impurities + phonons) diminish as T diminishes, leveling out to a constant value (residual resistivity) for T sufficiently small. In combination with these terms, ρ_K must then result in a resistivity minimum. The phonon contribution to the resistance varies at low temperatures as T^5, so that the temperature at which the resistance is a minimum is given by

$$T_{min} \propto (c|J\rho(\varepsilon_F)|)^{1/5} \qquad (18)$$

This prediction was the first explanation of the resistance minimum "problem" of MacDonald and the Ottawa group.[18] An example is the electrical resistivity of Ag:Mn as measured by Malm and Woods,[19]

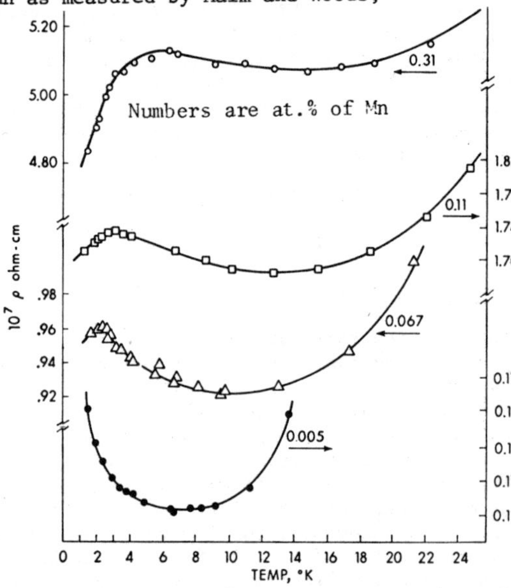

Fig. 11. The resistance as a function of temperature for dilute Ag:Mn alloys. From Ref. 19.

The weak concentration dependence of T_{min} ($c^{1/5}$) is apparent.

Likelwise, the magnetic susceptibility also experiences Kondo $\ln T$ terms. From Scalapino,[17]

$$\chi^{(0)} + \chi^{(2)} = \{(2\mu_B)^2S(S+1)/3k_BT\}$$
$$\times \{1 + (J\rho(\varepsilon_F))^2\ln(k_BT/D)\} \qquad (19)$$

where $\chi^{(0)}$ is the Curie term (first square brackets), and $\chi^{(2)}$ the second order correction in $J\rho(\varepsilon)$.

From (17) and (19), serious problems emerge. As T diminishes, it appears that ρ becomes larger without limit (and χ goes negative). This is silly - the dilute alloy will not become an insulator. Clearly, the simple perturbation theory is failing. Abrikosov[20] patched up this problem to some extent by summing the perturbation series to all orders in $J\rho(\varepsilon_F)$, keeping only the leading divergent terms in ℓnT. He found

$$\rho_K = \frac{\rho_M}{\{1 - J\rho(\varepsilon_F)\ell n(k_B T/D)\}^2} \qquad (20)$$

[Notice that (17) is a high temperature expansion of (20).] But now, ρ_K diverges at a finite temperature, i.e., when

$$1 = J\rho(\varepsilon_F)\ell n(k_B T/D) \qquad (21)$$

This defines the temperature T_K, the Kondo temperature,

$$T_K = (D/k_B)\exp\{1/J\rho(\varepsilon_F)\} \qquad (22)$$

It is the characteristic temperature at which perturbation theory appears to go wrong. The high temperature results, (17) and (19), are valid when the ℓnT correction is small, i.e., when $T >> T_K$. But something seriously wrong appears to occur for $T \lesssim T_K$. Because D is so large, the characteristic temperature T_K varies over a wide range. It is now "accepted" that

$$T_K(\text{Ag:Mn}) = 10^{-16}\ \text{K}; \quad T_K(\text{Cu:Mn}) = 8 \times 10^{-3}\ \text{K};$$
$$T_K(\text{Cu:Fe}) = 20\ \text{K} \qquad (23)$$

Over a broader temperature range,[21]

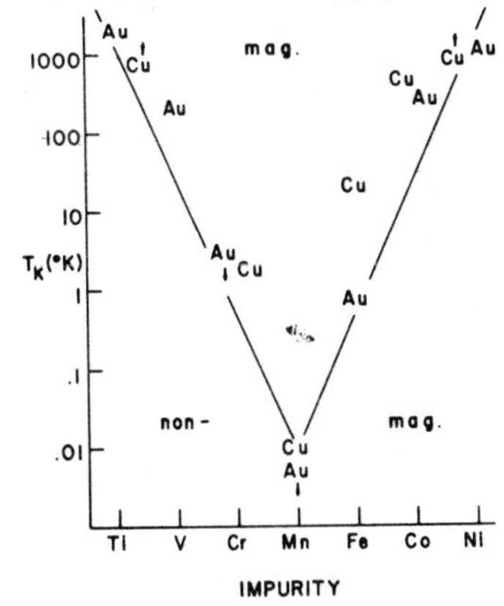

Fig. 12. A plot of Kondo temperatures for transition metal impurities in noble metal hosts. From Ref. 21.

Hamann extended the resistivity calculation, including potential scattering, to find

$$\rho_K = \frac{mc}{\pi Z N e^2 \hbar \rho(\varepsilon_F)} \{1 - \cos(2\delta_V) \frac{\tilde{\alpha}(T)}{[\tilde{\alpha}^2(T) + 4b]^{1/2}}\} \qquad (24)$$

where b is a constant, and

$$\tilde{\alpha}(T) = 1 - \{J\rho(\varepsilon_F)\cos^2\delta_V\}\ell n(k_B T/D)$$

Here, δ_V is the phase shift from potential scattering. No divergence takes place near T_K. The expression for ρ_K varies smoothly through the region of temperature near T_K.

Suhl and Wong[22] used dispersion theory to obtain approximate solutions for the resistivity, susceptibility, and thermopower over the full temperature range. The comparison between theory and experiment for the (scaled) electrical resistivity

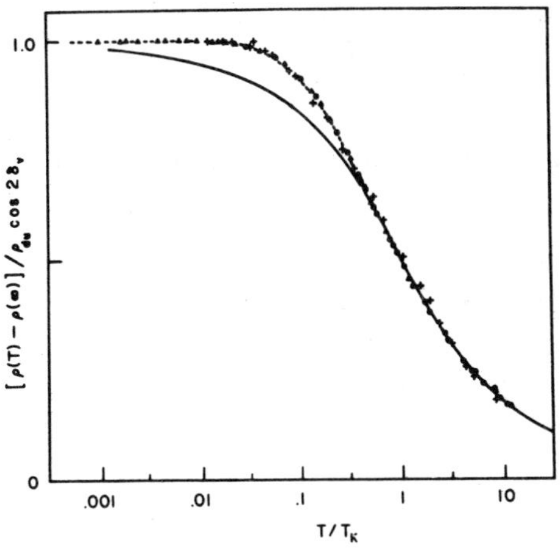

Fig. 13. Comparison of the Suhl-Wong theory with experiment for the (scaled) resistivity versus T/T_K for a variety of alloys. From Ref. 21.

is adequate at and above T_K, but poor below T_K. Both Hamann, and Suhl and Wong, predict departures from the $T = 0$ value (or unitarity limit) of the resistivity in powers of $\ell n(k_B T/d)$. The actual departure appears to be more gentle (see Fig. 13). The differences are strikingly exhibited in the susceptibility[21]:

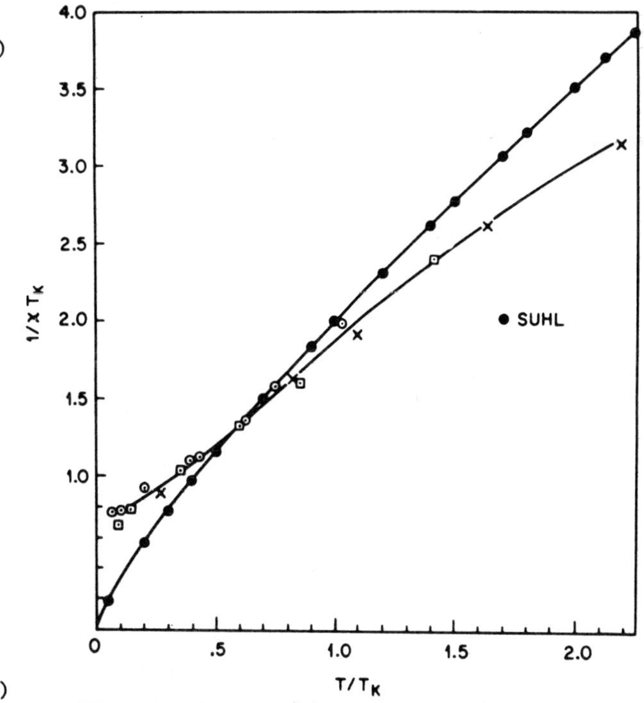

Fig. 14. A plot of inverse susceptibility versus temperature for the Suhl-Wong theory and the computer solutions of Schotte and Schotte.

In addition to Suhl and Wong's expression, the computer simulation results of Schotte and Schotte (essentially exact, apart from corrections due to the finite number of spins) are plotted in Fig. 14. Whereas Suhl and Wong predict an infinity in the zero temperature susceptibility, Schotte and Schotte predict a finite behavior. Clearly, something is wrong with the perturbation results.

The solution to this dilemma (we have omitted many other solutions which exhibit a wide variety of "crazy" behaviors) was pioneered by Anderson and Yuval,[23] and by Anderson, Yuval, and Hamann.[24] In a most significant paper, Wilson[25] achieved the exact zero temperature solution, and began finite temperature correction calculations. Most recently, Nozieres[26] has provided a simple physical model with which it appears that finite temperature properties can be calculated rather directly.

The technique which these authors used has its roots in "scaling theory."[27] It is based on the physical observations:

1. The characteristic temperature T_K determines the energy scale for the Kondo problem.
2. The high energy cut-off (D) cannot be important for the dynamics of the Kondo problem.
3. One may be able to scale the intractable problem into a tractable one. The solution can then be "unscaled" and returned to the physical system.

The first observation allows one to separate energy and temperature regimes. At low temperatures (by low, we now say precisely $T << T_K$), the local moment and conduction electrons (the Kondo "system") are locked into the ground state. Thermal fluctuations are insufficient to excite the system. The Kondo energy, k_BT_K, is therefore a measure of the "binding energy" for the system. For $T >> T_K$, the system unlocks, and the simple high temperature (now a precise meaning) perturbation theory leading to (17) and (19) suffices.

The second observation allows us to obtain the character of the condensed state. If it is true that T_K is a fixed measure of the energy scale, we should be allowed to change the bandwidth D (e.g., reduce it) and not affect the character of the low energy ($<< k_BT_K$) excitations of the alloy. We must obtain the same physical properties of our system if we let $D \rightarrow D + \delta D$ in our substance. But this would correspond to a different material. Scaling allows us to map the original material onto this new material if we preserve T_K. T_K sets the fixed energy scale - all activation energies are measured relative to it.

The expression for T_K is:

$$1 = J\rho(\varepsilon_F)\ln(k_BT/D) \tag{25}$$

Let $D \rightarrow D + \delta D$. Then T_K will remain the same only if $J \rightarrow J + \delta J$ where, from (25),

$$\delta D/D = \{1/J^2\rho(\varepsilon_F)\}\delta J \tag{26}$$

Thus, if we take $\delta D < 0$ (narrow the band - high energy excitations cannot be important) $\delta J < 0$. But for antiferromagnetic exchange $J < 0$. Hence, narrowing the bandwidth so that only low lying excitations need be considered can be accomplished only at the price of increasing the magnitude of the exchange. We obtain the correct energy scale in the ersatz material for $D \rightarrow 0$ only if $J \rightarrow -\infty$. In Wilson's language,[25] this is the stable fixed point of the Kondo problem for antiferromagnetic exchange. For ferromagnetic exchange ($J > 0$), δJ remains < 0 as D diminishes, and therefore $J \rightarrow 0$ is the stable fixed point.

These two limits are quite remarkable.[28] For antiferromagnetic exchange, the ground state of the system ($J \rightarrow -\infty$) must be a singlet with completely correlated localized moment and conduction electron spins. Because the energy scale has been preserved throughout this process, at $T << T_K$ this state can never be broken by energy conserving transitions (the spin-flip scattering rate must vanish): (See Fig. 15)

For the ferromagnetic case ($J \rightarrow 0$), the local spin and conduction electrons are not locked together at all. There are spatial correlations, but no pairing. The local moment of the "complex" is fully developed.

We are therefore faced with the striking result that the ground states of the two cases are completely

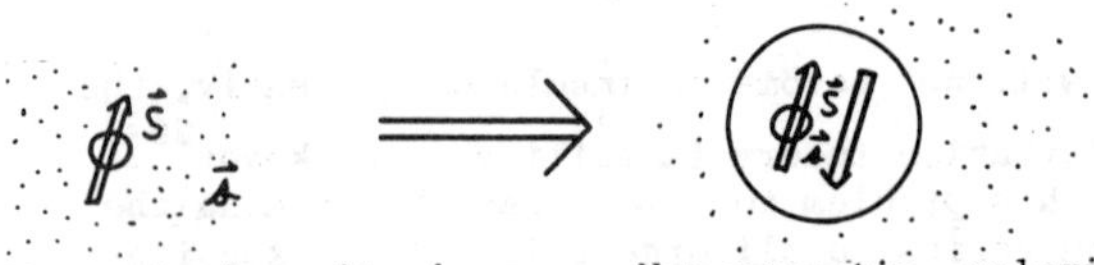

Magnetic impurity in a non-interacting electron gas

Non-magnetic, polarizable singlet with a localized interaction (repulsive) between conduction electrons

Fig. 15. Schematic of the Wilson description of the transition to the Kondo ground state.

different. The material has undergone a phase transition with a finite magnetization jump at $J = 0$.[25,26]

The low temperature antiferromagnetic Kondo problem thereby reduces to an estimate of the response of the electron gas to the bound singlet. This enormous simplification is the cause for Anderson's title,[1] and his expressed hope for a full solution. The subsequent perturbation theoretic approach of Nozieres[26] is the cause for the choice of title of this paper. Nozieres uses the fact that at $J \rightarrow -\infty$, the singlet cannot be broken with finite energy excitations. For J away from the $-\infty$ limit (i.e., finite), virtual excitations into 0 or 2 electron states on the impurity site become possible.

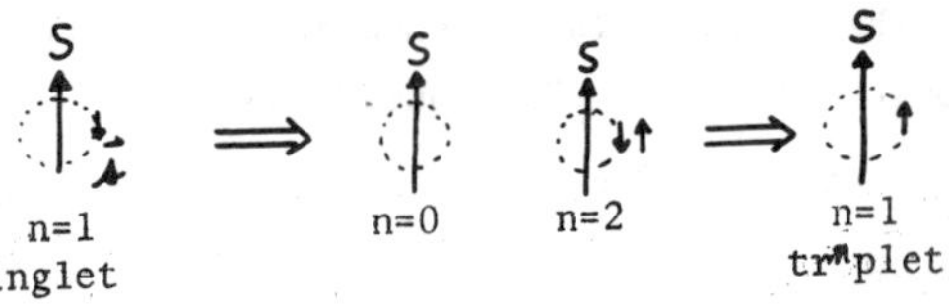

Fig. 16. Schematic of the polarization process of Nozieres (Ref. 26).

This allows for polarization of the singlet, and thence to a repulsive interaction between conduction electrons. This repulsive electron-electron interaction through the intermediary of the polarizable singlet is analogous to the B.C.S. attractive electron-electron interaction through virtual phonon exchange. The difference is that no phase transition amongst the conduction electrons can occur for repulsive interactions.[29] They can scatter only weakly off the repulsive potential. This situation is ideally suited to Fermi liquid theory. Nozieres[26] uses the kinetic (or site transfer) term in the electronic Hamiltonian as a perturbation, and calculates the leading term in the repulsive interaction (between electrons of opposite spin direction) to be of order D^4/J^3. Fermi liquid theory then predicts the form for $\rho(T)$ and χ. Everywhere that T_F would occur, one now inserts T_K, because T_K determines the scale for energy excitations in the Kondo problem. The Fermi liquid approach is only valid at $T << T_K$, because only in that temperature regime is the singlet ground state a correct description. This is because its binding energy is of order k_BT_K, and at temperatures comparable to T_K the use of a singlet state with which to calculate electronic interactions would not be valid.

Nozieres[26] (and Wilson[25]) find for the conductivity

$$\sigma(T) = 1/\rho(T) = \sigma(0)\{1 + (\pi^2/3)(T/T_K)^2\} \quad ; \tag{27}$$

We have taken Noziere's A [his Eqs. (12) and (22), $0 < A < 1$], equal to unity. This is equivalent to setting the energy scale for variation of the phase shift (his α) equal to $1/T_K$. From Fermi liquid theory (using localized spin fluctuations as a guide[30]) the susceptibility can be written as

$$\chi(T) = \chi(0)\{1 - \xi(\pi^2/3)(T/T_K)^2\} \quad . \tag{28}$$

ξ is a constant of order unity. The zero temperature conductivity $\sigma(0)$ is that of the unitarity limit, corresponding to a bound state exactly at ε_F, equivalent to a phase shift of $\pi/2$. The zero temperature result for χ, $\chi(0)$, was calculated by Wilson[25] [Abrikosov and Migdal[31] calculated the first term]

$$\chi(0) = \frac{0.103 \pm 0.003}{D} \{J\rho(\varepsilon_F)\}^{-1/2}$$

$$\times \exp \{\frac{1}{|J\rho(\varepsilon_F)} - 1.58|J\rho(\varepsilon_F)| + \ldots\}$$

$$= C/T_K \quad \text{where} \tag{29}$$

$$k_BT_K = D \exp\{-(1/|J\rho(\varepsilon_F)|) - 1.58|J\rho(\varepsilon_F)| + \ldots\}$$

From Wilson,[25] the ratio of the low temperature impurity specific heat to the low temperature impurity magnetic susceptibility is given by

$$\lim_{T\to 0} [c_V(T)/T\chi(T)] = \frac{2\pi^2}{3} (1 \pm 0.03)k_B^2 \tag{30}$$

More succinctly, the ratio of the relative change in susceptibility to the relative change in specific heat equals[26]

$$\frac{\Delta\chi/\chi}{\Delta C_V/C_V} = 1 + \text{universal dimensionless constant independent of } T_K$$

$$\simeq 2 \text{ (non-interacting electron gas would give 1)} \tag{31}$$

the local moment having been locked into the singlet state.

In conclusion, at low temperatures ($T << T_K$):
For antiferromagnetic exchange ($J < 0$)

1. Magnetic impurities give no energy dependences more singular than ε^2 (equivalently, T^2).
2. $\chi(0)$ is finite.
3. The magnetic impurity behaves as a true singlet (the Nagaoka[32] conjecture).
4. Susceptibility (or other experimental) results exhibiting singular behavior near $T = 0$ are caused by impurity-impurity interaction effects.

For ferromagnetic exchange ($J > 0$)

1. Finite spin S on the magnetic impurity at $T = 0$.
2. Logarithmic correlation functions.

The low temperature properties for both signs can now be analyzed within a theoretical framework which yields exact results for $T = 0$. The method can be developed, using Fermi liquid theory, to yield quantitative relationships at finite temperatures. The transition region ($T \sim T_K$) appears to be presently somewhat inaccessible, and is in need of further analysis. In the interim, extrapolation formulae[21] are available which smoothly connect the low and high temperature regimes. A remarkably rich physical problem appears to have been solved.

REFERENCES

* Supported in part by the National Science Foundation and the U. S. Office of Naval Research, Contract No. N00014-69-0200-4032.

1. P. W. Anderson, Comments on Solid State Physics 5, 73 (1973).
2. N. Abragam and B. Bleaney, Electron Paramagnetic Resonance of Transition Ions (Clarendon Press, Oxford, 1970), p. 404.
3. A. Abragam and B. Bleaney, ibid., p. 445.
4. A. M. Clogston, B. T. Matthias, M. Peter, H. J. Williams, E. Corenzwit, and R. C. Sherwood, Phys. Rev. 125, 541 (1962).
5. A. J. Heeger, Solid State Phys. 23, 283 (1969).
6. Charlotte E. Moore, Atomic Energy Levels (U. S. Government Printing Office, Washington, D. C., 1952), Volume II.
7. P. W. Anderson and A. M. Clogston, Bull. Am. Phys. Soc. 6, 124 (1961).
8. J. Kondo, Progr. Theor. Phys. 28, 846 (1962); ibid., 32, 37 (1964).
9. P. de Faget de Casteljau and J. Friedel, J. Phys. Radium 17, 27 (1956); J. Friedel, Can. J. Phys. 34, 1190 (1956); J. Phys. Radium 19, 573 (1958); Suppl. Nuovo Cimento VII, 287 (1958); A. Blandin and J. Friedel, J. Phys. Radium 19, 573 (1958).
10. P. W. Anderson, Phys. Rev. 124, 41 (1961).
11. This question is discussed at some length by A. Blandin, Magnetism, Vol. 5, Ed. by H. Suhl (Academic Press, N. Y., 1973), p. 58.
12. There are further subtleties. V. Jaccarino and L. R. Walker [Phys. Rev. Lett. 15, 258 (1965)] have shown that in some instances it is not the band properties [$\rho(\varepsilon_d)$ through Δ] which determine the criteria for magnetization, but rather the character of the local environment (number of nearest neighbors of a given species). There have been many confirmed examples of this behavior.
13. J. R. Schrieffer and P. A. Wolff, Phys. Rev. 149, 491 (1966).
14. T. Moriya, Theory of Magnetism in Transition Metals, Ed. by W. Marshall, Enrico Fermi Course 37 (Academic Press, N. Y., 1967), p. 206.
15. D. Davidov, C. Rettori, R. Orbach, A. Dixon, and E. P. Chock, Phys. Rev., accepted for publication, Spring, 1975.
16. H. Hasegawa, Progr. Theor. Phys. (Kyoto) 21, 483 (1959).
17. D. J. Scalapino, Phys. Rev. Lett. 16, 937 (1966).
18. D. K. C. MacDonald, W. B. Pearson, and I. M. Templeton, Proc. Roy. Soc. (London) A266, 161 (1962).
19. H. L. Malm and S. B. Woods, Can. J. Phys. 44, 2293 (1966).
20. A. A. Abrikosov, Physics 2, 5 (1965).
21. M. Daybell, Magnetism, Ed. by H. Suhl, Vol. 5 (Academic Press, N. Y., 1973), p. 121.
22. H. Suhl and D. Wong, Physics 3, 17 (1967).
23. P. W. Anderson and G. Yuval, Phys. Rev. B1, 1522 (1970).
24. P. W. Anderson, G. Yuval, and D. R. Hamann, Phys. Rev. B1, 4464 (1970).
25. K. G. Wilson, Collective Properties of Physical Systems, Novel Symposium 24 (Academic Press, N. Y., 1974), p. 68.
26. P. Nozieres, J. Low Temp. Phys. 17, 31 (1974).
27. These ideas were initially developed for phase transitions - K. G. Wilson and M. E. Fisher, Phys. Rev. Lett. 28, 240 (1972) - and applied most lucidly to the Kondo problem by M. Fowler, Phys. Rev. B6, 3422 (1972).
28. Note that the change of D and J is an artifact. Scaling allows one to devise a "new" material, where the ground state problem can be solved more easily than in the original material. This new material must be so designed, however, that the energy excitation scale is the same as that originally. After a solution is obtained, one scales the solution back to the original material. The entire process is reminiscent of conformal mapping, and its use to solve complicated electrostatics problems by mapping them onto simple shapes.
29. K. Sawada, Phys. Rev. 116, 1344 (1959).
30. N. Rivier and M. J. Zuckermann, Phys. Rev. Lett. 21, 904 (1968).
31. A. A. Abrikosov and A. A. Migdal, J. Low Temp. Phys. 3, 519 (1970).
32. Y. Nagaoka, Phys. Rev. A138, 1112 (1965).

PHASE TRANSITIONS AND CRITICAL PHENOMENA IN MAGNETISM*

M. BLUME
Brookhaven National Laboratory, Upton, New York 11973
and
State University of New York at Stony Brook,
Stony Brook, New York 11794

ABSTRACT

Much of the theoretical and experimental work on phase transitions in the past decade has been concerned with, stimulated by, or tested on magnetic systems, and much of this work has been reported at sessions of this Conference. A review of the development of this field was given, designed primarily for those workers in magnetism who might have been present at conferences on magnetism and magnetic materials, but who did not attend the sessions on critical phenomena.

Recent work has been done against the background of the classical theories of Curie, Weiss, Ornstein, Zernike, and Landau. As a result of Onsager's 1944 solution of the two-dimensional Ising model, it was realized that the singular behavior of different thermodynamic functions such as magnetization, susceptibility, and specific heat, was not necessarily as predicted in the classical theories. Tentative numerical work in the fifties led to the consideration of various critical indices, the exponents which characterize the nature of divergences in the thermodynamic quantities near T_c. Experimental measurements on these indices showed that they agreed much more closely with results of analysis of series expansions than with classical theories. Theoretical developments of the sixties, which were presented at CMMM sessions, included inequalities and equalities relating the indices - the scaling relations. Experimental tests of these relations on magnetic systems were also reported. The most recent developments have included the theoretical calculation of these indices by the renormalization group theories of Wilson and Wilson and Fisher. These theories have shed considerable light on the nature of the singularities near critical points and have given some indication of the factors which affect the singularities.

Attention at these conferences in the last few years has been turned to more complex types of critical points (e.g. tricritical points) which can be found in antiferromagnets in external fields, and we can expect close experimental and theoretical examination of the phase diagrams of magnetic systems in external fields.

It should be emphasized that the relationship between the two fields of study, critical phenomena and magnetism, is symbiotic and not parasitic. There has been and will continue to be much beneficial cross-fertilization between them, as advances in one stimulate developments (both theoretical and practical) in the other, and vice versa.

* Work at Brookhaven National Laboratory supported by U. S. Atomic Energy Commission, and at Stony Brook by National Science Foundation.

THE CONFIGURATION-BASED APPROACH TO MAGNETIC IMPURITIES IN METALS+

L. L. Hirst

Institut für Theoretische Physik der Universität Frankfurt,
6 Frankfurt a.M., Germany

+Project of the Sonderforschungsbereich für Festkörperspektroskopie
Darmstadt-Frankfurt, supported by Deutsche Forschungsgemeinschaft.

ABSTRACT

In the configuration-based approach to 3d or 4f impurities in metals, the bare impurity states are taken as conventional ionic-type several-electron states, corresponding to well-defined configurations $3d^n$ or $4f^n$ with intraconfigurational splitting into L-S terms, crystal-field levels, etc. The condition for the impurity to remain "magnetic" as the interaction with the conduction electrons is switched on is that the stabilization energy of the bare-impurity ground state, relative to levels belonging to higher configurations, be large compared to the mixing width. Under this condition, the ionic-type intraconfigurational structure of the impurity also continues to be defined. Although a different historical line of development hindered its recognition, there is good evidence for such ionic-type structure in 3d as well as 4f impurities. Our conclusion is that a configuration-based description is the most physical one for a magnetic impurity. This conclusion has interesting implications for the theory of magnetism in concentrated metallic systems.

I. INTRODUCTION

A potentially magnetic impurity in a metal may reasonably be described by a generalized Anderson model of the form [1,2,3]

$$H = H_{ion} + H_{cond} + H_{mix} \tag{1}$$

Here H_{ion} describes the localized interactions at the impurity site; H_{cond} is the energy of non-interacting conduction electrons; and H_{mix} is a one-electron mixing interaction.

$$H_{ion} = -V n + U n(n-1)/2 + H_{intra} \tag{2}$$

Here n is an operator for the total number of 3d (or 4f) electrons. V is a binding potential at the impurity site, yielding an energy proportional to n. U is the leading part of the 3d-3d Coulomb interaction and yields an energy proportional to the number of 3d pairs. H_{intra} symbolizes the intra-configurational splitting energies, which are responsible for formation of L-S terms, crystalline electric field (CEF) levels, spin-orbit levels, etc. H_{mix} may be written as

$$H_{mix} = V_{mix} \sum_{km\sigma} a^{+}_{m\sigma} c_{km\sigma} + h.c. \tag{3}$$

where a^{+} and c are creation and annihilation operators for 3d and conduction electrons respectively. We use a partial-wave description for the conduction electrons, and the mixing interactions conserve orbital angular momentum [2,3].

In the above Hamiltonian, H_{ion} describes a tendency toward formation of ionic-type localized states, belonging to definite configurations, $3d^n$. The mixing interaction tends to break up the configurations and mix the 3d electrons into the conduction band. The basic physics of the impurity problem is a competition between these two tendencies. A natural measure of the mixing strength is

$$\Delta = \pi |V_{mix}|^2 \rho(\epsilon) \tag{4}$$

where $\rho(\epsilon)$ is the density of conduction-electron states. The Coulomb energy U may be taken as a measure of the strength of H_{ion}. For the potentially-magnetic impurities in which we are interested, U is larger than Δ. The physically appropriate approach in such a case is to deal with H_{ion} before considering mixing effects. That is, we take several-electron bare impurity eigenstates, which belong to definite configurations and include appropriate intraconfigurational structure according to a shell-model description, as our starting point, both computationally and conceptually.

II. CONFIGURATIONAL ENERGIES

The configuration-based energy-level scheme, in the absence of mixing interactions, is indicated in Fig. 1. We assume the configuration n to be stable and show in addition the neighboring configurations n±1, with intraconfigurational splittings indicated schematically. On the right we show the conduction-electron states, which are occupied up to the Fermi energy. The conduction states are one-electron states obeying Fermi statistics, whereas the impurity states are several-electron states obeying Boltzmann statistics (one and only one impurity state is occupied at any time). This causes the energy scheme of Fig. 1 to have a few unconventional properties, which may be slightly confusing when first encountered. The main point to be emphasized is that only the differences between energies of impurity levels, not the absolute energies, can have physical significance, since the number of impurities is conserved. Thus, in Fig. 1 the position of the impurity's ground level relative to the Fermi level is arbitrary and without physical significance.

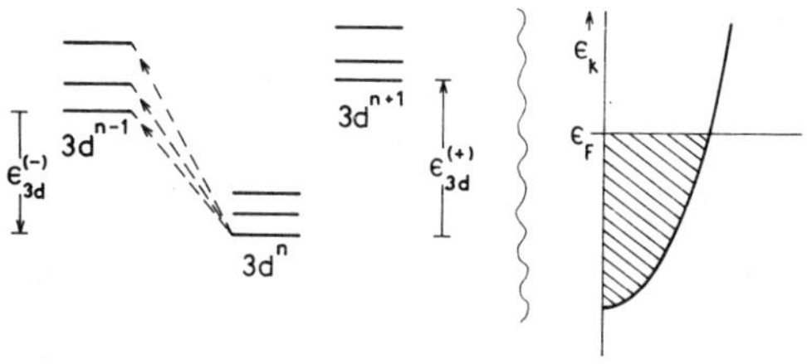

Fig. 1. Configuration-based energy levels of an ion with stable configuration n . Dashed lines are XPS transitions.

An important recent development is the use of X-ray photoemission spectroscopy (XPS) to directly measure the separation between impurity levels belonging to different configurations [4-6]. In Fig. 1 the ion is initially in the intraconfigurational ground level of its stable configuration, n . A high energy X-ray photon knocks an electron out of the 3d or 4f shell, leaving the ion in any of the levels of the configuration n-1 . The possible transitions between impurity states are indicated by dashed lines in Fig. 1. By catching the emitted photoelectron and measuring its energy, the corresponding impurity-level separations may be determined. Measurements performed on 4f materials show well-resolved lines corresponding to the various intraconfigurational levels in the final state.

Assuming the impurity initially to be in the lowest level of configuration n , the minimum interconfiguration excitation energy needed to reach the configuration n-1 is obtained when the 3d (or 4f) electron removed from the impurity is put into the conduction band at the Fermi energy. The required energy is

$$E^{(-)}_{exc} = \epsilon_F - [E_n - E_{n-1}] \approx \epsilon_F - [-V + (n-1)U] \tag{5}$$

where E_n or E_{n-1} denotes the energy of the lowest level within each configuration, and where the approximate expression at the right is obtained by neglecting the intraconfigurational energies. Similarly, the minimum energy needed to add a 3d electron is

$$E^{(+)}_{exc} = [E_{n+1} - E_n] - \epsilon_F \approx [-V + nU] - \epsilon_F \tag{6}$$

In the zero-mixing limit, the condition for the configuration n to be stable is that both these energies be positive.

If we start from a situation in which n is stable, by decreasing V we eventually reach n-1 as the stable configuration. The configuration crossover, where the two configurations are equally stable, is of particular interest. In Fig. 2 we show the energy-level scheme and the allowed XPS transi-

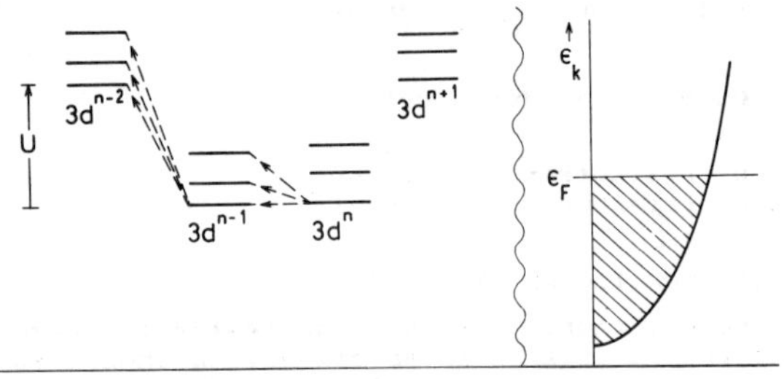

Fig. 2. Configuration-based energy levels of an ion at the crossover between configurations n and n-1. Dashed lines are XPS transitions. (This diagram, like Fig. 1, presumes the Fermi energy to have been chosen as the zero of one-electron energies.)

tions for this case. Since the incoming photon may find the ion in the ground level of either configuration, one expects two families of lines, corresponding to the transitions $n \to n-1$ and $n-1 \to n-2$. Such spectra have in fact been observed in 4f systems [6]. Measurements of XPS at the configuration crossover are very important because they permit a direct determination of the Coulomb interaction U, which is so important for impurity theory.

III. EFFECTS OF MIXING

We stated above that the quantity Δ, eq.(4), is an appropriate measure of the mixing strength. More precisely, Δ is a measure of the rate at which mixing transitions occur when the configurational energetics do not interfere. Consider again the impurity of Fig. 1, and suppose that it has been excited into the unstable configuration n-1, for example by a high-energy photon in an XPS experiment. The impurity will then spontaneously absorb a conduction electron via the mixing interaction to return to the stable configuration n. The rate at which this occurs is $\sim 2\Delta/\hbar$, and the corresponding lifetime width (HWHM) is $\sim\Delta$. This is the ideal linewidth for an XPS experiment, although in practice the instrumental widths are dominant. (A more detailed discussion yields differences among the widths of various excited levels, which perhaps can be measured in the future.)

By contrast, if the impurity of Fig. 1 is initially in its ground state, it cannot make a spontaneous mixing transition to an excited-configuration level, for reasons of energy conservation. The lifetime broadening $\sim\Delta$ therefore applies only to those impurity levels which are energetically unstable relative to levels belonging to other configurations. This includes all levels belonging to excited configurations, but may also apply to the higher intraconfigurational levels of the ground configuration. It might be possible to detect the latter effect in optical absorption experiments on 4f substances near the configuration crossover.

Within the lowest levels of a stable configuration the lowest-order mixing processes are "frozen out" by energy-conservation requirements. However, there remain higher-order processes in which the impurity reaches another ground-configuration state via virtual transitions to the excited configurations. These may be treated by a generalized Schrieffer-Wolff transformation [7,8]. The result is an effective coupling between impurity and conduction electrons which is analogous to an exchange coupling, but in the realistic case of a 3d or 4f impurity contains orbital as well as spin coupling terms. The effective coupling strength is

$$I = |V_{mix}|^2 / E_{exc} \qquad (7)$$

where

$$E_{exc} = \left[\left(E^{(+)}_{exc}\right)^{-1} + \left(E^{(-)}_{exc}\right)^{-1} \right]^{-1} \qquad (8)$$

The effective widths of the low-lying levels of a stable impurity configuration are given by expressions which are 2nd order or higher in I. For example, the generalized Korringa width [9,10] is $\sim |I\rho(\epsilon_F)|^2\, kT = (\Delta/\pi E_{exc})^2\, kT$. Such widths are many orders of magnitude smaller than Δ, which is the reason why magnetic resonances can be observed in such impurities [3,9,10].

Let us now define more carefully what is meant by configurational stability in the presence of a finite mixing strength. The essential point is that spontaneous mixing transitions are suppressed by energy-conservation requirements. In order to achieve this effect, the interconfiguration excitation energy has to be not merely positive but larger than the mixing strength Δ. We may write this condition as

$$\Delta < E_{exc} \qquad (9)$$

The stable configuration of the isolated impurity ion may or may not be stable in the sense of eq.(9). If it is not, then no single configuration is predominant in the many-body ground state of the total system.

It is useful to consider a configurational "phase" diagram of impurity behavior, Fig. 3, obtained by plotting eq.(9) in the plane of binding potential V and mixing strength Δ. For $\Delta=0$, corresponding to the isolated impurity, each configuration is stable over a certain range of V, with the crossovers occurring approximately at V=0, U, 2U,... For finite Δ, the configuration-stability condition (9) is most easily satisfied for a V value lying at the middle of the stability range of the isolated impurity, in which eq.(9) becomes $\Delta < U/4$. As one approaches a V value corresponding to configuration crossover of the isolated ion, the maximum allowed Δ approaches zero.

Since eq.(9) has only a qualitative significance, the "phase" boundaries cannot be taken as implying sharp transitions in impurity behavior. Indeed, we know from uncertainty-principle arguments that sharply-defined phase transitions cannot occur in a microscopic system. The "phase" boundaries do however represent a useful nominal dividing line between local-moment behavior and other types of behavior. as will now be discussed.

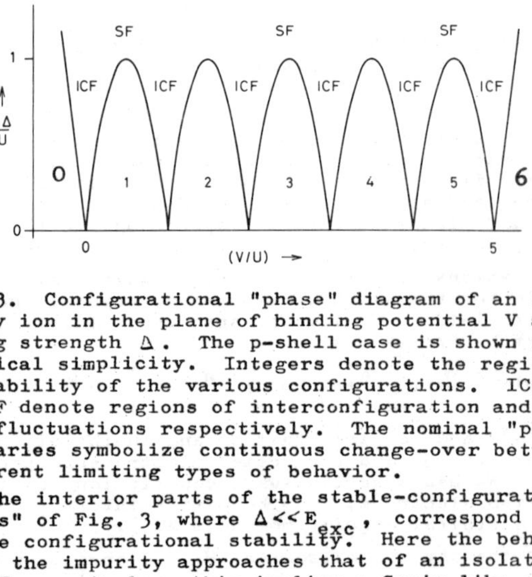

Fig. 3. Configurational "phase" diagram of an impurity ion in the plane of binding potential V and mixing strength Δ. The p-shell case is shown for graphical simplicity. Integers denote the regions of stability of the various configurations. ICF and SF denote regions of interconfiguration and spin fluctuations respectively. The nominal "phase" boundaries symbolize continuous change-over between different limiting types of behavior.

The interior parts of the stable-configuration "phases" of Fig. 3, where $\Delta \ll E_{exc}$, correspond to extreme configurational stability. Here the behavior of the impurity approaches that of an isolated ion. In particular, this implies a Curie-like susceptibility (except for the full- or empty-shell configurations, and aside from special effects of CEF or spin-orbit interactions in a few 4f cases.) Consequently, we may say that stabilization of a single configuration is the condition for local-moment behavior of an impurity.

(It should be emphasized that the Hund couplings, which tend to align the 3d or 4f electrons to produce maximum L and S, are not essential for the qualitative appearance of a local moment. It is easy to show that an isolated ion with a stable configuration but zero intraconfigurational splitting has a Curie susceptibility with an effective moment comparable to that obtained when Hund coupling is included.)

Outside the regions of configurational stability we distinguish two cases. In the upper part of the "phase" diagram, say for $\Delta > U/4$, the mixing is so strong as to effectively break down the configurational energy-level scheme. We may call this the spin-fluctuation (SF) regime, since here it is reasonable to use a SF theory in the sense of [11], in which the Coulomb and other intra-ionic interactions are treated in perturbation theory. In the regions lying between the stable-configuration "phases", we have the interconfiguration fluctuation regime. Here, because we are near the crossover between two configurations (say n and n-1), we have spontaneous fluctuations between the two, driven by the mixing interaction. However, the mixing strength Δ is not large compared to U, so the various ionic configurations remain well-defined, and so does the intraconfigurational level structure on a scale coarser than Δ. The physically appropriate approximation in this case would be to project away all impurity states belonging to the configurations other than n and n-1. A full quantitative theory along these lines has not been found, mainly because the Boltzmann statistics of the configuration-based impurity states and the Fermi statistics of the conduction-electron states tend to resist simultaneous treatment. However, one can make a semi-quantitative treatment of such ICF systems by taking theoretical expressions for the isolated ion, averaging over the two configurations with appropriate weightings, and inserting the mixing width Δ "by hand" where physical arguments indicate it should appear [12,13]. For example, for an ICF impurity we expect a Curie susceptibility C/T to be replaced approximately by $C/(T+T_{ICF})$, where $T_{ICF} \sim \Delta$. In the case of 4f impurities, $\Delta \sim .01$ eV ≈ 100 K is small enough to permit this temperature dependence, as well as more subtle dependences resulting from the intraconfigurational splittings, to be observed. For 3d impurities, on the other hand, we expect $\Delta \sim 1$ eV $\approx 10{,}000$ K, implying a very small susceptibility without appreciable temperature dependence for the regions where an ICF (or also SF) theory is appropriate. Therefore, when Curie-Weiss fits for 3d impurities indicate characteristic temperatures in the range of hundreds of degrees or less, this indicates that the impurity is in the range where a Kondo theory is appropriate, and the characteristic temperature should, in our view, be interpreted as a Kondo temperature.

The Kondo problem contemplates an impurity in the interior of a stable-configuration "phase", for which a Schrieffer-Wolff effective k-d coupling may be used. An estimate for the ratio of Kondo temperature to Fermi temperature is

$$-\ln(T_K/T_F) \sim 1/I\rho(\epsilon_F) = \pi E_{exc}/\Delta \qquad (10)$$

The right-hand form indicates that the Kondo temperature becomes large as one moves from the interior toward the boundary of the stable-configuration "phase". However, eq.(10) becomes inapplicable near the boundary, where the Schrieffer-Wolff transformation is poorly justified. As the boundary is approached vertically or laterally, we expect the Kondo temperature to go over continuously into a spin or interconfiguration-fluctuation temperature, T_{SF} or T_{ICF}, which in either case is $\sim\Delta$. (The present question of how the Kondo temperature varies within a given stable-configuration "phase" should be distinguished from the question of how it depends upon the atomic number of different impurities in a given host, for which reliable estimates are much more difficult 2,14 .)

Where do real impurities lie in the configuration "phase" diagram? For 4f impurities, where $\Delta/U \sim 10^{-3}$, one is very near the V axis. Fig. 3 indicates that configuration stability is by far the most probable, but configuration crossover with ICF behavior is also possible [15]. For 3d impurities, it appears that Δ/U can sometimes be large enough to put the system into the SF region, where practically non-magnetic behavior results. This is probably the case for 3d impurities in such hosts as Al, where the large density of states is expected to yield a large Δ. However, cases such as Mn in noble-metal hosts, where near-ideal local-moment behavior is observed, show that Δ/U can be small enough to put a 3d impurity well into a stable-configuration "phase". For the remaining 3d impurities in the noble metals, one observes a gradual diminution of local-moment behavior as the end of the transition series is approached on either side of Mn [16,17]. On the basis of EPR and other evidence, the configurations of the strongly-magnetic impurities can be identified as those corresponding to the divalent ions [9,10]. However, Cu metal (and hence a Cu "impurity" in a Cu "host") is believed to be essentially monovalent. This suggests that the Mn impurities are located near the center of the n=5 stable-configuration "phase", and that the sequence Fe, Co, Ni, Cu moves toward and through the configuration crossover, so that Ni lies in the 8+9 crossover region and Cu lies in the n=10 "phase". We find this explanation of the 3d systematics more plausible than an alternative one in which the Kondo temperatures are estimated under the assumption that Δ and E_{exc} are the same for all impurities in a given host [2,14].

Such interpretations are necessarily somewhat speculative at present. Perhaps in the future it will be possible to test them by means of XPS or by Mössbauer isomer shifts.

IV. VIRTUAL BOUND STATES

The first theoretical discussions of magnetic impurities in metals approached the problem from the point of view of the virtual bound state (v.b.s.) [1,18,19]. In the present section we shall briefly discuss the relation between Anderson's Hartree-Fock (HF) version of the v.b.s. theory [1] and the configuration-based approach. For reasons which will become clear, we consider an orbitally degenerate 3d impurity, as opposed to the more familiar hypothetical s-shell impurity. The HF solutions for the orbitally-degenerate case are rather complicated, but much of the complication can be removed by omitting all intraconfigurational splittings. This is a reasonable thing to do, since as noted above, and as emphasized by Anderson, these splittings are not essential to local-moment formation.

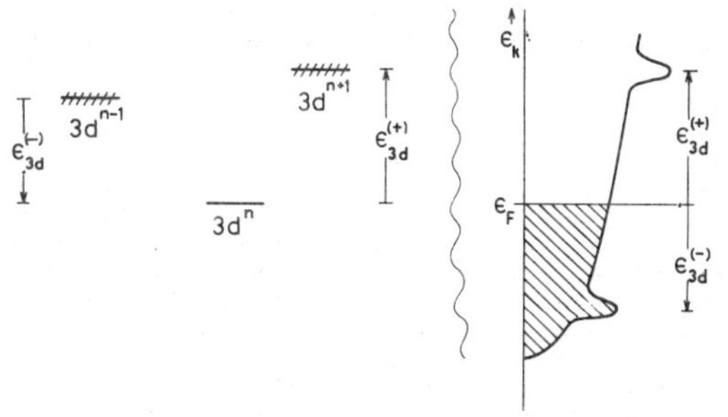

Fig. 4. Comparison between configuration-based and virtual-bound-state levels for a strongly-magnetic 3d impurity. The humps below and above the Fermi surface on the right symbolize n and 10-n degenerate v.b.s. respectively.

The HF equations yield 10 v.b.s. energies, $E_{m\sigma}$. When Δ/U is sufficiently small, the absolutely stable solutions are found [20] to be of the type shown in Fig. 4, where there is a spontaneous HF splitting between n v.b.s. at a lower energy and 10-n v.b.s. at a higher energy. Such a HF solution type will be called a protoconfiguration n, since it resembles the ionic configuration $3d^n$ and reduces to it in the limit $\Delta = 0$. Any permutation of the indices $m\sigma$ yields a solution belonging to the same protoconfiguration and having the same energy. This can be interpreted as Curie paramagnetism, where the distinct solutions obtained by permutations of $m\sigma$ correspond to different orientations of the local moment. In addition to these magnetic solutions, there are non-magnetic solutions, in which all 10 v.b.s. energies are equal, which become stable at larger Δ

The regions of the V, Δ plane in which the various HF solution types are absolutely stable are shown in Fig. 5, which may be compared to Fig. 3. The HF theory works well in the outer parts of the plane, where it yields non-magnetic solutions corresponding to the spin fluctuations of Fig. 3. In the stable-configuration regions, the HF solution is only partially successful, since it exhibits a physically correct tendency toward configuration stabilization but has other shortcomings to be discussed later. In the regions of configuration crossover, the HF solution fails completely, since instead of the ICF behavior described above it yields an unphysical, discontinuous jump from one protoconfiguration to the next. This failure of the HF theory suggests that the chances of extracting a satisfactory theory of ICF from a higher-order fermion-based perturbation development are slim. It also means that the oft-quoted Anderson criterion for existence of a local moment, $\pi\Delta < U$, should be replaced by the more stringent condition (9). The latter agrees approximately with the Anderson condition at V values corresponding to maximum configurational stability, but shows that moment formation becomes increasingly difficult as a crossover is approached. These difficulties of the HF solution near the configuration crossover are not seen in calculations for a hypothetical s-shell impurity, where there is only one non-trivial configuration.

In [1] it was remarked that under certain circumstances the HF solutions for an orbitally-degenerate impurity yield no orbital magnetization. This remark has been widely misunderstood as implying that the v.b.s. theory contains some intrinsic mechanism for quenching orbital magnetization. Actually, the effect in question is a secondary and relatively weak effect of the intraconfigurational Hund splittings, which arises only near the onset of magnetization and (as a more careful analysis shows [20]) not necessarily even there. Since this onset zone is precisely where the HF theory is least reliable, the effect should not be taken very seriously. There is actually no qualitative conflict between the configuration-based and v.b.s. theories as concerns orbital magnetization: both agree that it can exist in the strongly-magnetic, stable-configuration regime. In our view there is clearcut experimental evidence that some strongly-magnetic 3d impurities do carry orbital magnetization, as expected [9,10].

One of the early apparent successes of the v.b.s. theory was an explanation [18,19] of the residual resistivities of strongly-magnetic 3d impurities, which seemingly followed a double-peaked curve as a function of atomic number. More recently, doubt has been expressed as to whether this should be regarded as the true experimental behavior [17]. Regardless of this, and regardless of possible shortcomings of the HF approximation, we must point out that the v.b.s. theory does not in general yield such behavior, at least if one accepts [1] as being the correct quantification of it. The explanation of the resistivities was based on the idea of v.b.s. with an exchange splitting according to spin orientation, with 5 v.b.s. of (say) spin up lying below 5 v.b.s. of spin down, and with the 10 levels moving as a whole toward lower energies with increasing atomic number so as to accommodate more 3d electrons. The HF solutions of Fig. 5 show a quite different splitting scheme, which cannot be put into correspondence with spin orientations. As a function of increasing V (equivalent to increasing atomic number), the v.b.s. fall one by one from above to below the Fermi level, by discontinuous jumps occurring at the boundaries between different protoconfigurations.

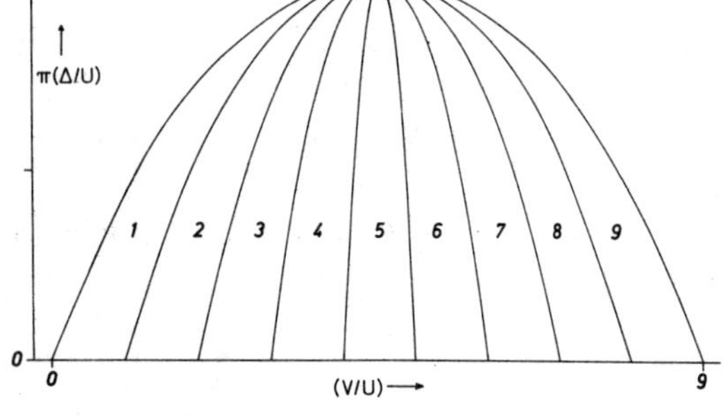

Fig. 5. "Phase" diagram for 3d impurity without intraconfigurational splitting, according to HF v.b.s. theory. Integers denote solutions of protoconfiguration type, which are Curie paramagnetic. Elsewhere unsplit, non-magnetic solutions are stable. All boundaries correspond to discontinuous changes in the effective moment and other properties.

Although the v.b.s. theory gives a qualitatively correct indication of regions of strong configuration stabilization, it does not provide a satisfactory framework for more detailed discussions of impurity properties within these regions. This is because the fermion-based v.b.s. theory corresponds to a perturbation scheme which is physically inappropriate in such regions. The comparison in Fig. 4 shows that the v.b.s. correspond to fermion addition or removal excitations between different impurity configurations. The width $\sim\Delta$ of these excitations really corresponds to a width of the excited-configuration levels, with the ground-configuration levels remaining sharp; but it is simply impossible to translate this important distinction into v.b.s. language. It is also impossible to translate intraconfigurational splittings into the v.b.s. description in a satisfactory way. Rather than trying to patch up the v.b.s. picture with heuristic arguments, one should recognize the fundamental nature of these shortcomings and learn instead to think in terms of configurational levels.

V. MAGNETISM IN CONCENTRATED SYSTEMS

When magnetic impurities first came into style, it was often said that they were being studied as a step toward the understanding of magnetism in transition metals. The impurity problem has proved a hard nut to crack, and one does not hear this remark so often any more. Nevertheless, we believe that the configuration-based approach to impurities does yield insights which should be valuable in future work on transition metals.

Many have discussed the tendency for a few configurations to be favored in a 3d ion in a metal [21-23], but few are willing to believe that this tendency could go so far as to permit a single configuration to become predominant. But according to the analysis we have given above, the experimentally observed strongly magnetic behavior of 3d impurities in certain metals shows precisely that in these cases a single configuration does predominate. This, in our opinion, is the important message which impurity studies bring to the transition-metal problem. This encourages us to take a new look at magnetism in concentrated systems from the configuration-based point of view. As in the impurity case, we start from the limit of zero mixing (and hopping) strength.

The configurational energy balance has the following new feature for concentrated ions as opposed to impurities: in order to complete the configuration crossover from n to $n-1$, it is necessary to transfer a finite concentration of electrons to the conduction band. This raises the Fermi level by a finite amount, which by eq.(5) means that a finite change in V is necessary to complete the crossover. If we assume for simplicity a constant density D of conduction-electron states, then a change of $1/D$ in V is necessary to complete the crossover, while each configuration is stable over a range U. This simple model suffices for our present purposes, although the energetics of the crossover can be more complicated in reality [24].

We now add interactions capable of transferring electrons to or from the magnetic ions, characterized by a total strength $W = \Delta + |h|$ which is the sum of mixing and hopping strengths. The discussion of configurational stability in the presence of finite W is similar to the impurity case, and yields the configurational "phase" diagram shown in Fig. 6. As in the impurity case, the "phase" boundaries are of qualitative significance and do not in general correspond to sharp transitions in the behavior of the system.

Of special interest are the regions of configuration crossover, which occur when W/U is small and V lies in an appropriate crossover range. It can be shown [24] that the filling of the conduction band proceeds in such a way that the interconfiguration excitation energy remains zero, so that spontaneous ICF between the two competing configurations occur with a rate $\sim W/\hbar$. A comparison of Fig. 6 to Fig. 3 suggests that ICF behavior should be found more often in concentrated as compared to dilute systems, and indeed many examples have been found. For 4f examples see e.g. [5], [12], and several papers at this conference. The basic mechanism of the metal-insulator transition in 3d oxides appears to be configuration crossover, although a different terminology is usual [25].

It has long been recognized that in 3d systems there is a competition between local and non-local behavior [23]. This is reflected in discussions of "local" vs. "itinerant" magnetism, and also in discussions of the metal-insulator transition. Our configurational "phase" diagram amounts essentially to a classification of systems according to the amount of localization they have, as will now be discussed.

The stable-configuration regime corresponds to extreme local behavior, in which each ion has a conventional ionic-type energy-level structure. This typically corresponds to a well-defined but freely-orientable local moment on each ion. Magnetic order arises as an alignment of these local moments, but the moments themselves continue to be well-defined in the paramagnetic regime.

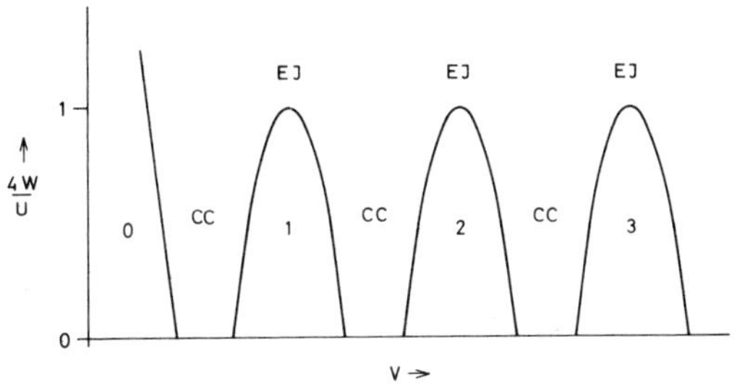

Fig. 6. Configurational "phase" diagram for concentrated 3d or 4f system, in plane of binding potential V and mixing + hopping strength W. Integers denote regions of configurational stability, CC denotes configuration crossover, EI denotes extreme itinerant. Diagram continues symmetrically further to the right. Breadths of stable-configuration and CC regions at W=0 are U and 1/D, assuming density D of conduction-electron states to be constant. The "phase" boundaries symbolize continuous changeover between different limiting types of behavior.

The extreme itinerant regime corresponds to the spin-fluctuation regime in the impurity case. Here the transfer interactions are so strong that the configuration-type local energy-level structure is effectively broken down. The physically appropriate approach is to take itinerant 3d electrons as the starting point and treat the local intra-ionic interactions as perturbations. Extreme itinerant models can yield ferromagnetism; but we do not believe that such models, in the ferromagnetic range, provide a realistic description of the tight-binding magnetic materials actually occurring in nature. Even the most optimistic extreme itinerant approach, the Stoner theory, requires $U \gtrsim W$ for ferromagnetism to occur. But this implies that one is getting into the area where the configurational tendencies are important.

The configuration crossover regime yields an interesting combination of local and itinerant behavior. The ionic configurations, and also the intraconfigurational splitting on a scale coarser than W, continue to be well defined; but there are 3d-like fermion excitations lying at at the Fermi energy, implying itineracy. As in the impurity case, the presence of ICF implies that well-defined local moments do not exist. Such a system can still show spontaneous magnetization, but this must occur as an intrinsically collective, non-local effect, not merely as an ordering of pre-existing local moments.

Configuration crossover provides a third simple limiting regime in addition to the extreme itinerant and extreme local regimes. The existence of this regime invalidates many previous phenomenological arguments as to the value of W/U, and suggests that this ratio could be smaller than generally believed. We believe that Ni and other demonstrably itinerant 3d ferromagnets can best be classified as configuration-crossover systems. Although results are not yet available, quantitative calculations for this regime seem feasible. We expect such calculations to yield a picture of 3d metals, and in particular of their fermion excitation spectra, considerably different from the conventional one.

Having earlier insisted that 3d ions can show magnetism of extreme local type even in a metallic environment, we want now to emphasize that concentrated 4f systems may sometimes show magnetism of an itinerant type. This could occur for configuration-crossover 4f systems, provided that effects of well-isolated singlet intraconfigurational ground states do not interfere. Despite the difference of two orders of magnitude in the interaction strengths W, and hence in the typical ordering temperatures, such systems would have considerable similarity with itinerant 3d metals. Their investigation is therefore of particular interest.

References

1. P. W. Anderson, Phys. Rev. 124, 41 (1961).
2. J. R. Schrieffer, J. appl. Phys. 38, 1143 (1967).
3. L. L. Hirst, Phys. kondens. Materie 11, 255 (1970)
4. S. B. M. Hagström, P. O. Hedén, and H. Löfgren, Solid State Comm. 8, 1245 (1970).
5. M. Campagna, E. Bucher, G. K. Wertheim, D. N. E. Buchanan, and L. D. Longinotti, Phys. Rev. Letters 32, 885 (1974).
6. M. Campagna, E. Bucher, and G. K. Wertheim, this conference.
7. B. Coqblin and J. R. Schrieffer, Phys. Rev. 185, 847 (1969).
8. L. L. Hirst, Z. Physik 244, 230 (1971).
9. L. L. Hirst, Adv. Phys. 21, 759 (1972).
10. L. L. Hirst, Archives des Sciences 27, 136 (1974)
11. P. Lederer and D. L. Mills, Phys. Rev. Letters 20, 1036 (1968).
12. M. B. Maple and D. Wohlleben, Phys. Rev. Letters 27, 511 (1971).
13. D. Wohlleben and B. R. Coles, in "Magnetism" (Ed. H. Suhl), Vol. V, Academic Press, New York, 1973.
14. L. L. Hirst, Intern. J. Magnetism 2, 213 (1972).
15. V. Allali, P. Donzé, and A. Treyvaud, Solid State Comm. 7, 1241 (1969).
16. M. D. Daybell and W. A. Steyert, Rev. Mod. Phys. 40, 380 (1968).
17. M. D. Daybell and W. A. Steyert, J. appl. Phys. 40, 1056 (1969).
18. J. Friedel, Nuovo Cimento Suppl. 7, 287 (1958).
19. A. Blandin and J. Friedel, J. Phys. Rad. 20, 160 (1959).
20. L. L. Hirst, unpublished.
21. J. H. van Vleck, Rev. Mod. Phys. 25, 220 (1964).
22. N. H. Mott, Adv. Phys. 13, 326 (1964).
23. C. Herring, "Magnetism" (Ed. G. T. Rado and H. Suhl), Vol. IV, Academic Press, New York, 1966.
24. L. L. Hirst, J. Phys. Chem. Solids 35, 1285 (1974)
25. L. M. Falicov, C. E. T. Goncalves da Silva, and B. A. Huberman, Solid State Comm. 10, 455 (1972) and references therein.

ELECTRONIC STRUCTURE AND LATTICE CONSTANT OF INVAR ALLOYS

Junjiro Kanamori, Yoshihiro Teraoka and Takeo Jo
Department of Physics, Faculty of Science, Osaka University, Toyonaka 560, Japan

ABSTRACT

A theory of the volume striction in ferromagnetic transition metal alloys is developed on the basis of the Alexander-Anderson-Moriya theory of the exchange interaction. The volume derivative of the interaction between a pair of atoms is calculated under the assumption that a volume change affects the state density of the s band and 3d atomic levels. In fcc Ni-Fe the number of 3d minority spin electrons of a Fe atom increases with increasing Fe concentration. The volume derivative of the interaction between two Fe atoms shows an anomalous dependence on the Fe concentration through the change of the number of 3d minority spin electrons which explains the large spontaneous volume striction, lattice constant anomaly and negative thermal expansion at low temperatures in invar alloys. The possibility of finding Fe atoms with magnetic moment antiparallel to bulk magnetization and its relevance to the lattice anomalies are discussed also.

I. INTRODUCTION

The invar alloys, in particular, Ni-Fe and Pt-Fe with about 70% of Fe are known to have a large positive spontaneous volume striction in the ferromagnetic phase[1,2] The lattice constant of fcc $Ni_{1-x}Fe_x$ decreases with increasing x in this invar region, whereas it increases linearly with increasing x in the range $x<0.5$[3] The thermal expansion coefficient is negative at low temperatures for $x>0.5$; its absolue magnitude at a given temperature reaches maximum at $x=0.65$ with varying x[4] There are many theoretical discussions about the origin of these anomalies based either on the localized moment picture[5,6,7] or on the rigid band approximation[8,9] In this paper we start with the Anderson model to show that these anomalies arise from the interaction between neighboring Fe atoms. The interaction energy, which we calculate by extending the theory of the interaction between two impurity atoms given by Alexander and Anderson[10] and Moriya;[11] comprises both the exchange energy and the band structure energy of the d band. In this calculation we assume that all Fe atoms are in the same magnetic state with magnetic moment parallel to bulk magnetization. We then examine the possibility of more-than-one states of Fe atoms proposed first by Weiss[6] by use of an extension of the coherent potential approximation(CPA). A preliminary numerical calculation indicates the possibility that the deviation of the saturation magnetization from the Slater-Pauling curve for $x>0.5$ in Ni-Fe alloys might be caused by the appearance of Fe atoms with antiparallel moment. Implications of the result are discussed briefly.

In section II we derive an expression of the interaction between two atoms on the basis of the pseudo Greenian formalism[12] Then we derive an expression of the total free energy of a given system. In section III we calculate the pressure arising from the interaction. We discuss the application of the theory to invar alloys in section IV. Section V deals with comparison of the calculation to the experimental data. Section VI is devoted to the discussion of the possibility of more-than-one states of Fe atoms. A discussion on the condition of obtaining 'invar' alloys is given in section VII.

II. THE TOTAL FREE ENERGY

We start with five d orbitals of each atom and the s band corresponding to the OPW states of a given alloy. Taking into account the s-d mixing, we can derive the following approximate expression for the density of states, $\rho(\varepsilon)$, for a given energy ε[12]

$$\rho(\varepsilon) = \rho_s(\varepsilon) + \rho_d(\varepsilon) \text{ with} \tag{1}$$

$$\rho_d(\varepsilon) = -(1/\pi)\mathrm{Im}(\partial/\partial\varepsilon)\log\det(W + T), \tag{2}$$

where $\rho_s(\varepsilon)$ is the density of states of the s band; W and T are matrices whose rows and columns refer to the 3d states of constituent atoms. W is defined to be a diagonal matrix with the elements given by

$$W_{j\mu\sigma} = \varepsilon - \varepsilon_{j\sigma} + i\Delta , \tag{3}$$

where μ specifies one of the 3d orbitals of the j-th atom; σ denotes the spin state; $\varepsilon_{j\sigma}$ represents the atomic energy level determined in the Hartree-Fock approximation; Δ is the half-width of the virtual bound level due to the s-d mixing; we assume, for simplicity, that $\varepsilon_{j\sigma}$ and Δ are independent of μ, omitting the subscript μ in $W_{j\mu\sigma}$ hereafter. T represents a non-diagonal matrix which has novanishing matrix elements between a pair of d orbitals of atoms at different sites with the same spin. Expanding $\rho_d(\varepsilon)$ in powers of T, we obtain up to the second power of T,

$$\rho_d(\varepsilon) = 5\textstyle\sum_{j\sigma}\rho_{j\sigma}(\varepsilon) + \rho_{int}(\varepsilon) \tag{4}$$

with $\rho_{j\sigma}(\varepsilon)$ and $\rho_{int}(\varepsilon)$ defined by

$$\rho_{j\sigma}(\varepsilon) = (\Delta/\pi)/\{(\varepsilon - \varepsilon_{j\sigma})^2 + \Delta^2\} \quad \text{and} \tag{5}$$

$$\begin{aligned}\rho_{int}(\varepsilon) &= 25(1/\pi)\mathrm{Im}(\partial/\partial\varepsilon)\textstyle\sum_{<jk>,\,\sigma}T_{jk}W_{k\sigma}T_{kj}W_{j\sigma} \\ &= -25(\partial/\partial\varepsilon)\textstyle\sum_{<jk>,\sigma}T_{jk}T_{kj}(\rho_{j\sigma}-\rho_{k\sigma})/(\varepsilon_{j\sigma}-\varepsilon_{k\sigma}),\end{aligned} \tag{6}$$

where $\rho_{j\sigma}(\varepsilon)$ is obtained from $W_{j\mu\sigma}$ defined by Eq.(3) with the assumption that Δ is independent of the energy ε; T_{jk} and T_{kj} are the average transfer matrix elements between a pair of 3d orbitals of the j-th and k-th atoms; <jk> indicates that the sum in $\rho_{int}(\varepsilon)$ is taken over pairs of atoms. Though T_{jk} and T_{kj} are not complex conjugate to each other according to the original derivation of Eq.(2)[12] we assume that $T_{jk}T_{kj}$ is a real positive quantity, since the nonhermite part of T is expected to be of minor importance.

Adopting the bottom of the s band as the origin of the energy variable ε, we approximate $\rho_s(\varepsilon)$ by the free electron state density given by

$$\rho_s(\varepsilon) = \Omega(1/2\pi^2)(2m/\hbar^2)^{3/2}\varepsilon^{1/2}, \tag{7}$$

where Ω is the volume. We assume that $\varepsilon_{j\sigma}$ can be expressed as

$$\varepsilon_{j\sigma} = \varepsilon_j - \varepsilon_s + K_j(n_{j\sigma} + n_{j-\sigma}) + U_j n_{j-\sigma}, \tag{8}$$

whre ε_j represents the 3d level in the absence of the intra-atomic interactions between d electrons; ε_s is the energy of the bottom of the s band which enters into the expression because of the above-mentioned choice of the origin of ε; we assume for simplicity that ε_j and ε_s are spin independent ; $n_{j\sigma}$ is the number of electrons occupying one of the 3d orbitals with spin σ of the j-th atom; K_j and U_j correspond to 4(U-J) and (U+4J), respectively, in Moriya's notation[11] with U and J representing the intra-atomic coulomb and exchange integral. $n_{j\sigma}$ and $n_{j-\sigma}$ are determined self-consistently by use of the equations,

$$5n_{j\sigma} = (\partial/\partial\varepsilon_{j\sigma})\int_{-\infty}^{\infty} g(\varepsilon;\mu)\{5\textstyle\sum_{j\sigma}\rho_{j\sigma}(\varepsilon) + \rho_{int}(\varepsilon)\}d\varepsilon \tag{9}$$

with $g(\varepsilon;\mu)$ defined by

$$g(\varepsilon;\mu) = -kT\log[1 + \exp\{(\mu-\varepsilon)/kT\}]. \tag{10}$$

We now write down the expression of the total free energy $F(T,\Omega)$ for a given temperature T and a given volume Ω. We assume

$$F(T,\Omega) = \int_{-\infty}^{\infty} g(\varepsilon;\mu)\{\rho_s(\varepsilon) + \rho_d(\varepsilon)\}d\varepsilon + (\mu+\varepsilon_s)(N_s+N_d) - 5\sum_j\{K_j(n_{j\sigma} + n_{j-\sigma})^2/2 + U_j n_{j\sigma}n_{j-\sigma}\} - F_{dd} - F_{sd} - F_{ss} + F_{core} + F_v , \quad (11)$$

where N_s and N_d are the total numbers of s and d electrons, respectively; F_{dd} is the interaction energy between d electrons belonging to different atoms which is counted twice in $\varepsilon_{j\sigma}$'s; F_{sd} is the interaction energy between s and d electrons; F_{ss} is the interaction between s electrons; F_{core} and F_v are the interaction energy between ion-cores and the free energy of lattice vibration, respectively. We assume that ε_j, ε_s, F_{dd}, F_{sd}, F_{ss}, F_{core} and F_v are functions of N_s and $n_{j\sigma}$'s, though we do not specify explicit expressions of these functions in the following discussions. We may then require that $F(T,\Omega)$ satisfies the extremum conditions,

$$\partial F/\partial N_s = 0,\ \partial F/\partial n_{j\sigma} = 0 \text{ and } \partial F/\partial\mu = 0. \quad (12),(13),(14)$$

When combined with Eq.(9), the first two conditions yield the equations which should be satisfied by the adopted expressions of ε_j, ε_s and F_{dd}, etc. as functions of N_s and $n_{j\sigma}$'s. The last condition yields

$$N_s + N_d = \int f(\varepsilon;\mu)\{\rho_s + \rho_d\}d\varepsilon, \quad (15)$$

where $f(\varepsilon;\mu)$ is the Fermi distribution function.

III. THE VOLUME STRICTION DUE TO THE INTERACTION

By use of the conditions (12), (13) and (14) we can show that the change of the total free energy due to the electron transfer between atoms is given by

$$\delta F(T,\Omega) = \int g(\varepsilon;\mu)\rho_{int}(\varepsilon)d\varepsilon \quad (16)$$

to the second order of the transfer integrals, T_{jk}'s. Since only this effective interaction energy depends on the relative orientations of the atomic magnetic moments, the pressure δP defined by

$$\delta P = \partial\delta F(T,\Omega)/\partial\Omega \quad (17)$$

is the origin of the spontaneous volume striction in the ferromagnetic phase. Since the condition $\partial F/\partial\Omega = 0$ determines the equilibrium volume, the equation,

$$\delta\Omega(\partial^2F/\partial\Omega^2) + \delta P = 0, \quad (18)$$

determines the change of the volume due to the interaction; $(\partial^2F/\partial\Omega^2)$ can be estimated by use of the observed value of the compressibility $\kappa = \Omega/(\partial^2F/\partial\Omega^2)$. The spontaneous volume striction is defined to be the difference of $\delta\Omega$ between the ferromagnetic and paramagnetic phases.

In order to obtain a rough estimate of δP, we calculate the volume derivative of $F(T,\Omega)$ by use of the conditions (12), (13) and (14) to derive

$$\partial F/\partial\Omega = (1/\Omega)\int g(\varepsilon;\mu)\rho_s d\varepsilon + N_s(\partial\varepsilon_s/\partial\Omega) + \sum_{j\sigma}n_{j\sigma}(\partial\varepsilon_j/\partial\Omega) + (\partial/\partial\Omega)(-F_{dd}-F_{sd}-F_{ss}+F_{core}+F_v), \quad (19)$$

where we neglect the volume dependence of K_j, U_j, Δ and T_{jk}; the volume dependence of T_{jk} will be discussed later; the first term in the r.h.s. of Eq.(19) is the pressure arising from the kinetic energy of the s electrons and other terms correspond to the pressure arising from other contributions to the cohesive energy. It is to be noted that N_s and $n_{j\sigma}$'s can be treated as if they are volume independent in calculating $\partial F/\partial\Omega$ because of the conditions (12) and (13). The electron transfer contributes to $\partial F/\partial\Omega$ through the changes of N_s and $n_{j\sigma}$'s caused by the perturbation. In fact we can show that the following expression of δP is equivalent to Eq.(17): we obtain from Eq.(19)

$$\delta P = \delta N_s(\partial P_1/\partial N_s) + \delta N_s(\partial P_2/\partial N_s) + \sum_{j\sigma}\delta n_{j\sigma}(\partial P_2/\partial n_{j\sigma}), \quad (20)$$

where P_1 is the first term in the r.h.s. of Eq.(19) and P_2 comprises the remaining terms; δN_s and $\delta n_{j\sigma}$'s are the changes of N_s and $n_{j\sigma}$'s due to the electron transfer.

Since P_1 is proportional to $N_s^{5/3}$ in the free electron approximation, the first term in the r.h.s. of Eq.(20) may be written as

$$\delta P_1 = \delta N_s(\partial P_1/\partial N_s) = (5/3)(\delta N_s/N_s)P_1. \quad (21)$$

We note that Eq.(19) corresponds to the pressure in the model assumed by Fuchs[13] in his theory of elastic properties of Cu. He divides P_2 into two parts, P_a and P_r: P_a is calculated by use of the Wigner-Seitz method with the assumption that d electrons are a part of core electrons and P_r arises from the repulsive interaction between overlapping d electron clouds of neighboring atoms. In order to obtain a rough estimate of the variation of P_2, we assume that P_a and P_r are functions of the total numbers of s and d electrons, N_s and N_d. In the classical limit P_a is proportional to the square of the net charge of the ion-core in which d electrons are included. Noting $\delta N_s = -\delta N_d$, we assume, therefore,

$$\delta P_a = 2(\delta N_s/N_s)P_a. \quad (22)$$

Since P_r will be roughly proportional to N_d^2 we obtain

$$\delta P_r = 2(\delta N_d/N_d)P_r = -2(\delta N_s/N_d)P_r. \quad (23)$$

From Eqs.(21), (22) and (23) we obtain

$$\delta P = (\delta N_s/N_s)\{(5/3)P_1 + 2P_a - (N_s/N_d)P_r\}, \quad (24)$$

which can be rewritten by use of $P_1 + P_a + P_r = 0$ as

$$\delta P = -(\delta N_s/N_s)[\{2+(N_s/N_d)\}P_r+(1/3)P_1] = -A(\delta N_s/N_s)P_1. \quad (25)$$

In the last expression of δP we assume $P_1=P_r$, since they are of the same order in Fuchs' calculation; the numerical factor A will be of the order of 2. Since P_1 can be calculated in the free electron approximation, our problem is now reduced to the calculation of δN_s.

In the following we derive general expressions of δN_s and also $\delta n_{j\sigma}$. We define $\rho_s[\mu]$ and $\rho_{j\sigma}[\mu]$ by

$$\rho_{s \text{ or } j\sigma}[\mu] = \int\{\partial f(\varepsilon;\mu)/\partial\mu\}\rho_{s \text{ or } j\sigma}(\varepsilon)\, d\varepsilon, \quad (26)$$

where the integral is equal to the state density at $\varepsilon=\mu$ when T=0. We note that the shift of the Fermi energy μ, $\delta\mu$, is related to δN_s by

$$\delta\mu = \delta N_s/\rho_s[\mu]. \quad (27)$$

From Eq.(9) we obtain

$$5\delta n_{j\sigma} = \sum_{k\tau} 5(1 + I)^{-1}{}_{j\sigma,k\tau}\rho_{k\tau}[\mu]\delta\mu + (1 + I)^{-1}{}_{j\sigma,k\tau}\int g(\varepsilon;\mu)(\partial\rho_{int}/\partial\varepsilon_{k\tau})d\varepsilon, \quad (28)$$

where the matrix I is defined by

$$I_{j\sigma,k\tau} = \rho_{j\sigma}[\mu]\ (\partial\varepsilon_{j\sigma}/\partial n_{k\tau}). \quad (29)$$

Eq.(28) leads us to

$$\delta N_s = -\frac{\rho_s[\mu]\sum_{j\sigma,k\tau}(1 + I)^{-1}{}_{j\sigma,k\tau}\int g(\partial\rho_{int}/\partial\varepsilon_{k\tau})d\varepsilon}{\rho_s[\mu] + 5\sum_{j\sigma,k\tau}(1 + I)^{-1}{}_{j\sigma,k\tau}\rho_{k\tau}[\mu]}, \quad (30)$$

which together with Eqs.(27) and (28) gives us an expression of $\delta n_{j\sigma}$ as a functional of $\rho_{int}(\varepsilon)$. Thus once the dependence of $\varepsilon_{j\sigma}$ on $n_{k\tau}$'s is given, Eqs.(28), (29) and (30) determine δN_s and $\delta n_{j\sigma}$'s.

IV. THE CALCULATION OF δP IN Ni-Fe Alloys

As will be discussed later, the contribution of the interaction between two Fe atoms to δP is predominantly large even when the Fe concentration, x, is small. We derive, therefore, the expression of δP of this origin more explicitly. Assuming that all Fe atoms are in the same magnetic state, we obtain $\rho_{int}(\varepsilon)$ of the Fe-Fe interaction by taking the limit, $\varepsilon_{j\sigma}^{int}-\varepsilon_{k\sigma}\to 0$, in Eq.(6) as

$$\rho_{Fe-Fe}(\varepsilon) = 5NzV^2(x^2/2)\sum_\sigma(\partial^2\rho_{Fe\sigma}(\varepsilon)/\partial\varepsilon^2), \qquad (31)$$

where z is the number of the nearest neighbors in the fcc lattice (z=12); N is the total number of lattice sites of a given sample; V^2 is defined by

$$V^2 = (5/z)\sum_k T_{jk}T_{kj}, \qquad (32)$$

where the sum is taken over all neighbors of the j-th site. $\rho_{Fe\sigma}(\varepsilon)$ in the r.h.s. of Eq.(31) is defined by Eq.(5) with the atomic energy $\varepsilon_{Fe\sigma}$. For the calculation of the matrix I defined by Eq.(29), we assume that $\varepsilon_{Fe} - \varepsilon_s$ appearing in the r.h.s. of Eq.(8) is independent of $n_{Fe\sigma}$'s, or in other words, we assume that K_{Fe} and U_{Fe} are the effective intra-atomic integrals including the contributions made by the interatomic interactions which should be incorporated in $\varepsilon_{Fe} - \varepsilon_s$ in the present formalism. Then we obtain from Eqs.(25), (30) and (31)

$$\delta P=5NzV^2(x^2/2)A(P_1/N_s)[\rho_s/(\rho_s+Nx\rho_{effFe}+N(1-x)\rho_{effNi})]$$
$$\times \left[\frac{\rho_\uparrow' + \rho_\downarrow' - U(\rho_\uparrow'\rho_\downarrow + \rho_\downarrow'\rho_\uparrow)}{1-U^2\rho_\uparrow\rho_\downarrow + K(\rho_\uparrow + \rho_\downarrow-2U\rho_\uparrow\rho_\downarrow)} \right], \qquad (33)$$

where we denote $\rho_s[\mu]$, $\rho_{Fe\sigma}[\mu]$ and $\partial\rho_{Fe\sigma}[\mu]/\partial\mu$ defined by Eq.(26) by ρ_s, ρ_σ and ρ_σ', respectively; we drop also the subscript Fe of K_{Fe} and U_{Fe}, for simplicity; ρ_{effFe} and ρ_{effNi} are defined by

$$\rho_{effj} = (\rho_{j\uparrow} + \rho_{j\downarrow} - 2U_j\rho_{j\uparrow}\rho_{j\downarrow})$$
$$\times 1/[1 - U_j^2\rho_{j\uparrow}\rho_{j\downarrow}+K_j(\rho_{j\uparrow}+\rho_{j\downarrow}-2U_j\rho_{j\uparrow}\rho_{j\downarrow})], \qquad (34)$$

where j stands for either Fe or Ni. ρ_{eff} is the effective state density at the Fermi level obtained by taking into account the redistribution of up and down spin electrons in the Hartree-Fock solution for a shift of the Fermi energy.

Since we start with the Lorentzian state density, $\rho_{Ni\sigma}[\mu]$ and also $\partial\rho_{Ni\sigma}[\mu]/\partial\mu$ are considerably small compared to the corresponding values of minority spin d electrons of a Fe atom. Thus the contribution of the Ni-Ni interaction to δP is concluded to be negligibly small in the present scheme of the calculation. In fact the spontaneous volume striction in pure Ni seems to be very small[1] Apart from this fact we can argue that the small state density at the Fermi level of a Ni atom is not very unreasonable, since the CPA calculation by Hasegawa and Kanamori[14] shows that the high density peak in the d band of Ni collapses with a rather small concentration of Fe in Ni-Fe alloys. As for the interaction between Ni and Fe atoms, the term proportional to $\partial\rho_{Fe\sigma}/\partial\varepsilon$ in ρ_{int} given by Eq.(6) can be taken into account as a shift of $\varepsilon_{Fe\sigma}$ given by $\delta\varepsilon_{Fe\sigma}=25\sum_k T_{\sigma k}T_{k\sigma}/(\varepsilon_{Fe} - \varepsilon_{Ni\sigma})$ in the state density $\rho_{Fe\sigma}$, where the sum is taken over neighboring Ni atoms of a Fe atom at the origin. The remaining term in ρ_{int} turns out to be negligibly small. With these considerations we concentrate on δP arising from the Fe-Fe interaction hereafter.

Since an ab initio calculation is difficult, we adopt a rather phenomenological approach in the calculation. With given values of U_{Fe} and K_{Fe} we can calculate δP if we assume that n_{Fe} of minority spin (called down spin hereafter) is a given quantity. The neutron diffuse scattering experiment on Ni-Fe alloys[15] shows that the the magnetic moment of a Fe atom decreases with increasing x to a considerable extent. The CPA calculation[14] which is in good agreement with the neutron data shows that $5n_{Fe\downarrow}$ increases with x from $5n_{Fe\downarrow}= 1.9$ at x=0 to $5n_{Fe\downarrow}=2.3$ at x=0.5, while $n_{Fe\uparrow}$ remains to be 1 in this range of x. Apart from the electron transfer between Fe atoms, two mechanisms are conceivable for the increase of $n_{Fe\downarrow}$.

Fe atoms will lose some electrons to Ni atoms when x is small if the Fermi level in pure Ni is lower than the value necessary to keep them. Also the upward shift of $\varepsilon_{Fe\downarrow}$ due to the Fe-Ni electron transfer will decrease $n_{Fe\downarrow}$. These effects are expected to decrease with increasing x, causing a relative increase of $n_{Fe\downarrow}$. For x>0.5 where the saturation magnetization deviates from the Slater-Pauling curve, we may expect a more rapid increase of $n_{Fe\downarrow}$, since some up spin states lose electrons to pour them into down spin states of a Fe atom. Considering these mechanisms, we choose rather arbitrarily several $n_{Fe\downarrow}$ vs. x relations as input information. Fig.1 shows examples of the choices. The curve A is obtained by interpolating the values of $n_{Fe\downarrow}$ at x=0 and x=0.5 of the CPA calculation[14] We extrapolate it to x>0.5. In the curves B and C we assume a slower increase of $n_{Fe\downarrow}$ for x<0.5, since the CPA calculation includes the effect of the Fe-Fe transfer which may not be included in the unperturbed value of $n_{Fe\downarrow}$. For x>0.5 we assume a rapid increase rather arbitrarily. The value of $n_{Fe\downarrow}$ at x=0 of the curve B was determined by fitting the magnitude of the atomic magnetic moment of Fe in the Hartree-Fock approximation of the Anderson model to the CPA calculation.

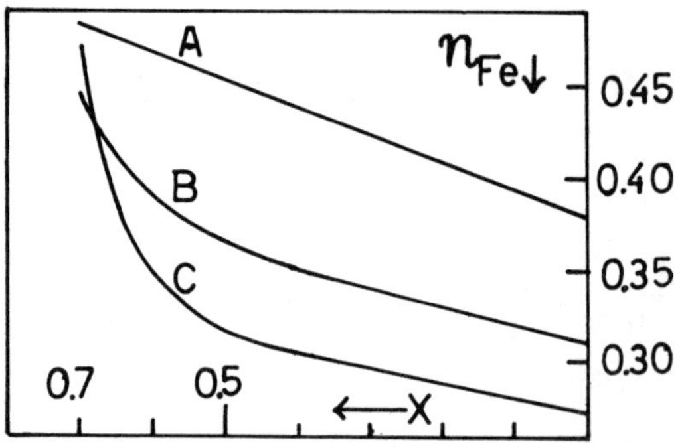

Fig. 1 $n_{Fe\downarrow}$ vs. x curves assumed in the calculation. x is the concentration of Fe in Ni-Fe alloys($Ni_{1-x}Fe_x$).

In Fig.2 we plot the calculated δP against x in the unit of $A(V/\Delta)^2$ which should be of the order of 1 if we take Moriya's choice of the parameter[11] In the calculation we assume Δ=0.05 and $U_{Fe}=K_{Fe}$=0.5 in ryd. The choice of (U/Δ) = 10 is again close to Moriya's choice[11] We assume also ρ_s=1.25 and ρ_{effNi}=2.5 per atom per ryd. The value of ρ_s is calculated in the free electron approximation with N_s/N=0.6. ρ_{effNi} is obtained from the Lorentzian state density with 9.4 d electrons in the nonmagnetic state. Calculations based on a somewhat larger ρ_{effNi} are also shown in Fig.2. The calculated δN_s is of the order of 0.01 per atom with negative sign.

V. COMPARISON WITH EXPERIMENT

As was shown by Moriya,[11] the interaction between two Fe atoms with antiparallel magnetic moments of the

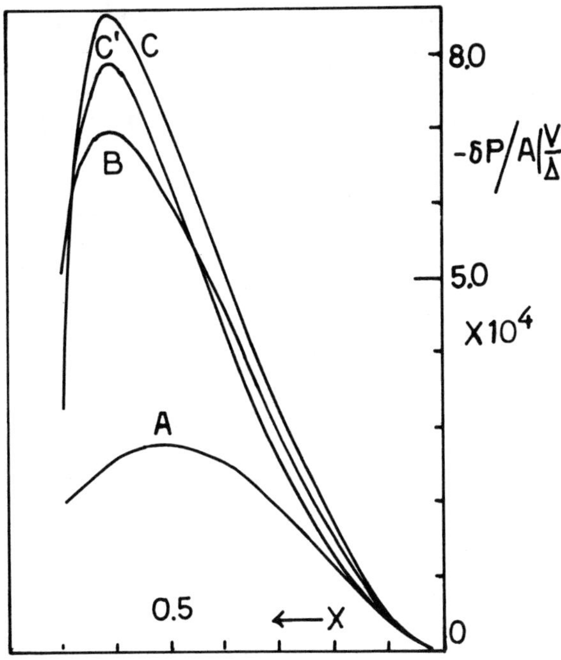

Fig. 2 δP vs. x curves. δP is given in the unit of ryd./(Bohr radius)^3 times $A(V/\Delta)^2$. Curves A, B and C are calculated with the curves A, B and C given in Fig. 1, respectively, with ρ_{effNi} = 2.5 per ryd. per atom; curve C' is a modification of the curve C with ρ_{effNi} = 5 in the same unit.

same magnitude does not depend on the number of electrons, making thus no contribution to δP. This fact leads us to a rough estimate of the spontaneous volume striction, $\omega = -\kappa\delta P/2$, since the number of pairs of Fe atoms with parallel moments in the paramagnetic state will be about a half of that in the ferromagnetic state. In Fig.3 we show the values of ω deduced by Hayase, Shiga and Nakamura[1] from the thermal expansion data, which should be compared to $|\delta P|$ in Fig.2. Since the compressibility of pure Ni is about 70 in (Bohr radius)^3/ryd., agreement is satisfactory as far as the order of magnitude is concerned. There are ambiguities both on theoretical and experimental sides. First of all the temperature effect on δP is neglected in the above estimate. Also the deduction of ω from thermal expansion data involves an ambiguity.

Next we calculate the linear thermal expansion coefficient at low temperatures by expanding every quantity depending on temperature in the r.h.s. of Eq.(33) in powers of T up to T^2; in the calculation the Hartree-Fock self-consistent condition is satisfied also to the order of T^2. The calculated thermal expansion coefficient α is proportional to T. We plot α/T against x in Fig.4. The experimental data on Ni-Fe alloys measured by Meincke and Graham[4] and White[16] show that α below 5 K is proportional to T with α/T decreasing rapidly with increasing x from α/T=0 at x=0.5 to $\alpha/T = -1\times10^{-7}$ at x=0.65; α/T for x<0.5 has small positive values decreasing slowly with increasing x. The thermal expansion coefficient of Pt-Fe alloys with about 70% of Fe is reported to be also negative at low temperatures.[17] Our calculation takes into account the Stoner type excitation only. The spin wave contribution, however, is proportional to $T^{3/2}$, which may be neglected at low temperatures.

Finally we discuss the concentration dependence of the lattice constant in Ni-Fe alloys. As is shown in Fig.3, the lattice constant shows an anomalous decrease with x in the invar region. Since our δP passes a maximum in this region, we may expect that it explains also the lattice parameter anomaly. For the calculation, however, we need the knowledge of the lattice constant in the unperturbed state, i.e., in the state before switching on the electron transfer between

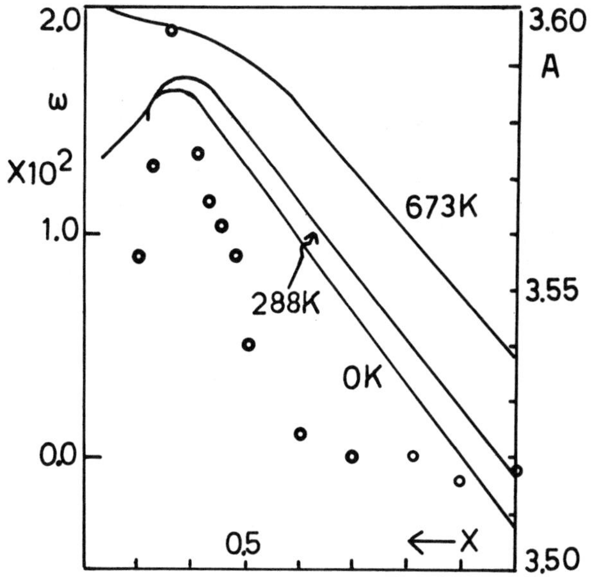

Fig.3 The lattice constant and the spontaneous volume striction ω at absolute zero. The curves are the lattice parameter at various temperatures of Ni-Fe alloys.[3] The circles are the values of ω estimated from the thermal expansion data of Ni-Fe alloys.[1]

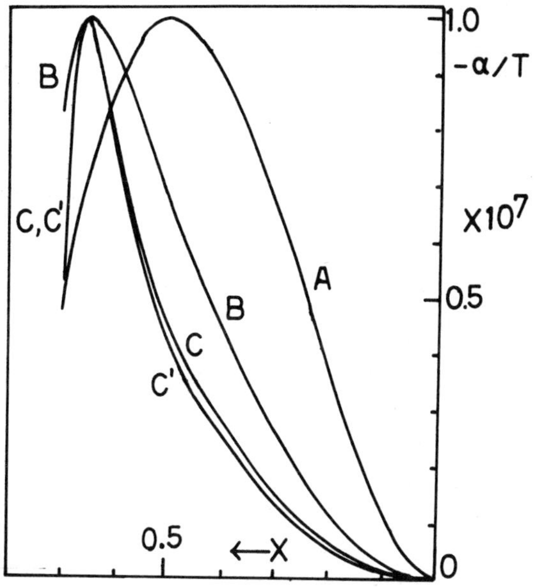

Fig.4 The calculated linear thermal expansion coefficient divided by T, α/T. The definitions of A,B,C and C' are the same as those in Fig.2. The maximum value of $-\alpha/T$ is normalized to 1×10^{-7}deg^{-2} which corresponds to the observed maximum value at x=0.65. The value of $A(V/\Delta)^2$ to yield the maximum value is 2.38 for the curve A, 1.08 for B, 1.03 for C and 1.09 for C'.

Fe atoms. Though it is expected that the unperturbed lattice constant is a monotonic function of x, we can not determine it without an ab initio calculation. Fig.5 shows several arbitrarily chosen unperturbed lattice constant vs. x curves together with the calculated lattice constants including the perturbing effect of δP just to demonstrate the possibility of obtaining a good agreement with the data. Though there is an argument supporting a linear dependence on x of the unperturbed lattice constant, i.e., the Vegard rule,[7] we have no reason to exclude a convex curve such as shown in Fig.5 for the unperturbed lattice constant.

Considering various crude approximations involved in our calculation, we may conclude that the overall agreement of our calculation with experiment is

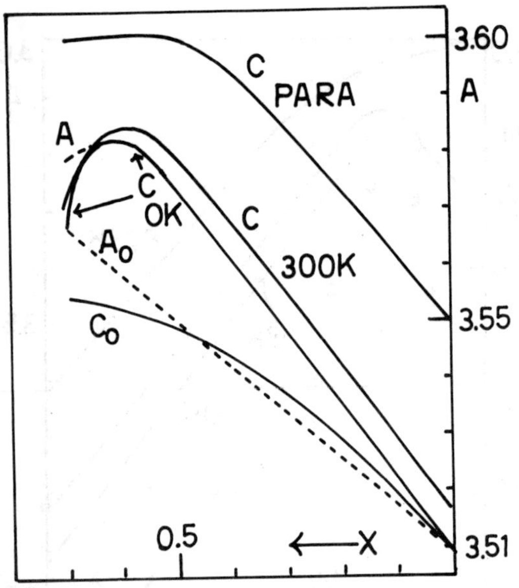

Fig. 5 The calculated lattice parameter. Curves C at 0K and 300K are calculated with the curve C of δP with $A(V/\Delta)^2 = 0.5$ and the curve C_o assumed to represent the lattice parameter of the unperturbed state; at 300K a constant is added to the curve C_o. The curve C with PARA represents the lattice parameter at high temperatures calculated by adding $-\kappa\delta P/6$ and an arbitrary constant to the curve C_o. The dotted curve A which overlaps with the curve C for $x<0.65$ is the calculation at 0K with the curve A of δP and the curve A_o for the unperturbed lattice parameter; $A(V/\Delta)^2$ is taken to be 1.

satisfactory. So far we have neglected the volume dependence of the transfer integral T_{jk}. We can show that its contribution to δP is proportional to $\rho_{Fe\uparrow}[\mu] + \rho_{Fe\downarrow}[\mu]$ which does not give rise to anomalies in the volume striction and the thermal expansion coefficient for $n_{Fe\downarrow}$ vs. x relations assumed in Fig.1. Moreover, the contribution to δP is of positive sign which is opposite to the above calculation, if $\partial V^2/\partial \Omega$ is assumed to be of negative sign. The contribution to α/T is also of opposite sign, being positive. Since this mechanism is equivalent to that arising from the volume dependence of the d band width, it is operating in other transition metals. According to the experimental data,[16] α/T in the cases of the positive sign is usually small compared to the absolute magnitude of α/T in invar alloys. We neglect it, therefore, in our calculation.

VI. MORE-THAN-ONE STATES OF Fe ATOMS

We have assumed so far that all Fe atoms are in the same state with magnetic moment parallel to bulk magnetization. It is argued by many authors, however, that a Fe atom surrounded by a sufficient number of Fe nearest neighbors may have a magnetic moment antiparallel to bulk magnetization.[5,19] We present here a discussion on this possibility. According to the CPA calculation, Ni-Fe alloys with x smaller than 0.5 is in the state of the 'strong' ferromagnetism, which is defined to be the state in which the up spin d states are fully occupied.[14] At x=0.5 the edge of the up spin band crosses the Fermi level. As is well known, the CPA treats a Fe atom as an impurity embedded in the medium described by the coherent potential. We can find in a wide range of input parameters two stable Hartree-Fock solutions for the atomic state of Fe, one with a parallel moment and another with an antiparallel moment.[20] Since the CPA calculation[14] precludes the possibility of the latter solution, the above-mentioned picture of the instability of the 'strong' ferromagnetism should be examined more carefully. As a first step we extend the CPA calculation to allow two states of Fe atoms, FeI and FeII. We treat, therefore, a Ni-Fe alloy as a ternary alloy of FeI, FeII and Ni. Adopting the same state density function of pure metal as that used in the previous CPA calculation, we solve the impurity problem in the medium described by a trial coherent potential to find generally two solutions, one with parallel moment(FeI) and the other with antiparallel moment(FeII). Assuming that a certain fraction of Fe atoms are FeII, we determine the coherent potential. Once the coherent potential is determined, we compare the energies of the two solutions by use of an energy expression.[20] We readjust then the assumed number of FeII until the energies of FeI and FeII become equal to each other. The atomic energy parameters are determined by the same procedure as that adopted in ref.14. Assuming $U_{Fe}=U_{Ni}= 1.4$ and $K_{Fe}=K_{Ni}=1.6$ in the unit of a half of the band width, we find that FeII appears for $x>0.5$ and its fraction amounts to 2% at $x=0.65$. The calculated saturation magnetization is shown in Fig.6 together with the experimental data. It turns out that the number of d electrons of FeII is generally smaller than that of FeI by the amount of $0.4\sim0.5$ per atom when nonvanishing fraction of FeII is obtained.

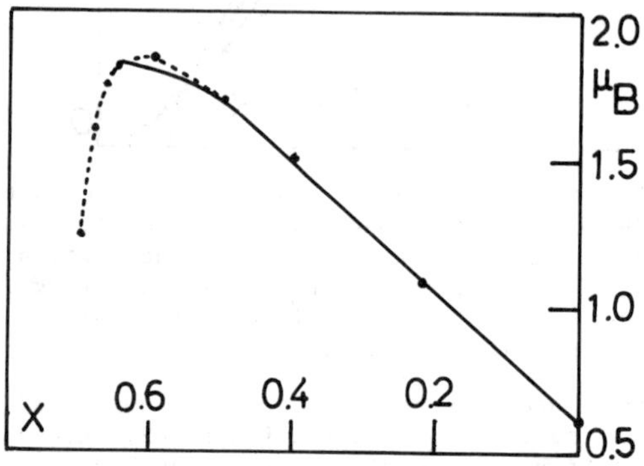

Fig. 6 The saturation magnetization per atom at absolute zero of Ni-Fe alloys. The full curve is the calculation(see text) and the points and the dotted curve are the experimental data.[18]

If we preclude the presence of FeII, the CPA calculation with the same parameters yields the 'strong' ferromagnetism up to $x\sim0.7$. Thus the result indicates the possibility that the instability of the 'strong' ferromagnetism may be caused by the appearance of FeII atoms rather than the dipping of the Fermi level into the up spin band, though we need to examine other choices of parameters before concluding definitely. We have two possible interpretations of the appearance of FeII atoms. In the first interpretation the appearance of FeII implies the onset of the dynamical fluctuation of the atomic moment. Since we assume the condition of equal energy, Fe atom may change its state between FeI and FeII, costing no energy within the limit of the Hartree-Fock approximation.

The second interpretation is to suppose that a Fe atom surrounded by a sufficiently large number of Fe neighbors has a static antiparallel magnetic moment. Then the ternary alloy approach is regarded as an approximation which takes into account the environment effect on the average. In order to get an insight into the environment effect, we apply the formalism developed by Miwa[21] to calculate the state of a Fe atom surrounded by a given number of Fe nearest neighbors in the tight binding model; in the calculation farther neighbors are replaced by the medium described by the coherent potential obtained by the ternary alloy approach. Since this calculation also has not been completed yet, we can say at this moment only that the calculated magnetic moment of the antiparallel solution decreases to a considerable extent with increasing number of Fe nearest neighbors. It suggests that the antiparallel solution may become almost nonmagnetic or even attain a small 'parallel' moment when surrounded by twelve Fe nearest neighbors. Since there is a computational difficulty, we have not made yet an energy comparison between the parallel and antiparallel solutions.

So far no experimental evidence has been obtained for the presence of Fe atoms with antiparallel moments in Ni-Fe alloys with $x \leq 0.65$.[22] Since the number of **Fe** atoms with antiparallel moments seems to be still small in that range of x, this may not exclude the possibility of the static antiparallel moment. It is apparent that we need a more refined theory to discuss the fluctuation of the atomic moment. Going back to our theory of the volume striction, we may say that our calculation of δP at absolute zero will not be affected much by the appearance of FeII either dynamical or static for $x \leq 0.65$, where the number of FeII atoms seems to be still small. Hoever, the fluctuation of the atomic moment might enhance the absolute magnitude of the thermal expansion coefficient at low temperatures which we calculated by taking into account the Stoner type excitations only. In fact we need a somewhat larger value of $A(V/\Delta)^2$ to fit the calculation to the experimental value of α/T than in the cases of the spontaneous volume striction and the lattice parameter (see Figs. 2-5). In Pt-Fe[23] and Pd-Fe[24] the saturation magnetization follows approximately the Slater-Pauling curve up to the 'invar' concentration. It is desirable to have the experimental data of the concentration dependence of α/T and other quantities of these alloys to compare with the data on Ni-Fe alloys. As for the $n_{Fe\downarrow}$ vs. x relation, our preliminary calculation described above shows that $n_{Fe\downarrow}$ of FeI is increased by the appearance of FeII, though we need furthe investigation to confirm quantitatively such a rapid increase of $n_{Fe\downarrow}$ as in the curves B and C in Fig. 1.

VII DISCUSSIONS

In our opinion, the properties which characterize the 'invar' alloys are a large spontaneous volume expansion in the ferromagnetic phase and a large negative α/T at low temperatures. In the present theory these properties are closely related to the position of the Fermi level in the Lorentzian state density. In the r.h.s. of Eq.(33) the factors $(x^2/2)$ and $\rho_\downarrow'$ play important roles in producing the maximum of the spontaneous volume expansion in invar alloys. In the case of α/T it is the third derivative of $\rho_\downarrow$ which is dominant in determining the sign and magnitude. Though details of the analysis will be reported in future publication, we can show that the 'invar' properties may appear only in a limited range of $n_\downarrow$ such as shown in Fig. 1 when down spin states are mainly responsible for the effect. With the assumed value of $U/\Delta = 10$ we find that a Fe atom is the best candidate for producing the 'invar' effect. Somewhat smaller values of $n_\downarrow$ can produce the effect; with such $n_\downarrow$ we obtain the total electron number rather close to a Mn atom. In this case, however, the antiferromagnetic coupling is more favored than the ferromagnetic coupling if we calculate the exchange energy defined by Moriya.[11] Though we cannot exclude other possibilities completely for different choices of the U/Δ value, we conclude that the invar properties have a chance to appear in those ferromagnetic alloys which contain enough interacting Fe-Fe pairs. Since $n_\downarrow$ to produce the maximum effect seems to be close to the critical value for the onset of the instability of the 'strong' ferromagnetism, we may say also that the system should be close to the boundary of the ferromagnetic phase in phase diagram. Fe atoms, however, are required to keep a magnetic moment of sufficient magnitude to have the most favorable value of $n_\downarrow$. The role of other elements in the alloys is to bring $n_\downarrow$ to that value by alloying. We conclude also that the crystal structure is of secondary importance. We hope that 'invar' alloys of different crystal structures will be found in future to confirm this conclusion. It has been found recently that $Zr_{0.7}Nb_{0.3}Fe_2$ with the Laves structure has a large spontaneous volume expansion in the ferromagnetic phase.[25]

Our theory can be applied to alloys other than invar alloys. We shall report several applications in future publication.

The authors would like to express their sincere thanks to Professor H.Miwa for valuable discussions and to Mrs.A.Egusa for her efforts in preparing the manuscript.

REFERENCES

1. M. Hayase, M. Shiga and Y. Nakamura, J. Phys. Soc. Japan 34, 925(1973)
2. M. Hayase, M. Shiga and Y. Nakamura, Phys. Satus Solid. B46, K117(1971); K. Sumiyama, M. Shiga and Y. Nakamura, to be published in J. Phys. Soc. Japan
3. W. B. Pearson, A Handbook of Lattice Spacings and Structure of Metals and Alloys, 638, Pergamon Press (1958).
4. P. P. M. Meincke and G. M. Graham, Proc. 8th Internl. Conf. Low Temp. Phys. 339(1963).
5. E. I. Kondorsky and V. L. Sedov, J. Appl. Phys. 31, 331S(1960).
6. R. J. Weiss, Proc. Phys. Soc. A82, 281(1963).
7. M. Shiga, AIP Conf. Proc. No. 18, 19th Magnetism & Mag. Mat., 463(1973).
8. E. P. Wohlfarth, Phys. Letters 31A, 525(1970)
9. K. Terao and A. Katsuki, J. Phys. Soc. Japan 37, 828(1974).
10. S. Alexander and P. W. Anderson, Phys. Rev. 133A, 1594(1964).
11. T. Moriya, Progr. Theor. Phys. 33, 157(1965).
12. J. Kanamori, K. Terakura and Y. Yamada, Progr. Theor. Phys. Supplement 46, 221(1970).
13. K. Fuchs, Proc. Roy. Soc. A151, 585(1935).
14. H. Hasegawa and J. Kanamori, J. Phys. Soc. Japan 33, 1599(1972).
15. C. G. Shull and M. K. Wilkinson, Phys. Rev. 97, 304(1955).
16. G. K. White, Proc. Phys. Soc. 86, 159(1965).
17. K. Sumiyama, Ph.D. Thesis, Kyoto Univ.(1974)
18. J. Crangle and G. C. Hallem, Proc. Roy. Soc. A272, 119(1963).
19. Mn in Ni-Mn with antiparallel moment was concluded by neutron diffraction experiment: J. W. Cable and H. R. Child, J. de Phys. 32, C1-67(1971).
20. J. Kanamori, J. de Phys. 35, C4-131(1974).
21. H. Miwa, Progr. Theor. Phys. 52, 1(1974).
22. B. Window, J. Phys. F 4, 329(1974).
23. A. Kussmann and G. G. V. Rittberg, Ann. Physik 7, 173(1950).
24. A. Kussmann and K. Jessen, J. Phys. Soc. Japan 17 Supplement B-1, 136(1962).
25. M. Shiga, private communication.

SPECTROSCOPY OF INTERCONFIGURATIONAL FLUCTUATIONS IN RARE EARTH COMPOUNDS

M. Campagna and G. K. Wertheim
Bell Laboratories, Murray Hill, New Jersey 07974
and E. Bucher
University of Konstanz, 775 Konstanz, West Germany

ABSTRACT

After a brief discussion on the application of X-ray photoemission (XPS) to the study of multiplet structures and intensities in 4f-photoemission, we present XPS spectra of 4f-electrons in rare earth compounds with intermediate valence state. This technique provides an instantaneous picture of the electronic configuration, and in each case shows the presence of both divalent and trivalent ions, which are identified by their multiplet structure and chemical shifts. These results allow a direct estimate of the Coulomb correlation energy U_{eff} within 4f-shells and demonstrate that Hund's rule correlation is well preserved within a "nonmagnetic", delocalized 4f-shell. We review recent temperature dependent XPS-studies on two distinct systems, the Tm monochalcogenides and the "chemically collapsed" (golden) phases of SmS. We analyze their relation to other physical properties, especially lattice constant. We also discuss the implications of these observations for the understanding of the semiconductor-to-metal transition in these compounds and the related peculiar magnetic properties.

INTRODUCTION

Rare Earth (RE) compounds with intermediate valence states have generated in the last few years increasing interest among theorists working on the problem of local magnetic moment formation and metal-insulator transition. From simple arguments 4f-shells should always show "magnetic", ionic behaviour, the ground state corresponding to a well defined integral number, n, of f-electrons. In contrast the physical properties of many RE-compounds and alloys cannot be understood without assuming, for example, contributions of f-functions to the Fermi surface. It is now accepted that the transition from localized to "delocalized" f-electrons is responsible for the occurrence of such anomalous physical properties as the very large linear values of the specific heat at low temperatures, and intermediate values of quantities like lattice constants, susceptibilities or isomer shifts as we shall briefly discuss below. For recent reviews of the status of theory and experiment we refer the reader to the articles by Wohlleben and Coles [1] and by Maple and Wohlleben [2]. These authors pointed out the advantages of RE-compounds and alloys over those involving 3d-elements for studying phenomena related to the existence of local moments. The following points are cited because of their importance in our considerations:

(1) due to the strong atomic-like character of 4f-shells the ionic approximation in the magnetic limit allows the identification of the Hund's rule ground state;

(2) even in the "non-magnetic" or "delocalized" limit the Hund's rule correlation is expected to be conserved, because no direct overlap of 4f-shells of neighboring cells is possible and the mixing interaction, V_{kf}, of f-states with the conduction electrons is small compared to the intra-atomic Coulomb correlation energy U_{eff} of f-shells. As pointed out by Coqblin and Blandin [3] experiments with concentrated RE-systems can be considered in some sense analogous to those involving isolated, very dilute impurities of 3d-elements in a metallic matrix.

It is the purpose of this contribution to show that the above factors turn out to be especially favorable for the application of spectroscopical methods like X-ray photoemission (XPS) to the study of these problems.

2. SPECTROSCOPY OF 4f-ELECTRONS

From the work of Eastman and Kutznietz [4], Wertheim et al. [5] and Heden et al [6] it became clear about 4 years ago that 4f-electrons in solids can be suitably studied only at higher photon energies ($h\nu \geqslant 40$ eV) and that the spectra of 4f-photoemission have to be understood in terms of the spectra of the hole state left behind the photoemitted electron. A recent systematic study of the multiplet structure in 4f-photoemission from cleaved single crystals of the RE-Sb using monochromatized

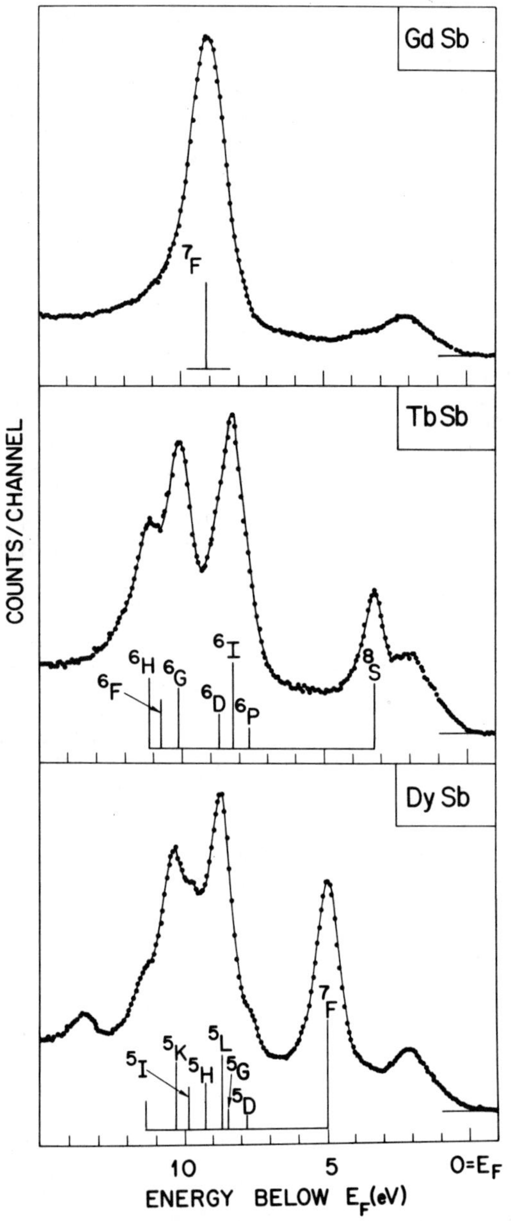

Figure 1. XPS-spectra of valence band and 4f-electrons of GdSb, TbSb and DySb.

Al K_α radiation and a total energy resolution of about 0.5 eV [7] revealed a gratifying agreement with detailed predictions of Cox et al [8] based on the method of the coefficients of fractional parentage. The basic procedure underlying these calculations had been developed by Racah[9] in the 40's. The many-electron initial state wave function of the $4f^n$-shell is developed in terms of products of (n-1)- and one-electron functions. The coefficients of fractional parentage represent the amplitudes of well-defined configurations contained in the expansion. The one-electron function represents the outgoing photo-electron while the (n-1)-configurations represent the possible final hole-states. The probabilities of occupying these hole states are proportional to the squares of the coefficients of fractional parentage. The different energies of the hole states are known from optical spectra of RE-ions[10]. As an example we show in Fig. 1 the X-ray photoemission spectra of valence and 4f- photoelectrons of GdSb, TbSb and DySb. In the case of GdSb, where the $4f^6$-final state peak (^7F) is energetically well separated from the valence band region, it is furthermore easy to obtain an experimental estimate of the relative photoionization cross-sections of localized $4f^7$ and (5p,d)-valence band electrons (6 electrons formula unit: $\sigma_{4f^n}/\sigma_{V.B.} \approx 10$ for n = 7 and $h\nu$ = 1486 eV. Comparison of these spectra with results of Eastman et al.[11] on in situ cleaved RE-sulfides obtained with energy resolution of 0.4 eV demonstrates that because of transition probabilities and escape depth effects spectra of 4f-electrons are best studied with photon energies $h\nu \geq 70$ eV. Herbst et al.[12] recently reported a systematic investigation of the electronic properties of RE-metals based on the renormalized-atom-scheme. An interesting feature of their work is a many-electron calculation of a parameter $\Delta_- = E(4f^n \rightarrow 4f^{n-1})$. Δ_- can be considered as the energy necessary to excite an f-electron into the conduction band at the Fermi level taking into account screening and relaxation. In Fig. 2 we have plotted the experimental and theoretical Δ_- values. The data for the metals are obtained from results by Baer and Busch [13], those for metallic RE-Sb

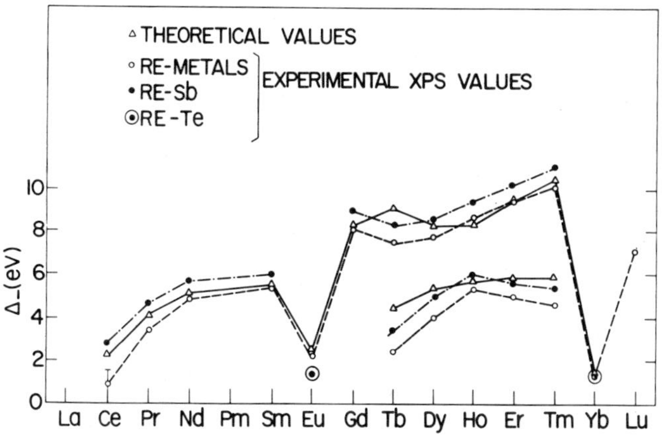

Figure 2. Comparison of photoionization energies of 4f-electrons relative to E_F calculated by Herbst et al. [12] using the renormalized-atom scheme with experimental values of Baer and Busch for the metals and from Ref. 7 for the RE-Sb (-Te).
$\Delta_-(\uparrow)$: majority -spin photoexcitation
$\Delta_-(\downarrow)$: minority-spin photoexcitation (final state in Hund's rule ground state).

from Ref. 7. The agreement between theory and experiment is encouraging. It shows that the renormalized-atom scheme is a promising technique for treating correlated many-electron shells. The fact that we are now able to understand in detail the spectra in 4f-photoemission makes XPS a very powerful tool for studying electronic phase transitions involving 4f-electrons. The state of a RE-ion can be unambiguously identified through its 4f-multiplet structure.

3. INTERMEDIATE VALENCE STATE OF RARE EARTH IONS IN SOLIDS

At the beginning, in the middle and at the end of the RE-series the energies of the $4f^n$- and $4f^{n-1}d$-configurations are not very different, according to the spectroscopic data. In solids the 5d (and 6s) wave functions of neighboring RE-atoms overlap and form bands. On the other hand, the atomic nature of the 4f-states, which lie spatially inside the $5s^2$ and $5p^6$ closed shells is usually not perturbed. If the Fermi level E_F lies between the $4f^n$-states and the empty 5d(6s)-bands then the compound is semi-conducting (insulating). The RE-ion is considered in this case to be formally divalent. An increase of the crystal field splitting between $5dt_{2g}$ and $5de_g$-bands is equivalent to a reduction of the excitation energy E_{exc} necessary for the transition $4f^n \rightarrow 4f^{n-1}5d$ (formally the trivalent state); this can be obtained experimentally by applying pressure to the crystal, as verified by Jayaraman et al.[14] When $E_{exc} \approx 0$ both configurations can coexist. In this case the 4f-states mix with the conduction band and many new anomalous physical properties can be observed. The mixed valence state is first of all characterized by formally intermediate value of the ionic radius of the RE-ion, between that of the divalent and that of the trivalent state. This is reflected in an intermediate value of the lattice constant. Among further signs that a RE-ion is in an intermediate valence state we note:

(1) anomalously large linear specific heat at low temperature[15]
(2) intermediate value of the magnetic susceptibility between those of the two adjacent ionic configurations[1,2], and
(3) intermediate value of the isomer shift between those of the divalent and trivalent states[16].

Various phenomenological models have recently been proposed to explain the occurrence of intermediate valence states of RE-ions in solids[17]. An appealing simple picture due to Hirst[18] is based on a modification on the Friedel-Anderson model of localized moment formation. Here, for $E_{exc} \approx 0$, the Anderson mixing interaction V_{kf} causes the individual ions to undergo fast spontaneous interconfigurational fluctuations (ICF) by emission and absorption of conduction electrons. For a width $\Delta \cong |V_{kf}|^2\rho(E_F)$ of the f-levels and $E_{exc} \leq \Delta$ the ions fluctuate at a rate $1/\tau = \Delta/\hbar$ between the two adjacent ionic configurations. In a temporal picture this means that one has a spatially homogeneous system and that the time average occupation of the f-shell is identical on each ion. The temporal mixture is of course fundamentally different from the spatial one. In the following we discuss briefly a few arguments that should support our point of view concerning the validity of the temporal picture and its relevance to the XPS-studies.

4. TEMPORAL VERSUS SPATIAL ADMIXTURE, RELEVANCE TO XPS-STUDIES

Estimates of Δ in metallic systems[3] involving f-shells lead typically to values of a few hundredths of an eV($\tau \approx 10^{-12}$-10^{-13} sec). This implies that for frequencies $\nu \ll 1/\tau$ the ICF-state is equivalent to a state of width Δ having the two ionic configurations I_1 and I_2 occupied in such a way as to yield a well defined time average. In Mössbauer measurements in

the limit of fast fluctuations defined by $\tau\Delta E/\hbar \ll 1$, where ΔE is the isomer shift separation of the two ionic configurations, a single line with width in excess of the natural width by $\tau\Delta E^2/4\hbar$ is obtained. This is indeed what is observed in many Eu-compounds[16]. The position of the absorption line is temperature dependent, showing that the mixture is temperature dependent, while its width is constant within experimental accuracy. From this an upper limit for the fluctuations can be estimated: $\tau \leq 10^{-11}$ sec. From their beautiful experiments Bauminger et al.[16] concluded that the Eu ion in $EuCu_2Si_2$, even at very low temperatures, is most probably not in a pure f^6 state. This situation is different from that in Eu_3S_4, where a hopping model is appropriate and a distribution of inequivalent lattice sites is known to exist[19]. A crystallographic distortion at 168°K has been related to a static charge ordering[20]. Below this temperature two well defined isomer shifts, corresponding to Eu^{2+} and Eu^{3+}, are indeed observed. Well defined upper limits for the fluctuation time are difficult to infer from magnetic, thermal or transport properties because the interpretation of the data usually involves too many adjustable parameters. Inelastic neutron scattering studies on paramagnetic trivalent TmSb using unpolarized and polarized neutrons[21] were extremely successful in determining crystal field parameters and the magnetic form factor. In contrast, in TmSe down to 1°K, i.e., below the magnetic phase transition temperature of about 3°K, neutron studies detected neither crystal field splittings nor long range magnetic order[22]. An analogous situation is present in $CeAl_3$[23], another system with intermediate valence state. It is clear that if the fluctuations between two valence states are comparable or slightly faster than the characteristic inverse frequency of thermal neutrons (10^{-12} sec) no splittings can be detected.
The time characteristic of XPS is thought to be 10^{-17} to 10^{-18} sec, i.e. much smaller than the time of the fluctuations. We therefore expect to be able to detect an instantaneous picture of the ions in the two valence states. The unambiguous identification of the two states is made possible by the following facts: (1) knowledge of the multiplet stucture of each f^n configuration, (2) conservation of the ionic structure of the f-shells in the mixed valence state, (3) separation by the intraconfiguration energy $U_{eff} \approx$ 5-10 eV of the two f^n and f^{n-1} configurations in the photoemission spectra, despite the small interconfiguration separation E_{exc}. The experimental XPS-results obtained so far on RE-systems in the intermediate valence state invariably show the presence of both divalent and trivalent ions. In the temporal picture this is reasonable and we can therefore set following limits for the ICF time: 10^{-11} sec $\leq \tau_{ICF} \leq 10^{-15}$ sec. Values of τ smaller than 10^{-15} sec would result in a detectable lifetime-broadening of the XPS lineshape of the 4f-multiplets.

5. XPS ON ICF-SYSTEMS

The main advantage of this spectroscopic method is its "universality". In fact, in contrast for example to the Mössbauer technique or neutron scattering, XPS can be applied to any material. ICF phenomena are expected to occur in semiconducting compounds (like TmTe) with E_{exc} of the order of a few hundred degrees, but more frequently, and under normal ambient conditions, in metallic systems. So far XPS has been applied successfully only to compounds with the simple NaCl-structure. This limitation arises largely from the need to prepare clean surfaces in situ in UHV by cleaving. However, since extremely clean conditions in XPS-spectrometers can now also be reached and maintained over relatively long periods of time[13], one should be able to prepare clean surfaces of more reactive materials, like the extremely interesting alloys $YbAl_2$ or $EuCu_2Si_2$. In the following we describe briefly the kind of information which one may obtain by using XPS on ICF systems.

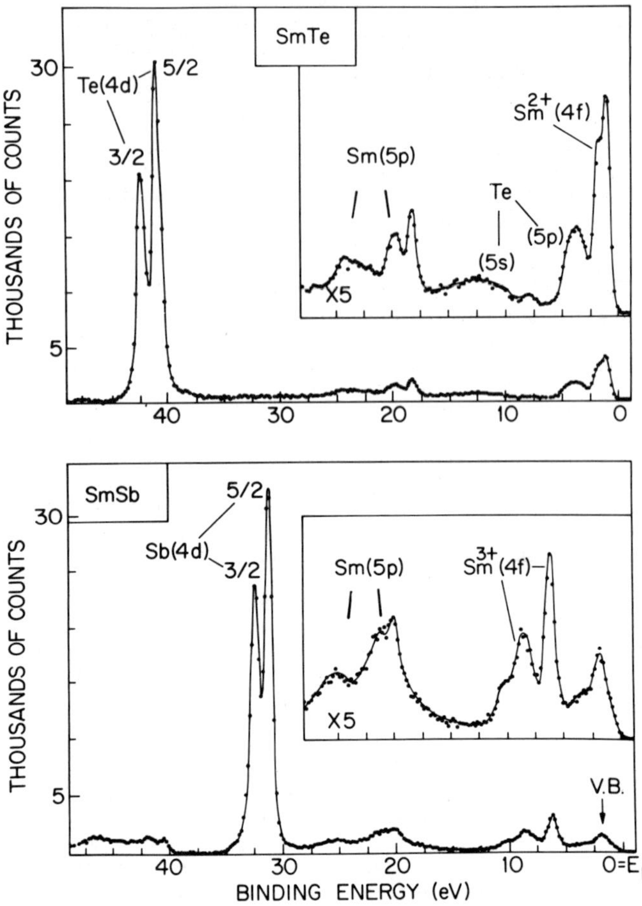

Figure 3. Overview of occupied energy levels of SmTe and SmSb in the range 0 - 45 eV below E_F.

5.1. Pressure induced changes in the f^n configurations.

SmTe and SmSb are formally divalent and trivalent compounds respectively. The corresponding f- configurations of Sm are $^7F_0$ and $^6H_{5/2}$ in the ground states. An overview of the occupied energy levels in the binding energy range from 0 to 45 eV below E_F is shown in Fig. 3. Of special interest for the present discussion is the relative position of the 4f-spectra. The spectrum corresponding to the transition $f^6 \rightarrow f^5$ (SmTe) is located in the binding energy region 0-4.5 eV, while that of the transition $f^5 \rightarrow f^4$ (SmSb) appears in the range 5-11 eV. For Sm^{2+} the total spin in the ground state is S = 3, for Sm^{3+}: S = 5/2. This is reflected in the structure of the Sm(5p)-core levels, the spin orbit doublet $5p_{3/2}$- $5p_{1/2}$ is masked by the coupling of the p-hole state with the spin in the f-shell[24]. This fact can be used as an additional indication for identifying the valence state of a RE-ion. The structure of the 5p-final state will be discussed in more detail in a forthcoming publication[25]. Continuous phase transitions from $Sm^{2+}Te$ to $Sm^{3+}Te$ ($f^6 \rightarrow f^5d^1$) can be obtained by applying external pressure (up to 60 kb)[14]. In SmS this transition is first order and occurs at the remarkably low pressure of 6.5 kbar. The effect of the external pressure can be simulated by substituting smaller cations for Sm^{2+} [26]. Susceptibility and lattice constant measurements show that the "chemically collapsed," metallic phase has an intermediate valence state, with valence of the Sm of about 2.7. Figure 4 shows two examples

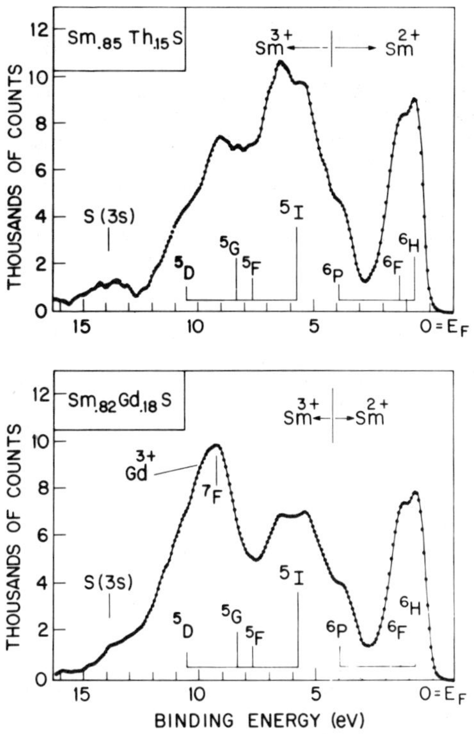

Figure 4. Coexistence of divalent and trivalent Sm-ions in chemically collapsed phases of SmS, as detected by the XPS-method.

of typical XPS results on chemically collapsed SmS. In the case of $Sm_{.85}Th_{.15}S$, it is easier to identify the contribution of the 2^+ and 3^+ initial states of Sm to the spectrum. Below the experimental results we have plotted the expected intensities[8], as calculated from the coefficients of fractional parentage. In an instantaneous picture of these ICF systems, both types ions are seen to coexist[27]. From the XPS-spectra it is possible to estimate also the mixing ratio $r = Sm^{3+}/Sm^{2+}$, by calculating the area under the corresponding multiplets. An accurate estimate requires a knowledge of contributions from electron states other than 4f. This could be achieved by "calibrating" the spectrum with an analogous material without f-electrons. However, since for $n = 7$, $\sigma_{4f^n} = 10\times\sigma_{V.B}$, as we have seen, an estimate with about 10%-20% accuracy can be obtained even by neglecting these contributions. In the specific case of $Sm_{.85}Th_{.15}S$ $r_{XPS} = 2.6$, to be compared with $r = 2.3$ from the lattice constant systematics of trivalent RE-sulfides[1]. The separation between the ^5I- and ^6H-multiplets gives an experimental measure of U_{eff}. We obtain from Fig. 4 for the metallic phase SmS: $U_{eff} = 5$-6 eV. This value makes it clear why the ionic structure of the f-shell is not destroyed by the interaction V_{kf} with the conductions electrons.

5.2. ICF-systems at normal ambient conditions.

In Fig. 5 we present results obtained with the singlet ground state paramagnet, TmSb, and with TmTe. TmTe behaves like a semiconductor, at least as far as resistivity is concerned, but its lattice constant ($a \simeq 6.33$ Å) deviates by about 5-15% from the extrapolated value obtained from the systematics of the lattice constants of divalent semiconducting RE-tellurides ($a = 6.37$ Å). We have observed that the value of the mixing ratio r_{XPS} in TmTe, as detected by XPS, depend strongly on vacuum conditions. Even after cleaving in a low 10^{-9} Torr vacuum in a baked system a steady conversion of Tm^{2+} to Tm^{3+} near the surface was detected. This is not the case for SmS, demonstrating further the inherent instability of the divalent state of the Tm ion. More recent photoemission and optical reflectivity work by Freeouf, Güntherodt and Eastman[29] on single crystals cleaved

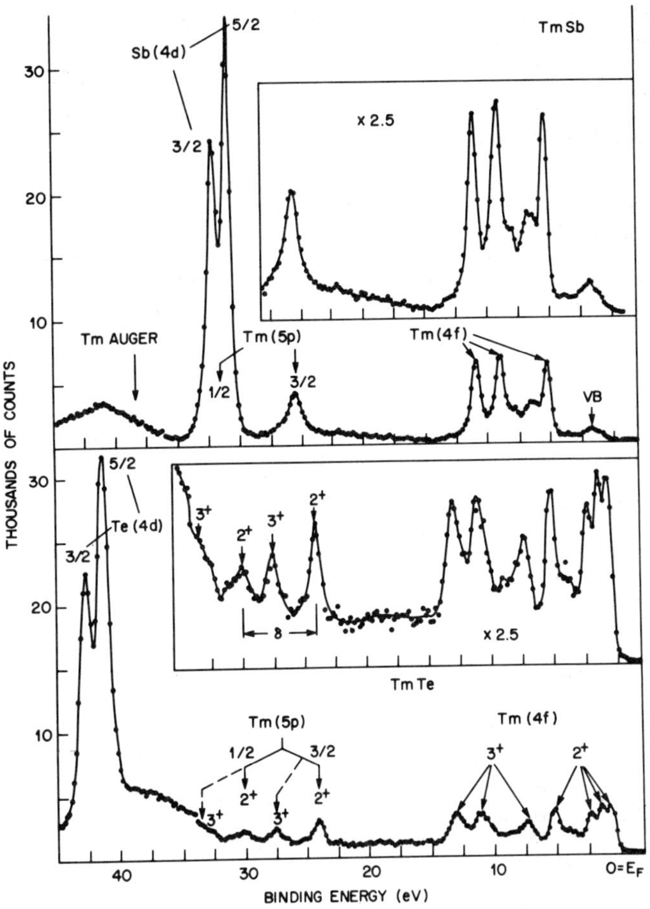

Figure 5. Like Fig. 3, but for TmSb and TmTe.

in situ, in the low 10^{-10} Torr range, shows that also for TmTe r_{XPS} is related to the value of r_a obtained from the systematics of the lattice constants. Although "semiconducting", TmTe shows a resistivity behavior that does not allow the estimate of a unique energy gap ΔE, since the resistivity does not follow a simple exponential law, although it changes by 5 orders of magnitude in the temperature range 100-300°K[28]. The structure observed in the spectra of Fig. 5 have been discussed in a recent letter[30]. Of relevance to the present discussion are peaks labeled Tm(5p) in TmTe; in going from Tm^{2+} to Tm^{3+} the doublet $5p_{3/2}$ is shifted toward higher binding energies by about 3.5 eV. This "chemical shift" is quite accurate because both charge states are measured at

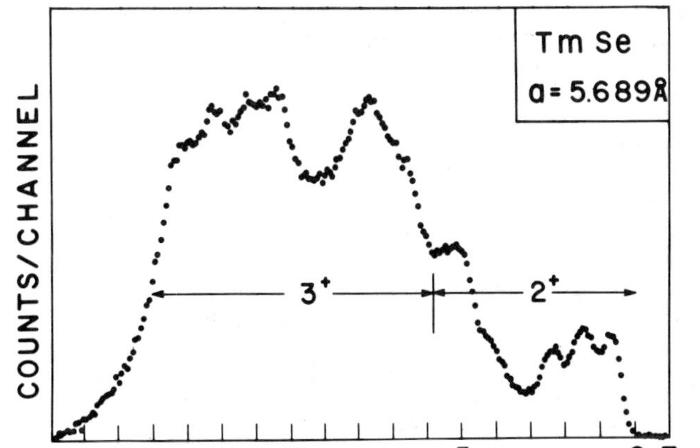

Figure 6. Like Fig. 4, but for the pure compound TmSe.

the same time on the same sample, avoiding therefore complications due to work function changes and relaxation. The doublet structure of the 5p-spectrum is not strongly perturbed by coupling between hole-state and the 4f-spin S. This is so because the 5p-spin-orbit splitting $\delta \cong 6$ eV is quite large and S = 1 and 1/2 for Tm^{3+} and Tm^{2+}, respectively[24].

In Fig. 6 we show the 4f multiplet structure of TmSe. The lattice constant of this compound deviates by about 15-25% from the one expected for trivalent TmSe (a = 5.64 Å). In TmSe the spectrum corresponding to Tm^{3+} is not as well resolved as the one of Tm^{2+}. The reason for this is not yet known. For TmSe, as is the case for every metallic ICF-system investigated so far, reasonable agreement is found between r_{XPS} and r_a inferred from the lattice constant systematics.

For TmSe $r_{XPS} \cong 4$. The Tm-monochalcogenides TmSe and TmTe show the interesting novel feature of a magnetic phase transition in an intermediate valence state, in strong contrast to the other ICF-materials involving Ce, Sm, Eu or Yb ($EuIr_2$ might also be an exception) The mechanism leading to magnetic ordering is not clarified. Mössbauer studies[31] down to 51 m°K in the cubic TmSe show surprisingly a fully resolved magnetic hyperfine structure in the absence of large quadrupole interactions. But the ratio $r = Tm^{3+}/Tm^{2+}$ obtained from these Mössbauer data is in conflict from that expected from the lattice constant systematics. The complete understanding of the physical phenomena occurring in these compounds, which are in the mixed valence state at normal ambient conditions like the "chemically collapsed" phases of SmS, will help to clarify the origin of the "nonmagnetic" behavior in other ICF-systems. XPS at very low temperatures would provide valuable information in this direction.

5.3. Temperature dependence of ICF

Again, from lattice constant measurements as a function of temperature[26,31] it can be shown that in many chemically collapsed phases of SmS $r = Sm^{3+}/Sm^{2+}$ decreases on cooling the samples below room temperature or by heating at 900°K. This reversible volume change, related to reversible variations in the mixing $4f^n/4f^{n-1}5d^1$, can also be detected by XPS[32,33]. It demonstrates more convincingly that the ICF-state is indeed responsible for the peculiar properties of these compounds.

6. CONCLUSIONS

XPS has been shown to be an ideal technique for studying the physics of ICF-systems. With this method information can be obtained on important parameters like the mixing ratio $r = RE^{2+}/RE^{3+}$, its temperature dependence and the Coulomb correlation energy U_{eff}. Effects of the ICF-state on core level binding energies as well on transition probabilities from f-shells can also be studied. Future work will involve investigations of the dependence of binding energies of f-levels on local environment and its relation to the semiconductor-metal transition of divalent RE-compounds.

ACKNOWLEDGMENT

We are indebted to many colleagues including K. Andres, R. J. Birgenau, S. T. Chui, R. L. Cohen, B. Coqblin, R. E. Dietz, L. L. Hirst, A. Jayaraman, I Nowik and C. Varma for helpful discussions. We also thank D. N. E. Buchanan and L. D. Longinotti for technical assistance, A. S. Cooper and P. Dernier for taking X-ray data.

REFERENCES

1. D. K. Wohlleben and B. R. Coles, Magnetism V, Ed. H. Suhl, Academic Press, 1973.
2. M. B. Maple and D. K. Wohlleben, AIP Conf. Proc. 18, 447, 1974.
3. B. Coqblin and A. Blandin, Adv. Phys. 17, 281, 1968.
4. D. E. Eastman and M. Kutznietz, J. Appl. Phys. 42, 1396, 1971.
5. G. K. Wertheim, A. Rosencwaig, R. L. Cohen and H. J. Guggenheim, Phys. Rev. Letters 27, 505, 1971.
6. P. O. Heden, L. Loefgren and S. B. Hagstroem, Phys. Rev. Letters 26, 432, 1971.
7. M. Campagna, E. Bucher, G. K. Wertheim, D. N. E. Buchanan and L. D. Longinotti, Proc. 11th Rare Earth Conf., p. 810, Traverse City, Mi., 1974.
8. P. A. Cox, Y. Baer and C. K. Jørgensen, Chem. Phys. Letters 22, 443, 1973.
9. G. Racah, Phys. Rev. 176, 1352, 1949.
10. G. H. Diecke and H. M. Crosswhite, Appl. Optics 2, 675, 1963.
11. D. E. Eastman, F. Holtzberg, J. L. Freeouf and M. Erbudak, AIP Conf. Proc. 18, 1030, 1973.
12. J. F. Herbst, D. N. Lowy and R. E. Watson, Phys. Rev. B6, 1913, 1972.
13. Y. Baer and G. Busch, Proc. Intl. Conf. on Electron Spectroscopy, Namur, April 15-19, 1974.
14. A. Jayaraman, V. Narayanamurti, E. Bucher and L. D. Longinotti, Phys. Rev. Letters 25, 1430, 1970.
15. S. D. Bader, N. E. Phillips and D. B. McWhan, Phys. Rev. B7, 4686, 1973. J. V. Mahoney, V. Rao, W. E. Wallace, R. S. Craig and N. G. Nereson, Phys. Rev. B9, 154, 1974.
16. E. R. Bauminger, I. Felner, D. Froindlich, D. Levron I. Novik, S. Ofer and R. Yanovsky, to be published in J. de Phys. (Paris).
17. P. W. Anderson and S. T. Chui, Phys. Rev. B9, 3229, 1974. C. M. Varma and V. Heine, to be published. J. R. Iglesias Sicardi, A. K. Bhattacharje, R. Jullien and B. Coqblin, Proc. 11th Rare Earth Conf., p. 593, Traverse City, Mi. 1974. B. Alascio and A. Lopez, Sol. St. Comm. 14, 321, 1974. H. S. Wio, B. Alascio and A. Lopez, to appear in Sol. St. Comm.
18. L. L. Hirst, Phys. Kond. Mat. 11, 255, 1971, and J. Phys. Chem. Solids 35, 1285, 1974.
19. C. Berkooz, M. Malamud and S. Strikman, Solid St. Comm. 6, 185, 1968.
20. H. H. Davis, I. Bransky, N. M. Tallan, J. Less. Comm. Metals 22, 193, 1970.
21. R. J. Birgenau, E. Bucher, L. Passel and C. K. Tuberfield, Phys. Rev. B7, 718, 1971. G. H. Lander, T. O. Brun and O. Vogt, Phys. Rev. B7, 1988, 1971.
22. D. E. Cox, L. Passell, R. J. Birgenau and E. Bucher, unpublished. H. A. Mook and H. L. Davis, private communication.
23. A. Furrer, private communication.
24. For a general review of multiplet splittings in XPS-spectra of open shell ions see: G. K. Wertheim, Proc. Nato Adv. Studies on "Electronic Structure of Inorganic Materials", Oxford, Sept. 8-18, 1974.
25. M. Campagna, E. Bucher, G. K. Wertheim, Y. Yafet, to be published.
26. A. Jayaraman, E. Bucher, P. D. Dernier, L. D. Longinotti, Phys. Rev. Letters 31, 700, 1973.
27. M. Campagna, E. Bucher, G. K. Wertheim and L. D. Longinotti, Phys. Rev. Letters 33, 165, 1974.
28. E. Bucher, K. Andres, F. J. di Salvo, J. P. Maita, A. C. Gossard, A. S. Cooper, and G. W. Hull, Jr. to be published in Phys. Rev. B, Jan. 1975.
29. J. L. Freeouf, G. Güntherodt and D. Eastman, private communication and to be published.
30. M. Campagna, E. Bucher, G. K. Wertheim, D. N. E. Buchanan and L. D. Longinotti, Phys. Rev. Letters 32, 885, 1974.
31. B. B. Triplett, N. S. Dixon, P. Boolchand and S. S. Hanna, Proc. Intl. Conf. on "Applications of the Mössbauer effect", Bendor, Sept. 2-6, 1974.
32. A. Jayaraman and P. Dernier, private communication. R. A. Pollak, F. Holtzberg, J. L. Freeouf and D. E. Eastman, Phys. Rev. Letters 33, 820, 1974.
33. M. Campagna and P. Dernier, to be published.

EXCITED-STATE SPIN WAVES, SOFT MODES AND CRYSTAL FIELD EFFECTS

W.J.L. Buyers
Atomic Energy of Canada Limited, Chalk River, Ontario, Canada, KOJ 1JO

ABSTRACT

There are several branches of the spin-wave spectrum in materials where the exchange and crystalline field are comparable. Spin waves out of excited states can play an important role by mixing and interacting with the conventional spin waves out of the ground state. These effects are illustrated by neutron scattering results for TbSb, where several branches were observed, for $TbAl_2$, where an excited-state spin wave has been directly observed, and for Pr_3Tl where the mode-mode interaction has been found to be responsible in part for the surprising lack of temperature dependence through T_c of the spin-wave peak. A theory of the dynamical susceptibility has been developed that includes, in RPA, all spin waves allowed within the ground multiplet, and is found to be in good agreement with experiment in contrast to the singlet-singlet or singlet-triplet models. For Pr_3Tl as $T \to T_c$ the frequencies of the singlet-triplet modes decrease but remain finite. The modes within the triplets do tend to zero frequency at T_c, but do not drive the transition since they are entirely transverse. The soft "mode" is a zero-frequency mode that appears in the scattering as a central peak that diverges at the phase transition. The central peak, which arises from localized spin fluctuations, has probably been observed in TbSb.

INTRODUCTION

A great deal can be learned about phase transitions by studying the spin dynamics of magnetic crystals. In contrast to the situation for phonons, especially ferroelectrics, the magnetic interactions are simple, yet the result can be a spectrum of considerable richness. This arises because each ion has a set of localized states, determined by the average exchange field, the crystal field and the spin-orbit interaction, between which transitions can take place. The exchange coupling between sites leads to the formation of as many bands of crystal excitations as there are transitions[1]. Because the dominant exchange contribution is bilinear in the spins, and because the excitation spectrum is customarily observed with a magnetically dipolar probe, namely neutron scattering or infrared absorption, it is usual to restrict attention to those transitions, the spin-wave transitions, that are magnetic dipole in character.

The nature of the phase transition when there are many degrees of freedom at each site is by no means clear despite extensive work. The canonical example, where the problem is clearest but the difficulties still acute, is the rare-earth system with a singlet ground state[2,3,4]. When the ratio of exchange to crystal field splitting, Δ, exceeds a critical value, the material undergoes a phase transition at a finite temperature T_c. Following the work of Cochran[5] one expects that, if the transition is second order, some "mode" must go soft at the phase transition and that this mode must be of the appropriate symmetry to make the longitudinal susceptibility diverge at T_c.

Two years ago at this conference R.J. Birgeneau[6] presented neutron results for the singlet-ground-state ferromagnet Pr_3Tl (T_c=11.6 K). He showed that, although only finite wave vectors could be studied, it was most unlikely that the frequencies of the zone-centre spin waves approached zero. Indeed his results showed that the frequencies at Q=0.6Å^{-1} remained approximately independent of temperature from 4.4 K to ~60 K (Fig.1). This unexpected result, contrasting strongly with the renormalization approximately with $\langle S_z \rangle$ found in pure spin systems[7], appeared to deal a vital blow to most

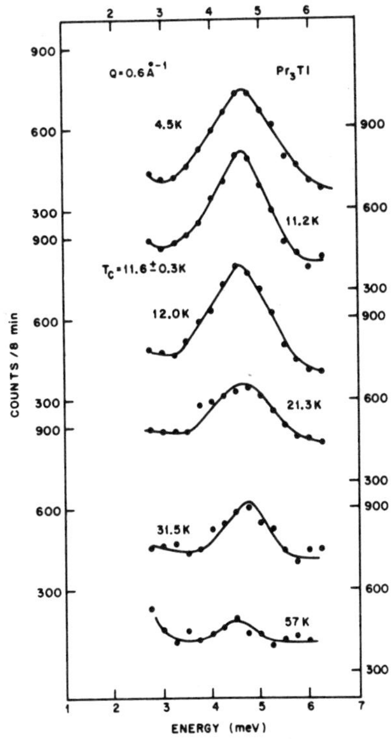

Fig. 1. Temperature dependence of the peak observed in the neutron scattering from Pr_3Tl (ref. 6).

simple theories of the phase transition. It was later shown[8] that the weak temperature dependence is in fact consistent with a many-level theory in the random phase approximation.

In another experiment[9], carried out on a single crystal of the single-ground-state antiferromagnet TbSb (T_N=14.9 K), the spin waves near $\vec{q}$=0 were examined in detail. Their frequencies were observed to

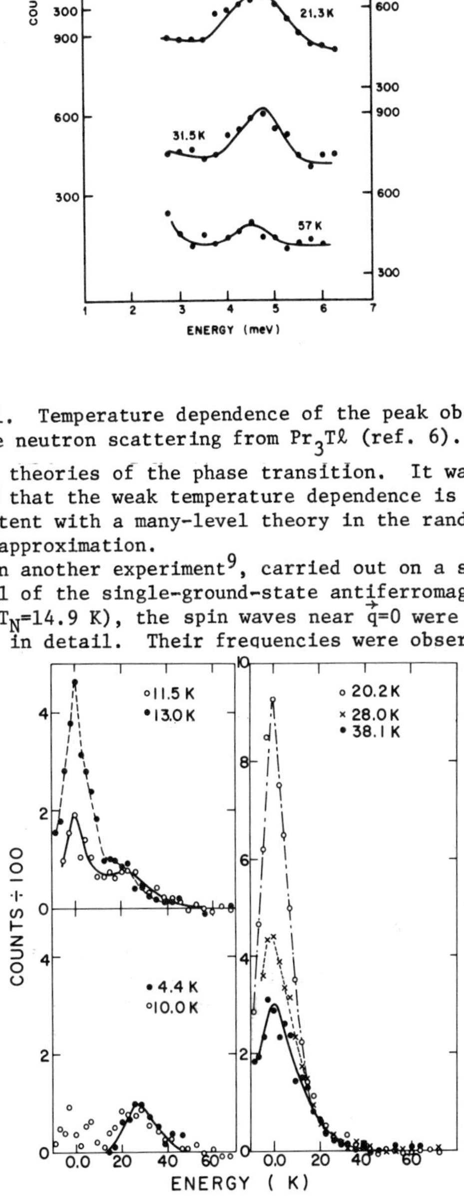

Fig. 2. Temperature dependence of the spin wave and central mode observed in TbSb (ref. 9) at $\vec{q}$ = (0.05,0.05,0.05)2π/a.

fall as $T \to T_N$ but that, before reaching zero frequency, they were obscured by the wings of a very intense peak centred on zero energy (Fig.2). The intensity of this peak appeared to diverge at T_N and was reminiscent of critical scattering. Although the ratio of exchange to crystal field in TbSb is more than three times greater than the ratio for $Pr_3T\ell$ the systems are similar in other respects. Both have a Γ_1 singlet ground state and a Γ_4 first excited triplet. It is likely that the central mode is common to both systems, although it could not be studied near $\vec{q}=0$ in $Pr_3T\ell$ since, in neutron measurements on polycrystals made in the (000) Brillouin zone, the smallest $\vec{q}$ is set by the smallest angle that the spectrometer can reach without running into the incident beam.

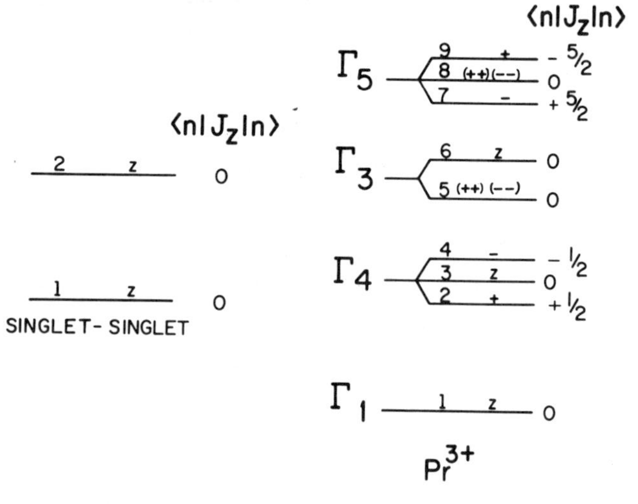

Fig. 3. Symmetries and moments of states for the singlet-singlet problem and for Pr^{3+} in a cubic field. Allowed transitions involving J_z are between states with the same labelling. Allowed transitions of J_+ can be found from knowing that J_+ acting on a state changes the labels as follows, $z \to +$, $- \to z$, $(++)(--) \to -$. J_- acting on a state changes the labels as follows, $z \to -$, $+ \to z$, $(++)(--) \to +$. All other transitions are absent.

In neither of these experiments was it possible to identify the symmetry of the mode or modes responsible for the phase transition. Figure 3 shows the symmetries of the energy levels of the Pr^{3+} ion in a cubic crystal field. Transitions involving matrix elements of J_+ or J_- are called transverse, while those involving matrix elements of J_z are longitudinal. We shall also distinguish scattering of two types, the scattering caused by inelastic transitions between different levels of the ion or spin-wave scattering, and the elastic† scattering within the same level of the ion. The two processes require non-zero matrix elements of the spin between and within the levels respectively. The elastic† scattering is entirely longitudinal.

The states of the lowest triplet are connected to the ground state by matrix elements of J_+, J_z and J_-. Within the triplet there are other inelastic transitions, both transverse, but there are no inelastic longitudinal transitions. This suggests that as $T \to T_c$ from below the transverse modes may soften as the triplet closes. This behaviour of the intra-triplet modes was confirmed by the many-level theory[10], but it is not sufficient to explain the phase transition.

What is required is a divergence of not the transverse but the longitudinal susceptibility associated with the appearance below T_c of the ordered moment $\langle J_z \rangle$. The only inelastic transition that will do is the longitudinal $1 \to 3$ transition. We will show, however, that this mode does not in fact go soft. We are therefore led to examine the divergence of the longitudinal elastic scattering associated with the matrix elements $\langle n|J_z|n\rangle$. Although the modes causing the scattering are not excitations in the conventional sense, these zero-frequency modes, first discussed by Smith[11] for the singlet-triplet problem, are valid degrees of freedom of the system with the correct longitudinal symmetry.

Much of the early work on singlet-ground-state systems was based on the singlet-singlet model[2]. The only non-zero matrix element is that of J_z between the states as shown in Fig.3. The model is likely too simple to show the behaviour found in real crystals. Thus two of the Γ_4 triplet states for Pr^{3+} carry a moment but neither of the states in the singlet-singlet system are magnetic. It is for this reason that the frequency of the longitudinal spin-wave mode falls to zero at T_c in the singlet-singlet model; there is no other degree of freedom that will lead to a divergence of the longitudinal susceptibility. In the singlet-triplet system or in a real crystal it is not necessary for the frequency of any longitudinal spin-wave to fall to zero.

In what follows we shall review current theories that can describe the soft-mode behaviour, the mode-mode interaction, and the two types of magnetic response. The discussion will be specialized to a rare earth ion in a cubic crystalline field but the conclusions regarding the nature of the phase transition are widely applicable. The reader is referred to refs. 8 and 10 for many-level calculations for the Pr^{3+} ion and, for a clear explanation of the behaviour of the singlet-triplet model, to the letter by Lines[12].

THEORY OF MANY-LEVEL SPIN WAVES

The moment induced in a crystal by a field $h_\beta(\vec{q}\,\omega)$ in the β direction varying sinusoidally in space and time with wave vector, $\vec{q}$, and frequency, ω, is given by

$$\delta J_\alpha(\vec{q}\,\omega) = G^{\alpha\beta}(\vec{q}\,\omega)\, h_\beta(\vec{q}\,\omega) \qquad (1)$$

where the change in moment is $\delta\vec{J} = \vec{J}(\vec{q}\,\omega) - \langle\vec{J}\rangle$. The field h is measured in units $(-g_J\mu_B)$ for rare earth ions of total angular momentum $\vec{J}$. The dynamic susceptibility, G, is directly related to the spectral properties of the crystal. For example the neutron scattering from the longitudinal spin components is given by

$$S(\vec{q}\,\omega) = \pi^{-1}[1-\exp(-\omega/kT)]^{-1}\ \mathrm{Im}\, G^{zz}(\vec{q}\,\omega)\ . \qquad (2)$$

As $\vec{q} \to 0$ and $\omega \to 0$ the same susceptibility gives the ordinary static susceptibility, $\chi^{zz} = g_J^2\mu_B^2\, G^{zz}(\vec{0}\,0)$, whose divergence provides the experimental evidence for a phase transition.

The susceptibility is defined in terms of correlation functions as the time and space fourier transform of

$$G^{\alpha\beta}(ij,t) = i\theta(t)\langle[\delta J_\alpha(i,t),\ \delta J_\beta(j,0)]\rangle \qquad (3)$$

where the triangular brackets denote the thermal average and $\delta J_\alpha(i,t)$ is the spin deviation at site i at time t.

The Hamiltonian for the spins consists of a crystal field part, V_{CF}, and an exchange part:

$$H = \sum_i V_{CF}(i) - \sum_{ij} \mathcal{J}(ij)\vec{J}(i).\vec{J}(j). \qquad (4)$$

A single-ion part may be extracted,

$$H_1 = \sum_i V_{CF}(i) - 2\sum_i J_z(i)\tilde{\mathcal{J}}(\vec{0})\langle J_z\rangle - \sum_i\sum_\alpha \mathcal{J}(\vec{0})p_\alpha J_\alpha(i)^2 \qquad (5)$$

where $\tilde{\mathcal{J}}(0) = \mathcal{J}(\vec{q}=0)(1-p_z)$. If the correlation parameter p_α of the correlated effective field (CEF) introduced by Lines[12] is neglected, then Eq.(5) is the

Hamiltonian for an ion in the molecular field approximation (MFA). In either case H_1 may be diagonalized exactly to yield a set of energies, ω_n, and states, $|n\rangle$, created by the operator C_n^+, so that H_1 may be written

$$H_1 = \sum_i \sum_n \omega_n C_n^+(i) C_n(i). \tag{6}$$

The appropriate ordered moment $\langle J_z \rangle$ differs in the MFA and CEF but the form of (6) does not.

Any operator can now be written[10] in terms of the operators $C_m^+ C_n$ that take an ion from the state $|n\rangle$ to the state $|m\rangle$. Thus

$$J_\alpha(i) = \sum_{mn} J_{\alpha mn} C_m^+(i) C_n(i) \tag{7}$$

where $J_{\alpha mn} \equiv \langle m|J_\alpha|n\rangle$. It then may be shown[10] from the Heisenberg equation of motion for all the possible $C_m^+ C_n$, with m,n running over the ground multiplet, that the transverse spin waves are given by the poles of

$$G^{+-}(\vec{q}\,\omega) = g^{+-}(\omega)/[1 - \mathcal{J}(\vec{q}) g^{+-}(\omega)] \tag{8}$$

with a similar expression for $G^{-+}(\vec{q}\,\omega)$, and the longitudinal spin waves by the poles of

$$G^{zz}(\vec{q}\,\omega) = g^{zz}(\omega)/[1 - 2\mathcal{J}(\vec{q}) g^{zz}(\omega)]. \tag{9}$$

The notation here follows that of Buyers and Holden[10]. Similar ideas have been developed by Peschel, Klenin and Fulde[13] for the paramagnetic phase, by Klenin and Peschel[14], and by Bak[15] for the ordered phase, and by Haley and Erdös[16] in a more formal manner. The correlated effective field approximation of Lines[12] in the $\alpha\alpha^*$ susceptibility is obtained by replacing $\mathcal{J}(\vec{q})$ by $\mathcal{J}(\vec{q}) - p_\alpha \mathcal{J}(\vec{0})$. The non-interacting susceptibility is

$$g^{\alpha\alpha^*}(\omega) = \sum_{mn} |\delta J_{\alpha mn}|^2 (f_n - f_m)/(\omega - \omega_{nm}), \tag{10}$$

where $\omega_{nm} \equiv \omega_n - \omega_m$ and f_n is the probability that the single-ion state $|n\rangle$ is occupied, $\exp(-\omega_n/kT)/Z$, where Z is the partition function.

Equations (8) and (9) are simple in appearance but contain a great deal of the physics. The temperature dependence above T_c is contained in the f_n and below T_c, where the state $|n\rangle$ depends on $|J_z\rangle$, also on the ω_{mn} and $J_{\alpha mn}$. Because the single-ion susceptibility, Eq.(10), describes transitions between all levels of the ground multiplet, the theory can describe spin waves corresponding to transitions out of excited states as well as the more usual spin waves out of the ground state. The extra spin-wave branch observed in $TbAl_2$ by Purwins et al.[17] is an example of an excited-state spin wave. Its intensity is enhanced by interaction with the strong spin wave out of the ground state. Mode-mode interaction also plays a part in producing the unusually weak temperature dependence of the neutron peak observed by Birgeneau et al.[6] in $Pr_3T\ell$ as we shall see shortly.

SPIN WAVES IN $Pr_3T\ell$

The theory of the last section leads to detailed predictions of the temperature dependence of all the spin waves of the ground multiplet. For $Pr_3T\ell$ the crystal field and exchange parameters are known from the work of Birgeneau[6] and are chosen to reproduce the dispersion relation measured at the lowest temperatures as shown in Fig.4. Calculations have also been made for the dilute alloy $(Pr_{1-x}La_x)_3T\ell$ by replacing $\mathcal{J}(\vec{q})$ by $(1-x)\mathcal{J}(\vec{q})$. The spin waves depend only on the inelastic transitions, $m \neq n$ in the $g(\omega)$ of Eq.(10), as there is no contribution to $g(\omega)$ at any finite frequency from the elastic or diagonal transitions with m=n. The energy gap at $\vec{q}=0$ in Fig.4 is characteristic

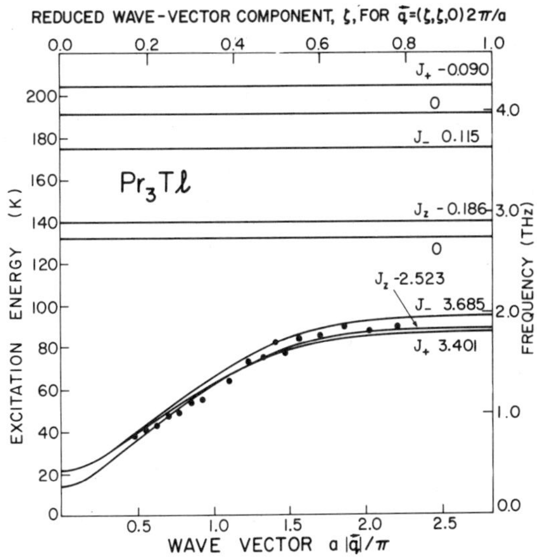

Fig. 4. Dispersion relations for many-level spin waves in $Pr_3T\ell$ from the dynamical susceptibility theory (lines, ref. 10) and neutron scattering (points, ref. 6).

of the spin waves in crystals where the crystalline field is important. The singlet-triplet model cannot predict a gap as it has a transverse Goldstone mode at all temperatures[18].

As the temperature is raised the exchange-split Γ_4 triplet becomes populated and permits transitions from Γ_4 to Γ_3 to take place. These lie within the

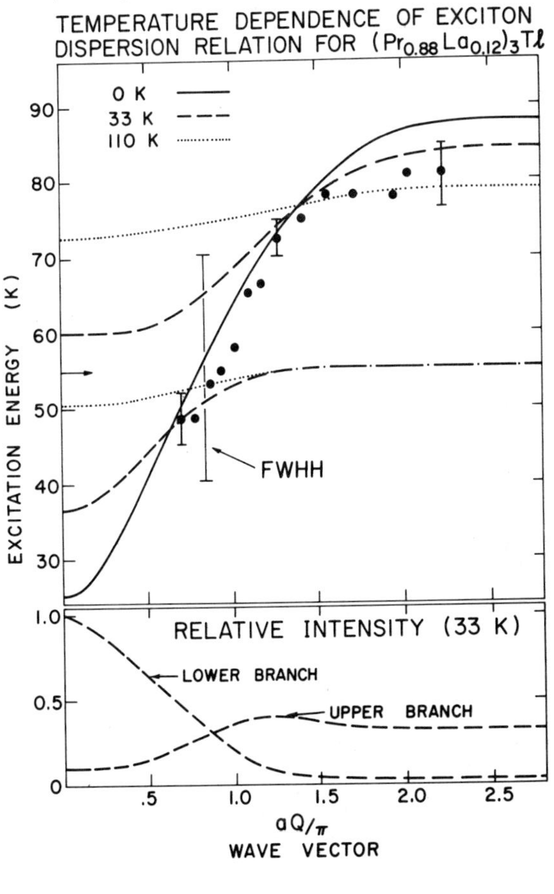

Fig. 5. Mode-mode interaction in $(Pr_{0.88}La_{0.12})_3T\ell$ (theory from ref. 8 and experimental points at 4.4 K from ref. 6).

band of Γ_1-Γ_4 spin waves and, since they have the same symmetry, mode-mode repulsion occurs. The behaviour shown in Fig.5 is a typical example[8] of anticrossing effects in this case for non-magnetic $(Pr_{0.88}La_{0.12})_3Tl$. The plot of the intensity at the foot shows how the intensity transfers smoothly from the lower to the upper branch as the wave vector increases. In Fig.5 the splitting between the two modes increases with temperature as expected from the theoretical prediction for two isolated modes that the splitting is proportional to $\exp(-\Delta/kT)$ at low temperatures, where Δ is the energy of the excited level giving rise to the interacting spin waves. Thus as $T\rightarrow 0$ the range of wave vectors where anticrossing effects are appreciable becomes vanishingly small.

From the full width at half height of the peak in the neutron scattering observed by Birgeneau et al.[6] (vertical bar in Fig.5) it is clear that the spectrometer samples both modes in the region of the crossover. The appropriate weighted average over the modes was computed in reference 8 and is shown in Fig.6. The many-level dynamical susceptibility theory is in excellent agreement with experiment. The temperature dependence is weak because the wave vector is relatively large and thus far from the zone centre where the most dramatic changes with temperature take place. Further, as the temperature is raised, some of the modes under the neutron peak increase in frequency while others decrease. The singlet-triplet model does not contain the latter effect since excited-state spin waves are not included. The main observable change is a decrease in peak intensity with temperature resulting from the depopulation of the ground state (lower part of Fig.6). Similar results have been obtained[10] for pure Pr_3Tl through its ordering temperature, and for other dilute alloys.

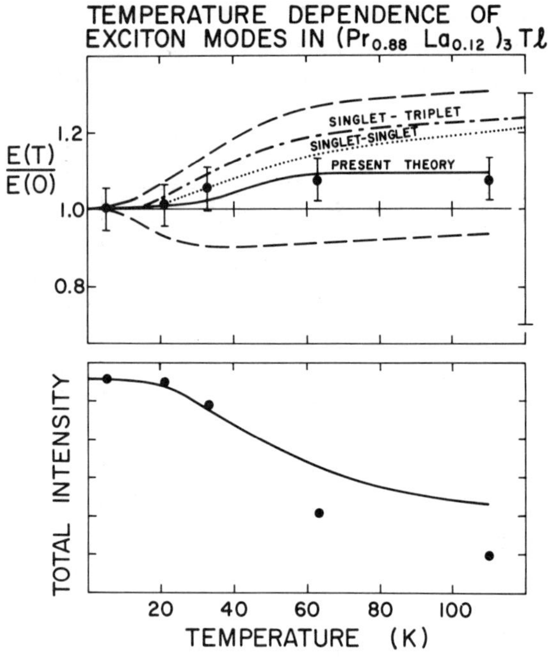

Fig. 6. Temperature dependence of normalized frequency E(T)/E(0), and of total intensity in arbitrary units, of the peak in the neutron scattering. The experimental points are from ref. 6.

SOFT MODE BEHAVIOUR

We have seen that the theory of the spin waves based on the many-level dynamic susceptibility gives a good description of what experimental results are available. The predictions of the theory are shown in Fig.7 for the behaviour of the spin waves at the zone centre, a region that was inaccessible in the experiment. The frequencies of all three members of the Γ_1-Γ_4 branch remain finite at the phase transition. The results show that two very low frequency transverse spin waves, associated with J_- inelastic transitions (see Fig.3) within the Γ_4 triplet, appear with appreciable intensity for $T \gtrsim 2/3\ T_c$ and finally go soft as $T\rightarrow T_c$. Two additional J_- modes within the Γ_5 triplet are not shown in Fig.7 as they are extremely weak. Above T_c the triplet modes have zero frequency while the main Γ_1-Γ_4 branch rises and tends to the crystal field splitting, Δ.

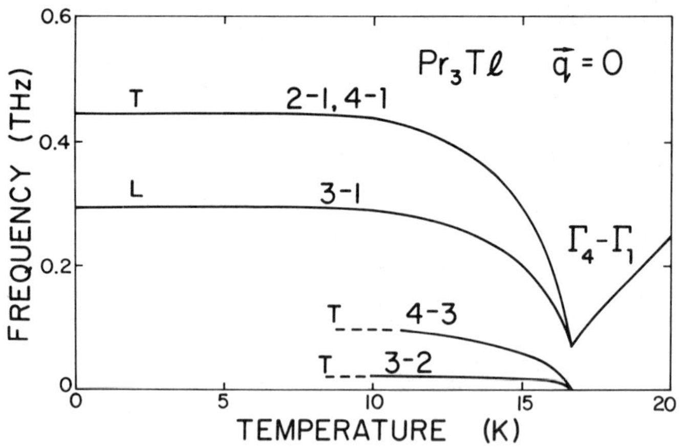

Fig. 7. Temperature dependence of the zone-centre spin waves in Pr_3Tl (ref. 10).

Although the triplet modes show soft-mode behaviour they tell only part of the story about the phase transition. To begin with they are entirely transverse. There are no soft inelastic spin waves of longitudinal symmetry. One can easily verify that, if only the inelastic transitions are included, the transverse static susceptibility diverges at T_c because of the soft transverse modes, but the longitudinal susceptibility does not. The additional part of the static susceptibility that is necessary for the divergence at T_c, is a contribution that has been ignored until now since it does not affect the spin dynamics directly. This contribution comes from the elastic transitions m=n in $g^{zz}(\omega)$. This suggests that the transition is driven by a set of zero frequency modes whose intensity diverges as $T\rightarrow T_c$. This account of the transition was originally proposed by Smith[11] and has been developed more fully by Lines[12] for the singlet-triplet model. We generalize Lines' treatment in what follows for an arbitrary ion with many levels.

To discuss the soft-mode behaviour it is useful to write the longitudinal dynamic susceptibility (henceforth we omit the zz suffixes) in terms of the eigenfrequencies, $\omega(\vec{q}\,j)$, as

$$G(\vec{q}\,\omega) = -\sum_j R(\vec{q}\,j) / [\omega - \omega(\vec{q}\,j)]\,. \tag{11}$$

The mode strengths are given by

$$1/R(\vec{q}\,j) = 4\,\mathcal{J}(\vec{q})^2\, dg/d\omega|_{\omega=\omega(\vec{q}\,j)}\,, \tag{12}$$

and the eigenfrequencies by the solutions of

$$g(\omega,T) = 1/(2\,\mathcal{J}(\vec{q}))\,. \tag{13}$$

The argument, T, has been inserted in the $g(\omega)$ of Eq.(10) to emphasize its temperature dependence. The non-interacting static susceptibility, divided into a diagonal part and a crystal field or inelastic part, can be written

$$g(0,T) = \langle \delta J_{zd}^{\,2}\rangle_T/kT + g_{CF}(0,T),$$

where the diagonal spin fluctuation is

$$\langle \delta J_{zd}^{2}\rangle_T = \sum_n |J_{znn}|^2 f_n - \langle J_z\rangle^2 .$$

The transition temperature, T_c, is given by

$$\langle \delta J_{zd}^{2}\rangle_{T_c}/kT_c + g_{CF}(0,T_c) = 1/(2\mathcal{J}(\vec{0})) \qquad (14)$$

In the limit $T\to 0$ the diagonal spin fluctuation vanishes, but it increases greatly when $T\to T_c$ from below since $\langle J_z\rangle$ is decreasing rapidly. For Pr_3Tl the diagonal term is only a small though essential fraction of the non-interacting susceptibility, $g(0,T)$, even near T_c; at 16K we find $g_{CF}(0,T) = 8.0334\ \mathrm{THz}^{-1}$, while $\langle\delta J_{zd}^2\rangle_T/kT$ is only $0.0175\ \mathrm{THz}^{-1}$, but the enhancement gives a large static susceptibility of $2886.2\ \mathrm{THz}^{-1}$.

From the mode analysis of Eq.(11) we see that the static susceptibility at the ordering wave vector,

$$G(\vec{0},0) = \sum_j R(\vec{0},j)/\omega(\vec{0},j), \qquad (15)$$

can only diverge if one of the frequencies tends to zero. This is the basic soft-mode concept developed by Cochran[5] for ferroelectrics, but it is valid for any second-order phase transition. Since none of the inelastic longitudinal spin-wave modes are soft we fix attention on the "modes" arising from the diagonal terms in the susceptibility. To obtain the equivalent mode frequency we take the limit $m\to n$ in (10), set $\omega_{nm} = \pm\varepsilon$ in a symmetrical way, where ε is a small quantity, and find that the frequencies of the diagonal modes are given by $\omega = \lambda_{\vec{q}}\varepsilon$, where

$$(1/kT)\langle\delta J_{zd}^{2}\rangle_T/(1-\lambda_{\vec{q}}^2) + g_{CF}(0,T) = 1/(2\mathcal{J}(\vec{q})). \qquad (16)$$

Comparison with Eq.(14) shows that $\lambda_{\vec{0}} \to 0$ as $T\to T_c$. Although the strength of the mode $\omega = \lambda_{\vec{q}}\varepsilon$ in $G(\vec{q}\,\omega)$ is vanishingly small, being proportional to ε, the neutron scattering from the mode has a well-defined limit independent of ε since the temperature factor in Eq.(2) brings in a cancelling factor kT/ε. The neutron scattering appears as a central mode whose integrated intensity is

$$S_0^{zz}(\vec{q}) \equiv \int S^{zz}(\vec{q}\,\omega)\,d\omega = \frac{\langle\delta J_{zd}^{2}\rangle_T/[1-2\mathcal{J}(\vec{q})g_{CF}(0,T)]}{1-2\mathcal{J}(\vec{q})g(0,T)} . \qquad (17)$$

Since the denominator for $\vec{q}=0$ vanishes at T_c as shown by (14), this shows that the intensity of the zone-centre central mode diverges at the phase transition.

The temperature dependence of the intensity, apart from the slowly varying non-resonant factors, is the same as that of the static susceptibility. The result (17) can also be obtained in the classical limit as the difference between the total scattering and the spin-wave scattering by using the Kramers-Kronig relations to show $S(\vec{q}) = kT\,G(\vec{0},0)$.

We conclude that the transition is driven by the catastrophic growth in intensity of the elastic modes of the system. The result for Pr_3Tl is shown in Fig.8. Very close to T_c the central mode peaks up and takes all the intensity. For the spin-wave contribution we have plotted $kT\,G^{zz}(\vec{0},0_+)$ where $\omega = 0_+$ is just large enough to avoid the central mode, and for the total $kT\,G^{zz}(\vec{0},0)$ is plotted. The central modes are entirely responsible for the phase transition in pure-spin magnets, as may be seen by setting $g_{CF} = 0$, in (17) and are observed as longitudinal critical scattering. For any cubic magnet the transverse susceptibility must also diverge at T_c and hence there must also be a soft transverse inelastic spin wave. For cubic magnets with crystal field effects, the transition is effected partially by mode-softening away from T_c, but close to T_c the central mode dominates the neutron scattering.

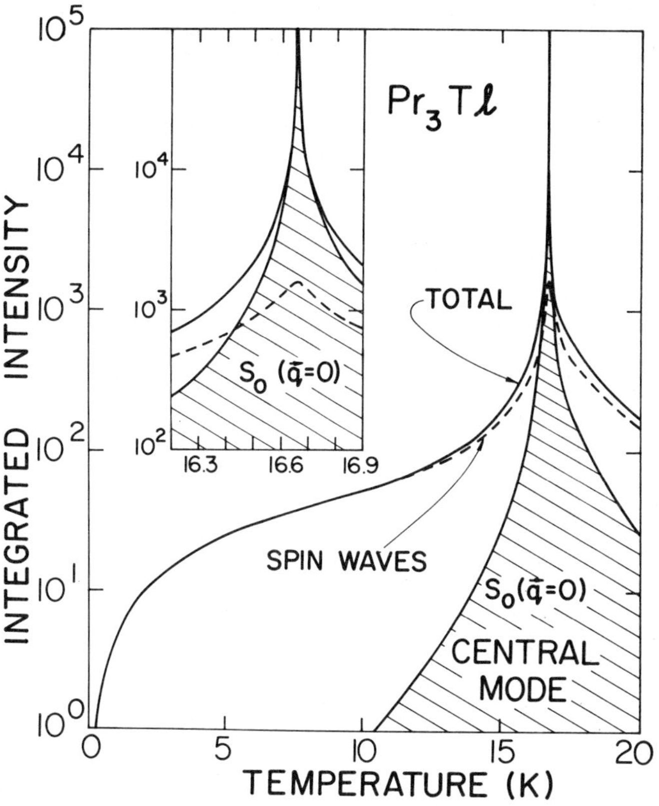

Fig. 8. Contributions to the neutron scattering in Pr_3Tl. The spin-wave and total contributions have been evaluated in the classical limit but give the qualitative behaviour near T_c.

At the opposite extreme to the pure-spin magnet is the singlet-singlet model which has no central mode since $\langle\delta J_{zd}^2\rangle_T = 0$. Any transition must then proceed by forcing the inelastic spin wave down to zero frequency. The singlet-singlet model is a highly artificial system, it has no transverse scattering and it is mainly of heuristic value. For real crystals having two non-magnetic singlets connected by J_z there are generally higher magnetic states present so that $\langle\delta J_{zd}^2\rangle_T \neq 0$. We conclude that, for most systems of physical interest, there will always be a divergent central longitudinal mode sufficiently close to the transition provided only there are magnetic states within the multiplet. Further, for cubic systems, the divergent central mode will be accompanied by a soft transverse mode.

The experimental evidence for a central mode in singlet-ground-state crystals is inconclusive. The best evidence is probably the divergent central peak observed[9] in TbSb (Fig.2), but no simultaneous saturation of the frequency of the $\vec{q}=0$ spin wave could be observed since it was obscured by the quasielastic peak near T_c. The experiments on Pr_3Tl were carried out at a large wave vector where any elastic scattering that is observed need not diverge.

To probe further the nature of the phase transition requires improvements in both the experiments and the theory. Experiments on single crystals that order weakly are necessary so that the zone centre modes can be studied, but this is not possible for Pr_3Tl. If one is thereby forced to study crystals like TbSb that order strongly it will be necessary to include the effect of the exchange fluctuations in limiting the lifetime and shifting the frequencies of the spin waves at finite temperatures. Initial calculations (A. Bahurmaz, M.J. Zuckermann and W. Buyers, private communication) appear to show that the fluctuations weaken still further the temperature dependence of the singlet-triplet mode. This suggests that when spin-wave interactions are included the zero-frequency mode must play an even greater part in driving the phase transition than is suggested by the simple calculations outlined above.

Valuable discussions were held with T.M. Holden. Martin Elliott helped with some of the calculations.

REFERENCES

1. For an example of the study of the many-branch spin-wave spectrum with neutron scattering see W.J.L. Buyers, T.M. Holden, E.C. Svensson, R.A. Cowley and M.T. Hutchings, J. Phys. C 4, 2139 (1971).
2. Y.L. Wang and B.R. Cooper, Phys. Rev. 185, 696 (1969).
3. D.A. Pink, J. Phys. C 1, 1246 (1968).
4. Y.Y. Hsieh and M. Blume, Phys. Rev. B 6, 2684 (1972).
5. W. Cochran, Advances in Physics 9, 387 (1960).
6. R.J. Birgeneau, AIP Conference Proceedings 10, 1664 (1973); R.J. Birgeneau, J. Als-Nielsen and E. Bucher, Phys. Rev. B 6, 2724 (1972).
7. S.V. Tyablikov, "Methods in the Quantum Theory of Magnetism" (Plenum, New York, 1967).
8. T.M. Holden and W.J.L. Buyers, Phys. Rev. B 9, 3797 (1974).
9. T.M. Holden, E.C. Svensson, W.J.L. Buyers and O. Vogt, Neutron Inelastic Scattering (IAEA, Vienna, 1972) p. 553, and Phys. Rev. B (to be published, November 1974).
10. W.J.L. Buyers and T.M. Holden, Phys. Rev. (to be published, January 1975).
11. S.R.P. Smith, J. Phys. C 5, L157 (1972).
12. M.E. Lines, J. Phys. C 7, L287 (1974).
13. I. Peschel, M. Klenin and P. Fulde, J. Phys. C 5, L194 (1972).
14. M. Klenin and I. Peschel, Phys. Condens. Materie 16, 219 (1973).
15. P. Bak, Risø Report No. 312 (1974).
16. S.B. Haley and P. Erdös, Phys. Rev. B 5, 1106 (1972).
17. H.-G. Purwins, J.G. Houmann, P. Bak and E. Walker, Phys. Rev. Lett. 31, 1585 (1973).
18. M. Blume and R.J. Birgeneau, J. Phys. C 7, L282 (1974).

†Elastic refers only to the scattering within the context of the single-ion levels. The interion coupling introduces fluctuating fields which lead to relaxation and quasielastic scattering.

VALENCE TRANSITIONS OF Sm IN $Sm_{1-x}R_xS$ (R = Y, La, OR Gd)

Lung-jo Tao
IBM Research Center, Yorktown Heights, NY 10598

ABSTRACT

Magnetic susceptibility and lattice parameter measurements[1,2] indicate that Sm in $Sm_{1-x}R_xS$ (R = Y, La, or Gd) undergoes valence transitions as a function of composition and of temperature, similar to SmS under pressure.[3] In the $Sm_{1-x}Y_xS$ system, the samples are black colored for x $<$ .15 and gold colored for x $>$ .3. The valence of Sm increases from divalent at x = 0 to $\simeq$ 2.3+ at x = .14, and to $\simeq$ 2.7+ for the entire range .3 $<$ x $<$ 1. No phase transition as a function of temperature is observed in these two composition ranges for 4.2 $<$ T $<$ 900 K. The magnetic susceptibility is independent of temperature for T $<$ 100 K, similar to collapsed SmS under a pressure larger than 7.5 Kbar.[4] Note that trivalent Sm with J = 5/2 as ground state in general exhibits Curie-like behavior at low temperature. For .15 $<$ x $<$ .3, samples show remarkable temperature-induced phase transitions. At x = .19, for example, the lattice parameter of the sample contracts on warming at T = 190 K but expands again at T > 600 K. Accompany the lattice variation, the color of the sample changes on warming from black to gold at T = 190 K and to black again at T > 600 K. The valence of Sm is estimated to change from ≃2.4+ to ≃2.7+ at T = 190 K and back to become more divalent at T > 600 K. We thus obtain a T vs x phase diagram for this system in which the valence of Sm changes smoothly from divalent at x = 0 to more trivalent as x increases at both low (T < 100 K) and high (T > 600 K) temperatures. At some intermediate temperature, a sharp transition from black phase (more divalent) to gold phase (more trivalent) occurs in the composition range .15 < x < .3. Similar effects are observed in the $Sm_{1-x}La_xS$ and the $Sm_{1-x}Gd_xS$ systems.

An interconfiguration fluctuation (ICF) model has been suggested for the intermediate valence state of Sm and other rare-earth ions and their common nonmagnetic state at low temperature.[4] According to this model, the Sm ion in its intermediate valence state switches from one integral valence state (2+) with 6 local electrons on the 4f shell to the other state (3+) with 5 local electrons on the 4f shell and the extra electron in the conduction (5d) band. If this fluctuation rate is faster than the relaxation rate of the Zeeman levels in the $4f^5$ configuration, proper Boltzmann distribution will not be achieved and the Curie behavior will be suppressed. By including crystal field effects and multilpe energy levels into the ICF model, the temperature-induced transitions between valence states can be explained.[2] The change of valence is a result of the relative change of the energy levels of the $4f^6$ configuration (with J = 1 as the first excited state about 400 K above the J = 0 ground state) and the $4f^5 5d$ configuration (with J = 5/2 as the ground state for $4f^5$ shell and t_{2g} for the 5d electron). At T > 600 K, we propose that the energy levels of the two configurations are such that the J = 5/2 level is between the J = 1 and the J = 0 levels for Sm in an intermediate valence state. Further, the J = 5/2 level splits into Γ_8 and Γ_7 states in cubic crystal. At T > 600 K, all these states are populated at one instant or another as the Sm ion switches between the two configurations. As the temperature lowers, the J = 1 state becomes depopulated. This will increase the population of the $4f^5 5d$ configuration at the expense of the $4f^6$ configuration and cause the lattice to contract. In addition, the system may find that at a certain temperature its total energy can be lowered if the energy separation between the Γ_7 and the Γ_8 levels is widened and that the Γ_7 level is more populated than the Γ_8 level which has higher energy. The separation between the two levels can be increased if the crystal field strength is increased. This in turn can be achieved by contraction of the lattice which is happening now as a result of the depopulation of the J = 1 level. Further, the t_{2g} band is lowered as the separation of the e_g and the t_{2g} branches widens as a result of the increase of the crystal field strength. These effects will further depopulate the J = 1 state and a spotaneous transition may result under favorable conditions. As a consequence of this transition, Sm becomes more trivalent, the number of the 5d electrons increases, and the color of the sample changes from black to gold. At an even lower temperature, however, the Γ_7 state will start to depopulate and the J = 0 level becomes more populated. This will increase the relative population of the $4f^6$ configuration and cause the lattice to expand. This in turn weakens the crystal field strength and decrease the relative energy level of the $4f^6$ configuration. This will further populate the J = 0 state and again, under a certain condition, a sharp spontaneous transition may occur which returns the system to a more divalent state, decreases the number of the 5d electrons, and changes the color of the sample back to black.

REFERENCES

1. F. Holtzberg, AIP Conf. Proc. 18, 478 (1974).
2. Lung-jo Tao and F. Holtzberg, Phys. Rev. to appear.
3. A. Jayaraman, V. Narayanamurti, E. Bucher, and R. G. Maines, Phys. Rev. Lett. 25, 1430 (1970).
4. M. B. Maple and D. Wohlleben, Phys. Rev. Lett. 27, 511 (1971).

SPIN RESONANCE OBSERVATION OF Sm^{3+} IN SEMICONDUCTING SmS

W. M. Walsh, Jr., E. Bucher, L. W. Rupp, Jr. and L. D. Longinotti
Bell Laboratories, Murray Hill, New Jersey 07974

ABSTRACT

Trivalent samarium ions have been observed via electron spin resonance as dilute "impurities" in the Val Vleck paramagnetic semiconductor SmS at 12 and 17.4 GHz in the temperature range 1.4 to 20 K. The spectra indicate the Sm^{3+} ions to reside in stable, bulk sites of tetragonal symmetry and to have a Γ_7 doublet ground state. The concentration of these ions appears to be ~0.5-5% in those crystals in which the signals are observable, roughly one third of those examined. These crystals do not exhibit appreciable conductive losses at low temperatures so the ionized electrons are apparently not mobile. These observations indicate that the trivalent state is not an intrinsic property of semiconducting SmS but is induced by the presence of a lattice defect such as a sulfur vacancy which can trap an electron from one of the neighboring Sm^{2+} ions. No control over the Sm^{3+} concentration has been achieved despite various modifications of the crystal growth conditions. The isotropically averaged experimental g-value $\bar{g} = 0.70 \pm 0.02$ is significantly enhanced with respect to the value 0.476 expected for the Γ_7 doublet. This is attributed to exchange coupling of the Sm^{3+} moment with the neighboring singlet Sm^{2+} ions. The magnitude and even the sign of the effect appear anomalous, however, in relation to earlier monochalcogenides. Rapid line broadening above 10K may result from spin relaxation as the trivalent state becomes locally itinerant in the Sm shell nearest to the stabilizing lattice defect.

INTRODUCTION

The rocksalt cubic samarium monochalcogenides have attracted considerable attention since they undergo isostructural "electronic collapse" when subjected to hydrostatic compression.[1,2] The effect is particularly striking in the case of SmS where the application of 6.5 kbar causes the black semiconducting material to suddenly become a golden metal.[3] Lattice parameter studies support a model in which the initially divalent samarium $4f^6$ ions release a third electron to the 5d, 6s conduction band. The electronic properties of the collapsed metallic phase are anomalous, however, implicating rapid valence fluctuations between the di- and trivalent configurations.[4] It has even been suggested that an appreciable admixture of the trivalent configuration occurs in the semiconducting state at atmospheric pressure[5] though this has been disputed.[6] We wish to report on electron spin resonance (ESR) studies of SmS in which Sm^{3+} ions have been identified at temperatures below 20K. While low concentrations of trivalent samarium ions are found in some crystals they do not appear to be an intrinsic property of the material but are the result of lattice defects or chemical contamination. The spectra are of intrinsic interest, however, as evidence is obtained for appreciable exchange coupling to the divalent "host" samarium ions which may be compared with results of earlier ESR studies of S-state ions in the semiconducting samarium monochalcogenides.[7]

EXPERIMENTAL RESULTS

ESR signals attributed to Sm^{3+} have been observed in several single-crystal samples of SmS in the temperature range from 1.4 to 20K. The experiments were performed at 12.0 and 17.4 GHz using conventional homodyne spectrometers. The most intense and best resolved 12 GHz spectra are shown in Fig. 1. While hyperfine identification is rather marginal due to poor resolution, the low-field group for H|| ⟨100⟩ appears to be consistent with a central peak accounting for ~70% of the total intensity flanked by 16 hyperfine lines of ~2% intensity each, as expected for Sm. The anisotropy of the spectrum, shown for the (110) and (100) planes in Fig. 2, is that of three equivalent anisotropic $S = 1/2$ resonances having ⟨100⟩ axial symmetry ($g_{||} = 0.51 \pm 0.02$, $g_{\perp} = 0.77 \pm 0.02$). The ground state of Sm^{3+} ($4f^5$, $^6H_{5/2}$, $g_J = 2/7$) in octahedral coordination is expected to be a Γ_7 doublet with a g-value $g = 5g_J/3 = 0.476$ in cubic symmetry. While this differs significantly from the isotropic average of the observed g-values ($\bar{g} = g_{\langle 100\rangle} = 0.70 \pm 0.02$) we can suggest no other plausible paramagnetic ion. As will be discussed below, such a shift of the average g-value is not unexpected since appreciable exchange couplings have been previously observed in the samarium monochalcogenides. There is evident in Fig. 2 a splitting of the signal in the vicinity of the ⟨100⟩ axes which we are unable to explain by crystalline field or other arguments.

The Sm^{3+} spectrum proved to be observable in about 1/3 of the SmS crystals which we have examined Concentrations estimated from the ESR intensities as compared with known concentrations of Eu^{2+} and Gd^{3+} were in the range 0.5-5%. No evidence of Sm^{3+} could

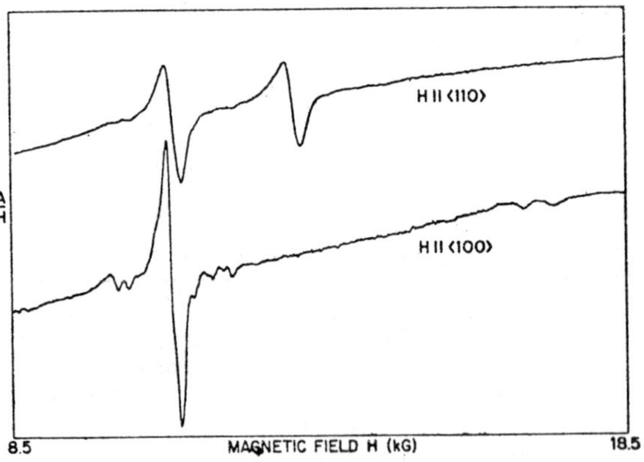

Figure 1. Spectra of Sm^{3+} trace "impurities" in semiconducting SmS taken at 4.2K and 12.0GHz in the ⟨110⟩ and ⟨100⟩ directions.

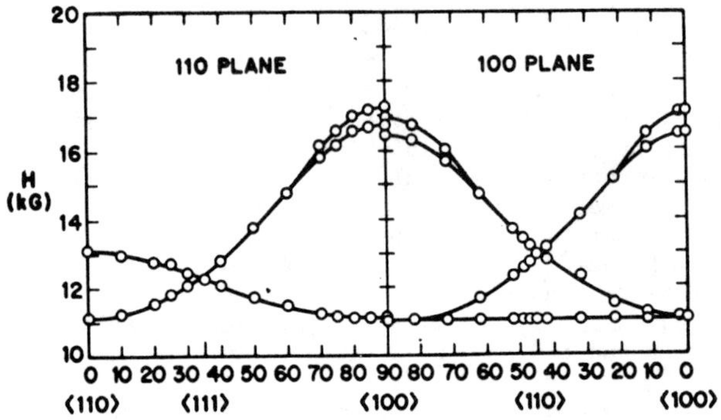

Figure 2. Anisotropy plots of the Sm^{3+} resonance spectrum taken under the conditions of Fig. 1. The splitting observed near the ⟨100⟩ axes is not understood. The other features are consistent with three equivalent S=1/2 ions in tetragonal sites oriented along ⟨100⟩ axes.

be found in either SmSe or SmTe. It was hoped that by raising the sample temperature it might be possible to observe thermally excited Sm^{3+} in SmS but all our attempts merely led to rapid line broadening of the tetragonal spectrum. Since this spectrum appeared to be unique but erratically observable we attempted to control its presence by deviations from stoichiometry, quenching, annealing and even oxygen contamination. These efforts have thus far proved fruitless. We are led to conclude that the presence of Sm^{3+} is not at all an intrinsic property of semiconducting SmS but results from a crystalline defect or contamination of unknown origin. It may prove interesting to study the spectrum while subjecting the crystal to hydrostatic pressure since the approach of the electronic phase transition could conceivably lead to appearance of intrinsic trivalent states.

DISCUSSION

Despite our inability to identify the origin of the extrinsic Sm^{3+}, the tetragonal site symmetry and the absence of appreciable microwave conductivity in the samples which exhibited the spectrum suggest a model of a point defect at an anion site (e.g., a sulfur vacancy) which is able to ionize one electron from one of the six neighboring Sm^{2+} ions and to stabilize that particular trivalent state by a slight distortion of that ion's position. The resulting Sm^{3+} would be subject to a superposition of a cubic crystal potential plus an axial component directed along a $\langle 100 \rangle$ axis thus producing the observed anisotropy. One might expect a magnetic resonance from the trapped electron in analogy with color centers in insulating crystals but we have no evidence of any additional resonance accompanying the Sm^{3+} spectrum. This need hardly serve to reject the model, however, as no resonance has been detected due to the "extra" electron which must be present in crystals intentionally doped with trivalent gadolinium.[7]

The enhancement of the observed average g-value relative to that expected theoretically ($3\bar{g}/5g_J = 1.47$) cannot result simply from the lack of cubic symmetry.[8] In view of our earlier studies of S-state ions present as substitutional impurities in the samarium monochalcogenides it is likely that the enhancement results from exchange coupling to the neighboring Sm^{2+} ions.[7] While the assumption of isotropic exchange was valid for the S-state ions it is unlikely to remain so in the tetragonal symmetry of the present problem. An even more fundamental difficulty or inconsistancy with the S-state ion results is the large magnitude and apparent ferromagnetic sign of the exchange coupling (the Sm^{3+} "feels" a larger field than we apply). The sign of the effect might at first appear consistent with the ferromagnetic coupling found for Eu^{2+} and Gd^{3+} in the SmS lattice, as well as with the ferromagnetic Sm-Sm coupling of the host lattice,[7] but the large orbital contribution to the Sm^{3+} magnetic moment must be considered. If exchange forces act only on the spin component of the total moment then the exchange coupling must be antiferromagnetic since the spin moment is opposed to the dominant orbital moment of this $L = 5$, $S = 5/2$, $J = 5/2$ ion. The anomolous sign and large magnitude of the apparent exchange for Sm^{3+} in SmS remain quite puzzling at this point.

Finally we recall the rapid line broadening above 10K mentioned above. The linewidth of the low-field resonance for $H \| \langle 100 \rangle$ increases from 300 G to 1800 G in the 10-20K interval. While this may result from Orbach relaxation via the excited Γ_8 state the nature of our model suggests an alternative possibility: If one regards the trivalent state as localized at low temperature on a particular samarium ion neighbor to a given defect only because of a slight distortion in its position it is conceivable that this situation becomes unstable above some characteristic temperature. The trivalent configuration would then become locally mobile in the Sm shell surrounding the particular electron trapping site. Such random hopping of the magnetic moment should be a particularly effective mechanism for spin-lattice relaxation.

We wish to acknowledge several helpful discussions with Dr. R. J. Birgeneau and with Prof. D. Davidov.

REFERENCES

1. A. Jayaraman, V. Narayanamurti, E. Bucher and R. G. Maines, Phys. Rev. Letters 25, 1430-33 (1970).

2. E. Bucher, V. Narayanamurti and A. Jayaraman, J. Appl. Phys. 42, 1741-45 (1971).

3. J. L. Kirk, K. Vedam, V. Narayanamurti, A. Jayaraman and E. Bucher, Phys. Rev. B6, 3023-26 (1972).

4. M. B. Maple and D. K. Wohlleben, Phys. Rev. Letters 27, 511-14 (1971).

5. J. L. Freeouf, D. E. Eastman, W. D. Grobman, F. Holtzberg and J. B. Torrance Phys. Rev. Letters 33, 161-164 (1974).

6. M. Campagna, E. Bucher, G. K. Wertheim and L. D. Longinotti, Phys. Rev. Letters 33, 165-168 (1974).

7. R. J. Birgeneau, E. Bucher, L. W. Rupp, Jr. and W. M. Walsh, Jr. Phys. Rev. B5, 3412-18 (1972).

8. R. J. Birgeneau (private communication).

OPTICAL STUDIES OF SEMICONDUCTOR TO METAL TRANSITION IN $Sm_{1-x}Y_xS$

G. Guntherodt and F. Holtzberg
IBM Thomas J. Watson Research Center, Yorktown Heights, New York 10598

ABSTRACT

$Sm_{1-x}Y_xS$ undergoes a semiconductor-metal transition as a function of the composition x. We have measured the reflectivity of cleaved single crystals as a function of composition and temperature for photon energies from 0.1 to 6eV. The prominent structure of the reflectivity of semiconducting SmS is assigned to transitions from the $4f^6$ level of Sm^{2+} into the crystal field split 5d states. The observed decrease in the intensity of the excited $4f^5$ multiplet structure with increasing doping is assumed to be indicative of an increased admixture of $4f^5 5d$ states into the $4f^6$ ground state, the Sm ion being in a state of mixed valence. The dc scattering time of the conduction electrons at 300K shows a much stronger scattering for samples in the collapsed metallic phase, indicating a degeneracy of the hybridized 4f level with the Fermi level. For these samples the plasma edge exhibits an unusually large shift to lower energies upon cooling due to a localization of conduction electrons in the admixed 4f level. The concurrent valence transformation of the admixed 4f level manifests itself in an increase in lattice constant, scattering time and in the oscillator strength of the $4f^6$-$4f^5 5d$ transitions.

INTRODUCTION

The divalent rare earth chalcogenides have the intriguing feature that a change in valence of the cation manifests itself in a drastic lattice contraction. This is due to the much smaller ionic radius of the nominally trivalent ion. For instance in SmS this valence change occurs at a pressure of 6.5 kbar, resulting in a first order isostructural semiconductor-metal transition and a nonmagnetic mixed valence state of the Sm ion[1,2]. This pressure induced phase transition can also be simulated chemically by substituting smaller cations for Sm or larger anions for S[3,4]. Thus due to the lattice pressure the semiconductor-metal transition occurs at atmospheric pressure as a function of the concentration of the substituent.

In this work we are concerned with the optical properties of SmS-YS alloys as a function of composition and temperature. The aim is to study optical inter- and intraband transitions and thus to gain information about their ground state, the unoccupied states involved in the phase transition and finally about the transport parameters of the conduction electrons.

EXPERIMENTAL RESULTS

$Sm_{1-x}Y_xS$ shows as a function of x[4-6] a discontinuity for $x \cong 0.15$ in the room temperature lattice constant, its collapsed value of ~ 5.69 Å being equivalent to that of SmS under 10 kbar[1]. Thus we have a composition-dependent electronic phase transition in going from the small gap semiconductor SmS[7] to the metal YS, which has one conduction electron per Y^{3+} ion. The lattice constant as well as the X-ray photoemission have confirmed a mixed valence state of the Sm ion[4-6,8,9]. In our optical investigation we have measured the reflectivity for photon energies from 0.1 to 6 eV on cleaved single crystals as grown from the melt. In order to compare the reflectivity spectra on an absolute scale for $0 \leq x \leq 1$, samples with $.15 \lesssim x \lesssim .7$ have been polished because of nonplanarity of their surfaces after cleaving.

Fig. 1 shows the room temperature reflectivity of a cleaved single crystal of $Sm_{.95}Y_{.05}S$ and YS and a polished $Sm_{.82}Y_{.18}S$ sample. The structure found in the

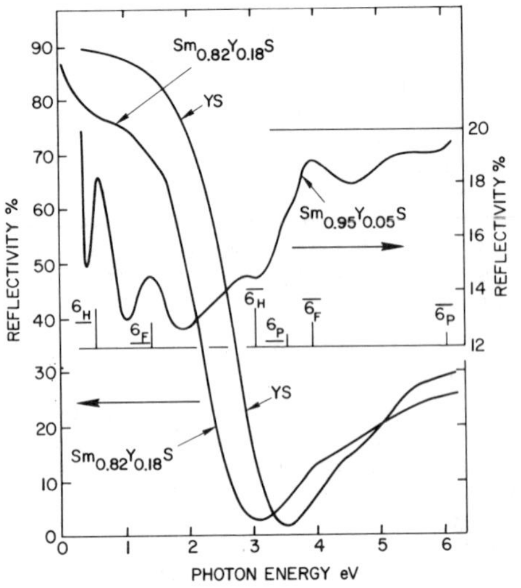

Figure 1. Reflectivity of $Sm_{.95}Y_{.05}S$ and YS (cleaved) and of $Sm_{.82}Y_{.18}S$ (polished) at 300K.

reflectivity spectrum of semiconducting SmS[10] can be understood on the basis[11,12] of optical transitions from the $4f^6$ level of Sm^{2+} into the crystal field split 5d states: $4f^6 \to 4f^5(^6H, ^6F, ^6P)\, 5dt_{2g}$ and $4f^6 \to 4f^5(^6H, ^6F, ^6P)\, 5de_g$. This assignment for SmS, including the calculated intensities of the $4f^5$ multiplet levels[13], is indicated for comparison in Fig. 1. For 5% Y in SmS we see a plasma edge below 0.4 eV. Although the $4f^5$ multiplet structure is somewhat weakened with respect to pure SmS, it still can be readily identified. The increasing absorption background above 3 eV in $Sm_{.95}Y_{.05}S$ as well as in the other samples in Fig. 1 and 2 is due to valence-to conduction band transitions as deduced from X-ray photoemission on $Sm_{1-x}Y_xS$ for $0 \leq x \leq 1$[8].

The position in energy of the plasma edge for a sample of the collapsed phase with x = 0.18 is very similar to that of metallic SmS under hydrostatic pressure[14]. The weak structure on top of the plasma edge of $Sm_{.82}Y_{.18}S$ around 1 and 1.5 eV and the shoulder around 4 eV is also found for a cleaved $Sm_{.75}Y_{.25}S$ sample at 300K in Fig. 2. In addition, the latter sample shows two weak shoulders at 2.7 and 3.5 eV presumably due to its better surface quality compared to a polished one. The detailed structure in the reflectivity of $Sm_{.82}Y_{.18}S$ and $Sm_{.75}Y_{.25}S$ might still be partly associated with the $4f^5$ multiplet structure, since we do not see any indication of it on a cleaved YS sample (Fig. 1). YS shows a structureless Drude like plasma edge below 3.5eV and an onset of p valence- to d, s conduction band transitions at about 4 eV as in LaS[15].

A most striking effect on the optical properties of metallic collapsed samples occurs upon cooling as shown in Fig. 2 for a cleaved $Sm_{.75}Y_{.25}S$ sample. We observe an unusually large shift of the plasma edge of 0.5 eV from 300K to 80K. Further cooling to 20K results in a very small additional shift of less than 0.1 eV. The structure at 300K, reminiscent of the $4f^5$ multiplet structure, becomes even more pronounced upon cooling. We resolve two shoulders at about 0.8 and 1.5 eV and a very prominent peak around 4 eV compared with a broad shoulder at 300K.

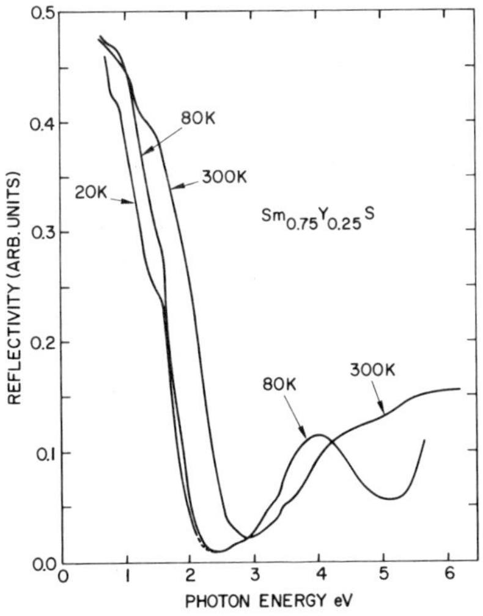

Figure 2. Reflectivity of $Sm_{.75}Y_{.25}S$ (cleaved) for various temperatures.

DISCUSSION

The driving mechanism of the first order phase transition in SmS under pressure is assumed to be the closing of the gap between the $4f^6$ level and the conduction band[7,16]. Assuming for small x in $Sm_{1-x}Y_xS$ the same compressibility as in SmS, the corresponding volume change for a closing of the gap is attained for $x \sim 0.05$. But the phase transition in $Sm_{1-x}Y_xS$ occurs for $x \simeq 0.15$. Hence we conclude that in the $Sm_{1-x}Y_xS$ system for $0.05 \lesssim x \lesssim 0.15$ the 4f level is already degenerate in energy with the conduction band. This degeneracy presumably results in an increased admixture of $4f^55d$ states into the $4f^6$ ground state and hence a weakening of the $4f^5$ multiplet structure in addition to the dilution of the Sm ions by Y doping.

Information about the position of the admixed 4f level with respect to the Fermi level E_F is obtained by calculating the scattering time τ_{dc} at zero frequency from the dc conductivity σ_{dc} and the plasma frequency ω_p of the conduction electrons. σ_{dc} has been measured[6] and ω_p has been determined from a KK analysis of the reflectivity[10] by taking into account the screening of the plasmons due to interband transitions. For $0 < x < 0.15$ τ_{dc} is almost constant ($\sim 4 \cdot 10^{-15}$sec), but drops by almost a factor of 4 for $x \simeq .15$ and then increases linearly towards the value of $7 \cdot 10^{-15}$sec for YS. Since τ_{dc} accounts for the scattering of electrons in a region of $\pm$ kT around E_F, the constant value for $x < 0.15$ indicates that the 4f level can be degenerate in energy with the conduction band as stated above, but not with its Fermi level. It thus can retain, due to a configurational admixture, a valence of up to 2.2 for $x \lesssim 0.15$ as derived from the lattice constant[6,8]. The strong decrease in τ for $x \gtrsim 0.15$ signifies that the admixed 4f level must be degenerate with E_F, resulting in a strong scattering of electrons at E_F. This degeneracy and the much smaller lattice constant in the collapsed metallic phase favor a much stronger configurational mixing of the 4f level, which then appears to be broadened compared to its width for $x < 0.15$. Hence the Sm ion is in a mixed valence state of 2.7 as derived from the lattice constant[6,8] for $x > 0.15$. The linear increase of τ as a function of increasing x ($x > 0.15$) is then due to the dilution across the solid solution system.

The effect of cooling for samples of the collapsed metallic phase results in an increasing localization of conduction electrons in the broadened 4f level and hence a shift of the plasma edge towards lower energies as seen in Fig. 2. At the same time this localization results in a smaller $4f^55d$ admixture into the ground state of the optical transitions as concluded from the strengthening of the $4f^5$ multiplet structure at low temperatures in Fig. 2. Particularly the peak around 4 eV at 80K in Fig. 2 is due to an increased oscillator strength of the $4f^6$-$5de_g$ transition. The background valence-to conduction band transitions are supposedly only weakly temperature dependent. The scattering time increases upon cooling to 20K to the same value as for samples with $x < 0.15$ at 300K. This shows clearly that the admixed 4f level is no longer degenerate with E_F and has a similar configuration to samples with $x < 0.15$ at 300K. The valence transformation of the admixed 4f level from 2.7 at 300K to about 2.5 at 80K[6] provides a change in the conduction electron concentration which can only partly account for the shift of the plasma edge upon cooling in Fig. 2. Thus part of this shift is also due to changes in the optical dielectric constant and the effective mass.

We would like to thank J. L. Freeouf, T. Penney, J. C. Tsang and S. von Molnár for many valuable discussions. The technical assistance of P. Fernandez and P. G. Lockwood is gratefully acknowledged.

LITERATURE

1. A. Jayaraman, V. Narayanamurti, E. Bucher and R. G. Maines, Phys. Rev. L, 25, 1430 (1970).
2. M. B. Maple and D. Wohlleben, Phys. Rev. L. 27, 511 (1971).
3. A. Jayaraman, E. Bucher, P. D. Dernier and L. D. Longinotti, Phys. Rev. L. 31, 700 (1973).
4. F. Holtzberg, AIP Conf. Proc. 18, 478 (1974).
5. L. J. Tao and F. Holtzberg, to be published.
6. T. Penney and F. Holtzberg, this conf. and to be published.
7. E. Kaldis and P. Wachter, Solid State Commun. 11, 907 (1972).
8. R. A. Pollak, F. Holtzberg, J. L. Freeouf and D. E. Eastman, Phys. Rev. L. 33, 820 (1974).
9. M. Campagna, E. Bucher, G. K. Wertheim and L. D. Longinotti, Phys. Rev. L. 33, 165 (1974).
10. G. Guntherodt and F. Holtzberg, to be published.
11. R. Suryanarayanan, C. Paparoditis and J. Ferré, Rare Earths and Actinides, Durham, England, Conf. Digest No. 3, 1971.
12. F. Holtzberg and J. B. Torrance, AIP Conf. Proc. No. 5, 860 (1971).
13. P. A. Cox, Y. Baer and C. K. Jorgensen, Chem. Phys. L. 22, 433 (1973).
14. J. L. Kirk, K. Vedam, V. Narayanamurti, A. Jayaraman and E. Bucher, Phys. Rev. B 6, 3023 (1972).
15. G. Guntherodt and P. Wachter, AIP Conf. Proc. 18, 1034 (1974).
16. B. Batlogg, J. Schoenes and P. Wachter, Phys. L. 49A, 13 (1974).

SEMICONDUCTOR-METAL TRANSITION IN SmS - A ^{149}Sm MOSSBAUER STUDY

J.M.D. COEY and S.K. GHATAK
Groupe des Transitions de Phases, Centre National de la Recherche Scientifique, B.P. 166, 38 042. Grenoble-Cedex, France

and F. HOLTZBERG
I.B.M. Thomas J. Watson Research Center, Yorktown Heights, New York 10598, USA.

ABSTRACT

The isomer shifts of ^{149}Sm have been measured in both the semiconducting and collapsed phases of samarium sulphide at room temperature. The isomer shift for SmS at atmospheric pressure (the semiconducting phase) is -0.72 ± 0.04 mm/sec relative to a source of ^{149}Eu in EuF_3, with a linewidth of 2.20 mm/sec. The isomer shift in the collapsed phase was measured both in an Yttrium-doped sample and in SmS as a function of pressure. The value in $Sm_{0.77}Y_{0.23}S$ was -0.37 ± 0.04 mm/sec, whereas that measured under a uniaxial stress of 11 kbar applied in a clamp device was -0.21 ± 0.05 mm/sec. These results confirm the mixed valence state proposed for the non-magnetic, metallic phase, and from the linewidth, 2.1 mm/sec, it is deduced that the $4f^6\ 5d \leftrightarrow 4f^6$ fluctuations are more rapid than 10^{-9} s. The decreasing isomer shift in the series SmX, X = S, Se, Te is inconsistent with the suggestion that there is configurational mixing in the semiconducting state of SmS which is absent in Sm Se.

INTRODUCTION

Recently many compounds of several rare earth metals have been studied experimentally because of their interesting electrical, magnetic and elastic properties,[1-3,12] the samarium monochalcogenides [1-3] and SmB_6 in particular. Pure SmS undergoes a sharp pressure-induced semiconductor-to-metal transition at 6.5 kbar at room temperature. This transition is accompanied by a large volume contraction (∿ 16 %) but no change in NaCl structure. These unusual properties of the collapsed phase of SmS have been attributed to fluctuations between the $4f^6$ and $4f^5\ 5d$ configurations. This interconfigurational fluctuation can also be produced either by 'chemical pressure', i.e by doping SmS with a trivalent cation (e.g. Y^{3+}, La^{3+}, Gd^{3+}, Th^{3+}) [2,7] or else by anion substitution (e.g. As)[2]. It occurs when the two states are energically near-degenerate. In that case, the interconfiguration energy is the energy necessary to take an electron from the $4f^6$ shell and to put it into the conduction band leaving the samarium in the $4f^5$ configuration. According to Hirst,[4] this energy goes to zero at the transition. Recent UV photoemission results [5] on SmS suggested that the semiconducting, zero pressure phase of SmS also contains an admixture of the $4f^5\ 5d$ (Sm^{3+}) configuration and the amount of admixture was found to be of the order of 15 %. Based on this result, Mott [6] suggested that ground state of SmS is an excitonic insulator and that a discontinuous transition from an excitonic insulating state to a metallic state is possible as pressure increases. On the other hand X-ray photoemission (XPS) study [7] on SmS predicted that any admixture is less than 3 %. The amount of admixture in the compounds $Sm_{.81}Y_{.19}S$ [8], $Sm_{.82}Gd_{.18}S$ [7] and $Sm_{.85}Th_{.15}S$ [7] has also been obtained both from XPS and from lattice constant measurements, and the configuration mixing ratio (Sm^{3+}/Sm^{2+}) at room temperature is found to be 1.4 for $Sm_{.81}Y_{.19}S$ and ∿ 4 for Gd and Th doped samples. This ratio is also observed to depend on temperature [8], showing that thermal fluctuations can influence configurational fluctuations.

Here we report the results obtained by Mössbauer spectroscopy on the configuration mixing in SmS and $Sm_{.77}Y_{.23}S$. We study the isomer shift, δ, in the mixed crystal, and in SmS both at zero and high pressure. This technique has been used before in SmB_6 [9] to estimate the mixing ratio, which was found to be 2.3.

EXPERIMENTAL RESULTS

The 22.5 keV 5/2 → 7/2 transition of ^{149}Sm was used. The source, 6 m Ci of ^{149}Sm in EuF_3 prepared by the reaction ^{149}Sm (p,n) by New England Nuclear, gave a linewidth of 2.20 mm/sec with an SmF_3 absorber. The natural linewidth is 1.71 mm/sec, but part of the difference may be due to an unresolved quadrupole interaction as both the ground and excited states of the ^{149}Sm nucleus possess quadrupole moments.

All spectra were collected at room temperature using a constant acceleration spectrometer and a Si Li detector. Absorbers were made from powdered crystals of SmS and $Sm_{0.77}Y_{0.23}S$ containing ∿ 16 mg/cm^2 of natural samarium which gives an absorption dip of ∿ 4 %. The data under pressure were obtained with a simple clamp device. The SmS was moulded into a plastic disc 0.5 mm thick and 12 mm in diameter which was clamped between two B_4C cylinders 12 mm long and 12 mm in diameter in a steel frame which allowed a clear passage for the γ-rays 8 mm in diameter. The count rate through the clamp was typically 40 counts/second.

Some spectra are shown in figure 1. a and c were collected at zero pressure whereas b was collected with the clamp, counting for 5-6 days. All the spectra were well fitted by a single Lorentzian peak by the method of least squares, and the isomer shifts and linewidths are listed in table I.

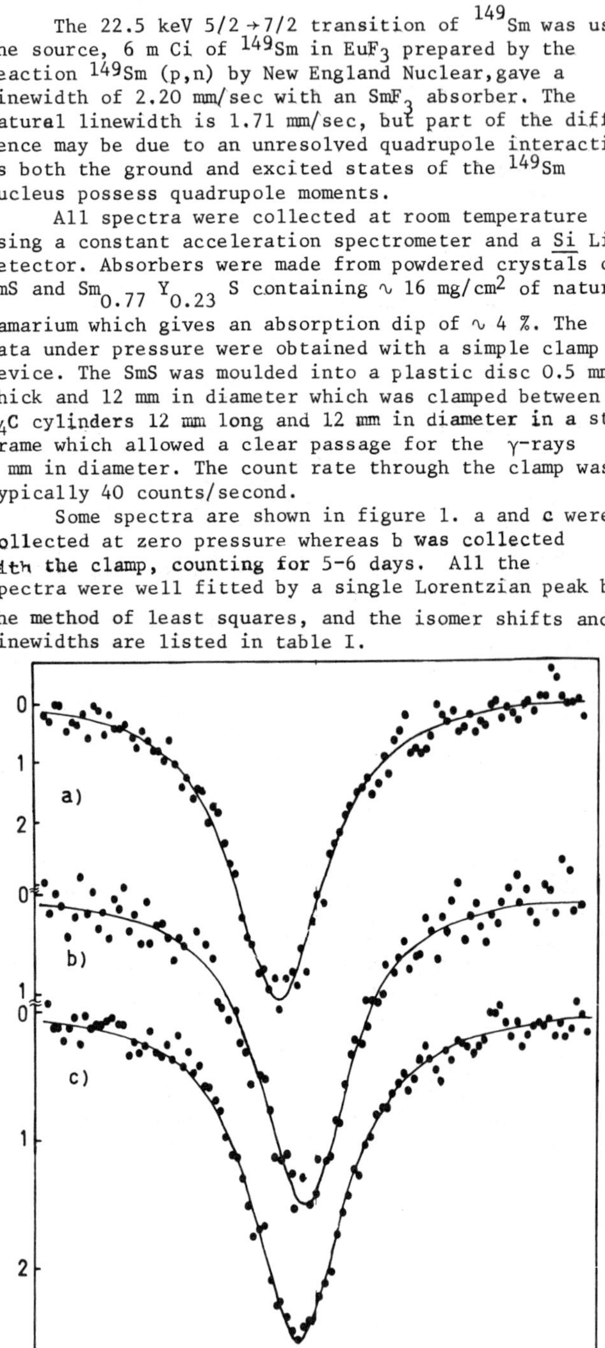

Fig. 1 : Room-temperature Mössbauer spectra of a) SmS at atmospheric pressure b) SmS at 11kbars and c) $Sm_{0.77}Y_{0.23}S$ at atmosphere pressure.

TABLE I

	p k bars	δ mm/s	Γ mm/s	n
SmS	0	−0.72 (0.04)	2.20 (0.10)	6.0
	4	−0.70 (0.06)	2.10 (0.12)	5.97 (0.08)
	7	−0.30 (0.06)	2.16 (0.15)	5.44 (0.08)
	11	−0.21 (0.05)	2.06 (0.12)	5.32 (0.07)
$Sm_{.77}Y_{.23}S$	0	−0.37 (0.04)	**2.26 (0.10)**	5.53 (0.05)

DISCUSSION

The isomer shift is a measure of electron charge density at the site of the nucleus ; Eibschütz et al.[9] found a difference between SmF_2 (−0.9 mm/sec) and SmF_3 (0.0 mm/sec). The difference in the electron density at the nucleus arises from the different shielding effect on the s-shells of the $4f^6$ and $4f^5$ configurations in divalent and trivalent compounds respectively. The addition of 5d/6s electrons in Sm metal only increases the IS to + 0.2mm /sec, that is 0.07 mm/sec per electron.

If we suppose that the configuration of SmS at zero pressure is $4f^6$ and that of Sm_2Sm_3 is $4f^5$, we can estimate the intermediate configurations by interpolating between the two isomer shifts (−0.72 ± 0.04 and −0.03 ± 0.04 mm/sec). This assumes that the contributions to the isomer shift due to covalency and overlap distortion are unaffected by n, the 4f occupation in sulphides. We arrive thus at the values of n for the configurations $4f^n$ listed in table I, where the values of isomer shift and linewidth (Γ) are also given. There is a sharp increase of the isomer shift of 0.4 mm/sec at about 6 Kbars, and a configuration n = 5.38 is derived for the high-pressure phase. The corresponding mixing ratio is 1.6. From the lattice constant measurement, n = 5.23 was deduced [2]. The configuration deduced for $Sm_{.77}Y_{.23}S$, n = 5.53 ±0.05 may be compared with those derived for $Sm_{.81}Y_{.19}S$ from XPS (5.42) and lattice constant [8] (5.31) measurements. Lower values of n, ∿ 5.2 were found for $Sm_{.82}Gd_{.18}S$ and $Sm_{.85}Th_{.15}S$ [7].

Eibschütz et al. found a small, systematic increase of isomer shift from SmS to SmSe to SmTe (from −0.73 to −0.69 to −0.60 mm/sec) which may be explained by increasing 3p →6s covalent transfer. The suggestion of a 15% $4f^5$ admixture in zero pressure SmS which is absent in SmSe[5] would mean a positive 4f contribution to the SmS isomer shift of 0.11 mm/sec for which there does not seem to be any evidence.

On the basis of the configuration fluctuation model the $4f^{5+x}$ configuration in collapsed phase of SmS can be looked on as the time average of x $4f^6$ and (1−x) $4f^5$ configurations. On a time scale short compared to the fluctuation time, one expects that the 4f-shell is either in a $4f^6$ or $4f^5$ configuration. In the Mössbauer experiment on $Sm_{.77}Y_{.23}S$ a narrow line corresponding to the average configuration $4f^{5.53}$ is observed. An attempt to fit the spectrum to two lines with isomer shifts of −0.72 and ± 0.03 mm/sec led to an unacceptably high value of χ^2, which means that the configuration fluctuation time is shorter than the time scale in the Mössbauer experiment (∿ 10^{-9} sec). But in XPS spectra [7,8] well resolved lines in the 'chemically collapsed' phase of SmS corresponding to $4f^6$ and $4f^5$ were observed and indicate that the time scale of fluctuation is larger than 10^{-16} sec which is time scale in the XPS experiment. Therefore, it appears from these considerations that the fluctuation time scale lies between 10^{-9} and 10^{-16} sec. In this connection the electronic specific heat measurements of Bader et al on the collapsed phase of SmS are of interest [11]. These authors find γ = 145 mJ/mole, which has been interpreted by Mott[6] in terms of an unenhanced f band. The width of such a band would be ∿ 0.1 eV, which corresponds to a charge fluctuation time ∿ 10^{-14} sec.

REFERENCES

1. A. Chatterjee, A.K. Singh, and A. Jayaraman, Phys. Rev. B 6, 2285-91 (1972)
2. F. Holtzberg, in Magnetism and Magnetic Materials - 1973 A.I.P. Conference Proceedings 18, 478 (1974)
3. M.B. Maple and D. Wohleben, Phys. Rev. Lett. 27, 511 (1971)
4. L.L. Hirst, J. Phys. Chem. Solids, 35, 1285-96 (1974)
5. J.L. Freeouf, D.E. Eastman, W.D. Grobman, F. Holtzberg, and J.B. Torrance, Phys. Rev. Lett. 33, 161-164 (1974)
6. N.F. Mott, Phil. Mag. 30, 403-16 (1974)
7. M. Campagna, E. Bucher, G.K. Wertheim, and L.D. Longinotti, Phys. Rev. Lett. 33, 165-68 (1974)
8. R.A. Pollak, F. Holtzberg, J.L. Freeouf, and D.E. Eastman, Phys. Rev. Lett. 33, 820-23 (1974)
9. R.L. Cohen, M. Eibschütz, and K.W. West, Phys. Rev. Lett., 24, 383-86 (1970)
10. M. Eibschütz, R.L. Cohen, E. Buehler, and J.H. Hernick, Phys. Rev. B 6, 18-23 (1972)
11. S.D. Bader, N.E. Phillips and D.B. McWhan, Phys. Rev. B 7, 4686-8 (1973)
12. B. Coqblin and A. Blandin, Adv. Phys. 17, 281-366 (1968)

LOW TEMPERATURE MAGNETIC PHASE TRANSITION IN TmSe

H.R. Ott
Laboratorium für Festkörperphysik, ETH Zürich,
8049 Zürich, Switzerland,
K. Andres and E. Bucher
Bell Telephone Laboratories Inc., Murray Hill,
New Jersey 07974

ABSTRACT

We have investigated the magnetically ordered state of TmSe below 3 K by means of thermal expansion and magnetostriction measurements (down to 1.5 K and up to 15 kOe) as well as by a.c. susceptibility measurements in different magnetic fields. From these data it is possible to construct a diagram of the magnetic phases at low temperatures. The ordering is antiferromagnetic in nature but we find 3 distinctly different phases below the Néel temperature T_N. The Néel temperature remains constant in external magnetic fields up to 5 kOe. For higher fields an unambiguous interpretation of the experimental features is difficult.

INTRODUCTION

Recent investigations of several physical properties (lattice constant, electrical resistivity, magnetic susceptibility [1], Mössbauer spectra [2]) of TmSe indicate a nonintegral valence of Tm in this compound. In addition it was concluded from room temperature ESCA-spectra [3] that temporal fluctuations between the divalent and the trivalent state of the Tm-ion are existent in TmSe and TmTe.

In earlier work on local moment formation it has been pointed out that experimentally observed nonmagnetic behaviour in rare earth compounds are associated with valence fluctuations in the 4f-shell [4-6]. Specific heat and magnetisation measurements of Tm-monochalcogenides[7], however, indicated a magnetic ordering of these materials at low temperatures, while neutron diffraction data did not reveal a long range order below the ordering temperature [8]. It is the intention of this work to study the nature of the magnetic phenomena of TmSe below 10 K in order to obtain a better understanding of the mechanisms behind the low temperature behaviour of this material and possibly the general features of substances with alleged valence fluctuations.

EXPERIMENTS AND RESULTS

The preparation of the crystals has been described earlier [7]. We have measured the linear thermal expansion of TmSe single crystals along the (100)-axis between 1.5 K and 10 K and in various external magnetic fields up to 15 kOe. In addition we have determined the magnetostriction at fixed temperatures in fields parallel and perpendicular to the direction in which the length changes were monitored. The length change measurements were made on crystals of approximately 2 mm length with a capacitance dilatometer. a.c. susceptibility measurements were made for varying temperature in different external fields.

In fig. 1 we show the linear thermal expansion coefficient as a function of temperature and in various magnetic fields. In this case the magnetic field is applied parallel to the direction of the length change measurements. In zero field we observe a well defined peak at 2.85 K. For increasing magnetic fields the peak maximum is also increasing but appears at the same temperature. A tendency of broadening on the low temperature side may be seen. Above 5 kOe the peak begins to shift towards lower temperatures with increasing fields and the peak maximum is drastically reduced. For field strengths of more than 6 kOe the peak vanishes completely and a broad

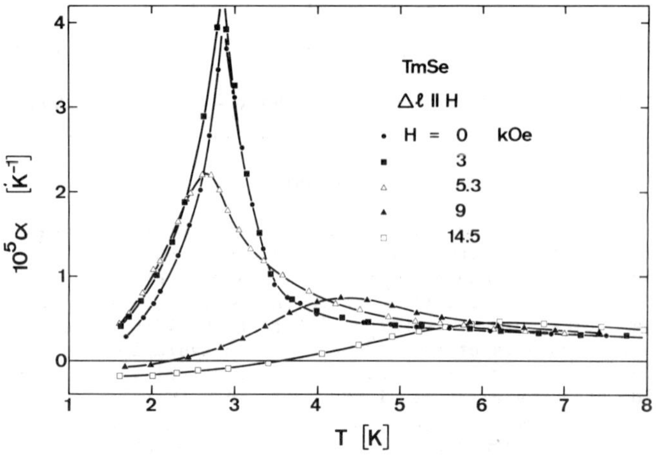

Fig. 1: Linear thermal expansion coefficient of TmSe parallel to the (100)-axis. H is parallel to the strain.

shoulder appears at higher temperatures being shifted towards higher temperatures with still increasing magnetic field. For strong magnetic fields the expansion coefficient becomes even negative at low temperatures.

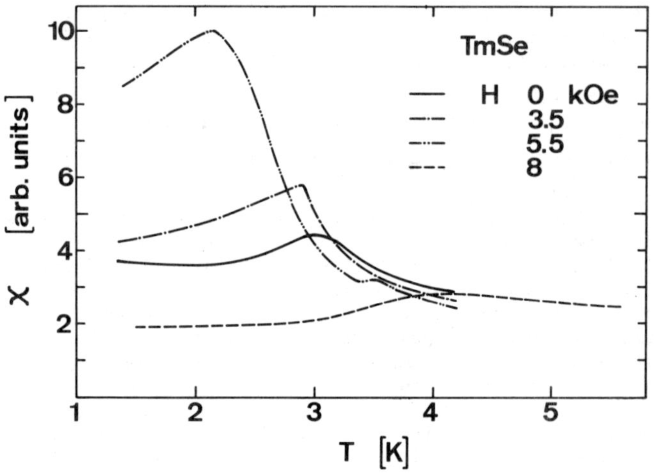

Fig. 2: a.c. susceptibility as a function of temperature in various magnetic fields.

Fig. 2 shows the temperature and field dependence of the a.c. susceptibility which bear a resemblance to the results on $\alpha(T)$ shown in fig. 1. For increasing fields we also observe the appearance of a broad shoulder whose maximum is shifted towards higher temperatures with raising strength of H.

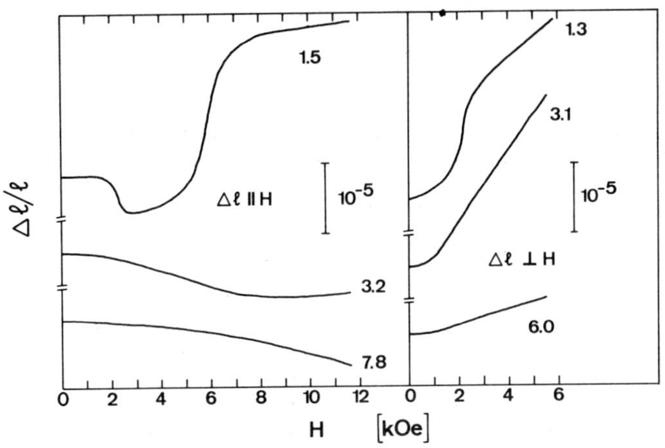

Fig. 3: Typical recorder tracings of magnetostriction measurements in parallel and perpendicular magnetic fields. The curves are shown for increasing fields only.

Selected recorder tracings of magnetostriction measurements at different temperatures are given in fig. 3. For the case of the strain being parallel to the field direction we observe two distinct critical fields at temperatures below T_N, where T_N is the temperature of the maximum in the thermal expansion anomaly in zero field. We find a discontinuous contraction of the crystal at approximately 2 kOe and a much larger expansion at approximately 5 kOe. Above 3 K the crystal shows a negative strain at low fields and a positive strain at higher fields. The change of slope is shifted towards higher fields with increasing temperature. In fields perpendicular to the strain measuring direction (only 6 kOe were available) we observe only one critical field at approximately 2 kOe, where the crystal shows an expansion comparable to the dilatation in the parallel case. In the paramagnetic region above T_N a much stronger deformation upon application of a transverse field is observed than in the case of a parallel field.

DISCUSSION AND CONCLUSION

From the results given above we obtain a tentative diagram of the magnetic phases of TmSe which is shown in fig. 4. The transition occurring when passing the lines I, II or III are presumably of first order since they are accompanied by a macroscopic volume change. When crossing the phase boundaries I and III no clear hysteresis effects could be detected, whereas a field sweep through line II leaves the phase between line II and III frozen in even at zero field. The virgin state can be obtained by heating the sample above T_N. Little can be said on the actual structure of ordering but if line II corresponds to a spin flop transition, this would call for a very small anisotropy energy.

Line IV derives from the broad maxima of the thermal expansion coefficient and the a.c. susceptibility. It is also compatible with a change of slope in the magnetostriction. Below line IV we observe the typical paramagnetic H^2-dependence whereas above line III and IV the slope has a different sign and the strain varies linearly with H.

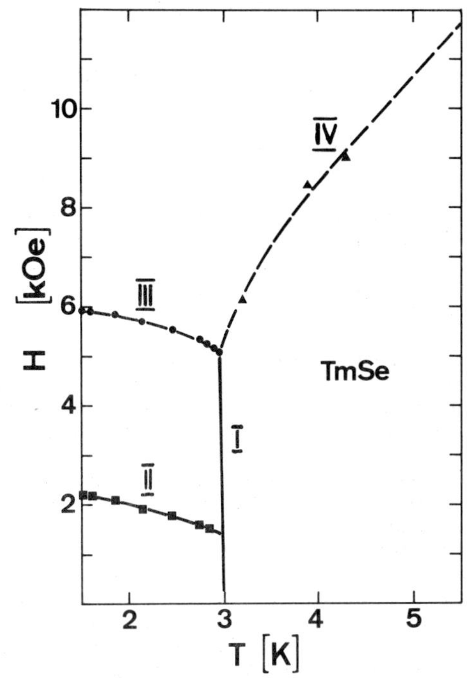

Fig. 4: Magnetic phase diagram of TmSe at low temperatures.

The discontinuous and hysteretic volume changes connected with the magnetic phase transition demonstrate that the ordering is of long range nature and it is not clear why this fact could not have been observed with neutron scattering experiments. There is also the problem of bringing in accordance the long range order with the very short time scale on which the valence fluctuations of the Tm-ions are reported to occur.

If we accept the concept of temporal valence fluctuations there remains the question whether the order is due to the Tm^{3+} or the Tm^{2+} ions. The ground state of Tm^{3+} would be a Γ_1 singlet and the order would be of the induced moment type. Tm^{2+} can have either a Γ_6 or a Γ_7 doublet ground state. For an ordering temperature of 3 K molecular field theory gives the following molecular field exchange constants λ for the different ground states: 2, 4.5 and 2.7 moles/emu for the Γ_1, Γ_6 and Γ_7 ground state respectively. It is therefore not easy to decide which level can be identified as the magnetic ground state. A more detailed analysis of this work is in preparation.

REFERENCES

1. E. Bucher, K. Andres, F.J. di Salvo, J.P. Maita, A.C. Gossard, A.S. Cooper and G.W. Hull Jr., to be published
2. B.B. Triplet, N.S. Dixon, P. Boolchand, S.S. Hanna and E. Bucher,
3. M. Campagna, E. Bucher, G.K. Wertheim, D.N.E. Buchanan and L.D. Longinotti, Phys. Rev. Letters 32, 885 (1974)
4. D. Wohlleben, J.G. Huber and M.B. Maple, AIP Conf. Proc. 2, 1478 (1971)
5. M.B. Maple and D. Wohlleben, AIP Conf. Proc. 18, 447 (1973)
6. D. Wohlleben and B.R. Coles, Magnetism, ed. G.T. Rado and H. Suhl (Academic Press, New York 1974), vol. V, p.3
7. E. Bucher, A.C. Gossard, K. Andres, J.P. Maita and A.S. Cooper, Proc. 8th Rare Earth Res. Conf. (USGPO Washington D.C., 1970) p. 74
8. D.E. Cox, L. Passell, R.J. Birgeneau and E. Bucher, unpublished.

MAGNETORESISTANCE DUE TO CONDUCTION BAND SPLITTING IN MAGNETIC SEMICONDUCTORS

R. L. Kautz and Y. Shapira, Francis Bitter National Magnet Laboratory*
Massachusetts Institute of Technology, Cambridge, Massachusetts 02139

ABSTRACT

In magnetic semiconductors an external magnetic field causes a spin splitting of the conduction band. This leads to a change not only in the resistivity due to spin disorder but also in the resistivity due to nonmagnetic scattering such as scattering from ionized impurities.

In a magnetic semiconductor it is useful to distinguish two electron populations, the localized _f_ electrons which are primarily responsible for magnetic properties and the itinerant _s_ or _d_ electrons which are responsible for electrical conduction. For the present purpose we model these two electron populations by a lattice of localized spins and a single conduction band with a fixed number of carriers. The _s-f_ or _d-f_ exchange interaction between localized spins and conduction electrons affects the conduction electrons in two ways. First, spatial fluctuations in the spin alignment scatter conduction electrons and contribute to the electrical resistivity. Such spin-disorder scattering results in the well-know resistivity peak associated with large spin fluctuations near the Curie temperature.[1] Second, a net magnetization $\vec{M}$ of the localized spins, whether spontaneous or induced by an external magnetic field $\vec{H}$, splits the conduction band into spin (+) and spin (-) subbands separated by an energy δ which is approximately proportional to M. When δ is field induced, it can be many times larger than the splitting $g\mu_B H$ that results from the direct interaction between the conduction electrons and H. As δ increases from zero electrons are transferred from the higher-to the lower-energy subband and for the large splittings obtainable in magnetic semiconductors this transfer can be nearly complete.

Near the magnetic ordering temperature, where spin-disorder resistivity is large, an external magnetic field produces a large decrease in the resistivity. This occurs for two reasons.[1] First, the field acts to suppress fluctuations in the alignment of localized spins, thereby reducing the scattering potential. Second, the spin splitting induced by the field acts to prevent scattering between the spin subbands. This second effect is a result of the fact that spin-disorder scattering is nearly elastic. As the subbands are split, the number of opposite-spin states into which scattering can occur is reduced for the majority of electrons.

Well above the ordering temperature, scattering due to mechanisms other than spin disorder becomes comparable to spin-disorder scattering and, at the same time, the negative magnetoresistance associated with spin-disorder scattering is often reduced to a few percent at typical fields. It then becomes necessary to take into account the effect of spin splitting on other scattering mechanisms. Such effects can be seen most clearly for the case of degenerate statistics. In this limit the resistivity depends only on the scattering times for electrons at the Fermi surface. Because the relaxation times for most scattering mechanisms are energy dependent and the Fermi energy of each subband shifts as electrons are redistributed, the resistivity is a function of spin splitting. It should be noted that spin-disorder, acoustical-phonon, and ionized-impurity scattering all have energy dependent relaxation times and are therefore affected by field-induced electron redistribution.

The case of ionized-impurity scattering demands special attention because the relaxation time depends not only on energy but also on the screening radius for the ionized-impurity Coulomb potential.[2] This screening radius r_o changes as electrons are redistributed between the subbands. Within the Thomas-Fermi approximation r_o is given by

$$r_o^{-2} = \frac{4\pi e^2}{\kappa}\left[\frac{dn^+}{d\zeta^+} + \frac{dn^-}{d\zeta^-}\right], \qquad (1)$$

where κ is the dielectric constant, n^+ and n^- are the densities of spin (+) and (-) electrons and ζ^+ and ζ^- are the Fermi energies for the respective subbands. The behavior of r_o as a function of the electron transfer parameter $\Delta = (n^+-n^-)/(n^++n^-)$ and degeneracy parameter $D = \zeta_o/kT$ (where ζ_o is the Fermi energy for $\Delta = 0$, $T = 0$) is shown in Fig. 1. The limit $D \gg 1$ corresponds to degenerate statistics and $D \ll 1$ corresponds to classical statistics. Because r_o measures the range of the scattering potential, it is evident that the ionized impurity relaxation time at a given energy decreases with increasing Δ.

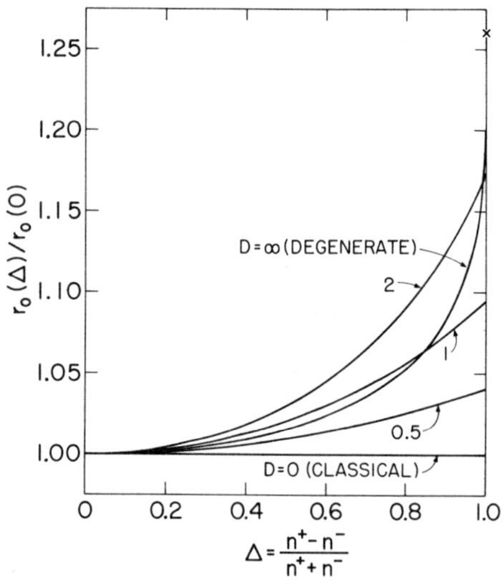

Fig. 1. Screening radius vs electron transfer for several values of the degeneracy parameter $D = \zeta_o/kT$. The point marked x indicates the endpoint of the $D = \infty$ curve at $\Delta = 1$.

To see the net effect of spin splitting on ionized impurity scattering it is useful to consider the Fourier transform of the screened impurity potential

$$v(k) = \frac{4\pi e^2/\kappa}{k^2 + r_o^{-2}}. \qquad (2)$$

For degenerate statistics the most important scattering events are those which involve values of k comparable to the Fermi momentum k_F. In going from $\Delta = 0$ to $\Delta = 1$ the situation changes from one with the conduction electrons equally divided between two subbands with a common Fermi momentum $k_F(\Delta = 0)$, to one with all electrons in the lower subband and a Fermi momentum $2^{1/3} k_F(\Delta = 0)$. Thus, as Δ changes from 0 to 1, the important scattering events shift to larger k. From Eq. (2), the increase in Fermi momentum would lead to a reduction in the scattering amplitude if r_o were constant. As has been argued, however, r_o increases with Δ and causes increased scattering for any given k. Thus, there are two competing mechanisms. Whether the resistivity increases or decreases with Δ will depend on whether $k_F(\Delta = 0)$ is smaller or larger than approximately $1/r_o(\Delta = 0)$. In Fig. 2 we show the ratio of the ionized

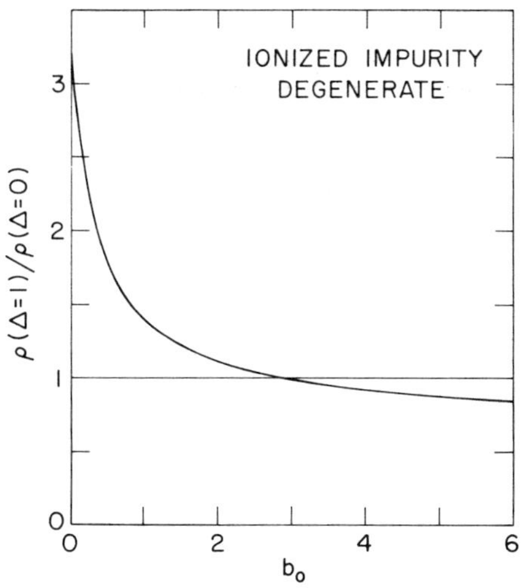

Fig. 2. Ratio of the resistivities at $\Delta = 1$ and $\Delta = 0$ as a function of $b_o = (2k_F r_o)^2{}_{\Delta = 0}$ for degenerate statistics.

impurity resistivities at $\Delta = 1$ and $\Delta = 0$ as a function of $b_o = (2\ k_F r_o)^2{}_{\Delta = 0}$.

We now consider the effect of spin splitting for classical statistics. In this limit the resistivity depends on an average of the relaxation time over energies within roughly kT of the band edge. For all processes having relaxation times which are not dependent on Δ, this average does not change with spin splitting and magnetoresistance due to electron redistribution disappears. Moreover, although the ionized-impurity relaxation time depends on Δ through r_o, Fig. 1 shows that r_o is independent of Δ for classical statistics. Thus, in the classical limit only spin-disorder scattering is affected by spin splitting and this because it involves scattering between subbands.

A detailed account of the effect of spin splitting on electrical conduction and experimental observations of these effects in EuTe, EuSe, and EuS will be published elsewhere.[3]

REFERENCES

1. C. Haas, CRC Crit. Rev. Solid State Sci. 1, 47 (1970).
2. V.I. Fistul, Heavily Doped Semiconductors (Plenum, New York, 1969), Chap. 3.
3. Y. Shapira and R.L. Kautz, Phys. Rev. B; Y. Shapira, S. Foner, N.F. Oliveira, Jr., and T.B. Reed, Phys. Rev. B; Y. Shapira, R.L. Kautz and T.B. Reed, in Proceedings of 12th Intl. Conf. on the Physics of Semiconductors, Stuttgart, 1974 (to be published).

*Supported by the National Science Foundation.

MAGNETIC POLARIZATION IN THE RAMIREZ-FALICOV MODEL OF VALENCY-CHANGE TRANSITIONS; APPLICATION TO CERIUM*

P.F. DE CHATEL and A.J.T. GRIMBERG**
Natuurkundig Laboratorium, University of Amsterdam, Amsterdam, The Netherlands

ABSTRACT

We have extended the Ramirez-Falicov model of the α-γ phase transition of cerium, allowing for a field-induced or spontaneous splitting of the bands and of the six-fold degenerate 4f level. To this end, we had to introduce two new parameters in the theory to take account of conduction electron-f electron and conduction electron-conduction electron exchange. The value of the latter can be estimated from the low--temperature susceptibility of α cerium, for the former an upper limit can be found on the basis of the phase diagram. We find the effect of presently available magnetic fields on the α-γ transition temperature in cerium to be negligible according to this theory. With higher values (not appropriate to cerium) for the conduction electron-f electron exchange parameter, the model shows first and second order transitions to a ferromagnetic γ phase. The relative stability of this phase with respect to the α phase is higher than that of the paramagnetic γ phase; under certain circumstances there is no direct paramagnetic γ→α transition.

* To be published in Phys.Rev.B, 15 November 1975.

** Present address: Koninklijke/Shell Exploratie en Produktie Laboratorium, Volmerlaan 6, Rijswijk, The Netherlands.

THE EFFECT OF CONDUCTIVITY ON THE Eu^{153} HYPERFINE INTERACTIONS IN EuTe*

K. Raj
Fordham University, Bronx, New York 10458

J. I. Budnick and T. J. Burch
University of Connecticut, Storrs, Connecticut 06268

ABSTRACT

The hyperfine interactions of Eu^{153} in two well characterized samples of EuTe have been measured at liquid helium temperatures by spin-echo NMR. One of the samples is a conductor and the other one essentially a nonconductor. Our studies indicate that the Eu^{153} hyperfine interactions are sensitive to the conductivity of the samples. Published magnetization measurements on these samples are used to interpret our NMR results. A tentative identification of the $Te^{123,125}$ resonances is made.

Europium telluride is the only member of the europium chalcogenides that is antiferromagnetic. Its Neél temperature is 9.64 K.[1] Neutron diffraction work[2] has shown that the ordering in EuTe is of the second kind with Eu^{2+} moments aligned parallel in (111) planes with alternate planes of spins aligned antiparallel. EuTe becomes ferromagnetic in an applied magnetic field of 84 KG at 1.5K.[3]

We have measured the spin-echo spectra in two samples of EuTe which had been used in magnetization,[4] resistivity and Hall effect[5] and ultrasonic[6] measurements. One sample, No. 102, is conducting with room temperature resistivity of 10^{-2} Ω-cm and has a saturation moment of 6.9μ_B/Eu^{2+}. This sample is in powder form. The other sample, No. 113, is nonconducting with room temperature resistivity and saturation moment values of 10^{6} Ω-cms and 6.6μ_B/Eu^{2+} respectively. This sample is in polycrystalline form.

A spectrum was taken of the conducting sample, No. 102, in a field to optimize the signal intensity. The Eu resonances in EuTe as in EuO, EuS[7] and EuSe[8] have a short frequency dependent total echo decay time, T_2^*, making correction of the raw data necessary to produce a plot of echo amplitude which we take to be proportional to the number of nuclei, versus frequency. The correction procedure has been described elsewhere.[7] All the spectra discussed here have received this correction unless it is specifically excluded. In this spectrum echoes were observed over the frequency ranges 79 to 145 MHz and 264.2 to 265.7 MHz. Two very intense peaks occur at 118.6 and 264.7 MHz. In the lower frequency range additional peaks of very low intensity at 115.1, 103.8, 98.0, 81.5 and 79.8 MHz are also observed.

We identify the peaks at 118.6 and 264.7 MHz as the Eu^{153} and Eu^{151} resonances respectively, since the ratio of the frequencies of the resonances is nearly the same as the ratio of the gyromagnetic ratios of the isotopes ($\gamma^{151}/\gamma^{153} = 2.25$).

We note that although echoes are observed over a wide frequency range (79 to 145 MHz) about the Eu^{153} line, no corresponding intensity is found about the Eu^{151} line. The additional intensity about the Eu^{153} spectrum must be associated with the nuclei other than the europium. We tentatively assign the peaks at 115.1 and 98.0 MHz as due to the Te^{125} and Te^{123} resonances respectively since the ratio of the frequencies of the lines is very close to the ratio of the gyromagnetic ratios of the isotopes ($\gamma^{125}/\gamma^{123} = 1.20$) and the intensities of the lines are roughly in the same proportion as the natural abundances of the isotopes. If these are the Te resonances, the hyperfine field at the tellurium nuclei would then be about 86 kG. Recent Mossbauer work[9] on EuTe indicates that the Te spectrum is broad with a tentative hyperfine field value between 80 and 100 kG. The origin of the lines occurring at 103.8, 81.5 and 79.8 MHz is not known.

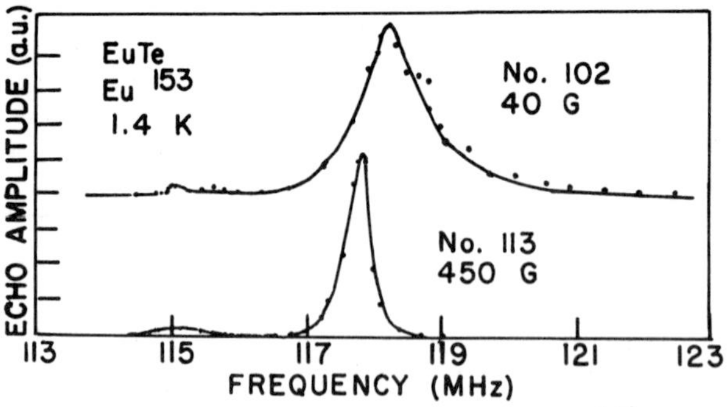

Fig. 1. Corrected spin-echo spectra of Eu^{153} in conducting (No. 102) and nonconducting (No. 113) samples of EuTe at 1.4K.

Figure 1 shows the corrected Eu^{153} spin-echo spectrum of the conducting sample No. 102 in a field of 40 G and the nonconducting sample No. 113 in a field of 450 G at 1.4 K. The conditions necessary to excite a signal differ drastically in the two samples. In the conducting sample, No. 102, echoes could be seen in zero external field and could be excited with low power. The echo amplitude was also found to be very sensitive to the power level. In the nonconducting sample, No. 113, the echoes were observed only in applied magnetic fields greater than about 400G and higher power was required to excite the echo. An extrapolation of the power versus external field curve indicates that very large power (unavailable at present) would be required to excite the echo in this sample in zero applied field.

These differences between the samples suggest that the signals in sample No. 102 are greatly wall enhanced and thus possibly this sample is not purely antiferromagnetic. This conclusion is consistent with magnetization data[4] which show that the conducting sample has a spontaneous moment and possibly a canted spin structure at H=0; and the nonconducting sample has no spontaneous moment and is antiferromagnetic at H=0.

It is possible to compare the spectra of Fig. 1 since in both samples the Eu^{153} NMR frequency and half width are independent of applied field. The maximum of the resonances are at 118.3 ± 0.2 and 117.9 ± 0.1 MHz in the conducting and nonconducting samples respectively. The half widths of the lines are 0.7 ± 0.2 and 0.3 ± 0.1 MHz respectively. These differences are thought to be associated with differences in conduction electron polarization of the two samples.

The Eu^{153} echo decay as a function of pulse separation is frequency dependent for all values of H studied (0-20 KG for sample No. 102 and 0.4-20 KG for sample No. 113). Non-exponential echo decay is observed in sample No. 102. This decay can be decomposed into two exponential components; a short and a long component. The values of these relaxation times at the center of the line (118.3 MHz) are $(T_2^*)_{short} = 6 \pm 2\mu S$ and $(T_2^*)_{long} = 18 \pm 3\mu S$ and are independent of external magnetic field.

The effective spin-spin relaxation time in the nonconducting sample can be described by a single exponential at all fields. At the center of the Eu^{153} resonance (117.9 MHz) in a field of 450 G T_2^* is about 2μS. T_2^* increases rapidly with external magnetic field.

The temperature dependence of the Eu hyperfine field between 4.2 and 1.2 K has a T^2 behavior for both samples. Figure 2 shows this dependence for the Eu^{153}

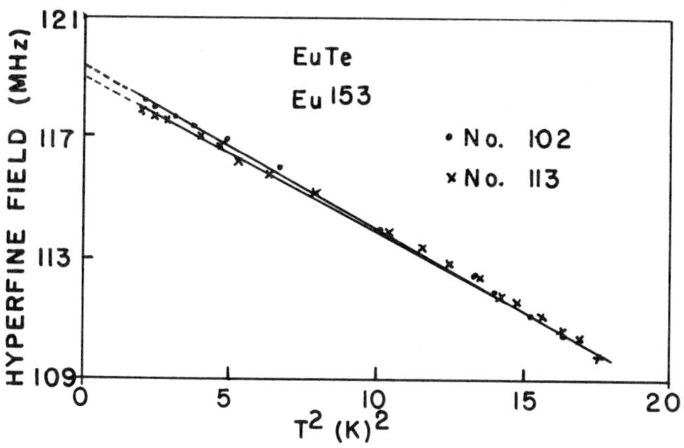

Fig. 2. The T^2 dependence of Eu^{153} resonance in EuTe samples Nos. 102 and 113 in the temperature range of 1.2 to 4.2K.

NMR in an external field of 40 G for sample No. 102 and in a field of 450 G for sample No. 113. A least square analysis yields values of the saturation hyperfine field at T=0K of 119.42 $\pm$ 0.18 and 119.00 $\pm$ 0.20 MHz respectively. Our values agree closely with that obtained by Hihara et al.[10] (118.7 MHz) but are much higher than the Mössbauer result (113.4 MHz).[11]

If we assume that the sign of the Eu hyperfine field is negative, as it is in the other europium chalcogenides, the conduction electrons in sample No. 102 produce a small negative spin density at the nucleus. The evidence for the conduction electron polarization is also supported by the increased value of the paramagnetic curie temperature θ in the conducting sample. θ for samples No. 102 and No. 113 are 4 $\pm$ 1 and -1 $\pm$ 2 K respectively. A similar relationship between θ and the hyperfine field has been found in $Eu_{1-x}Gd_xSe$.[8]

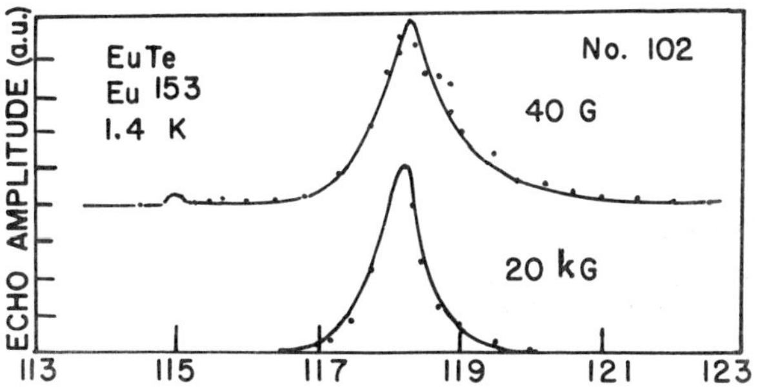

Fig. 3. Corrected Eu^{153} spectra in EuTe (No. 102) in external magnetic fields of 40 G and 20 kG.

Figure 3 shows the Eu^{153} spectra in sample No. 102 at 1.4 K in a 40 G field and in a 20 kG field. The frequency of the line has not changed at the higher field and also its width is roughly the same. Similar results are found in the data from sample No. 113. Hihara et al.,[9] however, have reported a field dependence of both the Eu^{153} frequency and line width in europium telluride.

Figure 4 shows the field dependence of the echo amplitude of Eu^{153} in the two samples. The echo amplitude was measured at the peak frequency and the power has been varied to provide low and constant turn angle conditions at each field. No T_2^* correction was necessary for sample No. 102 because T_2^* was relatively constant up to 9 KG. In sample No. 113 T_2^* showed a strong field dependence. Below 2 KG T_2^* was too short to provide a satisfactory correction. Both an uncorrected and a portion of the corrected curve is shown for this sample. Hihara et al.[10] also found the

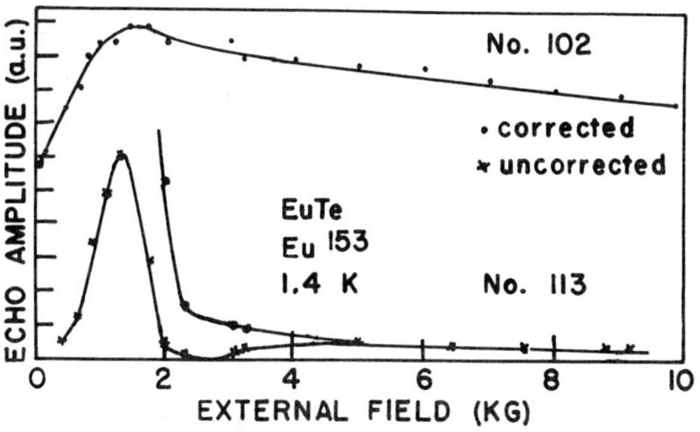

Fig. 4. Magnetic field dependence of Eu^{153} spin-echo amplitude in conducting (No. 102) and nonconducting (No. 113). In the nonconducting sample the corrected amplitude is not fully shown because of its very large relative value and uncertainty in the peak due to short T_2^*.

intensity of the resonance field dependent.

In both samples, the echo amplitude first rises steeply, becoming a maximum around 1 KG and then begins to decrease. This maximum agrees with the applied field, 770 $\pm$ 200 G, at which Battles[12] found a spin-flop transition in EuTe.

Magnetization measurements[4] on sample No. 113 show a large increase in the field dependent differential susceptibility, $d\sigma/dH$, peaking sharply at about 1 kG and then decreasing. The echo amplitude in this sample shows a similar behavior. In sample No. 102, $d\sigma/dH$ peaks at about 200 G (1.2K). The difference between the spin-echo and magnetization measurements[4] for this sample is probably due to demagnetization effects. Magnetization data on sample No. 102 was taken on a single crystal.

The power necessary to excite an echo in the nonconducting sample decreases as the magnetic field is increased above zero consistent with a change of magnetic spin structure to a canted phase enabling power to couple with the Eu spins. The power reaches a minimum at about 1 kG and then increases. In the conducting sample the power increases smoothly with field.

In summary the spin-echo excitation characteristics for the conducting and nonconducting EuTe samples are different. The addition of conduction electrons to EuTe does not significantly influence the magnitude of the Eu hyperfine field. However, it does increase the line width and the effective spin-spin relaxation time as a result of magnetic spin structure changes.

ACKNOWLEDGEMENT

We are grateful to Dr. T. B. Reed (MIT) for supplying us the EuTe samples and enlightening discussions and to N. Blum (Applied Physics Lab.) for discussions of his Te Mossbauer results prior to publication.

REFERENCES

* Supported in part by the University of Connecticut Research Foundation.

1. G. Busch, P. Junod, R. G. Morris and J. Muheim, Helv. Phys. Acta 37, 637 (1964).
2. G. Will, S. J. Pickart, H. A. Alperin and R. Nathans, Phys. Chem. Solids 24, 1679 (1963).
3. C. R. Pidgeon, J. Feinleib and T. B. Reed, Solid State Commun. 8, 1711 (1970).
4. N. F. Oliveira, Jr., S. Foner, Y. Shapira and T. B. Reed, J. Appl. Phys. 42, 1783 (1971); Phys. Rev. B5, 2634 (1972).
5. Y. Shapira, S. Foner, N. F. Oliveira, Jr. and T. B. Reed, Phys. Rev. B5, 2647 (1972).
6. Y. Shapira and T. B. Reed, Phys. Rev. B5, 2657 (1972).

7. K. Raj, T. J. Burch and J. I. Budnick, Int. J. Magnetism 3, 355 (1972).
8. K. Raj, T. J. Burch and J. I. Budnick, AIP Conf. Proc. Part II 10, 1564 (1972).
9. N. Blum, Applied Physics Lab., Maryland; Private Communication.
10. T. Hihara, T. Komaru and Y. Koi, International Conf. on Ferrites held in July 1970 in Kyoto, Japan.
11. G. K. Shenoy, F. Holtzberg, G. M. Kalvius and B. D. Dunlap (to be published).
12. J. W. Battles, Report No. NWC TP 4851, Naval Weapons Center, China Lake, California.

VIRTUAL BOUND STATES AND CONFIGURATIONAL MIXING IN $Sm_{1-x}Y_xS$ ALLOYS

T. Penney and F. Holtzberg, IBM Thomas J. Watson Research Center, Yorktown Heights, New York 10598

ABSTRACT

Resistivity and Hall effect measurements have been made on $Sm_{1-x}Y_xS$ alloys which exhibit black to gold phase transitions in temperature and composition, similar to SmS under pressure.

SUMMARY

We have made measurements of resistivity and Hall effect in $Sm_{1-x}Y_xS$ alloys which undergo changes in electronic configuration as a function of composition and temperature.[1] The room temperature lattice constant data for the $Sm_{1-x}Y_xS$ system shows a discontinuity at $x \sim 0.15$ which divides the system into an expanded "black" phase and a collapsed "gold" phase.

Using this lattice constant data we calculate the Sm valence v and find $2.0 < v < 2.6$ for the "black" phase and $v = 2.7$ for the "gold". These non-integral valence values result from the presence of both the divalent f^6 and trivalent f^5d configurations. We would expect that the Hall effect would measure the d electrons from the Sm f^5d configurations as well as the Y d electrons.

Surprisingly, the experimental one band Hall densities $N_H \equiv 1/R_He$, for the "black" phase samples corresponds to the density of Y d electrons only. Apparently, in the "black" phase, the Sm d electrons from the f^5d configuration do not appear as free electrons in the d band.

We suggest a model of a virtual bound state, $B(\varepsilon)$ of mixed configuration, $(1-\varepsilon)f^6 + \varepsilon f^5d$, which is degenerate with the d conduction band but below the Fermi level, E_F, where it is undetected by the Hall effect.

For samples which collapse upon warming from low T ($x \gtrsim 0.15$) there is an increase in ρ and a decrease in R_H. In the virtual band state model, the lattice collapse causes a lowering of the d band energy so that E_F now lies within $B(\varepsilon)$. The result is an increase in the d scattering and the resistivity. The Hall effect is small due to a cancellation between the electrons in the d band and holes at the top of $B(\varepsilon)$.

1) T. Penney and F. Holtzberg, to be published.

ELECTRICAL RESISTIVITY OF Gd AND La MONOCHALCOGENIDES

R. Hauger, E. Kaldis, G. von Schulthess, P. Wachter, C. Zürcher
Labor. für Festkörperphysik, ETHZ, 8049 Zürich, Switzerland.

ABSTRACT

The Gd monochalcogenides form a very interesting system of ionic and antiferromagnetic metals ($40\ K < T_N < 60\ K$) which permit a wide variation in stoichiometry without changing crystal symmetry or phase purity. The paramagnetic Curie temperature and the carrier concentration can be varied drastically with stoichiometry, resulting in a spectacular change in color [1,2,3]. In this paper we present electrical resistivity data of GdS, GdSe,GdTe and LaS, LaSe,LaTe single crystals, which have been measured between 4 K and 300 K by an ac four probe technique for samples with varying stoichiometry. The La monochalcogenides have been chosen as the nonmagnetic counterparts to investigate the influence of alone the stoichiometry variation on the resistivity. It is observed that in the Gd chalcogenides a kink in the resistivity at T_N for excess metal develops into a maximum at T_N for stoichiometric samples. For these compounds the Langer-Fisher theory of spin disorder scattering sufficiently explains the observed results [4]. For metal deficiency GdS and GdSe exhibit a maximum far below T_N and in GdSe and GdTe the off-stoichiomtry can be driven so far that a steady increase of the resistivity below about 80 K is found. In contrast to this, the La-chalcogenides with similar metal deficiencies did not show an increase of resistivity with decreasing temperature, in fact at low enough temperatures a superconducting transition is observed. This comparison shows that it is not a simple electronic defect scattering mechanism which causes the anomalous resistance increase in the Gd chalcogenides at low temperatures. In this case of metal deficiency conventional scattering theories are unable to explain the observed behavior. Therefore new models will be proposed based on magnetic defect scattering.
A complete report will be given elsewhere.

[1] F. Holtzberg, D.C. Cronemeyer, T.R. McGuire and S.von Molnar: NBS Spec.Publ.364, Proc.of 5th Mat. Res.Symp., 1972 p.637
[2] W.Beckenbaugh, G.Güntherodt, R. Hauger, E. Kaldis J.P.Kopp and P.Wachter:AIP Conf.Proc. 18, 540 (1973)
[3] G.Güntherodt and P.Wachter:AIP Conf.Proc.18, 1034 (1973)
[4] M.E. Fisher and J.S.Langer:Phys.Rev.Letters 20, 665 (1968)

TEMPERATURE DEPENDENCE OF EPR LINEWIDTH IN EuSe: FIELD SUPPRESSION OF FLUCTUATION SCATTERING PROCESSES*

P. Streit†, Atsuo Kasuya††, and Glen E. Everett
Physics Department, University of California, Riverside, CA 92502

ABSTRACT

The linewidth versus temperature of EuSe (T_N=4.6K) at 24 GHz exhibits a broad maximum at ~12K. Additional results at 9.05, 18.26, and 35.25 GHz have been obtained. At 35.25 GHz no linewidth peak is observed while at 9.05 GHz the peak is strongest and occurs at ~T_N. The behavior is qualitatively interpreted following Tomita and Kawasaki. Measured linewidths versus temperature are proportional to the product of the lifetime and the density of fluctuations. Increasing the field (frequency) at a fixed temperature decreases the fluctuation density and decreases the linewidth. In large fields, the linewidth dependence on the critical divergence in fluctuation lifetimes is completely obscured, but at low fields, dominates the observed temperature dependence. The maximum linewidth with increasing T at $T > T_N$ occurs when the increasing density of fluctuations compensates the decreasing lifetime. At 9.05 GHz, the excess linewidth due to fluctuations can be described by $\Delta H-\Delta H(T=\infty)\sim(T-T_N)^{\delta}$ with δ=-1.33 and does not agree with the value $\delta \cong -2/3$ predicted by T-K.

The linewidth in EuSe has been measured versus temperature in single crystal-spherical samples at different frequencies. The previously reported peak in the linewidth at T=12K (T_N=4.6K in zero field)[1] has been partially explained as originating in the interplay between the expected divergence in the critical fluctuation lifetimes as the Nèel temperature is approached and the number of spin fluctuations excitable at a given temperature in the presence of a magnetic field. This interpretation has been given previously by us for earlier results in EuSe as a single frequency[1] and by Campbell and Lawson for their results in $CrCl_3$[2].

The role of the field in suppressing the number of excitations is expected to have an exponential character reflecting the field induced energy gap. To test these ideas the measurements have been extended to higher and lower frequencies. The results are shown in Fig. 1 together with those of Ketcham and Everett[1]. At the frequency 35.25 GHz, the field for resonance is 12.7 kOe. Relative to previous results at 24.0 GHz[1], it is seen that the linewidth

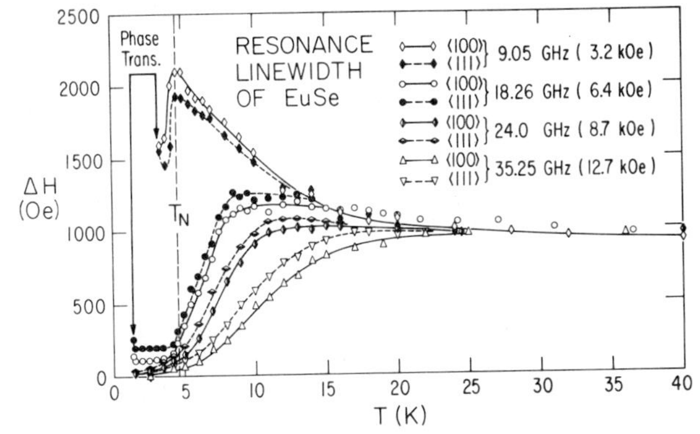

Fig. 1. Temperature dependence of the linewidth in EuSe determined at different frequencies. At each frequency results are shown for the field applied along crystallographic ⟨100⟩ and ⟨111⟩ directions representing extremal values of ΔH. The resonance field for $T > T_N$ is indicated for each frequency.

is suppressed and has the appearance of being displaced to higher temperature with no peak persisting. The data at the intermediate frequency, 18.26 GHz, is, by contrast, shifted to lower temperatures. The peak observed is more pronounced and occurs at ~9K. The behavior at 9.05 GHz, the lowest frequency measured, shows the behavior expected from a critical divergence of the fluctuation lifetimes. The peak in linewidth occurs at T_N dropping rapidly for $T < T_N$.

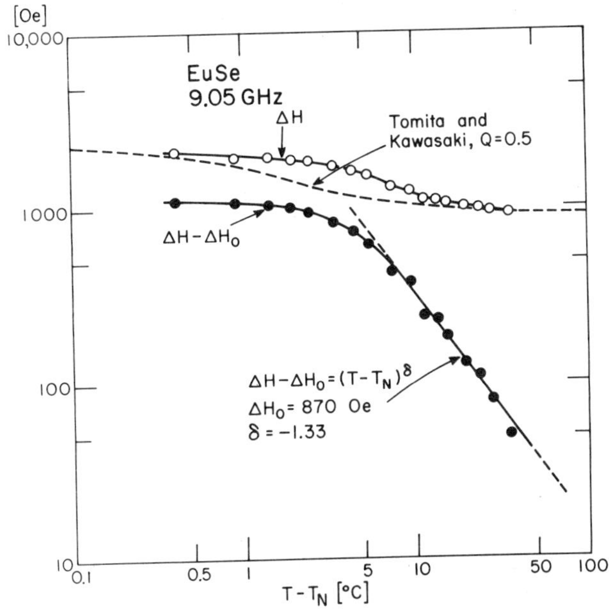

Fig. 2. The total linewidth (open circles) plotted versus temperature. The dashed curve is the theoretical result of Tomita and Kawasaki[3]. The excess linewidth (solid circles) obtained by subtracting the high temperature dipolar contribution is also shown.

A comparison of these results with the theory of Tomita and Kawasaki[3] is shown in Fig. 2. Also included for comparison, is a plot of the theoretical results of T-K with Q=0.5. The experimental results are similar, however, no detailed fit using Q as an adjustable parameter is justified. The parameter Q is related to an anisotropy energy. Previous anisotropy measurements[4] give $Q \ll 1$ while an improved fit to the linewidth would require $Q > 1$. Also shown in Fig. 2 is a log-log plot of the excess linewidth, $\Delta H-\Delta H_o$, versus $T-T_N$. The parameter, ΔH_o, was chosen to give the best fit to the data and represents the limiting high temperature linewidth due to dipolar interactions. The result, expressed as

$$\Delta H - \Delta H_o = (T - T_N)^{\gamma}$$

corresponds to the value $\gamma = -(1.33 \pm 0.05)$ for $(T - T_N) > 10$. This value is different from that predicted by T-K, $\gamma \cong -2/3$, but is close to the value $\gamma = -5/3$ obtained by Huber[5]. The departure from this relation at $(T-T_N) < 10$ is interpreted as a suppression of the effect of fluctuations due to the finite field, H=3.2 kOe. The value of ΔH_o=(870±30)G is in good agreement with previous measurements at room temperature[6] and the theoretically expected value from dipolar broadening of 840G.

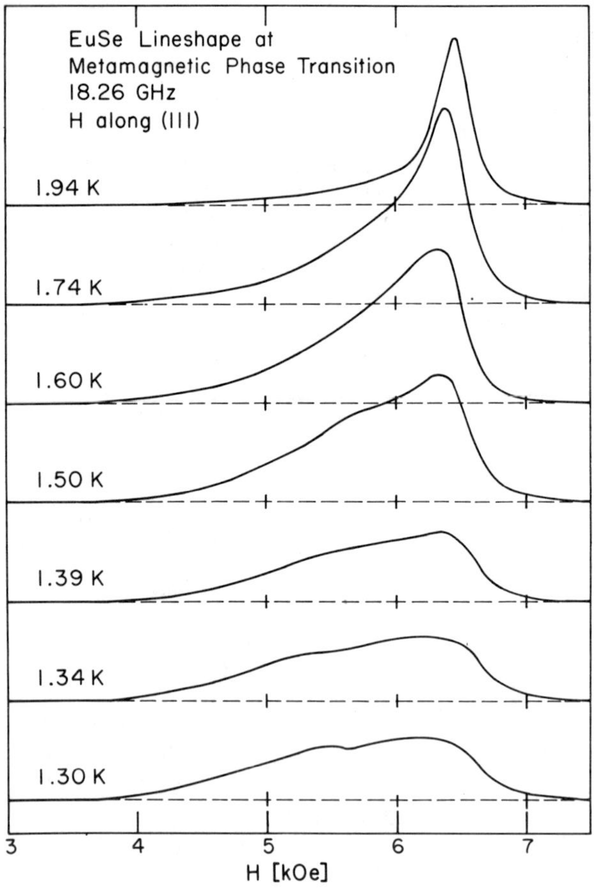

Fig. 3. The absorption lineshape at different temperatures for $T < T_N$, obtained at fixed frequency as a function of the applied field. The metamagnetic transition in this temperature interval is first order. As the applied field is varied, the relative proportions of the two coexisting magnetic phases changes continuously giving rise to the apparent splitting of the line as well as causing broadening.

The region below T_N is strongly influenced by the metamagnetic transitions which exist in this material. For a spherical sample, the resonance field (neglecting magnetic anisotropy) is independent of temperature but the internal field increases with decreasing temperature in constant applied field. The swept field operation of the microwave spectrometer, particularly with the large linewidths occurring at low frequencies, causes distortion of the measured absorption line due to the proximity of the metamagnetic transition. This effect is shown at 18.26 GHz for a series of temperatures, Fig. 3, displaying an increasing asymetry in the linewidth with decreasing temperature. At 9 GHz, this effect is much larger having the appearance of the lines being "cut off" below a critical field. This distortion of the line for $T < T_N$ makes impractical any interpretation in this region. It would be of interest to extend these measurements to lower frequencies to determine if the "critical divergence" becomes stronger. However, the large

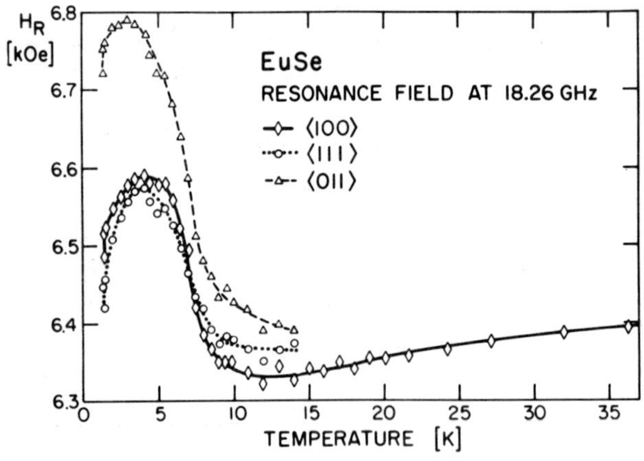

Fig. 4. The dependence of the resonance field on temperature for three directions of the applied field. Previous results at 24.0 GHz[(4)] have been interpreted in terms of the temperature dependence of the cubic anisotropy constants.

linewidth and the sensitivity of the magnetic free energy to changes in the internal field indicate that meaningful results could only be expected from measurement at constant field in a swept frequency spectrometer. With data obtained in this way, it might be possible to interpret reuults for $T < T_N$. For completeness, we note that there has been interest in the nature of the anisotropy constants in this material[(7)]. We show in Fig. 4, the temperature dependence of the resonance field, measured at 18.26 GHz for the field in the directions indicated. The temperature dependence of the cubic anisotropy constants, K_1 and K_2, were determined previously[(4)] from an analysis of the angular dependence of the resonance field at 24.0 GHz. Uncertainties regarding the nature of the high field phase for $T < T_N$ casts doubt on this interpretation and a similar analysis of these data has not been undertaken.

*Research supported by the U.S. Atomic Energy Commission. This constitutes AEC Report No. 34 P77-46.

† Present address: Brown, Boveri and Cie. BADEN, Switzerland.

†† Present address: The Research Institute for Iron, Steel, and Other Metals, Tohoku University, Sendai, Japan.

REFERENCES

1. R.A. Ketcham and Glen E. Everett, AIP Conf. Proc., 10, 102 (1972).
2. T.G. Campbell and A.W. Lawson, AIP Conf. Proc. 18, 759 (1974), (Magnetism and Magnetic Materials-1973).
3. K. Tomita and T. Kawasaki, Prog. Theo. Phys. 44, 1173 (1970).
4. Glen E. Everett and Ray C. Jones, Phys. Rev. B5, 1561 (1971).
5. D.L. Huber, Phys. Rev. B6, 3180 (1972).
6. G. Busch, B. Natterer, and H.R. Neukomm, Phys. Lett. 23, 190 (1966).
7. Atsuo Kasuya and M. Tachiki, Phys. Rev. B8, 5298 (1973).

PRESSURE DEPENDENCE OF THE RARE EARTH KNIGHT SHIFT IN THE SINGLET GROUND STATE SYSTEMS PrP AND TmP*

H. T. Weaver and J. E. Schirber
Sandia Laboratories
Albuquerque, New Mexico 87115

ABSTRACT

We have measured the pressure dependence of the Knight shift (K) for ^{141}Pr in PrP and ^{169}Tm in TmP at 4 K and find d ℓn K/dP = 0.83% and 0.48%/kbar, respectively. Both compounds are paramagnetic but exhibit a temperature independent (Van Vleck) susceptibility near T = 0 due to the crystal field induced singlet electronic ground state. The rare earth Knight shift provides a direct measurement of the susceptibility (χ_{VV}), from which the crystal field splitting (Δ) may be determined, i.e., $\chi_{VV} \propto \Delta^{-1}$. Thus, our result of d ℓn K/dP > 0 is in qualitative disagreement with the usually successful effective point charge model. Such a model implies that Δ varies as some inverse power of the lattice constant and, sonsequently, that the presure derivative (d ℓn K/dP) is negative. We checked for anomalous conduction electron-nuclear coupling effects by measuring the pressure dependence of the ^{31}P Knight shift in PrP. However, the observed d ℓn K/dP (1.6% kbar) excludes this possibility and probably results from the same susceptibility change that is responsible for the rare earth resonance behavior.

The magnitude of the effect of hydrostatic pressure on hyperfine coupling, as observed by nuclear magnetic resonance (NMR) techniques, has been found to be comparable in magnitude to the material compressibility (B^{-1}). Typical materials exhibit compressibilities of order 0.1%/kbar; thus, in order to study pressure induced changes in Knight shifts (typically K are of order 1%) we must measure changes in the resonance position of roughly 10^{-5} of the applied field. In many cases this is much smaller than the line width. As a consequence, NMR-pressure work has been confined, in the main, to highly compressible materials[1] or to specific situations in which electronic transitions are driven by modest pressures.[2]

An alternative approach to the study of pressure effects of hyperfine coupling is to investigate systems with Knight shifts sufficiently large that even with normal compressibilities, pressure-induced changes of the resonance line position are easily detectable. The largest Knight shifts that have been observed[3] to date in non-magnetic materials are exhibited by ^{141}Pr and ^{169}Tm contained in intermetallic compounds with the pnictides. Here the RE Knight shifts range from 500 to 10,000%. Well defined resonance lines with Knight shifts of such magnitudes owe their existence to the occurrence of a singlet electronic ground state for the f-electrons of the RE. The singlet is produced by crystal field splitting of the lowest total angular momentum (J) state. However, the conduction electron screening limits the energy splitting (Δ) to the range of order 100°K. Thus, the ground state singlet, with $\langle S \rangle = 0$, drastically reduces the RKKY broadening of the resonance, and the low lying excited states give rise to a large Van Vleck susceptibility or orbital Knight shift.

The RE-pnictide intermetallic compounds have been thoroughly characterized in terms of their structures, magnetic properties, and crystal field interactions.[4] Some problems still remain concerning (1) the transferred hyperfine mechanism and (2) the nature of the crystal field interaction. Our data have relevance to both questions.

TABLE I

Summary of Knight shift pressure derivatives in PrP and TmP. The number in parentheses represents the error in the last quoted digit.

	d ℓn K/dP (%/kbar)
^{31}P in PrP	1.6(4)
^{141}Pr in PrP	0.83(5)
^{169}Tm in TmP	0.48(4)

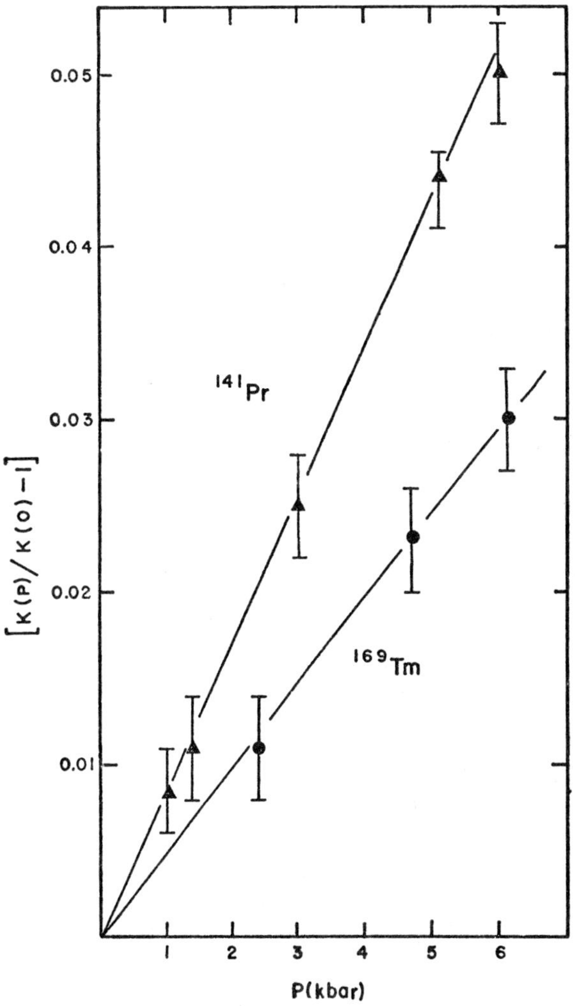

Figure 1. Plot of pressure dependence at 4°K for the ^{141}Pr and ^{169}Tm Knight shifts in PrP and TmP, respectively.

We have made NMR measurements of the resonance position of ^{141}Pr and ^{31}P in PrP and ^{169}Tm in TmP as a function of hydrostatic pressure to 6 kbar at 4°K. A conventional marginal oscillator was used to detect the resonance and solid helium pressure methods were used. A more detailed explanation of the procedures is contained in Ref. 2.

The RE Knight shifts are plotted as a function of pressure in Fig. 1 and the pressure derivatives for all three nuclei are listed in Table I. Note that the ^{31}P pressure derivative is roughly an order of magnitude smaller than that for the RE. The phosphorus shift is probably in the range of the material compressibilities (which have not yet been measured) with the RE derivatives exhibiting a much larger value.

There are three contributions to the RE Knight shift: (1) the orbital (K_{orb}), (2) core polarization (K_{cp}), and (3) conduction electron spin polarization (K_{ce}). Jones has decomposed the observed shift and finds K_{orb} much larger than the other two. Thus, any pressure changes must either be due predominantly to changes in K_{orb}, or enormous changes in K_{cp} and K_{ce} are implied. Since K_{cp} reflects core electron interaction, it is highly unlikely that small pressure changes could drastically affect its value. It is possible that the s-f exchange could be appreciably altered by pressure, due to the behavior of the conduction electrons. The phosphorus shift is a sensitive monitor of this interaction and the observed small pressure derivative for ^{31}P is good evidence that s-f exchange has not been radically altered. Thus, we attribute our observed RE pressure derivative to changes in K_{orb}.

Orbital and spin-dipolar interactions give rise to K_{orb}. Following Jones we write

$$K_{orb} = 2\langle r_f^{-3}\rangle\langle J||N||J\rangle\ \chi/Ng_j \quad , \qquad (1)$$

where r_f is the 4f-electron radius, χ the susceptibility and $\langle J||N||J\rangle$ is the operator-equivalent factor tabulated by Elliot and Stevens.[5] For $T \ll \Delta$

$$\chi \approx \frac{360}{7}\frac{Ng^2\mu_B^2}{\Delta} \quad . \qquad (2)$$

We see that pressure changes in K must be due to changes in either $\langle r_f^{-3}\rangle$ or Δ. The f-electron radius is much smaller than the atomic radius of the RE. As a consequence, the change in r_f quite likely is much less than the compressibility of the material. Thus, our experimental $d\ \ln K/dP$ is approximately equal to $-d\ \ln \Delta/dP$ and we have the surprising result that

$$d\ \ln \Delta/dP < 0. \qquad (3)$$

That is, Δ decreases as the lattice is compressed.

Due to screening effects on interatomic potentials in metals, no comprehensive theory for crystal field interactions has been developed. Usually an effective point charge model is applied, which for a predominantly fourth order interaction[6,7] yields $\Delta \propto Ze^2\langle r_f^4\rangle/R^5$, where Ze is the effective charge and R is the lattice constant. Good agreement between this model and neutron diffraction experiments has been achieved for RE lighter than Tb using $Ze = -1.2e$. A somewhat larger $|Ze|$ was needed for the heavier RE, indicating a breakdown of the model. Our data for Pr, as well as Tm, i.e., both light and heavy RE, require that Ze be pressure dependent. Such an effect is outside the scope of the effective point charge model.

As a final item we consider the ^{31}P Knight shift. The origin of the ^{31}P shift has commonly been attributed to s-f exchange-produced conduction electron polarization. For uniform polarization,[9]

$$K = K_o\left[1 + \Gamma\frac{(g_J - 1)}{2Ng_J\beta^2}\right]\chi \quad , \qquad (4)$$

where K_o is the temperature independent shift and Γ is the s-f exchange energy. Assuming $d\ \ln K(Pr)/dP = d\ \ln \chi/dP$ and $K \gg K_o$ we have

$$\frac{d\ \ln \Gamma}{dP} = \frac{d\ \ln K(P)}{dP} - \frac{d\ \ln K(Pr)}{dP} \approx 0.8\%/\text{kbar} \ . \qquad (5)$$

Thus, we have an indication of a large pressure deviation for Γ, although our uncertainity is large.

ACKNOWLEDGMENTS

The samples were prepared by Dr. R. K. Quinn. We acknowledge the technical assistance of R. L. White and B. R. Hansen.

REFERENCES

* Work supported by the U. S. Atomic Energy Commission.
1. G. B. Benedek and T. Kushida, J. Phys. Chem. Solids 5, 241 (1958); H. T. Weaver and A. Narath, Phys. Rev. B 1, 973 (1970).
2. H. T. Weaver, J. E. Schirber, and A. Narath, Phys. Rev. B 8, 5443 (1973).
3. E. D. Jones, Phys. Rev. Letters 19, 432 (1967).
4. See E. D. Jones, Phys. Rev. 180, 455 (1969) and references contained therein.
5. B. Bleaney in Magnetic Properties of Rare Earth Materials, ed. R. J. Elliot (Plenum, New York, 1972).
6. R. J. Birgeneau, E. Bucher, J. P. Maita, L. Passell, and K. C. Tuberfield, Phys. Rev. B 8, 5345 (1973).
7. P. Junod, A. Menth, O. Vogt, Phys. Letters 23, 626 (1966).
8. Pressure could change the ratio (x) of fourth order to sixth order crystal field interactions. Since the interaction is predominantly fourth order[6,7] at atmospheric pressure we believe an increase in sixth order is the most likely change in this ratio, if any at all. Such a change tends to increase Δ in PrP and produces a very slight decrease in TmP. (See K. R. Lee, M. J. M. Leask, and W. P. Wolf, J. Phys. Chem. Solids 23, 1381 (1962).) Thus, large differences in the pressure dependence of x between PrP and TmP are required for experimental agreement. This is considered very unlikely.
9. V. Jaccarino, J. Appl. Phys. 32, 1025 (1961).

MAGNETIC MOMENT DISTRIBUTION IN URANIUM DIOXIDE

John Faber, Jr.

Argonne National Laboratory, Argonne, Illinois 60439

ABSTRACT

The magnetic form factor of uranium in UO_2 has been examined by means of neutron diffraction. Uranium dioxide (CaF_2 structure) orders antiferromagnetically at 30.8 K with the type I magnetic structure. The intensities of 1250 magnetic Bragg reflections (reducing to 71 independent hkl values and giving full three-dimensional data for sin $\theta/\lambda \leq 0.83$ Å^{-1}) have been measured with a well-characterized stoichiometric single crystal at 5°K. The values of (μf) plotted versus sin θ/λ show considerable asphericity for sin $\theta/\lambda > 0.5$ Å^{-1}. Comparisons with free-ion (+ crystal field) calculations give an experimental magnetic moment of 1.78 (4) μ_B and indicate that the general shape of the form factor is in agreement with theory for low scattering angles, but that the nonspherical contributions at higher angles are at least a factor of five greater than expected. The significance of these results is discussed in terms of the extended nature of the 5*f* wavefunctions in UO_2.

INTRODUCTION

Measurements of the magnetic form factor yield direct information on the radial extent of the unpaired electrons. This information is of particular value in the actinides since, in contrast to 3*d* and 4*f* systems, the nature of magnetism in 5*f* systems appears to be strongly correlated with the extent of overlap between neighboring 5*f* ions. The form factor is also sensitive to the electronic ground state of the ion - a problem especially difficult to resolve in actinide systems. Since theoretical calculations have suggested that the $5f^2$: $^3H_4$ ground state is appropriate for UO_2, a measurement of the magnetic form factor should be able to provide precise information on the spatial extent of the 5*f* electrons. In addition, the antiferromagnetic structure of UO_2 is simple type I with an ordered moment $\approx 1.8\ \mu_B$ / U atom,[1] which are features that tend to simplify both the experiment and interpretation of the data.

EXPERIMENTAL

A UO_2 single crystal (1.6 x 2.4 x 5.1 mm) was heated for 6 h at 1950°C in a 10^{-6} to 5 x 10^{-7} torr vacuum to ensure the stoichiometric composition.[2] The room temperature lattice constant, a=5.4702(2)Å, agrees well with values found in the literature.[3] The neutron diffraction experiments were performed at the CP-5 reactor using an instrument capable of collecting 3-dimensional crystallographic data at low temperature.[4] The neutron wavelength was λ=0.992Å. Specimen characterization as well as parameters needed for processing the magnetic intensities were obtained at 80 and 5°K by measuring the nuclear Bragg reflections, with the results summarized in Table I. Extinction corrections were ~5% for the weak nuclear reflections, whereas for the magnetic data only the (100) and (110) reflections required significant corrections (4 and 2% respectively). Azimuthal (Σ) scans on four magnetic reflections at 5°K suggested that the effects of multiple scattering are negligible. The first-order phase transformation characterized on this sample by T_N 30.7(2)°K is in excellent agreement with the literature.[1]

Table I. UO_2 nuclear reflections parameter summary at 80 and 5°K.

Parameter	T=80°K	T=5°K
Scale constant	50.4(2)	50.3(2)
Extinction parameter, g	1110(46)	1155(43)
B_u (Å^2)	0.10(1)	0.07(1)
B_{ox} (Å^2)	0.24(1)	0.22(1)
χ^2	1.4	1.2
Residual (weighted)	0.9%	0.6%
# observations	594	540
# indep. hkl's	32	32

The intensities of 1250 magnetic Bragg reflections (reducing to 71 independent hkl values for sin $\theta/\lambda \leq 0.83$Å^{-1}) were collected at 5°K from a multidomain UO_2 single crystal. The magnetic intensites were consistent with UO_2 being a type I antiferromagnet with a <001> propagation direction and moments in the XY plane. The domain populations were equal (within experimental error), thus precluding a unique determination of the moment direction in the XY plane. A <100> or <110> direction for the moments appears most probable,[5] and the magnetic unit cell symmetry is then orthorhombic. The experimentally determined product of the magnetic moment (μ, in Bohr magnetons) and the form factor (f) is shown in Fig. 1. Three points of interest are: (a) the slow fall-off of the form factor with sin θ/λ suggests a $5f^2$ rather than $6d^2$ outer-electron configuration, (b) the U^{4+} effective moment and the magnitude of the form factor anisotropy for sin $\theta/\lambda > 0.5$ Å^{-1} are in agreement with the limited zone data of Frazer et al.,[1] (c) the anisotropy in the form factor for sin $\theta/\lambda > 0.5$ Å^{-1} is clearly outside experimental error.

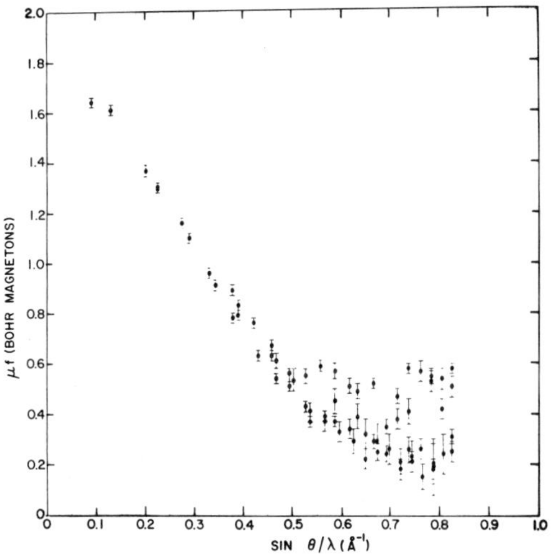

Fig. 1. Experimental values for μf in antiferromagnetic UO_2.

DISCUSSION

The magnetic form factor can be considered as arising from two fundamental properties of the uranium ion: (a) the one-electron radial wavefunctions of the 5*f* shell, and (b) the electronic ground state of the U^{4+} ions in the lattice. In this model anisotropy is a result of the unquenched orbital angular momentum and crystal-field effects. For (a) we have integrated relativistic Dirac-Fock wavefunctions $U^2_{5f}(r)$, supplied by Freeman and Desclaux[6], to obtain the $\langle j_i(\vec{K})\rangle$ functions where

$$\langle j_i(\vec{K})\rangle = \int_0^\infty U^2_{5f}(r)\ j_i\ (Kr)\ dr$$

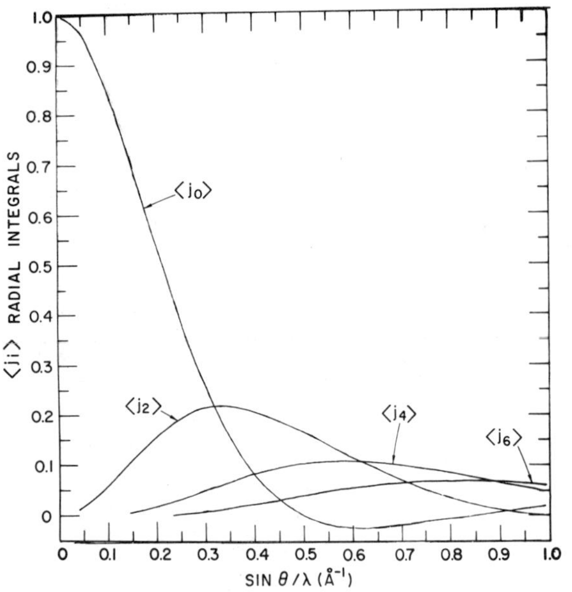

Fig. 2. Radial integrals calculated from Dirac-Fock one-electron wavefunctions for a $5\underline{f}^2$ ground state.

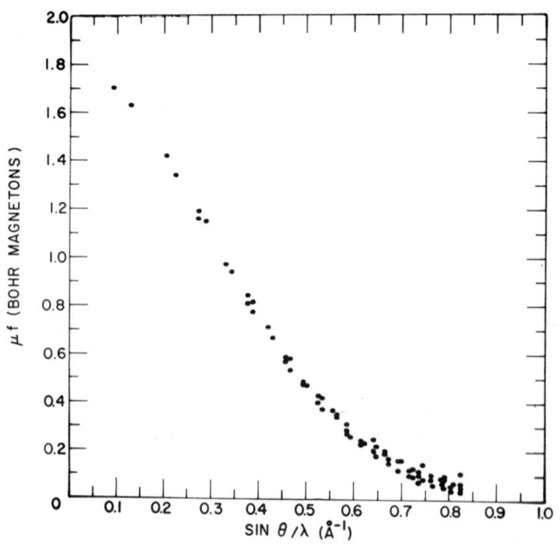

Fig. 3. Calculated values for μf with moments along a <100> direction, with $|M_J = 3>$.

in which $j_i(Kr)$ is the spherical Bessel function. The radial integrals, $<j_i(K)>$ are shown in Fig. 2. The $<j_0>$ component of the wavefunctions contains terms that are spherically symmetric, whereas terms in $<j_2>$ etc., contain aspherical contributions. Of particular interest is the angular dependence of $<j_0>$ in comparison with the observed data in Fig. 1. The large scatter of points about a smooth curve in Fig. 1 is a measure of the anisotropy in the form factor of UO_2 and occurs predominantly for $\sin\theta/\lambda > 0.5$ Å^{-1}, in which region the spherical contributions, $<j_0>$ in Fig. 2, have passed through zero and remain small. For (b), the electronic ground state, we choose a model that considers UO_2 an insulator with a localized $5\underline{f}^2$ configuration. Within the framework of Russell-Saunders coupling this gives a $^3H_4$ state with a magnetic moment of $3.2\mu_B$. The wavefunctions of the free-ion model may then be characterized as $|M_J = 4>$. Crystal-field calculations have been reported on UO_2 and suggest that the Γ_5 triplet is the ground state[7]. In this model an exchange field polarizes the triplet to yield the magnetic moment of $1.8\mu_B$ at low temperature. The tensor-operator method[8,9] provides a simple method of handling the magnetic scattering from ions with unquenched orbital moments and gives the cross section in the form

$$q^2\mu f(\vec{k}) = \Sigma_{i=0,2,4,6}\ a_i <j_i(\vec{K})>$$

where $f(\vec{K})$ is the form factor, q^2 the square of the magnetic interaction vector, μ the magnetic moment, and a_i are coefficients.

For H || <100> the wave functions are predominantly $|M_J = 3>$ whereas for H || <110> the $|M_J = 4>$ state is the majority component of the wavefunction. As a first approximation form factors calculated for $|M_J = 3>$, and $|M_J = 4>$ were fit to the experimental data with a weighting scheme based on $\sigma_{obs}(\mu f)$. To match the calculated and experimental values of (μf) for a given M_J state one parameter, the magnetic moment, was allowed to vary. The best fit gave $\mu = 1.78(4)\mu_B$. The (μf) values for $|M_J = 3>$ are shown in Fig. 3. As expected, the calculated form factor (either M_J state) fits the data quite well for $\sin\theta/\lambda < 0.5$ Å^{-1}, but neither calculation can account for the magnitude of the anisotropy at higher angles. For $\sin\theta/\lambda > 0.5$ Å^{-1} a comparison between Fig.1 and 3 shows that the magnitude of the anisotropy is at least a factor of five greater than that predicted on the basis of the unquenched orbital moment.

Fourier transform techniques can yield the magnetization density throughout the unit cell and the data in Fig. 1 as well as the calculated form factor given in Fig. 3 were used to generate such magnetization maps. Several features have emerged from the study of the magnetization maps. First, the radial dependence of the magnetization density around the U^{4+} ion at the origin was found to be consistent with $|M_J = 3>$ and incorrect for $|M_J = 4>$. This is consistent with the preliminary crystal-field calculations. Second, integration techniques[10] were used to calculate the magnetic moment per uranium ion from the experimental data, yielding $\mu = 1.74\mu_B$. In addition a non-zero density at the octahedral interstices (1/2,1/2,1/2, etc.) was found to represent a moment of $\sim 0.1\mu_B$. However, these results must be viewed with caution since no data were obtained for $\sin\theta/\lambda > 0.83$Å^{-1} and series termination effects are probably important.

A few concluding remarks appear appropriate since the experimental data presented here tend to ask (rather than answer) more questions concerning our understanding of the physics of actinide magnetism, particularly for UO_2. We plan to do a full crystal-field calculation and hence remove some of the constraints associated with the free-ion model. These calculations may account for some of the observed anisotropy. Invoking covalency, where unpaired spin remains on the ligand ion, does not appear too promising, as pointed-out by Lovesey,[11] because of the symmetry associated with the cubical environment of the U^{4+} ions. Finally we plan to use polarized neutron experiments to examine the induced moment for UO_2 and compare it with the antiferromagnetic form factor of Fig. 1.

REFERENCES

1. B.C. Frazer, G. Shirane, D.E. Cox, C.E. Olsen, Phys. Rev., 140, A1448 (1965)
2. R.K. Edwards, M.S. Chandrasekhardiah, P.M. Davidson, High Temperature Sci., 1, 98 (1969)
3. L. Lynds, J. Inorg. Nucl. Chem., 24, 1007 (1962)
4. L. Heaton, M.H. Mueller, M.F. Adam, R.L. Hitterman, J. Appl. Cryst., 3 289 (1970)
5. S. J. Allen, Jr., Phys. Rev., 166, 530 (1967)
6. A. J. Freeman, J.P. Desclaux, Private Comminication
7. H. U. Rahman, W.A. Runciman, J. Phys. Chem. Solids, 27, 1833 (1966)
8. S.W. Lovesey, D.E. Rimmer, Rept. Progr. Phys. (London) 32, 333 (1969)
9. G.H. Lander, T.O. Brun, J. Chem. Phys., 53, 1387 (1970)
10. R.M. Moon, Intern. J. Magnetism, 1, 219 (1971)
11. S.W. Lovesey, Private Communication

MAGNETIC STRUCTURES AND DIMERIZATION IN ANTIFERROMAGNET $La_2Fe_2S_5$

R. Plumier, M. Sougi, A. Miedan-Gros, M. Lecomte
Service de Physique du Solide et de Résonance Magnétique
CEN-Saclay. BP n°2, 91190 Gif-sur-Yvette France

ABSTRACT

The magnetic structure of orthorhombic $La_2Fe_2S_5$ previously observed at 4.2K is studied in the whole range of temperatures up to $T_N = 88K$. From T_N down to 30 K a different magnetic structure is observed with magnetic group $P2_1'a'b$ in an (a,b,2c) magnetic cell. At lower temperatures, the magnetic structure previously given is confirmed. However a dimerization of Fe ions antiferromagnetically coupled through superexchange interactions along [001] is observed along this direction, with a consequent lowering of the nuclear cell symmetry. At 17K and down to 4.2K, an additional lowering of the nuclear cell symmetry, which becomes monoclinic and which we ascribe to further dimerization along [100] is observed jointly with the appearance of weak ferromagnetism in that direction.

The magnetic structure of orthorhombic[1] $La_2Fe_2S_5$ has recently been given[2]. Two kinds of magnetic Fe^{2+} ions (octahedral and tetrahedral) are found on separate chains parallel to [001], and are antiferromagnetically coupled on each chain through superexchange interactions, with nearest neighbour antiferromagnetic chains being uncoupled. It was also shown that, for tetrahedral Fe, next-nearest neighbour chains are coupled through super-superexchange interactions of the type Fe-S-S-Fe.

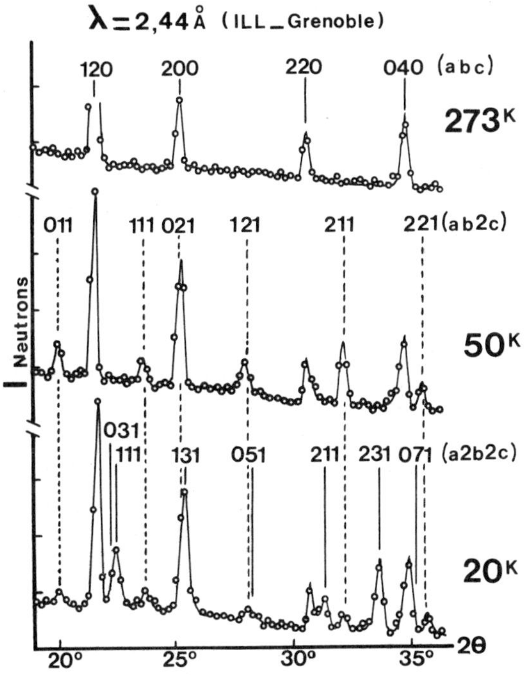

Fig.1 - Part of neutron spectra at various temperatures

Further detailed neutron diffraction investigations, a summary of which is given below, have recently been performed at EL3 Saclay and I.L.L. Grenoble at various temperatures (Fig.1). Mossbauer effect[3] and specific heat measurements indicate an ordering temperature of about 88K, different from the temperature (T= 30K) where the set of magnetic reflections observed at 4.2K[2] disappears.

a) In the temperature range $180K > T > 90K$, in the paramagnetic region, drastic changes (Fig.2) are noticed in the intensities of various nuclear reflections, whereas no noticeable changes are observed in the cell parameters and symmetry $A2_1am$ (Fig.4a). We interpret these modifications as a result of coherent shifts, in the (X-Y) plane, of tetrahedral Fe as well as those sulfur ions, 1 and 2 (Fig.3), which ensure antiferromagnetic superexchange interactions for octahedral Fe. Although in this range of temperatures the antiferromagnetic chains remain magnetically uncoupled, those

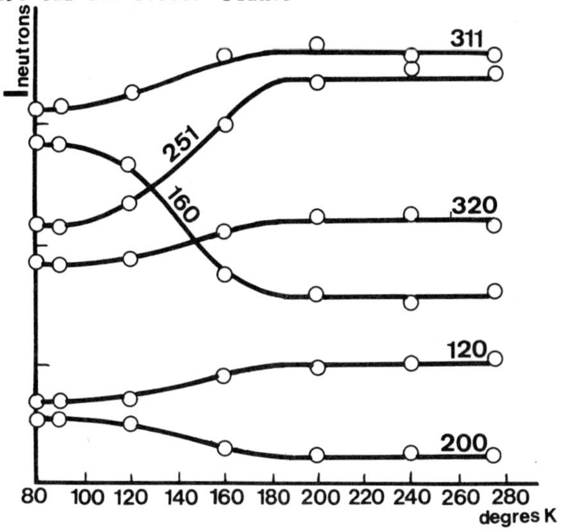

Fig.2 - Evolution of some nuclear reflections between 180 K and 90 K.

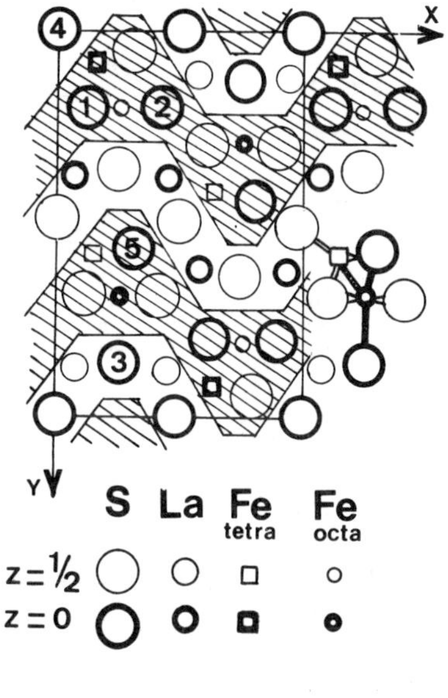

Fig.3 - (X,Y) projection of the nuclear structure of $La_2Fe_2S_5$; light areas refer to the rigid part of the lattice.

shifts which occur through an exchange-striction mechenism are elastically coupled, La and sulfur ions 3 and 4 (Fig.3) ensuring the stability of the whole lattice. At lower temperatures, no further change of the nuclear reflection intensities is observed.

b) Starting at 88K and down to 30K additional reflections we attribute to the appearance of long range magnetic ordering are observed (Fig.1). With moments pointing along [001], the corresponding magnetic structure may be described in an (a,b,2c) orthorhombic magnetic cell which belongs to the magnetic group $P2_1'a'b$ (Fig.4b).

c) Beneath 30K the general aspect of the spectrum (Fig.1) is analogous to the one previously obtained at EL3 Saclay[2]. However, thanks to the good resolution of the spectrometer used at I.L.L.Grenoble, a detailed examination of neutron spectra obtained in that range of temperatures leads us to a more complex description of both nuclear and magnetic structures. In addition to the magnetic reflections observed in the past[2], which may be described in an (a,2b,2c) magnetic cell, we

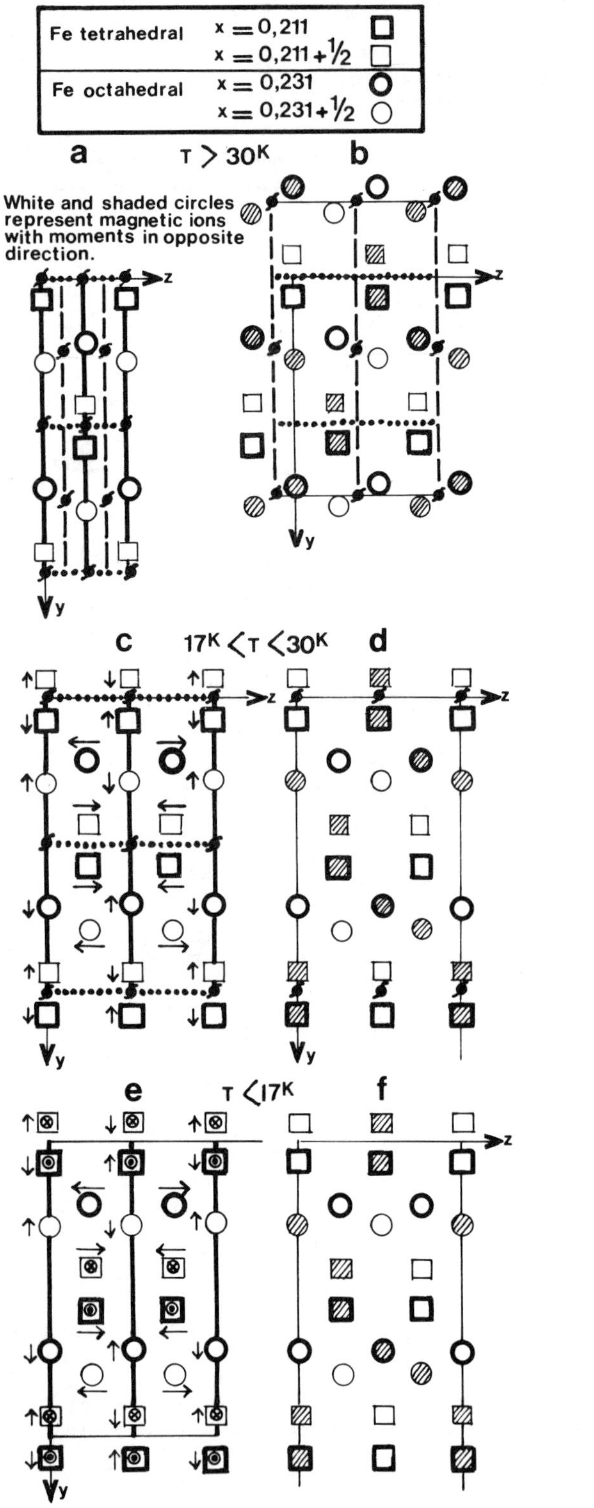

Fig.4 - (Y,Z) projections of the nuclear and magnetic structures at various temperatures.

notice that, although they are much weaker and have different intensities ratios, additional reflections occurring at the same positions as the ones of the magnetic cell (a,b,2c) observed in the temperature range 88K >T > 30K are still there. Beneath 30K , we consider that set of additional reflections to be superstructure lines arising from a lowering of the symmetry of the nuclear cell. The best agreement between observed and calculated intensities is obtained for the nuclear cell (a,b,2c) depicted in Fig.4c . Such a superstructure contains a dimerization of half the Fe^{2+} ions along [001], the other half, although possibly shifted in the (X,Y) plane, remaining at even distances. The symmetry of the (a,b,2c) nuclear cell is $P2_1am$, and

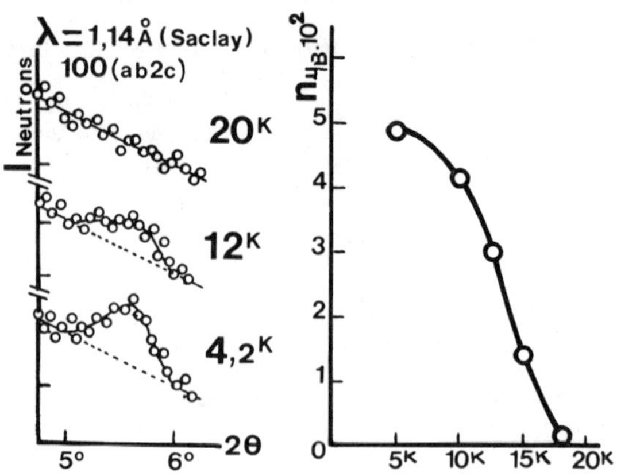

Fig.5 - Thermal evolution of weak ferromagnetism and (100) nuclear reflection.

the corresponding magnetic structure in that range of temperatures is described in an (a,2b,2c) monoclinic cell belonging to the magnetic group $P2_1'$ (Fig.4d), moments being aligned along [001] .

d) Starting at 17K and down to 4.2K an additional (100) reflection (Fig.5) is observed which we attribute to a further lowering of the nuclear symmetry. With ⊗ and ⊙ having their usual meaning, fair agreement between observed and calculated intensities is reached with the nuclear cell depicted in Fig.4e, which contains a further dimerization along [100]. In that range of temperatures, the symmetry of the (a,b,2c) nuclear cell is Pm, and the corresponding magnetic structure is described in an (a,2b,2c) triclinic cell with P1 magnetic symmetry (Fig.4f) and with antiferromagnetic moment components aligned along [001]. By allowing free rotation of the crystallites, we find from neutron diffraction that the weak ferromagnetic moment (Fig.5) is aligned along [100], whereas at higher temperatures, the easy direction of magnetization is along [010]. This set of results is related to the fairly large spin-orbit coupling of Fe^{2+} ions on both tetrahedral and octahedral sites.

A summary of the various nuclear and magnetic cell symmetries at all temperatures may be found in the following table, where letters a to f refers to Fig.4.

T		88K = T_N	30K	17K
M		ab2c (b) $P2_1'a'b$	a2b2c (d) $P2_1'$	a2b2c (f) P_1
N	abc (a) $A2_1am$		ab2c (c) $P2_1am$	ab2c (e) Pm

We think $La_2Fe_2S_5$ offers another example of the dimerization of magnetic chains theoretically studied by various authors[4,5] and recently observed in $V_{1-x}Cr_xO_2$ [6]. When dimerization comes into play, not only do the magnetic exchange interactions take new values, but their number is increased. If dimerization occurs at temperatures where long range magnetic ordering already exists, another magnetic structure may very well be stabilized. This is probably what is happening in $La_2Fe_2S_5$ at 30K and 17K . Let us mention that neutron diffraction experiments performed around 30K indicate hysteresis in the appearance of the two competing magnetic structures $P2_1'a'b$ and $P2_1'$, suggesting a first order transition. No such hysteresis is found around 17K .

References
1. G. Collin, P. Laruelle, Bull. Soc. Fr. Cristallogr. 94, 113 (1971).
2. R. Plumier, M. Sougi, G. Collin, Solid State Commun. 14, 971 (1974).
3. F. Varret, private communication.
4. G. Beni, P. Pincus, J. Chem. Phys. 57, 3531 (1972).
5. J.Y. Dubois, J.P. Carton, J. Physique 35, 371 (1974).
6. J.P. Pouget et al., Phys. Rev. B10, 1801 (1974).

MAGNETIC SUSCEPTIBILITY AND MÖSSBAUER STUDY OF SINGLE CRYSTAL FeS

J.R. GOSSELIN, J.L. HORWOOD, M.G. TOWNSEND, R.J. TREMBLAY and A.H. WEBSTER, Dept. of Energy, Mines & Resources, Mines Branch, Ottawa, Ontario, Canada K1A 0G1.

ABSTRACT

Stoichiometric FeS is characterized by two phase transitions, one structural at T_α and the other magnetic at T_s.

We have obtained magnetic susceptibility and Mössbauer data on a single crystal of very nearly stoichiometric FeS; the (macroscopic) magnetic susceptibility data show the two transitions well separated [$T_\alpha \doteq 420°K$, $T_s \doteq 445°K$], whereas the (microscopic) Mössbauer data show that at room temperature the 2C structure, with spins ||c, coexists with a few percent 1C structure with spins ⊥c. The concentration of the latter increases with temperature; at T_α the 2C structure transforms to 1C with spins ∥ c; however, in the range $T_\alpha<T<T_s$, a fraction of 1C with spins ⊥c is present. Above T_s the spins are ⊥ c in the 1C structure.

The susceptibility curves of FeS suggest a localised moment, in contrast to NiS which is now generally accepted to be an itinerant antiferromagnet.

INTRODUCTION

Stoichiometric iron sulphide is antiferromagnetic and is characterized by two phase transitions. The α-transition which we observe near 420°K (T_α) is a transition from an NiAs-type structure at high temperature to a closely related 2C superstructure below T_α[1,2]. Neutron diffraction on powdered nearly stoichiometric FeS showed that a spin-flop transition occurs at a temperature T_s higher than T_α ($T_s \simeq 450°K$)[3,4]. T_s is very sensitive to composition[4], and earlier magnetic studies on single crystals of nominally stoichiometric FeS indicated that T_α and T_s coincided[5].

In order to aid further understanding of the nature of the transitions at T_α and T_s in $Fe_{1-x}S$, magnetic susceptibility and Mössbauer effect measurements have been made in single crystals of composition $Fe_{0.996}S$ and $Fe_{0.988}S$.

EXPERIMENTAL

Crystals of $Fe_{0.996}S$ and $Fe_{0.988}S$ were grown by the Bridgman method in sealed evacuated silica capsules. The capsule containing the charge was lowered slowly (6mm/hr) from the hot zone of a furnace. Crystals were annealed at 850°C for 10 days; their compositions were determined from powder X-ray measurements of the 102 interplanar spacings of the NiAs-type cell (114 interplanar spacings of the 2C superlattice (troilite) cell). For Mössbauer measurements, oriented plates cut perpendicular to the c-axis were cemented to glass slides and were thinned to from 0.002- to 0.001-in. thickness. For magnetic measurements (made by the Faraday method in fields up to 8,000 oersteds) oriented pieces from the same boule were sealed into small evacuated silica capsules.

RESULTS AND DISCUSSION

The magnetic susceptibilities parallel and perpendicular to the c-axis in $Fe_{0.996}S$ are shown in Figure 1. The α-transition at about 420°K coincides with the endothermic peak observed on heating in our differential thermal analysis DTA curves and is well separated from the spin rotation at $T_s \sim 445°K$. However, with a small increase in metal deficiency, in $Fe_{0.988}S$ the two transitions appear to be superimposed. Earlier measurements[5] were made on material that was not sufficiently close to stoichiometry; the

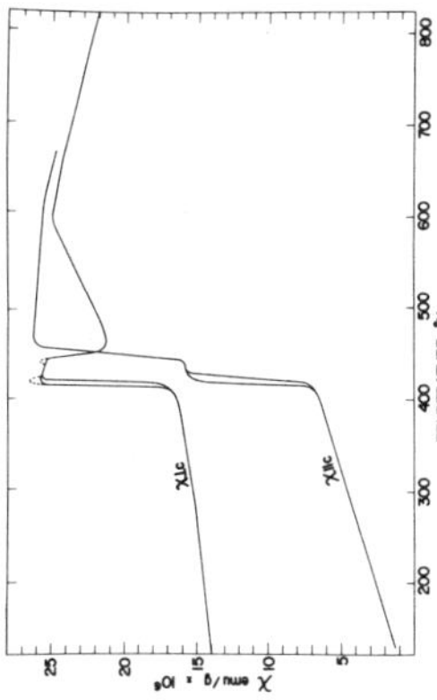

Fig. 1. Magnetic susceptibility of single crystal $Fe_{0.996}S$.

distinction between T_s and T_α was not clear and this le d to the erroneous conclusion that the spin rotation resulted from the abrupt change in c/a at T_α[7].

Measurements of the anisotropy of the susceptibility lead to the conclusion that the spins rotate only in a narrow range of temperatures around $T_s(x)$, ∿445°K, Figure 1. Furthermore the spins do not rotate gradually with increasing temperature as suggested earlier[3], but rather owing to a small spread in composition of the crystal, spins in slightly different environments rotate from || to ⊥c at different transition temperatures, $T_s(x)$.

The abrupt increase in $\chi_\perp$ on heating through T_α reflects a decrease in the exchange energy. From the expression $\chi_\perp = \frac{g^2 \mu_B^2}{2ZS(S+1)J_3}$ [8] where Z is taken as 2 and J_3 is the intersublattice magnetic exchange interaction in the c direction, we obtain $J_3 = -0.0039$ eV below T_α and -0.0023 eV above T_α, giving a decrease in $|J_3|$ at T_α of 1·6 meV.

Above the Neel temperature, at about 598°K, the susceptibility follows the Curie-Weiss law; the effective moment $5.5\mu_B$, is compatible with localised Fe(II) having an orbital contribution in the ground state; $\theta_p = +1160°K$.

Recently, Takahashi[6] has reported magnetic susceptibility data on nearly stoichiometric FeS, that are broadly in agreement with those reported here for $Fe_{0.996}S$.

To first order, the relative intensities of a Mössbauer magnetic hyperfine spectrum are (3:0:1) for spins || the γ-ray direction and (3:4:1) for spins ⊥ the γ-ray direction. The Mössbauer spectrum

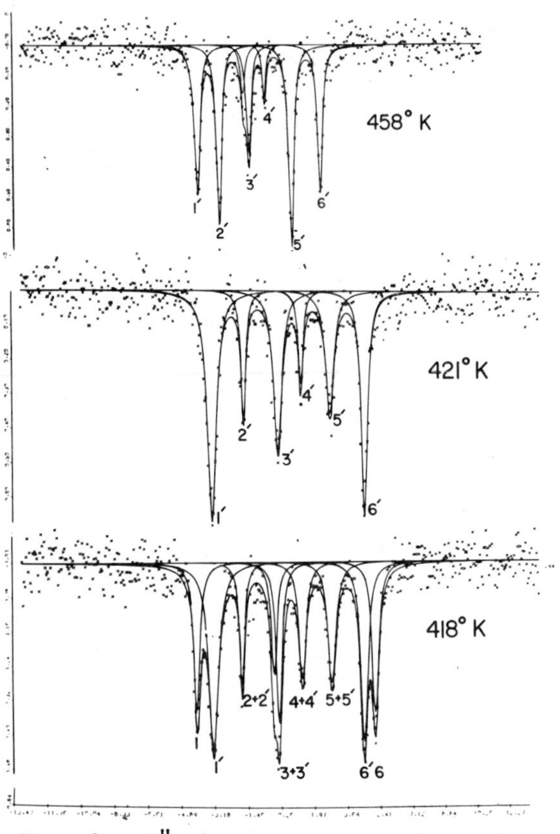

Fig. 2. Mössbauer spectra of single crystal $Fe_{0.996}S$ for c|| the γ-ray direction.

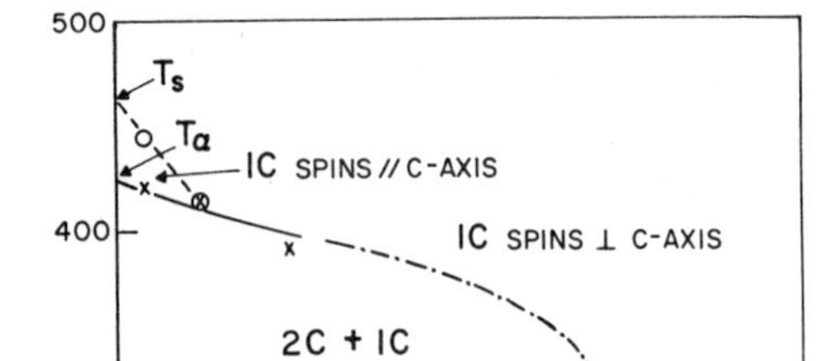

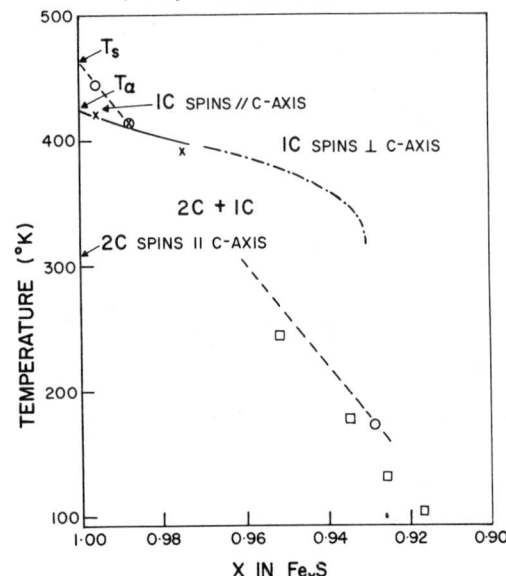

Fig. 3. Magnetic phase diagram of $Fe_{1-x}S$ x-T_α from our DTA, o-T_s from our magnetic susceptibility and Mössbauer results, □-T_s from ref. (9), -·-·- from ref. (10), ---$T_s(x)$.

at 300°K of $Fe_{0.996}S$, oriented with the c-axis || the γ-ray direction, shows weak lines in addition to the four strong lines anticipated for spins ||c. These result from a slight misorientation of the crystal, imperfectly collimated γ-rays and from a small concentration of high-temperature phase. Figure 2 shows that just below T_α the concentration of the high-temperature phase, as indicated by lines (1′-6′), increases dramatically at the expense of the low temperature phase, lines (1-6). At 421°K, all the 2C structure appears to have transformed to the high-temperature phase. At temperatures between 300° and 400°K, the relative intensities of lines 1′ and 6′ to lines 2′ and 5′ indicate that the spins in the high-temperature 1C phase are ⊥c coexisting with spins ||c in the 2C phase. For $T_\alpha<T<T_s$, the relative intensities of lines 1 and 6 to lines 2′ and 5′ indicate thatthere is a mixture of two magnetic phases in the high-temperature phase; one with spins ||c and the other with spins ⊥c. This occurs because of a small spread in composition and the fact that T_s decreases rapidly with increasing metal deficiency, Figure 3. As the temperature is increased above T_α, the fraction with spins ⊥c increases until above T_s only this phase remains, Figure 2.

A localised moment is suggested for Fe in FeS both by the Curie-Weiss behaviour above T_N and because the experimental parameters are fitted quite well by the molecular field theory; the value of T_N calculated from $\chi_\perp$ by molecular field theory or spin wave theory using a Heisenberg model is of the same order as the experimental value. This contrasts with NiS where the moment is generally accepted to be delocalised[8,11]. One difference between the two materials probably lies in the position of the metal 3d band relative to the sulphur 3p band. In NiS the Ni 3d band is superimposed on the S 3p band[12,13]; this reduces the 3d coulombic interaction in Ni[13]. In FeS, the Fe 3d band appears to be above the S 3p band; such a shift might be expected[14], for decreasing atomic number. Our conclusions are not consistent with the model[13] of FeS as an itinerant antiferromagnet.

REFERENCES

1. Haraldsen, H., Z. anorg. allgem. Chem. 231, 78 (1937).
2. Bertaut, E.F., Bull. Soc. Franc. Minér. Crist. 79, 276 (1956).
3. Sparks, J.T., Mead, W. and Komoto, T., J. Phys. Soc. Japan 17, Suppl.B-1, 249, (1962).
4. Andresen, A.F. and Torbo, P., Acta. Chem. Scand. 21, 2841 (1967).
5. Hirahara, E. and Murakami, M., J. Phys. Chem. Solids, 7, 281 (1958).
6. Takahashi, T., Solid State Communications 13, 1335 (1973).
7. Goodenough, J.B., 'Magnetism and the chemical bond', Interscience (1963).
8. Briggs, G.A., Duffill, C., Hutchings, M.T.H., Lowde, R.D., Satya-Murthy, N.S., Saunderson, D.H., Stringfellow, M.W., Waeber W.B. and Windsor, C.G., in Proceedings of the Fifth IAEA Symposium on Neutron Inelastic Scattering, Grenoble 1972.
9. Hiharo, T., J. Sci. Hiroshima U. 24A, 31, (1960).
10. Yund, R.A. and Hall H.T., Mat. Res. Bull., 3, 779 (1968).
11. Coey, J.M.D., Brusetti, R., Kallel, A., Schweizer, J. and Fuess, H., Phys. Rev. Letts. 32, 1257 (1974).
12. Townsend, M.G., Tremblay, R.J., Horwood, J.L. and Ripley, L.G., J. Phys. C: Solid St. Phys., 4, 598 (1971).
13. White, R.M. and Mott, N.F., Phil. Mag. 24, 845 (1971).
14. Heine, V. and Mattheiss, L.F., J. Phys. C., 4, L191 (1971).

MAGNETIC PROPERTIES OF IRON IMPURITIES IN A HOST OF METALLIC NICKEL SULPHIDE

H.ROUX-BUISSON and J.M.D.COEY,
Groupe des Transitions de Phases, Centre National de la Recherche Scientifique,
B.P. 166, 38 042. Grenoble-Cedex, France,

and P. HAEN,
Centre de Recherches des Très Basses Températures, Centre National de la Recherche Scientifique, B.P. 166, 38 042. Grenoble-Cedex, France.

ABSTRACT

$Ni_{0.95}S$ is a Pauli-paramagnetic d-band metal which never orders magnetically. Susceptibility, conductivity and Mössbauer spectra have been measured to determine the magnetic behaviour of iron impurities substituted into an $Ni_{0.95}S$ matrix in concentrations of 0.1-2.0 %. The extra susceptibility due to the iron is proportional to its concentration and follows a Curie-Weiss law with $\theta = 6$ K and $p_{eff} = 1.2\ \mu_B$. No sign of spin-glass order is found in the iron-doped samples down to 1.5 K. The extra residual resistivity is also proportional to iron concentration, and a resistivity minimum occurs between 7 and 20 K. These observations suggest that the iron impurities are in a non-magnetic state at low temperatures.

INTRODUCTION

Forms of NiS and FeS exist with the nickel-arsenide structure. Their magnetic properties are strikingly different. In the paramagnetic state, nickel sulphide is a metal with a small Pauli susceptibility which actually increases slightly with temperature [1]. Iron sulphide however has a Curie-Weiss susceptibility with an effective moment, $p_{eff} = 5.2\ \mu_B$ per iron [2], close to the ionic ferrous value of 4.9 μ_B. Differences are also found in the antiferromagnetic state. Iron sulphide shows normal antiferromagnetic behaviour with $\chi_\perp$ approximately independent of temperature and $\chi_\parallel$ falling sharply with decreasing temperature below the Néel point, $T_N \simeq 600$ K [2]. For nickel sulphide, both $\chi_\parallel$ and $\chi_\perp$ are very small and independent of temperature below the first order magnetic transition at $T_c = 265$ K [1]. We have suggested that this result, together with the strong dependence of the ordered moment on lattice parameter and the T^2 variation of the ^{61}Ni hyperfine field [3] shows that the magnetically -ordered phase of NiS is an itinerant-electron antiferromagnet [4]. Altogether, the magnetic properties indicate that the intra-atomic correlations are strong in FeS compared with the d-bandwidth giving a local moment on the iron,whereas in NiS the correlations are weak, and itinerant d-electron behaviour results.

The antiferromagnetic-paramagnetic transition in nickel sulphide may be supressed by external pressure [5], or by nickel vacancies [6] which have a similar effect in reducing the lattice parameters. 20 kbars or 4 % vacancies suffice to supress the transition completely, so that the paramagnetic-metal phase is conserved down to 0 K. The electronic specific heat coefficient for $Ni_{0.95}S$ is 7 mJ/mole K^2 [7] which suggests that it is a normal d-band metal with a bandwith of 1-2 eV and no strong electronic correlations. This interpretation is strongly corroborated by the magnitude of the Pauli susceptibility [1] and Seebeck coefficient [8], the lack of diffuse magnetic neutron scattering [4,6] and the recent band-structure calculations for the paramagnetic-metal phase of $Ni_{1.00}S$ [9], above T_c.

The volume of the unit cell of metallic $Ni_{.95}S$ is 10 % smaller than that of FeS, most of the difference being due to the much shorter c-axis. The purpose of the present work was to find out what happens to the iron moment when it is substituted into a host of $Ni_{.95}S$. It seems that it remains localised, but that there is evidence for a spin-compensated state or Kondo effect at low temperatures.

EXPERIMENTAL RESULTS

We have measured the magnetic susceptibility and electrical resistivity of six samples of $(Ni_{1-c}Fe_c)_{.95}S$ with c = 0, 0.1, 0.2, 0.4, 1.0 and 2.0 % and ^{57}Fe Mossbauer spectra for samples with c = 1.0 and 2.0 %. The samples were prepared from a 1 g mixture of the elements in sealed, evacuated quartz tubes . None of the samples showed any sign of an antiferromagnetic phase transition, and Debye-Scherrer X-ray photographs indicated that they were all single-phase, with the NiAs structure. The lattice parameters were found to increase linearly with iron concentration as expected for a random distribution of impurities.

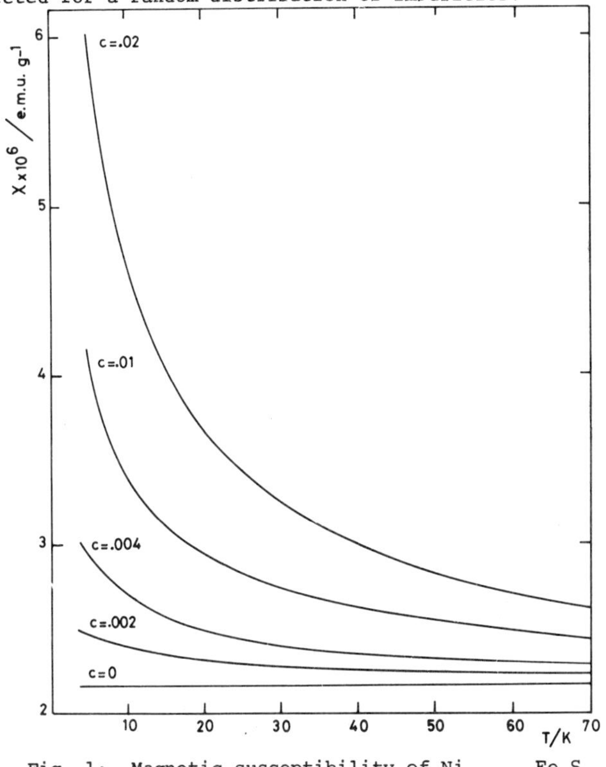

Fig. 1: Magnetic susceptibility of $Ni_{0.95-c}Fe_cS$

The magnetic susceptibility was measured on ∿ 400 mg samples of the powdered materials as a function of temperature using a vibrating sample magnetometer. The data are shown in figure 1. $Ni_{0.95}S$ has a small, almost constant susceptibility whose value is similar to that of $Ni_{1.00}$ above T_c. The materials containing iron have a quite different, temperature-dependent susceptibility. At a fixed temperature, χ is linear in c, so we may define the susceptibility of an iron ion in $Ni_{0.95}S$ as the ratio of excess susceptibility to iron concentration. In the range 4.2 - 70 K the iron susceptibility follows a Curie-Weiss law with a paramagnetic Curie temperature $\theta = -6 \pm 1$ K and an effective moment $p_{eff} = 1.2 \pm 0.1\ \mu_B$. θ depends somewhat on the temperature range from which the extrapolation is made. Not only is the moment much smaller than the ionic value, but there is no sign of the magnetical-

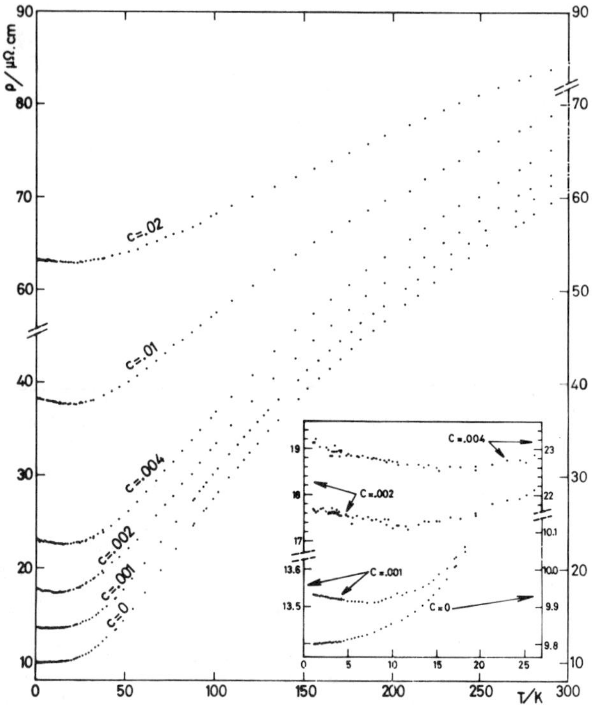

Fig. 2: Electrical resistivity of $Ni_{0.95-c}Fe_cS$

ly-ordered state anticipated if θ could be interpreted simply in terms of an exchange interaction. Mössbauer spectra and magnetic susceptibility of the compound with c = 1 % showed no sign of magnetic order down to 1.5 K, so if spin-glass ordering occurs at all, it is below 1.5 K.

The electrical resistivity of the materials was measured on small bars with dimensions ∿ 4 x 1 x 1 mm using the four probe method [10]. Data collected from 1.2 K to 300 K is shown in figure 2. The undoped sample shows typical metallic behaviour with a monotonic increase of resistivity with temperature. Below 40 K it may be fitted to $\rho = \rho_o + a T^2 + b T^3$. The rather high value of the residual resistivity ρ_o = 9.8 μΩ cm, may be due to the large concentration of defects in the sample. Besides the nickel vacancies due to non-stoichiometry, nickel sulphide is thought to contain a few percent of unoccupied sites of both kinds [11]. Like the susceptibility, the resistivity at 4.2 K increases nearly linearly with iron content (∿ 26 μΩ cm per %), suggesting that the iron is behaving like a collection of independent impurities. Most notable however are the shallow resistivity minima increasing in temperature from 7 K for c = 0.1 % to 20 K for c = 2 %. The excess resistivity $\Delta\rho = \rho\,(\underline{Ni}_{.95}\,\underline{S}\,Fe) - \rho\,(\underline{Ni}_{.95}\,\underline{S})$ was best fitted to the function $A + B\,\ell n \sqrt{1 + (T/\theta_c)^2}$ [12]. Values of θ_c between 20 and 50 K were found for the three highest iron concentrations.

DISCUSSION

We have emphasised that the paramagnetic phase of nickel sulphide behaves like a normal metal. According to the band-structure calculations [9] of Kasowski and Mattheiss, the conduction band has predominantly nickel 3d character though there is considerable admixture of sulphur 3p character. A bandwidth of 1-2 eV is consistent with the small, nearly constant Pauli susceptibility if there are no strong electron correlations. The T^2 term in the resistivity which is predominant below 10 K may be due to electron-electron scattering, whereas the T^3 term might be due to interband scattering, as in the transition metals.

Most of the work so far on magnetic impurities in non-magnetic hosts has been concerned with noble metal hosts (e.g. AuFe, CuMn). There the magnetic impurities only behave independently for concentrations ≤ 500 ppm. At higher concentrations the impurities interact magnetically via the long-range RKKY interaction and a spin glass forms at a temperature proportional to their concentration [13]. A peculiarity of the $\underline{Ni}_{.95}\underline{S}$ Fe system is that the excess susceptibility and excess residual resistivity are proportional to iron content up to the remarkably high value of 2 %, and no sign of spin glass order was found in the present system down to 1.5 K. It is possible that the RKKY interaction is weakened by the large vacancy concentration which is several times the impurity concentration.

The resistivity minima together with the behaviour of the susceptibility suggest a spin-compensated state or Kondo effect. At temperature near the value of θ in the Curie-Weiss law (6 K) the excess resistivity is rather independent of temperature. It is only above about 50 K that Δρ varies logarithmically with temperature suggesting a Kondo temperature ∿ 100 K. It proved impossible to fit the excess resistivity to the Hamann formula, which is only valid above T_K [14]. Hence the formula $\Delta\rho / c = A + B\,\ell n \sqrt{1 + (T/\theta_c)^2}$ was used which reproduces both the logarithmic behaviour at high temperatures and a T^2 behaviour well below θ_c. We have no explanation at present for the discrepancy between θ_c and the Curie Weiss θ.

CONCLUSION

The results of the measurements discussed above amount to a Kondo syndrome for the $\underline{Ni}_{.95}\underline{S}$ Fe system, despite the large impurity concentrations involved. So far as we are aware, such behaviour has not previously been observed in a host which has usually been regarded as a covalent compound.

1. R.F. Koehler and R.L. White, J. Appl. Phys. 44, 1682-6 (1973) ; J.M.D. Coey, J.C. Bruyère, H. Roux-Buisson and R. Brusetti, Proceedings of the International Conference on Magnetism, Moscow 1973 (in the press)
2. E. Hirahara and M. Murakami, J. Phys. Chem. Solids 7, 281-9 (1958)
3. J. Fink, G. Czjzek, H. Schmidt, K. Ruebenbauer, J.M.D. Coey and R. Brusetti, J. Physique, Colloque C 6, 35 (1974) (in the press)
4. J.M.D. Coey, R. Brusetti, A. Kallel, J. Schweizer and H. Fuess, Phys. Rev. Lett. 32, 1257-60 (1974)
5. D.B. McWhan, M. Marezio, J.P. Remeika and P.D. Dernier, Phys. Rev. B 5, 2552-5 (1972)
6. J.T. Sparks and T. Komoto, Rev. Mod. Phys. 40, 752-4 (1968)
7. J.M.D. Coey and R. Brusetti, Phys. Rev. B 10 (1974) (in the press)
8. T. Ohtani, V. Kosuge and S. Kachi, J. Phys. Soc. Japan, 29, 521 (1970).
9. R.V. Kasowski, Phys. Rev. B 8, 1378-82 (1973) ; L.F. Mattheiss, Phys. Rev. B 10, 995-1005 (1974)
10. P. Haen and J. Teixeira, Rev. Phys. Appliquée, 9, 879-93 (1974)
11. M. Lafitte, Bull. Soc. Chim. Franc. 1959, 1211-21 (1959)
12. R.M. Roshko and G. Williams, Phys. Rev. B 9, 4945-53 (1974) ; J.W. Loram, R.J. White and A.D.C. Graslie, Phys. Rev. B 5, 3659-69 (1972)
13. J.L. Tholence and R. Tournier, J. Physique Colloq. 35, C 4 - 229-35 (1974)
14. D.R. Hamann, Phys. Rev. 158, 570-80 (1967)

Acknowledgement. We wish to thank J. Souletie for some helpful discussions.

SHEAR STRAINS AND MAGNETIC STRUCTURE IN $FeCl_2$ UNDER MAGNETIC FIELD, $\vec{H}$, PERPENDICULAR TO THE TERNARY AXIS, $\underline{c}$, OF THE CRYSTAL. THEORETICAL MODEL AND MÖSSBAUER EXPERIMENTS.

J.A. Nasser
DPh-G/PSRM, CEN-Saclay, 91190 Gif-sur-Yvette, France
and
F. Varret
Centre Universitaire, 72000 Le Mans, France

ABSTRACT

We present a model describing magneto-elastic coupling in $FeCl_2$. Shear strains and orientation-dependent magnetic effects are predicted when a magnetic field perpendicular to $\underline{c}$ is applied. Mössbauer experiments at 4.2 K provide evidence for such orientation-dependent magnetic effects.

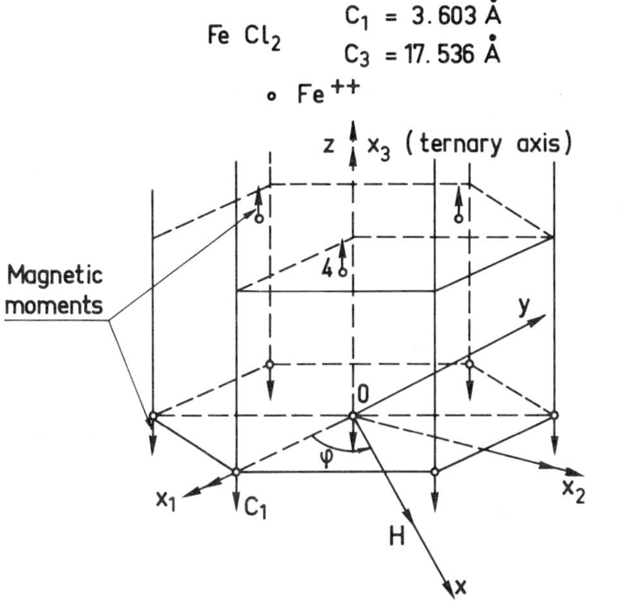

Fig. 1 – Magnetic and crystallographic structure of $FeCl_2$ at 0°K

$FeCl_2$ is a two-sublattices antiferromagnet ($T_N \sim$ 24 K) in which the ternary crystalline axis $\underline{c}$ is a magnetic easy axis. Its rhombohedral ($R\bar{3}m$) crystalline structure results from a packing of planes perpendicular to $\underline{c}$, containing successively Fe^{2+}, Cl^-, Cl^-, Fe^{2+} ions (Fig.1). The model of Ôno et al.[1] was successfully used for interpreting previous magnetic experiments[2]. The following Hamiltonian was used :

$$\mathscr{H}_m = \sum_i \left\{ \Delta(\ell_{zi}^2 - \tfrac{2}{3}) - \lambda \vec{\ell}_i . \vec{S}_i \right\} - \sum_{i \neq j} 2J_{ij} \vec{S}_i . \vec{S}_j + \sum_i \mu_B \vec{H}(-\vec{\ell}_i + 2\vec{S}_i),$$

where the quantization axis Oz is the ternary axis $\underline{c}$; $\mathscr{H}_m$ has axial symmetry around $\underline{c}$. The values of Δ, λ, and exchange coupling were determined as : $\Delta/\lambda \sim 1.25$, $\lambda \sim -95\ cm^{-1}$, ferromagnetic coupling $\sim 6\ cm^{-1}$, antiferromagnetic coupling $\sim -0.7\ cm^{-1}$ (ref.1,2).

When a magnetic field $\vec{H}$ is applied along Ox (perpendicular to $\underline{c}$). $\mathscr{H}_m$ leads to the following predictions at O K :

i) for $H < H_c \sim 120$ kOe, a canted antiferromagnetic structure is obtained (Fig.2a) ; the spins S_1 and S_2 of the two sublattices are symmetrical with respect to the $\vec{H}$-direction :

$$\langle S_x \rangle_1 = \langle S_x \rangle_2 = x(H)$$
$$\langle S_y \rangle_1 = \langle S_y \rangle_2 = 0$$
$$\langle S_z \rangle_1 = -\langle S_z \rangle_2 = z(H) \quad ,$$

where x(H) and z(H) can be deduced by using molecular field calculations.

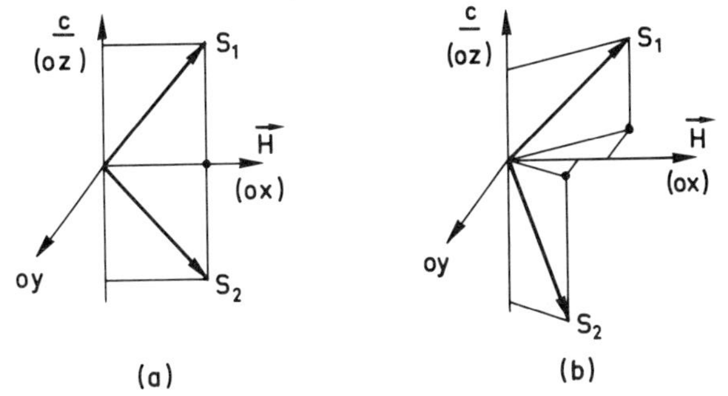

Fig. 2 – Sublattices 1 and 2
a) Ôno's model.
b) Our model.

ii) for $H \geq H_c$ the saturated paramagnetic phase is obtained where $\vec{S}_1$ and $\vec{S}_2$ are parallel to the $\vec{H}$-direction :

$$\langle S_z \rangle_1 = \langle S_z \rangle_2 = 0 \quad .$$

We introduce now the magnetoelastic couplings. Let ε_k (k = 1,6) be the components of the macroscopic strains of the crystal, expressed in the Oxyz axes bound to the H-direction (Fig.1). Then we replace $\mathscr{H}_m$ by [3] :

$$\mathscr{H}_{me} = \sum_i \left\{ \Delta(\varepsilon_\alpha)(\ell_{zi}^2 - \tfrac{2}{3}) - \lambda \vec{\ell}_i . \vec{S}_i \right\} - \sum_{i \neq j} 2J_{ij}(\varepsilon_\alpha)\, \vec{S}_i\, \vec{S}_j + \sum_i \mu_B \vec{H}\,(-\vec{\ell}_i + 2\vec{S}_i) + \sum_i \delta W_i \quad .$$

In the above expression, $\Delta(\varepsilon_\alpha)$ and $J_{ij}(\varepsilon_\alpha)$ are linear functions of $(\varepsilon_1 + \varepsilon_2)$ and ε_3 , and are responsible for magnetostriction in volume[3]; δW_i , originating from crystal-field terms, is related to shear strains.

The study of $\mathscr{H}_{me}$ has been carried out in ref.3, using the following assumptions:

i) magnetoelastic couplings are perturbations before $\mathscr{H}_m$;

ii) only the lowest three states of each ion are considered ; so a fictitious spin $\vec{s}$ (s = 1) may be used ;

iii) exchange effects are treated by the molecular field theory.

Within these assumptions, δW was calculated as :

$$\delta W = A\,[\varepsilon_5(s_x s_z + s_z s_x) + \varepsilon_4(s_y s_z + s_z s_y)] + iB\,(\varepsilon_5 \cos 3\varphi - \varepsilon_4 \sin 3\varphi)(s_+^2 - s_-^2) - B\,(\varepsilon_5 \sin 3\varphi + \varepsilon_4 \cos 3\varphi)(s_+^2 + s_-^2) \quad ,$$

where A and B are real constants associated with the magnetoelastic couplings, and φ the angle between H and the crystal axis Ox_1 (Fig.1).

By minimizing the total free energy of the crystal, we determine the strain components ε_k^{eq} corresponding to the thermodynamical equilibrium : these components are expressed as follows :

$$\varepsilon_6^{eq} = 0 \qquad \varepsilon_3^{eq} \propto K(H)$$
$$\varepsilon_5^{eq} \propto G(H)\, B \sin 3\varphi \qquad \varepsilon_1^{eq} + \varepsilon_2^{eq} \propto K'(H) \qquad \text{(R1)}$$
$$\varepsilon_4^{eq} \propto G(H)\, B \cos 3\varphi \qquad \varepsilon_1^{eq} - \varepsilon_2^{eq} \propto G(H)\, B$$

(The computed values of G(H), K(H), K'(H) are given in ref.3).

From (R1) it must be pointed out that the crystalline symmetry is lowered ; the predicted shear strains show an angular dependence upon φ , with a $2\pi/3$ periodicity which agrees with the initial three-fold symmetry of the crystal.

By using the calculated ε_k^{eq} , we finally obtain the spin orientations in sublattices 1 and 2 ; the most striking results are :

$$\langle S_x\rangle_1 - \langle S_x\rangle_2 \propto M_x(H)\ AB\ \sin 3\varphi$$

$$\langle S_y\rangle_1 = -\langle S_y\rangle_2 \propto M_y(H)\ AB\ \cos 3\varphi \qquad (R2)$$

$$\langle S_z\rangle_1 + \langle S_z\rangle_2 \propto M_z(H)\ AB\ \sin 3\varphi$$

(The computed values of $M_{x,y,z}(H)$ are given in ref.3).

From (R2) it must be pointed out that the symmetry of the magnetic sublattices with respect to the H-direction is destroyed in the antiferromagnetic phase (Fig.2b); in the saturated paramagnetic phase, all spins are parallel, but shear strains and magnetization along $\underline{c}$ are still predicted.

In order to check the magnetic predictions of our model, we must emphasize that :

i) the total magnetization along H (Ox) shows no angular dependence[3], and differs little from the predictions of Ôno et al.'s model ;

ii) the total magnetization along Oy is nul[3] ;

iii) the total magnetization along Oz, although depending upon φ, is likely to be cancelled out at a macroscopic scale by the onset of magnetic domains similar to ferromagnetic domains.

Consequently, we are left with methods allowing the separate observation of the two sublattices, such as neutron diffraction or resonance methods (in addition to lattice studies which might check the relations (R1)).

In the present paper we describe a Mössbauer experiment which gives evidence for the non-symmetrical behaviour of the two sublattices, as well as for the $\sin 3\varphi$ angular dependence.

The sample was a single crystal sheet (0.3 mm thick), perpendicular to $\underline{c}$, and had a circular shape (8 mm diameter) ; the γ-beam orientation was along $\underline{c}$. By the use of a Helmholtz-type superconductive magnet, a 60 kOe magnetic field perpendicular to $\underline{c}$ was applied, the temperature being 4.2 K . Various values of the φ-angle were obtained by rotating the sample around the $\underline{c}$ axis ; because of this convenient geometry, all the other parameters such as thickness effects, field inhomogeneity, demagnetizing field, γ-beam orientation, crystalline axes orientations, remained the same throughout the complete set of experiments. Consequently, any difference between the Mössbauer spectra could be related to the expected angular dependence of the magnetic effects.

Indeed, the Mössbauer spectra obtained show clearly a good angular dependence (Fig.3). The angular periodicity of the spectra was found to be $\pi/3$, since the spectra recorded for $\varphi = 0°$, $60°$, $180°$, $0°$ (Fig.3, top) and $\varphi = 30°$, $90°$, $150°$ (Fig.3, bottom), respectively, exhibited very similar lineshapes. This $\pi/3$ periodicity agrees very well with the predicted $\sin 3\varphi$ dependence (periodicity $2\pi/3$) : since the Mössbauer experiment cannot decide whether one given spin belongs to sublattice 1 or 2, the experimental angular dependence is actually $|\sin 3\varphi|$ which indeed has a $\pi/3$ periodicity.

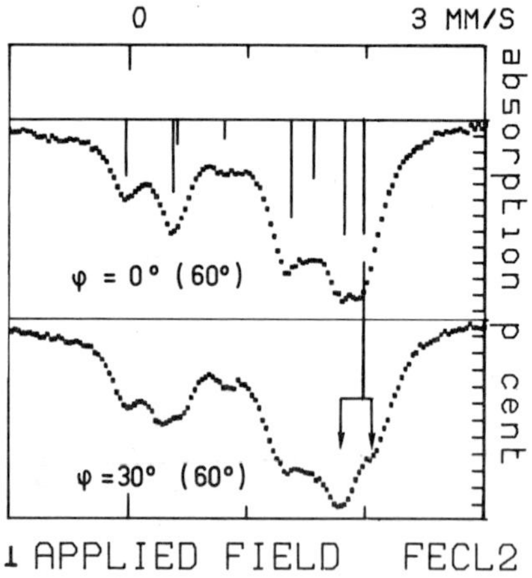

Fig.3 – Mössbauer spectra

A really remarkable fact is that the top spectrum of Fig.3 ($\varphi = 0°$) may be roughly described by one single hyperfine pattern ; the spins S_1 and S_2 are equivalent from the Mössbauer point of view. This was expected since relations (R2) for $\varphi = 0$ lead to spins symmetrical with respect to Ox, i.e. the direction of the Mössbauer observation.

On the contrary, for $\varphi = 30°$, the Mössbauer spectrum splits into two hyperfine patterns, corresponding to spins S_1 and S_2 which are no longer symmetrical with respect to Ox. The splitting of one line is indicated on Fig.3. At the present time, we have not completely interpreted these spectra which involve a large number of unknown hyperfine parameters (due to induced-field effects on the electric field gradient as well as on the hyperfine field). As a first approach, let us remark that :

i) the hyperfine field tensor along Oz is so small[1] that any angular dependence of the Oz component of the hyperfine field is expected to be small ;

ii) the Oy components of the hyperfine field are expected to have the same value , because of (R2).

Consequently, the splitting of the Mössbauer pattern is mainly related to the $\langle S_x\rangle_1 - \langle S_x\rangle_2$ difference ; according to preliminary fittings of the spectra, this difference might be sizeable :

$$|\langle S_x\rangle_1 - \langle S_x\rangle_2| \sim 0.2\ |\langle S_x\rangle| \quad \text{for} \quad \varphi = 30° .$$

(a detailed interpretation of the Mössbauer spectra is under process).

As a conclusion, we claim that shear strains due to magnetoelastic coupling may greatly alter the magnetic structure when a magnetic field is applied. Experimental evidence for such a situation may be given by the Mössbauer effect when the ferrous ion is concerned.

References

1. K. Ôno, A. Ito and T. Fujita, J. Phys. Soc. Jap. 19, 2119 (1964).
2. J.A. Nasser, J. Physique 34, 891 (1973), and references therein.
3. J.A. Nasser, J. Physique (to be published).

CONTROL OF THE SPIN REORIENTATION WITH IMPURITY MODIFICATION IN $RFeO_3$ and Fe_3BO_6

N. Koshizuka, M. Hirano, T. Okuda, S. Nakamura*, H. Hiruma*, and T. Tsushima
Electrotechnical Laboratory, Tanashi, Tokyo 188, Japan

ABSTRACT

Spin reorientation phenomena were studied on impurity modified crystals of rare earth orthoferrite and iron borate. Faraday rotation in the series of (Er,Sm) FeO_3 crystals with incident light parallel to the optical axis was observed to measure the variation of spin reorientation temperature(T_{SR}) with the composition of the rare earths. Changes of Faraday rotation with temperature and applied magnetic field reflect the spin configuration of Fe ions to be G_ZF_X and G_XF_Z below and above T_{SR}, respectively.

Spin configurations in an undoped Fe_3BO_6 crystal were determined as G_ZF_X and G_XF_Z below and above T_{SR}, respectively, by magnetic measurements and a crystal symmetry consideration. Controlling of the spin reorientation in Fe_3BO_6 system was tried, and T_{SR} was found to be lowered from 416°K to 0°K on simultaneous doping of Co and Ti.

A new spin structure assigned as G_Y type appeared with further modification of Fe with Co and Ti than 6 atm%, and a magnetic phase diagram was drawn in the system of Fe_3BO_6:(Co,Ti).

INTRODUCTION

Recently, studies on the physical properties of the spin reorientation phenomenon have been considerably developed in many kinds of weak ferromagnets[1][2]. In this paper the phenomenon was studied on the impurity modified single crystals of rare earth orthoferrite $RFeO_3$ and iron borate Fe_3BO_6 not only by usual magnetic measurements but also by a magnetooptical method. Both of them have higher T_{SR} than room temperature, and are thought to be useful for device applications in future, and it is thought interesting to see some effects of impurity modifications upon the phenomenon.

$RFeO_3$ shows various magnetic properties, but there have been performed few magnetooptical investigations. It may be due to the complexity caused by the birefringent properties associated with the orthorhombic crystal symmetry. However, large Faraday rotation, not elliptic birefringence, can be observed for incident light along any optical axes in such biaxial crystals[3][4].

Spin reorientation phenomenon in the system of $(Er_xSm_{1-x})FeO_3$ single crystals and the variation of T_{SR} against the composition of the rare earths were observed through Faraday rotation for the first time. And the spin configuration in an undoped crystal of Fe_3BO_6[5] was investigated by means of magnetic measurements and by a crystal symmetry consideration. Also examined was a controlling of the spin reorientation in the Fe_3BO_6 system by a modification of Fe with some kinds of impurity atoms.

EXPERIMENTAL

Single crystals of mixed orthoferrites $Er_xSm_{1-x}FeO_3$ were grown by a floating zone technique[6] using an infrared heating-type furnace(where x=0,0.3,0.35,0.5,0.7, and 1.0), and the single crystals of pure and impurity modified Fe_3BO_6 were prepared by a more improved flux method based on the previously reported one[5]. Two types of modification were examined:one is a simple doping with single element such as Ga,Al,Cr, or Co, and the other is a coupled doping of Co and Ti or of Zn and Ti.

Compositions of the rare earths in $Er_xSm_{1-x}FeO_3$ were determined by means of interpolation of the lattice parameters which showed a good agreement with the nominal values, and impurity concentrations in the Fe_3BO_6 system were determined by means of electron probe microanalysis.

Magnetic properties were measured with a vibrating-sample magnetometer, a torquemeter, and a balance, and Faraday rotation was observed with a recording-type optical rotation spectrometer(JASCO MOE-7).

Optical axes in some of rare earth orthoferrites are known to lie in the a-plane making angles of 51°∿52° with the c-axis at the wavelength of 0.6μm[3][4], and the specimens of the orthoferrites for the present experiment were cut at 51° from the c-axis in the a-plane. They were mechanically polished for the optical measurements.

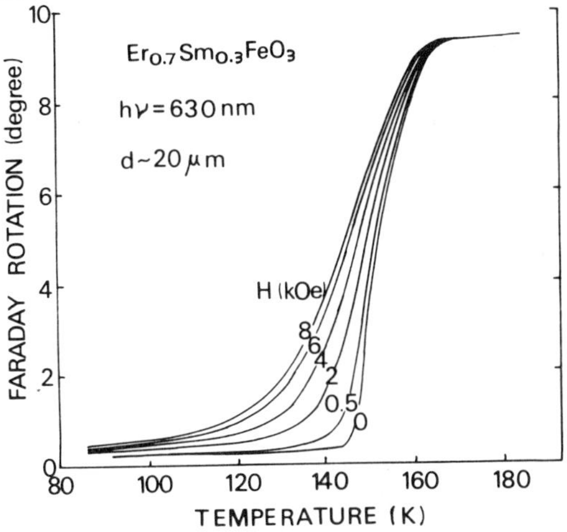

Fig.1 Sharp change of Faraday rotation near T_{SR}. H is applied magnetic field.

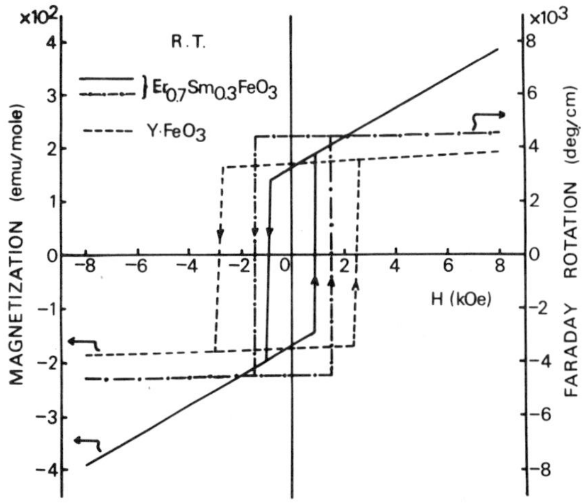

Fig.2 Magnetic field dependences of magnetization for $Er_{0.7}Sm_{0.3}FeO_3$ and $YFeO_3$ and that of Faraday rotation for $Er_{0.7}Sm_{0.3}FeO_3$ at room temperature. The directions of applied field and incident light are parallel to the optical axis.

RESULTS

1) Spin Reorientation and Faraday Rotation in $Er_xSm_{1-x}FeO_3$

From the measurements of Faraday rotation and optical absorption of $Er_{0.7}Sm_{0.3}FeO_3$ in the wavelength region from 0.6 to 2.0μm at room temperature, it turned out that the anomalies in the optical rotation at 0.96,1.26, 1.40, and 1.52μm were originated from the absorption of Fe^{3+}, Sm^{3+}, and Er^{3+} ions. Temperature variations of the Faraday rotation in $Er_{0.7}Sm_{0.3}FeO_3$ with incident light of 0.63μm are shown in Fig.1 for various magnetic fields. The Faraday rotation decreases below 160°K and approaches to zero near 145°K at zero magnetic field which corresponds to the variation of the magnetic structure of Fe^{3+} ions in $Er_{0.7}Sm_{0.3}FeO_3$: Γ_4 (G_XF_Z) type above T_{SR} and Γ_2 (G_ZF_X) type below T_{SR} in the magnetically ordered state. Comparing the magnetic field dependence of Faraday rotation in $Er_{0.7}Sm_{0.3}FeO_3$ with that of the magnetization curves of the specimens of the same composition and of $YFeO_3$ in Fig.2, it is confirmed that the magnitude of Faraday rotation is proportional to the sum of each component of the Fe^{3+} sublattice moments along the optical axis. We obtained the same composition dependence of T_{SR} with that by Sherwood et al.[7] in the present magnetooptical measurements for the other x's.

2) Spin Configuration in Pure Fe_3BO_6

Fe_3BO_6 belongs to space group Pnma[8] and three types of spin configurations compatible with its crystal structure are listed in Table I, where G^+G is dominant. In Fig.3 are shown the temperature dependences of weak ferromagnetic moment and magnetic susceptibility for the three principal axes. It is deduced that the sublattice moments are parallel to [001] with a canting along [100] below T_{SR}, and they are parallel to [100] with a canting along [001] above T_{SR}. Thus it is concluded that the spin configurations in Fe_3BO_6 below and above T_{SR} are assigned as Γ_2 and Γ_4, respectively[9].

Table I. Spin configurations of Fe_3BO_6.

	8d			4c	
Γ_1	C_x^+	G_y^+	A_z^+	G_y	
Γ_2	F_x^+	A_y^+	G_z^+	F_x	G_z
Γ_4	G_x^+	C_y^+	F_z^+	G_x	F_z

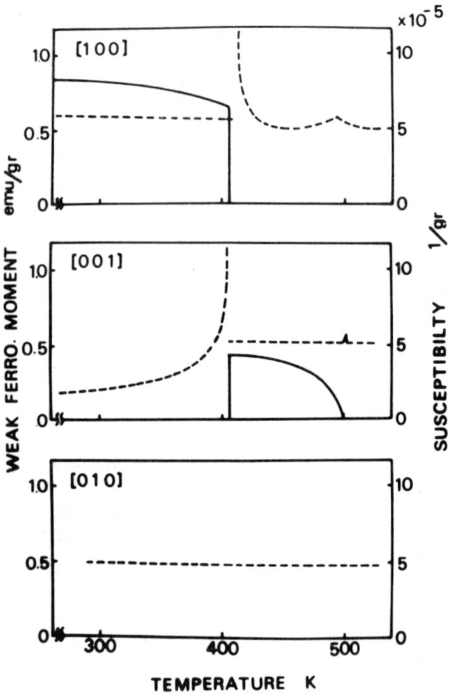

Fig.3 Temperature dependences of weak ferromagnetic moment and susceptibilities of Fe_3BO_6.

3) Impurity Effects on Fe_3BO_6

Co^{2+} is easily doped into the Fe_3BO_6 host if Ti^{4+} is doped simultaneously, and the reason may be considered as the charge compensation. Fig.4 indicates the spin configurations of the system of Fe_3BO_6:(Co^{2+},Ti^{4+}) by taking temperature and Co^{2+} concentration as the parameters.

It is remarkable that T_{SR} is lowered from 416°K to 0°K with increasing modification of Fe^{3+} up to 6 atm% with Co^{2+} and Ti^{4+}, and that a new antiferromagnetic phase Γ_1 with the spin configuration G_y is produced with the further substitution of Co^{2+} as seen in Fig.4.

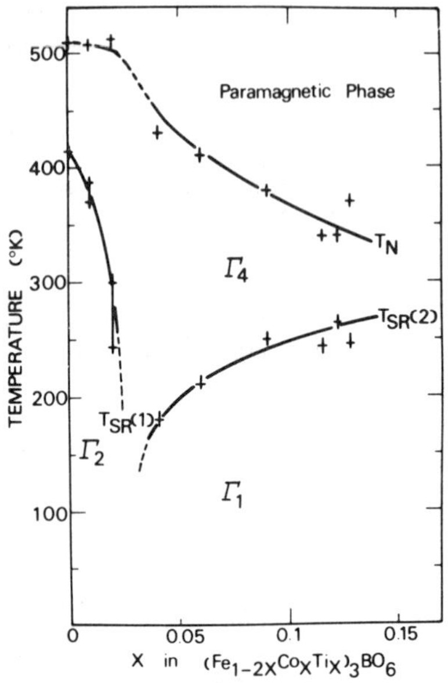

Fig.4 Magnetic Phase diagram of Co and Ti doped Fe_3BO_6 with respect to Co concentration and temperature.

CONCLUSIONS

1) Spin reorientation in $(Er_xSm_{1-x})FeO_3$ crystals was observed by Faraday rotation, and the variation of T_{SR} with x was obtained.
2) Spin configurations in Fe_3BO_6 were determined as G_ZF_X and G_XF_Z below and above T_{SR}, respectively.
3) A magnetic phase diagram in the modified system of Fe_3BO_6:(Co,Ti) was drawn, and a new antiferromagnetic phase was produced.

*Research fellow from Tokyo Science Univ., Shinjuku-ku, Tokyo.

REFERENCES

1) R.L.White, J.Appl.Phys. 40, 1061 (1969)
2) K.Tsushima, Kotai Butsuri 7, 592 (1972) (in Japanese)
3) W.J.Tabor, A.W.Anderson, and L.G.Van Uitert, J.Appl. Phys. 41, 3018 (1970)
4) M.V.Chetkin, J.U.S.Didosjan, and A.I.Akhutkina, IEEE Trans.Mag. Sept., 401 (1971)
5) R.Wolfe, R.D.Pierce, M.Eibshütz and J.W.Nielsen, Solid State Commun. 7, 949 (1969)
6) T.Okada, K.Matsumi, and H.Makino, Proc.International Conference on Ferrites, Kyoto, 372 (1970)
7) R.C.Sherwood, L.G.Van Uitert, R.Wolfe, and R.C.LeCraw, Phys.Letters 25A, 297 (1967)
8) J.G.White, A.Miller and R.E.Nielsen, Acta Crystal. 19, 1060 (1965)
9) M.Hirano, S.Umemura, S.Nakamura, K.Kohn, T.Okuda, and T.Tsushima, Solid State Commun. 15, 1129 (1974)

DIRECT OBSERVATION OF THE SPIN-REORIENTATION TRANSITION AT Yb SITES IN $YbFeO_3$

G. R. Davidson* and B. D. Dunlap*
Argonne National Laboratory, Argonne, Illinois 60439

M. Eibschütz and L. G. van Uitert
Bell Laboratories, Murray Hill, New Jersey 07974

ABSTRACT

We report a direct observation of a 90° reorientation of Yb spins in $YbFeO_3$ by Mössbauer spectroscopy of single crystals. Our spectra are consistent with a continuous rotation of the Yb spins in the region $6.6\ K \leq T \leq 7.8\ K$ with second-order phase transitions at the end points. Our results agree with qualitative predictions by Horner and Varma and by Yamaguchi, but the quantitative predictions of Yamaguchi are found not to apply to $YbFeO_3$.

INTRODUCTION

When cooled, several of the rare earth orthoferrites ($RFeO_3$) undergo a spontaneous spin reorientation in which the weak ferromagnetic moment rotates from the c to the a axis.[1] Proposed theoretical models for this transition[2,3] indicate that the reorientation involves either an abrupt first-order change or a rotation that is continuous over a temperature interval $T_1 \leq T \leq T_2$ with second-order phase transitions at T_1 and T_2.

In the case of $YbFeO_3$ bulk measurements have indicated that a continuous rotation occurs. A specific heat experiment[4] yielded $T_1 = 6.55$ K and $T_2 = 7.83$ K, while pyromagnetic response data[5] yielded 6.31 K and 7.59 K. The results of both were qualitatively consistent with the phenomenological free energy model of Horner and Varma,[2] but neither provided quantitative verification.

In the present work we have investigated the nature and origin of the transition by means of a direct local measurement of the magnitudes and orientations of the Yb spins in the vicinity of the reorientation. This information was obtained from ^{170}Yb Mössbauer absorption spectra of $YbFeO_3$ single crystals between 4.2 and 10 K.

BACKGROUND[1]

$YbFeO_3$ is an orthorhombically distorted perovskite (space group Pbnm). At $T_N \approx 630$ K, the Fe^{3+} ions order nearly antiferromagnetically along the a axis with a slight canting ($\sim 0.5^o$ [1,6]) to produce a weak ferromagnetic component along c. Each unit cell has four Yb sites (commonly labelled 5, 6, 7, and 8). The conventional magnetic symmetry assignments[7] (obtained by assuming that the magnetic and chemical unit cells are identical) require the four spins to be along the c axis above T_2 and to form two pairs below T_2 with one pair obtainable from the other by reflection in the ac plane (Fig. 1). The orientations of the two pairs may be specified by the opening angle 2Φ between the pairs and the rotation angle γ of the net Yb moment away from c. (See Fig. 1). Below T_1 the two pairs lie in the ab plane (i.e., $\gamma = 90^o$).[1]

EXPERIMENT

The Mössbauer measurements involved the 84.3 keV $0 \rightarrow 2$ transition of ^{170}Yb, a standard transmission geometry, and a source of ^{170}Tm in $TmAl_2$. We studied two single-crystal absorbers, a 0.6 mm ab platelet and a 0.4 mm bc platelet, each cut from a flux-grown single crystal. The orientation of each was determined by morphological considerations.[8]

Temperature was measured to $\sim \pm 0.1$ K and controlled to ± 0.05 K during each measurement by use of a carbon resistor thermometer.

RESULTS

Spectra were taken above, within, and below the reorientation region $T_1 \leq T \leq T_2$. At each temperature we measured the spectra of both platelets and made a simultaneous least-squares fit to the two sets of data. With the specific heat values of T_1 and T_2 adopted,[9] the hyperfine parameters of the fits were restricted in each temperature range to values consistent with the conventional crystallographic and magnetic symmetry assignments. In addition, within the reorientation region three parameters found to be only slightly temperature-dependent were held fixed at the averages of their values above and below the transition.[10] These procedures were required for obtaining well-defined parameters from the fits. All fits assumed a linewidth of 2.8 mm/sec, a value comparable to that obtained with the same source and a single-line absorber. This linewidth indicates a coherent rotation process with equivalent hyperfine parameters at all Yb sites, a finding consistent with symmetry requirements.

The determination of the Yb magnetic moment $\vec{\mu}$ from the spectra is based on the proportionality between $\vec{\mu}$ and $\vec{H}_{hf}$. Relativistic free ion calculations indicate that a moment of 4 μ_B corresponds to a hyperfine field of 4232 kOe.[11] The orientation of $\vec{\mu}$ with respect to crystallographic axes a, b, c is obtained from the angular hyperfine parameters specifying the orientation of $\vec{H}_{hf}$ in the principal axis system (PAS) of the electric field gradient (EFG) acting at the Yb nucleus and from the angular parameters specifying

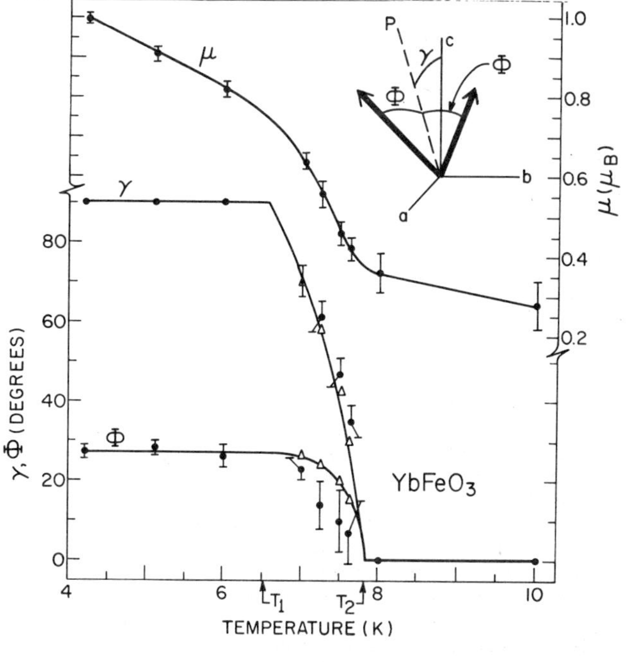

Figure 1. Upper right corner: configuration of the two Yb spin pairs in the reorientation region. Φ is the angle between a spin and its projection P on the ac plane; γ is the angle between P and the c axis. Below: temperature dependence of Φ, γ, and Yb magnetic moment magnitude $\vec{\mu}$. The triangular symbols are values obtained with $\vec{\mu}$ restricted to rotate in the plane determined by its end positions at T_1 and T_2. The lines are drawn to help the eye follow the data.

the orientation of the EFG PAS relative to the crystallographic axes. The latter angles are determined partially by symmetry and partially by fitting our single-crystal spectra. The dots in Fig. 1 show our results for μ, Φ, and γ. The angular parameters have values close to those of the triangular symbols of Fig. 1. These were obtained by imposing the additional restriction that $\vec{\mu}$ rotates in the plane defined by its end positions at T_1 and T_2.

Given the magnitudes and directions of the magnetic moments of the Yb ions, one can easily calculate the bulk Yb magnetization $\vec{\sigma}^{Yb}$. In Fig. 2 we compare the resulting values of $\vec{\sigma}^{Yb}$ with measured values[12] of the bulk spontaneous magnetization $\vec{\sigma}$. The major point to be noted is that the Yb magnetization is substantially larger than the total magnetization for temperatures $T < T_1$. Since this difference must be due to the Fe magnetization ($\vec{\sigma}^{Fe} = \vec{\sigma} - \vec{\sigma}^{Yb}$), we conclude that the net Fe and net Yb moments are antiparallel.[13] (At 4.2 K, $\sigma_a^{Fe} = -3.0 \pm 0.4$ emu/gm, indicating an Fe canting angle of $1.7 \pm 0.2°$.) As the net Fe and net Yb moments are parallel above T_2, they must rotate in opposite directions during the reorientation.

DISCUSSION

The phenomenological free energy model of Horner and Varma[2] deals with the rotation angle θ of the bulk magnetization away from the c axis. It predicts that θ is continuous between T_1 and T_2 and that $\partial\theta/\partial T$ is infinite at T_1 and T_2. We have measured not θ, but the net Yb moment rotation angle γ. Our data show that γ is continuous between T_1 and T_2 and are not inconsistent with $\partial\gamma/\partial T$ being infinite at T_1 and T_2. These results are plausible (though not required) consequences of the Horner-Varma model.

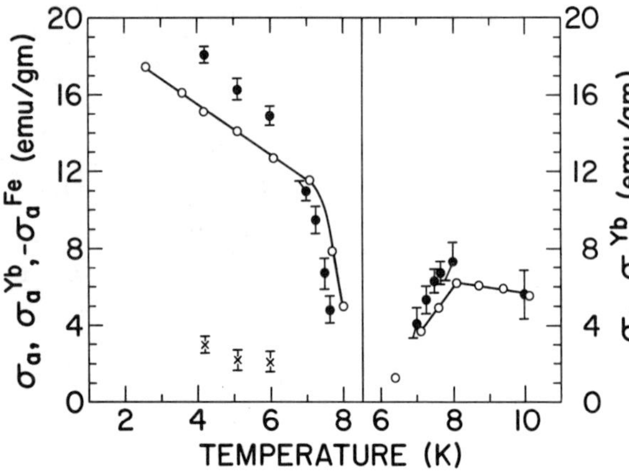

Figure 2. Temperature dependence of various components of spontaneous bulk magnetization. Left: total magnetization σ_a from Ref. 12 (o); Yb contribution σ_a^{Yb} obtained from Mössbauer data (•); Fe contribution σ_a^{Fe} obtained by subtraction (x). Right: σ_c from Ref. 12 (o); σ_c^{Yb} from Mössbauer data (•). The lines are drawn to help the eye follow the total magnetization.

The microscopic interaction treatment of Yamaguchi[3] deals explicitly with the parameters Φ and γ. His calculations are based on a very general Hamiltonian which includes all possible bilinear Fe-Fe and Fe-rare earth exchange interactions as well as Fe anisotropy terms. Our results are consistent with his predictions that Φ and γ are continuous between T_1 and T_2 and that $\partial\Phi/\partial T$ and $\partial\gamma/\partial T$ have infinite discontinuities at T_2 and at T_1 and T_2, respectively. He also predicts that in the transition region

$$\mu \cos \Phi = \mu_o \quad ,$$

where μ_o is the value of μ at T_2. We find, however, $\mu_o = 0.37\ \mu_B$, but $\mu(T_1)\cos\Phi(T_1) = 0.67\ \mu_B$. A possible cause of this large discrepancy is Yamaguchi's assumption that the net Fe and rare earth moments are collinear during the reorientation, an assumption invalid for $YbFeO_3$. His neglect of rare earth anisotropy may also be a poor approximation for this case.

SUMMARY

We find a continuous rotation of the Yb spins between T_1 and T_2, with the net Yb moment rotating counter to the net Fe moment. Our results are consistent with the occurrence of second-order phase transitions at T_1 and T_2 such as those predicted by Horner and Varma and by Yamaguchi, but the quantitative predictions of Yamaguchi are found not to apply.

ACKNOWLEDGEMENTS

We are grateful to W. F. Flood for cutting the single crystal platelets and to E. G. Avery for calibration of the carbon resistance thermometer.

REFERENCES

1. For reviews of orthoferrites see D. Treves, J. Appl. Phys. 36, 1033-9 (1965) and R. L. White, J. Appl. Phys. 40, 1061-9 (1969).
2. H. Horner and C. M. Varma, Phys. Rev. Letters 20, 845-6 (1968).
3. T. Yamaguchi, J. Phys. Chem. Solids 35, 479-500 (1974).
4. M. R. Moldover, G. Sjolander, and W. Weyhmann, Phys. Rev. Letters 26, 1257-9 (1971).
5. W. J. Schaffer, R. W. Bené, and R. M. Walser, Phys. Rev. B 10, 255-64 (1974).
6. M. Eibschütz, S. Shtrikman, and D. Treves, Phys. Rev. 156, 562-77 (1967).
7. E. F. Bertaut, in Magnetism, Vol. III, edited by G. T. Rado and H. Suhl (Academic Press, New York, 1963), Ch. 4.
8. S. Shtrikman, B. M. Wanklyn, and I. Yaeger, Intern. J. Magnetism 1, 327-31 (1971).
9. The pyromagnetic response value of T_2 was found inconsistent with spectra at 7.63 K and 7.74 K.
10. The three parameters were the quadrupole interaction constant e^2qQ, the asymmetry parameter η, and the angle between a and the z principal axis of the EFG at the Yb nucleus. The spectra and further details of the fitting process will be published elsewhere (G. R. Davidson, B. Dunlap, M. Eibschütz, and L. G. van Uitert, in preparation).
11. B. D. Dunlap, in Mössbauer Effect Methodology, Vol. 7, edited by I. J. Gruverman (Plenum Press, New York, 1971). pp. 123-47.
12. Data for $\vec{\sigma}$ are reported by T. Beaulieu, Microwave Laboratory Rept. No. 1530, Stanford University (1967).
13. Previously, this point was unsettled. White (Ref. 1) indicates that the net Yb and Fe moments are antiparallel below T_1, whereas Beaulieu (Ref. 12, p. 73) concludes that they are parallel.

*Work at Argonne National Laboratory performed under the auspices of the USAEC.

ANTIFERROMAGNETIC ORDERING IN $ErPO_4$

B. W. Mangum and D. B. Utton
National Bureau of Standards
Washington, D.C. 20234

ABSTRACT

The magnetic susceptibility and dM/dB as a function of field have been measured for $ErPO_4$ from 25 mK to 4.2 K. They show that $ErPO_4$ orders antiferromagnetically with $T_N = 100 \pm 2$ mK with the spins along the tetragonal a-axes. The magnetic phase diagram is presented which includes a spin-flop phase in the basal plane. It is proposed that dipolar interactions may be responsible for the magnetic ordering.

INTRODUCTION

The rare-earth phosphates, arsenates, and vanadates, which have the tetragonal zircon structure[1] at room temperature, exhibit a variety of interesting magnetic properties at low temperatures. Er^{3+} has a ground state multiplet of $^4I_{15/2}$ which is split by a crystalline tetragonal field into eight Kramers doublets and the energies of the five lowest crystal field states of Er^{3+} in YPO_4 are known from optical absorption spectra.[2] The first excited state lies 32 cm^{-1} (46 K) above the ground state doublet, whose g-values have been determined by paramagnetic reasonance to be $g_{\parallel} = 6.42$ and $g_{\perp} = 4.81$.[3] It is expected that approximately the same values exist in pure $ErPO_4$.

$ErPO_4$ has the zircon structure[1]; the symmetry is tetragonal, the space group is D_{4h}^{19} ($I4_1$/amd) and the point symmetry at each Er^{3+} ion is D_{2d}. The unit cell dimensions are a = 6.863 Å and c = 6.007 Å. There are four Er^{3+} ions per unit cell, all magnetically equivalent.

The magnetic susceptibility of $ErPO_4$ between 1.5 and 300 K has been reported[4] with no evidence of magnetic ordering. We report measurements of the magnetic susceptibility of $ErPO_4$ between 25 mK and 4.2 K and its magnetic field dependence below 0.1 K. These measurements have enabled us to construct the magnetic phase diagram.

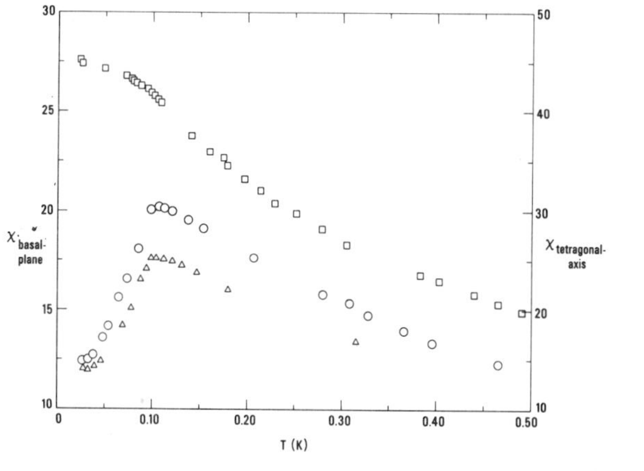

Fig. 1. The magnetic susceptibility of $ErPO_4$ measured along the crystallagraphic axes. Squares represent the data obtained along the c-axis, circles represent those obtained along one of of the a-axes and triangles represent the data obtained along the other a-axis. Arbitrary units are used for the susceptibility.

RESULTS AND DISCUSSION

The magnetic susceptibility of $ErPO_4$ was measured along the c-axis and the two a-axes. The measurements below 0.5 K are presented in Fig. 1. These data indicate a transition to an antiferromagnetic state below T_N = 100 mK with the spins lying in the basal plane. At temperatures above 1 K, the difference between the susceptibility measured along the c-axis and that measured along the a-axes is in good agreement with the published g-values.[3] The difference between the susceptibilities measured below 1 K along the two a-axes can be attributed to demagnetizing effects. We shall discuss later the nonvanishing susceptibility in the basal plane at T=0. The increase in the susceptibility measured along the c-axis below 100 mK is greater than is normally observed for antiferromagnetic systems in which exchange is the dominant interaction and it can probably be attributed to dipolar interactions, which for $ErPO_4$ appear to be dominant. The assumption that dipolar interactions are dominant is supported by (1) Meijer's calculation of the heat capacity[5], which

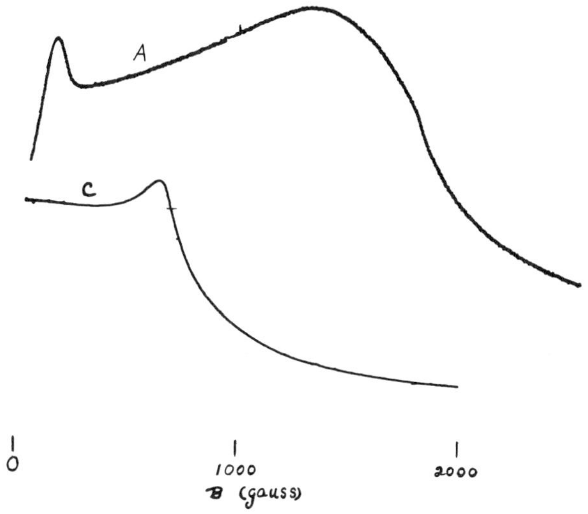

Fig. 2. Typical dM/dB curves of $ErPO_4$ as a function of applied magnetic field at a fixed temperature below 0.1 K. The scale along the ordinate is arbitrary. Curve A was obtained with the dc field applied along one of the a-axes. Curve C was obtained with the field applied along the c-axis.

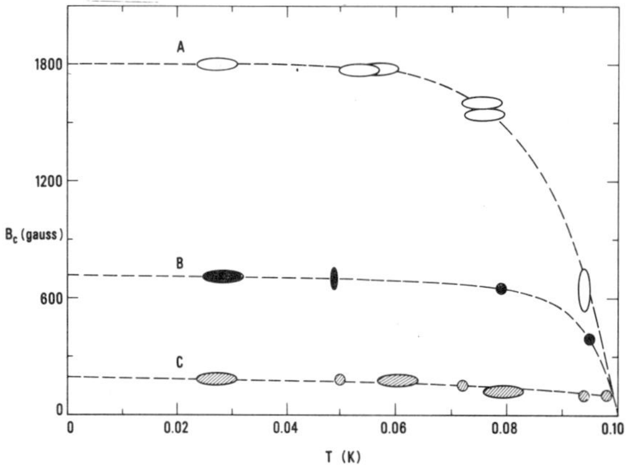

Fig. 3. The magnetic phase diagram of $ErPO_4$. The heavily screened symbols represent the critical field curve separating the antiferromagnetic and the paramagnetic phases when the field is applied along the c-axis. The symbols containing the diagonal lines represent the critical field curve separating the antiferromagnetic and the spin-flop phases when the external field is applied along either of the a-axes. The open symbols represent the critical field curve separating the spin-flop and the paramagnetic phases when the external field is applied along either of the a-axes.

predicts that dipolar interactions will cause $ErPO_4$ to order antiferromagnetically at $\sim$0.2 K and, (2) the EPR pair spectra observed by Hillmer[6].

We are unable to determine a Weiss constant since the data departed progressively in an antiferromagnetic way from a Curie-Weiss law over the temperature region of our measurements. This is to be contrasted with the results of Will, Lugscheider, Zinn, and Patscheke[4] who found that their susceptibility departed in a ferromagnetic way from the Curie-Weiss law. They reported a paramagnetic Curie temperature of $\theta = -1$ K.

In order to determine the magnetic phase diagram, we measured dM/dB as a function of magnetic field applied along the c-axis and along the two a-axes at various temperatures. Typical curves obtained at temperatures below 100 mK are shown in Fig. 2. From such a family of curves we obtained the magnetic phase diagram which is shown in Fig. 3. The field measurements confirmed an antiferromagnetic transition temperature of $T_N = 100 \pm 2$ mK.

For the case of dM/dB as a function of field directed along the c-axis, we interpret the peak in curve C of Fig. 2 as due to a transition from an antiferromagnetically ordered state, with the spins lying along the a-axes, to a paramagnetic state. Curve A of Fig. 2 shows typical behavior of dM/dB at temperatures below 100 mK when the field is applied along either of the a-axes. We attribute the fairly sharp peak at low fields to the transition from the antiferromagnetic state, with the spins aligned along the direction of the field, to the spin-flopped state in which the spins are essentially normal to the field. The abrupt decrease in dM/dB on the high field side of the broad high-field peak is associated with the transition from the spin-flopped state to the paramagnetic state. The position of the maximum of the broad peak decreases with increasing temperature but is nevertheless still evident well above T_N. Such behavior above T_N is probably due to antiferromagnetic short-range order effects but is certainly associated with the paramagnetic region. Since the dominant interaction probably is dipolar, the situation at temperatures below T_N is not as simple as it would be if exchange interactions dominated and it is thought that this more complicated situation accounts for the existence of the broad maximum at temperatures below T_N. The precise mechanism, however, is unclear.

Angular dependence of the critical field in the basal plane showed that the system is anisotropic but that the two a-axes are equivalent directions of the ordered spins. Three comments can be made concerning the alignment of the spins along the a-axes. (1) If the crystal is tetragonal, then the dipolar field at a given Er^{3+} site due to the spins of all the surrounding Er^{3+} ions is anisotropic and energetically favors alignment along the a-axes. Hence, the dipolar field of the tetragonal system can be the source of the observed anisotropy. (2) If the crystals undergo a distortion from tetragonal to orthorhombic symmetry, then the a and b directions, corresponding to the tetragonal a-axes, are no longer equivalent. This is a disagreement with our measurements and, hence, would appear to be excluded as a possibility. (3) If a cooperative Jahn-Teller distortion from tetragonal to orthorhombic symmetry occurred, then the tetragonal a-directions would still be magnetically equivalent but there would be anisotropy in the basal plane. The susceptibility above T_N, however, did not indicate that such a distortion had taken place, although it is possible that it occurred at a higher temperature.

If the system has tetragonal symmetry with the a-axes being the directions of spin alignment or if there is a cooperative Jahn-Teller distortion present, then the zero field susceptibility measured along the tetragonal a-axes at T=0 should be one-half the value at T_N. At the very lowest temperatures attained the susceptibility was approaching a constant value approximately 65% of the value at T_N. There is some evidence, however, that thermal contact to the $ErPO_4$ sample became rather poor at the lowest temperatures and that the temperature of the sample was higher than that indicated by the CMN thermometer. Assuming this to be the case, the susceptibility can readily be extrapolated to a value at T=0 equal to one-half the value at T_N.

Calculations, using the tetragonal structure and dipolar interactions only, reveal that an antiferromagnetic state with the spins aligned along an a-axis has the lowest energy, in agreement with our measurements. The proposed magnetic structure contains ferromagnetic sheets in the ac-plane and is the same as that observed in $DyVO_4$ and $DyAsO_4$[7] rather than that of $DyPO_4$[8] which has spins aligned along the c-direction only. If dipolar interactions are the only interactions present, then all field effects can be calculated directly. Such calculations are in progress.

ACKNOWLEDGMENTS

We are grateful to Drs. J. C. Wright, J. N. Lee, and H. W. Moos of the John Hopkins University for providing us with the $ErPO_4$ single crystals used in this investigation.

REFERENCES

1. R. W. G. Wyckoff, Crystal Structures, 2nd edition, Vol. 3, Interscience Publishers, New York, 1965, Chap. VIII.
2. D. Kuse, Z. Phys. 203, 49-58, (1967).
3. M. Dzionara, H. G. Kahle, and F. Schedewie, phys. stat. sol. (b) 47, 135-136 (1971).
4. G. Will, W. Lugscheider, W. Zinn, and E. Patscheke, phys. stat. sol. (b) 46, 597-601 (1971).
5. P. H. E. Meijer, private communication.
6. W. Hillmer, phys. stat. sol. (b) 55, 305-314 (1973).
7. H. Göbel and G. Will, phys. stat. sol. (b) 50, 147-154 (1972).
8. J. C. Wright, H. W. Moos, J. H. Colwell, B. W. Mangum, and D. D. Thornton, Phys. Rev. B3, 843-858 (1971).

THE X-AXIS METAMAGNETIC TRANSITION OF THE TERBIUM ORTHOFERRITE

R. Bidaux, J.E. Bourée, J. Hammann
Service de Physique du Solide et de Résonance Magnétique
CEN-Saclay, BP n°2, 91190 Gif-sur-Yvette
France

ABSTRACT

Magnetization curves at 1 K with a magnetic field applied along the x direction are reported for two typical single crystals of $TbFeO_3$. The different behaviors which have been observed, are ascribed to a partial quenching of domains in one of the samples. This quenching is at the origin of the important temperature dependent discrepancies between the two samples[2].

INTRODUCTION

The Terbium orthoferrite $TbFeO_3$ undergoes at $T_N = 651$ K a transition to a weak ferromagnetic structure (F_zG_x) which only involves the iron system. The small anisotropy of the Fe system in the xOz plane, which is mainly due to dipolar interactions, as deduced from the study of $YFeO_3$, makes it possible to rotate the ferromagnetic moment from the z to the x direction, and thus to induce a coupling between the Fe and the Tb ions (the Tb moments are located in the xOy plane). This possibility is observed in a set of samples (samples I) which show a reordering temperature at 6.2K corresponding to the onset of the ferromagnetic structure of Fig.1. The same set exhibits a second reordering at 3.1 K into the antiferromagnetic configuration (Fig.1) where the Fe moments rock back to their high temperature configuration.[1] But another set of samples (samples II) simply display one transition at 3.4 K directly to the antiferromagnetic structure, involving now only the Tb system[2].

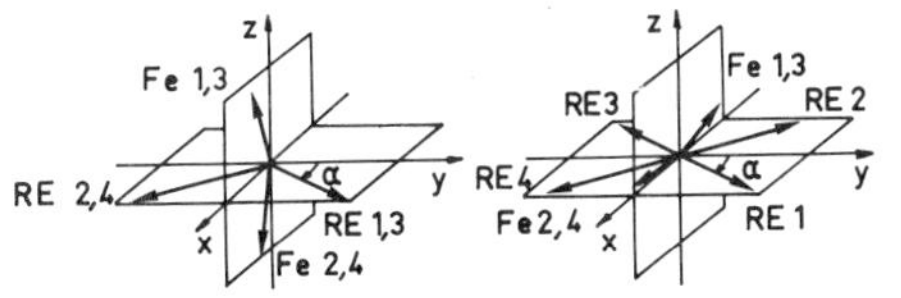

FIG.1_FERROMAGNETIC AND ANTIFERROMAGNETIC STRUCTURES_

These different behaviors have already been analysed in Ref.3, taking into account the dipolar interactions together with isotropic Fe-Tb and Tb-Tb exchange interactions : the presence of dipolar interactions stabilizes the ferromagnetic configuration , as far as the onset of an appropriate domain structure can occur. Therefore we assume that such a domain structure is present in samples I, whereas it is hindered in samples II ; this assumption is supported by the relaxation measurements reported in ref. 2 and should be compatible with the x-axis metamagnetic behaviors.

EXPERIMENTAL RESULTS

Fig.2 reports the results of magnetization measurements at 1.15 K for sample I which was a spherical single crystal of diameter 2mm. The well defined linear part in the curve which has been obtained, indicates a first order metamagnetic transition. The slope of this linear part corresponds quite well to the demagnetizing factor of the sample : $N = 1/3$. This is related to the possibility of mixing the ferromagnetic and antiferromagnetic phases inside the single crystal, in such a manner that the ferromagnetic domains minimize the dipolar energy. In particular, the transition should start at increasing fields with the creation of thin ferromagnetic needles whose effective demagnetizing factor is zero.

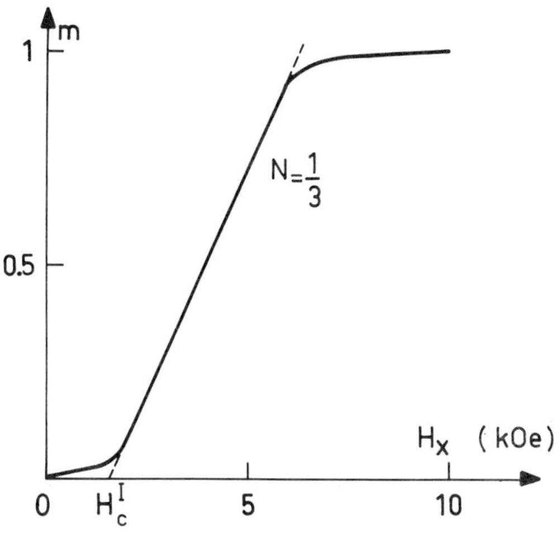

FIG.2_EXPERIMENTAL MAGNETIZATION CURVE OF SAMPLE I AT T = 1.15 K

In this sample, the internal critical field can be defined, by the extrapolation of the linear part of the magnetization curve, which leads to $H_c^I = 1.52$ kOe. This value should characterize the transition in an infinite needle shaped sample where the whole volume passes abruptly from the antiferromagnetic to the ferromagnetic phase, because there can appear no domain with a smaller demagnetizing field than that corresponding to the overall shape of the sample.

From this description of the metamagnetic transition, the behavior of sample II, shown on Fig.3, can now be well understood. It must first be noted that the sample II which has been used was an irregular thin slab (approximate diameter $\simeq$ 3mm, thickness $\simeq$ 0.4mm), for which no uniform demagnetizing factor could be defined.

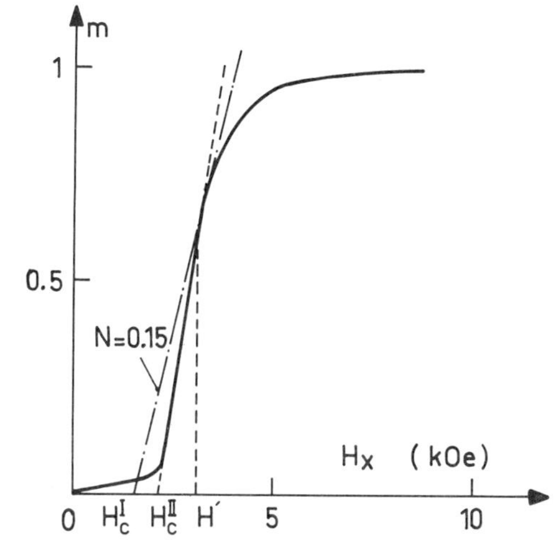

FIG.3_EXPERIMENTAL MAGNETIZATION CURVE OF SAMPLE II AT T = 1.15 K

Only an estimate of the average value of this factor can be made : $0.15 \leq N \leq 0.2$. Nevertheless it is visible on Fig.3 that the slope of the approximately linear part in the magnetization curve does not correspond to this demagnetizing factor. This behavior explains the value of $H_c^{II} = 2.12$ kOe which is larger than H_c^I .

If sample II could have an appropriate domain

structure, it should display a critical internal field equal to H_c^I and a behavior corresponding to the straight line with a slope defined by $N \simeq 0.15$, as shown on Fig.3. If no domain at all could appear, the transition should occur at $H' \simeq 3.1$ kOe with an abrupt jump of the magnetization.

DISCUSSION

The theoretical magnetization curves at T = OK of Fig.4 have been determined for a demagnetizing factor of N = 0 . The method of calculation will be briefly described in an other publication[4] . Both curves $s \neq 0$ (s = antiferromagnetic order parameter) and $s = 0$ have two parts corresponding to the situations $\theta \neq 0$ and $\theta = 0$ where θ is the angle between the weak ferromagnetic moment and the x-axis.

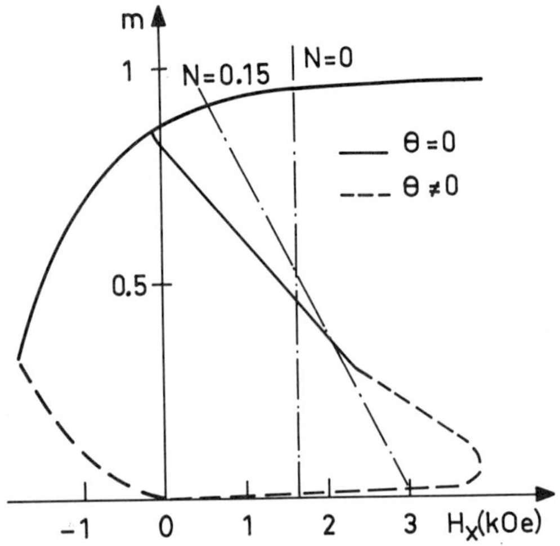

FIG.4 _ THEORETICAL MAGNETIZATION CURVES AT T = 0K

These curves can only describe the case where no domains appear during the transition. Now the behavior of sample I, when reported as a function of the internal field, is always equivalent to that of a N = 0 single domain. H_c^I has thus to be compared with the theoretical critical field obtained by applying the Maxwell rule on Fig.4, which leads to $H_c = 1.65$ kOe.

The properties of sample II can be compared with that of a single domain of $N \simeq 0.15$ which is obtained on Fig.4 through the transformation $H' = H + NM$ (M = magnetization, H = applied field). The Maxwell rule now applied in the oblique axes gives $H_c' \simeq 3.1$ kOe which fits the experimental value of H'.

As a conclusion, it is thus quite clear that the x - axis metamagnetic behavior of both types of samples is well understood if the quenching of the ferromagnetic domains is taken into account : no change in the other physical parameters is needed, which supports the assumption that all the observed temperature dependent differences between the samples are accounted for by the shape effects due to the dipolar interactions.

References

1. E.F. Bertaut, J. Chappert, J. Mareschal, J.P. Rebouillat, J. Sivardière, Solid State Comm. 5, 293 (1967)
2. J.E. Bourée, J. Hammann, J. Physique (to be published)
3. R. Bidaux, J.E. Bourée, J. Hammann, J.Phys.Chem.Solids (to be published)
4. R. Bidaux, J.E. Bourée, J. Hammann, (submitted for publication to J. Physique).

MAGNETIC STRUCTURE OF HEXAGONAL Ba_2NiReO_6

C.P. Khattak* and D.E.Cox*
Brookhaven National Laboratory, Upton, New York 11973
and
F.F.Y. Wang
State University of New York at Stony Brook, New York 11794

ABSTRACT

The compound Ba_2NiReO_6 (prepared by the hydrothermal method) has been reported[1] to have a cubic, ordered perovskite type structure. Solid state synthesis of the polycrystalline compound has yielded a six layered hexagonal structure. Long wavelength neutron powder data (λ=2.359Å) revealed a contribution from magnetic scattering which disappeared around 80°K. The magnetic structure could be analyzed on the basis of doubling of the C axis of the hexagonal chemical cell with the Ni moments on adjacent (001) layers coupled antiparallel. The Ni moment had a value of $(1.6 \pm 0.3)\mu_B$ and is directed along the C axis. The x-ray powder data are consistent with the structure of C_{3v}^1 (P3m) type, with Ni and Re ordered on alternate (001) layers.

REFERENCES

*Work supported by U. S. Atomic Energy Commission.
1. A. W. Sleight and J. F. Weiher, J. Phys. Chem. Solids 33, 679 (1972).

MAGNETIZATION PROCESS IN $DyCrO_3$

K. TSUSHIMA, T. TAMAKI, and Y. YAMAGUCHI*
Broadcasting Science Research Laboratories of Japan
Broadcasting Corp., (NHK) Setagaya-ku, Tokyo 157
*Electro-Technical Laboratory, Tanashi, Tokyo 188 JAPAN

ABSTRACT

The magnetic structure of the weak ferromagnet $DyCrO_3$ below the second critical temperature, T_{N2}=2.01K, has been given as $\Gamma 25$ (F_x, C_y, G_z: F_x^R, C_y^R, G_x^R, A_y^R) in zero field. Dysprosium is a good candidate for the Ising-type of spin and it is confined in the orthorhombic a-b plane, directed ± 27° from the b-axis. Chromium spins, on the other hand, have an antiferromagnetic arrangement, G_z, together with a weak ferromagnetic component, F_x, along the a-axis and a small hidden canting component, C_y, along the b-axis. Magnetization measurements down to 0.48 K revealed a two-step magnetization curve along the a-axis with discontinuous jumps at 3.5 kOe. and 14.3 kOe. This is quite different from $DyAlO_3$ in which only one jump has been observed. The change in magnetization at each jump amounts to 2.2 μB which corresponds to the flip of a dysprosium spin from the +(-)27° direction to a -(+)27° direction, measured from the b-axis in the a-b plane. This means that the unit cell must be taken to be twice as large as the original one along the a- and b-axes when the field is applied along the a-axis.

INTRODUCTION

Rare earth orthochromites,[1.] as well as rare earth orthoferrites, have been known as weak-ferromagnetic crystals which have a spin reorientation transition, in general, between the temperatures, T_2 and T_1, below the first Neel temperature, T_{N1}. They also have another transition temperature, T_{N2}, below which rare earth spins begin to reorder by the magnetic interaction between rare earth ions.
Among them, dysprosium orthochromite is of particular interest[2.] as the dysprosium ion is a good example of an Ising spin in the sense that it has an extremely large anisotropy : that is, the g-value along the anisotropy axis is as large as 18, which is determined to be along the 27° from the orthorhombic b-axis in the a-b plane, whereas the g-value along the perpendicular direction, the c-axis, is nearly zero[3.]

The aim of the present experiment is to measure the magnetization along the weak-ferromagnetic moment, the a-axis, as well as other directions such as the 27° direction from the b-axis and the b- and the c-axes below T_{N2}=2.01K.

EXPERIMENT

In our experiment, the magnetization was measured by the method of sample extraction from a pick-up coil. The lowest temperature down to 0.4 K was made available by using liquid He^3. The field was applied by a superconducting solenoid. The measurement of the susceptibility was done by the magnetic balance-type of magnetometer.
Single crystals of $DyCrO_3$ were grown by the flux method, a mixture of PbF_2 and PbO being the flux.

EXPERIMENTAL RESULTS

The spin structure of $DyCrO_3$ in zero field below T_{N2} is given schematically in Figure 1. Here arrows, 1, 2, 3 and 4 represent Cr^{3+} spins, arrows, 5, 6, 7 and 8 represent the Dy^{3+} spins. A short one along the a-axis is the weak-ferromagnetic moment resulting from the overt canting of the four chromium ions.
Figure 2 shows the two sharp increases in the magnetization, which are observed at the lowest temperatures, when the magnetic field is applied along the a-axis. Experimentally, the feature of the sharp two-step-wise magnetization is unchanged up to 0.9 K. Above T_{N2}= 2.01 K, the step disappears as is seen from the 4.2 K data in the same figure. In these cases, the chromium spins have been known to show a little change of the magnetization[4.] that amounts nearly 10^{-5} μB.Oe^{-1} and it arises from the field-induced increase of the weak-ferromagnetic moment along this axis.
The most part of the change in magnetization, therefore, can be accounted for to come from dysprosium spins. The amount of the increased moment at the two transition fields, 3.5 and 14.3 kOe, is 2.2 μB and is the same in both cases. The two transitions also show a time hysteresis of the order of a few tens of seconds in both increasing and decreasing fields at the two transitions.
In Figure 3, the magnetization curves along various directions are shown at T=0.48 K as well the one along the a-axis. In contrast to the one along the a-axis, one step jumps are observed along the b- and the 27° directions. For the θ=63° direction, in which two dysprosium spins, namely, the 7-th and the 8-th arrows in Figure 1, are perpendicular to the applied field, the magnetization shows a kink at 4.8 kOe.

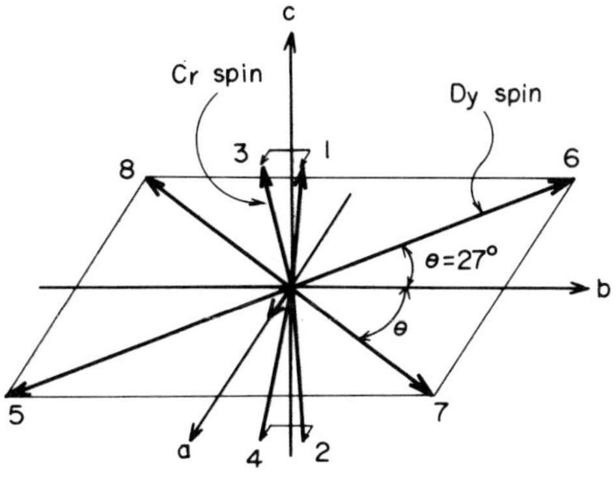

Figure 1. Schematic spin structure of $DyCrO_3$ below T_{N2}=2.01 K.

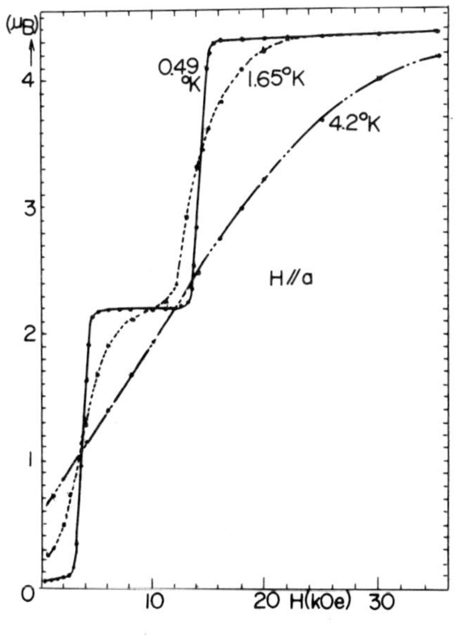

Figure 2. The magnetization curves along the a-axis in $DyCrO_3$

This demonstrates the inequivalence of the 5-th spin with the 6-th one in the unit cell. For the c-axis, no field-induced spin reorientation of Cr^{3+} spin is observed. A precise determination of T_{N2} in the crystal is done by measurement of the susceptibility along the 27° direction from the b-axis and it appeared to be 2.01 ±0.002 K, which is seen in Figure 4.
The weak-ferromagnetic moment $\sigma(a)$, and the susceptibility, χa, along the a-axis are also measured as a function of temperature, and they are shown to drop abruptly below the onset of the reordering of Dy^{3+} spins at T_{N2}, as shown in Figure 5.

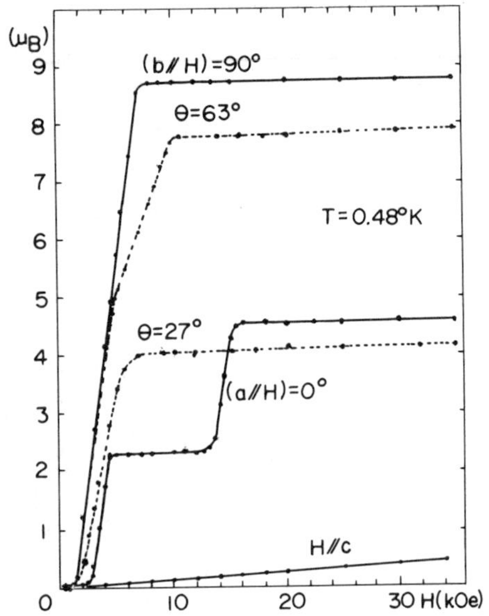

Figure 3. Magnetization curves at T=0.48 K in $DyCrO_3$

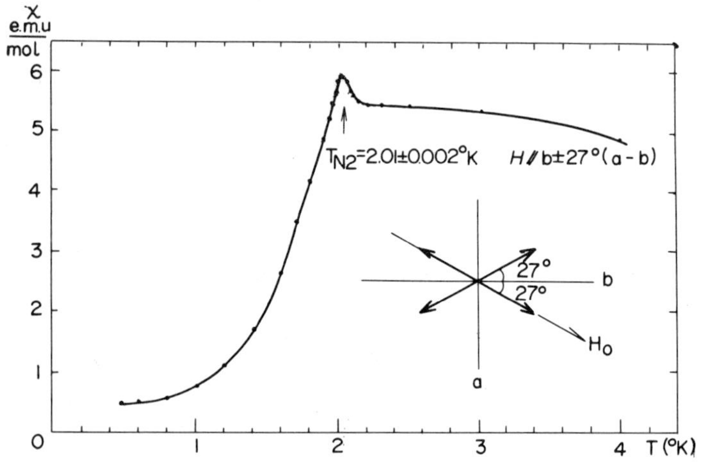

Figure 4. Precise measurement of χ along H// b±27° in the a-b plane near T_{N2} in $DyCrO_3$.

DISCUSSION

In an isomorphous compound, $DyAlO_3$, Holmes et al[5.] have observed the magnetization curve in the a-b plane below the ordering temperature of Dy^{3+} spins. In their case, magnetization in all directions reveals only one jump, even along the a-axis.
The main difference between $DyAlO_3$ and $DyCrO_3$ is that the Al^{3+} ion is non-magnetic in the former, but Cr^{3+} ion is magnetic in the latter. From the results along

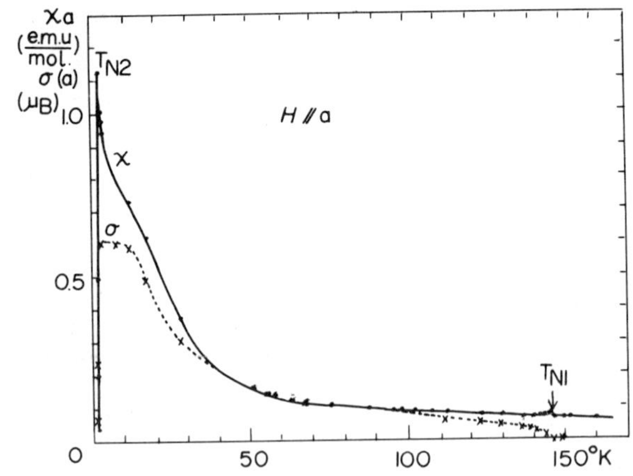

Figure 5. Temperature dependence of the weak-ferromagnetic moment, $\sigma(a)$, and the susceptibility χa along the a-axis in $DyCrO_3$.

the a-axis in the preceeding section below T_{N2}, it is obvious that the magnetic unit cell in the a-b plane must be doubled along the a- and the b-axes as compared with the one above T_{N2}.
Furthermore, the fact that the two transitions at 3.5 kOe and 14.3 kOe exhibit hysteresis means these transitions are of first order.
The presence of the magnetic chromium ions is essential for making the magnetization two-stepwise along the a-axis below T_{N2} in $DyCrO_3$.
The indirect superexchange interaction between crystallographically adjacent dysprosium ions is induced through the Cr^{3+} - Dy^{3+} superexchange interaction. The strength of this interaction and the crystallographic topology are crucial in determining the observed magnetization process[6.] It is basically analogous to the treatment for the metamagnetic $CoCl_2 \cdot 2H_2O$ by Kanamori[7.].
The magnetic symmetry of $DyCrO_3$[8.] at 0 Kelvin in zero field is given as $\Gamma 25$ (F_x, C_y, G_z: F_x^R, C_y^R, G_x^R), where the spin notation symbols are used after Bertaut [9.] and R in the superscript denotes the dysprosium spins.

The one above H=14.3 kOe is also described as Γ_2 (F_x, C_y, G_z: F_x^R, C_y^R), that means dysprosium is ferromagnetic along the a-axis and antiferromagnetic, C-type, along the b-axis.
In the intermediate field region, 3.5 ≤ H ≤ 14.3 kOe, the magnetic symmetry has not uniquely been assigned. It could be determined, for instance, by experiments using the magneto-electric effect.

REFERENCES

1. K. Tsushima, Kotai Butsuri 7, 592 (1972) (in Japanese)
2. K. Tsushima, T. Tamaki and R. Yamaura, Proc. Int. Conf. Magnetism, 73, 5 (1974) : to be published
3. Y. Uesaka, I. Tsujikawa, K. Aoyagi, K. Tsushima and S. Sugano, J. Phys. Soc. Japan, 31, 1380 (1971)
4. I.S. Jacobs, H.F. Burne and L.M. Levinson, J. Appl. Phys., 42, 1631 (1971)
5. L.M. Holmes, L.G. Van Uitert, R.R. Hecker and G. W.Hull, Phys. Rev., B5, 138 (1972)
6. T. Yamaguchi, to be published in J. Phys. Soc. Japan
7. J. Kanamori, Prog. Theor. Phys., 35, 16 (1966)
8. T. Yamaguchi and K. Tsushima, Phys. Rev., B8, 5187 (1973)
9. E.F. Bertaut, in Magnetism III, edited by G.T. Rado and H.Suhl (Academic Press, New York), 149 (1963)

MAGNETIC AND TRANSPORT PROPERTIES OF $(SnTe)_{1-x}(MnTe)_x$

R.W. Cochrane, F.T. Hedgcock, A.W. Lightstone and J.O. Strðm-Olsen
Eaton Electronics Laboratory, McGill University, Montreal, Quebec H3C 3G1

ABSTRACT

In order to extend our earlier investigations[1] of $(GeTe)_{1-x}(MnTe)_x$ to lower carrier concentrations we have begun a study of the SnTe based alloys with $N_c=10^{20}/cm^3$. At small concentrations these alloys order ferromagnetically with a T_c of approximately 2K/at.% Mn as inferred from E.S.R., magnetization and resistivity measurements. ESR spectra showing a single line for which $g = 2.00 \pm 0.02$ have been observed up to 25K. The magnetization derived from these data are in agreement with the magnetization measured directly on a vibrating sample magnetometer and both are consistent with a Mn spin of $2.5\mu_B$ and an s-d exchange interaction, J_{sd}, of order 1 eV. The zero field residual resistivity increases linearly with x up to at least 5 at.% Mn. The magnetoresistance is negative as is usual for scattering from local moments. Finally, in contrast to the GeTe system, a weak resistance minimum is observed in the SnTe alloys suggesting that $J_{sd}<0$.

Previously reported experiments on GeMnTe have shown that the system orders ferromagnetically through an RKKY interaction with an s-d exchange constant of approximately 1 eV in magnitude. In order to investigate the influence of varying the range and strength of the interaction in a comparative system we have begun an extensive investigation of the properties of MnTe dissolved in SnTe where the charge carrier concentration is a factor of ten smaller. We present here studies on a sample containing 2 at.% Mn with a charge carrier density of $1.0 \times 10^{20}/cm^3$ which orders ferromagnetically at 4^oK. In the SnMnTe system it is possible, in contrast to the GeMnTe system, to observe a well resolved spin resonance line up to 25^oK. As well as ESR data, included in this study are magneto-resistance, magnetization and resistivity data for the same sample.

Shown in figure 1 are the linewidth and inverse magnetization as derived from the directly measured areas under the absorption curve both plotted as a function of temperature. The ESR spectra show a single line with a g value of 2.00 ± 0.02 and a zero temperature linewidth of order 200 gauss. Also shown in fig. 1 is the inverse magnetization at 3.2kG measured directly on a vibrating sample magnetometer and it can be seen that there is a good qualitative correlation indicating the observed resonance is due to the manganese ions. If the linewidth is controlled by the spin system relaxing directly to the lattice then it would be expected that the temperature dependence of the linewidth should be similar to that of the resistivity. This general feature can be observed by comparing the resistivity data in figure 1(a) with that of the linewidth.

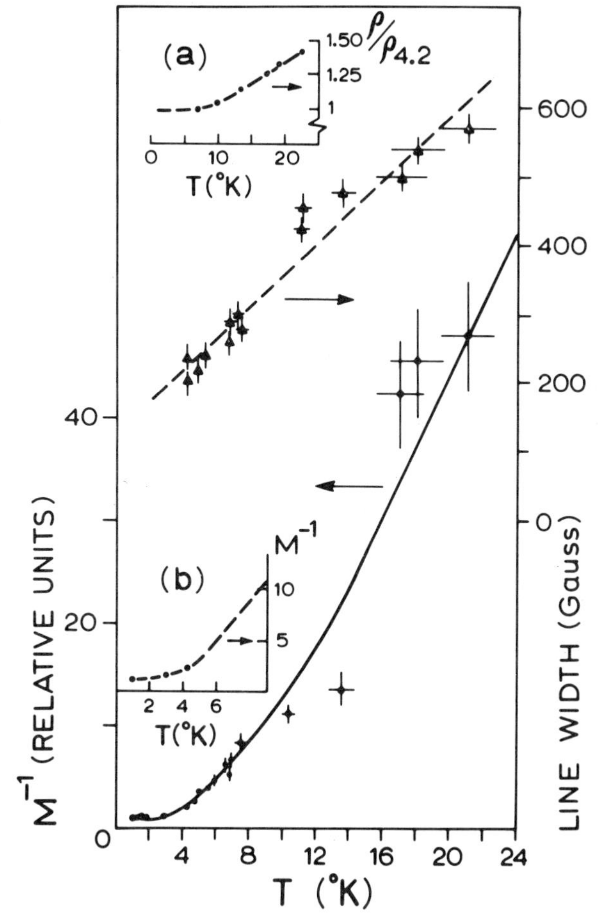

Fig. 1: ESR linewidth and inverse magnetization as a function of temperature for a sample $(SnTe)_{98}(MnTe)_2$. Inset (a) is resistivity data for a typical sample. Inset (b) is taken from directly measured magnetization data. The relative magnetization is measured relative to the saturation value. The linewidth is defined by the peak to peak separation of the derivative of the absorption signal.

The inverse susceptibility as a function of temperature obeys a Curie-Weiss law with a manganese moment of $2.5\mu_B$. The value of the Curie-Weiss temperature indicates a value of approximately 1 eV for the s-d exchange constant which is comparable to that previously reported for the GeMnTe system. That the residual resistivity is a linear function of concentration (see inset fig. 2) suggests that the manganese enter as

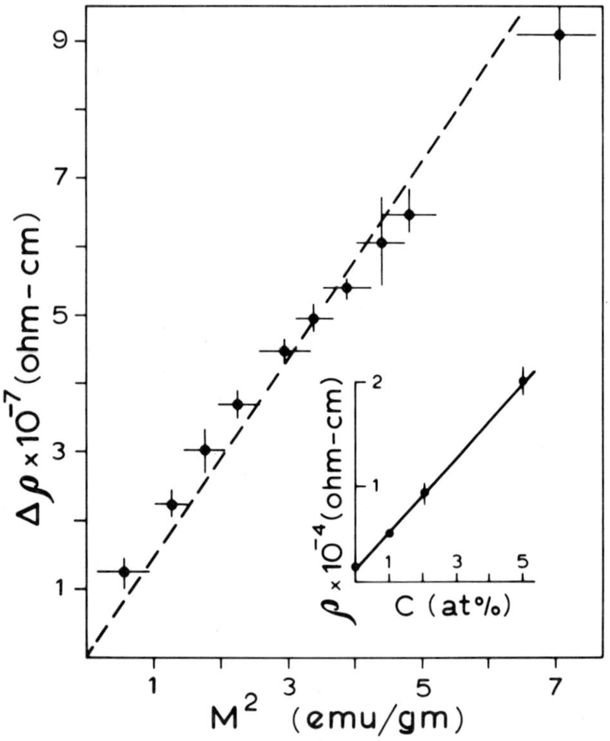

Fig. 2: Correlation of magneto-resistance at 4.2°K with the square of the magnetization at 4.2°K for a sample $(SnTe)_{98}(MnTe)_2$. The inset shows the resistivity at 4.2°K as a function of concentration of manganese.

single impurities, whose d levels do not overlap to form a band; the ferromagnetic ordering in the 5% sample does not affect this argument since magnetic scattering represents only 1% of the total scattering from the manganese. If the magneto-resistance is governed by spin-flip scattering then a second order perturbation calculation yields the fact that $\Delta\rho$, the change in resistivity due to the application of a magnetic field, should vary as the square of the magnetization. As is illustrated in fig. 2 this is indeed the case within the experimental error. Since the second order perturbation calculation depends on the square of the s-d exchange constant (J_{sd}) the sign can be either positive or negative. It was argued since no Kondo effect was observed in the GeMnTe system that J_{sd} was positive. However, preliminary results do indicate that in the SnMnTe system a Kondo effect, though small in magnitude, is indeed present.

REFERENCE

1. R.W. Cochrane et al., Phys. Rev. B8, 4262 (1973). See also R.W. Cochrane, Phys. Rev. B9, 3013 (1974).

JASIFERROMAGNETIC BEHAVIOR OF A CANTED ANTIFERROMAGNET-$NdCrO_3$

R.M. Hornreich, Y. Komet, R. Nolan, I. Yaeger,
Department of Electronics, The Weizmann Institute of Science, Rehovot, Israel and B.M. Wanklyn,
Clarendon Laboratory, Oxford, England

ABSTRACT

Single crystals of $NdCrO_3$ were studied by means of bulk magnetization and susceptibility measurements and optical absorption spectroscopy. This compound, in conformity with other rare earth orthochromites, exhibits canted antiferromagnetic behavior. It orders at T_N=214°K in a $\Gamma_2(F_x)$ spin structure which persists down to T_R=35°K. At T_R a first order phase transition occurs and, down to below 4.2°K, the spins are in a $\Gamma_1(0)$ mode. The temperature dependence of the spontaneous magnetization is significantly different from that of other rare earth orthochromites and is reminiscent of that found in many ferrimagnetic materials. Using a single ion model in the molecular field approximation, it is shown that this behavior is due to magnetization contributions from the two lowest Kramers' doublets of the Nd^{3+} ground multiplet. The calculated Stark splitting is 114±10°K, in excellent agreement with the absorption spectroscopy measurements. Above T_R, the ground doublet splitting is essentially due to the NdCr interaction while below T_R the additional polarization due to Nd-Nd coupling becomes significant.

EXPERIMENTAL STUDIES OF THE ELECTRICAL CONDUCTIVITY AND PHASE TRANSITION IN Fe_3O_4*

B. J. Evans†
Department of Geology and Mineralogy
University of Michigan
Ann Arbor, Michigan 48104

ABSTRACT

Above the Verwey transition the conduction electrons in Fe_3O_4 are described best by a band model. Those experimental data interpreted as indicating localized hopping conduction are either not susceptible to definitive interpretations or have been incorrectly interpreted. A band model is also appropriate above the Néel temperature and the influence of magnetic order on the conduction mechanism is only of minor significance. The Verwey transition is complex and involves both electronic and structural aspects. The temperatures of these transitions may be different or identical depending upon the purity of the Fe_3O_4. Good correlations exist between Mössbauer effect and thermal properties measurements concerning the complexity and qualitative characteristics of the Verwey transition. Resistivity measurements appear to confirm the Mössbauer effect and thermal properties measurements.

INTRODUCTION

There have been several timely reviews[1-3] of electrical conduction mechanisms in transition metal oxides, including Fe_3O_4, whose electrical conductivities undergo dramatic changes in a small temperature interval. And while one review has been devoted exclusively to Fe_3O_4[4], a comprehensive and critical appraisal of most of the available electrical, magnetic and thermal properties data of Fe_3O_4 has been lacking. Nonetheless, the previous reviews[1-3] do provide an adequate appraisal of the electrical properties measurements that were available as of 4 to 6 years ago. Since these earlier reviews, however, there has been a great increase in investigations, both experimentally and theoretically, of Fe_3O_4; and consequently, considerably more facts and different ideas have been generated regarding the electrical properties and low temperature polymorphism in Fe_3O_4. These new results will be the primary concern of this inquiry. Even though these new data when simply juxtaposed are rather confusing and contradictory, it will be shown in the following that a reasonably consistent picture as to the electrical conduction mechanism and the low temperature polymorphism emerges from a *critical analysis* of these data, especially when compared with the results of some very recent studies.

It is encouraging that most of the data resulting from earlier studies of the electrical and magnetic properties of Fe_3O_4 above and in the region of the Verwey transition[5,6], T_V, have been confirmed by recent investigations[7,8]. Therefore, our consideration of data obtained in the temperature interval $T_V<T<T_N$, (T_N = Néel temperature) will be directed primarily at resolving several alternative interpretations of the conduction mechanism rather than a detailed consideration of the data thermselves. Ideas concerning the structure of the low temperature phase[9], below T_V, have, however, been substantially altered by more recent studies[10]. Thermal property measurements in the region of the Verwey transition have also indicated a complexity beyond that previously reported[11]. The second major objective of this paper is to establish that the low temperature phase transition is a complex one. In this instance, data only recently available must be considered; these data result from different property measurements on identical samples and the complexity of the phenomena appears to be firmly established.

The principal conclusions of this investigation are the following. For the temperature interval $T_V<T<T_N$, a band description is appropriate for the conduction mechanism in Fe_3O_4. Previous data cited as indicating a hopping mechanism are found to be either incorrectly interpreted or to have resulted from poor samples. The conclusion reached for the temperature interval $T_V<T<T_N$ receives additional support from Mössbauer and resistivity measurements for $T>T_N$. For $T<T_V$, a band description of the conduction mechanism is also supported by the available data, but this conclusion is less definitive than that reached for $T_V<T<T_N$. In the region of the Verwey transition, there appear to be two phase transitional phenomena: one related in the limiting case to purely electronic phenomena and another related to purely structural (crystal symmetry, site symmetry and atomic positional coordinates) phenomena. The phase transitional phenomena associated with the electronic state of the conduction electrons are more sensitive to the presence of impurities and defects than the structural transition. As expected for a structural transition in which there is a substantial lowering of the symmetry of the structure and changes in the atomic positional coordinates of most, if not all, of the ions, impurities on either the A or B sites have significant influences on temperatures and energy changes associated with the transition.

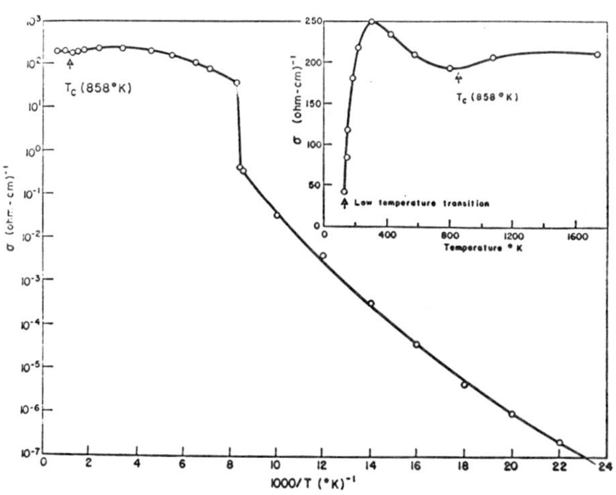

Fig. 1. Temperature dependence of the DC conductivity of single-crystal magnetite (ref. 6).

TEMPERATURE INTERVAL $T_V<T<T_N$

As might be expected, most of the available data of any sort has been obtained in this temperature interval[6]. The most important result in this region is the temperature dependence of the resistivity[6] shown in Fig. 1. Recent investigations have confirmed the early measurements[7], and provided more precise Hall effect and thermopower data[12]. The central question regarding the electrical conductivity in this region has been whether to describe it in terms of a localized hopping model or in terms of a band model. Until relatively recently, the description of the conduction mechanism in terms of a hopping model was rather qualitative and appeared to be based almost wholly on the conductivity being activated, apparently, and on the early suggestion[5] that the high electrical conductivity resulted from rapid electron interchange between the B site Fe^{2+} and Fe^{3+} ions. Fe^{57} nuclear gamma-ray resonance[13] measurements soon after the discovery of the Mössbauer effect demonstrated the absence of distinct Fe^{2+} and Fe^{3+} ions on the octrahedral site in Fe_3O_4. This result was interpreted in terms of a hopping model but is equally consistent with a band model in which there is significant screening of the ion cores. That the isomer shift and hyperfine field of the ^{57}Fe B site pattern are intermediate to those of an isolated Fe^{3+} and Fe^{2+} ion offers no *unique* support of a hopping model. Similar hyperfine fields and isomer shifts can be produced by appropriate screening. In addition, the hyperfine field and isomer shift of a hypothetical Fe^{2+} ion in Fe_3O_4 are of rather uncertain magnitudes and the magnetic hyperfine splitting of what may be considered an Fe^{2+} ion below T_V[14,16] corresponds to a magnetic hyperfine field somewhat larger than that observed in other spinels that contain Fe^{2+} ions[17].

The crucial experiment in establishing support for a hopping model was the measurement of the temperature dependence of the linewidth of the ^{57}Fe B site Mössbauer spectrum[18]. It has been confirmed in almost every Mössbauer study of Fe_3O_4, that the B site pattern has larger linewidths, Γ_B, than those of the A site pattern, Γ_A. On the assumption that there was no broadening of the A site linewidth, some function of the difference between Γ_A and Γ_B, given by Eq. 1, could be taken as measure of the $Fe^{3+}(B) - Fe^{2+}(B)$ electron hopping relaxation time

$$\Delta\Gamma = \tau_s \Delta^2/2 \qquad (1)$$

where $\Delta\Gamma = \Gamma_B - \Gamma_A$; Δ is the difference in the *assumed* frequencies of the *unrelaxed* Fe^{2+} and Fe^{3+} lines associated with Γ_B of the relaxed line and τ is the electron hopping relaxation time. Questionable assumptions regarding the hyperfine field isomer shift of an hypothetical Fe^{2+} ion in Fe_3O_4 figured prominently in this interpretation of the data. At any rate, the quantity $(\Gamma_B^2 - \Gamma_A^2)$ increased with decreasing temperature and at 298 K a hopping time of about 1 nanosecond was deduced[18]. The conductivity calculated on the basis of this measurement of the relaxation time for the hopping mechanism was found to be two orders of magnitude less than the experimentally measured conductivity. This first linewidth study

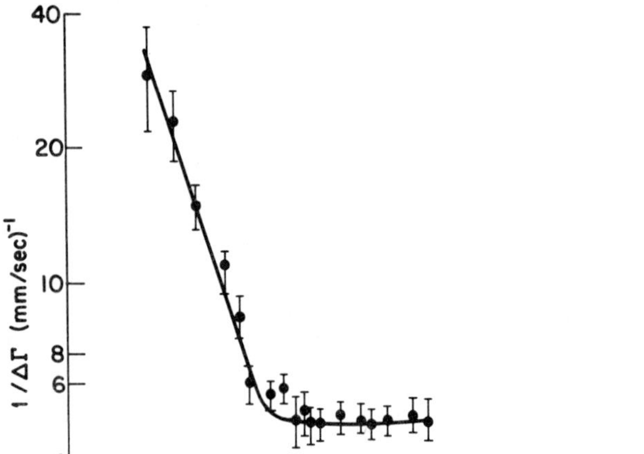

Fig. 2. Temperature dependence of the linewidth difference, $\Delta\Gamma$, of the A and B site patterns of Fe_3O_4, determined from lines 1A and 1B of the Mössbauer spectrum, as reported in ref. 19.

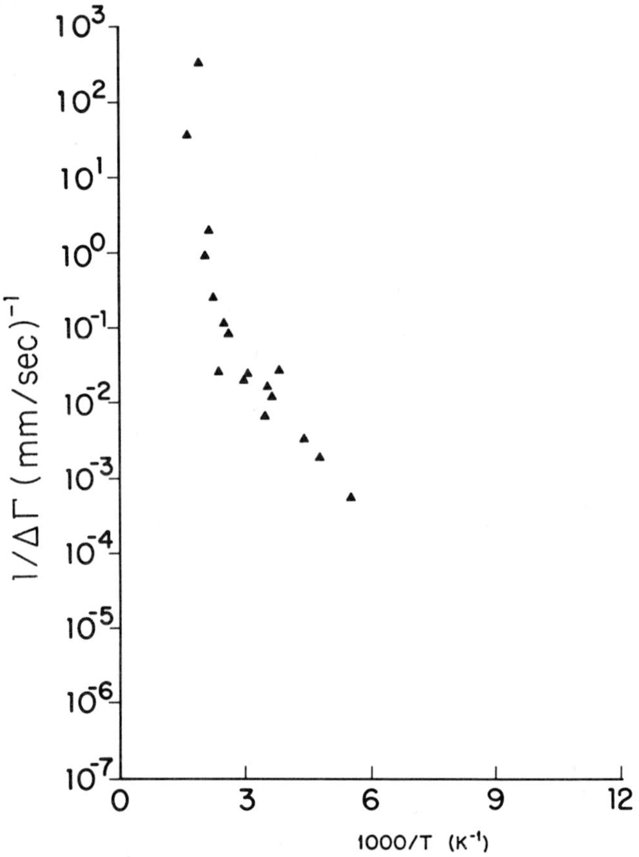

Fig. 3. Temperature dependence of the magnitude of the linewidth difference, $\Delta\Gamma$, of the A and B site patterns of Fe_3O_4, determined from lines 1A and 1B of the Mössbauer spectrum. This study.

was conducted only for temperatures below 300 K. A later study[19] extended these measurements over a larger temperature range; and in contrast to the earlier study[18], the difference between Γ_{1A} and Γ_{1B} was found to be independent of temperature below 250 K (Fig. 2). However, it was just in this temperature interval that the temperature dependence of $(\Gamma_B^2 - \Gamma_A^2)$ was used to deduce the relaxation time in the earlier study[18]. Disagreement between the temperature dependence of $(\Gamma_{1B} - \Gamma_{1A})$ and the temperature dependence of the conductivity was also noted in this latter study. Our measurements of the temperature dependence of the linewidths of the A and B site ^{57}Fe Mössbauer patterns reveals the temperature dependence shown in Fig. 3. These results are in contrast to those reported in Reference 19 inasmuch as $\Gamma_{1B} - \Gamma_{1A}$ is found to be temperature dependent over the entire temperature range from 800 K to temperatures just above the Verwey transition. However, for $T>300$ K, the decrease in $(\Gamma_{1A} - \Gamma_{1B})$ is not due to a decreasing Γ_B as demanded by a hopping model but rather to an increase in Γ_A (cf. Fig. 4). In another recent study[20] the decrease in $(\Gamma_A - \Gamma_B)$ with increasing temperature above 300 K was also found to be due more to an increase in Γ_A than to a decrease in Γ_B. For example, at 296 K $\Gamma_{1A} = 0.349$ mm/sec and $\Gamma_{1B} = 0.392$ mm/sec and at 703 K $\Gamma_{1A} = 0.376$ mm/sec and $\Gamma_{1B} = 0.371$ mm/sec[20]. Therefore, in the temperature interval $300<T<T_N$, the decrease in $(\Gamma_{1A} - \Gamma_{1B})$ is not simply related to electron hopping among the octahedral sites; and it is reasonable to conclude that the temperature dependence of $\Gamma_B - \Gamma_A$ is not indicative of electron hopping.

Further, it has been found that Γ_B is larger than Γ_A even in spinels that have very low electrical conductivities[21,22]. Then too, there is rather straightforward interpretation for Γ_B being larger at some temperatures than Γ_A. Evans[23] was the first to suggest that the apparent broadening of the B site lines

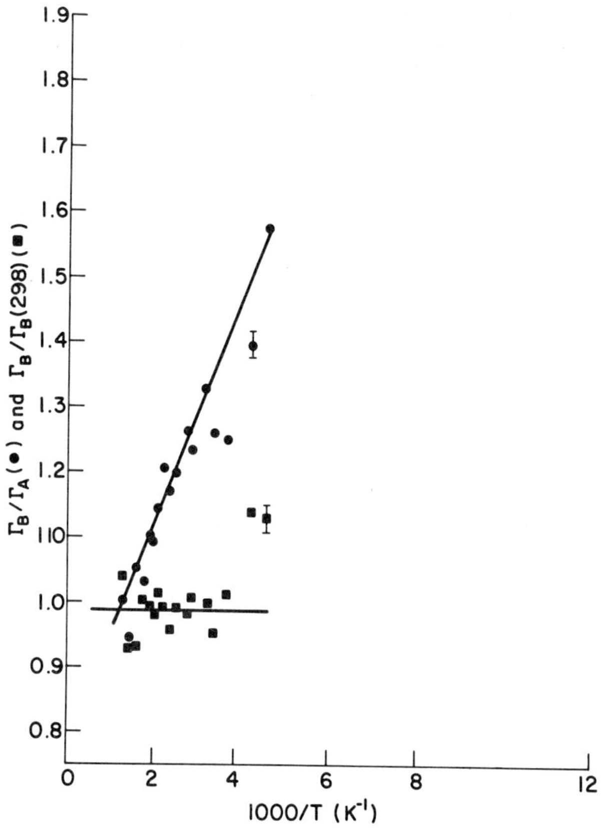

Fig. 4. Temperature dependence of Γ_{1A} (●) and Γ_{1B} (■), normalized to Γ_{1B} at 298 K. This study.

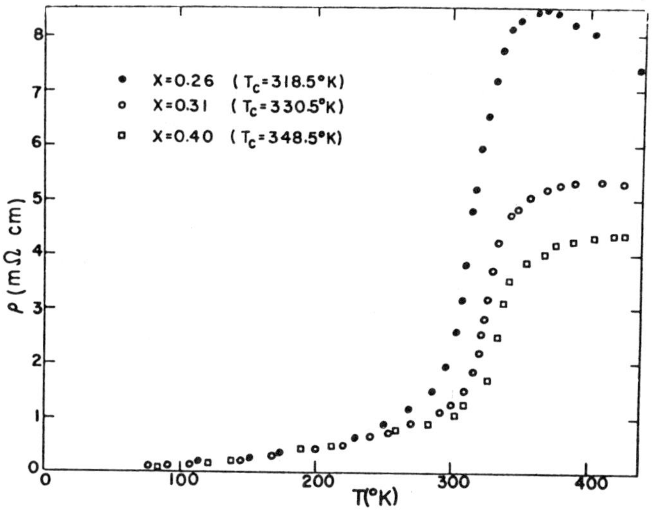

Fig. 5. Temperature dependence of the resistivity of $La_{1-x}Pb_xMnO_3$, ref. 27. Note the increase in ρ by almost an order of magnitude at T_C for x = 0.26.

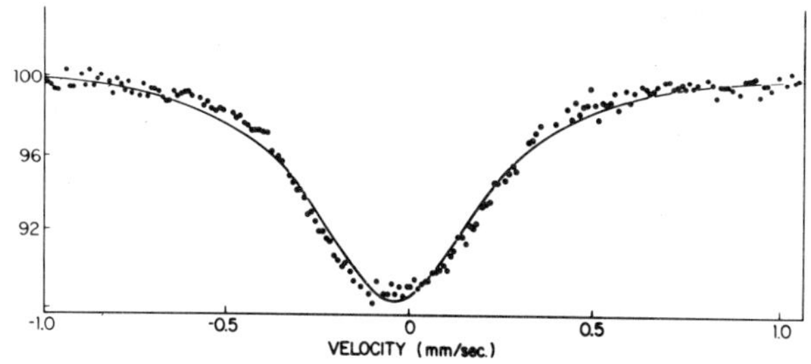

Fig. 6. ^{57}Fe Mössbauer spectrum of Fe_3O_4 above T_N, ref. 28.

was not due to electron hopping but rather resulted from the different quadrupole splittings at the B site as a consequence of the different angles between the principal axis of the electric field gradient tensor and the magnetic hyperfine field. That is to say, the shift of the absorption lines due to the electric quadrupole interaction is given by

$$\varepsilon_{m_I} = (-1)^{(|m_I| + 1/2)} e^2qQ(3\cos^2\theta-1)/8 \qquad (2)$$

where θ is the angle between the magnetic field and the principal axis of the electric field gradient tensor; and for spinel ferrites with the easy axis of magnetization being [111], there are two values of θ, θ = 0 and θ = 70°54", occurring with a relative frequency of 1 and 3, respectively. Therefore, there will be two patterns for the B site iron ions. A similar analysis would also apply for any anisotropy in the magnetic hyperfine field. Susequently, another Mössbauer investigation of Fe_3O_4[24] has led to a similar conclusion to that of Evans[23] regarding the broadening of the B site lines.

Two B site patterns have also been resolved using NMR[16,25] but in these instances the splitting of the B site pattern is observed to rise from the anisotropy in the magnetic hyperfine field. The results of the two measurement techniques are, however, in accord; the lattice sums that lead to the anisotropy in the magnetic hyperfine field are also the ones that lead to a non-vanishing electric field gradient tensor. The different interpretations are due, on the one hand, to the much higher resolution of the NMR measurements, and on the other hand, to the fact that the electric quadrupole interaction is not detected in the NMR measurements. There is in all probability a splitting of the B site pattern due to both an anisotropic magnetic hyperfine field and a non-zero electric quadrupole interaction.

As has been recognized previously[26] if the conduction mechanism could be described by a localized hopping mechanism, the magnetic double exchange mechanism would also be operative. The available data, which is rather definitive, indicate that the double exchange mechanism is not essential to the electron transport in Fe_3O_4. In materials in which the double exchange mechanism is operative, the conductivity decreases vary rapidly at the Curie temperature as shown in Fig. 5 for some $La_{1-x}Pb_x MnO_3$ perovskites[27]. As shown in Fig. 1, this is not the case in Fe_3O_4; the decrease in the conductivity for Fe_3O_4 at T_N is barely discernible. Also if the double exchange mechanism, and therefore localized hopping, was essential to the electron transport, one would expect the electron hopping relaxation time to be greater above T_N than below T_N. In which case, the Fe^{2+} and Fe^{3+} B site ions would give rise to more or less distinct hyperfine patterns. This is not the case, Fig. 6. These data which were reported earlier[28] have been verified by more recent unpublished studies. The surprising aspect of these data, however, is that not only are there no distinct Fe^{2+} and Fe^{3+} B site patterns above T_N but the A site Fe^{3+} pattern is also unresolved.

Finally, recent resistivity, Hall effect and thermopower measurements all lead to the conclusion that the conduction mechanism in Fe_3O_4 does not involve localized hopping[12]. The conduction electrons are best described by band states, albeit of rather narrow widths. The apparent activation of the conductivity and the broad maximum in the conductivity can be understood in terms of a degenerate semiconductor. Theoretical studies also lead to the same conclusions[29].

The neglect in the above discussion of much of the data on so-called "non-stoichiometric magnetites" has been deliberate. First of all, several investigators have demonstrated the doubtful character of the materials used in these studies[30]. It has also been shown that these "non-stoichiometric magnetites" are frequently multiphase materials and/or chemically inhomogeneous[30,31]. Microscopic measurements such as the Mössbauer effect can be understood in terms of the different phases present but macroscopic measurements such as electrical conductivity represent intractably complex convolutions of phenomena and their interactions. Many new ideas regarding conduction mechanisms in such non-stoichiometric materials resulted from these studies but it is not certain that these ideas are particularly germane to the *intrinsic* conductivity of Fe_3O_4. As we shall see in what follows, the technique of doping Fe_3O_4 with small quantities of other elements whose local crystal chemistries in spinel oxides are well understood and simple leads to considerably more tractable phenomena than non-stoichiometric materials.

$T \simeq T_V$

Associated with the decrease in the conductivity at the Verwey transition is also a λ-type anomaly in the heat capacity[32] and changes in unit cell volume and crystal symmetry[33,9]. More recent measurements have, however, indicated considerably more complexity in phenomena occurring at the transition than previously thought. First of all, instead of one λ-type anomaly in the heat capacity versus T curve, two anomalies have been observed[11]. Both the temperature and heat content of these anomalies depend on the amounts and kinds of impurties in Fe_3O_4[34]. Secondly, the lattice of the low temperature phase is of lower symmetry than the orthorhombic or rhombohedral lattices deduced from earlier studies[10]. Finally, resistivity measurements for very slow heating or cooling rates in the region of the Verwey transition also exhibit two maxima[35].

The main objective of the following discussion is establish the complexity of the Verwey transition and to present a possible interpretation of the systematics of the available data.

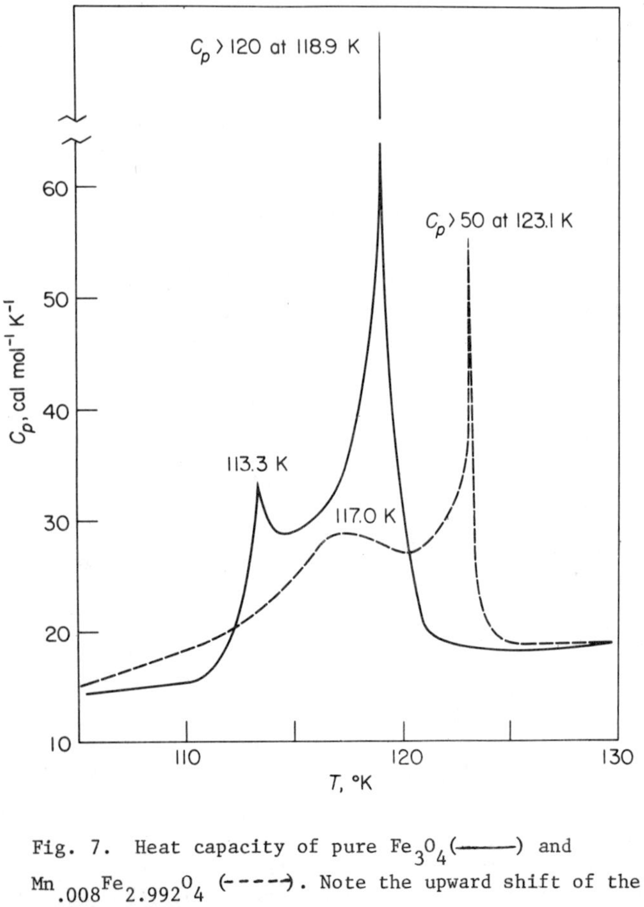

Fig. 7. Heat capacity of pure Fe_3O_4 (———) and $Mn_{.008}Fe_{2.992}O_4$ (- - - -). Note the upward shift of the high temperature anomaly and the smearing of the lower one for $Mn_{.008}Fe_{2.992}O_4$.

The C_p versus T curves of Fe_3O_4 and $Mn_{0.009}Fe_{2.99}O_4$ in the vicinity of Verwey transition are shown in Fig. 7[11]. The bifurcated nature of the peaks and the dependence of the peak shapes on composition are noteworthy features. The enthalpies, line profiles and temperatures of the anomalies are seen to depend sensitively upon the presence of impurities: For $Mn_{0.008}Fe_{2.992}O_4$ the temperature of the maximum C_p in the high temperature anomaly is shifted 4 K from that of Fe_3O_4[31], and the enthalpy content is 71 cal/mol compared to 98 cal/mol for Fe_3O_4[36]. The low temperature anomaly is less sensitive to impurities; at *low dopant levels* its enthalpy is essentially independent of temperature but the temperature of the maximum in C_p increases[34]. With these new thermal data it may be concluded that (1) substitutions of closed shell ions on the A site are not innocuous; (2) the increase in temperature of the maxima in the heat capacity with decrease in Fe^{2+} content and/or increase in lattice constant holds for Zn^{2+} and Cd^{2+} dopant level less than 0.1 atom percent; for $Cd_{0.005}$, it appears that the two anomalies occur at the same temperature; (3) at dopant levels much above 0.1 atom percent, the magnitudes of the heat capacity anomalies decrease rapidly, i.e., both anomalies are absent in a material with the composition $Zn_{0.066}Fe_{2.934}O_4$.

There is a body of literature[37] that ascribed such furcation of λ-type anomalies to inhomogeneities; and this as a possible interpretation of the new thermal properties data cannot be rejected out of hand. The samples used in our studies have, however, been prepared with sufficient care to insure against inhomogeneities. Further, the furcation persists for samples prepared by a variety of different techniques and exhibits a cogent dependence on the systematics of the crystal chemistry of the dopants[38].

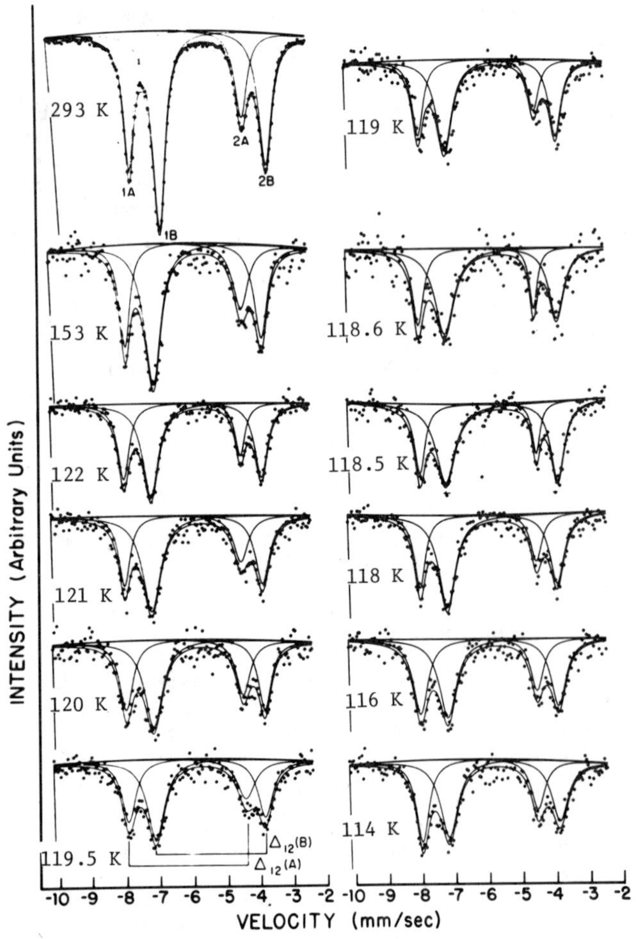

Fig. 8a. First two A and B site lines of the ^{57}Fe Mössbauer spectrum of $Zn_{.005}Fe_{2.995}O_4$ as a function of temperature.

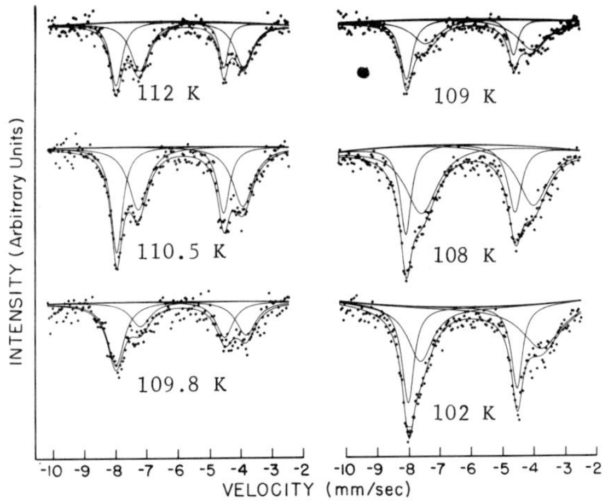

Fig. 8b. First two A and B site lines of the ^{57}Fe Mossbauer spectrum of $Zn_{.005}Fe_{2.995}O_4$ as a function of temperature.

The thermal properties measurements have proven to be crucial in delineating the complexity of the Verwey transition in Fe_3O_4, but by themselves, they provide little phenomenological insight into the crystal physics of the thermal properties variations. Therefore, ^{57}Fe Mössbauer measurements have been made in the region of the Verwey transition on the pure Fe_3O_4 and $Zn_{0.005}Fe_{2.995}O_4$ samples used in the thermal properties measurements[11,35]. The most significant changes in the ^{57}Fe Mössbauer spectrum occur for the $\Delta m = 1$, $+ 1/2 \rightarrow + 3/2$ and $\Delta m = 0$, $+ 1/2 \rightarrow + 1/2$ transitions whose absorption lines occur in negative velocity region, Fig. 8. These four lines have been used previously to monitor the Verwey transition[30] and changes in their positions and profiles have been used to confirm the complexity of the Verwey transition[11]. Some additional studies on non-stoichiometric magnetites have led to the conclusion that these lines are not reliable indicators of the Verwey transition[39] but conclusions drawn from measurements on non-stoichiometric magnetites are to be accepted with great caution. At any rate, it is demonstrated below that a good correlation exists between the temperature of the λ-anomalies and changes in the Mössbauer spectrum for the above mentioned lines in the negative velocity region. A weak absorption line between +2mm/sec and +4mm/sec, which is believed to be associated with more or less distinct Fe^{2+} species[18], has not been used in these measurements because of the long counting times necessary for good statistics and consequent stringent requirements on temperature stability.

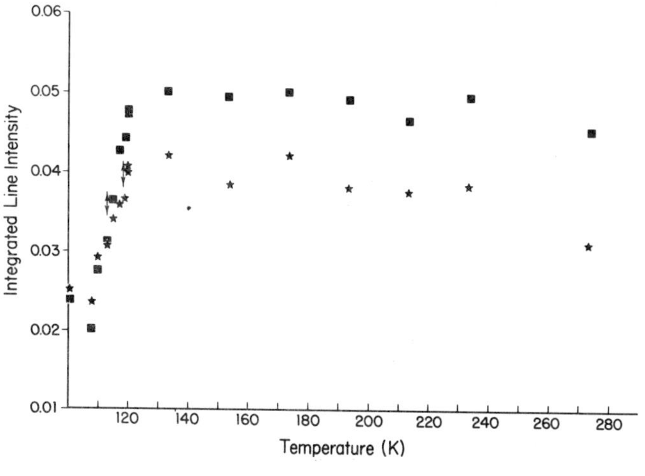

Fig. 9. Temperature dependence of the integrated intensity of lines 1 (■) and 2 (★) of the B site ^{57}Fe Mössbauer spectrum of pure Fe_3O_4. Arrows indicate the temperatures of the λ-anomalies in C_p.

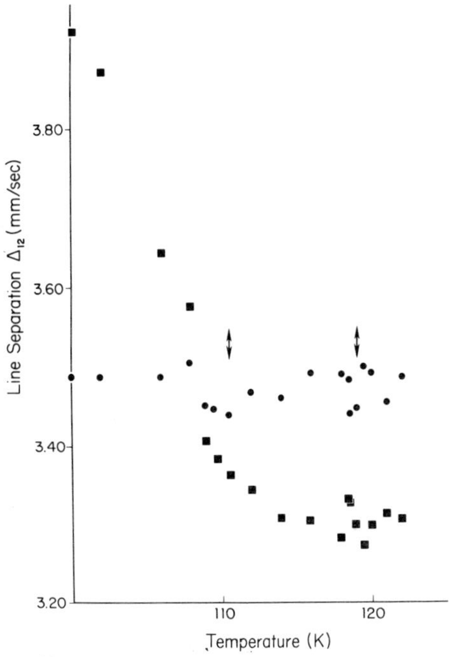

Fig. 10. Temperature dependence of the line separation, Δ_{12}, (cf.Fig. 8a) of the B site (■) and A site (●) Mössbauer patterns of $Zn_{.005}Fe_{2.995}O_4$. Arrows indicate the temperatures of the λ-anomalies in C_p.

As shown in Figs. 9 and 10, the intensities and splittings Δ_{12} of the four lines show fundamental changes in the vicinity of the Verwey transition. At temperatures corresponding to the high temperature λ-anomaly the relative intensities of the lines undergo rapid change whereas the splittings, Δ_{12} while showing a greater scatter of values, has nearly the same value above and below the temperature corresponding to this first λ-anomaly. At the lower temperature heat capacity anomaly, however, Δ_{12} and the relative line intensities, both, undergo very rapid changes. There is, thus a 1:1 correspondence between dramatic changes in the relative line intensities, the splittings, Δ_{12}, or both and the anomalies in the heat capacity. It is possible to advance a tentative explanation of these results which are consistent with and lend support to a band description of the conduction electrons above and below the Verwey transition.

If the Verwey transition is a semimetal or degenerate semiconductor→nondegenerate semiconductor transition[12], then the differences in the electronic structure above and below T_V are primarily quantitative and not qualitative in nature. Therefore, this transition is to be associated with the change in the relative intensity of the lines in the Mössbauer pattern at a temperature roughly the same as that of the high temperature heat capacity anomaly. The extreme sensitivity of the temperature and line profile of this anomaly to impurities is consistent with its being associated with the degenerate-nondegenerate semiconductor transition since this transition is expected to be rather sensitive to changes in the electron concentration.

The changes in the splittings, Δ_{12}, of the B site pattern results either from a change in the electric quadrupole splitting, the magnetic hyperfine field, or both and are indicative of major changes in the overall structure and local site symmetries since changes in local site symmetry are expected to influence substantially the orbital angular momentum contribution to the magnetic hyperfine field and the electric quadrupole interaction. It is concluded that the low temperature heat capacity anomaly and the changes in the Mössbauer spectrum are associated with major structural changes. The relative magnitudes of the maxima observed in resistivity measurements[32] are consistent with this interpretation of the heat capacity anomalies and the

Mössbauer spectra.

SUMMARY

A reasonably consistent description of the electrical conduction mechanism and Verwey transition in Fe_3O_4 seems to be the following: Above the Verwey transition Fe_3O_4 is a semimetal or degenerate semiconductor with the conduction electrons being appropriately described by a band model. There are no data that *uniquely* support a localized hopping model. There is a semimetal or degenerate semiconductor→nondegenerate semiconductor transition at ≈120 K that is accompanied by a small increase in the resistivity. This transition may be related in its origin to a Mott-Wigner metal-nonmetal transition. Previous concerns[26] about a Mott-Wigner description of the transition were based on measurements of impure magnetites that were in fact *too impure* to provide a test of the effect of a decrease in electron density on the temperature of the electronic transition. Impurity levels sufficiently low that they can be regarded as small perturbations on the lattice and electron concentration yield an increase in the transition temperature as expected for a Mott-Wigner transition.

At a lower temperature, there is a phase transition that is accompanied by major structural changes, as well as changes in the electrical properties. The temperature interval between these two transitions is affected by the amounts and kinds of impurities. Hence, the two transitions may occur at the same temperature for some impure samples. Below the Verwey transition the conduction electrons are in band states and distinct Fe^{2+} and Fe^{3+} oxidation states for *all of the iron species* do not obtain. Whether the absence of distinct Fe^{2+} and Fe^{3+} states for all of the iron atoms is due to the complexity of the structure of the low temperature phase or to charge density waves[40] is one of the questions to be resolved by further study.

Further measurements on samples with A site impurity levels slightly below and above 0.1 atom percent would be useful. The kind of dopant should also be extended to include those that occupy the octrahedral sites. A particularly important measurement would be the determination of the resistivity at very slow heating and cooling rates in the vicinity of the Verwey transition. Lastly, further conductivity and Mössbauer measurements on pure and doped Fe_3O_4 samples above the Néel temperature are desiderata; the ^{57}Fe Mössbauer spectrum above T_N is consistent with a band theoretical description of the conduction process but the spectrum is sufficiently perplexing and suggestive of extensive electron delocalization over both the A and B sites to warrant further studies.

ACKNOWLEDGMENT

Support of the work by the National Science Foundation, Grant GH-41419 is gratefully acknowledged. Additional support from the Sloan Foundation is also acknowledged.

REFERENCES

+ Alfred P. Sloan Research Fellow.
1. D. Adler, Solid State Physics, Vol. 21, F. Seitz, D. Turnbull, and H. Ehrenreich, eds., (Academic Press, New York, 1968) pp. 1-113.
2. J. B. Goodenough, Progress in Solid State Chemistry, Vol. 4, H. Reiss, ed. (Pergamon, New York, 1972), pp. 145-399.
3. N. F. Mott, Phil. Mag. 20, 163 (1969).
4. E. A. Callen, Phys. Rev. 150, 367 (1966).
5. E. J. W. Verwey and P. W. Haayman, Physica 8, 979 (1941).
6. P. A. Miles, W. B. Westphal, and A. von Hippel, Rev. Mod. Phys. 29, 279 (1957) and references therein.
7. J. R. Drabble, T. D. Whyte, and R. M. Hooper, Solid State Commun. 9, 275 (1971).
8. Yu. D. Tropin and A. A. Lepishev, Sov. Phys.-Solid State 14, 2654 (1973).
9. W. C. Hamilton, Phys. Rev. 110, 1050 (1958).
10. T. Yamada, K. Suzuki and S. Chikazumi, Appl. Phys. Letters 13, 172 (1968).
11. B. J. Evans and E. F. Westrum, Jr., Phys. Rev. B5, 3791 (1972) and references therein.
12. W. J. Siemons, I.B.M. J. Res. Develop. 14, 245 (1970).
13. R. Bauminger, S. G. Cohen, S. Ofer and E. Segal, Phys. Rev. 122, 1447 (1961).
14. B. J. Evans and S. S. Hafner, J. Appl. Phys. 40, 1411 (1969).
15. R. S. Hargove and W. Kundig, Solid State Commun. 8, 303 (1970).
16. M. Rubinstein, G. A. Stauss and F. Bruni, AIP Conf. Proc. 10, 1384 (1973).
17. F. Hartmann-Boutron and P. Imbert, J. Appl. Phys. 39, 775 (1968).
18. W. Kundig and R. S. Hargove, Solid State Commun. 7, 223 (1969).
19. G. A. Sawatzky, J. M. D. Coey and A. H. Morrish, J. Appl. Phys. 40, 1402 (1969).
20. H. Topsøe, Ph.D. Thesis, Stanford University (University Microfilms, Ann Arbor, 1974), p. 137.
21. B. J. Evans and S. S. Hafner, J. Phys. Chem. Solids 29, 1573 (1968).
22. B. J. Evans and L. J. Swartzendruber, J. Appl. Phys. 42, 1628 (1971).
23. B. J. Evans, AIP Conf. Proc. 5, 296 (1971).
24. C. I. Nistor, Rev. Roum. Phys. 18, 867 (1973).
25. N. M. Kovtun and A. A. Shamyakov, Solid State Commun. 13, 1345 (1973).
26. A. Rosencwaig, Canad. J. Phys. 47, 2309 (1969).
27. C. W. Searle and S. T. Wang, Canad. J. Phys. 48 2023 (1970).
28. B. J. Evans, AIP Conf. Proc. 10, 1398 (1972).
29. D. Ihle and B. Lorenz, Phys. Stat. Sol. 58, 79 (1973) and references therein.
30. H.-P. Weber, Ph.D. Thesis, University of Chicago, 1972, p. 62.
31. H. Topsøe, J. A. Dumesic, and M. Boudart, J. de Phys. (Paris), In Press.
32. R. W. Millar, J. Amer. Chem. Soc. 51, 215 (1929).
33. L. R. Bickford, Jr., Rev. Mod. Phys. 25, 75 (1953).
34. J. J. Bartel and E. F. Westrum, Jr., AIP Conf. Proc. 10, 1393 (1974).
35. S. Iida, M. Yamamoto, and S. Umemura, AIP Conf. Proc. 18, 913 (1974).
36. J. J. Bartel, Ph.D. Thesis, University of Michigan, 1974.
37. B. Ya. Sukarevskii, A. V. Alapina, Yu. A. Dushechkin, T. N. Kharchenko, and I. S. Shchetkin, Sov. Phys.-JETP 31, 820 (1970).
38. J. J. Bartel and E. F. Westrum, Jr., These Proceedings.
39. V. P. Romanov, V. D. Checherskii, and V. V. Eremenko, Phys. Stat. Sol. 9, 713 (1972).
40. J. B. Sokoloff, Phys. Rev. 5, 4496 (1972).

CHARGE ORDERING AND LATTICE INSTABILITY IN MAGNETITE

Y. Yamada*
IBM Thomas J. Watson Research Center
Yorktown Heights, New York 10598 USA

ABSTRACT

We propose a new type charge ordering scheme in Fe_3O_4 which is basically different from that postulated by Verwey. Recent observation of neutron critical scattering[1] above the 119°K transition has revealed that the internal lattice mode with wave vector 1/2 C* of the cubic reciprocal lattice becomes unstable at the transition. This suggests that the electron-phonon interaction is playing an important role to initiate this phase transition, and that the transition would be understood as the condensation of a coupled charge density-phonon mode. Based on this standpoint, the behavior of the phonons and the charge density waves along [00ξ] direction (Δ-points) and of the neutron scattering cross sections due to these fluctuations is studied. The analysis of the observed intensity distributions of the critical scattering shows that the condensing phonon mode belongs to the two dimensional irreducible representation Δ_5. When combined with the strong electron-phonon interaction, this leads to the condensation of the charge density wave with the same symmetry Δ_5, which is basically different from Verwey's model. The scattered neutron spectra are also discussed in connection with the hopping relaxation time of the electrons.

INTRODUCTION

The 119°K (T_v) phase transition of magnetite, Fe_3O_4, has been the subject of extensive investigations in connection with charge ordering as originally proposed by Verwey.[2] Recent observations on this phase transition by electron diffraction,[3] neutron diffraction,[4] Mössbauer effect,[5,6] and magnetic resonance[6,7] cast doubt on the validity of the Verwey type charge ordering scheme.

The electron diffraction pattern[2] exhibits some super structure lines indicating that the unit cell in the low temperature phase is doubled along the cubic principal axis. Samuelson et al[4] also confirmed the super structure by neutron diffraction.

Moreover, they concluded that the doubling of the unit cell is mainly caused by small displacements of atoms.

This strongly suggests that some internal lattice mode, rather than uniform bulk distortion, is associated with this phase transition. Using the prevailing concept of the lattice instability at the structural phase transition, we may consider that the unstable (soft) phonon mode will exist above T_v, and that below T_v this mode "condenses" giving rise to the static displacement field which corresponds to the eigenvectors of the soft mode.

When combined with the existence of the charge ordering below T_v, these considerations naturally lead to the assumption that the interaction between electrons and phonons is playing the important role in this phase transition. Cullen and Callen[8] developed a theory of Verwey transition based on electron-electron correlation interaction. As such it does not explicitly explain the lattice deformation coexisting with the charge ordering. The importance of the electron-phonon interaction has been already pointed out by several authors[9,10,11,12,13] in connection with the uniform bulk distortion from the cubic to the orthorhombic system.

On the contrary, our stand point is that the phonon mode which is primarily associated with the phase transition is the internal mode with the wave length twice the cubic unit cell dimension, and the uniform deformation is considered as an induced effect.

The important implication of the above assumption is that the charge density fluctuation which is coupled to the condensing phonon mode due to the electron-phonon interaction is also an "internal" mode modulated by the same wave length as the phonon mode. This possibility of a modulated charge ordering has been pointed out by Sokoloff.[10] Mössbauer and NMR observations which revealed multiple internal fields within the octahedral sites can be easily understood by assuming a modulated charge density wave.

In this paper, we try to understand the 119°K transition in Fe_3O_4 on the basis that this phase transition is viewed as the condensation of a coupled charge density-phonon mode with short wavelength.

The leading electron-phonon coupling is assumed to be bilinear with respect to the charge-density and the phonon amplitude. This implies a close relation between the present problem and the cooperative Jahn-Teller phase transition. In fact, it will be shown that the effective Hamiltonian to describe our system is formally equivalent to that formulated for the Jahn-Teller system.

The most direct experimental evidence for the existence of the postulated coupled mode with short wavelength will be given by neutron scattering measurements. In the particular case of Fe_3O_4, where spin degeneracy of 3d-electrons is lifted due to magnetic ordering,[+] the neutron scattering serves as a very powerful probe because both the phonons and the charge (spin) density modes can be detected as the nuclear scattering and the magnetic scattering respectively.

Recently, an extensive observation of the neutron critical scattering above T_v has been carried out by Fujii and Shirane[1] at Brookhaven National Laboratory. They have observed strong critical scatterings due to large amplitude atomic fluctuations with short wavelengths.

From the analysis of their experimental results based on the present model, we conclude that the condensing charge density mode, and hence the static charge ordering scheme is basically different from that proposed by Verwey. We postulate a new type charge ordering scheme.

Another important result of their experiment is that the neutron spectra of the critical scattering showed a "central peak", implying the large atomic fluctuations are mainly a diffusive mode. These dynamical aspects will also be discussed in connection with the electron hopping relaxation time.

As for the electronic state of the 3d-electrons on the octahedral sites, we may take either the localized electron picture or the band picture. In this paper, we take the former standpoint, assuming that the coupling energy between electrons and phonons is greater than the electron transfer energy. It will be shown that the opposite extreme case, namely the wide band width case, is inconsistent with the results of the neutron experiments.

In the course of the analysis of the phonon modes, the symmetry properties of the structure of Fe_3O_4 have

* On leave from Osaka University, Toyonaka, Osaka, Japan.

+ $T_c \approx 850°K$

been effectively utilized. To facilitate the description of the phonon properties such as eigenvectors, the structure data of the cubic phase are summarized in the Appendix.

SYMMETRY PROPERTY OF THE PHONON FIELD AND THE CHARGE DENSITY FIELD

The observed neutron critical scattering has shown strong peaks at [H,0,L ± 1/2] reciprocal lattice points accompanied by ridge-like streaks running along [001] direction.[1] This suggests that the relevant phonon modes which become unstable at the phase transition have the wave vector $\vec{k} = (00\xi)$ (specified as Δ-points in the reciprocal space). Especially, the mode with $\vec{k} = (00\ 1/2)$ will become condensed in the low temperature phase as is confirmed by the electron diffraction and the neutron diffraction measurements. Therefore we begin with examining the symmetry property of the phonon field as well as the charge density field at Δ-points in the cubic reciprocal space.

Phonon Field

The little group of $\vec{k} = (00\xi)$ of space group O_h^7 has five irreducible representations denoted by Δ_1, Δ_2, Δ_3, Δ_4 and Δ_5. Among these, Δ_5 is a two dimensional representation and the others are all one dimensional, Δ_1 being the identity representation. The 42 phonon branches are characterized by the following symmetry properties,

$$\Delta_\nu = 7\Delta_1 + 3\Delta_2 + 3\Delta_3 + 7\Delta_4 + 11\Delta_5. \tag{1}$$

Among these phonon branches, the Δ_4-modes and the Δ_5-modes are particularly important for the later discussions. The type of the eigenvectors, or the pattern of the atomic displacements of these two types of modes are listed in Table I. It should be pointed out that all the singlet modes share the common feature in the pattern of displacement in that the atom pair which is related by two fold axis along [001] (for instance, oxygen (3) and oxygen (4), Fe (11) and Fe (12), etc.) moves out of phase, while in the Δ_5-mode, they move in phase as is seen in Fig. 3. It should be also mentioned that for an arbitrary value of $\vec{k} = (00\xi)$, the atomic displacement is modulated along [001] direction with the corresponding wave length a/ξ. Thus, for the particular case of $\vec{k} = (00\ 1/2)$, the wavelength is twice the cubic unit cell dimension along the c-axis.

Charge Density Field

The electrons which are associated with the 119°K phase transition are a pair of excess 3d-electrons coming from two Fe^{2+} ions per primitive cell located at octahedral sites. Since the magnetic ordering has been attained at this temperature, the spin degeneracy is already lifted.[14] Therefore, there are four orbitals per primitive cell to accommodate these two electrons.

As for the electronic states of these electrons above T_v, there is evidence[15,16,17] that these electrons are localized at each site rather than forming bands, thus giving rise to a dynamic disorder in charge density at the octahedral sites. We take this standpoint and consider the symmetry property of the localized charge density fluctuations. The comparison with the results based on the band model will be discussed in the last section.

We define $\Delta\rho_{\ell m}$ as the deviation of the charge density from the "averaged" density, $\bar{\rho} = 1/2$, at the m'th octahedral site in the ℓ'th primitive unit cell.

TABLE 1. The displacement of ions for the phonon modes with the symmetry Δ_4 and Δ_5. The numbers in the parenthesis following the parameters of the Δ_5-mode are the displacements used for the tentative calculation of F_d^Δs.

mode	Δ_4			$\Delta_5^{(1)}$			$\Delta_5^{(2)}$		
atom	x	y	z*	x	y	z	x	y	z
1	0	0	w_1	u_1	$-u_1$	0	u_2	u_2	0
2	0	0	w_1	u_2	$-u_2$	0	u_1	u_1	0
3	u_3	$-u_3$	w_3	u_3	$-u_3$	w_3	u_7	u_7	0
4	$-u_3$	u_3	w_3	u_3	$-u_3$	$-w_3$	u_7	u_7	0
5	u_5	u_5	w_5	u_5	$-u_5$	0	u_9	u_9	$-w_9$
6	$-u_5$	$-u_5$	w_5	u_5	$-u_5$	0	u_9	u_9	w_9
7	u_3	u_3	$-w_3$	u_7	$-u_7$	0	u_3	u_3	$-w_3$
8	$-u_3$	$-u_3$	$-w_3$	u_7	$-u_7$	0	u_3	u_3	w_3
9	$+u_5$	$-u_5$	$-w_5$	u_9	$-u_9$	w_9	u_5	u_5	0
10	$-u_5$	$+u_5$	$-w_5$	u_9	$-u_9$	$-w_9$	u_5	u_5	0
11	u_{11}	$-u_{11}$	w_{11}	u_{11}	$-u_{11}$	w_{11}	u_{13}	u_{13}	0
12	$-u_{11}$	u_{11}	$+w_{11}$	u_{11}	$-u_{11}$	$-w_{11}$	u_{13}	u_{13}	0
13	u_{11}	u_{11}	$-w_{11}$	u_{13}	$-u_{13}$	0	u_{11}	u_{11}	$-w_{11}$
14	$-u_{11}$	$-u_{11}$	$-w_{11}$	u_{13}	$-u_{13}$	0	u_{11}	u_{11}	w_{11}

* z is taken along the direction of $\vec{k}$. Independent parameters for the Δ_4-mode: u_3, u_5, u_{11}, w_1, w_3, w_5, w_{11}. Independent parameters for the Δ_5-modes: $u_1(-0.5)$, $u_2(0.5)$, $u_3(0.5)$, $u_5(0.5)$, $u_7(1.0)$, $u_9(1.0)$, $u_{11}(0.25)$, $u_{13}(0.25)$, $w_3(0.5)$, $w_9(0.5)$, $w_{11}(0.25)$. The phase factors associated with wave vector $\vec{k}$ are omitted in the Table.

The charge density wave with the wave vector $\vec{k}$ are then defined as

$$\Delta\rho_{(\vec{k},m)} = \frac{1}{\sqrt{N}} \sum_{\ell,m} \Delta\rho_{\ell,m} e^{i\vec{k}\cdot\vec{r}_\ell}, \tag{2}$$

$$m = 1,\ 2,\ 3,\ 4$$

where $\vec{r}_\ell$ is the position vector of the origin of the ℓ'th primitive cell, and m = 1, 2, 3, 4 specifies the octahedral site occupied by Fe (11), Fe (12), Fe (13), and Fe (14) respectively. (See Fig. A-1.) N is the number of the unit cell. The scalar field of $\Delta\rho_{(\vec{k},m)}$'s will also be symmetrized so that the linear combination of $\Delta\rho_{(\vec{k},m)}$'s will form the basis of the irreducible representations of space group O_h^7.

It is shown that the four scalar quantities associated with each octahedral sites is symmetrized as follows,

$$\Delta_s = \Delta_1 + \Delta_4 + \Delta_5, \tag{3}$$

and the basis vectors of each representation are explicitly given by,

$$\Delta_1: \ \Delta\rho_{(\vec{k},1)} + \Delta\rho_{(\vec{k},2)} + \Delta\rho_{(\vec{k},3)} + \Delta\rho_{(\vec{k},4)},$$

$$\Delta_4: \ (\Delta\rho_{(\vec{k},1)} + \Delta\rho_{(\vec{k},2)}) - (\Delta\rho_{(\vec{k},3)} + \Delta\rho_{(\vec{k},4)}),$$

$$\Delta_5: \begin{cases} \Delta\rho_{(\vec{k},1)} - \Delta\rho_{(\vec{k},2)}, \ (\Delta_5^{(1)}), \\ \Delta\rho_{(\vec{k},3)} - \Delta\rho_{(\vec{k},4)}, \ (\Delta_5^{(2)}). \end{cases} \quad (4)$$

Among these, a large fluctuation with symmetry Δ_1 will be ruled out because in such fluctuations, especially with long wavelengths, the charge density piles up within the unit cell and breaks the local charge neutrality condition (Anderson's condition). Therefore we are left with the charge density modes with the symmetry Δ_4 and Δ_5.

It is noticeable that the charge density field with the symmetry Δ_4 corresponds to Verwey ordering except that it is modulated with an arbitrary wave vector $\vec{k} = (00\xi)$. (See Fig. 1-a.) For the uniform mode, it exactly corresponds to the model proposed by Verwey. On the other hand, the Δ_5-modes give new charge ordering schemes. In one of the degenerate Δ_5-modes, Fe^{2+}-Fe^{3+} charge ordering is attained between Fe (11) and Fe (12) while Fe (13) and Fe (14) remain disordered. The other Δ_5-mode, $\Delta_5^{(2)}$, in turn gives the charge ordering within Fe (13) and Fe (14) leaving Fe (11) and Fe (12) disordered. (See Fig. 1-b.)

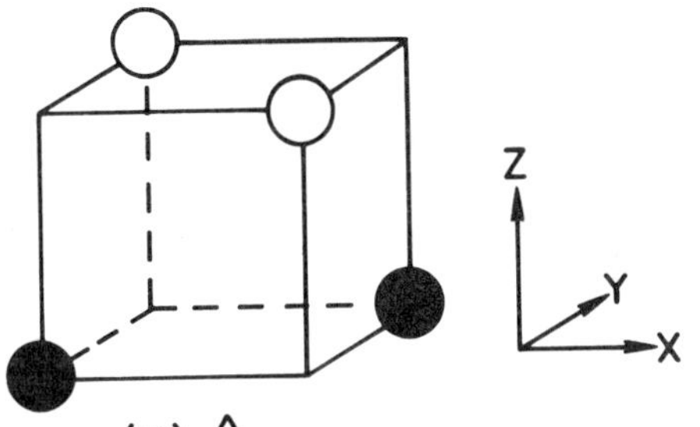

(a) Δ_4

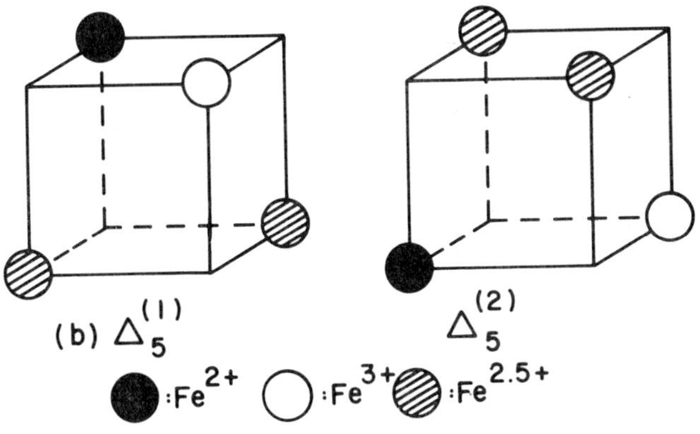

Figure 1-a. The charge density mode with the symmetry Δ_4. For an arbitrary value of k, this pattern is modulated by the corresponding wavelength. The uniform mode (k = 0) corresponds to Verwey ordering.

Figure 1-b. The charge density mode with the symmetry Δ_5. The charge ordering is attained in one of the atomic planes perpendicular to k, while the other is left disordered.

MODEL HAMILTONIAN AND THE NEUTRON SCATTERING CROSS SECTION

The experimental results suggest that both the charge density and the amplitude of the atomic displacement are playing the role of the order parameter in the low temperature phase. Whence, it is reasonable to assume that the bilinear coupling between the phonon field and the charge density field is important in the ordering process. We will retain only the lowest order bilinear coupling term hereafter.

The interaction Hamiltonian, $\mathscr{H}_{ep}$, is then essentially the same as that used to establish small polaron model[18,19] except that we explicitly specify four equivalent sites in the primitive cell;

$$\mathscr{H}_{ep} = \sum_{\substack{k,s \\ \ell,m}} \frac{g_{\vec{k},s}}{\sqrt{N}} (Q_{\vec{k},s} + Q^*_{\vec{k},s}) \, C^+_{\ell m} C_{\ell m} e^{i\vec{k}\cdot\vec{r}_\ell}, \quad (5)$$

where $C^+_{\ell m}$ and $C_{\ell m}$ are the creation and the annihilation operators of the electron localized at site (ℓ, m), while $Q_{\vec{k},s}$ and $Q^*_{\vec{k},s}$ refer to the amplitude of the phonons specified by the wave vector $\vec{k}$ and branch s, with the characteristic frequency $\omega_{k,s}$. To be consistent with the description in the preceding section, we redefine the local charge density fluctuation operator $\Delta\rho_{\ell m}$ as

$$\Delta\rho_{\ell m} = -\frac{1}{2} + C^+_{\ell m} C_{\ell m}. \quad (6)$$

Then the effective coupling energy will be expressed as

$$\mathscr{H}_{ep} = \sum_{k,s} \sum_{m} \frac{g_{\vec{k},s}}{\sqrt{N}} (Q_{\vec{k},s} + Q^*_{\vec{k},s}) \, \Delta\rho_{(\vec{k},m)}, \quad (7)$$

where $\Delta\rho_{(\vec{k},m)}$ is the spacial Fourier transform of $\Delta\rho_{\ell m}$.

Since the Hamiltonian itself should be totally symmetric, the charge density wave which couples to a particular phonon mode should belong to the same irreducible representation of the group of $\vec{k}$ as does the phonon mode. Making use of the results obtained in the preceeding section, we obtain the following expressions when $\vec{k} = [00\xi]$,

$$_{ep} = \sum_{k,s} \frac{g_{\vec{k},s}}{\sqrt{N}} (Q_{\vec{k},s} + Q_{\vec{k},s}) \, \Delta\rho_{k,s}, \quad (8)$$

$$s: \ \Delta_4, \Delta_5.$$

Here, s runs only the modes belonging to Δ_4 and Δ_5. The charge density operators with symmetry Δ_4 and Δ_5 are explicitly given by

$$\Delta_4: \ \rho_{\vec{k},\Delta_4} = C^+_{(\vec{k},1)} C_{(\vec{k},1)} + C^+_{(\vec{k},2)} C_{(\vec{k},2)} - (C^+_{(k,3)} C_{(\vec{k},3)} + C^+_{(\vec{k},4)} C_{(\vec{k},4)}),$$

$$\Delta_5: \begin{cases} \rho_{\vec{k},\Delta_5^{(1)}} = C^+_{(\vec{k},1)} C_{(\vec{k},1)} - C^+_{(\vec{k},2)} C_{(\vec{k},2)}, \\ \rho_{\vec{k},\Delta_5^{(2)}} = C^+_{(\vec{k},3)} C_{(\vec{k},3)} - C^+_{(\vec{k},4)} C_{(\vec{k},4)}. \end{cases} \quad (9)$$

Only the phonon modes belonging to Δ_4 and Δ_5 will couple to these two charge density waves respectively.

It is convenient to define three independent classical Ising operators to specify those symmetrized charge density waves as follows:

$\sigma_\ell^{(1)}$ is associated with $\Delta\rho_\ell, \Delta_5^{(1)}$ by

$$\sigma_\ell^{(1)} = \begin{cases} +1, \text{ when site (11) is } Fe^{2+} \text{ and site (12) is } Fe^{3+} \\ -1, \text{ when site (11) is } Fe^{3+} \text{ and site (12) is } Fe^{2+} \end{cases}$$

$\sigma_\ell^{(2)}$ is associated with $\Delta\rho_\ell, \Delta_5^{(2)}$ by

$$\sigma_\ell^{(2)} = \begin{cases} +1, \text{ when site (13) is } Fe^{2+} \text{ and site (14) is } Fe^{3+} \\ -1, \text{ when site (13) is } Fe^{3+} \text{ and site (14) is } Fe^{2+}, \end{cases}$$

$\sigma_\ell^{(3)}$ is associated with $\Delta\rho_\ell, \Delta_4$ by

$$\sigma_\ell^{(3)} = \begin{cases} +1, \text{ when site (11) and site (12) are } Fe^{2+} \\ \quad \text{and site (13) and site (14) are } Fe^{3+}, \\ -1, \text{ when site (11) and site (12) are } Fe^{3+} \\ \quad \text{and site (13) and site (14) are } Fe^{2+}. \end{cases}$$

Then the system is described by a coupled pseudospin-phonon system with the effective interaction between pseudospins and phonons given by

$$\mathcal{H}_{ep} = \sum_k g_{\vec{k},\Delta_4} (Q_{\vec{k},\Delta_4} + Q^*_{\vec{k},\Delta_4}) \, \sigma^{(3)}_{(\vec{k})}$$
$$+ \sum_{\vec{k}} g_{\vec{k},\Delta_5} \{ (Q_{\vec{k},\Delta_5}(1) + Q^*_{\vec{k},\Delta_5}(1)) \, \sigma^{(1)}_{(\vec{k})}$$
$$+ (Q_{\vec{k},\Delta_5}(2) + Q^*_{\vec{k},\Delta_5}(2)) \, \sigma^{(2)}_{(\vec{k})} \}. \quad (10)$$

The property of such a coupled system is well known in connection with cooperative Jahn-Teller effect etc.[20,21,22,23] That is, by a suitable choice of the coupling parameter $g_{\vec{k},s}$, the system undergoes a phase transition, below which the ordering of pseudospins (the charge density, in our case) and the condensation of the phonons take place at the same time. Moreover, large amplitude fluctuations of the charge densities as well as of the atomic displacements are predicted in the critical region of the phase transition.

Let us consider the effect of these critical fluctuations on the neutron scattering cross sections. This observable quantity is particularly interesting for the following reason. As previously stated, in the temperature region in question the spin degeneracy of the 3d electrons at Fe-sites is already lifted due to magnetic ordering. Hence, the charge density fluctuations are always accompanied by the equivalent fluctuations of the magnetic moments. Therefore, the atomic and the charge density fluctuations are, in principle, detectable by neutrons both in nuclear and the magnetic diffuse scattering. These scattering cross sections are expressed as

$$\left(\frac{d^2\sigma}{d\Omega d\omega}\right)_{nuc} = N \sum_s |F_d^s(\vec{k})| \, \phi_{QQ} (\vec{k},\omega), \quad (11)$$

$$\left(\frac{d^2\sigma}{d\Omega d\omega}\right)_{mag} = N \left(\frac{e^2\gamma}{mc^2}\right)^2 \sum_s F_m^s(\vec{K}) \, \phi^s_{\sigma\sigma} (\vec{k},\omega), \quad (12)$$

$$\vec{K} = \vec{K}_h + \vec{k},$$

where $F_d(\vec{K})$ is the dynamical structure factor associated with the condensing phonon mode and $F_m^s(\vec{K})$ is the dynamical magnetic structure factor of the charge density wave with symmetry s. $\vec{K}$ is the scattering vector and K_h is the reciprocal lattice vector. The quantities $\phi^s_{QQ}(\vec{k},\omega)$ and $\phi^s_{\sigma\sigma}(\vec{k},\omega)$ are the spectral densities defined by

$$\phi^s_{QQ}(\vec{k},\omega) = \int \langle Q(\vec{k},s,0)Q(-\vec{k},s,t)\rangle \, e^{i\omega t} dt. \quad (13)$$

$$\phi^s_{\sigma\sigma}(\vec{k},\omega) = \int \langle \sigma^s(\vec{k},o)\sigma^s(-\vec{k},t)\rangle \, e^{i\omega t} dt. \quad (14)$$

The property of these spectral densities, especially of $\phi^s_{QQ}(\vec{k},\omega)$, will be discussed in the last section.

We will momentarily concentrate on the dynamical structure factor $F_d^s(\vec{K})$. It is well known that $F_d^s(\vec{K})$ is explicitly given in terms of the phonon eigenvectors as follows,

$$F_d^s(\vec{K}) = \sum_j \frac{b_j}{\sqrt{M_j}} (\vec{K}\cdot\vec{e}^{\,j}_{k,s}) \, e^{i\vec{K}\cdot\vec{r}_j}, \quad (15)$$

where $\vec{e}^{\,j}_{k,s}$ is the eigenvector and b_j is the nuclear scattering length of the j'th atom. By comparing the observed diffuse scattering intensities around various reciprocal lattice points (H,0,L $\pm$ 1/2), we can obtain information on the eigenvectors of the condensing modes. Particularly important is that since we know the basic difference between the eigenvectors of the Δ_4 modes and of the Δ_5 modes, we should be able to determine the symmetry of the condensing phonon mode, and hence the type of the charge ordering which will be stabilized in the low temperature phase.

CONDENSING PHONON MODE AND THE CHARGE ORDERING SCHEME

The atomic displacements for the Δ_4- and the Δ_5-phonon modes are tabulated in Table I. The displacements of the Δ_4 mode contains seven independent parameters while the Δ_5 modes contains eleven parameters in accordance with the number of respective branches. In spite of the large number of parameters to be determined, the following characteristic of the eigenvectors help decisively to determine which is the condensing mode. In the Δ_4-mode the atom pairs which are related by the symmetry operation of two fold axis along the z-direction move out of phase, while in the Δ_5 mode these atom pairs move together. This gives rise to a systematic tendency in the neutron scattering pattern for the Δ_4 mode to give strong intensities at the lattice point (H,0,L $\pm$ 1/2) with H,L = 4n $\pm$ 2, while the Δ_5-modes give strong intensities at H,L = 4n. On the other hand, the observed intensity distribution of the critical scattering shows that the indices giving stronger intensities are (8,0 $\pm$ 1/2), 8,0,4 $\pm$ 1/2) (4,0, $\pm$ 1/2), (4,0,4 $\pm$ 1/2), and (4,0,8 $\pm$ 1/2). This fact definitely indicates that the condensing mode belongs to Δ_5. In fact, even by a tentative set of 11 parameters given in Table I, we obtain fairly good agreement between $|F_d|^2_{obs}$ and $|F_d|^2_{cal}$ as is shown in Fig. 2. Once the symmetry of the condensing phonon modes is established as Δ_5, the charge density waves which couple with these phonon modes are determined as $\Delta\rho(\vec{k},\Delta_5^{(1)})$ and $\Delta\rho(\vec{k},\Delta_5^{(2)})$. Namely, it is expected that in the critical temperature region, there exists a coupled charge density - atomic displacement fluctuation whose pattern is depicted in Fig. 3-b. In principle, these charge density fluctuations will give rise to the magnetic critical scattering of neutrons. The magnetic dynamical structure factor of the $\Delta_5^{(1)}$ type charge density wave is given by

$$F_m^{\Delta_5^{(1)}}(\vec{K}) = \Delta S_\perp f(K)(e^{i\vec{K}\cdot\vec{r}_{11}} - e^{i\vec{K}\cdot\vec{r}_{12}}), \quad (16)$$

where $\Delta s_\perp$ is the amplitude of the spin fluctuations perpendicular to $\vec{K}$; f(K) is the magnetic form factor. Finite scattering cross sections are expected at [H,0,L ± 1/2] with H = 4n + 2. However, no appreciable intensities are experimentally observed even at the optimum lattice point of (2,0 ± 1/2).

Nevertheless, the important implication of the Δ_5-phonon condensation is that at least in the low temperature phase, the static charge ordering with symmetry Δ_5 will be attained accompanied by the lattice distortion with the same symmetry property. This charge ordering scheme is quite different from the currently accepted Verwey ordering, which belongs to the Δ_4 symmetry. (See Fig. 3-a.) We may characterize the Δ_4-type and the Δ_5-type charge ordering scheme as the "longitudinal" and the "transverse" ordering respectively, because in the former, the dipole moment produced by the displacement of charges from the disordered state is parallel to the direction of the modulation ([001]), while in the latter, it is perpendicular.

As for the static charge ordering, in the low temperature phase, a particular complication arises from the fact that the condensing mode is doubly degenerate. Any linear combination of the $\Delta_5^{(1)}$ and the $\Delta_5^{(2)}$ modes is equally probable within our approximation. Namely, the relative phase difference as well as the relative amplitude between these two modes should be known in order to determine the low temperature charge ordering uniquely. These interesting aspects are outside our scope at the present stage where only the experimental data above the transition is available. A neutron scattering experiment in the low temperature phase is certainly very important to confirm the postulated new type charge ordering scheme.

DISCUSSION

There are 11 phonon branches with the symmetry Δ_5. Among these, we infer that the lowest lying Δ_5 mode becomes dominantly unstable at the transition, partly because the transition temperature is given by

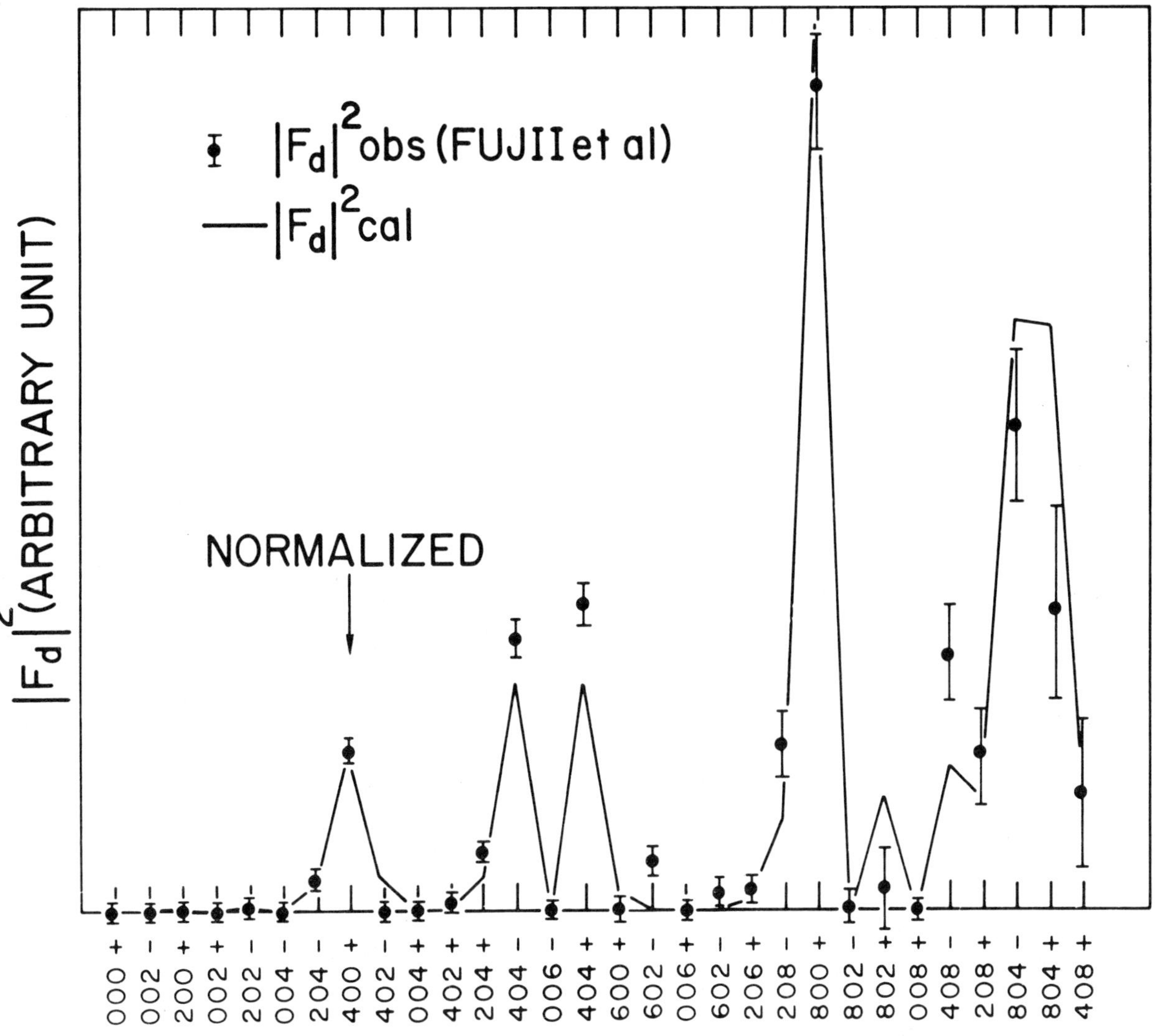

Figure 2. The comparison of the observed relative intensities at various reciprocal lattice points, (H,0,L ± 1/2), and the calculated values, $\frac{1}{2}\{|F_d^{\Delta_5^{(1)}}|^2 + |F_d^{\Delta_5^{(2)}}|^2\}$, where the condensing mode is taken as the Δ_5-mode. The numbers followed by "+" or "−" means the indices $(H,0,L+\frac{1}{2})$ and $(H,0,L-\frac{1}{2})$ respectively.

$$kT_v = |g_{k,s}|^2/\omega^2_{k,s}. \qquad (17)$$

The transverse acoustic branch is identified as the lowest Δ_5-mode. The temperature dependence of the frequencies of this branch has been carefully observed to check the softening of the frequencies around

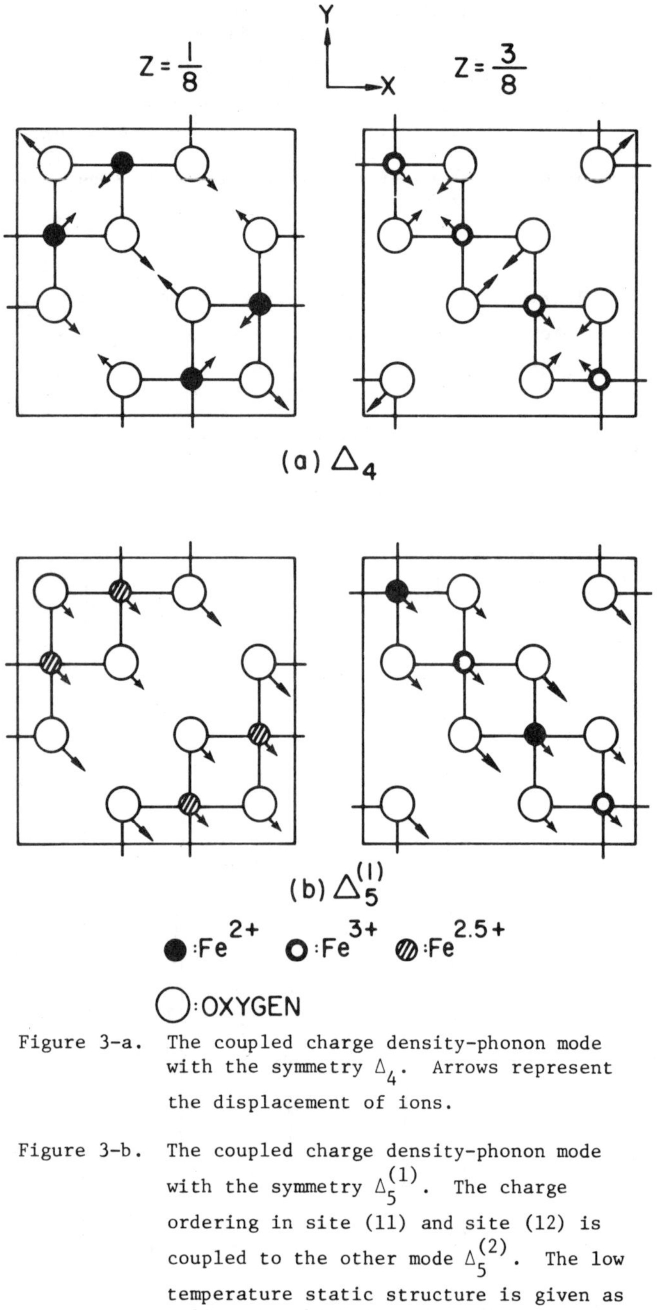

Figure 3-a. The coupled charge density-phonon mode with the symmetry Δ_4. Arrows represent the displacement of ions.

Figure 3-b. The coupled charge density-phonon mode with the symmetry $\Delta_5^{(1)}$. The charge ordering in site (11) and site (12) is coupled to the other mode $\Delta_5^{(2)}$. The low temperature static structure is given as a linear combination of these two degenerate modes.

$k \sim (00\ 1/2)$. No softening or anomalies in the line shape has been observed.[1] Instead, the large response due to critical fluctuations is concentrated at $\omega = 0$, (that is the spectrum shows the central peak), implying that the critical fluctuation is a relaxational mode.

Recently, Yamada et al[24] studied dynamical behavior of the coupled pseudospin-phonon system which is described by the same Hamiltonian as Eq. (10). They showed that the characteristic of the correlation functions $\phi^s_{QQ}(\vec{k},\omega)$ and $\phi^s_{\sigma\sigma}(\vec{k},\omega)$, exhibits wide variation depending on the relative magnitude of $\omega_{\vec{k},s}$ and τ_σ, the relaxation time of the flipping motion of the pseudospin. Namely, in the limit of $\tau_\sigma\omega_{\vec{k}s} \ll 1$ (the fast relaxation case), the spectrum of $\phi^s_{QQ}(\vec{k},\omega)$ shows critical softening, while in the opposite case, $\tau_s\omega_{\vec{k},s} \gg 1$ (the slow relaxation case), the spectrum gives "three peak" structure and only the central (relaxation) component shows critical behavior leaving phonon side bands unchanged.

When compared with the observed spectra of $(d^2\sigma/d\Omega d\omega)_{nuc}$ which is proportional to $\phi_{QQ}(\vec{k},\omega)$, we conclude that the present case falls into the category of the "slow relaxation" case. In our case, the pseudospin operator corresponds to the electron configurations as is illustrated in the previous section. Accordingly, "flipping" of the pseudospin from the state of $\sigma = 1$ to that of $\sigma = -1$ simply means electron hopping from one site to the neighboring site. In principle, the relaxation time of the hopping motion can be estimated from the width of the central peak of the observed neutron spectra. Unfortunately, however, the width due to the experimental resolution exceeds the intrinsic width due to dynamical behavior of the fluctuations. Hence, we can only set the minimum value of τ_σ as

$$\tau_\sigma > 10^{-10}\text{sec.}$$

From Mössbauer measurement, Kündig and Hargrove[13] gave the value for the electron hopping relaxation time at room temperature as

$$\tau = (1.1 \pm 0.2) \times 10^{-9}\text{sec.}$$

The above lower limit on τ_σ is at least consistent with their results.

So far, we have treated the electrons as localized at each octahedral sites, assuming that the electron-phonon interaction energy exceeds the energy associated with electron transfer. The transport properties of Fe_3O_4 as well as other transition metal oxides seem to be consistent with this picture.[25]

It is interesting, however, to examine the results based on the other picture. Namely, the electrons are considered as itinerant whose band width and band gap are far wider than the electron-phonon coupling energy. Starting from this electronic state, we can also discuss the phase transition based on the band Jahn-Teller effect. It is well known that when there is a non-filled degenerate band, the lattice tends to distort spontaneously to split the band (Peierls instability of the lattice). It is easily shown that there is in fact a doublet band along the [001] direction, which may, by accommodating only N-electrons, be a suitable band scheme. Therefore, we assume this doublet band is responsible for the phase transition.

Following a similar procedure to the localized electron case described in the preceeding section, we arrive at the effective Hamiltonian for this case given by

$$\mathscr{H}_{ep} = \sum_{k,\ell} \frac{g_{\vec{k},\Delta_4}}{\sqrt{N}} (Q_{\vec{k},\Delta_4} + Q^*_{\vec{k},\Delta_4})\, \sigma^z_\ell\, e^{ik\cdot r_\ell}$$

$$+ \sum_{k,\ell} \frac{g_{\vec{k},\Delta_3}}{\sqrt{N}} (Q_{\vec{k},\Delta_3} + Q_{\vec{k},\Delta_3})\, \sigma^x_\ell\, e^{ik\cdot r_\ell}, \qquad (18)$$

where σ^z_ℓ and σ^x_ℓ are Pauli spin operators associated with ℓ'th unit cell.

It is interesting to see that in this case the symmetry of the condensing mode does not include Δ_5. The difference arises simply from the fact that we have only taken two particular electronic states forming the double degenerate band, neglecting all the interband interaction terms. At least within the framework described above we conclude that the band model does not give the correct condensing mode as is observed by the experiment.

APPENDIX

The crystal structure data in the cubic phase (above T_v) is summarized in the following:

Space group: O_h^7 - Fd3m

Lattice constant: a = 8.394 (300°K)

TABLE A-1. Equivalent sites, site symmetries, ionic species, ion labels, and coordinates, referred to the cubic cell, of ions in the primitive cell.

Equivalent Position	Site Symmetry	Ion	Ion Label	x,	y,	z
8a	$\bar{4}3m$	Fe^{3+}	1	0,	0,	0,
			2	1/4,	1/4,	1/4
32e	3m	O^{2-}	3	1/4-u,	1/4+u,	1/4+u
			4	1/4+u,	1/4-u,	1/4+u
			5	1/4-u,	1/4-u,	1/4-u
			6	1/4+u,	1/4+u,	1/4-u
			7	u,	u,	u
			8	-u,	-u,	u
			9	u,	-u,	-u
			10	-u,	u,	-u
16d	$\bar{3}2/m$	$Fe^{2.5+}$	11	-1/8,	-3/8,	-1/8
			12	-3/8,	-1/8,	-1/8
			13	-1/8,	-1/8,	-3/8
			14	-3/8,	-3/8,	-3/8

$u = 3/8 + \delta$

Ideal Spinel: $\delta = 0$ Fe_3O_4: $\delta = 0.004$

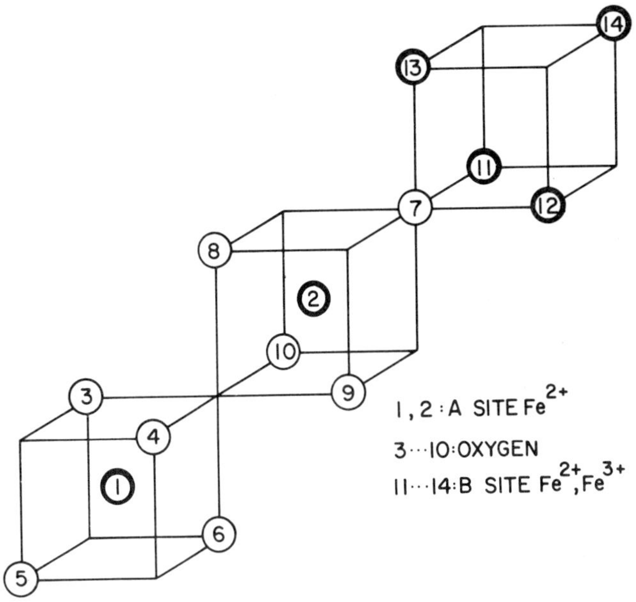

Figure A-1. The arrangement of ions in the primitive cell. The numbers represent the ion labels sited in the column 4 of Table 4.

The ionic labels in the column 4 of Table A-1 are used throughout this paper to specify the positions of the ions. The following figure provides the relationship between ion labels and the spacial arrangement of the ions.

ACKNOWLEDGEMENTS

The author wishes to express his thanks to G. Shirane and Y. Fujii for showing their experimental data prior to publication. He is indebted to S. LaPlaca for his help in the process of data analysis. He also expresses his thanks to E. Pytte, J. B. Torrance, M. Blume and J. D. Axe for their illuminating discussions.

REFERENCES

1. Y. Fujii, G. Shirane, and Y. Yamada, to be published.
2. E. J. Verwey and P. W. Haayman, Physica 8, 979 (1941); E. J. Verwey, P. W. Haayman, and F. C. Romeijn, J. Chem. Phys. 15, 181 (1947).
3. T. Yamada, K. Suzuki, and S. Chikazumi, Appl. Phys. Lett. 13, 172 (1968).
4. J. Samuelson, E. J. Bleeker, L. Doborzynski, and T. Riste, J. Appl. Phys. 39, 1114 (1968).
5. R. S. Hargrove and W. Kundig, Solid State Commun. 8, 303 (1970).
6. M. Rubinstein, and D. W. Forester, Solid State Commun. 9, 1675 (1971).
7. M. Rubinstein, G. H. Strauss, and F. J. Bruni, AIP Conf. Proceedings No. 10, Magnetism and Magnetic Materials 18th Annual Conf. - Denver, 1972, p. 1384.
8. J. R. Cullen, and E. Callen, J. Appl. Phys. 41, 879 (1970).
9. B. K. Chakraverty, Solid State Commun. 15, 1271 (1974).
10. J. Sokoloff, Phys. Rev. B 5, 4496 (1972).
11. J. L. Verble, private communication.
12. I. G. Austinand and N. F. Mott, Adv. Phys. 18, 42 (1969).
13. J. B. Goodenough, Prog. Solid State Chem. 5, 308 (1971).
14. K. Yoshida and M. Tachiki, Progress Theor. Phys. (Kyoto) 17, 331 (1957).
15. W. Kündig and R. S. Hargrove, Solid State Commun. 7, 223 (1969).
16. A. A. Samokhvalov, N. M. Tutikov, and G. P. Skovnyakov, Soviet Phys. Solid State 10, 2172 (1969).
17. D. L. Camphausen, Solid State Commun. 11, 99 (1972).
18. H. G. Reik, Phys. Lett. 5, 236 (1963).
19. H. G. Reik and D. Heese, J. Phys. Chem. Solids 28, 581 (1967).
20. J. Kanamori, J. Appl. Phys. 39, 688 (1968).
21. E. Pytte, Phys. Rev. B 3, 3503 (1971).
22. K. K. Kobayashi, J. Phys. Soc. Japan 24, 497 (1968).
23. J. Feder and E. Pytte, Phys. Rev. B 8, 3978 (1973).
24. Y. Yamada, H. Takatera, and D. L. Huber, J. Phys. Soc. Japan 36, 641 (1974).
25. J. Yamashita and T. Kurosawa, J. Phys. Chem. Solids 5, 34 (1958).

Fe_3O_4 VERWEY TRANSITION. CALORIMETRIC DETERMINATION OF THE EFFECTS OF CONCENTRATION OF Zn AND Cd DOPANTS*

James J. Bartel and Edgar F. Westrum, Jr., Department of Chemistry, University of Michigan, Ann Arbor, Michigan 48104.

ABSTRACT

To determine the effect of increase of the lattice constant and decrease in electron concentration on the Verwey transition in Fe_3O_4, heat capacity measurements have been made on Zn- and Cd-doped Fe_3O_4 from 5 to 350 K. Bifurcation of the anomaly depends sensitively on the nature and concentration of dopant. Although high dopant levels (e.g., Cd(0.010)) occasion the diminuition of the transitional enthalpy increment, for lower concentration (e.g., Cd(0.005)) the bifurcation has disappeared and both anomalies are merged into a single peak. At even higher dopant levels (e.g., Zn(0.066)) virtually no anomaly is evident over the transition region but an excess heat capacity in the region 25 to 150 K can be discerned. The doping occasions upward migration of the higher-temperature anomaly and downward migration of the lower-temperature anomaly. The former is consistent with the prediction of the Wigner crystallization model and the latter with the usual behavior of an order-disorder transition.

INTRODUCTION

The electronic transition of Fe_3O_4 at ~120 K is a well-known but poorly understood phenomenon. Previous experiments by other techniques [1] suggest that a more systematic confirmation of the mechanism of the bifurcation [2] of the Verwey transition of Fe_3O_4 as a Wigner-type electron crystallization was urgent especially since the expected increase in the transition temperature with the decreasing electron concentration had not been observed until recently by magnetic and electrical studies. Our recent measurements on a Mn-doped Fe_3O_4 sample revealed an upward migration in a heat-capacity anomaly associated with the Verwey transition. This finding is of sufficient difference from previous experience and of such theoretical significance that it warranted further systematic confirmation. To this end Zn- and Cd-doped Fe_3O_4 samples have been synthesized and their thermal properties investigated especially in the region of the Verwey transition. These dopants decrease the electron concentration without disturbing the octrahedral sublattice and therefore to a first approximation their effect is simply to decrease the electron concentration. They should provide good model systems for confirming whether the Verwey transitions have a marked Wigner nature. Both Zn and Cd have a clear preference for spinel tetrahedral A-sites, completely filled orbitals and have four-coordinate, Shannon and Prewitt radii of 0.50 and 0.95 A respectively. The former value is near that Fe^{3+} (0.60 A); therefore, at low dopant levels the zinc reduces electron concentration, as will cadmium; however, the latter will increase the unit cell size substantially. The reduction of octahedral Fe^{2+} is directly proportional to the amount of tetrahedral dopant consistent with the following formula: $(M^{2+}_x Fe^{3+}_{1-x})(Fe^{2+}_{1-x} Fe^{3+}_{1+x})O_4$.
We report results for M = Cd, x = 0.010, 0.005, and for M = Zn, x = 0.066, 0.005.

EXPERIMENTAL

Reactants were Johnson-Matthey Grade I "Specpure", 99.998% pure with respect to both cations and anions. Amounts of Fe_2O_3, Fe, and dopant oxide, MO, consistent with the reaction:

$$(MO) + (1-\frac{x}{3})Fe + (4-\frac{x}{3})(Fe_2O_3) = 3M_{\frac{x}{3}}Fe_{3-\frac{x}{3}}O_4$$

were dry mixed in an agate mortar, loaded into a quartz ampoule which was then evacuated to 10^{-6} min. and sealed off. The ampoule and contents were fired at 800 C for 8 to 14 hours, and the furnace was cooled to room temperature. After grinding and sizing (to 200 mesh or less), the sample was compressed in a steel die, the resulting pellet was loaded into a special quartz ampoule packed with alumina wool to prevent movement of the pellet, and the bottom of the ampoule was fused shut. The then evacuated and sealed ampoule and contents were fired at 985 C for 24 hours and furnace cooled. Surface layers were removed and the pellet was crushed.

X-ray powder patterns at 300 K using a Guinier-Hage camera and $CuK\alpha_1$ radiation revealed a linear variation of the a_0 lattice parameter with dopant concentration. Chemical analysis confirmed the stoichiometries within the limits of analytical accuracy.

Heat capacity measurements were made from 5 to 350 K in the Mark II adiabatic cryostat. Temperature increments as small as 50 mK were used to evaluate the details of the C_p anomalies. The entire transition region was traversed several times with single enthalpy increments to determine the total enthalpy and entropy of transition.

RESULTS

Transitional thermophysical properties of the doped-Fe_3O_4 samples including those previously published are presented in Table I. It will be observed that not only are the peaks in the heat capacity displaced by addition of the dopants in a significant manner, but that at low dopant levels the enthalpy of transition (ΔH_t) is almost independent of dopant level and this suggests that the phenomena occurring in the transition region change only slightly at low dopant levels. Moreover, it is observed that the temperature of the higher-temperature anomaly migrates upwards for low dopant levels of ions whose radii are not significantly different from those of Fe^{3+} and Fe^{2+}. For larger ions, such as Cd^{++}, the effects on the thermal parameters are more dramatic. Even in the most dilute Cd sample examined Cd(0.005) the bifurcation of the anomaly is no longer present whereas in the corresponding Zn-doped sample two peaks were still observed. For higher dopant levels of either ion not only is the bifurcation no longer present but the transition itself is suppressed in size and smeared-out over a large temperature range.

Somewhat similar migration effects have also been noted for vanadium oxides [4] and parallels exist with Ti_4O_7's Verwey transition.[5] This study has, however, quantified the energetics as noted in Table I. In addition, subtle changes occur in the heat capacities and entropies in regions remote from the transition especially at relatively high dopant levels. All doped samples clearly had finite heat-capacity maxima and achieved equilibria rapidly (15-20 min) without hysteresis near the maxima.

The lower-temperature transition is seen to occur at temperatures more elevated than those for pure Fe_3O_4 in several samples in table I; yet in all Cd and Zn samples and Millar's sample a significant decrease was observed. Further study will be needed at lower dopant levels to establish the initial sense and rate of change. In intermediate dopant levels e.g., Cd(0.005), the magnitude of the observed

Table I. SELECTED TRANSITIONAL PROPERTIES*

Compound	T_ℓ/K	T_h/K	ΔH_t	ΔH_ℓ	ΔH_h	ΔS_t	$C_{p,\ell}$ max	$C_{p,h}$ max	Ref.
Fe_3O_4	113.3	118.88	164	66	98	1.4	33.2	117	(2)
Hyd-Fe_3O_4§	117.0	123.0	139	68	71	1.8	27.2	(46.8)	(1)
Xtal-Fe_3O_4¶	117.30	(119)	140	140	--	1.2	109.6	28.8	(1)
Zn(0.005)	110.57	119.09	142	126	16	1.1	53.7	36.1	
Cd(0.005)	114.66	---	155	155	--	1.3	101.5	----	
Millar	114.2	---	108	108	--	1.0	40.	----	(3)
Cd(0.010)	105.7	---	104	104	--	0.98	31.1	----	
Zn(0.066)	80	---	70	70	--	0.87	10.4	----	

*ΔH's are enthalpy increments in cal mol^{-1}. ΔS's are entropy increments and Cp's are heat capacities; both in Cal K^{-1} mol^{-1}. Subscripts on thermodynamic quantities--t, ℓ, and h--refer to total lower temperature, and higher temperature transitions.

§Corresponds to Mn(0.008)

¶Two large magnetite crystals.

ΔH_t and the single transition imply the coincidence of both transitions. At still higher concentrations further suppression, downward migration, and smearing-out of the transitions are found.

DISCUSSION

This study confirms the reality of the bifurcation originally observed in synthetic Fe_3O_4 [2] on samples prepared in different laboratories, by different techniques beyond any reasonably doubt. The profiles and the ΔH_t's of the two anomalies permit us to draw tentative conclusions about the mechanism underlying the transitions. For example the upward migration of the high-temperature anomaly at low-dopant levels for decreases in electron concentration is consistent with its description as an essentially pure electron crystallization of the Wigner-type. The relative insensitivity of the low-temperature peak to impurities as well as its gradual suppression with increased impurity levels and its downward migration is consistent with ascription to a phenomenon of the more usual order-disorder type with a second order character and its relation to a structural transition. Similar systematics occur in VO_2 and Ti_4O_7. [4,5]

Bifurcation is seen to be affected both by identity and quantity of dopant. The results of this study indicate that most other measurements on Fe_3O_4 have been made on relatively impure materials. The effects of impurities have been mistakenly attributed as being those of a pure sample. Millar's [3] impurity level, for example, must approximate that of Cd(0.005) since both show a single anomaly. In addition, the much greater influence of Cd^{2+} doping on the thermal anomalies relative to that of Zn^{2+} is consistent with these systematics inasmuch as Cd causes a greater lattice expansion than does Zn at an equivalent dopant concentration. At higher levels it is no longer correct to treat the dopant as a minor perturbation on the lattice system because of the real differences in the parameters of the ions concerned. The model thus becomes inapplicable and at sufficiently high dopant levels the distortions predicted by molecular-field approximations become relevant.

Samara [6] observed a decrease in the transition temperature of Fe_3O_4 with increasing pressure. The observed increase in the temperature of the higher-temperature transition at low dopant levels observed in this study is entirely consistent with the results of Samara if one describes the transition as a Wigner crystallization. These data are also totally consistent with the observations of Miyahara, [5] who found consistent decreases in the magnetization transition temperature of Al, Co, Ni, Cu, Zn, and Ti dopants at concentrations greater than 0.05.

REFERENCES

* Supported by the National Science Foundation.
1. J. J. Bartel and E. F. Westrum, Jr., AIP Conf. Proc. 10, 1393 (1973).
2. E. F. Westrum, Jr. and F. Grønvold, J. Chem. Thermodynamics 1, 543 (1969).
3. R. W. Millar, J. Amer. Chem. Soc. 51, 215 (1929).
4. J. M. Reyes, J. R. Marks, and M. Sayer, Solid State Communications 13, 1953 (1973).
5. C. Schlenker, S. Lakkis, J. M. D. Casdy, and M. Marezio, Phys. Rev. Lett. 32, 191 (1974).
6. G. A. Samara, Phys. Rev. Lett. 21, 795 (1968).
7. Y. Miyahara, J. Phys. Soc. Japan 32, 629 (1972).

SPIN POLARIZED PHOTOEMISSION FROM Fe_3O_4

W. Eib, F. Meier, D.T. Pierce, K. Sattler,
H.C. Siegmann, Laboratory of Solid State Physics,
Swiss Federal Institute of Technology,
CH-8049 Zürich, Switzerland.

ABSTRACT

Ferrimagnetic Fe_3O_4 (magnetite), being the prototype of the ferrites, is of both fundamental and technical importance. Yet the electronic structure is still uncertain mainly due to the difficulties arising from simultaneous effects of exchange and crystal field splittings at the tetrahedral (A) and octahedral (B) lattice sites.
We report the first measurements of the photoelectric yield and photoelectron spin polarization on surfaces of natural single crystals obtained by cleaving in UHV. The experiment yields the width and position of highly spin polarized d-bands with respect to the Fermi level. Interesting structure is resolved in the spectra of spin polarization, that is related to electron emission from Fe-ions at A and B sites. Near photothreshold, the predominant direction of the photoelectron magnetic moment is antiparallel to the magnetization. Our results support the one electron energy diagram proposed by Camphausen, Coey and Charkraverty.

While the structural, electronic transport, and magnetic properties of magnetite, Fe_3O_4, have been widely investigated, little is known about the electronic structure. Magnetite has the inverse spinel structure where the magnetic moments of the 8 Fe^{+++} ions on tetrahedral A sites are aligned antiparallel to the moments of the remaining 8 Fe^{+++} and the 8 Fe^{++} ions occupying octahedral B sites. At the Verwey temperature, T_V = 119 K, a first order phase transition occurs where the conductivity decreases by a factor of 100 and the crystal structure changes from cubic inverse spinel to one of lower symmetry. With regard to the electronic structure, however, it is not even known where the localized or narrow band d-states lie energetically with respect to the 2p(O) and 4s(Fe) bands. On the one hand, it has been suggested that all of the occupied d-states lie in the gap between the 2p(O) and 4s(Fe) bands[1], while, on the other hand, it has been suggested that all but one of the occupied d-states lie below the top of the 2p(O) bands[2]. In this paper we report spin polarized photoemission measurements from Fe_3O_4 which locate directly the magnetic states in an energy level scheme.

In the spin polarized photoemission experiment electrons are photoemitted by ultraviolet light and then accelerated to 100 keV where the spin is measured by Mott scattering from a gold foil[3]. A magnetic field is applied perpendicular to the sample surface to align the domains. The polarization P is defined as $P = (N\uparrow - N\downarrow)/(N\uparrow + N\downarrow)$ where $N\uparrow$ ($N\downarrow$) is the number of electrons with magnetic moments parallel (antiparallel) to the applied field. By measuring P as function of the photon energy $h\nu$, one obtains a spectrum of spin polarization $P(h\nu)$. The d electrons are polarized, whereas electrons from doubly occupied 2p(O) states are not.

Measurements were made on natural magnetite crystals from Zermatt, Switzerland. The crystals were cleaved and measured in a vacuum of 3×10^{-10} Torr. No change in the polarization was observed over a period of 15 hours. The measurements of the spectrum $P(h\nu)$ extended from near photothreshold to 11.2 eV. A McPherson 225 vacuum ultraviolet monochromator was used with a Hg-Xe high pressure lamp with suprasil envelope for $h\nu \leq 6.4$ eV and with a H_2 discharge in a modified hot filament Hinteregger lamp[4] for $h\nu > 6.4$ eV. In the measurements presented here, no attempt was made to measure P in the neighbourhood of T_V where, in contrast to the conductivity, the magnetic properties change relatively little.

A typical spectrum $P(h\nu)$ for magnetite is shown in Fig.1. The temperature was held constant at 4.2 K and the applied magnetic field at ~ 8.5 kG. The height of the rectangular fields represents the statistical uncertainty in the electron counting (plus or minus one standard deviation) and the width the monochromator pass band. The photothreshold Φ denoted by the arrow was determined to be $\Phi = 4.05 \pm 0.05$ eV.

The striking feature of this curve is the large negative polarization near photothreshold, the first ever observed for a magnetic material. For $h\nu$ somewhat above 5 eV the emission of electrons with moments parallel to the magnetic field becomes dominant. The polarization changes dramatically from -30% to +50% as the photon energy changes only 1 eV.

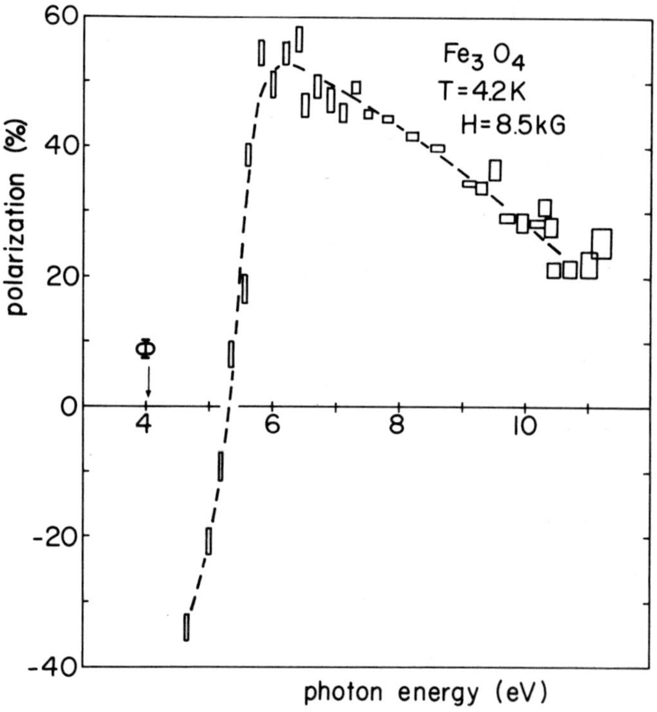

Fig. 1. Spectrum of photoelectron spin polarization from magnetite.

The spectrum of Fig.1 is an integral measurement in that all of the electrons emitted at a given photon energy are measured without further selection according to the electron kinetic energy. In the case of excitation from the localized d-state (flat bands) of magnetite, the influence of selection rules and final state density is expected to be constant so that a density of states approximation is valid. Then, the changes in polarization with $h\nu$ can be attributed to the new energetically deeper lying states excited as the photon energy is increased. Formally the photoelectron currents are given by

$$N_{\uparrow,\downarrow} = \int_{E_F + \varphi - h\nu}^{E_F} n_{\uparrow,\downarrow}(E)\, T(E) dE \qquad (1)$$

where $n_\uparrow$ ($n_\downarrow$) is the density of states of electrons with magnetic moments parallel (antiparallel) to the magnetic field. T(E) is the escape function and φ is the work function. If the density of states is known, for example from an x-ray photoemission measurement, we can analyze $P(h\nu)$ to determine separately $n_\uparrow(E)$ and $n_\downarrow(E)$.

Even without a knowledge of the density of states, we can draw several important conclusions from the spectrum of spin polarization. First, the polarization of all the d electrons taken together is the number of Bohr magnetons per formula unit ($4\mu_B$) divided by the number of d electrons (16) which gives 25%. The polarization above $h\nu = 10$ eV is approximately 25%, suggesting strongly that photoemission is possible from all the d-states but not yet from the p-states. We cannot absolutely rule out a tailing of p-states into the d-states, but we can expect that at the onset of transitions from p-states there will be a rapid decrease in the polarization due to the higher optical absorption cross section of the p-states relative to the d-states at these photon energies[5]. Second, antiparallel electrons, such as the extra electron beyond the half filled d shell of the Fe^{++} ion, occupy the states near the Fermi level. If the density of states near E_F is higher than that corresponding to this one extra electron, then some of the negative polarization must also be attributed to antiparallel electrons on A sites. Third, there is a large density of states of majority electrons on B sites starting about 1 eV below E_F which gives rise to the rapid increase in the polarization. Fourth, the gradual decrease in P beginning at $h\nu \sim 6.5$ eV indicates emission of antiparallel electrons from A sites beginning ~ 2.5 eV below E_F.

These results are in general agreement with the one electron energy diagram proposed by Camphausen, Coey and Chakraverty[1] based on indirect experimental evidence. With a further density of states analysis we hope to make these four conclusions more quantitative and obtain estimates of the crystal field splitting and exchange splitting of the electron orbitals.

ACKNOWLEDGEMENTS

We wish to thank U. Buchenau for giving us the magnetite crystals. The encouragement and support of Prof.Dr.G. Busch is greatly appreciated. The financial support of the Schweizerische Nationalfonds is gratefully acknowledged.

REFERENCES

1. D.L. Camphausen, J.M.D. Coey and B.K.Chakraverty, Phys.Rev.Lett.29, 657 (1972).
2. I. Balberg and J.I. Pankove, Phys.Rev.Lett.20, 1371 (1971).
3. M. Campagna, D.T. Pierce, K. Sattler and H.C.Siegmann, J.de Physique 34, C6/87 (1971).
4. D.E. Eastman and J.J. Donelon, Rev.Sci.Instr.41, 1648 (1970).
5. D.E. Eastman and M. Kuznietz, J.Appl.Phys.42, 1396 (1971).

THEORY OF THE ORDER-DISORDER AND INSULATOR-METAL TRANSFORMATIONS IN MAGNETITE

Tomoyasu Tanaka and Charles C. Chen
Department of Physics, Ohio University, Athens, Ohio 45701

ABSTRACT

Electronic energy bands for Fe_3O_4 are calculated in the neighborhood of the Verwey transition temperature using a Hartree-Fock decoupled Hamiltonian. Parameters appearing in the H-F Hamiltonian are determined by minimizing the free energy in configuration space in the pair approximation of the cluster variation method. Four sublattice construction of the spinel structure, charge density of 1/2 electron per site, and the Verwey charge ordering along the c-axis are assumed. The Verwey order parameter, effective exchange fields and the specific heat are calculated as functions of temperature. Those results are substituted into the H-F Hamiltonian and four energy bands are obtained. At sufficiently low temperatures, those bands are split into two, almost doubly degenerate, bands separated by a large energy gap, corresponding to an insulator phase. As temperature approaches the Verwey transition point, the band width increases rapidly, accompanied by a narrowing of the energy gap. The lower filled bands and upper vacant bands eventually touch each other, changing the system into a metallic state. It is found that the insulator-metal transition takes place at a temperature where the Verwey order is still finite, although very small. As long as the Verwey transition remains second order, the theory predicts two different transition temperatures, one for the insulator-metal transition and the other for the order-disorder transition.

DISTRIBUTION OF MAGNON MODES IN AMORPHOUS TWO DIMENSIONAL FERROMAGNETS AND ANTIFERROMAGNETS*

D. L. Huber and R. P. Siemann
Physics Dept., Univ. of Wisconsin, Madison, Wis., 53706

ABSTRACT

The distribution of linearized magnon modes in "lattice" models of amorphous, two dimensional ferromagnets and antiferromagnets is investigated. Disorder is introduced by postulating that the exchange integral connecting sites i and j is of the form $J_{ij} = \bar{J}(1-\frac{1}{2}\Delta+\Delta Y)$, where Y is a random number between 0 and 1. Computer calculations of the distribution of modes in 30x30 arrays are compared with the predictions of theories based on the bond coherent potential approximation. With the ferromagnet there is good agreement between experiment and theory; the antiferromagnet case is slightly less satisfactory.

INTRODUCTION

In this paper we report the results of computer studies of the distribution of magnon modes in "lattice" models of amorphous two dimensional ferromagnets and antiferromagnets. The computer calculations, which are exact results for finite arrays, are compared with the predictions of a theory based on the bond coherent potential approximation.[1] The goal of this work has been to test the validity of approximate calculations against exact results obtained from the same Hamiltonian. Such comparisons are important since the model is in itself a rather severe approximation to the Hamiltonian of a real magnet. The accuracy of the approximate calculations on the model Hamiltonian must be established before meaningful comparisons can be made with data from real materials.

The model system consists of a lattice of spins, $\vec{S}$, interacting through a nearest-neighbor isotropic exchange interaction, $J_{ij}\vec{S}_i\cdot\vec{S}_j$. Disorder is introduced by postulating that the exchange integral, J_{ij}, connecting sites i and j is of the form $J_{ij} = \bar{J}(1-\frac{1}{2}\Delta+\Delta Y)$ where Y is a random number uniformly distributed between 0 and 1, with no correlation between different bonds. We are interested in the low lying excitations or magnons and consider both ferromagnets and antiferromagnets, the latter being made up of interpenetrating sublattices with oppositely directed spins. The calculations were carried out in the linear approximation, valid at low temperatures, where S_z^k is replaced by $\pm S$ in the equations of motion. Although we treat only two dimensional arrays, our conclusions about the accuracy of the approximations are expected to apply to three dimensional systems as well.

A square lattice was assumed with S=1 and $\bar{J}$=0.25 so that in the absence of disorder the magnon frequencies take the form

$$\omega_q = 1-\tfrac{1}{2}(\cos q_x + \cos q_y) , \qquad (1)$$

for ferromagnets and

$$\omega_q = (1-\tfrac{1}{4}(\cos q_x + \cos q_y)^2)^{\frac{1}{2}} , \qquad (2)$$

for antiferromagnets. In the computer studies calculations were carried out on 30x30 arrays with free surface boundary conditions. Using numerical methods developed by Dean and Bacon[2] the distribution of modes was obtained from the dynamical matrix associated with the (linearized) equations of motion for the operators S_+^j ($S_+ = S_x + iS_y$). Assuming a harmonic time dependence with frequency ω, we obtain equations of the form[3]

$$\omega S_+^j = (\Sigma_\delta J_{j,j+\delta})S_+^j - \Sigma_\delta J_{j,j+\delta}S_+^{j+\delta} , \qquad (3)$$

for ferromagnets and

$$\omega(-1)^{n_j}S_+^j = (\Sigma_\delta J_{j,j+\delta})S_+^j + \Sigma_\delta J_{j,j+\delta}S_+^{j+\delta} , \qquad (4)$$

for antiferromagnets, where j+δ refers to the nearest neighbors of site j. In Eq. (4) n_j = 0(1) if the jth spin is on the up (down) sublattice. Note that we have implicitly assumed that there is a small anisotropy field present to insure ordering. However its influence on the magnon spectrum is neglected.

FERROMAGNETS

The results of computer runs with Δ=1.0 and Δ=2.0 are shown in Fig. 1, which is a histogram of the number of modes of the 900 spin array with frequencies between ω-0.004 and ω. Each curve is the output from a single run which took approximately twelve minutes on a Univac 1110 computer. It is seen that the effect of disorder is to round off the peak in the distribution at ω=1.0 and to induce tailing beyond 2.0, the upper limit to the distribution for the ideal lattice.

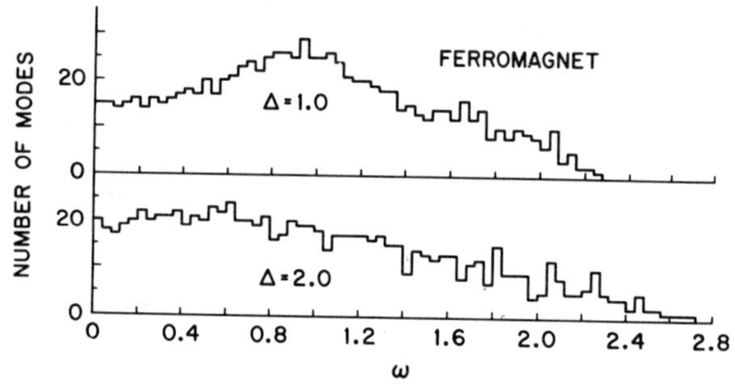

Fig. 1 Distribution of magnon modes in an amorphous ferromagnet. Shown are the number of modes in a 30 x 30 array with frequencies between ω and ω-0.04. Δ is the relative width of the distribution of exchange integrals.

The data in Fig. 1 are to be compared with a calculation of the density of states carried out in the bond coherent potential approximation of Foo and Bose.[1] In Fig. 2 we show the density of states,

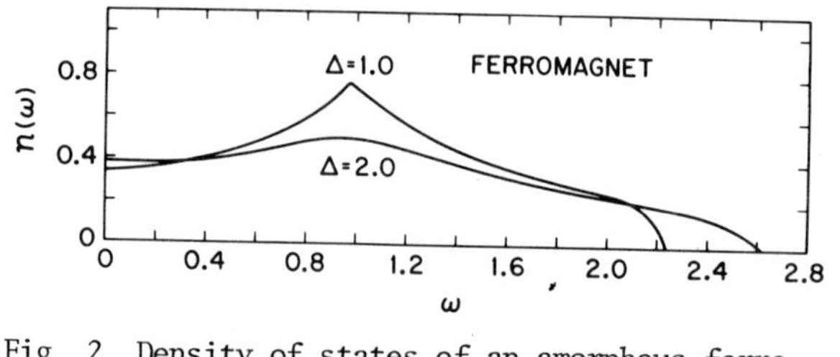

Fig. 2 Density of states of an amorphous ferromagnet in the bond coherent potential approximation. Δ is the relative width of the distribution of exchange integrals.

$n(\omega)$, obtained when the formalism of Ref. 1 is applied to a two dimensional ferromagnet. It is apparent that the theory is in remarkably good overall agreement with the computer generated data. Such agreement is not surprising in view of earlier work on one dimensional systems.[4,5] The data obtained with the uniform distribution of exchange integrals are similar to the data obtained with a Gaussian distribution.[3] The only significant difference is that in the latter case there are modes with frequencies less than zero whereas in the uniform distribution all modes have frequencies greater than or equal to zero. The negative frequencies do not appear in Fig. 1 since we have considered only values of Δ such that $J_{ij} \geq 0$.

ANTIFERROMAGNETS

The data for the antiferromagnetic arrays are shown as histograms in Fig. 3 for the cases Δ=1.0 and Δ=2.0. Since the dynamical matrix has both positive and negative eigenvalues we have added together the modes with frequencies between -ω and -ω+0.025 and ω-0.025 and to obtain a distribution of the magnitudes of the frequencies. It is apparent that the principal effect of disorder is to smear out the peak in the distribution for the ideal lattice which is at ω=1.0.

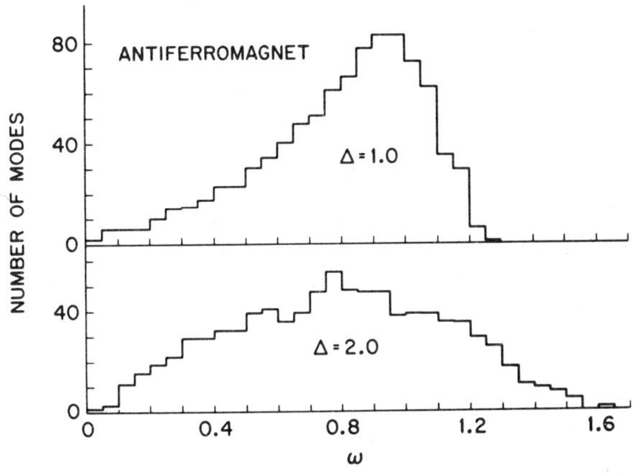

Fig. 3 Distribution of magnon modes in an amorphous antiferromagnet. Shown are the number of modes in a 30 x 30 array with frequencies between ω and ω-0.025. Δ is the relative width of the distribution of exchange integrals.

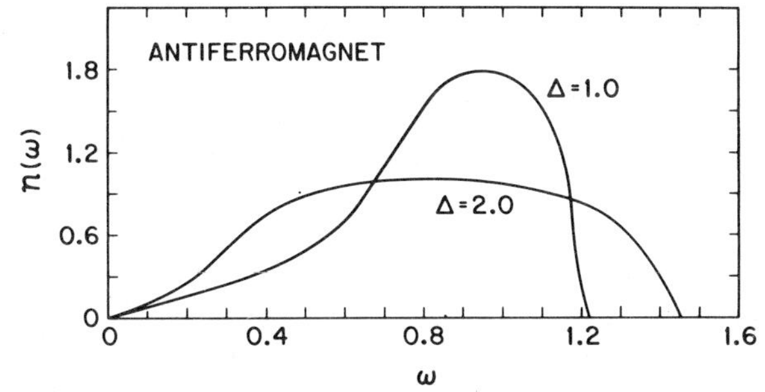

Fig. 4 Density of states of an amorphous antiferromagnet in the bond coherent potential approximation. Δ is the relative width of the distribution of exchange integrals.

In Fig. 4 we display the density of states obtained from a calculation based on the bond coherent potential approximation for an antiferromagnetic lattice. It is similar to the calculation for the ferromagnetic lattice outlined in Refs. 1 and 5, the only difference being that an ideal lattice Green's function appropriate to an antiferromagnetic array is used in place of the corresponding ferromagnetic function.[6] Comparing Fig. 4 with Fig. 3 it can be concluded that the theory again is in good qualitative agreement with the computer data. The only significant quantitative discrepancy is in the fall-off of the distribution beyond ω=1.0. For Δ=1.0 the computer generated data extend to ω=1.3, while the predicted upper limit is 1.23. When Δ=2.0 the disagreement is even more severe with the data extending to ω=1.7 whereas the predicted limit is 1.45. In comparison, for the ferromagnetic lattice the corresponding numbers are Δ=1.0, upper limit = 2.3 (expt.), 2.24 (theory); Δ=2.0, upper limit = 2.7 (expt.), 2.62 (theory).

CONCLUSION

The principal conclusion to be drawn from this work is that the bond coherent potential approximation provides a realistic description of the magnon distribution in the lattice model of an amorphous magnet. The theory appears to work somewhat better for ferromagnets, but is still satisfactory as a first approximation for an antiferromagnet.

REFERENCES

*Research supported by the National Science Foundation

1. E.-N. Foo and S. M. Bose, Phys. Rev. B 9, 347-350 (1974).
2. P. Dean and M. D. Bacon, Proc. Roy. Soc. A283, 64-82 (1965).
3. D. L. Huber, Solid State Comm. 14, 1153-1155 (1974).
4. D. L. Huber, Phys. Rev. B 8, 2124-2129 (1973).
5. E. N. Foo and S. M. Bose, Phys. Rev. B 9, 3944-3945 (1974).
6. The same transcription occurs in Tahir-Kheli's theory of the dilute bond antiferromagnet (R. A. Tahir-Kheli, Phys. Rev. B 6, 2826-2837 (1972)).

SPIN-WAVE DENSITY OF STATES IN LINEAR CHAINS WITH RANDOM LONGITUDINAL AND TRANSVERSE EXCHANGE*

Raza A. Tahir-Kheli
Department of Physics, Temple University
Philadelphia, Pa. 19122

ABSTRACT

Using Dean's numerical procedures, Huber[1] has recently evaluated the spin-wave density of states in disordered magnetic chains. His work is specific to the case where the Heisenberg exchange bonds are spatially isotropic. In realistic amorphous systems, even if exchange bond isotropy were to obtain 'on the average', it is expected not to hold at the microscopic level(see, for example, the discussion in Ref.2). We have,therefore, generalized Huber's computations to include independently fluctuating longitudinal and transverse exchange bonds. Further, to test the accuracy of our CPA for doubly fluctuating bonds, we have compared these one dimensional results, for the total density of states, with those obtained by the use of the formulation given in Ref.2. Finally, we have analyzed the predictions of our DFCPA in three dimensions. Results for the spin-wave stiffness and the frequency-wave-vector dependent density of states are presented for the b.c.c. lattice.

References

1 D.L. Huber, Phys. Rev. B8,(1973)2173.; also, to be published.
2 R.A. Tahir-Kheli, phys. stat. sol. (b) 63, (1974)K17.

(DFCPA)

* Supported by the NSF Grant # GH 39023

MAGNETIC PROPERTIES OF $MnO \cdot Al_2O_3 \cdot SiO_2$ GLASSES

R. W. Kline,* R. A. Verhelst,* and A. M. de Graaf,† Wayne State Univ., Detroit, MI 48202, and H. O. Hooper,* Univ of Maine, Orono, ME 04473

ABSTRACT

The magnetic susceptibility of concentrated Mn aluminosilicate glasses has been measured from 1.5 to 300 K. The susceptibilities above ∿50 K follow a Curie-Weiss behavior characterized by a large negative paramagnetic Curie temperature, indicating strong antiferromagnetic interactions among the manganese ions. Below ∿50 K the susceptibilities exhibit a superparamagnetic type behavior which terminates in a sharp peak at 3.3 to 6.5 K. In general with increasing Mn concentration the peak shifts to a higher temperature while the peak susceptibility decreases. Although the sharp peaks in the susceptibility seem to indicate the onset of long range antiferromagnetic order, we show that this behavior can be explained quantitatively by postulating that these glasses contain small (∿50 Å) manganese rich monodomains (which order antiferromagnetically well above the peak temperature) separated by manganese poor paramagnetic regions, and applying Néel's theory of small antiferromagnetic particles. The crucial parameters of this theory are the anisotropy energy, K, the volume, V, and the magnetic moment μ, of the monodomains. For very reasonable values of these parameters ($KV \sim 10^{-15}$ erg, $\mu \sim 200\ \mu_B$) quantitative agreement with the experimental data is obtained.

INTRODUCTION

In recent years significant attention has been paid to the properties of magnetic glasses. These include the manganese borates,[1] the chromium phosphates,[2] as well as the iron, manganese, and vanadium phosphate glass systems.[3] In this study we have measured the magnetic susceptibility of manganese aluminosilicate glasses and have found general agreement with the behavior of the systems mentioned above. However, the susceptibility of this system exhibits a relatively sharp peak at low temperatures. This phenomenon is unique among the glasses studied so far, and resembles closely the sharp peaks observed in the so-called spin glasses.[4]

EXPERIMENTAL

The glass samples for this study were prepared by firing appropriate mixtures of reagent grade maganese carbonate and aluminum oxide with pure silica sand in an arc-image furnace. In the heating process the carbonate is converted to manganese oxide by dissociation. The molten material was quenched by pooring into an aluminum mold and allowing it to cool to room temperature. The compositions which were prepared as seemingly good glasses and which were subsequently studied are listed in Table I. Here the compositions are those determined from the batch powder mixtures. Portions of the samples were examined by powder x-ray diffraction and electron microscopy. No crystallinity was detected in any of the samples by these techniques. Specimens for the magnetic measurements were cut from the cast samples in the form of regular parallelopipeds of dimensions about 10 x 4 x 4 mm.

TABLE I

Batch Compositions of Samples

Molar %			Atomic %
MnO	Al_2O_3	SiO_2	Mn
35	10	55	12.3
40	10	50	14.3
50	5	45	19.2
60	10	30	23.1
70	5	25	29.2
80	5	15	34.8
90	3	7	41.7

The magnetic susceptibilities were measured at 500 Hz over the temperature range 1.5 to 300 K using a low field (∿5 G) mutual inductance bridge technique. The high temperature susceptibilities follow a Curie-Weiss behavior with a large negative paramagnetic Curie temperature, indicating strong antiferromagnetic interactions among the manganese ions. From ∿50 K to ∿10 K the susceptibilities are markedly enhanced, resembling a superparamagnetic type behavior. Between 3.3 and 6.5 K the susceptibility exhibits a sharp peak for the various samples. In general, with increasing Mn concentration the peak shifts to a higher temperature, broadens, and decreases in magnitude.

From the high temperature susceptibility the effective ionic moment, μ_{eff}, and the paramagnetic Curie temperature, Θ, are determined. These parameters, and also the temperature of the susceptibility maximum, T_{max}, are listed in Table II for each of the samples. The range and magnitudes of the effective moments observed in these glasses are not fully understood at present. Values of μ_{eff} above ∿6.0 μ_B for manganese ions have not previously been reported in the literature.

TABLE II

Magnetic Susceptibility Data

At.% Mn	$\mu_{eff}(\mu_B)$	Θ(K)	T_{max}(K)
12.3	6.96	-194	3.3
14.3	6.89	-194	4.6
19.2	6.01	-245	5.55
23.1	4.69	-226	5.8
29.2	4.85	-210	6.15
34.8	5.20	-236	5.1
41.7	5.37	-224	6.5

DISCUSSION

The presence of the sharp low temperature peaks in the susceptibility might indicate the onset of long range antiferromagnetic order. However, rather than pursuing the question of whether or not an ordered magnetic state is possible at all in a structurally disordered glass, we have chosen to capitalize on the similarity between our experimental susceptibility curves and the predictions of Néel's theory of small magnetic particles.[5]

We shall consider the behavior of small particles which order antiferromagnetically at a temperature T_c. Néel has shown that below its ordering temperature such a particle develops a net magnetic moment due to an imperfect compensation of the magnetic sublattices. This net moment can be on the order of a hundred times the moment of a single ion. An important consideration in Néel's theory is the anisotropy energy of the particle. For simplicity it is assumed that the anisotropy is uniaxial, with an energy density K. Then the energy of a particle with net moment $\vec{\mu}$ in the presence of an applied field $\vec{H}$ is given by

$$E = KV \sin^2\beta - \mu H \cos\theta \tag{1}$$

where V is the volume of the particle, β is the angle between $\vec{\mu}$ and the anisotropy direction, and θ is the angle between $\vec{\mu}$ and $\vec{H}$.

Another essential quantity in this theory is the relaxation time

$$\tau = \tau_o \exp(KV/kT), \tag{2}$$

which represents the time required for the magnetic moment of the particle to reach thermal equilibrium. In the case that τ becomes greater than the time of measurement the moment will not be able to reach thermal equilibrium and will freeze in the direction of the anisotropy below a temperature on the order of KV/k. As a consequence of this freezing of the moment the susceptibility of the particle will have a maximum.

To calculate the susceptibility of such a particle the classical partition function, Z, is formed using Eq. 1. In thermal equilibrium the susceptibility of a particle with a large magnetic moment would be given by

$$\chi_{\text{equil.}} = \frac{d}{dH}\left\{ \frac{M(0)}{Z} \int \exp(-E/kT) \cos\theta \, d\Omega \right\}, \quad (3)$$

where M(0) is the magnetization at T = 0 K. In order to take into account the freezing of the magnetic moment we must restrict the motion of $\vec{\mu}$ and require it to point nearly in the direction of the anisotropy at all times. To that purpose we divide the integral in Eq. 3 into two parts, each an integration over a hemisphere about the anisotropy axis. We have evaluated χ numerically for various values of KV/k. An additional integration was performed to take into account the randomness of the anisotropy direction for a system of such particles. The results of this calculation are shown in Fig. 1a.

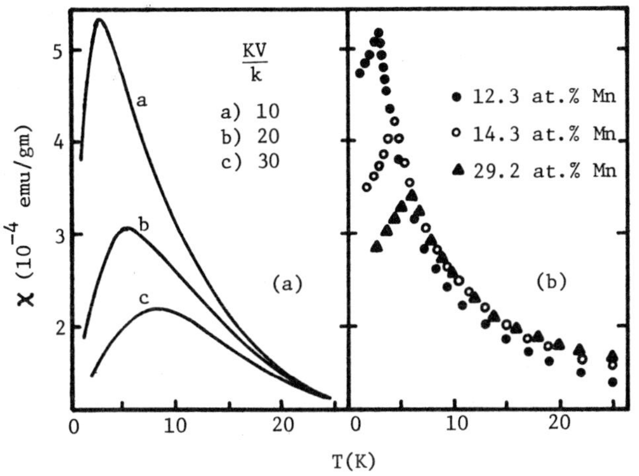

Fig. 1 Susceptibility vs. temperature: a) theoretical curves generated for different values of the parameter KV/k, b) experimental maxima of several of the samples.

The striking similarity between the curves in Fig. 1a and our experimental data, displayed in Fig. 1b, has led us to the monodomain model of the manganese aluminosilicate system. This model assumes that these glasses consist of regions (i.e., monodomains) of relatively high manganese concentration separated from one another by areas of lesser manganese content. The magnetic moments within the monodomains are assumed to order antiferromagnetically (at $\sim$50 K) well above the blocking temperature, due to strong intradomain superexchange interactions. A mechanism of anisotropy acting in each monodomain might have its source in magnetocrystalline-type effects or in the local strains which are invariably present in glassy materials.[6]

Employing this model, it is seen that above the ordering temperature of the monodomains the susceptibility will follow a Curie-Weiss behavior characterized by antiferromagnetic interactions among the manganese ions. Below the ordering temperature the susceptibility is superparamagnetic due to the large net moment of each monodomain. To quantitatively examine the susceptibility in the region of the maximum we neglect the contribution of the interdomain areas and consider each sample as a collection of small antiferromagnetically ordered particles. As seen in Fig. 1, the concentration dependence of the susceptibility peaks corresponds to a scaling of KV with manganese concentration. This is reasonable since it is likely that increased concentration results in larger monodomains.

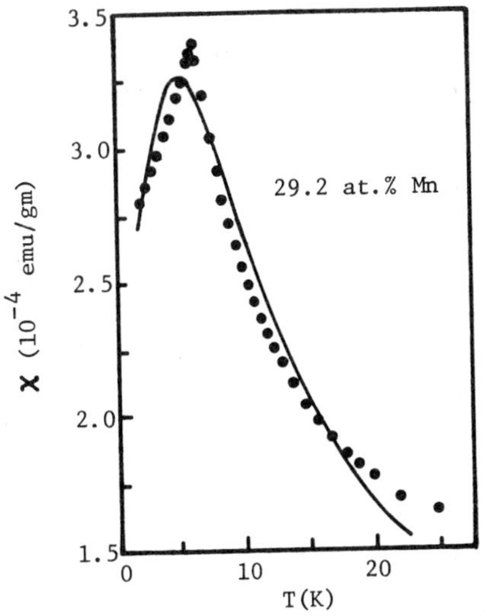

Fig. 2 Susceptibility vs. temperature for a 29.2 at.% Mn glass. Solid curve is the theoretical fit.

The theoretical susceptibility (averaged over random anisotropy directions) was used to obtain a least squares fit to the experimental data. The parameters adjusted in the fitting routine were μ and the product KV. A uniform distribution of K and V values was assumed as it had been determined that such a distribution provided the best fit. The results of this procedure are shown in Fig. 2 for the 29.2 at.% Mn glass. The parameters required to achieve this fit were KV/k = 19.4 Kelvin and $\mu = 230\ \mu_B$. If we take for $K \sim 10^6$ erg/cm^3 (which is a value typical for crystalline manganese compounds), we find $V^{1/3} \sim 15$ Å. This distance does not represent the monodomain size but rather the extent of local atomic order. In order to obtain the actual size of the monodomains one must use the value for μ. Utilizing the fact that $\mu = 230\ \mu_B \simeq 4.85\ \mu_B \sqrt{N}$, where N is the number of manganese ions in a monodomain,[5] we obtain a size of $\sim$50 Å. This length is a measure of the extent of magnetic order in the glass.

We can achieve even better agreement between the theory and the experimental data by allowing the anisotropy constant to be weakly temperature dependent. We have determined that only a 10% variation of K about its value at the peak temperature would yield perfect agreement with experiment. A temperature dependence of K is not at all unreasonable since antiferromagnetic materials usually do show such dependence. Since our model is only approximate we are, however, not able to determine the precise temperature dependence of K.

*Supported in part by the Air Force Office of Scientific Research under Grant #AFOSR-71-2002.

†Supported in part by the National Science Foundation.

1. C. J. Schinkel and G.W. Rathenau, Physics of Non-Crystalline Solids, North Holland, Amsterdam, 1965, 215-9.
2. R.J. Landry, J.T. Fournier and C.G. Young, J. Chem. Phys. 46, 1285-90, 1967.
3. E.J. Friebele, L.K. Wilson and D.L. Kinser, Amorphous Magnetism, edited by H.O. Hooper and A.M. de Graaf, Plenum Press, New York, 1973, 27-45.
T. Egami et al., J. Phys. C 5, L261-5, 1972.
4. V. Cannella and J. Mydosh, Phys. Rev. B 6, 4220-37, 1972.
5. L. Néel, Low Temperature Physics (Les Houches, 1961) edited by C. de Witt, B. Dreyfus and P.G. de Gennes Gordon and Breach, New York, 1962, 413-40.

ANOMALOUS MAGNETIC BEHAVIOR OF A Pt-1at.%Fe ALLOY

N. C. Koon and D. U. Gubser
Naval Research Laboratory, Washington, D.C. 20375

ABSTRACT

Detailed magnetization and susceptibility measurements have been performed on an alloy of Pt-1at.%Fe, which is slightly above the percolation concentration for ferromagnetic ordering. A well defined Curie temperature was observed near 4 K. The extrapolated zero field magnetization dropped off with increasing temperature much more rapidly than predicted by conventional spin wave or molecular field theory, in common with other systems having a broad exchange field distribution.[1] Above 1 K there was no detectable hysteresis. Below 1 K, however, the coercive force increased very rapidly as the temperature dropped, reaching 350 Oe at 40 mK. Both the square shape of the hysteresis loop and the temperature dependence of the coercive force are similar to those recently reported for amorphous $TbFe_2$.[2] We suggest that similar mechanisms are operative in both cases, except that in Pt-Fe the random uniaxial anisotropy is due to dipolar forces rather than magnetocrystalline anisotropy.

[1] N. C. Koon and A. I. Schindler, Phys. Rev. 8, 5257-5262 (1973).

[2] J. J. Rhyne, J. H. Schelleng, and N. C. Koon, Phys. Rev., to be published.

THE MICTOMAGNETIC TRANSITION IN Cu-Mn*

H. Claus and J. S. Kouvel
Department of Physics, University of Illinois,
Chicago, Illinois 60680

ABSTRACT

Detailed magnetic measurements on an alloy of composition $Cu_{81}Mn_{19}$ reveal that the temperature T_m at which the low-field susceptibility has a sharp maximum (as previously reported for this alloy system[1]) marks a qualitative change in the magnetization behavior at higher fields. Below T_m (∿85°K), the magnetization M for increasing field H rises linearly from zero, accelerates to an inflection point (located at ∿25 kOe at 4.2°K), then slowly approaches saturation. If the field is lowered isothermally from any value H' above the initial linear region, hysteresis and remanence are exhibited and grow with increasing H'. The initial reversible region and the amount of hysteresis developed at higher fields both reduce to zero as the temperature is raised to T_m. Above T_m, M vs H is everywhere reversible and concave-downward. All these effects indicate the onset of a strong anisotropy at this characteristic temperature. They are thus consistent with the observation that the changes in magnetization produced by cooling the alloy in a field (which typify a mictomagnetic state[2]) disappear upon warming to this same temperature.

*Supported in part by the National Science Foundation.

1. V. Cannella, in "Amorphous Magnetism", H. O. Hooper and A. M. de Graaf, eds (Plenum Press, New York, 1973) pp. 195-206.
2. P. A. Beck, in "Magnetism in Alloys", P. A. Beck and J. T. Waber, eds (TMS-AIME, New York, 1972) pp. 211-230.

IMPURE CLASSICAL HEISENBERG CHAIN*

T. Tonegawa† and P. Pincus
Department of Physics
University of California
Los Angeles, California 90024

and

H. Shiba
Department of Physics
Osaka University
Toyonaka 560, Japan

The thermodynamic properties (specific heat, spin-spin correlation function, magnetic susceptibility, and density-density correlation function) of an impure one-dimensional classical Heisenberg chain with nearest neighbor exchange are calculated exactly in the thermodynamic limit. We consider both bond and site impurities and consider the quenched and annealed limits for each of these models. The present theory is an extension of Fisher's work[1] for the pure case. In the bond model, the annealed and quenched limits lead to the same results. In the site model, the difference between the annealed and quenched limits is predominant at low temperatures. For various combinations of the exchange constants (both ferro- and antiferromagnetic) we examine analytically how the low-temperature behavior of the zero field susceptibility varies with concentration. Numerical results are given as functions of temperature and concentration. It is found that in the annealed limit of the site model the specific heat versus temperature curve has a maximum at a finite temperature. This maximum comes from the short-range ordering of the constituent ions. The results of the calculation will be applicable to alloys of $(CH_3)_4NMnCl_3$ (TMMC) and its isomorphs.

Details of this work will be published in Physical Review.

*Supported in part by the National Science Foundation and the Office of Naval Research.

†Permanent address: Department of Physics, Kobe University, Rokkodai, Kobe 657, Japan.

1. M.E. Fisher, Am. J. Phys. 32, 343 (1964).

MICTOMAGNETIC TO FERROMAGNETIC TRANSITION IN Cr-Fe ALLOYS

R. D. Shull and Paul A. Beck
Materials Research Laboratory and Metallurgy Department, University of Illinois, Urbana, Ill. 61801

ABSTRACT

The b.c.c. Cr-Fe solid solutions are mictomagnetic in the composition range 9 to 23 at.% Fe. The change to ferromagnetism is gradual, but alloy $Cr_{75}Fe_{25}$ is essentially ferromagnetic. By analyzing the paramagnetic data, the average giant moment of the magnetic clusters was determined as a function of temperature and composition. The discrepancy between the Curie temperature of $Cr_{75}Fe_{25}$ and the "kinkpoint" temperature measured in a steady 113 Oe field is interpreted in terms of a relaxation effect.

INTRODUCTION

Cr-Fe b.c.c. solid solution alloys containing more than 30 at.% Fe are ferromagnetic, with Curie temperatures near or above room temperature[1]. Straight line extrapolation of the Curie temperatures gives 0°K at 18 at.% Fe. Alloys containing less than 5 at.% Fe are antiferromagnetic. Their Néel temperatures (T_N), as determined by neutron diffraction studies[2,3] coincide with the anomalous electrical resistivity minima[4,5,6]. Extrapolation of T_N to 0°K is not inconsistent with a composition of 18 at.% Fe. However, saturation magnetization measurements[7] on alloys containing more than 20 at.% Fe give mean atomic moments, which extrapolate to zero at a composition as low as approx. 8 at.% Fe. Moreover, remanent magnetization was found not only at 18 to 20 at.% Fe[8], but also at 15 at.% Fe and, at very low temperatures, even at 9 at.% Fe[9]. The apparent overlapping of the antiferromagnetic and ferromagnetic composition ranges between 5 and 18 at.% Fe suggests the possibility of mictomagnetism in these alloys.

SPECIMEN PREPARATION

In order to search for more definite indications of mictomagnetism, we prepared 4 alloys (using 99.87% pure Fe and 99.96% pure Cr), containing 19, 21, 23, and 25 at.% Fe, by arc melting in 1 atm. of argon. The alloy specimens were homogenized at 1200°C for seven days in 1 atm. of argon and water quenched. After grinding them to spherical shape, we annealed them again at 1200°C for 1 hour and quenched them into cold water.

EXPERIMENTAL RESULTS AND DISCUSSION

Magnetization measurements by the Faraday method were made as a function of field up to 12.6 kOe and as a function of temperature from 1.6°K to room temperature. Measuring and cooling the 19 at.% Fe alloy in a field of 113 Oe showed a broad maximum of the magnetization at 22°K. When cooling the specimen in zero field to 4.2°K and making the measurements in a field of 113 Oe, with the temperature increasing, the magnetization was considerably lower at temperatures below the maximum (Fig.1). This thermomagnetic history effect is characteristic for mictomagnetic alloys[10,11].

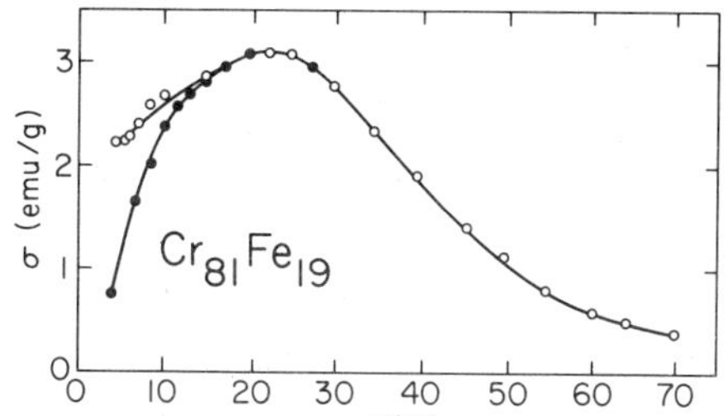

Figure 1

Magnetization at H=113 Oe vs. temperature for $Cr_{81}Fe_{19}$, field-cooled ○ and zero-field-cooled ●.

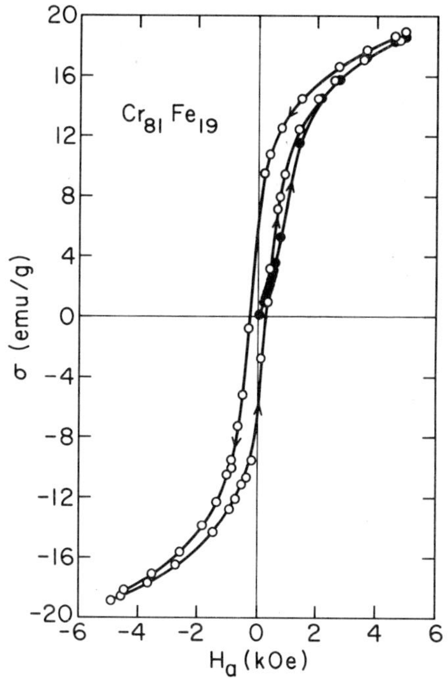

Figure 2

Hysteresis loop at 1.6°K after field cooling at H_a=12.7 kOe ○ and a 1.6°K isotherm measured after zero-field cooling ●.

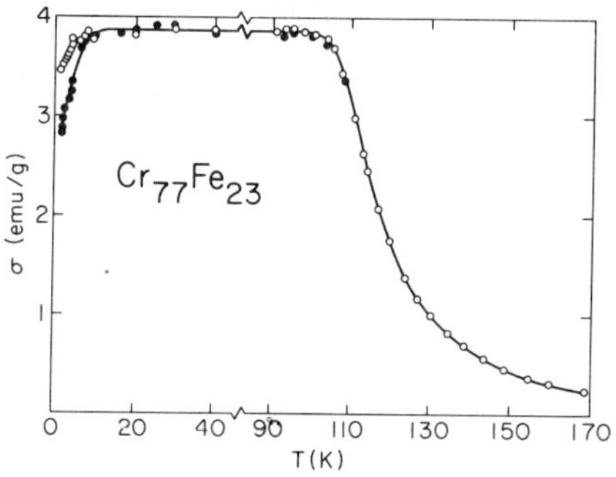

Figure 3

Magnetization at H=113 Oe v. temperature for $Cr_{77}Fe_{23}$, field-cooled ○ and zero-field-cooled ●.

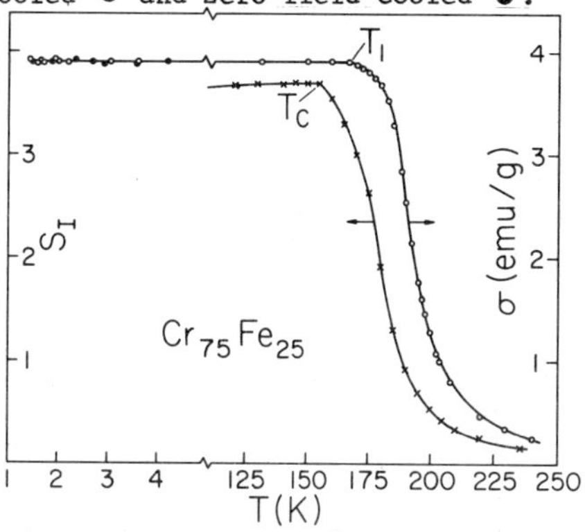

Figure 4

Magnetization at H=113 Oe vs. temperature for $Cr_{75}Fe_{25}$, field-cooled ○, zero-field cooled ●, and susceptibility in alternating 4 Oe field vs. temperature x.

The hysteresis loop for $Cr_{81}Fe_{19}$, measured at 1.6°K after field cooling (H=12.7 kOe), does not show any displacement (Fig. 2). However, the isotherm measured at 1.6°K with increasing field, after cooling the specimen to 1.6°K in zero field, lies largely outside the hysteresis loop. This phenomenon has been observed with mictomagnetic alloys even at temperatures somewhat higher than those at which loop displacement occurs[12].

Alternating low field susceptibility measurements on $Cr_{81}Fe_{19}$ show a relatively broad peak, corresponding to a matrix freezing temperature range at around 35°K (regardless of the field, within the range of 4 mOe to 4 Oe RMS). The lack of sharpness of the peak even at the lowest field used indicates that, despite the homogenizing treatment of 7 days at 1200°C, this alloy specimen still has some residual inhomogeneity. Above the matrix freezing temperature range, the specimen is paramagnetic.

A difference in the magnetization at low temperatures between the field-cooled and the zero-field-cooled states was also noted for $Cr_{79}Fe_{21}$ and $Cr_{77}Fe_{23}$ (Fig. 3), but for $Cr_{75}Fe_{25}$, no measurable thermomagnetic history effect was observed down to 1.6°K (Fig. 4).

The results indicate that, as previously reported for $MnAu_4$ with increasing long range order[12], the change from mictomagnetism to ferromagnetism in Cr-Fe alloys with increasing Fe-content is a gradual one. The alloy $Cr_{77}Fe_{23}$ is not yet a "good" ferromagnet, but $Cr_{75}Fe_{25}$ may be regarded as definitely ferromagnetic.

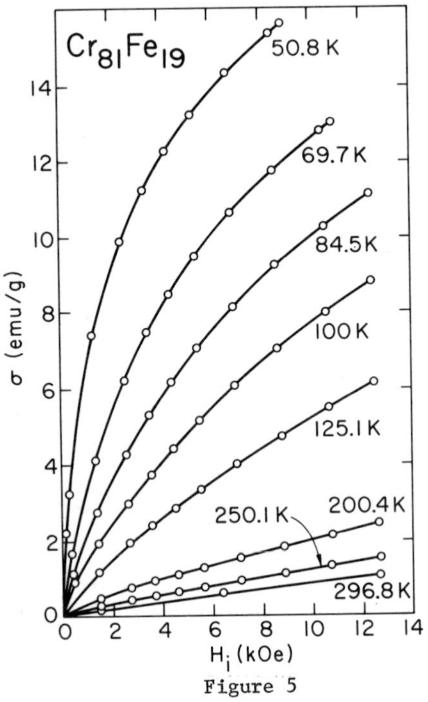

Figure 5

Magnetization vs. field for $Cr_{81}Fe_{19}$ at temperatures indicated.

The inverse initial susceptibility vs. temperature graphs for all four alloys are curved in the entire temperature range up to 300°K, suggesting that the magnetic structure here is a function of temperature. Therefore, the field dependence of the magnetization (shown for $Cr_{81}Fe_{19}$ in Fig. 5) was analyzed for pairs of successive isotherms above the matrix freezing temperature, neglecting the change in the parameters in the temperature interval between the two isotherms. The analysis was made in terms of a field-independent susceptibility (χ'), an average magnetic cluster moment (μ) and dipole concentration (c) according to the following equation:

$$\sigma = \chi' H + \mu c B\left[\mu, \frac{\mu\{H+\lambda(\sigma-\chi' H)\}}{kT}\right]$$

where H is the internal field, λ is the molecular field coefficient, and $B\left[\mu, \frac{\mu\{H+\lambda(-\chi' H)\}}{kT}\right]$ is the Brillouin function. It is assumed here that the orbital moment is quenched, so that μ=gS, that g=2, and that S does not need to be an integral multiple of ½. The results of this analysis indicate that the average magnetic cluster moment for $Cr_{81}Fe_{19}$ decreases from $81\mu_B$ (for the 70° and 84° K isotherms) to $17\mu_B$ (for the 250° and 300° K isotherms). The corresponding molecular field coefficients are 1610 and 302071, and the χ' values are 6.8 and 6.9 (in 10^{-5} emu/g Oe). With increasing Fe content, the average moment increases to $29\mu_B$ for the 25 at.% Fe alloy (270° and 300° K isotherms). In this case, λ=10016 and χ'=4.7 in the same units.

It is worth noting that, for $Cr_{75}Fe_{25}$, the temperature at which the magnetization begins to drop in the 113 Oe steady field measurements (the "kinkpoint"[13] T_1 in Fig. 4) is approximately 13° higher than the corresponding temperature in the alternating 4 Oe field measurements (T_C in Fig. 4). It was found that decreasing the alternating field to 4 mOe RMS has no effect on the temperature of the kinkpoint. Clearly, T_C is the Curie temperature. At T_1, $Cr_{75}Fe_{25}$ is in the superparamagnetic state. At lower temperatures, the magnetization remains constant while the internal field is zero, the susceptibility having reached a "critical" value large as compared with the reciprocal demagnetization factor. The present results show that the 113 Oe steady field is able to turn at T_1 the giant moments of the magnetic clusters sufficiently to achieve the critical value of the susceptibility. This susceptibility value is reached with the 4 Oe alternating field only at T_C. Steady field measurements using a field of 20 Oe give T_1=173°K, i.e., an increase of 6°K with decreasing field. Thus, the difference between T_C and T_1 is evidently not due to the difference in the field values used, but rather due to the alternating, as opposed to the steady, field. Apparently, the relaxation time of the magnetic clusters is longer than the half cycle of the 200 Hz alternating field. Such long relaxation times have been observed with Cu-Mn alloys[14,15]. For $Cr_{81}Fe_{19}$, a mictomagnet, the alternating low field susceptibility peak occurs typically at a temperature higher than the corresponding maximum of the magnetization measured in a 113 Oe steady field. For this alloy, a value of the susceptibility, critical in the above sense, is not reached at any temperature in either type of measurement. The temperature of the alternating field susceptibility peak here is determined by the freezing of the matrix, and it is essentially independent of frequency.

This work was supported by the U. S. Atomic Energy Commission.

REFERENCES

1. F.Adcock, J.Iron and Steel Inst.124, 129 (1931).
2. A.Arrott and S.A.Werner, Phys.Rev.153, 624 (1967).
3. Y.Ishikawa, S.Hoshino, and Y.Endoh, J.Phys.Soc.Jap. 22, 1221 (1967).
4. N.S.Rajan, R.M.Waterstrat, and P.A.Beck, J.App.Phys. 31, 731 (1960).
5. S.Arajs and G.R.Dunmyre, J.App.Phys.37, 1017 (1966).
6. M.M.Newman and K.W.H.Stevens, Proc.Phys.Soc.74, 290 (1959).
7. M.J.Besnus and A.J.P.Meyer, Phys.Rev.2, B2999 (1970).
8. M.V.Nevitt and A.T.Aldred, J.App.Phys.34, 463 (1963).
9. Y.Ishikawa, R.Tournier, and J.Filippi, J.Phys.Chem. Sol.26, 1727 (1965).
10. P.A.Beck, Met.Trans.2, 2015 (1971).
11. P.A.Beck, in Magnetism in Alloys, edited by P.A. Beck and J.T.Waber, TMS, AIME (1972) pp.211-230.
12. D.J.Chakrabarti and P.A.Beck, Int.J.Mag.3, 319 (1972).
13. The "kinkpoint," theoretically related to critical phenomena (P.J.Wojtowicz and M.Rayl, Phys.Rev.Lett. 28, 1489 (1968), and M.Rayl and P.J.Wojtowicz, Phys. Lett.28, A142 (1969)) was used to determine the Curie temperature of iron (S.Arajs and R.V.Colvin, J.App.Phys.35, 2424 (1964)).
14. P.A.Beck, J.Less-Common Met.28, 193 (1972).
15. A.Mukhopadhyay and P.A.Beck, "Magnetic Clusters in Mictomagnetic Cu-Mn Alloys," to be published.

SPIN-STIFFNESS COEFFICIENT AND CURIE TEMPERATURE OF DISORDERED FERROMAGNETIC BINARY ALLOYS

J.Bernasconi and H.J.Wiesmann
Brown Boveri Research Center, CH-5401 Baden, Switzerland

ABSTRACT

A relation between the effective static exchange integral J_{eff} of a disordered Heisenberg ferromagnet and the effective conductivity σ_{eff} of the corresponding disordered resistance network is used to investigate the concentration dependence of the spin-stiffness coefficient and the Curie temperature of disordered ferromagnetic alloys. Calculations of σ_{eff} are performed within the scope of various Effective-Medium Theories (EMT's) that take into account at least part of the correlations between the values of exchange integrals involving a common atom. In particular, we discuss a certain straightforward cluster extension of the Single Bond EMT and develop the method of "disturbed neighborhoods". Comparison is made with exact results for low concentrations and with Monte Carlo calculations. We show that our approximate methods in general give a good description of the Monte Carlo data, although the convergence (with increasing cluster-size) to the exact low concentration results is very slow. This latter deficiency turns out to be serious only for dilute or "nearly dilute" ferromagnets, for which a "low concentration cluster-EMT" gives much better results over the whole concentration range.

Disordered classical resistance networks have recently been studied extensively by various techniques, especially in connection with the electrical conductivity of amorphous systems[1]. There exists, however, also an interesting connection between the conduction in such resistance networks (random conductances σ_{ij} between sites i and j of a regular lattice) and the propagation of spin-excitations in disordered ferromagnetic Heisenberg-systems (random exchange integrals J_{ij})[1,2]. If we treat our Heisenberg system in the spin-wave approximation, and assume all magnetic atoms to have the same spin S, the excitation energy of spin-waves with small wavenumbers k is given by

$$E(\underset{\sim}{k}) = Dk^2 \equiv 2SJ_{eff}\, k^2 \qquad (1)$$

Here we have introduced the so-called spin-stiffness coefficient D and the effective static exchange integral J_{eff} of our disordered magnetic system. Based on calculations of Brenig et al.[3] and Kirkpatrick[1,2] we can now derive a relation between this J_{eff} and the effective conductivity σ_{eff} of the corresponding disordered resistance network (exchange integrals J_{ij} replaced by conductances σ_{ij} with the same relative values). To be definite, we consider the case of a binary random system of A-sites (concentration p) and B-sites (concentration 1-p), and restrict ourselves to 2-dimensional square and 3-dimensional simple cubic lattices with only nearest-neighbor interactions. (To ensure a ferromagnetic groundstate, we add an effective magnetic field term to the Hamiltonian of the 2-dimensional magnetic system). The relation between σ_{eff} and J_{eff} of corresponding systems (σ_{AA}, σ_{AB}, σ_{BB} $\leftrightarrow$ J_{AA}, J_{AB}, J_{BB}) can then be written as follows:

$$\sigma_{eff}(p)/\sigma_{eff}(1) = P(p)\; J_{eff}(p)/J_{eff}(1) \qquad (2)$$

where $P(p) = 1$ if all J_{ij} are positive. For a dilute ferromagnet ($J_{AB} = J_{BB} = 0$), P(p) is the fraction of sites which lie in the infinite cluster connected by J_{AA}-bonds[1,2] and is known for all important lattices[4]. Using relation (2), and making the usual assumption that the Curie temperature T_c is proportional to J_{eff}, we can now study the concentration dependence of T_c and of the spin-stiffness coefficient D(= $2SJ_{eff}$) by investigating the conductivity σ_{eff} of binary site-disordered ($\sigma_{ij} = \sigma_{AA}$, σ_{AB} or σ_{BB}) resistance networks.

In the following, we shall briefly discuss several Effective-Medium Theories (EMT's) for the conductivity of such site-disordered resistance networks. These EMT's are all based on a certain (exact or approximate) treatment of small clusters of bonds and take into account at least part of the correlations between the values of adjacent conductances. These correlations are entirely neglected in the usual Single Bond EMT[1,5,6] which, in general, gives quite unsatisfactory results for our site-disordered systems. For low concentrations ($p \to 0$ and $p \to 1$) we can compare our approximate results with exact calculations based on the work of Izyumov[7]. We have also carried out Monte Carlo calculations for several 2-dimensional systems. Two examples are shown in Figs.1 and 2, where we make comparison with the different approximations ($\Delta\sigma$ denotes the deviation with respect to the results of the Single Bond EMT).

In our Effective-Medium approximations we use the following selfconsistent procedure for the determination of σ_{eff}[8]. We consider a cluster of conductances σ_{ij}^c embedded in the effective medium (all conductances equal to σ_{eff}) and calculate the total resistance R_c between two opposite boundary points of the cluster. The value of σ_{eff} is then determined by the requirement that

$$\langle R_c \rangle = R_{eff} \qquad (3)$$

where <....> denotes an averaging-procedure that depends on the treatment of the specified cluster. R_{eff} is the value of R_c when the conductances in the cluster are also put equal to σ_{eff}. The calculation of R_c is performed by Fourier-transforming the appropriate Kirchhoff equations (similar to the procedure described in Appendix B of Ref.6).

We begin with the discussion of a <u>Cluster-EMT for low concentrations</u>, analogous to that for the site-percolation conductivity[9,10]. For small concentrations ($p \to 0$ or $p \to 1$) almost all σ_{ij}'s of the disordered system have the same value (σ_{BB} or σ_{AA}), with the exception of a few isolated σ_{AB}-clusters consisting of the nearest-neighbor connections of a given site. In this limit, it seems therefore appropriate to construct an EMT by considering such clusters of equal conductances (σ_{AA}, σ_{AB} or σ_{BB}, with probabilities p^2, $2p(1-p)$ and $(1-p)^2$, respectively). This low concentration cluster-EMT, denoted by ① in the examples of Figs.1 and 2, indeed becomes exact in the limits $p \to 0$ and $p \to 1$, and for general cases (Fig.1) it only gives a good description near these limits. For

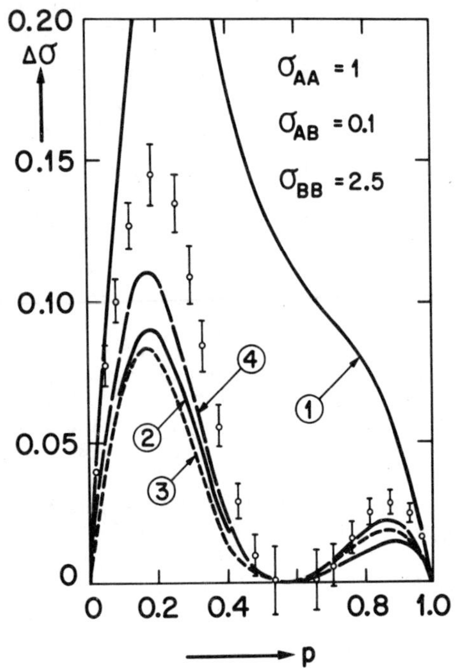

Fig.1 Comparison of Monte Carlo calculations (circles) and different Effective-Medium approximations (described in the text) for a general 2-dimensional system. $\Delta\sigma$ denotes the deviation with respect to the results of the Single Bond EMT.

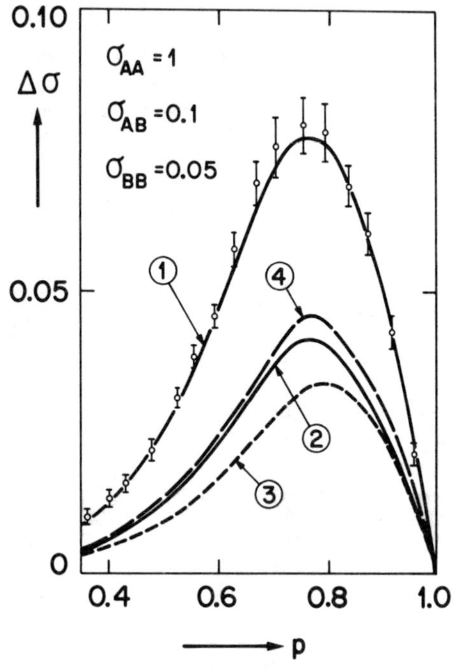

Fig.2 Same as in Fig.1, but for a "nearly dilute" system ($\sigma_{AA} >> \sigma_{AB}$, σ_{BB}).

dilute ($J_{AB} = J_{BB} = 0$) or "nearly dilute" ($J_{AA} >> J_{AB}$ and J_{BB}) systems, however, the agreement turns out to be excellent over the whole range of concentrations (see Fig.2), except in critical regions (dilute systems).

A straightforward Cluster-Extension of the Single Bond EMT is obtained by treating clusters of bonds exactly, instead of only a single bond. In Figs. 1 and 2 the results of such calculations, based on clusters consisting of all nearest-neighbor bonds of a given site, are denoted by ②. The convergence (with increasing cluster-size) to the exact low concentration results turns out to be very slow. Empirically we find that the deviations from the exact slopes at $p = 0$ and $p = 1$ decrease about linearly with the inverse of the number of bonds in the cluster. This deficiency, however, turns out to be serious only for dilute and "nearly dilute" systems (Fig.2), whereas in general cases (Fig.1) the agreement with the Monte Carlo results is remarkably good already for our small clusters.

Finally, we derive a new class of Effective-Medium approximations by introducing the method of "disturbed neighborhoods". Again a small cluster of bonds is treated exactly, but now its neighborhood is distinguished from the rest of the lattice. The adjacent conductances are determined by a consistent averaging-procedure that takes into account the presence of the exactly treated cluster, and are not simply put equal to σ_{eff}. In general they now become functionals of σ_{eff}. We have carried out two such calculations, corresponding to disturbed neighborhoods of a single site and a single bond, respectively. The results are denoted by ③ and ④, respectively, in the figures, and are comparable to those of exactly treated clusters of equivalent size (e.g. curves ②). This is quite a remarkable result if we observe that the "disturbed-neighborhood" treatment is much simpler than the exact treatment of an equivalent cluster.

REFERENCES

1. For a review see S.Kirkpatrick, Rev.Mod.Phys.45, 574 (1973).
2. S.Kirkpatrick, Solid State Commun. 12, 1279 (1973).
3. W.Brenig, G.Döhler and P.Wölfle, Z.Phys. 246, 1 (1971).
4. V.K.S.Shante and S.Kirkpatrick, Advan.Phys. 20, 325 (1971).
5. J.Bernasconi, Phys.Rev. B7, 2252 (1973).
6. J.Bernasconi, Phys.Rev. B9, 4575 (1974).
7. Yu.Izyumov, Proc.Phys.Soc.(London) 87, 505 (1966).
8. A similar formulation of Effective-Medium approximations is used by P.Erdös and S.B.Haley (preprint).
9. B.P.Watson and P.L.Leath, Phys.Rev. B9, 4893 (1974).
10. J.Bernasconi and H.J.Wiesmann, Spring Meeting of the Swiss Physical Society, Berne, 26-27 April 1974 (to be published).

STATIC PROPERTIES OF DILUTE FERRI- AND ANTIFERROMAGNETS

S. KIRKPATRICK, IBM T. J. Watson Research Center, Yorktown Heights, N.Y. 10598 and
A. B. HARRIS, Dept. of Physics, University of Pennsylvania, Philadelphia, Pa. 19104

ABSTRACT

We report microscopic calculations of the static magnetic and spin-wave properties of dilute antiferromagnets (e.g. $KMn_{1-x}Zn_xF_3$) and of dilute ferrimagnetic garnets. The results are valid at low temperatures but are not restricted to low concentration of dilutents.

MACROSCOPIC THEORY

The energies of long wavelength spin waves can be expressed in terms of static quantities which can then be calculated either analytically or by Monte Carlo simulation. For ferromagnets, for example, the relation

$$\omega(q) = 2\gamma(A/M)q^2 \equiv Dq^2, \tag{1}$$

where A is the exchange stiffness,[1] M is the magnetization per unit volume, and γ is the gyromagnetic ratio, can be derived from hydrodynamic arguments,[2] irrespective of the microscopic details of the system. The result derived for dilute ferromagnets in Ref. 3 by a microscopic calculation is a special case of Eq. (1).

The hydrodynamic[2] result for antiferromagnets is

$$\omega(q) = \gamma(2A/\chi^{\perp})^{1/2}q \equiv Cq, \tag{2}$$

where $\chi^{\perp}$ is the susceptibility in the plane perpendicular to the staggered magnetization.

Spin excitations in ferrimagnets with oppositely oriented sublattice magnetizations, M_A and M_B, can also be treated by a classical continuum theory. The details of the calculation will be described later in a fuller paper. The main result is

$$\chi^{\perp}(\omega/\gamma)^2 + (M_A-M_B)\omega/\gamma - 2Aq^2 = 0. \tag{3}$$

Here $\chi^{\perp}$ is a generalized perpendicular susceptibility, defined as the response, disregarding overall rotations of the magnetization, to magnetic fields, H_A and H_B, applied to the two sublattices, with the ratio of H_A to H_B such that no net torque results. Equation (3) has two roots given by

$$\omega(q) = [\omega_{opt}{}^2/4 + 2\gamma^2Aq^2/\chi^{\perp}]^{1/2} \pm \omega_{opt}/2, \tag{4}$$

where

$$\omega_{opt} = \gamma|M_A - M_B|/\chi^{\perp}$$

is the frequency of the simplest optical mode at $q = 0$. When ω_{opt} is large, the smaller root reduces to the quadratic form of Eq. (1) with $M = |M_A-M_B|$. At compensation, $\omega_{opt} = 0$, and Eq. (4) reduces, as it must, to the linear antiferromagnetic result of Eq. (2). Near compensation, the dispersion relation has both quadratic and linear parts:

$$\omega(q) \approx 2\gamma(A/M)q^2 \quad \text{if } q \ll q^* , \tag{5a}$$

$$\approx \gamma(2A/\chi^{\perp})^{1/2}q - 1/2\,\omega_{opt} \quad \text{if } q \gg q^* , \tag{5b}$$

where $q^* = M/(2A\chi^{\perp})^{1/2}$.

MICROSCOPIC CALCULATION

The second step of the treatment is the calculation of the macroscopic quantities A and $\chi^{\perp}$ from the microscopic Hamiltonian, H, describing the randomly diluted magnet. We take H to be

$$H = -\sum_{ij} J_{ij}\vec{S}_i \cdot \vec{S}_j, \tag{6}$$

where i and j are summed only over magnetically occupied sites and $\vec{S}_i$ is treated classically. We have used direct computer simulation, solving for the linear response of samples of 5000 or more spins to an applied field (for $\chi^{\perp}$) or to a weak spatial variation in the direction of magnetization (for A). Exact expressions at low dilution were obtained for all the simple lattices studied, in order to check the simulations.

To find A, we calculate the energy, E, needed to rotate the direction of magnetization through an angle θ ($\theta \ll 1$) over a distance L. Minimizing the total energy leads to the condition

$$\sum_j J_{ij}S^2\,(\theta_i - \theta_j) = 0 \tag{7}$$

from which the equilibrium angle of the ith spin, θ_i, can be found. Then A is found from the relation,

$$E = 1/2\sum_{ij} J_{ij}S^2\,(\theta_i - \theta_j)^2 = A\theta^2\Omega L^{-2}, \tag{8}$$

where Ω is the volume. Equation (7) is Kirchoff's law for a network of conductances in which we identify the θ_i as voltages and the $J_{ij}S^2$ as conductances. With this identification, equation (8) sums the power dissipation, IV, in each conductance. Thus, apart from units, A is the macroscopic conductivity of the equivalent disordered network.[3]

The determination of $\chi^{\perp}$ is similar. If the sublattice magnetizations are exactly equal, we may write the conditions on the equilibrium angular deflections in the presence of a perpendicular field, H, as

$$2\sum_{ij}(\theta_i + \theta_j)J_{ij}S^2 = (\gamma hS)H. \tag{9}$$

Then $\chi^{\perp}$ is determined by the analog of Eq. (8):

$$E = -1/2\gamma hS\sum_i \theta_i H = -1/2\,H^2\Omega\chi^{\perp}. \tag{10}$$

The above derivations are not limited to nearest neighbor interactions, but require only that the Neel state be a good approximation to the true ground state. Then the sublattice magnetization is proportional to the fraction of sites in the "infinite" occupied cluster. (For ferromagnetic Bravais lattices $M(x)/M(0)$ is equal to the percolation probability, $P(x)$.)

Our results for the simple cubic lattice are shown in Fig. 1. For $x \sim x_p$, we find $P(x) \propto (x_p - x)^{0.3 \pm 0.1}$, $A(x) \propto (x_p - x)^{1.6 \pm 0.2}$, and $\chi^{\perp}$ appears to diverge as $(x_p - x)^{-0.6}$. The divergence is cut off close to x_p in our calculations by effects of finite sample size. We attribute the divergence in $\chi^{\perp}$ to the occurrence of large regions of incompletely compensated magnetization. Fig. 1 shows that the effect of disorder on D and C is very similar. However, $D(x)/D(0)$ and $C(x)/C(0)$ have different slopes near $x = 0$, so they cannot be exactly equal.

Since $T_c(x)$ is readily measured, many authors have sought to establish an empirical relation between $t \equiv T_c(x)/T_c(0)$ and either $d \equiv D(x)/D(0)$ or $a \equiv A(x)/A(0)$.

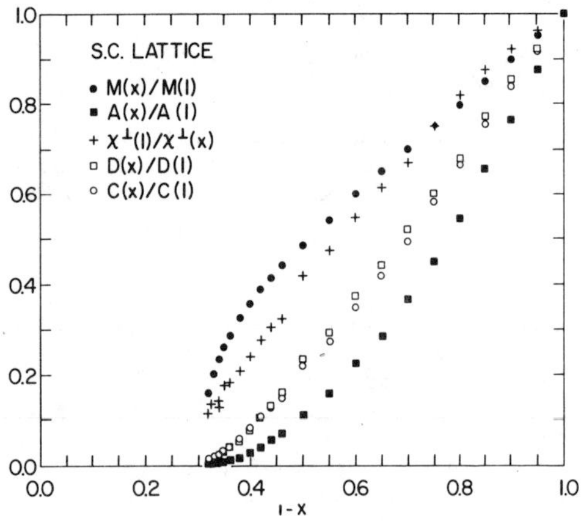

For small x the result for the sc lattice is[4] $d(x)=1-1.53x$, $a(x)=1-2.53x$. High temperature series calculations of $T_c(x)$ give $t(x)=1-1.37x$ for the sc lattice. The coefficients obtained for other 3D lattices are within .1 of these values. Clearly t and a are unrelated, while the equality between t and d is at best qualitative.

SUBSTITUTED GARNETS

We have performed similar calculations for substituted garnets. Second neighbor interactions are small but nonnegligible in the garnets. In YIG, for example, van der Ziel et al[5] find $J_{ad}=-20\text{cm}^{-1}$, $J_{dd}=-2\text{cm}^{-1}$, and $J_{aa}=0$. However, for moderate doping or for substitution primarily into the majority sublattice, we find that the nearest neighbor approximation is accurate, as is shown by the two calculations of A in Fig. 2. Well above the percolation threshold, our results for a garnet with composition $Y_3Fe_{3(1-x)}Fe_{2(1-y)}M_{(3x+2y)}O_{12}$ may

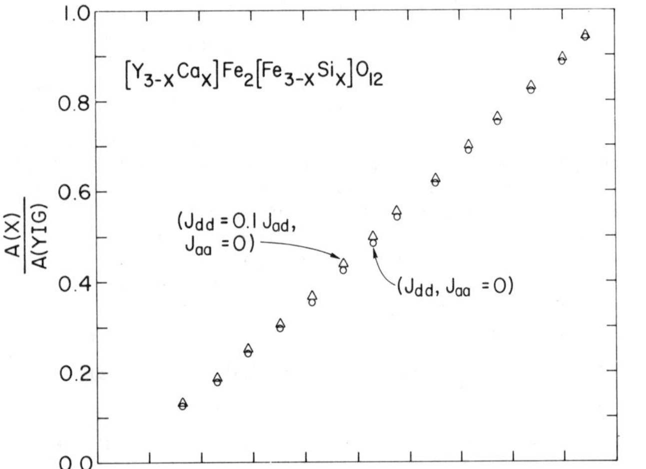

2. Calculated exchange stiffness, A, of YIG diluted on the majority sublattice only, with and without next-nearest neighbor interactions. The data (as well as those in Figs. 3-5) were obtained on Monte Carlo samples of 5x5x5 cubic unit cells, or 5000 Fe sites.

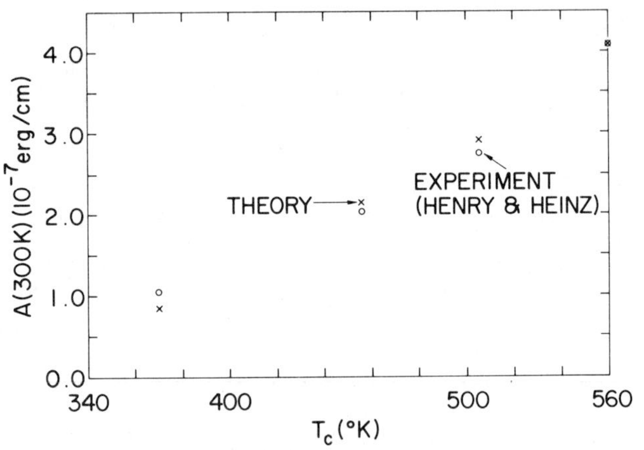

3. Comparison of calculated exchange stiffness for Ga-doped YIG with room temperature data of Ref. 6.

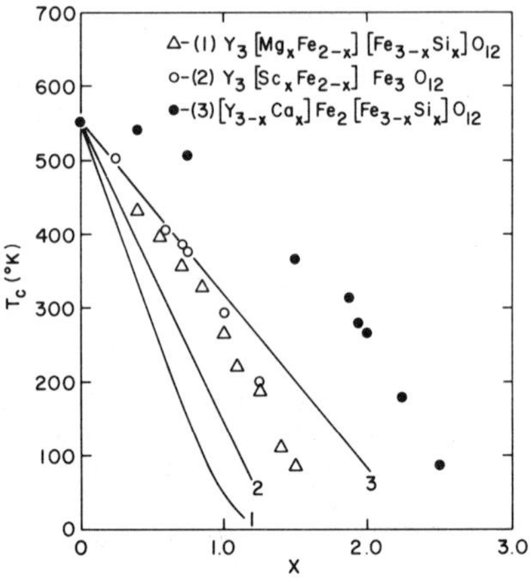

4. Comparison of measured (Ref. 7) T_c for three substituted YIG series and the values T_c would take (solid lines) if it were proportional to A.

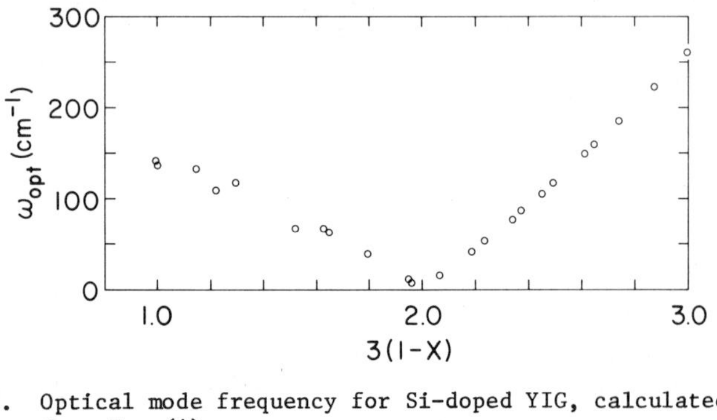

5. Optical mode frequency for Si-doped YIG, calculated from Eq. (4).

be summarized by the relation

$$A(x,y)/A(0,0)=1-1.28x-1.45y+1.5xy. \tag{11}$$

In the absence of low-temperature measurements of A on substituted garnets, we compare Eq. (11) with the recent room-temperature measurements of Henry and Heinz.[6] For this comparison we use the RPA result,

$$A(T)/A(0) = 4\langle S_a^z\rangle\langle S_d^z\rangle/25, \tag{13}$$

where <> denotes a thermal average. To obtain the theory points in Fig. 3, we estimated the compositions not stated in Ref. 6 by comparison with the measurements of Geller et al.[7], and calculated $\langle S_a^z\rangle$ and $\langle S_d^z\rangle$ in a mean field approximation, with J_{ad} adjusted to give the observed ordering temperature, T_c. Agreement with experiment, as shown in Fig. 3, is excellent, but due to the approximations made, it is hardly definitive.

For the dilute garnets, it is clear that $T_c(x)$ cannot be directly related to D(x). Several authors have suggested that t(x)=a(x). In Fig. 4, we test this relation by comparing the observed T_c's to the quantity a(x) for three families of diluted YIG. In no case is there qualitative agreement. In all cases, a(x) decreases more rapidly than t.

The generalized susceptibility needed to discuss the low-lying modes of Eq. (4) was calculated for Si-doped YIG. The resulting optical mode frequency, which should be experimentally observable, is plotted in Fig. 5.

REFERENCES

1. Chikazumi, S., Physics of Magnetism, (J. Wiley, New York 1964) p. 189.
2. Halperin, B.I. and Hohenberg, P.C., Phys. Rev. 188, 898-918 (1969).
3. Kirkpatrick, S., Rev. Mod. Phys. 45, 574-588 (1973).
4. Morgan, D.J. and Rushbrooke, G.S., Mol. Phys. 4, 291-303 (1961).
5. van der Ziel, J.P., Dillon, J.F. Jr., and Remeika, J.P., A.I.P. Conf. Proc. 5, 254-258 (1971).
6. Henry, R.D. and Heinz, D.M., A.I.P. Conf. Proc. 18, 194-198 (1973).
7. Geller, S., Williams, H.J., Espinosa, G.P., and Sherwood, R.C., Bell Syst. Tech. J. 43, 565-623 (1964).

EXCITATIONS IN RANDOMLY DILUTED FERROMAGNET

Alba Theumann*
Laboratoire de Physique des Solides, Universite Paris-Sud, Centre d'Orsay, France

and

Raza A. Tahir-Kheli†
Department of Physics, Temple University, Philadelphia, Pennsylvania 19122

ABSTRACT

Dynamics of randomly diluted, quenched Heisenberg ferromagnet has recently been analyzed by several authors. Within an effective medium type of approach, the correct site aspect of this problem is best given by the recent work of Harris et al. Their theory, which introduces spurious degrees of freedom for the non-magnetic vacancies and then projects them out by the use of an appropriate pseudopotential, however, is applicable at low and intermediate concentrations of the nonmagnetic vacancies. For large vacancy concentrations, their results for the magnetic response leak over into the negative frequency region. Also in their theory the spin-wave stiffness becomes complex for relative vacancy concentrations of order 49%. Here we present an alternative effective medium approach to the study of this problem. We avoid the use of additional degrees of freedom for the vacancies by working directly with the equations of motion for the magnetic spins. Effective medium ansatz is introduced through the use of a generalization of the path CPA approach introduced by Brouers et al in their study of random electronic alloys. For low and intermediate vacancy concentrations, our results are found to be of comparable quality to those given by Harris et al. Moreover, the first three frequency moments of the response are preserved exactly in our work. On comparison with 'exact' results--obtained via Padé procedures making use of numerically computed frequency moments--we find that our theory continues to yield qualitatively reasonable results even when the vacancy concentration is large, e.g. 40% in a simple cubic lattice.

*Present address: Facultad de Ciencias Exactas y Naturales Departmento de Fisica, Ciudad Universitaria (Nunez) Buenos Aires, Argentina

†Supported by NSF Grant #GH39023

SPIN-WAVES IN DILUTE ANTIFERROMAGNETS

W.K. Holcomb*, Oxford University, Oxford, England and
A.B. Harris**, University of Pennsylvania, Philadelphia, Pa. 19174

ABSTRACT

The effect of dilution on spin waves in isotropic Heisenberg antiferromagnets is studied. The model includes only nearest-neighbor interactions for a bcc lattice and spin-wave interactions are neglected, i.e. the results are correct in the limit $s\to\infty$. The dynamical susceptibility $\chi(\vec{k},\omega)$ and inelastic neutron cross section are obtained for arrays of 8192 sites randomly occupied by a concentration c of magnetic ions. For a given array the calculation is done by inverting the dynamical matrix and thus is essentially exact. Our results are as follows. For large k we find that Ising-like resonances corresponding to different numbers of occupied neighboring sites become increasingly prominent as c is decreased. The envelope of these resonances agrees with previous results using the coherent potential approximation where fluctuations in environment are suppressed. For small k we find a single spin-wave resonance broadened by the random dilution. The application of these results to $Mn_cZn_{1-c}F_2$ is discussed.

A large amount of effort has been spent studying the elementary excitations of random systems. Much of the effort has concerned itself with amorphous electronic problems, but real systems are difficult to describe in terms of models that are sufficiently simple that theoretical studies can be made. Any comparison between theory and experiment involves the difficult task of determining whether the model used is sufficiently accurate to place any real test on the theoretical approximations used. Similar comments are true for phonons in substitutional alloys. Thus much recent work has concentrated on magnetic alloys where simple models can be expected to describe the excitations accurately. The randomly diluted magnet is especially well suited for study since all the parameters in the model Hamiltonian can be determined from experiments on pure systems.

In this paper we study the randomly diluted isotropic Heisenberg antiferromagnet on a bcc lattice with lattice constant a for the purpose of clarifying the nature of the spin-wave dynamics. Explicitly, we consider the Hamiltonian

$$H = s\Sigma_{ij}p_ip_jJ_{ij}[a_i^\dagger a_i + a_j^\dagger a_j + a_i^\dagger a_j^\dagger + a_i a_j] \quad (1)$$

where $J_{ij} = J$ if i and j are neighbor sites and $J_{ij} = 0$ otherwise, $p_i = 1$ or 0 depending on whether or not the site i is occupied by a magnetic atom and finally the operator a_i creates a spin deviation at the site i. In the pure case (all $p_i = 1$) many phenomena, in particular, those that do not selectively pick out long wavelength behavior are well described by the much simplier Ising model. In these cases it is the diagonal terms of (1) that are the most important. This suggests that in the dilute case one might expect to find peaks in the response function at the Ising energies corresponding to different numbers of occupied neighboring sites. Such structure has not been observed experimentally.[1,2] Previous theoretical work[1,3] has not analyzed this situation in an unbiased way. For example, the single site coherent potential approximation (CPA) for the system[3] neglects fluctuations in the number of neighbors and perforce predicts a singly peaked response. On the other hand, cluster theories[4] assume such fluctuations are the dominant effect and treat the off diagonal terms of 1 in a more approximate way and thus predict a highly structured response . In any event we expect the response at long wavelengths to have a single peak broadened and shifted from the perfect crystal value. This is because the long wavelength response samples the system in such a way that it is relatively insensative to the fluctuations.

In order to resolve the question of the existence of Ising-like structure in the dynamics we have studied the response of finite systems (8192 sites) using the technique developed by Harris[5]. We generate each system by occupying each site randomly with a nominal probability of c. The quantity of interest is the susceptibility

$$\chi(\vec{k},\omega) = \Sigma_{ij}\, e^{i\vec{k}\cdot(\vec{r}_i-\vec{r}_j)}[\omega \underset{\approx}{1} - \underset{\approx}{K}]^{-1}_{ij}, \quad (2)$$

where $\underset{\approx}{1}$ is the unit matrix and $\underset{\approx}{K}$ is the dynamical matrix of the system using periodic boundary conditions. Here and below the indices i and j are restricted to magnetically occupied sites. The neutron scattering cross section is given essentially by

$$I(\vec{k},\omega) = \mathrm{Im}\, \chi(\vec{k},\omega - i0^+)/N, \quad (3)$$

where N is the total number of sites and 0^+ is a positive infinitesimal. We find $\chi(k,\omega)=\Sigma_j x_j e^{-i\vec{k}\cdot\vec{r}_j}$ by solving the equation

$$\Sigma_j\left[(\omega - i\epsilon)\underset{\approx}{1} - \underset{\approx}{K}\right]_{ij} x_j = e^{i\vec{k}\cdot\vec{r}_i} \quad (4)$$

for x_j. For each value of the parameters $\epsilon,\vec{k},\omega$ we study several samples and compute an average $\chi(\vec{k},\omega)$. To use this to represent the response on the infinite system we proceed as follows. We evaluate $\chi(\vec{k},\omega)$ only on the largest cluster and for most of the calculations ϵ was taken to be 0.1 (in units where 2Js = 1). This second step is necessary since the response for the finite system consists of a series of poles on the real axis (ϵ= 0). We believe that by drawing a smooth curve through these points we have a resonable representation of the response of the infinite cluster in an infinite system. This is because besides the finite clusters which can be corrected for later, the only sharp states are the "blocked" states described by Eggarter and Kirkpatrick[6] and these are statistically rare. A plot of $I(k,\omega)$ for $\vec{k}$ at the zone boundary, $\vec{k}$ = (1,0,0), (We measure $\vec{k}$ in units of π/a.) and ϵ = 0.1 is given in Fig. 1. We can extrapolate back to the real axis (ϵ= 0) by taking

$$I(\vec{k},\omega) = \mathrm{Im}\left\{\chi(\vec{k},\omega - i\epsilon) + i\epsilon\, d\chi(\vec{k},\omega - i\epsilon)/d\omega\right\} \quad (5)$$

and the second term can be estimated from our evaluation of Re $\chi(\vec{k},\omega - i\epsilon)$. The results of such extrapolations are shown in Fig. 1. The importance of this extrapolation may be judged from the curves of Fig. 1. The

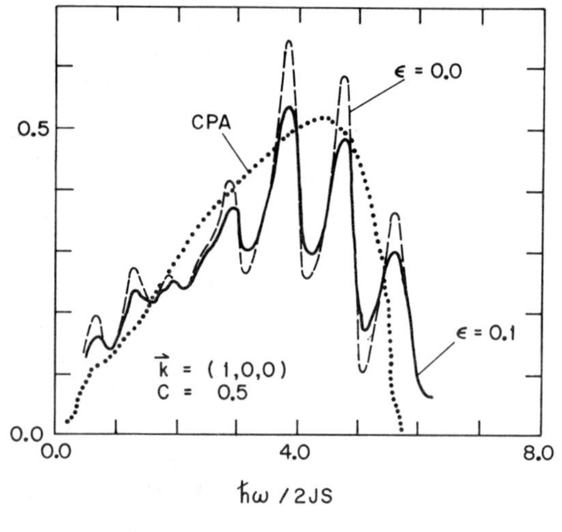

1. Response at $\vec{k}$ = (1, 0, 0) , c = 0.5 for ϵ = 0.1, solid curve; ϵ = 0.0 dashed curve; and the single site CPA also at ϵ = 0, dotted curve.

curves shown in Fig. 2 and 3 are also the results of this procedure, i.e. $I(k,\omega)$ averaged over several samples ($\varepsilon = 0.1$) and then extrapolated to the real axis ($\varepsilon = 0.0$). Corrections for the response due to finite clusters could be made by enumerating the smallest clusters and adding their response to that calculated above. This procedure will work except for concentrations too near the percolation concentration where large clusters become important.

The dependence on wavevector k is shown in Fig. 2 and the dependence on c is shown in Fig. 3. We see that for large k there is a sharp Ising-like structure. It is interesting to note that the peaks do not fall exactly at the Ising-energies of 1.0, 2.0, 3.0 ...(2Js). This is easy to understand as follows. If we treat the off diagonal terms of (1) as a perturbation then in second order the peak of the local density of states is shifted from the Ising-energy n (in units of 2Js) by an amount $\Delta E = -n/(n + zc)$, where z is the number of nearest neighbors. This is in agreement with our results: namely ΔE decreases as either c increases or n decreases. We also see that at small k the response goes over to a single peak as expected. This is even more pronounced in the results for k = (1/8, 0, 0) (not reproduced here) which is the smallest k we can examine for our sample sizes.

To determine if this structure is observable we must estimate the magnitude of spin-wave interactions. The Ising splitting is of order H_E/z where $H_E = 2Jsz$. Spin-wave interactions lead to a shift of order H_E/zs. If we concentrate on the case of $Mn_cZn_{1-c}F_2$ the Mn^{++} has s = 5/2 so this is not serious.[1] Also in $Mn_cZn_{1-c}F_2$ we should consider the role of other neighbor interactions. If the exchange constant for the first ferromagnetic neighbor is denoted J' we have $J'/J \cong 0.2$. This could lead at worst to a splitting into peaks having relative energy shifts of 0,0.2, and 0.4 (in units of 2Js = 1). This tends to increase the width of our peaks by 0.2 but the peaks should remain discreet. The resolution of current neutron scattering experiments[2] on $(MnZn)F_2$ is not sufficiently good to see this structure. The neutron scattering data is not inconsistent with our results. In fact the single site CPA which accurately describes the neutron scattering data[3] also accurately describes the envelope of the Ising peaks we calculate, as can be seen from the CPA curve given in Fig. 1.

Finally we note that our technique is in principle applicable to the binary alloy. The main limitations are the time consuming nature of the calculation and the finite mesh size in space ($\Delta k = \pi/8$).

We wish to thank the University of Oxford and the University of London computer centers for providing the extensive machine time used in this calculation. Also we gratefully acknowledge the many fruitful discussions with Prof. R.J. Elliott and thank Dr. R.J. Birgeneau for pointing out the importance of further neighbor interactions. Finally one of us (WKH) thanks the Science Research Council for the support provided while this study was conducted.

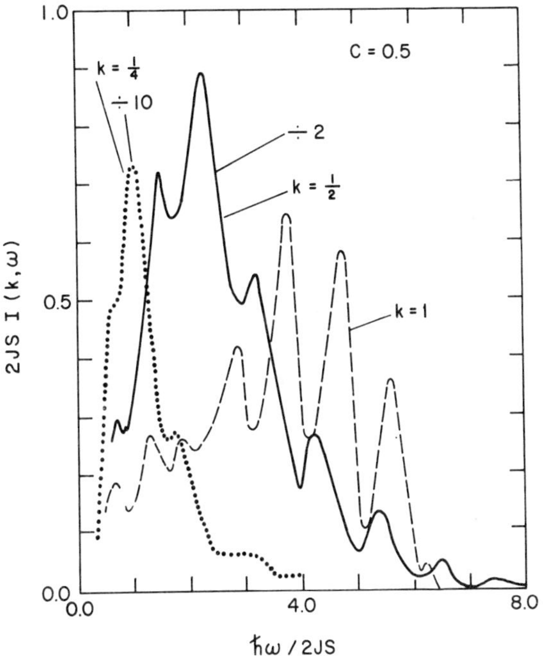

2. $I(k,\omega)$ for c = 0.50 ε = 0.0 and $\vec{k}$ = (1,0,0) dashed; $\vec{k}$ = ($\frac{1}{2}$, 0,0) solid; k = ($\frac{1}{4}$, 0, 0) dotted.

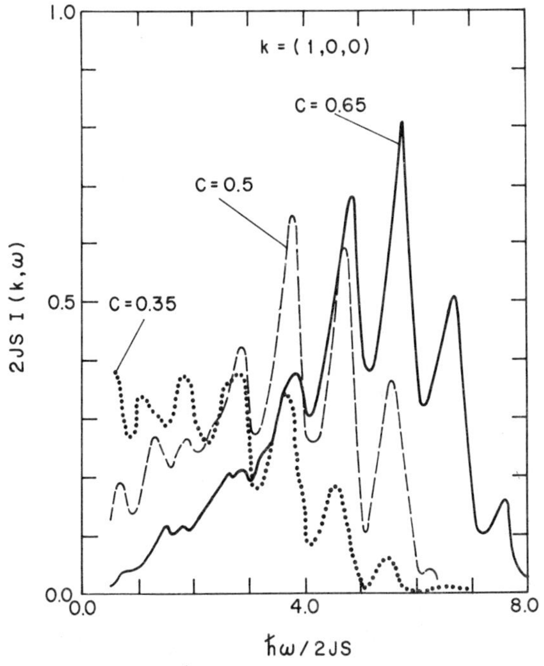

3. $I(k,\omega)$ for ε = 0 $\vec{k}$ = (1, 0, 0) and c = 0.65 solid; 0.5 dashed; and 0.35 dotted.

REFERENCES

*Address for 1975: Department of Physics, University of Pennsylvania, Philadelphia, Pa. 19174.

**Partially supported by the National Science Foundation.

1. Cowley, R.A., Buyers, W.J.L., Rev. Mod Phys. 44, 406-450 (1972).
2. Coombs, G.J., Cowley, R.A., Svensson, E.C., Holden, T.M., Jones D.A., Buyers, W.J.L., 1974 to be published.
3. Holcomb, W.K., J Phys C7, to be published (1974).
4. Buyers, W.J.L., Pepper, D.E., Elliott, R.J., J. Phys C6 1933-52 (1973).
5. Harris, A.B. to be published.
6. Kirkpatrick, S., Eggarter, T.P., Phys Rev B6 3598 (1972).

MAGNETIC AND SPATIAL SHORT RANGE ORDER IN ANNEALED A-B PARAMAGNETS

G. Bruce Taggart*
Department of Physics, Virginia Commonwealth University, Richmond, VA 23284
and
Raza A. Tahir-Kheli+
Department of Physics, Temple University, Philadelphia, PA 19122

ABSTRACT

In a paramagnetic A-B alloy, exchange interactions as well as interatomic potentials contribute to its magnetic and spatial short range order. We have studied an annealed Ising paramagnet with arbitrary concentrations of spins S_A and S_B, exchange interactions J^{AA}, J^{BB}, and J^{AB}, and interatomic potentials V^{AA}, V^{BB}, and V^{AB}. High temperature expansions for the spatial (i.e. the X-ray) and the magnetic (i.e. the neutron scattering) structure factors are carried out in the inverse temperature. Special features of the magnetic correlation function arising from the presence of the interatomic potentials and those of the X-ray structure factor, caused by the exchange interactions, are examined. Following a procedure which is similar to the Opechowski and Ornstein-Zernike procedures, estimates of the spatial and magnetic order-disorder temperatures are also made.

The study of Ising ferromagnets with magnetic, or non-magnetic, impurities has long been an area of active investigation[1]. Likewise, the similar problem of order-disorder phenomena in multicomponent alloys[2] has occupied the attention of physicists and metallurgists for a long time. However, studies of the interaction between order-disorder phenomena and magnetic phenomena[3] have been somewhat scarce in comparison.

Using techniques developed earlier[2], we investigate the properties of a binary Ising paramagnet with arbitrary concentrations of magnetic atoms having arbitrary spins. The appropriate correlation functions, i.e., magnetic and spatial, are calculated in terms of the concentrations, spin magnitudes and the magnetic and atomic interactions. Having done this, we then indicate how these correlation functions can be related to their corresponding ordering temperatures. These, in turn, tell us how the atomic interactions influence the magnetic behavior of the system, and, of course, how the magnetic interactions influence the spatial behavior.

Consider a binary alloy with an arbitrary concentration of Ising spins of magnitudes S_A and S_B. The Hamiltonian for this system is given by $H = H_{MAG} + H_{OD}$, where,

$$H_{MAG}=-\sum_{i,j}J^{AA}_{(ij)}S^z_{iA}S^z_{jA}\sigma^A_i\sigma^A_j-\sum_{i,j}J^{BB}_{(ij)}S^z_{iB}S^z_{jB}\sigma^B_i\sigma^B_j-\sum_{i,j}J^{AB}_{(ij)}\left\{S^z_{iA}S^z_{jB}\sigma^A_i\sigma^B_j+S^z_{iB}S^z_{jA}\sigma^B_i\sigma^A_j\right\}-h^A\sum_i S^z_{iA}\sigma^A_i-h^B\sum_i S^z_{iB}\sigma^B_i, \qquad (1a)$$

and,

$$H_{OD}=\tfrac{1}{2}\sum_{i,j}\left\{V^{AA}_{(ij)}\sigma^A_i\sigma^A_j+V^{BB}_{(ij)}\sigma^B_i\sigma^B_j+V^{AB}_{(ij)}\left(\sigma^A_i\sigma^B_j+\sigma^A_j\sigma^B_i\right)\right\}. \qquad (1b)$$

$S^z_{i\nu}$ is the z-component of a spin of type ν at site i; h^A and h^B are proportional to the Zeeman fields; σ^λ_i is an occupation operator such that $\sigma^\lambda_i = 1$ if a λ-type spin is at site i, and is zero otherwise. The magnetic interactions between spins of types λ and ν, at sites i and j respectively, are given by $J^{\lambda\nu}(ij)$, while the comparable atomic interactions are given by $V^{\lambda\nu}(ij)$.

This Hamiltonian can be transformed to an equivalent grand canonical Hamiltonian by replacing the occupation operators by pseudo-spin operators, c^z_i, $[\sigma^A_i = \tfrac{1}{2} + c^z_i;\ \sigma^B_i = \tfrac{1}{2} - c^z_i]$ which also specify the configuration of the spins, and by introducing an arbitrary chemical potential μ, to be determined such that

$$\langle c^z_i\rangle \equiv m = \tfrac{1}{2}(c_A - c_B). \qquad (2)$$

c_ν is the concentration of ν-type spins ($c_A + c_B = 1$) and the angular brackets represent a thermal averaging. The grand canonical Hamiltonian thus becomes,

$$=\tfrac{1}{2}\sum_{i,j}J(ij)c^z_ic^z_j-\mu\sum_i c^z_i-\tfrac{1}{4}\sum_{i,j}\left\{J^{AA}_{(ij)}S^z_{iA}S^z_{jA}+J^{BB}_{(ij)}S^z_{iB}S^z_{jB}+2J^{AB}_{(ij)}S^z_{iA}S^z_{jB}\right\}-\frac{h^A}{2}\sum_i S^z_{iA}-\frac{h^B}{2}\cdot\sum_i S^z_{iB}-\sum_{i,j}\left\{J^{AA}_{(ij)}S^z_{iA}S^z_{jA}c^z_i-J^{BB}_{(ij)}S^z_{iB}S^z_{jB}c_i+J^{AB}_{(ij)}\cdot S^z_{iA}S^z_{jB}c^z_i-J^{AB}_{(ij)}S^z_{iA}S^z_{jB}c^z_j\right\}-h^A\sum_i S^z_{iA}c^z_i+h^B\sum_i S^z_{iB}c^z_i-\sum_{i,j}\left\{J^{AA}_{(ij)}S^z_{iA}S^z_{jA}+J^{BB}_{(ij)}S^z_{iB}S^z_{jB}-2J^{AB}_{(ij)}S^z_{iA}S^z_{jB}\right\}\cdot c^z_ic^z_j, \qquad (3)$$

where $J(ij) = V^{AA}_{(ij)} + V^{BB}_{(ij)} - 2V^{AB}_{(ij)}$.

The correlation functions are given by,

$$\langle A\rangle = \frac{\mathrm{Tr}_{c,s}\,e^{-\beta\mathcal{H}}A}{\mathrm{Tr}_{c,s}\,e^{-\beta\mathcal{H}}}, \qquad (4)$$

where A is a product of configuration (c^z_i) and spin (S^z_{iA}, S^z_{jB}) operators. The trace is taken over spin variables first, and then over configuration variables, and corresponds to the annealed alloy . The correlation functions of interest here are the "two-site" functions, i.e. the X-ray short range order function, $\langle c^z_ic^z_j\rangle$, and the spin correlation functions, $\langle S^z_{iA}S^z_{jA}\rangle$, $\langle S^z_{iB}S^z_{jB}\rangle$, and $\langle S^z_{iA}S^z_{jB}\rangle$.

Appropriate Green's function equations can be generated, using the grand canonical Hamiltonian, in order to evaluate the correlation functions in (4). This has been done and, since we are interested in the binary paramagnet, a high temperature expansion has been performed on the correlation functions generated in this manner. As a test of this equation of motion method, a direct expansion of (4) has been performed which substantiates the results obtained by the Green's function method. For the Ising paramagnet with $h^A=h^B=0$, we have $\langle S^z_{iA}\rangle = \langle S^z_{iB}\rangle = 0$, and the notation,

$$\langle (S^z_{i\nu})^2 \rangle \equiv x_\nu = \tfrac{1}{3} S_\nu (S_\nu + 1), \tag{5}$$

has been used.

The results for the configuration correlation, i.e. spatial short range order, are:

$$\langle c^z_g c^z_p \rangle = m^2 - \beta c^2_A c^2_B J(gp) + 2\beta^2 m^2 c^2_A c^2_B J(gp)^2 \cdot + \beta^2 c^3_A c^3_B \sum_f J(gf) J(pf) + 2\beta^2 x^2_A c^2_A c^2_B J^{AA}(gp)^2 + 2\beta^2 \cdot$$
$$\cdot x^2_B c^2_A c^2_B J^{BB}(gp)^2 - 4\beta^2 x_A x_B c^2_A c^2_B J^{AB}(gp)^2 + \delta_{gp}\left\{c_A c_B - \beta^2 c^3_A c^3_B \sum_f J(gf)^2\right\} + O(\beta^3). \tag{6}$$

The magnetic correlation function, or magnetic short range order is given by:

$$\langle S^z_{gA} S^z_{pA} \rangle = x_A \delta_{gp} + 2\beta x^2_A c^2_A J^{AA}(gp) + 4\beta^2 \cdot \cdot x^3_A c^3_A \sum_f J^{AA}(gf) J^{AA}(pf) + 4\beta^2 x^2_A x_B c^2_A c_B \sum_f J^{AB}(gf) \cdot J^{AB}(pf) - 2\beta^2 x^2_A c^2_A c^2_B J^{AA}(gp) J(gp) - \beta^2 \delta_{gp} \cdot \cdot \left\{4 x^3_A c^3_A \sum_f J^{AA}(gf)^2 + 4 x^2_A x_B c^2_A c_B \sum_f J^{AB}(gf)^2\right\} + O(\beta^3), \tag{7}$$

[$\langle S^z_{gB}\, S^z_{pB} \rangle$ is obtained by interchanging A ↔ B]. The magnetic short range order between unlike species of spins is given by,

$$\langle S^z_{gA} S^z_{pB} \rangle = 2\beta x_A x_B c_A c_B J^{AB}(gp) + 4\beta^2 x^2_A x_B c^2_A c_B \cdot \cdot \sum_f J^{AA}(gf) J^{AB}(pf) + 4\beta^2 x_A x^2_B c_A c^2_B \sum_f J^{BB}(gf) J^{AB}(pf) + 2\beta^2 x_A x_B c^2_A c^2_B J^{AB}(gp) J(gp) - \delta_{gp} \beta^2 \left\{4 x^2_A x_B c^2_A c_B \cdot \cdot \sum_f J^{AA}(gf) J^{AB}(gf) + 4 x_A x^2_B c_A c^2_B \sum_f J^{BB}(gf) J^{AB}(gf)\right\} + O(\beta^3). \tag{8}$$

The spatial (Warren-Cowley) short range order parameter, $\alpha_c(gp)$, is related to the spatial correlation function by,

$$c_A c_B \alpha_c(gp) = \langle c^z_g c^z_p \rangle - m^2 \tag{9}$$

As the temperature of the binary paramagnet approaches the critical temperature for spatial ordering, or phase separation, the Fourier transform of $\alpha_c(gp)$ approaches infinity, indicating long range correlations. An appropriate representation of this behavior[2] is

$$\alpha_c(\vec{k}) = D(1 - X(\vec{k}))^{-1} \tag{10}$$

where D is a normalization constant, and $X(\vec{k})$ is obtained by inverting (6), along with the definition (9). Values of the critical temperature for spatial ordering of phase separation are thus obtained by solving the equation, $1 - X_c(\vec{k}_m) = 0$, where $X_c(\vec{k}_m)$ is a polynomial in β_c. These solutions for β_c, the order-disorder transition temperature, are functions of the spin magnitudes and of the magnetic interactions, as well as functions of the concentrations and atomic interactions.

In a similar way the magnetic correlation functions can be related to the susceptibility, and singularities in the susceptibility lead to expressions for the magnetic ordering temperature which are dependent on the concentration of spins and on the atomic interactions, in addition to the magnetic interactions, and spin magnitudes. A more detailed analysis of this work will be published elsewhere.

References

[1] D.C. Rapaport, J. Phys. C: Solid State Physics 5 (1972), 1830.

[2] G.B. Taggart and R.A. Tahir-Kheli, Physica 68 (1973).

[3] T.W. McDaniel and C.L. Foiles, Solid State Comm. 14 (1974), 835.

[4] R. Brout, Phys. Rev. 115 (1959), 824.

* Part of this work was done at Temple University under NSF Grant #GH 39023.

+ Supported by NSF Grant #GH 39023.

HIGH TEMPERATURE SUSCEPTIBILITIES IN QUENCHED MAGNETIC ALLOYS

Tatuo Kawasaki†
Physics Department, Temple University
Philadelphia, Pennsylvania 19122

ABSTRACT

The method of high temperature series expansion is applied to study the dependence of the Curie temperature T_C and the susceptibility exponent γ on the concentration p_A of A atoms using the following Hamiltonian

$$\mathcal{H} = -\sum_{\substack{\lambda\lambda' \\ ij}} \{J_L^{\lambda\lambda'}(ij)S^z_{i\lambda}S^z_{j\lambda'} + J_T^{\lambda\lambda'}(ij)(S^x_{i\lambda}S^x_{j\lambda'} + S^y_{i\lambda}S^y_{j\lambda'})\}\sigma_i^{\lambda}\sigma_j^{\lambda'}$$

$$- H\sum_{i\lambda} u^{\lambda} S^z_{i\lambda}\sigma^{\lambda} .$$

λ stands for A or B and σ's are the occupation operators. Judging from the results, the case of spin one-half in the Heisenberg model seems to be exceptional to the others in several respects. The value γ is, as is expected, rather insensitive to the concentration or the anisotropy of coupling constants J_T/J_L and J^{AB}/J^{AA}, except at their limiting values. The critical concentration in the A-B alloy has been estimated to be $p_A = 0.11$ in the classical limit($S=\infty$).

INTRODUCTION

For randomly diluted ferromagnet models, including the site as well as the bond cases, much work has been done using the high-temperature series procedures[1-3]. In this paper we apply this method to a random magnetic alloy with two magnetic species, A and B, whose Hamiltonian is as given in the abstract above[4]. The occupation operator σ_i^λ takes the value 1 or 0 according to whether the site is occupied by a λ-atom or not.

By calculating the magnetization first by the formula

$$< \sigma_0^A S_{0A}^z > = (\mathrm{Tr}\ e^{-\beta\mathcal{H}})^{-1}\ \mathrm{Tr}\ \sigma_0^A S_{0A}^z\ e^{-\beta\mathcal{H}}$$

in powers of $J^{AA}/k_BT(\equiv K)$, and then averaging over the randomness[5], the reduced total susceptibility χ is obtained up to order K^5 in a simple cubic lattice;

$$\chi = \sum_n a_n(S_A, S_B, p_A, J_L^{AA}, J_L^{AB}, J_L^{BB}, J_T^{AA}, J_T^{AB}, J_T^{BB})\ K^n$$

where a_n's are rather lengthy expressions depending on the parameters indicated. For simplicity in the following discussion, only five parameters in the a_n's are explicitly retained; S_A, S_B, p_A, J^{AB}/J^{AA}, and the anisotropy of the exchange constant J_T/J_L for the sake of avoiding unnecessary complication. The Padé approximant is applied to the derivative of log χ with respect to K. Owing to the shortness of our series, only [1,1], [2,1] and [1,2] approximants are usable in the present evaluation of T_c and γ. [1,2] will be mainly used hereafter which has two poles. [M,N] denotes the Padé approximant using a function $\sum_m b_m z^m / \sum_n c_n z^n (m_{max}=M,\ n_{max}=N)$.

RESULTS

Although the whole expression for χ can not be presented here within a limited space, the formula has been tested in its limiting cases by using those of perfect and dilute crystals. Various symmetric properties are also used as an important check.

Curie temperature $\Theta_c(S_A, S_B, p_A, J^{AB}, J_T/J_L)$

The first figure shows the result of the normalised Curie temperature $\Theta_c(p_A)/\Theta_c(0)$ for a model with $S_A=S_B=1$. The qualitative profile of graphs is almost the same as that in the molecular field theory(MF), which is obtainable by using only the first two coefficients of the series. Near $p_A = 0$ the values of $\Theta_c(p_A)$ seem to vary linearly with p_A as in the dilute cases[1-3]. For spins bigger than 1, there appears no significant change in their behaviors. In Fig. 2 the abscissa is scaled by J^{AB}/J^{AA}. As the magnitude of spin increases, curves both in Ising and Heisenberg approach to those of the molecular field theory, which is linear in J^{AB}.

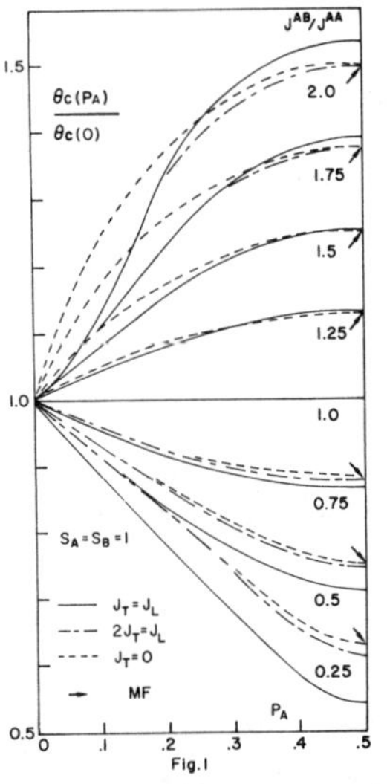

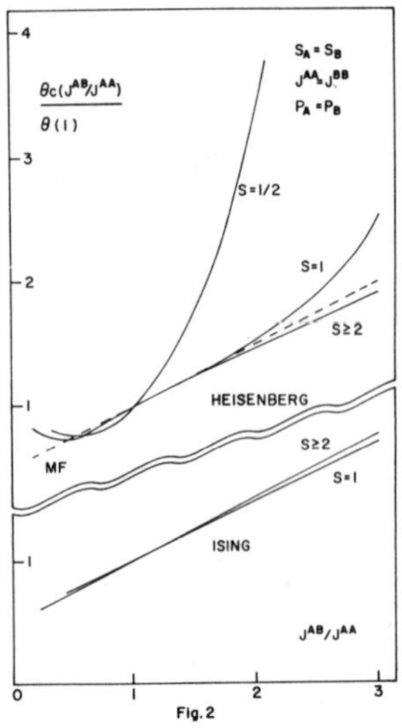

Fig. 1 Curie Temperature as a function of p_A. The arrows indicate values obtained by the molecular field theory.

Fig. 2 Curie Temperature as a function of J^{AB}. Dashed lines show results given by the molecular field theory

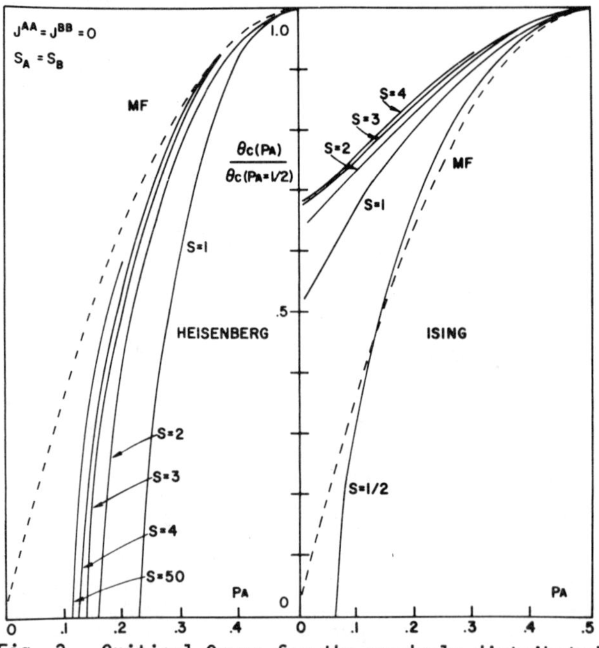

Fig. 3 Critical Curve for the randomly distributed AB-magnetic alloy with $J^{AA} = J^{BB} = 0$, and $J^{AB} = 0$. The dashed line shows the result in the molecular field theory.

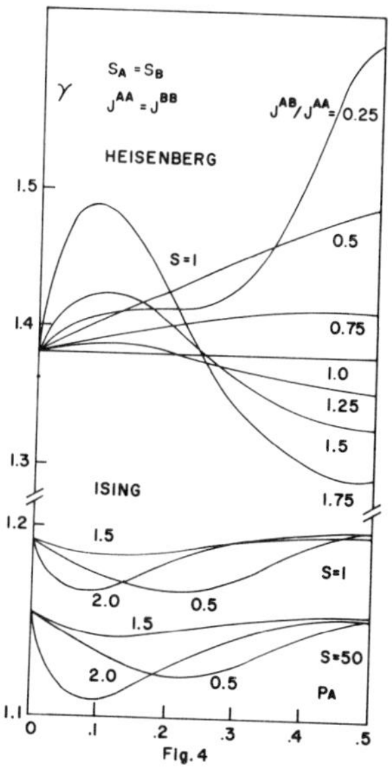

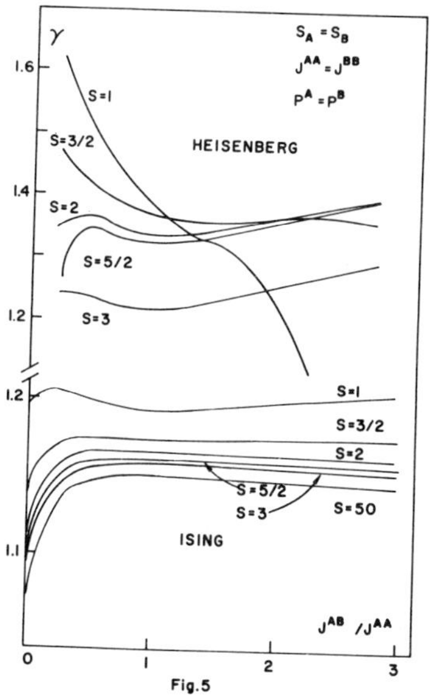

Fig. 4 Susceptibility Exponent as a function of p_A.

Fig. 5 Susceptibility Exponent as a function of J^{AB}.

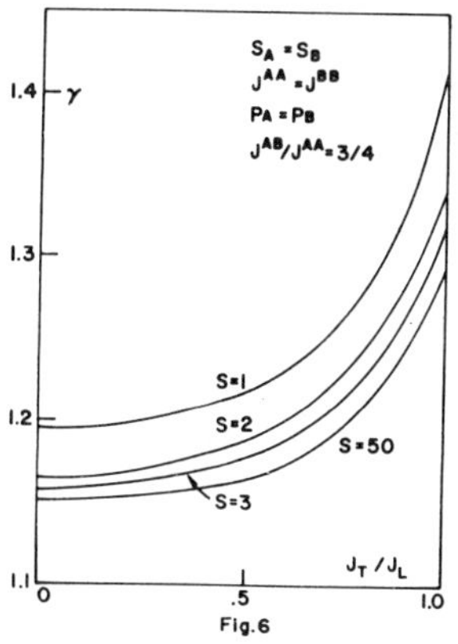

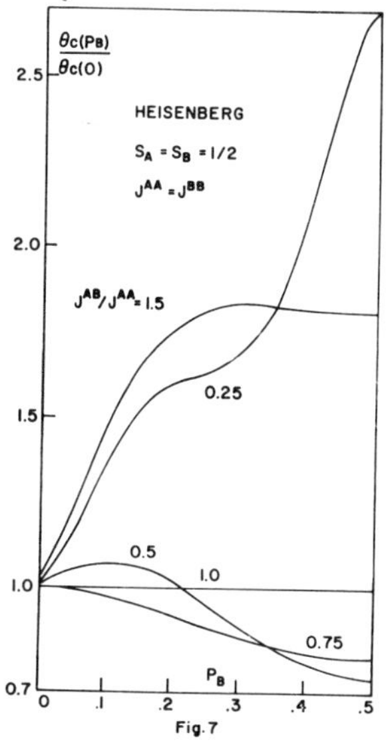

Fig. 6 Susceptibility Exponent as a function of J_T/J_L.

Fig. 7 Curie Temperature as a function of p_A for the Heisenberg model with S = 1/2.

Critical Concentration p_c

In a special case where $J^{AA}=J^{BB}=0$ and $J^{AB}\neq 0$ in the binary magnet, there may exist some sort of a critical concentration p_c as in the dilute magnet. In the molecular field theory $\Theta_c(p_A) = 2\Theta_c(1/2)\sqrt{p_A p_B}$ which has no critical concentration, while in the BPW method we may have a critical concentration $p_c=0.042(\sqrt{p_A p_B} = (z-1)^{-1})$ in the simple cubic lattice. Fig.3 shows how Θ_c varys with the parameter J_T/J_L. In the present analysis, the p_c for the Heisenberg and the Ising model are found to be different from each other, which should be unsatisfactory.

Critical Exponent γ

Since in the present approximation the value of γ (S=1/2) is 0.498(0.5 inMF) in the pure crystal, the detailed, quantitative discussion on γ seems to be inadequate. Fig. 4 shows variation of γ with p_A. Except near the region $J^{AB}=0$, the index γ approaches a steady value, although the absolute value is not definitely determined. As the system becomes Ising-like($J_T \approx 0$), γ approaches a steady value in a wider range of J^{AB}, which is shown in Fig. 5. Fig. 6 may show that γ will be constant in the region $0 \leq J_T/J_L \leq 1$ and suddenly or very sharply change its value at or near $J_T/J_L=1$.

DISCUSSION

After looking at all the data, including those not recorded here, the case of spin one-half in the Heisenberg model gives us considerably unsatisfactory results both for Θ_c and γ. One of the examples is seen in Fig. 7. In some range of S and J^{AB}, Θ_c becomes sometimes complex even though the coefficients of series seem to be fairly monotonic at a glance. It is also noted that the Padé approximant lower than [1,2] which we used does not give us no critical concentration even in the Heisenberg case. The preliminary calculation about the percoration probability shows that p_c is 0.31 for a square lattice with 4096 atoms and is less than 0.30 for a simple cubic lattice. Further discussion will be given elswhere, with the whole expression of χ.

The author would like to thank Prof. R.Tahir-Kheli for his cooperation in this work.

† Supported by NSF Grant #GH 39023.

On leave from College of Liberal Arts and Science, Kyoto University, Kyoto, Japan 606.

1 G.S.Rushbrooke and D.J.Morgan, Mol.Phys. 4 1(1961)
2 D.J.Morgan and G.S.Rushbrooke, ibid. 4 291(1961).
3 E.Brown, J.W.Essam and C.M.Place (to be published).
4 R.A.Tahir-Kheli, Phys.Rev. B6 2808(1972).
5 R.Brout, Phys.Rev. 115 824(1959).

SHORT RANGE ORDER EFFECTS IN RANDOMLY DILUTED QUENCHED HEISENBERG PARAMAGNET

Kazuko Kawasaki† and Raza A. Tahir-Kheli*
Department of Physics, Temple University
Philadelphia, Pennsylvania 19122

ABSTRACT

The temperature and concentration dependence of second and fourth moments of the spectral line shape are investigated for the randomly diluted Heisenberg paramagnet. We also study the first four terms of high temperature expansion of the wave-dependent susceptibility and the spin correlation function. The results are compared with those of Collins[1] for the limiting case, i.e. when the non-magnetic impurity concentration is vanishing. Finally, using phenomenological procedures based on the use of the first four moments[2], we construct the neutron scattering lineshapes. The results for the lineshape are compared and contrasted with those recently obtained by us for the case of infinite temperature.[3]

1. M. F. Collins, Phys. Rev. 2, 4552 (1970); 4, 1588 (1971).
2. R. A. Tahir-Kheli and D. G. McFadden, Phys. Rev. 182, 604 (1969).
3. R. A. Tahir-Kheli and Kazuko Kawasaki, Physical Rev. (to be published)

†Permanent address: Department of Physics, Nara Women's University, Nara, 630 Japan.
*Supported by the U. S. National Science Foundation Grant #GH 39023.

MAGNETISM IN RARE EARTH-TRANSITION METAL AMORPHOUS ALLOY FILMS

Kenneth Lee and Neil Heiman
IBM Research Laboratory, San Jose, California 95193

ABSTRACT

A number of rare earth (RE) transition metal (TM) amorphous alloy films (RE = Gd, Ho, Tb, Y; TM = Co, Fe, Ni, Mn) have been prepared across a broad compositional range by thermal evaporation. The samples were examined by polar Faraday rotation, polar Kerr rotation and magnetometer techniques. Results show that the major differences between amorphous and crystalline alloys are due to the altered state of the TM. Specifically the ordering temperature of amorphous RE-Ni and RE-Co alloys is higher than in the corresponding crystalline alloys, whereas the reverse is true for RE-Fe alloys; at the same time, the TM "sublattice" magnetization is larger in the amorphous state for both Fe and Co compounds. Furthermore amorphous RE-Mn alloys are not ordered. These facts tend to support the general concepts of the charge transfer model. Other properties such as coercivity and anisotropy are strongly dependent on the RE constituent.

INTRODUCTION

In order to determine the systematics of the magnetic interactions in amorphous rare earth (RE) - transition metal (TM) alloy films and to test theoretical models[1,2] of those interactions, we have prepared and studied several different binary RE-TM alloy systems across broad compositional ranges. To isolate the effects due to the TM, we have varied the TM constituent systematically with Mn, Fe, Co and Ni for a given RE. Similarly keeping each TM fixed, we have varied the RE constituent with Y, Gd, Ho and Tb. Although Y is not strictly a RE element, it has been traditionally found to be a good RE substitute. Thus one might expect the Y-TM alloys to provide direct evidence of the TM behavior. Gd, being the most magnetic of the RE metals and a spherically symmetric ion, should provide information on the effect of the RE magnetism without the complications of orbital moments or local anisotropy. Ho, on the other hand having the largest orbital moment, can provide information on such effects. Tb is the most anisotropic of the heavy RE atoms and should most strongly exhibit the effects of local anisotropy.

EXPERIMENT

Amorphous RE-TM films were prepared by two source thermal evaporation in a UHV system. Film thicknesses ranged from 0.07 to 1.0 μm and chemical analysis was monitored by X-ray fluorescence and electron microprobe analysis on a number of samples to insure the composition determined by the rate monitors.

The polar Faraday rotation or polar Kerr rotation of the films were measured from about 100K (the lower limit of our cryostat) to about 600K (the temperature at which recrystallization and oxidation become a consideration). The magneto-optic signals at λ = 0.6328μm were generally large and exhibited abrupt sign reversal at the compensation temperature (T_{comp}), thus providing an excellent determination of T_{comp}, the coercive field (H_c) and the Curie temperature (T_c). An important point is that the magneto-optic data, particularly in regard to T_{comp} determinations, are not subject to questions of saturation, sample density, and other technical difficulties normally associated with magnetometer measurements. For completeness vibrating sample magnetometer measurements were made from 4.2K to 300K using a superconducting magnetometer capable of 70 kOe. With the exception of some Tb-Fe and Tb-Co compositions, no difficulties were encountered in saturating our samples.

TABLE I

	T_c (K)		T_{COMP} (K)		μ(4.2K) (μ_B/F.U.)	
ALLOY	XTAL	AMORP	XTAL	AMORP	XTAL	AMORP
$GdCo_2$	409	550	Gd Dom.	510	5.0	4.2
$GdCo_3$	612	750	Gd Dom.	400	2.2	2.8
Gd_2Co_7	775	>500	410	300	2.4	4.2
$GdCo_5$	1008	>500	–	80	1.2	0
$Gd_{57}Fe_{43}$	–	350	–	Gd Dom.	–	–
$Gd_{40}Fe_{60}$	–	>500	–	Gd Dom.	–	–
$GdFe_2$	785	490	Gd Dom.	450	2.8	–
$GdFe_3$	728	460	–	150	1.6	–
Gd_6Fe_{23}	659	420	–	≈100	–	–
$GdNi_2$	85	38			7.1	8.7
$GdMn_2$	(AF) Tn=86	NOT MAG.	–	–	–	–
Gd_6Mn_{23}	473	NOT MAG.	–	–	–	–

TABLE II

	T_c (K)		T_{COMP} (K)		μ(4.2K) (μ_B/F.U.)	
ALLOY	XTAL	AMORP.	XTAL	AMORP.	XTAL	AMORP.
$Ho_{45}Co_{55}$	–	375	–	Ho Dom.	–	–
$Ho_{40}Co_{60}$	–	600	–	325	–	–
$HoCo_2$	85	>600	Ho Dom.	270	7.8	4.8
$HoCo_3$	418	>600	350	150	5.6	4.3
$HoCo_5$	1000	>600	80	Co Dom.	1.1	0.2
$Ho_{40}Fe_{60}$	–	250	–	180	–	–
$HoFe_2$	612	260	Ho Dom.	120	5.5	2.0
$HoFe_3$	567	290	400	50	4.6	2.8
Ho_6Fe_{23}	501	300	40	Fe Dom.	–	–
Ho_2Ni_{17}	162	>400			12.2	–
$HoNi_5$	10	400			8.1	–
$HoNi_2$	22	15			8.4	4.0*
$HoMn_2$	–	NOT MAG.			–	0
Ho_6Mn_{23}	434	NOT MAG.			49.8	0

* Still very temperature dependent at 4.2K

TABLE III

	T_c (K)		T_{COMP} (K)		μ(4.2K) μ_B/F.U.	
ALLOY	XTAL	AMORP.	XTAL	AMORP.	XTAL	AMORP.
$TbCo_2$	256	>600	Tb Dom.	500	6.7	4.3
$TbCo_3$	506	>600	Tb Dom.	250	3.4	2.5
$TbCo_5$	980	>600	100K	Co Dom.	0.5	2.4
$TbFe_2$	711	390	Tb Dom.	Tb Dom.	4.7	>4.0

TABLE IV

ALLOY	T_c (K) XTAL	T_c (K) AMORP.	μ(4.2K) μ_B/T.M. ATOM XTAL	μ(4.2K) μ_B/T.M. ATOM AMORP.
Y_4Co_3	13	450	0.03	.2
Y Co	–	550	–	.6
$Y Co_2$	NOT MAG.	>600	0	1.6
$Y Co_3$	301	>600	0.5	–
$Y Co_5$	977	>600	1.4	1.5
Y Fe	–	NOT MAG.	–	0
$Y Fe_2$	550	400	1.4	.2
$Y Fe_3$	–	>450	–	.8
$Y_6 Fe_{23}$	484	–	1.6	–
$Y Fe_5$†	–	>600	–	1.9

† Showed evidence of bcc Fe microcrystalline phase

RESULTS

Although many different alloy compositions were investigated, the difference between amorphous and crystalline RE-TM alloys are best seen by examining the data summarized in Tables I-IV. In these tables the data for crystalline alloys were obtained from the literature.[3] Also the data on amorphous Gd-Co alloys of Tao et al.[4] are included for completeness.

With some exceptions to be discussed, certain general trends occur in the data for all amorphous RE-TM alloys.

1) T_c of Co and Ni compounds are higher for amorphous than crystalline alloys.
2) T_c of Fe alloys are lower for amorphous than crystalline alloys.
3) Amorphous Mn compounds do not appear to be magnetically ordered.
4) The 4.2K magnetic moments of the amorphous alloys (with the notable exception of Y-Fe) are more TM dominated than in crystalline alloys.
5) T_{comp} of all amorphous alloys are lower than T_{comp} of the crystalline materials.
6) Just as in the case of crystalline materials, amorphous Gd alloys have higher T_c than Tb alloys which in turn have higher T_c than Ho alloys.

The trend described in item 6 is broken with Y compounds which appear to be anomalous in a number of ways (as also noted in item 4). The Y-Co alloys seem to behave as other amorphous RE-Co alloys in that T_c and the magnetic moment are increased; however, T_c are higher than for Gd-Co alloys. Amorphous Y-Fe is more anomalous in that the Fe moment drops sharply as Y content increases beyond 20%. Similar behavior for amorphous YFe_2 has been alluded to by Rhyne et al.[5] Furthermore, T_c and magnetic moment increase linearly with increasing Fe content as if Y were a pure dilutent of bcc Fe. The X-ray diffraction pattern and room temperature Fe^{57} Mössbauer spectrum of YFe_5 show the presence of microcrystalline bcc Fe, whereas X-ray and Mössbauer effect data show amorphous behavior for all other RE TM_5 films. Clearly in the case of amorphous alloys, unlike the crystalline alloys, Y must not be treated simply as a non-magnetic RE but rather must be handled as a separate case altogether.

Thus excluding Y-TM films, the trends in the data outlined above can best be explained in terms of a picture for amorphous RE-TM films in which:

1) The TM magnetic moments at 4.2K are larger than in the crystalline compounds.
2) The RE magnetic moments at 4.2K are near their crystalline value.
3) Co-Co and Ni-Ni exchange is enhanced but Fe-Fe (and Mn-Mn) exchange is reduced.
4) RE-RE exchange and/or RE-TM exchange is reduced.

This picture rules out the local anisotropy model proposed by Harris et al.[2] and to some extent supports the concept of reduced charge transfer suggested by Tao et al.[1] as we have reported in a previous paper.[6] However, the lack of magnetic ordering in amorphous RE-Mn compounds and the anomalous behavior of amorphous Y-TM alloys imply that the absence of specific crystal structures is also an important consideration and that perhaps for these amorphous alloys atomic spacing and coordination plays a significant role. Cargill[7] showed by radial distribution analysis of amorphous Gd Fe_2 that relative to the crystalline Laves phase compound, the Fe-Fe distance decreased from 2.60Å to 2.54Å, the Gd-Gd distance increased from 3.18Å to 3.47Å and the coordination numbers for Fe-Gd neighbors were reduced by half. It is interesting to note that in crystalline alloys with Fe-Fe spacings less than 2.6Å, the Fe magnetic moments couple antiferromagnetically.[8] This, together with Cargill's results, suggest that the effects resulting from altered atomic dimensions and coordination can not be neglected. Similar radial distribution function results for other amorphous RE-TM alloys would support the hypothesis that the altered spacing and coordination reduce Fe-Fe ferromagnetic exchange, enhance Co-Co and Ni-Ni exchange and reduce RE-TM and RE-RE exchange.

Finally, although the anisotropy and orbital moment of the RE does not play a significant role in the altered exchange in amorphous RE-TM alloys, it strongly affects H_c. Above room temperature all amorphous alloys exhibit H_c on the order of a few Oe. As the temperature is reduced to room temperature or slightly below Tb-TM alloys develop H_c values greater than 20 kOe. At very low temperatures Ho-Fe compounds also develop large H_c.

ACKNOWLEDGEMENTS

We wish to thank S. L. Lawrence for his technical assistance and R. Geiss for X-ray fluorescence analysis. We particularly wish to thank J. Smit and J. C. Suits for their suggestions and helpful discussions.

REFERENCES

1. L. J. Tao, S. Kirkpatrick, R. J. Gambino and J. J. Cuomo, Solid State Commun. 13, 1491 (1973).
2. R. Harris, M. Plischke, and M. J. Zuckermann, Phys. Rev. Letters 31, 160 (1973).
3. K. N. R. Taylor, Advan. Phys. 20, 551 (1971), and references cited therein.
4. L. J. Tao, R. J. Gambino, S. Kirkpatrick, J. J. Cuomo, and H. Lilienthal, in Magnetism and Magnetic Materials-1973, AIP Conference Proceedings No. 18, edited by C. D. Graham, Jr. and J. J. Rhyne, p 641 (1974).
5. S. J. Pickart, J. J. Rhyne and H. A. Alperin, Phys. Rev. Letters 33, 424 (1974).
6. N. Heiman and K. Lee, Phys. Rev. Letters 33, 778 (1974).
7. G. S. Cargill III, in Magnetism and Magnetic Materials-1973, AIP Conference Proceedings No. 18, edited by C. D. Graham, Jr. and J. J. Rhyne, p 631 (1974).
8. R. Forrer, J. Physique 7, 605 (1952).

TEMPERATURE DEPENDENCE OF MAGNETIZATION IN AMORPHOUS Gd-Co-Mo FILMS

R. Hasegawa, B. E. Argyle and L-J. Tao
IBM Thomas J. Watson Research Center, Yorktown Heights, New York 10598

ABSTRACT

The temperature dependence of the magnetization of amorphous ferrimagnetic $(Gd_{1-x}Co_x)_{1-y}Mo_y$ films prepared by rf sputtering with x=0.79–0.88 and y=0.08–0.16 was studied in the temperature range 4.2-500°K. The data, taken using a vibrating sample magnetometer and the polar Kerr effect, were found to be satisfactorily described by Néel's two sublattice model.

INTRODUCTION

Metallic rare-earth cobalt amorphous films have received considerable attention largely because of their potential application in magnetic devices.[1-5] The magnetization for these films, however, has a strong temperature dependence at room temperature in the vicinity of which lies the magnetization compensation temperature. To modify this unfavorable temperature dependence of the magnetization, molybdenum is added to form a ternary system of composition $(Gd_{1-x}Co_x)_{1-y}Mo_y$. The choice of Mo as a third element is based on the following reasons: (i) Mo has a low density of states at the Fermi level and behaves like a non-magnetic metal as evidence by the Kondo effect in the Mo host dilute magnetic alloys.[6] (ii) The saturation magnetic moment on Co is reduced at a rapid rate of about $6.5\mu_B$ per fraction of Mo atom in the Co-Mo binary system.[7] We present in the following a simple molecular field analysis of the magnetization data for the present amorphous alloys. This may provide a basis for further predicting the magnetization behavior for different compositions of the present alloy system.

EXPERIMENTAL TECHNIQUES AND METHOD OF DATA ANALYSIS

The amorphous films, 1-2μm thick, were deposited by the sputtering techniques described elsewhere[1] on glass substrate from an arc-melted target of a composition close to $(Gd_{.25}Co_{.75})_{.85}Mo_{.15}$. The deposited film composition was determined by electron microprobe analysis with an absolute and relative accuracy of ±5 and ±2% respectively in the weight fractions measured, and ranges as $0.79\leq x<0.90$ and $0.078\leq y<0.16$. The magnetization was measured using a vibrating sample magnetometer (VSM) between 4.2 and 500°K with an applied field of 10 kOe and also from the field for saturation or the initial susceptibility determined from a polar Kerr hysteresis loop, a method mentioned in Ref. 4. The magnetization data thus obtained were fitted to a set of coupled Brillouin functions, one representing the Gd sublattice and the other the Co sublattice. In using the two sublattice model, we assume that the role of Mo is to fill the 3d band of Co. Thus the total magnetization per atom is given by

$$\bar{S} = |(1-y)(1-x)\bar{S}_1 + (1-y)\,x\,\bar{S}_2| \qquad (1)$$

where $\bar{S}_1$ and $\bar{S}_2$ are the Gd and Co sublattice magnetization respectively, and x and y are the fractions of Co and Mo atoms respectively in the system $(Gd_{1-x}Co_x)_{1-y}Mo_y$. The magnetization $\bar{S}_i(i=1,2)$ may be expressed by

$$\bar{S}_i = S_i B_{S_i}(g_i\mu_B H_i S_i/k_B T) \qquad (2)$$

where 1=Gd and 2=Co, S_i is the spin value for the atom i, μ_B is the Bohr magneton, k_B is the Boltzmann constant and we take $g_i=g=2$. The effective fields H_i can be written as

$$\left.\begin{aligned} H_1 &= 2J_{11}z_{11}\bar{S}_1/g\mu_B + 2J_{12}z_{12}\bar{S}_2/g\mu_B + H_a \\ H_2 &= 2J_{21}z_{21}\bar{S}_1/g\mu_B + 2J_{22}z_{22}\bar{S}_2/g\mu_B + H_a \end{aligned}\right\} \qquad (3)$$

where J_{11}, $J_{12}=J_{21}$ and J_{22} are the exchange constants for the Gd-Gd, Gd-Co and Co-Co interactions respectively, z_{ij} is the number of nearest neighbors of the atom j (j=1,2) for the atom i(i=1,2) and H_a is an applied field. We assume that atoms are randomly distributed over the non-crystalline atomic sites with an average coordination number of 12 as suggested from x-ray data for a similar amorphous system.[8] Then we have $z_{11}=12(1-y)(1-x)=z_{21}$ and $z_{22}=12x(1-y)=z_{12}$. The justification for using Eq. (2) to describe the magnetizaiton for the two sublattices is in order. It has been reported[9] that the magnetization of a single crystal Gd as a function of temperature can be expressed by a form similar to Eq. (2). This may be expected in view of the highly localized nature of the 4f band and also the fact L=0 for Gd. These facts further warrant that we may take $S_1=7/2$. Despite the spread of the 3d band $\bar{S}$ for 3d metals is known to obey the form of Eq. (2) relatively well. Thus we tentatively assume that the Co sublattice magnetization can be expressed by Eq. (2).

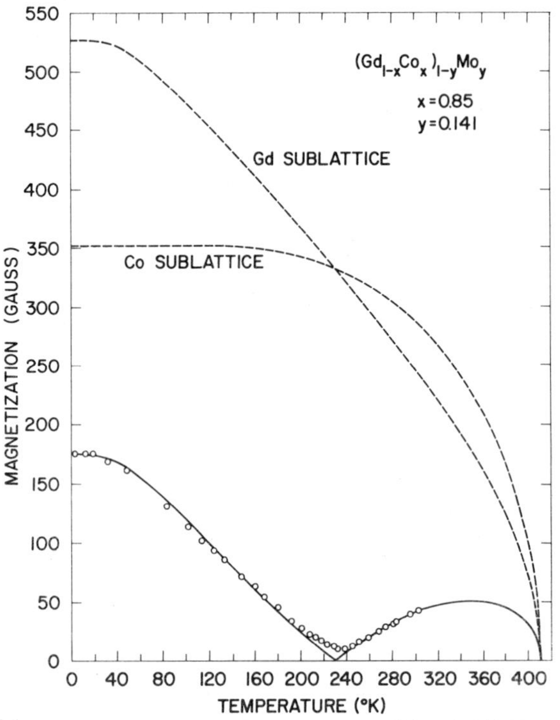

Fig. 1. Total and sublattice magnetizations as a function of temperature determined through Eq. (1)-(3) from the experimental data shown by open circles.

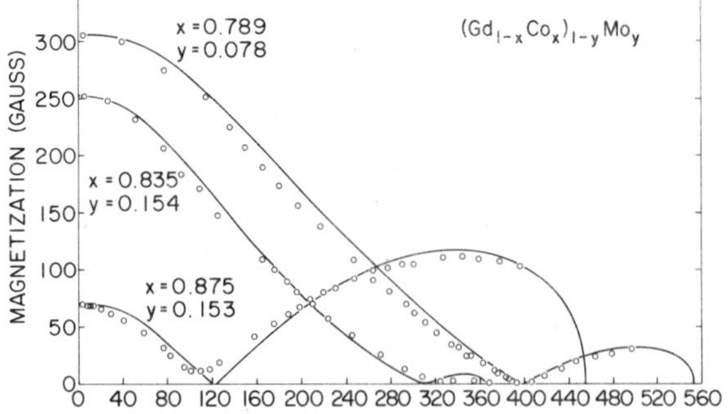

Fig. 2. Calculated (solid line) and experimental (open circle) magnetization versus temperature for various combinations of x and y.

RESULTS AND DISCUSSION

We show in Figs. 1 and 2 typical examples of the magnetization versus temperature VSM data fitted to Eq. (1) with the fitting parameters given in Table 1. We assume that the total number of atoms per cm^3 is given by $N=6.3\cdot10^{22}$ as previously.[4] It is noted that

TABLE I. Fitting parameters S_1, S_2, $J_{11}(10^{-16}erg)$, $J_{12}(10^{-15}erg)$ and $J_{22}(10^{-14}erg)$ for different values of x and y.

x	y	S_1	S_2	J_{11}	J_{12}	J_{22}
.798	.078	3.5	.577	2.0	-2.4	1.25
.835	.154	3.5	.387	2.0	-2.4	1.45
.85	.141	3.5	.413	2.0	-2.3	1.5
.875	.153	3.5	.419	2.0	-2.2	1.68

both J_{11} and J_{12} are fairly constant at 2.0 10^{-16} and $-(2.3\pm1)10^{-15}$ erg respectively. The relatively large value for the Gd-Co exchange constant is reflected in the temperature dependence of the Gd sublattice magnetization shown in Fig. 1. Especially in the vicinity T=0, the magnetization of Gd does not decrease rapidly with temperature which results in the net magnetization resembling that of a 3d metal for $T<<T_{comp}$ where T_{comp} is the magnetization compensation temperature. The exchange constant J_{22} for the Co-Co interaction was found to vary as $J_{22}=(5.1x-2.8\pm0.05)\cdot10^{-14}$ erg for the range of x covered in this study. The increase of J_{22} with the cobalt concentration may be expected and can be further clarified by a detailed structure study of the present alloy system. The saturation magnetic

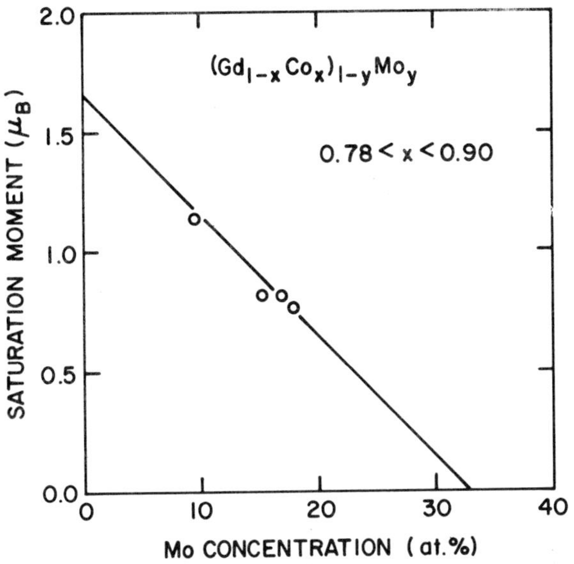

Fig. 3. Saturation moment for Co per atom versus fraction of Mo (y/[y+(1-y)x]).

moment for Co, μ_{Co}, given by gS_2 is plotted in Fig. 3 against the fraction of Mo with respect to Co concentration. Thus we have

$$\mu_{Co} = 1.65-5.0y/[y+(1-y)x] \ (\mu_B/\text{atom}). \quad (4)$$

The value of μ_{Co} extrapolated to the moment axis (y=0) corresponds to the average saturation moment of Co for the binary crystalline $Gd_{1-x}Co_x$ alloy in the present concentration range in which μ_{Co} is relatively constant at $1.65\mu_B$.[10] Furthermore μ_{Co} given in Eq. (4) is not very different from the similar relationship for the binary Co-Mo system.[7] These two findings, therefore, support the supposition that the role of Mo is essentially to fill the d-band of Co. By using the present exchange constant J_{11}, J_{12} and J_{22} and Eq. (4) with $S_1=7/2$ in Eqs. (1)-(3), we can predict compositional and temperature dependence of the magnetization within the concentration range considered here. From these magnetization curves we can further predict the compensation (T_{comp}) and the Curie temperature (T_c), which are shown in Fig. 4. We find the predicted T_c and T_{comp} and the corresponding experimental values agree within about 10%, as shown in Table II.

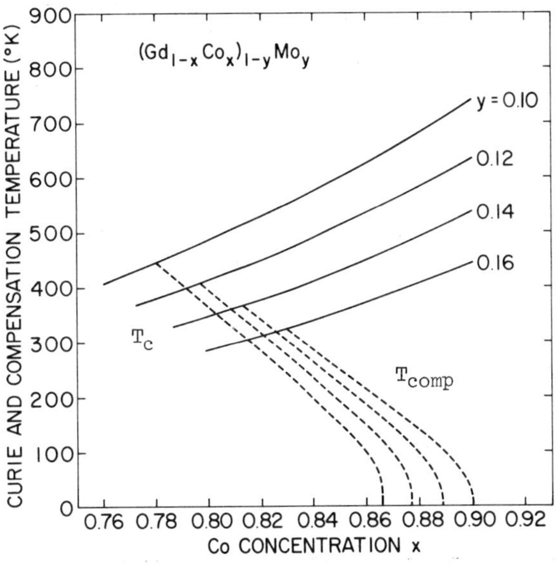

Fig. 4. Calculated Curie and compensation temperatures as a function of x with y as a parameter.

TABLE II. Calculated and observed values of T_c and T_{comp} for various combinations of x and y. The asterisks refer to the polar Kerr effect data.

x	y	Experimental T_{comp}	Experimental T_c	Calculated T_{comp}	Calculated T_c
.80	.144	none	335±5°K	none	335°K
.86	.16	238±1	383±1	210	364
*.85	.15	266±1	390±5	245	390
*.83	.11	238±1	---	250	515

Additional information extracted from the present analysis is the temperature dependence of the Co sublattice magnetization, which can be directly compared with the polar Kerr effect data. As we see in Ref. 11, agreement between the predicted and experimental Kerr rotation angle which is proportional to the Co sublattice magnetization is satisfactory.

ACKNOWLEDGEMENTS

One of the authors (R. H.) wishes to thank J. C. Slonczewski for his helpful discussion and advice, and would like to acknowledge A. A. Levi for instructing him in APL programming. R. J. Gambino and P. Chaudhari were helpful in various aspects of the present work.

REFERENCES

1. P. Chaudhari, R. J. Gambino and J. J. Cuomo, Appl. Phys. Lett. 22, 337 (1973).
2. M. H. Kryder and H. L. Hu, AIP Conf. Proc. 18, 213 (1974).
3. D. C. Cronemeyer, AIP Conf. Proc. 18, 85 (1974).
4. R. Hasegawa, J. Appl. Phys. 45, 3109 (1974).
5. R. Hasegawa, R. J. Gambino, J. J. Cuomo and J. F. Ziegler, J. Appl. Phys. 45, 4036 (1974).
6. A. Narath, K. C. Brog and W. H. Jones, Jr., Phys. Rev. B2, 2618 (1970).
7. R. M. Bozorth, Ferromagnetism (D. van Nostrand Co., N.Y. 1951) p. 292.

8. G. S. Cargill, III, AIP Conf. Proc. 18, 631 (1974).
9. H. E. Nigh, S. Legvold, F. H. Spedding, Phys. Rev. 132, 1092 (1963).
10. See for example, E. Burzo, Phys. Rev. B6, 2882 (1972).
11. B. E. Argyle, R. J. Gambino and K. Y. Ahn, (this conference).

MAGNETIC EXCITATIONS IN AMORPHOUS Co_4P

H. A. Mook
Solid State Division, Oak Ridge National Laboratory,* Oak Ridge, Tennessee 37830

D. Pan
Harvard University, Cambridge, Massachusetts 02138

J. D. Axe† and L. Passell†
Brookhaven National Laboratory, Upton, New York 11973

ABSTRACT

Neutron scattering techniques have been used to study the magnetic excitations in the amorphous ferromagnet Co_4P. Distinct spin waves with a narrow line width have been measured near momentum transfer $Q = 0$ using a three axis spectrometer. The measurements around $Q = 0$ can be performed only over a very restricted range in Q since the spin wave branch is quite steeply rising and at momentum transfers larger than 0.03 $Å^{-1}$ the spin wave energy is too high to conserve energy in the scattering process so that the scattering triangle cannot be closed. However, measurements were possible over a range of 0.005 to 0.03 $Å^{-1}$. These measurements could be well fit with a quadratic dispersion law $E = DQ^2$ if D has a value of about 185 meV $Å^2$.

Polarized neutron time-of-flight measurements were able to show additional low-lying spin wave excitations near the Q equal to the first peak in the static structure factor S(Q). The polarized beam technique is especially valuable for these measurements as only spin flip excitations are measured and the time-of-flight technique permits measurements at a large number of momentum transfers simultaneously. The spin wave excitations at these larger values of Q again closely follow a quadratic dispersion law and can be reasonably well fit by the relation $E = \Delta + Dq^2$ where q is measured from the Q equal to the first peak in S(Q), D is kept at 185 meV $Å^2$ and an energy gap Δ of about 35 meV is introduced. The gap is a result of the positional disorder of the system and the excitation spectrum appears in some respects similar to the phonon excitation spectrum for liquid ^4He, the higher Q excitations being analogous to the roton excitations in the liquid.

*Operated by Union Carbide Corporation for the USAEC.

†Research sponsored by the U. S. Atomic Energy Commission.

GALVANOMAGNETIC EFFECTS IN Gd-Co SPUTTERED FILMS

K. OKAMOTO, T. SHIRAKAWA, S. MATSUSHITA, AND Y. SAKURAI
Department of Control Engineering, University of Osaka,
Toyonaka, Osaka, Japan

ABSTRACT

The transverse Hall effect(THE) and the transverse magnetoresistance effect(TME) are measured simultaneously in Gd-Co sputtered films which have the uniaxial magnetic anisotropy perpendicular to the film plane. The results show that the THE loop is related to the magnetization curve, and that the Hall output voltage, V_H, is proportional to the total area of reversed magnetic domains in the sample. It is also found that the fractional change of resistance, $\Delta R/R$, depends on the number of magnetic domain walls with which the control current interacts. The magnitude of $\Delta R/R$ is of the order of 10^{-4} or less.

INTRODUCTION

It has been reported that Gd-Co sputtered films with magnetization perpendicular to the film plane exhibit a remarkable extraordinary Hall effect, and that the hysteresis loop of the Hall voltage agrees well with those observed by the Kerr magneto-optic effect and also is similar to the B-H curve obtained using the vibrating-specimen magnetometer.[1] The transverse magnetoresistance effect(TME) is also interesting because one can possibly acquire some information about the relationship between the domain configuration and the value of resistance(or resistivity) for such films. We therefore measured the THE and TME simultaneously in Gd-Co sputtered films.

EXPERIMENTS

The Gd-Co films are prepared by the rf-sputtering method using glass substrates in argon atmosphere. The composition is about 25 at. percent Gd.[2,3] The samples used are made from the as-sputtered films. Each sample is circular in shape with a diameter of a few millimeters and is provided with two pairs of terminals(one for the control current I and the other for the Hall output). The electrical contacts are made by conductive paint. In order to investigate the dependence of the THE and TME on the magnetic domain configuration, two types of samples are examined. That is, one type of samples has random domains(Type-1) and the other has parallel domains(Type-2). In regard to the Type-2 samples, measurements are performed in the two cases, namely; (A) when the control current I is passed across the domain walls and (B) when I is passed along the domain walls, as shown in Fig.1.

Samples are placed in the air gap of an electromagnet. A constant current source is connected to the current terminals. The hysteresis loops of the Hall voltage, V_H, and the fractional change of resistance, $\Delta R/R$, are traced on an X-Y recorder simultaneously. The values of I are from 10 to 30 mA. The driving magnetic field has a triangular waveform with frequency of a few Hz or less.

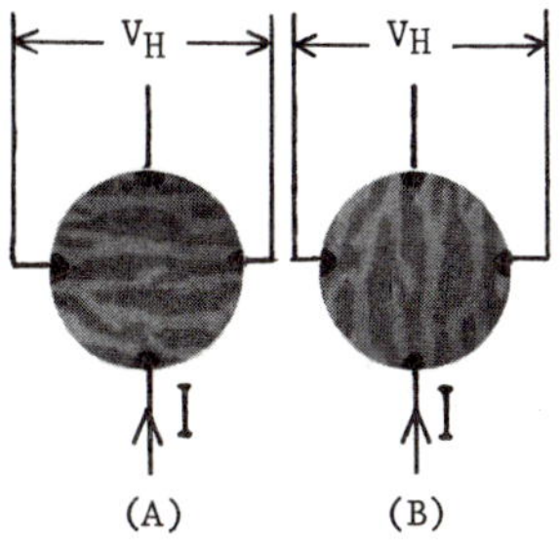

Fig.1. Arrangement of a sample for the measurement in the two cases; (A) control current I is vertical to domain walls, (B) I is parallel to domain walls.

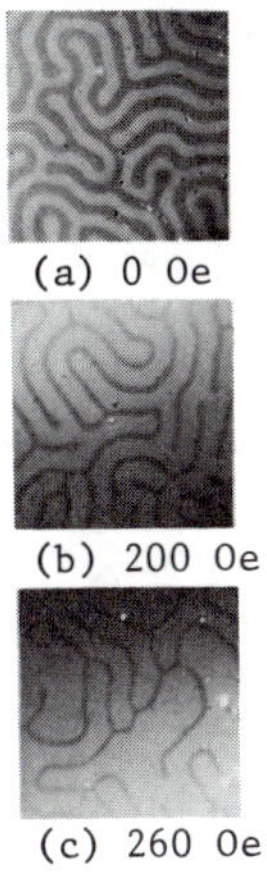

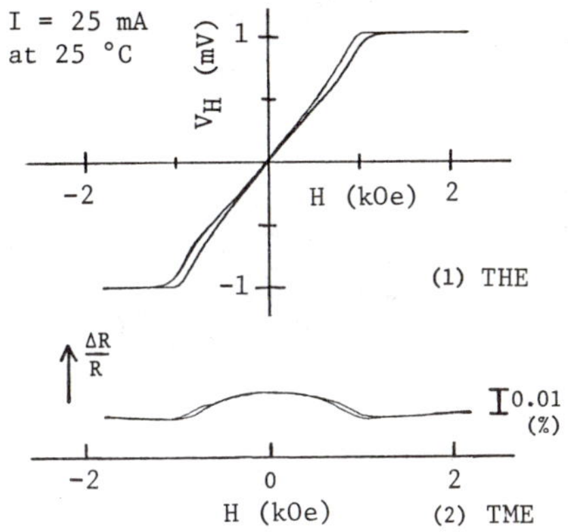

Fig. 2.(Left) Maze domain pattern and its variation under magnetic fields observed in a Type-1 sample.

Fig. 3.(Right) Hysteresis loops of the THE and TME for the sample described in Fig. 2.

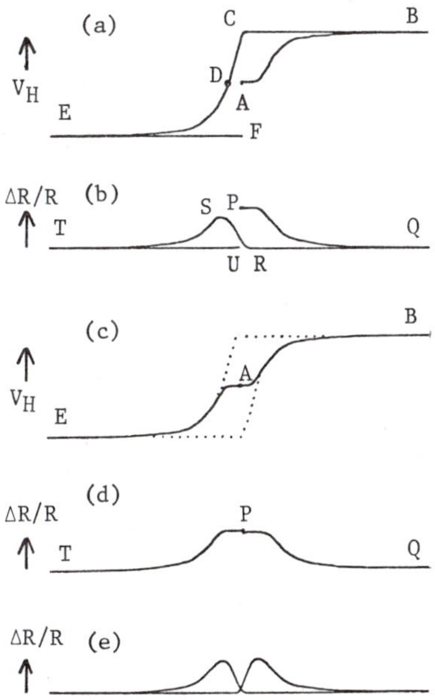

Fig.4. Correspondence between a THE curve and a TME curve traced simultaneously. Where, the points from A to F correspond to those from P to U respectively.
(a) THE curve start from the demagnetized state A.
(b) TME curve correspond to the THE curve in (a).
(c) THE initial curves from A to B and from A to E.
(d) TME curves correspond to the THE curve in (c).
(e) Major loop of the TME.
(The dotted curve in (c) is the major loop of THE.)

RESULTS AND DISCUSSION

Fig.2(a) shows a typical maze domain pattern observed in a Type-1 sample, and (b) and (c) are its variations under magnetic fields. Fig.3 indicates the hysteresis loops of THE and the corresponding TME measured in the sample shown in Fig.2. Comparing the two loops in Fig.3, it is obvious that V_H changes in the vicinity of H = 0, whereas $\Delta R/R$ does not change there. This seems to indicate that V_H is related to the area of reversed domains, while $\Delta R/R$ relates to the number of domain walls. The two loops do not change when the input terminals and the output terminals are interchanged. This would be reasonable because the THE and TME are isotropic in such a film that has a random domain pattern.

The measurements of THE and TME were also performed on a film which has a rectangular magnetization characteristic. The results are shown in Fig.4. In this figure the points from A to F correspond to those from P to U, respectively. Fine multidomains are visually observed in state A, and wide multidomains

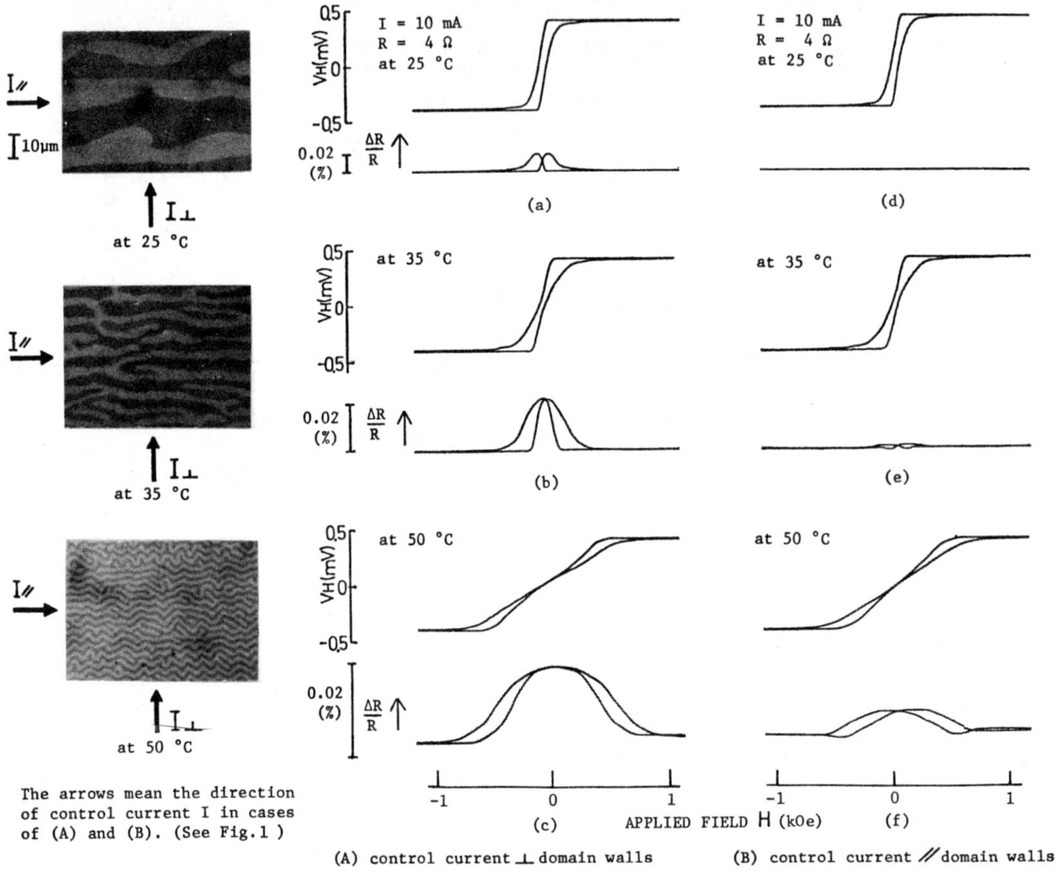

Fig.5. Comparison of the THE and TME loops with respect to the current direction measured in a Type-2 sample and their variation at three temperatures; 25°, 35° and 50°C. (a),(b) and (c) correspond to Fig.(A), (d),(e) and (f) correspond to Fig.(B). The photographs exhibit the domain patterns in the ac-demagnetized state at the three temperatures.

are observed in the state D. Fig.4(a) is the THE curve started from the ac-demagnetized state A, and Fig.4(b) the TME curve correspond to Fig.4(a). It is evident from Fig.4(b) that the value of ΔR/R is a maximum at point P and a minimum at the saturation regions where the film has a single domain. The discrepancy in level between points P and S may be due to the difference in the domain configuration. Fig.4(c) represents the initial THE curves(from A to B and from A to E), and Fig.4(d) is the TME curves correspond to (c). Fig.4(e) is the complete major loop of TME.

From these experimental facts described above, we may conclude that ΔR/R is dependent on the number of domain walls.

Fig.5 illustrates the change of the THE and TME loops with respect to the current direction measured in a Type-2 sample, and its variation at three temperatures. The control current, I, is vertical to the domain walls in (a),(b)&(c), and parallel to the domain walls in (d),(e)&(f), as shown in Fig.1. The photographs indicate the domain patterns in ac-demagnetized states. It is apparent that the TME in the case of (B) is much smaller than that of (A). In particular, the TME is not detected at 25 °C in this measurement. Therefore, we may conclude further that ΔR/R is a function of the number of domain walls which lie in the path of current flow. The THE loop, however, does not change between the two cases. This fact suggests that the transverse Hall voltage depends on the area of reversed domains in the sample and is not affected by the direction of the current.

It may seem unnatural that there is no change in the maximum value of V_H in Fig.5. But indeed, it has been found that the maximum value of V_H is nearly constant in the temperature range from 77 K to the Curie temperature except in the neighborhood of the compensation point and the Curie temperature. Further, it is also found that the shape of the THE loop is rectangular(e.g. like Fig.5(a)) in the region near the compensation point and linear(e.g. like Fig.5(c)) in the other temperature regions.

CONCLUSION

The transverse Hall effect and the transverse magnetoresistance effect are investigated in the Gd-Co sputtered films. After examining the experimental results, the followings became evident: (1) the transverse Hall voltage is the function of the total area of reversed magnetic domain, and not affected by the direction of the control current in the sample, (2) the fractional change of resistance, ΔR/R, is dependent on the number of magnetic domain walls with which the control current interacts.

REFERENCES

1. K.Okamoto, T.Shirakawa, S.Matsushita, and Y.Sakurai: IEEE Trans. on Mag., MAG-10 No.3, 799, 1974.
2. P.Chaudhari, J.J.Cuomo, and R.J.Gambino: IBM J. Res.Dev., 17, 66, 1973.
3. T.Shirakawa, K.Okamoto, K.Onishi, S.Matsushita, and Y.Sakurai: IEEE Trans. on Mag., MAG-10 No.3, 795, 1974.

MAGNETIC HYPERFINE STRUCTURE IN AMORPHOUS $DyFe_2$

D.W. Forester, Naval Research Laboratory, Washington, D.C. 20375
R. Abbundi† and R. Segnan†, American University, Washington, D.C. 20016
D. Sweger, National Bureau of Standards, Washington, D.C. 20234

ABSTRACT

In this paper we summarize the results of a Mössbauer study of the amorphous rare earth-transition metal (RE-TM) alloy $DyFe_2$ in the magnetically ordered state. Both the ^{57}Fe and ^{161}Dy spectra were investigated as a function of temperature and reveal a broad, though measurable, distribution of Fe hyperfine fields (H_{eff}) and quadrupole splittings (QS). There is a distribution similar in magnitude for the Dy hyperfine fields but the observed quadrupole splittings are narrowly distributed. The temperature variation of the ^{161}Dy spectra indicates a large variation of local exchange splittings at the Dy sites. These results are discussed in terms of a model in which the Dy spins are more strongly coupled to their random-direction anisotropy axes than to the neighboring Fe spins (which also experience a weaker random-direction anisotropy.)

INTRODUCTION

Amorphous, magnetic, RE-TM metal alloys have now been prepared at several laboratories and x-ray[1] and neutron scattering[2] studies in sputtered $TbFe_2$ and $GdFe_2$ reveal a short-range atomic order significantly different from that in their crystalline (Laves phase) counterparts. Magnetization studies[3] of amorphous $TbFe_2$ have been described in terms of the model mentioned above which also accounts for the reduced magnetic moment per formula unit compared with the corresponding crystalline alloy (RE and TM antialigned). Previous Mössbauer studies[4] have concentrated on the well-defined electric field gradient (EFG) observed in the paramagnetic state of amorphous $REFe_2$ alloys which can be explained in terms of a dense random packed sphere model. This model[4] yields an EFG at TM sites and a magnetic anisotropy at RE sites both of which are nearly constant in magnitude but random in direction.

EXPERIMENTAL

Samples of amorphous $DyFe_2$ were cut from a bulk plate (2.5×0.1 cm disk) prepared at Battelle Pacific Northwest Laboratories by d.c. krypton sputtering from a pressed powder target. (We are grateful to Dr. A.E. Clark for making the sample available to us.) Clark[5] has measured the low temperature magnetic properties of this sample and finds them to be quite similar to amorphous $TbFe_2$, but there are no other published data. Our own x-ray analysis shows a characteristic amorphous diffraction pattern and no crystalline reflections were found either before or after the material was prepared as a Mössbauer absorber. X-ray fluorescence analysis shows the stoichiometry to be correct to within ± 2 percent. ^{57}Fe backscattering spectra taken at room temperature on the original sputtered disk are the same as transmission spectra taken after the Mössbauer absorbers were prepared. A typical absorber was prepared by powdering 100-200 mg of $DyFe_2$ alloy in a dry, inert-gas atmosphere. The sample was immediately mixed with an organic binder and pressed into a 0.5 inch diameter pellet. None of the Mössbauer spectra taken in this study revealed evidence for either Dy or Fe oxide phases.

A) ^{57}Fe Spectra. Although previous Mössbauer studies[4] of ^{57}Fe in amorphous $(RE)Fe_2$ alloys, including $DyFe_2$, reported an absence of resolved hyperfine structure (hfs) below the Curie temperature (T_c), by optimizing the absorber thickness and by taking more than 3×10^6 counts per channel, the present work revealed a broad but resolved distribution of hyperfine fields. ^{57}Fe spectra taken at several temperatures are shown in Fig. 1. These data are quite similar to those obtained for the amorphous ferromagnet $Fe_{44}Pd_{36}P_{20}$.[6] Since the outer lines are much broader than the inner lines, there clearly must be a large distribution of hyperfine fields which is estimated to have a width of ≃90 kOe. The skewed shapes and asymmetrical intensities of the central lines are reminiscent of powder-patterns found in polycrystalline EPR and NMR spectra. As can be seen, during the temperature dependent collapse of the Fe^{57} hyperfine field (H_{eff}), the asymmetry of the center two lines changes with increasing temperature. The Curie temperature was determined to be 299±3°K from a thermal scan of the transmission at zero velocity. At low temperatures the mean value of $H_{eff}(Fe)$ is 225 kOe, virtually the same as for crystalline $DyFe_2$ (228 kOe)[7] although much smaller than for elemental iron presumably due to electron transfer from the RE to the Fe d-band.

B) ^{161}Dy Spectra. There have been no previous Mössbauer studies of either H_{eff} or QS for RE atoms in the amorphous $REFe_2$ alloys. In Fig. 2 are shown our ^{161}Dy spectra in polycrystalline $DyFe_2$ at T=4.2K and 300K and amorphous $DyFe_2$ at 4.2°K. The magnetic hyperfine splitting in the amorphous sample is completely collapsed at room temperature. At 4.2K the average H_{eff}, 6.3 MOe, is about half-way between the value for Dy metal and crystalline $DyFe_2$.[8] The QS ($e^2qQ/2$) is 6.3 cm/sec. The outer lines of the amorphous spectrum are considerably broadened while the central lines are not. This requires a distribution of width $\Delta H_{eff}(^{161}Dy)\cong100$ kOe at 4.2K.

DISCUSSION

We find that the Fe^{57} spectra may be described in terms of the model of Rhyne et al.[3] mentioned above. The Fe spins are pictured as aligned by exchange coupling over a spatial region or "domain". As a simple starting point we assume both an axial EFG and pseudo-dipolar magnetic field, H_D, varying as $(3\cos^2\theta-1)$, where θ is measured with respect to the "domain" magnetization direction. Also, over the spatial region of the "domain" the EFG and H_D are allowed to be randomly distributed in direction. The expected spectrum is then that of a powder pattern and a typical computer generated spectrum is shown above the T=14K spectrum in Fig. 1. For reference, the bar diagram shows the T=4.2K line positions for ^{57}Fe in crystalline $DyFe_2$. The generated spectrum used the measured 300K QS($e^2Qq/2$)=0.51 mm/sec, an estimated H_D=20 kOe based on H_{eff} anisotropy in the Laves phases[9] and the previously mentioned symmetric distribution of isotropic hyperfine fields. Qualitatively the T=14K spectral features are reproduced well. The change in asymmetry of the data with increasing temperature does not follow immediately from our model, presumably because only a single powder pattern distribution was used.

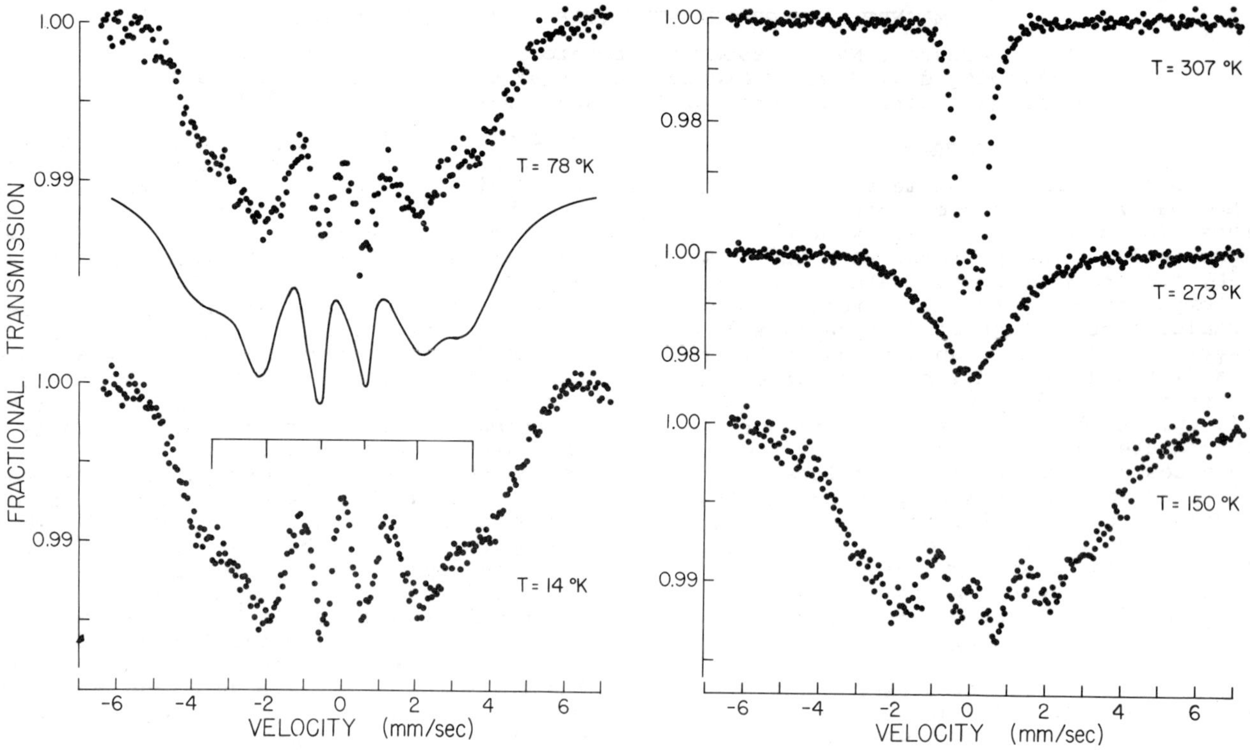

Fig. 1. ^{57}Fe Mössbauer spectra in amorphous $DyFe_2$. See text for description of theoretical curve. The bar graph gives line positions for crystalline $DyFe_2$ at T=4.2K.

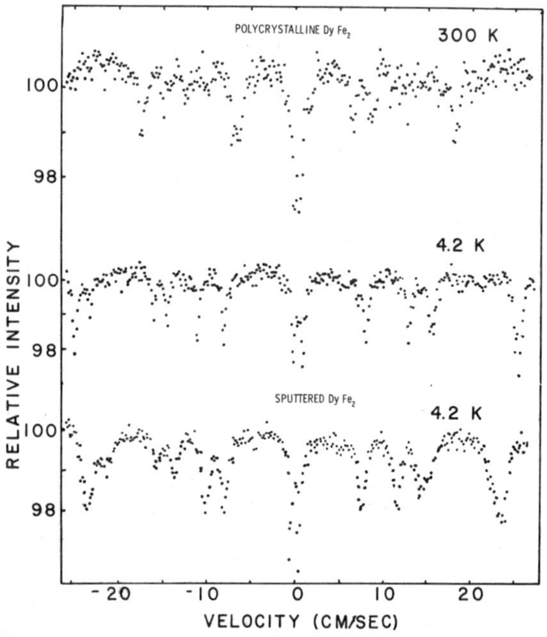

Fig. 2. ^{161}Dy Mössbauer spectra in polycrystalline and amorphous $DyFe_2$.

In contrast to the ^{57}Fe data, the ^{161}Dy hyperfine fields (not shown in the Figures) were found to collapse very rapidly with T/T_c and the spectral lines broadened by orders of magnitude indicating a distribution of local exchange.

We note that, whereas 12 Fe nearest neighbors surround RE atoms in crystalline $ReFe_2$ this number is reduced, on the average, to about 6.7 in the amorphous state.[1] From our T=4.2K measurements we find a reduction in $H_{eff}(^{161}Dy)$ of 0.5 MOe and a reduction in $e^2qQ/2$ of about 2 cm/sec in going from the crystalline to the amorphous state. If we attribute the decreases in H_{eff} to a change in the number of nearest neighbors, we find that H_{eff} is reduced by 75 kOe for each Fe neighbor removed. On the other hand, if the decrease is due to a reduction in $\langle J_z \rangle$, then our results would imply a possible 6 percent decrease in the Dy moment.

1. G.S. Cargill, AIP Conf. Proc. Series XVIII, 631 (1974).
2. J.J. Rhyne, S.J. Pickart and H.A. Alperin, Phys. Rev. Letters 29, 1562 (1972).
3. J.J. Rhyne, J.H. Schelleng and N.C. Koon (To be published) Phys. Rev. B, Dec. 2 (1974).
4. D. Sarkar, R.Segnan, E.K. Cornell, E. Callen, R. Harris, M. Plischke and M.J. Zuckerman, Phys. Rev. Letters 32, 542 (1974)(and references contained therein.)
5. Arthur E. Clark, Appl. Phys. Letters 23, 642 (1973).
6. T.E. Sharon and C.C. Tsuei, Phys. Rev. B5, 1047 (1972).
7. This study and G.K. Wertheim and J.H. Wernick, Phys. Rev. 125, 1937 (1962).
8. G.J. Bowden, D.St.P. Bunbury, A.P. Guimares and R.E. Snyder , J. Phys. C. (Proc. Phys. Soc.)1, 1376 (1968).
9. M.P. Dariel, U. Atzmony and D. Lebenbaum, phys. stat. sol. B59, 615 (1973)(and references contained therein.)

† Work supported in part by ONR Contract No. N0014-68-A-0245-0002.

"CRITICAL" NEUTRON SCATTERING IN AMORPHOUS $HoFe_2$ AND $GdFe_2$*

S. J. Pickart,[†] J. J. Rhyne and H. A. Alperin
Naval Ordnance Laboratory, White Oak, Maryland 20910 and
National Bureau of Standards, Washington, DC 20234

ABSTRACT

Small angle neutron scattering measurements were made on amorphous $HoFe_2$ and $GdFe_2$. For $HoFe_2$ the intensity for $T \gtrsim T_c$ (195 K) is Lorentzian in form and can be analyzed for an inverse correlation length κ_1 which remains finite at T_c. Below T_c a giant anomalous magnetic component appears, which follows a $q^{-2.4}$ dependence and can be analyzed in terms of magnetic inhomogeneities of dimension comparable to κ_1^{-1} at T_c. Limited measurements at fixed q indicate the absence of this component in $GdFe_2$, suggesting an origin based on local crystalline field effects.

INTRODUCTION

Recent small angle scattering measurements[1] in amorphous $TbFe_2$ gave anomalous results in that the inverse correlation length κ_1 remains finite at T_c and a large magnetic component related to magnetic inhomogeneities exists below T_c. In an effort to determine how general these phenomena are in amorphous systems and whether they are related to the presence of local magnetocrystalline anisotropy, we have extended these measurements to sputter-deposited, amorphous $HoFe_2$ and the S-state-ion amorphous alloy $GdFe_2$. The Curie temperature for $HoFe_2$ is 195K as determined from an Arrott plot of the magnetization. The spontaneous moment vanishes at a higher temperature near 220K also suggesting that the phase transition is not well defined. T_c for amorphous $GdFe_2$ is 500K[2] as obtained by either method.

EXPERIMENTAL

The measurements were taken at the NBS reactor with an incident wavelength of 2.43 A, obtained from a curved graphite monochromator and filtered with 1-3/4 inches of graphite. With a 20'-10'-10' collimator sequence, this allowed measurements down to a wavevector transfer q of .02A^{-1} before the background became excessive.

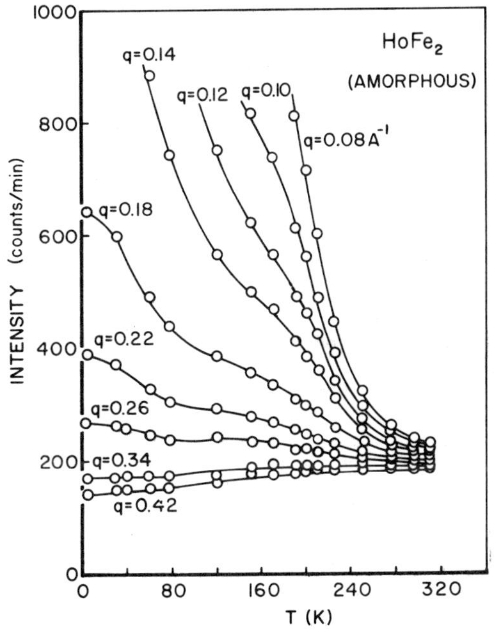

Figure 1 - Scattered neutron intensity from $HoFe_2$ as a function of temperature and wavevector.

The samples were prepared by rapid sputtering at Battelle Pacific Northwest Laboratories. The $HoFe_2$ was of 1 mm thickness and of irregular shape, cut from a 2.5 cm dia. disk. The $GdFe_2$ was powdered and dispersed in a Cu powder matrix to provide approximately 1/e absorption. Careful scans of the strong (111) magnetic peak position of the Laves phase showed no evidence of recrystallization.

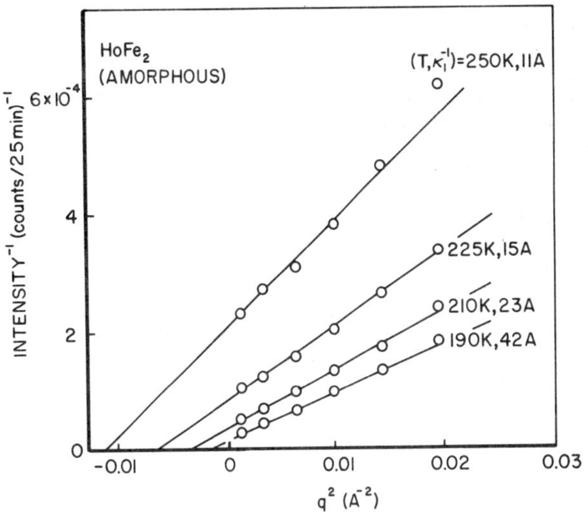

Figure 2 - Inverse intensity plotted vs. q^2 in $HoFe_2$. The coherence lengths (κ_1^{-1}) have been corrected for resolution.

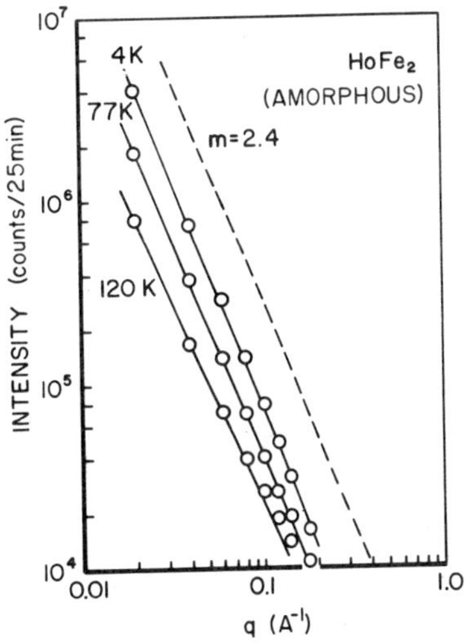

Figure 3 - Log-log plot of intensity vs. q for several temperatures in $HoFe_2$. Dashed line shows slope of -2.4.

RESULTS AND DISCUSSION

The scattering observed from $HoFe_2$, corrected for absorption and background, is plotted as a function of temperature for various q in Fig. 1. These curves are reminiscent of the observations[1] in $TbFe_2$ with the exception that the broad "hump" observed in the region of the Curie point is less pronounced; for $q \lesssim .08A^{-1}$ it is completely swamped by the anomalous small angle component.

If the high temperature asymptotic level for $q \leq 0.14$ A^{-1} is subtracted, the scattering in Fig. 1 has a Lorentzian line-shape, $I \propto (\kappa_1^2 + q^2)^{-1}$ for temperatures above T_c which also persists down to 40° below T_c.

The scattered intensity isotherms for $T > T_c$ are shown in Fig. 2, along with the correlation lengths κ_1^{-1} obtained after applying corrections for vertical and horizontal resolution. Lower temperature data show κ_1 approaching zero near 150 K; however this interpretation is complicated by scattering from magnetic inhomogenieties as discussed below. Because of the uncertainty in determining T_c, it is not feasible to establish the power law dependence of κ_1 on $T - T_c$.

At temperatures below 150 K, the wavevector dependence goes over to $q^{-2.4}$ (see Fig. 3) exactly as in the case of $TbFe_2$ with a temperature dependence for small q proportional to the square of the macroscopic magnetization. The asymptotic formula[3] for small q scattering from a distribution of spherical particles has a q^{-4} dependence; which, after allowing for our resolution, becomes $q^{-2.4}$. Interestingly enough, the specific particle surface area appearing in this formula, when evaluated using the 4K, $q = .02 Å^{-1}$ data corresponds to a spherical particle of mean radius within a factor of two of the coherence length at T_c.

Because of the very high absorption and consequent low signal, data were taken on amorphous $GdFe_2$ at only a single fixed angle, $q = .04 Å^{-1}$. Two separate runs showed only a broad hump in the region of T_c (∿500K) with no evidence of the anomalous low temperature component below T_c observed in both the $TbFe_2$ and $HoFe_2$. In spite of the low signal intensity our previous experience with magnetic scattering in Gd compounds[4] would lead us to believe that a component of such an unusual magnitude would have been visible.

CONCLUSIONS

The observations reported here for $HoFe_2$ are in every detail similar to those for $TbFe_2$, while $GdFe_2$ shows no anomalous small angle scattering below T_c nor evidence for sharp critical scattering at T_c. Subject to confirmation of the latter negative results with better statistical accuracy which could only be provided by an isotopic Gd composition, we conclude that the Curie point, as reflected by finite fluctuations down to a wavevector scale of $q = .02 Å^{-1}$, is not a well defined quantity in these amorphous alloys, although above T_c (as determined from magnetization measurements) the correlation length behaves qualitatively like that of a normal paramagnet. At low temperatures, magnetic inhomogeneities are present in the non-S-state ion rare earth systems suggesting that they are stabilized by local crystalline field effects.[2] For temperatures just below T_c, due to the rapidly increasing $q^{-2.4}$ dependent scattering from the inhomogeneities, one cannot determine if the correlation length remains finite to lower temperatues or diverges at about 150K (some 40° below T_c). The inelasticity of the scattering may influence some of the details and further experiments are in progress to investigate these effects.

REFERENCES

*Work supported by the Office of Naval Research, the Naval Sea Systems Command, and the Naval Surface Weapons Center Independent Research Fund.

Present address: Department of Physics, University of Rhode Island, Kingston, RI 02881

1. S. J. Pickart, J. J. Rhyne and H. A. Alperin, Phys. Rev. Letters 33 424 (1974).

2. J. J. Rhyne, J. H. Schelleng and N. C. Koon, Phys. Rev. B 10, 4672 (1974).

3. A. Guinier and G. Fournet, Small Angle Scattering of X-Rays, Wiley & Sons, New York, 1955, p. 17 and 156-160.

4. T. McGuire, R. Gambino, S. Pickart and H. Alperin, J. Appl. Phys, 40, 1009 (1969).

TEMPERATURE DEPENDENCE OF NEUTRON FERROMAGNETIC DISORDER SCATTERING FROM Si AND Mn IMPURITIES IN Fe*

H. R. Child and J. W. Cable
Solid State Division, Oak Ridge National Laboratory, Oak Ridge, Tennessee 37830

ABSTRACT

We report measurements of the ferromagnetic disorder scattering of 4.4 Å neutrons from 3 at.% Si and 2 at.% Mn impurities in bcc Fe as a function of temperature for $0.3 \le T/T_C \le 0.8$. Molecular field theory predicts that this type of scattering should depend on T/T_C. This work is part of the first experimental observation of this temperature dependent effect. Our results show that the Mn moment is $0.6 \pm 0.3\ \mu_B$ at ambient temperature ($T/T_C = 0.3$) and decreases more rapidly than the host moment as T/T_C increases. The disturbance in the Fe neighbors of the Mn impurity increases in magnitude and range at larger T/T_C. These effects are consistent with the assumption that the Mn is weakly coupled to the host. The disturbance around the Si impurity also increases with temperature and appears to change sign above $T/T_C \simeq 0.6$.

*Research sponsored by the U. S. Atomic Energy Commission under contract with Union Carbide Corporation.

SPIN WAVES IN AN AMORPHOUS METALLIC FERROMAGNET*

J. D. Axe and L. Passell
Brookhaven National Laboratory, Upton, New York 11973
and
C. C. Tsuei
IBM Watson Research Center, Yorktown Heights, New York 10598

ABSTRACT

Low angle inelastic neutron scattering experiments have been carried out on the amorphous ferromagnet $Fe_{75}P_{15}C_{10}$, with a Curie temperature $T_c = 597°K$. At reduced temperatures, $t = T/T_c$, between 0.5 and 0.83 well-defined spin waves have been observed with propagation vectors $|q| = 2\pi/\lambda < 0.18\ Å^{-1}$ with a ferromagnetic dispersion relation $\hbar\omega = D(T)q^2 + E(T)q^4$. Inability to close the scattering triangle rather than intrinsic broadening prevents extension of the measurements to even shorter wavelengths. At room temperature $D = 149.5(\pm 3)meV\text{-}A^2$ and $E = 680(\pm 200)\ meV\text{-}A^4$.

INTRODUCTION

Over certain ranges of composition alloys involving a transition metal and lesser amounts of C, Si, or P can be induced to form metastable amorphous metallic phases, which can be characterized structurally by a dense random packing of hard spheres.[1] Many of these amorphous metals exhibit ferromagnetic long range order below a well-defined Curie temperature, T_c, and there have been many recent studies of the structural and bulk magnetic properties of these materials.[1-4] In particular, Cochrane and Cargill[4] find the behavior of the spontaneous magnetization of Co-P alloys at low temperatures is consistent with excitation of spin waves with the familiar Dq^2 dispersion relation.

Neutron scattering measurements have provided valuable information concerning the spatial correlations of spins in $TbFe_2$, an amorphous ferrimagnet with the same dense random packed structure.[5] However, inelastic neutron scattering studies of this same material have thus far revealed only broad featureless magnetic scattering rather than sharp spin-wave excitations. This paper is concerned with the long wavelength dynamics of spin fluctuations in amorphous magnets and reports inelastic neutron scattering experiments which have successfully observed sharp spin-wave excitations in amorphous $Fe_{75}P_{15}C_{10}$. (The subscripts denote the atomic concentration of the elements.)

EXPERIMENTAL

The amorphous alloy $Fe_{75}P_{15}C_{10}$ was obtained by rapid quenching from the liquid state using the piston and anvil technique.[6] The details of the alloy preparation have been described in Ref. 7. The resulting quenched samples are foils approximately 2 cm in diameter and 40-50 microns thick. Since the actual cooling rate may vary from sample to sample, each foil was carefully checked using an x-ray diffractometer to insure that only the broad bands indicative of amorphous structure were present. Quenched samples which showed any deviation from a smooth broad diffraction band were rejected for use in this work. A total of 85 amorphous foils were used in the neutron scattering experiment.

The neutron experiments were performed using a triple-axis spectrometer in the "constant-Q" mode at the Brookhaven High Flux Beam Reactor. In crystalline materials spin-wave measurements are conventionally performed at a momentum transfer $\vec{Q} = \vec{G} + \vec{q}$, where $\vec{G}$ is a reciprocal lattice vector, thereby allowing spin waves with small propagation vectors, $\vec{q}$, to be studied at conveniently large scattering angles. As there are no nonzero reciprocal lattice vectors in an amorphous material, the present measurements had to be performed about the origin of reciprocal space, which is to say at unusually small scattering angles ($2\theta_s$ between 0.3° and 1.5°). This technique has been used successfully in the study of spin waves in polycrystalline samples.[8] It is very important to minimize spurious low angle scattering from the nearby incident beam, and in particular in the present experiments it was crucial to position the pyrolytic graphite filters used to remove order contamination in the incoming beam upstream as far as possible from the sample.

A closely related and ultimately more restrictive difficulty encountered in experiments of this sort results from the scattering kinematics, which forbids transfer of more than a small fraction of the incident neutron energy, while maintaining the necessarily small momentum transfer. In the present instance the spin waves are rather stiff, so the incident neutron energy necessary to close the scattering triangle is quickly driven beyond the peak of the reactor spectrum, with concomitant losses not only in intensity but in the resolution of the spectrometer. (For example it was necessary to use an incident energy of 130 meV in order to transfer 5 meV to the sample.) In these experiments a filtered 14.5 meV beam using the (002) reflection from pyrolytic graphite monochromator was used at the smallest energies. Higher incoming energy beams were obtained from a (113) germanium monochromator. 10 or 20 minute collimators were used, depending upon the intensity of the scattering.

Spin waves were observed with wave vectors ranging between 0.04 $Å^{-1}$ and 0.18 $Å^{-1}$, at temperatures between 296°K and 471°K. Typical low q data are shown in Fig.1. At the lowest temperature the high energy portion of

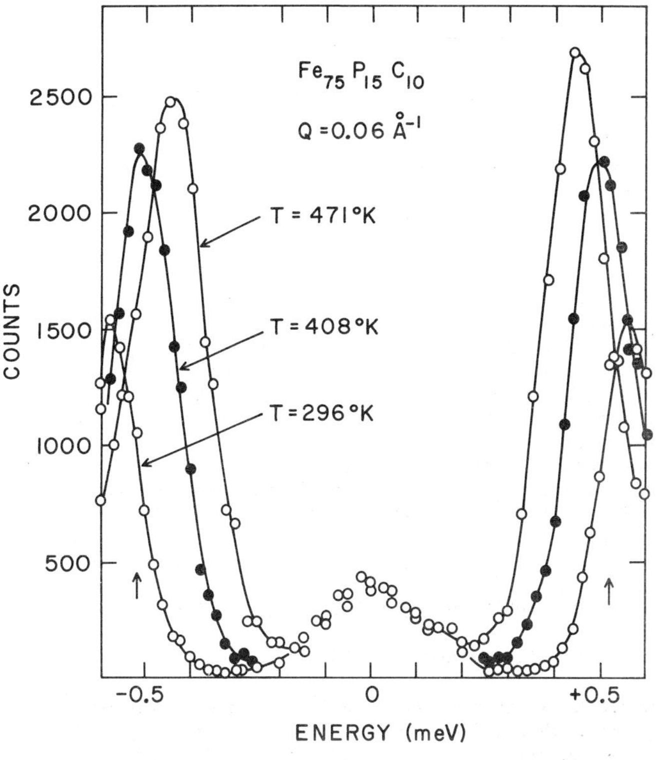

Fig. 1. Inelastic neutron scattering spectrum at constant momentum transfer, q = 0.06 $Å^{-1}$. The upper limit of energy transfer is limited by kinematic considerations (see text). The arrow indicates the value of $D_{eff}q^2$ deduced from the 296°K data and shows the magnitude of the resolution corrections.

the peaks is truncated for the reasons discussed above. The small central elastic component is nearly temperature independent but it is not yet clear whether it is primarily magnetic or structural in origin.

The data were analyzed by fitting the observed peaks to calculated line shapes obtained by convoluting the known spectrometer resolution function with a trial dispersion relation, $\hbar\omega = D_{eff}q^2$ and adjusting the effective spin-wave stiffness constant $D_{eff}(q,T)$. The importance of these corrections can be seen from the shift of the observed peak from the nominal value, $D_{eff}q^2$, indicated by the arrows in Fig. 1. The resulting corrected spin-wave energies plotted as D_{eff}/q^2 vs q^2 for three temperatures are shown in Fig. 2.

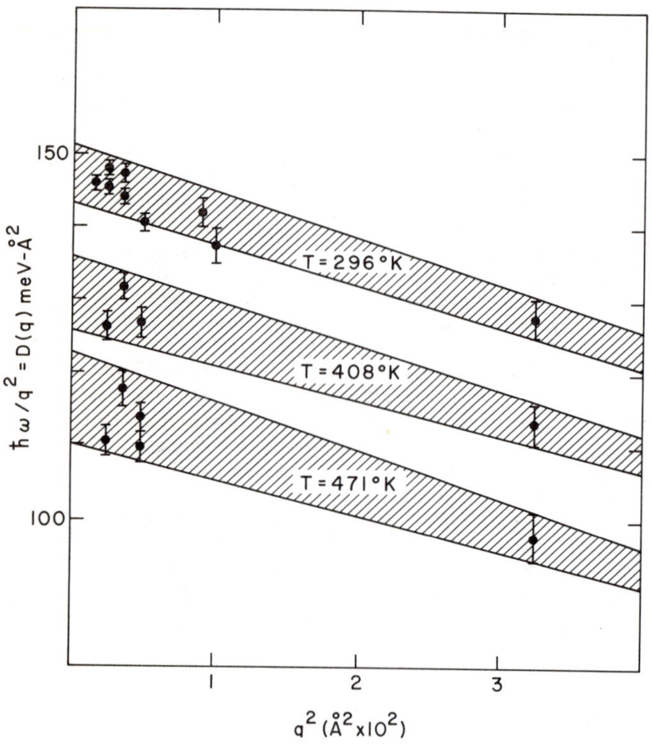

Fig. 2. Resolution corrected spin-wave energies, represented as $\omega(q) = D_{eff}/q^2$ plotted vs q^2.

DISCUSSION

The results of the type shown in Fig. 1 and summarized in Fig. 2 show that very well-defined spin-wave excitations exist in amorphous $Fe_{75}P_{15}C_{10}$, over the entire range of propagation vectors accessible for study. The spin-wave dispersion is proportional to q^2 for small q, but show a definite deviation from this behavior at high q. Within the limits of accuracy of our measurements all of our data can be satisfactorily fit to a dispersion relation of the form $\hbar\omega(q,T) = D(T)q^2 + E(T)q^4$. Specifically

$$T = 296°K \quad \hbar\omega(q) = 149.5(\pm 3)q^2 - 680(\pm 200)q^4$$

$$T = 408°K \quad \hbar\omega(q) = 129.5(\pm^{7}_{3})q^2 - 500(\pm 200)q^4$$

$$T = 113°K \quad \hbar\omega(q) = 113.0(\pm^{10}_{3})q^2 - 480(\pm 200)q^4$$

where $\hbar\omega$ is in meV and q in Å^{-1}. (Note by way of comparison that Collins et al[7] find for pure bcc Fe at room temperature that D = 281 (±10) meV Å^2 and E = -270(±35) meV Å^4.) The large ratio of E/D implies appreciable exchange from longer than nearest neighbors, and is typical of itinerant ferromagnetism.

The temperature dependent renormalization of the spin-wave energies has not been studied over a sufficient range of temperature to be definitively characterized, but the expression

$$D(T) = D(0)(1 - \lambda T^x) \qquad \text{with } 2 \leqslant x \leqslant 3$$

fits the existing data very satisfactorily. In a crystalline itinerant ferromagnet a $T^{5/2}$ term is expected for magnon-magnon interaction at low temperature, whereas a T^2 term is predicted for magnon-electron interactions.[9] The width of all of the spin waves studied can be accounted for satisfactorily by instrumental resolution effects alone. The instrumental line widths are acceptably small for all except for the q = 0.18 data, where it is ∿ 2 meV.

The existence of propagating spin waves in an amorphous magnetic medium is no more surprising than is that of sound waves in an amorphous elastic medium. They can perhaps be discussed quantitatively most directly as quantized excitations of a magnetic continuum so long as the wavelength is very large compared with the characteristic atomic structure, as is certainly true in the present case. This has the distinct advantage that the microscopic details of both the magnetic interactions and the structure can be left unspecified. It is worthwhile to point out however that the familiar lowest order representation of the exchange energy density[9], $\mathcal{H}_{ex} = C \sum_{ij}(\partial M_i/\partial X_j)^2$ gives rise to a strictly quadratic spin-wave dispersion, and (as a direct consequence) no magnon-magnon renormalization of the spin-wave energies within the usual Hartree-Fock factorization.

It seems clearly worthwhile to attempt to extend these rather preliminary results in two obvious directions. The first is to increase the temperature range, concentrating in particular on the behavior of the critical region near $T = T_c$. This is impossible with the present sample composition because recrystallization occurs (slowly) below T_c. Secondly, it is of interest to continue these measurements to larger momentum transfer where the microscopic random nature of the material will begin to manifest itself in the spin dynamics. It is possible that both of these problems can be alleviated in a material with somewhat less magnetic stiffness and we are currently attempting to synthesize such materials.

We would like to thank G. S. Cargill III, C. J. Glinka, Y. Ishikawa, V. J. Minkiewicz, and G. Shirane for useful discussions of these results and Pol Duwez for permission to prepare the specimens at Cal Tech.

REFERENCES

* Work performed in part under the auspices of the U. S. Atomic Energy Commission.

1. See, for example, Amorphous Magnetism, edited by H. O. Hooper and A. M. deGraaf, Plenum Press, New York (1973).
2. G. S. Cargill III, J. Appl. Phys. 41, 12 and 2249 (1970).
3. C. C. Tsuei, G. Longworth and S. C. H. Lin, Phys. Rev. 70, 603 (1968).
4. R. W. Cochrane and G. S. Cargill III, Phys. Rev. Lett. 32, 476 (1974).
5. J. J. Rhyne, S. J. Pickart and H. A. Alperin, in "Magnetism and Magnetic Materials - 1973," AIP Conference Proceedings no. 18, edited by C. D. Graham, Jr. and J. J. Rhyne (AIP, New York, 1974); S. J. Pickart, J. J. Rhyne and H. A. Alperin, Phys. Rev. Lett. 33, 424 (1974).
6. P. Pietrokowsky, Rev. Sci. Inst. 34, 445 (1962).
7. Pol Duwez and S. C. H. Lin, J. Appl. Phys. 38, 4096 (1967).
8. L. Passell, O. W. Dietrich, and J. Als-Nielsen, "Magnetism and Magnetic Materials - 1971," AIP Conference Proceedings no. 5, edited by C. D. Graham, Jr. and J. J. Rhyne (AIP, New York, 1972), pp. 1251-55.
9. M. F. Collins, V. J. Minkiewicz, R. Nathans, L. Passell, and G. Shirane, Phys. Rev. 179, 417 (1969).
10. See, for example, F. Keffer in "Handbuch der Physik" Vol. 18, edited by H. P. J. Wijn, Springer, Berlin (1966), pp. 1-268.

INELASTIC MAGNETIC SCATTERING FROM AMORPHOUS $TbFe_2$

J. J. Rhyne, Naval Ordnance Laboratory, White Oak, Maryland 20910 and National Bureau of Standards, Washington, DC 20234, D. L. Price, Argonne National Laboratory, Argonne, Illinois 60439, and H. A. Mook, Oak Ridge National Laboratory, Oak Ridge, Tenn. 37830.

ABSTRACT

Inelastic neutron scattering measurements have been made on sputter-deposited amorphous $Tb_{33}Fe_{67}$ using time-of-flight spectrometers at Argonne and at Oak Ridge. At 433K (above T_c) the magnetic scattering function $S(Q,\omega)$ exhibited an approximate Gaussian dependence on energy transfer characteristic of a paramagnet. Below T_c, at 299K, a broad distribution of magnetic inelastic scattering is present centered approximately at ΔE = 11 meV. As expected this distribution was significantly broadened and was shifted to a lower energy than the equivalent polycrystalline compound $TbFe_2$ which showed a single inelastic peak near 18 meV. Polarized beam time-of-flight data taken at Oak Ridge showed evidence for discrete spin wave excitations for $Q \lesssim 1.0$ $Å^{-1}$ in the amorphous sample.

INTRODUCTION

The rare earth alloy of composition $TbFe_2$ prepared by sputtering has been shown to be crystallographically amorphous[1] with an atomic stacking sequence consistent with a dense-random-packed model,[2] and also to exhibit long-range magnetic order[3] with a Curie temperature of approximately 390 K. Low angle neutron scattering measurements[4] near T_c have indicated an absence of conventional critical scattering and the existence of a finite value of the spin correlation length at the bulk T_c.

In light of these features, the question of the nature of spin wave excitations in this material arises. Preliminary time-of-flight (TOF) inelastic data[5,1] taken at the National Bureau of Standards Reactor at two scattering angles indicated only a broad distribution of inelastic scattering. Due to the nature of time of flight experiments, data at a single scattering angle are not representative of a single wave-vector transfer Q, rather Q changes with increasing energy transfer (e.g. from Q = .5 $Å^{-1}$ for ΔE = 0 to 1.1 $Å^{-1}$ for ΔE = 12 meV).

In the interest of eliminating this difficulty, time-of-flight experiments were performed on the multi-detector time-of-flight facility[6] at the CP-5 reactor at Argonne National Laboratory. In the configuration used for these experiments, an incident neutron energy of 35 meV was obtained from a double monochromator, pulsed (chopped) and after scattering from the sample, the neutrons were counted by 44 detectors arranged into 16 subgroups. Two hundred fifty-six time channels were available for each subgroup. This allowed 16 independent time-of-flight spectra to be accumulated simultaneously, each corresponding to a given scattering angle. The angular range covered was from 2.4 to 50.4 degrees corresponding to an elastic $Q = 4\pi \sin\theta/\lambda$ of 0.17 $Å^{-1}$ to 3.4 $Å^{-1}$. Constant Q spectra could then be obtained by an interpolation procedure through the 16 constant angle data sets.

RESULTS AND DISCUSSION

The uncorrected TOF data obtained at 433 K (above T_c) and at 295 K are shown in Fig. 1 and correspond to an elastic Q of 1.35 $Å^{-1}$. The spectrum shows both neutron energy gain (positive ΔE) and energy loss processes on a non-linear scale as produced by the time of flight method. The elastic peak at ΔE = 0 is a measure of the structure factor of the amorphous material.

In Fig. 1 above T_c strong quasi-elastic scattering is observed characteristic of a paramagnet. Below T_c the quasi-elastic component is absent (energy breadth is that of the instrumental resolution) and is replaced

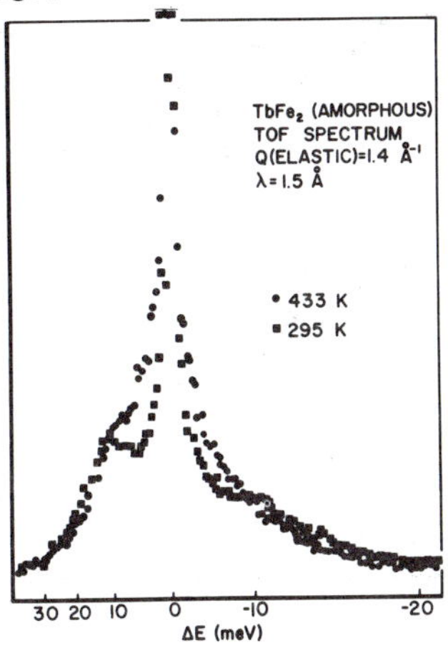

Figure 1 - Time of flight scattered intensity above and below the magnetic ordering temperature as a function of channel number. Corresonding energy transfer is indicated for reference.

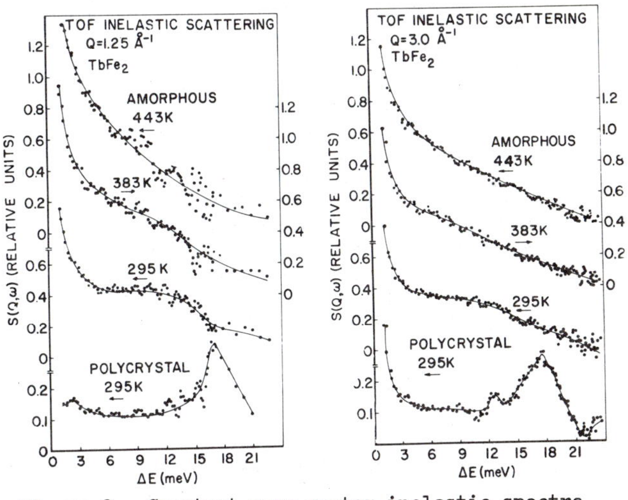

Figure 2 - Constant wave-vector inelastic spectra for amorphous and polycrystalline $TbFe_2$ at temperatures indicated.

by a broad peak on the neutron energy gain side and a corresponding shoulder in the energy loss spectrum.

Figure 2 shows the results of two constant Q sections through the accumulated spectra. The data have been corrected for instrumental factors, detailed balance, and the time of flight scale and include both energy gain and energy loss events. The result is then equivalent to the scattering function $S(Q,\omega)$ which in this material is expected to be dominated by the magnetic scattering. At Q = 1.25 $Å^{-1}$ the high temperature data show an approximate Gaussian energy dependence as expected for a paramagnet with exchange.[7] At T_c (390 K) and below, the spectrum consists of a broad inelastic distribution centered about 11 meV. At larger Q the magnitude of the distribution is suppressed (see Fig. 2 right side), consistent with the magnetic form factor.

The lack of well defined excitations was not unexpected, since the wave-vector Q is essentially not a valid description of an excitation in a highly disordered structure except at very small Q. The result at large Q is that the observed distribution is representative of the spin wave density of states. Discrete spin wave excitations would also be expected to be strongly broadened by life time effects.

Calculations by a number of authors[8] have shown that the magnon density of states should shift to lower energy in an amorphous material relative to

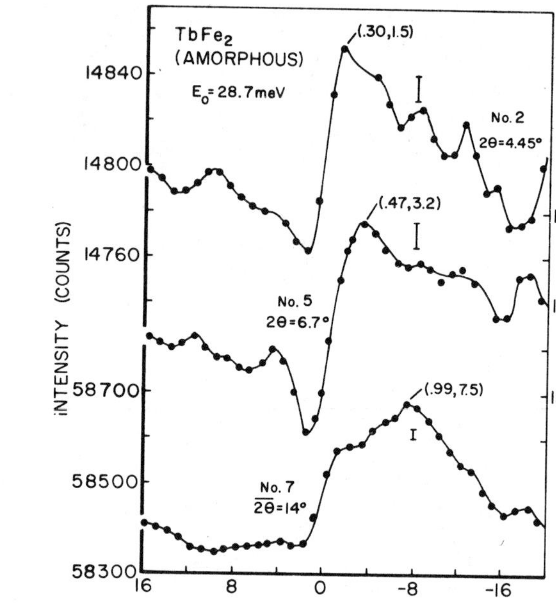

Figure 3 - Room temperature polarized-beam time-of-flight data at scattering angles indicated. The number refers to the detector group. The Q and ΔE values of the major peaks are given in parenthesis. The data have been corrected for the time of flight scale and instrumental factors.

the crystalline analogue. The predicted shift in magnon energies in the amorphous state is indicated in Fig. 2 by comparison to the inelastic scattering observed in polycrystalline $TbFe_2$. In this case the distribution centered about ΔE = 18 meV is broadened by polycrystalline averaging of Q vectors but does show the presence of significantly higher energy excitations than the amorphous alloy. It should be noted that the magnitude of the shift in the distribution may be affected somewhat by spin wave renormalization as the two room temperature spectra do not represent the same fraction of the Curie temperature.

The existence of discrete spin wave modes at low Q has been examined also using a triple axis spectrometer at NBSR. The limited range of energy transfer accessible at low Q can severly limit the observations of spin waves with large stiffness constants, or modes for which the rare earth local anisotropy interaction may introduce an appreciable energy gap. Both of these difficulties can be circumvented by measurements near T_c where the renormalization should bring the spin wave energies into the observable range dictated by energy and momentum conservation. The results of such measurements were reported in part in reference 4 and showed that in the range from 30 K below T_c to 10 K above T_c and in the Q range from .06 A^{-1} to greater than 1.0 A^{-1}, no discrete excitations could be observed. At these temperatures, however, the spin waves may be severly broadened. Below T_c an additional complication is the large elastic low angle component[4] proportioned to the square of the magnetization and associated with the magnetic inhomogeneities[3] which may again obscure the observation of spin waves.

In the interest of eliminating this elastic component, very recently time-of-flight measurements were made at Oak Ridge using a polarized beam correlation technique[9] which isolates the neutron spin-flip scattering cross-section. Room temperature results were taken at 15 distinct scattering angles, three of which are shown in Fig. 3. In this technique energy loss processes result in positive peaks with respect to the elastic channel while energy gain events result in negative peaks with broadened resolution. The spectra show evidence for several statistically significantly discrete excitations at the smaller scattering angles, with a broad distribution appearing again for larger Q.

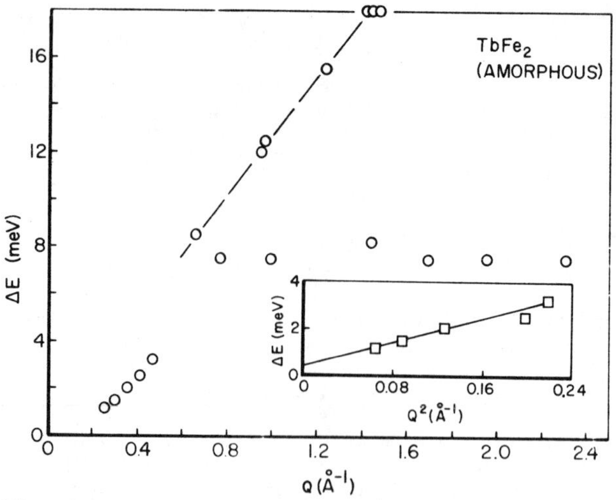

Figure 4 - Plot of observed peak energies in TOF energy loss spectra as a function of Q. The inset represents the energies of the low lying branch plotted versus Q^2. The data near 8 meV are the maxima in the broad inelastic distribution observed at the higher scattering angles (e.g. #7 of Fig. 3).

Figure 4 shows a ΔE vs Q plot of these excitations and provides evidence for a low Q dispersive spin-wave branch and also portions of a rapidly rising branch at higher energy similar to that observed for crystalline $Ho_{.88}Tb_{.12}Fe_2$[10]. The low Q data yield a relatively small spin-wave stiffness constant D = 15 meV-A^2 and evidence of an anisotropy gap of order 0.25 meV (see insert Q^2 plot in Fig. 4) both with considerable uncertainty due to the limited data. The significant reduction in the values of D for this branch dominated by the rare earth-iron exchange coupling is not unexpected for the amorphous phase. The higher Q data (>1.0 A^{-1}) show a broadened distribution centered near 8 meV. The discrepancy between this energy and the higher 11 meV distribution observed previously suggests that the unpolarized data may contain a higher energy contribution from a magneto-vibrational component. Additional details and refinements of these data will be reported later.

ACKNOWLEDGEMENT

It is a pleasure to acknowledge many discussions with S. J. Pickart, H. A. Alperin and J. R. Cullen concerning spin waves in amorphous alloys.

REFERENCES

1. J. J. Rhyne, S. J. Pickart and H. A. Alperin, Phys. Rev. Letters 29, 1562 (1972); J. J. Rhyne, S. J. Pickart and H. A. Alperin, AIP Conf. Proc. Series 18, 562 (1974).
2. D. E. Polk, J. Non-Cryst. Solids 11, 381 (1973); G. S. Cargill, AIP Conf. Proc. Series 18, 631 (1974).
3. J. J. Rhyne, J. H. Schelleng and N. C. Koon, Phys. Rev. B 10, 4672 (1974).
4. S. J. Pickart, J. J. Rhyne and H. A. Alperin, Phys. Rev. Letters 33, 424 (1974).
5. H. A. Alperin, J. J. Rhyne and S. J. Pickart, Proc. Int. Conf. on Magnetism 1973, Vol. IV, Nauka, Moscow, pp. 358.
6. R. Kleb, G. E. Ostrowski, D. L. Price, and J. M. Rowe, Nucl. Instrum. and Meth. 106, 221 (1973).
7. C. G. Windsor, Proc. Phys. Soc. (London) 89, 825 (1966).
8. R. Tahir Kheli, Amorphous Magnetism, Hooper and De Graaf, Plenum Press, New York, 1973, pp. 393; J. E. Gubernatis and P. L. Taylor, ibid, pp. 405; R. M. Stubbs and C. G. Montgomery, ibid pp. 413.
9. H. A. Mook (to be published).
10. R. M. Nicklow, N. C. Koon, C. M. Williams and J. B. Milstein, (this conference).

NON-CRYSTALLINE $Y_3Fe_5O_{12}$ STUDIED BY MOSSBAUER EFFECT AND MAGNETIZATION

Th.J.A. Popma and A.M. van Diepen
Philips Research Laboratories, Eindhoven, The Netherlands

ABSTRACT

Non-crystalline $Y_3Fe_5O_{12}$ is studied by X-ray, electron microscope, BET surface, magnetic susceptibility, and Mössbauer effect techniques. The material appears to consist of X-ray amorphous platelets of several microns, which are agglomerations of particles of about 200 Å. Magnetization measurements above 50 K are consistent with a superparamagnetic behavior of particles with a size of about 200 Å. Mössbauer spectra at 300 K show a paramagnetic doublet, indicative of Fe being in a tetrahedral surrounding, while at 5 K hyperfine splitting for both octahedral and tetrahedral sites is found. In the transition range the spectra reveal magnetic relaxation effects in accordance with particles of ferrimagnetic amorphous $Y_3Fe_5O_{12}$ of ~200 Å.

Non-crystalline $Y_3Fe_5O_{12}$ was prepared[1] from a solution of Y-nitrate and Fe(III)-citrate to which an excess of citric acid was added. The solution was dehydrated by heating at 120°C and the material was pyrolized at 400°C in an atmosphere of 95% N_2 and 5% O_2. Of the obtained product portions were refired in air at temperatures T_a between 400 and 800°C. Samples heated below 680°C are X-ray amorphous while heating above 680°C results in a diffraction pattern of YIG (See Fig.1). Electron micrographs of the materials

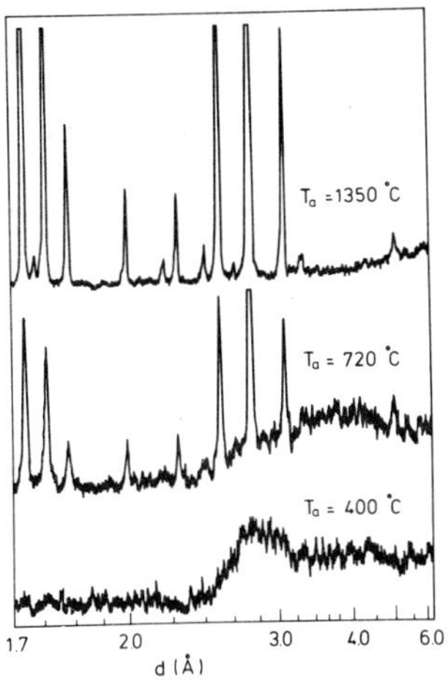

Fig.1. X-ray diffractograms of samples fired at 400, 720, and 1350°C.

heated below 680°C show platelets with in-plane dimensions of several microns. These dimensions do not change if $T_a < 800$°C. Materials prepared below 650°C also show no electron diffraction, indicating the absence of crystallographic order over a range longer than about 30 Å. The BET-surface amounts 54 m^2/g for T_a=400°C and decreases with increasing T_a and more prolongated heating (See Table 1). The magnetization as a function of applied magnetic field was determined above 300 K with a recording null-coil pendulum magnetometer. The X-ray amorphous samples exhibit hysteresis only below ~50 K. (Coercive field and remanence at 4 K are ~400 Oe and ~2 emu/gram, respectively). The magnetic moment at 4 K in a field of 10 kOe amounts to about 30% (~13 emu/g) of that of well-crystallized YIG. Above 50 K the magnetization in a constant magnetic field decreases almost linearly with temperature up to about 800-900 K depending on T_a.[1] Extrapolation of this linear behavior to zero magnetic moment yields an "apparent magnetic ordering temperature" for individual particles T_o (See Table 1). Samples heated above 680°C show magnetic properties close to those of YIG.

The ^{57}Fe Mössbauer spectra were taken at several temperatures by means of a conventional constant-acceleration spectrometer with a fixed absorber and a moving ^{57}Co in Pd source. For the low temperature

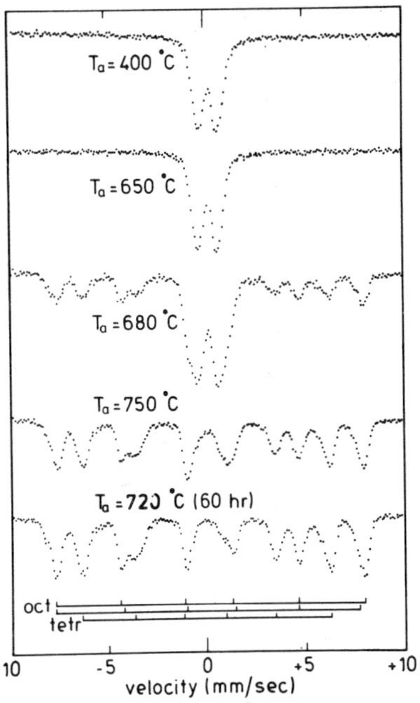

Fig.2. Mössbauer spectra at 300 K for samples fired at various temperatures for 4 hours. Positions for octahedral and tetrahedral subspectra of crystalline $Y_3Fe_5O_{12}$ are indicated.

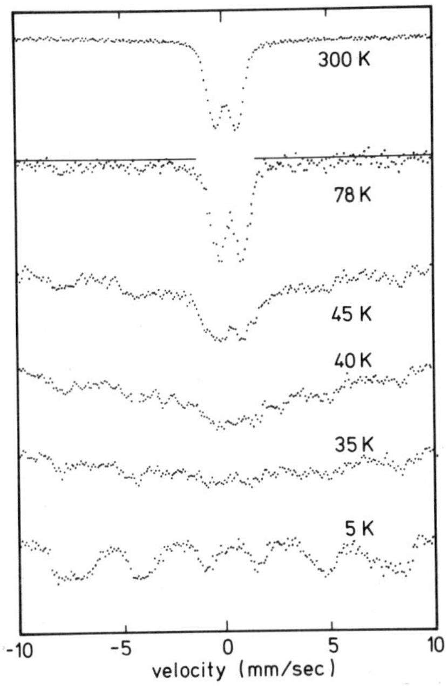

Fig.3. Mössbauer spectra of the sample fired at 400°C measured at various temperatures.

spectra a cold-finger cryostat was used. Spectra are given in Figs.2 and 3.

The magnetization data of the samples prepared below 680°C can be described within the model of superparamagnetism, except for the low field data for $T <$ 50 K (hysteresis region). In a superparamagnetic system the magnetic moment per particle can reach its average value in a magnetic field within the measuring time by virtue of the small magnetic energy content if the particles are sufficiently small. As a consequence no hysteresis is observed and for sufficiently low anisotropy the magnetization can be described by a Langevin function (See e.g. Ref.2):

$$M = Nm(T)\left[\text{cotanh}\ m(T)/kT - kT/m(T)H\right], \qquad (1)$$

where M is the magnetization, N the number of particles per unit volume, m(T) the magnetic moment per particle at temperature T, and H the applied magnetic field.

The plotting of M vs H at constant temperature indeed results in a curve very similar to a Langevin function for materials prepared at temperatures below 680°C. The magnetization data of these samples can be represented in a single curve, in accordance with a Langevin-like behavior, by plotting $M(M_{350}/M_T)$ vs $(M_T/M_{350})H/T$ (See Fig.4) in which M_T is the saturation magnetization at temperature T. Deviations from the Langevin function may be caused by a distribution in particle size. From the fitting of the magnetic data to a Langevin curve we obtain the average magnetic moment per particle m and the effective number of particles N. The results are given in Table 1. From these data we calculate a particle size of about 200 Å The surface calculated from N is about twice the value of the experimentally determined BET surface. The difference can be explained by the fact that according to electron micrographs we are dealing with platelets of several microns, apparently agglomerations of small particles.

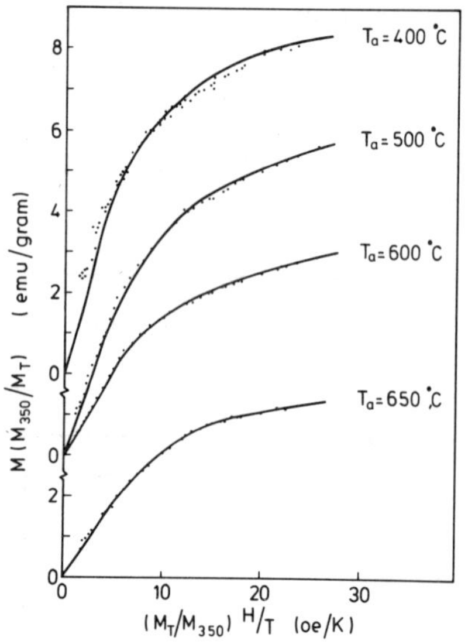

Fig.4. Reduced magnetization versus reduced M_TH/T for different values of T_a. Drawn curves are Langevin functions.

At 300 K the material shows a paramagnetic doublet in the Mössbauer spectrum, with a quadrupole splitting of about 1.0 mm/sec, about the same value as for tetrahedral iron in garnets. An octahedral component is not very clearly present and probably smeared out, indicating that in the amorphous state octahedra are more disturbed than tetrahedra. As was observed with the other techniques also in the Mössbauer spectra we see that heating below about 650°C does not change the properties very much, whereas heating at higher temperatures induces crystalline properties (Fig.2). The particle size remains virtually unchanged by heating at 680°C (See BET-surfaces in Table 1). This may be an indication for a transformation from the amorphous state into the crystalline state. The spectra of Fig.2 were all taken after the samples were cooled down to room temperature.

The temperature dependence of the Mössbauer spectra of non-crystalline samples is shown in Fig.3. At 5 K a magnetically split spectrum is observed, with roughly two subspectra, indicating octahedrally and tetrahedrally surrounded Fe (hyperfine fields 506±20 and 452±15 kOe), as in crystalline YIG, but with much broader lines. The temperature dependence of the relaxation time τ_o of superparamagnetic particles is given by:[2] $\tau_o^{-1} = 2f_o\exp(-KV/kT)$, with f_o the Larmorfrequency in the anisotropy field H_A=2K/M_s, K the anisotropy constant ($\approx 6\times10^3$ erg/cm^3 for crystalline YIG), and V the particle volume. From a comparison of our spectra with theoretical ones[3,4] we estimate that $\tau_o \approx 5\times10^{-9}$ sec at T ≈ 40 K. The size of the particles is then derived to be 150-200 Å. It should be noted that $\tau_o^{-1}/2f_o$ becomes close to 1/e for $\tau_o \approx 5\times10^{-9}$sec, so that for temperatures higher than 40K there is a rather slow variation of τ_o, while it varies much faster for lower temperatures. This makes that the outer peaks in the Mössbauer spectra remain present up to rather high temperatures (e.g. T = 78 K), and that the derived hyperfine field does not follow a Brillouin-type temperature dependence as is normally observed when going through a magnetic ordering point in crystalline materials or amorphous ferromagnetic materials.[5]

It is concluded that the present material can be described in a superparamagnetic model with amorphous ferrimagnetic particles of about 200 Å. The particles become crystalline by heating in the temperature region ~650-700°C, while higher temperatures are required for particle growth. Below ~40 K the relaxation time of the magnetic moment of the amorphous particles becomes larger than ~5x10-9sec, the characteristic Mössbauer time.

1. T.J.A. Popma and A.M. van Diepen, Mat.Res.Bull. 9, 1119 (1974).
2. E. Kneller, Theorie der Magnetisierungskurve kleiner Kristalle. Handbuch der Physik XVIII/2, Springer, Berlin (1966).
3. M. Blume and J.A. Tjon, Phys.Rev. 165, 446 (1968).
4. H.H. Wickman and G.K. Wertheim, in Chemical Applications of Mössbauer Spectroscopy (Academic Press, New York, 1968) p. 548 ff.
5. See e.g. T.E. Sharon and C.C. Tsuei, Phys.Rev. B5, 1047 (1972).

TABLE 1. Properties of $Y_3Fe_5O_{12}$ annealed at the temperature T_a for t hours. Saturation magnetization M_{350} and average magnetic moment per particle m(350) have been derived from the fitting to a Langevin function (See Fig.4). The number of particles N has been derived from M_{350} and m(350); the surface from N and the X-ray density of YIG (5.17 g/cm^3) assuming spherical particles. T_o is the apparent magnetic ordering temperature for a single particle. Estimated errors in units of the last decimal place are in parentheses.

T_a °C	t hours	M_{350} emu/gram	m(350) emu	N 1/gram	surface m^2/gram	BET-surface m^2/gram	T_o K
400	4	9.40(10)	$3.9(4)\times10^{-17}$	$2.4(3)\times10^{17}$	100(12)	53.7(5)	920
500	4	8.95(10)	$3.6(4)\times10^{-17}$	$2.5(3)\times10^{17}$	103(12)		900
600	4	5.85(6)	$3.2(3)\times10^{-17}$	$1.8(2)\times10^{17}$	91(10)	45.4(5)	870
650	4	5.45(6)	$3.0(3)\times10^{-17}$	$1.8(2)\times10^{17}$	91(10)		780
680	4					39.2(4)	
720	60					3.9(5)	

EFFECTS OF AN AMORPHOUS YIG SURFACE LAYER ON THE FERROMAGNETIC RESONANCES OF A THIN YIG FILM

J. P. Omaggio and P. E. Wigen
Department of Physics, The Ohio State
University, Columbus, Ohio 43210

ABSTRACT

Studies[1] in thin YIG films, argon implanted, report that an amorphous layer is produced and a surface spinwave is generated at the perpendicular orientation. Earlier studies of YIG films have reported surface modes at the parallel orientation only.[2]

In this paper the effect of such an amorphous layer on the spinwave spectra was observed at 9 GHz as a function of temperature (from 20° - 300°K) and orientation, and at 23 and 34 GHz as a function of orientation at room temperature. A critical angle is observed at all temperatures and frequencies. The deviations of the low order spinwave modes from a quadratic dependence and the presence of the surface mode in the perpendicular orientation are consistent with the amorphous layer having a larger magnetization than the bulk region of the film. As the temperature is decreased, the localization of the surface mode is seen to go from 106 Oe at room temperature to 425 Oe at 85°K. This indicates that in this region the amorphous magnetization is increasing at a faster rate than the bulk magnetization as the temperature decreases. Below 85°K the trend is reversed.

INTRODUCTION

The spinwave resonances of two single crystal thin YIG films of thicknesses 0.63μm and 0.51μm have been studied. The films were grown epitaxially on gadolinum gallium garnet substrates with the <111> orientation perpendicular to the plane of the film. The films were then implanted[3] with 75 KeV argon ions to a dosage of 2×10^{15} atoms/cm^2.

A new type of surface mode, appearing at the perpendicular orientation is observed only after implantation.[1] These modes are not seen at the parallel orientation and upon rotation of the applied field, a uniform mode, where only one strong mode is excited, is observed.

EXPERIMENTAL RESULTS

At X band, f ~ 9.2 GHz, the resonance spectra of the films were studied from room temperature to 25°K. A surface mode is present at the perpendicular orientation in both samples and at all temperatures. The localization is given in fig. 1. This temperature dependence is due to the different temperature dependences of the magnetization or internal fields between the bulk and the implanted regions. A study of the orientational dependence of the modes indicates that a uniform mode can be generated at some proper applied field angle at all temperatures studied. The surface mode is not observed at angles beyond this "critical angle".

Studies of the body mode spectra in both films show a significant deviation from the quadratic law for low order modes. This deviation indicates an inhomogeneity in the internal fields across the film.[4]

ION IMPLANTATION OF FILM

A recent study by Johnson, North and Wolfe[5] on implantation of single crystal magnetic garnets, indicates a damaged or polycrystalline region is produced by implantation. Fig. 6 in their publication indicates the damage concentrations produced as a function of film depth for 100 KeV protons implanted at various dosages. Fig. 7 of the Johnson et. al publication shows quite clearly that damage concentration rises dramatically with ion mass.

Using more massive ions, comparable dosages and similar targets in this study, it is assumed for purposes of developing a model that the damage profile of our films has a shape similar to that given by fig. 3A. It is believed that the implantation of the films has caused damages at the amorphous level and that the width of this region is about 300Å.[3]

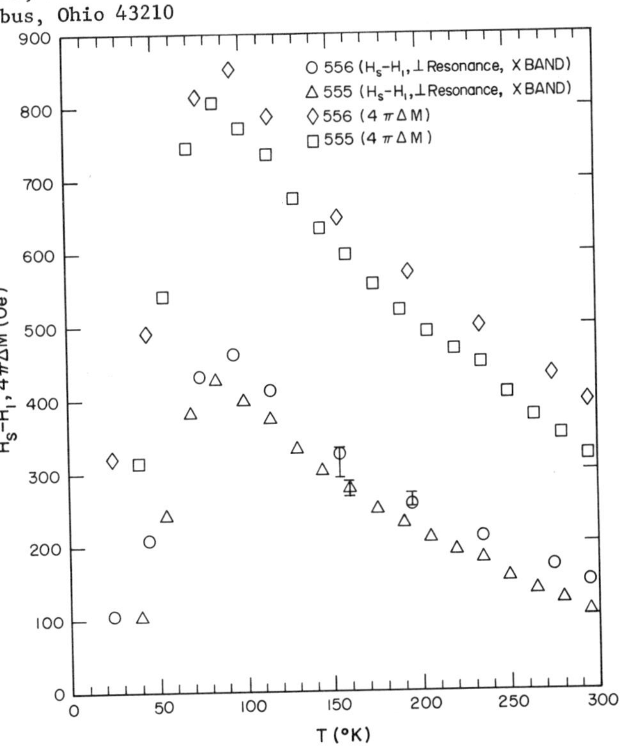

fig 1. The difference in field position between the surface mode and highest field body mode as a function of temperature. The mode positions are for the perpendicular configuration taken at X band (9 GHz). Also the values of 4πΔM as a function of temperature.

PHENOMENOLOGICAL MODEL

If exchange coupled spinwaves in a thin ferromagnetic film are excited with the applied field perpendicular to the plane of the film, the equation of motion for the circular component of the magnetization is given by:[5]

$$-\frac{2A}{M}\frac{d^2m}{dz^2} + \left[H - 4\pi M + \frac{2A}{M}\frac{d^2M}{dz^2}\right] m = \frac{\omega}{\gamma} m \qquad (1)$$

where $M_x + iM_y \equiv M^+ = me^{-i\omega t}$.

Also, H is the applied field, M is the saturation magnetization, A is the exchange stiffness constant and γ is the gyromagnetic ratio for the material.

In the model, it is assumed that the magnetization of the amorphous layer is larger than that in the bulk region as shown in fig. 3A. Not knowing the details of the profile of the magnetization in the interface region the details of the d^2M/dz^2 term will be ignored, and the resultant field is shown in fig. 3B.

The spinwave spectrum expected from such a model can be compared with the experimental observations. Eqn. (1) has the same form as Schrodinger's equation for a one particle, one dimensional system with potential $H - 4\pi M$. If the function m satisfies certain boundary conditions at the film surfaces only "eigenfunctions" m_i with certain "eigenvalues" ω_i/γ will be allowed. Depending on the thickness, L_1, and depth $4\pi\Delta M \equiv 4\pi M_A - 4\pi M_B$ (M_A is the magnetization in the amorphous region and M_B is the magnetization in the single crystal region), of the well in the amorphous region, there can exist zero, one or more "surface"

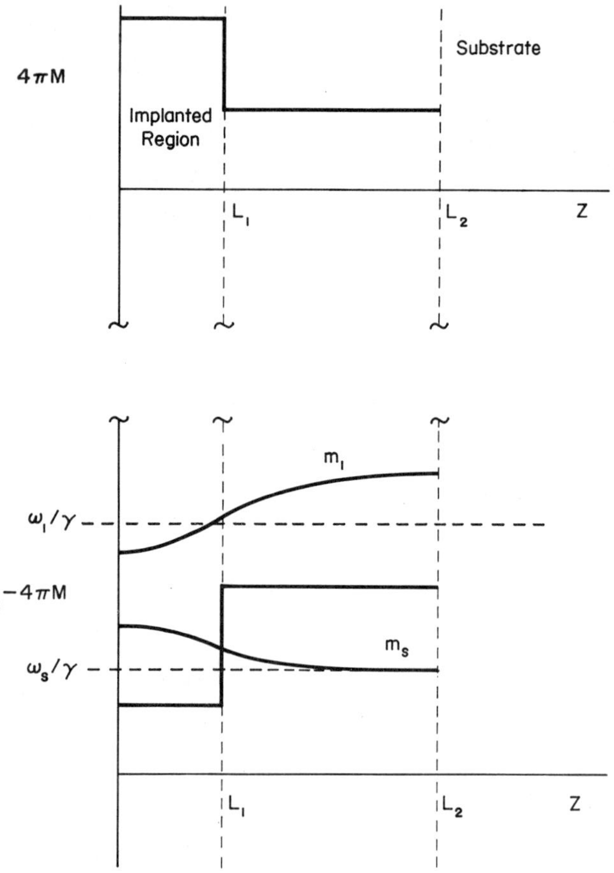

fig 2. A (top): The proposed special dependence of $4\pi M$ across the film thickness. The amorphous region is $0 < z < L_1$ and the single crystal region is $L_1 < z < L_2$.

B (bottom): The resulting special dependence of the effective potential, $H_a - 4\pi M$. m_s denotes the qualitative dependence of the lowest energy, ω_s/γ surface mode. Similarly m_1 shows the dependence of the lowest energy, ω_1/γ, body mode.

modes where $\omega/\gamma < H - 4\pi M_B$. "Body" modes are allowed for $\omega/\gamma > H - 4\pi M_B$. Experimentally, a weaker high field mode is observed in addition to the body spinwave modes, at the perpendicular orientation at all temperatures studied. As discussed below, the position of this mode is observed to be much more sensitive to the amorphous region well depth than are the positions of the body modes. This is consistent with this mode being a surface mode in our model.

At the parallel orientation the lack of cylindrical symmetry complicates the form of the equation of motion. The result changes the sign of the demagnetization field term resulting in an effective potential that is larger in the surface layer than in the bulk region. As a result, a surface mode would not be allowed at the parallel orientation.

As a function of the angle of the magnetization between the perpendicular and parallel orientations, the depth of the "well step" in the potential varies continuously between the two extremes. Thus at some "critical angle", θ_c, the effective potential or equivalently, the internal field, will be homogeneous across the film thickness.

The experimental behavior of the spinwave spectra is consistent with this model. Upon rotation from the perpendicular orientation towards the critical angle, the surface mode is observed to increase in energy relative to the body modes which are relatively uneffected. At θ_c, the surface mode is transformed into the uniform mode and is the only mode excited. This angle is consistent with the position where a uniform potential is predicted to exist across the film and therefore suggests boundary conditions given by

$$\frac{dm}{dz} = 0 \text{ at } z = 0,\ z = L_2 \qquad (2)$$

where L_2 is the film thickness. Upon rotation past θ_c, surface modes are no longer observed but the body modes are again excited.

Due to the lack of knowledge of the behavior of $M(z)$ near the amorphous -bulk boundary, the $\frac{2A}{M^2}\frac{d^2M}{dz^2}$ term in the model is neglected. For films with heavily damaged implanted regions, d^2M/dz^2 will be large only near the amorphous -bulk interface. Since this term is present at all orientations, if it were of significant magnitude it would introduce inhomogeneities in the effective internal field which forbids a uniform mode solution. The fact that only one intense mode is observed at the critical angle is consistent with the assumptions that this term is small compared to $H - 4\pi M$ and that the surface spins are unpinned.

QUANTITATIVE CALCULATIONS

For surface modes, m has a propagation constant K in the amorphous region and an exponential dependence with decay constant α in the bulk region. From the equation of motion:

$$DK^2 = \frac{\omega_s}{\gamma} - H - 4\pi M_a \qquad (3)$$

$$D\alpha^2 = H - 4\pi M_B - \omega_s/\gamma \qquad (4)$$

where D is the exchange constant for YIG, assumed constant in both regions. Requiring m_s and its first derivative to be continuous at the amorphous-bulk interface and demanding that m obeys the boundary conditions given by eqn. 2, yields the condition:

$$0 = \exp(\alpha(a - L)) \cdot (\alpha^2\cos(Ka) + K\alpha \sin(Ka)) \qquad (5)$$
$$+ \exp(\alpha(L - a)) \cdot (K\alpha \sin(Ka) - \alpha^2\cos(Ka))$$

where a is the amorphous layer thickness and L is the total film thickness. Specifying a, L, D, ω, $4\pi\Delta M$ and $4\pi M_B$, a computer program was written to generate the allowed values of K, α, and H_s. By comparing the computer generated values of H_s with those determined experimentally, the values of $4\pi\Delta M$ were determined for both samples as a function of temperature. The results of this calculation are shown in fig. 1.

The value of D used was $.516 \times 10^{-8}$ $(\text{Oe} - \text{cm}^2)$[6] and the values of $4\pi M_B$ used for each temperature were determined from the experimental spectrum. It should be noted that the values of $4\pi M_B$ and the values of

$4\pi M_A = 4\pi M_B + 4\pi\Delta M$ include all internal fields (other than the applied field), including the crystalline anisotropy and magnetostrictive effects.

CONCLUSION

The magnetization of a surface layer of argon implanted YIG has been determined using ferromagnetic resonance. The temperature dependence of the surface like mode localized in the amorphous region and the first body mode in the bulk region indicate that the magnetization of the amorphous layer is larger between 25K and 300K and has a temperature dependence that differs considerably from the bulk materials.

REFERENCES AND FOOTNOTES

1. R. D. Henry, P. J. Besser, D. M. Heinz and J. E. Mee, IEEE MAG vol. 9, pp. 535 - 37 (1973).
2. S. D. Brown, R. D. Henry, P. E. Wigen and P. J. Besser Sol. St. Comm. vol. 11, pp. 1179 - 82 (1972).
3. The films were implanted and supplied by Rockwell International at Anaheim, Cal.
4. A. M. Portis Appl. Phys. Lett. vol. 2 pp. 69 - 71 (1965).
5. W. A. Johnson, J. C. North and R. Wolfe J. Appl. Phys. vol. 44 pp. 4753 - 57 (1973).
6. R. C. LeCraw and L. R. Walker J. Appl. Phys. vol. 32 pp. 1675 - 1685 (1961).

OBSERVATION OF rf-INDUCED SIDEBAND EFFECTS IN AN AMORPHOUS MAGNETIC MATERIAL*

C. L. Chien and J. C. Walker
The Johns Hopkins University, Baltimore, Maryland 21218

R. Hasegawa
IBM Research Center, Yorktown Heights, New York 10598

ABSTRACT

We report here the first observations of the rf sideband effect in amorphous magnetic material. The sideband effect in amorphous $Fe_{75}P_{15}C_{10}$ can be easily observed using the Mössbauer effect with a pulsed rf field. The sideband intensities are greater than those observed for an Fe foil of similar dimensions.

The observation of r-f sidebands by Mössbauer spectroscopy was first reported by Heiman et al.[1] Subsequent experiments[2-5] have clearly demonstrated that this frequency modulation effect of the γ-rays is caused by acoustic vibrations driven by magnetostriction when the ferromagnetic sample is subjected to radio frequency magnetic fields. It is interesting to investigate the possibility of observing sideband effects in an amorphous ferromagnetic material, where macroscopic magnetostriction might be expected to be quite small due to the absence of any long range crystal structure.

An amorphous sample of $Fe_{75}P_{15}C_{10}$ was chosen for the present work. The amorphous Fe-P-C system has been extensively studied in the past few years by standard magnetic techniques,[6] Mössbauer spectroscopy,[6] Lorentz electron microscopy,[7] and other methods.[8] Tsuei et al.[6] have shown that a very similar system, $Fe_{80}P_{12.5}C_{7.5}$, is ferromagnetic below 586 K. The high ordering temperature makes it unnecessary to cool the sample from room temperature. When heated above 690 K, the amorphous sample rapidly transforms to its crystalline state.

The amorphous $Fe_{75}P_{15}C_{10}$ sample was made by rapid quenching from the liquid state. It is about 30 μm thick and 1.5 cm in diameter. X-ray and electron microscopy studies verify the amorphism and indicate that there are no crystals larger than 15 Å.

Conventional Mössbauer spectroscopy was used to study the sample maintained at room temperature. The sample was sandwiched between two pieces of microscope cover glass to ensure mechanical stability. This arrangement has been shown to approximate a free hanging foil.[5] The r-f field was applied parallel to the sample face by means of an eight-turn loosely wound coil around the sample. The r-f power was pulsed at a rate of 2.5 kHz with 50% duty cycle to avoid sample heating.

A Mössbauer spectrum of amorphous $Fe_{75}P_{15}C_{10}$ with zero r-f field is shown in Fig. 1 (top spectrum). A rather poorly-resolved spectrum is commonly experienced in amorphous samples due to the existence of many sites. The symmetric pattern with the absence of any apparent quadrupole effects can possibly be explained by the random angles that the electric field gradient axes make with respect to the hyperfine field axis.[6] A correct fit to this spectrum would require a complicated distribution of hyperfine fields. However, the spectrum can also be reasonably well fit by assuming a six-line pattern with different widths for the three pairs of lines as shown by the solid curve in Fig. 1. This simplification is particularly necessary for fitting the resulting spectrum when one turns on the r-f field because additional lines may appear for each parent line.

Mössbauer spectra taken with an 11 MHz r-f field are also shown in Fig. 1. The frequency of 11 MHz is chosen because it provides additional lines roughly half-way between the parent lines. The fit assumes

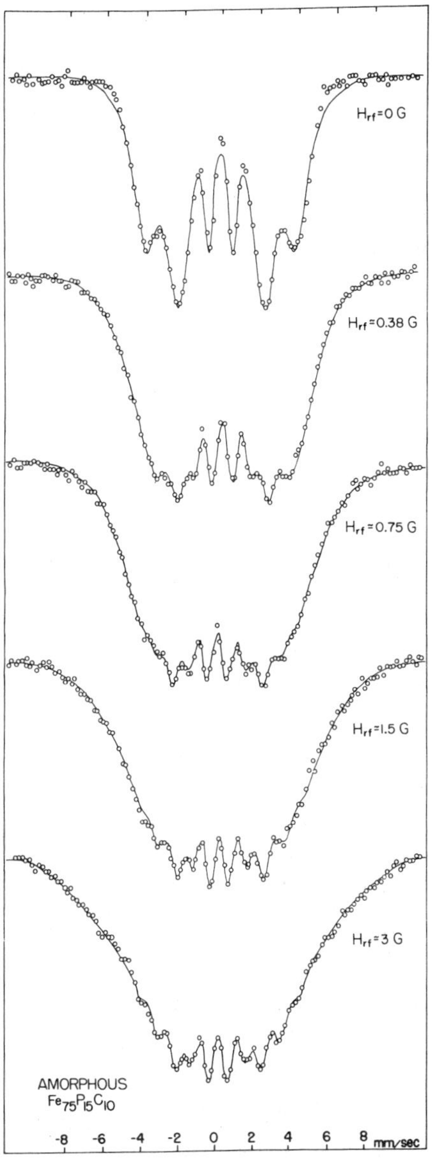

Fig. 1. Mössbauer spectra of amorphous $Fe_{75}P_{15}C_{10}$ at 300 K with and without r-f field. The r-f field strength is denoted as H_{rf}.

six parent lines as mentioned above. The sideband intensities are assumed to vary according to $e^{-m^2} I_n(m^2)$, where I_n is the modified Bessel function of the first kind and m is the modulation index.[2] Because of the poor resolution only a few sideband lines can be clearly resolved. Nevertheless, the change of the pattern clearly demonstrated the sideband effect. A typical value of m is 1.9 for $H_{rf} \simeq 1.5$ G. Due to the relatively poor resolution of the spectrum, the measured values of m are probably not particularly reliable.

It should be noted that the r-f field strength needed to observe the sideband effect in amorphous $Fe_{75}P_{15}C_{10}$ is rather small - much smaller than that required to produce a comparable effect at 11 MHz in a metallic Fe coil of similar dimensions. As is shown in Fig. 2, for a 25 μm thick Fe foil sidebands begin to appear at an r-f field strength ≳1.5 G. This is at least five times larger than the field required to

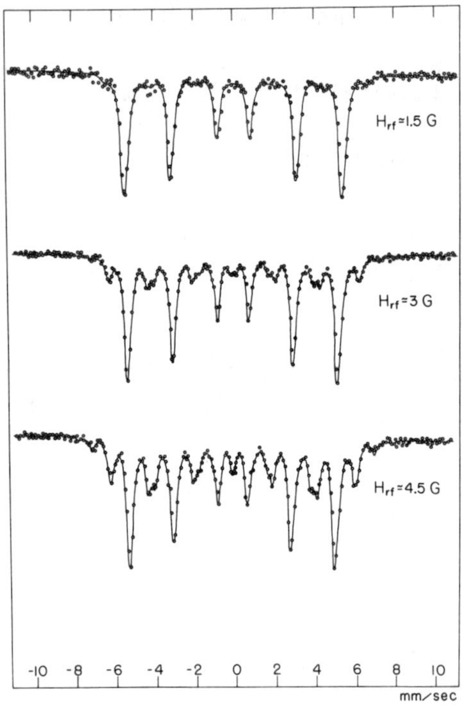

Fig. 2. Mössbauer sideband spectra of 25 μm thick metallic Fe at 300 K. The r-f field strength is denoted as H_{rf}.

just show sidebands in $Fe_{75}P_{15}C_{10}$.

The appearance of r-f induced sidebands depends not only on the existence of magnetostriction, but also on the acoustic Q, and how efficiently acoustic vibrations are scattered off the sample face.[2] These factors can be very different for different materials. Therefore, direct comparison of sideband intensities (or modulation indices) for different ferromagnetic materials does not necessarily reflect the difference in their magnetostrictions. However, under the presently accepted interpretation,[2] the fact that sidebands exist does indicate that significant magnetostriction is present in the sample.

Magnetostriction in an amorphous sample is expected to be significantly reduced from its crystalline value. There is experimental evidence in some amorphous systems which supports this expectation - most notably the rare-earth iron system.[9] For $Fe_{75}P_{15}C_{10}$, no measurement of magnetostriction has yet been conclusively made, partly due to the fact that the sample preparation method restricts the available sample thickness. Our results indicate that the magnetostriction in $Fe_{75}P_{15}C_{10}$ is quite sizable. The r-f sideband technique can provide an additional method for determining magnetostriction in thin amorphous samples. Since sideband effects have been observed in the past both by direct Mössbauer spectroscopy and by recoilless Rayleigh scattering,[4] sideband techniques for detecting magnetostriction can be applied to all ferromagnetic amorphous materials whether or not they contain iron. It is, however, necessary to determine the acoustic Q of these materials. Work is in progress to measure Q for amorphous $Fe_{75}P_{15}C_{10}$.

We are indebted to Professor P. Duwez for kindly providing the amorphous samples.

REFERENCES

*Work supported by the National Science Foundation.

1. N. D. Heiman, L. Pfeiffer and J. C. Walker, Phys. Rev. Letters 21, 93 (1968).

2. L. Pfeiffer, N. D. Heiman and J. C. Walker, Phys. Rev. B6, 74 (1972).

3. R. F. Heile and J. C. Walker, Bull. Am. Phys. Soc. 18, 85 (1973).

4. R. F. Heile and J. C. Walker, Bull. Am. Phys. Soc. 17, 546 (1972).

5. N. D. Heiman, R. K. Hester and S. P. Weeks, Phys. Rev. 8, 3145 (1973).

6. C. C. Tsuei, G. Longworth and S. C. H. Lin, Phys. Rev. 170, 603 (1968).

7. D. I. Paul, J. Marti and L. Valadez, AIP Conf. Proceedings No. 18, 1377 (1974).

8. S. C. H. Lin and P. Duwez, Phys. Stat. Sol. 34, 469 (1969).

9. A. E. Clark, AIP Conf. Proceedings, No. 18, 1015 (1974).

PROPERTIES OF THE AMORPHOUS MAGNET: MgF_2/Fe

Z. Shanfield, P. H. Barrett, P. A. Montano
Department of Physics, University of California, Santa Barbara, Calif. 93106

ABSTRACT

The properties of thick films of simultaneously deposited MgF_2 and Fe have been studied as functions of concentration and temperature. X-ray diffraction patterns display no lines assignable to α-Fe, γ-Fe or Fe compounds; lines assignable to the substrates (teflon or aluminum) and MgF_2 were observed. The highly ionic MgF_2 matrix perturbs the electronic configuration of the neutral Fe atom as demonstrated by the Mössbauer effect isomer shift (+ 0.46 mm/sec with respect to Fe metal at room temperature). The Mössbauer spectra linewidths had weak temperature dependence, while both quadrupole splitting and linewidths showed weak concentration dependence. Mössbauer spectroscopy also showed that long range magnetic ordering did not occur above 1.2°K in the concentration range studied (10 at. % Fe to 45 at. % Fe). The Mössbauer spectra in an external magnetic field were typical of those of materials with paramagnetic relaxation processes. The relaxation times and hyperfine fields are concentration dependent.

The Mössbauer effect has been used to study the amorphous magnet, MgF_2/Fe, over a range of temperatures and concentrations. Thick films were prepared in a vacuum evaporator by simultaneously depositing Fe atoms and MgF_2 molecules onto aluminum foil or teflon tape substrates kept at room temperature (RT). Since it was desired to study the suppression of magnetic order in an amorphous system, only those three samples that showed a single broad Mössbauer effect line at RT were selected for comprehensive measurements.

Electron microprobe analysis was used to determine sample concentrations (10 ± 6, 22 ± 3 and 45 ± 8% at. Fe) and to verify that the samples were uniform on the scale of microns. X-ray diffraction patterns displayed no lines assignable to α-Fe, γ-Fe, or Fe compounds; lines assignable to the substrates and MgF_2 were observed.

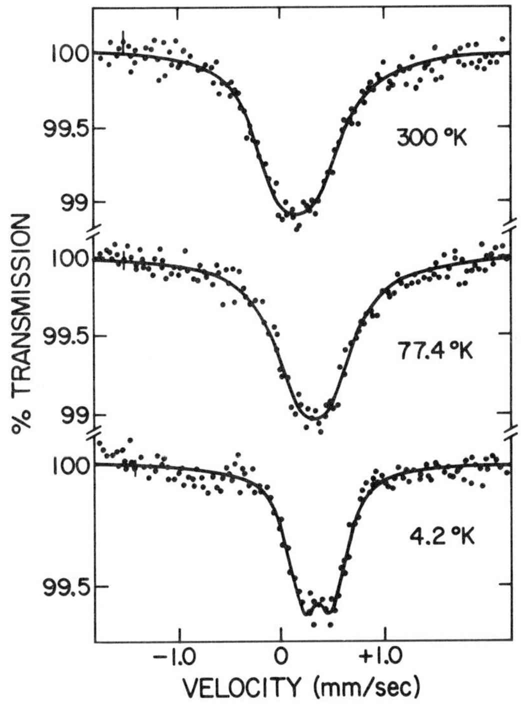

Fig. 1. Mössbauer absorption spectra for 45% at. Fe sample of MgF_2/Fe at various temperatures. Doppler velocity relative to Fe metal at room temperature.

Mössbauer spectra taken between room temperature and 1.2°K displayed a single, broad line that was least squares fitted to a quadrupole doublet (Fig. 1). Because of the steepness of the dip in the observed spectra, attempts to fit to a single, broad Lorentzian line were unsuccessful. The quadrupole splitting decreased slightly with increasing concentration from 0.38 ± 0.05 mm/sec for the 10% sample to 0.32 ± 0.05 mm/sec and 0.30 ± 0.05 mm/sec for the 22% and 45% samples respectively. Linewidths were of the order of 0.6 mm/sec at room temperature decreasing to around 0.4 mm/sec at 4.2°K, the decrease being concentration dependent. The temperature dependence of the central shift was fitted to an Einstein lattice model[1] giving, within the experimental uncertainty, a concentration independent isomer shift of =0.46 ±0.03 mm/sec with respect to Fe metal at RT. An Einstein temperature, θ_E = 274 ± 20°K, was also found (Fig 2). Using the $\Delta\langle r^2\rangle$ value of Micklitz and Barrett[2] and the electron densities and quadratic interpolation scheme of Blomquist, et al.,[3] an electronic configuration of $3d^{6.45}\,4s^{1.55}$ was determined for neutral Fe. One interpretation of these fractional electron assignments is that a strong mixing of the free atom $3d^6\,4s^2$ ground state configuration and the $3d^7\,4s^1$ first excited state configuration occurs due to the strong perturbation of the MgF_2 matrix. The areas of the Mössbauer spectra show very strong, abnormal dependence on temperature; the areas decreased as the temperature decreased, the decrease being greater for higher concentrations.

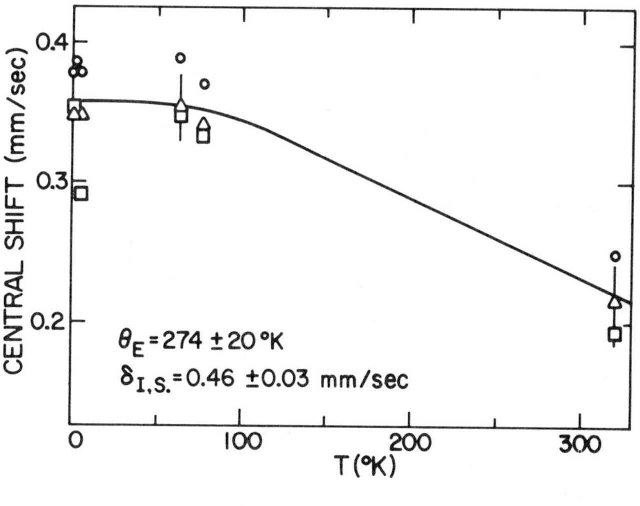

Fig. 2. Central shift versus temperature, ○, 10% at Fe. Δ, 22% at. Fe. □, 45% at. Fe.

Mössbauer spectra were also taken at 4.2°K in external magnetic fields of 15.2 KOe and 30.5 KOe applied along the direction of propagation of the γray (Fig.3). These spectra were analyzed using the relaxation theory of Wegener,[4] giving maximum hyperfine fields of - 440 ± 60 KOe,- 245 ±100 KOe, and - 100 ±60 KOe for the 45%,22% and 10% samples respectively. Also obtained were relaxation times of the order of 2.0×10^{-7} sec and 1.0×10^{-9} sec for the 45% and 22% samples. The relaxation time for the 10% sample was much faster than 10^{-9} sec. A strong increase of the resolved Mossbauer spectra area with applied field was

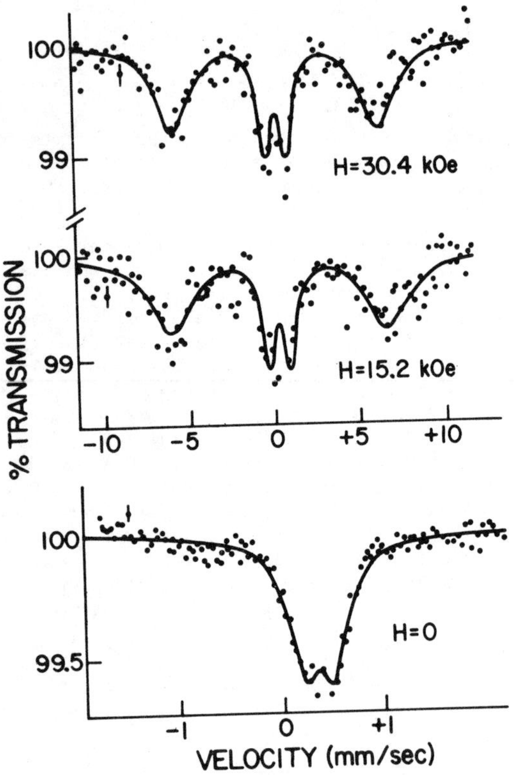

Fig. 3. Mössbauer absorption spectra for 45% at. Fe sample at. 4.20°K at various applied magnetic fields.

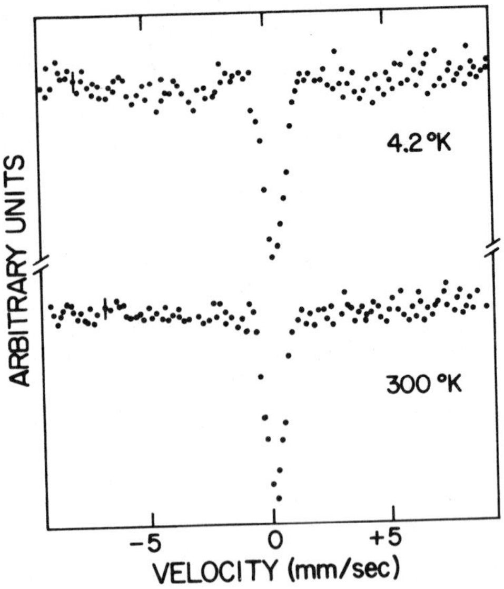

Fig. 4. Mössbauer absorption spectra for 45% at. Fe sample of MgF_2/Fe at 4.2°K and room termperature for H=0 kOe. Note that the velocity scan is greater than in Fig. 1.

also observed; the change from zero field to 15.2 KOe for the 45% sample was larger than the increase in area for the 22% sample.

The behavior of the spectral area with temperature and field can be explained using the following model. The magnitude of the effective hyperfine field at the nucleus is a function of the atomic relaxation time and atomic parameters such as spin, etc.[4] Since these samples are crystallographically disordered, there is a variation of the atomic parameters especially exchange. As the temperature is lowered those spins whose exchange coupling is sufficiently strong will correlate, producing short-range-ordered "magnetic clusters."[5]. The relaxation times will lengthen, causing the appearance of an effective magnetic field at the iron nucleus. Due to the distribution of magnetic interactions in these "clusters," the Mössbauer lines will produce a broad unresolved background. In Figure 4 Mössbauer spectra, with a much larger velocity range than in Fig. 1, are shown for the 45% at. Fe sample. At 300°K the spectrum shows a broad, single line and a flat background. At 4.2°K, however, there is a central line plus a suggestion of a broad unresolved background possibly caused by the "clusters" and which could account for the missing area. With the application of an external field the relaxation time of the "cluster" atoms will decrease and the background lines will coalesce, increasing the observed resolved area. In the 45% sample, a 30 KOe field nearly saturates the magnetization[6] causing the lines to coalesce, increasing the observed area. Magnetization measurements have been carried out and will be reported in the near future.

REFERENCES

[1] V. I. Goldanskii and R. H. Herber, Chemical Applications of Mössbauer Spectroscopy (Academic, New York, 1968) Chapt. 1.

[2] H. Micklitz and P. H. Barrett, Phys. Rev. Letters 28, 1547-1550 (1972).

[3] J. Blomquist, B. Roos, and M. Sundbom, J. Chem. Physics 55, 141-145 (1971).

[4] H. Wegener, Z. Physik 186, 498-511 (1965).

[5] P. A. Beck, Met. Trans. 2, 2015-2024 (1971).

[6] Z. Shanfield, Dissertation, Univ. of Calif., 1974 unpublished.

SPIN GLASSES AND MICTOMAGNETS

J.A. Mydosh
Institut für Festkörperforschung der Kernforschungsanlage, 517 Jülich, Germany

ABSTRACT

Over the past few years two new descriptive terms have entered the lexicon of magnetism: spin glass and mictomagnet. Both of these terms apply to non-dilute magnetic alloys where there is a random freezing of impurity moments at a distinct temperature which gives rise to a sharp peak in the low field susceptibility and a sudden splitting of the Mössbauer spectrum, along with remanence, irreversibility and relaxation. Spin glass refers mainly to the lower concentration regime where the long range indirect RKKY interaction is responsible for the cooperative freezing of the spins. Mictomagnetism is more appropriate in describing the magnetic clusters which form in the spin glass matrix at higher concentrations, due to local (first or second nearest neighbor) exchange, or when there is a tendency towards short range atomic ordering. The present paper will attempt to illustrate this type of random magnetism: (1) by presenting a variety of known (and conjectured) alloy systems which exhibit such characteristics, (2) by reviewing the salient experimental features of the susceptibility, magnetization, Mössbauer effect, neutron scattering, resistivity, thermopower and the specific heat, and (3) by surveying the development of theoretical models to treat these effects.

INTRODUCTION

The term spin (or magnetic) glass was first suggested by B.R. Coles to be applied to the strange magnetic behavior of the "weak" moment AuCo system /1/. Simultaneously, this expression appeared, again at Coles' instigation, in a treatise from P.W. Anderson /2/ linking localization theory with the long standing CuMn problem. At about the same time P.A. Beck, in discussions with J.S. Kouvel over this very problem, originated the word mictomagnetism which stems from the Greek prefix meaning mixed. A description of this concept was originally presented in Beck's survey article on recent results in the magnetism of alloys /3/. It was also in 1971 that the observance /4/ of rather sharp peaks in the low field magnetic susceptibility for a series of low concentration AuFe alloys was reported. Last year yet another term, speromagnetism /5/, was coined with respect to the random spin freezing of an amorphous ferric gel. This lead the editors of Nature to request a taxonomist for magnetism /6/. In the past few years there has been an increasing usage of the terms spin glass and mictomagnetism. While it is certainly premature to attempt a definitive review of this area, nevertheless some advantage might now be gained by surveying these random phenomena in magnetic alloys.

The first problem which arises is to obtain a working understanding for the physical concepts behind the "catch-words" spin glass and mictomagnet. We attempt to represent schematically a non-dilute, 2-dimensional magnetic alloy in Fig. 1. The lattice is crystalline and has a random (or quasi random - allowing for the possible existence of short range atomic order) distribution of localized magnetic spins, here $\approx$10 at.%. For very large impurity separations, i.e. at the lowest concentrations, the impurities are mainly non-interacting, and we have the "single impurity" or dilute limit which exhibits experimental properties describable within the framework of the Kondo effect /7/. By increasing the concentration, the local moments begin to interact via the RKKY mechanism /8/ - indirect exchange mediated by the conduction electrons. As the temperature is lowered in zero external field, the impurity spins "freeze out" or become "locked" in random directions, i.e. the vector average of all local moments gives no net macroscopic moment, and there is no long range magnetic order. Yet the freezing occurs at a well-defined temperature, T_o. For $T \ll T_o$, local or diffuse excitations may exist within the frozen matrix. The concentration, c, at which impurity correlations become important can be estimated by setting the freezing temperature $T_o(c)$ equal to T_K, the Kondo temperature /9/. For this regime, the experimental parameters scale as universal functions of T/c and H_e/c, and the freezing temperature, T_o, is proportional to c /10/. Furthermore, the magnetization processes reflect the "viscous" freezing at T_o by exhibiting irreversibility, thermomagnetic and isothermal remanence, and time dependent effects. Tholence and Tournier /9/ have proposed a model for this spin glass state where the alloy for $T \lesssim T_o$ spontaneously divides itself into regions ($\approx$500 impurities) within which the individual moments are coupled by the RKKY interaction. The resulting moment of each region is frozen in the direction of a random anisotropy axis.

As the concentration is further increased,

Fig.1 SPIN GLASS WITH MICTOMAGNETIC CLUSTERS ≈ 10 at.% IMPURITIES.

there is a breaking down of this concentration scaling, since at a few thousand ppm short range correlations, characteristic of the particular impurity, (a local pairing, tripling, etc.) are becoming important. Now small groups of impurities - clusters, form at one temperature, interact with each other via the long range RKKY, and then cooperatively freeze at the lower freezing temperature. For concentrations over a few at.%, large clusters are present with effective moments $\approx$ 20-20,000μ_B, and these dominate the magnetic behavior /11/. The cluster moments can freely rotate or superparamagnetically respond to an external field until the temperature is decreased to T_o. Then the spin glass matrix freezes and the clusters are blocked due to an anisotropy energy barrier in random directions. Fig. 1 illustrates these mictomagnetic clusters embedded within the spin glass matrix. With greater growth of the clusters, a magnetic percolation limit is reached for long range order. For the ferromagnetic case of an fcc-lattice, an aligned path of spins reaches from one end of the crystal to the other. This occurs at about 15at.% /12/. When the spins are antiferromagnetically coupled, enough of them are needed to build a long range two sub-lattice structure in fcc space. A much larger concentration is therefore required $\approx$45 at.% /13/. Above these concentrations, an inhomogeneous long range magnetic order exists.

Therefore, a spin glass may be described as a cooperative or collective, random freezing of the impurity spin below a characteristic temperature. Under this freezing temperature, the viscous metastable nature of the "locking-in" process produces the remanence, irreversibility and relaxation. Many of the spin glass properties are related to the ferro- and antiferromagnetic monodomains introduced long ago by Néel /14/. The term mictomagnet, which describes effects originally observed by Kouvel /15/ and considered in his "ensemble of domains" model, is particularly appropriate for the cases: a) At higher concentrations $c \gtrsim 10$ at.%, where purely from statistics there are large probabilities of nearest (or next nearest) neighbor magnetic impurities, and b) in those systems, e.g. CuMn, where there is a tendency towards short range atomic ordering. In both of these cases, the experimental properties will reflect the high temperature, a few hundred Kelvin, magnetic cluster formation. Then at T_0 with the freezing of the spin glass matrix, these cluster moments are no longer free to rotate and become fixed in random directions. There are also low temperature excitations for the spin glass and within or on the periphery of the clusters.

Noble Metal - Transition Metal

Host \ Impurity →	V	Cr	Mn	Fe	Co	Ni	
Cu	X-S	X-S	good	X-S+T*	X-S	X-T	Zn Mn
Ag	X-S	X-S	good	X-S	X-S	X-S	and
Au	X-T	good	good	good	X-S+T*	X-T	Al_{30} Fe_{70}

*See text

Transition Metal - Transition Metal

Host \ Impurity →		Cr	Mn	Fe	Co	
Mo	simple	?	good	good	X-S+T	V Mn
Rh	simple	X-T	good	X-T ■	X-T ■	$Ni_{75}Mn_{25}$
Pd	exchange enhanced	good	both LRferroO & S.G.	LRferroO	LRferroO	$Pd_{75}Mn_{25}$ / (Pd Ag) Fe or Mn
Pt	exchange enhanced	X-T	good	LRferroO	good	(Pd Fe or Mn) H

■ See text

Fig.2 Spin Glass/Mictomagnet Combinations. X means the alloy is not a particularly favorable system because of a limited solubility, X-S, or a weak moment(T_K too large),X-T. LRferroO refers to the preference for long range ferromagnetic order.

SOME SPIN GLASS/MICTOMAGNETIC SYSTEMS

We begin with the well-studied noble metal-3d transition metal alloys, and invoke two criteria to find the simplest spin glass behavior: (a) "good moment" systems, i.e. the Kondo temperature, T_K, should be less than $\approx$1K, so that no complications are encountered with the weakening of the local moments at low temperatures, and (b) a favorable solubility such that at least 10 at.% of the 3d metal may be dissolved in the noble metal host. This latter criterion in conjunction with a proper homogenization process provides for a random distribution of impurities and eliminates difficulties with chemical precipitation. Besides the archetypal examples of CuMn and AuFe, there are only three other uncomplicated spin glass systems, shown by a "good" in Fig. 2: AuCr, AuMn and AgMn. The low field susceptibility /16/ and the electrical resistivity /17/ exhibit very similar properties for these five alloy systems. By referring to the listings in Fig. 2, problems are encountered for many of the other combinations, with either the Kondo temperature or especially the solubility limit. However, a number of special cases exist, as for example with CuFe which possesses a rather poor solubility and a $T_K \approx 30$ K, but which nonetheless shows the magnetization characteristics of a mictomagnet /18/. The deviation from randomness (precipitation of Fe for $c \gtrsim 1$ at.%) is reflected in the low field susceptibility by the sharp peaks at T_0 changing into broad maxima /16/. Also in this more complex category, we would include AuCo for which the Co solubility is poor, but not as bad as CuFe. The difficulty here is with the very large - few hundred Kelvin - Kondo (or spin fluctuation) temperature which is strongly dependent upon the number of Co nearest neighbors. For this system an effective magnetic concentration /16/ should be used instead of the actual one. The specific heat /19/ and magnetic susceptibility /20/ of ZnMn also show the characteristics of a spin glass. Although the solubility of Mn in Zn is rather limited (in the thousand ppm region), the hexagonal crystal structure may help to elucidate the anisotropy mechanisms present. Finally, displaced hysteresis loops and torque measurements indicate mictomagnetism and exchange anisotropy in an $Fe_{0.7}Al_{0.3}$ alloy /21/.

We now turn out attention to binary combinations of transition metals. A collection of various possibilities is given in Fig. 2. For the "simple" hosts Mo and Rh, a number of clearly favorable systems exist. The special cases of RhFe /22/ and RhCo /23/ require a sufficiently large concentration of magnetic impurity ($\approx$1 at.% Fe and 20 at.% Co), before spin glass behavior manifests itself. Below this concentration paramagnetic spin fluctuations - "weak moments" - are present, while for c large enough there is a transition to ferromagnetic order. Thus, within a triangularly shaped region of a T-c phase diagram, spin glass behavior occurs.

For the exchange enhanced hosts, Pd and Pt, a strong tendency towards "giant moment" ferromagnetism exists even at the lowest concentrations. However, there is a keen competition between this long range order and the mixed or antiferromagnetic correlations of the random alloy. Hence, particularly for Mn or Cr impurities and their inclination for antiparallel couplings, the spin glass state may overcome the ferromagnetism at certain concentrations. An especially interesting situation is with PdMn alloys where for $c \lesssim 4$ at.% Mn the giant moment ferromagnetism prevails. But upon further increasing the Mn concentration (bringing the Mn atoms closer together), T_C falls, and the spin glass freezing appears as $c \gtrsim 5$ at.% Mn /24,25,26/.

We should at this point also mention the weak, but giant moment, systems PdNi, RhNi, VFe and VMn. (CuNi would additionally be counted among these). The word "weak" refers to the model stipulation that a sufficient number of Ni, Fe or Mn nearest neighbors are present to produce the local magnetic moment which is *per se* a cluster. For all these systems with the exception of VMn there seems to be a change of behavior, with increasing concentration, directly from paramagnetic (or spin fluctuations) to ferromagnetic ordering without any intervening spin glass or mictomagnetic regime /22/. A critical-like concentration is used to describe the onset of ferromagnetism, namely at Pd + 2.3 at.% Ni, $Rh_{37}Ni_{63}$, $V_{75}Fe_{25}$ and $Cu_{58}Ni_{42}$ /22,27/. Nonetheless, the onset of long range magnetic order in a random solid solution remains one of the least understood areas of the magnetic alloy problem. Further efforts would be required to develop better experimental criteria /28/ for describing such systems around their "critical" concentrations. A theoretical framework has recently been proposed to treat this situation /29/. The exception, VMn, seems to show mictomagnetism above about 50 at.% Mn /30/, but this behavior would be in conflict with the Sato-Kikuchi /13/ antiferromagnetic percolation limit for a bcc lattice of 21 at.%. A "weak" Mn moment and a strongly preferred local order might account for this discrepancy.

A number of well ordered ferro- or antiferromagnets, e.g. Ni_3Mn /31/ or Pd_3Mn /26/, become mictomagnetic upon disturbing the atomic order. Then there are certain elegant methods to destroy or reduce the large exchange enhancement of Pd, thereby forming a spin glass state at low concentrations of Fe and Mn. One example is the ternary alloy (PdAg) Fe (or Mn) /32/; another would be the hydrogenation of PdFe or PdMn /33,34/.

Due to the lack of space we will be unable to review the rare earth metal impurities which would include two topics of current interest: Kondo/spin glass - superconductors, and the coexistence of magnetism and superconductivity. We also omit any discussion of the disordered lattice spin glasses.

SUMMARY OF EXPERIMENTAL PROPERTIES

In this section we limit ourselves to the noble metal - 3d metal systems and present some experimental highlights of the spin glass/mictomagnetic behavior. First of all, we treat the completely reversible, low field, magnetic susceptibility. This susceptibility, χ, exhibits the very sharp peaks at T_o, and thus uniquely defines the freezing temperature. Some typical $\chi(T)$ measurements are presented in Fig. 3 /35/. Below the freezing temperature, for $0.3T_o \lesssim T \lesssim 0.8T_o$, $\chi(T)$ may be simply described by $\chi = a + bT^n$ where n is approximately 2. As $T \rightarrow 0$, $\chi(0)$ extrapolates to a finite value, and the ratio $\chi(0)/\chi(T_o)$ is nearly constant (≈ 0.6), for not too large impurity concentrations ($c \lesssim 5$ at.%). Above T_o, the behavior is Curie-Weiss-like with values of the effective moments, p_{eff}, for the lower concentrations (≈ 0.1 to 1.0 at.%), consistent with those found in the dilute or Kondo limit. There is an increase of p_{eff} with larger c which is attributed to the growing importance of short range magnetic correlations. The magnitude of the paramagnetic - θ is close to zero when short range correlations are unimportant, but $|\theta|$ increases to small values as the short range correlations and p_{eff} grow /36/. Recently, Mukhopadhyay and Beck /37/ have extended susceptibility measurements to room temperature for Cu + 9 and 25at.% Mn. At ≈ 200 - 300 K, a simple Curie-Weiss law is obeyed with p_{eff} that of the dilute or single impurity limit, and $\theta \approx 100$ K indicating the presence of interactions. Then as the temperature is lowered below 200 K, there is a gradual

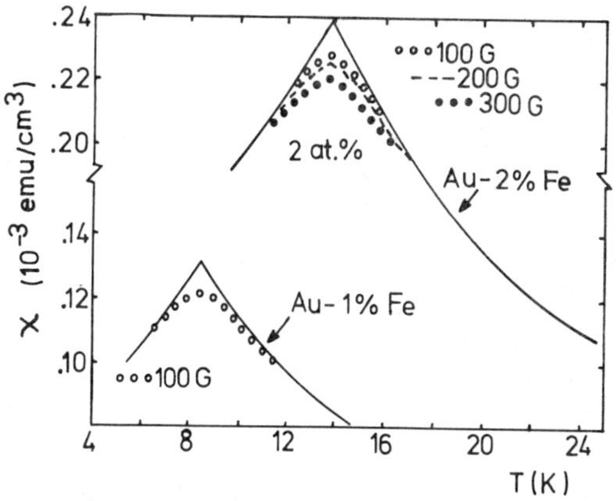

Fig.3 Typical low field susceptibility with increasing external field. The solid lines represent the zero-field limit-χ.

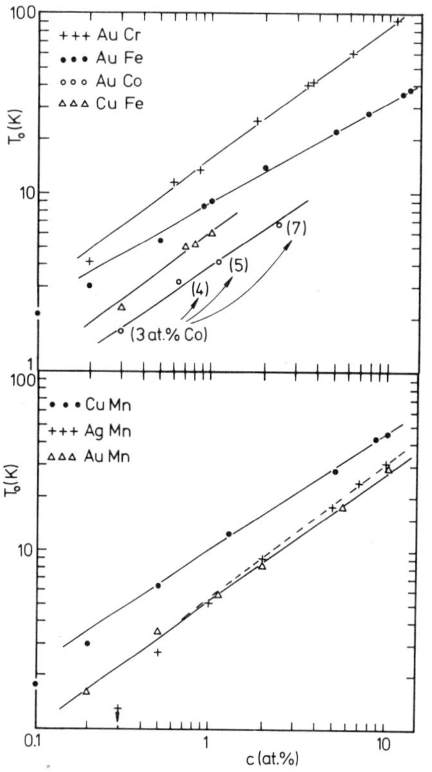

Fig.4 Freezing temperatures for the noble metal based Spin Glasses. (See Ref. 16)

diminishing of the $1/\chi$ versus T slope meaning an increase of p_{eff} or cluster growth. These clusters which form at about 150 - 200 K, then freeze at a much lower T_o (≈ 40 K and ≈ 100 K for the two concentrations). In the concentration region ≈ 0.1 to 10 at.%, $T_o \propto c^m$ where m is between 0.55 and 0.75. Fig. 4 shows a collection of freezing temperatures plotted as a function of the concentration for the various noble metal - 3d impurity systems /16/. There seems to be a roughly linear concentration dependence, $T_o \propto c$, for the regions both above and below these 0.1 to 10 at.% limits.

A more quantitative estimate of the cluster formation and growth for $T > T_o$ is given by Brillouin function fitting of magnetization data /3,38,39/: $M = \mu_c NB[\mu_c H/k_B T]$. Depending upon concentration, heat treatment and temperature, the concentration of superparamagnetic clusters, N, and their average moment, μ_c, can be determined. The magnetization measurements for $T \lesssim T_o$ depend very strongly on the external field present as the specimen is cooled through

T_o. The zero-field cooled, low field susceptibility $\chi \equiv (M/H)|H \rightarrow 0$, discussed in the previous paragraph, loses its reversibility and becomes greatly smeared /35,40,41/ when measured in an external field, H_e, greater than a threshold value $(H_e)_c$ (see Fig. 3). In conjunction with this, hysteresis losses and a isothermal remanent magnetization (IRM) are present /9,42/. In Fig. 5 we illustrate these magnetization processes. Part (a) is the zero-field cooled M-H character; the inner portion being reversible, with the small hysteresis and IRM developing after $(H_e)_c$ is exceeded. Part (b) represents a field cooled, displaced hysteresis loop /15,42,43/. This displacement is a dramatic illustration of the unidirectional exchange anisotropy caused by the field cooling. For comparison, a typical ferromagnetic loop with its virgin curve is given in part (c). One can also measure the thermal remanence magnetization (TRM) in zero field after cooling in field H_e and define an irreversible susceptibility $\chi_{irr} \equiv TRM(H)/H|H \rightarrow 0$. For $T \leq T_o$, the total susceptibility, $\chi_{TOT} = \chi + \chi_{irr}$, seems to become independent of the temperature at least for AuFe /9/. The remanent magnetizations decay with characteristic relaxation times τ_i(c, H_e, T). A single τ does not seem to be sufficient to represent the data which would mean that various thermal activation processes are responsible for the relaxation effects in a spin glass /43/.

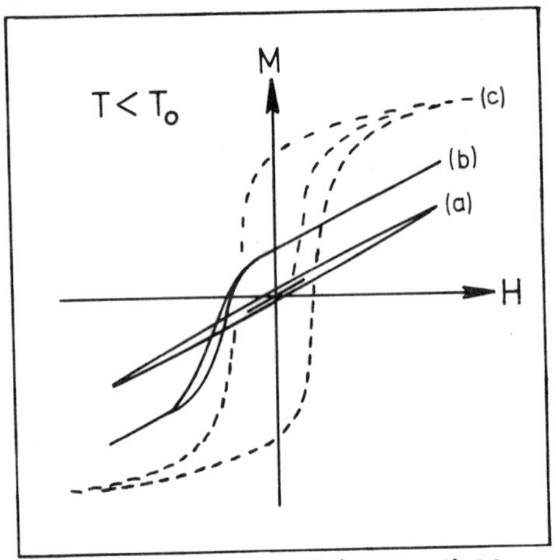

Fig.5 Hysteresis loops for (a) zero-field cooled Spin Glass, (b) field cooled Spin Glass, and (c) ferromagnet.

Mössbauer measurements, both for Fe^{57} and Sn^{119} doped alloys, give a clear confirmation of the freezing temperature /44-47/. For, at T_o, there is a sudden splitting of the Mössbauer spectra from a single (paramagnetic) line into the six fingered structure. This onset of a hyperfine field, assumed proportional to the atomic magnetization, is usually associated with magnetic ordering. However, upon careful analysis of the spectra, the intensity ratios indicate a random alignment of spin /48,49/. The long relaxation times of the frozen spins and the collective nature of the freezing process thus create the coherent splitting at T_o. An external field can produce a ferromagnetic component consistent with the magnetization results. Borg /49,43/ has recently proposed a simple model based upon a random freezing of spins to account for the moment discrepancies between the Mössbauer and the magnetization measurements. And Window /50/ has found some evidence from his Mössbauer investigations of certain easy directions for the alignments of spins in a cluster.

Long ago, by analyzing his neutron scattering data, Arrott pointed out the absence of long range magnetic order in these types of alloys /51/. Recently, however, a series of neutron diffraction measurements /52/ have detected the existence, even at a few hundred ^oC, of short range atomic order over a few unit cells. For CuMn (15 to 25 at.% Mn), this in turn leads to the corresponding short range type of magnetic coupling - a giant ferromagnetic moment. Moreover, this ferromagnetic clustering is perfectly consistent with what was determined from the susceptibility and the magnetization /37/. Polarized neutron studies /53/, which clearly give the magnetic scattering cross section, show that the bulk static susceptibility of CuMn alloys at low enough temperature is definitely less than the neutron scattering cross section $\propto \chi(\kappa)$ in the limit $\kappa \rightarrow 0$. This indicates that below a certain temperature a static disordered structure has been formed and that the correlations are simply frozen "in a manner analogous to the formation of a glass". Further, as the concentration of Mn is reduced $c \lesssim 5$ at.%, the tendency towards complete randomness is increased, and hence, the statistical fluctuations would then be responsible for the cluster formation rather than the short range order. Most probably the key to the microscopic understanding of the spin glasses lies with these neutron investigations. A series of such studies would be suggested to probe the formation of magnetic correlations at higher temperatures, the freezing region around T_o, and the dynamics or excitations as $T \rightarrow 0$.

Although electrical resistivity measurements have been performed for many years on spin glass/mictomagnetic alloys (especially in the dilute Kondo limit), only now is a good experimental - theoretical comprehension evolving. Systematic studies of the five most favorable noble metal - 3d impurity systems show a generally similar overall behavior /17/ which is schematically illustrated in Fig. 6. $\Delta\rho$ represents an estimate of the pure magnetic or spin resistivity, since the host contribution has been subtracted, viz. $\Delta\rho = \rho(\text{alloy}) - \rho(\text{host})$. By starting from the limit $T \rightarrow 0$, there is a large residual component $\Delta\rho_o$ due to the "disorder" scattering from the randomly frozen spin glass matrix and mictomagnetic cluster. As the temperature is increased, $\Delta\rho = \Delta\rho_o + A(c)T^{3/2}$ where A(c) is a slowly decreasing function of the concentration, e.g. $A(c) \propto - \ln c$ or $c^{-1/5}$. Some evidence has recently been offered for the appearance of a T^2 term at the very lowest temperatures $\lesssim 0.3$ K (when $T \approx T_K$) for these systems /54/. As the temperature is increased a strong linearly dependent $\Delta\rho \propto T$ is found in the temperature region about T_o. A good correlation between the temperature of the maximum in the temperature coefficient of the resistivity, $d(\Delta\rho)/dT$, and T_o exists for AuFe /55/, but for the Mn and Cr systems this relationship breaks down /56/. Nonetheless, T_o and the "unfreezing" of the spin glass are marked by this steeply rising linear resistivity contribution. For yet higher temperatures, a much slower $\Delta\rho(T)$ variation gradually leads to a maximum in $\Delta\rho$ at T_m, followed by a continuing fall-off as room temperature is reached. This decreasing $\Delta\rho$ with increasing T is surprisingly steep for AuCr with $\Delta\rho(300) \lesssim \Delta\rho_o$. Very high temperature measurements /57/ suggest the above inequality to be generally true. The maximum in $\Delta\rho$ would indicate the approximate temperature region where correlations between the impurities first appear and then the cluster begin to form. However, at such high temperatures there are definite complications present as for example with deviations from Matthiessen's rule. Hence it would be an oversimplification to completely attribute the negative slope of $\Delta\rho(T)$ to a single impurity Kondo-like mechanism. Presently, studies /58/ of the resistivity for such systems are being carried out at very high pressures ≈ 100 kbar, in order to investigate the nature of these impurity interactions when the Kondo temperature is varied with the pressure. In addition, a new theory by Rivier and Adkins /59/ has ascribed the low temperature $\Delta\rho$ behavior to low wave vector elementary excitations which are diffusive in character. If all the diffusion modes $0 \leq q \leq q_{max}$ can be excited, $\Delta\rho(T)$ = const. $c\,(T/\Lambda)^{3/2}$ where Λ is a diffusion constant which varies as c times some function of the electronic mean free path. Finally we remark that another transport property, the thermoelectric power exhibits strikingly dissimilar behavior among the various spin glass systems /17/. A more extensive experimental study, along with some additional theoretical guidance are at the moment lacking.

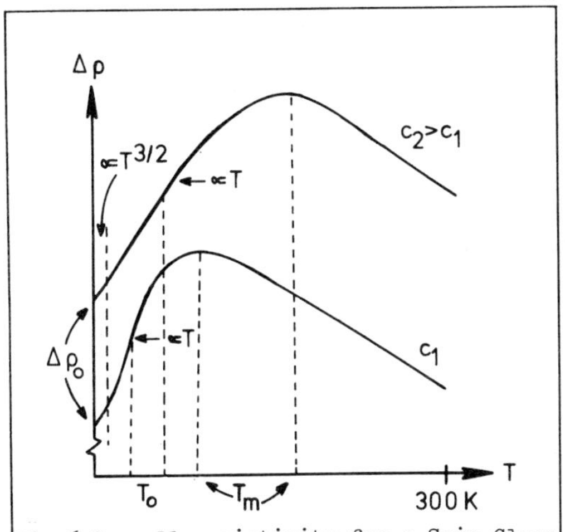

Fig.6 Overall resistivity for a Spin Glass

Specific heat measurements are an especially important method to examine the spin glass for ordering, freezing or excitations. To extract the magnetic portion of the specific heat, C_m, a particularly precise experiment is required along with a secure knowledge of the γT and βT^3 terms /60/. At the lowest temperature C_m has a large, anomalous, concentration independent, linear in T, contribution /61,62,63, 19,10/ which seems to be a general property of disordered materials - amorphous glasses or spin glasses /64/. A maximum in C_m near T_0 is present, followed by a 1/T fall off at higher temperatures. With the lower concentration samples, the scaling of C_m/c with T/c holds rather well. Very recently, Window /65/ has discussed an anisotropy - cluster model to account for the concentration independent linear term. However, the description of the maximum in C_m near T_0 remains incomplete when considered from either the experimental or the theoretical end. Exactly how sharp (equivalent to χ?) and coincident with T_0 is this maximum, and what do the theories which consider the peak in have to say about C_m? The experimental question might easily be resolved by employing the latest pulse methods for specific heat measurement, and we will briefly return to the theoretical aspects in the next section.

SURVEY OF THEORETICAL MODELS

The usual theoretical approach /66/ to such non-dilute magnetic alloys has been the mean random molecular field approximation /67-70/. Here the statistically distributed impurities, indirectly coupled via the RKKY interaction produce a probability distribution of the random internal field and of the temperature, P(H,T). At high temperatures P(H) is a single-peaked, symmetric Lorentzian distribution. But for lower temperatures where the spin correlations are very important, there is a greater probability of higher fields, so that P(H) changes into a double-peaked Lorentzian with a smaller but finite $P(0) \propto 1/c$. In addition, the contributions to P(H) at a given point, r_0, may be divided into a correlated region around r_0 and an uncorrelated remainder beyond a certain correlation length. When this P(H,T) is convoluted with a Brillouin function to determine the susceptibility (and similarly for the other physical quantities), a broad maximum in $\chi(T)$ at a temperature directly proportional to the concentration is obtained. The magnitude of $\chi(T)$ is approximately independent of concentration and external field. Also due to the RKKY interaction being proportional to $1/r^3$, there is an invariance of the product (cr^3). This naturally results in the "scaling" where such measurable quantities as the magnetization M/c, the specific heat C_m/c, etc. are universal functions of T/c and H_e/c /10/. Thus, $\chi = f(T/T_0)$ where $T_0 \propto c$ for the scaling spin glass regime.

In another approach /71,72/ to the problem of interacting magnetic impurities, a virial expansion of the free energy in terms of c has been employed to calculate the various thermodynamic functions and the resistivity. A scaling in H_e/c and T/c is also found here. This method seems to be especially valid at higher temperatures $T>T_0$, and not too large concentrations where it agrees nicely with specific heat measurements /19/, and predicts, consistent with experiment, a simple Curie-Weiss law for the susceptibility with the paramagnetic-θ = $(\text{const})cV_0$. V_0, the only adjustable parameter of the theory, is the RKKY coupling constant. For the resistivity with $T>T_0$, $\Delta\rho = c\rho_{so}[1-K(cV_0/T)]$. ($\rho_{so}$ is the constant spin disorder scattering and K is a constant for given impurity spin.) This result would correspond to the slowly rising experimental resistivity roughly between T_0 and T_m - see Fig.6.

In an attempt to directly treat the sharp maximum in the susceptibility at T_0, three theories have very recently been proposed. The first of these from Adkins and Rivier /73/ generalizes the local molecular field theory by calculating P(H) from first principles, using the analogue with the random walk problem. The theory then assumes a local order parameter (a local magnetization) m(T) which gives a definite orientation of the spins within a correlation length. m(T) is a collective feature of the entire ensemble of spins due to the long range nature of the RKKY interaction. By seeking a self-consistent solution for the coupled m(T) and P(H) equations, a nontrivial temperature dependence for m(T) admits finite values below a certain temperature, T_0. This results in a shift of P(H) to higher fields, and, when the zero field susceptibility is calculated, a cusp at T_0. The cusp in χ is associated with the sudden onset of short range order for $T \leq T_0$. Additionally, a Curie-Weiss dependence is found at $T>T_0$ (with some deviations near T_0), and χ remains finite in the limit $T \rightarrow 0$. A small external magnetic field causes a smearing in m(T) around T_o, and thus the "sharpness" in χ is washed out.

In the second approach to the theory of spin glasses, Edwards and Anderson /74/ have constructed a time dependent correlation function of the spin at a single site i: $q = \langle \vec{n}_i(1) \cdot \vec{n}_i(2) \rangle$, where n_i is the probability of finding a particular spin orientation, and the (1) and (2) represent "today" and "tomorrow". They then proceed to develop a formalism for the ensemble free energy F. Here the exchange coupling energy between a pair of spins J has a Gaussian probability distribution P(J). By performing a series of mathematical manipulations for the ensemble integrals, an expression for F(q) is obtained. The susceptibility has the general form $\chi = (C/T)[1-q]$ which has the following limits: $\chi = C/T$ for $T > T_o$ and $\chi = C/T_o$ at T_o, $\chi = (C/T_o)\,[1+1/2(1-T_o^2/T^2)] \approx C/T_o - \mathrm{const}(T_o-T)^2$ for $T \lesssim T_o$, and $\chi = (C/T_o)(2/3\pi)^{1/2}$ for $T \rightarrow 0$. Therefore, the ratio $\chi(0)/\chi(T_o) = (2/3\pi)^{1/2} = 0.46$. The specific heat is linear in temperature at the lowest temperature, approaches T_o from under $(T^- \rightarrow T_o)$ with zero slope and from over $(T \gtrsim T_o)$ as $C_m \propto T_o^2/T^2$. Thus, there is maximum (semi-cusp) in C_m at T_o. However, the mean field theory usually does not correctly predict the shapes at a singularity, and a more symmetric form would be expected.

The final theoretical model from D.A. Smith /75/ utilizes the competition between the long range RKKY interactions, which seek to build correlated spin clusters, and the thermal disorder energy, which interferes with these correlations. At low enough temperatures, the RKKY coupling must win and an "infinite cluster" is formed at T_o which spontaneously freezes (an infinite relaxation time). The specific heat may be calculated from this model in the limit of very low temperatures as arising from loose spins in the infinite cluster, here $C_m \propto T$. For $T \gg T_o$, a high temperature expansion similar to that of Larkin et al. /70,71/ yields $C_m \propto 1/T$.

ACKNOWLEDGEMENTS

I wish to acknowledge my most fruitful and long-standing collaboration with V. Cannella and P.J. Ford. In addition, I have greatly benefited from discussions with P.A. Beck, H. Claus, K. Fischer and N. Rivier.

REFERENCES

1. M.H.Bancroft,Phys.Rev.B2,2597(1970).
2. P.W.Anderson,Mat.Res.Bull.5,549(1970).
3. P.A.Beck,Met.Trans.2,2015(1971).
4. V.Cannella,J.A.Mydosh,and J.I.Budnick,J. Appl.Phys.42,1689(1971).
5. J.M.D.Coey and P.W.Readman,Nature 246,476 (1973).
6. See Nature 246,445(1973).
7. See the recent reviews of the Kondo effect by M.D.Daybell,in Magnetism V,editors G.T. Rado and H.Suhl (Academic,N.Y.,1973)pg.121, and C.Rizzuto,Rep.Prog.Phys.37,147(1974).
8. M.A.Ruderman and C.Kittel,Phys.Rev.96,99 (1954);T.Kasuya,Progr.Theoret.Phys.(Kyoto) 16,45(1956);K.Yosida,Phys.Rev.106,893(1957).
9. J.L.Tholence and R.Tournier,J.Phys.(Paris) 35 C4,229(1974).
10. J.Souletie and R.Tournier,J.Low Temp.Phys.1, 95(1969).
11. P.J.Beck,J.Less Common Metals 28,193(1972).
12. K.Duff and V.Cannella, in Amorphous Magnetism editors H.O.Hooper and A.M.de Graaf(Plenum, N.Y.,1973)pg.207,and AIP Conf.Proc.10,541 (1973).
13. H.Sato and R.Kikuchi,AIP Conf.Proc.18,605 (1974).
14. L.Néel,Ann.Géoph.5,99(1949),and in Physique des Basses Températures (Gordon and Breach, N.Y., 1961).
15. J.S.Kouvel,J.Phys.Chem.Sol.24,795(1963),and experimental references therein.
16. V.Cannella and J.A.Mydosh,AIP Conf.Proc.18, 651(1974),and in Proc.Int.Conf.on Magnetism 2,74(Publ.House Nauka,Moscow,1974).
17. P.J.Ford and J.A.Mydosh,J.Phys.(Paris)35 C4, 241(1974),and in Proc.Int.Conf.on Magnetism 2,79(Publ.House Nauka,Moscow,1974).
18. S.Mishra and P.A.Beck,Phys.Stat.Sol.(a)19, 267(1973).
19. F.W.Smith,Solid State Commun.13,1267(1973), and Phys.Rev.B9,942(1974).
20. F.W.Smith,Phys.Rev.B10,2034(1974).
21. J.S.Kouvel,J.Appl.Phys.30,313S(1959).
22. B.R.Coles in Proc.of the Michigan State University Summer School on Alloys-1972,pg.183, and R.Rusby,J.Phys.F4,1265(1974).
23. B.R.Coles,A.Tari and H.C.Jamieson in Low Temperature Phys.-LT13,editors K.D.Timmerhaus, W.J.O'Sullivan and E.F.Hammel(Plenum,N.Y., 1974)Vol.2,pg.414.
24. B.R.Coles,J.Phys.(Paris)35 C4,203(1974);B.R. Coles,H.Jamieson,R.H.Taylor and A.Tari,to be published in J.Phys.F.
25. H.A.Zweers and G.J.Van den Berg, to be published in J.Phys.F. See also G.J.Nieuwenhuys,Thesis,University of Leiden (1974).
26. D.J.Chakrabarti,Int.J.Magn.6,305(1975).
27. P.Pataud,J.P.Perrier and R.Tournier,J.Phys. (Paris)35 C4,189(1974).
28. See for example the recent magnetization measurements on AuFe by A.P.Murani,J.Phys. (Paris)35 C4,181(1974)and J.Phys.F4,757(1974) and on VFe by H.Claus,Phys.Rev.Lett.34,26 (1975).
29. D.Sherrington and K.Mihill,J.Phys.(Paris)35 C4,199(1974).
30. S.Chakravorty,P.Panigrahy and P.A.Beck,J. Appl.Phys.42,1698(1971).
31. J.S.Kouvel and C.D.Graham,J.Appl.Phys.30, 312S(1959)and J.Phys.Chem.Solids11,220(1959)
32. J.I.Budnick,V.Cannella and T.J.Burch,AIP Conf.Proc.18,307(1974).
33. J.A.Mydosh,Phys.Rev.Letters33,1562(1974).
34. J.P.Burger and D.S.McLachlan,Solid State Commun.13,1563(1973).
35. V.Cannella and J.A.Mydosh,Phys.Rev.B6, 4220(1972).
36. A full account of these susceptibility measurements will shortly be forthcoming. V. Cannella and J.A.Mydosh,to be published.
37. A.Mukhopadhyay and P.A.Beck,to be published.
38. P.A.Beck in Magn.in Alloys,editors P.A.Beck and J.T.Waber,(TMS-AIME,N.Y.,1972)pg.211.
39. B.de Mayo,AIP Conf.Proc.5,492(1972);J.Phys. Chem.Solids 35,1525(1974).
40. O.S.Lutes and J.L.Schmit,Phys.Rev.125,433 (1962)and 134,A676(1964).
41. J.Owen,M.E.Brown,V.Arp and A.K.Kip,J.Phys. Chem.Solids 2,85(1957).
42. J.S.Kouvel,J.Phys.Chem.Solids 21,57(1961).
43. R.J.Borg and T.A.Kitchens,J.Phys.Chem.Solids 34,1323(1973).
44. R.J.Borg,R.Booth and C.E.Violet,Phys.Rev. Letters 11,464(1963)and C.E.Violet and R.J. Borg,Phys.Rev.149,56(1966).
45. V.Gonser,R.W.Grant,C.J.Meecham,A.M.Muir,and H.Wiedersich,J.Appl.Phys.36,2124(1965).
46. B.Window,Phys.Letters24A,659(1967).
47. A.P.Jain and T.E.Cranshaw,Phys.Letters 25A, 425(1967).

48. P.P.Craig and W.A.Steyert,Phys.Rev.Letters 13,802(1964).
49. R.J.Borg,Phys.Rev.B1,349(1970).
50. B.Window,Phys.Rev.B6,2013(1972).
51. A.Arrott,J.Appl.Phys.36,1093(1965)and in Magnetism IIB,editors B.T.Rado and H.Suhl (Academic,N.Y.,1966)pg.295.
52. S.A.Werner,H.Sato and M.Yessik,AIP Conf. Proc.10,679(1973)and H.Sato,S.A.Werner and R.Kikuchi, J.Phys.(Paris)35 C4,23(1974).
53. N.Ahmed,S.J.Campbell and T.J.Hicks,in Proc. of the Int.Conf.on Magn. 4,91(Publ.House, Nauka,Moscow,1974),and N.Ahmed and T.J.Hicks, Solid State Commun.15,415(1974).
54. O.Laborde and P.Radhakrishna,J.Phys.F3,1731 (1973),and in Proc.of the Int.Conf.on Magn. 4,82(Publ.House Nauka,Moscow,1974).
55. J.A.Mydosh,P.J.Ford,M.P.Kawatra and T.E.Whall, Phys.Rev.B10,2845(1974).
56. A complete description of the resistivity measurements will shortly be forthcoming. P.J.Ford and J.A.Mydosh,to be published.
57. C.Domenicali and E.L.Christenson,J.Appl.Phys. 32,2450(1961).
58. J.Schilling,J.Crone,P.J.Ford,S.Methfessel, and J.A.Mydosh,J.Phys.F4,L116(1974);and J. Schilling,P.J.Ford,and J.A.Mydosh,to be published.
59. N.Rivier and K.Adkins,to be publ.in J.Phys.F.
60. For an extensive review of magnetic alloy specific heats, see N.E.Phillips,Crit.Rev. Solid State Sci.2,467(1971).
61. J.E.Zimmermann and F.E.Hoare,J.Phys.Chem. Solids17,52(1960).
62. J.De Nobel and F.J.Du Chatenier,Physica 25, 969(1959);F.J.Du Chatenier et al.Physica 32, 403,561,1097(1966).
63. B.Dreyfus,J.Souletie,R.Tournier and L.Weil, Compt.Rend.259,4266(1964).
64. P.W.Anderson in Amor.Magn.,editors H.O.Hooper and A.M.de Graaf (Plenum,N.Y.,1973)pg.1.
65. B.Window,Phys.Rev.Letters 32,665(1974).
66. For a more detailed theoretical description along with an excellent review of spin glass properties,see K.Adkins,Thesis,University of London (1974).
67. W.Marshall,Phys.Rev.118,1519(1960).
68. M.W.Klein and R.Brout,Phys.Rev.132,2412(1963)
69. M.W.Klein,Phys.Rev.136,A1156(1964);173,522 (1968);188,933(1969).
70. S.H.Liu,Phys.Rev.157,411(1967).
71. A.I.Larkin and D.E.Khmel'nitskii,Zh.Eksp. Teor.Fiz.58,1789(1970).[English transl. JETP 31,958(1970)].
72. A.I.Larkin,V.I.Mel'nikov and D.E.Khmel'nitskii,Zh.Eksp.Teor.Fiz.60,846(1971).[English transl. JETP 33,458(1971)].
73. K.Adkins and N.Rivier,J.Phys.(Paris)35 C4, 237(1974);N.Rivier,to be published.
74. S.F.Edwards and P.W.Anderson,to be published.
75. D.A.Smith,to be published.

FERROMAGNETISM IN AMORPHOUS SOLIDS*

G. S. Cargill III
Department of Engineering and Applied Science
Yale University, New Haven, Connecticut 06520

ABSTRACT

Experimental results on ferromagnetism in amorphous metals and alloys are reviewed, together with models for atomic arrangements in these materials. Consequences of chemical and structural disorder are discussed, with particular emphasis on magnetic moments, spin waves, and the temperature dependence of magnetization in amorphous ferromagnets.

INTRODUCTION

Some amorphous solids are magnetically ordered, although they are structurally disordered, i.e. they lack the long range structural periodicities which characterize crystalline solids. Coexistence of ferro- and ferrimagnetic order with structural disorder has been studied in many amorphous materials. Results of some of these studies are summarized in this review, which focusses on effects of structural and alloy (chemical) disorder on magnetic moments, temperature dependence of magnetization, spin waves, magnetic anisotropy, and hysteresis in amorphous ferromagnets.

ATOMIC ARRANGEMENTS[1]

Most experimental studies of ferromagnetism in amorphous solids have employed alloys of transition metals [Fe, Co, Ni] with elements of Groups IIIA, IVA, or VA [B, C, Si, P], containing approximately 20 at.% of the latter. These "metal-metalloid" alloys have been produced as amorphous solids by rapid quenching from the liquid state ("splat cooling" or "roller quenching"), by evaporation in vacuum and condensation on cooled substrates, by electrodeposition from aqueous solutions, and by "electroless" (chemical) deposition. Most of these amorphous alloys are metastable at room temperature and revert to their more stable crystalline forms only when heated. Crystallization temperatures are typically in the range 250°C - 450°C. Interpreting magnetic properties of these materials is complicated by their being alloys. Their equilibrium crystalline forms are usually mixtures of two or more phases, which in most cases rules out comparisons of properties between amorphous and (single phase) crystalline materials of the same composition. This makes it difficult to separate effects of chemical disorder and composition from those of structural disorder.

Information on atomic arrangements in amorphous solids have come from x-ray, electron, and neutron scattering experiments and from density measurements. Amorphous solids produce diffuse scattering patterns with broad, overlapping peaks. Scattering measurements can be used to obtain radial distribution functions (RDF's), which describe correlations among atomic positions.

RDF's for metal-metalloid alloys are very similar, as shown in Fig. 1. These distribution functions are dominated by contributions from metal-metal atom pairs, because the metal atoms are more numerous and also have larger scattering factors than the metalloid atoms. These RDF's illustrate several differences between atomic arrangements in amorphous and crystalline alloys:

(1) Long range structural periodicity is absent in the amorphous alloys; there are only weak correlations between atomic positions separated by more than four or five atom diameters.

(2) There is no unique nearest neighbor distance in these amorphous alloys. Widths of the nearest neighbor maximum in the RDF's are between 0.4Å and 0.5Å FWHM (full width at half maximum) after correction for termination effects. Radial distribution functions for crystalline Fe, Co, or Ni would have much narrower nearest neighbor maxima (~0.2Å FWHM).

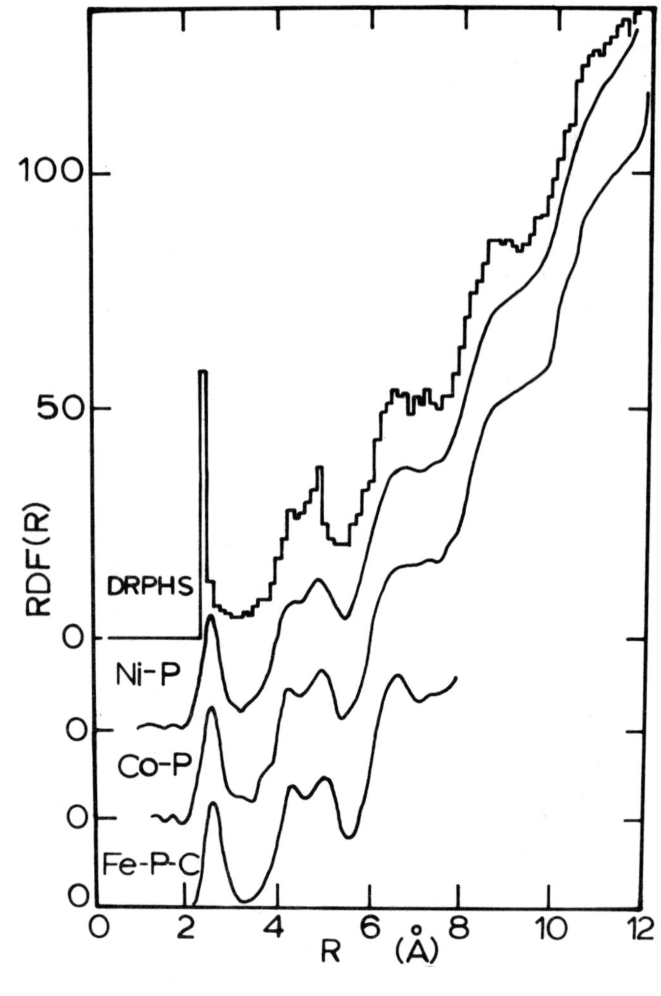

Fig. 1. RDF's for amorphous metal-metalloid alloys: $Ni_{76}P_{24}$, $Co_{78}P_{22}$, and $Fe_{80}P_{13}C_7$, and for dense random packing of hard spheres (DRPHS) structural model.[1]

(3) Average metal-metal near neighbor separations in the Fe, Co, and Ni based amorphous alloys are only 1% - 3% larger than distances of closest approach in the close packed (fcc or hcp) crystalline forms of these elements, but the average Fe-Fe separation in amorphous $Fe_{80}P_{13}C_7$ is 7% larger than that in bcc Fe.

Density measurements indicate that atoms in amorphous metal-metalloid alloys are quite densely packed together. Crystallization of these alloys increases their densities by only 0.5% - 1.0%. RDF's and density measurements provide statistical descriptions of atomic arrangements in amorphous alloys and serve as critical tests for three dimensional structural models. Dense random packing hard sphere (DRPHS) models[2] agree well with measured distribution functions of the amorphous metal-metalloid alloys,[3] as illustrated in Fig. 1. Hand built and computer generated DRPHS models have been used in theoretical studies of vibrational[4] and elastic properties[5] of these materials.

Experiments with amorphous films of Fe-Au alloys (3-55 at.% Au) and of Fe and Co, with controlled impurity additions as small as 1 at.%, have been particularly useful in separating effects of chemical and structural disorder on magnetic properties.[6] These amorphous films, prepared by evaporation in vacuum

(<10^{-6} Torr) onto substrates cooled to 20°K, transform to simple, single phase crystal structures if heated to room temperature. Magnetic properties of these films were studied by in situ measurements. It is unfortunate that no density measurements or RDF's are available for these materials. Their characterization as amorphous was based on diffuseness of x-ray scattering patterns and sharp, irreversible decreases in electrical resistivity on heating.

Amorphous films of Co and of Fe, with no intentional impurity additions, have been prepared by vacuum evaporation at 4°K and 10^{-7} - 10^{-10} Torr.[7] These films crystallize at approximately 50°K. Kwan and Hoffman[8] have reported Mössbauer spectra for Co films, both amorphous and crystalline. Ichikawa[9] and Leung and Wright[7] used in situ electron scattering measurements to obtain RDF's for amorphous Co and Fe films. Their results are very similar to the distribution functions shown in Fig. 1; DRPHS structural models may be appropriate for amorphous transition metal films as well as for amorphous metal-metalloid alloys. Average nearest neighbor separations for the amorphous Co and Fe films were within 1% of the distances of closest approach in corresponding close packed crystalline phases (hcp and fcc Co, fcc not bcc Fe), but the distributions of nearest neighbor separations were similar to those for amorphous metal-metalloid alloys.

Amorphous rare earth-transition metal (RE-TM) alloys which are metastable at room temperature have been made by vacuum evaporation and by d.c. and r.f. sputtering. Magnetization measurements indicate that amorphous alloys in the Tb-Fe, Gd-Fe, and Gd-Co systems are ferrimagnetic, with RE and TM moments antiferromagnetically coupled.[10,12] Although amorphous RE-TM alloys have been obtained over wide composition ranges[10,11], RDF's have been determined only for $Tb_{33}Fe_{67}$, $Gd_{36}Fe_{64}$, and $Gd_{18}Co_{82}$.[13-15,53] These compositions are approximately those of the equilibrium crystalline phases $TbFe_2$, $GdFe_2$, and $GdCo_5$, but short range order in the amorphous alloys differs significantly from that in these crystalline compounds. Average RE-RE near neighbor separations in the amorphous alloys are 10% - 15% larger than those in crystalline (Laves phases) $TbFe_2$, $GdFe_2$, and $GdCo_2$, but they are only 1% - 7% smaller than distances of closest approach in the pure crystalline RE metals. Densities of the amorphous RE-TM alloys (∿67 at.% TM) are approximately 10% lower than those of the corresponding RE TM_2 crystalline phases.[11,13,16] Computer generated binary DRPHS models are consistent with observed nearest neighbor coordinations of amorphous RE-TM alloys, but experimentally observed structure in RDF's for larger r-values have not yet been reproduced in these models.[1,15]

MAGNETIC MOMENTS

The saturation magnetization of a magnetically ordered material, measured at low temperatures and extrapolated to T = 0°K, is one of its basic magnetic properties. Results of such measurements, often expressed as the average magnetic moment per atom or per "magnetic atom", in units of Bohr magnetons (μ_B), are available for several ferromagnetic and ferrimagnetic amorphous solids.

To what extent is the magnetic moment per atom affected by structural disorder? Most experiments provide ambiguous answers, because both chemical and structural disorder are present. Magnetic moments per transition metal atom in amorphous Fe-P-C and Co-P alloys are smaller than those in the pure, crystalline transition metals (bcc Fe, hcp or fcc Co),[3,17-20] but this decrease may be due mainly to chemical, rather than structural disorder. Data on effects of chemical disorder in many crystalline ferromagnetic alloys are limited to dilute solute concentrations because of restricted solid solubilities. Available data[21] on solutes Si and Al in crystalline Fe (bcc) and Co (fcc) are compared with measurements on amorphous Fe-P-C and Co-P alloys in Fig. 2.

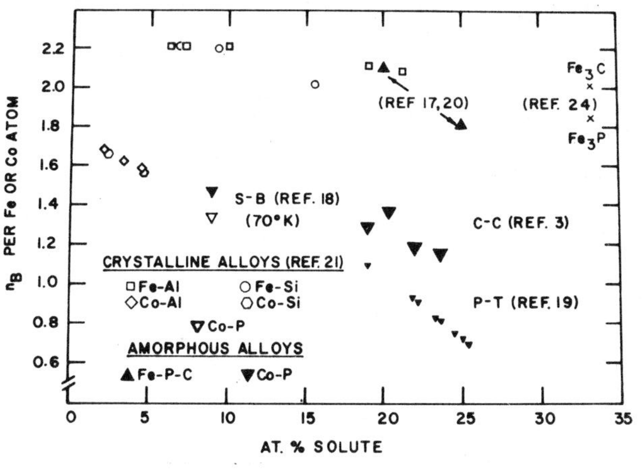

Fig. 2. Magnetic moment per "magnetic atom" for ferromagnetic crystalline and amorphous alloys.

Additions of Al or Si to bcc Fe reduce the moment per Fe atom only slightly; these solutes in fcc Co produce much larger magnetic moment reductions. Magnetic moments in amorphous Fe-P-C and Co-P alloys show trends similar to those observed for moments in dilute crystalline alloys based on these transition metals; reduced moments in the amorphous alloys may be due to chemical effects than to structural disorder.

Simpson and Brambley[18] reported that a chemically deposited Co-P alloy of 9 at.% P was both amorphous and ferromagnetic, but that annealing at 278°C for 30 min. transformed the alloy to a (non-equilibrium) hcp solid solution of P in Co and decreased the magnetic moments per Co atom by 10%.

Magnetization measurements have been reported for other amorphous metal-metalloid alloys, which contain more than one type of metal atom. The results of Mizoguchi, et al.[22], on $(Fe_{100-x}M_x)_{80}B_{10}P_{10}$ alloys, with M = Ni, Co, Mn, Cr, V and x = 0 - 100, are particularly interesting. These amorphous alloys were prepared by rapid quenching and were used to obtain a "Slater-Pauling" curve for magnetic moments per metal atom in these amorphous magnetic systems. For amorphous alloys containing Fe-Ni and Fe-Co, the dependence of magnetic moment per metal atom on "average outer electron concentration" for metal atoms was similar to that reported for crystalline binary alloys of these transition metals. The main differences were that (1) the amorphous alloys did not show the "invar" anomaly and (2) magnetic moments per metal atom for the amorphous alloys were uniformly about 0.5μ_B smaller than those for the corresponding binary crystalline alloys. However, additions of Mn, Cr, or V to the amorphous Fe-based alloys produced more rapid decreases in the magnetic moment per metal atom than for the binary crystalline alloys.

Magnetization measurements on more dilute transition metal-metalloid alloy films, in both amorphous and single phase crystalline forms, were reported by Felsch.[6] He found that the magnetization at 20°K of amorphous Co-Si alloy films (1-6 at.% Si) did not change by more than 3% when the films transformed to the crystalline (hcp) form. The composition dependence of magnetic moment (20°K) agreed with other measurements on bulk Co-Si alloys, as shown in Fig. 3(a). Similar measurements on Fe-Si alloy films (1-8 at.% Si) indicated that incorporation of the Si produced no detectable change in the moment per Fe atom, in either amorphous or crystalline films, but that amorphous films with less than 1 at.% Si had lower magnetizations per Fe atom, as shown in Fig. 3(b). Qualitatively similar results were obtained for Fe-Ge alloy films. Reasons for the "anomalous" behavior of very dilute Fe-based amorphous alloys are not known.

Felsch[6] also reported magnetization measurements on amorphous Fe-Au alloys (3-55 at.% Au). Over most of this composition range (7-55 at.% Au), the amor-

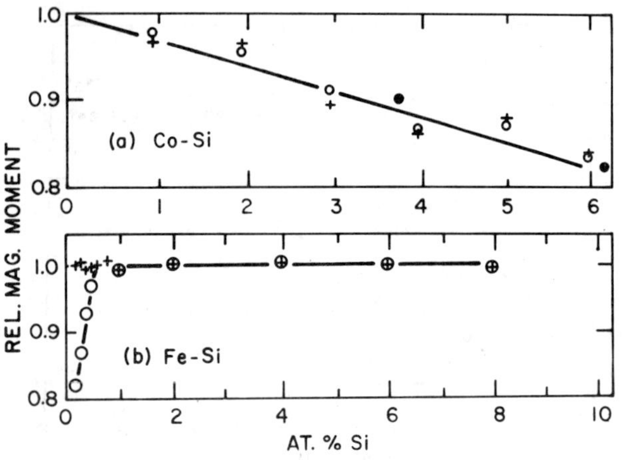

Fig. 3. Relative magnetic moments (20°K) for amorphous (O) and crystalline (+) alloy thin films and bulk crystalline alloys (●): (a) per atom for Co-Si alloys and (b) per Fe atom for Fe-Si alloys (from Felsch[6]).

phous films had magnetic moments of $2.9\mu_B$ (±10%) per Fe atom. Amorphous alloys of 40 to 55 at.% Au crystallized to (non-equilibrium) fcc solid solutions, with no detectable change (± 3%) in magnetization, but amorphous alloys of 7-40 at.% Au crystallized to (non-equilibrium) bcc solid solutions, with magnetic moments between $2.2\mu_B$ and $2.4\mu_B$ per Fe atom. Magnetic moments per Fe atom in amorphous Fe-Au alloys of less than 7 at.% Au decreased with decreasing Au content.

These very different results for the moment per Fe atom in rather dilute Fe-Au amorphous alloys ($2.9\mu_B$/Fe atom for 7 at.% Au) and in Fe-Si and Fe-Ge amorphous alloys ($2.2\mu_B$/Fe atom for 8 at.% Si or 7 at.% Ge) suggest that these amorphous alloys have very different atomic scale structures and/or that Au and Si or Ge, even in these low concentrations, produce striking changes in electronic structures of these amorphous alloys.

Accurate magnetic moment measurements for elemental amorphous ferromagnets are not yet available. However, Kwan and Hoffman [8] have measured Mössbauer spectra from films of Co, in both amorphous and crystalline forms, at temperatures between 4°K and 55°K. Crystallization of the amorphous Co films increased the average hyperfine field by only 1.5%. This suggests that the magnetic moment per Co atom is not strongly affected by structural disorder.

Structural effects on magnetic moments in amorphous ferrimagnetic rare earth-transition metal alloys have been studied for several alloy systems. The observed ($T \rightarrow 0°K$) magnetic moment per formula unit for amorphous $GdCo_2$ ($4.16 \pm 0.3\mu_B$) is significantly reduced from that in the corresponding crystalline compound ($4.95\mu_B$).[11] This change in moment has been interpreted by Tao, et al.,[11] in terms of an increase in the Co magnetic moment in the amorphous alloy ($1.4 \pm 0.15\mu_B$) relative to that in the crystalline Laves phase ($1.02\mu_B$), assuming that the antiparallel alignment of Co and Gd spins of the Laves phase is maintained in the amorphous alloy and that the Gd moment has the same value in both forms of $GdCo_2$.

Decrease of the Co-moment in crystalline $GdCo_2$ below that of elemental Co ($1.7\mu_B$) has been attributed to charge transfer from rare earth ions to the transition metal ions. The smaller decrease of Co-moment in amorphous $GdCo_2$ suggests a smaller charge transfer than in crystalline $GdCo_2$. If the charge transfer in crystalline $GdCo_2$ is associated with compression of Gd-Gd separations, less transfer in the amorphous form would be consistent with the experimental observations of lower density[11] in amorphous $GdCo_2$ and of larger Gd-Gd separations in amorphous $Gd_{80}Co_{20}$.[1,15]

Similar differences in density[16] and in Gd-Gd separations[14] have been observed for crystalline and amorphous $GdFe_2$, but the Fe moment is apparently $1.55\mu_B$ in both forms of $GdFe_2$.[12] This different behavior for Co and Fe has not yet been explained, but differences might have been anticipated from data shown in Fig. 2 for amorphous and crystalline Co and Fe based alloys.

TEMPERATURE DEPENDENCE

Curie Temperatures. The strengths of magnetic interactions in ferromagnets are indicated by their Curie temperatures. Measurements of T_c have been made for several amorphous metal-metalloid alloys based on Fe and Co.[3,17,19,20,23] Results of these measurements are shown in Fig. 4, together with results for crystalline Fe- and Co-based alloys with Si and Al[21] and for the intermetallic crystalline compounds Fe_3C and Fe_3P.[24] Values of T_c for the amorphous Co-P alloys fall on a reasonable extrapolation of data for dilute Co-Al and Co-Si crystalline alloys. Very different results are found for the amorphous Fe-P-C alloys, for which observed Curie temperatures are well below extrapolations based on crystalline Fe-Al and Fe-Si alloys, falling between those of the crystalline compounds Fe_3P and Fe_3C and increasing slightly with increasing metalloid content.

As with the magnetic moment data, these measurements do not provide unambiguous answers concerning effects of structural disorder on Curie temperatures. Chemical effects may be mainly responsible for Curie temperatures of amorphous metal-metalloid ferromagnets being lower than those of the pure crystalline transition metals. Crystallization well below T_c prevents Curie temperature measurements in very dilute amorphous alloys.

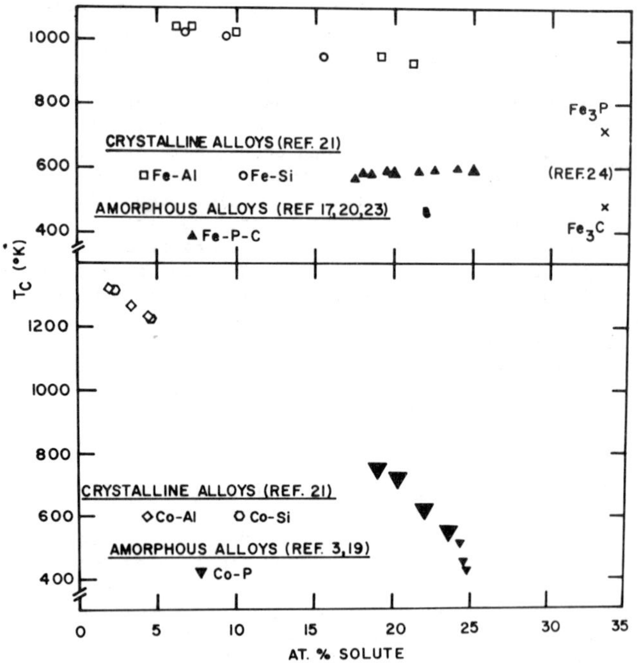

Fig. 4. Curie temperatures for ferromagnetic crystalline and amorphous alloys.

Over-all Temperature Dependence. Experimental results on the temperature dependence of spontaneous magnetization are available for several amorphous ferromagnetic alloys: $Fe_{80}P_{12.5}C_{7.5}$ (splat cooled),[17] $Co_{100-x}P_x$ with x = 19 - 24 (electrodeposited),[25] $Co_{75}P_{25}$ (electrodeposited),[19] and $Fe_{44}Pd_{36}P_{20}$ (splat cooled).[26] These results, from either direct magne-

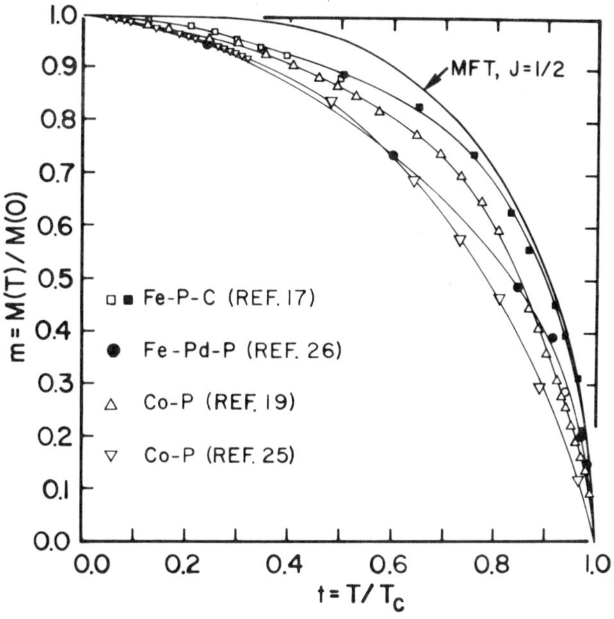

Fig. 5. Over all temperature dependence of magnetization for MFT, $Fe_{80}P_{12.5}C_{7.5}$, $Fe_{44}Pd_{36}P_{20}$, $Co_{75}P_{25}$ and $Co_{78}P_{22}$. Open symbols represent direct magnetization measurements; filled symbols, Mössbauer hyperfine field measurements.

tization measurements[19,25] or from Mössbauer hyperfine field measurements,[17,26] are shown as m = M(T)/M(0) versus t = T/T_c in Fig. 5; also shown is m(t) from mean field theory (MFT) with J = $\frac{1}{2}$. The m(t) curve reported for amorphous Fe-P-C falls slightly below that for crystalline Fe, but curves for the other amorphous alloys fall considerably below those for Fe-P-C. Reasons for the different m(t) results for the various amorphous alloys and the extents to which these curves are affected by structural and by chemical disorder are unknown. Although m(t) curves for crystalline ferromagnetic elements Ni, Fe, and Co are very similar to one another, some crystalline alloys based on these elements have m(t) curves which fall well below those of the pure elements.[21,27,28]

Theoretical investigations of temperature dependence of magnetization in amorphous ferromagnets have involved Heisenberg models and either MFT or approximate Green's function methods.[29] Most theoretical treatments have employed periodic atomic arrangements with assumed exchange fluctuations. The m(t) behaviors predicted by these theories vary in their dependence on the magnitude of exchange fluctuations; however, most of the theoretical results suggest that m(t) curves for amorphous ferromagnets should be flatter, i.e. should fall below, those expected for simple crystalline materials with unique exchange parameters. Assessing the accuracy of such theoretical predictions by comparison with experiments is complicated by not having an independent measure of exchange parameter fluctuation. Furthermore, relevance of the presently available theories to amorphous ferromagnetic metals is questionable. There is little evidence to support using Heisenberg-localized moment models for this class of materials.[30,31] Amorphous magnetic insulators would certainly provide better tests--or subjects--for currently available theories than do the amorphous metallic alloys. Another possible deficiency of these theories is their failure to incorporate realistic models for atomic arrangements in amorphous ferromagnets.

Spin Waves. The behavior of spin waves in amorphous ferromagnetic materials has been inferred from low temperature magnetization measurements[20,25] and from inelastic neutron scattering measurements.[32,33] Neutron scattering measurements on an amorphous electrodeposited Co-P alloy (20 at.% P), to be reported by Mook, et al.,[32] at this conference, indicate a value of 200 mev $Å^2$ for the spin-wave dispersion coefficient D, assuming a dispersion relation of the form E(k) = D k^2. Cargill and Cochrane[3] have deduced values of D from their low temperature magnetization measurements on amorphous electrodeposited Co-P alloys with 19 - 24 at.% P. For these samples, the demagnetization was dominated by a $T^{3/2}$ temperature dependence up to 2/10 of T_C. For a 20.3 at.% P sample they found that $B = 2.6 \times 10^{-5}$ $deg^{-3/2}$, with

$$\Delta M(T) = M(T) - M(0) = - M(0)\, B\, T^{3/2}. \qquad (1)$$

Demagnetization at low temperatures in crystalline ferromagnets is attributed to thermal excitation of long wavelength spin waves; D and B should be related by the following equation

$$B = \xi(3/2)\, \frac{g\mu_B}{M_o} \left(\frac{k_B}{4\pi D}\right)^{3/2} \qquad (2)$$

where g is the g-factor, μ_B is the Bohr magneton in emu, M_o is the magnetization in emu/$Å^3$, and k_B is Boltzmann's constant. $\xi(3/2)$ is the Zeta function given by 0.0587 $(4\pi)^{3/2}$. Although Eq. 2 is most easily derived using the Heisenberg model and a definite, simple crystal structure, Herring and Kittel[34] and Keffer[35] have argued that its validity is much more general, and that it can be derived by a field theoretical treatment of the ferromagnetic material as a continuous medium in which densities of the components of spin are regarded as amplitudes of a quantized vector field. Use of Eq. 2 with the value of B obtained by Cochrane and Cargill[25] (g = 2.1, M_o = 900 emu/cm^3) indicates that D = 115 mev $Å^2$, which is considerably smaller than the value obtained by neutron scattering[32] on a sample of nearly the same composition (200 mev $Å^2$).

For crystalline ferromagnets there appears to be good agreement between values of B and D obtained from low temperature magnetization measurements and from neutron scattering measurements. There is an almost linear correlation between $T_C{}^{3/2}$ and values of 1/B for some fcc and bcc crystalline ferromagnetic metals and alloys, i.e. "corresponding states" behavior in the very low temperature region. Published data[36-40] are shown in Fig. 6, together with values of 1/B for amorphous Co-P alloys from low temperature magnetization measurements[25] and for amorphous $Co_{80}P_{20}$

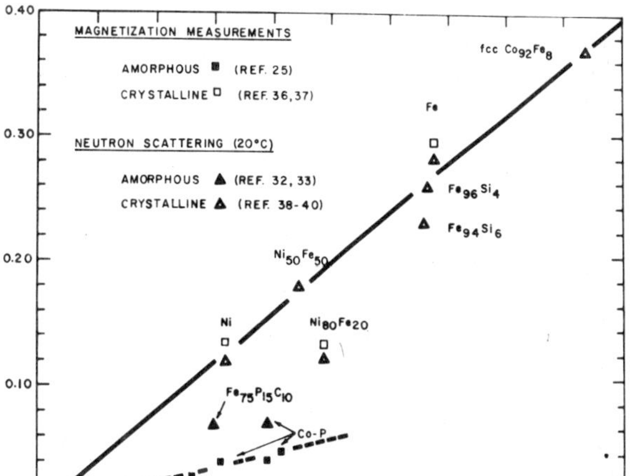

Fig. 6. Experimental correlation between the coefficient B of the $T^{3/2}$ term in low temperature demagnetization and the Curie temperature T_c for ferromagnetic metals and alloys.

and $Fe_{75}P_{15}C_{10}$ alloys from inelastic neutron scattering measurements (D = 200 and 149 mev Å^2).[32,33]

Values of 1/B from magnetization measurements for the amorphous Co-P alloys are also nearly proportional to $T_c^{3/2}$, but these data fall well below those for the crystalline ferromagnets. Preliminary low temperature magnetization measurements[20] on amorphous $Fe_{75}P_{15}C_{10}$ (T_c = 597°K) suggest that the 1/B value for this alloy would fall among those of the Co-P alloys. The two values of 1/B obtained with Eq. 2 from inelastic neutron scattering measurements[32,33] are significantly larger than those from magnetization measurements. Data shown in Fig. 6 suggest that Eq. 2 may not be valid for amorphous ferromagnetic alloys, and that relations between low temperature magnetic excitations and the complete break down of magnetic long range order at T_c differ substantially in the amorphous ferromagnetic alloys and in some crystalline ferromagnetic elements and alloys. Additional measurements on both amorphous and crystalline ferromagnetic alloys are needed to test these suggestions.

Critical Behavior. Do amorphous ferromagnets have a well defined Curie temperature? Mitzoguchi and Yamauchi[41] have addressed this question in their careful measurements of the magnetization of an amorphous, splat cooled $Co_{70}P_{20}B_{10}$ alloy near its Curie temperature. They used "Arrot" plots to evaluate spontaneous magnetization, paramagnetic susceptibility, and their temperature dependences.

Their results indicate that this amorphous ferromagnet has a definite Curie temperature as the critical point of the second order phase transition, in spite of chemical and structural disorder. Magnetization measurements,[19] Mössbauer measurements,[17,26] and magnetic specific heat measurements[23] on other amorphous ferromagnetic alloys also indicate that the magnetic transition is sharp, with a definite Curie temperature. Exceptions are found for some amorphous ferrimagnetic alloys.[53]

MAGNETIC ANISOTROPY

Although atomic arrangements in amorphous solids have commonly been assumed to be macroscopically isotropic, some amorphous ferrimagnetic and ferromagnetic alloys are magnetically anisotropic. Ferromagnetic resonance measurements have confirmed that some sputtered amorphous ferrimagnetic Gd-Co and Gd-Fe alloy films have perpendicular easy axis anisotropy, with anisotropy fields H_K larger than $4\pi M_s$.[42,43] Heiman, et al.,[44] report at this conference that many amorphous evaporated films of Gd, Tb, and Ho, alloyed with Fe and Co, also have strong uniaxial magnetic anisotropy.

Occurrence of magnetic anisotropy has received much less attention for amorphous ferromagnetic materials, although the first experimental suggestion of anisotropy in an amorphous material was probably made by Mader and Nowick[45] in their early paper on ferromagnetism in amorphous Co-Au alloys. More recently, Cargill, et al.,[46] found that amorphous electrodeposited Co-P and Co-Ni-P alloy films have weak perpendicular easy-axis anisotropy, with $H_K/4\pi M_s$ = 0.01 - 0.1. Sherwood, et al.,[47] at this conference report evidence for long range magnetic anisotropy in an amorphous ferromagnetic metal-metalloid alloy, quenched from the liquid by a "roller quenching" technique.

Most amorphous ferromagnetic solids probably have non-zero magnetostrictions. Substrate induced strains in sputtered or evaporated films are obvious sources for magnetic anisotropy. Anisotropic microstructures, involving density or composition fluctuations, with accompanying spatial variations in magnetization, can also produce effective magnetic anisotropies (internal shape effect). However, perpendicular easy-axis anisotropies produced in films by this mechanism can have H_K no larger than $4\pi M_s$. The presence of such anisotropic microstructures, having scales of 10Å - 10,000 Å, might be established by small angle x-ray or electron scattering.[48,49]

Some amorphous materials which are not subjected to external stresses apparently also have large magnetic anisotropies, i.e. $H_K > 4\pi M_s$. The origin of this magnetic anisotropy is thought to involve anisotropies in short range structural or compositional ordering. Gambino, et al.,[50] have proposed that an excess of Co-Co near neighbor pairs in the film plane is responsible for perpendicular anisotropy in amorphous Gd-Co alloy films. They estimated the excess number of in-plane pairs to be of the order of 10^{20}-10^{21}, which corresponds to 0.1% - 1.0% more pairs "in-plane" than "out-of-plane".

Data of the type shown in Fig. 7 could provide direct experimental evidence for structural anisotropies, if they were many times larger than those estimated by Gambino, et al.[50] Near neighbor portions of two RDF's for an amorphous sputtered $Gd_{18}Co_{82}$ alloy are shown.[15] One was obtained from x-ray scattering data taken with the scattering vector perpendicular to the film surface; the other, with the scattering vector parallel to the film surface. Results from these two types of scattering measurements should be identical if local atomic arrangements are macroscopically isotropic. Arrows in Fig. 7 indicate average atomic separations for Co-Co, Co-Gd, and Gd-Gd nearest neighbor pairs; these data indicate that the number of Co-Co pairs aligned perpendicular to the film is very close to the number aligned parallel to any in-plane direction. The difference is certainly less than 5%. However, the data do suggest that there is a deficiency of "in-plane" Gd-Gd near neighbor pairs. This composition is quite near the compensation composition (79.5 at.% Co), at which the perpendicular anisotropy energy dips to a low value.[43] Following Gambino's arguments,[50] this film is a poor candidate for seeking evidence for anisotropic Co-Co nearest neighbor distributions. Similar measurements on high anisotropy films may provide direct evidence for the atomistic origin of magnetic anisotropy, but expected experimental uncertainties are of the order of 3%. At this conference Heiman, et al.,[44] are to report evidence for "preferential substructure" in amorphous Gd-Co and Ho-Co films associated with magnetic anisotropy in these materials.

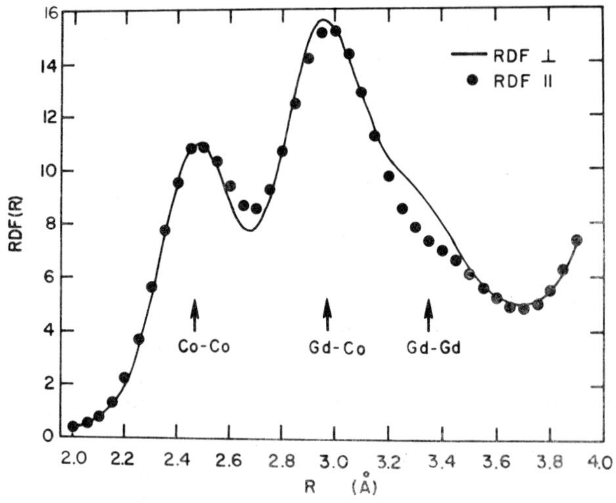

Fig. 7. Comparison of RDF's from scattering measurements with scattering vector K_o perpendicular and parallel to the surface (K_{max} = 16 Å^{-1}) of an amorphous, sputtered $Gd_{18}Co_{82}$ film (22μm thick).[15]

HYSTERESIS

Hysteresis loss, coercivity, and remanence are important for many technical applications of magnetic materials, and these aspects of the magnetic behavior of amorphous metallic alloys are receiving increasing attention. Papers at this conference by Egami, et al.,[51] and by Sherwood et al.,[47] discuss experimental results for amorphous metal-metalloid alloys based on combinations of Fe, Co, and Ni, prepared by rapid quenching.

What are the consequences of structural disorder for these magnetic properties? M-H loops (60 Hz) for amorphous $Fe_{75}P_{15}C_{10}$ (splat cooled) and $Co_{78}P_{22}$ (electrodeposited) are shown in Fig. 8, together with loops for the crystallized alloys. The nearly linear approach to saturation for the amorphous Co-P foil is associated with weak perpendicular magnetic anisotropy.[46] Coercivity and hysteresis losses increase for both materials when crystallization has occurred. As with many other property changes for these alloys,

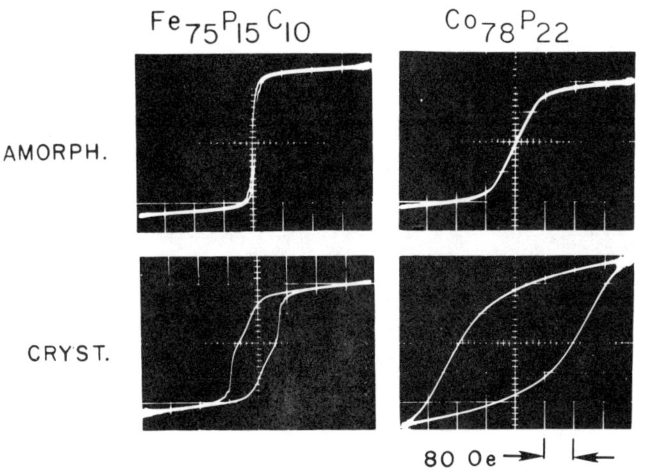

Fig. 8. M-H loops (60 Hz) for a splat cooled $Fe_{75}P_{15}C_{10}$ foil (∿ 40μm thick, supplied by C. C. Tsuei) and for an electrodeposited $Co_{78}P_{22}$ foil (250μm thick), before and after crystallization.

this comparison is complicated because the crystalline material consists of a mixture of phases.

Felsch's studies of amorphous evaporated films of Fe-Si, Fe-Ge, Fe-Au, and Co-Si alloys also included coercivity measurements on the films, both before and after crystallization.[6] Typical coercivity values, at 20°K for films approximately 500 Å thick, were 10 Oe and 40 Oe, for amorphous and single phase crystalline films respectively of Fe-based alloys, and 15 Oe and 140 Oe for Co-Si alloys.

Available data indicate that structural disorder in amorphous ferromagnets reduces hysteresis loss, coercivity, and remanence. The magnetic "softness" of amorphous ferromagnets probably results from the absence of grain boundaries and of other clearly defined structural defects in these materials. High electrical resistivities (∿ 150μΩcm) of amorphous alloys also reduce eddy current losses.

Major exceptions to these observations have been reported by Rhyne, et al.,[12] for amorphous sputtered samples of $TbFe_2$, which has a high coercive force (∿100 Oe) at 20°C compared with other amorphous ferromagnetic and ferrimagnetic (Gd-Fe, Gd-Co) alloys; this coercivity increases sharply at lower temperatures, to 30 kOe at 4°K. Similar behavior has also been reported for amorphous sputtered samples of $DyFe_2$ and $SmFe_2$.[52] The high coercivities of these amorphous Tb, Dy, and Sm based alloys seems to be associated with large local crystal field interactions in these materials.[12]

CONCLUSIONS

This short review on ferromagnetism in amorphous solids is necessarily incomplete. Material was selected to illustrate most clearly differences and similarities between amorphous and crystalline ferromagnets.

Short range atomic order in these materials may be similar, although differences in long range order have been clearly established. Magnetic moments and Curie temperatures of amorphous ferromagnets appear to be more closely linked with alloy composition than with structural disorder. Experimental measurements on amorphous Fe-Au films and on amorphous transition metal films with several at.% Ge or Si permit direct comparison of single phase amorphous and crystalline ferromagnets,[6] and additional studies of these materials would be very useful. Spin-wave experiments on amorphous ferromagnets provide several puzzling results which deserve further attention, both experimental and theoretical.

Further understanding of amorphous ferromagnetic materials requires additional, careful experiments, with proper attention to both chemical and structural characterization, and development of theoretical models which reflect more realistically the nature of magnetic interactions and the chemical and structural features of this class of materials.

ACKNOWLEDGMENTS

R. W. Cochrane's collaboration in studies of magnetism in amorphous Co-P alloys, S. Kirkpatrick's collaboration on atomic arrangements in amorphous RE-TM alloys, and C. C. Tsuei's providing unpublished results on amorphous $Fe_{75}P_{15}C_{10}$ are gratefully acknowledged. In preparing this review, I have also benefited from discussions with D. Pan, J. Logan, H. A. Mook, J. D. Axe, J. J. Rhyne, and G. C. Chi.

REFERENCES

* Research supported by N. S. F.

1. A comprehensive review will be published in Solid State Physics (F. Seitz, D. Turnbull, and H. Ehrenreich, eds.), Academic Press, New York, Vol. 30, 1975.
2. J. L. Finney, Proc. Roy. Soc. Lond. A 319, 479 (1970).
3. G. S. Cargill III and R. W. Cochrane, J. de Physique 35, C4-269 (1974).
4. L. v. Heimendahl and M. F. Thorpe, to be published.
5. D. Weaire, M. F. Ashby, J. Logan, and M. J. Weins, Acta Met. 19, 779 (1971).
6. W. Felsch, Z. Phys. 219, 280 (1969); Z. Angew. Phys. 29, 217 and 30, 275 (1970).
7. P. K. Leung and J. G. Wright, Phil. Mag. 30 , 185 (1974), and to be published.
8. M. Kwan and R. W. Hoffman, to be published, J. Appl. Phys. (Japan).
9. T. Ichikawa, Phys. Stat. Sol. (a) 19, 707 (1973).
10. J. Orehotsky and K. Schroder, J. Appl. Phys. 43, 2413 (1972).
11. L. J. Tao, R. J. Gambino, S. Kirkpatrick, J. J. Cuomo, and H. Lilienthal, AIP Conf. Proc. 18, 641 (1974); Solid State Comm. 13, 1491 (1973).
12. J. J. Rhyne, J. H. Schelleng, and N. C. Koon, to be published, Phys. Rev. B, Dec. (1974).
13. J. J. Rhyne, S. J. Pickart, and H. A. Alperin, AIP Conf. Proc. 18, 563 (1974).
14. G. S. Cargill III, AIP Conf. Proc. 18, 631 (1974).
15. G. S. Cargill III and S. Kirkpatrick, to be published.
16. J. J. Rhyne, unpublished.
17. C. C. Tsuei, G. Longworth, and S. C. H. Lin, Phys. Rev. 170, 603 (1968).
18. A. W. Simpson and D. R. Brambley, Phys. Stat. Sol. (b) 43, 291 (1971).
19. D. Pan and D. Turnbull, J. Appl. Phys. 45, 1406 (1974).
20. C. C. Tsuei, unpublished.
21. D. Parsons, W. Sucksmith, and J. E. Thompson, Phil. Mag. 3, 1174 (1958).
22. T. Mizoguchi, K. Yamauchi, and H. Miyajima, in Amorphous Magnetism (H. O. Hooper and A. M. deGraaf, eds.), Plenum Press, New York, 1973, p. 325.
23. H. S. Chen, Phys. Stat. Sol. (a) 17, 561 (1973).
24. A.I.P. Handbook, 3rd ed., 1972, p. 5-146.

25. R. W. Cochrane and G. S. Cargill III, Phys. Rev. Letters 32, 476 (1974).
26. T. E. Sharon and C. C. Tsuei, Phys. Rev. B5, 1047 (1972).
27. S. A. Ahern, M. J. C. Martin, and W. Sucksmith, Proc. Roy. Soc. Lond. 248, 145 (1958).
28. J. Crangle and G. C. Hallam, Proc. Roy. Soc. Lond. 272, 119 (1963).
29. K. Handrich, Phys. Stat. Sol. 32, K55 (1969). Several other theoretical treatments are given in Amorphous Magnetism (H. O. Hooper and A. M. deGraaf, eds.), Plenum Press, New York, 1973.
30. C. Herring, J. Appl. Phys. 31, 3S (1960).
31. N. F. Mott, Adv. in Phys. 13, 325 (1964).
32. H. A. Mook, D. Pan, J. D. Axe, and L. Passell, these proceedings (4B-7).
33. J. D. Axe, L. Passell, and C. C. Tsuei, these proceedings (4B-6).
34. C. Herring and C. Kittel, Phys. Rev. 81, 869 (1951).
35. F. Keffer, in Handbuch der Physik, Vol. XVIII/2 (S. Flugge, ed.), 1966, p. 1.
36. B. E. Argyle, S. H. Charap, and E. W. Pugh, Phys. Rev. 132, 2051 (1963).
37. B. E. Argyle and S. H. Charap, J. Appl. Phys. 35, 802 (1964).
38. G. Shirane, V. J. Minkiewicz, and R. Nathans, J. Appl. Phys. 39, 383 (1968).
39. F. Menzinger, G. Caglioti, G. Shirane, and R. Nathans, J. Appl. Phys. 39, 455 (1968).
40. S. J. Pickart, H. A. Alperin, V. J. Minkiewicz, R. Nathans, G. Shirane, and O. Steinvoll, Phys. Rev. 156, 623 (1967).
41. T. Mizoguchi and K. Yamauchi, J. de Physique 35, C4-287 (1974).
42. P. Chaudhari, J. J. Cuomo, and R. J. Gambino, IBM J. Res. Dev. 17, 66 (1973).
43. D. C. Cronemeyer, AIP Conf. Proc. 18, 85 (1974).
44. N. Heiman, A. Onton, D. F. Kyser, K. Lee, and C. R. Guarnieri, these proceedings (4D-5).
45. S. Mader and A. S. Nowick, Appl. Phys. Letters 7, 57 (1965).
46. G. S. Cargill III, R. J. Gambino, and J. J. Cuomo, IEEE Trans. Mag. MAG-10, 803 (1974).
47. R. C. Sherwood, E. M. Gyorgy, H. S. Chen, S. D. Ferris, G. Norman, and H. J. Leamy, these proceedings (8E-4).
48. R. H. Wade and J. Silcox, Phys. Stat. Sol. 19, 57 (1967).
49. G. S. Cargill III, Phys. Rev. Letters 28, 1372 (1972).
50. R. J. Gambino, P. Chaudhari, and J. J. Cuomo, AIP Conf. Proc. 18, 578 (1974).
51. T. Egami, P. J. Flanders, and C. D. Graham Jr., these proceedings (4A-4).
52. A. E. Clark, Appl. Phys. Letters 23, 642 (1973).
53. See also S. J. Pickart, J. J. Rhyne, and H. A. Alperin, Phys. Rev. Letters 33, 424 (1974), for comments on RDF for amorphous YFe_2 and on anomalous critical scattering in amorphous ferrimagnetic $TbFe_2$.

SPIN WAVES AND MAGNETIC INTERACTIONS IN THE FERROMAGNETIC HEUSLER ALLOYS

Y. Ishikawa*† and Y. Noda
Physics Department, Tohoku University, Sendai, Japan 980 and
*Institut Laue-Langevin, Grenoble Cedex 38042

ABSTRACT

The magnetic interactions in the ferromagnetic Heusler alloys X_2MnY (Pd_2MnSn, Ni_2MnSn and Cu_2MnAl) have been investigated using neutron scattering. The spin wave dispersion relations along the [100], [011] and [111] directions have been determined across the whole magnetic Brillouin zone. The results have been analyzed in terms of spin wave theory with the assumption that only Mn atoms have a localized magnetic moment. It has been found that interactions between more than sixth neighbors (R > 11 Å) should be taken into account to reproduce the observations with a theoretical relation $\hbar\omega_{\vec{q}} = 2S(J(0) - J(\vec{q}))$. The exchange integrals J_R at large distances have an oscillatory character which can be interpreted by the s-d type interaction based on the nearly free electron model. However the n.n and n.n.n interactions have opposite signs to those expected from the model. These interactions depend strongly on the kind of X atom and Cu atoms have the largest effect. The spin wave renormalization found in Pd_2MnSn could also be interpreted by the spin wave theory with dynamical interactions. Neutron paramagnetic scattering for Pd_2MnSn gives an energy spectrum of Gaussian shape with the second momentum in good agreement with the value calculated using the exchange parameters determined from the spin wave study. All these facts suggest that the Heusler alloys can be classified as typical Heisenberg systems with long range interactions.

1. INTRODUCTION

This paper summarizes recent neutron scattering results of several different studies on the Heusler alloys. Heusler alloys with a chemical formula X_2MnY are ordered 3d compounds with the $L2_1$ structure. The atoms X (Cu, Ni, Co and Pd) occupy the corner sites of the b.c.c. structure (A, C sites), while the Mn and Y (Al, In, Ge, Sn, Sb, Te) atoms occupy alternate body center sites (B and D sites, respectively). The crystal structure is shown in Fig. 1. As modifications of this structure, Mn and X atoms occupy randomly the B

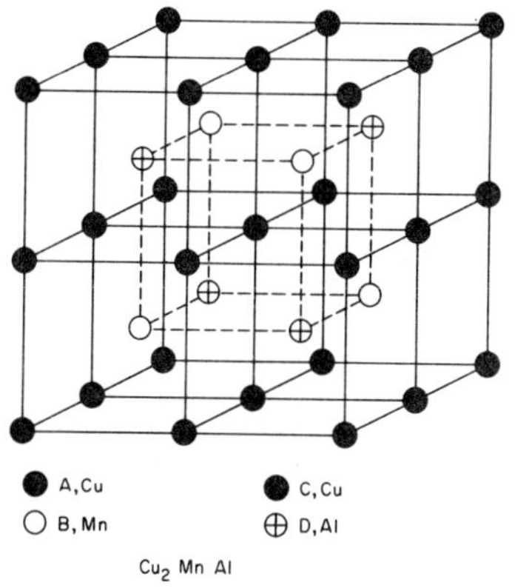

Fig. 1. Crystal structure of Heusler alloy with $L2_1$ structure.

and D sites in the B2 structure (Pd_2MnAl), while in the $C1_b$ structure, which is realized in XMnY as PdMnSb, the C sites in the $L2_1$ structure are vacant. The magnetic properties of the Heusler type alloys which are pertinent to the present study are summarized in Table 1. Except for the particular case where X is Co,[6,8] the main carriers of the magnetic moment are Mn atoms[9]

TABLE 1. MAGNETIC PROPERTIES OF HEUSLER TYPE ALLOYS

Type	Compounds	a (A)	$T_C(T_N)$ (K)	r_{MM} (A)	μ/Mn (μ_B)	n_c	Ref.
$L2_1$	Cu_2MnAl	5.949	630	4.21	3.8	5.8	(1)
$L2_1$	Cu_2MnIn	6.206	520	4.39	4.0	6.0	(1)
$L2_1$	Cu_2MnSn	6.173	∿530	4.36	4.1	7.1	(1)
$C1_b$	Cu MnSb	6.095	(55)*	4.36	4.56	7.5	(2)
$L2_1$	Ni_2MnGa	5.825	379	4.12	4.17	4.2	(1)
$L2_1$	Ni_2MnIn	6.068	323	4.29	4.40	4.4	(1)
$L2_1$	Ni_2MnSn	6.052	344	4.28	4.05	5.1	(1)
$L2_1$	Ni_2MnSb	6.001	334	4.24	4.31	6.3	(3)
$C1_b$	Ni MnSb	5.922	750	4.18	4.0	6.0	(2)
B2	Pd_2MnAl	6.165	(240)*	<u>3.08</u>	4.4	4.4	(4)
$L2_1$	Pd_2MnIn	6.373	(142)*	4.50	4.3	4.3	(5)
$L2_1$	Pd_2MnSn	6.383	189	4.51	4.23	5.2	(5)
$L2_1$	Pd_2MnSb	6.424	247	4.54	4.40	5.4	(5)
$L2_1$(P)	Pd_2MnGe	6.182	260	<u>2.68</u>	4.0	5.0	(6)
$C1_b$	Pd MnSb	6.246	500	4.42	3.95	4.95	(2)
$C1_b$	Pd MnTe	6.271	(23)*	4.43	4.8	4.8	(7)

( )*: Néel temperature, $L2_1$(P): partially disordered

r_{MM} = Mn-Mn n-n distance,

μ/Mn = Magnetic moment per Mn atom,

The number of conduction electrons per formula was estimated that 3d or 4d bands of the X atoms is completely filled.

which are widely separated by non-magnetic X and Y atoms in the $L2_1$ and $C1_b$ structure. Therefore the Mn atoms have been considered to have a definite localized moment and they are coupled via the conduction electrons; either by the s-d interaction of R-K-K-Y type[10] or by the double resonance type interaction proposed by Caroli and Blandin (C-B).[11] Although many of the magnetic properties, such as the Curie temperature[2,11,12] or the hyperfine fields on the X and Y atoms,[11-16] have been analyzed based on an assumption that these s-d interactions play a main role in the ferro or antiferro magnetism of the Heusler alloys, no definitive experimental evidence has ever been provided to verify this assumption. Actually it has been recognized that the simple s-d type interaction gives a negative Curie temperature, if it is calculated using appropriate Fermi wave vector k_F and phase shift ϕ of the conduction electrons.[2,12] Furthermore, recent studies suggest that the hyperfine field on the Y atom depends more strongly on the local character of this atom due to the charge screening than on the polarization by the environment. Therefore the information on the magnetic interaction cannot easily be deduced from the measurement of the hyperfine fields.

The neutron spin wave scattering is the most direct method of determining the magnetic interactions. We have studied by this technique the spin wave dispersions in Pd_2MnSn (Ishikawa, Noda[21]), Ni_2MnSn (Noda, Ishikawa) and Cu_2MnAl (Ishikawa, Tajima, Stringfellow, Webster and Tocchetti[22]) and the results are presented in Part I of this paper together with discussions on the mechanism of the magnetic interactions in the Heusler alloys.

Another important aim of the present investigation is to examine whether the dynamical behavior of the localized moment in the Heusler alloys is the same as that found in the insulators (the Heisenberg system). Unlike the localized moment in the rare earth metals, where the 4f electrons are well screened by outer shell electrons, the magnetic moments in these 3d compounds possibly have a strong coupling with the conduction electrons by s-d mixing. Therefore some differences in the dynamical properties would be expected between the 3d compounds and the Heisenberg system for which the theory of spin dynamics has been well developed. The spin wave theory with dynamical interactions has been found to explain the renormalization of the spin waves in the Heisenberg systems (MnF_2,[23] MnO,[24] EuO[25]) even at a temperature T = 0.9 T_c. Such an agreement has not been obtained for the ferromagnetic 3d metals.[26,27] The spin dynamics in the paramagnetic region can also give a good interpretation to the neutron paramagnetic scattering from the Heisenberg system.[28] Therefore we have studied the temperature dependence of the spin wave dispersions in Pd_2MnSn (Noda, Ishikawa) as well as the neutron paramagnetic scattering from Pd_2MnSn (Watanabe, Ishikawa, Tsuzuki[29]) and the results were analyzed in terms of the Heisenberg Hamiltonian, which are described in Part II. This kind of analysis would be a starting point of understanding the dynamical magnetic properties such as the spin waves of the 3d metals where the concept of the localized moment and the utility of the Heisenberg type interactions are quite ambiguous.[26,27]

PART I - SPIN WAVE STUDIES

2. EXPERIMENTAL TECHNIQUES AND RESULTS

Neutron spin wave scattering has been measured with the conventional triple axis spectrometers either of Tohoku University [TUNS] (Pd_2MnSn, Ni_2MnSn) or of ILL [IN1] and of Harwell (Cu_2MnAl). Single crystals of Pd_2MnSn and Ni_2MnSn used in the experiments were about 3 cc in volume. They were prepared by the Bridgman method and were annealed to have the maximum ordering in silica tubes filled with argon and furnace cooled.[5,15] Pd_2MnSn is reported to keep the ordered structure even if it is quenched from high temperatures.[4,6] A single crystal of Cu_2MnAl about 8 cc in volume was provided by Dr. J. Schweizer (I.L.L.). The crystal was also heat treated to have the maximum ordering.[30]

The neutron scattering cross sections at the scattering vector $\vec{Q}$ for creation of one magnon, with the energy $\varepsilon(\vec{q})$ and the momentum $\vec{q}$ have the form[31]

$$\frac{d^2\sigma}{d\Omega dE} \propto \sum_{\vec{q}} |f(\vec{Q})|^2 (\langle n_{\vec{q}}\rangle+1) T_{\vec{q}}(\vec{Q})\,\delta(\vec{Q}-\vec{\tau}-\vec{q})\,\delta(E-\varepsilon(\vec{q})), \quad (1)$$

where $\langle n_{\vec{q}}\rangle$ is the magnon population factor given by

$$\langle n_{\vec{q}}\rangle = \{\exp(\hbar\omega_{\vec{q}}(T)/kT)-1\}^{-1} \quad (2)$$

and $T_{\vec{q}}(\vec{Q})$ is the structure factor for inelastic scattering (dynamical structure factor).

The magnon dispersion relations were determined for three principal symmetry directions [001], [110] and [111]. The main experimental data were collected in two magnetic Brillouin zones centered at the (111) and (002) reciprocal lattice points by using both the constant Q and constant E modes of operation. It is remarked that the acoustic phonon has a small dynamical structure factor near the center of the magnetic Brillouin zones. Therefore this alloy has a great advantage that the magnon dispersion relation can be determined without being influenced by the acoustic phonons, in sharp contrast to the usual ferromagnet.

The magnon dispersion relations in Pd_2MnSn determined at 50 K (T/T_c = 0.26) are shown in Fig. 2. The center of each mark corresponds to the peak position of the observed magnon group, while the size of the mark represents approximately the observed line width. The value at the center (Γ) was estimated from the anisotropy constant K_1 (∿1·10^4erg/cc) determined by the static torque measurement. The dispersion relations are plotted against the reduced wave vectors ζ, but the scales in the three directions are changed so that the absolute magnitude of the magnon wave vector q has the same scale factor in these three directions. The dispersion relations are characterized by a small hump in the [001] direction and a rather large dip around the K point in the [110] direction just as is found in the magnon dispersions in the c-direction in the rare earth metals.[32] The dispersion in the [111] direction increases monotonically to the zone boundary (L).

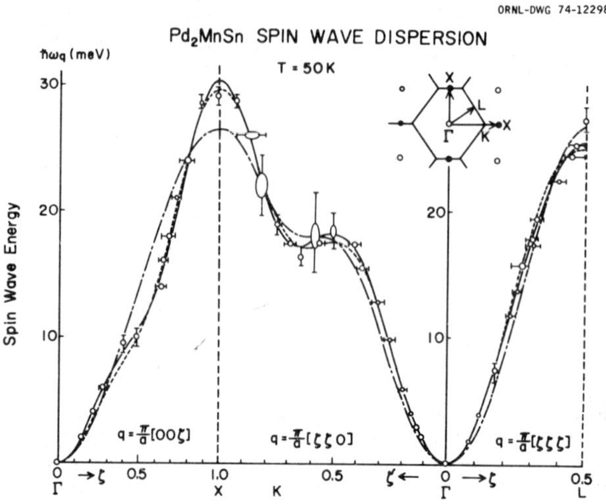

Fig. 2. Spin wave dispersions along [001], [110] and [111] directions in Pd_2MnSn measured at 50°K. Solid (——), broken (---) and chain (-·-·-) lines are calculated using respectively eight, six, and four exchange parameters determined by least squares fit to the data.

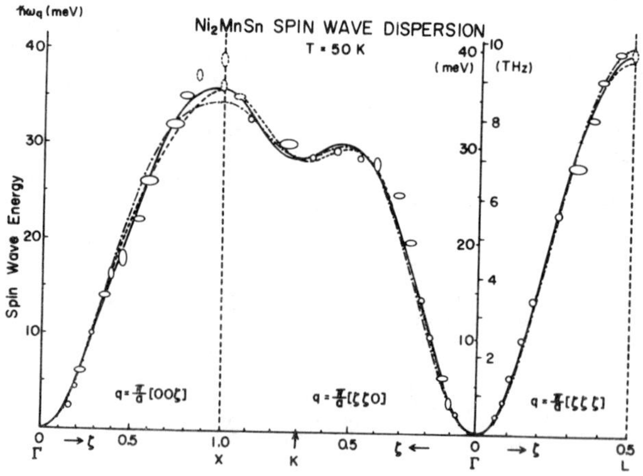

Fig. 3. Spin wave dispersions along [001], [110] and [111] directions in Ni_2MnSn measured at 50°K. Solid, broken and chain lines have the same meaning as those in Fig. 2.

The magnon dispersion relations in Ni_2MnSn observed at 50 K (T/T_c = 0.15) is shown in Fig. 3. The dispersion relations are quite similar to those in Pd_2MnSn, but the anomaly found in the [001] direction is less pronounced. It is remarked that the magnon energies at the zone boundaries do not differ as much as might be expected from the difference in the Curie temperature between these two substances ($T_c(Ni_2MnSn)/T_c(Pd_2MnSn)$ = 1.8).

The magnon dispersion relations in Cu_2MnAl were determined at 4.2°K.[22] They have characteristics very similar to those in Ni_2MnSn. The magnon energy increases monotonically to the zone boundaries in both [001] and [111] directions, but some structure similar to that found in Pd_2MnSn exists around the K point in the [110] direction. The zone boundary magnon energies are 94 ± 2 meV and 86 ± 2 meV at the X and L points, respectively. The structure in the [110] direction which may exist in the momentum region ($0.95 > \zeta > 0.55$) has not yet been determined. Well defined magnon groups could not be detected in this range. It is not certain whether this is due to some physical reason or to lack of energy and momentum resolution in the measurements. A detailed study is now in progress. The magnon energies at the zone boundaries are reasonable compared with those of Pd_2MnSn ($T_c(Cu_2MnAl)/T_c(Pd_2MnSn) = 3.3$).

The exchange stiffness constants $D(= \hbar\omega_{\vec{q}}/q^2)$ were determined from the average value of the initial slope of the dispersion relations in the three directions They are 90 ± 10 meV A^2 for Pd_2MnSn, 140 ± 10 meV A^2 for Ni_2MnSn and 170 ± 10 meV A for Cu_2MnAl. The error in the value is mainly due to the difference in slope in the different directions.

3. ANALYSIS OF DATA

The dispersion relations at the lowest temperature determined in the whole range of the magnetic Brillouin zone are analyzed in terms of the Heisenberg Hamiltonian $H = -2\sum J_{ij} S_i S_j$. The anisotropy term which is found to be quite small was neglected in the first approximation. If we assume that only Mn atoms with the f.c.c. structure have localized moments of $gS\mu_B$ with $g = 2$,[33] the spin wave dispersion relation is given in a simple form

$$\hbar\omega_{\vec{q}} = 2S[J(0) - J(\vec{q})], \quad (3)$$

where $J(\vec{q})$ is the Fourier transform of the exchange integrals $J_{\vec{R}}$ $(= \sum_{\vec{R}} J_R e^{i\vec{q}\vec{R}})$. The solid lines in Figs. 2 and 3 are calculated using eight parameters determined by a least squares fit to the data. The agreement between the observations and the calculations is satisfactory for both samples and the exchange parameters determined are tabulated in Table 2, together with the results of analysis for Cu_2MnAl. The exchange parameters in Cu_2MnAl are not definitive because of the ambiguity in the dispersion relations, but the nearest neighbor interaction J_1 would not be modified by more than 10%.

In the case of Pd_2MnSn, eight parameters were found to be definitely necessary to reproduce the observations by using Eq. (3).[21] If eleven parameters are used in the analysis, we could get better agreement between the theory and experiment, but the magnetic interactions between further than eighth neighbors are quite small and the exchange parameters determined for the first eight neighbors are qualitatively the same as those determined by the eight parameter fit.[21] However if we used only six parameters, we could not reproduce the small hump of the dispersion curve in the [001] direction as shown in broken lines in Fig. 2. The difference in the eight parameter fit and the six parameter fit is rather small in the case of Ni_2MnSn. The exchange parameters J_R thus determined for Pd_2MnSn and Ni_2MnSn are plotted in Fig. 4 as a function of the distance R. The range of each exchange parameter shown in the figure indicates the range of variation of values of the exchange parameter due to changing the number of parameters in the analysis. The exchange parameter decreases monotonically to the fourth neighbors, then it starts to oscillate with distance. The fact that the four parameter fit fails to reproduce the observations (the chain lines in Figs. 2 and 3) suggests that the oscillatory nature of the exchange parameters is absolutely important to explain the spin wave dispersions in Pd_2MnSn and Ni_2MnSn. It is noted that the oscillatory part is quite similar in both substances except for the last parameter where the termination error is important. In the case of Cu_2MnAl the nearest neighbor interaction J_1 is particularly large, while the other interactions are similar to those in the other Heusler alloys, as shown in Table 2. In this table, the Fourier transform $J(0)$ as well as the paramagnetic Curie temperature $k\theta_p$ calculated using the molecular field approximation ($k\theta_p = 2/3\, S(S+1)J(0)$) are also tabulated. The value for Pd_2MnSn (205.0 K) is in excellent agreement with the observed paramagnetic Curie temperature of 201 ± 3 K. Although no data are available for the paramagnetic Curie temperature of Ni_2MnSn, the calculated value (337.2°K) is also reasonable compared with the Curie temperature of 340 K. However the value for Cu_2MnAl is slightly higher compared with the reported Curie temperature of 630°K. The exact Curie temperature of Cu_2MnAl would be higher than the reported value, because Cu_2MnAl starts to transform into another nonmagnetic phase above 600°K.[30,34]

The exchange stiffness constants D_o are also calculated using a formula $D_o = 1/3\, S R_1^2 J^{(2)}$, where $J^{(n)} = \sum_i Z_i J_i (R_i/R_1)^n$ and R_1, the nearest neighbor distance. They are also tabulated in Table 2. The agreement between the observed value and the calculation is excellent for Pd_2MnSn indicating that the parameters determined are accurate. The agreement is not satisfactory for the other Heuslers. This would be because of the inaccuracy of J_R at large distances with which D varies sensitively.

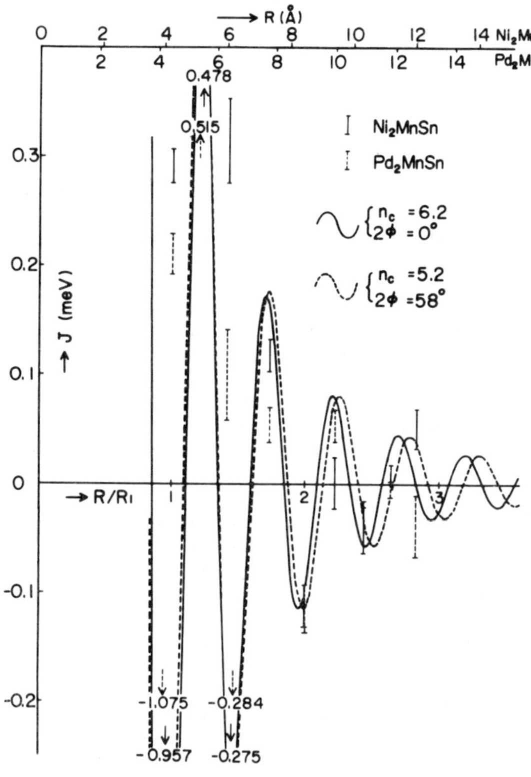

Fig. 4. Exchange parameters determined for Pd_2MnSn (---) and Ni_2MnSn (——) by least squares fit to data plotted against distance R between pairs normalized by n.n. distance R_1. Solid and broken lines are calculated using s-d type interactions with parameters nc and 2ϕ given in the figure.

4. MAGNETIC INTERACTIONS IN THE HEUSLER ALLOYS

We have found that the magnetic interactions in the Heusler alloys are long range (> 12 Å) and they have an oscillatory nature at large distances which is a characteristic of the s-d type interaction. According to the simple R-K-K-Y theory based on the free electron model, the exchange integral at large distances is given by[10]

$$J_R \sim -\frac{9\pi}{8}\frac{n^2 I^2}{E_F}\frac{\cos(2k_F R)}{(k_F R)^3}, \quad (4)$$

while that by the C-B theory of the double resonance scattering is expressed as[11]

TABLE 2. EXCHANGE PARAMETERS J_i(meV) FOR Pd_2MnSn, Ni_2MnSn AND Cu_2MnAl OBTAINED BY LEAST SQUARES FIT TO DATA USING EIGHT PARAMETERS. The Fourier transform J(0), the paramagnetic Curie temperature θ_p(K) as well as the exchange stiffness constant D_o(meV A^2) calculated using these exchange parameters are also tabulated.

Substance	S	J_1	J_2	J_3	J_4	J_5	J_6	J_7	J_8	J(0)	θ_p	D_o
Pd_2MnSn	2.1	0.195	0.130	0.066	-0.133	0.042	-0.057	0.011	-0.017	4.069	205.0	106.1
Ni_2MnSn	2.0	0.270	0.356	0.128	-0.126	-0.020	-0.059	0.015	0.068	7.103	337.2	168.0
Cu_2MnAl	1.9	0.97	0.16	0.30	-0.19	0.036	0.084	-0.050	-0.10	16.1	688	144

$$J_R \sim -\frac{25}{4\pi}\frac{E_F}{S^2}\sin^2\phi^- \frac{\cos(2k_F R + 2\phi^-)}{(k_F R)^3}, \qquad (5)$$

where n is the number of conduction electrons per atom, I is the s-d exchange constant, E_F and k_F are the Fermi energy and wave vector, respectively. ϕ^- is the phase shift of the scattered wave function of down spin electrons from the potential of Mn atoms and is related to the number of down spin electrons Z^- or the magnetic moment $m\mu_B$ of the Mn atoms by

$$\phi^- = (\pi/5)Z^- = (\pi/5)(5 - m).^{11}$$

We have then used a formula $J_R = C\cos(2k_F R + 2\phi)/(k_F R)^3$ to analyze the oscillatory part of the exchange parameters given in Fig. 4. Two parameters k_F and ϕ were determined by a least squares fit to the exchange parameters further than the next nearest neighbors. The number of conduction electrons per chemical formula nc was estimated from the Fermi wave vector k_F based on the free electron model. The result of analysis indicates that the oscillatory character can be interpreted with any number of conduction electrons between 5.3 and 8.3[21] if the parameter ϕ is determined to minimize the root mean square (RMS) of the difference between the calculated values and those plotted in Fig. 4. The RMS varies only slightly for these values of nc. In the figure, the calculated values are plotted for two sets of parameters (nc,ϕ). One set (6.2, 0) has values determined to fit the data for Pd_2MnSn, while another set (5.2, 24°) is that obtained from the fit to the data of Ni_2MnSn. The effective numbers of conduction electrons nc (cf. Table 1) as well as the phase shifts ϕ^- estimated for Pd_2MnSn and Ni_2MnSn are (nc = 5.2, ϕ^- = 36°) and (nc = 5.1, ϕ^- = 28°), respectively. The effective value of nc determined from the observed oscillatory behavior should be larger than this value, because the free electron Fermi surface drawn in the Heusler alloys almost touches the nuclear Brillouin zone boundary in the [011] direction and thus the actual effective Fermi wave vector will be somewhat increased from the value of the free electron model. Therefore the value of nc between 5.2 and 6.2 would be appropriate for these alloys. We also evaluated the values of E_F and I in Eqs. (4) and (5) using the normalization constant C determined to fit the experimental data. The values obtained are I(eV) $\sim 0.24\sqrt{E_F(eV)}$ for the R-K-K-Y interaction and $E_F \sim 1.5$ eV for the C-B interaction. These values are considered to be reasonable at least as an order of magnitude, if they are compared with the values of I = 0.2 eV and E_F = 7 eV expected for the dilute MnCu alloys.[11] Therefore we can conclude that the magnetic interactions in the Heusler alloys at large distances (R > 8 Å) can well be interpreted in terms of the s-d type interaction based on the nearly free electron model. The experimental fact that the values of the exchange parameters between the pairs further than the third neighbors in Pd_2MnSn and Ni_2MnSn agree with each other gives a good support for it, because both substances have almost the same number of conduction electrons. The energy gap which is expected to exist at the nuclear Brillouin zone boundary to stabilize the b.c.c. structure[35] does not seem to play an important role in these interactions.

Another more important conclusion for the present study is that the exchange parameters between the first two near neighbors which are the largest of all interactions have opposite signs to those expected from the simple s-d type interaction given in Eqs. (4) and (5) as shown in Fig. 4. Therefore these interactions cannot be explained by this type of interaction. The situation is also the same for Cu_2MnAl. The exchange parameters due to the simple s-d type interaction are also calculated for Cu_2MnAl using appropriate values of nc and ϕ^- and the same normalization constant C as used in Pd_2MnSn and the values for the first four neighbors are presented in Table 3. The result indicates that the simple s-d type interaction gives also a negative sign for the first two near neighbors, contrary to the observation. On the other hand, in the case of Pd_2MnIn which is antiferromagnetic of the second kind, the near neighbor interactions were evaluated in a simple molecular field approximation to be $J_1 + J_3 = 0.34$ meV and $J_2 = -0.508$ meV respectively.[36] If these values are compared with those calculated for the simple s-d type interaction which are also listed in Table 3, we find that the exchange parameters between the first two nearest neighbors have always the sign opposite to that expected by the simple s-d type interaction.

TABLE 3. EXCHANGE PARAMETERS DUE TO THF SIMPLE s-d TYPE INTERACTION GIVEN BY $C\cos(2k_F R + 2\phi)/(k_F R)^3$

	nc	ϕ	J_1(meV)	J_2	J_3	J_4
Cu_2MnAl	5.8	43	-0.224	-0.489	+0.120	-0.044
Pd_2MnIn	4.3	25	-0.704	+0.1118	0.003	0.009

The same normalization constant C as used for Pd_2MnSn is used.)

These short range interactions can also be estimated for other ferromagnetic Heusler alloys from their Curie temperatures, because our analysis of the experiments has shown that the first two nearest neighbor interactions contribute more than 75% to the Fourier transform J(0). Therefore by making reference to the Curie temperatures of various Heuslers listed in Table I, we can summarize the general features of these short range interactions as follows: (1) J_1 and J_2 always have opposite signs to those expected by the simple s-d type interaction. (2) The magnitude of J_1 depends strongly on the kind or electronic state of X atoms, but is not sensitive to the number of conduction electrons. (3) The magnitude of J_1 decreases in the sequence Cu-Ni-Pd, and J_1 in Cu_2MnAl is more than four times greater than J_1 in Pd_2MnSn. (4) J_2 varies only a little with the X atoms.

Two mechanisms can be considered to explain these ferromagnetic interactions in the Heusler alloys. One mechanism stands always on the s-d type interaction but it considers that the preasymptotic form at small distances has opposite signs to those of the asymptotic form given in Eqs. (4) and (5). Another mechanism attributes the ferromagnetic interactions to a different type of interaction.[36] As one example of the former approach, Alloul discussed[37] the spin polarization due to magnetic impurity in dilute alloys based on the Anderson Hamiltonian and showed that the preasymptotic correction becomes quite important if the distance R from the impurity becomes smaller than a critical distance $k_F R_C \sim E_F/\Delta \sin\phi$, where Δ is the width of the virtual bound state. The spin polarization changes signs to the asymptotic limit for $R < R_C$. $k_F R_C$ is approximately 2π in his case. The main origin of the deviation is due to the d character of the impurity scattering. A similar result has also been reported by Geldart.[38] Since the next nearest neighbor distance $k_F R_2$ in the Heusler alloys is just 2π, the effect of the preasymptotic correction of this kind would also be important, but it is not clear that the sign of J_2 changes by this mechanism. Kasuya took account of the k dependence of the s-d exchange constant I in Eq. (4) by considering the charge density fluctuation of the conduction electrons in the actual crystal.[39] In the case of the Heusler alloy, it is expected that the charge density is located more on the Y site near the bottom of the conduction band. Near to and above the Fermi level, the density on the Mn site increases. Therefore the s-d exchange interaction on the Mn site should become important above some critical energy of the conduction electron. This effect was evaluated[39] at first by assuming the analytic form for the k dependent I

$$I(k) = I \frac{k^{2n}}{k^{2n} + k_0^{2n}}, \tag{6}$$

where k_0 indicates the position of the sharpest change and n the width of the change. By introducing this form of I in the R-K-K-Y type interaction, he could get the correct sign for the first two nearest neighbor interactions with parameters $n = 6$ and $k_0 = 1.5\ k_F$. The interactions for further neighbors are not so much affected by introducing a k dependent I. Therefore this mechanism can explain qualitatively the magnetic interactions in the Heusler alloys. It is, however, quite uncertain that this mechanism is really most important in the Heusler alloys, because the parameters used in the calculation are quite unrealistic.[39] It is noted that the k dependence of the exchange constant I has been calculated for paramagnetic Gd metal by Harmon and Freeman[40] using the nonrelativistic APW wave functions. The strong angular dependence of I was found due to band crossings, which may explain the anomaly of $J(\vec{q})$ found in the c-direction of Gd.[41] All these studies suggest that the s-d type interaction would account for the magnetic interactions in the Heusler alloy, if the realistic situations in the crystal are taken into account to calculate J_R.

As another approach to explain the ferromagnetic interaction, Kasuya has proposed[36] a new mechanism of virtual double exchange interaction in which the d-electron with the down spin vitually excited in the unoccupied d-band of the Mn atoms couple to the spin on the adjacent atoms by a double exchange interaction (the d-d transfer mechanism). He showed that this interaction becomes important when the number of 4d holes in the X atom decreases, because the charge transfer from the X atom to the $3d^-$ band of the Mn atoms prevents a lowering of the $3d^-$ band by this mechanism. Although the prediction is in good agreement with the observation (2) and (3), it does not explain the fact (1); the antiferromagnetic interaction in Pd_2MnIn. Therefore the preasymptotic modification of the simple s-d type interaction would always be responsible for the next nearest interaction.

It is remarked that the antiferromagnetism of Pd_2MnAl with the B2 structure is due to the antiferromagnetic coupling between the Mn atom on the B and D sites.[36] The Mn-Mn distance is small (∿3 Å) and the direct exchange interaction of Alexander-Anderson-Moriya[42] type which predicts the negative coupling should be predominant.[36]

PART II - HIGH TEMPERATURE SPIN DYNAMICS IN Pd_2MnSn

5. TEMPERATURE DEPENDENCE OF SPIN WAVES

In the theory which takes into account the dynamical interactions between pairs of spin waves in the Heisenberg system, the spin wave dispersion at a finite temperature T is given by[43]

$$\hbar\omega_{\vec{q}}(T) = 2S[J(0)-J(\vec{q})] - \frac{1}{N}\sum_{\vec{k}} \langle n_{\vec{k}} \rangle [J(0)-J(\vec{q})+J(\vec{k}-\vec{q})-J(\vec{k})] \tag{7}$$

In the long wave limit, $\hbar\omega_{\vec{q}}$ is reduced to $\hbar\omega_{\vec{q}} = D(T)q^2$ in the cubic crystal and D(T) is given by

$$D(T) = D(0)\left(1 - C(T/T_c)^{5/2} + \alpha(T/T_c)^{7/2}\ldots\right) \tag{8}$$

with

$$C = \frac{\pi v_o}{S} \frac{R_1^2 J^{(4)}}{J^{(2)}} \left(\frac{k_B T_c}{4\pi D_o}\right)^{5/2} \zeta(\tfrac{5}{2}), \tag{9}$$

where $J^{(n)}$ is defined by $J^{(n)} = \sum_R J_R (R/R_1)^n$ and R_1 is the nearest neighbor distance. We have made a selfconsistent calculation of the spin wave dispersion at finite temperature $\hbar\omega_{\vec{q}}(T)$ using Eqs. (7) and (2) and eight parameters given in Table 2. The results are displayed in Fig. 5 together with the experimental observation made at 158°K ($T/T_c = 0.83$). The agreement between the observation and the calculation is rather satisfactory, if we consider that 158°K is almost the limiting temperature to have a selfconsistent solution. The temperature dependence of the exchange stiffness constant obtained from the measurement along the [00ζ] direction is plotted in Fig. 6 against $(T/T_c)^{5/2}$. A linear relation holds up to 0.83 T_c. The temperature dependence was also calculated using Eqs. (8) and (9)

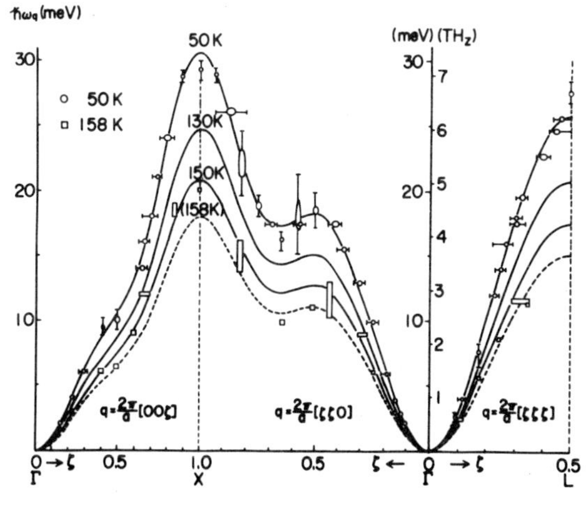

Fig. 5. Temperature dependences of spin wave dispersions in Pd_2MnSn calculated by Eqs. (7) and (2) and comparison with experiments at 158°K.

and the experimentally determined exchange parameters, and the results are plotted by a broken line in the figure. Although the calculated value of D(0) agrees well with the observation at the lowest temperature, the calculated temperature variation is much smaller than the observation. However the self consistent calculation using Eq. (7) gives the correct decrease of the exchange stiffness constant as shown by a solid line in the figure. This means that the higher order terms in a development of D(T) with respect to temperature are quite important for this substance because of the long range of the interactions. The temperature variation of the static magnetization can also be well explained up to 0.7 T_c by the spin wave theory with dynamical interactions.

The kinematic interaction between the spin waves is another important factor affecting the magnons at high temperatures.[44] In the case of the Heisenberg ferromagnet, this interaction has no effects on the renormalization of long wave length magnons but it affects the neutron spin wave scattering cross section

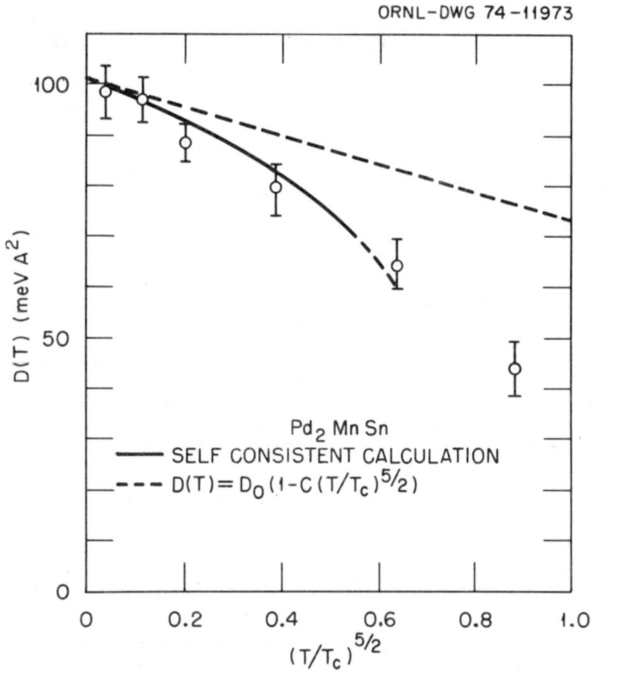

Fig. 6. Temperature dependences of exchange stiffness constants D(T) in Pd_2MnSn and comparison with calculations. A broken line was calculated using Eqs. (8) and (9), while a solid line was the result of selfconsistent calculations based on Eqs. (7) and (2).

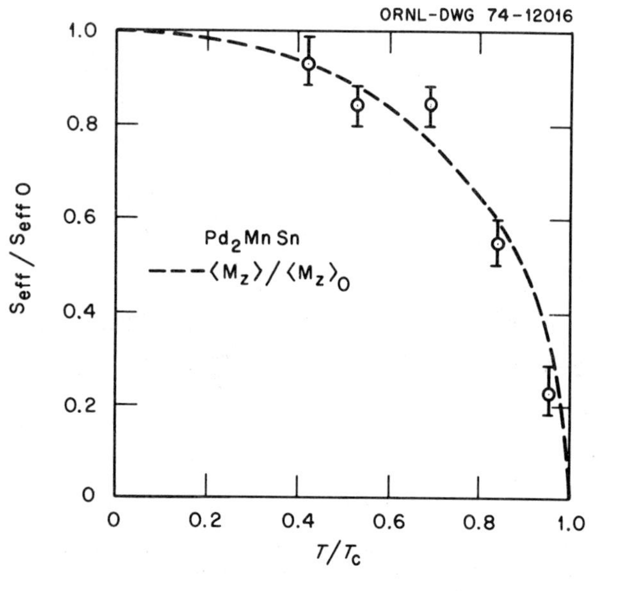

Fig. 7. Temperature dependence of integrated intensity of magnon groups with q = 0.15 A in Pd_2MnSn corrected for magnon population factor $\langle n_{\vec{q}} \rangle$. A broken line is static magnetization of Pd_2MnSn.

so that it varies with temperature as the magnetization $\langle M_z \rangle$ for small spin wave momenta.[45] Therefore an accurate measurement of the temperature dependence of the intensity is particularly of interest in the case of ferromagnetic metals, because the existence of other kinds of excitation such as the Stoner excitation would affect greatly the neutron cross section. Actually, in the cases of Fe, Ni[26,27] and Fe_3Pt (Invar alloy),[45] the decrease in the neutron spin wave scattering cross section for small momentum transfer is found to be much faster than the magnetization.

The temperature dependence of the intensity of the spin wave scattering was studied for Pd_2MnSn. The magnon group with momentum transfer q = 0.15 $Å^{-1}$ ($\hbar\omega_q \sim 3$ meV) were measured at various temperatures by constant Q scans. The integrated intensity was corrected for the magnon population factor $\langle n_{\vec{q}} \rangle$ and its temperature variation was compared with that of the magnetization in Fig. 7. Although we have not yet made the resolution correction on the intensity, the agreement between the scattering intensity and the static magnetization $\langle M_z \rangle$ is quite satisfactory. This is the first experimental result which showed clearly the existence of the kinematic interaction between the spin waves in the Heisenberg ferromagnet.

6. PARAMAGNETIC SCATTERING FROM Pd_2MnSn[29]

Neutron paramagnetic scattering is the most direct method of examining the existence of a localized moment in the system. The energy spectrum of the neutron scattering is quite different between a Heisenberg system and an itinerant electron sytem.[28] The neutron paramagnetic scattering cross section of the Heisenberg system is given by

$$\frac{d^2\sigma}{d\omega d\Omega} \propto \frac{2}{3} N|f(\vec{Q})|^2 S(S+1)\left(\frac{1}{2\pi\bar{\omega}^2}\right)^{\frac{1}{2}} \exp\left(-\frac{\omega^2}{\bar{\omega}^2}\right) \quad (10)$$

for large momentum transfer $\vec{q}$ (= $\vec{\tau}-\vec{Q}$),[28,46] namely the energy spectrum of the scattering is of Gaussian type with an energy spread given by the second moment $\hbar^2\bar{\omega}^2$ defined by

$$\hbar^2\bar{\omega}^2 = \frac{8}{3} S(S+1) \sum_{\vec{R}} J_R^2(1-\cos\vec{Q}\vec{R}). \quad (11)$$

The neutron paramagnetic scattering from a single crystal of Pd_2MnSn was measured at room temperature (T/T_c = 1.6) with a time-of-flight spectrometer installed at a pulsed neutron source produced by an electron LINAC of Tohoku University.[29] The same crystal as used in the spin wave measurements was employed. It was found that the energy spectrum at the zone boundary is really of Gaussian form with the second moment $(\hbar\bar{\omega})^2$ = 23.3 ± 0.2 $(meV)^2$ which is in reasonable agreement with the calculated value of 18.4 $(meV)^2$ using the exchange parameters in Table 2. Therefore it is definitive that the alloys keep a localized moment even in the paramagnetic state and the system can be considered as a Heisenberg system.

7. CONCLUSION

The most important conclusion we have obtained from the present study is that the 3d electrons in the Heusler alloys are well localized on the Mn atoms and they behave just like the localized moment in the insulator, and the theory developed for the Heisenberg system can successfully explain the spin dynamics of this system. Therefore the Heusler alloys can be considered as an ideal Heisenberg ferromagnet with long range interactions. The detailed investigations of the intensity of neutron magnon scattering as well as the magnon life time of this system are quite useful to test the validity of the theory. The neutron critical scattering of this system is also quite interesting in this connection.

The spin wave studies have shown that the magnetic interactions in the Heusler alloys are long range, extending to more than eighth neighbors (R ∿ 14 Å) and the interactions further than the third nearest Mn-Mn interactions (R > 8 Å) can be reasonably well explained by the s-d type interaction based on the nearly free electron model. However in all of the Heusler alloys we studied, the first two nearest neighbor interactions are ferromagnetic and the Curie point is mainly determined by them. These interactions are of indirect type and the electronic state of the X atoms (Cu, Ni, Pd) plays an important role in determining the magnitude of the interactions. The virtual double exchange interaction or the preasymptotic correction of the s-d type interaction which takes into account the d band of the X atoms may be possible origins of the interaction. Therefore band theoretical calculations of J_R of this system are highly required to understand the mechanism of the magnetic interaction.

ACKNOWLEDGMENT

The authors would like to thank Drs. N. Watanabe, K. Tajima, P. Radhakrishna, D. Tocchetti, M. W. Stringfellow and J. Webster with whom we enjoyed collaborating on the investigations which became the basis of this paper. They are indebted to Prof. T. Kasuya for crucial discussions on the mechanism of the magnetic interaction, which promoted our research. Their thanks are also due to Dr. J. Schweizer for providing the crystal and to Mrs. M. Kohgi and M. Onodera and Drs. A. Boeuf, S. Wilson, and A. Hartman for assistance in the measurement; to Drs. J. Cooke, H. Alloul and I. A. Campbell for discussions and to Dr. J. W. Cable for reading the manuscript. One of the authors (Y.I.) gratefully acknowledges many colleagues in Institut Laue-Langevin and Solid State Division in the Oak Ridge National Laboratory for their hospitality during his stay in these laboratories. A part of the experiment was carried out in the former laboratory, while the paper was prepared in the latter laboratory.

REFERENCES

1. P. J. Webster, Contemp. Phys. 10, 559 (1969).
2. K. Endo, J. Phys. Soc. Japan 29, 643 (1970).
3. T. Shinohara, J. Phys. Soc. Japan 28, 313 (1970).
4. P. J. Webster and R. S. Tebble, J. Appl. Phys. 39, 471 (1968).
5. R. J. Webster and R. S. Tebble, Phil. Mag. 16, 347 (1967).
6. M. G. Natera, M. R. L. N. Murthy, R. J. Begum and N. S. Satya Murthy, phys. stat. sol. (a) 3, 959 (1970).
7. H. Masumoto, K. Watanabe and S. Ohnuma, J. Phys. Soc. Japan 32, 570 (1972).
8. P. J. Webster, J. Phys. Chem. Solids 32, 1221 (1971).
9. G. P. Felcher, J. W. Cable and M. K. Wilkinson, J. Phys. Chem. Solids 24, 1663 (1963).
10. K. Yoshida, Phys. Rev. 106, 893 (1957).
11. B. Caroli and A. Blaindin, J. Phys. Chem. Solids 27, 503 (1966).
12. D. J. W. Geldart and P. Ganguly, Phys. Rev. B 1, 310 (1970).
13. S. Ogawa and J. Smit, J. Phys. Chem. Solids 30, 657 (1969).
14. T. Shinohara, J. Phys. Soc. Japan 28, 313 (1970).
15. W. Leiper, D. J. W. Geldhart and P. J. Pothier, Phys. Rev. B 3, 1637 (1971).
16. J. M. Williams, J. de Physique 32, C1-790 (1971).
17. L. J. Swartzendruber and B. J. Evans, AIP Conf. Proc. 5, 539 (1972).
18. L. J. Swartzendruber and B. J. Evans, AIP Conf. Proc. 10, 1369 (1973).
19. A. Blandin and I. A. Campbell, Phys. Rev. Lett. 31, 51 (1973).
20. C. C. M. Campbell and W. Leiper, AIP Conf. Proc. 18, 319 (1974).
21. Y. Ishikawa and Y. Noda, Solid State Comm. 15, 833 (1974).
22. The work done in ILL and in Harwell, to be published.
23. G. G. Low, Proc. Symp. on Inelastic Scatt. of Neutrons in Solids and Liquids, 1964 (IAEA, 1965).
24. M. Kohgi, Y. Ishikawa, I. Harada, and K. Mochizuki, J. Phys. Soc. Japan 36, 112 (1974).
25. C. J. Glinka, V. J. Minkiewicz, L. Passell and M. W. Shafer, AIP Conf. Proc. 18, 1060 (1974).
26. M. W. Stringfellow, J. Phys. C 1, 950 (1968).
27. H. A. Mook, J. W. Lynn and R. M. Nicklow, Phys. Rev. Lett. 30, 556 (1973).
28. M. F. Collins, J. Appl. Phys. 39, 533 (1968).
29. N. Watanabe, Y. Ishikawa and T. Tsuzuki, Nucl. Instrum. Methods 120, 293 (1974).
30. A. Delapalme, J. Schweizer, G. Couderchon, R. Perrier de la Bathie, Nucl. Instrum. Methods 95, 589 (1971).
31. R. D. Lowde, J. Appl. Phys. 36, 884 (1965).
32. H. B. Møller, J. C. Gylden Houman and A. R. Mackintosh, J.A.P. 39, 807 (1968).
33. G. G. Scott, J. Phys. Soc. Japan 17, Suppl. B-1, 372 (1962).
34. We are indebted to Dr. Rebouillat (CNRS) for the measurement.
35. H. Jones, Proc. Roy. Soc. (London) A144, 255 (1934)
36. T. Kasuya, Solid State Commun. 15, 1119 (1974).
37. H. Alloul, J. Phys. F (1974), in press.
38. D. J. W. Geldart, Phys. Lett. 38A, 25 (1972).
39. T. Kasuya, private communication, to be published.
40. B. N. Harmon and J. A. Freeman, to be published in Phys. Rev.
41. W. C. Koehler, H. R. Child, R. M. Nicklow, H. G. Smith, R. M. Moon and J. W. Cable, Phys. Rev. Lett 24, 16 (1970).
42. T. Moriya, Theory of Magnetism in Transition Metals, ed. by W. Marshall (Academic Press, 1967), p. 206.
43. W. Marshall and S. W. Lovesey, Theory of Thermal Neutron Scattering (Oxford U. P., Oxford, England, 1971), p. 265.
44. W. Marshall and G. Murray, J. Phys. C 2, 539 (1969
45. Y. Ishikawa and K. Tajima, Rep. Phys. Soc. Meet. Japan (1973) Oct.
46. W. Marshall and S. W. Lovesey, Theory of Thermal Neutron Scattering (Oxford U. P., Oxford, England, 1971), p. 493.

†A temporary stay in Oak Ridge National Laboratory as a guest scientist.

MAGNETIC EXCITATIONS IN RARE EARTH Al_2 COMPOUNDS

Per Bak*
AEC Research Establishment Risø, 4000
Roskilde, Denmark

ABSTRACT

In the rare earth intermetallic compounds, the crystal field is often comparable to the exchange interaction between the magnetic ions. The magnetic excitations (excitons) in such systems are transitions between single-ion atomic levels, propagating through the lattice via the exchange interaction. This article reviews recent experimental and theoretical work on rare earth Al_2 compounds. The excitons in $NdAl_2$ and $TbAl_2$ have been studied by inelastic neutron scattering. The dispersion relations are analysed within the random phase approximation using a Hamiltonian including 4th and 6th order crystal field terms and an isotropic exchange interaction. The spectrum of $NdAl_2$ at 5 K comprises both transverse and longitudinal modes, corresponding to transitions between the ground state and excited mean-field states. The bulk magnetic properties of $NdAl_2$ can be understood quantitatively using the crystal field parameters derived from the analysis of the exciton spectrum. At 4 K the magnetic excitations in $TbAl_2$ are spin waves. At temperatures above 35 K the neutron scattering measurements show a double peak structure. This splitting of the spin wave branch is the result of an interaction between the spin-waves and excitations associated with higher-lying mean-field states, a magnon-exciton interaction.

INTRODUCTION

The magnetic properties of rare earth systems, in which the exchange energies and the crystal field nearly balance, differ substantially from those of conventional magnets.[1] In particular, the magnetic excitations are fundamentally different from the spin-waves encountered in materials where the exchange interaction dominates the crystal field, such as the heavy rare earth elements. Instead the magnetic excitations, or magnetic excitons, can be described as transitions between single-ion states which propagate through the lattice via the exchange interactions.[2-5] In the paramagnetic regime, the excitons are pure crystal field transitions which may have a $\vec{q}$ dependence due to two-ion interactions.[2] In the ordered regime we shall see that the excitons can be thought of as transitions between mean-field states.

The elementary magnetic excitations can in principle be studied directly via inelastic neutron scattering techniques. In conventional magnets, the main effect of the crystal field is simply to create energy gaps in the spin wave spectrum, and it is difficult to get reliable information on the crystal fields from the dispersion relations. In this case a more precise determination can be obtained from single crystal magnetisation studies,[6] or from inelastic neutron scattering[7] or susceptibility measurements[8] on diluted samples. For systems where crystal field and exchange energies are comparable, an analysis of the exciton spectrum may yield information on both the crystal field parameters and the wave vector dependence of the exchange interaction between the rare earth ions.

In this paper a recent study of the magnetic excitons in some $REAl_2$ (RE = rare earth) compounds will be reviewed.[9,10] The local surroundings of the rare earth ions are cubic, which limits the number of independent crystal field parameters to two and makes these compounds suitable for detailed numerical calculations. We shall see that the crystal field creates a variety of interesting dynamical effects, and that a microscopic description of the magnetic properties can be given in terms of the crystal field and the exchange parameters derived from an analysis of the exciton spectrum.

The format of the paper is as follows. In Sec. II we set up the theory for the dispersion relations for magnetic excitons in complicated many-level systems to be used in the analysis. This RPA-theory includes not only excitations from the ground state, but also excitations out of excited states "excited state spin waves", which are important at elevated temperatures. Section III is devoted to a detailed discussion of $NdAl_2$ where several modes of different nature have been observed. In Sec. IV we discuss the temperature dependence of the spin wave spectrum of $TbAl_2$. In this compound the excited state spin waves have been directly observed through an interaction with the main spin-wave mode. Finally, Sec. V contains a general discussion.

II. THEORY OF MAGNETIC EXCITATIONS

In this section, a method will be presented for the study of collective excitations in a lattice of identical magnetic ions with discrete energy levels. The excitation spectrum will be calculated directly by considering the equations of motion of operators creating transitions between single-ion states. Using the double-time Green's function techniques, the equations of motion will be developed in the random-phase-approximation (RPA). This theory gives a clear picture of the physical nature of the magnetic excitations in contrast to theories working in artificial pseudo-spin spaces.[12]

We consider the Hamiltonian

$$\mathcal{H} = \sum_i V_{ci} - \sum_{ij} \mathcal{J}_{ij}\vec{J}_i\cdot\vec{J}_j \;. \qquad (1)$$

V_{ci} is the crystal field potential on the ion at site i which may conveniently be expressed as a linear combination of Steven's operators.[11] $\mathcal{J}_{ij}$ is the exchange integral between the ions i and j. For metals the $\mathcal{J}_{ij}$ originates mainly from the indirect coupling via the conducting electrons, the RKKY interaction.[13] However, the present theory can be extended to systems with any two-ion interaction, both in the paramagnetic and the ordered phase without introducing essential complications. In the ordered phase, the Hamiltonian is further separated into a two-ion term $\mathcal{H}_1$ and a single-ion term $\mathcal{H}_0$ in the following way:

$$\mathcal{H} = \mathcal{H}_0 + \mathcal{H}_1 + N\,\mathcal{J}(0) < J^z >^2 \qquad (2)$$

$$\mathcal{H}_0 = \sum_i V_{ci} - 2\mathcal{J}(0) < J^z > \sum_i J_i^z \;, \qquad (3)$$

$$\mathcal{H}_1 = -\sum_{ij} \mathcal{J}_{ij}\, \vec{j}_i\cdot\vec{j}_j \;, \qquad (4)$$

$$\vec{j}_i = \vec{J}_i - <J> \vec{\epsilon}_z \;, \qquad \mathcal{J}(0) = \sum_i \mathcal{J}_{ij} \qquad (5)$$

$\mathcal{H}_0$ is the molecular field Hamiltonian and $\mathcal{H}_1$ is the interacting part of the Hamiltonian giving rise to dispersion of collective excitations. The statistical average $< J^z >$ must be calculated self-consistently. ϵ_z is a unit vector in the magnetization direction. The eigenstates of the single-ion Hamiltonian $\mathcal{H}_0$ are denoted $|n>_i$ with energies E_n. To proceed further we transform the total Hamiltonian into a representation where the single-ion term is diagonal, that is to say we project any operator O onto the single-ion states $|n>$

$$O_i = \sum_{mn} < n|O_i|m > (|n>< m|)_i \ . \tag{6}$$

The operators $(|n> < m|)_i$, which create transitions between the single ion states $|m>_i$ and $|n>_i$, are denoted a_{nm}^i. The operators $a_{0n}^i = (|0>< n|)_i$ and $a_{n0}^i = (|n>< 0|)_i$, where $|0>$ is the ground state, correspond to Grover's pseudoboson operators b_n and b_n^+.[14] In a singlet-singlet system these operators are identical to the pseudo-spin operators introduced by Wang and Cooper.[15]

We obtain:

$$\mathcal{H} = \sum_{in} E_n a_{nn}^i - \sum_{ij} \sum_{\substack{mn \\ m'n'}} \mathcal{J}_{ij}\, a_{mn}^i \times a_{m'n'}^j$$

$$\times \sum_{\alpha} < m|j^{\alpha}|n > < m'|j^{\alpha}|n' > \ . \tag{7}$$

α denotes the cartesian coordinates. This transformation is, of course, exact. We now introduce the retarded double time dependent Green's functions

$$G^{ij}_{pr,p'r'}(t) = \langle\langle a^i_{pr}(t);\, a^j_{p'r'}(0) \rangle\rangle$$

$$= -i\theta(t) < [a^i_{pr}(t),\, a^j_{p'r'}(0)] > \tag{8}$$

as defined by Zubarev.[16] $\theta(t)$ is the unit step function. The equations of motion of the Green's functions are evaluated using the equation $i\,\dot{a}^i_{pr} = [a^i_{pr}, \mathcal{H}]$. For the higher order Green's functions involving operators on three different sites we apply the simplest random phase (RPA) decoupling.

$$\text{RPA:} \ \langle\langle a^i_{pr}(t)\, a^{i'}_{p'r'}(t);\, a^{i''}_{p''r''}(0) \rangle\rangle$$

$$= < a_{pr} > \langle\langle a^{i'}_{p'r'}(t);\, a^{i''}_{p''r''}(0) \rangle\rangle$$

$$+ < a_{p'r'} > \langle\langle a^{i}_{pr}(t);\, a^{i''}_{p''r''}(0) \rangle\rangle \ . \tag{9}$$

The thermal averages $< a_{pr} >$ must, in principle, be determined in a self-consistent way. However, to facilitate the calculations we insert the molecular-field values calculated from the single-ion Hamiltonian $\mathcal{H}_0$:

$$< a_{pr} > = < a_{pp} > \delta_{pr} = n_p \delta_{pr} \tag{10}$$

n_p is the occupation number of level p. Because of time and translational invariance in a simple Bravais lattice, we can Fourier transform our Green's functions with respect to time and space.

The doubly Fourier-transformed equations of motion turn out to be:

$$G_{pr,p'r'}(\vec{q},\omega)(E_r - E_p - \omega)$$

$$- \sum_{mn} G_{mn,p'r'}(\vec{q},\omega)(n_p - n_r)$$

$$\times 2\mathcal{J}(\vec{q}) \sum_{\alpha} < m|j^{\alpha}|n>< r|j^{\alpha}|p >$$

$$= -\frac{1}{2\pi}(n_p - n_r)\delta_{pr'}\delta_{rp'} \tag{11}$$

The Green's functions can thus be determined by solving this coupled system of linear equations. The energies of the magnetic excitations are the poles of the Green's functions, which are the roots of the characteristic determinant of the system of equations for fixed values of p' and r'. The determinant is independent of p' and r', such that the excitation frequencies are unequivocally determined.

For rare earth ions in a cubic environment, at least two of the matrix elements $< n|J^+|m >$, $< n|J^-|m >$, and $< n|J^z|m >$ are zero. This means that the equations (11) determining the exciton dispersion relations can be separated into two set of equations. One set involves matrix elements of J^+ and J^- only. The excitation branches determined by this part of the linear system of equations are denoted transverse. The other set of equations, involving matrix elements of J^z only, determines the longitudinal excitations. At zero temperature this approach is equivalent to the pseudo-Boson method introduced by Grover,[14] and the many level spinwave theory of Buyers et al.[17] used to interpret the exciton spectrum of TbSb.[4]

The complex frequency dependent susceptibility can be expressed by the Green's functions:

$$\chi^{\alpha\beta}(\vec{q},\omega) = 2\pi\, i\, \text{F.T.}\,\theta(t) < [J^{\alpha}_i(t),\, J^{\beta}_j(0)] >$$

$$= -2\pi \sum_{\substack{pr \\ p'r'}} G_{pr,p'r'}(\vec{q},\omega) < p|J^{\alpha}|r>< p'|J^{\beta}|r' > . \tag{12}$$

F.T. denotes the Fourier transformation with respect to time and lattice. By inserting (12) into (11) we obtain

$$\chi^{\alpha\beta}(q,\omega) = \chi_0^{\alpha\beta}(\omega) + 2\mathcal{J}(\vec{q}) \sum_{\gamma} \chi_0^{\alpha\gamma} \chi^{\gamma\beta}(q,\omega) \tag{13}$$

where

$$\chi_0^{\alpha\beta}(\omega) = \sum_{pr} \frac{< p|J^{\alpha}|r > < r|J^{\beta}|p > (n_p - n_r)}{E_r - E_p - \omega} \ . \tag{14}$$

χ_0 is the single-ion susceptibility. Equation (13) is the usual molecular-field equation to determine the complex susceptibility. The excitations determined by Eq. (11) are thus in agreement with the excitations calculated by the method of Peschel et al.[18]

For a neutron scattering experiment it is important to know the strength of the different lines. The cross section can be expressed as the imaginary part of the dynamic susceptibility[19] using the fluctuation-dissipation theorem

$$\frac{d^2\sigma}{d\Omega d\omega} = \left(\frac{1.91\, e^2 g F(\vec{\kappa})}{2mc^2}\right)^2 \frac{k_f}{k_i}\, T \tag{15}$$

where

$$T = [1 - \exp(-\beta\omega)]^{-1} \sum_{\alpha\beta} (\delta_{\alpha\beta} - \hat{\kappa}_{\alpha}\hat{\kappa}_{\beta})\, \text{Im}\chi^{\alpha\beta}(\vec{\kappa},\omega) \tag{16}$$

The intensities can thus be calculated from the residues of the dynamics susceptibility. For a cubic multidomain sample, the factor $(\delta_{\alpha\beta} - \hat{\kappa}_{\alpha}\hat{\kappa}_{\beta})$ must be summed over equivalent directions, which effectively removes the $\hat{\kappa}$ dependence.

The unit cell of the rare earth Al_2 compounds contains two magnetic ions. The spectrum will therefore consist of both acoustic and optical modes. The theory can easily be extended to this case. One simply has to replace $\mathcal{J}(\vec{q})$ by $\mathcal{J}(\vec{q})+|\mathcal{J}'(\vec{q})|$ for the acoustic branches and by $\mathcal{J}(\vec{q})-|\mathcal{J}'(\vec{q})|$ for the optical branches in the equations determining the energies and neutron intensities. $\mathcal{J}(\vec{q})$ is the Fourier transform within one sublattice and $\mathcal{J}'(\vec{q})$ is the Fourier transform between the two sublattices. However, the magnetic structure factor $1 \pm \cos(\vec{\tau}\cdot\vec{\rho}+\Phi)$, must be taken into account.[19] The plus sign is valid for the acoustic modes, the minus sign for the optical modes. This factor makes it possible to distinguish between optical and acoustic modes. Φ is a phase angle given by $\exp(i\Phi) = \mathcal{J}'(\vec{q})/|\mathcal{J}'(\vec{q})|$.

III. MAGNETIC EXCITATIONS IN $NdAl_2$

$NdAl_2$ is an example of a magnetic system in which the crystal field and exchange energies are of the same order of magnitude. $NdAl_2$ orders in a simple ferromagnetic structure for $T < 65K$ with magnetic moments pointing along $< 100 >$ directions. The magnetic moment at 4.2K is $2.5\mu_B \pm 0.1\mu_B$,[20] which is significantly lower than the free ion value $3.28\mu_B$ of the $^4I_{9/2}$ ground state multiplet of the Nd^{3+} ion. The crystal field is thus strong enough to produce considerable quenching, and we expect it therefore to have strong effects on the magnetic exciton spectrum.

An inelastic neutron scattering study of the excitons in the ordered state has recently been reported by Houmann, Bak, Purwins and Walker.[9] The experiment was performed on a triple axis spectrometer at Riso. The experimental results for momentum transfer in the $< 110 >$ direction are given in Fig. 1.

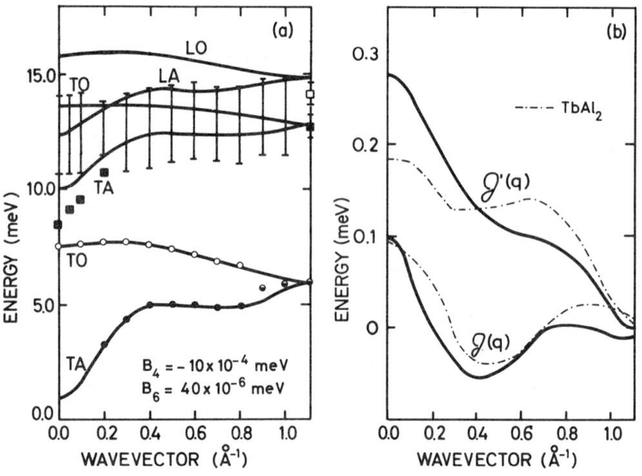

Fig. 1. (a) Dispersion relations for magnetic excitations in $NdAl_2$ for momentum transfer in the $<110>$ direction. The lines are the results of the analysis described in the text. (b) $\mathcal{J}(\vec{q})$ and $\mathcal{J}'(\vec{q})$ deduced from the fit. TA: Transverse acoustic modes; TO: Transverse optical branches; LA: Longitudinal acoustic branch; LO: Longitudinal optical branch. Data taken from Ref. 9.

Two well-resolved low-lying branches were observed, one acoustic and one optical. It was not possible to resolve the acoustic branch below $q = 0.2Å^{-1}$. However, the feature of the spectrum which makes it fundamentally different from an ordinary spin wave spectrum is the existence of a band of scattered neutron intensity at higher energies ~ 10-15 meV. This part of the spectrum is completely resolved into two peaks only near the zone boundary, with intensities corresponding to the low-lying branches. Near the zone centre, one acoustic branch with a marked wave vector dependence is clearly resolved. The optical branches could not be traced separately in this energy region for experimental reasons. The existence of three well-defined excitons at the zone boundary, where optical and acoustic modes are degenerate, implies six modes for a general point in the Brillouin zone. We shall now see that the features of this spectrum can be understood using the theory described in Sec. II

For the cubic symmetry appropriate to the rare earth Al_2 compounds, the most general crystal field Hamiltonian describing any single ion interactions inside a manifold of given J is simply

$$V_{ci} = B_4[O_{4i}^0 + 5\,O_{4i}^4] + B_6[O_{6i}^0 - 21\,O_{6i}^4] \qquad (17)$$

The coordinate axes have been chosen along the cube edges. In the analysis we shall assume that the Hamiltonian can be written on the form (1) with this crystal field potential. All anisotropic and higher order exchange interactions have thus been omitted together with magnetoelastic effects. The exciton spectrum was calculated for a wide range of parameters B_4' and B_6. The mean-field constant $\mathcal{J}(0)+\mathcal{J}'(0)$ was determined to yield a Curie temperature of 65K in agreement with experiment. $\mathcal{H}_0$ was diagonalized self-consistently to yield single-ion wavefunctions and energies, and the exciton energies were computed directly from Eq. (11). All ten single ion levels were taken into account. The exchange interactions $\mathcal{J}(\vec{q}) \pm \mathcal{J}'(\vec{q})$ can be determined for $\vec{q} \neq 0$ by fitting the dispersion relations calculated by this procedure to the two low-lying well-defined optical and acoustic branches. By means of an iteration process following these lines, B_4 and B_6 were determined as the parameters giving the best agreement with the upper branches. Generally spoken, one can say that the lower branches give information on the exchange interactions, and the positions of the upper branches yield information on the crystal field parameters. Once the lower branches are calculated, the dispersion of the upper branches is fixed without further fitting of the exchange. It turns out that the analysis can be performed in an unique way because of the small number of independent crystal field parameters. The resulting values of the crystal field parameters obtained in this way are $B_4 = -1.0\cdot10^{-3}$ meV and $B_6 = 4.0\cdot10^{-5}$ meV. The theoretical spectrum consists of six strong modes for a wide range of parameters; however, even relative small deviations (~ 15%) from the best values will shift the positions of the exciton branches severely

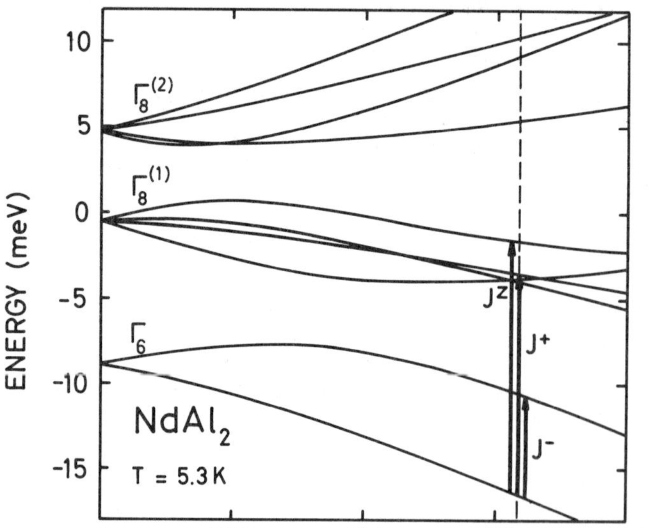

Fig. 2 Single-ion level scheme of $NdAl_2$ as a function of molecular field. The vertical line indicates the actual field obtained by the self consistent diagonalisation of $\mathcal{H}_0$. (Ref. 9)

and destroy any agreement with experiment. Thus Houmann et al.[9] have shown that an analysis of the magnetic excitons in the ferromagnetic regime for systems with large crystal field may give detailed information on both the exchange interaction and the crystal field parameters.

Figure 2 shows the single-ion level scheme as a function of the molecular field. The crystal-field-only scheme has a doublet, Γ_6, as the ground state and two quartets, $\Gamma_8^{(1)}$ and $\Gamma_8^{(2)}$, as excited states. The magnetic moment associated with the Γ_6 state is $1.33\mu_B$. In the exchange field, the degeneracies of these states are lifted and the excited levels are admixed into the ground state. The magnetic moment of the mean-field ground state is $2.6\mu_B$ in agreement with the value $2.5\mu_B$ determined by neutron diffraction.[20] The quenching of the free ion moment of $3.28\mu_B$ can thus be explained as a pure crystal field effect.

We denote the mean field states $|1\rangle$, $|2\rangle$... ...$|10\rangle$ according to increasing energy. The most important matrix elements creating excitations between the levels at low temperature are

$$\langle 2|J^-|1\rangle = 3.25 \qquad \langle 5|J^+|1\rangle = 2.28$$
$$\langle 6|J^z|1\rangle = -1.71$$

The first two matrix elements give rise to two optical and two acoustic transverse modes, and the third matrix element creates two longitudinal branches. The two lowest exciton branches originate mainly from the J^- coupling between the two states associated with the Γ_6 crystal field state. Since the corresponding wavefunctions are approximately $|J,J\rangle$ and $|J,J-1\rangle$, these modes correspond to the spin waves in magnetic systems with strong exchange interaction. The remaining four branches of longitudinal and transverse character are associated mainly with transitions between the ground state and excited states originating from the $\Gamma_8^{(1)}$ quartet. Matrix elements between the ground state and higher lying states are much smaller, and consequently the excitons associated with these states are not observable.

$\mathcal{J}(\vec q)$ and $\mathcal{J}'(\vec q)$ are given in Fig. 1 as full lines. The broken line shows the corresponding functions for $TbAl_2$ as derived from spin wave measurements by Bührer, Godet, Purwins and Walker.[21] The similarity indicates that the exchange interactions are of the same nature in the two compounds. By Fourier transforming $\mathcal{J}(\vec q)$ and $\mathcal{J}'(\vec q)$, Bührer et al. showed that the exchange in $\vec r$ space is of long range and oscillatory, with a dominant contribution from the interaction between nearest neighbour RE-ions. The results thus support the RKKY model of the exchange interaction. However, it must be stressed, that the exchange does not scale with the de Gennes factor $(g-1)^2$, indicating that orbital contributions to the two-ion interactions are important, as already pointed out by Levy[22] from an analysis of the Curie temperatures.

The magnetic excitons in $NdAl_2$ can thus be understood within the framework of the RPA theory using a Hamiltonian including crystal field and isotropic Heisenberg exchange interaction terms. Further measurements are suggested to obtain more detailed experimental information. In particular one may hope to measure the dispersion of the crystal field transitions in the paramagnetic regime and thus determine the crystal field and exchange-parameters in an independent way. Furthermore a better resolution of the upper exciton branches should be possible. The smearing out of these modes seems to indicate that linewidth considerations are important.

We now discuss the macroscopic magnetic properties of $NdAl_2$. We assume that the Nd^{3+} ion in an externally applied magnetic field, $\vec H$, can be described by the single ion Hamiltonian

$$\mathcal{H}_0 = B_4[O_4^0 + 5\,O_4^4] + B_6[O_6^0 - 21\,O_6^4] - g\mu_B(\vec H + \vec H_{MF})\cdot\vec J \qquad (18)$$

The mean field, $\vec H_{MF}$, obeys the self consistency condition

$$\vec H_{MF} = g\mu_B\lambda\langle\vec J\rangle,\ \lambda = \frac{2[\mathcal{J}(0) + \mathcal{J}'(0)]}{g^2\mu_B^2} \qquad (19)$$

The crystal field parameters B_4 and B_6 and the mean-field constant λ were determined from the analysis of the exciton experiment. We are then in possession of all the parameters necessary to calculate the bulk magnetic properties of $NdAl_2$. This gives us an opportunity of testing the reliability of the crystal field parameters derived by Houmann et al.

The magnetisation as a function of an external magnetic field in high symmetry directions can be calculated by solving Eqs. (18) and (19). When the magnetic field is in the $\langle 100\rangle$ direction, the magnetic moment remains in this direction, and the problem is reduced to one dimension. When the external magnetic field is applied along other symmetry directions, the analysis is more complicated. In this case we solve the equations self-consistently in the plane given by the direction of the magnetic field and the $\langle 100\rangle$ direction, implicitly assuming that the magnetic moment is confined to this plane. The computations can be carried out using an iteration method.[23] This procedure allows us to calculate $\vec M$ in an external magnetic field without introducing new phenomenological, temperature-dependent anisotropy parameters.

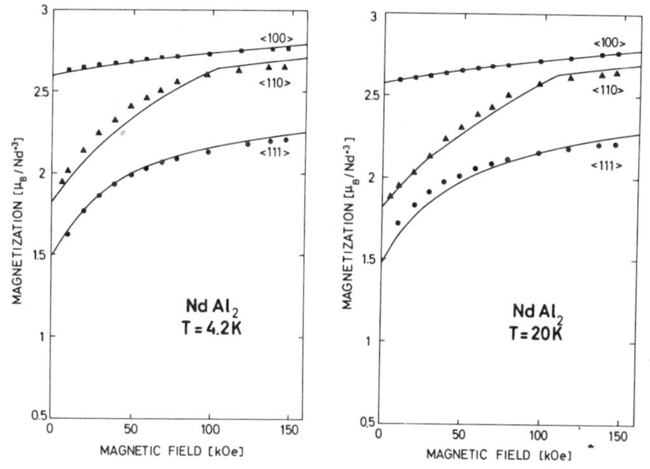

Fig. 3 The magnetisation as a function of the external magnetic field in the three main symmetry directions. The moments shown are the components along the applied magnetic field. The measured points have been augmented by 5% to normalise to the calculated zero field moment at 4K. Data taken from Refs. 23 and 25.

Theoretical single-crystal magnetisation curves are shown in Fig. 3 together with experimental points at 4.2 and 20K measured by Cock, Roeland, Purwins, Walker and Furrer.[25] To facilitate the detailed comparison with experiment and to avoid confusing the drawing, we have normalised the experimental points to agree with the theoretical zero field magnetisation at 4K. This involves an enhancement of the observed points by 5%. The normalised experimental points are seen to be in detailed agreement with theory. The kinks on the theoretical magnetisation curves indicate the critical magnetic fields where the magnetisation vector is aligned along the field. At 4.2K the kink occurs at 100 kOe. The 5% discrepancy is certainly within what could be expected from a mean-field theory without any fitting parameters, further it is well within the uncertainties associated with the determination of the crystal field parameters and the mean-field constant from the exciton experiment. The measurements thus provide strong support for the crystal field parameters extracted from the exciton dispersion curves.

The present data give no justification for changing the simple crystal field model of $NdAl_2$ by introducing additional effects, such as anisotropic exchange interactions or magnetoelastic effects. It seems rather unlikely that a consistent picture could be obtained, if such effects were important.

IV. OBSERVATION OF EXCITED STATE SPIN WAVES IN $TbAl_2$

It has been pointed out by Peschel, Klenin and Fulde,[18] that in the paramagnetic regime magnetic excitations from the ground state of a lattice of rare earth ions can interact with excitations out of excited magnetic states with very low population, provided that the energies of the non-interacting excitations are comparable. This interaction leads to a splitting of the ground state exciton mode, which in principle can be observed by inelastic neutron scattering. In this section we shall discuss the system $TbAl_2$, where a splitting of the same nature has been directly observed in the ordered phase by Purwins, Houmann, Bak and Walker.[10] This was the first observation and theoretical interpretation of an interaction of this kind.

$TbAl_2$ orders ferromagnetically below 105K[26] with magnetic moments in the < 111 > direction. The Curie temperature is high compared with other $REAl_2$ compounds. Accordingly the quenching of the low temperature spontaneous magnetic moment is very small, and the magnetic excitations are ordinary spin waves with an energy gap at $\vec{q} = 0$ due to crystal field effects.[21] However, we shall see that when the temperature is raised, the excitation spectrum behaves in a most unusual way.

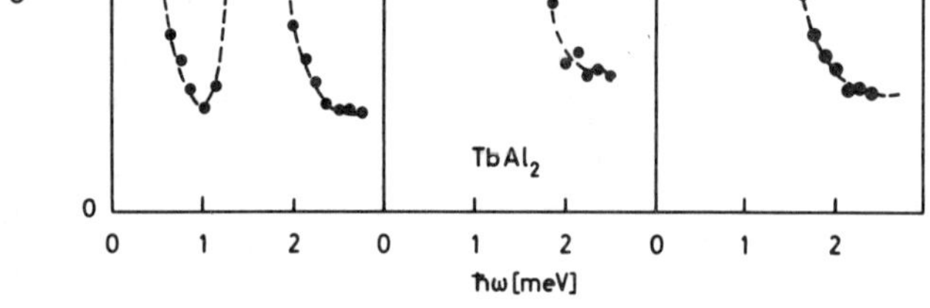

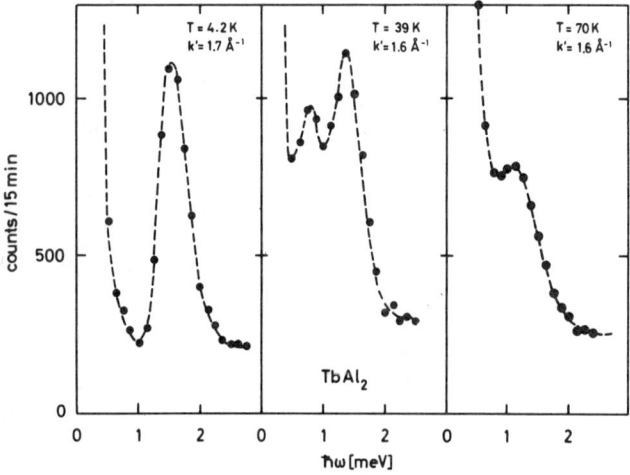

Fig. 4. Typical neutron spectra for $TbAl_2$ as a function of temperature (Ref. 10).

The neutron scattering experiment on the temperature dependence of the spin-waves was performed on a triple axis spectrometer at the DR 3 reactor at Riso. Some typical experimental results are shown in Fig. 4. At 4.2K a well-defined magnon peak at 1.58 meV is observed. As the temperature is raised to about 35K, a new peak arises in the neutron spectrum with lower energy than the original magnon peak. The intensity of the new peak increases

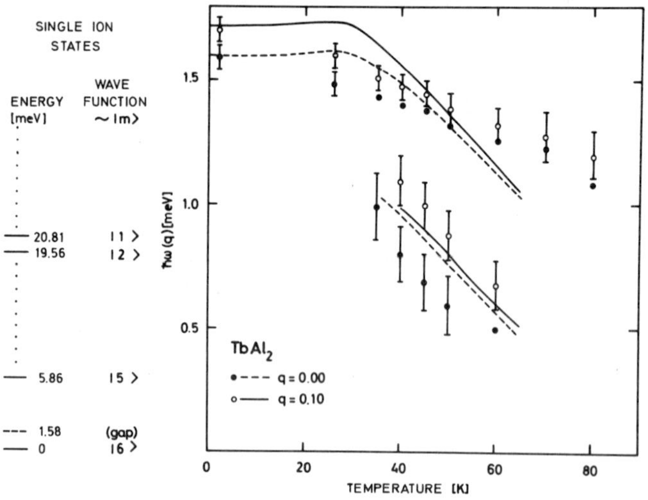

Fig. 5. Temperature dependence of the magnetic excitation energies in $TbAl_2$. Full and dotted lines: theory. Points: experiment. The left hand side of the figure shows the single ion states of Tb^{3+} at 4K. Data taken from Ref. 10.

with increasing temperature while the intensity of the upper peak decreases. Above 60K the low-energy peak, which has taken over most of the intensity, disappears into the background. Figure 5 shows the experimental points for all measured wave vector transfers and temperatures. We shall interpret this double peak structure as an interaction between the magnon mode and a magnetic exciton, an "excited state spin wave" originating from a transitions between some higher-lying excited magnetic states.

The strong interactions between the magnetic ions and the corresponding small quenching of the magnetic moment implies that the eigenfunctions of the magnetic ions are simple $|J,M_J>$ states. The $J_i^- J_j^+$ part of the isotropic exchange interaction between the ions i and j has large matrix elements between neighbouring levels, whereas the matrix elements of J^z can be ignored. Only transverse magnetic excitations (like spinwaves) are expected in this case. The crystal field part of the Hamiltonian acts as a perturbation shifting the mean-field levels without changing the wavefunctions substantially. The shift between the two lowest lying levels determines the energy gap associated with the low-temperature spin-waves. In general the temperature dependence of the lowest excited levels combined with a change in population of the mean field levels gives rise to the temperature-renormalisation of the spin wave energies and the energy gap. However, if the energy difference between two succeeding excited levels becomes comparable to the spin wave energy, the coupling between the corresponding excited state magnon (exciton) and the spin waves via the two-ion matrix elements of the exchange interaction becomes important, i.e., an interaction between the modes takes place. This resonance gives rise to a mixing of the magnon mode and the exciton mode, and hence to the observed anti-crossing effects, analogous to the magnon-phonon interaction, causing similar splittings of the magnon branches.[28] However, in the latter case, the phonon represents a transition from the boson ground state of the lattice at low temperatures, and therefore the interaction can be observed even at 0 K. In the actual case the excited states corresponding to the exciton mode are not populated at low temperature, and consequently there will be no interactions. However, when the temperature increases, the excited states become populated, and the interaction becomes possible.

The qualitative physical picture described here can be confirmed quantitatively using the Green's function theory described in Sec. II. The Hamiltonian for the Tb ions is written as the sum of crystalline field terms and isotropic exchange terms on the form (1) with

$$V_{ci} = -\frac{2}{3} B_4 [O_4^0 - 20\,(2)^{\frac{1}{2}}\, O_4^3] + \frac{16}{9} B_6 [O_6^0 + \frac{35}{4}\,(2)^{\frac{1}{2}}\, O_6^3 + \frac{77}{8}\, O_6^6] \qquad (20)$$

This potential is essentially the same as the potential (17) which was used to describe $NdAl_2$. However, the quantisation axis is chosen along the direction of magnetisation (< 111 >).

$\mathcal{H}_0$ is diagonalised self-consistently as a function of temperature to yield mean-field eigenfunctions and energies to serve as a basis for calculating the exciton spectrum. The calculations of the acoustic modes determined by $\mathcal{J}_{eff}(\vec{q}) = \mathcal{J}'(\vec{q}) + \mathcal{J}(\vec{q})$ are carried out for a wide range of crystal parameters B_4 and B_6 in the same way as for $NdAl_2$. The macroscopic magnetic properties calculated on the basis of these parameters must be in qualitative agreement with the experimental facts, that the easy direction of magnetisation is the < 111 > direction, the Curie temperature is 105K, the low temperature magnetic moment is near to the free ion value $9\mu_B$,[24] and the spin wave energy gap at 4K is 1.6 meV. These conditions impose severe restrictions on the parameters. There is only very little freedom left to vary the two parameters to explain the detailed behaviour of the excitation spectrum as a function of temperature.

The excitation energies are the roots of the secular determinant of (11), including transitions between all 13 levels. The roots of the diagonal elements of the characteristic matrix of Eqs. (11),

$$\omega_{diag.}(q) = E_r - E_p - 2(n_p - n_r)\,\mathcal{J}(\vec{q}) \sum_{\alpha} |\langle r\,|j^{\alpha}|\,p\rangle|^2 \qquad (21)$$

are the non-interacting excitation frequencies. For a simple ferromagnet this term yields the low-temperature spin wave theory, if $|r\rangle$ and $|p\rangle$ are chosen as the two lower states. The dispersion of the excited states is small because of the Boltzmann factors n_p and n_r. The off-diagonal elements due to two-ion interactions give rise to perturbations of the non-interacting energies. According to general perturbation theory, the energy shifts are largest, when the corresponding diagonal coefficients approach each other. This effect can thus be described as a resonance (or interaction) between two exciton modes. The best agreement with experiment was obtained using the parameters $B_4 = 7.0 \cdot 10^{-4}$ meV and $B_6 = 2.15 \cdot 10^{-6}$ meV. The calculated energies of the modes with observable neutron intensities are given in Fig. 5 together with the single ion level scheme at 4K. The parameters used in the calculation give T_c = 105K and nearly pure J^z single ion states. We therefore denote these states by the corresponding J^z quantum numbers. B_4 is of the same order of magnitude as the parameter used to describe magnetisation data of $TbAl_2$.[6] However, a large discrepancy exists for the parameter B_6, although the theoretical magnetisation curves calculated on the basis of the present parameters are in fair agreement with experiment.

At low temperatures the calculation shows only one acoustic excitation branch, which can be interpreted as an ordinary magnon branch with an energy gap, since it is mainly due to a $|J\rangle \rightarrow |J-1\rangle$ transition. As the temperature is raised the upper states become slightly populated. The renormalised magnon energy approaches and crosses the energy difference between the excited states $|2\rangle$ and $|1\rangle$, so that an interaction between the $|2\rangle \rightarrow |1\rangle$ excited state spin waves and the main magnon mode can take place. In the temperature range where the measurements were carried out, the population of the excited states corresponding to the bare exciton is very small. Consequently the neutron scattering cross section for the exciton is at least two orders of magnitude smaller than that for the magnon. The excited states can thus only be observed through the magnon-exciton interaction. The experimental ratios between the intensities of the low-lying and the high-lying excitation is roughly 1 to 4 at 35K and 1 to 1 at 45K, compared with the theoretical prediction (Eq. 16) of 1 to 5.2 and 1 to 1.3 respectively.

We conclude that interactions between magnetic excitations may occur if the energies of two modes are comparable, and this resonance may be directly measured by inelastic neutron scattering. The anti-crossing effects that were observed in $TbAl_2$ may well occur in other rare earth compounds. Recently Holden and Buyers[29] have suggested that the weak temperature dependence of the magnetic excitons in Pr_3Tl[3] originates from a mode-mode interaction of the same nature.

V. DISCUSSION

The analysis of the magnetic excitations in $TbAl_2$ and, in particular, in $NdAl_2$, has given detailed information on the fundamental microscopic magnetic interactions in rare earth intermetallic compounds. The exciton spectra can be reproduced using a simple Hamiltonian including two crystal field parameters and an isotropic RKKY-like exchange interaction. Since the analysis yields numerical values of the parameters B_4 and B_6, it is interesting to see how these parameters are related to parameters obtained for other $REAl_2$ compounds and by means of other kinds of measurements. Such a comparison has been performed by Purwins *et al.*[24] Assuming a simple effective point charge model, like the one successfully applied to the rare earth monopnictides by Birgeneau *et al.*,[30] the parameters for the whole series of $REAl_2$ compounds are found to be identical within 30%, when scaling with the appropriate reduced matrix elements are taken into account. This indicates that the effective charge distributions around the magnetic ions are independent of the particular rare earth constituent.

Finally we shall mention, that besides influencing the magnetic properties, the crystal field levels may play a crucial role in determining the superconducting properties of diluted magnetic systems.[31] The nonmagnetic compound $LaAl_2$ is superconducting below 3.3K. As the simple crystal field model seems to be valid also for $RE_x\,La_{(1-x)}Al_2$ compounds, the effects on the superconducting properties may be calculated using the crystal field parameters derived from neutron inelastic scattering and magnetisation work. Measurements of T_c as a function of Tb concentration in $LaAl_2$ clearly demonstrates the influence of the crystal field levels.[32]

REFERENCES

* Present address: Brookhaven National Laboratory, Upton, New York 11973.

1. R.J. Birgeneau, "Magnetism and Magnetic Materials 1972" (AIP Conference Proceedings, No. 10, ed. by C.D. Graham and J.J. Rhyne) p. 1664-88.
2. B.D. Rainford and J.C.G. Houmann, Phys. Rev. Letters 26, 1254-56 (1971) and J.C.G., Houmann, et al., to be published.
3. R.J. Birgeneau, J. Als-Nielsen and E. Bucher, Phys. Rev. Letters 27, 1530-33 (1971); Phys. Rev. B6, 2724-29 (1972).
4. T.M. Holden, E.C. Svensson, W.J.L. Buyers, and O. Vogt, "Neutron Inelastic Scattering (International Atomic Energy Agency, Vienna), p. 553-561 (1972).
5. K.C. Turberfield, L. Passell, R.J. Birgeneau and E. Bucher, Phys. Rev. Letters 25, 753-55 (1970); J. Appl. Phys. 42, 1746-54 (1971).
6. J.G. Purwins, E. Walker, B. Barbara, M.J. Rossignol and P. Bak, J. Phys. C7, (1974) 3573-81.
7. O. Rathmann, J. Als-Nielsen, P. Bak, J. Hog and P. Touborg, Phys. Rev. B 10 (1974) 3983-88.
8. J. Hog and P. Touborg, Phys. Rev. B9, 2920 (1973); P. Touborg and J. Hog, Phys. Rev. Letters 33, 775-78 (1974).
9. J.G. Houmann, P. Bak, H.-G. Purwins and E. Walker, J. Phys. C7 (1974) 2691-96.
10. H.-G. Purwins, J.G. Houmann, P. Bak and E. Walker, Phys. Rev. Letters 31, 1585-87 (1973).
11. K.W.H. Stevens, Proc. Phys. Soc. A65, 209-15 (1952).
12. Y.Y. Hsieh and M. Blume, Phys. Rev. B6, 2685-2709 (1972); D.A. Pink, J. Phys. C1, 1246-57 (1968).
13. M.A. Ruderman and C. Kittel, Phys. Rev. 96, 99-102 (1954).
14. B. Grover, Phys. Rev. 140A, 1944-51 (1965).
15. Y.L. Wang and B.R. Cooper, Phys. Rev. 172, 539-51 (1968).
16. D.N. Zubarev, Sov. Phys. Usp. 3, 320-44 (1960).
17. W.J.L. Buyers, T.M. Holden, E.C. Svensson and M.T. Hutchings, J. Phys. C4, 2139-59 (1971).
18. I. Peschel, M. Klenin and P. Fulde, J. Phys. C5, L194-96 (1972).
19. W. Marshall and S.W. Lovesey, Theory of Thermal Neutron Scattering (Clarendon Press, Oxford, 1971) p. 230-52.
20. N. Nereson, C. Olsen and G.J. Arnold, Appl. Phys. 37, 4575-80 (1966).
21. W. Bührer, M. Godet, H.-G. Purwins and E. Walker, Sol. State Commun. 13, 881-84 (1973).
22. P.M. Levy, J. Appl. Phys. 41, 902-04 (1970).
23. P. Bak, J. Phys. C7, in press.
24. H.-G. Purwins, E. Walker, B. Barbara, M.J. Rossignol and P. Bak, J. Phys. C7, in press.
25. G.J. Cock, L.W. Roeland, H.-G. Purwins, E. Walker and A. Furrer, to be published.
26. A.H. Millhouse, H.-G. Purwins and E. Walker, Sol. State Commun. 11, 707-12 (1972).
27. H.-G. Purwins, E. Walker, B. Barbara and M.F. Rossignol, Phys. Letters 45A, 427-28 (1973).
28. J. Jensen and J.C.G. Houmann, Phys. Rev. to bc published.
29. T.M. Holden and W.J.L. Buyers, Phys. Rev. B9, 3797-3805 (1973).
30. R.J. Birgeneau, E. Bucher, J.P. Maita, L. Passell and K.C. Turberfield, Phys. Rev. B8, 5345-47 (1973).
31. J. Keller and P. Fulde, J. Low Temp. Phys. 4, 289-98 (1971).
32. G. Pepperl, T. Umlauf, A. Meyer and J. Keller, Sol. State Commun. 14, 161-65 (1974).

TEMPERATURE DEPENDENCE OF MAGNETIC EXCITATIONS IN Pd_3Fe

T.M. Holden
Atomic Energy of Canada Limited, Chalk River, Ontario, Canada

ABSTRACT

Measurements of the magnetic excitation spectrum of ferromagnetic Pd_3Fe (T_C = 499 K) have been made at 293,373,423,473 and 573 K by neutron inelastic scattering techniques. Constant energy and constant momentum transfer modes of operation were employed with reduced wavevectors between 0.11 and 0.40 propagating in the [ζ00] direction. At 293 K one branch of the dispersion relation for magnetic excitations is observed, which is characterized by an exchange stiffness D of 52.7±2.7 THzÅ^2 and by neutron groups which have maxima below ~6 THz. The intrinsic linewidth is a strong function of wavevector and varies from 0.0±0.5 THz at ζ = 0.13 to 2.5±0.5 THz at ζ = 0.3. As the temperature is raised the excitation frequencies fall and the linewidths increase. Between 293 and 473 K at ζ = 0.3 the frequency falls by a factor of 1.8 and the linewidth increases by the same factor, nevertheless a broad inelastic response is still observed. At 473 K at wavevectors, less than ζ = 0.13, the spin wave is overdamped and the magnetic scattering has the form of a peak of appreciable width centred on zero frequency.

INTRODUCTION

The study of magnetic excitations in ferromagnetic transition metals such as Ni[1,2] and Fe[3] is leading to a better understanding of exchange effects in metals[4]. Studies of the temperature dependence of the excitations should likewise refine our presently unreliable ideas on how the exchange splitting of the bands alters with temperature. In order to obtain the frequencies and widths associated with the spin waves in an itinerant ferromagnet over a wider wavevector region than is possible in pure materials, neutron inelastic scattering studies of the chemically ordered ferromagnetic metal Pd_3Fe have been carried out at 293,373,423,473 and 573 K (T_C = 499 K). The principal advantage of studying Pd_3Fe is that the exchange interactions are weaker than in the pure materials and hence the frequencies are lower and accessible to high resolution studies.

In previous studies[5] of Pd_3Fe at 293 K the scattering cross-section became too weak to measure well before the zone boundary was reached in the [ζ00] direction, but the present results show that in this wavevector region the spin-wave lifetime is markedly reduced and this partly accounts for the loss of intensity. In Ni at elevated temperatures (1.1 T_C) the magnetic excitations, as observed by means of constant-energy scans[6], remain sharp although both the frequency and intensity are less than at low temperatures. In Pd_3Fe, however, the temperature induced broadening of the spin waves even below T_C is sufficient to make the excitations overdamped over a much larger wavevector region than for Ni, and leave very broad spin waves at higher wavevectors. In constant energy scans, groups with a sharp profile at 293 K had completely disappeared at 473 K.

THE EXPERIMENT

The experiment was performed using the triple axis crystal spectrometer at the DIDO reactor, U.K.A.E.A. Harwell which operates in the fixed incident frequency (E_O = 21.1 THz) mode. The sample of Pd_3Fe, which was close to being chemically ordered, was used in the previous studies of the excitation spectrum[5]. The crystal was mounted with the [010] axis vertical and measurements were made of spin waves propagating in the [ζ00] direction in the Brillouin zone containing the (100) superlattice reflection. The sample was mounted within a cylindrical Ta foil heating element inside an evacuated water-cooled furnace and the temperature was controlled and measured to within ±2 K by a chromel-alumel thermocouple attached to the crystal.

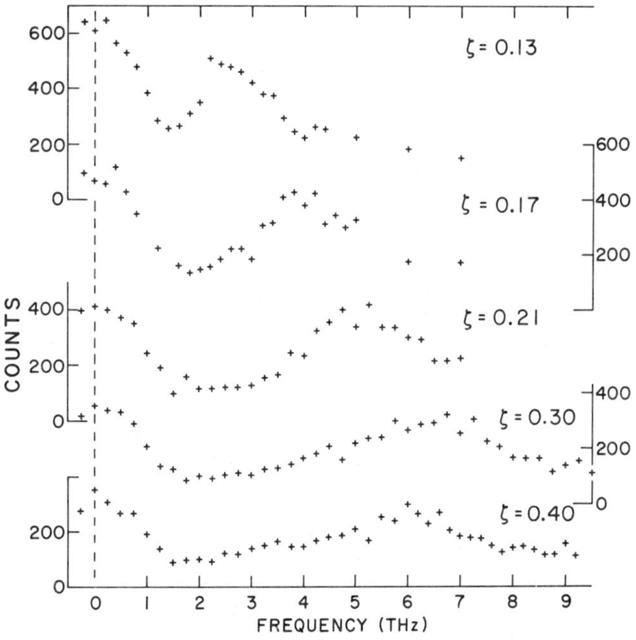

Fig. 1. Constant-Q scans in Pd_3Fe at 293 K normalized to the same number of incident neutrons for several reduced wavevectors ζ (in units of 2π/a) in the [ζ00] direction.

The distribution in frequency of scattered neutrons for several reduced wavevectors at 293 K, after reduction to a form where it is proportional[7] to S(Q,ω), is shown in Fig. 1. The spectra consist of two parts, an elastic incoherent peak arising from static fluctuations in the long range order, and the magnetic inelastic scattering. Below ζ=0.11 (units of 2π/a) the spin wave scattering merges with the elastic scattering and beyond 0.40 the intensity was too weak to measure and there was also the possibility of extra phonon scattering. The lineshape of the inelastic group at ζ=0.13 is determined completely by instrumental resolution effects but at ζ=0.30 the major part of the width comes from an inherent broadening mechanism. While it is possible that this broadening is caused by interactions between spin waves and single particle modes, there are other possible mechanisms at ~0.59 T_C which may be important; further experiments at lower temperatures will be necessary to check this conjecture.

In order to follow the temperature dependence of the magnetic scattering it is necessary to subtract the temperature independent elastic incoherent peak from the data. A gaussian was found to reproduce the form of the elastic peak accurately and this was scaled to the average intensity at zero frequency for temperatures of 293, 373 and 423 at each wavevector and then subtracted from the experimental scans. The temperature dependence of the scattering at 293, 473 and 573 for ζ=0.13 and 0.30 is shown in Fig. 2. The results illustrate the magnetic origin of the scattering since the neutron group has shifted in each case on raising the temperature. In addition to the lowered frequencies there is a thermally induced broadening. At ζ=0.13 (q=0.21Å^{-1}) the excitation is overdamped at 473 but at ζ=0.3 a broad inelastic response is still observed.

To obtain the frequencies and widths of the excitations from the measured groups resolution corrections were made. These were obtained by computing from a Heisenberg model[8], which approximately fitted the measured dispersion for each temperature, the lineshape[9] expected from the instrumental parameters. The

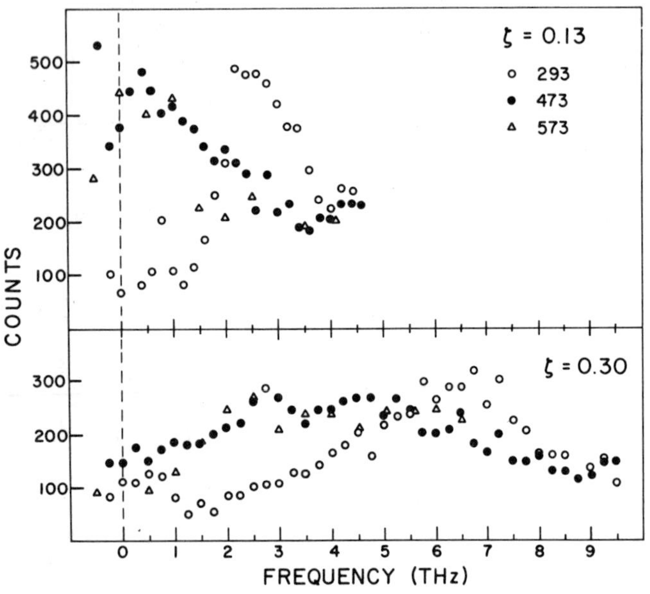

Fig. 2. Temperature dependence of the neutron scattering observed at ζ=0.13 and 0.30 in Pd_3Fe.

shift of the Heisenberg model frequency from the peak of the calculated lineshape for each case was then used to obtain the true frequency from the experimental lineshape and the procedure repeated to achieve consistency. The widths were obtained by assuming that both the intrinsic broadening and the instrumental broadening had a Lorentzian form.

The effective exchange stiffness, computed from Heisenberg model[8] fits to the corrected frequencies, is reduced by a factor 2 over the temperature range 293 (0.59 T_c) to 473 K (0.96 T_c), while the frequency at ζ=0.3, for example, is reduced by a factor 1.8 ± 0.3 in this temperature range. As an illustration of the increase of the spin wave linewidth it is found that the ratio of the full width at half maximum to the frequency at ζ=0.3 increases from 0.4 to 1.3 in the temperature range 293 to 473 K.

DISCUSSION

The reduction in frequency over a similar temperature range in Ni (0.47 to 1.1 T_c) is close to that for Pd_3Fe. For Ni the frequency at a reduced wavevector (0.13,0.13,0.13) ($q \sim 0.4 Å^{-1}$) falls by a factor 1.47 and the exchange stiffness by a factor 1.54. Near T_c, however, the region of reciprocal space over which the modes are overdamped in Ni is smaller than for Pd_3Fe ($q \lesssim 0.125\ Å^{-1}$ in Ni compared with $q \leqslant 0.21\ Å^{-1}$, $\zeta \leqslant 0.13$ in Pd_3Fe). The crucial difference between Ni and Pd_3Fe thus appears to be that the linebroadening is much more severe in the alloy and is moreover strongly temperature dependent and this damps out a sharp response just below and above T_c.

The exchange stiffness of a ferromagnetic metal is expected[10] to vary with temperature according to the expression

$$D(T) = D_0 + D_1T^2 + D_2T^{5/2}...$$

where D_1 is governed by magnon-single-particle interactions and D_2 is governed by magnon-magnon interactions. Mathon and Wohlfarth[10] estimate, for a simple closed bandshape, that $D_1/D_0 \sim -10^{-6}\ K^{-2}$. If all the temperature dependence of the stiffness is assumed to come from a T^2 term the ratio D_1/D_0 for Pd_3Fe is found to be $(-2.6\pm0.4)\times10^{-6}\ K^{-2}$. For Fe, Stringfellow[11] found $D_1/D_0 = -5.4\times10^{-6}\ K^{-2}$. The temperature dependence of the stiffness of Pd_3Fe thus appears to be of the same magnitude as for other ferromagnetic metals and is consistent with estimates based on an itinerant electron model.

ACKNOWLEDGEMENTS

The author acknowledges the experimental assistance of G. Briggs, N. Clarke and J. Stanton and the invaluable advice of Drs. D.H. Saunderson, C.G. Windsor, M.W. Stringfellow and M.J. Zuckermann. He wishes to thank Dr. R.A. Cowley for many profitable discussions and for the loan of the crystal. The author was supported in part during the experiment by a Science Research Council (U.K.) Visiting Fellowship and the initial analysis of the results was performed while the author was a guest at the Department of Physics, University of Edinburgh.

REFERENCES

1. R.D. Lowde and C.G. Windsor, Advan. Phys. 19, 813-909 (1970).
2. H.A. Mook, R.M. Nicklow, E.D. Thompson and M.K. Wilkinson, J. Appl. Phys. 40, 1450-1 (1969).
3. R.M. Nicklow and H.A. Mook, Phys. Rev. B7, 336-42 (1973).
4. J.F. Cooke and H.L. Davis, AIP Conf. Proc. 10, 1218-37 (1973).
5. W.G. Stirling and R.A. Cowley, Solid State Commun. 11, 271-4 (1972).
6. H.A. Mook, J.W. Lynn and R.M. Nicklow, Phys. Rev. Letters 30, 556-9 (1973).
7. G. Dolling and V.F. Sears, Nucl. Instrum. Meth. 106, 419-28 (1973).
8. F. Leoni and C. Natoli, Nuovo Cimento 55B, 21-39 (1968).
9. M.J. Cooper and R. Nathans, Acta. Cryst. 23, 357-67 (1967).
10. J. Mathon and E.P. Wohlfarth, Proc. Roy. Soc. (London) A 302, 409-18 (1968).
11. M.W. Stringfellow, J. Phys. C, Ser. 2, 1, 950-65 (1968).

SINGLE-ION RESONANT STATES IN TWO SPIN WAVE SPECTRA*

S.T. Chiu-Tsao, C. Paulson and P.M. Levy
New York University,
New York, New York 10003

Quadrupole interactions affect the two spin wave spectrum of magnetic materials.[1] In addition to changing the binding energy of the exchange bound states these interactions give rise to a single-ion resonant state inside the two spin wave continuum. This resonance can be seen in the Raman two spin wave density of states. We have calculated the density of states curves for an isotropic spin-one ferromagnet for several ratios of quadrupole to dipole interaction constants for a linear chain and simple cubic lattice. As the ratio increases for ferroquadrupolar coupling we find that the resonance becomes sharper and more prominent in the density of states curve. When the quadrupole and dipole couplings are of equal strength, i.e. the spin-one Schrödinger Hamiltonian, the single-ion state has infinite lifetime. For sufficiently negative quadrupolar coupling bound states are found above the two spin wave continuum. At the zone center the weight of the bound state and its splitting from the continuum increases as the quadrupole coupling becomes increasingly negative.

*Research supported in part by the National Science Foundation under Grant No. GH 32422.

[1] D.A. Pink and P. Tremblay, Can. J. Phys. 50, 1728 (1972); D.A. Pink and R. Ballard, Can. J. Phys. 52, 33 (1974).

MAGNETIC EXCITATIONS IN Mn-DOPED CoF_2

E.C. Svensson, S.M. Kim and W.J.L. Buyers
Atomic Energy of Canada Limited, Chalk River, Ontario, Canada, K0J 1J0
S. Rolandson
AB Atomenergi, Studsvik, Fack, S-61101 Nyköping 1, Sweden
R.A. Cowley
University of Edinburgh, Edinburgh, Scotland
D.A. Jones
University of Aberdeen, Aberdeen, Scotland

ABSTRACT

The magnetic excitation spectra of $Co_xMn_{1-x}F_2$ single crystals having x=0.9 and 0.95 have been studied by neutron inelastic scattering. In both cases there is a branch of strongly perturbed "cobalt" excitations and a branch of lower frequency "manganese" excitations whose intensity is very strongly peaked at the zone centre. For x=0.9 (0.95) the zone-centre frequencies for the Co and Mn modes are 1.46±0.05 THz (1.23±0.02 THz) and 0.67±0.04 THz (0.81±0.02 THz) respectively. For x=0.95, separate shell modes associated with the Co neighbours of Mn impurities have been observed at frequencies ~2.25 THz near zone boundaries. Low concentration Green function theory, which adequately describes the behavior for low cobalt concentration, does not even qualitatively describe the current results. It predicts flat branches of undamped excitations for the manganese and shell modes whereas both have finite intrinsic widths and the manganese modes exhibit a marked frequency variation. Calculations based on the CPA theory of Buyers, Pepper and Elliott do, however, give a rather good description of the results.

EXPERIMENTAL RESULTS

In this paper we present the results of neutron-scattering measurements on specimens of CoF_2 doped with 5% and 10% MnF_2. The measurements on the 5% and 10% specimens were carried out respectively at Studsvik and Chalk River using triple-axis crystal spectrometers operating in the constant-momentum-transfer mode. In both cases the rutile-structure single-crystal specimens were oriented with a (010) axis vertical and excitations propagating along the a and c directions were studied in the antiferromagnetic phase at 4.2 K.

Scattered-neutron distributions observed at zone-centre and zone-boundary positions are shown in Fig.1. Much higher experimental resolution was used for $Co_{0.95}Mn_{0.05}F_2$ than for $Co_{0.9}Mn_{0.1}F_2$. The lower (higher) frequency peaks in the zone-centre distributions (B and C) are associated mainly with manganese (cobalt) ions. Note that the relative intensities of the two peaks depends critically on composition. For both materials, the intensity of the lower frequency peak falls off very rapidly as one goes away from the zone centre, and in $Co_{0.95}Mn_{0.05}F_2$ we were only able to follow the lower branch about half way to the zone boundary (see Fig.2).

Near zone boundaries in $Co_{0.95}Mn_{0.05}F_2$ we were also able to observe scattering by the shell modes associated with the cobalt neighbors to manganese impurities as illustrated by the weak higher frequency peak in distribution A of Fig.1. The stronger peak corresponds to the cobalt band mode. The very long counting time (≈3 hours/point, 2½ days total) and the large incoherent scattering by the cobalt ions accounts for the large background intensity. The resolution width was sufficiently small that distribution

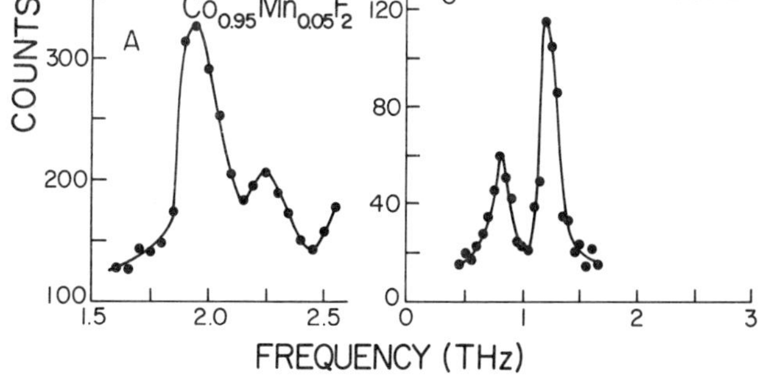

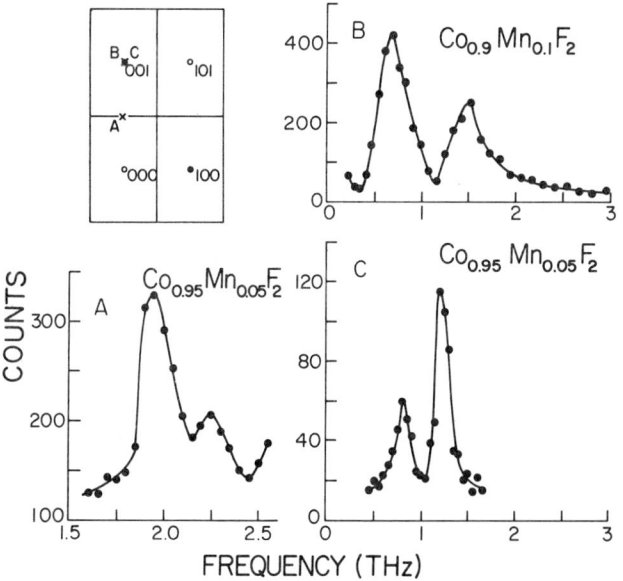

Fig. 1. Scattered-neutron distributions for $Co_{0.9}Mn_{0.1}F_2$ and $Co_{0.95}Mn_{0.05}F_2$ observed at zone-boundary (A) and zone-centre (B and C) positions as indicated on the reciprocal-lattice diagram. The curves are simply a guide to the eye.

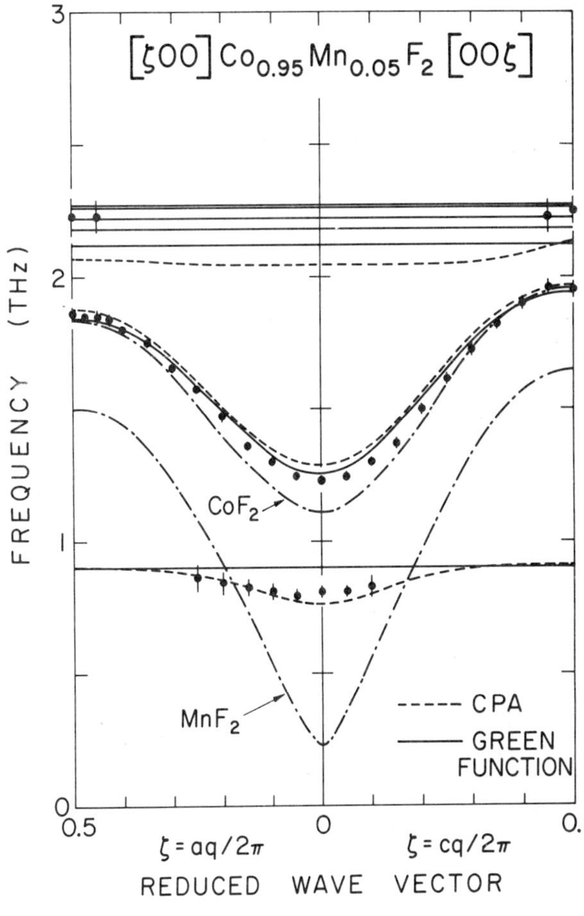

Fig. 2. Experimental frequencies (solid circles) of magnetic excitations in $Co_{0.95}Mn_{0.05}F_2$ and the predictions of CPA and Green-function theories. Dash-dot curves indicate the dispersion relations for CoF_2 (Ref.1) and MnF_2 (Ref.2).

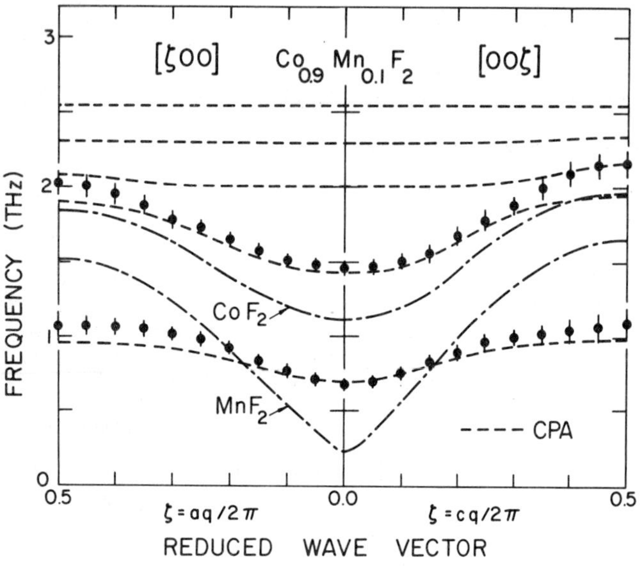

Fig. 3. Experimental frequencies (solid circles) of magnetic excitations in $Co_{0.9}Mn_{0.1}F_2$ and the predictions of CPA theory. Dash-dot curves indicate the dispersion relations for CoF_2 (Ref.1) and MnF_2 (Ref.2).

A is close to the real lineshape. Separate shell modes were not observed for $Co_{0.9}Mn_{0.1}F_2$. The higher frequency peaks for this material do, however, have long high-frequency tails and may correspond to scattering by both shell and band modes which could not be separated by the resolution employed.

The experimental results for $Co_{0.95}Mn_{0.05}F_2$ are summarized in Fig.2. For comparison we also show the dispersion relations[1,2] for CoF_2 and MnF_2. The lower branch of manganese excitations appears to have an appreciable bandwidth in contrast to the flat branch of localized cobalt excitations observed[3] at 3.57 THz for 5% CoF_2 in MnF_2. In Mn-doped CoF_2 the defect modes interact more strongly with the host band modes than in Co-doped MnF_2 because they lie closer together in frequency[4]. Hence the modes associated with manganese impurities in CoF_2 are less localized and overlap to form a band at a much lower impurity concentration than do the highly localized modes associated with cobalt impurities in MnF_2. As expected, we see in Fig.2 that those cobalt band modes which lie closest to the manganese modes, i.e. the modes near the zone centre, are the most strongly perturbed.

The results for $Co_{0.9}Mn_{0.1}F_2$ are summarized in Fig.3. As expected, there is a larger perturbation of the cobalt band modes than for $Co_{0.95}Mn_{0.05}F_2$ and the manganese modes have a substantially larger bandwidth.

THEORY AND DISCUSSION

We have compared two theories with the experimental results. One is first-order Green-function theory[4] which is exact in the limit of low impurity concentration. The other theory is valid for all concentrations and is based on the coherent-potential approximation (CPA) developed by Buyers et al.[5] for antiferromagnets. This theory has previously been shown[5] to give a reasonably good description of the experimental results[6,7] for $Co_{0.7}Mn_{0.3}F_2$. The only free parameter in the theories is the Co-Mn antiferromagnetic exchange interaction and this is chosen so that each theory correctly predicts the experimentally determined[8,9] frequency of the local mode in Co-doped MnF_2 in the dilute limit. Since the orbital angular momentum of the cobalt ion is not quenched, it is necessary to calculate the mixing of the cobalt levels by the composition-dependent exchange field of the surrounding cobalt and manganese ions.

First-order Green-function theory, which gave an excellent description[4,7] of the results[3] for $Co_{0.05}Mn_{0.95}F_2$, is not even qualitatively correct for $Co_{0.95}Mn_{0.05}F_2$ and $Co_{0.9}Mn_{0.1}F_2$. It predicts flat branches (see Fig.2) of localized δ-function excitations for the manganese and shell modes whereas both types of modes are observed to exhibit finite intrinsic widths and the manganese modes have a finite bandwidth. Note that this theory gives all the s,p,d and f shell modes expected from symmetry (see Fig.2 and Ref.4). Both theories give frequencies for the cobalt band modes in $Co_{0.95}Mn_{0.05}F_2$ that are slightly too high near the zone centre (see Fig.2), but this may not be significant since comparison with the AFMR results of Enders et al.[10] suggests that the manganese concentration in our specimen may be somewhat less than the nominal 5%.

The main success of the CPA theory is in predicting approximately the correct bandwidth for the manganese modes for both 5% (Fig.2) and 10% (Fig.3) manganese concentration. The CPA theory predicts only one s-like shell mode. For $Co_{0.95}Mn_{0.05}F_2$ this mode has a finite intrinsic width as observed but has a frequency somewhat lower than observed (Fig.2). The shell mode is only one of a series of resonances predicted by the CPA theory for cobalt ions surrounded by n=1,2,3...8 antiferromagnetically-coupled manganese neighbors. By convention, the term shell mode is used for the n=1 resonance. The dispersion curves for n=1,2 and 3 are shown in Fig.3 whereas only the curve for n=1 is shown in Fig.2. The theory in fact predicts that at the c-axis zone-boundary position in $Co_{0.9}Mn_{0.1}F_2$ the n=2 mode is as strong as the cobalt band mode and the n=1 mode is more than twice as strong. No evidence for this multiple-peaked structure was obtained in the experiment. Measurements at much higher experimental resolution are required to determine whether or not the multiple-peaked structure predicted by the CPA theory and other theories[11] really exists.

ACKNOWLEDGMENTS

We would like to thank R. Campbell, H. Nieman, M. Potter, D. Tennant, K.O. Isaxon and S. Sandell for valuable technical assistance. One of us (ECS) is grateful to the staff of AB Atomenergi, in particular R. Stedman and R. Pauli, for their hospitality and assistance during his stay at Studsvik.

REFERENCES

1. P. Martel, R.A. Cowley and R.W.H. Stevenson, Can. J. Phys. 46, 1355-70 (1968).
2. G.G. Low, A. Okazaki, R.W.H. Stevenson and K.C. Turberfield, J. Appl. Phys. 35, 998-9 (1964).
3. W.J.L. Buyers, R.A. Cowley, T.M. Holden and R.W.H. Stevenson, J. Appl. Phys. 39, 1118-9 (1968); T.M. Holden, R.A. Cowley, W.J.L. Buyers and R.W.H. Stevenson, Solid State Commun. 6, 145-7 (1968).
4. For a detailed discussion of the application of Green function theory to magnetic-impurity problems see R.A. Cowley and W.J.L. Buyers, Rev. Mod. Phys. 44, 406-50 (1972).
5. W.J.L. Buyers, D.E. Pepper and R.J. Elliott, J. Phys. C 5, 2611-28 (1972).
6. W.J.L. Buyers, T.M. Holden, E.C. Svensson, R.A. Cowley and R.W.H. Stevenson, Phys. Rev. Lett. 27, 1442-5 (1971).
7. E.C. Svensson, W.J.L. Buyers, T.M. Holden, R.A. Cowley and R.W.H. Stevenson, AIP Conf. Proc. No.5, pp. 1315-33 (1972).
8. R. Weber, J. Appl. Phys. 40, 995-6 (1969).
9. G. Parisot, S.J. Allen, Jr., R.E. Dietz, H.J. Guggenheim, R. Moyal, P. Moch and C. Dugautier, J. Appl. Phys. 41, 890-1 (1970).
10. B. Enders, P.L. Richards, W.E. Tennant and E. Catalano, AIP Conf. Proc. No.10, pp. 179-83 (1973).
11. R.J. Elliott, J.A. Krumhansl and P.L. Leath, Rev. Mod. Phys. 46, 465-543 (1974).

STATISTICAL THEORY OF MAGNETIC EXCITATIONS IN HEISENBERG PARAMAGNETS

S. H. Liu and P. A. Swanson
Iowa State University, Ames, Iowa 50010

ABSTRACT

We propose a statistical method to calculate the dynamical correlation function $G(\vec{q},t) = \langle \vec{S}_{\vec{q}}(t) \cdot \vec{S}_{-\vec{q}}(0) \rangle$ for the Heisenberg model above the ordering temperature. The function may be expanded into a power series of the time and the coefficients are expressed in terms of static multi-spin correlation functions. In the limit $\lambda > q^{-1} > r$, where λ is the spin correlation length and r is the range of interaction, the dominant contributions to the coefficients come from spins within correlated clusters. Therefore, we treat the dynamics of such a cluster by the spin-wave theory for an infinite crystal, then weigh the contribution of each cluster to $G(\vec{q},t)$ according to its probability. The result is that the excitations in the range $q\lambda > 1$ are magnon-like modes broadened by a mean-free-path measured by λ. In addition, there is a central peak at zero frequency, which is the diffused remnant of the magnetic Bragg peak. The theory compares favorably with the neutron scattering data on TMMC and nickel.

There exists a wealth of inelastic neutron scattering data on the magnetic excitations above the critical temperature of magnetic systems.[1-5] The nature of the results was first described qualitatively by Marshall.[6] In the long wavelength case, i.e. $q^{-1} \gg \lambda$, the magnetic excitation appears as a peak in scattering cross-section around zero energy transfer. The spin diffusion theory[7] and the more refined dynamical scaling theory[8] give a good description of the line shape. In the short wavelength case the observed excitations are broadened magnon-like modes. There has been a great deal of theoretical interest in this problem.[9-11]

In this paper we present a statistical theory for the magnetic excitations in the short wavelength region. The inelastic neutron scattering cross-section is determined by the Fourier transform of the dynamical correlation function[12]

$$G(\vec{q},t) = \langle \vec{S}_{-\vec{q}}(t) \cdot \vec{S}_{\vec{q}}(0) \rangle = \sum_j \langle \vec{S}_i(t) \cdot \vec{S}_j(0) \rangle e^{i\vec{q}\cdot\vec{R}_{ij}} \quad (1)$$

The two spin correlation function, which we denote by $G_{ij}(t)$, may be expanded into a power series of t, i.e.

$$G_{ij}(t) = \sum_{n=0}^{\infty} G_{ij}^{(n)} (-it)^n/n!, \quad (2)$$

where the coefficients are

$$G_{ij}^{(0)} = \langle \vec{S}_i \cdot \vec{S}_j \rangle,$$

$$G_{ij}^{(1)} = \langle [\vec{S}_i, H] \cdot \vec{S}_j \rangle,$$

$$G_{ij}^{(2)} = \langle [\vec{S}_i, H] \cdot [H, \vec{S}_j] \rangle, \quad (3)$$

etc., and $H = -\sum_{i,j} J_{ij} \vec{S}_i \cdot \vec{S}_j$ is the Heisenberg Hamiltonian. The commutator of $\vec{S}_i$ with H brings in a group of spins $\vec{S}_\ell$ that are coupled to $\vec{S}_i$ by the interaction $J_{i\ell}$. Thus, the coefficients involve static two, three, four and more spin correlation functions. From Eq. (1) one can see that the important contribution to $G(\vec{q},t)$ comes from spins i, j that are less than a wavelength apart. In the case $\lambda > q^{-1} > r$, where r is the range of J_{ij}, all the spins in the leading coefficients are within a correlated cluster of spins. Therefore, the much used pair-wise decoupling scheme, which assumes negligible correlation between pairs of spins, is not a good method to approximate these static multi-spin correlation functions.

We suggest a more physical approximation as follows. Let $\rho(i,j)$ be the probability that $\vec{S}_i$ and $\vec{S}_j$ are correlated, and $1 - \rho(i,j)$ the probability that they are not correlated. We may divide $G_{ij}^{(n)}$ into two parts.

$$G_{ij}^{(n)} = G_{ij}^{(n)} \rho(i,j) + G_{ij}^{(n)} [1 - \rho(i,j)] = G_{ij}^{(n)}(1) + G_{ij}^{(n)}(2). \quad (4)$$

The probability function may be identified as the normalized twp spin static correlation function

$$\rho(i,j) = \langle \vec{S}_i \cdot \vec{S}_j \rangle / S(S+1). \quad (5)$$

In $G_{ij}^{(n)}(1)$ all the spins are correlated, so we approximate the multi-spin correlation function by the spin-wave theory. In $G_{ij}^{(n)}(2)$ the spins $\vec{S}_i$ and $\vec{S}_j$ are uncorrelated, so we apply the pair-wise decoupling sheme. By doing a few model calculations we were able to show that the second term makes a very small contribution in the short wavelength region. Thus, we approximate the entire coefficient $G_{ij}^{(n)}$ by the first term. The calculation is very easy to carry out, for example, for a simple ferromagnet

$$G_{ij}^{(1)} = \frac{S}{N}\sum_{\vec{q}'} [\omega_{q'}(n_{q'}+1) - \omega_{q'} n_{q'}] e^{i\vec{q}'\cdot\vec{R}_{ij}} \rho(i,j)$$

$$+ \frac{1}{2N^2} \sum_{\vec{q}'\vec{q}''} [(\omega_{q'} - \omega_{q''}) n_{q''}(n_{q'}+1)$$

$$+ (\omega_{q''} - \omega_{q'}) n_{q'}(n_{q''}+1)] e^{i(\vec{q}' - \vec{q}'')\cdot\vec{R}_{ij}} \rho(i,j). \quad (6)$$

In the above equation we adopt the notation $\omega_q = 2S[J(0) - J(\vec{q})]$, $J(\vec{q}) = \sum_j J_{ij} e^{i\vec{q}\cdot\vec{R}_{ij}}$ and $n_q = [e^{\beta\omega_q} - 1]^{-1}$. The group of terms involving one sum on the wavevector come from the transverse part of the correlation function, and those involving double sums come from the longitudinal part. These may be recognized as one-magnon and two-magnon terms. There are also three and more magnon terms, but they are neglected. In this manner all the coefficients in Eq. (2) may be calculated. After resummation and Fourier transformation, we obtain

$$G(\vec{q},\omega) = 2\pi[S^2 - \frac{2S}{N}\sum_{\vec{q}'} n_{q'}] \rho(\vec{q})\, \delta(\omega)$$

$$+ \frac{2\pi S}{N} \sum_{\vec{q}'} [n_{q'} \delta(\omega + \omega_{q'}) + (n_{q'}+1)\, \delta(\omega - \omega_{q'})] \rho(\vec{q} - \vec{q}')$$

$$+ \frac{\pi}{N^2} \sum_{\vec{q}'\vec{q}''} [n_{q'}(n_{q''}+1)\, \delta(\omega + \omega_{q'} - \omega_{q''})$$

$$+ n_{q''}(n_{q'}+1)\, \delta(\omega - \omega_{q'} + \omega_{q''})] \rho(\vec{q} - \vec{q}' + \vec{q}'') \quad (7)$$

where $\rho(\vec{q})$ is the Fourier transform of $\rho(i,j)$.

The result in Eq. (6) has a simple meaning. Since $\rho(i,j)$ is the probability of finding a cluster of spins of size R_{ij}, the result is an average of contributions from all clusters, each weighted by its probability. The contribution from each cluster is estimated by the spin-wave theory for an infinite crystal.

The first term in Eq. (7) is a central peak at $\omega = 0$. It comes from the longitudinal correlation function, and it is simply the remnant of the elastic Bragg peak which is diffused in space due to spin disorder. In our approximation scheme the peak has zero width in frequency. The one and two-magnon terms are all broadened by convolution with $\rho(\vec{q})$. Physically the width of the central peak measures the decay time of the local order parameter.[9] We may append the theory at this point by adding a spin diffusion factor $\exp(-\Lambda q^2 t)$ to the central peak term. This amounts to a separation of the long time behavior (spin diffusion) from the short time behavior (magnon oscillation) of the system. In our simple theory we deal successfully with the latter effect at the expense of the former. We will discuss this point in detail in another publication.

There are a few inherent uncertainties in our theory. 1. With each succeeding commutation of $\vec{S}_i$ with H, a ring of more distant spins is brought into the correlation function. So the approximation that all spins in a correlation function are within a correlation length must break down at a high enough order of $G_{ij}^{(n)}$. The problem is most acute when $q\lambda \cong 1$. 2. The temperature within a cluster is taken to be the same as that of the whole crystal. This is not justified when the clusters are small and the local temperature fluctuation is large. 3. The magnon-magnon interaction is ignored. This means that as soon as T_c is reached the magnon linewidth becomes zero. It is hoped that all these defects may be removed by more careful formulations.

Another consequence of our approximation scheme is that the theory breaks down at the infinite temperature limit.[12,13] We regard this feature as a strength of our theory rather than a weakness. For instance, as shown by McLean and Blume,[9] a downward extrapolation of the infinite temperature theory failed to explain the behavior of the one-dimensional antiferromagnet TMMC, but an extrapolation from the T = 0 theory worked quite well.

Similar to the other theories above T_c, our theory requires a detailed knowledge of the static two-spin correlation. For Ni we used the Gaussian form $\rho(\vec{q}) \propto \exp(-\lambda^2 q^2)$ where $\lambda^2 = 2D/k(T-T_c)$, D is the spin-wave stiffness. This resulted in distinctly resolved peaks with linewidths plotted in Fig. 1. On the same graph we show the experimental data with the instrumental resolution subtracted out.[4] The theoretical line has the same slope as the experimental curve near T_c. The rise of linewidth below T_c is not reproduced by reason discussed earlier. The saturation of linewidth above 800K arises when $\lambda \le r$. We will show in another publication that a simple correction of our theory accounts for this effect.

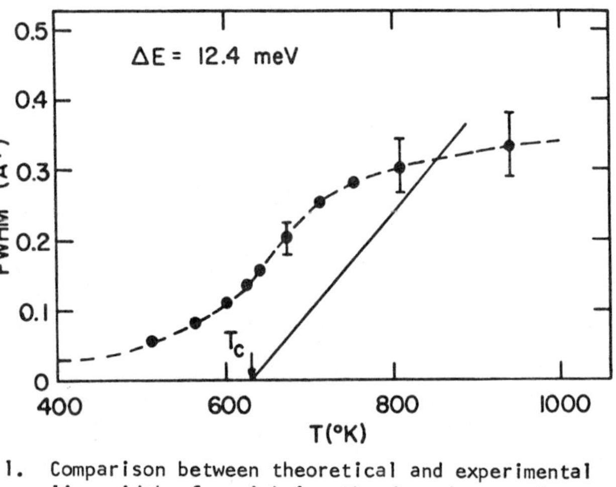

Fig. 1. Comparison between theoretical and experimental line widths for nickel. The dotted curve is drawn through the data points. The straight line through T_c is the theoretical result. The instrumental resolution has been subtracted from the data.

For one-dimensional systems the critical temperature is suppressed to 0 K by geometry. So our assumption of perfect order at T_c is justified. Furthermore, the two-spin correlation function is known for the classical model,[15] and the result works well for TMMC.[3] The one-magnon part of the dynamical correlation function for an antiferromagnetic chain is

$$G(q,\omega) = \frac{2\pi S}{N} \sum_{q'} [(n_{q'}+1)\delta(\omega_{q'}-\omega) + n_{q'}\delta(\omega_{q'}+\omega)] \times [A_{q'}^{(+)}\rho(q+q'-\pi) + A_{q'}^{(-)}\rho(q+q')], \quad (8)$$

where $A_q^{(\pm)} = \cot q \pm \csc q$, the wave vector q is measured in units of $(c)^{-1}$, c being the lattice parameter. The sum on q' is carried out in the magnetic Brillouin zone. In Fig. 2 we show the theoretical line shape with the experimental data. The theoretical curves were calculated from Eq. (6) and convoluted with a resolution function of width 0.3 meV.[3] For the experimental value of $q = 0.1(\pi/c)$ the area under the central peak was estimated to be negligible. One can see that the observed line shapes are better reproduced than previous theories.[9,10,16]

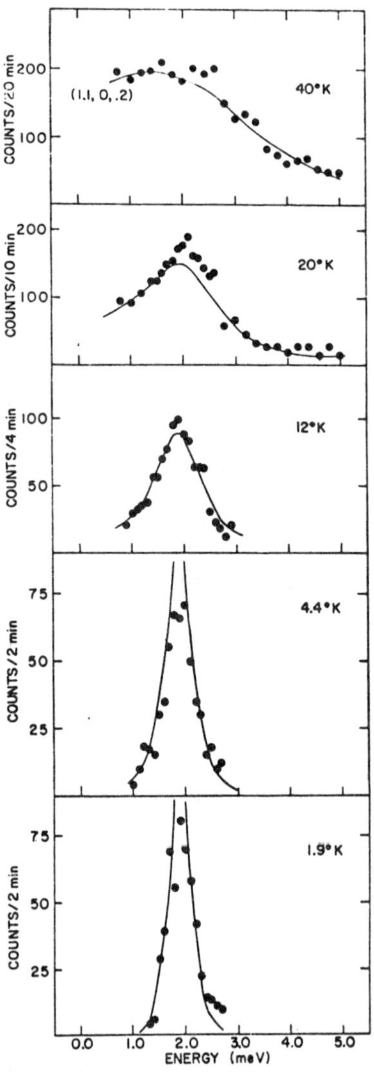

Fig. 2. Comparison between theoretical and experimental line shapes for TMMC. The solid curves represent theoretical line shapes convoluted with a resolution function of 0.3 meV wide. The momentum transfer is $q = 0.1$ (π/c).

REFERENCES

1. Minkiewicz, Collins, Nathan, and Shirane, Phys. Rev. 182, 624 (1969), and references cited therein.
2. Nathan, Menzinger and Pickart, J. Appl. Phys. 39, 1237 (1968).
3. Hutchings, Shirane, Birgeneau, and Holt, Phys. Rev. B5, 1999 (1972).
4. Mook, Lynn and Nicklow, Phys. Rev. Lett. 30, 556 (1973).
5. Mook, Lynn and Nicklow, AIP Conf. Proc. No. 18, 781 (1974).
6. Marshall, Conference on Critical Phenomena, (NBS Misc. Publications 273, 1965), p. 135.
7. Mori and Kawasaki, Prog. Theor. Phys. 27, 529 (1962).
8. Halperin and Hohenberg, Phys. Rev. 177, 952 (1969).
9. McLean and Blume, Phys. Rev. B7, 1149 (1973), and references cited therein.
10. Tomita and Hashiyama, Prog. Theor. Phys. 48, 1133 (1972), and references cited therein.
11. Lovesey and Meserve, J. Phys. C6, 79 (1972).
12. van Hove, Phys. Rev. 95, 249 (1954).
13. Reiter, Phys. Rev. B7, 3325 (1973).
14. Myles and Fedder, Phys. Rev. B9, 4872 (1974).
15. Fisher, Am. J. Phys. 32, 343 (1964).
16. Scales and Gersch, Phys. Rev. Lett. 28, 917 (1972).

SPIN WAVES IN THE LAVES COMPOUND $Ho_{.88}Tb_{.12}Fe_2$

R. M. Nicklow
Solid State Division, Oak Ridge National Laboratory,* Oak Ridge, Tennessee 37830

N. C. Koon, C. M. Williams, and J. B. Milstein
Naval Research Laboratory, Washington, D. C. 37830

ABSTRACT

The energies of spin waves in the ferrimagnetic cubic Laves compound $Ho_{.88}Tb_{.12}Fe_2$ (T_C = 610°K) have been investigated by neutron spectrometry. The measurements were carried out at room temperature for wave vectors $\vec{q}$ in the [100], [110], and [111] principal symmetry directions. At this temperature the alloy studied exhibits small magnetic anisotropy. Thus the acoustic branch of the dispersion relation is characteristic of that expected for a cubic ferrimagnet with only isotropic Heisenberg exchange interactions, i.e. it possesses a negligible energy gap (< 0.2 meV) and for small q it is isotropic and shows nearly quadratic dispersion, $\hbar\omega \cong Dq^2$ with D = 84 meV Å^{-2}. The lowest energy optic branch at about 7.5 meV shows very little dispersion and is nearly degenerate with the acoustic branch at the zone boundaries. Qualitative considerations of the scattering structure factors suggest that this optic branch is primarily associated with the rare-earth sublattice, while the remaining four optic branches, all above 15 meV, are primarily associated with the iron sublattice. One of these optic branches is particularly interesting in that it shows extremely strong dispersion, similar to that exhibited by the spin waves in pure iron. Preliminary calculations based on a simple near-neighbor exchange model indicate that such a model is inadequate to describe the experimental results.

*Operated by Union Carbide Corporation for the U.S.A.E.C.

THEORETICAL MAGNON DISPERSION CURVES FOR Gd*

B. N. Harmon
Ames Lab.-USERDA, Iowa State Univ., Ames, Iowa 50010
P.-A. Lindgård
DAEC, Risø, 4000 Roskilde, Denmark
A. J. Freeman
Northwestern Univ., Evanston, Ill. 60201 and
Argonne National Lab., Argonne, Ill. 60439
J. Rath
Physics Department, Northwestern Univ., Evanston, Ill. 60201

ABSTRACT

The indirect (RKKY) exchange matrix elements $j(\bar{k},\bar{k}')$ between the conduction electrons and the local 4f moments have been calculated for Gd using nonrelativistic APW wavefunctions and assuming an unscreened Coulomb interaction. The matrix elements exhibit[1] a great deal of structure as a function of $\bar{k}$ and $\bar{k}'$ and are largest for d-like conduction electrons. The matrix elements and energy eigenvalues are used to calculate the magnon spectrum of Gd along the Γ-A symmetry direction. The calculation demonstrates: (1) the need to include the $\bar{k}$ and $\bar{k}'$ dependence of the matrix elements, and (2) the importance of including transitions from all bands below the Fermi energy to bands well above. The dispersion curves are found to be rather insensitive to fine details in the band structure. Comparison of the theoretical results with various experiments indicates agreement is obtained if the matrix elements are reduced by a constant scale factor of about two, a reduction which may be due to screening effects not taken into account in our calculations.

[1]B. N. Harmon, A. J. Freeman, Phys. Rev. B15 (to be published).

*Supported in part by AEC, DAEC, NSF, and AFOSR.

ANTIFERROMAGNETIC RESONANCE STUDIES OF $NaMnF_3$ +

Geoffrey O. White and Robert L. White
Stanford Electronics Laboratories
Stanford University, Stanford, California

ABSTRACT

Antiferromagnetic resonance experiments have been conducted on $NaMnF_3$ at frequencies from 31 to 37 GHz in magnetic fields up to 18 K Oe, and at temperatures ranging from 4.2°K to 67°K. The principal observed resonances may be qualitatively explained by the simple antiferromagnetic model as modified by the addition of antisymmetric exchange and/or crystalline anisotropic canting terms. In addition to antiferromagnetic resonances, several highly structured resonances (probably not antiferromagnetic) were observed for temperatures below 10°K.

INTRODUCTION

$NaMnF_3$ is an orthorhombic perovskite with a_0 = 5.57 Å, b_0 = 5.74 Å and c_0 = 7.99 Å(1). It has been determined to be a canted antiferromagnet with a Néel temperature of 66.7°K and a canting angle of 0.18°(2), approximately one-third that of $YFeO_3$ with which it is believed to be isostructural and isoelectronic. Our interest in the antiferromagnetic resonance of $NaMnF_3$ derives from a desire to have an experimental determination of the spin wave spectrum of a canted antiferromagnet because we are interested in the role such spin waves play in domain wall relaxation. The low Néel temperature and small anisotropy of $NaMnF_3$ mean that its antiferromagnetic resonance spectrum falls in an experimentally accessible region of the microwave spectrum, in contrast with that of $YFeO_3$ which probably falls in the near IR. We propose both to scale the $NaMnF_3$ results to $YFeO_3$ and to perform domain wall velocity measurements on $NaMnF_3$.

EXPERIMENTAL RESULTS

The resonance experiments were conducted at fixed frequencies of 31.7, 35.17 and 37.25 GHz in a magnetic field which could be swept from 0 to 18 K Oe. X-ray oriented single crystal $NaMnF_3$ spheres were placed in a TE-011 microwave cavity such that the rf magnetic field was perpendicular to the biasing quasi-static field. The temperature of the cavity could be raised from 4.2°K to above the Néel temperature and controlled to within 0.1°K using a Au-Fe vs Cu thermocouple in conjunction with a feedback heater system.

Data was taken with the DC field in the a-b plane (rf h along c) and with the DC fields in the a-c plane (rf h along b). The data at several temperatures for H in the a-c plane are shown in Figure 1. The data for H along a at two frequencies are plotted from 4.2°K to the Néel temperature in Figure 2. Since the positions of the resonances were strongly temperature-dependent, it was necessary to run the experiments at very low

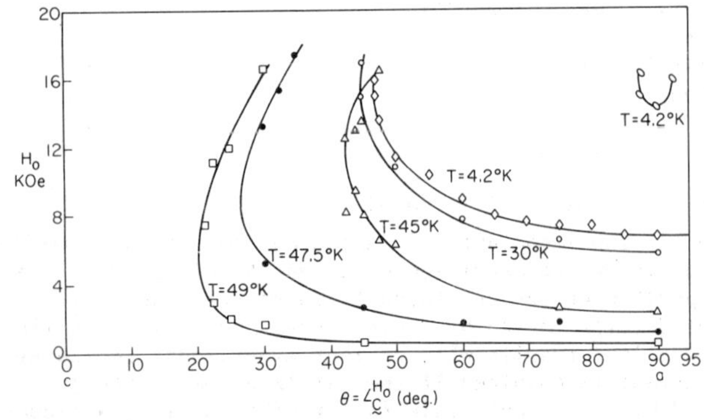

Figure 1. DC bias field (Ho) required for antiferromagnetic resonances in $NaMnF_3$ at35.17 GHz with Ho in the a-c plane, and with the rf magnetic field (h) in the b direction.

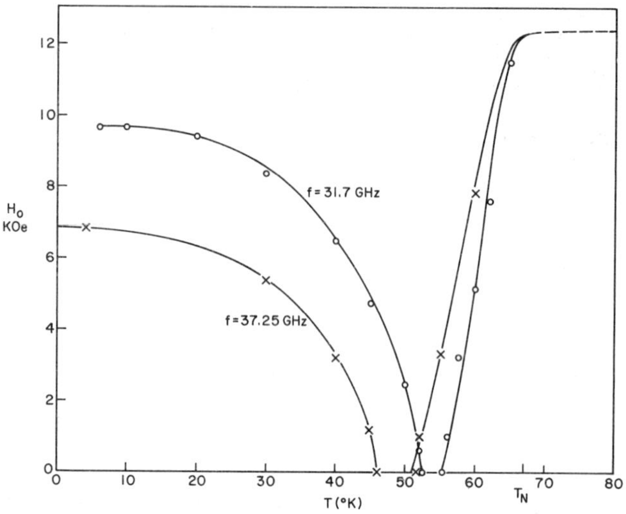

Figure 2. DC bias field (Ho) required for antiferromagnetic resonance in $NaMnF_3$ at the frequencies shown, with Ho in the a direction and with the rf magnetic field (h) in the b direction.

power levels to prevent sample heating by the microwave power absorbed. The data plotted are the dominant strong absorptions. In addition to these, we have, in some samples, observed some very weak, highly structured (five to seven component) lines. These lines decreased in strength and resolution as the temperature was increased from 4°K, finally disappeaing altogether at about 10°K. Although we cannot presently account for these lines, we feel that their presence is not relevant to our explanation of the strong resonance behavior, and will not discuss them further in this communication.

DISCUSSION

The data can be qualitatively understood by reference to Figure 3 which plots the resonant frequencies of the system versus the magnitude of applied field along the a axis. Figure 3 is based on curves by Cinader(3) for a similar but simpler situation. The curves resemble those for a simple antiferromagnet except that the antisymmetric canting fields and the canting anisotropy fields lift the twofold degeneracy

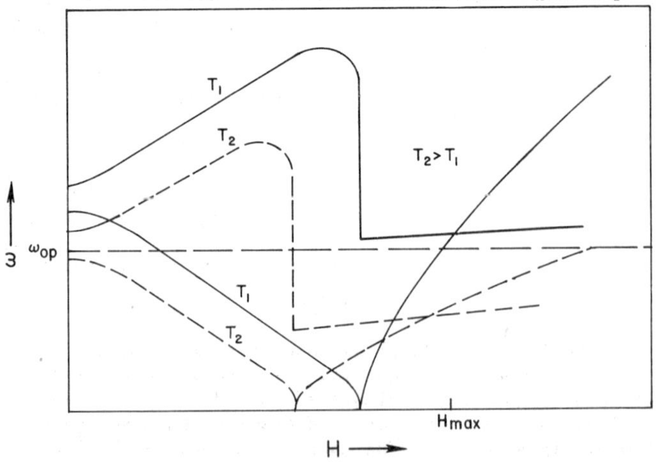

Figure 3. Proposed dependence of antiferromagnetic resonance frequency (ω) on DC bias field (H) in $NaMnF_3$ with Ho applied in the easy (a) direction.

of the zero-field resonance and also convert the discontinuous spin-flop of the antiferromagnetic case to a continuous rotation in the canted case.

For H along the b and c axes both resonances are expected to tune upward in frequency[4] with applied field and will be observed only when the zero-field frequency is below the observation (microwave) frequency. We believe the data to be understood as follows:

For temperatures below approximately 50°K both zero-field resonances are above the observation frequency, and absorptions are observed only for applied fields near the a axis and only the downward tuning branch of the resonance is observed, perhaps twice, once before and once after the spin reorientation (the somewhat speculative downward dip shown in Figure 3 of the higher frequency mode near the reorientation was never observed). For a short temperature interval near 50°K, with the exact position of the interval depending on the observation frequency, the observation frequency is between the two modes and no resonance is observed over the range of available fields. Above about 52°K the zero-field frequencies of both modes fall below the observation frequency, and the upward tuning higher mode is observed, essentially at all field orientations.

We are currently engaged in a theoretical calculation of the resonance frequencies of the canted antiferromagnet with H applied along the a direction using the forms of the crystalline anisotropy suggested by Herrmann[4]

$$-F_K = A_{XX}(X_1^2+X_2^2) + A_{YY}(Y_1^2+Y_2^2) + A_{ZZ}(Z_1^2+Z_2^2) + A_{XZ}(X_1Z_1-X_2Z_2)$$

where $(X_i, Y_i, Z_i) \equiv \underline{M}_i/M$.

The equations are such that computer solution is necessary, but the final result will be a quantitative and appropriate version of Figure 3, which may then be compared quantitatively with these and further experiments.

We wish to thank Robert Raymakers for growing the crystals used in these experiments and Dr. G. F. Herrmann for illuminating discussions of our results.

REFERENCES

+ This work supported by the Air Force Office of Scientific Research through Grant AFOSR-72-2260C and by the National Science Foundation through the Center for Materials Research at Stanford University.

1. Shane, J. R., Lyons, D. H., Kestigan, M. JAP, Vol. 38 Number 3, pp. 1280-1282, 1967.

2. Maartense, I., Intern. J. Magnetism, Vol. 2, pp.117-122, 1972.

3. Cinader, G., Phys. Rev., Vol. 155, #2, pp. 453-457, 1967.

4. Herrmann, G. F., J. Phys. Chem. Solids, Vol. 24, pp. 597-606, 1963.

AFMR VERSUS ORIENTATION IN WEAKLY FERROMAGNETIC $BaMnF_4$*

E. L. Venturini† and F. R. Morgenthaler
Department of Electrical Engineering and Center for Materials Science and Engineering
Massachusetts Institute of Technology, Cambridge, Massachusetts 02139

ABSTRACT

The angular dependence of the antiferromagnetic resonance (AFMR) spectrum in barium manganese fluoride has been studied at 4.2 °K. The data reveal a splitting of one AFMR mode into two distinct modes as the applied static magnetic field is rotated away from the easy (b) axis in the (100) plane. No splitting is observed for magnetic fields in the (001) plane.

This behavior has been explained using a two-magnetic-sublattice model which includes a Dzyaloshinsky-Moriya (D-M) field along the crystal a axis. This field causes a canting of the otherwise antiparallel sublattices producing a weak ferromagnetic moment along the c axis. This canting may be masked in static magnetization measurements by the presence of domains with oppositely directed weak moments. The AFMR frequency differs in these two types of domains when the static magnetic field has a component along the c axis. This frequency splitting offers a new and sensitive measure of the strength of the D-M field.

MAGNETIC FREE ENERGY DENSITY

Transparent barium manganese fluoride $BaMnF_4$ is orthorhombic at room temperature (point group 2mm) and is strongly piezoelectric.[1] It has low microwave frequency acoustic attenuation except near a crystallographic phase transition at 255 °K (the symmetry of the low temperature phase is not known).[2] It orders antiferromagnetically at 25 °K with the crystal b axis as the magnetic easy direction.[3]

We have studied the antiferromagnetic resonance (AFMR) spectrum of $BaMnF_4$ at 4.2 °K to determine its magnetic properties as a first step in understanding elastic and magnetic wave interactions in this unique material. A previous AFMR study using magnetic fields applied along the three crystal axes concluded that $BaMnF_4$ is an orthorhombic antiferromagnet below its ordering temperature.[4] The present study which includes the angular dependence of the AFMR spectrum indicates that $BaMnF_4$ is actually a weak ferromagnet with a domain structure masking the net moment.

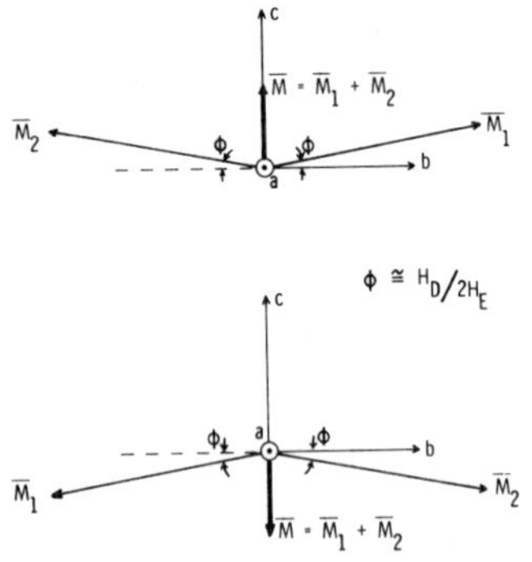

Figure 1. Energetically equivalent sublattice configurations for zero applied magnetic field.

The AFMR data led us to choose a magnetic free energy density which has the two equilibrium magnetic sublattice configurations shown in Fig. 1. The two magnetizations are nearly antiparallel in zero applied field but are slightly canted away from the b axis toward the c axis in the (100) plane (angle ϕ is three milliradians in $BaMnF_4$ at 4.2 °K). The free energy density F_m (appropriate to magnetic point group 2m'm') to second order in the magnetizations is given by[5]

$$F_m M_o = H_E \vec{M}_1 \cdot \vec{M}_2 - H_{A1} M_{1a} M_{2a} - H_{A2} M_{1c} M_{2c} - H_D \hat{a} \cdot (\vec{M}_1 \times \vec{M}_2)$$

where M_o is the magnitude of the two magnetizations $\vec{M}_1$ and $\vec{M}_2$ at zero applied field (their magnitudes are allowed to differ in nonzero fields through a parallel magnetic susceptibility). H_E is the isotropic exchange field, H_{A1} and H_{A2} are the anisotropy fields along the a and c axes, respectively, and H_D is the Dzyaloshinsky-Moriya (D-M) field which points along the a axis. This latter field produces the two energetically equivalent canted configurations shown in Fig. 1. $BaMnF_4$ is assumed to contain domains corresponding to both configurations.

AFMR IN A WEAK FERROMAGNET

Fig. 2 shows the measured and calculated AFMR spectrum (the lowest frequency mode) in $BaMnF_4$ at 4.2 °K when the magnetic field is applied along the easy (b) axis. The two branches correspond to resonance below and above the spin flop transition at 10 kOe.

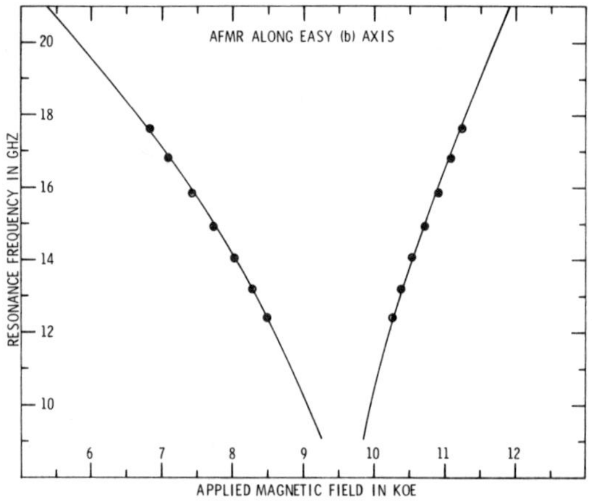

Figure 2. AFMR in $BaMnF_4$ along (010) axis at 4.2 °K.

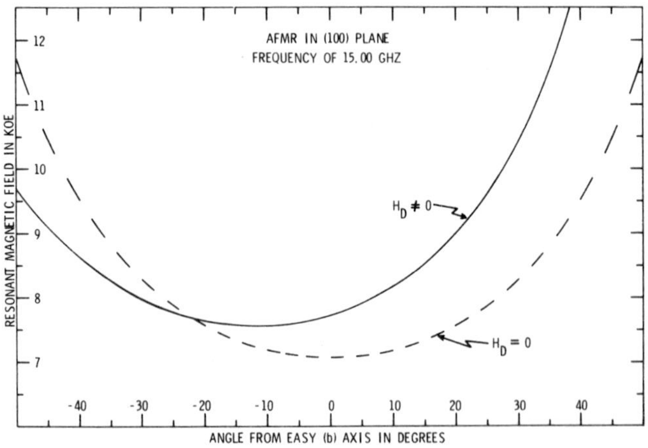

Figure 3. Effect of D-M field on AFMR in (100) plane.

If the frequency is fixed at 15 GHz and the crystal is rotated about its a axis, the magnetic field for resonance has the angular dependence shown in Fig. 3 using the free energy density given above. The dashed curve depicts the behavior in an orthorhombic antiferromagnet with the same anisotropy fields as $BaMnF_4$ but with $H_D = 0$. Note the symmetry about the b axis (zero degrees). This symmetry is broken if $H_D \neq 0$, and the minimum field for resonance occurs at a nonzero angle. The solid curve corresponds to AFMR in a domain with its net moment along the +c axis. A domain with its net moment along -c has the same response with the sign of the angle on the horizontal axis reversed.

AFMR data in the (100) plane of $BaMnF_4$ at 12.40 and 17.62 GHz are shown in Figs. 4 and 5. The solid lines were calculated using the free energy density above and the two sublattice configurations in Fig. 1 (the applied magnetic field alters these configurations slightly). The splitting of the lowest AFMR mode in the (100) plane provides a new and sensitive measure of the D-M interaction in a weak ferromagnet. The configuration with the higher field for resonance in the split mode has a lower free energy; the energy

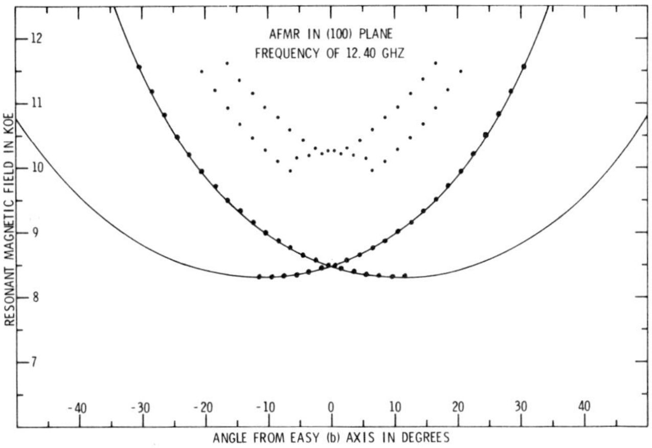

Figure 4. Theory and experimental data in (100) plane at 12.40 GHz and 4.2 °K.

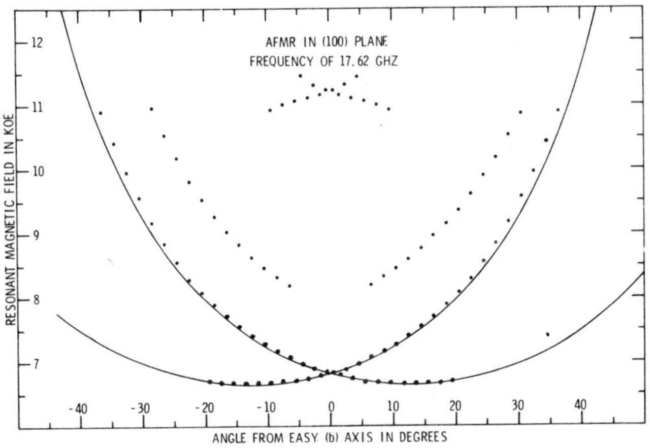

Figure 5. Theory and experimental data in (100) plane at 17.62 GHz and 4.2 °K.

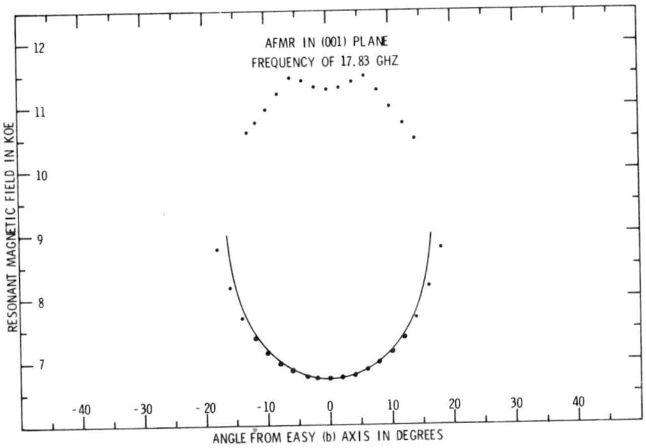

Figure 6. Theory and experimental data in (001) plane at 17.83 GHz and 4.2 °K.

difference depends on the component of the applied field along the c axis. As this component increases, the AFMR mode with the lowest resonance field weakens and then disappears, presumably due to domain collapse. The data at the highest fields in Figs. 4 and 5 also show a mode splitting and correspond to AFMR above spin flop. No calculations were made for off-axis fields in this region of the spectrum. The central mode in these two figures has not been explained, but it weakens and disappears as the applied field approaches the b axis.

For comparison the AFMR spectrum in the (001) plane at 17.83 GHz is shown in Fig. 6. Note the symmetry of the lower mode about the b axis and the absence of splitting; the minimum field for resonance occurs at zero degrees. The calculated behavior shown as a solid line is identical for the two configurations in Fig. 1. The interesting dip in the upper field data near the b axis has not been explained, but qualitatively similar behavior has been reported for the orthorhombic antiferromagnet $CuCl_2 \cdot 2H_2O$.[6]

Molecular field values used in the calculations for Figs. 2 through 6 depend on the value of the exchange field H_E determined from static magnetization data.[3] We estimate H_E = 475 kOe at 4.2 °K. The remaining fields found as the "best fit" to the AFMR data are H_{A1} = 82 Oe, H_{A2} = 780 Oe, H_D = 3150 Oe, and the ratio of parallel to perpendicular magnetic susceptibilities = 0.13. For comparison we have calculated the classical dipolar contribution to the magnetic anisotropy using the manganese ion positions reported at room temperature.[1] The dipolar anisotropy fields along the a and c axes are 83 and 1010 Oe.

CONCLUSIONS

We have measured the angular dependence of the AFMR spectrum in $BaMnF_4$ at 4.2 °K, and we find a unique mode splitting in the (100) plane. We have shown that this mode splitting is expected for a weakly ferromagnetic structure produced by a Dzyaloshinsky-Moriya field along the a axis. This effect offers a new and sensitive determination of the strength of the Dzyaloshinsky-Moriya interaction.

ACKNOWLEDGEMENTS

The authors thank E. G. Spencer of Bell Labs for sharing his unpublished AFMR and paramagnetic resonance data on $BaMnF_4$. The crystals used in this study were grown by D. Gabbe, A. Linz, and R. Mills of the Crystal Physics Lab, Center for Materials Science, M.I.T.

REFERENCES

*Based on a thesis submitted by one of the authors (ELV) in partial fulfillment of the requirements for the degree of Doctor of Philosophy in Electrical Engineering at the Massachusetts Institute of Technology, June, 1974. Research supported by the NSF under Contract GH-33635.

†NSF Graduate Fellow, 1971-1973. Present address: Sandia Laboratories, Org. 5132, Albuquerque, NM 87115

1. E. T. Keve et al., J. Chem. Phys., 51, 4928 (1969) and references therein.
2. E. G. Spencer et al., Appl. Phys. Lett., 17, 300 (1970); J. F. Ryan and J. F. Scott, Solid State Commun., 14, 5 (1974).
3. L. Holmes et al., Solid State Commun., 7, 973 (1969).
4. S. V. Petrov et al., Soviet Phys. JETP, 35, 981 (1972).
5. Additional terms are allowed by symmetry and have been included in a more complete treatment.
6. H. Yamazaki and M. Date, J. Phys. Soc. Japan, 21, 1615 (1966).

NEW RESONANCE EFFECTS IN ANTIFERROMAGNETIC MnF_2

W. E. Tennant
Rockwell International Science Center
Thousand Oaks, Ca.

and

R. B. Bailey and P. L. Richards
Department of Physics, University of California
and Inorganic Materials Research Division,
Lawrence Berkeley Laboratory, Berkeley, Ca 94720

ABSTRACT

A series of weak absorption lines of unknown origin has been observed below the 8.7 cm^{-1} antiferromagnetic resonance (AFMR) in the far infrared transmission spectra of MnF_2 powders. These extra modes appear in powders from different sources but are not present in the single crystal samples from which powders were prepared. Impurities are therefore unlikely to be the explanation. Conventional magnetostatic effects cannot account for sharp lines as far from the AFMR as those observed in the infrared spectra. The strongest mode, at 8.3 cm^{-1}, shows the same magnetic splittings as the AFMR in low fields, yet is apparently not present in 5 cm^{-1} microwave spectra measured at high field values. The integrated strengths of five lines observed between 7.4 cm^{-1} and 8.3 cm^{-1} increase as the particle size is reduced. Surface magnon modes have been predicted in this frequency range. However, the measured dependence on surface area of the 8.3 cm^{-1} mode intensity is not rapid enough to support this interpretation. Microwave transmission spectra show a number of lines below the AFMR field with strengths comparable to that expected for surface modes. No correlation with surface area or with the infrared mode positions has yet been found.

INTRODUCTION

Manganese flouride has been widely studied as an example of a simple, uniaxial antiferromagnet.[1,2] Its antiferromagnetic resonance (AFMR) occurs at 8.7 cm^{-1} and is accessible to both optical and microwave spectroscopic techniques. In this paper we report the discovery of additional resonances in the far infrared transmission spectrum of MnF_2 powders. A number of possible explanations are considered including the presence of surface spin wave modes. These have been predicted for several crystal structures but not yet observed in any material.

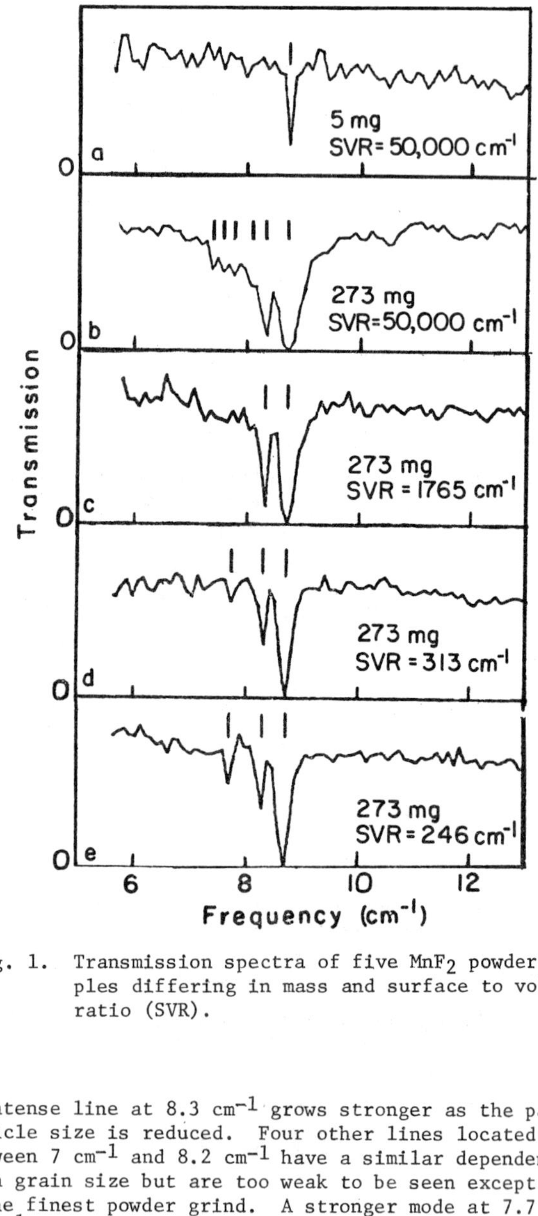

Fig. 1. Transmission spectra of five MnF_2 powder samples differing in mass and surface to volume ratio (SVR).

INFRARED MEASUREMENTS

Far infrared transmission spectra were obtained using the techniques of fourier transform spectroscopy. The output of a Michelson interferometer was coupled into a cryostat containing the sample, a high field superconducting solenoid, and a doped germanium bolometer. Powder samples were mounted in cylindrical light pipe sections and saturated in wax to preserve random orientation in magnetic fields. They were immersed in the 4.2 K helium bath.

Figure 1 shows the spectra of five powder samples ground from a single crystal and sieved to segregate grain sizes. The AFMR at 8.7 cm^{-1} appears broad in the 273 mg samples due to magnetostatic modes and to strong saturation of the absorption. In the 5 mg sample the line is approaching saturation with a 0.15 cm^{-1} linewidth comparable to the instrumental resolution of 0.1 cm^{-1}. Mode strengths for the extra absorptions observed below the AFMR depend strongly on the size of the powder grains. The most intense line at 8.3 cm^{-1} grows stronger as the particle size is reduced. Four other lines located between 7 cm^{-1} and 8.2 cm^{-1} have a similar dependence on grain size but are too weak to be seen except in the finest powder grind. A stronger mode at 7.7 cm^{-1} has a more complicated dependence on particle size. The approximate strengths of these extra modes range from 1/500 to 1/20 of the AFMR strength based upon comparison with the AFMR intensity in unsaturated 1 mg, 5 mg, and 10 mg samples.

The AFMR exhibits a linear magnetic field splitting when the magnetization axis is parallel to the applied field and a smaller quadratic splitting when it is perpendicular. The resulting spectrum for powder samples in an applied field is a broadened absorption with sharp edges on the low and high frequency sides where crystals aligned with the field experience the maximum splitting. For fields up to 14 kOe our spectra show that the splitting pattern of the 8.3 cm^{-1} mode is the same as that of the AFMR indicating its probable antiferromagnetic origin. At higher fields its absorption is too broad to be detected. Studies of the alloy systems (Co,Mn)F_2 and (Fe,Mn)F_2 also show that the 8.3 cm^{-1} mode is closely related to the AFMR.[3,4] The two lines merge and the absorption shifts to higher frequencies as the iron or cobalt concentration increases. They are separately visible only for impurity concentrations less than 2 mole percent.

So far no explanation for the extra absorptions has been found which satisfactorily accounts for their position, shape, field dependence, and the variation of absorption strength with particle size. Substitutional impurities in certain antiferromagnetic crystals do cause sharp satellite lines near the AFMR.[3,5] The modes we observe however appear in powder samples from three different sources but are not present in the single crystals from which the powders were prepared. Impurity effects in the original crystals are therefore ruled out. Surface contamination introduced during the grinding process is a possibility but would probably cause broader structure with a different field dependence.

Several other surface and size effects have been eliminated as possible sources of the 8.3 cm^{-1} mode. Interaction of sample crystals with the mounting wax or with each other has been ruled out by preparing a sample without the wax and one with the MnF_2 grains diluted in 10 times the volume of ZnF_2 powder. Both showed the 8.3 cm^{-1} absorption line with the expected intensity. Annealing a powder sample also produced no change in the 8.3 cm^{-1} mode. This makes strains and dislocations introduced during powder grinding an unlikely source of the extra modes. In addition to the 8.7 cm^{-1} AFMR, which corresponds to uniform precession of all spins in a crystal, there exist shape dependent magnetostatic modes[6,7] displaced from the AFMR by as much as 0.13 cm^{-1}. These, along with damping effects due to surface scattering, account for the linewidth of the AFMR but cannot explain a sharp line at 8.3 cm^{-1}.

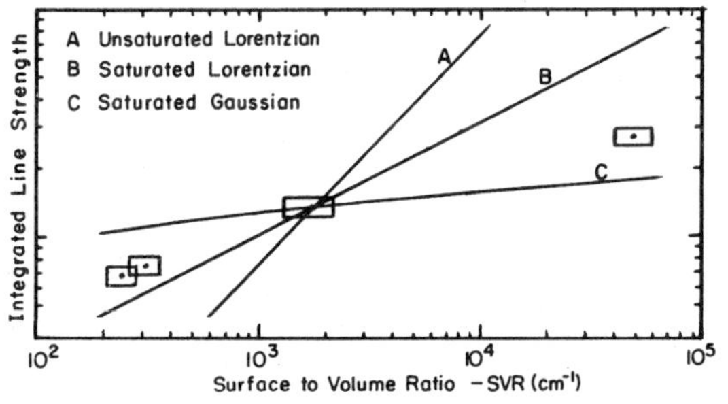

Fig. 2. Logarithmic plot of integrated absorption strength vs surface to volume ratio for the 8.3 cm^{-1} line. The solid lines show the expected dependence for both Lorentzian and Gaussian lineshapes.

Finally we consider surface spin waves as a possible source of the modes detected between 7 cm^{-1} and 8.3 cm^{-1}. A free surface in an antiferromagnetic crystal reduces the number of neighboring spins and therefore the exchange forces felt by spins on surface sites. Calculations for different crystal structures and surfaces[8,9] show that the resulting surface magnon frequencies lie in the range 6.2 cm^{-1} to 8.7 cm^{-1} for MnF_2. Predicted penetration depths are typically between 5 and 50 atomic layers and are largest for modes nearest the AFMR. Figure 2 plots the integrated strengths for the 8.3 cm^{-1} line in Fig. 1 against the measured surface to volume ratios (SVR) for four 273 mg powder samples. A surface mode's integrated strength should vary linearly with SVR for very small fractional absorptions. Assuming a lorentzian line shape this should change to a $\sqrt{SVR}$ dependence for a saturated absorption. The measured variation with surface area is not as rapid as we would expect for a surface magnon mode even in the limit of a fully saturated line. The relative strengths of the surface and bulk modes is the ratio of the crystal volumes participating in the two resonances. This is found by multiplying the SVR of a powder sample by the penetration depth. Figure 1e, where the 8.3 cm^{-1} mode has about 1/50 the AFMR strength and the SVR is 246, requires a penetration depth of at least one micron (2000 lattice constants) for the 8.3 cm^{-1} mode. This is much larger than theory predicts. However the weaker modes appearing below 8.3 cm^{-1} in Fig. 1b do have the intensity expected for surface modes with penetration depth of approximately 10 lattice constants. Optical techniques are not sensitive enough to study the magnetic field or SVR dependence of these lines.

MICROWAVE MEASUREMENTS

Field swept transmission spectra using a 5 cm^{-1} microwave source also show reproducible structure below the AFMR field in powder samples. However, none of it has the expected dependence on surface area and none can be identified with the lines seen in the infrared spectra. For powders, the microwave experiment detects the linear absorption edge and achieves about the same sensitivity as the infrared data. Surface modes penetrating at least 10 lattice constants are visible in the spectra. Even so the strong 8.3 cm^{-1} mode is not present. This provides evidence that the field dependence observed below 14 kOe does not persist to the 50 kOe field values used to observe the microwave resonance. Microwave measurements on single crystal samples were less sensitive to surface magnons than the powder measurements due to the small surface areas of single crystals.

CONCLUSIONS

None of the possibilities we have considered satisfactorily accounts for the strong 8.3 cm^{-1} absorption detected in MnF_2 powder samples. The weaker lines, visible only in the finest powder, could not be studied as thoroughly. Their intensity, frequency range, and dependence on particle size do, however, suggest surface spin waves as a possible cause. Conclusive proof of this possibility requires more sensitive single crystal spectra in order to identify each mode with a particular crystal surface. Measurements of this type are in progress.

REFERENCES

1. F. M. Johnson and A. H. Nethercott, Phys. Rev. 114, 705 (1959).
2. S. Foner, in Magnetism I, G. T. Rado and H. Suhl, eds., (Academic Press, N.Y., 1963), p. 383.
3. B. Enders, P. L. Richards, W. E. Tennant, and E. Catalano, Magnetism and Magnetic Materials 1972, C. D. Graham, Jr. and J. J. Rhyne, eds. (AIP, N.Y., 1972) p. 179.
4. W. E. Tennant, Ph.D. Thesis, University of California, Berkeley, 1974.
5. R. A. Cowley and W. J. L. Buyers, Rev. Mod. Phys. 44, 406 (1972).
6. J. P. Kotthaus and V. Jaccarino, Phys. Rev. Letters 28, 1649 (1972).
7. J. P. Kotthaus and V. Jaccarino, Magnetism and Magnetic Materials 1972 (AIP, N.Y., 1972).
8. D. L. Mills and W. M. Saslow, Phys. Rev. 171, 488 (1968).
9. T. Wolfram and R. E. DeWames, Phys. Rev. 185, 762 (1969).

THEORY OF MULTIMAGNON RELAXATION IN ANTIFERROMAGNETS

R. M. White
Xerox Palo Alto Research Center, Palo Alto, California 94304

S. M. Rezende and L.C.M. Miranda
Instituto de Física,† Universidade Federal de Pernambuco, Recife, Brazil

ABSTRACT

The general form of the AFMR relaxation rate associated with n/2 antiferromagnetic magnons scattering into n/2 magnons is presented. In the low temperature limit this process gives a temperature dependence of the form T^{3n-8}, while at high temperatures this becomes T^{n-2}. These results are applied to the relaxation of the antiferromagnetic resonance in MnF_2. It is found that, except at low temperatures ($T < T_N/4$), higher order spinwave processes quickly become very important.

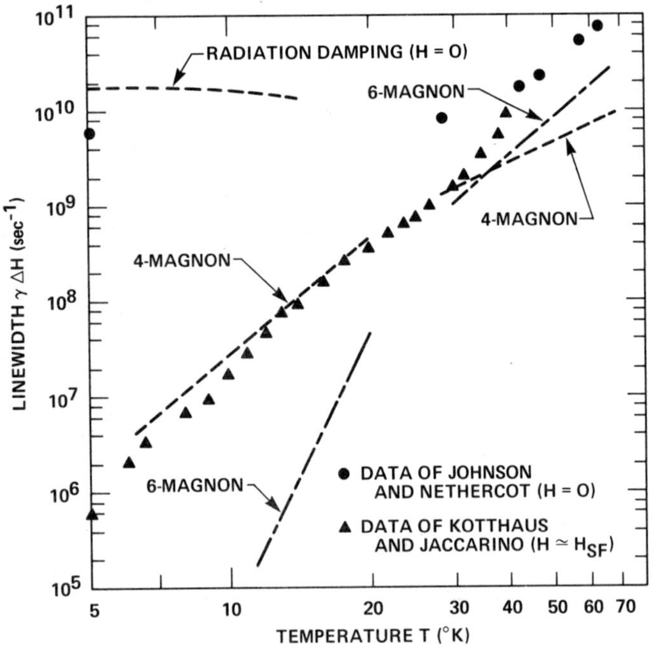

Fig. 1 Comparison of theoretical and experimental[1,8] linewidths in MnF_2 ($T_N = 68°K$).

Recent measurements[1] of the linewidth of the antiferromagnetic resonance in MnF_2, in which care has been taken to avoid radiation damping,[2] have been made as a function of temperature. The results are indicated by the triangles in Fig. 1. At low temperatures the data follow a T^4 behavior. White, Freedman, and Woolsey[3] (WFW) have carried out a theoretical calculation based on a four-magnon scattering process which does in fact give a T^4 dependence at low temperatures. Furthermore, since it also agrees quantitatively with the data in this region without any adjustable parameters, we believe this to be the correct relaxation mechanism at low temperatures. However, above about 15°K the experimental data maintains approximately the same T^4 dependence whereas the WFW result assumes a lower T^2 dependence. Furthermore, in order to include all the aspects of this process, WFW were forced to evaluate their result numerically. Guided by this result, we present here an approximate, but analytical, approach to magnon-magnon relaxation which can be extended to higher order processes.

† Work partially supported by CNPq, BNDE, and CAPES (Brazilian Government).

It was suggested in WFW that the high temperature relaxation might be associated with higher order magnon scattering processes. For the uniaxial antiferromagnet considered by WFW, one finds that the n-magnon terms all involve n/2 input magnons scattering into n/2 output magnons. The relaxation rate for magnon 1 associated with this process is

$$\eta_1^{(n)} = \frac{2\pi}{\hbar^2}\left[e^{\beta\hbar\omega_1}-1\right]\sum_{\vec{k}_2}\sum_{\vec{k}_3}\cdots\sum_{\vec{k}_n}\left|C^{(n)}(\vec{k}_1,\vec{k}_2\ldots\vec{k}_n)\right|^2 \cdot e^{\beta\hbar(\omega_2+\ldots\omega_{n/2})}\,\bar{n}_2\bar{n}_3\ldots\bar{n}_n\,\delta(\omega)\Delta(\vec{k}) , \qquad (1)$$

where

$$\delta(\omega) \equiv \delta(\omega_1+\omega_2\ldots+\omega_{n/2}-\omega_{n/2+1}\ldots-\omega_n) ,$$

$$\Delta(\vec{k}) \equiv \Delta(\vec{k}_1+\vec{k}_2\ldots+\vec{k}_{n/2}-\vec{k}_{n/2+1}\ldots-\vec{k}_n) ,$$

and $C^{(n)}(\vec{k}_1,\vec{k}_2\ldots\vec{k}_n)$ is an appropriate coupling coefficient. $C^{(4)}$, for example, is given by Eqs. (13) and (15) in Ref. 3. The total relaxation rate is the sum over the different relaxation channels associated with the different values of n. For large values of n the evaluation of Eq. (1) becomes extremely difficult. A similar problem was faced by Sparks and Sham[4] in dealing with multiphonon absorption. However, in their case they have one virtual phonon of frequency ω decaying into n phonons. These n phonons were all assumed to have the same frequency ω_g, which is the frequency at which the density of states is largest. In our case the situation is more complicated because we have additional input particles whose frequencies depend upon the temperature as well as the density of states. Our first step therefore was to develop an approximation scheme for the four-magnon process that compared favorably to the numerical result of WFW. This scheme involves the following steps: (a) Since $\beta\hbar\omega_1$ is small, $[\exp(\beta\hbar\omega_1)-1]$ is approximated by $\beta\hbar\omega_1$; (b) the sums in Eq. (1) are converted to integrals in k-space; (c) the magnon dispersion relation is approximated by $\omega = \omega_0 + \bar{v}k$. Because of these linear dispersion relations, energy and momentum conservation require that $\vec{k}_2$, $\vec{k}_3$, and $\vec{k}_4$ be colinear. One can then show that the energy delta function becomes $\delta(\omega) = (k_2-k_3)\delta(\cos\theta-1)/\,\bar{v}k_2k_3$, where θ is the angle between $\vec{k}_2$ and $\vec{k}_3$. The condition that $k_2 \geq k_3$ also follows from energy conservation; (d) since $k_1 = 0$ and ω_1 is small, energy-momentum conservation makes the joint density of states largest when one of the output magnons has $k \simeq 0$ and the other two magnons involved have large k. Furthermore, numerical investigations show that $C^{(4)}(0,k,k,0)$ is nearly independent of k. Therefore we approximate the coupling coefficient by $\bar{C}^{(4)} \equiv C^{(4)}(0,k_{BZ},k_{BZ},0) + C^{(4)}(k_{BZ},0,k_{BZ},0) + C^{(4)}(0,k_{BZ},0,k_{BZ}) + C^{(4)}(k_{BZ},0,0k_{BZ})$ where k_{BZ} is the value of k at the Brillouin zone boundary. This coupling constant may then be taken outside the integrals.

The resulting expression for the relaxation rate then becomes

$$\eta_1^{(4)} = \frac{2\pi}{\hbar^2\bar{v}}\left(\frac{\hbar\omega_1}{k_BT}\right)\frac{|\bar{C}^{(4)}|^2}{2}\,\frac{1}{2}\left(\frac{N\Omega}{2\pi^2}\right)^2\left(\frac{k_BT}{\hbar\bar{v}}\right)^5 I^{(4)}(T) , \qquad (2)$$

where

$$I^{(4)}(T) = \int_0^{x_{max}} dx \int_0^{x} dy \frac{xe^x}{e^x-1} \frac{y}{e^y-1} \frac{x-y}{e^{x-y}-1} , \quad (3)$$

and $x_{max} = \frac{\hbar \bar{v} k_{max}}{k_B T}$, $k_{max} = \left(\frac{6\pi^2}{\Omega}\right)^{1/3}$, where Ω is the volume of the unit cell. The $C^{(n)}$ that appears in Eq. (1) is obtained by permuting all the input and output wavevectors. The factor of 2 dividing $|\bar{C}^{(n)}|^2$ corrects for the fact that the two output magnons are equivalent. From Ref. 3 we find $\bar{C}^{(n)} = (1.8\ z_2 J_2 - 13.5\ K)/2N$. For MnF_2, $J_2 = 1.76°K$ is the next near neighbor exchange, $z_2 = 8$ is the number of next nearest neighbors, and $K = 0.41°K$ is the uniaxial anisotropy constant. The cell volume is $\Omega = (4.87)^2(3.3)\ Å^3$, so $k_{max} = 0.91\times10^8\ cm^{-1}$. The integral $I^{(4)}(T)$ was evaluated numerically. As a function of x_{max} this was found to separate into two regions: a low temperature region ($T \le T_N/4$) where $I^{(4)}(T)$ is independent of temperature and equal to 8.7; and a high temperature region ($T > T_N/4$) where $I = x_{max}^2/2$. The value of $\bar{v}$ is an important parameter in this theory because it enters to high powers. At low temperatures we take for $\bar{v}$ the slope of the spinwave dispersion relation in the long wavelength region. This gives $\bar{v} = 1.3\times10^5$ cm/sec. At high temperatures spinwaves near the Brillouin zone become more important. Since the group velocity of these excitations is small, $\bar{v}$ is reduced. We therefore take $\bar{v} = 1.2\times10^5$ cm/sec. for $T > T_N/4$. The four-magnon relaxation rate becomes

$$\eta_1^{(4)} = \begin{cases} 3.1\times10^3\ T^4 & T < T_N/4 \\ 2.0\times10^6\ T^2 & T > T_N/4\ . \end{cases} \quad (4)$$

This is plotted as the dashed line in Fig. 1. In the low temperature region this agrees very well with the calculation of WFW. Therefore we conclude that the approximations made in arriving at this result are valid. Pincus[5] has also shown that the relaxation rate should vary as T^2 at high temperatures. Harris et al.,[6] however, using a Dyson-Maleev boson formulation, found that the low temperature rate varies as T^3.

Let us now consider the case of the general n-magnon process. The first two steps, (a) and (b), indicated above are still valid. The manner in which energy and momentum are now conserved, however, is more complicated. The wavevectors, for example, are not necessarily colinear. This may be seen by considering the number of degrees of freedom entering the wavevector sums. In general, these sums may be converted to integrals in the usual fashion. There are n-1 such $\vec{k}$-space integrations. However, momentum conservation removes one of these. Thus, there remain (n-2) ϕ integrations and (n-2) θ integrations. But one of the θ integrations may be eliminated by energy conservation. We shall now simplify this angular integration problem by assuming that all the wavevectors are colinear. This introduces a factor of $(2\pi)^{n-2}$ from the ϕ integrations and a factor of $(2)^{n-3}$ from the θ integrations. The n-order relaxation rate then becomes

$$\eta_1^{(n)} = \alpha^{(n)}\ T^{3n-8}\ I^{(n)}\ (T)\ , \quad (5)$$

where

$$\alpha^{(n)} = 2\pi\omega_0 |\bar{C}^{(n)}|^2\ 2^{n-3} \left(\frac{N\Omega}{4\pi^2}\right)^{n-2} \frac{(k_B)^{3n-8}}{(\hbar\bar{v})^3 (n-2)} , \quad (6)$$

and $I^{(n)}(T)$ is the generalized version of Eq. (3).

Let us now use this general result to evaluate the contribution from six-magnon scattering. When the antiferromagnetic next-near neighbor exchange interaction is expanded in spinwave amplitudes the coefficient of the six-magnon term is

$$C^{(ex)}_{123456} = (J_2 z_2/16N^2S)(u_1v_2v_3v_4v_5v_6\gamma_1 + v_1v_2v_3v_4v_5u_6\gamma_6 + v_1u_2u_3u_4u_5u_6\gamma_1 + u_1u_2u_3u_4u_5v_6\gamma_6 - 2u_1v_2v_3u_4u_5v_6\gamma_{6-2-3} - 2u_1u_2v_3v_4v_5u_6\gamma_{4+5-3})\ . \quad (7)$$

As in the four-magnon case, all permutations (36) of the input and output wavevectors must be included in C, but then the relaxation rate must be corrected for those magnons that are identical. This introduces a factor of 1/12. Thus the effective coupling constant becomes $\bar{C}^{(6)} = (z_2J_2/16N^2S)(24u_0v_0/\sqrt{12})$. The integral $I^{(6)}(T)$ is evaluated in the low temperature limit by replacing $(e^y-1)^{-1}$ by e^{-y} and extending the limit x_{max} to infinity. The result gives 672. In the high temperature region $(e^y-1)^{-1}$ is replaced by y^{-1} which leads to a value of $(17/72)x_{max}^6$ for $I^6(T)$. The six-magnon relaxation rate becomes

$$\eta_1^{(6)} = \begin{cases} 1.2\times10^{-5}\ T^{10} & sec^{-1} & T < T_N/4 \\ 1.3\times10^3\ T^4 & sec^{-1} & T > T_N/4\ . \end{cases} \quad (8)$$

This is plotted in Fig. 1. We see that the six-magnon rate exceeds the four-magnon rate above 35°K. In general, the temperature at which the (n+2)-magnon rate exceeds the n-magnon rate increases with n. Therefore the higher order processes become increasingly more important as one approaches the Néel temperature.

Data have also been published[7] on the antiferromagnet $GdAlO_3$. When this data, which extends from 0.25 T_N up to 0.75 T_N, is plotted on a log-log plot it appears to fit a T^5 law. However, it is interesting to note that on such a plot it is not possible to distinguish this from $H = 3.3\ T^4 + 1.7\ T^6$, which would correspond to six- plus eight-magnon scattering.

Finally, a word about phonons. One might think that since phonons are also characterized by a linear dispersion relation, their contribution would be indistinguishable from the magnons. However, the phonon amplitudes are proportional to $\omega_q^{-1/2}$. This introduces a factor of q^{-1} in the relaxation rate which reduces the temperature dependence. But, perhaps more importantly, is the fact that the Debye temperature now enters the denominator of the relaxation rate in place of the Néel temperature. Thus the magnon-phonon coupling must be strong enough to overcome this factor. Since Mn^{2+} and Gd^{3+} are S-state ions this coupling is very weak. However, in FeF_2 there is evidence that a phonon mechanism may be operative.

REFERENCES

1. J. P. Kotthaus and V. Jaccarino, Phys. Lett. A 42, 361 (1973).
2. R. W. Sanders, D. Paquette, V. Jaccarino, and S. M. Rezende, Phys. Rev. B 10, 132 (1974).
3. R. M. White, R. Freedman, and Roy B. Woolsey, Phys. Rev. B 10, 1039 (1974).
4. M. Sparks and L. Sham, Solid State Comm. 11, 1451 (1972); Phys. Rev. Lett. 31, 714 (1973).
5. P. Pincus, J. Phys. Radium 23, 536 (1962).
6. A. B. Harris, D. Kumar, B. Halperin, and P. Hohenberg, Phys. Rev. B 3, 961 (1971).
7. P. Doussineau and B. Ferry, Phys. Lett. 46A, 135 (1973).
8. F. M. Johnson and A. H. Nethercot, Phys. Rev. 114, 705 (1959).

THERMAL TRANSPORT BY MAGNETOELASTIC MODES IN $CoCl_2 \cdot 6H_2O$+

J. E. Rives and S. N. Bhatia
Department of Physics and Astronomy, University of Georgia, Athens, Georgia 30602

ABSTRACT

The thermal conductivity of $CoCl_2 \cdot 6H_2O$ has been measured between 0.35 and 4.2 K. For $T<T_N$ the conductivity has been analyzed by assuming the heat carriers are coupled magnetoelastic modes. The uncoupled phonon contribution to the conductivity can be determined by applying a magnetic field of sufficient strength to depopulate the spin states. Using a Debye model, we show that the only important phonon scattering, under these conditions, is that due to boundaries and point defects. The treatment of the magnon-phonon interaction is restricted to one-magnon-one-phonon processes with coupling between the transverse phonons and the two magnon branches. The total conductivity is assumed to be the sum of contributions from the magnetoelastic branches and the longitudinal phonon branch. With reasonable approximations for the magnon-phonon coupling strength and the magnon-magnon lifetimes a satisfactory fit is obtained for the zero field data. There is evidence for strong critical scattering of phonons near T_N. For $T>T_N$ the observed effect of the magnetic scattering on the conductivity is in qualitative agreement with recent theories of ultrasonic attenuation near magnetic critical points, however, the observed critical exponent is surprisingly large for an anisotropic Heisenberg antiferromagnet.

The thermal conductivity in magnetic insulators is affected by the spin system in two ways. For temperatures well below the ordering temperature, magnons can be effective heat carriers adding to the phonon heat current. Near the ordering temperature, phonon damping due to the spin-phonon interaction is important. We have measured the thermal conductivity of $CoCl_2 \cdot 6H_2O$ (T_N=2.29K) in the presence of magnetic fields up to 55 kOe between 0.35 and 4.2 K.

Within the relaxation time approximation the Boltzmann equation gives for the thermal conductivity[1]

$$K = (1/6\pi^2) \int_0^{q_{max}} \hbar\omega[\partial N(q)/\partial T]\ell(q)(\partial\omega/\partial q)q^2 dq \, , \quad (1)$$

where the sum over the modes has been replaced by an integral over the wavevector q. N(q) is the distribution function and $\ell(q)$ is the mean free path. The inverse relaxation time $\tau^{-1}(q) = (\partial\omega/\partial q)\ell^{-1}(q) = \sum_i \tau_i^{-1}(q)$, where the sum is over all contributing scattering processes including magnetic scattering due to the spin system.

It is possible to eliminate the magnetic scattering by applying a magnetic field of sufficient strength to completely saturate the spin system. Assuming the Debye model for phonons, in the high field limit Eq. 1 can be expressed in the form

$$K_{DIA} = K(\mu H/k_B T>>1) = AT^3 \int_0^{\Theta/T} x^4 e^x/(e^x-1)^2 \tau_{DIA}^{-1} \, dx, \quad (2)$$

where τ_{DIA}^{-1} includes all but the magnetic scattering. At lower fields

$$K_H = AT^3 \int_0^{\Theta/T} x^4 e^x/(e^x-1)^2 (\tau_{DIA}^{-1} + \tau_m^{-1}) \, dx, \quad (3)$$

where τ_m^{-1} is the inverse relaxation time for magnetic scattering. $\Delta K/K_H = (K_{DIA}-K_H)/K_H$ is just the ratio $\tau_m^{-1}/\tau_{DIA}^{-1}$ averaged over the thermal distribution of phonons.[2] Since it is possible to determine τ_{DIA}^{-1} from a measurement of K_{DIA}, the effect of the magnetic scattering can be isolated.

For $T<T_N$ it is necessary to include the heat carried by the magnons and the effect of the spin-phonon interaction on both carriers. The magnon conductivity can be calculated using Eq. 1.

The magnon-phonon interaction has been studied by White et al.[3] For coupling between the magnon and the transverse phonon branches the energies of the coupled modes in terms of the uncoupled phonon and magnon frequencies Ω_q and ω_q are given by[4]

$$E_{1,3} = (\hbar/2)(\Omega_q+\omega_q) \pm (\tfrac{1}{2})[\hbar^2(\Omega_q-\omega_q)^2+8D_q^2]^{\frac{1}{2}} \quad (4)$$

and

$$E_2 = \hbar\omega_q \, . \quad (5)$$

The relaxation times of the coupled modes are determined using the method of Kittel[5] which takes account of the scattering by attributing an imaginary part to the energy of both phonons and magnons. Thus, $\hbar\tilde{\Omega}_q = \hbar \times (\Omega_q+i\tau_{ph}^{-1})$ and $\hbar\tilde{\omega}_q = \hbar(\omega_q+i\tau_{mag}^{-1})$. The resulting complex coupled mode energies are of the form $\tilde{E} = E+i\hbar\tau_{me}^{-1}$. E and τ_{me}^{-1} are used in Eq. 1 to calculate the conductivity.

In Fig. 1 we show the thermal conductivity data as a function of temperature at zero external field and in a field of 50 kOe applied along the crystallographic b-axis. Up to about 1.0 K the data is independent of field at 50 kOe, however, at higher temperatures complete saturation is not obtained at the highest fields. The solid line is the best fit to the high field data up to 1.0 K using Eq. 2 including boundary and point defect scattering only. The dashed line is the expected result for boundary scattering alone.

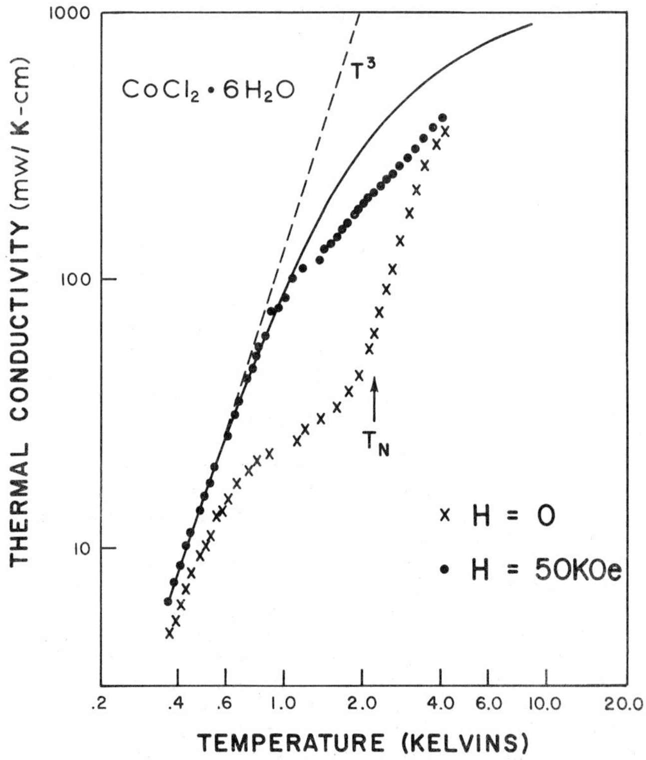

Fig. 1. Thermal conductivity of $CoCl_2 \cdot 6H_2O$. ● ✖ data; ---- boundary scattering; —— boundary plus point defect scattering.

The field dependence of the conductivity was measured at several temperatures. At 0.55 K the conductivity at 30 kOe exceeded the high field saturated value by approximately 10 percent, giving direct evidence of a substantial magnon conductivity. The conductivity decreases sharply in the vicinity of the antiferromagnetic to paramagnetic phase transition at 43 kOe.

The behavior of the thermal conductivity suggests that a strong magnetoelastic interaction plays an important role in the conduction process. This is consistent with specific heat[6] and magnetization[7] studies which also show evidence of strong magnetoelastic coupling.[8]

The expected behavior of the zero field conductivity for $T<T_N$ was calculated by using the coupled mode energies given by Eqs. 4 and 5. The magnon frequencies were taken from the orthorhombic model calculations of Keffer[9] for an external field perpendicular to the preferred direction. The high field data were used to determine τ_{ph}^{-1}.

For the magnon scattering we consider boundary, magnetic point defect, and magnon-magnon scattering to give

$$\ell^{-1}_{mag}(q) = L_B^{-1} + AN_D q^4 + BT^{5/2} q^2 [e^{-E_g/kT} + e^{-(E_g+\Delta)/kT}]. \quad (6)$$

The boundary mean free path L_B is chosen equal to the value found for phonons L_B=0.132 cm. A is given by Callaway et al.[10], and N_D, the density of defects is taken equal to the density of defects determined from the phonon scattering. The last two terms for magnon-magnon scattering are calculated following the treatment of Lawrence and Petitgrand[4], simplified for the present case of large energy gap E_g and small dispersion 2Δ. The value of the constant B is chosen to give a magnon conductivity in agreement with the excess conductivity observed at 0.55 K at 30 kOe. The interaction constant D_q is taken as the value used by Kimura[8] in the interpretation of the specific heat results of Donaldson and Edmonds.[6]

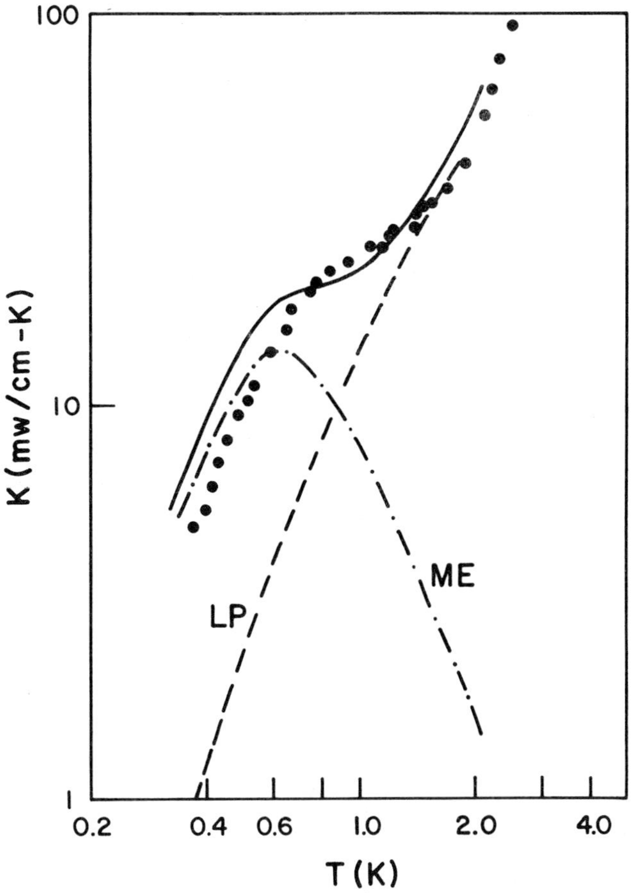

Fig. 2. Analysis of zero field conductivity of $CoCl_2 \cdot 6H_2O$. ● data; — · — magnetoelastic mode contribution; — — longitudinal phonon contribution; —— total conductivity.

In Fig. 2 we show the comparison between the calculated and experimental conductivity. When the magnetoelastic contribution is added to that of the longitudinal phonons, a remarkably good fit to the data is obtained. The ratio of the longitudinal to transverse phonon velocities was used as a final adjustable parameter.

This treatment of the thermal conductivity breaks

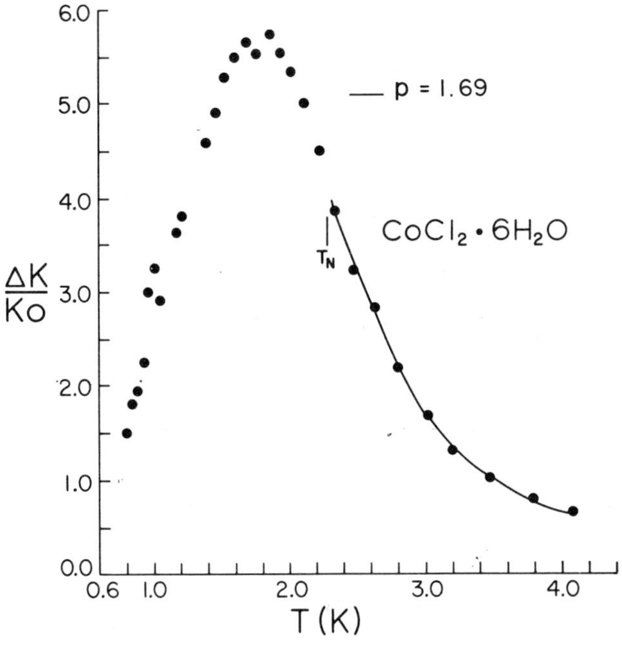

Fig. 3. $(K_{DIA} - K_0)/K_0$ versus temperature for $CoCl_2 \cdot 6H_2O$. ● data; —— best fit to the data for $T>T_N$ from theory of critical scattering.

down as T_N is approached. Above T_N the phonon conductivity is reduced by magnetic scattering. The effect of the magnetic scattering is isolated by plotting $(K_{DIA}-K_H)/K_H$ versus T. This is shown in Fig. 3 where it is seen that the magnetic scattering appears to be critically enhanced as T_N is approached from above.

For critically enhanced phonon scattering the ultrasonic attenuation is expected to diverge as T_N is approached.[11] The inverse relaxation time is simply related to the ultrasonic attenuation, and is also expected to diverge. Microscopic fluctuations are always present to prevent a real divergence. A form for τ_m^{-1} which satisfies both these conditions, and which has been successful in fitting similar data on several ferromagnets[2] can be written as

$$\tau_m^{-1}(\omega) = C\omega^2/(\varepsilon^p + D\omega) , \quad (7)$$

where p is the critical exponent associated with the critical scattering, and $D\omega$ is included to prevent τ_m^{-1} from diverging.

The solid line in Fig. 3 represents the best fit to the data for $T>T_N$ using the above form for τ_m^{-1}. The observed value of the critical exponent p=1.69 ± 0.05 can be compared with the predicted value of Laramore and Kadanoff[11] for a uniaxial Heisenberg antiferromagnet of $p = \gamma - \alpha$. Since γ for this model is approximately 1.38, and α is positive, the experimental value of p is not consistent with the predicted value.

+Research sponsored by the National Science Foundation

REFERENCES

1. P. Carruthers, Rev. Mod. Phys. 33, 92 (1961).
2. G.S. Dixon, Phys. Rev., accepted for publication.
3. R.M. White, M. Sparks, and I. Ortenburger, Phys. Rev. 139, A450 (1965).
4. G. Lawrence and J.B. Petitgrand, Phys. Rev. B 8, 2130 (1973).
5. C. Kittel, Phys. Rev. 110, 836 (1958).
6. R.H. Donaldson and D.T. Edmonds, Proc. Phys. Soc. 85, 1269 (1965).
7. J.P. Renard, J. de Phys. 29, 767 (1968).
8. I. Kimura, J. Phys. Soc. Japan 28, 1182 (1970).
9. F. Keffer, Handbuch der Physik (Springer-Verlag, New York, 1966) Vol. 18.
10. J. Callaway and R. Boyd, Phys. Rev. 134, A1655 (1964).
11. G.E. Laramore and L. Kadanoff, Phys. Rev. 187, 619 (1969).

MAGNON HEAT CONDUCTION AND MAGNON LIFETIMES IN THE METALLIC FERROMAGNET $Fe_{68}Co_{32}$ AT LOW TEMPERATURES*

Y. Hsu and L. Berger
Physics Department, Carnegie-Mellon University, Pittsburgh, Pennsylvania 15213

ABSTRACT

The thermal and electrical conductivities of the alloy $Fe_{68}Co_{32}$ have been measured with high accuracy between 1.2K and 4.5K, in fields up to 5.2T. The results show that magnons contribute about 10% of the heat conduction at 4K. Magnon lifetime is 7×10^{-10} sec. at 4K, and varies like $T^{-1.2}$.

INTRODUCTION

Heat transport by magnons was observed recently[1] in two nickel-rich Ni-Fe alloys, between 1.2 and 4.5K. This was the first such observation in transition metals. The results were roughly consistent with a kinetic model for magnon thermal conduction which assumes s-d exchange scattering of magnons by conduction electrons. Magnon lifetime was about 1.5×10^{-10} sec. at 4K, and varied roughly like $T^{-0.6}$ and T^{-1} respectively in the two alloys.

In the present work, we have measured with high accuracy the thermal and electrical conductivities of the alloy $Fe_{68}Co_{32}$ between 1.2K and 4.5K, in magnetic fields up to 5.2T. The results show that at 4K the magnon contribution is about 10% of the total heat conductivity. The magnon conductivity is again consistent with magnon-electron scattering. Magnon lifetimes are larger than that of the nickel-iron alloys,[1] by a factor of at least four at 4K. These lifetimes vary like T^{-1} above 2K.

KINETIC MODELS FOR MAGNON THERMAL CONDUCTIVITY

The thermal conductivity of a magnon gas is given by:

$$\kappa_m = \frac{1}{3}\sum c\, v^2\, \tau_m \quad , \tag{1}$$

where c is the specific heat, v the group velocity, and τ_m the lifetime of a magnon mode. The dipole-dipole interaction changes[2] the dispersion relation of a magnon, and therefore influences c and v. This influence was taken into account only to first order in the saturation magnetization M_s in the work[1] on Fe-Ni. Due to the larger M_s of Fe-Co, it is advisable to treat the dipole-dipole interaction exactly. Then Eq.(1) becomes:

$$\kappa_m = \frac{k_B^{7/2}T^{5/2}}{8\pi^2\hbar^2 D^{1/2}}\int_{x=0}^{\infty}\int_{a=-1}^{1}\tau_m\frac{e^{\ell}}{(e^{\ell}-1)^2}\{2(x+z)x - jz(1-a^2)\}^2 a^2 da\,\frac{dx}{x^{1/2}} \quad , \tag{2}$$

where $x = Dq^2/k_BT$, q is the magnon wave number and D the exchange stiffness for the alloy, $z = g\mu_B B_E/k_BT$, B_E is the external magnetic field, $j = g\mu_B M_s/k_BT$, $\ell = \{(x+z)^2 + j(x+z)(1-a^2)\}^{1/2}$, $a = \cos\theta$ and θ is the angle between $\vec{q}$ and $\vec{M}_s$.

If magnon-electron scattering mediated by the s-d exchange interaction is assumed to be dominant,[1] and if the "dirty" limit $\Lambda_e q \ll 1$ holds, where Λ_e is the electron mean free path, then one can easily show[1,3] that $\tau_m \propto \omega^{-1}$ or $\tau_m\,\omega = Q$, where Q is a constant. Thus:

$$\kappa_m(M_s,B_E,T) = \frac{Qk_B^{5/2}T^{3/2}}{8\pi^2\hbar D^{1/2}} J_D(j,z) \quad , \tag{3}$$

where $J_D(j,z) = \int_{x=0}^{\infty}\int_{a=-1}^{1}\frac{e^{\ell}}{\ell(e^{\ell}-1)^2}\{2(x+z)x - jz(1-a^2)\}^2 a^2 da\,\frac{dx}{x^{1/2}}$.

For $B_E=M_s=0$, $J_D(0,0) = (8/3)\Gamma(7/2)\zeta(5/2) = 11.883$.

On the other hand, if the "clean" limit $\Lambda_e q \gg 1$ holds, then[4] the s-d exchange theory predicts $1/\tau_m = b\omega/q$ where b is a constant. Therefore:

$$\kappa_m(M_s,B_E,T) = \frac{k_B^3 T^2}{8\pi^2\hbar b D} J_C(j,z) \quad , \tag{4}$$

where $J_C(j,z) = \int_{x=0}^{\infty}\int_{a=-1}^{1}\frac{e^{\ell}}{\ell(e^{\ell}-1)^2}\{2(x+z)x - jz(1-a^2)\}^2 a^2 da\, dx$.

For $B_E=M_s=0$, $J_C(0,0) = (8/3)\Gamma(4)\zeta(3) = 19.232$ and $\tau_m = b^{-1}(\hbar/D)^{-1/2}\omega^{-1/2}$. When z=j=0, the integrand of $J_D(j,z)$ peaks at $1.8\ k_BT$ and that of $J_C(j,z)$ at $2.6\ k_BT$. The integrals J_D and of J_C have been evaluated for various values of z and j, with the help of a computer.

Using Eqs.(3) or (4), the variation of the magnon thermal conductivity between two field values, B_{E1} and B_{E2}, can be written as:

$$\frac{\kappa_m(B_{E2},M_s,T)-\kappa_m(B_{E1},M_s,T)}{\kappa_m(0,0,T)} = \frac{J(z_2,j)-J(z_1,j)}{J(0,0)} \quad . \tag{5}$$

where J may be J_D or J_C, depending on conditions. Actually, the choice between J_D and J_C affects the right side of Eq.(5) only by a few percents. If we substitute experimental values of $\{\kappa_m(B_{E2},M_s,T)-\kappa_m(B_{E1},M_s,T)\}$ into Eq.(5), we can solve for the values of $\kappa_m(0,0,T)$.

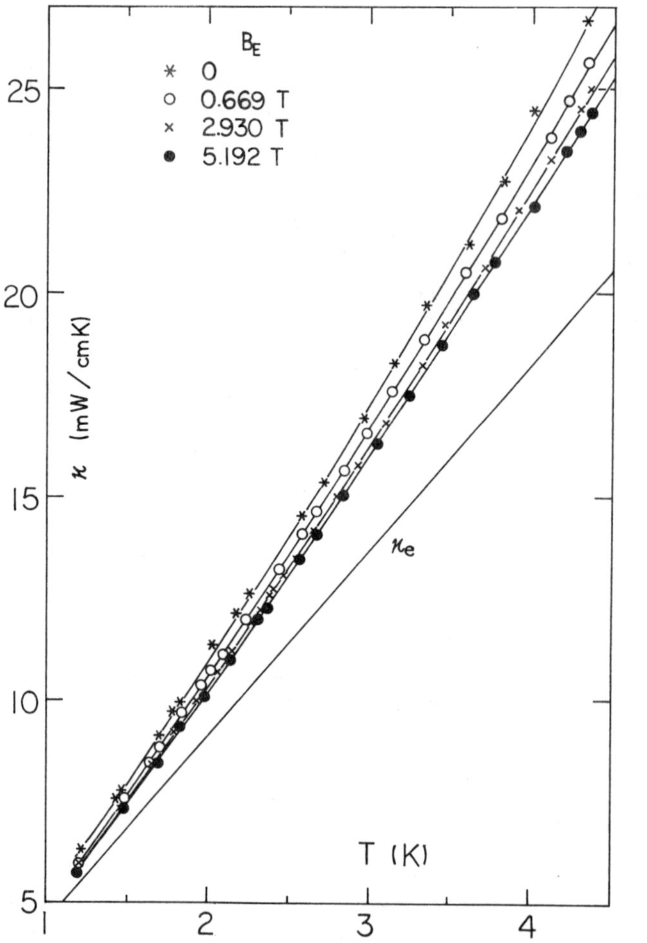

Fig. 1. Measured thermal conductivity κ at various fields, and calculated electronic thermal conductivity κ_e at saturation.

EXPERIMENTAL

The sample was prepared from high purity Johnson-Mathey metals by levitation induction melting, and was homogenized for 48 hours in dry hydrogen at 1385 C. The dimensions are 0.47×0.50×6.5 cm.

The apparatus used in thermal conductivity measurements is an improved version of the one described in Ref. 1. The method for data reduction is also similar to the one of Ref. 1. The external magnetic field is parallel to the direction of heat current, and takes the values 0, 0.669, 2.930, and 5.192 T. The three last field values are sufficient to saturate the sample. Three runs were made at the second and third field values. Six runs were made at the highest field, because the dispersion of data points was the largest at that field.

The electrical resistivity was measured at 4.2K and 1.9K in a direction parallel to the external magnetic field, between 0 and 5.3 T. To the experimental accuracy of 0.1%, it was the same at the two temperatures.

RESULTS AND DISCUSSION

The electrical resistivity ρ of the sample varies slowly above saturation, and is described by $\rho = 5.503 + 0.00476B_E$ in units of 10^{-8} Ωm. This formula is used to calculate the electronic part κ_e of the thermal conductivity, employing the Wiedemann-Franz law.

The measured thermal conductivity κ and the calculated κ_e are plotted in Fig. 1. For κ only one run is shown at each field. The thermal conductivity κ is seen to decrease for increasing fields, above saturation, while κ_e is almost constant. The variation of conductivity $\Delta\kappa$ between the two field values 0.669T and 5.192T can be evaluated by using the smooth curves which result from fitting the κ data of Fig. 1 to third-degree polynomials. The small variation $\Delta\kappa_e$ between the same two fields, is calculated from the Wiedemann-Franz law. Then, assuming $\Delta\kappa = \Delta\kappa_e + \Delta\kappa_m$, we find $\Delta\kappa_m$. Substituting this value into the numerator on the left side of Eq.(5), we derive the magnon conductivity $\kappa_m(0,0,T)$, shown on Fig. 2. At 4K it amounts to about 10% of the total conductivity. Figure 2 indicates that $\kappa_m(0,0,T)$ varies like $T^{1.3}$ on the average over the range 1.5 to 4.5K. This is rather close to the $T^{1.5}$ variation predicted by Eq.(3) for the dirty limit. The agreement is even better if we consider only the data between 2K and 4.5K, where $\kappa_m \propto T^{1.5}$ on the average (Fig. 2).

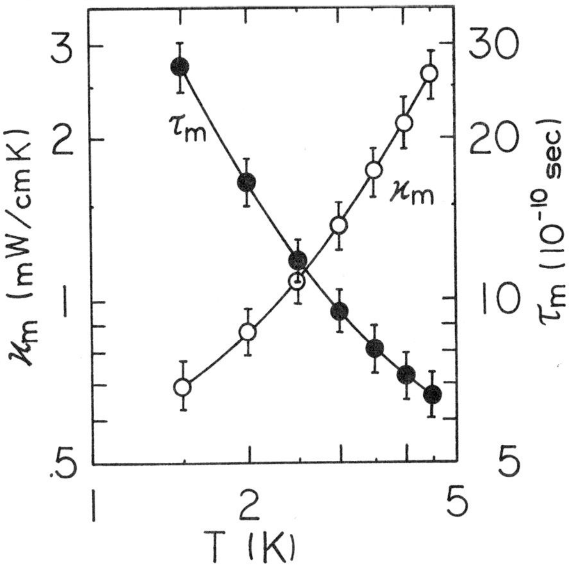

Fig. 2. Magnon thermal conductivity κ_m and magnon lifetime τ_m at $B_E = M_s = 0$.

Substituting these $\kappa_m(0,0,T)$ values into Eq.(3), we can solve for Q at any given temperature. We assume $M_s = 2.62$T (Ref. 5), $g = 2.09$ (Ref. 6), $D = 6.84\times10^{-40}$ Jm^2 (Ref. 7). Then we define an average magnon lifetime $\bar{\tau}_m = Q/\bar{\omega}$, where we choose $\bar{\omega} = 1.8\ k_BT/h$ for the average magnon frequency since the integrand in J_D peaks at $x = 1.8$. The resulting values of $\bar{\tau}_m$ are shown on Fig.2. We find $\bar{\tau}_m \propto T^{-1.2} \propto (\bar{\omega})^{-1.2}$ on the average over the range between 1.5K and 4.5K, in rough agreement with the prediction $\tau_m \propto \omega^{-1}$ of the s-d exchange model in the dirty limit. The agreement is even better between 2 and 4.5K, where our data (Fig.2) give $\bar{\tau}_m \propto (\bar{\omega})^{-1}$ on the average. At 4K, $\bar{\tau}_m \simeq 7.3\times10^{-10}$ sec, which is at least four times the values found in nickel-rich Ni-Fe alloys.[1]

These large τ_m values in $Fe_{68}Co_{32}$ may be related to the very large and well defined momentum gap[1] present in this material, and to the very small electronic density of states at the Fermi level.

*Work supported by the U. S. National Science Foundation.

REFERENCES

1. W. B. Yelon and L. Berger, Phys. Rev. B6, 1974 (1972).
2. T. Holstein and H. Primakoff, Phys.Rev.58, 1098(1940).
3. G. N. Grannemann and L. Berger, to be published; B. Heinrich, D. Fraitova and V. Kambersky, Phys. Status Solidi 23, 501 (1967).
4. A. H. Mitchell, Phys. Rev. 105, 1439 (1957).
5. R. M. Bozorth, Ferromagnetism (Van Nostrand, New York, 1951), p.194.
6. G. G. Scott and H. W. Sturner, Phys. Rev. 184, 490 (1969).
7. R. D. Lowde, et al., Phys. Rev. Letters 14, 698 (1965).

EFFECT OF SPIN WAVE AMPLIFICATION ON THE MAGNETORESISTANCE OF FERROMAGNETIC SEMICONDUCTORS

M. D. Coutinho, F°
Departamento de Fisica, Universidade Federal de Pernambuco, 50000 Recife, Pe., Brasil and Laboratory of Atomic and Solid State Physics, Cornell University, Ithaca, N.Y. 14853
and
L. C. M. Miranda and S. M. Rezende
Departamento de Fisica, Universidade Federal de Pernambuco, 50000 Recife, Pe., Brasil

ABSTRACT

Very recently[1,2], the present authors have investigated the conditions under which the charge carrier-magnon interactions can lead to spin wave (magnon) amplification in ferromagnetic semiconductors. We have concluded that in the most suitable ferromagnetic semiconductor known for these experiments, namely $CdCr_2Se_4$, the d-spin-s-orbit interaction between the carriers and the localized spins gives the most important contribution to the amplification. The purpose of this communication is to analyze the effect of magnon amplification on the magnetoresistance of ferromagnetic semiconductors, with application to $CdCr_2Se_4$. This entails relating the transport coefficients to the characteristics of the magnon system. A simple model for the non-linear relaxation of magnons is proposed which can account for the observed[7] positive electric-field dependent magnetoresistance of $CdCr_2Se_4$.

INTRODUCTION

In previous papers[1,2] we have investigated both in the classical and quantum regimes the conditions under which the charge carrier-magnon interactions can lead to amplification in the magnon population in ferromagnetic semiconductors. It has been shown that this state of instability is reached when the drift velocity of the carrier due to an applied dc electric field exceeds the phase velocity of the spin waves. A net amplification then occurs when the resulting rate of creation of magnons is greater than the magnetic losses due to other effects. This problem has been also analyzed by a number of authors[3] in the framework of phenomenological theories. We have concluded[2] that in the most suitable ferromagnetic semiconductor known for these experiments, namely p-type $CdCr_2Se_4$, the spin-orbit (or spin-current) interaction between the carriers and the localized spins[4] gives the most important contribution for the amplification. In this material, because of the fact that the holes are weakly coupled by exchange to the localized moments[5] the s-d interaction is not important for the amplification. We have also shown that the plasmon assisted processes contribute negligibly to the amplification. Another information, which results from an analysis of the classical and quantum spin-orbit amplification coefficients,[1,6] is the fact that the state of instability can be reached more easily in the classical regime $k\ell < 1$, i.e., when the magnon wavelength is longer than the electron mean free path. This is not surprising because the only electron-magnon interaction which corresponds to the classical situation is the spin-current interaction mentioned above.

It has been recently suggested by Balberg and Pinch[7] that spin wave amplification due to carrier interaction can account for the observed positive electric-field dependent magnetoresistance of $CdCr_2Se_4$. Those authors made a qualitative fit of their experimental results to the classical theory of spin wave amplification imposing the additional constraint that the magnon population reaches a saturation value with increasing carrier drift velocity.

The purpose of this paper is to analyze the effect of magnon amplification on the magnetoresistance of ferromagnetic semiconductors, with application to $CdCr_2Se_4$. Contrary to the main assumption of Balberg and Pinch[7], we shall be interested in the situation when the magnon population has not reached its saturated value with increasing carrier drift velocity.

MAGNETORESISTANCE AND DRIFT VELOCITY

The equation of motion for the density of carrier momentum is

$$\frac{d\vec{P}}{dt} = n_o e \vec{E}_o - \frac{\vec{P}}{\tau} - \vec{P}_M, \quad (1)$$

where τ and n_o are respectively the relaxation time and the density of the carriers, and $\vec{E}_o$ is the d.c. applied electric field. $\vec{P} = mn_o\vec{v}_d$, and

$$\vec{P}_M = \frac{1}{V}\sum_{\vec{k}} \hbar\vec{k}\,\frac{dN(\vec{k})}{dt} \quad (2)$$

is the density of carrier momentum transferred to the magnons ($dN/dt > 0$), or received from the magnons ($dN/dt < 0$), per unit time, with m being the carrier effective mass. In equilibrium we should have $dP/dt = 0$, from which using (1) we can express the carrier drift velocity in the form $\vec{V}_o - \vec{V}_M$, where $\vec{V}_o = (e\tau/m)\vec{E}_o$ and

$$\vec{V}_M = \frac{\tau}{Vmn_o}\sum_{\vec{k}} \hbar\vec{k}\,\gamma(\vec{k})\,\overline{N}(\vec{k}) \quad (3)$$

Here $\vec{V}_o$ and $\vec{V}_M$ are respectively the ohmic drift velocity and the magnetoelectric velocity, $\gamma(\vec{k})$ is the classical spin-orbit amplification coefficient and $\overline{N}(\vec{k})$ is the magnon population in the steady state regime. Under the experimental conditions assumed[7] we have[1,2]

$$\gamma(\vec{k}) = \frac{\omega_M\omega_p^2}{c^2k^2}(\vec{k}\cdot\vec{V}_d - \omega_k) \quad (4)$$

where $\omega_p^2 = 4\pi n_o e^2/m$ is the plasma frequency, $\omega_M = 4\pi\gamma M_o$ ($\gamma = g\mu_B/\hbar$ is the gyromagnetic ratio), ω_k is the magnon frequency and M_o is saturation magnetization. $\vec{V}_d$ is then determined selfconsistently from the relation $\vec{V}_d = \vec{V}_o - \vec{V}_M$.

Now, when there is spin wave amplification this effective magnetoelectric velocity will lead to an effective increase of the crystal resistivity. Under small-signal conditions, the change in resistivity can be expressed as[8]

$$\frac{\rho-\rho_o}{\rho_o} = \frac{V_o - V_d}{V_o} = \frac{V_{MZ}}{V_o} \quad (5)$$

Here $\vec{V}_o$ was assumed in the $\hat{z}$ direction and V_{MZ} means $\vec{V}_M\cdot\hat{z}$. In order to evaluate (5) it is necessary to estimate the magnon population in the steady state regime.

The rate equation for the magnon population is given by

$$\frac{dN(\vec{k})}{dt} = \left[\gamma(\vec{k}) - \nu(\vec{k})\right]N(\vec{k}) + \left.\frac{dN(\vec{k})}{dt}\right|_{NL}, \quad (6)$$

where $\nu(\vec{k})$ represents the magnetic losses due to linear processes and the last term expresses the losses due to the nonlinear processes. For defining the region of instability in k-space $\left(\gamma(\vec{k}) > \nu(\vec{k})\right)$, we have taken[9] $\nu \sim 10^7 sec^{-1}$. For the non-linear losses we shall make the assumption that there exists one dominant non-linear process such that

$$\frac{dN(\vec{k})}{dt}\bigg|_{NL} \simeq -\frac{N^2(\vec{k})}{\tau_{NL}(\vec{k})}, \qquad (7)$$

where $\tau_{NL}(\vec{k})$ is the relaxation time associated with the process. Equations (6) and (7) determine the amplified magnon population in the steady state, which reads

$$\overline{N}(\vec{k}) \simeq \tau_{NL}(\vec{k})\left[\gamma(\vec{k}) - \nu(\vec{k})\right] \qquad (8)$$

We expect this model to be a reasonable approximation when the width of the region of instability in k-space and the width of the cone of amplification (defined by the angle between $\vec{k}$ and $\vec{V}_d$) are both very small. In this circumstance the amplified magnons have essentially the same wave vector $\vec{k}$ mainly in the $\hat{z}$ direction.

The next problem is to search the dominant nonlinear process which would lead the magnon system to a steady state regime. Among the various possible decay processes involving two unstable magnons, we have concluded[6] that the most efficient one is 2 magnons→1 phonon due to the magnetoelastic interaction involving only the magnetic $\tilde{Cr}$ ions which had a charge change because of Ag doping[7]. These Cr ions have nonzero angular momentum and therefore can provide a strong spin-lattice interaction. To describe this localized interaction, the magnetoelastic Hamiltonian can be modeled by[10]

$$H_{ME} = N'\int d\vec{x}\, e^{-\frac{x^2}{a^2}} \left[\sum_{i,j,\alpha,\beta} b_{ij\alpha\beta} M_\alpha(\vec{x}) M_\beta(\vec{x}) \epsilon_{ij}(\vec{x})\right] \qquad (9)$$

where N' is the number of magnetic impurity ions, a is the lattice parameter, $b_{ij\alpha\beta}$ are the magnetoelastic constants, $M_\alpha(\vec{x})$ are the components of the magnetization and $\epsilon_{ij}(\vec{x})$ are the components of the strain tensor. The choice of a gaussian factor to localize the interaction is just a matter of mathematical convenience. Its presence prevents momentum conservation for processes taking place around the magnetic impurity ions, such as, 2 unstable magnons→1 phonon. The relaxation time defined in (7) is then easily evaluated if one uses the magnon and phonon representation for the magnetization and strain tensor, respectively. The final result is

$$\frac{1}{\tau_{NL}} \simeq \frac{(\pi\tilde{b}\tilde{c}g\mu_B M_o)^2 \omega_k^3}{\hbar\eta V_{ph}}, \qquad (10)$$

where $\tilde{b}$ is the magnetoelastic constant which takes into account all kinematic possibilities, $\tilde{c} = N'/N$ is the magnetic impurity concentration, η is the crystal density and V_{ph} is the phonon velocity. According to (10) and (8), one notices that variations in $\vec{V}_d$ imply variations of $N(\vec{k})$ in the same direction. This result can account for the rapid increase in the magnetoresistance observed[7] for electric fields between 30 and 300 V cm^{-1}.

Since there is no available data for the magnetoelastic constant $\tilde{b}$ for the experimental case under study, quantitatively we may at best make some order of magnitude estimative of the validity of the theory. Taking $V_d \sim 6 \times 10^5$ cm sec^{-1}, $\omega_k \sim 3 \times 10^{10}$ sec^{-1}, $\omega_p^2 \sim \pi\, 10^{28}$ sec^{-2}, $\omega_M \sim 6.7 \times 10^{10}$ sec^{-1}, $V_{ph} \sim 10^5$ cm sec^{-1}, $\tau \sim 10^{-12}$ sec^{-1}, $m \sim 10^{-28}$ gr, $T \sim 10^2$ °K, $\tilde{c} \sim .4\%$, $\eta \sim 1$ gr cm^{-3}, the change in resistivity[7] $\Delta\rho/\rho_o \sim 2\%$ and using (3), (4), (5), (8) and (10), we obtain

$$\tilde{b} \sim 10^2 \text{ erg cm}^{-3}\text{ G}^{-2} \qquad (11)$$

This value of $\tilde{b}$ is of the same order as the magnetoelastic constant in typical magnetic insulators.[11]

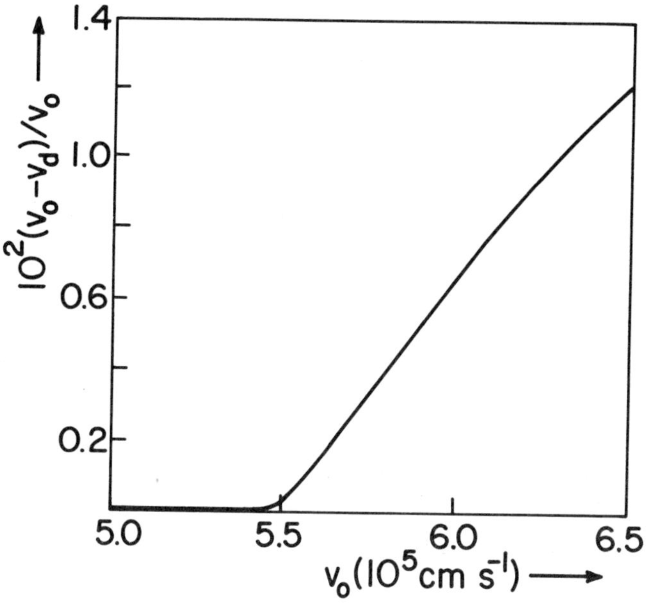

Fig. 1 Fractional change in resistivity as a function of the ohmic drift velocity

This result seems to be a reasonable check for the theory taking into account its simplicity and the uncertainties in some of the parameters involved. Using the value (11) for $\tilde{b}$ Fig. 1 shows the variation of the change in resistivity (5) with V_o.

In conclusion we remark that the result of the theory is in qualitative agreement with the experimental ones, in the region of relatively small electric fields. However, in the region of more intense electric fields, where the magnetoresistance begins to saturate, the present model fails. In this region, one should have other relaxation mechanisms, which were not considered here, whose effectiveness should lead the magnon population to a definite saturated value.

REFERENCES

1. M. D. Coutinho, Fº, L. C. M. Miranda and S. M. Rezende, phys. stat. sol. (b) 57, 85 (1973).
2. M. D. Coutinho, Fº, L. C. M. Miranda and S. M. Rezende, phys. stat. sol. (b) to be published.
3. A. I. Akhiezer, V. G. Bar'yakhtar and S. V. Peletminskii, Phys. Letters 4, 129 (1963); B. Vural, J. Appl. Phys. 37, 1030 (1966); H. N. Spector, Solid State Comm. 6, 811 (1968).
4. E. Abrahams, Phys. Rev. 98, 387 (1955); E. A. Turov, in Ferromagnetic Resonance (ed. by S. V. Vonsovskii, Pergamon Press, Oxford, 1966), chap. V.
5. C. Haas, Crit. Rev. Solid State Sci., 1, 47 (1970).
6. M. D. Coutinho, Fº, L. C. M. Miranda and S. M. Rezende, phys. stat. sol. (b), to be published.
7. I. Balberg and H. L. Pinch, Phys. Rev. Letters 21, 909 (1972).
8. A. Many and I. Balberg, in Electronic Structures in Solids (ed. by E. D. Haidemenakis, Plenum Press, N.Y., 1969).
9. J. S. Bartokvwski, J. S. Page and R. C. Le Craw, J. Appl. Phys. 39, 1071 (1968).
10. R. M. White and M. Sparks, Phys. Rev. 130, 632 (1963).
11. See, e.g., F. Keffer, in Handbuch der Physik (ed. by S. Flügge, Springer-Verlag, Berlin, 1966), vol. XVIII/2.

EFFECT OF SINGLE-ION ANISOTROPY ON TWO-MAGNON RAMAN SPECTRA OF HEISENBERG FERROMAGNETS*

B. J. Choudhury and P. D. Loly
Department of Physics, University of Manitoba, Winnipeg, Manitoba, Canada R3T 2N2

ABSTRACT

Two-magnon Raman spectra exhibit magnon pairing effects for zero total wavevector ($\vec{K} = 0$) as resonances within, or bound states below, the two-magnon continuum. Calculations of the appropriate spectral function describing the excitation of two spin deviations on one site are reported for nearest neighbor s.c. and f.c.c. Heisenberg ferromagnets with single-ion anisotropy as a variable parameter. Since our approach probes effects both within and outside the continuum, the results go considerably beyond conventional bound state analyses. As the relative strength (D/J) of the single-ion anisotropy to the exchange is increased, the resonance found in the isotropic case moves lower in the two-magnon band, which itself is elevated by the anisotropy gap. The line profile varies in a complicated fashion, narrowing at first, then broadening before weakening as the single-ion resonance moves out of the bottom of the band to become a bound state. The single-ion bound state has been located through the real part of the spectral function and is confirmed by a sum rule analysis.

During the last decade two-magnon bound states have received much attention but their relationship to the resonances occurring in two-magnon Raman spectra has remained largely unprobed. This, despite the fact that these resonances probably represent the most observable consequences of magnon pairing interactions. In fact the bound states have only been found at large values of the total wavevector ($\vec{K}$) of the magnon pair in s.c.[1] and b.c.c.[2] ferromagnets, they do exist throughout the entire Brillouin zone for large enough exchange[1] or single-ion[3] anisotropies in s.c. ferromagnets. Although Boyd and Callaway[4] showed that the bound states initially broaden into resonances when they enter the two-magnon continuum at non-zero $\vec{K}$, it has not so far proven possible to follow the resonances all the way to the Raman axis at $\vec{K} = 0$.

It is the purpose of this investigation to examine the conversion of the Raman resonance into a bound state at $\vec{K} = 0$ as the anisotropy is increased and this is done with the case of single-ion anisotropy and the excitation of two spin deviations on the same site. We should note that although the resonances were originally predicted[5] in the case of antiferromagnets, for which they were subsequently observed[6], a recent study[7] for isotropic ferromagnets shows a qualitatively similar resonance phenomenon. Any results obtained for ferromagnets will shed light on the less tractable antiferromagnetic problem.

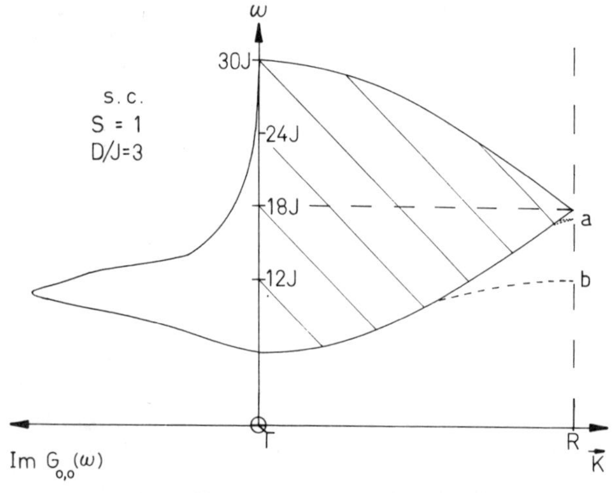

Fig. 1. The two-magnon continuum (hatched) for S = 1 and D = 3J as a function of total wavevector $\vec{K}$, including the exchange (a) and single-ion (b) bound states as located by Silberglitt and Torrance[1], compared with the present calculations of the two-magnon Raman spectrum, which is proportional to Im $G(o,o,\omega)$.

Using a ferromagnet with the Hamiltonian

$$H = -\frac{1}{2} J \sum_{i,\delta} \vec{S}_i \cdot \vec{S}_{i+\delta} - D \sum_i (S_{iz})^2 \quad (1)$$

with the nearest neighbor exchange (J) and the single-ion anisotropy (D) as positive parameters, we work with zero-temperature spin Green functions $G(\delta,\delta')$,

$$G(\delta,\delta') = \frac{J}{4SN} \langle\langle A^+_\delta ; A_{\delta'} \rangle\rangle \text{ where } A_\delta = \sum_i S^-_i S^-_{i+\delta}. \quad (2)$$

The two-magnon Raman spectrum for the creation of two-spin deviations on a single site via a spin-orbit coupling mechanism has been shown by Thorpe[8] to be given by Im $G(o,o,\omega)$. One then proceeds via the equation of motion of the $G(\delta,\delta')$ to construct Dyson equations which specify $G(o,o)$ in a manner which is a straightforward extension of the isotropic problem. As in the isotropic case the formulation involves no approximations because of the use of zero-temperature and thus fully incorporates the effects of magnon-magnon interactions with the result,

$$G(o,o) = [1 - 1/(2S)]\, K\, [1 + DK/(JS)]^{-1} \quad (3)$$

where $$K = [g_0 - g_1/(2S)][1 - g_1\tilde{\omega}/(4S^2J)]^{-1} \quad (4)$$

with $$g_\delta = \frac{2SJ}{N} \sum_k \exp(i\vec{k}\cdot\vec{\delta})(\tilde{\omega} - 2\omega_k)^{-1}, \quad (5)$$

$$\tilde{\omega} = \omega - 2(2S - 1)D, \quad (6)$$

and $$\omega_k = S[J(o) - J(k)]. \quad (7)$$

In (4), g_0 and g_1 are lattice Green functions for $\delta = 0$ and for nearest neighbor $\vec{\delta}$ respectively, and are related by an identity

$$2SJ = (\tilde{\omega} - 2SJZ)g_0 + 2SJZ\, g_1 \quad (8)$$

where Z is the number of nearest neighbors. The real and imaginary parts of g_0 required for the s.c. case have been obtained from a tabulation of Hone, Callen and Walker,[9] and those for $G(o,o)$ were then obtained by using the complex arithmetic facility of FORTRAN. The results were analysed with the aid of sum rules for the zeroth and first moments of the spectrum which were derived in the same manner[7,8] as in the isotropic case (and will be written down explicitly in the next section).

Our calculated two-magnon Raman spectra, proportional to $-\text{Im}\, G(o,o,\omega)$, are shown in Fig. 1 and 2 for spin one. In Fig. 1 we show the Raman resonance ($\vec{K} = 0$) side by side with the bound state information derived by Silberglitt and Torrance[3] for the (1,1,1) $\vec{K}$-direction. The region of the Brillouin zone in the (1,1,1) direction over which the bound state is below the two-magnon continuum extends from a critical total wavevector ($\vec{K}_c$) up to the zone corner. For D = 0 the single-ion bound state is not present, leaving only the s- and d-wave exchange bound states.[1] Silberglitt and Torrance[3] showed that $\vec{K}_c$ went to zero for D/J between 3 and 10.

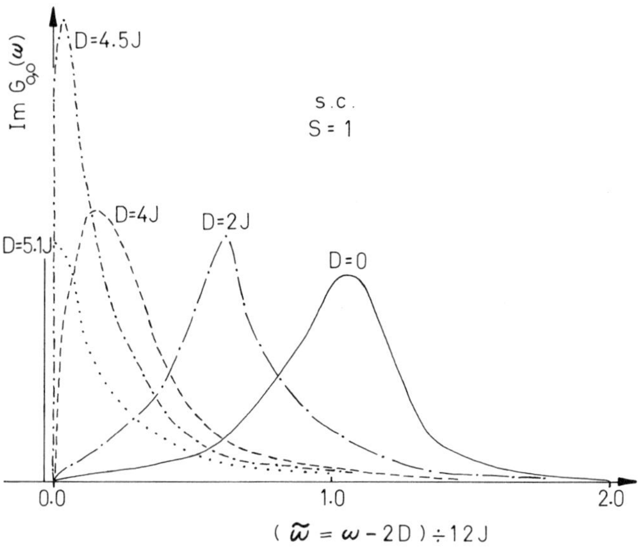

Fig. 2. Calculated spectra for a range of uniaxial anisotropy parameters, D, including the delta-function bound state for D = 5.1 J.

Fig. 2 shows the variation of the Raman spectra for a representative range of values of D/J including the isotropic case (D = 0). It is clear that the resonance peak is moving lower in the two-magnon band, changing shape but maintaining a constant integrated intensity in accordance with the sum rule

$$(-) \int \text{Im } G(o,o,\tilde{\omega})d\tilde{\omega} = \pi J(2S - 1) \qquad (9)$$

for D/J = 0,1,2, and 4.5. A dramatic drop in integrated intensity within the band occurs between D/J equal to 4.5 and 5.1, signalling the appearance of a single-ion bound state below the two-magnon continuum in the latter case. The weight in this bound state is just that required to satisfy (9) and its position can then be determined with the aid of the first moment sum rule

$$(-) \int \text{Im } G(o,o,\tilde{\omega})\tilde{\omega}d\tilde{\omega} = \pi J(2S - 1)(2SJZ - 2D). \qquad (10)$$

Although the delta function lineshape of the bound state will not be seen directly in Im G(o,o), it will appear as a divergent discontinuity in Re G(o,o,ω) and this provides corroboration for the bound state.

Finally in Fig. 3 we plot the frequencies of the band maximum and bound state against the single-ion anisotropy parameter D. The limits of the two-magnon continuum serve to show how the band maximum moves before and after the bound state separates from the band, while the constant frequency (for S = 1) of the single-ion bound state at the zone corner (as calculated for the Ising situation[10]) serves as an upper bound on the $\vec{K}$ = 0 single-ion bound state.

We have shown how the Raman lineshape varies with the amount of single-ion anisotropy and also how single-ion bound states emerge below the two-magnon-continuum above a critical value of D/J. It is expected that a similar behaviour will result for exchange anisotropy increases which narrow the two-magnon continuum and facilitate the eventual emergence of exchange bound states (which correspond to spin deviations on neighboring sites in the Ising limit) on the Raman axis.

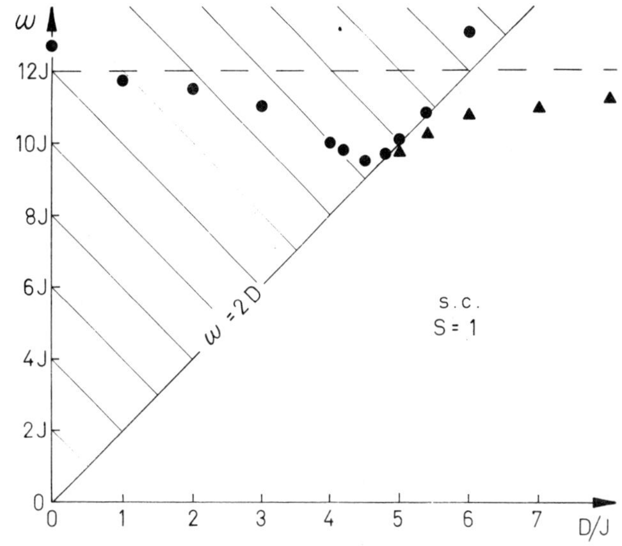

Fig. 3. Frequencies of the band maximum (circles), and, when appropriate, the single-ion bound state at $\vec{K}$ = 0 (triangles), together with the lower portion of the two-magnon continuum (hatched) and the position of the bound state at the zone corner (dashed line).

In addition we have found a similar behaviour for the f.c.c. case as D/J increases. There, the principal difference is a more complex resonance lineshape which at one point develops a double peak before the separation of the bound state from the band. It is perhaps worth noting that there are no exchange bound states in the nearest neighbor f.c.c. case because the two-magnon continuum does not narrow to a point as in the s.c. and b.c.c. cases. As a consequence the single-ion bound state can not appear even for finite $\vec{K}$ until a critical value of D/J is attained. In both s.c. and f.c.c. cases a rich structure develops in the remnants of the band for very large D/J when the bound state is a long way below its lower edge. At the present time we do not understand the reasons for this, nor indeed, for some of the lineshape variation before the emergence of the bound state.

The spectral probe approach used in this investigation yields considerably more information for the $\vec{K}$ = 0 spectra than does a conventional bound state study[3] and it is clear that an analogous approach to the finite $\vec{K}$ problem would enable one to follow completely the transition between the Raman resonance and bound states for cases where the bound states do not exist at $\vec{K}$ = 0.

REFERENCES

1. M. Wortis, Phys. Rev. 132, 85-97 (1965).
2. A. M. Bonnot and J. Hanus, Phys. Rev. B 7, 2207-2209 (1973).
3. R. Silberglitt and J. B. Torrance, Phys. Rev. B 2, 772-778 (1970).
4. R. G. Boyd and J. Callaway, Phys. Rev. 138A, 1621-1629 (1965).
5. R. J. Elliott et al., Phys. Rev. Letters 21, 147-150 (1968); R. J. Elliott and M. F. Thorpe, J. Phys. C 2, 1630-1643 (1969).
6. P. A. Fleury, Phys. Rev. Letters 21, 151-153 (1968).
7. P. D. Loly, B. J. Choudhury and W. R. Fehlner, submitted.
8. M. F. Thorpe, Phys. Rev. B 4, 1608-1613 (1971).
9. D. Hone, H. B. Callen and L. R. Walker, Phys. Rev. 144, 283-295 (1966).
10. R. J. Elliott and A. J. Smith, J. de Physique 32, C1-585-C1-589 (1971).

*Research supported in part by the National Research Council of Canada.

MAGNETIC CIRCULAR DICHROISM IN Si AND Sn DOPED GARNETS

B. Antonini, A. Paoletti†, P. Paroli, A. Tucciarone
Laboratorio Elettronica Stato Solido C.N.R., Via Cineto Romano, 42, Roma (Italy)
J. F. Dillon, Jr., E. M. Gyorgy and J. P. Remeika
Bell Laboratories, Murray Hill, N. J. 07974

ABSTRACT

Both theoretical and experimental evidence have recently indicated that magnetic circular dichroism in Si doped YIG is a manifestation of the same "impurity centers" responsible for the photoinduced effects in this material. In this paper we report measurements of magnetic circular dichroism in (011) plates of Si and Sn doped crystals at 4.2°K. Magnetic circular dichroism was measured in the wavelength range 1.1-1.6μm. Results for the two different dopants are compared and discussed. It appears that magnetic circular dichroism occurs both for Si which prefers tetrahedral sites and for Sn which prefers octahedral sites. However, the detailed dependence on wavelength is markedly different for the two dopants. To our knowledge this is the first demonstration that the properties of the Fe^{++} ion depend on the identity of the compensating ion.

It is generally assumed that the substitution of tetravalent dopants, such as Si^{4+} or Ti^{4+}, in yttrium iron garnet (YIG), gives rise to Fe^{2+} ions in these systems, as a result of charge compensation. In the case of YIG (Si) it has been extensively established experimentally that these Fe^{2+} impurities are the active centers in a mechanism that allows many properties of the material (such as coercive force, initial permeability, static anisotropy and linear dichroism) to be changed by means of irradiation with near infrared light(1-6). An effective technique for investigating these Fe^{2+} centers consists in irradiating first the material with linearly polarized white light and then measuring the photoinduced linear dichroism(4,7). One difficult with this method is that photoinduced linear dichroism is essentially a "second-order" effect since its magnitude is the result of two subsequen processes: the "irradiating" one and the "sensing" one. This particular character of photoinduced linear dichroism not only has the consequence of making the effect often too small to measure, but, in addition, makes its analysis quite difficult most of the times, because of the occurrence in the system, after the "irradiating" process, of relaxation mechanisms tending to obliterate its effect altogether. Recently, however, it has been suggested that the Fe^{2+} impurities in the garnets should give rise to straight circular dichroism, a first-order effect, that could thus constitute a powerful direct probe in the study of these centers(8). Subsequently, first-order circular dichroism (1st order CD) has been measured in YIG(Si) at several temperatures and experimental evidence has been given supporting the view that, indeed, it is a manifestation of the Fe^{2+} centers(9). Furthermore, it has been shown that the relative amount of Fe^{2+} centers in several plates of YIG(Si) can be directly determined by means of a simple and nondestructive CD measuremen at room temperature. In this paper we report measurements of first-order circular dichroism in (110) plates of Si and Sn doped crystals at 4.2°K. It is shown that also in the Sn-doped specimens 1st order CD provides a simple direct measurement of the relative amounts of Fe^{2+} centers. In general, however, the detailed dependence on wavelength of 1st order CD is markedly different in the case of Si and Sn dopan thus giving direct evidence of the fact that the quadrivalent dopant determines not only the existence but also the behavior of the Fe^{2+} centers.

TABLE I

List of samples with their dopant contents in atoms/formula: the third column gives the dopant amount as it was measured; the fourth the dopant amount in excess to the one in the "purest" sample with the same dopant, as explained in the text.

Sample No.	Dopant	x	$x-x_o$
1	Si	0.017	0.011
2	Si	0.045	0.039
3	Sn	0.045	0.026
4	Sn	0.077	0.058

Measurements were made on (110) single crystal plates which were grown by slow cooling from a PbO, B_2O_3 flux. In Table I we list our samples giving the respective Si and Sn contents, x, in atoms per formula unit. The silicon content was determined by measuring the optical absorption at 1.06μ and using the Wood and Remeika calibration curve(10). Following these authors we assume that some amount of Si in the garnets goes to compensate for existing impurities. This amount is $x_o = 0.006$ and we take it as the true zero in the x scale. In the case of tin we applied the same procedure using the calibration curve of Fig. 1, measured by

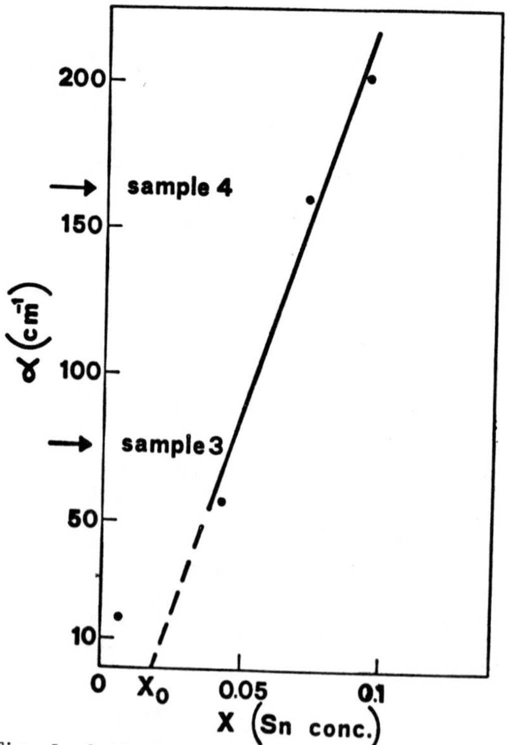

Fig. 1 Optical absorption at λ = 1.20μ for Sn doped YIG. The straight line is best fitted to the experimental points by the lower one. The value of x_o, obtained by extrapolation of the line to α=0, is assumed to be the Sn concentration of the "purest" sample, as explained in the text.

us at $\lambda = 1.2\mu$. The value of x_o was determined by linear extrapolation of the curve at the higher tin concentrations and was found equal to 0.019. The site distribution of the dopants is, according to Geller et al.[11,12], as follows: Si^{4+} ions prefer tetrahedral sites and Sn^{4+} ions octahedral sites.

The circular dichroism measurements were carried out in the wavelength range 1.1-1.6μ. The sample was placed inside a liquid helium cryostat with optical windows and cooling was always carried out in the dark, with an applied field of 4 kOe along the [110] axis. The sensing beam from a monochromator was circularly polarized by means of a polarizer and a Babinet-Soleil compensator, then passed through the sample and was finally detected by a PbS cell, the output of which was fed into a lock-in amplifier. The circular dichroism Δ was determined by the relation $\Delta = (1/t)(\ln I^+ - \ln I^-)$, where t is the sample thickness, and I^+, I^- are the intensities transmitted by the sample when the applied magnetic field is parallel and antiparallel to the spin of the photons respectively.

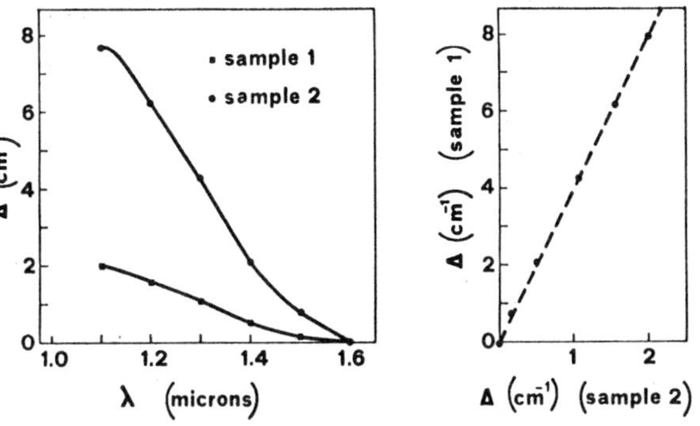

Fig. 2(a) Circular dichroism spectra of Si-doped samples. (b) Circular dichroism of sample 1 vs circular dichroism of sample 2.

Figure 2(a) shows the circular dichroism spectra $\Delta(\lambda)$ at liquid-helium temperature for the Si-doped samples. The spectra are the same in shape, since, as shown in Reference 9, they are associated with the silicon-induced impurity centers that do not change in character with increasing Si concentration, although they change in number. The universality of the shape is best demonstrated by the plot of Fig. 2(b) which shows that the ratio of the Δs of the two samples is a constant, with changing wavelength. This constant gives the ratio between the numbers of impurity centers in the two specimens and is equal to 3.7. Figure 3 shows the same plots as Fig. 2, referred to the Sn-doped specimens. In particular, Fig. 3(b) shows that also the Sn-doped samples, as the Si-doped ones, have spectra $\Delta(\lambda)$ that are all the same in shape. The constant ratio between the two spectra, giving the relative amount of Fe^{2+} ions in the two

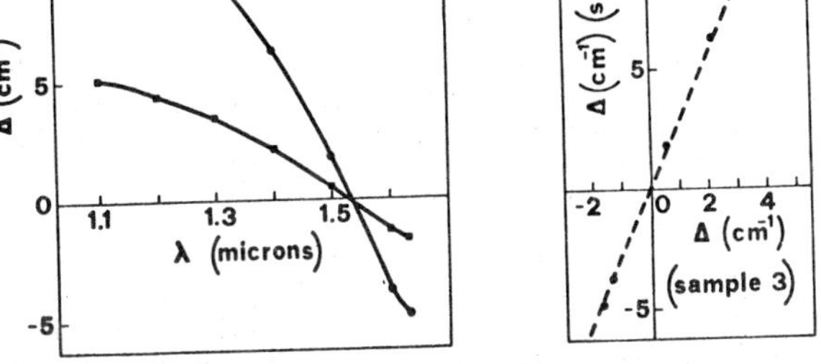

Fig. 3(a) Circular dichroism spectra of Sn-doped samples. (b) Circular dichroism of sample 4 vs circular dichroism of sample 3.

Sn-doped systems, is equal to 2.8. The $\Delta(\lambda)$ associated with the Sn doping are definitely different spectra from the ones associated with the Si doping. This is best demonstrated by plotting the two dichroisms, one versus the other at the same wavelength, as in Fig. 4. To our knowledge, the difference in the two spectra is the first demonstration that the role of the quadrivalent dopant goes well beyond the mere creation of the Fe^{2+} centers since it determines also their character.

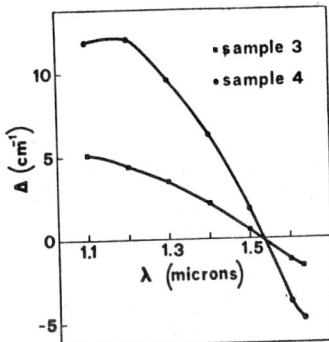

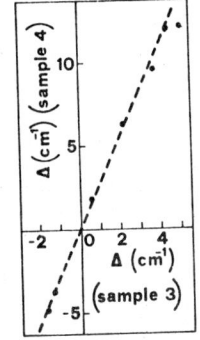

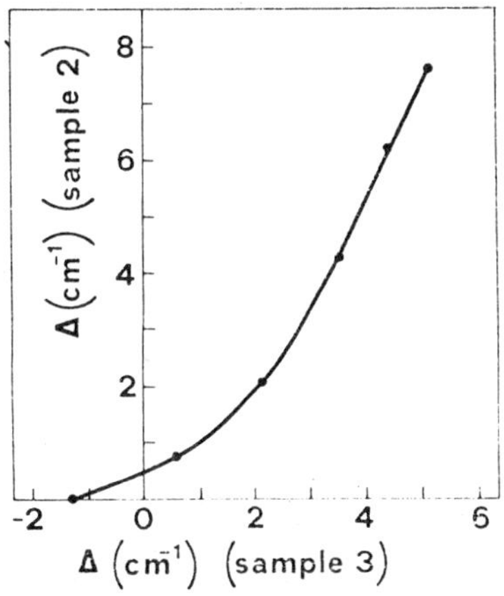

Fig. 4 Circular dichroism of sample 2 (Si-doped) vs circular dichroism of sample 3 (Sn-doped).

ACKNOWLEDGMENT

The authors wish to thank Mr. F. Scarinci for expert technical assistance, and Mrs. Ghislaine A. Pasteur for determining the optical absorption coefficients of Fig. 1.

REFERENCES

†also: Istituto di Fisica, Facolta di Ingegneria, Universita di Roma

1. Teale, R. W. and Temple, D. W., Phys. Rev. Lett. 19, 904 (1967).
2. Enz, U. and Van Der Heide, H., Solid State Comm. 6, 347 (1968).
3. Pearson, R. F., Annis, A. D. and Kompfner, P., Phys. Rev. Lett. 21, 1805 (1968).
4. Dillon, J. F., Gyorgy, E. M. and Remeika, J. P., Phys. Rev. Lett. 22, 643 (1969).
5. Holtzwijk, T. H., Lems, W., Verhulst, A. G. H. and Enz, U., I.E.E.E. Trans. Mag. 6, 853 (1970).
6. Enz, U., Metselaar, R. and Rijnierse, P. J., J. Phys. 32, C1-704 (1971).
7. Alben, R., Gyorgy, E. M., Dillon, J. F., Remeika, J. P., Phys. Rev. B 5, 2560 (1972).
8. Tucciarone, A., Lett. Nuovo Cimento 6, 20 (1973).
9. Antonini, B., Paoletti, A., Paroli, P., Tucciarone, A., Dillon, J. F., Gyorgy, E. M., Remeika, J. P. (to be published).
10. Wood, D. L., Remeika, J. P., J. Appl. Phys. 37, 1232 (1966).
11. Geller, S., Williams, H. J., Sherwood, R. C., Espinosa, G. P., J. Phys. Chem. Solids 23, 1525 (1962).
12. Geller, S., Bozorth, R. M., Gilleo, M. A., Miller, C. E., J. Phys. Chem. Solids 12, 111 (1959).
13. This method enhances the photoinduced difference in population distribution from one irradiation to the other and will be discussed in detail in a forthcoming communication.

OBSERVATION OF THE TWIN-WALLS MOVEMENT IN $KNiF_3$ BY DICHROISM

R.H. PETIT, J. FERRÉ and J. BADOZ, Laboratoire d'Optique Physique,
EPCI, 10 rue Vauquelin, 75231 PARIS CEDEX 05

ABSTRACT

In a cubic antiferromagnet it has been predicted by Néel that the reversible movement of the twin-walls under a magnetic field H or a stress σ below T_N precedes the spin rotation inside a monodomain. Magnetic circular dichroism (MCD) and stress linear dichroism (SLD) measurements on exciton and exciton-phonon bands in $KNiF_3$ have allowed us to follow accurately these twin-walls movements versus H and σ for the first time. The modification of the shape of the MCD spectrum and the variation of the SLD amplitude due to the reorientation of the spins during this process is explained. Untwinned crystals have been obtained above low critical values $H_{cr} \sim 20$ kOe and $\sigma_{cr} \sim 50$ kg/cm^2.

Néel has predicted (1) that the twin-walls movement under a magnetic field H or a stress σ in a cubic antiferromagnet such as $KNiF_3$ below $T_N = 246$ K, precedes the well-known spin-flopping inside a single domain which occurs in uniaxial antiferromagnets. In this paper we confirm this suggestion by using sensitive optical methods : the magnetic circular dichroism (MCD) and the stress induced linear dichroism (SLD) on $KNiF_3$. We show the spectra of the $^3A_2 \rightarrow {}^1E$ transition but the same modifications have been observed on the other absorption bands in the visible range.

Without perturbation three kinds of domains (d_x , d_y , d_z) exist the spins of which are oriented along the fourfold axis. A magnetic field H_z favours d_x and d_y domains and a stress σ_x stabilises d_x domains (4).

MCD EXPERIMENTS

In these experiments the magnetic field was oriented along the light beam direction z. In a low magnetic field the spin up and spin down sublattices are energetically inequivalent for d_z domains. Then, for narrow lines the MCD ΔA_c looks like the first derivative of the absorption curve $df(\bar{\nu})/d\bar{\nu}$. Inside d_x or d_y domains the spins tilt a little towards the magnetic field direction, which give rise to a term proportional to $(H/H_E)\ f(\bar{\nu})$ (5). In the more general case, if N_z represents the proportion of spins inside d_z domains the MCD is expressed as :

$$\frac{\Delta A_c}{A_m} = \mu_B H \left[N_z\ a\ \frac{df(\bar{\nu})}{d\bar{\nu}} + (1 - N_z)\ \frac{c}{g\mu_B H_E}\ f(\bar{\nu}) \right] \qquad (1)$$

A_m stands for the maximum absorbance of the line in the abscence of magnetic field and the a and c coefficients depend on the optical transition (5).

The spectra for the $^3A_2 \rightarrow {}^1E$ transition are reported on fig. 1 for a thin crystal plate (0.195 mm). In order to be sure that no linear dichroism appears, we have verified that the MCD spectrum is symmetric with respect to the base line when reversing H. Linear dichroism would appear for example if there remains inequivalency between d_x and d_y domains. When H is increased, the shape of the MCD spectrum varies drastically (fig. 1b, 2). Moreover the signal is found to be reversible when decreasing or increasing H.

This modification arises from the N_z variation in Eq. (1).

We have considered a displacement x of the wall between a d_z domain and d_x or d_y domains. It corresponds to a decrease of the width of the d_z domain. We have minimized the variation of the total free energy during this process with respect to x, including the energy of the restoring force $kx^2/2$. This leads to $x = (\chi_\perp - \chi_\parallel)\ H^2/2k$. This result indicates that the twin-walls move gradually with H and that d_z domains disappear at $H_{cr} = \sqrt{2kl/(\chi_\perp - \chi_\parallel)}$, where l is the domain width.

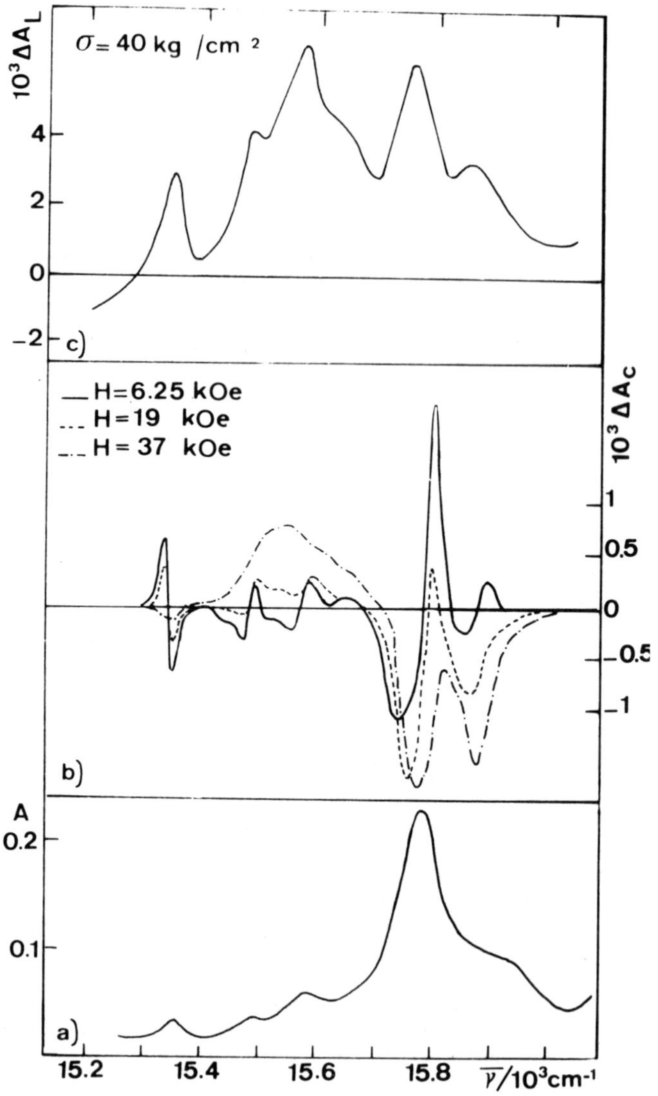

Figure 1 - Absorption (a) and MCD (b) for a 0.195 mm thick crystal of $KNiF_3$. SLD for a 0.5 mm thick crystal (c). All experiments are made at 15 K on the $^3A_2 \rightarrow {}^1E$ transition (bandwidth 3 cm^{-1}).

We have compared the field dependence of the MCD term which varies as $df(\bar{\nu})/d\bar{\nu}$ to the expression $N_z H \sim H(1-H^2/H_{cr}^2)$ (fig. 2a), and that of the term varying as $f(\bar{\nu})$ to $H(H^2/H_{cr}^2)$ (fig. 2b).

In figure 2b, we observe that below 12 kOe, there is no variation of the MCD term. Thus we introduce a parameter H_1 such that for $H < H_1$ the crystals remains untwinned along z and that for $H > H_1$: $x = l(H^2 - H_1^2)/H_{cr}^2 - H_1^2)$.

The experimental results fit well with $H_1 = 12$ kOe, $H_{cr} = 20$ kOe (fig. 2). The H_1 value depends on the crystal and may be related to a difference of the anisotropy between z and x or y directions. For an ideal crystal ($H_1 = 0$) the critical field is about 10 kOe. The facts that H_{cr} varies with the sample and is diffe-

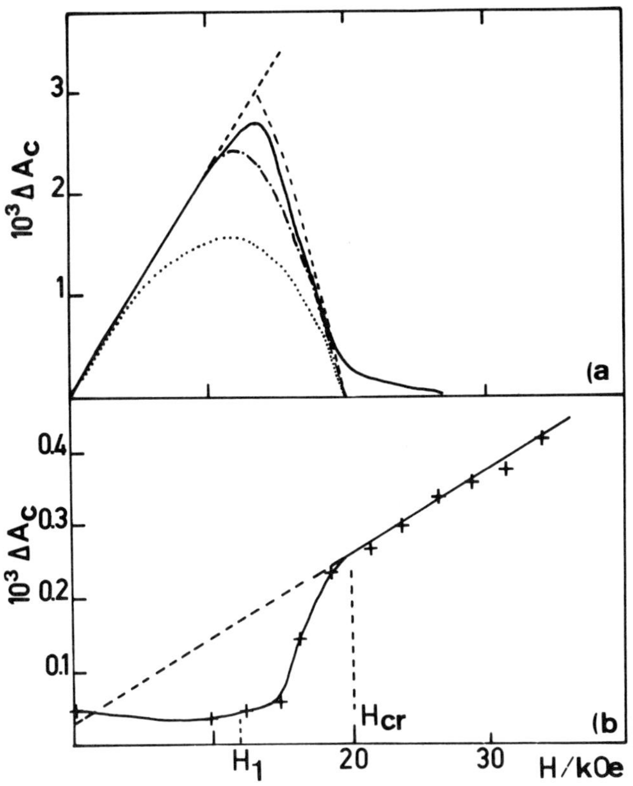

Figure 2 - (a) Experimental (——) and calculated for H_{cr} = 20 kOe, H_1 = 0 (...), H_1 = 11,6 kOe, (-.-), H_1 = 14 kOe (---) field dependence of the MCD $df(\bar{\nu})/d\bar{\nu}$ term for the magnetic dipole $^3A_2 \rightarrow {}^1E$ transition ($\bar{\nu}$ = 15347 cm^{-1}).
(b) Field dependence of the MCD $f(\bar{\nu})$ term for the peak located at 15485 cm^{-1}.

rent from the frequency of the one-magnon Raman line at 4.5 cm^{-1} (6) supports Néel's idea.

The sign and amplitude of the parameters a (Eq. 1) for the exciton line (15 347 cm^{-1}) and for the two first exciton-phonon (the latter being at $\vec{k}$ = 0, with the symmetry t_{1u}), are in good agreement with the calculations (7). At high field ($H > H_{cr}$) the observed MCD c terms may be well related to a net ferromagnetism along $\vec{H}$ (5).

SLD EXPERIMENTS

We have measured the difference in absorption for light polarised parallel (y) and perpendicular (x) to the stress direction. SLD experiments have been carried out on a 0.5 mm thick sample.

We have previously (2, 3) demonstrated that the SLD has a magnetic origin, i.e., is due to the spin reorientation along the stress direction. Then we have calculated the sign and amplitude of the SLD for exciton and exciton-phonon (the latter at $\vec{k}$ = 0) bands, which are those of the magnetic linear dichroism (MLD) for a single d_y domain. We have extended the paramagnetic MLD results (5) to the antiferromagnetic phase by substituting the exchange field for the magnetic field. This gives $f(\bar{\nu})$ line shape terms for each spin orbit component.

We have observed a reversible variation of $f(\bar{\nu})$ SLD terms versus σ (fig. 1c, 3). This variation, which is linear under σ_{cr} = 50 kg/cm^2, saturates above this value.

For a stressed crystal the SLD should be proportional to the difference $N_y - N_x$ of the number of the spins inside d_y and d_x domains. This quantity is proportional to the wall displacement x between d_y and d_x. By minimizing the total free energy variation we have found that the twin-walls move linearly until an untwinned crystal is obtained for $\sigma_{cr} = 2kl/3\lambda_{100}$, where λ_{100} stands for the fourfold axis magnetostrictive coefficient.

The reversible SLD variation versus the stress (fig. 3) reveals this behaviour. We get an untwinned crystal at $\sigma_{cr} \sim 50$ kg/cm^2 and above this value the signal saturates. As H_{cr}, σ_{cr} depends on the shape of the sample because it is related to k and l. As it is expected, the variations of the free energy deduced from σ_{cr} and H_{cr} experimental values are comparable.

We have related the modification of the MCD and SLD spectra to the twin-walls movement in $KNiF_3$. This general phenomenon for cubic antiferromagnets must always be taken into account so as to interpret experiments under various external properly perturbations.

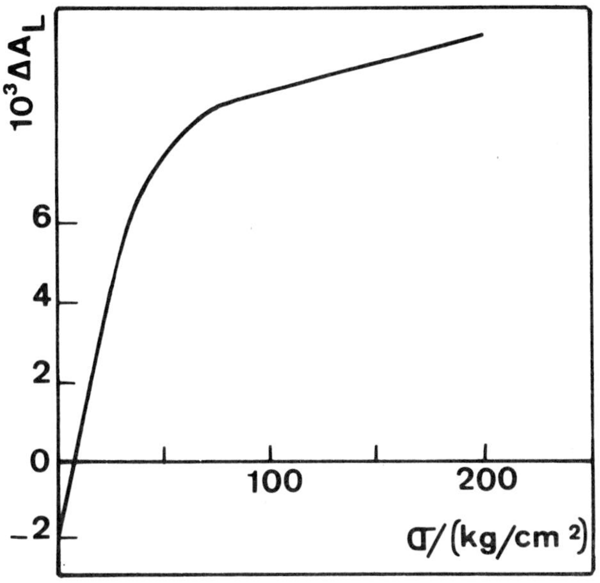

Figure 3 - Stress dependence of the SLD signal for the peak located at 15575 cm^{-1}.

REFERENCES

(1) L. Néel, Proc. of the Kyoto Conf. on theoretical Physics, 1953. Science Council of Japan, Tokyo, 701-14, 1954.
(2) R.V. Pisarev, J. Ferré, J. Duran and J. Badoz, Sol. State Commun., 11, 913-7, 1972
(3) J. Ferré, B. Briat, R.V. Pisarev and J. Nouet, Intern. Conf. Mag. Moscow, 1973.
(4) K. Hirakawa, T. Hashimoto and K. Hirakawa, J. Phys. Soc. Jap., 16, 1934-9, 1961.
(5) R.H. Petit, J. Ferré, unpublished results.
(6) P. Moch and C. Dugautier, Intern. Conf. Mag. Moscow, 1973.
(7) M. J. Harding, S. F. Mason, D. J. Robbins and A. J. Thomson, J. Chem. Soc., A, 3047-62, 1971.

A NEW CLASS OF ROOM-TEMPERATURE MAGNETO-OPTIC INSULATORS: THE COBALT FERRITES

R. K. Ahrenkiel, T. J. Coburn, D. Pearlman,
E. Carnall, Jr., T. W. Martin, and S. L. Lyu
Research Laboratories
Eastman Kodak Company, Rochester, New York 14650

ABSTRACT

Large magneto-optical activity has recently been associated with the crystal field transitions of tetrahedral Co^{++} in a magnetic medium. Most of the materials, which have been reported, are sulfo-spinels with transition temperature below room temperature. The ferrite $CoFe_2O_4$ has a transition temperature of 790°K but it is an inverse spinel and has fairly small magneto-optic activity.[1] The addition of certain trivalent ions such as Rh^{3+}, Cr^{3+} and Al^{3+} convert the spinel to a normal configuration with the distribution $Co[FeT]O_4$. Here T is one of the above trivalent ions. This substitution has produced large magneto-optic effects in the range of the Co^{++} (T_d) crystal field transitions. Peak Faraday effects of about 10^5 deg/cm are seen in the region of the $^4A_2(F) \longrightarrow {}^4T_1(P)$ transition at 15,800 cm^{-1} (0.63 μm). These effects are seen at room temperature as the transition temperatures of the above compounds are in the range 350°K to 500°K. Thus, the substituted cobalt ferrites are a new family of room temperature compounds with large magneto-optical activity. Possible applications of these materials are modulators, isolators, and optical memories.

Large magneto-optical effects have recently been found to originate from the ligand field transitions of tetrahedral Co^{++} in a magnetic medium.[1-4] This magneto-optical activity is dependent upon the exchange splitting of the cobalt $^4A_2(F)$ ground state in the ferromagnetically ordered cobalt sublattice. The spin-orbit components of the principal crystal field transitions ($^4A_2(F) \longrightarrow {}^4T_1(F)$, $^4T_1(P)$) are predominately circularly polarized leading to a large circular dichroism in all of the optical constants.

The materials studied above are primarily sulfo-spinels with transition temperatures below room temperature. The cooling requirements of these materials limit their applicability. However, the proposed model[2] of this phenomenon suggests that a wide range of cobalt compounds may show large magneto-optic effects. The principal criteria for such effects are:

(1) Divalent cobalt in a tetrahedral lattice site.
(2) Ferromagnetic ordering of the cobalt sublattice.

If these requirements are fulfilled, the model predicts large magneto-optic effects in the spectral region of the crystal field transitions. A large number of cobalt compounds then are candidates for magneto-optic materials. Many of these cobalt compounds have transition temperatures above ambient.

Reflectance-circular-dichroism (RCD) studies of $CoFe_2O_4$ (T_c = 790°K) have been reported.[3] However, the magneto-optical activity of the $^4A_2(F) \longrightarrow {}^4T_1(F)$ transition is much smaller than is observed in the sulfo-spinel $CoCr_2S_4$.[1] We observed a 1.2 percent effect in $CoFe_2O_4$ at 293°K compared to 30 percent RCD in $CoCr_2S_4$ at 80°K. This difference was attributed to $CoFe_2O_4$ being predominantly an inverse spinel. The Co^{++} that occupied octahedral sites does not contribute significantly to the magneto-optical activity.

Blasse[5] has shown that $CoFe_2O_4$ transforms into a normal spinel by the substitution of certain trivalent metals for trivalent iron. Examples of such compounds are $Co[Rh_xFe_{2-x}]O_4$, $Co[Cr_xFe_{2-x}]O_4$ and $Co[Al_xFe_{2-x}]O_4$ with x greater than a certain critical value. The elements in the brackets indicate the B-site species. The Curie temperatures of the rhodium, chromium, and aluminum compounds with x = 1.0 are 355°K, 430°K, and 420°K, respectively. These compounds comprise a new class of room temperature magneto-optic compounds.

Here, we will describe in detail magneto-optic studies of the series $Co[Rh_xFe_{2-x}]O_4$. In a study of the magnetic susceptibility of this series, Blasse found that the spinel became normal for values of x greater than about 0.5. Also, the Curie temperature decreases rapidly with x.

We will describe here the compositions x = 0.5, 1.0, and 1.5 which have transition temperatures of 580°K, 350°K, and 160°K, respectively. Powders were prepared according to techniques which have been described in the literature.[5] Analysis of the diffuse reflectance of the x = 1.0 powder using the Kubelka-Munk function[6] showed strong absorption bands centered at about 0.7- and 1.5 μm wavelength. These absorption bands correspond closely to the $^4A_2(F) \longrightarrow {}^4T_1(P)$ and $^4A_2(F) \longrightarrow {}^4T_1(F)$ crystal field transitions of Co^{++} in ZnO[7] and are blue shifted relative to the corresponding transitions in the sulfide. The powder spectrum indicates the onset of broad and strong absorption at the short-wavelength side of the 0.7 μm transition. This is probably due to the semiconductor band edge.

Polycrystalline samples were prepared by hot-pressing at 1000°C and 50,000 psi. The sample surface was prepared by polishing with successively finer abrasives and a specular surface was obtained. Specular reflectance measurements of the surface showed barely resolvable, 2-3 percent peaks associated with the two crystal field transitions at 0.7 μm and 1.5 μm.

A Kramers-Kronig analysis of this spectrum indicates an absorption coefficient of about $5 \cdot 10^4$ cm^{-1} at 0.7 μm. As these samples can be mechanically ground to a minimum thickness of about 50 μm, transmission data could not be obtained.

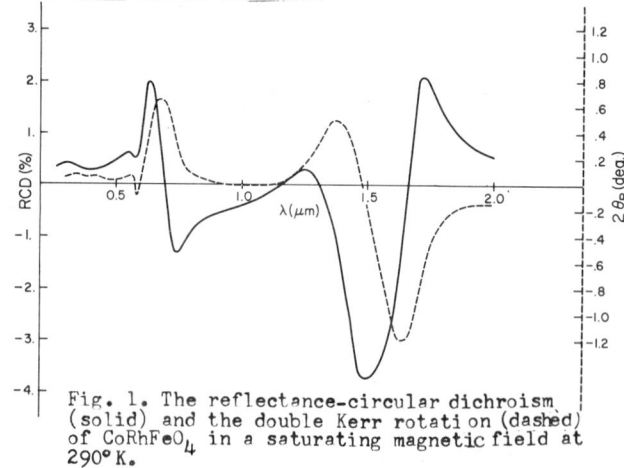

Fig. 1. The reflectance-circular dichroism (solid) and the double Kerr rotation (dashed) of $CoRhFeO_4$ in a saturating magnetic field at 290°K.

The RCD of the x = 1.0 sample at 290°K and a saturating magnetic field is shown in Fig. 1. The characteristic RCD pattern[2] of Co^{++} (T_d) is fairly prominent and the RCD peak at 1.5 μm has a value of 3.75 percent. The RCD of an x = 0.5 sample indicated weaker features, characteristic of $CoFe_2O_4$ with the peak at 1.5 μm being about 1.0 percent. Based upon saturation moments and the Néel model, Blasse[5] finds this composition to partially inverse. These studies support that conclusion as the smaller magneto-optical activity indicates a lower proportion of tetrahedral cobalt. The RCD of an x = 1.5 sample was measured at temperatures lower than its magnetic transition temperature of 160°K. The RCD signal is somewhat weaker than in the x = 1.0 sample. As the Rh^{++} ion is a low spin, nonmagnetic ion, it probably does not exchange couple into the A and B sublattice magnetization. Therefore, in a random distribution of the two B-site species about each A-site, some cobalt sites will have a larger number of RH^{+++} ions and may not effectively exchange couple with the B-site magnetization. Such ions will not be magneto-optically active and will not contribute to the RCD signal.

A Kramers-Kronig (KK) analysis of the RCD[4] gives the Kerr rotation. Such determinations of the Kerr rotation have been shown to be as accurate as a direct measurement.[4] Fig. 1 shows also the KK-calculated double Kerr rotation at room temperature. A peak rotation of 1.2 degrees at 1.65 μm is observed. These magneto-optic effects are comparable in magnitude to the magneto-optic coefficients in ferromagnetic metals[8] and MnBi.[9]

A knowledge of the RCD and Kerr rotation allows us to calculate[3] the Faraday rotation. However, when the extinction coefficient K is much less than n, the index of refraction,[10] the Faraday rotation is very accurately given by:

$$\theta(\text{deg/cm}) \cong 4.5 \times 10^5 \frac{n^2-1}{\lambda(\mu m)} \quad \text{RCD}$$

As n here is about 3.0 and K at 0.7 μm is about 0.25, the above expression should be very accurate and avoids a Kramer-Kronig transform. Values of the Faraday rotation, calculated by either of the above techniques, were found to agree to within several percent. These calculated values of the Faraday rotation are plotted in Fig. 2.

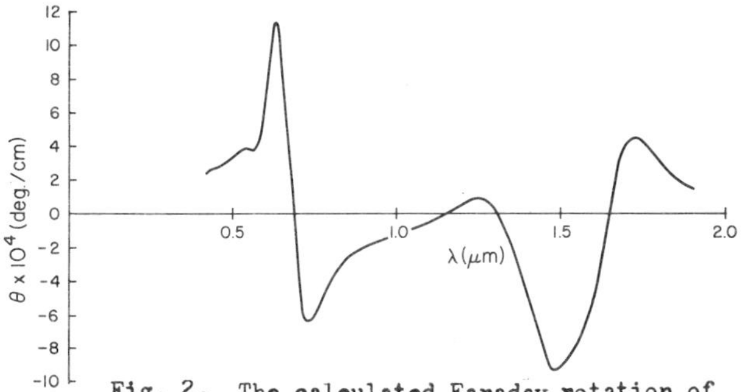

Fig. 2. The calculated Faraday rotation of $CoRhFeO_4$ at 290°K.

A peak Faraday effect of about 1.5×10^5 deg/cm is observed at 0.63 μm, a fortuitous matching with the He-Ne laser output. Taking[4] the absorption coefficient at 0.63 μm as 5×10^4 cm^{-1}, the figure of merit, $2\theta/\alpha$, is about 6 degrees. This is comparable to MnBi (l.t.p.) which has a figure of merit of 3.6 degrees[11] at 0.63 μm. However, the absorption coefficient here is about an order of magnitude lower.

In the spectral range of the $^4T_1(F)$ transition, the absorption band is almost completely circularly polarized. We write the polarized absorption coefficient as

$$\alpha_+(\alpha_-) = CNfF_+(\lambda) \text{ for } T < T_c.$$

Here N is the number of tetrahedral cobalt ions $Co^{++}(T_d)$, per unit volume; f is the oscillator strength per ion; and $F(\lambda)$ is the shape factor of the absorption band normalized to unity. C is a constant that depends on the particular absorption band. The spin-orbit interaction splits $^4T_1(F)$ into a triplet of double group states transforming like G, $E_{1/2}$ and a degenerate pair G and $E_{3/2}$. The oscillator strength to these components is almost completely circularly polarized, either (+) or (-). The primary temperature effects for $T < T_c$ will be a "sharpening" of $F(\lambda)$ as the vibronic interaction decreases.

The RCD is linearly related[10] to the polarized absorption and therefore RCD peaks are proportional to the peak values of $F(\lambda)$. Cooling to 78°K allows us to compare the peak RCD values associated with the $^4A_2(F) \longrightarrow {}^4T_1(F)$ transitions of $CoCr_2S_4$ and $CoRhFeO_4$. For $CoRhFeO_4$ at 78°K, the peak RCD at 1.5 μm increases to 8.6 percent compared to 30 percent for $CoCr_2S_4$-(λ_p = 1.7 μm) at the same temperature. Thus, the magnetic-optical effects are about 3.5 times larger in the sulfide. As $F(\lambda)$ is about the same in the two materials at 78°K, we find that N_sf_s/N_of_o equals about 3.5, where the subscripts refer to the sulfide (s) and oxide (o), respectively. Also, as both the sulfide and oxide are normal spinels, then N_s equals approximately N_o. Thus, we reach the conclusion that $f_s/f_o \cong 3.5$ and the oscillator strength is larger in the sulfide, producing a larger magneto-optic effect. Weakliem[7] has measured the oscillator strengths of the $Co^{++}(T_d)$ transition in ZnS and ZnO and found them to be 2.9×10^{-3} and 7.8×10^{-4}, respectively, so that a ratio $f_s/f_o = 3.7$ is obtained. We may tentatively conclude that smaller magneto-optic coefficients in the oxide are a result of an intrinsically smaller oscillator strength of the ligand field transitions. The lower magneto-optical coefficients reported for $CoFe_2O_4$[3] are a result of the lower oscillator strength as well as partial spinel inversion.

$CoRhFeO_4$, like many previous hot-pressed materials,[4] possesses normal remanence of the RCD exceeding 90 percent. This property is useful for optical memory applications. As the material is an insulator, the thermal diffusion of the heated spot may be less severe than in metals. Also, the adjustable Curie temperature may be an advantage in minimizing the writing power.

In the spectral range from 2.0-10.0 μm, one would expect a figure of merit comparable to $CoCr_2S_4$.[12] As the Faraday rotation is about proportional to the cobalt absorption coefficient in this range,[12] then $2\theta/\alpha$ should be independent of α. Thus, $CoRhFeO_4$ should have a large figure of merit in the near infrared. As many of the efficient gas lasers like the Co and Co_2 laser operate in this wavelength range, infrared magneto-optic devices utilizing these materials may be useful in modulation and isolation.

In conclusion, $CoRhFeO_4$ is the first room temperature material to demonstrate the large magneto-optic effects associated with divalent cobalt. The compound is a magnetic insulator with a large magneto-optic figure of merit. This material suggests a new family of magnetic insulators with a variety of device applications.

Properties of other members of this new family of room temperature materials are currently under investigation and will be presented in the near future.

REFERENCES

1. R. K. Ahrenkiel and T. J. Coburn, Appl. Phys. Lett. 22, 340 (1973).
2. R. K. Ahrenkiel, T. H. Lee, S. L. Lyu, and F. Moser, Solid State Commun. 12, 1113 (1973).
3. T. Coburn, R. Ahrenkiel, E. Carnall, and D. Pearlman, Magnetism and Magnetic Materials-1973, Ed. by C. D. Graham and J. J. Rhyne (AIP, New York, 1974) p. 1118.
4. R. K. Ahrenkiel, T. J. Coburn, and E. Carnall, Jr., IEEE Trans. on Magnetics 10, 2 (1974).
5. G. Blasse, Philips Res. Reports, Suppl. No. 3, 72 (1964).
6. W. W. Wendlandt and H. C. Hecht, Reflectance Spectroscopy, Interscience, New York, (1966).
7. H. A. Weakliem, J. Chem. Phys. 36, 2117 (1962).
8. K. H. Clemens and J. Taumann, Z. Physik 173, 135 (1963).
9. D. Chen, J. F. Ready, and E. Bernal, J. Appl. Phys. 39, 3916 (1968).
10. R. K. Ahrenkiel, S. L. Lyu, and T. J. Coburn, J. Appl. Phys. (in press).
11. D. Chen, Appl. Apt. 13, 767 (1974).
12. E. Carnall, Jr., D. Pearlman, T. J. Coburn, F. Moser, and T. Martin, Mat. Res. Bull. 1, 1361 (1972).

MAGNETOOPTICAL AND MAGNETIC PROPERTIES OF TERBIUM IRON GARNET AT LOW TEMPERATURES.

Tran Khanh Vien, H. Le Gall, A. Lepaillier-Malécot and D. Minella.
Laboratoire de Magnétisme et d'Optique des Solides, CNRS,
92190 Meudon-Bellevue, France.

M. Guillot
Laboratoire de Magnétisme, CNRS, 38 Grenoble, France.

ABSTRACT

The specific circular ϕ_F and linear ϕ_{CM} magnetic birefringences of $Tb_3 Fe_5 O_{12}$ single-crystal are reported as a function of the d.c. field H_a and from liquid helium temperature to room temperature. The sign, amplitude and slope of the Cotton-Mouton effect are strongly dependent on T, H_a and on the direction of this field with respect to the crystal axis. Well defined maxima of the slope $\Delta\phi_F/\Delta H_a$ and $\Delta\phi_{CM}/\Delta H_a$, observed near T=45°K, are discussed both from the paramagnetic evolution of the Tb^{3+} ions in the anisotropic exchange and crystal fields which induce a non-colinear magnetic structure.

INTRODUCTION

Being proportional respectively to the magnetic long-range order parameter S and the square of this quantity S^2, the circular and linear magnetic birefringences may provide interesting information on the evolution of magnetic state and magnetic phase transitions in ferri- weak ferro- and antiferromagnets. Recently the LMB has been used to investigate magnetic phase transition in cubic rare-earth (R.E.) ferrimagnetic garnets, in noncubic ferrimagnets and weak ferromagnets |1 2| and in antiferromagnets as MnF_2, FeF_2, CoF_2 |3 4|. As pointed out by Pisarev et al |1| the specific Cotton-Mouton phase-shift $\phi_{CM}=(q_{/\!/} - q_{\perp})$ may be comparable to the specific Faraday rotation $\phi_F=1/2\ (q_L-q_R)$ and presents a strong anisotropy even in "cubic" isotropic crystal as $Y_3 Fe_5 O_{12}$. In fact we observe in $Tb_3 Fe_5 O_{12}$ a LMB three times higher than the CMB at low temperatures (10°K) and with d-c fields large enough (=20kOe) to saturate the sample and avoid a domain structure. The origin of the large LMB in garnets is not clear at the present time. Many contributions have been introduced to explain this property such as a Cotton-Mouton effect induced through a first-order perturbation of the exchange interaction between the excited orbital states of a pair of magnetic ions |1 5|, or such as the spontaneous magnetostriction which produces a distorsion of the crystallographic symetry and a corresponding modification of the on-diagonal element of the permittivity tensor. This later contribution has been invoked by Dillon et al |2| to explain the LMB of rare-earth (R.E.) ferrimagnetic garnets and by Jahn and Dachs in three dimensional antiferromagnets Mn F_2, Fe F_2, Co F_2 and more recently by Kleemann and Farge in the 2-dimensional Heisenberg ferromagnet K_2 Cu F_4 |6|. In the present work we present magnetooptical measurements on TbIG at low temperature. Our investigation is concentrated on the d.c. field evolution of the dispersive part of the CMB and the LMB at low temperatures where an anomaly of the cristallographic and magnetic structures of R.E. garnets have been recently discussed by Bertaut, Sayetat and Tcheou |7|.

Measurements: In all experiments, two single-crystal plates of TbIG cut perpendicularly to the |111| direction (CMB) and the $|1\bar{1}0|$ direction (LMB) have been used. Typical evolution of the LMB are given on the figure 1 for light propagation parallel to a $|1\bar{1}0|$ direction and the applied d.c. field parallel either to the easy magnetization axis |111| or the hard magnetization axis |001|. In the former case ($H_a /\!/ |111|$) the d.c. field evolution of the LMB is simpler than in the latter. For T higher than the magnetic compensation temperature (T_{comp}= 249°K) the LMB of TbIG is small and is comparable to the LMB in YIG. For T lower than T_{comp} a strong increase of the absolute value of the LMB is observed. It is to be noted first: the small d.c. field necessary to saturate the crystal by suppressing the domain structure (H_S < 10 KOe) even at low temperatures, second: the linear increase with H_a of the LMB for $H_a > H_S$ and third: the high amplitude of LMB which at T=18°K is three times (ϕ_{CM}=11 000°/cm) higher than the corresponding specific Faraday rotation observed when light propagation and d.c. field are parallel to the "easy-axis". ($\phi_F \simeq$ 3 000°/cm). The slope of the linear evolution of the LMB depends on T and will be discussed later. In all the cases of propagation of light parallel to the |100| |110| and |111| direction of the d.c. field evolution of ϕ_F is similar to the evolution of LMB in the case of H_a // |111|.

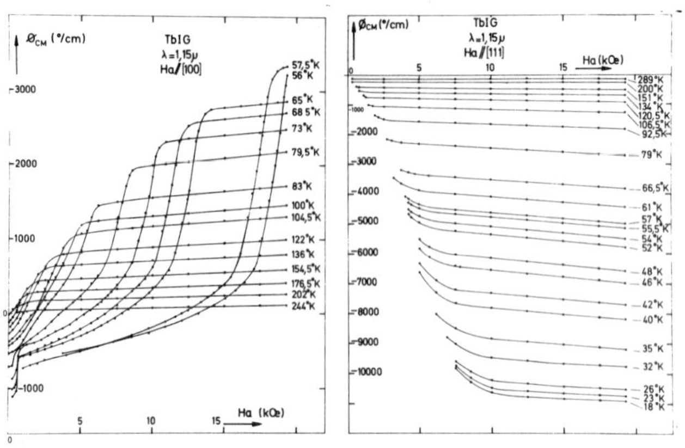

Fig. 1. d.c. field evolution of LMB of $Tb_3Fe_5O_{12}$

The d.c. evolution of the LMB for H_a parallel to |001| is more complicated due to the strong magnetocrystalline anisotropy of TbIG at low temperatures (Fig. 1-a).

DISCUSSION

The anomalous behaviour of the optical anisotropy as shown from the LMB on the figure 1, is correlated to the occurence of magnetic ordering in the R.E. sublattice at low temperatures. Therefore a parallel evolution of the LMB with the square of the Tb^{3+} magnetization, is expected and verified from experiment. As indicated above, the LMB is increasing linearly with the d.c. field for values of H_a higher than the saturation value necessary for the iron sublattice. That shows an appreciable evolution of the magnetic state associated with the paramagnetic evolution of the rare-earth moment in the effective field $H_{eff}=H_a+H^{c-d}_{exch}$ where H^{c-d}_{exch} is the exchange field as seen by the R.E. ions from the high exchange interaction between the c and d (tetrahedral sites. We observed such a paramagnetic process from the curves $\phi_F(H_a,T)$ and $\phi_{CM}(H_a,T)$ even when H_a is applied along the easy magnetization axis. The evolution of the slope $\Delta\phi_{CM}/\Delta H_a$ as a function of temperature is described on the figure 2. This slope increases at low temperature, shows a maximum near T=45°K then goes to a low finite value near the liquid helium temperature. We observe a simi-

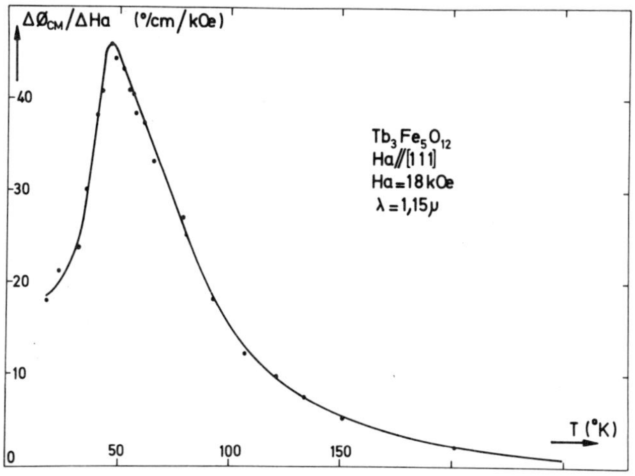

Fig. 2. Temperature dependence of the derivative of the LMB in $Tb_3Fe_5O_{12}$

lar evolution for $\Delta\phi_F/\Delta H_a$ as deduced from the curves ϕ_F (H_a,T). Pauthenet first studied the paramagnetic process of the R.E. ions in ferrimagnetic garnets and verified that the corresponding susceptibility χ_c of the site c follows the law:

$$\chi_c = \nu C_c/(T-\theta_p)$$

The paramagnetic Curie temperature θ_p is given by $\theta_p=\nu C_c\, n_{cc}$ where ν, C_c and n_{cc} are respectively the proportion of R.E. ions in the magnetic cell (=3/8), the Curie constant and the molecular field coefficient associated with the exchange between two R.E. ions. Since $\Delta\phi_F$ corresponds to the evolution of the rare-earth moment, $\Delta\phi_F/\Delta H_a$ is proportional to χ_c. Therefore it is expected an evolution of $(\Delta\phi_F/H_a)^{-1}$ proportional to χ_c^{-1} or proportional to T. We verified such a proportionality from which are deduced the following values: θ_p=-40°K and n_{cc}=-9. These values are five times higher than the values previously obtained from polycristalline TbIG |8| .

The anomalous maximum of $\Delta\phi_{CM}/\Delta H_a$ or $\Delta\phi_F/\Delta H_a$ must be correlated to the corresponding maxima of χ_c observed in most of R.E. garnet by Pauthenet and the maximum of the derivative of the forced magnetostriction $d\lambda_n/dH_a$ as studied more recently by Belov et al |9|. Without making any microscopic hypothesis, the later author attributed such maxima to the occurence of a "sharp change of the long-range magnetic order in the rare-earth sublattice" which appears near a low temperature point T_ℓ where the interaction energy of the R.E. ions with the exchange field H^{c-d}_{exch} is comparable with the thermal energy:

$$T_\ell = 2\mu_B\, S_{Tb}\, H^{c-d}_{exch}/\, k_B$$

With $H^{c-d}_{exch} \simeq$ 220 kOe for TbIG this relation predicts a maximum near T=60°K. A sharp change of the long range order in the c sublattice can be attributed to occurence of magnetic and crystallographic distorsions as observed from elastic neutron scatterings |10|, and X-rays experiments |7|. The direction of the Tb^{3+} moments depends on the spin-orbit coupling of the R.E. and on the anisotropy of the crystal field of the c-site which tends to align the spins along one |100| direction, and depends on the anisotropy of the c-d exchange interaction which tends to align the spins along one |111| direction. At high temperature, the anisotropy of the crystal field is much lower than the anisotropy of the exchange field, so a pure ferrimagnetic arrangement is observed.

On the other hand, at low temperature an opposite situation is obtained and a non-collinear magnetic ("umbrella" type) structure appears for the Tb^{3+} spins around the |111| direction. The evolution of this structure is maximal near T=45°K and can explain the corresponding evolution of ϕ_{CM}.

The evolution of the magnetic structure induces a distorsion of the crystallographic structure (cubic → rhombohedral) through the strong magnetostriction of the Tb^{3+} ions. As discussed by many authors, the LMB must be correlated to the spontaneous magnetostriction. Similarly the d.c. field evolution of the LMB can be associated in part to the corresponding forced magnetostriction which induces a crystallographic distor - sion.

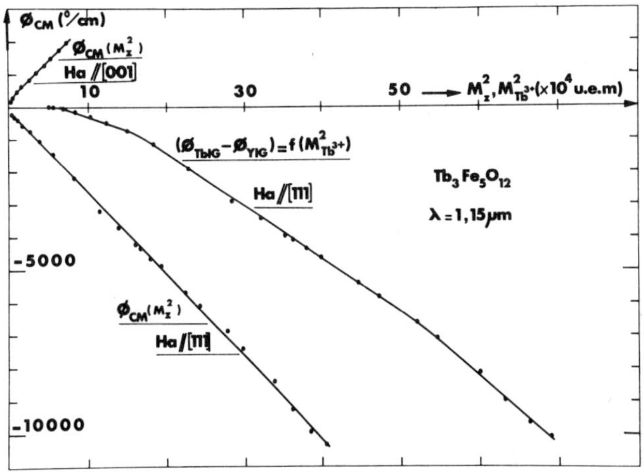

Fig. 3 - Quadratic evolution of the LMB with the magnetization in TbIG.

On the figure 3 we have reported the evolution of the LMB of TbIG as a function of the square of the total magnetization M_z^2 and the evolution of the LMB of the Tb^{3+} ions as a function of the square of the terbium magnetization $M^2(Tb^{3+})$. The later curve, which is obtained from the difference ϕ(TbIG)-ϕ(YIG), shows, for $T < T_{comp}$, three linear portions which verify a quadratic evolution of $\phi(Tb^{3+})$ with the R.E. sublattice magnetization. The changes of the slope of this curve appear first near T=100°K and next near T=45°K where the magnetic and crystallographic distorsions become important.

REFERENCES

|1| Pisarev R.V., Sinii I.G. and Smolensky G.A. ZhETF Pis. Red. 9, 112, 1969.

|2| Dillon J.F., Remeika J.P. and Staton C.R., J. Appl. Phys., 41, 4613, 1970.

|3| Borovik-Romanov A.S., Kreines N.M., Talalaef M.A. Sov. Phys. J. exp. theor. Phys. Pis., 13, 80, 1971.

|4| Jahn I.R. and Dachs H., Sol. Stat. Com., 9, 1617, 1971
and
Jahn J.R., Phys. Stat. Solidi (b), 57, 681, 1973.

|5| Le Gall H., Journ. Phys., C,32, 590, 1970.

|6| Kleeman W. and Farge Y., Journ. Phys. Lettres, 35, L135, 1974.

|7| Bertaut E.F., Sayetat F., and Tcheou F., Solid State Com., 8, 239, 1970.
and Sayetat F., Thèse, Grenoble, 1974.

|8| Pauthenet R., Ann. Phys., 3, 424, 1958.

|9| Belov K.P. and Sokolov V.I., Izv. Ann. SSSR, Ser. fiz. 30, 1073, 1966.

|10| Herpin A., Boucher B., Meriel P., and Plumier R., Bull. Soc. Sci. Bretagne, 39, 59, 1964.

|11| Tchéou, Thèse, Grenoble, 1972.

MAGNETO-OPTICAL STUDIES OF FERROMAGNETIC METALS*

J. L. Erskine†

Department of Physics
University of Washington
Seattle, Washington 98055

ABSTRACT

Magneto-optic Kerr effect (MOKE) spectroscopy is one of several experimental techniques that can be used to study the electronic structure of ferromagnets. Specifically, MOKE spectra contain information related to the electron spin-polarization (ESP), exchange splitting, widths and shapes of minority and majority spin bands. MOKE spectra are therefore related not only to optical and photoemission results, but also to other experiments which measure ESP. These latter experiments, which include spin-polarized photoemission and field emission, and spin-dependent tunneling have produced some results which appear to be in conflict with various theoretical predictions. Magneto-optical experiments can be used to study these discrepancies and may be helpful in resolving some of them. Magneto-optical spectroscopy of transition metal and rare earth ferromagnets is discussed. The d-band widths of Fe, Co, and Ni determined from MOKE spectroscopy are wider than corresponding widths measured by photoemission, and the sign of ESP near the Fermi level in Ni is found to be negative as predicted by band theory. Several interesting future applications of MOKE spectroscopy are suggested by these results.

INTRODUCTION

Magneto-optical effects have provided a useful probe of the electronic structure of atomic, molecular and solid state systems.[1] The greatest utility has been in application to simple systems where a direct relationship between the observed effects and their microscopic origins exists. In metals, the relationship is considerably more complicated than in simple systems, which perhaps is part of the reason why magneto-optical spectroscopy of metals has received limited use as an electronic structure probe.

The primary objective of this paper is to discuss applications of magneto-optical spectroscopy to ferromagnetic metals for the purpose of obtaining electronic structure information. It is shown that certain features of the magneto-optical spectra of ferromagnetic metals have direct interpretation in terms of their electronic structure. The magneto-optical spectra of several ferromagnetic metals are analyzed and the main electronic structure features of their magnetic bands are obtained. The results are discussed in relation to other relevant experimental work.

The magneto-optical effect associated with ferromagnetic metals is the magneto-optic Kerr effect (MOKE) named after its discoverer.[2] The effect is observed when plane polarized light reflects from a magnetized ferromagnet: The reflected light emerges elliptically polarized having its major axis rotated with respect to the incident polarization axis. The rotation and ellipticity are proportional to the magnetization and are found to be typically on the order of 1 min. of arc at saturation magnetization for visible wavelengths. Although the MOKE is a relatively small effect, it can be measured with sufficient precision to resolve spectral structure, and therefore clearly has potential application as an electronic structure probe.

The microscopic origin of MOKE is spin-orbit coupling acting in conjunction with the net spin-polarization that exists in ferromagnets.[3] The structure in MOKE spectra results from interband absorption processes similar to those occurring in nonmagnetic metals with the added feature that the electron spin is coupled to orbital motion of electron states resulting in an asymmetry in the interaction of right and left circularly polarized (RCP and LCP) light with the magnetized metal. MOKE spectra therefore have bearing on the results of recent spin-polarized photoemission,[4] spin-polarized field emission,[5] and spin-dependent tunneling[6] experiments. These investigations have aroused a significant amount of attention which has focused on the disturbing fact that their predictions appear to be in conflict with various theoretical models.[7] A good summary of the experimental observations as they relate to band models of magnetism has been given by Gutzwiller.[8]

THE CONDUCTIVITY OF MAGNETIC METALS

The conductivity of a solid with cubic or higher symmetry magnetized along its z-direction is described by the tensor

$$\tilde{\sigma} = \begin{pmatrix} \sigma_{xx} & \sigma_{xy} & 0 \\ -\sigma_{xy} & \sigma_{xx} & 0 \\ 0 & 0 & \sigma_{zz} \end{pmatrix} . \qquad (1)$$

In Eq.(1), diagonal terms describe ordinary optical properties and are equal ($\sigma_{xx}=\sigma_{zz}$) and independent of the magnetization, M, to first order. The off-diagonal terms describe the MOKE and are proportional to M. Both diagonal and off-diagonal terms are complex having real and imaginary components $\sigma_{ij} = \sigma_{ij}^{(1)}+i\sigma_{ij}^{(2)}$. The absorptive components, $\sigma_{xx}^{(1)}$ and $\sigma_{xy}^{(2)}$, are related to absorption of RCP and LCP light propagating along the z-direction in the sample by

$$\sigma_{xx}^{(1)} \propto A_{LCP} + A_{RCP} ,$$
$$\sigma_{xy}^{(2)} \propto A_{LCP} - A_{RCP} . \qquad (2)$$

The asymmetry in absorption of RCP and LCP light produced by spin-orbit coupling and the spin-polarization makes MOKE spectroscopy a particularly useful probe for electronic structure of magnetic metals, mainly because structure in the spectra can be unambiguously associated with magnetic bands.

In analogy with ordinary optical absorption in metals magneto-optical absorption contains both interband and intraband contributions. The intraband contribution can be characterized by[9]

$$\sigma_{xy}^{(2)}(\omega) = \frac{\eta}{\omega\tau} , \qquad (3)$$

where τ is the same lifetime associated with the Drude formula, and η is a parameter that is proportional to the spin-orbit coupling strength and spin-polarization of the conduction band. This term does not contribute sharp structure to MOKE absorption, and numerical estimates[9] show that it is typically small for energies above the infrared range where interband effects dominate.

The interband contribution to MOKE absorption is expressed microscopically by[10]

$$\sigma_{xy}^{(2)}(\omega) = \frac{e^2}{2m^2\hbar}\sum_{\alpha}\sum_{\beta}\left\{\frac{(\pi^-_{\alpha\beta})^2}{(\omega^-_{\alpha\beta})^2-\omega^2} - \frac{(\pi^+_{\alpha\beta})^2}{(\omega^+_{\alpha\beta})^2-\omega^2}\right\}, \quad (4)$$

where $\pi^{\pm} = \pi_x \pm i\pi_y$, and $\pi = p + \frac{e}{c}A + \frac{1}{2mc^2}\sigma x\nabla$ is the momentum operator including the spin-orbit contribution. The sums extend over all occupied states β and empty states α, and matrix elements, $\pi^{\pm}_{\alpha\beta}$, are between wavefunctions which include spin-orbit coupling. The parameters $\omega^{\pm}_{\alpha\beta}$ correspond to the energy differences between the states α and β for RCP and LCP light including spin-orbit coupling effects.

Application of Eq.(4) to a real metal is not straight forward, and unless one is prepared to proceed directly using a band structure computer program,[11] an alternate approach is needed. The model calculation described in the next section provides a useful method.

MODEL CALCULATION OF MAGNETO-OPTICAL ABSORPTION

Magneto-optical absorption in ferromagnetic metals can be produced by two distinct mechanisms. One mechanism contributes through the energy denominator in Eq.(4) via spin-orbit coupling produced shifts in $\omega^{\pm}_{\alpha\beta}$. The other mechanism contributes through the matrix elements $\pi^{\pm}_{\alpha\beta}$ via spin-orbit perturbation of the wavefunctions α and β. The wavefunction mechanism dominates when orbital quenching is strong; the energy denominator mechanism dominates when one of the states is atomic like, as in the case of 4f-levels. Both mechanisms produce magneto-optical absorption of similar strength having a definite sign that depends on whether majority or minority spin transitions dominate the absorption processes. A model calculation approach based on the energy denominator mechanism is used in this section to illustrate the basic features of magneto-optical absorption from a microscopic viewpoint.

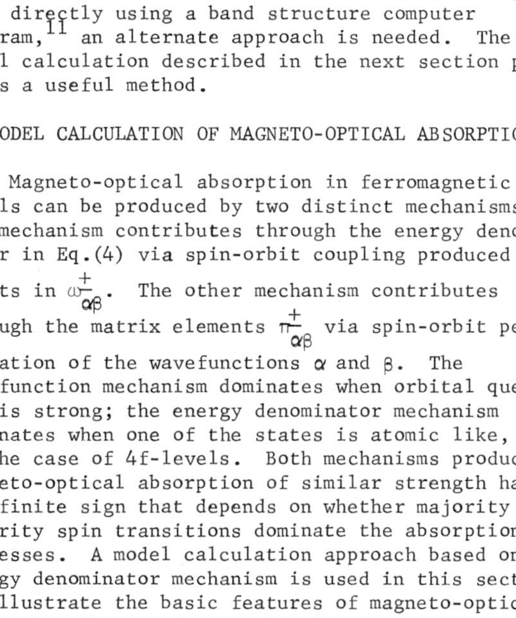

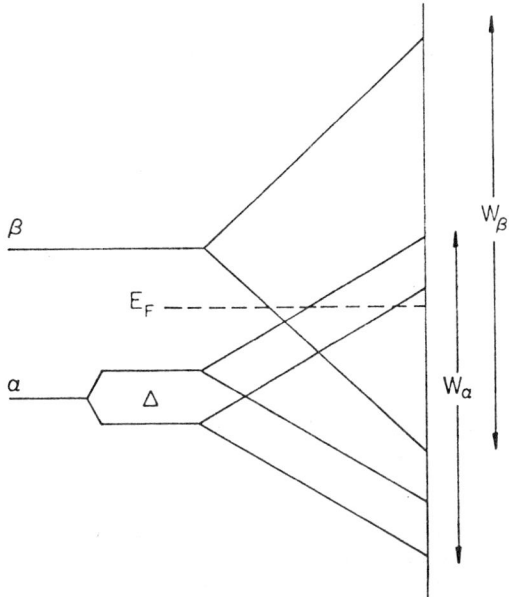

Fig. 1: Two band model of a metal formed from atomic states α and β. Δ is the spin-orbit splitting of the atomic state α. Angular components of α and β are assumed to be distributed uniformly over the bandwidths W_α and W_β.

The model calculation described here provides a description of the gross spectral dependence and an amplitude estimate of magneto-optical absorption resulting from interband transitions between a pair of bands. The calculations are based on the model shown in Fig. 1 which illustrates the formation of metallic bands from corresponding atomic levels. Several assumptions are made to equate the oscillator strengths calculated based on the atomic picture with interband absorption in the metal.

Inherent in Fig. 1 is the assumption that appreciable mixing of higher atomic levels into the bands is not important. This assumption, together with the f-sum rule[12] assures equality of the total oscillator strength in the atomic system and solid. This assumption is not strictly valid for magnetic metals: the conduction bands of the transition metals as well as the rare-earth metals contain s-character electrons in addition to the d- and p-character states used for model predictions. However, this does not impose a serious limitation on the model.

Additional assumptions are that band broadening preserves the average spin-orbit splitting of atomic levels,[13] and that all atomic components are spread uniformly throughout the band. The first of these enables one to make absolute quantitative estimates of the amplitude of MOKE absorption. The second is needed in order to associate structure in MOKE absorption with variation in the joint density of states as opposed to changes in oscillator strength within the band. This is a common approximation used in the interpretation of optical data, and its validity is supported by band calculations.

It is straightforward to show by direct calculation that in the atomic system, all first order effects associated with spin-orbit coupling are taken into account by the splitting Δ. In this case, $\pi^{\pm}_{\alpha\beta} = i\omega_{\alpha\beta} m|x \pm iy|_{\alpha\beta}$, and the matrix elements become electric dipole type between unperturbed atomic wavefunctions. The radial component can be evaluated by numerical integration of tabulated wavefunctions,[14] and the weighting associated with the various angular components are given by standard texts on atomic spectra.[15] To incorporate a net spin-polarization, m_j states with spin up components are weighted by n_1 and and those with spin-down components by n_2 with $n_1+n_2 = n_t$ and $n_1-n_2 = \mu_B/n_t$, where n_t represents the number of band electrons and μ_B is the magnetic moment per atom associated with the band.

Figure 2 illustrates the result of computing the difference in absorption for two cases, $n_1 = n_2$ and $n_1 \neq n_2$. The atomic oscillator strengths are shown broadened into bands, and the shape represents the joint density of states. In an unmagnetized solid, a detailed cancellation of oscillator strengths occurs, and $\sigma_{xy}^{(2)}(\omega) \equiv 0$. When $n_1 \neq n_2$, RCP and LCP oscillator strengths are shifted relative to each other due to spin-orbit splitting, and the cancellation no longer occurs. The energy shift is given by

$$\Delta E = \frac{1}{n_1+n_2}\sum_{LCP} n_{\alpha\beta}\Delta_{\alpha\beta} - \sum_{RCP} n_{\alpha\beta}\Delta_{\alpha\beta}, \quad (5)$$

where $n_{\alpha\beta}$ is the relative weighting including the angular matrix element and weighting due to spin-polarization, and $\Delta_{\alpha\beta}$ is the relative spin-orbit shift equal to $+\frac{1}{2}\ell$ for $j = \ell + \frac{1}{2}$ and $-\frac{1}{2}(\ell+1)$ for $j = \ell - \frac{1}{2}$. For example, consider transitions from an S level to a $P_{1/2}P_{3/2}$ doublet. The relative oscillator strengths[16] to the $P_{3/2}$ levels are 1 and 3, and to the $P_{1/2}$ level, 2. Weighting the spin-up and spin-down m_j P states with n_1 and n_2, Eq.(5) yields $\Delta E = 2\Delta(n_1-n_2)/3(n_1+n_2)$. Application is easily extended to P→D and D→F transitions.

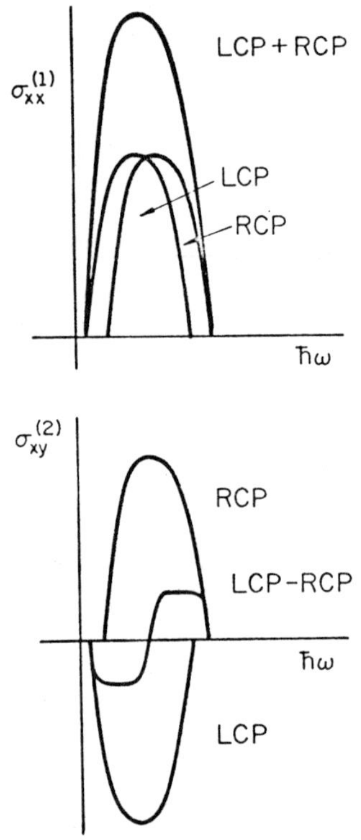

Fig. 2: Optical (sum of LCP and RCP) and magneto-optical (difference of LCP and RCP) absorption. The shift in center of gravity for LCP and RCP absorption is proportional to the net spin-polarization and spin-orbit coupling strength.

The model has the following features as can be verified by referring to Eq.(5) and Fig. 2: 1) The width W of the positive and negative structure corresponds to the width of the joint density of states; 2) positive and negative areas are equal (if the whole bandwidth is considered); 3) the weight of the MOKE absorption (first peak) $\langle \sigma_{xy}^{(2)} \rangle$ is equal to $\frac{1}{2}\,\frac{\Delta E}{W} \times \langle \sigma_{xx}^{(1)} \rangle$; and, 4) the initial sign of the structure depends on the sign of the spin polarization.

EXPERIMENTAL DATA

The optical[17] and MOKE[18] absorption spectra for Ni, Co, and Fe, shown in Figs. 3 and 4 clearly illustrate the enhanced structure characteristic of differential absorption spectroscopy. Concentrating on the optical absorption spectra of Ni, a well-defined peak is observed centered around 4.5 eV and a weaker peak occurs near 1 eV. The MOKE absorption exhibits enhanced structure consistent with these two optical absorption peaks. This can be seen by referring to Fig. 2, which shows the expected shape of MOKE absorption for a single peak in optical absorption. The apparent difference in the ratios of absorption at 1 and 4.5 eV for optical and MOKE absorption primarily reflects the differences in the sp-contribution[13] to the optical absorption peaks that does not contribute to MOKE absorption process.

The MOKE absorption spectra of Ni leads to the density of states model shown in Fig. 5. Transitions from d-bands to p-character bands above E_F produce the

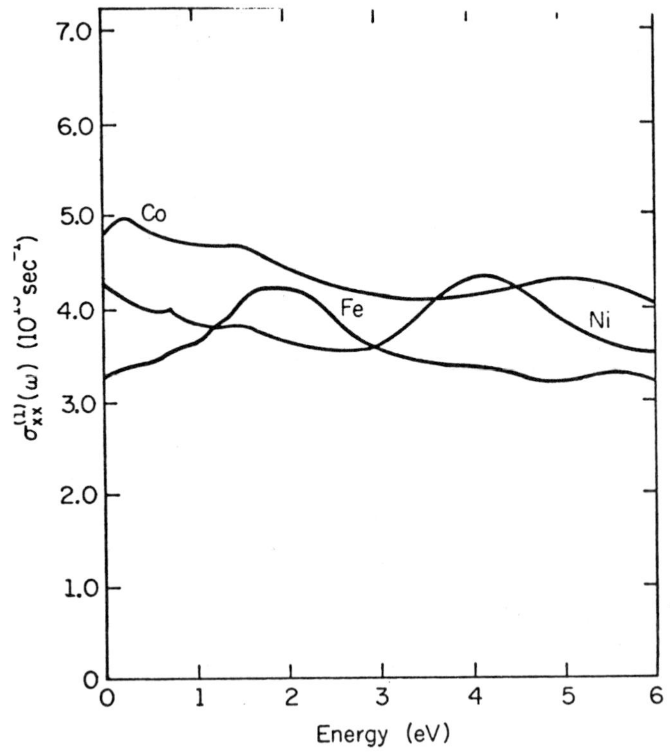

Fig. 3: Absorptive component of the optical conductivity of Ni, Co, and Fe measured by Johnson and Christy, Ref. 17.

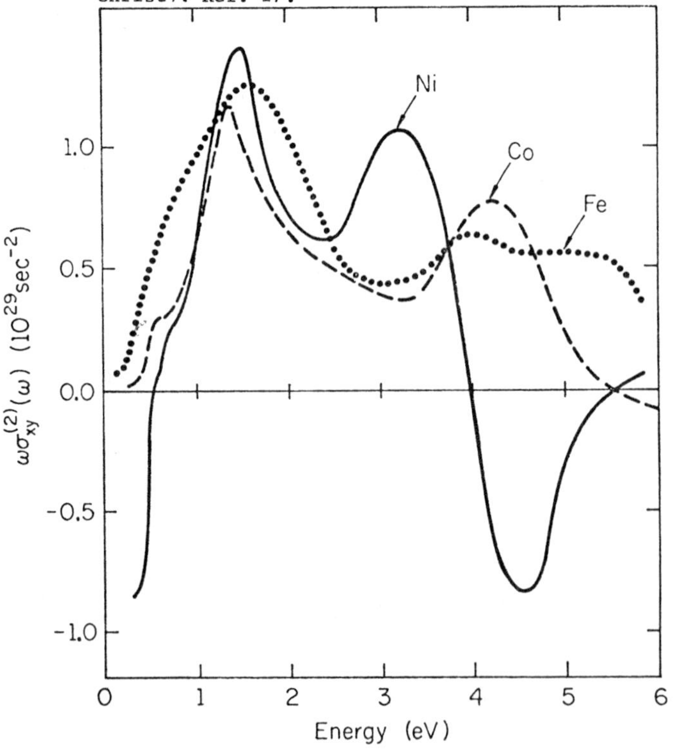

Fig. 4: Absorptive component of the magneto-optical conductivity of Ni, Co, and Fe measured by Krinchik and Artem'ev (Ref. 18).

MOKE absorption, and some broadening due to the spread in final states is expected. However, the sharpness of the structure suggests that this is not more than 1 eV. The predicted density of states consists of two peaks of approximately equal area, the peak nearest E_F having a slightly narrower bandwidth. This model is in good qualitative agreement with density of state curves given by band calculations.[19] The total d-band width of Ni predicted by MOKE absorption is about 4.5

eV, a value considerably wider than 3.3 eV predicted by photoemission.[20]

The initial negative sign of MOKE absorption in Ni is indicative of minority spin transitions. The data extends to 0.22 eV, however, no sign change is expected at lower energies.[21] The spin-polarized photoemission experiments[4] and spin-dependent tunneling experiments[6] measure a net positive (majority) spin-polarization for (d) electrons originating from within less than a tenth of an eV from the Fermi level. This is hard to reconcile with the band model shown in Fig. 5. Although many explanations[7,8] have been offered,

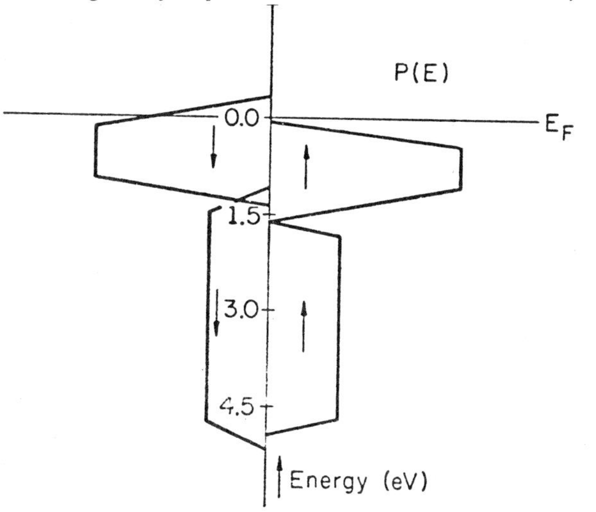

Fig. 5: Density of states model for d-bands of Ni predicted by MOKE absorption.

none appear to be consistent with all of the experimental data. It has been suggested that the apparent contradiction may arise from oversimplified interpretation of some of the experiments.[22] In the tunneling, photoemission and field emission experiments, both sp- and d electrons are involved. In the interpretation, it is assumed the d-electrons dominate, yet this is not obviously true. The density of states is clearly dominated by d-electrons, but the processes also depend on matrix elements which typically favor s-p electrons. The s-band polarization in transition metals may have either sign,[23] and this provides an alternate explanation for the observed phenomena. The MOKE absorption resulting from sp-bands with a net spin-polarization would still be dominated by absorption due to d-electrons because their spin-orbit coupling strength is much stronger.

Density of states predictions for the whole band similar to that derived for Ni in Fig. 5 cannot be obtained from the MOKE spectra of Fe and Co because of insufficient spectral range coverage. However, the data clearly suggests that the d-bands of Co and Fe are wider than that of Ni, and that majority spin electrons dominate minority spin transitions at low energies, implying a dominance of majority spin electron states near E_F in these metals. The d-band widths for Co and Fe measured by photoemission[20] are 3.6 and 3.8 eV, respectively, which are also considerably narrower than the widths indicated by MOKE absorption.

The optical[24] and magneto-optical[25] absorption spectra of the three rare-earth metal ferromagnets shown in Figs. 6 and 7, also exhibit the increased sensitivity of MOKE spectroscopy. The pronounced double peak in the MOKE spectra of Gd is consistent with the two major peaks separated by a deep minima in d-states above E_F predicted by band calculations.[26] Model calculations[9] of the weight $\langle \sigma_{xy}^{(2)} \rangle$ have shown that this structure is due to transitions from occupied p-bands into unfilled d-bands. Interpretation of the MOKE spectra of Tb and Dy is complicated by the fact that transitions involving 4f-electrons are expected to occur in this range. Extension of the wavelength coverage of experimental data and guidance from more accurate band models now available should provide enough information to make specific assignments for the origin of the structure.[25]

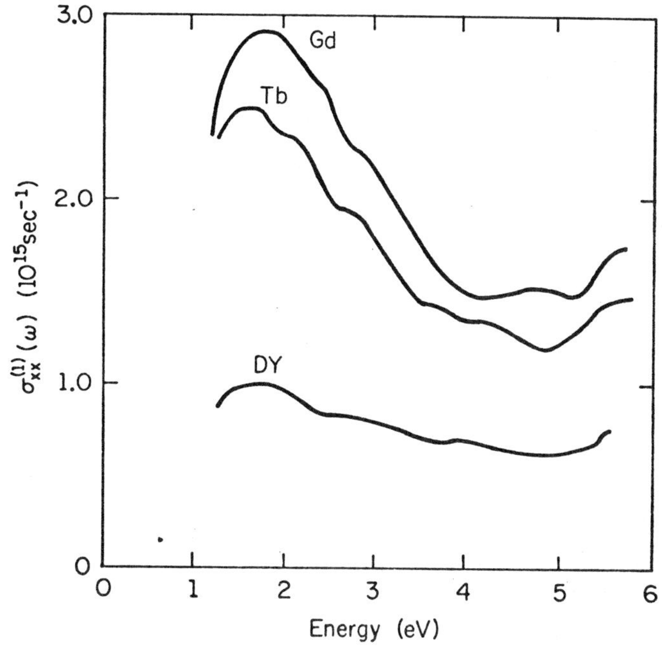

Fig. 6: Absorptive component of the optical conductivity of Gd, Tb, and Dy measured by Erskine, Blake, and Flaten (Ref. 24).

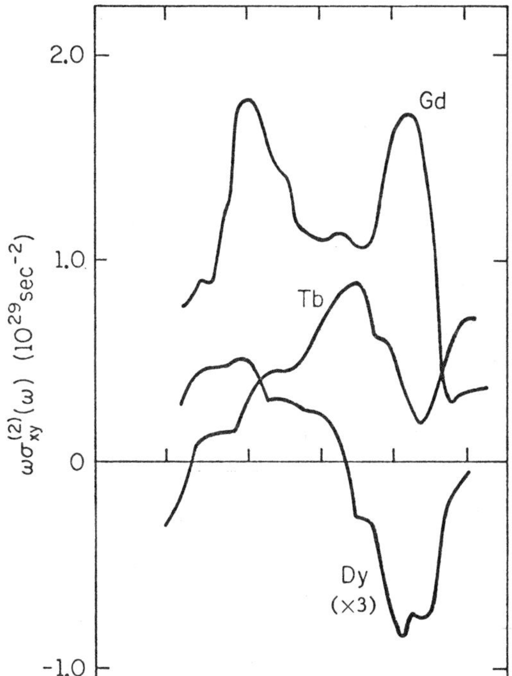

Fig. 7: Absorptive component of the magneto-optical conductivity of Gd, Tb, and Dy measured by Erskine and Stern, (Ref. 9, 25).

CONCLUSIONS

Magneto-optical spectroscopy provides a useful probe for studying electronic structure of ferromagnetic metals. Advantages offered by this technique are increased sensitivity to d- and f-electrons in proportion to their net spin-polarization and spin-orbit coupling strength. Magneto-optical techniques are less sensitive to surface effects than electron emission and tunneling techniques since optical penetration depths are typically hundreds of Angstroms compared with tens of Angstroms or less which characterize electron escape depths. In addition, the work function does not enter into interpretation of optical data. Certain characteristics of the MOKE absorption spectra are directly related to electronic structure parameters. The initial sign of the structure determines the spin-polarization associated with the onset of transitions, and the amplitude and width roughly characterize the joint density of states. Calculations based on an atomic model yield a fairly accurate description of the weight of magneto-optical absorption, which is helpful in analyzing experimental data.

There are several interesting opportunities for future work using MOKE techniques. Magnetic metals have not been studied in the vacuum uv using magneto-optical techniques. The rare-earth metals are particularly attractive candidates for such work. The 4f-electrons responsible for their magnetic properties are located 6-10 eV below E_F and are expected to produce large magneto-optical effects. These atomic like levels provide a possible means of probing the unfilled d-bands in the rare earths. The discussion in this paper has centered around the gross features of MOKE absorption. Fine structure is also observed in the MOKE spectra. With improved experimental techniques, and the capability to compute MOKE absorption directly from band calculations recently demonstrated,[11] this fine structure could provide a very precise check on calculated band models.

*Research supported by the Air Force Office of Scientific Research.

†Present address: Department of Physics, University of Illinois, Urbana, Illinois 61801.

REFERENCES

1. A complete bibliography of magneto-optics of solids (to 1967) is given by E. D. Palik and B. W. Henvis, Appl. Optics 6, 4, 603 (1967).
2. Kerr, J., Phil. Mag. 3, 321 (1877); and Phil. Mag. 5, 161 (1878).
3. Hulm, H. R., Proc. Roy. Soc. (London) A135, 237 (1932); Kittel, C., Phys. Rev. 83, 208(a)(1951); Argyres, P. N., Phys. Rev. 97, 334 (1955).
4. Busch, G., Campagna, M., Pierce, D. T., and Siegmann, H. C., Phys. Rev. Letters 28, 611 (1972); Busch, G., Campagna, M., and Siegmann, H. C., Solid State Comm. 7, 755 (1969) and J. App. Phy. 41, 1044 (1970); Busch, G. Campagna, M., Cotti, P., and Siegmann, H. C., Phys. Rev. Letters 22, 597 (1969); Busch, G., Campagna, M., and Siegmann, H. C., Phys. Rev. B 4, 746 (1971); Baenninger, V., Busch, G., Campagna, M., and H. C. Siegmann, Phys. Rev. Letters 25, 585 (1970).
5. Regenfus, G. and Helwig, R., Verh Deut. Phys. Ges. 5, 181 (1970); Gleich, W., Regenfus, G. and Sizmann, R., Phys. Rev. Letters 27, 1066 (1971); Chrobok, G., Hoffmann, M., and Regenfus, G., Phys. Letters 26A, 11, 551 (1968).
6. Tedrow, P. M. and Meservey, R., Phys. Rev. Letters 26, 192 (1971) and Phys. Rev. B7, 1, 318 (1973).
7. Wohlfarth, E. P., Phys. Letters 36A, 2, 131 (1971); Anderson, P. W., Phil. Mag. 24, 203 (1971); Smith, N. V. and Traum, M. M., Phys. Rev. Letters 27, 30, 1388 (1971); Politzer, Beverly A. and Cutler, P. H., Phys. Rev. Letters 28, 20, 1330 (1972).
8. Gutzwiller, M. C., AIP Conf. Proc. No.10, Part 2, 1197 (1973).
9. Erskine, J. L. and Stern, E. A., Phys. Rev. B8, 3, 1239 (1973).
10. Bennett, H. S. and Stern, E. A., Phys. Rev. 137, A448 (1965).
11. Wang, C. S. and Callaway, J, to appear in Phys. Rev. B, Nov. (1974).
12. Wilson, A. H., The Theory of Metals (Cambridge University Press, Cambridge, England, 1953), p. 47.
13. The spin-orbit splitting in metallic nickel is computed by Wang and Callaway, Ref. 11, to be slightly larger than the atomic value.
14. Herman, F. and Skillman, S., Atomic Structure Calculations (Prentice-Hall, Englewood Cliffs, New Jersey, 1965).
15. Condon, E. V. and Shortley, G. H., The Theory of Atomic Spectra (Cambridge University Press, Cambridge, England, 1970).
16. Schiff, L. I., Quantum Mechanics, (McGraw-Hill, New York, 1955), p.292.
17. Johnson, P. B. and Christy, R. W., Phys. Rev. B9, 12, 5056 (1974).
18. Krinchik, G. S. and Artem'ev, V. A., Soviet Phys. JETP 26, 6, 1080 (1968); also see Yoshino, T. and Tanaka, S., Optics Comm. 1, 149 (1969).
19. Zornberg, E. I., Phys. Rev. B1, 1, 244 (1971); Callaway, J. and Zhang, H. M., Phys. Rev. B 1, 1, 305 (1970); also see Ref. 11.
20. Eastman, D. E., J. Phys. (Paris), Colloq. 32, C1-293 (1971).
21. Wang and Callaway, Rev. 11, have calculated $\sigma_{xy}^{(2)}(\omega)$ for Ni directly from their band calculation code. Their results are in reasonably good agreement with Fig. 4 if a slight change in energy scale is made. The calculated spectra extends to zero energy with no change in sign. Similar success has been achieved by Singh, Wang, and Callaway (to be published) in computing $\sigma_{xy}^{(2)}(\omega)$ for Fe.
22. Erskine, J. L. and Stern, E. A., Phys. Rev. Letters 30, 26, 1329 (1973).
23. Mott, N. F., Adv. in Physics, 13, 52, 325 (1964).
24. Erskine, J. L., Blake, G. A., and Flaten, C. J., to appear in J. Oct. Soc. America.
25. Erskine, J. L. and Stern, E. A., to be published.
26. Dimmock, O. J., Freeman, A. J., and Watson, R. E. Optical Properties and Electronic Structure of Metals and Alloys, F. Abeles, Ed. (North-Holland, Amsterdam, 1966), p.237; Harmon, B. N. and Freeman, A. J., AIP Conf. Proc. No.10, Part 2, 1309, (1973); Harmon, B. N., private communication.

MAGNETIC TRICRITICAL BEHAVIOR OF $FeCl_2$

J. A. Griffin† and S. E. Schnatterly
Joseph Henry Laboratories
Princeton University
Princeton, New Jersey 08540

ABSTRACT

Magneto-optic measurements of the tricritical behavior of $FeCl_2$ are presented. These measurements utilize the magnetic circular dichroism of a near infrared absorption of the Fe^{++} ions in order to determine the total magnetization as a function of temperature and magnetic field near the tricritical point. The resultant description of the magnetic tricritical behavior is consistent with most predictions of a classical Landau analysis, and is found to exhibit a remarkable similarity with the tricritical behavior of ^3He-^4He mixtures.

INTRODUCTION

Optical techniques have frequently been utilized as a probe for studying many aspects of magnetism and magnetic materials. One aspect which has recently received considerable interest is the field-dependent behavior of highly anisotropic antiferromagnetic systems and their resultant tricritical cooperative phenomena. In this paper we will describe the application of magneto-optic techniques to the study of the magnetic tricritical behavior in $FeCl_2$. Although we have observed the effects of magnetic ordering in measurements of optical absorption and of magnetic circular dichroism from the near ultraviolet to the near infrared, we will concentrate here only on the infrared circular dichroism[1]. These measurements, which utilized the dichroism of an absorption band of the Fe^{++} ion, are used to monitor the temperature and magnetic field dependence of the total magnetization. The optical data is then used to construct magnetic phase diagrams of total magnetization versus temperature and of internal magnetic field versus temperature, and to extract tricritical point exponents associated with the total magnetization. The resultant tricritical behavior of $FeCl_2$ exhibits a striking similarity with that of ^3He-^4He mixtures, and is found to agree in most respects with the predictions of a classical Landau analysis.

METAMAGNETIC BEHAVIOR

Metamagnetic $FeCl_2$ is a highly anisotropic antiferromagnet which consists of hexagonal layers of Fe^{++} magnetic moments.[2] Within these layers there is a strong ferromagnetic exchange coupling which aligns the magnetic moments in a parallel arrangement, and between the layers there is an antiferromagnetic coupling. The overall magnetic order is that of antiferromagnetically aligned ferromagnetic layers. The magnetic moments are strongly confined to the hexagonal c axis (perpendicular to the layers) by an anisotropy energy that exceeds both the ferromagnetic and antiferromagnetic exchange energies. The fact that this anisotropy exceeds the antiferromagnetic exchange and that there is ferromagnetic coupling within the layers results in the rather interesting field-dependent behavior.

In the presence of a sufficiently strong magnetic field applied perpendicular to the layers, there is an antiferromagnetic to paramagnetic phase transition. Due to the strong anisotropy this field-induced transition involves a spin-flip in which the initially antiparallel planes flip directly to the paramagnetic configuration, without the spin-flop phase present in many antiferromagnets. Due to the ferromagnetic coupling within the layers this spin-flip transition is first order at finite temperature, and it results in a finite discontinuity in the total magnetization. The first order nature of the transition persists up to the tricritical temperature T_t, above which the transition is second order up to the Neel temperature T_N. The point in the metamagnetic phase diagram at which this field-induced transition changes from first to second order has been labeled a tricritical point[3], and it is the cooperative behavior associated with this special point that is responsible for the current interest in metamagnetic systems.

EXPERIMENTAL TECHNIQUE

Magnetic circular dichroism is the field-induced difference in absorption coefficients for left and right circularly-polarized light for a geometry in which the magnetic field and incident beam are parallel.[4] Measurements of the infrared dichroism of $FeCl_2$ were made by placing the sample in the bore of a superconducting solenoid with optical access parallel to the magnetic field. The sample, a flat plate, was mounted so that its hexagonal c axis was parallel to the magnetic field and was masked in order that the incident beam was limited to the central portion of the plate, across which the internal magnetic field is homogeneous. The incident monochromatic beam was passed first through a linear polarizer, next through a rotating phase plate, next through a mechanical chopper, and finally through the sample. The suitably analyzed intensity at the detector is proportional to

$$\frac{e^{-2\alpha_L d} - e^{-2\alpha_R d}}{e^{-2\alpha_L d} + e^{-2\alpha_R d}} \qquad (1)$$

from which the circular dichroism

$$MCD(\lambda,T,H) \equiv \alpha_L - \alpha_R \qquad (2)$$

is obtained. The resolution of the optical system used in the experiment was approximately 0.2% of the saturated paramagnetic dichroism, and this resolution has been extended to approximately 0.03% using optical bridge techniques.

The temperature of the sample was carefully controlled. For this purpose the sample was placed in a cylindrical holder equipped with optical windows on each end and which was filled with ^4He exchange gas. No effects due to sample heating by the incident beam were found to occur. The temperature of the holder was measured by a calibrated four-wire germanium resistance thermometer and an ac resistance bridge[5] before and after each magnetic field scan. In order to control the sample holder temperature during the field scan, a magnetic field-independent $SrTiO_3$ capacitance thermometer was used in an ac capacitance bridge temperature controller.[6] This technique eliminates the coupling of temperature and magnetic field which occurs due to the magnetoresistance present in germanium or carbon resistance thermometers. Using this scheme, the sample temperature was constant to approximately 3 mK along each isothermal field scan.

EXPERIMENTAL RESULTS

Fig. 1 illustrates the optical absorption of $FeCl_2$ in zero magnetic field at four temperatures. The lowest optical transition occurs at approximately 1.45 microns, and is a phonon-assisted electric dipole transition between the cubic field split components, the lower $^5T_{2g}$ and the upper ^5E_g, of the free ion ^5D term. The doublet structure arises from the Jahn-Teller effect in which the two-fold orbital degeneracy of the ^5E_g level is lifted by even-parity phonons.[7] This transition is spin-allowed, and shows no temperature dependence at low temperatures (T < 88 K). At higher energies there are several spin-forbidden, quintet-to-triplet transitions. The three bands between 1.5 eV and 2.3 eV are broad and structureless, and exhibit no temperature dependence. At higher energies the spectrum consists of narrow lines which are strongly temperature-dependent below the Neel temperature.

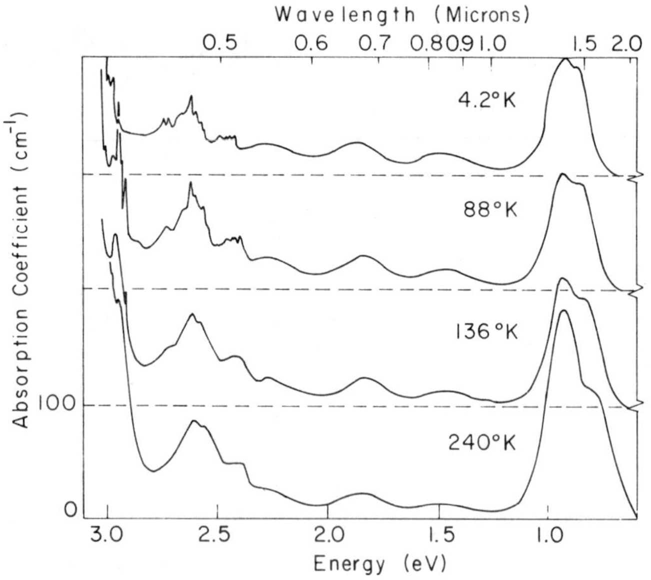

Fig. 1: Absorption spectrum of $FeCl_2$ at four temperatures.

The data on which we will concentrate here is the magnetic circular dichroism of the absorption at 1.45 microns. The intensity of this dichroism exhibits the same temperature and field dependence as the total magnetization, and it is this dichroism which we will utilize to discuss the tricritical behavior of $FeCl_2$.

Although magnetic circular dichroism is not in general proportional to the magnetization, due to the fact that optical transitions involve a final state as well as the ground state, there are many examples in which dichroism and magnetization are proportional[8]. Due to the complexity of the phonon-assisted optical transition[7] this proportionality has not been theoretically verified in $FeCl_2$. Perturbation theory estimates of mechanisms that could preclude this proportionality indicate that the deviation from proportionality has an upper limit of approximately 0.1%.

The dichroism measurements were made at 1.45 microns as a function of external magnetic field along thirty isotherms. Fig. 2 illustrates this dichroism, normalized by the saturated paramagnetic value, as a function of external magnetic field for nineteen isotherms near the tricritical point. The effect of the demagnetization field NM, where N is the classical demagnetization factor and M is the total magnetization, is seen in this figure. At the first order transition the presence of the demagnetization field results in the spontaneous production of saturated paramagnetic domains which are spatially distributed throughout the otherwise antiferromagnetically ordered crystal. These domains then grow with increasing external field in such a manner that the magnetization in this mixed phase is linear in external magnetic field with a slope of 1/N.[9] This behavior is well illustrated by the dichroism at 16.94 K. The antiferromagnetic phase occurs up to a magnetic field of approximately 9.6 kG, and is characterized by a small susceptibility. At this field the mixed phase is entered, and the dichroism exhibits a linear dependence in external magnetic field up to 15.3 kG, at which the saturated paramagnetic phase is entered. As the temperature is increased, thus increasing the thermal disorder, the range of the mixed phase is decreased, but the linear field behavior persists with the same slope. This behavior is illustrated by the isotherm at 20.39 K, for example. As the temperature is increased above T_t, the spin-flip transition is second order, and there is no longer any linear field dependence characteristic of

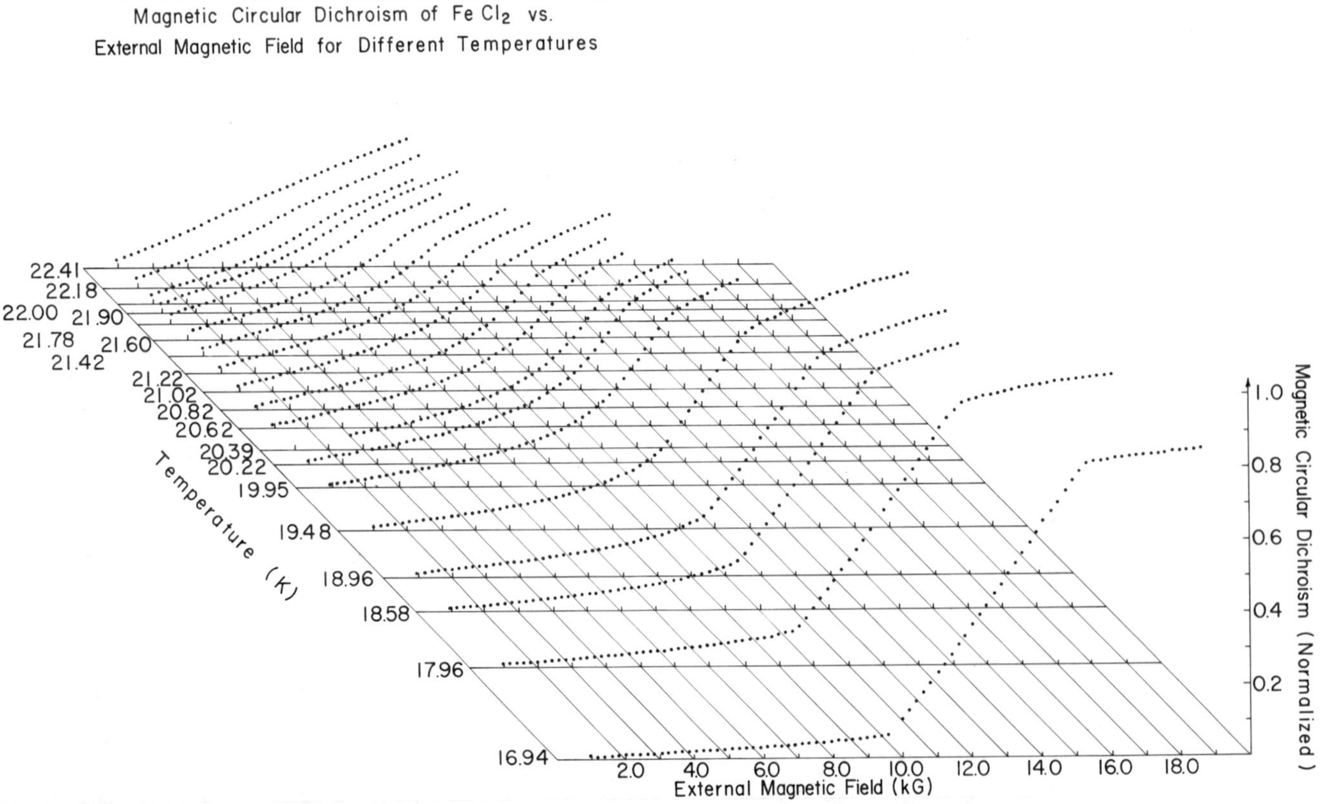

Fig. 2: Normalized magnetic circular dichroism versus external magnetic field for nineteen isotherms near the tricritical point.

the first order transition. The second order transition is located by a maximum in the susceptibility at the critical external field.

The behavior depicted in Fig. 2 is accentuated in Fig. 3, which illustrates the dichroism as a function of internal magnetic field for the same nineteen isotherms. The transformation to internal magnetic field was accomplished by least squares fitting the dichroism in the mixed phase in Fig. 2 to determine the demagnetization factor, and numerically computing the internal magnetic field at each data point according to

$$H = H_E - NM \tag{3}$$

where H is the internal magnetic field, and H_E is the external magnetic field. The demagnetization factor was determined for eleven temperatures between 9.95 K and 19.48 K, and was found to deviate from the average by at most 2%, thus illustrating that the factor of proportionality between dichroism and magnetization is temperature independent. As a function of temperature the internal field dependence of the dichroism illustrates the physical behavior of the tricritical system. At 16.94 K the simultaneous spin-flip of the large blocks of magnetic moments in the antiparallel layers is manifested by a discontinuity in the dichroism at the critical internal field. As the temperature is increased, thus reducing the ferromagnetic coupling within the layers, the discontinuity in the dichroism becomes smaller. This behavior is illustrated by isotherms at 18.58 K, 19.48 K, and 20.22 K. At temperatures beginning with 20.82 K the transition is second order, for the finite discontinuity occurring at the critical field at low temperatures is replaced by a continuous dichroism. This continuous behavior above T_t is associated with the flipping one-by-one of the magnetic moments in the antiparallel layers. The line of these second order transitions is traced out by the susceptibility maximum at the critical internal field.

Fig. 4 illustrates the normalized critical magnetic circular dichroism versus temperature. In the first order region the two data points for each temperature depict the dichroism at the onset and termination of the vertical discontinuity depicted in Fig. 3. The two branches are denoted by M^+ and M^-. In the second order region the single data point at each temperature denotes the value of the dichroism on the λ line, and this branch is labeled by M_λ. The tricritical point occurs at $T_t = 20.79 \pm 0.11$ K and $M_t = 0.38 \pm 0.01$. There are several interesting features of the data in Fig. 4. First, both branches in the first order region appear to approach the tricritical point linearly. The tricritical point exponents associated with the total magnetization are described by $(M^+ - M_t)/M_t = A_+((T_t - T)/T_t)^{\beta_+}$ and $(M_t - M^-)/M_t = ((T_t - T)/T_t)^{\beta_-}$.[10] From the data in Fig. 4 we obtain $\beta_+ = 1.03 \pm 0.05$ for $8.3 \times 10^{-3} \leq (T_t - T)/T_t \leq 1.06 \times 10^{-1}$ and $\beta_- = 1.13 \pm 0.14$ for $8.3 \times 10^{-3} \leq (T_t - T)/T_t \leq 6.3 \times 10^{-2}$. According to a Landau analysis these exponents are given by $\beta_+ = \beta_- = 1.00$. Secondly, the slopes of the branches M^+ and M_λ at the tricritical point are not equal. We obtain $d(M^+/M_t)/d(T/T_t) = 7.0 \pm 0.3$ and $d(M_\lambda/M_t)/d(T/T_t) = 4.1 \pm 0.2$, whereas a Landau analysis predicts that the slope of the total magnetization on the upper branch of the first order transition should be equal to that on the λ line at the tricritical point.

The internal critical magnetic field versus temperature phase diagram is illustrated in Fig. 5. This figure is a plot of the critical internal field at each temperature at which the spin-flip transition occurs in Fig. 3. This data illustrates that at the tricritical point the first order branch of this phase diagram intersects the λ branch with a continuous slope. A Landau description predicts that the slope and magnitude of these branches should be continuous at the tricritical point, and such behavior appears to be the case in Fig. 5.

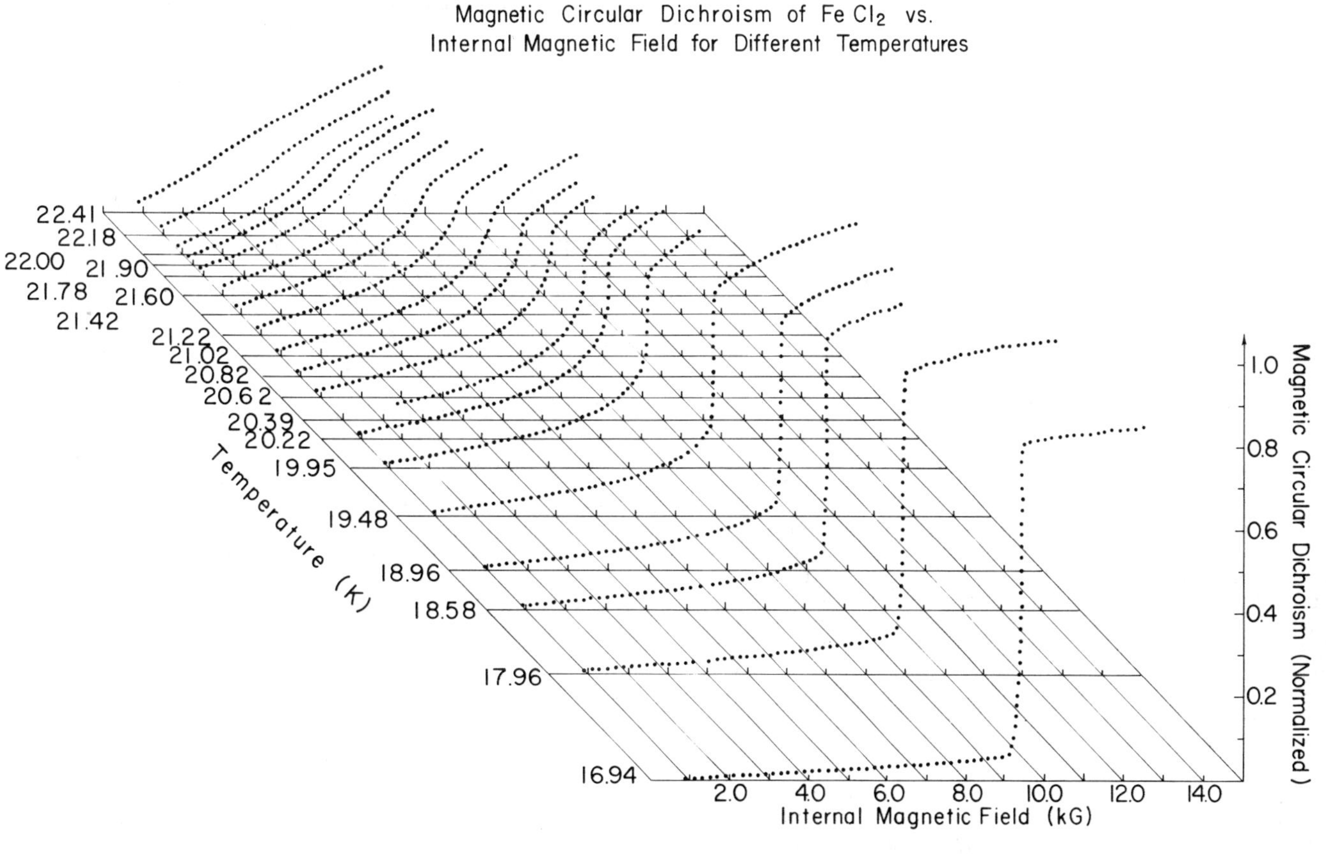

Fig. 3: Normalized magnetic circular dichroism versus internal magnetic field for nineteen isotherms near the tricritical point.

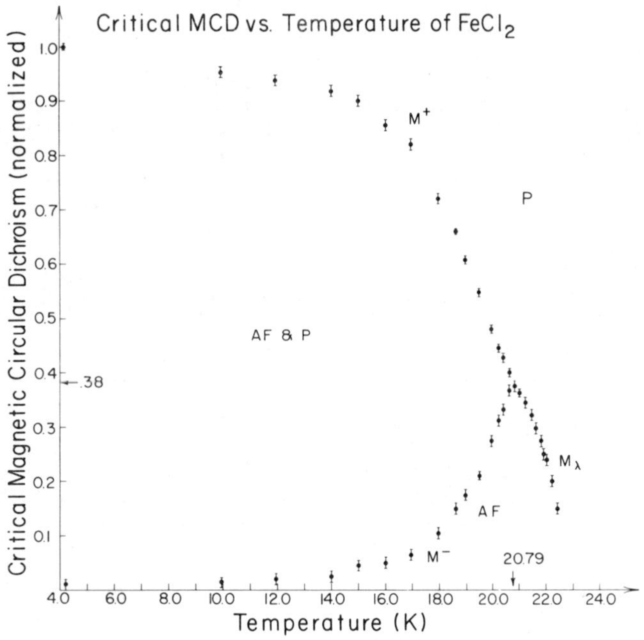

Fig. 4: Critical magnetic circular dichroism versus temperature phase diagram of $FeCl_2$.

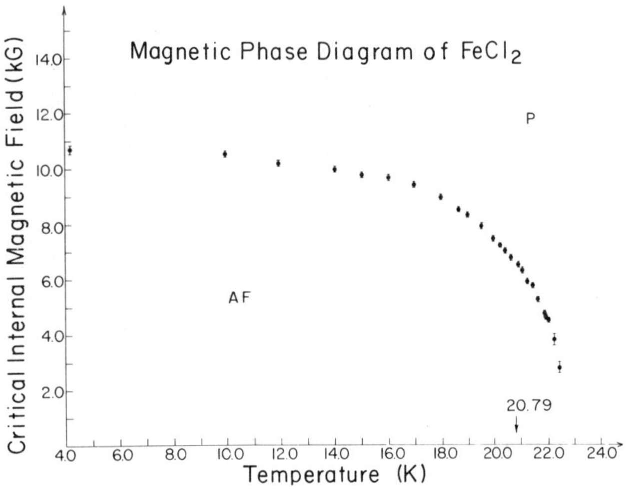

Fig. 5: Critical internal magnetic field versus temperature phase diagram of $FeCl_2$.

DISCUSSION

Cooperative behavior at a tricritical point is expected to differ from that found at an ordinary critical point. Renormalization group techniques have demonstrated that ordinary critical point phenomena behave nonclassically in dimensions less than four, and these techniques have also predicted numerical values of critical point exponents in three dimensions that quantitatively agree with experimental results.[11] These same techniques have been recently applied to tricritical phenomena in ^{3}He-^{4}He mixtures and metamagnetic systems.[12] It was found that tricritical point phenomena behave classically in dimensions less than four, and that, apart from logarithmic corrections, classical exponents are expected for dimensions equal to three. A Landau analysis has also been applied to tricritical phenomena.[13] This classical approach, which is quantitatively incorrect at ordinary critical points, predicts numerical values of exponents and describes shapes of phase diagrams in the tricritical region.

^{3}He-^{4}He mixtures have provided the first empirical test of the classical description of a tricritical point.[14] In this system the ^{4}He superfluid to normal transition is known to decrease in temperature with the addition of ^{3}He down to 866 mK and 67% ^{3}He. Below this point in the phase diagram the ^{3}He-^{4}He mixture exhibits a first order separation into two phases, one a normal ^{3}He-rich fluid and the other a ^{4}He-rich superfluid. Numerous experiments have essentially demonstrated that a classical description of the ^{3}He-^{4}He tricritical point is quantitatively valid, with the exception of a slope discontinuity in the X_3 (fraction of ^{3}He) versus T phase diagram. Magnetic tricritical phenomena are providing the second empirical test of the cooperative behavior of a tricritical system, and $FeCl_2$ is the first metamagnetic system on which detailed tricritical measurements have been made. We have presented here a description of the magnetic tricritical behavior of $FeCl_2$ based on measurements of the magnetic circular dichroism. The tricritical point exponents and the phase diagrams so determined are, within experimental uncertainties, consistent with the classical description of a tricritical point, with the exception of the slope discontinuity in the M versus T phase diagram. This discontinuity, which is analogous to the discontinuity found in the X_3 versus T phase diagram of ^{3}He-^{4}He mixtures, represents in this experiment the only discrepancy between the empirically observed behavior and that which is predicted by a Landau theory.

In conclusion, we have demonstrated the utility of magneto-optic techniques in the study of magnetic tricritical behavior by using the magnetic circular dichroism of a near infrared absorption band as a monitor of the total magnetization. Although there are several additional tricritical exponents that we have not determined in the present experiment, we have presented optical data which strongly imply the validity of a classical description of the magnetic tricritical behavior of $FeCl_2$ and which illustrates the striking similarity found between the tricritical behavior of magnetic $FeCl_2$ and ^{3}He-^{4}He mixtures.

We are pleased to acknowledge helpful discussions with Y. Farge, F. A. Ferrone, M. P. Fontana, J. J. Hopfield, J. M. Kincaid and S. R. Nagel.

REFERENCES

†Present Address: Department of Physics and Center for Materials Science and Engineering, Massachusetts Institute of Technology, Cambridge, Mass. 02139

1. S. E. Schnatterly and M. Fontana, J. Phys. (Paris) 33, 691-697 (1972); J. A. Griffin, S. E. Schnatterly, Y. Farge, M. Regis and M. P. Fontana, Phys. Rev. B10, 1960-1966 (1974); J. A. Griffin and S. E. Schnatterly, Phys. Rev. Lett., to be published.
2. M. K. Wilkinson, J. W. Cable, E. O. Wollan and W. C. Koehler, Phys. Rev. 113, 497-507 (1959); I. S. Jacobs and P. E. Lawrence, Phys. Rev. 164, 866-878 (1967); R. J. Birgeneau, W. B. Yelon, E. Cohen and J. Makovsky, Phys. Rev. B5, 2607-2615 (1972); W. B. Yelon and R. J. Birgeneau, Phys. Rev. B5, 2615-2621 (1972); C. Vettier, H. L. Alberts and D. Bloch, Phys. Rev. Lett. 31, 1414-1417 (1973) E. Yi Chen, J. F. Dillon, Jr. and H. J. Guggenheim in Proceedings of the Conference on Magnetism and Magnetic Materials, Boston (1973), to be published.
3. R. B. Griffiths, Phys. Rev. Lett. 24, 715-718 (1970).
4. N. V. Starostin and P. P. Feofilov, Soviet Phys.-Uspekhi 12, 252-270 (1969); P. N. Schnatz and A. J. McCaffrey, Quarterly Rev. 23, 552-584 (1969); P. J. Stephens, J. Chem. Phys. 52, 3489-3516 (1970).

5. L. G. Rubin and Y. Golanhy, Rev. Sci. Instr. 43, 1758-1762 (1972).
6. J. A. Griffin, Rev. Sci. Instr., to be published.
7. G. D. Jones, Phys. Rev. 155, 259-261 (1967).
8. S. Geschwind, Electron Paramagnetic Resonance, ed. by S. Geschwind, Plenum Press, New York (1972), p. 389.
9. A. F. G. Wyatt, J. Phys. C1, 684-686 (1968).
10. R. B. Griffiths, Phys. Rev. B7, 545-551 (1973).
11. K. G. Wilson and J. Kogut, Physics Reports C12, 75-200 (1974).
12. E. K. Riedel and F. J. Wegner, Phys. Rev. Lett. 29, 349-352 (1972); D. R. Nelson and M. E. Fisher, to be published.
13. M. Blume, V. J. Emery and R. B. Griffiths, Phys. Rev. A4, 1071-1077 (1971); R. Bausch, Z. Phys. 254, 81-88 (1972); J. M. Kincaid, Ph.D. Thesis, The Rockefeller University (1974).
14. G. Ahlers, The Physics of Liquid and Solid Helium, ed. by K. H. Benneman and J. B. Ketterson, John Wiley, New York, Vol. I, to be published.

MAGNETOOPTICAL STUDIES OF METAMAGNETS

J. F. Dillon, Jr., E. Yi Chen and H. J. Guggenheim
Bell Laboratories, Murray Hill, New Jersey 07974

ABSTRACT

Techniques based on magnetooptical rotation have recently been of great use in the study of metamagnetic crystals. Under achievable conditions this rotation is a measure of M which may be measured automatically as H or T vary. The distinctly different rotations of coexisting anti- and paramagnetic phases at the first order metamagnetic transition make possible direct examination of the mixed state with a polarizing microscope. Furthermore, many properties of mixtures of the two time reversed antiferromagnetic states may be deduced from their appearance in applied fields. Two metamagnets have been studied:

$FeCl_2$: $T_N \cong 23.5$ K, layer structure with competing ferromagnetic intra- and antiferromagnetic interlayer interactions.

$Dy_3Al_5O_{12}$(DAG): $T_N \cong 2.54$ K, garnet structure rich in symmetry properties. There are magnetooptical manifestations of coupling between the antiferromagnetic order parameter and fields along [111].

Moving pictures will be shown on the metamagnetic transition in DAG and $FeCl_2$ as well as that between time reversed antiferromagnetic states in DAG.

INTRODUCTION

Experimental techniques based on the magnetooptical Faraday rotation have recently proved useful in studying Ising-like antiferromagnetic crystals.[1,2,3] This rotation which may be plotted automatically as field and temperature vary is a sensitive convenient measure of magnetization. Furthermore, when distinctively different rotations are associated with the two magnetic phases at a first order transition, it is possible to examine directly the coexisting states with a polarizing microscope. The simple Ising-like antiferromagnets are of special interest because fundamental theory of their properties is often tractable. Part of this interest derives from the fact that the phase relationships for these antiferromagnets are closely analogous to those of another much studied simple system, He^3-He^4 mixtures.[4] It seems important to characterize the magnetic systems and to make detailed comparisons.

In this paper we shall briefly describe techniques used to measure the variation of magnetooptical rotation with applied field, internal field, and temperature. The experimental behavior of two Ising-like antiferromagnets will be discussed in terms of broad features of their magnetic phase diagrams. With this background in hand we will present polarizing microscope studies of mixed magnetic phases in these simple systems.

Micrographic moving pictures of first order magnetic phase transitions in both $FeCl_2$ and $Dy_3Fe_5O_{12}$ will be shown with the oral version of this paper.

Both the rotation studies and the microscope studies have been performed with two antiferromagnetic crystals: $FeCl_2$ in fields along [0001] and $Dy_3Fe_5O_{12}$(DAG) in fields along [111]. The properties of $FeCl_2$ in the neighborhood of its tricritical point have been discussed in two papers at this conference.[5,6] Similarly, a broad exposition of recent work on DAG has been given.[7] Much of the material in those three papers serves as introduction to the descriptions given here. In Table I we list for reference some properties of the two crystals. Note that in both cases the work discussed in this paper only refers to the field directions specified. An indicator of the relative complexity of the two structures may be seen in the number of magnetic ions in the primitive chemical cell.

TABLE I

	$FeCl_2$	DAG
In fields along	[0001]	[111]
Structure	Hexagonal sheets between Cl layers	Cubic garnet structure
Mag. ions/prim. chem. cell	1	12
Growth Techniques	$Fe+Cl_2 \rightarrow FeCl_2$, Bridgman	Czochralski, flux
Crystals	Sublimation flakes	Sound 3-D; may be sawn, polished
Néel Temp.	24.5 K	2.54 K
A→P transition is first order to (T,H_i)	(21.8 K,7.2 kOe)	(1.66 K,3.35 kOe)
Moment confined to	[0001]	100
Saturation induction at 1.3 K, $4\pi M$	7.58 kOe	10.85 kOe
H_i for coexistence at 1.3 K	10.60 kOe	3.72 kOe

The garnet is a complicated three dimensional structure, and $FeCl_2$ a simple hexagonal layer structure. The $FeCl_2$ specimens we have used were grown by the reaction of Cl_2 gas with Fe sponge at about 950 C. The soft sublimation flakes thus obtained were merely trimmed around the edges to fit into our holders. Conventional sample preparation and polishing techniques could be used to make suitable samples from the large hard Czochralski grown crystals of DAG.

Let us first review Fig. 1(a), the magnetic phase diagram of an Ising antiferromagnet in the space of temperature and the two intensive fields, H_i and H_s. This follows the approach of R. B. Griffiths[4,8]. The staggered field H_s is the hypothetical field distribution which would produce a pure staggered magnetization, one sublattice up and the other down. Such a field can have either sign, and is conjugate to the two time reversed [TR] antiferromagnetic states which we might designate A^+ and A^-. The other field variable H_i is the field distribution which would produce a uniform magnetization on the two sublattices. In the case of $FeCl_2$ it corresponds to a properly demagnetized external field. As we shall see later, DAG is a more complicated case. The vertically shaded area in H_s=0 represents the coexistence surface of the two antiferromagnetic states. If one could prepare a sample by approaching this surface from very large positive H_s and H_i=0 the spin system would all be in the positive antiferromagnetic state. A sample cooled through $(T_N,0,0)$ is as likely to be in one state as the other, and would presumably be in a 1:1 mixture of the two TR states. The dashed line joining $(T^t,H_i^t,0)$ and $(T^t,-H_i^t,0)$ contains the Néel point, and is a line of critical points along which the two antiferromagnetic states cease to be distinct. Above this line the crystal is paramagnetic. However at temperatures below T^t the transition from antiferromagnetic to positive and negative paramagnetic is first order. In

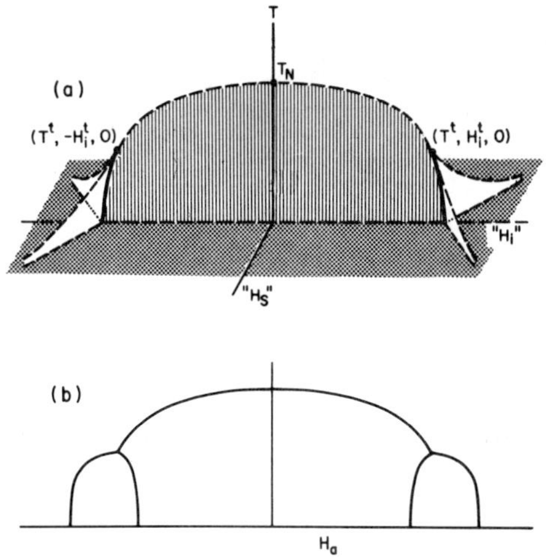

Fig. 1(a) Phase diagram of an Ising antiferromagnet in the coordinates T-H_i-H_s. H_i and H_s are discussed in the text. (b) The corresponding phase diagram in the plane T-H_a for a case such as $FeCl_2$ in which applied field H_a differs from H_i by a demagnetizing correction, and does not contribute to H_s.

the H_s=0 plane these are shown as the solid line segments extending down from $(T^t,\pm H_i,0)$ to the T=0 plane. Looking in the dimension H_s we see that the antiferromagnetic coexistence surface splits along this line, and that "wings" extend out into positive and negative H_s. These wings, white in our sketch, are again coexistence surfaces between A^+ and P^+, between A^- and P^+, etc. Along their high temperature edges these wings are also bounded by critical lines. The tricritical points $(T^t,\pm H_i^t,0)$ each stand at the intersection of three lines of critical points. As indicated in Fig. 1(b), when the H_s=0 section of this phase diagram is transformed into a plot of T versus applied field H_a, the A-P coexistence line opens out into a coexistence area because of demagnetizing effects.

ROTATION MEASUREMENTS

We have measured specific magnetooptical rotation ρ as field and temperature vary, and take it to be a measure of the magnetization. That this proportionality is valid when the sample is in a single magnetic phase is clear. It can also hold in a mixed state if most of the light passing through the specimen encounters the two phases in the average proportion and if the rotation contrast is not too large. As will be seen below, we have an experimental check of the proportionality ρ/M.

The apparatus used to plot rotation automatically is illustrated in Fig. 2. Linearly polarized monochromatic light is passed through a specimen held at the desired temperature. Below 4.2 K this may be maintained by immersion in liquid He, above about 3 K by contact with the thermostated copper block of a variable temperature Dewar. The sample is also subject to the field of a large electromagnet which lies normal to its major faces and along the axis of the optical system. After passing through the sample, the light encounters a linear polarizer rotated at about 50 Hz. Light reaching the detector is thereby modulated at about 100 Hz, and the rotation which it has undergone is contained in the phase of that signal. We have used a commercial phase computer to yield a voltage proportional to the phase angle between the detected signal and a reference. From a Hall probe taped to the pole piece of the magnet we obtain a direct measure of the applied field. The two voltages allow us to produce plots of rotation versus H_a, the upper recorder in Fig. 2.

Figure 3(a) is an example of a $\rho(H_a)$ plot obtained with this apparatus. It is for an $FeCl_2$ sample at 1.3 K. The field segment marked A with small rotation

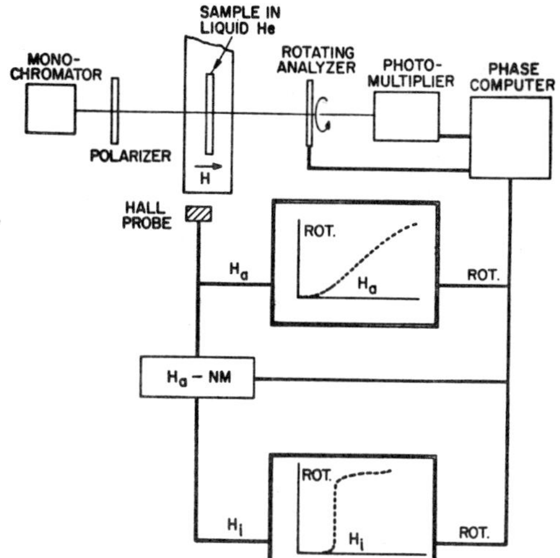

Fig. 2 Sketch of apparatus for automatically plotting of rotation versus H_a and H_i. Above liquid He temperatures a variable temperature Dewar is used.

corresponds to the range in which the sample is antiferromagnetic. The long straight portion of the curve in the region A+P is the mixed phase region in which the spin system is divided into discrete antiferromagnetic and paramagnetic volumes. In terms of the magnetizations M_{anti} and M_{para} of the two states the fraction f which is paramagnetic is determined by the requirement that the interfaces between the two states be subject to a net field equal to the thermodynamic coexistence field H_{co}. In the familiar expression for demagnetized field we may set $H_{co} = H_a - NM = H_a - N[(1-f)M_{anti} + fM_{para}]$ as f goes from 0 to 1. It follows that $dM/dH_a = 1/N$, a geometric quantity. If $\rho(H_a)$ is linear in this range we have demonstrated $\rho \propto M$. If $\rho(H_a)$ is not quite linear we may alter the experiment slightly to make it so, e.g. use a longer wavelength at which rotation is smaller, or resort to a thinner specimen. Finally, in the segment designated P the spins are all in the paramagnetic state and rotation and magnetization approach saturation in the highest fields.

Since rotation is equivalent to magnetization, we may multiply the rotation voltage by an appropriate factor to demagnetize the applied field, and obtain a voltage proportional to internal field H_i. These operations may be performed with operational amplifier circuits. In Fig. 2 we have shown a second recorder which plots rotation against H_i. The choice of the "appropriate factor" is made by adjusting a feedback potentiometer so that the mixed phase region corresponds to a vertical line on the plot of $\rho(H_i)$. Once made for a given sample and wavelength this adjustment should be independent of temperature, and indeed is found to be so.

Directly recorded plots of rotation against internal field have been of particular value in the study of DAG where a completely unexpected hysteresis was encountered. In Fig. 3(b) is shown the A and part of the A+P ranges of the rotation curve against H_i for DAG at 1.3K. As discussed above, H_i does not change in the mixed phase region. The approach to the mixed phase was found to take place along different paths depending on the past history of the sample. If the spin system was last exposed to large negative fields, the curve followed the path marked I (for initial). If last exposed to large positive fields the curve followed the path S (for second). We designated the phenomenon I-S hysteresis, and found it to be a very real effect quite outside the then accepted magnetic phase diagram.[1] The two curves imply that there are two different antiferromagnetic states. Its observation was a demonstration of the utility of the optical techniques, for the phenomenon had not been seen in the course of conventional magnetization measurements by a number of workers.

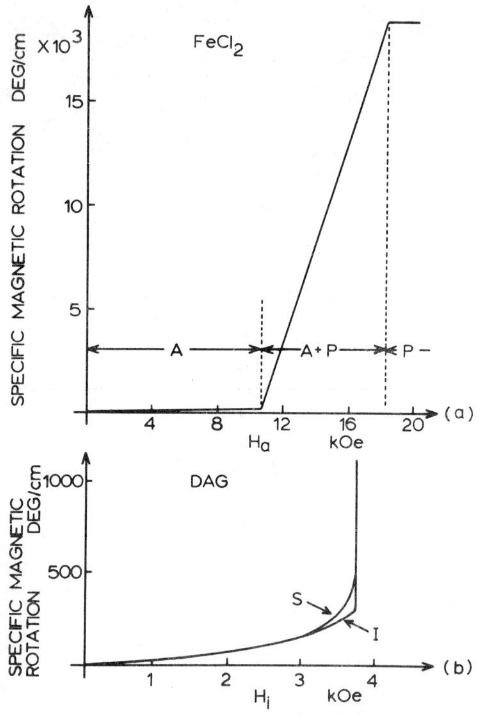

Fig. 3 (a) Plot of rotation vs H_a for a thin sheet of $FeCl_2$ at 1.3 K. In segment A the sample is antiferromagnetic; in P it is paramagnetic; in A+P a mixture of the two. (b) Detail of rotation vs H_i along [111] for DAG at 1.3 K showing the two paths produced depending on the past magnetic history of the spin system.

Our present understanding of these two antiferromagnetic states, indeed of the DAG phase diagram, follows from the insight presented in a recent letter by Blume, Corliss, Hastings and Schiller.[9] They showed that the symmetry of the garnet structures allows a coupling between the staggered magnetization and a physical field along [111], i.e. an externally applied field couples to the staggered magnetization. This means that in the phase diagram of Fig. 1 for DAG with fields along [111] both of the field coordinates must be fictitious. An "H_i" which varies from site to site is required to produce a purely uniform magnetization, and another combination of local fields "H_s" is required to produce a purely staggered magnetization with no uniform component. A physical field has components of both, and thus moves us out of the "H_s"= 0 plane. The surface T-H_a in which we are constrained to take data furls through the diagram of Fig. 1, crossing the A^+-A^- coexistence surface along the T-axis. The second order line joining the two tricritical points does not lie in the T-H_a surface except at H_a=0. For positive H_a the physical plane intersects one of the wing coexistence surfaces for the antiferromagnetic and paramagnetic states. In Fig. 4(a) we show the DAG T-H_a

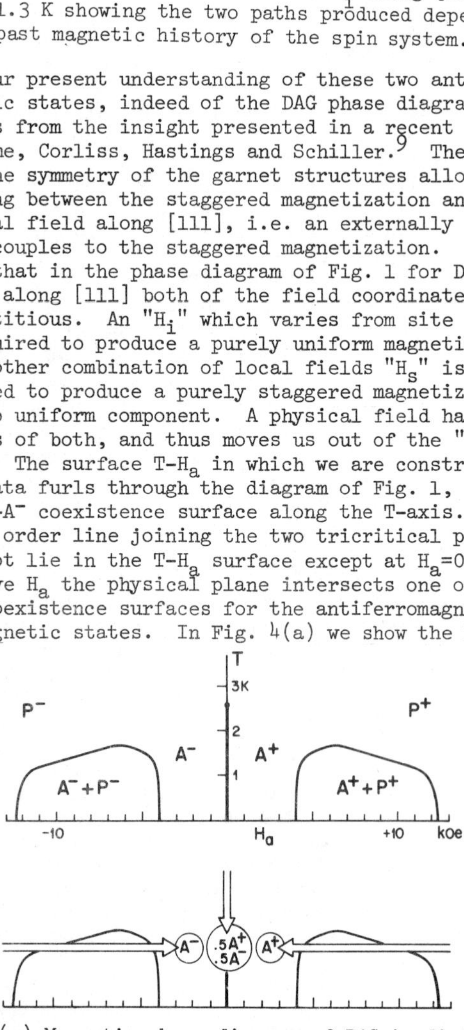

Fig. 4 (a) Magnetic phase diagram of DAG in fields along [111]. As discussed in the text A^+ and A^- are the two time reversed antiferromagnetic states, P^+ and P^- the paramagnetic states. (b) Field-temperature treatments which will produce A^+, A^- and a 1:1 mixture of the two.

diagram which may be compared with Fig. 1(b). Since positive H_a couples to one sign of the staggered magnetization and negative to the other, the line H_a=0 is a coexistence line for the transition between the two antiferromagnetic states. It is apparent from our experimental work that the energy between these two states is exceedingly small in much of the antiferromagnetic region of the phase diagram. Let us define as A^+ the antiferromagnetic state which is stable in positive H_a at 1.3 K, and A^- that which is stable in negative H_a at 1.3 K. Because the energy difference is small, A^- is metastable over much of the accessible positive field range, and A^+ is metastable over much of the negative range. Since A^+ is stable in positive H_a, if we enter the antiferromagnetic region along the path 1 of Fig. 4(b) we will produce a spin system essentially all in A^+. Similarly, path 2 will produce A^-. If the sample is cooled through the Néel point in zero field, path 3 , we expect and do find it to be in a 1:1 mixture. It is possible to store any of these states or mixtures of states indefinitely by reducing the field to zero and maintaining a low temperature. At 1.3 K A^- is metastable in positive fields up to the point where P^+ appears. At a temperature such as 1.8 K the conversion $A^- \rightarrow A^+$ takes place in hundreds of seconds at applied fields of about 3.6 kOe.

Against the background of Ising antiferromagnets we have described techniques for the measurement of magnetooptical rotation. Much of this has been a useful preliminary to the description of microscope studies of these phase transitions. Though not emphasized here the techniques and their variations may be used to explore many aspects of the magnetic phase diagrams of such compounds.

MICROSCOPY OF MIXED PHASES

Using a polarizing microscope we can actually see many details of mixed phases in these metamagnetic crystals. This includes the nucleation and growth of the new phase at either edge of the A-P coexistence range as well as characteristics of the manner in which the two phases are disposed in each other throughout the mixed phase. In DAG in fields along [111] we can provoke a rotation difference between A^+ and A^- by the application of such a field, and thereby study their behavior. In both $FeCl_2$ and DAG we find that it is possible to see antiferromagnetic domain walls by a decoration technique.

The great body of Faraday rotation microscopy on insulating ferro- and ferrimagnets is prologue to this work, and many of the phenomena may be understood readily because of that background. An earlier example of microscopy in antiferromagnets was reported by King and Paquette.[10] These authors studied the first order antiferromagnetic to spin flop transition in MnF_2 at 4.2K which occurs over the field range 92.5 to 93.2 kOe. They achieved optical density contrast between the two states by using a specific wavelength which was absorbed much more strongly in the spin flop state than in the antiferromagnetic.

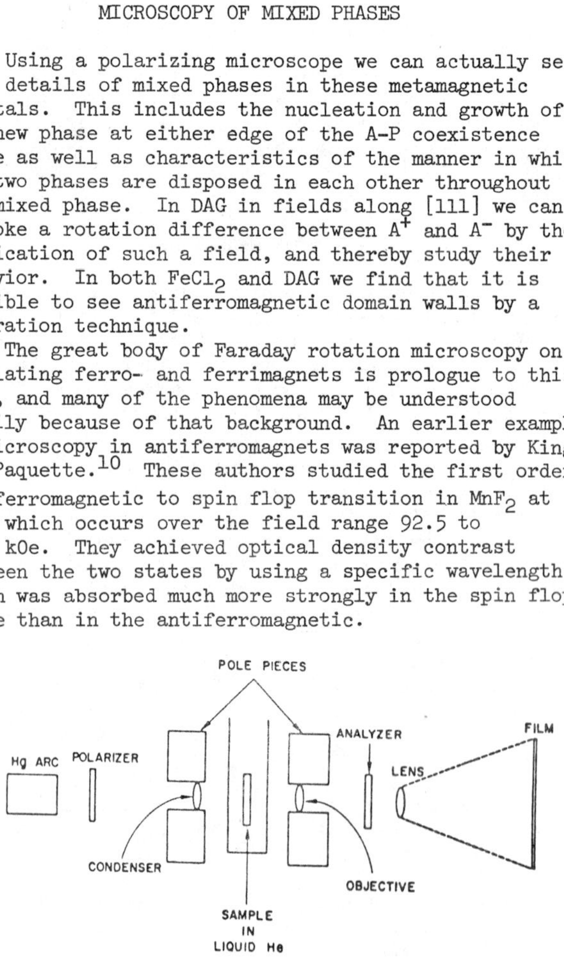

Fig. 5 Sketch of polarizing microscope used to observe mixed magnetic phases. Above 3 K the glass Dewar may be replaced by a variable temeprature Dewar.

The microscope used is sketched in Fig. 5. It is unusual only in that its optic axis passes through the pole pieces of a large electromagnet and the specimen is situated in one of the Dewars mentioned in connection with the rotation measurements. Matching long focal length microscope objectives serve as condenser and objective lenses. The rudimentary camera shown in the sketch has at various times been replaced by a moving picture camera, a television camera, and most often by an ordinary eyepiece and an observer. Contrast between the two phases depends on their differing rotations. This will be discussed briefly after the shape of the objects has been ascertained.

$FeCl_2$

Consider a thin clean sublimation flake of $FeCl_2$ perhaps 50 μm thick by 5 mm diameter which has been cooled to the antiferromagnetic state in zero field. Examining this in the microscope of Fig. 5 we find that image structure appears immediately on entering the mixed A-P phase. In fact we have been able to observe structure all along the antiferromagnetic edge of the A-P region from the lowest accessible temperatures to within about 0.1 K of the tricritical point at ∿21.8 K. We begin by describing the nucleation and development at the lowest temperature available, 1.3 K.[11] In order to provide a dark background against which the new phase will appear bright, we set H_a near the upper edge of the antiferromagnetic region and adjust the analyzer to extinguish the whole field of view. On gently entering the mixed phase a short bright strip typically appears somewhere within the field of view. Further very slight increases in applied field will cause this strip to lengthen. It may grow to extend across the whole field of view; it may join other segments which have grown in a similar fashion. Figure 6(a) illustrates the ribbon at this stage. If we decrease the field so as to leave the mixed phase slowly the bright strip does not simply roll back, but instead homogeneously dims until it is no longer visible. On subsequent reentry, the whole strip homogeneously brightens at the original position. It is apparent that the spin system now contains a line or ribbon that acts as a nucleus for the paramagnetic strip.

A deeper penetration into the mixed phase requires the fraction in the paramagnetic state to increase. At 1.3 K this occurs by convolution or meandering of the existing ribbon. Figures 6(b)(c)(d) show three steps in the process. If after taking any of these photographs we decrease the field, the whole meandering ribbon homogeneously dims to extinction. Reentering the mixed phase the whole ribbon homogeneously brightens in the same configuration. We have managed to convolute not only the paramagnetic ribbon but also the underlying nucleus.

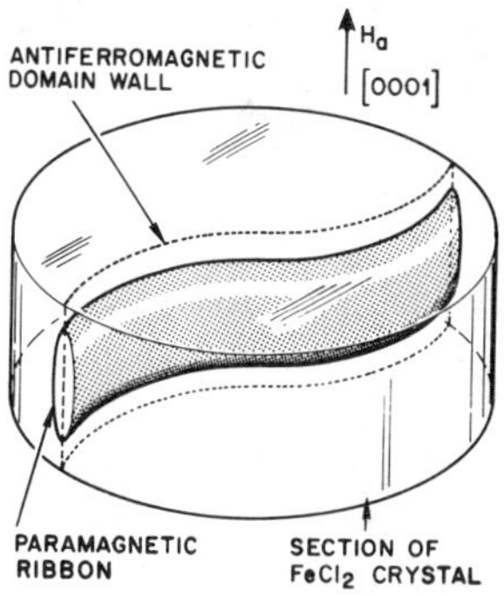

Fig. 7 Interpretive sketch of paramagnetic ribbon nucleated on an antiferromagnetic domain wall in a section of an $FeCl_2$ sheet

We identify the line nucleus of the two preceding paragraphs as an antiferromagnetic domain wall, the transition layer between the two time reversed antiferromagnetic states. Figure 7 shows an interpretive sketch of this exceedingly thin sheet extending from one surface of the crystal to the other. Nucleated on it, perhaps not extending the full thickness, is an elliptical cross section ribbon of spins in the paramagnetic state. This is the object which we view on edge. Judging from the fact that the convoluting ribbon seems to drag the nuceus with it, there is an obvious coupling between the two. Clearly the antiferromagnetic wall can only act as a nucleus if it contains the [0001] direction. When the spin system was cooled through the Néel temperature both time reversed phases may have grown from a number of nucleation sites resulting in antiferromagnetic walls which wander through the crystal without particular orientation. We suggest that the gradual lengthening from a short segment to the situation of Fig. 6(a) represents the rectification of the domain wall so that it comes to contain [0001]. Once it is so rectified, and the crystal brought back to the antiferromagnetic state, no forces other than surface tension act on the wall and it maintains its configuration.

At higher temperatures the antiferromagnetic domain walls continue to act as nuclei Fig. 8(a). However, the subsequent evolution goes very differently. At about 4.2 K branches grow very easily on the original ribbon gradually filling the crystal with a completely connected pattern of ribbons as in Fig. 8(b). On decreasing the field from this point, the ribbons roll back to those associated with antiferromagnetic domain walls and these dim homogeneously. At, say, 9 K and above, the wall still acts as the first nucleus Fig. 8(c), but other point nuclei also seem effective. On increasing the field these appear to extend into ribbons which then fragment to leave paramagnetic

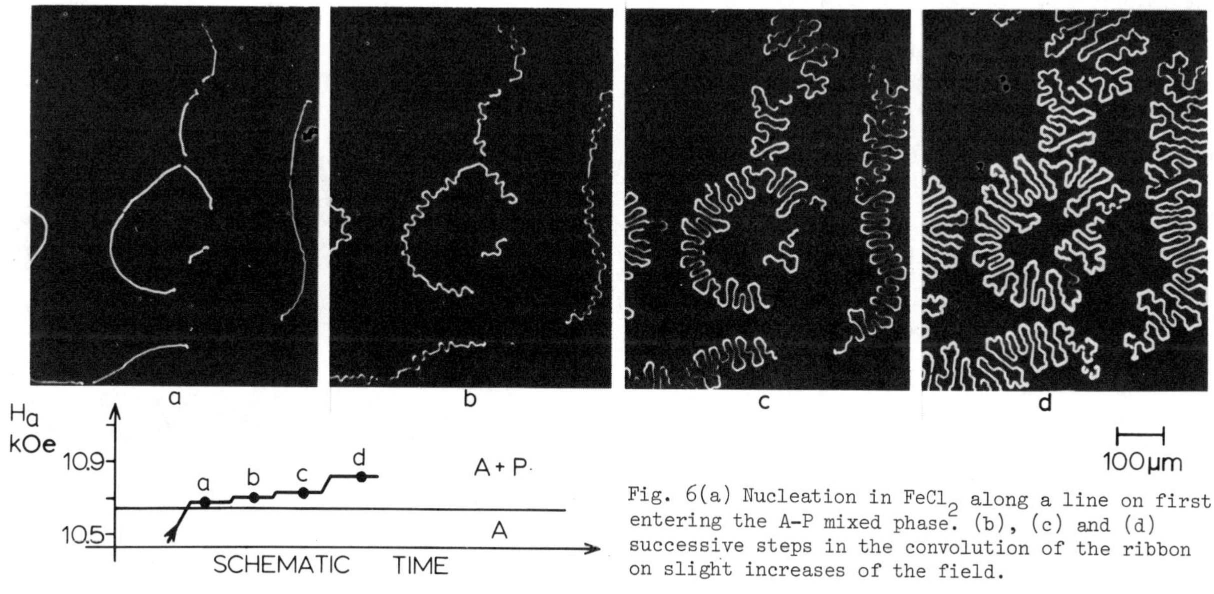

Fig. 6(a) Nucleation in $FeCl_2$ along a line on first entering the A-P mixed phase. (b), (c) and (d) successive steps in the convolution of the ribbon on slight increases of the field.

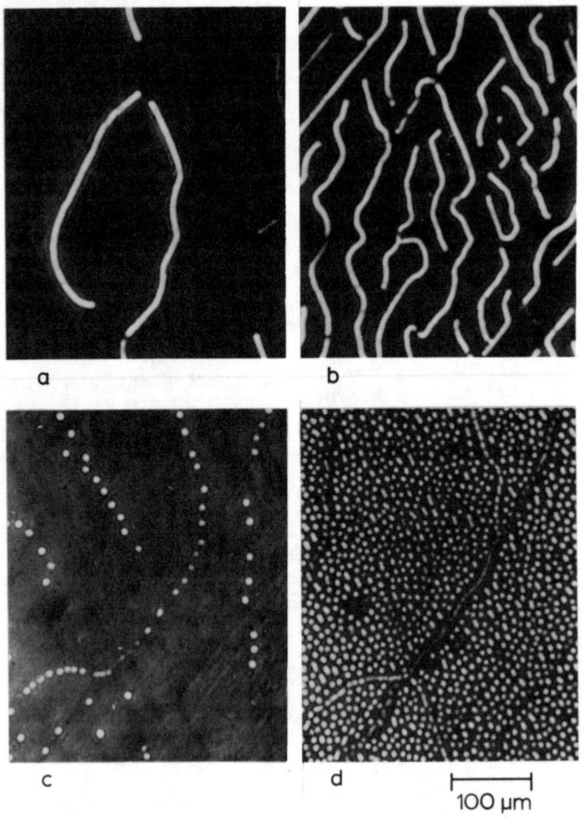

Fig. 8 (a) At 4.2 K a ribbon nucleates on antiferromagnetic domain wall. (b) On increasing the field ribbon grows by extension to a completely connected pattern. (c) At 18 K the first nucleation is again along the antiferromagnetic domain walls. (d) On increase of field the paramagnetic phase grows to a pattern of disconnected bubbles.

"bubbles" scattered over the field of view as in Fig. 8(d). However, the process takes place so rapidly that we cannot follow it by eye.

Though they must be treated elsewhere, several other aspects of this microscopy of the A-P phase of $FeCl_2$ should be mentioned. On approaching the mixed phase from the high field side, there are no entities corresponding to antiferromagnetic domain walls which can serve as line nuclei for the new phase. Thus the low temperature process is very different from that described above. At higher temperature, however, extension and fragmentation of ribbons of the antiferromagnetic phase are seen which are very similar to that of Fig. 8. This difference between behavior at the high and low field bounds of the A-P region is an important factor in the metamagnetic hysteresis encountered by workers using pulse techniques.[12] The phase structure seen in the middle of the mixed phase is a complicated interleaving of thin ribbon-like volumes of the two components. The dimensions are small compared to the thickness of the sample; they approach the resolution of the microscope; and detailed interpretation of the photographs is difficult.

DAG

Consider a (111) sheet of DAG about 200 μm in thickness x 5 x 5 mm^2 which has been brought to the state A^+ at 1.3 K from P^+ through the first order transition as discussed above. As before, the analyzer is set to extinguish A^+, and the field is slowly brought to a value perhaps 20 oe inside the mixed A^+-P^+ phase.[1] The observer sees a number of light objects form, both spots and lines. In the course of 10 or 20 seconds these reach some maximum brightness. Such a field of view is shown in the top frame of Fig. 9. If the experiment is repeated, or if entry is made from the paramagnetic edge, the same points and lines act as

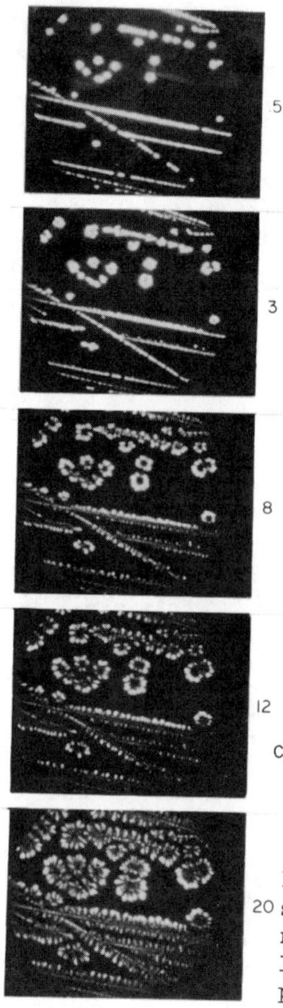

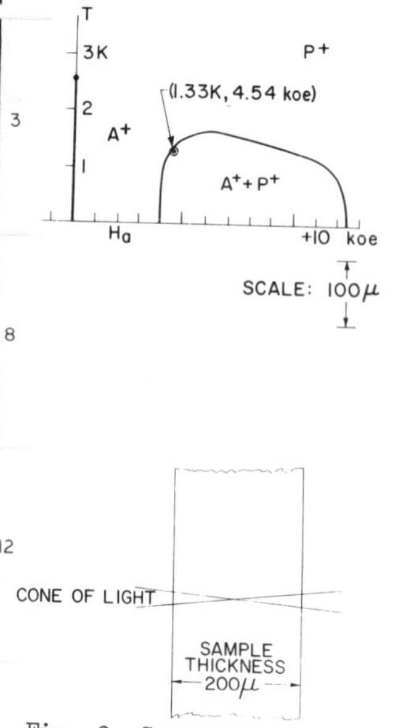

Fig. 9 Sequence of photographs showing the evolution of paramagnetic nuclei in DAG at 1.3 K at the low field edge of the mixed A-P phase.

nuclei. However, the situation of this top photograph is not stable. In the course of minutes the bright circles develop "petals", and some of them come to look very much like stylized daisies. In the 20 minutes covered by these photographs, the petals have moved out and left behind them a much dimmer diffuse structure which fills a large part of the picture. A preliminary interpretation of this series is given in Fig. 10. This depicts a cross section of the crystal in which one of the initial bright spots appears. The brightening of the spot is thought to represent the growth of paramagnetic material from a surface nucleus into a cylinder of paramagnetic material extending from one surface to the other. It is this process which accounts for the initial dim appearance and rapid brightening of the object. Apparently this configuration is unstable, and gradually gives way to a collection of ellipsoidal needles roughly on the center plane of the sheet specimen. These needles, parallel and strongly magnetized, repel each other and gradually move apart. The process by which the cylinder breaks

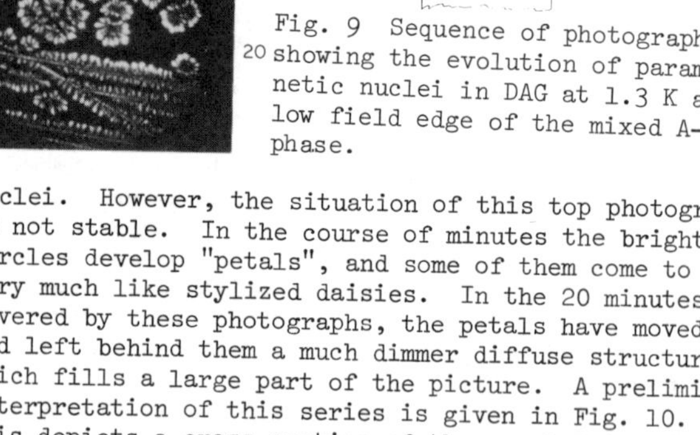

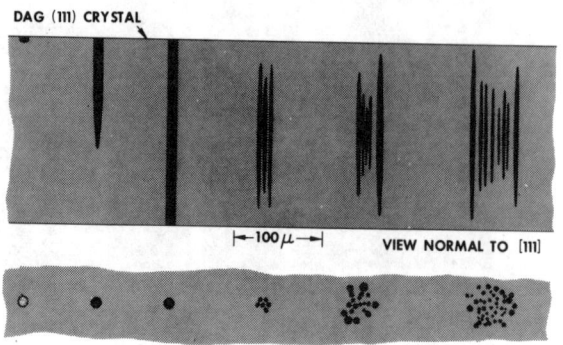

Fig. 10 Sketch of the initial growth and subsequent evolution of nuclei seen in Fig. 9. The first three steps take place in about 20 seconds, the last three in about 30 minutes.

up and the retreating needles leave a trail of smaller objects in their wake is by no means clear. However, the equilibrium state toward which the phase distribution tends clearly has a large number of small paramagnetic needles. It takes a very long time to approach this state. The evolution of the mixed phase after entry from the high field edge is very similar to that of Fig. 9.

In DAG we can distinguish A^+ and A^-, and observe the conversion from one to the other. Figure 11 presents some examples. In (a) the sample was prepared in the A^- state. Though this is metastable at 1.3 K it is easily brought just into the positive mixed state. Nuclei of P^+ form, and are distinguishable as bright rods, P^+ in A^-. If the field is reduced, P^+ converts by first order transition to the stable A^+, and we have A^+ in A^-. This situation is not stable; the A^+ volume grows at the expense of the A^- volume. The photograph (c) was made after about 10 minutes. Left at this field the conversion would go to completion. However, the field was increased to reenter the mixed A-P phase. Ribbons of bright paramagnetic material are seen to form on the A^+-A^- walls. These rapidly break up to needles, and conversion of P^+ to A^+ and A^- to A^+ proceeds. The photograph Fig. 11(d) shows volumes of A^+ in a predominantly A^- sample with needles of P^+ strung along the antiferromagnetic domain walls. Again this is unstable, all the A^- will disappear, and the P^+ in A^+ distribution will relax to some distribution such as that toward which the system of Fig. 9 is tending.

MICROSCOPY AND MOVING PICTURES

Relative to the microscopy itself, the contrast mechanism and some properties of the image should be mentioned briefly. Figures 7 and 10 suggest the shape of some of the spin objects as we interpret them. A cone of linearly polarized light, shown in the inset of Fig. 9, illuminates these, and is collected by the objective. The rotation associated with an individual ray is determined by the path length within the object projected on the magnetization. For the paramagnetic phase of DAG the specific rotation is about $0.6°/\mu m$, for $FeCl_2$ $1.5°/\mu m$, both measured at 1.3 K, .546 µm wavelength. It is clear that the maximum rotations are very large, often over 100°, but the polarization axis of most of the light defining the image is rotated considerably less than this. Thus, images of small objects are made up of light which is depolarized rather than light whose polarization is rotated by some fixed amount. By properly setting the analyzer the background can be extinguished and the object made bright. At least for small objects the contrast cannot be reversed as typically possible with the images of domains in ferrimagnetic films and crystal sheets. The image is complicated by the fact that the object extends along the line of sight so that we have to deal with out of focus images both above and below the focal plane. The flat glass windows of the Dewar contribute to the spherical aberration the overall optical system. The spread in rotation, the out of focus images, the spherical aberration all combine with the resolution of the microscope to make difficult the detailed interpretation of the shape of the spin objects seen in the A-P mixed state.

MOVING PICTURES

Many aspects of the first order phase transitions seen here are difficult to describe even with the liberal use of still photographs. Moving pictures are exceedingly valuable in conveying an understanding of the way in which the mixed phases nucleate and evolve. In presenting this paper at the Conference we display a short moving picture illustrating some of these phenomena. The film includes segments dealing with the following:

$FeCl_2$

1.3 K	Initial nucleation on an antiferromagnetic domain wall (AFDW), growth by convolution, homogeneous dimming, nucleation and growth without an AFDW.
4.2 K	Nucleation on AFDW and growth by extension.
9 K	Nucleation on AFDW, extension and fragmentation.

DAG

1.3 K	Traversal of the mixed A-P phase, P^+ in A^+ nucleation and time lapse photos of development, distribution after rapid entry, A^+ in P^+.
1.8 K	$A^- \rightarrow A^+$ conversion.
1.3 K	Precipitation of P^+ on A^+-A^- walls and sequelae.

ACKNOWLEDGMENTS

We are happy to acknowledge a great deal of valuable interaction with W. P. Wolf during the course of this work. We also thank R. Alben, R. J. Birgeneau, M. Blume, L. M. Corliss, N. Giordano, J. M. Hastings and L. R. Walker for stimulating discussions.

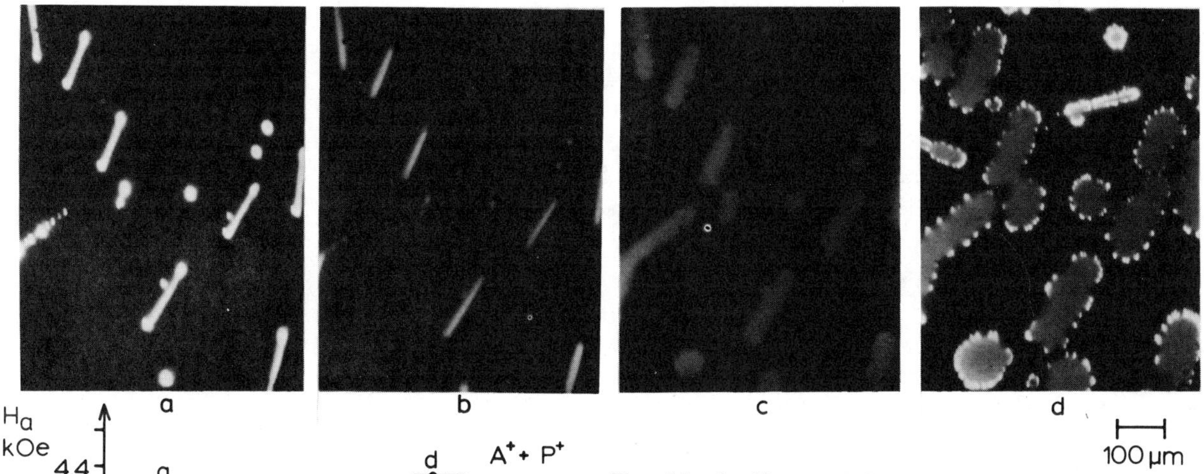

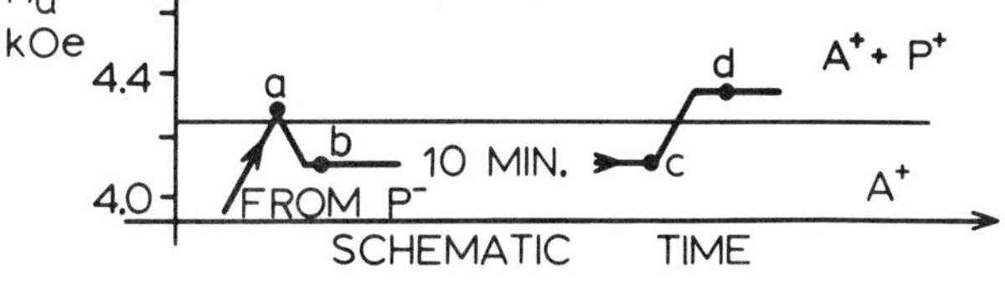

Fig. 11 As discussed in the text, a DAG crystal at 1.3 K was prepared in the state A^-. (a) On entering the mixed phase P^+ nuclei appeared. (b) On decreasing the field P^+ is replaced by A^+. (c) After 10 minutes the A^+ volume has grown at the expense of A^-. (d) On reentering the mixed phase P^+ nucleates on A^+-A^- walls.

REFERENCES

1. Dillon, J. F., Jr., Chen, E. Yi and Wolf, W. P., Proceedings of the International Conference on Magnetism, Moscow 22-28 September, 1973 (In press).
2. Chen, E. Yi, Dillon, J. F., Jr. and Guggenheim, H.J. AIP Conference Proceedings, No. 18, Magnetism and Magnetic Materials 1973, p. 329.
3. Dillon, J. F., Jr., Chen, E. Yi, Giordano, N. and Wolf, W. P., Phys. Rev. Letters 32, 98-101 (1974).
4. Griffiths, R. B., Phys. Rev. Letters 24, 715-717 (1970).
5. Birgeneau, R. J., Shirane, S., Blume, M. and Koehler, W. C., Proceedings of this Conference.
6. Griffin, J. A. and Schnatterly, S. E., Proceedings of this Conference.
7. Wolf, W. P., Proceedings of this Conference.
8. Griffiths, R. B. in "Critical Phenomena in Alloys, Magnets, and Superconductors" ed. by Mills, Ascher and Jaffe, McGraw Hill (1971) p. 377.
9. Blume, M., Corliss, L. M., Hastings, J. M. and Schiller, E., Phys. Rev. Letters 32, 544-547 (1974).
10. King, A. R. and Paquette, E., Phys. Rev. Letters, 30, 662-666 (1973).
11. Dillon, J. F., Jr., Chen, E. Yi and Guggenheim, H.J. Solid State Communications (In press).
12. Jacobs, I. S. and Lawrence, P. E., Phys. Rev. 164, 866-878 (1967).

MULTIPLET SPLITTING OF X-RAY PHOTOEMISSION SPECTRA CORE LEVELS IN MAGNETIC METALS*

S.P. Kowalczyk, F.R. McFeely, L. Ley,† and
D.A. Shirley
Department of Chemistry
Lawrence Berkeley Laboratory, University of California
Berkeley, California 94720

ABSTRACT

The results of high resolution x-ray photoemission studies of the multiplet splitting of the 3s core levels in the 3d transition metals and the 4s, 5s, and 4d core-levels in the lanthanide metals are reported.

INTRODUCTION

Measurements of core-level binding energies and valence band density-of-states by x-ray photoemission spectroscopy (XPS), has proved very valuable to the understanding of the electronic structure of solids. For systems with unpaired spin, information about the initial state spin S can in principle be obtained by correlation with the measured splitting of a core level caused by the phenomenon of multiplet splitting.[1] The basis of obtaining S is the use of Van Vleck's Theorem,[2]

$$\Delta E_{n\ell} = \frac{2S+1}{2\ell+1} G^{n\ell}(n\ell, n'\ell'), \qquad (1)$$

where $\Delta E_{n\ell}$ is the splitting of the $n\ell$ level, $G^{n\ell}$ is the appropriate atomic exchange integral, and n and n' are the principle quantum numbers of the level measured and the level with the unpaired spin respectively. Unfortunately the situation is not so straightforward. Using (1) overestimates $\Delta E_{n\ell}$ by a factor of ~2 when n = n'. It is now understood that this discrepancy is due to intra-shell electron correlations.[3,4] It will be shown in this paper that by using systems where the correlation effects are nearly the same, measured ΔE_{ns} can still be used to obtain S.

EXPERIMENTAL

High resolution XPS spectra were obtained from the 3d metals Sc through Zn and the rare earth metals La through Lu (excluding Pm). The samples were prepared in situ under ultra high vacuum conditions ($<5 \times 10^{-10}$ torr) by either vapor or sputter deposition.[5]

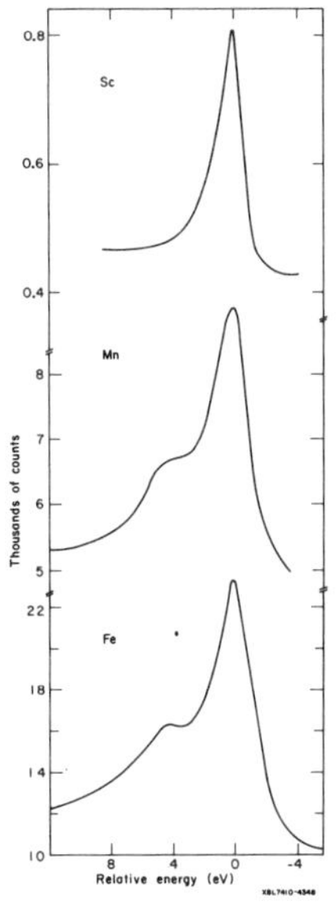

Fig. 1.

XPS 3s spectra of Sc, Mn, and Fe.

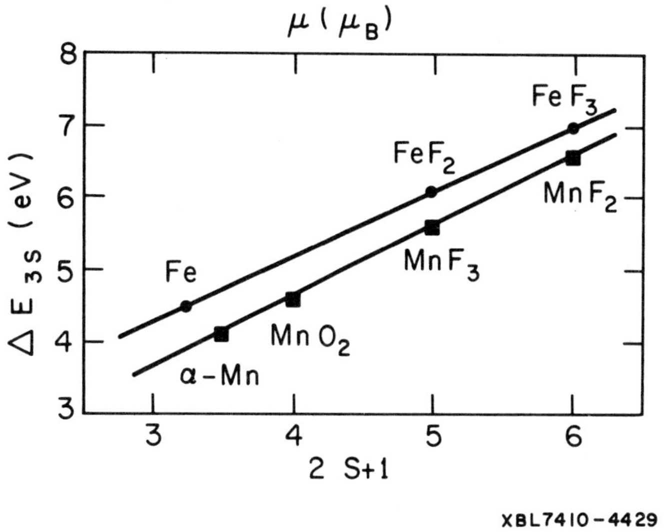

Fig. 2.

Calibration curves for Fe and Mn. Fe, α-Mn, FeF_2 and MnF_2 from data of the authors. FeF_3 and MnF_3 are taken from J.C. Carver, G.K. Schweitzer, and T.A. Carlson, J. Chem. Phys. 57, 973-982 (1972).

RESULTS AND DISCUSSION

The 3d metals can be divided into 3 classes according to their magnetic properties: paramagnetic (Sc, Ti, V), antiferromagnetic (Cr, Mn), and ferromagnetic (Fe, Co, Ni). Figure 1 shows a typical 3s spectrum of a metal from each of these classes. The Sc spectrum shows a single peak as expected from a band theory explanation of Pauli paramagnetism. Paramagnetic Ti and V exhibit 3s spectra similar to Sc (i.e. no splitting). Antiferromagnetic Cr and α-Mn show sizeable 3s splittings (2.8 and 4.1 eV, respectively). Likewise ferromagnetic Fe, Co and Ni show multiplet structure. Unfortunately the structure can not be resolved in the cases of Co and Ni due to correlation effects (see Ref. 5 for detailed discussion). Cu and Zn with filled 3d-shells exhibit no multiplet structure in their 3s spectra.

Figure 2 demonstrates how S can be obtained from measured ΔE_{ns} despite large correlation effects. The integral in (1) should be dominated by the atomic core region and be relatively insensitive to bonding effects outside the atomic core. Thus for an atom in various environments, ΔE_{ns} should still be proportional to 2S+1. By plotting measured ΔE_{ns} vs 2S+1 of ionic systems, where S is well defined, one obtains in effect a "calibration curve" on which one can place an observed ΔE_{ns} of a metal or an atom in an alloy or a compound and get a value for 2S+1 or the localized moment. Using this calibration procedure for Fe with an observed $\Delta E_{3s} = 4.5$ eV yields a magnetic moment $\mu = 2.2$ μB which agrees quite well with the generally accepted value of 2.22 μ_B.[6] This agreement suggests the correlation factor is roughly constant for Fe atoms in different surroundings. Our α-Mn measurements were obtained at room temperature and from the above procedure implied a localized moment with an average value of ~2.5 μ_B. This is at odds with neutron diffraction measurements above 100°K, which set an upper limit of 0.5μ_B per atom.[7] This is due to the fact the XPS time scale (10^{-15} sec) is

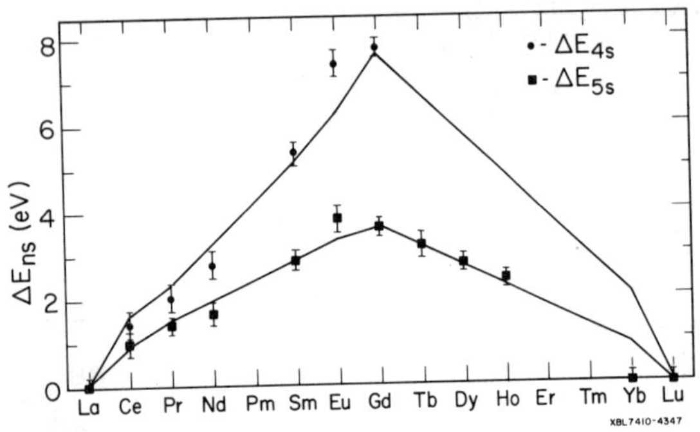

Fig. 3.

Observed 4s (●) and 5s (■) multiplet splitting in the rare earth metals. The solid curves are estimates based on Eq. 1. The 4s estimates have been reduced by a factor 0.55

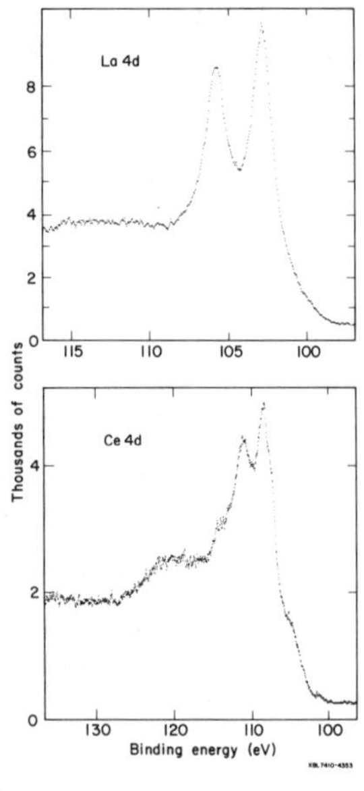

Fig. 4.

La and Ce 4d spectra.

several orders of magnitude faster than neutron diffraction.[5] The XPS observation of two well defined peaks indicates that all the Mn atoms have a moment close to the average value.

Figure 3 summarizes the results for the 4s and 5s measured splittings for the rare earth metals. The solid lines are Van Vleck's theorem estimates for the metals in their trivalent state. The estimates for ΔE_{4s} have been scaled by 0.55 to account for correlations. The observed values follow the theoretical curve quite well which supports the idea of the near constancy of correlation. Only Eu and Yb significantly deviate and this is due to their divalent character. This suggests that ΔE_{ns} measurements could be applied to the problem of determining valency ratios in mixed valency rare earth compounds. Levels with nonvanishing orbital angular momentum often possess very complicated multiplet structures.[4b, 8] This structure can be useful as a valency fingerprint. The 4d spectra of La($4f^0$) and γ-Ce($4f^1$) metals are displayed in Fig. 4. One could for instance determine if Ce was tetravalent ($4f^0$) or trivalent ($4f^1$) in a particular compound or alloy. A trivalent Ce 4d spectrum would resemble that of γ-Ce, while tetravalent Ce would exhibit a 4d spectrum similar to La metal which just shows spin-orbit splitting.

FOOTNOTES AND REFERENCES

*Work done under the auspices of the U.S. Atomic Energy Commission.

†IBM Fellow.

1. C.S. Fadley and D.A. Shirley, Phys. Rev. A 2, 1109-1120 (1970).
2. J.H. Van Vleck, Phys. Rev. 45, 405-419 (1934).
3. P.S. Bagus, A.J. Freeman, and F. Sasaki, Phys. Rev. Lett 30, 850-853 (1973).
4. a) S.P. Kowalczyk, L. Ley, R.A. Pollak, F.R. McFeely, and D.A. Shirley, Phys. Rev. B 7, 4009-4011 (1973).
   b) S. P. Kowalczyk, L. Ley, F.R. McFeely, and D.A. Shirley, Phys. Rev. B, in press.
5. For more detailed experimental information, see L. Ley, S. P. Kowalczyk, F.R. McFeely, and D.A. Shirley, Phys. Rev. B, submitted for publication.
6. C. Kittel, Introduction to Solid State Physics, 3rd Edition (John Wiley & Sons, Inc., New York, 1968) p. 461.
7. C.G. Shull and M. K. Wilkinson, Rev. Mod. Phys. 25, 100-107 (1953).
8. S. P. Kowalczyk, N. Edelstein, F. R. McFeely, L. Ley, and D. A. Shirley, Chem. Phys. Lett. in press.

COVALENCY EFFECTS ON TETRAHEDRAL AND OCTAHEDRAL Fe^{3+} SITES IN YIG: CHARGE AND SPIN DENSITIES AND NEUTRON FORM FACTORS*

E. Byrom, A. J. Freeman and D. E. Ellis
Physics Department, Northwestern University, Evanston, Illinois 60201

ABSTRACT

Polarized neutron studies of YIG by Bonnet et al.[1] show the form factor for the Fe^{3+} ion in the tetrahedral [d] site to be reduced below that of the octahedral [a] site. Spin unrestricted HFS model potential calculations for the tetrahedral $(FeO_4)^{5-}$ and octahedral $(FeO_6)^{9-}$ molecular clusters show that when the model potential includes the effect of the YIG crystalline environment on the clusters, there is a more pronounced covalency between the magnetic 3d orbitals and the ligand 2p orbitals at the [d] than at the [a] site. This greater covalency leads to a reduction of the neutron scattering form factor at the [d] site below that at the [a] site and to correct signs for the Mössbauer magnetic hyperfine and isomer shifts between the two sites. The magnetic moments transferred to the oxygen ions are different in the two clusters, implying a net moment on the oxygen ions in YIG.

INTRODUCTION

The yttrium iron garnet (YIG) crystal has a structure described by the space group O_h^{10}; a bcc Bravais lattice with four formula units $Y_3Fe_2(FeO_4)_3$ per primitive unit cell. There are two distinct Fe sites: the [a] site in which the Fe^{3+} ion is surrounded by six O^{2-} ions arranged in a slightly irregular octahedron; and the [d] sites in which the Fe^{3+} ion is surrounded by a slightly irregular tetrahedron of O^{2-} ions. The neutron scattering form factors of the two Fe sites have recently been shown[1] to be different. The form factor for Fe [a] agrees with the calculated Hartree-Fock (HF) Fe^{3+} free ion value[2] while that for Fe [d] falls below it in the range $\sin\theta/\lambda < 0.8 Å^{-1}$. The same work shows a magnetic moment density near the O site. The conclusion drawn from these results is that there are different degrees of covalent mixing between O orbitals and Fe orbitals at the different sites.

A quantitative investigation of this conclusion has been attempted, in which the molecular clusters $(FeO_4)^{5-}$ and $(FeO_6)^{9-}$, approximating the two environments of the Fe^{3+} ion in YIG, are treated by the Discrete Variational Method (DVM).

DISCRETE VARIATIONAL METHOD

This method has been discussed in detail elsewhere.[3] In the Hartree-Fock-Slater (HFS) approximation, an effective Hamiltonian

$$H = \frac{-\nabla^2}{2} + V_{nuclear} + V_{coulomb} + V_{exchange} \tag{1}$$

is used, the exchange term being replaced by a local potential,

$$V_{exchange,\sigma}(\underline{r}) = -3\alpha\left(\frac{3}{4\pi}\rho_\sigma(\underline{r})\right)^{1/3} \tag{2}$$

where σ refers to the spin state.[4] In these calculations α is set equal to 1. One electron HFS eigenfunctions are obtained in terms of a linear combination of atomic orbitals (LCAO) basis set

$$\psi_i(\underline{r}) = \sum_j A_j(\underline{r})c_{ji} \tag{3}$$

An error for eigenfunction i at point r is defined

$$\delta_i(\underline{r}) = (H-\epsilon_i)\psi_i(\underline{r}) \tag{4}$$

and the coefficients c_{ji} are determined by minimizing weighted averages over a discrete set of sample points. A secular equation is derived, identical to the Rayleigh-Ritz equation

$$\underline{\underline{H}}\,\underline{c} = \underline{\underline{S}}\,\underline{c}\,E \tag{5}$$

but with matrix elements given in discrete form. The choice of sample points corresponds to a Diophantine procedure of numerical integration.

In these calculations the $A_j(\underline{r})$ are Slater-type orbitals (STO's) centered on both O and Fe sites, giving a basis set of double zeta quality. Matrix elements H_{ij} and S_{ij} of the Hamiltonian and overlap matrices are evaluated by the Diophantine procedure over a grid of 3000 points and the HFS equations are thus solved in a model potential. A part of the effect of the crystal environment on the cluster is included in the model potential, which is generated from the superposed charge densities of all the ions in a 'giant cluster' of the ~ 100 ions closest to the central Fe site, together with a sum of point ion potentials. To simplify the calculations, the O sites in the clusters are arranged in regular polyhedra about the Fe site.

RESULTS

The spin-polarized Fe^{3+} ion has the nominal configuration $(3d\uparrow)^5(3d\downarrow)^0$ with inner-shell (doubly occupied) orbitals having slightly differing energies and radial wavefunctions due to the exchange potential. The 3d orbitals, which of course dominate the neutron form factor, are allowed by symmetry to mix with oxygen-2p orbitals in the e_g and t_{2g} representations of the octahedral cluster (O_h point group), and in the e and t_2 representations of the tetrahedral cluster (T_d point group). In addition, the oxygen-2s orbital mixes in e_g and t_2 representations, which accounts for the observable transferred hyperfine field in the simplest models. Iron p-d mixing is also allowed in the T_d symmetry, leading to interesting consequences for the nuclear quadrupole splitting. Because of space limitations we will not discuss orbitals and densities found for the nominally closed shell representations (a_{1g}, t_{1u}, t_{2u}, etc.).

For the O_h cluster the ground state MO configuration was taken to be...$(4e_g\uparrow)^2(4e_g\downarrow)^0(2t_{2g}\uparrow)^3(2t_{2g}\downarrow)^0$; for the T_d cluster it is...$(2e\uparrow)^2(2e\downarrow)^0(7t_2\uparrow)^3(7t_2\downarrow)^0$. While the unoccupied magnetic orbitals, e.g. $(4e_g\downarrow)$, display a well defined predominant iron-3d character, the occupied 3d character is distributed over several MO's, mixed with ligand basis functions. Thus while the total spin density distribution can be simply described in terms of 3d → oxygen-2s,2p transfer, discussion of individual orbital covalency parameters does not appear to be particularly useful. The spin density in the vicinity of the 3d maximum is found to be reduced by ~ 10% compared to the free ion; this "iron component" of the density is remarkably similar along bond directions for both [a] and [d] sites. The transferred spin density, appearing in the vicinity of the ligands, is considerably greater in magnitude for the $(FeO_4)^{5-}$ cluster; this is achieved by "sweeping up" contributions from non-bonding molecular regions. As the Fe [a] and [d] sites form magnetic sublattices of opposite alignment, a small (< 0.1 μ_B) net magnetic moment is found on the O sites.

From the greater spatial extension of the spin density in the clusters, compared to the free ion, it follows that the neutron scattering form factors decrease more rapidly in the region of small $\sin\theta/\lambda$, as shown in Figure 1. The form factor of $(FeO_4)^{5-}$ falls below that of $(FeO_6)^{9-}$, the maximum difference between the values of $f(\sin\theta/\lambda)$ being 0.08, in agreement with

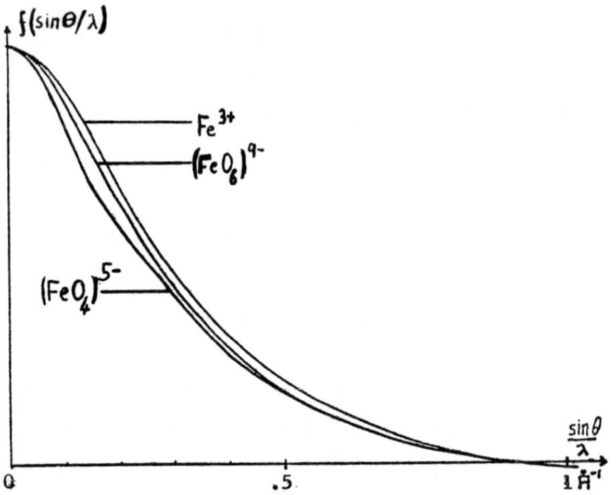

Figure 1 Neutron scattering form factors of the Fe^{3+} free ion, and the $(FeO_6)^{9-}$ and $(FeO_4)^{5-}$ clusters.

the results of Bonnet et al.[1], although the calculated maximum difference occurs at too low a value of $\sin\theta/\lambda$. In comparing the cluster results to the Fe^{3+} free ion form factor[2], the effects of covalency should be separated from those of the exchange approximation. The HFS form factor falls off more rapidly than the HF for small $\sin\theta/\lambda$ because of the large effect on the radial densitites of the differing exchange approximations. The maximum difference between them, 0.07, is comparable to the covalency effect. An HFS-DVM calculation was performed for the Fe^{3+} free ion, and yielded the third curve in Figure 1. Because of the covalency present in the $(FeO_6)^{9-}$ cluster, its form factor lies below this free ion result.

Since these calculations do not include the self-consistent effects of relaxation of the core orbitals in the expanded spin density, they do not give accurate Mössbauer magnetic hyperfine or isomer shift results. However, it should be noted that the difference between the charge densities at the Fe nucleus in the two cases has the correct sign,[5] and that the spin density at the Fe nucleus, or contact hyperfine field, has a greater (negative) magnitude in the $(FeO_6)^{9-}$ case, again in agreement with experiment.[6]

REFERENCES

*Research supported by the NSF and the AFOSR.

[1]M. Bonnet, A. Delapalme, F. Tcheou and H. Fuess, Proceedings ICM-73, IV, 251-7 (1974).

[2]R. E. Watson and A. J. Freeman, Acta. Cryst. 14, 27 (1961).

[3]D. E. Ellis and G. S. Painter, Phys. Rev. B2, 2887, (1970); T. Parameswaran and D. E. Ellis, J. Chem. Phys. 58, 2088 (1973).

[4]For a discussion of the local exchange potential, see J. C. Slater, "Quantum Theory of Molecules and Solids" Vol. IV, McGraw-Hill, New York, (1974). While the HF approximation for ground state properties is generally preferable, the computationally simpler HFS approximation has proved quantitatively useful in a variety of molecular and solid state applications. The exchange scaling parameter α may be chosen, either by a particular scheme of exchange averaging, or by optimising the fit to experimental data. The original Slater averaging scheme (α=1) is used in this paper, but the value of α is not critical at the level of approximation implied by a non-self-consistent calculation.

[5]G. N. Belozerskii, V. N. Gittsovich, A. N. Murin, Yu. P. Khimich, and Yu. M. Yakoulev, Sov. Phys. Solid State 12, 2323 (1971) and references therein.

[6]R. Bauminger, S. G. Cohen, A. Marinov and S. Ofer, Phys. Rev. 122, 743 (1961).

COVALENCY OF HYDRATED COMPLEXES OF Co^{2+} FROM ^{17}O ENDOR*

J. W. Culvahouse, H. Glotfelty, and W. P. Unruh
Department of Physics, The University of Kansas, Lawrence, Kansas 66045

ABSTRACT

ENDOR spectra of ^{17}O in water molecules of the complexes $(Co\cdot 6H_2O)^{2+}$ in the two non-equivalent sites of double nitrate crystals $[La_2Zn_3(NO_3)_{12}\cdot 24H_2O]$ have been utilized to obtain transferred hyperfine interactions. The X site exhibits very little trigonal distortion and the interaction is an isotropic part, A_s arising from the spin density at the O nuclei plus an anisotropic part with axial symmetry about the Co-H_2O bond. For the two inequivalent O nuclei in this site, we find -24.0 ±0.4 MHz and -23.3 ±0.4 MHz for A_s. Interpreted as transferred spin fractions, f_s, for the 2s orbitals of O, we find 0.0099 and 0.0096. Andriessen[1] has obtained wavefunctions for $(Mn\cdot 6H_2O)^{2+}$ which yield spin densities in remarkable agreement with these results. Expressing the calculated densities as $f_{se}\times|\psi_{2sO}(0)|^2$, one finds f_{se} to be 0.0113 and 0.0108 for the two oxygen nuclei. These wavefunctions show little t_{2g} bonding, suggesting little difference for Mn and Co, and in good accord with the relative strengths of the spin-spin interactions between Mn, Co, and Ni ions in the X and Y sites. Neglecting t_{2g} bonding, the transferred hyperfine interactions for O in the water molecules of the Y site may be approximated as a part arising from the spin density on the O nuclei which has the same anisotropy as g_s (the part of the g tensor arising from the electronic spin of Co^{2+}); and a part which has the same form as the point dipole interaction, but with g_s replacing the total g tensor. In this model, A_{zz}, where z is the C_3 axis of the complex, arises entirely from the contact part of the hyperfine interaction and corresponds to f_s=0.0100. From the wavefunctions, one finds f_{se}=0.0113. The measured values of the interactions attributed to spin densities in the 2p orbitals of oxygen are about twice as large as those calculated from the wavefunctions. However, the wavefunctions combined with accepted descriptions of the hydrogen bonds which bridge the X and Y sites give an accurate account of the magnetic interactions between 3d ions in these sites.

*Research supported in part by National Science Foundation Grant GH-34582. Helium gas supplied by Office of Naval Research Contract No. Nonr-2775 (00).

[1]J. Andriessen, J. Mol. Phys., 23, 1103 (1972).

A SYSTEMATIC STUDY OF THE ORBITALLY DEPENDENT RARE EARTH - Gd^{3+} EXCHANGE PARAMETERS IN $GdCl_3$*

R.S. Meltzer and D.K. Braswell
Department of Physics and Astronomy, University of Georgia, Athens, Georgia 30602

and

R.L. Cone
Department of Physics and Astronomy, University of Georgia, Athens, Georgia 30602
and
Department of Physics, Montana State University, Bozeman, Montana 59715†

ABSTRACT

Recent studies have shown that the two-electron tensor operator method can provide a unified description of isotropic and anisotropic rare earth exchange interactions in ground and excited states. The present study demonstrates that such experimentally determined exchange parameters are applicable to other rare earth ions in related systems. An analysis of optically measured exchange splittings of Nd^{3+}, Gd^{3+}, and Er^{3+} ions in ferromagnetic $GdCl_3$ demonstrates that the two-electron exchange parameters have a smooth systematic variation over two thirds of the lanthanide series and indicates a significant ability of this approach to predict exchange effects in a variety of systems.

INTRODUCTION

The two-electron spherical tensor operator description of the interionic exchange interaction has recently been successfully used to describe both static[1,2] and dynamic[3] exchange effects involving rare earth ions. Moreover, these studies have demonstrated that this approach has the important ability to relate exchange effects for different levels within a $4f^n$ configuration. This success results from directly accounting for the orbital dependence of the interaction by summing over the interactions of pairs of electrons on the two centers.

The purpose of the present paper is to demonstrate that the "two electron" exchange parameters resulting from such an analysis are also applicable to other rare earth ions with different electronic configurations. In particular, we compare Er^{3+}-Gd^{3+} and Nd^{3+}-Gd^{3+} parameters determined from exchange splittings of impurity levels in ferromagnetic $GdCl_3$ with the previously-determined Gd^{3+}-Gd^{3+} parameters[3] for the pure material.

DETERMINATION OF THE EXCHANGE SPLITTINGS

Both Er^{3+} with configuration $4f^{11}$ and Nd^{3+} with configuration $4f^3$ have a large number of optically observable energy levels which are Kramers doublets in the absence of interionic interactions. The exchange splittings of these levels in $GdCl_3$ provide a direct means for measuring the exchange parameters of interest. To avoid the problem of accounting for Gd^{3+} spin deviations, it is desirable to measure the splittings in an environment of perfectly aligned Gd^{3+} ions; however, the $GdCl_3$ magnetization is only about 85% of saturation at 1.2K, the lowest temperature attainable in our experiments (T_c=2.2K). Large external magnetic fields were thus used to substantially reduce the population of Gd^{3+} spin deviations. Extrapolation of measurements at 3 kG increments between 12 and 30 kG by a linear least squares procedure gave values of both the zero field splittings and magnetic splitting factors for each observed impurity doublet. These fields maintained the magnetization above 95% of saturation; moreover, the absence of an observable temperature dependence in the spectra at the higher fields provided further evidence that the results represent accurate values for the condition of saturation.

Table I. Observed energies and Kramers doublet exchange splittings for Er^{3+}:$GdCl_3$. All energies are in cm^{-1}.

State	$(2\mu)^a$	$E_{av}{}^b$	$\Delta E_{obs}{}^c$	$\Delta E_{dip}{}^d$	$\Delta E_{exch}{}^e$ obs	$\Delta E_{exch}{}^e$ calc
$^4I_{15/2}$	(5)	0	−0.36±.05	−0.30	−0.06±.10	−0.17
	(3)	38.22	−0.45±.13	−0.12	−0.33±.27	−0.22
	(1)	94.9	−0.78±.16	−0.27	−0.51±.20	−0.33
$^4I_{13/2}$	(3)	17923.75^f	0.47±.03	0.33	0.14±.08	0.16
	(1)	17916.43^f	0.63±.09	0.03	0.60±.12	0.51
	(1)	17900.97^f	1.99±.03	2.10	−0.12±.24	0.61
	(3)	17864.75^f	−2.22±.10	−1.39	−0.82±.31	−0.17
$^4I_{9/2}$	(3)	12390.66	−0.09±.10	−0.35	0.26±.15	0.38
	(1)	12485.24	0.05±.14	0.15	−0.10±.16	−0.20
$^4F_{9/2}$	(1)	15265.44	0.19±.07	0.23	−0.04±.10	−0.05
	(3)	15294.92	0.90±.11	0.53	0.37±.19	0.20
$^4S_{3/2}$	(3)	18368.71	2.62±.06	0.92	1.70±.20	1.73
	(1)	18397.53	0.79±.06	0.31	0.47±.11	0.45

a J manifold and μ quantum number (irreducible representation of the site group C_{3h}).
b $E_{av}=[E(\mu) + E(-\mu)]/2$, where $E(\mu)$ and $E(-\mu)$ indicate the energies of the two components of a doublet level.
c $\Delta E_{obs}=E(\mu) - E(-\mu)$.
d $\Delta E_{dip}=-2\mu_z H_{dip}$.
e $\Delta E_{exch}=\Delta E_{obs} - \Delta E_{dip}$.
f Energies are those of fluorescence from $^2H_{11/2}$, μ=5/2.

The average zero field energies E_{av} and zero field splittings ΔE_{obs} for thirteen Er^{3+} levels have been determined from optical absorption and fluorescence experiments and are given in Table I. A crystal field analysis including the effects of intermediate coupling and J mixing[4] has been carried out for these levels along with the observed levels of Dieke,[5] and good agreement results for both crystal field energies (2.2 cm^{-1} rms deviation) and magnetic splitting factors for all levels shown in Table I. We thus conclude that the resulting crystal field eigenfunctions provide a good basis for the analysis of the exchange splittings.

Absorption measurements on Nd^{3+}:$GdCl_3$ have been complicated by over absorption, which must be overcome by preparing thinner samples; however, accurate measurements have been obtained for several levels. Table II gives the average zero field energies and zero field splittings for two $^4S_{3/2}$ levels which, as we point out below, are sufficient to determine the isotropic Nd^{3+}-Gd^{3+} exchange parameter.

Table II. Observed and calculated exchange splittings for Nd^{3+}: $GdCl_3$. All energies are in cm^{-1}. (For explanation of symbols, refer to Table I.)

State	(2μ)	$\Delta E_{exch(obs)}$	$\Delta E_{exch(calc)}$
$^4S_{3/2}$	(3)	2.6±0.5	2.4
	(1)	0.7±0.6	0.8

For both Er^{3+} and Nd^{3+}, the Kramers doublet splittings are predominantly a first order effect;[1] hence, the exchange contribution ΔE_{exch} and magnetic dipolar contribution ΔE_{dip} are simply additive. The electric multipole interactions give negligible contributions. Since the dipolar contributions could be accurately calculated using the measured Er^{3+} and Nd^{3+} splitting factors and the known dipolar field for the samples, the exchange contributions were readily determined. The resulting values of ΔE_{dip} and ΔE_{exch} are given in Tables I and II.

DETERMINATION OF THE EXCHANGE PARAMETERS

Previous studies[1-3] have shown that the exchange Hamiltonian of Elliott and Thorpe[6] and Levy[7] adequately describes rare-earth earth exchange interactions.

$$\mathcal{H}_{exch} = \sum_{i=1}^{n} \sum_{j=1}^{n'} \sum_{\substack{kk'\\qq'}} -\Gamma^{kk'}_{qq'}\ u^{(k)}_{q}(i)\ u^{(k')}_{q'}(j)\{\tfrac{1}{2} + 2\vec{s}_i\cdot\vec{s}_j\}, \quad (1)$$

where the $\Gamma^{kk'}_{qq'}$ are parameters, the $u^{(k)}_{q}$ are spherical tensor operators acting on the orbital angular momentum of individual electrons, and $\vec{s}_i$ and $\vec{s}_j$ are the spin operators for individual electrons on each ion. The Γ^{00}_{00} term represents the familiar isotropic exchange which is usually written as $-2J\vec{S}\cdot\vec{S}$. ($\Gamma^{00}_{00}=7J$ from the definition of $u^{(0)}_{0}$.)[3] Terms with $k\neq0$ are anisotropic.

Dynamic or transfer of energy exchange terms[3] are not involved in the present analysis since none of the impurity levels are energetically near Gd^{3+} levels. Since all Gd^{3+} neighbors are "frozen" in essentially pure $^8S_{7/2}$ states, k' q'=0. For f electrons, $k\leq6$, and the C_{3h} site symmetry allows only five terms to contribute: Γ^{00}_{00}, Γ^{20}_{00}, Γ^{40}_{00}, Γ^{60}_{00}, and $\Gamma^{60}_{\pm60}$. It is thus clearly feasible to determine the above parameters from an analysis of the exchange splittings. All Gd^{3+} neighbors contribute, and the experimental parameters involve a sum over neighbors. The two nearest neighbors and six next nearest neighbors are expected to give important contributions.

Since the splittings are a first order effect, they depend linearly on the exchange parameters. A standard least squares fitting procedure was thus used. Good agreement between theory and experiment was obtained for the thirteen observed Er^{3+} levels with five parameters. The calculated splittings are given in Table I, and the fitted parameters are given in Table III. The anisotropic terms gave individual contributions as large as 0.65 cm^{-1}; hence, they must be included to obtain a proper description of such exchange splittings.

The reported Nd^{3+} splittings gave a good determination of the Γ^{00}_{00} term since the $^4S_{3/2}$ state of Nd^{3+} is 95% pure ^4S (vs. only 69% purity for the corresponding Er^{3+} state).[8] Calculated Nd^{3+} splittings are compared with experiment in Table II, and the fitted parameter is given in Table III. Anisotropic parameters for Nd^{3+} will be available as soon as further measurements on thinner samples are completed.

Table III. Experimental exchange parameters for Er^{3+}, Gd^{3+}, and Nd^{3+} in $GdCl_3$. Units are cm^{-1}, and summation is over neighboring Gd^{3+} sites.

	$\{\Sigma\Gamma^{00}_{00}\}$	$\{\Sigma\Gamma^{20}_{00}\}$	$\{\Sigma\Gamma^{40}_{00}\}$	$\{\Sigma\Gamma^{60}_{00}\}$	$\{\Sigma\Gamma^{60}_{\pm60}\}$
Er^{3+}-Gd^{3+}	0.8±.1	-0.2±.2	1.7±.9	-0.5±.2	-0.4±.3
Gd^{3+}-Gd^{3+}	0.9±.1	-	-	-	-
Nd^{3+}-Gd^{3+}	1.0±.2	-	-	-	-

The Γ^{00}_{00} term for Gd^{3+} splittings has been determined earlier[2] and is given in Table III for comparison. Unfortunately the Γ^{20}_{00} term is not well determined by ^6P_J Gd^{3+} level splittings.

Comparison of the Γ^{00}_{00} terms in Table III indicates that the "two-electron" isotropic exchange parameter has essentially the same value for Er^{3+}-Gd^{3+}, Gd^{3+}-Gd^{3+}, and Nd^{3+}-Gd^{3+}. The apparent increase for lighter ions would be consistent with a simple interpretation involving less tightly-bound electrons due to lower effective nuclear charge. The smooth variation is reminiscent of that occurring for crystal field parameters and "free ion" parameters.[4]

CONCLUSION

The three ions studied here span over two thirds of the lanthanide series, yet their two-electron isotropic exchange parameters are essentially identical and vary smoothly from $4f^3$ to $4f^7$ to $4f^{11}$. This provides strong evidence for a significant ability of the above approach to predict exchange effects in a wide variety of systems based on simple measurements. Extension to anisotropic terms is being completed.

ACKNOWLEDGEMENTS

The authors wish to thank Dr. H.M. Crosswhite for making doped $GdCl_3$ samples available for these experiments.

REFERENCES

*Work supported by National Science Foundation Grant GH-43934.

†Present address.

1. R.L. Cone, and W.P. Wolf, "Magnetism and Magnetic Materials - 1972" AIP Conf. Proc., No. 10, Ed. by C.D. Graham and J.J. Rhyne (AIP, NY, 1973) pp. 1039-1043.

2. R.S. Meltzer and R.L. Cone, "Magnetism and Magnetic Materials - 1973" AIP Conf. Proc., No. 18, Ed. by C.D. Graham and J.J. Rhyne (AIP, NY, 1974) p. 545-549.

3. R.L. Cone and R.S. Meltzer, Phys. Rev. Lett. 30, 859 (1973).

4. R.L. Cone, J. Chem. Phys. 57, 4893 (1972).

5. G.H. Dieke, "Spectra and Energy Levels of Rare Earth Ions in Crystals" (Interscience, NY, 1968) p. 298.

6. R.J. Elliott and M.F. Thorpe, J. Appl. Phys. 39, 802 (1968).

7. P.M. Levy, Phys. Rev. 177, 509 (1969).

8. K. Rajnak, J. Chem. Phys. 43, 847 (1965).

EXCHANGE INTERACTIONS IN ISOLATED TRINUCLEAR CLUSTERS OF FE^{3+}: MAGNETIC AND MOSSBAUER STUDIES

L.N. Mulay and G.H. Ziegenfuss
Materials Research Laboratory and Material Sciences Dept.
The Pennsylvania State University, University Park, Pa. 16802

ABSTRACT

The magnetic susceptibility data obtained for the following typical solids over a wide temperature range were fitted to the Van Vleck-Kambe equation:

(A) $[Fe_3(C_6H_5COO)_5(OH)_3](C_6H_5COO)\cdot H_2O$

(B) $[OFe_3(C_6H_5COO)_6\cdot 3H_2O]ClO_4\cdot H_2O$

(C) $[OFe_3(CH_2CNCOO)_6\cdot 3H_2O]ClO_4\cdot 6H_2O$

A computer-aided least squares analysis yielded the following antiferromagnetic exchange interactions (J in°K) (A)-22.8 (B)-5.14 and (C)-27.28 and the corresponding "g" values 1.05, 0.79, and 1.81. Despite the equivalence of Fe^{3+} (high spin, 5/2) ions with an equilateral triangular configuration inferred from Mossbauer spectroscopy, the (extrapolated) magnetic results suggest the possibility of non-equivalent exchange interactions in the very low temperature region. These observations and the relatively large values of the temperature independent paramagnetic terms and the Weiss constants are interpreted in terms of a novel "spin pairing" model and concepts of resonant structures. The relevance and applicability of this approach to certain transitions in ionic solids are pointed out.

INTRODUCTION

Molecular complexes of transition metal ions provide fascinating systems for elucidating exchange interactions between spins isolated within clusters[1]. Our interest in this area stemmed partly from the need to explore the nature of the so-called "intramolecular" antiferromagnetism" in clusters present in molecular solids and to extend these concepts to somewhat similar clusters in ionic solids such as the Magneli phases of titanium oxides (Ti_n-O_{2n-1}), which exhibit semiconductor to metal transitions. This approach by Mulay et al.[2] has been indeed very successful and led them to postulate the presence of "constrained antiferromagnetism" within two and four ion clusters of Ti^{3+} ($3d^1$) in the low temperature, semiconducting regions of these oxides.

Within the scope and limitations of this paper we report the salient features of our recent studies on the rather diverse "Welo" type[3] trinuclear complexes of iron. Of the eleven complexes investigated, results and discussion on three typical solids shown above and key references are presented.

EXPERIMENTAL

The above complexes were synthesized and purified according to the method of Weinland[3] and were analyzed for Fe as well as for C and H. A close agreement (±2%) in general was found with the expected chemical composition. The composition for B and C is that suggested by Figgis et al.[4]. The Fe^{57} Mossbauer spectra were obtained on polycrystalline solids (∿0.5 mm thick) at room temperature with a Nuclear Science and Engineering Spectrometer using a nuclear data multichannel analyzer and a Co^{57} in Pd source.

The per gram magnetic susceptibilities (χ) were measured with a Faraday magnetic balance and a cryostat (70°-500°K) described extensively by Mulay[5]. Measurements at different fields (up to 8000 Oe) showed no presence of ferromagnetic impurities. One important aspect of our work involved making ten measurements of χ at each temperature (using two different samples of each complex) to provide statistically significant information.

RESULTS AND DISCUSSION

The isomer shift (δ, mm/sec) and the quadrupole

TABLE II

	$N\alpha(x10^{-6})$	$S_e\ (x10^{-6})$	g	J(°K)
A	2016	12.1	1.05	-22.8
B	2629	17.9	0.79	-5.1
C	120	18.9	1.81	-27.3

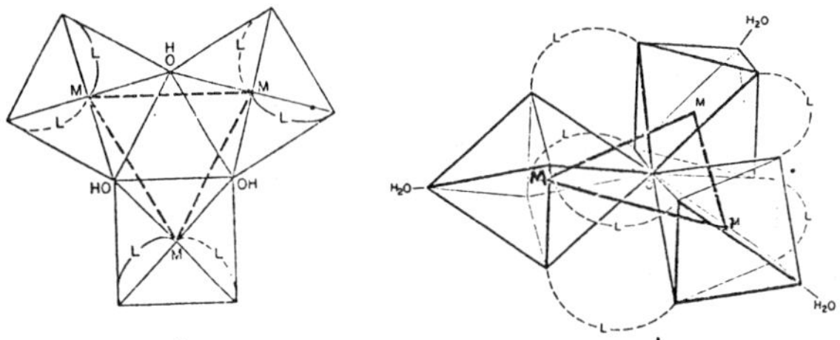

a b

Fig. 1(a) Plan view of structures for complexes A, here L = C_6H_5COO. (b) Structure for B and C (L = C_6H_5COO or CH_2CNCOO) [In all cases M = Fe, which is at the center of the octahedron.]

splitting (ΔE_Q, mm/sec), measured relative to a standard NBS sodium nitroprusside absorber are given in Table I.

All parameters(including linewidths) compared with the WWJ etc plots indicate the presence of high-spin Fe (III) ions[6]. The presence of a quadruply split doublet in all cases shows that each Fe(III) is at an equivalent site in the above complexes. These observations strongly support the equilateral triangular configuration[7,8] proposed for B and C. Figure 1(b) shows an isometric view of this arrangement.

The situation for complex A, however, is quite different. In the absence of any x-ray structural information this complex could be represented by the structure shown in Figure 1(a). Here all the benzoic acid groups would be included within the complex as suggested previously[9]. However, instead of only one oxygen being shared by three Fe ions (as in B and C), three oxygens are involved. This structural information has been used for the following discussion on the magnetic susceptibility results.

Since Van Vleck[10] has shown the importance of temperature-independent paramagnetic term ($N\alpha$) we have used a modified Curie-Weiss law containing this term:

$$\chi_{Fe} = \frac{C}{T-\theta} + N\alpha \quad \text{---- I}$$

The values of temperature and molar susceptibility (χ_m, corrected for the diamagnetism of ligands, etc.) were fitted by means of a computer-aided nonlinear least squares analysis[11] to Eqn 1 and values of C, θ, and $N\alpha$ were computed. The effective magnetic moments (μ_{eff} Bohr Magnetons), calculated from $\mu_{eff} = 2.828 \times \sqrt{(\chi_{Fe}-N\alpha)(T-\theta)}$ were consequently subnormal in a and B.

Values of C, θ (°K) and $N\alpha$ ($x10^{-6}$) cgs units) per Fe are given in Table I.

The exchange interactions are best described by the following Van Vleck-Kambe equation with the inclusion of the $N\alpha$ term, and the assumption of a single J to describe the equivalent interactions among all three spins ($S_1 = S_2 = S_3 = 5/2$).

$$\chi_{Fe} = (Ng^2\beta^2/12KT)[\Sigma A_i \exp(-B_i/KT)/\Sigma C_i \exp(-B_i/KT)] \quad \text{- II}$$

TABLE I

	δ	ΔE_Q	C	$-\theta$	$N\alpha$	μ_{eff}
A	0.70±0.05	0.44±0.02	1.737	499	1725	3.73
B	0.71±0.05	0.51±0.03	0.817	100	2452	2.56
C	0.70±0.05	0.57±0.03	7.153	740	-1765	7.56

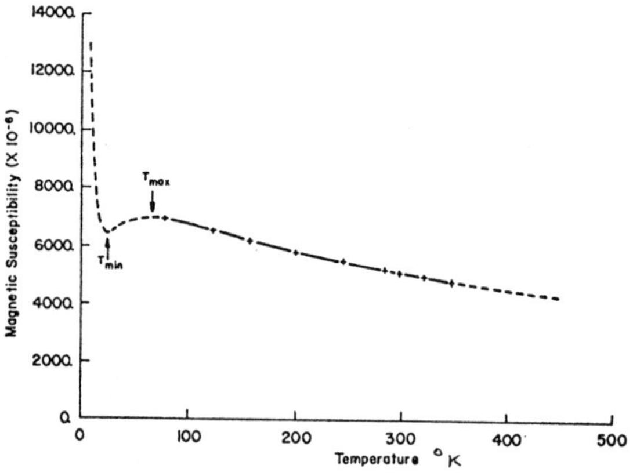

Fig. 2 Typical plots (χ) of magnetic susceptibility versus temp for complex C and the theoretically fitted curve of Eqn II (dashed line).

Here B_i refers to the J terms ranging from zero to 63J and other terms have their usual significance[12].

Due to limitations of space a typical plot of observed values of χ_m versus T and the theoretically fitted (nonlinear least-squares)[11] curve are shown for complex C only (Figure 2). Important derived magnetic parameters are given in Table II. The standard errors of estimate (S_e) which represent one standard deviation tolerance on each side of the theoretical curve are also given.

We wish to stress the following important observations with regard to our new approach to elucidate magnetic interactions.

(i) The application of the modified Curie-Weiss Law (Eqn I) shows large negative values of θ, which only qualitatively suggest strong antiferromagnetic coupling between spins. It also yields large values for Van-Vleck paramagnetism ($N\alpha$), and in complex C yields a negative value, to which no physical significance can be attached at present. Hence, fitting experimental data to Eqn I should be undertaken with caution.

(ii) Fitting our data to the Van Vleck-Kambe Eqn II gives positive but large values for $N\alpha$ in all cases. The $N\alpha$ term is generally believed to arise from the "mixing in" of high energy levels into the ground state. This concept appears to be compatible with the spin-manifold for each Fe in the trinuclear iron complexes, and the separation of their energy levels with respect to KT.

(iii) Eqn II yields negative values for J comparable to those observed in other trinuclear iron complexes and quantitatively state the strong antiferromagnetic coupling between spins. J in effect is seen to determine the separation of the higher energy levels from the ground state; hence, higher the value of J, the lower is the $N\alpha$ term. It should be noted that in all previous studies[1] most workers seemed to arbitrarily assume a value of $N\alpha \sim 600 \times 10^6$ cgs units and have missed an interesting approach to magnetic investigation.

(iv) These workers have also assumed g = 2 in all types of complexes, including the binuclear and trinuclear systems. (Unfortunately we could not detect any EPR signals for our complexes; hence we could not get any independent corroboration for our g value.) Frequently g values are typical of single ion states. New sets of J values are now being evaluated using the usual values of g in place of the curve-fitted values. The results will be reported separately.

(v) A direct exchange resulting from the overlap of d-orbitals of Fe ions is ruled out because of the large Fe-Fe distances involved (these are ∿3.29Å in the acetate analog)[8]. Hence, super-exchange is proposed via oxygen anions. In complex A, three oxygen atoms are involved, one between each pair of Fe atoms. In B and C only one oxygen is involved. Thus we postulate a novel mechanism in which one oxygen may provide a path for superexchange for three Fe ions "simultaneously" or in "pairs at a time" leaving the third spin "free." We designate this a "resonant" superexchange mechanism. Although the Fe-O-Fe angles involved in A, B, and C do not correspond to the ideal situation (180°), superexchange via the actual configuration is still plausible.

(vi) Figure 2 shows an unusual behavior for χ_m predicted by the Van Vleck-Kambe formulation at lower temperatures. A maximum at $T_{max} \simeq 2J$ and a minimum at $T_{min} \sim J$ is observed. Further decrease in temperature shows an increase in χ_m, going to almost infinity as $T \rightarrow 0$. We propose to ascribe the following physical significance to this behavior, which appears to be compatible with our picture of the aforesaid "resonant" superexchange. Because of the odd number of spins involved here, "pairs" of spin at a time (S_1S_2, S_2S_3, S_3S_1) would dominate the antiferromagnetic coupling giving a maximum at T_{max}. Such coupling would leave the third spin free (S_3, S_1 and S_2) when the other pairs are formed. This is believed to give rise to the increasing paramagnetism, expected at very low temperatures. Analysis of this curve has given satisfactory values for magnetic parameters which are compatible with a free spin. A combination of all such effects would produce the curve shown in Fig. 2.

Further details of this work will be published elsewhere. We thank the Matthey Bishop Inc. and the National Science Foundation for providing a fellowship and traineeship to one of us (G.H.Z.) during the course of this work. We are grateful to Professor W.E. Hatfield for helpful discussions.

REFERENCES

1. P. Day, "Electronic Structure and Magnetism of Inorganic Compounds," Specialist Periodical Reports, Chemical Society of London (1972-1974). These summarize work on bi- and tri-nuclear complexes.
2. L.N. Mulay and W.J. Danley, J.Appl. Phys. 41, 877 (1970); Mat. Res. Bull. 7, 739 (1972); L.N. Mulay and J.F. Houlihan, Inorg. Chem. 13, 745 (1974); Phys. Stat. Sol. (b) 61, 647 (1974); (b) 54, 247 (1972).
3. R. Weinland and O. Loebich, Z. Anorg. Allgem Chem. 151, 271 (1926).
4. B.N. Figgis and G.B. Robertson, Nature 205, 694 (1965).
5. L.N. Mulay, "Magnetic Susceptibility" Repring Monograph, Interscience-Wiley, New York (1966); "Magnetic Susceptibility Techniques" in "Physical Methods of Chemistry" A. Weissberger and B.W. Rossiter, Eds., Intersci.-Wiley, New York (1972).
6. N. Greenwood and T.C. Gibb, "Mossbauer Spectroscopy" Chapman Hall Ltd., London (1971).
7. L. Orgel, Nature 187, 4504 (1960). See also Ref. 4.
8. K. Anzenhofer and J.J. DeBoer, Recueil des Trav. Chim. Pays-Bas, 88(3), 286 (1969).
9. A. Earnshaw, B.N. Figgis and J. Lewis, J. Chem.Soc. 12, 1656 (1966).
10. J.H. Van Vleck, "The Theory of Electric and Magnetic Susceptibilities." Oxford Univ. Press, London (1932).
11. D.W. Marquart, "Lease Squares Estimation of Non-Linear Parameters" IBM Share Library (Distr. No. 3094) Yorktown Heights, N.Y. (1964).
12. A. Earnshaw "Introduction to Magnetochemistry" Academic Press, N.Y. (1968). See also K. Kambe, J. Phys. Soc. Japan 5, 48 (1950).

MAGNETIC HYPERFINE STRUCTURE AND RELAXATION EFFECTS IN BROMOBIS(MORPHOLYLDITHIOCARBAMATO)IRON(III)*

J. M. Grow and H. H. Wickman
Department of Chemistry
Oregon State University
Corvallis, Oregon 97331

ABSTRACT

Mössbauer resonance hfs show magnetic order and relaxation effects in the ^4A ground state of bromobis (morphylyldithiocarbamate)iron(III), $(S_2CNC_4H_8O)_2FeI$. Interpretation of the data required an extension of earlier relaxation calculations to include general orientations of the following inputs: axial and rhombic crystal field terms, intermolecular spin-dipolar interactions, electric field gradient, and an intermolecular Weiss field. Polarizations of electronic levels were converted to internal effective magnetic fields (only core polarization mhfs is present). Relaxation processes were calculated quantum mechanically using the Liouville operator formalism of Blume or Gabriel. Below T_C, agreement between theory and the major spectral region is satisfactory. However, evidence of a spread in transition temperatures and possible spin canting is also present.

INTRODUCTION

Halo(bisdithiocarbamate)iron(III) complexes are molecular solids, several of which exhibit collective magnetic transitions in the helium temperature range.[1,2] The ground state of the iron is an orbital-singlet, spin-quartet term with typical zero field splitting (zfs) in the range 2-30 cm^{-1}.[3,4] We report here Mössbauer spectra of the bromo(bismorphylyldithiocarbamate)iron(III) complex (denoted MBr). This complex orders at 3.5°K; this is the highest ordering temperature observed thus far in bis-dtcs. We have also observed magnetic transitions in the corresponding chloro and iodo complexes.[5] Unlike previously studied complexes,[1] the morphylyl derivatives exhibit variable, complex spectra. Consistent with such spectra, the complexes show polymorphism, solvent adducts, and possibly crystallite size effects.

A second complication, typical of low symmetry complexes, is the potential non-coincidence of the principal axes of the crystal field, exchange and hfs interactions. By analogy with known structures, the iron in MBr is assumed to occupy the centroid of a rectangular pyramid formed by a plane of four sulfur atoms, and an apical halide. Since these complexes show magnetic relaxation effects in their hfs below T_C, it has been necessary to generalize an earlier relaxation calculation to include orientation effects. Spectra calculated using the Liouville matrix method of Gabriel[6] and Blume[7] have been used to fit observed data.

EXPERIMENTAL

The MBr was prepared by standard methods, and recrystallized from CH_2Cl_2. Chemical analysis showed the crystals were of composition $MBr\cdot\frac{1}{4}CH_2Cl_2$. Mossbauer spectra were obtained using a constant acceleration spectrometer. Figure 1 shows data for several temperatures below 4.2°K. Temperature measurement and control employed vapor pressure thermometry and a calibrated germanium resistor at the sample. The solid curves through the spectra for temperatures below 3.5°K are fits to the data using the model described below. The calculations involve variation of several parameters and require significant machine time. As a result, we have not yet attempted least squares fitting methods. The computer work used a CDC-7600 at the Lawrence Berkeley Laboratory.

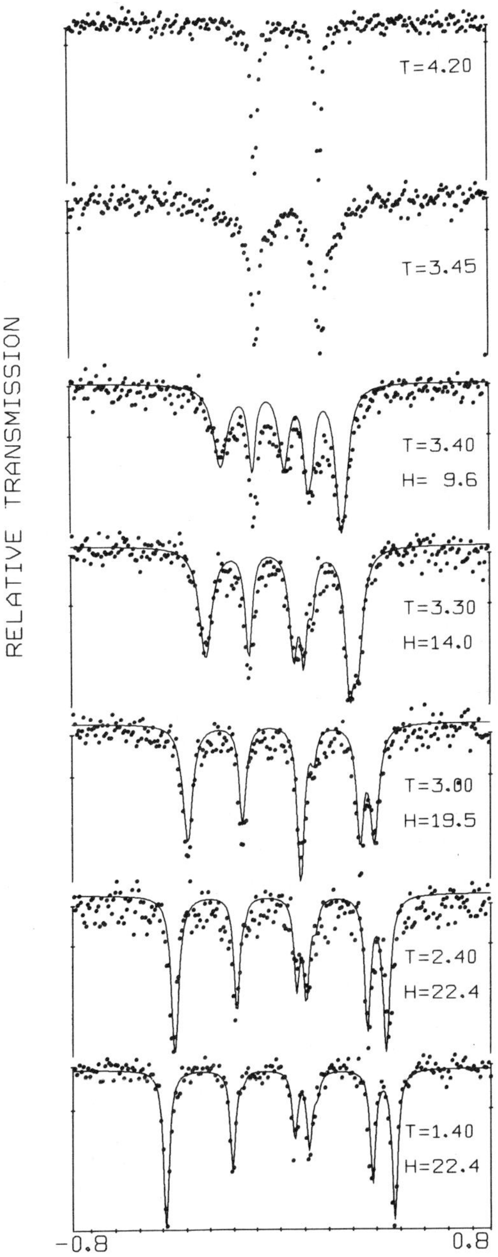

Fig. 1

DISCUSSION

A magnetic transition at T_C = 3.50 ± 0.05°K was derived from spectra showing the onset of magnetic hfs. Above 3.5°K, a nearly temperature independent quadrupole doublet is observed. The interactions within the ground electronic term (S = 3/2) are

$$\mathcal{H} = \beta\vec{H}_e\cdot g\cdot\vec{S} + D[S_z^2 - 5/4] + E(S_x^2 - S_y^2) + \vec{I}\cdot A\cdot\vec{S} + (e^2qQ/4)[3I_{z'}^2 - 15/4 + \eta(I_{x'}^2 - I_{y'}^2)] + \mathcal{H}'_{dip} \quad (1)$$

The first term represents the Weiss field approximation; the last term is the usual two-spin, spin-dipolar interaction which is used to compute relaxation rates (see below). The electronic g-tensor and the core polarization hyperfine tensor were isotropic in these calculations. Recent magnetic measurements of Mitra et al., in related bis-dtcs have shown only a small (~5%) anisotropy in the g-tensor.[4]

Because magnetic transitions in bis-dtcs occur at low temperatures, the hfs and relaxation effects are dominated by the magnetic character of the ground Kramers level. When this state is Ising-like, the nucleus sees an effective field, and relaxation calculations are reasonably straightforward.[8] In the present case, D is positive, the ground doublet has $g_x \sim g_y \sim 4$, $g_z \sim 2$ (effective spin $S = \frac{1}{2}$), and the effective field approximation fails. A complete stochastic calculation requires a Liouville matrix of dimention $(2S + 1)^2(2I_1 + 1)(2I_0 + 1) = 128$.[9] The situation simplifies when a magnetic field is present with sufficient strength to dominate hyperfine interaction. The electronic levels are polarized and may be represented at the nucleus by classical fields along the direction of $\langle \vec{S} \rangle$ in the four levels. This reduces the dimension of the matrix to 32. In MBr with $T<T_C$, the exchange field produces the necessary polarization.

A calculated pattern is obtained as follows. The magnetic hyperfine constant was taken as 220 kOe/<S>.[1] The quadrupole parameters are derived from the 1.4°K spectra. The principal axis system of the efg is the preferred coordinate system. In the ordered state, the orientations and magnitudes of $\vec{H}_e$, D, and E determine the four effective fields among which relaxation occurs. Relaxation is fixed by the parameters $g^2\beta^2/r^3$, θ and ϕ appearing in $\mathcal{H}_{dip}$. The foregoing parameters are specified as input to a program which uses the Liouville matrix and spectral expression given in Eq. (22) of reference 7. The calculated spectra in Figure 1 correspond to the following input parameters to Eq. (1): $(e^2gQ/2) = 0.272$ cm/sec, $\eta = 0.16$; $D = 8$°K, $E = 0.4$°K, $\theta_{CF} = 90°$, $\phi_{CF} = 10°$; $g^2\beta^2/r^3 = 2.2$ cm/sec, $\theta_{dip} = 90°$, $\phi_{dip} = 0°$; $\theta_{H_e} = 90$; $\phi_{H_e} = 90°$; values of $\vec{H}_e(T)$ are given in Fig. 1.

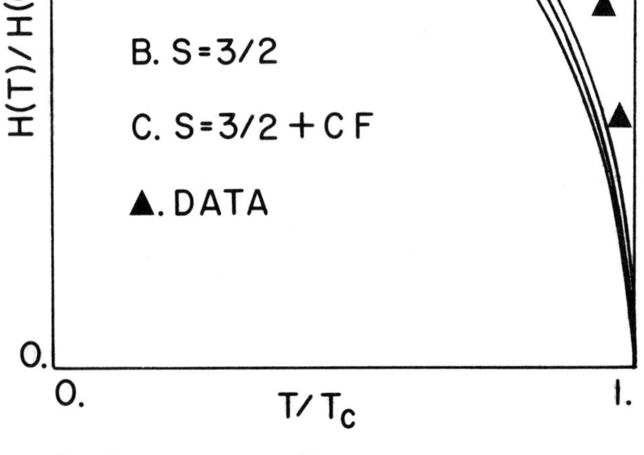

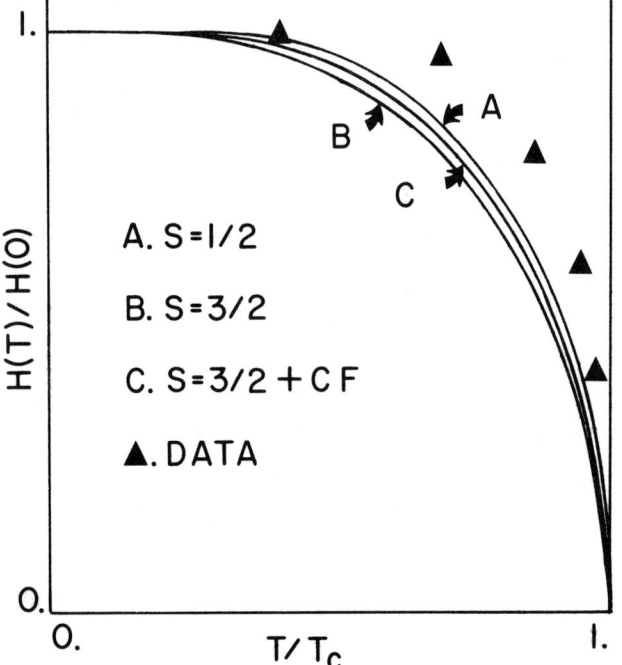

Figure 2. Exchange field temperature dependence in MBr.

In previous work, the temperature dependence of the internal field was well approximated by a Brillouin function, S = 3/2.[8] This is not true in MBr. Figure 2 shows the field dependence predicted for $S = \frac{1}{2}$ (isolated ground doublet), S = 3/2 (negligible crystal field splittings), and curve C is a self-consistent calculation which includes the crystal field parameters. The onset of the magnetism, as shown by the data points, is clearly much faster than is predicted by molecular field calculations. A complete explanation of this observation is not yet available. It is likely that spin canting occurs in MBr although the orientation of $\vec{H}_e$ with T has not been varied. That is, if the orientation of <S> or H_e were allowed to vary with temperature (canting) a self-consistent crystal field calculation of $H_e(T)$ might lead to better agreement between theory and experiment. Because of the many parameters appearing in the calculation, a resolution of this point must await more detailed measurement below T_C. A second distinction between the hfs of Fig. 1, and spectra reported in other bis-dtcs,[1,2,9] is the fact that over a substantial range of temperatures, the patterns correspond to a fast electronic relaxation so there is an effective field approximately given by $\langle S \rangle_{ion}$. (In the other dtcs relaxations rates are intermediate to slow compared with the nuclear precession frequency.) At the lowest temperatures, population effects produce a slower relaxation rate. The polarization in the ground state is <S> = 1.23; this results from a substantial mixing of the M_S states by the combined crystal field and molecular field.

A discrepancy between calculated and observed spectra is seen in the 3.4°K pattern, a portion of which is composed of an unsplit quadrupole doublet. This is indicative of a breadth of transition temperatures in the sample. This effect, which is variable, depending on sample preparation, is also observed in the iodo derivatives.[5,10] A number of explanations of such behavior is possible, but discussion of this point is deferred until crystal structure data are complete. We note, however, that the influence of preparation may be illustrated by the behavior of samples of MBr which are obtained by recrystallization from CH_2Cl_2/toluene solution. In such cases, the MBr does not exhibit a magnetic transition to 1.40K. Barring destructive crystallographic changes, such behavior illustrates that solvent effects may play an important role in stabilizing crystal structures with different magnetic properties.

*Supported by the National Science Foundation

REFERENCES

1. G. E. Chapps, S. W. McCann, H. H. Wickman, and R. C. Sherwood, J. Chem. Phys., 60, 990 (1974).
2. A. Kostikas, D. Petrides, A. Simopoulos, and M. Pasternak, Solid State Comm., 13, 1661 (1973).
3. G. C. Brackett, P. L. Richards, and W. S. Caughy, J. Chem. Phys. 54, 4383 (1971).
4. P. Ganguli, V. R. Mitra, S. Mitra, and R. L. Martin, Chem. Phys. Letters, in press.
5. J. M. Grow and H. H. Wickman, to be published.
6. H. Gabriel, J. Bosse, and K. Rander, phys. stat. sol. 27, 301 (1968).
7. M. Blume, Phys. Rev. 174, 351 (1968).
8. H. H. Wickman and C. F. Wagner, J. Chem. Phys. 51, 435 (1969).
9. M. J. Clauser and M. Blume, Phys. Rev. B3, 581 (1971).
10. A. Kostikas, private communication

TEMPERATURE INDEPENDENT ^{55}Mn MAGNETIC HYPERFINE FIELDS IN $ErMn_2$ AND $TmMn_2$

R. G. Barnes and B. K. Lunde[+]
Ames Laboratory, USAEC, and Physics Department
Iowa State University, Ames, Iowa. 50010.

ABSTRACT

Analysis of the complex NMR spectra of ^{55}Mn in the hexagonal Laves phase compounds $ErMn_2$ and $TmMn_2$ reveals only a very slight temperature dependence of the ^{55}Mn Knight shift K at either of the two inequivalent lattice sites, in contrast to the Curie-Weiss behavior of the susceptibility of these compounds and their known ferromagnetic ordering properties. In addition, K differs negligibly from that in isostructural $ScMn_2$ which does not order magnetically. Both neutron diffraction and NMR data indicate that Mn carries no local moment, and the Knight shift measurements suggest that K results essentially entirely from the orbital contribution, K_{orb}. Unlike the case of the $LnAl_2$ cubic Laves phase compounds, for example, the net contribution to K resulting from indirect interaction between ^{55}Mn and the rare-earth moments is relatively small.

INTRODUCTION

Nuclear hyperfine fields have been studied by magnetic resonance (NMR) methods in a number of sequences of intermetallic compounds involving paramagnetic metal ions. The Knight shifts of the nuclei of the non-magnetic ions in such compounds, especially those involving rare earth ions, have usually been found to be proportional to the magnetic susceptibility. In general, this reflects the indirect coupling between rare earth spins brought about by the conduction electrons according to the RKKY mechanism,[1] and the shift behavior has been discussed both in terms of a uniform polarization model and more sophisticated ones based on Friedel-type oscillations in the conduction electron spin density.[2]

The rare earth compounds of the present work, $ErMn_2$ and $TmMn_2$, have previously been the object of investigation by Mössbauer effect,[3] magnetization and susceptibility measurements,[4,5] and neutron diffraction.[4] These previous investigations had shown that the Mn ion carried little if any magnetic moment and that the effective charge on the Mn ion was very nearly zero. The compounds have the hexagonal Laves phase structure (C14, $P6_3/mmc$). In this case, the B-element (Mn) occurs in two crystallographically inequivalent sites, one possessing axial symmetry and the other not. The rare earth ion occurs only in an axially symmetric site. The compounds are formed in structurally very well-ordered condition, so that the NMR spectra are well-resolved. The compound $ScMn_2$ which has the same structure serves as a reference compound with which to compare the shift behavior of the Mn NMR.

EXPERIMENTAL

Samples were prepared in the Metallurgy Division of the Ames Laboratory using Ames Laboratory rare earth metals and Foote Mineral Co. electrolytic manganese. Arc-melted buttons were turned and remelted five times to ensure homogeneity, and were then annealed at approximately 900°C for two weeks. For the NMR work the buttons were crushed in a ceramic mortar and sieved to -325 mesh.

Standard crossed-coil nuclear induction methods were employed for the NMR measurements. Signal averaging and magnetic field strengths to 25 kOe aided in improving the sensitivity and spectral resolution. Magnetic susceptibility measurements were made with a Faraday system.

RESULTS

Despite the relatively high quality of the NMR spectra, analysis of the ^{55}Mn spectra is complicated by the occurrence of two inequivalent lattice sites and the strong quadrupole interaction. Manganese ions in site I have point symmetry $\bar{3}m$ and hence axially symmetric Knight shift and quadrupole interaction tensors, whereas those in site II have point symmetry mm and non-axially symmetric hyperfine interactions. At both sites the spectrum is dominated by the quadrupole interaction and the isotropic Knight shift, the non-isotropic components of the shift tensor being unresolved.

The NMR results at 300 K are summarized in Table I. A more detailed account of the analysis of the $ScMn_2$ spectrum, as well as a representative experimental trace, has appeared elsewhere.[6] The Mn shifts were measured with respect to the revised ^{55}Mn effective gyromagnetic ratio of 1.0500 MHz/kOe.[7] Apart from small increases in the quadrupole couplings, the

Table I. Summary of isotropic Knight shift and quadrupole coupling parameters at 300 K in hexagonal RMn_2 compounds. Point symmetry of Mn site I is $\bar{3}m$ and of site II is mm. The manganese shifts are measured with respect to the revised ^{55}Mn effective gyromagnetic ratio 1.0500 MHz/kOe.

Compound	Nucleus and site	K_{iso} (%)	e^2qQ/h (MHz)	η
$ScMn_2$	^{45}Sc	-0.09 ± 0.01	0.777 ± 0.004	0
	$^{55}Mn_I$	$+0.9 \pm 0.2$	21.3 ± 1.3	0
	$^{55}Mn_{II}$	$+0.69 \pm 0.05$	10.58 ± 0.09	0.71 ± 0.02
$ErMn_2$	$^{55}Mn_I$	$+0.58 \pm 0.06$	18.4 ± 0.3	0
	$^{55}Mn_{II}$	$+0.34 \pm 0.2$	13.3 ± 1.2	0.36 ± 0.10
$TmMn_2$	$^{55}Mn_I$	$+0.9 \pm 0.2$	18.8 ± 1.3	0
	$^{55}Mn_{II}$	$+0.56 \pm 0.1$	14.9 ± 1.5	0.39 ± 0.14

parameters are unchanged at 77 K to within the stated uncertainties in the shift parameters.

Results of the susceptibility measurements are summarized as follows: for $ErMn_2$, $\mu_{eff} = 9.56 \pm 0.03\ \mu_B$, $\theta = 14 \pm 1$ K; for $TmMn_2$, $\mu_{eff} = 7.64 \pm 0.05\ \mu_B$, $\theta = 6 \pm 3$ K.

DISCUSSION

The Curie-Weiss behavior of both compounds is we well-established. Previous magnetization measurements indicated a ferromagnetic ordering temperature of 25 K and an extrapolated value of 7.86 μ_B/molecule for $ErMn_2$ and of 12 ± 2 K and 5.44 μ_B/molecule for $TmMn_2$.[4] In addition, other more recent susceptibility measurements on ErMn2 obtained $\mu_{eff} = 9.6\ \mu_B$ and $\theta = 30$ K.[5] The μ_{eff} values are in excellent agreement with the free ion values for the rare-earth ions (9.59 μ_B for Er^{3+} and 7.57 μ_B for Tm^{3+}). This, as well as the neutron diffraction results, indicates that the Mn ion carries no moment. The observability of the ^{55}Mn NMR in all three compounds in the temperature range 77-300 K, with linewidths characteristic of nuclear dipolar broadening only, also supports this conclusion. A previous Mössbauer effect study of ^{169}Tm in hexagonal Laves phase compounds showed that the experimental crystal-line electric field (CEF) parameters for the Tm^{3+} ion were best accounted for by taking the effective Mn charge to be zero.[3] This, coupled with the apparent zero magnetic moment, could be accounted for in terms of a highly localized band picture for the Mn ion in which no net spin polarization of the band occurs.

As is evident from Table I, correspondences between the shift and quadrupolar parameters in the three compounds are not altogether total. Although the ^{55}Mn Knight shifts in $ScMn_2$ and $TmMn_2$ are very similar, the quadrupole asymmetry parameter values differ by a factor of two. Similarly, the ^{55}Mn shifts in $ErMn_2$ and $TmMn_2$ differ substantially. The ^{55}Mn quadrupole couplings are however fairly constant throughout. The ^{55}Mn shifts are somewhat larger than the positive shifts measured in both α and β - Mn,[8,9] but are smaller than those measured for dilute Mn in vanadium.[10] In the latter case, in particular, a reasonable partitioning of the Knight shift data was possible, reaching the conclusion that the large positive shift value was attributable to the orbital susceptibility. The same conclusion would appear to apply in the present case, at least qualitatively.

REFERENCES

+Present address: Metallurgy Division, Ames Laboratory, USAEC, Iowa State University, Ames, Iowa 50010.

1 K. Yosida, Phys. Rev. 106, 893 (1957).
2 V. Jaccarino, J. Appl. Phys. 32, 102S (1961).
3 D. L. Uhrich et al., Phys. Rev. 166, 261 (1968).
4 G. P. Felcher et al., J. Appl. Phys. 36, 1001 (1965).
5 H. Oesterreicher, J. Less-Common Metals 23, 7 (1971).
6 R. G. Barnes and B. K. Lunde, J. Phys. Soc. Japan 28, 408 (1970).
7 W. B. Mins et al., Phys. Letters 24A, 481 (1967).
8 L. O. Andersson, Phys. Letters 26A, 279 (1968).
9 L. E. Drain, Proc. Phys. Soc. 88, 111 (1966).
10 E. von Meerwall and D. S. Schreiber, Phys. Rev. B3, 1 (1971).

ON SUPEREXCHANGE THEORY†

R.S. Tu and T.A. Kaplan
Department of Physics, Michigan State University,
East Lansing, Michigan 48824

ABSTRACT

A major presumption of Anderson's theory of superexchange[1] is that the perturbation theory defined in terms of "the exact" Wannier functions would be rapidly convergent. (The small perturbation parameter is the nearest neighbor overlap S.) His Hartree-Fock (HF) definition of these exact Wannier states was shown by Silva and Kaplan[2] to be unsatisfactory; in particular these states do not satisfy his requirement that they be non-magnetic (non-magnetic wave functions are defined as products $w(r)\sigma$ of spatial and spin functions in which the spatial functions are independent of spin σ). We nevertheless investigate the presumption of rapidity of convergence using a different variational definition, namely the non-magnetic localized solutions in the thermal single-determinant approximation (TSDA).[3,4]

Our investigation is made within a previously studied[5] 4-electron 3-orbital 3-site model of 180° superexchange. We calculated the exchange integral J exactly to $O(S^4)$, this calculation involving perturbation theory through 4th order, with the following results. (1) To leading order the TSDA Wannier functions are the same as those derived by GT by a different method (their method also differs from Anderson's HF theory although they use the term "Hartree-Fock" to describe their approach). (2) The size of the contributions to J from each order in perturbation theory is very sensitive to the choice of Wannier functions. (3) Using the TSDA choice, J is given exactly to order S^4 by 1st- plus 2nd-order perturbation theory.[6]

† Work supported by NSF Grant GH-34565

1. P.A. Anderson, Solid State Physics 14, 99 (1963)
2. N.P. Silva, T.A. Kaplan, AIP Conf. Proc. No. 18, Magnetism and Magnetic Mat'ls (1973) p. 656
3. T.A. Kaplan, Petros N. Argyres, Phys. Rev. B1, 2457 (1970)
4. In Ref. 2 another variational approach was studied. However, the present one is simpler and adequate to our more limited purposes.
5. Ken-Ichiro Gondaira, Tukito Tanabe, J. Phys. Soc. Japan 21, 1527 (1966), referred to as GT.
6. GT also carried out perturbation calculations using these basis functions, but they did not discuss the corresponding 3rd and 4th order perturbation-theoretic terms.

MÖSSBAUER EFFECT STUDY OF $Co^{57}Cl_2.2H_2O$ and $Co^{57}Cl_2.6H_2O$

B. A. Alavi, D. F. Goble and W. Leiper
Dept. of Physics, Dalhousie University, Halifax, Nova Scotia, Canada

There has been some discussion in the literature regarding the hyperfine spectra of $Co^{57}Cl_2.2H_2O$ and $Co^{57}Cl_2.6H_2O$. [1-3] Ingalls and de Pasquali [1] claim to have observed charge states up to Fe^{4+} in hydrated cobalt salts and what was interpreted as quadrupole doublets arising from Fe^{2+} and Fe^{3+} states in $CoCl_2.2H_2O$. By contrast, Mullen and Ok [2] interpreted their measurements on $CoCl_2.2H_2O$ in terms of anisotropy in the source (i.e. the Goldanskii-Karyagin effect) and on $CoCl_2.6H_2O$ in terms of a local rearrangement of the lattice to a configuration close to that of the dihydrate. However, in neither of these investigations was the crystal structure of the samples verified by x-ray crystallography. Friedt and co-workers [3] have observed Fe^{2+} and Fe^{3+} states in both $Co^{57}Cl_2.2H_2O$ and $Co^{57}Cl_2.6H_2O$, which they ascribe to self-radiolysis of the water ligands in these compounds. We have studied carefully prepared samples of $Co^{57}Cl_2.2H_2O$ and $Co^{57}Cl_2.6H_2O$, the structure of the samples having been verified both by x-ray diffraction and by gravimetric analysis. The Mössbauer spectra of both compounds at 77K were satisfactorily fitted to two doublets, and at 4.2K (below the Neel temperature) $CoCl_2.2H_2O$ gave a complex spectrum which suggests the presence of an Fe^{3+} and Fe^{2+} doublet. The x-ray and gravimetric analysis preclude the possibility that the second doublet arises from the presence of any of the other hydrates of cobalt chloride. We conclude, therefore, that Auger after-effects are significant in hydrated cobaltous halides.

1. Ingalls, R. and De Pasquali, G., Phys. Lett. 15 (1965) 262.
2. Mullen, J. G. and Ok, H. N., Phys. Rev. Lett. 17 (1966) 287.
3. Friedt, J. M., Shenoy, G. K., Abstreiter, G. and Poinsot, R., J. of Chem. Phys. 59 (1973) 3831.

RESONANCE ENERGY TRANSFER IN $YCrO_3$

C. Satoko
Institute for Solid State Physics, Tokyo University, Minato-ku, Tokyo, 106

S. Washimiya
Broadcasting Science Research Laboratories of Nippon Hôsô Kyôkai, Setagaya-ku, Tokyo, 157

ABSTRACT

It has been shown that the magnetic-field dependence of the energy-transfer matrix elements in 1-2 and 1-4 Cr^{3+}-pairs in $YCrO_3$ obtained by analysing the observed spectral changes of the exciton absorption lines during the field-induced spin reorientation can well be explained empirically on the basis of the direct energy-transfer model proposed by Sugano, Aoyagi and Tsushima.

The purpose of the present paper is to derive expressions for these matrix elements of the resonance energy transfer, based on this model, and to discuss its magnetic-field dependence. Theoretical expressions for v_{12} and v_{14} are expressed in terms of the ground- and the excited-state wavefunctions and the operator equivalent to the real Hamiltonian. These expressions give the observed magnetic-field dependences of $|v_{12}|$ and $|v_{14}|$ and the estimated order of them is in fair agreement with the experiment. However, this model is not applicable to the resonance energy-transfer v_{13} and v_{42}. It would be necessary to take into account indirect transfer via $1\to2\to3$ and $2\to3\to4$ paths for these transfer.

EXCHANGE STIFFNESS CONSTANTS IN Ni-Co ALLOYS

H. YAMAUCHI, I. MAEDA and H. WATANABE, The Research Institute for Iron, Steel and Other Metals, Tohoku University, Sendai, Japan.

ABSTRACT

Magnetization measurements of f.c.c. Ni-Co alloys using pendulum-type magnetometer have been made from liquid He temperature to room temperature. Magnetization vs. temperature curves have been analysed on the basis of the spin wave theory. The exchange integral J and the exchange stiffness constant D are estimated. The composition dependence of J and D are not consistent with that obtained by Hinoul et al. from spin wave resonance experiments nor with that obtained by Wakoh from the calculation on the basis of itinerant electron model. It is suggested that the contributions to the spin wave dispersion relation from the itinerant and localized characters of electrons are additive as shown by Englert et al. and Yamada et al. The role of the inter-atomic exchange interaction in Ni-Co alloys seems to be important in order to explain the observed values of D.

CALORIMETRIC EVIDENCE FOR SPIN FLUCTUATIONS IN UAl_2*

R. J. Trainor and M. B. Brodsky
Argonne National Laboratory, Argonne, IL 60439
and L. L. Isaacs
Argonne National Laboratory and Dept. of Chem.
Engr., The City College of C.U.N.Y., New York, NY 10031

ABSTRACT

We present results of heat capacity measurements on UAl_2 between 1.8 and 400K. The data are compared with recent resistivity and susceptibility measurements which indicate the existence of localized spin fluctuations in a narrow 5f band. Below about 50K the electronic contribution to the heat capacity becomes large, equivalent to $\gamma \approx 70$ mJ/mole-K^2. Below 6K there is an upturn in C/T which is proportional to $T^2 \log (T/T_{SF})$, where T_{SF} = 10.6K is identified as the spin fluctuation temperature. Extrapolation of this term to zero temperature yields $m^*/m \sim 2$ for the spin-fluctuation mass enhancement. At 300K, UAl_2 exhibits more typical metallic behavior, with $\gamma \approx 15$ mJ/mole-K^2. Data are also presented for nonmagnetic URh_3; at low temperatures $C = \gamma T + \beta T^3$, with γ = 14.5 mJ/mole-K^2 and β corresponding to θ_D = 336K.

INTRODUCTION

In a large number of actinide intermetallic compounds which do not order magnetically, the existence of localized spin fluctuations (l.s.f.) has been proposed to explain resistivities which have T^2 dependences at low temperatures.[1,2] In support of this proposal are susceptibility data which often show magnetic (i.e., temperature dependent) behavior at high temperature, but level off or become weakly temperature dependent at lower temperatures. Two compounds which display these properties and which have attracted considerable attention recently are UAl_2 and $PuAl_2$.[3]

If strong spin fluctuation effects are active in materials like UAl_2 and $PuAl_2$, they may be expected to manifest themselves in the specific heat, as they do in He^3 at ultralow temperatures.[4] Such effects had not been previously observed in any pure compound.

In this paper the results of measurements on UAl_2 between 1.8 and 400K are presented and are considered in terms of the l.s.f. model. We also present heat capacity data for the compound URh_3, in which the 5f band has been broadened by 5f-6d hybridization to such an extent that it lies well into the nonmagnetic region when interpreted in terms of a Friedel-Anderson model.[5]

EXPERIMENTAL PROCEDURE

The samples were prepared by arc-melting 99.99% U with the appropriate amounts of either high purity Al or Rh. X-ray patterns verified that the samples were all single phase (cubic Laves phase for UAl_2; fcc $AuCu_3$ for URh_3). Two UAl_2 samples were prepared, weighing 1.5 and 5 grams. The URh_3 sample weighed 14 grams.

Measurements were made in a strong heat-leak calorimeter, designed to measure heat capacities of small self-heating samples (e.g., Np and Pu) by essentially a d.c. heat-pulse technique. Details of the technique will be discussed elsewhere.[6] Measurements on 15 grams of high purity copper showed deviations from accepted values of 1% or less below 10K, increasing to 1 1/2% near 40K. The overall accuracy is estimated to be about 1%. Above 50K a differential technique, discussed previously, was used.[7]

*Work supported by the U. S. Atomic Energy Commission.

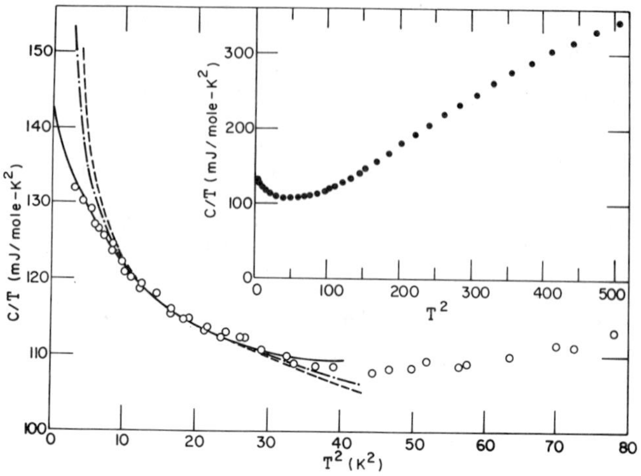

Figure 1. C/T versus T^2 for UAl_2. Fits using $C = AT + BT^3 + Df(T)$ are shown, where $f(T) = 1/T^2$ (dashed curve), $1/T$ (dashed-dot), and $T^3 \log T$ (solid curve). The inset shows the same data over a larger temperature range (MSD Negative 60450).

RESULTS

Figure 1 shows a plot of C/T versus T^2 for UAl_2. The striking feature of the data is the magnitude and nonlinearity of C/T at the lowest temperatures. We have assumed that below about 6K, where the Debye T^3 law is expected to characterize the lattice contribution, the total specific heat may be written as

$$C = AT + BT^3 + Df(T) \qquad (1)$$

where the latter term is the contribution responsible for the deviation from simple metallic behavior. Least-squares fits using $f(T) = 1/T$ and $f(T) = 1/T^2$ were clearly incompatible with the data (see Figure 1) and would require negative lattice terms. A third fit using $f(T) = T^3 \log T$, which characterizes a spin fluctuation system well below the spin flucutation temperature T_{SF},[8] gives an excellent description of C below 6K.

If the spin fluctuation model is invoked, the coefficients in Equation 1 will have the following approximate physical meanings:[9]

$$A = (m^*/m)\gamma$$

$$B = \beta - (\alpha\gamma/T_{SF}^2) \log T_{SF}$$

$$D = \alpha\gamma/T_{SF}^2$$

where m^*/m is an electronic mass enhancement for the electrons responsible for spin fluctuations, γ is the electronic specific heat coefficient, and β is the coefficient of the lattice term. The quantity α is proportional to $U_o^2\, N(E_f)^2 S$, where U_o is the interspin Coulomb repulsion, $N(E_f)$ is the density of states at the Fermi level, and S is the Stoner exchange enhancement. The coefficients obtained from the fit are A = 143, B =-4.38, and D = 1.94, when energy is in millijoule units. Assuming a reasonable value for β of 0.2 $mJ/mole\text{-}K^4$ (corresponding to $\theta_D \sim 300K$) comparison of B and D yields $T_{SF} = 10.6K$.

No attempt was made to precisely fit the data above 6K because of expected deviations from Debye theory as well as a possible temperature dependence of γ. However, these data are roughly consistent with $\gamma \simeq 70$ $mJ/mole\text{-}K^2$ (meaning $m^*/m \simeq 2$) and $\theta_D \simeq 275K$. An approximate value of $\gamma \simeq 88$ $mJ/mole\text{-}K^2$ has been previously reported, but θ_D was not given.[10]

Measurements on UAl_2 were extended up to 400K in order to obtain the unenhanced electronic coefficient γ_o and the high-temperature Debye temperature θ_∞. If we ignore anharmonic phonon contributions, the total heat capacity is given by

$$C_v = \gamma_o T + 3R\left[1 - \frac{1}{20}(\theta_\infty/T)^2\right]$$

where the second term describes the lattice for $0.7\,\theta_\infty < T < 1.3\,\theta_\infty$.[11] Over this temperature range a plot of $(C-3R)/T$ versus $1/T^3$ should then yield a straight line whose intercept is γ_o and whose slope is proportional to θ_∞^2. For UAl_2 such a plot is linear and gives $\gamma_o \simeq 15$ $mJ/mole\text{-}K^2$ and $\theta_\infty \simeq 300K$. We note that these values are only approximate, since we have neglected anharmonicity (which is expected to contribute an additional but small linear term) and dilation effects (no thermal expansion data exists).

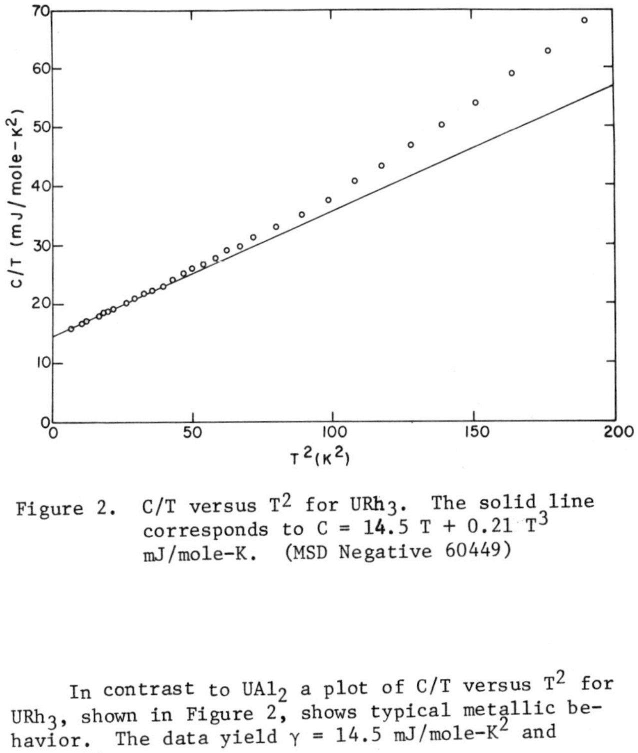

Figure 2. C/T versus T^2 for URh_3. The solid line corresponds to $C = 14.5\,T + 0.21\,T^3$ mJ/mole-K. (MSD Negative 60449)

In contrast to UAl_2 a plot of C/T versus T^2 for URh_3, shown in Figure 2, shows typical metallic behavior. The data yield $\gamma = 14.5$ $mJ/mole\text{-}K^2$ and $\theta_D = 336K$.

DISCUSSION

Gossard et al.[10] proposed that UAl_2 is characterized by a narrow 5f band near the Fermi level. This would give rise to a high density of states at the Fermi energy and would be reflected by large values for the low temperature susceptibility and electronic heat capacity coefficient. A low-temperature resistivity proportional to T^2 in UAl_2 has been interpreted in terms of Doniach's model[12] as arising from spin fluctations in the narrow f band.[13] This idea receives further support from susceptibility data which show a Curie-Weiss-like behavior above about 80K, leveling off below that temperature.[10,13,14] Nuclear spin-lattice relaxation measurements on the Al^{27} nucleus in UAl_2 and $PuAl_2$ point to the existence of spin fluctuations in $PuAl_2$, but remain somewhat inconclusive for UAl_2, since weaker interband coupling in UAl_2 causes the Al relaxation rate to be more insensitive to their presence.[3]

The $T^3 \log T$ term associated with spin fluctuation contributions to the heat capacity has not been observed previously in any pure compound. In fact, this term has been unambiguously identified only in He^3 below 100 mK,[4] but low-temperature upturns in C/T of some alloys are believed to be manifestations of l.s.f. effects (e.g., $LaPd_3$:Ce[15] and Pd:U[16]).

There are a number of ways to determine T_{SF} from experimental measurements but the values obtained for any one system can vary greatly. These discrepancies are associated both with theoretical and experimental uncertainties, as has been discussed previously.[2] The value obtained from these data ($T_{SF} = 10.6K$), is thus in quite good agreement with the value obtained from the T^2 region of the resistivity data ($T_{SF} = 16K$), but is considerably below that obtained any other way.[2] (We note, however, that the other values for T_{SF} are in considerable mutual disagreement.)

Other possible explanations for the low temperature upturn in UAl_2 exist. Fradin[17] has shown that the effect of sharp structure in the density of states on interband phonon scattering is capable of quantitatively explaining why some metals have highly temperature-dependent resistivities at high temperatures, yet exhibit slowly changing susceptibilities. In some cases a temperature-dependent γ is also predicted. The model could conceivably explain UAl_2, but this remains to be seen. On the other hand, it is possible that the upturn in C/T is the precursor of a magnetic transition below 1.2K. Heat capacity measurements down to 0.4K are underway to explore this possibility, but at present there exists no evidence that ordering may occur in UAl_2. Less likely explanations involve magnetic-impurity effects and atomic disorder. We note that data were taken on two different samples, prepared from different batches of U; the low temperature behavior is reproducible from sample to sample. NMR studies on ternary $U_{1-x}Pu_xAl_2$ (for $x = 0.1$ and greater) do show effects of chemical disorder, enhanced by radiation self-damage of the Pu, but there is minimal evidence of such problems in pure UAl_2.[3]

We conclude that the l.s.f. model provides the most plausible explanation for the low-temperature heat capacity, especially when previous resistivity and susceptibility results are considered.

In contrast to the low-temperature behavior of UAl_2 we point to the more typical situation represented by URh_3 and similar compounds, where either l.s.f. behavior is apparently unimportant ($\rho \propto T^3$ for URh_3[5]) or T_{SF} is too high (as deduced from the T^2 resistivity[14]) to make an observable nonlinear contribution to C.[18]

REFERENCES

1. For a general review of the magnetic properties of actinide elements and compounds see The Actinides: A. J. Freeman and J. B. Darby, eds. (Academic, New York, 1974).
2. M. B. Brodsky, Phys. Rev. B9, 1381-87 (1974).
3. F. Y. Fradin, M. B. Brodsky, and A. J. Arko, AIP Conf. Proc. 10, 192-6 (1973); A. J. Arko, F. Y. Fradin, and M. B. Brodsky, Phys. Rev. B8, 4104-18 (1973).

4. W. R. Abel, A. C. Anderson, W. C. Black, and J. C. Wheatley, Phys. Rev. 147, 111 (1966).
5. W. J. Nellis, A. R. Harvey, and M. B. Brodsky, AIP Conf. Proc. 10, 1076-80 (1973).
6. R. J. Trainor, M. B. Brodsky, G. S. Knapp, and R. B. Snyder (to be published).
7. R. W. Jones, G. S. Knapp, and B. W. Veal, Rev. Sci. Instr. 44, 807-10 (1973).
8. S. Doniach and S. Engelsberg, Phys, Rev. Letters 17, 750-53 (1966); N. Berk and J. R. Schrieffer, Phys. Rev. Letters 17 433 (1966); W. F. Brinkman and S. Engelsberg, Phys. Rev. 169, 417-31 (1968).
9. R. L. Chan, Ph.D thesis, Univ. of Surrey, Guildford, 1974 (unpublished).
10. A. C. Gossard, V. Jaccarino, and J. H. Wernick, Phys. Rev. 128, 1038-43 (1962).
11. D. C. Wallace, Thermodynamics of Crystals (Wiley, New York, 1972) p. 54
12. S. Doniach, in Ref. 1, Vol. 2, Chap. 2, pp. 51-72.
13. A. J. Arko, M. B. Brodsky, W. J. Nellis, Phys. Rev. B5, 4564-69 (1972).
14. K. H. J. Buschow and H. J. van Daal, AIP Conf. Proc. 5, 1464 (1971).
15. R. D. Hutchens, V. U. S. Rao, J. E. Greedan, and R. S. Craig, J. Phys. Soc. Japan 32, 451-54 (1972).
16. W. J. Nellis, M. B. Brodsky, H. Montgomery G. P. Pells, Phys. Rev. B2, 4590-96 (1970).
17. F. Y. Fradin, Phys. Rev. Letters 33, 158-61 (1974).
18. M. H. van Maaren, H. J. van Daal, K. H. J. Buschow, and C. J. Schinkel, Solid State Comm. 14, 145-7 (1974).

THE HEAT CAPACITY OF USn_3*

S. D. Bader, G. S. Knapp, and H. V. Culbert
Argonne National Laboratory, Argonne, Illinois 60439

ABSTRACT

The heat capacity of USn_3 was measured between 0.6 and 420 K. An unusually large linear term in the low-temperature heat capacity was observed. The coefficient of this linear term was found to be 171 mJ/°K^2-mole, in excellent agreement with a recently reported calorimetric measurement above 1.5 K. An anomalous leveling off of C/T versus T^2 below 3.5 K reported in an earlier study[1] was not found and, hence, is not an intrinsic property of USn_3. Below 1.5 K, small anomalies were found in the heat capacity of two samples with different impurity levels. At present, no auxiliary physical-property measurements below 4 K are available to clarify the origin of the low-temperature anomalies.

Since $C_p \simeq 3R[1 - 1/20(\theta_\infty/T)^2] + \gamma T$ at high temperatures, a value of the Debye temperature θ_∞ of 196 K was determined from the slope of a $(C_p-3R)/T$ versus T^{-3} plot. The $T^{-3} = 0$ intercept of this plot indicates that at high temperatures the electronic heat-capacity coefficient drops to $\sim$20 mJ/°K^2-mole, a value more typically associated with a conduction electron heat-capacity contribution. The electronic entropy in excess of 20T mJ/°K-mole that is removed at low temperatures is 1.05 Rln2. The accuracy to which the excess electronic entropy is known is quite sensitive to the model used to describe the lattice heat capacity. The one-parameter harmonic representation used can introduce errors of the order of $\pm$25% in the 1.05 Rln2 value. A rapid drop of γ from 171 to $\sim$20 by 80 K indicates the presence of unusual structure in the electronic density of states at the Fermi energy. It is proposed that USn_3 is an interconfigurational-fluctuation system.

1. M. H. van Maaren, H. J. van Daal, and K. H. J. Buschow, Solid State Commun. 14, 145-7 (1974).

---

*Work supported by the U. S. Atomic Energy Commission.

ROLE OF ITINERANT 5f STATES ON THE FERMI SURFACE OF Th AND POSSIBLE IMPURITY LOCAL MOMENT FORMATION*

D. D. Koelling
Argonne National Laboratory, Argonne, Illinois 60439
and
A. J. Freeman
Physics Department, Northwestern University, Evanston, Illinois 60201 and
Argonne National Laboratory, Argonne, Illinois 60439

ABSTRACT

The nature (itinerant vs. localized) of the 5f electrons in the actinides determines their magnetic properties. Extensive optical and de Haas-van Alphen data which exist only for Th, the simplest actinide with no occupied 5f states, allow detailed comparisons to be made between theory and experiment. The 5f bands were found, by means of ab initio RAPW energy band calculations, to lie above the Fermi energy, to be itinerant in character and to significantly affect the reflectivity measurements of Veal, et al. We find that the 5f bands when included in the RAPW calculations, give excellent agreement with the de Haas-van Alphen results of Boyle and Gold, for all sections of the Lungs, Dumbell and Superegg Fermi surfaces, in contrast with calculations based on artificially removing the 5f resonance. Simple overlap charge density impurity scattering potential model calculations are used to discuss the possible occurrence of impurity local moment formation in Th based alloys.

INTRODUCTION

Theoretical studies of the actinide metals have shown that their magnetic and other properties are determined by the nature of their 5f elections--itinerant for the light actinides (Th-Pu) and localized for the heavy actinides (Am and beyond).[1] Unfortunately, the low temperature phases of the light actinides are structurally complex making extremely difficult both the band calculations and interpretations of experimental data. Thorium, at the beginning of the series, with no occupied 5f states, has the fcc structure as its low temperature phase and thus appears to be an ideal candidate for theoretical and experimental investigations. The availability of high purity single crystals has permitted accurate de Haas-van Alphen (dHvA) measurements to be made.

Gupta and Loucks[2] artifically removed the 5f bands from their RAPW calculations, in order to overcoment the difficulties found by Keeton and Loucks[3] in their full Slater exchange (α = 1) calculations, and obtained reasonable agreement with the dHvA measurements. By contrast, Koelling and Freeman[1,4] argued that the 5f electrons in the lighter actinides are sufficiently delocalized to significantly overlap neighboring atoms and to hybridize strongly with the 6d and 7s bands. As in the case of the lighter actinides, the f electrons have small Coulomb correlation energy (relative to the now large effective band width) and hence must be considered itinerant; i.e., are to be included in and described by band calculations unlike the case of the 4f electrons in the rare earths.

Recently normal incidence reflectivity measurements on thorium by Veal, et al.,[5] showed the existence of band-like f states 1-5 eV above E_F. Because of the importance of this result as providing strong confirming evidence for the prediction that the occupied 5f states in the lighter actinide metals are itinerant, a highly precise determination of the FS of Th appears necessary as a severe test of the itinerant model.

We have found that the itinerant model (f states included) gives Fermi surface dimensions in excellent agreement with the dHvA measurements unlike the predictions of the localized (f state removed) model. In addition, we have used the overlapping charge density impurity potential scattering model to calculate for the dilute impurity in Th problem some information about the possible f resonance states on the impurities and to discuss the possibility of impurity local moment formation.

INFLUENCE OF UNOCCUPIED 5f STATES ON THE FERMI SURFACE

The relativistic energy band structure of fcc Th consists of s and d-like states as in any high Z fcc transition metal.[1-4] For the itinerant model, there are in addition f bands located in the range of 1 to 5 eV above E_F. In both models the Fermi Surface (FS) consists of three distinct pieces: (1) a hole surface at the center of the Brillouin zone (BZ) shaped like a rounded cube called "superegg"; (2) electron surfaces on symmetry lines ΓK<110> shaped like pairs of "lungs"; and (3) the hole surfaces on the symmetry lines ΓL<111> shaped like "dumbbells" with triangular ends. In comparing their model with experiment,[6] Gupta and Loucks[2] concluded that their model was "only in qualitative agreement with the experimental results". In many ways, the agreement obtained is rather remarkable but may be understood in terms of the potential insensitivity of the underlying transition-element band structure found by us from comparisons of the Fermi surfaces resulting from four different calculations: (1) the Gupta-Loucks (GL) calculation (d^2s^2 α = 1, f states removed); (2) the calculation for the d^2s^2, α = 2/3 potential with the f states removed; (3) the calculation for the d^2s^2, α = 2/3 potential with the f states included; and (4) the calculation for the d^3s^1, α = 2/3 potential with f states included.

The FS showed a very clear separation between the calculations with the f states included and those with the f states excluded. The two FS resulting from the calculations (3) and (4) with the f states included and with the configuration varied were close to identical. (This tends to give one confidence that these results will not differ greatly from a fully self-consistent calculation.) The two FS for the f states removed calculations give some index of the small but observable sensitivity to exchange approximation of the underlying transition-element band structure. Further, we find that the effect of changing α is to move the f states relative to the remaining bands. Thus the parameter α mainly affects these other bands and their Fermi surface via the f states and their hybridization with these bands.

Having (briefly) described the effects of self-consistency (through configuration) and exchange approximation, we examine the effect on the FS of including or omitting the f states. This effect was found to be larger than that due to change of configuration or exchange. The effect on the hole surfaces was to move volume from the superegg to the dumbbell. The superegg shows somewhat more structure in addition to being smaller. The dumbbell increased in size primarily on the ball portion, the rod section toward L remaining relatively unchanged. The electron surface (the lungs) is not centered on a high-symmetry point but on a line and thus can change its position. The inclusion of the f states moves this surface slightly in the direction of Γ. The tube which connects the two lobes across the Σ line is reduced in size. The lobes themselves are taller and narrower. Qualitatively, these changes are all in the correct direction to improve the agreement between theory and experiment.

However, in order to obtain a more quantitative comparison, one must actually perform the area integrals. The results of these calculations give dHvA areas in excellent agreement with the experiment for all the experimental orbits observed by Boyle and Gold[6] (provided their orbits c and d are interchanged). The effect of including the f-states was to reduce the largest discrepancies on the hole surfaces from over 100% to less than 15%. In addition, comparing the calculated values of m* with the few experimental numbers give reasonable enhancement factors (0.4 to 0.7)--again provided orbits c and d are interchanged. This excellent agreement with experiment adds strong confirmation to the itinerant nature of 5f states in the lighter actinides.

POSSIBLE LOCAL MOMENTS OF DILUTE IMPURITIES IN Th

A number of striking effects are observed if dilute impurities are alloyed with Th metal.[7-10] As an ideal BCS superconductor with a transition temperature of 1.36°K, Th has been used as a host for a number of alloy experiments which focus on the magnetic (pair-breaking) properties of the impurities since their dramatic effect on superconducting properties gives a sensitive test for magnetic moment formation. Thus the experimental results have been compared with Kaiser's[11] theory for the effect of nonmagnetic resonant impurity states on the superconductivity of the hot metal and/or with the Abrikosov-Gorkôv (AG) theory for the pairing breaking effect of a localized magnetic moment. It is found that: Ce behaves as a nonmagnetic impurity and fits the Kaiser theory; Gd, as a magnetic impurity fits the AG theory; and U is an intermediate case which is associated with localized spin fluctuations.

In order to understand the results of these experiments, we have examined the overlapping charge density model for a single impurity in a Th host. Within this model (which is not necessarily self-consistent), we can then hope to see the nature of any f-resonances which might form, and relate them to the possibility of local moment formation. Thus we superpose (overlap) the impurity charge density and the host atom charge densities, construct a muffin tin (MT) potential and solve the radial Dirac equation to determine the (logarithmic derivatives and the) scattered-wave phase shifts. To define the quantities of interest we note that this procedure applied to a pure metal (and excluding hybridization) would yield ε^j_{min} and ε^j_{max} (relative to E_F) for which the maximum of the wave function and its zero (which would be at infinity for the free atom) occur at the MT radius. These would be upper and lower bounds for the states with a given j-character and so give some idea of the energy range over which f resonances will have an appreciable effect. The other two parameter ε^j_r and Γ^j are obtained from the expansion in energy

$$\cot \delta_j = \frac{\varepsilon^j_r - \varepsilon}{\Gamma^j} + \ldots \qquad (1)$$

of the cotangent of the phase shift near the energy for which it is zero. This expansion has been used extensively in deriving combined basis representations from the Green's function formalism;[12] ε^j_r is the resonant energy (center of the band for a collection of such potentials) and Γ^j is the Born transition probability. [To establish the scale of Γ, we note that $\Gamma = 0.044$ Ry for the d-states in bcc Fe.] Clearly it will be the difference of the impurity values from the Th values which are significant since if the impurity scattered like Th it would have no effect on the electron states.

Our results for Ce, Gd and U are given in Table I, for j = 5/2 and 7/2 of each impurity. The results for Gd show quite clearly that since the transition probability Γ is small and the resonance lies well below the E_F one will obtain a local moment on the impurity site. (Because of the large Coulomb correlation energy, U_{eff}, an occupied f^8 configuration would move far above the Fermi energy; thus these one-electron energies must be used with care when the states are beginning to localize.) The results show quite clearly that Ce will not support any f states but that one will get considerable scattering. Since the "clear-cut" cases of Gd and Ce are seen to fit well into our model, we now consider the case of U.

It is interesting to note that U ($f^3d^2s^1$) yields nearly the same scattering parameters as Th. Thus, if U were to remain in this configuration, it would have very little effect on the Th host. However, as this would require the occupied f-states to remain in their energy states well above the Fermi energy, this cannot be the ground state of the system. If, instead, we look at U ($f^2d^3s^1$), the f levels are lowered by 0.16 Ry (which gives an estimate of the Coulomb correlation energy) and can be occupied. Here we notice that the widths $\Delta = (\varepsilon^j_{max} - \varepsilon^j_{min})$ and/or transition probabilities Γ^j are reduced to 40% of their U ($f^3d^2s^1$) values because the atomic f states have contracted by roughly 10%. Hence, for the case of U impurities, the f states are located at E_F as required by the specific heat data[10] and self-consistency requirements, are much more localized than with the trivalent configuration, and strongly correlated ($U_{eff} \approx \Delta$).

Table I. f-state parameters (in Ry) for impurities in Th

Impurity	$\varepsilon^{5/2}_{min}$	$\varepsilon^{5/2}_{max}$	$\varepsilon^{5/2}_{r}$	$\Gamma^{5/2}$	$\varepsilon^{7/2}_{min}$	$\varepsilon^{7/2}_{max}$	$\varepsilon^{7/2}_{r}$	$\Gamma^{7/2}$
Th(d^3s^1)	0.06	0.30	0.19	0.031	0.09	0.37	0.24	0.038
Gd($f^7d^2s^1$)	-0.13	-0.09	-0.10	0.001	-0.07	-0.04	-0.06	0.002
Ce($f^1d^2s^1$)	0.11	0.20	0.15	0.009	0.12	0.22	0.18	0.012
Ce(d^3s^1)	0.10	0.20	0.15	0.011	0.12	0.22	0.17	0.012
U($f^3d^2s^1$)	0.10	0.30	0.20	0.025	0.15	0.36	0.26	0.032
U($f^2d^3s^1$)	-0.06	0.09	0.02	0.010	-0.01	0.16	0.08	0.015

REFERENCES

*Supported by the AEC, the NSF and the AFOSR.

1. See the review article by A. J. Freeman and D. D. Koelling, in the Actinides: Electronic Structure and Properties, A. J. Freeman and J. B. Darby, Jr., Eds. (Academic Press, N.Y., 1974), Vol. I, p. 51.
2. R. P. Gupta and T. L. Loucks, Phys. Rev. Letters 22, 458 (1969).
3. S. C. Keeton and T. L. Loucks, Phys. Rev. 146, 429 (1966).
4. D. D. Koelling and A. J. Freeman, Solid State Commun. 9, 1369 (1971).
5. B. W. Veal, D. D. Koelling, and A. J. Freeman, Phys. Rev. Letters 30, 1061 (1973).
6. D. J. Boyle and A. V. Gold, Phys. Rev. Letters 22, 461 (1969).
7. W. R. Decker and D. K. Finnemore, Phys. Rev. 172, 430 (1968).
8. J. G. Huber and M. B. Maple, J. Low Temp. Phys. 3, 537 (1970).
9. M. B. Maple, J. G. Huber, B. R. Coles, and A. C. Lawson, J. Low Temp. Phys. 3, 137 (1970); H. L. Watson, D. T. Peterson, and D. K. Finnemore, Low Temperature Physics-LT 13, K. D. Timmerhaus, W. J. O'Sullivan, and E. F. Hammel, Eds. (Plenum Press, N.Y., 1974), Vol. II, p. 590.
10. C. A. Luengo, J. M. Cotignola, J. Sereni, A. R. Sweedler, and M. B. Maple, Low Temperature Physics-LT 13, K. D. Timmerhause, W. J. O'Sullivan, and E. F. Hammel, Eds. (Academic Press, N.Y., 1974), Vol. II, p. 585.
11. A. B. Kaiser, J. Phys. C 3, 409 (1970).
12. D. G. Pettifor, J. Phys. C 3, 367 (1970).

PRESSURE EFFECTS ON THE BAND STRUCTURE OF GADOLINIUM

M I Darby and N Richardson
Department of Pure and Applied Physics,
University of Salford, Salford M5 4WT, England

ABSTRACT

APW calculations of the band structure of gadolinium have been made to investigate the effects of decrease in lattice parameters coupled with an increase in c/a ratio. It is found that compared with normal pressure results decreasing the lattice constants lowers the bottom of the band, and increasing c/a modifies the detailed band structure, particularly near Γ. The changes in the density of states have been computed and it is found that the Fermi energy is now much closer to the levels near K. It is estimated that the density of states at the Fermi energy, and magnetic properties depending on it, are reduced by 15% from normal.

INTRODUCTION

Several experiments have been performed to determine the effects of hydrostatic pressure on the properties of gadolinium. It is found that increasing pressure lowers the Curie temperature[1-3], and at 25 kbar there occurs a structural transition from the hcp to Sm-type crystal structure[3-4], above which pressure the magnetic order is not simple, probably being antiferromagnetic[3]. Since the exchange interaction is an indirect one the changes in magnetic properties are assumed to be associated with those in the electronic properties.

The band structure of Gd under normal conditions is well known[5-6]. It would be expected that a decrease in lattice spacing produced by compression would lower the bottom of the conduction band and generally increase the separations of the energy levels[7-8]. On the other hand increasing pressure has the effect of increasing the ratio of the lattice constants, c/a, towards the ideal value, which might be expected to favour higher degeneracies at symmetry points. It is therefore of interest to investigate what effect a combination of these factors, and the latter in particular, has upon the band structure and related properties of gadolinium.

THE BAND STRUCTURE

The non-relativistic form of the augmented plane wave method has been employed with a muffin tin potential constructed in a standard way[9], including the Slater exchange term, from the wavefunctions of Herman and Skillman[10] for the configuration $4f^7 5d^1 6s^2$. Differences in the band structure obtained using the relativistic method are small for gadolinium[6]. The lattice constants of the hcp structure under normal pressure have been taken as a = 6.871 (a.u.), c = 10.930 (a.u.), giving c/a = 1.59, and the computed bands are illustrated in Fig.1 for the symmetry direction ΓK of the Brillouin zone. While these are similar to previous results[5] it is necessary to reproduce them to compare with the high pressure case in order to avoid differences arising from the choice of potential.

Conflicting values have appeared in the literature[3,11] for the c/a ratio of the hcp phase at high pressure, and earlier calculations[12] have assumed that the ratio is little altered by pressure. The temperature variation of the lattice constants at normal pressure[13] indicates that as the value of a decreases, corresponding to a decrease in temperature, the ratio c/a increases.

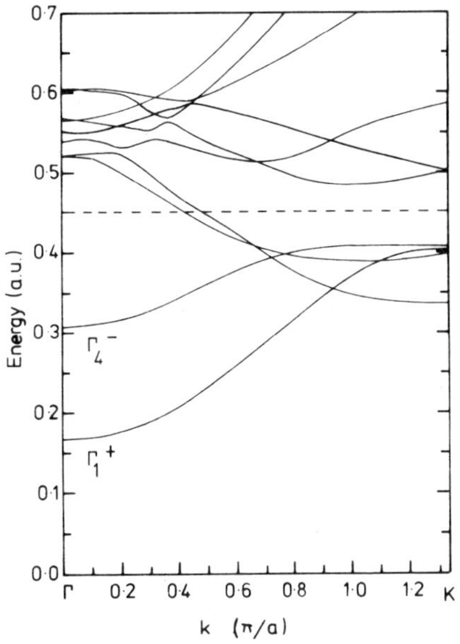

Figure 1. Band Structure of gadolinium in the direction ΓK. The broken line is the Fermi energy.

It is assumed here that c/a behaves in the same way when a is decreased by pressure. To illustrate the effects upon the band structure of changes in the ratio c/a, results have been obtained for the ideal value 1.63. The lattice parameters were estimated by extrapolating the data for the temperature variation of c/a as a function of a, and were taken to be a = 6.708(a.u) and c = 10.930(a.u). The qualitative features of the band structure are independent of the exact values chosen. Results for the ΓK direction are shown in Fig.2.

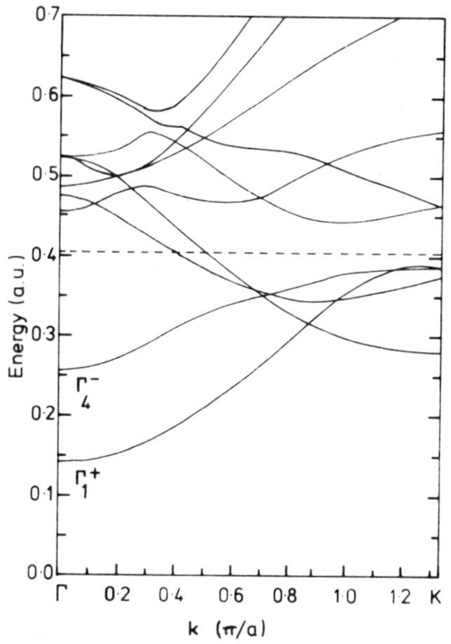

Figure 2. Band Structure at high pressure. The broken line is the Fermi energy.

DISCUSSION

Comparison of the bands for the compressed lattice employing the higher c/a ratio with those of Fig.1 shows that the bottom of the band Γ_1^+ is shifted downwards, the next higher band Γ_4^- is shifted more so, and the detailed structure of the levels is modified, particularly near Γ at energies ~0.5 (a.u.). In the latter region there is evidence of increased degeneracies.

The density of states has been computed employing 105 points in 1/24th of the Brillouin zone. Compared with the normal pressure results, which are similar to those of Dimmock and Freeman[5], at high pressure there are more states at lower energies and the heights of the peaks are reduced. One consequence is that the flat, d-like bands near the point K are now much closer to the Fermi energy, being 0.02 (a.u.) below, compared with the normal 0.04 (a.u.). However the Fermi surface is almost unaltered. The density of states at the Fermi energy is found to be 15% lower than the normal pressure value, and hence there will be a corresponding reduction in the electronic contributions to the saturation magnetization and specific heat, and a lowering of the Curie temperature[14]. Assuming that the results are applicable at 25 kbar and that there is a linear variation with pressure, the change in the polarization contribution to the moment per ion is $-3.3 \times 10^{-3}\ \mu_B\ \text{kbar}^{-1}$, which is in approximate agreement with experimental[2] and other calculated values[12], as can be seen from Table I.

Table I. Pressure dependence of the electronic contribution to the moment per ion μ_c. At normal pressure $\mu_c = 0.55\ \mu_B$. The assumed values of the lattice constants at high pressures (20-25 kbar) are also shown.

	a (a.u.)	c (a.u.)	$d\mu_c/dP$ (x $10^3\mu_B$ kbar^{-1})
Present	6.708	10.930	-3.3
Ref.12	6.745	10.737	-3.2
Experiment[2]			-2.2

CONCLUSION

While the detailed band structure is sensitive to the values of lattice parameters and their ratio c/a chosen to represent gadolinium under pressure, the density of states at the Fermi surface, and related magnetic properties, is less sensitive, and is found to be reduced by 15% from the normal pressure value.

REFERENCES

1. P R Patrick, Phys.Rev. 93, 384-392, 1954.
2. D Bloch and R Pauthenet, in Proceedings of the International Conference on Magnetism, Nottingham, 1964 (The Institute of Physics and The Physical Society, London, 1965) p 255-259.
3. D B McWhan and A L Stevens, Phys.Rev. 139, A682-689, 1965.
4. A Jayaraman and R C Sherwood, Phys.Rev.Letters 11, 22-23, 1964.
5. J O Dimmock and A J Freeman, Phys.Rev.Letters 13, 750-752, 1964.
6. S C Keeton and T L Loucks, Phys.Rev. 168, 672-678, 1968.
7. J M Ziman, Proc.Phys.Soc. 91, 701, 1967.
8. N D Lang and H Ehrenreich, Phys.Rev. 168, 605-622, 1968.
9. T L Loucks, Augmented Plane Wave Method, 1967 (W A Benjamin Inc.,New York).
10. F Herman and S Skillman, Atomic Structure Calculations, 1963 (Prentice Hall Inc., New Jersey).
11. D B McWhan and A L Stevens, Phys.Rev. 154, 438-445, 1967.
12. G S Fleming and S H Liu, Phys.Rev. B2, 164-166, 1970.
13. W B Pearson, Handbook of Lattice Spacings and Structure of Metals and Alloys, Vol.1, 1964 (Pergamon Press, New York), p 673.
14. S H Liu, Phys.Rev. 127, 1889, 1962.

ELECTRONIC STRUCTURE OF MnBi

W. Stutius, R. M. White, and Tu Chen
Xerox Palo Alto Research Center, Palo Alto, California 94304

and

G. R. Stewart*
W. W. Hansen Laboratory of Physics, Stanford University, Stanford, California 94305

ABSTRACT

Heat capacity and electrical conductivity measurements have been carried out on single crystals of low temperature phase (LTP) MnBi and high temperature phase (HTP) $Mn_{1.08}Bi$ single crystals. These measurements are combined with previous measurements to argue that MnBi is a defect semiconductor.

I. INTRODUCTION

MnBi is a room temperature ferromagnet which has been of great practical interest for many years. This interest stems from the fact that MnBi has a very large uniaxial anisotropy constant as well as a reasonably large magnetooptical coefficient. The large anisotropy leads to a large coercivity in fine particles and gives MnBi a high figure of merit as a permanent magnet. The large magnetooptic coefficient makes MnBi attractive as a recording medium for a beam addressable optical memory. The purpose of this paper is to present measurements of the heat capacity and resistivity on single crystals of MnBi which, together with previous data on other properties, enable us to suggest a model for the electronic structure of this material.

MnBi has the NiAs structure and exhibits a first order phase transition at 628 K. This corresponds to a phase separation identified as[1] MnBi $\rightarrow$ $Mn_{1.08}Bi$ + Bi. Although this transition is not a Curie point in the usual sense, it is accompanied by a discontinuous loss of magnetic order. Prior to the identification of this phase separation the terms "low-temperature" phase (LTP) and "high-temperature" phase (HTP) were used to characterize these two phases. We shall continue to employ this description.

II. TRANSPORT

A. Electrical Resistivity

The electrical resistivity of single crystals of low temperature phase (LTP) MnBi and quenched high temperature phase (QHTP) $Mn_{1.08}Bi$ has been measured between 4.2 K and the decomposition temperature of the high temperature phase at 720 K. The measurements were performed in zero external magnetic field on samples with linear dimensions of about 2x1x7 mm. A constant rms ac-current of low frequency (between 20 and 75 Hz) was applied and the voltage drop across the sample was monitored as a function of temperature. The samples selected had no visible bismuth inclusions, cracks or other inhomogeneities and therefore the absolute accuracy of the measurements (±10%) is mainly given by the accuracy with which the spacing of the voltage probes could be determined. Pressure contacts were used for the current and the voltage leads.

Figure 1 shows the resistivity of LTP MnBi as a function of temperature. The room temperature value is 55 μΩ-cm. At the transition at 628 K the resistivity increases about 60% and is essentially constant up to 720 K. This constant resistivity is characteristic for the QHTP $Mn_{1.08}Bi$ in its paramagnetic state. The change of resistivity at the spin-flip transition at around 90 K is very small and agrees with previously published results.[2]

In all the LTP MnBi samples investigated the resistivity can be described by $\rho = \rho_0 + A \cdot T^2$ over a large temperature range between 5 K and 30 K. The coefficient A was found to have values between 0.9×10^{-9} and 1.3×10^{-9} Ω-cm. ρ_0 is rather different from sample to sample, varying between 0.05 and 0.5 μΩ-cm, but does not seem to be related to the magnitude of A.

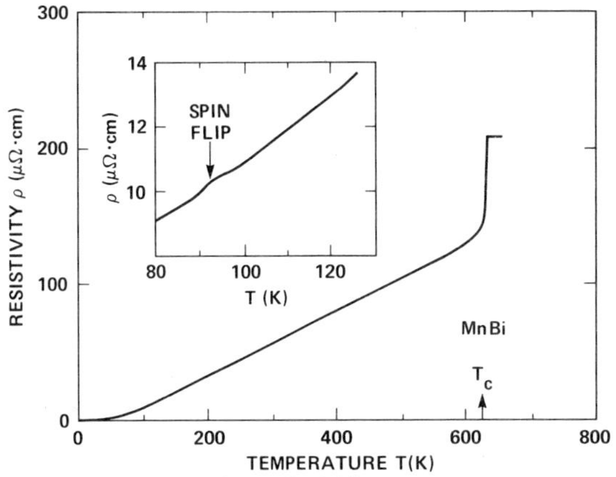

Fig. 1 Resistivity of MnBi vs. temperature.

Such a T^2-law is characteristic of transition metals and is generally attributed to electron-electron, or Baber scattering. The coefficient A is proportional to the <u>difference</u> in the effective masses of these electrons. The relatively large (10^{-9} Ω-cm-deg^{-2}) coefficient of T^2 compared to transition metals (10^{-11} Ω-cm-deg^{-2}) would therefore require the coexistence of two very different Fermi surfaces for which there is no strong supporting evidence.

Both magnon and paramagnon scattering can also give rise to a resistivity that varies as T^2. However, since we are far below the estimated Curie temperature (~700 K) we feel these mechanisms will not be important. At the end of Section IIIA we suggest an alternative mechanism for such a T^2 dependence.

B. Hall Effect and Thermoelectric Power

Hall effect measurements on oriented LTP MnBi films have been reported by Chen et al.[3] The ordinary Hall constant R_0 is positive and, if one assumes a simple one-band model, it leads to a hole density of 10^{21} cm^{-3} which leads to mean-free-path of 200 Å at room temperature. If the d-electrons also contribute to the Fermi surface then the interpretation of a Hall measurement is very complicated. However, it is interesting to note that using the density of holes given above to calculate the electronic contribution to the specific heat gives a value for γ of 1.3 millijoules/(mole-deg^2) which is consistent with the small measured value (see below).

We have measured the thermopower of a single crystal of LTP MnBi at room temperature with its c-axis oriented perpendicular to the temperature gradient. The thermoelectric voltage was determined with respect to a copper reference. The absolute thermopower is negative and about -20 μV/deg (±20%).

This negative value is not consistent with a simple parabolic hole band. However, the thermopower depends strongly on the details of the Fermi surface and is, even when its temperature dependence is known, a cumbersome electronic transport quantity to interpret.

III. BULK PROPERTIES

A. Magnetic

The magnetic properties of MnBi have been well documented[1] and we shall merely recall those aspects relevant to our band model. One of the most important magnetic parameters is the saturation moment. Both neutron scattering[4] and magnetometer measurements[1] show that the moment at low temperatures is approximately 3.8 μ_B. The magnetometer measurement is clearly a volume average. Neutron scattering on the other hand, provides information on the atomic moments. However, this is not unambiguous, but depends upon how well one can fit a magnetic form factor and an assumed moment arrangement to the scattering intensities. In MnBi, for example, the moment of 3.8 μ_B could arise from a ferrimagnetic arrangment in which 2% of the manganese ions are in interstitial sites with the divalent ^6A ground state while the major contribution is from the Mn^{3+} ions with a ^5E ground state.

Another important magnetic aspect of MnBi and $Mn_{1.08}Bi$ are their magnetic anisotropies. The anisotropy constant of MnBi is negative at low temperatures, increasing as T^2 in the region below 30 K. It reaches zero at 90 K and continues to a large positive value at about 500 K. This change in sign at 90 K is accompanied by a rotation of the magnetic moments from a direction in the hexagonal plane at low temperatures to a direction parallel to the c-axis at high temperatures.

The anisotropy constant of the QHTP at low temperatures is twice as large as the largest value for the LTP, but decreases with increasing temperature.

One of the major origins of magnetic anisotropy is the spin-orbit interaction, $\lambda\vec{\ell}\cdot\vec{s}$. The spin-orbit coupling associated with the 6p electron of Bi^{2+} is 13,859 cm^{-1} compared with 542 cm^{-1} for the 3d electron on Mn^{6+}. Thus a large overlap of the bismuth and manganese wavefunctions would enable the manganese to "utilize" this large spin-orbit interaction. We shall see below that the nuclear specific heat does, in fact, suggest a large overlap of the bismuth and manganese wavefunctions. The fact that in the QHTP about 15% of the Mn ions reside in interstitial sites[4] suggests that these may be more effective in "transmitting" the spin-orbit coupling. Therefore if we assume that the temperature dependence of the anisotropy in the LTP is connected with the presence of interstitials then these interstitials will also provide temperature-dependent impurity-like scattering centers for the holes. This will give rise to a resistivity with the same temperature dependence of the anisotropy, namely T^2.

B. Specific Heat

The heat capacity of LTP MnBi and QHTP $Mn_{1.08}Bi$ was measured in the temperature range between 1.2 K and 10 K using a calorimeter described elsewhere.[6] The method is based upon determining the thermal-relaxation time constant $\tau_1 = C/k$, where C is the heat capacity of the sample and k the thermal conductance linking the sample to the heat reservoir. The results are shown in Fig. 2.

There is a great deal of information contained in such data. This information is generally extracted by fitting the data to a function of the form $C_p = A/T^2 + \gamma T + \beta T^3 + C_{magnon}$. The term varying as T^{-2} is the first term in a high-temperature expansion of the Schottky anomaly associated with the nuclear degrees of freedom. This contains contributions from both the magnetic hyperfine interaction as well as the electric quadrapole interaction. We have made estimates of the electric field gradients at the Bi and Mn nuclei and found that they are small and may be neglected. The hyperfine contribution to the coefficient A is given by $(\mu_n H_{eff}/k_B)^2(I+1)/3I$. Since both Mn^{55} and Bi^{209} possess nuclear moments they contribute additively to the nuclear specific heat. The hyperfine field at the Mn nucleus in the LTP has been measured by nuclear resonance[2] and was found to be -224.5 kOe at T = 0 K. From our specific heat data we therefore obtain a hyperfine field at the Bi nucleus of 1.8 MOe in the LTP and 1.2 MOe in the QHTP. These large fields are consistent with the trend as one goes from MnAs to MnSb. This field arises from the unpairing of the bismuth 6-s shell produced by the manganese d-electrons. The fact that this effect is large implies a large 3d-6s overlap.

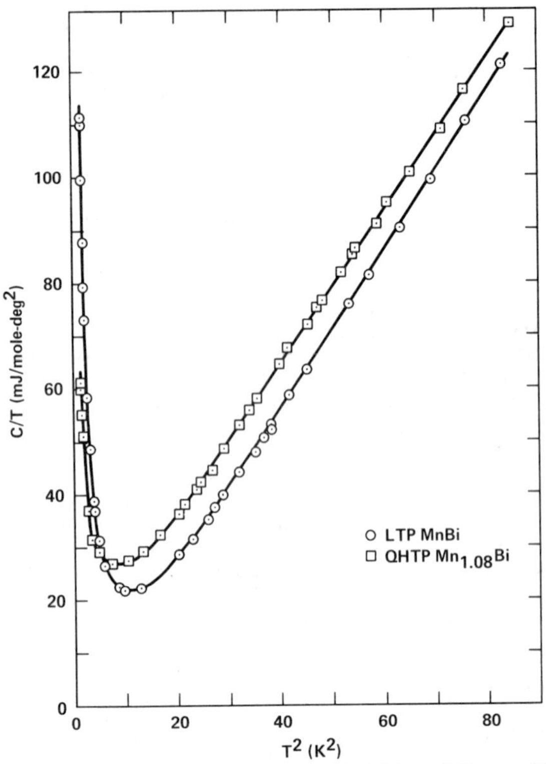

Fig. 2 Specific heat, C, of MnBi and $Mn_{1.08}Bi$.

The second term in the expression for C_p is the electronic contribution. From Fig. 2 we see that the value of γ for the LTP is very small which is consistent with the Hall measurement assuming a simple one-band model. There have been suggestions that the NiAs-type materials may involve itinerant d-electrons. This small value of γ presents a challenge to such a model. The value γ = 7 mJ/mole-K^2 for the QHTP is much larger. This could be due to the fact that the additional Mn ions in the interstitial sites are now providing enough direct Mn-Mn overlap to produce a d-band.

The third term is the phonon contribution. From the slope β we find that the Debye temperatures of the LTP and the QHTP are approximately the same, 145 K.

Finally, the last term in C_p is the magnon contribution. Using the dispersion relations appropriate for easy-plane and easy axis anisotropies with an effective spinwave stiffness of 55 K we find that this contribution is small in this temperature region.

IV. OPTICAL PROPERTIES

The optical properties of MnBi are not particularly illuminating due to their rather featureless nature. The absorption, reflectivity and Faraday and Kerr rotations have been measured[5] in the visible. From 2.5 to about 3.0 eV these quantities are what one would expect from Drude-like intraband transitions. There is an increase in the absorption at 3.4 eV which could be the onset of the Bi 6p-to-Mn 4s interband, charge-transfer transition.

V. CONCLUSION

In discussing each of the physical properties of MnBi we have tried to adopt a simple model in which the manganese states are localized. It appears that this model is not inconsistent with any of the data presented above. Thus we suggest that the conductivity in MnBi arises from a small pocket of holes at the top of the Bi p-band. If each Mn ion were trivalent there would be no such holes. However, we suggest that 2 to 5% of the Mn ions move statistically into the interstitial sites in the Ni-As structure where they assume the divalent configuration. Thus MnBi is, in fact, a defect semiconductor with the Fermi level lying 0.3 eV below the top of the Bi p-band which is 3.5 eV below the bottom of the Mn 4s-band. It is still an interesting question as to why MnBi prefers to create interstitials rather than fill the Bi band or for that matter, why the HTP accepts an additional 8% of Mn into interstitials.

ACKNOWLEDGEMENTS

The authors gratefully acknowledge helpful discussions with Prof. T. Geballe and Dr. J. Allen.

* Supported by the Air Force Office of Scientific Research, Air Force Systems Command, USAF, under Grant No. AFOSR 73-2435A.

REFERENCES

1. Tu Chen and W. Stutius, IEEE Trans. MAG-10, 581 (1974).
2. T. Hihara and Y. Koi, J. Phys. Soc. Japan 29, 343 (1970).
3. D. Chen, Y. Gondo, and M. D. Blue, J. Appl. Phys. 36, 1261 (1965).
4. A. Andresen, W. Hälg, P. Fischer, and E. Stoll, Acta Chem. Scand. 21, 1543 (1967).
5. D. Chen, J. Ready, and E. Bernal, J. Appl. Phys. 39, 3916 (1968).
6. R. Bachmann et al., Rev. Sci. Inst. 43, 205 (1972).

NEUTRON SCATTERING DETERMINATION OF THE CRYSTAL FIELD SPLITTINGS IN TmN*

H. L. Davis and H. A. Mook
Solid State Division, Oak Ridge National Laboratory
Oak Ridge, Tennessee 37830

ABSTRACT

Inelastic neutron scattering experiments have been performed at 4.2, 78, and 293°K on the paramagnetic compound TmN by utilizing the ORNL magnetically-pulsed time-of-flight spectrometer. After suitable theoretical analysis, the data from these experiments has provided a complete specification of the octahedral crystal field present at the $Tm^{3+}(4f^{12})$ sites in TmN. Values obtained for the crystal field parameters at 78°K are $A_4\langle r^4\rangle = 25.9 \pm 0.8$ meV and $A_6\langle r^6\rangle = 1.08 \pm 0.16$ meV. Although these parameters specify a crystal field producing the same relative ordering of the energy levels, Γ_i, in TmN as in previously studied TmAs, TmSb, and TmBi, the quantitative values of the parameters indicate a major difference between the crystal field of TmN and the crystal fields of the other three thulium compounds. This difference cannot be completely accounted for by the variation in lattice parameter between the compounds; thus, it must be due to differences in the electronic structures possessed by the compounds. These differences in electronic structure will be discussed, and indications given as how their consideration might provide clues to the microscopic origins of the crystal fields in these and other NaCl-structured rare-earth compounds.

*Research sponsored by the U. S. Atomic Energy Commission under contract with the Union Carbide Corporation.

X-RAY PHOTOEMISSION STUDY OF THE DENSITY OF STATES OF THE TRANSITION METALS

L. LEY, S. P. KOWALCZYK, F. R. MCFEELY and D. A. Shirley,
Department of Chemistry and Lawrence Berkeley Laboratory,
University of California, Berkeley, California 94720.

The entire 3d transition metal series was studied by x-ray photoemission spectroscopy (XPS) under ultra high vacuum conditions. The total valence band density of states (DOS) was obtained and properties such as the d bandwidth, total bandwidth and the relative density of states at the Fermi level E_F were derived. These properties are considered important for understanding the magnetic properties of these metals. Also from detailed comparisons with theoretical band structure calculations, the energies of certain symmetry points were assigned.
Hcp Sc and Ti and bcc V and Cr illustrates the effects of rigidity on the DOS. In Cr, a low DOS at E_F was observed for the first time by a spectroscopic method. The experimental DOS of Fe was compared in detail with the expectations of the itinerant band model of ferromagnetism. In particular the observed features of a low DOS at E_F and a peak just below E_F in bcc Fe was shown to be obtainable from two sets of bcc Cr bands shifted by the appropriate exchange splitting. Likewise Co and Ni were found to be in good agreement with predictions of the itinerant band model.

MAGNON-PHONON COUPLING IN METAMAGNETIC SYSTEMS: EVIDENCE FOR ONE PHONON-TWO MAGNON INTERACTIONS

R. S. Silberglitt, National Science Foundation, Washington, D.C. 20550, K. L. Ngai, Naval Research Laboratory, Washington, D.C. 20375, and E. N. Economou and J. Ruvalds, University of Virginia, Charlottesville, Virginia 22901

ABSTRACT

Spin-phonon interactions in the linear chain compounds $FeCl_2 \cdot 2H_2O$, $CoCl_2 \cdot 2H_2O$, and $CoBr_2 \cdot 2H_2O$ are considered. It is shown that the recent data of Torrance and Hay on $CoBr_2 \cdot 2H_2O$ provides strong evidence for the existence of coupling terms involving one phonon and two spin operators. A semi-quantitative microscopic discussion is given of the spin-phonon Hamiltonian, within which existing experimental data on all three systems is correlated. It is shown that a consistent picture of the spin-phonon coupling in these materials can be generated, and additional experiments are suggested to further verify this picture and further probe the spin-phonon interaction.

INTRODUCTION

The metamagnetic systems $FeCl_2 \cdot 2H_2O$ (FC2), $CoCl_2 \cdot 2H_2O$ (CC2), and $CoBr_2 \cdot 2H_2O$ (CB2) have been extensively studied[1]. These systems have proven to be of great interest because of the wealth of physical effects which have been observed in the magnetic field and temperature dependence of their excitation spectra. Both the observation of magnon bound states and the coupling of the magnons to a phonon mode corresponding to vibrations of the waters of hydration has been made in FC2 and CC2[1]. Excellent agreement has been obtained between the results of these experiments and theory based on either Heisenberg or Ising Hamiltonians for the spin system and a phenomenological spin-phonon interaction[1]. A microscopic derivation of the spin-phonon coupling has been made for FC2 by Torrance and Slonczewski[2], and this is also in excellent agreement with experiment.

Recently, Torrance and Hay have reported observations of the far infrared absorption spectrum of CB2 which provide strong evidence for the direct observation of a magnon-phonon pair[3], and have argued that this combined magnon-phonon excitation may be in a bound state, with a binding energy of about $2cm^{-1}$. Their data is shown in Fig.1. They also suggested a mechanism for the excitation of this combined mode, utilizing the same bilinear spin-phonon interaction term responsible for the hybridization in FC2 and CC2. However, in an attempt to explain the binding energy, Ngai, Ruvalds and Economou[4] invoked higher order terms, in particular, scattering of the magnons by a phonon. It is the purpose of this paper to show that a consistent picture of the data on FC2, CC2, and CB2 can be generated and that the experimental data strongly suggest the presence of spin-phonon coupling terms involving one phonon and two magnon operators. We will also discuss the microscopic calculation of the spin-phonon Hamiltonian, describe the symmetry requirements on the types of terms allowed, and use these results to correlate the existing experimental observations. Finally, we suggest additional experiments to verify the picture and further probe the spin-phonon interaction in these materials.

DISCUSSION OF DATA

The far infrared absorption spectrum of CB2 shown in Fig.1 has been analyzed[3] in a similar manner to that of FC2 and CC2, with excellent agreement obtained for the single magnon lines labelled a_1, b_1, c_1, d_1, and e_1, and the two and three magnon bound states e_2, and e_3. In contrast to the other systems, there is no phonon observed to hybridize with the magnons, and there is observed a line with a g-factor corresponding to a single magnon but a much higher energy. A broad field-independent line (not shown) is also observed at 40-43 cm^{-1}. Torrance and Hay[3] have fitted this higher energy data with a magnon excited at $k = \pi/b$ and a field independent excitation (presumably a phonon) with $\omega = 18cm^{-1}$. The assumption that the 40-43 cm^{-1} line

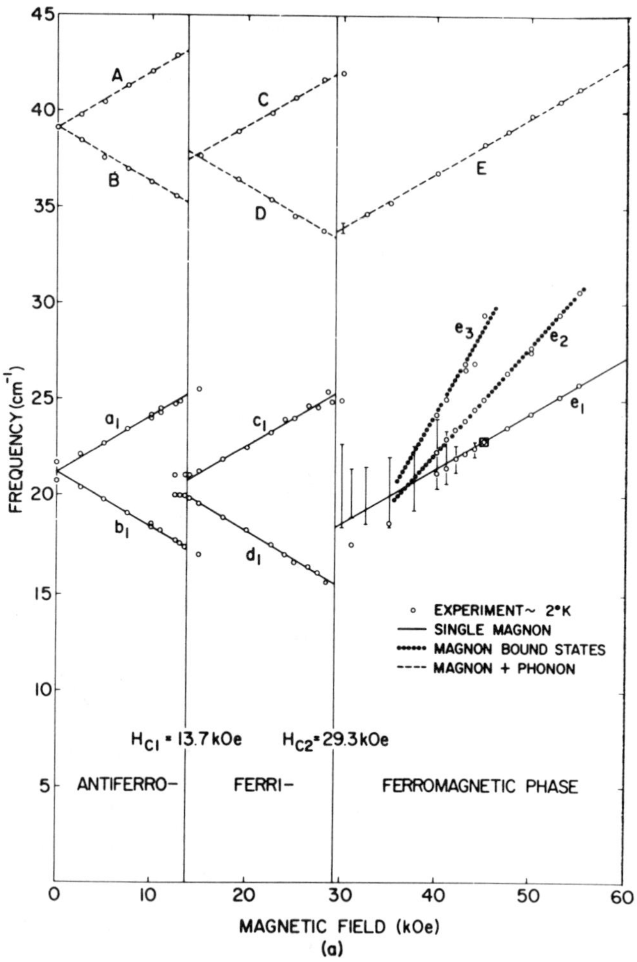

Fig. 1. The far infrared absorption spectrum, by Torrance and Hay, of $CoBr_2 \cdot 2H_2O$, after Ref.3.

corresponds to the excitation of two phonons yield a binding energy of about $2cm^{-1}$ for this magnon-phonon pair.

In order to interpret these results, we must consider the form of the spin-phonon coupling. Fig. 2 shows the lowest order magnon-phonon interaction terms.

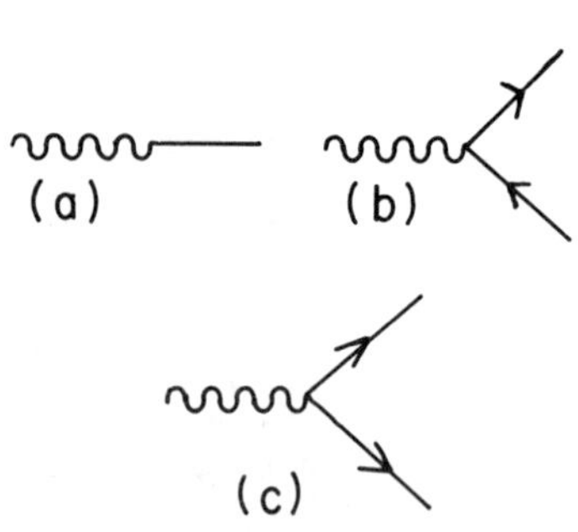

Fig. 2. Lowest order magnon-phonon scattering process Wavy lines represent phonons and solid lines represent magnons.

2a leads to the linear hybridization observed in FC2 and CC2 and 2b has been shown[4] to provide an attractive interaction between magnon-phonon pairs. We will discuss 2c later. The bilinear term of Fig 2a is suggested in Ref.3 as a mechanism for the excitation of the magnon-phonon pair. However, the following argument rules out the presence of this term for the phonon under consideration: if this term were present, then the two-phonon line at 40-43 cm^{-1} would hybridize with the magnon-phonon line itself. No such hybridization is observed. Moreover, as discussed in Ref.4, in order to generate the bound magnon-phonon state, the one phonon-two magnon term of Fig.2b is required.

A possible explanation of the mechanism for the excitation of the pair mode is also supplied by the presence of this type of interaction, as shown in Fig. 3. Fig. 3a corresponds to the usual excitation of a

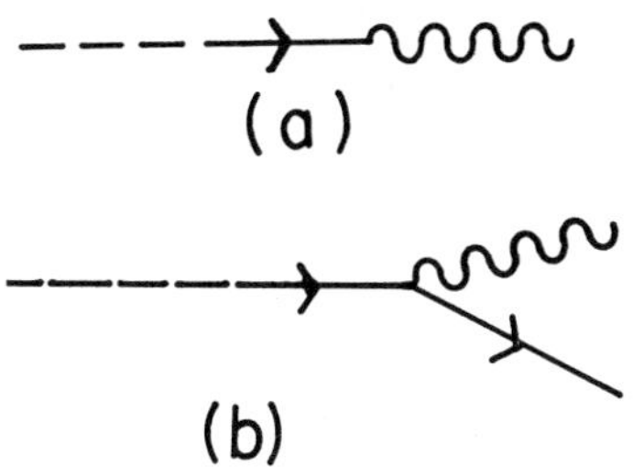

Fig. 3. Mechanisms for excitation of non-infrared active modes by mixing with the single magnon state. Dashed line represents a photon. 3a describes excitation of a phonon, while 3b describes excitation of a magnon-phonon pair.

phonon, and 3b to the excitation of a magnon-phonon pair. We note that for a binding energy of $2cm^{-1}$, as observed, the interaction strength for the diagram of Fig.3b must be of the order of 6-7 $cm^{-1}(g^2/[2\omega_M-\omega_{PH}] \sim 2cm^{-1})$. This predicts that the intensity of the magnon-phonon line should be about 10 percent of the intensity of the magnon mode itself, which is in qualitative agreement with experiment. Thus the far infrared absorption data on CB2 cannot be explained with the bilinear term, and a consistent phenomenological explanation can be generated with the one phonon-two magnon process of Fig. 2b.

SPIN-PHONON INTERACTION

In this section we outline qualitatively the microscopic calculation of the spin-phonon interaction for the case of CC2 or CB2. The results of quantitative calculations will be given in a fuller paper. This derivation is similar to the earlier work of Allen and Guggenheim[5] on CoF_2, Torrance and Slonczewski[2] on FC2, and Lovesey[6] on FeF_2. We wish here to emphasize the presence of one phonon-two spin operator terms, and the symmetry requirements on these terms. The Co^{++} case differs from the Fe^{++} case in that the latter has weak spin-orbit coupling and the lowest multiplet remains S=2[1].

In order to obtain an effective magnon-phonon Hamiltonian, we must consider the lowest energy levels of the Co^{++} ion in the presence of the crystal field of the CB2 (CC2) lattice. The lowest free ion level is a ^4F state, and under the influence of a cubic crystal field this is split into Γ_2, Γ_5, and Γ_4 levels with the Γ_4 triplet lying lowest[7]. The rhombic component of the crystal field removes the degeneracy of these orbital levels, leaving only the four-fold spin degeneracy. Thus we must treat the three Γ_4 orbital states, each multiplied by an S= 3/2 spin function. The mechanism for the spin-phonon coupling is as follows: (a) The phonon modulates the Coulomb potential between the f-electrons and the waters of hydration. (b) The potential gradient has non-zero matrix elements within the Γ_4 manifold. (c) the Γ_4 states are also mixed by the spin-orbit interaction, so that the resulting lowest lying levels can be written in terms of a combined spin and phonon Hamiltonian. If we use a point charge approximation for the Coulomb potential, the gradient takes the form:

$$\Delta V_u \sim \frac{Z'e^2}{R^4} \quad f(x,y,z)\ u \qquad (1)$$

where Z' is the effective charge on the waters of hydration, R is the distance between the Co^{++} ion and the water of hydration, u is the phonon amplitude, and the function $f(x,y,z)$ describes the transformation properties of the phonon. This introduces selection rules for the coupling. For example (taking x,y,z as the a,b,c axes) the phonons transforming as $x^2(y^2,z^2)$ produce only diagonal matrix elements of ΔV within Γ_4 and hence couple to spin operators like S_α^2 whereas those phonons transforming like xy(yz,xz) yield only off-diagonal matrix elements and couple to $S_\alpha S_\beta$. Furthermore, the spin-orbit coupling is strong enough to split off the lowest Kramers doublet well below all other levels, so that one should transform from S=3/2 to an effective spin σ =1/2, as discussed by Lines[8]. The final result is:

$$H_{M-P} = Au_{xz}\ \sigma_x + Bu_{yz}\ \sigma_y + Cu_z^2\ \sigma_z \qquad (2)$$

In Eq. (2), z is the magnetization axis[9], Δ_1 and Δ_2 are the energies of the two lowest lying orbital levels, as measured from the ground state, $A,B \sim (\lambda^2/\Delta_1\Delta_2)$, and $C \sim (\lambda^2/\Delta_1^2)$. This suggests that the diagrams of 2b and 2c are always larger than that of 2a (2c does not exist for CB2 (CC2) because σ =1/2, but is nonzero for FC2). Unfortunately, the splittings Δ_1 and Δ_2 are not known for CB2. In CC2[10], $\Delta_2 \sim 2\Delta_1$, and in FC2, $\Delta_2 \sim 5\Delta_1$. Note also that the symmetries of the phonons which couple via 2a and 2b, c are different.

where (Δ_1, Δ_2 are the separations of the lowest orbi levels): A, B $\sim$ ($\lambda^2/\Delta_1\ \Delta_2$), but C $\sim$ ($\lambda^2/\ \Delta_1^2$) This suggests that the diagrams of 2b and 2c are always larger than that of 2a (2c does not exist for CB2(CC2) b cause σ=1/2, but is nonzero for FC2). Unfortunately, splittings Δ_1 and Δ_2 are not known for CB2. In CC2[10] $\Delta_2 \sim 2\Delta_1$, and in FC2, $\Delta_2 \sim 5\Delta_1$. Note also that the symmetries of the phonons which couple via 2a and 2b are different.

SUMMARY AND CONCLUSIONS

We have shown that the data on CB2 provides strong evidence for the presence of phonon-magnon scattering terms. Microscopic considerations suggest these terms occur only for the phonons with $x^2(y^2,z^2)$ symmetry, and that the ratio of the amplitude of scattering processes of the type 2b,c to those of the type 2a is Δ_2/Δ_1. From the known values of Δ_1, Δ_2 for FC2 and CC2, and the observations on CB2, we have then constructed a consistent picture of the experimental results on all three materials, as follows: (1) The 30 cm^{-1} phonon in FC2 and CC2 has proper symmetry for 2a and its interaction strength is $\sim$ 1-2 cm^{-1}. (2) The 20 cm^{-1} phonon in CB2 has proper symmetry for 2b and its interaction strength is $\sim$ 6-7 cm^{-1}. These considerations suggest, in agreement with physical intuition, that the phonons in CB2 are somewhat different in frequency from those in FC2 and CC2 (due to the different lattice properties), but that the magnitude of the spin-phonon interaction is very similar in all three materials. In order to further verify the picture, either Raman or neutron scattering experiments should be performed. Of particular interest is the process 2c, which will have an interaction strength of about $5cm^{-1}$ (Δ_2/Δ_1 for FC2), and will cause a hybridization of the $x^2(y^2,z^2)$ phonon with magnon pairs. The effects of this kind of interaction have been previously treated[11] for the case of a Heisenberg spin Hamiltonian, with the spin-phonon interaction derived microscopically from exchange modulation. The present work complements this calculation in that the crystal field modulation is the important effect here.

ACKNOWLEDGEMENT

The authors are greatly indebted to Dr. J.B. Torrance, Jr. for many helpful discussions.

REFERENCES

1. J. B. Torrance and M. Tinkham, Phys. Rev. 187, 587; 595 (1969) and references contained therein.
2. J. B. Torrance and J. C. Slonczewski, Phys. Rev. B5, 4648 (1972).
3. J. B. Torrance and K. A. Hay, Phys. Rev. Letters 31, 163 (1973).
4. K. L. Ngai, J. Ruvalds, and E. N. Economou, Phys. Rev. Letters 31, 166 (1973).
5. S. J. Allen, Jr. and H. J. Guggenheim, Phys. Rev. B4, 937; 950 (1971). These authors explicitly treat the four lowest Co^{++} levels, and show that this is necessary for CoF_2.
6. S. W. Lovesey, J. Phys. C5, 2769 (1972).
7. A. Abragam and M.H.L. Pryce, Proc. Roy. Soc., Ser. A 206, 173 (1951).
8. M. E. Lines, Phys. Rev. 137, A982 (1965).
9. For CC2 and CB2, the magnetization direction is along the b-axis, while for FC2 it is in the a-c plane[1].
10. A. Narath, Phys. Rev. 140 A552 (1965).
11. R. Silberglitt, Phys. Rev. 188, 786 (1969).

OBSERVATION OF VIRTUAL PHONON EXCHANGE IN A MAGNON-LIKE EXCITON OF $Tb(OH)_3$*

R. L. Cone+ and R. S. Meltzer
Department of Physics and Astronomy, University of Georgia, Athens, Georgia 30602

ABSTRACT

We report strong evidence for a large contribution of virtual phonon exchange to the energy dispersion of a magnon-like exciton in the ferromagnet $Tb(OH)_3$. The energy dispersion and resulting interionic energy transfer interactions for the nearer neighbors have been determined by a comparison of the observed line shapes with calculations of the band-to-band transitions in fluorescence from the $^5D_4$ manifold to one of the crystal field components of the $^7F_6$ (ground) and $^7F_5$ (first excited) manifolds. A consideration of the simple nature of the single-ion states, an examination of the selection rules on the various energy transfer mechanisms and an analysis of the range dependence of these mechanisms has enabled us to conclude that 1) the exchange plays a major role in both the $^7F_5$ and $^7F_6$ bands, 2) the electric multipole interaction does not make the dominant contribution to the dispersion in either manifold, 3) the magnetic dipole-dipole interaction plays a small but significant role particularly in the $^7F_6$ manifold, and 4) the virtual phonon exchange plays a role comparable to that of exchange in the $^7F_6$ band, a result which is of special interest in view of the few cases in which its importance has been clearly demonstrated.

There have been only a few instances[1] in which the role of virtual phonon exchange (VPE) has been clearly identified; however Birgeneau et al[2] have recently shown it to be important in the soft mode structural phase transition in $PrAlO_3$. We show here strong evidence for its contribution to energy transfer in a magnon-like exciton at 118 cm^{-1} in the ferromagnet $Tb(OH)_3$.

The single-ion states of Tb^{+3} in $Tb(OH)_3$ are well known.[3] The exciton nature of the crystal states is described by the interionic interaction Hamiltonian $H_{i(p)j(q)}$ summed over all ion pairs, where p and q label the two sublattices. Sublattice excitons are defined in the normal manner, and the interionic interaction energy matrix is constructed in this representation. Two exciton branches result from each single-ion state with eigenvalues

$$E^{\mu}_{\vec{k}}(\pm) = E^{\mu}_{0} + H^{\mu}_{11}(\vec{k}) \pm |H^{\mu}_{12}(\vec{k})| \quad (1)$$

where E^{μ}_{0} is the single-ion energy and $H^{\mu}_{11}(\vec{k})$ and $H^{\mu}_{12}(\vec{k})$ are the matrix elements of the interionic interaction between sublattice excitons on the same and opposite sublattices, respectively.

The emission line shapes of the band-to-band exciton transitions are calculated assuming a single-ion transition mechanism. The energy distribution of the emission resulting from the dispersive nature of the excitations is given by the line shape function

$$\alpha^{\mu\mu'}(h\nu)=C\sum_{\vec{k}}\sum_{tt'}|\vec{M}^{\mu\mu'}_{tt'}(\vec{k})|^2\delta[E^{\mu'}_{\vec{k}}(t')-E^{\mu}_{\vec{k}}(t)-h\nu]\langle n^{\mu'}_{\vec{k}}(t')\rangle \quad (2)$$

where the constant C depends on the multipole of the single-ion transition mechanism, t′ and t label the exciton branches in the upper(μ′) and lower(μ) exciton bands, respectively, and $\langle n^{\mu'}_{\vec{k}}(t')\rangle$ is the upper exciton occupation number which, assuming quasi-thermodynamic equilibrium, is a Boltzmann distribution whose energy is measured with respect to the bottom of the upper exciton band. The $\vec{k}$-dependent transition moments $\vec{M}^{\mu\mu'}_{tt'}(\vec{k})$ are determined from the eigenvectors of the energy matrix and the symmetry properties of the single-ion transition moments.

In order to determine the individual ion pair transfer-of-energy matrix elements (TEME) it is convenient to break them up into long range contributions such as magnetic dipole-dipole (MDD) and electric dipole-dipole (EDD) interaction and short range interactions which include exchange, higher order electric multipole (EMI), and VPE, here limited to the two nearest neighbors (nn) and six next-nearest neighbors (nnn). The $\vec{k}$ dependence of the short range contributions to $H^{\mu}_{pq}(\vec{k})$ in terms of the ion pair TEME v^{μ}_1 and v^{μ}_2 for nn and nnn, respectively, has previously been given for this lattice structure.[4] The long range MDD contributions are calculated exactly using the Ewald technique for evaluating the lattice sums.[5]

The various mechanisms which contribute to v^{μ}_1 and v^{μ}_2 are isolated by taking advantage of the simple nature of the single-ion groundstate and terminal states of the fluorescence. The ground state is essentially pure $|^7F_6\ J_z=-6\rangle$ and has a nearly pure S_z and L_z composition $|S=3,S_z=-3;L=3,L_z=-3\rangle$. The terminal states of the fluorescence which show large energy transfer effects are the μ=1 levels of the $^7F_6$ and $^7F_5$ manifolds at 118 cm^{-1} and 2170 cm^{-1}, respectively, which are nearly pure $J_z=-5$. In the $|S,S_z;L,L_z\rangle$ representation these are

$$|^7F_{6[5]},J_z=-5\rangle = 2^{-\frac{1}{2}}\,(|3,-3;3,-2\rangle+[-]|3,-2,3,-3\rangle) \quad (3)$$

where the bracketed symbols refer to the J=5 state. The fluorescing states are the nearly degenerate $\mu=3^+$ and μ=2 components of the $^5D_4$ manifold at 20,500 cm^{-1}.

We briefly consider the selection rules for the various contributions to the TEME. Recently the anisotropy of the dynamic electronic exchange interaction has been successfully described using an effective operator expression consisting of products of single electron spherical tensor operators and spin operators for electron pairs on the two ions.[6] The selection rules for this operator for multi-electron states of the f^N configuration are $|\Delta S|\le 1$, $|\Delta L|\le 6$, and $|\Delta J|\le 7$, and are satisfied for both the ^7F_J and $^5D_4$ manifolds. Since other mechanisms are not allowed for the $^5D_4$ states, exchange is almost certainly the dominant mechanism of energy transfer there.

The EMI can be written as an effective operator consisting of products of single-electron spherical harmonics[7] with the resulting selection rules $|\Delta S|\le 0$, $|\Delta L|\le 6$, and $|\Delta J|\le 6$. Thus the EMI cannot transfer excitations between $^7F_6$ and $^5D_4$ except by virtue of the small admixture of other spin states. Since all states of interest belong to $4f^8$, only even parity EMI terms should be important (EDD is rigorously zero for μ=1 levels). If we assume that covalency and overlap effects do not invalidate the electric multipole expansion, then the range and coordination angle dependence of the various terms in the EMI are known, making it possible to predict the ratios of the contributions of different near neighbor pairs to the TEME. These predicted ratios can be compared with the experimentally determined TEME ratios to ascertain the importance of this mechanism.

VPE is an additional induced EMI arising from the exchange of virtual phonons. The general form and selection rules are identical to those of the static EMI discussed above, however significant differences occur in the range dependence and general magnitudes of the various terms. In addition, it contains an energy denominator not present in the static EMI which arises in the second order perturbation model and which leads to an inverse proportionality to the excitation energy being transferred, greatly favoring its contribution in the $^7F_6$ manifold. Differences in the TEME for the two μ=1 states of Eq. (3) could arise from the cross terms, which have opposite signs; however, these are zero for the exchange, EMI, and VPE. Thus, only the energy denominators in the VPE contributions can lead to different contributions to v_1^μ and v_2^μ for the two states, and it is possible to isolate the VPE contribution to the energy transfer process by comparing their band-to-band fluorescence line shapes.

Line shapes were calculated using Eq. (2) by varying v_1^μ and v_2^μ so as to obtain a best fit to the observed line shape. The MDD contribution was explicitly included leading to a contribution to the energy dispersion of 0.9 cm^{-1} for the $^7F_6$ exciton and 0.1 cm^{-1} for the $^7F_5$ exciton. The line shapes were calculated on a computer by randomly selecting 4000 points in the first Brillouin zone. A Gaussian broadening of half width 0.67 cm^{-1} was given to the contribution to the line shape of each point to account for inhomogeneous broadening.

The π-polarized line shape at 1.3 K of the transition from the $^5D_4$ $\mu'=2$ exciton to the $^7F_5$ $\mu=1$ exciton bands in a magnetic field of 20 kG along the c axis is compared with some calculated line shapes in Fig. 1. A best fit, shown by the solid curve, is obtained by setting the nnn interactions to zero in both the $^5D_4$ and $^7F_5$ states. The resulting nn parameters are $v_1^{\mu=1} = -1.84(1.79)\pm0.15$ cm^{-1} and $v_1^{\mu'=2} = 0.10(-0.10)\pm0.03$ cm^{-1} for the terminal $^7F_5$ and fluorescing $^5D_4$ states, respectively. The values in parenthesis are an alternative choice leading to identical line shapes.

The absence of a nnn interaction indicates that EMI is not the dominant energy transfer mechanism. Within the assumptions discussed above, the electric quadrupole-quadrupole (EQQ) contributions to the TEME for nn and nnn pairs should be in the ratio $v_2^\mu/v_1^\mu = -0.103$. A line shape calculation based on this ratio produces a poor fit to the observed line shape, shown by the dashed curve in Fig. 1. By varying v_2^μ, we estimate $v_2^\mu/v_1^\mu \leq 0.05$. It thus appears that EQQ alone cannot account for the observed TEME in the $^7F_5$ exciton. Similar arguments apply to the higher order EMI.

The VPE is isolated by comparing the line shape of the fluorescence to the $^7F_5$ exciton with that to the $^7F_6$ exciton. The latter is shown for two magnetic fields in Fig. 2. The fluorescence to $^7F_6$ is two orders of magnitude weaker so that the data are not nearly as good. The possibility also exists that some of the intensity may arise from weak non-intrinsic emission. We conclude, however, from the magnetic field dependence of the spectrum, that the peaks labeled (2) and (3) in Fig. 2 are the intrinsic emission of interest. The peaks are separated by 12 cm^{-1}, whereas, as discussed above, their separation should be identical to that of fluorescence to $^7F_5$ aside from an additional 0.8 cm^{-1} contribution due to MDD. Given the nature of these single-ion states, only the VPE energy transfer mechanism can explain the additional dispersion in this magnon-like exciton.

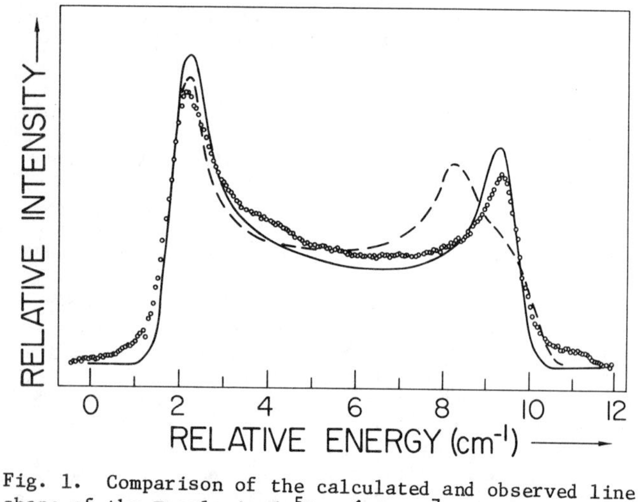

Fig. 1. Comparison of the calculated and observed line shape of the π-polarized $^5D_4$ $\mu'=2$ to $^7F_5$ $\mu=1$ transition at 1.3 K and in a magnetic field of 20 kG. Solid curve- $v_1^{\mu=1} = -1.84$ cm^{-1}, $v_1^{\mu'=2} = 0.10$ cm^{-1} and all $v_2^\mu = 0$. Dashed curve-calculation with $v_1^{\mu=1} = -1.84$ cm^{-1}, $v_2^{\mu=1} = 0.19$ cm^{-1}, $v_1^{\mu'=2} = v_2^{\mu'=2} = 0$.

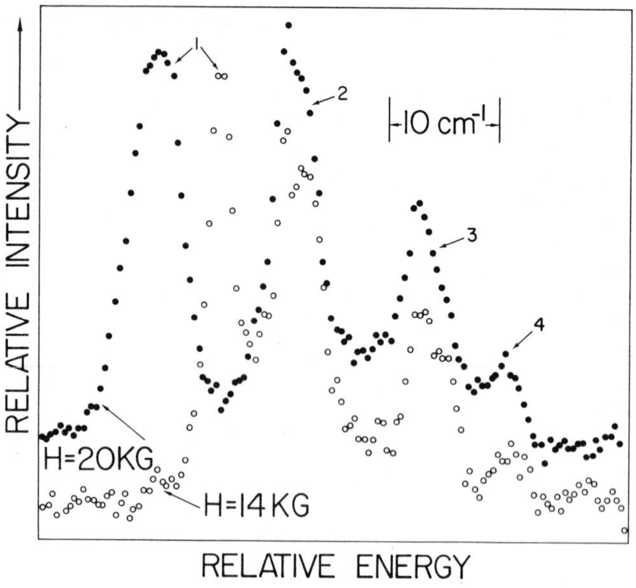

Fig. 2. Observed line shape of the transition from $^5D_4$ $\mu=3^+$ to $^7F_6$ $\mu=\pm1$ at 2.0 K in a magnetic field of 14 and 20 kG.

ACKNOWLEDGEMENTS

We express our gratitude to the Bell Telephone Laboratories for the gift of the 1.8 meter Jarrell Ash Spectrometer used in these experiments and to Professor W. P. Wolf and S. Mroczkowski for the $Tb(OH)_3$ samples.

*Work supported by NSF Grant GH-43934
+Present address, Department of Physics, Montana State University, Bozeman, Montana 59715

REFERENCES

1. For references, see J. M. Baker, Rep. Progr. Phys. 34, 109 (1971).
2. R. J. Birgeneau, J. K. Kjems, G. Shirane, and L. G. Van Uitert, Phys. Rev. B 10, 2512 (1974).
3. P. D. Scott, H. E. Meissner, and H. M. Crosswhite, Physics Letters 28A, 489 (1969), P. D. Scott and W. P. Wolf, J. Appl. Phys. 40, 1031 (1969), and P. D. Scott, Ph.D. Thesis, Yale Univ. (1970).
4. R. S. Meltzer and H. W. Moos, Phys. Rev. B 6, 364 (1972).
5. C. D. Marquard, Proc. Phys. Soc. (London) 92, 650 (1967).
6. R. L. Cone and R. S. Meltzer, Phys. Rev. Lett. 30, 859 (1973).
7. W. P. Wolf and R. J. Birgeneau, Phys. Rev. 166, 376 (1968).

X-RAY PHOTOEMISSION STUDY OF METAL-INSULATOR TRANSITIONS IN VO_2, V_2O_3 AND NiS

G. K. Wertheim, M. Campagna, H. J. Guggenheim,
J. P. Remeika, and D. N. E. Buchanan
Bell Laboratories, Murray Hill, New Jersey 07974

ABSTRACT

The X-ray photoemission density of states has been obtained for a number of transition metal compounds both above and below the metal-insulator transition. In metallic VO_2 the valence band exhibits a typical metallic Fermi edge. In the insulating state the data show a ~0.7 eV gap below the Fermi energy. Comparison with a recent band structure calculation for the metallic state is generally satisfactory when the effects of covalent mixing are taken into account. Data for NiS confirm earlier work at lower resolution but reveal some additional structure in the d-band. No change was resolved on going through the metal-insulator transition. At room temperature in the metallic state V_2O_3 also has a well-resolved Fermi edge.

INTRODUCTION

The electronic structure in the immediate vicinity of the Fermi energy has been studied by X-ray photoemission spectroscopy (XPS) at the metal-semiconductor (M-S) transition of VO_2, V_2O_3 and NiS. In VO_2 [1] the M-S transition is accompanied by a substantial lattice distortion from the high-temperature rutile structure to a low-temperature monoclinic phase in which the V atoms pair up along the c-axis.[2] In the other compounds crystallographic changes are much smaller.

The samples of VO_2 were grown in a melt of V_2O_5 by controlled decomposition. The surface for the X-ray photoemission experiment was prepared in a vacuum of 5×10^{-9} Torr by cleaving or scraping in the sample preparation chamber of a HP 5950A spectrometer. The vacuum in the spectrometer itself is characteristically in the high 10^{-10} Torr range. The data were obtained with an instrumental resolution of 0.55 eV FWHM, determined from measurements of the Fermi edge of silver.

Earlier studies of samples cleaved in air invariably gave evidence of surface oxidation. Samples prepared by etching with aqueous HF had partially hydrated surfaces. The O 1s and V $2p_{3/2}$ lines provide the best indication of surface chemistry. In samples exposed to air the O 1s line has additional structure at higher binding energy indicative of adsorbed water, and the V $2p_{3/2}$ line has a clearly resolved component at higher binding energy indicative of surface oxidation. These features were not observed after vacuum cleaving or scraping.

RESULTS AND DISCUSSION

A set of data for the valence band region above and below the 341°K M-S transition is shown in Fig. 1. The metallic conduction band of the high temperature data is also shown magnified vertically in order to exhibit the cut-off at the Fermi energy more clearly. The observed shape is in accord with the convolution of the known resolution function with the Fermi function, which at this point spacing resembles a step function. In the semiconducting state the occupied density of states has clearly moved away from the Fermi energy by an amount consistent with an energy gap of ~0.7 eV. In spite of the increasing localization of the d electrons implied by this gap, there is no evidence for the formation of a magnetic state in susceptibility, nuclear magnetic resonance, or Mössbauer effect.[2] This is ascribed to the pairing of vanadium species in the monoclinic phase.[2] The appearance of ^{119}Sn nuclear magnetic hyperfine structure in Mössbauer experiments on insulating $V_{1-x}Sn_xO_2$ is apparently a direct consequence of the substitutional Sn which breaks V-V bonds resulting in unpaired electrons.

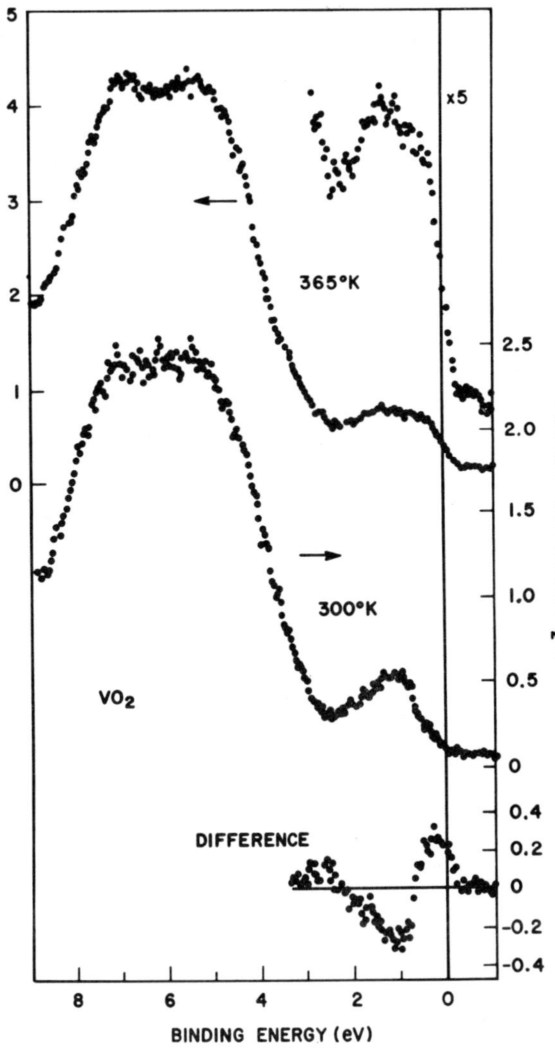

Figure 1. XPS spectra of the valence band region of VO_2 at 365°K in the metallic state and at 300°K in the semiconducting state.

Also shown in Fig. 1 is the difference between the two spectra normalized to equal p-band height. This curve indicates a transfer of density of states from the Fermi energy to a region 1 eV below E_F at the transition. The p-band, which appears between 3 and 8 eV is strikingly similar in the two spectra. The general characteristics which emerge support the conclusion regarding the band structure drawn from optical studies.[4-6]

With this information we re-examine earlier UV photoemission data,[7] and confirm that the major feature observed there is the p-band beginning ~2.5 eV below E_F. It is clear, however, that the work function cut-off did not allow the full p band to be observed, even at a photon energy of 11.2 eV. Neither the conduction band nor the Fermi edge is resolved in the UV data at 100°C, but the room temperature results do exhibit a shoulder which can now be identified as the 3d band.

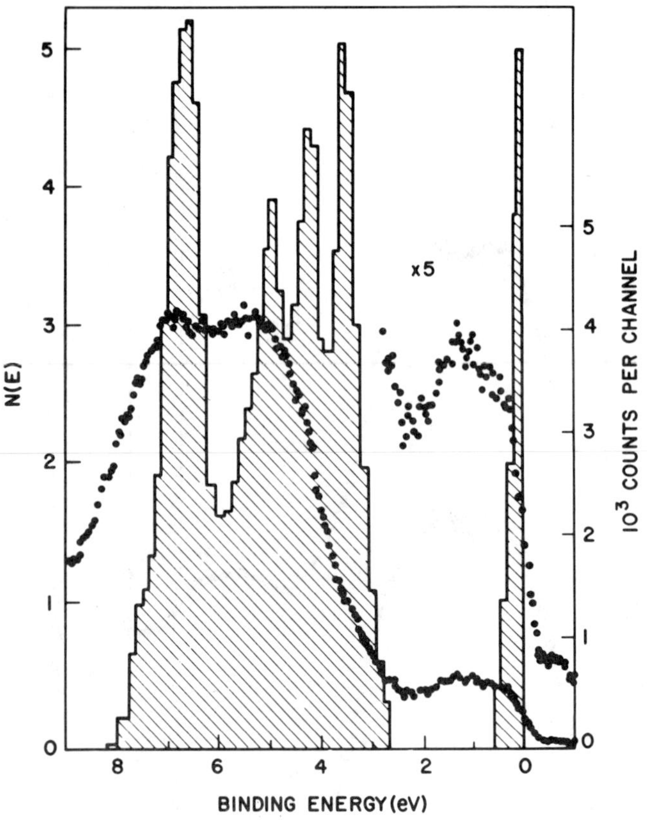

Figure 2. Comparison of the data for the metallic state with a band structure calculation from Ref. 7.

For a fuller interpretation of the data we turn to a recent band structure calculation for metallic VO_2.[7] The data are superposed with arbitrary vertical scale on the calculated density of states in Fig. 2. Two discrepancies are apparent. One arises from the fact that a direct comparison of the total density of states with the XPS data is not justified.[9] This rests on the fact that the cross sections for photoemission of oxygen 2p and metal ion d electrons are not the same. In fact, the former have a cross section so much smaller at 1.5 keV in ReO_3 that the XPS spectrum of the p band consists largely of response due to the covalent admixture of rhenium d states into the oxygen p-band.[9] To a lesser extent this is also true in VO_2. The band structure calculations show that the upper nonbonding part of the p-band, between 2.7 and 4.8 eV, has very little 3d admixture. This region appears only weakly in XPS. Admixture is substantial, however, in the lower part of the band which corresponds to the major XPS response. This is quite analogous to the behavior in ReO_3, and suggests that XPS is again dominated by d admixture into the p band. The upper half of the p-band does appear strongly in UV photoemission where the ratio of d to p cross section is much smaller. As a result one cannot bring theory into agreement with the XPS data by proposing a modification of the calculation to shift the p-band toward greater binding energy.

The second problem concerns the XPS density of states peak at 1.3 eV, which falls into the gap of the calculated band structure. This peak, which was observed in both vacuum cleaved and scraped material, agrees in position with the d-band peak in the semiconducting state. Since XPS samples a thin, typically 15 Å thick, surface layer, this raises the question whether the density of states near the surface is like that of the bulk. This is a particularly important question in the case of VO_2 where a lowering of symmetry, at the M-S transition, produces a dramatic change in the density of states. A possible interpretation of the 1.3 eV peak is that it represents the d-band of a surface layer which remains nonmetallic above the M-S transition by virtue of the lower symmetry and crystallographic distortion at the surface. Further work will be required to resolve this problem.

A comparison of the data for the insulating state with the corresponding band structure calculation[10] is less satisfactory. This is immediately apparent from the fact that the calculated band structures above and below the transition differ greatly throughout the valence band region, while the data differ significantly only in the immediate vicinity of the Fermi energy.

The core levels of V and O were also examined by XPS to facilitate comparison with the results of an X-ray spectroscopic study of VO_2.[11] In comparing XPS with L_{III} X-ray emission data one must remember that the two techniques do not detect all parts of the density of states with equal sensitivity. The selectivity of XPS for the d admixture in the p-band has already been cited. Similarly L_{III}-emission corresponds largely to the metal ion admixture into O 2s and O 2p band.[12] A detailed comparison shows good agreement with the X-ray emission work insofar as the location of core and valence levels is concerned. Only the O 2s emission at 488 and 491.8 eV identified with the $1a_{1g}$ and $1e_g$ orbitals[12] fall at energies quite distinct from that obtained from the XPS data (494.2 eV). Since XPS is here not sensitive to d admixture or selection rule effects, the latter value is probably representative of the position of the major O 2s states density.

Data on a V_2O_3 single crystal surface etched in aqueous HF exhibit a 2.5 eV wide conduction band with a Fermi cut-off in the metallic state.

Measurements on vacuum cleaved NiS gave results very similar to those obtained earlier[13] and did not show any resolvable change in the density of states at the Fermi energy on passing through the M-S transition This is not surprising in view of the small, 0.14 eV, gap found in optical studies[14] and the absence of a major crystallographic distortion at the M-S transition. It also supports the view that NiS is an itinerant electron antiferromagnet.[15]

CONCLUSIONS

The present XPS measurements of the valence band region of transition metal compounds above and below the M-S transition provide a direct demonstration of the opening of a gap at the Fermi energy in VO_2 and confirm the general conclusions regarding the band structures drawn from optical and UV photoelectron measurements.

NOTES AND REFERENCES

1. F. J. Morin, Phys. Rev. Letters 3, 34 (1959).
2. See for example, D. Adler, Rev. Mod. Phys. 41, 714 (1969) or J. B. Goodenough in "Progress in Solid State Chemistry," edited by H. Reiss (Pergamon, New York, 1971) Vol. 5, p. 145. For a discussion specific to VO_2 see J. P. Pouget, H. Launois, T. M. Rice, P. Dernier, A. C. Gossard, G. Villeneuve and P. Hagenmuller (to be published).
3. P. B. Fabritchnyi, M. Bayard, M. Pouchard, and P. Hagenmuller, Solid State Commun. 14, 603 (1974).
4. A. S. Barker, Jr., H. W. Verleur, and H. J. Guggenheim, Phys. Rev. Letters 17, 1286 (1966).
5. H. W. Verleur, A. S. Barker, Jr. and C. N. Berglund, Phys. Rev. 172, 788 (1968).
6. C. N. Berglund, and H. J. Guggenheim, Phys. Rev. 185, 1022 (1969).
7. R. J. Powell, C. N. Berglund and W. E. Spicer, Phys. Rev. 178, 1410 (1969).
8. E. Caruthers, L. Kleinman, and H. I. Zhang, Phys. Rev. B7, 3753 (1973).
9. G. K. Wertheim, L. F. Mattheiss, M. Campagna, and T. P. Pearsall, Phys. Rev. Letters 32, 997 (1974).
10. E. Caruthers, and L. Kleinman, Phys. Rev. B7, 3760 (1973).
11. D. W. Fischer, J. Appl. Phys. 40, 4151 (1969). The change in the density of states at the Fermi energy was not detected in this work.
12. D. W. Fischer, J. Appl. Phys. 41, 3561 (1970).
13. S. Hüfner , and G. K. Wertheim, Phys. Letters 44A, 133 (1973).
14. A. S. Barker (unpublished) has obtained a gap of 0.14 eV in NiS.
15. J. D. M. Coey, R. Brusetti, A. Kallel, J. Schweitzer, and H. Fuess, Phys. Rev. Letters 32, 1257 (1974).

MAGNETIC FIELD INDUCED INFRARED ACTIVITY OF PHONONS IN $TbPO_4$

G. A. Prinz and J. F. L. Lewis
Naval Research Laboratory, Washington, D.C. 20375

ABSTRACT

Many of the rare-earth ortho-arsenates, -phosphates and -vanadates which possess a high temperature zircon structure (space group D_{4h}^{19}) undergo a spontaneous crystallographic distortion as the temperature is lowered. Previous work indicates that $TbPO_4$ distorts at 3.5°K, that the distortion can be induced by an applied perpendicular magnetic field, and that unlike the other materials it may have a crystallographic symmetry lower than orthorhobic in the distorted phase. To learn more about the distortion behavior of $TbPO_4$ we have investigated the infrared activity of normally IR-inactive phonons in magnetic fields from 0 to 100 kOe at temperatures from 4.2 to 1.6°K, with light polarized σ ($E \perp C$) and π ($E \parallel C$), in the frequency region 80-180 cm^{-1} (formerly silent). The intensities grow with field but the frequencies remain fixed. For $H \parallel C$, only the E_g phonon at 131 cm^{-1} becomes IR-active, in π only, and its strength and frequency increase with field. The implications of these results regarding the crystallographic distortions in $TbPO_4$ will be discussed.

HYPERFINE ENHANCED NUCLEAR COOLING TO 1.3 mK IN $PrNi_5$

K. Andres, P. H. Schmidt and S. Darack
Bell Laboratories, Murray Hill, New Jersey 07974

ABSTRACT

We have grown single crystals of $PrNi_5$, which crystallizes in the hexagonal $CaCu_5$ structure, and have observed that the well known maximum in the susceptibility occurs only normal to the c-axis, contrary to previous theoretical crystal field predictions. Below 4.2K, in the Van Vleck paramagnetic regime, the hyperfine enhancement of the local field at the Pr-nuclei over the external applied field is 7.0 along the c-axis and 14.5 normal to it. Hyperfine enhanced nuclear adiabatic demagnetization experiments starting from 25 mK and 26 kOe yield end temperatures as low as 1.3 mK. The nuclear specific heat at 1.8 mK is still relatively low, confirming that exchange interactions are small and suggesting that cooling to even lower temperatures should be possible.

INTRODUCTION

It was pointed out already in 1962 by Nesbitt et al.[1] that in the hexagonal $CaCu_5$ structure-compound $PrNi_5$ the d-states of the Ni-ions are filled and hence non-magnetic. The relatively weak maximum observed in the susceptibility around 16K was for some time associated with an antiferromagnetic transition.[2] Recent specific heat measurements by Craig et al.[3] however showed conclusively that this maximum is a pure crystal field effect and that the ground state of Pr^{3+} in $PrNi_5$ is actually a Γ_4 non-magnetic singlet state. The latter authors were able to fit their specific heat and (polycrystalline) susceptibility data reasonably well with a crystal field calculation which predicted that the maximum in χ was caused mainly by the c-axis susceptibility ($\chi_{||}$) and that in the Van Vleck paramagnetic regime (below ~4.2K) there was almost no anisotropy left ($\chi_{||} \simeq \chi_{\perp}$).

SINGLE CRYSTAL RESULTS

Van Vleck paramagnetic Praseodymium compounds can be useful for nuclear magnetic cooling experiments if both exchange interactions and the nuclear pseudoquadrupole splitting are small,[4] the latter being proportional to the anisotropy of the Van Vleck susceptibility. We have grown single crystals of $PrNi_5$ in an argon tri-arc furnace equipped with a rotating pull rod by means of which we can pull crystals from the molten button. The hexagonal crystals have a tendency to grow in their c-direction. Susceptibility measurements both along and normal to the c-axis are shown in Fig. 1 and reveal that the maximum around 16K occurs only normal to the c-axis ($\chi_{\perp}$) and that $\chi_{\perp}$ is always larger than $\chi_{||}$ (at 4.2K, $\chi_{||}$ = .037 emu/mole, $\chi_{\perp}$ = .077 emu/mole). This is opposite to the behavior predicted by the calculations of Ref. 3, which indicated that the maximum in χ occurs along the c-axis. We should mention here that we have also grown single crystals of the isostructural compounds $PrCu_5$ and $PrPt_5$, which are both Van Vleck paramagnetic. None of these show a maximum in χ, and the anisotropy at 4.2K for $PrPt_5$ ($\chi_{||}$ = .067 emu/mole, $\chi_{\perp}$ = .21 emu/mole) has the same sign as the one for $PrNi_5$. For $PrCu_5$ ($\chi_{||}$ = 1.05 emu/mole, $\chi_{\perp}$ = .15 emu/mole) the anisotropy is opposite and we are able to explain the susceptibility with a crystal field calculation (taking into account a ferromagnetic exchange enhancement). For $PrNi_5$ and $PrPt_5$ however, our attempts of finding ratios of second, fourth and sixth order crystal field amplitudes which produces a level scheme that would yield the observed susceptibilities have failed so far.

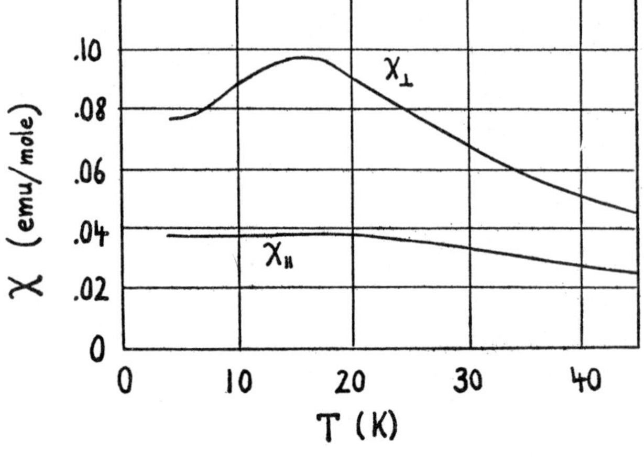

Figure 1. Molar susceptibility of $PrNi_5$ in H = 1 kOe along ($\chi_{||}$) and normal ($\chi_{\perp}$) to the c-axis.

MAGNETIC COOLING RESULTS

The Van Vleck susceptibility below 4.2K leads to hyperfine enhancements of the local nuclear field of 7 in the direction of the c-axis and of 14.5 normal to it. For the nuclear pseudoquadrupole splitting $P[I_z^2 - \frac{1}{3} I(I+1)]$ we obtain

$$P/k_B = -(A^2/2g^2\mu_B^2k_B)(\chi_{||}-\chi_{\perp}) = .23 \text{ mK}$$

so that the overall splitting is only 1.36 mK. Nuclear adiabatic demagnetizations from 25 mK and 26 kOe down to 367 Oe yield end-temperatures of 1.3 mK, as determined by an independent $AuIn_2$ nuclear susceptibility thermometer.[5] Nuclear specific heat measurements in 367 Oe above 1.8 mK are shown in Fig. 2. The dashed line in Fig. 2 is a simple superposition of the expected specific heat contribution from the "ideal" nuclear Zeeman splitting in 367 Oe and from the pseudoquadrupole splitting mentioned above. While at higher

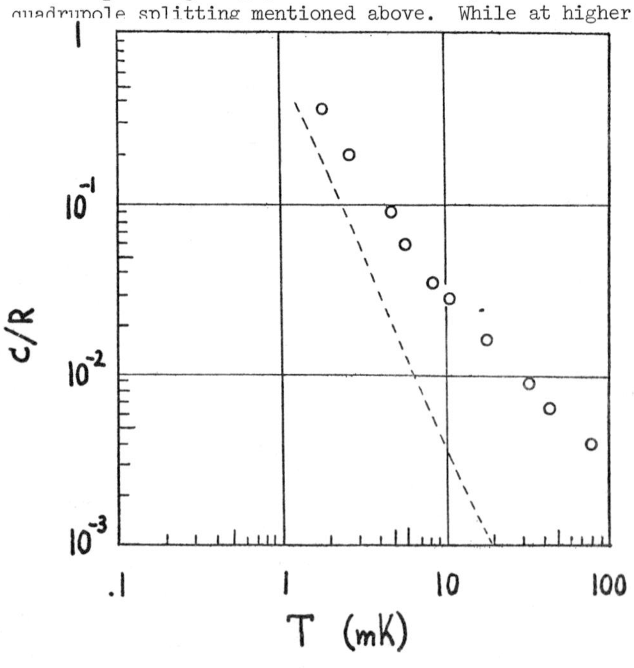

Figure 2. Molar specific heat (in units of R) of $PrNi_5$ in H = 367 Oe at very low temperatures.

temperatures the observed specific heat deviates substantially from the "ideal" value, the deviation diminishes at lower temperatures. Exchange interactions must indeed be small in $PrNi_5$, since down to 1.3 mK there is no sign yet of any onsetting magnetic order. Demagnetization from lower starting temperatures and higher fields to zero field should yield even lower end temperatures than 1.3 mK.

REFERENCES

1. E. A. Nesbitt, A. J. Williams, J. H. Wernick and R. C. Sherwood, J. Appl. Phys. 33, 1674-78 (1962).
2. W. E. Wallace and M. Aoyagi, Mh. Chem. 102, 1455 (1971).
3. R. S. Craig, S. G. Sankar, N. Marzouk, V. U. S. Rao, W. E. Wallace and E. Segal, J. Phys. Chem. Solids 33, 2267 (1972).
4. K. Andres and E. Bucher, J. Low Temp. Phys. 9, 267-289 (1972).
5. K. Andres and J. H. Wernick, Rev. Sci. Instrum. 44, 1186-88 (1973).

MAGNETIC EXCITONS IN PRASEODYMIUM

J. G. HOUMANN and A. R. MACKINTOSH, Research Establishment Risø, Denmark, B. D. RAINFORD, Imperial College, University of London, O. D. McMASTERS and K. A. GSCHNEIDNER JR., Iowa State University.

ABSTRACT

Using a large single crystal, we have performed further neutron scattering measurements on paramagnetic Pr. The exciton dispersion relations corresponding to the transition from the ground state J = |0> to the J = |1a, 1s> excited crystal field levels on the hexagonal sites have been measured at 6 K along the major symmetry directions. The results generally agree with those reported earlier[1] but are considerably more detailed and precise. Except along ΓA and the point K the two-fold degeneracy of the modes is removed by anisotropic exchange whose magnitude is comparable to that of the isotropic component. A careful study of the energies and the intensities of the neutron groups allow an explicit determination of the form of the anisotropic exchange. The minimum in the dispersion relation occurs along ΓM, at a point in k space which corresponds to the periodicity of the magnetic ordering in Pr-5% Nd[2]. The energy of this mode decreases from 1.75 meV to 1 meV as the temperature is lowered from 20 K to 4 K, but remains almost constant when the temperature is further decreased to 0.3 K. The lifetime also increased rapidly with decreasing temperature. The temperature dependence observed is in very good agreement with an RPA calculation and it is found that the exchange is approximately 90% of that which would be required to produce magnetic ordering in Pr. Neutron diffraction experiments reveal no long-range magnetic order in this crystal at 0.3K[2]. The effect of an applied magnetic field on the exciton dispersion has been studied in fields up to 44 kOe. When the field is applied in either the <100> or the <110> direction it is found to have a large effect on the exciton energies with relative energy shifts of up to 100%. In addition strong field dependent interactions with the phonons are observed.

1. B. D. Rainford and J. G. Houmann, Phys. Rev. Lett. 26, 1254 (1971) and 27, 223 (E).
2. B. Lebech and K. A. McEwen, to be published.

MICROWAVE RESONANCE EFFECTS IN PARAMAGNETIC SINGLET-GROUND-STATE SYSTEMS

K. Sugawara and C. Y. Huang
Case Western Reserve U.*, Cleveland, Ohio 44106
and B. R. Cooper**
General Electric Res. & Dev. Ctr., Schenectady, N. Y. 12301
and West Virginia U., Morgantown, W. Va. 26506

ABSTRACT

EPR in the excited $\Gamma_5^{(2)}$ triplet state of Tm^{3+} in cubic TmP has been studied. From the temperature dependence of the g-value, it was found that the fourth-order term dominates in the crystal-field, and the exchange interaction between Tm^{3+} ions is weakly antiferromagnetic. The most striking feature of the observed behavior is the presence of an anomalous maximum in the g-factor temperature dependence occurring at about 40K, where the intensity and line widths also have maxima. Analysis of the EPR intensity gives the separation between the excited $\Gamma_5^{(2)}$ state and the singlet Γ_1 ground state as about 60K. Brief mention is made of the results of EPR observations in the excited Γ_5 level of Pr^{3+} in PrP and for Gd^{3+} as a dilute impurity in TmVA and PrVA (VA = P, As, Sb, Bi).

Recently considerable theoretical and experimental attention has been given to rare earth-Group VA intermetallic compounds where the crystal-field ground state of the rare earth ion is a singlet.[1-3] The value of EPR observations in an excited crystal-field level of the rare earth ion was pointed out by Cooper et al[4], who observed the EPR in the excited $\Gamma_5^{(2)}$ level of Tm^{3+} in TmN. We report here the EPR in the excited $\Gamma_5^{(2)}$ level of Tm^{3+} in TmP and the information gained about the exchange and crystal-field interactions.

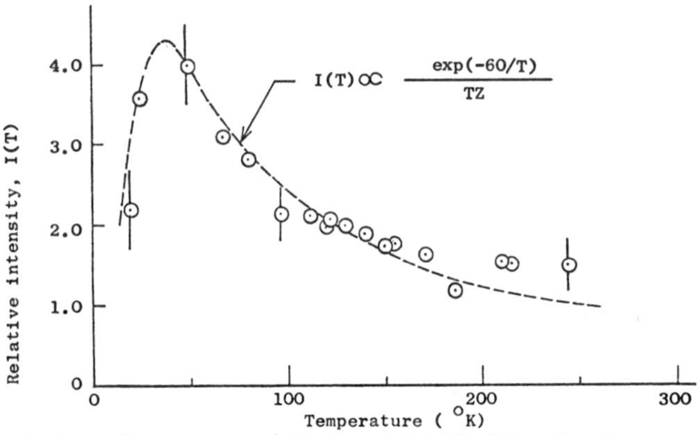

Fig.1. Temperature dependence of the EPR intensity of the $\Gamma_5^{(2)}$ level of Tm^{3+} in TmP. The solid circles are the experimental data, and the dashed line is the theoretical intensity.

The sample was made using the method described by Jones.[5] EPR was performed for a powdered sample at 9 GHz. The identification that the EPR observed originated from the excited $\Gamma_5^{(2)}$ level of Tm^{3+} was made from the temperature dependence of the EPR intensity shown in Fig. 1. In an octahedral crystal-field as occurs in the NaCl-structure rare earth monopnictides, Tm^{3+} has two Γ_5 levels. The $\Gamma_5^{(2)}$ level is the lower of these.[6] The variation of g expected as the ratio of fourth to sixth-order crystal-field contributions varies as shown in Fig. 2 of Ref. 4.

Fig. 2 and Fig. 3 show the experimentally observed g-values and line widths respectively. The g-value approaches 2.20 with increasing temperature indicating that the crystal-field is predominatly fourth-order. The most striking experimental feature is the anomalous maximum in the g-value occurring at about 40K, where the intensity and line widths also have maxima.

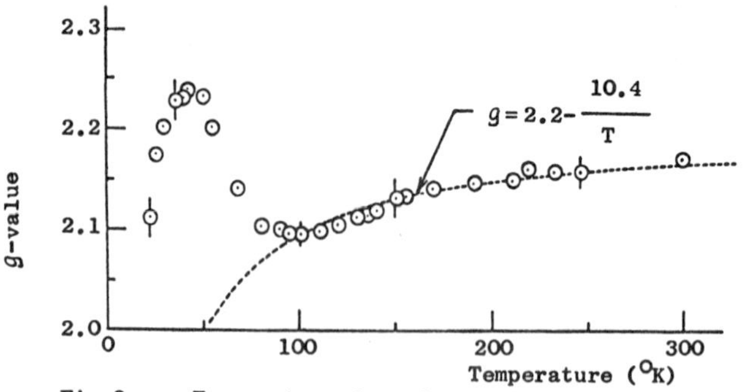

Fig.2. Temperature dependence of the g-value of the excited $\Gamma_5^{(2)}$ level of Tm^{3+} in TmP.

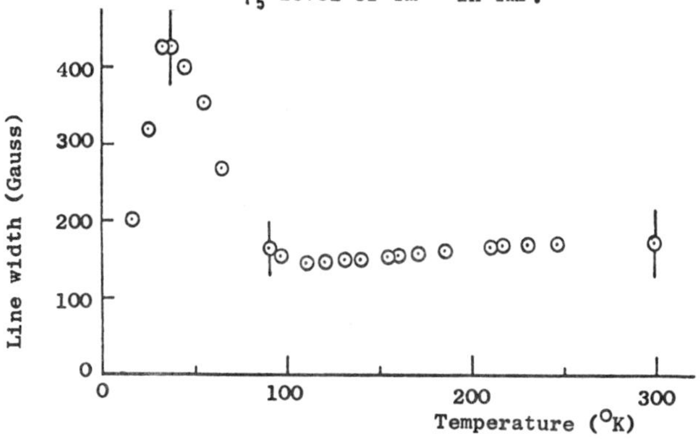

Fig.3. Temperature dependence of the EPR line width of the excited $\Gamma_5^{(2)}$ level of Tm^{3+} in TmP.

For isotropic exchange small compared to crystal-field effects, the g-value of the excited $\Gamma_5^{(2)}$ level is given in Ref. 4 by (using the notation of Ref. 4),

$$g = g_5^{(2)} + \frac{2J(0)\chi}{\lambda\beta^2} \langle\Gamma_{5a}^{(2)}|J_z|\Gamma_{5a}^{(2)}\rangle \qquad (1)$$

$$-(2J(0)\lambda)/(kTZ)\langle\Gamma_{5a}^{(2)}|J_z|\Gamma_{5a}^{(2)}\rangle\langle\Gamma_{5b}^{(2)}|J^-|\Gamma_{5a}^{(2)}\rangle^2$$

$$\times \exp(-E(\Gamma_5^{(2)})/kT)$$

$g_5^{(2)}$ is the crystal-field-only g-value. The second term is proportional to the magnetic susceptibility. The last term comes from the "flip-flop" part of the exchange interaction. $J(0)$ is the total exchange coupling of any ion with all neighbors. The matrix elements in (1) can be calculated from the crystal-field-only wave functions.[6] The experimentally observed g-values above 100K fit.

$$g = 2.2 - 10.4/T \qquad (2)$$

Eq. (2) together with Eq. (1) yields

$$J(0) \simeq -0.17K \qquad (3)$$

$J(0)$ can also be calculated from the magnetic susceptibility at zero temperature. Using Eq. (44) of Ref. 4 together with the experimental results of Busch et al[7], we get

$$J(0) \approx -0.11K \qquad (4)$$

where the value used for $E(\Gamma_4)-E(\Gamma_1)$ is 27K which is implied by the intensity behavior of Fig. 1 for fourth-order crystal-field anistoropy. This value also agrees with that estimated from an analysis of the linewidth of Gd^{3+} ions diluted in TmP. The agreement between

J(0) in (3) and (4) is quite reasonable considering the bases of analysis.

The anomalous g-value behavior of Fig. 2 is similar to that found[4] in TmN, although the TmN behavior was followed down only to 77K. While the last term of (1) has temperature dependence that might contribute to the anomalous behavior, it is much too small[4] (i.e. amounting to perhaps 5% of the anomalous increase) and too broad in its temperature dependence to account for the anomalous behavior. The anomalous behavior may have its origin in anisotropic exchange effects and may also be related to the magnetic exciton dispersion[1,3] pertinent to the paramagnetic regime.

We have also observed the EPR in the excited Γ_5 level, at about 400K above the Γ_1 ground state of Pr^{3+} in PrP for temperatures between 270 and 500K. There is no anomaly in the g-value. However, considering the lack of reliable data below 270K and the slight variation of intensity expected and observed over the temperature range studied, this lack of anomalous g-value behavior may be of no particular significance.

In addition to the experiments on the pure singlet-ground-state materials TmP and PrP, the EPR has been studied from 4.2K to 80K for Gd^{3+} between 0.1 and 10 at %, diluted in TmVA and PrVA (VA = P, As, Sb, Bi). The g-shift of Gd^{3+} serves as a probe of the isothermal susceptibility of the host materials and measures the strength of the Gd-Tm or Gd-Pr exchange interaction.

REFERENCE

* Supported by NSF grant DMR74-08033.

** Present address: Dept. of Physics, West Virginia University, Morgantown, W. Va. 26506.

1. B. R. Cooper and O. Vogt, J. Phys. (Paris) 32, C 1-958 (1971); R. J. Birgeneau, AIP Conf. Proc. No. 10, 1664 (1973).
2. T. M. Holden and W. J. L. Buyers, Phys. Rev. B 9, 3797 (1974).
3. For a recent review see B. R. Cooper, Lectures at the Eleventh Annual Winter School for Theoretical Physics, U. of Wroclaw, Poland, 1974 (to be published).
4. B. R. Cooper, R. C. Fedder, and D. P. Schumacher, Phys. Rev. 163, 506 (1967).
5. E. D. Jones, Phys. Rev. 180, 455 (1969).
6. K. R. Lea, M. J. M. Leask, and W. P. Wolf, J. Phys. Chem. Solids 23, 1381 (1962).
7. G. Busch, A. Menth, O. Vogt, and F. Hulliger, Phys. Lett. 19, 622 (1966).

LATTICE EFFECTS IN SINGLET GROUND STATE SYSTEMS

E. BUCHER, P. D. DERNIER, J. P. MAITA and L. D. LONGINOTTI, Bell Laboratories, Murray Hill, New Jersey 07974 and B. LUTHI and P. S. WANG, Physics Department, Rutgers University, New Brunswick, New Jersey 08903

ABSTRACT

The influence of phonons on Singlet Ground State dynamics is studied. Three systems were investigated: 1) TbP: Here a softening of the c_{44}-mode above T_N is observed, which can be quantitatively explained as an induced cooperative Jahn-Teller effect. 2) $PrSn_3$: In this system the elastic constants show little temperature dependance for $T>T_N$ = 7.3°K despite the fact that the Γ_5 level is only about 9°K above Γ_1. We observe a 4% softening in the c_{44} mode. 3) Pr_3X_4 with X = S, Se, Te: X-ray and elastic constant measurements revealed a structural transition above T_C for X = S, Se. In Pr_3Se_4 T_C = 13°K, T_a = 40°K. This is not of the coop. J-T type because La_3Se_4 also exhibits a cubic-tetragonal transition at T_a = 62°K. These structural transitions are probably induced by similar mechanisms as those occuring in other metal alloys (e.g. In-Tℓ). Pr_3Te_4 exhibits no structural transition, but very strong magnetic field dependent effects. Although all these systems exhibit interesting features (phonon softening and structural instabilities), there is no apparent strong coupling between the elastic waves and the soft magnetic mode. The reason is that they couple different crystal field levels ($\Gamma_1 - \Gamma_3$, $\Gamma_1 - \Gamma_5$ for phonons and $\Gamma_1 - \Gamma_4$ for magnetic excitations in cubic symmetry). A full report will be published elsewhere.

FLUCTUATIONS IN SINGLET-GROUND-STATE MAGNETS*

M.A. Klenin, SUNY, Stony Brook, New York 11974 and
Brookhaven National Laboratory, Upton, New York 11973 and
J.A. Hertz†, University of Chicago, Chicago, Illinois 60637

ABSTRACT

We discuss several effects of coupling between magnetic excitations in singlet-ground-state ferromagnets above T_c. The mode-mode coupling is described in a Hartree approximation analogous to the one we used earlier to describe fluctuations in itinerant ferromagnetism - every magnetic exciton is dressed by arbitrary numbers of other excitons. This theory, which includes terms of all orders in the exchange interaction, is most naturally expressed in the functional integral formalism. The result can be expressed as a correction to the effective single-ion susceptibility, obtained by averaging it over a distribution of external fields. The consequent form of the excitation spectrum has two features not found in RPA: a continuum part with a threshold at the crystal field splitting, and a zero-frequency peak which causes the transition to occur before the magnetic excitation has gone soft.

In view of the recent interest in the magnetic excitations of singlet-ground-state materials, we have undertaken a study of fluctuations in a simple model of such systems, treating the mode-mode coupling in a generalized Hartree fashion. Even in this simple approximation, the resulting excitation spectra have interesting features not found in RPA treatments, which we feel call for further experimental and theoretical investigation. This preliminary report will be supplemented by a longer paper in preparation.[1]

The simplest model of this sort of system approximates the single-ion crystal field level scheme by a pair of singlets, $|0\rangle$ and $|1\rangle$, and postulates that the only nonvanishing matrix elements of the total angular momentum are $\langle 0|j^z|1\rangle = \langle 1|j^z|0\rangle = M$. In terms of Pauli matrices in the space spanned by $|0\rangle$ and $|1\rangle$, the angular momentum on the ith ion, j_i^z, is then just $M\sigma_i^x$, and

$$H = H_{cf} + H_{ex} = \tfrac{1}{2}\Delta\sum_i \sigma_i^z - \tfrac{1}{2}M^2\sum_{ij} J_{ij}\sigma_i^x\sigma_j^x \tag{1}$$

Using standard techniques[2], the partition function may then be represented by a functional integral in which Z is the average, over a normal distribution of time-and space-dependent magnetic fields, of partition functions describing noninteracting ions in the presence of these fields:

$$Z=\int D\xi \exp[-\sum_i \frac{1}{\beta}\int_0^\beta d\tau\, \xi^2(\tau)]\, Z\{\xi\} \equiv \int D\xi\, \exp(-\psi[\xi]) \tag{2}$$

where

$$Z\{\xi\} = \mathrm{tr}\,\mathbf{T}\exp\,[-\int_0^\beta d\tau\, (H_{cf}(\tau) + \sum_{ij}\xi_i(\tau)A_{ij}\sigma_j^x(\tau))] \tag{3}$$

$\mathbf{T}$ is the time-ordering operator and the matrix A_{ij} is defined by $(A^2)_{ij} = 2M^2J_{ij}/\beta$.

The integrand in (2) may be expanded in powers of the random field ξ, generating a diagrammatic representation of the interactions between magnetic excitons. We pursue this approach elsewhere[1]; here we deal directly with approximations in the functional integral. The functional $\psi[\xi]$ (for which formally exact expressions can be written) is a generalized Landau-Ginzburg free energy functional, including terms of all orders in the order parameter field $\xi_i(\tau)$. If this expansion ended at second order, Z could be evaluated quite simply, since the functional integration would reduce to a product of simple Gaussian integrals. The RPA, in fact is obtained by truncating the expansion at second order in ξ. Our approximation is to replace the true functional by the best (in a variational sense) quadratic one, as Mühlschlegel and Zittartz did for the Ising model[3] and we did for the Hubbard model[4]. We showed quite generally there that the optimal approximate functional is

$$\psi_o[\xi] = \tfrac{1}{2}\sum_{\vec{q}m}\left\langle \frac{\partial^2\psi[\xi]}{\partial\xi_{\vec{q}m}\partial\xi_{-\vec{q},-m}}\right\rangle |\xi_{\vec{q}m}|^2 \tag{4}$$

where the ξ_{qm} are Fourier components of $\xi_i(\tau)$(m labels the Matsubara frequencies $2\pi m/\beta$)and the average is over the approximate distribution $\exp(-\psi[\xi])$. As was the case in the Hubbard model,using(4) is equivalent to a Hartree approximation in treating the interactions between excitation modes[1]. Using (3) we find

$$\frac{\delta^2 \ln Z\{\xi\}}{\delta\xi_i(\tau)\delta\xi_j(\tau')} = \sum_{\ell m} A_{i\ell}A_{jm}[\langle T\,\sigma_\ell^x(\tau)\sigma_m^x(\tau')\rangle_\xi - \langle\sigma_\ell^x(\tau)\rangle_\xi\langle\sigma_m^x(\tau)\rangle_\xi \tag{5}$$

The subscript ξ on the averages means that these expectation values are in the presence of the field $h_j(\tau) = \sum_i \xi_i(\tau)A_{ij}$. These of course cannot be done exactly,[1] so as in our previous work, we make the further approximation of evaluating the average as if h were static and uniform. We have discussed the implications of this kind of approximation fully in the context of the Hubbard model; this situation is quite analogous. Here we only remark that it is equivalent to ignoring most of the nonlocality in the mode-mode interaction vertices. Then (4) becomes

$$\psi_o[\xi] = \sum_{\vec{q}m}[1 - J(\vec{q})\phi(i\omega_m)]\,|\xi_{\vec{q}m}|^2 \tag{6}$$

with $\phi(i\omega_m) = \langle\chi_o^x(i\omega_m,h)\rangle_h$, the single-ion susceptibility in a static field h, averaged over h. Notice that RPA is obtained by ignoring the dependence of ϕ on $\vec{q}$.

To evaluate $\chi_o^x(i\omega_m,h)$, it is simplest to rotate coordinate axes in the pseudospin space through an angle $\theta = \tan^{-1}(2h/\Delta)$ around the y axis, to diagonalize H.

$$\tilde{H} = \tfrac{1}{2}\tilde{\Delta}\tilde{\sigma}^z, \text{with } \tilde{\Delta}^2 = \Delta^2 + (2h)^2 \tag{7}$$

Then in terms of the transformed-spin susceptibilities

$$\chi_o^x = \cos^2\theta\,\tilde{\chi}_o^x + \sin^2\theta\,\tilde{\chi}_o^z \tag{8}$$

The z component is easy. Since H is proportional to $\tilde{\sigma}^z$, the perturbation commutes with H, and a finite χ is obtained only for a static perturbation. We may use the classical formula

$$\tilde{\chi}_o^z(i\omega_m) = M^2\beta[\langle(\tilde{\sigma}^z)^2\rangle - \langle\tilde{\sigma}^z\rangle^2]\,\delta_{mo} = M^2\beta[1 - \tanh^2(\tfrac{1}{2}\beta\tilde{\Delta})]\,\delta_{mo} \tag{9}$$

(This is just a Curie law modified to account for the finite field.) χ_o^x is not as trivial, since there is a finite response at finite frequency. It can, however, be evaluated directly from the exact spectral representation of χ_o: (Here $Z = \sum e^{-\beta\epsilon_n}$; $\epsilon_n = \pm\tilde{\Delta}/2$.)

$$\tilde{\chi}_o^x(i\omega_m) = \frac{1}{Z}\sum_{\ell n}(e^{-\beta\epsilon_\ell} - e^{-\beta\epsilon_n})\frac{|\langle\ell|M\tilde{\sigma}^x|n\rangle|^2}{\epsilon_n - \epsilon_\ell - i\omega_m} \tag{10}$$

leading to

$$\tilde{\chi}_o^x(i\omega_m) = \frac{2\tilde{\Delta}M^2}{\tilde{\Delta}^2 - (i\omega_m)^2}\tanh(\tfrac{1}{2}\beta\tilde{\Delta}) \tag{11}$$

Using the results (9) and (11) leads to

$$\chi_o^x(i\omega_m,h) = \frac{2M^2\Delta}{\sqrt{(\Delta^2+4h^2)}}\,\frac{\tanh[\tfrac{1}{2}\beta\sqrt{\Delta^2+4h^2}]}{\Delta^2+4h^2-(i\omega_m)^2} + \delta_{mo}\frac{4M^2\beta h^2}{\Delta^2+4h^2}\,\mathrm{sech}^2[\tfrac{1}{2}\beta\sqrt{\Delta^2+4h^2}] \tag{12}$$

We now need to average eqn.(12) over the distribution of ξ, or, equivalently, over the distribution of single-ion fields h. This is straightforward because of our hypothesized form of $\psi_o[\xi]$. The ξ_{qm} in (4) are normally and independently distributed, hence so are the Fourier components of the field $h_{qm} = A(q)\xi_{qm}$, and the local fields h_i are consequently normally (although not independently) distributed, with variance

$$\sigma^2 = \frac{1}{N}\sum_{\vec{q}m} A^2(q)\langle|\xi_{\vec{q}m}|^2\rangle = \frac{M^2}{\beta N}\sum_{\vec{q}m} \frac{J(\vec{q})}{[1-J(\vec{q})\phi(i\omega_m)]} \qquad (13)$$

by virtue of the central limit theorem. The self-consistency condition on the distribution arises from the fact that calculating σ^2 in (13) requires the distribution as input in evaluating $\phi(i\omega_m)$ on the right-hand side.

The averaging gives a spectral weight function for ϕ of

$$\mathrm{Im}\,\phi(\omega) = \theta(|\omega|-\Delta)\sqrt{\frac{\pi}{2\sigma^2}}\;\frac{M^2\Delta^2\tanh(\frac{1}{2}\beta\omega)}{|\omega|\sqrt{\omega^2-\Delta^2}}\;\exp\left(\frac{-\omega^2+\Delta^2}{8\sigma^2}\right) \qquad (14)$$

The delta functions in the RPA Im ϕ are asymmetrically broadened into continua with singular thresholds at $\pm\Delta$, and the pole singularities in Re ϕ reduced to divergences like $|\omega-\Delta|^{-\frac{1}{2}}$.

The dynamical susceptibility can be obtained from the mean square fluctuations of ξ:

$$\chi(\vec{q},i\omega_m) = \frac{2}{J(\vec{q})}\left(\langle|\xi_{\vec{q}m}|^2\rangle - \frac{1}{2}\right) = \frac{\phi(i\omega_m)}{1 - J(\vec{q})\phi(i\omega_m)} \qquad (15)$$

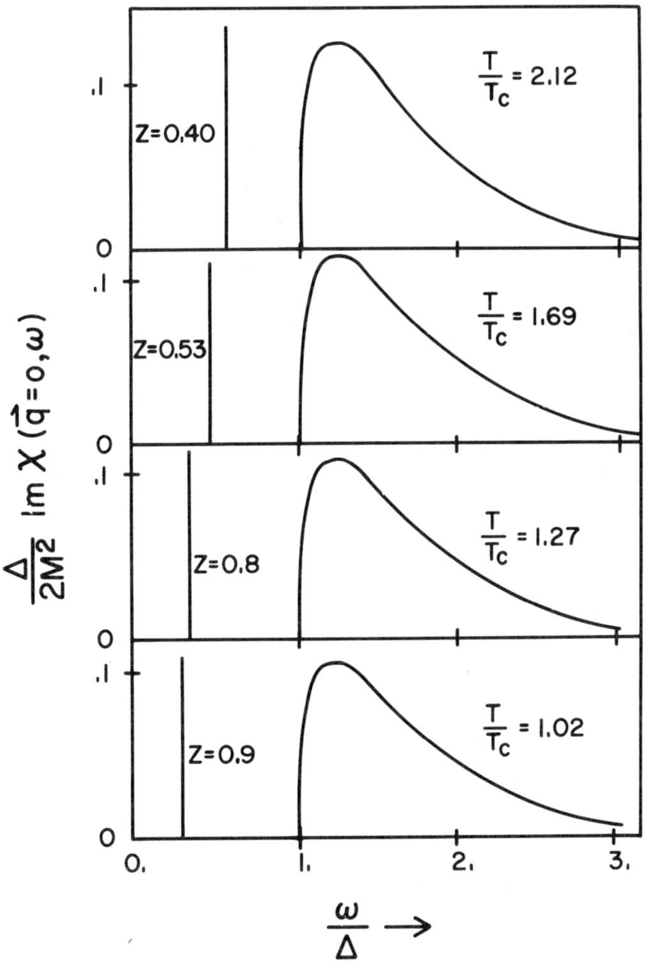

Figure Caption: The excitation spectrum given by Eqns. 14 and 15 for $2M^2J_o/\Delta = 2.10$ $T_c = 0.236\ \Delta$ at the values of T shown. The position and integrated spectral weights z of the RPA-like spike are given (in the same units) for comparison.

On the real axis, then, χ can have singularities of two sorts - exciton poles, at energies $\omega(q)$ where the denominator of (15) vanishes, and branch cuts in the regions $|\omega|>\Delta$ where Im $\phi(\omega)$ is nonvanishing. Fig. 1 shows plots of Im $\chi(0,\omega)$ for several values of $T>T_c$. We have chosen a case of moderately strong coupling: $M^2J(0) = 1.05\Delta$. In these calculations, the mean square field (13) was evaluated approximately as follows: Anisotropy was completely ignored, and the finite range of the exchange simulated by taking $J(q) = J(0)(1-q^2/q_o^2)$ in a sphere of radius q_o at the center of the Brillouin zone and $J(q) = -J(0)(1 - (q-q_m)^2/q_o^2)$ in a sphere of the same size about the point $q_m = (\frac{\pi}{a}, \frac{\pi}{a}, \frac{\pi}{a})$. This agrees with the expansion to lowest order in q for a nearest neighbor interaction in a cubic crystal. Furthermore, terms with $\omega_m \neq 0$ were ignored, an approximation we can expect to be valid when T is greater than typical $\omega(q)$'s.

As T is lowered, χ will diverge before the exciton softens completely, because of the elastic (δ_{m0}) term in ϕ, the average of the second term in (12). $T_c = \beta_c^{-1}$ (always $< T_c^{rpa}$) is determined by

$$1 = J(0)M^2\int_{-\infty}^{\infty}\frac{dx}{\sqrt{8\pi\sigma^2}}\,e^{-\Delta^2x^2/8\sigma^2}\left[\frac{2\tanh(\frac{1}{2}\beta_c\Delta\sqrt{1+x^2})}{(1+x^2)^{3/2}} + \frac{\beta_c\Delta x^2\,\mathrm{sech}^2(\frac{1}{2}\beta_c\Delta\sqrt{1+x^2})}{1+x^2}\right] \qquad (16)$$

while the softening of the exciton (if it occurred) would be determined by (16) without the second term in the integrand. We can estimate $\omega(0)$ at T_c by setting $J(0)\phi(\omega(0)) = 1$ and using (16). Then to lowest order in $\omega(0)/\Delta$

$$\frac{\omega^2(o)}{\Delta^2} \simeq \left(\frac{T_c}{\Delta}\right)^{\frac{1}{2}} e^{-\Delta/T_c} \qquad (17)$$

This final approximate expression is appropriate to the large fluctuation case.

We expect that the zero-frequency anomaly in $\chi(\omega)$ will be broadened into a central peak of finite width by effects, whose origin may be either extrinsic or intrinsic, which we have omitted here. Such a peak would appear in neutron scattering spectra. Although we can say little about the width of this peak, the strength of the extra elastic part of χ, as calculated here, should give a reasonable estimate of the total integrated intensity in it.

In summary, the magnetic excitation spectra of singlet-ground-state systems above T_c may exhibit features absent from RPA theories. The central peak, the continuum part of Im χ above Δ, and the finite $\omega(0)$ at T_c all call for further experimental and theoretical investigation.

*Supported by N.S.F. and N.S.F.-M.R.L.
†Alfred P. Sloan Foundation Fellow.

REFERENCES

1) M.A. Klenin and J.A. Hertz, to be published.
2) J. Hubbard, Phys. Rev. Letters 3 77(1959); S.Q. Wang, W.E. Evenson and J.R. Schrieffer, Phys. Rev. Letters 23 92 (1969); J.R. Schrieffer CAP Summer School notes, Banff, 1969 (unpublished); D.R. Hamann, Phys. Rev. B2 1373 (1970).
3) B. Mühlschlegel and H. Zittartz, Z. Phys. 175 553 (1963).
4) J.A. Hertz and M.A. Klenin, Phys. Rev. B10 1084 (1974).

FIELD-INDUCED TRANSITIONS IN DySb*

T.O. Brun, G.H. Lander, F.W. Korty† and J.S. Kouvel†
Argonne National Laboratory, Argonne, Illinois 60439 and
University of Illinois, Chicago, Illinois 60680

ABSTRACT

The NaCl-structured compound DySb, which in zero field transforms abruptly at $T_N \approx 9.5$ K to a Type-II antiferromagnetic (A) state with a nearly tetragonal lattice distortion, was previously found to exhibit rapid field-induced changes in magnetization at 1.5 K. The field-induced transitions in a DySb crystal have been studied by neutron diffraction and magnetization measurements in fields up to ∿ 60 kOe applied parallel to each of the principal axes. In the <100> case, the transition from the A to an intermediate ferrimagnetic (Q) state is first-order at 4.2 K (critical field $H_c \approx$ 21 kOe) but is continuous from ∿ 6 K up to T_N, as $H_c \to 0$. The Q-to-paramagnetic (P) transition is rapid but continuous at 4.2 K ($H_c \approx$ 40 kOe) and becomes broad as T_N is approached. In the <110> case the A-to-Q transition remains essentially first-order from 4.2 K ($H_c \approx$ 15 kOe) up to T_N; above T_N rapid P-to-Q transitions occur at very high fields. The magnetic structure of the Q state is found to be that of HoP.

INTRODUCTION

In zero magnetic field, the compound DySb exhibits a first order magnetic transition[1,2] at $T_N \approx$ 9.5 K. Above T_N DySb has the cubic NaCl-structure. Below T_N the compound orders antiferromagnetically in the type-II structure that consists of ferromagnetic (111) planes stacked antiferromagnetically along the [111] axis. The first order transition is accompanied by a predominantly tetragonal lattice distortion (c/a = .993). The ordered magnetic moment at 6 K is 9.5 μ_B per Dy, almost the saturation moment of 10 μ_B for Dy^{3+}, and is parallel to the tetragonal [001] axis. Elastic as well as magneto-thermal data for DySb have been analyzed in a molecular field approximation,[3,4] and the analysis shows that biquadratic pair interactions between the Dy^{3+} ions may be as important as the bilinear exchange.

Previous magnetization measurements[5] made on a single crystal of DySb at 1.5 K in fields up to 60 kOe show that a ferrimagnetic state exists. For all three principal directions of the field the values of the magnetization in this state are consistent with the magnetic structure of DySb being that of HoP,[6] in which the moments within each (111) plane of Dy atoms are ferromagnetically aligned with 10 μ_B per Dy and oscillate between two perpendicular <100> directions in alternating (111) planes.

RESULTS

<u>Magnetization</u> measurements were performed on single crystals of DySb using a vibrating-sample magnetometer. Magnetic fields of up to 56 kOe were applied parallel to the <100>, <110> and <111> crystallographic directions. The fields have been corrected for demagnetization. The results for the <100> direction are shown in Fig. 1 as isotherms M vs H. Below T_N we observe two critical fields at which the magnetization increases very rapidly. At 4.2 K the magnetization changes discontinuously at $H_{c1}^{<100>}$ = 21.6 kOe and very rapidly at $H_{c2}^{<100>}$ = 40.6 kOe. Both critical fields decrease with increasing temperature, both transitions becoming broader as the temperature approaches T_N. At 4.2 K the magnetization is almost 5 μ_B per Dy atom just above $H_{c1}^{<100>}$ and approaches 10 μ_B per Dy atom above $H_{c2}^{<100>}$.

In Fig. 2 the M vs H isotherms are shown for the field along <110>. The isotherms show one transition which remains abrupt while its critical field, $H_{c1}^{<110>}$, decreases from 15.6 kOe at 4.2 K to zero at T_N. The magnetization at 4.2 K approaches a value of approximately 7.2 μ_B at high fields. In addition, the

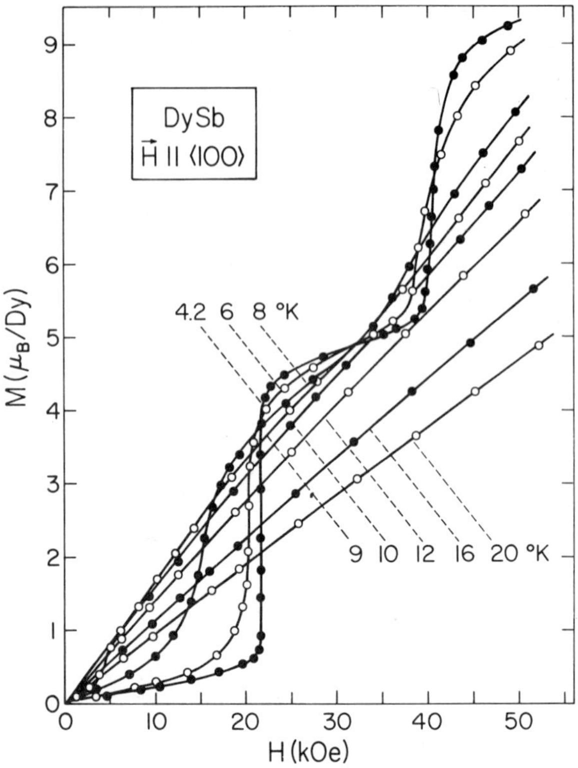

Figure 1. Magnetization of DySb vs. field applied along <100> at constant temperature.

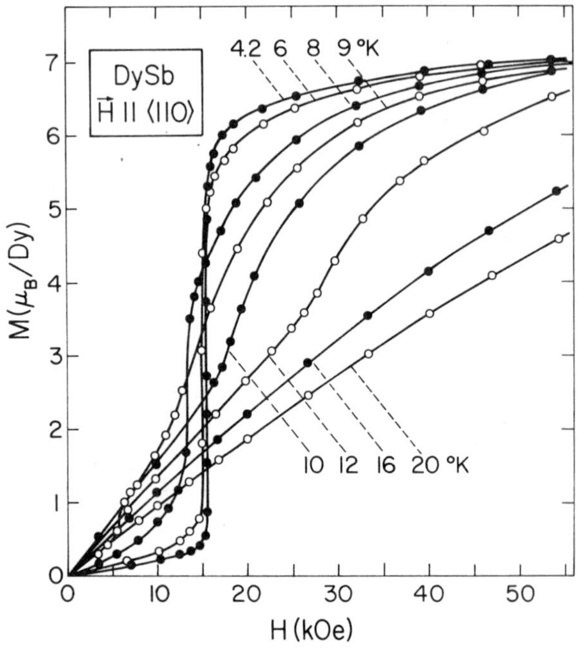

Figure 2. Same as Fig. 1 for field along <100>.

~~isotherm~~ for 9 K shows a very rapid increase ~~in~~ dM/dH at fields just above $H_{c1}^{<110>}$. The 10 K and 12 K isotherms above T_N show a similar change in dM/dH at fields just above 15 kOe and 26 kOe, respectively.

To determine the ranges of temperature and field for the various magnetic states of DySb, we have converted our data to field vs temperature curves of constant magnetization. The phase boundaries have been identified from the rapid changes in dH/dT. Fig. 3 shows the magnetic phase diagram for $\vec{H}$ parallel to

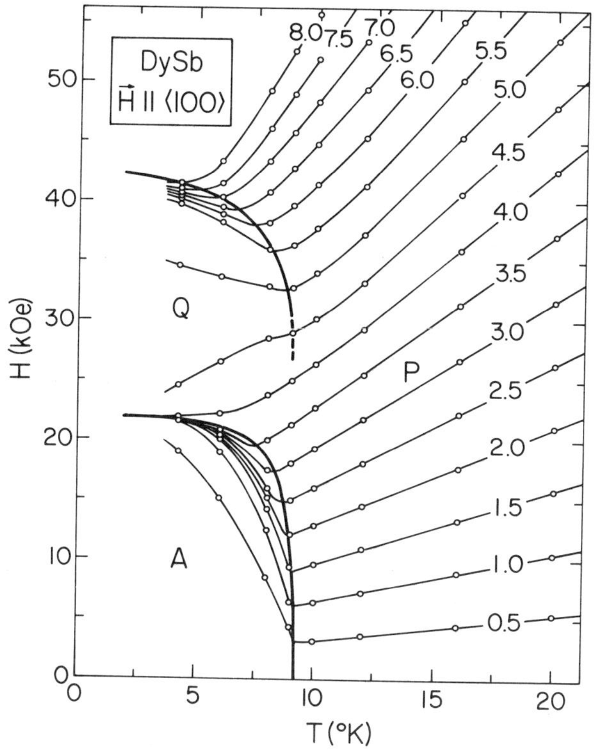

Figure 3. Field along <100> vs. temperature at constant magnetization (in μ_B/Dy) for DySb.

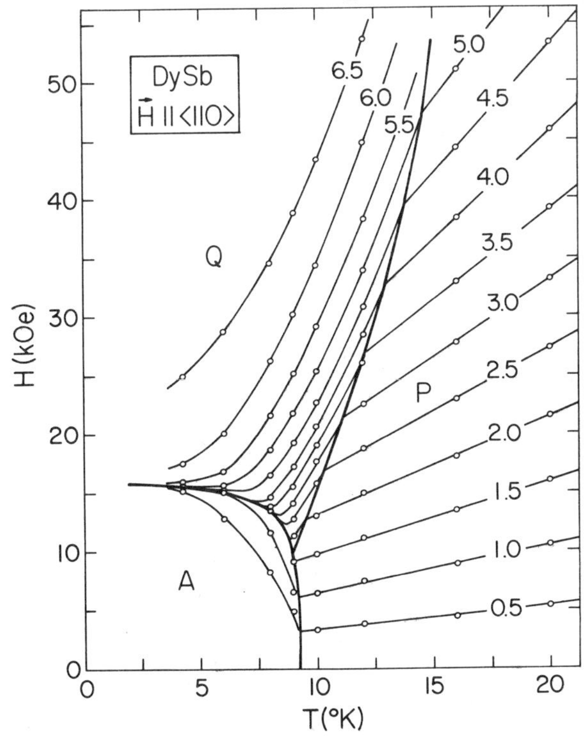

Figure 4. Same as Fig. 3 for field along <110>.

<100>. The phase marked P is paramagnetic, the A phase is the Type-II antiferromagnetic state, and the Q-phase is the intermediate ferrimagnetic state.

The magnetic phase diagram for $\vec{H}$ parallel to <110> is shown in Fig. 4. The P, A, and Q phases are the same as for H parallel to <100>. An unusual feature of this diagram is that the boundary between the Q and P phases extends up to 15 K at the maximum field.

The magnetization data for $\vec{H}\|$<111> are not shown, but the magnetic phase diagram is similar to that for $\vec{H}\|$<110>. The critical field, $H_{c1}^{\langle 111\rangle}$, is 19 kOe at 4.2 K, and the value of the magnetization in the Q phase is $\sim 5.8\ \mu_B$ per Dy atom.

Neutron diffraction measurements have been performed on a single crystal of DySb with $\vec{H}\|$ [110] to determine the structure of the Q-phase. We have measured the integrated intensities of the $(\frac{1}{2}\,\frac{1}{2}\,\frac{1}{2})$ and the $(\frac{5}{2}\,\frac{3}{2}\,\frac{1}{2})$ antiferromagnetic reflections at 4.2 K and 11 K and for H = 0 and 30 kOe. **The observed intensities for the Q-phase are in good agreement with 10 μ_B per Dy parallel to [100] and [010] in alternate** (111) planes. At 11 K no magnetic reflections are present in zero field, but by increasing the field the $(\frac{1}{2}\,\frac{1}{2}\,\frac{1}{2})$-reflection appears at the P to Q boundary, with the same intensity as at 4.2 K. Preliminary data for the lattice distortions in the Q phase indicate that the lattice expands in the [001] direction, normal to the plane of the moments.

DISCUSSION

The results of the present magnetization measurements are in good agreement with the results of Busch and Vogt.[5] As pointed out by these authors the value of the magnetization in the Q phase for each of the three principal field directions is consistent with the HoP structure. Furthermore if the three Q phases have identical magnetic structures with the same exchange and anisotropy energies, the differences between the Zeeman energies of the three Q-phases predict that $H_{c1}^{\langle 100\rangle} = \sqrt{2}\, H_{c1}^{\langle 110\rangle} = 2/\sqrt{3}\, H_{c1}^{\langle 111\rangle}$. The experimental values of the three critical fields are in good agreement with this prediction. Both magnetization and neutron diffraction results for the Q-phase of DySb show therefore that this phase has the HoP structure with 10 μ_B per Dy atom.

For $\vec{H}\|$<100> the HoP structure cannot be stabilized by bilinear exchange interactions alone; additional couplings such as biquadratic pair interactions are needed. Stevens and Pytte[7] have suggested that the dominant contribution to this biquadratic interaction comes from the couplings between the ion and the lattice. In their model cooperative shifts of the Sb ions relative to the Dy ions create local distortions stabilizing the HoP structure.

Further neutron diffraction measurements are required to verify the HoP structure of the Q phases for other direction of the field and to test the model of Stevens and Pytte.[7] Our magnetization data for DySb above T_N are being compared in detail with crystal-field calculations of magnetization vs effective field in order to expose any evidence for biquadratic or any other higher-order coupling, as was done in a recent investigation of PrAg.[8]

*Based on work performed under the auspices of the U. S. Atomic Energy Commission.
†Supported in part by the Office of Naval Research.

REFERENCES

1. E. Bucher, R. J. Birgenau, J. P. Maita, G. P. Felcher and T. O. Brun, Phys. Rev. Letters 28, 746 (1972).
2. G. P. Felcher, T. O. Brun, R. J. Gambino and M. Kuznietz, Phys. Rev. B8, 260 (1973).
3. T. J. Moran, R. L. Thomas, P. M. Levy and H. H. Chen, Phys. Rev. B7, 3238 (1973).
4. L. F. Uffer, P. M. Levy and H. H. Chen, AIP Conf. Proc. 10, 553 (1973).
5. G. Busch and O. Vogt, J. Appl. Phys. 39, 1334 (1968).
6. H. R. Child, M. K. Wilkinson, J. W. Cable, W. C. Koehler and E. O. Wollan, Phys. Rev. 131, 922 (1963).
7. K. W. H. Stevens and E. Pytte, Solid State Comm. 13, 101 (1973).
8. T. O. Brun, J. S. Kouvel, G. H. Lander and R. Aitken, Solid State Comm. 15, 1157 (1974).

EFFECTS OF HYBRIDIZATION ON ELECTRON LOCALIZATION TRANSITIONS IN SOLIDS*

J. M. Robinson
Purdue University, Fort Wayne Campus, Fort Wayne, Indiana 46805

ABSTRACT

We consider a model of an actinide or rare earth compound in which a localized non-degenerate 5f or 4f state on each metal ion is overlapped by and hybridized with a band of itinerant states. There is also a short range Coulomb interaction between the band and f electrons. The thermodynamic properties are calculated at T=0K and at finite temperature with a Green's function method. As T is increased, the electronic occupation probability of the f-levels (and of the band) may either increase or decrease discontinuously, but for a large hybridization the size of the discontinuity is significantly less than unity. The model may be applied to electron localization or valence-change transitions.

I. INTRODUCTION

Recently, L. M. Falicov and others[1-3] proposed a theory of metal-insulator transitions in iron-group oxides and rare earth materials in which approximately one electron per metal ion "jumps" suddenly from the conduction band to a localized 3d or 4f level (or *vice versa*) as the temperature T is increased. We call such transitions electron localization (or delocalization) transitions. P. Erdos and the present author[4,5] generalized this model to take into account magnetic ordering and crystal field effects and obtained quantitative explanations of magnetic and thermodynamic properties associated with first-order phase transitions observed in some metallic actinide compounds. In these theories, the hybridization, or quantum mechanical admixture of the localized and itinerant states, is neglected. This approximation is probably poor for many actinide materials, where the 5f electrons are thought to form narrow hybridized bands.[6,7] This motivates the present work, which includes hybridization in the above theories. A hybridized model of actinide metals was proposed by Jullien and Coqblin[7] and by Doniach[8], but first-order transitions were not considered by these authors.

II. THE HYBRIDIZED MODEL

In the previous theory[4,5] the electron localization transitions occur at low temperature in the magnetically ordered state, where a strong exchange field removes the degeneracy of the localized f-levels. This has the result that the thermodynamic properties near the transition temperature can be obtained under the assumption of a single localized state per metal ion having a fully aligned magnetic moment. The model Hamiltonian for such a situation is written as follows:

$$H=\Delta'\sum_i a^+_{i\uparrow} a_{i\uparrow} + N^{-\frac{1}{2}}\sum_{i,\vec{k}}(V_{\vec{k}}\exp(i\vec{k}\cdot\vec{R}_i)a^+_{i\uparrow} b_{\vec{k}\uparrow} + \text{h.c.}) + + GN^{-1}\sum_{i,\vec{k},s} a^+_{i\uparrow} a_{i\uparrow} b^+_{\vec{k}s} b_{\vec{k}s} + \sum_{\vec{k},s} \varepsilon_{\vec{k}} b^+_{\vec{k}s} b_{\vec{k}s} \; . \quad (1)$$

The operator $a_{i\uparrow}$ destroys an electron with spin "up" in a state having energy Δ' and localized on lattice site $\vec{R}_i$. The number of lattice sites (assumed to be equivalent) is N, and for simplicity only the ferromagnetic case is treated here. The operator $b_{\vec{k}s}$ destroys an electron in a band state of wavevector $\vec{k}$, spin s, and unperturbed energy $\varepsilon_{\vec{k}}$. The hybridization between a band and a localized state is $V_{\vec{k}}\exp(i\vec{k}\cdot\vec{R}_i)$, and h.c. denotes Hermitian conjugate of the preceding term. The quantity G is the Coulomb interaction between a band and a localized electron in the same atomic cell. The interaction of the f-electrons with the exchange field may in the following treatment be subsumed within the parameters Δ' and G. There are assumed to be N(1+z) electrons available, of which at most one can occupy the localized level at site i. The corresponding localization probability (1-p) is defined by the equation $\langle a^+_{i\uparrow} a_{i\uparrow}\rangle = 1-p$, where the brackets denote quantum statistical averaging.

The Hamiltonian is treated by the Green's function equation of motion technique as described by Zubarev[9], and the details will be omitted here. Decoupling in the Hartree approximation, we find the following quasiparticle dispersion curves:

$$2E^{\pm}_{\vec{k}} = E_0 + e_{\vec{k}} \pm ((e_{\vec{k}}-E_0)^2+4|V_{\vec{k}}|^2)^{\frac{1}{2}} \; ; \quad (2)$$

$$e_{\vec{k}} = \varepsilon_{\vec{k}} + G(1-p). \quad (3)$$

In Eq.(2) E_0 is defined by the expression $E_0=\Delta'+G(p+z)$. The delocalization probability p and the Fermi level ζ' are solutions of the two self-consistent equations

$$(1-p) = (2\pi)^{-3}v\int d\vec{k}(r^+_{\vec{k}}f(E^+_{\vec{k}}) + r^-_{\vec{k}}f(E^-_{\vec{k}})) \; ; \quad (4)$$

$$1+z = (2\pi)^{-3}v\int d\vec{k}(f(E^+_{\vec{k}}) + f(E^-_{\vec{k}}) + f(e_{\vec{k}})), \quad (5)$$

where

$$r^{\pm}_{\vec{k}} = (\pm e_{\vec{k}} \mp E^{\pm}_{\vec{k}})(E^-_{\vec{k}} - E^+_{\vec{k}})^{-1}. \quad (6)$$

In Eqs.(4) and (5), f(E) is the Fermi-Dirac function, and v is the volume of a Wigner-Seitz cell. The integrations are over the Brillouin zone. The internal energy $U=\langle H\rangle$ in thermal equilibrium is found to be

$$UN^{-1} = \Delta'(1-p) + G(1-p)(p+z) + v(2\pi)^{-3}\int d\vec{k}(\varepsilon_{\vec{k}}f(\varepsilon_{\vec{k}}) + (r^-_{\vec{k}}\varepsilon_{\vec{k}}+2t_{\vec{k}})f(E^+_{\vec{k}}) + (r^+_{\vec{k}}\varepsilon_{\vec{k}}-2t_{\vec{k}})f(E^-_{\vec{k}})), \quad (7)$$

where $t_{\vec{k}}=|V_{\vec{k}}|^2(E^+_{\vec{k}}-E^-_{\vec{k}})^{-1}$. The entropy S is as follows:

$$SN^{-1} = -v(2\pi)^{-3}\int d\vec{k}(q(E^-_{\vec{k}}) + q(E^+_{\vec{k}}) + q(\varepsilon_{\vec{k}})); \quad (8)$$

$$q(E) = f(E)\ln(f(E)) + (1-f(E))\ln(1-f(E)). \quad (9)$$

The free energy F is now determined from the equation F=U-TS.

III. THERMODYNAMICS OF THE MODEL

The solutions to Eqs. (4) and (5) have been studied under the following restrictions: (a) a $\vec{k}$-independent hybridization ($V_{\vec{k}}=V$), (b) dispersion $\varepsilon_{\vec{k}}=(\hbar^2/2m^*)k^2$ for the conduction electrons, and (c) a spherical Brillouin zone of radius k_0, where $4\pi k_0^3=(2\pi)^3/v$.

The parameters of the model are chosen to be z, G, V, W, and Δ, where $W=\varepsilon(k_0)$, and $\Delta=\Delta'+G(1+z)$. The change of variable $\zeta=\zeta'-G(1-p)$ is also made in Eqs. (4) and (5).

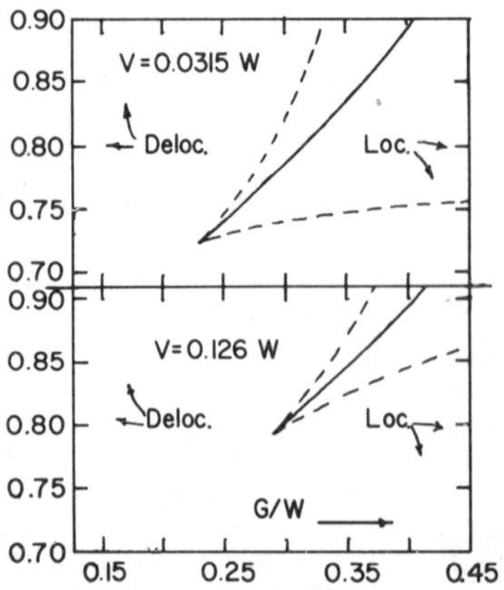

Fig. 1. First-order transition lines (solid curves) separating regions of localization (loc.) and delocalization (deloc.) for z=0.2 and T=0K.

A numerical program was written which finds all of the solutions of the self-consistent equations for given values of the temperature T and of the model parameters. Each solution is a pair of values of p and ζ and represents a possible physical state or phase. The most stable phase is that having the lowest free energy at the given temperature.

In Fig. 1 the regions of stability of the phases at T=0K are shown as functions of Δ and G for two values of the hybridization V. Inside the wedge-shaped regions there are two competing stable solutions corresponding to the localized ($p \cong 0$) and delocalized ($p \cong 1$)

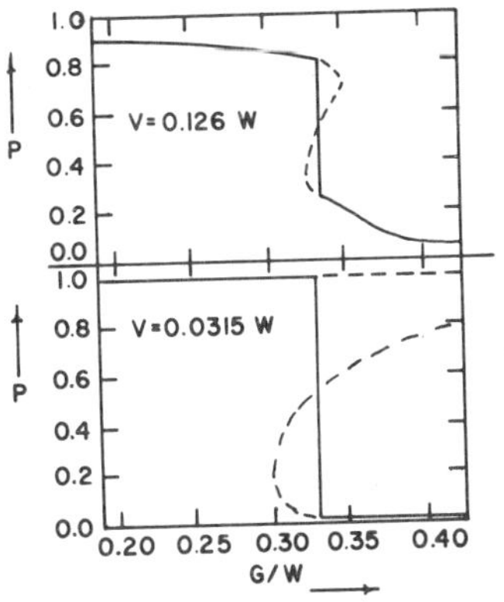

Fig. 2. The delocalization probability p as a function of G (solid curves) for the case Δ=0.819W, z=0.2, and T=0K. The dashed curves pertain to phases of higher free energy.

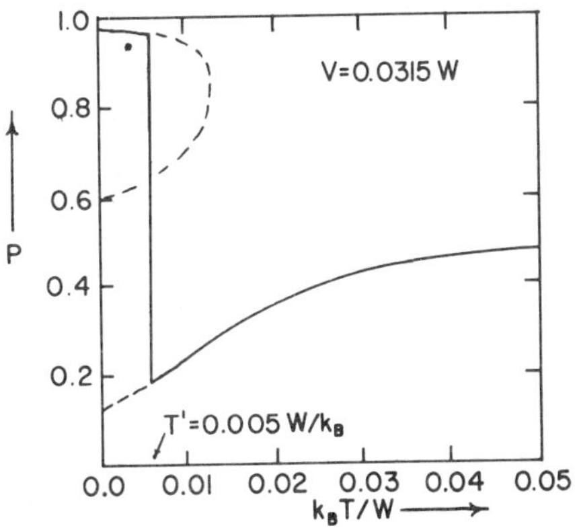

Fig. 3. The delocalization probability p (solid curve) of the most stable phase as a function of T for the case Δ=0.756W, G=0.267W, and z=0.2. The dashed curves show p(T) in phases of higher free energy.

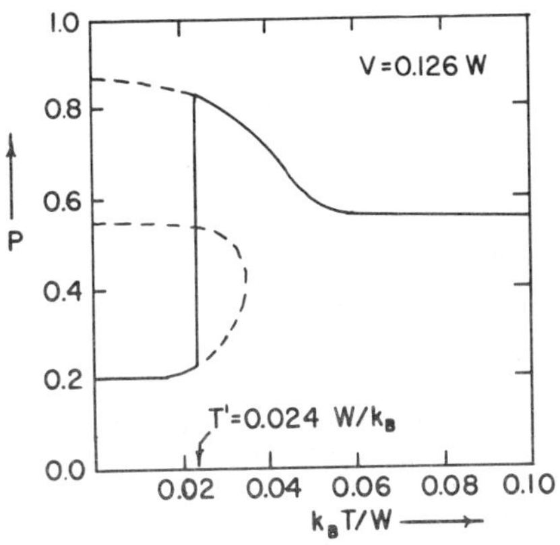

Fig. 4. The delocalization probability p(T) of the most stable phase (solid curve) for the case Δ=0.850W, G=0.356W, and z=0.2. The dashed curves show p(T) in phases of higher free energy.

phases. The free energies of the two phases are equal along the solid curve, which ends in a critical point, and there occurs a discontinuous transition between the two phases upon crossing the solid curve (See also Fig. 2). For finite V the metal ions have no well-defined valence, and the discontinuity in p does not represent a complete valence change. Outside of the wedge-shaped regions in Fig. 1 there is only one phase, and p varies continuously with Δ and G. Figure 3 shows the temperature dependence of p in the most stable phase for a small value of V. The sudden localization of electrons as T is increased to a critical temperature T' is similar to the transition discussed in Refs. 4 and 5. A more surprising result is seen in Fig. 4 for a larger value of V, for which there occurs a sudden delocalization transition. Note also the smaller discontinuity in p. The physical cause of the transitions in Figs. 3 and 4 is the electronic entropy resulting from the location of the Fermi level near a peak in the self-consistent density of states in the high temperature phase.

IV. DISCUSSION

We have investigated the effects of hybridization on first-order electronic phase transitions, using a model Hamiltonian valid at temperatures well below the magnetic ordering temperature. The small discontinuity in the band occupation (Figs. 2 and 4) resulting from the hybridization may explain why the changes in the electrical resistivity observed at the "moment-jump" transitions in UP and UAs are smaller than those expected on the basis of the previous theory[4,5], which neglected hybridization. The present calculations also support an earlier suggestion[2] that hybridization may account for the fractional valence change of 0.61 observed at the α-γ phase transition in Ce metal.

In using the simple one-body mixing term in Eq. (1), we have adopted a single particle (band) description of the electronic states. Intraatomic correlations should be considered in treating the localized states. The case of a k-dependent hybridization as proposed by Jullien and Coqblin[7] is presently being considered, and the preliminary results indicate no significant change in the nature of the first-order transitions.

REFERENCES

*This work was supported by the Purdue Research Foundation with a 1974 Summer Faculty XL Grant.

1. R. Ramirez, L. Falicov, and J. Kimball, Phys. Rev. B2, 3383 (1970).
2. R. Ramirez and L. Falicov, Phys. Rev. B3, 2425 (1971)
3. L. Falicov, C. Goncalves da Silva, and B. Huberman, Solid State Commun. 10, 455 (1972).
4. J. Robinson and P. Erdös, Phys. Rev. B8, 4333 (1973); B9, 2187 (1974).
5. P. Erdös and J. Robinson, A.I.P. Conf. Proc. 10, 1070 (1973).
6. A. Arko et al, Phys. Rev. B5, 4564 (1972).
7. R. Jullien and B. Coqblin, Phys. Rev. B8, 5263 (1973)
8. S. Doniach, discourse, International Symposium on the Electronic Structure of the Actinides, Argonne National Laboratory, Argonne, Illinois, 1974.
9. D. Zubarev, Soviet Physics-Uspekhi 3, 320 (1960).

EXCITED STATE RESONANCE STUDY OF CRITICAL BEHAVIOR IN SINGLET-GROUND- STATE ANTIFERROMAGNETS

C. Y. Huang and K. Sugawara
Case Western Reserve U*, Cleveland, Ohio 44106
and B. R. Cooper**
General Electric Res. & Dev. Ctr., Schenectady, N. Y. 12301
and West Virginia U., Morgantown, W. Va. 26506

ABSTRACT

The EPR of the excited $\Gamma_5^{(2)}$ level of Tb^{3+} in TbVA (VA = P, As, Sb) intermetallic compounds has been studied. The temperature dependence of the EPR intensity for TbP serves to confirm that the EPR is originating from the excited $\Gamma_5^{(2)}$ state. For TbP and TbAs the g-shift from the crystal-field only value can be expressed as $A/(T-T_N)^n$ where A = 300 (TbP), 350 (TbAs) and n = 1.3 ± 0.3 (for both TbP and TbAs). These results are interpreted by assuming that the g-shift is proportional to the staggered magnetic field. The line widths of TbP and TbAs do not broaden when the temperature approaches T_N, while that of TbSb increases. These results may be explained by the Kawasaki-Huber and Krueger theory for the line width in antiferromagnets in the vicinity of the Néel temperature. The temperature dependence of the line widths of TbP and TbAs can be accounted for under the assumption that the line width is proportional to the auto-correlation function of the spin fluctuations. From these analyses the energies of the Γ_4 and $\Gamma_5^{(2)}$ levels of Tb^{3+} in TbP and TbAs are about 16K and 35K respectively.

The EPR of the excited $\Gamma_5^{(2)}$ triplet state[1] of Tb^{3+} in cubic TbVA (VA = P, As, Sb) has been studied in the temperature range between 50K and the Néel temperatures (8K, 10.5K, and 16.5K respectively[2]). All samples were made using the method described by Jones.[3] The average particle size of the powdered samples was in the range between 10 microns and 70 microns. This is comparable with the microwave skin depth[4] at the x-band frequencies (≈ 9GHz) employed in these experiments. The first derivative of power absorbed has been recorded, and the typical shape is shown in Fig. 1. The EPR signal for TbP and TbAs becomes flat at high magnetic fields ($H \gtrsim 2000$ Gauss); hence the flat tail section is adopted as the base line in defining the resonance field. The EPR signal for TbSb does not become flat even when the magnetic field is greater than 7000 Gauss. This together with the much broader absorption width for TbSb forced the analysis of our results for TbSb to be more restricted than that for TbP and TbAs. Analysis has shown that the

EPR signal of TbAs at 13.0K

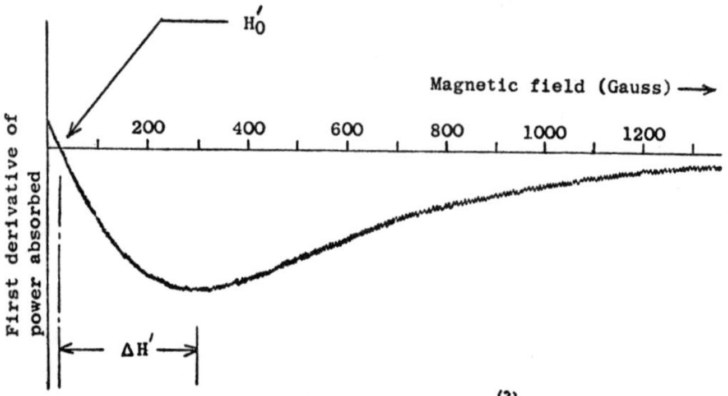

Fig.1. Shape of the EPR of the excited $\Gamma_5^{(2)}$ level of Tb^{3+} in TbAs.

EPR signals for TbP and TbAs are very close to Lorentzian. Therefore H_o', defined in Fig. 1, is taken as the resonant field. The "line width" $\Delta H'$ is also defined in Fig. 1.

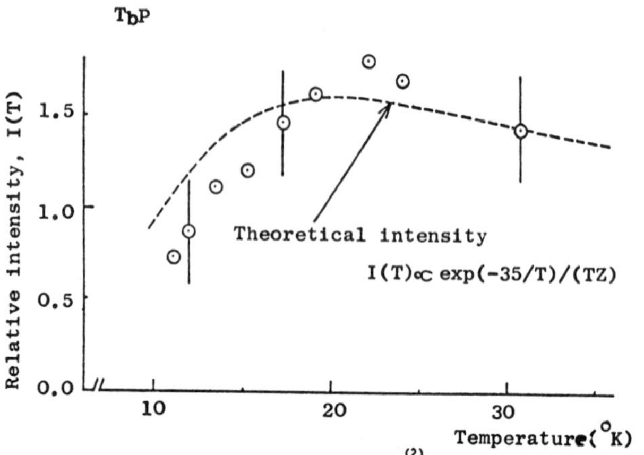

Fig.2. EPR intensity of the $\Gamma_5^{(2)}$ level of Tb^{3+} in TbP. The solid circles are the experimental data, and the dashed line is the theoretical intensity. [For $E(\Gamma_5^{(2)}) - E(\Gamma_1) = 35K$]

The variation of EPR intensity with temperature for TbP is shown in Fig. 2. The dashed curve gives the theoretical intensity if the fourth-order term dominates the crystal-field, and the $\Gamma_5^{(2)}$ level is 35K above the Γ_1 singlet ground state. (In the expression for the theoretical intensity in Fig. 2, Z is the crystal-field-only partition function of a single Tb^{3+} ion.) The value of 35K compares reasonably to the value of 25.3K obtained for the same energy splitting in TbSb using the crystal-field parameters found by Cooper and Vogt.[5] Also, as discussed below, the value of 35K leads to theoretical expected line width behavior in agreement with experiment. Thus the close agreement between the experimental and theoretical thermal behavior of the intensity in Fig. 2 serves to confirm that the EPR is originating from the excited $\Gamma_5^{(2)}$ state.

For TbP and TbAs the g-values are found from the ratio of the signal frequency to the resonant field H_o' defined in Fig. 1. In the limit of dominant fourth-order crystal-field anisotropy, the crystal-field-only g-value[6] is 2.8. The shift of the experimental g-values from this crystal-field-only value are shown in Fig. 3 plotted against $T-T_N$, where T_N is the Néel temperature. These shifts in g-value can be expressed as

$$\Delta g = A/(T - T_N)^n \quad (1)$$

with A = 300, n = 1.3 ± 0.3, T_N = 8K for TbP and A = 350, n = 1.3 ± 0.3, T_N = 10.5K for TbAs. These results are interpreted by assuming that the g-shift is proportional to the staggered susceptibility. It is somewhat difficult to determine the exponent n accurately because of the relatively large error bars in the g-shifts near the Néel temperature. The Mean-Field-Approximation predicts n = 1.0; however, our result is close to the value 1.238 obtained for the critical index of the longitudinal staggered susceptibilty in the neutron experiments of Schulhof et al[7] on MnF_2.

The line width of TbP and TbAs does not increase when the temperature approaches T_N, while that of

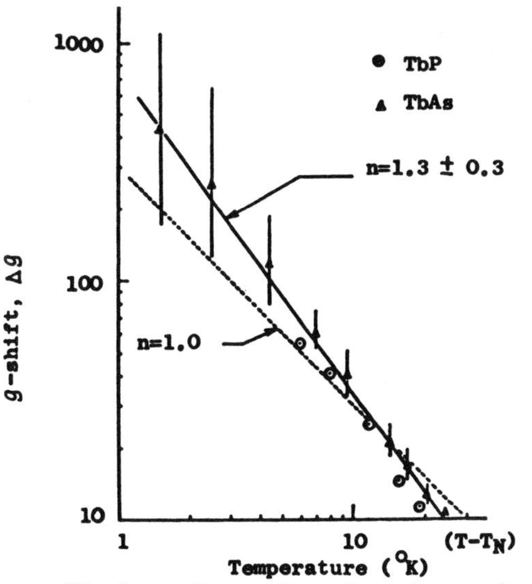

Fig.3. Temperature dependence of the g-shifts of Tb^{3+} in TbP and TbAs.

TbSb does. These results may be explained by the Kawasaki[8]-Huber and Krueger[9] theory for the line width in antiferromagnets in the vicinity of the Neel temperature. Their theories predict that the line width would not increase in the vicinity of the Neel temperature if the damping constant, $\Gamma_{\vec{K}_o}$, of the spin correlation function $(S_{\vec{K}_o}(t), S_{\vec{K}_o})$ is smaller than the microwave frequency employed in the EPR experiment.[9a] The appropriate value of the damping constant $\Gamma_{\vec{K}_o}$ for TbVA can be calculated following the paper by Huber and Krueger[8],

$$\Gamma_{\vec{K}_o} \simeq J(\chi_o T_N)^{1/2}(\kappa a)^{3/2} \qquad (2)$$

where we have used the same notation as in Ref. 8. The exchange interaction coefficient J can be found from the paper by Busch et al[2], and the uniform magnetic susceptibility, χ_o, in the paper by Cooper and Vogt.[5] Here κ is the inverse correlation length. Eq. (2) gives

$$\Gamma_{\vec{K}_o} \simeq 7 \text{ GHz at } T = T_N + 1 \text{ for TbP, TbAs}$$
$$\Gamma_{\vec{K}_o} \simeq 11 \text{ GHz at } T = T_N + 1 \text{ for TbSb} \qquad (3)$$

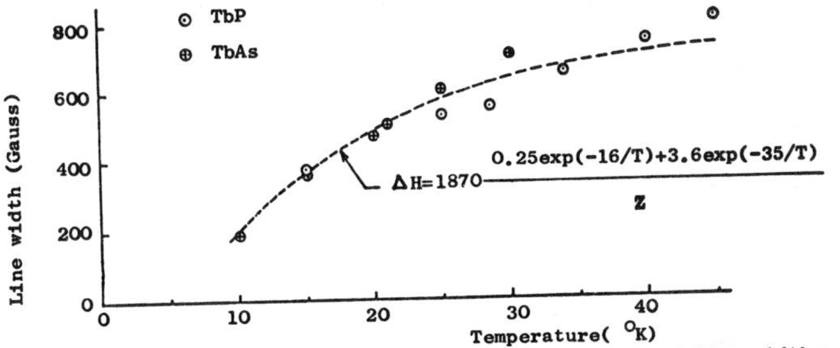

Fig.4. Comparision between the experimentally observed line widths of TbP (⊙), TbAs(⊕) and the theoretical line width(the dashed line).

The experimental temperature dependence of the line width, $\Delta H'$, for TbP and TbAs is shown in Fig. 4. This temperature dependence can be accounted for under the assumption that the line width is proportional to the auto-correlation function of the spin fluctuations.[10] The theoretical estimate for the line width on this basis is (assuming the $\Gamma_5^{(2)}$ state is at 35K and the Γ_4 state is at the corresponding value of 16K implied in the limit of fourth-order-only crystal field anisotropy),

$$\Delta H \simeq 1100 \frac{0.25 \exp.(-16/T) + 3.6 \exp.(-35/T)}{Z} \qquad (4)$$

As shown in Fig. 4, this gives a satisfactory fit to the observed line widths of TbP and TbAs if we multiply the right side of Eq. (4) by 1.7. This result indicates that the energy levels of the Γ_4 and $\Gamma_5^{(2)}$ triplet crystal-field states of Tb^{3+} in TbP and TbAs are indeed about 16K and 35K respectively above the singlet ground state Γ_1. This is consistent with the intensity behavior of Fig. 2.

REFERENCES

* Supported by NSF grant DMR 74-08033

** Present address: Dept. of Physics, West Virginia University, Morgantown, W. Va. 26506

1. The notation used for the crystal-field levels is that of K. R. Lea, M. J. M. Leask, and W. E. Wolf, J. Phys. Chem. Solids 23, 1381 (1962).
2. G. Busch, P. Schwob, O. Vogt, and F. Hulliger, Phys. Letters 11, 100 (1964); G. Busch, O. Vogt, and F. Hulliger, Phys. Letters 15, 301 (1965); G. Busch, O. Vogt, O. Marincek, and A. Menth, Phys. Letters 14, 262 (1965).
3. E. D. Jones, Phys. Rev. 180, 455 (1969).
4. L. H. Brixner, Inorg.& Nuclear Chem 15, 199 (1960).
5. B. R. Cooper and O. Vogt, Phys. Rev. B1, 1218 (1970).
6. The internal resonance field "seen" by the ionic moments corresponds to this g-value, and therefore at the signal frequency of 9.1 GHz is always 2322 Gauss, even though the "observed" external resonance field, H_o', is considerably lower.
7. M. P. Schulhof, P. Heller, R. Nathans, and A. Linz, Phys. Rev. B1, 2304 (1970).
8. K. Kawasaki, Progr. Theoret. Phys. 39, 285 (1967).
9. D. L. Huber and D. A. Krueger, Phys. Rev. Letters 24, 111 (1970).

9a. We should point out that the theories of Refs. 8 and 9 were developed to explain the line widths in isotropic antiferromagnets. Significant modifications may be required for compounds, such as those discussed here, with large crystal-field effects. Thus the quantitative estimates discussed here on the basis of these theories should be treated with caution.

10. T. Moriya and Y. Obata, J. Phys. Soc. Japan 13, 1333 (1958).

CORRELATION MODEL OF A BINARY INDUCED MOMENT CRYSTAL IN A MEAN FIELD APPROXIMATION. SINGLET-TRIPLET SYSTEM.

E. Shiles
Department of Physics, Virginia Commonwealth University
Richmond, Virginia 23284

ABSTRACT

The magnetization of a two-component induced moment crystal is examined in a mean-field model that includes correlations in the occupancy of the nearest neighbor shell. The mixing is treated exactly in the limit of complete randomness. Ions with singlet ground state and triplet lowest excited state are considered. The magnetization and the critical temperature for magnetic ordering are found to be considerably larger than that from earlier work on a similar mean field model without correlations.

In an earlier paper[1] we examined the magnetization and ordering temperature of a two-component induced moment crystal with ions having singlet ground state and triplet lowest excited state. In this model, referred to herein as MFM, the magnetic interactions were treated using a mean-field approximation, and the mixing of the components was treated exactly in the limit of complete randomness. The present calculation extends this earlier model by including correlations in the occupancy of the nearest neighbor sites of a chosen ion. The correlations are included in a manner similar to that of Richards[2], who examined the magnetization of a random ferromagnet.

We have examined the effect of nearest neighbor correlations in a binary induced moment crystal in a singlet-singlet model[3], and found the magnetization and the ordering temperature to be larger than that for the MFM model without correlations. This work is extended here to the singlet-triplet model, as this model describes crystals available for experimental measurements. The procedure of Hsieh and Blume[4], where two spin one-half operators are used to describe the singlet-triplet system, is used to find the mean field states at each ion. These states are then used to calculate the magnetization, and the critical temperature is obtained by extrapolating the result to zero magnetization. The model is limited to describing a mixture of two singlet-triplet systems, both of which can be described by the Hsieh and Blume representation, or the case where one of the components is non-magnetic.

The mean-field Hamiltonian for a mixture of type A and type B ions is, with no external magnetic field,

$$H = H^A + H^B \tag{1}$$

where (ν=A,B; τ=A,B; $\tau\neq\nu$),

$$H^\nu = \sum_i V_{ci}^\nu \sigma_i^\nu - \sum_{ij} \mathfrak{J}^{\nu\nu}(ij)\sigma_i^\nu\sigma_j^\nu\mu_i^\nu J_j^{\nu z} - \sum_{ij}\mathfrak{J}^{\nu\tau}(ij)\sigma_i^\nu\sigma_j^\tau\mu_i^\nu J_j^{\tau z}. \tag{2}$$

i and j are summed over all sites in the crystal, V_{ci}^ν is a single-ion crystal field operator at a ν-type ion, which gives a singlet ground state and triplet lowest excited state, $\mathfrak{J}^{\nu\tau}(ij)$ is the exchange constant between ions of types ν and τ at sites i and j, $\mu_i^\nu = \langle J_i^{\nu z}\rangle$ is the thermal expectation value of the z-component of angular momentum $\vec{J}_i^\nu$, and σ_i^ν is is an occupation operator equal to unity if a ν-type ion is at site i and zero otherwise. In this work we will assume that the crystal field at an ion is independent of the composition of that ion's surroundings, and also, for simplicity, consider exchange coupling with only nearest neighbors.

Following Hsieh and Blume we take, where $\vec{S}$ and $\vec{T}$ are two pseudospin-½ operators,

$$J_i^{\nu z} = a_\nu S_i^{\nu z} + b_\nu T_i^{\nu z}. \tag{3}$$

In this formalism the crystal field only Hamiltonian is represented by $\sum_i \Delta^\nu \vec{S}_i^\nu \cdot \vec{T}_i^\nu$, where Δ^ν is the splitting between the singlet ground state and triplet lowest excited state of an isolated ion of type ν. The a_ν and b_ν must be chosen for the particular crystal under study. For example, ions Tb^{3+} and Tm^{3+} (J=6), in an octahedral crystal field, require $a_\nu=\frac{1}{2}(1+2\alpha_\nu)$ and $b_\nu=\frac{1}{2}(1-2\alpha_\nu)$, where $\alpha_\nu=\sqrt{14}$ is the matrix element between the ground state and one of the excited states (from the states Γ_1 (singlet) and Γ_4 (triplet) given by Lea, Leask, and Wolf[5]). The calculation of the magnetic moments is the same as in reference 1, where we diagonalized the Hamiltonian and used the new states to determine the thermal expectation value μ_i^ν. We obtain for the magnetization at a central site i = 0, which is a ν-type ion surrounded by n^ν ν-type ions on the exchange coupled nearest neighbor sites,

$$\mu_o^\nu(n^\nu) = \frac{1}{Z_o^\nu}\Big\{(a_\nu+b_\nu)e^{-\beta\Delta^\nu}\sinh\left[\beta(a_\nu+b_\nu)\mathfrak{J}_o^\nu(n^\nu)\right] + 2\mathfrak{J}_o^\nu(n^\nu)(a_\nu-b_\nu)^2 e^{-\frac{1}{2}\beta\Delta^\nu}\sinh\left[\tfrac{1}{2}\beta\Delta^\nu\mathcal{g}_o^\nu\right]/\Delta^\nu\mathcal{g}_o^\nu\Big\} \tag{4}$$

where $\mathfrak{J}_o^\nu(n^\nu) = \mathfrak{J}^{\nu\nu}\sum_\delta \mu_\delta^\nu\sigma_\delta^\nu + \mathfrak{J}^{\nu\tau}\sum_\delta\mu_\delta^\tau\sigma_\delta^\tau$ ($\tau\neq\nu$, δ summed over the nearest neighbors) is the exchange strength, $\mathcal{g}_o^\nu=(1+(2(a_\nu-b_\nu)\mathfrak{J}_o^\nu(n^\nu)/\Delta^\nu)^2)^{\frac{1}{2}}$, and $Z_o^\nu=2e^{-\beta\Delta^\nu}\cosh(\beta(a_\nu+b_\nu)\mathfrak{J}_o^\nu(n^\nu)) + 2e^{-\frac{1}{2}\beta\Delta^\nu}\cosh(\frac{1}{2}\beta\Delta^\nu\mathcal{g}_o^\nu)$ is the partition function. A similar expression is obtained for $\mu_\delta^\nu(p^\nu)$, the magnetization on a neighbor site δ on which a ν-type ion resides and which is surrounded by p^ν ν-type nearest neighbors. For this case the exchange strength has the expression

$$\mathfrak{J}_\delta^\nu(p^\nu) = \mathfrak{J}^{\nu\nu}\sum_{\delta'}\mu_{\delta'}^\nu\sigma_{\delta'}^\nu + \mathfrak{J}^{\nu\tau}\sum_{\delta'}\mu_{\delta'}^\tau\sigma_{\delta'}^\tau, \tag{5}$$

where δ' is summed over the neighbors of the site δ. The magnetization on the τ-type nearest neighbors is expressed similarly.

We want to calculate $\overline{\mu}_o^\nu$, the magnetization at the central site averaged over all possible configurations of the z nearest neighbors, from the expression

$$\bar{\mu}_o^\nu = \sum_{n^\nu=o}^{z} \binom{z}{n^\nu} (m^\nu)^{n^\nu}(1-m^\nu)^{z-n^\nu}\mu_o^\nu(n^\nu), \quad (6)$$

where m^ν is the fractional concentration of ν-type ions ($m^\nu + m^\tau = 1$). In the MFM we excluded correlations by taking $\mu_\delta^\nu = \bar{\mu}_o^\nu$; in the present calculation we include these correlations by calculating μ_δ^ν for each of the numbers n^ν. We close the set of equations (restrict the exact consideration to the central site and its z nearest neighbors) by a suitable approximation on the sums in (5). These sums can be expressed

$$\sum_{\delta'} \mu_{\delta'}^\nu \sigma_{\delta'}^\nu = \mu_o^\nu\sigma_o^\nu + \sum''_{\delta'}\mu_{\delta'}^\nu\sigma_{\delta'}^\nu + \sum'_{\delta'}\mu_{\delta'}^\nu\sigma_{\delta'}^\nu . \quad (7)$$

The first term involves site zero, the double primed sum involves those nearest neighbors of δ that are also nearest neighbors of the central site (forming nearest neighbor triangles), and the single primed sum includes all other nearest neighbors of δ. A similar expression exists for the sum with τ superscripts. To consider only correlations in the nearest neighbor shell we approximate the single primed sum by an average value $\sum'_{\delta'}\mu_{\delta'}^\nu\sigma_{\delta'}^\nu \approx z'm^\nu\bar{\mu}_o^\nu$, where z' is the number of such neighbors. For calculational simplicity we make one more approximation; the double primed sum is approximated by the simple average $\sum''_{\delta'}\sigma_{\delta'}^\nu\mu_{\delta'}^\nu \approx (z''/z)n^\nu\mu_\delta^\nu$, where z'' is the number of nearest neighbors of δ that are also nearest neighbors of the central site. For a crystal structure with no nearest neighbor triangles, this term is zero. In this approximation all μ_δ^ν are the same and all μ_δ^τ are the same, for a particular n^ν. The exchange term (5) becomes

$$\mathcal{J}_\delta^\nu = \mathcal{J}^{\nu\nu}\mu_o^\nu\sigma_o^\nu + (z''/z)n^\nu\mathcal{J}^{\nu\nu}\mu_\delta^\nu + z'm^\nu\mathcal{J}^{\nu\nu}\bar{\mu}_o^\nu + (z''/z)(z-n^\nu)\mathcal{J}^{\nu\tau}\mu_\delta^\tau + z'(1-m^\nu)\mathcal{J}^{\nu\tau}\bar{\mu}_o^\tau \quad (\nu\neq\tau) \quad (8)$$

and the exchange term $\mathcal{J}_o^\nu(n^\nu) = n^\nu\mathcal{J}^{\nu\nu}\mu_\delta^\nu + (z-n^\nu)\mathcal{J}^{\nu\tau}\mu_\delta^\tau$. We now have a closed set of equations, from which we can solve for $\bar{\mu}_o^\nu$ and $\bar{\mu}_o^\tau$; this calculation must be done numerically and self-consistently.

For a numerical comparison of the correlation model with the earlier MFM, we choose a binary system consisting of induced moment ions (type A) and non-magnetic ions (type B). An example of such a system is $Tb_{m^A}Y_{1-m^A}Sb$. Cooper and Vogt[6] have determined, for this crystal, splitting $\Delta^A = 11.9°K$, and ordering temperature for pure TbSb, $T_c = 15.1°K$. Using this splitting, using $T_c = 15.1°K$ to obtain the exchange constant $\mathcal{J}^{AA}$ between magnetic ions, and using the values of a_A and b_A given earlier for Tb^{3+}, we obtain the curves of figure 1 for $\bar{\mu}_o^A$ as a function of temperature. A crystal structure with z=6, z'=5, and z''=0 (simple cubic) is assumed. Curve I is for m^A=1.0, which is the same for MFM and the present correlation model, since the difference is in the way the mixing is treated. For curves II, m^A=0.5. The correlation model shows a considerably larger magnetization at this magnetic concentration than does the model MFM (dashed curve). Figure 2 shows the magnetic ordering temperature T_c as a function of magnetic concentration, obtained by extrapolating the magnetization curves to zero magnetization. Again the dashed curve is the MFM result. The correlation model shows a considerably larger critical temperature. The experimental measurements of Cooper and Vogt are also shown in this figure. The large disagreement is probably due to the exchange coupling in the crystal being more complicated than the simple nearest neighbor scheme utilized in our model.

We plan to examine this correlation model for other crystal structures to see the effects on the magnetization and ordering temperature as z, z', and z'' are varied.

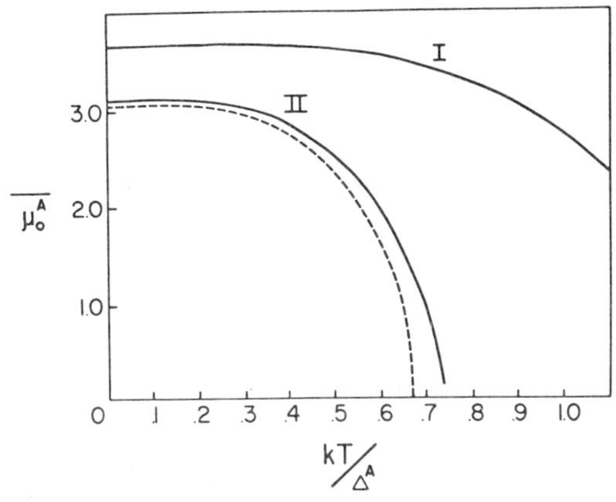

Fig. 1. Magnetization as a function of temperature. Curve I is for m^A = 1.0 and II for m^A = 0.5. The dashed curve is the MFM result.

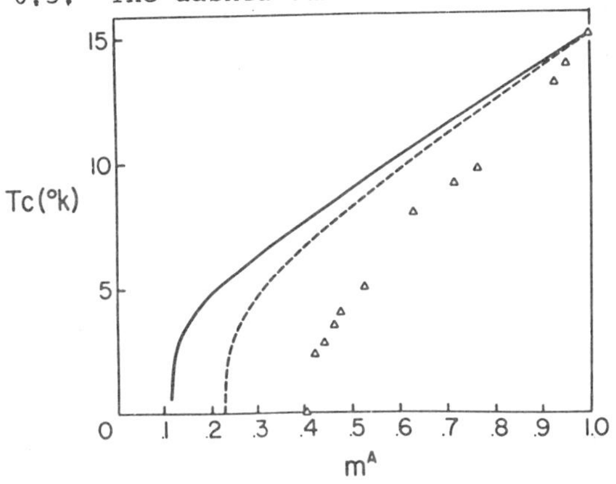

Fig. 2. Critical Temperature as a function of magnetic concentration. The dashed curve shows the MFM result and the symbols (Δ) are the experimental results on $Tb_{m^A}Y_{1-m^A}Sb$[6].

1. E. Shiles, R.A. Tahir-Kheli, and G.B. Taggart, J. Phys. C (London) 7, 2487(1974).
2. P.M. Richards, Phys. Letters 44A, 389(1973; A.I.P. Conference Proceedings No. 18, ed. by C.D. Graham and J.J. Rhyne, A.I.P., New York, 600(1974).
3. E. Shiles and G.B. Taggart, to be published in Solid State Communications.
4. Y.Y. Hsieh and M. Blume, Phys. Rev. B6, 2684(1972).
5. K.R. Lea, M.J.M. Leask, and W.P. Wolf, J. Phys. Chem. Solids 23, 1381(1962).
6. B.R. Cooper and O. Vogt, Phys. Rev. Bl, 1218(1970).

MAGNETIC FIELD DEPENDENCE OF MOLECULAR ORIENTATIONS IN LIQUID CRYSTAL DROPLETS

Murray J. Press and Anthony S. Arrott
Simon Fraser University, Burnaby, British Columbia, Canada V5A 1S6

ABSTRACT

Calculations of the behavior of liquid crystal droplets in magnetic fields applied along the axis of cylindrical symmetry have been carried out using relaxation methods on the finite difference approximation to the Euler-Lagrange equations derived from the elastic strain energy. The nature of the solutions and how they change with field is discussed. The application of these methods to problems in magnetism is suggested.

It has been said that the similarities between the science and technology of magnetism and that of liquid crystals are sufficient to entice a magnetician into the field of liquid crystal research but are not sufficient to do him much good after he is there. Yet it seems that a case can be made for the inverse effect of feedback from liquid crystal research to magnetism. There is a long standing problem in magnetism of finding the lowest energy configurations for the magnetization patterns in various geometries when the only terms in the energy are the isotropic exchange energy and the magnetostatic interactions. There is a similar problem in liquid crystals on which some progress has been made recently. The methods used in the liquid crystal system suggest an approach to the magnetism problem. The work reported here illustrates these methods for the treatment of the magnetic field dependence of the configuration of the molecular alignment in a liquid crystal droplet.

The molecular alignment is described by a director field $\hat{n}$ which is quite analogous to the unit vector $\vec{M}/|M_s|$. In a cylindrical coordinate system

$$\hat{n}(\rho,z) = \cos\theta(\rho,z)\ \hat{z} + \sin\theta(\rho,z)\ \sin\varphi(\rho,z)\ \hat{\Phi} + \sin\theta(\rho,z)\ \cos\varphi(\rho,z)\ \hat{\rho} \qquad (1)$$

where θ and φ are polar angles referred to the $\hat{z}$ axis and the local $\hat{\rho}$ axis respectively. Cylindrical symmetry has been assumed. The energy[1] is written as

$$E = \pi\int \rho d\rho dz \left\{S\ \mathrm{div}^2\hat{n} + T(\hat{n}\cdot\nabla\times\hat{n})^2 + B(\hat{n}\times\nabla\times\hat{n})^2\right\} \qquad (2)$$

where the coefficients S, T, and B stand for Splay, Twist, and Bend. If S=B=T=2A one can write for the energy

$$E = 2\pi A\int \rho d\rho dz \left\{\mathrm{div}^2\hat{n} + \mathrm{curl}^2\hat{n}\right\} \qquad (3)$$

and this is the same form as the isotropic exchange energy in magnetism[2]. In magnetism the magnetostatic interactions tend to eliminate div $\vec{M}$ and one thus sees that the exchange problem is one of minimizing $\int \mathrm{curl}^2\vec{M}\,dv$ with the constraint div$\vec{M}$=0. In this formulation of the problem some things are clearer than if, as usual, the exchange is written as $(\nabla\alpha)^2 + (\nabla\beta)^2 + (\nabla\gamma)^2$. The liquid crystals of interest here are nematic and the order parameter is quadrapolar rather than dipolar as in magnetism. The energy as written above assumes that $\hat{n}$ is slowly varying on an atomic scale, as does the expression for the exchange energy.

In many liquid crystal problems it is necessary to take account of the naturally occurring inequality of the elastic constants which typically have B>S>T[3]. Here the problem is simplified for the sake of discussion by taking S=T=B even though many of the actual calculations that have been carried out are for specific examples of the general case. Including the effect of a magnetic field along the z axis upon the anisotropic susceptibility ($\Delta\chi$) of the liquid crystal molecules one finds for the energy[4]

$$2\pi A\int \rho d\rho dz \left[\sin^2\theta\left[\frac{1}{\rho} + \rho\,\varphi_\rho^{\ 2} + \rho\,\varphi_z^{\ 2} - 2\cos\varphi\ \theta_z + 2\rho\sin\varphi\,(\theta_z\varphi_\rho - \theta_\rho\varphi_z) + \frac{\Delta\chi}{2A}H^2\right] + \rho\,\theta_\rho^{\ 2} + \rho\,\theta_z^{\ 2} + 2\sin\theta\cos\theta\ \theta_\rho\right] \qquad (4)$$

where a subscript signifies a partial derivative.

This energy has a term $\cos\varphi\ \theta_z$ which can be used to decrease the energy through a most effective mechanism which has been termed "splay cancelling". In cylindrical coordinates

$$\mathrm{div}\hat{n} = \sin\theta\left[\frac{\cos\varphi}{\rho} - \theta_z - \sin\varphi\ \varphi_\rho\right] + \cos\theta\ \cos\varphi\ \theta_\rho \qquad (5)$$

and if the boundary conditions dictate a change of θ with z a suitable choice of $\cos\varphi$ leads to a reduction in div$\hat{n}$. This point is emphasized in Fig. 1 where the similarly appearing configurations 1a and 1b differ greatly in energy because of splay cancelling in case 1b.

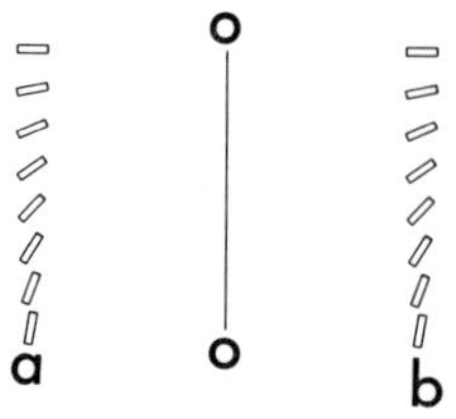

Fig. 1 Compare the two different patterns generated by rotation of configuration a or b about the central axis 00. For a the first two terms in div$\hat{n}$ add and for b they cancel.

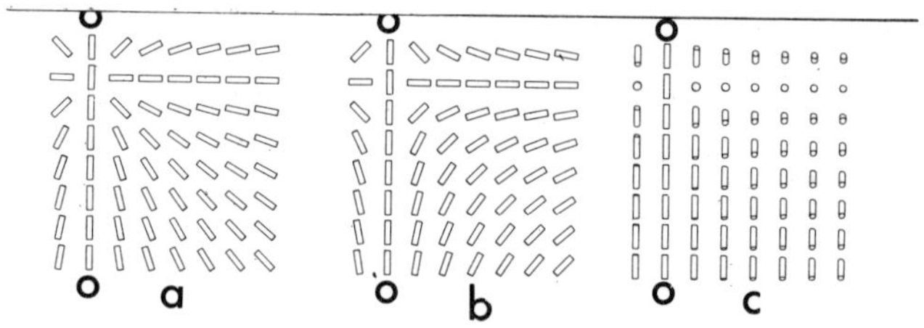

Fig. 2 Molecular alignment about a point singularity. Each pattern has rotational symmetry about the vertical axis 00. $\theta(\rho,z)$ is the same in each pattern. In a $\cos\varphi = -1$, in b $\cos\varphi = +1$, and in c $\cos\varphi = 0$.

Point singularities with cylindrical symmetry are shown in Fig. 2. These are described by

$$\tan\theta = \frac{\rho}{z} \qquad (6)$$

with $\varphi = \pi$ for a, $\varphi = 0$ for b, and $\varphi = \pi/2$ for c. The energy in a sphere of radius R about the singularity is $16\pi AR$ for 2a, $16\pi AR(1/3)$ for 2b and $16\pi AR(2/3)$ for c. These configurations are solutions of the torque equations found by the calculus of variations from the energy of Eq. (4):

$$\theta_{\rho\rho} + \frac{1}{\rho}\,\theta_\rho + \theta_{zz} = \cos\theta\ \sin\theta\left\{\frac{1}{\rho^2} + \varphi_\rho^2 + \varphi_z^2 + \frac{\Delta\chi}{2A}H^2\right\} \qquad (7a)$$

$$\varphi_{\rho\rho} + \frac{1}{\rho}\,\varphi_\rho + \varphi_{zz} = -2\cot\theta\ \left\{\varphi_\rho\theta_\rho + \varphi_z\theta_z\right\} \qquad (7b)$$

In fact $\tan\theta = \rho/z$ satisfies these equations for any constant value of φ as long as $H^2 = 0$. The torque equations contain no knowledge about splay cancelling. Its effect is noticed only if the energy is evaluated or if one looks at the boundary conditions that arise from the Euler-Lagrange equations (Eq. (7)).

The experimental system most studied was that of the nematic liquid crystal methoxybenzilidene butylaniline (MBBA) at an interface between water (at top) and air (at bottom). This produces lens-shaped droplets because of the effects of surface tension. The full treatment of a droplet necessary to give details of the internal molecular alignment has been beyond present ambitions. More restricted calculations have been carried out using a geometry which simulates only the central portion of a droplet. The top and bottom surfaces are taken as flat and the side boundary as cylindrical. The modeling of the droplet behavior is then achieved through the choice of boundary conditions at this cylindrical surface. From the integration by parts one obtains

$$\varphi_\rho + \sin\varphi\ \theta_z = 0 \tag{8a}$$

and

$$\sin^2\theta\ \sin\varphi\ \varphi_z - \theta_\rho = \frac{\sin\theta\ \cos\theta}{\rho} \tag{8b}$$

as the conditions to be satisfied if θ and φ are not pinned at the cylindrical surface. For MBBA at the water-air interface one has the experimentally determined values of $\theta=\pi/2$ on the top surface and $\theta=\pi/12$ on the bottom surface. The suitable elastic constants are $B/S=5/4$, and $T/S=5/8$. In this case the torque equations (Eq. (7)) are increased in complexity to 41 terms in each, yet these have been treated for various ratios of elastic constants and for various applied fields[4]. Here the discussion centers on the artificial case S=T=B with $\theta=\pi/2$ at the top and $\theta=\pi/12$ on the bottom as boundary conditions. Note that these boundary conditions do not pin φ on either surface. A similar pinning condition occurs in magnetism where the component normal to a surface is determined by the charge density required to give a demagnetizing field sufficient to (almost) cancel the applied field. The components in the plane of the surface are not pinned. An initial set of calculations compares the configurations and energies for a range of field parameters that correspond to 0-20 kG for a typical MBBA droplet 20μ thick and 150μ in diameter. The differential equations are replaced by difference equations and solutions computed using relaxation methods on a grid of 30 by 40 points.

With no field present the lowest energy configuration is characterized by $\varphi=0$ everywhere and θ increasing monotonically from the bottom to the top surfaces except from radii less than 7 microns as shown in Fig. 3a. As the magnetic field increases the molecules become oriented more closely to the axis of symmetry of the droplets. Above 3 kG θ decreases from both surfaces. For magnetic fields less than 10 kG the molecules continue to align more closely to the axis of symmetry but there is no change in their azimuthal angles. Above 10 kG the energy is reduced if the molecules near the lower surface rotate about an axis parallel to the axis of symmetry. For 12 kG the azimuthal angles near the lower surface vary from about $\pi/2$ for small radii to about π for large radii while along the upper surface φ remains at about 0. The driving force for this rotation is the effect of splay cancellation. When the magnetic field is small θ increases from the bottom surface to the top surface which from Eq. (8a) produces a negative value for $\varphi_\rho/\sin\varphi$ (if $\sin\varphi\neq0$). Any set of values for φ other than 0 (or π which is a much higher energy metastable solution) is therefore unstable and decays to a uniform value for φ of zero everywhere. When the magnetic field is increased above 3 kG and θ_z becomes negative near the lower surface $\varphi_\rho/\sin\varphi$ in this region becomes positive and any deviation of φ from zero near

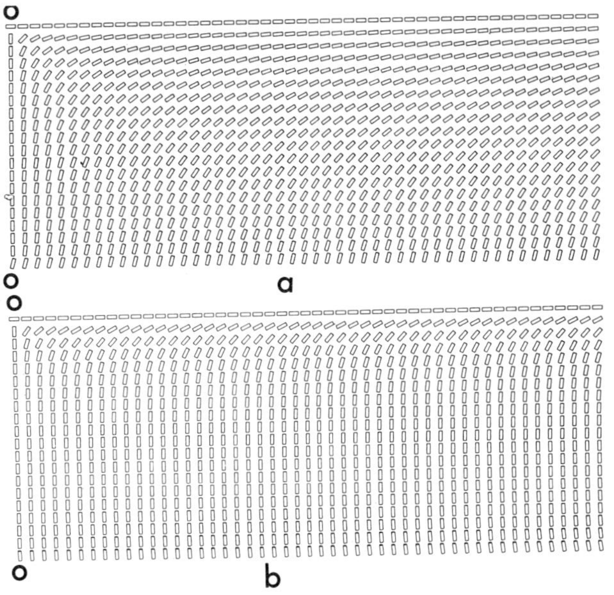

Fig. 3 The calculated molecular configuration in the ρ-z plane of "MBBA" (S=T=B, $\Delta\chi/S=.15$ cgs) between air and water. In a there is no magnetic field present while in b a field of 12 kG parallel to the symmetry axis has been applied.

the bottom surface might therefore grow. However near the upper surface $\varphi_\rho/\sin\varphi$ is still negative which requires φ near the upper surface to be zero. These two regions could be joined by a transition region in which φ varies from some finite value ($\sim\pi$) near the bottom surface to 0 near the upper surface. This transition region can occur where θ_z is zero. However, the energy (Eq. (4)) contains the term $\rho\ \sin^2\theta\ \varphi_z^2$ which is large if the transition from $\varphi\sim\pi$ to $\varphi\sim0$ occurs at non-negligible values of θ. Thus as long as the magnetic field is sufficiently weak so that θ (at large radii) never becomes close to zero the transition cannot occur and φ remains 0 everywhere. At 10 kG θ is sufficiently close to zero far from the surfaces that the $\varphi=0$ solution becomes metastable and the molecules can change their azimuthal angles. The driving force determining φ at 10 kG is very small and the exact nature of the transition has not been determined. By the time the field has reached 12 kG the transition has occurred and the azimuthal angles are between $\pi/2$ and π along the lower surface and 0 along the upper surface separated by a transition region where θ is approximately zero: the effect of the magnetic field has been to divide the droplet into two independent regions.

With realistic values of the elastic constants the phase transition occurs at lower fields for the same sized drop. Experimentally no discontinuities are observed when droplets are viewed under crossed polarizers in vertical magnetic fields up to 10 kG. The color phenomena in these droplets in magnetic fields are considered spectacular.

REFERENCES

1. This is referred to as the Frank elastic energy. For entre to the field of liquid crystals see P.G. Gennes Liquid Crystals, Oxford University Press, Great Britain, 1974.
2. A.S. Arrott, B. Heinrich and D.S. Bloomberg, IEEE Trans. on Magnetics, Vol. Mag-10, No. 3, 950-3, 1974.
3. M.J. Press and A.S. Arrott, Phys. Rev. Letters, Vol. 33, No. 7, 403-6, 1974.
4. M.J. Press and A.S. Arrott, J. Physique, Jan. 1975 (in press); M.J. Press thesis, Simon Fraser University, 1974.

MAGNETIC INTERACTIONS IN Ni^{++} SALTS WITH THE $NiSnCl_6 \cdot 6H_2O$ STRUCTURE*

M. Karnezos and S. A. Friedberg
Physics Department, Carnegie-Mellon University, Pittsburgh, Pennsylvania 15213

ABSTRACT

A He^3 -He^4 dilution cryostat has been used to extend our magnetic susceptibility measurements on single crystal $NiSnCl_6 \cdot 6H_2O$ and several isostructural Ni^{++} salts down to ~0.05K. These crystals are trigonal and contain one axially distorted $[Ni(H_2O)_6]^{++}$ complex per unit cell. Earlier data[1] between 0.4 and 4.2K are well described by the Hamiltonian $=DS_z^2 + g\,\mu_B\,\underline{S}\cdot\underline{H}_{eff}$ with S=1, g≃2.3, $\underline{H}_{eff} = \underline{H}_o + n\underline{M}$, where n is a mean field constant and $\underline{M}$ the magnetization. For $NiSnCl_6 \cdot 6H_2O$, D/k ≈ +0.6K and n = -0.02 mole/emu, i.e. the ground state is a singlet and spin coupling is weakly antiferromagnetic. According to mean field theory this interaction is subcritical and no spontaneous ordering is expected for $\underline{H}_o=0$. None is observed down to ~0.05K although discrepancies with mean field theory are seen. For $NiZrF_6 \cdot 6H_2O$ (D/k = -2.8K and n = +0.09 mole/emu) and $NiTiF_6 \cdot 6H_2O$ (D/k = -1.7K and n = +0.08 mole/emu) the ground state is a doublet and interaction is ferromagnetic. We detect ferromagnetism in these salts below T_c ≈ 0.19K ($NiZrF_6 \cdot 6H_2O$) and T_c ≈ 0.18K ($NiTiF_6 \cdot 6H_2O$) as compared with mean field estimates of T_c = 0.22K and 0.19K, respectively. Of particular interest is the series $NiPtCl_6 \cdot 6H_2O$ (D/k = +0.66K), $NiPtBr_6 \cdot 6H_2O$ (D/k ≈ +0.32K) and $NiPtI_6 \cdot 6H_2O$ (D/k ≤ +0.07K) in which the singlet-doublet separation decreases rapidly. Spin interactions, which are subcritical for the chloroplatinate, may be approaching critical size for the iodoplatinate. A full account of the results on these and other compounds with the $NiSnCl_6 \cdot 6H_2O$ structure will be published elsewhere.

*Work supported by the NSF and the ONR.

[1]M. Karnezos and S. A. Friedberg, Bull. Am. Phys. Soc. 19, 369 (1974).

FIELD INDUCED MAGNETIC PHASE TRANSITIONS IN DAG†

W. P. Wolf
Department of Engineering and Applied Science, Yale University, New Haven, Connecticut 06520

ABSTRACT

Dysprosium aluminum garnet (DAG) is a much studied antiferromagnet, and in this paper we review its present status as a model system for understanding several kinds of field induced phase transitions. Until redently, the behavior in fields along [111] seemed to be quite simple and generally consistent with that of a two sublattice Ising model, although there were some indications of discrepancies in detailed comparisons. Recent optical and neutron scattering experiments have shown that there is in fact an extra dimension previously neglected in this system, in that an applied field can couple not only to the magnetization but also to the staggered magnetization. This is equivalent to an induced staggered magnetic field and it leads to significant changes in the nature of the phase diagram. General considerations of symmetry and the microscopic mechanisms responsible for the coupling indicate that similar effects should not influence the phase diagrams for special directions such as [001] and [110]. However, for these directions it is necessary to take into account the multi-sublattice structure of DAG, and this results in the somewhat more complicated situation of coupled order parameters which vanish simultaneously as the phase transition is approached.

INTRODUCTION

Dysprosium aluminum garnet, $Dy_3Al_5O_{12}$, (DAG), has been studied by many techniques since the discovery of its unusual antiferromagnetic properties in 1962.[1,2] The principal feature of this material is an extremely strong local anisotropy which constrains the Dy^{3+} moments to one of three axes, so that the ordered state involves three sets of orthogonal pairs of sublattices parallel to ±X, ±Y, and ±Z. In the presence of a field along [111] these three directions become equivalent and the system can be closely approximated by a two sublattice Ising model. For this reason, almost all experiments on DAG have been made with fields along [111] and we shall here limit most of our discussion to this case. However, in the final section we shall also consider H||[001] and H||[110], two cases which provide specially interesting situations.

A detailed discussion of the correspondence between DAG and the Ising model has been given in several papers and the most complete summaries may be found in Refs. 3 and 4. It seems clear that the Ising model should be an extremely good approximation, with non-Ising terms about two orders of magnitude smaller. Recent developments have shown that non-Ising terms even as small as this may produce some unusual effects under special conditions, and we shall discuss these below. However, it would appear that almost all of the present experimental results for DAG can in fact be explained in terms of an appropriate Ising model.

Estimates of the strengths of the interactions between nearest, next nearest, etc. neighbors have been made,[5,6] and the results have been summarized in Ref. 4. The main features include a number of competing near neighbor interactions, resulting from the rather complex garnet crystal structure, and relatively large contributions from long range magnetic dipole forces. The combined effect of all the interactions leads to a Néel temperature of about 2.5 K. The corresponding energy of the antiferromagnetic (A) state relative to the paramagnetic (P) state in an applied field results in critical fields for A-P phase transitions in the region of a few kOe. This makes DAG a very attractive material for the study of field induced phase transitions, and such studies have revealed a number of interesting effects.

For temperatures below about 1.66 K the observed phase transitions were clearly of first order, though the interpretation of the experiments was complicated by rather large demagnetizing effects. Making allowances for these, the expected discontinuities in magnetization and entropy were found.[2]

In the region between 1.66 K and T_N the behavior was not so clear. For a simple Ising model one would expect second order singularities, whereas both the specific heat and the differential susceptibility showed maxima which were smooth and finite to an extent which could not be readily ascribed to experimental "rounding." The analysis was again complicated by demagnetizing effects, but even after allowance was made for these it was difficult to reconcile the results with the expected behavior.

This problem was discussed by Boccara[7] and by Benguigui[8] who pointed out that it is in fact possible for a system such as DAG to exhibit non-singular behavior in its macroscopic thermodynamic functions while still undergoing a second order phase transition with respect to its antiferromagnetic order parameter. Such behavior would be quite unusual since the singularity in the variation of the order parameter at a second order transition is generally reflected also in other derivatives of the free energy, but it is a possibility which can not be excluded without additional information.

Recently, there have been reports of three independent series of measurements which provide some further insight into the phase transitions in DAG, and they show that DAG is indeed a most unusual system, with features which have so far not been observed for any other material.

RECENT EXPERIMENTS

1. Tricritical Point Scaling. If DAG were a normal antiferromagnet, one would expect the point at which the field induced phase transition ceases to be first order to be a tricitical point, in the sense defined by Griffiths.[9] Following this assumption, Tuthill et al.[10] attempted to reduce the available information on the magnetization as a function of field and temperature in this region to a single scaling function and they fitted the appropriate critical exponents, $\beta_u = 0.65$ and $\delta_u = 3$. This attempt was subsequently followed by a similar analysis based on more accurate magnetization measurements[11] but the results were essentially the same ($\beta_u = 0.5\pm0.2$, $\delta_u = 3.2\pm1.0$). Both studies showed that the fitted exponents were quite far from the values one might expect for a classical tricritical point ($\beta_u = 1$, $\delta_u = 2$) and there was also some evidence of systematic deviations from a simple scaling relation. It seemed clear that some additional insight into the nature of the transition was required.

2. Magneto-optical Experiments. Further evidence of unusual behavior in DAG was provided by a series of magneto-optical experiments by Dillon et al.[12-14] This work is being reviewed in another paper in this conference[15] and we will here mention only the most significant result. This involved the observation of two "essentially distinct antiferromagnetic states" depending on the past magnetic history of the sample, and a corresponding hysteresis as the field was cycled about zero. These states subsequently turned out to be the two time reversed conjugate ordered states, ("up-down" and "down-up"), which have different energies in the presence of a magnetic field. The origin of this unexpected difference was explained by Blume et al.,[16] in connection with some similarly puzzling neutron scattering experiments.

3. Neutron Scattering with H||[111]. Blume et al.[16] had attempted to measure the field dependece of the antiferromagnetic order parameter as the field approached the region in which the thermodynamic measurements had indicated a second (or higher) order transition. Since neutrons measure the order parameter directly, it was expected that the elastic Bragg scattering would go to zero at some field, corresponding to the disappearance of long range antiferromagnetic order. In fact, however, no such disappearance was found and Bragg scattering, indicating long range order was observed well into the region previously thought to be paramagnetic.

DISCUSSION

An explanation for the neutron scattering effect as well as for the anomalous optical and magnetic results was provided by Blume et al.,[16] who noted that the symmetry of DAG permits a coupling between the antiferromagnetic order parameter, η, and an applied field, so that there is in fact no temperature at which $\eta = 0$ in the presence of a field. Such a coupling is well known in systems in which a spontaneous magnetic moment is allowed, i.e. for canted antiferromagnets,[17] but it had never been considered for systems such as DAG in which a spontaneous moment is not allowed by symmetry. For systems with a spontaneous moment, the coupling is linear in H, i.e. there is a term in the free energy proportional to $-\eta H$, while for DAG the allowed term turns out to have the form $\eta H_x H_y H_z$.

Such a coupling is equivalent to an induced staggered field, H_s, which would by itself lead to a term in the free energy of the form $-\eta H_s$. The induced H_s is thus proportional to H^3 and we might therefore expect the effect to be quite small at low fields. An analysis of the observed hysteresis has shown[18] that $H_s \approx 1$ Oe for an applied field of 3 kOe at T = 1.3 K, but this apparently is quite sufficient to distinguish between the two time reversed states and to destroy the second order phase transition. The point at which phase transition ceases to be first order may now be identified as a "wing" critical point,[16] close to but distinct from the true tricritical point, which would correspond to $H_s = 0$.

Two alternative microscopic mechanisms have been proposed to account for the coupling between η and H. One involves the small but non-zero g-values, g_x and g_y, which can lead directly to a difference in the coupling between the field and the spins at two different types of sites in the lattice.[16] This mechanism would imply a deviation from the ideal Ising model and its effect can be calculated readily in terms of the known g-values. It turns out that the contribution to the energy is quite small, at least in the regions in which the measurements were made, and it would appear that this mechanism is not able to account quantitatively for the observed hysteresis or for the neutron scattering results.

The second mechanism is based on an unusual topological difference in the interaction linkages between spins at different sites in the lattice. This effect is discussed in some detail in another paper at this conference[18] and we shall state here merely that it would appear to account for all the observed effects in terms of the previously determined interactions, without having to invoke any deviations from an ideal Ising model. The special feature of the DAG lattice which gives rise to such an unusual effect is related to the nature of the interactions between the various near neighbors, which when defined relative to one particular set of axes show a characteristic variation of sign from pair to pair. As a result, thermal excitations on certain sites occur preferentially over those on other sites in the presence of a field. Details of this mechanism are discussed elsewhere.[18]

Both the general theory based on symmetry arguments and the two microscopic mechanisms indicate that all the unusual effects in DAG should occur only if the field has components along all three cubic axes, i.e. $H_x H_y H_z \neq 0$. For special symmetry directions for which this product vanishes, one would expect no coupling between the antiferromagnetic order parameter and an applied field and consequently a more normal magnetic behavior. Two directions which satisfy this condition are [001] and [110] and it is of interest to review the available information for these cases.

EXPERIMENTAL RESULTS FOR H||[001] AND H||[110]

Since the special character of these directions was not recognized until recently there are relatively few experimental results, and in particular there are no detailed studies of the phase transition. Early magnetization[19,6] and specific heat measurements[20] led to phase diagrams[20] which are qualitatively similar to that deduced for H||[111],[2] at least down to 1 K. However, it is important to recognize one significant factor whenever H is not parallel to [111], that the three sets of perpendicular sublattices will then behave differently, so that the system will not be described by just one order parameter η.

A discussion of the various possible order parameters classified according to the symmetry of the garnet lattice has been given by Mukamel and Blume.[21] Since the unit cell contains 24 spins there is a very large number of possible terms in the free energy, and in the presence of an arbitrary applied field many different order parameters may be coupled by the field. A finite antiferromagnetic η will therefore generally imply also a finite value of one or more other types of order. At a second order phase boundary, coupled order parameters will vanish together and the system should behave essentially in a normal manner. However, it seems clear that the presence of more than one order parameter will generally complicate the quantitative analysis of such a phase transition.

There is at present no detailed information to indicate whether the low field phase transitions for H||[001] and H||[110] are in fact second order and correspondingly, whether the onset of the observed first order transitions at higher field corresponds to a true tricritical point. Further experiments are clearly called for.

One interesting feature of the phase diagrams for H||[001] and H||[110] which is unrelated to any of the above questions but which deserves comment and further experimental study, concerns a second phase transition which is observed[20] in each case at high fields $H_i > 5$ kOe and low temperatures ($T \approx 1$ K for H||[001] and $T \approx 0.6$ K for H||[110]). These transitions can be understood in terms of an ordering of the spins which are perpendicular to the applied field and which therefore do not interact directly with the field. In the high field paramagnetic phase these spins also do not couple to the spins parallel to the field due to the symmetry of the garnet lattice and they are therefore free to order according to their own mutual interactions. These interactions are relatively weak and the rather low ordering temperatures are consistent with the known details of the interactions. Discussions of these rather unusual phase transitions have been given in Refs. 6, 20 and 22.

CONCLUSIONS

In this brief review we have aimed primarily to introduce the wide variety of unusual effects which have been found in DAG and we must refer the reader to the cited references for additional details. The intriguing feature of DAG is that it is in some respects very simple, closely approximating a classical Ising model, while its relatively complex three dimensional structure leads to some unusual features not found in more conventional Ising-like systems. The striking effects observed in DAG which result from the coupling between the antiferromagnetic order parameter and the field through a term in the free energy of the form $\eta H_x H_y H_z$ will certainly be present also in other systems.

Indeed a recent group theoretical study by Alben et al.[23] has shown that of the 59 magnetic point groups which correspond to compensated antiferromagnetic states, no fewer than 27 should show similar effects, while another 11 should show coupling effects in fifth, seventh or ninth order in the field. The effects first observed in DAG with H||[111] should therefore be far from rare, though they might be masked by other complications in systems which can not be analyzed as unambiguously as DAG. It is clear that further experiments remain to be done, both for DAG and for other similar systems.

ACKNOWLEDGEMENTS

The author would like to acknowledge many helpful and stimulating discussions with R. Alben, M. Blume, L.M. Corliss, J.F. Dillon Jr., N. Giordano, J.M. Hastings and D. Mukamel. He would also like to thank the Physics Department of Brookhaven National Lab for its hospitality during the summer of 1974, when some of the ideas in this paper were first formulated.

REFERENCES

†This work was supported in part by the National Science Foundation, and in part by the Army Research Office, Durham.

1. M. Ball, M.T. Hutchings, M.J.M. Leask, and W.P. Wolf, Proc. 8th Int. Conf. on Low Temp. Physics (Butterworth, London) 248 (1963).
2. D.P. Landau, B.E. Keen, B. Schneider, and W.P. Wolf, Phys. Rev. B 3, 2310 (1971). This paper also contains an extensive list of references to earlier work.
3. W.P. Wolf, D.P. Landau, B.E. Keen, and B. Schneider, Phys. Rev. B 5, 4472 (1972).
4. J.C. Norvell, W.P. Wolf, L.M. Corliss, J.M. Hastings, and R. Nathans, Phys. Rev. 186, 557 (1969).
5. B. Schneider, D.P. Landau, B.E. Keen, and W.P. Wolf, Phys. Lett. 23, 210 (1966).
6. R. Bidaux, P. Carrara and B. Vivet, J. Physique 29, 357 (1968).
7. N. Boccara, Solid State Comm. 7, 331 (1968).
8. L. Benguigui, Phys. Rev. B 8, 412 (1973).
9. R.B. Griffiths, Phys. Rev. Lett. 24, 715 (1970).
10. G.F. Tuthill, F. Harbus, and H.E. Stanley, Phys. Rev. Lett. 31, 527 (1973).
11. A.T. Skjeltorp, R. Alben, and W.P. Wolf, AIP Conf. Proc. 18, 770 (1974).
12. J.F. Dillon Jr., E. Yi Chen, and W.P. Wolf, Proc. Int. Conf. on Magnetism, Moscow, 1973 (in press).
13. J.F. Dillon Jr., E. Yi Chen, and W.P. Wolf, AIP Conf. Proc. 18, 334 (1974).
14. J.F. Dillon Jr., E. Yi Chen, N. Giordano, and W.P. Wolf, Phys. Rev. Lett. 33, 98 (1974).
15. J.F. Dillon Jr., E. Yi Chen, and H.J. Guggenheim, AIP Conf. Proc. 1975, to be published.
16. M. Blume, L.M. Corliss, J.M. Hastings, and E. Schiller, Phys. Rev. Lett. 32, 544 (1974).
17. A good review of previous work on canted antiferromagnetic systems has been given by A.S. Borovik-Romanov, in Elements of Theoretical Magnetism, ed. by S. Krupicka and J. Sternberk, translated by W.S. Bardo (CRC Press, Cleveland, Ohio, 1968) p. 193.
18. N. Giordano and W.P. Wolf, AIP Conf. Proc. 1975 (to be published). A more extensive discussion will be published later.
19. A.F.G. Wyatt, Thesis, Oxford University, 1963 (unpublished).
20. B.E. Keen, D.P. Landau, and W.P. Wolf, Phys. Lett. 23, 202 (1966) and to be published.
21. D. Mukamel and M. Blume (private communication, to be published).
22. R. Bidaux and B. Vivet, J. Physique 29, 57 (1967).
23. R. Alben, M. Blume, L.M. Corliss, and J.M. Hastings, Phys. Rev. B (in press).

CRITICAL AND TRICRITICAL BEHAVIOR IN $FeCl_2$ - A SUMMARY

Robert J. Birgeneau
Bell Laboratories, Murray Hill, N. J. 07974

EXTENDED ABSTRACT

In the last decade there has been considerable progress in our experimental characterization and theoretical understanding of thermodynamic behavior around ordinary critical points.[1] As a result of these advances it is now both possible and indeed important to direct our attention towards multicomponent systems[2] in which there are multiple order parameters each with its corresponding conjugate field. Critical points then evolve into lines, surfaces etc. of phase transitions. This, in turn, may give rise to totally new critical behavior[3]. Perhaps the simplest non-trivial variant on an ordinary critical point occurs in a two component system in which, as a function of the non-ordering density conjugate field, the line of second order transitions ends suddenly and becomes a line of first order transitions (in fact, a line of triple points). This situation has been discussed extensively by Griffiths[4] in the context of He^3-He^4 mixtures. Griffiths labelled this junction point a tricritical point due to the fact that it actually occurs at the intersection of three lines of second order transitions, the other two branching out into the ordering-density-conjugate-field $\gtrless 0$ directions.

At the tricritical point both the ordering and nonordering densities go critical so that totally new critical behavior is expected. Such behavior was already well known in He^3-He^4 mixtures[5] and Griffiths suggested that similar behavior should be observed in certain other systems including metamagnets such as $FeCl_2$. In this case the ordering and nonordering densities are the sublattice magnetization M_s and the bulk magnetization M respectively while the corresponding conjugate fields are the staggered field H_s and internal field Hint. Since Griffiths' work two theoretical advances of particular importance to this talk have occurred:

(1) Riedel and Wegner[6] using renormalization group techniques for a simplified model of He^3-He^4 mixtures have shown that at the tricritical point in three dimensions the critical exponents should be those of a Landau or mean field theory up to logarithmic correction terms
(2) Nelson and Fisher[3,7] have domonstrated this explicitly for the $FeCl_2$-type metamagnet, thus suggesting at least limited universality for tricritical points.

Supporting evidence for mean field behavior in the metamagnet also comes from the series expansion work of Harbus and Stanley[8].

In this talk I discuss the results of a series of neutron diffraction experiments by Gen Shirane, Martin Blume, Wally Koehler and myself[9,10] on the critical behavior of $FeCl_2$ in a magnetic field. $FeCl_2$ is a simple and now much-explored[11-17] anisotropic two-sublattice antiferromagnet. From the vantage point of critical phenomena[13], it may be viewed as being made up of hexagonal sheets of ferromagnetically coupled S = 1 Ising spins with successive planes weakly coupled antiferromagnetically. This planar structure gives rise to metamagnetic behavior in an applied field, as discussed by Landau[11] in 1933. The modern era of research in $FeCl_2$ began with the experiments of Jacobs and Lawrence[12]. They showed that at low temperatures as a function of increasing internal field Hint (I will assume that all fields are applied along the crystalline c-axis) $FeCl_2$ undergoes a first order transition from an antiferromagnetic (A/f) to a paramagnetic (para) state. Above a critical temperature of $\sim$ 21K, however, the A/f - para transition appears to become continuous; the Néel temperature in zero field is $\sim$ 23.6 K. It should be emphasized that in all experiments it is the applied field, Happ which is varied. Happ and Hint are related by

$$\text{Hint} = \text{Happ} - 4\pi \text{NM}(\text{Hint}, T) \qquad (1)$$

where M(Hint, T) is the bulk magnetization and N is the demagnetizing factor. Clearly in a phase transition experiment it is essential to have N as uniform as possible.

In our neutron diffraction experiments we have measured both M and M_s as a function of applied field and temperature. The magnetization was determined from the flipping ratio of polarized neutrons at the (300) reflection while the sublattice magnetization could be measured directly from the intensity at the (201) superlattice position. Where possible, measurements were carried out with sufficient precision to yield the appropriate critical exponents. The principal information given so far by these experiments is

(1) the phase diagram in the Happ - T plane
(2) the phase diagram in the M - T plane.

By combining (1) and (2) with Eq. (1) one obtains

(3) the phase diagram in the Hint - T plane
(4) the critical behavior of M_s along the A/f side of the first order line approaching the tricritical point (T_t, Hint_t)
(5) the critical behavior of M_s for several (T_N, Happ_N) points along the λ-line up to the tricritical point as a function of T at fixed Happ and vice versa.

I will briefly summarize the results of these experiments.

For reference purposes I show below explicitly the phase diagram for our sample in the Happ - T plane. These results are in general accord with those of Jacobs and Lawrence[12] and later workers[14,15] although, of course, they are rather more detailed. At about T = 21.15 K Happ = 10,200G the phase transition appears to crossover from second to first order. The explicit location of the tricritical point may be inferred from the concomitant kink in the upper phase boundary. However, since (T_t, Happ_t) plays a pivotal role in all of our analysis, it is clearly desirable to develop an independent technique for locating it. In the neutron scattering experiment it turns out that this may be effected quite readily by monitoring the strength of the A/f critical fluctuations along the upper phase boundary. As discussed in Ref. 9, 10, the A/f critical scattering is found to vary smoothly along the λ-line until T_t = 21.15 ± 0.1K, at which point there is a sudden break and the critical scattering rapidly vanishes below T_t. This is a clear signature of a crossover from a second to a first order transition and hence it gives us an unambiguous value for (T_t, Happ_t). Before leaving the Happ - T phase diagram I should point out that a detailed discussion of domain effects in the mixed phase region are given by Dillon et al[17] in a separate paper in this conference. Also I should note that both T_N and T_t are found to be somewhat sample dependent, quite possible due to varying Fe^{3+} impurity concentrations.

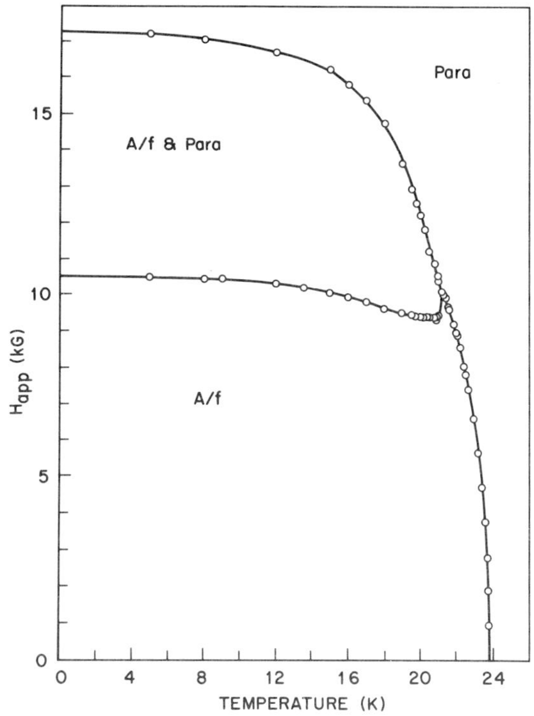

Fig. 1 Phase diagram in the Happ - T plane for $FeCl_2$. Data taken from Ref. 10.

The M - T phase diagram may be immediately derived by measuring the magnetization along the three phase boundaries shown above; these results are discussed in Ref. 9. The magnetization curves along both the λ-line and the para first order line are found to approach (T_t, M_t) linearly, that is $\beta = 1$, in agreement with theory. There is, however, a discontinuity in the slope at T_t, in disagreement with the simplest Landau theory but consistent with a more general scaling description[4]. Both of these features are also found in He^3-He^4 mixtures in the concentration - temperature plane[5]. The results along the A/f first order line are, however, rather anomalous. Over the reduced temperature range $\sim 10^{-1} < 1 - T/T_t < 10^{-2}$, an exponent $\beta\bar{u} \sim 0.36$ rather than 1 seems to be appropriate. Thus we have an apparent gross contridiction with the theory. I will discuss this in more detail below.

As noted above, by combining the Happ - T phase diagram with the M - T results and using Eq. (1), one automatically obtains the Hint - T diagram. These results are again discussed in Ref. 9. By definition the upper and lower boundaries of the mixed phase region in the Happ - T diagram collapse onto a single line in the Hint - T plane. This line is found to be continuous with the λ-line in agreement with the Landau theory and also the corresponding experimental results in the chemical potential - temperature plane in He^3-He^4 mixtures. The shape of the Hint - T curve near (T_t, $Hint_t$) is consistent with a cross-over[6] exponent of $\varphi_t = 1/2$ although the data are not precise enough to determine this exponent accurately.

I now discuss the critical behavior of the sublattice magnetization M_s along various paths. For Happ = 0, M_s exhibits conventional critical behavior with $M_s(T) = 1.47(1-T/T_N)^{0.29}$ for $10^{-1} < 1-T/T_N < 10^{-3}$. This was first demonstrated by Yelon and Birgeneau[13]. One may also look at the discontinuity ΔM_s in the sublattice magnetization across the first order line, M_s being finite in the A/f phase and, of course, zero in the para phase. Over the reduced temperature range $2 \times 10^{-1} < 1 - T/T_t < \sim 5 \times 10^{-3}$ we find $(\Delta M_s/M_s(0))^2 = 1.5 (1-T/T_t)^{0.38}$, that is $2\beta_1 = 0.38$, compared with the mean field prediction of $2\beta_1 = 1$. Recall that along the same path $M_t - M^- \sim (1-T/T_t)^{0.36}$ so that $\beta\bar{u} = 0.36$ rather than 1. Thus we have the identical discrepancy with the Landau theory occuring for both M and M_s along the A/f first order line. If these exponents truly represent the asymptotic behavior then we have a major contradiction with both the mean field and scaling descriptions. It should be emphasized, however, that in He^3-He^4 mixtures the asymptotic exponents along this particular direction are only achieved for $1 - T/T_t < 10^{-2}$ and large deviations occur for $1 - T/T_t > 10^{-2}$. Thus the above discrepancy may only reflect an asymmetry in the size of the critical region depending on the direction of approach. I should also note that Griffin and Schnatterly[16] have determined the M - T phase diagram in $FeCl_2$ using an optical technique. Their results are in general in good agreement with ours. Somewhat surprisingly, however, they find an approximate linear variation of $M_t - M^-$ for $10^{-1} < 1 - T/T_t < 10^{-2}$ in contradiction with our findings. Clearly further experiments are required to resolve this apparent conflict. The optimal experiment would be one involving simultaneous optical and neutron scattering measurements, a difficult but not impossible task.

Finally, I consider the critical behavior of M_s for various paths across the λ-line. Ideally, one would like to perform such experiments as a function of Hint and T. However, due to the fact that M(Hint, T) itself is expected to exhibit critical behavior this is not possible in our experiments (cf Eq.(1)) and instead one is forced to perform such measurements as a function of Happ and T. As pointed out to us by Griffiths, however, this simply means than one measures $\beta eff = \beta/1-\alpha$ rather than β itself; that is, the exponent is renormalized in the Fisher sense. Hence along the λ-line away from T_t one should measure $\beta eff \simeq 0.33$ since $\beta \simeq 0.29$ (experiment) and $\alpha \simeq 0.12$ (theory). Further, according to universality this exponent should be independent both of the position on the λ-line and the direction of approach provided that the path is not asymptotically parallel to the λ-line. These experiments should, of course, provide an experimental test of these ideas. Near T_t, βeff should crossover to a value of $\beta eff = 0.50$ since in the Landau theory $\beta_t = 1/4$, $\alpha_t = 1/2$. The tricritical exponent may, however, be affected by logarithmic correction terms[6].

Measurements were performed relative to the following positions on the λ-line: (Happ, T) = (8010, 22.45), (9140, 21.93) (9900, 21.45) and (10,200, 21.17) $\simeq$ (Happ, T_t). The first two experiments were performed at fixed T as a function of Happ and vice versa while for the latter two only measurements at fixed T could be performed. These results are discussed in detail in Ref. 10. In all cases, single power law behavior is observed for $2 \times 10^{-1} < 1 - T/T_N < 10^{-3}$, $2 \times 10^{-1} < 1 - H/Happ_N < 10^{-2}$. The exponents so-determined for each of the above four points are βeff = 0.34 ± 0.04, 0.35 ± 0.04, 0.36 ± 0.04, 0.37 ± .03 respectively, independent of the direction of approach. The λ-line exponents have the value expected to within the quoted errors; however βeff does not seem to crossover to 0.50 at T_t. As discussed in Ref. 10, this apparent discrepancy may simply originate in the logarithmic correction terms predicted by Wegner and Riedel[6], that is, for $10^{-1} < \varepsilon < 10^{-3}$ $(\varepsilon \log \varepsilon)^x$ may simulate a simple power law with reduced x. Confirmation of this hypothesis, however, must await explicit theoretical calculations of amplitudes.

In conclusion, then, these experiments have firmly established $FeCl_2$ as a prototypical metamagnetic material exhibiting interesting critical and tricritical behavior. In general, our results are consistent with current theoretical ideas for multicomponent systems; in addition, they confirm a number of detailed predictions. There is, however, one serious caveat; along the A/f side of the first order line for $2 \times 10^{-1} \lesssim 1 - T/T_t < \sim 5 \times 10^{-3}$ both $M_t - M^-$ and $(\Delta M_s)^2$ have effective exponents of ~ 0.37 compared with the mean field value of 1. Additional experiments, hopefully extending much closer to T_t, are required to probe this discrepancy further. Clearly also experiments on the critical fluctuations associated with M_s and M would be of great interest.

REFERENCES

1. H. E. Stanley, Introduction to Phase Transitions and Critical Phenomena, (Oxford U. P. 1971)
2. R. B. Griffiths and J. C. Wheeler, Phys. Rev. A2, 1047, (1970).
3. M. E. Fisher, this conference
4. R. B. Griffiths, Phys. Rev. Lett. 24, 715 (1970).
5. For a review see G. Ahlers in The Physics of Liquid and Solid Helium edited by K. H. Benneman and J. B. Ketterson (Wiley, New York, to be published) Vol. I.
6. E. K. Riedel and F. J. Wegner, Phys. Rev. Lett. 29, 349 (1972); F. J. Wegner and E. K. Riedel Phys. Rev. B7, 248 (1973).
7. D. F. Nelson and M. E. Fisher, Phys. Rev. B, 1975 (in press)
8. F. Harbus and H. E. Stanley, Phys. Rev. Lett. 29, 58 (1972)
9. R. J. Birgeneau, G. Shirane, M. Blume and W. C. Koehler, Phys. Rev. Lett. 33, 1098 (1974).
10. R. J. Birgeneau, G. Shirane, M. Blume and W. C. Koehler (to be submitted to Phys. Rev. B).
11. L. Landau, Z. Physik Sowjetunion 4, 675 (1933).
12. I. S. Jacobs and P. E. Lawrence, Phys. Rev. 164, 866 (1967).
13. R. J. Birgeneau, W. B. Yelon, E. Cohen and J. Makovsky, Phys. Rev. B5, 2607 (1972); W. B. Yelon and R. J. Birgeneau Phys. Rev. B5 2615 (1972).
14. C. Vettier, H. L. Alberts and D. Bloch, Phys. Rev. Lett. 31, 1414 (1973).
15. J. A. Griffin, S. E. Schnatterly, Y. Farge M. Regis and M. Fontana, Phys. Rev. B10, 1960 (1974).
16. J. A. Griffin and S. E. Schnatterly, Phys. Rev. Lett. 33, 1576 (1974); J. A. Griffin (this conference).
17. J. F. Dillon Jr., E. Yi Chen and H. J. Guggenheim (this conference).

PARAMAGNETIC RESONANCE LINEBROADENING AND SPIN-SPIN RELAXATION NEAR MAGNETIC CRITICAL POINTS *

M. S. Seehra†
Physics Department, West Virginia University, Morgantown, W. Va. 26506

and

D. L. Huber
Physics Department, University of Wisconsin, Madison, Wisconsin 53706

ABSTRACT

Recent studies of electron paramagnetic resonance (EPR) linewidths near critical points in magnetic insulators are reviewed. The behavior of representative three-dimensional antiferromagnets $RbMnF_3$, MnF_2, $NiCl_2$, MnO and MnS and ferromagnets EuO and $CrBr_3$ is analyzed. The connection between the EPR linewidth and the relaxation rate of the fluctuations in the total magnetization in zero field is established. Symmetry arguments are used to confirm the hypothesis that the anomalies in the EPR linewidth arise from a coupling between the fluctuations in the total magnetization and the long wavelength fluctuations in the order parameter. This coupling is induced by anisotropic terms in the Hamiltonian. The dependence of the linewidth on temperature and on the direction and magnitude of the applied field is discussed. The contribution to the linewidth coming from the spin-phonon interaction is also considered.

INTRODUCTION

In this paper we review the present status, both theoretical and experimental, of the anomalies in the electron paramagnetic resonance (EPR) linewidth and the spin-spin relaxation rate near magnetic critical points. We have restricted the discussion to the three-dimensional magnetic insulators since EPR spin dynamics in one and two-dimensional systems and in non - insulators is considerably different. Even in the case of three - dimensional magnetic insulators we have chosen only a few model systems for discussion purposes. In this we are motivated by the fact that experiments in these model systems lend themselves to a clear comparison with theoretical models. References to other materials where EPR linewidth anomalies have been observed can be found at appropriate places in this review or in the references cited. It is shown that the symmetry of the magnetic lattice plays a very significant role in the nature of the linewidth anomaly near the critical point. Therefore the discussion throughout the review usually has been presented separately for cubic and uniaxial systems.

The arrangement of the material in the review is along the following lines. In section II we present a summary of the experimental results in the model systems. The theory for the zero-field relaxation rate is outlined in section III. In the following section the relation between the zero-field relaxation rate and the EPR linewidth is presented along with a discussion of the anomalies in the EPR and NMR linewidths. Section V deals with a comparison between theory and experiment. In the last section we outline the remaining problems which need to be looked at in the future.

II. SURVEY OF EXPERIMENTAL RESULTS

Here we review the experimental results on the temperature dependence of the EPR linewidth ΔH in a number of prototype antiferromagnets and ferromagnets. For reasons of symmetry, which will be shown to play an important role in the critical point anomalies of the linewidth and the relaxation rate, a further subdivision into uniaxial and cubic systems is made.

A. Antiferromagnets.

(i) Uniaxial systems: The most prominent of the uniaxial systems on which detailed measurements have been made is MnF_2. Its Neel temperature T_N= 67.3°K with the c - axis as the easy axis. Fig. 1 shows the linewidth ΔH (normalized to $(\Delta H)_\infty$= 290 Oe) vs. $\varepsilon = (T - T_N)/T_N$ for the applied field $\vec{H}_0 \parallel$ c-axis.[1] An order of magnitude increase in ΔH is observed in the critical region. For $0.01 < \varepsilon < 0.3$, $(\Delta H)_T = \Delta H - \Delta H_\infty$ was found to vary nearly as $\varepsilon^{-1.2}$. For $\vec{H}_0 \perp$ c-axis an anomaly is also observed such that the ratio $R = (\Delta H)_T[\vec{H}_0 \parallel c]/(\Delta H)_T[\vec{H}_0 \perp c]$ is nearly equal to two in the critical region[2], as shown in Fig. 2. At higher operating frequencies of 24 GHz and 35 GHz, the linewidth anomaly was observed to be suppressed.[3] Recently the zero - field spin-spin relaxation rate in MnF_2 has been measured by Verbeek et al.[4] They observed that the parallel relaxation rate $\Gamma_\parallel$ shows no singularity in the critical region. Furthermore, considerable agreement was noted between the measurements and the relaxation rate derived from the earlier linewidth measurements.[1] This is evident from a plot given in Fig. 3.[5] Another uniaxial system in which a detailed EPR linewidth study has been reported is $NiC\ell_2$ ($T_N \simeq 49.5$ °K).[6] In this material the spins order in a plane normal to the hexagonal c-axis. The behavior of the linewidth vs. ε, taken from Ref. 6, is shown in Fig. 4. Here again as in MnF_2, the ratio $R \simeq 2$ throughout much of the critical region. The critical exponent of 0.7 is somewhat smaller than that in MnF_2. For $\varepsilon > 2$, there is evidence of a considerable contribution to the linewidth from the spin-phonon interaction.

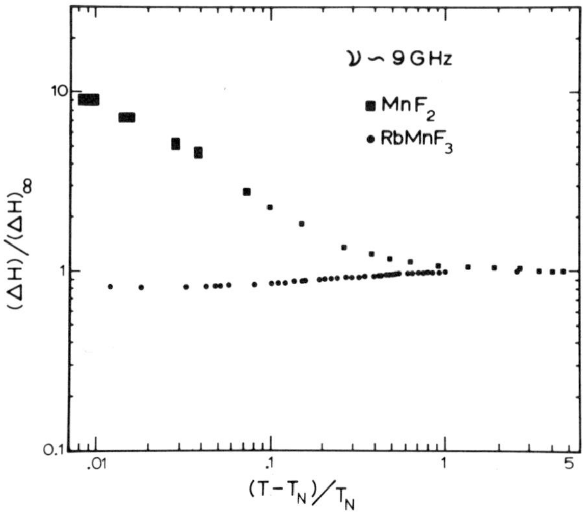

Fig. 1 Reduced linewidth vs. temperature for the magnetic field along the easy axis of MnF_2 (Ref. 1) and $RbMnF_3$ (Ref. 7).

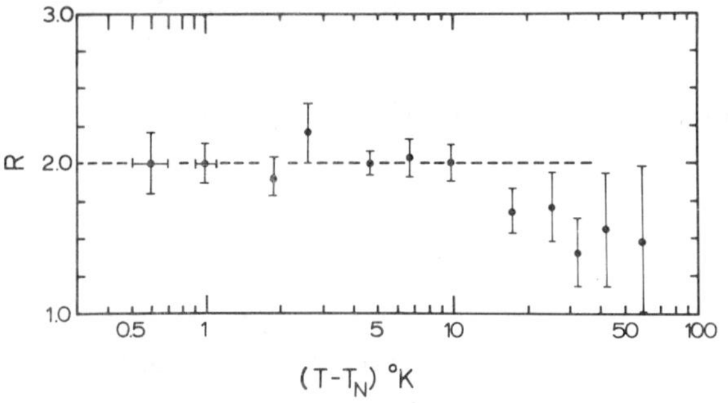

Fig. 2 The ratio $R = (\Delta H)_T(\vec{H}_0 \parallel c)/(\Delta H)_T(H_0 \perp c)$ for MnF_2 plotted against $(T-T_N)$. The graph is taken from Ref. 2.

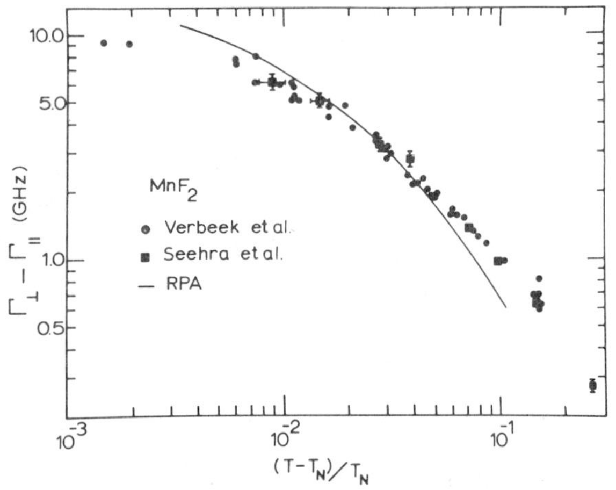

Fig. 3 The critical part $\Gamma_\perp - \Gamma_\parallel$ of the relaxation rate in MnF_2 vs. (T/T_N)-1. The circles are the data of Verbeek et.al. (Ref. 4). The squares denote $(\Delta H)_T$ scaled to coincide with the zero field data at $(T/T_N)-1=0.028$. The solid curve is a theoretical estimate based on the RPA (Ref. 5). $\Gamma_\perp$ and $\Gamma_\parallel$ correspond to $1/T_{2\perp}$ and $1/T_{2\parallel}$ of the text, respectively.

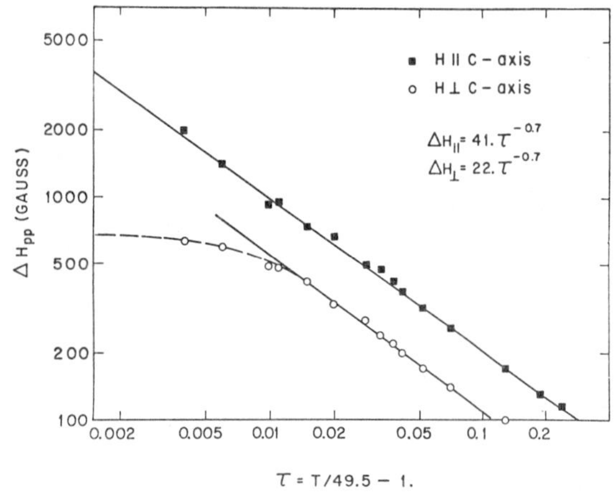

Fig. 4 Log-log plot of the peak-to-peak linewidth $(\Delta H)_{pp}$ vs. reduced temperature in $NiCl_2$ (Ref. 6) $\nu \sim 17.6$GHz.

(ii) Cubic Systems: $RbMnF_3(T_N \simeq 83\ °K)$ has a simple cubic structure, and is a nearly ideal example of a Heisenberg antiferromagnet. The measured linewidth (in reduced units with $\Delta H_\infty = 58$ Oe) for $\vec{H}_0 \parallel [111]$, the easy axis, is shown in Fig. 1. In contrast to the anomaly in MnF_2,

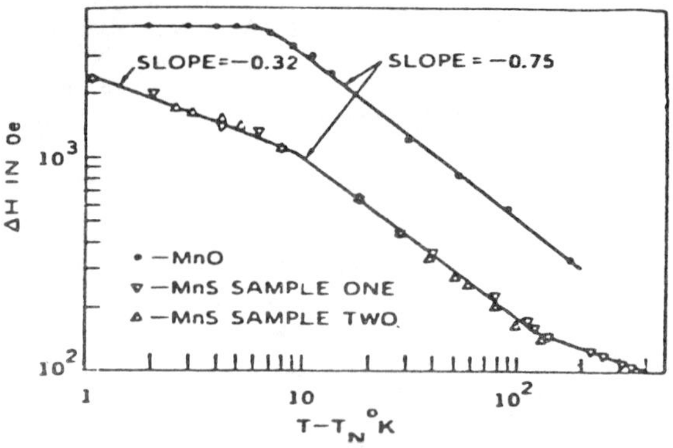

Fig. 5 EPR linewidth of MnS and MnO vs. $(T-T_N)$. The magnetic field is parallel to a [112] direction in sample one and to a [100] direction in sample two. (Ref. 9). $\nu \sim 35$GHz.

the linewidth in $RbMnF_3$ remains nearly constant in the critical region. Similar behavior was observed in the cubic antiferromagnet $KMnF_3$.[8]

MnO ($T_N \simeq 118$ °K) and α - MnS ($T_N \simeq 151$ °K) are face-centered cubic antiferromagnets in which the spins are known to lie in the (111) plane. Unlike $RbMnF_3$ and $KMnF_3$ there is a considerable increase in the linewidth in both cases as $T \to T_N^+$. The measurements by Battles[9] at 35 GHz are shown in Fig. 5. According to Battles no angular dependence was observed in the critical region.

B. Ferromagnets

(i) Uniaxial system: The best known example of a uniaxial ferromagnetic insulator is $CrBr_3$. Below the Curie temperature $T_C \simeq 32.5$ °K, the moments align along the hexagonal c-axis. Measurements of the EPR linewidth in $CrBr_3$ at 9 GHz for $\vec{H}_0 \parallel$ c-axis and $\vec{H}_0 \perp$ c-axis have recently been reported by Seehra and Gupta[10] for 20 - 550 °K, with particular attention to the critical region. Their data near T_C are shown in Fig. 6. The most significant features of the data are (i) a general narrowing of the line as T is lowered from the high temperature side, (ii) increasing anisotropy between the two directions and (iii) observation of a minimum in $(\Delta H)_\parallel$ near 50 °K. Between 50 and 34 °K $(\Delta H)_\parallel$ increases reaching a maximum value at $T_m \simeq 34$-35 °K. Note also that at T_m, $(\Delta H)_\parallel / (\Delta H)_\perp \simeq 2$. Below T_m, $(\Delta H)_\parallel$ decreases by an order of magnitude within few degrees. In the critical region the behavior of ΔH at 9 GHz differs substantially from the ΔH measured at 24 GHz.[10,11]

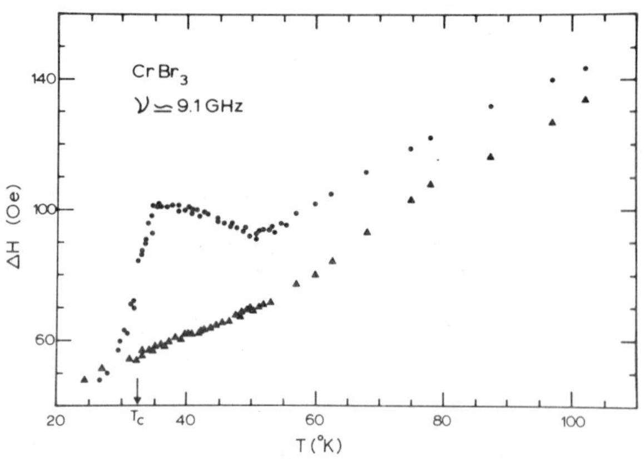

Fig. 6 EPR linewidth of $CrBr_3$ vs temperature for $\vec{H}_0 \parallel$ c-axis (circles) and $\vec{H}_0 \perp$ c-axis (triangles). The graph is taken from Ref. 10.

(ii) Cubic System: The best known prototype of the Heisenberg ferromagnet is EuO with $T_C \simeq 69.6$ °K. The earlier linewidth measurements of Eastman at 25 GHz[12] along with the recent data of Seehra and Sturm at 9 GHz[13] are shown in Fig. 7. Again, as in $CrBr_3$[10], there are substantial differences in the critical region between the linewidth measured at 9 GHz and that measured at 25 GHz. Several features of the data at 9 GHz are similar (although not as dramatic) to those observed in $CrBr_3$ in Fig. 6. However measurements in EuO at 9 GHz showed no anisotropy in the linewidth either at room temperature or in the critical region.

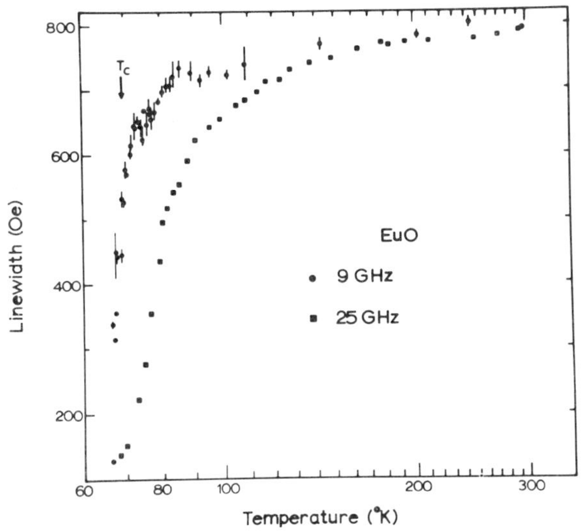

Fig. 7 EPR linewidth of EuO at 9GHz (Ref. 13) and 25 GHz (Ref. 12) vs. temperature. The two measurements are normalized at room temperature.

III. ZERO FIELD RELAXATION RATE

Rather than deal directly with the paramagnetic resonance linewidth it is advantageous first to analyze the related problem of the relaxation of the fluctuations of the total spin in the absence of a field. For the systems considered in Sec. II the magnetic Hamiltonian can be written in the form

$$H = -\frac{1}{N}\Sigma_q J(q)\vec{S}(q)\cdot\vec{S}(-q) + \frac{1}{N}\Sigma_q \vec{S}(q):\overleftrightarrow{D}(q):\vec{S}(-q) \qquad (1)$$

where N is the number of spins and the sum on q is over the Brillouin zone associated with the magnetic lattice. Also $\vec{S}(q)$ is the Fourier transform of the spin operators, while J(q) and $\overleftrightarrow{D}(q)$ are the transforms of the exchange interaction and anisotropy tensor, respectively.

The relaxation rate, $1/T_{2\alpha}$, for the fluctuations of the α component of the total spin, $S_\alpha(0)$, is given by the integral[14,15,16]

$$\frac{1}{T_{2\alpha}} = \frac{g^2\mu^2}{\chi_\alpha(0)}\int_0^\infty dt\ (\dot{S}_\alpha(0,t),\ \dot{S}_\alpha(0)) \qquad (2)$$

where $\dot{S} = (i/\hbar)[S,H]$. and (A,B) denotes a relaxation function defined, for example, in Ref. 14. Likewise g is the electronic g-factor, μ is the Bohr magneton and $\chi_\alpha(0)$ is the uniform field susceptibility in the α-direction.

Since S(0) commutes with the isotropic exchange interaction a finite value for $1/T_2$ indicates the presence of anisotropic terms in the Hamiltonian. At very high temperatures all terms in the Fourier expansion of the anisotropic interaction make roughly comparable contributions to $1/T_2$ (for three dimensional lattices). However as T approaches the critical temperature, T_c, the integrand in (2) will be dominated by contributions from the critical modes (i.e. the long wavelength fluctuations in the order parameter). Since these modes are characterized by divergent correlation lengths and undergo thermodynamic slowing down[17] their contribution becomes anomalously large, leading to a rapid increase in $1/T_2$.[15,16]

Information about the dependence of the anomalies on lattice symmetry and crystallographic direction can be inferred by singling out the contribution of the critical modes.[18] Apart from the special case of dipolar anisotropy in cubic ferromagnets, we can define that part of the Hamiltonian that is directly responsible for the anomalies by

$$H' = \sum_{|q-K_0|\lesssim\kappa} \vec{S}(q):\overleftrightarrow{D}(\vec{K}_0):\vec{S}(-q) \ . \qquad (3)$$

where κ ($\sim\varepsilon^{2/3}$) is the inverse correlation length while $K_0 = 0$ for ferromagnets and in antiferromagnets is the magnetic superlattice vector. In the case of simple cubic and body centered cubic antiferromagnets $\overleftrightarrow{D}(K_0)$ is proportional to the unit tensor so that $H' \sim \vec{S}(q)\cdot\vec{S}(-q)$ with the consequence that $[S_\alpha(0), H'] = 0$. As a result no anomaly in $1/T_2$ is predicted for these systems. On the other hand in a face centered cubic antiferromagnet with type II ordering and dipolar anisotropy, e.g. MnO and MnS, $\overleftrightarrow{D}(K_0)$ is of the form[19,20]

$$\begin{pmatrix} 0 & c & c \\ c & 0 & c \\ c & c & 0 \end{pmatrix},$$

relative to the [100] axes, so that anomalies are possible except for α along [111].

In the case of uniaxial symmetry $\overleftrightarrow{D}(K_0)$ can be written

$$\begin{pmatrix} D_\perp & 0 & 0 \\ 0 & D_\perp & 0 \\ 0 & 0 & D_\parallel \end{pmatrix},$$

for both ferromagnets and antiferromagnets. Since $[S_\parallel(0), H'] = 0$ while $[S_\perp(0), H'] \neq 0$ anomalies are predicted for $1/T_{2\perp}$ but not $1/T_{2\parallel}$. In systems of lower symmetry anomalous behavior is possible for fluctuations along any one of the three principal directions.

As noted, the approximation of replacing $\overleftrightarrow{D}(q)$ by $\overleftrightarrow{D}(K_0)$ can not be made in the case of cubic ferromagnets with dipolar anisotropy. This happens because of the singular behavior in $\overleftrightarrow{D}(q)$ at small wave vectors. Were this not the case we would have $\overleftrightarrow{D}(0)$ proportional to the unit tensor so that there would be no anomaly in $1/T_2$. The presence of the linewidth anomaly in EuO is thus a direct consequence of the long range character of the dipolar forces.

The temperature dependence of the anomalies in $1/T_2$ is a complicated problem. Generally speaking, the stronger the anisotropy the weaker the anomaly, relative to the relaxation rate at high temperatures. Furthermore the anomalous increases do not persist indefinitely but in most cases saturate. However in some instances $1/T_2$ is predicted first to increase, peak, and then go to zero at T_c.

The influence of the anisotropy on the temperature dependence comes about because the critical fluctuations, themselves, are influenced by the anisotropy. In the case of uniaxial, easy-axis systems the anomalous part of $1/T_2$ involves a four-spin function of the form $(S_\parallel(t)S_\perp(t), S_\parallel(0)S_\perp(0))$. If the anisotropy is sufficiently weak there will be a region near T_c where the critical fluctuations are approximately isotropic. At this point $1/T_2$ will show an apparent divergence. At lower temperatures only the parallel fluctuations show true critical behavior. In this region the linewidth will begin to saturate since the parallel contribution to $(S_{\parallel}(t)S_{\perp}(t),\ S_{\parallel}(0)S_{\perp}(0))$ is suppressed by the perpendicular modes as shown, for example in Fig. 3. In the case of α along an easy axis of a weakly anisotropic orthorhombic ferromagnet saturation will occur in the numerator of Eq. (2). However $\chi_\alpha(0)$ will continue to diverge so that $1/T_2$ is predicted ultimately to vary as $\chi_\alpha(0)^{-1}$ in the limit $T \to T_c^+$. Similar behavior, but with a variation as $\chi_\alpha(0)^{-1/2}$ has also been predicted for cubic ferromagnets with dipolar anisotropy.[21]

Two points should be emphasized here. First it may happen that the anisotropy is sufficiently strong relative to the isotropic exchange that the anomalies in $1/T_2$ and the linewidth are completely suppressed throughout the entire critical region. The second point pertains to the form of Eq. (2). In the case of antiferromagnets the susceptibilities vary slowly near T_c so that the anomalies reflect only the behavior of the integral in the numerator. In the case of ferromagnets the $\chi_\alpha(0)$ increase near T_c. Divergent behavior in $\chi_\alpha(0)$ will tend to mask the contribution from the numerator. In order to remove this effect, as is desirable in comparisons between ferromagnets and antiferromagnets, the analysis should be made in terms of the function $T\chi_\alpha(0)(1/T_{2\alpha})$.

Theoretical studies of the temperature dependence of $1/T_2$ near T_c have largely been based on the random phase or independent mode approximation where the four-spin functions are factored into products of two-spin functions. If the further approximation is made of using functional forms for the two-spin functions which are appropriate for a completely isotropic spin system then the following behavior is predicted[15,22]

$$\frac{1}{T_2} \propto \varepsilon^{-5/3} \tag{4}$$

for antiferromagnets and[16,22]

$$\frac{T\chi(0)}{T_2} \propto \varepsilon^{-7/3} \tag{5}$$

for ferromagnets. It should be kept in mind that even if the random phase approximation were exact the "asymptotic" behavior displayed in Eqs. (4) and (5) would not be maintained arbitrarily close to T_c. Ultimately the anisotropy must be incorporated into the two-spin functions. As noted the anisotropy in the critical fluctuations removes the divergence in $1/T_2$. At a finite temperature above T_c anisotropy in the two-spin function leads to an effective "softening" in the "critical" exponent of $1/T_2$.

IV. CONNECTION BETWEEN THE ZERO FIELD RELAXATION RATE AND THE EPR LINEWIDTH

The electron paramagnetic resonance experiments provide information about the dynamic susceptibility perpendicular to an applied field. The analysis of the linewidth is similar to the analysis of the zero field relaxation rate with two exceptions. First, in Eq. (2) a factor of $\exp[-i\omega_0 t]$ is introduced in the integrand. In the magnetic systems considered here the decay rate of $(S_\alpha(0,t),\ S_\alpha(0))$ is almost always large in comparison with the Larmor frequency, ω_0 so that the presence of this factor has little effect on the value of the integral. Second, $(S_\alpha(0,t),\ S_\alpha(0))$ itself has an intrinsic field dependence. In antiferromagnets like MnF_2 the dependence on field is probably unimportant for fields on the order of three kilogauss (see below), although it may be important for those systems which order at very low temperatures. In ferromagnets the field effects are likely to be much more significant. Generally speaking the external field acts to suppress the anomalies in the linewidth relative to the corresponding zero field relaxation rates, as shown for example, in Fig. 7.

In situations where the field dependence of the linewidth is unimportant there is a direct connection between the EPR linewidth and the zero field relaxation rate. In this case the EPR linewidth is identified with the angular average of the zero field relaxation rate in the plane perpendicular to the applied field. In uniaxial systems this leads to an equation relating the linewidth, ΔH, to T_2 of the form[15]

$$g\mu\,\Delta H(\theta) = \frac{\hbar}{T_{2\perp}}\cos^2\theta + \frac{\hbar}{2}\left(\frac{1}{T_{2\parallel}} + \frac{1}{T_{2\perp}}\right)\sin^2\theta \tag{6}$$

where θ is the angle between the applied field and the c-axis. More generally, if α,β,γ refer to the principal directions of the anisotropy tensor, with the static field along α, we have[23]

$$g_\alpha \mu \Delta H_\alpha = \frac{\hbar}{2}\left[\frac{1}{T_{2\beta}} + \frac{1}{T_{2\gamma}}\right] \tag{7}$$

From the above analysis it is apparent that the EPR measurements provide information about the dynamics of the spin fluctuations at the center of the Brillouin zone. This is in contrast to neutron scattering which probes only the dynamics at finite wave vector,[24] and nuclear magnetic resonance (NMR) which (in the paramagnetic phase) provides information only about the time integral of the auto and near-neighbor spin correlation functions.[25] In particular, the EPR linewidth anomalies, although similar to the linewidth anomalies reported in nuclear magnetic resonance studies, represent distinctly different aspects of critical behavior.

V. COMPARISON BETWEEN EXPERIMENT AND THEORY

A. Antiferromagnets:

(i) MnF_2: Discussion in section III has shown that for a system with uniaxial anisotropy like MnF_2, $1/T_{2\parallel}$ should show no anomaly in the critical region. Zero field measurements of $1/T_{2\parallel}$ reported in Ref. 4 confirm this. Another important consequence of this, using Eqs. (6) and (7), is that the ratio $R = (\Delta H)_T(\vec{H}_0 \parallel c)/(\Delta H)_T(\vec{H}_0 \perp c)$, discussed in section II, is equal to two. As shown in Fig. 2, this is found to be true experimentally in the critical region of MnF_2. The deviation from this behavior in the precritical region possibly results from the fact that the contribution of the spin fluctuations near $q = K_0$ to the relaxation rate is no longer dominant at these temperatures.[2]

The relationship between the zero-field spin-spin relaxation rates and the EPR linewidth, as given by Eqs. (6) and (7), can be further checked by making a direct comparison between the two measurements. Such a comparison in MnF_2 is shown in Fig. 3. It is observed that the relaxation rates derived from the 9 GHz linewidth data show a similar temperature dependence as the direct measurements of Verbeek et al..[4] From this remarkable agreement one can also infer that the magnetic field effects are negligible in MnF_2 for the X-band measurements.

The softening of the critical exponent for the linewidth and the relaxation rates near T_N in MnF_2 is evident in the data shown in Figs.1 and 3. As discussed in section III, this can be understood as a result of the effect of the anisotropy since in the presence of the anisotropy only the parallel fluctuations show the true critical behavior. The solid curve in Fig. 3 comes from a calculation using the random phase approximation (RPA) in which the effect of the anisotropy has been included.[5]

The critical exponent of 5/3 (Eq. (4), which is based on the RPA approximation of decoupling the four-spin functions into products of two-spin functions and which assumes isotropy of the two-spin functions, is somewhat larger than the measured value in MnF_2. For $.03<\varepsilon<0.3$, an exponent of 1.20 was observed[21], with lower values for smaller ε. Possible reasons for this discrepancy are discussed

in the following section.
(ii) $NiCl_2$: The linewidth measurements of Birgeneau et al.[6] in $NiCl_2$ near T_N are shown in Fig. 4. The ratio R defined above is found to be equal to two throughout most of the critical region as expected from the theory discussed in section III. (The increase in R below $\varepsilon = 0.01$ is attributed to the suppression of the non-secular terms in the dipolar interaction by the magnetic field.[6,15]) Furthermore it is observed that for this system $\Delta H \sim \varepsilon^{-0.7}$ for $0.004<\varepsilon<0.25$. Thus the critical exponent of 0.7 is considerably smaller than the predicted value of 5/3. In Ref. 6, it was argued that $NiCl_2$ was more nearly isotropic than MnF_2 so that the observed critical exponent was characteristic of a region where $\kappa a \ll 1$ yet the spin fluctuations are approximately isotropic. Therefore it was concluded that the decoupling of the four-spin functions using the random phase approximation was the main culprit for the lower exponent. On the other hand for MnF_2 an exponent of about 0.7 is observed in a region where the spin functions are becoming increasingly anisotropic with decreasing temperatures. Further discussion of this apparent discrepancy is given in section VI.
(iii) $RbMnF_3$, MnO and MnS: Discussion in section III shows that for antiferromagnets with simple cubic and body centered cubic magnetic lattices there should be no anomaly in the relaxation rates and the linewidth in the critical region. The linewidth data of Ref. 7 on $RbMnF_3$ are shown in Fig. 1. Only a very small narrowing, on the order of few percent, is observed in the critical region. This agreement between theory and experiment assures us that the basic assumption of the theory may be assumed to be correct. Since $S_\alpha(0)$ does not commute with $\vec{S}(q):\overleftrightarrow{D}(q):\vec{S}(-q)$ at arbitrary points in the Brillouin zone the absence of an anomaly confirms the hypothesis that the increase in ΔH comes from a coupling of the fluctuations in the total spin (or total magnetization) to spin fluctuations with $q \approx K_o$.

Another prediction of the theory[15] is that in antiferromagnets with a face-centered cubic magnetic lattice, an increase in the linewidth should be observed as T approaches T_N^+. Earlier measurements of Battles[9] in fcc MnO and MnS are shown in Fig. 5. The observed critical exponent of 0.75 with softening near T_N is somewhat similar to the behavior observed in MnF_2. The predicted angular dependence of the linewidth in fcc lattices however remains to be observed. Recently Dormann et al.[26] have shown that the linewidth behavior in MnO and MnS in the pre-critical region is also consistent with the theory discussed in section III.
(iv) Two other three-dimensional antiferromagnets where careful linewidth measurements on single crystals have been reported are $CuCl_2 \cdot 2H_2O$[27] and $CuF_2 \cdot 2H_2O$.[28] In both cases the observed behavior of $(\Delta H)_T = \Delta H - \Delta H_\infty$ is similar to that of MnF_2 i.e. a softening of the critical exponent as $T \to T_N^+$. $CuCl_2 \cdot 2H_2O$ is an orthorhombic antiferromagnet with $T_N = 4.33$°K, whereas $CuF_2 \cdot 2H_2O$ is a monoclinic antiferromagnet with $T_N = 10.9$°K.

B. Ferromagnets

(i) $CrBr_3$: Measurements of the EPR linewidth of $CrBr_3$ near T_C and at 9 GHz are shown in Fig. 6. For $H_o \parallel$ c-axis, a minimum in the linewidth is observed near $\varepsilon \sim 0.5$. For lower temperatures the linewidth increases, reaching a maximum at T_m near $\varepsilon \approx 0.1$ and then the linewidth sharply decreases as T_C is approached. A behavior qualitatively similar to this has been predicted in the discussion and the references noted in section III. At $T = T_m$, $(\Delta H)_\parallel / (\Delta H)_\perp \simeq 2$. This observed anisotropy can be understood on the basis of the uniaxial anisotropy of $CrBr_3$.[10] The basis of this anisotropy is the same as in the case of the uniaxial antiferromagnet viz. that the parallel relaxation rate $1/T_{2\parallel}$ should show no anomaly in the critical region.

The linewidth is predicted to vary as ε^{-1} for $0.1 \lesssim \varepsilon \lesssim 0.5$[16,29] and as $\chi_\perp^{-1}$ for $0 < \varepsilon < 0.1$. The observed strength of the anomalies in $CrBr_3$ is considerably weaker than what is predicted. It is possible that the applied magnetic field, which is necessary to observe the resonance, and which directly couples to the order parameter in ferromagnets is the source of at least part of this discrepancy.[10] A quantitative understanding of this effect however is lacking at this time. Zero-field measurements of the relaxation rate, similar to the measurements of Verbeek et al. in MnF_2, or EPR linewidth measurements at a frequency lower than 9 GHz are needed to settle this question.

According to Eq. (5) $T\chi(0)/T_2$ should vary as $\varepsilon^{-7/3}$ in the critical region. Since in cases where magnetic field effects can be neglected $(\Delta H) \propto 1/T_2$, a plot of $(\Delta H)_\parallel T \chi_\perp$[30] is shown in Fig. 8. Unfortunately the magnetic susceptibility data in the region 34 - 47 °K in $CrBr_3$ are not available so that the plot could not be extended to the region near T_C. However the increase of $(\Delta H)_\parallel T \chi_\perp$ as T is lowered towards T_C is in qualitative agreement with the prediction. A better test of Eq. (5) is provided by the data on EuO discussed in the following section. Incidently the increase of $(\Delta H)_\parallel T \chi_\perp$ with increasing T for $T > 3T_C$ has been interpreted to mean that the spin-phonon interaction makes a substantial contribution to the EPR linewidth of $CrBr_3$ for $T > 3T_C$.[30] However this mechanism is predicted to have no significant effect on the critical behavior discussed above.

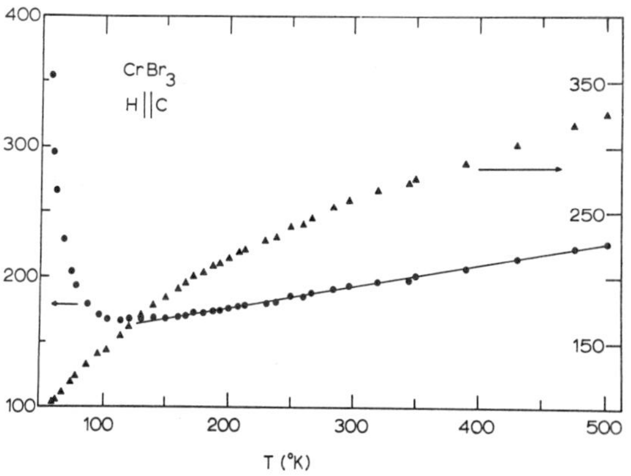

Fig. 8 Plot of $(\Delta H)_\parallel T\chi_\perp$ vs. temperature in $CrBr_3$. The linewidth $(\Delta H)_\parallel$ is also plotted for purposes of comparison. The graph is taken from Ref. 30. $\chi_\perp$ is the uniform field susceptibility perpendicular to the c-axis.

(ii) EuO: EPR linewidth measurements in EuO are shown in Fig. 7. As noted in section II, the qualitative behavior of the linewidth at 9 GHz as a function of temperature is similar to the observations in $CrBr_3$ for $H_o \parallel$ c-axis. However the strength of the predicted anomalies is even weaker as compared to the observations in $CrBr_3$. From a comparison of the data at 9 GHz and 25 GHz, both in $CrBr_3$[1] and EuO[13] (Fig. 7), it is clear that the magnetic field has a pronounced effect on the spin dynamics in ferromagnets near T_C. Because of coupling between the total magnetization and the magnetic field in a ferromagnet it is unlikely that the measurements at 9 GHz ($H_o \simeq 3$ kOe) give essentially the zero field behavior as was found to be true in the case of MnF_2.

A plot of $\Delta H \chi_o T$ versus T using the susceptibility data of Menyuk et al.[31] is shown in Fig. 9. The linewidth data are also shown for comparison purposes on the same log-log plot. As was evident from the linewidth data in Fig. 7, a weaker exponent of about 6/5, as compared to the predicted value of 7/3 for $0.1 < \varepsilon < 0.5$, is observed. The observations for $\varepsilon < 0.1$ though in qualitative agreement with the predictions also show a much weaker temperature dependence of ΔH. Comparison of the plots in Figs. 8 and 9 suggests that there is no evidence for a contribution to the EPR linewidth in EuO coming from the spin-phonon interaction.

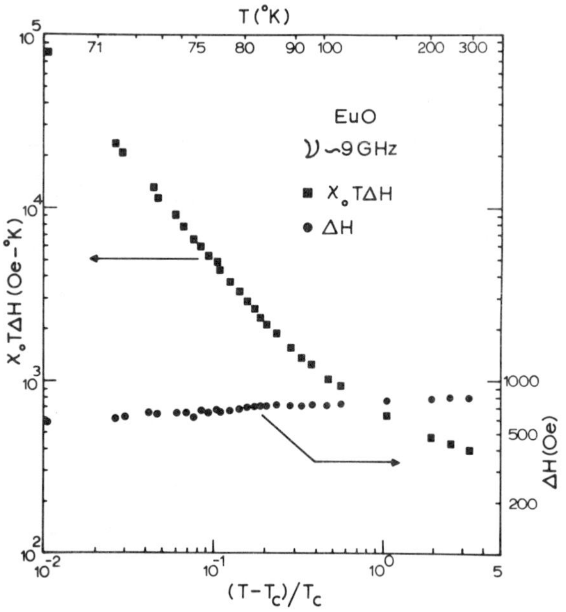

Fig. 9 Log-log plot of $\chi_0 T\Delta H$ vs. reduced temperature in EuO. The linewidth ΔH is also plotted for comparison purposes. The data is from Ref. 13. χ_0 is the uniform field susceptibility.

VI. REMAINING PROBLEMS

A. Magnitude of Critical Exponent.

Comparison of the experimental data with the theory in section V has shown that the critical exponents for the anomalies in the linewidth near T_c, both for antiferromagnets and ferromagnets, are smaller than the predicted values. A possible explanation, which has been convincingly raised by the work on $NiC\ell_2$, is that the decoupling of the four-spin functions into products of two-spin functions using the random phase approximation is inadequate. If the anisotropy in $NiC\ell_2$ is as weak as Ref. 6 suggests then this decoupling may be responsible, at least in part, for the higher predicted value of the exponent. The calculations on MnF_2[5] based on the above decoupling suggest that a softening of the exponent is obtained if the effect of the anisotropy on the temperature dependence of the two-spin function is included. The crossover occurs when the system goes from nearly isotropic to easy-axis type in spin space. However this does not explain the observations in MnF_2 completely since the observed exponents in both regions, (isotropic as well as easy-axis type) are still lower than the predicted values (see Fig. 3). Therefore it seems that further theoretical work on approximating the four-spin functions is needed.[32]

B. Magnetic Field Dependence

From the discussion presented in sections IV and V, it is evident that for MnF_2, the linewidth data at 9 GHz give nearly the zero field behavior. Similar conclusions can be reached for $RbMnF_3$ since a recent ΔH measurements at 5 GHz[33] agree exactly with the data at 9 GHz. Thus magnetic field effects for antiferromagnets with $T_c \gtrsim 100$°K are likely to be unimportant at X-band frequencies.

As pointed out in section IV, the magnetic field dependence in ferromagnets is likely to be more important. Measurements in $CrBr_3$[10] and EuO[13] (Fig. 7) at 9 and 25 GHz strongly suggest this. If the zero-field relaxation rates were available, they would provide direct evidence of any effects due to magnetic fields at 9 GHz. However from the analysis in MnF_2 one can say that the major effect at 9 GHz in these compounds is likely to be of intrinsic nature i.e. the temperature dependence of the relaxation function $(\vec{S}_\alpha(0,t), \vec{S}_\alpha(0))$ in Eq. (2) for ferromagnets is likely to be suppressed due to the magnetic field. Some discussion related to this point is given in Ref. 10. Thus the weaker nature of the anomalies observed in ferromagnets as compared with the predictions may be partly due to the effect of the applied field and partly due to inadequate decoupling of the four-spin functions. A quantitative understanding of both these effects is lacking at present.

C. Spin-Phonon Interaction and High Temperature Behavior.

A comparison of the plots in Figs. 8 and 9 shows that the high temperature behavior of the linewidth for $CrBr_3$ and EuO is different. In particular an increase of $\Delta HT\chi$ with increasing temperatures in $CrBr_3$ suggests that the spin-phonon interaction makes a significant contribution to the linewidth of $CrBr_3$ for $T > 3T_C$. However as pointed out in a recent paper[30], the contribution of the spin-phonon interaction to the EPR linewidth in magnetic insulators such as $CrBr_3$ and $NiC\ell_2$ can not yet be estimated quantitatively. On the other hand, systems with S-state ions like Mn^{++} and Eu^{++} are unlikely to have any significant contribution from the spin-phonon interaction, since in these cases the orbit-lattice interaction is very weak.

REFERENCES

*Supported in part by the National Science Foundation.
†A. P. Sloan Research Fellow.

1. M. S. Seehra and T. G. Castner, Jr., Solid State Commun. 8, 787 (1970).
2. M. S. Seehra, Phys. Rev. B 6, 3186 (1972).
3. M. S. Seehra, J. Appl. Phys. 42, 1290 (1971).
4. P. W. Verbeek, J. C. Verstelle and J. A. Tjon, Physica 66, 545 (1973). Similar results have been recently obtained by A. M. Gottlieb, M. Feldman, M. Littman and P. Heller, AIP Conf. Proc. 18, 764 (1974).
5. D. L. Huber, M. S. Seehra and P. W. Verbeek, Phys. Rev. B 9, 4988 (1974).
6. R. J. Birgeneau, L. W. Rupp, Jr., H. J. Guggenheim, P. A. Lindgard and D. L. Huber, Phys. Rev. Letters 30, 1252 (1973).
7. R. P. Gupta and M. S. Seehra, Phys. Letters A 33, 347 (1970).
8. R. P. Gupta, M. S. Seehra, and W. E. Vehse, Phys. Rev. B 5, 92 (1972).
9. J. W. Battles, J. Appl. Phys. 42, 1286 (1971).
10. M. S. Seehra and R. P. Gupta, Phys. Rev. B 9, 197 (1974).
11. J. F. Dillon, Jr. and J. P. Remeika, Proceedings of the Eleventh Colloque Ampere, Endhoven, edited by J. Smit (North-Holland, Amstordam 1962).p.480.
12. D. E. Eastman, Phys. Letters A 28, 136 (1968).
13. M. S. Seehra and D. W. Sturm (to be published).
14. H. Mori and K. Kawasaki, Progr. Theor. Phys. 27, 529 (1962); 28, 971 (1962).
15. D. L. Huber, Phys. Rev. B6, 3180 (1972).
16. D. L. Huber, J. Phys. Chem. Solids 32, 2145 (1971).
17. L. Van Hove, Phys. Rev. 95, 1374 (1954).
18. S. Maekawa, J. Phys. Soc. Japan 33, 573 (1972).
19. J. S. Smart, Effective Field Theories of Magnetism (Saunders, Philadelphia, 1966).
20. M. H. Cohen and F. Keffer, Phys. Rev. 99, 1128 (1965).

21. S. V. Maleev, Phys. Letters A47, 111 (1974).
22. K. Kawasaki, Prog. Theor. Phys. 39, 285 (1968).
23. W. M. deJong and J. C. Verstelle, Phys. Letters A42, 297 (1972).
24. W. Marshall and R. D. Lowde, Rep. Prog. Phys. 31, 705 (1968).
25. T. Moriya, Prog. Theor. Phys. 28, 371 (1962).
26. E. Dormann and V. Jaccarino, Phys. Letters A 48, 81 (1974).
27. N. J. Zimmerman, F. P. Van der Mark and I. Van den Handel, Physica 46, 204 (1970).
28. K. Nagata and M. Date, J. Phys. Soc. Japan 19, 1823 (1964).
29. K. Tomita and T. Kawasaki, Prog. Theor. Phys. 44, 1173. (1970).
30. D. L. Huber and M. S. Seehra, J. Phys. Chem. Solids (to be published).
31. T. Menyuk, K. Dwight and T. B. Reed, Phys. Rev. B 3, 1689 (1971).
32. An alternative theory of the linewidth exponent, proposed by K. Kawasaki (Phys. Lett. 26A, 543 (1968), yields a value $\approx$ 3/4 for nearly isotropic antiferromagnets. Full details of the theory have not yet been published so it is not possible to consider its applicability in detail.
33. D. W. Olson, private communication.

STATIC, DYNAMIC AND CRITICAL EFFECTS AT THE SPINFLOP TRANSITION OF ANTIFERROMAGNETS

H. Rohrer*
Department of Physics, University of California, Santa Barbara, CA 93106

ABSTRACT

It is shown that the competition of intralattice and interlattice anisotropy gives rise to a variety of spinflop transitions in antiferromagnets with orthorhombic anisotropy. Experimental results are presented for the extension of the ordinary spinflop transition and the width of the domain state and the metastable region associated with it. Close to the bicritical point, critical effects are observed and compared with predictions of extended scaling. The paramagnetic phase boundaries are compatible with the predicted anisotropy crossover exponent ϕ; magnetization discontinuity across the spinflop transition and the divergence of the susceptibility, however, favour a smaller value of ϕ.

INTRODUCTION

Magnetic field induced transitions in antiferromagnets have been investigated extensively both theoretically(1) and experimentally(2). They show a variety of different types of transitions and critical points which make them also attractive for studies of critical effects. Here we discuss some aspects of the spinflop transition which separates the antiferromagnetic (AF) state from the spinflop (SF) state for the case of orthorhombic anisotropy. In Section I the different types of SF transitions, intermediate (IF) phases replacing them, and the possibility of higher order critical points are explored for different regimes of anisotropy. In Section II experimental results on the extension of the SF transition and its associated domain state and metastability region are presented for $GdAlO_3$. Section III deals with critical effects observed near the bicritical point.

I TYPES OF TRANSITIONS

The different types of transitions have been mainly investigated at T = 0 and for fields along the easy axis and the uniaxial case has been summarized by Yamashita(1). Finite temperatures and fields in the principal planes usually require numerical calculations.

For the following discussion, we write the MFA free energy in the form

$$F = NS^2\{J_1 \underset{\sim}{\sigma}_A \cdot \underset{\sim}{\sigma}_B + 1/2\, J_2 (\sigma_A^2 + \sigma_B^2) + \underset{\sim}{\sigma}_A \cdot \underset{\approx}{K}_1 \cdot \underset{\sim}{\sigma}_B + 1/2 (\underset{\sim}{\sigma}_A \cdot \underset{\approx}{K}_2 \cdot \underset{\sim}{\sigma}_A + \underset{\sim}{\sigma}_B \cdot \underset{\approx}{K}_2 \cdot \underset{\sim}{\sigma}_B) - (g\mu_B/S)\, \underset{\sim}{H} \cdot (\underset{\sim}{\sigma}_A + \underset{\sim}{\sigma}_B) - (kT/S^2)(\eta(\sigma_A) + \eta(\sigma_B)) \quad (1)$$

σ_A and σ_B are the average moments of sublattice A and B, respectively ($\sigma_A = \langle S_A \rangle / S$), J_1 and $\underset{\approx}{K}_1$ interlattice coupling, J_2 and $\underset{\approx}{K}_2$ the intralattice coupling constants, and $\eta(\sigma)$ is the entropy per site. The anisotropy has the components ΔK, 0, K along the principal axis denoted by x,y,z. For simplicity, the single ion anisotropy is incorporated in K_2; its more complicated σ dependence does not change the results qualitatively. The condition for the z-axis to be the easy axis is

$$0 < K_1 - K_2 > \Delta K_1 - \Delta K_2 \quad (2)$$

For H along z, the paramagnetic (PM) state becomes unstable with respect to a staggered mode in the y direction provided

$$\Delta K_1 - \Delta K_2 < 0 \quad (3)$$

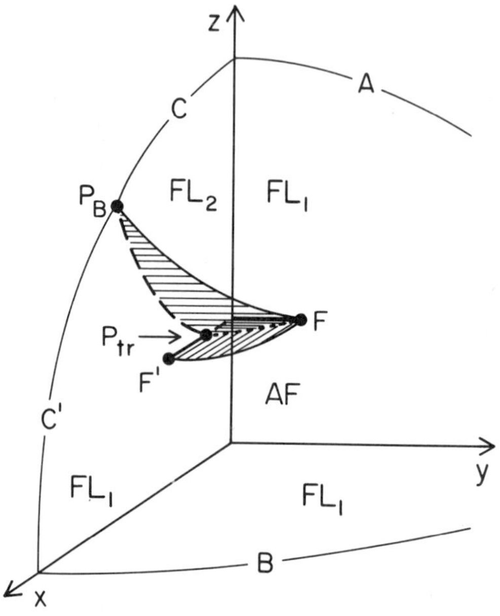

Fig. 1. Phase diagram of the orthorhombic antiferromagnet. Heavy lines: in plane SF transitions; dashed heavy lines: out of plane SF transition; shaded areas: surface of first order transition. The dotted line connecting P_{tr} and F is a line of triple points.

This defines the x-axis as the hard axis. In the following K is called the uniaxial and ΔK the orthorhombic component of the anisotropy. The phase diagram for the simple orthorhombic antiferromagnet is shown in Fig. 1 at T = 0. The instability of the PM phase is always a perpendicular staggered mode $\delta\sigma^\dagger$ determining the symmetry of the ordered phase. (We do not consider here the case of a first order PM-AF transition possible when $K_2 < -J_1(2J_1+K_1)/(3J_1+2K_1)$). Along A and B, the instability occurs by an in plane mode $\delta\sigma^\dagger_{zy}$ and $\delta\sigma^\dagger_z$ (in plane refers to the plane containing the easy axis and applied field, or, for $H \parallel z$, to the z-y plane). In the z-x plane the instability changes from an out of plane mode ($\delta\sigma^\dagger_y$) along C to an in plane mode ($\delta\sigma^\dagger_{zx}$) along C'. The two phases therefore have to be separated by either a first order transition or by a third phase. K_2 will determine which of the two cases applies. For the moment we assume a first order transition, which then connects to the PM boundary in a bicritical(3) point P_B. In the z-y plane, the SF transition does not connect to the PM boundary but ends in a critical point F. In the following in plane transitions and in plane phases are denoted by subscript 1; out of plane by subscript 2. AF phases are always in plane phases.

The SF_2 transition has been shown to be

*On leave from IBM Zurich Research Laboratory CH 8803 Ruschlikon, Switzerland

a hyperbola[4]. However, this is only the case not too close to the z-axis. For $\Delta K_1-\Delta K_2 \to 0$ P_B moves to the z-axis. In the special case $\Delta K_1-\Delta K_2 = 0$ uniaxial symmetry is restored and an SF_1 type transition must exist in the z-x plane. This transition cuts the hyperbola near its apex and changes at the triple point P_{tr} from SF_2 type to SF_1 type. Thus two first order spinflop transitions exist for a range of angles ψ between easy axis and applied field.

As a function of temperature and H along z, the PM instability changes from a perpendicular mode for $T < T_B$ to a parallel mode for $T > T_B$. (The case of an intermediate piece of a first order AF-PM transition between the two different second order PM boundaries[5] is not considered here). Again the two phases are separated by a first order transition or an intermediate phase depending on K_2. In the case of a first order transition, T_B is a bicritical point, and critical effects in its vicinity will be considered in Section III. For fields in the z-x plane, the first order SF transition always connects to the PM boundary and ends in a bicritical point. For fields in the z-y plane, however, the SF transition ends in a critical point.

We turn now to the case of intermediate phases. Considering instability in the z-y plane only, the stability limits of the AF and SF phases at T=0 and for H along z are given by[2,6]

$$H_{sh} = \sqrt{(K_1-K_2)(2J_1+K_1-K_2)} \quad (4a)$$

$$H_{sc} = \sqrt{(K_1-K_2)(2J_1+K_1+K_2)/(2J_1+K_1-K_2)} \quad (4b)$$

where H_{sh} is the stability limit of the AF phase or superheating field and H_{sc} the stability limit of FL phase or supercooling field. For

$$K_2 > 0 \quad (5)$$

$H_{sh} < H_{sc}$ and an intermediate (IF_1) bounded by two second order transitions (IA and IS) appears. This intermediate or canted phase with the sublattices tilted from the z direction by

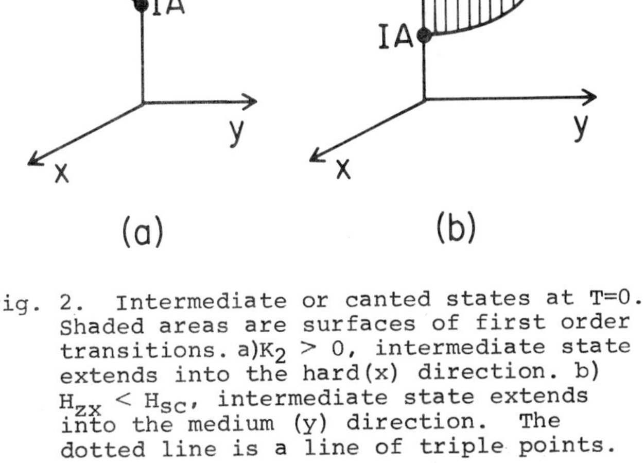

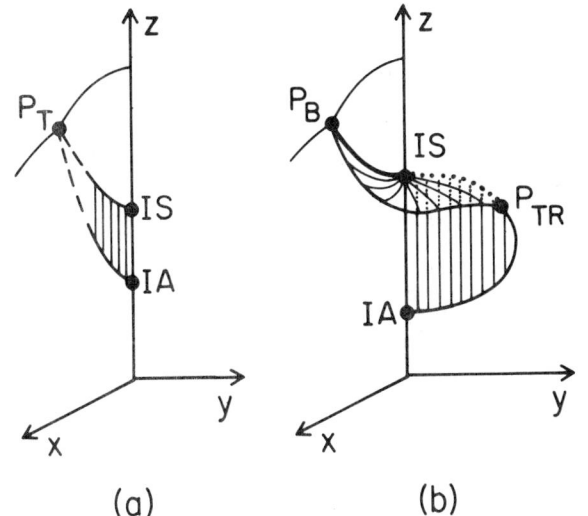

Fig. 2. Intermediate or canted states at T=0. Shaded areas are surfaces of first order transitions. a)$K_2 > 0$, intermediate state extends into the hard(x) direction. b) $H_{zx} < H_{sc}$, intermediate state extends into the medium (y) direction. The dotted line is a line of triple points.

different angles extends only into the z-x plane. The two second order transitions IA and IS are simply endpoints of a surface of first order transition lying in the z-x plane (see Fig. 2a). This IF phase does not, however, necessarily connect to the PM boundary in a tetracritical point[3,7] at all temperatures. Preliminary numerical calculations show that for small values of K_2 and with $H \parallel z$, the IF state vanishes at a temperature T_I and a first order SF transition appears for $T > T_I$. In other words, the surface of first order transition changes from parallel to z at $T < T_I$ to perpendicular to z for $T > T_I$. The point T_I is a peculiar point insofar as it is the intersection of four lines of critical points (boundaries of surfaces of first order transitions) but the magnetization discontinuity remains finite. A similar situation exists at T=0, where the stability limits coincide for $K_2=0$, but ΔM is finite.

Finally we consider different regimes of ΔK at T = 0. For $\Delta K_1-\Delta K_2 < 0$, $K2 < 0$, and $H \parallel z$, the FL state is an in plane state (FL_1) and its stability mode an in plane mode and thus independent of the value of $\Delta K_1-\Delta K_2$. The AF phase, however, becomes unstable with respect to $\delta\sigma_{zx}$ at

$$H_{zx} = \sqrt{(2J_1+K_1-K_2+\Delta K_1+\Delta K_2)(K_1-K_2+\Delta K_2-\Delta K_1)} \quad (6)$$

The hyperbolas $H_{zx} = H_{sh}$ and $H_{zx} = H_{sc}$ bound regions of intermediate states. $H_{zx} > H_{sh}$ is the case of orthorhombic anisotropy discussed above. For $H_{zx} < H_{sc}$, an intermediate phase has to appear. It is similar to that with $K_2 > 0$ but with the sublattices canted in the z-x plane (IF_2). Its boundary to the FL_1 phase is therefore of first order. The transition to the AF phase on the other hand is of second order. The two transitions IA and IS again are the endpoints of a surface of first order transition lying in the z-y plane (see Fig. 2b), but IS is now a triple point and the line IS-P_{TR} is a line of triple points. The point P_{TR} is the intersection of three lines of critical points and reminds one of a tricritical point.

II THE SPINFLOP TRANSITION IN $GdAlO_3$

The Neel temperature $T_N = 3.87$ K, bicritical point $T_B = 3.126$ K, SF-field $H_c \approx$ 11 kOe, and a relatively large perpendicular extension of the SF-transition ($\psi_c(T=0) \sim 4°$) make $GdAlO_3$ convenient for experimental studies[2,8]. $GdAlO_3$ has orthorhombic anisotropy and the following pertains to the z-y plane part of the SF transition only.

A. Extension of the SF Transition

The maximum angle between easy axis and internal applied field under which a SF transition exsits, decreases from its T=0 value[9]

$$tg|\psi_c^i| = -K_2/(2J_1+K_1-K_2) \quad (7)$$

to zero at T_B. Thus only the K_2 type anisotropy is important for the critical angle. Experimentally it is easier to determine T_c, the endpoint of the SF transition, for fixed ψ. The results are shown in Fig. 3 for different sample shapes. In the case of the long cylinder and the disc with disc plane parallel to the easy axis, T_c was located by the adiabatic susceptibility[2]. There one makes use of the fact that at the SF transition, adiabatic modulation sweeps the sample along the coexistence line rather than across it, resulting in a sharp reduction of the susceptibility maximum at the

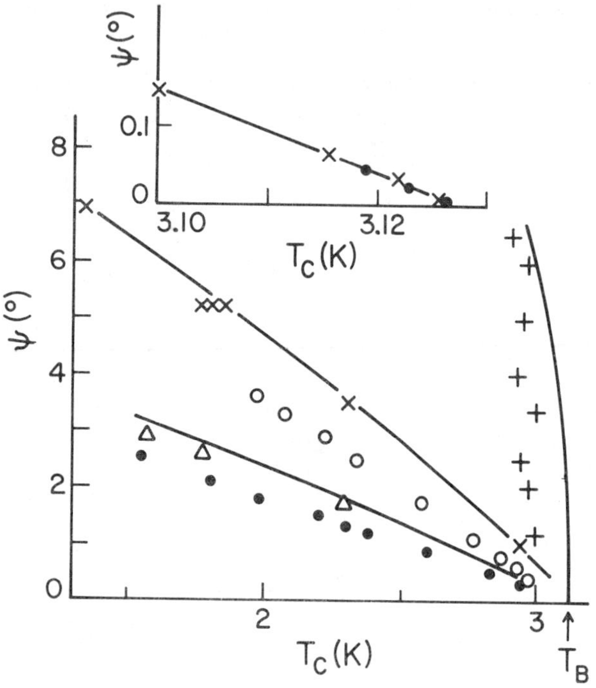

Fig. 3. Extension of the ordinary spinflop state perpendicular to the easy axis for long slab (crosses), sphere (open circles) and disc with plane ∥z (triangles) and ⊥z (filled circles). The drawn curves are MFA calculations. Insert: in situ alignment of slab.

SF transition. Since the relevant relaxation time for the spin system at these low temperatures is the thermal relaxation of the whole sample, frequencies in excess of 1 kc can be considered adiabatic for samples with shortest dimension of 0.5 mm. For the sphere and the disc with plane perpendicular to z, the onset of the domain state consisting of AF and SF domains (see II B) determined T_c. The agreement with MFA, with the coupling constants derived from the SF and FL-PM transition at low temperatures and properly redefined for demagnetization, is quite satisfactory. The insert of Fig. 3 shows the results of an in situ alignment, where the applied field was tipped through the z-axis (x: $\psi>0$, •: $\psi<0$) with an additional perpendicular component. This verifies the linear relation between ψ and T_c required by MFA very close to T_B. Finally, also the approach of T_c to T_B in the z-x plane is shown. Although T_c is nearly independent of ψ, as required by MFA, a small misalignment of H in the y direction pushes T_c below T_B.

B. Domain State

At the first order spinflop transition, demagnetization favours a domain state consisting of AF and FL domains. Direct experimental evidence of such a domain state was first presented by King and Paquette in MnF_2 [10]. They observed both NMR and optical absorption lines characteristic of each phase in a field region corresponding to the expected width of the domain state. Close to T_c, the two domains become similar and cannot easily be distinguished. We have measured $\tilde{\chi}_{zy}$, the susceptibility perpendicular to the applied field in order to determine the onset and width of the domain state. The latter determines how the magnetization discontinuity vanishes at T_c (see Section III). Above T_c, the sublattice

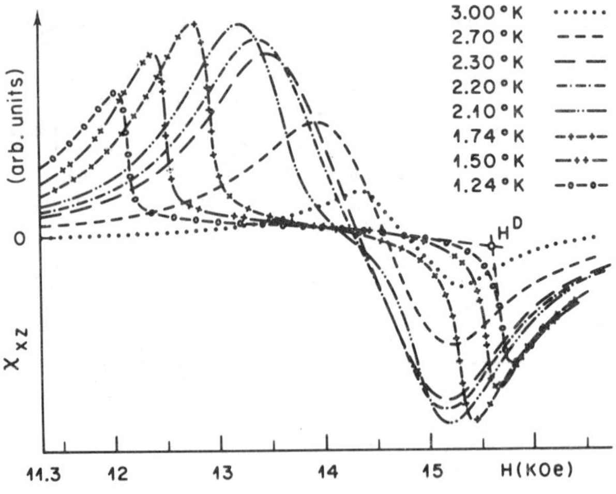

Fig. 4. $\tilde{\chi}_{zy}$ vs H for different temperatures for a disc with disc plane perpendicular to the easy axis and the applied field at an angle of 1.4° from the easy axis. The middle portion of $\tilde{\chi}_{zy}$ shows the domain state for $T \leq 2.2$ K.

magnetizations rotate in a narrow field interval from an AF like configuration to an FL like phase, causing the magnetization to tip away from the applied field. This results in a dispersion like curve of $\tilde{\chi}_{zy}$. In the domain state, however, $M_y \approx$ const. since $M_y(AF) \approx M_y(FL)$ at the SF transition. Thus $\tilde{\chi}_{zy} \approx 0$ in the domain state region and a break in the middle protion of the dispersion like $\tilde{\chi}_{zy}$ is observed. Fig. 4 shows $\tilde{\chi}_{zy}$ for a disc of $GdAlO_3$ with the disc plane perpendicular to z and with the field applied at an angle $\psi = 1.4°$ from the easy axis above and below T_c.

If the frequency of the modulation field, ν_m, becomes too high, the finite mobility of the domain walls prevents compensation of the modulation field by demagnetization and the dispersion like shape of $\tilde{\chi}_{zy}$ is restored in the domain state. For $\nu_m < 10$ kHz the domain walls move freely whereas for $\nu_m > 100$ kHz the domain structure stays rigid.

C. Metastability

Associated with the first order SF transition are metastable regions causing hysteresis. In contrast to ferromagnets and metamagnets, no hysteresis has yet been observed in antiferromagnets at the transition between the AF and SF state. Keffer and Chow [11] have invoked surface states to explain the lack of hysteresis. In their model, a surface layer of the other phase appears before the thermodynamic transition field H_c is reached, becoming unstable at H_c and driving the sample into the other phase. However such a surface state has not yet been observed.

The SF transition is a transition from a state of low anisotropy to a state of high anisotropy. The potential barrier separating the two states is of the order $\frac{1}{4}H_a/2H_{ex}$ compared to H_a separating the phases in ferromagnets and metamagnets. This greatly reduced potential barrier facilitates the transition and might well be the reason for the lack of hysteresis.

D. Antiferromagnetic Resonance (AFMR)

The frequency of the low frequency AFMR mode vanishes at the stability limit of the AF phase [2]. Although the antiferromagnet cannot be brought into the metastable region, the branch in the stable region is expected to reflect the instability field. At T=0, the

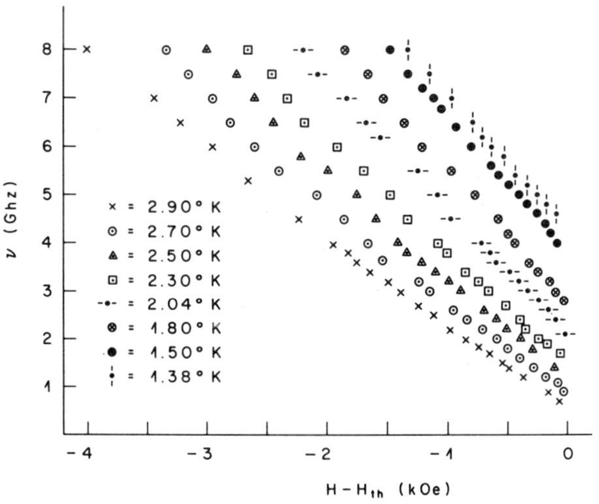

Fig. 5. Low frequency antiferromagnetic resonance branch vs H at different temperatures.

width of the superheated metastability region is

$$\Delta H_{sh} = H_{sh}-H_{th} \approx -K_2\sqrt{(K_1-K_2)/(2J_1+K_1)} \approx \Delta H_y^m \qquad (8)$$

where ΔH_y^m is the maximal perpendicular extension of the SF transition given by Eq. (7). ΔH_{sh} depends in the same way on temperature and demagnetization as ΔH_y(max). Fig 5 shows the low frequency AFMR at different temperatures for a thin slab of $GdAlO_3$ with field along the easy axis. Extrapolation to zero frequency gives ΔH_{sh} slightly larger than ΔH_y^m determined above. This is due to the orthorhombic component ΔK present in $GdAlO_3$. In the case of orthorhombic anisotropy (intrinsic or shape), the AFMR is pushed to a higher frequency by ΔK; close to H_{sh}, however, the precession mode degenerates into an oscillatory mode in the z-y plane, ΔK becomes ineffective and the frequency drops rapidly[13]. For finite angles ψ between easy axis and applied field, the stability limit of the AF phase approaches

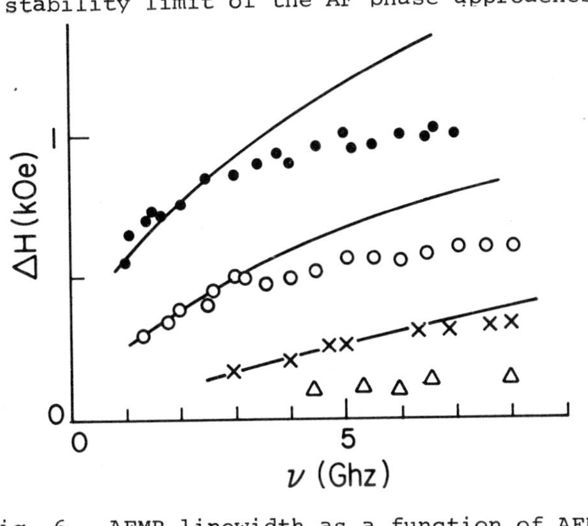

Fig. 6. AFMR linewidth as a function of AFMR frequency for T = 2.9 K (filled circles), T = 2.5 K (open circles), T = 1.9 (crosses), and T = 1.45 (triangles). The lines are calculations described in the text with frequency independent relaxation times ($\tau \propto T^{-2.7}$) giving $\Delta H \propto T^4$ at ν = 3 GHz.

rapidly the transition field H_{th}[9]. This collapse of the metastability region, however, cannot be observed by AFMR because the precession mode is no longer the soft mode for ψ finite and its frequency is nearly independent of ψ in the stable region[14].

The temperature dependence of the AFMR linewidth is still controversial[15]. In MnF_2, experiments indicate a T^4 law due to 4 magnon scattering at low and 6 magnon scattering at high temperatures[16]. In $GdAlO_3$, the linewidth inferred from antiferroaccoustic resonance was reported to follow a T^5 law[17]. The linewidth of AFMR of long cylinders with a considerably smaller temperature independent part (<100 Oe), however, favours a T^4 law, with the low frequency (ν <2 GHz) linewidth decreasing with $T^{4.5\pm0.5}$ and the high frequency linewidth ($\nu \approx$ 7 GHz) with $T^{3.5\pm0.5}$. Using a relaxation type equation of motion[18], the frequency dependence of the linewidth can be explained by a frequency independent relaxation time at low temperatures. At high temperatures, however, the agreement is not good and the linewidth becomes independent of frequency at high frequencies. (See Fig. 6).

III CRITICAL EFFECTS NEAR THE BICRITICAL POINT

Recently, Fisher and Nelson[2] have shown that the critical behaviour of antiferromagnets near the bicritical point is governed by the anisotropy crossover exponent. We present here experimental results on the PM boundaries and the magnetization discontinuity ΔM across the PM transition for $GdAlO_3$. The width of the transitionless region $(T_B - T_C)/T_B$ = 0.0096 indicates a misalignment of 0.08°.

The PM boundaries are predicted to follow

$$(t-Bg)^{\phi} = \pm Ag \qquad (9)$$

where t is the reduced temperature $(T_B-T)/T_B$,

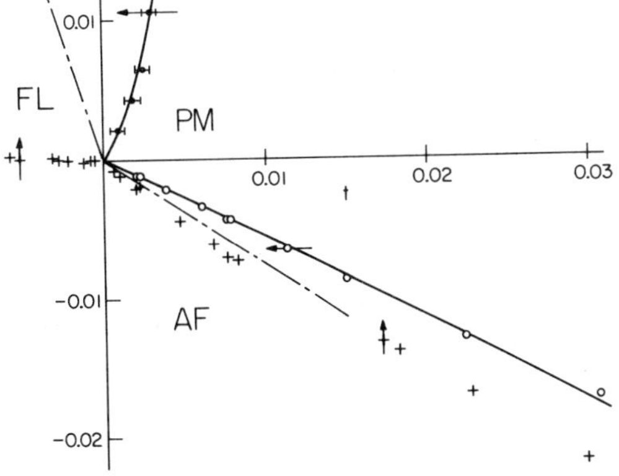

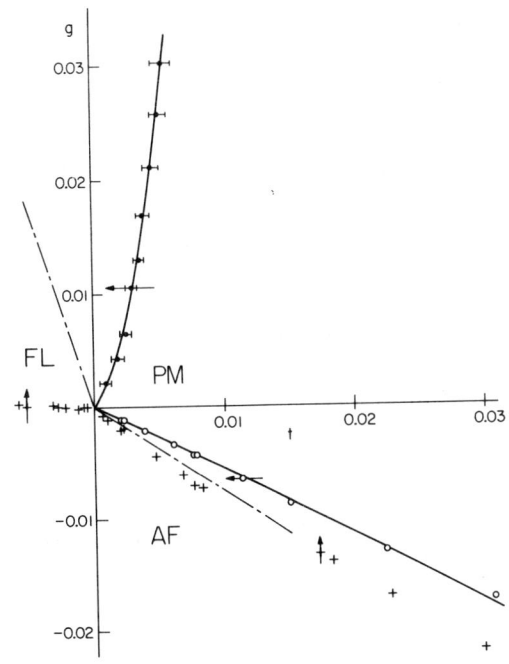

Fig. 7. PM boundaries as a function of reduced field g and reduced temperatures t. The arrows indicate sweep directions. Solid lines: Eq. (9) with the parameters given in the text; dashed-dotted lines: MFA boundaries.

g the reduced field $(H-H_B)/kT_B + j_B t$ where $j_B = (dH_{SF}/dT)T_B$, $\phi(n=2) = 1.18$ the anisotropy crossover exponent, and the ± stands for the PM-FL and PM-AF transition, respectively. Since $GdAlO_3$ is orthorhombic, a n=2 component Heisenberg system appropriately describes its critical behaviour[2]. Eq. (9) requires the PM boundaries to meet tangentially at T_B; however, this behaviour will partially be masked by the analytic term Bg. Fig. 7 shows the PM boundaries determined by the maximum of the parallel susceptibility χ. Although the PM-FL boundary is distinct from its MFA behaviour, no unambiguous fit to Eq. (9) can be made without knowing B. If we assume for B its MFA value $B_{FL} = -0.28 \pm 0.05$, a least square fit gives $\phi = 1.24 \pm 0.04$ and $A = 0.26 \pm 0.05$, (curve drawn in Fig. 7), in reasonable agreement with the theoretical value $\phi = 1.18$. In the case of the PM-AF boundary, the analytic term is dominant and a fit of the experiments to Eq. (7) does not give significant results. The curve in Fig. 7 is Eq. (9) with $\phi = 1.18$, $B_{FL}(MFA) = -1.33$, and $A = 0.23$). Further, the boundaries determined by field sweep and temperature sweep are different but converge at T_B.

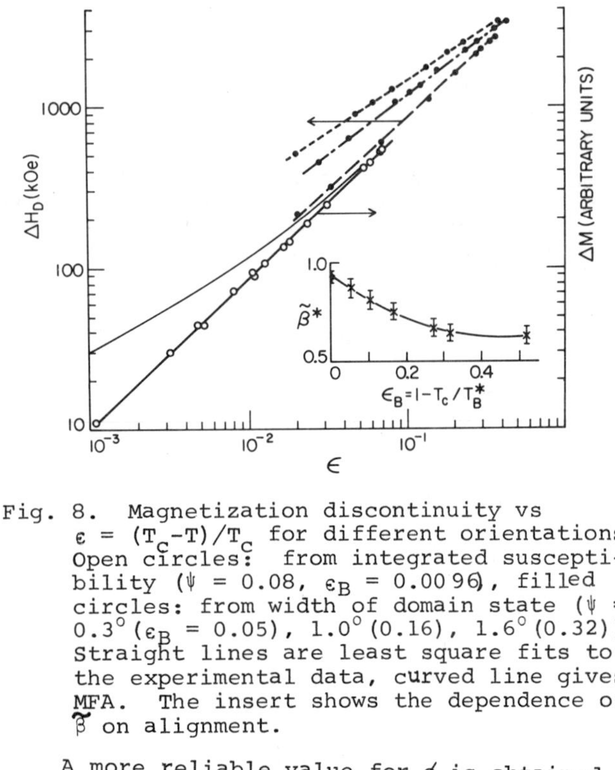

Fig. 8. Magnetization discontinuity vs $\varepsilon = (T_c-T)/T_c$ for different orientations. Open circles: from integrated susceptibility ($\psi = 0.08$, $\varepsilon_B = 0.0096$), filled circles: from width of domain state ($\psi = 0.3°$ ($\varepsilon_B = 0.05$), $1.0°$ (0.16), $1.6°$ (0.32)). Straight lines are least square fits to the experimental data, curved line gives MFA. The insert shows the dependence of $\tilde{\beta}$ on alignment.

A more reliable value for ϕ is obtained from the magnetization discontinuity across the SF transition. Extended scaling predicts that ΔM vanishes with power

$$\tilde{\beta} = 2 - \phi - \alpha \qquad (10)$$

where $\alpha(n=2) = -0.02$ is the exponent of the specific heat along the line g = 0. ΔM has been obtained from the width of the domain state and, more accurately, from the isothermal susceptibility integrated over a narrow field interval with the corresponding area at T_c subtracted. The second method largely eliminates uncertainties in ΔM due to rounding effects of M(H) near the SF transition. The results are shown in Fig. 8 for different alignments. ΔM is seen to vanish with a power law but with the exponent dependent on $\varepsilon_B = (T_B-T_c)/T_B$ (which is a measure of the alignment). Extrapolation to perfect alignment gives $\tilde{\beta} = 0.93 \pm 0.03$, with Eq. (10) $\phi = 1.09 \pm 0.03$, a value considerably smaller than predicted. A similar value of ϕ is inferred from $\tilde{\gamma}$, the divergence of the direct susceptibility along g = 0. Experiment gives $\tilde{\gamma} = 0.15$,[19] a divergence considerably weaker than predicted and with the scaling relation $\tilde{\gamma} = 2\phi + \alpha - 2$ one obtains $\phi = 1.09$. Although the PM boundaries are compatible with the predicted value of ϕ (depending on the choice of the analytic term) the present experiment appears to favour a smaller value. Better internal consistency can be obtained by choosing a larger value of $|B|$ in Eq. (9). A procedure to this effect (although the origin of B is not necessarily an analytic term in the free energy) is outlined in the following paper by Fisher.

ACKNOWLEDGEMENT

Discussions with Dr. A. King, Dr. K.A. Muller and Dr. H. Thomas, the technical assistance of Ch. Gerber and the growth of single crystals by H.J. Schell are gratefully acknowledged. We also would like to thank Professor Fisher and Professor Aharony for making their work available to us prior to publication.

REFERENCES

1. For ref. see N. Yamashita, J. Phys. Soc. Japan, 32, 610 (1972).
2. For ref. see K.W. Blazey, R. Webster and H. Rohrer, Phys. Rev. B 4, 2287 (1971).
3. M.E. Fisher and D.R. Nelson, Phys. Rev. Letters 32, 1350 (1974).
4. C.J. Gorter and J. Haantjes, Physica 18, 285 (1952); T. Nagamiya, Prog. Theor. Phys. 11, 309 (1954).
5. C.J. Gorter and T. van Peski-Tinbergen, Physica 22, 273 (1956).
6. C.W. Fairall and J.A. Corsen, Phys. Rev. B2, 4636 (1970).
7. A. Bruce and A. Aharony, to be published.
8. J.D. Cashion, A.H. Cooke, T.L. Thorp, and M.R. Wells, Proc. Roy. Soc. (London) A318, 473 (1970).
9. H. Rohrer and H. Thomas, J. Appl. Phys. 40, 1025 (1969); M.J.Kaganov and G.K. Chepurnykh, Sov. Phys. Solid State 4, 745 (1970).
10. A.R. King and D. Paquette, Phys. Rev. Letters 30, 662 (1973).
11. K. Keffer and H. Chow, Phys. Rev. Letters 31, 1061 (1973).
12. F.B. Anderson and H.B. Callen, Phys. Rev. 136, A1068 (1964); J. Feder and E. Pytte, Phys. Rev. 168, 640 (1968); K.W. Blazey, K.A. Muller, M. Ondris and H. Rohrer, Phys. Rev. Letters 24, 105 (1970).
13. H. Rohrer, Magnetic Resonance and Related Phenomena, ed. G.I. Hovi (North-Holland), 1973.
14. S. Foner in Magnetism I, ed. G.T. Rado and H. Suhl (Academic Press), New York, 1968.
15. For refs. see J.P. Kotthaus and V. Jaccarino, AIP Conf. Proc. 10, 57 (1972).
16. R.M. White, S.M. Rezende and L.C.M. Miranda, Conf. on Magnetism and Magnetic Materials, 1974.
17. P. Doussineau and B. Ferry, Phys. Letters 46A, 135 (1973).
18. H. Thomas, Proc. of the Chania International Conf. on Magnetism, Crete, 1969 (Gordon and Breach, New York); K.W. Blazey, M. Ondris, H. Rohrer and H. Thomas, J. Phys. 32, 1020 (1971).
19. H. Rohrer, to be published.

THEORY OF MULTICRITICAL TRANSITIONS AND THE SPIN-FLOP BICRITICAL POINT

Michael E. Fisher

Baker Laboratory, Cornell University, Ithaca, New York 14853

ABSTRACT

Some recent developments in the theory of multicritical points are reviewed with emphasis on the spin-flop bicritical point (T_b, H_b) in anistropic antiferromagnets, such as MnF_2. An extended scaling theory[1] is outlined with stress on the existence of optimal scaling axes. Renormalization group calculations for competing (perpendicular and parallel) order parameters justify the bicritical scaling theory and yield values for the corresponding exponents. Specific predictions follow for the vanishing of the magnetization discontinuity across the spin-flop line as $T \to T_b-$, for the $(M_\parallel, H_\parallel)$ bicritical isotherm, for the asymptotic variation of the susceptibility, specific heat, and scattering intensity, and for the shape of the paramagnetic phase boundaries near the bicritical point. The spin-flop phase diagram in $(T, H_\parallel, H_\perp)$ space is analyzed, and the universality of magnetic tricritical points is discussed briefly.

I. INTRODUCTION

In this talk I will review some recent theoretical work on magnetic multicritical points[1-6], specifically on tricritical points[1,2] and bicritical points.[3-5] The researches have been performed in close association with D. R. Nelson and J. M. Kosterlitz and involve scaling analyses[7] and renormalization group calculations utilizing the $\epsilon = 4-d$, dimensionality expansion.[8,9]

The first point will be to define a "multicritical point". This will be done in the context of anisotropic antiferromagnets which exhibit various types of multicritical points. In particular, the spin flop transition will be characterized as terminating at a spin flop bicritical point.[3] In writing a scaling hypothesis for multicritical points, emphasis will be placed on the optimal choice of (linear) scaling axes. The unknown exponents required in the scaling hypotheses may be found by renormalization group analysis combined, for greater accuracy, in $d = 3$ dimensions, with series expansion calculations.[10,11] The renormalization group calculations confirm a fairly wide degree of tricritical universality for magnetic systems leading, for $d = 3$, to the same classical exponents modified by logarithmic factors, as found originally by Riedel and Wegner[12,13] for a simpler Hamiltonian appropriate to helium three-four mixtures. Explicit renormalization group predictions are obtained, and will be presented, for the exponents determining behavior in the vicinity of the spin-flop bicritical point. In particular, the shape of the phase boundaries (or lambda lines) as they come into the bicritical point in real systems will be discussed. It will be seen that it is crucial to identify the orientation of both scaling axes if experimental data are to be properly analyzed. Finally, predictions for the shape of the low field phase boundary of an ideal isotropic antiferromagnet will be described.

II. MULTICRITICAL POINTS

To define a multicritical point in a nutshell, one may say it is a point on a line of critical points (or a lambda line) at which the basic characteristics of the transition including, normally, the various critical exponents, change abruptly. Different multicritical points correspond to different types of change and to different sorts of behavior "beyond" the multicritical point. To make this definition more concrete and to place it in a magnetic setting, consider a uniaxial or orthorhombic anisotropic antiferromagnet. In zero external field this exhibits a Néel point at $T_c^\parallel(0) = T_N$, below which the spins align antiferromagnetically, but parallel to the easy axis (as denoted by the superscript $\parallel$). If now, as illustrated in Fig. 1(a), a magnetic field $H\ (\equiv H_\parallel)$ is imposed (parallel to the easy axis) the Néel point is drawn out into a lambda line, $T_c^\parallel(H)$. (Note that we will always refer to the internal magnetic field, which must be derived from experiments made as a function of applied field, by making appropriate demagnetization corrections.) Owing to the anisotropy, which singles out just one component of the spins, the critical behavior all along this lambda line should be Ising-like, with exponents $\alpha = \alpha_I(d)$, $\beta = \beta_I(d)$, $\gamma = \gamma_I(d), \ldots$ appropriate to d dimensions. Now, in a metamagnet such as $FeCl_2$,[14] which is strongly anisotropic, this critical line in the (T, H) plane ends suddenly and becomes a line of first order transitions at a point (T_t, H_t). Following the lead of Griffiths[15] this point is called a tricritical point.

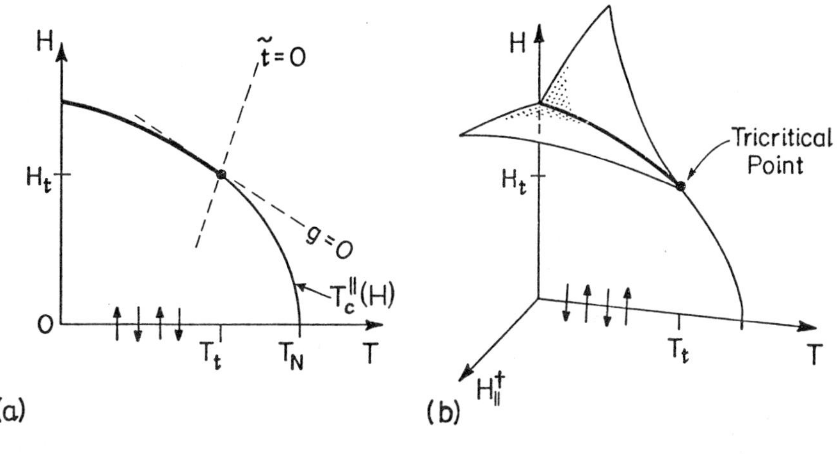

Fig. 1 (a) Phase diagram for an anisotropic antiferromagnet in a parallel field H illustrating a tricritical point at (T_t, H_t); the bold line denotes a first order transition. (b) Corresponding phase diagram in the presence of a staggered, ordering field $H^\dagger$.

Fig. 2 (a) Phase diagram for a weakly anisotropic antiferromagnet exhibiting a first order spin flop line (bold curve) and a bicritical point at (T_b, H_b). (b) Corresponding phase diagram in the presence of a staggered, ordering field, $H_{\parallel}^{\dagger}$, conjugate to the uniaxial phase only.

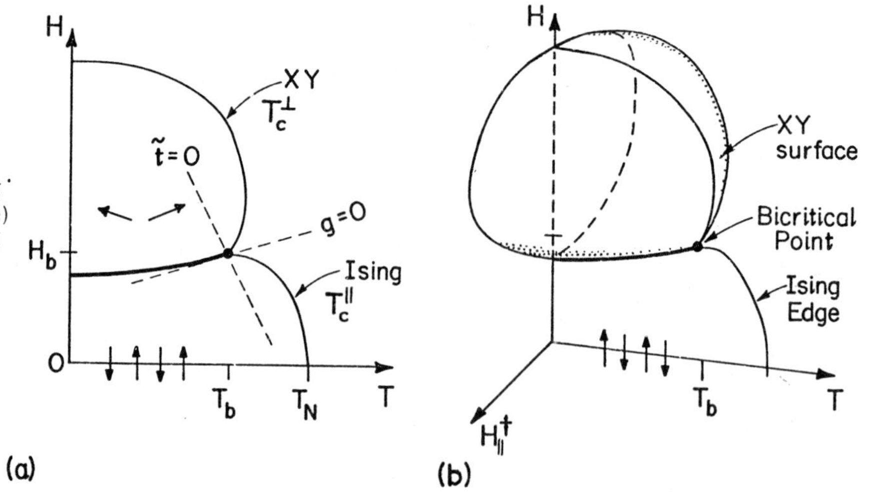

On the otherhand, in weakly anisotropic antiferromagnets, such as MnF_2 which is uniaxial,[16] and $GdAlO_3$ which is orthorhombic,[17,18] the behavior is different. As illustrated in Fig. 2(a) the critical line $T_c^{\parallel}(H)$ meets a second, distinct critical line, $T_c^{\perp}(H)$, at the point (T_b, H_b); at temperatures below this multicritical point a line of first order transitions emerges. This first order line, $H=H_{\varphi}(T)$, is just the spin-flop line at which the spins switch over from their low-field alignment parallel to the easy axis into a "perpendicular" alignment transverse to the field. In the uniaxial case this perpendicular ordering lies in the easy plane and hence is XY-like. The corresponding lambda line, $T_c^{\perp}(H)$, should thus be characterized by XY-like critical behavior with exponents $\alpha = \alpha_{XY}(d), \ldots,$ as indicated in Fig. 2(a). This very direct prediction for the high field critical behavior has not yet been tested experimentally. Evidently the multicritical point corresponds to a competition between two distinct sorts of ordering, namely parallel and perpendicular. For this reason the word bicritical point was coined[3] to describe it.

In the uniaxial case (as in MnF_2) a total of n=3 separate spin components become critical at the bicritical point. By contrast, in orthorhombic cases (like $GdAlO_3$), the spin flop occurs into an alignment along the second, or intermediate magnetic axis. This ordering is again Ising-like so that exponents across the perpendicular lambda line, $T_c^{\perp}(H)$, should still be $\alpha = \alpha_I(d)$, etc. Similarly, the total number of spin components with divergent fluctuations at the bicritical point is only n=2 for orthorhombic systems.

As stressed by Griffiths,[15] it is instructive to consider the effect of the corresponding ordering field on the tricritical phase diagram. For an antiferromagnet in low fields this is the staggered magnetic field $H^{\dagger}$, which acts oppositely on complementary sublattices. The appearance of the $(T, H, H^{\dagger})$ diagram for a tricritical system is exhibited in Fig. 1(b) and is now well known.[15] It reveals that the first order boundary, $H_{\tau}(T)$, below T_t is actually a line of triple points; in addition three distinct lines of critical points are seen to converge and meet at the tricritical point. This last observation provided the original motivation for the name "tricritical". By the same token, two lines of critical points [in the (T, H) plane] meet at a bicritical point. The unwary should be warned, however, that this nomenclature is not systematic. But until a complete classification of multicritical points is achieved (which may not even be possible) it seems reasonable to proceed in an ad hoc fashion. For most simple antiferromagnets a nonzero staggered field cannot be produced in the laboratory (although its response function, $\hat{\chi}^{\dagger}(\vec{k})$, can be studied by neutron diffraction). However, it has now been realized[19] that in many of the experiments on the well known antiferromagnet dysprosium aluminum garnet, one cannot, in fact, avoid imposing a finite staggered field. Thus the full phase diagram of Fig.2(b) must be considered, and one can actually study the "wing critical points" which lie out of the $H_{\parallel}^{\dagger} = 0$ plane.

Introduction of the ordering fields for a bicritical system is more complex since there are two of them namely, for an antiferromagnet $H_{\parallel}^{\dagger}$ and also $\vec{H}_{\perp}^{\dagger}$. Since, as regards the perpendicular (or spin-flopped) phase, $H_{\parallel}^{\dagger}$ is merely a nonordering field, one anticipates that the $(T, H, H_{\parallel}^{\dagger})$ diagram is as shown in Fig. 2(b). The perpendicular phase is now bounded by a critical surface which balloons out of the $H_{\parallel}^{\dagger} = 0$ plane (and is probably of XY character for the uniaxial, n=3 case illustrated).

III. SCALING THEORY

We may, initially, consider bicritical and tricritical points in a unified fashion in constructing a scaling theory. The first step is to introduce the deviations

$$\Delta T = T - T_m, \qquad T_m = T_b, T_t, \tag{3.1}$$

$$\Delta H = H_{\parallel} - H_m, \qquad H_m = H_b, H_t, \tag{3.2}$$

and thence the reduced temperature and ordering field

$$t = \Delta T / T_m, \qquad \vec{h} = m\vec{H}^{\dagger}/k_B T_m, \tag{3.3}$$

where m is the magnetic moment per spin. Now, owing to the slope $(dH_{\mu}/dT)_m = p/T_m$ $(\mu = \tau, \varphi)$ of the first order line as it meets the multicritical point, it is natural and reasonable, to introduce ΔH in the combination

$$g = \Delta H - pt . \tag{3.4}$$

Indeed, this is imperative for a proper scaling description since both variables t and ΔH otherwise carry one away from the multicritical point in an arbitrary direction and, hence, are equivalent for scaling purposes.

On the otherhand, variation of t for $g=0$, carries one (asymptotically) along the phase boundary (below T_m), while variation of g then carries one away from the boundary.

We may now state the (extended) scaling hypothesis for the singular part of the free energy as

$$F_{sing.}(T,\vec{H}^{\dagger},H) \approx t^{2-\alpha} Y\left(\frac{h_{\|}}{t^{\Delta_{\|}}}, \frac{\vec{h}_{\perp}}{t^{\Delta_{\perp}}}, \frac{g}{t^{\phi}}\right). \quad (3.5)$$

To allow for the two distinct phases meeting at a bicritical point we have decomposed $\vec{h}$ into its $\|$ and $\perp$ components and allowed them to scale separately. Only $h_{\|}$ and $\Delta_{\|}$ are relevant at a tricritical point. The exponents $\alpha, \Delta_{\|}, \Delta_{\perp}$, and the crossover exponent ϕ are, of course, characteristic of the multicritical point and need not bear any relation to the exponents for the lambda lines. The values of these exponents according to the classical phenomenological theories are[12, 13, 15]

$$\text{Tricritical: } \alpha=\alpha_u=-1, \quad \Delta=\phi\Delta_t=\frac{5}{2}, \quad \phi=\Delta_u=2, \quad (3.6)$$

and[20]

$$\text{Bicritical: } \alpha=0 \text{ (discon.)}, \quad \Delta_{\|}=\Delta_{\perp}=\frac{3}{2}, \quad \phi=1, \quad (3.7)$$

The symbols α_u, Δ_u, and Δ_t serve to display the relation to Griffiths' notation.[15]

At this point we observe that the variable t is not obviously the best variable with which to scale. For a simple critical point in zero field, symmetry dictates that ΔH and ΔT are independent scaling variables, but one knows from soluble examples[21] that, more generally, one should expect the existence of a second (linear) scaling axis, of slope say $(dT/dH)_m = -qT_m$, corresponding to the optimal scaling variable

$$\tilde{t} = t + q\,\Delta H. \quad (3.8)$$

This conclusion is borne out also by the renormalization group approach.[8-10] Thus everywhere in the scaling hypothesis (3.5) we should replace t by $\tilde{t}$. It is easy to check that when the crossover exponent satisfies $\phi > 1$ this change has no effect on the leading asymptotic behavior as $t, g \to 0$; but, as we will see below, it has a drastic effect on the range over which the leading asymptotic terms describe the behavior satisfactorily.

IV. RENORMALIZATION GROUP THEORY

To determine the critical exponents appropriate to magnetic tricritical and bicritical behavior, renormalization group theory has been applied[1-5] to the anisotropic antiferromagnetic Hamiltonian

$$\mathcal{H} = -\tfrac{1}{2}\sum_{\underset{\sim}{R},\underset{\sim}{R}'} [J(\underset{\sim}{R}-\underset{\sim}{R}')\vec{S}_{\underset{\sim}{R}} \cdot \vec{S}_{\underset{\sim}{R}'} + D(\underset{\sim}{R}-\underset{\sim}{R}')S^{\|}_{\underset{\sim}{R}} S^{\|}_{\underset{\sim}{R}'}]$$
$$-\sum_{\underset{\sim}{R}} [H_{\|} S^{\|}_{\underset{\sim}{R}} + \vec{H}_{\perp} \cdot \vec{S}^{\perp}_{\underset{\sim}{R}} + e^{i\underset{\sim}{k}_0 \cdot \underset{\sim}{R}} \vec{H}^{\dagger} \cdot \vec{S}_{\underset{\sim}{R}}], \quad (4.1)$$

in which $J(\underset{\sim}{R})$ is the isotropic exchange coupling, while $D(\underset{\sim}{R})$ represents the uniaxial anisotropy energy. (Note that $\frac{1}{2}D(\underset{\sim}{0})$ is just the single-ion, quadratic anisotropy.) As usual,[9] the spins S_R at sites $\underset{\sim}{R}$ are taken to be classical vectors of n components whose lengths, $|S_R|$, may vary continuously but subject to a weighting factor of the form $\exp[-\frac{1}{2}|S|^2 - \tilde{u}_4|S|^4 - \tilde{u}_6|S|^6 - \ldots]$.

I will summarize first the results for large anisotropy, where one may effectively take $n=1$, obtaining Ising spins with coupling $J_{\|}(\underset{\sim}{R}) = J(\underset{\sim}{R}) + D(\underset{\sim}{R})$. The simplest model expected to exhibit a tricritical point is the layered metamagnet in which ferromagnetically coupled sublattice layers, with $J_{\|} = J_{aa} = J_{bb} > 0$, interact via weak, intersublattice antiferromagnetic couplings, with $J_{\|} = J_{ab} < 0$. Owing to the competition between antiferromagnetic, layered ordering and totally ferromagnetic ordering, which develops as $H_{\|}$ and $H^{\dagger}$ increase, a convincing theory must allow for both sorts of order. This has been achieved[1] by decomposing the spin field $S_{\underset{\sim}{R}}$ into two distinct fields $s_a(\underset{\sim}{R})$ and $s_b(\underset{\sim}{R})$, for the two sublattices, with corresponding reduced Brillouin zones in momentum space. After transformation, the different spin fields must be renormalized in distinct ways.[1, 9] Finally, one can show how the renormalized Hamiltonian effectively reduces to a simpler (single-field) form, originally discussed by Riedel and Wegner[12] in connection with He^3-He^4 mixtures.

Following Riedel and Wegner's analysis[12] one then concludes that metamagnetic tricritical points should be described by the scaling hypothesis (3.5) together with, for three dimensions, the classical exponent values (3.6). This prediction seems to be quite well borne out by the experiments on $FeCl_2$.[14] As a matter of fact, renormalization group theory[12] indicates the presence of additional logarithmic factors so that e.g., one expects $\chi^{\dagger}(T) \sim t^{-1}|\ln t^{-1}|^{\frac{1}{3}}$. It may, however, be very hard to detect such factors experimentally, even once theory can specify their relative magnitude.

The prediction of classical exponent values is also confirmed by series expansions[22] and by Monte Carlo studies[23] on the metamagnetic layered Ising model which yield $\delta = \delta_u = 2$ for the tricritical M vs H isotherm. This agrees with (3.6) via the scaling relation $\tilde{\delta} = \phi/(2-\alpha-\phi)$. But, for a next nearest neighbor simple cubic Ising model[22, 23] with $J_{\|} = J_{nn} < 0$ and $J_{\|} = J_{nnn} > 0$, tricritical points were found numerically[22, 23] with, apparently, $\tilde{\delta} = 4 \pm 1$. The renormalization group analysis of the layered model may, however, be taken over straightforwardly,[2] for the nnn model by making the identifications $J_{nnn} \Rightarrow J_{aa}$ and $J_{nn} \Rightarrow J_{ab}$. Theory again predicts classical exponent values (for $d=3$) and, indeed, indicates a rather large degree of tricritical universality for magnetic systems. Following Wortis,[24] we thus believe that numerical difficulties in locating T_t reliably, have led to misestimates of $\tilde{\delta}$ in the series and Monte Carlo studies; a slightly revised value for T_t seems consistent with the renormalization group exponent predictions.[24]

V. BICRITICAL EXPONENTS AND BEHAVIOR

The Hamiltonian (4.1) in the case of weak anisotropy is expected to yield spin flop and bicritical behavior; a renormalization group attack using the $\epsilon = 4-d$ dimensionality expansion[9] and different renormalizations for four or more independent spin fields[3-5] can then determine the character of the corresponding fixed point.[9] The situation for general n and d is found to be quite complicated[3-5] (including "biconical" behavior and various tetracritical possibilities.[6] However, for n below a borderline value $n^{x}(d)$, where $n^{x}(3) \simeq 3.13$, one finds that the bicritical exponents entering the scaling relation (3.5) are the same as those for the corresponding, fully isotropic n-component system. For d=3 the Heisenberg and XY values are known most accurately from series expansion studies[10, 11, 25] which thus yield the exponent values:

Bicritical (d = 3)

$$n=3 \text{ (uniaxial)}: \quad \alpha \simeq -0.10, \quad \Delta_{\parallel}=\Delta_{\perp}\simeq 1.74, \quad \phi \simeq 1.25, \qquad (5.1)$$

$$n=2 \text{ (orthorhombic)}: \quad \alpha \simeq -0.02, \quad \Delta_{\parallel}=\Delta_{\perp}\simeq 1, \quad \phi \simeq 1.175. \qquad (5.2)$$

which replace the classical values (3.7). It should be noted that these exponents are all subject to extrapolation uncertainties in the range ± 0.015 to ± 0.05.

The scaling hypothesis may now be used to make concrete predictions for bicritical behavior. As we will see the crossover exponent ϕ, which previously has been almost unobservable, will play a dominant role. The principal predictions are:

(A) The specific heat C_M on the locus $g=0$ (or $M=M_b$) varies as $C_c - At^{|\alpha|}$ i.e., with a sharp (but finite) cusp, in contrast to the divergent behavior on the $T_c^{\parallel}$ lambda line (where $\alpha_{\parallel} \simeq +0.13$).

(B) The magnetization discontinuity $\Delta M_{\varphi}(T)$ across the spin flop line below T_b vanishes as $|t|^{\beta}$ with exponent

$$\beta = 2 - \alpha - \phi$$
$$= 0.85 \pm 0.07 \ (n=3), \qquad 0.84 \pm 0.05 \ (n=2), \qquad (5.3)$$

which deviates clearly from the classical value $\tilde{\beta}=1$. The first experiments to observe this exponent, by Rohrer[18] in $GdAlO_3$, do indicate $\tilde{\beta} < 1$, but suggest a value slightly larger than (5.3) with n=2.

(C) The standard susceptibility, $\tilde{\chi} = (dM_{\parallel}/dH_{\parallel})_T$, for $g=0$ should diverge as $t^{-\tilde{\gamma}}$ with

$$\tilde{\gamma} = 2\phi + \alpha - 2$$
$$= 0.40 \pm 0.09 \ (n=3), \qquad 0.33 \pm 0.06 \ (n=2). \qquad (5.4)$$

This implies an appreciably sharper singularity than the very weak divergence or cusp expected in $\tilde{\chi}$ across the lambda lines (where it should mirror the corresponding specific heats but with a smaller amplitude relative to the background[26]).

(D) Along the locus $g=0$ both the ordering susceptibilities $\chi_{\parallel}$ and $\chi_{\perp}$ should be observed, in scattering experiments, to diverge with the same exponent γ which takes the values 1.38 ± 0.02 and 1.315 ± 0.02 for n = 3 and 2, respectively (as at the corresponding critical points).

(E) On the bicritical isotherm $T=T_b$ (or, better, on the locus $\tilde{t}=0$, see below) the field deviation ΔH varies with the magnetization as $|M_{\parallel} - M_b|^{\delta}$ where

$$\tilde{\delta} = \phi/\tilde{\beta} = \phi/(2-\alpha-\phi),$$
$$= 1.47 \pm 0.15 \ (n=3), \qquad 1.40 \pm 0.11 \ (n=2), \qquad (5.5)$$

which should not be too hard to distinguish from the classical value $\tilde{\delta} = 1$.

(F) The critical lines, $M_c^{\perp}(T)$ and $M_c^{\parallel}(T)$, in the $(T, M_{\parallel})$ plane should come into the critical point as $\pm t^{\tilde{\beta}}$ and hence, with a vertical tangent.

(G) This last prediction follows from the fact that within the extended scaling hypothesis (3.5), the $\perp$ and $\parallel$ critical lines must be given simply by

$$g/\tilde{t}^{\phi} = +w_{\perp}, \quad -w_{\parallel}, \qquad (5.6)$$

where $w_{\perp}$ and $w_{\parallel}$ are positive constants. This result together with $\phi > 1$, finally implies that in the original $(T, H_{\parallel})$ plane both lambda lines come in tangent to the spin flop line $[g=0$ or $H \approx H_{\varphi}(T)]$ at the bicritical point. This predicted tangency is illustrated schematically in Fig. 3, which is a plot in the (T, H^2) plane; the reason for using H^2 in place of H $(\equiv H_{\parallel})$ will be explained below. At first sight, the experimental data points for the corresponding critical lines of MnF_2, shown in Fig. 4, do not appear to be meeting the flop line tangentially; indeed, the $T_c^{\perp}$ line appears to be almost orthogonal to it! Let us discuss this problem in more detail.

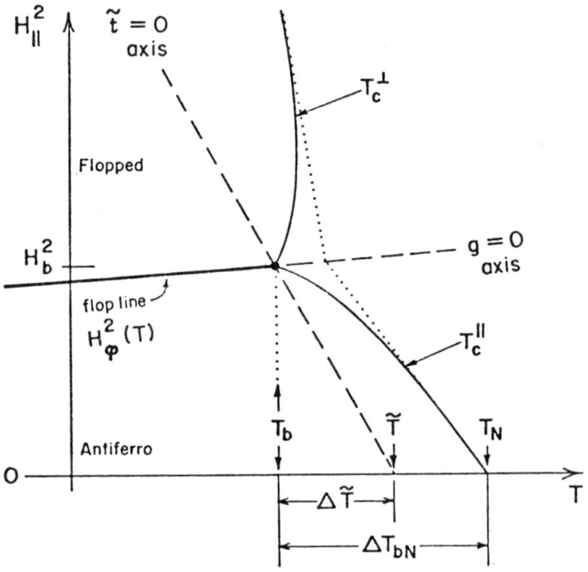

Fig. 3 Schematic phase diagram of a spin flop bicritical point in the (T, H^2) plane (with $H \equiv H_{\parallel}$) illustrating the expected geometry of the spin flop line, $H^2 = H_{\varphi}^2(T)$, and the critical lines $T_c^{\parallel}(H)$ and $T_c^{\perp}(H)$. Note that $T=\tilde{T}$ is the intercept of the $\tilde{t}=0$ axis with the zero field axis.

VI. THE GEOMETRY OF THE CRITICAL LINES

The renormalization group theory[3-5] not only shows that the bicritical exponents should be Heisenberg-like, but also indicates that the (T, H^2) spin flop phase diagram should map onto the (t, g_1) diagram for the simpler uniaxial (Heisenberg/Ising or XY/Ising) ferromagnetic Hamiltonian

$$\mathcal{H}(g_1) = -\sum_{(\underset{\sim}{R}, \underset{\sim}{R}')} J(\underset{\sim}{R} - \underset{\sim}{R}')[\vec{S}_{\underset{\sim}{R}} \cdot \vec{S}_{\underset{\sim}{R}'} - g_1 \mathcal{Q}_1(\underset{\sim}{R}, \underset{\sim}{R}')], \qquad (6.1)$$

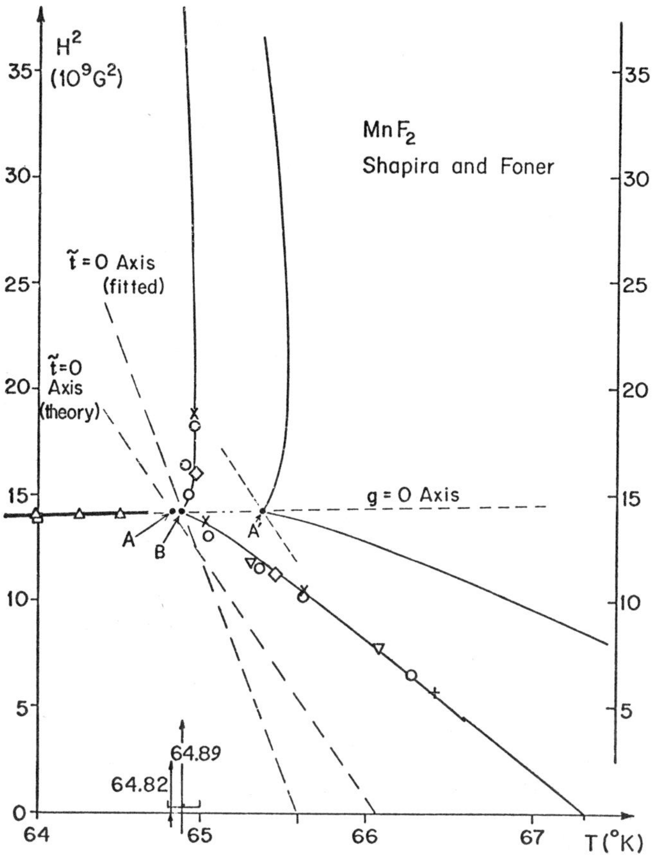

Fig. 4 Phase diagram in the (T, H^2) plane for MnF_2 (after Shapira and Foner, Ref. 16). Note that data points on the $T_c^{\perp}$ line above $H^2 = 20 \times 10^{12} G^2$ have not been shown. The significance of the various fitting lines is explained in the text.

where, for a correct scaling theory, one must write the anisotropic coupling, for general n, in the symmetry adapted form[10,27,28]

$$\mathcal{Q}_1(\underset{\sim}{R}, \underset{\sim}{R}') = S^{(1)}_{\underset{\sim}{R}} S^{(1)}_{\underset{\sim}{R}'} - \frac{1}{n} S_{\underset{\sim}{R}} \cdot S_{\underset{\sim}{R}'}$$

$$\propto S^{(1)}_{\underset{\sim}{R}} S^{(1)}_{\underset{\sim}{R}'} - \frac{1}{n-1} \sum_{\mu=2}^{n} S^{(\mu)}_{\underset{\sim}{R}} S^{(\mu)}_{\underset{\sim}{R}'} , \qquad (6.2)$$

for which $\langle \mathcal{Q}_1 \rangle_{g_1=0} \equiv 0$. Series expansion studies for the cubic behavior for the cubic lattices[11,28] confirm that the critical behavior for $\mathcal{H}(g_1)$ crosses over to Ising-like form for $g_1 < 0$ (and, for n=2, is XY-like for $g_1 > 0$); the critical lines are found to be determined by

$$g_1 \approx A_{\perp} t^{\phi}, \quad \text{and} \quad -A_{\parallel} t^{\phi}, \qquad (6.3)$$

where ϕ is the crossover exponent of (5.1) and (5.2). These results, which correspond to (5.6), are plotted for n=3 in Fig. 5. The amplitude ratio

$$Q_n = A_{\perp}(n)/A_{\parallel}(n), \qquad (6.4)$$

should have a universal value; the series analyses[10,29] for n=3 confirm this and yield $Q_3 \simeq 2.51$, as used in the figure.

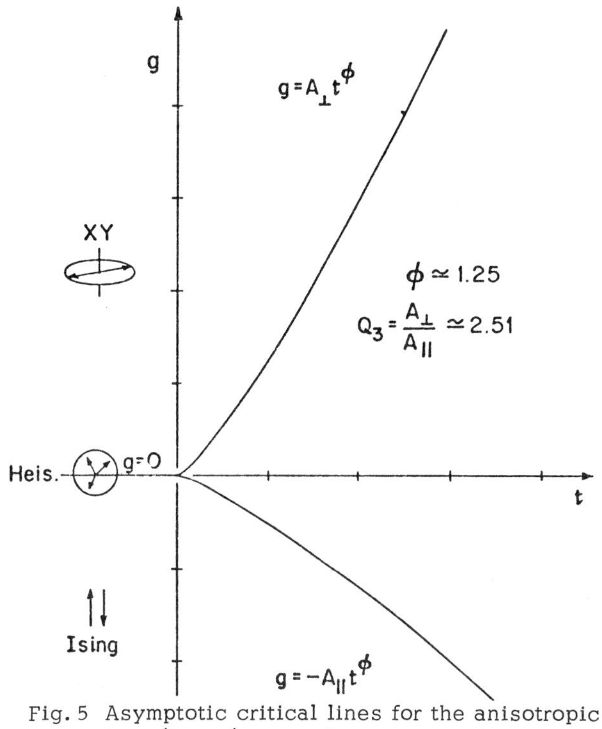

Fig. 5 Asymptotic critical lines for the anisotropic Heisenberg/Ising/XY Hamiltonian according to series expansion calculations (Refs. 11 and 29).

In leading order in $\epsilon = 4-d$ one may determine Q_n simply by using (6.2) and the mean field critical temperatures, which are evidently

$$T_0^{\parallel}(g_1) = T_0(1 + |g_1|) \quad \text{for} \quad g_1 \le 0,$$

$$T_0^{\perp}(g_1) = T_0[1 + g_1/(n-1)] \quad \text{for} \quad g_1 \ge 0, \qquad (6.5)$$

where the $g_1 = 0$ mean field critical point is

$$T_0 = \hat{J}(0)/k_B = k_B^{-1} \sum_{\underset{\sim}{R}} J(\underset{\sim}{R}) . \qquad (6.6)$$

On introducing $t_{\epsilon=0} = (T - T_0)/T_0$ and noting that[27,28] $\phi(\epsilon = 0) = 1$ one immediately finds, from (6.2) to (6.5),

$$Q_n = n - 1 + O(\epsilon) . \qquad (6.7)$$

But note that for n=2 one must have $Q_2 = 1$ for all ϵ by symmetry.

One feature of Fig. 5 is that the asymptotic tangency as $t \to 0$ is not really visible even though it is rigorously present. Indeed, since $\phi - 1 = 0.25$ is quite small, the plots appear almost straight! To obtain the spin flop phase diagram one must implement the mapping

$$g_1 \Rightarrow g = \Delta(H^2) - \bar{p} t , \qquad (6.8)$$

$$t \Rightarrow \tilde{t} = t + \bar{q} \Delta(H^2), \qquad (6.9)$$

where $\Delta H^2 = H^2 - H_b^2$, and $\bar{p}$ and $\bar{q}$ accordingly replace the original (T, H) slopes, p and q. To do this one must know the slope $\bar{p}$ of the g=0 axis, i.e. the tangent to the spin flop line $H_{\varphi}^2(T)$; but this is easily found from experimental data [see Figs. 3 and 4]. However, one also needs the slope $\bar{q}$ of the $\tilde{t} = 0$ axis, which is not immediately obvious, and which is also a nonuniversal parameter.

Lacking this information, one may adopt the theoretical values of ϕ and of $Q_n = A_\perp / A_\parallel = w_\perp / w_\parallel$, and treat $w_\parallel$ [in (5.6)] and $\bar{q}$ as fitting parameters. If experimental data near (T_b, H_b^2) are sparse, as in Fig. 4, one must also allow T_b to vary somewhat. We have performed a rough analysis for MnF_2 along these lines taking $\phi = 1.25$ and $w_\perp/w_\parallel = 2.51$ (n=3), and finding directly from the data[16]

$$H_b \simeq 119\,\text{kG}, \qquad T_b \simeq 64.89\,°\text{K},$$

$$\bar{p} = T_b (dH_\varphi^2/dT)_b \simeq 1.28 \times 10^{10}\,G^2 . \qquad (6.10)$$

Fitting the critical lines to the form (5.6) with $\tilde{t}$ and g defined as in (6.8) and (6.9), then yields

$$\bar{q} = \Delta\tilde{T}/T_b H_b^2 \simeq 7.6 \times 10^{-13}\,G^{-2}, \quad w_\parallel \simeq 1.46 \times 10^{12}\,G^2 . \qquad (6.11)$$

The parameter $\Delta\tilde{T} = \tilde{T} - T_b$, specifies the slope of the $\tilde{t} = 0$ axis via its intersection, $\tilde{T}$, with the $H^2 = 0$ axis (see Fig. 3). The resulting fit (shown as the two solid curves springing from the point B in Fig. 4) is surprizingly good (many additional data points for $H^2 > 20 \times 10^9\,G^2$ have been ommitted); but until further and more precise data become available, too much significance should not be attached to it. Nevertheless, it shows dramatically how the almost vertical $T_c^\perp$ line may, in fact, be asymptotically tangent to the spin flop line!

From a theoretical viewpoint one must now ask: "Is the value $\Delta\tilde{T}_{fit} \simeq 0.70\,°K$, reasonable?" It turns out that even though the value of $\Delta\tilde{T}$ is non-universal, one can actually estimate $\Delta\tilde{T}$ to leading order in the anisotropy and in $\epsilon = 4-d$. The analysis follows the lines used in deriving (6.7); indeed, for small anisotropy, the mean field critical curves (6.5) correspond to the dotted straight lines shown in Fig. 3. The straightness provides the motivation for plotting vs $H_\parallel^2$ in place of $H_\parallel$. For small anisotropy, the slopes $dT_c^\parallel/d(H_\parallel^2)$ and $dT_c^\perp/d(H_\parallel^2)$ are, according to mean field theory[17] (which is appropriate for $\epsilon = 0$), in the ratio 3:1. This then leads[30] to the simple result

$$\Delta\tilde{T} = \tilde{T} - T_b \simeq \frac{n+2}{3n}\,\Delta T_{bN}, \qquad (6.12)$$

where $\Delta T_{bN} = T_N - T_b$ and T_N is the zero-field Néel temperature, as shown in Fig. 3. The accuracy of this formula can be increased by renormalizing ΔT_{bN} approximately by subtracting $\Delta T_b^0 \simeq T_b^0 - T_b$, where T_b^0 is the effective mean field bicritical point estimated by a linear extrapolation from low (and high) fields as indicated by the dotted lines in Fig. 3.

If one fits the MnF_2 data with a $\tilde{t} = 0$ axis determined by (6.12), one is lead to adopt the slightly lower bicritical point estimate $T_b \simeq 64.82\,°K$ (which is still within the $64.9 \pm 0.1\,°K$ range quoted by Shapira and Foner[16]). This is indicated in Fig. 4 by the point A. Together with $T_N = 67.33\,°K$ and $\Delta T_b^0 \simeq 0.26\,°K$ the theory then indicates $\Delta\tilde{T} \simeq 1.25\,°K$. We conclude that both the sign and magnitude of $\Delta\tilde{T}_{fit}$ ($\simeq 0.70\,°K$) are perfectly reasonable.

The $\tilde{t} = 0$ axis following from $\Delta\tilde{T} = 1.25\,°K$ is labelled "theory" in Fig. 4. Adopting this, one can vary $w_\parallel$ to fit the two critical lines. To avoid crowding the figure, the results for one choice have been displaced along the g=0 axis to a shifted bicritical point A′; although the fit to $T_c^\perp$ is acceptable over a fair range, the curve for $T_c^\parallel$ rather quickly deviates above the data points. Since $\Delta\tilde{T}_{fit}/\Delta\tilde{T}_{th.} \simeq 0.56$ this is not very surprizing: the deviation in $\Delta\tilde{T}$ must probably be attributed to the full renormalizing effects of the bicritical fluctuations in d=3 dimensions, but more extensive experimental data in the bicritical region would be valuable.

Finally, I may mention that very recent and precise measurements of the $T_c^\perp$ line for $GdAlO_3$ by Rohrer[18] give a rather clear indication of $\phi > 1$. Furthermore, preliminary fits to $\phi \simeq 1.18$, $Q_2 = 1$ and $\Delta\tilde{T} \simeq \frac{2}{3}\Delta T_{bN}$ (for n=2) are rather encouraging.

VII. OTHER PHASE DIAGRAMS

It is instructive to consider the antiferromagnetic Hamiltonian (4.1) in the isotropic limit $D(R) \equiv 0$, which might be realizable experimentally to fairly high precision. For $H \equiv |\vec{H}| = 0$ the critical behavior at $T_N = T_c(0)$ is now isotropic, or Heisenberg-like with corresponding exponents. But imposition of a field must lead to an instantaneous spin flop, as indicated schematically in Fig. 6(a). Thus the lambda line $T_c(H)$ should (for n=3) be XY-like in character; renormalization group analysis[5] confirms this. Mean field or phenomenological theories would also indicate that $T_c(H) - T_N$ should vary as H^2 for small H, as drawn in Fig. 6(a).

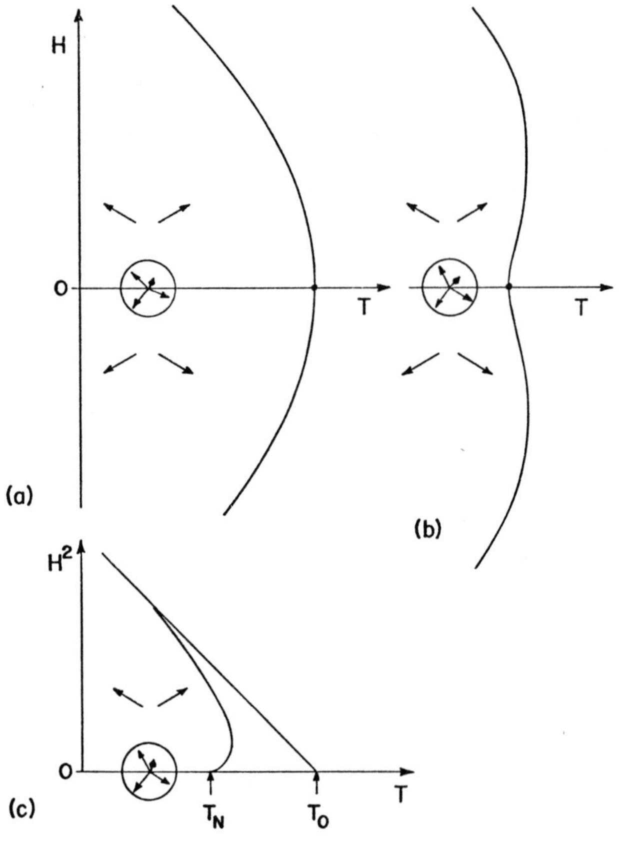

Fig. 6 Critical lines for a fully isotropic antiferromagnet in a field H according to (a) mean field or phenomenological theories, (b) degenerate bicritical scaling theory with $\tilde{\psi} = 1.60$, and (c) in the (T, H^2) plane. Note that the critical behavior for nonzero field is (n-1)- or XY-like.

However, since as we now see, the Néel point is really a multicritical point (which might be termed a "degenerate bicritical point"), this conclusion is suspect! Indeed, the renormalization group calculations[5] show that (6.3) still applies if g is given by (6.8) with by symmetry, $H_b = 0$ and $\overline{p} = 0$, so that $g \propto H^2$. The relation (6.9) remains valid and thus we find

$$T_c(H) - T_N \approx BH^{\tilde{\psi}} - \overline{\overline{q}}H^2 + \ldots \qquad (7.1)$$

with

$$\tilde{\psi} = 2/\varphi \simeq 1.60 \ (n=3) \quad 1.70 \ (n=2). \qquad (7.2)$$

Botn B and $\overline{\overline{q}}$ are expected to be positive, so (7.1) leads to the re-entrant or bow-shaped critical lines illustrated in Fig. 6(b) and, vs H^2, in Fig. 6(c), where the straight line contrasts the mean field prediction. The isotropic multicritical point may be regarded as an umbilical point on the full $T_c(\vec{H})$ surface since although the surface is smooth there, the curvature diverges. It would be interesting to attempt an experimental test of (7.1) for n=2 and 3.

To observe a bicritical point in a real system, it is important[17, 18] to align the total external field $\vec{H} = (H_\parallel, H_\perp)$ precisely parallel to the easy axis. Our previous discussion has assumed this tacitly although experimentally special procedures must be used.[18] However, the theory can be adapted[5] to study the full $(T, \vec{H}_\parallel, \vec{H}_\perp)$ space. One finds[5] that the transverse field $H_\perp$ also scales as t^ϕ near the bicritical point but, as in the discussion just presented of the fully isotropic antiferromagnet, the (n-1) or XY-like character of the perpendicular lambda line, $T_c^\perp$, is destroyed by nonzero $\vec{H}_\perp$. Thus, as illustrated in Fig. 7, the critical surface should, normally,[5] have an overall Ising-like character but, for $n \geq 3$, display an XY-like "seam" (for $\vec{H}_\perp = 0$, $H > H_b$), and a sharp, Heisenberg-like, bicritical "umbilicus". The XY- or (n-1) - like seam should lie in a shallow "furrow" characterized, as in (7.1), by an exponent $\tilde{\psi}(n-1)$.

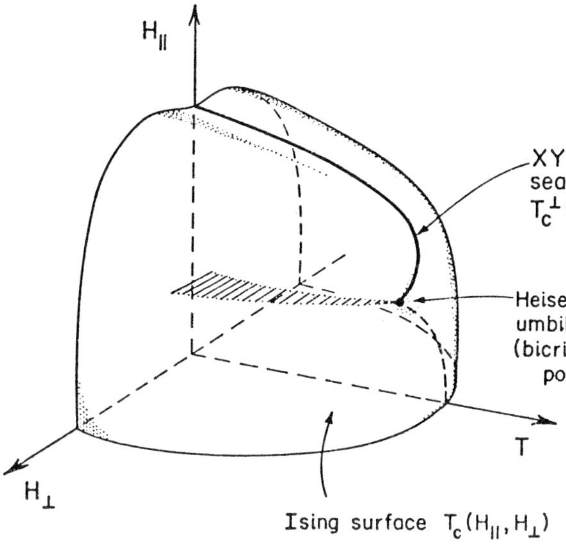

Fig. 7 Phase diagram of an anisotropic antiferromagnet in a general field $\vec{H} = (H_\parallel, H_\perp)$, illustrating (for n=3) the XY seam, lying in a furrow, and leading to the Heisenberg, bicritical umbilicus.

Mean field theory and the detailed experiments[17, 18] on $GdAlO_3$ show that, inside the critical surface, the spin flop line is drawn out into a rather narrow, pointed "shelf" of first order transitions, as shown schematically in Fig. 7. The bicritical behavior of the width of this shelf as $T \to T_b-$, and the nature of the lines of critical points bounding its edges beneath T_b, remain fascinating theoretical problems awaiting attack.

ACKNOWLEDGEMENTS

It is a pleasure to acknowledge strong interactions with Dr. J. M. Kosterlitz and Dr. D. R. Nelson, and stimulating comments from Dr. A. Aharony and Dr. A. D. Bruce. Professor D. Jasnow kindly made available the results of his series expansion work with Dr. S. Singh. The researches reported here have been supported by the National Science Foundation, in part through the Materials Science Center at Cornell University.

REFERENCES

1. D. R. Nelson and M. E. Fisher, "Renormalization Group Analysis of Metamagnetic Critical Behavior", Phys. Rev. B. (1975) [in press].
2. M. E. Fisher and D. R. Nelson, "Universality of Magnetic Tricritical Points", (to be published).
3. M. E. Fisher and D. R. Nelson, Phys. Rev. Lett. 32 1350 (1974), Note the values $\alpha(n=2) \simeq -0.02$ and $\tilde{\gamma}(n=2) \simeq 0.33$ should replace those (mis)printed.
4. D. R. Nelson, J. M. Kosterlitz, and M. E. Fisher, Phys. Rev. Lett. 33, 813 (1974).
5. J. M. Kosterlitz, D. R. Nelson, and M. E. Fisher, "Bicritical and Tetracritical Points in Anisotropic Antiferromagnetic Systems", (to be published).
6. See also A. Aharony and A. D. Bruce, Phys. Rev. Lett. 33, 427 (1974).
7. See e.g. the articles by M. E. Fisher, by L. P. Kadanoff, and by H. E. Stanley, A. Hankey and M. H. Lee, in "Critical Phenomena" Proc. Internat. School of Physics "Enrico Fermi" Course 51, edited by M. S. Green (Academic Press, New York, 1971).
8. K. G. Wilson and M. E. Fisher, Phys. Rev. Lett. 28, 548 (1972).
9. For reviews of the $\epsilon = 4-d$ expansion and the renormalization group approach, see K. G. Wilson and J. Kogut, Physics Reports 12C, 75 (1974), and M. E. Fisher, Rev. Mod. Phys. 46, 597 (1974).
10. P. Pfeuty and M. E. Fisher, "Magnetism and Magnetic Materials - 1972" AIP Conf. Proc. No. 10 p. 817 (1973).
11. P. Pfeuty, D. Jasnow, and M. E. Fisher, Phys. Rev. B 10, 2088 (1974).
12. E. K. Riedel and F. J. Wegner, Phys. Rev. Lett. 29, 349 (1972).
13. F. J. Wegner and E. K. Riedel, Phys. Rev. B 7, 248 (1973).
14. See the reviews of measurements on $FeCl_2$ presented at this Conference by R. J. Birgenau, G. Shirane, M. Blume and W. Koehler, by J. A. Griffith and S. E. Schnatterly, and by J. F. Dillon Jr., E. Yi. Chen and H. J. Guggenheim.
15. R. B. Griffiths, Phys. Rev. Lett. 24, 715 (1970); Phys. Rev. B 7, 545 (1973).

16. The phase diagram of MnF_2 has been studied particularly by Y. Shapira, S. Foner, and A. Misetich, Phys. Rev. Lett. 23, 98 (1969), and Y. Shapira and S. Foner, Phys. Rev. B 1, 3083 (1970).
17. K. W. Blazey, H. Rohrer, and R. Webster, Phys. Rev. B4, 2287 (1971).
18. See the invited paper at this Conference by H. Rohrer.
19. See the review by W. P. Wolf at this Conference, and M. Blume et al. Phys. Rev. Lett. 32, 544 (1974), and J. F. Dillon et al. Phys. Rev. Lett. 33. 98 (1974).
20. See the analysis of K.-S. Liu and M. E. Fisher, J. Low Temp. Phys. 10, 655 (1973).
21. M. E. Fisher and B. U. Felderhof, Ann. Phys. 58, 217 (1970), Secs. 8 and 9.
22. F. Harbus and H. E. Stanley, Phys. Rev. Lett. 29, 58 (1972); Phys. Rev. B8, 1141, 1156 (1973).
23. D. P. Landau, Phys. Rev. Lett. 28, 449 (1972); B. L. Arora and D. P. Landau, AIP Conf. Proc. 10, 870 (1973).
24. M. Wortis, comment at the Conference on Critical Phenomena in Multicomponent Systems, April 15-17, 1974, Athens, Georgia, and private communication.
25. We have also used the exponent compilation of M. E. Fisher and D. Jasnow (to be published).
26. See. M. E. Fisher, Phil. Mag. 7, 1731 (1962).
27. M. E. Fisher and P. Pfeuty, Phys. Rev. B 6, 1889 (1972).
28. F. J. Wegner, Phys. Rev. B 6, 1891 (1972).
29. D. Jasnow and S. Singh (to be published).
30. Details will be published elsewhere.

μ^+ STUDIES OF CRITICAL SPIN FLUCTUATIONS IN Ni*

B. D. Patterson, K. Nagamine**
Lawrence Berkeley Laboratory, University of California, Berkeley, California 94720

C. A. Bucci†, and A. M. Portis††
University of California, Berkeley, California 94720

ABSTRACT

Critical fluctuations above the Curie point in nickel have been studied by observing the relaxation of the free precession of positive muons. This work was prompted by the apparent discrepancy in nickel spin correlation time as determined by neutron scattering and by perturbed angular correlation (PAC) experiments. Up to 20K above the Curie point we obtain a temperature dependence in agreement with that determined from PAC. At higher temperatures the computed nickel spin correlation time becomes independent of temperature. The observed high temperature muon relaxation time of 16 μsec is very much shorter than expected from either the observed contact field at the low temperature muon site or from classical dipolar coupling to the muon nearest neighbors. The increased relaxation rate may arise from an enhanced pseudodipolar interaction.

INTRODUCTION

In this communication, we report the observation of critical spin fluctuations in nickel using positive muons. The implanted positive muon, as has been described in the literature,[1,2,3] is a convenient probe of local magnetic fields in solids in general and in magnetic materials in particular. In Ni at temperatures just above the Curie temperature (T_C = 630K), the muon spin is relaxed by spin fluctuations; our experiment consists of measuring the relaxation time T_2 as a function of $(T-T_C)$ and externally applied magnetic field (see Fig. 1). In most of what follows, it will be assumed that the site of the implanted muon is the octahedral interstitial site in the fcc Ni lattice as is indicated from studies below T_C[2,3].

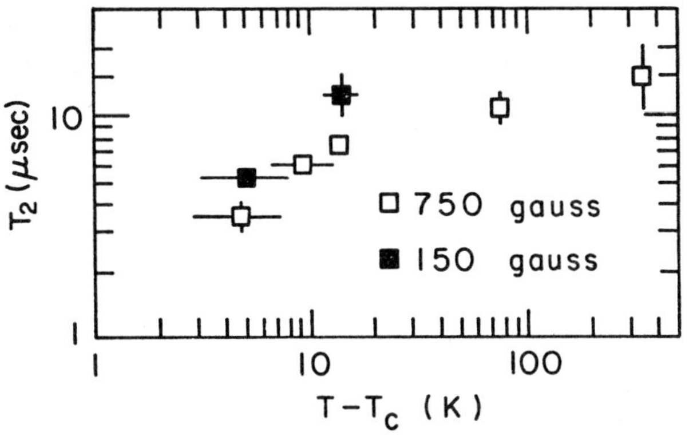

Fig. 1: The experimental dependence of T_2 on temperature and external field.

RESULTS AND DISCUSSION

For values of $(T-T_C)$ greater than 70K, we find that $1/T_2$ is approximately constant and has the value of $\sim 10^5\ \text{sec}^{-1}$ (see Fig. 1). Microwave studies have established[4] that at these temperatures effectively all correlation between Ni spins is destroyed and the spins fluctuate with a correlation time $\tau_c \sim 1/\omega_e \sim \hbar/J \sim 2 \times 10^{-14}$ sec. Here ω_e is the exchange frequency and J is the exchange energy. If each of N mutually uncorrelated spins produces a magnetic field of strength ω/γ_μ at the μ^+, fluctuations of correlation time τ_c will relax the μ^+ at a rate given approximately by:

$$\frac{1}{T_2} \simeq N\omega^2\tau_c \tag{1}$$

for $\omega\tau_c \ll 1$. Two magnetic interactions responsible for ω suggest themselves immediately: (1) the isotropic contact interaction of the μ^+ with its screening cloud of (polarized) electrons, and (2) the classical dipolar interaction with the neighboring Ni cores.

The strength of the contact interaction may be inferred from the measured[2] hyperfine field at the μ^+ in the ordered state, B_{hf} = −.66kG. The octahedral site has N=6 nearest neighbors, and we make the assumption that they share equally in the interaction, each contributing −.11kG. Thus,

$$\omega_{cont} = \frac{\gamma_\mu B_{hf}}{N} \simeq 10^7\ \text{sec}^{-1}, \tag{2}$$

where γ_μ is the gyromagnetic ratio of the muon. This implies a relaxation rate

$$\frac{1}{T_{2_{cont}}} \simeq N\omega_{cont}^2\ \tau_c \simeq 10\ \text{sec}^{-1}, \tag{3}$$

which is well below the observed rate. Clearly, the isotropic contact interaction is too weak to account for the observed relaxation.

Although the dipolar fields from neighboring Ni cores vanish by symmetry at the octahedral site in the ordered state, dipolar fields from fluctuating spins can cause relaxation when cubic symmetry is destroyed by the disappearance of spin correlation. In this case, ω is given by (for nearest neighbors):

$$\omega_{dip} = \frac{\gamma_\mu \mu_{Ni}}{(a/2)^3} \simeq 10^8\ \text{sec}^{-1}, \tag{4}$$

where μ_{Ni} = .6 μ_B is the magnetic moment of a Ni core and a is the lattice constant. Since ω_{dip}^2 for neighbors further removed from the μ^+ drops off rapidly with distance, we consider only the nearest neighbor contribution:

$$\frac{1}{T_{2_{dip}}} \simeq N\omega_{dip}^2\ \tau_c \simeq 10^3\ \text{sec}^{-1} \tag{5}$$

which is too small by a factor of ∿100. This implies that the dipolar interaction is too weak by a factor of 10.

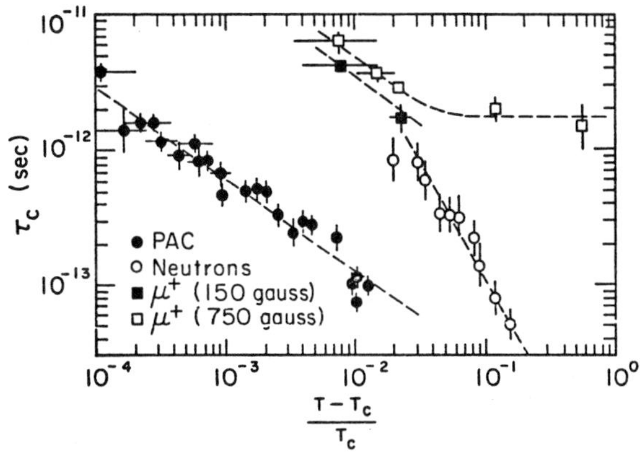

Fig. 2: A comparison of τ_c as measured by various techniques.

It is often the case that in addition to the isotropic contact interaction between the spins of a magnetic material, there exists a contact interaction with dipolar symmetry. This "pseudo-dipolar" enhancement of the classical dipolar field is known[4] to be the source of the ferromagnetic resonance linewidth and the magnetic anisotropy in Ni. We believe that pseudo-dipolar enhancement by a factor of 10 may not be unreasonable for the case of μ^+ in Ni.

Figure 2 compares the muon data with existing[5] perturbed angular correlation and neutron scattering data. From each experiment, a spin correlation time τ_c is extracted and plotted versus $(T-T_C)/T_C$. For the PAC and muon techniques, τ_c is derived from T_2 via the relation $1/T_2 \propto \omega^2\tau_c$. The known (hyperfine) interaction strength is used for PAC data, and for purposes of illustration, the dipolar field (without pseudo-dipolar enhancement) is used for the muon data. The extraction of τ_c from the neutron data is more involved and requires fairly restrictive assumptions[5]. We see that for all three probes, as the temperature is lowered toward T_C, the correlation time increases according to the power law:

$$\tau_c \propto \left[\frac{T-Tc}{T_c}\right]^{-n} \qquad (6)$$

where n for PAC and μ^+ is .7 and is 1.4 for neutrons. The fact that the neutron data disagree with both the PAC and the μ^+ data may be due to a failure of the assumptions made[5] in extracting τ_c from the neutron data, or it may possibly be inherent in the manner in which the different probes sample the momentum spectrum of the critical fluctuations.

As T approaches T_C several things happen: (1) the product $\omega\tau_c$ is no longer small compared to one, and the above treatment breaks down. It is believed that this will happen only in a temperature region inaccessibly close to T_C. (2) The nearest neighbors to the muon will begin to correlate. When this occurs, the dipolar (and pseudo-dipolar) fields will cancel by symmetry, and the interaction will be solely via the weaker contact term. It is interesting to note that the shift from a predominantly dipolar interaction to a contact interaction would change the way in which the muon samples spin fluctuations. This is because the μ^+ preferentially sees fluctuations of small wave number via the contact interaction while the dipole interaction weights fluctuations of higher wave number---as they are more likely to destroy the cubic symmetry near the μ^+.

A potential complication would arise if the muon were diffusing through the Ni lattice instead of being well localized at the octahedral site. It is known that hydrogen is diffusing rapidly in Ni at these temperatures[6]. Evidence so far[2] points to the fact that if the muon is diffusing, it spends by far the greatest amount of time in one type of site (presumably the octahedral site). The question then becomes, under what conditions does a muon diffuse a distance of the order of a fluctuation wavelength in a time short compared to the correlation time of that fluctuation. Rough considerations suggest that this criterion is difficult to satisfy, and that the motion of the muon may be ignored.

As seen in Figure 2, the effect of an external field is to increase the correlation time. This is the anticipated dependence since an external field is expected to stabilize the ordered state and hence to increase spin correlation. Since the PAC and neutron scattering experiments were performed in zero external field, a quantitative comparison with μ^+ measurements requires an extension of this work to zero field.

We are indebted to Professors K. M. Crowe and T. Yamazaki, Drs. J. H. Brewer, F. N. Gygax, and Y. Ishikawa, and Mr. R. F. Johnson for fruitful discussions and assistance.

REFERENCES

1. M. L. G. Foy, N. Heiman, W. J. Kossler, and C. E. Stronach, Phys. Rev. Lett. 30, 1064 (1973).
2. B. D. Patterson, K. M. Crowe, F. N. Gygax, R. F. Johnson, A. M. Portis, and J. H. Brewer, Phys. Lett. 46A, 453 (1974).
3. B. D. Patterson and L. M. Falicov, Solid State Comm., 15, 1509 (1974).
4. M. B. Salamon, Phys. Rev. 155, 224 (1967).
5. A. M. Gottlieb and C. Hohenemser, Phys. Rev. Lett. 31, 1222 (1973).
6. Y. Ebisuzaki, W. J. Kass, and M. O'Keeffe, J. Chem. Phys. 46, 1373 (1967).

*Work done under the auspices of the U.S. Atomic Energy Commission and the National Science Foundation.
†NATO Fellow, University of Parma, Italy.
**Permanent address: University of Tokyo. Supported by the Japanese Society for Science Foundation.
††Supported by the U.S. National Science Foundation, GH-32817.

CRITICAL BEHAVIOR OF THE SUBLATTICE MAGNETIZATION OF Dy NEAR THE NEÉL POINT*

C. L. Chien and J. C. Walker
The Johns Hopkins University, Baltimore, Maryland 21218

E. Loh
Towson State College, Baltimore, Maryland 21204

ABSTRACT

We report here the measurement of the temperature dependence of sublattice magnetization of a high quality single crystal of dysprosium metal using the Mössbauer effect. Dy-metal is anisotropic with hexagonal close-packed structure. Below the Neél point of about 180 K, it is antiferromagnetically ordered. No relaxation effects were observed in the spectra and it was reasonable to assume a proportionality between the effective field $H_{eff}(T)$ and the sublattice magnetization M(T). The data were then found to satisfy the relation,

$$\frac{M(T)}{M(0)} = D(1-\frac{T}{T_N})^\beta$$

within the temperature range $0.001 < 1 - \frac{T}{T_N} < 0.3$. A careful analysis of the data yields the following values for the parameters in the above relation: $T_N = 180.4$ K, $D = 1.07 \pm 0.01$ and $\beta = 0.335 \pm 0.01$. In terms of its magnetism dysprosium metal can be considered a three-dimensional system with two degrees of freedom for the spins. Recent model calculations seem to be particularly applicable to this system.

*Work supported by the National Science Foundation.

SMALL-ANGLE NEUTRON SCATTERING FROM COBALT IN THE CRITICAL REGION

C. J. Glinka*† and V. J. Minkiewicz*†,
University of Maryland, College Park, Maryland 20742
and
L. Passell
Brookhaven National Laboratory††, Upton, New York 11973

ABSTRACT

The temperature dependence of the small-angle critical scattering of neutrons from cobalt has been measured both above and below its Curie temperature, T_c. Below T_c the scattering is dominated by well-defined spin wave modes which exhibit nearly quadratic dispersion. As T_c is approached, the spin waves renormalize and broaden but no evidence of the longitudinal component of the susceptibility is observed. Above T_c, the exponents $\gamma = 1.23 \pm 0.05$ and $\nu = 0.65 \pm 0.04$, describing the power law dependences of the static susceptibility and the inverse correlation range, respectively, have been obtained from the small-angle scans after taking full account of the inelasticity of the scattering. The energy widths of the scattering above T_c have been measured and are well described by a dynamical scaling function over a range of temperature and wave vector extending throughout the critical region and well into the hydrodynamic regime.

RESULTS

Among the transition metal ferromagnets, the critical properties of iron and nickel have been studied in some detail with a variety of techniques. The experimental results regarding cobalt are, however, far less complete and those available have dealt solely with the static critical properties.[1,2,3] By performing both elastic and inelastic, small-angle neutron scattering measurements on polycrystalline cobalt, we have been able to investigate dynamic as well as static features of the critical behavior.

Below T_c the critical scattering is dominated by the spin wave modes. The extreme steepness of the spin wave dispersion in cobalt, together with the limited range of energy transfer available in a small-angle experiment, restricted our spin wave measurements to temperatures very near T_c where the spin wave energies are greatly reduced. Well-defined, though broadened, spin waves exhibiting nearly quadratic dispersion were observed for wave vectors $0.03 \lesssim q \lesssim 0.08$ Å^{-1} over the temperature range $0.002 \lesssim 1 - T/T_c \lesssim 0.03$. The observed spin wave line shapes were completely describable in terms of a two-peaked function which gradually coalesce into a single peak at T_c, as can be seen in Fig. 1. In this respect cobalt is like other previously investigated isotropic ferromagnets which have all failed to exhibit a central diffusive peak below T_c corresponding to the longitudinal component of the susceptibility.

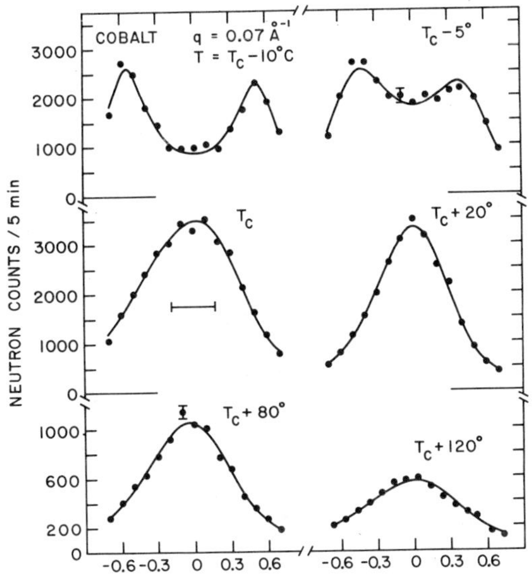

Fig. 1. Energy distribution of the scattering at $q = 0.07$ Å^{-1} on passing through the transition temperature. The curves in the figure were obtained by convoluting the neutron cross section with the instrument resolution. The double-ended bar below the curve at T_c indicates the width of the instrument resolution.

Above T_c, two complementary series of measurements were undertaken to examine both static and dynamic aspects of the critical scattering. In the first of these, the angular distribution of the scattering was studied by operating the spectrometer in the double-axis mode which allows one to observe processes whereby neutrons incident with wave vector $\vec{k}_i$ are scattered through a fixed angle θ. The quantity of interest here is the static, wave-vector-dependent susceptibility $\chi(q)$ whose Fourier transform yields the instantaneous spatial correlation of spin fluctuations. $\chi(q)$ may be measured directly provided the total scattering at a fixed angle θ can be associated with a particular wave vector q. If there is some inelasticity to the scattering, however, so that the scattered neutrons have a spread of wave vectors $\vec{k}_f$, then many momentum transfers are including through the relation $\vec{q} = \vec{k}_i - \vec{k}_f$. As a result, the cross section which is directly measured becomes[4]

$$\sigma(\theta) \propto \int_{-\infty}^{\hbar k_i^2/2m} \{k_i^2 - (2m/\hbar)\omega\}^{1/2} \{\beta\hbar\omega/(1-e^{-\beta\hbar\omega})\} \times \{K_1{}^2 + q^2\}^{-1} F(q,\omega)\, d\omega \qquad (1)$$

where $\hbar\omega$ is the energy lost by the neutron in the scattering process and the integration is done over a path in q-ω space corresponding to a fixed scattering angle θ. The first factor in the integrand arises because $\sigma(q,\omega)$ is proportional to k_f/k_i while the second factor is the detailed-balance factor with $\beta = 1/k_BT$. Also included in Eq. (1) are the Ornstein-Zernike expression for $\chi(q)$ and the spectral weight function $F(q,\omega)$ whose Fourier transform yields the temporal development of spin fluctuations. Eq. (1) implies that unless $F(q,\omega)$ is a narrow function of ω, i.e. the scattering is nearly elastic, knowledge of $F(q,\omega)$ is necessary to infer $\chi(q)$.

We were able to characterize $F(q,\omega)$ by carrying out a separate series of triple-axis measurements in which the frequency distribution of the scattering at fixed wave vectors was determined. Scans were taken for wave vectors in the range $0.03 \lesssim q \lesssim 0.10$ Å^{-1} at temperatures up to 150°C above T_c. This information was then used to numerically evaluate the integral in Eq. (1) for an estimated value of the inverse correlation range K_1, taken to be the measured half-width of the critical scattering. The intensities measured in the double-axis experiment were then corrected for inelasticity by multiplying by $\sigma(q)/\sigma(\theta)$, where $\sigma(q)$ is essentially the Ornstein-Zernike expression, obtained from Eq. (1) by replacing $F(q,\omega)$ with a delta-function. This procedure was then repeated until the value of K_1 used to compute the inelasticity correction agreed with that derived from the corrected data.

The success of this treatment of our two-axis data is illustrated in Fig. 2. There the reciprocal of the intensities measured in a typical angular scan at 35°C above T_c are plotted as a function of q^2. Also shown in the figure are these same reciprocal intensities corrected for inelasticity effects. That the corrected reciprocal intensities lie along a straight

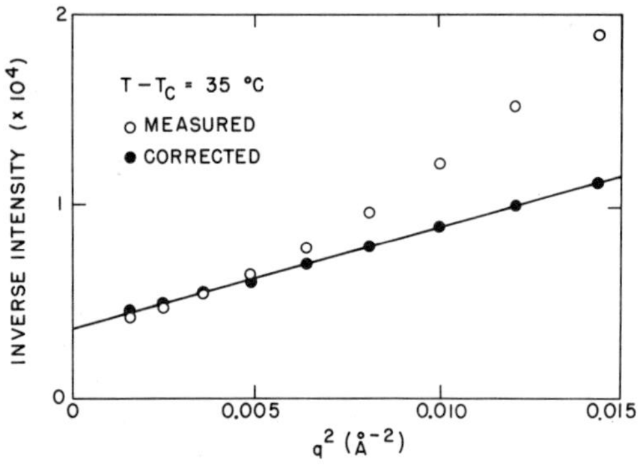

Fig. 2. Inverse intensities of the angular distribution of the scattering at 35°C above T_c plotted versus q^2. The filled circles are obtained from the measured values by correcting for inelasticity as described in the text.

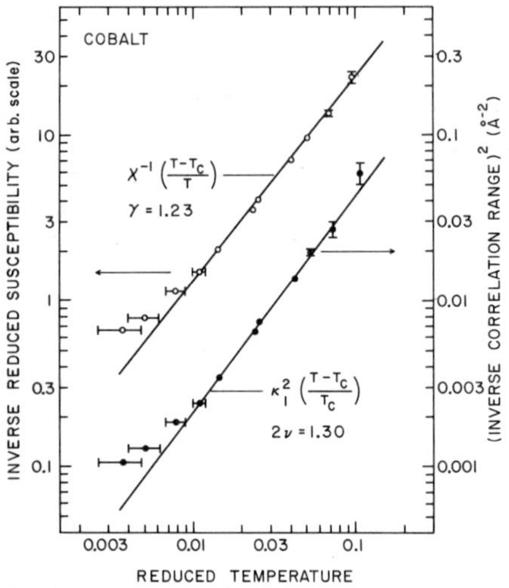

Fig. 3. Temperature dependence of the square of the inverse correlation range and the inverse reduced susceptibility.

line, in accord with the Ornstein-Zernike form for the reduced susceptibility $\chi(q)$, serves to demonstrate the internal consistency between the measured data and the cross-sections used in the analysis.

Values for K_1^2 and the inverse macroscopic susceptibility $\chi^{-1}(q=o)$ obtained from the corrected two-axis data are plotted versus reduced temperature on a log-log scale in Fig. 3. The exponents $2^\nu = 1.30 \pm .08$ and $\gamma = 1.23 \pm .05$ are obtained for the power law dependences of these two quantities, respectively. The exponent γ agrees with macroscopic measurements[2,3] as well as with previous neutron scattering results[1].

The energy widths of the critical scattering determined from our triple-axis measurements have been compared with Halperin and Hohenberg's prediction of the functional form for the characteristic frequency $\Gamma(q,K_1)$ of the spectral weight function $F(q,\omega)$. Using dynamical scaling arguments, they suggest[5] that for a ferromagnet

$$\Gamma(q,K_1) = C\, q^{5/2}\, f(K_1/q) \qquad (2)$$

where the homogeneous function $f(K_1/q)$ is normalized to unity for $K_1/q = 0$ and has the limiting behavior $f(K_1/q) \propto (K_1/q)^{1/2}$ for $K_1 >> q$ in accord with Van Hove's result[6] for the hydrodynamic regime.

Eq. (2) indicates that at T_c, where $K_1 = 0$, the energy widths of the scattering should vary as $q^{5/2}$. We find that our measured line widths at T_c are consistent with a five-halfs power law with a coefficient $C = 300$ meV-$Å^{5/2}$. As the temperature is increased above T_c, the distinctive feature of the line widths is their pronounced narrowing and subsequent broadening upon approaching the hydrodynamic regime. This behavior can be seen in the line shapes shown in Fig. 1.

By forming the ratios, $\Gamma(q,K_1)/C\, q^{5/2}$, of the line widths above T_c to those at T_c, and plotting these as a function of K_1/q, as shown in Fig. 4, the form of the dynamical scaling function $f(K_1/q)$ could be deduced from our line width data. Indeed, the line width ratios when plotted in this way do lie on a single curve as anticipated by dynamical scaling. The dashed curve in Fig. 4 represents a calculation by Resibois and Piette[7] of the scaling function $f(K_1/q)$ for a Heisenberg ferromagnet. The solid curve in the figure was obtained by varying the parameters of the Resibois-Piette calculation in a weighted least-squares fit to the data. The data in the figure, while in overall qualitative agreement with the Resibois-Piette calculation, show a deeper minimum followed by a more rapid broadening than can be accounted for by their calculation, even by varying the parameters of the calculation.

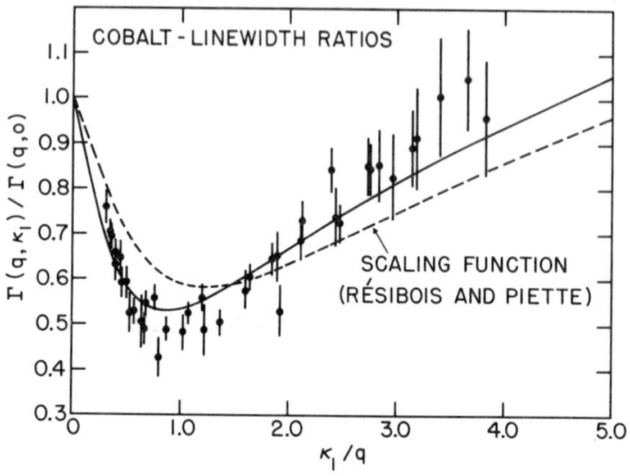

Fig. 4. Ratios of the line widths above T_c to those at T_c plotted as a function of κ_1/q.

CONCLUSIONS

Cobalt provides an example of a magnetic system whose critical scattering exhibits a high degree of inelasticity. The line widths we observe in cobalt are, for example, more than twice as broad as those in iron[8]. The effect of the inelasticity is to cause the angular distributions of the critical scattering to deviate significantly from the Ornstein-Zernike expressions for the reduced susceptibility $\chi(q)$. However, by empirically determining the form of the dynamical scaling function which characterizes the width of the spectral weight function $F(q,\omega)$, the deviations from Ornstein-Zernike susceptibility can be satisfactorily accounted for. While the magnitude of the inelasticity corrections required for cobalt preclude the determination of small modifications to the Ornstein Zernike expression[9], the ability to make such corrections provides assurance that both the static and dynamic cross-sectional forms used in the analysis are essentially correct.

* Work supported by National Science Foundation
† Guest scientist at Brookhaven National Laboratory
†† Work at Brookhaven supported by U. S. AEC

1. D. Bally, M. Popovici, M. Totia, B. Grabcev, and A. M. Lungu, Proc. Fourth Symposium on Neutron Inelastic Scattering, Copenhagen, 1968, Vol.II, p.75.
2. R. V. Colvin and S. Arajs, J. Phys.Chem. Solids 26 435 (1965).
3. K. K. Geissler & H. Lange, Z. angew. Phys. 21, 357 (1966).
4. J. Als-Nielsen, Phys. Rev. Lett. 25, 730 (1970).
5. B. I. Halperin and P. C. Hohenberg, Phys. Rev. 177, 952 (1969).
6. L. Van Hove, Phys. Rev. 93, 268 (1954) & 95, 249 (1954).
7. P. Resibois & C. Piette, Phys. Rev. Lett. 24, 514 (1970).
8. M. F. Collins, V. J. Minkiewicz, R. Nathans, L. Passell & G. Shirane, Phys. Rev. 179, 417 (1969).
9. M. E. Fisher & R. Burford, Phys.Rev. 156 583 (1967).

CRITICAL EXPONENTS OF $FeBO_3$

D. M. Wilson* and S. Broersma
University of Oklahoma, Norman, OK 73069

ABSTRACT

A sample of $FeBO_3$ containing 1-5 μm diameter platelets was prepared. Magnetic measurements were made using a vibrating coil magnetometer and fields of 0.1 to 5 kOe. Data analysis included the effect of crystallite alignment and corrections for the small demagnetizing field. The Curie temperature was found to be 347.85 ±0.2 K. The critical exponents and the temperature ranges used for their determination are as follows:

$\gamma = 1.38 \pm 0.08$,	$1.002 < T/Tc < 1.02$
$\gamma' = 1.0 \pm 0.1$,	$0.98 < T/Tc < 0.999$
$\beta = 0.350 \pm 0.007$,	$0.95 < T/Tc < 0.9998$
$\delta = 3.9 \pm 0.4$,	$\lvert T-Tc \rvert < 0.03$ K

INTRODUCTION

Iron borate, $FeBO_3$, is a canted antiferromagnet [1,2] having a low anisotropy field in its basal plane. Scaling theory describes magnetic behavior near the Curie temperature T_c using critical exponents. Two critical exponents for $FeBO_3$ have been reported[3,4]. These exponents were remeasured and two additional ones were obtained. The theory of weak ferromagnetism was modified to make it correspond more closely to scaling theory and to make it apply to a powdered sample.

APPARATUS AND SAMPLE

Magnetization was measured as a function of temperature and magnetic field with a vibrating double coil magnetometer. The coils were driven perpendicularly to the applied field by a synchronous motor. The 15 Hz vibration had a displacement of 2.54 cm. A 6-inch magnet varied the field from 0.1 to 5 kOe.

The sample, contained in a Be-Cu holder, was placed in a dewar. Liquid from a heat bath was circulated to control the sample temperature. A thermocouple calibrated to 0.2°C measured the temperature of the liquid around the sample. A thermistor located in one end of the Be-Cu holder was used to detect temperature changes of 0.01°C. The temperature ranged from 20°C to 100°C.

A 10.19 g sample of $FeBO_3$ filled a cylindrical cavity 1.41 cm in diameter and 3.3 cm in length. The demagnetizing factor is 1.7π. The demagnetizing field was always less than 1/20 of the applied field.

A method for making powdered $FeBO_3$ was published in 1968 by Joubert, Shirk, White, and Roy[2]. Our sample was prepared as follows. A 5.6:1 molar ratio of B_2O_3 to Fe_2O_3 was ball milled and dried. Then the powdered mixture was placed in a quartz tube, which was evacuated and heated 18 hours at 600°C followed by 120 hours at 800°C. Excess B_2O_3 was removed with hot water. Quartz and $FeBO_3$ were separated by a magnet. A microscopic examination showed that the sample consisted of transparent green platelets 1 to 5 μm in diameter. Chemical analysis determined the iron and boron content. The remaining material was assumed to be oxygen. The result can be expressed as $FeB_{.96}O_{3.13}$, which could be $FeBO_3$ within the errors of the chemical analysis. An X-ray diffraction scan revealed only $FeBO_3$ peaks.

THEORY

The theory will be based on Dzialoshinskii[5] weak ferromagnetism as given by Borovik-Romanov and Ozhogin[6] and Moriya[7]. Using sublattice magnetizations $\vec{M}_1$ and $\vec{M}_2$, we define

$$2M_0 \vec{\lambda}(T,H) = \vec{\ell} = \vec{M}_1 - \vec{M}_2 , \quad (1)$$

$$2M_0 \vec{\mu}(T,H) = \vec{m} = \vec{M}_1 + \vec{M}_2 , \quad (2)$$

where M_0 is a sublattice magnetization at 0 K. The magnetic part of the free energy can be written as

$$\tilde{\Phi} = 2M_0 \{\tfrac{1}{2}A\lambda^2 + \tfrac{1}{2}B\mu^2 + \tfrac{1}{4}C\lambda^4 + \tfrac{1}{2}a\lambda_z^2 - D(\lambda_y\mu_x - \lambda_x\mu_y) - \vec{\mu}\cdot\vec{H}\} . \quad (3)$$

The z-axis is perpendicular to the plane of easy magnetization (basal plane). A(T), B(T), C(T), and a(T) characterize the exchange interactions and D(T), the Dzialoshinskii interaction. Because the anisotropy field in the basal plane is less than one oersted[8], the magnetization lies in the H-z plane. We choose the x-axis in this same plane. The anisotropy field perpendicular to the basal plane as characterized by a(T) is large and positive[8]. With these conditions the minimum value of the free energy occurs when

$$H_x = \frac{B^2}{D^2}\left(A - \frac{D^2}{B}\right)\left(\mu_x - \frac{H_x}{B}\right) + \frac{CB^4}{D^4}\left(\mu_x - \frac{H_x}{B}\right)^3 \quad (4)$$

$$H_z = B\mu_z \quad . \quad (5)$$

Equation (4) is similar to equations describing ferromagnets near T_c; here the variable is $\mu_x - H_x/B$.

Robert Brout[9] gives for the critical region the equation of state

$$H = \varepsilon^\gamma R f(R^2/\varepsilon^{2\beta}) \quad . \quad (6)$$

R is the magnetic moment and $\varepsilon \equiv (T-T_c)/T_c$.
If the f-function can be approximated by a power series, equation (6) becomes

$$H = k_1\varepsilon^\gamma R + k_2\varepsilon^{\gamma-2\beta}R^3 \; . \quad (7)$$

Identifying $(\mu_x - H_x/B)$ in Eq. (4) with R in Eq. (7) one is led to the temperature dependence above T_c

$$A - D^2/B = \nu\varepsilon^\gamma \quad , \quad C = C_0\varepsilon^{\gamma-2\beta} \quad , \quad T > T_c \quad (8)$$

Below T_c we use γ', β', and $|\varepsilon|$ in Eq. (7) so that

$$A - D^2/B = -\nu'|\varepsilon|^{\gamma'} \quad , \quad C = C_0' \; |\varepsilon|^{\gamma'-2\beta'} , \quad T < T_c \quad (9)$$

Here the constants ν, ν', C_0, and C_0' are positive. At T_c, $f(R^2/\varepsilon^{2\beta})$ must be proportional to $(R^2/\varepsilon^{2\beta})^{\gamma/2\beta}$ to cancel the ε singularity for a non-zero H. Thus

$$H = k_3R^{1+\gamma/\beta} = k_3R^\delta \quad . \quad (10)$$

Uniqueness of δ requires that $\gamma'/\beta' = \gamma/\beta$. Because only β below T_c will be used, we drop the prime.

The magnetization of the powdered sample will now be related to the single crystal magnetization. The magnetization component m_H of a general crystallite shown in Fig. 1 is

$$m_H = m_x\sin\theta + m_z\cos\theta \quad . \quad (11)$$

In principle, Eq. (4) can be inverted to give

$$m_x = (2M_0/B)H\sin\theta + 2M_0 g(H\sin\theta) \quad (12)$$

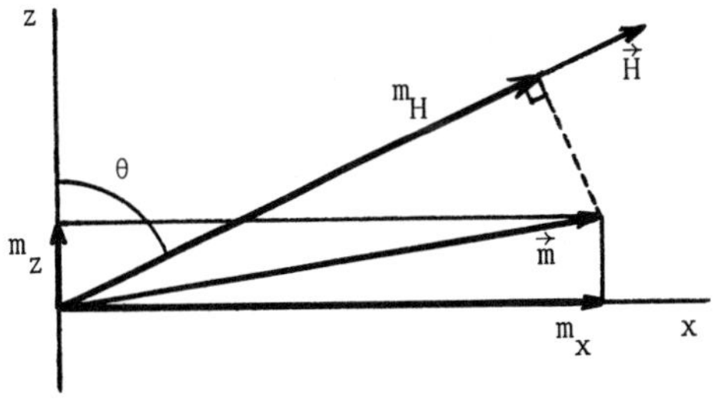

Fig. 1. Crystallite Coordinate System.

where g is a function. Using Eqs. (5) and (12) we get

$$m_H \doteq (2M_0/B)H + 2M_0 g(H\sin\theta)\sin\theta \ . \qquad (13)$$

We now assume the crystallite orientations to have a probability distribution $P(\theta)$. This can include effects caused by a partial alignment of the crystallites during the initial application of the magnetic field. The average magnetization is parallel to H and is given by

$$\langle m\rangle_H = \frac{2M_0 H}{B} + 2M_0 \int_0^{\pi} g(H\sin\theta)\ \sin\theta\ P(\theta)d\theta \qquad (14)$$

where $P(\theta)$ is normalized. We define

$$s_n = \int_0^{\pi} \sin^n\theta\ P(\theta)d\theta \ . \qquad (15)$$

For a randomly oriented sample $P(\theta) = \frac{1}{2}\sin\theta$, thus $s_1 = \pi/4$ and $s_2 = 2/3$. For fully aligned crystals all $s_n = 1$.

Below T_c the inversion technique of Borovik-Romanov and Ozhogin is applied to Eq. (4). The spontaneous magnetization $\langle m_s\rangle_H$ and differential susceptibility $\langle\chi\rangle_H$ as H goes to zero are

$$\langle m_s\rangle_H = \frac{2M_0 s_1 D}{B}\left(\frac{\nu'}{C_0'}\right)^{\frac{1}{2}}\left(\frac{T_c - T}{T_c}\right)^{\beta} , \qquad (16)$$

$$\langle\chi\rangle_H - \frac{2M_0}{B} = \frac{M_0 D^2 s_2}{B^2\nu'}\left(\frac{T_c}{T_c - T}\right)^{\gamma'} . \qquad (17)$$

Above T_c Eq. (4) is inverted using Newton's method. The susceptibility $\langle\chi\rangle_H$ in the limit as H goes to zero is, from Eq. (14),

$$\langle\chi\rangle_H - \frac{2M_0}{B} = \frac{2M_0 D^2 s_2}{B^2\nu}\left(\frac{T_c}{T - T_c}\right)^{\gamma} . \qquad (18)$$

At T_c we find from Eqs. (4), (10) and (14)

$$\langle m\rangle_H - (2M_0/B)H = 2M_0 (H/k_3)^{1/\delta}\ s_{(1+1/\delta)} \ . \qquad (19)$$

RESULTS

The above equations apply to the magnetization and susceptibilities extrapolated to zero field. Below T_c interfering effects such as domains, strain, etc., are overcome by linear extrapolation from high field values. As T_c is approached extrapolation of the data has to be limited to progressively lower fields so that the approximations used apply.

The magnetization below T_c is expressed as the sum $\langle m_s\rangle_H + \langle\chi\rangle_H H$. The Curie temperature is found when the extrapolated value of $\langle m_s\rangle_H$ goes to zero using an initial value for β. The best value for β is then found from a log-log plot of Eq. (16).

Before γ' can be obtained from a log-log plot of Eq. (17), $2M_0/B$ must be determined. It is obtained from interpolation of the limiting susceptibility for data near 20° C and near 100° C. Above T_c Newton's method gives the magnetization as $\langle\chi\rangle_H H - k H^3$. Using the initial susceptibility in Eq. (18), we determined γ in the same manner as γ'. At T_c, δ is found from a log-log plot of Eq. (19).

Our results along with published values are shown in Table I. The values of T_c differ slightly; those for β agree well. The values of δ disagree.

Table I

Parameter	Temperature Range	Source
$T_c = 347.85 \pm 0.2$ K	333 - 348 K	
$T_c = 348.35 \pm 0.2$ K	303 - 348 K	Ref. (3)
$\gamma = 1.38 \pm 0.08$	$1.002 < T/T_c < 1.02$	
$\gamma' = 1.0 \pm 0.1$	$0.98 < T/T_c < 0.999$	
$\beta = 0.350 \pm 0.007$	$0.95 < T/T_c < 0.9998$	
$\beta = 0.354 \pm 0.005$	$0.94 < T/T_c < 0.9996$	Ref. (3)
$\beta = 0.353$	$0.95 < T/T_c < 0.998$	Ref. (4)
$\delta = 3.9 \pm 0.4$	$\lvert T-T_c\rvert < 0.05$ K	
$\delta = 3$		Ref. (4)

If the Curie temperature had been determined from spontaneous magnetizations obtained solely by linear extrapolation from a high field or from the maximum differential susceptibility in a high field, it would have been 0.5 K higher.

Using a log-log plot, for T near T_c, as suggested by Eq. (19), we find that the data points approach a straight line, but from different directions depending whether the temperature was above or below T_c. According to Eq. (19) the straight line occurs only at the critical temperature; this can be a criterion for determination of T_c.

Because the critical exponents are obtained using log-log plots, the value of s_n does not affect the exponents. For the spontaneous magnetization of the powder we find $4\pi\langle m_s\rangle = 90$ G at 299 K. Literature values for single crystals are 115 G at 297 K[8] and 122 G at 300 K[10]. These imply that s_1 is 0.78 or 0.74, close to the random value $\pi/4 = 0.79$.

According to Eq. (10) γ' should equal $\beta(\delta-1)$. Our experimental exponents are consistent with this relation well within the stated accuracy.

*Present address: Pfizer, Easton, PA. 18042

REFERENCES

1. Bernal, Struck, White, Acta Cryst. 16, 849-50 (1963).
2. Joubert, Shirk, White, Roy, Mat. Res. Bull. 3, 671-6 (1968).
3. Eibschütz, Pfeiffer, Nielsen, J. Appl. Phys. 41, 1276-7 (1970).
4. Yakimov, Ozhogin, Gamlitskii, Cherepanov, Pudkov, Phys. Lett. 39A, 421-3 (1972).
5. Dzialoshinskii, JETP 32, 1547-62 (1957), Soviet Phys. JETP 5, 1259-72 (1957).
6. Borovik-Romanov, Ozhogin, JETP 39, 27-36 (1960), Soviet Phys. JETP 12, 18-24 (1961).
7. Moriya, Magnetism Vol. I, (Academic Press, N.Y., 1963), 85-125.
8. Wolfe, Kurtzig, LeCraw, J. Appl. Phys. 41, 1218-24 (1970).
9. Brout, Statistical Mechanics, ed. Rice et al, (U. of Chicago Press, 1972), 279-98.
10. Petrov, Smolenskii, Paugurt, Kizhaev, Chizhov, Soviet Phys.-Solid State 14, 87-91 (1972).

CRITICAL PHENOMENA IN THE A.C. SUSCEPTIBILITY OF IRON WHISKERS

B. Heinrich and A.S. Arrott
Simon Fraser University, Burnaby, British Columbia, Canada V5A 1S6

The dependence of the outphase component of the a.c. susceptibility upon frequency, temperature and d.c. bias field has been investigated for temperatures from T_c-100 mdeg C to T_c+200 mdeg C for iron whiskers. Specific results are given for a large whisker first as grown and then after electropolishing to approximate an ellipsoid of revolution. A superanomalous region from T_c-30 mdeg C to T_c+70 mdeg C is identified. The behavior may be associated with critical slowing down.

A.c. susceptibility measurements are used to obtain information about the magnetization processes within iron whiskers for temperatures in the immediate vicinity of T_c. This information is obtained mostly from the dependence of the outphase component $\chi''(\omega)$ upon frequency, temperature and d.c. bias field. Second harmonic generation which amplifies subtle effects is also informative. The inphase component is primarily governed by the demagnetizing field H_D below T_c and by the combination of intrinsic susceptibility χ_i and H_D above T_c. In the limit of zero frequency

$$\frac{1}{\chi'(0,z)} = \frac{1}{\chi_i} + 4\pi D(z) \qquad (1)$$

where the z dependence indicates that the response varies along the whisker as has been previously discussed in detail[1]. Below T_c, $1/\chi_i \ll 4\pi D$ and above T_c it seems to work[2] to use the formula for critical behavior

$$\chi_i = ((T-T_c)/T_1)^{-\gamma}. \qquad (2)$$

Within a few hundredths of a degree of T_c there is some problem about how to extrapolate $\chi'(\omega)$ to $\chi'(0)$. The losses are such that the phase $\varphi = \chi''/\chi'$ is still 0.1 at T_c for the lowest frequencies used. The frequency dependence of χ' does slightly reflect the losses, but it is more direct to look at χ'' and from that deduce changes in magnetization processes.

Several tenths of a degree below T_c there is a change from the Landau domain structure to a curling pattern in which one may think of the domain wall extending throughout the sample[2]. This change is accompanied by a decrease in χ'' by a factor of 3 to 4 depending upon the whisker. The decrease in loss is accounted for by comparing the eddy current loss for a single wall in motion with that for a more homogeneous rotation of magnetization as occurs in the curling pattern[3].

The most striking effects occur even closer to T_c. These are shown in Fig. 1 for a whisker with dimensions 230μ × 230μ × 16.2mm and in Fig. 2 for the same whisker after electropolishing to remove the corners and edges to approximate an ellipsoid of revolution (140μ × 13.2mm). χ' and χ'' have been normalized to the low frequency limit of χ' in the region of the curling pattern where[4] $\chi'=1/4\pi\bar{D}$.

In the region above T_c-0.4 deg C the data are interpreted as showing a curling pattern. This region ends quite abruptly. For as-grown whiskers there is a strange decrease in loss, most noticeable at higher frequencies, over a range of 5 mdeg C about 35 mdeg C below T_c. This is followed by a rapid increase in loss reaching a peak very close to T_c and then falling rapidly down. The losses appear linear in frequency in the curling region and again in the region above T_c + 100 mdeg C. There are difficulties with phase in these measurements and it is necessary to subtract out a small contribution from the inphase signal. The instrumental origin of this effect has not been isolated. The data in Fig. 1 and Fig. 2

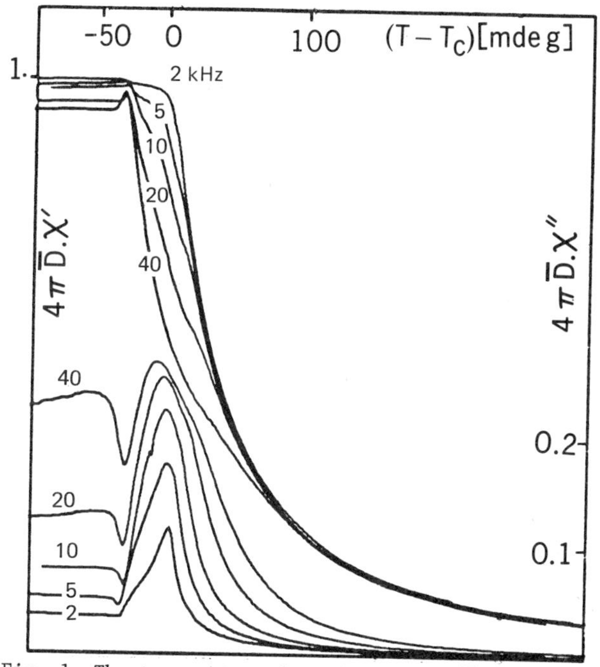

Fig. 1 The temperature dependence of $4\pi\bar{D}\chi'(\omega)$ and $4\pi\bar{D}\chi''(\omega)$ for frequencies 2, 5, 10, 20, and 40 kHz for an iron whisker 230μ × 230μ × 16.2mm. $4\pi\bar{D}$ is estimated to be .0083. The lines are drawn from data point to data point in this and all succeeding graphs. Typically the data is in mdeg C steps.

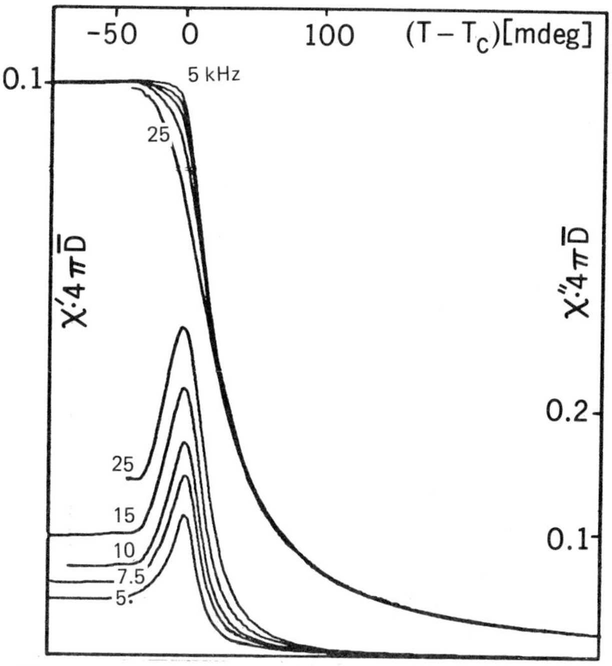

Fig. 2 The temperature dependence of $4\pi\bar{D}\chi'(\omega)$ and $4\pi\bar{D}\chi''(\omega)$ for frequencies 5, 7.5, 10, 15, 25 kHz for the same whisker as in Fig. 1, but after electropolishing to a figure of revolution of maximum diameter 140μ and length 13.2mm. $4\pi\bar{D}$ is estimated to be .005.

have not been corrected. In calculating the loss coefficients as discussed below corrections have been applied. At T_c it appears that χ'' varies approximately as $f^{1/2}$.

The behavior of $\chi'(\omega)$ and $\chi''(\omega)$ with d.c. bias field along the axis shows some anomalies that are particularly pronounced for the as-grown whisker. These are thought to be connected to the behavior in the corners of the square cross-section. Just above T_c there is a strange spike[5] in χ'' that occurs over the narrow range of d.c. fields: ± 0.01 Oe. To investigate this carefully it is necessary to reduce the amplitude of the a.c. driving field to this order from the usual 0.05 Oe value. This spike in the field dependence shows also as a second peak in the temperature dependence of χ'', that is in addition to the usual peak near T_c there is one at T_c + 5 mdeg. The region of rapid rise in the loss is characterized by the appearance of hysteresis with change of d.c. bias. This is seen most clearly in the second harmonic. It is not present in the region of the curling pattern and goes away at T_c. The second harmonic hysteresis is present also in the polished whisker but on a reduced scale. The spike above T_c and the sharp decrease in χ'' between the curling region and the region of rapid rise are not seen in the polished specimen. In addition this sample shows a much smoother temperature dependence of χ' near T_c.

To compare these results with a specific model one can calculate χ'' for a uniformly magnetized sample with eddy current losses in which case

$$\chi'' = K\sigma R^2\omega\chi'^2 \tag{3}$$

where K is a geometrical factor, R is the diameter or thickness of the sample and σ is the conductivity. A loss coefficient is defined as

$$\beta \equiv \frac{\chi''}{\omega(\chi'^2 + \chi''^2)} \tag{4}$$

where the inclusion of the small correction χ''^2 in the denominator takes into account the contribution of the outphase component to the flux change. If the losses are calculated for an inhomogeneous process χ'' is of the same form as eq (3) but the value of K is 3 to 4 times larger[3]. Thus an increase in β is interpreted as a change toward a less homogeneous magnetization process. The temperature dependence of the loss parameter $\beta(\omega)$ is shown in Fig. 3 and 4. The data has been corrected for phase mixing[6].

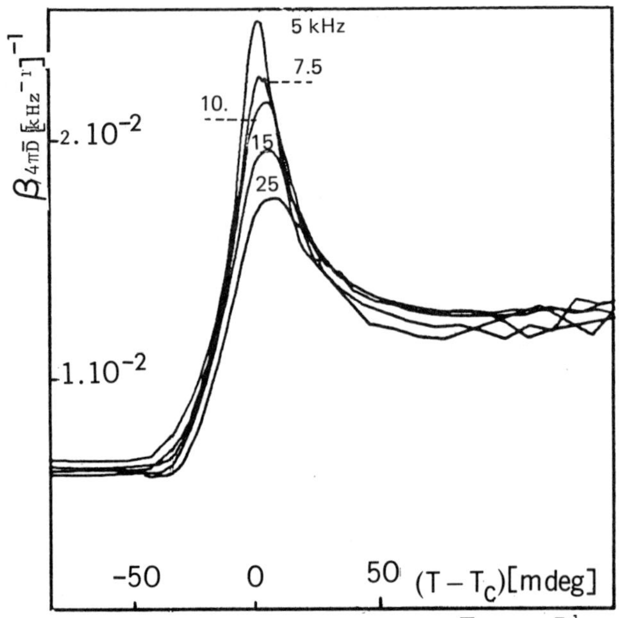

Fig. 4 The loss coefficient $\beta/4\pi\bar{D}$ in kHz^{-1} calculated from the data of Fig. 2.

The region between T_c-30 mdeg C and T_c+70 mdeg C shows a super-anomalous loss coefficient dependent upon frequency and temperature and quite large in magnitude. The region above T_c+70 mdeg C shows a loss coefficient which is independent of frequency and temperature but still anomalous in that it is too large by a factor of 3 to be ascribed to a homogeneous magnetization process. It is in fact comparable to the loss coefficient for the Landau structure.

T_c can be assigned from extrapolation of $\chi'(\omega)$ to zero frequency using Eqs. (1) and (2). This T_c corresponds to the low frequency maxima in β, φ and χ'' within 2 mdeg C. With increasing frequency the maximum in χ'' shifts to lower temperatures, the maximum in β shifts to higher temperatures, and the peak in φ remains within millidegrees of T_c.

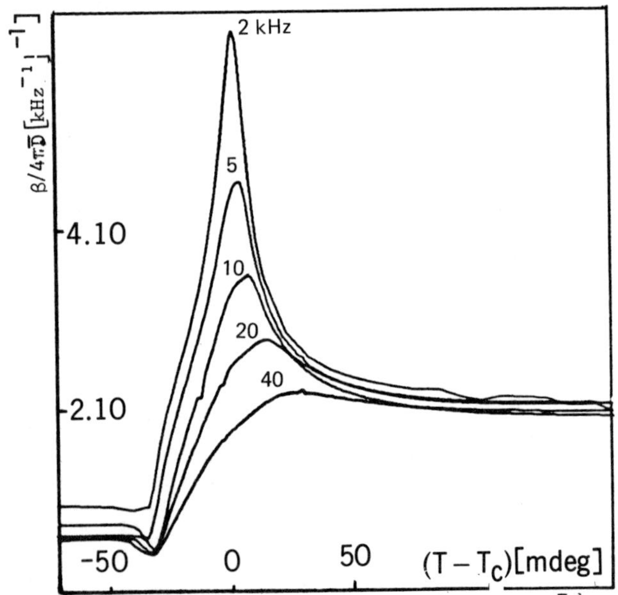

Fig. 3 The loss coefficient $\beta/4\pi D$ in kHz^{-1} calculated from the data of Fig. 1.

It seems clear that fluctuations are the origin of the inhomogeneity of the processes in the super-anomalous and the anomalous regions. In the latter region the eddy currents generated by fluctuation as they grow and decay in a changing field add to the loss. As long as the frequency of the most important fluctuations are high compared to the driving frequency the loss is linear in f and β is frequency independent. In the superanomalous region the fluctuations have frequencies comparable to the driving frequency. At T_c the fluctuations should last long enough to be reversed by the driving field. If the amplitude of the driving field is large compared to the field necessary for such reversal, then by a general argument due to Becker[7], a case can be made for an $f^{1/2}$ frequency dependence. Unfortunately there is no formulation of the above ideas into a calculable model. Furthermore none of this gives a suggestion as to why the rapid increase in losses starts so abruptly from 30 mdeg C below T_c. Yet if there is anything to the above ideas, the critical slowing down for wave lengths comparable to the size of the whisker diameter is in the range of kH_z for $-3\times10^{-6}<(T-T_c/T_c<7\times10^{-6}$.

REFERENCES

1. A.S. Arrott, B. Heinrich, and D.S. Bloomberg, A.I.P. Conf. Proc. 6 941 (1973), D.S. Bloomberg, thesis, Simon Fraser University 1973.
2. B. Heinrich and A.S. Arrott, Proc. Int. Conf. of Magnetism, ICM-73, vol. IV, 556 NAUKA, Moscow 1974.
3. See H.J. Williams, W. Shockley and C. Kittel, Phys. Rev. 80 1090 (1950), S. Chikazumi, Physics of Magnetism, John Wiley, Inc., New York 1964, p 321-3 and ref. 2 above for several different examples.
4. The bar indicates an average over D(z) suitable to the detector coil used in the induction measurement. For the coils used in these experiments this should approximate the magnetometric average.
5. See ref. 1 p. 958.
6. It is assumed that χ'' is proportional to χ'^2 in the higher temperature range to find the magnitude of the phase mixing by writing $\chi'' = a+b\chi'+c\chi'^2$. The values found for b correspond to a phase offset of 1 deg. The term in a is very small. The coefficient c is proportional to the loss coefficient β at these temperatures.
7. J.J. Becker, J. App. Phys. 34 1327 (1963) (see p. 1330).

CORRECTION TO MEAN FIELD DESCRIPTION OF TRICRITICAL SYSTEMS*

C. Paulson and P. M. Levy
New York University,
New York, New York 10003

ABSTRACT

Attempts to include fluctuations of the order parameter near the critical point of a three-dimensional system lead to inconsistencies. However, for tricritical systems there is no inconsistency,[1] and we can include fluctuations of the order parameter to correct the mean field description of the thermodynamic properties of these systems. We account for fluctuations of the order parameter by including a fluctuating component to the internal field.[2] These fluctuating fields are weighted according to the free energy of the fluctuations. The coupling between the fluctuations and the fluctuating field is left as a parameter; it is determined by using the fluctuation theorem.[3]

We have applied the above approach to calculate the free energy of holmium antimonide. By including fluctuations we are able to determine the singular part of the specific heat for $T > T_n$ and we find that the critical temperature is lowered by about 10% from its mean field value. As a check on our procedure we calculate the transition temperature for the tricritical point of the Blume-Emery-Griffiths model and compare it with the temperature determined from high temperature series.

*Research supported in part by National Science Foundation under Grant No. GH 32422.

[1]R. Bausch, Z. Physik 254, 81 (1972).
[2]A. Furrer and H. Herr, Phys. Rev. Lett. 31, 1350 (1973).
[3]M.E. Lines, Phys. Rev. B9, 3927 (1974).

TRICRITICALITY OF CUBIC RARE-EARTH COMPOUNDS*

P.M. Levy and L.F. Uffer
New York University
New York, New York 10003

In appropriate cubic fields rare-earth ions have six-fold degenerate ground states. When the angular momentum of the rare-earth is large the six levels are characterized by states that are directed along the cube edges.[1] Within these states the angular momentum operators J_x J_y and J_z have particularly simple matrix representations.

We calculate the free energy for a system with a Heisenberg exchange interaction between rare-earths in these six-fold degenerate states. In the mean field approximation such a system undergoes a phase transition which is tricritical-like. To the extent that the mean field approximation is an accurate guide we predict that there are many cubic rare-earth compounds which exhibit tricritical-like behavior.

In addition we calculate the free energy for a system with pure quadrupole coupling between rare-earths in these six-fold degenerate states. By using the mean field approximation we find that such systems have a first-order phase transition which is close to the critical point at the end of the coexistence line. A small single-ion anisotropy is sufficient to make the system critical with a specific heat exponent $\alpha = 2/3$. The validity of these mean field results is discussed.

*Research supported in part by the National Science Foundation under Grant No. GH 32422.

[1]G.T. Trammell, Phys. Rev. 131, 932 (1963).

ULTRASONIC PROPAGATION NEAR THE CURIE POINT OF MnP

B. Ferry* and B. Golding
Bell Laboratories, Murray Hill, N.J. 07974

ABSTRACT

Ultrasonic attenuation and phase velocity measurements in the frequency range 10-500 MHz have been performed in the vicinity of the 290 K paramagnetic-ferromagnetic transition of MnP. Longitudinal waves propagating along the principal axes exhibit critical attenuation and dispersion and detailed measurements along the a-axis reveal that the temperatures of the attenuation and velocity extrema depend on the ultrasonic frequency. Below T_C, transverse waves exhibit non-critical attenuation which depends on the particular sample studied and on its thermal history.

INTRODUCTION

MnP is an orthorhombic metal which undergoes a paramagnetic to ferromagnetic transition near 290 K.[1] The magnetocrystalline anisotropy is uniaxial with easy direction along the c-axis, where the axes are identified by the lattice parameter convention $a > b > c$. From previous measurements of the uniaxial stress dependence of T_c[2], we expect a variation of magnetostrictive coupling from a to c axes of at least ~10 to 100, as previously reported, and as observed in this report in ultrasonic experiments. In the present investigation, the propagation of longitudinal and transverse sound waves in MnP is studied in zero applied magnetic field in the temperature region $T_c \pm 15$ K for frequencies from 10 to 500 MHz. Recently a report of longitudinal wave attenuation along the b-axis in applied magnetic fields at a single frequency has been presented.[4] Our study is generally confined to the critical region near T_c in the absence of applied fields and emphasizes different aspects of the critical acoustic behavior of MnP.

EXPERIMENTAL

Conventional ultrasonic methods were utilized throughout. Single-ended pulsed and double-ended cw resonance techniques were used, with the latter confined to the lowest frequency range (~10 MHz). Quartz or tourmaline resonant transducers were bonded to the MnP crystals with Nonaq grease. Specimens were contained within an evacuated isothermal enclosure and temperature measured and controlled to within a few mK by a platinum resistance thermometer.

RESULTS

a-axis. The magnetostrictive coupling is weakest in MnP along the a direction (although comparable with most "normal" magnets) and has permitted detailed study of the transition region attenuation and velocity dispersion over the largest frequency range. Figure 1 shows the attenuation of longitudinal sound near T_c at 65, 225, and 520 MHz plotted on a semi-log scale. The most striking feature of these data is the appearance of an attenuation maximum which shifts to lower temperatures as the frequency increases. This phenomenon is qualitatively similar to behavior previously observed in liquid He[5] near the λ-point and in Ni[6] near its Curie temperature.

The longitudinal sound velocities have been measured from 10 to 70 MHz. Figure 2 shows the relative change in velocity vs. temperature at 10 MHz (right scale). The change $(\Delta v/v)_a$ is approximately 1×10^{-3}. At higher frequencies, substantial dispersion is evident, with $(\Delta v/v)_a$ decreasing by about a factor of 2 at 70 MHz, and with a shift of the minimum by -0.2 K. We observed a difference in the 10 MHz $(\Delta v/v)_a$ measured by a pulsed method (pulse superposition) and a cw resonant method. We believe that pulse distortion in the highly dispersive region near T_c may make cw methods preferable.

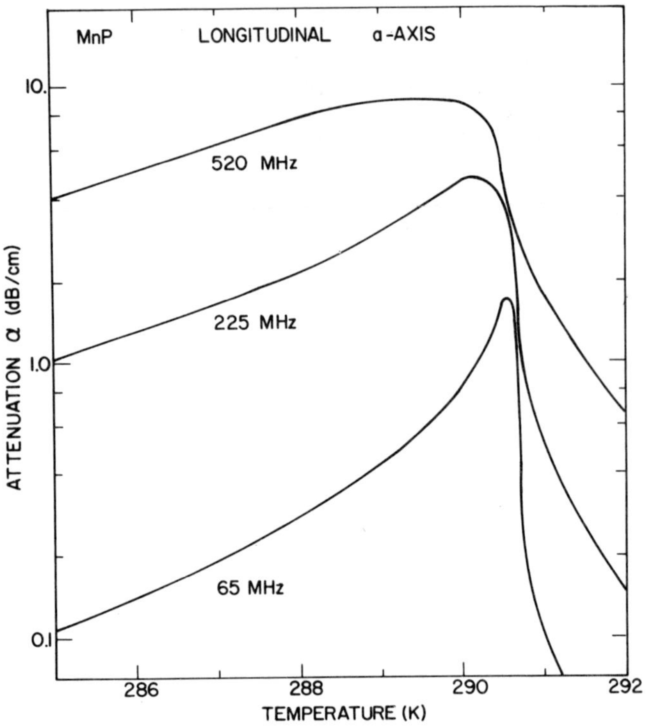

Fig. 1. Attenuation of longitudinal sound waves propagating along the a-axis in MnP near T_c (≈ 290.70K for this sample). The attenuation maximum shifts to lower temperatures as the frequency increases.

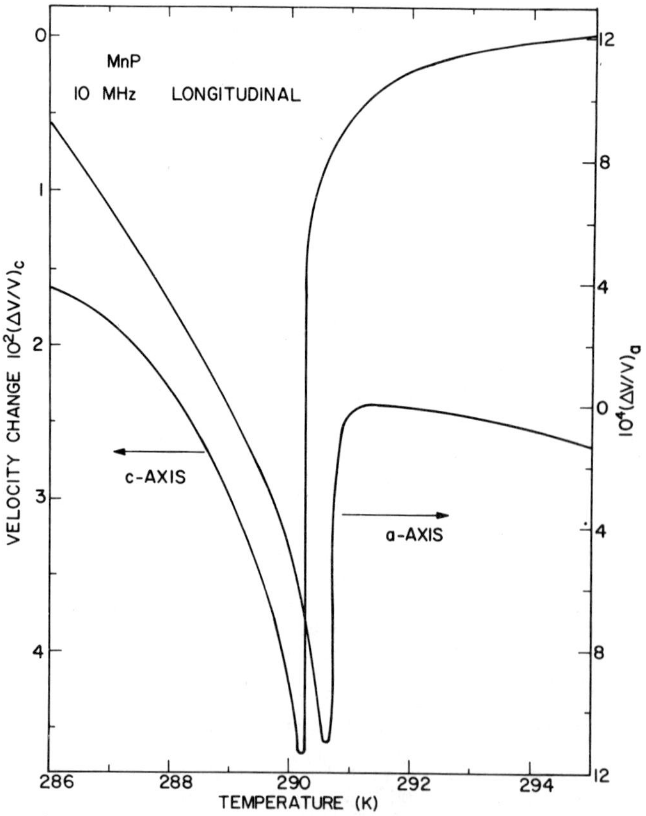

Fig. 2. Velocity changes near T_c for 10 MHz longitudinal sound waves propagating along the c and a axes of MnP. Note the different scale factors for Δv/v for the two axes.

b-axis. We have not conducted detailed experiments for this direction except to confirm that the attenuation maxima occur at the same temperature as the a-axis maxima for the same frequency. The attenuation is larger along the b-axis than the a-axis by a factor of ≈ 5 and its magnitude is in agreement with Ref. 4.

c-axis. The magnetostrictive coupling along the c-axis is anomalously large as evidenced by $(\Delta v/v)_c \approx 5\times10^{-2}$ as shown in Fig. 2 (left scale). These data were taken on a thin (~1 mm) plate resonated near 10 MHz since the high attenuation made pulsed measurements impossible near T_c. This crystal originated from a different boule and had a slightly lower transition temperature as can be seen by comparing the velocity minima in Fig. 2.

Attempts to measure the longitudinal attenuation were thwarted by the unusual mode velocities, namely, the three pure modes along ⟨001⟩ have approximately the same velocity. In fact, the b polarized transverse mode is faster than the longitudinal mode (see Table I). We found it impossible to generate ⟨001⟩ longitudinal waves without generating signals which propagated at the two ⟨001⟩ transverse velocities, even though it was possible to generate a pure transverse mode with no detectable orthogonal polarization. This effect was observed in three different crystals and we have been unable to determine whether the "transverse" signals are generated in the crystal or by the transducer. The nearly coincident acoustic echos exhibit phase interference effects which, because they depend on the relative velocities of the modes, create amplitude oscillations as the crystal is cooled, making reliable attenuation measurements difficult.

The propagation of transverse waves along the c-axis was studied for two samples and an attenuation maximum and velocity minimum was observed a few tenths of a degree below T_c. In one sample, an attenuation change of about 5 dB/cm and a maximum velocity decrease of about 5×10^{-4} were observed for a-axis polarized waves. The magnitude of the attenuation on heating through this region depended on the thermal history of the sample i.e. hysteresis was observed. On cooling from the paramagnetic state the attenuation was always reproducible but on heating the attenuation depended on the lowest temperature attained and was lower than the attenuation on cooling. On yet another sample, the maximum attenuation was about 1.5 dB/cm and no hysteresis was observed. For both samples, the attenuation was nearly frequency independent between 10 and 50 MHz. Similar results were observed for b-axis polarized waves.

We tentatively interpret these phenomena as due to the interaction of transverse waves with ferromagnetic domains which exist below T_c in MnP. The detailed domain configuration is expected to be sample dependent and could also be a function of the thermal history of the sample. We feel that the transverse wave phenomena are not attributable to critical scattering of sound by fluctuation phenomena.

$\vec{\varepsilon}$ \ $\vec{k}$	⟨100⟩	⟨010⟩	⟨001⟩
⟨100⟩	6.92	5.03	4.21
⟨010⟩	5.02	7.10	4.87
⟨001⟩	4.27	4.86	4.80

Table I. Sound velocities at 20 MHz in units of 10^5 cm/sec along the principal axes of MnP at T = 296 K. $\vec{k}$ and $\vec{\varepsilon}$ refer to the sound propagation and polarization vectors, respectively. Within the experimental error of ±1%, the matrix is symmetric, as expected for the orthorhombic system.

Ferromagnetic domains were directly observed under the optical microscope on c-axis faces by means of a Bitter-pattern technique. The patterns were parallel lines which extended from one lateral face to another in the b direction (the b-axis has anisotropy intermediate to the a and c-axes)[1] and whose width was 10^{-3} to 10^{-2} cm. Since this dimension is roughly that of an acoustic wavelength, simple acoustic behavior resulting from sound wave-domain interaction should not be expected.

CONCLUSIONS

We have presented a number of diverse acoustic properties of MnP near its ferromagnetic Curie point. For reference, we list in Table I the pure mode sound velocities for the three principal axes. We call attention to the unusual behavior for phonons propagating along the c-axis, namely, that a transverse mode propagates faster than the longitudinal mode. A better perspective might be to view the c-axis longitudinal mode as being unusually "soft", at least with respect to the other longitudinal modes. It would be of some interest to have knowledge of the phonon dispersion relations along ⟨001⟩.

A more detailed description of the critical attenuation and dispersion of ⟨100⟩ sound waves will be presented elsewhere.

ACKNOWLEDGMENTS

We are indebted to E. Hirahara for supplying the MnP crystals. We thank R. C. Sherwood for assistance in viewing the domain patterns and P. A. Busch for orienting and polishing the crystals.

REFERENCES

*Present address: Laboratoire D'Ultrasons, Université Paris VI, Paris, France.

1. E. E. Huber and D. H. Ridgley, Phys. Rev. 135, A1033-40 (1964).
2. E. Hirahara, T. Suzuki, and Y. Matsumura, J. Appl. Phys. 39, 713 (1968). As indicated by Iwata (Ref. 3), the stress derivatives appear to have been mislabelled and should be $|\partial T_c/\partial\sigma_c| > |\partial T_c/\partial\sigma_b| > |\partial T_c/\partial\sigma_a|$.
3. N. Iwata, J. Sci. Hiroshima Univ. 33, 1-21 (1969).
4. T. Komatsubara, A. Ishizaki, S. Kusaka, and E. Hirahara, Sol. St. Commun. 14, 741-5 (1974).
5. C. E. Chase, Phys. Fluids 1, 193 (1958); R. D. Williams and I. Rudnick, Phys. Rev. Lett. 25, 276-80 (1970).
6. B. Golding and M. Barmatz, Phys. Rev. Lett. 23, 223-6 (1969).

THERMAL EXPANSIVITY OF MnP NEAR THE FERROMAGNETIC CRITICAL POINT

B. Golding and C. A. Helms*
Bell Laboratories, Murray Hill, N.J. 07974

ABSTRACT

The thermal expansivity of metallic MnP has been measured in the vicinity of its paramagnetic to ferromagnetic transition near 290 K. Data are presented for the linear expansivities along the b and c axes, as measured by capacitive dilatometry. Several power law functions are fit to the data and it is concluded that the data cannot be described well by a simple generalized power law. It is suggested that a crossover may exist in the region accessible to experiment.

INTRODUCTION

Manganese phosphide is an orthorhombic metallic compound which undergoes a paramagnetic to simple ferromagnetic transition near 290 K and transforms to several complex magnetic structures below 50 K.[1-4] Although greater attention has been focussed on the low temperature phases, an inelastic neutron scattering study[5] of the 290 K transition has indicated that the spin dynamics are in accord with certain predictions of the dynamical scaling theory.[6] The critical behavior of the thermal expansivity of an anisotropic magnet such as MnP is of interest for two principal reasons. The measurement of the thermal expansivity along the principal axes allows the determination of the critical exponents α and α' which, because of widely differing spin-lattice coupling along the different axes, might conceivably show different behavior. Secondly, the magnitude of the uniaxial anisotropy suggests the existence of a crossover from 3-d Heisenberg-like to Ising-like behavior at a reduced temperature $t \equiv 1-T/T_c$ near 5×10^{-3} as T_c is approached. Thus α might be expected to change from $\alpha_H \approx -0.1$ to $\alpha_I \approx +0.1$ near T_c and this relatively large change in exponent could, in principle, be detected.

EXPERIMENTAL

The MnP sample was a right parallelopiped with sides between 4 and 7 mm, with each face oriented to within a degree of the principal axes ($a > b > c$). The specimen was bonded with GE varnish to a copper capacitive dilatometer which was suspended in a vacuum chamber. A radiation shield surrounding the capacitance cell was maintained at the same temperature as the cell, and the vacuum can was held at fixed temperature by an ice bath. Cell temperatures were measured with a platinum resistor operated in an ac bridge circuit.

A capacitance datum was taken, generally, after thermal equilibrium was attained throughout the cell, usually after about 15 min. However, a continuous (drift) method was initially employed for a c-axis run very near T_c to check for evidence of a first order transition. A hysteresis of less than 10 mK in the expansivity minimum was observed on consecutive heating and cooling sweeps. This difference was consistent with the measured thermal equilibration times and the $\sim$1 mK/min drift rate so we conclude that any first order character must be on much smaller temperature scale.

Expansivities were calculated by fitting four consecutive data points (4/15 of a decade in $|t|$) to a second degree polynomial and computing analytic derivatives.

RESULTS

The linear thermal expansivities along the b and c axes, α_b and α_c, respectively, are shown in Fig. 1. The magnitude of α_c is unusually large and negative. The expansivity along the a-axis is not presented because extremely long time constants prevented an accurate measurement. However, its magnitude near T_c is somewhat smaller than α_b and is positive. These results are in qualitative agreement with those of Iwata[7], whose measurements extended over a wider temperature range but not particularly near T_c.

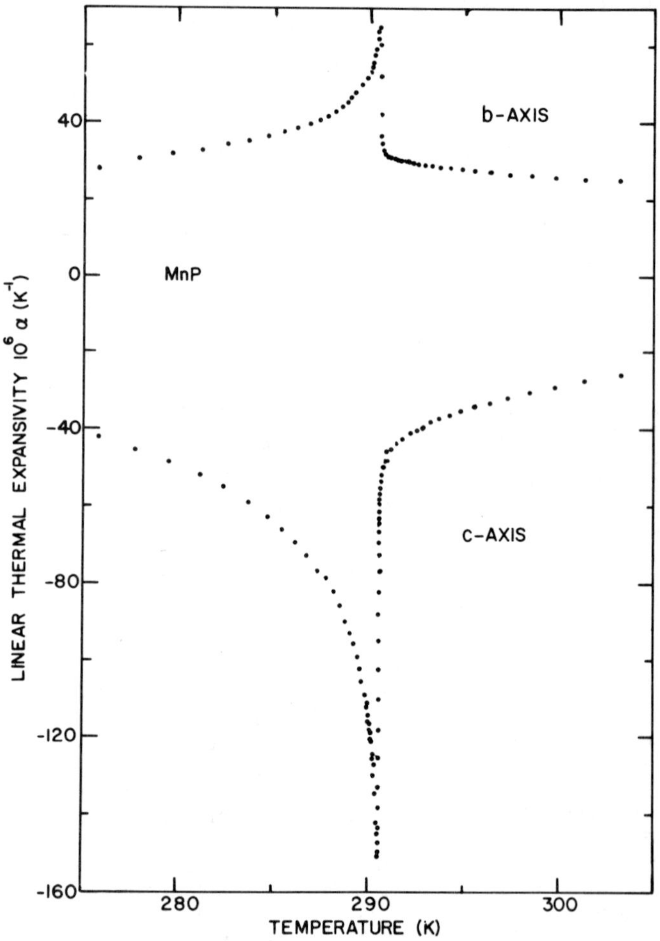

Fig. 1. Thermal expansivity of MnP in the b and c directions as a function of temperature.

A plot of α_b and α_c vs. $\log|t|$ is shown in Fig. 2. A transition temperature of 290.600 K was arbitrarily chosen. Note that this choice for T_c causes the most rapid change in expansivity to occur for $t < 5\times10^{-4}$.

In an attempt to determine critical exponents the following function was fit to the data, generally over the range $5\times10^{-4} < |t| < 5\times10^{-2}$, with a non-linear least-squares routine:

$$\alpha(T>T_c) = A|t|^{-\alpha}(1+D|t|^{x}) + B + Et \qquad (1a)$$

$$\alpha(T<T_c) = A'|t|^{-\alpha'}(1+D'|t|^{x'}) + B' + E't \qquad (1b)$$

Initially, Eqs. (1) were used with constraints $\alpha = \alpha'$, $T_c = T_c'$, $D = D' = 0$, and $E = E'$. The results for the b and c-axes fell within the range $\alpha = \alpha' = -0.08 \pm .04$, $T_c = T_c' = 290.60 \pm .02$, $A/A' = .16 \pm .04$ $(B-B')/A = 3.5 \pm 1$. Although α and α' are in agreement with predictions for the 3-d Heisenberg model, the amplitude ratio A/A' is much smaller than $A/A' \gtrsim 1$ observed for 3-d Heisenberg magnets.[8,9] In addition, since $B \neq B'$, an expansivity discontinuity exists at T_c, so that the leading singularity is actually $\alpha = \alpha' = 0$.[10]

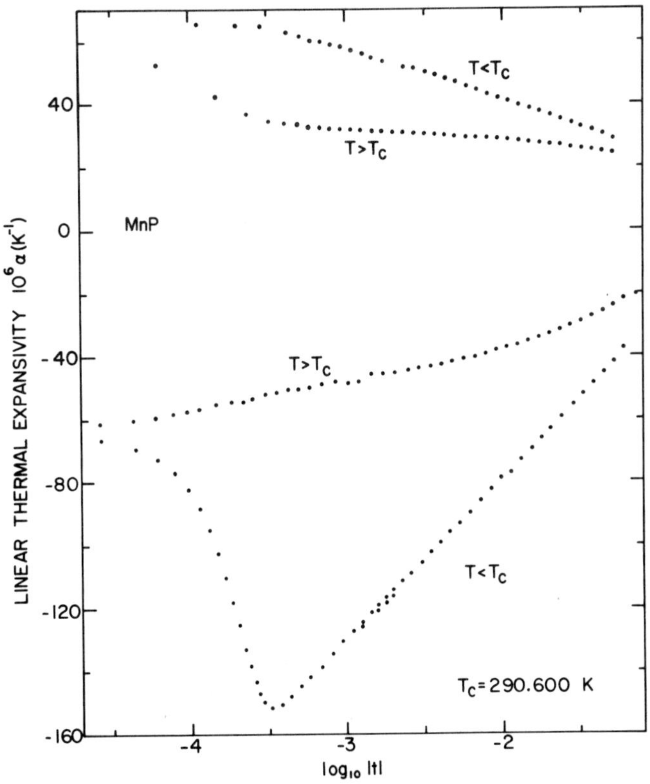

Fig. 2. Thermal expansivity of MnP along the b and c axes vs. reduced temperature $|t| \equiv |1-T/T_c|$. T_c is chosen arbitrarily as 290.600 K, which is about 50 mK above the expansivity extrema.

In order to examine the effect of removing the discontinuity at T_c, a fit was performed with the same previous constraints plus $B = B'$. The results were $\alpha = \alpha' = +0.16$, $T_c = T_c' = 290.63$, and $A/A' = 0.40$. However, the rms error increased by a factor of four indicating the function did not represent the data well. Next, the constraint $B = B'$ was maintained and D and D′ allowed to be non-zero, with $0.2 < x < 1.0$. The fit improved greatly with $\alpha = \alpha' = +0.13 \pm .1$, but $A/A \approx 0.1$, and $D \gtrsim 10$, the latter result suggesting that Eq. (1) was still not appropriate.

CONCLUSIONS

We believe that the thermal expansivity in the range $5\times10^{-4} < |t| < 10^{-1}$ is not expressible by the generalized power law, Eq. (1). Our analysis does not permit compatibility with a 3-d Heisenberg model for which $\alpha = \alpha' \cong -0.1$, $A/A \gtrsim 1$, and $B = B'$, nor does it allow comparison with the 3-d Ising model for which $\alpha = \alpha' \approx +0.1$, $A/A' \approx 0.5$, and probably $B = B'$.[11,12]

One possibility for understanding the results is that a crossover from Heisenberg-like to Ising-like behavior occurs within the measurement range as suggested by the ratio $VK/Nk_BT_c = 5\times10^{-3} \approx t_{crossover}$, where K is the magnetocrystalline anisotropy energy density and V is the molar volume. If a crossover is occurring, the behavior in the crossover region is more complex than one's expectation of a smooth and monotonic variation in exponents and amplitude ratios.

Another possibility for understanding these results is that there is a strain-induced renormalization of the critical behavior brought about by the relatively large volume change near T_c. Also, it is possible that the transition is actually first order, although hysteresis was not detectable experimentally.

ACKNOWLEDGMENTS

We thank K. Binder and P. C. Hohenberg for their helpful comments. We are indebted to E. Hirahara for supplying the MnP crystal and to P. Busch for his technical assistance.

REFERENCES

*Present address: Department of Physics and Astrophysics, University of Colorado, Boulder, Colorado 80302.
1. E. E. Huber and D. H. Ridgley, Phys. Rev. 135, A1033-40 (1964).
2. J. B. Forsyth, S. J. Pickart, and P. J. Brown, Proc. Phys. Soc. 88, 333-9 (1966).
3. Y. Ishikawa, T. Komatsubara, and E. Hirahara, Phys. Rev. Letts. 23, 532-4 (1969).
4. A. Ishizaki, T. Komatsubara, and E. Hirahara, J. Phys. Soc. Japan 30, 292 (1971).
5. V. J. Minkiewicz, K. Gesi, and E. Hirahara, J. Appl. Phys. 42, 1374-5 (1971).
6. B. I. Halperin and P. C. Hohenberg, Phys. Rev. 177, 952-978 (1969).
7. N. Iwata, J. Sci. Hiroshima Univ. 33, 1-21 (1969).
8. A. Kornblit and G. Ahlers, Phys. Rev. B8, 5163-74 (1973).
9. F. L. Lederman, M. B. Salamon, L. W. Shacklette, Phys. Rev. B9, 2981-88 (1974).
10. M. E. Fisher in Reports on Progress in Physics, Vol. XXX, Part II (Institute of Physics and the Physical Society, London, 1967), p. 623.
11. K. G. Wilson and M. E. Fisher, Phys. Rev. Letts. 28, 240 (1972); K. G. Wilson, Phys. Rev. Letts. 28, 548 (1972).
12. C. Domb, Adv. Phys. 19, 339-70 (1970).

SPIN 3/2 ISING MODEL FOR TRICRITICAL POINTS IN TERNARY FLUID MIXTURES*

S. Krinsky and D. Mukamel
Brookhaven National Laboratory, Upton, New York 11973

ABSTRACT

We present a lattice-gas model of ternary fluid mixtures. Within the mean field approximation, we study a non-symmetric tricritical point in this model. We compare our results to the experimental observations on the system ethanol-water-carbon dioxide.

In the course of our work, we have studied a Landau theory describing the neighborhood of a fourth-order-critical point. Also, we have noted that mean field theory indicates the existence of a fourth-order-critical point in the spin 1 model of Blume, Emery, and Griffiths, corresponding to $K < 0$, $J + K > 0$.

* Work supported by U. S. Atomic Energy Commission. (This work will be published in Phys. Rev. B 1, Jan. 1975).

AN EXACTLY SOLVABLE MODEL FOR TRICRITICAL PHENOMENA*

V. J. Emery
Brookhaven National Laboratory, Upton, New York 11973

ABSTRACT

An n-component continuous spin model with a tricritical point is solved exactly in the limit $n \to \infty$, and its properties are described.

For the discussion of tricritical phenomena, it is clearly desirable to have an exactly solvable model to compare with experiments and the various approximate calculations. In this paper we describe an n-component continuous-spin model which may be solved exactly in the limit $n \to \infty$, and has a tricritical point.

The model has n spins $s_{i\alpha}$ with $\alpha = 1,2,\dots,n$ at each lattice site i and the Hamiltonian is

$$\mathcal{H} = \sum_{\alpha}^{n}\sum_{i,j}^{N} J_{ij}s_{i\alpha}s_{j\alpha} + \sum_{i}^{N}\left[n\phi(\Delta, n^{-1}\sum_{\alpha}^{n} s_{i\alpha}^2) + H\sum_{\alpha}^{n} s_{i\alpha}\right] \tag{1}$$

The exchange coupling J_{ij} is a function of i-j and it is assumed that its Fourier transform $\mathcal{J}_q$ vanishes for q=0 and satisfies $\mathcal{J}_q > 0$, otherwise. ($\mathcal{J}_0=0$ can always be arranged by subtracting a diagonal part from the exchange integral and adding it into ϕ). The principal assumption for $\phi(\Delta,\tau)$ is that it is well-behaved and analytic so that all singularities may be attributed to critical phenomena. The magnetic field is H and the temperature T has been absorbed into J_{ij} and (possibly) ϕ. The factors n associated with ϕ ensure that the limit $n \to \infty$ is well defined. The parameter Δ plays the role of the crystal field splitting in the BEG model[1] and variation of Δ leads to a tricritical point.

In the limit $n \to \infty$, the free energy F per site per degree of freedom is given exactly by a generalization of the Hartree approximation:

$$F = \Phi(\kappa) + \phi(\Delta, \langle s^2 \rangle) - \kappa^2 \langle s^2 \rangle \tag{2}$$

with self-consistency conditions

$$\kappa^2 = \left[\frac{\partial\phi(\Delta,\tau)}{\partial\tau}\right]_{\tau=\langle s^2\rangle} \tag{3}$$

$$\langle s^2 \rangle = \frac{\partial\Phi}{\partial\kappa^2} \tag{4}$$

In these equations Φ is the free energy of the Gaussian model obtained by replacing $\phi(\Delta,\tau)$ by $\kappa^2\tau$ in Eq. (1):

$$\Phi(\kappa) = \frac{1}{N}\sum_q \ln(\mathcal{J}_q + \kappa^2) - H^2/4\kappa^2 - \frac{1}{2}\ln\pi , \tag{5}$$

Also, $\langle s^2 \rangle$ is the mean value of any of the $s_{i\alpha}^2$, and κ is the inverse coherence length in the Gaussian model. Equations (2)-(5) may be derived[2] by making an integral representation of $\exp[-n\phi(\Delta, n^{-1}\sum_{\alpha}^{n} s_{i\alpha}^2)]$ for each i, after which the $s_{i\alpha}$ integrals become Gaussian and the representation integrals may be evaluated by the method of steepest descents as $n \to \infty$. More rigorously, it may be shown[3] that F on Eq. (2) is a lower bound for the free energy for all n and that the thermodynamic variation principle[4] gives an upper bound which approaches F as $n \to \infty$. When ϕ is of fourth order in the $s_{i\alpha}$, Eqs. (2)-(5) are analogous to the usual Hartree approximation in Quantum Mechanics.[5] The magnetization per site per degree of freedom is $m = -\partial F/\partial H$ and, using Eqs. (2)-(4),

$$m = H/2\kappa^2 \tag{6}$$

The equation of state is obtained by solving Eqs. (3) and (4) and substituting for κ^2 in Eq. (6).

The properties of the model will be considered as functions of (Δ,T) or of (x,T) where $x = -(\partial F/\partial\Delta)_T$ is the analog of the concentration in helium mixtures. For brevity the discussion will be restricted to the case of three space dimensions. The following results are obtained:

A The critical exponents are spherical along its λ-line and there is a tricritical point at which the exponents are Gaussian for constant Δ.

B The concentration susceptibility $(\partial x/\partial\Delta)_T$ diverges at the tricritical point with an exponent of 1/2 for constant Δ and an exponent of 1 along the coexistence curve or for constant x.

C The phase boundary has a discontinuous slope at the tricritical point in the (x,T) plane.

D The tricritical exponents remain Gaussian to order n^{-1}.

Of these results, A and B have been obtained for related models by renormalization group arguments[6,7] or by series analysis[8] and the exponents for the divergence of $(\partial x/\partial\Delta)_T$ are in agreement with experiment on helium mixtures.[9] C has been obtained in series calculations[8] and experiments on helium mixtures[9] and $FeCl_2$[10] but has not been found in mean field theory or other relevant analytic approximations.

We now summarize the method of deriving those results. Equation (6) shows that, in the ordered phase, $\kappa \to 0$ as $H \to 0$, since m remains finite. Across the λ-line, the spontaneous magnetization appears continuously, so it is the line $m=0$, $\kappa=0$ or, from Eq. (3), $\phi_0' = 0$. (We use the notation

$$\phi_0^{(n)} \equiv \frac{\partial^n\phi(\Delta,\tau_0)}{\partial\tau_0^n} , \quad \tau_0 = \langle s^2 \rangle \text{ evaluated at } m=0,$$

$\kappa=0$). This is a line in the (Δ,T) plane. Near to the λ-line, m and κ are small and, from Eqs. (4), (5) and (6),

$$\langle s^2 \rangle = \tau_0 + m^2 - \gamma\kappa \tag{7}$$

where γ is a constant. Equation (3) may then be expanded to give

$$\kappa^2 = \phi_0' + \phi_0''(m^2-\gamma\kappa) + \frac{1}{2}\phi_0'''(m^2-\gamma\kappa)^2 + \dots \tag{8}$$

For $T \gtrsim T_c$, $m=0$ and $\phi_0' \sim (T-T_c)$. Then if $\phi_0''>0$, it follows that $\kappa \sim (T-T_c)$ near to the λ-line, as in the spherical model. The other critical exponents may be shown to be spherical in a similar way. Now suppose that, as we move along the line $\phi_0'=0$, we come to a region in which $\phi_0'' < 0$ and $\phi_0''' > 0$. Then in addition to $m=0$, $\kappa=0$, Eq. (8) has additional solutions, one with $m=0$, $\kappa \neq 0$, the other with $\kappa=0$, $m \neq 0$. Of these, the latter solution $(m \neq 0)$ has the lowest free energy and hence the transition has taken place above the line $\phi_0'=0$. It is then of first order since $m=0$, $\kappa=0$ simultaneously only when $\phi_0'=0$. Thus at $\phi_0''=0$, the transition changes from second order to first order and it is a tricritical point. $(T=T_t)$ Just above that point both ϕ_0' and ϕ_0'' are linear in $T-T_t$ and hence Eq. (8) gives $\kappa \sim (T-T_t)^{1/2}$ which is a Gaussian result. All other exponents are Gaussian at fixed Δ.

To derive B, we use Eq. (2) to obtain the concentration susceptibility $\left(\frac{\partial x}{\partial\Delta}\right)_T = -\left(\frac{\partial^2 F}{\partial\Delta^2}\right)_T$. The divergent part is $\left(\frac{\partial\langle s^2\rangle}{\partial\Delta}\right)_T$ which may be obtained from Eqs. (3) and (7)

$$\left(\frac{\partial < s^2 >}{\partial \Delta}\right)_T = \left[\frac{-\gamma \frac{\partial^2 \phi}{\partial \Delta \partial \tau}}{2\kappa + \gamma \frac{\partial^2 \phi}{\partial \tau^2}}\right]_{\tau = < s^2 >} \qquad (9)$$

For Δ = constant, $\kappa \sim (T-T_t)^{1/2}$, $\frac{\partial^2 F}{d\tau^2} \sim (T-T_t)$ near to the threshold temperature T_t, and hence $(\frac{\partial x}{\partial T})_T \sim (T - T_t)^{-\frac{1}{2}}$.

On the other hand, along the coexistence curve and for x = constant, the Gaussian model is constrained and the exponents are renormalized[11] to their spherical model values. Then $\kappa \sim (T-T_t)$ and, from Eq. (9), $(\frac{\partial x}{\partial \Delta})_T \sim (T-T_t)^{-1}$.

For C, we need to consider the phase boundary. Its slope turns out to be continuous in the (Δ,T) plane, but in the (x,T) plane it is not. Suppose the phase boundary is a curve $x = x_0(T)$ in the (x,T) plane. From Eqs. (2), (3) and (4), $x = -(\frac{\partial F}{\partial \Delta})_T = -(\frac{\partial \phi}{\partial \Delta})_T$ and this depends upon $< s^2 >$ which, according to Eqs. (4) and (5) is a function of κ for the phase with $m = 0$. Thus to evaluate $\frac{dx_0}{d\Delta}$, it is necessary to know $\frac{d\kappa}{d\Delta}$ along the phase boundary. Along the λ-line $\kappa = 0$, so $\frac{d\kappa}{d\Delta} = 0$. Along the coexistence curve κ varies and $\frac{d\kappa}{d\Delta} \neq 0$. Thus $\frac{dx_0}{d\Delta}$ is discontinuous at the tricritical point. The value of the discontinuity depends upon the form of Φ.

Finally we mention that it is possible to evaluate the order n^{-1} corrections to F. They change the critical exponents along the λ-line[5] but at the tricritical point where $\phi_0'' = 0$, the entire order n^{-1} correction vanishes and the tricritical exponents are unchanged.

REFERENCES

1. M. Blume, V.J. Emery and R.B. Griffiths, Phys. Rev. A4, 1071 (1971).
2. V.J. Emery, Phys. Rev. B (1975).
3. V.J. Emery, to be published.
4. B. Mühlschlegel and H. Zittartz, Z. Phys. 175, 553 (1963).
5. R.A. Ferrel and D.J. Scalapino, Phys. Rev. Letters, 29, 413 (1972); Phys. Letters 41A, 371 (1972).
6. D.J. Amit and C.T. de Dominicis, Phys. Letters 45A, 195 (1973).
7. E.K. Riedel and F.J. Wegner, Phys. Rev. Letters 29, 349 (1972).
8. D.M. Saul, M. Wortis and D. Stauffer, Phys. Rev. B9, 4964 (1974).
9. E.H. Graf, D.M. Lee and J.D. Reppy, Phys. Rev. Letters 19, 417 (1967).
10. R.J. Birgeneau, G. Shirane, M. Blume and W.C. Koehler, to be published.
11. M.E. Fisher, Phys. Rev. 176, 257 (1968).

SYMMETRY BREAKING "IRRELEVANT" OPERATORS AND TETRACRITICAL POINTS IN ANISOTROPIC MAGNETS

Amnon Aharony *
Lyman Laboratory of Physics, Harvard University, Cambridge, Ma. 02138
and
Alastair D. Bruce
Physics Department, University of Edinburgh, Edinburgh, U.K.

ABSTRACT

The phase diagram of a cubic magnet, with a variable uniaxial anisotropy, is studied. For a range of values of the cubic anisotropy energy, it exhibits a tetracritical point, at which three ordered phases and the disordered phase coexist. These are separated by four second-order lines. The shapes of these lines, as functions of the uniaxial anisotropy, are calculated using scaling arguments and expansions in $\varepsilon = 4-d$ and in $1/n$.

INTRODUCTION

Many magnetic materials have single-ion interactions which reflect the symmetry of the underlying lattice. For cubic lattices, the lowest order of these have the form

$$\mathcal{H}_v = v\sum_{\vec{x}}\sum_{\alpha=1}^{n} S_\alpha^4(\vec{x}), \qquad (1)$$

where $S_\alpha(\vec{x})$ $(\alpha=1,2,\ldots,n)$ is the α component of the spin vector $\vec{S}$ at the lattice site $\vec{x}$. The sign of v determines whether the spins tend to align along a cubic axis, e.g. $(1,0,\ldots,0)$ $(v<0)$ or along a diagonal, e.g. $(1,1,\ldots,1)$ direction $(v>0)$.

Various conditions may lead to a lower, uniaxial, symmetry, which is reflected in terms like

$$\mathcal{H}_g = \frac{1}{2}g\sum_{\vec{x}}\left[S_1^2(\vec{x}) - \frac{1}{n-1}\sum_{\alpha=2}^{n}S_\alpha^2(\vec{x})\right]. \qquad (2)$$

Such terms arise from a uniaxial symmetry in the underlying lattice, which may be caused by an external anisotropic stress along the $(1,0,\ldots,0)$ direction. The Hamiltonian (2) leads to an alignment of the spins along $(1,0,\ldots,0)$ $(g<0)$ or perpendicular to $(1,0,\ldots,0)$ $(g>0)$.

For $v<0$, the Hamiltonians (1) and (2) both lead to an ordering in which the spins align along an axis, and there is no competition between them. Thus, the line $g=0$ (in the g-T plane) represents a first-order spin flop line, at which the spins flop from the $(1,0,\ldots,0)$ direction to the direction of one of the perpendicular axes. This line ends at a bicritical point, at which two second-order lines [paramagnetic to $(1,0,\ldots,0)$-ordering or paramagnetic to $(0,1,0,\ldots,0)$, $(0,0,1,\ldots,0)$ etc. ordering] also meet. The phase diagram is described in Fig. 1, and its details were studied by Fisher, Nelson, and Kosterlitz.[1]

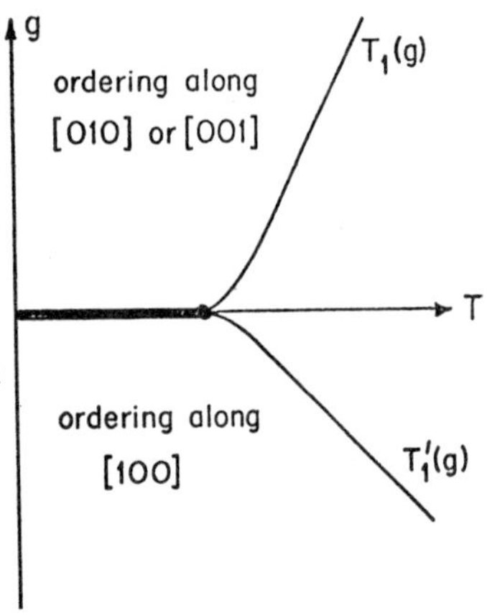

Figure 1

Phase diagram for $v<0$ and $n=3$, displaying a bicritical point and a spin-flop line (the bold line).

For $v>0$, the Hamiltonians (1) and (2) prefer different directions of spin alignment, and the competition leads to the appearance of a new, "intermediate" phase, separated by two second-order lines from the two ordered phases mentioned above. The phase diagram now assumes the form shown in Fig. 2.

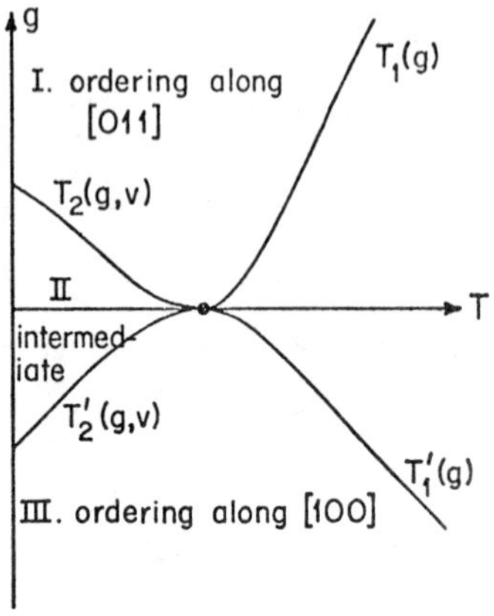

Figure 2

Phase diagram for $v>0$ and $n=3$, displaying a tetracritical point, and an intermediate ordered phase.

The four second-order critical lines meet at a tetracritical point. It is this situation that we wish to describe here. A more detailed study will be published elsewhere.[2]

CUBIC CORRECTIONS TO SCALING AND EASY AXES

In mean field theory, one replaces $\vec{S}$ in (1) and (2) by its average $\vec{M}$, and minimizes the resulting free energy with respect to the direction of the vector $\vec{M}$. A straightforward algebra leads to the phase diagram of Figs. 1 $(v\leq 0)$ and 2 $(v>0)$, where all second order lines are straight lines. In the intermediate phase, the vector $\vec{M}$ rotates continuously from the $(0,1,\ldots 1)$ direction to the $(1,0,\ldots,0)$ direction.

A renormalization group study[3] at $g=0$ shows that for $n\leq n_c$ [$\simeq 3.13$ at $d=3$], v is an irrelevant parameter, and that therefore the tetracritical point is an isotropic Heisenberg-like fixed point. Being interested in $n=3$, we start with a discussion of this case. Although v is "irrelevant", it leads to corrections to the leading scaling behavior, of order $v|t|^{-\phi_v}$, with $t=(T-T_c)/T_c$, and

$$\phi_v = \frac{n-4}{2(n+8)}\varepsilon + \frac{n^3+16n^2+4n+240}{2(n+8)^3}\varepsilon^2 + O(\varepsilon^3), \qquad (3)$$

$(\varepsilon = 4-d)$ being the cubic crossover exponent. Because of these corrections, the system has easy axes at any finite (T_c-T), and these are along diagonals for $v>0$. Thus, one must not ignore the "irrelevant" variable v when discussing ordered phases with particular directions of the magnetization $\vec{M}$.

SCALING THEORY AND DIAGRAMMATIC EXPANSIONS

The free energy at the tetracritical point is now assumed to have the scaling form

$$F(t,g,v) \approx |t|^{2-\alpha}\mathfrak{F}(g|t|^{-\phi_g}, v|t|^{-\phi_v}), \qquad (4)$$

where ϕ_g is the spin-anisotropy crossover exponent.[4] This function is singular on each of the four second-order lines. On the lines $T_1(g)$ and $T_1'(g)$, the transitions become XY-like or Ising-like.[4] The singularity is at $g|t_1|^{-\phi_g}$ = const., and hence the shift in the critical temperature is of the form[1]

$$t_1 \sim T_1(g) - T_c \sim |g|^{1/\psi_1}, \quad \psi_1 = \phi_g . \tag{5}$$

On the lines $T_2(g,v)$ and $T_2'(g,v)$, the singularity is of the form

$$g|t_2|^{-\phi_g} \sim (v|t_2|^{-\phi_v})^{\theta} . \tag{6}$$

This form is dictated by the demand that $t_2 = 0$ when $v = 0$. To obtain explicit values of the exponent θ, diagrammatic expansions were used. To illustrate these, we summarize the calculation of $T_2'(g,v)$ (see Fig. 2). The transition at $T_1'(g)$ is Ising-like, into a phase in which only the first spin component is ordered. In this ordered phase, we can calculate the susceptibilities transverse to the direction of this spin component. All of these are equal, and diverge at $T_2'(g,v)$, when the other spin components start to order, presumably in an XY-like [(n-1)-like] transition. The diagrammatic calculation of the inverse transverse susceptibility, r_T, is simplified using a Ward identity. The final result has the form[2]

$$r_T = -A_v v \overline{M}^{\sigma_v} - A_g \frac{ng}{n-1} \overline{M}^{\sigma_g} = 0 , \tag{7}$$

where A_v and A_g are constants, $\overline{M}$ is the magnetization (of the first spin component) on the line T_2', and where σ_g and σ_v are found, both to order ε^2 and to order $1/n$, to be related to the appropriate crossover exponents ϕ_g and ϕ_v via

$$\sigma = (\gamma - \phi)/\beta , \tag{8}$$

where β and γ are the magnetization and susceptibility exponents at the n-component isotropic Heisenberg-like tetracritical point. Using $\overline{M} \sim |t_2|^{\beta}$ we immediately find that

$$T_c - T_2'(g,v) \sim |t_2| \sim |g|^{1/\psi_2} , \tag{9}$$

with

$$\psi_2 = \phi_g - \phi_v , \tag{10}$$

so that $\theta = 1$ in (6).

For $n < n_c$, $\phi_v < 0$, and therefore $\psi_2 > \phi_g$. Hence, the lines T_2 and T_2' approach the axis $g = 0$ faster than the lines T_1 and T_1'. Asymptotically close to the tetracritical point, T_2 and T_2' may seem to coincide into a single first-order line (which will still differ from the flop line, due to the divergence of the transverse susceptibilities on it). However, ϕ_v is very small (the ε-expansion of ϕ_v is very poorly convergent at $\varepsilon = 1$, but $n = 3$ seems to be very close to n_c, at which $\phi_v = 0$), and thus this asymptotic region may never be reached in practice.

If $n > n_c$, then the tetracritical point becomes the cubic fixed point,[3] and one has[5]

$$\psi_1 = \psi_2 = \phi_g^C = 1 + \frac{1}{6}\varepsilon - \frac{5n^2 - 49n + 53}{81n^2}\varepsilon + O(\varepsilon^3) ,$$

$$= 1/(1 - \alpha_I) + O(1/n) \tag{11}$$

(α_I is the Ising model specific heat exponent). In this case, the lines T_2 and T_2' are distinct all the way into the asymptotic critical region.

CONCLUSIONS

We expect cubic magnets, with a variable uniaxial anisotropy (caused e.g. by a uniaxial stress) to exhibit bicritical or tetracritical points, with various interesting features. An experimental verification of the existence of an intermediate phase in a magnetic material (it seems to have been seen in mixed magnetic crystals, when $n = 2$ and g represents the concentration[6]) and a determination of the exponents describing the phase lines bounding it will be of great interest.

ACKNOWLEDGEMENTS

Discussions with M. E. Fisher and D. R. Nelson were very helpful. This work was started when the authors were at Cornell University, where the research was supported in part by the National Science Foundation, partly through the Materials Science Center at Cornell University.

REFERENCES

* Work supported in part by the National Science Foundation under Grant No. 32774.

1. M. E. Fisher and D. R. Nelson, Phys. Rev. Lett. 32, 1350 (1974). See also D. R. Nelson, J. M. Kosterlitz and M. E. Fisher, Phys. Rev. Lett. 33, 813 (1974) and A. Aharony and A. D. Bruce, Phys. Rev. Lett. 33, 427 (1974).
2. A. D. Bruce and A. Aharony, Phys. Rev. B (in press).
3. A. Aharony, Phys. Rev. B8, 4270 (1973).
4. M. E. Fisher and P. Pfeuty, Phys. Rev. B6, 1889 (1973); F. J. Wegner, Phys. Rev. B6, 1891 (1973).
5. A. Aharony, Phys. Lett. 49A, 221 (1974).
6. F. J. Wegner, Solid State Commun. 12, 785 (1973).

ZERO- AND LOW-TEMPERATURE PHASE TRANSITIONS AND THE RENORMALIZATION GROUP*

J. A. Hertz †
The University of Chicago, Chicago, Illinois 60637

ABSTRACT

At temperatures low compared with characteristic atomic frequencies, classical free energy functionals where the order parameter field ψ depends only on position are inadequate to describe fluctuation phenomena; the noncommutativity of quantum-mechanical operators leads to a ψ which also depends on time. We show how to generalize the renormalization group to such functionals. For phase transitions at T=0 caused by variation of a coupling parameter, the boundary between classical and nonclassical exponent behavior is at 4-z, not 4 dimensions, where $z \geq 1$ depends on the dynamics of the system: z=1 for van Vleck ferromagnets, z=2 for itinerant antiferromagnets, z=3 for clean and z=4 for dirty itinerant ferromagnets. For small T_c there is a crossover region where the renormalization group equations go from their T=0 form to a Wilson form, implying a large precritical region where classical exponents are correct and a small true critical region. One also has an explicit construction of the Wilson free energy functional valid in the critical region.

The highly successful descriptions of both equilibrium and dynamical critical phenomena in terms of the Wilson renormalization group[1,2] are all accomplished within the bounds of classical statistical mechanics. For classical systems, the dynamics and statics can be separated, in the sense that the latter can be solved ignoring the former, given only knowledge of the Landau-Ginzburg-Wilson (LGW) free energy functional $\Phi[\psi(x)]$. In quantum statistics, however, no such separation is possible. What one does have is a formally defined generalized free energy functional where the order parameter ψ depends on both space and (imaginary) time, the latter in the interval $[0,-i\beta]$, where β is the inverse temperature. The aim of this paper is to investigate some aspects of the scaling procedure (the quantum renormalization group) used on these generalized functionals. The results are consequential for an understanding of phase transitions whenever T_c is much less than characteristic energies in the microscopic Hamiltonian, and for zero-temperature transitions caused by changes in a coupling parameter.

As an example, consider spin fluctuations near a ferromagnetic instability in a Hubbard model.[3] The techniques first introduced by Hubbard allow one to express the partition function as a functional integral,[4] and the exponent in the integrand of this functional integral is the exact generalized LGW functional for this system. We approximate it here, expanding the quadratic coefficient only to lowest nonzero order in q and ω, ignoring the q and ω dependence of the quartic coefficient, and neglecting higher-order terms altogether. The justification for this truncation has been discussed elsewhere.[5] The result is a natural generalization of the LGW static model functional:

$$\Phi[\psi] = \beta^{-1}\sum_{q,\omega} (1-UN(E_F) + \mu^2 q^2 + \lambda|\omega|/q)|\psi(q,\omega)|^2$$
$$+\beta^{-3}\sum_{q_1\omega_1}\sum_{q_2\omega_2}\sum_{q_3\omega_3} \tfrac{1}{12}U^2 N''(E_F)\psi(q_1,\omega_1)\times$$
$$\times\psi(q_2,\omega_2)\psi(q_3,\omega_3)\psi(-q_1-q_2-q_3,-\omega_1-\omega_2-\omega_3)\,. \qquad (1)$$

U is the intraatomic Coulomb repulsion, and the parameters λ and μ depend on details of the band structure. One can choose units so that Φ can be written as[6]

$$\Phi[\psi] = \beta^{-1}\sum_{q\omega}(r_o + q^2 + |\omega|/q)|\psi(q,\omega)|^2$$
$$+ u_o\beta^{-3}\sum_{q_1\omega_1}\sum_{q_2\omega_2}\sum_{q_3\omega_3}\psi(q_1,\omega_1)\psi(q_2,\omega_2)\times$$
$$\times\psi(q_3,\omega_3)\psi(-q_1-q_2-q_3,-\omega_1-\omega_2-\omega_3)\,. \qquad (2)$$

When T is much less that the natural cutoff in ω in these sums, the Matsubara frequencies can be treated as a continuum, and it is this case with which we deal here.

Beal-Monod[6] proposed analyzing the functional (2) with the Wilson renormalization group, and asserted that the critical behavior would simply be that of a (d+1)-dimensional system. The conclusions reached here are somewhat different, and the difference can be traced to the $|\omega|/q$ term in the quadratic coefficient. This unusual anisotropic coupling in the extra dimension prevents one from applying the Wilson analysis straightforwardly.

The scaling operation of (2) proceeds in three steps: (a) Terms with q and $|\omega|$ in an "outer shell" are integrated out of the functional integral, leaving a functional which, when its logarithm is expanded in powers of ψ, produces changes in the coefficients r_o and u_o. (b) The integrations over the remaining wavenumbers and frequencies, which now only run up to a cutoff <1, are converted to integrations over new variables which run up to unity. (c) The fields ψ are rescaled, so that in terms of the new fields and new wavenumbers and frequencies, the q- and ω-dependent quadratic terms look just like those in the original functional (2). Suppose step (a) is carried out to lowest order in u by integrating out ψ's with $e^{-\ell}<q,|\omega|<1$. Then after rescaling to new $q' = qe^{\ell}$ and $\omega' = \omega e^{\ell}$, the quadratic term in (2) looks like

$$Q' = e^{-(d+1)\ell}\int_0^1 \frac{d^dq'\,d\omega'}{(2\pi)^{d+1}}(\bar r_o + q'^2 e^{-2\ell} + |\omega'|/q')\times$$
$$\times|\psi(q'e^{-\ell}\omega'e^{-\ell})|^2 \qquad (3)$$

where $\bar r_o \neq r_o$ because of step (a). It is evident that one cannot rescale the ψ's so that the quadratic coefficient has the form $(r_o' + q'^2 + |\omega'|/q')$ because the factor $e^{-2\ell}$ multiplies the q'^2 term but not the $|\omega'|/q'$ in (3). (We make the ansatz that this form remains and that neither term disappears in the $\ell\to\infty$ limit.)

The solution lies in an anisotropic renormalization procedure. Values of q between $e^{-\ell}$ and unity are integrated out, but the range of $|\omega|$ removed is $e^{-z\ell}$ to 1. The value of z is determined by the condition that the coefficient of $|\omega'|/q'$ in the new

$$Q' = e^{-(d+z)\ell}\int_0^1 \frac{d^dq'\,d\omega'}{(2\pi)^{d+1}}(\bar r_o + q'^2 e^{-2\ell}$$
$$+ |\omega'|e^{-z\ell}/q'e^{-\ell})|\psi(q'e^{-\ell},\omega'e^{-z\ell})|^2 \qquad (4)$$

be $e^{-2\ell}$, the same as the coefficient of q'^2. For this model, clearly, one obtains z=3.

In general, z will depend on the form of the frequency-dependent term in Φ. If it describes diffusive dynamics (such as one would obtain by introducing random scattering centers into this system[7]) the $|\omega|/q$ term becomes $|\omega|/D_oq^2$, and z=4. In an antiferromagnet, the dynamics above T_c would be relaxational, and ω would enter in the form $|\omega|/\Gamma_o$, leading to z=2. Finally, in an induced-moment system above T_c, one finds a Φ with a quadratic term of the form $r_o + q^2 + |\omega|^2$, describing propagation of the triplet excitons.[8] In this case z=1, and Beal-Monod's analysis applies.

Whether one obtains classical or nonclassical exponents will depend on how the quartic term behaves under this rescaling, in which $\psi'(q',\omega') = \exp[-(d+z+2)\ell/2]\psi(q e^{-\ell}, \omega e^{-z\ell})$. Because of the rescaling, it is multiplied by a factor of e^W, where W = $-3(d+z)\ell$ (from the three q and ω integrations) + $4 \cdot (d+z+2)\ell/2$ (from the four ψ fields) = $(4-z-d)\ell$.

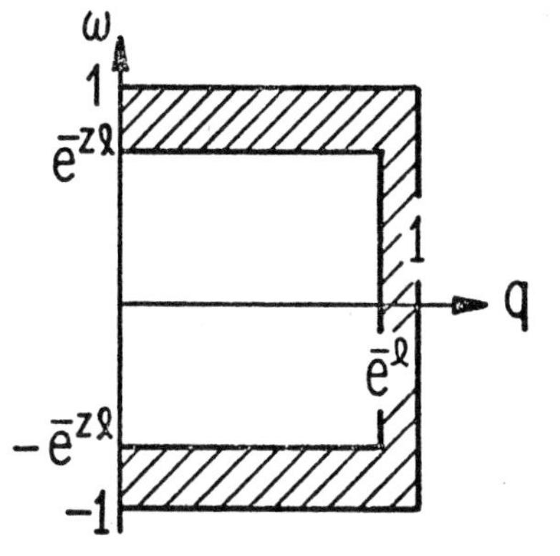

Figure 1. The region of (q,ω) space whose ψ fields are integrated out of the functional integral to obtain the renormalization group equations.

To derive the renormalization group equations, we also have to carry out step (a), as shown in fig. 1. For the functional (2) the change in Φ, to first order in u, is

$$\delta\Phi = \frac{1}{2}\left(\frac{2\ell z}{2\pi}\right)\int_o^1 \frac{\Omega_d}{(2\pi)^d} q^{d-1} dq\, \ell n(r + q^2 + \frac{1}{q} + 6u\beta^{-1}\sum_{k,\nu} |\psi(k,\nu)|^2) + \frac{1}{2}\left(\frac{\Omega_d \ell}{(2\pi)^d}\right)\int_o^1 \frac{2d\omega}{2\pi}\, \ell n(r + 1 + \omega + 6u\beta^{-1}\sum_{k,\nu} |\psi(k,\nu)|^2) \quad (5)$$

where Ω_d is the area of the d-dimensional hypersphere. The first term is the $|\omega| \approx 1$ strips (horizontal in fig. 1); the second, the $q \approx 1$ strip (vertical). Expanding $\delta\Phi$ to second order in u gives the changes in r and u. Explicitly, the renormalization group equations take the form

$$\frac{dr}{d\ell} = 2r + \frac{6u\Omega_d}{(2\pi)^{d+1}}\left[\ell n\left(\frac{r+2}{r+1}\right) + z\int_o^1 \frac{x^d dx}{x^3 + rx + 1}\right] \quad (6a)$$

$$\frac{du}{d\ell} = \epsilon u - \frac{18u^2\Omega_d}{(2\pi)^{d+1}}\left[\frac{1}{(r+1)(r+2)} + z\int_o^1 \frac{x^{d+1} dx}{(x^3 + rx + 1)^2}\right] \quad (6b)$$

where ϵ = 4-d-z. The second term in each of these equations comes from integrating out the large q and ω fields, and the first comes from the rescaling of the remaining fields and their arguments. The explicit form of the second terms in (6) depends on the ω-dependence of the functional (2), but the second term in (6b) will always be negative. Therefore the fixed point of these equations will be Gaussian ($u^*=0$) when $d+z \geq 4$. For three dimensions, all the models mentioned above satisfy this criterion, except the induced-moment system, which is on the borderline. It can be expected to have logarithmic corrections to classical exponent behavior; all the others will be completely classical in this respect.

This analysis is valid as long as the Matsubara frequencies continue to be effectively continuously distributed. But as we rescale ω to $\omega e^{z\ell}$, the spacing between them also increases by a factor of $e^{z\ell}$. The scaling must stop, then, when this spacing is comparable to the highest frequency in the sum. This happens at $\ell = \bar{\ell} = (1/z)\ell n(\omega_o/T)$, where the natural cutoff ω_o has been explicitly restored. ($\omega_o \approx W$ in the case of the Hubbard model.) At this point, $q' = q^{(1/z)}$, so fluctuations with $q \geq q_c = (T/\omega_o)^{1/z}$ are describable by the quantum renormalization group, and, for three-dimensional systems, RPA theories are qualitatively valid. Fluctuations of smaller q must be described by the classical (Wilson) renormalization group acting on the remaining static functional, leading to nonclassical exponent behavior. That is, there will be a crossover from classical to nonclassical exponents at the temperature where the correlation length becomes as large as q_c^{-1}.

Note also that the renormalization procedure up to the crossover provides an explicit derivation of the classical LGW functional valid in the critical region. The parameters r and u are changed from their original values by the effect of dressing by the modes of nonthermal energies. Such effects were studied by Murata in the spin fluctuation problem,[9] although not in a scale-invariant fashion. His results are qualitatively similar to these.

REFERENCES

*Work supported by NSF and NSF-MRL.
†Alfred P. Sloan Foundation Fellow.

1. K. G. Wilson, Phys. Rev. B4, 3184 (1971); K. G. Wilson and J. Kogut, Physics Reports (to be published); S.-K. Ma, Rev. Mod. Phys. 45, 589 (1973); F. J. Wegner and A. Houghton, Phys. Rev. A8, 401 (1973).
2. B. I. Halperin, P. C. Hohenberg, and S.-K. Ma, Phys. Rev. B10, 139 (1974); B. I. Halperin, P. C. Hohenberg, and E. D. Siggia, Phys. Rev. Letters 32, 1289 (1974); S.-K. Ma and G. Mazenko, preprint.
3. J. Hubbard, Proc. Roy. Soc. A276, 238 (1963).
4. J. Hubbard, Phys. Rev. Letters 3, 77 (1959).
5. J. A. Hertz and M. A. Klenin, Phys. Rev. B10, 1084 (1974).
6. M. T. Beal-Monod, Solid State Comm. 14, 677 (1974).
7. P. Fulde and A. Luther, Phys. Rev. 170, 570 (1968).
8. M. A. Klenin and J. Hertz (to be published).
9. K. Murata, AIP Conference Proceedings 18, 692 (1974).

MAGNETIC ORDERING AND CRITICAL BEHAVIOR NEAR SURFACES

P. C. Hohenberg, TU München, Garching, W. Germany, and Bell Laboratories, Murray Hill, N. J. 07974 and K. Binder, Universität Saarbrücken, W. Germany.

ABSTRACT

Films and semi-infinite systems are studied for Ising and Heisenberg models, using mean-field theory, scaling, series expansions, and Monte Carlo simulations. Predictions are made for the magnetization as a function of temperature and distance from the surface, and for the critical exponents. Possible experimental consequences of the theory are briefly reviewed.

INTRODUCTION

The physical properties of magnetic surfaces are interesting from several points of view. Firstly, real systems are necessarily finite, and it is important to assess the effect of surfaces on measurements of "bulk" properties, especially near a phase transition, where correlations become long-ranged. Secondly, surfaces are interesting in their own right, since a number of chemical and metallurgical phenomena are quite sensitive to magnetic surface properties, and experimental techniques can be devised to study these properties in detail (e.g. by spin-polarization[1] or magneto-optic[2] measurements, or by electron[3] and neutron[4] scattering). Finally, from a theoretical standpoint, our recent understanding of bulk critical behavior[5] highlights the importance of the spatial dimensionality, and systems with surfaces present an interesting challenge to the theory, since they do not possess a well-defined dimensionality. A complete study of realistic models of magnetic surfaces is clearly a formidable task, since surfaces can have a drastic effect on the microscopic structure of a material. For example, the crystal structure and electronic band structure may be different on the surface and in the bulk, so that the exchange interactions and magnetic order may be severely modified near the surface.[6]

As a first step toward a theoretical understanding of surface magnetism, we have studied a simple model, consisting of a semi-infinite lattice of localized magnetic moments, interacting via exchange interactions which may be different for surface and bulk atoms. The main physical effect of the surface is twofold: 1) An atom on the surface has fewer neighbors than in the bulk, and 2) Translational invariance is broken in the direction normal to the surface. The theoretical methods we shall use to study this model are the same as the ones which have been applied to the bulk:[5] a phenomenological mean-field theory,[7-9] a scaling approach[9-11] (which is consistent with exact results[12]), and numerical techniques, such as series expansions,[9,13,14] and Monte Carlo calculations.[14]

MEAN-FIELD THEORY

We consider first the case where the nature of the ordering at the surface is the same as in the bulk. Then we may describe it by one local order parameter $m(z)$, where z is the distance from the surface (Fig. 1). If T is near T_c, one expands the local free energy in terms of $m(z)$ and $\nabla m(z)$, to get the total magnetic free energy of the system[7-9]

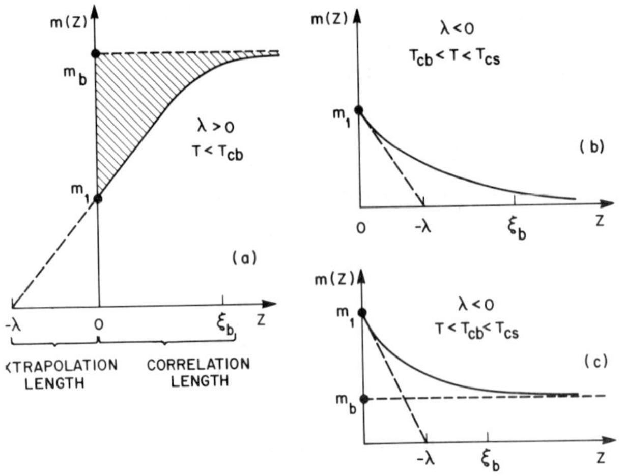

Fig. 1: Mean-field magnetization profile near a surface for $\lambda > 0$ (a), and $\lambda < 0$ (b),(c) (schematic).

$$\Delta F = \int_0^\infty dz \left\{ a'\left(\frac{T}{T_{cb}}-1\right)m^2(z) + bm^4(z) + c\left(\frac{dm}{dz}\right)^2 \right\} + c\lambda^{-1}m^2(0), \tag{1}$$

where T_{cb} is the bulk transition temperature,[15] and a', b, c and λ are phenomenological constants. The only difference between (1) and the bulk expression is the presence of the last term, proportional to the inverse length λ^{-1}. As in the bulk, symmetry requires that only even powers of m enter the expression, and only the lowest order term has been retained. The magnetization profile $m(z)$ follows from the "Ginzburg-Landau" equation[8,9]

$$-m(z)\,a'\left(1-\frac{T}{T_{cb}}\right) + 2bm^3(z) - c\left(\frac{d^2m}{dz^2}\right) = 0\ , \tag{2a}$$

$$\frac{1}{m}\left(\frac{dm}{dz}\right)_{z=0} = \lambda^{-1}\ . \tag{2b}$$

From (2) it is evident that λ plays the role of an extrapolation length (Fig. 1). It also follows that $m(z)$ is appreciably different from the bulk value[15] $m_b = m(\infty)$, only within a correlation distance

$\xi_b = \sqrt{2c/a'}\,|1-T/T_c|^{-\frac{1}{2}}$ from the surface.

The physical properties of the system depend crucially on the sign[7-9] of the parameter λ. For $\lambda > 0$, (Fig. 1a), $m(z)$ is always smaller than m_b, and $m_1 \equiv m(0)$ vanishes at the bulk transition temperature T_{cb} with an exponent[15] $\beta_1 = \beta_b + \nu_b = 1$. For $\lambda < 0$, $m(z)$ is larger than m_b, and surface ordering sets in at a temperature $T_{cs} > T_{cb}$ (Figs. 1b, 1c), with $\beta_1 = \frac{1}{2}$. If one considers a film of thickness $L \gg \xi_b$ (with two free surfaces at z=0 and z=L), one is also interested in the deviations of the average magnetization from m_b, which defines a "surface magnetization" m_s:

$$L^{-1}\int_0^L dz\; m(z) \equiv m_b + 2L^{-1} m_s \; , \qquad (3)$$

with an exponent β_s. Similarly, in the presence of a uniform field H, and a field H_1 which couples to the surface spins only, one can write the free energy of the film as

$$F_L(T,H,H_1) = F_b(T,H) + 2L^{-1}F_s(T,H,H_1) + \ldots, \qquad (4)$$

where F_b is the bulk free energy, and the terms omitted in (4) are of higher order in L^{-1}, for large L. It may be shown[9,11] that $m_1 = (\partial F_s/\partial H_1)$ and $m_s = (\partial F_s/\partial H)$, which suggests the introduction of various susceptibilities, by further differentiation, e.g. $\chi_s = \partial^2 F_s/\partial H^2$, $\chi_1 = \partial^2 F_s/\partial H \partial H_1$, $\chi_{1,1} = \partial^2 F_s/\partial H_1^2$. The associated exponents[9] γ_s, γ_1, and $\gamma_{1,1}$ may be calculated in mean-field theory. It should also be mentioned that the quantities introduced above are all measurable, at least in principle: m_s and χ_s by careful thin-film measurements, and the quantities m_1, χ_1, and $\chi_{1,1}$ by use of a local probe, such as nuclear resonance, or LEED.[3] In the bulk[5] it is well known that mean-field theory is not quantitatively accurate near T_c, and there is no reason to believe that it will work better for surface properties. In fact, an exact calculation[12] for an Ising half-plane (d=2), yields $\beta_1 = \frac{1}{2}$, which disagrees with the mean-field answer $\beta_1 = 1$, and with the expression $\beta_1 = \beta_b + \nu_b = 9/8$, which results from the assumption of a temperature-independent extrapolation length λ.[3]

SCALING THEORY

Just as in the bulk,[5] it is possible to construct scaling expression for the free energy F_s, of the form[9,11,14,15]

$$F_s(T,H,H_1) = |t|^{2-\alpha_b-\nu_b} f_s(H\cdot|t|^{-\Delta_b},\; H_1\cdot|t|^{-\Delta_1}), \qquad (5)$$

where $t = (T-T_c)/T_c$, $\Delta_b = \beta_b\delta_b$, and Δ_1 is a new exponent. The above expression implies a number of scaling relations,[5] for example[9,11,14,15]

$$\beta_s = \beta_b - \nu_b \; ,$$

$$\beta_1 = 2-\alpha_b - \Delta_1 = \beta_b + \gamma_b - \gamma_1 = \tfrac{1}{2}(2-\alpha_b - \nu_1 - \gamma_{1,1}) \; . \qquad (6)$$

By a related scaling argument[10] one can also evaluate the shift in T_c of a film of thickness L, as

$$T_c(L) - T_c(\infty) \sim L^{-1/\nu_b} \; , \; L \to \infty . \qquad (7)$$

In addition to the above "thermodynamic scaling", a scaling assumption may also be introduced[9] for the anisotropic correlation functions of spins near the surface. It turns out[9] that there are two new correlation exponents $\eta_\perp$ and $\eta_\parallel$ (which are related by $\eta_\parallel = 2\eta_\perp - \eta_b$), whereas the correlation length ξ is isotropic, and has the bulk exponent ν_b. It is interesting to note[9] that the scaling theory is consistent with mean-field theory in four dimensions, and also with the exact results[12] in two dimensions ($\beta_1 = \frac{1}{2}, \eta_\parallel = 1, \gamma_{1,1} = 0$). In the following section we shall compare the predictions of scaling with series results and with Monte Carlo calculations in three dimensions.

SURFACE EXPONENTS FOR THE ISING MODEL: SERIES EXPANSIONS AND MONTE-CARLO STUDIES

High temperature series for Ising half-spaces have been carried out[9] in two and three dimensions, and the exponents estimated. For d=2, the results were compatible with the exact exponents, and for d=3, we found[9]

$$d=3: \;\; \gamma_1 = 7/8, \; -1/8 < \gamma_{1,1} < +1/8 \; . \qquad (8)$$

The most interesting consequence of these results is the prediction,[9] via Eq. (6), that $\beta_1 \approx 2/3$. Since, however, this value seemed inconsistent with a LEED experiment[3] on NiO, and with a preliminary series evaluation by Barber,[11] both of which suggested $\beta_1 \approx 1$, this problem was investigated by means of a Monte Carlo calculation.[14,16]

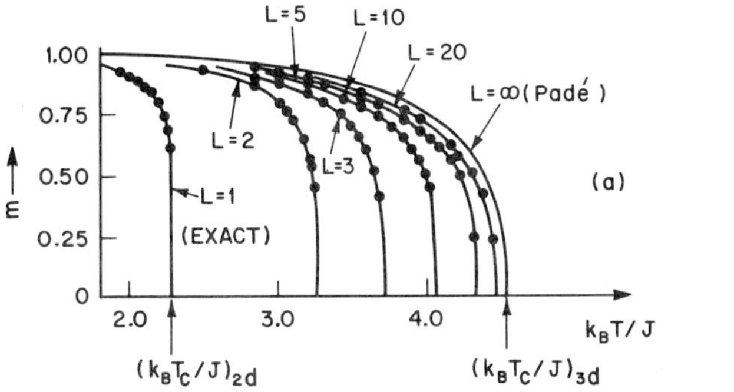

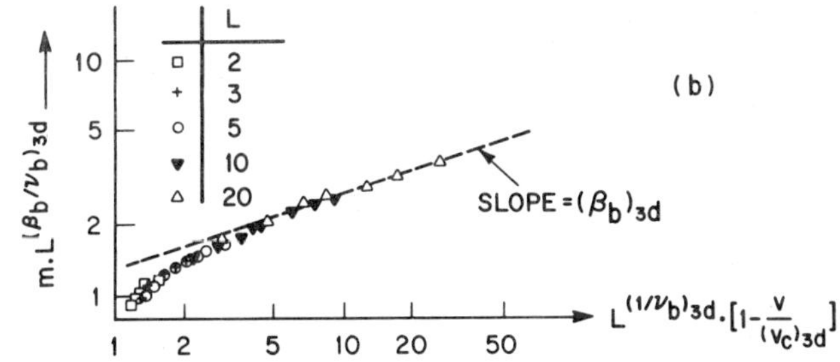

Fig. 2: Mean magnetization of Ising films with L layers, plotted vs. temperature (a), and replotted in scaled form (b). The points are the Monte Carlo results of Ref. 16; $v \equiv \tanh(J/k_BT)$

The systems studied were NxNxL lattices, with two free NxN surfaces, and otherwise periodic boundary conditions. For the chosen values of T and L(L=1,2,5,10, and 20), the effects due to the finiteness of N (=55) were negligible. Figure 2a shows the mean magnetization of these films versus temperature, with a striking changeover from two-dimensional critical behavior for L=1, to three-dimensional behavior for large L. This changeover is displayed in a scaled form in Fig. (2b), where all the data are seen to fit a single curve, with no adjustable parameters (the bulk 3d exponents were taken from series estimates[5]). The critical temperatures implied in Fig. (2a), together with series estimates of Allan and Fisher,[17,10] are plotted in Fig. 3, and show remarkable agreement with the scaling relation (7). It is also straightforward to obtain the magnetization profile m(z) from the computer experiments, as shown in Fig. 4a. Since for the thicker films m(z)

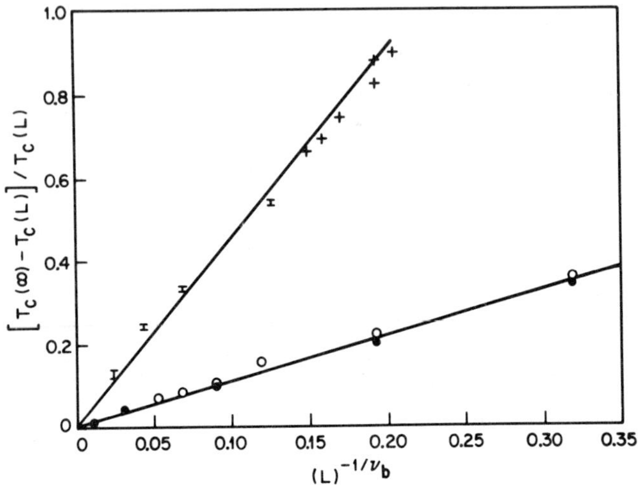

Fig. 3: Monte Carlo (•) and series (o) estimates for the shift of T_c of a film plotted vs. L^{-1/ν_b}; also included are data for finite cubes (I,+). After Ref. 14.

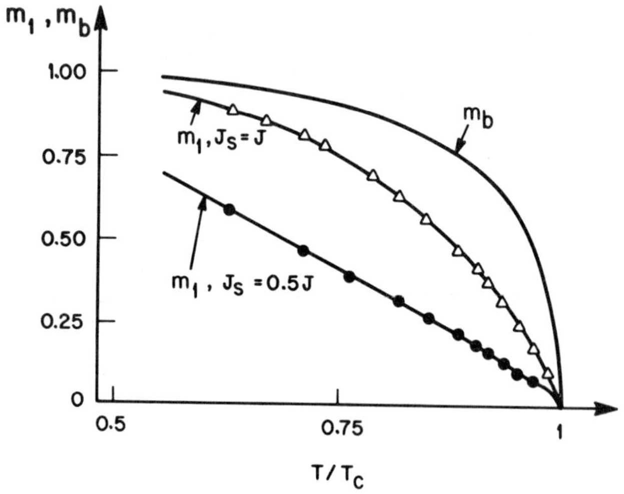

Fig. 5: Bulk and surface-layer magnetization plotted vs. temperature. After Ref. 14.

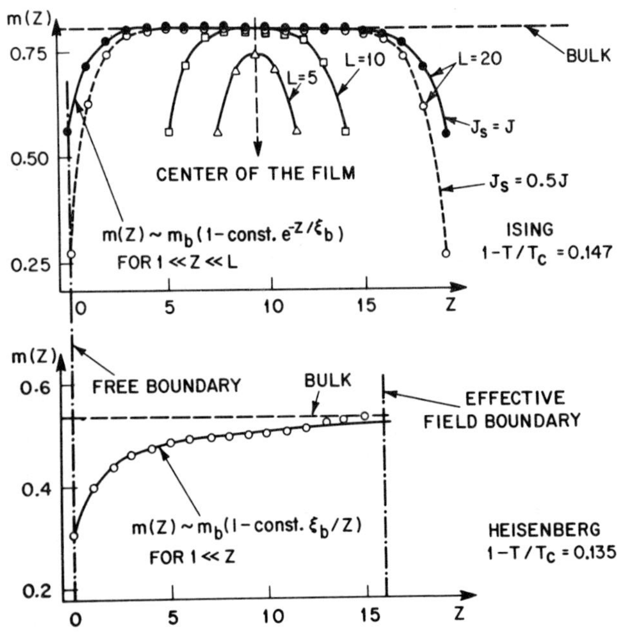

Fig. 4: Magnetization profile for Ising films of thickness L = 5,10,20 and for a Heisenberg "halfspace". After Ref. 14.

reaches m_b in the center of the film, the value of m at the film surface can be taken as an accurate representation of m_1 for a semi-infinite system. Also shown in Fig. 4a, is a calculation[14] in which the exchange constant J_s between surface spins has the value $J_s = 0.5J$, in which case m(z) is strongly depressed. The temperature dependence of m_1, shown in Fig. 5, yields an exponent $\beta_1 \approx 2/3$ over a wide temperature range, for $J_s = J$, in agreement with the scaling prediction.[9] Moreover, an evaluation[13] of the low temperature series for a different lattice than in Ref. 11 also turned out to be compatible with this value of β_1. In the case $J_s = 0.5J$, on the other hand, the expected "universal" exponent $\beta_1 = 2/3$ is at best only observable extremely close to T_c, and the data are seen to fit an exponent $\beta_1 \approx 1$, especially if a small shift in T_c is permitted. This effect could account for the experimental results[3] on NiO.

The above results on Ising models are quite consistent with the surface scaling theory.[9-11, 14] Moreover, a recent renormalization-group calculation by Lubensky and Rubin[18] also yields a partial confirmation of scaling, and shows in addition that the speculation of Fisher[11] that $\Delta_1 = 1/2$ for all d, is violated in linear order in $\varepsilon = 4-d$.

HEISENBERG MODEL WITH SURFACES

In order to apply the Monte Carlo method to the Heisenberg model, a semi-infinite system was simulated,[14] by using the "self-consistent effective field" boundary condition[19] on the back surface, placed at z=16 (Fig. 4b). Such a device is necessary since the magnetization tends to its bulk value according to a power law in the Heisenberg model (Fig. 4b), in contrast to the Ising case, where the behavior is exponential (Fig. 4a). Such a power law behavior is correctly predicted by spin-wave theory,[20] which agrees very well with the computer simulations[14] at low temperatures. Near T_c, the latter yield exponents[14] $\beta_1 \approx 3/4$, and[21] $\delta_1 = 2.3$, which are close, but presumably not identical, to the Ising values.

THE NATURE OF MAGNETIC SURFACE ORDERING

The scaling theory obtained so far is restricted to the case in which the magnetic order at the surface has the same form as in the bulk. Of course, this is not the most general case, since it was already found in mean-field theory that for $\lambda < 0$ the surface orders at a higher temperature than the bulk. In fact, the nature of surface ordering may be investigated[14], using the molecular field approximation, for arbitrary values of the surface exchange J_s. The results for a nearest-neighbor cubic lattice are depicted in Fig. 6, where it is seen that surface ferromagnetism occurs above the bulk T_c for $J_s/J > 1.25$, whereas surface antiferromagnetism is found for $J_s/J < -1.5$. [Also shown in Fig. 6 are the results of a more accurate calculation using high-temperature series, which

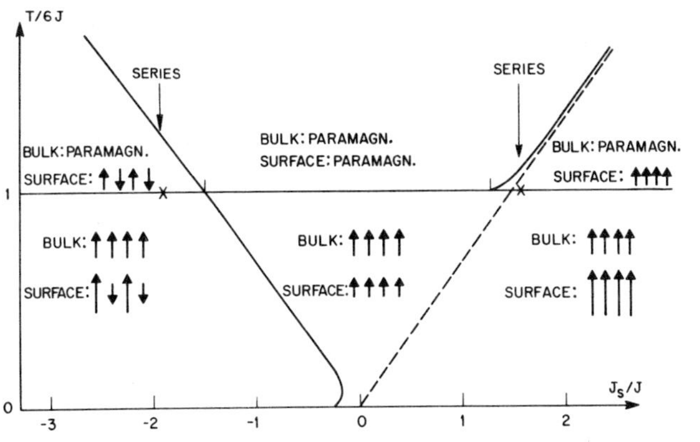

Fig. 6: Molecular-field phase diagram for a system with surface exchange J_s and bulk exchange J, showing the different states of magnetic order. After Ref. 14.

yields surface ferromagnetism for $J_s/J > 1.6$, and antiferromagnetism for $J_s/J < -1.9$]. Below the bulk T_c there is a regime where the surface has both a total magnetization, and a staggered magnetization.

Finally, let us remark that for Ising models with purely ferromagnetic interactions, it can be shown rigorously that "dead layers" do not occur, i.e. situations in which the bulk is ordered but the surface is disordered are excluded. Indeed, using Griffiths inequalities[5] one finds,

$$\langle\mu(\vec{\rho}_1,0)\mu(\vec{\rho}_2,0)\rangle \geq \langle\mu(\vec{\rho}_1,0)\mu(\vec{\rho}_1,z)\rangle^2 \langle\mu(\vec{\rho}_1,z)\mu(\vec{\rho}_2,z)\rangle \tag{9}$$

where $\mu(\vec{\rho}_1,z)$ is the spin at the site $\vec{\rho}_1$ in the z'th layer, with the surface at z=0. Choosing $|\vec{\rho}_1-\vec{\rho}_2|$ to be large, Eq. (9) implies $m_1^2 \geq G_\perp^2(z)m^2(z)$, where $G_\perp(z) \neq 0$ is a correlation function perpendicular to the surface, which cannot vanish unless some interaction constants are zero. Thus, since $m^2(z) > 0$, m_1 cannot vanish. The phase diagram in Fig. 6 is expected to be qualitatively correct for any uniaxial system. For a Heisenberg model, the non-existence of long-range order in two dimensions[22] precludes a state of surface order above T_{cb},[14,23] in contradiction to an earlier suggestion.[24]

CONCLUSION: EXPERIMENTAL CONSEQUENCES

As mentioned in the Introduction, a variety of experimental methods may yield information on surface properties of real materials. We showed already that the LEED experiments[3] on NiO could be interpreted by our theory, assuming a reduced exchange at the surface. Similarly, the results for the shift of T_c in films, reported by Lutz et al.[25] for Ni, are consistent with the exponent $1/\nu_b$ of Eq. (7), although the data are not sufficiently precise to exclude rather different values. In both the above cases, of course, one is dealing with itinerant systems, for which our models are only very crude approximations, even for ideal samples with perfectly defined geometry. We must therefore recognize that our ignorance of the microscopic structure of real surfaces will necessarily make the experimental verification of the theory very difficult. In fact, we believe that the situation may be turned around: as a result of the theoretical work described briefly in this paper, we believe that the surface magnetic behavior of simple models is rather well understood, and the methods can be further refined to answer specific questions. Real magnetic surfaces, on the other hand, contain considerable complexity, and the main problem in understanding these is to find models which will adequately represent their properties. It is hoped that by making simultaneous measurements of various quantities on the same sample (e.g. m_b, m_1, χ_b, χ_1, $\chi_{1,1}$, etc.) and by comparing the results to model calculations, one can narrow down the uncertainty in the parameters, and thus use the theory of critical phenomena as a tool in elucidating the microscopic structure of surfaces!

REFERENCES AND FOOTNOTES

1. U. Banninger, G. Busch, M. Campagna and H. C. Siegmann, Phys. Rev. Letters 25, 585 (1970).
2. G. S. Krinchik, A. P. Khrebtov, A. A. Askochenskii, and V. E. Zubov, JETP Letters, 17, 335 (1973).
3. T. Wolfram, R. E. DeWames, W. F. Hall, and P. W. Palmberg, Surface Sci. 28, 45 (1971).
4. I. I. Gurevich and L. V. Tarasov, Low Energy Neutron Physics, (North-Holland, Amsterdam, 1968).
5. M. E. Fisher, Rept. Progr. Phys. 30, 615 (1967); see also C. Domb and M. S. Green, Phase Transitions and Critical Phenomena, Vols. I-V (Academic, N.Y., 1972-75).
6. See, e.g., H. Takayama, K. Baker, and P. Fulde, Phys. Rev. B10, 2022 (1974).
7. D. L. Mills, Phys. Rev. B3, 3887 (1971); see also, A. Corciovei, G. Costache, and D. Vamanu, Solid State Physics 27, 237 (1972) and references therein.
8. M. I. Kaganov and A. N. Omelyanchuk, JETP 34, 895 (1972).
9. K. Binder and P. C. Hohenberg, Phys. Rev. B6, 3461 (1972).
10. M. E. Fisher, in Critical Phenomena, M. S. Green, ed. (Academic, N.Y., 1971).
11. M. N. Barber, Phys. Rev. B8, 407 (1973); M. E. Fisher, J. Vac. Sci. Technol. 10, 665 (1973).
12. B. M. McCoy and T. T. Wu, Phys. Rev. 162, 436 (1967).
13. M. N. Barber, J. Phys. C. 6, L262 (1973).
14. K. Binder and P. C. Hohenberg, Phys. Rev. B9, 2194 (1974).
15. We shall consistently use a subscript b to denote bulk quantities, and s for surface quantities.
16. K. Binder, Thin Solid Films 20, 367 (1974); see also, K. Binder, in Ref. 5.
17. G. A. T. Allan, Phys. Rev. B1, 352 (1970).
18. T. C. Lubensky and M. H. Rubin, Phys. Rev. Letters 31, 1469 (1973); and (to be published).
19. H. Müller-Krumbhaar and K. Binder, Z. Physik 254, 269 (1972).
20. D. L. Mills and A. A. Maradudin, J. Phys. Chem. Solids 28, 1855 (1967); and Corciovei et al., Ref. 7 above.
21. H. Müller-Krumbhaar (to be published).
22. N. D. Mermin and H. Wagner, Phys. Rev. Letters 17, 1133 (1966); P. C. Hohenberg, Phys. Rev. 158, 383 (1967).
23. R. A. Weiner (preprint).
24. R. A. Weiner, Phys. Rev. Letters 31, 1588 (1973).
25. H. Lutz, J. D. Gunton, H. K. Schurmann, J. E. Crow, and T. Mihalisin, Solid State Comm. 14, 1075 (1974).

FINITE SIZE BEHAVIOR OF THE ISING SQUARE LATTICE WITH FREE EDGES+

D. P. Landau
University of Georgia, Athens, Georgia 30602

ABSTRACT

Using an importance sampling Monte Carlo technique we have studied NxN Ising square lattices with free edges for $3 \leq N \leq 60$. Both the shift in "ordering temperature" $T_c(N)$ as well as the broadening of the transition due to finite N are greater than for identical lattices with periodic boundary conditions (p.b.c.). We find that $1-T_c(N)/T_c(\infty) = a/N$ with $a=1.24 \pm 0.04$ and that $(C/R)_{max} \approx 0.33 \ln N + 0.2$. For $N \geq 30$ the specific heat begins to approach the same asymptotic size dependence as found for much smaller lattices with p.b.c. Similarly, the finite size dependence of the order parameter is much more dramatic for free edges than p.b.c. and the limiting behavior is not reached until $N \geq 30$.

INTRODUCTION

Although the properties of the infinite Ising square lattice are quite well understood, the behavior of finite lattices is less clear. Exact calculations for the specific heat and internal energy of NxN lattices with periodic boundary conditions (p.b.c.) have been made by Ferdinand and Fisher[1], and we have used the Monte Carlo method to study the magnetic properties of these lattices[2,3]. By and large these results are consistent with a simple theory[4] of finite-size effects and with finite-size scaling. Our knowledge of the finite-size behavior of Ising lattices with free edges is much less complete. Ferdinand and Fisher also considered a few small lattices ($N \leq 4$) with free edges, but these systems were too small to allow determination of the asymptotic large N behavior. In addition, in a Monte Carlo study of the simple cubic Ising lattice with free edges, Binder[5] showed that the size variation of the "transition temperature" for small lattices did not agree with the theoretical prediction.[4] Later work by Binder[6] and Binder and Hohenberg[7] showed that the disagreement was probably due to the non-asymptotic nature of the data. We would hope that our investigation would not only demonstrate the effect of free edges vs p.b.c. for the square lattice but would also lend further support to the explanation of the simple cubic lattice behavior.

MONTE CARLO METHOD

The calculations were made using an importance sampling Monte Carlo method similar to that described by Fosdick[8]. In principle the technique estimates the properties of the system by generating a relatively small carefully chosen fraction of the states of the system. Details of the calculation are given elsewhere[3,9]. We have examined NxN square lattices with free edges for $3 \leq N \leq 60$. Spin-spin interactions are limited to ferromagnetic nearest-neighbor coupling of magnitude K_{nn}. For each lattice, configurations were chosen from different starting conditions and then averages were taken over the different samples of states. For $N \leq 10$, as many as 5×10^4 configurations were generated at each temperature. With increasing lattice size the number of states generated was reduced and for the largest system (N=60) a maximum of 4×10^3 configurations were included in the averages.

RESULTS

One effect of the free edge boundary conditions is to make the ground-state energy size dependent (due to the "dangling bonds" at the surface). The size dependence of the internal energy is complicated by a change in the scale factor in addition to the change in the "pseudo-critical" behavior. This difficulty is eliminated in the specific heat, and for this reason we shall use the specific heat data, even though they are less precise, for studying thermal properties. The results for several different lattices are shown in

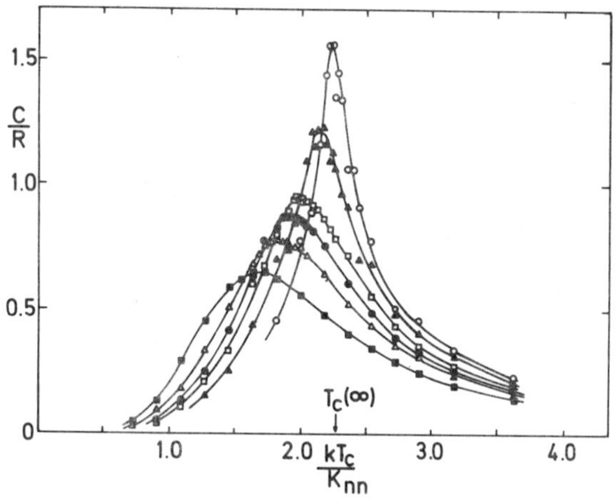

FIG. 1 Specific heat of NxN lattices with free edges: N=4,■; N=6,△ ; N=8,● ; N=10,□ ; N=20,▲ ; N=60, o.

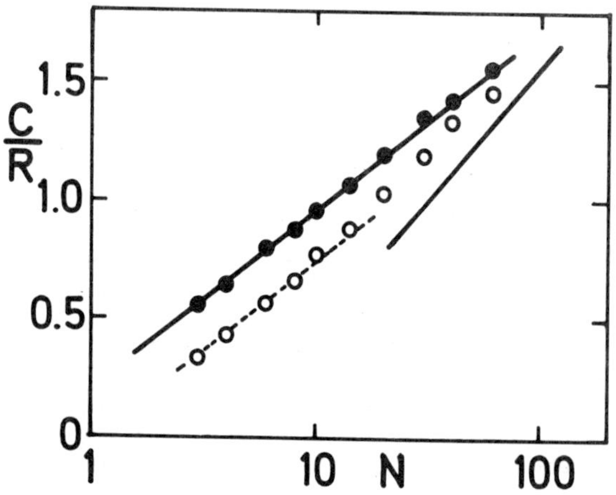

FIG. 2 Specific heat of NxN lattices with free edges. Maximum value,● ; value at $T_c(\infty)$, o. The solid line to the right has slope $= A_o = 0.494$.

Fig. 1. As the size of the lattice is reduced the specific heat peak is broadened and shifted to lower temperatures. We shall take the position of the peak to define an N-dependent pseudo-critical temperature $T_c(N)$. These data were compared with the exact curves of Ferdinand and Fisher and excellent agreement was found. Small random differences of about ±1% were found for N=3 and ±2% for N=4. The maximum specific heat values vary as $A_m \ell n\ N$ for $3 \lesssim N \lesssim 60$ (see Fig. 2) although the coefficient A_m is only 2/3 of $A_o = 0.494$ which was found for the p.b.c. case. The specific heat at $T = T_c(\infty)$ also has the same size dependence for small N, but for $N \gtrsim 20$ the data begin to bend upwards and the last few points are consistent with a slope $= A_o \simeq 0.494$, as predicted by scaling theory.[4]

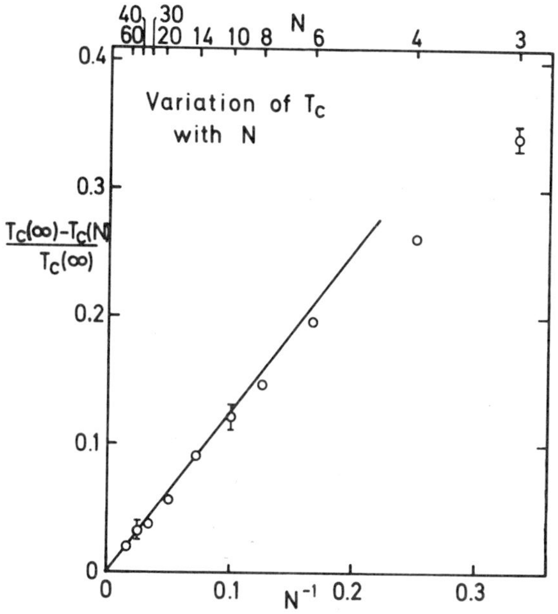

FIG. 3 Variation of T_c with lattice size. The asymptotic behavior as $N \to \infty$ is given by the solid line with slope 1.24 ± 0.04.

The size dependence of $T_c(N)$ was also analyzed. A log-log plot of $T_c(\infty) - T_c(N)$ vs N yields a slope of 0.96 ± 0.03 if data for all the lattices are used and 1.01 ± 0.02 if only data points for $N \gtrsim 10$. This is strong evidence that T_c varies as a/N. In Fig. 3 we show a plot of $[T_c(\infty) - T_c(N)]/T_c(\infty)$ vs N^{-1} which yields a = 1.24 ± 0.04. For comparison we note that Fisher and Ferdinand estimated[1,4] a $\simeq$ 1.35 ± 0.08 from results on small NxN lattices and a $\simeq$ 0.90 for $N \times \infty$ strips. For the case of p.b.c. they obtained a = -0.36. A comparison of the widths at half peak height indicates that both cases vary as $N^{-1/2}$ but the amplitude for the free edge case is ~1.5 times that for p.b.c.

Because the spontaneous magnetization of a finite lattice can easily reverse itself, we have chosen to define the order parameter M as the absolute value of the spontaneous magnetization. The data for the order parameter are shown in Fig. 4. Although these results are qualitatively similar to those obtained with p.b.c. there are substantial

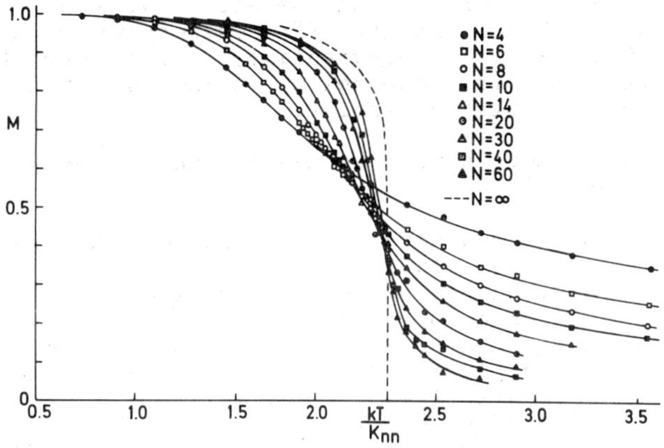

FIG. 4 Order parameter of NxN lattices with free edges: N=4, ●; N=6, ■; N=8, ○; N=10, ■; N=14, ▲; N=20, ⊙; N=30, ▲; N=40, ⊡; N=60, ▲. The infinite lattice result[10] is given by the dashed line.

quantitative differences. Whereas the p.b.c. results show little size dependence for $N \gtrsim 10$ until $T \gtrsim 0.95\ T_c(\infty)$, the data taken with free edges vary significantly all the way down to $0.6\ T_c(\infty)$. Here, even for N=60, substantial differences are found in comparison with the infinite lattice result. The size dependence of the order parameter can be given by a scaling relation for $T < T_c$:

$$M = N^{-\beta/\nu} f(\varepsilon N^{1/\nu}) \qquad (1)$$

where β and ν are the infinite lattice critical exponents and f is a function of $x = \varepsilon N^{1/\nu}$ only. As $x \to \infty$ (i.e. $N \to \infty$) $f \simeq B\, x^{\beta}$ asymptotically in order to give the known critical form. A plot of $y = MN^{\beta/\nu}$ vs x then defines the function f(x). For free edges the data appear to be entering the asymptotic region for $x \gtrsim 6$ whereas the p.b.c. data obey the asymptotic form for $x \gtrsim 0.25$. Taken together with the specific heat results this implies that the asymptotic size dependence is obtained only for much larger lattices when free edge boundary conditions are used. We feel that it is therefore almost certain that the simple cubic lattices studied by Binder were too small to show the asymptotic size behavior. (The variation of T_c shown in Fig. 6 of Ref. 5 was limited to a maximum lattice size of 8x8x8 although a 12x12x12 lattice was also studied.)

Although our results on the square lattice cannot differentiate between a N^{-1} and $N^{-1/\nu}$ size dependence (since ν=1) they do suggest that Fisher's theory can be tested using simple cubic lattices and that the disagreement between theory and "experiment" will disappear when larger lattices are considered. Because the region of asymptotic size dependence is reached for rather small lattices it is clear that p.b.c. are much more useful in extracting the infinite lattice behavior. Free edges, however, are more physical in terms of describing the behavior of small grains. In real systems with long range (dipolar interactions) one might also expect that significantly larger lattices must be used before the asymptotic size dependence is seen. The properties of very fine grains may therefore seem to have no simple size dependence.

We wish to thank Drs. R. Swendsen and H. Horner for helpful comments. We also wish to thank the Institut für Festkörperforschung of the Kernforschungsanlage, Jülich, West Germany, for its hospitality during the preparation of this manuscript.

+Supported in part by the National Science Foundation

1. A.E. Ferdinand and M.E. Fisher, Phys. Rev. 185, 832 (1969).
2. D.P. Landau, Phys. Letters 47A, 41 (1974).
3. D.P. Landau (to be published).
4. M.E. Fisher, Proc. Int. School of Physics Enrico Fermi, Course 51, Varenna, 1970 (Academic Press, N.Y., 1971).
5. K. Binder, Physica 62, 508 (1972).
6. K. Binder, Thin Solid Films 20, 367 (1974).
7. K. Binder and P.C. Hohenberg, Phys. Rev. B9, 2194 (1974).
8. L. D. Fosdick, Methods in Computational Phys. 1, 245 (1963).
9. D.P. Landau, AIP Conf. Proc. 18, 819 (1974).
10. L. Onsager, Nuovo Cimento 6 Suppl, 261 (1949).

HYPERFINE FIELDS AND CRITICAL POINT EXPONENTS OF SOME TWO-DIMENSIONALLY LAYERED Fe(2+) SALTS

J. L. Schurter[+] and R. G. Barnes
Ames Laboratory, USAEC, and Physics Department,
Iowa State University, Ames, Iowa 50010, and
R. D. Willett,
Department of Chemistry, Washington State
University, Pullman, Wash. 99163

ABSTRACT

The nuclear hyperfine properties of the two-dimensionally layered antiferromagnetic ferrous chloride complexes, methylammonium ferrous chloride (MA_2FeCl_4) and benzylammonium ferrous chloride (BA_2FeCl_4), have been studied by (57)Fe Mössbauer spectroscopy. Temperature dependence of the effective magnetic hyperfine fields, as well as isomer shifts and quadrupole couplings, were determined by transmission experiments on powders. Saturation hyperfine fields and Neel temperatures were found to be: for MA_2FeCl_4, H(0) = 280 kOe and T_N = 95.5 K, and for BA_2FeCl_4, H(0) = 265 Oe and T_N = 75.6 K. Analysis of the hyperfine fields in terms of critical point behavior, $H(T)/H(0) = B(1 - T/T_N)^{\beta}$ yields for MA_2FeCl_4, B = 1.10 and β = 0.23, and for BA_2FeCl_4, B = 1.04 and β = 0.28. These values are compared with the predictions of the two-dimensional Ising model, B = 1.22 and β = 0.125, and with the exponents γ and γ' obtained previously from susceptibility measurements.

INTRODUCTION

Two-dimensional ({2})* magnetic systems may be divided into two basic groups: (a) those in which some cancellation of interlayer forces occurs so that the principal magnetic interaction is intralayer, and (b) those in which the layers are physically separated to such an extent that the intralayer interaction dominates.

The antiferromagnetic divalent transition metal fluorides such as K_2CoF_4,[1] Rb_2MnF_4, and others[2] are good examples of the first type. Examples of the second kind which have been investigated, such as the copper (II) chloride complexes of the type $(RNH_3)_2CuCl_4$,[3] rely on the length of the substituted ammounium cation for the magnetic layer separation. The compounds reported on here, methylammonium ferrous chloride, $(CH_3NH_3)_2$ $FeCl_4$, or $(MA)_2$ $FeCl_4$, and benzylammonium ferrous chloride, $(C_6H_5CH_2NH_3)_2$ $FeCl_4$, or $(BA)_2$ $FeCl_4$, are believed to be of this second kind. These compounds crystallize in a tetragonal unit cell; magnetic iron atoms are located in {2} metal-halogen sheets, the separation of which is determined by the length of the ammonium cation. The magnetic susceptibility of powder samples has been measured and showed maxima at 95 K and 73 K for the methyl and benzyl compounds, respectively.[4] The compounds appear to be antiferromagnetic since no field dependence was observed below the ordering temperature. The critical indices γ and γ' were determined from the susceptibility.[4]

EXPERIMENTAL

Crystals of these compounds were prepared as previously described.[5] Since, due to the oxidation of Fe^{2+} to Fe^{3+}, the compounds quickly deteriorate if exposed to air, they were handled only under dry argon. Appropriate Mössbauer absorbers were made by sealing approximately 30 mg in a plexiglass holder using a low temperature varnish as adhesive. The γ-ray source used was a commercially prepared 50 mC ^{57}Co source in a copper matrix. With this source at 300 K the full width at half maximum intensity of the resonance lines of Fe foil was 0.22 mm/sec, as compared to the natural linewidth for folded source and absorber of 0.19 mm/sec, hence assuring good resolution of relatively complex spectra.

RESULTS

Mössbauer spectra were obtained for $(MA)_2FeCl_4$ starting at 8.8 K and at regular intervals to 91.5 K, as well as at 300 K. Typical spectra of $(MA)_2FeCl_4$ are shown in Fig. 1. At 300 K the typical two-line quadrupolar pattern of ferrous compounds appears, with a splitting of 2.21 ± 0.03 mm/sec and a positive isomer shift of 1.11 ± 0.03 mm/sec relative to magnetic iron. A third line seen in the 300 K spectrum, belonging to an unidentified impurity, occurred in all spectra and remained unsplit at all temperatures. Below T_N combined magnetic dipole and electric quadrupole hyperfine splitting appeared. Also detectable in this temperature range were the six lines of a quite dilute impurity, for which H(0) = 474 kOe at 8.8 K. This impurity is believed to be Fe_3O_4 since its H(0) is reported variously as 473 kOe or 507 and 450 kOe at two non-identical sites.[6]

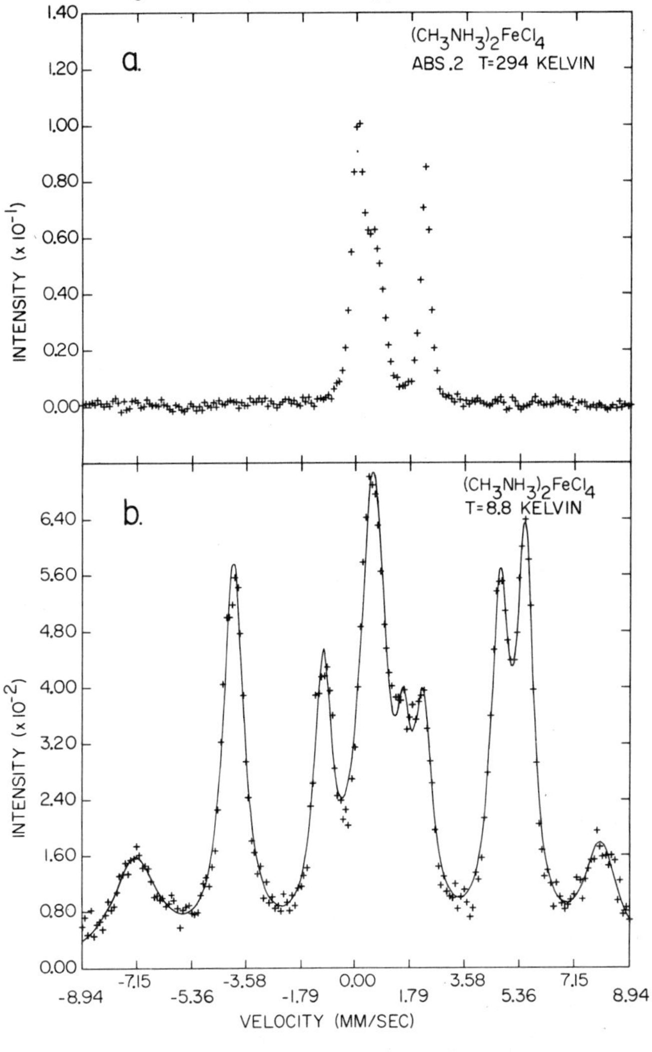

Fig. 1. Mössbauer effect spectrum of ^{57}Fe in $(CH_3NH_3)_2$ $FeCl_4$ at (a) 294 K, and (b) 8.8 K.

Theoretical spectra were computer-fit to the experimental data points after the two impurity spectra had been subtracted.[7] The value of H(0) = 280 kOe was determined by extrapolation of the low temperature points, and the critical temperature was obtained by fitting the data points near the critical temperature to the {2} Ising model expression,[8]

$$H(T)/H(0) = B(1 - T/T_N)^{\beta} \; . \qquad (1)$$

This resulted in values of $T_N = 95.5 \pm 0.6$ K, B = 1.1, and $\beta = 0.23$ for $(MA)_2FeCl_4$. The critical parameter β was also determined from a log-log plot of H(T)/H(0) vs $(1 - T/T_N)$, as shown in Fig. 2, yeilding $\beta = 0.23$ in agreement with the direct fit result. The results for $(BA)_2FeCl_4$, obtained in the same manner, were H(0) = 265 kOe, $T_N = 75.6$ K, 1.04, and $\beta = 0.28$.

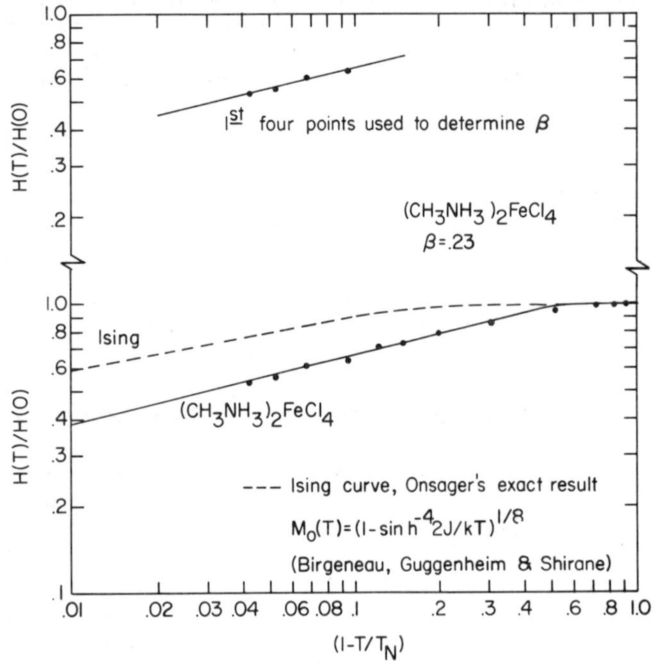

Fig. 2. Log-log plot of reduced H(T) data for $(CH_3NH_3)_2FeCl_4$, showing comparison with the exact Ising model calculation.

Due to spectral line broadening, meaningful data could not be obtained at temperatures closer to T_N than $0.96T_N$ for $(MA)_2FeCl_4$ and $0.97T_N$ for $(BA)_2FeCl_4$, at which points $H_{eff} = 0.531H(0)$ and $0.328H(0)$, respectively. At least four possible explanations for this behavior suggest themselves. (1) A distribution of T_N values could result from the presence of slippage or stacking vaults in the crystals. (2) A distribution of temperatures within the sample would result in a distribution of hyperfine fields. Near T_N small temperature gradients give rise to relatively large H_{eff} deviations. (3) If the temperature of the entire sample were to fluctuate, line broadening would result. (4) It is possible that relaxation effects may become important near T_N.

DISCUSSION

The critical exponent β is of particular interest. The data were shown to fit the curve, $H(T)/H(0) = B(1 - T/T_N)^{\beta}$, in the range $0.6 < T/T_N \leq 0.97$ (see Fig. 2). The meaning of β, however, is questionable except in the immediate neighborhood of T_N. Whereas β has been successfully determined by neutron scattering experiments[1,2] for {2} magnetic structures in which the magnetic interplanar interaction cancels (Type I compounds), there are no reported values of β in compounds in which the {2}character arises from large magnetic plane separation (Type II compounds).[4,5,10] For Type I compounds the reported β values vary from 0.14 to 0.23, in relatively good agreement with the {2} Ising model prediction of 0.125.[8] Since these Type I materials were studied by neutron scattering, the {2} character of the crystals is indicated not only by the β values, but also by the presence of {2} correlations. For several samples, the magnetization data were shown to fit $M(T)/M(0) = B(1 - T/T_N)^{\beta}$ over a wide temperature range, a behavior not common when {3} ordering occurs.

Since crystal quality of the Type II compounds generally does not permit neutron scattering, the Mössbauer effect furnishes a useful means by which some of these compounds may be examined. Although less sensitive than neutron scattering, it does provide sufficient information to suggest several conclusions. The fact that the H_{eff} data for both $(MA)_2FeCl_4$ and $(BA)_2FeCl_4$ fit Eq. (1) over a wide temperature range suggests {2} ordering. The β value obtained for $(MA)_2FeCl_4$, 0.23, agrees reasonably well with that expected theoretically (0.125);[8] the value obtained for $(BA)_2FeCl_4$, 0.28, is in poorer agreement. The effects of the inexact nature of the Type II crystal structure on the critical exponent values are not well understood. It is expected, however, that crystal asymmetry will cause the β values to be increased, as is indicated by the present results.

The results of these measurements of the ^{57}Fe hyperfine field are consistent with previous measurements of the susceptibility of these two compounds.[4] For $(MA)_2FeCl_4$, the susceptibility in the vicinity of T_N was found to vary as $(T - T_N)^{-\gamma}$ above T_N and as $|T - T_N|^{-\gamma'}$ below T_N, with $T_N = 94.9$ K, $\gamma = 1.67$, and $\gamma' = 1.60$. These are to be compared with the expected values for the {2}Ising model, $\gamma = \gamma' = 1.75$.[8] In summary, the measurements reported here further substantiate the conclusion that the behavior of the compounds $(CH_3NH_3)_2FeCl_4$ and $(C_6H_5CH_2NH_3)_2FeCl_4$ approximates that of a {2}Ising model antiferromagnet.

REFERENCES

+ Present address: MacMurray College, Jacksonville, Illinois 62650.

* {d} means d-dimensional.

1 E. J. Samuelsen, J. Phys. Chem. Solids 35, 785 (1974).

2 R. J. Birgeneau et al., Phys. Rev. B1, 2211 (1970).

3 L. J. de Jongh et al., J. Appl. Phys. 40, 1363 (1969).

4 R. D. Willett and B. C. Gerstein, Phys. Letters 44A, 153 (1973).

5 M. F. Mostafa and R. D. Willett, Phys. Rev. B4, 2213 (1971).

6 A. H. Muir et al., Mössbauer Effect Data Index 1958-1965, Interscience, New York (1966).

7 W. Kündig, Nucl. Instr. and Methods 48, 219 (1967).

8 H. E. Stanley and T. A. Kaplan, Phys. Rev. Letters 17, 913 (1966).

9 E. M. Peterson and R. D. Willett, J. Chem. Phys. 56, 1879 (1972).

CRITICAL BEHAVIOR OF THE SUSCEPTIBILITY IN A TWO-DIMENSIONAL FERROMAGNET

M. Aïn and J. Hammann
Service de Physique du Solide et de Résonance Magnétique
CEN-Saclay, BP n°2, 91190 Gif-sur-Yvette
France

ABSTRACT

Magnetic susceptibility measurements on a single crystal layer type ferromagnet $CuCl_4(CH_3NH_3)_2$ have been plotted as Log χ versus Log$((T-T_c)/T_c)$ for $10^{-3} < (T-T_c)/T_c < 1$ the curve shows a crossover behavior with exponents $\gamma_0 \simeq 1.42$ and $\gamma_1 \simeq 2.3$. The nature of the crossover is found to be of dimensionality of space type. The exchange between copper planes is 5.10^{-5} to 5.10^{-4} times smaller than the exchange in the copper planes.

INTRODUCTION

During the last few years a great attention has been payed, mainly by de Jongh, Miedema and coworkers[1] to the copper compounds of general formula $CuCl_4(C_nH_{2n+1}NH_3)$. Their two-dimensional characters have been well-established by specific heat and magnetic susceptibility experiments. Among those compounds the first one, with $n = 1$, has its Cu^{2+}, $s = 1/2$ copper ions disposed in a nearly quadratic array within planes that are very weakly ferromagnetically coupled in such a manner that $R = j/J \sim 5.5\ 10^{-4}$ [1], where j is the interplanar exchange factor and J is the intraplanar exchange factor which is also ferromagnetic and corresponds to an intraplanar exchange field $H_E \sim 2.8\ 10^5$ Oe. Below the critical temperature the magnetic moments are in the planes.

During this same time great theoretical progress has been made in the field of critical phenomena leading to a better understanding of the magnetic materials behavior near their critical temperature T_c . Following these theories we will write an expression for the initial susceptibility as :

$$\chi^{-1}\ \alpha \left(\frac{T-T_c}{T_c}\right)^{\gamma} = \varepsilon^{\gamma} \quad ,$$

where γ is sensitive only to very general parameters such as d the dimensionality of space and n the dimensionality of spins, leaving to the proportionality factor the possibility of varying with more specific parameters such as crystal structure or exchange coupling.

EXPERIMENTS

We have measured the initial complex susceptibility $\chi_m = \chi'_m - i\chi''_m$ with a mutual inductance bridge at a frequency of 17 Hz. The a.c. measuring field was in the copper ion planes, with a peak value smaller than 1.5 Oe. The sample[2] of $CuCl_4(CH_3NH_3)_2$ used was a single crystal shaped into a disk of .2 mm thickness and 3.4 mm diameter. We assimilated this disk to an oblate ellipsoïd and we introduced a constant N in order to take into account the effect of the demagnetizing field

$$\chi_e = \frac{\chi_m}{1 - N\ \chi_m}$$

where χ_e and χ_m are respectively the effective and measured susceptibilities. The complex part of χ_m was negligible above T_c ; in what follows we will write χ_m for χ'_m .

Figure 1 shows the curves of $\chi^{/\!/}_m$ and $\chi^{\perp}_m$: the parallel and perpendicular susceptibilities. Since $\chi^{/\!/}_e$ diverges at $T_c = 8.779$ K , the demagnetizing factor was taken equal to $1/(\chi^{/\!/}_m \max)$, the inverse of the maximum value achieved by $\chi^{/\!/}_m$ near T_c . Notice that $\chi^{/\!/}_m$ and $\chi^{\perp}_m$ are superposed at high temperature and then split into two near T_c showing that there is an anisotropy in the copper planes which prevent $\chi^{\perp}_e$ from diverging. Also noteworthy is the fact that $\chi^{/\!/}_m$ keeps a constant value $\chi^{/\!/}_m$ max just below T_c as long as the domain walls can move freely, and then begins to decrease with temperature indicating that the walls are being quenched.

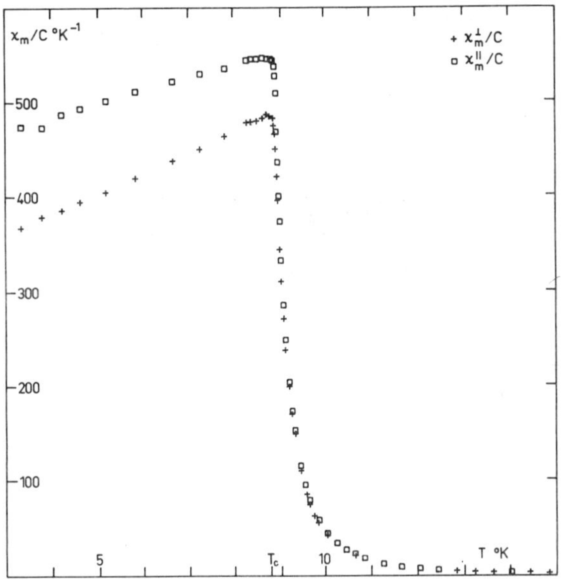

Fig.1 : Plot of $x^{\perp}_m/C$ and $x^{\parallel}_m/C$ versus T

In order to estimate the anisotropies of this $CuCl_4(CH_3NH_3)_2$ compound we performed two spin-flop experiments on our sample : with an applied field in the copper planes perpendicular to the direction of magnetization and with an applied field perpendicular to the copper planes. This gave $H^{in}_A \simeq 50$ Oe for the anisotropy in the planes and $H^{out}_A \simeq 1000$ Oe for the anisotropy out of the planes.

DISCUSSION

While this work was in progress, de Jongh and Miedema have published a plot of $\chi T/C$ versus $(T-T_c)/T$ for two compounds $CuCl_4(CH_3NH_3)_2$ and $CuCl_4(C_{10}H_{21}NH_3)_2$ (see ref.1 on page 214). For $CuCl_4(CH_3NH_3)_2$ our $T_c = 8.779$ is smaller than their $T_c = 8.91$, this discrepancy could be attributed to the dependence of the critical temperature on the magnitude of the measuring

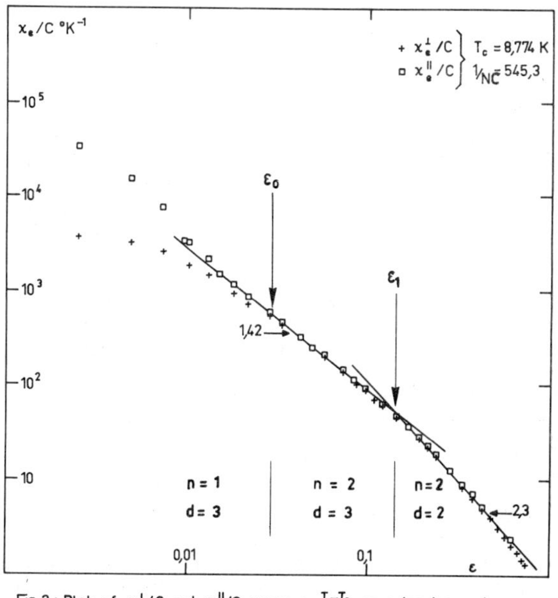

Fig.2 : Plot of $x^{\perp}_e/C$ and $x^{\parallel}_e/C$ versus $\varepsilon = \frac{T-T_c}{T_c}$ on a log-log scale

field. Otherwise a plot of χ versus ε is equivalent to the plot of $\chi T/C$ versus $(T-T_c)/T$. When $\varepsilon < 5\ 10^{-2}$ or whenever the system is two-dimensional and Ising like, still it gives a lower value of γ for temperatures high above T_c (see Fig.3, curve IV).

Figure 2 represents a plot of $\chi_e^{/\!/}$ and $\chi_e^{\perp}$ versus ε, on Fig.3, curve I is the same as $\chi_m^{/\!/}$ on Fig.2, curve II shows the effect of an upward shift of 1‰ on T_c only and curve III the effect of a variation of 2.5% on N only. These variations being of the order of our experimental uncertainties on T_c and N, we have not tried to estimate the value of γ in the last decade of ε. Curve IV is a plot of $\chi_m^{/\!/} T/T_c$ versus $(T-T_c)/T$.

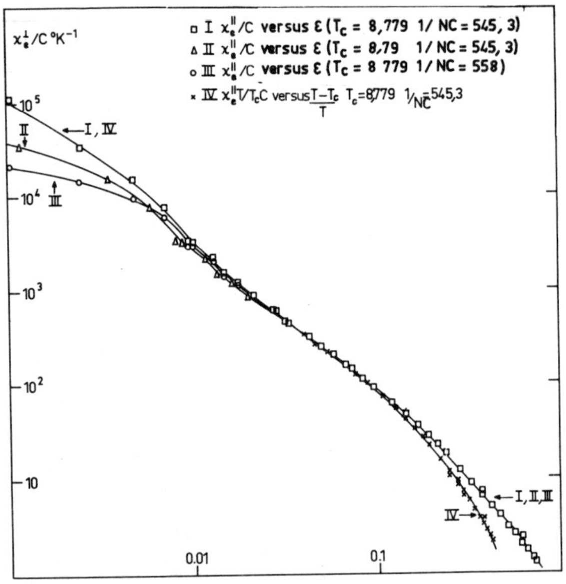

Fig.3 : Curve I is a plot of $\chi_e^{\parallel}/C$ versus ε as on fig.2. Curve II shows the dependence of Curve I on T_c, Curve III shows its dependence on N and Curve IV is a plot of $\chi_e^{\parallel} T/T_c C$ versus $T-T_c/T$.

Returning to Fig.2 one can see that the critical region of $CuCl_4(CH_3NH_3)_2$ extend on an unusually wide region above T_c. It displays a crossover behavior near $\varepsilon_1 \simeq 0.1$, from $\gamma_0 \simeq 1.42$ to $\gamma_1 \simeq 2.3$. Now the problem is to know what kind of crossover it is. To answer this question let us examine Fig.2, there we find one reliable event which is the split between $\chi_e^{/\!/}$ and $\chi_e^{\perp}$ near $\varepsilon_0 = 5.10^{-2}$, in this region we should be able to see a crossover in anisotropy (under the effect of $H_A^{in} = 50$ Oe) from an Ising model $(n=1)$ to an XY $(n=2)$; if our curves were more precise below ε_0. Nevertheless above ε_0 we can trust our exponents to 5%, and we think that $\gamma_0 \simeq 1.42$ stands for the γ of a three dimensional XY system (calculation for $d=3$, $n=2$ gives $\gamma = 1.35 \pm 0.03$ [6], note that it also gives $\gamma = 1.44 \pm 0.01$ [7] for $d=3$, $n=3$ but a three dimensional Heisenberg system would not be consistent with the sequel in our demonstration). Following $\chi_e^{/\!/}$ with ε we reach ε_1, the second crossover region which we identify to be a crossover in dimensionality $(d=3\ ,\ n=2) \rightarrow (d=2\ ,\ n=2)$ for two reasons : first the exponent $\gamma_2 = 2.3$ is comparable to the values obtained by series expansion techniques for a two-dimensional system ; second this crossover cannot be attributed to the anisotropy H_A^{out}; it occurs much too close to the first one considering the perpendicular anisotropy $H_A^{out} \sim 1000$ Oe.

We can now calculate an order of magnitude for R. We start with the expression of Liu and Stanley[4] :

$$\chi_{d+1} = \chi_d \left[1 + 2\frac{RJ}{k}\frac{\chi_d}{C} + \left(2\frac{RJ}{k}\frac{\chi_d}{C}\right)^2 f\left(\frac{RJ}{k}\right) \right] , \qquad \text{(a)}$$

C is the Curie constant and $1 < f(\ ) < 2$: where χ_{d+1} corresponds to our curve $\chi_e^{/\!/}$; χ_d is the susceptibility for the two-dimensional XY system which can be approximated by fitting our experimental points above ε_1 to a law such as

$$\frac{C}{\chi_d} \simeq A \left(\frac{T-T_{co}}{T_{co}} \right)^{\gamma} . \qquad \text{(b)}$$

We found $A = 1.1$ K $T_{co} = 8.37$ K $\gamma = 2.368$. Note that $T_{co} < T_c$ in accordance with the scaling approximate result $T_c - T_{co} \propto R^{1/\gamma}$ where T_c is the critical temperature of the three dimensional system and T_{co} the critical temperature of the two dimensional system. Relations (a) and (b) gave values for R spread between 5.10^{-5} and 5.10^{-4} for several temperatures in the crossover region. These values are of the same order of magnitude than those found in the literature[1,5].

CONCLUSION

It seems worthwhile to perform the measurement of $\chi_{\perp}$ perpendicularly to the copper planes in order to locate the crossover region due to the perpendicular anisotropy H_A^{out}. This amounts to a new way of checking our present interpretation.

We wish to thank Drs. Sarma, Toulouse and Pfeuty for very fruitfull suggestions.

References

1. de Jongh, Miedema, Advances in Physics 23, n°1 (1974)
2. The sample has been kindly given to us by J.P. Renard, Lab. d'Electronique, Faculté des Sciences Paris-Sud, 91405 Orsay, France.
3. Actually $\chi(T)^{-1} \propto T(1 - T_c/T)^{7/4}$ for the magnetic susceptibility of a plane Ising model, see M.E. Fisher, Physics 25, 521 (1959).
4. Yamasaki, Proc. Int. Conf. Magnetism, Moscow (1973).
5. L.L. Liu, H.E. Stanley, Phys.Rev. B 8, 2279 (1973).
6. D.D. Betts, C.J. Elliot, M.H. Lee, Can. J. Phys. 48, 1566 (1970).
7. G.A. Baker Jr. H.E. Gilbert, J. Eve and G. Rushbrooke, Phys. Rev. 164, 800 (1967).

EFFECT OF RANDOMNESS ON CRITICAL BEHAVIOR OF SPIN MODELS*

T. C. Lubensky and A. B. Harris
Department of Physics, University of Pennsylvania, Philadelphia, Pa. 19174

ABSTRACT

Renormalization group methods are used to analyze the critical behavior of random Ising models. The Wilson-Fischer ϵ-expansion for the recursion relations for n-component continuous spin models are developed for randomly inhomogeneous systems. In addition to the usual variables for a homogeneous system there appears a variable which in essence describes local fluctuations in T_c. From the structure and stability of the fixed points we conclude that critical exponents are unaffected by randomness for $n > 4$ but are renormalized by randomness for $1 < n < 4$. In both cases $\alpha < 0$, as expected from a simple physical argument.

It is well known that uniform magnetic systems undergo sharp phase transitions with divergent susceptibilities. If, however, the system is randomly diluted, or if the interactions between spins are randomized, the situation is less clear. Is the transition sharp or smeared? If the transition is sharp, are the exponents the same as for the homogeneous system or are they renormalized? High temperature expansions[1] seem unable to answer these questions. An exact solution[2] for a special two-dimensional random Ising model predicts a smeared transition. However, in view of the long range correlations in the randomness of this special model, it is not clear whether the results represent behavior typical of local randomness. In view of these uncertainties it is natural to try to clarify the situation using renormalization group techniques which have been so successful in calculating critical properties of homogeneous systems.[3] Two formulations of the renormalization group suitable for this purpose are the cluster expansion for discrete spins given by Niemeyer and van Leeuwen[4] and the ϵ-expansion for continuous spins of Wilson and Fisher.[3] Previously[5] we outlined the general scheme for applying the renormalization group to systems with random potentials. Since most of that discussion described the discrete-spin method, we will confine the present discussion to the continuous spin technique. Results to first order in ϵ will be given here; higher order results will be presented elsewhere.

We begin with the reduced Hamiltonian

$$\mathcal{H} = \tfrac{1}{2}\int V_2(\vec{q}_1,\vec{q}_2)\,\vec{S}_1\cdot\vec{S}_2\,dq_1\,dq_2 + \int V_4\,\vec{S}_1\cdot\vec{S}_2\,\vec{S}_3\cdot\vec{S}_4\,dq_1\,dq_2\,dq_3\,dq_4, \qquad (1)$$

where $\vec{S}_n = \vec{S}(\vec{q}_n)$ is the Fourier transform of an n-component vector field, $\int dq \equiv (2\pi)^{-d}\int d^d\vec{q}$, where the integration is over a sphere of radius Λ, and V_2 and $V_4 \equiv V_4(q_1,q_2,q_3,q_4)$ are arbitrary random potentials for an inhomogeneous system governed by a probability distribution P. We then develop recursion relations for the inhomogeneous potentials in the standard way:[6]

$$\{V'_\ell\} = R\{V_\ell\} \equiv R_s\,R_b\,\{V_\ell\}, \qquad (2)$$

where R_b represents the removal of all spin degrees of freedom with $b^{-1}\Lambda < |\vec{q}| < \Lambda$ and R_s represents a scale change $\vec{q} \to b\vec{q}$ and a spin renormalization $\vec{s} \to \zeta\vec{s}$. As shown in Fig. 1, R_b can be developed diagrammatically as in the homogeneous case. As discussed in Ref. 4, Eq. (2) gives rise to recursion relations for the probability distribution:

$$P'(\{V'_\ell\}) = \int \delta(\{V'_\ell\} - R\{V_\ell\})P(\{V_\ell\})d\{V_\ell\} \qquad (3)$$

where the integral is over all degrees of freedom in $\{V_\ell\}$. Thus in the random problem, one seeks a fixed point for the probability distribution $P(\{V_\ell\})$ rather than for the potentials.

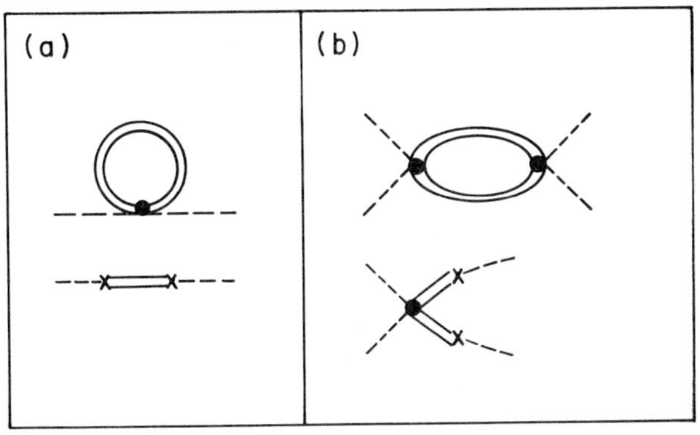

1. Diagrams for V_2' (a) and V_4' (b) to leading order in V_4 and δV_2. The double line represents the Gaussian propagator for the inhomogeneous system $[V_2(q,q')]^{-1}$. To obtain Eqs. (6) use $V_2^{-1} = \langle V_2\rangle^{-1} - \langle V_2\rangle^{-1}\,\delta V_2\langle V_2\rangle^{-1}$ ... in these diagrams and average the resulting equations using $\langle \delta V_2\,\delta V_2\rangle = \Delta(q_1+q_2+q_3+q_4)$.

It is obvious that $P(\{V_n\}) = \delta(\{V_n\} - \{V_n^*\})$ is a fixed point of Eq. (3) if $\{V_n^*\}$ is the fixed point value of $\{V_n\}$ for the homogeneous system. To study systems with narrow probability distributions, we develop recursion relations for the cumulants, $\langle V_\ell\rangle$ $\langle V_\ell V_m\rangle - \langle V_\ell\rangle\langle V_m\rangle$ etc. of P. The averaging process restores translational invariance, so we can write $V_2(\vec{q}, \vec{q}') = \langle V_2(\vec{q}, \vec{q}')\rangle + \delta V_2(\vec{q}, \vec{q}')$, where

$$\langle V_2(\vec{q}, \vec{q}')\rangle = (r + q^2)\,\delta^d(\vec{q} + \vec{q}'). \qquad (4)$$

The spin renormalization coefficient ζ is then chosen so that the coefficient of q^2 in Eq. (4) remains unity after each iteration (i.e. $\zeta = b^{1+\frac{1}{2}d-\frac{1}{2}\eta}$). In the long wavelength limit, we can also write

$$\langle V_4(\vec{q}_1,\vec{q}_2,\vec{q}_3,\vec{q}_4)\rangle = u\,\delta^d(\vec{q}_1+\vec{q}_2+\vec{q}_3+\vec{q}_4) \qquad (5a)$$

$$\langle \delta V_2(\vec{q}_1,\vec{q}_2)\,\delta V_2(\vec{q}_3,\vec{q}_4)\rangle = \Delta\delta^d(\vec{q}_1+\vec{q}_2+\vec{q}_3+\vec{q}_4). \qquad (5b)$$

If there are no long range correlations in the random potentials, Δ will be a constant in the long wavelength limit. Thus Δ behaves like a four-spin potential and must be treated on the same level as u in the recursion relation. All other cumulants and momentum dependences are irrelevant variables near four dimension for the same reason that u_6 and q-dependent corrections to u are irrelevant in the homogeneous case.[6]

To first order in $\epsilon = 4-d$ the recursion relations are

$$r' = b^{2-\eta}\{r - A(r)[4(n+2)u - \Delta]\} \qquad (6a)$$

$$u' = b^{\epsilon-2\eta}\{u - K\ln b[4(n+8)u^2 - 6u\Delta]\} \qquad (6b)$$

$$\Delta' = b^{\epsilon-2\eta}\{\Delta - K\ln b[8(n+2)u\Delta - 4\Delta^2]\}, \qquad (6c)$$

where $K^{-1} = 2^{d-1}\pi^{\frac{1}{2}d}\Gamma(\tfrac{1}{2}d)$ and $A(r) = \int_{\Lambda/b}^{\Lambda}(q^2+r)^{-1}d^d\vec{q}$. The analysis of these equations proceeds exactly as for a spin system with a hypercubic potential.[7] There are four fixed points to first order in ϵ, and the flow diagram showing their stability is given in Fig. 2. They are

1) a Gaussian fixed point with $u^* = \Delta^* = 0$;
2) a Heisenberg fixed point with $u^* = \epsilon/[4K(n+8)]$, $\Delta^*=0$, $\lambda_\Delta = \epsilon(4-n)/(n+8)$, $\alpha = \frac{1}{2}\epsilon(4-n)/(n+8)$;
3) an unphysical fixed point with $u^*=0$, $\Delta^* = -\frac{1}{4}K^{-1}\epsilon$, $\lambda_\Delta = -\epsilon$, $\lambda_u = -\frac{1}{2}\epsilon$, $2\nu = 1+\epsilon/8$;

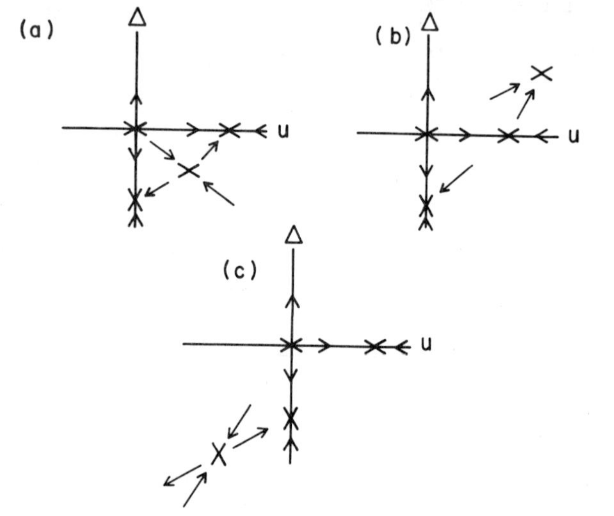

2. Fixed point flow diagrams for a) $n > 4$, b) $1<n<4$, and c) $n < 1$.

4) a randomness dominated fixed point with $u^* = \epsilon/[16K(n-1)]$, $\Delta^* = \epsilon(4-n)/[8K(n-1)]$, $\lambda_1 = -\epsilon$, $\lambda_2 = \frac{1}{4}(n-4)\epsilon/(n-1)$, $2\nu = 1+3n\epsilon/[16(n-1)]$, $\alpha = \epsilon(n-4)/[8(n-1)]$.

Our conclusions are therefore: A) the third fixed point is always stable but can never be reached, since physically Δ must be positive. B) If $n > 4$, the Heisenberg fixed point is stable, in particular, with respect to turning on a small amount of randomness. We interpret this to mean that for $n > 4$, there is a sharp phase transition in the random system with the same exponents as in the homogeneous system. C) For $1 < n < 4$, the random fixed point is stable. At this fixed point Δ is non zero and the exponents differ from those of the homogeneous system. D) for $n < 1$, there is no stable fixed point with u and Δ positive. This presumably corresponds to a transition which is different from the usual second order one. The behavior for n near unity is not well understood yet.

A heuristic argument by one of us[8] predicts that there can be a sharp transition only if the specific heat exponent α is negative. Note that conclusions B and C are in accord with this argument inasmuch as α is negative in both cases. Intuitively, making n large decreases the effect of randomness because the number of degrees of freedom is increased.

A second order (in ϵ) calculation of the stability of the Heisenberg fixed point gives

$$\lambda_\Delta = [(4-n)/(n+8)]\epsilon - [(n+2)(13n+44)/(n+8)^3]\epsilon^2 \quad (7)$$

and to order ϵ^2 we may write this as $\lambda_\Delta = \alpha/\nu$.[9] Thus the Heisenberg fixed point is never stable with respect to randomness when α is positive in agreement with the heuristic argument.

The above results can also be obtained by a formulation due to Emery.[10] In his method one studies the free energy, F_R, of the random model with a Hamiltonian

$$\mathcal{H}_R = \Sigma_{rs} J_{rs} \Sigma_\alpha S_{r\alpha} S_{s\alpha} + v\Sigma_r \Sigma_{\alpha,\beta} S_{r\alpha}^2 S_{r\beta}^2 - \Sigma_r \psi_r\, g(S_{r1}, S_{r2}, \dots S_{rn}), \quad (8)$$

where r and s are spatial indices, α and β are component labels and are summed from 1 to n, and ψ_r is a random variable governed by the distribution function $P(\psi_r)$. Emery[10] shows that F_R is the same as the free energy F_e associated with the Hamiltonian

$$\mathcal{H}_e = \Sigma_{rs} \Sigma_\alpha \Sigma_k J_{rs}\, x_{r\alpha k}\, x_{s\alpha k} + v\Sigma_r \Sigma_{\alpha,\beta} \Sigma_k\, x_{r\alpha k}\, x_{r\alpha k}\, x_{r\beta k}\, x_{r\beta k} + \Sigma_r\, f[i\Sigma_m\, g(x_{r1k}, x_{r2k}, \dots x_{rnk})], \quad (9)$$

where $\vec{x}_r$ is a vector variable with components $x_{r\alpha k}$, $f(s) = -\ln[\int P(z) e^{-isz}\, dz]$, the component label k is summed from 1 to m, and the limit $m \to 0$ is taken. We take $g(S_{r1}, S_{r2}, \dots S_{rn}) = \sum_\alpha (S_{r\alpha})^2$. We now develop recursion relations for Hamiltonians of the form $\mathcal{H}_e$. Since terms of sixth order in x are irrelevant[7], we replace $f(s)$ by its expansion up to order s^2: $f(s) = \frac{1}{2}\overline{\psi^2} s^2$, where $\overline{\psi^2} > 0$ is the average value of ψ^2. Thus the last term in Eq. (9) is of the form $w\Sigma_r(\Sigma_{k\alpha} x_{r\alpha k}^2)^2$. Then the recursion relations for the generalized hypercubic model[11] follow:

$$r' = b^{2-\eta}\{r - A(r)[(4n+8)v + (4mn + 8)w]\} \quad (10a)$$

$$v' = b^{\epsilon-2\eta}\{v - K \ln b[4(n+8)v^2 + 48\, vw]\} \quad (10b)$$

$$w' = b^{\epsilon-2\eta}\{w - K \ln b[8(n+2)vw + 4(nm+8)w^2]\}. \quad (10c)$$

In the limit $m \to 0$, these relations reproduce Eq. (6), if the identifications $v = u$ and $w = -\Delta/8$ are made.[12]

If $V_2(x,x')$, where x is a position coordinate, is constrained to be constant within a p-dimensional subspace, then Δ in Eq. (5b) will be proportional to $\delta^P(\vec{k}_1 + \vec{k}_2)$, where $\vec{k}_1$ is the projection of $\vec{q}_1$ onto the p-dimensional subspace. In this case the recursion relations yield $\Delta' \sim b^{\epsilon+p}\Delta$ and all fixed points are unstable with respect to randomness within the ϵ expansion. This may explain why the "striped" randomness treated in Ref. 2 leads to a broadened transition, whereas the renormalization group treatment given elsewhere[5] suggests a sharp transition. This result also suggests that the transition for $n < 1$ (see D above) may be a broadened one.

REFERENCES

* Supported in part by the National Science Foundation.
1. Rapaport, D. C. J. Phys. C 5, 1830-1858 (1972).
2. McCoy, B. M. and Wu, T. T. Phys. Rev. Lett. 23, 383-386 (1968).
3. Wilson, K. G., Phys. Rev. B4, 3174-3205 (1971); Wilson, K. G. and Fisher, M. E., Phys. Rev. Lett. 28, 240-243 (1972).
4. Niemeyer, T. and van Leeuwen, J. M. J., Physica 71, 17-40 (1974).
5. Harris, A. B. and Lubensky, T. C., Phys. Rev. Lett. (in press).
6. Wilson, K. G. and Kogut, J., (in press).
7. Aharony, A., Phys. Rev. B 8, 4270-4273 (1973).
8. Harris, A. B., J. Phys. C 7, 1671-1692 (1974).
9. We would like to thank Dr. A. Aharony for clarifying this relation.
10. Emery, V., to be published.
11. Brezin, E., Guillou, J. C. and Zinn-Justin, J. Phys. Rev. B 10, 892-900 (1974).
12. Grinstein, G. and Luther, A. (to be published) have also obtained Eq. (10).

CRITICAL BEHAVIOR OF A RANDOM m-VECTOR MODEL

G. Grinstein*
Gordon McKay Laboratory
Harvard University, Cambridge, Mass. 02138

ABSTRACT

The critical behavior of an m-vector model with random impurities is investigated through the use of renormalization group techniques near four dimensions. For almost all values of m, this disordered model has two stable fixed points and a single nontrivial, unstable fixed point. The model exhibits sharp phase transitions with exponents which do not depend continuously on the concentration of impurities. The familiar m-component fixed point, which is normally assumed to characterize the critical behavior of the pure m-vector model, is unstable whenever m<4.

THE EFFECTIVE HAMILTONIAN: FIXED POINTS

The problem of magnetic phase transitions in alloys is a formidable one both experimentally and theoretically. Neither the conditions under which disordered systems exhibit sharp phase transitions,[1,2] nor the dependence of the exponents on impurity concentration has been rigorously elucidated.

The most powerful and exciting new method for handling problems of critical phenomena in pure systems is renormalization group analysis.[3,4] In an attempt to bring this technique to bear upon the random problem we investigate a simple, disordered m-vector model from the renormalization group point of view.

Consider a randomly dilute m-vector model, defined by allowing each site of a lattice in d-dimensions to be either occupied by a classical, m-component spin, or empty.[5] The probability of occupation of any site is given by the number p. We consider ferromagnetic interactions between nearest neighbor spins. In a separate publication,[6] we show how to construct a translationally invariant, "effective" Hamiltonian, $H\{\phi\}$, whose critical behavior is formally identical to that of the disordered model. This effective Hamiltonian has the familiar "continuous spin", or field theoretic, form.[3] Not surprisingly, it contains many-body interactions of arbitrarily high order, with coefficients which are polynomial functions of p. Fortunately, all of these interactions are explicitly short-ranged (in fact point) interactions. This fact, together with its translational invariance, makes H directly amenable to renormalization group analysis.

Only terms up to ϕ^4 are "relevant"[3] for determining the fixed points of the effective Hamiltonian in 4-ε dimensions. These terms can be written

$$H\{\phi\} = \int d^dx\{\frac{1}{2}\sum_{\alpha\beta} [(\nabla\phi^\alpha_\beta)^2 + r_o(\phi^\alpha_\beta)^2] + wp\sum_{\substack{\alpha\alpha'\\ \beta}} (\phi^\alpha_\beta)^2(\phi^{\alpha'}_\beta)^2 - up(1-p)\sum_{\substack{\alpha\alpha'\\ \beta\beta'}} (\phi^\alpha_\beta)^2(\phi^{\alpha'}_{\beta'})^2\} \quad (1)$$

where u and w are positive quantities which depend (in a simple, computable manner) on the temperature. Note that the "order parameter", ϕ^α_β, has two separate component labels. Upper labels (α) are to be summed from 1 to m, lower labels (β) from 1 to n. In order to preserve the equivalence between the $H\{\phi\}$ and the original, dilute m-vector model, we must set the parameter n equal to zero. This constraint[7,8] is to be understood as a calculational prescription: Compute correlation functions, critical exponents, etc., from $H\{\phi\}$ as a function of m and n. Then set n equal to zero in order to obtain results for the disordered m-component system.

Hamiltonian (1) is a rather familiar one. Brézin et al.[9] have applied renormalization group techniques to the continuum field theory form of this Hamiltonian in 4-ε dimensions. They find four fixed points whose properties are summarized in Table I.

TABLE I: FIXED POINTS OF THE EFFECTIVE HAMILTONIAN

Fixed Point	Exponents	Region of Stability
Gaussian	Mean field	Never
m-component	m-vector model exponents	$m \geq 4$
mn-component	mn-vector model exponents	$mn < 4$
"Extra"	$\gamma = 1 + \frac{3m(n-1)\varepsilon}{m^2n+8mn-16m+16}$	$\frac{(4-mn)(m-4)}{m^2n+8mn-16m+16} > 0$

The first two nontrivial fixed points in Table I give rise to well known critical exponents which have been calculated to $O(\varepsilon^3)$.[10] The last, or "extra", fixed point gives rise to a less familiar set of indices. Brézin et al.[9] have pointed out that for each pair of positive integral values of m and n, only one of the four fixed points is stable. Of course, only the n=0 case is of direct interest to us here. It is clear from Table I that this case is rather unique. Two fixed points are stable for almost all values of m. The most striking feature of Table I is that the m-component fixed point, which characterizes the critical behavior of the pure m-vector model, is unstable when n=0 and m<4. The exponents of the Ising, XY, and Heisenberg models are therefore altered by the presence of impurities. The clear presumption[11] is that the critical behavior of the dilute m-component system is governed by the m-component fixed point for $m \geq 4$ and the "extra" fixed point for 1<m<4.

We have yet to face the thorny global question of which (if any) of the available fixed points is actually attained by the system for a given value of p. A rigorous analytic solution of the global problem seems out of the question, even for the grossly truncated effective Hamiltonian of equation (1). Nonetheless, one can make some qualitative statements about the special case for which n=0 and m=1. This case, the random Ising model, is unique in that the "extra" fixed point does not occur at all. Only the mn-component fixed point is stable; this constitutes a considerable simplification. We therefore restrict our discussion to n=0 and m=1 from here on.

The u term of Hamiltonian (1) is easily seen to be negative for all values of p. The w term, on the other hand, is proportional to p and is always positive. For values of p sufficiently close to unity, the w term is large enough to ensure that the partition function arising from H is well-defined. So although the negative sign of u is not disastrous, it does suggest that first order and tricritical behavior may occur. It is natural to include a hitherto omitted ϕ^6 term in H to consider this possibility. We write the resulting Hamiltonian as:

$$\mathcal{H} = \int d^dx\{\frac{1}{2}\sum_\beta[(\nabla\phi_\beta)^2 + r_o(\phi_\beta)^2] + w_o\sum_\beta(\phi_\beta)^4 + u_o\sum_{\beta\neq\beta'}(\phi_\beta)^2(\phi_{\beta'})^2 + v_o\sum_{\beta\beta'\beta''}(\phi_\beta)^2(\phi_{\beta'})^2(\phi_{\beta''})^2\} \quad (2)$$

where we have suppressed the upper component labels on the fields since m=1. It can furthermore be shown[6] that v_o is proportional to p(1-p)(2p-1).

When u_o and v_o are zero (as is the case for p=1) Hamiltonian (2) gives rise to critical behavior characterized by the Ising fixed point. With respect to the Ising fixed point, one can imagine (in the terminology of Riedel and Wegner[12]) computing the scaling densities and determining the corresponding scaling fields.

Since the Ising fixed point is known[4] to be unstable with respect to operators of the form $(\phi^2)^2$ in n-component theories, two of these scaling fields will be relevant. (In the presence of a magnetic field there would be three relevant scaling fields.) Hence the Ising fixed point is a tricritical fixed point[13] of the system. The occurrence of this non-Gaussian tricritical point is somewhat unusual, although indications of such behavior can be found in the literature.[4,14] It is interesting to note that there are three relevant scaling fields with respect to the Gaussian fixed point in the present model. As will be discussed shortly, however, the Gaussian fixed point is inaccessible to the true random Ising system.

The vanishing of two relevant scaling fields determines the location of the Ising tricritical point. As usual,[12] this requirement places two conditions on the three parameters r_o, u_o, and v_o. One of these conditions merely sets the temperature to its critical value; the other is a relation between u_o and v_o of the form

$$u_o = f(v_o) \quad . \tag{3a}$$

When both u_o and v_o are small (which is true for p near unity) the relation is linear:

$$u_o + \alpha v_o = 0 \quad . \tag{3b}$$

Here α is a positive constant which is readily calculable to $O(\varepsilon)$. If $(u_o+\alpha v_o)$ is positive, the system is in the regime of the critical line and undergoes a second order phase transition characterized by the mn-component fixed point. Since n is zero, the critical exponents have their n=0 values along this line. If $(u_o+\alpha v_o)$ is negative, the system presumably undergoes a first order phase transition.

Recall now that r_o,u_o, and v_o are not arbitrary, but are known functions of the concentration, p, and the temperature, T. The two conditions which determine the Ising tricritical point thus fix tricritical values (T_t,p_t) for these parameters. (It is for this reason that the Gaussian fixed point is inaccessible to the random system. There one must fulfill three conditions on p and T, an obvious impossibility.) One has no guarantee that p_t will fall in the physically meaningful region: $p\leq 1$. The quantities u_o and v_o are both nonlinear functions of p and T. Furthermore, the critical temperature is undoubtedly a nonlinear function of p. All told, then, equation (3a) is a complicated, highly nonlinear equation for p. P=1 is a trivial solution of this equation. We have been unable to find a second solution for p very near unity, where equation (3b) is valid. This does not eliminate the possibility that (3a) has a solution for some value of p less than one. If such a solution does exist, the phase diagram of the random Ising system would have the qualitative behavior shown in figure 1. If not, we conclude that the phase transition is first order for p<1. We have been unable to construct a convincing argument in favor of one or the other of these two alternatives. Although on physical grounds it seems rather unlikely, one cannot even rule out the possibility that there is more than one solution for p<1.

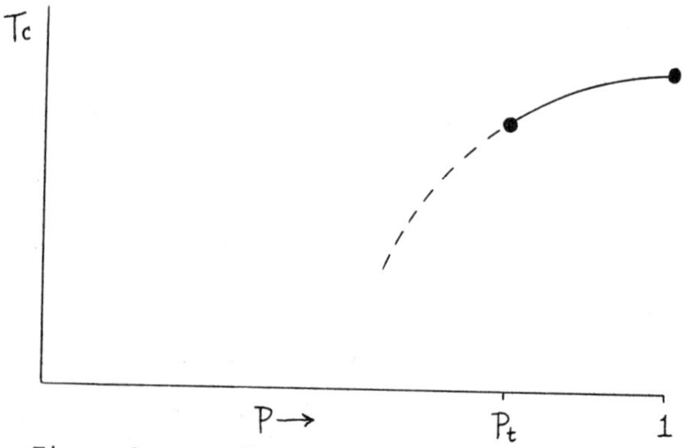

Figure 1. Possible phase diagram of the dilute Ising model. Heavy dots and solid line indicate second order phase transitions with Ising and "n=0" exponents, respectively. Dotted line shows first order transition.

The foregoing discussion has been based on the simple Hamiltonian (2). The complete effective Hamiltonian actually contains an infinite number of higher order terms which have been summarily omitted from $\mathcal{H}$. While their presence does not alter the fixed points of the theory, it will undoubtedly affect the precise location of the tricritical point (or points) as a function of p. It is therefore somewhat futile to try to sharpen our highly qualitative remarks about the nature of the random Ising model phase diagram. The details depend on too many high order operators.

CONCLUSIONS

The effective Hamiltonian obtained for the randomly dilute m-vector model exhibits many-body interactions which are explicitly short-ranged. (Such effective Hamiltonians are characteristic[6,8] of disordered systems wherein there is no long range correlation between the random variables associated with different lattice sites.) Except for the n=0 constraint, therefore, this Hamiltonian looks very much like Hamiltonians which are used to describe pure systems. Not surprisingly, it gives rise to sharp phase transitions and familiar critical exponents. Critical indices do not depend continuously on impurity concentration. Qualitatively, then, the dilute m-vector model behaves very much like models of pure magnets. We believe that similar statements apply to a rather broad class of disordered systems.

ACKNOWLEDGMENTS

I would like to thank Professor A.Luther for invaluable contributions to this work. Numerous conversations with Professor P.C. Martin and Dr. V.J.Emery have been tremendously helpful. The financial support of the National Research Council of Canada is gratefully acknowledged.

* Present Address: Department of Physics,University of Illinois, Urbana, Illinois 61801.

REFERENCES

1. For an example of a random model which undergoes a sharp phase transition, see F. Matsubara, Prog. Theor. Phys. 51, 378 (1974).
2. That such rounding does occur in some random models has been explicitly demonstrated by B. M. McCoy and T. T. Wu, Phys. Rev. 176, 631 (1968).
3. K. G. Wilson and J. Kogut, Phys. Reports [in press] (1974).
4. K. G. Wilson and M. E. Fisher, Phys. Rev. Lett. 28, 240 (1972).
5. See R. B. Griffiths and J. L. Lebowitz, J. Math. Phys. 9, 1284 (1968).
6. G. Grinstein and A. Luther (to be published).
7. The n=0 limit was first used by P. G. de Gennes, Phys. Lett. 38A, 339 (1972).
8. The relevance of the n=0 limit to very general random systems has been elucidated by V. J. Emery (to be published).
9. E. Brézin, J. C. Le Guillou, and J. Zinn-Justin, Phys. Rev. B10, 892 (1974).
10. E. Brézin, J. C. Le Guillou, J. Zinn-Justin, and B. G. Nickel, Phys. Lett. 44A, 227 (1973).
11. This conclusion has been reached independently by A. B. Harris and T. C. Lubensky (to be published).
12. E. K. Riedel and F. J. Wegner, Phys. Rev. Lett. 29, 349 (1972).
13. Here (and subsequently) we define hypercritical points solely in terms of the number of relevant scaling fields. See, for example, E. K. Riedel, AIP Conf. Proc. 10, 865 (1973).
14. F. Harbus and H. E. Stanley, Phys. Rev. Lett. 29, 58 (1972).

RENORMALIZATION-GROUP CALCULATION OF CRITICAL EXPONENTS IN THREE DIMENSIONS*

G. R. Golner and E. K. Riedel
Department of Physics, Duke University, Durham, North Carolina 27706

ABSTRACT

A scaling-field representation of Wilson's exact renormalization-group equation is derived and used to calculate critical exponents for the continuous-spin Ising model in three dimensions. Here the results for two successive truncations of the infinite hierarchy of scaling-field equations are presented. The truncation that retains only the two most relevant scaling fields yields the critical exponents $\nu \approx 0.55$ and $\eta = 0$; the truncation retaining the four most relevant fields yields the improved values $\nu \approx 0.60$ and $\eta \approx 0.06$. Approximations to the full hierarchy of scaling-field equations introduce a weak dependence of the critical exponents on redundant parameters in the renormalization-group Hamiltonian. That question is also discussed.

IDEA

Wilson's renormalization group approach[1,2] provides a framework for remarkably simple analytical calculations of critical phenomena near the molecular-field limit (ε-expansion about d = 4 dimensions) and the spherical-model limit (1/n expansion).[3] However, for a precise understanding of critical behavior in three dimensions or, for example, an accurate renormalization-group calculation of Ising model critical exponents, the methods are of limited usefulness due to the extrapolations they require for d = 3 or n = 1. In this paper we present preliminary results on what promises to be a convergent method for studying critical behavior in three dimensions. The idea is to transform Wilson's exact functional equation into a more tractable hierarchy of ordinary differential equations for scaling fields.[4] Then approximations are generated by truncating the hierarchy of equations according to the degree of relevance of the fields to be retained. Two successive truncations are presented below.

METHOD

The eigenfunctionals of the Gaussian tricritical fixed point in three dimensions[2,5] provide an operator basis Q_i in the space of effective renormalization-group Hamiltonians H_ℓ, which allows the expansion[6]

$$H_\ell = H^* + \sum_i \mu_i(\ell) Q_i . \qquad (1)$$

Hence, the evolution of H_ℓ as a function of ℓ can be equivalently described by the evolution of the expansion coefficients $\mu_i(\ell)$. In Eq. (1) and in the following the subscript i denotes the pair of indices $i = \{m,p\}$ which, following Wilson's notation,[2] characterize the Gaussian basis operator having canonical index $y_{m,p} = 3 - (\frac{1}{2} m + p)$. By substituting the expansion (1) into the exact Wilson equation [Eq. (XI. 17) of Ref. 2] and projecting all terms onto the basis Q_i via an operator product expansion the Wilson equation is transformed into an infinite hierarchy of coupled differential equations,[7]

$$\frac{d\mu_i}{d\ell} = y_i\mu_i + \sum_{jk} a_{i,jk}\, \mu_j\mu_k + \Delta \{ \sum_j a_{i,j}\, \mu_j + a_i \}. \qquad (2)$$

The quantity Δ allows the proper normalization of each fixed point Hamiltonian and determines the critical exponent η associated with that fixed point. The projection coefficients a in Eq. (2) are ℓ-independent constants that are expressed as closed form integrals using the operator product expansion and evaluated numerically.[7] Approximations to the hierarchy of equations (2) are generated by truncating as described above. At this point the implicit assumption is made that the expansion of Eq. (1) converges and that the sequence in relevance of the critical operators does not drastically differ from that of the Gaussian basis. We remark on this in the following section.

RESULTS

We have analyzed the fixed point properties, and associated critical exponents, of two truncated sets of scaling-field equations: (i) the coupled equations for the two most relevant scaling fields μ_{20}, μ_{40}, and (ii) the coupled equations for the four most relevant scaling fields μ_{20}, μ_{40}, μ_{60}, μ_{22}. Some of the results of these investigations are shown in Figs. 1 and 2.

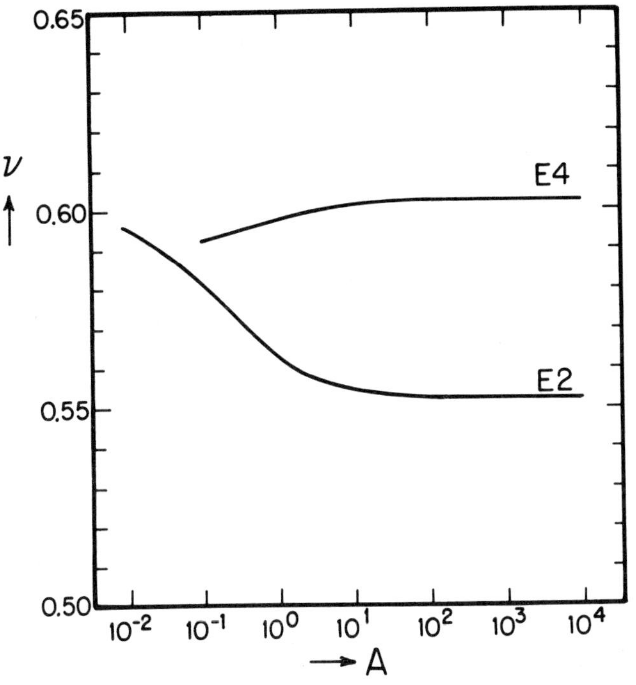

Figure 1. Critical exponent ν as a function of A for the sets of two scaling-field equations (E2) and four scaling field equations (E4).

The variable A denotes the normalization parameter of the kinetic energy term in the initial Hamiltonian

$$H_0 = -\frac{1}{2} A \int d^3q\, q^2 \sigma_q \sigma_{-q} + \cdots \qquad (3)$$

The normalization A enters Eq. (2) via the projection coefficients a. The scaling-field equations yield consequently lines of fixed points parametrized by A.

The critical exponent ν as a function of A is shown in Fig. 1 for the sets of two equations (lower curve) and four equations (upper curve). First, the two successive truncations improve the value of the exponent ν compared to the mean-field result, ν = 0.5, giving ν ≈ 0.55 and 0.60, respectively. Second, over the interval $10^{-1} \leq A \leq 10^4$ the exponent ν varies by 11% for the first truncation but only by 1.7% for the second truncation. Therefore, the exponent calculated from the truncated equations depends on the "redundant" parameter A, but only weakly.[8]

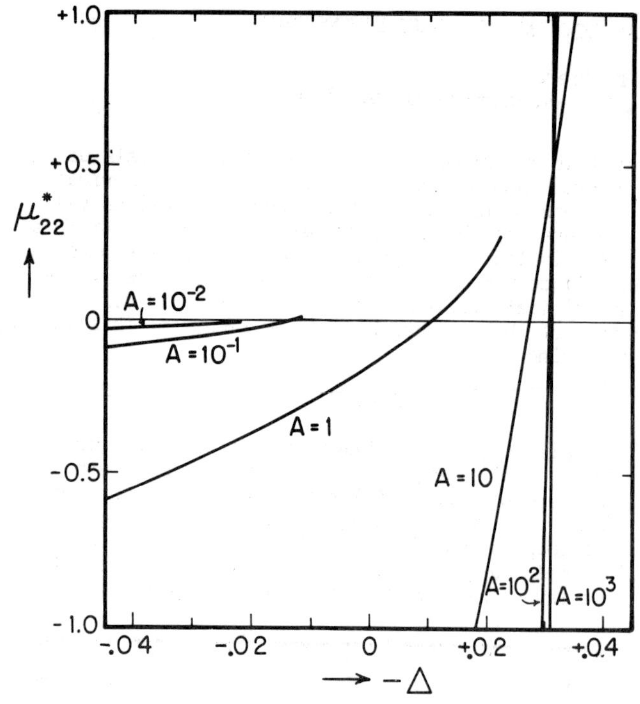

Figure 2. Fixed point coordinate μ^*_{22} as a function of $(-\Delta)$ for six values of A. The lines of fixed points for $A = 10^{-2}$, 10^{-1}, and 1 terminate as shown.

The determination of the critical exponent η for the four coupled scaling-field equations is shown in Fig. 2. We find a critical fixed point for a range of values for the parameter Δ. The figure exhibits the fixed point coordinate μ^*_{22} as a function of $-\Delta$ for six values of A. The fixed point value $\Delta = \Delta^*$ is determined by fixing the normalization of the critical fixed point Hamiltonian as in Eq. (3). That implies the fixed point condition $\mu^*_{22}(\Delta^*) = 0$, which, since[2]

$$\eta = -2\Delta^* , \qquad (4)$$

determines the critical exponent η. We note that with increasing A the solutions of $\mu^*_{22}(\Delta^*) = 0$ converge towards a common value $\Delta^* \approx -0.031$, which according to Eq. (4) yields $\eta \approx 0.062$. For small A the value of η varies considerably as a function of A, and no solution exists for $A \lesssim 10^{-2}$. We have tested the consistency of our value of η by calculating it also from the critical exponent, $y_{10} = \frac{1}{2}(d + 2 - \eta)$, that we determine from the equations (2) for the three most relevant odd scaling fields μ_{10}, μ_{30}, μ_{50} by linearizing about the critical fixed point $\mu^*_i(\Delta^*)$. The result agrees exactly with the value obtained using Eq. (4) above. For the case of two coupled scaling field equations no condition for Δ^* exists and we choose $\Delta^* = 0$ consistent with the Gaussian fixed point value. We note that the exponents ν plotted in Fig. 1 are those of the fixed points with $\Delta = \Delta^*$ and $\Delta = 0$, respectively. The accuracy of the calculations presented here is to three significant figures.

Although it is too early, in terms of the number of scaling fields considered, to do more than speculate about the convergence of this approach, we find the behavior of the equations as A becomes large very encouraging, for in this limit the solutions approach their ideal behavior, i.e., independence of A and uniqueness of Δ^*.[2] If the expansion is convergent, we expect this limiting behavior to be approached for smaller values of A as the truncation is removed.

SIGNIFICANCE

Obviously, the next step is to study the equations taking more irrelevant scaling fields into account. This, given the operator product expansion, is now a straightforward program.

Already, in addition to the above results, the scaling-field equations permit an exact calculation of logarithmic corrections to Gaussian critical and tricritical behavior, verifying the results of Wegner and Riedel's treatment based on Wilson's approximate recursion formula.[9] Furthermore, the derivation of the scaling field equations puts the Riedel-Wegner discussion of crossover phenomena[10] on a microscopic basis, making possible the calculation of scaling functions, including those for crossover phenomena. Much work is in progress.

ACKNOWLEDGEMENT

We have benefited from a discussion with Professor K. G. Wilson on the numerical determination of fixed point solutions of the scaling-field equations.

REFERENCES

* Work supported in part by the National Science Foundation through Grant No. GH-36882.

1. K. G. Wilson, Phys. Rev. B4, 3184 (1971).
2. K. G. Wilson and J. Kogut, Physics Reports 12C, 75 (1974)
3. For references see the Proceedings of the Conference on the Renormalization Group, Temple University, Philadelphia, Pa., 29-31 May 1973, and the recent review by K. G. Wilson and J. Kogut (Ref. 2).
4. A brief account of this aspect of the work has been presented at the Conference on Critical Phenomena in Multicomponent Systems, University of Georgia, Athens, Ga., 15-17 April 1974. Full details of this work will be published elsewhere.
5. E. K. Riedel and F. J. Wegner, Phys. Rev. Letters 29, 349 (1972).
6. F. J. Wegner, Phys. Rev. B5, 4529 (1972).
7. G. R. Golner and E. K. Riedel, to be published.
8. F. J. Wegner, J. Phys. C 7, 2098 (1974).
9. F. J. Wegner and E. K. Riedel, Phys. Rev. B7, 248 (1973).
10. E. K. Riedel and F. J. Wegner, Phys. Rev. B9, 924 (1974).

GLOBAL NONLINEAR RENORMALIZATION GROUP ANALYSIS FOR MAGNETIC SYSTEMS

J.F. Nicoll, T.S. Chang, and H.E. Stanley

Physics Department, Massachusetts Institute of Technology, Cambridge, Ma. 02139

ABSTRACT

Recent applications of the renormalization group to critical phenomena in magnetic systems have been based mainly on local linear arguments. It has been implicitly assumed that the global nonlinear effects are important only in crossover effects and that the behavior asymptotically close to the critical point is determined by the stablest fixed point alone. We have given a nonlinear analysis which incorporates the crossover between the Wilson-Fisher and mean field behavior. We point out in this paper that this competition expresses itself in globally valid solutions which can upset the dominance presumed from the linear stability analysis unless certain regularity conditions are imposed.

In a recent paper[1], we gave a solution describing the crossover between Gaussian and Wilson-Fisher[2] (WF) critical behavior of a set of renormalization group equations within an ε expansion ($\varepsilon \equiv 4-d$). Briefly we considered a reduced Hamiltonian density of the form (for an n-component spin $\vec{s}$)

$$H = |\nabla\vec{s}|^2 + r\vec{s}^2 + u(\vec{s}^2)^2 \tag{1}$$

Applying an approximate form[3] of the differential generator of Wegner and Houghton[4] we obtained nonlinear differential equations for the renormalization behavior of the parameters r and u. We found, that nonlinear scaling fields can be written to $O(\varepsilon)$ as

$$S_G = x\bar{G}^{-(n+2)/(n+8)}(1+r) \tag{2a}$$

$$S_{WF} = xy^{-(n+2)/(n+8)}(1+r)^{(4-n)/(n+8)} \tag{2b}$$

where $x \equiv r/(1+r) + u/(1+r)^2((n+2)/n)$ and $\varepsilon y \equiv u/(1+r)^2(2(n+8)/n)$. The function $\bar{G}$ is given by

$$\bar{G} = (1-y/\phi)\exp\{((n+2)/(n+8))\varepsilon xy/\phi\} \tag{2c}$$

where $y=\phi(x)$ is the equation of the separatrix connecting the WF to the infinite Gaussian (high temperature) fixed point (cf. Fig. 1).

Many thermodynamic functions can now be expressed as generalized homogeneous functions of the scaling fields given in (2). These scaling fields are not unique since a generalized homogeneous function of any pair of scaling fields is again a suitable scaling field. On the other hand, if we require that the scaling fields be proportional to the variable x for small $x (x \sim T-T_c)$, the scaling fields given in (2) are unique up to multipication by arbitrary functions of the renormalization invariant.

$$I = x^{\varepsilon}\bar{G}^{2-\varepsilon(n+2)/(n+8)}/y^2(1+r)^d \tag{3}$$

which are nonzero at I=0. Note that I vanishes on the separatrix $y=\phi(x)$ as well as on the line x=0; $I=\infty$ on the pure Gaussian trajectory y=0.

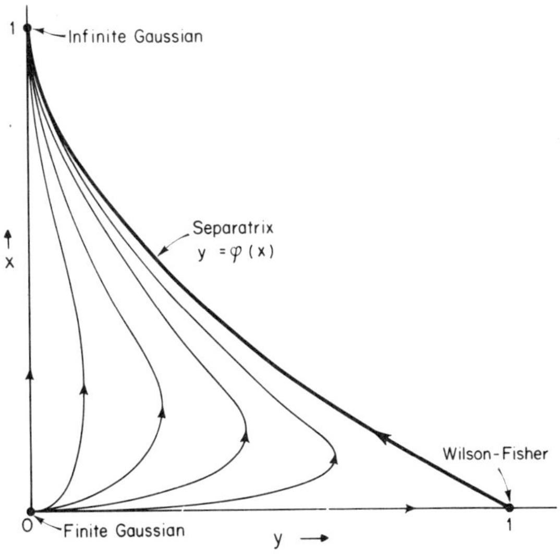

Figure 1 Nonlinear Renormalization Group Trajectories in the x-y Plane

We may, however, use the scaling fields (2) without loss of generality since we have not specified the form of any generalized homogeneous function. For the zero field correlation length we can choose as an example the simple form

$$\xi = S_G^{-1/a_G} + S_{WF}^{-1/a_{WF}} \tag{4}$$

where the scaling powers of S_G and S_{WF} are $a_G=2$ and $a_{WF}=2 \doteq \varepsilon(n+2)/(n+8)$. Since the two scaling fields appear symmetrically in (4), this form has the virtue of reducing to the appropriate linear solution as either of the two fixed points is approached. For x fixed and $y \to 0$, $S_{WF} \to \infty$, and the Wilson-Fisher term vanishes. Similarly, as the separatrix is approached, $S_G \to \infty$, and the Gaussian term vanishes. For intermediate values of y, both singularities contribute, giving the expected nonlinear crossover.

A more complicated behavior is exhibited by the Gibbs potential G. In addition to the spin-dependent terms, an additive constant v_o in the Hamiltonian density also contributes. This spin-independent term grows at the rate of $e^{d\ell}$. To first order in ε, we find

$$G(h,S_G,S_{WF}) = e^{-d\ell}G(he^{a_h\ell},S_Ge^{a_G\ell},S_{WF}e^{a_{WF}\ell}) + (dn/2)\int_0^{\ell}d\ell'\,\ell n\{1+r(\ell')\}e^{-d\ell'} \tag{5a}$$

where h is the ordering field and $a_h=1+d/2$. As the renormalization average proceeds ($\ell \to \infty$), information about the Gibbs potential passes from the first term on the right hand side of (5a) to the second term[2]. In some circumstances, (in particular, zero magnetization) it may be possible to take the limit $\ell \to \infty$ and consider only the second term. This method has been utilized by some authors[5,6,7] to calculate an approximate Gibbs potential. For our case, the result would be

$$G = (n/2)\ell n\{1+r(0)\} + n\int_0^{\infty}e^{-d\ell}x(\ell)d\ell \tag{5b}$$

However, for fixed ℓ, the second term contains information that should be unimportant for critical behavior. Accordingly, we will deal with the homogenous term only in our discussions. Thus, when discussing the Gibbs potential and its temperature like derivatives, we will confine our attention to $x \to o$, even though the solutions for the nonlinear scaling fields are valid for all $x<1$. The difficulty does not arise when studying the derivatives of the Gibbs potential with respect to the ordering field h (such as the magnetization and susceptibility) since the second term in (5a) is not dependent on h and does not contribute. We could, therefore, phrase our discussion of crossover in terms of these functions; we will discuss the Gibbs potential to allow the closest connection between this work and other phenomenological discussions of crossover.

In general, $G(h, S_G, S_{WF})$ will generate critical point exponents that do not satisfy exponent inequalities as equalities. This is to be expected since G depends on three distinct scaling powers. The usual scaling equalities which relate three exponents are satisfied because there are only two independent scaling powers. An example of a Gibbs potential which is a nonscaling global solution of the renormalization group equations is given by

$$G = G_G(h,S_G) + G_{WF}(h,S_{WF}) \qquad (6)$$

where G_G and G_{WF} are separately generalized homogeneous functions. Each piece of the Gibbs potential generates its own singularities with exponents that satisfy equalities. However, since $\gamma_G<\gamma_{WF}$ it follows that $\alpha_G>\alpha_{WF}$. The measured exponents would be γ_{WF} and α_G. Therefore, $\alpha+2\beta+\gamma>2$.

A solution of the form (6) cannot, however, be matched to the expected linear solutions near the two fixed points since as the scaling fields diverge the Gibbs potential becomes infinite. To show this more explicitly, we consider the h=0 potential. We may write it in two ways,

$$G = \{S_G\}^{2-\alpha_G} f_G(I) \qquad (7a)$$

or

$$G = \{S_{WF}\}^{2-\alpha_{WF}} f_{WF}(I) \qquad (7b)$$

If the asymptotically valid value of α were α_G, then it follows that $f_G(0)$ would be a finite constant. By (3), $f_G(I)$ is also well behaved as the separatrix $y=\phi(x)$ is approached. Since S_G is singular on the separatrix, the Gibbs potential would be singular there as well. On the other hand, if the asymptotically valid value of α is α_{WF}, then $f_{WF}(0)$ is finite. It is also finite on the separatrix. However, as $y\to 0$, the invariant $I\to\infty$. Therefore, the divergence in S_{WF} as $y\to 0$ may be cancelled by an appropriate behavior of f_{WF}. An example which has this property is given by

$$G = G_G(h,S_G)G_{WF}(h,S_{WF})/(G_G+G_{WF}) \qquad (8)$$

It is easy to check that (8) generates exponents that agree with those of G_{WF} alone; the linear analysis is thereby justified by the global results. If $\varepsilon<0$, the arguments given above are precisely reversed and the Gaussian fixed point (in this case the stabler) determines the asymptotically valid critical point exponents.

The analysis presented above gives a theoretical understanding of a mechanism for possible nonscaling critical behavior as global renormalization group solutions with singularities on portions of the boundary of the solution region. This possibility and further nonlinear analysis will be explored in a separate paper.[7]

We acknowledge interesting discussions with Professor M.E. Fisher and Dr. D.R. Nelson. This research is sponsored by NSF, AFOSR, and ONR.

1. J.F. Nicoll, T.S. Chang, and H.E. Stanley, Phys. Rev. Lett. 32, 1446-1449 (1974).

2. K.G. Wilson and M.E. Fisher, Phys. Rev. Lett. 28, 240-243 (1972).

3. J.F. Nicoll, T.S. Chang, and H.E. Stanley, Phys. Rev. Lett. 33, 540-543 (1974).

4. F.J. Wegner and A. Houghton, Phys. Rev. A8, 401-412 (1973).

5. M. Nauenburg and B. Nienhuis, Phys. Rev. Lett. 33, 1598 (1975).

6. D.R. Nelson, A.I.P. Conference Proceedings (this volume).

7. J.F. Nicoll, T.S. Chang, and H.E. Stanley, Phys. Rev. (to be published).

8. T.S. Chang, A. Hankey, and H.E. Stanley, Phys. Rev. B8, 1446 (1973).

CROSSOVER SCALING FUNCTIONS FROM RENORMALIZATION GROUP TRAJECTORY INTEGRALS

David R. Nelson
Baker Laboratory and Materials Science Center
Cornell University, Ithaca, New York 14853

ABSTRACT

We express the free energy of a system in its critical region as a line integral along a renormalization group trajectory. The kernal associated with the trajectory integral is just the spin independent part of the Hamiltonian generated by each renormalization group iteration. If the trajectory integral is parametrized in terms of nonlinear scaling fields, a closed expression for the crossover scaling function is obtained. Crossover scaling functions associated with free energy are calculated explicitly for recursion relation models treated by Riedel and Wegner, and the relevance to ε- expansions near four dimensions is discussed. A breakdown of hyperscaling above the borderline dimensions ($d = 4$ for critical points, and $d = 3$ for tricritical points) arises naturally in the formalism.

SUMMARY[1]

Phenomenological scaling ideas[2] have been successful in describing the thermodynamic singularities associated with critical points in ferromagnets and fluids. Further understanding of these concepts has come from explicit calculations of scaling functions[3] and associated critical exponents[4] within the framework of the renormalization group epsilon expansion ($\varepsilon = 4-d$, where d is the spatial dimensionality).

Riedel and Wegner[5] have formulated a scaling theory which incorporates the phenomenon of crossover from one type of critical behavior to another. Their scaling hypothesis has been useful in describing behavior near critical points of systems with tricritical points,[6] saytems with spin flop transitions,[7] etc.

We have developed a formalism which expresses the free energy of a thermodynamic system as a trajectory integral along a renormalization group flowline.[8] This yields a method for calculating (by ε- expansion or other means) the crossover scaling functions associated with these phenomenological theories. The kernal of the trajectory integral is related to the spin-independent or "constant" term generated at each renormalization group iteration.

When treating these more complicated systems by renormalization group techniques, one must deal simultaneously with at least two fixed points,[9] characterized by distinct critical exponents. The importance of a multiple fixed point situation emerges clearly from the trajectory integral analysis. Singularities in crossover scaling functions are seen to arise from the character of the Hamiltonian flows, rather than from singularities in the kernal of the trajectory integral itself.

Our analysis makes repeated use of the concept of nonlinear scaling fields.[10] These are nonlinear functions[11] $g_t(t, h)$ and $g_h(t, h)$ of the "fundamental" thermodynamic parameters of the system t and h which exhibit exact scaling properties under the action of the renormalization group. The parameter t is taken to be a temperature-like variable. When described in terms of these scaling fields, the free energy $F(g_t, g_h)$ may be written as the Laplace transform,

$$F(g_t, g_h) = \int_0^{\infty} e^{-d\ell} G_0(e^{\lambda_t \ell} g_t, e^{\lambda_h \ell} g_h) d\ell \,. \qquad (1)$$

Here, d is the dimensionality and $G_0(g_t, g_h)$ is the contribution to the free energy resulting from a renormalization group transformation in the limit that the spatial rescaling factor $b \equiv e^{\delta}$ goes to unity. It is easy to show from the work of Wegner and Houghton[12] on differential renormalization groups that, for the usual[4] continuous spin models,

$$G_0(g_t, g_h) = \tfrac{1}{2}\eta(g_t, g_h) + \ln[1 + r(g_t, g_h)], \qquad (2)$$

where the function $\eta(g_t, g_h)$ reduces to the critical exponent η at a fixed point and $1 + r(g_t, g_h)$ is the propagator evaluated at the momentum cutoff.

The expression (1) may be used to explicitly seperate the free energy into a regular and singular part and to obtain a closed expression for the crossover scaling function which fits the predictions of the phenomenological theories.[5,13] The regular part is seen to allow a breakdown of the dimensionality-dependent hyperscaling relationships above a borderline dimensionality ($d = 4$ for critical points, $d = 3$ for tricritical points). At the borderline dimensionalities, the regular and singular parts conspire to give logarithms.

The formalism has been used to calculate crossover scaling functions for the "recursion relation models" discussed by Riedel and Wegner.[14] The method avoids their "matching" technique, and yields the specific heat crossover scaling functions. The crossover from the Gaussian to Heisenberg fixed points induced by the us^4 term in continuous spin models has also been analyzed in detail. Results for the singular part of the specific heat C to first order in ε evaluated

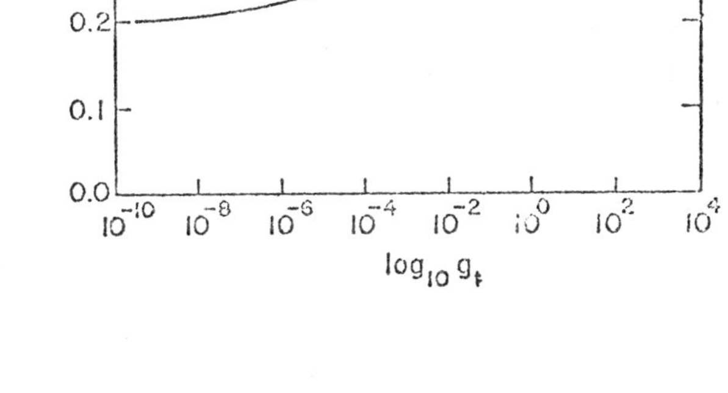

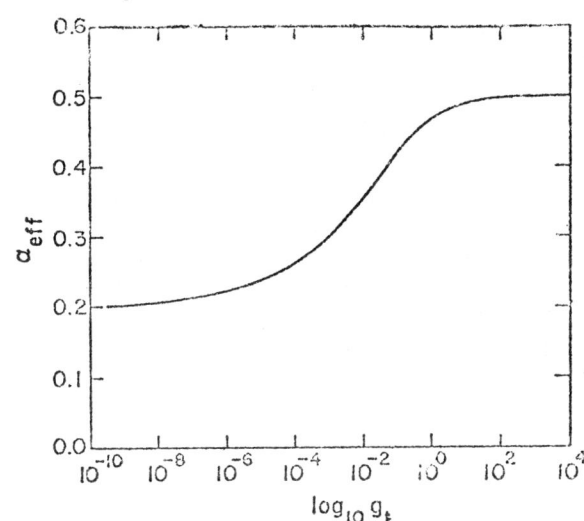

FIG. 1 Plot of the effective specific heat index α_{eff} vs. the logarithm of g_t.

at $\epsilon = 1$ are shown in Fig. 1. The effective specific heat index

$$\alpha_{eff} = (\partial \ln C / \partial \ln g_t)_{g_h} \quad (3)$$

is plotted against the logarithm of the temperature-like variable g_t for a fixed value of g_h. The exponent α_{eff} clearly crosses over from a "tricritical" value $\alpha_{eff} = .5$ to an Ising-like value $\alpha_{eff} = .2$. The "correct" Ising exponent to first order in ϵ is $\alpha = 1/6$. We have employed a different procedure here for truncating the epsilon expansion in order to evaluate the scaling function numerically, which results in $\alpha = 1/5$.

REFERENCES

1. A full account of this work will be published elsewhere.
2. B. Widom, J. Chem. Phys., 43, 3898-3905 (1965). For a review and other references to the original work, see M. E. Fisher in Proc. of the Enrico Fermi Summer School of Physics, Varenna, 1970, edited by M. S. Green (Academic, New York, 1971).
3. E. Brézin, D. J. Wallace, and K. G. Wilson, Phys. Rev. Lett. 29, 591-594 (1972), Phys. Rev. B7, 232 (1973), and E. Brézin and D. J. Wallace, Phys. Rev. B7, 232-239 (1973).
4. For a review see K. G. Wilson and J. Kogut, Physics Reports, 12 C, 77-199 (1974).
5. E. Riedel and F. Wegner, Z. Phys. 225, 195-215, (1960), and Phys. Rev. Lett. 24, 730-733, 930 (E) (1970).
6. E. K. Riedel, Phys. Rev. Lett. 28, 675-678 (1972).
7. M. E. Fisher and D. R. Nelson, Phys. Rev. Lett. 32, 1350-1353 (1974).
8. A similar formalism has been applied to ordinary critical points by J. Rudnick, Phys. Rev. B (to be published), by M. Nauenberg and B. Nienuis, Utrecht preprint, and by M. Nauenberg, Max-Planck Institut preprint.
9. For a review, see A. Aharony, AIP Conf. Proc. No. 18, Magnetism and Magnetic Materials 1973, 863-864 (AIP, 1974), and M. E. Fisher, Rev. Mod. Phys. 46, Oct. 1974.
10. F. J. Wegner, Phys. Rev. B5, 4529-4536 (1972).
11. D. R. Nelson and M. E. Fisher, AIP Conf. Proc. No. 18, Magnetism and Magnetic Materials 1973, 888-890, (AIP, 1974), and to be published.
12. F. J. Wegner and A. Houghton, Phys. Rev. A8, 401-412 (1973).
13. It is convenient to work with the extended crossover scaling hypothesis of Fisher and Jasnow, see e. g., P. Pfeuty, M. E. Fisher, and D. Jasnow, AIP Conf. Proc. No. 10, Magnetism and Magnetic Materials 1572, 817, 821 (AIP, 1973) and Phys. Rev. B 10, 2088-2126 (1974).
14. E. K. Riedel and F. J. Wegner, Phys. Rev. B9, 294-315 (1974).

RENORMALIZATION GROUP CALCULATION OF HEISENBERG-TO-DIPOLAR CROSSOVER SCALING FUNCTIONS

Alastair D. Bruce*, J. M. Kosterlitz†, and David R. Nelson
Dept. of Physics and Materials Science Center
Cornell University, Ithaca, New York 14853

ABSTRACT

Renormalization group analysis of an isotropic ferromagnet with dipole-dipole interactions reveals crossover Hamiltonian flows from an unstable Heisenberg fixed point to a fixed point associated with dipolar critical behavior. This crossover is analyzed with the aid of a formalism which expresses the free energy as a trajectory integral along the flow lines. We solve the appropriate differential recursion relations to first order in $\varepsilon = 4-d$, obtaining (to this order) the non-linear scaling fields, and the crossover scaling functions for the specific heat and the susceptibility. We evaluate these numerically, and show analytically that the amplitudes exhibit the expected double-scaling form in the asymptotic critical region.

SUMMARY[1]

Recently, a formalism has been developed[2] which allows the calculation of crossover scaling functions by means of renormalization group trajectory integrals. This technique has been applied to the model recursion relations discussed by Riedel and Wegner[3] and to the Gaussian-to-Heisenberg crossover problem to first order in $\varepsilon = 4-d$. It is of interest to calculate crossover scaling functions for problems of more immediate physical relevance.

One class of systems to which trajectory integral methods can be applied are isotropic ferromagnets with dipole-dipole interactions. Renormalization group analysis of these systems has revealed crossover Hamiltonian flows from an unstable Heisenberg fixed point to a fixed point associated with dipolar critical behavior.[4] By expressing various thermodynamic functions as trajectory integrals along renormalization group flowlines, we have calculated, to first order in ε, the specific heat and susceptibility crossover scaling functions for systems with dipolar interactions.

Previous work on trajectory integrals concentrated on the free energy.[2] Thus, by integrating up one of the differential renormalization group equations derived by Wegner and Houghton[5], the free energy was expressed as a trajectory integral whose kernal was related to the logarithm of the quadratic spin term in the usual continuous spin Hamiltonian.[6] Although the susceptibility could in principle be evaluated in this way by introducing a symmetry breaking field and then differentiating, the analysis is complicated. We have found it convenient to extend the trajectory integral formalism by analyzing the Wegner-Houghton differential equation[5] for the quadratic spin coupling $v_2(\ell)$,

$$\frac{\partial v_2(\ell)}{\partial \ell} = [2-\eta(\ell)]v_2(\ell) + (3d/2)v_4(\ell)/v_2(\ell). \quad (1)$$

Here, ℓ is related to the spatial rescaling factor b through $b=e^{\ell}$, $\eta(\ell)$ reduces to the critical exponent η at fixed points, and $v_4(\ell)$ is the four-spin coupling constant. Integrating this equation up one obtains a closed form expression for the susceptibility crossover scaling function without the necessity of introducing a symmetry breaking field. Equation (1) pertains to an Ising system. It is easily generalized to n-component dipolar systems,[4] where the propagator assumes a tensorial character.

Various refinements of the previously developed techniques[2] are employed in the calculations. There are essentially three variables of interest in the dipolar problem: these are the coefficients (r, u, g) of the quadratic spin term, of the four spin coupling, and of the dipolar interactions respectively.[4] Loosely speaking, u is an irrelevant variable at both the Heisenberg and dipolar fixed points. We account for this third irrelevant variable by constructing a two-dimensional manifold in the three-dimensional parameter space on which crossover scaling holds exactly. This manifold is the analogue of the critical value of the four spin coupling u_{0c} introduced by Wilson,[7] for which ordinary critical point scaling is exact. The appropriate differential equations restricted to this manifold, are solved to first order in ε. Thus, we obtain the nonlinear scaling fields,[8] which allow immediate calculation of the crossover scaling functions.

These scaling functions may be evaluated numercally, and can be shown to exhibit the expected crossover behavior. In addition, the possibilities and limitations of a direct Feynman graph analysis of crossover have been studied, and the necessity for resorting to some sort of global renormalization group analysis demonstrated. The trajectory integral approach yields an analytic demonstration of the Fisher-Jasnow extended scaling[9] behavior.

* Present address: Dept. of Physics, University of Edinburgh, Edinburgh EH9 3JZ, Scotland.
† Present address: Dept. of Mathematical Physics, The University of Birmingham, Birmingham B15 2TT, England.

REFERENCES

1. A full account of this work will be published elsewhere.
2. D. R. Nelson, see the proceedings of this conference, and to be published.
3. E. K. Riedel and F. J. Wegner, Phys. Rev. B 9, 294-315 (1974).
4. M. E. Fisher and A. Aharony, Phys. Rev. Lett. 30, 559-562 (1973), and A. Aharony and M. E. Fisher, Phys. Rev. B 8, 3323-3341 (1973).
5. F. J. Wegner and A. Houghton, Phys. Rev. A 8, 401-412 (1973).
6. For a review, see K. G. Wilson and J. Kogut, Physics Reports 12 C, 77-199 (1974).
7. Wilson, K. G., Phys. Rev. Lett. 28, 548-551 (1972).
8. F. J. Wegner, Phys. Rev. B 5 4529-4536 (1972).
9. This follows from the extended crossover scaling hypothesis of Fisher and Jasnow, see P. Pfeuty, M. E. Fisher, and D. Jasnow, Proc. of the 18th Annual Conference on Magnetism and Magnetic Materials, AIP Conf. Proc. 10, 817-821 (1972) and Phys. Rev. B (in press).

HIGH-TEMPERATURE SERIES STUDIES OF ISING-LIKE WILSON MODELS IN THREE DIMENSIONS

William J. Camp and J. P. Van Dyke
Sandia Laboratories
Albuquerque, New Mexico 87115

ABSTRACT

We have studied the susceptibility of continuous-spin Landau-Wilson models on the FCC lattice using exact series expansions through 10-th order in the inverse temperature. Both order-disorder (double-well) and displacive (single-well) models have been considered. The exponent γ of the dominant singularity is found to be 1.25 for both types of model. The confluent singularity arising from corrections to scaling is found to have the universal value $\delta = 0.5$. We discuss in detail the crossover from Gaussian to Ising behavior as a function of the quartic coupling constant, and find a crossover exponent $\phi = 1$ in all dimensions.

In this work we consider corrections to scaling in the critical behavior in three dimensions (d=3) of Landau-Wilson Hamiltonians of the form

$$H = -\mathcal{H}/kT = \Sigma_{\vec{r}}\, P_{2n}(Q(\vec{r})) + \tfrac{1}{2}K\Sigma_{\vec{r},\vec{\delta}}\, Q(\vec{r})\cdot Q(\vec{r}+\vec{\delta}) \; , \qquad (1)$$

where $P_{2n}(x)$ is a polynomial of degree 2n in $|x|$, $Q(\vec{r})$ is a continuous variable with range $-\infty < Q(\vec{r}) < \infty$, and $K(=J/kT)$ is a reduced interaction strength. Such models have proven interesting for at least two reasons. First, they (and their n-vector equivalents) provide the simplest basis for a discussion of the renormalization group[1] and its realization within the ϵ[1,2] and 1/n expansions.[1,3] Second, more practically, with a proper choice of $P_{2n}(x)$ they form a reasonable zeroth-order model for displacive and order-disorder structural phase transitions.[4,5] To date these models [with Q scalar] have only been studied to low order in $\epsilon(= 4-d)$,[1,2] or within the context of molecular-field and self-consistent phonon approximations.[4,5] The detailed applicability of the ϵ-expansion results in three dimensions $[\epsilon \equiv 4-d = 1]$ is an open question; and the self-consistent approximations are known to fail in the critical region.[1,2,4]

To study the critical behavior of Landau-Wilson models in three dimensions, we have formed the series for the zero-field susceptibility, $\chi = \Sigma_{\vec{r}}\langle Q(\vec{o})Q(\vec{r})\rangle$, through 10-th order in $K(= J/kT)$ on the FCC net. The series coefficients have been obtained exactly in terms of the bare vertex weights, $I_{2\ell}$, defined by

$$I_{2\ell} = \int_{-\infty}^{\infty} dx\, x^{2\ell} \exp\{P_{2n}(x)\} \Big/ \int_{-\infty}^{\infty} dx\, \exp\{P_{2n}(x)\} \; . \qquad (2)$$

The series are too involved to present here. The general series for the triangular net through 8-th order has been presented elsewhere;[6] and detailed 10-th order results for all important lattices are forthcoming.[6]

For simplicity we have specialized to the cases $P_{2n}(x) = r_2 x^2 - u_n|x|^n$ with n = 3, 4, 5, 6, 7, 8, 9 and 10. In all that follows u_n is non negative. For $r_2 > 0$ P_{2n} describes a double-well potential (order disorder case[4,5]), while with $r_2 \le 0$ P_{2n} describes a single-well potential (displacive case[4,5]). With $r_2 < 0$ and $u_n = 0$, the model reduces to the Gaussian model which exhibits mean-field behavior in all dimensions.[1] Our primary concerns in studying these models have been (i) the exponent, γ, of the dominant singularity defined by $\chi \approx \chi_0|T-T_c|^{-\gamma}$, (ii) corrections to this dominant behavior of the form $\chi_1|T-T_c|^{-\gamma+\delta}$, and (iii) the nature of the crossover to Gaussian behavior as $u_n \to 0$ at fixed $r_2 < 0$.

Corrections to scaling of the form $\chi_1|T-T_c|^{-\gamma+\delta}$ have been predicted using renormalization group theory by Wegner[1b] and explicitly demonstrated for the case of the spin-S Ising model using extended series-expansion methods.[7] In particular it has been shown[7] that, for the spin-S Ising model,

$$\chi(T) \approx \chi_0[T-T_c(S)]^{-\gamma} + \chi_1[T-T_c(S)]^{-\gamma+\delta} + \ldots \qquad (3)$$

where $\chi_1(S)$ vanishes at S=1/2, and the exponents have the universal values $\gamma = 1.250 \pm 0.003$ and $\delta = 0.50 \pm 0.08$. In studying the Landau-Wilson models we have used both series analysis extended to treat the confluent correction terms in Eq. (3),[7] and straightforward end-shift ratio analysis[8] appropriate to Eq. (3) when the correction terms are both weak and analytic $(\delta=1, |\chi_1| \ll |\chi_0|)$.

The double well potential $r_2\{x^2 - 2|x|^n/n\}$ describes, as r_2 tends to infinity, a two state $(x=\pm 1)$ model, i.e. a spin-half Ising model.[9] We have studied such models (with n=3, 4 and 6) for r_2 = 1, 2, . . .10 and 20. For $r_2 \sim 1$ the apparent exponent from endshifts is $\bar{\gamma} \simeq$ 1.23 - 1.24, while extended analysis appropriate to Eq. (3)[7] yields $\gamma = 1.25$ and $\delta \simeq 0.5 \pm 0.1$. As r_2 increases the apparent exponent $\bar{\gamma}$ from endshifts rapidly approaches 1.25, while the amplitude, χ_1, of the correction term in Eq. (3) decreases -- this decrease being accompanied by increased scatter in the estimates for δ. For values of r_2 greater than 6 for n=3, 4 for n=4 and 3 for n=6, the amplitude χ_1 is found to be zero to the accuracy of our extrapolation schemes.[7] For all values of n and r_2 considered, the extended analysis predicts $\gamma = 1.25$. These results are similar to those we obtained for the spin-S Ising model.[7]

The Landau-Wilson model with $r_2 = 0$ and $u_n = 1$ approaches the spin-infinity Ising model as $n \to \infty$. [To see this note that $|x|^n$ tends to zero (infinity) as $n \to \infty$ if $|x|<1$ ($|x|>1$).] To study the approach to the spin-infinity limit we have analyzed the series for $P(x) = -|x|^n$ with n = 3, 4,...10, 15 and 20. In Table 1, we list the endshift estimates $\bar{\gamma}$ for the exponent, the estimates for the correction exponent δ(assuming γ=5/4) and the estimates for γ obtained by fixing δ=1/2, for various values of n.[10] The results in Table 1 are clearly consistent with γ=1.25 and $\delta \simeq 0.5$ <u>independent of n.</u> For the S = ∞ Ising model we have previously found[7] an apparent exponent $\bar{\gamma} = 1.232 \pm 0.001$, and deduced [from extended analysis appropriate to Eq. (3)] the results $\gamma = 1.250 \pm 0.001$ and $\delta = 0.50 \pm 0.08$ -- all of which are consistent with the results of Table 1. Additionally the critical point $K_c(n)$ and the amplitudes $\chi_0(n)$ and $\chi_1(n)$ for this model are found to smoothly approach their respective values for the S=∞ Ising model.[7]

Table 1

n	$\bar{\gamma}$	$\gamma(\delta=1/2)$	$\delta(\gamma=5/4)$
3	1.205	1.24±0.01	0.40±0.10
4	1.212	1.25±0.01	0.46±0.10
5	1.217	1.25±0.00	0.48±0.05
6	1.222	1.25±0.00	0.48±0.05
7	1.225	1.25±0.00	0.49±0.05
8	1.226	1.25±0.00	0.48±0.05
9	1.227	1.25±0.01	0.48±0.05
10	1.228	1.25±0.01	0.49±0.05
15	1.230	1.25±0.01	0.48±0.05
20	1.231	1.25±0.01	0.48±0.05

We note that the exponents estimated for this single-well potential are the same as those found above in the double-well case. It is interesting to pursue this matter of double versus single-well potentials in more detail -- especially since the two cases have been thought of as quite distinct in the context of structural phase transitions.[5]

Table 2

r_2	n=3 $\bar{\gamma}$	n=3 $\gamma(\frac{1}{2})$	n=3 $\delta(\frac{5}{4})$	n=4 $\bar{\gamma}$	n=4 $\gamma(\frac{1}{2})$	n=4 $\delta(\frac{5}{4})$	n=6 $\bar{\gamma}$	n=6 $\gamma(\frac{1}{2})$	n=6 $\delta(\frac{5}{4})$
-2.0	1.14	1.16	0.22	1.18	1.21	0.31	1.20	1.23	0.37
-1.0	1.17	1.20	0.29	1.20	1.24	0.38	1.21	1.24	0.43
-0.5	1.19	1.22	0.33	1.21	1.24	0.42	1.22	1.25	0.46
+0.5	1.22	1.25	0.45	1.23	1.25	0.49	1.23	1.25	0.50
+1.0	1.23	1.25	0.50	1.23	1.25	0.51	1.24	1.25	0.50
+2.0	1.24	1.24+	0.49	1.24	1.24+	0.48	1.24	1.25	0.68

We have considered the Landau-Wilson models with $P(x) = r_2 x^2 - |x|^n$ for n = 3, 4 and 6. The series were analyzed for a large number of values of r_2 in the range $-2 < r_2 < 2$. In Table 2 we present the analysis for 6 representative values, $r_2 = \pm\frac{1}{2}$, ±1 and ±2. Two aspects of these results are striking: (i) for fixed r_2 the apparent exponent $\bar{\gamma}$ from endshifts, and the exponents $\gamma(\frac{1}{2})$ and $\delta(5/4)$ obtained by assuming Eq. (3) are all increasing functions of n(=3, 4 and 6); (ii) for fixed n all three exponents $\bar{\gamma}$, $\gamma(\frac{1}{2})$ and $\delta(5/4)$ increase as r_2 increases from -2 to +2. For $r_2 > 0$ the results are completely consistent with $\gamma = 5/4$ and $\delta = \frac{1}{2}$. For $r_2 < 0$, especially for n=3 there are significant apparent deviations from these values. However these deviations cannot be caused by the change from a double-well to a single-well potential, since as we saw in Table 1 the single-well potential $P(x) = -u|x|^n$ with $n \geq 4$ yields estimates $\gamma = 1.25 \pm 0.01$ and $\delta = 0.49 \pm 0.05$, in good agreement with the double-well cases (for example with the model for which $P(x) = \frac{1}{2}x^2 - x^4$).[11] Rather the apparent r_2 dependence seen in Table 2 for $r_2 < 0$ is due to a crossover effect,[12] namely the crossover over from Gaussian behavior ($P(x) = r_2 x^2 - u_n|x|^n$ with $u_n/r_2 \to 0$) to Ising behavior.

To clarify the nature of this crossover to Gaussian behavior we have studied the Landau-Wilson models with $P(x) = -x^2 - u_n|x|^n$ for n=3, 4 and 6. Now however the interaction strength is varied exponentially in the range $10^{-6} < 10^{-1}$. For u_n identically zero, the Gaussian model has susceptibility $\chi = \chi_0/[1-K/Kc]$ with <u>no</u> correction terms. That is, $\gamma = 1$ and δ is undefined. For any finite value of u_n, the model exhibits Ising-like behavior, namely $\gamma = 5/4$ and $\delta \simeq 1/2$. In the language of renormalization group theory,[1b] the coupling constant u-u* (where u*, the fixed-point value of u, is zero for the Gaussian fixed point and finite for the Ising fixed point) is a relevant variable with respect to the Gaussian fixed point and an irrelevant variable with respect to the Ising fixed point in three dimensions.[1,2] This means that the Gaussian fixed point is unstable with respect to u, and for small u we find a typical crossover behavior.[12] It is characteristic of such situations that the deviations from the unstable-fixed-point exponent do not begin to appear in the apparent exponents obtained from the analysis of the first N terms of a high-temperature series until N is of order u^{-1}. More specifically, for $u \ll 1$ we expect[7,12] that the value of γ extrapolated from N-th order series will be $\gamma \simeq 1 + \alpha N u$ where $\alpha \sim 1$.

In Table 3 we exhibit the estimates of the critical parameters obtained using 10-th order series for $u_4 = 10^{-1}, \ldots 10^{-6}$. Listed are the apparent exponent, $\bar{\gamma}$, from endshifts, the value of δ estimated with γ fixed at $\gamma = 5/4$, and a parameter <u>A</u> whose significance is clarified below. First note that $\alpha \approx 0.7$, and that δ decreases with u_4 like $\delta = 0.8\, u_4$. We can understand this easily. A straightforward way of estimating δ is afforded by noting that the ratios R_n of successive coefficients χ_n in the high-temperature expansion for χ in Eq. (3) behave like $R_n \equiv \chi_n/\chi_{n-1} \approx (kT_c/J)[1 + (\gamma-1)/n + A/n^{1+\delta} + \cdots]$.[7] Thus A in Table 3 is the amplitude of the $n^{-(1+\delta)}$ term in R_n. The fact that $\delta \simeq 0$

Table 3

u_4	$\bar{\gamma}$	$\delta(\gamma=\frac{5}{4})$	A
10^{-1}	1.150	0.24	-0.248
10^{-2}	1.050	0.064	-0.254
10^{-3}	1.007	0.008	-0.251
10^{-4}	1.0007	0.0008	-0.2501
10^{-5}	1.00007	0.00008	-0.25001
10^{-6}	1.000007	0.000008	-0.250001

for small enough u_4 simply reflects the fact that the series still looks Gaussian through 10-th order for small u_4. Indeed with $\delta = 0$, the quantity <u>A</u> becomes a shift in γ. Since <u>A</u> and δ were obtained assuming $\gamma = 5/4$, $\delta = 0$ and $A = -1/4$ imply $\gamma = 1$, in agreement with straightforward endshift analysis. Thus, the apparent decrease in δ as r_2 becomes increasingly negative in Table 2 simply signals the approach to the crossover region. Furthermore the fact that δ increases toward 0.5 as we increase n holding the magnitudes of r_2 and u_n fixed follows from the fact that the model with $P(x) = -r_2 x^2 - u|x|^n$ becomes farther from the Gaussian fixed point as n increases with u fixed.

To conclude we assume that for small u_{2n} we can write a scaling form for χ of the form $\chi = t^{-1} X(u_{2n}/t^{\phi})$,[12] where $t = 1-K/K_c$ and ϕ is the crossover exponent.[1,2,10] To estimate ϕ note that according to the scaling assumption

$$F_{2n}(t) \equiv \frac{1}{\chi} \frac{\partial \chi}{\partial u_{2n}} = \tilde{F}_0 t^{-\phi}. \quad (u_{2n} \to 0). \quad (4)$$

One can exactly obtain the $u_{2n} = 0$ limit of $F_{2n}(t)$ for all dimensions, d, and all $n \geq 2$. The crossover exponent φ is found to be unity <u>independent of n and d.</u> However the detailed form of F_{2n} is interesting because it exhibits strong corrections to scaling.[1b,7] We find $F_{2n}(t) \sim t^{-1}[-f(t)]^{n-1}$ where with d=2 $f(t) \approx \ln[t/(t + Q^2)]$, with d=3 $f(t) \approx (Q - \frac{\pi}{2}\sqrt{t})$ and with d=4 $f(t) \approx Q^2 + \frac{t}{2}\ln[t/(t + Q^2)]$. Here Q is a high momentum cutoff due to the finite lattice spacing.

REFERENCES

1. S. Ma, Rev. Mod. Phys. <u>45</u>, 589 (1973), and references cited. (b) F. J. Wegner, Phys. Rev. B <u>5</u>, 4529 (1972).
2. K. G. Wilson and M. E. Fisher, Phys. Rev. Lett. <u>28</u>, 240 (1972).
3. S. Ma, Phys. Rev. A <u>7</u>, 2172 (1973).
4. W. J. Camp, Bull. Am. Phys. Soc. <u>17</u>, 673 (1972), and unpublished work.
5. N. S. Gillis and T. R. Koehler, Phys. Rev. Lett. <u>29</u>, 369 (1972).
6. J. P. VanDyke and W. J. Camp, AIP Conf. Proc. <u>18</u>, 878 (1974); and work in preparation.
7. W. J. Camp and J. P. VanDyke, Phys. Rev. B to be published. D. Saul, M. Wortis and D. Jasnow, Phys. Rev. B to be published.

8. J. P. VanDyke and W. J. Camp, Phys. Rev. B 9, 3121 (1974).
9. The S = 1 Ising model is described by the $A \to \infty$ limit of the triple well potential, $P(x) = -Ax^2(x^2-1)^2$, $+ (\ln 2)X^2$, while the spin-S model is described by the $A \to \infty$ limit of the (2S+1) well potential, $P(x) = -A(x^{4S+2} + \Sigma_{\ell=0}^{2S} a_\ell x^{2\ell})$, with an appropriate choice of $\{a_\ell\}$.
10. As discussed in Ref. (7) if one attempts to analyze the susceptibility, assuming Eq. (3) with both γ and δ unspecified the scatter in the estimates for γ and δ increases inordinately.
11. Thus, the essential differences between double-well and single-well potentials are in the nature of the corrections to the dominant behavior. This result is in contrast to the results of approximate theories which find distinct behavior in the two cases. See Ref. (5).
12. P. Pfeuty, D. Jasnow and M. E. Fisher, Phys. Rev. B 10, 2088(1974). J. P. VanDyke and W. J. Camp, unpublished.

THEORY OF ITINERANT FERROMAGNETISM†

R. E. Prange and V. Korenman
Department of Physics and Astronomy,
and Center for Theoretical Physics,
University of Maryland, College Park, Maryland 20742

ABSTRACT

A "fluctuating band" picture of ferromagnetism is presented which agrees qualitatively with the entire range of phenomena in itinerant ferromagnets at energies comparable to or less than thermal, and at temperatures up to the Curie point and above, as well as providing a unified picture in which band theory clearly is correct at low temperature and magnetic fluctuations dominate the thermodynamics in the neighborhood of the Curie point. These phenomena include the insensitivity of the magnetic features of the band structure (the exchange splitting) to the temperature and phase transitions as found by Mook et al.[1] and Lonzarich and Gold[2]. The basic idea used is that the bands are for the most part locally determined, and therefore may be calculated in the presence of fairly long wavelength and slow fluctuations in the magnetisation before averaging thermodynamically over these fluctuations. The effect of the magnetic fluctuations on the single electron states in the band is calculated by a perturbation procedure.

It is firmly established by experiments done at low temperature, notably measurements on the de Haas-van Alphen effect, that the itinerant ferromagnets (IF) have spin-split Fermi surfaces with the properties usually encountered in good metals. Further, these Fermi surfaces are in reasonably good accord with semi-phenomenological band structure calculations.

There are also measurements which demonstrate the existence of spin waves or magnons in the IF. Their properties can be studied qualitatively by use of the random phase approximation. It is clearly an extremely difficult numerical problem to calculate from first principles all the various renormalizations necessary to correct the Hartree-Fock and random phase approximations, but the form of the result is quite clear. That is, we expect to have a Landau theory of Fermi liquids generalized to include magnons. Such a theory has been written down by Izuyama and Kubo[3], and exploited recently by Edwards[4].

While a theory of this type is phenomenologically able to explain all the low temperature data, [even though it contributes little to understanding the ground state energetics] it is not clear what to do for temperatures comparable to or above the Curie temperature. It is difficult or impossible to fit with a Stoner type of theory, the observed magnetisation vs. temperature curves, the magnetic specific heat and entropy, and the susceptibility. All of these quantities are suspiciously similar to Heisenberg model results and indicate that the remnants of magnetism persist locally above the transition temperature. They also do not encourage the idea that the mechanism for magnetism is distinct in different IF. Adding in the spin waves and renormalizing may produce correct answers but such an approach is very difficult and does not yield much insight.

Two recent experiments have indicated more direct temperature anomalies. The first shows that the exchange splitting at low temperature is nearly constant as compared with the temperature variation of the magnetisation.[1] This indicates that the exchange splitting should not be regarded as the central symmetry breaking parameter as in the Stoner model. Although Fermi liquid theory explains this[4], it is not apparent how to extend this result to the temperature range for which the net magnetisation is small. The other is the neutron measurement[2] which shows that spin waves continue to exist at and even well above T_c, for wave numbers $q > Q$ where $Q \sim .2\ \text{A}^{\circ -1}$ in Ni at T_c. If the exchange splitting vanished, direct decay of spin waves into Stoner excitations via strong Coulomb matrix elements could proceed, and it would be difficult to understand the well-defined existence of the spin waves, and their similarity to those observed at low temperatures. In addition, it is found that the maximum observed spin wave momentum is temperature independent, and it is argued that this maximum should be identified with the intersection of the spin wave dispersion relation with the Stoner continuum.

The easiest way to interpret these results is to note that the properties of spin waves of wavelength λ can be determined from the local properties of a region of size $d \gtrsim \lambda$, since we could make a wave packet substantially confining the excitation to the chosen region. The weak dependence of spin wave dispersion on T for $q > Q$ therefore implies that over a region of size $2\pi/Q$ the state of the IF is not much different from what it is at low temperature, where we have seen that the Fermi-liquid theory and band structure works very well.

The phase transition in our model is not due to thermal excitation of Stoner pairs, but rather to the softening of the long wavelength spin fluctuations. This softening occurs well below the Stoner transition, and of course is responsible for the phase transition in a Heisenberg magnet as well.

This situation is a well studied one in the modern theory of phase transitions. We thus break up the partition function (following the renormalization group idea)[5] into

$$Z = \mathrm{Tr}\, e^{-\beta H} = \mathrm{Tr}_{q<Q}\, \mathrm{Tr}_{q>Q}\, e^{-\beta H} = \mathrm{Tr}_{q<Q}\, e^{-\beta F_Q[M]}$$
$$= \int_{q<Q} \mathcal{D}M_q \; e^{-\beta F_Q[M]} \tag{1}$$

where the restricted free energy $F_Q[M]$ is defined for an arbitrary fixed static $M(\vec{r})$ with no fourier components of wave number larger than Q. The notation $\mathrm{Tr}_{q>Q}$ means that only fluctuations with wave numbers obeying the inequality are included.

The cutoff Q is chosen as before, which has the following advantages. First, we can find F_Q since we can carry out the trace for the short wavelength fluctuations. For these wavelengths the short range order dominates. Second, because spin wave energies for $q<Q$ are much less than kT, such spin waves will be excited to high quantum numbers and a classical treatment is appropriate. This means that we do not have to worry about the failure of different components of M to commute. It also justifies the use of static magnetisation fluctuations.

To calculate $F_Q[M]$ we may note that fluctuations in the magnitude of M are less serious than fluctuations in the direction of M. Let us neglect the magnitude fluctuation initially, and choose a typical $\hat{M}(\vec{r})$. Now we go to a spin coordinate system for which the z-axis is in the direction $\vec{M}(\vec{r})$. The Hamiltonian H when expressed in this system becomes H'=H+V, where V vanishes with Q and will be treated in perturbation theory. Such calculations are already well known in the theory of the Bloch wall, for example.[6] A similar approach has been applied by H. Capellmann[7] to the Hubbard model in the Hartree-Fock approximation.

Solving the posed problem is comparable in difficulty to the low temperature problem. Clearly we must choose the unperturbed solution (with V zero) for which the electron spins are quantized in the local $\hat{z}$ direction. Temperature effects will come in through Stoner excitations, and through spin wave excitations for $q>Q$. The main effect reducing the net local magnetisation will be the latter, but this is not too serious. Using the measured spin wave dispersion leads to a net reduction in the local magnetisation at T_c of about 25% in both iron and nickel. The spin wave dispersion will be weakly temperature dependent as Landau theory (or magnon-magnon interactions in the Heisenberg model) predicts.

Doing second order perturbation theory in V leads to an energy shift of order Q^2. This calculation is comparable in difficulty to a calculation of the spin wave dispersion Dq^2, and in fact the result may be related to this quantity in a well known way . A by-product is that the modification of the local band structure due to long wavelength spin fluctuations can be estimated as well.

We summarize our theory as follows: Because of the considerable short range order characterized by the small magnitude of Q, we are enabled to separate the problem of itinerant ferromagnetism into three numerically difficult but physically clear subproblems. These are the calculation of a local band structure and local short wavelength spin waves, the (second order) perturbation corrections proportional to Q, and the final thermodynamic average over long wavelength fluctuations.

One may attempt to find local band structure in other situations with sufficient short range order. We may remark however, that the nearest neighbor Heisenberg model (in three dimensions but not in two) has little short range order above T_c, and consequently well defined spin waves do not exist there. The existence of well defined short range order in the itinerant case is doubtless a consequence of the range of the exchange forces. It is perhaps no accident that this range is comparable with $2\pi/Q$.

It may finally be worth while to remark that Q has no relationship to ξ, the coherence length, which specifies the long-range order. In particular, extrapolation of ξ vs. T into a region for which the long range order is shorter than the short range order should be avoided.

We finally remark that our approach has some similarity to that of Murata and Doniach[8] who study weak IF. It differs in that we assume spin waves are the most important type of thermodynamic fluctuation, which requires a vector order parameter in expressions such as (1). It also emphasizes the role of short range order. We are in agreement however, in believing that the Stoner transition temperature is considerably above the actual Curie point, and in our use of expressions like (1) to study the thermodynamics.

[1]H. A. Mook, J. W. Lynn, and R. M. Nicklow, Phys. Rev. Letters 30, 556 (1973).

[2]G. Lonzarich and A. V. Gold, Can. J. Phys. 52, 694 (1974)

[3]T. Izuyama and R. Kubo, J. Appl. Phys. 35, 1074 (1964).

[4]D. M. Edwards, Can. J. Phys. 52, 704 (1974).

[5]K. G. Wilson, Phys. Rev. B 4, 3174, 3184 (1971).

[6]C. Herring, Magnetism, Vol. IV, G. Rado and H. Suhl, Eds. (Academic Press, N. Y., 1960).

[7]H. Capellmann, J. Phys. F. 4, 1112,(1974) and to be published.

[8]K. K. Murata and S. Doniach, Phys. Rev. Letters 29 285 (1972).

†Work supported in part by National Science Foundation

PHONON-MODULATED SUSCEPTIBILITY OF A NARROW-BAND HUBBARD MODEL*

Mustansir Barma and Robert A. Bari
State University of New York at Stony Brook
Stony Brook, New York 11794

The effects of the coupling between lattice vibrations and electrons on the magnetic properties of a narrow-band Hubbard model are investigated. The most important consequence of the coupling is to produce modulation of the one-electron transfer integrals. The magnetic susceptibility is calculated using a high temperature expansion. Over a certain range of temperature, we find that the susceptibility χ follows a renormalized Curie-Weiss Law:

$$\frac{1}{\chi} = \frac{R}{C}\,(T+R\theta)$$

Here C and θ are the values of the Curie constant and the Néel temperature, respectively, in the absence of phonons and

$$\frac{2R\theta}{R-1} = T_{Debye}\left(\frac{\text{Decay length of local elec. w.f.}}{\ell_{Lattice}}\right)^2$$

where $\ell_{Lattice}$ is a characteristic lattice-vibrational length, defined by

$$(\ell_{Lattice})^2 = \frac{2\hbar}{(\text{Mass of ion})\times(\text{Debye freq.})}.$$

R is the ratio by which the Néel temperature is raised and the square of the effective magnetic moment is reduced. The relation of our results to NMP-TCNQ is discussed. Further details will be published elsewhere.[1]

* Worked supported by the National Science Foundation Grant No. H040884.

[1] Mustansir Barma and Robert A. Bari, to appear in Phys. Rev. B, February, 1975.

GENERALIZED SUSCEPTIBILITY OF Sc METAL*

J. Rath and R. P. Gupta
Magnetic Theory Group, Physics Department, Northwestern University, Evanston, Illinois 60201

A. J. Freeman
Physics Department, Northwestern University, Evanston, Illinois 60201
and
Argonne National Laboratory, Argonne, Illinois 60439

ABSTRACT

The generalized magnetic susceptibility $\chi(\vec{q})$ of Sc metal has been determined from accurate Augmented Plane Wave (APW) method calculations of their energy band structure using a novel computational scheme developed as an extension of the work of Lehmann and Taut and Jepsen and Andersen on the density of states problem. The procedure yields simple analytic expressions for the $\chi(\vec{q})$ integral inside a tetrahedral microzone of the Brillouin zone which depends only on the volume of the tetrahedron and the differences of the energies only at its corners. Constant matrix element results have been obtained for Sc which show very good agreement with the results of Liu, Gupta and Sinha and with a first maxima in $\chi(\vec{q})$ at $(0,0,0.31)2\pi/c$ vs $(0,0,0.35)2\pi/c$ obtained by Liu et.al. which relates very well to dilute rare-earth alloy magnetic ordering data $\vec{q}_m=(0,0,0.28)2\pi/c$ and to kink in the LA phonon dispersion curve at $(0,0,0.27)$ $2\pi/c$.

INTRODUCTION

The real part of the frequency and wave vector dependent spin susceptibility function in RPA is given by the well known expressions

$$\mathrm{Re}\chi(\vec{q},\omega)=\frac{1}{N}P\sum_{\substack{n,n'\\ \vec{k}}}|M^{n,n'}(\vec{k}+\vec{q}+\vec{K}_o,\vec{k})|^2\frac{f(E_n(\vec{k}))\,[1-f(E_{n'}(\vec{k}+\vec{q}+\vec{K}_o)]}{E_{n'}(\vec{k}+\vec{q}+\vec{K}_o)-E_n(\vec{k})-\omega} \tag{1}$$

where $E_n(\vec{k})$ is the energy at the point $\vec{k}$ of the Brillouin zone (BZ) in the band n and $M^{nn'}$ is the oscillator strength matrix element. At T=0, the Fermi function $f(E_n(\vec{k}))$ has the value 0 or 1 depending whether $E_n(\vec{k})$ is above or below the Fermi energy, E_F. The reciprocal lattice vector $\vec{K}_o$ keeps the point $\vec{k}+\vec{q}$ in the first Brillouin Zone. We shall drop the ω dependence from consideration as this can easily be inserted at a later point.

Extensive computational efforts have been directed recently towards the problems of the accurate calculation of $\chi(\vec{q})$, which gives the response of the magnetization of the electron gas in a metal to a spatially varying field of wave vector $\vec{q}$. Its importance lies in the well known fact that structure in $\chi(\vec{q})$ has been related to the magnetic interaction energy of a real metal and hence to the possibility of magnetic ordering with wave vector $\vec{q}$ and to phonon soft modes and Peierls instabilities. Similarly, observed magnon dispersion may also be related (in a constant matrix element approximation) to $\chi(\vec{q})$. For real transition or rare-earth metals recent efforts have focused on including the actual band structure into the calculations and relating maxima in $\chi(\vec{q})$ to certain nesting features in the Fermi surface. Recent calculations have emphasized the need to include matrix elements in addition to the realistic energy eigenvalues.[1-3]

A basic problem, however, for any of these studies has been the need for a highly accurate, rapid and efficient computational method. Major advances have been made by a number of authors[4] towards performing the required summations (integrations) of Eq. (1). We have developed a novel computational scheme for the calculation of $\chi(\vec{q})$ which is derived as an extension of the work of Jepsen and Andersen[5] (JA) and Lehmann and Taut[6] (LT) on the density of states, N(E), problem.[7]

Using tetrahedrons as microzones with which to divide the BZ, a geometrical analysis is made of the occupied and unoccupied regions of any tetrahedron which reduces the problem of calculating $\chi(\vec{q})$ essentially to the problem of performing a volume integral over a tetrahedron with a linearized energy denominator. This procedure yields simple analytic expressions for the BZ integral, which depend only on the volume of the tetrahedron and the differences of energies, $\Delta E_n(\vec{k})$, at its corners. The result is a computational scheme which is highly accurate and, because of its simplicity, rapid to perform.[7]

As a first application of this method to a real metal, we present here a detailed study of the generalized susceptibility, $\chi(\vec{q})$, of Sc metal determined from an accurate Augmented Plane Wave (APW) method calculation of its energy band structure in the constant matrix element approximation. The results show good agreement with the work of Liu et.al.;[1] the first maxima in $\chi(\vec{q})$ and relates very well to the dilute rare-earth alloy magnetic ordering data and to the kink in the LA phonon dispersion curve.[8]

ANALYTIC TETRAHEDRON LINEAR ENERGY METHOD FOR THE CALCULATION OF $\chi(\vec{q})$

As in the case of the density of states calculation, our purpose is to develop a suitable analytic integration scheme for the Brillouin zone integration of Eq. (1) taking into account the principal value nature of the integral. As in the work of LT and JA, the scheme proceeds by dividing the entire BZ (or irreducible part of it) into non-overlapping tetrahedra. Although not necessary we assume that all tetrahedra are of the same size as this saves a certain amount of computational effort. Since we make a linear approximation to the energy bands inside the tetrahedron, the surface of constant energy is simply a plane.[4] Now, the fractional volume of a given tetrahedron that contributes to the susceptibility function is determined by the intersection of constant energy surfaces (planes) corresponding to $E_n(\vec{k})$ and $E_{n'}(\vec{k}+\vec{q})$. Inside this fractional volume of the tetrahedron the product of $f(E_n(\vec{k}))[1-f(E_{n'}(\vec{k}+\vec{q}))]$ has the value unity. It turns out that this fractional volume is either a tetrahedron or the sum of (at most nine) tetrahedrons. Thus after taking care of the Fermi factors it is then only necessary to be able to perform a volume integration over a tetrahedron with a linearized energy denominator.

Assuming the matrix elements to be constant inside the tetrahedron we have to perform the following integral

$$I_{nn'}(\vec{q}) = \int_{\mathrm{tetra}} \frac{d^3k}{E_{n'}(\vec{k}+\vec{q})-E_n(\vec{k})} \tag{2}$$

It is convenient to arrange the energy at the corners of the tetrahedrons in increasing or decreasing order. Let $\vec{k}_i (i=1,2,3,4)$ be the coordinates of the 4 corners of the tetrahedron and denoting $E_n(\vec{k}_i)$ by E_i, we obtain the analytic expression[7]

$$I_{nn'} = 3\Omega\Big[\frac{V_1^2}{D_1}\ln\Big|\frac{V_1}{V_4}\Big| + \frac{V_2^2}{D_2}\ln\Big|\frac{V_2}{V_4}\Big| + \frac{V_3^2}{D_3}\ln\Big|\frac{V_3}{V_4}\Big|\,\Big] \tag{3}$$

where Ω=volume of the tetrahedron and $D_i=\prod_{j\neq i}(V_i-V_j)$

$$V_i = E_{n'}(\vec{k}_i+\vec{q}) - E_n(\vec{k}_i)$$

At this stage we must consider how to include the matrix elements in our integration scheme. Although a large number of calculations have been performed assuming constant matrix elements,(1) recent calculations(2) have shown that matrix elements play a very important role in determining the location and magnitude of structures in $\chi(\vec{q})$. The inclusion of matrix elements in our scheme is simply done by calculating the matrix elements at the center of each tetrahedron and assuming it to be constant throughout the tetrahedron. This perhaps is not too bad an approximation if the tetrahedrons are small enough.

$\chi(\vec{q})$ for Scandium Metal

Interest in the electronic structure of Sc metal is high because its unusual properties as a light transition metal and because many of its physical properties resemble those of the heavy rare-earth metals which also have hcp structure. Although not magnetic like the rare-earths, knowledge about the $\chi(\vec{q})$ of Sc can be instructive about the magnetic structures of its alloys with magnetic rare-earths.

The electronic energy band structure was determined using the Augmented Plane Wave (APW) method. The warped muffin tin crystal potential was derived as a superposition of Hartree-Fock-Slater neutral atomic charge densities but for the configuration $3d^2 4s^1$ of Sc. Exchange was treated by Slater's $\rho^{1/3}$ approximation with the exchange parameter $\alpha=1$.

We have calculated $\chi(\vec{q})$, in the constant matrix element approximation, for $\vec{q}$ along the Γ-A direction using 1536 tetrahedrons in $1/24^{th}$ of the irreducible BZ. Only the 3rd and 4th bands were included in computing $\chi(\vec{q})$ since these two bands, which intersect the Fermi energy in Sc, determine the structure.

A plot of $\chi(\vec{q})$ vs $\vec{q}$ along Γ-A, shown in Fig. 1, is similar to that shown by Liu et.al.(1) However, we find two peaks in the curve situated at $\vec{q}=(0,0,0.31)2\pi/c$ and $\vec{q}=(0,0,0.72)2\pi/c$ rather than the three broad peaks(8) shown by Liu et.al.(14) at approximately $\vec{q}=(0,0,0.35)2\pi/c$, $(0,0,0.57)2\pi/c$ and at $(0,0,0.77)2\pi/c$. We also do not find the structure (oscillations) shown in this first and (especially) the third peak but these small differences may be entirely due to the small differences between our respective Fermi surfaces. It is gratifying to note that the first peak in $\chi(\vec{q})$ in Fig. 1 corresponds very closely to the experimental value of the magnetic wave vector $\vec{q}_m=(0,0,0.28)2\pi/c$ found for the dilute alloys of Sc with rare-earths.(9) One expects a sharp maximum in $\chi(\vec{q})$ to manifest itself also as an anomaly in the phonon dispersion curve. For Sc metal, Wakabayashi et al(8) have observed a kink in the longitudinal acoustic (LA) branch at a wave vector, $\vec{q}=(0,0,0.27)2\pi/c$. This $\vec{q}$ value is close to the observed magnetic wave vector, $\vec{q}_m$, and close to our first peak in $\chi(\vec{q})$.

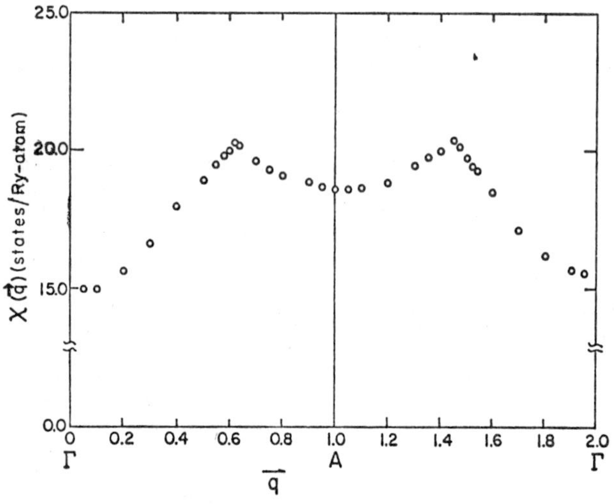

Fig. 1. $\chi(\vec{q})$ for Sc along the Γ-A-Γ direction.

Finally, as noted by Liu et.al,(1) although yttrium metal has a very similar electronic structure, and shows similar peaks in the theoretical(1) $\chi(\vec{q})$, to that of Sc, no phonon anomalies were found by Sinha(10) et. al. near the first peak in $\chi(\vec{q})$. Instead, two sharp dips in the longitudinal optic (LO) branch at $(0,0,0.625)2\pi/c$ and $(0,0,0.775)2\pi/c$ were observed. In view of the similarities in the electronic structures of Sc and Y, one expected the Kohn anomalies in both the metals to be observed roughly at the same positions. Even though the susceptibility matrix plays a very complex role in the dynamical matrix for phonon spectra it is possible that the $\chi(\vec{q})$ determined from matrix elements using accurate solid state wave functions will help in resolving this question. Such calculations have become increasingly important since it is now widely recognized that the matrix elements do play an important role in determining the peaks and the general structure. Calculations are now underway for both Sc and Y which utilize the methods developed here and APW wave functions for the calculation of matrix elements and $\chi(\vec{q})$ according to Eq. (1).

REFERENCES

[1]W. E. Evenson and S. H. Liu, Phys. Rev. 178, 783 (1969).

[2]R. P. Gupta and S. K. Sinha, Phys. Rev. B3, 2401 (1971).

[3]S. Liu, R. P. Gupta and S. K. Sinha, Phys. Rev. B4, 1100 (1971).

[4]The review by G. Gilat [J. Comp. Phys. 10, 432 (1972)] contains a comparison of methods and extensive references.

[5]O. Jepsen and O. K. Andersen, Sol. State. Comm. 9, 1763 (1971). See also the recent extension of the tetrahedron method by P. A. Lingård (Risø-M-1701, Library of the Danish AEC and Lectures at the XI Annual Winter School for Theoretical Physics, Poland, 1974, to be published).

[6]G. Lehmann, P. Rennert, M. Taut and H. Wonn, Phys. Stat. Sol. 37, K27 (1970); G. Lehmann and M. Taut, Op. Cit. 54, 469 (1972).

[7]The work reported here is based on the detailed work of J. Rath and A. J. Freeman, submitted to Phys. Rev.

[8]N. Wakabayashi [Ph.D. thesis, Iowa State University, Ames, Iowa, 1969 (unpublished)]. The observed phonon spectra is reported by N. Wakabayashi, S. K. Sinha, and F. H. Spedding, Phys. Rev. B4, 2398 (1971).

[9]H. R. Child and W. C. Koehler, Phys. Rev. 174, 562 (1968) and references contained therein.

[10]S. K. Sinha, T. O. Brun, L. D. Muhlestein and J. Sakurai, Phys. Rev. B1, 2430 (1970).

*Supported by the NSF, the AFOSR and the AEC.

DYNAMIC SUSCEPTIBILITY CALCULATIONS IN FERROMAGNETIC IRON*

J. F. Cooke, J. W. Lynn†, and H. L. Davis
Solid State Division, Oak Ridge National Laboratory
Oak Ridge, Tennessee 37830

ABSTRACT

Calculations of the low temperature dynamic susceptibility of ferromagnetic iron have been carried out within the framework of an itinerant electron model. The formalism incorporates band and wave vector dependent matrix element effects and is essentially the same as that used in previous work on nickel.[1] The spin wave dispersion curves calculated along the three principal symmetry directions are found to be in excellent agreement with neutron scattering experiments. In addition the calculations predict correctly the experimentally observed spin wave disappearance. These results indicate that low temperature spin waves in both nickel and iron can be adequately described in terms of an itinerant electron model without recourse to ad hoc assumptions about local moment behavior.

1. INTRODUCTION

The dynamic susceptibility, $\chi(\vec{q},\omega)$, is a quantity of fundamental importance in the study of magnetism in the transition metals. One important aspect of $\chi(\vec{q},\omega)$ is that it can be directly measured by inelastic magnetic neutron scattering techniques. Thus, from a theoretical point of view a calculation of $\chi(\vec{q},\omega)$ not only leads to a detailed description of the dynamic properties of a magnetic system but also provides a means of testing various models of magnetism. For example, in a recent paper a zero temperature result for $\chi(\vec{q},\omega)$ was obtained for ferromagnetic nickel within the framework of an itinerant electron model.[1] The calculation was based on an approximation of $\chi(\vec{q},\omega)$ suggested by Cooke which incorporates both band and wave vector dependence of matrix elements which appear in the theory.[2] Results from this calculation were found to be in good agreement with neutron scattering experiments, which supports the view that the itinerant model is correct for nickel at low temperatures.

This apparent success of the itinerant model for nickel does not necessarily imply that it is applicable to the rest of the transition metal series. In particular, it has been argued by some that the itinerant model must be modified to include the possibility of localized electron behavior and of strong Hund's rule coupling which might occur when two or more unpaired spin electrons occupy the same site.[3-6] If this argument is correct one might expect that the itinerant model would work for nickel since, on the average, there is less than one unpaired-spin electron per site. However, for iron, which averages about two unpaired-spin electrons per site, one would certainly expect poor agreement with the itinerant model.

The purpose of this paper is to present some results from a calculation of $\chi(\vec{q},\omega)$ for ferromagnetic iron. These results tend to support the view that the magnetic properties of iron can be adequately described in terms of an itinerant model without recourse to ad hoc assumptions about strong Hund's rule coupling and local moment behavior. The remainder of this paper is divided into three sections. Section 2 contains a generalization of the results obtained in reference (2) which form the basis of the calculation, section 3 presents the results of the calculation of $\chi(\vec{q},\omega)$, and section 4 contains a comparison with experiment and discussion of conclusions.

2. APPROXIMATION FOR $\chi(\vec{q},\omega)$

In the itinerant model one assumes that the magnetic electrons occupy energy bands. If one assumes that these bands are narrow enough it is tempting to suppose that Wannier functions, which are "localized" functions formed from linear combinations of Bloch functions, do not overlap appreciably from site to site. If this is the case the many body problem can be greatly simplified by using the Wannier representation and neglecting electron-electron interaction terms which involve the overlap of Wannier functions.[7] Although no definitive statement can be made at present, there is some evidence which indicates that Wannier functions are not well localized in the transition metal series.[1,8]

For cases where the Wannier functions are spread out it appears to be simpler to work within a Bloch representation. As shown in reference (2) a closed form expression for $\chi(\vec{q},\omega)$ based on a generalized random phase approximation can be obtained for this representation which incorporates both the band and wave vector dependence of matrix elements in a self-consistent (approximate) way. We have generalized these results to include the possibility that the screened coulomb matrix element parameter, U^{d-d}_{eff}, defined in reference (2) might be different for e_g and t_{2g} symmetries.[9] This generalization follows directly from the theory and does not alter the fact that all of the information needed to calculate $\chi(\vec{q},\omega)$ can be obtained from a self-consistent solution of the energy band equations.

The results for iron presented in this paper were obtained from a numerical evaluation of this generalization of the theory. The results for nickel referred to above can be thought of as coming from the same (generalized) theory but with the restriction that U^{d-d}_{eff} is the same for e_g and t_{2g} symmetries.

3. NUMERICAL RESULTS FOR $\chi(\vec{q},\omega)$

The first step in evaluating $\chi(\vec{q},\omega)$ is to generate a reasonable ferromagnetic band structure. The method we use is to first develop a "paramagnetic" crystal potential from a $3d^8$ atomic configuration according to a prescription suggested by Mattheiss. A second neighbor, s-p-d Slater-Koster interpolation scheme is then used to generate the "paramagnetic" bands. The 27 Slater-Koster parameters are determined from a least-squares fit to results obtained from a KKR calculation of the "paramagnetic" bands. Given the "paramagnetic" bands and values for U^{d-d}_{eff} the ferromagnetic band structure can then be obtained from a self consistent solution of the energy band equations.

Our choice of values of U^{d-d}_{eff} for t_{2g} and e_g symmetry was determined by requiring that the moment as well as the ratio of e_g to t_{2g} character in the moment agreed with experiment. Spin splitting of s-states was ignored. The resulting ferromagnetic band structure can quite simply be described in terms of a rigid splitting of 1.94 eV for d-states and a strong wave vector dependent splitting of other states due to s-d hybridization effects. A rather strong wave vector dependence of the matrix elements was also found which indicates that Wannier functions overlap considerably for the band structure we have adopted.

The numerical procedure for evaluating the relevant integrals is based on the Gilat-Raubenheimer (GR) integration scheme[10] and is the same as that used in reference (1). Six spin up and six spin down bands were incorporated in this calculation. The traditional method of replacing $(f_{\vec{n}\vec{k}\downarrow} - f_{\vec{m}\vec{k}+\vec{q}\uparrow})$ by one or zero in each of the small cubes used in the (GR) integration scheme was found to cause some convergence problems in the calculation. This type of problem will always be present in the (GR) scheme but it is amplified in the iron calculation because the Fermi level falls in the middle of the d-bands. This particular problem can be avoided by adopting an integration procedure proposed by Lehmann and Taut[11] which breaks the irreducible Brillouin zone up into tetrahedrons instead of cubes.

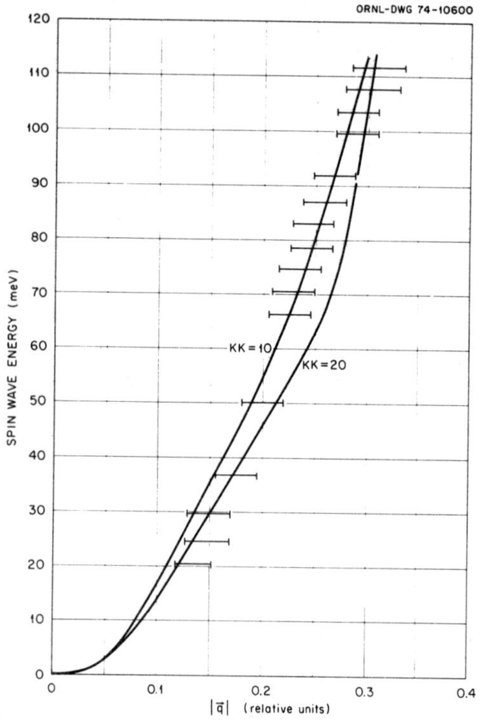

Fig. 1. Spin wave dispersion curve for iron. The wave vector, $\vec{q}$, is measured in units of $2\pi/a_o$. Solid curves labeled KK = 10 and KK = 20 were obtained using a (GR) integration mesh of 440 and 3080 cubes respectively in the irreducible zone. The bars represent neutron scattering results of Mook and Nicklow.[12]

In order to overcome this problem within the (GR) scheme one must go to a relatively large number of cubes in the irreducible zone. The results presented in this paper were obtained using up to 3080 cubes in the irreducible zone. We feel that the numerical results based on 3080 cubes, while not exact, do provide a reasonably good description of $\chi(\vec{q},\omega)$.

In order to compare with neutron scattering experiments we obtained numerical results for Im $\chi(\vec{q},\omega-i\varepsilon)$, which are directly related to the neutron scattering cross-section, as a function of neutron energy loss, ω, for fixed values of the momentum transfer, $\vec{q}$, along the principal symmetry directions. The results were found to be nearly independent of the direction of $\vec{q}$. Well defined spin wave peaks were found at low $|\vec{q}|$. As $|\vec{q}|$ was increased these peaks shifted to higher energies, broadened somewhat, and eventually died out completely. The spin wave peak was found to completely vanish for $|\vec{q}|$ somewhere in the range $.35 \leq |\vec{q}| \leq .4$, where $|\vec{q}|$ is measured in units of $2\pi/a_o$ (a_o = lattice parameter). The spin wave dispersion curve obtained from the position of the spin wave peaks was found to be virtually isotropic and is shown in Fig. 1 plotted as a function of $|\vec{q}|$. The curves labeled KK = 10 and KK = 20 refer to the use of 440 and 3080 cubes respectively in the irreducible zone. The bars represent room temperature ($T/T_c \cong .3$) neutron scattering results obtained by Mook and Nicklow.[12]

4. COMPARISON WITH EXPERIMENT AND CONCLUSIONS

The results presented in Fig. 1 should be viewed as a demonstration that the itinerant model is capable of providing an adequate description of the low temperature spin wave dispersion curve in iron. The isotropic character of the calculated dispersion curve is a non-trivial result of the calculation and is in agreement with experiment.

In order to compare the theoretical and experimental spin wave scattering intensities it must be remembered that the experiment was performed at constant energy (variable $\vec{q}$) while the calculation was made for constant $\vec{q}$ (variable energy). Generally speaking this would not cause any serious problems but, in this case, due to computer time limitations the calculations can not be made at enough $\vec{q}$ points to allow us to fold in the experimental resolution function and calculate the spin wave scattering intensity which one would expect from a constant energy experiment. It is possible to infer, however, that because of the steepness of the dispersion curve the broadening in energy which was found to occur in the calculated spin wave peaks as $|\vec{q}|$ increases is not sufficient to cause any significant broadening in $\vec{q}$ beyond that expected from resolution effects. In other words the calculation would predict no significant broadening of the neutron groups in a constant energy experiment (apart from that expected from resolution effects), which is in agreement with the neutron measurements.

Finally, the disappearance of the spin wave somewhere in the range $.35 \leq |\vec{q}| \leq .40$, which corresponds to an energy range of about 120 to 130 meV, is in excellent agreement with experiment. This result does not appear to be an artifact of the numerics since the same result was obtained using 440 and 3080 cubes in the integration procedure.

In summary then, it appears that the itinerant model is capable of producing good overall quantitative agreement with the low temperature neutron scattering results in iron. Thus, the results of this calculation tend to support the view that the itinerant model is adequate for describing the low temperature magnetic properties of transition metals without recourse to ad hoc assumptions about local moment behavior.

REFERENCES

*Research sponsored by the U. S. Atomic Energy Commission under contract with Union Carbide Corp.
†Present address, Department of Physics, Brookhaven National Laboratory, Upton, New York 11973.

1. J. F. Cooke and H. L. Davis, AIP Conf. Proc. No. 10, 1218 (1972).
2. J. F. Cooke, Phys. Rev. B 7, 1108 (1973).
3. C. Herring, in Magnetism, edited by G. T. Rado and H. Suhl (Academic, New York, 1966).
4. T. Arai and M. Parrinello, Phys. Rev. Lett. 27, 1226 (1971).
5. L. C. Bartel, Phys. Rev. B 7, 3153 (1973).
6. M. B. Stearns, Phys. Rev. B 8, 4383 (1973).
7. J. Hubbard, Proc. Roy. Soc., A 276, 239 (1963).
8. D. A. Goodings and R. Harris, Phys. Rev. 178, 1189 (1969).
9. J. F. Cooke, to be published.
10. G. Gilat and L. J. Raubenheimer, Phys. Rev. 144, 390 (1966).
11. G. Lehmann and M. Taut, Phys. Stat. Sol. 54, 469 (1972).
12. H. A. Mook and R. M. Nicklow, Phys. Rev. B 7, 336 (1973).

THE INFLUENCE OF MANY-ELECTRON EXCHANGE EFFECTS ON THE GROUND STATE TOTAL SPIN IN FERROMAGNETICS

M.POPOVIĆ-BOŽIĆ
Institute of Physics, 11000 Beograd, P.O.Box 57

ABSTRACT

In the new approximate Hamiltonian[1] for exchange interaction of two Hund's rule atoms (which has been derived from first principles with the aid of Girardeau's second quantization representation for composite particles[2]) the first and the second term associated with the exchange of one and two pairs of electrons respectively, are studied in the case of atomic spin one. The aim of the study was to investigate whether two-pair exchange effects can affect the value of the two-atom ground state spin. The answer to the above question is affirmative. In fact, in the ground state of two Hund's rule atoms, each having spin one, apart from the values two and zero (these are the only possible values in the ground state of two atoms in the Heisenberg theory which comprises one-pair exchange effects only) the value one is available in the ground state if two-pair exchange effects can not be neglected. It has been suggested that this result may be used to account for the fact that the mean magnetic moment per atom in magnetic 3d metals is less than the spin magnetic moment of an individual atom.

INTRODUCTION

The concept of exchange coupling of the spins of two or more magnetic atoms, formulated in terms of an effective spin Hamiltonian - the so-called Heisenberg model - has dominated in the theory of magnetism. In the set of assumptions of this concept the following one appears to be crucial: The effects of a simultaneous exchange of two or more pairs of electrons between a pair of neighbouring atoms can be neglected.

As a consequence of this assumption it is concluded that if spin-orbit interaction can be neglected, then in the ground state of two magnetic atoms each having spin S^g the only possible values of the total spin are $2S^g$ and 0.

Hence in order that in a certain crystal a spontaneous magnetization may occur it is necessary and sufficient that the exchange integral be positive at least for certain pairs of neighbouring atoms.

In other words, in the Heisenberg concept of exchange coupling, either the exchange integral is negative and there is no magnetic ordering or the exchange integral is positive and the macroscopic spin of the ordered set of atoms has to be as high as possible (the mean magnetic moment per atom, $\bar{\mu}$, has to be equal to the spin magnetic moment of free atoms, $\mu_S=2S^g\mu_B$).

However, the mean magnetic moment $\bar{\mu}$ in metals and alloys of all 3d transition elements is less than $2S^g\mu_B$ despite the fact that the atomic spin S^g, determined by applying Hund's rule, can be taken to be good quantum number, even for atoms in a crystal. On the other hand, in many calculations[3,4] of the exchange integral in magnetically ordered 3d transition metals, it has been found to be negative, which is in contradiction with Heisenberg's condition for magnetic ordering.

Among other numerous unexplained magnetic phenomena the above two belong to the fundamental ones.

In the theory exposed in the present paper it will be shown that the above two facts, which appear to be contradictory in the Heisenberg model, are no longer contradictory if we do not neglect two-pair exchange effects.

1. EXCHANGE INTERACTION OPERATORS ASSOCIATED WITH THE EXCHANGE OF ONE AND TWO PAIRS OF ELECTRONS

In the previous paper[1] the laws of the two-atom exchange coupling have been expressed in terms of an effective Hamiltonian $H^{ex}_{nm}(N)$ operating in spin space only, but involving one, two, ..., N pair exchange effects (N is the number of electrons in the unfilled atomic shell). There it was pointed out that the assumptions we have made in order to obtain this new model for the exchange interaction, are applicable mainly in the case of 3d metals where eigenvalues of the square of the operator of atomic spin and of its z-projection can be taken to be good quantum numbers.

For two atoms with less than half-filled shells, fixed at points n and m, exchange interaction operator involving effects of exchange of one and two pairs of electrons has the form[1]

$$H^{ex}_{nm}(N)=H^{(1)}_{nm}(N)+H^{(2)}_{nm}(N) \tag{1.1}$$

$$H^{(1)}_{nm}(N)=-N^2\cdot Q^{(1)}_{nm}(N)\sum_{\substack{S_1S_2S_3S_4=-S^g\\ S_1+S_2=S_3+S_4}}^{S^g} {}^{N}I_1(S_1S_2S_3S_4)a^+_{nS_1}a_{nS_3}a^+_{mS_2}a_{mS_4} \tag{1.2}$$

$$H^{(2)}_{nm}(N)=\binom{N}{2}^2\cdot Q^{(2)}_{nm}(N)\sum_{\substack{S_1S_2S_3S_4=-S^g\\ S_1+S_2=S_3+S_4}}^{S^g} {}^{N}I_2(S_1S_2S_3S_4)a^+_{nS_1}a_{nS_3}a^+_{mS_2}a_{mS_4} \tag{1.3}$$

where S_1,S_2,S_3,S_4 stand for atomic spin z-component. The normal product of atomic creation and annihilation operators $a^+_{nS_1}a_{nS_3}$ operates in the spin space of the atom in site n in the following way:

$$a^+_{nS_1}a_{nS_3}|nS^gS_n\rangle=\delta_{S_3S_n}|nS^gS_1\rangle \tag{1.4}$$

$Q^{(j)}_{nm}(N)$ and $I_j(S_1S_2S_3S_4)$, j=1, 2, are the matrix elements of exchange interaction defined by

$$Q^{(j)}_{nm}(N)=\int u^*_{ng}(r_1\ldots r_N)u^*_{mg}(r'_1\ldots r'_N)\Big[2E_g(N)+\frac{Z^2e^2}{|n-m|}+\sum_{i,k=1}^{N}\frac{e^2}{|r_i-r'_k+n-m|}-\sum_{i=1}^{N}\frac{Ze^2}{|r'_i-n+m|}-\sum_{i=1}^{N}\frac{Ze^2}{|r_i+n-m|}\Big]$$
$$u_{ng}(r'_1+m-n,\ldots,r'_j+m-n,r_{j+1}\ldots r_N)\,u_{mg}(r_1+n-m,\ldots,r_j+n-m,r'_{j+1}\ldots r'_N)dr_1\ldots dr_N dr'_1\ldots dr'_N$$
$$j=1,2 \tag{1.5}$$

$$I_j(S_1S_2S_3S_4)=\sum_{\substack{\sigma_1\ldots\sigma_N\\ \sigma'_1\ldots\sigma'_N}}\chi^*_{S^g,S_1}(\sigma_1\ldots\sigma_N)\,\chi^*_{S^g,S_2}(\sigma'_1\ldots\sigma'_N)\times$$
$$\times\chi_{S^g,S_3}(\sigma'_1\ldots\sigma'_j,\sigma_{j+1}\ldots\sigma_N)\times$$
$$\times\chi_{S^g,S_4}(\sigma_1\ldots\sigma_j,\sigma'_{j+1}\ldots\sigma'_N)$$
$$j=1,2 \tag{1.6}$$

$u_{ng}(r_1\ldots r_N)$ and $\chi_{S^g,S_k}(\sigma_1\ldots\sigma_N)$ are the coordinate and spin part of atomic ground state wave function in the coordinate system centered at nucleus. $E_g(N)$ is the ground state energy of the atom.

Eigenstates of the operator $H^{ex}_{nm}(2)$ are simultaneously the eigenstates of the square of the two-atom spin and of the operator of the z-component of the two-atom spin.

2. EXCHANGE INTERACTION OF TWO ATOMS HAVING SPIN ONE AND TWO 3d ELECTRONS

In this Section we will study the operators $H^{(1)}_{nm}(2)$ and $H^{(2)}_{nm}(2)$.

$(S_1S_2S_3S_4)$	$^{aa}I_1(S_1S_2S_3S_4)$	$^{\alpha\alpha}I_2(S_1S_2S_3S_4)$
(1111)	1	1
(1010)	1/2	0
(1001)	1/2	1
(0000)	1/2	1
(1-11-1)	0	0
(1-1-11)	0	1
(001-1)	1/2	0

Table 1. The values of spin factors in first and second order exchange interaction matrix elements for atoms with N=2 and S^g=1.

The values of coefficients $^2I_1(S_1S_2S_3S_4)$ and $^2I_2(S_1S_2S_3S_4)$ for different combinations of numbers S_1,S_2S_3,S_4 are given in Table 1, whereas we will take the coefficients $Q^{(j)}_{nm}(2)$, j=1,2 as unknown parameters.

With the aid of the square of the two-atom spin and of the operator of the z-component of the two-atom spin the following results for the eigenvalues of $H^{ex}_{nm}(2)$ are obtained.

$$E^{ex}_2(2,nm)=-4\cdot Q^{(1)}_{nm}(2)+Q^{(2)}_{nm}(2)$$
$$E^{ex}_1(2,nm)= \quad -Q^{(2)}_{nm}(2)$$
$$E^{ex}_0(2,nm)=2\cdot Q^{(1)}_{nm}(2)+Q^{(2)}_{nm}(2) \qquad (2.1)$$

Let us investigate the dependence of the order of energy levels $E^{ex}_2(2,nm)$, $E^{ex}_1(2,nm)$, $E^{ex}_0(2,nm)$ on the relation between coefficients $Q^{(1)}_{nm}(2)$ and $Q^{(2)}_{nm}(2)$. We have to solve six sets of inequalities. One may proceed by taking one inequality from each of the following three pairs:

$$E^{ex}_2(2,nm)\gtrless E^{ex}_1(2,nm) \Leftrightarrow Q^{(1)}_{nm}(2)\lessgtr\frac{1}{2}\cdot Q^{(2)}_{nm}(2)$$
$$E^{ex}_2(2,nm)\gtrless E^{ex}_0(2,nm) \Leftrightarrow Q^{(1)}_{nm}(2)\lessgtr 0$$
$$E^{ex}_1(2,nm)\gtrless E^{ex}_0(2,nm) \Leftrightarrow Q^{(1)}_{nm}(2)\lessgtr -Q^{(2)}_{nm}(2) \qquad (2.2)$$

In Fig. 1 the plane $Q^{(1)}_{nm}(2)$, $Q^{(2)}_{nm}(2)$ is divided into six separated regions. The coefficients $Q^{(1)}_{nm}(2)$ and $Q^{(2)}_{nm}(2)$ in these regions satisfy one of the above six sets of inequalities, so that each of the six possible orders is associated with one of the regions.

The direct inspection shows:

If $Q^{(2)}_{nm}(2)=0$, we have the same conditions for the ground state value of the total spin as in Heisenberg's theory. In this particular case, if $Q^{(1)}_{nm}(2)$ is greater (less) than zero, the two-atom spin in the ground state equals two (zero).

If $Q^{(2)}_{nm}(2)\neq 0$, the test for the ground state value of the spin is altered quantitatively and qualitatively. In order that the total spin may be equal to two it is not sufficient that $Q^{(1)}_{nm}(2)>0$ but also $Q^{(1)}_{nm}(2)>\frac{1}{2}Q^{(2)}_{nm}(2)$ must hold. The most interesting effect is the possibility that the two-atom spin be equal to one in the ground state. This value, as a ground state value, is not allowed in the Heisenberg approximation. In our model the two-atom spin in the ground state may be equal to one if

$$Q^{(2)}_{nm}(2)>0 \text{ and } -Q^{(2)}_{nm}(2)<Q^{(1)}_{nm}(2)<Q^{(2)}_{nm}(2)/2.$$

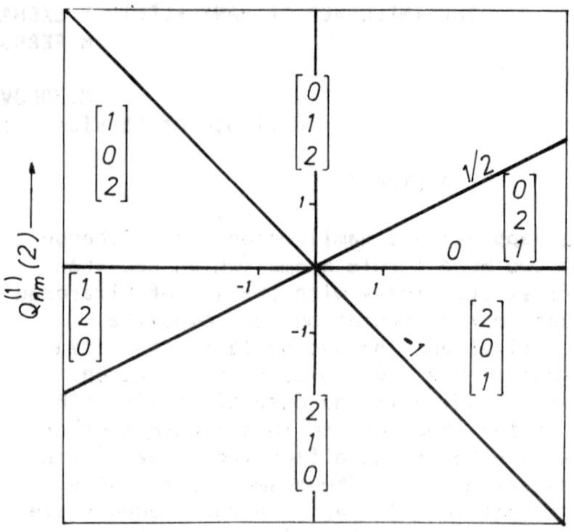

Figure 1. The illustration of the solutions of six sets of inequalities obtained from the inequalities (2.2) so that each of the sets corresponds to one of six possible orders of energy levels. The symbol $\begin{bmatrix}2\\0\\1\end{bmatrix}$ means that $E^{ex}_1(2,nm)<E^{ex}_0(2,nm)<E^{ex}_2(2,nm)$. The other similer symbols have the analogous meaning. The number on each line equals to the tg. of the angle between that line and the $Q^{(2)}_{nm}(2)$ axes.

CONCLUSION

In the new model, involving one and two-pair exchange effects, Heisenberg's conclusion that the total two-atom ground state spin is either $2S^g$ or 0 does not hold. The two pair exchange effects alter the total ground state spin, and in a particular case, when they can be neglected with respect to the one-pair exchange effects, Heisenberg's result is valid.

Accordingly, two-pair exchange effects can influence the value of the mean magnetic moment per atom in such a manner as to make it to be $0<\bar{\mu}<2S^g\mu_B$. (The importance of two-pair exchange effects is discussed more in the papers[5].) A precise calculation of the total spin i.e., of the mean magnetic moment in monocrystals is a problem which has not yet been solved. For the time being we assert that the fact that an average magnetic moment per atom is not exactly equal to $2S^g\mu_B$ can be explained without assuming that d-electrons are itinerant or that s-d interaction gives a negative contribution to the magnetization, but by taking into account higher order effects of electronic exchange.

REFERENCES

1. M. Popović-Božić, phys. stat. sol. (b) 65 (1974), 305.
2. M.D. Girardeau, J. Math. Phys. 4 (1963), 1096; 11 (1970), 681; 12 (1971) 1799.
3. A.Y. Freeman, R.E. Watson, Phys. Rev. 120 (1960), 1439.
4. A.Y. Freeman, R.K. Nesbet, R.E. Watson, Phys. Rev. 125 (1962), 1978.
5. Z. Marić, M. Popović-Božić, A New Model for the Ground State Spin Determination in Ferromagnetics, I, II, in press.

MULTIPLE SPIN CORRELATION EFFECTS

N. Giordano and W. P. Wolf
Department of Engineering and Applied Science, Yale University, New Haven, Connecticut 06520

ABSTRACT

A simple two dimensional Ising model with a particular arrangement of ferromagnetic and antiferromagnetic nearest neighbor interactions is used to illustrate some of the unusual effects which may be expected in systems in which multiple spin correlation effects are important. These systems are unusual in that many of the most common statistical approximations do not give even qualitatively correct descriptions of their behavior. The results provide the basis for a microscopic explanation of some unexpected optical and neutron scattering measurements recently reported for dysprosium aluminum garnet.

I. INTRODUCTION

There have been many studies of Ising models with different kinds of lattices and various combinations of interactions, and the results have been used to explain and predict phenomena in real systems. Among the materials which have been thought of as Ising-like is the antiferromagnet dysprosium aluminum garnet (DAG),[1] and the recent discoveries of some unexpected magneto-optical[2] and neutron scattering[3] effects in fields along [111] were therefore interpreted initially as due to complications in the Hamiltonian, not represented by the Ising approximation. We shall demonstrate that the observed effects may in fact be explained in terms of an Ising model, but one which belongs to a class which is not normally considered. These are systems in which the topology of the lattice is important, with the result that spin correlations involving several sites become essential in determining qualitative aspects of the macroscopic behavior.

In Section II we will introduce and derive the properties of a simple two dimensional model which exhibits all the important features of this class of systems, without the complications of a three dimensional lattice or the long range interactions which must be included in the discussion of real materials. Then in Section III we discuss the unusual properties of our model and indicate how these properties relate to those of DAG. Finally in Section IV we shall discuss some of the more general implications of our findings for other systems.

II. THE MODEL

The model we shall consider is shown in Fig. 1. It is a Kagomé net with Ising spins at each vertex, in which half of the bonds are ferromagnetic and half antiferromagnetic. Each spin has a similar number of nearest neighbor (nn) interactions, but the topology of the lattice is such that the two sublattices are clearly inequivalent. In particular, the ferromagnetically coupled nn of the a type spins are also ferromagnetically coupled nn of each other, while the ferromagnetically coupled nn of the b type spins are not even nn of each other.

In zero field the model is equivalent to the ferromagnetic Kagomé net, and from the exact result[5] we know that both models exhibit a second order phase transition. The behavior in a field is, of course, not known exactly but we can get a qualitative understanding from the following general arguments.

At low temperatures and high fields, the system will be in its ground state with all spins aligned parallel to H, and the lowest energy excitations will correspond to states with either one a type spin or one b type spin flipped. These excitations have equal energies and degeneracies and hence occur with equal probability. Similarly, excitations in which two spins are flipped also occur with equal probability on the two sublattices. However, if we now consider excitations of three spins, we see that they are no longer symmetrical with respect to the two sublattices. In particular, there can be an excitation in which three a type spins which are all nn of each other are flipped, but there is no such excitation involving b type spins. Thus the probabilities of spins on the two sublattices being excited are not the same, and therefore the sublattice magnetizations will not be equal. Hence we see that in the high field paramagnetic state there will be a non-zero staggered magnetization.

We can also investigate the low T, low H properties by considering excitations from the appropriate ground state. Because of the inequivalence of the two sublattices, there will now be two possible ground states which we will denote as A^+ and A^-, where A^+ is the ground state with the a sublattice antiparallel to the field and A^- is the time reversed conjugate of A^+. Just as in the high field case, excitations involving one or two spins occur with equal probability from either ground state, but excitations of three (or more) spins do not. Thus the two antiferromagnetic states will have different properties and, except in zero field, only one will be stable.

We have calculated the properties of this model over a range of fields and temperatures using the Monte Carlo method,[6] and Fig. 2 shows the results for a temperature $T = 0.75\ T_N$, where T_N is the Néel temperature.[7] At high and low fields the Monte Carlo results could be checked against the leading terms of exact series expansions, and it can be seen in Fig. 2 that the agreement is quite satisfactory. The series expansions were also used to calculate the behavior of the non-equilibrium states near H = 0. The results for other temperatures below T_N were all qualitatively similar.

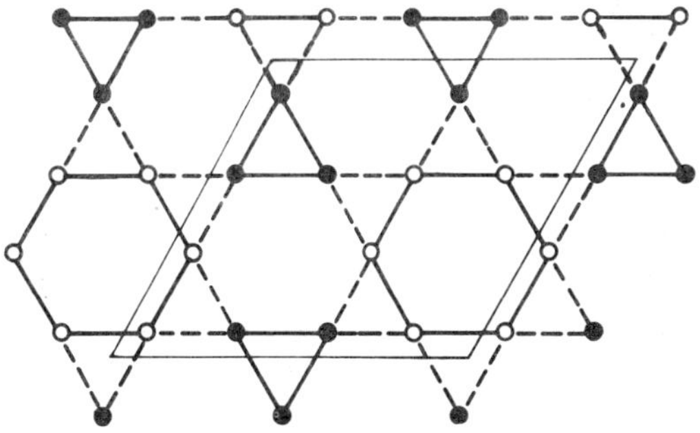

Fig. 1 Two dimensional net with ferromagnetic —— and antiferromagnetic ---- interactions. Ising spins $\sigma = \pm 1$ are denoted by ● (a-sites) and o (b-sites). The unit cell (shown by the faint line) contains 12 spins.

III. DISCUSSION

Figure 2 shows that our model has several properties not found in "normal" antiferromagnets. First, the staggered magnetization is non-zero for all finite values of H. This implies that there will be no phase transition in the presence of a field, in marked contrast to "normal" antiferromagnets which generally undergo second order phase transitions over wide ranges of field.

Second, the two antiferromagnetic states are found to have different properties in the presence of a field and only one of the states will therefore be stable for any particular field. This behavior is also in sharp contrast to that of a "normal" antiferromagnet, for which the two time reversed antiferromagnetic states coexist in equilibrium for all values of H and T for which there is long range order.

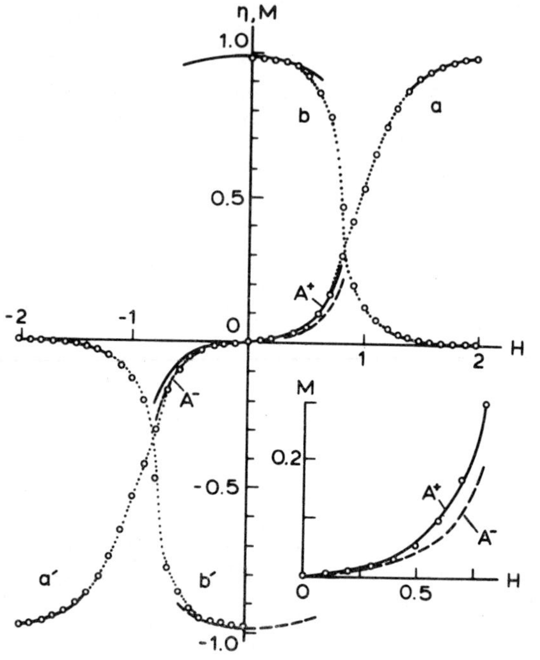

Fig. 2 Variation of magnetization M (curves a and a') and staggered magnetization η (curves b and b') with magnetic field for the net shown in Fig. 1. The curves ... were obtained by extrapolation of Monte Carlo calculations, the sizes of the o's roughly indicating the estimated uncertainties in the extrapolations. The solid —— and broken --- curves were obtained from low temperature series expansions. Near H = 0 the solid curves correspond to the A^+ state (stable for H > 0), the broken curves correspond to the A^- state (stable for H < 0). η and M are expressed in units of their values at T = 0 K, and H in units of the critical field at T = 0 K.

In low magnetic fields, the energy difference between the two states will not be very large, and since a transition from one state to the other would involve the coherent reversal of almost all the spins, we may expect real systems of this kind to exhibit hysteresis effects as the field is cycled through zero.[4]

In the above discussion of excitations in both low and high fields, we noted that the unusual effects entered into the calculations only when simultaneous excitations of three or more nn spins were considered. This feature will persist also at intermediate fields and at higher temperatures, and we therefore conclude that it is the multiple spin correlations reflecting the topology of the lattice which play a crucial role in determining the qualitative behavior of these kinds of systems. Mean field theory and other approximations which ignore correlation effects would therefore not predict any of the properties discussed above. This failure is somewhat unusual, as mean field theory is generally considered to be a rather reliable guide to the qualitative features of the magnetic behavior.

Recent experiments have shown that the behavior of DAG is similar to that of our type of Kagomé net. More specifically, the different magnetizations of the two antiferromagnetic states have been found,[2,8] and the corresponding hysteresis in the response to an applied field has been observed.[2,9] Also, neutron scattering experiments[3] have shown that a significant amount of long range antiferromagnetic order is present in regions of the phase diagram which other measurements clearly identify as belonging to the paramagnetic state. This behavior is discussed more fully in Ref. 4, where it is shown that all of the experimental results can be described quantitatively within the accepted Ising approximation for DAG, provided that multiple spin correlation effects are taken into account.

IV. CONCLUSIONS

We have considered a specific model to illustrate a microscopic process in its simplest form, but the effects which we have found are quite general. Alben et al.[10] have recently discussed the symmetry considerations which show whether coupling between the antiferromagnetic order parameter and the field is allowed for other types of lattices. It turns out that for some lattices there are no unusual effects until terms of ninth degree in the field are considered, and it seems clear that quite subtle approximations would be required to treat such cases properly. So far, DAG is the only real system in which these effects have been observed experimentally,[2,3,9] but one may now be able to find other examples.

In this connection we should note that superficially similar effects have in fact been known for some time in systems in which a weak ferromagnetic moment is allowed by symmetry.[11,12] However, in such systems, the physical mechanisms which provide the coupling between the antiferromagnetic order parameter and the field are quite different, and in particular they arise from specific terms in the microscopic Hamiltonian which reflect the particular symmetry.[12] Correspondingly, there is a first order coupling between the field and the antiferromagnetic order, which manifests itself as a spontaneous moment for all temperatures below T_N, even in the absence of a field.

For both DAG and our simple Kagomé net, there is no spontaneous moment in zero field and there are no special terms in the Hamiltonian. Both systems can be described in terms of the usual Ising model interactions and the special features arise only from the topological arrangement of the individual interactions.

In more complicated systems it is of course possible to have both multiple spin-correlation effects and the effects of additional terms in the Hamiltonian, and one would then have to consider competition between them. Some of these possible complications will be discussed elsewhere.[4]

ACKNOWLEDGEMENTS

We would like to thank R. Alben, M. Blume, J.F. Dillon Jr., E. Yi Chen, and M.F. Thorpe for a number of helpful discussions. One of us (N.G.) gratefully acknowledges the support of an NSF graduate fellowship. This work was supported in part by AROD and by NSF.

REFERENCES

1. See for example W.P. Wolf, B. Schneider, D.P. Landau and B.E. Keen, Phys. Rev. B5, 4472 (1972).
2. J.F. Dillon Jr., E. Yi Chen and W.P. Wolf, Proc. Int. Conf. on Magnetism, Moscow, 1973 (in press).
3. M. Blume, L.M. Corliss, J.M. Hastings and E. Schiller, Phys. Rev. Lett. 32, 544 (1974).
4. N. Giordano and W.P. Wolf (to be published).
5. S. Naya, Prog. Theor. Phys., Japan 11, 53 (1954).
6. J.R. Ehrman, L.D. Fosdick and D.C. Handscomb, J. Math. Phys. 1, 547 (1960).
7. From Ref. 5, $T_N = 2.14\ J/k_B$ where J is the exchange constant and k_B is Boltzmann's constant.
8. J.F. Dillon Jr., E. Yi Chen, N. Giordano and W.P. Wolf, Phys. Rev. Lett. 33, 98 (1974).
9. J.F. Dillon Jr., E. Yi Chen and W.P. Wolf, AIP Conf Proc. 10, 344 (1974).
10. R. Alben, M. Blume, L.M. Corliss and J.M. Hastings, Phys. Rev. (in press).
11. I.E. Dzialoshinsky, J. Phys. Chem. Solids 4, 241 (1958).
12. T. Moriya, Phys. Rev. 117, 635 (1960); ibid 120, 91 (1960).

THE SPECTRAL WEIGHT OF LOW-LYING EXCITATIONS IN THE ONE-DIMENSIONAL SPIN ½ HEISENBERG ANTIFERROMAGNET*

J. C. Bonner† and B. Sutherland
University of Utah, Salt Lake City, Utah 84112
Peter M. Richards
Sandia Laboratories, Albuquerque, New Mexico 87115

ABSTRACT

We have examined numerical calculations on a spin ½ one-dimensional (1D) Heisenberg antiferromagnet for N = 8 spins, both for the states and their spectral weights. This work is of interest because of the recent neutron scattering experiments in $CuCl_2 \cdot 2NC_5D_5$ which show the spectral weight to be concentrated at the des Cloizeaux and Pearson (dCP) spin wave frequencies. The states form a two-parameter continuum between the dCP value $E_1(k) = \pi|J||\sin k|$ and the upper limit $E_2(k) = 2\pi|J||\sin(k/2)|$. Though most of the spectral weight is at the dCP frequency for N = 8, about 10% of it is found near E_2 for $k = \pi$. There is no numerical evidence for the existence of singlet bound states as predicted by Ovchinnikov.

INTRODUCTION

Recent experimental determinations of the neutron scattering cross section as a function of wave-vector k and frequency (or energy) ω[1] have been carried out on $CuCl_2 \cdot 2NC_5D_5$ which has become the archetypal experimental realization of an $S=\frac{1}{2}$ one dimensional antiferromagnet (AF). It appears that the spectral weight is concentrated at the des Cloizeaux and Pearson (dCP) excitation frequencies[2] over the whole of the Brillouin zone. In the light of conventional spin wave theory and some recent work,[3,4] this result may not seem too surprising. However exact calculations on the linear S-½ XY model, whose spectrum of elementary excitations shows a substantial resemblance to that of the isotropic antiferromagnet, show a very different behavior in terms of the spectral weight.[5]

We examine here the nature of the low-lying excitations and the spectral weight of the corresponding correlation functions from a predominantly numerical point of view.[6] We conclude that although most of the spectral weight is at the dCP value, there is some weight at higher frequencies, at least for finite chains. We find no evidence for the existence of singlet bound states, as predicted by Ovchinnikov.[7,8] Limitations on machine calculations have restricted us in this preliminary study to rings no larger than 8 or 9 spins.

LOW-LYING SPECTRAL EXCITATIONS

The lowest-lying states of the spectrum for an N=8 linear antiferromagnet are shown in Fig. 1 and will be discussed in the light of expectations for the $N \to \infty$ limit. A prescription scheme which indicates the wave vector k, the type of state, whether singlet, triplet, quintet, etc., and, roughly, the excitation energy has been employed to characterize the various states. This prescription is an extension of the Yang picture of hole-particle excitations from a filled Fermi sea. In Fig. 1 we observe that the primary (or lowest-lying) dispersion branch φ contains only a single state, the singlet AF ground state. Branch φ will have the dCP dispersion relation

$$E_1(k) = \pi|J||\sin k| \tag{1}$$

(where energies are measured upwards from the ground state and the Hamiltonian is given as $\mathcal{H} = 2|J| \sum_{i=1}^{N} \vec{S}_i \cdot \vec{S}_{i+1}$, $N + 1 \equiv 1$). The next important branch is branch α containing only triplet states. As $N \to \infty$ branch α approaches the limiting dCP dispersion value (1) as 1/N. However, these states are not the only single particle/hole excitations. There are altogether a total of $N(N+2)/8$ low-lying triplets, related to the α states and forming a triplet spin wave continuum (two-parameter continuum) whose lower limit is the dCP dispersion curve (1), and whose upper limit is the envelope dispersion curve

$$E_2(k) = 2\pi|J||\sin(k/2)| \tag{2}$$

which clearly has twice the amplitude and half the periodicity of (1). (It is interesting to note that this scheme has strong similarities to the two-overturned-spin continuum in the ferromagnetic Heisenberg linear chain.[9])

The excitation scheme so far is in accordance with the analytical work of Yamada,[10] which discusses only triplet excitations. Ovchinnikov,[8] however, also by an analytic calculation, comes to the conclusion that two types of excitation are dominant. One type is the dCP triplets and the second type corresponds to bound state singlets whose dispersion curve is (2), which we obtain as an envelope of the spin wave continuum, except that the Ovchinnikov dispersion curve has a termination point at $k = \pm \pi/2$, whereas our envelope continues on to $k = \pm \pi$. Although we do see a continua of singlet states (one can show that for $N \to \infty$ the singlet and triplet continua are identical) arising out of the dispersion branch γ, there is no evidence from calculations on up and including N = 10 for a dispersion branch of the Ovchinnikov type. The apparent lack of bound states in the AF region of the spectrum is in accordance with analytic work of Johnson et al.,[11] who discuss the evolution of bound states as a parameter in the Hamiltonian varies between XY and Heisenberg limits, and conclude that all bound states appear in the ferromagnetic region of the spectrum. We feel, therefore, that the question of the existence of bound excitations in antiferromagnets (and also in Hubbard models[8]) should be carefully reexamined. It is also interesting to note that if the spin wave excitations form a two-parameter band with upper and lower limits given by (2) and (1), respectively, the double periodicity of the single parameter dCP dispersion curve, which echoes the double periodicity of the spin wave AF is lost, and the Brillouin zone should be regarded as extending not from, say, $k = +\pi/2$ to $k = -\pi/2$ but from $k = \pi$ to $k = -\pi$, as for the linear ferromagnet.

In addition to the singlet and triplet excitations, quintet states, which result from two particle-two hole excitation in the Yang prescription, appear (see dispersion branch δ), and are comparable in energy with states in the spin wave continuum. The higher-order excitations become more apparent at relatively lower energies as N increases, and can hardly be ignored, as has been the case in previous theoretical work. An important consideration is whether these higher-order states carry any spectral weight, which will be examined, numerically, in the next section.

SPECTRAL WEIGHT OF THE VARIOUS EXCITATIONS

In Fig. 2 we show the spectral weight $\langle S_k S_{-k} \rangle_\omega$ (i.e. the Fourier components of the time-dependent correlation function) for the states of Fig. 1 at $k = 3\pi/4$ and π for an 8-spin ring. Since we are dealing with a finite system we employ a histogram representation at a temperature $k_B T/|J| = 0.1$. All weights not shown on the graph are less than those shown by at least a factor of 10^{-4}. The upper and lower limits of the triplet (and singlet) continua [Eqs. (2) and (1)] are indicated by the arrows U and L, respectively. So that spectral weight may be correlated with the individual states of Fig. 1, these are shown on the ω axes of Fig. 2, using the same symbolism as in Fig. 1. From these histograms we conclude that the singlet and

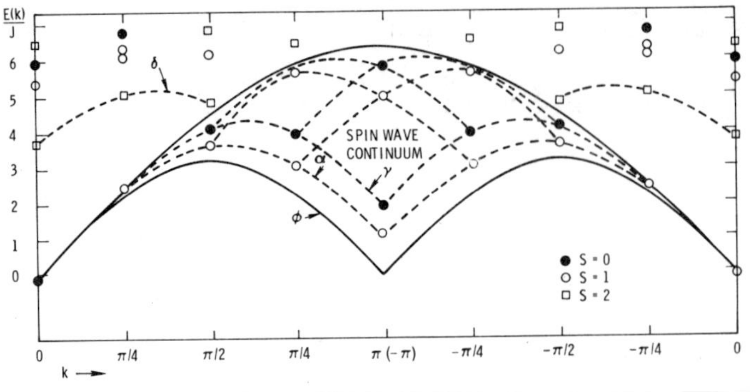

Fig. 1. Low-lying states of the N = 8 spin $\frac{1}{2}$ antiferromagnetic Heisenberg linear chain.

quintet states carry negligible spectral weight, so that previous theoretical work which has discussed only the spin wave triplets[3] may yield reasonable results for the spin dynamics. For $k = \pi/4$ and $\pi/2$ (not shown) the lowest triplet contains all the weight. However, the $k = 3\pi/4$ and $k = \pi$ results show that the higher triplet excitation does contribute a measurable amount to the spectral density. Calculations on a sequence of longer spin rings should test whether an appreciable tail of spectral density above the dCP spin wave states is likely to persist as $N \to \infty$, and give some idea of the form of the true continuum spectral density function. At present one may conclude that the N = 8 results are consistent with the Hartree-Fock calculations,[3] and suggest that further study, numerical, analytical and experimental, might be worthwhile.

INTEGRATED INTENSITY AS A FUNCTION OF WAVE VECTOR

In Fig. 3 we show the integrated spectral density (intensity) as a function of k for several models. Curve A is the result of an ideal spin wave model and is given by the expression, normalized to unity at $k = \pi/2$,

$$I = \frac{\langle S_k\, S_{-k}\rangle}{\langle S_{\pi/2}\, S_{-\pi/2}\rangle} = \frac{1 - \cos k}{\sin k} \propto \frac{1 - \cos k}{\omega_k} \tag{3}$$

which is valid as $T \to 0$ and clearly diverges as $k \to \pi$.

Curve B is the equivalent expression obtained numerically for the N = 8 ring. It is evidently rising steeply for $k > 3\pi/4$, suggesting the possibility of a divergence in the large N limit (only for $N \to \infty$ is the dCP frequency $\omega_k = 0$ at $k = \pi$). At $k = \pi$ curve B is about 50% below $(1-\cos k)/\omega_k$ with ω_k the dCP frequency for N = 8. Curve C is an exact result for ZZ correlations in the linear XY model, which is characterized by a low lying two-parameter spin wave continuum very much like that of the Heisenberg AF. Specifically, the lower and upper limits of the continuum are given by equations (1) and (2), respectively, with the difference of a factor $\pi/2$. The intensity varies as $k/2$, and this result would be true also for the Heisenberg AF if all states within the continuum contributed equally, as is the case for the ZZ correlations for the XY model. However, it is implicit in the work of McCoy[12] that for the XX and YY correlations for the XY model, the staggered susceptibility, and also intensity, diverge as $k \to \pi$. There is evidence from the work of Barber and Baxter[13] that this should happen also for the 1D Heisenberg AF. This is consistent with a two parameter dispersion continuum.

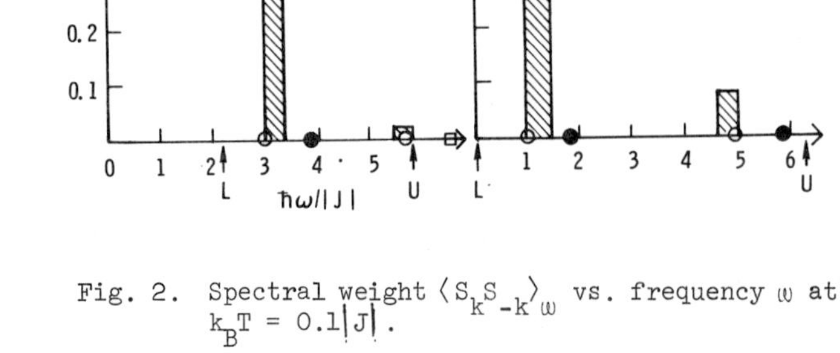

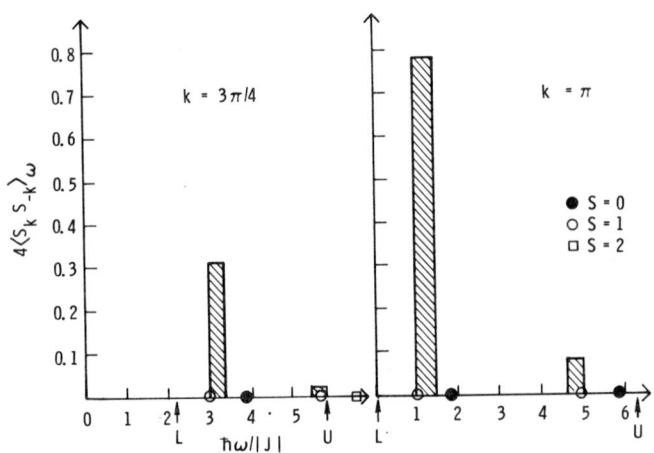

Fig. 2. Spectral weight $\langle S_k S_{-k}\rangle_\omega$ vs. frequency ω at $k_B T = 0.1|J|$.

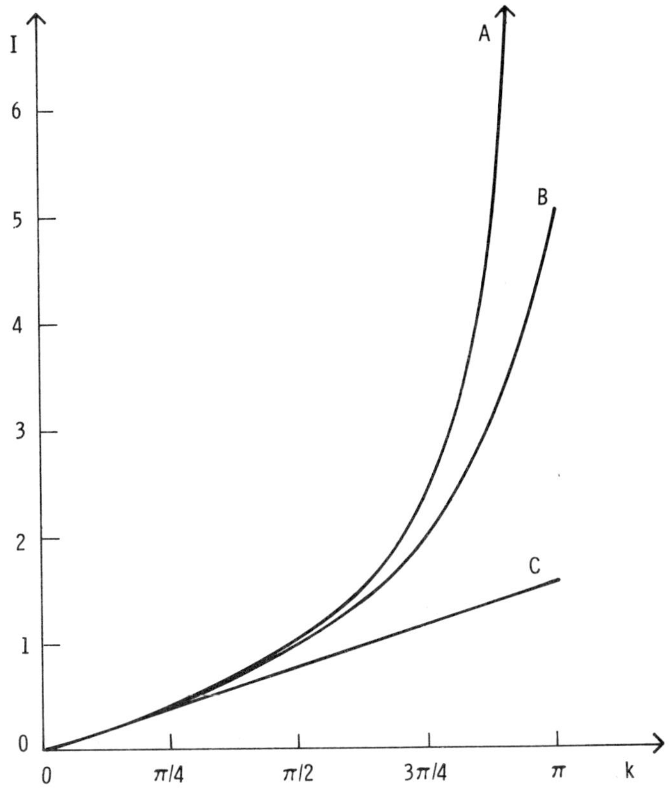

Fig. 3. Integrated intensity, or static correlation $\langle S_k S_{-k} \rangle$, for three models, as explained in text.

* Supported by N.S.F. Grant GP-38705 and by the E.R.D.A. at Brookhaven National Lab. and Sandia Labs.

† Present address: Dept. of Applied Mathematics, Brookhaven National Laboratory, where part of this work was performed.

REFERENCES

1. Y. Endoh, G. Shirane, R. J. Birgeneau, P.M. Richards and S. L. Holt, Phys. Rev. Lett. 32, 170 (1974).
2. J. des Cloizeaux and J. J. Pearson, Phys. Rev. 128, 2131 (1962).
3. T. Todani and K. Kawasaki, Progr. Theor. Phys. (Japan) 50, 1216 (1973).
4. P. C. Hohenberg and W. F. Brinkman, Phys. Rev. B 10, 128 (1974).
5. R. K. G. Liem, Thesis, Carnegie-Mellon University, (1969). Unpublished.
6. J. C. Bonner and M. E. Fisher, Phys. Rev. 135, A640 (1964); P. M. Richards and F. Carboni, Phys. Rev. B 5, 2014 (1972). We also make use here of previously unpublished calculations of F. Carboni.
7. A. A. Ovchinnikov, Sov. Phys. JETP 29, 727 (1969).
8. A. A. Ovchinnikov, Sov. Phys. JETP 30, 1160 (1970).
9. See, for example, J. C. Bonner, thesis, University of London (1968), unpublished.
10. T. Yamada, Progr. Theor. Phys. (Japan), 41, 880 (1969).
11. J. D. Johnson, S. Krinsky and B. M. McCoy, Phys. Rev. 8, 2526 (1973).
12. B. M. McCoy, Phys. Rev. 173, 531 (1968).
13. M. Barber and R. Baxter, J. Phys. C 6, 2913 (1973).

DYNAMICAL SPIN CORRELATION FUNCTIONS IN A SYSTEM OF RANDOMLY DISTRIBUTED SPINS WITH r^{-n} INTERACTIONS

Peter A. Fedders*
Physics Department, Washington University, St. Louis, Missouri 63130

Charles W. Myles and C. Ebner
Battelle Memorial Institute, 505 King Avenue, Columbus, Ohio 43201

ABSTRACT

A theory is proposed for the self-consistent calculation of dynamical two-point spin correlation functions in a system of randomly distributed spins with r^{-n} interactions. It is applicable at all spin concentrations and in the high temperature limit. The formalism employs the previously described "bubble" approximation for the spin self energy while the impurity average is accomplished by a self-consistent effective medium approximation. The results of applying this theory to spin systems interacting via dipole-dipole (r^{-3}) forces are presented and compared with experiment.

I. FORMALISM

There have been a number of calculations of the spectral functions for a system of randomly distributed impurity spins with r^{-n} interactions.[1-7] Many of these treatments are semiphenomenological in that they assume a functional form for a spin correlation function which is then averaged. Also, many of them refer exclusively to either the high or low concentration limit. In this paper we present a first principles method, applicable at all concentrations and in the high temperature limit, which is based on a rigorously formulated theory of interacting spins that has previously been used successfully to calculate spectral functions for perfect lattices of spins.[8] Our method is thus a generalization of this diagrammatic technique and involves no lineshape assumptions or adjustable parameters.

We start by defining a set of two point correlation functions

$$G_\alpha(i,j;t) = \langle A_\alpha(i,t)A_\alpha^\dagger(j,0)\rangle\theta(t) \qquad (1)$$

where $\theta(t)$ is the step function, $\langle\ \rangle$ denotes the ensemble average, and i,j denote lattice sites containing spins. The irreducible multipole operators $A_\alpha(i)$ where $\alpha=(\ell,m)$ form a complete set of spin operators at the site i; their relationship to the usual vector spin operators is discussed elsewhere.[8,9] We assume the anisotropic part of the spin-spin interactions is negligible. Although this restriction is not necessary, it allows us to neglect the off-diagonal Green's functions.

In the high temperature limit, we express the Green's function in terms of a self energy $\Sigma_\alpha(i,j;t)$ defined by[8]

$$\left(i\frac{\partial}{\partial t} - m\,\omega_o\right)G_\alpha(i,j;t) - \sum_k \int d\bar{t}\ \Sigma_\alpha(i,k;t-\bar{t})G_\alpha(k,j;\bar{t}) = i\,\delta_{i,j}\delta(t). \qquad (2)$$

Here and subsequently, a lattice sum is understood to be taken only over occupied sites. In this paper we consider the Hamiltonian

$$H = -\hbar\,\omega_o \sum_i S_z(i) + \tfrac{1}{2}\sum_{i,j,\alpha,\beta} A_\alpha(i)\,J_{\alpha\beta}(i,j)A_\beta(j), \qquad (3)$$

where $\omega_o=\gamma H_o$, γ is the gyromagnetic ratio, H_o is an external magnetic field whose direction defines the z-axis, and the choice of J specifies the interaction.

The derivation of the basic equations is lengthy; we defer the details to a later paper and present only the essential features here. In earlier work[8] we have used the "bubble" approximation for the self energy. This is a lowest order approximation in a hierarchy of self consistent approximations. We have also previously[8] discussed a "local" approximation whereby, within the "bubble" approximation, both the correlation function $G_\alpha(i,j;t)$ and its self energy $\Sigma_\alpha(i,j;t)$ are replaced by the local functions $G_\alpha(i,t)\delta_{i,j}$ and $\Sigma_\alpha(i,t)\delta_{i,j}$. After these have been found, $G_\alpha(i,t)\delta_{i,j}$ is substituted into the non-local "bubble" expression for Σ to give

$$\Sigma_\alpha(i,i';t) = i\sum_{\beta,\gamma,j}|\Omega_{\alpha\beta\gamma}(i,j)|^2 G_\beta(i,t)G_\gamma(j,t)\delta_{i,i'} + i\sum_{\beta,\gamma}\Omega_{\alpha\beta\gamma}(i,i')\Omega^*_{\alpha\gamma\beta}(i',i)G_\beta(i,t)G_\gamma(i',t), \qquad (4)$$

where $\Omega_{\alpha\beta\gamma}=\Sigma_\delta C(\alpha,\delta,\beta)J_{\delta\gamma}/\hbar$, and $[A_\alpha,A_\beta]=\Sigma_\gamma C(\alpha,\beta,\gamma)A_\gamma$. The equations for the local Green's functions are similar to Eqs.(2) and (4), the main difference being the absence of the last term in Eq.(4). Note that $G_\alpha(i,t)$ is not itself a good approximation to $G_\alpha(i,j;t)$ but it can be used to generate a good one.

Although formally both the "bubble" and "local" approximations are best when each spin has a large number Z of nearest neighbors, the constraint is somewhat misleading. For example, the second spectral moment obtained using the full "bubble" approximation is exact for any distribution of spins. We contend that the spectral lineshape at all c is dominated by the interactions of a given spin with a group of other spins and not with a single other spin. This is the basic idea upon which our next approximation is based.

Given a random distribution of spins, $G_\alpha(i,t)$ depends upon its local environment. In order to solve for the local and non-local G_α's we replace $G_\gamma(j,t)$ in any expression where j is summed over by its average $\bar{G}_\gamma(t)=N^{-1}\Sigma_j G_\gamma(j,t)$, where N is the number of spins. This is an effective medium approximation (EMA) reminiscent of the CPA.[10] Further, most experiments are done at zero wavevector and thus measure the spatial average, $\bar{F}_\alpha(t)$, of $F_\alpha(i,t)\equiv\Sigma_j G_\alpha(i,j,t)$; $\bar{F}_\alpha(t)$ is defined in analogy with $\bar{G}_\alpha(t)$. By the use of the EMA in the equation of motion for $G_\alpha(i,t)$ one can solve for $G_\alpha(i,t)$ and $\bar{G}_\alpha(t)$. Similarly, by summing Eq.(2) over j, combining that result with Eq.(4), and using the EMA twice, one can obtain an equation involving only $F_\alpha(i,t)$, $\bar{F}_\alpha(t)$, $G_\alpha(i,t)$, and $\bar{G}_\alpha(t)$.

For numerical computation it is convenient to express the equations in terms of a distribution function. For some interactions (e.g. dipolar), $\Sigma_j|\Omega_{\alpha\beta\gamma}(i,j)|^2 = B_{\alpha\beta\gamma}\Sigma_j f(i,j) = B_{\alpha\beta\gamma}f(i)$ and $\Sigma_j\Omega_{\alpha\beta\gamma}(i,j)\Omega_{\alpha\beta\gamma}(j,i) = B'_{\alpha\beta\gamma}f(i)$. If we let $p(f)df$ be the probability that $f\le f(i)\le f+df$, then, after changing from variable i to f, the local equation can be written

$$\left(i\frac{\partial}{\partial t} - m\,\omega_o\right)G_\alpha(f,t) - i\int d\bar{t}\sum_{\beta\gamma}B_{\alpha\beta\gamma}f \times G_\beta(f,\bar{t})\bar{G}_\gamma(\bar{t})G_\alpha(f,t-\bar{t}) = i\,\delta(t), \qquad (5)$$

while the nonlocal equation takes the form

$$\left(i\frac{\partial}{\partial t} - m\omega_o\right)F_\alpha(f,t) - i\int d\bar{t}\,\sum_{\beta\gamma}\Big\{B_{\alpha\beta\gamma}\,f\,G_\beta(f,t-\bar{t})\bar{G}_\gamma(t-\bar{t}) \tag{6}$$

$$\times F_\alpha(f,\bar{t}) + B'_{\alpha\beta\gamma}\,f\,G_\beta(f,t-\bar{t})\bar{G}_\gamma(t-\bar{t})\bar{F}_\alpha(\bar{t})\Big\} = i\,\delta(t),$$

where now $\bar{G}_\gamma(t) = \int df\,p(f)\,G_\gamma(f,t)$ with an analogous definition for $\bar{F}_\alpha(t)$.

We complete the presentation of the formalism with Eqs.(5) and (6). Given P(f), they may be solved self-consistently. There are two basic approximations in addition to the "bubble" approximation. These are $Z \gg 1$ and the EMA. Both work well as $c \to 1$; as $c \to 0$ the situation is not as clear. However, if one thinks of a "super-lattice" with the spin spacings increased to give the correct c, then the approximations are independent of c. Of course, this picture ignores clustering effects. We note however that if a spin's environment is dominated by one very near neighbor, then the spectral weight of these spins is pushed into the wings of the lineshape. Thus the approximations are best for those spins which contribute to the center of the line. Furthermore, the numerical results presented below tend to support the view that these approximations are at least qualitatively correct as $c \to 0$.

II. RESULTS

In this section, we treat only the dipolar interaction in a s.c. lattice. Further, because of space limitations, numerical solutions are given only for $c = 1$ and $c \to 0$, although the method is applicable at all c. For the dipolar interaction the only correlation functions of interest are those with $\alpha = (1,m)$. Thus, in what follows, whenever a quantity has a Greek index, that index will be replaced by m. Further, it turns out that the correlation functions with $m = \pm 1$ are equivalent and that $F_o(f,t) = \bar{F}_o(t) = 1$, so that it is only necessary to compute $\bar{F}_1(t)$.

Under quite general conditions, it can be shown that[11]

$$P(f) = \int_{-\infty}^{\infty} \frac{du}{2\pi}\, e^{iuf}\, \prod_j \left[1 - c\left(1 - e^{-iuf(k,j)}\right)\right], \tag{7}$$

where the product is over all lattice sites. For arbitrary c, P(f) must be computed numerically; however for $c \to 1$, $P(f) = \delta\big(f - cf(i)\big)$, while for $c \to 0$, $P(f) = \left(8\pi/3\sqrt{3}\,\pi\; c/27f^{2}\right)^{\frac{3}{2}} \exp(-64\pi^3 c^2/243f)$. The $c \to 1$ form is valid independent of the interaction, while the $c \to 0$ form is valid only for dipolar forces in a s.c. lattice.

For $c = 1$, $\bar{G}_\alpha(t) = G_\alpha(f_d,t)$ and $\bar{F}_\alpha(t) = F_\alpha(f_d,t)$, where $f_d = f(i)$ for the dense lattice. The solution for $\bar{F}_1(t)$ is plotted as a function of $\tau = \omega_d t$ in Fig.1; $\omega_d = \gamma^2\hbar\sqrt{s(s+1)}\,/\,\sqrt{12}\,a^3$, a is the lattice spacing, and s is the spin quantum number. The experimental free induction decay (fid) curve[12] for F^{19} in CaF_2 with $H_o \parallel [100]$ is also shown. We believe that our theoretical fid curve is the first entirely first principles calculation for the dipolar lattice. Others[13] have obtained curves which are in better agreement with experiment, but they have used semi-phenomenological theories which contain adjustable parameters. As is evident from the figure, the best agreement with experiment occurs at short times, before the first zero. The discrepancy between theory and experiment with regard to the position of the zeros ranges from 5.65% for the first zero to 22.6% for the eighth zero.[12] It should be re-emphasized that the theoretical curve was obtained from a calculation which is only expected to be accurate to order 1/Z. In light of this and the fact that the results were obtained from a first principles calculation, we regard the calculation as being in good agreement with experiment. Furthermore, we believe that our approximation can be systematically improved.

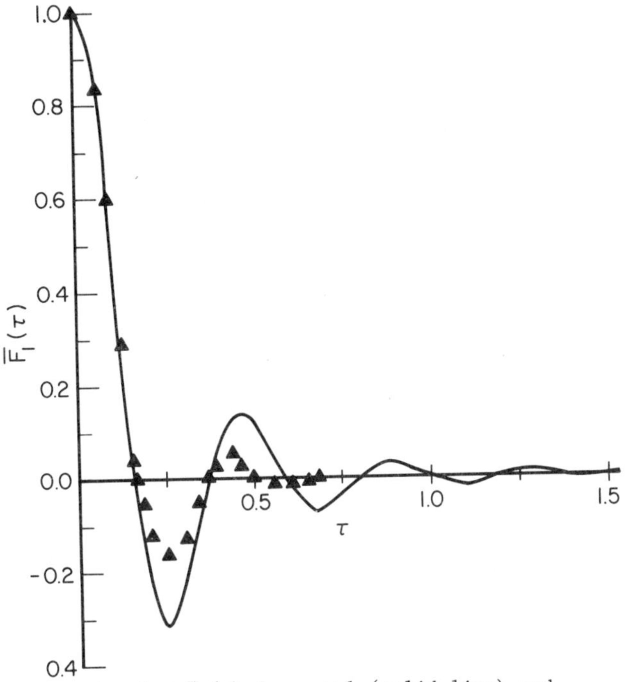

Fig. 1. $F_1(\tau)$ for $c = 1$ (solid line) and experimental fid curve for F^{19} in CaF_2 (triangles) vs. τ.

We have also solved Eqs.(5) and (6) for $c \to 0$, in which case they scale with c and so need only be solved once. The Fourier transform $\bar{F}_1(\omega)$ has Lorentzian-like behavior at small ω, and the linewidth, $1/\bar{F}(\omega = 0)$ is proportional to c. Both features are in agreement with EPR experiments on Cr^{+3} in ruby[3] and with other theories.

In summary, the above theory for computing spin correlation functions in systems of randomly distributed spins with r^{-n} interactions is applicable at all c for $T \to \infty$. It contains no lineshape assumptions or adjustable parameters but rather approaches the problem from an entirely first principles point of view. We have shown that the theory yields magnetic resonance lineshapes for interacting dipolar systems which are in reasonable agreement with experiment in both the $c \to 1$ and $c \to 0$ limits. We are now in the process of using this theory to obtain results on these systems for the intermediate c range and on other systems with r^{-n} interactions.

REFERENCES

* Supported in part by the N.S.F.

1. P. W. Anderson, Phys. Rev. B 2, 842 (1951).
2. C. Kittel and E. Abrahams, Phys. Rev. 90, 238 (1953).
3. W. J. C. Grant and M. W. Strandberg, Phys. Rev. 135, A715 (1964); 135, A727 (1964).
4. C. C. Sung, Phys. Rev. 167, 271 (1968).
5. A. B. Harris, Phys. Rev. B 2, 3495 (1970).
6. J. Hama, T. Inuzuka, and T. Nakamura, Prog. Theor. Phys. 48, 1769 (1972).
7. C. C. Sung and L. G. Arnold, Phys. Rev. B 7, 2095 (1973); C. Ebner and C. C. Sung, Phys. Rev. B 8, 5226 (1973).
8. C. W. Myles and P. A. Fedders, Phys. Rev. B 9, 4872 (1974); C. W. Myles and C. Ebner, to be published in Phys. Rev. B.
9. G. F. Reiter, Phys. Rev. B 5, 222 (1972).
10. B. Velický, S. Kirkpatrick, and H. Ehrenreich, Phys. Rev. 175, 747 (1968).
11. M. H. Cohen and F. Reif, in Solid State Physics, Vol.5, 321 (Ed. by F. Sectz and D. Turnbull), 1957.
12. M. Englesberg and I. J. Lowe, Phys. Rev. B 10, 822 (1974).
13. F. C. Barreto and G. F. Reiter, Phys. Rev. B 6, 2555 (1972).

APPLICATION OF PERCOLATION THEORY TO MAGNETIZATION IN 1D-3D SYSTEMS*

I. Riess and M. Pollak
Department of Physics, University of California
Riverside, CA 92502 U.S.A.

ABSTRACT

We consider magnetization in the Heisenberg model through a modified percolation approach assuming that not any infinite spin cluster may be magnetic as in the Ising model but that also correlations between the various sites must exist which impose restrictions on the topology of the magnetic cluster. An approximation of the correlations between nearest-neighbors (n.n.) is considered (valid at high temperatures, $T > T_c$, for a n.n. interaction, $S = \frac{1}{2}$ and disordered systems), and is first applied to 2D Heisenberg systems where $T > T_c$ for any finite T. It is found that this approximation is a sufficient improvement to exclude the 2D honeycomb and square lattice from being magnetic. In 3D if we assume that the magnetization set in at a critical concentration x_c which is higher than the site percolation critical concentration x_c^o, our result is then self consistent and x_c is found to be quite a bit higher than x_c^o. The result in 3D is then in contradiction with the spin wave approximation of Kumar and Harris where $x_c = x_c^o$, which is also a self consistent one. In particular we find that the diamond structure is not magnetic. While this may indicate a breakdown in our approximation it still cannot be rejected offhand since the exact calculation of Mermin and Wagner does not exclude that possibility. The origin for this result may lie in the contribution to the density of states from all the low energy levels of the many spin deviations states, contributions which are neglected in the spin wave approximation.

---

*Work supported in part by NSF Grant No. GH-40035.

INSULATING STATES AND METAL-NONMETAL TRANSITION IN THE HUBBARD MODEL *

Tadashi Arai
Argonne National Laboratory, Argonne, Illinois 60439

ABSTRACT

If a band is narrow and is not nearly empty or nearly full, the Hubbard lattice is an insulator. A nearly half-filled lattice is a Hubbard-type insulator with a gap but if the number of electrons decreases or increases beyond a certain limit, the lattice becomes a "gapless" insulator. As the bandwidth increases, these insulating states will turn metallic. The mechanism involved in the transition is presented.

INTRODUCTION

In the Hubbard model,[1] a band splits into two and, if the lower band is filled and the upper band empty, the lattice is an insulator, while, if the lower (or upper) band is partly filled, the lattice is metallic. According to Herring,[2] however, the ratio between the Fermi-surface volume and the number of electrons deviates from the value predicted for a normal metal by as much as a factor of 2, in contradiction to Luttinger's theorem[3] that the Fermi-surface volume is unchanged by electron interaction. As the density of electrons per volume increases, the energy gap disappears, yielding a metallic state with overlapping bands. However, this state again exhibits the abnormally large Fermi-surface volume.

EXISTENCE OF A GAPLESS INSULATOR

Previously,[4] we have calculated an improved Green's function for the Hubbard model by using the functional derivative method[5] and showed that, if a band is narrow and is not nearly empty or nearly full, the Hubbard lattice remains insulating. The lattice containing about one electron per atom will be a Hubbard-type insulator with gap. If the number of electrons decreases or increases beyond a certain range, the gap is no longer at the Fermi surface, but an attractive interaction appears between an electron and the lattice, yielding a finite threshold E_A for the electron to be excited and making the lattice a gapless insulator.[6]

Physically, the appearance of the attractive interaction may be explained as follows. In the Hubbard lower band, electrons avoid each other and two electrons with opposite spins will never occupy a site at the same time. Consequently, the energy spectrum of the lower band obtained in Hubbard I, $(1 - n_{\bar{\sigma}})\varepsilon_k + O(\varepsilon_k/I)$, does not contain a term linear in the interaction I, and the kinetic energy ε_k is reduced by the factor $(1 - n_{\bar{\sigma}})$ where $n_{\bar{\sigma}} \equiv \langle C^{\dagger}_{R\bar{\sigma}} C_{R\bar{\sigma}} \rangle$. The Hubbard solution may be improved by perturbation and the leading term, being the second order perturbation, is always attractive even though I is repulsive.

In this calculation, higher order Green's functions involved in the two equations of motion for basic Green's functions considered in Hubbard I are reduced to functional derivatives of the one-electron Green's function G. The derivatives and the self-energy correction $\Sigma^{(1)}$ are then calculated by replacing G by the Hubbard I solution G^o. The result is similar to the expression in Eq. (1) except that $B(\omega)$ and $C(\omega)$ are replaced by $[\omega - (1 - n_{\bar{\sigma}})I]^2$. Note that if the electron Green's function G_{el} is being calculated, $0^{\pm} = 0^{+}$ and $iG_{RR\sigma}(0^{\pm}) \equiv i\langle\langle C^{\dagger}_{R\sigma}(0^{\pm}) C_{R\sigma}(0)\rangle\rangle$ is equal to $1 - n_{\sigma}$ and, for the hole Green's function G_{hole}, $iG_{RR\sigma}(0^{-}) = - n_{\sigma}$. Therefore, the energy $\omega_{el}(k)$ calculated by G_{el} becomes lower than the corresponding energy $\omega_{hole}(k)$ obtained by G_{hole} by the amount E_A. Since the attractive interaction between the electron and the lattice is included in G_{el} but not in G_{hole}, E_A calculated above is the activation energy needed to remove the electron from the lattice.

IMPROVED GREEN'S FUNCTION FOR A FINITE BAND

Since $\Sigma^{(1)}$ lacks some linear terms in hopping motion ε, the result cannot be extended to a metallic state. In fact, $-\delta\Sigma^{(1)}/\delta\varepsilon$ contains the same types of terms as those in $\delta(G^o)^{-1}/\delta\varepsilon$ and the second order self-energy correction $\Sigma^{(2)}$ obtained from $-\delta\Sigma^{(1)}/\delta\varepsilon$ involves, in addition to terms quadratic in ε, linear terms. However, terms linear in ε appearing in successive approximations are limited and the calculation can be extended up to infinite order.[7] The result is

$$2\pi G^{-1}(k\sigma, \pm\omega) = \frac{(\omega-\omega_1)(\omega-\omega_2)}{\omega-(1-n_{\bar{\sigma}})I} - \frac{I^2 n_{\bar{\sigma}}(1-n_{\bar{\sigma}})\varepsilon_k}{C(\omega)} - \frac{I^2[iG_{RR\sigma}(0^{\pm})d-d']}{B(\omega)}, \quad (1)$$

where

$$d = \sum_{R'\neq R} \varepsilon_{RR'}\langle C^{\dagger}_{R\bar{\sigma}} C_{R'\bar{\sigma}}\rangle,$$

$$d' = \sum_{R'\neq R} \varepsilon_{RR'}\langle C^{\dagger}_{R\bar{\sigma}} C_{R'\bar{\sigma}}\rangle\langle C_{R\sigma} C^{\dagger}_{R'\sigma}\rangle \exp[-ik(R-R')],$$

$$B(\omega) = [2\omega - (\tfrac{3}{2} - n_{\bar{\sigma}})I]^2 - (I/2)^2\chi(0^{\pm}),$$

$$C(\omega) = [\omega - (1 - n_{\bar{\sigma}})I]^2 - I^2 n_{\bar{\sigma}}(1 - n_{\bar{\sigma}}) - I^2 iG_{RR\sigma}(0^{\pm})d[\omega - (1 - n_{\bar{\sigma}})I][B(\omega)]^{-1},$$

$$\chi(0^{\pm}) = [iG_{RR\sigma}(0^{\pm})]^2 - \langle C^{\dagger}_{R\sigma} C_{R'\sigma}\rangle\langle C^{\dagger}_{R'\sigma} C_{R\sigma}\rangle. \quad (2)$$

Here ω refers to an electron and $-\omega$ to a hole; ω_1 and ω_2, respectively, are the Hubbard I solutions for the lower and upper bands; $\varepsilon_{RR'}$ and ε_k are the hopping matrix element and its Fourier transform. Here, only nearest neighbor hoppings are considered regarding $\chi(0^{\pm})$ as constant, and terms proportional to d' and of higher orders are neglected in $C(\omega)$. Otherwise, the above expression is exact up to terms linear in ε and, in addition, involves a higher order term (the term involved in $C(\omega)$ and proportional to $[B(\omega)]^{-1}$).

Poles of $G(k\sigma,\omega)$ may be estimated graphically. If $\varepsilon_k = 0$, the second term vanishes, and the solutions $\omega'_{1,2}$ and $\omega'_{\pm\chi}$ are given by intersects of the two curves

$$h = (\omega - \omega_1)(\omega - \omega_2)B(\omega)/4,$$

$$g = (I/2)^2 iG_{RR\sigma}(0^{\pm})d[\omega - (1 - n_{\bar{\sigma}})I], \quad (3)$$

as is illustrated in Fig. 1. Here $\omega_{\pm\chi}$ are the solutions of $B(\omega) = 0$, and the sign of g is opposite for an electron and a hole. The solution of $C(\omega) = 0$ can be calculated similarly by replacing, in Eq. (3), ω_1 and ω_2 by ω_a and ω_b, respectively, and may be labelled as $\omega'_{a,b}$ and $\omega'_{\pm\chi'}$. Here ω_a and ω_b are the solutions of $[\omega - (1 - n_{\bar{\sigma}})I]^2 - I^2 n_{\bar{\sigma}}(1 - n_{\bar{\sigma}}) = 0$.

If $\varepsilon_k \neq 0$, the poles of $G(k\sigma,\omega)$ are calculated from the intersects of the two curves:

*Work performed under the auspices of the USAEC.

$$h' = (\omega''-\omega_1')(\omega''-\omega_2')(\omega''-\omega_a')(\omega''-\omega_b')(\omega''-\omega_{-\chi}')$$

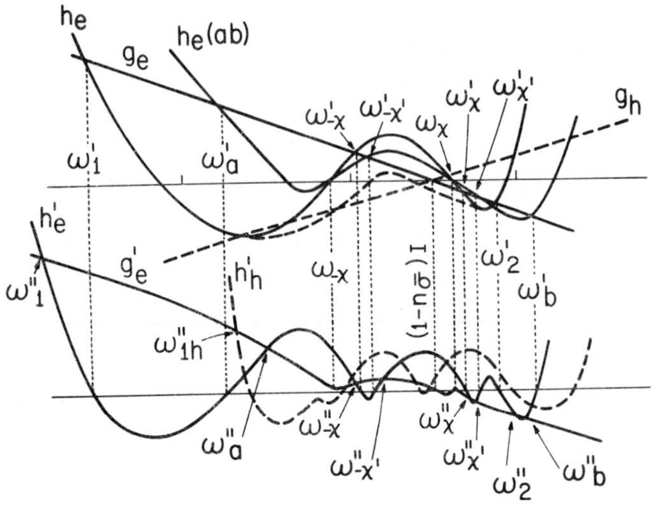

Fig. 1 Solutions of Eqs. (3) and (4) for an electron. Intersects of h_e and g_e give $\omega_{1,2}'$ and $\omega_{\pm\chi}'$, while intersects of $h_e(ab)$ and g_e give $\omega_{a,b}'$ and $\omega_{\pm\chi'}$. The complete solutions ω'' are given by intersects of h_e' and g_e'. For a hole, the h and g curves will be modified as illustrated by broken lines. Note the drastic shift from h_e' for an electron to h_h' for a hole. g_h' for a hole will not be so different from g_e'.

$$\times\ (\omega''-\omega'_{-\chi'})(\omega''-\omega_\chi')(\omega''-\omega_{\chi'}') \quad ,$$

$$g' = I^2\, n_\sigma^-(1-n_\sigma^-)\varepsilon_k[\omega''-(1-n_\sigma^-)I](\omega''-\omega_{-\chi})^2(\omega''-\omega_\chi)^2. \tag{4}$$

Since ω''s are the solutions of Eq. (3), the complete solutions $\omega''_{1,2}$, $\omega''_{a,b}$, $\omega''_{\pm\chi'}$ are calculated in two steps as is shown in Fig. 1.

The sign of g' is the same for an electron and a hole and, if $\varepsilon_k < 0$, it is as illustrated in Fig. 1; but, if $\varepsilon_k > 0$, it will be reversed. The difference in the sign of g makes ω_1'' for an electron lower than the corresponding ω''_1 for a hole, confirming the previous conclusion that the activation energy is needed to excite an electron and that the lattice is an insulator. As was discussed previously, d is negative and remains a fraction of the bandwidth 2D. For a majority of electrons, therefore, $|\varepsilon_k| > |d|$ and g' dominates over g. If $\varepsilon_k << 0$, therefore, ω_2'' and ω_b'' (and also ω_χ'' and $\omega_{\chi'}''$ for an electron and ω_a'' and $\omega''_{-\chi'}$ for a hole) become complex.

The spectral weight $A(\omega_1'')$ may be calculated as before.[4] Here we note that, if $\varepsilon_k < d < 0$ and if the value of $|\varepsilon_k|$ happens to be small, the result given by Eqs. (1) and (2) becomes poor since ω_b'' is real and $A(\omega_b'')$ is negative. If higher order hopping processes are added to Eqs. (1) and (2), however, the result will become parallel to the result for $\varepsilon_k << 0$, yielding complex ω_2'' and ω_b'' of the form $\omega_0 \mp i\Delta$. Then a single energy state $\omega_0 \mp i\Delta$ with a positive but reduced spectral weight $A(\omega_0 \mp i\Delta) = A(\omega_2'') + A(\omega_b'')$ will appear, where $-i\Delta$ refers to an electron and $+i\Delta$ refers to a hole. Similarly ω_χ'' and $\omega_{\chi'}''$, for an electron and ω_a'' and $\omega''_{-\chi'}$ for a hole yield single energy states with positive but reduced A's, while the other ω'' remain real with positive A's.

CONCLUSION: THE METAL-NONMETAL TRANSITION

Let us now consider a less-than-half-filled lattice and calculate the energy spectrum of an electron emitted from the lattice. In the narrow band region where $A(\omega_1'') \approx 1 - n_\sigma^-$, the lowest hole band ω_1'' is filled up to the state k_{hole}^{max} where k_{hole}^{max} is greater than the Fermi momentum $k_F{}^o$ expected in the weakly interacting limit. As 2D increases, four solutions ω_a'', $\omega''_{-\chi'}$, ω_2'' and ω_b'' become complex with vanishingly small A. Although $\omega''_{\pm\chi}$ and $\omega''_{\chi'}$ are real, these solutions are bonded in narrow regions such that $\omega_{-\chi} < \omega''_{-\chi} < \omega'_{-\chi}$ and $\omega_\chi' < \omega_\chi''$, $\omega_{\chi'}'' < \omega_{\chi'}'$, yielding negligibly small A. As $|\varepsilon_k|$ increases, therefore, $A(\omega_1'')$ of the lowest band should increase to one, doubly-filling the lowest band up to $k_F{}^o$. Hence an electron emitted from the lattice should behave like a normal metallic electron, except that a few electrons near the Fermi level satisfying $|\varepsilon_k| < |d|$ will exhibit complex ω_1'' with a finite width Δ.

The energy spectrum of the state to which an electron is excited is calculated by G_{el}. If $\varepsilon_k << d < 0$, ω_2'', ω_b'', ω_χ'' and $\omega_{\chi'}''$, for an electron turn complex with small A's. The lower four solutions are real and the A's of $\omega''_{-\chi}$ and $\omega''_{-\chi'}$ remain negligible. However, $A(\omega_a'')$ increases and $A(\omega_1'')$ of the lowest band will never approach one. Instead, $A(\omega_1'')$ decreases as $A_H(\omega_1'')(1-x) \approx (1-x)$ while $A(\omega_a'') \to x$, where $x \equiv (\omega_a''-\omega_a')/(\omega_a''-\omega_1'')$. If $\varepsilon_k > |d|$, the sign of g' is reversed, yielding complex ω_1'' with a vanishingly small A.

In the narrow band limit, therefore, the electron band ω_1'' is filled up to the state k_{el}^{max} ($> k_{hole}^{max}$). As 2D increases, $A(\omega_1'')$ decreases and electrons have to occupy higher states, increasing the difference $k_{el}^{max} - k_{hole}^{max}$ as well as the activation energy. If k_{el}^{max} reaches the zone boundary, however, electrons have to occupy the second band ω_a'', and the energy required to add an electron increases suddenly to ω_a'', the value equal to the energy (ω_1'' for a hole) required to remove it within the uncertainty Δ of ω_1'' for a hole. Then the lattice discontinuously becomes metallic with a normal Fermi surface $k_F{}^o$.

Since, for a nearly half-filled case ($n_\sigma \approx 1/2$), the two electron bands ω_1'' and ω_a'' lie below the hole bands ω_1'' and ω_a'', only after higher bands $\omega''_{\mp\chi}$ and $\omega''_{\mp\chi'}$ begin being occupied, does the lattice turn metallic. As 2D increases, the gap between $\omega''_{-\chi}$ and ω''_χ will decrease, thus effectively eliminating the Hubbard gap completely.

REFERENCES

1. J. Hubbard, Proc. Roy. Soc. (London) A276, 238 (1963); A281, 401 (1964).
2. C. Herring, in Magnetism, edited by G. T. Rado and H. Suhl (Academic, New York, 1966).
3. J. M. Luttinger, Phys. Rev. 119, 1153 (1960).
4. T. Arai, Phys. Rev. Letters 33, 486 (1974).
5. T. Arai and M. P. Tosi, Solid State Commun. 14, 947 (1974).
6. The conductivity calculated by the Hubbard I solution is suppressed by the narrowing of the band, but remains metallic. See, R. A. Bari, D. Adler, and R. V. Lange, Phys. Rev. B2, 2898 (1970).
7. T. Arai and M. P. Tosi, submitted to Phys. Rev. B.

COMPARISON OF THE MAGNETIC PROPERTIES OF THE HUBBARD HAMILTONIAN IN THE WEAK CORRELATION LIMIT TO THE HFA

L. C. Bartel*
Sandia Laboratories, Albuquerque, New Mexico 87115

H. S. Jarrett
E. I. duPont deNemours and Company, Wilmington, Delaware 19898

ABSTRACT

Resonance broadening is included along with spin-disorder scattering to study the magnetic properties of the Hubbard Hamiltonian in the weak correlation limit. The calculated susceptibility is compared to the exchange enhanced Hartree-Fock approximation (HFA) susceptibility, and limits are set on the applicability of the HFA. For a semicircular density of states, HFA appears to be applicable for $I/\Delta < 0.5$ and a half-filled band, but for low electron (or hole) densities the range of applicability in I/Δ decreases.

INTRODUCTION

In band structure calculations, a Hartree-Fock approximation (HFA) is usually made. The HFA assumes that an electron moves in an average field created by the other electrons and in the main neglects electron correlation effects. The HFA allows one to make a one-electron approximation and thus allows a tractable solution. However, for narrow bands, the HFA is inappropriate in some instances. In this paper we establish limits on the apparent applicability of the HFA when correlation is present for a particular single-particle density of states.

We recently studied the properties of the Hubbard Hamiltonian using a locator technique[1] where we include both spin-disorder scattering and the two resonance-broadening terms: electron-hole and spin-flip scattering. In previous work,[1] we restricted our investigation to the strong correlation limit. Here we discuss the approach of the density of states, magnetic susceptibility and its associated enhancement factor, and the single site spin-spin correlation function to the weak correlation limit where the HFA should be applicable.

RESULTS AND DISCUSSION

The pseudoparticle density of states $\rho_\sigma(E_F)$, given by the imaginary part of the single-particle Green's function is compared in Fig. 1 with the semicircular single-particle densities of states (DOS) for $I/\Delta = 0.2$ and for selected n, the average number of electrons per site (n = 1 corresponds to a half-filled band). Recall in the atomic limit[2] that the energy of one electron on an otherwise empty nondegenerate site is T_0, while for two electrons of opposite spin on the same site the energy is T_0+I, where I is the Coulomb repulsion between the two electrons. For finite hopping rate between sites, these two atomic states broaden into bands separated by the Mott-Hubbard gap of approximate width $I-\Delta$, where Δ is the single-particle band width. For a semicircular single-particle DOS, the bands merge for $I/\Delta < \frac{1}{2}\sqrt{3}$. In the limit $I/\Delta \ll \frac{1}{2}\sqrt{3}$, motional band narrowing takes place as seen by the decrease in density in the wings of the DOS curves of Fig. 1. Incidentally, those states shown by the cross-hatched region are localized in the Anderson[3] sense, where the Economou-Cohen[4] Yonezawa-Watabe[5] criterion defines the regions of localization. Motional band narrowing occurs because the $-\sigma$ spin electron hops rapidly on and off a site causing the σ spin electron "to see" only the average energy $E_\sigma = In_{-\sigma}$, where $n_{-\sigma}$ is the average number of $-\sigma$ spin electrons per site.

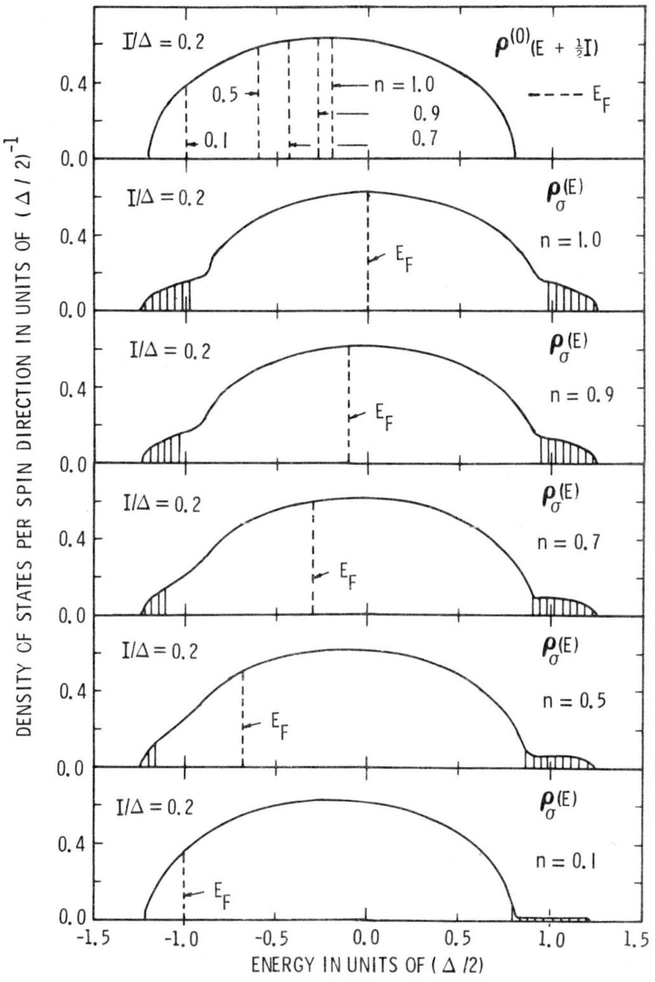

Figure 1. Density of States vs Energy

The reduced magnetic susceptibility[1] at T = 0 is

$$\chi_r = 2\rho(E_F)\,[1 + K]^{-1} \qquad (1)$$

where $\chi = \mu_B^2 N\chi_r$ is shown in Fig. 2 as a function of I/Δ. The HFA susceptibility is

$$\chi_{HFA} = 2\rho^{(0)}(E_F)\,[1-I\rho^{(0)}(E_F)]^{-1}\,, \qquad (2)$$

where $\rho^{(0)}(E_F)$ is the single-particle DOS. The approach of χ_r, to the HFA limit Eq. (2) (dashed lines) is evident. Note that the HFA approximation improves near the half-filled band case. The reason for this behavior can be seen by considering the enhancement factor K in Eq. 1. Integration of Eq. 2.12 of Fukuyama and Ehrenreich[6] gives

$$K = \pi^{-1}\,\mathrm{Im}G_\sigma(E_F)\left(\frac{\partial E_\sigma}{\partial n_{-\sigma}}\right)_{E_F} \,. \qquad (3)$$

In the HFA limit of rapid electron hopping, $(\partial E_\sigma/\partial n_{-\sigma})_{E_F}$ I and Eq. 1 approaches Eq. 2. For large n ($E_F > I$), K approaches the HFA value of $-I\rho^{(0)}(E_F)$, but for small n ($E_F < I$), the deviation of K from the HFA value is greatest. This effect arises because the probability for double occupancy of a site is highest when $E_F > I$.

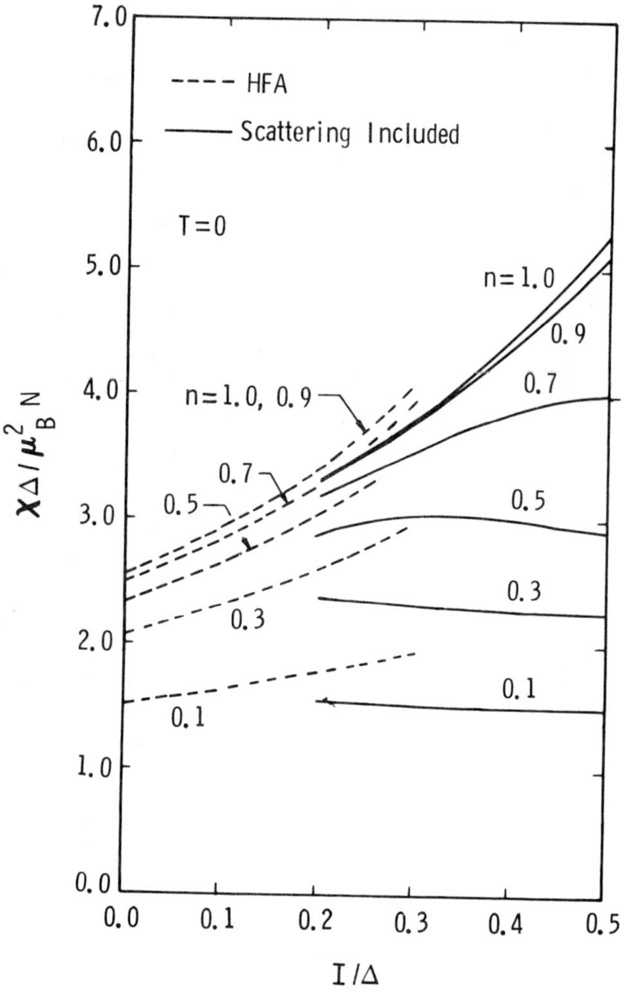

Figure 2. Susceptibility vs I/Δ

The electron therefore "sees" only a motionally averaged repulsion energy (HFA limit). However, for $E_F<I$ many states are available and single occupancy of a site is energetically more favorable than overcoming the Coulomb repulsion by double occupancy. Electron correlation is then greater and $(\partial E_\sigma/\partial n_{-\sigma})_{E_F}<I$.[7]

This behavior is also manifested in the spin-spin correlation function $L_o = (3/2N)\ \Sigma_i \langle n_{i\sigma}(1-n_{i-\sigma})\rangle$ shown in Fig. 3. Deviation of L_o from the correlated value of $(3/4)n$ at large I/Δ occurs as a distinct "knee" at a value of I/Δ for which $E_F(n)$ becomes less than I. For $E_F>I$, double occupancy destroys some electron correlation, $\langle n_{i\sigma}n_{i-\sigma}\rangle \to \langle n_{i\sigma}\rangle\langle n_{i-\sigma}\rangle$, and $L_o \to \frac{3}{4}n(1-\frac{n}{2})$ the HFA value. For $E_F(n)<I$, single occupancy is more likely, $\langle n_{i\sigma}n_{i-\sigma}\rangle \to 0$, and the correlated value of L_o persists to smaller I/Δ.

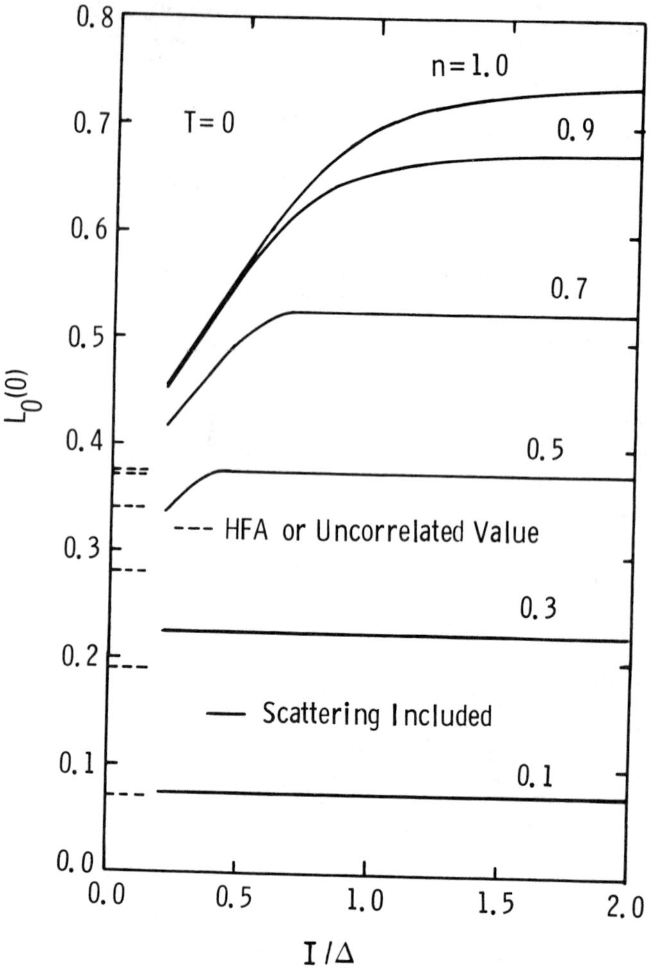

Figure 3. Single-site spin-spin correlation function vs I/Δ

SUMMARY

We have shown that the HFA best approximates bands in which $E_F>I$ and that low electron density does not arbitrarily imply HFA is applicable. This result is consistent with Wigner condensation at low electron density where correlation can cause a super lattice of ordered singly occupied sites. These results contrast with those of Bari & Kaplan[8] who found that in the absence of electron scattering (Hubbard I), the weak correlation limit does not approach the HFA limit. However we have shown that it is just these scattering terms that are necessary to give the HFA limit and one is not surprised by their results.

REFERENCES

* Work at Sandia Laboratories supported by the U. S. Atomic Energy Commission.

1. L. C. Bartel and H. S. Jarrett, Phys. Rev. B 10, 946 (1974).
2. J. Hubbard, Proc. R. Soc. A 281, 401 (1964).
3. P. W. Anderson, Phys. Rev. 109, 1492 (1958).
4. E. N. Economou and M. H. Cohen, Phys. B 5, 2931 (1972).
5. F. Yonezawa and M. Watabe, Phys. Rev. B 8, 4540 (1973); F. Yonezawa and M. Watabe, Solid State Commun. 11, 1667 (1972).
6. H. Fukuyama and H. Ehrenreich, Phys. Rev. B 7, 3266 (1973).
7. J. Kanamori, Prog. Theor. Phys. 30, 275 (1963).
8. R. A. Bari and T. A. Kaplan, Phys. Letters A 33, 400 (1970).

APPLICATION OF THE NARROW BAND MODEL FOR DISORDERED FERROMAGNETIC ALLOYS TO THE SOLID SOLUTIONS $Fe_{1-x}Co_xS_2$ AND $Ni_{1-x}Co_xS_2$*

G.F. ABITO, University of Iowa, Iowa City, Iowa 52242

J.W. SCHWEITZER, University of Iowa and
Argonne National Laboratory, Argonne, Illinois 60439

ABSTRACT

The narrow band model of Hubbard, generalized for random disordered alloys, is used to discuss the ferromagnetism in the pyrite solid solutions $Fe_{1-x}Co_xS_2$ and $Ni_{1-x}Co_xS_2$. This discussion is made within the coherent potential approximation for treating disorder.

INTRODUCTION

The solid solutions of transition metal disulfides appear to be exemplary systems for studying the magnetic properties of narrow band electrons.[1] For FeS_2 through ZnS_2 one pictures the conduction band arising from the $3d$ orbitals of e_g symmetry without overlapping wide bands. It is believed that this two-fold degenerate e_g band is being filled as one proceeds across the series with Fe, Co, Ni, Cu, and Zn contributing 0, 1, 2, 3, and 4 electrons respectively to the band. Consistent with the above picture is the fact that FeS_2 is a nonmagnetic semiconductor, CoS_2 is metallic and ferromagnetic ($Tc \simeq 120°K$), and NiS_2 is semiconducting, presumably due to strong intra-atomic Coulomb correlations, and is probably antiferromagnetic. CuS_2 shows metallic conductivity and is paramagnetic, and ZnS_2 is an extrinsic semiconductor as the e_g band is completely filled with 4 electrons.

In the solid solutions $Fe_{1-x}Co_xS_2$ ferromagnetism exists for $x \geqslant 0.05$. For $0.15 \leqslant x \leqslant 0.95$ the saturation magnetization increases linearly with Co concentration and is equal to one Bohr magneton per Co atom. At $x \simeq 0.95$ this magnetization begins to decrease and for CoS_2 one finds $\mu_S \simeq 0.9\mu_B$. With the addition of Ni to CoS_2 the magnetization decreases rapidly and ferromagnetism only exists in $Ni_{1-x}Co_xS_2$ for $x \geqslant 0.9$. Here there is probably a transition to antiferromagnetic behavior. The experimental results of Jarrett et al.[1] are shown in **Fig. 1.** Throughout this wide range of solid solutions one observes metallic conductivity. Hence, these solid solutions provide an example where one sees itinerant ferromagnetism over a range of electron concentrations in the e_g band between 0.15 and 1.1 if one assumed the Fe, Co, and Ni introduce 0, 1, and 2 electrons respectively into the band. In the present paper we show that this observed ferromagnetism can be semi-quatitatively understood using the Hubbard[2] model generalized for disorder and treating this disorder within the framework of the coherent potential approximation (CPA).[3]

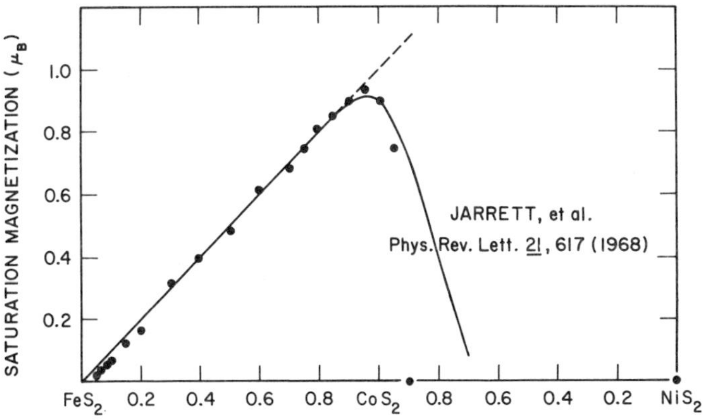

Fig. 1. Experimental results of Jarrett et al. for the saturation magnetization of ferromagnetic $Fe_{1-x}Co_xS_2$ and $Ni_{1-x}Co_xS_2$. The solid curve is the result of the present calculations.

MODEL AND CALCULATIONS

We assume for simplicity the nondegenerate narrow-band model of Hubbard[2] generalized to include site-diagonal disorder.[4,5] This model is described by

$$H = \sum_{i,j,\sigma} t_{ij}\, C^{+}_{i\sigma} C_{j\sigma} + \sum_{i,\sigma} \epsilon_i\, n_{i\sigma} + \frac{1}{2}\sum_{i,\sigma} I_i\, n_{i\sigma}\, n_{i-\sigma}$$

The atomic level energies ϵ_i and the intra-atomic Coulomb repulsion parameters I_i take on values depending on the type of atom occupying the i-th site while the transfer term t_{ij} is independent of the disorder.

Clearly we have neglected the two-fold degeneracy of the e_g band we wish to model. A minimum account of this degeneracy is obtained by scaling the e_g electron concentrations by a factor of one-half. This amounts to assuming the two bands are completely decoupled which ignores the interband intra-atomic Coulomb repulsion and exchange effects. Although it is difficult to see why these effects might be small, there is some indication that their neglect may not be too serious a shortcoming of the model. Only NiS_2, which has a half-filled band, shows evidence of strong correlation effects resulting in a split band. Both CoS_2 and CuS_2 with 1 and 3 e_g electrons respectively are metallic.

To complete the specification of the model we assume a semi-elliptical form $\frac{2}{\pi}(1 - \epsilon^2)^{\frac{1}{2}}$ for the density of states $\rho^o(\epsilon)$ associated with the transfer terms. According to our picture of the disulfides the scaled electron concentration is $(1/2)\,x$ for $Fe_{1-x}Co_xS_2$ and $1 - (1/2)x$ for $Ni_{1-x}Co_xS_2$. The atomic level energies are chosen such that $\epsilon_{Fe} > \epsilon_{Co} > \epsilon_{Ni}$, and I_{Co} is chosen to give the experimental value for the saturation magnetization of CoS_2. Furthermore, in order to reduce the number of parameters we set $I_{Fe} = I_{Ni} = I_{Co}$. This leaves $\epsilon_{Fe} - \epsilon_{Co}$ and $\epsilon_{Co} - \epsilon_{Ni}$ as adjustable parameters.

If one adopts a Hartree-Fock (H-F) approximation for the Coulomb term one obtains an alloy generalization of the Stoner theory. Using our $\rho^o(\epsilon)$ one finds that the CoS_2 saturation magnetization of 0.9 μ_B requires $I_{Co} = 1.86$ (bandwidth = 2.0). Now it is interesting to note that a rigid band treatment of these alloys yields a saturation magnetization as a function of alloy composition which is qualitatively different from that observed. Using $I = 1.86$ to fit CoS_2, the rigid band calculation gives a magnetization for $Fe_{1-x}Co_xS_2$ which decreases rapidly with Fe concentration such that the magnetization is zero for $x \leqslant 0.33$. Furthermore, it predicts a magnetization which increases with Ni concentration in $Ni_{1-x}Co_xS_2$. These results depend, of course, on our choice for $\rho^o(\epsilon)$, however it would take a most unrealistic choice for $\rho^o(\epsilon)$ in order to reproduce the experimental results.

It is a simple matter to improve on the treatment of disorder by using the CPA, particularly if we retain the H-F approximation. The H-F plus CPA treatment of this model is well discussed in the literature.[4,5,6] Our calculations show that one reproduces the magnetization data very well when $\epsilon_{Fe} - \epsilon_{Co}$ and $\epsilon_{Co} - \epsilon_{Ni}$ are chosen to be approximately equal to the bandwidth. Such calculations yield similar results to the solid curve in Fig. 1 when $\epsilon_{Fe} - \epsilon_{Co} = \epsilon_{Co} - \epsilon_{Ni} = 2.0$. If $\epsilon_{Co} - \epsilon_{Ni}$ is chosen still larger one finds a faster decrease in magnetization with increasing Ni concentration; however, in general these results are not very sensitive to the choice for these disorder parameters.

One shortcoming of these calculations is their prediction that the ferromagnetism in $Fe_{1-x}Co_xS_2$ persists as x goes to zero with a saturation moment proportional to the Co concentration. This is an obvious defect of one-electron theories where there is no distinction between ferromagnetism and local moment behavior. However, since this defect only occurs at low concentrations, it does not seriously detract from the overall success of the CPA treatment.

Much more serious is the fact that in order to produce the rapid decrease in magnetization with Ni concentration the H-F plus CPA calculation puts nearly 4 electrons on each Ni site. Clearly there should not be large deviations from the electronic configuration of Ni as assigned in the pure NiS_2. In $Ni_{1-x}Co_xS_2$ the Ni component band is lower than the Co component so there is nothing to prevent it from filling when the Ni concentration is low. This difficulty is due to the neglect of the strong intra-atomic Coulomb correlations which are evidenced by the fact that NiS_2 is a semiconductor.

We have previously studied[6] this problem of Coulomb correlations in the Hubbard model with disorder; and although considerable uncertainties exist, the following picture seems to emerge. If the electron concentration in a component band is small compared to the half-filled value, the result that one obtains for the magnetization is similar to the H-F result with the Coulomb repulsion parameter I renormalized to a value which is of the order of the bandwidth. Note that this suggests our H-F plus CPA treatment of $Fe_{1-x}Co_xS_2$, with $I_{Fe} = I_{Co} = 1.86 \approx$ bandwidth, gives a reasonably accurate description since the scaled concentrations are approximately 0 and 1/2 for Fe and Co respectively. However, when the electron concentration in a component band is near the half-filled value, one expects to see the effects of band splitting and narrowing when the intra-atomic Coulomb repulsion is strong.

These nontrivial correlation effects can be included in a not unreasonable manner by introducing the energy-dependent effective atomic level energy[6]

$$\epsilon_{Ni,\sigma}(\epsilon) = \left[1 - \langle n_{Ni,\bar{\sigma}}\rangle\right]^{-1} \left[\epsilon_{Ni} + \left(I^{eff}_{Ni} - \epsilon\right)\langle n_{Ni,\bar{\sigma}}\rangle\right]$$

in place of the Hartree-Fock expression. In this way we describe the lower portion of the band which is split due to the intra-atomic Coulomb correlations. This lower portion contains only one state.

Using this treatment of correlations at the Ni sites together with the CPA our calculations produce the curve in Fig. 1 for $Ni_{1-x}Co_xS_2$. Here $\epsilon_{Co} - \epsilon_{Ni} = 1.5$ and $\langle n_{Ni}\rangle$ is found to have a value just slightly less than 2. The full character of our results can be seen in Fig. 2 where the component densities of states are shown for two cases. In Fig. 2(a), where the Fe concentration is small, there is a narrow Fe band above the Fermi energy ϵ_F such that $\langle n_{Fe}\rangle \simeq 0$. As the Co concentration decreases the Co band narrows but the Fermi energy remains fixed near the middle of the up-spin Co band. For $x_{Co} \leqslant 0.9$ the Fermi energy is below the bottom of the Co down-spin band, which yields a magnetization 1 μ_B per Co atom. When the Ni concentration is dilute as in Fig. 2(b), the Ni enters in a narrow down-spin band below the bottom of the Co band. Hence each Ni acts like a localized S = 1 spin state which is polarized opposite to the Co magnetization. This produces the rapid decrease of the total magnetization with Ni concentration. Presumably at some critical concentration of Ni, an antiferromagnetic phase appears; but this is outside the scope of the present considerations.

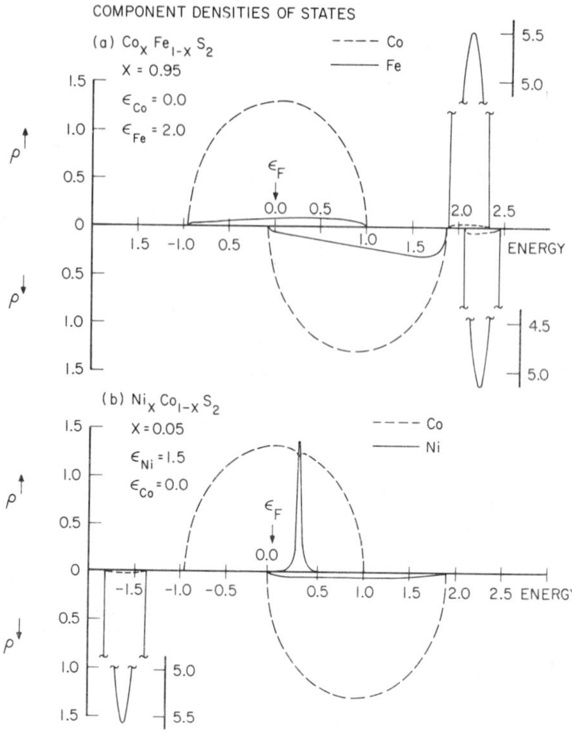

Fig. 2. Component densities of states for up and down spin states.

In summary, the present treatment seems to describe quite well the ferromagnetism of these solid solutions. However, the full role of the degeneracy of the e_g band needs to be clarified.

*Supported in part through the auspices of the USAEC.

REFERENCES

1. Jarrett, H.S., Cloud, W.H., Bouchard, R.J., Butler, S.R., Frederick, C.G., and Gillison, J.L., Phys. Rev. Letter 21, 617 (1968). Chandler, R.N, and Bene, R.W., Phys. Rev. B, 8, 4979 (1973).
2. Hubbard, J., Proc. Roy. Soc. A276, 238 (1963).
3. Soven, P., Phys. Rev. 156, 809 (1967); 178, 1136 (1969); Velicky, B., Kirkpatrick, S., and Ehrenreich, H., Phys. Rev. 175, 747 (1968).
4. Roth, L.M., Phys. Letters 31A, 440 (1970).
5. Hasegawa, H., and Kanamori, J., J. Phys. Soc. Japan 31, 382, (1971).
6. Abito, G.F. and Schweitzer, J.W., Phys. Rev. B. Jan. (1975).

MAGNETIC PROPERTIES OF THE NEPTUNIUM LAVES PHASES: $NpMn_2$, $NpFe_2$, $NpCo_2$, $NpNi_2$

A. T. Aldred, B. D. Dunlap, D. J. Lam, G. H. Lander, M. H. Mueller and I. Nowik
Argonne National Laboratory, Argonne, Illinois 60439

ABSTRACT

The magnetic properties of the cubic Laves phases (C 15 structure) $NpMn_2$, $NpFe_2$, $NpCo_2$, and $NpNi_2$ have been studied from 4 to 300 K by means of magnetization, neutron-diffraction, and nuclear-gamma-ray resonance (Mössbauer) measurements. The respective ordering temperatures (and types of ordering) are 18 K (ferro), ∿500 K (ferro), 15 K (antiferro), and 32 K (ferro). Magnetic moments are present on the neptunium atom in all compounds and on the Fe atom in $NpFe_2$. Certain magnetic properties such as the large anisotropy and the relationship between the magnetic moment and the hyperfine field are best interpreted in terms of localized 5*f* electrons, whereas other properties suggest hybridization between the 5*f* and 3*d* electrons.

INTRODUCTION

We have recently reported experimental investigations of the Np*v* (*v* = N, P, As, and Sb) and NpM_2 (M = Al, Os, Ir, and Ru) compounds.[1,2] We have now extended these measurements to a series of neptunium intermetallic compounds with the 3*d* transition metals. In a companion paper we present magnetic measurements on the uranium and plutonium Laves phases with the 3*d* transition metals.[3]

EXPERIMENTAL

The sample preparation techniques have been described previously.[2] To characterize the polycrystalline samples, x-ray and neutron-diffraction patterns were taken at room temperature. Additional diffraction lines (identified as arising from NpO_2) were present only for $NpMn_2$. Except for $NpCo_2$, the intensities of the neutron-diffraction lines are in good agreement with those predicted from the stoichiometric compounds. The difficulties with $NpCo_2$ are discussed below.

To determine the magnetic properties of these compounds a series of neutron experiments have been performed with both polarized and unpolarized neutrons. In particular, the use of polarized neutrons together with a 60 kOe superconducting magnet allows us to study details of the magnetic behavior that would be impossible with unpolarized neutrons and such small samples. The experimental techniques used in the magnetization and NGR experiments were described previously. For $NpFe_2$, $NpNi_2$, and $NpCo_2$, the NGR spectra are similar to those given by Gal et al.[4] The hyperfine coupling parameters are given in Table I.

RESULTS

$NpFe_2$

The compound $NpFe_2$ is ferromagnetic at room temperature and the magnetization curves at 5 and 300 K are very similar. The large field dependence (the magnetization increases by ∿26% between 3 and 13 kOe) presumably reflects the presence of magnetocrystalline anisotropy, although the approach to saturation is much slower than the H^{-2} dependence that is valid for a normal polycrystalline ferromagnet.[5] This slow approach is typical of actinide ferromagnets,[2] and makes a reliable extrapolation to saturation rather uncertain. To determine the individual magnetic moments we have measured the spin dependent cross section of polarized neutrons reflected from a number of Bragg planes. Since the Np and 3*d* atoms are on different sites they contribute differently to the various Bragg reflections and a unique assignment of the magnetic moments is possible. We have used the magnetic form factor for neptunium as reported earlier.[6] The values of the magnetic moments are given in Fig. 1 as a function of the applied magnetic field. A difficulty in these experiments is to know the extent of neutron depolarization as the neutron beam traverses the loosely packed polycrystalline sample. Previous experiments on actinide ferromagnets have shown that this is a substantial effect at low fields,[7] but that the depolarization decreases exponentially with the approach to saturation and is negligible for H > 30 kOe. As a further test, experiments were conducted on a polycrystalline sample of US (solid triangles in Fig. 1). The open triangles correspond to magnetic measurements on polycrystalline US.[8] The agreement

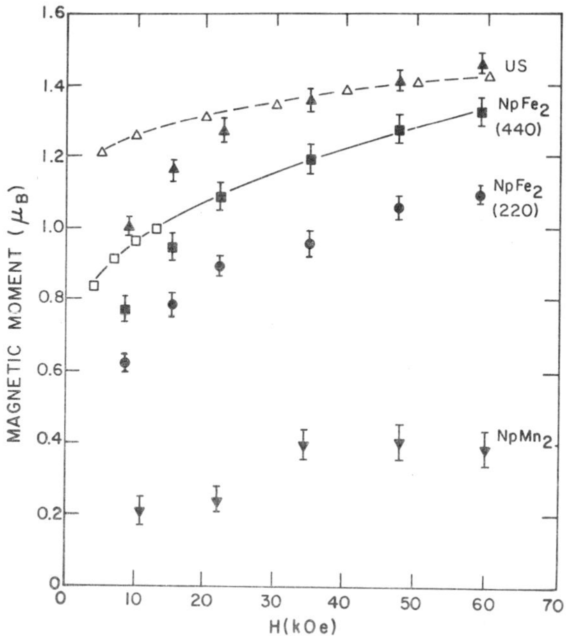

Fig. 1. Magnetic moments as a function of field for polycrystalline samples at 5 K. The closed symbols represent neutron-diffraction data and the open symbols were obtained from the magnetization results. All neutron data points except the $NpFe_2$ (440) refer to the actinide moments only. For the (440) reflection, the quantity plotted is $\mu_{Np}f_{Np} + 2\mu_{Fe}f_{Fe}$.

for H > 30 kOe is good and allows us to proceed with an analysis of the results for the neptunium compounds. Since $\mu_{Np} \simeq \mu_{Fe} = \bar{\mu}/3$, where $\bar{\mu}$ is the magnetic moment per mole (measured by the magnetization experiments), the values of $\bar{\mu}$ can be compared with the high-field neutron data and are shown as open squares in Fig. 1. In fitting the combined data of Fig. 1 we have used the expression

$$\sigma = \sigma_{sat}(1 - a/H^n) + \chi H$$

where σ_{sat} is related to the magnetic moment per mole at saturation μ_{sat}, a and n are constants, and χ is the high-field susceptibility. For soft magnetic materials, such as iron and nickel, n = 2. For US, the best fit is with n = 1/2 and $\chi = (5 \pm 1) \times 10^{-3}$ emu/mole, in agreement with the value of 3.8×10^{-3} obtained by magnetization measurements on single crystals.[9] For $NpFe_2$ a somewhat smaller value of n = 1/3 gives the best fit and the parameters are given in Table I. Because there is no *a priori* reason for choosing a particular value of n, the results should be treated cautiously.

TABLE I

Magnetic properties of the Np Laves phases. μ_{tm} represents the magnetic moment at the transition metal site. H_{hf} is the magnetic hyperfine field at the Np site.

	$NpFe_2$	$NpNi_2$	$NpMn_2$	$NpCo_2$
Unit cell, a_o	7.144	7.098	7.230	7.043
Magnetism	F	F	F	AF
Order temp (°K)	∿500	32(1)	18(2)	15(1)
H_{hf} (kOe) ± 100	1670	2350	∿400	980
μ_{Np} (μ_B)	1.0(1)	1.20(15)	∿0.3	0.5(1)
μ_{tm} (μ_B)	1.05(10)	<0.3	∿0.2	<0.15
$\bar{\mu}_{sat}$ (B/mole)	3.10(10)	1.10(5)	0.40(5)	--
χ (10^{-3} emu/mole)	60(5)	18(5)	24(3)	--

Standard deviations in parens refer to the least significant digit.

$NpNi_2$

This compound is ferromagnetic below 32 K. In the ordered state, the field dependence of the magnetization is similar to that of $NpFe_2$ and suggests a large anisotropy. No polarized-neutron experiments were performed on $NpNi_2$. The magnetic intensities, obtained by determining the difference between 80 and 5 K scans with unpolarized neutrons, are consistent with a ferromagnetic arrangement of Np moments. The value of μ_{Np} is in good agreement with the estimate of 1.2 μ_B from the magnetic hyperfine field. An approach-to-saturation analysis also gave a similar magnetic moment.

$NpMn_2$

$NpMn_2$ is ferromagnetic with $T_C \simeq 18$ K. The field dependence of the magnetic moment on the Np site as determined with polarized neutrons is shown in Fig. 1. The results of the data analyses are given in Table I.

$NpCo_2$

The susceptibility shows a sharp maximum at 15 ± 1 K, suggesting antiferromagnetic ordering. The RT neutron scans did not indicate sample stoichiometry in the case of $NpCo_2$. However, a number of attempts to understand the compositional problem (e.g., by comparing the data with those obtained from a UCo_2 sample) were unsuccessful. We believe that possible metallurgical problems also contribute to the severe line broadening observed at low temperature in the NGR spectra.[6] Additional reflections arising from the antiferromagnetic ordering were not observed in the low-temperature neutron patterns, but this is not surprising in view of the small sample (2 g.) and the small Np moment (∿0.5 μ_B) expected on the basis of the hyperfine-field result. However, a sudden increase in the neutron signal at 25 ± 2 kOe indicated the sample was metamagnetic at 5 K; the increase corresponded to 0.5 ± 0.1 μ_B/Np atom. Taken together with the NGR and susceptibility data, both of which suggest the existence of an ordered phase below 15 K, the neutron experiments are compatible with antiferromagnetic ordering in zero field. A ferromagnetic state is induced with a field of 25 kOe.

DISCUSSION

Previous experiments on neptunium intermetallics have been confined to compounds with nonmagnetic partners.[1,2] By extending these studies to intermetallics with the 3*d* transition metals we can anticipate more complicated behavior. However, except for $NpFe_2$, the neutron experiments show that the 3*d* metal atom does not carry a significant magnetic moment. (In the case of $NpMn_2$ the value $\mu_{Mn} \simeq 0.2\ \mu_B$ is observed in a 60 kOe applied field and is probably a field-induced effect.) In similar lanthanide compounds both iron and cobalt atoms carry magnetic moments of between 1 and 2 μ_B.[10] In $NpFe_2$ the iron and neptunium moments are aligned parallel. This arrangement is similar to that observed in the light lanthanide (Pr, Nd) compounds and is regarded as a consequence of the long-range exchange interaction mediated by the conduction electrons.

An interesting property of the present Np compounds is that the iron compound orders near 500 K, whereas the others order magnetically below 40 K. We assume, by analogy with the lanthanide compounds, that the high ordering temperature of $NpFe_2$ is a result of the strong *d* - *d* exchange. Evidence for localized 5*f* behavior is deduced from a consideration of the large anisotropy as well as from the linear relationship between μ_{Np} and H_{hf} (Table I).[11] However, an important difference between the lanthanide and actinide Laves phases concerns the variation of μ_{Np} from compound to compound. The small values of μ_{Np} (relative for example, to the values found in the Np*v* series[1]) are difficult to interpret in terms of a strictly localized configuration. The apparent large values of the high-field susceptibility (χ in Table I) are also characteristic of itinerant magnetism. Further experiments to investigate this property of the Np Laves phases are now in progress.

A more detailed account of this work will be published in the Physical Review B (1 Jan 1975).

We would like to express our thanks to A. W. Mitchell and J. F. Reddy for sample preparation.

REFERENCES

1. A. T. Aldred, B. D. Dunlap, A. R. Harvey, D. J. Lam, G. H. Lander, and M. H. Mueller, Phys. Rev. B 9, 3766 (1974).
2. A. T. Aldred, B. D. Dunlap, D. J. Lam, and I. Nowik, Phys. Rev. B 10, 1011 (1974).
3. D. J. Lam and A. T. Aldred, Paper of this Conf.
4. J. Gal, Z. Hadari, U. Atzmony, E. R. Bauminger, I. Nowik, and S. Ofer, Phys. Rev. B 8, 1901 (1973).
5. See, for example, R. M. Bozorth, Ferromagnetism (Van Nostrand, New York, 1951), pp. 484 et seq.
6. G. H. Lander, B. D. Dunlap, M. H. Mueller, I. Nowik, and J. F. Reddy, Int. J. Magnetism 4, 99 (1973).
7. G. H. Lander and M. H. Mueller, Phys. Rev. B 10, 1994 (1974).
8. W. Suski, T. Palewski, and T. Mydlarz, Int. J. Magnetism 4, 305 (1973), and private communication.
9. F. A. Wedgwood, J. Phys. C 5, 2427 (1972) and references therein.
10. K. N. R. Taylor, Advances in Physics 20, 551 (1971).
11. B. D. Dunlap and G. H. Lander, Phys. Rev. Letters 33, 1046 (1974).

MAGNETIC PROPERTIES OF URANIUM AND PLUTONIUM LAVES PHASES WITH 3*d* TRANSITION ELEMENTS*

D. J. Lam and A. T. Aldred
Argonne National Laboratory, Argonne, Illinois 60439

ABSTRACT

We have measured the magnetization of UMn_2, UFe_2, UCo_2, UNi_2, $PuMn_2$, $PuFe_2$, $PuCo_2$, and $PuNi_2$ from 4-300 K in fields up to 14 kOe. The susceptibility of UMn_2 shows a small maximum near 240 K (which may indicate an antiferromagnetic transition) in agreement with previous results. Our data for UFe_2, UCo_2 and UNi_2 are not in good agreement with earlier work; the ferromagnetic ordering temperature (158 K) of a single crystal sample of UFe_2 is lower than any reported value, UNi_2 orders ferromagnetically at 30 K, and UCo_2 may order below 5 K. In contrast, $PuMn_2$, $PuCo_2$ and $PuNi_2$ have small weakly-temperature-dependent susceptibilities. $PuFe_2$ is ferromagnetic at room temperature, in agreement with previous Mössbauer results, and has a saturation moment of $\sim 2.6\ \mu_B$/mole.

INTRODUCTION

In the past year an extensive study of the magnetic properties of neptunium-based C-15 Laves phases has been undertaken at Argonne National Laboratory.[1-3] The magnetic characteristics of the compounds where the second component is not a 3*d* transition element show a regular dependence on neptunium-neptunium distance.[1,2] However, the Laves phases of neptunium with the 3*d* transition metals (Mn, Fe, Co, Ni) show no such simple correlations, although they all order magnetically at low temperatures.[3]

To further study the magnetism of actinide Laves phases, we have made magnetization measurements from 4 to 300 K in applied fields up to 14 kOe on the uranium and plutonium Laves phases with Mn, Fe, Co, and Ni. All the compounds have the C-15 structure except for UNi_2 which has the hexagonal C-14 Laves phase structure.

The samples were prepared by arc melting (weight losses were negligible except for the Mn compounds) and were polycrystalline in nature. In the case of UFe_2, the single-crystal used in the neutron-diffraction experiments of Yessik[4] was available.

RESULTS

The procedures for data analysis and determination of magnetic susceptibilities are given in Ref. 1. In particular, the magnetization of the Np compounds in the paramagnetic regime showed a small ferromagnetic component (determined in the data analysis) which had a substantial temperature dependence and could not be attributed to an impurity. Similar effects, of greater or lesser magnitude, were observed in the present investigation. Metallographic examination of the UMn_2, UCo_2, and UNi_2 samples showed no evidence (<1%) of any second phase. It is, again, difficult to rationalize this behavior in terms of the phase equilibria or possible impurities.

UMn_2

The temperature dependence of the reciprocal molar susceptibility χ_M^{-1} (Fig. 1) is in good agreement with the previous study of Lin and Kaufman.[5] In particular, the small susceptibility maximum near 240 K is reproduced. Recent neutron diffraction experiments at Argonne show that at $\sim$240 K a structural transition occurs in UMn_2. No evidence for magnetic ordering is observed at 4.2 K.

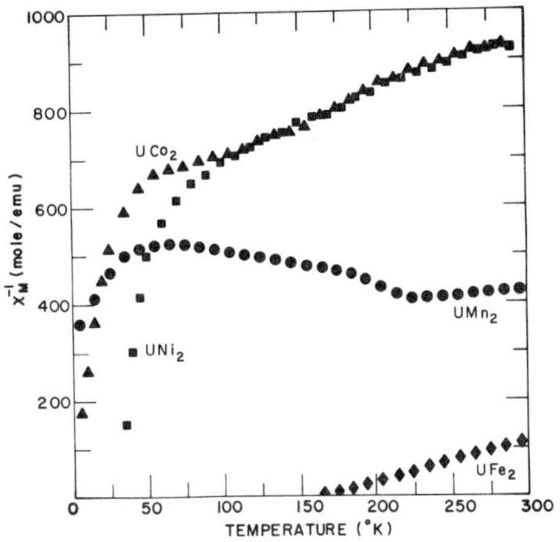

Fig. 1. Reciprocal molar susceptibility as a function of temperature for UMn_2, UFe_2, UCo_2, and UNi_2.

UFe_2

Previous magnetic measurements on polycrystalline material[6] had shown a range of Curie temperatures from 172 to 195 K, and a saturation magnetization equivalent to $\sim 1\mu_B$/mole; neutron-diffraction studies[4] indicated that the moment is associated almost entirely with the Fe atoms. The present work on UFe_2, which will be presented more fully in a separate publication, shows that a sample with the <111> easy axis parallel to the applied field has a spontaneous moment at 4 K of 1.05 μ_B/mole and a Curie temperature of 158 K. The susceptibility above the Curie temperature is given in Fig. 1.

UCo_2

The susceptibility shows approximately Curie-Weiss behavior between 50 and 300 K (Fig. 1). The rapid increase in susceptibility below 50 K may presage the onset of magnetic ordering at very low temperatures (below 5 K). These results are not in agreement with the sketchy published data.[6]

UNi_2

Previous results[6] are again incomplete. The present molar susceptibility is almost identical with that of UCo_2 down to $\sim$100 K (Fig. 1). Below 100 K there is again a rapid increase in susceptibility, similar to but more pronounced than that in UCo_2, and the compound orders ferromagnetically (as indicated by Arrott plots) at 30 $\pm$ 1 K. The maximum moment at 5 K and $\sim$13 kOe is 0.12 μ_B/mole, and the relative increase in magnetization with field is $\sim$11% from 3 to 13 kOe. An approach to saturation fit[3] yielded μ_{sat} = 0.13 μ_B/mole; the high-field susceptibility was zero (within the statistical uncertainty).

$PuMn_2$

This compound shows an almost temperature independent susceptibility between 5 and 300 K (Fig. 2).

$PuFe_2$

Previous Mössbauer measurements[7] had shown that $PuFe_2$ is ferromagnetic at 300 K with a hyperfine field at the Fe site. The present work confirms this result, and the $\sim$7% change in magnetization between 5 and 300 K suggests a Curie temperature in the neighborhood of 600 K. The $\sim$17% increase in magnetization from 3 to 13 kOe (independent of temperature) suggests a large magnetocrystalline anisotropy. The maximum moment at 5 K and $\sim$13 kOe. was 2.09 μ_B/mole. An approach to saturation analysis yielded μ_{sat} = 2.6 μ /mole and again a zero high-field susceptibility term (in sharp contrast to $NpFe_2$).

TABLE I.

Magnetic properties of uranium and plutonium Laves phases with Mn, Fe, Co, and Ni.

	UMn_2	UFe_2	UCo_2	UNi_2	$PuMn_2$	$PuFe_2$	$PuCo_2$	$PuNi_2$
Magnetism [a]	AF(?)	F	P	F	P	F	P	P
Ordering Temp. (°K)		158		30		∿600		
$\bar{\mu}_{sat}$ (μ_B/mole)		1.05		0.13		∿2.6		
CW range (°K)			50-290	100-290				50-300
μ_{eff} (μ_B/mole)			2.61	2.50				4.1
θ (°K)			-505	-445				-1200
χ_M (300 K) (10^{-3} emu/mole)	2.4	8.2	1.0	1.0	3.0		2.3	1.4
χ_M(5 K) (10^{-3} emu/mole)	2.8	-	5.7	-	3.0		2.9	2.9

[a]AF - antiferromagnetism, F - ferromagnetism, P - paramagnetism.

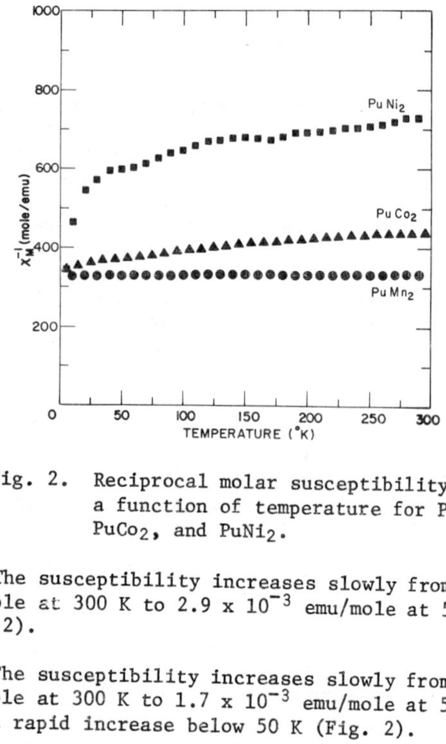

Fig. 2. Reciprocal molar susceptibility as a function of temperature for $PuMn_2$, $PuCo_2$, and $PuNi_2$.

$\underline{PuCo_2}$

The susceptibility increases slowly from 2.3 x 10^{-3} emu/mole at 300 K to 2.9 x 10^{-3} emu/mole at 5 K (Fig. 2).

$\underline{PuNi_2}$

The susceptibility increases slowly from 1.4 x 10^{-3} emu/mole at 300 K to 1.7 x 10^{-3} emu/mole at 50 K, with a more rapid increase below 50 K (Fig. 2).

DISCUSSION

The magnetic properties of the uranium and plutonium Laves phases studied here are summarized in Table I. For the compounds where there is a substantial linear χ^{-1} vs T region, Curie-Weiss constants have been determined and are reported in Table I. These results may be compared with the more detailed data available for the corresponding neptunium Laves phases.[3] The most striking feature is the greater magnitude of the ordering temperatures of the Fe compounds relative to the other compounds -- presumably a reflection of the strong *d - d* exchange. Similar behavior is observed in the corresponding lanthanide compounds.[8] The uranium atoms in UFe_2 carry almost no moment, whereas the Np atoms have a moment of 1 μ_B in $NpFe_2$, and the Pu atoms appear to have little or no moment in $PuFe_2$. (The hyperfine field at the iron site in $PuFe_2$ is a factor of two larger than that in $NpFe_2$.)[7] When the behavior of the Mn, Co, and Ni compounds are considered in this context, a trend becomes evident. It appears that the uranium atoms have little or no moment in any of the *3d* compounds, the neptunium atoms in all the *3d* compounds carry moments, and the plutonium atoms again are nonmagnetic in all these compounds, except possibly $PuFe_2$. This behavior is in contrast to the corresponding lanthanide compounds[8] and presumably indicates strong *f - d* hybridization. We may speculate that a substantial transfer of *f* electrons from the actinide atoms to the *d* bands leaves essentially no localized *f* electrons at the uranium atoms in these compounds. The corresponding data for the neptunium compounds[3] are not inconsistent with one electron per neptunium site. If there were two localized *f* electrons per plutonium site, we would again expect a nonmagnetic crystal-field ground state[9] and this is consistent with the present data. However, a more complete understanding of these materials will await a more detailed atomistic study of their magnetic behavior.

REFERENCES

1. A. T. Aldred, B. D. Dunlap, D. J. Lam, and I. Nowik, Phys. Rev. B 10, 1011 (1974).
2. A. T. Aldred, D. J. Lam, A. R. Harvey, and B. D. Dunlap, Phys. Rev., to be published.
3. A. T. Aldred, B. D. Dunlap, D. J. Lam, G. H. Lander, M. H. Mueller, and I. Nowik, Phys. Rev. (B1, Jan 1975); see also paper of this conference.
4. M. Yessik, J. Appl. Phys. 40, 1133 (1969).
5. S. T. Lin and A. R. Kaufman, Phys. Rev. 108, 1171 (1957).
6. D. J. Lam and A. T. Aldred in *The Actinides: Electronic Structure and Related Properties*, (J. B. Darby, Jr., and A. J. Freeman, eds., Academic Press) Vol. I, Chap. 3.
7. J. Gal, Z. Hadari, E. R. Bauminger, I. Nowik, S. Ofer, and M. Perkal, Phys. Lett. 31A, 511 (1970).
8. K. N. R. Taylor, Advances in Physics, 20, 551 (1971).
9. B. Bleaney, Proc. Roy. Soc. A276, 28 (1963).

*Work supported by the U. S. Atomic Energy Commission.

HIGH-FIELD SUSCEPTIBILITY IN FERROMAGNETIC $NpOs_2$*

B. D. Dunlap, A. T. Aldred, D. J. Lam and G. R. Davidson
Argonne National Laboratory, Argonne, Illinois 60439

ABSTRACT

$NpOs_2$ is known to be a ferromagnet with a Curie temperature of 7.5 K. Previous bulk magnetization measurements indicated a field induced magnetization even well below the transition temperature. We have extended this by a measurement of the local high-field susceptibility, using the Mössbauer effect in ^{237}Np. At 1.6 K, a susceptibility of $(1.2 \pm 0.2) \times 10^{-2}$ emu/mole is obtained, in general agreement with the bulk measurement. Such a large susceptibility is best understood by a model of itinerant magnetism, although other properties of the material indicate localized behavior.

INTRODUCTION

We have previously reported[1] a study of the magnetic properties of a number of cubic neptunium Laves phases (C-15 structure). These materials are of interest in that they show a systematic variation of magnetic parameters which can be interpreted as a transition from localized to itinerant magnetism as the lattice constant is decreased. $NpOs_2$, which has a ferromagnetic transition at 7.5 K, lies intermediate between the two extremes. An interesting feature of the $NpOs_2$ investigation was an apparent increase of the bulk magnetization with external field, even well below the magnetic ordering temperature. This was interpreted as a field-induced band magnetism, and discussed in terms of a model given by Murata and Doniach.[2]

In general, magnetization data taken for actinide systems in a magnetically ordered state are difficult to interpret because the large magnetic anisotropy makes magnetic saturation difficult with the applied fields available. One is then forced to make assumptions concerning the field dependence of the approach to saturation. While some procedures are available,[3] at this stage the situation is sufficiently complicated that such data reductions should be viewed with some caution, at least until higher field data become available. For this reason, magnetic hyperfine field measurements such as obtained by the Mössbauer effect can be very useful. The hyperfine interaction provides a local measurement of the electronic properties without the need for extrapolation. While the intensities of some lines in the spectrum will reflect the extent of domain alignment, the line positions (which give directly the hyperfine field) are not affected by the degree of alignment. In addition, investigation of a number of Np intermetallics has shown[4] a constant relationship between the hyperfine field H_{hf} obtained by the Mössbauer effect and the Np magnetic moment μ measured by neutron diffraction experiments:

$$H_{hf} = (1915 \text{ kOe}/\mu_B)\mu \quad (1)$$

If we assume this relation applies to other less well studied systems, then the hyperfine field can be quantitatively converted to a local magnetization. Such a method therefore provides a measure of the magnetization within individual domains which is not sensitive to the magnetic saturation problem.

RESULTS

In order to clarify the magnetic behavior of $NpOs_2$, we have obtained hyperfine field measurements as a function of temperature and external field, using the 59.6 keV Mössbauer resonance of ^{237}Np. Data were taken at 4.2 K ($T/T_c = 0.56$) and 1.5 K ($T/T = 0.2$) in external fields H_{ext} up to 34 kOe. Typical spectra, obtained at 1.5 K, are shown in Fig. 1. Inspection of the data reveals several important features: (i) The solid line in Fig. 1 for $H_{ext} = 0$ is a computer fit to the data using intensities which assume a random orientation of the domains. For $H_{ext} = 34$ kOe, the intensities indicate essentially total alignment of the moments in the external field. (ii) On increasing H_{ext}, the hyperfine pattern expands considerably, indicating an increase in the local magnetization with external field. (iii) The extrapolated hyperfine field for $H_{ext} = 0$ and $T = 0$ is 800 kOe. At 1.5 K, H_{hf} is reduced slightly to 780 kOe. However, application of a 34 kOe field at 1.5 K increases the hyperfine field to 950 kOe, considerably in excess of the saturation value obtained in exchange field alone.

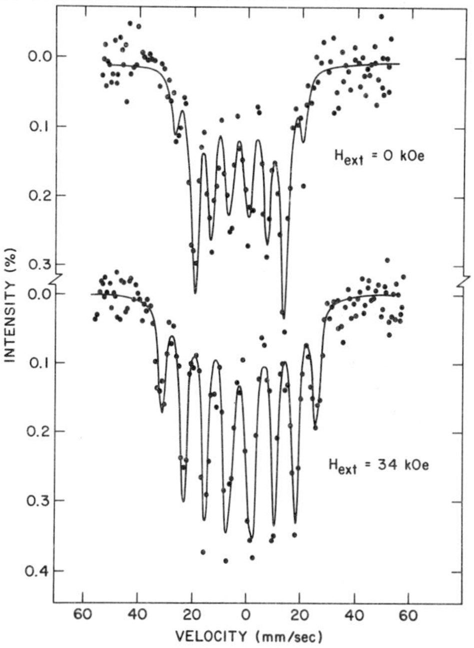

Figure 1. Hyperfine spectra in $NpOs_2$ at 1.5 K in external fields of 0 and 34 kOe.

Results of the hyperfine measurements are summarized in Fig. 2. The hyperfine data have been converted to magnetization on the right axis by assuming the relationship given in Eq. 1. Comparison with the highest field bulk magnetization data ($H_{ext} \approx 10$ kOe) shows good agreement within the uncertainties of this conversion procedure and of the data analysis methods. From the hyperfine field data,

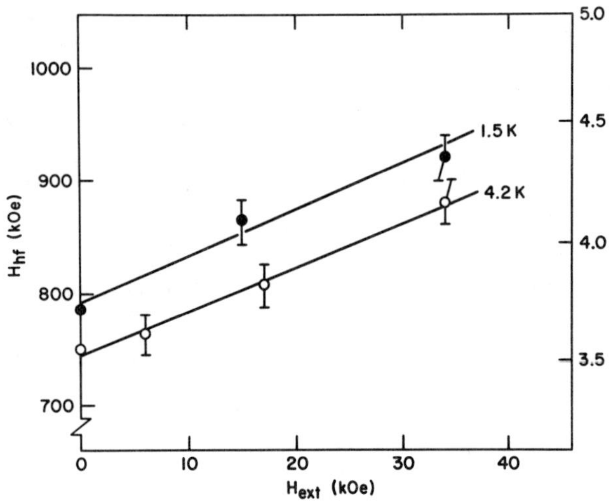

Figure 2. Dependence of the hyperfine field on external field at 1.5 K and 4.2 K.

one obtains a high-field susceptibility of $(1.2 \pm 0.2) \times 10^{-2}$ emu/mole, independent of temperature between 4.2 and 1.5 K. The zero field saturation value of 800 kOe indicates a saturation moment μ_{sat} of $(0.41 \pm 0.05)\ \mu_B$. We have attempted to describe the bulk magnetization data by including terms describing both the approach to saturation plus a high-field susceptibility χ in the form[3]

$$\sigma = \sigma_{sat}\ (1 - a/H^n_{ext}) + \chi H_{ext} \qquad (2)$$

For data taken at 1.6 K, a value of n = 1/2 gives[4] $\chi = 1.5 \times 10^{-2}$ emu/mole and $\mu_{sat} = 0.5\ \mu_B$. In view of the uncertainties involved in all these procedures, this may be considered to be good agreement between the local and the bulk magnetic properties.

DISCUSSION

It is not simple to resolve these various pieces of data in a self-consistent way. From some viewpoints, an approach based on localized electrons seems necessary, while from other viewpoints this assumption seems quite inadequate. The agreement between the measured local and bulk properties implies that Eq. 1 is essentially correct for this system. In considering the systematics of a large number of Np intermetallics and how their observed properties relate to calculated free-ion expressions, we have argued that this expression implies a large degree of localization for the 5f electrons.[5] Thus, it would seem that the electrons contributing to the magnetic properties of $NpOs_2$ can be considered to be localized. On the other hand, it seems quite difficult to understand the high field susceptibility from a simple localized point of view. Within a molecular field framework, the application of an external field can generally not provide a magnetization larger than that obtained in zero field at T = 0. An external field can induce small effects, such as an admixing of crystal field levels, provided the external field is comparable to the exchange field H_{exch}. If we estimate the exchange field by $\mu H_{exch} \approx kT_c$, then we find for the present case $H_{exch} \approx 300$ kOe. Thus it seems unlikely that the application of fields of the size used here will have a significant effect by such a mechanism.

Perhaps the simplest approach to understanding the susceptibility alone is to adopt a band picture. Such results have been seen in the past. For 1% Co in Pt, measurements[6] in fields as high as 150 kOe gave a high field susceptibility of 3.4×10^{-4} emu/mole. However, the present result is two orders of magnitude higher than this. As previously discussed, the approach of Murata and Doniach at least provide a framework in which the measurements can be considered. By using the obtained low temperature susceptibility, the transition temperature and the Curie constant immediately above the transition temperature, one finds that $NpOs_2$ fulfills the Murata and Doniach requirement for a weak, itinerant ferromagnet.[1]

In summary, it appears that neither the localized nor the itinerant electron approach alone can provide an adequate description of the observed results, although each describes some aspect of the magnetic and electronic properties. This emphasizes once again our feeling of the need for a theoretical approach which is truly intermediate between the two extremes if one is to understand the magnetic properties of the actinide intermetallics.

REFERENCES

1. A. T. Aldred, B. D. Dunlap, D. J. Lam and I. Nowik, Phys. Rev. B 10, 1011-1019 (1974).
2. K. Murata and S. Doniach, Phys. Rev. Letters 29, 285-288 (1972).
3. A. T. Aldred, B. D. Dunlap, D. J. Lam, G. H. Lander, M. H. Mueller and I. Nowik, this conference.
4. A. T. Aldred, D. J. Lam, A. R. Harvey, and B. D. Dunlap, Phys. Rev., in press.
5. B. D. Dunlap and G. H. Lander, Phys. Rev. Letters, 33, 1046 (1974).
6. S. Foner, E. J. McNiff Jr., and R. P. Guertin, Phys. Letters, 31A 466-467 (1970).

*Based on work performed under the auspices of the United State Atomic Energy Commission.

MAGNETIC SUSCEPTIBILITY MEASUREMENTS ON ACTINIDE-Be_{13} INTERMETALLIC COMPOUNDS*

M. B. Brodsky and R. J. Friddle
Argonne National Laboratory, Argonne, Illinois 60439

ABSTRACT

Magnetic susceptibilities have been measured between 2° and 300°K for the compounds $AnBe_{13}$ (An = Th, U, Np, Pu), all of which form the $NaZn_{13}$ structure. The susceptibility of $ThBe_{13}$ is temperature-independent. The data for UBe_{13} and $NpBe_{13}$ may follow two Curie-Weiss regions with the breaks occurring at 200° and 70°K respectively. The Curie-Weiss parameters are not substantially different for the two regimes, p_{eff} = 3.16 and 2.99 μ_B for UBe_{13}, and p_{eff} = 2.35 and 2.76 μ_B for $NpBe_{13}$. For $PuBe_{13}$, the Curie-Weiss fit requires a temperature independent term and yields $\chi_o = 1.20 \times 10^{-6}$ emu/g, $p_{eff} = 0.74\ \mu_B$/Pu, and $\theta = +11$°K. The intermediate coupling scheme yields $5f^3$, $5f^4$, $5f^5$ configurations for the U, Np, and Pu compounds, respectively. Thus, each actinide ion is trivalent if the f-electrons are considered to be localized. Of these compounds, only $PuBe_{13}$ gives evidence for long-range magnetic order, becoming antiferromagnetic at T_N = 11.5°K.

INTRODUCTION

It is well known that magnetism often occurs in actinide materials due to the unfilled 5f shell. The type of magnetic behavior observed may be long-range ferromagnetism or antiferromagnetism described in terms of localized moments;[1] itinerant ferromagnetism;[2] or localized spin fluctuations.[3] One of the major factors determining the presence of magnetism is the actinide-actinide distance.[4] If this distance is too small, ∿3.4 Å, direct 5f-5f overlap causes broadening into a narrow band, which may prevent magnetic order. In the pure metals the 5f level is above the Fermi surface for Th, is at or near the Fermi level in Pa, U, Np, and Pu, and may be thought to be fully below the Fermi level only by Am.[5] It has not been possible to follow the development of magnetic moments and long-range order in too many actinide systems because there are not many systems of common structure type among the actinides. Two such studies were Pd-An solid solutions (An = Th, U, Np, Pu)[6] and $AnRh_3$ intermetallic compounds.[7,8] In the latter case, the 5f level is not seen in $ThRh_3$; it forms a transition metal-like broad band in URh_3 due to 5f-6d hybridization; behaves as a spin-fluctuation system in $NpRh_3$; and orders antiferromagnetically in $PuRh_3$.

Another structure type which is common among the actinides is the $NaZn_{13}$ structure formed with Be.[9] In this system, there should be no complications due to the presence of a transition metal "B" element, and the large actinide-actinide distance, ∿5.1 Å, should prevent direct 5f-5f overlap. Furthermore, this is the only intermediate compound formed in these systems. The only study reported on an actinide beryllide is that of Troc, et al. on UBe_{13}.[10]

EXPERIMENTAL

Samples about 1 gram each were arc-melted with an initial excess of Be inside the Pu glovebox system. Arc melting continued until the total weight reached the expected weight for the amount of actinide used. All samples were arc melted at least 5 times. X-ray diffraction patterns indicated no additional lines, and the lattice parameters were consistent with published values (see Table I). Susceptibility samples, 0.2-0.6 grams were electrolytically machined, and sealed in aluminum for measurements in a Faraday apparatus similar to one described previously.[11] All measurements were taken at successively high temperatures, and were made at applied fields of 3, 4, 6, 9, 12, and 14.5 kOe.

RESULTS

The susceptibility-temperature curve for each material is shown in Fig. 1. Note the scale differences. Although the data for $ThBe_{13}$ have a slight temperature dependence there is no evidence for 5f behavior. The slight upturn at low temperatures is probably due to impurities. The data for the other three compounds all show strong temperature dependences, and their Curie-Weiss behavior is seen in Fig. 2 (the data for $PuBe_{13}$ are for $\chi - 1.2 \times 10^{-6}$ emu/g).

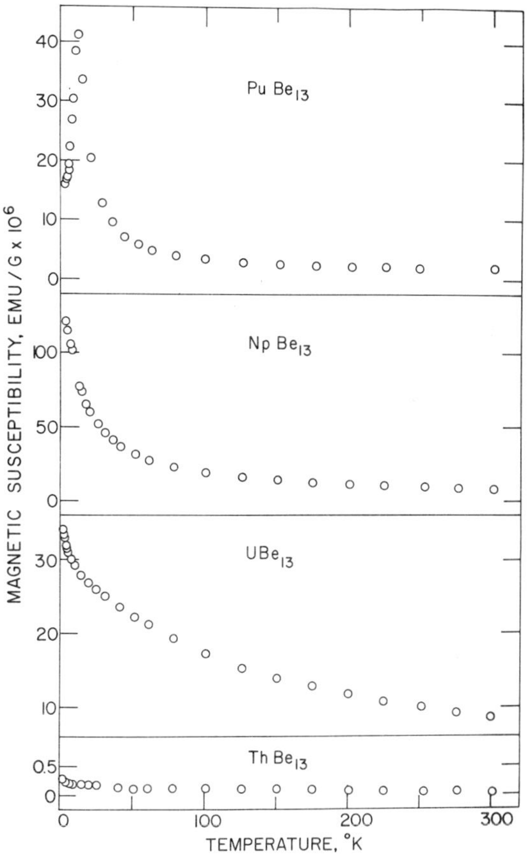

Fig. 1. Susceptibility-temperature curves for the $AnBe_{13}$ compounds. (MSD Neg. # 60447).

There is an apparent break in the curves for UBe_{13} at 200° and for $NpBe_{13}$ at 70°K. This behavior was also reported by Troc, et al., with the break occurring at 260°K in their UBe_{13} sample. The parameters of these curves are also listed in Table I. It is seen that the parameters for the high and low temperature regions are not drastically different. The higher temperature θ for UBe_{13} is nearly the same as that reported by Troc, et al. (-70°K), but the effective moment is lower than their value of 3.4μ_B. Since the data of Ref. 10 were taken over the temperature range 4.2-1000°K, their Curie-Weiss result is undoubtedly more accurate. The present data for UBe_{13} also show a small deviation at lowest temperatures in the direction of high magnetization, and may be due to impurity effects.

The results for $PuBe_{13}$ are simpler. Although a temperature independent term is needed in the Curie-Weiss curve fitting, there is no discontinuity in the curve. The sharp maximum at 11.5 ± 0.5°K is undoubtedly due to antiferromagnetic order. The data drop below the Neel temperature to 35% of the maximum value.

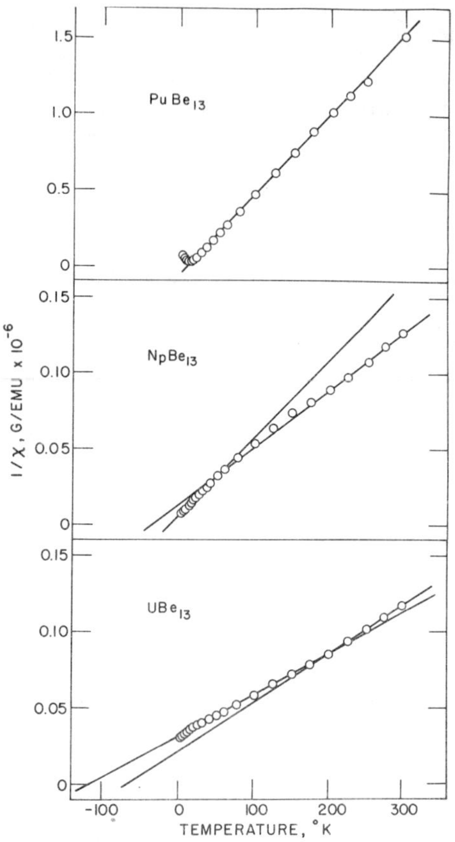

Fig. 2. Curie-Weiss low dependences. The data for $PuBe_{13}$ are for $\chi-\chi_o$. (MSD Neg. # 60448).

Table I. Data for the Various $AnBe_{13}$ Compounds

Actinide	Th	U	Np	Pu
a_o, Å (a)	10.393	10.255	10.254	10.273
p_{eff}, μ_B	- - -	2.99 (b) 3.16	2.76 2.35	0.74
θ, °K	- - -	-68 -110	-42 -12	+11
χ_o, 10^{-6}emu/g	∿0	0	0	1.20

(a) The actinide-actinide distance is $a_o/2$.
(b) The first number listed is for the higher temperature region.

DISCUSSION

The results of this study are in agreement with some of the expectations. No magnetic behavior is seen in the $ThBe_{13}$ sample. Each of the U, Np, and Pu compounds show local moment paramagnetism, as was expected from the large inter-actinide spacing. The lattice parameters also show evidence for the f-series beginning at or near U. The parameter decreases by over 1% between Th and U, and then increases slowly, rising by only 0.2% between U and Pu.

As has been pointed out by Chan and Lam,[12] it is not appropriate to use straight-forward L-S or J-J coupling schemes for the actinides. It is necessary to use intermediate coupling schemes in which many low-lying J-manifold states are allowed to mix. In the absence of a complete crystal field calculation for the $NaZn_{13}$ structure, the results of Chan and Lam for NaCl structure actinide compounds will be used. This is not truly appropriate since the 24 nearest neighbor Be atoms form a snub cube, but there is a six-fold actinide-only coordination cage in both cases. The results of Chan and Lam[12] show that for $5f^2$, $5f^3$, $5f^4$, $5f^5$, and $5f^6$ configurations, the effective paramagnetic moments are 0, 3.33, 2.59, < 1.41, and 0, respectively. Matching these to the values for the present compounds yields $5f^3$ for UBe_{13}, $5f^4$ for $NpBe_{13}$, and $5f^5$ for $PuBe_{13}$. Although Russell-Saunders coupling could yield the same conclusion for $NpBe_{13}$ and $PuBe_{13}$, it would also predict $5f^2$ for UBe_{13}, whereas the intermediate coupling scheme yields zero moment for $5f^2$.

These configurations would indicate that each of these three actinide ions is in the trivalent state. It should be noted, however, that although the magnetic moments may be localized, a straight-forward description of these light actinides in terms of localized 5f electrons is probably not correct.[5] Probably there is some f-character in a band made up of s-f or d-f hybridized states.

Since $PuBe_{13}$ is the only compound to order netically, the rather large Weiss constant for UBe_{13} and even that for $NpBe_{13}$ are not to be taken as indicative of the molecular field. In view of the large inter-actinide spacing, it is likely that the exchange in $PuBe_{13}$ is primarily RKKY type. Therefore it would be expected that the U and Np compounds should order since S^2 for $5f^5$ is only ∿1.5 times as large as for $5f^4$ (Np^{3+}).[13] It is likely that the non-equal actinide spacing comes into play, also.

The discontinuties in the curves for UBe_{13} and $NpBe_{13}$ in Fig. 2 do not seem sharp enough to be explained by "valence" shifts or lattice distortions.[14] It is possible that the break is really a curvature due to crystal field effects. It is possible to obtain a good fit to the susceptibility-temperature data for $NpBe_{13}$ with Chan and Lam's calculations.[15] However, the temperature dependences for UBe_{13} and $PuBe_{13}$ cannot be reproduced by the calculations. Further work using the full $NaZn_{13}$ crystal structure in the crystal field calculations, and additional experimental data on these materials is necessary to explain the source of the discontinuities.

The authors wish to thank D. J. Lam for use of his crystal field calculations and for stimulating discussions of this work; and B. R. Coles for suggesting this work on the beryllides.

REFERENCES

1. D. J. Lam and A. T. Aldred in The Actinides; Electronic Structure and Related Properties, ed. A. J. Freeman and J. B. Darby, Jr. (Academic Press, New York, 1974), Vol. I, Ch. 3.
2. A. R. Harvey, M. B. Brodsky, and W. J. Nellis, Phys. Rev. B7, 4137-4142 (1973).
3. M. B. Brodsky, Phys. Rev. B9, 1381-1387 (1974).
4. H. H. Hill, Nuclear Metallurgy 17, 2-19 (1970).
5. A. J. Freeman and D.D. Koelling, in The Actinides; Electronic Structure and Related Properties, ed. A. J. Freeman and J. B. Darby, Jr. (Academic Press, New York, 1974), Vol. I, Ch. 2.
6. W. J. Nellis and M. B. Brodsky, Phys. Rev. B4, 1594-1601 (1971).
7. W. J. Nellis, A. R. Harvey, and M. B. Brodsky, A. I. P. Conf. Proc. 10, 1076-1080 (1973).
8. G. P. Sykora, A. J. Arko, and D. O. Van Ostenburg, this conference.
9. M. Hansen, Constitution of Binary Alloys, 2nd, ed (McGraw-Hill, New York, 1958), pp. 291-299.
10. R. Troc, W. Trzebiatowski, and K. Piprek, Bull. Acad. Polon. Sci., ser. sci. chim. 19, 427-432 (1971).
11. J. W. Ross and D. J. Lam, Phys. Rev. 165, 617-620 (1968).
12. S. K. Chan and D. J. Lam, Nuclear Metallurgy 17, 219-232 (1970). S. K. Chan and D. J. Lam, in The Actinides; Electronic Structure and Related Properties, ed. A. J. Freeman and J. B. Darby, Jr. (Academic Press, New York, 1974), Vol. I, Ch. 1.
13. S. Chikazumi, Physics of Magnetism, Engl. ed. (John Wiley and Sons, New York, 1964), pp. 440-458.
14. A. T. Aldred, B. D. Dunlap, D. J. Lam, and I. Nowik, Phys. Rev. B10, 1011-1019 (1974).
15. D. J. Lam, private communication.

*Work supported by the U. S. Atomic Energy Commission.

MAGNETIC PROPERTIES OF DONOR-ACCEPTOR COMPOUNDS OF TETRATHIOFULVALENE WITH BIS-DITHIOLENE METAL COMPLEXES *

I. S. Jacobs, L. V. Interrante and H. R. Hart, Jr.
General Electric Research and Development Center, PO Box 8, Schenectady, N.Y. 12301

ABSTRACT

Donor-acceptor compounds of tetrathiofulvalene (TTF) with bis-dithiolene (BDT) metal complexes are studied as flexible analogues to (TTF)-tetracyano-quinodimethane. For BDT complexes, $MS_4C_4X_4$, the $X=CF_3$ derivatives with M = Ni and Pt, and the X=H derivative with M=Ni, have been examined magnetically for susceptibility and magnetization behavior. The 1:1 (TTF)(BDT) $PtCF_3$ compound follows a Curie-Weiss law with θ=16K for dominant ferromagnetic interactions, but the susceptibility peaks near 12K indicating anti-ferromagnetic coupling. Its ferromagnetic subsystems are probably chain-or layer-like. The μ_{eff} value of $2.24\mu_B$ is consistent with two S =1/2 spins per formula. The 1:1 $NiCF_3$ compound has a similar μ_{eff} ($2.27\mu_B$), but exhibits dominant antiferromagnetic interactions with θ = -18K down to 40K. At lower temperature its Curie constant is reduced to that of a single spin with an indicated ferromagnetic intercept $\theta \approx +5K$. Below 10K the magnetization-field behavior tends toward saturation and a Curie point near 2.18K is suggested. A model of two nearly independent magnetic systems, analogous to $CuSO_4 \cdot 5H_2O$ is offered. The 2:1 X=H derivative $(TTF)_2(NiS_4C_4H_4)$ is essentially paramagnetic ($|\theta| < 2K$) with one unpaired S = 1/2 spin per formula.

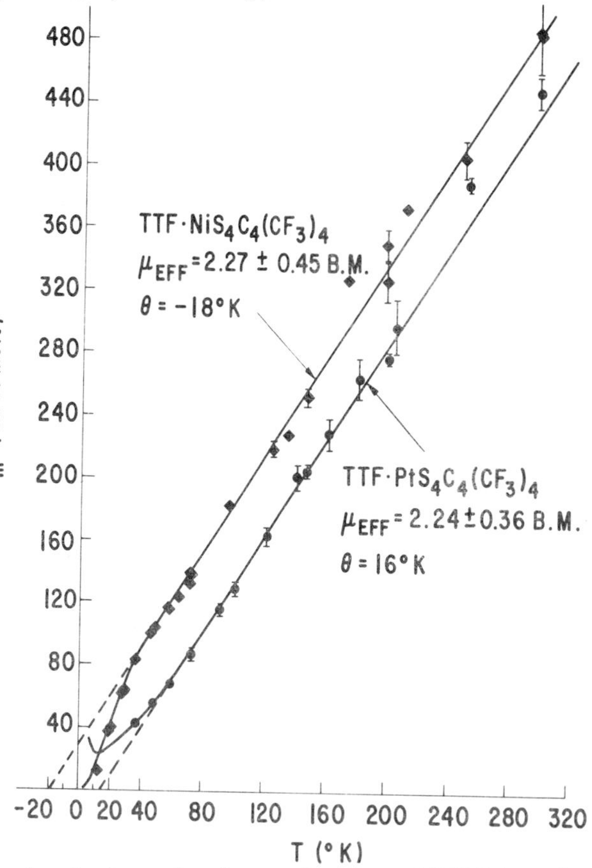

Fig. 1. Reciprocal molar powder susceptibilities of $(TTF)(MS_4C_4X_4)$ where $X=CF_3$ and M=Ni and Pt.

Considerable interest has been directed recently to various organic materials on account of their physical behavior including significant metallic conductivity, metal-insulator transitions and low dimensional electrical and/or magnetic properties. Compounds such as tetrathiofulvalene-tetracyanoquino-dimethane (TTF)(TCNQ)[1] and N-methyl-phenazinium-(TCNQ)[2] are examples receiving much attention. We have studied several donor-acceptor compounds[3] of TTF, formally analogous to (TTF)(TCNQ) but where (TCNQ) is replaced as the acceptor by planar bis-dithiolene (BDT) metal complexes. This choice is stimulated by analogies[4] between BDT complexes and TCNQ in molecular and electronic structure and reversible electron transfer behavior. Another attraction is the system flexibility inherent in BDT complexes $MS_4C_4X_4$, where M=Ni, Pt, Pd, or other metals and X=H, CH_3, CN, CF_3, etc.

We report here the magnetic properties of three compounds in this family[3], i.e. the $X = CF_3$ derivatives, $(TTF)[MS_4C_4(CF_3)_4]$ with M=Ni, Pt; and one derivative with X = H, $(TTF)_2(NiS_4C_4H_4)$. It should be noted that the neutral constituents (TTF) or (BDT), either in solution or as molecular crystals are diamagnetic. The presence of a transition metal in the BDT does not guarantee an unpaired spin.

Magnetization and susceptibility measurements were made on powder samples of approx. 100 mg. using a Foner vibrating sample magnetometer in the range $4K \leq T \leq 300K$, up to 18 kOe. In addition, for the two Ni-BDT compounds collections of oriented crystals weighing 2 mg were measured down to 2.2K with a sensitive Faraday balance based on a Cahn Electrobalance and a Nb-Ti solenoid (H up to 40 kOe) allowing independent variation of H and dH/dz.

The more interesting magnetic properties are found in the 1:1 CF_3 derivatives for which the powder reciprocal susceptibilities are shown in Fig. 1. At high temperatures each follows a Curie-Weiss behavior $\chi^{-1}=(T-\theta)/C$ with nearly identical slopes, where $C=N\mu^2_{eff}/3k$. This temperature dependent paramagnetism, along with results of solution conductivity and electronic absorption spectra, serves to identify the ionic charge transfer character as $TTF^+MS_4C_4(CF_3)_4^-$ in which one electron has been transferred producing two electronic spins per formula. These spins may be strongly coupled as a triplet S=1 or essentially uncoupled as two S=1/2 spins. The effective moment, $\mu_{eff}=g\mu_B[zS(S+1)]^{1/2}$, using g = 2 from epr[5], and where z=1 or 2 according to the model, would be 2.83 or 2.45, respectively. Comparison with the observed values of μ_{eff} = 2.27 for M=Ni and 2.24 for M=Pt suggests that the latter (uncoupled spins) model is correct. The Weiss temperature θ projected from the high temperature behavior of these CF_3 derivatives suggest a dominance of ferromagnetic (F) interactions for the Pt compound ($\theta \approx 16K$), and antiferromagnetic (AF) interactions for the Ni compound ($\theta \approx -18K$). At lower temperatures additional magnetic interactions come into evidence.

For the Pt compound, the reciprocal susceptibility has a minimum near 12K and is still rising at 4.2K. This behavior indicates the presence of AF interactions and probably long range AF order. Such a combination of F and AF interactions is associated with lower dimensional character, i.e. the F subsystems are layers or chains which are subject to a weaker AF inter-subsystem coupling.[6] This is supported by a crystal structure analysis[7] which suggests that each subsystem is a chain-like stack of alternating TTF^+ and BDT^- ions.

The $NiCF_3$ compound contrasts markedly with the Pt analogue. Below about 35K the reciprocal susceptibility departs from its AF Curie-Weiss high temperature behavior. For a limited range it may be approximated with $\theta \approx +5K$ (i.e. F interactions) and $\mu_{eff} \approx 1.7\ u_B$ (a single S = 1/2 spin/formula). At still lower temperatures the magnetization vs. field behavior (Fig. 2) shows marked curvature toward ferro- or paramagnetic saturation, reaching a moment of $0.6\mu_B$/formula at the limits of our measurement. We employ the Belov-Goryaga[8] plot (M^2 vs. H/M) and the χ_o^{-1} values

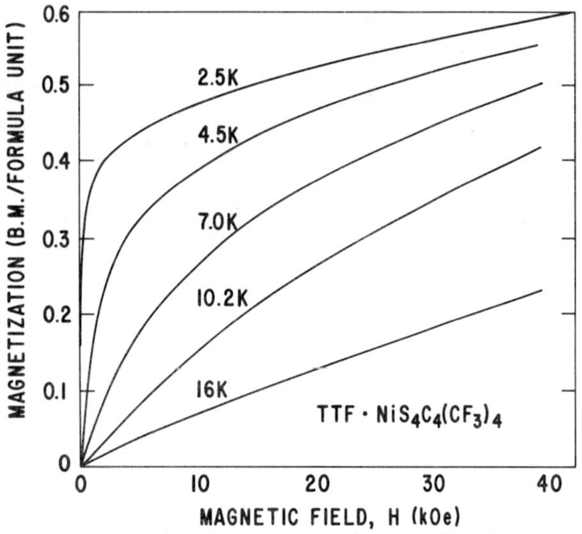

Fig. 2. Magnetization curves of several oriented crystals of (TTF)$(NiS_4C_4(CF_3)_4)$; H ∥ needle axis.

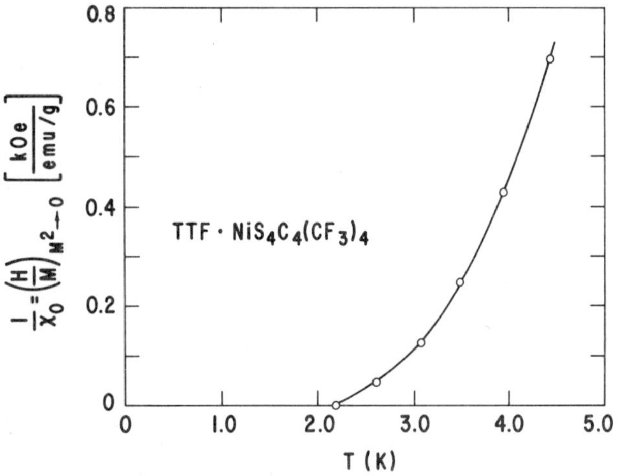

Fig. 3. Reciprocal initial susceptibility of oriented crystals of (TTF)$(NiS_4C_4(CF_3)_4)$, derived from Belov-Goryaga plot at low temperatures.

derived therefrom to ascertain if long range ferromagnetism has been achieved. From Fig. 3 it appears that a Curie point is reached near 2.18K.

The curve of χ^{-1} vs. T for this $NiCF_3$ compound recalls a novel model previously employed for $CuSO_4 \cdot 5H_2O$,[9] (following a suggestion of Geballe and Giauque). We propose two nearly independent (S=1/2) spin systems; one with relatively stronger AF interactions, the other with weaker F interactions. We cannot presently draw any conclusions on the dominant dimensionality of the subsystems, but we suggest that it is less than 3. In Fig. 4 we compare the observed data to this model, using chain subsystems with a) Heisenberg exchange (S=∞ scaled to finite spin)[10] and b) Ising exchange (S=1/2)[11]. The agreement is adequate to justify the basic concept of the model. Other models, e.g. long period ferrimagnetism, may also be invoked. Final choice awaits further investigation.

The third compound $(TTF)_2(NiS_4C_4H_4)$ is to a good approximation a Curie paramagnet ($|\Theta| < 2K$) with $\mu_{eff} \simeq 1.7\mu_B$ (S=1/2, g=2). The epr g-values are close to 2.0[5]. The problem of how only one unpaired spin can appear in a charge transfer between diamagnetic complexes is discussed elsewhere[3] in the context of a complete three dimensional crystal structure analysis.

A further interpretation of the results on all these compounds will be developed with information obtained by other measurements. We acknowledge helpful conversations with W. P. Wolf and B. R. Cooper as well as the use of unpublished results of our colleagues, J. S. Kasper and G. D. Watkins.

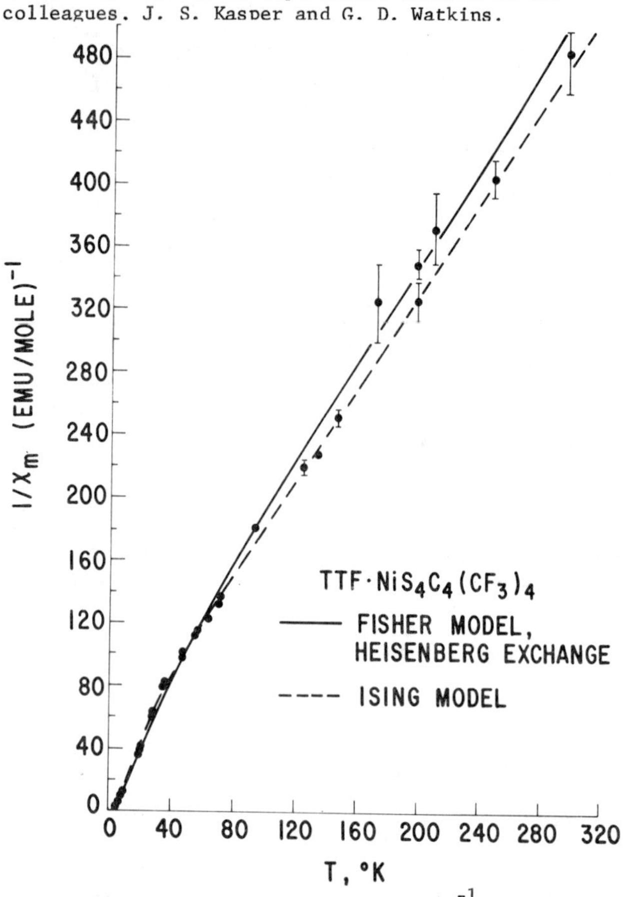

Fig. 4. Comparison between observed χ_M^{-1} and model of non-interacting (chain) subsystems: Heisenberg-Fisher, $|J_A|=6.7J_F=67k$; Ising, $|J_A| = 7.2J_F=124k$.

REFERENCES

*Work supported in part by AFOSR, under contract F-44620-71-C-0129

1. L. B. Coleman, M. J. Cohen, D. J. Sandman, F. G. Yamagishi, A. F. Garito, and A. J. Heeger, Solid State Comm. 12, 1125-1132, (1973).
2. A. J. Epstein, S. Etemad, A. F. Garito and A. J. Heeger, Phys. Rev. 5, 952-977 (1972).
3. L. V. Interrante, K. W. Browall, H. R. Hart, Jr., I. S. Jacobs, J. S. Kasper, C. A. Secaur, G. D. Watkins and S. H. Wee, submitted for publication.
4. K. W. Browall and L. V. Interrante, J. Coord. Chem. 3, 27-38 (1973).
5. R. D. Schmitt and A. H. Maki, J. Am. Chem. Soc. 90, 2288-2292 (1968), Ni anion in frozen glass; G. D. Watkins, private communication, this compound.
6. L. D. Landau, Physik. Z. Sowjetunion 4, 675-679 (1933); J. S. Smart in "Magnetism" G. T. Rado and H. Suhl, eds., (Academic Press, New York, 1963) Vol. III, Ch 2; L. J. de Jongh and A. R. Miedema, Adv. Physics. 23, 1-260 (1974).
7. J. S. Kasper, private communication.
8. K. P. Belov and A. N. Goryaga, Fiz. Met. Metallov. 2, no. 1, 3-9 (1956); J. S. Kouvel and M. E. Fisher 136, A1626-32 (1964).
9. A. R. Miedema, H. Van Kempen, T. Haseda and W. J. Huiskamp, Physica 28, 119-130 (1962).
10. M. E. Fisher, Am. J. Phys. 32, 343-346 (1964); T. Smith and S. Friedberg, Phys. Rev. 176, 660-5, (1968).
11. H. A. Kramers and G. Wannier, Phys. Rev. 60, 252-262 (1941); S. Katsura, Phys. Rev. 127, 1508-1518 (1962).

ANTIFERROMAGNETIC ORDER IN AMPHIBOLE ASBESTOS+

J.C. Eisenstein,* M.F. Taragin, and D.D. Thornton
The George Washington University, Washington D.C. 20009

ABSTRACT

We have made magnetic susceptibility, electron spin resonance, and Mössbauer measurements on two different amphiboles, (Fe, Mg)$_7$Si$_8$O$_{22}$(OH)$_2$, and crocidolite, Na$_2$Fe$_5$Si$_8$O$_{22}$(OH)$_2$, which indicate that these materials are ordered antiferromagnetically at liquid helium temperatures. Amphibole minerals are characterized structurally by long double chains of linked silicate tetrahedra. The metal anions are 'sandwiched' between pairs of double silicate chains to form long narrow ribbons relatively well separated from each other. In amosite the only magnetic species present is Fe^{2+}. This has been confirmed for our samples by the Mössbauer measurements. The ac magnetic susceptibility measurements on oriented fibers indicate that there are two antiferromagnetic transitions in amosite; one around 22 K associated with an easy axis perpendicular to the fiber axis, and one around 7 K associated with an easy axis parallel with the fiber axis. The Mössbauer spectra support this interpretation. Both Mössbauer and spin resonance indicate that any magnetic impurities must be less than one percent.

INTRODUCTION

The fibrous silicate minerals known collectively as asbestos fall into two major crystallographic groups: crysotile and amphibole. In a survey of the low temperature magnetism of these minerals we observed antiferromagnetic behavior in crocidolite and amosite.[1] In this paper we report further results which indicate the possibility of two distinct antiferromagnetic transitions in amosite.

CRYSTAL STRUCTURE

Both amosite and crocidolite are monoclinic amphiboles. The most general chemical formula for the amphiboles can be written as $X_{2-3}Y_5Z_8O_{22}(OH,F)_2$[2,3]. In the naturally occurring minerals Z is usually silicon and OH is preferred to F. The X ions may be any of a variety of mono- or di-valent cations (eg. Na, Mg, Mn), while the Y ions may be di- or tri-valent cations (eg. Mg, Fe, Mn, Al). The amphibole structure[2] is based on "infinite" double chains of silicate tetrahedra with a base unit Si_4O_{11}. The Y cations and the OH anions are sandwiched between two of these chains so that the Y ions occupy three octahedrally coordinated M_4 sites along the edges of these long strips and serve to bind them into the overall crystal structure.

In amosite, most of the metal cation sites are occupied by divalent iron[4] with magnesium, calcium and aluminum as minor substitutional constituents The unit cell dimensions of amosite have been determined[5] as a = 0.989, b = 1.826, c = 0.530 nm., $\beta = 70^\circ$. The space group is given as I2/m. The details of the crystal structure of amosite are not known, except that it is very similar to the structure of Bolivian crocidolite, which has been determined.[6] Crocidolite has the unit cell a = 0.989, b = 1.795, c = 0.531 nm, $\beta = 72\frac{1}{2}^\circ$.

THE FERROUS ION

The ferrous ion has a $3d^6$ configuration with a ^5D ground term. In a distorted octahedral crystal field such as we have at the M_1, M_2, M_3 sites in amosite, we would expect the angular momentum to be quenched and the five fold spin degeneracy to be lifted by the spin orbit coupling. In compounds with similar coordination the overall splitting of the spin quintuplet varies from 10 cm^{-1} to 100 cm^{-1}.[7] The resulting structure is usually an upper triplet and a lower doublet, each of which may be split further. The doublet is very anisotropic. The crystollographic situation is less well defined for the M_4 site.

EXPERIMENTAL

We measured the ac magnetic susceptibility of a powder sample (χ_p) of finely milled amosite (U.I.C.C.) over the temperature range from 1 K to 30 K. The data are plotted as circles in figure 1. In addition the ac susceptibility of oriented fibers of Malipsdrift amosite (from which the U.I.C.C. amosite is milled) was measured both parallel (χ_c) and perpendicular (χ_{ab}) to the fiber axis. These data are plotted in figure 1 as closed circles (ellipses) and

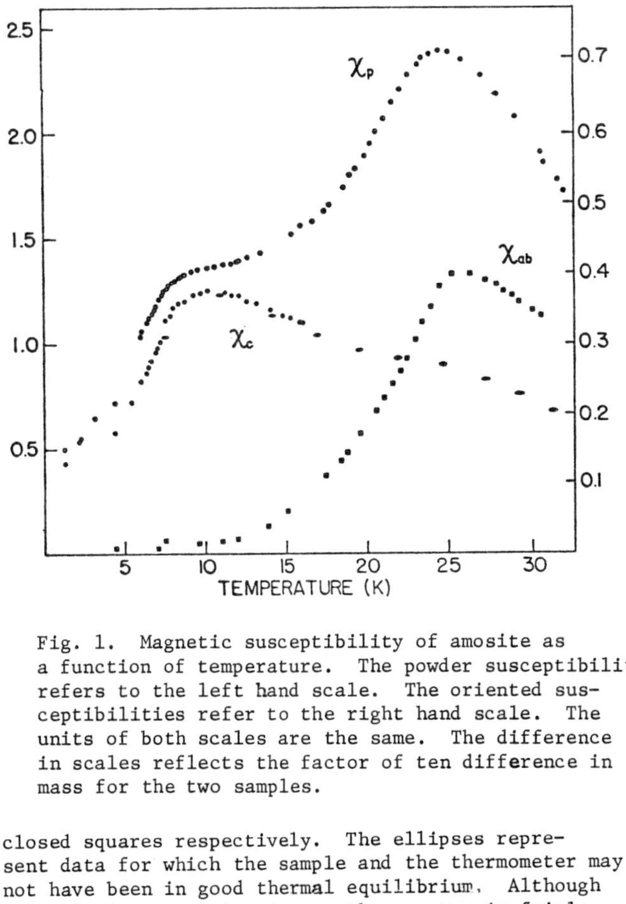

Fig. 1. Magnetic susceptibility of amosite as a function of temperature. The powder susceptibility refers to the left hand scale. The oriented susceptibilities refer to the right hand scale. The units of both scales are the same. The difference in scales reflects the factor of ten difference in mass for the two samples.

closed squares respectively. The ellipses represent data for which the sample and the thermometer may not have been in good thermal equilibrium. Although the relative precision in our thermometry is fairly good, difficulties in obtaining a proper calibration in the high temperature range (10 K to 30 K) force us to assign a progressive uncertainty in absolute temperature amounting to 2 K at the highest temperatures of the susceptibility measurements.

Mössbauer spectra obtained from a sample of milled U.I.C.C. amosite at several temperatures are shown in figure 2. Additional measurements include the magnetization of the fiber sample in the direction of the fiber axis in fields up to 2.0 Tesla, at temperatures from 1 K to 300 K; electron spin resonance at 9 GHz. The esr was a null experiment in that no resonances were observed in fields up to 2.0 Tesla at any temperature. This certainly precludes the presence of any significant amount of ferric impurity.

DISCUSSION

The room temperature Mössbauer spectrum consists of two pairs of peaks corresponding to two groups of ferrous ions with different quadrupole splittings. The relative intensities indicate that the proportion of ions in the two groups is approximately 5 to 2. This is consistent with assigning the pair with the larger quadrupole splitting to the octahedrally coordinated M_1,M_2,M_3 ions and the smaller splitting to the eight fold coordinated M_4 sites. These results are in agreement with previous work.[4,8] The spectra at 77 K and 30 K are essentially the same as at room temperature, showing only small temperature dependence of the isomer shift, quadrupole splittings and line widths. At 15 K the outer doublet is still easily discernible, though the lines are quite a bit broader. In place of the inner doublet we see a broad magnetic hyperfine structure. At 11 K the huperfine structure is more clearly defined and the lines of the outer quadrupole doublet are broadened further. At 4.2 K both groups of ions show magnetic hyperfine structure.

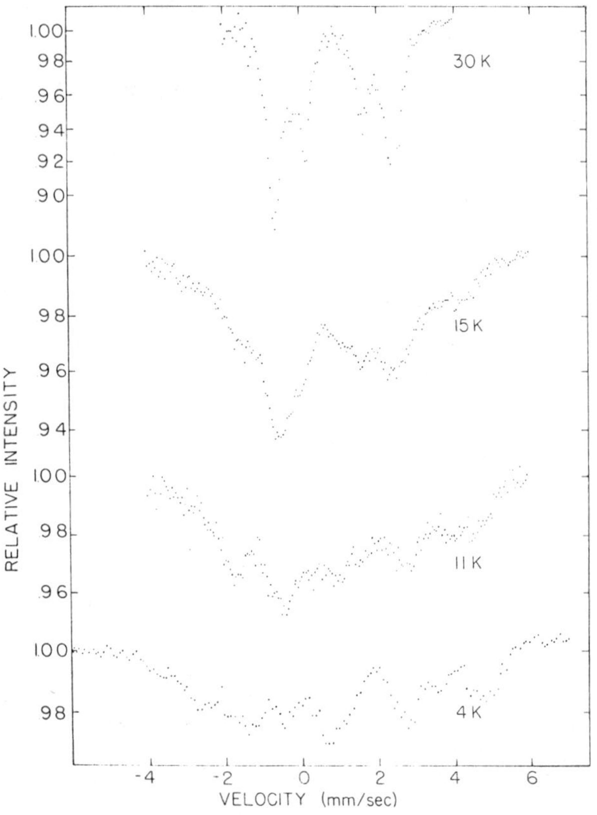

Fig. 2. Mössbauer spectra of amosite

The powder susceptibility shows evidence of two peaks, one at 25 K, the other at about 10 K. In the fiber susceptibility, χ_{ab} passes through a maximum at 25 K and decreases essentially to zero by about 11 K. Except for a small hump which arises from the presence of some misaligned fibers in the sample, χ_c shows no strong departure from paramagnetic behavior down to about 10 K, where it also goes through a maximum. Again with the exception of a small effect arising from misaligned fibers, χ_{ab} does not change significantly below 11 K. It is striking that one component of the susceptibility can exhibit the kind of temperature dependence normally associated with an antiferromagnetic transition while the perpendicular component gives no indication of any change in the magnetic state.

Since the Mössbauer spectra indicate that the M_4 ions order between 25 K and 10 K, the high temperature peak in χ_{ab} must be associated with the M_4 ions. This indicates that the ferrous ions in the M_4 sites are indeed highly anisotropic, that their easy axis is perpendicular to the fiber axis, and that they form a system which orders antiferromagnetically below 25 K. Similarly, the 10 K peak in χ_c must be associated with the ferrous ions in the M_1, M_2, M_3 sites indicating that these ions also have a highly anisotropic ground state whose easy axis is parallel to the fiber axis, and that they form a system which orders antiferromagnetically below 10 K. At this point it is not possible to draw any conclusions regarding the nature of the order of either system, nor do the data absolutely exclude alternative interpretations.

We are indebted to Dr. Vernon Timbrell who introduced us to asbestos and has generously supplied us with the samples used in these studies. We would also like to acknowledge the hospitatlity of Dr. B.W. Mangum and the Cryogenic Physics Section of the National Bureau of Standards where some of the experiments were conducted.

REFERENCES

* Supported in part by N.S.F. grant GP 29520.
+ Supported in part by N.S.F. grant GU-3287.
1 D.D. Thornton, M.F. Taragin and J.C. Eisenstein, Physics Letters, to be published.
2 E.J.W. Whittaker, Acta Cryst. 13, 291 (1960).
3 B.E. Warren, Z. Kristallogr. 72, 493 (1930).
4 T.C. Gibb and N.N. Greenwood, Trans Faraday Soc. 61, 1317 (1965).
5 R.I. Garrod and C.S. Rann, Acta Cryst. 5, 285 (1952).
6 E.J.W. Whittaker, Acta Cryst. 2, 312 (1949).
7 See for example

A STUDY OF THE MAGNETIC BEHAVIOR OF IRON IMPURITIES IN THE LINEAR ANTIFERROMAGNET $CsNiCl_3$

P. A. Montano
Dept. of Physics, University of California,
Santa Barbara, Calif. 93106

ABSTRACT

The behavior of iron impurities in the quasi-one dimensional antiferromagnet $CsNiCl_3$ has been studied using susceptibility and Mössbauer measurements. Samples of $CsNiCl_3$ containing 4% Fe were used. A magnetic field region from 1 to 16 kOe was investigated and no field dependent susceptibility was observed. From the absence of any temperature dependent anisotropy in the susceptibility as well as no observable hyperfine magnetic splitting at the ^{57}Fe nucleus at 4.2K, a lowering of the transition temperature (4.8K) from pure $CsNiCl_3$ was inferred. In $CsNiCl_3$ the Fe^{2+} impurity has a negative single-ion anisotropy and seems to be antiferromagnetically coupled to the neighboring Ni ions. By constrast in $CsFeCl_3$[1] the Fe^{2+} has a positive single ion anisotropy and a ferromagnetic exchange interaction between the Fe^{2+} ion along the chain.

INTRODUCTION

It is the purpose of this work to study the influence of iron impurities in the quasi one dimensional antiferromagnet $CsNiCl_3$[2] (T_N = 4.8 K).[3] A single crystal of $CsNiCl_3$ containing 4% Fe was studied[4] using Mössbauer and susceptibility measurements.

The susceptibility measurements were carried out using the Curie-Faraday method. These measurements were carried out on powdered as well as single crystal samples. There was a linear relationship between the field and the magnetization down to temperatures of 2K and up to 16 kOe fields. The Mössbauer measurements were carried out down to 1.8 K on a single crystal sample. Figure 1 shows a typical spectrum at room temperature (RT).

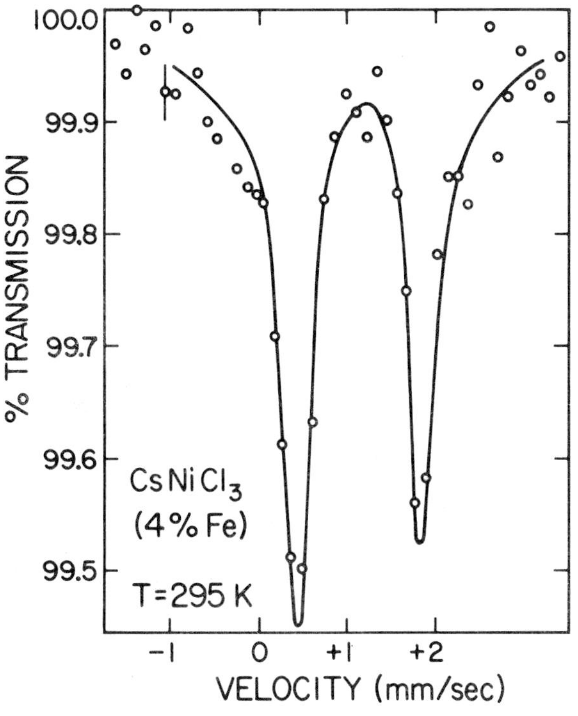

Fig. 1. Mössbauer spectrum of $CsNiCl_3$ (4% Fe).

EXPERIMENTAL RESULTS AND DISCUSSION

The Fe^{2+} ion is subjected to the crystal field produced chiefly by the distorted octahedron of neighboring Cl^- ions. The ^5D term of Fe^{2+} free ion is then split by the cubic component of the crystalline field into an orbital ^5E_g doublet and a $^5T_{2g}$ orbital triplet where the latter is the ground state. There is a further splitting caused by the spin-orbit interaction $-\lambda\vec{L}\cdot\vec{S}(L = 2, S = 2)$ and the residual trigonal component of the crystal field $V_T = -\Delta(L_z^2 - 2)$, where Δ is the crystal field strength and the sign of Δ depends upon the nature of the distortion. In the temperature range of this experiment only the $^5T_{2g}$ level will be populated. The sign of Δ can be determined from Mössbauer measurements on single crystals. This was carried out on a $CsNiCl_3$ 4% Fe single crystal sample with the propagation axis of the γ-ray at an angle of 90° with the symmetry axis of the crystal (c-axis). From these measurements it was inferred that $\frac{1}{2}e^2qQ > 0$, where Q is the quadrupole moment of the I = 3/2 level of ^{57}Fe and eq is the electric field gradient (EFG). The above result is consistent with $\Delta > 0$ and a ground state doublet. By contrast in $CsFeCl_3$ $\Delta < 0$ and consequently $e^2qQ < 0$.[1] From the temperature dependence of the QS a value of $\Delta/\lambda \simeq 1.5$ was found, $\lambda = 75\ cm^{-1}$.[1] Due to the different lattice parameters along the c-axis in $CsNiCl_3$(Ni - Ni = 2.96Å)[2] and $CsFeCl_3$(Fe-Fe = 3.03Å) as well as the larger ionic radius of Fe^{2+} one will expect a different type of crystalline distortion for Fe^{2+} impurities in $CsNiCl_3$. At low temperatures the Fe^{2+} can be described by an effective spin S = 1 with single ion anisotropy $D = -32 \pm 5 cm^{-1}$ and $J_\parallel \simeq 2.5 J_\perp$.[1] The QS at low temperatures shows evidence of some perturbation by the magnetic interaction of the neighboring Ni ions (QS = 1.44 mm/sec at RT to 2.00 ± 0.05 at 4.2 K). At 1.8 K it is possible to observe the appearance of some hyperfine magnetic splitting on the Fe^{2+} indicating the onset of magnetic order or slow electronic relaxation time. By contrast, no hf magnetic splitting was observed at 4.2 K.

The Ni^{2+} ion has an orbital ground state singlet in cubic octahedral symmetry with S = 1.

In pure $CsNiCl_3$ the exchange and single ion anisotropy D has been evaluated from the analysis of the susceptibility using a classical spin field model.[5] One obtains $D = -0.3\ cm^{-1}$ and $J = -11\ cm^{-1}$ with $g_\parallel$ = 2.24 and $g_\perp$ = 2.28 in good agreement with values reported in the literature.[2]

From 300°K down to 100 K the susceptibility in single crystal and powdered samples of $CsNiCl_3$ 4% Fe could be fitted using a molecular field approximation with $\theta_p = -68 \pm 2K$ almost identical to pure $CsNiCl_3$ ($\theta_p = -69K$).[2] In Fig. 2 the experimental results are shown. The measurements have been corrected for Van Vleck temperature independent paramagnetism of the Ni^{2+}. At low temperature (below 80 K) the results of measurements on single crystal are shown in Fig. 3 (At

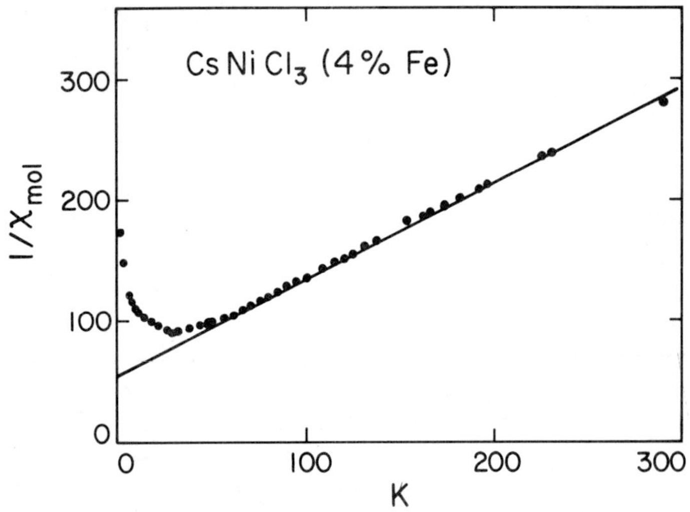

Fig. 2. Inverse susceptibility of a powdered sample of $CsNiCl_3$ (4% Fe) vs. T.

higher temperature there are no differences between single crystal and powdered sample measurements). One observes that $\chi_\perp > \chi_\parallel$ and that the difference between $\chi_\perp - \chi_\parallel$ remains almost constant down to 2K. By constant, in pure $CsNiCl_3$ there is a strong anisotropy $\chi_\perp - \chi_\parallel$ below 15 K where this anisotropy changes by a factor of 5 down to 4K.[2] This change coincides with the onset of long range order. From the behavior of susceptibilities $\chi_\perp$ and $\chi_\parallel$ at low temperature one can infer that the Fe^{2+} is weakly coupled to the neighbors and probably with an antiferromagnetic interaction; a ferromagnetic interaction should produce a different behavior of the susceptibility at low temperature.

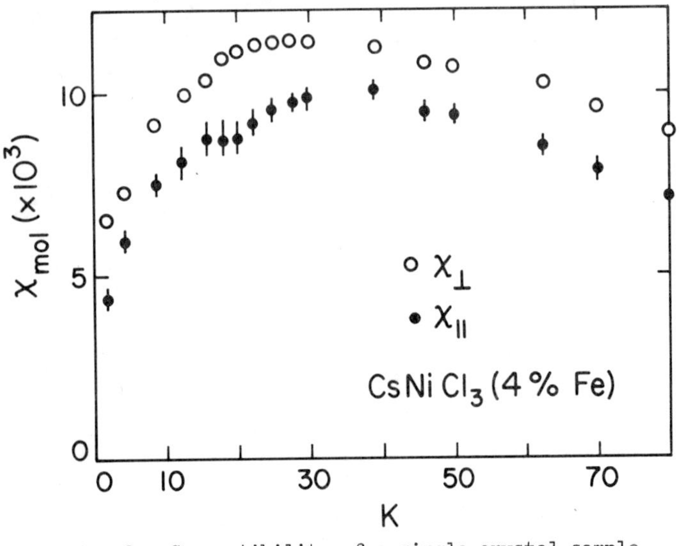

Fig. 3. Susceptibility of a single crystal sample of $CsNiCl_3$ (4% Fe) below 80K.

The absence of a temperature dependent anisotropy $\chi_\perp - \chi_\parallel$ in $CsNiCl_3$ 4% Fe as well as the lack of hf magnetic splitting in the Mössbauer spectrum at 4.2 K suggests a lowering of the transition point. Indeed such an effect is not unexpected in one dimensional systems. A simple argument using a classical spin approach and the results of Scalapino, Imry and Pincus[6] can explain such a change. Let us suppose that the impurities produce a break in the antiferromagnetic linear chain of Ni^{2+} (for simplicity the magnetic interaction with the impurity will be neglected). Then the susceptibility for a finite number of spins (n+1) is given by (staggered susceptibility for an antiferromagnet)

$$\chi_n = \frac{Ng^2\mu_B^2}{(n+1)3kT}\left\{(n+1)\frac{1+u}{1-u} - 2u\,\frac{1-u^{n+1}}{(1-u)^2}\right\} \qquad (1)$$

$$u = \coth K - 1/K \quad K = 2J\ell/kT \, .$$

One can use the relation[6] $1 = 2ZJ'\bar{\chi}(T_c)^*$ where Z is the number of neighboring chains and J' the interchain interaction. When one compares the values of χ for infinite chains and for finite chains (assuming J' is equal) one obtains $\chi(T_c) = \chi_n(T_c')$. Since $\chi_\infty > \chi_n$ for the same temperature (staggered susceptibility) one obtains for $n \approx 20$ spins a T_c reduction of about 20%. One-dimensional systems are extremely sensitive to the presence of impurities as demonstrated by the large perturbations they produce on the three-dimensional ordering temperature.

The author acknowledges helpful discussions with Drs. P. H. Barrett, Y. Imry and D. Hone.

REFERENCES

[1] P. A. Montano, Hanan Shechter, E. Cohen and J. Makovsky, Phys. Rev. B9, 1066 (1974).

[2] N. Achiwa, J. of the Phys. of Jap. 27, 561 (1969).

[3] W. B. Yelon and D. E. Cox, Phys. Rev. B7, 2024 (1973).

[4] The samples used in this experiment were kindly supplied by Dr. Hanan Shechter.

[5] A. McGurn and D. J. Scalapino, to be published.

[6] D. J. Scalapino, Y. Imry and P. Pincus, to be published.

* Susceptibility per spin.

MAGNETIC TRANSITIONS IN TWO DIMENSIONAL SYSTEMS AS A FUNCTION OF INTERLAYER SEPARATION: BIS-PROPYLAMMONIUM TETRACHLOROMANGANATE(II) AND BIS-PENTADECYLAMMONIUM TETRACHLOROMANGANATE(II)

B. C. Gerstein and Chee Chow
Ames Laboratory-USAEC and Department of Chemistry
Iowa State University, Ames, Iowa 50010
and
Ruth Caputo and Roger Willett
Department of Chemistry, Washington State University
Pullman, Washington 99163

ABSTRACT

The initial susceptibilities of the two dimensional systems $(C_3H_7NH_3)_2MnCl_4$ ($PrAMnCl_4$) and $(C_{15}H_{31}NH_3)_2MnCl_4$ ($PDAMnCl_4$) have been measured from 4 - 200K. Both salts exhibit sharp transitions in the neighborhood of 42K. In the latter case, the transition is found to be field dependent, with the magnitude of the real and imaginary susceptibilities increasing with decreasing fields, with ac fields in the range 13 - 30 oe. This behavior is similar to that found in the 2-D system $(CH_3NH_3)_2MnCl_4$ which exhibits ordering at 45.3K. The magnetic losses are found to be as sensitive an indicator of the transitions as are the dispersion for both salts. The fact that the transition temperatures for all three salts are in the neighborhood of 40 K indicates that the transition is independent of interlayer separation.

INTRODUCTION

Two dimensional magnets with nearest neighbor interactions are the simplest examples of systems which can exhibit co-operative magnetic interactions. The two dimensional spin 1/2 Ising model has been solved by Onsager,[1] and has been found to exhibit an ordered ground state at temperatures above T = 0. For the classical Heisenberg model, Mermin and Wagner[2] have shown that no ordering can occur above T = 0. On the other hand, Stanley and Kaplan,[3] Stanley,[4] Moore,[5] and Betts, Elliott, and Ditzian[6] have found a value of T_c greater than zero for the 2-D classical Heisenberg model. It is therefore of interest to realize real physical systems which may exhibit two dimensional interactions such that the above discrepancy, if real, can be resolved. In past work,[7] we have found that the layer system bis-methylammonium tetrachloromanganate(II), $(CH_3NH_3)_2MnCl_4$ ($MAMnCl_4$) exhibited a peak in the initial susceptibility, both absorption and dispersion, centered at 45.3 K.

$PrAMnCl_4$ and $PDAMnCl_4$ are layer compounds with essentially the same structure[8] as $MAMnCl_4$. The structure consists of corner shared octahedra of chloride co-ordinated Mn^{++} ions forming an infinite 2-D network. Ammonium group hydrogens of the alkylammonium ions hydrogen bond to chloride ions in the layers. The methyl "heads" of the alkylammonium chains face each other in the interlayer space. The number of carbon atoms in the chain therefore determines the interlayer spacing. The interlayer spacing for $PrAMnCl_4$ is 12.97 Å, and that for $PDAMnCl_4$ is ~ 30 Å (compared to a spacing of 9.5Å for $MAMnCl_4$. Since superexchange interactions are found to vary[9] as r^{-10}, the behavior of the methylammonium complex compared to that of the propyl- and pentadecylammonium complexes should offer a reasonable test of the two dimensionality of the interactions thus found.

EXPERIMENTAL

The bis-alkylammonium tetrachloromanganate(II) compounds were prepared by the method of Foster and Gill[10] as previously described.

Initial susceptibility measurements were made at 33 Hz on $PDAMnCl_4$, and 45 Hz on $PrAMnCl_4$. No significance is to be attached to the difference between these frequencies, other than convenience of performing the measurement on a given sample at a given time on a particular inductance apparatus. Measurements on the $PDAMnCl_4$ were performed as a function of ac field. Field dependent measurements were not made on $PrAMnCl_4$. The thermocouple and temperature scale for both samples were the same as for those used in measurements on $MAMnCl_4$. Accuracy of the temperature scale above 20 K was 0.5 K. This value was inferred from calibration at the melting point of CS_2, as well as reproducibility of measurements herein reported (vide infra).

RESULTS AND DISCUSSION

The initial susceptibility χ' and its reciprocal, of $PrAMnCl_4$ and $PDAMnCl_4$, respectively are shown in Figs. 1 and 2. A sharp peak in both the absorption, χ'', and the dispersion, χ', is found in the neighborhood of 40 K for both salts.

The absorption and dispersion of $PrAMnCl_4$ at an ac field of 19.5 oe, in the neighborhood of

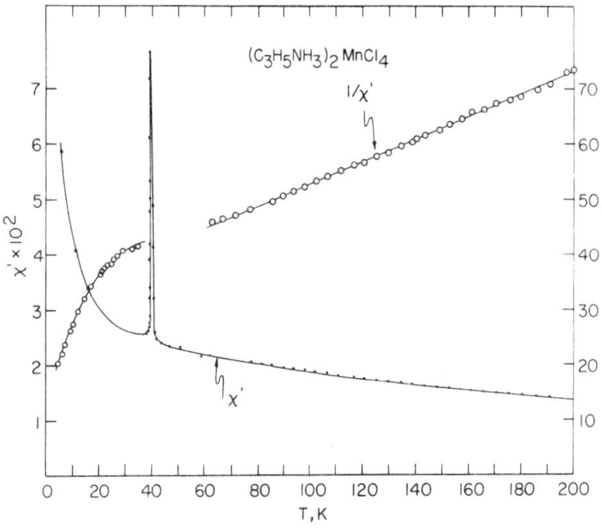

Fig. 1. Initial susceptibility, χ', and reciprocal susceptibility, $1/\chi'$, of $PrAMnCl_4$.

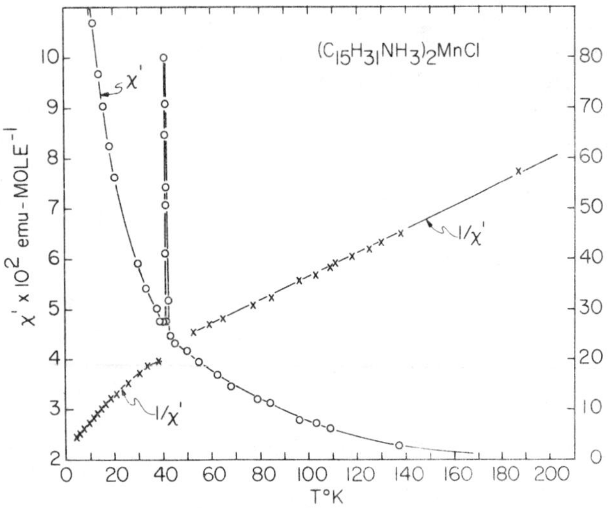

Fig. 2. Initial susceptibility, χ', and reciprocal susceptibility, $1/\chi'$, of $PDAMnCl_4$.

the peak, is shown in Fig. 3. The two dispersion curves were measured on different days, and indicate the accuracy, and reproducibility, of the temperature scale. It is interesting to note that the peak in the absorption occurs, to within experimental accuracy, where the slope of the dispersion changes sign on the low temperature side.[11]

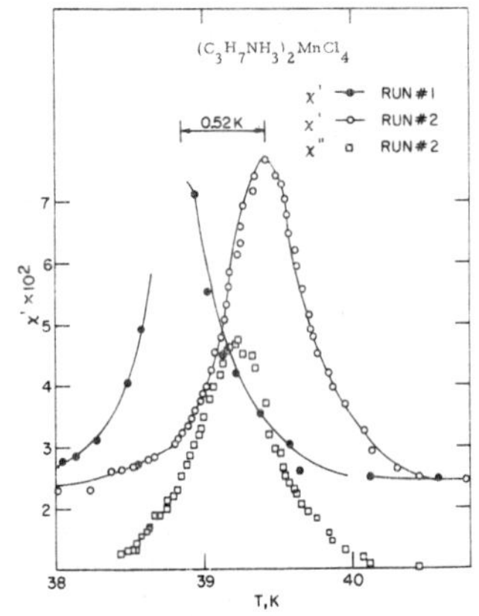

Fig. 3. Absorption, X'', and Dispersion, X', of $PrAMnCl_4$ in the neighborhood of T_c. The ac field is 19.5 oe.

The absorption and dispersion as a function of field for $PDAMnCl_4$ are given in Figs. 4 and 5, respectively, for fields of 12 and 30 oe. The magnitude of both are found to increase with decreasing field. Any attempt to extract critical point exponents from such behavior, therefore, is subject to question, unless there is found a field

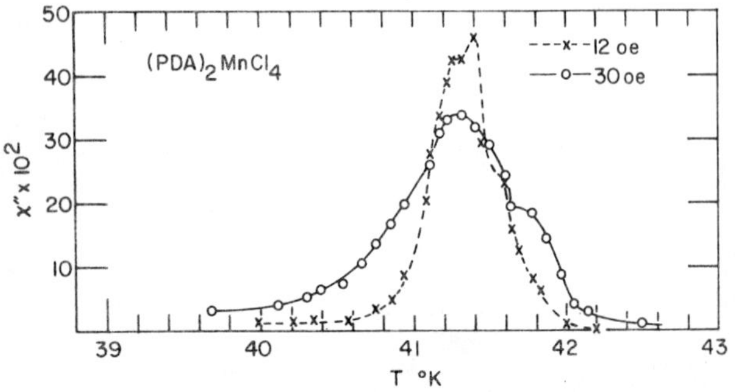

Fig. 4. Absorption, χ'' of $PDAMnCl_4$, as a function of ac field in the neighbornood of the peak in the susceptibility.

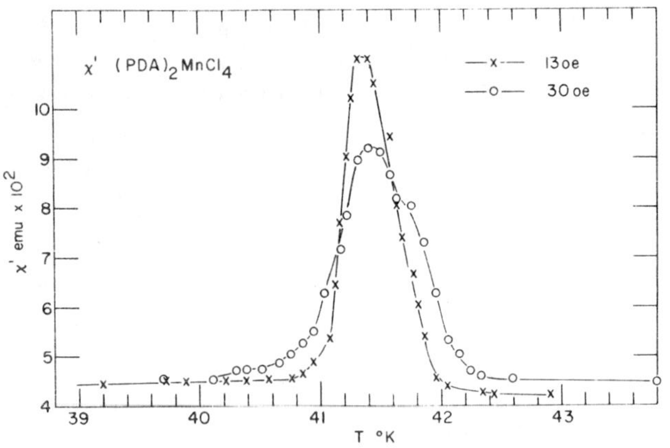

Fig. 5. Dispersion, χ' of $PDAMnCl_4$ as a function of ac field in the neighborhood of the peak in the susceptibility.

below which reproducible behavior exists. Under any circumstances, one must be sure that it is the isothermal susceptibility from which one extracts such exponents,[12] and that this is the case can only be determined by a study as a function of frequency. This type of study is impossible with present equipment in this laboratory.

The fact that the susceptibilities of both $PrAMnCl_4$ and $PDAMnCl_4$ deviate positively from Curie-Weiss behavior below the sharp peak in susceptibility, in contrast to behavior for $MAMnCl_4$ needs explanation. The primary magnetic interlayer interaction must be dipole-dipole coupling. We note that the ground state of planes of dipoles aligned perpendicular to the plane is ferromagnetic, whereas that of planes of dipoles with moments in the plane is antiferromagnetic.

We infer that the sharp peak in the susceptibilities found, essentially at a temperature completely independent of interlayer spacing, for such spacing varying from 9.5 Å to 30 Å is indicative solely of intralayer interactions.

REFERENCES

1. L. Onsager, Phys. Rev., 65, 117 (1944).
2. N. D. Mermin and H. Wagner, Phys. Rev. Lett. 17, 1133 (1966).
3. H. E. Stanley and T. A. Kaplan, Phys. Rev. Lett., 17, 913 (1966); J. Appl. Phys., 38, 975 (1967).
4. H. E. Stanley, Phys. Rev., 164, 709 (1967); Phys. Rev. Lett., 20, 150 and 589 (1968); J. Appl. Phys., 40, 1272 (1969).
5. M. A. Moore, Phys. Rev. Lett., 23, 861 (1969).
6. D. D. Betts, C. J. Elliott and R. V. Ditzian, Can. J. Phys., 49, 1327 (1971).
7. B. C. Gerstein, Kun Chang and R. D. Willett, J. Chem. Phys., 60, 3454 (1974).
8. E. R. Peterson and R. D. Willett, J. Chem. Phys., 56, 1879 (1972).
9. D. Bloch, J. Phys. Chem. Solids, 27, 881 (1966); M. T. Hutchings, R. J. Birgeneau, and W. P. Wolf, Phys. Rev., 168, 1026 (1968).
10. J. J. Foster and N. W. Gill, J. Chem. Soc., A 1968, 269.
11. M. E. Fisher, Phil. Mag., 7, 1731 (1962).
12. A. J. Van Duyheveldt, J. Soeteman and L. J. deJongh, "Differential Susceptibility Measurements in Antiferromagnetic $MnCl_2 \cdot 4H_2O$ and $MnBr_2 \cdot 4H_2O$ as a Function of Field, Frequency, and Temperature" (to be published).

LINEAR CHAIN COMPOUNDS: METAMAGNETISM IN $Co(pyr)_2Cl_2$, $Fe(pyr)_2Cl_2$, $Fe(pyr)_2(NCS)_2$ AND $Ni(pyr)_2Cl_2$

S. Foner, R.B. Frankel, E.J. McNiff, Jr.
Francis Bitter National Magnet Laboratory,* Massachusetts Institute of Technology, Cambridge, Massachusetts 02139

W.M. Reiff†
Northeastern University, Boston, Massachusetts 02115

B.F. Little and G.J. Long††
University of Missouri, Rolla, Missouri

ABSTRACT

We report the observation of low field metamagnetic behavior in powder samples of $Co(pyr)_2Cl_2$, $Fe(pyr)_2Cl_2$, $Fe(pyr)_2(NCS)_2$ and $Ni(pyr)_2Cl_2$ where pyr=pyridine. These materials have linear chain structures with strong ferromagnetic interactions along the chains, and relatively weak antiferromagnetic interactions between chains. In all four compounds the transitions were observed as a rapid increase in the magnetic moment with increasing magnetic field. In $Fe(pyr)_2Cl_2$, $Fe(pyr)_2(NCS)_2$, and $Ni(pyr)_2Cl_2$, the transitions at 4.2 K were observed at 0.7 kG, 1.1 kG and 2.7 kG respectively. In the Co compound the transition (at ~ 0.7 kG) was observed for $T < 3\,K(T_N=3.17\,K)$. Differential magnetic moment measurements in $Fe(pyr)_2Cl_2$ and $Fe(pyr)_2(NCS)_2$ are presented. Above the low field transitions the magnetic moment continues to increase and saturation is not achieved for applied fields up to 200 kG. High field Mössbauer data consistent with these results are discussed.

INTRODUCTION

The magnetic properties of $Co(pyr)_2Cl_2$, where pyr=pyridine, have the characteristics of Ising linear chains.(1) Because of the current interest in nearly one-dimensional systems, studies of the susceptibility, specific heat, crystal structure and related properties have recently been reported for this and similar linear chain magnetic structures. In this work we discuss the magnetic properties of $Fe(pyr)_2Cl_2$, $Fe(pyr)_2(NCS)_2$, $Ni(pyr)_2Cl_2$ and $Co(pyr)_2Cl_2$. These compounds have strong ferromagnetic interactions along the chains and relatively weak antiferromagnetic interactions between the chains. We will show that their magnetic properties are quite similar.

EXPERIMENTAL DETAILS

The experimental results reported here involve static magnetic moment measurements made in fields to 60 kG with a vibrating sample magnetometer(2) adapted to a superconducting solenoid and measurements in dc fields to 210 kG in water-cooled Bitter solenoids with a very low frequency VSM.(3) Differential magnetic moment (DMM) measurements were made in a small balanced coil system at zero field and low fields with small amounts of material (~ 1 mg) as described elsewhere.(4) The materials examined were powders; single crystals were not available.

EXPERIMENTAL RESULTS

A. Low Field Transitions

In all the compounds we find that below the ordering temperature, T_N, the magnetic moment remains small at very low field, shows a rapid increase at relatively low fields and above a few kG shows a gradual increase in moment up to the highest dc fields we have available (see Figs. 1 and 2). This behavior is consistent with the occurrence of a low field metamagnetic phase transition when the antiferromagnetic coupling between chains is overcome.(5)

Generally measurements have been made in such systems at zero field so that a large anomaly in susceptibility is observed. When conventional susceptibility measurements are made one should observe a relatively large value of suscep-

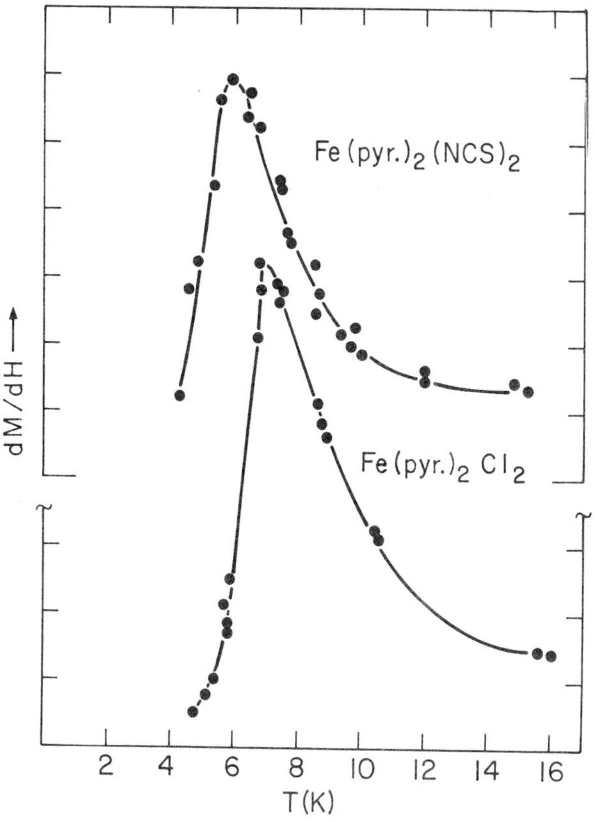

Fig. 1. Differential magnetic moment versus temperature, T, at $B_0=0$ for $Fe(pyr)_2Cl_2$ and $Fe(pyr)_2(NCS)_2$. The ac field is less than 10 G.

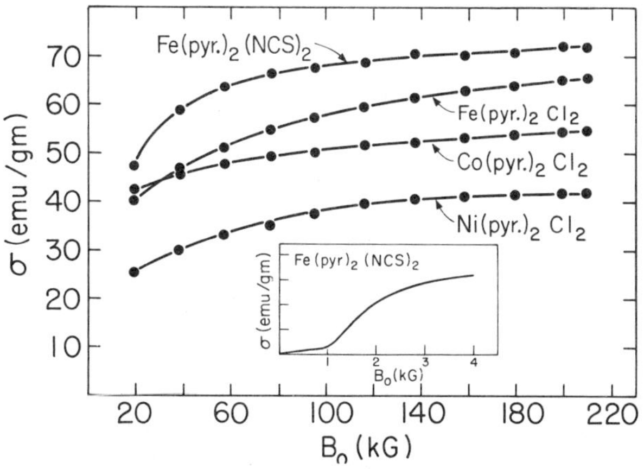

Fig. 2. Magnetic moment, σ, versus applied field, B_0, for several pyridine compounds. The insert shows example of low field transition for $Fe(pyr)_2Cl_2$.

tibility below T_N because the applied fields are usually much greater than the transition fields. Examples of DMM results for $Fe(pyr)_2Cl_2$ and $Fe(pyr)_2(NCS)_2$ are shown in Fig. 1. It should be noted that the transitions are quite sharp. The accuracy of the data is estimated to be better than 1 K, and the position of the transition depends on whether one defines it as the maximum or the most rapid increase in χ vs T. The value of $T_N = (7 \pm 0.5)$ K for the $Fe(pyr)_2Cl_2$ compares to 10.5 K reported by Sanchez et al.[6] For the $Fe(pyr)_2(NCS)_2$, $T_N = (6 \pm 0.5)$ K. It should be noted that for both of these materials the metamagnetic behavior is observed at 4.2 K.

The evidence for apparent metamagnetic behavior for these systems is not unambiguous because these are powder samples. There is thus the possibility that we are observing one of several possible low field transitions, e.g., a spin-flop transition. However, for a spin-flop transition we would expect that the transition field H_c would decrease rapidly with decreasing temperature because $\chi_{\parallel}(T)$ is rapidly decreasing. We observe a slight increase in H_c as T decreases, consistent with metamagnetic behavior. Furthermore, the systems are similar to $FeCl_2 \cdot 2H_2O$ (the pyr are replaced by H_2O) where single crystal data shows two metamagnetic transitions for this six-sublattice system.[7] For the pyr materials $T_N \leqslant 7$ K whereas for the dihydrate materials $T_N \sim 17$ K so that the interchain antiferromagnetic exchange field is smaller in the former. In addition, the spin-flop transition occurs when $H_A \ll H_{ex}^{AF}$ whereas metamagnetic transitions are expected for highly anisotropic systems (such as expected for these linear chain systems) where $H_A > H_{ex}^{AF}$. The magnitude of the moment change at the low field transition does not correspond to complete alignment of the atomic moments, however, as expected for a polycrystalline aggregate. If these systems are similar to the dihydrate material we might expect two transitions versus field. Possible reasons for not observing two are: a) the second transition occurs at very low fields in which case it would be observed at low temperatures, b) it occurs at higher fields and may be averaged out; c) the relative exchange constants are such that the pyridine materials are effectively two-sublattice systems.

The transition fields at 4.2 K for $Fe(pyr)_2Cl_2$, $Fe(pyr)_2(NCS)_2$, and $Ni(pyr)_2Cl_2$ are ~0.7 kG, 1.1 kG, and 2.7 kG respectively; and for $Co(pyr)_2Cl_2$ at 1.2 K the transition field is ~0.7 kG. Here the transition fields are defined as the extrapolation of the linear portion of the σ versus B_0 curve (see insert of Fig. 2) to the $\sigma = 0$ axis.

B. High Field Results

The high field approach to saturation for these materials is shown in Fig. 2. It should be noted that: a) the $Co(pyr)_2Cl_2$ compound is not ordered at 4.2 K ($T_N \simeq 3.17$ K[1]) so that the σ versus B_0 data reflects paramagnetic behavior; b) none of the systems are completely saturated at 200 kG. This is consistent with high anisotropy as expected. The moments at 200 kG are given in Table I. If the low field transition reflected a spin-flop transition, we would expect a nearly linear increase in σ versus B_0 up to $\simeq 2H_{ex}^{AF}$ at which point the system would be saturated. For the low T_N's of these materials this would occur at $\leqslant 100$ kG (see, e.g., Ref.4). Extrapolation from the low field transition would indicate that saturation should occur at $\leqslant 20$ kG if this transition were a spin-flop transition. Thus the high field behavior is inconsistent with a low field spin-flop transition.

Table I. High Field Magnetic Moments of Pyridine Compounds

Compound	200 kG Moment (μ_B)	Spin-Only Moment (μ_B)
$Fe(pyr)_2Cl_2$	3.3	4
$Fe(pyr)_2(NCS)_2$	4.3*	4
$Co(pyr)_2Cl_2$	2.8	3
$Ni(pyr)_2Cl_2$	2.2	2

*Mössbauer measurements showed evidence of some Fe^{3+} so that this value may be slightly high.

C. Mössbauer Measurements

Mössbauer measurements have been reported in $Fe(pyr)_2Cl_2$[6,8] and $Fe(pyr)_2(NCS)_2$.[9] The spectrum of the latter consists of a quadrupole doublet with splitting $\Delta E_Q = 3.03$ mm/sec above T_N, and a magnetic hyperfine pattern below T_N, with $H_n = -280$ kOe. Measurements in longitudinal external magnetic fields up to 80 kOe show that the electronic moments tend to align parallel to the external field, but the polarization is incomplete at the highest field, as is also seen in the high field magnetization measurements. The low field metamagnetic transition cannot be seen in powder samples by Mössbauer effect.

$Fe(pyr)_2Cl_2$ is isomorphous with $Co(pyr)_2Cl_2$ at room temperature and undergoes a transition at 250 K[6] which has been postulated[8] to be a form which is isomorphous with $Cu(pyr)_2Cl_2$ below 250 K. For $250\text{ K} > T > T_N$, the spectrum consists of a quadrupole doublet with $\Delta E_Q = 1.26$ mm/sec. Below T_N, a magnetic hyperfine field is observed. External magnetic field measurements up to 80 kOe show that the iron moments tend to polarize along the field direction, but that the polarization is incomplete at 80 kOe, again consistent with the high field magnetization results. The overall splitting of the spectrum increases with increasing B_0, indicating that the hyperfine field has a positive sign. The smaller quadrupole splitting and effective magnetic field below T_N in the chloride compared to the thiocyanate suggest less complete quenching of the orbital angular momentum in the chloride, and hence considerable orbital admixture in the ground state. This should be reflected in larger magnetic anisotropy of the chloride compared to the thiocyanate, and is indeed consistent with the data presented in Fig. 2 and Table I.

CONCLUSION

We have shown that the linear chain materials $Fe(pyr)_2Cl_2$, $Fe(pyr)_2(NCS)_2$, $Co(pyr)_2Cl_2$ and $Ni(pyr)_2Cl_2$ are antiferromagnetic at low temperature and undergo metamagnetic transitions in low external fields. The high anisotropy of these materials is demonstrated in the high field magnetization and Mössbauer data. Further elucidation of the magnetic properties of these materials will be possible when single crystals become available. More detailed results will be presented elsewhere.

REFERENCES

1. K. Takeda, S. Matsukawa and T. Haseda, J. Phys. Soc. Japan, 30, 1330 (1971).
2. S. Foner, Rev. Sci. Instr. 30, 548 (1959).
3. S. Foner and E.J. McNiff, Jr., Rev. Sci. Instr. 39, 171 (1968).
4. N.F. Oliveira, Jr., S. Foner, and Y. Shapira, Phys. Rev. 5, B 2634 (1972).
5. A preliminary report of the low field behavior is given by S. Foner, R.B. Frankel, W.M. Reiff, B.F. Little and G.J. Long, Solid State Comm. (in press).
6. J.P. Sanchez, I. Asch, and J.M. Freidt, Chem. Phys. Letters 18, 250 (1973).
7. A. Narath, Phys. Rev. 139, A 1221 (1965).
8. W.M. Reiff, R.B. Frankel, B.F. Little, and G.J. Long, Chem. Phys. Letters 28, 68 (1974).
9. W.M. Reiff, R.B. Frankel, B.F. Little, and G.J. Long, Inorganic Chem. 13, 2153 (1974).

* Supported by the National Science Foundation.
† Supported by National Science Foundation Grant No. GH39010.
†† Supported by National Science Foundation Grant No. GP8653.

MAGNETIC ORDERING IN SEVERAL Fe-CHAIN SILICATE COMPOUNDS*

R. J. Borg
Lawrence Livermore Laboratory
University of California
Livermore, CA 94550
and
F. R. Szofran, W. L. Burmester, and D. J. Sellmyer
Behlen Laboratory of Physics
University of Nebraska
Lincoln, NE 68508

ABSTRACT

Magnetic ordering in selected Fe-chain silicate compounds has been studied with Mössbauer, high-field magnetization, and susceptibility measurements. The compounds have the monoclinic crystal structure and the general formula

$(Na,Ca)_{2-3}(Fe^{+2},Fe^{+3},Mg,Al)_5Si_8O_{22}(OH,F)_2$.

The stacking sequence of the Fe atoms forms a one dimensional chain along the [001] axis. The appearance of magnetic hyperfine splitting has been used to determine magnetic ordering temperatures which range from 6.2 to 26.0 K. In one of the compounds, the Fe^{+3} spins have been shown to be canted at 29° from [001] and the susceptibility has shown the ordering to be antiferromagnetic. Magnetization evidence for high field spin transitions to the paramagnetic state has been obtained. Evidence for the presence of ferromagnetic and antiferromagnetic interactions is discussed.

INTRODUCTION

There is considerable interest in the influence of dimensionality on critical phenomena in magnetic materials. We have investigated the onset of magnetic ordering in four Fe-chain silicate compounds which have the monoclinic crystal structure. The Fe atoms can occupy three distinct crystallographic sites in the structure and these sites are arranged in a stacking sequence which forms a one-dimensional chain parallel to the c axis. Each chain is separated from adjacent ones by linked SiO_4 tetrahedra so that one might anticipate strong magnetic coupling between the Fe atoms within each chain and relatively weak interchain coupling.

EXPERIMENTAL RESULTS

The magnetic properties of these compounds were studied with apparatus of conventional design: a constant acceleration Mössbauer spectrometer using 25-35 mCi of ^{57}Co in Cu as a source, a vibrating sample magnetometer in an 80 kOe solenoid, and a Faraday-method susceptibility system employing a conventional electromagnet.

Mössbauer spectra for certain of the compounds show the development of magnetic hyperfine splitting (mhfs) at low temperatures. Of the four compounds studied thus far, three develop such indications of magnetic ordering at temperatures ranging from 6.2 to 26.0 K. The major difference in the compounds is in their iron concentrations with the higher concentration samples having the higher ordering temperatures T_o. Chemical analyses and some preliminary Mössbauer data for the three compounds which order magnetically are given in earlier work,[1,2] wherein it is also shown that the Mössbauer effect is capable of determining Fe^{+2}/Fe^{+3} ratios from the magnetic hyperfine spectra of Fe^{+3}. Because of the crystallographic and expected magnetic similarities of all of the compounds, we concentrate in this report on details of the temperature, orientation, and field dependences observed in the compound designated SP-1. This compound essentially has the chemical formula $Na_2Fe_5Si_8O_{22}(OH,F)_2$.

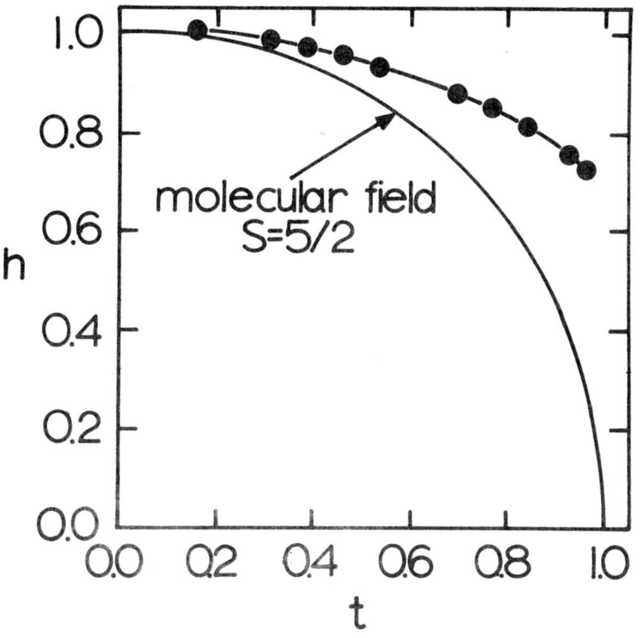

Fig. 1. Temperature dependence of reduced field at the nucleus as a function of reduced temperature $t = T/T_o$; T_o = 26.0 K. Sample SP-1.

Figure 1 shows the temperature dependence of the reduced field at the Fe^{+3} nucleus, $h \equiv H_n(T)/H_n(4.6)$. It is clearly quite different from the molecular field curve calculated for S = 5/2, and this is not surprising in view of similar results on other highly anisotropic (e.g., quasi-two dimensional) magnetic systems.[3]

Angular orientation Mössbauer measurements on

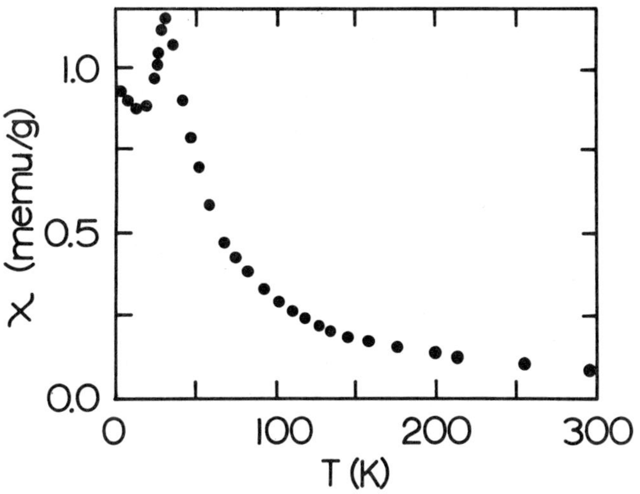

Fig. 2. Temperature dependence of the susceptibility for powdered SP-1 at H = 3.2 kOe.

single crystals of compound SP-1 have shown that the atomic spins associated with the ferric ions are canted at an angle of 29° from the [001] axis and one nearly perpendicular to (121) and (1̄21). The Mössbauer data also indicates the Fe^{+2} moments to be approximately parallel to the Fe^{+3} moments.

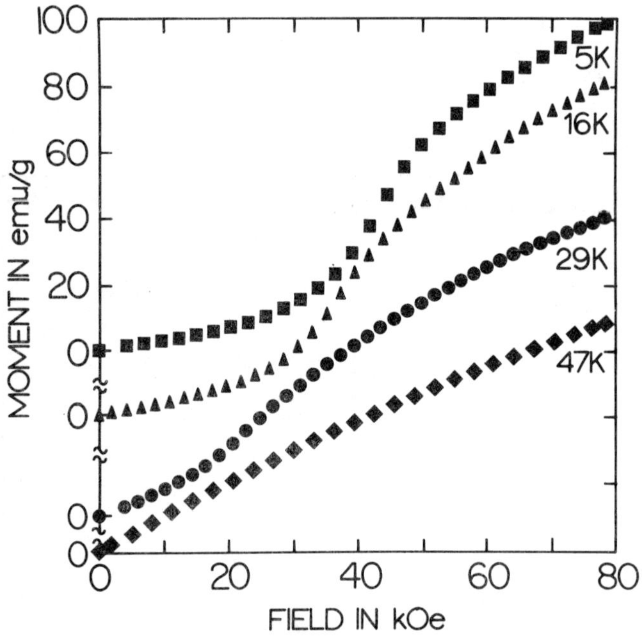

Fig. 3. Dependence of the magnetization on applied field at several temperatures. $\vec{H}||[001]$. Note the shift of the curves along the ordinate. The sample is a rectangular parallepiped.

Figure 2 shows the magnetic susceptibility $\chi(T)$ for a powdered sample of SP-1. There is a sharp peak at T_{max} = 31.5 K which suggests an antiferromagnetic transition. In Figure 3 is shown the field dependence of the magnetization M(H) for a single crystal with $\vec{H}||[001]$. The curves are shifted along the ordinate for clarity. The data suggest the presence of a spin-flop or metamagnetic transition which gradually disappears as T becomes greater than T_{max}. No evidence for hysteresis or remanence was observed. For $\vec{H}\perp(110)$, a cleavage plane, no anomalous field dependence was observed; at 5 K M(H) is qualitatively similar to the 47 K curve for $\vec{H}||[001]$. The field dependence of the differential susceptibility (dM/dH) has been determined from the data shown in Figure 3. Below T_{max} the curves show fairly sharp maxima which can be used to define a critical field H_c. The data show that H_c decreases with temperature but, as yet, there is insufficient data to carefully map out a phase diagram.

DISCUSSION

As a first step towards an understanding of the data we have fit the susceptibility to the expression

$$\chi(T) = \chi_o + C/(T - \theta), \qquad (1)$$

for $T \gtrsim 40$ K. The results gave χ_o = 3.6 μemu/g, C = 2.26 x 10^{-2} K emu/g, and θ = +22.4 K. If one ignores for the moment the fact that there are two different ionic states present, one calculates from the Curie constant that p_{eff} = 5.8 which, reasonably, is bracketed by the values expected for the states $^6S_{5/2}$ for Fe^{+3} and $^5D_4$ for Fe^{+2}, namely, 5.92 and 4.90, respectively.[4] In Figure 3 it appears that M(H) at 5 K is approaching saturation at the highest fields. If one roughly takes $M_{sat} \approx 120$ emu/g, then the saturation moment per iron atom is $\mu_s = 4.1\ \mu_B$. Since the Fe^{+3}/Fe^{+2} ratio is 0.38/0.62,[1] and assuming spin values of 5/2 and 2 and a g value of 2, the weighted value for μ_s is 4.4 μ_B. This approximate agreement indicates that at the highest fields the antiferromagnetic interactions have been broken up leading to the resulting nearly saturated paramagnetic state.

From a microscopic viewpoint the Fe ions occupy three distinct octahedrally coordinated sites M_I, M_{II}, and M_{III}. The sites form a chain of two or three atoms in width which are aligned along the [001] axis in the following stacking sequence: (II,III,II)(I,I)(II,III,II)(I,I)···. The Fe^{+2} ion predominantly occupies sites I and III whereas site II is occupied by Fe^{+3}. Each chain consists of the above sequence of nearest-neighbor Fe sites, but is separated from nearest-neighbor chains by linked SiO_4 tetrahedra. As mentioned above the Curie-Weiss fit leads to a positive value for θ which, in view of the apparent antiferromagnetic ordering, suggests a mixture of ferromagnetic and antiferromagnetic couplings. Either the Fe^{+3} and Fe^{+2} spins would each have to form an intrachain sublattice which itself is antiferromagnetic so the two types of spin species sum to zero net spin, or it may be possible that the intrachain interactions may be predominantly ferromagnetic while the interchain ones are antiferromagnetic. Defining ε as $(1 - T/T_o)$, the temperature dependence of the hyperfine field follows within the experimental uncertainty (∿2%) the expression $h(\varepsilon) \propto \varepsilon^\beta$ with T_o = 26.0 K, β = 0.13 ± 0.05, and $0.03 \lesssim \varepsilon \lesssim 0.05$. The relatively large error estimate for β was derived by choosing various T_o values which differ by ≃ 2 K. The β value obtained is close to the Ising value[3] for two-dimensional coupling (β = 1/8) but, unfortunately, we cannot obtain data for ε values smaller than ≃ 0.03. Thus it is not possible to state unequivocally that evidence exists for two-dimensional Ising behavior.

Several points, including the origin of the upturn in $\chi(T)$ for $T \lesssim 12$ K, and the nature of the high-field transition, remain to be resolved. The apparent discrepancy between T_o and T_{max} is explainable in view of the fact that T_{max} occurs at the temperature corresponding to the maximum rate of change of short range spin order, whereas T_o occurs at the lower temperature at which the spin-lattice relaxation becomes commensurate with the nuclear Larmor precession time. Further investigations on all of these compounds are continuing, and comprehensive results will be reported in a future publication.

We thank Dr. John Goodenough and Professor M. B. Salamon for helpful discussions.

REFERENCES

1. R. J. Borg, D. Y. Lai, and I. Y. Borg, Nature, Phys. Sci. 246, 46 (1973).
2. R. J. Borg and I. Y. Borg, in Proc. Int. Conf. on Application of Mössbauer Effect, Bendor, France, and to be published.
3. L. J. de Jongh and A. R. Miedema, Adv. in Physics 23, 1 (1974).
4. C. Kittel, Introduction to Solid State Physics, 4th ed. (John Wiley, New York, 1971). p. 510.

*Research supported at the University of California by the Atomic Energy Commission and at the University of Nebraska by the National Science Foundation.

THE ONSET OF THE 5*f* ENERGY BAND IN $Th_{1-x}U_xRh_3$ INTERMETALLIC COMPOUNDS*

G. P. Sykora and A. J. Arko
Argonne National Laboratory, Argonne, Illinois 60439

and

D. O. Van Ostenburg
DePaul University, Chicago, Illinois 60604

ABSTRACT

The cubic $AuCu_3$-type intermetallic compounds $AnRh_3$ (where An = Th, U, Np and Pu) have been previously studied and were found to exhibit 5*f* electron character similar to pure actinide metals; i.e., no 5*f* band in $ThRh_3$ and a broad hybridized 5*f*-6*d* band in URh_3 which becomes progressively narrower as the actinide component is substituted across the actinide series. The purpose of the present investigation is to determine the U concentration x in the $U_xTh_{1-x}Rh_3$ system which first manifests evidence for an occupied hybridized 5*f*-6*d* band. The room-temperature magnetic susceptibilities, electrical resistivities, and lattice parameters were measured for x = 0, 0.1, 0.3, 0.5, 0.7, 0.9 and 1.0. A sharp break in the plot of room temperature parameters vs x is observed at $x \approx 0.4$ for susceptibility and $x \approx 0.55$ for resistivity with both parameters increasing much more rapidly at higher values of x. The sharp rise is attributed to the introduction of an electron band with a high density of states at the Fermi level. It is believed that this is a hybridized 5*f*-6*d* band. The discrepancy between the resistivity and susceptibility results is not understood, but magnetic susceptibility is judged to be more closely related to the density of states and hence a better indicator of the introduction of 5*f* electrons.

*Work supported by the U. S. Atomic Energy Commission.

A NEW SERIES OF CUBIC ROOM-TEMPERATURE-MAGNETIC OXIDES

B. Bochu*, J. Chenavas**, A. Collomb**, M.N. Deschizeaux*, J.C. Joubert*, M. Marezio**
* Laboratoire de Génie Physique, INPG, BP 15 - Centre de Tri, 38040 Grenoble Cedex (France)
** Laboratoire des Rayons X, CNRS, BP 166 - Centre de Tri, 38042 Grenoble Cedex (France)

ABSTRACT

A new series of cubic room-temperature-magnetic oxides with the general formula $CCu_3(MM')_4O_{12}$ has been synthesized under high pressure - high temperature conditions. The structure is of distorted perovskite type. Magnetization curves as a function of temperature have been determined for the compounds $CaCu_3Mn_4^{4+}O_{12}$, $YCu_3Mn^{3+}Mn_3^{4+}O_{12}$ and $ThCu_3Mn_2^{3+}Mn_2^{4+}O_{12}$.

A new large series of magnetic oxides has been synthesized under high pressure and high temperature conditions. The prototype of the series is $CaCu_3Mn_4^{4+}O_{12}$, whose structure has been refined from single crystal data (1) and found to be isostructural with $NaMn_7O_{12}$ (2). The arrangement is perovskite-like with the Ca^{2+} and Cu^{2+} cations ordered on the large A-sites, while the octahedral B-sites are occupied by the Mn^{4+} cations. The space group is Im3. The cation sublattice is not distorted with respect to the ideal perovskite structure. The doubling of the lattice parameter is due to the ordering of Ca^{2+} and Cu^{2+} cations on the A-sites as well as to the distortion of the oxygen sublattice. Because of this distortion the MnO_6 octahedra form zig-zag chains along the three [100] directions, with Mn-O-Mn angles of 142° instead of 180° as in the ideal perovskite. The main consequence of this octahedral tilting is that the Cu^{2+} sites are highly distorted with four close neighbors at 1.942 Å forming almost a square and four other at 2.707 Å located on a rectangle perpendicular to the square. There are four other oxygen atoms at 3.181 Å, but they must be considered as second-nearest neighbors. The Ca^{2+} cations are surrounded by twelve equidistant oxygen atoms at 2.562 Å forming a nearly regular icosahedron. The high pressure is needed for the synthesis of this compound because some of the cations must occupy sites with higher coordination than usual namely, eight for Cu^{2+} and twelve for Ca^{2+}. The distortion of the oxygen polyhedron around the Cu^{2+} cations is due to a Jahn-Teller effect, which seems to be responsable for the stability of the structure. A schematic representation of the structure is shown in fig. 1.

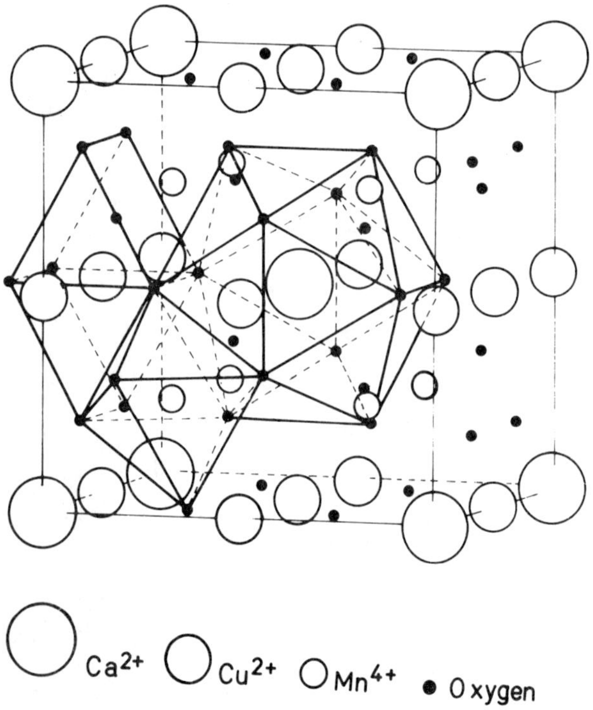

Ca^{2+} Cu^{2+} Mn^{4+} • Oxygen

Figure I : The three kinds of coordination polyedra in the $CaCu_3Mn_4O_{12}$ structure type.

Samples were prepared by thoroughly mixing $Ca(OH)_2$, 3CuO and $4MnO_2$ and subjecting the mixture to 50 kbar and 1000°C for one hour. The quenched materials consisted of well crystallized black powders containing small brilliant cuboctahedral single crystals. If one keeps the Cu^{2+} cation as a constituent, a large series of new isostructural compounds can be synthesized by performing partial or complete substitutions of Mn^{4+} and Ca^{2+} cations on the octahedral and icosahedral sites, respectively. Table 1 gives the formulae, the lattice parameters, the synthesis conditions and the magnetic behavior at room temperature for the members of the series which have been obtained so far as pure phases.

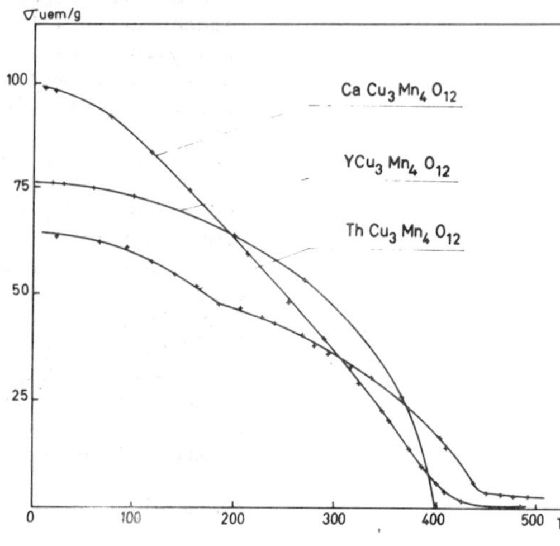

Figure II :
a - magnetization curve of $ThCu_3Mn_4O_{12}$ determined with a vibrating sample magnetometer in a permanent field of 9560 Oe.
b and c - saturation curves of $YCu_3Mn_4O_{12}$ and $CaCu_3Mn_4O_{12}$ obtained by extrapolation at H = 0 of isotherm curves M = f(H) saturated in a field of 24,000 Oe.

Fig. 2 shows the magnetization curves as a function of temperature for the compounds $CaCu_3Mn_4^{4+}O_{12}$, $YCu_3Mn_3^{4+}Mn^{3+}O_{12}$ and $Th^{4+}Cu_3Mn_2^{4+}Mn_2^{3+}O_{12}$. The values of the spontaneous saturation moments at liquid-helium temperature obtained from isotherm curves $\mu(H)$ determined up to a field of 25 kgauss are 11.5 μ_B (100 u.e.m./g), 9.4 μ_B (75 u.e.m./g) and 12.9 μ_B (86 u.e.m./g) for the three compounds, respectively. The values corresponding to the two last compounds indicate a ferrimagnetic ordering between the Cu and Mn sublattices. On the other hand the value of 11.5 μ_B found for $CaCu_3Mn_4O_{12}$ seems to indicate that the two sublattices are not directly opposite and that a ferromagnetic coupling would be possible. In order

TABLE I : SYNTHESIS CONDITIONS OF $CaCu_3Mn_4O_{12}$ TYPE COMPOUNDS

Formula	Starting materials	Synthesis conditions			Cell Parameter	Magnetic behaviour
		P (Kbar)	T°C	time	(Å)	at room temperature
$Ca\ Cu_3\ Mn_4^{4+}\ O_{12}$	$Ca(OH)_2$,CuO,MnO_2,	40	1000	1 hr	a = 7,241	magnetic
$Cd\ Cu_3\ Mn_4^{4+}\ O_{12}$	$Cd(OH)_2$,CuO,MnO_2,	40	1000	1 hr	a = 7,237	magnetic
$Sr\ Cu_3\ Mn_4^{4+}\ O_{12}$	$Sr(OH)_2$,CuO,MnO_2,	50	1000	1 hr	a = 7,336	magnetic
$Y\ Cu_3\ Mn_3^{4+}Mn^{3+}O_{12}$	Y_2O_3 ,CuO,MnO_2,	80	1400	1 hr	a = 7,252	magnetic T_C = 394 K
$Ho\ Cu_3\ Mn_3^{4+}Mn^{3+}O_{12}$	Ho_2O_3 ,CuO,MnO_2,	80	1400	1 hr	a = 7,256	magnetic $T_C \simeq$ 400 K
$La\ Cu_3\ Mn_3^{4+}Mn^{3+}O_{12}$	La_2O_3 ,CuO,MnO_2,	50	1200	1 hr	a = 7,322	magnetic
$Th^{4+}Cu_3\ Mn_2^{4+}Mn_2^{3+}O_{12}$	ThO_2 ,CuO,MnO_2,	40	1000	1 hr	a = 7,361	magnetic $T_C \simeq$ 450 K
$Th\ Cu_3\ Mn_2^{4+}Fe_2^{3+}O_{12}$	ThO_2 ,CuO,MnO_2,Fe_2O_3	80	1000	1 hr	a = 7,403	non magnetic
$Th\ Cu_3\ Mn_2^{4+}Cr_2^{3+}O_{12}$	ThO_2 ,CuO,MnO_2,Cr_2O_3	40	1000	1 hr	a = 7,320	magnetic
$Ca\ Cu_3\ Cr_4^{4+}\ O_{12}$	$CaCrO_4$,$CuCrO_4$,	80	1000	1 hr	a = 7,265	non magnetic
$Na\ Mn_3^{3+}Mn_2^{4+}Mn_2^{3+}O_{12}$	NaOH ,Mn_2O_3, MnO_2	80	1000	1 hr	a = 7,308	non magnetic

to determine the spin configurations, neutron diffraction experiments are presently being performed.

The isostructural $[NaMn_3^{3+}](Mn_2^{3+}Mn_2^{4+})O_{12}$ compound undergoes a crystallographic phase transition at 185 K. Since the crystal symmetry lowers from cubic to monoclinic this transition seems to be due to an ordering which takes place between the Mn^{3+} and Mn^{4+} cations on the octahedral sublattice, with a cooperative Jahn-Teller distortion of the Mn^{3+} sites (3). The anomaly observed on the $ThCu_3Mn_2^{3+}Mn_2^{4+}O_{12}$ magnetization curve at 190 K could correspond to a similar crystallographic transition because in these two compounds the octahedral sites are occupied by the same cations ($2Mn^{3+}$ and $2Mn^{4+}$). The low crystallographic ordering temperatures can be explained by an electron transfer process.

REFERENCES

(1) J. Chenavas, J.C. Joubert, M. Marezio and B. Bochu J. Solid State Chem., (1975)

(2) M. Marezio, P.D. Dernier, J. Chenavas and J.C. Joubert J. Solid State Chem., 6, 16 (1973)

(3) J. Chenavas, F. Sayetat, A. Collomb, J.C. Joubert and M. Marezio (to be published)

INFLUENCE OF ANNEALING ON THE ABSORPTION OF Bi-BASED IRON GARNET FILMS

A. Akselrad, R. E. Novak, and D. L. Patterson
RCA Laboratories Inc., Princeton, New Jersey 08540

ABSTRACT

Annealing of LPE-grown iron garnet films in reducing atmosphere changes their optical absorption and their electrical conductivity. In Bi-based films where the dominant impurity, lead, reaches 0.35 mole%, both the conductivity and the absorption below 2.4 eV decrease after annealing.

Annealing experiments were performed on films approximated as $(Bi_{.6}Pb_{.06}Tm_{2.34})(Fe_4Ga_1)O_{12}$. The maximum observed decrease in absorption was of the order of 40%, while the conductivity decreased by 2 - 3 orders of magnitude. The decrease in the absorption is independent of the accompanying redistribution of Ga ions and its final value depends mostly on the Pb content and on the annealing conditions. Experimental results lead to a model where oxygen vacancies diffusing into the crystal ($D \sim 10^{-8}$ cm^2sec^{-1} at $870^{\circ}C$) replace Fe^{4+} as charge-compensators for the Pb^{2+} impurity.

The high Faraday Rotation observed in Bi-based iron garnet films[1,2] is usually accompanied by an absorption higher than the absorption of pure $Y_3Fe_5O_{12}$. This is a study of the origin of this dopant induced absorption.

Below 2.4 eV the absorption spectrum of iron garnets is strongly affected by dopants.[3] The most commonly invoked example is the increase in the absorption which is observed when a non trivalent cation-dopant, such as Ge^{4+} or Ca^{2+}, enters the garnet substitutionally. Although recent studies[4] have demonstrated the presence of Fe^{2+} in $Y_3Fe_{5-x}Si_xO_{12}$ (YIG:Si), little is known about acceptor levels in garnets or about the effect of the doping level on the defect distribution in LPE films. The second mechanism through which a given dopant might influence the absorption of its host-garnet depends on interatomic interactions. The evidence of such influences is contained in the spectral studies of YIG:Ga, Sc,[3,5] and it can be inferred from the work on Bi-induced FR in Bi-doped iron garnets.[6,7]

In this paper we discuss the dopant-induced absorption in LPE films of $(BiTm)_3(FeGa)_5O_{12}$ grown from PbO flux. Specifically we distinguish between the effects related to the valence of Pb impurity and a possible enhancement of electronic transitions of YIG, due to strong interatomic interactions with Bi or Pb.

In an elegant experiment Le Craw et. al.[8] have shown that microscopic redistribution of Ga between the octohedral (a) and the tetrahedral (d) sites in $(YEu)_3(FeGa)_5O_{12}$ is facilitated when these films are annealed in a reducing Si-environment. The increased diffusion rate of Ga was ascribed to oxygen vacancies, introduced into the crystal during annealing. We have decided to use this technique to alter the stoichiometry of our films and to look for the resulting changes in their absorption and in their electrical conductivity.

EXPERIMENTAL RESULTS AND DISCUSSIONS

The growth procedures and the magnetic and magneto-optic properties of films of $(Bi_xPb_yTm_{3-x-y})(Fe_{5-z}Ga_z)O_{12}$† with $x \sim .6$, $y \sim .06$, $z \sim 1$, were described previously.[1] For samples to be annealed in Si-environment, Si layers 5000 Å thick, were evaporated onto the surface of the magnetic films. After annealing, Si was removed and the absorption spectrum of the sample was compared with that of the as-grown film. Such comparison is shown in Fig. 1. Throughout the investigated frequency range the absorption coefficient (α) decreases as a result of annealing. The magnitude of the effect $|\Delta\alpha|$, increases with frequency, while $\Delta\alpha/\alpha$ increases towards the infra-red (where the intrinsic absorption of garnets is low).

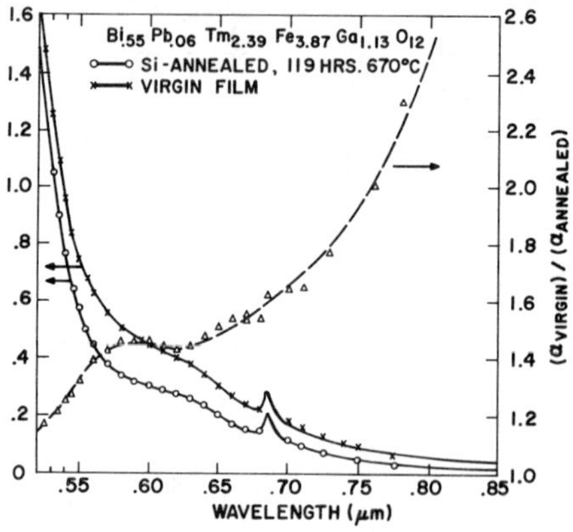

FIG.1 ABSORPTION SPECTRUM BEFORE AND AFTER ANNEALING.

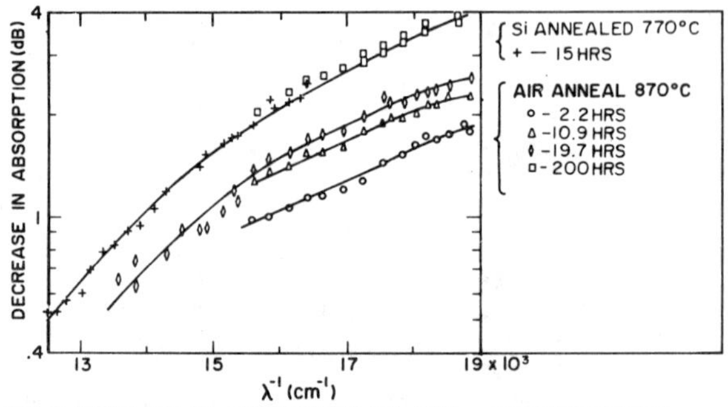

FIG.2. DECREASE IN ABSORPTION AS FCN. OF FREQUENCY MEASURED AFTER ANNEALING IN AIR OR IN Si-ENVIRONMENT. ℓ = 23.3μm.

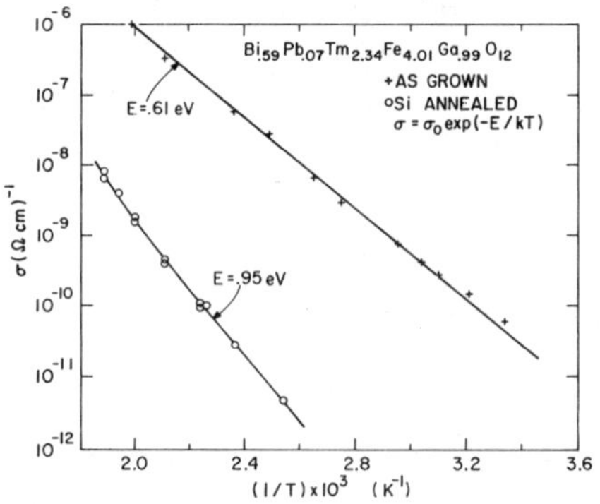

FIG.3 ELECTRICAL CONDUCTIVITY AS FCN. OF TEMPERATURE BEFORE AND AFTER ANNEALING.

In the sample most altered by annealing, $\Delta\alpha/\alpha$ is $\sim 40\%$ and $\sim 60\%$ at 5900 and 7400 Å respectively. No changes in Faraday Rotation were observed. Quantitative results obtained with Si-annealing are erratic and depend on the garnet-Si interface. Moreover, Si constitutes a sink for the oxygen from the garnet, and Si-annealing ends with deterioration of the garnet. We have found that for Bi-based films air itself constitutes a lightly reducing atmosphere. Therefore, for quantitative measurements presented below, BiTm films were annealed in air. Figure 2 is the spectrum of $(\Delta\alpha)\ell$ where ℓ is the thickness of the film. The parallel curves of Fig. 2 imply that the line shape of $\Delta\alpha$ remains constant throughout the annealing process; i.e. we are dealing with a specific set of transitions. Moreover, annealing in Si is equivalent to annealing in air since the same set of transitions is affected by either. Except for the magnitude of $\Delta\alpha$ all of our films yielded spectra identical to those in Fig. 2. Between 12500 cm^{-1} and 17240 cm^{-1} $\Delta\alpha$ increases almost exponentially with frequency and tends towards a broad peak at ~ 18200 cm^{-1}. Although an unequivocal identification of the absorption is not attempted here, it is noted that a similar spectrum with a peak around 5800 Å was identified by B. Faughnan as a charge transfer absorption of Fe^{4+} in $SrTiO_3$.[9]

Figure 3 shows the effects of annealing on the electrical conductivity (σ), as measured, using the apparatus described in Ref. 10. Between $30^{\circ}C$ and $300^{\circ}C$ the simple exponential law: $\sigma = \sigma_o \exp(-E/kT)$ is obeyed before and after annealing and the activation energy, E = 0.6 eV, measured in the virgin film, coincides with E for p-type σ in YIG:Ca.[3] After annealing, σ drops by a factor of $\sim 10^3$, and E increases to $\sim$ 0.95 eV.

The changes observed in the optical absorption and in the electrical conductivity are most easily explained in terms of color centers. In the virgin film the defect distribution is dominated by the high level of Pb^{2+} which is charge compensated with Fe^{4+}. Absorption having a line-shape of $\Delta\alpha$ is due to strong transitions involving Fe^{4+} and the film has p-type conductivity. Our annealing experiments indicate that isothermal LPE growth does not lead to an equilibrium distribution of defects for $600^{\circ}C < T < 1100^{\circ}C$ and $p(O_2) \sim .2$ torr. Annealing in Si or in air leads to a partial reduction of the material. Oxygen vacancies, V_o^*, diffuse into the crystal and the resulting donor provides an alternative charge compensation mechanism.

$$2Pb^{2+} + 2Fe^{4+} + V_o^* \rightarrow 2Pb^{2+} + 2Fe^{3+} + V_o^{2+} \qquad (1)$$

Replacement of $(Pb^{2+} + Fe^{4+})$ acceptors with deeper lying levels $(Pb^{2+} + 1/2V_o^{2+})$ results in an increase of E-average and in a decrease of α and of σ associated with Fe^{4+}, as observed in Figs. 1,2 and 3.

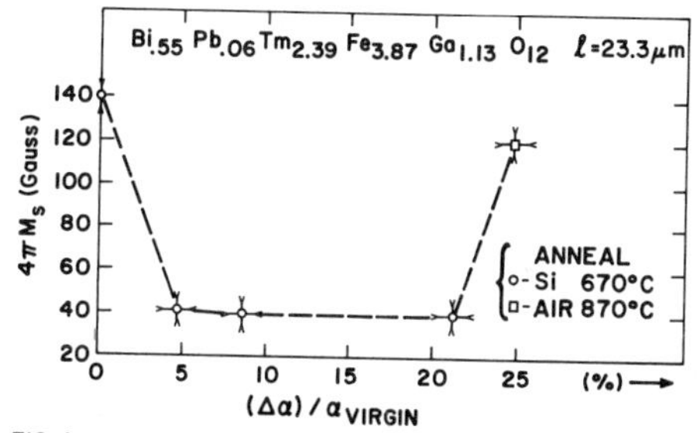

FIG.4. CHANGE IN ABSORPTION vs $4\pi M_s$ AT VARIOUS STAGES OF ANNEAL.

Although the above explanation is consistent with our results, it is not a priori impossible that changes in α and in σ are independent of each other and that $\Delta\alpha$ is related to the change in the saturation magnetization ($4\pi M_s$).[†] It can be shown that in a BiTm film annealed at $700^{\circ}C$, approximately 3% of Ga ions will move from (a) to (d) sites; i.e. the absorption of (Fe^{3+}) d should decrease and that of (Fe^{3+}) a should increase, by 3% respectively. Using a film of $(YEu)_3(FeGa)_5O_{12}$ as a standard we were unable to detect any change in α which could be associated with Ga redistribution. One can still argue that since the details of the orbital overlap of Bi^{3+} 6p with O^{2-} 2p or with Fe^{3+} 3d orbitals are unknown, the effects of Ga redistribution on α can be different in garnets containing Bi or Pb. Figure 4 shows that although annealing leads to both effects: a decrease in α and a change in $4\pi M_s$, the former is not the result of the latter. To obtain the results of Fig. 4, 3 samples were Si-annealed at $670^{\circ}C$, for 4 hrs., 10 hrs., and 119 hrs. respectively. The corresponding changes in $4\pi M_s$ and in α (at 5900 Å) are shown as circles of Fig. 4. After 4 hour anneal $\Delta\alpha/\alpha$ was 5% and $4\pi M_s$ dropped from 140 to 40 Gauss. Afterwards, $4\pi M_s$ remained constant while $\Delta\alpha/\alpha$ kept increasing and reached 21% after 119 hrs. The square in Fig. 4 shows that when the 3rd sample was re-annealed in air at $870^{\circ}C$ so that its $4\pi M_s$ rose back to 120 Gauss $\Delta\alpha/\alpha$ of that sample did not increase.

Having decided that: (i) For a given sample (at a given λ), $\Delta\alpha$ obtained after annealing for time t is proportional to the number of V_o^* introduced into the film during t. (ii) For air-annealing $(\Delta\alpha)_t/(\Delta\alpha)_\infty$ measures the approach to the equilibrium concentration of V_o^*, the diffusion constant of V_o^* can be obtained from measurements of $\Delta\alpha$ as a function of t. Fitting the experimental data to the diffusion equations (derived for a solute contained in a slab and evaporating from its surface[11]) we obtain 6×10^{-8} $cm^2 sec^{-1}$ as a tentative value of the diffusion constant at $870^{\circ}C$.

The results of the annealing experiments presented above, indicate that the dopant-induced absorption[†††] of Bi-based films containing $\sim$ 0.3 mole% of Pb consists of at least 2 components. Below 16700 cm^{-1} transitions involving Fe^{4+} (probably charge transfer) account for the entire extrinsic absorption and this contribution is eliminated through partial reduction. At higher frequencies α extrinsic has an additional component which does not add to the electrical conductivity and which is probably due to an increase in the oscillator strength of some neighboring Fe^{3+} lines.

ACKNOWLEDGMENTS

The authors are deeply indebted to R. Shahbender for helpful discussions and continuous encouragment, to D. Hoffman for her help with Si deopsition, to D. Staebler for the use of his experimental equipment to measure the electrical conductivity, to T. Ward for his unfailing help in the experimental work and to E. P. Bertin for performing electron-probe analysis of the samples. Helpful discussions with I. Gordon and experimental help of C. Horak and R. Noack are gratefully acknowledged. In addition, one of us is especially grateful to her husband for taking care of "everything else" when absolutely necessary

REFERENCES

1. A.Akselrad, R.E.Novak, D.L.Patterson, AIP Conf.Proc. no.18, Magnetism and Magnetic Mat'ls (AIP-NY 1974) p.949.
2. S.Wittekoek, et.al., Ref. 1, p.944.
3. D.L.Wood, J.P.Remeika, J.Appl.Phys.37, 1232,(1966).
4. R.W.Teale, et.al., J.Appl.Phys.40, 1435,(1969). V.Enz, et.al., J.de Phys., Suppl.32, C1-703(1971). D.J.Epstein, AFML-TR-70-210, Final Report, Aug. 1970.
5. S.H.Wemple, et.al., Phys.Rev.B9, 2134, (1974).
6. S.Wittekoek, D.E.Lacklison, Phys.Rev.Lett.28,740(1972)
7. A.Akselrad, AIP Conf.Proc.,no. 5, Magnetism and Magnetic Mat'ls, (AIP-NY 1972) p.249.
8. R.C.LeCraw, P.A.Byrnes,Jr., W.A.Johnson, H.J.Levinstein, J.W.Nielsen, R.R.Spiwak and R. Wolfe, IEEE Trans. Magn., MAG-9, 422 (1973)
9. B.W.Faughnan, Phys.Rev.B4, 3623, (1971).
10. J.Blanc, D.L.Staebler, Phys.Rev.B4, 3598, (1971).
11. J.Crank, The Mathematics of Diffusion, Oxford 1956, p. 56.

† abbreviated below as BiTm films.
†† in a charge-compensated garnet we measured $\sigma \simeq 1.6 \times 10^{-}$ $(\Omega\text{-cm})^{-}$ and $E \simeq 1.16$ eV which is $\simeq 1/2$ of the band-gap.
††† defined as the excess α of a BiTm film over α of its Bi and Pb free counterpart.

INCREASED CURIE TEMPERATURE AND SUPEREXCHANGE INTERACTION GEOMETRY IN BISMUTH AND VANADIUM SUBSTITUTED YIG*

S. Geller
Department of Electrical Engineering
University of Colorado, Boulder, CO 80302

A. A. Colville
Department of Geology
California State University at Los Angeles, Los Angeles, CA 90032

ABSTRACT

An unexpectedly large Bi^{3+} for Y^{3+} substitution has been made in YIG; single crystals have the formula $\{Y_{1.12}Bi_{1.88}\}[Fe_2](Fe_3)O_{12}$. The structures of this garnet and of single crystal $\{Bi_{0.34}Ca_{2.66}\}[Fe_2](Fe_{1.67}V_{1.33})O_{12}$ have been refined, the purpose being to determine whether there is a relation between the relatively high Curie temperatures of such garnets and the superexchange interaction geometry. The results indicate mainly an increase in the [Fe]-O distances and a decrease in the (Fe)-O distances as compared with the analogous distances in YIG. This may imply that d-p wave-function overlap between (Fe^{3+}) and the intervening oxygen is more important than between $[Fe^{3+}]$ and the intervening oxygen in determining J_{ad}.

INTRODUCTION

There exist two cases in which a diamagnetic ion when introduced into iron garnets causes a large increase in Curie temperature. One of these is V^{5+} substitution for tetrahedral Fe^{3+}, which is remarkable in that the increase in J_{ad} as a result of V^{5+} substitution more than compensates for the decrease in the number of interactions.[1,2] The other case is the substitution of Bi^{3+} for Y^{3+} in YIG.[3] It has been shown that 0.45 La^{3+} substitution increases the Curie temperature by about 4°K,[4] but Bi^{3+} is about four times as effective as this, i.e. the increase in Curie temperature is 38°/Bi/formula unit.[3] The size of the ion alone is therefore not the cause of this increase in T_c; in fact, the increase in T_c for some of the smaller rare earth ions is greater[4] than it is for La^{3+}. It has been hypothesized[1-3] that the increase in J_{ad} resulting from the V^{5+} or Bi^{3+} substitution is very likely associated with a more favorable [Fe]-O-(Fe) interaction geometry.

In considering the superexchange interaction between two magnetic ions through an intervening oxygen ion, both interatomic angle and interatomic distance appear to play a role.[5] However, in an iron garnet a large change in the $Fe^{3+}(a)$-O^{2-}-$Fe^{3+}(d)$ angle is unlikely because the Fe^{3+}-O^{2-} distances are constrained to lie within narrow limits about 2.01Å and 1.88Å for the octahedral and tetrahedral distances respectively.[6,7] The garnet structure is such that a substantial increase in angle is not possible without substantial decrease in at least the sum of these distances.

The advantage of studying the crystal structure of a Bi-substituted YIG is that Bi substitution may be used as a probe to determine the effect on the $Fe^{3+}(a)$-O^{2-}-$Fe^{3+}(d)$ interaction geometry free of interference by substitution for the Fe^{3+}. With this in mind, we decided to grow crystals with high Bi content.

Earlier,[3] the maximum Bi substitution was estimated to be 1.5±0.05 out of 3. However, in our more recent experiments, we succeeded in obtaining a specimen with formula $\{Y_{1.12}Bi_{1.88}\}[Fe_2](Fe_3)O_{12}$, with lattice constant 12.531Å; the maximum lattice constant for an iron garnet has been predicted to be 12.540Å.[8] Not only does the lattice constant indicate the correctness of the formula, but electron microprobe analysis as well as the least squares calculations with the X-ray data confirm it. Such high Bi substitution in YIG single crystals should be of interest because of the increased Faraday notation with increased Bi substitution.[9]

We have also grown crystals of $\{Bi_{0.34}Ca_{2.66}\}[Fe_2](Fe_{1.67}V_{1.33})O_{12}$,[10] with lattice constant equal to 12.489Å, and refined its structure as well as that of $\{Y_{1.12}Bi_{1.88}\}[Fe_2](Fe_3)O_{12}$ to see what comparisons could be made.

DIFFRACTION DATA

A crystal of $\{Y_{1.12}Bi_{1.88}\}[Fe_2](Fe_3)O_{12}$(I) will absorb very highly all X-ray wavelengths. For AgKα radiation, used to collect the data, the linear absorption coefficient, μ, is 283.0 cm^{-1}. (For MoKα, the linear absorption coefficient is 532.1 cm^{-1}.) The crystal, ground to a sphere of radius R, 0.06 mm, had μR=1.7. The data were dollected (at the University of Colorado) with a Buerger-Supper single crystal diffractometer automated by a Nova 1200 computer. Scan rate was 1°/min; scan interval was (1.5+0.5Lp) degrees (Lp is the Lorentz-polarization-Tunell factor); background, before and after peak scan, was counted at 1/6 the scan time interval; Pd and Mo balanced filters were used; reflections in the range $10 \leq 2\theta \leq 50°$ were collected.

Data on $\{Bi_{0.34}Ca_{2.66}\}[Fe_2](Fe_{1.67}V_{1.33})O_{12}$(II) were collected (at California State University, Los Angeles) with an automated four-circle goniometer (Syntex $P2_1$) and graphite crystal monochromatized MoKα radiation. The crystal was a ground sphere with radius 0.134 mm; linear absorption coefficient for MoKα is 139.7 cm^{-1} and μR=1.9. Intensities were collected in the range $5° \leq 2\theta \leq 65°$ using variable scan rate and scan interval: 2°/min scan rate for the weakest, 20°/min for the strongest reflections. Background counts were taken on both sides of each peak with total background time/scan time = 1.0. Intensities were corrected for deadtime and the scaled net count was output on a 1°/min basis.

STRUCTURE REFINEMENT

For crystal I, all intensities were first used in least squares[11] calculations to establish scale and thermal parameters of the cations. The thermal parameters of the cations were then held constant. The data used in the final refinement were those most sensitive to oxygen positional parameters.[12] There were 208 of these in the structure factor list, 115 of which were above threshold. The discrepancy factor $\Sigma||F_o|-|F_c||/\Sigma|F_o|$ was 0.06. For crystal II, all data above threshold, 169 structure amplitudes, were used in the refinement. The discrepancy factor was 0.04.

Table I Oxygen ion parameters

	(I) $\{Y_{1.12}Bi_{1.88}\}[Fe_2](Fe_3)O_{12}$	(II) $\{Bi_{0.34}Ca_{2.66}\}[Fe_2](Fe_{1.67}V_{1.33})O_{12}$
x	-0.0279(7)	-0.03339(18)
y	0.0548(7)	0.05197(22)
z	0.1571(5)	0.15189(20)
β_{11}	0.0021(5)	0.00047(12)
β_{22}	0.0001(4)	0.00110(15)
β_{33}	0.0021(4)	0.00058(12)
β_{12}	-0.0003(4)	-0.00018(12)
β_{13}	-0.0004(4)	-0.00018(12)
β_{23}	0.0000(3)	0.00077(12)

Note: Numbers in parentheses are the standard errors applied to the last digits.

The final oxygen ion parameters for the two crystals are given in Table I.

DISCUSSION

The interionic distances calculated[13] from the oxygen parameters given in Table I and from the special cation positions (c, a, d of the space group Ia3d) are given in Tables II and III. Unfortunately, the estimated standard errors for crystal I are three to four times those of crystal II. Nevertheless some important conclusions may be drawn.

Table II Interionic distances and angles $\{Y_{1.12}Bi_{1.88}\}[Fe_2](Fe_3)O_{12}$

FeO_4 tetrahedron		
Fe-O	(4)1.863(9)Å*	
O-O	(2)2.821(18)Å,	(4)3.146(16)Å
∠O-(Fe)-O	(2)98.4(6)°,	(4)115.3(3)°
FeO_6 octahedron		
Fe-O	(6)2.051(10)Å	
O-O	(6)2.760(17)Å,	(6)3.036(16)Å
∠O-[Fe]-O	(6)84.5(4)°,	(6)95.5(4)°
{Y,Bi}-O_8 dodecahedron		
{Y,Bi}-O	(4)2.379(9)Å,	(4)2.493(9)Å
O-O	(2)2.821(18)Å,	(4)2.760(17)Å
	(4)2.846(18)Å,	(2)3.039(18)Å
	(4)3.618(7)Å,	(1)3.909(18)Å
	(1)4.010(17)Å	
∠O-{Y,Bi}-O	(4)68.9(5)°,	(2)71.4(4)°
	(4)72.7(4)°,	(2)75.1(4)°
Others		
{Y,Bi}-[Fe]	(4)3.503Å	
{Y,Bi}-(Fe)	(2)3.133Å,	(4)3.837Å
[Fe]-(Fe)	(6)3.503Å	
∠[Fe]-O-(Fe)	(6)126.9(5)°	
[Fe]-O-{Y,Bi}	(6)100.4(3)°,	(6)104.2(4)°
(Fe)-O-{Y,Bi}	(4)94.4(4)°,	(4)122.8(5)°
{Y,Bi}-O-{Y,Bi}	(8)103.9(3)°	

*(Frequency) distance or angle (standard error).

Table III Interionic distances and angles $\{Bi_{0.34}Ca_{2.66}\}[Fe_2](Fe_{1.67}V_{1.33})O_{12}$

$(Fe,V)O_4$ tetrahedron		
(Fe,V)-O	(4)1.798(3)Å	
O-O	(2)2.773(5)Å,	(4)3.014(4)Å
∠O-(Fe,V)-O	(2)100.95(14)°,	(4)113.87(7)°
FeO_6 octahedron		
Fe-O	(6)2.048(3)Å	
O-O	(6)2.837(4)Å,	(6)2.954(5)Å
∠O-[Fe]-O	(6)87.68(10)°,	(6)92.32(10)°
{Bi,Ca}-O_8 dodecahedron		
{Bi,Ca}-O	(4)2.416(2)Å,	(4)2.530(3)Å
O-O	(2)2.773(5)Å,	(4)2.837(4)Å
	(4)2.953(5)Å,	(2)2.984(5)Å
	(4)3.619(2)Å,	(1)4.039(5)Å
	(1)4.142(5)Å	
∠O-{Bi,Ca}-O	(4)69.95(12)°,	(2)73.28(10)°
	(4)70.05(12)°,	(2)72.27(10)°
Others		
{Bi,Ca}-[Fe]	(4)3.491Å	
{Bi,Ca}-(Fe)	(2)3.122Å,	(4)3.824Å
[Fe]-(Fe)	(6)3.491Å	
∠[Fe]-O-(Fe)	(6)129.73(14)°	
[Fe]-O-{Bi,Ca}	(6)98.82(10)°,	(6)102.59(10)°
(Fe)-O-{Bi,Ca}	(4)94.50(10)°,	(4)123.06(12)°
{Bi,Ca}-O-{Bi,Ca}	(8)101.24(9)°	

In YIG, the [Fe]-O-(Fe) interaction angle is 126.6° while in crystal I it is 126.9°. The standard errors in both determinations render this difference statistically insignificant. The [Fe]-O-(Fe) distances are constrained to be close to their values in YIG.[6,7] The sum of these distances is not apt to differ substantially from (2.01+1.88)Å=3.89Å. A substantial increase in angle, however, would require a decrease in this sum; for example, if [Fe]-O were 2.00Å and (Fe)-O were 1.84Å, the [Fe]-O-(Fe) angle would be 131.6° in crystal I.

The structure analysis indicates an increase in [Fe]-O and a decrease in (Fe)-O distances in crystal I. The sum of the two is 3.91Å, not significantly different from the analogous sum in YIG.

For crystal II, we find good agreement of distances, but not of angle, with some structural results published earlier.[14] For this crystal, there is no doubt that the [Fe]-O distance is significantly larger than in YIG; this gives credence to the [Fe]-O distance found in crystal I.

The (Fe,V)-O distance and [Fe]-O-(Fe,V) angle are average values; they do not tell us directly what the (Fe)-O distance and [Fe]-O-(Fe) angle are. The results of a neutron diffraction study[15] of $\{NaCa_2\}[Cu_2](V_3)O_{12}$ give a value of 1.77Å for the (V)-O distance, implying an (Fe)-O distance of 1.82Å and an [Fe]-O-(Fe) angle of 128.9° in crystal II.

In crystal I, all distances are really averages also. An oxygen is at the corner of four polyhedra[6]: a tetrahedron, an octahedron and two dodecahedra, the last involving two different {M}-O distances. If the {Y}-O distances in crystal I are close to what they are in YIG, namely 2.37 and 2.43Å and the [Fe]-O and (Fe)-O distances associated with them are close to what they are in YIG, then those [Fe]-O and (Fe)-O distances associated with the {Bi}-O distances must be larger and smaller respectively than the averages. The values would be 2.08 and 1.85Å respectively for [Fe]-O and (Fe)-O. These values would give an even smaller [Fe]-O-(Fe) angle, namely 126°.

In summary, the results of the structure refinements of the crystals of $\{Y_{1.12}Bi_{1.88}\}[Fe_2](Fe_3)O_{12}$ and $\{Bi_{0.34}Ca_{2.66}\}[Fe_2](Fe_{1.67}V_{1.33})O_{12}$ indicate an increase in the [Fe]-O distance and a decrease in the (Fe)-O distance. The implication is that a small increase in d-p wave-function overlap between the tetrahedral Fe^{3+} ion and the intervening oxygen ion is far more effective in increasing J_{ad} than a much larger decrease in the d-p wave-function overlap of the octahedral Fe^{3+} ion with the intervening oxygen is in decreasing J_{ad}.

REFERENCES

*The crystals used in this work were grown in 1971 while the authors were at the Science Center/Rockwell International.

1. S.Geller, G.P.Espinosa, H.J.Williams, R.C.Sherwood and E.A.Nesbitt, Appl. Phys. Lett. 3, 60-61 (1963).
2. S.Geller, G.P.Espinosa, H.J.Williams, R.C.Sherwood and E.A.Nesbitt, J. Appl. Phys. 35, 570-572 (1964).
3. S.Geller, H.J.Williams, G.P.Espinosa, R.C.Sherwood and M.A.Gilleo, Appl. Phys. Lett. 3, 21-22 (1963).
4. J.Loriers and G.Villers, Compt. Rend. 252, 1590-1592 (1961).
5. P.W.Anderson, "Theory of Exchange in Insulators" in "Solid State Physics" (Academic Press, New York, 1963), F.Seitz and D.Turnbull, eds., Vol. 14, pp. 99-214 and pertinent references therein.
6. S.Geller and M.A.Gilleo, J. Phys. Chem. Solids 3, 30-36 (1957); 9, 235-237 (1959).
7. S.Geller, Z. Krist. 125, 1-47 (1967) and pertinent references therein.
8. S.Geller, H.J.Williams and R.C.Sherwood, Phys. Rev. 123, 1692-1699 (1961); G.P.Espinosa, J. Chem. Phys. 37, 2344-2347 (1962).
9. C.F.Buhrer, J. Appl. Phys. 40, 4500-4502 (1969).
10. G.P.Espinosa and S.Geller, J. Appl. Phys. 35, 2551-2552 (1964).
11. W.R.Busing, K.O.Martin and H.A.Levy, Oak Ridge National Lab. Report ORNL-TM-305 (1962).
12. M.D.Lind and S.Geller, Z.Krist. 129, 427-434 (1969).
13. W.R.Busing, K.O.Martin and H.A.Levy, Oak Ridge National Lab. Report ORNL-TM-306 (1964).
14. E.L.Dukhovskaya and Yu.G.Samsonov, Soviet Phys.-Solid State 13, 181-183 (1971).
15. Yu.V.Lipin and Yu.Z.Nozik, Latv. PSR Zinat. Akad. Vestis Fiz. Tech. Zinat. Ser., 1971, pp. 123-4. This information obtained from Chem. Abstr. 75, 256 (abstr. 92181j) (1971).

CATION DEFICIENCY IN Si-SUBSTITUTED YIG*

R. M. Housley
Science Center/Rockwell International
Thousand Oaks, California 91360

S. Geller
Department of Electrical Engineering, University of Colorado
Boulder, Colorado 80302

G. P. Espinosa
Science Center/Rockwell International
Thousand Oaks, California 91360

ABSTRACT

The results of Mössbauer and crystal chemical experiments on specimens of silicon-substituted YIG prepared by solid state reaction indicate that the Fe^{2+} ion content is substantially lower than the Si^{4+} content resulting in large concentrations of cation vacancies. These vacancies can occur in all three types of cation sites. It is highly probable that some of the small number of Fe^{2+} ions present are in dodecahedral sites.

INTRODUCTION

In this communication, we present Mössbauer effect and crystal chemical evidence for extensive cation deficiency in silicon-substituted yttrium iron garnet (YIG) samples prepared by solid state reaction in air. It has previously been shown[1] that the maximum possible x in the nominal formula $\{Y_3\}[Fe^{2+}_xFe_{2-x}](Fe_{3-x}Si_x)O_{12}$ is 0.45. Attempts to force Fe^{2+} ion into dodecahedral sites met with very limited success; it seemed that some Fe^{2+} ion would enter dodecahedral sites only when substantial Fe^{2+} ion entered octahedral sites as in $\{Y_{2.9}Fe^{2+}_{0.1}\}[Fe^{2+}_{0.3}Fe_{1.7}](Si_{0.4}Fe_{2.6})O_{12}$. It thus appeared that the Fe^{2+} ions preferred the octahedral sites when substituted into YIG.

Lattice constant measurements at room temperature[1] and especially results of magnetic measurements at 1.4K seemed to confirm the above formulas and the prior speculation by Wickersheim et al.[2] that Fe^{2+} ions electrostatically compensate for Si^{4+} substitution in YIG crystals. Many papers have since been published concerned with magnetic anneal and photoinduced magnetic anisotropy effects in silicon-substituted YIG.

RESULTS AND DISCUSSION

The first specimen examined in the present study was that with nominal formula $Y_3Fe_{4.6}Si_{0.4}O_{12}$ with exactly the same lattice constant as the specimen made for the Geller et al. 1962 paper.[1] The Mössbauer spectrum taken above the magnetic ordering temperature at 300°C (Fig. 1) shows no distinct component corresponding to Fe^{2+}. If we assume that Fe^{2+} is present but that electron hopping with a relaxation time short compared with about 10^{-8} sec prevents the appearance of a distinct Fe^{2+} spectral component, then we would expect a shift in the center of gravity of the spectrum proportional to the Fe^{2+} fraction. Indeed comparison of the spectrum of the substituted sample with that of pure YIG shows a small but definite shift of 0.010 to 0.015 mm/sec. Shift values for Fe^{2+} in octahedral coordination given in the literature are narrowly clustered at a value 0.8 mm sec^{-1} more positive than the weighted average of the garnet shifts. Using this value and the observed shift, the Fe^{2+} content is constrained to be in the range 0.05 to 0.10 atom per formula unit. This upper limit is considerably lower than the initially expected 0.40 atom per formula unit. Measurements were also made on a specimen with nominal composition $\{Y_3\}[Fe^{2+}_{0.3}Fe^{3+}_{1.7}](Fe_{2.7}Si_{0.3})O_{12}$; this is exactly the same specimen of the Geller et al. 1962 paper.[1] Again less than 0.10 atom per formula unit of Fe^{2+} was found.

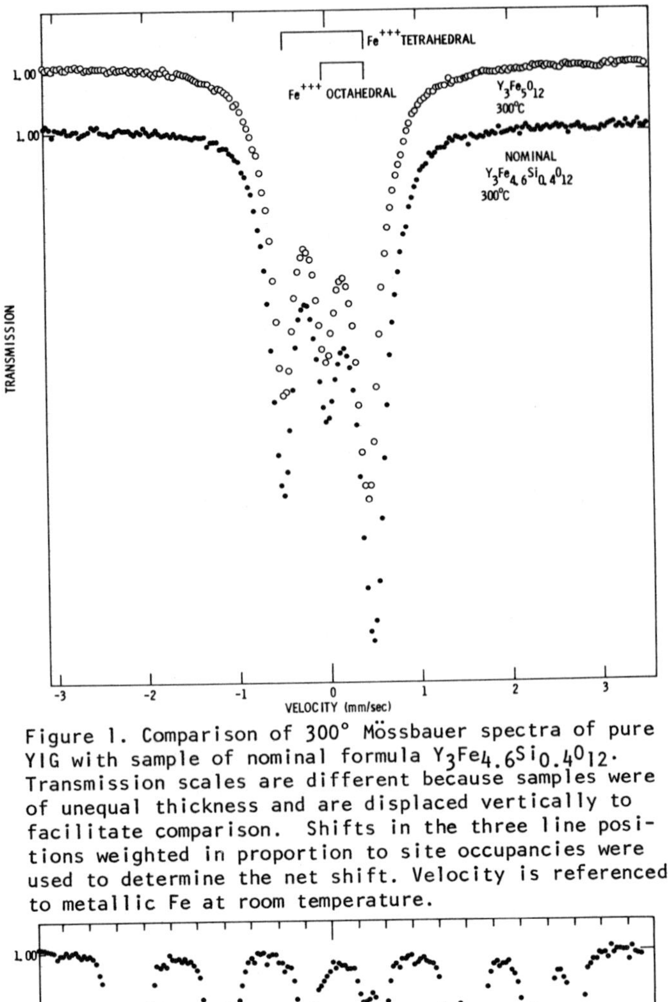

Figure 1. Comparison of 300° Mössbauer spectra of pure YIG with sample of nominal formula $Y_3Fe_{4.6}Si_{0.4}O_{12}$. Transmission scales are different because samples were of unequal thickness and are displaced vertically to facilitate comparison. Shifts in the three line positions weighted in proportion to site occupancies were used to determine the net shift. Velocity is referenced to metallic Fe at room temperature.

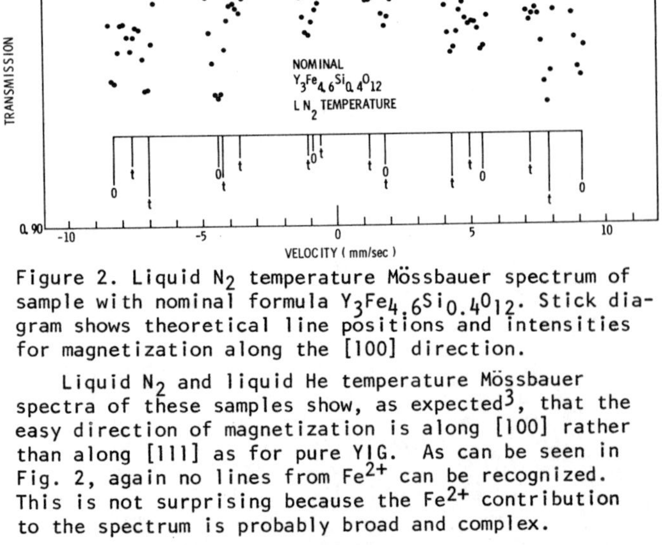

Figure 2. Liquid N_2 temperature Mössbauer spectrum of sample with nominal formula $Y_3Fe_{4.6}Si_{0.4}O_{12}$. Stick diagram shows theoretical line positions and intensities for magnetization along the [100] direction.

Liquid N_2 and liquid He temperature Mössbauer spectra of these samples show, as expected[3], that the easy direction of magnetization is along [100] rather than along [111] as for pure YIG. As can be seen in Fig. 2, again no lines from Fe^{2+} can be recognized. This is not surprising because the Fe^{2+} contribution to the spectrum is probably broad and complex.

A Mössbauer spectrum, taken at liquid N_2 temperature, of the specimen with nominal formula $\{Y_3\}[Fe^{2+}_{0.1}Fe^{3+}_{1.9}](Fe^{3+}_{2.9}Si_{0.1})O_{12}$ shows no change from the YIG [111] easy direction. If we assume that the ratio of Fe^{2+} to Si^{4+} in these samples, which were similarly prepared,[1] is constant we may conclude that between 0.01 and 0.075 Fe^{2+} ion per formula unit is necessary to change the easy axis of magnetization at liquid N_2 temperature.

We are convinced that the above samples are single phase and that the Mössbauer results demonstrate substantial cation deficiency. No evidence of extraneous phases could be found in the x-ray powder photographs. Both high and low temperature Mössbauer spectra were consistent with those expected from garnet.

To substantiate the above conclusion further, and in the hope of slowing electron hopping to the point where recognizable Fe^{2+} spectra could be obtained we attempted to prepare $Y_3Fe_{0.3}Ga_{4.4}Si_{0.3}O_{12}$. It was expected that such a material would be paramagnetic down to low temperatures. Divalent Ga is unknown, and monovalent Ga is very unlikely to occur in this compound. The material prepared was not precisely single phase as determined from the x-ray powder photograph. However, examination in a scanning electron microscope with energy dispersive analysis showed that most of the Si and Fe were contained in the major phase.

The room temperature Mössbauer spectrum (Fig. 3) shows a broad high velocity tail, but not a sharp Fe^{2+} component. A graphical determination of the maximum shift consistent with the data leads to the conclusion that at least 60% of the Fe is trivalent. Although distributed over both octahedral and tetrahedral sites, it prefers the octahedral sites.[4] This again requires cation deficiency. Surprisingly almost no change occurred in the spectrum on cooling to liquid He temperature suggesting that the broadening of the Fe^{2+} contribution is not a relaxation effect.

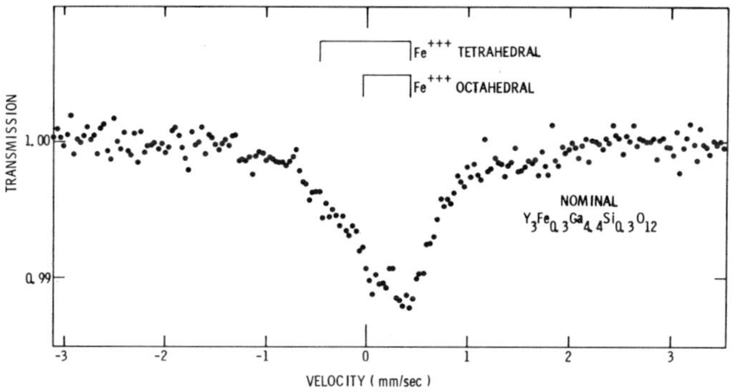

Figure 3. Room temperature Mössbauer spectrum of sample with nominal composition $Y_3Fe_{0.3}Ga_{4.4}Si_{0.3}O_{12}$. Note dominance of Fe^{3+} distributed over both octahedral and tetrahedral sites.

If we assumed that the iron in the sample with nominal formula $Y_3Fe_{4.6}Si_{0.4}O_{12}$ were all trivalent, then normalizing to twelve oxygens per formula unit, the formula would be $Y_{2.95}Fe_{4.52}Si_{0.39}O_{12}$, implying the presence of 0.14 cation vacancies per formula unit. (This neglects the possibility of oxygen vacancies, which if present would probably be small in number.) If, on the other hand, we assume that the specimen contains 0.1 Fe^{2+} per formula unit, the maximum allowed by the Mössbauer data, recalling that the 0°K magnetization per old formula unit[1] is $3.20\mu_B$ or $3.17\mu_B$ per new formula unit, and using $M(Fe^{2+})=4.2\mu_B$, the distribution of ions ranges between $\{Y_{2.97}Fe^{2+}_{0.03}\}[Fe^{2+}_{0.07}Fe^{3+}_{1.86}](Fe^{3+}_{2.55}Si_{0.4})O_{12}$ and $\{Y_{2.97}\}[Fe^{2+}_{0.1}Fe^{3+}_{1.86}](Fe^{3+}_{2.58}Si_{0.4})O_{12}$. This implies that there are definitely vacancies in both octahedral and tetrahedral sites.

To gain further insight regarding the location of the vacancies, we prepared several other specimens. While it is unlikely that Y^{3+} can be made to substitute for Fe^{3+} in the octahedral sites, we nevertheless tried to prepare a specimen with formula $Y_{3.1}Fe_{4.5}Si_{0.3}O_{12}$.[5] Several different heat treatments did not produce a single phase specimen. A specimen with vacancies only in the a and/or d sites, having formula $Y_3Fe_{4.47}Si_{0.40}O_{12}$ also proved unattainable as a single phase. We did, however, succeed in making a single phase specimen with formula $Y_{2.867}Fe_{4.6}Si_{0.4}O_{12}$, seemingly having only dodecahedral vacancies. The lattice constant, 12.337±0.003Å, of this specimen is quite different from that, 12.350±0.003Å, of the original 0.4 Si specimen.

Because a Mössbauer effect spectrum at LN_2 temperature indicated that the specimen $Y_{2.867}Fe_{4.6}Si_{0.4}O_{12}$ has a [100] easy direction, it was necessary to reconsider this formula; the change in easy direction implies the presence of at least 0.0125 to 0.10 Fe^{2+} ion per formula unit. Thus readjusting the oxygen content and again assuming the maximum possible Fe^{2+} ion content (in accord with the Mössbauer effect data) of 0.1 Fe^{2+} ion per formula unit, we may write the more nearly correct formula for this specimen: $\{Y_{2.88}Fe^{2+}_{0.02}\}[Fe^{2+}_{0.08}Fe^{3+}_{1.92}](Fe^{3+}_{2.60}Si_{0.4})O_{12}$. However, more or all the Fe^{2+} could be in dodecahedral sites and the garnet could have vacancies in the octahedral and/or tetrahedral sites also. Regardless of how many Fe^{2+} ions this specimen contains, it can be shown that some of them must be in dodecahedral sites. (The magnetization of this specimen has not yet been measured; such a measurement would give a more definitive result as to the distribution of ions in this specimen.)

In summary, we have shown that the Fe^{2+} ion content in silicon-substituted YIG is substantially less than the silicon content and that this results in a large concentration of vacancies; that vacancies may occur in any one or combination of two or three of the c, a, d cation sites; that some of the small number of Fe^{2+} present may occupy dodecahedral sites. These features may play a role in the magnetic anneal and photoinduced effects in these materials.

ACKNOWLEDGMENTS

Work done recently in collaboration with A. Tucciarone, B. Antonini and P. Paroli at the Laboratorio di Elettronica dello Stato Solido del CNR in Rome on our silicon-substituted YIG crystals has helped to clarify the results of our experiments. Discussions with R. W. Grant and A. Paoletti are gratefully acknowledged.

REFERENCES

*The experimental part of this work was carried out prior to August 1971, when S.G. was still at the Science Center.

1. S. Geller, H.J. Williams, R.C. Sherwood and G.P. Espinosa, J. Phys. Chem. Solids 23, 1525-1540(1962)
2. K.A. Wickersheim, R.A. Lefever and B.M. Hanking, J. Chem. Phys. 32, 271-276 (1960). See also R.A. Lefever and A.B. Chase, J. Chem. Phys. 32, 1575-1576 (1960).
3. R.P. Hunt, J. Appl. Phys. 38, 2826-2836 (1967).
4. M.A. Gilleo and S. Geller, Phys. Rev. 110, 73-78 (1958); S. Geller, J.A. Cape, G.P. Espinosa and D.H. Leslie, 148, 522-524 (1966) and appropriate references therein.
5. This formula was based on the possibility of the existence of a solid solution between 0.9 $Y_3Fe_2Fe_3O_{12}$ and 0.1 $Y_4Si_3O_{12}$. For $Y_4Si_3O_{12}$, see N.A.Toporov, I.A. Bondahr and M.M. Piryutko, Dokl. Akad. Nauk, SSR 156, 619-621 (1964).

PREPARATION AND PROPERTIES OF $CdCr_2Se_4$ THIN FILMS*

D. I. Tchernev and A. J. Syllaios
The University of Texas at Austin, Austin, Texas 78712

ABSTRACT

$CdCr_2Se_4$ thin films were prepared by RF sputtering of stoichiometric mixtures of CdSe and Cr_2Se_3 on sapphire substrates. An amorphous mixture of the constituent elements, highly deficient in Se but with the correct Cd to Cr ratio, was thus deposited on the substrate. A subsequent annealing in Se atmosphere was then carried out and $CdCr_2Se_4$ films were grown according to the reaction

$$Cd + 2Cr + 4Se \rightarrow CdCr_2Se_4 \quad (500^oC, 1\ h).$$

X-ray analysis revealed no other phases present. The lattice constant at 10.75 A° is in good agreement with the bulk value.

The temperature dependence of the magnetization and the g-factor is investigated using Ferromagnetic Resonance. It is found that the g-factor is constant between 77°K and room temperature at 1.98 and in good agreement with the bulk value. The magnetization estimated from the resonance equations at 24 GHz is an effective one and is smaller than the bulk saturation magnetization. This may be due to magnetostriction and the porosity of the films. The linewidth is of the order of 100 Oe compared to the reported bulk value of 20 Oe at 77°K, and it increases with temperature. Spin wave resonance has not been observed.

PREPARATION OF $CdCr_2Se_4$ THIN FILMS

Cadmium Chromium Selenide belongs to a varied family of Transition Metal Chalcogenide Spinels of composition AB_2X_4 (A = transition metal, Mg, Zn, Cd, Hg; B = transition metal, Ga, Al, In; X = S, Se, Te). These compounds can in general be prepared by reaction of the constituent elements or the binary chalcogenides.[1] Single crystals of $CdCr_2Se_4$ are grown by closed tube chemical transport reaction with $CdCl_2$ or Cl_2 as a transport agent. However, the preparation of $CdCr_2Se_4$ thin films has not been reported. Attempts to grow thin films of this material by Chemical Vapor Deposition methods apparently were not successful.[2]

The preparation of $CdCr_2Se_4$ thin films is carried out in two separate steps. First, a stoichiometric mixture of the constituent elements is deposited on a substrate by RF sputtering of CdSe and Cr_2Se_3. Then the deposited mixture is reacted on the substrate at elevated temperatures to form $CdCr_2Se_4$.

The sputtering is carried out in an MRC Model No. 8620 RF sputtering module placed over an oil diffusion vacuum system with an ultimate pressure of 5×10^{-7} Torr. The sputtering gas is Argon purified by passing through an MRC Model No. V4-198 inert gas purifier with a titanium getter filament before entering the system. The Argon pressure is maintained at 15×10^{-3} Torr.

The target was prepared from CdSe and Cr_2Se_3 powders thoroughly mixed and pressed to a disc 5 cm in diameter and 0.3 cm thick. The disc was then attached to the cathode electrode with conductive epoxy. Its composition by weight is 35.95% CdSe and 64.05% Cr_2Se_3 as dictated by the $CdCr_2Se_4$ composition in order to yield stoichiometric films. The substrates were single crystal polished sapphire although single crystal NaCl and glass slides were also used in preliminary experiments. The cathode electrode and the substrate support were water cooled.

The composition of films sputtered under various conditions was measured by electron probe microanalysis. In the bias sputtering mode and for 1% to 25% of the RF power applied on the substrate, very little deposition was observed on glass and sapphire substrates even for prolonged sputtering (several hours) for target powers up to 300 W. Films deposited on NaCl were highly enriched in Chromium. For bias levels above 10% in particular there was hardly any Selenium detectable by the microprobe. Nearly stoichiometric films were deposited under the following conditions:

No Bias, RF Power = 250 W, Argon Pressure = 15 microns, Sapphire Substrates. (1)

Sputtering at lower powers yields films deficient in Cr whereas at high power the films are Cr enriched.

Films deposited on sapphire under conditions (1), though stoichiometric in Cd and Cr, (within the microprobe analysis experimental errors), are deficient in Se in comparison with the $CdCr_2Se_4$ composition. Furthermore, they are amorphous. In view of this fact the sputtered films are subsequently annealed in Selenium atmosphere in closed quartz tubes.

The ampoules are heated in a horizontal furnace to 520 °C for 2 hours and then they are furnace cooled. X-ray analysis of films reacted at other temperatures showed that below 400 °C the films were amorphous. For temperatures in the range of 400 °C to 480 °C the films consisted of mixtures of CdSe, $CdCr_2Se_4$, Se and possibly CrSe. Films consisting of only $CdCr_2Se_4$ were formed at 520 °C. For higher temperatures, depending on the film thickness, the films were discontinuous. The appearance of cracks is attributed to tensile stresses present on the film due to thermal mismatch between film and substrate.

It is presumed that $CdCr_2Se_4$ is formed on the substrate through the following solid state chemical reaction

$$Cd + 2\ Cr + x\ Se + (4-x)Se \xrightarrow[2\ hours]{520\ ^oC} CdCr_2Se_4 \quad (2)$$

with x the amount of Se existing in the sputtered film and the balance (4-x) provided by Se vapors condensing on the substrate.

Films thus grown have the crystallographic structure of the normal spinel $CdCr_2Se_4$. The lattice constant was found to be 10.75 A° which is in good agreement with the bulk value.[1]

FERROMAGNETIC RESONANCE

Ferromagnetic resonance measurements were performed at 24 GHz. Films of approximate area of 2 mm x 1 mm and 4500 A° thick were placed at the center of a copper cylindrical cavity along its axis. The cavity-sample temperature is controlled by a resistor heater in contact with the cavity and monitored by a Cu-Constantan thermocouple. The entire assembly is mounted in a conventional glass dewar system.

The cavity operates at the TE_{011} mode and the transmitted rf power is monitored with a Schottky barrier diode. An AFC circuit locks the klystron on the resonant frequency of the cavity.

The magnetic field is provided by a six-inch electromagnet external to the dewar. The magnet can be rotated about the axis of the cavity so that the applied magnetic field and the rf field remain perpendicular to each other. Fields are measured with a rotating coil gauss meter calibrated against the EPR line of DPPH. The magnetic field is automatically swept from 0 to 14 KOe at a constant rate via a feedback control circuit.

The external magnetic field is modulated by adding to it a small, low frequency (220 Hz), alternating field produced by two Helmholtz coils, so that the field derivative of the absorption line is obtained. Phase sensitive detection is accomplished through the use of a PAR lock-in amplifier.

Perpendicular and parallel FMR measurements at different temperatures yield the temperature variation of the magnetization M and the g-factor through the equations:[3]

$$\frac{f(GHz)}{1.4\,g} = H_{\perp}(KOe) - 4\pi\,M(KGauss) \qquad (3)$$

$$\left(\frac{f}{1.4\,g}\right)^2 = H_{\parallel}^2 + 4\pi\,H_{\parallel}M \qquad (4)$$

The linewidth is obtained from the field difference between the points of maximum slope of the resonance line.

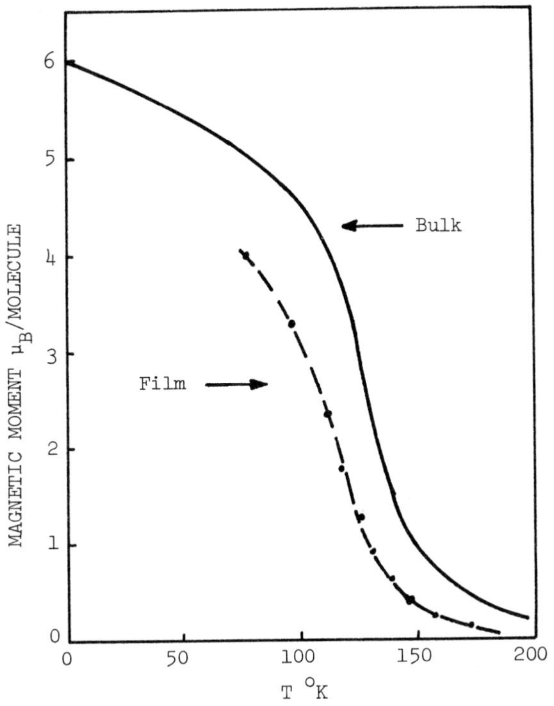

Figure 1. The Magnetic Moment Obtained from the FMR at 24 GHz. The Solid Curve is the Moment of the Bulk from Ref. 4.

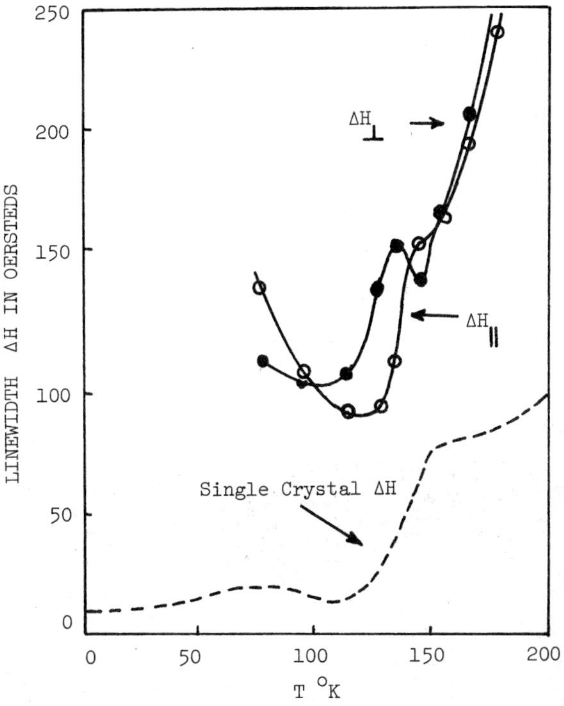

Figure 2. FMR Linewidth at 24 GHz. The Dashed Curve is the Single Crystal Linewidth from Ref. 4.

It was found that the g-factor is constant at 1.98 in the whole temperature range. This value is in good agreement with the reported one for single crystals[4] of 1.983 at 4.2 °K. Figure 1 shows the temperature dependence of the magnetic moment per molecule, as compared to the reported moment for single crystalline bulk material[4] determined by static methods.

The perpendicular $\Delta H_{\perp}$ and parallel $\Delta H_{\parallel}$ linewidth temperature variations are shown in Figure 2. These are not monotonic functions of temperature, but as in the case of the bulk material[4] they show two maxima. The temperatures where the maxima occur are not the same for the cases of the bulk and perpendicular film linewidths.

It was also found that the linewidth varies upon annealing the films in vacuum after they have been grown in Se atmosphere. In particular the linewidth (both perpendicular and parallel) decreases from values of the order of 500 Oe for the case of no vacuum annealing to about 100 Oe for annealing temperature 230 °C. For higher annealing temperatures the linewidth increases again. There is no change in the resonance field, however, resulting from this heat treatment of the films. The linewidths plotted in Figure 2 are for films vacuum annealed at 230 °C having the smaller linewidth.

Spin wave resonance has not been observed in any of these films. The absence of observable distinct spin wave resonance lines is attributed to the large linewidth exhibited by the films. The spin wave structure is obscured in the main resonance peak, but the resulting apparent single line is slightly skewed in the low field side by the spin wave substructure. The larger linewidth of the films as compared to that of single crystals in Figure 2 is attributed to the polycrystallinity and the presence of nonmagnetic voids (porosity) and nonmagnetic impurities in the films. These would account for the smaller than the bulk magnetization.

On the other hand, because the films are grown in Se atmosphere, a layer of Se is condensed on the film surface. Annealing the films in vacuum removes most of the Se, but some of it may still be present in the form of small islands on the surface or in the grain boundaries. These surface conditions may seriously affect the pinning of the magnetic moments at the film surface and thus suppress spin wave resonance.

CONCLUSIONS

The preparation of thin films of the ferromagnetic semiconductor $CdCr_2Se_4$ has been described. The essential factor in this process is the deposition of Cd and Cr in relative concentrations reflecting the $CdCr_2Se_4$ composition. The balance in Se is provided by Se vapors condensing on the substrate during the post deposition annealing in Se atmosphere.

A drawback of this process is the condensation of Se on the films which affects their magnetic (and electrical) properties. Annealing the films in vacuum has proved to be inadequate for the total removal of the condensed Se between the grain boundaries.

It should be noted that sputtering cannot be substituted by vacuum evaporation to obtain stoichiometric deposition. It was found experimentally that vacuum evaporation of CdSe and Cr_2Se_3 resulted in deposited films highly deficient in chromium. The large vapor pressure difference between Se and Cr leads to fractionation of Cr_2Se_3.

*This research was supported by the Joint Services Electronics Program under Contract F44620-71-C-0091 and National Science Foundation Grant DMR73-02416.

REFERENCES

1. P. K. Baltzer, P. J. Wojtowicz, M. Robbins, and E. Lopatin, Phys. Rev., 151, 151 (1966).
2. H. L. Pinch and L. Ekstrom, RCA Review, 31, 692 (1970).
3. H. Nosé, J. Phys. Soc. Japan, 16, 2475 (1961).
4. R. C. LeCraw, H. von Philipsborn, and M. D. Sturge, J. Appl. Phys., 38, 965 (1967).

PREPARATION AND MICROCHEMICAL CHARACTERISATION OF MAGNETIC SEMICONDUCTING $EuCr_2Se_4$ SINGLE CRYSTALS

Pink, H.

Siemens AG, Forschungslaboratorien, 8000 München 80, Balanstraße 73, W-Germany

ABSTRACT

Crystals of $EuCr_2Se_4$ have been grown and analyzed. This substance crystallizes in a hexagonal system. Attempts to grow single crystals resulted in very thin, about 12 mm long needles of a few microgramms in weight each. The amount of Eu, Cr and Se in the crystals was determined. The results of the analysis show that crystals with the full stoichiometric composition have never been grown.

INTRODUCTION

The compounds $EuCr_2S_4$ and $EuCr_2Se_4$ are isostructural with $PbCr_2S_4$, $SrCr_2S_4$ and $BaCr_2S_4$ which crystallize in a hexagonal structure [1]. They are of interest for research because of the high magnetic moment of $7/u_B$ which is caused by Eu^{2+}. In a former [2] work they have been characterized as magnetic semiconductors with Curietemperatures of 68 K and 153 K, respectively, with the magnetic and electric properties being very dependent on deviations from stoichiometric composition, i.e. on the preparation conditions. Therefore, we have improved crystal growth and have elaborated a microchemical method for determining the composition of the crystals.

PREPARATION

We prepared powder samples by means of solid state reactions starting with the binary compounds EuS + Cr_2S_3 and EuSe + Cr_2Se_3, respectively. The heating of mixtures of these couples in evacuated, sealed quartz ampoules to about 1000 °C for 7 days, resulted in products, that were single phased as detected by X-ray analysis. By the chemical analysis it was shown that a very small part of the powder samples was soluble in diluted HCl, while for the most part only concentrated HNO_3 effected a clear solution. The HCl solutions contained only small amounts of Eu ions and no Cr or Se, whereas in HNO_3, Eu, Cr and Se were dissolved. By macrochemical analysis it was shown that in no case compounds with the wanted stoichiometric composition $EuCr_2S_4$ or $EuCr_2Se_4$ have been prepared.

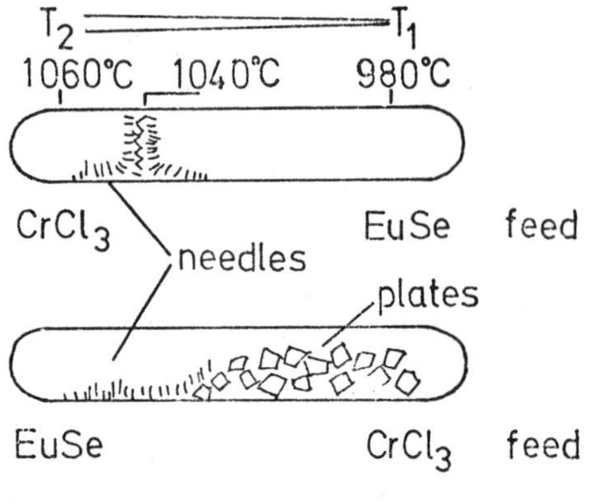

Fig. 1 Experiments for growing crystals of $EuCr_2Se_4$

We started in growing single crystals of these substances to prove whether these are stoichiometric in composition or not. For the present we concentrated in growing the selenides. Preliminary experiments have shown that the Europium chalcogenides react with Chromium chloride by vapor transport if they are heated in evacuated, sealed quartz ampoules to temperatures of 1060 °C at one end and 980 °C at the other end of the ampoule (that is to say with a temperature gradient of 80 K along the ampoules). When starting the experiment with the Europium chalcogenide at the cooler end (T_1) of the ampoule and the Chromium chloride at the end of higher temperature (T_2) (Fig. 1a), the ampoule cross section at T = 1040 °C was filled with very thin needles of about 2 mm length after a reaction time of seven days. With the reverse arrangement of the reactants but almost the same conditions (Fig. 1b), the experiment resulted in growing only a few needles - again in the temperature zone of 1040 °C - and large plates of Cr_2Se_3 between this zone and the cooler end that was fed with the Chromium chloride.

Knowing these results, we started our next experiments with a homogeneous mixture of Europium selenide and Chromium chloride in a molar ratio of 2 : 1 according to the equation

$$4\ EuSe + 2\ CrCl_3 \longrightarrow EuCr_2Se_4 + 3\ EuCl_2 \quad (1)$$

This mixture was distributed over the whole length of the evacuated and sealed ampoule, and the ampoule was heated to 1040 °C for seven days without any gradient in temperature. With this arrangement of the reactants, needle shaped crystals were grown looking like a brush with extremely thin hairs of 12 mm in length (Fig. 2).

X-ray analysis showed that the needles are single phased as the powder samples prepared by solid state reaction.

MICROCHEMICAL ANALYSIS

Unfortunately, the weight of an as grown crystal was so small that macrochemical methods for the detection of elements must fail and even a microchemical one needs about 20 - 30 needles to result in reproducible values with an error less than 0.5 %. Therefore we elaborated a method for analyzing a

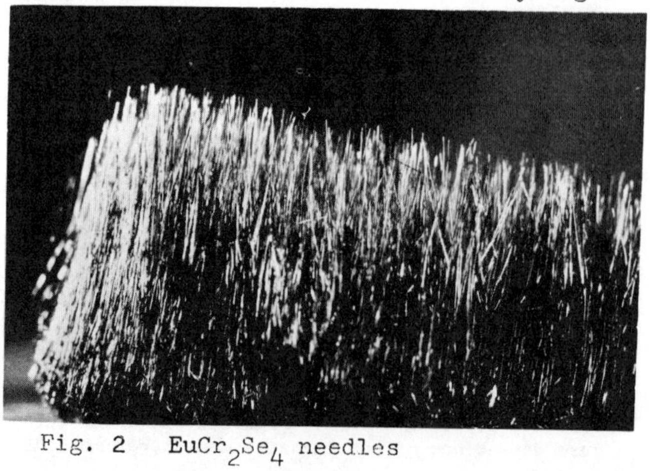

Fig. 2 $EuCr_2Se_4$ needles

sample of 3 mg - 30 mg. The quantitative dissolution of such small amounts succeeded with concentrated HNO_3 (65 %). The ions in the solutions were Eu^{3+}, Cr^{3+} (some Cr^{6+}) and SeO_3^{--}. They could be detected in aliquote parts of the solution by means of colorimetry. Eu^{3+} reacts with the organic reagent Arsenazo at a pH-value of 8.0 to a violet colored complex which has a maximum of absorption at 570 nm. As is shown by a calibration curve determined for this method the amount of Eu is proportional to the extinction at 570 nm in the range 0 - 100 µg Eu. The measurement is not disturbed by the presence of Chromium- and Selenit ions.
Cr reacts, after its oxydation to the Cr^{6+}- ion in acidic solution with Diphenylcarbazid to a complex, redviolet in color with a maximum of adsorption at 545 nm. Measuring the extinction at this wavelength allows the determination of Cr within 0 - 25 µg Cr. The presence of Eu and Selenit ions does not disturb.
Finally, Se can be determined in the range 0 - 1000 µg Se by reducing SeO_3^{--} ions with Ascorbinic acid to the element. The precipitate arising in this reaction is stabilized by adding a solution of gelatin (5 %). The maximum of absorption occurs at 530 nm. At this wavelength, measuring of extinction gives satisfactory values especially in the concentration range 500 - 1000 µg Se, as is seen from a calibration curve. The presence of Eu and Cr ions does not disturb.

CONCLUSIONS

Using the methods described above, we have analyzed needles from several preparations. We have found that samples from one run did not differ in composition within the error of detection (0.5 %). Samples from different runs, however, deviated from one another. Typical formulas, calculated from the measured values for different runs, are

1. $Eu_{0.72}Cr_{2.0}Se_{3.35}$
2. $Eu_{0.73}Cr_{2.0}Se_{3.40}$
3. $Eu_{0.69}Cr_{2.0}Se_{3.30}$

The results show that not only powder samples, grown by solid state reactions, but also crystals deviate from the stoichiometric composition $EuCr_2Se_4$. This means it is important for the understanding of deviations that occur sometimes at the measurements of physical data in a material to determine its composition first.

REFERENCES

[1] Omloo, W.P.F.A.M., Jellinek, F.: Rec. Trav. Chim. Pays-Bas T. 87, 545 (1968)

[2] Lugscheider, W., Pink, H., Weber, K., Zinn, W.: "Darstellung und Eigenschaften von ACr_2X_4-Verbindungen mit hexagonaler Struktur (A = Eu^{2+}, Sr,Ba X = S,Se)", Zeitschrift für angewandte Physik, 32 (1971) 80-83

THE STABILIZATION OF THE HIGH TEMPERATURE PHASE OF MnBi BY THE ADDITION OF RHODIUM OR RUTHENIUM

Kenneth Lee, J. C. Suits and G. B. Street
IBM Research Laboratory, San Jose, California 95193

ABSTRACT

Manganese bismuth has long been considered as a material for use in thermomagnetic recording. It exists as a low temperature phase α and high temperature phase β, of which the β phase is more attractive for thermomagnetic applications as it requires only about 1/4 of the writing energy required for the α phase.[1] The principal problem associated with the β phase is that it tends to convert to the more stable α phase on thermal cycling. Stabilization is therefore an important problem which has received much attention. It has been found that rhodium and ruthenium are effective in stabilizing both thin films and bulk samples of β-MnBi. Addition of 1 at.% Rh in bulk MnBi increases the minimum time constant for the $\beta \rightarrow \alpha$ transformation by five orders of magnitude and lowers the transformation temperature from 360°C to 326°C. Films of nominal composition $Mn_{0.89}Rh_{0.16}Bi$ showed no evidence of transformation for temperatures up to 250°C even after 10^6 secs.

REFERENCES

1. R. L. Aagard, F. M. Schmit, T. S. Liu and D. Chen IEEE Trans. on Magnetics MAG-9, 463 (1973)

THE MAGNETIC PROPERTIES OF V_5Se_8

B. G. Silbernagel, A. H. Thompson and F. R. Gamble
Corporate Research Laboratories
Exxon Research and Engineering Company
Linden, New Jersey 07036

ABSTRACT

Magnetic susceptibility and wideline NMR observations on V_5Se_8 demonstrate the coexistence of vanadium atoms with localized and delocalized d electrons. A transition at 150K is observed which is electronic and associated with a change in the character of the localized moment. Application of strong magnetic fields suppresses the onset of antiferromagnetic order with a field dependence of the form $T_N(H) = T_N(0)[1-(H/H_c)^4]$.

INTRODUCTION

V_5Se_8 is one of a series of layered materials containing two dimensional arrays of localized magnetic species. Its structure consists of alternating layers of (hexagonally arranged) sulfur and vanadium atoms. In every second vanadium layer, only one quarter of the sites are occupied (the other vanadium layers are completely filled). The occupied sites in the partially filled layers are isolated from one another.[1] Magnetic susceptibility and NMR observations in isostructural V_5S_8 show that two d electrons on these isolated atoms are localized,(while the filled vanadium layers are metallic).[2,3] In this present study, we find localized moments on the isolated atoms in V_5Se_8 and a magnetic susceptibility generally similar to V_5S_8. However, NMR and susceptibility data suggest that at high temperatures these moments have a different electronic character than those of V_5S_8 and that a change in the character of the moment occurs near 150K. Further, the magnetic ordering temperature in V_5Se_8 is highly field dependent and varies as the fourth power of the magnetic field, a phenomenon not observed in V_5S_8. We suggest that these differences reflect the more covalent character of the V-Se bond.

EXPERIMENTAL PROCEDURE

The samples were prepared by heating mixtures of the reactants as indicated in Ref. 1. X-ray examination of the resulting materials indicates a pure V_5Se_8 product. NMR observations were performed with a Varian WL-112 spectrometer between 90K and 573K. Susceptibility data were obtained between 2K and 270K using a PAR vibrating sample magnetometer.

The magnetic susceptibilities of powder samples of V_5S_8 and V_5Se_8, shown in Fig. 1, exhibit generally similar high temperature behavior and are distinctly not Curie-Weiss like. Cusps in the low field χ data at $T_N = 28K$ for V_5S_8 and $T_N = 25.5K$ for V_5Se_8 suggest antiferromagnetic order. Observations on single crystal samples establish that the easy axis of magnetization is perpendicular to the layers in both cases. The fact that $\chi_{||}$ approaches zero as $T \rightarrow 0$ and that $\chi_{\perp}$ remains constant below T_N confirms the antiferromagnetic character of the low temperature state.

High field susceptibility observations on the powders reveal a significant difference in the ordered state. Fig. 2 shows the field dependence of T_N for V_5S_8 and V_5Se_8, where T_N is defined from the cusp in the $1/\chi$ vs. T curve. While T_N is field independent in V_5S_8 to 70KG, there is a pronounced depression of T_N for V_5Se_8, which is described by the relationship $T_N(H) = T_N(0)[1-(H/H_c)^4]$, where $T_N(0) = 25.5K$, $H_c = 75.6KG$. The depression of T_N is also reflected in the magnetization curves (M vs. H) in the ordered state, which exhibit an initial positive curvature at low H, followed by a break at the field where order is destroyed, and linear behavior at higher fields. The magnetization data confirm the H^4 field dependence in V_5Se_8.

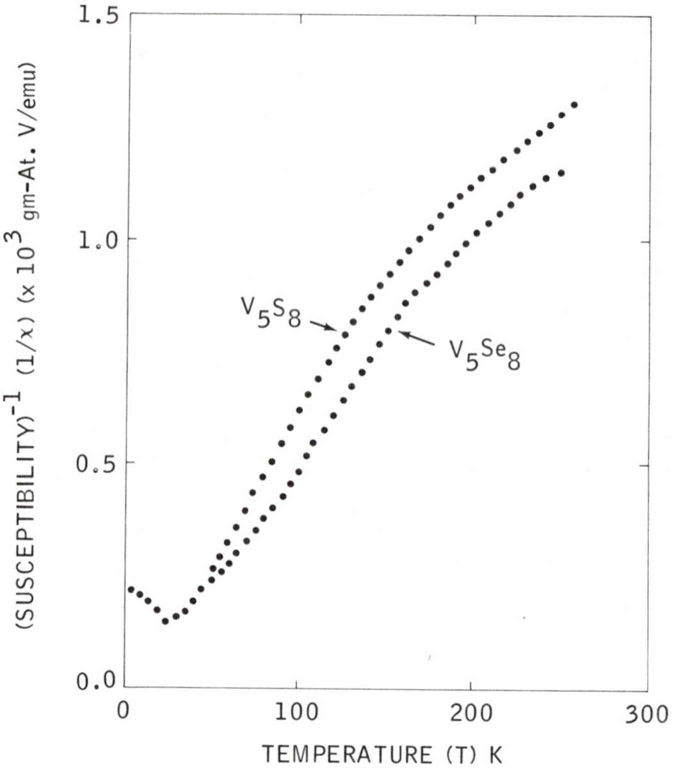

Fig. 1. Susceptibility vs. T

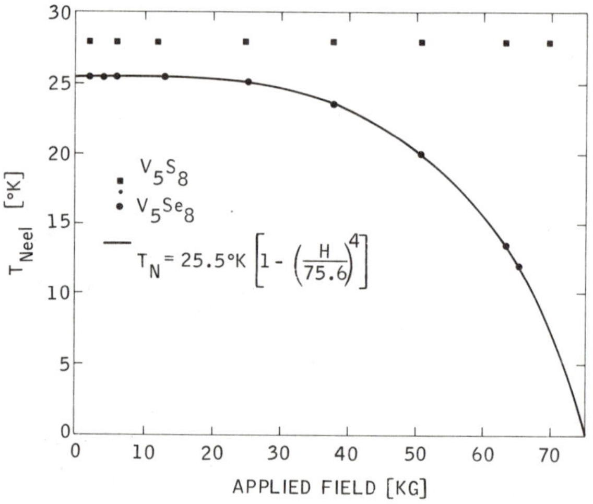

Fig. 2. T_{Neel} vs. H

As in V_5S_8,[3] three distinct NMR absorbtions are observed: a broad, highly shifted absorbtion from nuclei on atoms in the partially filled layers having localized moments (called Type I), a second from atoms in the filled layer with a neighboring moment in the adjacent, partially filled layer (Type II), and third from atoms in the filled layer with no neighboring moment (Type III). The temperature dependent paramagnetic shifts of these lines are plotted as a function of χ in Fig. 3. The Type I shift is large and diamagnetic, while Type II and III shifts are smaller and paramagnetic. The principal feature of the data of Fig. 3 is a break in the K vs. χ curves for all three sites which occurs at ~150K, in contrast to the

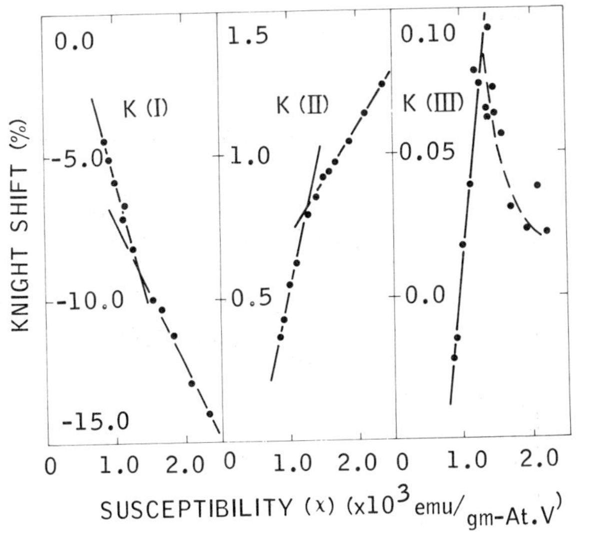

Fig. 3. Knight Shift vs. Susceptibility

linear variations observed in V_5S_8.[3] Low temperature line shapes reveal an axial anisotropy to the paramagnetic shift at the Type II site and non-axial anisotropy at the Type III site. The Type II and III resonances exhibit quadrupolar splitting, but the powder pattern spectra observed are of lower quality than in V_5S_8. Values of the quadrupolar coupling are estimated to be $e^2qQ/h = 7.2$ MHz for Type II and 6.6 MHz for Type III.

Two aspects of the V_5Se_8 data deserve detailed consideration: The apparent change of electronic character at 150K and the suppression of T_N by applied fields. The significance of the former can be assessed by analyzing the moment-induced hyperfine fields at each site which have the form $H_{hf}(i)=N_I\mu_B(\Delta K(i)/\Delta\chi)$, N_I being the number of localized moments per gm. at. of vanadium. The results for V_5S_8 and V_5Se_8, shown in Table I, indicate reductions of $H_{hf}(i)$ for V_5Se_8 at all three sites below 150K. The low temperature values of $H_{hf}(i)$ for V_5Se_8 are very similar to those for V_5S_8. Inferences about the character of the moments can be drawn from a semi-quantitative estimate of $H_{hf}(I)$.[4] If the orbital angular momentum of the moment is completely quenched, the spin-only contribution to $H_{hf}(I)$ should be $\sim$-125KG/μ_B, nearly independent of valence.[3] Orbital contributions are of opposite sign, leading to smaller values of H_{hf} in the case of incomplete orbital quenching. The high temperature value, $H_{hf}(I) = -109$KG/μ_B, implies relatively small orbital contributions. The significantly smaller value of $H_{hf}(I)$ at low temperatures reflects a change in the orbital (and hence spatial) character of the moment. In contrast, the observations in V_5S_8 suggest that this transition occurs at higher temperatures. A detailed examination of Fig. 1 suggests a break in the χ vs. T curve for V_5Se_8 at around 150K. In V_5S_8, examination of the powder quadrupole spectrum of the Type II resonance indicates that the electric field gradient is non-axial, suggesting a crystal distortion. One possibility is that this transition at 150K is associated with the establishment of a charge density wave in the system. In fact, recent systematic studies on the dichalcogenides,[6] when extended to V_5Se_8, suggest that a transition should occur at approximately 150K. The anomalous behavior of the Type III shifts below 150K may result from the establishment of a concomitant spin density wave.

The H^4 field dependence observed for the suppression of T_N is not presently understood since the usual molecular field analysis would suggest an H^2 dependence.[7] However, the differences between the way V_5S_8 and V_5Se_8 respond to the applied field may be quantitative in nature and reflect the greater covalency of the V-Se bond. For example, if H_c for V_5S_8 were 2 times that of V_5Se_8, then the T_N depression at 70KG for V_5S_8 would be less than 5% of $T_N(0)$. The antiferromagnetic exchange interaction is probably not responsible for the different values of H_c since the values of $T_N(0)$ are so similar. However, increased covalency will lead to a larger (ferromagnetic) intrasublattice coupling in V_5Se_8. This conjecture is supported by the values of θ/T_N (θ being extracted from the linear region of the $1/\chi$ vs. T plot between 50K and 150K) of 0.31 for V_5Se_8 and 0.75 for V_5S_8 which, in the molecular field picture, implies an intrasublattice coupling nearly four times larger in V_5Se_8. However, since χ is not Curie-Weiss like, such arguments must be regarded as suggestive.

Table I
Vanadium Atom Hyperfine Fields (in KG/μ_B)

	Type I	Type II	Type III
V_5S_8 (Ref. 3)	- 66	+ 6.7	0.67
V_5Se_8 (High T)	-109	+11.0	3.80
V_5Se_8 (Low T)	- 59	+ 4.00	--

We thank H. M. McConnell and J. R. Schrieffer for helpful comments and L. G. Falls for assistance in collecting the experimental data. T. Zatwarnicki has been most helpful in preparation of the text.

REFERENCES

1. S. Brunie and M. Chevreton, C. R. Acad. Sci. Paris, 258, 5847 (1964).
2. A. B. deVries and C. Haas, J. Phys. Chem. Solids, 34 541 (1973).
3. B. G. Silbernagel, R. B. Levy and F. R. Gamble, Bull. Am. Phys. Soc. 19, 253 (1974), and to be published.
4. A. J. Freeman and R. E. Watson, in Magnetism v. IIA (H. Suhl and G. Rado, eds.) (Academic Press, New York, 1964), p. 268.
5. A. H. Thompson, submitted to Physical Review Letters.
6. P. M. Williams, G. S. Parry, and C. B. Scruby, Phil. Mag., 29 695 (1974) and J. A. Wilson, F. J. DiSalvo, and S. Mahajan, Phys. Rev. Lett., 32, 882 (1974).
7. T. Nagamiya, K. Yosida and R. Kubo, Advan. Phys. 4, 1 (1955).

MAGNETISM IN NEPTUNIUM BORIDES*

J. L. Smith and H. H. Hill
University of California, Los Alamos Scientific Laboratory, Los Alamos, New Mexico 87544

ABSTRACT

Magnetization measurements were performed on NpB_2, NpB_4, and NpB_{12} over the temperature range 2-180 K. NpB_2 is ferromagnetic below $T_C = 99.5 \pm 2$ K; NpB_4 antiferromagnetic below $T_N = 52.5 \pm 2$ K. These compounds exhibit approximate Curie-Weiss behavior above their ordering temperatures. The corresponding Pu compounds as well as PuB_6 were found not to exhibit magnetic ordering above 2 K. The susceptibility of NpB_{12} is relatively independent of temperature. Attempts to prepare NpB_6 in the arc furnace were unsuccessful. The ferromagnetism of NpB_2 is apparently itinerant in nature, the moment being unsaturable at 52 kOe and 2 K and having magnitude ~ 0.3 μ_B per Np atom. The very occurrence of magnetism in the compound is somewhat surprising considering the relatively small separation between nearest-neighbor Np atoms in the compound, $d_{Np-Np} = 3.16$ Å.

INTRODUCTION

Following the discovery[1] that nonmagnetic UB_4 could be made ferromagnetic by alloying with nonmagnetic rare earth tetraborides (YB_4, LaB_4, LuB_4), it became of interest to investigate the magnetic properties of the other actinide tetraborides NpB_4 and PuB_4. We report here the results of studies on these compounds and also studies on NpB_2, PuB_2, PuB_6, and NpB_{12}. The diboride compounds are hexagonal, with alternating layers of actinide (An) and boron atoms; the tetraborides are tetragonal, with an atomic arrangement intermediate between the layered hexagonal diborides and the cubic hexaborides, the latter possessing a CsCl lattice of An atoms and boron octahedra. The dodecaboride structure consists of a NaCl arrangement of An atoms and boron cubo-octahedra.[2] We were unable to prepare NpB_6 by standard arc-melting techniques.[3] No attempt was made to prepare PuB_{12}.

EXPERIMENTAL

The alloys were prepared by melting together in a He-Ar arc furnace appropriate amounts of An metal and boron pellets. Magnetic measurements were made before and after annealing in vacuum at 1300°C for 2 hrs. The results obtained were qualitatively similar, but sometimes differed slightly quantitatively. The data employed were those obtained after annealing.

X-ray powder diffraction analyses revealed that all but the tetraboride samples consisted of a major and a minor phase even after heat treatment. Intensity comparisons indicated that the relative amounts of the minor phases (given in parentheses) were as follows: NpB_2 (5% NpO_2), PuB_2 (10% PuB_4), PuB_6 (25% PuB_4), NpB_{12} (10% NpB_4). These values were used to ascertain the magnetic strengths appropriate to the major phases of interest.

The magnetic studies were performed in a Foner-type vibrating sample magnetometer incorporating a superconducting solenoid capable of fields up to 52 kOe.[4]

In the NpB_2 and NpB_4 samples (and perhaps to a very slight extent in NpB_{12}), there was found at all temperatures investigated a weak, easily saturated ferromagnetic component apparently not directly related to the magnetic properties of the phases of interest. This parasitic ferromagnetism has been observed previously in other Np compounds and has been dubbed "σ_o."[5] The origin of σ_o is not known. In our studies a possible source could be NpC,

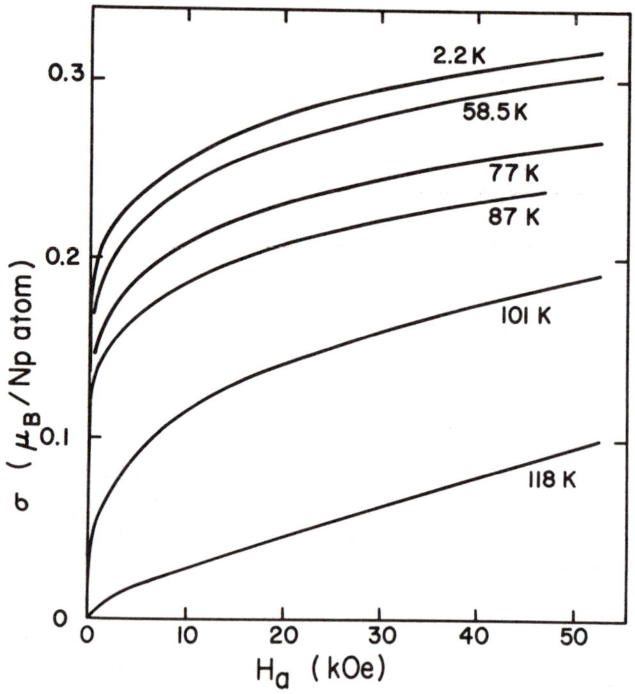

Fig. I. Magnetization of NpB_2 at various temperatures.

since our Np metal contained 125-340 ppm C by weight.[4] NpC is ferromagnetic below 200 K[6], and, were the C in our starting material present in the alloys as NpC, it could account for the σ_o values observed by us ($\lesssim$40 emu/mole Np). In other alloys, however, σ_o has been observed to persist to room temperature,[5] so, were C at fault in those cases, it would have to exist in a more complex form.

Corrections for demagnetizing effects were small in all the alloys studied and were ignored.

RESULTS

$\underline{NpB_2}$. The magnetization of NpB_2 as a function of applied field--determined at several representative temperatures between 2 and 180 K--is shown in Fig. 1. The curves generated are qualitatively similar to those obtained for the traditional itinerant ferromagnets $ZrZn_2$ and Sc_3In and the apparently itinerant $NpOs_2$.[5] To be noted are 1) the finite slope of M vs. H even at high fields and low temperatures, and 2) the small value of the induced moment (~ 0.3 μ_B per Np atom) even at H = 52 kOe and T = 2 K. It seems likely that NpB_2 is an itinerant ferromagnet, although an unequal local-moment ferrimagnetic or similarly complex configuration would not seem necessarily ruled out by the present data. We indicate below that itinerancy is probably more compatible with the interatomic spacing between Np atoms in the compound.

Generally, magnetizing and demagnetizing curves were almost indistinguishable above a few kOe of applied field. The remanence of ~ 0.14 μ_B per Np atom for the 2.2 K trace is illustrated in Fig. 1.

In the paramagnetic region $1/\chi$ was approximately linear with T between 140 and 180 K, yielding the tentative Curie-Weiss parameters $\mu_{eff} \sim 1.7$ μ_B per Np atom and $\theta_p \sim 82$ K. The parasitic σ_o was subtracted out in this region by a conventional χ vs. $1/H$ plot, treating σ_o as a ferromagnetic impurity.

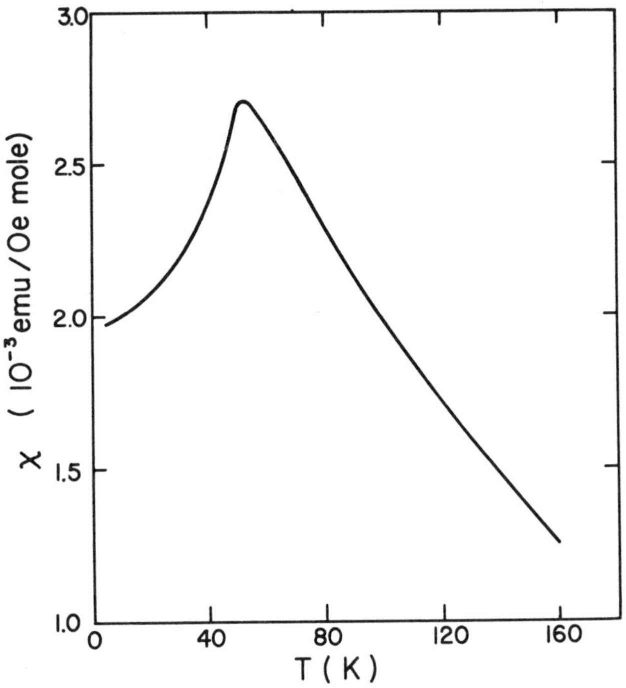

Fig. 2. Magnetic susceptibility of NpB_4.

In the ferromagnetic region, σ_o was eliminated by the following approximate method. The sample was cooled in zero field to the desired temperature, and the observed spontaneous moment subtracted as a field-independent quantity from the ensuing magnetization curve. This technique yielded Arrott plots of reasonable linearity in the vicinity of T_C, which was determined to be 99.5 ± 2 K.

<u>NpB_4.</u> The susceptibility of NpB_4 is shown as a function of temperature in Fig. 2. The peak in χ at 52.5 K is characteristic of a transition into an antiferromagnetic state. Between 80 and 140 K, a Curie-Weiss fit to the data yielded $\mu_{eff} \sim 1.4\ \mu_B$ per Np atom and $\theta_p \sim -30$ K. σ_o was subtracted from the data as described above for the paramagnetic region of NpB_2.

<u>PuB_2</u>, <u>PuB_4</u>, <u>PuB_6</u>, and <u>NpB_{12}</u>. The magnetic susceptibilities of these compounds were found to be relatively independent of temperature and of magnitudes (x 10^{-4} emu/Oe mole): PuB_2: 4.2, PuB_4: 7.2, PuB_6: 15.9, NpB_{12}: 9.8. Because of the technique used to determine the amounts of each phase in the two-phase samples, the numbers for all but PuB_4 should be regarded as accurate to no more than ± 25%. Also, because of the limited sensitivity of the magnetometer, the "temperature independence" of χ for these compounds is meant to imply a variation of no more than ± 20% between 2 and 180 K.

DISCUSSION

The magnetism of NpB_2 and NpB_4 is apparently the first observation of such in binary An-B systems. In discussing the magnetism of these compounds, the most useful reference point is probably UB_4. The suggested critical (minimum) interatomic An-An spacings for magnetism in U, Np, and Pu compounds are 3.4-3.6, ~3.25, and ~3.4 Å, respectively.[7] With a U-U spacing of 3.71 Å, UB_4 must nevertheless be diluted to be made magnetic.[1] As we have learned, dilution is not necessary for NpB_4 (d_{Np-Np} = 3.72 Å) perhaps because the suggested critical spacing for Np compounds is smaller than that for U compounds. Spacing considerations alone do not explain the absence of magnetism in PuB_4, however, and further studies, including the formation of ternary alloys, are planned to further elucidate the nature of the magnetism in these systems.

The Np-Np spacing in NpB_{12} is very large (5.29 Å), and one would be inclined to associate its magnetism with local moment Van Vleck behavior. Troć et al, on the other hand, have suggested a band description for the isomorphous UB_{12}.[8]

With an interatomic Np-Np spacing of only 3.16 Å, magnetism (albeit itinerant) was not expected in NpB_2. Such a small spacing is very close to that in α-Np itself (d_{Np-Np} = 3.11 Å, ave.).[7,9] Perhaps only the increased 5f-5f overlap in α-Np due to a larger coordination number (CN 11-13[9] compared to CN 6 for NpB_2[2]) and/or the presence of four very close neighbors in the α-Np structure (d < 2.7 Å[9]) prevents the occurrence in the pure metal of magnetism similar to that in NpB_2.

It is implicit in the analysis of actinide superconductivity and magnetism on the basis of An-An interatomic spacing[7] that alloys/compounds with spacings near the critical value for a given element may be expected to posses a marginal character (UPt, e.g.[10]); and the apparently itinerant nature of NpB_2 (and $NpOs_2$[5]) are consistent with this description, but the magnitude (~ 100 K) of the Curie temperature of NpB_2 is quite surprising. We are cognizant of the moderately impure nature of our Np starting material, but several consistency aspects of the results obtained make difficult any conclusion other than that the ferromagnetism observed is indeed due to NpB_2. It would seem that Np metal and Np alloys/compounds are more prone to magnetism than had been expected.

ACKNOWLEDGEMENTS

We wish to thank R. B. Roof and his staff for performing the X-ray phase identification analyses and V. O. Struebing for preparing the samples.

REFERENCES

1. A. L. Giorgi, E. G. Szklarz, R. W. White, and H. H. Hill, J. Less-Common Metals 34, 348 (1974); H. H. Hill, A. L. Giorgi, E. G. Szklarz, and J. L. Smith, ibid 38, 239 (1974).
2. B. Post, Boron, Metallo-Boron Compounds and Boranes, R. M. Adams, Ed. (Interscience, New York, 1964), p. 301.
3. H. A. Eick and R. N. R. Mulford, J. inorg. nucl. Chem. 31, 371 (1969).
4. J. L. Smith and H. H. Hill, phys. stat. sol. (b) 64, 343 (1974).
5. A. T. Aldred, B. D. Dunlap, and D. J. Lam, Magnetism and Magnetic Materials-1973, C. D. Graham, Jr. and J. J. Rhyne, Eds., AIP Conf. Proc. No. 18 (AIP, New York, 1974), p. 366; A. T. Aldred, B. D. Dunlap, D. J. Lam, and I. Nowik, Phys. Rev. B 10, 1011 (1974).
6. J. W. Ross and D. J. Lam, J. Appl. Phys. 38, 1451 (1967).
7. H. H. Hill, Plutonium 1970 and Other Actinides, W. N. Miner, Ed., Nucl. Metall., v. 17 (1970), p. 2.
8. R. Troć, W. Trzebiatowski, and K. Piprek, Bull. Acad. Polon. Sci., Ser. Chim. 19, 427 (1971).
9. W. H. Zachariasen, Acta Cryst. 5, 660 (1952).
10. J. G. Huber, M. B. Maple, and D. Wohlleben, Magnetism and Magnetic Materials-1972, C. D. Graham, Jr. and J. J. Rhyne, Eds., AIP Conf. Proc. No. 10 (AIP, New York, 1973), p. 1075 and to be published.

*Work performed under the auspices of the U. S. Atomic Energy Commission.

NONFERROMAGNETIC Cr-Si-Mn INVAR ALLOYS

K. Fukamichi and H. Saito
The Research Institute for Iron, Steel and Other Metals,
Tohoku University, Sendai, Japan

ABSTRACT

To obtain nonferromagnetic Invar alloys, the physical properties of some Cr-base dilute alloys are investigated. Cr is an antiferromagnetic metal and its physical properties around the Néel temperature T_N are sensibly affected by the addition of solute atoms.

It has been found that the antiferromagnetic transition in Cr-Si alloys is of the first kind. The thermal expansion coefficients of the alloys below T_N become very small, showing the Invar characteristic; the Invar region is observed below room temperature and the region occurs at lower temperatures with increasing Si concentration. In Cr-Mn alloys, T_N increases with Mn concentration and the thermal expansion coefficients below T_N become smaller, but their T_N are far above room temperature.

Taking these results into consideration, the thermal expansivity and the magnetic properties of Cr-Si-Mn primary solid solution alloys have been measured. It was found that the alloys are antiferromagnetic showing the Invar characteristic, and the magnetic susceptibility of the alloys is very small; their magnetism is practically negligible.

INTRODUCTION

Invar alloys have wide applications in control and electromagnetic devices and precision instruments, but these practical alloys are all ferromagnetic as in the case of Fe-Ni and Co-Fe-Cr Invar alloys[1]. The ferromagnetic alloys exhibit a large magnetostriction in a static magnetic field and a magnetostrictive oscillation in an alternating field. These phenomena restrict the applications of the ferromagnetic Invar alloys. Therefore, considerable effort is being devoted to the development of nonferromagnetic Invar alloys.

Cr is an antiferromagnetic metal, and its physical properties can be explained by the spin density wave (SDW) theory proposed by Overhauser[2]. Physical properties of Cr such as thermal expansivity and electrical resistivity around the Néel temperature are sensibly modified by the addition of other elements[3~5]. The additional elements used in research are generally transition metals which occupy positions adjacent to Cr in the periodic table. The physical properties of these alloys have been explained by the theory of the two band model, which was proposed by Lomer and modified by Fedders and Martin[6~7].

Recently, studies of Cr-base dilute alloys have been made not only with alloys containing transition metals, but also with alloys containing non-transition metal such as Al, Sn or Si[8~10].

We have investigated the physical properties of Cr-Si and Cr-Mn primary solid solution alloys. According to those experimental results, we expected that Cr-Si-Mn ternary alloys which are of the antiferromagnetic Invar type can be obtained by a proper mixing of binary alloys of Cr-Si and Cr-Mn.

EXPERIMENTAL METHOD

Cr(99.99%), Si(99.99%) and Mn(99,99%) were used as starting materials for the alloys. The alloys were prepared by melting in an arc furnace under an argon atmosphere. The specimens were vacuum-sealed in quartz ampoules and annealed for 3 days at 800°C for homogenization.

The thermal expansion was automatically recorded with a dilatometer of the differential transformer type, the electrical resistivity was measured by the 4-terminal method, and the magnetic susceptibility with a magnetic balance.

RESULTS AND DISCUSSION

The electrical resistivity of Cr and its dilute alloys exhibits a minimum value at the Néel temperature T_N because of a fractional truncation of the Fermi surface by the appearance of the antiferromagnetism. The Néel temperature T_N of Cr-base dilute alloys with Si or Mn can be determined from the tempera-

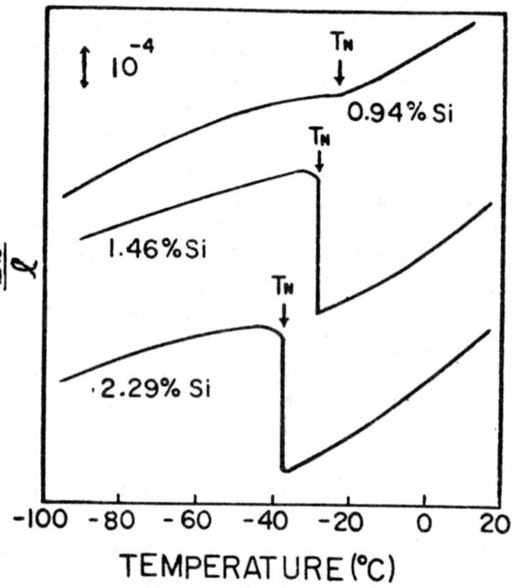

Fig. 1. Temperature dependence of the thermal expansivity of Cr-Si dilute alloys.

ture dependence of the electrical resistivity. In Cr-Si dilute alloys, T_N decreases with Si concentration. Discontinuous volume change is observed at T_N showing that the transition is first-order in Cr-1.46%Si and Cr-2.29%Si alloys(Fig. 1). The thermal expansion coefficient of these alloys becomes very small showing the Invar characteristic below T_N, but the Invar region is below the room temperature and the region shifts to lower temperatures with increasing Si concentration. In Cr-Mn dilute alloys, T_N increases with Mn concentration, and the thermal expansion coefficient below T_N becomes smaller, but the temperature range of small coefficient is above room temperature.

Thus, both alloy systems are known to have a temperature range below T_N where the thermal expansion coefficient becomes smaller, and T_N decreases with the addition of Si and increases

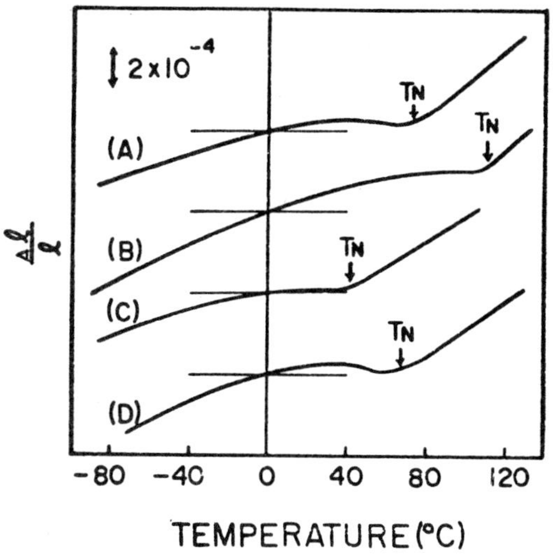

Fig. 2. Thermal expansion curves for several Cr-Si-Mn ternary alloys.

(A) Cr-2.15%Si-0.37%Mn (B) Cr-2.12%Si-0.58%Mn
(C) Cr-2.95%Si-0.53%Mn (D) Cr-1.43%Si-0.21%Mn

with the addition of Mn. Therefore, it is expected that Cr-Si-Mn ternary alloys which exhibit the Invar characteristic in the vicinity of room temperature can be obtained by a proper mixing of Cr-Si and Cr-Mn alloys to control the thermal expansion and T_N.

Fig. 2 shows the thermal expansion curves for Cr-Si-Mn ternary alloys. As expected, Invar type expansion curves are revealed over a wide temperature range below T_N(determined from the temperature dependence of the electrical resistivity). It is also shown that with decreasing Si concentration or with increasing Mn concentration in the ternary alloys, the Invar region shifts to higher temperatures so that the desirable temperature range of the Invar region for applications can easily be obtained by an appropriate selection of the Si and Mn concentrations.

With increasing temperature, magnetic susceptibility χ diminishes gradually, showing a very small peak at T_N. The value of χ at room temperature for the alloys is of the order of 5×10^{-6} emu/g in a magnetic field of 1500 Oe. The above results of magnetic measurements show that these alloys are antiferromagnetic and their magnetic susceptibility is practically negligible.

Similar results with Cr-Fe-Mn, Cr-Fe-Sn and Cr-Co dilute alloys were already reported by present authors[11~14].

The addition of 4d and 5d transition metals Ru, Rh, Re, Os, Ir and Pt, which are to the right of Cr in the periodic table, increase the electron-to-atom ratio of Cr and raise T_N, as seen in the case of Mn. Therefore, it is predicted that by the addition of these elements to the Cr-Si system, T_N is increased and the thermal expansion coefficient below T_N made smaller. Fig. 3 shows examples of the thermal expansion curves of the Cr-Si-X ternary alloys, showing that the Invar characteristic can also be obtained in these ternary alloys.

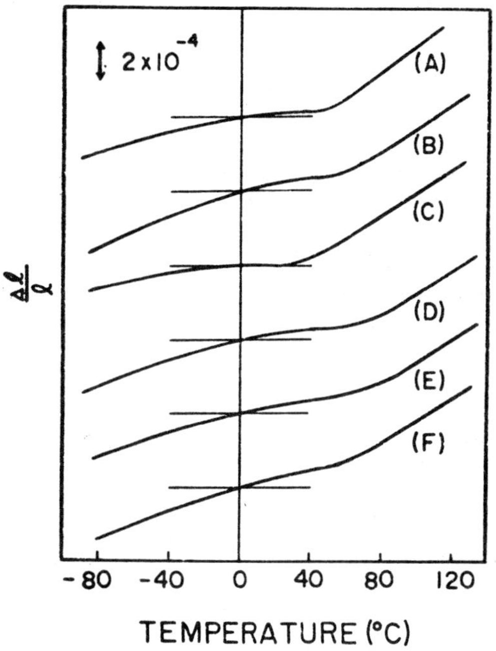

Fig. 3. Thermal expansion curves for some Cr-Si-X ternary alloys.

(A) Cr-3.0%Si-0.5%Ru (B) Cr-2.0%Si-0.5%Rh
(C) Cr-3.0%Si-0.5%Re (D) Cr-3.0%Si-0.5%Os
(E) Cr-2.0%Si-0.5%Ir (F) Cr-3.0%Si-0.5%Pt
(Nominal composition)

REFERENCES

1. S. Chikazumi, T. Mizoguchi, N. Yamaguchi and P. Beckwith, J. Appl. Phys., 39, 939(1968).
2. A. P. Overhauser, Phys. Rev. Lett., 4, 462(1960).
3. T. Suzuki, J. Phys. Soc. Japan, 21, 442(1966).
4. Y. Ishikawa, S. Hoshino and Y. Endoh, J. Phys. Soc. Japan, 22, 1221(1967).
5. Y. Endoh, Y. Ishikawa and H. Ohno, J. Phys. Soc. Japan, 24, 263(1968).
6. W. M. Lomer, Proc. Phys. Soc., 80, 489(1962).
7. P. A. Fedders and P. C. Martin, Phys. Rev., 143, 245(1966).
8. A. Kallel and F. De Bergevin, Solid State Commun., 5, 955(1967).
9. K. Fukamichi and H. Saito, J. Phys. Soc. Japan, 33, 1485(1972).
10. S. Arajs and Wm. E. Katzenmeyer, J. Phys. Soc. Japan, 23, 932(1967).
11. K. Fukamichi and H. Saito, Phys. Status solidi (a), 10, K129(1972).
12. H. Saito and K. Fukamichi, IEEE Trans. Mag., 8, 687(1972).
13. K. Fukamichi, Y. Suzuki and H. Saito, J. Japan. Inst. Metals, 37, 927(1973).
14. K. Fukamichi, N. Fukuda and H. Saito, J. Japan. Inst. Metals, 38, 327(1974).

CHARACTERIZATION OF FERROMAGNETIC PRECIPITATES IN GLASSES BY VARIABLE TEMPERATURE FERROMAGNETIC RESONANCE

E.J. Friebele, D.L. Griscom, and C.L. Marquardt
Naval Research Laboratory, Washington, D.C. 20375

ABSTRACT

The ferromagnetic phases which precipitate from various iron-containing silicate glasses have been identified and characterized by studying the integrated intensity of the ferromagnetic resonance absorption as a function of sample temperature between 4.2 and 570°K. The analysis makes use of the effect of microwave skin depth on the FMR intensity to estimate the average particle size for fine dispersions of metallic iron, and experimental results on well-characterized glasses are in good qualitative agreement with theory. Furthermore, by using the results of modelling experiments carried out on iron and magnetite precipitates produced independently in different glasses, it is possible to detect and discriminate between metallic iron and ferric iron spinel, even when both are co-present in an unknown sample. This result is important because spectra characterized by positive magnetocrystalline anisotropy constants have been observed in samples containing fine-grained magnetite.

INTRODUCTION

The nature of the ferromagnetic phases precipitated from iron-containing silicate glasses is often difficult to infer since magnetite is stable below ~600°C even under the most reducing conditions[1]. The characterization of these phases may also be ambiguous since fine-grained ferric iron spinels (FIS) may be indistinguishable from bcc metallic iron by both static thermomagnetic analysis[2] and ferromagnetic resonance lineshape[3]. In addition, up to 0.1 wt% large grained nonstoichiometric magnetite can escape detection by Mössbauer techniques[4], while superparamagnetic (SPM) particles of iron and magnetite both give rise to an excess area near zero velocity[5]. The present study was undertaken to provide a means of detecting fine-grained metallic iron and FIS precipitated in glass and discriminating between them.

EXPERIMENTAL

FMR spectra were obtained using a Varian E-9 spectrometer operating a X-band (9 GHz). Sample temperatures were varied between 90-573°K by means of a Varian variable temperature accessory and between 4.2-140°K by means of an Air Products LTD-3 Helitran apparatus.

A complete description of the samples used in this study and the preparation techniques involved has appeared elsewhere[6]. In general, the base glass composition modelled that of a lunar soil and contained both iron oxide and titanium oxide. The OKA-B magnetite was crushed mineral Fe_3O_4 dispersed in SiO_2, and 2.1-600-16 was a calcium aluminoborosilicate glass containing only iron oxide[7]. These glasses were given various heat treatments to produce iron and FIS (probably titanomagnetite) precipitates in the size ranges shown in the figures. Sample 2.1-600-16 was found by x-ray and electron diffraction to contain precipitated nonstoichiometric Fe_3O_4[7]. The size and morphology of the precipitates were studied by transmission and scanning electron microscopy.[6,7]

Since the first derivative of absorption is experimentally observed, a double numerical integration[8] of the spectra was performed to determine the FMR intensity. Broad resonances are observed for multidomain (MD)[9] (see Fig. 1a) and acicular[10,11] particles of iron. Narrow resonances are expected for single domain

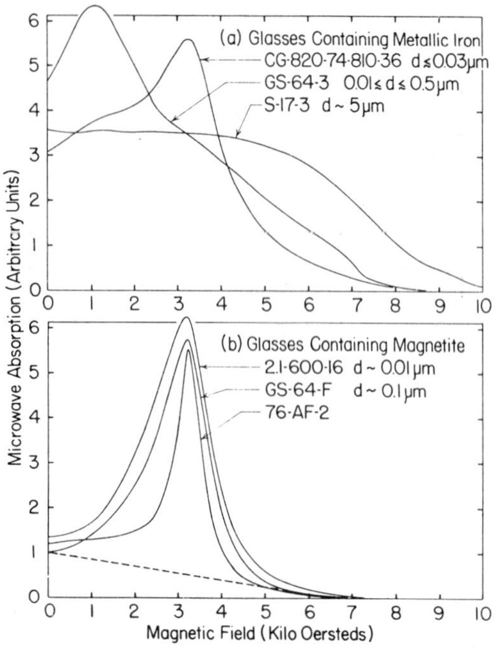

Fig. 1. X-band FMR absorption curves for (a) metallic iron and (b) ferric iron spinel precipitates in glass.

(SD) and SPM iron and observed for equant particles of FIS precipitates[12]. Under certain conditions the FMR absorption peak of MD iron samples may be shifted from g = 2 due to domain effects[9], e.g., Fig. 1a. As shown in Fig. 1b, the FMR absorption of the FIS-containing samples consists of both a narrow and broad part. This is further evidence that iron and FIS cannot be identified on the basis of FMR lineshape alone.

To separate the broad from the narrow parts of the resonance, it is possible to integrate just the narrow part, defined as the area under the absorption curve lying above a line intersecting the absorption curve near zero field and tangent to the curve at high field (5-6 kG) (e.g. dashed line in Fig. 1b).

RESULTS AND DISCUSSION

Although there have been few previous studies of the temperature dependence of the integrated FMR intensity, it might be intuitively assumed that the intensity would vary as the saturation magnetization M_s for particles smaller than the skin depth δ of the microwave radiation. For particles larger than δ, the intensity might be assumed to vary as the fractional particle volume which is penetrated by the microwaves and hence as δ. These assumptions are borne out in the present study.

The temperature dependence of M_s for iron is well-known[13]; the temperature dependence of $\delta = 1/\sqrt{\pi\sigma\mu\nu}$ will be primarily determined by the temperature dependence of the conductivity σ since the permeability μ is only slightly temperature dependent below 600°K. Thus, the FMR intensity of sufficiently large particles should be proportional to δ and hence to $1/\sqrt{\sigma}$. A normalized plot of $1/\sqrt{\sigma}$ vs T, where the values of σ have been taken from the literature[14], is shown in Fig. 2a as the curve "particles >> skin depth." The absolute value of δ in an FMR experiment is difficult to predict[6]. As shown in Fig. 2a reasonably good agreement between the results for 5μ parti-

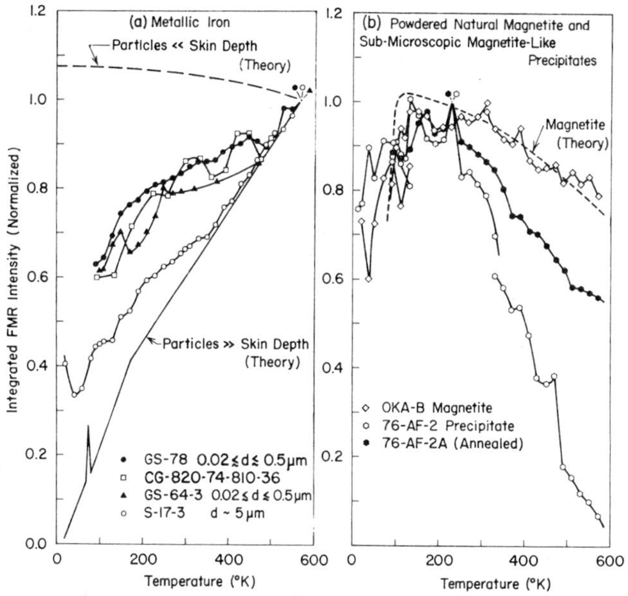

Fig. 2. Total FMR intensity vs. temperature for (a) metallic iron and (b) magnetite and ferric iron spinel precipitates in glass.

cles and the theoretical curve for $d \gg \delta$ is obtained, implying the $\delta \ll 5\mu$. Thus, the FMR intensity of iron particles is seen to be an increasing function of temperature, surprisingly even for particles as small as 0.015μ (CG-820-74-810-36).

Since the conductivity of magnetite is substantially less than that of iron, δ is relatively large ($\geq 5\mu$ over the temperature range of interest). Thus, it is to be expected that the FMR intensity will vary as M_S for particles $\ll 5\mu$ above the Verwey temperature ($\sim 120^oK$). Below this temperature, the intensity is expected to drop since the FMR intensity of magnetite is known to decrease drastically upon cooling between 119 and 105°K [15]. These expectations are incorporated into the "theory" curve shown in Fig. 2b.

The agreement between the theory and the experimental results for OKA-B magnetite is striking. The FMR intensity of the FIS precipitates decreases more rapidly than the Fe_3O_4 sample with increasing temperature, which is probably a result of a lower Curie temperature in the precipitates than in pure magnetite. However, all samples display the characteristic drop in intensity near the Verwey temperature. It should be noted that Bickford[15] observed the FMR intensity of single crystal Fe_3O_4 to approach zero when the external field was aligned along certain crystal axes. In the present (polycrystalline) case, the intensity does decrease below the Verwey temperature, but remains finite.

The data shown in Fig. 2 are derived from the total areas under the FMR absorption curves and provide a clear means for distinguishing between fine grained iron and fine grained magnetite. However, it is often easier to obtain the intensity of only the narrow part of a resonance. When the narrow integral can be obtained for an iron-containing sample (SD + SPM), the temperature dependence is similar to that of the total integral (Fig. 3a), and the narrow integrals of samples containing FIS precipitates display the characteristic drop in intensity below the Verwey temperature[6] (Fig. 3b). The broad absorption of the FIS samples gains intensity upon cooling in analogy with the increase in the low field (broad) absorption of Fe_3O_4 upon cooling through the Verwey temperature[15]. This causes the drop in the narrow integral of the FIS-containing sample of Fig. 3b to be greater than that of the broad integral.

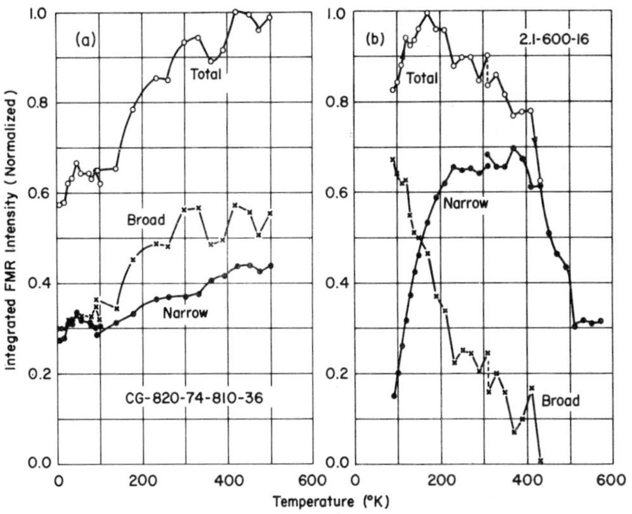

Fig. 3. FMR intensity vs. temperature for (a) metallic iron and (b) magnetite precipitates in glass.

In summary, the temperature dependence of the FMR intensity may be used to identify and distinguish between precipitated iron and/or ferric iron spinels (or magnetite) in glass. A sample which displays an inverted-U-shape FMR intensity vs. temperature behavior can be inferred to contain FIS precipitates whether or not metallic iron is present. As little as ~ 0.1 wt % FIS has been detected in the presence of 1.0 wt % metallic iron, detected by conventional spectroscopic techniques[6].

REFERENCES

1. R.J. Williams, and E.K. Gibson, Earth Planet.Sci. Lett. 17, 84-88 (1972).
2. D.L. Griscom, E.J. Friebele, C.L. Marquardt, N. Sugiura and T. Nagata, EOS, Trans.Am.Geophys. Union 55, 329 (1974).
3. D.L. Griscom, Geochim. Cosmochim. Acta 38, 1509-1519 (1974).
4. D.W. Forester, ibid. Suppl. 4, Vol. 3, 2697-2707 (1973).
5. R.M. Housley, R.W. Grant and M. Abdel-Gawad, ibid. Suppl. 4, Vol. 1, 1065-1076 (1972).
6. D.L. Griscom, C.L. Marquardt and E.J. Friebele, submitted to J. Geophys. Res.
7. M.P. O'Horo, Bull.Am.Ceram.Soc. 53, 324-356 (1974).
8. P.B. Ayscough, Electron Spin Resonance in Chemistry Methuen (London) 1967, p. 442.
9. D.L. Griscom, C.L. Marquardt, E.J. Friebele, and D.J. Dunlop, Earth Planet. Sci.Lett. 24, 78-86 (1974).
10. R.A. Weeks, J.L. Kolopus, D. Kline and A. Chatelain, op.sit. 3, Suppl. 3, Vol. 1, 797-817 (1970).
11. E.J. Friebele, D.L. Griscom, C.L. Marquardt, R.A. Weeks and D. Prestel, op.sit. 3, Suppl. 5, Vol. 3 (in press).
12. D.L. Griscom, E.J. Friebele, and C.L. Marquardt, op.sit. 3, Suppl. 4, Vol. 3, 2709-2727 (1973).
13. A.H. Morrish, The Physical Principles of Magnetism, John Wiley (New York) 1965.
14. Handbook of Chemistry and Physics (1956), C.D. Hodgman, ed. Chemical Rubber Publ. Co., p 2362.
15. L.R. Bickford, Jr., Phys.Rev. 78, 449-457 (1950).

FERROMAGNETISM IN THE METALLIC COMPOUND Fe_xTaS_2 ($x \simeq 1/3$)

M. Eibschütz, F. J. Di Salvo, and G. W. Hull, Jr.
Bell Laboratories, Murray Hill, New Jersey 07974

ABSTRACT

Magnetic susceptibility, electrical resistivity and Mössbauer effect (ME) measurements in single crystals of the metal intercalation layer compound Fe_xTaS_2 are reported. Crystals are grown by iodine vapor transport from $Fe_{1/3}TaS_2$ powder. The crystals contain less Fe than the powder with $x \simeq 0.28$. The resistivity (ρ) shows a metallic behavior, $\rho = 2.7\times10^{-5}$ ohm-cm at 300°K. Magnetization curves as a function of temperature, obtained cooling the sample in H = 10 kG, revealed that the compound is a ferromagnet with a Curie temperature (T_c) of (85 ± 3)°K. Above T_c the susceptibility obeys a Curie Weiss law with $\mu_{eff}^{\parallel} = \mu_{eff}^{\perp} = 4.5 \pm 0.1\ \mu_B$ and $\theta_{\perp} = 30 \pm 2$°K and $\theta_{\parallel} = 90 \pm 3$°K. Above T_c, μ_{eff} and θ are very similar for powder (x = 1/3) and single crystal ($x \simeq 0.28$) samples. At 4.2°K the magnetic moment is 3.5 μ_B per Fe and is parallel to the hexagonal c axis. A large coercive force of 50 kG is required to reverse the spins. A field of 60 kG is not enough to saturate the moment at low temperatures. In the paramagnetic state the ^{57}Fe ME spectra show two resonance lines due to quadrupole splitting;; $e^2qQ/2 = 0.64 \pm 0.03$ mm/sec at room temperature. The isomer shift with respect to iron metal is 0.83 ± 0.03 mm/sec at 300°K, and indicates that the iron atoms are in the Fe^{2+} state. In the ferromagnetic state a distribution of hyperfine fields have been observed for the Fe^{2+} atoms contrary to previous structure determinations (from powder diffraction data) where only one ordered Fe site (octahedral) was proposed. At 4.2°K the average hyperfine field is $H_{hf} = 185$ kG and $e^2qQ/2 = 0.85$ mm/sec. In light of the high coercive force, the high magnetic moment and the complexity of the Mössbauer spectrum (at 4.2°K), it appears that even in single crystals two-dimensional Fe clusters occur. These observations may explain the diversity of magnetic properties reported for the same 3d intercalation compounds of either $2H\text{-}NbS_2$ or $2H\text{-}TaS_2$.

SINGLE CRYSTAL GROWTH OF $Ho_{1-x}Tb_xFe_2$

J. B. Milstein, N. C. Koon, L. R. Johnson, and C. M. Williams
Naval Research Laboratory, Washington, D. C. 20375

ABSTRACT

Numerous magnetic measurements of cubic Laves phase $REFe_2$ (RE = rare earth) compounds have been reported in the literature. Because of the difficulty in single crystal synthesis, the majority of these measurements have been carried out on polycrystalline samples. We wish to report the growth of single crystals of a number of compositions in the system $Ho_{1-x}Tb_xFe_2$, where $0.1 \leq x \leq 0.2$, which exhibit the interesting properties of quite large magnetostriction (1,2) and low anisotropy (3,4).

The crystals were prepared from 99.9% rare earth ingot and 99.95% iron metal by the Czochralski method. An electric arc furnace similar to that of Reed and Pollard (5), powered by a stabilized DC arc welding power supply, was used. Typical growth conditions were a pull rate of 0.5 cm/hr, seed rotation of 40-80 rpm, and melt rotation of 50-75 rpm. Gettered argon served as a growth atmosphere, a slight over-pressure being maintained in the furnace at all times. Large single crystal specimens (typically 1 cm diameter by 1-2 cm in length) have been grown.

On the basis of a variety of measurements, including x-ray diffraction, neutron diffraction (6), electron microprobe, and metallographic examination, the crystals appeared to be macroscopically single phase and homogeneous, with the Ho-Tb ratio varying no more than approximately ±1% in a given specimen. Furthermore, the compounds apparently melt congruently in the composition region investigated.

REFERENCES

(1) N. C. Koon, A. I. Schindler, C. M. Williams, and F. L. Carter, J. Appl. Phys., 45, 5389 (1974).
(2) N. C. Koon, C. M. Williams, and J. B. Milstein, to be presented at Int. Conf. on Magnetics.
(3) C. M. Williams and N. C. Koon, Phys. Rev., to be published.
(4) C. M. Williams, N. C. Koon, and J. B. Milstein, unpublished.
(5) T. B. Reed and E. R. Pollard, J. Crystal Growth 2, 243 (1968).
(6) R. M. Nicklow, N. C. Koon, C. M. Williams, and J. B. Milstein, paper 1C3, this Conf.

HIGH FREQUENCY PYROMAGNETIC MEASUREMENTS OF $CoS_{2-x}Se_x$*

M. L. Malwah, R. W. Bené, and R. M. Walser
Department of Electrical Engineering and Electronics Research Center,
The University of Texas at Austin, Austin, Texas 78712

ABSTRACT

In this paper we present the results of the pyromagnetic measurements on $CoS_{2-x}Se_x$ single crystals in the frequency range of 100 Hz to 10 MHz. The high frequency response measurements reveal in these compounds the presence of a surface magnetic state with a Curie temperature higher than that of the bulk material. The surface Curie temperature, as well as the thickness of the surface region, depends on the level of selenium doping: the depth of the surface region increases with higher level of selenium doping and varies from about 0.5 to 25 microns. The cause for the existence of the surface magnetic state is believed to be a uniform selenium deficiency in a surface region.

INTRODUCTION

High frequency pyromagnetic measurements were made on single crystals of $CoS_{2-x}Se_x$ compounds. These single crystals were grown by the method of chlorine vapor transport[1] which resulted in typical crystals ranging from about 2 to 5 mm in linear dimensions. Static magnetic properties of $CoS_{2-x}Se_x$ single crystals have been discussed comprehensively in the literature.[2-5] The crystals have pyrite structure with each cation in the center of anion octahedron and each anion has a tetrahedral coordination of one anion and three cations. All of these compounds are metallic in nature, but their magnetic properties depend rather crucially on the level of selenium doping. CoS_2 is ferromagnetic with a saturation magnetic moment of $0.88\mu_B$ and a Curie temperature of about 120°K. In the compositional range $0 < x \leq 0.20$, the magnetic moment per cobalt ion does not change, but the Curie temperature decreases almost linearly with increasing x. Above x = 0.20, the moment decreases rapidly with increasing x and reaches zero for x = 0.25. For x > 0.25, the order changes its character from ferromagnetic to antiferromagnetic and for $0.30 \leq x \leq 2.0$, $CoS_{2-x}Se_x$ shows antiferromagnetic behavior with almost constant Neel temperature around 90°K.

In the course of a study of high frequency pyromagnetic effects in materials, pyromagnetic measurements were made on annealed single crystals of $CoS_{2-x}Se_x$ (for $0 < x \leq 0.20$) and some interesting and unexpected surface effects were observed. These effects are not likely to be detected by static magnetic measurements.

To illustrate how pyromagnetic measurements could be used to study the surface properties of magnetic materials, consider a magnetic sample inside a coil. Then, the magnetic flux density $B_z(\vec{r},t)$ along z-direction inside the coil is given by

$$B_z(\vec{r},t) = Ho + 4\pi\, m_z(\vec{r},t) \qquad (1)$$

where the externally applied d.c. magnetic field Ho, magnetization density $m_z(\vec{r},t)$ and the coil axis all point along the z-direction. Then pyromagnetic voltage is simply the voltage induced in the coil because of the thermally driven change in the magnetic moment of the sample. For a sample loosely coupled to the pick up coil, this voltage can be expressed as

$$V_p = \frac{-4\pi\lambda}{c} \int_v d^3r \frac{\partial m_z}{\partial t}(\vec{r},t) \qquad (2)$$

where λ is the coupling constant which depends upon the number of turns and cross-sectional area of the coil and the dimensions and geometrical shape of the coil. c is the velocity of light, and v is the volume of the sample. Upon re-expressing the time derivative of the magnetization density $m_z(\vec{r},t)$ because of the temperature fluctuation, Eq. (2) can be written as

$$V_p = \frac{-4\pi\lambda}{c} \left[\frac{\partial m_z}{\partial T}\bigg|_{T,Ho}\right] \int_v d^3r \frac{\partial T}{\partial t}(\vec{r},t) \qquad (3)$$

where only the first derivative in the Taylor series expansion of $m_z(T + \delta T(\vec{r},t), Ho)$ has been retained. This assumption is only true for small temperature fluctuations so that the saturation effects do not occur. Under these circumstances, the pyromagnetic coefficient $\frac{\partial m_z}{\partial T}\Big|_{T,Ho}$ depends only on the ambient temperature T and the applied magnetic field Ho.

The pyromagnetic response peaks near the Curie temperature Tc, because the slope $\frac{\partial m_z}{\partial T}\Big|_{T,Ho}$ maximizes. The integral in Eq. (3) can be evaluated using Fourier's law of heat conduction. It is known that if the temperature of a substance is modulated at its surface (z = 0), an exponentially decaying temperature wave propagates into the sample[6] whose amplitude is proportional to e^{-qz} where $q = (\omega/2D_T)^{1/2}$ and ω and D_T are the angular modulation frequency and the thermal diffusivity, respectively. At low frequencies, the thermal diffusion length 1/q is of the sample dimensions and the whole sample responds. But as ω increases, the thermal diffusion length starts decreasing and at high enough frequencies, only a very small fraction of the sample near the surface is responding pyromagnetically. This shows that the measurement of the frequency response of the pyromagnetic voltage can act as a thermomagnetic scanner for the sample.

EXPERIMENTAL ARRANGEMENT AND RESULTS

The beam intensity of a HeNe laser was amplitude modulated with the help of a mechanical chopper or an electro-optic modulator and was allowed to shine on the surface of a crystal. A small coil of roughly 50 turns surrounded the sample, and the pyromagnetic voltage induced in this coil was detected with the help of either an audio lock-in amplifier or a narrow band r.f. radio receiver, depending upon the frequen-

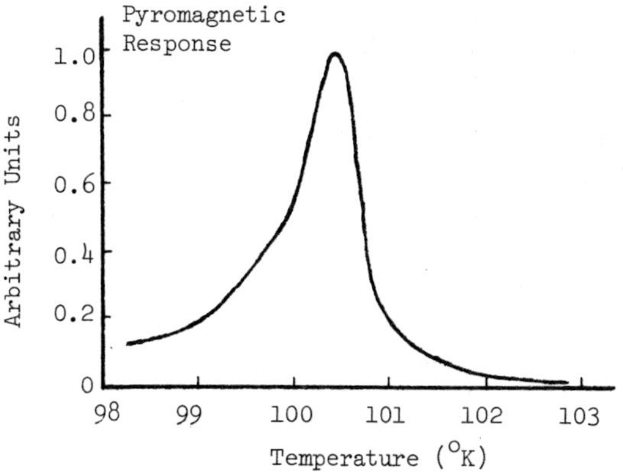

Figure 1. Temperature response behavior of pyromagnetic voltage of $CoS_{1.94}Se_{0.06}$ single crystal at 2 KHz in an applied magnetic field of 1 KOe.

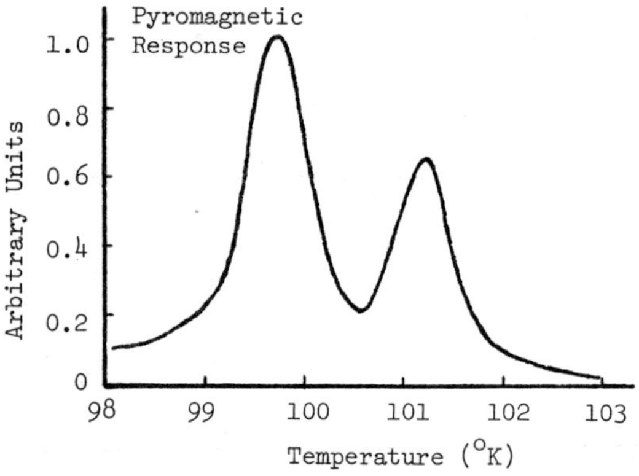

Figure 2. Temperature response behavior of pyromagnetic voltage of $CoS_{1.94}Se_{0.06}$ single crystal at 2.8 MHz in an applied magnetic field of 1 KOe.

cy. The Curie temperature of $CoS_{2-x}Se_x$ system is in the low temperature region (∿40°K to 120°K). The samples were cooled down to their respective Curie temperature using a cold finger in a closed cycle helium gas cryogenic system with a thermocouple to monitor the ambient temperature.

In the low frequency region, all of the selenium doped samples exhibited only one narrow pyromagnetic peak as the temperature was swept through their respective Curie temperatures as is shown in Figure 1. As the modulation frequency was increased, a second pyromagnetic peak began to appear at a temperature higher than that of the low frequency peak. For still higher frequencies, the second peak became more and more distinct (see Figure 2) and above a certain characteristic frequency, the higher temperature peak dominated the lower temperature peak. For what follows, the lower and the higher temperature peaks will be denoted by I and II respectively. Only the temperature difference between the two peaks was measured accurately and there was an indeterminacy of a degree Kelvin or so in measuring the absolute temperature for various temperature response curves at different frequencies which can be noticed in Figures 1 and 2, also. The characteristic frequency as well as the difference in temperature between the two peaks was dependent on the level of selenium doping as is shown in Table I. This type of frequency response behavior for selenium doped samples was in contrast to that of pure CoS_2 single crystals where only peak I was observed both at low and high frequencies.

Table I

x	Tc_1 °K	Tc_2 °K	$T = Tc_2 - Tc_1$ °K	Characteristic Frequency, KHz
0	118	-	-	-
0.06	100.3	101.8	1.3	8×10^3
0.14	69	75	6	25
0.17	62	85K	23	1

DISCUSSION

The appearance of peak II can be understood if it is assumed that the Curie temperature Tc_2 of a small volume fraction of selenium doped samples near the surface is higher than the bulk Curie temperature Tc, and further that the depth of the surface region depends upon the level of selenium doping. At low modulation frequencies, the thermal diffusion length is of the order of sample dimensions and pyromagnetic response from the bulk of the material is observed. The surface region will contribute negligibly, because the ratio of the volume of the surface region to the volume of the bulk is very small. At increased modulation frequencies, the thermal diffusion length becomes smaller and smaller and at a certain frequency (depending on the doping level x), the two volume fractions actively contributing to the pyromagnetic signal will become comparable and peak II corresponding to the transition temperature Tc_2 will begin to appear. For increased modulation frequencies, the thermal profile will tend to lie more and more in the surface region. Consequently, peak II will keep on growing whereas peak I will decrease in magnitude. (In the pyromagnetic process, the smaller net active region and smaller amplitude of modulation are compensated for by the increase due to increasing rate of modulation.)

To verify this model, one of the samples ($CoS_{1.81}Se_{0.19}$) was cut in the middle. In the frequency response measurements, the outside showed the presence of both the peaks, whereas only peak I was observed for the inside. In another test, one of the surfaces of $CoS_{1.94}Se_{0.06}$ sample was half polished with jeweler's rouge. The unpolished surface showed the presence of both the peaks, whereas only peak I was observed from the polished surface.

The thickness of the surface region was estimated from the knowledge of the modulation frequency at which the surface region became active and the thermal diffusivity D_T. Temperature wave method[7] was used to measure D_T and a numerical value of 5×10^{-2} cm^2/sec was obtained. Thickness of the surface region ranged between 0.5 to 25 microns depending upon the level of selenium doping. The depth was larger for higher doped samples. Since the two peaks were observed only in selenium doped samples, it is believed that the surface state is caused by the slight selenium deficiency near the surface. This conviction is further strengthened by the fact that when pyromagnetic measurements were made on single crystal $CoS_{1.81}Se_{0.19}$ annealed in selenium atmosphere, peak II did not appear. The fact that 2 sharp peaks occurred (rather than a broad single peak) indicated that there is a sharp boundary between the surface region and the bulk.

*Supported by Office of Naval Research, Contract No. N00014-67-A-0126-0014.

REFERENCES

1. R. J. Bouchard, J. Crystal Growth, 2, 40, 1968.
2. V. Johnson and A. Wold, J. Solid State Chem., 2, 40, 1970.
3. K. Adachi, K. Sato, and M. Takeda, J. Phys. Soc. (Japan), 26, 631, 1969.
4. J. B. Goodenough, J. Solid State Chem., 3, 26, 1971.
5. M. Hittori, K. Adachi, and H. Nakano, J. Phys. Soc. (Japan), 26, 642, 1969.
6. H. S. Carslaw and J. C. Jaeger, Conduction of Heat in Solids, 2nd Edition, Clarendon Press, Oxford, 1959, p. 65.
7. Y. Kagure and Y. Hiki, J. Appl. Phys. (Japan), 12, 814, 1972.

MAGNETIC HYPERFINE FIELD STRUCTURE OF IRON URUSHIBARA TYPE CATALYSTS

B. J. Evans[+]
University of Michigan
Ann Arbor, Michigan 48104

L. J. Swartzendruber
National Bureau of Standards
Gaithersburg, Maryland 20760

ABSTRACT

We have utilized the Mössbauer effect to study the hyperfine field structure of iron Urushibara catalysts. The Mössbauer spectra of those catalysts prepared using zinc show that they consist of a mixture of magnetic and non-magnetic Fe-Zn alloys. Both the magnetic field distribution in the magnetic phase and the relative amounts of magnetic and non-magnetic phases depend on the Fe-Zn ratio used in the preparation of the catalyst. This dependence on Fe-Zn ratio is in contrast to iron Urushibara catalysts prepared using aluminum (and Raney iron type catalysts) in which the active phase is almost pure α-Fe, irrespective of the Fe-Al ratio. The activity of zinc prepared iron Urushibara catalysts for certain hydrogenation reactions is known to be greater than that of aluminum prepared iron Urushibara catalysts and the above results suggest a relationship between activity and the modification of the iron catalyst by alloyed zinc. The alloying behavior of the Fe-Zn particles may be analogous to that of the so-called bimetallic clusters observed in other alloy systems.

INTRODUCTION

Relationships between the electronic and magnetic properties of solids and their activity as heterogeneous catalysts have been the subject of considerable discussion and controversy.[1] In particular, the fact that many transition metals and their alloys have high activity as hydrogenation catalysts is well known and yet the development of an unambiguous relationship between electronic properties and catalytic activity has been elusive. Most catalysts are prepared by techniques which give a material with high surface area. These preparation techniques also give a material with a complex structure which is difficult to determine due largely to the fact that the active material is in the form of small particles. One of the most difficult parameters to determine, and one which is of great importance in investigating relationships between electronic structure and catalytic properties, is the extent to which the metals used in preparing the catalysts are alloyed together. In this paper we utilize the ^{57}Fe Mössbauer effect to study the effect of preparation technique on the alloying and magnetic behavior of Urushibara type iron hydrogenation catalysts.

Urushibara Fe can be prepared[2] from either a ferrous chloride or ferric chloride solution. The resulting catalysts are denoted Urushibara-Fe(II) and Urushibara-Fe(III), respectively. Here we consider catalysts of the Urushibara-Fe(III) type prepared utilizing both Al and Zn and denoted Urushibara-Fe(III)-Al and Urushibara-Fe(III)-Zn, respectively. In the course of this study it has been discovered that those prepared using Zn form multi-phase Fe-Zn alloys in which the total Zn can be varied considerably. There appears to be very little alloying behavior for those prepared using aluminum. This is in contrast to previous work in which it was believed[2] that both were nearly pure α-Fe.

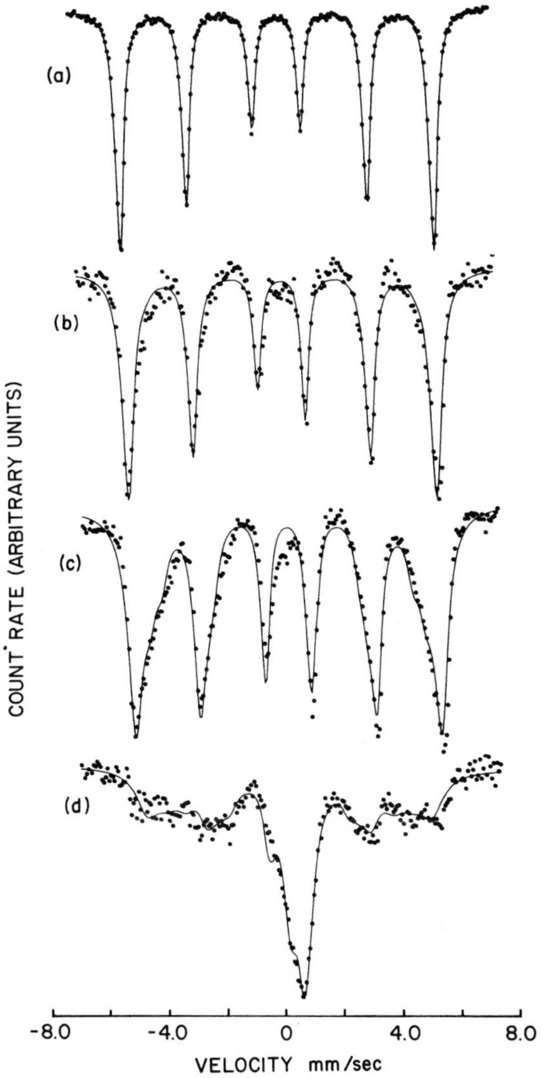

Fig. 1 Mössbauer spectra from catalysts prepared as described in the text. Circles are data points and solid line is the least squares fit. Zero velocity represents the center of a pure Fe spectrum. All spectra were taken at 298±2K. (a) Urushibara-Fe(III)-Al, (b) Urushibara-Fe(III)-Zn1, (c) Urushibara-Fe(III)-Zn2, (d) Urushibara-Fe(III)-Zn3.

EXPERIMENTAL

Samples were prepared according to the methods given in detail by Hata.[2] Hata[2] also gives a detailed discussion of chemical reactions, particle sizes, x-ray diffraction, etc. Urushibara-Fe(III)-Zn was prepared by adding iron chloride ($FeCl_3 \cdot 6HO$) to a mixture of 25g zinc dust in 8 ml water. The product was washed with water and the solid recovered by filtering. This material was then treated with acetic acid and the resulting catalyst collected on a sintered glass filter and washed with ethyl alcohol. A small amount of the catalyst was mounted in amyl acetate and Mössbauer spectra obtained at room temperature. Four samples are reported on in

Table I

Parameters obtained from least squares fitting the spectra of Fig. 1. Spectra were obtained and 298±2K and isomer shifts are relative to α-Fe at that temperature.

Parameter	SAMPLE U-Fe(III)- Al	Zn1	Zn2	Zn3
$\langle H_1\rangle$,kG	329	325	324	300
σ_{H_1},kG	-	12	16	39
H_1 rel. intensity	1.0	0.96	0.61	0.30
$\langle H_2\rangle$,kG	-	285	290	260
σ_{H_2},kG	-	12	24	40
H_2 rel. intensity	0	0.02	0.27	0.15
$\langle H_3\rangle$,kG	-	-	265	220
σ_{H_3},kG	-	-	35	40
H_3 rel. intensity	0	0	0.10	0.18
Line 1 Isomer Shift, mm/s	-	+0.82	+0.85	+0.60
Width, mm/s	-	0.30	0.28	0.57
relative intensity	-	0.02	0.02	0.17
Line 2 Isomer Shift	-	-	-	+0.12
Width	-	-	-	0.73
relative intensity	-	-	-	0.20

spectrum from Urushibara-Fe(III)-Al (Fig. 1a) was fitted to a single six-line magnetic hyperfine field pattern. The spectra from the Urushibara-Fe(III)-Zn samples (Figs. 1b, 1c, and 1d) were fitted to three separate six-line magnetic hyperfine field patterns with average field $\langle H\rangle$. Each of the three six-line patterns was constrained to have a 3:2:1:1:2:3 line intensity ratio and each was assumed to be broadened by a hyperfine field distribution with width σ_H. For Urushibara-Fe(III)-Zn1 and Urushibara-Fe(III)-Zn2 (Figs. 1b and 1c), one additional line was assumed responsible for the increased intensity evident in detail here: for Urushibara-Fe(III)-Zn1, 20 g of iron chloride were added to the zinc dust-water mixture; for Urushibara-Fe(III)-Zn2, 7.5 g were added; and for Urushibara-Fe(III)-Zn3, 2.5 g were added; a sample of Urushibara-Fe(III)-Al was prepared utilizing 10 g of iron chloride and 20 g of Al powder.

RESULTS

Mössbauer spectra from the four samples are shown in Fig. 1. The spectra were least squares fitted using a program described previously.[3] The one of the inner two lines of the magnetic pattern and was included in the least squares fit. For Urushibara-Fe(III)-Zn3 (Fig. 1d) two additional central lines were included in the fit. As can be seen from the Figure, this fitting procedure gave reasonable fits. These fits are considered adequate to reveal the main features of the spectra. Numerical results are displayed in Table I.

DISCUSSION

The spectrum for Urushibara-Fe(III)-Al (Fig. 1a) is, within our experimental error, indistinguishable from that obtained for pure α-Fe. For Urushibara-Fe(III)-Zn (Figs. 1b, 1c, and 1d) a two phase material is indicated, one phase being magnetic and the other non-magnetic. The possibility of incipient superparamagnetism was ruled out by applying a 1 kG field. This produced no detectable change in the observed spectra. The presence of iron oxides, hydroxides, etc. can be ruled out on the basis of known hyperfine fields and isomer shifts. For the magnetic phase, both the hyperfine field value and the broadened lines indicate an alloy which is a primary solid solution of Zn in α-Fe. Currently, no information concerning the Mössbauer spectra of iron-rich Fe-Zn alloys is available. However, in analogy with disordered Fe-Al, Fe-V, etc., one expects the hyperfine field, H_n, for an Fe atom surrounded by n nearest-neighbor Zn atoms to be given approximately by

$$H_n \approx H_o - n\Delta H. \qquad (1)$$

The effect of second, third, etc., nearest neighbor Zn atoms is, in first approximation, a reduction in hyperfine field value and a broadening of each line.

Using these simple considerations and assuming randomness, one can make a rough estimate of the concentration of Zn in the magnetic phase. These estimates are: 1/2% for Urushibara-Fe(III)-Zn1, 6% for Urushibara-Fe(III)-Zn2; and 8% for Urushibara-Fe(III)-Zn3. The accuracy of these results depend on the validity of the model used (especially the assumption of randomness) and, considering the rather large value of 40 kG found for ΔH, probably represent lower limits. The six and eight percent values are considerably in excess of the equilbrium solid solubility at room temperature (∿4%).

Both the considerable alloying of Fe and Zn and the absence of alloying of Fe and Al in these catalysts are contrary to what one might expect if only the bulk phase diagram were considered. It is possible that the thermodynamic stability of the small metallic particles is such than an Fe-Zn alloy is more stable than separate Fe and Zn particles, as has been demonstrated theoretically and experimentally in a number of other alloy systems.[4] For example, systems such as Cu-Os, which are almost immiscible in the bulk, have been shown to exhibit a much enhanced miscibility when in the form of extremely small particles, forming so-called[5] "bimetallic clusters." The enchanced solid solution of Zn in Fe observed in the present study may be due to a similar phenomenon.

The possibility of high local temperatures in the region of the embryonic Fe particles might also be considered as a source of the high degree of Fe-Zn alloying. However, magnetic iron oxides would also be expected to form under these conditions and we have no evidence for a magnetic Fe oxide being present. In addition, such local heating would also be present in the preparation of Urushibara-Fe(III)-Al, and Fe-Al alloys would also be expected to form, but there is no evidence for these alloys. The alloying that takes place thus appears due to chemical driving forces during the catalyst preparation.

The catalytic activities of Urushibara iron prepared using Zn or Al are known to be considerably different. For example, Urushibara-Fe(III)-Zn is active for the partial hydrogenation of 2-butyne-1,4-diol, whereas Urushibara-Fe(III)-Al is entirely inactive.[6] The Mössbauer results presented here suggest that this promotion effect of Zn could be due to the formation of Fe-Zn alloys. If this is the case, then there should be an optimum value of Zn content which maximizes the reaction rate.

REFERENCES

+ Partial support by the Petroleum Research Fund is acknowledged.

1. R.J.H. Voorheeve, AIP Conf. Proc. 18, 19 (1974).
2. K. Hata, "New Hydrogenating Catalysts," University of Tokyo Press, 1971.
3. W. Wilson and L.J. Swartzendruber, Computer Phys. Commun. 7, 151 (1974).
4. D.F. Ollis, J. Catalysis 23, 131 (1971).
5. J.H. Sinfelt, J. Catalysis 29, 308 (1973).
6. S. Taira, Bull. Chem. Soc. Japan 34, 1072 (1961).

SIZE EFFECT ON THE SPIN WAVE RESONANCE SPECTRA OF HEISENBERG ANTIFERROMAGNETIC FILMS*

S. T. Chiu-Tsao
S.U.N.Y. at Stony Brook,
Stony Brook, N.Y. 11790
and
New York University,†
New York, New York 10003

The spin wave resonance absorption spectra in simple cubic Heisenberg antiferromagnetic films of different thicknesses are calculated, the surfaces being (111) planes. By using spin wave theory, we obtain one surface spin wave resonance absorption line and a series of bulk spin wave resonance absorption lines. We find that the intensity of the surface spin wave line relative to those of the bulk increases as the film becomes thinner. For very thin films, the spectrum is quite sensitive to whether the number of layers is even or odd. For thick films, if the anisotropy field is much smaller than the exchange field, the surface spin wave frequency is lower than that of the lowest bulk spin wave frequency by approximately a factor of $\sqrt{2}$. The bulk spin wave frequencies obey the n^2 law and the intensities vary as $1/n^2$; this agrees with the work of Orbach and Pincus.[1]

*Research supported in part by the National Science Foundation under Grants No. GH32422 and 31695A.
†Present address.
[1]R. Orbach and P. Pincus, Phys. Rev. 113, 1213 (1959).

DIFFRACTION OF A THERMAL ATOMIC BEAM FROM A MAGNETIC SURFACE*

E. D. Thompson† and G. Felcher
Argonne National Laboratory, Argonne, Illinois 60439, U.S.A.

ABSTRACT

A thermal energy atomic beam is diffracted from a crystalline surface as from a corrugated hard wall. For an atomic hydrogen beam and a crystalline surface containing magnetic ions, the magnitude of the corrugations is spin dependent. The magnetic part depends on the relative size of the magnetic shell to the atomic radius. We calculate the scattering of a thermal energy atomic hydrogen beam from the (001) surface of MnO and find that the antiferromagnetic diffracted intensity is comparable to the specular intensity. Magnetic effects in the scattering from Eu^{++}-containing compounds are negligible. It is concluded that atomic beam scattering is potentially an important new tool for studying the magnetic state of the surface of insulating crystals.

INTRODUCTION

Thermal energy atomic beam scattering (TEAM[1]) shows great potential in determining the electronic state of atoms on the surface of crystalline materials. Recent developments[2] in the technology of production and detection of nearly monochromatic beams, as well as in the preparation of surfaces, promise large advances in this area in the coming years.

The atomic beam - solid surface interaction[3] is mainly short range and repulsive, and it arises from wavefunction overlap between electrons on the beam particle and electrons in the solid. The solid can be regarded as a hard wall to the thermal atomic beam with a wall roughness with characteristic wavelengths equal to the atomic spacing and to the deBroglie wavelength of the beam. Roughness of the crystalline surface gives rise to surface Bragg reflections in addition to the specularly-reflected peak. This model has been applied successfully in explaining the elastic scattering of He from LiF[4]. A somewhat different model, containing more fitting parameters, has also been employed in such calculations[5], including inelastic effects.

The scattering of a beam of particles possessing a magnetic moment, such as atomic hydrogen, is sensitive to the magnetic as well as the electronic state of the surface. The purpose of this paper is to present preliminary estimates of the magnetic scattering of a hydrogen atomic beam from surfaces containing magnetic ions such as the spherically symmetric Mn^{++} and Eu^{++} ions. It is concluded that large magnetic contributions exist in the scattering from solid surfaces containing 3d transition metal ions but that such contributions are negligible for the 4f rare earth ions.

INTERACTION OPERATOR

In the Born-Oppenheimer approximation, the interaction operator between an insulating crystal and an hydrogen atom is given by the form[6]

$$V(\vec{r}) = \sum_i [V_{di}(\vec{r}-\vec{R}_i) + V_{oic}(\vec{r}-\vec{R}_i) + V_{oie}(\vec{r}-\vec{R}_i)\ \vec{s}\cdot\vec{S}_i] \qquad (1)$$

where the sum is over the ions in the crystal, R_i and r are the positions of the ith ion and atom respectively, s and S_i the spin operators for the hydrogen atom and ith ion respectively, V_d is the van der Waal's attractive potential, and V_o is the strong repulsive potential due to wavefunction overlap. The classical magnetic dipole and electric polarization potential are neglected. Eq. (1) can be approximated by

$$V(\vec{r}) = \sum_i [-C_i/|\vec{r}-\vec{R}_i|^6 + B_{oic}e^{-\alpha_{ic}|\vec{r}-\vec{R}_i|} + B_{oie}e^{-\alpha_{ie}|\vec{r}-\vec{R}_i|}\ \vec{s}\cdot\vec{S}_i] \qquad (2)$$

where C_i is the van der Waal's constant and B and α are empirical parameters of the Buckingham potential.[3]

The van der Waal's constant and the α parameters are obtained by scaling from the known parameters for hydrogen and for the inert gases. This scaling is performed using the relative radii of theoretical atomic functions[7]. Heitler London calculations of the repulsive potential show that the largest contributions to B come from the overlap integral[8,9], and that B is proportional to the number of overlap integrals weighted by the magnitudes of the wavefunctions along the axis of the molecule. From this empirical observation, B is obtained from the known values of B for the inert gases. The obtained parameters are given in Table I. The calculated exchange interaction for H-Mn^{++} is in reasonable accord with Gardner and Karo's[10] preliminary molecular calculations.

TABLE I
PAIR INTERACTION POTENTIALS

Pair	C(°K-Å^6)	B(°K)	α(Å^{-1})
H-Mn^{++}	1.9×10^4		
Coulomb		2.2×10^7	4.5
Exchange		0.36×10^7	4.5
H-Eu^{++}	1.7×10^5		
Coulomb		8.2×10^6	3.46
Exchange		5.5×10^6	8.2

In calculating the van der Waal's contribution to the potential V, we have replaced the sum in (1) by an integral over the crystal which is assumed to occupy an infinite half space. This yields a potential contribution to V which depends only upon the distance from the crystalline surface (assumed to be the z co-ordinate as measured from the plane of surface atoms) and which gives no direct contribution to the non-specular scattering. On the other hand, this attractive contribution allows the hydrogen atom to penetrate closer to the crystalline surface, increasing the effective surface roughness.

When substituted into Eq. (1), the only repulsive terms that contribute to the sum when the atom is outside the crystalline surface are those terms from the ions on the surface and then only those terms nearest the impact point. To carry out a specific calculation, we consider the (001) surface of MnO below the Neél temperature. The surface is then a square lattice with ferromagnetic [01] rows of Mn^{++}, the rows being coupled antiferromagnetically (the [01] surface direction is the [$\bar{1}$10] crystal direction). Assuming a single domain with the spins directed in the [$\bar{1}$10] direction, we have calculated the potential V of Eq. (1) as a function of distance from the surface and position along the surface for an H beam polarized along the [01] direction. Clearly no antiferromagnetic lines are observable for the beam polarized in the [10] direction.

SCATTERING

The desired solution of the scattering equation has the asymptotic solution

$$\psi = \exp(i\vec{k}_i\cdot\vec{r}) + \Sigma_{\vec{G}} f_{\vec{G}} \exp(i(\vec{k}_r+\vec{G})\cdot\vec{r}) \qquad (3)$$

at large distances from the surface where the first term is the incident beam and the second term is summed over all surface reciprocal lattice vectors $\vec{G}$. The components of $\vec{k}_i$ and $\vec{k}_r$ parallel to the surface are equal and energy conservation requires $|\vec{k}_i| = |\vec{k}_r + \vec{G}|$. The specular scattered beam is the term with $\vec{G} = 0$, and the relative intensity of a particular surface Bragg peak is $|f_{\vec{G}}|^2$. For a finite beam size, the intensity must be corrected for the projected beam area.

We have solved the scattering equation approximately by extending the approximation procedure of Levi et al[4]. In this approximation, the repulsive potential gives hard wall scattering. The hard wall's position is a function of the surface co-ordinates (x,y) with a Fourier series representation. Levi kept the first term in the Fourier series and used the coefficient as an adjustable parameter in fitting experimental data.

In the case of an antiferromagnetic surface, we must keep at least two independent coefficients in the Fourier series expansion. Explicitly, the hard wall's position z is

$$z = \bar{z} + z_c[\cos\frac{2\pi x}{a} + \cos\frac{2\pi y}{a}] + z_m\cos\frac{\pi x}{a} \qquad (4)$$

where a is the square lattice spacing and z_c and z_m are the Fourier coefficients. To obtain z(x,y), we define z as the classical turning point, $V(z(x,y)) = E_\perp$ where $E_\perp$ is the normal component of the beam energy. For a normal beam energy of 300°K, the results yield $z_c = 0.2$ Å and $z_m = 0.067$ Å. The value of z_c compares favorably with that of Levi[4].

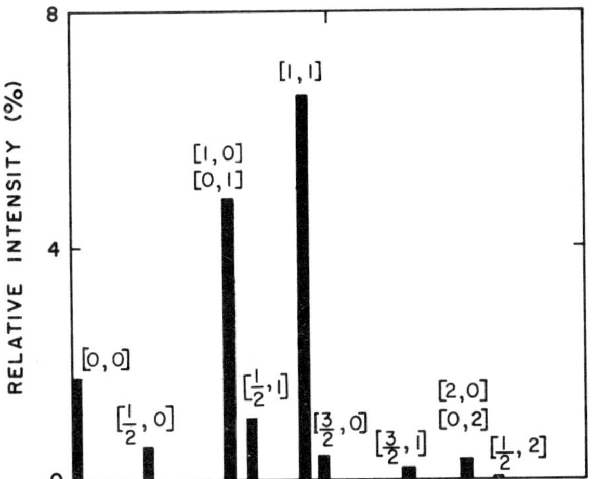

Fig. 1. The relative scattering intensity of a normally incident, 300°K, atomic hydrogen beam from the (001) surface of antiferromagnetic MnO. The scattered angle is measured relative to the surface normal and the surface indexing [1,0] is parallel to the crystalline [110] direction. A single antiferromagnetic domain with spins directed in the [$\bar{1}$10] direction and with the beam polarized in the same direction has been assumed. Half integer labelled peaks arise solely from the magnetic part of the interaction and the [0,0] peak is the specular scattering.

Using (4) and neglecting multiple scattering, the Bragg intensity is given by the relation

$$|f_{\vec{G}}|^2 = J_\ell^2\,(K_z z_c) \times |\sum_{m=-\infty}^{\infty} J_m(K_z z_c)\, J_{2s-2m}\,(K_z z_m) e^{-im\pi/2}|^2 \qquad (5)$$

where J is the Bessel function, ℓ and s are the indices for the Bragg peak, and K_z is the change of the z-component of the wave vector in the scattering event. ℓ is integer and s is half integer or integer corresponding respectively to the magnetic and non-magnetic Bragg scattering. Due to the smallness of the argument $K_z z_m$, only one or two terms contribute to the sum in (5). The calculated scattering is shown in Fig. 1.

DISCUSSION

The results in Fig. 1 are clear --- magnetic effects in the scattering of atomic hydrogen from MnO are large and comparable to the specular scattering. This is, of course, under the assumption that the surface of MnO is uncontaminated by non-magnetic adsorbed layers and that the surface is antiferromagnetic as in the bulk of the crystal. The calculation also assumes implicitly that competitive inelastic scattering processes due to phonons are negligible.[5] Experimental results for He scattered from LiF indicate increased inelastic scattering with surface temperature. Antiferromagnetic Bragg scattering, with peaks two orders of magnitude smaller than the specular peak, have been observed in LEED experiments on NiO[11]. Due to the penetration of the electron beam past the surface atomic layer into the interior of the solid, these results cannot be interpreted directly as properties of the atomic surface layer.

The calculations show that the atomic scattering is sensitive to the outermost part of the atomic wavefunctions. This means that TEAM serves as a potential probe to the electronic and magnetic state of the atom which is reflected in the outer part of the atom. Such is the case for Mn^{++} considered here and would also be the case for non-spherically symmetric 3d transition metal ions for which the crystal field symmetry at the surface differs from that of the bulk. On the other hand, the magnetic interaction H-Eu^{++} is quite small since the distance of closest approach of the beam atoms is dictated by the tails of the 5s-5p shells rather than that of the 4f (cf. Table I). Atomic scattering therefore is not sensitive to the electronic and magnetic structure in the interior of the atom.

Uncertainties in the calculation, uncertainties which we plan to eliminate in ongoing calculations, are small relative to the magnitude of the calculated result and to the experimentally observed ratio of the elastic Bragg to the background scattering of inert gas atoms from insulating ionic compounds. Let us end with a simple physical argument that the calculated surface roughness due to magnetic effects is reasonable. The distance of closest approach between the atom and ion is somewhat smaller than typical atom-atom spacings in the solid. The magnetic interaction strength should be comparable at this point to that found in magnetically active crystals -- of the order of 100°K. At this point, the repulsive potential is increasing rapidly and, whether we use a Lennard-Jones, Morse, or Buckingham potential, we find the rate of increase is 1000-2000°K/Å. The classical turning point variation with magnetic effects in order of magnitude wise as calculated.

REFERENCES

*Work performed under the auspices of the U.S. Atomic Energy Commission.

†Permanent address: Case Western Reserve University, Cleveland, Ohio, 44106.

1. J. N. Smith, Jr., Sur. Sci. 34, 613 (1973).
2. G. Boato and P. Cantini, lecture notes, Varenna summer school 1973.
3. D. M. Young and A. D. Crowell, Physical Adsorption of Gases, Butterworths, London, 1962, chap. 2.
4. G. Boato, P. Cantini, U. Garibaldi, A. C. Levi, L. Mattera, R. Spadacini, G. E. Tommei, J. Phys. C. 6, L394 (1973).
5. F. O. Goodman and W. K. Tan, J. Chem. Phys. 59, 1805 (1973).
6. H. Margenau and N. R. Kestner, Theory of Intermolecular Forces, Pergamon Press, New York, 1969.
7. J. T. Waber and D. T. Cromer, J. Chem. Phys. 42, 4116 (1965).
8. W. E. Bleick and J. E. Mayer, J. Chem. Phys. 2, 252 (1934).
9. M. Kunimune, J. Chem. Phys. 18, 754 (1950).
10. M. A. Gardner and A. M. Karo, unpublished. The authors thank Drs. Gardner and Karo for making their calculations available prior to publication.
11. P. W. Palmberg, R. E. DeWames, L. A. Vredevoe, Phys. Rev. Lett. 21, 682 (1968).

THE SURFACE OF A FERROMAGNETIC ELECTRON GAS

R.L. Kautz and Brian B. Schwartz
Francis Bitter National Magnet Laboratory,* Massachusetts Institute of Technology
Cambridge, Massachusetts 02139

ABSTRACT

The ferromagnetic electron gas has often served as a simple model for itinerant ferromagnetism. Linear-response functions for a ferromagnetic electron gas were first derived by Izuyama, Kim, and Kubo,(1) assuming that both the direct and exchange interactions are of delta-function form. Recently, Kim, Schwartz, and Praddaude(2) have shown that it is necessary to retain a Coulomb form for the direct interaction in order to assure charge conservation. This modification proves to be essential to the treatment of surfaces since an infinite surface energy results if charge neutrality is violated. In this paper we present an approximate solution for the surface of a ferromagnetic electron gas based on the linear-response functions of Kim et al. and a ferromagnetic extension of the Hohenberg-Kohn(3) density-functional formalism.

MODEL

The model we consider is a semi-infinite electron gas for which the neutralizing background charge is assumed to end abruptly at a plane surface. The electrons are treated in the Hartree-Fock approximation using the Coulomb (e^2/r) direct interaction and a delta-function exchange interaction ($\tilde{V}\delta(r)$). The Coulomb direct interaction assures charge neutrality while the delta-function exchange leads to a magnetic ground state described by the Stoner model.

The energy per unit volume in the bulk as a function of the bulk density n_o and relative magnetization $m_o = (n_{o+} - n_{o-})/n_o$ (where n_{o+} and n_{o-} are the majority and minority spin densities) is given by

$$\epsilon = \frac{3^{5/3}\pi^{4/3}}{20}\frac{\hbar^2}{m}\left[(1+m_o)^{5/3} + (1-m_o)^{5/3}\right]n_o^{5/3} - \frac{1}{8}\tilde{V}\left[(1+m_o)^2 + (1-m_o)^2\right]n_o^2. \quad (1)$$

The two terms in Eq. 1 represent the kinetic and exchange energies respectively. When m_o is chosen to minimize the energy, it is found that the bulk is partially magnetized ($0 < m_o < 1$) for $260 < n_o\tilde{V}^3 < 438$ (in atomic units) and fully magnetized ($m_o = 1$) for $n_o\tilde{V}^3 \geq 438$. Given this behavior for the bulk, we expect the surface to show a tendency toward paramagnetism since the density of electrons decreases at the surface.

METHOD

Our surface calculation makes use of the density functional formalism as extended to the ferromagnetic case by Rajagopal and Callaway.(4) This formalism is based on a theorem which states that the ground state energy of an electron gas is, for a given external charge density $\rho(r)$ and magnetic field $\vec{H}(r)$, a functional of the electron density $n(r)$ and magnetization $\vec{m}(r)$ and that this functional is minimized by the correct charge density and magnetization. The energy functional takes the form

$$E = \frac{1}{2}\int dr dr' \frac{(\rho(r) + en(r))(\rho(r') + en(r'))}{r - r'} - \int dr\, \vec{m}(r)\cdot\vec{H}(r) + G[n(r), \vec{m}(r)]$$

where G is a universal functional which does not depend on the external charge density or field. If G is known then the electron density and magnetization can be determined variationally from the stationarity of E.

For the present purpose we assume that G is a functional only of $n(r)$ and the magnitude of $\vec{m}(r)$, or equivalently of the majority and minority spin densities $n_+(r)$ and $n_-(r)$. In making this simplification we are assuming that the exchange energy dominates magnetic field energies and that the direction of magnetization relative to the surface is unimportant. Defining an energy density functional g such that

$$G[n_+, n_-] = \int dr\, g[n_+, n_-],$$

we proceed, following Hohenberg and Kohn, by expanding g in terms of the gradients of n_+ and n_-. Symmetry arguments can be used to reduce such an expansion to one of the form

$$g[n_+, n_-] = g_o(n_+, n_-) + g_2^+(n_+, n_-)|\vec{\nabla} n_+|^2 + g_2^-(n_+, n_-)|\vec{\nabla} n_-|^2 + g_2^{+-}(n_+, n_-)\vec{\nabla} n_+\cdot\vec{\nabla} n_- \quad (2)$$

where the coefficients g_o and g_2 are functions of n_+ and n_- which have been derived using the linear response functions of Kim et al.

SOLUTION

Following Smith,[5] who considered a paramagnetic surface, the above gradient expansion is now used to write an expression for the surface energy. Taking the surface as the $x=0$ plane and a positive background charge of the form $\rho(r) = -en_o\theta(-x)$, we obtain

$$S = \frac{e}{2}\int_{-\infty}^{\infty} dx\,\phi(x)\,(n(x) - n_o\,\theta(-x)) + \frac{3^{5/3}\pi^{4/3}}{2^{1/3}5}\frac{\hbar^2}{m}\int_{-\infty}^{\infty}\sum_{\sigma}(n_\sigma^{5/3}(x) - n_{o\sigma}^{5/3}\,\theta(-x))\,dx - \frac{\tilde{V}}{2}\int_{-\infty}^{\infty} dx\sum_{\sigma}(n_\sigma^2(x) - n_{o\sigma}^2\,\theta(-x)) + \frac{1}{72}\frac{\hbar^2}{m}\int_{-\infty}^{\infty} dx\sum_{\sigma}\frac{1}{n_\sigma}\left(\frac{dn_\sigma}{dx}\right)^2 \quad (3)$$

Here ϕ is the electrostatic potential and σ denotes spin direction. The four terms in Eq. 3 represent electrostatic, kinetic, exchange, and inhomogeneity energies. For the paramagnetic gas, Smith considered an electron density of the form

$$n = \begin{cases} n_o(1 - \frac{1}{2}e^{\beta x}) & x < 0 \\ n_o\frac{1}{2}e^{-\beta x} & x < 0 \end{cases}$$

and adjusted the parameter β to minimize the surface energy. By analogy one might parameterize the spin densities for the ferromagnetic gas by

$$n_\sigma = \begin{cases} n_{o\sigma}(1 - \frac{1}{2}e^{\beta x}) & x < 0 \\ n_{o\sigma}\frac{1}{2}e^{-\beta x} & x > 0 \end{cases} \quad (4)$$

This form was used by Pant and Rajagopal,(6) and it assumes a constant relative magnetization. In the present calculation, the following less restrictive form was used in which the half value points of the spin densities can be displaced from the surface.

$$n_\sigma = \begin{cases} n_{o\sigma}(1 - \frac{1}{2}e^{\beta_\sigma(x - x_\sigma)}) & x < x_\sigma \\ n_{o\sigma}\frac{1}{2}e^{-\beta_\sigma(x - x_\sigma)} & x > x_\sigma \end{cases} \quad (5)$$

The parameters β_+, β_-, x_+, and x_- were adjusted to obtain the minimum surface energy.

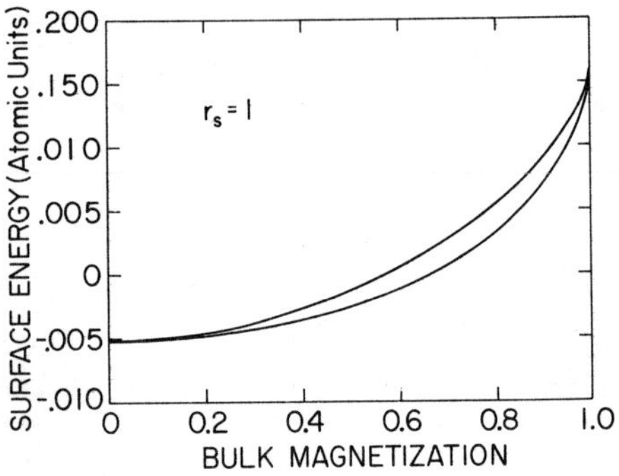

Fig. 1. Surface energy as a function of bulk magnetization m_0 for the parameterizations described by Eq. 4 (upper curve) and Eq. 5 (lower curve) for $r_s = 1$.

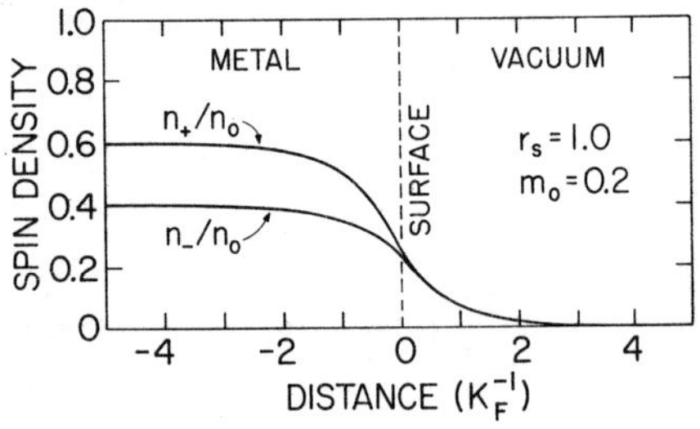

Fig. 2. Density of spin + and - electrons as a function of distance (measured in reciprocal Fermi wavenumbers) for $r_s = 1.0$ and $m_0 = 0.2$.

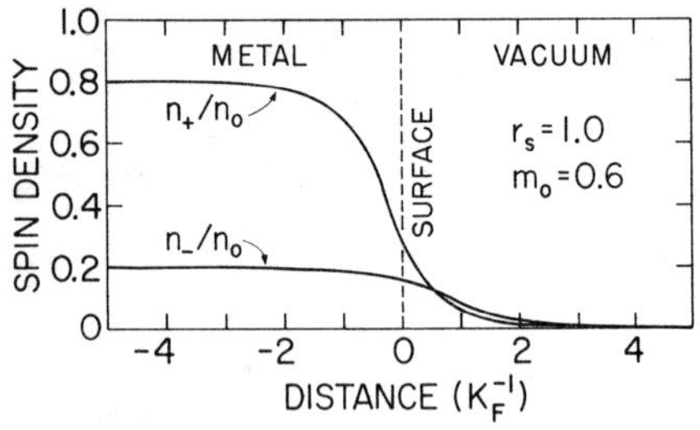

Fig. 3. Density of spin + and - electrons as a function of distance for $r_s = 1.0$ and $m_0 = 0.6$.

In Fig. 1 we compare the surface energy resulting from the parameterization (5) with that of (4) as a function of the bulk magnetization m_0 for a gas with a bulk density defined by $r_s = 1$. The two parameterizations are equivalent at $m_0 = 0$ and $m_0 = 1$ but for a partially magnetized bulk the surface energy of (5) falls below that of (4). The difference in surface energy between these two parameterizations is certainly significant, being comparable to the surface energies of simple metals.

Typical results for the spin densities as a function of position are shown in Figs. 2 and 3 for two values of the bulk magnetization. This tendency for the spin (+) and (-) densities to coalesce in the vacuum region, especially evident in Fig. 2, suggests that the surface energy is minimized by a paramagnetic surface. As has been stated, this tendency toward surface paramagnetism reflects the fact that the homogeneous gas is paramagnetic at low densities.

REFERENCES

1. T. Izuyama, D.J. Kim, and R. Kubo, J. Phys. Japan, 18, 1025 (1963).

2. D.J. Kim, B.B. Schwartz, and H.C. Praddaude, Phys. Rev. B, 7, 205 (1973).

3. P. Hohenberg and W. Kohn, Phys. Rev., 136, B864 (1964).

4. A.K. Rajagopal and J. Callaway, Phys. Rev. B 7, 1912 (1973).

5. J.R. Smith, Phys. Rev., 181, 522 (1969).

6. M.M. Pant and A.K. Rajagopal, Solid State Comm., 10, 1157 (1972).

* Supported by the National Science Foundation.

FIELD-EMISSION-ENERGY-DISTRIBUTION (FEED) CURVES OF Ni(100) BELOW AND ABOVE THE CURIE POINT

M. Campagna and T. Utsumi
Bell Laboratories, Murray Hill, New Jersey 07974

FEED-curves of Ni(100) have been measured at T = 78°K ($T/T_c \cong 0.12$, T_c = 631°K = Curie point of Ni) and 950°K ($T/T_c \cong 1.50$) with an energy resolution of about 50 meV. The tips were etched from a 99.999 pure Ni-rod and cleaned in vacuum by field evaporation using He as imaging gas. The results put an experimental upper bound of 10^{-1} to 10^{-2} to the transmission coefficient r of an electron of 3d-character to that of a free electron of the same energy. r is furthermore field independent. Since the FE current from Ni(100) is dominated by contributions from the free electron 4s-p band no significant change in the FEED is detected in going from the ferro- to the paramagnetic state.

INTRODUCTION

The trends in the theory of ferromagnetism of 3d-metals in the last decades followed two distinct pathes, characterized by the opposite limits of $\Delta/U_{eff} \gg\ll 1$, where Δ is the 3d-bandwidth and U_{eff} the effective Coulomb correlation energy of 3d-states. In the strong correlation limit little progress has been made despite major theoretical efforts.[1] Fundamental properties of d-states, like degeneracy, cannot be taken into account in realistic calculations. In contrast in the weak correlation limit well defined predictions of the behaviour of measurable quantities like susceptibility, specific heat or optical properties can be made. In this case, i.e. in the Stoner-Wohlfarth-Slater (SWS) band theory of ferromagnetism, experimental results should be able to provide tests for the validity of the model. The present status has been recently reviewed by Siegmann[2]. The most significant experiments involving spin-dependent superconducting tunneling[3], spinpolarization of photoelectrons[4], temperature dependence of the photoemission energy distribution curves[5,6] cannot be reconciled with the predictions of the band model. Spin polarized field emission experiments from Ni tips show. on the other hand, very intriguing results[7].

In Fig. 1 we show the accepted essential features of the band structure of ferromagnetic Ni ($T \ll T_c$). In the paramagnetic state the exchange splitting goes to zero [8,9].

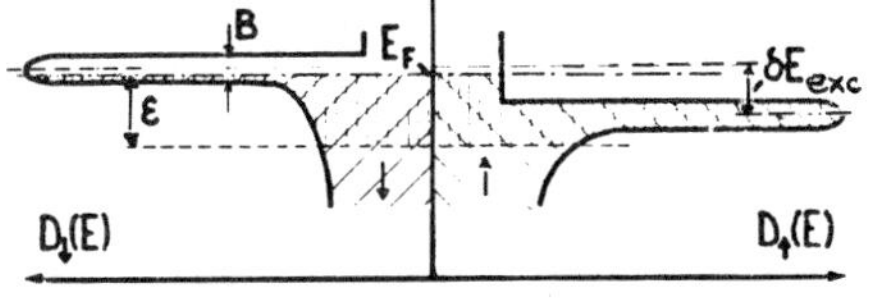

Figure 1. Band model of ferromagnetism. Essential features of the density of states near the Fermi level E_F, and the splitting into two spin subbands for a strong ferromagnet (Ni or Co). ε is the energy depth of interest for the FE-experiment (ε ≅ 600 meV), δE_{exc} the exchange splitting, B the width of the leading peak of d-character in the density of states (B ≥ 200 meV).

It has been recently recognized, mostly in research at the National Bureau of Standards [10] that field emission (FE) is not only capable of yielding information on the electronic properties of adsorbates on metal surfaces but can also be used to investigate the electronic structure of atomically clean surfaces. Penn and Plummer [10] concluded in their analysis of the applicability of F.E. to this specific question that in general the field emission current per unit energy, at energy ω, j(ω) will contain some weak structure due to the electronic properties of the substrate. Because of the tunneling probability $D^2(E_\perp)$, j(ω) measures the density of states at the classical turning point, restricted to those states with $k_\parallel \cong 0$($k_\parallel$ is the electron momentum parallel to the surface). The smaller the contribution of $k_\parallel$ is, the larger the influence of the one-dimensional density of states at the turning point (density of states perpendicular to the surface). The purpose of the present investigation is to measure FE energy distribution curves (FEED) of atomically clean Ni with the following objectives: (1) detect structure related to the presence of d-bands near E_F, (2) estimate experimentally the magnitude of the ratio of the transmission coefficient of an electron in a 3d state to that of a free electron (3) verify the detailed predictions of the calculations of Politzer and Cutler [11] regarding the FEED for the Ni(100) face (4) detect structural changes in the FEED in going through the Curie point and their relation to the disappearing of the Stoner splitting.

EXPERIMENTAL

The instrument used for this investigation is basically the same one used for recent experiments of field ion spectroscopy by Utsumi and Smith [12]. It consists of a field ion-electron microscope with a 0.9 mm probe hole on a 75 mm diameter fluorescent screen, a retarding lens, a double grid retarding energy analyzer and a channeltron. The overall energy resolution is estimated about 50 meV. The vacuum during the experiments is 1×10^{-9} Torr. Tip temperatures higher than the Curie point (631 °K) are reached by passing d.c. current through the Mo-

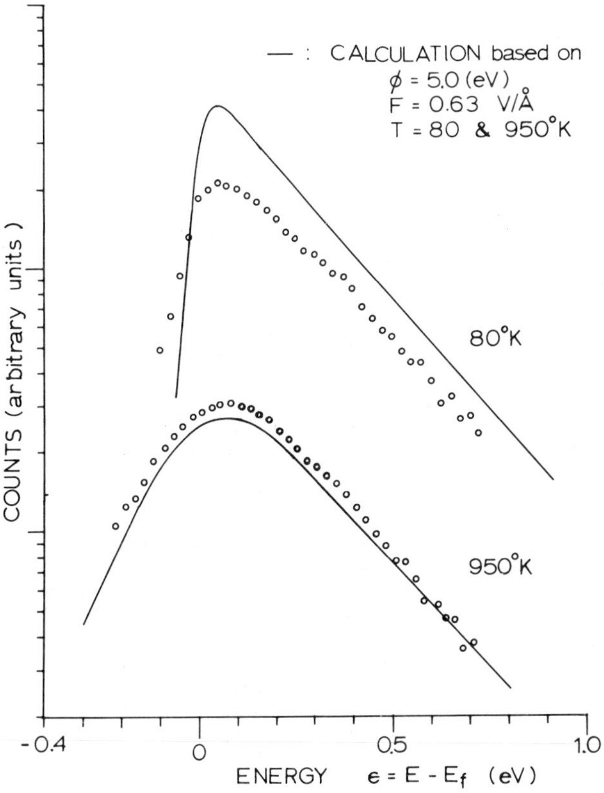

Figure 2. Field emission energy distribution of Ni(100)-crystal face at T = 80° ≅ 0.12 T_c and T = = 950° K ≅ 1.5 T_c compared with calculated Young's total energy distribution curves.

loop supporting the tip. The temperature is then measured by a pyrometer. The retarding potential is swept 10 cycle/sec and applied between the retarding grid and the Ni tip specimen itself. The specimen is etched from 10 mils oriented Ni-wire without preannealing. He- 1% H_2 FIM image of the tip exhibits (100) orientation along the wire axis. Tip diameters of 800-1000 Å were usually obtained.

RESULTS AND DISCUSSION

Measured FEED at 80 °K and 950 °K of the Ni(100) face are shown in Fig. 2, together with Young's[10] total energy distribution (TED) calculation based on a free electron model. Close similarity between the measurements and the calculated curves can be noted. No significant structure due to the presence of d-bands near E_F is detected at T = 80°K and no relevant structural change in the FEED is noted by crossing the Curie point. All this is rather striking, since as previously noted, a narrow 3d-peak in the DOS of Ni is present at a few 100 meV below E_F. Its intensity is estimated from calculations and photoemission experiments to be more than 10 times larger the one of a free electron conduction band[9]. At temperatures below T_c, by assuming a distance of the top of the majority spin peak to E_F of about 40-60 meV[8], one would expect some contribution from this peak to the FE current. This would be true even if the contributions from states with $k_{\parallel} \neq 0$ were negligable. In fact, from band structure calculations, we expect the d-states to show up also in the [001] direction, corresponding to the normal to the surface[9]. Since no significant structure is detected at T = 80°K, it is then not surprising that no major changes in the FEED are observed in going through T_c. This despite the fact that in the paramagnetic region the energy of both d-peaks of majority and minority spin electrons is equal.

Following Plummer and Penn[10] we tend to conclude that for the Ni(100)-crystallographic face the DOS at the classical turning point weighted by the barrier penetration probability with image-potential corrections is very near of being free electron like. This implies that the ratio r of the transmission coefficient of an electron of d-character at energy E near E_F to that of free electrons with the same energy is 10^{-1} to 10^{-2}. This experimental estimate and the overall results compare favorably with calculations of Politzer and Cutler[11]. These authors performed field emission calculations, taking into account band structure effects. They first evaluated r and found for Ni(100) at energies $E \cong E_F$: $6x10^{-3} \leq r \leq 1.5x10^{-1}$; the smaller limit is valid when the matching plane is located one lattice constant from the last layer of ion cores, the upper one when this distance is one-half the lattice spacing. They then calculated the FEED curve for the Ni(100)-face taking into account band structure effects. Their conclusion, strongly supported by the present data, is that the FE-current from the (100) face of Ni is largely dominated by the free-electron-like 4s,p-bands[13]. The motivation for such a detailed calculation were partially the spinpolarization (ESP) measurements on Ni[7]. The present data are consistent with the calculations of Politzer and Cutler, but resolve neither the controversy regarding the validity of the ESP data in field emission[2], nor provide tests of crucial features of the SWS theory for Ni. Experiments for verifying the predictions of Politzer and Cutler on the dependence of the transmission coefficient on the crystal planes and the ESP results are in progress.

REFERENCES

1. M. C. Gutzwiller, AIP Conf. Proc. 10, 1197, 1973.
2. H. C. Siegmann, to appear in Reports Progr. Phys.
3. P. M. Tedrow and R. Meservey, Phys. Rev. Letters 25, 1270, 1970.
4. H. Alder, M. Campagna and H. C. Siegmann, Phys. Rev. B8, 2075, 1973 and ref. cited therein.
5. D. T. Pierce and W. E. Spicer, Phys. Rev. B6, 1787, 1972.
6. J. E. Rowe and J. C. Tracy, Phys. Rev. Letters 27, 799, 1971.
7. W. Gleich, G. Regenfus and R. Sizman, Phys. Rev. Letters 27, 1066, 1971.
8. E. P. Wohlfarth, J. Appl. Phys. 41, 1205, 1970.
9. E. I. Zornberg, Phys. Rev. B1, 244, 1970; and Ref. cited therein.
10. For a review of the present status of field emission energy distributions studies see the excellent review: J. W. Gadzuk and E. W. Plummer, Rev. Mod. Phys. 45, 487, 1973; or D. R. Penn and E. W. Plummer, Phys. Rev. B9, 1216, 1974.
11. B. A. Politzer and P. H. Cutler, Phys. Rev. Letters 28, 1330, 1972; and Surf. Sci. 22, 277, 1970.
12. T. Utsumi and N. V. Smith, to be published in Phys. Rev. Letters.
13. The same conclusion has been reached by N. J. Dionne and T. N. Rhodin (Phys. Rev. Letters 32, 1311, 1974) in their extensive work on Pd and 5d-metals. Only in the case of Ir convincing evidence for the existence of the d-bands near E_F is present.

LARGE MAGNETIC EFFECT ON PHOTOTHRESHOLDS IN DOPED EuO

W. Eib, F. Meier, D.T. Pierce, P. Ruchti
Laboratory of Solid State Physics, Swiss Federal
Institute of Technology, CH-8049 Zürich, Switzerland

ABSTRACT

Generally, a magnetic phase transition has very little influence on photoemission. We have found that this does not apply to doped EuO, by measuring the photoelectric yield Y of pure and doped EuO single crystals up to 10 eV both above and below the ferromagnetic transition. For pure EuO the photothreshold changes only very little. In contrast, for EuO doped with ~ 1 % La, it shifts by as much as 0.5 eV on going through T_c. An interpretation of the results is attempted in terms of a change of the vacuum level of doped crystals at the magnetic phase transition.

The electronic level structure of EuO has been clarified by a number of workers and is now well established[1]. Also the influence of trivalent dopants as Gd and La on the magnetic ordering temperature T_c, optical and transport properties has been investigated[2]. In this note we report on a very pronounced effect, which seems to have escaped attention until now. It concerns the change of the quantum yield Y of La doped EuO when going through T_c.

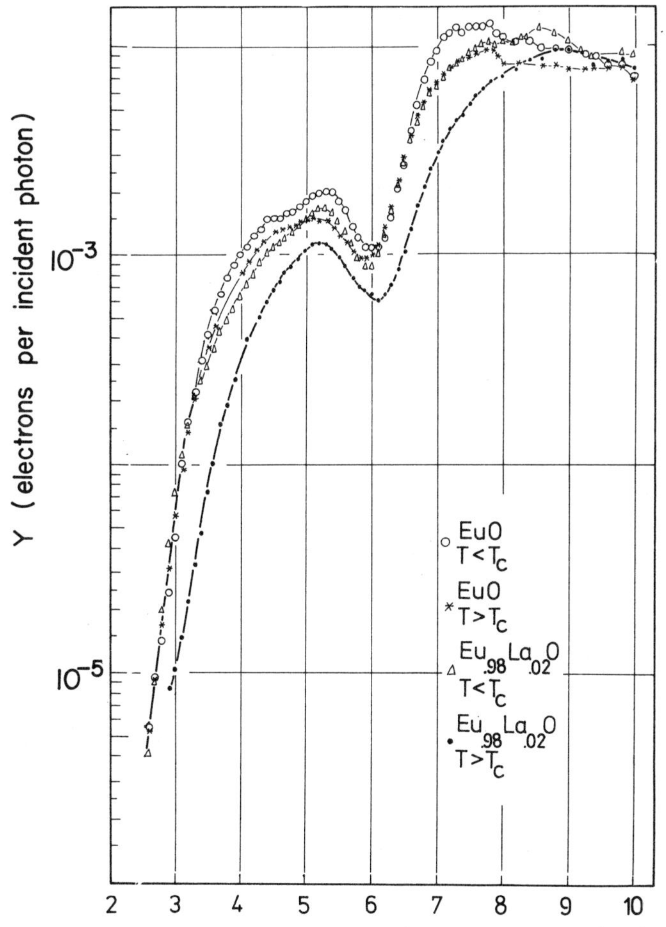

Fig.1. Quantum yield of pure and 2 % La - doped EuO above and below T_c.

Fig.1 shows Y - curves of pure and 2 % La doped EuO above and below T_c. They exhibit the following features:

(1) For pure EuO no significant difference exists between the yield curves at $T > T_c$ and $T < T_c$.

(2) For doped EuO there is a clear separation between the yield curves at $T > T_c$ and $T < T_c$.

(3) All curves have at 5.5 eV a local maximum and at 6 eV a local minimum.

(4) The yield of doped EuO at $T < T_c$ practically coincides with the yield of pure EuO.

In the following a tentative explanation of features (1)-(3) is given. Point (4) will not be discussed any further because at present there is no sufficiently accurate theory available to attack this problem.

The fact that the Y-shift occurs only for doped EuO led us to the conclusion that the extra electrons introduced by doping are responsible for the observed phenomenon. It is supposed that the Fermi energy coincides with the bottom of the conduction band, such that even at 4.2 K the semiconductor is degenerate. This assumption is corroborated by measurements of the electrical transport properties of highly doped Eu-Chalcogenides which seem to be interpretable within the framework of the free electron model of metals[3].

An important point to be clarified is whether the difference between the yields of the doped EuO is caused by a vertical shift on the Y-scale or a horizontal shift on the energy scale. In case of a vertical shift the <u>same</u> levels contribute to the photocurrent at given energy, but the quantum efficiency is smaller for $T > T_c$. In case of a horizontal shift there are levels which contribute to the photocurrent at $T < T_c$ <u>but not</u> at $T > T_c$. The simplest cause of such an effect is an increase of the energy of the vacuum level with respect to the filled bands of EuO.

The only possible reason for a vertical shift we can find would be a drastic change in optical reflectivity upon magnetic ordering. To the best of our knowledge no such effect has been reported up to now. Therefore, an explanation in terms of an energy shift is tried. It is not disturbing that the 5.5 eV maxima of the $T > T_c$ and $T < T_c$ curves do not reach the same value, as will be explained when feature (3) is discussed. On the other hand the fact that the Y-values of both curves reach the same value at $E > 9$ eV gives strong support to the horizontal shift assumption.

Our aim, therefore, is to show that an energy shift is understandable by making a simple model of doped EuO. This is as follows:

The conduction electrons are considered as an interacting, degenerate electron gas. The electron density for a 2 % - doped sample is about $10^{21} cm^{-3}$. The Eu^{++}, O^{--}, and La^{+++} ions are thought of as constituting a homogeneous positive background. The crudity of such a model needs no comment and obviously no quantitative results can be expected. Only sign and order of magnitudes of possible effects should be in agreement with experiment.

Kohn and Lang[4] calculated the work function of an interacting electron gas moving in a uniform positive, compensating charge distribution. The work function is given by

$$\Phi = e \cdot \Delta\phi - \mu \tag{1}$$

where $\Delta\phi$ is the change of the mean electrostatic potential across the surface and μ is the bulk chemical potential of the electrons relative to the mean electrostatic potential in the metal.

The origin of $\Delta\phi$ lies in the dipole layer formed at the surface. The electrons spread out of the metal, whose surface is determined by the discontinuous drop of the positive charge density from a constant value to zero. The characteristic length over which the electron charge extends beyond the surface is given by the Thomas-Fermi screening length and therefore depends on the electron concentration. As this dependence is of the form $\sim n^{1/6}$, the width of the layer is about the same for all metals. Consequently, one expects a small $\Delta\phi$ for low n metals and large $\Delta\phi$ for high n metals. This conclusion is confirmed by the exact quantum mechanical calculation[4]. In our case, where the electron concentration gives an electronic Wigner-Seitz radius of $r_s = 12$, $\Delta\phi$ is less than 0.01 eV and therefore negligible compared to Φ.

When evaluating the bulk part μ of the work function, exchange and correlation effects must be taken into account. For the electron gas, these expressions are well known[5]:

$$\mu = \mu_s + \frac{d}{dn}\left[n \cdot \epsilon_{xc}(n)\right] \qquad (2)$$

with $\epsilon_{xc}(n) = -096/r_s - 0.88/(r_s + 7,8)$ ry (3)

$\epsilon_{xc}(n)$ is the mean exchange and correlation energy per electron and $\mu_s = \hbar^2 k_F^2/2m$.

In our case, $r_s = 12$, the work function amounts to 1.76 eV.

The above calculation is a good approximation when the electron gas is unpolarized, i.e. all orbital states are doubly occupied. Below $T = T_c$, however, this is no longer true. In the ordered state the electrons are themselves polarized by the Eu^{++} - ions thereby acting as go-betweens in the indirect exchange between the magnetic ion cores. Sattler[6] showed that in EuO the polarization is 100 %: This unique phenomenon[7] brings about a remarkable change in Φ, when going from $T \gg T_c$ to $T \ll T_c$. Clearly, the problem is to calculate μ for the polarized gas. In a rough approximation, the correlation energy of the polarized gas is set to zero, so that $\mu = \mu_s + \mu_x$. One obtains for the work function $\Phi^{POL} = 1.19$ eV.

One notices that the absolute values of Φ are much too low. This is understandable in view of the simplicity of the model. However, for the difference $\Delta\Phi$, the errors committed might cancel each other to a large extent. Possibly for that reason the calculated $\Delta\Phi = 0.567$ eV compares favorably with the measured shift of about 0.5 eV.

Finally a remark concerning property (3) of the Y-curves. The reason for the decrease of Y between 5.5 and 6 eV is that in this region the 4f electrons can excite secondary electrons, but are afterwards no longer able to escape[8]. At the same time, the photons can excite 2p valence electrons: but also these are not energetic enough to leave the crystal. One notices that the decrease of Y is obviously due to the particular band structure of EuO, and is therefore not dependent on whether the conduction electrons are polarized or not. Now, in order to understand the measured Y curves imagine that this special absorption processes were not present. Then Y-curves were obtained which would look like the full lines of Fig.2a. In reality one has to superimpose the dip due to special absorption. This is indicated by dotted lines. As a result, the Y-curves obtained this way look very much like the measured ones, see Fig.2b.

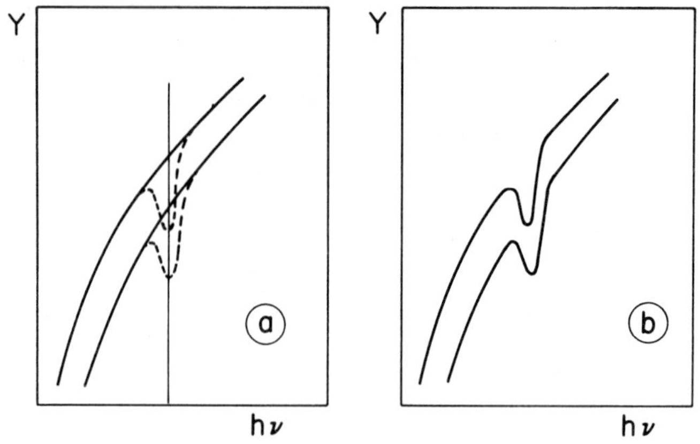

Fig.2. Schematic representation of the Y-structure between 5.5 and 6.0 eV.

ACKNOWLEDGEMENTS

We wish to thank E. Kaldis for supplying us with the EuO crystals.

REFERENCES

1. D.E. Eastman and M. Kuznietz, J.Appl.Phys.42, 1396 (1971).
2. S. Methfessel and D.C. Mattis, Encyclopedia of Physics (Springer-Verlag, Berlin, Vol.XVII, Part 1, pp. 389-562).
3. S.von Molnar, IBM J.Res.Develop.14, 269 (1970).
4. N.D. Lang and W. Kohn, Phys.Rev.B3, 1215 (1971).
5. See e.g. D. Pines, Elementary Excitations in Solids (W.A. Benjamin, Inc., New York 1963, chapter 3).
6. K. Sattler, thesis 1974, Swiss Federal Institute of Technology, CH-8049 Zürich, Switzerland.
7. This is e.g. not true in case of Fe, Co, Ni, and therefore no change of Φ occurs upon magnetic ordering.
8. D.E. Eastman, Phys.Rev.B8, 6027 (1973).

ON ELECTRON-SPIN-DEPOLARISATION BY INTERACTION BETWEEN ELECTRONS AND SPIN-WAVES

By G. Heber: Gesamthochschule Duisburg,
Fachbereich 6, Duisburg, West-Germany

ABSTRACT

We discuss the problem of depolarisation of the spin of photoelectrons from ferromagnetic materials. The degree of polarisation of such electrons should give information about details of electronic structure in such solids, however an understanding of the different mechanisms of depolarisation of the photoelectrons is needed. In this paper a brief discussion of the different possibilities of such depolarisation processes is given. More attention is drawn to the spin-flip-interaction of photoelectrons with spin-waves. The amplitude of such interaction with volume magnon-modes surely is very small, but the interaction with surface-magnon modes or other localized modes, living near the surface, is much more effective. Most effective in producing spin-flip should be some exchange interactions between photoelectronstates and spin-wave-states, beeing both localized insides of the same limited region of the volume of the crystal (near its surface, of course). Soft-magnon-modes, resulting from magnetic instability of the surface sheet of the crystal, in connection with surface-resonance-electron-states should give remarkable probability for spin-flip-processes.

INTRODUCTION

Since few some years, experiments are possible, measuring the degree of spin-polarisation of photo-electrons from ferromagnetic materials. There are measurements with samples of EuO[1], EuS[2] tungsten tips, coated with films of EuS[3] and Fe, Co, Ni, covered by a Cs-monolayer[4]. The intention is to draw from such experiments conclusions of the electronic structure of the spin up- and spin down-electrons separately.
The experimental findings however show for the itinerant ferromagnets Co and Ni, that there seems to be no simple connection between the degree of spin-polarisation of the photoelectrons and the degree of spin-polarisation of the electrons insides of the material, the photoelectrons are coming from. The sign of polarisation near threshold is opposite to that of majority-electrons in bulk material[4].

- 2 -

A strong dependence of the ESP on crystallographic direction[5] is expected. So, we must look for an understanding of the mechanism for changing the spin of the photoelectron, whilst traveling through the surface-sheets of the probe to the detector. We wish to mention, that there are two alternatives to such dynamical processes:

1) Photoelectrons stem from surface-sheets of the probe. Insides of this surface-sheet may exist completely different electronic states, also different spin-polarisation (surface-resonance-states, eg., with even opposite majority-spin than electrons in bulk material), compared to bulk material[6].
2) Our picture of electronic states insides of bulk ferromagnetic material might be not good enough to allow an understanding of these experiments. Attempts to improve this picture are published[7]. However, there are also doubts, if really the band model of ferromagnets is unable to explain the polarisation of photoelectrons[8].

OUR MODEL

Here we wish to discuss the influence of an exchange-interaction between one photo-electron and a spin-wave on the process of depolarisation of the photoelectron. Our model is as simple as possible, being defined by the Hamiltonian:

$$H = \frac{\hbar^2}{2m}\,\Delta - 2J\,\vec{S}_m(t)\cdot\vec{S}_{el} \qquad (1)$$

where:

first term: kinetic energy of photoelectron

S_{el} : Spin operator of photoelectron
S_m : Averaged spin-vector of magnon
J : exchange integral between photoelectron and the electrons, responsible for our magnon.

We shall assume for simplicity that $\vec{S}_m$ in the considered part of the sample doesn't depend on $\vec{r}$ (so-called homogeneous magnon-mode or long-wavelength). This assumption enables us to separate the eigenfuction ψ of the problem:

$$\psi(\vec{r},t,s) = \varphi(\vec{r},t)\,\chi(t) \qquad (2)$$

where $\varphi(\vec{r},t)$ describes only the orbital movement of the photoelectron, $\chi(t)$ gives the dynamics of spin of our electron.

- 3 -

Application of the usual separation procedures leads to the following time-dependent Schrödinger-equation for χ:

$$\frac{\hbar}{i}\dot{\chi}(t) = -2J\,\vec{S}_m(t)\cdot\vec{S}_{el}\cdot\chi(t) + \varepsilon\chi(t); \qquad (3)$$

here ε is some separation constant, which may be choosen arbitrarily and normalized by fixing some reference-system for our energy; in the following calculations we did set $\varepsilon = 0$. The equation for $\varphi(\vec{r},t)$, resulting from (1) and(2), we don't need for our discussion in this paper. Since we presently are not interested in the dynamics of the magnetic system himself, we may neglect the reaction of the quantities $S_m(t)$ on the interaction with the photoelectron. In that case, it should be a not too bad approximation, to treat $\vec{S}_m(t)$ like a classical vector-field (quasiclassical approximation for the spin-system), not as operator-quantity. There are many effects in spin-dynamics of magnetically ordered systems, for which this approximation is a quite good one, as is well-known.
In the spirit of this approximation, the existence of a spin-wave with frequency ω is described by

$$\vec{S}_m(t) = (S'\cos\omega t,\ S'\sin\omega t,\ S^0), \qquad (4)$$

where we have assumed, that the macroscopic magnetisation is in z-direction; the averaged spin (per atom) in z-direction is S^0, S' is measuring the number n of reversed spins; we have:

$$S' \approx \left(\frac{nS}{N}\right)^{1/2}, \quad S^0 \approx S - \frac{n}{N} \qquad (5)$$

where: S the spin-quantum-number of one atom, N the number of atoms contributing to this spin-wave, n the number of reversed spins ($\frac{n}{SN} \ll 1$ assumed). (4) is describing the well known spinprecession insides of a spin-wave. Now we remember, that mathematically exactly the same problem arises, if we consider the spin-dynamics of an electron in a constant magnetic field H_0 in z-direction and a rotation magnetic field $\vec{H}'$

$$\vec{H}' = (H_x'(t), H_y'(t), 0) = (H' \cos \omega t, H' \sin \omega t, 0) \qquad (6)$$

in x-y-plane with frequency ω.
This case may be solved exactly, see 9).
The Hamiltonian has the form

$$H = \mu_B (S_x H' + S_y H_y' + S_z H_0) \qquad (7)$$

- 4 -

The Schrödinger-equation $-\frac{\hbar}{i} \frac{\partial \chi}{\partial t} = H\chi$ (7a) is solved with the "Ansatz".

$$\chi(t) = u(t)\alpha + v(t)\beta,$$

$$\alpha = \binom{1}{0}, \beta = \binom{0}{1} \quad \text{: the Pauli-spinors.}$$

Asking for the solution with the "initial condition" $\chi(t=0) = \alpha$, one finds [6] :

$$\chi(t) = \left[\cos \Omega t - \frac{\omega_0 - \frac{1}{2}\omega}{\Omega} i \sin \Omega t\right] e^{-\frac{i\omega}{2}t} \alpha - \frac{\omega'}{\Omega} i \sin \Omega t \cdot e^{\frac{i\omega}{2}t} \cdot \beta, \qquad (8)$$

with $\Omega = \left[(\omega_0 - \frac{1}{2}\omega)^2 + \omega'^2\right]^{1/2}$;

$$\omega_0 = \frac{\mu_B H_0}{\hbar}; \quad \omega' = \frac{\mu_B H'}{\hbar}.$$

From (8) we get the probability P for spin-flip in the interval of time from $t=0$ to $t=\Delta t$:

$$P = \left(\frac{\omega'}{\Omega}\right)^2 \sin^2(\Omega \Delta t). \qquad (9)$$

Comparing (7), (7a) with (3) we get (using also (4) and (5)):

$$\omega' \approx \frac{2J}{\hbar}\left(\frac{Sn}{N}\right)^{1/2}, \text{ for } \frac{n}{N} \ll S; \; \omega_0 \approx \frac{2J}{\hbar}\left(S - \frac{n}{N}\right). \qquad (10)$$

DISCUSSION AND CONCLUSIONS

We discuss the order of magnitude of the probability P according to (9) in some limiting cases:

1. $\omega = 2\omega_0$ (resonance); in that case: $\Omega = \omega'$,
   $P = \sin^2(\omega'\Delta t) \approx (\omega'\Delta t)^2$ (for $\omega'\Delta t \ll 1$);
2. $\omega \ll \omega_0$; $\Omega \approx (\omega_0^2 + \omega'^2)^{1/2}$
   $P \approx (\omega'\Delta t)^2$ (for $\Omega \Delta t \ll 1$).

(we may see, that the dependence of P on ω is indeed very weak, as far as $\Omega \Delta t \ll 1$)

3. Taking $J = 10^{-1}$ eV and putting n=1 (one spin-wave only), we get

$$P \approx \left(10^{14} \frac{\Delta t}{\sec}\right)^2 \frac{1}{N} \qquad (11)$$

Δt is the time, during which there is some interaction between the photoelectron and the spin-waves; N the number of atoms, responsible for the spin-wave. From (11) we may conclude, that there will be no appreciable spin-flip-probability between one photoelectron and volume-spin-wave-modes, but there may be some appreciable effect, if there exist localized spin-wave-modes just insides of that part of the surface, in which the photoelectron has been excited. Another reason for some not-negligible P may arise, if you integrate (9) with respect to ω over the real spin-wave-spectrum (including local (surface)modes). Particulary if there are

- 5 -

soft magnon-modes at the surface (magnetic instability of surface), such modes should contribute much to P. We conclude, that exchange between photoelectrons and magnons may be of some importance for the spin-depolarisation, if there exists some high density of surface magnons, or some magnetic instability of the surface (see also 6) and 10)).

ACKNOWLEDGEMENTS

The author is greatly indebted to Prof. S. Methfessel, who gave him the possibility to work at his chair and encouraged the calculations by helpful discussions. Further we wish to thank Prof. S. Fulde (München/Stuttgart), Prof. D. Wagner (Bochum) and his staff, Prof. H. Hoffmann (Regensburg) and Prof. P. Wachter (Zürich) for illuminating conversations.

REFERENCES

1) Sattler, K. and Siegmann, H.C. Phys.Rev.Lett. 29, 1565 (72)
2) Busch, G., Campagna, M. and Siegmann HC Intern.Journal Magnetism, 4, 25 (73) Campagna, M. et al. AIP Conf.Proc. 18/2, 1388 (74)
3) Müller, N., Eckstein, W., Heiland W. and Zinn, W.,, Phys.Rev.Lett. 29, 1651 (72)
4) Busch, G., Campagna, M., Pierce, D.T. and Siegmann, H.C., Phys.Rev.Lett. 28, 611 (72), Alder, H. et al. Phys.Rev. B 8, 2075(73)
5) Politzer, B.A., Cuttler, P.H., Phys. Rev.Lett. 28, 1330 (72)
6) Fulde, P., Luther, A. and Watson, R.E. Phys.Rev. B 8, 440 (73) Takayama, H., Baker K. and Fulde P. preprint 1974 Levin, K., Liebsch, A. and Bennemann KH Phys.Rev. B 7, 3066 (73) Aoi, K. and Bennemann K.G. preprint 1974
7) Anderson, P.W., Phil.Mag. 24, 203 (71) Doniach, S., AJP - Conference- Proceedings, Vol. 5, 549 (72)
8) Wang, C.S. and Callaway, J., Phys.Rev. B 9, 4897 (74)
9) Flügge, S., Pracitical Quantum Mechanics, Vol. II, p. 20 (problem 138). Berlin, New-York, Springer (71)
10) Helmann, J.S. and Siegmann, H.C. Sol.State. Comm. 13, 891 (73)

OBSERVATIONS ON SURFACE SPIN WAVES IN NiO

Richard G. Schlecht
Engineering Science and Systems Department, University of Virginia, Charlottesville, Virginia 22901

ABSTRACT

In the past it has been difficult to make observations on the surface spin states of magnetic materials. Recently studies have been made on the antiferromagnetic material NiO by electron scattering. The temperature dependence of these studies have led to the prediction of a low lying surface spin wave in NiO. We have made a study of ultrafine particles of NiO by Raman scattering. Earlier reports of Raman scattering in NiO have shown second and fourth order bulk magnon scattering in this near cubic crystal where first order scattering is forbidden. In ultrafine particles we observe the first order bulk spin wave scattering which agrees with the infrared absorption results in NiO. We also observe a low lying mode which we attribute to the surface spin wave in this material. We will report on the size and temperature dependence of this first observation of surface spin waves by Raman scattering.

INTRODUCTORY REMARKS

The antiferromangnetic crystal NiO has received a great deal of attention in the past. It is an interesting material for experimental study in that it possesses a relatively simple structure and is quite stable. The bulk crystal has been studied extensively by several techniques but we will limit our discussion to optical and electron scattering results. In 1960 Kondoh[1] reported the first observation of the bulk spin wave in NiO by far-infrared absorbtion measurements. This was shortly followed by the observations of Sievers and Tinkham[2] which varified the results of Kondoh. The bulk spin wave at $\vec{k} = 0$ was observed at 36.5 cm^{-1} at low temperature and the mode softened and decreased in amplitude as the Néel temperature (523°K) was approached. Raman scattering on single crystals of NiO was observed by Dietz, Parisot and Meixner[3] who observed two magnon scattering at the zone edge. Thus all of the experimental work in the optical region on NiO has been done on bulk magnetic excitations. Several authors have reported work on the theory of surface spin waves in antiferromagnetic materials. Mills and Saslow[4] have calculated that the $\vec{k} = 0$ surface spin wave frequency for a body centered cubic structure in which the anisotropy field is much smaller than the exchange field is $\omega_o/\sqrt{2}$ where ω_o is the frequency of the bulk spin wave. Wolfram and DeWames[5] have also obtained this result. Palmberg, DeWames and Vredevoe[6] demonstrated the existence of magnetic order on the surface of NiO by low energy electron diffraction and DeWames and Wolfram[7] have predicted the existence of a low lying surface spin wave in NiO by the interpretation of the temperature dependence of exchange-scattered low energy electrons.

RESULTS

We have performed experiments on microcrystals of NiO by Raman scattering. The crystals were pressed into discs and scattering was observed at 90°. We have observed first order scattering from the $\vec{k} = 0$ magnon in spite of the fact that NiO is a nearly cubic crystal below the Néel temperature and thus first order Raman scattering is not allowed by symmetry considerations from the bulk excitations. The results agree with the infrared absorbtion results reported by Kondoh[1]. We have also observed the surface spin wave in the face-centered cubic material at 19.5 cm^{-1} at low temperature. At this temperature the bulk antiferromagnetic resonance was observed at 36 cm^{-1}. The bulk magnon and the surface spin wave were observed to have very different temperature characteristics. As the temperature was increased toward the Néel temperature for the bulk magnon the bulk magnon softened from 36 cm^{-1} to 22 cm^{-1} at which point the intensity became too weak to observe at higher temperature. For this same temperature scan the surface spin wave softened by only 3 cm^{-1}. It was also observed that the surface spin wave frequency is strongly dependent on particle size. At a particle size of 1400 Å the surface spin wave was observed at 19.5 cm^{-1} whereas at a particle size of 1700 Å it was observed at 14 cm^{-1}. Further work is necessary to determine the transition temperature of the surface spin waves. This was not possible with the particle sizes investigated. Smaller particle sizes where the signal to noise ratio would be better are needed to obtain this information.

1. H. Kondoh, J. Phys. Soc. (Japan) 15, 1970-1975 (1960)
2. A.J. Sievers and M. Tinkham, Phys. Rev. 129, 1566-1571 (1963)
3. R.E. Dietz, G.I. Parisot and A.E. Meixner, Phys. Rev. B4, 2302-2310 (1971)
4. D.L. Mills and W.M. Saslow, Phys. Rev. 171, 488-506 (1968)
5. T. Wolfram and R. E. DeWames, Prog. Sur. Science 2, 233-330 (1972)
6. P.W. Palmberg, R.E. DeWames and L.A. Vredevoe, Phys. Rev. Letters 21, 682-685 (1968)
7. R.E. DeWames and T. Wolfram, Phys. Rev. Letters 22, 137-139 (1969)

SPIN PINNING AT FERRITE-ORGANIC INTERFACES

A.E. Berkowitz, J.A. Lahut, I.S. Jacobs, L.M. Levinson, General Electric Corporate Res. & Dev., Schenectady, New York 12301, and D.W. Forester, Naval Res. Lab., Washington, D.C. 20375.

ABSTRACT

We have previously reported a drastic moment decrease in $NiFe_2O_4$ fine particles (~100Å dia.) coated with an organic surfactant such as oleic acid.[1] Removal of the organic coating restored the moment of the particles. Continued investigation has demonstrated that the apparent moment decrease is due to strong pinning of the spins of those ferrite cations that are bonded to the organic molecules. The evidence for this model includes: (1) Uncoated $NiFe_2O_4$ particles, prepared in an otherwise identical fashion and with the same size distribution showed no decrease in moment. This observation eliminates the possibility that defects or abnormal surface morphology are responsible for the decreased moment. (2) Low temperature Mössbauer measurements in zero field of coated and uncoated particles showed identical spectra associated with ordered fine particle $NiFe_2O_4$. (3) Mössbauer data taken on coated particles in a field of 68.5 kOe applied along the direction of γ-ray emission showed only a small decrease of the $\Delta m=0$ lines; data on uncoated particles showed the normal decrease. This is consistent with the existence of a substantial fraction of pinned or canted spins in coated particles. (4) Pulsed fields of 200 kOe were not sufficient to saturate coated particles.

(1) A.E. Berkowitz and J.A. Lahut, AIP Conf. Proc. 10, 966 (1973).

HELICAL SPIN ORDERING AND MAGNETIC CONSTANTS OF $(Sr_{0.8}Ba_{0.2})_2Zn_2Fe_{12}O_{22}$

MASARU MITA and NOBUYUKI MOMOZAWA
Department of Physics, Faculty of Science and Technology,
Science University of Tokyo, Noda, Chiba, Japan, 278.

ABSTRACT

On hexagonal ferrite $(Sr_{0.8}Ba_{0.2})_2Zn_2Fe_{12}O_{22}$ with proper-helix type spin ordering, the difference of Fourier transforms of exchange energy, $J(\vec{Q})-J(0)$ and $J(\vec{Q})-J(2\vec{Q})$, and uniaxial anisotropy constants, K_1 and K_2, are determined, where the $\vec{Q}$ is the propagation vector of helix. Nagamiya's parameters θ, ξ_1 and ξ_2 depending on the external magnetic field are defined.

The magnetization in the experiment is clearly explained in complete agreement with the theory in terms of θ, ξ_1 and ξ_2. Neutron diffraction pattern changing with field is attempted to analyze by the use of θ, ξ_1 and ξ_2.

§1 INTRODUCTION

Hexagonal ferrite $(Sr_{0.8}Ba_{0.2})_2Zn_2Fe_{12}O_{22}$ has a proper-helix type spin ordering[1,2,3] with the propagation vector $\vec{Q}$ directed to the crystallographic C axis. This fact was suggested by Enz[1] for the first time and was proved through detail study on neutron diffraction performed by Sizov and his co-workers.[2,3] They also carried out the experiments on magnetization processes and neutron diffraction in presence of the external magnetic field H. In this work, analyses of magnetization (§2) and of the diffraction pattern (§3) in presence of H are attempted by the use of three parameters θ, ξ_1 and ξ_2 introduced by Nagamiya's theory.[4]

§2 MAGNETIZATION PROCESS

The (M/M_S), the induced magnetization M divided by the saturation magnetization M_S, against H are shown in Fig.1. The black circles show the results for H parallel to the C axis and the black triangles for H perpendicular to the C axis. Sizov et al suggested the magnetization processes on the basis of neutron diffraction studies, but did not analyze their results correctly. Theoretically obtained magnetization for helical spin system by Nagamiya[4] and spin behavior are the following:

(i) In case of H=0, all the spins lie in the C plane. As spins form helix, spontaneous magnetization does not arise.

(ii) In case of H parallel to the C axis, spins are in the cone state represented by cone angle θ that is measured from the C axis. And the magnetization is expressed by

$$(M/M_S) = \cos\theta. \quad (1)$$

(iii) In case of H perpendicular to the C axis, spins lie in the C plane and are in slightly distorted helix state (sd-helix state) in weak field and in the fan state in high field. These phases are characterized by parameters ξ_1 and ξ_2, respectively. The (M/M_S), expressed by the use of ξ_1 or ξ_2, for the sd-helix state is equal to

$$(M/M_S) = -(1/2)\cdot\xi_1, \quad (2)$$

and for the fan state is equal to

$$(M/M_S) = 1-(1/4)\cdot\xi_2^2. \quad (3)$$

The θ, ξ_1 and ξ_2^2 are determined from the minimizing conditions of spin Hamiltonian, containing isotropic exchange, anisotropy, Zeeman and demagnetization energies;

$$4K_2\cdot\cos^3\theta+[U+2(K_1-2K_2)+(4\pi/3)\cdot M_S^2]\cdot\cos\theta-M_S\cdot H=0, \quad (4)$$

$$\xi_1 = -2M_S\cdot H/[U+V+(4\pi/3)\cdot M_S^2], \quad (5)$$

$$\xi_2^2 = 8[U-M_S\cdot H+(4\pi/3)\cdot M_S^2]/[3U+V-M_S\cdot H+4\pi M_S^2]. \quad (6)$$

Additional conditions are $U=J(\vec{Q})-J(0)>0$, $V=J(\vec{Q})-J(2\vec{Q})>0$ and $J(\vec{Q}) = NS^2\cdot\sum_n J_{nm}\cos\vec{Q}\cdot(\vec{R}_n-\vec{R}_m)$, where J_{nm} is exchange constant, $J(\vec{Q})$ Fourier transform of exchange energy. The S and N are thermally averaged magnitude of spin and numbers of spin in one period of helix, respectively. The K_1 and K_2 are uniaxial anisotropy energy constants corresponding to one period of helix.

By substituting (4), (5) and (6) into (1), (2) and (3), it is understood that (M/M_S)'s are expressed by the four magnetic constants; U, V, K_1 and K_2.

The fittest expressions to reproduce the experimental results are obtained for the first time; $U=J(\vec{Q})-J(0)=10300\cdot M_S-(4\pi/3)\cdot M_S^2$, $V=J(\vec{Q})-J(2\vec{Q})=750\cdot M_S$, $K_1=3000\cdot M_S$ and $K_2=1100\cdot M_S$. The coefficients of M_S still contain S and N, that is to say, 1100, the coefficient in K_2 is equal to $D_2S^3/g\mu_B$ as example, where D_2 is the uniaxial anisotropy energy constant for one ion, g spectroscopic splitting factor and μ_B Bohr magneton. The expressions of the four magnetic constants, however, are sufficient to explain magnetization processes. The M_S's cancel out each other in (4), (5) and (6). A good agreement is obtained between the calculated (M/M_S) and the experimental. (Fig.1)

The magnetic phases are found in the following forms: For H//C, Helix [H=0 Oe] → Cone-Helix [0<H<16300] → Ferro [H≥16300] and for H⊥C, Helix [H=0] → SD-Helix [0<H≲800] → Fan [1000≲H<10300] → Ferro [H≥10300]. Field H_f's for saturation of magnetization are $H_f=[U+2K_1+(4\pi/3)\cdot M_S^2]/M_S=16300$ Oe for H parallel to the C axis and $H_f=[U+(4\pi/3)\cdot M_S^2]/M_S=10300$ Oe for H perpendicular to the C axis. Consequently, the θ, ξ_1 and ξ_2 are determined as shown in Fig.2.

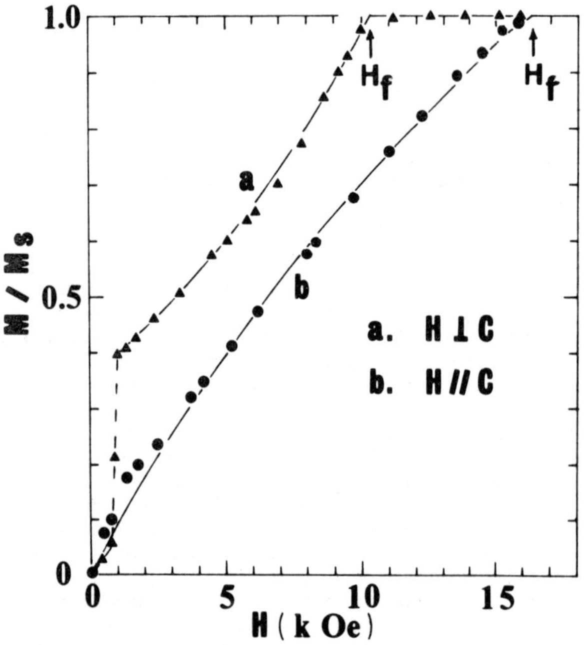

Fig. 1 Black circles and black triangles denote the experimentally obtained M/M_S (after T. M. Perekalina et al.[2]). Solid lines denote the theoretical results.

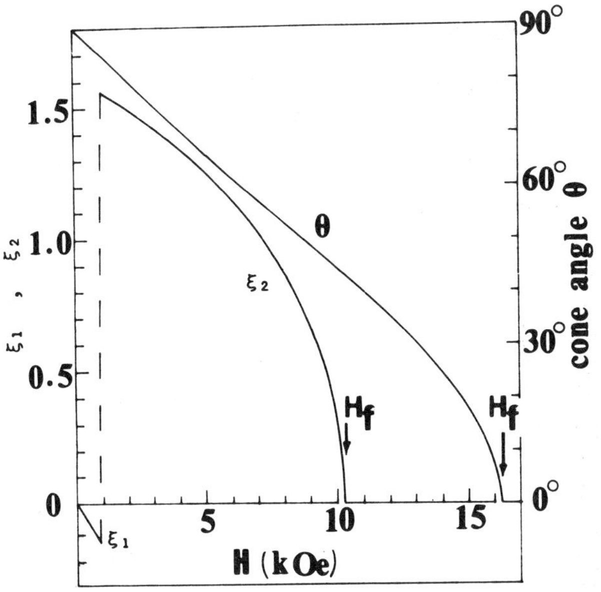

Fig. 2 θ, ξ_1 and ξ_2 against H.

The spin system in fan state in field region from near 1 kOe to 10.3 kOe smoothly changes with increase of H, i.e., with decrease of ξ_2 from near 1.6 to zero, and changes into the ferro state for H≳10.3 kOe. The abrupt destruction of the helical spin orderiong in 8 kOe, that was suggested by Sizov et al, does not arise from the present theory. It is expected, also, that the results of neutron diffraction by Sizov et al[2,3] corresponding to this field region would be interpreted by the use of such a smoothly changing parameters ξ_2.

§3 CROSS SECTIONS FOR UNPOLARIZED NEUTRON

The following characteristics of neutron experiments are mentioned by Sizov et al.[2,3]

(I) In case of H parallel to the C axis, a pair of satellite lines, which are accompanied by (00ℓ) nuclear diffraction lines, do not change their diffraction angles in spite of change of H. While the scattering intensity decreases with increase of H and vanishes for H≳16000 Oe.

(II) In case of H perpendicular to the C axis, the satellites moves slightly toward central peak (nuclear diffraction line) in H=800 Oe. But the satellites for H=3 kOe situate in angle more apart than that for H=0 Oe, and seem to remain up to H=8 kOe. For further increase of H, the satellite lines tend to vanish and to merge with the central peaks.

The differential cross section for the pure magnetic Bragg scattering is[5]

$$(d\sigma/d\Omega\cdot d\omega)=(1.91\cdot e/\hbar c)^2\cdot\delta(\omega)\langle\vec{M}_\perp(\vec{q})\rangle\cdot\langle\vec{M}_\perp(-\vec{q})\rangle, \quad (7)$$

where $\vec{q}$ is the incident vector $\vec{K}_0$ of neutron subtracted by the scattered vector $\vec{K}_1$. The symbol $\langle\,\rangle$ denotes an average over the initial states. The $\vec{M}_\perp(\vec{q})$ is normal compornent of Fourier transform of electron magnetization, $\vec{M}(\vec{q})$. The thermal average of n-th spin $\vec{S}_n$ is expressed by (S, θ_n, ψ_n) in spherical polar co-ordinates. The polar angle θ_n is measured from the C axis. The eq.(7) for the reflections from (00ℓ) planes is

$$(d\sigma/d\Omega\cdot d\omega)=\alpha\cdot\sum_{n,m} e^{-i\vec{q}(\vec{R}_n-\vec{R}_m)}\sin\theta_n\cdot\sin\theta_m\cdot\cos(\psi_n-\psi_m), \quad (8)$$
$$\alpha=(1.91\cdot e/\hbar c)^2\cdot(g\mu_B S)^2\cdot\delta(\omega).$$

In case of H=0 and $\vec{H}$ parallel to the C axis, the ψ_n is always equal to $\vec{Q}\cdot\vec{R}_n$. And the θ_n is $\pi/2$ and θ, respectively. The Bragg conditions are

$$\vec{K}_1-\vec{K}_0\pm\vec{Q} = 2\pi\cdot\vec{B}_{00\ell}, \quad (9)$$

where the $B_{00\ell}$ is the reciprocal lattice vector. And scattering intensity of satellite has $\sin^2\theta$ dependence. As the θ in Fig.2 varies from $\pi/2$ to zero with increase of H, the intensity decreases with increase of H and vanishes for H≳16300 Oe. The results are consistent with the experimental results by Sizov et al.[3]

The characteristics of (II) in this section would be explained in consistent with the explanation of the magnetization. In this case, θ_n is always equal to $\pi/2$. According to Nagamiya's proposal, ψ_n is $\vec{Q}\cdot\vec{R}_n+\xi_1\cdot\sin\vec{Q}\cdot\vec{R}_n$ for the sd-helix state and is $\xi_2\cdot\sin\vec{Q}\cdot\vec{R}_n$ for the fan state. The terms of $\exp[\pm i(1+\xi_1)\vec{Q}\cdot(\vec{R}_n-\vec{R}_m)]$ or of $\exp[\pm i\xi_2\vec{Q}\cdot(\vec{R}_n-\vec{R}_m)]$ obtained from substitution of ψ_n's into eq.(8) lead the following Bragg conditions:

$$\vec{K}_1-\vec{K}_0\pm(1+\xi_1)\cdot\vec{Q} = 2\pi\cdot\vec{B}_{00\ell} \quad (10)$$

for the sd-helix state, and

$$\vec{K}_1-\vec{K}_0\pm\xi_2\cdot\vec{Q} = 2\pi\cdot\vec{B}_{00\ell} \quad (11)$$

for the fan state. As the equation $\vec{Q}\cdot\vec{C}=2\pi$ is satisfied[3] at room temperature, the spacing $2\Delta\Theta$ between satellites is obtained to be $2\tan\Theta_0\cdot(1+\xi_1)/\ell$ [rad.] or $2\tan\Theta_0\cdot\xi_2/\ell$, where the Θ_0 is the diffraction angle of central peak and $\vec{C}$ is the lattice vector along the C axis. It is shown that all of the $2\Delta\Theta$'s in the papers by Sizov et al[2,3] are consistent with this interpretation of satellites. The criteria on scattering intensity, however, is not yet made in the present situation.

§4 SUMMARY

(1) Magnetic constants, $U=J(\vec{Q})-J(0)$, $V=J(\vec{Q})-J(2\vec{Q})$, K_1 and K_2,and Nagamiya's parameters, θ, ξ_1 and ξ_2, are determined.
(2) Magnetization processes are correctly explained by the use of Nagamiya's parameters θ, ξ_1 and ξ_2.
(3) The magnetic phases in the external magnetic field are clarified.
(4) In case of H parallel to the C acis, the neutron diffraction patterns are correctly explained by the use of θ.
(5) In case of H perpendicular to the C axis, the interpretations of diffraction angle of satellite are attempted by the use of ξ_1 and ξ_2. The present proposal on diffraction angles is in consistent with the experimental results by Sizov et al. The scattering intensity is still under discussion.

REFERENCES

1) U. Enz, J. Appl. Phys. Suppl.32, 22S (1961).
2) T. M. Perekalina, V. A. Sizov, R. A. Sizov, I. I. Yamzin and R. A. Vonskanyan, Soviet Phys. JETP. 25, 266 (1967) [English translation].
3) V. A. Sizov, R. A. Sizov and I. I. Yamzin, Soviet Phys. JETP. 26, 736 (1968), [English translation].
4) T. Nagamiya, Solid State Phys., Vol.20, P.305 (1967), Academic Press, New York and London.
5) P. G. De Gennes, Magnetism III (ed. by Rado and Suhl), P.115 (1963), Academic Press, New York and London.

STRAIN WAVES IN CHROMIUM AND ITS ALLOYS

Y. TSUNODA, M. MORI, Y. NAKAI and N. KUNITOMI
Faculty of Science, Osaka University, Toyonaka, Osaka 560, Japan

ABSTRACT

The properties of a periodic lattice distortion (strain wave) in Cr and its alloys have been studied in their incommensurate SDW state by X-ray and neutron diffractions. The period of the strain wave is a half of that of SDW. The neutron data analysis under the assumption of the rigid spin model shows that the lattice expansion takes place at the loop of SDW. The maximum amplitude of the distortion is estimated from the satellite intensity of X-ray as 0.25(±0.05)% of a lattice constant. Second and third harmonic components of the modulation are not observed within the experimental error. X-ray measurements were also made for Cr(Fe) and Cr(Mn) alloys and show that the strain wave can be observed whenever the incommensurate SDW exists. The discontinuous expansion of the lattice was observed at the incommensurate-commensurate transition temperature for each of these alloys. Both exchange striction and CDW models are the possible explanations of these properties.

INTRODUCTION

In the previous paper[1], we have reported the existence of a periodic lattice distortion (strain wave, hereafter we call it SW) in pure Cr below the Néel temperature. In this note, several properties of SW in Cr and its alloys observed by X-ray and neutron diffractions are reported. The experimental details of the X-ray measurements were described in the previous paper and neutron scattering measurement was made by a conventional type neutron diffractometer.

PROPERTIES OF STRAIN WAVE

(A) Wavelength of SW - The satellite peaks appear at positions 2η away from the Bragg point (η is the wave number of the sinusoidal modulation of SDW). This relation holds at any temperature below the Néel temperature though the absolute value of η decreases with increasing temperature. (It, of cource, means that the wavelength of SW is just a half of that of SDW.) The existence of SW was confirmed also by neutron diffraction around the 2 0 0 Bragg point.

(B) Amplitude - If we write the position vector of the l-th atom with SW as

$$\mathbf{r}_l = \mathbf{r}_l^0 + \Delta \sin 2\boldsymbol{\eta}\cdot\mathbf{r}_l^0 \qquad (1),$$

where Δ and $\mathbf{r}_l^0$ are the amplitude of the displacement and the position vector of the l-th atom without SW, respectively, the intensity of the satellite reflection due to SW is expressed as $(\mathbf{K}\cdot\Delta/2)^2/3 \times I_{Bragg}$ under the assumption of the uniform distribution of the magnetic domains. From the observed intensity ratio of the satellite to the Bragg reflections, the maximum strain Δ/a is estimated as $\Delta/a = 2.5(\pm0.5)\times 10^{-3}$ at T = 153 K. In order to confirm the orientation dependence caused by the $(\mathbf{K}\cdot\Delta)$ term in the expression of the satellite intensity, measurements were carried out around the 4 0 0 Bragg point in the directions parallel and perpendicular to $\mathbf{K}$. The reflection peak vanished in the perpendicular direction, indicating that the satellite reflection comes from a static lattice wave. The satellite intensities due to SW and due to SDW were studied as a function of the temperature by X-ray and neutron diffractions, respectively, for the same sample. As shown in Fig.1, the former decreased faster than the latter as the temperature was increased. In the further measurements of SW below spin flip temperature, the satellite intensity and its position did not show remarkable changes.

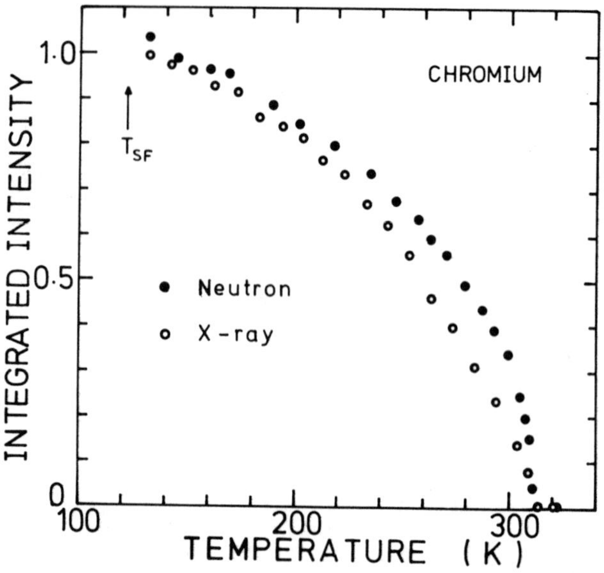

Figure 1. Temperature dependence of satellite intensities due to SW and that due to SDW measured by X-ray and neutron diffractions, respectively.

(C) Higher harmonics - Second and third harmonic components of SW were investigated in high precision. However, no peak was observed within an experimental error. Higher harmonic components are estimated to be less than 3.7% (0.14% in the intensity) of the primary one if there exist.

(D) Phase relation - The phase relation between SDW and SW has been studied by neutron diffraction, and analyzed under the assumption of the rigid spin model. When the atomic position with SW is given by eq. (1), the magnetic moment at site l can be written as $S_l = S_0 \sin(\mathbf{Q}\cdot\mathbf{r}_l^0)$ ----(a) if the displacement of the atomic position due to SW and the amplitude of the SDW are in phase, and $S_l = S_0 \cos(\mathbf{Q}\cdot\mathbf{r}_l^0)$ ----(b) if both are out of phase. (Where $\mathbf{Q}$ is equal to $2\pi/a - \boldsymbol{\eta}$.) Under these circumstances, the ratio of the magnetic satellite intensities of $1-\eta$ 0 0 and $1+\eta$ 0 0 is calculated as

$$I(1-\eta)\,/\,I(1+\eta) \simeq 1 \mp 2\,\mathbf{K}\cdot\Delta$$

where - sign corresponds to the case (a) and + sign to the case (b). Experimental value of this ratio is 1.044±0.003 at the room temperature after making the corrections of the extinction effect, magnetic form factor[2] and the Lorentz factor, indicating that the case (b) occurs in Cr. The results can be understood by the model that the lattice expands at the loop of the SDW. The phase relation between SDW and SW is illustrated schematically in Fig.2. The amplitude of the distortion is estimated as $\Delta/a = 3.5\times10^{-3}$ at room temperature. This value is slightly larger than that obtained by the X-ray diffraction experiment.

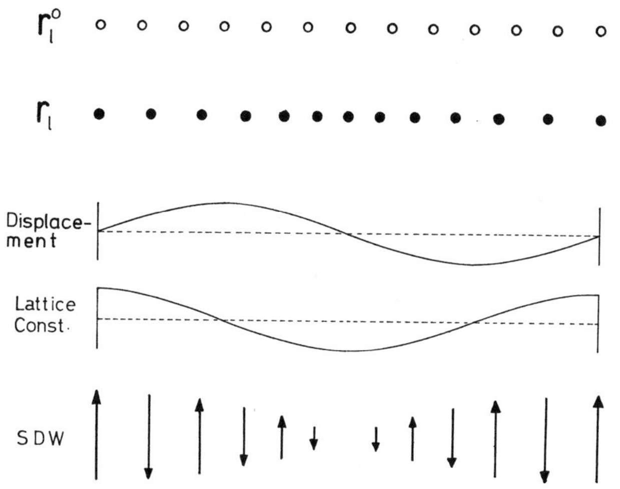

Figure 2. Schematic illustration of phase relation between SW and SDW.

Cr ALLOYS

The SDW state of Cr is sensitive to the contamination of a small amount of 3d-impurities. To obtain further knowledge about SW, X-ray measurements were performed for several Cr(Fe) and Cr(Mn) alloys in the temperature range from 135 K to 300 K.

(A) Cr(Fe) alloy - For specimens containing 0.4% and 1.5% Fe, satellite reflections caused by SW were found for temperatures between 135 K and T_N. On the other hand, no satellite was observed in the 3% Fe sample at any temperature. For the 2% Fe sample, a weak satellite peak was found in the temperature range from 240 K to 260 K, while no peak was observed below 240 K. The region in which the satellite peaks were observed just coincides with that of incommensurate SDW (ISDW) state determined from the neutron diffraction[3]. In other words, SW can be observed whenever ISDW exists. The satellite peak positions for these samples were plotted in Fig.3 as a function of the temperature. It is to be noted that the wavelength of the SW becomes shorter as the temperature is increased in the Fe 2% alloy in which the commensurate SDW (CSDW) is stable below 240 K. It was found from the measurement of the position of 2 0 0 Bragg reflection, that the lattice expands discontinuously by an amount of 2.5×10^{-4} at the incommensurate-commensurate transition temperature for 2% Fe alloy. Similar discontinuous expansion by an amount of 6×10^{-4} was found at the paramagnetic to CSDW transition temperature for the 3% Fe alloy.

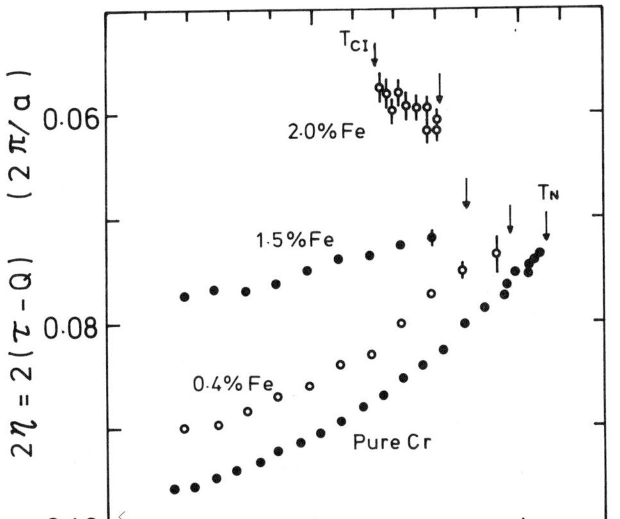

Figure 3. Temperature dependences of the SW satellite positions for Cr(Fe) alloys. Negative slope for Fe 2% alloy indicates a difference between Cr(Fe) system and Cr(Mn) system.

(B) Cr(Mn) alloy - In contrast with the case of the Cr(Fe) 2% alloy, ISDW state is stable in the Cr(Mn) system at low temperatures. In this measurement of the 0.5% Mn sample, satellite peak was observed below 250 K and vanished suddenly just above it. The maximum amplitude of SW is estimated to be $\Delta/a = 1.6\times 10^{-3}$ at 133 K, which is slightly smaller than that of pure Cr. The expansion of the lattice was observed at the ISDW tỏ CSDW transition temperature by an amount of $\delta a/a \sim 1.1\times10^{-3}$.

DISCUSSIONS

Our description of SW by eq.(1) does not contain the uniform component of the lattice expantion which was observed in pure Cr below T_N by Gordenko and Nikolaev[4]. Its magnitude is roughly estimated to be of the order of 10^{-4}. The jump of the lattice spacing accompanied by the commensurate-incommensurate transition observed in Cr(Fe) and Cr(Mn) alloys is also of the order of $10^{-3} \sim 10^{-4}$. The fact that the lattice expands in an antiferromagnetic phase is consistent with the result obtained in the phase relation between SW and SDW in Cr, where the maximum expansion of the lattice occurs at the place with the maximum antiferromagnetic coupling.

Theoretical explanations of SW are proposed by Kanamori and Teraoka[5] from the stand point of the exchange striction model and by Young and Sokoloff[6], Nakajima and Kurihara[7] and Kotani[8] from the stand point of an itinerant electron model. The latters lead first to the charge density wave (CDW) with even harmonics of SDW. Then, the formation of the static lattice distortion with sinusoidal modulation is discussed by introducing the strong electron phonon coupling. The amplitude of SW calculated from the itinerant electron theory seems to be rather small comparing with the experimental results.

The authors would like to thank Professor J. Kanamori for valuable discussions, and Professors J.B.Sokoloff and S. Nakajima and Dr. A. Kotani for informing their unpublished results.

REFERENCES

(1) Y. Tsunoda, M. Mori, N. Kunitomi, Y. Teraoka and J. Kanamori; Solid State Comm. 14 287 (1974).
(2) A.J. Freeman and R.E. Watson; Acta Crystal. 14 231 (1961).
(3) A. Arrott, S.A. Werner and H. Kendrick; Phys. Rev. 153 624 (1967).
(4) V.A. Gordenko and V.I. Nikolaev; JETP Letters 14 3 (1971).
(5) J. Kanamori and Y. Teraoka; to be presented at this Conference.
(6) C.Y. Young and J.B. Sokoloff; J. Phys. F: Metal Phys. 4 1304 (1974).
(7) S. Nakajima and Y. Kurihara; to be published in J. Phys. Soc. Japan.
(8) A. Kotani; to be published in J. Phys. Soc. Japan.

PHENOMENOLOGY OF THE SPIN FLIP TRANSITION IN CHROMIUM

J. W. Allen
Xerox Palo Alto Research Center, Palo Alto, California 94304

and

C. Y. Young
Northeastern University, Boston, Massachusetts 02115

ABSTRACT

The spin flip transition at T_{sf} = 123 K in chromium is described by assuming the existence of a uniaxial magnetic anisotropy and magnetoelastic interaction. The measured strains occurring at T_{sf} and measured elastic constants imply the following magnetoelastic tensor elements in units of 10^5 erg/cm^3: $(F_{11} - F_{13}) = 15.2$, $(F_{12} - F_{13}) = 1.8$, $(F_{31} - F_{33}) = -9.1$. For 95 K < T < T_{sf}, spin flop experiments imply that the anisotropy constant K is $K(T) = (.32\times10^3)(T - T_{sf})$ erg/cm^3. From estimates of $\chi_\perp$ and K at T = 0, the q = 0 spin wave energy is estimated to be about 9 cm^{-1}. The stress dependence of T_{sf} is calculated from the model.

Below T_N = 312 K chromium is antiferromagnetically ordered with an incommensurate linear spin density wave.[1] As the temperature is lowered through T_{sf} = 123.5 K, the polarization of the spin density wave changes abruptly from transverse (AF1 phase) to longitudinal (AF2 phase), and various lattice strains occur.[2] This paper presents a simple phenomenological description of the transition. The phenomenology is very similar to that used[3,4] to describe the spin flip (Morin) transition in the uniaxial antiferromagnetic insulator Fe_2O_3, and consists of including magnetic anisotropy and magnetoelastic terms of uniaxial symmetry in the Gibb's free energy. Chromium is cubic in its paramagnetic state and therefore it might be expected that the largest anisotropy and magnetoelastic terms would have cubic symmetry. However the existence of the spin flip transition, in which the magnetization but not the spin density wave Q-vector changes direction, suggests that in the single-Q magnetic state chromium should be regarded as uniaxial with the c-axis along the Q-vector. Indeed single-Q chromium is found to be distorted with uniaxial (tetragonal) symmetry in the AF2 phase.[2] The additional distortions to orthorhombic symmetry which occur as the magnetization rotates at the spin flip transition are much smaller,[2] and are here ascribed to magnetoelastic distortions superposed on the large uniaxial distortion associated with the presence of the incommensurate Q-vector. No attempt has been made to account for Q-switching effects, and the cubic magnetic anisotropy[5] has been neglected.

Following the preceding discussion the Gibb's free energy Φ is written as

$$\Phi = K(T)\cos^2\theta + \sum_{ij} \varepsilon_i F_{ij}\alpha_j + \frac{1}{2}\sum_{ij} \varepsilon_i C_{ij}\varepsilon_j \ . \qquad (1)$$

In Eq. (1) K is the uniaxial anisotropy constant, which would be zero for cubic symmetry. F, C, and ε are the magnetoelastic, elastic, and strain tensors, respectively, in Voigt notation. The α_i are quadratic products of the direction cosines of the magnetization axis, also in Voigt notation, and $\cos^2\theta = \alpha_3$. For the highest tetragonal symmetry, the nonzero elements of F_{ij} are[6] $F_{11} = F_{22}$, $F_{12} = F_{21}$, $F_{13} = F_{23}$, $F_{31} = F_{32}$, F_{33}, F_{44}, F_{66}. It is assumed that F is temperature independent in the range 0 to T_{sf}. C_{ij} appears to have cubic symmetry.[7] The slight temperature dependence[7] of C in the range 0 to T_{sf} and the small anomalies at T_{sf} are neglected.

To find the equilibrium state Eq. (1) should be minimized with respect to α and ε simultaneously, but if K dominates, θ is determined by the first term, and the equilibrium ε_i minimize the last two terms for fixed θ. If K varies monotonically from negative to positive, passing through zero at T_{sf}, θ is 0 for $T < T_{sf}$ and $\pi/2$ for $T > T_{sf}$, as in chromium. K can be found from measurements of $\Delta\chi \equiv (\chi_{AF1} - \chi_{AF2})$ and the spin flop field[8,9] $H_c = (-2K/\Delta\chi)^{1/2}$. For 95 K < T < T_{sf}, the experimental results are[1] $H_c^2 = 6.76\times10^{10}(1-T/T_{sf})$ G^2 and[10] $\Delta\chi = 14.4\times10^7$ emu/cm^3. K(T) is then determined as shown in Fig. 1, with the dashed lines showing extrapolation from experimental data. The extrapolation is based on the behavior of Fe_2O_3 and other magnetic systems. It is interesting to note that Fe_2O_3 also has a linear variation of K through the Morin temperature, suggesting an analytic behavior of Φ at T_{sf}.

From Eq. (1) it follows that the latent heat ℓ is $T_{sf}(\partial K/\partial T)_{T_{sf}} = 4.86\times10^4$ ergs/cm^3, as pointed out by Street et al.[1] The Clausius-Clapyron equation[11] connects the relative volume change $\Delta v = (v_{AF1} - v_{AF2})$ at T_{sf} and $\partial T_{sf}/\partial P$ with ℓ as $\ell = T_{sf}\Delta v/(\partial T_{sf}/\partial P)$. The quantity $(1/T_{sf})(\partial T_{sf}/\partial P)$ has been measured[12] as -4.63×10^{-11} dynes^{-1}cm^2. Steinitz et al.[2] measured Δv as -1.4×10^{-6} and pointed out that this leads to $\ell = 3.02\times10^4$ erg/cm^3, nearly the same as the value obtained from the temperature dependence of K.

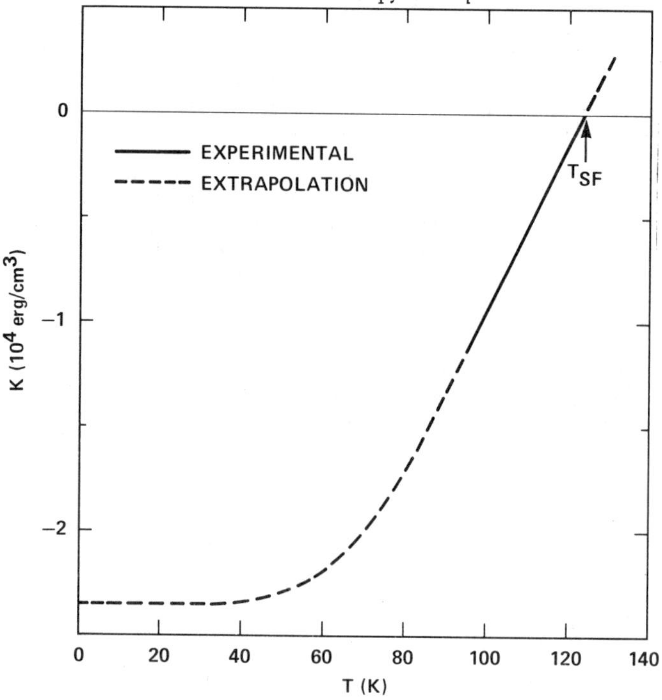

Fig. 1 Temperature dependence of uniaxial anisotropy constant of chromium.

For fixed α_k Eq. (1) is minimized by $\bar{\varepsilon}_i$ given by

$$\bar{\varepsilon}_i = -\sum_{ik} (C^{-1})_{ij} F_{jk}\alpha_k \ . \qquad (2)$$

For AF2, $\alpha_k = \delta_{k3}$, and for AF1, $\alpha_k = \delta_{k2}$. Steinitz et al. determined the individual strain changes giving rise to Δv to be $\overline{\Delta\varepsilon_1} = -.2\times10^{-6}$, $\overline{\Delta\varepsilon_2} = -4.7\times10^{-6}$, and $\overline{\Delta\varepsilon_3} = 3.5\times10^{-6}$, where $\overline{\Delta\varepsilon_i} = \bar{\varepsilon}_i(AF1) - \bar{\varepsilon}_i(AF2)$ at T_{sf}. Using these values and the measured C_{ij} in Eq. (2) then yields $(F_{11} - F_{13}) = 15.2$, $(F_{12} - F_{13}) = +1.8$, and $(F_{31} - F_{33}) = -9.1$, in units of 10^5 ergs/cm^3.

The effect of stresses T_k can be related to the magnetostrictive strains $\overline{\Delta\varepsilon_i}$ by substituting $\varepsilon_i = (C^{-1})_{ik} T_k$ in Eq. (1) (with the convention that negative T_k is compressive) and using Eq. (2). Since hydrostatic pressure ($P = T_1 = T_2 = T_3$) and uniaxial stress T_3 preserve the uniaxial symmetry, anisotropic terms due to P or T_3 can always be expressed as proportional to α_3 using the relation $\alpha_1 + \alpha_2 + \alpha_3 = 1$. It is readily found that $(\partial K/\partial P) = (\overline{\Delta\varepsilon_1} + \overline{\Delta\varepsilon_2} + \overline{\Delta\varepsilon_3}) = \Delta v$, and $(\partial K/\partial T_3) = \overline{\Delta\varepsilon_3}$. Since $K(T_{sf},X) = 0$, where X is T_3 or P, $(\partial T_{sf}/\partial X) = -(\partial K/\partial X)/(\partial K/\partial T)$, leading to $(1/T_{sf})(\partial T_{sf}/\partial P) = 2.88\times10^{-11}$, and $(1/T_{sf})(\partial T_{sf}/\partial T_3) = -7.2\times10^{-11}$. The former is, apart from a sign difference due to the convention that compressive $P < 0$, nearly the experimental value quoted previously.

Stresses T_1 and T_2 render the crystal orthorhombic and so it is convenient to separate out the stress-induced anisotropic part of Φ in terms of α_1 and α_2, giving

$$\Phi = \begin{cases} (-\overline{\Delta\varepsilon_2})T_1\,\alpha_1 + (-\overline{\Delta\varepsilon_1})T_1\,\alpha_2 + K\,\alpha_3 \\ (-\overline{\Delta\varepsilon_1})T_2\,\alpha_1 + (-\overline{\Delta\varepsilon_2})T_2\,\alpha_2 + K\,\alpha_3\,. \end{cases} \quad (3)$$

Since $K < 0$ for $T < T_{sf}$, and $\overline{\Delta\varepsilon_1}, \overline{\Delta\varepsilon_2} < 0$ with $|\overline{\Delta\varepsilon_2}| > |\overline{\Delta\varepsilon_1}|$, a compressive stress of $(-K/\overline{\Delta\varepsilon_2})$ along either the x or y axis should induce the spin to flop to that axis for $T < T_{sf}$.

For nonzero K, the $q = 0$ spin wave energy E is expected to be nonzero. The phenomenological spin wave theory, which is valid in the long wavelength limit and should apply to chromium as well as any other magnetic material, gives[8,9] $E = g(e\hbar/2mc)(-2K/\chi_\perp)^{1/2}$. $\chi_\perp$, the perpendicular susceptibility of the magnetic electrons only, can be estimated[10] as the fraction of the Pauli paramagnetic susceptibility implied by the decrease in specific heat between paramagnetic and antiferromagnetic chromium, which gives about 5×10^{-6} emu/cm^3. From Fig. 1, $K(T = 0)$ is estimated as about -2.35×10^4 erg/cm^3, with $(e\hbar/2mc) = 0.46\times10^{-4}$ cm^{-1}/G, and assuming $g = 2$, E is estimated to be 8.9 cm^{-1}.

At T_{sf} the difference in Φ for AF1 and AF2 must be zero. This difference is defined to be, per unit volume, $0 = (\Phi_1 - \Phi_2) = (u_1 - u_2) - T_{sf}(s_1 - s_2) + p\Delta v$, where u and s are, respectively, the internal energy and entropy per unit volume. Since $\ell = T_{sf}(s_1 - s_2)$ and $\Delta v = (1/T_{sf})(\partial T_{sf}/\partial P)$, $(u_1 - u_2) = \ell(1 - \delta)$ at T_{sf}, where $\delta = (P/T_{sf})(\partial T_{sf}/\partial P)$. For room pressure $\delta \cong 10^{-4}$, so $(u_1 - u_2) = \ell = (3.02$ to $4.86)\times10^4$ erg/cm^3, at T_{sf}. At $T = 0$, $(\Phi_1 - \Phi_2) = (u_1 - u_2) = -K(T = 0)$, again neglecting the volume difference term. The above estimate of $-K(T = 0) = 2.35\times10^4$ erg/cm^3 implies that $(u_1 - u_2)$ increases between $T = 0$ and T_{sf}, which suggests that the spin flip transition occurs because of the entropy difference developed between the two states at T_{sf}. An experimental determination of the $q = 0$ spin wave energy E would be useful in this connection because it would serve to establish $K(T = 0)$ more firmly.

If only the $K(T)\cos^2\theta$ term of Eq. (1) is present to determine θ, as in the approximation used above, then the transition at T_{sf} due to K passing through zero is not strictly first order because there is never more than a single minimum as a function of θ in $\Phi(\theta,T)$. At $T = T_{sf}$, θ is undetermined, since $\Phi(\theta,T_{sf}) = 0$. The fluctuations in θ as T passes through T_{sf} provide a measure of the abruptness of the transition in this model, and can be obtained from a classical statistical thermodynamics analysis. The probability $p(\theta)d\theta$ that θ lies between θ and $\theta + d\theta$ is given by[11]

$$p(\theta)d\theta = C\,\exp(-\Phi(\theta)/kT)d\theta = C\,\exp(-2Z\,\cos^2\theta)d\theta\ , \quad (4)$$

where $Z = -K/2kT$ and C is a normalization constant. By relating the integral for C to a modified Bessel function, and noting the relation $\langle\cos^2\theta\rangle = (C/2)\,d(1/C)/dZ$, it is fairly straightforward to obtain for $\beta^2 = \langle\sin^2\theta\rangle = 1 - \langle\cos^2\theta\rangle$

$$\beta = (1/\sqrt{2})[1 - I_1(|Z|)/I_0(|Z|)]^{1/2}\ , \quad (5)$$

where I_n is the modified Bessel function of order n. The properties of the Bessel functions imply that for $|Z|$ small, $\beta \cong (1/\sqrt{2})(1 - |Z|/2)^{1/2}$, and for $|Z|$ large $\beta \cong \sqrt{1/4|Z|}$. From the experimental results for K and the relation between K and ℓ given previously, it follows that for a large temperature interval near T_{sf} $|Z| = \ell\Delta T/2kTT_{sf}$ where $\Delta T = |T_{sf} - T|$. Putting this expression for $|Z|$ into the large-$|Z|$ result for β gives a formula reminiscent of mean field results for other systems. The small-$|Z|$ result for β shows that for $T = T_{sf}$, $\beta^2 = \frac{1}{2}$ as expected since $p(\theta)$ is uniform for T_{sf}. Examining the various formulas it is evident that large fluctuations will occur only if ΔT is very small (recall that $k = 1.38\times10^{-16}$ erg/K), so that $|Z| \cong \ell\Delta T/2k\,T_{sf}^2 \cong 10^{+11}\,\ell\Delta T$. For $\ell = 10^4$ erg/cm^3, as in chromium, in order to make $|Z|$ even as small as 10^4, ΔT must be $\sim10^{-11}$ for a cubic centimeter of material. Basically the conclusion is that if ℓ is measurably large, ΔT is inaccessibly small, and the transition appears to be sharp in this model.

To sum up, this paper has presented a simple but quite successful phenomenological description of the spin flip transition in chromium, and the associated stress and magnetoelastic effects. The description is based on the assumption of uniaxial symmetry. Other authors[13,14] have pointed out some of these things before, but there has not been a systematic, quantitative treatment of the type given here. This treatment provides no insight into the microscopic origin of the rather large (for a basically cubic crystal) uniaxial effects. Shimizu[14] has given a very interesting phenomenological description of chromium based on cubic symmetry but including anisotropy and magnetoelastic interactions involving space derivatives of the magnetization, which leads to uniaxial effects. But his analysis suffers from certain difficulties, such as not giving a latent heat at T_{sf}. It is interesting to speculate that the uniaxial anisotropy may be related to the presence of the recently observed strain wave[15] inasmuch as the strain wave may owe its existence to a weak charge density wave induced by the spin density wave.[16] It is hoped that the phenomenological analysis presented here may be useful as an intermediate link between the experimental results and a more microscopic theory.

REFERENCES

1. R. Street, B. C. Munday, B. Window, and I. R. Williams, J. Appl. Phys. 39, 1050 (1968).
2. M. O. Steinitz, L. H. Schwartz, J. A. Marcus, E. Fawcett, and W. A. Reed, Phys. Rev. Lett. 23, 979 (1969).
3. V. I. Ozhogin and V. G. Shapiro, Sov. Phys. - JETP 28, 915 (1969).
4. J. W. Allen, Phys. Rev. B 8, 3224 (1973).
5. R. A. Montalvo and J. A. Marcus, Phys. Lett. 8, 151 (1964).
6. W. P. Mason, Physical Acoustics and the Properties of Solids (Van Nostrand, Princeton, N.J., 1958).
7. S. B. Palmer and E. W. Lee, Phil. Mag. 24, 311 (1971).
8. S. Foner, in Magnetism, edited by G. T. Rado and H. Suhl, Vol. I, p.383 (Academic Press, New York, 1963).
9. A. I. Akhiezer, V. G. Bar'yakhtar, and M. I. Kaganov, Soviet Physics Uspekhi 3, 567 (1961).
10. M. O. Steinitz, E. Fawcett, C. E. Burleson, J. A. Schaefer, L. O. Frishman, and J. A. Marcus, Phys. Rev. B 5, 3675 (1972).
11. L. D. Landau and E. M. Lifshitz, Statistical Physics (Pergamon, London, 1958).
12. H. Umebayashi, G. Shirane, B. C. Frazer, and W. B. Daniels, J. Phys. Soc. Japan 24, 368 (1968).
13. B. C. Munday and R. Street, J. Phys. F: Metal Phys. 1, 498 (1971).
14. M. Shimizu, Progr. Theoret. Phys. (Kyoto) Suppl. 46, 310 (1970).
15. Y. Tsunoda, M. Mori, N. Konitomi, Y. Teraoka, and J. Kanamori, Solid State Commun. 14, 287 (1974).
16. C. Y. Young and J. B. Sokoloff, AIP Conference Proceedings, No. 18, p. 342 (1974).

MAGNETIC PROPERTIES OF Cr CONTAINING DILUTE CONCENTRATIONS OF Co IN THE NEIGHBORHOOD OF THE NÉEL TEMPERATURE

Sigurds Arajs, E. E. Anderson, Jack R. Kelly, and K. V. Rao
Department of Physics
Clarkson College of Technology
Potsdam, New York 13676

ABSTRACT

Magnetic susceptibility (χ) of Cr and Cr alloys containing 1.7, 2.2, 2.7, 4.4, and 6.2 at.% Co have been measured as a function of temperature (T) between 300 and 600 K. Each of the χ vs T curves exhibits a well-defined maximum at the Néel temperature (T_N). Assuming that χ of the matrix is unchanged by small additions of solute, the magnetic susceptibility of Co (χ_{Co}) in Cr has been determined over the above-mentioned T range. Above T_N (except close to T_N)χ_{Co} obeys the Curie-Weiss law. The magnetic moment per Co atom decreases with increasing Co concentration. It has been found the χ_{Co} just above T_N cannot be described by a power law with constant critical exponents.

INTRODUCTION

Previous experimental studies[1-3] of the magnetic properties of Cr containing Fe, Co, and Ni strongly suggest that there are localized magnetic moments on Fe and Co, but not on Ni atoms. Furthermore, it appears that below the Néel temperature, the localized moments on Co atoms strongly couple with the Cr lattice, while those on Fe atoms interact weakly with the average magnetic moment of Cr. The purpose of this paper is to present new and detailed studies of the magnetic susceptibility in the neighborhood of the Néel temperature and to discuss the significance of the experimental results.

EXPERIMENTAL CONSIDERATIONS

The Cr-Co samples used in this study, containing 1.7, 2.2, 2.7, 4.4, and 6.2 at.% Co, were cut from the specimens used previously[4] for the electrical resistivity investigations. The magnetic susceptibilities were determined using the Faraday method. The magnetic field was produced by a Magnion 7-inch electromagnet with tapered pole pieces in conjunction with a Magnion Model 1050B power supply. Force measurements were made with an Ainsworth Type 15 electrobalance. The signal output from the balance was read on a Hewlett-Packard 3439A digital voltmeter coupled to a 3443A range unit. The overall accuracy of the magnetic susceptibility measurements was ±0.5% with a relative precision of 1 part in 10^4. It was possible to maintain temperatures within ±0.1 K above about 500 K and within ±0.01K at lower temperatures.

RESULTS AND DISCUSSION

The mass magnetic susceptibility, χ, of Cr and Cr-Co alloys is shown as a function of the absolute temperature, T, in Fig. 1. The anomaly in χ for Cr at the Néel temperature, $T_N \approx 312$ K, is very small in comparison with the large peaks at T_N of Cr-Co alloys. Since the sample containing 1.7 at.% Co has T_N below 300 K, there is no maximum in the χ vs T curve as shown in Fig. 1. The usual method of interpretation of χ data on dilute alloys is to assume that χ of the bulk alloy consists of two components, the susceptibility of the matrix, χ_o, and the susceptibility of the solute impurity, χ_{Co}. Furthermore, it is usually assumed that χ_o is unchanged by small additions of Co. Then

$$\chi = w\chi_{Co}+(1-w)\chi_o,$$

where w is the weight fraction of Co. The T dependence of the quantities $1/\chi_{Co}$, calculated from the above equations are shown in Fig. 2. The straight line behavior, except very close to T_N, clearly confirms that there is a localized moment on Co atoms. Table I gives the values of T_N, determined from $d\chi/dT$ vs T plots, the effective paramagnetic moments, μ_{Co}, and the T range over which the Curie-Weiss law has been used for the determination of μ_{eff}.

Our values of T_N are in good agreement with those determined by Booth[3]. According to Suzuki[2] μ_{Co} is 2.9±0.1 μ_B for the Co concentrations between 1 to 6 at.%. Booth[3] finds that μ_{Co} decreases slightly from 2.3±0.4 μ_B at 1.9 at.% Co, and to 2.0±0.2 at 9.3 at.% Co. Our values of μ_{Co} are higher than those found by Suzuki and Booth. Our μ_{Co} values also show a strong decrease with increasing Co concentrations. It should be remarked that Lingelbach[1] expects μ_{Co} to be 3.87 μ_B, which is the approximate value of μ_{Co} if our data are extrapolated to low levels of Co content. Furthermore, according to Fallot,[5] μ_{Co} in metallic Co in the paramagnetic region is 3.1 μ_B. The reasons for the discrepancies between Suzuki, Booth and our results on μ_{Co} are not clear at the present time.

Lastly, we also have examined in detail the paramagnetic behavior of χ_{Co} in the neighborhood of T_N. Plots of $\ell n\chi_{Co}$ vs $\ell n(T-T_N)$ (which due to space limitations are not shown here) for the samples containing 2.2, 2.7, 4.4, and 6.2 at.% Co, are not straight lines but continuous curves with gradually increasing negative scope as ($T-T_N$) increases from 1 to 40 K. Thus it is not possible to assign a well-defined critical exponent for χ_{Co} above T_N.

ACKNOWLEDGMENTS

This work has been supported by grant number GH-37908 from the National Science Foundation and by the Office of Naval Research under grant number N0014-70-A-0311-0001.

Table I

T_N and μ_{Co} of Cr-Co Alloys

Co centration (at.%)	T_N(K)	$\mu_{Co}(\mu_B)$	Temperature Range (K)
1.7		3.6	304-425
2.2	312	3.5	348-520
2.7	336	3.4	380-582
4.4	323	3.0	340-584
6.2	317	2.9	340-595

REFERENCES

1. R. Lingelbach, Z. Phys. Chem. 14, 1 (1958).
2. T. Suzuki, J. Phys. Soc. Japan 21, 442 (1966).
3. J. G. Booth, J. Phys. Chem. Solids 27, 1639 (1966).
4. S. Arajs, G. R. Dunmyre, and S. J. Dechter, Phys. Rev. 154, 448 (1967).
5. M. Fallot, J. Phys. Radium 5, 153 (1944).

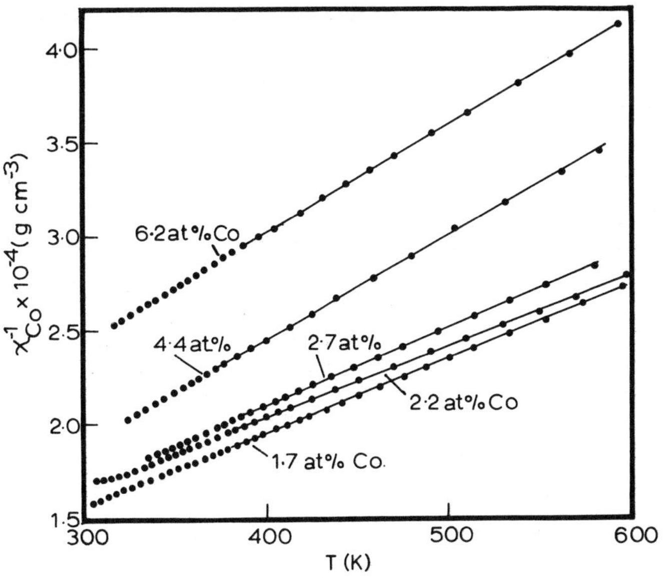

fig.2 χ_{Co} in Cr-Co alloys

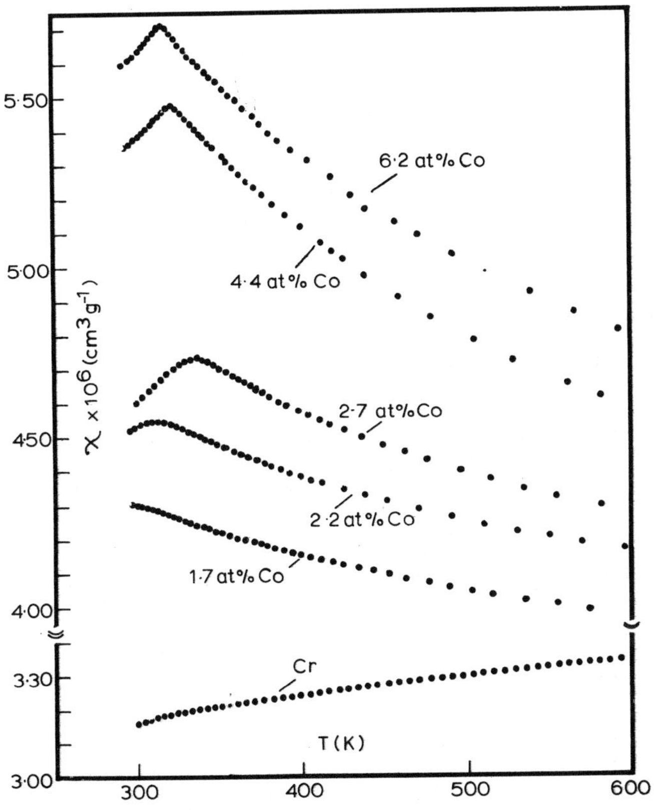

fig.1 χ of Cr-Co alloys

PRESSURE DEPENDENCE OF THE MAGNETIC ORDERING TEMPERATURES IN Cr-Fe ALLOYS*

L. R. Edwards and I. J. Fritz
Sandia Laboratories, Albuquerque, New Mexico 87115

ABSTRACT

The pressure dependences of the magnetic transition temperatures of Cr-2.5 and 2.8 at.% Fe single crystal alloys have been determined by electrical resistance and ultrasonic measurements. The results show that the 1 bar triple point (P,I,C) in the temperature concentration plane occurs at a concentration of ~ 2.5 at.% Fe in contrast to the previously published value of 4 at.% Fe. The high pressure data showed an abrupt change in the pressure dependence of the I-C transition at 3.3 kbar, from which it is speculated that an additional magnetic phase exists.

INTRODUCTION

It is well known that the Cr rich-transition metal alloys exhibit several different magnetic phases and that the transition temperatures between phases are very sensitive to hydrostatic pressure.[1,2] In these alloys three distinct magnetic phases have been observed: (1) paramagnetic (P), (2) incommensurate antiferromagnetic (I), and (3) commensurate antiferromagnetic (C). Among the various Cr-transition metal systems, the magnetic properties of the Cr-Fe alloy system are anomalous.[2,3] There is also some uncertainty in the magnetic phase diagram of this system, especially in the concentration range 2 to 4 at.% Fe.[4] Neutron diffraction measurements by Ishikawa et al[5] have indicated that in the 2 to 4 at.% Fe concentration range the ordering sequence at a pressure of 1 bar is P-I-C with decreasing temperature; however, Arrott et al[6] have suggested from their neutron diffraction results on a Cr-2.3 at.% Fe alloy that the appearance of the I-phase may be due to concentration inhomogeneities. Recently we have shown that well-defined phase transitions can be observed via electrical resistance and ultrasonic measurements on a pure, homogeneous single crystal alloy.[7] Specifically, on a Cr-2.8 at.% Fe sample, it was found that only one transition (P-C) occurred at a pressure of 1 bar and that the triple point among the P, I, and C phases exists in the positive pressure plane. The implication of these results is that the triple point among the P, I, and C phases in the temperature-concentration plane occurs at a concentration less than 2.8 at.% Fe.

The purpose of this work was to locate the P,I,C triple point in the temperature-concentration plane (at 1 bar) and to study the pressure dependences of the transitions. We report the results of electrical resistance and acoustic measurements on two single crystal alloys with concentrations of 2.5 and 2.8 at.% Fe. (The results presented here for 2.8 at.% Fe alloy are an extension to higher pressures of previous work.[7]) They show that the triple point in the temperature-concentration plane occurs very near 2.5 at.% Fe. The high pressure results have led to the speculation of the existence of a new magnetic phase, and it also appears that the I-C transition temperature may be driven to 0°K with easily accessible laboratory pressures (< 10 kbar).

EXPERIMENTAL RESULTS

The single crystals used in this study were grown by an arc-zone melting technique.[8] Both constituent elements contained less than 50 ppm of other transition metals and the compositions of the crystals were determined by atomic absorption spectroscopy to be 2.5 and 2.8 at.% Fe. The resistance and acoustic measurement techniques along with the high pressure helium gas vessel are described elsewhere.[7]

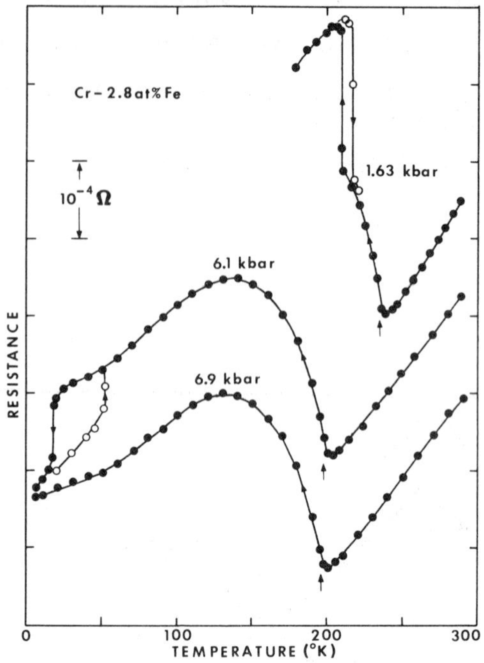

Fig. 1 Typical resistance isobars for the 2.8% alloy. The arrows indicate the location of the inflection points.

In Fig. 1 several typical resistance isobars are shown for the Cr-2.8 at.% Fe alloy. For the various isobars an inflection point is observed ~ 5°K below the resistance minimum. These inflection points correspond in temperature to a sharp "cusp" type of anomaly in the ultrasonic velocities and are interpreted as the P-I transitions.[7] For the 1.63 kbar isobar a second transition is observed, which is characterized by a discontinuous increase in resistance with decreasing temperature. This is the I-C transition and the hysteresis associated with this first order transition increases with increasing pressure. The magnitude of the discontinuous increase in resistance is found to decrease with increasing pressure and becomes very small over the pressure range 3 to 4 kbar. For pressures in excess of 4 kbar, the discontinuous increase in resistance is observed to change to a discontinuous decrease in resistance (cf. the 6.1 kbar isobar in Fig. 1). The magnitude of the discontinuous decrease in resistance increases with increasing pressure, which is opposite of the behavior observed for the lower pressure isobars. The possible significance of the change in the resistance discontinuity will be discussed in the next section. Finally, it is observed that the I-C transition temperature can be driven towards 0°K with the application of sufficient pressure (note that there is no I-C transition down to 4°K for the 6.9 kbar isobar in Fig. 1). Similar resistance anomalies are also observed for the Cr-2.5 at.% Fe alloy.

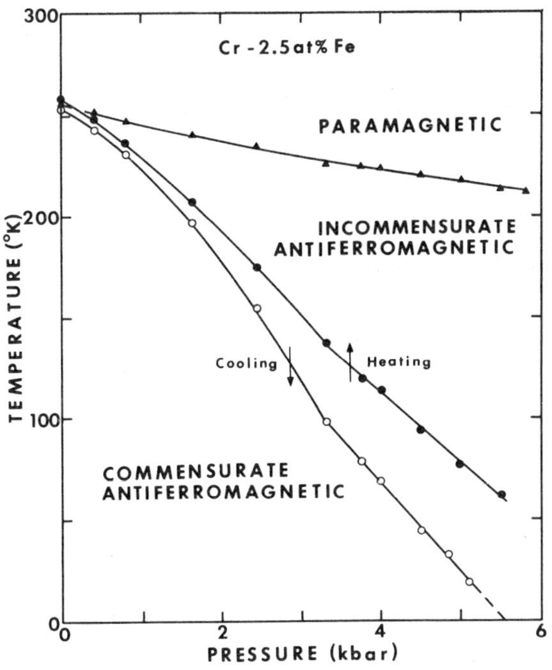

Fig. 2 The magnetic phase diagram for the 2.5% alloy. The arrows indicate the hysteresis.

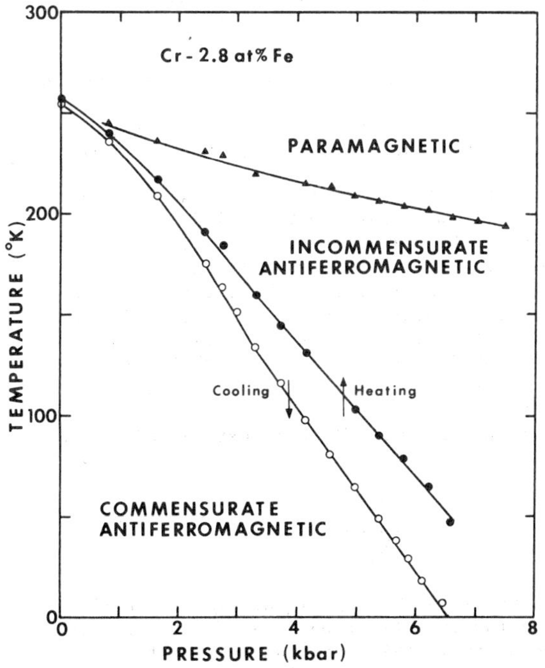

Fig. 3 The magnetic phase diagram for the 2.8% alloy. The arrows indicate the hysteresis.

The temperature-pressure magnetic phase diagrams for the 2.5 and 2.8 at.% Fe alloys are shown, respectively, in Figs. 2 and 3. Except for the low pressure region, the general features of the phase diagrams are similar for both alloys. The P-I transition temperature is observed to be relatively insensitive to pressure and to become progressively more insensitive to pressure with increasing pressure. In contrast, the I-C transition temperatures are very sensitive to pressure — decreasing quadratically with pressure up to ~ 3.3 kbar. At 3.3 kbar an abrupt change to a linear dependence is observed.

DISCUSSION

The various features of the phase diagrams (Figs. 2 and 3) will be discussed in order of increasing pressure. At low pressures for the 2.8% alloy, it has been previously shown[7] that only the P-C transition occurs at 1 bar and that the triple point (P,I,C) occurs at ~ 0.5 kbar. (There are actually two triple points because of the hysteresis associated with the I-C transition.) The current results, for the 2.5% alloy at 1 bar, show that with decreasing temperature there are two transitions (P-I and I-C) which are separated by only 2K°. On the other hand, with increasing temperature only one transition (C-P) is observed and the triple point (P,I,C) occurs in the positive pressure plane at ~0.25 kbar. The results for the cooling cycle show that the P-I and I-C boundaries extrapolate to an intersection point in the negative pressure plane (~.25 kbar). It is concluded from these observations that the triple point (P,I,C) at 1 bar in the temperature-concentration plane occurs at an Fe concentration very close to 2.5%. It should be noted that there will be two triple points in the temperature-concentration plane because of hysteresis effects.

In the intermediate pressure range an abrupt change in the pressure dependence of the I-C transition temperature is observed for both alloys. This change from a quadratic to a linear pressure dependence occurs at ~ 3.3 kbar and is independent of concentration and the direction of the temperature cycle. The temperature at which this change occurs, however, is dependent on concentration and the sense of the temperature cycle. The slope $(\partial T/\partial P)$ at 3.3 kbar was observed to change from 71°K/kbar to 44°K/kbar for the 2.5% alloy, and for the 2.8% alloy the slope change was 53°K/kbar to 41°K/kbar. In addition to the change in the functional pressure dependence of the I-C transition temperature at 3.3 kbar, the character of the resistance anomaly associated with this transition changes (cf. Fig. 1). These results seem to suggest the presence of an additional magnetic phase. A search was made by the present measuring techniques for a new phase boundary, and the results were null. Another technique, such as neutron diffraction measurements under pressure, will be necessary to establish the possible existence of a new magnetic phase.

In the high pressure limit the I-C transition temperature is observed to approach absolute zero linearly with pressure. For the cooling cycle data the I-C phase boundary extrapolates to 0°K at critical pressures of 5.5 and 6.5 kbar, respectively, for the 2.5 and 2.8% alloys. In fact at a pressure of 6.7 kbar there was no evidence of the I-C transition down to 4°K for the 2.8% alloy (cf. Fig. 1). This appears to be the first observation where the I-C transition temperature of a Cr alloy has been followed to liquid helium temperatures as a function of pressure.

The authors are grateful to Mr. F. A. Schmidt of the Ames Laboratory, Iowa State University, for the preparation of the single crystal alloys. Acknowledgment is also due to Ms. D. I. Miller for sample analysis, and to Mr. J. D. Pierce and J. Clickner for their expert technical assistance.

REFERENCES

* This work was supported by the U. S. Atomic Energy Commission.

1. T. M. Rice, A. Jayaraman and D. B. McWhan, J. Phys. (Paris) Colloq. 32, C 1-39 (1971).
2. R. Nityanada, A. S. Reshamwala and A. Jayaraman, Phys. Rev. Lett. 28, 1136 (1972).
3. A. Kotani, J. Phys. Soc. of Jap. 36, 103 (1974).
4. A. S. Barker and J. A. Ditzenberger, Phys. Rev. B1, 4378 (1970).
5. Y. Ishikawa, S. Hoshino and Y. Endoh, J. Phys. Soc. Jap. 22, 1221 (1967).
6. A. Arrott, S. A. Werner and H. Kendrick, Phys. Rev. 153, 624 (1967).
7. L. R. Edwards and I. J. Fritz, AIP Conf. Proc. 18, 401 (1974).
8. O. N. Carlson, F. A. Schmidt and W. M. Paulson, Trans. Am. Soc. Metals 57, 356 (1964).

MAGNETIC ENTROPY AND MAGNETIZATION OF $ZrZn_2$*

R. Viswanathan+ and J. R. Clinton
Department of Applied Physics and Information Science
University of California at San Diego
La Jolla, California

ABSTRACT

The previously published low temperature heat capacity data of four $ZrZn_2$ samples were analyzed to extract the magnetic entropy, $S_m(T)$. The well-ordered vapor deposited $ZrZn_2$ with $T_c \sim 22$ K showed a finite but small $S_m(T)$ of 110 mJ/gm mole K, indicating that it is very much an itinerant electron ferromagnet. However the dramatic increase of $S_m(T)$ with the increase of Curie temperature suggests that better the samples more the deviation from the Stoner-Wohlfarth band theory. The magnetic measurements, carried out on the same samples confirm the general applicability of the band theory and the sensitivity of the samples to the details of preparation and heat treatment.

INTRODUCTION

The metallic compound $ZrZn_2$ has been shown to be well-described, at least below its Curie temperature, T_c, by the itinerant-electron theory of ferromagnetism.[1,2] Many of its properties have however remained puzzling. Widely different Curie temperatures have been reported depending on the measurement technique and the details of the sample preparation.[3,4,5]

Early measurements of the heat capacity of $ZrZn_2$ failed to show any anomaly near the Curie temperature,[4,6] but our recent measurements using the steady state calorimetric technique have detected these anomalies in four different samples with different T_c's.[7] In this paper we analyze these published data to evaluate the magnetic contribution to the heat capacity and the magnetic entropy for each of these samples. The magnetic measurements, performed on the same samples, are compared with the heat capacity data to verify the predictions of the band theory and to confirm the sensitivity of the Curie temperature to the sample inhomogeneity.

RESULTS AND DISCUSSIONS

The measured heat capacity of a ferromagnetic metal is due to magnetic ordering as well as the usual lattice and electronic contributions. The separation of the magnetic contribution is generally complicated by an inability to determine the temperatures at which it begins and ends.

If one assumes that there is magnetic contribution to heat capacity only at temperatures $\sim 7° \lesssim T \lesssim T_c$, then γ and β could be deduced from below 7 K and the higher order terms from above Tc. This will give a base line corresponding to the nonmagnetic $ZrZn_2$. When this is done for sample #42, the magnetic entropy

$$S_m = \int_0^\infty \frac{c_m}{T}\, dT$$

derived is ~27 mJ/gm mole K. This value agrees very well with the 26 mJ/gm mole K obtained[8] by comparing the heat capacity data of $ZrZn_2$ to that of an isostructural nonmagnetic compound $ZrCu_{1.1}Al_{0.9}$. This suggests that such an analysis of the data is reasonably meaningful for taking out the magnetic heat capacity which is otherwise almost impossible. The results for all four samples are given in Fig. 1, the numbers within the triangular graphs being the respective magnetic entropy to within ±5%.

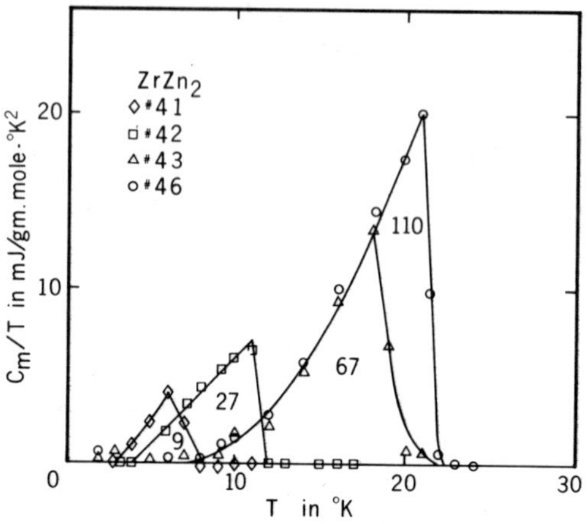

Fig. 1. Magnetic Heat Capacity for $ZrZn_2$ Samples

The magnetic entropy, $S_m(T)$ is $R \ln (2S+1)$ in the local moment model[9] and zero in the band model,[1] for $T \geq T_c$. The observed magnetic entropy for the well ordered $ZrZn_2$ (sample #46) is 110 mJ/gm mole K which is only $.02\ R \ln 2$, for a spin 1/2. This indicates that the magnetic moments in $ZrZn_2$ are not localized. The fact that the magnetic entropy has a definite nonzero value and also that it increases dramatically with the Curie temperature (and hence lattice order), suggests that for better samples, there will be greater deviation from the band description.

The band theory predicts a heat capacity jump, ΔC, at the T_c given by

$$\Delta C = \frac{M^2(0,0)}{\chi_o T_c} ,$$

where the values of $M(0,0)$, χ_o and T_c, are evaluated from the magnetization data.[10] For the most homogeneous sample #46, the observed ΔC of 420 mJ/gm mole K is somewhat larger than the predicted value of 280 mJ/gm mole K. Considering the extreme sensitivity of χ_o to sample condition, the agreement is considered to be good. The comparisons for the other samples are invalid due to their large inhomogeneity and unusually field-dependent magnetization.[10]

The Curie temperature T_c can be determined by any one of the following methods. It is the highest temperature at which (i) the heat capacity transition begins, (ii) the M^2 vs H/M plot goes through the origin, (iii) the low field susceptibility begins to diverge and (iv) the hysteresis of the material just disappears. The T_c's of the four $ZrZn_2$ samples in the present investigations, as defined above, are given in Table I.

Table I
T_c Data for $ZrZn_2$ Samples

Sample	From Spec. Heat	From Arrott Plot	From ac Susc.	From Hysteresis
	T_c in K			
#41	7	8.6	9	31
#42	12	12.8	14	24
#43	19	21.4	22.5	35
#46	22	20.8	22.2	24

The extreme sensitivity of the magnetic properties of $ZrZn_2$ to the details of sample preparation and heat treatment which has been already pointed out[4,7,11] is apparent in these samples. In a nonhomogeneous sample it has been suggested[4] that the hysteresis loop method may give a substantially higher T_c than that from the Arrott plot. This is clearly seen from Table I, for sample #41, 42 and 43, which show traces of ferromagnetism up to and above 24 K. Since these samples do not contain appreciable amounts of second phase-viz, $ZrZn_3$ - one could conclude that the variations in their magnetic properties and T_c's are due to differences in the degree of the respective lattice order. The vapor grown sample #46, is clearly crystallographically the most well-ordered and homogeneous and it shows the highest magnetization.[10] For such a homogeneous sample the hysteresis loop T_c agrees very well with those deduced from other methods.

ACKNOWLEDGMENT

We would like to thank Professors Mydosh and Bringer for interesting discussions; Professor Wohlfarth for critical reading of the manuscript, and Professor Luo for supplying the samples and for his continued interest.

REFERENCES

1. Wohlfarth, E. P., J. Appl. Phys. 39, 1061 (1968).
2. Knapp, G. S., J. Appl. Phys. 41, 1073 (1970).
3. Foner, S., McNiff, Jr., E. J. and Sadagopan, V., Phys. Rev. Lett. 19, 1233 (1967).
4. Ogawa, S. and Sakamoto, N., J. Phys. Soc. Jap. 22, 1214 (1967).
5. Blythe, H. J., J. Phys. C.1, 1604 (1968).
6. Hoare, F. E. and Wheeler, J. C. J., Phys. Lett. 23, 402 (1966).
7. Viswanathan, R., Luo, H. L. and Massetti, D. O., AIP Conf. Proc. on Magnetism and Magnetic Materials, No. 5, 1290 (1971).
8. Viswanathan, R., J. Physics F.; Metal Physics 4, L57 (1974).
9. Gopal, E. S. R., "Specific Heats at Low Temperatures," Plenum Press, N.Y. (1966).
10. Clinton, J. R. and Viswanathan, R. (to be published).
11. Matthias, B. T., Clogston, A. M., Williams, H. J., Corenzwit, E. and Sherwood, R. C., Phys. Rev. Lett. 7, 7 (1961).

*Work supported by U.S. Atomic Energy Commission under contract USAEC-AT-(04-3)34.
+Presently at Brookhaven National Lab., Upton, N.Y. 11973.

FERMI SURFACE "NESTING" IN Tb ALLOYS

P. Burgardt and S. Legvold
Ames Laboratory-USAEC and Department of Physics, Iowa State University, Ames, Iowa 50010.

ABSTRACT

Magnetization and electrical resistivity measurements from 4.2 to 300 K have been made on alloys of Tb with Th, Mg, and Yb in order to test valency effects on magnetic ordering in Tb. Tetravalent Th raises E_F and promotes ferromagnetism by reducing the "nesting" feature in the Tb Fermi surface. Divalent Mg and Yb lower E_F to enhance the "nesting" feature and promote the helical structure. The results are understood in terms of valency effects on the Fermi surface and impurity size effects.

INTRODUCTION

At present it is understood that magnetic order in the rare earth metals is governed by the Ruderman-Kittel-Kasuya-Yosida (RKKY) indirect exchange interaction. An appropriate way to consider the RKKY interaction is through a generalized susceptibility function $\chi(\vec{q})$ which describes polarization of the conduction electrons by the internal field. A peak in $\chi(\vec{q})$ represents a minimum in free energy and hence a stable magnetic structure of periodicity $\vec{q}$. Peaks in $\chi(\vec{q})$ are generated by parallel or "nested" regions in the Fermi surface where many states are separated by a characteristic $\vec{q}$. This dependence of magnetic structure on Fermi surface geometry was first suggested by Lomer[1] in relation to Cr and was later applied to the heavy rare earths by Evenson and Liu[2] using the band calculations of Keeton and Loucks[3].

The pair of bands which determine the shape of the Fermi surface in the heavy rare earths are quite flat along A to H such that Fermi surface "nesting" is very sensitive to changes of Fermi energy E_F. For a given set of bands, lowering of E_F is expected to increase the $\vec{q}$ at which the peak in $\chi(\vec{q})$ occurs and should produce a more prominent peak in $\chi(\vec{q})$. Hence, lowering of E_F should increase the dominance of the periodic magnetic ordering. On the other hand, raising of E_F can be expected to diminish the stability of such ordering by reducing the "nesting" feature.

These theoretical facts provided the motivation for a study by Mellon and Legvold[4] in which an attempt was made to induce a periodic magnetic structure in Gd by lowering E_F through alloying of the trivalent rare earth with divalent Mg. That study was inconclusive, presumably due to failure of rigid band ideas to extend to the more concentrated alloys. In this study we circumvent these difficulties by examining Fermi surface effects in terbium which exhibits anti-ferromagnetic (helical structure) ordering in the pure metal. We alloyed Tb with divalent Mg and Yb to lower E_F and with tetravalent Th to raise E_F and were able to obtain interesting, useful results in relatively dilute alloys.

EXPERIMENTAL PROCEDURE AND RESULTS

The single phase solid solution alloys used in this study were prepared from 99.9+% pure materials by melting together carefully weighed amounts of the constituents. The Tb + Th alloys were arc melted into buttons which were annealed at 1100 C for 3 days followed by an ice water quench. The Tb + Mg and Tb + Yb alloys were melted into the form of fingers inside Ar filled Ta crucibles and then annealed at 900 C for 3 days, again inside the crucibles, followed by an ice water quench. Samples were cut using a spark erosion apparatus and were hand lapped to the final dimensions. Single crystals of a usable size were not obtained so all results reported are for polycrystalline samples.

The magnetization measurements were performed on a Foner type vibrating sample magnetometer having a sensitivity of $\sim 10^{-5}$ emu. The sample temperature could be varied from 4.2-300 K in a helium cryostat using the vapor flow for cooling the samples. Available fields ranged from 0 to 30 kOe. The resistivity measurements were performed in the same cryostat using a four probe dc technique and a voltage sensitivity of 10^{-8}V.

Magnetization versus temperature results at a low field (50 Oe) were first obtained on a pure Tb sample which had been carried through the annealing procedure. The ordering temperatures obtained were 229.5 K and 220.8 K for the Néel temperature T_N and Curie temperature T_C respectively. These values agree nicely with the results of Hegland et al.[5].

Magnetization versus temperature experiments at low fields (50-100 Oe) gave the ordering temperatures shown in Fig. 1. The effectiveness of

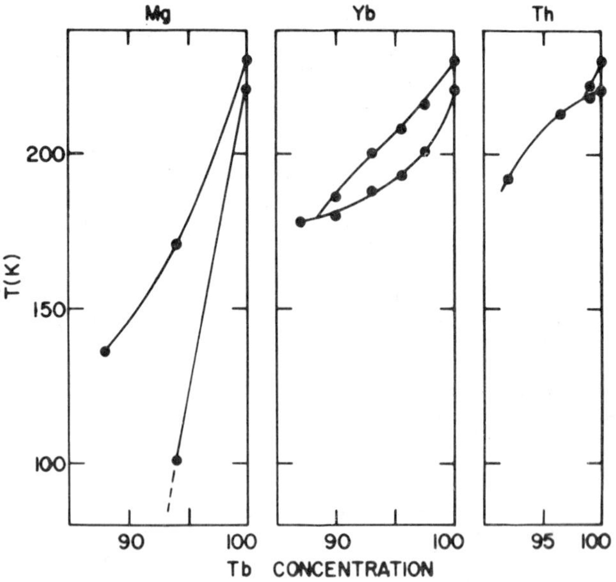

Figure 1: Ordering temperatures for the Tb with Mg, Yb, and Th alloys versus impurity concentration.

Mg in promoting the helical structure and the unusual ability of Th to promote ferromagnetism are particularly noticeable. It is also apparent that divalent Mg and Yb produce alloys having significantly different properties. The great difference between Mg and Yb must be due to the large size difference between the ions since they are otherwise similar non-magnetic diluents. Thus, in order to understand correctly the valence effects we must experimentally remove impurity size effects. We propose that an appropriate way to handle size effects is to use as a baseline the known results for a non-magnetic trivalent impurity having a size equal to that of the divalent or tetravalent impurity. When performing this operation we must be careful to account for the overall suppression of ordering temperatures due to dilution of the magnetic host: this is easily done by comparing alloys of equal impurity concentration. On the basis of atomic size we believe that Mg should be compared with Sc and that Yb and Th should be referenced to La. The actual ordering temperatures are not easily compared but the parameter T_N-T_C, which is an indicator of helical structure stability and of $\chi(\vec{q})$ peak height, has proven useful. From work by Child and Koehler[6], it was found that for a Tb+10%Sc alloy T_N-T_C = 41 K. For an equally dilute Tb+10%Mg alloy we estimate from Fig. 1 that T_N-T_C = 145 K. In this case, lowering E_F has apparently been very effective in promoting the helical structure. In a similar fashion, we use the La alloy results of Koehler et al.[7] in addition to a Tb+4%La alloy prepared for this study, which we found to have T_N = 208 K and T_C = 204 K, to determine that T_N-T_C becomes zero at ~5% La concentration. The Tb + Yb alloys gave T_N-T_C = 0 at ~12% Yb concentration (recall that the additional host dilution at 12% Yb content indicates an even greater contrast between the 5% for La and the 12% for Yb) which again demonstrates the effectiveness of lowering E_F in promoting the helical magnetic structure. The Tb + Th alloys have T_N-T_C = 0 at ~2% Th concentration to be compared with ~5% for La. In this case, raising E_F has been very effective in diminishing T_N-T_C, which we use as an indicator of the stability of the helical magnetic structure. Thus, shifting of E_F through alloying has produced the predicted results to confirm the dependence of magnetic ordering on Fermi surface "nesting".

The electrical resistivity results are shown in Fig. 2. These results can be used to complement the magnetization data in establishing the ordering temperatures since it is known that electrical resistivity is sensitive to magnetic ordering. In all cases, the ordering temperatures from both types of measurements correspond quite well. It can be seen that Yb and Th alloys exhibit resistivities which are what one might expect for non-magnetic impurities in Tb. The Mg alloys are very interesting in that they exhibit large residual resistivities. This effect may be due to removal of d electrons from the conduction process by resonant d bound states located on the Mg sites as proposed by Mellon and Legvold[4]. In addition, lattice strains due to the relatively small size of the Mg ions, may play a significant role. At this time we must conclude that the resistivities as a function of temperature are not well understood.

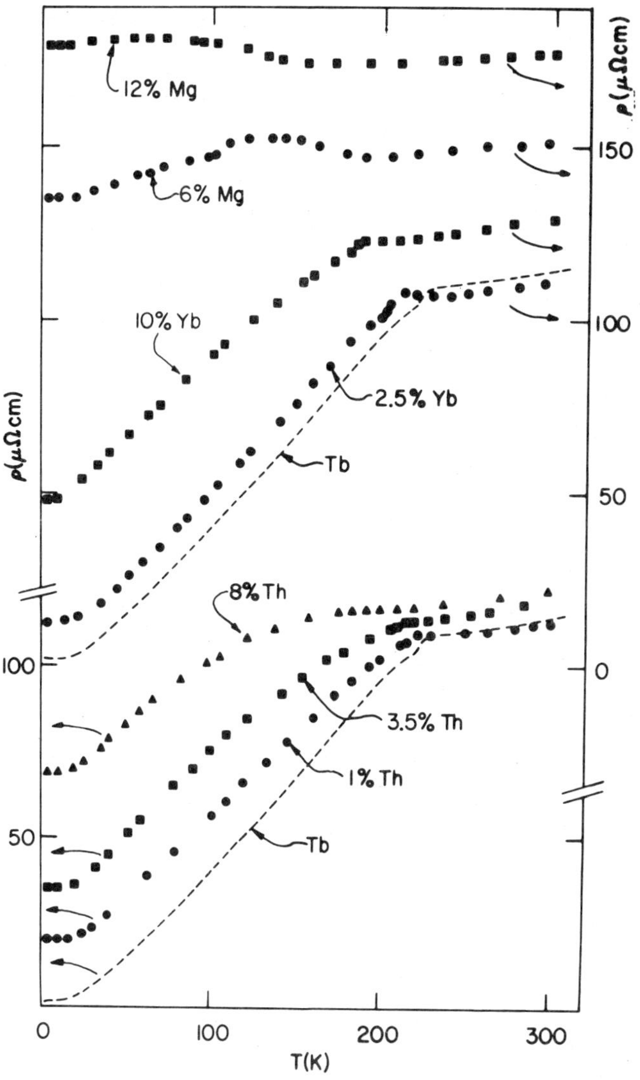

Figure 2: Electrical resistivity versus temperature results for the Tb with Mg, Yb, and Th alloys.

Helpful discussions with S. H. Liu and S. K. Sinha and the aid of B. Beaudry in sample preparation are gratefully acknowledged.

[1] W. M. Lomer, Proc. Phys. Soc. (London) 80, 489 (1962).

[2] W. E. Evenson and S. H. Liu, Phys. Rev. 178, 783 (1969).

[3] S. C. Keeton and T. L. Loucks, Phys. Rev. 168, 672 (1968).

[4] D. W. Mellon and S. Legvold, J. Appl. Phys. 42, 1295 (1971).

[5] D. E. Hegland, S. Legvold, and F. H. Spedding, Phys. Rev. 131, 158 (1963).

[6] H. R. Child and W. C. Koehler, J. Appl. Phys. 37, 1353 (1966).

[7] W. C. Koehler, J. W. Cable, H. R. Child, R. M. Moon, and E. D. Wollan, Proc. Int. Conf. on Magnetism, Nottingham (1964), p. 271.

MAGNETIC PROPERTIES OF SOME Au_3R COMPOUNDS

L. R. Sill and B. Prindiville
Northern Illinois University, DeKalb, Illinois 60115

ABSTRACT

The magnetization of a series of intermetallic compounds Au_3R, where R is Gd through Yb, has been investigated at temperatures from 2.5° to 300°K in applied fields up to 26 kOe. All the compounds studied exhibited the orthorhombic $TiCu_3$-Do_a type structure. For high temperatures the temperature dependence of the inverse susceptibility followed a Curie-Weiss law yielding effective paramagnetic moments in good agreement with the values calculated for free tripositive rare earth ions. At low temperatures deviations from Curie-Weiss behavior were observed in all cases. These deviations are ascribed to the influence of crystal-field and exchange interactions.

INTRODUCTION

The magnetic properties of rare earth intermetallic compounds are of fundamental interest and have been studied extensively in recent years.[1] The main focus of these studies has been on binary compounds of composition RX_n where X is a magnetic atom since these compounds usually exhibit ferri- or ferro-magnetic ordering and show no crystal field effects. In contrast, compounds of composition RX_n (n=1,2,3,4) where X is a nonmagnetic atom have been given much less attention. From existing magnetic studies it is found that only in some cases is the exchange interaction sufficiently strong to build up a cooperative phase on cooling. In all other cases no magnetic ordering occurs on cooling and the behavior of these compounds at low temperature can be understood in general by assuming a large interaction of the crystal field with the incomplete 4f shell which is stronger than the exchange interaction.

We have previously studied the static magnetic properties of the systems RAu_2[2] and RAu_4;[3] we now report the results of magnetic measurements on the series of compounds RAu_3 (R=Gd through Yb). Similar studies have been reported for the systems RAl_3,[4] RSn_3[5] and RPd_3.[6]

The crystal structures of the RAu_3 (R=Gd through Yb) compounds have been determined by Sadagopan et al[7] and confirmed by McMasters and Gschneider.[8] Sadagopan et al found that the structure of the Au_3R phase can be understood as a gradual transition from A_3-like (h.c.p. Mg-A_3) atomic arrangement to an A_2-like (b.c.c. W-A_2) one and can be described as a distortion of an A_3 structure type to an orthorhombic $TiCu_3$-Do_a structure type. McMasters and Gschneider found that the structure type $TiCu_3$ did not exist for any other lanthanide-metal compounds.

EXPERIMENTAL

The compounds were prepared by arc-melting a stoichiometric mixture of high purity constituents (Au 99.99%, R 99.9+%) under an argon atmosphere. To insure homogeniety each button was arc-melted three times, the buttons being flipped over between each melting. The weight losses were negligible. To insure equilibrium the buttons were homogenized in evacuated capsules for from 3 days to 3 weeks at temperatures from 600° to 850°C and furnace cooled. Debye-Sherrer powder patterns showed that the specimens had the proper structure and were free of a second phase.

Spheres approximately 1/8 inch in diameter were formed from the arc-melted buttons. Before making any measurements on the spheres they were reannealed in evacuated capsules for about 3 hours at a temperature of 600°C and furnace cooled. Isothermal magnetization measurements were made in applied fields up to 26 kOe at regular temperature intervals from 2.5° to 300°K using a commercial version of a vibrating sample magnetometer. All magnetic data were obtained from polycrystalline samples.

RESULTS and DISCUSSION

For the 7 compounds studied the reciprocal susceptibility, χ^{-1}, was observed to depend linearly on the temperature from 200° to 300°K. The values for the effective moments, μ_{eff}, and the asymptotic Curie temperatures, θ, have been derived from a least squares fit of a straight line to this part of the χ^{-1} vs T curves (Table I). The observed effective moments are in reasonable agreement with the calculated free tripositive ion values.

Table I. Measured properties of Au_3R compounds

R	θ (°K)	μ_{eff} (μ_β/formula unit)	$g\sqrt{J(J+1)}$ (μ_β/rare earth atom)
Gd	17.9	7.64	7.94
Tb	6.3	9.90	9.72
Dy	21.0	10.36	10.63
Ho	10.8	10.09	10.60
Er	23.7	8.99	9.60
Tm	26.3	7.21	7.60
Yb	33.5	3.95	4.54

An interesting characteristic of the Au_3R intermetallic compounds is the positive value of the asymptotic Curie temperature, θ, and the fact that there is no clear indication of magnetic ordering at low temperatures. Furthermore, θ does not show the expected systematic variation throughout the series of compounds. The dominant interaction responsible for magnetic ordering in intermetallic compounds of rare earths with other nonmagnetic metals is generally accepted to be the indirect exchange interaction between rare earth ions proceeding via the conduction electron spins. The mechanism for this interaction is best described by the RKKY theory.[9] DeGennes[10] showed that for the RKKY theory, using a molecular field approximation, the asymptotic Curie temperature should vary according to $(g-1)^2J(J+1)$ which also represents the strength of the RKKY interaction. In the present study the highest asymptotic Curie tempera-

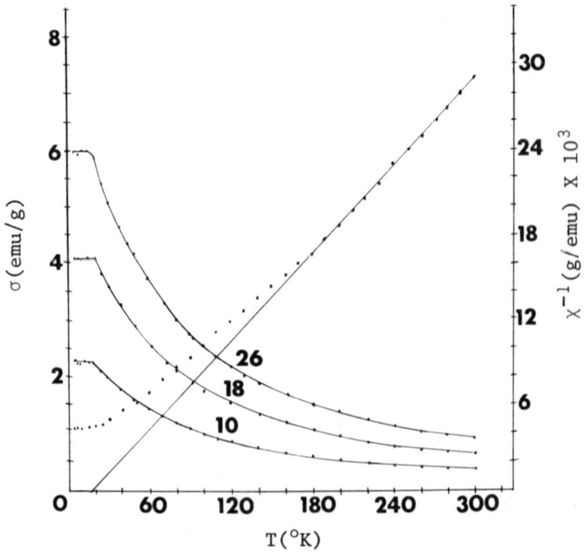

Fig. 1. Magnetic results for Au_3Gd

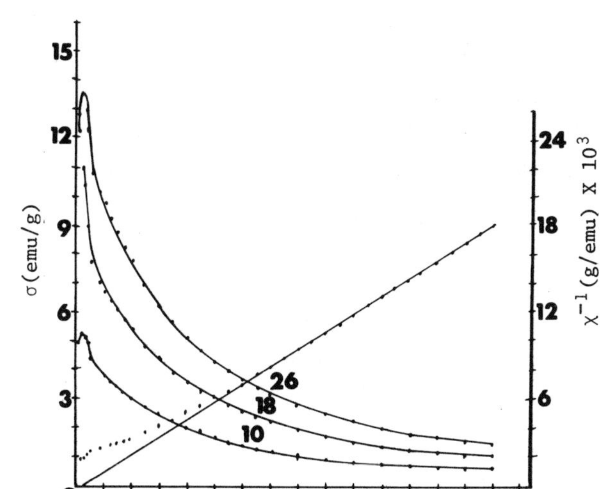

Fig. 2. Magnetic results for Au_3Tb

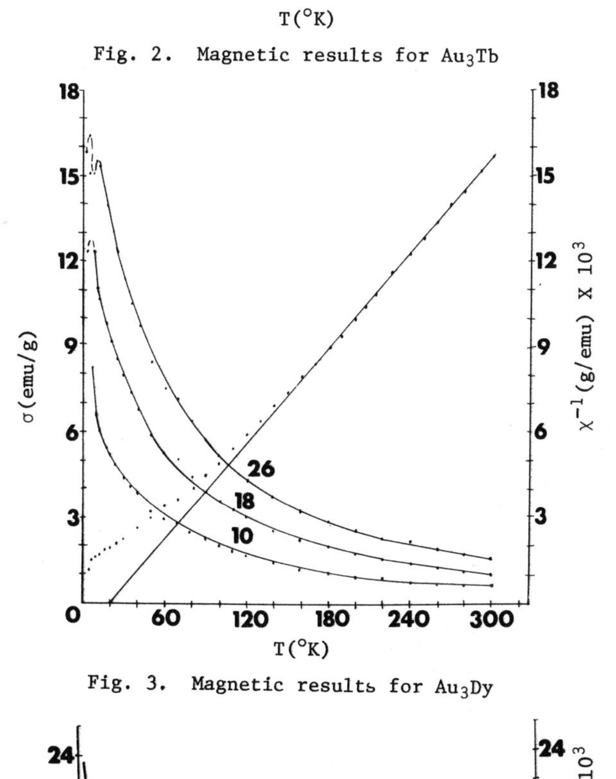

Fig. 3. Magnetic results for Au_3Dy

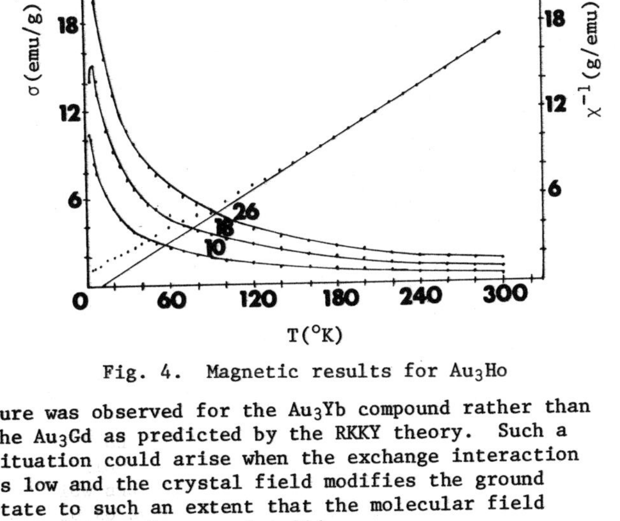

Fig. 4. Magnetic results for Au_3Ho

ture was observed for the Au_3Yb compound rather than the Au_3Gd as predicted by the RKKY theory. Such a situation could arise when the exchange interaction is low and the crystal field modifies the ground state to such an extent that the molecular field approximation becomes invalid.

The χ^{-1} vs T curves as well as σ vs T curves for applied fields of 10, 18, and 26 kOe are shown in Figs. 1-4 for Gd through Ho compounds. The curves for Er, Tm and Yb compounds are similar to the ones for the Ho compound except there are no peaks in the σ vs T curves. Inspection of the χ^{-1} vs T curves shows that at low temperatures there are significant deviations from the line drawn through the high temperature data. Again, these deviations are not well understood. On the one hand, this kind of behavior suggests that the effects of the crystal field on the 2J+1 fold degenerate ground levels of the rare earth ions are important. Furthermore, since the χ^{-1} vs T curves provide no clear indication of magnetic ordering above 2.5°K and since the asymptotic Curie temperatures, θ, are relatively small, it seems as though the indirect exchange interactions are relatively weak. On the other hand, the fact the deviation from Curie-Weiss behavior is a maximum for the Gd compound, where the crystal field effect is expected to be negligible since Gd^{3+} ion is in an all-spin state, suggests that the deviation of the magnetic behavior from the free-ion type must be due to the exchange interaction. When the exchange energy has approximately the same order of magnitude as the crystal field energy, the competition between these two effects gives rise to complex magnetic structures and to unusual magnetization processes. In particular, when crystal field effects are present, a value of θ is obtained which is related in part to exchange interactions and in part to the population of the higher state(s).

While the χ^{-1} vs T curves in Figs. 1-4 provide no clear evidence of the occurance of cooperative magnetic phenomena in any of these compounds, the peaks in the σ vs T curves in the same figures and the nonlinearity of the σ vs H_a curves at 4.2°K in Fig. 5 for all compounds except Au_3Gd are characteristic of interacting paramagnets. In addition, for Au_3Dy we observed a very small remanent moment at 4.2°K.

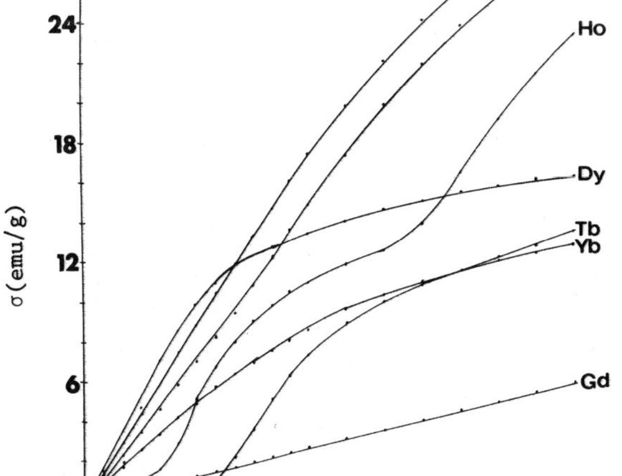

Fig. 5. Magnetic results for Au_3R at 4.2°K

The apparent saturation of the σ vs T curve for Au_3Gd below 20°K is hard to understand in the light of the linearity of the σ vs H_a curves for all temperatures. The peak in the σ vs T curves for the Au_3Tb, Au_3Dy and Au_3Ho and the steep rise in magnetization when the applied field exceeds a critical value for Au_3Tb and Au_3Ho strongly suggests that these compounds are weakly, antiferromagnetic. The appearance of critical field behavior as well as the nonlinearity of all but one of the σ vs H_a curves emphasizes the competitive nature of the crystal field and exchange field interactions as a result of their comparable magnitudes in these compounds.

It is a pleasure to thank A. E. Dwight for technical assistance in the course of this work.

REFERENCES

1. K.N.R. Taylor, Adv. in Phys., 20, 551 (1971). W.E. Wallace, "Rare Earth Intermetallics," Academic, New York, 1973.
2. L.R. Sill, S.R. Snow and A.J. Fedro, AIP Conf. Proc. 10, 1060 (1973).
3. L.R. Sill, W.J. Mass, A.J. Fedro, J.C. Shaffer and C.W. Kimball, J. Appl. Phys., 42, 1297 (1971).
4. K.H.J. Buschow and J.F. Fast, J. Phys. Chem. 50, 1 (1966).
5. K.H.J. Buschow, H.W. deWijn and A.M. vanDiepen, J. Chem. Phys. 50, 137 (1969).
6. W.E. Gardner, J. Penfold, T.F. Smith and I.R. Harris, J. Phys. F, 2, 133 (1972).
7. V. Sadagopan, B.C. Giessen and N.J. Grant, J. Less-Com. Met. 14, 279 (1968).
8. O.D. McMasters and K.A. Gschneider, Jr., J. Less-Com. Met. 25, 135 (1971).
9. M.A. Ruderman and C. Kittel, Phys. Rev., 96, 99 (1954). T. Kasuya, Progr. Theor. Phys. (Kyoto) 16, 45 (1956). K. Yosida, Phys. Rev., 106, 893 (1957).
10. P.C. deGennes, J. Phys. (Paris) 23, 510 (1962).

A NEUTRON INVESTIGATION OF THE SINUSOIDAL LATTICE WAVE IN PURE CHROMIUM

C. F. Eagen
Scientific Research Staff
Ford Motor Company
Dearborn, Michigan 48121

and

S. A. Werner
Scientific Research Staff, Ford Motor Company, Dearborn, Michigan 48121 and Department of Nuclear Engineering, University of Michigan, Ann Arbor, Michigan 48105

ABSTRACT

A neutron investigation on a chromium single crystal, which is 80% single-$\vec{Q}$, has confirmed the existence of the sinusoidal lattice displacement observed below the Néel temperature by Tsunoda et al[1] with x rays. The displacement wave was found to have its polarization parallel to the spin density wavevector, $\vec{Q}$, and to have wavevector $2\vec{Q}$ in both the longitudinally and transversely polarized spin density wave phases. Characteristic of a small sinusoidal lattice wave, the measured intensities of the $2\vec{Q}$ satellite peaks near the (1,1,0), (2,2,0) and (2,0,0) nuclear reflections were found to depend on the square of the scattering vector. By measuring the intensity of a given satellite peak on both sides of the spin flip temperature, the displacement amplitude was found to be independent of the direction of the magnetic polarization. Measurement of the intensities of the magnetic peaks near (1,0,0) and of the satellite peak below (1,1,0) at 130K, 248K, and 268K indicates that the displacement amplitude scales with the square of the magnetization as a function of temperature. Normalization of the various satellite peaks to the main nuclear and magnetic reflections yields a self-consistent value for the displacement amplitude, $\Delta a/a$, of $(0.17 \pm .02)\%$ at 130K. This value appears to be significantly lower than the $(0.25 \pm .05)\%$ value obtained in the x-ray investigation. One possible explanation for the discrepancy is that neutrons measure the periodic displacement of the nuclei while x rays are sensitive to both periodic displacements and periodic distortions of the atomic electron distribution.

[1] Y. Tsunoda, M. Mori, N. Kunitomi, Y. Teraoka, and J. Kanamori, Solid State Commun., 14, 287 (1974).

FIELD-INDUCED TRANSITIONS IN $DyAu_2$, $DyAg_2$, $TbAu_2$ and $TbAg_2$

T. KANEKO, M. OHASHI, S. MIURA, K. KAMIGAKI, A. HOSHI and H. YOSHIDA, The Research Institute for Iron, Steel and Other Metals, Tohoku University, Sendai, Japan

ABSTRACT

The magnetization of the compounds $DyAu_2$, $DyAg_2$, $TbAu_2$ and $TbAg_2$ were measured in pulsed magnetic fields up to 180 kOe and steady fields up to 90 kOe. These compounds exhibit two kinds of magnetically ordered structure, designated by the α-state and the β-state. Two kinds of abrupt change of the magnetization were observed at critical fields, which are considered to correspond to the transition fields from the β-state to the α-state and from the α-state to the metamagnetic state, respectively. The magnetization showed a saturation at higher fields. The characteristic feature of the magnetization curves of $DyAg_2$, $TbAu_2$ and $TbAg_2$ are somewhat different from that of $DyAu_2$. The saturation magnetic moment per Dy or Tb ions in these compounds are smaller than those expected for the 3^+ ions of Dy or Tb and such a depression of the magnetic moment is due to the effect of the crystalline field. On the basis of these experimental results, the H-T phase diagram of $DyAu_2$, $TbAu_2$ and $TbAg_2$ are proposed.

Rare-earth metals (R) form a series of intermetallic compounds, RAu_2 and RAg_2, all of which have the $MoSi_2$-type crystal structure.[1] The measurements of the magnetic properties of these intermetallic compounds are useful for studying the indirect exchange interaction between rare-earth ions embedded in a simple crystal together with ions of Au or Ag of relatively simple electronic configurations. The neutron diffraction studies by Atoji[2)~8)] have shown that most of these compounds have two states of different antiferromagnetic spin structure, that is, a layered antiferromagnetic spin structure (β-state) below T_N (β) and a transverse-sinusoidal spin structure between $T_N(\beta)$ and $T_N(\alpha)$. According to the measurements of susceptibility by the present authors,[9)~11)] the susceptibility versus temperature curves exhibit a maximum in susceptibility at T_N (α) upon cooling and an abrupt decrease at $T_N(\beta)$. These two kinds of magnetic transitions were also comfirmed bu the measurements of the electric resistivity and thermal expansion.[12), 13)]

In this paper the results of the measurements are reported on the magnetization of the compounds $DyAu_2$, $DyAg_2$, $TbAu_2$ and $TbAg_2$ in pulsed magnetic fields up to 180 kOe and steady fields up to 90 kOe.

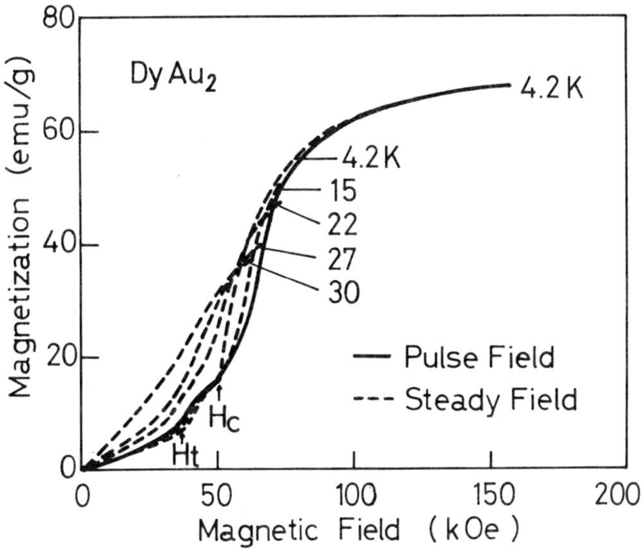

Fig. 1. Magnetization curves of $DyAu_2$, ——: under the pulsed field, ·····: under the steady field.

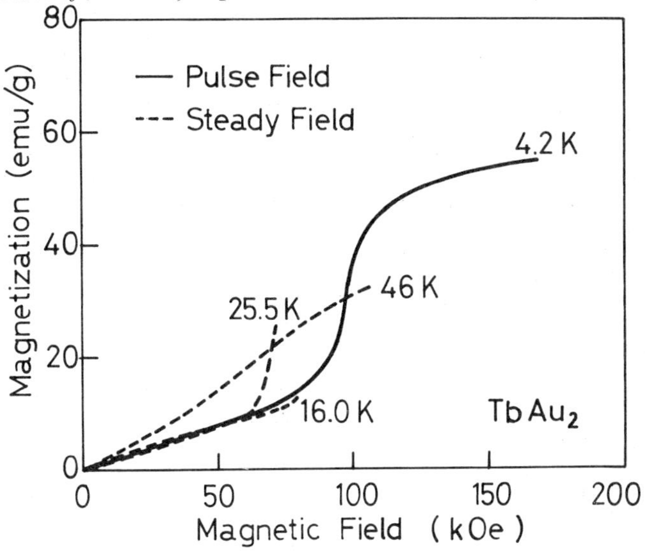

Fig. 2. Magnetization curves of $TbAu_2$, ——: under the pulsed field, ·····: under the steady field.

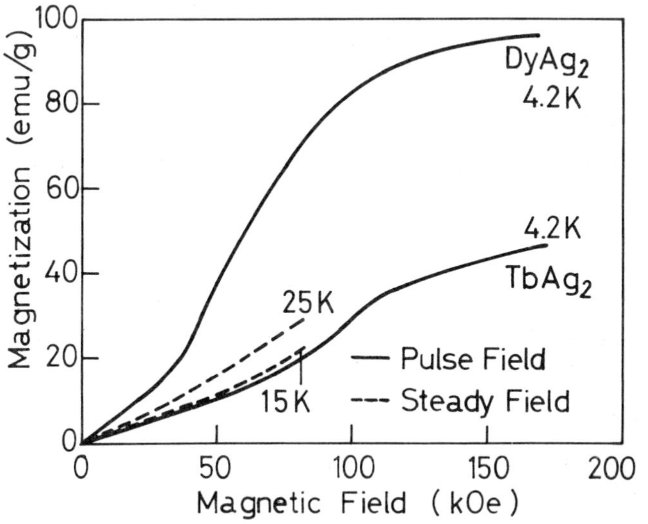

Fig. 3. Magnetization curves of $DyAg_2$ and $TbAg_2$, ——: under the pulsed field, ·····: under the steady field.

The specimens were prepared by arc-melting the mixtures of rare-earth metals with gold or silver (all the constituents are 99.9 % pure) of formula proportion in an argon atmosphere. The ingots were annealed at 500°C for 500 hours. X-ray analyses showed that all of the prepared specimens had the crystal structure of the $MoSi_2$-type with lattice paramenters in good agreement with those of Dwight.[1)]

The pulsed magnetic field was generated using a power source excited by 12-phase mercury rectifiers capable of generating 350 V and 40 kA for durations as long as 0.1 sec and applying current to the solenoid cooled in liquid nitrogen. The maximum field generated was 300 kOe. The use of such a long-duration pulsed field is effective for the measurement of magnetization of conductive substances, because the supression of the eddy current is perfect. The static magnetic field was generated by using the same power source and a water cooled solenoid of the Bitter-type. A maximum field of 100 kOe was attained for the desired duration. The results of the measurements of the magnetization are shown in Figs.1, 2 and 3. At temperatures below $T_N(\beta)$.

In $DyAu_2$ the magnetization curve shows a two stage transition at critical fields designated by H_t and H_c in the figure. The low field results obtained in the pulsed magnetic field show good agreement with those in the static field. The critical field H_t is the field where the transition from the antiferromagnetic β-state to the α-state occurs, and H_c is the transition field from the antiferromagnetic α-state to the metamagnetic state.
In both $DyAg_2$ and $TbAu_2$ there are also the α-state and the β-state, but in these compounds the two stage transition was not observed. The magnetization increases abruptly at a critical field H_c and then saturates with further increase of the magnetic field. The transitions observed in these compounds are considered to be direct transitions to the metamagnetic state from the antiferromagnetic β-state, from the fact that the magnetization tends to saturate with a high magnetization at higher fields subsequent to the abrupt increase at the critical field H_c. The saturation magnetic moment per magnetic ion calculated from the magnetization curves at 4.2 K by applying the law of approach to saturation, $\sigma = \sigma_0(1 - b/H^2) + \chi H$, for $DyAu_2$, $DyAg_2$ and $TbAu_2$ is shown in Table 1. The values are smaller than those expected for the free ions Dy^{3+} and Tb^{3+}, $10\mu_B$ and $9\mu_B$, respectively. The main origin of such a depression of the moment can be attributed to the effect of the crystalline field.

Table 1

	$T_N(\alpha)$ (K)	$T_N(\beta)$ (K)	θ_p (K)	$\mu_{eff.}$ (μ_B)	$\mu_{sat.}$ (μ_B)
$DyAu_2$	31.0	25.0	-13	10.6	7.2
$TbAu_2$	55.0	43.0	-21	9.6	6.0
$DyAg_2$	15.0	9.5	-21.5	10.5	6.9
$TbAg_2$	34.4		-31.5	9.7	4.3

The measurements of the neutron diffraction, the magnetic susceptibility and the electric resistivity show that there is no β-state in $TbAg_2$ in contrast with $DyAu_2$, $DyAg_2$ and $TbAu_2$. In $TbAg_2$, the magnetization increases also abruptly at a critical field H_c, and is hard to saturate. The saturation magnetic moment per Tb^{3+} ion derived in the same way mentioned above is 4.3 μ_B and this value is very small compared with that of a free ion Tb^{3+}. The origin of such a large depression of the magnetic moment is not clear at present.

According to the neutron diffraction measurements. the α-state has a transverse-sinusoidal spin structure. In the case of $DyAu_2$, the magnetic state in fields between H_t and H_c at 4.2 K and the state at temperatures between $T_N(\beta)$ and $T_N(\alpha)$ is the α-state. However, the value of the magnetic moment observed for the metamagnetic state is, in general, close to that of the maximum moment in the α-state rather than the averaged value of the magnetic moments of each lattice point in the α-state. Hence, the depressed magnetic moment of the α-state increased simultaneously with the flopping of the spin direction in the metamagnetic state. Such an increase of the atomic magnetic moment associated with spin flopping has not been observed before.

In Fig. 4 are shown the magnetic phase diagrams of $DyAu_2$, $TbAu_2$ and $TbAg_2$ determined by the present authors.

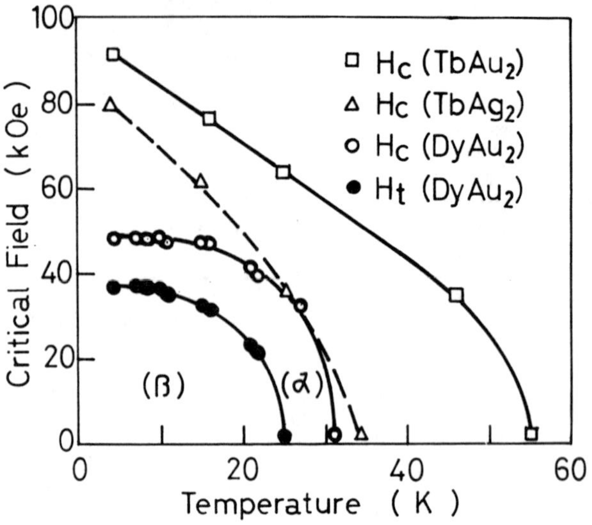

Fig. 4. Magnetic phase diagrams of $DyAu_2$, $TbAu_2$ and $TbAg_2$.

REFERENCES

1) A. E. Dwight et al.: Acta. Cryst. 22 (1967) 745.
2) M. Atoji: J. Chem. Phys. 48 (1968) 560.
3) M. Atoji: J. Chem. Phys. 48 (1968) 3380.
4) M. Atoji: J. Chem. Phys. 51 (1969) 3877.
5) M. Atoji: J. Chem. Phys. 51 (1969) 3882.
6) M. Atoji: J. Chem. Phys. 57 (1972) 851.
7) M. Atoji: J. Chem. Phys. 57 (1972) 2402.
8) M. Atoji: J. Chem. Phys. 57 (1972) 2407.
9) S. Miura, T. Kaneko, M. Ohashi and K. Kamigaki: J. Phys. (France) 32 (1971) C1-1124.
10) T. Kanako, S. Miura, M. Ohashi and H. Yamauchi: ICM-73 (Proc. Inter. Conf. of Magnetism, Moscow) vol. 5.
11) S. Miura, T. Kaneko, M. Ohashi and H. Yamauchi: J. Phys. Soc. Japan 37 (1974) 1464.
12) M. Ohashi, T. Kaneko, K. Kamigaki and S. Miura: J. Phys. Soc. Japan 34 (1973) 553.
13) M. Ohashi, T. Kaneko, K. Kamigaki and S. Miura: J. Phys. Soc. Japan (to be published)

MAGNETIC FORM FACTOR OF ELECTRONS IN 3d BANDS

R. M. Moon
Solid State Division, Oak Ridge National Laboratory,* Oak Ridge, Tennessee 37830

ABSTRACT

The formalism for calculating the spin density or magnetic form factor for tight-binding band electrons is considered in the case where nearest-neighbor overlap is not negligible. The formalism leads to a description of the system in terms of a two-component form factor, one of which is identical to the atomic form factor and the other is related to the diffuse overlap density. Numerical calculations for Fe and Ni show that the net overlap density is opposite in direction to the atomic spin. Comparison with experiment is presented for the case of Fe.

The polarized-neutron studies of Fe,[1] Co,[2] and Ni[3] led to a phenomenological model in which the total spin density was described as atomic 3d-like distributions sitting in a nearly constant sea of negative polarization. This model has since been used successfully to fit experimental results on a wide variety of 3d alloys. Unfortunately, the model has no sound theoretical basis, and the origin and physical significance of the negative polarization have been matters of some dispute.

In this paper we show that this phenomenological model does have some theoretical credibility when the crystal wave functions are written in the tight-binding form. The negative polarization may be associated with the overlap of the 3d atomic-like functions used in synthesizing the crystal wave functions.

Mostly, we will follow the notation of Hodges, Ehrenreich and Lang[4] (HEL) and carry their formalism one step further by explicitly including the effects of nearest-neighbor overlap of the atomic d functions. We consider hybridized d bands formed of linear combinations of atomic 3d orbitals and orthogonalized plane waves. The Bloch functions are written as

$$B_{kn\sigma}(\vec{r}) = \sum_{\mu} a_{n\mu\sigma}(\vec{k}) b_{k\mu\sigma}(\vec{r}) + \sum_{K} a_{nK\sigma}(\vec{k}) b_{kK\sigma}(\vec{r}), \qquad (1)$$

where n is a band index, μ is a 3d orbital symmetry index, σ is the electron spin index and $\vec{K}$ is a reciprocal lattice vector used in constructing the OPW $b_{kK\sigma}(\vec{r})$. It is assumed that the wave functions are known, either through a first-principles calculation or through an interpolation scheme. The 3d part of the wave function is given by

*Operated by Union Carbide Corporation for the USAEC.

$$b_{k\mu\sigma}(\vec{r}) = N^{-\frac{1}{2}} \sum_{\ell} e^{-i\vec{k}\cdot\vec{R}_\ell} \phi_{\mu\sigma}(\vec{r}-\vec{R}_\ell), \qquad (2)$$

where we have assumed a Bravais lattice defined by the set of position vectors $\vec{R}_\ell$ and N is the total number of unit cells. The functions $\phi_{\mu\sigma}(\vec{r}-\vec{R}_\ell)$ are constructed from atomic 3d symmetry functions, $\psi_{\mu\sigma}(\vec{r})$, to satisfy the orthogonality relationships,

$$\int \phi^*_{\mu\sigma}(\vec{r}-\vec{R}_\ell)\phi_{\nu\sigma}(\vec{r}-\vec{R}_m)d^3r = \delta_{\mu\nu}\,\delta_{\ell m}. \qquad (3)$$

The overlap integrals are defined as

$$\Delta^m_{\mu\nu\sigma} \equiv \int \psi^*_{\mu\sigma}(\vec{r})\psi_{\nu\sigma}(\vec{r}-\vec{R}_m)d^3r. \qquad (4)$$

We assume that for the 3d metals, only terms involving nearest-neighbor overlaps are significant, and further that the nearest-neighbor overlap is small enough that terms quadratic in the $\Delta^m_{\mu\nu\sigma}$ may be neglected. With these assumptions, our orthogonalized atomic functions are

$$\phi_{\mu\sigma}(\vec{r}) = \psi_{\mu\sigma}(\vec{r}) - \tfrac{1}{2}\sum_{m\nu} \Delta^m_{\mu\nu\sigma}\psi_{\nu\sigma}(\vec{r}-\vec{R}_m), \qquad (5)$$

where m runs over the nearest-neighbor sites.

We wish to evaluate the Fourier inversion of the z component of the spin density,

$$s_z(\vec{Q}) = \rho_\uparrow(\vec{Q}) - \rho_\downarrow(\vec{Q}), \qquad (6)$$

where

$$\rho_\sigma(\vec{Q}) = \int e^{-i\vec{Q}\cdot\vec{r}} \sum_{\substack{kn\\ \text{occ.}}} B^*_{kn\sigma}(\vec{r})B_{kn\sigma}(\vec{r})d^3r. \qquad (7)$$

In evaluating Eq. (7) there will be three types of terms: those involving the product of two d functions, of the product of d and OPW functions, and of two OPW functions. For the same reasons as set forth by HEL, we neglect the latter two types of terms and consider only the d-d contribution. Substituting Eq. (1) and (2) into Eq. (7) and rearranging terms, we find that

$$\rho_\sigma(\vec{Q}) = N^{-1}\sum_\ell e^{-i\vec{Q}\cdot\vec{R}_\ell} \sum_{\substack{kn\\ \text{occ.}}} \sum_{m\mu\nu} e^{i\vec{k}\cdot\vec{R}_m} a^*_{n\mu\sigma}(\vec{k})a_{n\nu\sigma}(\vec{k}) \times \int e^{-i\vec{Q}\cdot\vec{r}} \phi^*_{\mu\sigma}(\vec{r})\,\phi_{\nu\sigma}(\vec{r}-\vec{R}_m)d^3r. \qquad (8)$$

This may be written as

$$\rho_\sigma(\vec{Q}) = \rho^0_\sigma(\vec{Q}) + \rho^1_\sigma(\vec{Q}) + \ldots\rho^i_\sigma(\vec{Q}) + \ldots, \qquad (9)$$

where $\rho^i_\sigma(\vec{Q})$ stands for the collection of terms in Eq. (8) for the ith shell of neighbors with a common value of $|R_m|$. With our assumptions on the magnitude of the overlap integrals, we need to consider only $\rho^0_\sigma(\vec{Q})$ and $\rho^1_\sigma(\vec{Q})$.

To evaluate $\rho^0_\sigma(\vec{Q})$, we set $\vec{R}_m = 0$ in Eq. (8) and find that

$$\rho^0_\sigma(\vec{\tau}) = N\sum_\mu n_{\mu\sigma} f^0_{\mu\sigma}(\vec{\tau}), \qquad (10)$$

where $\vec{\tau}$ is a reciprocal lattice vector and

$$f^0_{\mu\sigma}(\vec{Q}) = \int e^{-i\vec{Q}\cdot\vec{r}}|\psi_{\mu\sigma}(\vec{r})|d^3r. \qquad (11)$$

In deriving Eq. (10) we have used the condition (see HEL) that

$$N^{-1}\sum_{\substack{kn\\ \text{occ.}}} a^*_{n\mu\sigma}(\vec{k})\,a_{n\nu\sigma}(\vec{k}) = n_{\mu\sigma}\,\delta_{\mu\nu}, \qquad (12)$$

and that

$$|\phi_{\mu\sigma}(\vec{r})|^2 = |\psi_{\mu\sigma}(\vec{r})|^2, \qquad (13)$$

which is consistent with Eq. (5) and our assumption that terms quadratic in the overlap integrals may be neglected. Equation (10), which was used by HEL in the case of Ni, is indistinguishable from that for a localized model. The real indication that we are dealing with Bloch wave functions is in the remaining contribution, $\rho^1_\sigma(\vec{Q})$, to the total density.

To evaluate this contribution we let $\vec{R}_m$ run over the nearest-neighbor sites in Eq. (8). Expanding the integral in Eq. (8) in terms of the atomic functions and keeping only the terms linear in the overlap integrals, we find that

$$\rho^1_\sigma(\vec{\tau}) = N\sum_{m\mu\nu} g^m_{\mu\nu\sigma}[\rho_{\mu\nu\sigma}(\vec{\tau}) - \sum_\eta \Delta^m_{\eta\nu\sigma}f^0_{\mu\eta\sigma}(\vec{\tau})], \qquad (14)$$

where

$$g^m_{\mu\nu\sigma} = N^{-1}\sum_{\substack{nk\\ \text{occ.}}} e^{i\vec{k}\cdot\vec{R}_m} a^*_{n\mu\sigma}(\vec{k})\,a_{n\nu\sigma}(\vec{k}), \qquad (15)$$

$$\rho^m_{\mu\nu\sigma}(\vec{Q}) = \int e^{-i\vec{Q}\cdot\vec{r}} \psi^*_{\mu\sigma}(\vec{r})\psi_{\nu\sigma}(\vec{r}-\vec{R}_m)d^3r, \qquad (16)$$

$$f^0_{\mu\nu\sigma}(\vec{Q}) = \int e^{-i\vec{Q}\cdot\vec{r}} \psi^*_{\mu\sigma}(\vec{r})\psi_{\nu\sigma}(\vec{r})d^3r. \qquad (17)$$

Note that $f^0_{\mu\eta\sigma}(0) = \delta_{\mu\eta}$ and $\rho^m_{\mu\nu\sigma}(0) = \Delta^m_{\mu\nu\sigma}$, so that Eq. (14) vanishes at $\vec{\tau} = 0$. Through symmetry arguments it can be shown that

$$\sum_{m\nu} g^m_{\mu\nu\sigma}\, \Delta^m_{\eta\nu\sigma} = \alpha_{\mu\sigma}\, \delta_{\mu\eta}, \tag{18}$$

so that

$$\rho^1_\sigma(\vec{\tau}) = N[\rho^{ov}_\sigma(\vec{\tau}) - \sum_\mu \alpha_{\mu\sigma}\, f^0_\mu(\vec{\tau})], \tag{19}$$

where the total overlap density is given by

$$\rho^{ov}_\sigma(\vec{\tau}) = \sum_{m\mu\nu} g^m_{\mu\nu\sigma}\, \rho^m_{\mu\nu\sigma}(\vec{\tau}). \tag{20}$$

Adding Eq. (10) and (20) to obtain the total density inversion, the total form factor is

$$f(\vec{\tau}) = \frac{\sum_\sigma P_\sigma[\rho^{ov}_\sigma(\vec{\tau}) + \sum_\mu (n_{\mu\sigma}-\alpha_{\mu\sigma}) f^0_{\mu\sigma}(\vec{\tau})]}{\sum_\sigma P_\sigma \sum_\mu n_{\mu\sigma}}, \tag{21}$$

where $P_\uparrow = +1$ and $P_\downarrow = -1$ (neutron case) or $P_\downarrow = +1$ (x-ray case). For cubic systems we use the symmetry functions defined by HEL and it can be shown that $\alpha_{1\sigma} = \alpha_{2\sigma} = \alpha_{3\sigma}$ and $\alpha_{4\sigma} = \alpha_{5\sigma}$ so that the analysis in terms of t_{2g} and e_g symmetry goes through as usual.

The 3d-like contribution to the form factor in Eq. (21) is augmented (or diminished depending on the sign of $\alpha_{\mu\sigma}$) over that obtained by counting the number of electrons of a given symmetry and spin. This extra scattering is exactly cancelled in the forward direction by the overlap contribution. However, the overlap density is very extended in real space so that $\rho^{ov}(\vec{\tau}) \cong 0$ for $\vec{\tau} \neq (000)$. The magnitude of this extra 3d scattering is determined in part by details of the band structure (through $g^m_{\mu\nu\sigma}$) and in part by the radial dependence of the atomic 3d functions (through $\Delta^m_{\mu\nu\sigma}$). Not only is the magnitude of the 3d-like scattering altered by the terms in $\alpha_{\mu\sigma}$, but the relative population of e_g and t_{2g} symmetry orbitals is affected.

We have made some numerical calculations to estimate the magnitude of the effects in Fe and Ni. Ferromagnetic energy bands have been calculated by Cooke and Davis[5] using an interpolation scheme with $\vec{k}$ dependent exchange. Using the values of $a_{\mu\sigma}(\vec{k})$ determined in their work, Cooke[6] has calculated $n_{\mu\sigma}$ and $g^m_{\mu\nu\sigma}$ for both Fe and Ni. For any atomic function of known radial dependence, it is then a simple matter to calculate the overlap integrals $\Delta^m_{\mu\nu\sigma}$, Eq. (4), and the quantities $\alpha_{\mu\sigma}$, Eq. (18). A convenient measure of the size and direction of the effects we have discussed is the quantity α, defined as

$$\alpha \equiv \frac{\sum_\mu(\alpha_{\mu\uparrow}-\alpha_{\mu\downarrow})}{\sum_\mu(n_{\mu\uparrow}-n_{\mu\downarrow})} = \frac{\rho^{ov}_\uparrow(0)-\rho^{ov}_\downarrow(0)}{\sum_\mu(n_{\mu\uparrow}-n_{\mu\downarrow})}. \tag{22}$$

This is the negative of the ratio of the apparent excess 3d-like spin to the total unpaired spin, or the ratio of the integrated overlap spin to the total unpaired spin.

TABLE I

ψ	α
Fe, $3d^6 4s^2$, Restricted[7]	-0.033
Fe, $3d^7 4s^1$, Restricted[8]	-0.073
Fe, $3d^8$, Unrestricted[9]	-0.142
Ni, $3d^8 4s^2$, Restricted[7]	-0.047
Ni^{+2}, $3d^8$, Unrestricted[10]	-0.047

Values of α for various atomic wave functions are shown in Table I, where the term restricted refers to a Hartree-Fock calculation in which the radial functions for both spin states are constrained to be identical, and the unrestricted calculations are not subject to this constraint. Note that all values of α are negative, indicating that the minority overlap density is dominant so that there is a diffuse negative polarization and a corresponding scale-up of the 3d-like spin density. Our earlier assertion that the overlap density would contribute very little to the Bragg reflections (except at the origin) was confirmed by calculating $[\rho^{ov}_\uparrow(\vec{Q}) - \rho^{ov}_\downarrow(\vec{Q})]$ along several symmetry directions. For example, for the case of Fe with the unrestricted $3d^8$ wave functions, the contribution of the overlap spin density to the form factor at the (110) Bragg position is 0.004. By comparing various calculations with the experimental results it was apparent that an unrestricted $3d^7 4s^1$ calculation might give good agreement for the case of Fe. A crude approximation to such a calculation was made by constructing a minority radial function slightly more expanded, and a majority radial function slightly more contracted, than that given by the restricted Hartree-Fock calculation. The form factor obtained with these radial functions is compared in Fig. 1 with the experimental results. Also shown, for the first few reflections, is the calculated form factor using only the $\rho^0(\vec{Q})$ contribution for both the restricted and approximate spin-polarized wave functions. Such good agreement has not yet been obtained for the Ni case, but it appears that an unrestricted calculation for a $3d^9$ configuration might be about right. For Ni, the neglected terms involving OPW functions are probably more important than in the case of Fe.

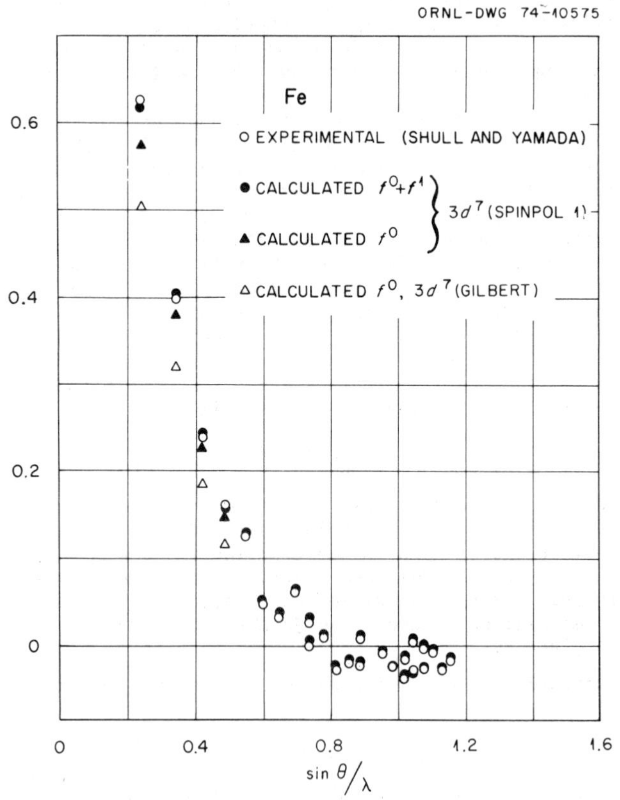

Fig. 1. Comparison of observed and calculated form factors for Fe. Orbital contribution to experimental data has been subtracted. f^0+f' is given by Eq. (21); f^0 is given by Eq. (21) with $\alpha_{\mu\sigma} = 0$ and $\rho^{ov}_\sigma = 0$. The SPINPOL 1 wave functions, different for the two spin states, were selected to give agreement with experiment, yet give an average density approximately equal to the restricted Hartree-Fock calculation of Gilbert (Ref. 8).

In summary, we have shown that a negative diffuse component of the spin density arises naturally when the wave functions are written in the tight-binding formalism with overlap effects explicitly considered. It seems likely that this formalism, combined with the proper spin-polarized atomic functions, can give good agreement with experimental results.

ACKNOWLEDGMENT

The author is indebted to J. F. Cooke for performing some of the numerical calculations and for many helpful discussions.

REFERENCES

1. C. G. Shull and Y. Yamada, J. Phys. Soc. Japan 17, Suppl. B-III, 1 (1962).
2. R. M. Moon, Phys. Rev. 136, A195 (1964).
3. H. A. Mook, Phys. Rev. 148, 495 (1966).
4. L. Hodges, H. Ehrenreich and N. D. Lang, Phys. Rev. 152, 505 (1966).
5. J. F. Cooke and H. L. Davis, AIP Conf. Proc. 10, 1218 (1973).
6. J. F. Cooke (private communication).
7. R. E. Watson, Phys. Rev. 119, 1934 (1960).
8. K. J. Duff and T. P. Das, Phys. Rev. B 3, 192 (1971).
9. R. E. Watson and A. J. Freeman, Phys. Rev. 123, 2027 (1961).
10. R. E. Watson and A. J. Freeman, Phys. Rev. 120, 1125 (1960).

SHORT-RANGE MAGNETIC ORDER IN Cu-Mn*

G. P. Felcher
Argonne National Laboratory, Argonne, Illinois 60439

S. A. Werner
Ford Motor Co., Dearborn, Michigan 48121

R. E. Majewski and J. S. Kouvel†
Argonne National Laboratory, Argonne, Illinois 60439
and University of Illinois, Chicago, Illinois 60680

ABSTRACT

Polarized-neutron scattering experiments were performed on a $Cu_{80}Mn_{20}$ alloy crystal at and above 4.2°K in a 100 kOe superconducting magnet, in a study of the relation between the magnetic and atomic short-range order. The atomic correlations give rise to diffuse diffraction peaks at $(1\,{}^1\!/_2\,0)$ and equivalent positions. The magnetic-nuclear interference cross-section $\Delta = [\sigma\uparrow(k) - \sigma\downarrow(k)]$ at these positions increases monotonically with applied magnetic field, whereas the sum cross-section $\Sigma = [\sigma\uparrow(k) + \sigma\downarrow(k)]$ is independent of both field and temperature below 78°K. A careful mapping was made of the cross sections in the [001] zone at T = 78°K and H_z = 52 kOe. Both Δ and Σ show broad maxima only around the $(1\,{}^1\!/_2\,0)$ positions; however, their shape is slightly different. The ratio Δ/Σ shows two maxima, approximately at the positions $(1\,{}^4\!/_{10}\,0)$, $(1\,{}^6\!/_{10}\,0)$. The inequivalence of the two cross sections indicates that the magnetic moments of manganese do not respond uniformily to the applied magnetic field, but instead they are modulated at the regions chemically more ordered, that have higher susceptibility.

*Performed under the auspices of the U. S. Atomic Energy Commission.

†Supported in part by the National Science Foundation.

NEUTRON SCATTERING AND MAGNETIZATION MEASUREMENTS ON THE KONDO COMPOUND $CeAl_3$*

A. S. Edelstein, T. O. Brun, and G. H. Lander
University of Illinois, Chicago, Illinois 60680 and
Argonne National Laboratory, Argonne, Illinois 60439

O. D. Mc Masters and K. A. Gschneidner, Jr.
Ames Laboratory, Ames, Iowa 50010

ABSTRACT

Previous measurements on the Kondo compound, $CeAl_3$ (hexagonal DO_{19}), indicate that with decreasing temperature it transforms gradually into a singlet state and does not magnetically order above 0.6 K. To determine the spatial extent of the magnetic moment we have measured the spin-dependent cross section of neutrons diffracted from the (201) planes of a polycrystalline sample in 50 kOe at 5, 19 and 81 K. At 50 kOe and these temperatures the total moment from magnetization measurements on the same sample is only 5-15% lower than the moment determined by the neutrons if one assumes a 4f form factor. This approximate agreement supports the hypothesis that the moment decreases without changing its spatial distribution. A similar spatially uniform reduction of the moment occurs also in dilute Kondo systems. From the field dependence of the magnetization M, the characteristic or Kondo temperature is estimated to be 5.1 K.

INTRODUCTION

Much interest in the Kondo problem has focussed on the spatial extent of the moment. Most of the present information concerning this distribution has been obtained for the system Cu:Fe for which NMR,[1] Mössbauer,[2] and neutron scattering[3] experiments indicate the absence of a large antiparallel conduction electron polarization cloud.

The intermetallic compound $CeAl_3$ has remarkable properties. The resistivity[4,5,6] $\rho(T)$ of $CeAl_3$ increases logarithmically with decreasing temperature below room temperature, reaches a broad maximum at 37 K and then decreases until at 0.3 K, $\rho(.3) \sim 0.03\rho(37)$. Specific heat measurements[7] indicate that essentially all the spin entropy has been removed by 1 K, i.e., the system is going into a singlet state. This cannot occur by crystal field effects alone, since, by Kramers theorem, a single 4f electron will be at least doubly degenerate in the presence of a crystalline field.

Further, susceptibility[6] and neutron diffraction measurements[7] indicate that no long-range magnetic ordering occurs above 1 K. The most likely explanation for this behavior is that $CeAl_3$ is a concentrated Kondo system.

RESULTS

To study the spatial extent and temperature dependence of the Ce moment we have performed magnetization and neutron experiments with polarized neutrons on a polycrystalline sample of $CeAl_3$. In the neutron scattering experiments the polarization ratio (or flipping ratio) of the (201) reflection at $s \equiv \sin\theta/\lambda = 0.207$ Å^{-1} has been measured for a magnetic field H = 50 kOe. Since the sample is polycrystalline the measurements yield the spin dependent cross section averaged over all directions of H in the (201) plane. This cross section is proportional to $\sin^2\phi \; M_{\|}/2 + (\cos^2\Phi + 1)M_{\perp}/2 = 0.362\, M_{\|} + .638\, M_{\perp}$ where $\Phi = 58.32°$ is the angle between the c axis and the normal to the (201) planes. The quantities $M_{\|}$ and $M_{\perp}$ are respectively the magnetization parallel and perpendicular to the c axis. Hence the average in the plane is approximately equal to the polycrystalline average $1/3M_{\|} + 2/3M_{\perp}$ obtained in magnetization measurements. The (201) reflection was the only reflection studied in detail. Other reflections have weaker intensities, and their average cross sections are proportional to quite different linear combinations of $M_{\|}$ and $M_{\perp}$.

Figure 1 is a plot of the average moment $\langle\mu\rangle$ in a field of 50 kOe and the reciprocal susceptibility χ^{-1} per mole as a function of temperature. The average moment determined by both magnetization and neutron scattering is shown. For the neutron scattering experiment we determine $\langle\mu\rangle$ from $\langle\mu\rangle = \langle\mu\rangle_s/f(s)$ where f(s) is the magnetic form factor. The experimental value of $\langle\mu\rangle_s$ is calculated from the polarization ratio using the value of 1.525×10^{-12} cm for the nuclear structure factor of the (201) reflection. We have used a theoretical value[8] for the Ce 4f form factor of f(0.207) = 0.85. The reciprocal susceptibility has been determined from χ^{-1} = const $H/\langle\mu\rangle$. The temperature dependence of χ^{-1} is approximately of a Curie-Weiss form in both cases with $\theta \approx 36$ K from the magnetization measurements and $\theta \approx 41$ K from the neutron measurements. If we interpret the magnitude of the effective moment using only crystal-field effects then the magnitude we have determined implies that the $m_J = \pm 1/2$ doublet is lowest. However, such an interpretation is incorrect, since other measurements[6,7] indicate that the system is gradually going into a singlet state and by 1 K this transition is nearly complete. The temperature dependence of the moment as determined by the neutron scattering experiments is essentially the same as that determined by magnetization measurements. Stassis and Shull[3] obtained similar results in their work on the Kondo system Cu:Fe.

To examine the temperature dependence of the spatial distribution of the moment more closely, the experimental value of the form factor $f(s) \equiv \langle\mu\rangle_s/\langle\mu\rangle$ is plotted in Fig. 2 for the (201) reflection for T = 5, 21, and 81 K. Here we have taken $\langle\mu\rangle$ as the moment determined from magnetization measurements. This is not strictly correct since different spatial averages occur in the two measurements but, as we shall see below, taking $\langle\mu\rangle$ from magnetization measurements leads to at most a 10% error. For s = 0.207 Å^{-1} the form factor f(s) is roughly independent of T. Though the uncertainty in f is large at T = 21 and 81 K, Fig. 2 suggests that f(0.207) decreases with decreasing temperature. Despite this possible tendency, the data shows that the moment is being reduced without drastically changing its spatial distribution. Also shown in Fig. 2 are the theoretical 5d and 4f atomic

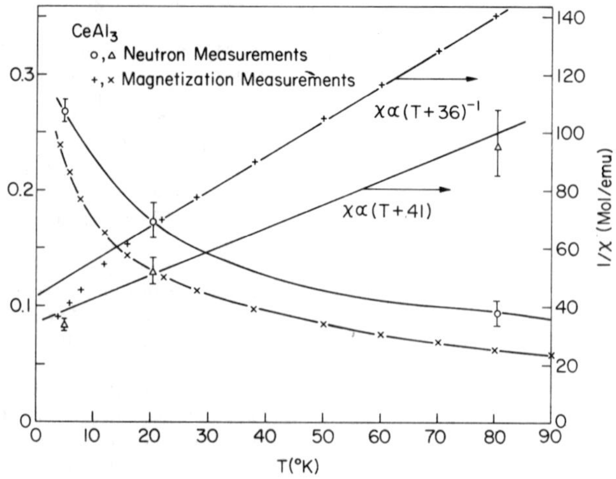

Figure 1. Temperature dependence of the average moment and inverse susceptibility as determined by magnetization and neutron scattering measurements.

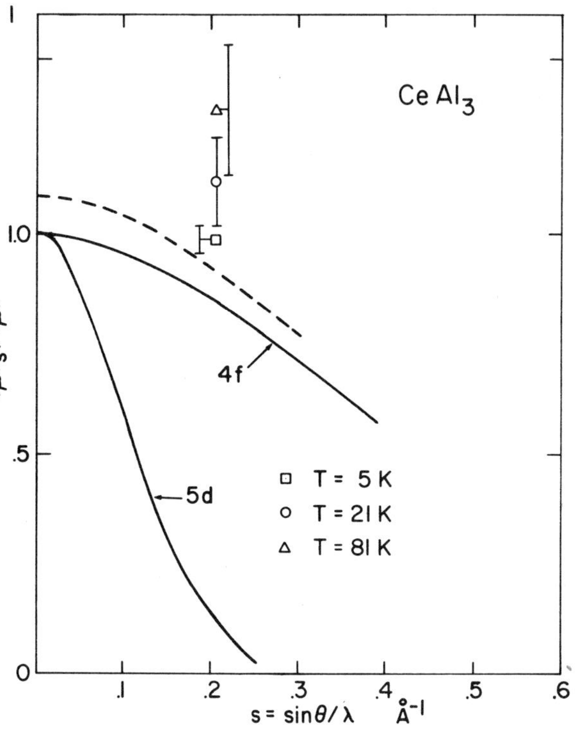

Figure 2. Plot of the form factor f as a function of s = sinθ/λ for T = 5, 21, and 81 K. The dashed curve is obtained by scaling the moment to allow for the maximum possible increase due to crystal-field effects.

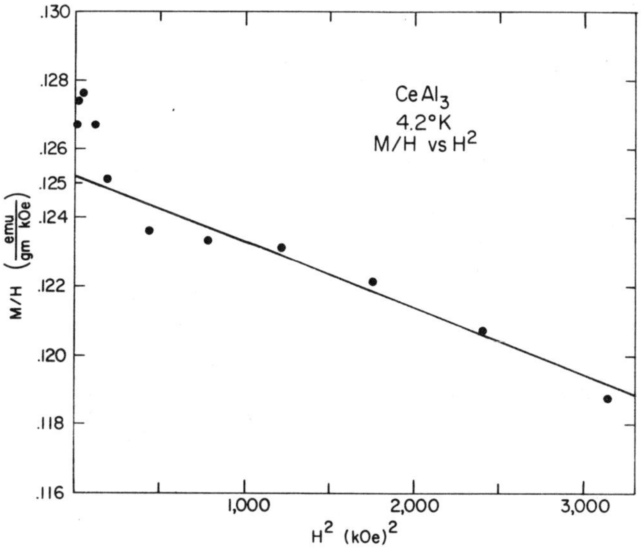

Figure 3. Plot of the magnetization divided by the magnetic field versus the square of the magnetic field at T = 4.2 K.

form factors.[8,9] The 5d form factor shown is that appropriate for Gd. This form factor is close enough to that appropriate for Ce that it can be used to illustrate the large difference between our measured value for f(0.207) and that expected for a 5d form factor. The dashed curve in Fig. 2 is the theoretical 4f prediction for Ce if one adjusts the normalization of the magnetic moment for the extreme case of $M_\perp = 0$. Thus, crystal-field effects cause at most a 10% change. One sees in Fig. 2 that the moment distribution is more localized than a 5d electron and resembles that of a 4f electron. Measurements[3] on Cu:Fe alloys also indicate that the average moment gradually goes to zero while keeping the same spatial distribution. The concentrated system $CeA\ell_3$ is in this sense acting like a dilute Kondo alloy.

Figs. 1 and 2 also illustrate that the 4f moment seen by neutrons is greater than the total moment measured by bulk magnetization. A possible explanation for this difference is that the observed moment is composed of two contributions. The first is a 4f contribution and the second is antiparallel to the 4f moment, approximately 15% of its magnitude, and not sufficiently localized to contribute to the (201) reflection.

Despite the fact that the magnetization M is nearly linear in H, we can use the slight curvature to make an estimate of the characteristic or Kondo temperature T_K. From a phenomenological model[10] for Kondo systems in the case of a doubly degenerate impurity, one has that

$$M = \mu \int_{-\infty}^{\infty} f(E)\,[N(E+\mu H) - N(E-\mu H)]\,dE \qquad (1)$$

where f(E) is the Fermi function and

$$N(x) = \frac{N_o}{x^2+(kT_K)^2} \qquad (2)$$

The low magnetic field form of M is a power series whose terms are of the form $A_{2n-1}H^{2n-1}$ (n=1,2,3,...). Our estimate of T_K is obtained by adjusting T_K so that the experimental and theoretical values of the ratio A_3/A_1 agree. We determine the experimental value of A_3/A_1 from the plot shown in Fig. 3 of M/H versus H^2. In making this plot a small low field contribution, which was probably due to impurities, was subtracted. In determining the theoretical value for the ratio A_3/A_1 we used $\mu = gJ\mu_B$ where the Lande g factor appropriate for Ce is 6/7 and we assumed the J value appropriate for the ground state doublet is 3/2. The estimate of T_K obtained as described above is 5.1 K.

In summary, we conclude that the moment $\langle\mu\rangle$ maintains a localized 4f-like spatial distribution even though its magnitude decreases with decreasing temperature. The additional contributions to the form factor and its possible temperature dependence are features that are not understood. The field dependence of M is consistent with a value of $T_K \sim 5.1$ K.

We wish to acknowledge the assistance provided by Dr. Helmut Claus.

REFERENCES

1. J. B. Boyce and C. P. Slichter, Phys. Rev. Letters 32, 61 (1974).
2. P. Steiner, W. V. Zdrojewski, D. Gumprecht, and S. Hufner, Phys. Rev. Letters 31, 355 (1973).
3. C. Stassis and C. G. Shull, Phys. Rev. B 5, 1040 (1972).
4. K. H. J. Buschow, H. H. van Daal, F. E. Maranzana, and P. B. van Aken, Phys. Rev. B3 , 1662 (1971).
5. A. Percheron, J. C. Achard, O. Gorochov, B. Cornut, D. Jerome, and B. Coqblin, Solid State Commun. 12, 1289 (1973).
6. A. S. Edelstein, C. J. Tranchita, O. D. McMasters, and K. A. Gschneidner, Jr. Solid State Comm. 15, 81 (1974).
7. J. V. Mahoney, V. U. S. Rao, W. E. Wallace, R. S. Craig, and N. G. Nereson, Phys. Rev. B 9, 154 (1974).
8. M. Blume, A. J. Freeman, and R. E. Watson, J. Chem. Phys. 37, 1245 (1962).
9. B. N. Harmon and A. J. Freeman, Phys. Rev. B 10, 1979 (1974).
10. A. S. Edelstein, Phys. Rev. Letters 29, 1522 (1972); A. S. Edelstein, L. R. Windmiller, J.B. Ketterson, and H. V. Culbert, AIP Conf. Proc. No. 5, 558 (1972).

*Based on work performed under the auspices of the U. S. Atomic Energy Commission.

MAGNETIC AND LATTICE PROPERTIES OF CeBi

G.H. Lander, M.H. Mueller,
Argonne National Laboratory, Argonne Illinois 60439

O. Vogt
Laboratory für Festkorperphysik, ETH, Zurich, Switzerland

ABSTRACT

The magnetic structures of CeBi have been examined as a function of field with neutron diffraction. At low fields unusual domain reorientation effects are observed. For H >18 kOe a 3+, 1- spin configuration is found. A first-order transition to induced ferromagnetic behavior occurs at 47.5 kOe. X-ray experiments at low temperature show that no lattice distortion occurs at either T_N = 25 K or the I - IA transition at 12.5 K. A volume discontinuity is observed at the low temperature transition.

INTRODUCTION

The compound CeBi (NaCl structure) is of interest because of its anomalous magnetization[1,2] and unusual magnetic structures in zero field.[3] At T_N (25 K) the magnetic structure is type I in which ferromagnetic (001) sheets are stacked in an alternating +,- sequence. The spin direction is parallel to [001]. However, at 12.5 K CeBi undergoes a first-order transition to the type IA magnetic structure, in which the spin direction remains unaltered but the ferromagnetic sheets are stacked in the ++-- sequence. The IA structure is unusual in that it is not predicted by molecular field theory, but has been observed in a number of uranium compounds.[4] Recently Cooper et al.[5] and Bartholin[6] have reported magnetization experiments on single crystals. These measurements have shown that the moments are parallel to the cube axis in all fields and that for 15 <H <45 kOe an intermediate phase exists in which the magnetization is 1/2 the value in the fully ordered state (= 2.1 μ_B/Ce atom).

<u>Neutron experiments</u>

The experiments were performed at the CP-5 Research Reactor with a 60 kOe superconducting magnet assembly. The field was applied parallel to [001] throughout. The following reflections (together with their interpretation) were examined as a function of field: (110) - type I modulation with the propagation direction $\tau \parallel H$, (210) - type I modulation $\tau \perp H$, (11 1/2) - type IA modulation $\tau \parallel H$, (2 1/2 0) - type IA modulation $\tau \perp H$, and the (111) and (200) nuclear reflections which are sensitive to a net ferromagnetic contribution and are equivalent to the magnetization. For (111) the nuclear contribution is $N(111)=b_{Ce} - b_{Bi} = -0.377 \times 10^{-12}$cm, and the magnetic contribution $M(111) = 0.27 \times f \times \mu \times 10^{-12}$cm, where f is the form factor (=0.91) from Blume et al.[7] and the dipole approximation.[8] Thus $M(111)=0.245 \times \mu$, where μ is the moment per Ce in Bohr magnetons. The polarized-neutron technique measures the ratio M/N so the (111) reflection is very sensitive to a ferromagnetic component.

a)Type IA phase, T = 5 K.

Figure 1 (left hand side) shows the field dependence of the (110) and (11 1/2) reflections together with the ferromagnetic component. Initially with H=0 only the type IA reflections are present. As H increases the (11 1/2) increases and the (2 1/2 0) decreases (not shown). This behavior corresponds to a domain reorientation such that preferred domains have $\tau \parallel H$, i.e. the longitudinal susceptibility is greater than the transverse, and was observed by Cable and Koehler.[3] For H ~15 kOe we believe fluctuations in the arrangement of the ferromagnetic (001) sheets leads to a lack of long-range coherence and the resulting loss

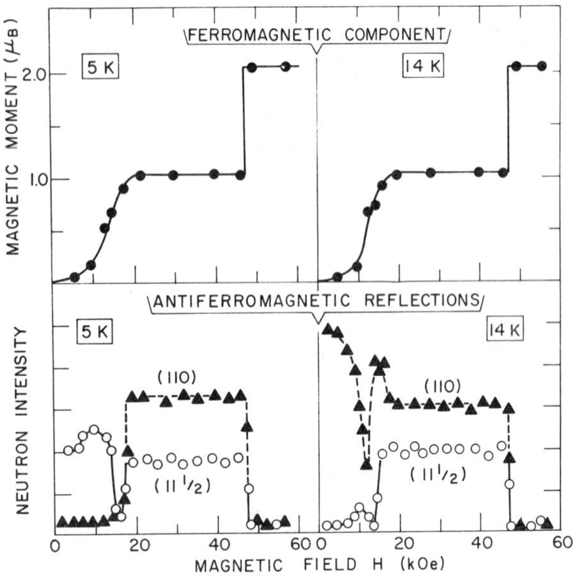

<u>Fig. 1.</u> Variation of ferromagnetic component (upper portion) and (110) and (11½) antiferromagnetic reflections (lower portion) in type IA (5 K) and type I (14 K) phases as a function of magnetic field.

of antiferromagnetic intensity. A sharp rise in the ferromagnetic component occurs at this field. For 18 <H <47.5 kOe both the (110) and (11 1/2) reflections are present, as well as a ferromagnetic component of 1.05 μ_B/Ce atom. No intensity was observed at the (210) or (2 1/2 0) positions. A combination of these modulations leads to the conclusion that the spin configuration is a 3+, 1- sequence of the ferromagnetic (001) sheets. For H=47.5 kOe the 3+, 1- structure collapses, leading to a totally induced ferromagnetic state with 2.1 μ_B/Ce atom.

b)Type I phase, 12.5 <T <25 K.

Figure 1 (right hand side) shows the field dependence of the (110) and (11 1/2) reflections, together with the ferromagnetic component at 14 K. In contrast to the low temperature behavior, the transverse susceptibility in the type I phase is greater than the longitudinal (as one expects for an antiferromagnet) and the (110) decreases as H increases. For H between 10 and 17 kOe the behavior is complex, and, we believe, represents both the domain reorientation effects and the inherent instability of the I and IA structures in the presence of an applied field. This behavior extends over a wider field range as the temperature is increased towards T_N (25 K). The lack of any magnetic intensity at the (210) or (2 1/2 0) positions eliminates the possibility of a canted spin arrangement. For 17.5 <H <47.5 kOe the 3+, 1- configuration is again present. The critical field for induced ferromagnetism shows little temperature dependence for T <20K, in agreement with magnetization results.[5,6]

<u>X-ray experiments</u>

We have examined the behavior of the lattice parameter of CeBi at low temperature with X-ray diffraction from the (800) planes of a single crystal. The results are shown in Fig.2. A careful

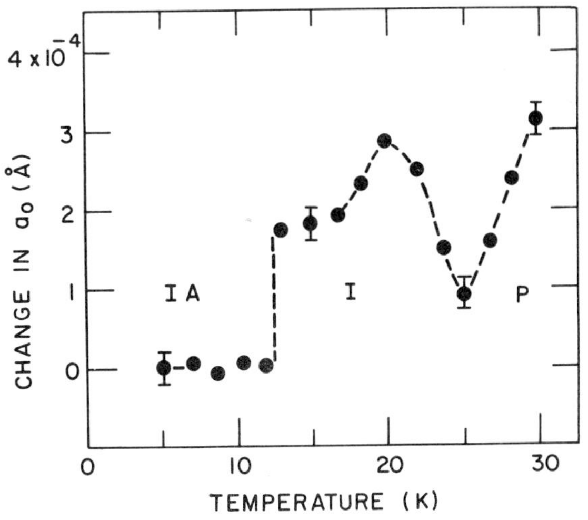

Fig. 2. Change of lattice parameter (relative to value of 6.4873 Å at 5 K) of CeBi as a function of temperature. I and IA refer to magnetic structures, P is paramagnetic regime.

search for a lattice distortion in the ordered state gave an upper limit of $|(c-a)/a|<1 \times 10^{-4}$ for any tetragonal distortion. As the figure shows, on warming through the IA-I transition (12.5K) the relative volume discontinuity associated with the first-order transition is $(8 \pm 1) \times 10^{-5}$. The thermal expansion in the type I magnetic phase is first positive (T <20 K) then negative for T >20 K. The lattice parameter at 5K is 6.4873 (1) Å. The change in the lattice parameter between 25 and 30K represents a linear expansion coefficient of 7×10^{-6}/K. [Note added: During the Conference H.R. Ott of ETH Zurich reported a measurement of a tetragonal distortion at T_N in CeBi. An initial comparison of these two results suggests that domain effects may be important, as is the case in magnetization experiments.[5,6]]

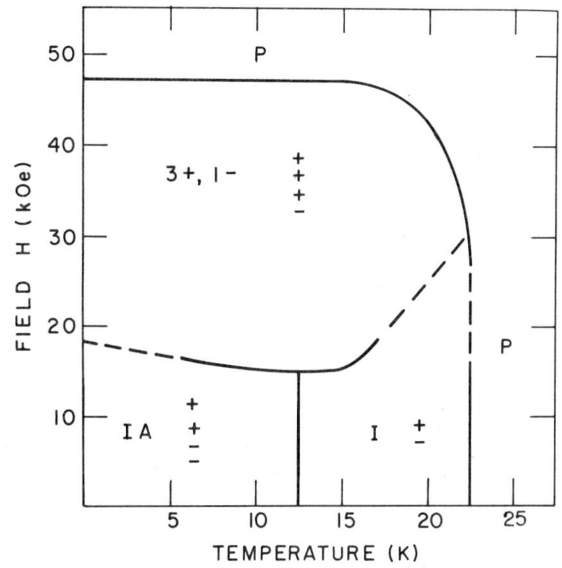

Fig. 3. Phase diagram for CeBi.

For symbols see Fig. 2 and text.

DISCUSSION

The experiments reported here confirm the 3+, 1- configuration for the intermediate field state in CeBi as first suggested by Tsuchida and Nakamura.[9] In fact their phase diagram is in excellent agreement with our results and we present a modified phase diagram with all the magnetic structures defined in Fig. 3. The dashed lines represent boundaries that have not been experimentally verified. The variations in the intensities of the (110) and (11 1/2) at low fields (see Fig.1) arise from domain reorientation effects and/or the instability of the I and IA phases to the application of a magnetic field and, as such, are not represented in the phase diagram. For H> 47.5 kOe and T< 25 K the correct description of the phase is paramagnetic rather than the "induced ferromagnetism" used in the literature.[9] However, the anisotropy in this state is considerable, and the spins cannot be rotated significantly away from the cube axis.[5,6] As discussed by Cooper et al.[5] the easy direction for the moments in CeBi should be <111> if the crystal field is the dominant interaction. Direct measurements of the crystal-field splittings in CeBi have found them to be small[10] so the exchange interactions probably define the easy axis of magnetization.

The mechanism that stabilizes the IA magnetic structure remains a mystery. A structural distortion, which would allow the introduction of an additional term in the free energy, is not observed at the I - IA transition at 12.5 K. The close analogy between the uranium compounds with the I and IA structures[4] (and their unusual behavior in a magnetic field) and CeBi is even more striking when one recalls that these uranium compounds, like CeBi, maintain cubic symmetry below their respective ordering temperatures.[11]

REFERENCES

1. Y. L. Wang and B. R. Cooper, Phys. Rev. B 2, 2607 (1970), and references therein.
2. T. Tsuchida, A. Hashimoto, and Y. Nakamura, J. Phys. Soc. Japan 36, 685 (1974)
3. J. W. Cable and W. C. Koehler, AIP Conf. Proc. 5, 1381 (1972)
4. G. H. Lander, M. H. Mueller, and J. F. Reddy, Phys. Rev. B 6, 1880 (1972); R. C. Maglic, G. H. Lander, M. H. Mueller, J. Crangle. and G. S. Williams, Phys. Rev. B 10, 1943 (1974)
5. B. R. Cooper, M. Landolt, and O. Vogt, Int. Conf. on Magnetism, Moscow, Aug. 1973, paper 27a-S4.
6. H. Bartholin, private communication.
7. M. Blume, A. J. Freeman, and R. E. Watson, J. Chem. Phys. 37, 1245 (1962)
8. G. H. Lander and T. O. Brun, ibid 53, 1387 (1970)
9. T. Tsuchida and Y. Nakamura, J. Phys. Soc. Japan 22, 942 (1967)
10. A. Furrer, W. Buhrer, H. Heer, W. Halg, J. Benes, and O. Vogt. Neutron Inelastic Scattering 1972 (IAEA, Vienna), 563, 1972
11. G. H. Lander and M. H. Mueller, Phys. Rev. B 10, 1994 (1974)

NEGATIVE MAGNETORESISTANCE IN HIGH-MAGNETIC-FIELD bcc TITANIUM ALLOY SUPERCONDUCTORS*

J.W. Lue, A.G. Montgomery, R.R. Hake
Indiana University, Bloomington, Indiana 47401

ABSTRACT

Weak normal-state negative magnetoresistance, and negative slope $d\rho/dT$ in the electrical resistivity ρ versus T, have been observed at temperatures $1.2 \leq T \leq 27K$ and applied magnetic fields $H \leq 140kG$ in very short electron-mean-free-path, superconducting, bcc Ti-base alloys such as $Ti_{92}Fe_8$, $Ti_{86}Mn_{14}$, $Ti_{92}Ru_8$, $Ti_{84}Mo_{16}$, $Ti_{92}Os_8$. This suggests that conduction electrons may interact with compensated or fluctuating localized spins, as might also account for the characteristic high ρ and negative $d\rho/dT$ at $\approx 10 < T < 300K$, H=0.

INTRODUCTION

Body-centered-cubic transition-metal alloys based upon Ti, Zr, and Hf form a prominent class of technologically important type-II superconductors. The unusual behavior of their normal-state electrical resistivity ρ has never been satisfactorily explained: (a) ρ values are quite large, in the 100 - 200 $\mu\Omega$ cm range for Ti-, Zr-, and Hf-rich alloys; (b) the highest-ρ alloys usually display an anomalous <u>negative</u> temperature slope[1] $(\partial\rho/\partial T)_{H=0}$ between about 300K and $\approx 2T_c$ [where apparent superconductive fluctuations result in a peak[2,3] in $\rho(T, H=0)$]; (c) alloying by addition of transition metal solutes to the right of Ti, Zr, and Hf in the periodic table, in concentrations c greater than those just sufficient to stabilize the bcc phase ($\approx$6-20 at. %), usually results in an anomalous <u>negative</u> concentration slope[1] $(\partial\rho/\partial c)_{T,\ H=0}$. In the present paper we report the observation[4] of yet a third anomalous <u>negative</u> slope $(\partial\rho/\partial H)_T$ [<u>n</u>egative <u>m</u>agnetoresistance (nm)].

EXPERIMENTAL RESULTS

Figures 1 and 2 show ρ at various temperatures in the 4 - 300K range for the bcc Ti alloys of the present study. A standard 4 - lead technique[1,5] was employed. The data show the typical negative high-T $(\partial\rho/\partial T)_{H=0}$ and typical negative $(\partial\rho/\partial c)_{T,\ H=0}$ (Fig. 1).

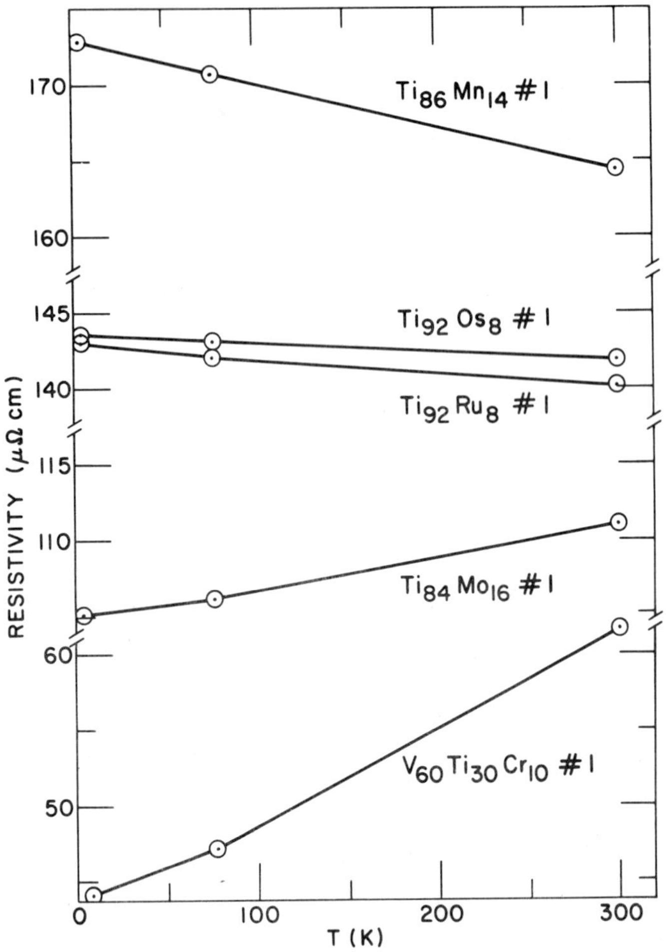

Fig. 2. Electrical resistivity versus T for various bcc alloys of the present study. All points are at H=0, except the 4.2K points where $H \approx 140$ kG.

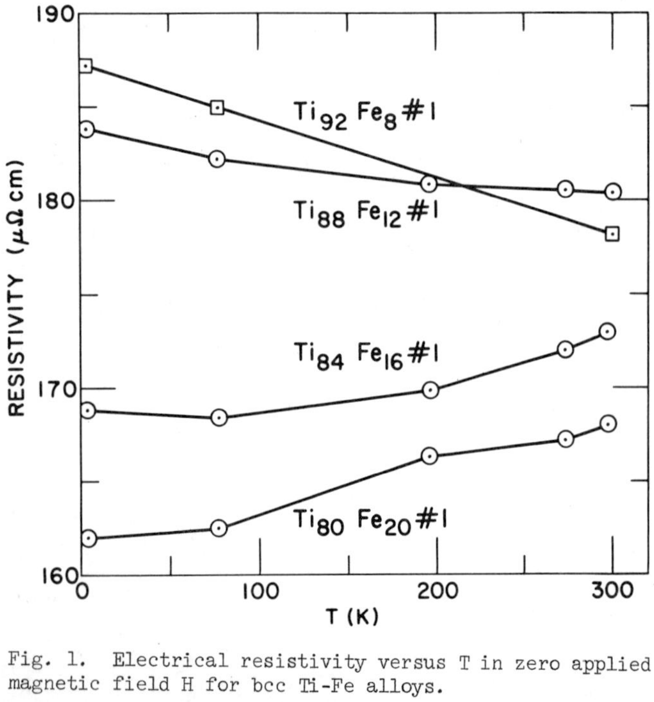

Fig. 1. Electrical resistivity versus T in zero applied magnetic field H for bcc Ti-Fe alloys.

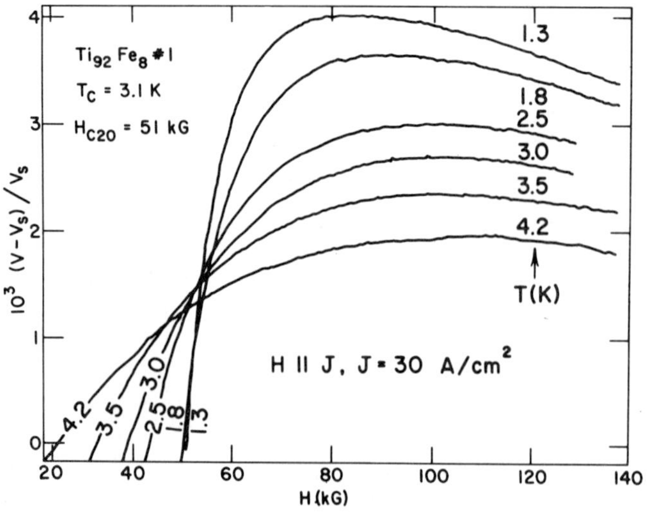

Fig. 3. Magnetoresistive curves (traced from an XY-recorder plot) for $Ti_{92}Fe_8$. V is the resistive voltage and V_s is a constant nulling voltage. H_{c20} is the resistively measured upper critical field as extrapolated to T = 0K.

Figure 3 shows magnetoresistive characteristics as measured for a typical bcc Ti-alloy superconductor $Ti_{92}Fe_8$, in the (H-T) region beyond both the upper and surface critical fields. Similar curves have been measured[6] for the other alloys of Figs. 1 and 2. At H< (80 - 100kG), $(\partial\rho/\partial H)_T$ is relatively large and positive and is attributed [2,6] to the H-quenching of superconductive fluctuations. At H > (80 - 100kG), $(\partial\rho/\partial H)_T$ is negative and increases in absolute value as T decreases.

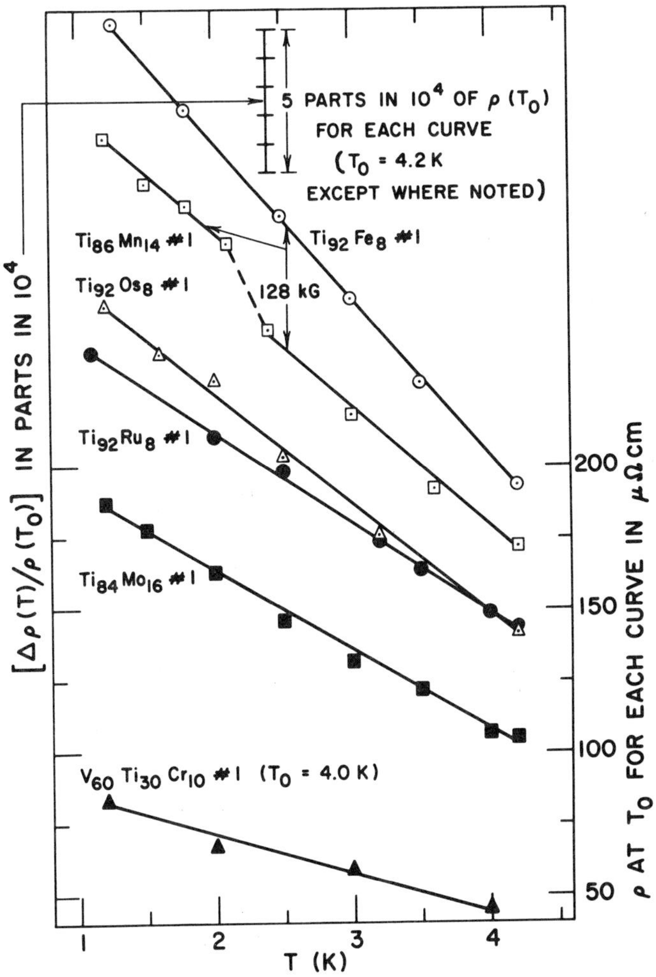

Fig. 4. Normalized resistivity changes versus T. The right-hand scale indicates the magnitude of ρ for each curve only at the arbitrary $T = T_o$. The break in the $Ti_{86}Mn_{14}$#1 curve is due to a slight shift in potential lead spacing between runs. H = 137-140 kG except where noted.

Figure 4 shows $\rho(T)$ at high constant H in the 1-4.2K range. A negative and constant $(\partial\rho/\partial T)_H$ is observed for each alloy, even for $Ti_{84}Mo_{16}$ and $V_{60}Ti_{30}Cr_{10}$, which display positive $(\partial\rho/\partial T)_{H=0}$ at T > 4.2K (where electron-phonon scattering is influential) as shown in Fig. 2.

Figure 5 shows magnetoresistive curves, all measured in the vicinity of T = 1.2K. A negative and constant $(\partial\rho/\partial H)_T$ is observed in the high-H range for most of the alloys with ρ (maximum) > 100 μΩ cm. We obtain, to within about $\pm 1 \times 10^{-9} G^{-1}$, $-\rho^{-1}(\partial\rho/\partial H)_T = 16, 15, 10, 5, 2 \times 10^{-9} G^{-1}$ for, respectively, $Ti_{92}Fe_8$, $Ti_{86}Mn_{14}$, $Ti_{92}Ru_8$, $Ti_{84}Mo_{16}$, and $Ti_{92}Os_8$. Magnetoresistive curves measured[6] for $Ti_{84}Mo_{16}$ and $Ti_{92}Ru_8$ at T = 27K, where superconductive fluctuation effects are very small, display very weak linear negative slopes (scaling roughly as 1/T from the lower temperature values) over the entire measured range $10 \le H \le 140$kG.

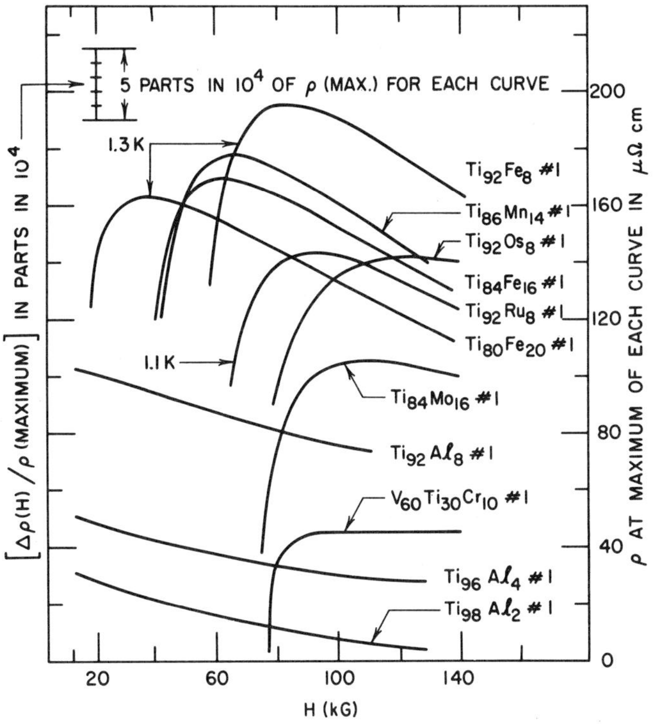

Fig. 5. Nomalized resitivity changes versus H. The right-hand scale indicates the magnitude of ρ only at the maximum ρ for each curve. T = 1.2 K except where noted.

The hcp Ti-Aℓ alloys of Fig. 5 are included to afford a comparison with the other alloys of Fig. 5 which are all bcc. Since dilute Mn additions to pure hcp Ti produce nm,[7] the negative $(\partial\rho/\partial H)_T$ in hcp Ti-Aℓ alloys may be due, at least in part, to magnetic moments localized on Mn impurities.

DISCUSSION

Low temperature negative magnetoresistance (nm) in metals and alloys is usually associated with the presence of localized magnetic moments.[8] However, standard localized-magnetic-moment behavior has never to our knowledge been observed in bcc alloys of the present type, even though calorimetric,[9] susceptibility,[10] magnetization,[5] and electrical resistance[1,2,5,11] measurements have been made on many different specimens, some containing more than 5 at.% of Mn, Fe, or Cr.

It seems likely that the present nm in bcc Ti alloys may be directly related to the anomalous negative temperature slope $(\partial\rho/\partial T)_H$ (nt) indicated in Fig.1,2,4. Spin-disorder scattering[1] and localized spin fluctuations (lsf)[3] have previously been advanced to explain nt.[12] It has also been suggested[13] that lsf might help to explain the magnitudes of T_c values in bcc Ti-V alloys. Despite possible interaction and local environment effects in the present alloys, it might still be useful to attempt a description in terms of very high lsf frequencies, or perhaps alternatively[14,15] very high[16] Kondo temperatures T_K and magnetic fields $H_K \approx k_B T_K/\mu_B$. In analogy with the behavior of dilute Cu-Cr[8,17] at $T \ll T_K$ one might then expect,[17,18] for the magnetic scattering contribution ρ_m to ρ, $\rho_m(H/H_K, T_o \ll T_K) \approx \rho_m(T/T_K, H_o \ll H_K)$ (where T_o, H_o = constant T, H of the measurement), since both H and T may act similarly to destroy the high-electron-scattering spin-compensated state. If we assume, from Figs. 1,2, $T_k > 300K$ and thus $H_K \approx k_B T_K/\mu_B > 4.6$ MG, then the linearity of $\rho(H)$ at $T_o/T_K < 0.004$ and $\rho(T)$ at $H_o/H_K < 0.03$ might reflect the usual[8,14,19] high-T linearity of $\rho_m(T)$ but extended to unusually low H/H_K and T/T_K by interaction[19] and/or local environment effects.

REFERENCES

1. See, e.g., R.R. Hake, D.H. Leslie, and T.G. Berlincourt, J. Phys. Chem. Solids 20, 177 (1961). The recent survey by J.H. Mooij, Phys. Stat. Solidi A17, 521 (1973) indicates that a negative $d\rho/dT$ at $\approx$ 300-350K is found in nearly all bulk and thin-film disordered transition metal alloys with $\rho > 130$ $\mu\Omega$ cm.
2. R.R. Hake, Phys. Rev. Letters 23, 1105 (1969).
3. V.A. Rassokhin, N.V. Volkenshtein, A.P. Romanov, A.F. Prekul, Zh. Exsp. Teor. Fiz. 66, 348 (1974).
4. J.W. Lue, A.G. Montgomery, and R.R. Hake, Bull. Am. Phys. Soc. 19, 349 (1974).
5. R.R. Hake, Phys. Rev. 158, 356 (1967).
6. J.W. Lue, A.G. Montgomery, R.R. Hake, submitted to Physical Review and unpublished data.
7. R.R. Hake, D.H. Leslie, and T.G. Berlincourt, Phys. Rev. 127, 170 (1962).
8. For a review see C. Rizzuto, Rep. Prog. Phys. 37, 147 (1974).
9. R.R. Hake and J.A. Cape, Phys. Rev. 135, A1151(1964); L.J. Barnes and R.R. Hake, Phys. Rev. 153, 435(1967).
10. J.A. Cape, Phys. Rev. 132, 1486 (1963).
11. T.S. Luhman, R. Taggart, D.H. Polonis, Scripta Met. 2, 169 (1968); V. Chandrasekaran, R. Taggart, D.H. Polonis, J. Mat. Sci. 9, 961 (1974).
12. Non-magnetic mechanisms have been suggested by E.W. Collings, Phys. Rev. B9, 3989 (1974).
13. K.H. Bennemann and J.W. Garland, A.I.P. Conf. Proc. No. 4, 1972, p. 103.
14. N. Rivier and V. Zlatic, J. Phys. F2, L87 (1972); ibid., L99.
15. K.D. Schotte, Z. Phys. 235, 155 (1970).
16. J.R. Schrieffer, J. Appl. Phys. 38, 1143 (1967).
17. M.D. Daybell and W.A. Steyert, Phys. Rev. Letters 20, 195 (1968). M.D. Daybell in Magnetism V, ed. by H. Suhl (Academic Press, 1973) p. 121.
18. G.S. Poo, J. Phys. F 4, L121 (1974); F.W. Fenton, Phys. Rev. B7, 3144 (1973).
19. E. Kovács-Csetényi, F.J. Kedves, L. Gergely, G. Grüner, J. Phys. F2, 499 (1972).

*Supported in part by NSF Grant GH33055.

INHOMOGENEITY AND INVAR EFFECT IN Fe-Pt ALLOYS

K. SUMIYAMA*, M. SHIGA and Y. NAKAMURA
Department of Metal Science and Technology,
Kyoto University, Kyoto, Japan
and
G. M. GRAHAM
Department of Physics, University of Toronto,
Toronto, Ontario, Canada M5S 1A7

In order to clarify the causality between the Invar effect and the inhomogeneity, which has been considered as an essential origin of Invar effect, we studied ordered state, A, and disordered state, B, of Fe-Pt Invar alloys, where A is thought as homogeneous and B as random. For $Fe_{72}Pt_{28}$ alloy, following results were obtained : At 4.2 K, magnetic moment is not sensitive to the atomic ordering (2.15 μ_B for A and 2.13 μ_B for B), while the hyperfine field at Fe nuclei is sensitive to the atomic ordering (342 kG for A and 363 kG for B) and distributes more widely in B, without superposition of two kinds of spectra. The Curie temperature, Tc, is 505 K for A and 371 K for B. At around Tc, negative thermal expansion coefficients with large magnitude have been obtained (-0.68×10^{-5} K^{-1} for A and -3.4×10^{-5} K^{-1} for B), and the magnetization changes against temperature more sharply for B. From these observations, we may safely conclude that the effect of inhomogeneities, i.e. atomic randomness, coexistence of two magnetic states and so forth, does not play an important role in the Invar effect.

* Presently at Department of Physics, University of Toronto, Toronto.

MAGNETORESISTANCE ANISOTROPY IN FERROMAGNETIC Ni-Co ALLOYS AT LOW TEMPERATURES

T. R. MC GUIRE, IBM Research Center, Yorktown Heights, New York 10598

ABSTRACT

The parallel ($\rho_{\parallel}$) and transverse ($\rho_{\perp}$) magneto-esistance has been measured in ferromagnetic alloys of $Ni_{1-x}Co_x$ with compositions $0 \leq x \leq .3$ and in fields up to 25kG at room temperatures, 77°K and 4.2°K. At each temperature the anisotropic ratio for magnetic saturation $\Delta\rho/\rho_o$ (where $\Delta\rho = \rho_{\parallel} - \rho_{\perp}$) increased with increasing cobalt content.

INTRODUCTION

The electrical resistance of several Ni-Co alloys have been measured at room temperature, 77°K and 4.2°K in magnetic fields up to 25kG. Resistivity was determined with the field applied parallel to the measuring current, $\rho_{\parallel}$, and transverse to the measuring current, $\rho_{\perp}$. Two effects are observed; the spontaneous anisotropic magnetoresistance due to the rotation of the magnetic domains of the specimen and the effect due to the applied field after the domains have been allígned. Both effects are well known and have been investigated in detail[1,2], but their physical origins are only partially understood[3].

In many respects the Ni-Co alloy system has ideal properties because of the close chemical and metallurgical similarity[4] of the two components. In agreement with this picture photoemission measurements[3] demonstrate that the Ni-Co system has a common d band in contrast for example to the Ni-Cu system. The above conditions are reflected in the average electrical resistivity, ρ_o , of Ni-Co, lower than other binary nickel alloys such as Ni-Cu, Ni-Fe and Ni-Mn.

Of the various nickel alloys, $Ni_{1-x}Co_x$ has the largest anisotropic magnetoresistance, $\Delta\rho = \rho_{\parallel} - \rho_{\perp}$, and also the largest ratio, $\Delta\rho / \rho_o$, where $\rho_o = 1/3\, \rho_{\parallel} + 2/3\, \rho_{\perp}$. However, for low Co concentration in Ni ($x < 0.1$) and at liquid helium temperatures over the complete range of compositions, Ni-Co has not been previously studied.

The understanding of the causes of anisotropic magnetoresistance is still developing. Theoretical models use spin-orbit coupling to account for the effect, showing that the scattering of electrons is greatest in the parallel case i.e. $\rho_{\parallel} > \rho_{\perp}$. However the subject is complicated by the fact that two types of scattering can be considered; that of s conduction electons by holes in the 3d band, and also by the Co impurity which may have a slightly different moment than its nickel neighbors. The study of dilute Co in Ni can give additional information on the scattering mechanisms.

SAMPLE PREPARATION & MEASUREMENTS

The samples were prepared by first arc melting Ni and Co in the desired proportions to make the richer cobalt alloys. A portion of these ingots were then used to produce the more dilute alloys. In each case the melting procedure was repeated about four times to insure initial homogeneity in the ingots. A rod about 2.5 cm long and 1 mm diameter was cut from the ingot. The rod specimens were then vacuum sealed in quartz tubes, annealed at 650°C for 18 hours, and then furnace cooled. A final step was to wash with acid.

In the following table are listed the samples studied including their composition and uniformity. This information was obtained by electron microprobe technique[5] and approximately four points were analyzed along each specimen.

A four probe resistivity measurement was made. The current leads were soldered to each end of the sample but the voltage leads were fine copper wires fixed by pressure and silver paint. A constant current power supply maintained a 500 milliamp current in the sample at a value which varied no more than ± 100 μa. The potential leads went to a digital voltmeter with 0.1 μ volt sensistivity.

Because small changes in temperature can cause changes in resistivity comparable with anisotropic magnetoresistive changes, measurements at low temperatures were made directly in liquid nitrogen and liquid helium. The sequence of measurements was to set the applied field, H, at a fixed value and rotate the sample so the cylinder axis alternated between parallel and perpendicular to H. The electric current was along the cylinder axis.

RESULTS

The resistivity measurements are shown in Figs. 1 through 4 and listed in the Table. For the composition, x=.005, measurements are given for three temperatures and for the remaining samples 4.2°K data are illustrated. Fig. 1 a, b, c show clearly the general behavior observed for all compositions. The spontaneous anisotropic effect occurs immediately upon saturation; for $\rho_{\parallel}$ the saturation occurs in fields less than 1 kG but for $\rho_{\perp}$ about 4kG is necessary in order to overcome demagnetization ($2\pi M_s$ perpendicular to the cylinder axis). At room temperature (Fig. 1a) as the applied field is increased up to 25kG both $\rho_{\parallel}$ and $\rho_{\perp}$ decrease in magnitude about the same amount. This decrease is a consequence of additional alignment of the magnetic spins above the M_s value. At 77°K the decrease in $\rho_{\parallel}$ and $\rho_{\perp}$ found at room temperature has vanished for two reasons; M_s is almost equal to M_o, the value at 0°K, and the ordinary magnetoresistance is beginning to show up. This is demonstrated clearly in the 4.2°K measurements, Fig. 1c, where the field dependence has the quadratic character typical of most metals and alloys. There are various problems connected with the interpretation of field dependence but in this paper we are

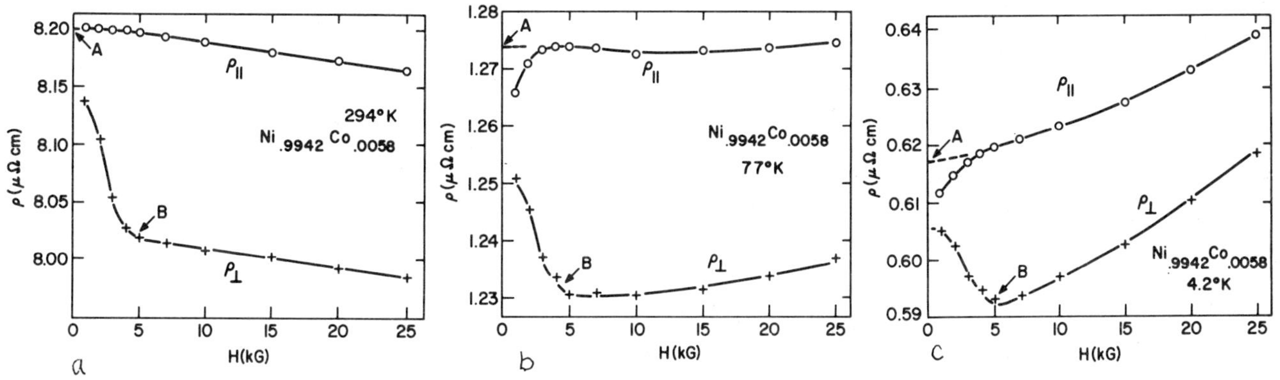

Figs. 1 a, b, c, Resistivity, ρ , in micro-ohm-cm. of $Ni_{.9942}Co_{.0058}$ as a function of magnetic field for a) room temperature, b) 77°K, and c) 4.2°K. The points A and B mark the selection of $\rho_{\parallel}$ and $\rho_{\perp}$ to determine the anisotropic magnetoresistance associated with the spontaneous moment.

primarily interested in thhe anisotropic effect, and the field effects are necessary only to the extent of obtaining the best value of $\Delta\rho$. On Fig. 1 the point marked A which is the extension of $\rho_{\parallel}$ to the zero field position and point B which represents $\rho_{\perp}$ at the field which demagnetization is overcome, represent the values used to determine $\Delta\rho/\rho_o$ as listed in the Table.

The measurements most difficult to interpret are those for pure nickel. This sample which had a resistance ratio, $\rho(297°K)/\rho(4.2°K)=83$, exhibited at 10kG and 4.2°K a cross over (Fig. 2) where $\rho_{\perp}$ became greater than $\rho_{\parallel}$. The condition $\rho_{\perp}>\rho_{\parallel}$ at low temperatures has been discussed by Smit[1] and van Elst[2] in terms of the normal magnetoresistance. Effects due to sample size and surface scattering are probably small as indicated by the work of Ehrlich and Rivier[10]. In addition, there is a strong initial increase in both $\rho_{\parallel}$ and $\rho_{\perp}$ at fields below 1 kG. The final choice for $\Delta\rho$ was the separation at 4kG where $\Delta\rho/\rho_o \simeq 2.2\%$. The values of $\Delta\rho$ and ρ_o for the remaining compositions presented no obvious problems.

The anisotropic ratio vs. composition is plotted in Fig. 4. At each temperature there is a gradual rise in magnitude of $\Delta\rho/\rho_o$ with increasing cobalt compositions. The functional behaviour illustrated here for the 4.2°K data is in disagreement with previously published values[6] for binary Ni alloys containing dilute amounts of Cr, V, Mn and Fe. For these alloys $\Delta\rho/\rho_o$ was reported independent of concentration showing a constant ratio with compositions as low as 1% impurity.

TABLE: $Ni_{1-x}Co_x$ with homogeneity, σ, and the magnetoresistance ratio at 4.2°K.

x	0	.0058	.025	.054	.107	.30
σ(%)	*	± 3.0	± 2.0	± 3.0	± 1.0	-
$\Delta\rho/\rho_o$(%)	2.2	3.8	8.7	10.6	16.4	26.7

*less than 50 ppm of Mn, Co, Fe, impurities

DISCUSSION

First it must be pointed out that the reason for the disagreement with previous measurements is not known. For the Ni-Co system the disagreement is based only on the one reported composition at $x \approx .02$ and only the ratio $\Delta\rho/\rho_o$ is given so a direct comparison of the measurements cannot be made. Never the less, it is of interest to examine the model which has been used to explain the composition independent data.

In a series of papers Campbell and Fert[7,8] and also Ferrell and Greig[9] have developed Mott's theory that the electrical resistivity of ferromagnetic transition metal alloys can be explained by a model in which spin up↑and spin down ↓ electrons conduct in parallel but with different conductivities. The basic approach is to compare resistivities to Matthiessen's rule and show the parallel conduction (two current model) explains the deviation that are found.

Campbell, Fert and Jaoul[6] related the two current model to anisotropic magnetoresistance by an induced resistivity transfer from ↑ to ↓ electrons due to spin-orbit interaction between conduction electrons and the ferromagnetic spin system. Next using the spin-orbit model of Smit[1] they conclude that the transfer process is stronger when conduction is parallel to the direction of magnetization, than when it is perpendicular. As a final step the anisotropic ratio for each binary alloy is given by the relation $\Delta\rho/\rho_o = (\alpha-1)\gamma$ where γ is a constant and α takes on a fixed value for each binary system.

From our measurements for $Ni_{1-x}Co_x$ we cannot justify the final step. Emperically the data shown in Fig. 4 fits a relation $\Delta\rho/\rho_o = .022 + x(1-x)C$ where C is a constant and x is the composition. It is of course possible that C is related to α but that cannot be determined here. What is of interest is that the expresion containing the term x(1-x) is identical to Nordheim's rule for resistivity of an alloy system. One interpretation is that the scattering is caused by a spin-disorder system made up of Ni and Co ions having small differences in moment. Development of this concept, which must include the temperature dependence behavoir of $\Delta\rho$ as well as the importance of the 3d band filling, in relation to the maximum value of $\Delta\rho$, will be considered in a future paper.

I wish to thank I. Kijac for sample preparation, F. Cardone for compositional analysis and H. Lilienthal for his assistance with the measurements.

REFERENCES

1. J. Smit, Physica, XVI, 612 (1951).
2. H. C. Van Elst, Physica, 25, 708 (1959).
3. T. R. McGuire, W. D. Grobman, and D. E. Eastman A.I.P. Conf. Proc. 18, 903 (1973).
4. F. H. Hayes, F. Muller, and O. Kubaschavski, J. Int. Metals. 96, 20 (1970).
5. W. Reuter, Sur. Sci. 25, 80 (1971).
6. I. A. Campbell, A. Fert, and O. Jaoul, J. Phys. C: Metals Phys. Suppl. No. 1, S95 (1970).
7. I. A. Campbell, A. Fert and A. R. Pomeroy Phil. Mag. 15, 977 (1967).
8. A. Fert and I. A. Campbell, Phys. Rev. Letters, 21, 1190 (1968).
9. T. Ferrell and D. Greig, J. Phys. C. Ser. 2, 1, 1359 (1968).
10. A. C. Ehrlich and D. Rivier, J. Phys. Chem. Solids 29, 1293 (1968).

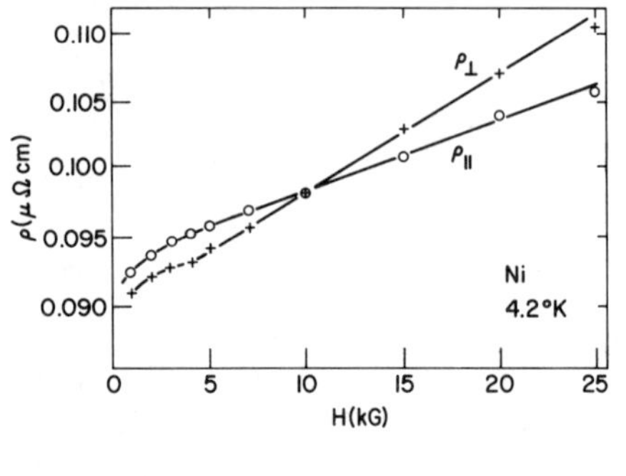

Fig. 2, Resistivity of pure Ni at 4.2°K as a function of H.

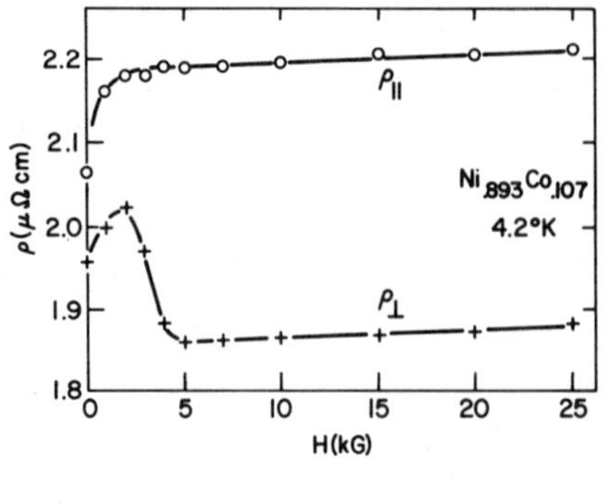

Fig. 3, Resistivity of $Ni_{.893}Co_{.107}$ at 4.2°K as a function of H.

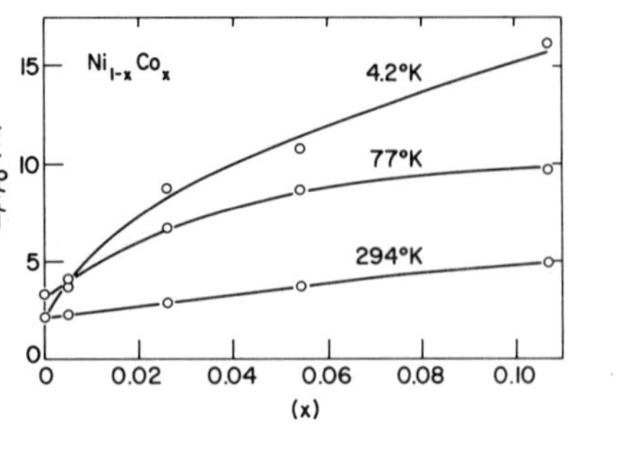

Fig. 4, Anisotropic magnetoresistance ratio vs. composition, x, in $Ni_{1-x}Co_x$ at three temperatures.

MAGNETIC PROPERTIES OF MnPt AND CoPt

C. W. Chen and R. W. Buttry
Ames Laboratory-USAEC and Department of Metallurgy, Iowa State University, Ames, Iowa 50010

ABSTRACT

Magnetization measurements were conducted on ordered and disordered polycrystalline specimens of MnPt and three Co-Pt alloys between 4.2 and 293°K. The magnetic data, while confirming the onset of antiferromagnetism in the ordered MnPt, reveal the presence of ferromagnetic regions in the disordered MnPt. Chemical order tends to exert a pronounced (∿30%) adverse effect on saturation magnetization ($M_{s,0}$) of the Co-Pt alloys at 0°K near the equiatomic composition. Meanwhile, the effect of Co concentration on $M_{s,0}$ of the Co-Pt alloys is shown to be appreciable only when the Co content exceeds 50%.

INTRODUCTION

Platinum and transition metals of the iron group form a series of isomorphous solid solutions at the equiatomic composition. In the disordered state, five NPt alloys, namely, CrPt, MnPt, FePt, CoPt, and NiPt crystallize into the face-centered cubic lattice. All these alloys tend to order chemically into the $L1_0$- or CuAu I-type superlattice, of which the c/a ratio is <1 based on the face-centered tetragonal structure, with room temperature values ranging from 0.915 for MnPt[1] to 0.969 for CoPt[2].

Investigation of the magnetic behavior of the NPt alloys was initiated by Kussmann[3] in 1935. Shortly thereafter[4], the ferromagnetic CoPt was found to display exceptionally high values of coercive force and $(BH)_{max}$, thus making the alloy an attractive permanent magnet. Besides the commercial potential of CoPt, interest in the NPt alloys lies in the prospect that, because of the striking resemblance in their crystal structure and their tendency to ordering, the alloys constitute a system within which the variation of magnetic properties with the species of magnetic component N and with the state of order could shed light on the distribution and interaction of magnetic moments in transition-metal alloys.

Despite the initial work by Kussmann and more recent studies by Pearson et al.[5] and by Krén et al.[6], magnetic data for the NPt alloys were still incomplete. For instance, new susceptibility data[7] for NiPt detected a change in the mode of spin coupling from ferromagnetic to antiferromagnetic as a result of the disorder→order transition. The latter result has motivated us to examine certain aspects of the magnetic behavior of MnPt and CoPt. The present report will reveal (1) the presence of ferromagnetic regions in the disordered MnPt, (2) a pronounced adverse effect of chemical order on the extrapolated value of saturation magnetization to 0°K ($M_{s,0}$) of CoPt, and (3) highly irregular variations of $M_{s,0}$ with Co concentration of the Co-Pt alloys near the equiatomic composition.

Table 1. Extrapolated Values of Saturation Magnetization to 0°K for Three Co-Pt Alloys.

	$Co_{0.54}Pt_{0.46}$		CoPt		$Co_{0.46}Pt_{0.54}$	
	Disord.	Order.	Disord.	Order.	Disord.	Order.
$M_{s,0}$, emu/g:	53.5	37.8	46.6	32.1	45.0	29.3

(A) Change in $M_{s,0}$ caused by ordering: $Co_{0.54}Pt_{0.46}$ 15.7(29%); CoPt 14.5(31%); $Co_{0.46}Pt_{0.54}$ 15.7(35%)

(B) Change in $M_{s,0}$ caused by variation in % Co;disord.: 6.9(15%) ($Co_{0.54}Pt_{0.46}$–CoPt); 1.6(3.4%) (CoPt–$Co_{0.46}Pt_{0.54}$)

(C) Change in $M_{s,0}$ caused by variation in % Co;ordered: 5.7(18%) ($Co_{0.54}Pt_{0.46}$–CoPt); 2.8(9%) (CoPt–$Co_{0.46}Pt_{0.54}$)

EXPERIMENTAL PROCEDURE

Four alloys (MnPt, $Co_{0.54}Pt_{0.46}$, CoPt, and $Co_{0.46}Pt_{0.54}$) were prepared from high purity (>99.98%) metals by arc melting. To attain composition homogeneity, the ingots were remelted several times on alternate sides and homogenized at appropriate temperatures for 12-46 hrs in an atmosphere of purified helium gas. Magnetic specimens were prepared in short cylinder form by machining. Disordered specimens were obtained by quenching into brine from 1125 and 950°C, respectively, for MnPt and Co-Pt alloys. Ordered specimens were obtained by prolonged annealing at several temperatures below the order-disorder transition temperature. A vibrating sample magnetometer equipped with a liquid helium metal dewar was used to measure magnetization under a field of maximum strength of 20 koe.

RESULTS AND DISCUSSION

(I) Kind of Magnetism in MnPt Below 293°K

Magnetic data obtained for the ordered MnPt between 4.2 and 293°K yield a linear relationship between magnetization (M) and magnetizing field (H). A typical linear M-H plot at 4.2°K is shown in Fig. 1. The deduced values for magnetic susceptibility (χ) show a slight, nonlinear decreasing tendency with decreasing temperature in the χ versus T plot. These results are consistent with the onset of antiferromagnetism established in the ordered MnPt by previous workers[5,6].

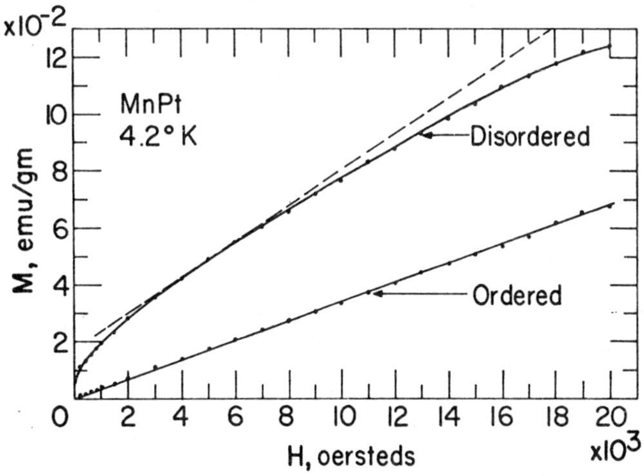

Fig. 1. Field dependence of magnetization in the disordered and ordered MnPt at 4.2°K.

Data for the disordered MnPt disclose a nonlinear M-H relationship in the same temperature range. Shown in Fig. 1 is a typical curve for the disordered MnPt at 4.2°K, of which the slope is positive, but decreasing progressively. The latter plot seems to imply the presence of ferromagnetic regions, possibly in the sense of mictomagnetism. To test this implication, we first measured M of the disordered MnPt between $+H_{max}$ and $-H_{max}$ at 4.2 and 293°K. A resolvable hysteresis loop

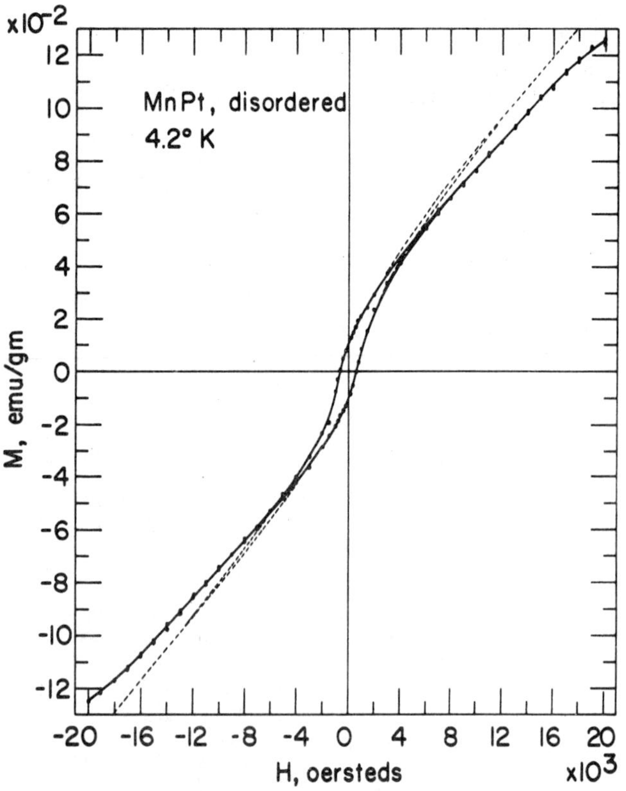

Fig. 2. Hysteresis loop displayed by the disordered MnPt at 4.2°K after cooling from room temperature in zero field (solid curves) and in a field of 20 koe (dashed curves).

emerged at both temperatures. The 4.2° loop is shown by the solid curves in Fig. 2. To see whether the emergence of this hysteresis loop is related to an unidirectional anisotropy induced by the exchange interaction between ferromagnetic and antiferromagnetic regions[8], we further cooled the specimen from room temperature to 4.2°K in a magnetic field of 20 koe. The hysteresis loop subsequently measured parallel to this field is shown by the dashed curves in Fig. 2. The magnetic cooling has neither introduced asymmetry with respect to the origin nor changed the main features (e.g., the coercive force and remanence) of the loop; it only opened up the loop toward the tips and slightly raised the maximum value of M at H_{max}. The results showing the insignificant effect of the magnetic cooling on the hysteresis loop suggest that the present case differs from that seen in the disordered Ni_3Mn[9], for which exchange anisotropy plays a vital role in the hysteresis behavior.

According to neutron diffraction data[6] for the ordered MnPt, only Mn atoms carry magnetic moments of $4.3 \pm 0.2\ \mu_B$, which are aligned antiferromagnetically in alternate (010) planes in the c-axis. Thus in terms of the Heisenberg theory, a negative exchange integral J exists between atom pairs of Mn in the nearest neighbors (NN) and a positive, but smaller, J exists between Mn-Mn pairs in the next nearest neighbors. Since the number of NN pairs of Mn is increased from 4 to 6 and the distance is decreased from 2.831 Å to 2.749 Å on going from the ordered to the disordered state, ferromagnetism is not expected in the disordered MnPt. The present results probably imply a nonuniform distribution of the constituent atoms, leading to a division of the crystals into regions, some rich in Co and others rich in Pt. The Pt-rich regions having compositions corresponding to 63-83 at % Pt-Mn give rise to the magnetic behavior shown in Figs. 1 and 2 because the latter alloys are known to be ferromagnetic.[6]

(II) The Effect of Chemical Order and Co Concentration on M_s of CoPt and Related Alloys

Although the ferromagnetic CoPt alloy attracted numerous studies in the 1950's[10], interests had been focused on structure-sensitive magnetic properties. In a more recent study, McCurrie and Gaunt[11] observed an 18% decrease in M_s of $Co_{0.48}Pt_{0.52}$ at 77°K as a result of chemical order. To further evaluate the effect of chemical order and Co content on M_s, we obtained the $M_{s,0}$ values at 0°K for CoPt and two related alloys by extrapolation first in the M vs 1/H plot and then in M_s vs T plot. The extrapolated values for the three alloys in both states of order are listed in Table 1. A persistent, 29-35% decrease is seen of $M_{s,0}$ in all three alloys upon ordering. Our data also reveal that the variations in $M_{s,0}$ of CoPt caused by a 4% change in Co content are larger on the Co-rich side than those on the Pt-rich side in both states of order.

The observed effect of chemical order and Co content on $M_{s,0}$ can be related to the crystal and magnetic structures. X-ray data for CoPt[2] gave a=3.751 Å in the disordered phase and a=3.793 Å and c=3.675 Å in the $L1_0$ superlattice at room temperature. From neutron data, van Lear[12] detected a total moment of $1.90 \pm 0.01\ \mu_B$ per formula CoPt. (Our $M_{s,0}$ values for CoPt give 2.12 and 1.43 μ_B per formula in the disordered and ordered states, respectively, indicating that van Lear's sample was only partially ordered.) The total moment was seen to split into $\lesssim 1.6$ and $\gtrsim 0.3\ \mu_B$ for the Co and Pt atom, respectively. In the disordered state, each atom is surrounded, on the average, by 6 Co and 6 Pt atoms at the same distance of d_{NN}=2.653 Å; whereas in the superlattice, each Co(or Pt) atom is surrounded by 8 Pt(or Co) atoms at d_{NN_1}=2.649 Å and by 4 Co(or Pt) atoms at d_{NN_2}=2.683 Å. In view of these environmental changes, we interpret our experimental results as implying that ordering causes a decrease in the moment of Co atoms that is considerably more than the possible increase in the moment induced at the Pt sites. The observed irregular effect of Co content on $M_{s,0}$ may be explained similarly in terms of changes in the number of Co-Co and Co-Pt pairs in the nearest neighbors.

REFERENCES

1. A. F. Andresen, A. Kjekshus, R. Mollerud, and W. B. Pearson, Phil. Mag. 11, 1245 (1965).
2. M. Hansen, Constitution of Binary Alloys (McGraw-Hill Book Co., New York, 1958) p. 492.
3. E. Friederich and A. Kussmann, Phys. Z. 36, 185 (1935).
4. W. Jellinghaus, Z. Tech. Phys. 17, 33 (1936); also H. Neumann, A. Buchner, and H. Reinboth, Z. Metallkde. 29, 173 (1937).
5. K. Brun, A. Kjekshus, and W. B. Pearson, Phil. Mag. 10, 291 (1964); A. F. Andresen, A. Kjekshus, R. Møllerud, and W. B. Pearson, Acta Chem. Scand. 20, 2529 (1966).
6. E. Krén, G. Kádár, L. Pál, J. Sólyom, P. Szabó, and T. Tarnóczi, Phys. Rev. 171, 574 (1968).
7. C. W. Chen, J. D. Greiner, and R. W. Buttry, AIP Conference Proceedings 18, 432 (1974).
8. W. H. Meiklejohn and C. P. Bean, Phys. Rev. 102, 1413 (1956) and 105, 904 (1957).
9. J. S. Kouvel and C. D. Graham, Jr., J. Appl. Phys. 30, 312S (1959).
10. E. P. Wohlfarth, Adv. in Phys. 8, 208 (1959).
11. R. A. McCurrie and P. Gaunt, Phil. Mag. 13, 567 (1966).
12. B. van Lear, J. Phys. (Paris) 25, 600 (1964).

VOLUME DEPENDENCE OF VBS PARAMETERS†

C. L. Foiles
Physics Department
Michigan State University
East Lansing, Michigan 48824

ABSTRACT

A simple VBS model is used to interpret the pressure dependence of residual resistivity in AgPd and CuNi alloys. For AgPd such an interpretation proves to have dubious significance. For CuNi the interpretation indicates the VBS recedes from the Fermi level as volume decreases and this result is qualitatively and quantitatively consistent with CPA calculations.

Several investigations of alloys involving Ce as an impurity have found a large, negative coefficient for the volume dependence of the residual resistivity.[1,2,3] Although these studies differ on a number of interpretative matters relating to detailed behavior, they concur on a virtual bound state (VBS) which moves toward the Fermi energy (E_F) as the volume decreases. This concurrence is sufficient to explain the above coefficient. Most alloy systems involving noble metal solvents and transition metal solutes also display a negative coefficient[4] and, since this is in direct contrast with the positive coefficient which is generally characteristic of non-transition metal solutes in these same solvents[4,5], it might be inferred that a VBS model is sufficient to explain the volume dependence of the former alloys. Two considerations present a major challenge to the preceding inference. First, some allowance must be made for the magnetic nature of the impurity which occurs in many of these systems. The importance of such an allowance can be demonstrated by considering the behavior of CuFe; the volume dependence of the impurity resistivity changes sign near 80°K.[6,7] This temperature is well above the Kondo temperature of CuFe and thus a unique value for the coefficient or a well defined separation of magnetic and non-magnetic contributions may not be possible. Second, if attention is limited to alloy systems where the impurity displays little or no magnetic character, then this subset has the smaller values for the coefficient and no sign pattern occurs. One striking example of the subset is CuNi. Although many properties of dilute CuNi are consistent with a VBS model having unique and well defined parameters,[8] the volume dependence of the residual resistivity has a sign and magnitude consistent with those for non-transition impurities.

The well documented ability of a VBS model to describe quantitatively many diverse properties for CuNi[8] makes this system an ideal candidate in pursuing the volume dependence of residual resistivity. The companion studies for AgPd with their greater scatter in VBS parameters[8] also provide insight into the problem and both systems are included in the present study. Using a single phase shift model and a VBS of Lorentzian shape (points which will be considered in greater detail later)

$$\alpha_V \equiv \frac{d\ell n\rho_o}{d\ell nV} = -\frac{d\ell nk_F}{d\ell nV} + 2\frac{d\ell n\Gamma}{d\ell nV} - \frac{2}{(E_d)^2+(\Gamma/2)^2}\{E_d^2\frac{d\ell nE_d}{d\ell nV} + (\Gamma/2)^2\frac{d\ell n\Gamma}{d\ell nV}\} \quad (1)$$

where ρ_o is the impurity resistivity per atomic % impurity, k_F is the Fermi wave-vector of the solvent, V is the volume, and Γ and E_d are, respectively, the half-width and the energy of the VBS. For a solvent density of states n(E) and a mixing term V_{kd}

$$\Gamma = 2\pi n(E)|V_{kd}|^2 \quad (2)$$

Using the standard free electron model for the solvent results in Γ being proportional to $V^{-1/3}$ and eqn. (1) becomes

$$\alpha_V = -\frac{1}{3} - \frac{2}{(E_d)^2+(\Gamma/2)^2}\{E_d^2\frac{d\ell nE_d}{d\ell nV} - \frac{1}{3}(\Gamma/2)^2\} \quad (3)$$

Prior to any discussion of results from eqn. (3), three features require more detailed consideration:

(i) Single phase shift model: This model is commonly used and it introduces negligible error for transport considerations in the present context. Use of a non-zero ℓ=0 phase shift produces an extra term in eqn. (3) but the logarithmic form generates weighting factors which make the effects of this term negligible. Use of a non-zero ℓ=1 phase shift complicates matters since an interference between ℓ=1 and ℓ=2 phase shifts occurs. However, numerical procedures indicate the errors are small as long as the latter phase shift, the resonant phase shift, dominates and this condition is met in the present circumstances.

(ii) VBS parameters: As indicated earlier, a unique set of parameters occurs for CuNi and the values Γ=0.5 eV and E_d= -0.75 eV are used in the present study. Greater uncertainty exists in the parameters for AgPd and table I summarizes the available data. Although E_d appears to have a consistent value,

Property	Γ(eV)	E_d(eV)	Ref.
C_e	0.6	-2.2	9
S_d	0.9	-3.1	10
Optical	1.8	-2.6	11
UPS	1(observed) 0.5 (with estimated spin orbit correction)	-2.1	12
XPS	0.6±0.2	-2.0±0.05	8
CPA	~ 0.6	~ -2.7	13

Table I. VBS parameters for AgPd.

Γ and E_d are defined in the text. These parameters were evaluated by using experimental data for electronic specific heat (C_e), diffusion thermopower (S_d), optical absorption (Optical), ultra-violet photoelectron spectroscopy (UPS), and x-ray photoelectron spectroscopy (XPS). A theoretical calculation using the coherential potential approximation (CPA) also provided values for the parameters.

the scatter in values for Γ is sufficient to cause problems. Of equal importance is the fact that a CPA calculation for AgPd indicates that Γ has a significant dependence upon impurity concentration[13] while a similar study for CuNi found no concentration dependence over the range of concentration involved in the present study.[14] The tentative values of Γ=1.0 eV and E_d= -2.0 eV are used for AgPd in this study.

(iii) Unique value for α_v: Measurements at 273 and 77.3°K have been made to detect any temperature effects in this parameter and the results for CuNi and AgPd are given in table II. The present results for CuNi are in reasonable agreement with previous results and there is no indication of any temperature effect. The result for AgPd is less precise. The agreement of various studies

Alloy	$\alpha_v \equiv \frac{d\ell n\rho_o}{d\ell nV}$ 77.3°K	~ 273°K	Ref
CuNi	1.2±0.1	1.2±0.1	
		0.9	4
		1.2	15
AgPd	-0.15±0.05	-0.21±0.05	
		-0.2	4

Table II. α_v at two temperatures for AgPd and CuNi.

near room temperature is good but the present study gives some indication of a temperature effect. The small values for α_v combine with the experimental uncertainty to prevent documentation of this dependence and the value of -0.20 is used for AgPd. The striking temperature dependence found in CuFe is clearly absent in the present systems.

With the preceding context, eqn. (3) and experimental data can be used to evaluate $d\ \ell n\ E_d/d\ \ell n\ V$. For AgPd, the resultant value is negative and has an absolute magnitude less than 0.1. In view of the near equality of terms in eqn. (3) and the various questions about a simple VBS model for this system, the significance of the final result is very dubious. However, the situation for CuNi is free from both objections and a straight-forward substitution yields $d\ \ell n\ E_d/d\ \ell n\ V = -0.8$.

The negative sign in this final result indicates the VBS moves further below the Fermi level as the volume decreases and such movement is the direct opposite of the movement which occurs in the alloys containing Ce. In these latter alloys pressure studies of superconducting properties provided confirming evidence for the movement of the VBS.[16] Clearly such confirmation is not possible for CuNi and instead the extensive CPA calculations for noble-transition metal alloys by Stocks and co-workers[13,14,17] are used. The parameter δ, a measure of the energy difference between constituent atomic orbitals which are used in the dd block of an interpolation Hamiltonian, plays an essential role in these calculations. Figure 9 in reference 14 depicts the formation of an impurity sub-band as δ becomes increasingly negative. Combining a dependence of E_d upon δ which can be obtained from this figure with a dependence of δ upon lattice parameter which CPA calculations as a function of lattice parameter revealed[13], yields the estimate $d\ \ell n\ E_d/d\ \ell n\ V = -0.7$.

CONCLUSIONS

The remarkable quantitative agreement between experimental and CPA estimates for $d\ \ell n\ E_d/d\ \ell n\ V$ in CuNi must be regarded as somewhat fortuitous. However, the existence of agreement is significant for both VBS and CPA models of this system and serves as one more confirmation of their ability to describe the observed properties. Both models indicate the VBS in CuNi recedes from the Fermi level as the volume decreases.

† Supported in part by a NSF grant.

1 K.S. Kim and M.B. Maple, Phys. Rev. B2, 4696 (1970).
2 W. Gey and E. Umlauf, Z. Phys. 242, 241 (1971).
3 M. Dietrich, W. Gey and E. Umlauf, Solid State Comm. 11, 655 (1972).
4 J.O. Linde, Arkiv for Fysik 39, 139 (1969).
5 J.S. Dugdale in Physics of Solids at High Pressures, edited by C.T. Tomizuka and R.M. Emrick (Academic Press, New York, 1965).
6 J.S. Dugdale and D. Gugan, Proc. Royal Soc. (London) A241, 397 (1957).
7 A closer inspection of refs. 4 and 6 indicates they differ on the sign of α_v for CuFe at room temperature. The latter reference had the behavior of Cu as its major concern and used only a single alloy concentration to determine the behavior of the Fe impurity. Such a procedure often produces serious errors and we have attempted to verify the result by using several concentrations below 0.1 at%. Our results agree with the sign of ref. 4 at 273°K and display a sign reversal by 77.3°K; thus, our basic results form a mirror image to those found in ref. 6 and confirm the fact of a sign reversal.
8 S. Hüfner, G.K. Wertheim and J.H. Wernick, Phys. Rev. B8, 4511 (1973).
9 Bengt Kjollerstrom, Solid State Comm. 7, 705 (1969).
10 Present study. Influence of a single phase shift model for transport upon the subsequent values deduced for VBS parameters can be obtained by comparing CuNi results with those obtained from a more complicated procedure; C.L. Foiles, Phys. Rev. 169, 471 (1968). There is very little effect upon Γ but a greater $|E_d|$ is deduced. This same trend is consistent with the entries in table II.
11 H.P. Myers, L. Walldén and A. Karlsson, Phil. Mag. 18, 725 (1968).
12 C. Norris and H.P. Myers, J. Phys. F, 1, 62 (1971).
13 G.M. Stocks, R.W. Williams and J.S. Faulkner, J. Phys. F. 3, 1688 (1973).
14 G.M. Stocks, R.W. Williams and J.S. Faulkner, Phys. Rev. B4, 4390 (1971).
15 P.W. Bridgman, The Physics of High Pressure (G. Bell and Sons, LTD, London, 1958) chapter IX.
16 M.B. Maple in Magnetism V, edited by H. Suhl (Academic Press, New York, 1973).
17 G.M. Stocks, Intl. J. Quantum Chem. 5, 533 (1971).

A CORRELATION BETWEEN ATOMIC VOLUME AND THE SQUARE OF THE ATOMIC MAGNETIC MOMENT FOR d GROUP TRANSITION METALS: APPLICATION

W. F. Schlosser, Dept. of Physics, Univ. of Manitoba, Winnipeg, Manitoba R3T 2N2, Canada

ABSTRACT

A simple magnetovolume relation previously checked on d group metals and alloys can be used to distinguish demagnetization due to a reduction in the magnitude of the atomic magnetic moments from that due to an increased disorientation of the moments. Such information is difficult, or impossible to obtain by other means, such as neutron diffraction.

The relation used is quite different from that proposed recently by Shiga.[9]

It appears that, although the interactions that determine the moment on an atom can have a strong interatomic contribution, the size (volume) of an atom is determined intra-atomically by its magnetic moment.

INTRODUCTION

The magnetic moment on an atom in a metal may change if the effective exchange coupling changes due to changes in temperature or atomic order, for instance. Changes in the magnitude of the atomic magnetic moment are very difficult to distinguish from changes in the relative alignment of these moments. One can, in principle, make this distinction with neutron diffraction, but only with great difficulty, and in a few special materials.

In this paper, a simple magnetovolume relation, previously demonstrated[1-4] to hold for a wide range of d group metals and alloys is used to show that volume effects should provide a ready and simple way to discriminate between changes in the magnitude and in the relative alignment of the magnetic moments.

The volume of a d group atom in a metal, and hence its contribution to the mean atomic volume of an alloy has been shown[1-4] to be

$$V = V_0 + k\mu^2 \tag{1}$$

where μ is the atomic magnetic moment corresponding to the spin polarization of the electrons in the Wigner-Seitz cell, and V_0 the atomic volume for $\mu = 0$. All non-magnetic thermal expansion effects etc. are included in V_0 which need not satisfy Zen's law (the mean atomic volume of an alloy varies linearly with concentration) though it usually does. Atomic volumes are taken as additive. $k = 0.108$ Å^3 μ_B^{-2} was found to be constant and not dependent on the element involved, the crystal structure, the near neighbour atomic environment, the type, or even the presence of magnetic or atomic order, or whether the alloy is substitutional or interstitial. Values of V_0 at 300 K[3-4] for Cr, Mn, Fe, Co and Ni, which can have large values of μ, are given in Table 1.

Relation (1) appears to hold from the weak paramagnet tungsten[1] to the strongest ferro- and antiferromagnets over a range of magnetic volume changes of $k\mu^2$ of 6×10^{10}.

TABLE 1

NON-MAGNETIC ATOMIC VOLUMES OF 3d ELEMENTS

Element	Atomic Volume $V_0 = V(\mu = 0)$, Å^3
Cr	12.000
Mn	12.160
Fe	11.248
Co	10.778
Ni	10.897

MAGNETIC MOMENT CHANGES, INVARS, AND MICTOMAGNETS

To illustrate the use of relation (1) in this paper, consider the possible behaviour of μ in a metal with magnetic order below T_C, (Fig. 1). Depending on the interactions producing the magnetic order, μ can remain constant (the classical mictomagnet), increase to a limit, or decrease to a limit not necessarily zero as T increases through T_C (Invar type behavior). For a ferromagnet, the magnetization M is not directly μ, but the time average of μ along the field direction. The temperature dependence of μ is thus intermediate between simple band models, where $\mu \to 0$ as $T \to T_C$ and simple localized models, where μ remains constant. By determining V_μ, the decrease of M as T increases can be resolved into contributions from the reduction in the magnitude of μ (simple band-like effects) and from the thermal disorientation of μ (simple disordered localized moment effects).

This paper is directly concerned with the application of this magnetovolume relation, to Invar type alloys and mictomagnets, and to show that these two types of materials can be clearly separated on the basis of the magnetovolume effect. The invar type metals have reduced magnetization, a reduced average value of V_μ, and a large magnetovolume increase as the magnetization increases. The mictomagnetic spin glasses such as Cu-Mn or atomically disordered Ni_3Mn have low magnetizations, but very small volume changes as the magnetization increases. Atomically disordered Ni_3Mn does not have long range magnetic order, while the atomically ordered Ni_3Mn is a very strong ferromagnet with $\mu_{Ni} = 0.3\ \mu_B$ and $\mu_{Mn} = 3.64\ \mu_B$[5], and yet there is almost no volume change on ordering.[6]

The Sidorov-Doroshenko model for these mictomagnetic systems[5,7] is based on atomic moments of fixed magnitude. The orientation of the moment on any Mn atom depends on the number of like atoms in the first neighbour shell. More than 3Mn atoms around a Mn atom and the central Mn atom has its moment opposed to that of its neighbours. The antiferromagnetic Mn-Mn interactions dominate the ferromagnetic Ni-Mn interactions. The magnitude of the atomic magnetic moment is not, however, a function of the relative orientation of the moments of an atom and its neighbours. Thus the demagnetization of Ni_3Mn takes place by the disorientation of moments of constant magnitude.

If one applies the model to invar-type alloys, the moments on the Fe atoms would not change, and there would be no temperature dependent magnetovolume anomaly. It is the dependence of the moment on an Fe atom (1.2 to 2.8 μ_B) on its near neighbour environment that produces the magnetovolume invar-type anomaly through relation (1). Thus, by determining the change in the magnetic contribution to the volume for an alloy, one can find the changes in the magnitude of the atomic magnetic moments independent of the demagnetization due to moment disorientation. The separation was carried out for pure Fe[3], the temperature dependence of V_μ and hence μ the magnitude of the moment derived and the results shown to be thermodynamically consistent with other data for Fe. This kind of data is difficult, if not impossible to obtain for these materials with other techniques such as neutron diffraction. It is very simply and readily obtained from the simple analysis at the magnetovolume effect used here and in previous papers. Conversely, a strict, conclusive proof of such a general relation is extremely difficult.

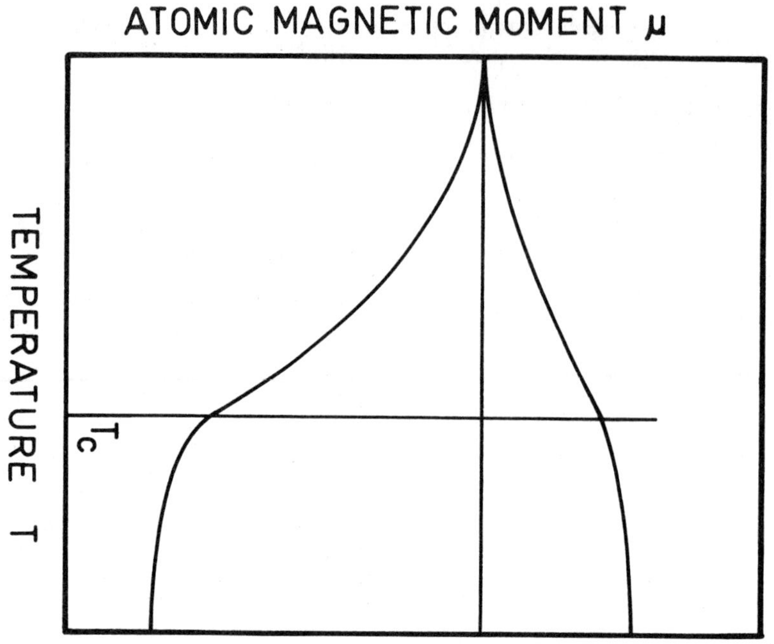

Fig. 1: The possible temperature dependences of the atomic magnetic moment μ in a metal with magnetic order below T_c.

SHIGA'S MAGNETOVOLUME RELATION

This treatment of the magnetovolume effect is simpler and more general than that suggested by Shiga.[9] Shiga proposed a linear dependence of the lattice parameter of a metal on the average of the magnitudes of the atomic moments, but found it difficult to accept such dependence, as theoretically and experimentally one expects a quadratic dependence on μ as in relation (1). The volume of an atom is a much better measure of its size than a lattice parameter, and can be readily used with any crystal structure. Both approaches agree on the necessity of considering demagnetization by a reduction in the magnitude as well as in the time averaged projection of μ.

CONSEQUENCES

The implications of the success of the magnetovolume relation (1) must be considered. The volume of an atom depends only on the moment μ on that atom, and only through the moment μ on the factors that determine μ. μ is determined primarily by interatomic interactions, whereas V_μ is essentially intra-atomic, depending only on the polarization of the electrons on the atom. Since this polarization is small near the boundaries of the Wigner-Seitz cell, interatomic effects are small. k is expected to be a constant characteristic of intra-atomic d electron interactions, and V_0 and V_μ to be additive, as found. The magnetovolume effect is thus an expansion of the atoms, not of the lattice.

The observed results cannot be accounted for by curves of the Bethe-Slater-Neel type (which cannot be generally determined anyway) or by calculations based on tight coupling of the magnetic and non-magnetic systems. The magnetovolume relations derived on such a basis are heavily dependent on such factors as the elements involved, the crystal structure, the value of the band splitting or the atomic moment itself, the band parameters and the type of magnetic order. For such tightly coupled systems, additivity of V_0 and V_μ is not expected.

CONCLUSIONS

The considerable success of the treatment of the magnetovolume effect used here, relation (1), emphasizes the need to reconsider current theoretical approaches to this subject. The simple form of relation (1) and the constancy of the parameter k found to be sufficient to account for the existing data suggests that a calculation of the spatial distribution of the spin polarization and the intra-atomic d electron interactions for a general d group atom should prove useful. Electron distributions are, however, apparently much harder to calculate than energy bands.

The absence of such theoretical, or strict experimental proof of the validity of the magnetovolume relation (1) does not however, prevent one from using the relation in practical cases where it can yield the only information available. The fact that the relation (1) does describe the data for a wide variety of materials lends considerable confidence to its use.

REFERENCES

1. W. F. Schlosser, Phys. Lett. A 42, 437-438 (1973).
2. W. F. Schlosser, Phys. Stat. Sol. (a) 17, 199-205 (1973).
3. W. F. Schlosser, Phys. Stat. Sol. (a) 22, K219-K222 (1974).
4. W. F. Schlosser, Int. J. Mag. (to be published).
5. S. K. Sidorov and A. V. Doroshenko, Phys. Met. Metal 20, 140-149 (1965).
6. M. C. Marcinkowski, Diss. Abs. 20, 257 (1959).
7. S. K. Sidorov and A. V. Doroshenko, Phys. Met. Metal 18, 6, 12-21 (1964).
8. S. K. Sidorov and A. V. Doroshenko, Phys. Met. Metal 19, 5, 132-134 (1965).
9. M. Shiga, Magnetism and Magnetic Materials, 1973 (19th Annual Conference - Boston) pp. 463-477, edited by C. D. Graham, Jr. and J. J. Rhyne, American Institute of Physics, New York (1974).

LOW TEMPERATURE MAGNETIC SUSCEPTIBILITY OF DILUTE Cu-Au(Fe) ALLOYS*

J. SHEU AND WM. R. SAVAGE
Department of Physics and Astronomy
University of Iowa
Iowa City, Iowa 52242

ABSTRACT

The magnetic susceptibility of a series of Cu-Au alloys with added Fe (< 760 ppm) has been measured at temperatures from 1.5 to 300 K and at six magnetic fields up to 13 kOe. The Au concentrations were 2.4, 4.8, 10, and 10.8 at.%. The results are analyzed on the basis of a model which separates the effects of single impurities and interaction effects of magnetic pairs of Fe.

At temperatures greater than 20 K the observed susceptibility was fitted to an expression which contained three terms: A constant term, a Curie-Weiss term, and a single Brillouin function with J = 3. When the constant term and Brillouin term obtained by this procedure were extended to low temperatures and subtracted from the measured susceptibility, the remainder agreed with the theory of Götze and Schlottmann for single impurities. The Brillouin function accounts for the field dependent susceptibility of our alloys. The characteristic temperature is observed to decrease with the addition of Au. A universal curve for the single impurity term $\chi_s(T)/\chi_s(0) = f(T/T_c)$ has been found where T_c is 13.6, 12.2, 10.6, 9.6, and 9.6 K for alloys with Au 0, 2.4, 4.8, 10.0, and 10.8 at.%, respectively.

I. INTRODUCTION

Considerable progress has been made in the understanding of metallic conductors that contain small amounts of magnetic transition element additions. Most theoretical[1] investigations have treated the effects of single impurities. The experimental studies show additional effects arising from correlations between the magnetic impurities. The correlations are most frequently attributed to magnetic pairs. The contributions of these pairs are more important at temperatures smaller than T_c, the characteristic temperature. The interpretation that the anomalous susceptibility of a dilute alloy could be in part due to magnetic pairs is known. The separation of the contribution of pairs from the one-impurity susceptibility has been a difficult problem in part because of the lack of the exact expression of the single-impurity effect. Attempts to separate these effects have been to use a series of Curie-Weiss terms[2], a series of Brillouin functions[3], or an expansion in terms of concentration dependences[4,5] to fit the measured susceptibility. Recently, an approximate theory for temperature-and field-dependent isolated-impurity susceptibility has been proposed by Götze and Schlottmann[6] that has agreed both qualitatively and quantitatively with the experiment on dilute Cu(Fe) system.[7]

In this study measurements were made on the magnetic susceptibility of dilute Fe in α-phase $Cu_{1-x}Au_x$ alloys in the temperature interval from 1.5 to 300 K and at six magnetic fields from 3.95 to 12.75 kOe. Measurements were also made on corresponding Fe-free $Cu_{1-x}Au_x$ host samples with similar Au concentrations. The latter measurements were subtracted from the former to evaluate the susceptibility of the Fe-bearing samples. The results have been analyzed by a simple procedure in which the single-impurity susceptibility can be separated from the field dependent susceptibility contributed by pairs.

II. EXPERIMENTAL

The samples used in this study are polycrystalline α-phase $Cu_{1-x}Au_x$ host alloys and dilute magnetic $Cu_{1-x}Au_x$(Fe) alloys. The samples used in the susceptibility experiment were cut from the same alloy bar used in a specific heat study by Delinger, et al.[8] The magnetic impurity content of the Fe-free samples was about 3-4 ppm as determined by a weak paramagnetic contribution to the susceptibility and the absence of a minimum in the electrical resistivity.[9]

A Faraday susceptibility apparatus was used for this magnetic susceptibility measurement that operates in the temperature range of 1.5 - 300 K with a precision of $\pm$ 4 x 10^{-10} cgs-emu/g. The susceptibility is obtained from a room temperature calibration that used 99.999+% zone refined ASARCO copper as the primary standard assumed to be 0.0858 x 10^{-6} cgs-emu/g.

Temperatures were determined to better that 0.1 K and the magnetic fields were determined to 2%. The chemical composition of the samples was measured by several independent methods including atomic absorption analysis.

III. RESULT AND DISCUSSION

Theoretical treatments and experimental measurements of the susceptibility of dilute alloys agree that the single-impurity susceptibility should follow a Curie-Weiss relationship at high temperatures and approach a constant value as T = 0. This indicates that a Curie-Weiss expression to describe the low temperature behavior of the single-impurity susceptibility may not be suitable. Our procedure for the analysis of the experimental data produce a universal curve that can be compared with the treatment by Götze and Schlottmann. The magnetic field dependence of the magnetization of the sample is fitted with a Brillouin function with J = 3. The high temperature (T > 20K) portion of the data is fit to the expression

$$\chi(H,T) = \chi_o + \frac{C}{T+\theta} + C_2 \frac{JB_{J=3}(H/T)}{H},$$

where χ_o is a constant, the second term is a Curie-Weiss term and the third term is the Brillouin function. The constants χ_o and C_2 resulting from this first step were used to obtain a field-independent susceptibility for the entire temperature range from the expression

$$\chi_s(T) = \chi(H,T) - \chi_o - C_2 \frac{JB_{J=3}(H/T)}{H}.$$

The value of $\chi_s(o)$ was obtained from the extrapolation of $\chi_s(T)$ to zero temperature. A characteristic temperature T_c was obtained from the expression $1.25T_c = \theta$, where θ is the parameter obtained from the high temperature data. The universal curve is a graph of the inverse of the reduced susceptibility, $\chi_s(T)/\chi_s(o)$ versus T/T_c. We are confident that this procedure separates the effects of single-impurities from that of pairs. Our experiment and that of others indicates a field-dependent susceptibility that could be described by parallel spin-coupled pairs each with an effective spin J = 3. The remaining susceptibility $\chi_s(T)$ is attributed to the Fe that acts as a single isolated impurity. We utilize the results of measurements of copper with Fe additions to establish the validity of the procedure. The result for the $\chi_s(T)$ term is in agreement within experimental errors with the Mössbauer results on a very dilute Cu(Fe) alloy[7] as shown in Fig. 1. The theoretical curve obtained by Götze and Schlottmann is also shown in this graph by the solid line. The difference in the T_c values for two different Cu(Fe) alloys may be due to the difference in the Fe concentration of two alloys. The results also show that the Brillouin function term can account for the field dependent part of the observed susceptibility. The estimated pair concentration

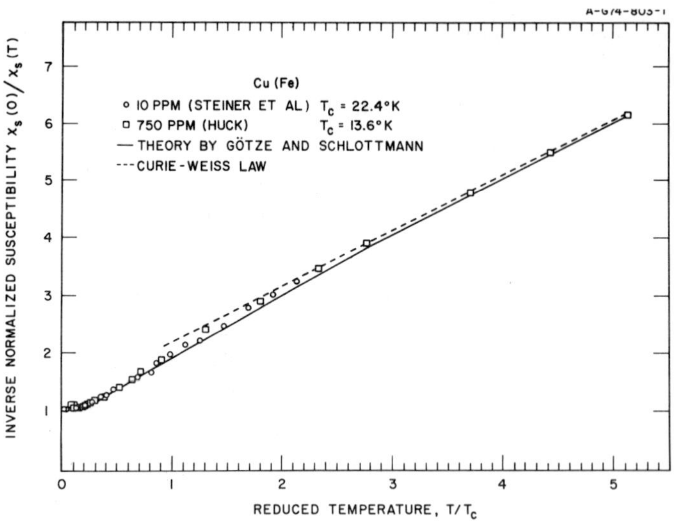

Fig. 1. Inverse normalized susceptibility due to Fe in Cu host as a function of the reduced temperature T/T_c. Where $\chi_s(o)$ is the zero temperature susceptibility. The solid curve is of the form suggested by Götze and Schlottmann with $\rho J = 0.2$, $\frac{kT_c}{D} = 1.38 \times 10^{-3}$. The dashed line represents high temperature Curie-Weiss law.

Table I

The Normalized Temperatures for CuAu(Fe) Alloys

Composition (at.% Au)	Normalized temperature (°K)			
	Present study	Delinger et al.[8]	Loram et al.[10]	Haddad and Sarachik[12]
2.4	12.2	14	——	——
4.8	10.6	12	13	15.8 (5 at.%)
10.8	9.6	8	8.6	10.4 (9 at.%)

and the effective moment of Fe, deduced from C and C_2 values in Eq. (1) for Cu(Fe) (0.07 at.%) alloy, are $3.4 \pm 0.1\ \mu_B$ and 68 C_{Fe}^2 respectively.

A similar analysis made on the results of measurements on a series of CuAu(Fe) alloys. The universal curve in the single-impurity susceptibility found[10] for the CuAu alloy system is the same as that for pure copper with the uncertainty of the data. We find the characteristic temperature T_c is 13.6, 12.2, 10.6, 9.6, and 9.6 K for alloys with Au concentration of 0, 2.4, 4.8, 10, and 10.8 at.% respectively. Table I lists the normalized temperatures for alloys of similar composition obtained from the present data and those found by Loram et al.,[11] Delinger et al.[8], and Haddad and Sarachik[12].
A more detailed comparison is not possible because the exact dependence of the normalized temperature upon Au and Fe concentrations is not known.[13]

We wish to thank Professor J.W. Schweitzer for his generous help during the analysis of this experiment.

*Supported in part by the National Science Foundation under Grant No. GH34359.

REFERENCES

1. Kondo, J., in Solid State Physics, edited by F. Seitz, D. Turnbull, and H. Ehrenreich (Academic, New York, 1969), Vol. 23, p. 183.
2. Ekström, H.E. and Myers, H.P., Phys. Kondens. Materie 14, 265(1972).
3. Huck, F.B., Savage, Wm.R., and Schweitzer, J.W., Phys. Rev. B, 8, 5213(1973).
4. Tholence, J.L. and Tournier, R., Phys. Rev. Letters 25, 867(1970).
5. Franz, J.M. and Sellmyer, D.J. in AIP Conference Proceedings No. 5, Magnetism and Magnetic Materials, (1971) p. 1150.
6. Götze, W. and Schlottmann, P., J. Low. Temp. Phys. 16, 84(1974).
7. Steiner, P., Zdrojewski, W.V., Gumprecht, D., and Hufner, S., Phys. Rev. Letters 31, 355(1973).
8. Delinger, W.G., Savage, W.R., and Schweitzer, J.W., Phys. Rev. B, 7, 1066(1973).
9. Sheu, J., M.S. Thesis (University of Iowa 1972).
10. Sheu, J. and Savage, W.R., (to be published).
11. Loram, J.W., Whall, T.E., and Ford, P.J., Phys. Rev. B2, 857(1970).
12. Haddad, J.B. and Sarachik, M.P., in AIP Conference Proceedings No. 18, Magnetism and Magnetic Materials (1974), p. 964, and Phys. Rev. (to be published).
13. Star, W.M., Basters, F.B., Nap, G.M., De Vroede, E., and Van Baarle, C., Physica 58, 585(1972).

LOW TEMPERATURE TRANSPORT AND SPECIFIC HEAT MEASUREMENTS IN Rh(Fe)

J. E. Graebner, J. J. Rubin, R. J. Schutz, F.S.L. Hsu, and W. A. Reed, Bell Laboratories, Murray Hill, New Jersey 07974, and R. J. Higgins, University of Oregon, Eugene, Oregon 97403.

ABSTRACT

Measurements of electrical resistivity of Rh(Fe) alloys from 20 mK to 7 K are consistent with a spin fluctuation temperature T_{sf} = 2 K, with impurity interactions still apparent at 0.1% Fe. Peaks are observed in both the thermoelectric power and heat capacity at T_{sf}. The heat capacity γT and βT^3 terms are strongly enhanced compared to pure Rh. Annealing the 0.1% Fe sample at room temperature for eight months resulted in a decrease in T_{sf} of 55%. The 1% Fe sample seems to display interaction - modified spin fluctuation characteristics and spin glass characteristics simultaneously.

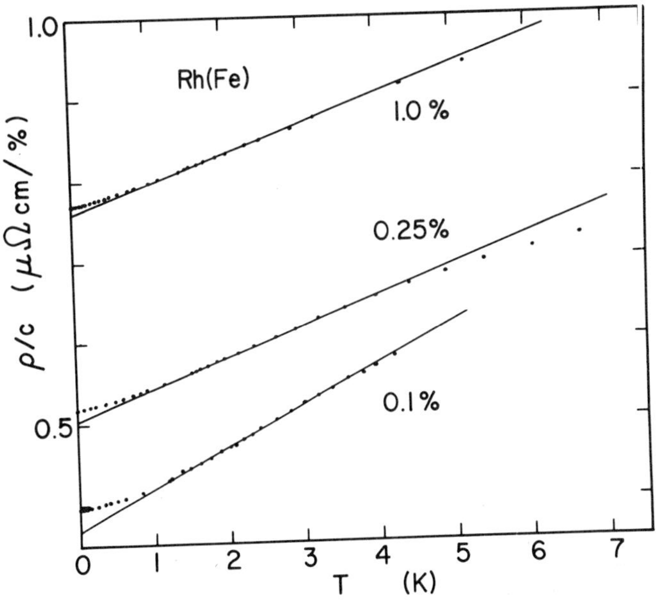

Fig. 1. Resistivity per atomic % vs. T.

Rh(Fe) was the first example[1] of a dilute alloy of a transition metal in a transition metal host which showed a positive temperature coefficient of electrical resistivity ρ down to low temperatures. Such behavior has been attributed to localized spin fluctuations,[2,3] which are predicted to give $\rho \propto T^2$ at the lowest temperature and T at higher temperature. Conflicting reports of only linear behavior[4] or both linear and quadratic behavior[5-7] have appeared. We present high resolution resistivity measurements from 20 mK to 7 K which exhibit both T^2 and T behavior and result in a spin fluctuation temperature $T_{sf} \approx 2.0$ K. We also report thermoelectric and heat capacity measurements which are consistent with impurity-impurity interactions being important down to 0.1% Fe, our lowest concentration.

The single crystal samples were prepared by electron-beam zone leveling[8] in a vacuum of 10^{-7} Torr. The ingots were spark-cut to 1 × 1 × 20 mm and showed concentration variations over their length no greater than 10% of the stated concentration.

The electrical resistance and thermoelectric ratio[9] are measured in a He^3-He^4 dilution refrigerator using a SQUID as null detector in a self-balancing voltmeter arrangement[10] with a sensitivity of 10^{-14} v. Maximum power dissipation in the sample is 10^{-12} w.

The heat capacity measurements from 50 mK to 2 K were made with a relaxation calorimeter[11] in a second dilution refrigerator, and from 1.5K to 40K in a pulse calorimeter.

Interaction effects are evident from the fact that ρ/c depends on c in Fig. 1. Nevertheless, a clearly defined linear region occurs for T ~ 1-4K. Below 1 K the data appear to deviate toward a T^2 law. Fig. 2 shows that a well-defined T^2 behavior is not reached until well below 0.1 K. A peculiar bump in the 0.25% sample occurs between 40-100 mK. We ascribe this to unidentified impurities in the starting material. The slight upturn below 30 mK for the 0.1% sample is likewise associated with some trace impurity, such as Cr,[12,13] which exhibits a resistance minimum in Rh and at these temperatures might be going as T^{-2}. The data of Figs. 1 and 2 for all samples were obtained after a high temperature anneal at 1400 C and 10^{-7} Torr for six hours followed by a quench in 3-4 sec. The heat treatment was performed when it was noticed that over a three-month period the value of ρ(T=0) for the 0.1% sample had increased by 2% while the slope in the linear region, s_1, had increased by 11%. The T^2 region

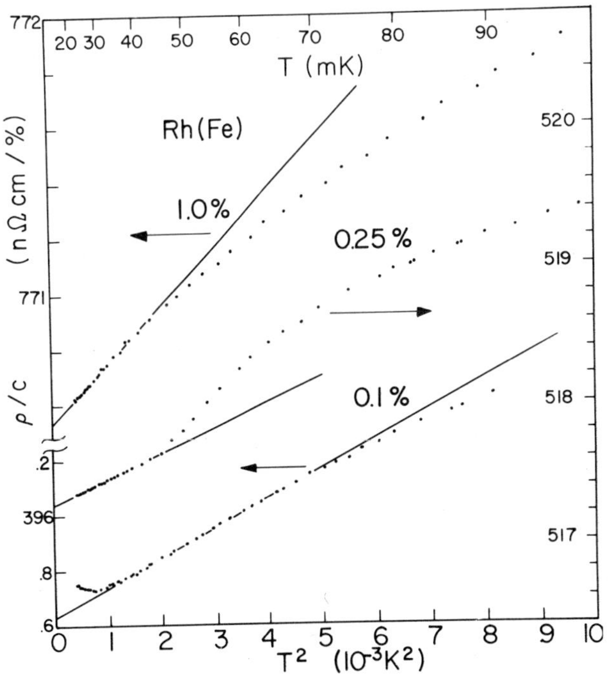

Fig. 2. Resistivity per atomic % vs. T^2 for T < 0.1 K.

was much less well defined before heat treatment, which together with the behavior of the 1% sample in Fig. 2, suggests that metallurgical problems as well as interaction effects[14] were responsible for the linear behavior observed[4] down to 85 mK. The aging effects we have observed suggest great caution in using this material as a resistance thermometer.

T_{sf} may be calculated[2] from the ratio of slope s_1 to the slope in the T^2 regime, s_2: $T_{sf}^{KD} = 2.094 \times s_1/s_2$. Fig. 3 exhibits the effect of aging on T_{sf}^{KD} for the 0.1% sample. The heat treatment seems to

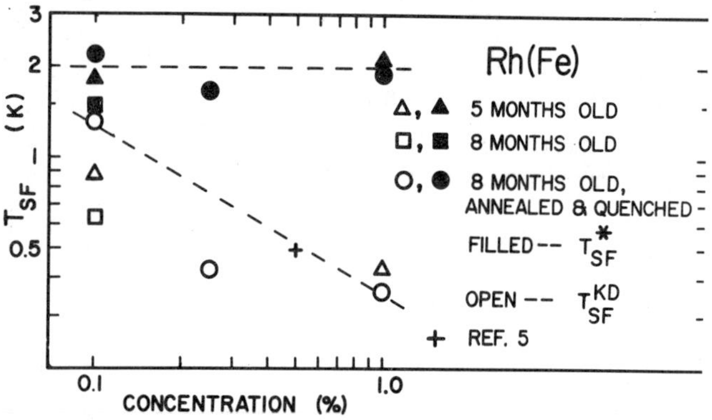

Fig. 3. T_{sf} vs. concentration in atomic %. The various symbols represent data taken after five months, after eight months, and finally after a high-temperature anneal an quench. T_{sf} is calculated from the data in two ways (see text).

restore T_{sf}^{KD} to what it may have been when the sample was first made, which was five months before the first measurement. We also see a definite dependence of T_{sf}^{KD} on concentration, probably due to interactions, and the dilute limit (c-independent) is not observed above 0.1%. We note that T_{sf}^{KD} of Ref. 5 for 0.5% Fe fits in well with our measurements.

We may, alternatively, calculate a characteristic temperature from the formula $\rho(T) = \rho(0)(1+(T/T_{sf})^2)$, analogous to that used[15] for more traditional resistance minimum alloys. We get $T_{sf} = (\rho(0)/s_2)^{1/2}$, which is also plotted in Fig. 3. T_{sf} displays sensitivity to aging and heat treatment but curiously almost no dependence on concentration, which suggests a limiting characteristic temperature of 2 K.

The fact that T_{sf}, determined only from very low temperature data, is insensitive to interactions while T_{sf}^{KD}, determined from both high and low T data, is not, suggests that the paramagnon spectral density function[2] $A(\omega) = \omega/(1+\omega^2)$ is distorted by the interactions. If the small-ω linear region is unaffected but the large-ω region near the peak is reduced by interactions, then the resistivity would be unaffected at very low T where only low frequency paramagnons are excited but would be decreased at higher T, as we observe.

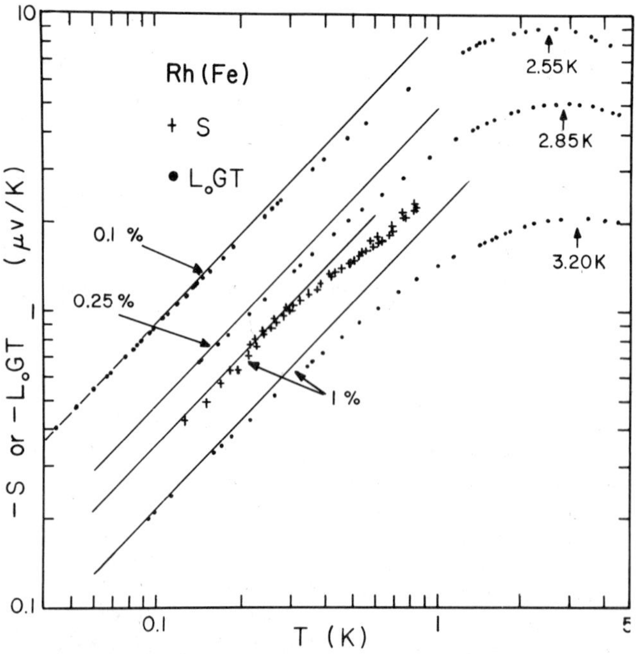

Fig. 4. Thermoelectric data vs. T. The straight lines indicate linear temperature dependence.

In Fig. 4, the thermoelectric power S for the 1% sample and the similar quantity[9] L_oGT for all samples (G is the thermoelectric ratio and L_o the Lorentz number) exhibit a peak[16] in the vicinity of 2-3 K and linear behavior below ~0.3 K. The data are insensitive to aging or heat treatment but the peak height and position depend on c. There are no theoretical estimates for the thermopower due to spin fluctuations except for an initial behavior[15] linear in T. Alternatively, one might view the peak as due to paramagnon drag, which should vary with temperature as the heat capacity of the paramagnons (see below). Touger[17] observes a similar though positive peak in Ir(Fe) at T_{sf}, and Foiles and Schindler[18] observe a positive peak in Pd(Ni) at $\sim T_{sf}/3$.

Heat capacity measurements of pure Rh and 0.1% Fe show the large values of $\Delta\gamma/\gamma\Delta c = 66$ and $\Delta\beta/\beta\Delta c = 126$ where γ and β are the coefficients of the linear and T^3 terms. We cannot say at this point whether γ and β change linearly in c. We also see a peak in ΔC_p of magnitude 1.5J/mole Fe·K centered at ~3 K and giving entropy of Rln(1.4±.2) per mole of Fe. The shape of the peak is very similar to the shape of the thermoelectric peaks. The 1% sample exhibits a large excess C_p of ~5 mJ/mole K from 1 K to ~10 K, which we associate with incipient spin-glass behavior.[7] Thus, the 1% sample seems to exhibit characteristics of an interacting spin-fluctuation system and a spin glass system simultaneously.

We thank T. Kometani for performing the atomic absorption spectroscopic analysis of the samples.

Note added in proof: Our resistivity measurements are in substantial agreement with the recently reported data of R. L. Rusby, J. Phys. F4, 1265 (1974).

References

1. B. R. Coles, Phys. Lett. 8, 243 (1964).
2. A. B. Kaiser and S. Doniach, Intern. J. Magnetism 1, 11 (1970).
3. N. Rivier and V. Zlatic, J. Phys. F2, L99 (1972).
4. N. F. Oliveira, Jr. and S. Foner, Phys. Lett. 34A, 15 (1971).
5. O. Laborde and P. Radhakrishna, Phys. Lett. 37A, 209 (1971).
6. R. Rusby, 1972 (to be published).
7. B. R. Coles, private communication.
8. 99.9999% Rh for 0.1% and 1% samples, 99.999% Rh for 0.25% sample. 99.999% Fe.
9. J. C. Garland, Appl. Phys. Lett. 22, 203 (1973) and to be published.
10. R. P. Giffard, R. A. Webb, and J. C. Wheatley, J. Low Temp. Phys. 6, 533 (1972).
11. R. J. Schutz, RSI 45, 548 (1974).
12. B. R. Coles, S. Mozumder, and R. Rusby, Proceedings of LT12, Kyoto, Japan (1970), p.737.
13. H. Nagasawa and N. Inone, ibid, p.741.
14. See also R. J. White and G. A. Swallow, Physics Lett. 35A, 427 (1971).
15. N. Rivier and M. J. Zuckermann, Phys. Rev. Letts. 21, 904 (1968).
16. H. Nagasawa, J. Phys. Soc. Japan 25, 691 (1968).
17. J. Touger, Ph.D. thesis, New York University, 1974.
18. C. L. Foiles and A. I. Schindler, Phys. Lett. 26A, 154 (1968).

MAGNETIZATION AND MAGNETORESISTANCE OF THE LOCALIZED SPIN FLUCTUATING SYSTEM PtCo

Gwyn Williams, Department of Physics, University of Manitoba, Winnipeg R3T 2N2, Canada, and
G. A. Swallow, A.D.C. Grassie and J. W. Loram, School of Mathematical and Physical Sciences, University of Sussex, Brighton, U.K.

ABSTRACT

The susceptibility, magnetization, resistivity and magnetoresistance of a Pt-0.061 at. % Co sample have been measured between 1.5 and 15 K in applied magnetic fields varying up to 59 kOe. The susceptibility data have been fitted to two phenomenological expressions, both of which yield a characteristic temperature of a few degrees. The magnetization data are complex, but they can be understood qualitatively by using the idea of an effective temperature. Finally, except at the highest field and lowest temperature, the magnetoresistance data is well represented by the predictions of localised spin fluctuation (ℓsf) theory, from which an ℓsf temperature of 0.7±0.1 K is estimated for this system.

INTRODUCTION

Currently many of the properties of alloys formed from first row transitional metal impurities in exchanged enhanced hosts such as Pd and Pt are being discussed in terms of ℓsf effects[1], with the estimated ℓsf temperature varying with impurity atomic number in much the same manner as the estimated Kondo temperature varied for the same impurities in noble metal hosts[2], viz. being low at the centre of the series (Mn) and high at its ends (V and Ni). In the case of the dilute PtCo system, recent susceptibility data[3] have been shown to fit a Curie-Weiss form above liquid helium temperatures, but at a temperature below an estimated characteristic temperature of 1.65 K it has been inferred that a rapid decrease in effective impurity moment occurs. Similar conclusions had been reached from prior analyses of nmr[4] and nuclear orientation[5] measurements.

RESULTS AND DISCUSSION

The susceptibility data, taken in a field of 4.6 kOe over the range 1.5 to 50 K, has been fitted to two phenomenological expressions. The first uses:

$$\chi(T) = \chi_{Pt}(T) + c\Delta\chi_e + \frac{cB}{(T+\theta)} \qquad (1)$$

where $\Delta\chi_e$ is a temperature independent solute contribution, c the impurity concentration, while the final term is of a Curie-Weiss form. We find $\theta \simeq 1.8\pm0.2$ K, μ_{eff} (from B) = 5.7±0.1 μ_B and $\Delta\chi_e = (4\pm0.1)\times10^{-6}$ emu. The second expression used was:

$$\chi(T) = \chi_{Pt}(T) + \frac{C}{T} + \frac{D}{(T+\theta)}. \qquad (2)$$

where the second term on the right is associated with stabilized moments and the third with unstable (fluctuating) moments. We obtain $\theta = 3.4\pm0.3$ K, μ_{eff} (both stable and unstable) = 5.6±0.1 μ_B, with 87% of unstable (i.e. fluctuating isolated) moments and 13% of stable interacting moments. Assuming that Co impurities interact "ferromagnetically", i.e. the latter stabilize individual moments via a reduction in their ℓsf temperature, then, if such a coupling is assumed to have a range of n atoms, it follows that $(1-c)^n \simeq 0.87$ given n ≃ 230 sites (quite close to the estimate[3] of 180 sites for this system). Figure 1 reproduces the magnetization data, taken

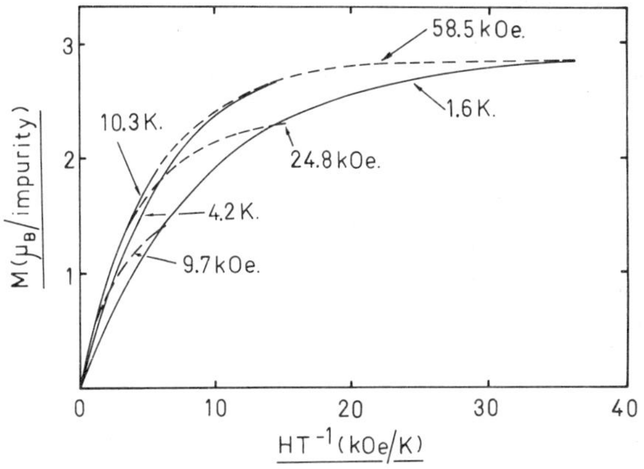

Fig. 1: The Magnetization M plotted against HT^{-1}. ---- constant field; —— constant temperature

both as a function of temperature in fixed field (dashed curves) and as a function of field at fixed temperature (solid curves). These curves can be understood qualitatively by introducing an effective temperature, T_{eff}, in the same manner as done by Suhl[6] in discussing interaction effects in the Kondo problem, with T_{eff} given by:

$$T_{eff}^2 = T^2 + \left(\frac{g\mu_B}{k_B}\right)^2 H^2 \qquad (3)$$

where g is the Lande factor and k_B Boltzmann's constant. μ_{eff} is assumed to increase with T_{eff}, climbing to its full "free spin" value in the limit $T_{eff}\to\infty$. This affords an explanation of why the magnetization at fixed values of HT^{-1} is larger for large H and T. Additionally such an idea explains why the magnetization taken at fixed temperature does not saturate (when plotted against HT^{-1}) at high HT^{-1} (increasing HT^{-1} at fixed T simply means increasing H and hence T_{eff}; thus μ_{eff} and consequently M continues to increase). Conversely the magnetization taken in fixed field does saturate with HT^{-1}; here increasing HT^{-1} implies decreasing T and hence T_{eff}; with the latter attaining a fixed value of $g\mu_B H k_B^{-1}$ as $T\to 0$.

In figure 2 the magnetoresistance data, taken as a function of temperature from 1.5 to 15 K in various fixed fields up to 59 kOe, is reproduced. This figure shows that except at the highest fields and lowest temperatures these data are well represented by:

$$\Delta\rho(T) = A + B\ell n\{[T^2 + \theta^2 + \left(\frac{g\mu_B H}{k_B}\right)^2]^{\frac{1}{2}}\} \qquad (4)$$

with A and B constants (the latter positive as predicted from phase shift considerations[7]). We find A increases from 0.0721 μΩ cm (H = 0) to 0.090 μΩ cm (H = 59 kOe) - due to the Kohler term, while B has the value 0.043 μΩ cm (at all fields examined). The ℓsf temperature θ is estimated at 0.7±0.1 K while g is found to have the value 1.5±0.1 (which we are currently not able to understand).

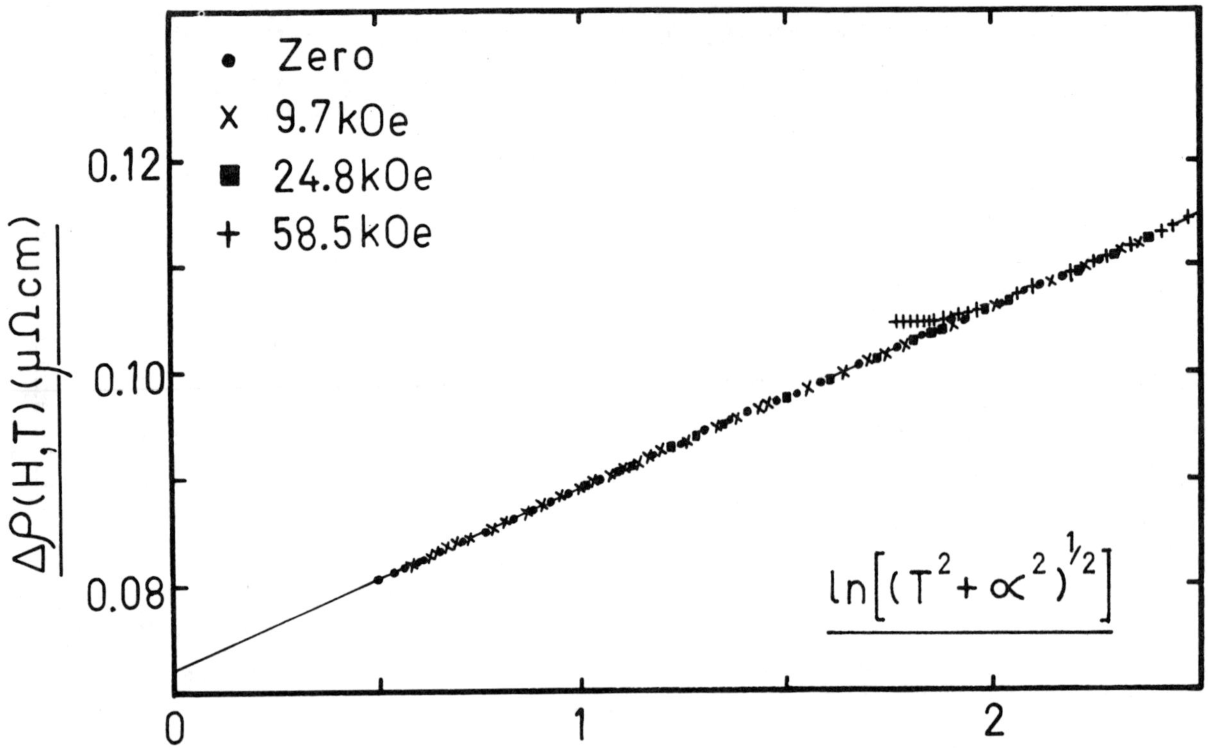

FIG. 2: THE INCREMENTAL RESISTIVITY $\Delta\rho(H,T)$ PLOTTED AGAINST $\ln[(T^2 + \alpha^2)^{1/2}]$ WITH α DERIVED FROM EQUATION (4).

ACKNOWLEDGEMENTS

Financial support for this work from the European Office of Aerospace Research, United States Air Force, the Defence Research Board and the National Research Council is acknowledged.

REFERENCES

1. C. Rizzuto, Rep. Prog. Phys. 14, 147 (1974).
2. M. D. Daybell and W. A. Steyert, Rev. Mod. Phys. 40, 308 (1968).
3. B. Tissier and R. Tournier, Solid State Communs. 11, 895 (1972).
4. L. D. Graham and D. S. Schreiber, J. Appl. Phys. 39, 963 (1968).
5. J. C. Gallop and I. A. Campbell, Solid State Communs. 6, 831 (1968).
6. H. Suhl, Phys. Rev. Letters 20, 656 (1968).
7. J. W. Loram, R. J. White and A.D.C. Grassie, Phys. Rev. B5, 3659 (1972).

INFLUENCE OF CRYSTAL FIELD SPLITTING OF Mn AND Cr ON THE ELECTRICAL RESISTIVITY OF ZINC

J. Kästner and E.F. Wassermann
II Physikalisches Institut, RWTH Aachen 1 Germany

F.T. Hedgcock and J.O. Strõm-Olsen
Eaton Electronics Laboratory, McGill University, Montreal, Quebec H3C 3G1

ABSTRACT

Investigations of the magnetic susceptibility indicate a possible crystal field splitting for both Mn and Cr in zinc with the Cr ground state supporting spin-flip scattering whereas the Mn ground state would not. Calculations indicate a qualitative difference in the low temperature electrical resistivity depending on whether the Kondo temperature is greater or smaller than the crystal field splitting. Measurements of the electrical resistivity of extremely dilute (<10 p.p.m.) zinc alloys containing manganese or chromium measured in the milli degree temperature range do not exhibit the detailed resistivity structure at least within the accuracy of 1 part in 10^4. However, below 0.1^oK the resistivity does show the expected T^2 law.

INTRODUCTION

It is well known that the Kondo effect can coexist with crystal field splittings in alloys containing rare earth impurities, provided the degeneracy of the rare earth ion is not completely removed.[1,2,3] Recent magnetic anisotropy measurements in ZnMn[4] and ZnCr[5] show that the same effect can occur with a 3d transition element. It was also found that the sign of the crystal field was opposite in these two cases, implying that the ground state of ZnCr could support spin-flip scattering whereas that of ZnMn could not. Calculations by Harris et al.[6] of the resistivity of a Kondo system indicate that a crystal field causes the usual lnT behaviour to give way to a T^2 behaviour; however, the sign of the crystal field only makes itself felt if the Kondo temperature, T_K, is much less than the crystal field temperature T_Δ, whereupon considerable structure is revealed. This theoretical behaviour is illustrated in fig. 1. Previous experiments[7] in these two systems have not shown any marked difference in the behaviour of ZnMn and ZnCr. However since the accuracy of these measurements ($\pm 0.1\%$) was relatively poor, we have reinvestigated the resistivity of very dilute ZnMn and ZnCr alloys in the mK range with an accuracy of 1 in 10^4.

EXPERIMENTAL

The concentration of the magnetic impurities is determined by measuring the slope of the resistivity $-\partial\rho/\partial \log T$ in the Kondo region. For ZnMn this slope is found to be 3.7μΩcm/at% dec., which is consistent with the value reported by Rizzuto[8]. For ZnCr the value $-\partial\rho/\partial \log T$ = 2.6μΩcm/at% dec. is used. The concentration data as taken from the slopes of the log T dependences results in the following impurity concentrations for the investigated alloys: ZnMn: 5.8 and 11.4 ppm Mn; ZnCr: 2.3 and 8.4 ppm Cr. This is consistent with the concentrations as determined by the residual resistivity (ZnMn: 2.8 at%$^{-1}$; ZnCr: 1.9 at%$^{-1}$)[8,9].

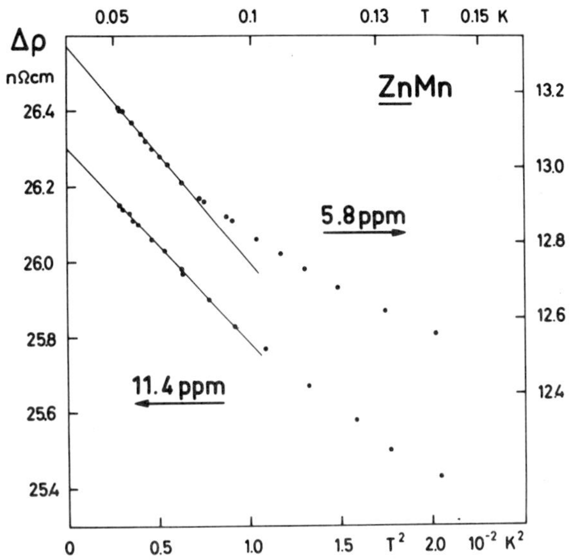

Fig. 2: Resistivity of ZnMn alloys as a function of T^2.

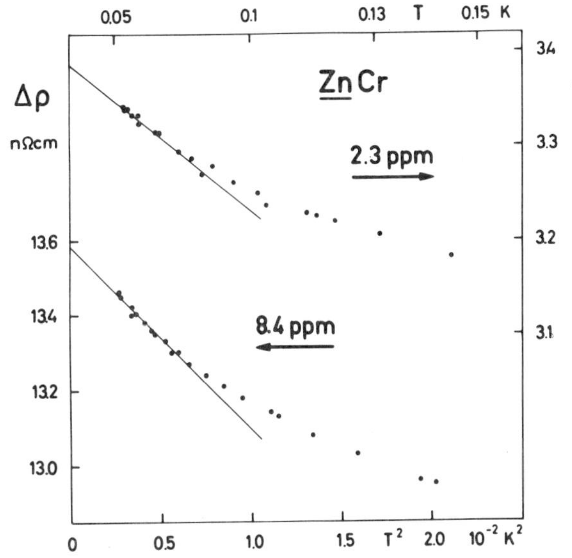

Fig. 3: Resistivity of ZnCr alloys as a function of T^2.

RESULTS AND DISCUSSION

Figs. 2 and 3 show the impurity resistivity Δρ plotted versus the square of the temperature on a logarithmic scale for the four samples with impurity concentrations as listed above. The data are fitted for both systems at 1K. At temperatures below 0.5^oK the data is fit to a T^2 low. This is

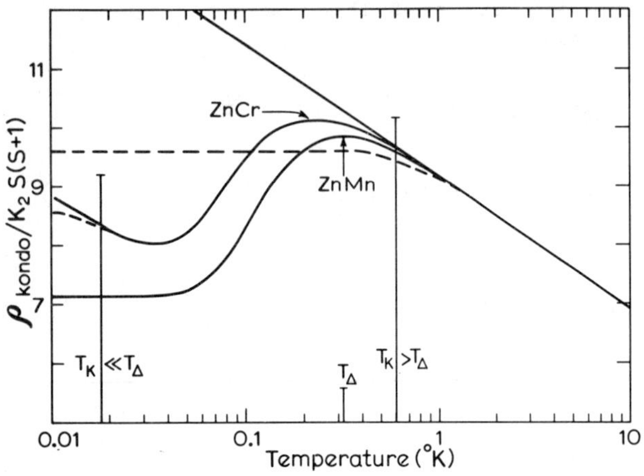

Fig. 1: The normalized Kondo resistivity as a function of temperature for ZnMn and ZnCr. The dashed line indicates schematically how the resistivity would saturate for both alloys if $T_k > T_\Delta$. T_Δ is the characteristic temperature related to the crystal field splitting.

shown in figures 2 and 3 for both alloy systems and results in a T^2 dependence of the form

$$\Delta\rho = \Delta\rho(0)\ \{1-\left(\frac{T}{\theta_R}\right)^2\}$$

The quantity θ_R is concentration dependent:

ZnMn: θ_R=0.71K (11.4ppm) θ_R=0.48K (5.8ppm)

ZnCr: θ_R=0.53K (8.4ppm) θ_R=0.46K (2.3ppm)

Values for the unitarity limit $\Delta\rho(0)$ are:

ZnMn: $\Delta\rho(0)$ = 21 μΩcm/at%
ZnCr: $\Delta\rho(0)$ = 14.5 μΩcm/at%

The experiments do not reveal the structure or difference in behaviour for the two systems as illustrated in figure 1. This is not surprising since the Kondo temperatures taken from a Hammon fit in the log T range are for ZnMn; T_k = 1°K (S = 3/2) and for ZnCr; T_k = 2°K (S = 1) which is in both cases larger than the crystal field splittings of roughly 0.30°K for Mn in zinc and of 0.16°K for Cr in zinc. However, the resistivity does show a T^2 behaviour as expected, although it should be noted that the agreement is not good quantitatively.

REFERENCES

1. A.S. Edelstein, L.R. Windmiller, J.B. Ketterson, G.W. Crabtree and S.P. Brown, Phys. Rev. Lett. 26, 516 (1971).
2. T. Sugawara and S. Yoshida, J. Low Temp. Phys. 4, 657 (1971).
3. S. deGennaro and E. Borchi, Phys. Rev. Lett. 30, 377 (1973).
4. P.L. Li, F.T. Hedgcock, W.B. Muir and J.O. Ström-Olsen, Phys. Rev. Lett. 31, 29 (1973).
5. F.T. Hedgcock, S. Lenis, P.L. Li, J.O. Ström-Olsen and E.F. Wassermann, Can. Jour. Phys., to be published.
6. R. Harris, F.T. Hedgcock, J.O. Ström-Olsen and M.J. Zuckermann, Can. Jour. Phys., to be published.
7. A. Pilot, R. Vaccarone and C. Rizzuto, Phys. Lett., 40A, 405 (1972).
8. C. Rizzuto, Reports on Progr. in Phys., 37, 147 (1974).
9. H.P. Falke, H.P. Jablowski and E.F. Wassermann, Z. Physik 269, 285 (1974).

LOCAL MOMENT-CONDUCTION BAND (s-d) EXCHANGE IN CuMn AND AgMn

R. E. Walstedt and L. R. Walker
Bell Laboratories, Murray Hill, New Jersey 07974

ABSTRACT

The derivation of s-d exchange parameters for dilute magnetic alloys from three experimental measurements is discussed. Mixing exchange parameters from appropriate data for CuMn are found to vary by a factor ~1.5. For AgMn available host NMR linewidth data are found to be inconsistent with exchange parameters from EPR studies.

INTRODUCTION

We discuss the derivation of the s-d exchange parameters of 3d magnetic impurities in metals from experimental data with particular reference to CuMn and AgMn. The s-d exchange in such systems is usually dominated by the "mixing" contribution treated by Kondo[1] and by Schrieffer and Wolff[2]. Our attention will be focused mainly on this contribution, but direct exchange will also be included. The treatment of the mixing comes from the foregoing sources and from the Anderson model[3,4].

The experimental effects we discuss are (a) long-range spin-density oscillations (SDO) caused by the impurity as reflected by host NMR line broadening, (b) high-temperature ^{55}Mn nuclear relaxation due to moment fluctuations which are in turn caused by exchange scattering of conduction electrons and (c) saturation moment (or susceptibility) of the impurity. A more thorough discussion of this material is given elsewhere[5].

THEORY OF s-d EXCHANGE PHENOMENA FOR DILUTE LOCAL MOMENTS

First, we present the basic formulae for the effects under consideration, beginning with the long-range SDO and associated NMR linewidth. The asymptotic long-range spin-density disturbance in an assumed spherical band may be written

$$\Delta n(r) = -\frac{\langle S_z\rangle \rho(E_F)\gamma(2k_F)^2}{4\pi r^3}\sum_\ell(-1)^\ell(2\ell+1)J'_\ell\cos(2k_Fr+2\delta_\ell) \quad (1)$$

where $\langle S_z\rangle$ is the expectation value of impurity spin, $\rho(E_F)$ is the bare density of states of the host band for one spin direction, $\gamma(2k_F)$ is the enhancement factor for the real part of the band susceptibility $\chi(k,\omega)$ at $\omega=0$ and $k=2k_F$, and $J'_\ell = J_\ell + J^m_\ell$ is the ℓ'th spherical component of the exchange coupling matrix $J_{\vec{k},\vec{k}'}$[6,7] where $|\vec{k}| = |\vec{k}'| = k_F$. J_ℓ is the direct and J^m_ℓ the mixing contribution. Whereas, in principle, the J_ℓ are all nonzero, the mixing part consists exclusively of J^m_2 for 3d moments in the approximation of a spherical impurity potential. The phase shifts δ_ℓ are subject to the condition imposed by the Friedel sum rule. A similar result to Eq. (1) neglecting phase shift effects was recently published by Davidov et al[7].

For the enhancement factor we use[8,9]

$$\gamma(k) = (1-I(k)\chi_o(k))^{-1} \quad (2)$$

$I(k)$ is a functional of the band exchange interaction.

The SDO amplitude is reflected by the host NMR line broadening. It has recently been shown[10] that the resulting lineshape is Lorentzian with half-width $\langle\Delta H\rangle$ proportional to SDO amplitude. The NMR linewidth data are characterized by a parameter

$$W = \langle\Delta H\rangle / c\,\langle S_z\rangle \quad (3)$$

which is independent of concentration, field and temperature. Defining a coefficient J_{sdo} (see Eq. (1)):

$$J_{sdo}\cos(2k_Fr+\varphi) = \sum_\ell(-1)^\ell(2\ell+1)J'_\ell\cos(2k_Fr+\delta_\ell), \quad (4)$$

J_{sdo} and W are then related by[5]

$$J_{sdo} = 6\mu_B W/K_s(1-\alpha)\gamma(2k_F)^2, \quad (5)$$

where K_s is the s-contact NMR shift and $\alpha = I(0)\chi_o(0)$.

Finally, the SDO mixing exchange parameter J^m_{sdo} ($=J^m_2$ defined earlier) is expressed in terms of Anderson model parameters as

$$J^m_{sdo} = \frac{\langle V^2_{kd}\rangle_{av}}{\langle S_z\rangle_{max}}\frac{(E^-_d-E^+_d)}{\left[(E_F-E^+_d)^2+\Delta^2\right]^{\frac{1}{2}}\left[(E_F-E_d)^2+\Delta\right]^{\frac{1}{2}}}, \quad (6)$$

where the $E^{\pm}_d$ represent the energy of the filled up-spin (+) and empty down-spin (-) d-states, $\Delta = \pi\rho(E_F)\langle V^2_{kd}\rangle_{av}$, and V^2_{kd} is averaged over the Fermi surface.

Turning to impurity-state susceptibility, one has, considering only ℓ = 0, 1, and 2 terms,[11]

$$\chi_{imp}(T)/\chi_S(T) \cong \left\{1+\tfrac{1}{2}\rho(E_F)\left[(J_0+3J_1+5J_2)/(1-\alpha)+5J^m_\chi\right]\right\}^2 \quad (7)$$

where $\chi_S(T)$ is the free-spin Curie susceptibility. The direct terms represent a physical polarization of the conduction band and are therefore enhanced $\times(1-\alpha)^{-1}$. The mixing terms, on the contrary, correspond to a reduction in impurity moment due to finite level width, with the band polarization small because of the compensation theorem.[3,12] The effect of band exchange on the mixing terms in Eq. (9) is therefore negligible to first order.[5,12] The mixing exchange parameter for susceptibility is

$$J^m_\chi = -(1/\pi\rho(E_F)S)\sin^{-1}\left\{\Delta(E^-_d-E^+_d)\times\left[(E^-_d-E_F)^2+\Delta^2\right]^{-\frac{1}{2}}\left[(E_F-E^+_d)^2+\Delta^2\right]^{-\frac{1}{2}}\right\}. \quad (8)$$

Finally, we consider high-temperature relaxation of the local moment nucleus as a third measure of s-d exchange. The nuclear relaxation rate is[13]

$$1/T_1 = \tfrac{2}{3}(\gamma_n\alpha_d\mu_B N_o p_{eff})^2 T_{1e}, \quad (9)$$

where α_d is the d-spin hyperfine coefficient, T^{-1}_{1e} is the (single impurity) EPR linewidth and

$$p_{eff} = g\sqrt{S(S+1)}\,\{\ \}, \quad (10)$$

where S = 5/2 for Mn in a nearly 5 d-electron configuration and { } is the corresponding expression on the right hand side of Eq. (7). The local-moment relaxation time can be expressed as[14,5]

$$T^{-1}_{1e} = \frac{\pi}{\hbar}\rho(E_F)^2k_BT(1-\alpha)^{-2}\sum_{\ell,\ell'}J'_\ell J'_{\ell'}\mathcal{K}_{\ell\ell'}(\alpha), \quad (11)$$

where

$$\frac{\mathcal{K}_{\ell\ell'}(\alpha)}{2(2\ell+1)(2\ell'+1)(1-\alpha)^2} = \int_0^1\frac{P_\ell(1-2x^2)P_{\ell'}(1-2x^2)x\,dx}{[1-\alpha\bar{I}(x)G(x)]^2} \quad (12)$$

and where $x = k/2k_F$, $\bar{I}(x) = I(x)/I(0)$, and $G(x)$ is the Lindhard function. For the dynamic exchange parameter due to mixing we take the expression[1,2]

$$J^m_{T_{1e}} = -S^{-1}\langle V^2_{kd}\rangle_{av}\left[(E^-_d-E_F)^{-1}+(E_F-E^+_d)^{-1}\right]. \quad (13)$$

Before turning to a discussion of specific data, we remark that the quantities $\langle S_z\rangle_{max}J^m_{sdo}/S$, J^m_χ, and $J^m_{T_{1e}}$ defined in Eqs. (6), (8) and (13), respectively, coincide in the limit of small Δ. For the cases discussed in the following sections, these quantities should agree within a few percent. It is this correspondence which provides an interesting experimental test of the consistency of the model theories which provide these results.

MEASURES OF s-d EXCHANGE FOR CuMn

In comparing the three measures of s-d exchange discussed above for the case of CuMn, we first present and discuss values of the basic parameters α and $\rho(E_F)$. For $\rho(E_F)$ we adopt the band structure density of states for Cu ($m^*/m = 1.27$) recently calculated by O'Sullivan et al.[15] This gives $\rho(E_F) = 0.135$ eV^{-1} per atom for each spin state. α is obtained from the de Haas-van Alphen g-factor measurements of Randles[16] which yield $\chi_s = 1.45\ \chi_{free\ electron}$ for Cu metal. Combining this with the m* value above leads to $\alpha = 0.12$. This rather small value has recently been reconciled[17] with the larger values of α suggested by earlier discussions of the Korringa product K^2T_1T.[18] In Ref. 17 it is suggested that the observed NMR shift K = 0.24% is composed of an s-contact shift $K_s = 0.18\%$ and a residual Van Vleck orbital shift $K_{VV} = 0.06\%$. We adopt this viewpoint here.

Susceptibility measurements give[19] $p_{eff} = 4.9$. Eqs. (10) and (7) then yield the relation

$$5|J^m_\chi| - 1.13\ (J_0+3J_1+5J_2) = 2.55\ \text{eV}. \qquad (14)$$

For the NMR linewidth measurements of SDO amplitude we have made a review of data from the literature.[5] Using values of $\langle S_z\rangle$ calculated from Eq. (9) with $p_{eff} = 4.9$ we find an average value $W = 9.8 \cdot 10^4$ Gauss ± 15%. The dipolar contribution to W is negligible.[5] In Eq. (5) we use the value of $\bar{I}(1) = 0.59$ obtained from the function I(k) in Ref. 9. Eq. (5) then gives $J^m_{sdo} = 2.00$ eV. It is clear that the J^m_{sdo} term dominates the right hand side of Eq. (4), since recent EPR work on AgMn[20] shows that the $J_\ell \lesssim 0.1$ eV. The right hand side of Eq. (4) may therefore be expanded to first order in the J_ℓ's, giving

$$5|J^m_{sdo}| - J_0\cos(2\delta_0)+3J_1\cos(2\delta_1)-5J_2 \cong 2.00\ \text{eV}, \qquad (15)$$

where we have taken $\delta_2^+ + \delta_2^- \cong \pi$ in Eq. (4).

The third source of s-d exchange data is high temperature ^{55}Mn nuclear relaxation measurements in CuMn[21] which we interpret with Eqs. (9)-(12). The $\mathcal{K}_{\ell\ell'}(\alpha)$'s have been calculated for $\alpha = 0.12$ again using $\bar{I}(x)$ as given by Shaw and Warren.[9] Using $T_1^{-1} = 2.3 \cdot 10^5$ sec^{-1} from Ref. 21 we find $T_{1e}^{-1} = 1.32 \cdot 10^{13}$ sec^{-1} at 1200K. Expanding Eq. (11) to first order in the J_ℓ's, one finds

$$5|J^m_{T1e}| - 0.02J_0 - 0.19J_1 - 5J_2 = 2.46\ \text{eV}. \qquad (16)$$

It is immediately clear from Eqs. (14)-(16) that the mixing terms dominate. Moreover, a closer examination reveals that the correspondence between $\langle S_z\rangle_{max} J^m_{sdo} / S, J^m_\chi$, and J^m_{T1e} noted earlier cannot be effected with reasonable values of the J_ℓ's. In Table I, values of the mixing parameters are displayed for three hypothetical sets of J_ℓ's assuming $2\delta_0 = 2\delta_1 = \pi/4$ corresponding to 5 d-electrons at the Mn site. Although it is difficult to establish error limits for these determinations, there appears to be a considerable discrepancy between these sources of exchange data, particularly between J^m_{sdo} and J^m_{T1e}. Improved data for these measurements are needed to sharpen this conclusion.

TABLE I

CuMn mixing exchange parameters for three hypothetical sets of J_ℓ parameters. All values are in eV.

Direct exchange			Corresponding mixing exchange		
J_0	J_1	J_2	$\langle S_z\rangle_{max}\lvert J^m_{sdo}\rvert/S$	$\lvert J^m_{T1e}\rvert$	$\lvert J^m_\chi\rvert$
0.10	0.05	0	0.36	0.49	0.57
0.09	0.06	0.03	0.38	0.52	0.60
0.05	0.05	0.05	0.40	0.54	0.60

We have also used these results to study the Anderson model parameters involved. Assuming 5 d-electrons at the Mn site and taking[22] $E_d^- - E_d^+ \sim 5$ eV one finds $\Delta \sim 0.45$ eV, in good agreement with estimates based on spin-flip scattering cross-section of 3d impurities in Cu[22] as well as band structure calculations of pure 3d metals as interpreted by Heine.[23]

One can also utilize Eq. (6) to establish an upper limit $n_d \lesssim 5.5$ on the number of d-electrons present at the impurity site.[5]

MEASURES OF s-d EXCHANGE FOR AgMn

The analysis of Ref. 20 gives the following exchange parameters for AgMn, based on EPR g-shift and linewidth data: $J_0 = 0.094$ eV, $J_1 = 0.06$ eV, and $J_2' = -0.081$ eV. Using Eq. (7) this corresponds to a minimum value $p_{eff} = 5.96$, i.e. a slight increase over the free-spin value $p_{eff} = 5.92$. This is within ~3% of the experimental values[19] and therefore in reasonable accord with susceptibility data.

The NMR linewidth data of Mizuno[24] present a more difficult situation to interpret. There it was found that high and low-temperature data correspond to values of W differing by nearly an order of magnitude, the high-temperature W value being larger. This dual-slope characteristic is inconsistent with the theory of RKKY line-broadening[9] and is considered to be evidence for more than one species of magnetic impurity.[5] Taking the values $\alpha = 0.47$[20] and $K_s = 0.53\%$[18] one finds that Eq. (5) gives $J_{sdo} = 1.51$ eV at high temperatures and an order-of-magnitude smaller than that at low temperatures. Neither value is in reasonable agreement with $J_{sdo} \sim 0.4$ eV obtained from Eq. (4) with the exchange parameters from Ref. 20. It is clear that further experimental host NMR linewidth studies are needed to resolve this matter.

REFERENCES

1. J. Kondo, Prog. Theor. Phys. 28, 846 (1962).
2. J. R. Schreiffer and P. A. Wolff, Phys. Rev. 149, 491 (1966).
3. P. W. Anderson, Phys. Rev. 124, 41 (1961).
4. T. Moriya, in Theory of Magnetism in Transition Metals, edited by W. Marshall, Academic Press, New York (1967).
5. R. E. Walstedt and L. R. Walker, submitted to Phys. Rev.
6. R. E. Watson, A. J. Freeman, and S. Koide, Phys. Rev. 186, 625 (1969).
7. D. Davidov, K. Maki, R. Orbach, C. Rettori, and E. P. Chock, Solid State Comm. 12, 621 (1973).
8. K. S. Singwi, A. Sjölander, M. P. Tosi, and R. H. Land, Phys. Rev. B 1, 1044 (1970).
9. R. Shaw and W. Warren, Phys. Rev. B 3, 1562 (1971).
10. R. E. Walstedt and L. R. Walker, Phys. Rev. B 9, 4857 (1974).
11. D. J. Scalapino, Phys. Rev. Lett. 16, 937 (1966).
12. T. Moriya, Prog. Theor. Phys. 34, 329 (1965).
13. R. E. Walstedt and A. Narath, Phys. Rev. B 6, 4118 (1972).
14. T. Moriya, Prog. Theor. Phys. 18, 516 (1963).
15. W. J. O'Sullivan, A. C. Switendick, and J. E. Schirber, in Electronic Density of States (edited by L. H. Bennett), Nat. Bur. Stand. (U.S.), Special Publication 323, 1971.
16. D. L. Randles, Proc. Roy. Soc. A 331, 85 (1972).
17. R. E. Walstedt and Y. Yafet, Sol. St. Comm. (to be published.
18. A. Narath and H. Weaver, Phys. Rev. 175, 373 (1968).
19. G. J. van den Berg, Progress in Low Temperature Physics, Vol. 4, edited by C. J. Gorter, Amsterdam (1964).
20. D. Davidov, et al., preprint.
21. R. E. Walstedt and W. W. Warren, Jr., Phys. Rev. Lett. 31, 365 (1973).
22. Y. Yafet, J. App. Phys., 39, 853 (1968).
23. V. Heine, Phys. Rev. 153, 673 (1967).
24. K. Mizuno, J. Phys. Soc. Japan 30, 742 (1971).

MAGNETIC BEHAVIOR OF Ni AND ITS DILUTE ALLOYS

Mary Beth Stearns
Scientific Research Staff
Ford Motor Company
Dearborn, Michigan 48121

ABSTRACT

The saturation magnetization, neutron scattering, and hyperfine field data of Ni and its dilute alloys are analyzed with the assumption that the behavior of Ni is largely dominated by the strong hybridization of the itinerant d electrons, d_i, with the localized d_ℓ and s electrons. This is in contrast with Fe where there is little interband mixing of the itinerant d_i's and thus its behavior is mainly determined by the d_i-d_ℓ Coulomb exchange interaction. It is found that the large host moment perturbation of Ni alloys can be very satisfactorily explained in terms of the d_i-d_ℓ interband-mixing interaction.

In the past Ni has usually been considered as representing the prototype behavior of an itinerant ferromagnet[1], whereas it now appears that on the contrary most ($\gtrsim 95\%$) of the Ni d electrons are localized.[2] The previous interpretation arose for reasons mostly connected with the small moment of Ni and the strong hybridization or interband-mixing of the outer electrons, i.e., the itinerant (d_i) and localized (d_ℓ) d-like and the 4s-like electrons. This is in contrast with the behavior of Fe where both band calculations[3] and Fermi surface measurements[4] show very little hybridization of the d_i electrons with the d_ℓ or 4s-like electrons. Since the moment of Ni is much smaller than that of Fe and they both have comparable numbers of d_i's, a greater fraction of the Ni moment is due to itinerant and hybridized d's. Thus some measurements have been erroneously interpreted as indicating that Ni is a more itinerant ferromagnet than Fe. The greater degree of hybridization of the d_i's with d_ℓ and s electrons in Ni causes Ni and Fe-based alloys to show strikingly different average saturation magnetization, neutron elastic-diffuse-scattering and hyperfine field spectra behavior.

Recently we have combined hyperfine field (hff) and saturation magnetization data on dilute Fe alloys to show that the moment perturbations of the host Fe atoms in alloys of Fe with transition element solute atoms are due to the polarization induced in the d_i's due to the Coulomb exchange interaction with the localized d_ℓ electrons.[5] The Fourier inversion of the moment perturbations[6] obtained from the hff data reproduced, very well, the varied curves obtained by neutron scattering experiments on dilute Fe alloys.[7]

Here we report on an analysis of the comparable experimental data for dilute Ni alloys.

I. SATURATION MAGNETIZATION DATA

Whereas alloys of Fe with non-transition elements show mainly simple dilution, with little if any host moment perturbations, those of Ni alloys show large moment decreases far exceeding that of simply removing a Ni moment.[8a] We attribute this difference in behavior as due to the strong interband-mixing of the Ni d_i's with its d_ℓ and s electrons. Thus upon alloying Ni with non-transition elements, the outer valence (s and p) electrons of the solute atoms intermix, through the s electrons, with the Ni d_i electrons. This s-d_i interaction changes the d_i population in the vicinity of the solute atom, which in turn causes moment changes on Ni atoms near the solute atom. We assume that the s-d_i hybridization in Ni changes the number of d_i electrons on Ni atoms in the vicinity of the solute atom in proportion to the excess number of valence electrons, ΔZ, introduced by the solute atom. That is, $\Delta Z = Z_{solute} - Z_{Ni}$, where $Z_{Ni} \simeq 0.6$. We then calculate the moment changes using the formulas, derived by Moriya[9] to describe the model of Anderson,[10] for the change in moment due to a localized moment intermixing with conduction electrons. In this case, however, the conduction electrons are taken as d_i electrons since the d_i-d_ℓ interaction is expected to be much stronger than the s-d_ℓ interaction.

Figure 1

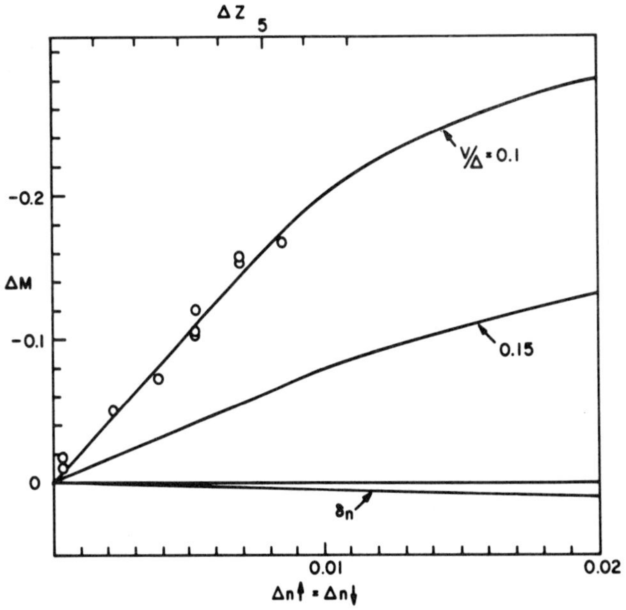

Fig. 1. Calculated moment change per Ni atom in the vicinity of a non-magnetic solute atom due to d_i-d_ℓ hybridization. The circles are experimental values for the moment change obtained as discussed in the text.

In the calculation of the moment changes, we use parameters in agreement with the Ni band calculations and other experimentally determined quantities. Thus, we assume that the six d electrons with T_{2g} symmetry at Γ are made up of about 0.4 d_i electrons and 5.6 localized electrons. These localized electrons are assumed to be low enough in the d_i conduction band that their moments do not change as the Fermi level varies due to the hybridization with the nearby solute atom. The rest of the d electrons having E_g symmetry at Γ are considered to be the localized electrons which interact with the d_i electrons. We then use the formulas in Ref. 9 to calculate the moment changes of the Ni atoms surrounding the solute atom. We indeed find that the moment changes are proportional to the increased number of d_i electrons and thus to ΔZ, as observed experimentally. This is shown in Fig. 1 where we give the calculated total moment change per Ni atom, ΔM, for various values of the d_i-d_ℓ interaction strength, V. V is in units of the width, Δ, of a parabolic d_i band. V should be comparable to the intraatomic interaction which is of the order of 1 ev. Thus $V/\Delta = 0.1$ corresponds to a band with an effective mass of about 2 m_e. Reasonable values for V/Δ are thus around 0.1 to 0.2. In Fig. 1 we also show the calculated moment change, δn, of a Ni atom due to the change in polarization of the d_i electrons. We see that this effect is negligible and practically all the moment change is due to the local moment decrease. This net moment decrease is mainly because the local moment spin-down band is very near the Fermi level and thus raising E_F slightly by increasing the number of d_i's causes a large increase in spin-down local moment.

Note from Fig. 1 that the smaller the value of V/Δ, the greater the net moment change for a given increase in number of d_i electrons. This arises because in this type of calculation the localized moment width comes entirely from the localized moment-conduction electron interaction. Thus, the smaller V/Δ is, the narrower the width, Γ, of the d_ℓ state. In reality there is a minimum value of Γ, due to other interactions, which determines the maximum moment change. A realistic value for this is obtainable from the Fe band calculations and is $\Gamma \simeq 0.5$ ev. This leads to the same moment change for all values of V/Δ smaller than about 0.1.

In Fig. 1 we also show the experimentally measured moment changes per Ni atom assuming that 30 Ni atoms surrounding the solute atom (e.g., equal loss for the first two neighbor shells and 1/2 that for the third shell) have their moments decreased. We see that for these realistic Ni parameters, a very small increase in the number of d_i electrons on a Ni atom, $\Delta n = \Delta n\uparrow + \Delta n\downarrow$, gives a rather large moment decrease. E.g., $\Delta n = 0.01$ decreases the moment per Ni atom by 0.1 μ_B for $V/\Delta = 0.1$. The ΔZ scale is assigned arbitrarily to fit the $V/\Delta = 0.1$ curve. We see that $\Delta Z = 1$ corresponds to a change in the number of d_i electrons on nearby Ni atoms of only about 0.003 electrons. This seems to be a very reasonable value for the change.

II. HYPERFINE FIELD DATA

We can obtain a rather good estimate of the summed s-like conduction electron polarization (sCEP) contribution to the hff from the surrounding Ni atoms, H_Σ, from the hff values at Cu, Ag and Au in Ni.[2] It is about -75 kG/μ_B, which is about the same as for Fe. A likely value for the nearest neighbor sCEP contribution is obtained from dilute CoNi alloys[11] to be about -10 kG/μ_B, again near the value for Fe. Unfortunately the structure in the spectra of the CoNi alloys in Ref. 11 does not vary properly with Co concentration. This is probably due to the experimental conditions and could be improved with present day understanding of ferromagnetic NMR behavior. The spectra of CoNi and perhaps FeNi alloys should be best for obtaining the sCEP curve of Ni since the d_i host perturbation effects are at a minimum in these alloys. Once the sCEP curve of Ni is known, it should then be possible to analyze the more complex spectra of other dilute Ni alloys to obtain the host moment perturbations. This was attempted on the Ni spectra of Aℓ, V and Cr alloys[11] but neither the sCEP nor Ni spectra are known well enough to obtain any conclusive results at this time. However, this should be a very fruitful line of research at present.

The values of moments on transition element solute atoms in Ni can be obtained in a way similar to that used to obtain moments in Fe.[2,5] However since the hff's are quite small in Ni, the method seems to be less accurate than in Fe. Probably because other contributions, e.g., orbital moments, become relatively more important. In Table I we list the moments obtained from the hff values at transition elements in Ni. We also list the moments as obtained from analyzing neutron scattering and saturation magnetization data.[8b] For the latter it is assumed that there are no host perturbations for the elements listed. This may not be so, especially for the higher transition series. We see that the various moments obtained for Mn and Fe are in good agreement but for some of the other solute atoms the agreement is very poor. In the case of V the neutron scattering data seems to definitely indicate a large negative moment, thus the hff value for V seems questionable. An orbital contribution could not resolve this discrepancy since it would lead to a more positive moment. The discrepancy for Co could be accounted for by a reasonable orbital contribution of $H_{orb} \simeq +50$ kG, however that of Rh would require an unreasonably large H_{orb}. If the host moments surrounding the Rh solute atoms had moment increases the various data on dilute RhNi alloys might be reconciled. Thus it is clear that much extension and improvement of the measurements of the hff's in dilute Ni alloys remains to be done.

Table I. Moments on transition element solute atoms in Ni as obtained from hff, neutron scattering and saturation magnetization data (in μ_B). The last column gives the average moment change per solute atom.

Element	hff	Neutron Scattering	Saturation Magnetization[8b]	$d\bar{u}/dc$
V	-0.1	-1.8		-5.2
Cr		-0.6		-4.5
Mn	3.1	2.5	3.0	2.4
Co	0.9		1.7	1.1
Fe	2.8	2.8	2.8	2.8
Nb	0.0			
Mo	0.0	0.0		-5.6
Ru	0.8	0.8		-3.0
Rh	0.8	2.0	2.6	2.0
Pd	0.7		0.6	0.0
W	0.0	0.0		-5.6
Re	-0.1	0.0		
Os	-0.1			-2.0
Ir	0.4		0.0	-0.6
Pt	0.1		0.0	-0.6

III. NEUTRON SCATTERING DATA

In examining the neutron scattering data for Ni alloys[7,12] we find that the data for non-transition solute atoms can be fit as well or better, with a Ni moment perturbation that has comparable decreases for the first two neighbor shells and falls off as $1/r^3$ for higher shells, than the $(e^{-ar})/r$ distribution which was previously used to fit the data.[12] The data for the transition element solute atoms V, Cr, Ru, Mo, W and Re were fit with the moments listed in column 3 of Table I plus a host perturbation given by moment decreases in the first few neighbor shells. In general, the shapes of all the neutron scattering curves with little or no moment on the solute atom are similar because the Ni has a small moment and the host moment perturbations are due to the d_i-d_ℓ interaction.

CONCLUSIONS

We find that we can satisfactorily fit the saturation magnetization, hyperfine field, and neutron scattering data of Ni and its dilute alloys with a model that assumes about 95% of the Ni d electrons are localized and that the itinerant d electrons are strongly hybridized with the d_ℓ and 4s-like electrons. Because some of the localized spin-down d_ℓ states are very close to the Fermi level, a slight change in the number of d_i electrons in the vicinity of a solute atom causes large host moment perturbations. Since the density of states of d_ℓ spin-down electrons near the Fermi level is even greater for Co than Ni, we might expect that the moment decrease per solute atom in Co alloys would be larger than that for Ni alloys. This appears to be so for Aℓ and Si solute atoms.[13]

REFERENCES

1. See e.g. N. F. Mott, Adv. in Phys. 13, 325 (1964); C. Herring in "Magnetism," Vol. IV edited by G. T. Rado and H. Suhl (Academic, New York, 1966).
2. M. B. Stearns, Phys. Rev. 8B, 4383 (1973).
3. Some band calculations strongly showing these features are: Ni; S. Wakoh, J. Phys. Soc. Japan 20, 1894 (1965); L. Hodges, H. Ehrenreich and N. D. Lang, Phys. Rev. 152, 505 (1966); J. Callaway and H. M. Zhang, Phys. Rev. B1, 305 (1970); E. I. Zornberg, Phys. Rev. B1, 244 (1970); J. Langlinais and J. Callaway, Phys. Rev. B5, 124 (1972): Fe; S. Wakoh and J. Yamashita, J. Phys. Soc. Jap. 21, 1712 (1966); K. J. Duff and T. P. Das, Phys. Rev. B3, 192 and 2294 (1971); R. A. Tawil and J. Callaway, Phys. Rev. 7B, 4242 (1973).
4. A. V. Gold, L. Hodges, P. T. Panousis, and D. R. Stone, Int. J. Magn. 2, 357 (1971).
5. M. B. Stearns, Phys. Rev. 9B, 2311 (1974) and Bull. of A.P.S. 19, 230 (1974), also submitted for publication in Phys. Rev.

6. M. B. Stearns and L. A. Feldkamp, submitted to Phys. Rev. and Bull. of A.P.S. 19, 230 (1974).
7. M. F. Collins and G. G. Low, Proc. Phys. Soc. 86, 535 (1965).
8. a) J. Crangle and M. J. C. Martin, Phil. Mag. 4, 1006 (1959); b) J. Crangle and D. Parsons, Proc. Roy. Soc. (London) A255, 509 (1960); also see J. Crangle, p. 51 in "Electronic Structure and Alloy Chemistry of the Transition Elements," edited by P. A. Beck, (Interscience Publ., New York, 1963).
9. T. Moriya, Prog. of Th. Phys. 33, 157 (1965).
10. P. W. Anderson, Phys. Rev. 124, 41 (1961).
11. R. L. Streever and G. A. Uriano, Phys. Rev. 139, A135 (1965); Phys. Rev. 149, 295 (1966).
12. J. B. Comly, T. M. Holden, and G. G. Low, J. Phys. C. (Proc. Phys. Soc.) 1, 458 (1968).
13. D. Parsons, W. Sucksmith, and J. E. Thompson, Phil. Mag. 3, 1174 (1958).

ELECTRICAL RESISTIVITY AND MAGNETO-RESISTIVITY OF VERY DILUTE Cu-Cr ALLOYS

S. Legvold, P. Burgardt, D. T. Peterson
Ames Laboratory-USAEC and Departments of Physics and Metallurgy
Iowa State University, Ames, Iowa 50010

T. A. Vyrostek
Johns Hopkins University Applied Physics Laboratory, Silver Spring, Maryland 20910

J. A. Schaefer
Physics Department, Loras College, Dubuque, Iowa 52001

ABSTRACT ONLY*

Samples containing 10, 19.6, 39.1 and 87.4 atomic ppm Cr were made by arc melting Cu and pieces of a master Cu-Cr alloy into fingers which were cold rolled, annealed at 800C for one day, quenched in ice water, spark cut and hand polished over emery cloth to about 1 x 1 x 36 mm. Four probe resistivity measurements were than made from 1.5-80 K. Because of the Kondo term, the residual resistivities were obtained from $\rho_{Cr}(25\ K) - \rho_{Cu}(25\ K)$ and these results turned out to be 1.02, 1.055, 1.095, and 0.992 n Ω cm/ppm Cr. These data lead to an equation $\rho_o/c = 1.065 - 0.0008c$ which works well up to c = 300 ppm Cr (ρ_o is in n Ω cm). All samples exhibited a resistivity minimum and these fell at 16, 17.8, 19.4, and 22.2 K respectively, giving $T_{min} \propto c^{0.15\pm0.02}$. The estimated depths of the minima were 5.8, 11.2, 22.0 and 41.7 n Ω cm respectively. The ρ versus log T plots for the samples showed linear behavior. The plot of the slope per ppm Cr versus Cr concentration was linear over the range spanned and fit the equation: slope/ppm = 5 - 0.014c with c in ppm.

The most interesting part of the study came from magnetoresistance data taken at 4.2 K in fields up to 100 kOe. In the main, the spin or Kondo magnetoresistance is theoretically expected to be negative because the scattering of the conduction electrons by the magnetic impurities should be suppressed in a strong magnetic field. A cautious approach was used in obtaining and treating the appropriate experimental data because it is necessary to eliminate the normal positive magnetoresistance of the host lattice. In order to obtain this positive part, the magnetoresistance of each sample was measured at 25 K. At this temperature the contribution from Cr spins was essentially negligible (resistivity minima occurred well below this temperature) and the phonon part was small so that, under the assumption it was like the phonon part of pure Cu, it was subtracted out. Thus, the Kondo spin magnetoresistance was obtained from $[\rho(4.2\ K, H) - \rho(25\ K, H)]_{Cr}$ $[\rho(4.2\ K, H) - \rho(25\ K, H)]_{Cu}$. Quite unexpectedly, the results showed that samples with Cr concentrations less than 100 ppm have positive magnetoresistance up to fairly large fields (~ 70 kOe). Since these observations were made above $T_K \sim 2K$, they fell in the intermediate regime with $g\mu_B H \sim k_B T \sim \mathcal{J}_{ij} \underset{\sim}{S}_i \cdot \underset{\sim}{S}_j$ where $\mathcal{J}_{ij}$ is the Cr impurity exchange constant and $\underset{\sim}{S}_i$ the Cr spin. It is believed that spin disorder scattering accounts for the positive magnetoresistance and that it is comparable to the Kondo spin effect in this regime. Apparently, conduction electrons are scattered from interacting Cr atoms (antiferromagnetically ordered clusters) and the applied field breaks down the Cr ordering thus enhancing the scattering and giving positive magnetoresistance under proper temperature and field conditions. At lower temperatures, the Kondo term dominates; e.g., the magnetoresistance of the 10 ppm Cr sample was negative at 1.9 K.

*To be published in Solid State Communications.

THE ELECTRICAL RESISTANCE DUE TO NONMAGNETIC IMPURITIES IN FERROMAGNETIC METALS

D.J. Kim
Department of Physics, Aoyama Gakuin University, Chitosedai, Setagaya-ku, Tokyo, Japan*
and
Francis Bitter National Magnet Laboratory,† Massachusetts Institute of Technology,
Cambridge, Massachusetts 02139

and

Brian B. Schwartz
Francis Bitter National Magnet Laboratory,† Massachusetts Institute of Technology,
Cambridge, Massachusetts 02139

ABSTRACT

We discuss the electrical resistance due to nonmagnetic impurities in ferromagnetic metals. The impurity resistance depends on the magnetization of the host metal since (1) the densities of states of ± spin electrons at the Fermi surface change with the magnetization of a metal, and (2) the screened impurity potential is different for the ± spin electrons and depends on the magnetization of the metal. The effect of the spin dependent screening on the electrical resistance in ferromagnetic metals was not fully considered previously. We choose a single band model for the metallic electrons and obtain an expression for the resistance which includes both of the above mechanisms, (1) and (2), simultaneously and self-consistently. We calculate the changes in resistivity due to nonmagnetic impurities in a ferromagnetic metal as a function of magnetization.

INTRODUCTION

The analysis of electrical resistance in ferromagnetic metals is not simple since there are many different mechanisms of electron scattering involved. In this paper we discuss the part of the electrical resistance due to the nonmagnetic impurities in the ferromagnetic state of metals. This problem was not discussed thoroughly in the past. In the paramagnetic state of metals the contribution of nonmagnetic impurities to the electrical resistance is temperature independent and accordingly the simplest to deal with. In the ferromagnetic state of metals, on the contrary, we find that the electrical resistance due to nonmagnetic impurities is strongly temperature-dependent in a not simple way. For certain situations, for instance, the electrical resistance first decreases and then increases, producing a resistance minimum, as we lower the temperature in the region below the Curie point.

ELECTRON SCATTERING DUE TO NONMAGNETIC IMPURITIES IN THE FERROMAGNETIC STATE OF METALS

In the present problem the interaction between electrons plays the following two roles. The first role is to produce the ferromagnetic state by spin splitting the electron energy bands and the second role is to screen the potential of the impurities. It is crucial to include these effects of electron interaction properly. A convenient formulation of the electron scattering problem as to include the above two roles of electron interaction simultaneously and self-consistently was recently carried out by using Green's function.[1] If we assume a single band for the electrons of a metal we obtain the following results for the Green's function describing the scattering of an electron from a state with wave number k and spin ± to another state with wave number k + q and spin ±,

$$G_{k+q,k\pm}(\omega) = \frac{1}{\omega-\epsilon_{k+q,\pm}} \frac{1}{\omega-\epsilon_{k,\pm}} U_{eff\pm}(q) \tag{1}$$

with

$$U_{eff\pm}(q) = \frac{U(q)}{\epsilon_{\pm}(q)} \tag{2}$$

where U(q) is the original potential of an impurity, $\epsilon_{\pm}(q)$ is the spin dependent dielectric constant given by[2]

$$\epsilon_{\pm}(q) = \left[1-\tilde{V}(q)F_{\pm}(q)\right]\left[1+V(q)\left\{\frac{F_{+}(q)}{1-\tilde{V}(q)F_{+}(q)} + \frac{F_{-}(q)}{1-\tilde{V}(q)F_{-}(q)}\right\}\right] \tag{3}$$

where $V(q)=4\pi e^2/q^2$ is the Coulomb interaction, $\tilde{V}(q)$ is the effective exchange interaction between electrons[2] and $F_{\pm}(q)$ are the Lindhard functions of ± spin electrons, and $\epsilon_{k\pm}$ is the one particle energy including the exchange self-energy. Clearly the effect of the spin splitting of the electron energy bands is incorporated in changing ϵ_k to $\epsilon_{k\pm}$ and the effect of screening is represented by replacing the scattering potential U(q) by $U_{eff\pm}(q)$.

CALCULATION OF ELECTRICAL RESISTANCE

The electrical resistance, ρ, can be calculated straightforwardly from the result in Eq. (1) by a standard method.[3] If we assume a point charge potential for the impurity, $U(q) = 4\pi Ze^2/q^2$, and a parabolic energy band for the electrons the specific electrical resistance is calculated explicitly in the zero temperature approximation as

$$\rho(M) = \rho_o \left[\frac{X_+^6}{\Theta_+} + \frac{X_-^6}{\Theta_-}\right]^{-1} \tag{4}$$

with

$$\rho_o \equiv \frac{N_i}{n} Z^2 e^2 (\hbar k_F)\left(\frac{2m^*}{\hbar^2 k_F^2}\right)^2 \frac{\pi}{4} \tag{5}$$

$$\Theta_{\pm} = 4\int_0^{\pi} \frac{1}{|\epsilon_{\pm}(2k_{F\pm}\sin\theta/2)|^2} \frac{\cos\theta/2}{\sin\theta/2} d\theta \tag{6}$$

where m^* is the effective mass of an electron, $X_{\pm} = k_{F\pm}/k_F = (1 \pm M)^{1/3}$, where $k_{F\pm}$ and k_F are the Fermi wave numbers of ± spin electrons respectively in the ferromagnetic and paramagnetic states and $M = (n_+ - n_-)/n$ is the relative magnetication of the electrons, where $n_{\pm}$ are the total numbers of ± spin electrons and $n = n_+ + n_-$.

In Fig. 1 we present an example of numerical calculation based on Eq. (4) for different relative magnetization corresponding to different choice of values for effective exchange interaction. The magnetization M at zero temperature and the effective exchange interaction is related by the Stoner model. The parameter C is introduced to characterize the range of the effective exchange interaction by the relation $\tilde{V}(q)=\tilde{V}(0)/(1+Cq^2/k_F^2)$. Note that the qualitative behavior of the electrical resistance depends rather sensitively on the range parameter C.

DISCUSSION

We may approximately replace the effect of a change in temperature by a change in the effective exchange interaction as to reproduce the same change in the magnetization. (As we lower temperature the relative magnetization increases from M = 0 toward M = 1.) In this approximation Fig. 1 can be viewed as to show the temperature dependence of the electrical resistance. Thus, as we lower the temperature below the Curie point, the electrical resistance due to nonmagnetic impurities would either monotonously increase (for C = 1) or have a resistance minimum (for C = 0) depending upon the range of the effective exchange interaction.

The dotted line in Fig. 1 is obtained by using an expression, $\epsilon(q) = 1 + V(q)\,[F_+(q) + F_-(q)]$, for the dielectric constant to screen the impurity potential in place of Eq. (3). We immediately notice the importance of using the proper dielectric constant in the present problem.

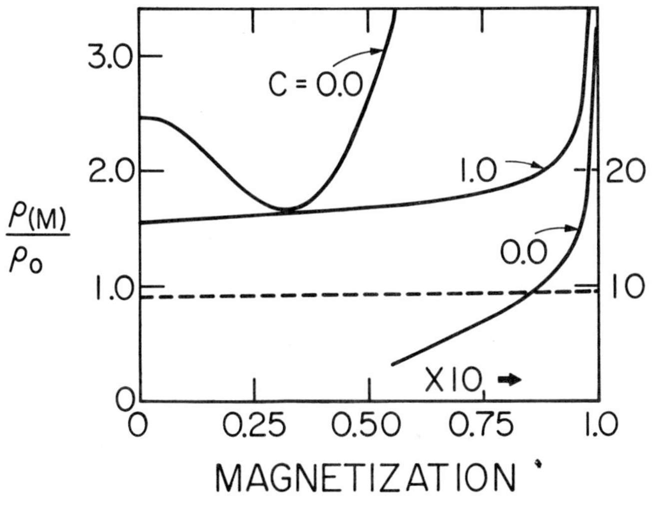

Fig. 1. A plot of $\rho(M)/\rho_0$ with m* equal to the free electron mass, $k_F = 10^8$ cm^{-1} and $\tilde{V}(q) = \frac{\tilde{V}(0)}{1+Cq^2/k_F^2}$. The dotted line corresponds to the use of simple Thomas-Fermi screening of the impurity potential.

REFERENCES

1. D.J. Kim, Phys. Letters 46A, 177 (1973).
2. D.J. Kim, B.B. Schwartz and H.C. Praddaude, Phys. Rev. B 7, 205 (1973).
3. N.F. Mott and H. Jones, The Theory of Properties of Metals and Alloys (Clarendon Press, Oxford, 1936).

* Permanent address.

† Supported by the National Science Foundation.

STUDY OF THE ANDERSON MODEL OF A DILUTE MAGNETIC ALLOY USING FEENBERG PERTURBATION THEORY

Samuel P. Bowen, University of Wisconsin, Madison, Wisconsin 53706

ABSTRACT

The Anderson Model of a dilute magnetic alloy has been studied by a generalization of Feenberg's perturbation theory applied to thermodynamic Green's functions. The method has the advantage of allowing the calculation of Green's functions which are not analytic functions of the coupling parameters as is the case here. The behavior of the impurity is found not to be determined by the Hartree Fock approximation nor by a narrow Abrikosow-Suhl resonance at the Fermi surface. Rather the thermodynamic properties are determined by the interplay of one electron host states with many electron correlated impurity states in a fashion quite analogous to the exactly soluble atomic limit. Self-consistent solutions for the thermodynamic Green's function give a susceptibility which is Curie-Weiss at high temperatures and Pauli-like at low temperatures, a resistivity which exhibits both a log(T) behavior at high temperatures and a T^2 behavior at low temperatures. The heat capacity exhibits both a Schottky anomaly and a change in the linear term of the heat capacity. The thermopower also exhibits the peak behavior as observed in experiment.

TEXT

Broadly speaking theoretical studies[1] of the dilute alloy have either been based on the Hartree Fock approximation and or standard field theoretic methods applied to Green's functions. It is generally the case that these methods fail to reproduce the atomic limit or give rise to logarithmic divergences. While these deficiencies have been somewhat overcome, agreement with experiment has not been complete in all temperature ranges.

In this paper is briefly described a theoretical method[2] for calculating Green's functions which avoids the logarithmic divergences, satisfies the atomic limit, and obtains the major experimental properties in a new unified description. The Hamiltonian is the single orbital model of Anderson[3] which characterizes the impurity by a localized orbital created by c^+_{ds} with a one electron energy ε_{ds} and an intra-atomic Coulomb interaction of U. The impurity resonance states are coupled to conduction band states $c^+_{\vec{k}s}$ with energy $\varepsilon_{\vec{k}s}$ by a hybridization matrix element $V_{\vec{k}d}$. The Hamiltonian is

$$H=\sum_s(\varepsilon_{ds}n_{ds}+\tfrac{1}{2}Un_{ds}n_{d\bar{s}})+\sum_{\vec{k}s} V_{\vec{k}d}(c^+_{\vec{k}s}c_{ds}+c^+_{ds}c_{\vec{k}s})+\sum_{\vec{k}s}\varepsilon_{\vec{k}s}n_{\vec{k}s} \quad (1)$$

The new method is based on a mathematical property[4] of thermodynamic Green's functions. This property is simply that Zubarev type Green's functions can be written as a certain type of matrix elements of the operator $(\omega-L)^{-1}$, where ω is a complex energy and L is a Hermitian operator acting on Heisenberg operators α defined as $L\alpha=[\alpha,H]$.

The matrix element is calculated using the inner product which determines the overlap between two Fermion operators A and B, each consisting of an odd number of Fermion operator factors, and defined by

$$(A,B) = \langle\{B,A^+\}\rangle\ , \quad (2)$$

where $\langle\ldots\rangle$ denotes the ensemble average and {,} is the anticommutator. In Zubarev's notation, the impurity Green's function is

$$G_{ds}(\omega) = \langle\langle c_{ds};c_{ds}\rangle\rangle_\omega = (c_{ds},(\omega-L)^{-1}c_{ds}). \quad (3)$$

The fact that the Green's functions are directly related to matrix elements of a special kind of resolvent means that we may use the standard matrix methods of band theory or renormalized perturbation theories like Feenberg's.[5] The, in principle, solution for the Green's function requires the determination of the secular matrix elements

$$S_{nn'} = \delta_{nn'}\omega - (\phi_n,L\phi_{n'}) \quad (4)$$

with respect to an orthonormal basis set of vectors ϕ_n which contains c_{ds}. The inverse of this matrix (4) yields the desired Green's functions.

The motivation behind this method is the same one which underlies the eigenvalue problem of standard quantum mechanics. The orthonormal basis ϕ_n, especially if it approximately diagonalizes L, corresponds to the fundamental physical states involved in the dynamics of the impurity. Put another way using the ϕ_n as intermediate states in perturbative calculations collects together more of the properties of the true equilibrium than the usual dressed propagator expansions of standard field theory.

An approximation scheme which is equivalent to the perturbation theory of E. Feenberg is to take a truncated secular matrix and seek its inverse either exactly or approximately, if there is some small parameter. In the following we exceed Feenberg's original proposal in that we invert a truncated, but infinite dimensional matrix.

The essential physics of the alloy is that the impurity is represented by atomic states which interact with the conduction electron states through the hybridization $V_{\vec{k}d}$. The atomic states of the impurity are of three types: a no electron state with energy 0, a two electron state with energy $2\varepsilon_d+U$, and the two one electron states with energy ε_{ds}. The coupling of these states to the conduction electrons gives rise to a broadening of each level and through screening a temperature dependent shift of the impurity resonances. It is this shift of the broadened resonances which manifests itself in the Kondo resistivity. As will be indicated below these results can be expanded to make contact with the Schrieffer-Wolf transformation and Kondo's perturbation calculation for the resistivity at high temperatures.

In the atomic limit ($V_{\vec{k}d}\to o$), the L secular equation is diagonalized by orthonormal operators

$$c_{\vec{k}s},\ \phi^{(r)}_{ds} = N^{(r)}_{d\bar{s}}\ c_{ds}/\langle N^{(r)}_{d\bar{s}}\rangle^{\frac{1}{2}}$$

where $r=\pm1$ and $N^{(-)}_{ds}=1-n_{ds}$ and $N^{(+)}_{ds}=n_{ds}$. The impurity green's function has simple resonances[6] at transition energies between 0 and 1 electron states with energy $E^{(-)}_{ds}=\varepsilon_{ds}$, and between 1 and 2 electron states with energy $E^{(+)}_{ds}=\varepsilon_{ds}+U$. The thermodynamics of this limit is determined by solving

$$\langle n_{ds}\rangle = (1-\langle n_{d\bar{s}}\rangle)\ f^{(-)}_{ds}+\langle n_{d\bar{s}}\rangle\ f^{(+)}_{ds} \quad (5)$$

where $f^{(r)}_{ds}=\langle\phi^{+(r)}_{ds}\phi^{(r)}_{ds}\rangle$, which in this limit are fermi functions of argument $E^{(r)}_s$.

When $V_{kd}\neq0$, the c_k rows of the L matrix are

$$Lc_{\vec{k}s} = \varepsilon_{ks}c_{\vec{k}s} + \Sigma_r\langle N^{(r)}_{d\bar{s}}\rangle^{\frac{1}{2}}V_{\vec{k}d}\phi^{(r)}_{ds}. \quad (6)$$

The impurity rows are $(r=\pm 1)$:

$$L\phi_{ds}^{(r)} = (E_s^{(r)} + \delta E_s^{(r)})\phi_{ds}^{(r)} + \Sigma_{\vec{k}}(M_{\vec{k}}^{(r)} c_{\vec{k}s} - M_{\vec{k}}^{(-r)} \phi_{d\vec{k}s})$$

$$+ \sum_{\vec{k}}(M_{1\vec{k}}\, \phi_{d\vec{k}ds} - M_{2\vec{k}}\, \phi_{\vec{k}dds}) \ . \qquad (7)$$

The orthonormal ϕ vectors, the impurity shift $\delta E_s^{(r)}$, and the matrix elements $M_1, M^{(r)}, M_2$ are determined by applying the Schmidt orthogonalization[7] to $L\phi_{ds}^{(r)}$. The small $V_{\vec{k}d}$ approximation of this study is to use a truncated L secular equation which is spanned by all the vectors of (6) and (7). The new feature of this approach is that the intermediate states, the ϕ's, are correlated many particle states and the matrix elements reflect this. The vector $\phi_{d\vec{k}s} \propto (n_{d\bar{s}} - \langle n_{d\bar{s}}\rangle)c_{\vec{k}s}$ displays impurity population fluctuations and a conduction electron excitation. The vectors $\phi_{d\vec{k}ds}$ and $\phi_{\vec{k}dds}$ which contain $c_{d\bar{s}}^{+} c_{\vec{k}\bar{s}} c_{ds}$ and $c_{\vec{k}\bar{s}}^{+} c_{d\bar{s}} c_{ds}$ respectively represent the excitation of a conduction hole or electron simultaneously with the zero or two electron impurity states. The most important aspect of this representation for present purposes is the shift $\delta E_s^{(r)}$ which arises from the orthogonality requirement. The bare resonance shift is defined by

$$\pi\delta E_s^{(r)} = -\Delta\Lambda_s/\langle N_{d\bar{s}}^{(r)}\rangle,$$

where

$$\Lambda_s = \langle N_{ds}^{(+)}\rangle^{\frac{1}{2}} \Sigma_{\vec{k}} V_{\vec{k}d} \langle c_{\vec{k}\bar{s}}^{+} \phi_{d\bar{s}}^{(+)}\rangle - \langle N_{ds}^{(-)}\rangle^{\frac{1}{2}} \sum_{\vec{k}} V_{kd} \langle c_{\vec{k}\bar{s}}^{+} \phi_{d\bar{s}}^{(-)}\rangle . \qquad (8)$$

The equal time averages $\langle c_{\vec{k}s}^{+} \phi_{ds}^{(r)}\rangle$ are determined using off-diagonal elements of the L-resolvent.

DISCUSSION

A very simple description of impurity green's function is gained if each impurity resonance is approximated by a Lorentzian quasi-particle resonance. Then the impurity resonances occur at $E_s^{(r)} - \Delta\Lambda'/\pi$ with a broadening Δ' and with a weighting factor of $\langle N_{d\bar{s}}^{(r)}\rangle$. The broadening Δ' is of order Δ and the renormalized shift function Λ' is much simpler than (8). The behavior of Λ' is controlled by two temperatures $T_- \sim |\varepsilon_{ds}|/\pi$, which is large $-\varepsilon_{ds} \sim 1$-2 eV, and $T_+ \sim \Delta'/\pi$, which can be quite small. Very small Δ' are found in the degenerate orbital model which will be discussed elsewhere. If T is large, $T > T_+, T_-$, then $\Lambda' = 0$, to order $1/D$, where D is the bandwidth. If T is between T_+ and T_-, Λ'_s is logarithmic in T,

$$\Lambda' = \ell n(3.55T_-/T) - 1.64\,(T/T_-)^2 . \qquad (9)$$

At low temperatures, $T < T_+$, we have

$$\Lambda' = \ell n(T_-/T_+) - 1.64(T/T_+)^2 \qquad (10)$$

These results are quoted for the case that the impurity resonance near $\varepsilon_{ds}+U>0$ is close to the Fermi energy. If this resonance energy is positive, the impurity is approximately singly occupied, the one electron doublet is the ground state and the impurity is magnetic. If this resonance were negative (below the Fermi energy), then the impurity would be approximately doubly occupied and the impurity ground state would be the two electron singlet. However, this double occupancy should cost an extra interatomic coulomb energy which has been left out of the model. Thus we must choose the value of U in this model to minimize the excess charge on the impurity. This charge neutrality condition can be approximately satisfied if we choose U so that the $E_s^{(+)}$ resonance is pulled down to the Fermi surface, but not below as $T \to 0$. This gives an equation for U which satisfies

$$(\varepsilon_{ds}+U)/\Delta = (2\pi)^{-1} \ell n(1 + (\varepsilon_{ds}/\Delta')^2) \ . \qquad (11)$$

This equation implies that U is slightly larger than $-\varepsilon_{ds}$ depending on Δ. For the ranges of ε_{ds} which arise in the renormalized atom picture[8] these values of U are like Herring's estimates.[9]

Let us now show that this picture gives the Kondo log T resistivity if $T > T_+$. At any temperature the resistivity is proportional to

$$\pi\rho_F \text{Imt} = \Delta\Sigma_{s,r} \langle N_{d\bar{s}}^{(r)}\rangle \Delta' / [(E_s^{(r)} - \Delta\Lambda'_s/\pi)^2 + \Delta'^2]$$

Expanding each denominator for small $\Lambda' \sim \ell n(T_-/T)$ yields a constant term of order Δ^2 and a logarithmic correction which is

$$\Delta\sum_{rs} \frac{\langle N_{d\bar{s}}^{(r)}\rangle \Delta' E_s^{(r)}}{[(E_s^{(r)})^2 + (\Delta')^2]^2} \frac{\Delta}{\pi} \ell n(\frac{T^{(-)}}{T})$$

This logarithmic correction term is clearly of order V^6 and, if Δ' factors in the denominator are neglected, gives a coefficient which is close to the Schrieffer-Wolff expression for the s-d exchange parameter J. While it is not immediate, if one examines the physical meaning of the operator matrix elements in this expansion, they can be seen to contain the same physics as Kondo's spin flip processes.

The physical significance of the $E^{(+)}$ resonance being pulled down to the Fermi energy is that the ground state of the impurity becomes strongly mixed with the two electron singlet impurity state. Without much violating the charge neutrality condition the singlet state could even become the ground state as $T \to 0$ if U is slightly smaller than in (11). This description of the spin compensation takes place almost exclusively within the impurity cell and is clearly in accord with recent experiments. Further, this picture makes the Kondo effect a strictly impurity effect in that D does not appear in the shift Λ' as in most s-d treatments.

The zero field impurity susceptibility of the impurity is Curie Weiss-like at high temperatures, $T > T_+$, and displays a reduced effective moment in accord with experiment.[10] At low temperatures the susceptibility is Pauli-like with T^2 corrections.

The heat capacity of the impurity displays at low temperature anomaly $T \sim 0.5T_+$ and contributes to the linear electronic heat capacity.

This description also displays negative peaks in the low temperature thermopower as seen in experiment.[10]

The detailed description of these effects and their comparison with experiment will be discussed elsewhere. The intended purpose of this note has been to briefly describe a new theoretical method, and to display the atomic-like description which this method yields for the dilute magnetic alloy. The ultimate value of the method and description is that it can be applied to the degenerate Anderson model for application to real transition metal impurities.

REFERENCES

1. For latest review, see A. Zawadowski and G. Gruner, Advances in Physics (1974) and references, therein.
2. Samuel P. Bowen, Phys. Rev., submitted 1974.
3. P. W. Anderson, Phys. Rev. 141, 41 (1962).
4. Samuel P. Bowen, AIP Conference Proc. 18, 683 (1973).
5. E. Feenberg, Phys. Rev. 74, 206 (1948).
6. J. Hubbard, Proc.Roy.Soc. A277, 237 (1963).
7. A. Messiah, Quantum Mechanics, Wiley, N.Y. 1962.
8. L. Hodges, H. Ehrenreich, R.E. Watson, Phys. Rev. B5, 3953 (1972); J.F. Herbst, D.N. Lowry, R.E. Watson, Phys.Rev. B6, 1913 (1972).
9. C. Herring, Magnetism IV, G.T. Rado and H. Suhl, eds. Academic Press, N.Y. 1966.
10. C. Rizzuto, Rep. Prog. Phys. 37, 147 (1974).

LATTICE LOCATION OF XE IN IRON

M. Van Rossum, G. Langouche, H. Pattyn, G. Dumont, J. Odeurs, A. Meykens, R. Coussement
University of Leuven, 3030 Heverlee, Belgium

and

P. Boolchand
University of Cincinnati, Cincinnati, Ohio 45221, U.S.A.

ABSTRACT

Mössbauer effect following ion implantation of 8.2 d ^{129m}Xe in metallic Fe has been systematically investigated as a function of implanted dose (10^{12}-10^{15} ions/cm^2). The significance of these experiments is threefold. (a) These have permitted a reliable measurement of the internal magnetic field H_{int} at the high field site, which is believed to be a substitutional Xe site in the b.c.c. lattice of the host. The H_{int} value was found to be 1480 ± 80 kOe and provides an important calibration point in the systematics of H_{int} at 5s-p impurities in Fe. (b) The general shape of the spectra were found to be sensitive to the implantation dose. They suggest the existence of a few magnetically inequivalent sites, the population of which are sensitive to implantation conditions such as host temperature and implanted dose. (c) Finally, values of H_{int} at the sites are of interest for purposes of structural identification. There is evidence that these sites may represent Xe lattice locations having one, two, three, etc. vacancies in the near neighbor shell.

INTRODUCTION

In order to study magnetic hyperfine interactions at radioactive nuclei, the presence of high magnetic fields is required. The ion implantation technique is a suitable tool for this purpose, as it allows one to make use of the large internal fields which nuclei experience when imbedded in a ferromagnetic lattice. The implanted atoms may not necessarily occupy purely substitutional lattice positions: the substitutional fraction may depend critically on such factors as implantation temperature and implanted dose. A careful analysis of the ion implantation behavior of the elements under study must then be performed before any significant conclusions can be drawn from the hyperfine interaction results. In this work, implantation of Xe in Fe has been studied by Mössbauer spectroscopy as a function of implanted dose. By this method, the various lattice locations of the Xe atoms can in principle be resolved, as they give rise to different Mössbauer absorption patterns. However, source strength requirements limit the usability of the Mössbauer effect (ME) to reasonably high doses. For a study of implantations at very low doses on the other hand, nuclear orientation (NO) may be more convenient.

EXPERIMENTAL RESULTS

Previous to this work, NO measurements were performed on ^{129m}Xe, ^{131m}Xe and ^{133m}Xe nuclei implanted in Fe.[1] In these experiments the implanted dose was varied between 3.5×10^{15} and 1.0×10^{11} atoms/cm^2. The analysis of the results was based on a two-site model: part of the implanted atoms experiencing the full field, $H_{\underline{XeFe}}$, (high-field fraction) and the rest experiencing a hyperfine field smaller than 0.2 $H_{\underline{XeFe}}$ (low-field fraction). For the high-field site, a value of $H_{\underline{XeFe}} = 1510 \pm 160$ kOe was obtained. Moreover, a clear dose-dependence of the high-field fraction was observed in these experiments

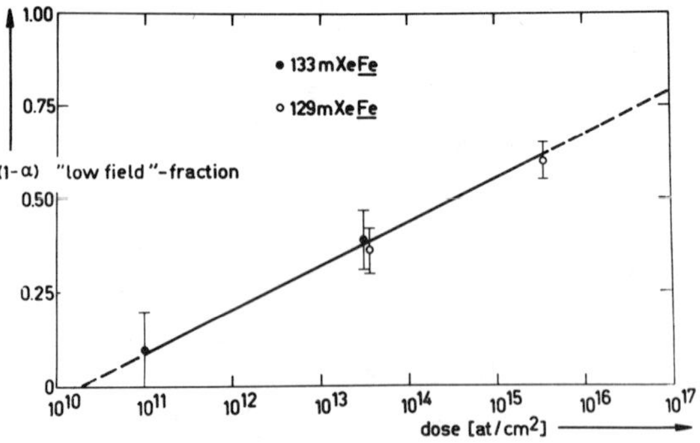

Fig. 1. Dose dependence of the low-field fraction of Xe implanted in Fe.

(Fig. 1). An important result of the NO experiments was that, at sufficiently low doses (1.0×10^{11} at/cm^2), these suggested that an almost purely substitutional implantation could be obtained.[2] However, these NO experiments did not allow a further distinction between the components contributing to the low-field fraction, nor did they yield a value for the magnetic field at those sites.

Therefore, a further investigation of the $\underline{XeFe}$ system using the ME became attractive to pursue. For this study, the 39.6 keV transition in the decay of ^{129m}Xe was chosen.[3] The use of this isomeric transition has the advantage of leaving the chemical state of the implanted atoms unchanged as a consequence of nuclear decay. The present activity was produced by thermal neutron irradiation of ^{127}I, yielding the reaction:

$$^{127}I \xrightarrow{n,\gamma} {}^{128}I \xrightarrow{\beta^-} {}^{128}Xe \xrightarrow{n,\gamma} {}^{129m}Xe$$

The activity was implanted into high-purity (99.99%) electropolished iron foils by the Leuven Isotope Separator at an energy of 75 keV. All implantations were done at room temperature. The ion current was measured at the target itself, the target being part of a Faraday cup. This method results in a reliable integrated dose measurement, the accuracy of which is close to 15%.

In the ME experiments, the source was moved by an electromechanical drive system in the constant acceleration mode. The unsplit compound $Na_4XeO_6 \cdot 2H_2O$ (obtained from Peninsular Chemical Research Inc., Gainesville, Florida) was used to make an absorber containing 20 mg Xe/cm^2. Both source and absorber were kept at liquid helium temperature.

Spectra of three sources, implanted to doses of 5.0×10^{12}, 5.0×10^{13} and 1.0×10^{15} at/cm^2 were studied. These showed definite evidence of different implantation sites. All three spectra contain a high-field component corresponding to a field value of $H_{high} = 1480 \pm 80$ kOe. This result is in good agreement with the values so far deduced from NO,[1] and ME[4] results on ^{131}Xe.

The spectra further showed that the so-called "low-field fraction" is not unique. An intermediate-field site with $H_{int} = 1200 \pm 100$ kOe and a low-field site with $H_{low} = 300 \pm 50$ kOe could be identified. In addition, the presence of a strong central absorption peak points towards the existence of a near-zero-field site.

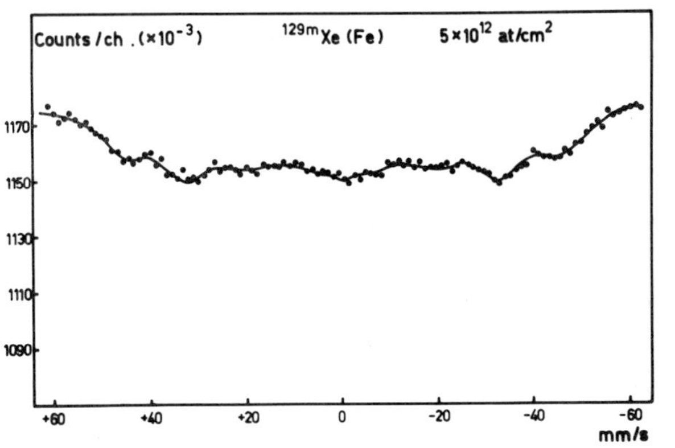

Fig. 2. Mössbauer spectrum at 4.2 K of ^{129m}Xe implanted in Fe at a dose of 5.0×10^{12} at/cm^2.

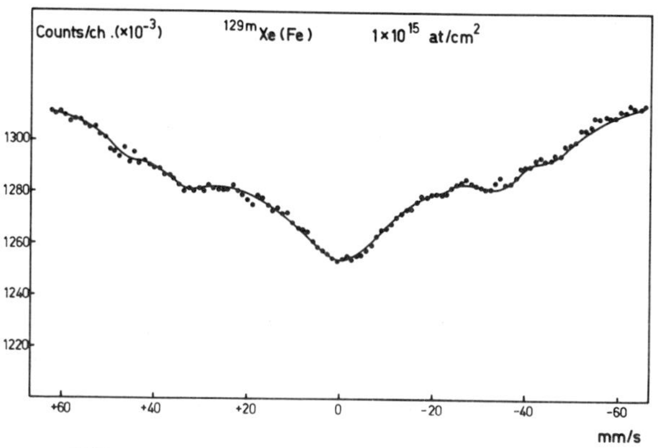

Fig. 3. Mössbauer spectrum at 4.2 K of ^{129m}Xe implanted in Fe at a dose of 1.0×10^{15} at/cm^2.

The dose dependence of the site populations is supported by the present ME measurements. This can be seen by comparing the spectra of Fig. 2 (dose of 5.0×10^{12}) and Fig. 3 (dose of 1.0×10^{15}). The relative intensity of the fraction of Xe atoms, not at a high-field site, is found to increase roughly by a factor of two. Information about the absolute populations of the different Xe sites can be deduced, only after the recoil-free fractions at these sites are reliably measured.

DISCUSSION

Our results yield a fairly accurate value for the H_{int} at Xe in Fe. The fields at 5s-p impurities in Fe as a function of the atomic number show a maximum for Xe. A knowledge of this field-value is important for studying the origin of the internal fields and the relative contribution of the different basic mechanisms.

Equaling significant, if not more, is the ion-implantation-information our measurements yield. The case of Xe implanted in Fe is an interesting one indeed, because the very big Xe atom (atomic radius 2.2 Å, vs. 1.25 Å for Fe) may be used as a probe-atom to study the interaction of the implanted species with the radiation damage caused by the implantation. That interaction is currently believed[5] to be the origin of the problems met in the use of ion implantation for hyperfine interaction studies, such as: different hyperfine fields found in the same system by different authors; the existence of several distinct fields in the same sample; and others.

a) The channeling experiments of Feldman and Murnick[6] utilized Xe implantations in Fe crystals and showed for implanted doses in the 10^{14} range a substitutional fraction of about 35 to 50%, while for an implanted dose of 10^{16}, a much lower substitutional fraction of 16 ± 3% was observed. The results of these experiments are directly related to the present ME experiments which also utilized Xe implantations in Fe. At lower implanted doses, the increase in the Xe high-field site fraction from the ME experiments on one hand, and the increase in the Xe substitutional site fraction from the channeling experiments on the other hand, suggests that the high field site may be identified with the substitutional one.

b) When one takes for granted that the substitutional fraction increases on lowering the dose, the reason why one sometimes has a "bad" implantation (i.e. where a considerable fraction of the implanted atoms are finally found on a site where they do not experience the full hyperfine field, this is a non-substitutional site) is to be found in the interaction of a part of the implanted atoms with the radiation damage (especially the vacancies) as produced in the collision cascade of the neighbor-atoms. This statement may be fully appreciated, when we relate it to the result of a NO-measurement at extremely low dose (1.0×10^{11} at/cm^2), where a high-field (substitutional) fraction of 0.90 ± 0.10 was found. This shows that the interaction of the Xe-atom with its own collision cascade ("correlated" damage) is, at least, very small.

We conclude that every model, describing these special implantation phenomena, has to be constructed on the basis of the interaction of the implanted atoms with the uncorrelated damage.

c) Internal magnetic fields found in the present experiments may be compared to these in ^{131}XeFe ME experiments recently reported by H. de Waard et al.[4] In both these experiments one is measuring internal fields at the same chemical element Xe in the same host, however, with the important difference that the implanted parents are different, in one case ^{131}I and in the other ^{129m}Xe. Although the internal fields at the different Xe sites in the two experiments appear to be the same, we believe that the population ratios in which these sites manifest themselves in the two experiments, depend on the chemistry of the implanted species and are in general different. There is clear evidence that apart from invoking a high-, intermediate- and low-field Xe site to account for the shape of the absorption spectrum, there is also the need to invoke a substantial near-zero-field site to account for the observed broad line centered in the vicinity of zero velocity. This latter feature appears to be unique to the Xe implantations in Fe. In our view this part of the spectrum should be attributed to a Xe site having three or more vacancies in the near neighbor shell, or conceivably such a site could also arise due to Xe atoms bunching together to provide a cubic, nonmagnetic environment as in solid Xe.

REFERENCES

1 H. Pattyn, R. Coussement, G. Dumont, E. Schoeters, R. E. Silverans, L. Vanneste, Phys. Lett. 45A, 131-139 (1973)
2 H. Pattyn, R. Coussement, G. Dumont, J. Odeurs, E. Schoeters, R. E. Silverans, L. Vanneste, to be published
3 M. Van Rossum, G. Langouche, H. Pattyn, G. Dumont, J. Odeurs, A. Meykens, P. Boolchand, R. Coussement, Proceedings of the International Conference on the Applications of the Mössbauer Effect, Bendor (France), to be published
4 H. de Waard, R. L. Cohen, S. R. Reintsema, S. A. Drentje, to be published
5 Meeting on Hyperfine Interactions and Radiation Damage, Dieppe (France) 6-7 June 1974
6 L. Feldman, D. Murnick, Phys. Rev. B5, 1-6 (1972)
7 H. de Waard, S. A. Drentje, Proc. Roy. Soc. A311, 139-150 (1969)

HYPERFINE FIELDS IN Nd_2Co_{17}, $NdCo_5$, AND $NdCo_3$

R. L. Streever
US Army Electronics Technology and Devices Laboratory (ECOM), Fort Monmouth, N.J. 07703

ABSTRACT

The nuclear magnetic resonances (NMR) of Co^{59}, Nd^{143}, and Nd^{145} have been studied at 4.2°K in the compounds Nd_2Co_{17}, $NdCo_5$, and $NdCo_3$. Co hyperfine fields in these alloys are found to depend on both the alloy and the Co site within a given alloy and vary from 30 kG to 220 kG. The Nd hyperfine fields are found to be in the range from 3.4-4.0 MG.

INTRODUCTION

The rare-earth cobalt intermetallic compounds have recently attracted considerable attention because of the large anisotropies and large saturation magnetizations which some of these alloys possess. In the present study we have investigated the NMR of three of these compounds Nd_2Co_{17}, $NdCo_5$, and $NdCo_3$. The $NdCo_5$ compound has the hexagonal $CaCu_5$ structure.[1] Nd_2Co_{17} has a rhombohedral structure[2] which can be derived from the $NdCo_5$ structure by ordered replacement of certain Nd atoms with pairs of Co atoms. The $NdCo_3$ structure which is isotypic with that of the $PuNi_3$ compound[3] can also be derived from the $NdCo_5$ structure by replacing certain Co atoms with Nd atoms.[4] In all of these compounds the Nd spin moments are coupled antiferromagnetically to the Co moments but, since the net Nd moment is opposed to its spin moment, the Nd moments add to those of the Co.

The samples were prepared by melting the appropriate amounts of Nd and Co metals. The ingots were then crushed or filed into approximately 50 micron powders. X-ray studies carried out on the powders established (to an accuracy of about 5%) that only the major phase was present.

The NMR spectra were obtained at 4.2°K with zero applied dc field by the usual technique of plotting the spin-echo amplitude as a function of frequency.

As a result of a problem involving samples the data given in the preliminary abstract was incorrect in that the results quoted for the $NdCo_5$ alloy should apply to the $NdCo_3$ alloy. The correct results are given. in the present paper.

RARE-EARTH NMR

For both Nd^{143} and Nd^{145} $I = 7/2$, so we expect the NMR of each isotope to be split by the quadrupole interaction into seven lines. The hyperfine field H_N can be obtained directly from the frequency of the central line. The frequency separation between adjacent lines Δ is just equal to $2|P|$ where P is defined by

$$P = 3e^2q\,Q/4I(2I-1)h \qquad (1)$$

In Table I we list the center frequencies of the Nd^{143} and Nd^{145} NMR observed in the three Nd-Co compounds along with the values of P obtained from the line splittings of the 143 isotope (due to a smaller value of Q the line splittings for the 145 isotope could not be resolved). Corresponding data[5] for Nd in Gd metal is also given. $NdCo_3$ has two rare-earth sites, however, a resonance from only one site was clearly identified. Further studies of $NdCo_3$ are planned.

Bleaney[6] has estimated P_{4f}, the contribution to P arising from the 4f electrons of the parent ion to be -5.3 MHz. The experimental values of P are probably negative and differences between these values and P_{4f} can probably be attributed to extra-ionic contributions arising from conduction-electron-enhanced crystal field effects.

Table I NMR center frequencies and quadrupole interaction parameters in various Nd compounds.

Compound	$\nu(Nd^{143})$ (MHz)	$\nu(Nd^{145})$ (MHz)	$P(Nd^{143})$ (MHz)
Nd_2Co_{17} [a]	905	563	± 3.5
$NdCo_3$ (one site)[a]	818	507	± 5
$NdCo_5$ [a]	795	494	(splitting unresolved)
Nd in Gd metal[b]	834	519	± 2.3

[a] this study [b] from Ref. 5

In Table II are listed values of H_N calculated from the NMR frequencies (taking $\mu_N(Nd^{143}) = -1.063$ nuclear magneton and $\mu_N(Nd^{143})/\mu_N(Nd^{145}) = 1.608$).[7,8] Also listed are values of H_N obtained from the Nd in Gd studies[5] and studies in Nd metal.[9] Bleaney[6] estimates the "free" ion value of H_N to be 4.30 MG for Nd^{+3}. The corresponding "free" ion moment is 3.27 μ_B. Magnetic moments derived from the H_N values by assuming a proportionality between H_N and μ are listed in Table II. Note that contributions to the Nd hyperfine field induced by moments on neighboring atoms should only be of the order of 0.1 MG (see Reference 6), and consequently, our assumed proportionality between H_N and μ should be relatively good. For $NdCo_3$ and $NdCo_5$ the moments derived from the H_N values agree well with moments of 2.6 and 2.5 μ_B, respectively, obtained from neutron diffraction.[4,10] Similar agreement has been pointed out for the Nd metal moments.[9]

Table II Hyperfine fields and derived magnetic moments (see text) in various Nd compounds.

Compound	H_N (MG)	$\mu(\mu_B)$
(Free ion)	4.30[a]	3.27
Nd_2Co_{17}	3.91[b]	3.0
Nd in Gd metal	3.60[c]	2.7
$NdCo_3$ (one site)	3.53[b]	2.7
$NdCo_5$	3.44[b]	2.6
Nd metal (hexagonal site)	3.29[d]	2.5
Nd metal (cubic site)	2.25[d]	1.7

[a] from Ref. 6 [c] from Ref. 5
[b] this study [d] from Ref. 9

COBALT NMR

The Co^{59} spectra in the three compounds are shown in Fig. 1. We show by arrows on the Nd_2Co_{17} spectra the positions of the four resonance lines observed by Desportes and Tsujimura[11] in a Y_2Co_{17} alloy. The R_2Co_{17} structure can take two closely related forms, a rhombohedral form and a hexagonal form. The Nd_2Co_{17} alloy takes only the rhombohedral form while Y_2Co_{17} can occur with both phases present. In any case, however, both forms have four sites available to the Co atoms and corresponding sites in the two structures have nearly the same environment. Consequently, we expect the spectra of the two alloys to differ primarily as a result of the different R atom.

We have tried to fit the Co hyperfine fields in these alloys to an expression of the following form

$$H_{Co} = a\mu_{Co} + bN_{Co}\mu_{Co} + cN_{Nd}\mu_{Nd}(S) \qquad (2)$$

The first term gives the hyperfine field induced by the moment of the parent atom. The second and third terms give the hyperfine fields induced by the Co moments and Nd spin moments of atoms in the immediate environment of a given Co site. N_{Co} and N_{Nd} are the numbers of Co and Nd neighbors respectively.

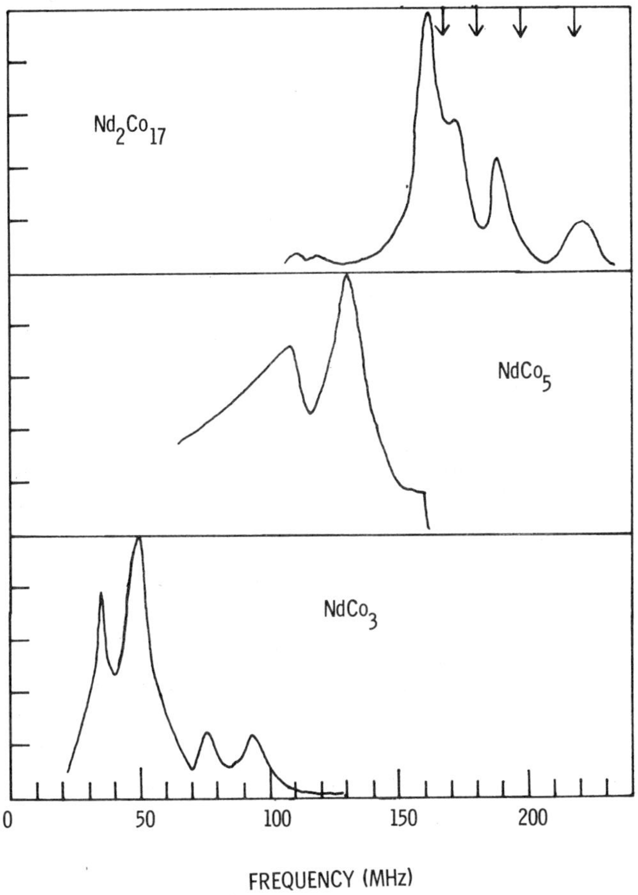

Fig. 1. Corrected relative Co^{59} spin-echo intensities at 4.2°K plotted as a function of frequency in Nd-Co compounds. The experimental echo amplitudes at each frequency ν have been divided by ν^2 to correct for the fact that the induced nuclear signal and the nuclear polarization are each proportional to ν.

The parameters a and b can be estimated from previous studies[12] of hyperfine field systematics to be -39.2 kG $(\mu_B)^{-1}$ and -7.32 kG $(\mu_B)^{-1}$ respectively. The value of $c\mu_{Nd}(S)$, the hyperfine contribution per Nd neighbor, has been taken to be +5 kG on the basis of the difference between the Nd_2Co_{17} and Y_2Co_{17} spectra. The Co moments have been taken to be 1.65, 1.5, and 1.0 μ_B in the Nd_2Co_{17}, $NdCo_5$, and $NdCo_3$ alloys respectively.[4,10]

Values of N_{Co} and N_{Nd} are listed in Table III for the different sites. We also compare values of H_{Co} calculated from Eq. (2) with those obtained experimentally from the line frequencies. The $NdCo_3$ peaks at 76 MHz and 94 MHz were assumed to belong to a single line centered at 85 MHz. The agreement between calculated and observed hyperfine fields is fairly good considering the simple model used. Line intensities are also approximately consistent with the relative numbers of sites per unit cell. Line frequencies and intensities observed[11] in Y_2Co_{17} are also consistent with the same model.

In $GdCo_2$ the hyperfine field is about 61 kG and is negative with respect to the Co magnetization.[13] The $GdCo_2$ results are, therefore, generally consistent with those of the present study if we consider that the Co moments are about 1.0 μ_B in both $GdCo_2$ and $NdCo_3$.

Table III Numbers of Co and Nd neighbors in the immediate environment of a given Co site. Values of H_{Co} calculated from Eq. (2) are compared with the experimental hyperfine fields (assumed to be negative) calculated from the line frequencies. Approximate relative experimental integrated line intensities I given in the last column are normalized to total to the total number of Co atoms per unit cell.

Compound	Sites	N_{Co}	N_{Nd}	H_{Co}(kG) (Eq. (2))	H_{Co}(kG) (exp.)	I (exp.)
Nd_2Co_{17}	6c	13	0	-222	-221	6
	18f	11	2	-188	-187	13
	9d	10	2	-176	-171	8
	18h	9	3	-158	-161	24
$NdCo_5$	2c	9	3	-143	-128	2
	3g	8	4	-127	-106	3
$NdCo_3$	6c	9	3	- 90	- 85	5
	18h	7	5	- 65	- 50	18
	3b	6	6	- 53	- 35	4

ACKNOWLEDGMENT

The author wishes to thank Professor A.E. Ray of the University of Dayton for providing the samples used in this study.

REFERENCES

1. J.H. Wernick and S. Geller, Acta Cryst. 12, 662-665 (1959).
2. G. Bouchet et al, C.R. Acad. Sci. Paris 262 B, 1227-1230 (1966).
3. D. Cromer and C.E. Olsen, Acta Cryst. 12, 689-694 (1959).
4. R. Lemaire, Cobalt, No. 32, 132-140 (1966); ibid, No. 33, 201-211 (1966).
5. S. Kobayashi et al, J. Phys. Soc. Japan 23, 474 (1967).
6. B. Bleaney, Magnetic Properties of Rare Earth Metals, Plenum Press, London and New York, 1972, Chapter 8.
7. K.F. Smith and P.J. Unsworth, Proc. Phys. Soc. 86, 1249-1257 (1965).
8. D. Halford, Phys. Rev. 127, 1940-1948 (1962).
9. A.C. Anderson et al, Phys. Rev. 183, 546-552 (1969).
10. R. Lemaire et al, J. Phys. Chem. Solids 28, 2471-2475 (1967).
11. J. Desportes and A. Tsujimura, C.R. Acad. Sci. Paris 277 B, 333-335 (1973).
12. J. Itoh et al, Proc. Int. Conf. Magnetism, Nottingham, 1964, Institute of Physics and the Physical Society, London, pp 382, 383; R.L. Streever and G.A. Uriano, Phys. Rev. 139, A135-A141 (1965).
13. K.N.R. Taylor and J.T. Christopher, J. Phys. C2, 2237-2245 (1969); I. Wang et al, AIP Conf. Proc., No. 18, 437 (1974).

MAGNETIC ORDERING IN PdGd: ELECTRICAL RESISTIVITY AND MAGNETIC SUSCEPTIBILITY

V. Cannella
Wayne State University, Detroit, Michigan 48202

T. J. Burch and J. I. Budnick
University of Connecticut, Storrs, Connecticut 06268

ABSTRACT

We have measured the temperature dependences of the electrical resistivity, $\rho(T)$, and the low field magnetic susceptibility, $\chi(T)$, for PdGd containing up to 5 at.% Gd. For alloys containing 2, 3, and 5 at.% Gd, $\rho(T)$ shows a sharp reduction in the spin disorder scattering at the ferromagnetic Curie temperature T_c, obtained from $\chi(T)$. Previous electrical resistivity measurements failed to exhibit the effect of magnetic ordering because the high field estimates of T_c used were significantly larger than the values we found. Resistivity data yield estimates of the conduction electron-local moment exchange interaction to be $\sim$.003eV.

INTRODUCTION

Alloys composed of low concentrations of Gd (or other magnetic rare earths)[1] dissolved in Pd do not exhibit the abnormally large magnetic moments per impurity atom which are found for the 3-d impurities Fe and Co in Pd. In PdGd the observation of a magnetic moment value close to the free Gd 3^+ ion value indicates that Gd impurities do not produce the strong ferromagnetic polarization of the Pd-electrons which cause the giant moments associated with Fe and Co impurities in Pd.[1,2,3,4] Nevertheless Crangle[5] found ferromagnetic ordering in PdGd alloys with concentrations C of Gd from 1 to 10 at.%. Crangle determined the magnetic Curie temperatures of the alloys using Arrott plots of H/σ vs. σ^2, where σ is the magnetization and H is the applied field. Crangle's low values of the saturated moment for Gd in the expected $^8S_{7/2}$ state were shown to be due to incomplete saturation by Guertin and co-workers[2] who found that saturation occurred only in fields much greater than 20 kG. Subsequently Sarachik and Shaltiel[6] reported the electrical resistivity $\rho(T)$ of PdGd for C up to 3 at.%. They found, however, no magnetic contribution to the resistivity near the T_c's determined by Crangle.

Measurements of the specific heat, C_p, were carried out by Bellarby and Crangle[7] on PdGd with C = 3, 5, and 7.7 at.% Gd. They found somewhat broadened peaks in C_p due to the magnetic phase transitions (similar to those found in PdFe[8]) but the temperatures of the peaks did not agree well with T_c found earlier by Crangle. The broadening of the peaks in C_p could be explained by a distribution of magnetic Curie temperatures due to the spatial variation of internal fields in the disordered alloys.[9] The discrepancies, however, between the temperatures of the maxima in C_p and the T_c's found by Crangle were not adequately accounted for.

We report here our measurements of the temperature dependences of the low field magnetic susceptibility, $\chi(T)$, and the electrical resistivity, $\rho(T)$. We find ordering temperatures significantly lower than those of Crangle[5], and our T_c's agree with the peaks in C_p of Bellarby and Crangle[7]. Furthermore $\rho(T)$ clearly shows the existence of spin disorder scattering which decreases below T_c as is expected for ferromagnetic alloys.

EXPERIMENT

$\chi(T)$ was measured between 1.3° and 10°K for ellipsoidal PdGd samples containing .5, 1, 2, 3, 5, and 7 at.% (nominal) Gd. The samples were annealed in a vacuum for over 24 hrs. at 950°, a procedure similar to that used by Crangle[5]. $\chi(T)$ was then measured using an ac mutual inductance bridge operating at a frequency of 50 Hz and a field of $\sim$5 gauss rms. $\rho(T)$ was measured between 1.2 and 4.2°K using a standard four terminal potentiometric arrangement on samples containing 2, 3, and 5 at.% Gd. These samples had been rolled, cut into thin strips, then annealed as above.

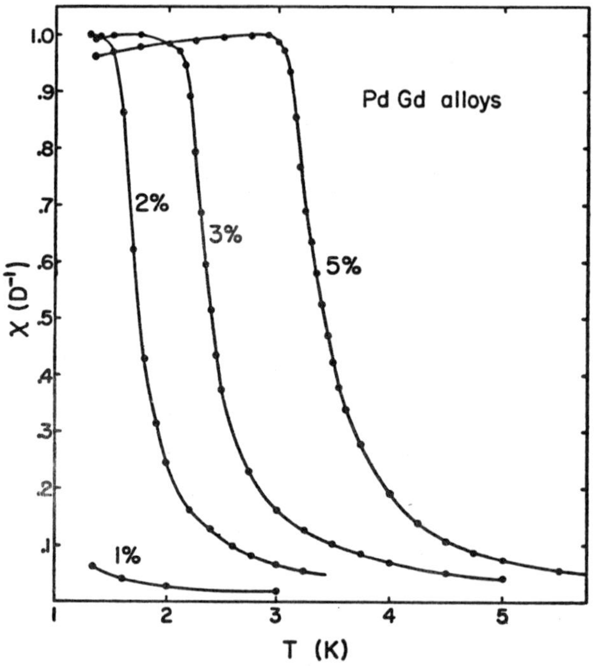

Figure 1. $\chi(T)$ is plotted in normalized units of D^{-1} where D is the demagnetizing factor of each sample.

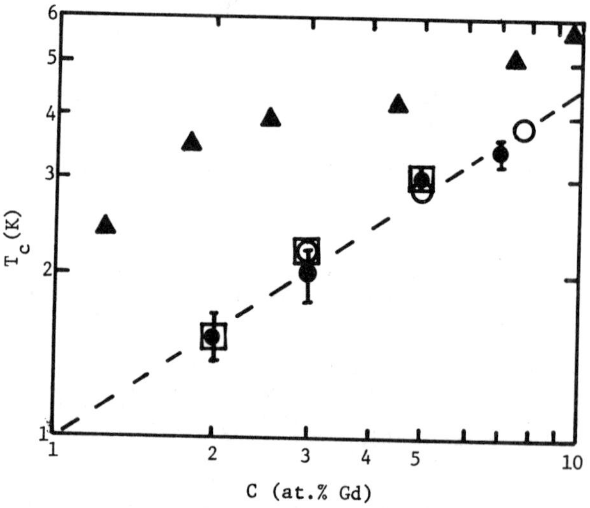

Figure 2. A log-log plot of T_c vs. concentration. Included are values ● from $\chi(T)$ and □ from $\rho(T)$ of our present work, values ○ corresponding to the temperature of $(C_p)_{max}$ of Bellarby and Crangle (reference 7) and values ▲ from the magnetization studies of Crangle (reference 5).

Plots of the measured $\chi(T)$ for C = 1,2,3, and 5% are shown in Fig. 1. For C = 2,3, and 5% the behavior of $\chi(T)$ is typical of ferromagnetic alloys: below T_c the real susceptibility diverges and the measured χ peaks at a value in emu/cm^3 equal to 1/D where D is the demagnetizing factor of the sample. Above T_c, $\chi(T)$ falls to a Curie-Weiss behavior which has been computer fitted well above T_c to obtain the effective moment per magnetic atom, p_{eff}, and the paramagnetic Θ. The rounding of the curves near T_c reflects the broadening of the magnetic transition expected in disordered dilute alloys. The resultant uncertainty in T_c is reflected in the error bars in Fig. 2 where T_c was taken to be between χ_{max} and $|d\chi/dT|_{max}$. Fig. 2 is a log-log plot of T_c vs C including values determined by the magnetization data of Crangle, by the peaks in C_p by Bellarby and Crangle, and from the "knee" in our $\rho(T)$ measurements (see the discussion below). Fig. 2 shows clearly the agreement between T_c determined from our $\chi(T)$ and $\rho(T)$, and the $(C_p)_{max}$ of Bellarby and Crangle. The T_c's from Crangle's magnetization measurements are clearly too high, which accounts for the inability of Sarachik and Shaltiel[6] to find a reduction in spin disorder scattering at those temperatures. From Fig. 2 we find that T_c varies more slowly than C in PdGd, following a power law $T_c = A\,C^m$ where m is between .6 and .7 and A is $\sim 1°K/(at.\%\ Gd)^m$. For alloys with C = .5 and 1 at.% Gd T_c was below 1.2°K and inaccessible to our apparatus, but the paramagnetic Θ's for .5 and 1% are consistent with T_c for higher concentrations. Table I lists the values found for T_c from our $\chi(T)$ and $\rho(T)$, along with the paramagnetic Θ and p_{eff} found from Curie-Weiss analysis of $\chi(T)$ well above T_c.

Table I

C	T_c(from χ)	T_c(from ρ)	Θ	p_{eff}
.5	——	——	.5 ± .15	6.5 ± .5
1	——	——	.8 ± .15	7.0 ± .5
2	1.5 ± .15	1.5 ± .1	1.7 ± .2	7.6 ± .7
3	2. ± .25	2.2 ± .1	2. ± .2	8 ± 1
5	3.0 ± .2	3.0 ± .1	3. ± .3	7 ± 1
7	3.4 ± .2	——	3. ± .3	8 ± .5

Table I. For various concentrations of Gd in Pd we present T_c found from $\chi(T)$, T_c found from $\rho(T)$, the paramagnetic Θ, and the effective moment, p_{eff}.

For our resistivity results, the resistivity of pure Pd[10] was subtracted, wherever significant, from that of the alloys to give $\Delta\rho = \rho_{alloy} - \rho_{Pd}$. $\Delta\rho(T)$ for C = 5 at.% is shown in Fig. 3. The curve clearly

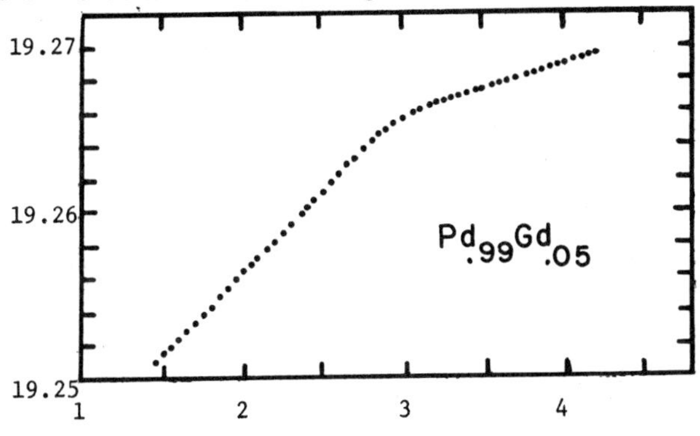

Figure 3. $\Delta\rho(T)$ for a PdGd sample with 5 at.% Gd.

shows a drop in $\Delta\rho$ below T_c which can be attributed to the decrease in spin disorder scattering in the ferromagnetic state. In these alloys $\Delta\rho$ varied nearly linearly with T just below T_c, so that T_c can be estimated from $\Delta\rho(T)$ to be at the "knee" in the curve. Considering electron-magnon scattering in disordered alloys, Long and Turner predict that the linear variation of $\Delta\rho(T)$ just below T_c should change to a $T^{3/2}$ behavior at lower temperatures.[11] Within our errors we did not find a $T^{3/2}$ dependence for $\Delta\rho$ further below T_c, but our data did not extend below 1.2°K. Since 1.2°K is a significant fraction of T_c in the 3% and 5% alloys, it seems reasonable that the $T^{3/2}$ dependence for $\Delta\rho$ may exist only below 1.2°K in these alloys. For C = 3 and 5 at.%, $\Delta\rho$ was extrapolated by extending the straight line below 1.2°K to produce an estimate of a minimum $\Delta\rho(T = 0)$, and we have taken $\Delta\rho(T = 1.2)$ as an estimate of a maximum $\Delta\rho(T = 0)$. These estimates of $\Delta\rho(T = 0)$ enable us to set limits on the values of J, the effective exchange interaction between conduction electrons and impurity spins, and V, the spin independent impurity potential. Following the procedure used by Williams and Loram in the limit of strong potential scattering, we find the lower limit of V to be .6eV which is comparable to the .57eV found for Fe in Pd.[12] Our estimates showed, however, that the effective exchange integral, J, for Gd in Pd is between .001 and .003eV, which is much smaller than the value of .06eV found for Fe in Pd.[12]

DISCUSSION

The agreement between the ordering temperatures determined from our $\chi(T)$ and $\rho(T)$, and the $(C_p)_{max}$ of Bellarby and Crangle shows that the ferromagnetism in PdGd alloys is quite similar in behavior to that found in other dilute alloy systems.[12,13] In particular, $\rho(T)$ behaves as expected for a ferromagnetic alloy showing a decrease in spin disorder scattering below T_c. The low values of T_c suggest that the exchange interactions between Gd spins are quite weak. The Gd atoms are expected to interact indirectly via a conduction electron exchange interaction, and the low value of J found from ρ is consistent with weak Gd-Gd interactions.

It is not immediately clear whether the indirect interactions via the conduction electrons would be expected to oscillate in sign, as the usual RKKY interaction. The absence of any contributions to the magnetic moment form the Pd d-electrons and the weak values of J estimated from ρ suggest that the range of the interactions is short. The work of Kim and Schwartz[14] suggests that if the range of interaction is short, then oscillations in the sign of the interaction will become more significant. This seems consistent with the high fields necessary to saturate Gd moments in Pd.[2]

REFERENCES

1. R. P. Guertin, H. C. Praddaude, S. Foner, E. J. McNiff, Jr., and B. Barsoumian, Phys. Rev. B 7, 274(1973).
2. R. P. Guertin, S. Foner, E. J. McNiff, Jr., and H. C. Praddaude, J. Appl. Phys. 42, 1550(1971).
3. J. Crangle and W. R. Scott, J. Appl. Phys. 36, 921(1965).
4. D. Shaltiel, J. A. Wernick, H. J. Williams, and M. Peter, Phys. Rev. 135, A1346(1964).
5. J. Crangle, Phys. Rev. Lett. 13, 569(1964).
6. M. Sarachik and D. Shaltiel, J. Appl. Phys. 38, 1155(1967).
7. P. W. Bellarby and J. Crangle, J. Phys. C 2, S362 (1970).
8. B. W. Veal and J. A. Rayne, Phys. Rev. 135, A442 (1964).
9. T. Takahashi and M. Shimazu, J. Phys. Soc. Japan 23, 945(1967).
10. A. I. Schinder and M. J. Rice, Phys. Res. 164, 759(1967).
11. P. D. Long and R. E. Turner, J. Phys. C 2, S127 (1970).
12. G. Williams and J. W. Loram, J. Phys. Chem. Solids 30, 1827(1969).
13. V. Cannella and J. A. Mydosh, Phys. Rev. 6, 4220 (1972).
14. D. J. Kim and B. B. Schwartz, Phys. Rev. Lett. 20, 201(1968); Phys. Rev. Lett. 21, 1744(1968).

SKEW SCATTERING AND QUADRUPOLE SCATTERING BY RARE EARTH IMPURITIES IN SILVER,GOLD AND ALUMINUM

A.FERT and A.FRIEDERICH

Laboratoire de Physique des Solides, Université de Paris-Sud, 91405, Orsay, France

ABSTRACT

We discuss the mechanisms of the skew scattering and of the quadrupole scattering which are shown by Hall effect and magnetoresistance measurements on silver, gold and aluminum containing rare earth impurities.

I INTRODUCTION

We discuss in this paper two sorts of asymmetric scattering of conduction electrons by magnetic impurities : skew scattering and quadrupole scattering.

The scattering of conduction electrons by a magnetic impurity is called skew if the deflection is different to the right and to the left (with respect to the direction of the impurity magnetic moment). When the magnetic impurities are polarized by a magnetic field, they all deflect the electrical current in the same direction and so contribute to the Hall effect. Skew scattering is known to be one of the mechanisms of the extraordinary Hall effect in the ferromagnetic metals and was first discussed by Smit (1). In recent years the problem of the skew scattering by isolated magnetic impurities in nonmagnetic metals has also started to attract attention (2) and Hall effect measurements have clearly shown the skew scattering by rare earth impurities(3). We shall summarize here a systematic study of the skew scattering by rare earth impurities in silver , gold and aluminum .

The interaction between conduction electrons and the electric quadrupole moment of a magnetic impurity (e.g.RE) makes the impurity cross section different according to whether the magnetic moment is parallel or perpendicular to the current. This results in an anisotropic magnetoresistance, the impurity resistivity being different in a magnetic field parallel or perpendicular to the electrical current. We shall present a study of the resistivity anisotropy of RE impurities in silver and gold (6) . The same sort of effect has been predicted by Kondo (5) for the magnetic disorder resistivity of the ferromagnetic metals.

Skew scattering and quadrupole scattering have in common that they cannot be explained by the standard interaction between conduction electrons and magnetic impurities

$$H = V(r) + J_{ex}(r)\ \vec{s}.\vec{S} \qquad (1)$$

or, for normal rare earth impurities that we shall consider from now:

$$H = V(r) + (g-1)\ \Gamma(r)\ \vec{s}.\vec{J} \qquad (2)$$

($\vec{s}$ = electron spin, $\vec{J}$ = total angular momentum of the f electrons, (g-1) = De Gennes factor). Skew scattering and quadrupole scattering can be only induced by additional terms to this spin independent + spin-spin interaction. In the case of an interaction between conduction and f electrons, these additional terms exist when the rare earth possesses an orbital angular momentum and have been expressed by several authors (18).We refer chiefly to the work of Kondo (5) who, limiting the plane waves of the conduction electrons to their s and p partial waves, derived a relatively simple form of the interaction (Coulomb + exchange). This interaction is given by the expression 2.45 of the Kondo paper (5). Keeping only three terms of this interaction, we write it :

$$H_{cf} = -\sum_{\vec{k}\,\vec{k}'} N^{-1} a^{*}_{\vec{k}'} a_{\vec{k}} \Big\{ (g-1)\ \Gamma(\vec{k},\vec{k}')\ \vec{J}.\vec{s} + \frac{i}{2}(2-g)\ F_2\ \vec{J}.(\vec{\kappa}\times\vec{\kappa}') + D\ \Big((\vec{J}.\vec{\kappa})(\vec{J}.\vec{\kappa}') - \frac{J(J+1)}{3}\ \vec{\kappa}.\vec{\kappa}' \Big) \Big\} \qquad (3)$$

where $\vec{\kappa}(\vec{\kappa}')$ denotes the unit vector in the direction of $\vec{k}(\vec{k}')$.

The first term of (3) is the usual spin-spin exchange interaction. The second term corresponds to a coupling between the orbital momentum $\vec{\ell}$ of the conduction electrons and $\vec{J}$ (indeed the matrix elements of the operator $\vec{\ell} = (\vec{r} \times \vec{p})\hbar^{-1}$ between plane waves $\vec{k}$ and $\vec{k}'$ is simply $12\pi i(\vec{\kappa}\times\vec{\kappa}')$ when the expansion of the plane waves is limited to partial wave $\ell = 1$ as in the Kondo paper (5)). We shall see in the next section that this second term in (3) can make the scattering skew as it changes sign when $\vec{k}$ and $\vec{k}'$ are interchanged (antisymmetric term).

The third term in (3) corresponds to the electrostatic interaction with the quadrupolar distribution of charge in the 4 f shell. We shall see in section III that it gives rise to an anisotropy of the impurity resistivity.

Actually the interaction 2.45 of the Kondo paper (5) includes also several other terms : two of them can give rise to resistivity anisotropy and one to skew scattering. But it can be shown that their contribution should not be significant and so we shall ignore them.

We shall see in the next section that, in silver and gold, skew scattering occurs even for a spherical 4 f shell (Gd^{+3},g=2). This skew scattering by Gd impurities cannot be explained by the interaction (3) and we have proposed an additional mechanism which is linked with the spin orbit coupling of the RE 5d states. On the contrary the resistivity anisotropy completely cancels out for the Gd impurities.

The interest of the study of the skew scattering and of the quadrupole scattering arises from the information provided on the antisymmetric and quadratic terms of the interaction between conduction electrons and magnetic impurities. As we shall see, this study can also be interesting for the determination of the crystal field splitting.

II SKEW SCATTERING

1) General: The scattering by a magnetic impurity is skew if the scattering probabilities between $\vec{k}$ and $\vec{k}'$ and between $\vec{k}'$ and $\vec{k}$ are different. Thus skew scattering can only arise if the scattering potential includes an antisymmetric part H^A:

$$< \vec{k}' \mid H^A \mid \vec{k} > \;=\; - < \vec{k} \mid H^A \mid \vec{k}' > \;=\; \text{imaginary}$$

Let us assume a very general scattering potential

$$H^{(S)} = \sum_{\vec{k}\vec{k}'} a^{*}_{\vec{k}'} a_{\vec{k}} (V_{\vec{k}\vec{k}'} + i\ U_{\vec{k}\vec{k}'})$$

where $V_{\vec{k}\vec{k}'}$ is real and symmetric and $U_{\vec{k}\vec{k}'}$ real and antisymmetric (this means antisymmetric with respect to the interchange of $\vec{k}$ and $\vec{k}'$). In the first Born approximation, the T matrix elements are

$$T^{(1)}_{\vec{k}\vec{k}'} = V_{\vec{k}\vec{k}'} + i\ U_{\vec{k}\vec{k}'}$$

and the scattering probability is symmetric (non-skew) :

$$W^{(1)}(\vec{k}\to\vec{k}') = \frac{2\pi}{\hbar} \mid T^{(1)}_{\vec{k}\vec{k}'} \mid^2 = \frac{2\pi}{\hbar} (V^2_{\vec{k}\vec{k}'} + U^2_{\vec{k}\vec{k}'})$$

Antisymmetric terms arise only if the T matrix is calculated to second Born approximation. Indeed the second order of $T_{\vec{k}\vec{k}'}$ includes terms such as

$$\sum_{\vec{k}''} \frac{V_{\vec{k}\vec{k}''}\quad V_{\vec{k}''\vec{k}'}}{\varepsilon_{\vec{k}} - \varepsilon_{\vec{k}''} + is} = -i\ \pi \sum_{\vec{k}''} V_{\vec{k}\vec{k}''} V_{\vec{k}''\vec{k}'}\ \delta(\varepsilon_k - \varepsilon_{k''}) + \text{real term}$$

It turns out that this imaginary symmetric term can interfere with the imaginary antisymmetric term $i\ U_{\vec{k}\vec{k}'}$, in $T^{(1)}_{\vec{k}\vec{k}'}$. The scattering probability then includes antisymmetric terms such as

$$-\frac{4\pi^2}{\hbar} \sum_{\vec{k}''} U_{\vec{k}\vec{k}'} V_{\vec{k}\vec{k}''} V_{\vec{k}''\vec{k}'}\ \delta(\varepsilon_k - \varepsilon_{k''})$$

We find that the lowest order skew terms of the scattering probability are of first order in U and second order in V. There are also skew terms of third order in U but they should generally be much smaller as U << V.

When the antisymmetric terms of the scattering probability are known, the contribution to the Hall resistivity can be found by introducing them in the Boltzmann equation. For elastic scattering one obtains (7)(8)

$$\rho^{\pm}_{xy} = -\left(\frac{\hbar}{8\pi^3 ne}\right)^2 \int \left(-\frac{\delta f^o}{\delta \varepsilon_k}\right)(\vec{k}.\vec{u})(\vec{k}'.\vec{v})\ W^a(\vec{k}\pm\to\vec{k}'\pm)\ d^3\vec{k}\ d^3\vec{k}'$$

where n is the number of electrons with a given spin direction, f^o the Fermi-Dirac distribution, $\vec{u}$ et $\vec{v}$ unit vectors in the directions x and y and $W^a(\vec{k}\pm\to\vec{k}'\pm)$ the antisymmetric part of the scattering probability. We point out that, on account of the factor $(\vec{k}.\vec{u})(\vec{k}'.\vec{v})$, the integral cancels out excepted if $W^a(\vec{k}\vec{k}')$ depends on $\vec{k}(\vec{k}')$ through p spherical harmonics (ℓ=1). This assesses that $W^a(\vec{k}\to\vec{k}')$ arises from products of two terms in T which depend on $\vec{k}(\vec{k}')$ through spherical harmonics ℓ and ℓ' with $\ell+\ell'$ = odd. This means that the scattering potential must act at least on two partial waves. There is no general rule for the dependence of ρ_{xy} on the magnetization of the magnetic impurities. However, in the case of RE impurities in silver or gold, our model (4) predicts that the Hall resistivity induced by skew scattering is proportional to $<J_z>$. Thus one expects in this case that the Hall resistivity should be the sum of two terms :

i) a term ρ^o_{xy} of ordinary Hall effect which is due to the Lorentz force. This term, on condition that $\omega_c\tau$ << 1 and phonon resistivity << impurity resistivity, is linear in the magnetic field H and independent of the temperature and of the concentration. Such a term is observed for silver with non magnetic impurities (9). In dilute magnetic alloys the "spin effect" (10) can raise the Hall resistivity at high field but is not observed in our alloys.

ii) a term ρ_{xy}^{E} due to skew scattering, proportional to the concentration of magnetic impurities and to their polarization (and thus to HT^{-1} at low field if the susceptibility obeys a Curie law). A contribution from side-jump scattering can be only observed at high concentration (11)(12).

2) Skew scattering by rare earth impurities in silver, gold and aluminum . Experimental results. We have studied the Hall effect of Ag RE, Au RE and Al RE alloys (RE =Pr, Nd, Sm, Gd, Tb, Dy, Ho, Er, Tm, Yb, Lu). The details on the preparation and the thermal treatment of the alloys are described elsewhere. The main metallurgical problem arises from the limited solubility of the RE in noble metals and aluminum . For heavy RE in silver or gold, a water-quenching from 800°C is sufficient to obtain concentrations of about 0.5 at %, sometimes 1 or 2 at %. For light RE in Ag or Au, one obtains about 0.1 at % by the same method. For the aluminum based alloys, 0.1 at % or a little more is obtained by splat-cooling but only for the heavy RE.

Considering first the silver and gold based alloys, several of them, in particular the alloys with light RE, contain some clusters of RE. However these clustered atoms contribute weakly to the normal and skew scattering : we have purposely clustered the RE of some alloys by annealing without quenching and then observed an almost complete cancellation of the resistivity and of the skew scattering Hall resistivity. This is also in agreement with the saturation of the resistivity which becomes almost independent of the concentration above the limit of solubility. Thus it turns out that, in spite of the possible existence of clusters, the normal and skew scattering are due mainly to non clustered impurities, at least in Ag and Au based alloys. This is less sure in the Al based alloys.

The Hall effect measurements have been performed by an a.c.thechnique in a superconductive coil up to 40 kG and between 1.2°K and 77°K.

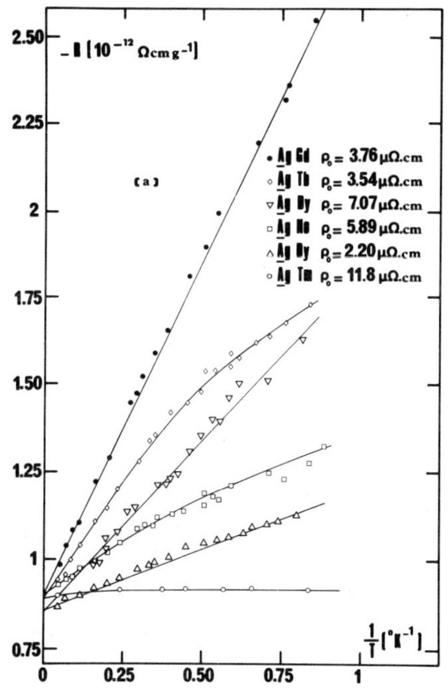

Figure 1 : The initial Hall coefficient (R = limit of ρ_{xy}/H for H → 0) vs T^{-1} for Ag Gd, Ag Tb, Ag Dy, Ag Tm, Ag Ho alloys.

Most of the measurements have been done on Ag and Au based alloys and we report first the experimental data for these systems. The skew scattering shows up first in the temperature dependences of the Hall effect. For example, the figure (1) shows the low field Hall coefficient R (R = limit of ρ_{xy}/H when H → 0) versus T^{-1} for series of Ag RE alloys. For Ag Gd on figure (1) , R can be decomposed in the sum of a constant term R_o (the ordinary Hall coefficient of the Ag Gd alloy) and of a term AT^{-1} that we ascribe to skew scattering by the Gd impurities :

$$R = R_o + AT^{-1} \qquad (4)$$

A skew scattering term in T^{-1} is just what is expected for Gd impurities whose the initial susceptibility obeys a Curie law. A constant value of the Hall coefficient is quite normal as the impurity scattering remains much stronger than the phonon scattering throughout the temperature range.

For the other silver, gold or aluminium based alloys (4) (e.g Ag Tb on figure 1) a variation $R = R_o + AT^{-1}$ is observed at high enough temperature. At lower temperature one observes deviations from T^{-1} which can be explained by crystal electric field (CEF) effects. Actually the initial magnetic susceptibility of most of the alloys also shows (13)(14) at low temperature similar deviations from the Curie law observed at high temperature. On figure (2) we have plotted at the same time the inverse of the magnetic susceptibility (from Murani ref14) and $(R-R_o)^{-1}$ as a function of the temperature for a Au Dy alloy. It appears that we can make the curves coincide and this means that the skew scattering term $(R-R_o)$ has the temperature dependence of the susceptibility.

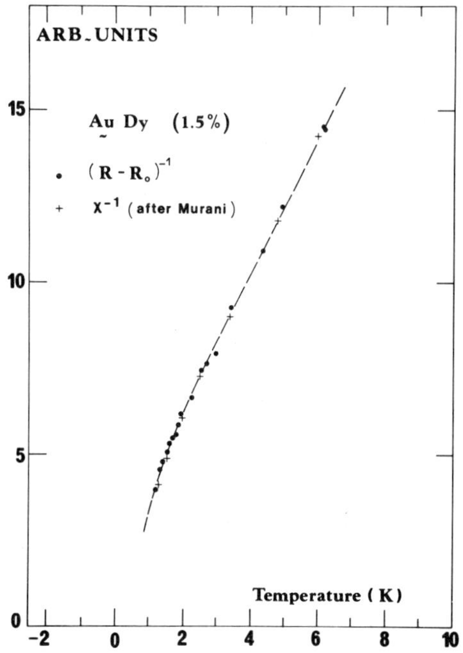

Figure 2 : The inverse of the magnetic susceptibility (from Murani (14)) and $(R - R_o)^{-1}$ (R = initial Hall coefficient, R_o = ordinary Hall coefficient) vs T for Au + 1.5 at % Dy.

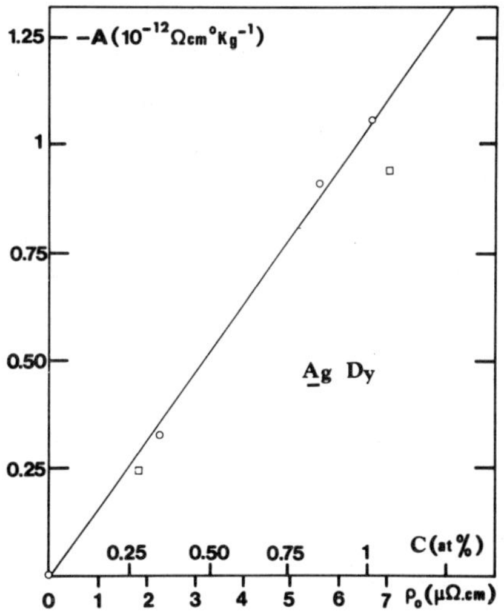

Figure 3 : A vs the residual resistivity and the impurity concentration for Ag Dy alloys (A is defined by (4)).

In some alloys with a very small skew scattering effect, the analysis of the temperature dependence of the Hall coefficient is less simple. We can say that the coefficient A of (4) is accurately determined for Gd, Tb, Dy, Ho in Ag and Au, Er and Tm in Ag but that the analysis of the experimental data is more uncertain for Er, Tm and Yb in Au and the light RE in Au and Ag.

The Hall effect induced by skew scattering is also expected to be proportional to the impurity concentration. We have determined the coefficient A in (4) for several Ag Dy alloys. On figure (3) A is plotted as a function of the concentration c and of the impurity resistivity ρ_o. It can be seen that A is proportional to c or ρ_o. This is also observed for several other alloys. So, in the high temperature limit, the experimental Hall coefficient can be written

$$R = R_o + \underline{a}\ \rho_o\ T^{-1} \qquad (5)$$

We characterize the magnitude of the skew scattering by a given impurity in the high temperature limit by the coefficient $\underline{a}$.

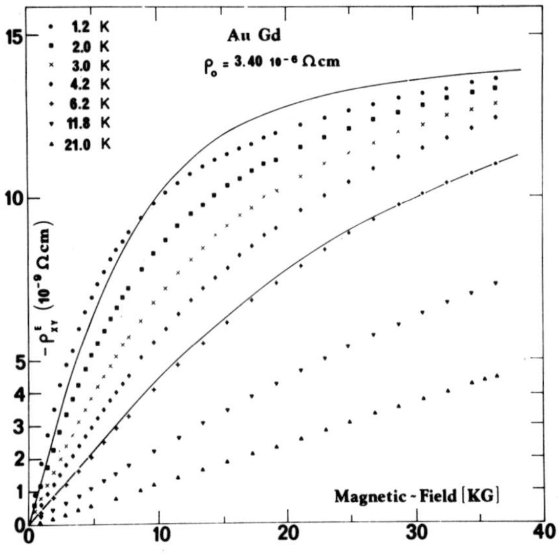

Figure 4 : The extraordinary Hall resistivity ρ^E_{xy} of a Au Gd alloy vs the magnetic field at several temperatures. The solid lines represent Brillouin functions for J = 7/2, $\mathbf{g}$ = 2, T = 6.2°K and T = 2°K.

On figure (4) the Hall resistivity ρ^E_{xy} induced by skew scattering in Au Gd ($\rho^E_{xy} = \rho_{xy} - R_o H$) is plotted as a function of the magnetic field at several temperatures. At 6.2°K the experimental curve nearly coincides with the Brillouin function $B_{7/2}$ (solid line). So ρ^E_{xy} seems proportional to $< J_z >$. At 2°K, the experimental curve departs somewhat from the Brillouin functions. This could be due, either to interaction effects, or to higher order terms in $< J^3_z >$. For non-S ions, the field dependence of the Hall resistivity is sometimes very different from that of a Brillouin function, certainly on account of crystal field effects.

We shall not concentrate on the CEF effects observed at low temperature but we shall chiefly discuss the data in the experimental range where the magnetic susceptibility nearly reaches its free ion value. In this range the skew scattering is characterized by the coefficient $\underline{a}$ of (5). Practically, the coefficient $\underline{a}$ is determined from experimental data up to 40°K. The magnetic measurements of Williams and Hirst (13) on Ag RE alloys and of Murani (14) on Au RE alloys show that, down to 20°K, the susceptibility deviates little (15) from the Curie law observed at higher temperature. The experimental value of the coefficient $\underline{a}$ for the RE impurities in Ag and Au are collected in figures (5) and (6).

We report finally the experimental results obtained for the Al RE alloys. The temperature dependence of the Hall coefficient at "high temperature" can also be fitted to expression (5) and we have determined the coefficient $\underline{a}$ for Tb, Dy, Ho and Er (figure 7). On account of the uncertainty of the metallurgy of the alloys, the accuracy on $\underline{a}$ is rather poor. We chiefly retain the sign of $\underline{a}$ and its rough variation in the RE series.

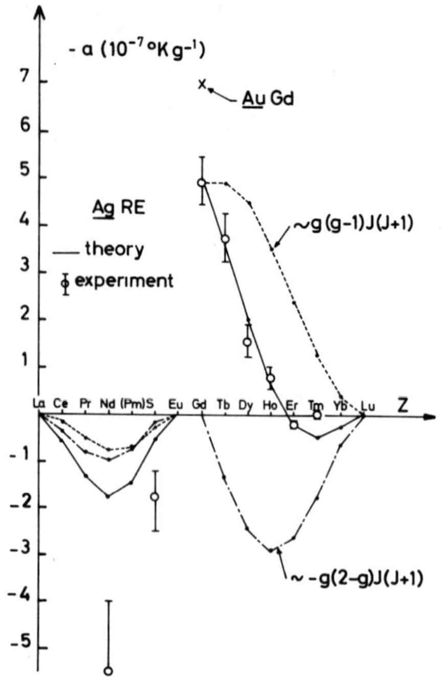

Figure 5 : The coefficient $\underline{a}$ which characterizes the skew scattering by each RE impurity in Ag is plotted throughout the RE series ($\underline{a}$ is defined by (5)). The solid, dotted-dashed and dashed lines respectively represent the values of $\underline{a}$, $\underline{a}_1$, $\underline{a}_2$ calculated from (11) ($\underline{a} = \underline{a}_1 + \underline{a}_2$).

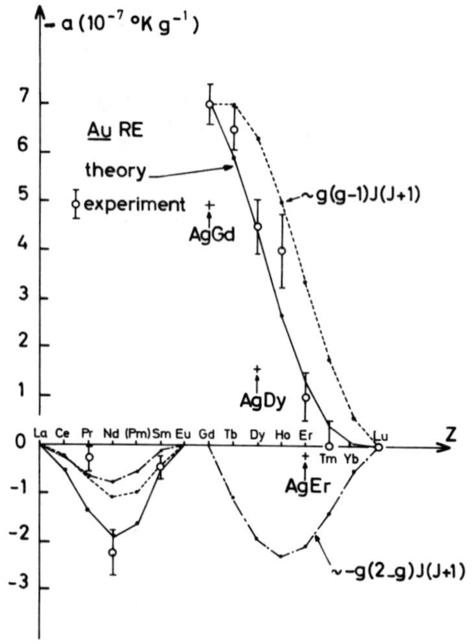

Figure 6 : The coefficient $\underline{a}$ which characterizes the skew scattering by each RE impurity in Au is plotted throughout the RE series ($\underline{a}$ is defined by (5)). The solid, dotted-dashed and dashed lines respectively represent the values of $\underline{a}_1$, $\underline{a}_1$ and $\underline{a}_2$ calculated from (12)

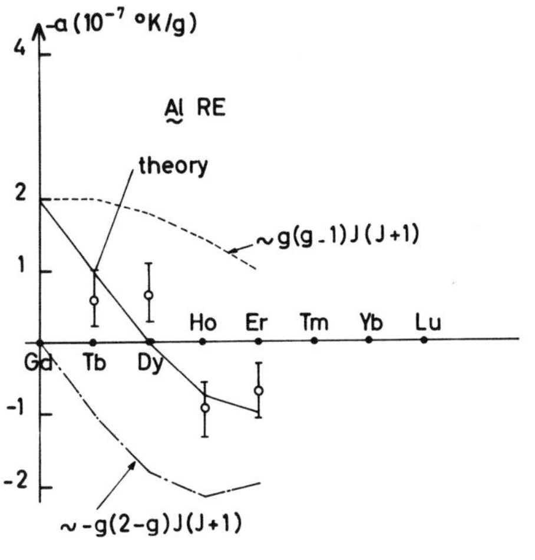

Figure 7 : The coefficient a which characterizes the skew scattering by each RE impurity in Al is plotted throughout the RE series. The solid, dotted-dashed and dashed lines respectively represent the values of a_1, a_1, a_2 calculated from (13).

3-Skew scattering by rare earth impurities in silver gold and aluminum - Theoretical results and discussion.

A model for the skew scattering by RE impurities in silver, gold and aluminum has been developed (5) from the following interaction between the conduction elec- trons and the RE impurities

$$H' = H^{(1)} + H^{(2)} + H^{(3)} + H^{(4)}$$

where

i) $H^{(1)}$ is the usual spin-spin exchange interaction (first term of the expression (3))

$$H^{(1)} = -\sum_{\vec{k}\vec{k}'} N^{-1} (g-1)\,\Gamma(\vec{k},\vec{k}')\,\frac{1}{2}\Big((a^*_{\vec{k}'+}a_{\vec{k}+} - a^*_{\vec{k}'-}a_{\vec{k}-})J_z + a^*_{\vec{k}'+}a_{\vec{k}-}J^- + a^*_{\vec{k}'-}a_{\vec{k}+}J^+\Big) \qquad (6)$$

with $\Gamma(\vec{k}\vec{k}') = 4\pi \sum_{\ell\leq 3,m} Y^*_{\ell m}(\Omega_{\vec{k}})\,Y_{\ell m}(\Omega_{\vec{k}'})\,\Gamma^{(\ell)}$

ii) $H^{(2)}$ is the skew term of the expression (3)

$$H^{(2)} = -\sum_{\vec{k}\vec{k}'} N^{-1}\frac{i}{2}(2-g)\,F_2(\vec{\kappa}\times\vec{\kappa}').\vec{J}a^*_{\vec{k}'}a_{\vec{k}} \qquad (7)$$

iii) $H^{(3)} + H^{(4)}$ expresses in (5) the screening effect around a RE ion. In silver and gold about two electrons are attrac- ted by the trivalent RE ion. It is generally admitted that the screen is mainly built with s and 5d electrons (5d vir- tual bound state), the main phase shifts being then η_o and η_2 with $2\eta_o + 10\eta_2 \sim 2\pi$. The existence of a 5d vbs for RE in Al has been also suggested (16), but, in this case, the re- pulsion of s and p electrons should balance the attraction of 5d electrons. $H^{(3)}$ also takes into account the spin orbit coupling of the 5d electrons and this introduces another skew term in the scattering probability.

From the scattering potential H', one derives the Hall resistivity induced by skew scattering and, in the tempera- ture range (kT > crystal field) where the magnetic suscepti- bility reaches the free ion susceptibility, one obtains the following low field Hall coefficient

$$R = R_o + (a_1 + a_2)\,\rho_o T^{-1} \qquad (8)$$

where R_o = ordinary Hall coefficient

$$a_1 = \frac{\pi\mu_B}{9k}\,\frac{\sin^2\eta_o - \sin^2\eta_2}{\sin^2\eta_o + 5\sin^2\eta_2}\,n(E_F)\,F_2\,g(2-g)\,J(J+1) \qquad (9)$$

$$a_2 = \frac{\pi\mu_B}{2k}\,\frac{\lambda}{\Delta}\,\frac{\sin^2\eta_2 \times \sin 2\eta_2}{\sin^2\eta_o + 5\sin^2\eta_2}\,n(E_F)\,(\Gamma^{(1)} - \Gamma^{(3)})\,g(g-1)\,J(J+1) \qquad (10)$$

ρ_o = impurity resistivity

λ = spin orbit constant of 5d electrons, Δ = width of the 5d virtual bound state, $n(E_F)$ = density of states per atom and spin direction).

a_1 arises from the antisymmetric interaction $H^{(2)}$ and a_2 from the spin orbit coupling of the 5d electrons. In a given host, η_o, η_2, λ, Δ, F_2, $\Gamma^{(3)}$ and $\Gamma^{(1)}$ should be nearly independent of the RE element. Then a_1 should vary throughout the RE series like the factor $g(2-g)J(J+1)$ and a_2 like $g(g-1)J(J+1)$.

We tried to fit the experimental variation of a throughout the RE series with a linear combination of $g(2-g)J(J+1)$ and $g(g-1)J(J+1)$.

For the AgRE alloys a good agreement (figure 5) is obtained with

$$a = 0.43g(2-g)J(J+1) - 1.57g(g-1)J(J+1) \qquad (11)$$

$$(a \text{ in } 10^{-8}\ {}^{o}K/g)$$

The competition between the terms a_1 and a_2 explains that a, maximum for Gd, decreases more rapidly than $g(g-1)J(J+1)$ for Tb, Dy, etc., and changes sign for Er.One obtains also the good sign of a for the light RE. However the quantitati- ve agreement is better for the heavy RE.

For the AuRE alloys, the fit of figure 6 is obtained with :

$$a = 0.34g(2-g)J(J-1) - 2.26g(g-1)J(J+1) \qquad (12)$$

$$(a \text{ in } 10^{-8}\ {}^{o}K/g)$$

A term a_2 larger than in the AgRE alloys explains that a is larger in Au and does not changes sign in the heavy RE series.

For the AlRE alloys the fit on figure (7) is obtai- ned with :

$$a = 0.32g(2-g)J(J+1) - 0.63g(g-1)J(J+1) \qquad (13)$$

$$(a \text{ in } 10^{-8}\ {}^{o}K/g)$$

We have identified the term in $g(2-g)J(J+1)$ with (9). Taking, for the AgRE and AuRE alloys, $Z_s = 1$, $Z_p = 0$, $Z_d = 1$ (this gives $\eta_o = \pi/2$, $\eta_2 = \pi/10$ and a residual resistivity $\rho_o = 6.42$ µΩcm/at.% in agreement with the experimental values) $n^o(E_F) = 0.15\ ev^{-1}$, one obtains

$$F_2 = 2\ 10^{-3} ev \text{ in Ag},\quad F_2 = 1.6\ 10^{-3} ev \text{ in Au}.$$

For the AlRE, F_2 appears to be of the same order of magnitude.

Thus we find that the coupling between the orbital angu- lar momentum of the conduction and f electrons is rather small, about 50 times smaller than the constant of the spin-spin cou- pling $\Gamma^{(o)}$. It turns out, for example, that the contribution of such a coupling to the relaxation of RE impurities in EPR should be negligible.

The magnitude of the term a_2 corresponds to reasonable values of the several parameters (λ = 0.1 ev, Δ = 0.25 ev(17), $Z_s = Z_p = 1$, $\Gamma^{(3)} - \Gamma^{(1)} = 0.06$ ev for the AgRE alloys).We have found that the term a_2 is smaller in Ag than in Au and much smaller in Al. This can be explained in terms of a 5d vbs wider in Ag than in Au and much wider in Al, in agreement with the de- creasing magnetic character of transition impurities from Au to Al based alloys.

III QUADRUPOLE SCATTERING

1 - Theory : We consider in the interaction between conduc- tion electrons and rare earth impurities in noble metals the the terms :

$$H'' = \sum_{\vec{k}\vec{k}'} N^{-1}\Big\{4\pi\sum_{\ell\leq 2,m} V_\ell\,Y^*_{\ell,m}(\Omega_{\vec{k}})\,Y_{\ell,m}(\Omega_{\vec{k}'}) - \frac{D}{k_F^2}\Big[(\vec{J}.\vec{k})(\vec{J}.\vec{k}') - J(\frac{J+1}{3})\vec{k}.\vec{k}'\Big]\Big\}\,a^+_{\vec{k}'}a_{\vec{k}} \qquad (13)$$

The first term is the isotropic and spin independent part of the potential (screening effect). The second term corresponds to the interaction with the quadrupolar distri- bution of charge in the 4f shell (cf. expression (3)) and scatters differently electrons with wave vector parallel or perpendicular to $\vec{J}$. It can be shown easily that, for RE im- purities in noble metal ($V_\ell >> \Gamma$),these two terms must provide the main contribution to the resistivity anisotropy. We have not included in the interaction the exchange cou- pling and the other terms of expression (3). The exchange coupling induces a well-known negative magnetoresistivity which is isotropic (that is independent of the relative direction of the current and of the field) and adds to the anisotropic magnetoresistivity calculated here.

The impurity resistivity for a current in the direc- tion of the unit vector $\vec{u}$ is :

$$\rho_{\vec{u}} = \frac{1}{2}\Big(\frac{\hbar}{8\pi^3 ne}\Big)^2\int\Big(-\frac{\delta f^o}{\delta\varepsilon_k}\Big)(\vec{k}.\vec{u})\big((\vec{k}-\vec{k}').\vec{u}\big)W_{\vec{k}\vec{k}'}\,d^3\vec{k}\,d^3\vec{k}' \qquad (14)$$

with $W_{\vec{k}\vec{k}'} = \frac{2\pi}{\hbar}\,|\langle\vec{k}|H''|\vec{k}'\rangle|^2\,\delta(\varepsilon_{k'} - \varepsilon_k)$

to first Born approximation (n is the number of conduction electrons with a given spin direction per unit volume). A straightforward calculation provides the resistivity $\rho^Q_{\parallel}$ and $\rho^Q_{\perp}$ induced by the quadrupole for a current parallel or perpendicular to the direction Oz of the field. In the case of potential terms V_ℓ much larger than D, one only retains in $|<\vec{k}\ |\ H''\ |\ \vec{k}'>|^2$ the cross terms in $V_\ell D$ and one obtains for $\rho^Q_{/\!/}$, $\rho^Q_\perp$ and for the resulting anisotropy of the impurity resistivity, the following expressions :

$$\rho^Q_{/\!/} = -2\,\rho^Q_\perp = \frac{2D(V_o + \frac{1}{5}V_2)}{3(V_o^2 + 5V_2^2)}\left(< J_z^2 > - \frac{J(J+1)}{3}\right)\rho_o \qquad (15)$$

$$\frac{\rho_{/\!/}(H) - \rho_\perp(H)}{\rho_o} = \frac{D(V_o + \frac{1}{5}V_2)}{V_o^2 + 5V_2^2}\left(< J_z^2 > - \frac{J(J+1)}{3}\right) \qquad (16)$$

where $\rho_o = \dfrac{2\pi\, m\, n(E_F)\, c}{z\, e^2\, \hbar}\,(V_o^2 + 5\,V_2^2)$ is the impurity resistivity at zero field (V_1 does not contribute to $\rho^Q_{/\!/}$ and $\rho^Q_\perp$ and has been neglected in ρ_o, z is the number of electrons per atom, $< J_z^2 >$' is the mean value of J_z^2). One actually knows that large values of the phase shifts η_o and η_2 ($2\eta_o + 10\eta_2 \sim 2\pi$) do not justify the calculation in first Born approximation. A more rigourous calculation provides :

$$\frac{\rho_{/\!/}(H) - \rho_\perp(H)}{\rho_o} = -\frac{\pi\, n(E_F)\, D(\sin 2\eta_o + \frac{1}{5}\sin 2\eta_2)}{2(\sin^2\eta_o + 5\sin^2\eta_2)} \times \left(< J_z^2 > - \frac{J(J+1)}{3}\right) \qquad (17)$$

At low field ($gJ\mu_B H << kT$) and without crystal field or interactions (17) becomes :

$$\frac{\rho_{/\!/}(H) - \rho_\perp(H)}{\rho_o} = -(J - \tfrac{1}{2})(2J^2 + 5J + 3)\,\frac{\pi\, n(E_F)\, D(\sin 2\eta_o + \frac{1}{5}\sin 2\eta_2)}{90(\sin^2\eta_o + 5\sin^2\eta_2)}\left(\frac{gJ\mu_B H}{kT}\right)^2 \qquad (18)$$

At low temperature the anisotropy should however be greatly dependent on the crystal field structure, as we shall see below.

2) Experimental results : We have studied the magnetoresistance of series of $Au_{1-x}RE_x$ and $Ag_{1-x}RE_x$ alloys (RE = Gd, Tb, Dy, Ho, Er, Tm, Yb, $5.10^{-3} < x < 2.10^{-2}$). The metallurgic and experimental problems have been briefly discussed in section II.

Figure 8a shows the magnetoresistance of a AuHo alloy for a field parallel or perpendicular to the current. It appears first that there is a clear magnetoresistance linked with the polarization of the impurities (and so vanishing at high temperature) and secondly that the curves show up an effect of resistivity anisotropy much stronger than the effect of isotropic negative magnetoresistance ($\Delta\rho = \rho_{/\!/} - \rho_\perp$ is much stronger than $\delta\rho = \rho_o - \frac{1}{3}(\rho_{/\!/} + 2\rho_\perp)$).

The magnetoresistance of AuGd is quite different. Figure 8b shows that $\rho_{/\!/}(H)$ and $\rho_\perp(H)$ coincide very nearly. One observes only the usual isotropic negative magnetoresistance of a dilute magnetic alloy. This is obviously related to the spherical symmetry of a S state

Anisotropy of the impurity resistivity is observed for all the other alloys (excepted AgGd and AuGd). According to the impurity type one observes, $\rho_{/\!/} > \rho_\perp$ (like for AuHo) or $\rho_{/\!/} < \rho_\perp$.

Table 1 compares the sign of ($\rho_{/\!/} - \rho_\perp$) throughout the heavy RE series to the sign of D predicted by Kondo (5). We see that the sign of the anisotropy is always the sign of D, the sign of ($\rho_{/\!/} - \rho_\perp$) and D being positive at the beginning of the series and negative for Er, Tm, Yb. This result confirms that the resistivity anisotropy of the RE impurities is due to the interaction with the quadrupolar moment of the 4f shell.

We shall now consider the effect of the crystal field splitting on the resistivity anisotropy and we shall discuss the different behaviors according to the degeneracy of the ground state.

For a free RE ion, the expressions (17) and (18) predicts that ($\rho_{/\!/} - \rho_\perp$) is proportional to H^2T^{-2} at low field ($g\,\mu_B H << kT$) and then becomes saturated for $g\,\mu_B H \gtrsim kT$. The same behavior can be predicted if the ground state is a triplet or a quartet. This type of behavior is approximately observed for AuHo on figure 8(a): the anisotropy is almost saturated at 1.2°K and 40 kG. This can be explained in terms of a $\Gamma_5^{(2)}$ triplet just few tenths of degrees above the non magnetic ground state and so populated, even at 1.2°K.

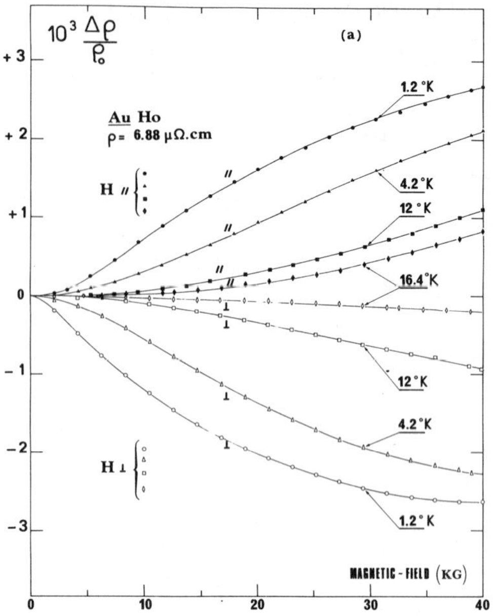

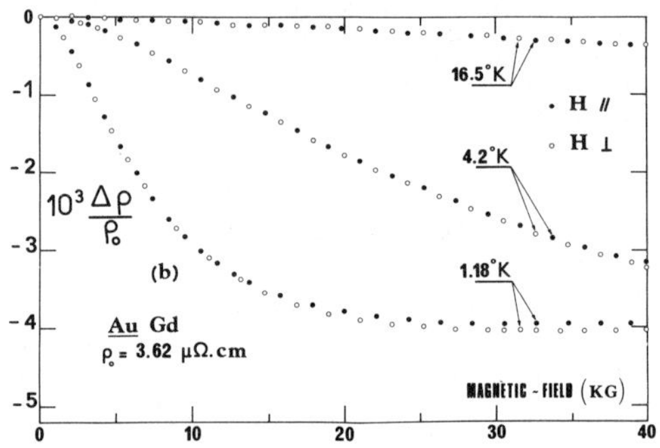

Figure 8 : The fractional longitudinal and transverse magnetoresistance of a 1 at. % AuHo alloy (a) and of a 0.5 at. % AuGd alloy (b) at several temperatures. The different behaviour of the magnetoresistance for a non-S state (Ho) and a S state (Gd) is obvious : for AuHo the main effect appears to result from the anisotropy of the impurity resistivity ($\Delta\rho_{/\!/} > 0$ and $\Delta\rho_\perp < 0$) while for AuGd one observes only an isotropic negative magnetoresistance. Note also that for both of the alloys the normal magnetoresistance (**independent of the temperature**) is small in comparison to the magnetoresistance linked with the impurity magnetization.

RE	$\rho_{/\!/} - \rho_\perp$		D
	AgRE	AuRE	
Gd	0	0	0
Tb	+	+	+
Dy	+	+	+
Ho	+	+	+
Er	—	—	—
Tm	—	—	—
Yb	non magnetic	—	—

Table I : Sign of $\rho_{/\!/} - \rho_\perp$ for RE impurities in Ag and Au and sign of the constant D of the coupling with the quadrupole (D from expression 2.45 and column C_1 in table I of a Kondo paper (5) ; we have assumed $10\,R_1 > F_3$; if not, all the signs in the column D should be changed).

If the ground state is a Kramers doublet, the rise of ($<J_z^2> - J(J+1)/3$) cannot be induced by the polarization of the doublet but can only be due to the mixing with excited states by the magnetic field. It can be shown easily that, at very low field ($\mu_B H << kT << \Delta$, where Δ is the separation from the excited states) the anisotropy $\Delta\rho$ is proportional to the mixing and also to the polarization of the doublet and so to :

$$\frac{\mu_B H}{kT} \times \frac{\mu_B H}{\Delta}$$

At a higher field ($\mu_B H \gtrsim kT$ but $\mu_B H << \Delta$) the doublet is almost completely polarized and one obtains $\Delta\rho \sim \mu_B H/\Delta$. Finally the saturation of the anisotropy can be only reached at high field: $\mu_B H \simeq \Delta$. This behavior is characterized chiefly by the replacement of the saturation for $\mu_B H \simeq kT$ ($H \simeq 20$ kG at 1.2°K) by an increase linear in H and independent of the temperature. This behavior is nearly observed for AgEr on figure 9: in the field and temperature range (say: 20-40kG, 1.2-4.2°K) in which the doublet ground state is expected to be almost polarized, ($\rho_\perp - \rho_\parallel$) is nearly linear in H and independent of T (the slight deviation from a linear variation above 30 kG seems to indicate a small separation from the excited states).

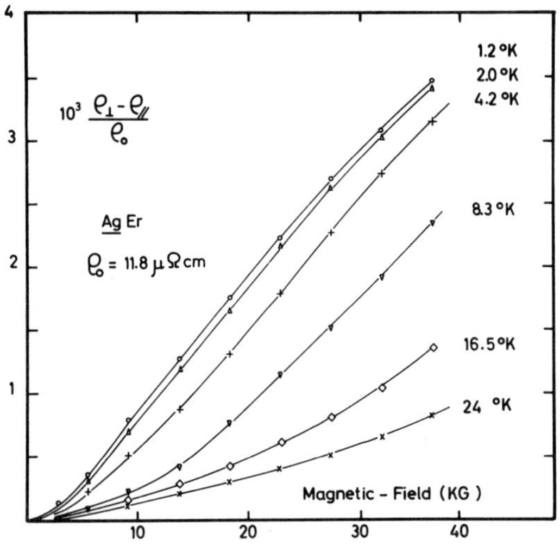

Figure 9 : The resistivity anisotropy of a AgEr alloys vs the magnetic field at several temperatures.

For a singlet ground state one finds that the anisotropy should be proportional to $H^2\Delta^{-2}$ if $\mu_B H << \Delta$ and $kT << \Delta$ (Δ = separation from the excited states). This sort of behavior is observed for AuTm alloys.

We have presented a very qualitative description of the CEF effects. An accurate analysis of the CEF effects on both the resistivity anisotropy and the skew scattering Hall resistivity ($\sim <J_z>$) is in progress and should also yield an accurate value of the coupling constant D for each impurity. A rough analysis of the data indicates that D is surprisingly large :

$|D| J^2 \simeq 0.05 \rightarrow 0.1$ eV ($DJ^2 \simeq$ energy of the interaction with the quadrupole).

This energy is much smaller than the exchange energy at the beginning of the heavy RE series ($E_{ex} \simeq \Gamma(g-1)J/2 \sim 0.3$ ev for Tb if $\Gamma \simeq 0.2$ ev) but not for elements at the end of the series ($E_{ex} \simeq \Gamma(g-1)J/2 \simeq 0.1$ ev for Tm). It seems that the strength of this electron-quadrupole interaction (and also of the resulting quadrupole-quadrupole interaction) has been underestimated up to now.

REFERENCES

(1) J.Smit, Physica 24, 39 (1958)
(2) A.Fert and O.Jaoul, Phys.Rev.letters 28,303 (1972)
Giovannini B, Phys Letters 42 A, 256 (1972)
Giovannini B, Low Temperature Phys.11 489 (1973)
(3) A.Fert and A.Friederich and J.Sierro, Solid State Commun 13, 997 (1973)
(4) A.Fert and A.Friederich, to be published.
(5) J.Kondo, Prog.Theor.Phys.27, 772 (1962)
(6) Preliminary results in A.Friederich and A.Fert, Phys. Rev. Letters 33, 1214 (1974)
(7) A.Fert, J.Phys F 3, 2126 (1973)
(8) The problem is more complicated for inelastic collisions. See B.Giovannini and G.C.Cohen, to be published.
(9) J.E.A. Alderson and C.M.Hurd, Can J.Phys.48, 2162 (1970)
(10) M.T. Beal-Monod and R.A.Weiner, Phys.Rev 185, 897 (1969)
(11) L.Berger, Phys.Rev. B 2, 4559 (1970)
(12) A.Fert, J.Phys Lettres (Paris) 35, L 107 (1974)
(13) G.Williams and LL.Hirst, Phys.Rev.186, 625 (1969)
(14) A.P. Murani, J.Phys. C 2, S 153 (1970)
(15) excepted for Au Yb
(16) C. Rettori, D.Davidov, R.Orbach, E.P.Chock and B.Ricks Solid State Communications 12, 621 (1973)
(17) A.B. Callender and S.E. Schnatterly, Phys.Rev.B 7, 4385 (1973)
(18) T.Kasuya and D.H.Lyons, J.Phys.Soc.Japan 21 287 (1966)
P.M.Levy, Solid State Commun.7, 1813 (1969)

BGS RELAXATION IN THE DILUTE ALLOYS La:Ln

D. S. SCHREIBER and R. F. WADE

Physics Department, University of Illinois, Chicago, Il. 60680

ABSTRACT

The enhanced relaxation rates $1/T_1$ of La^{139} in dilute alloys La:Ln, Ln = Ce, Pr, Nd, Gd, or Lu have been measured and the electronic moment life times have been deduced.

INTRODUCTION

The relaxation data to be presented in this paper spans a number of impurity ions, wide ranges of concentration, temperatures from 1.4 to 4.2°K, magnetic fields from 15.6 to 50 koe, as well as two different host lanthanum crystal structures. The dependences of the relaxation rate enhancements on each of these variables will be exploited to identify the responsible mechanism and to derive information about the dynamics of the impurity spin. These dependences, which follow from the general treatment of Giovanini et al. (ref. 1) are explicitly shown here.

$$T_1^{-1}|_{GH} \propto T_1^{-1}|_{Korr} C \frac{<S_z>}{H}, \qquad (1)$$

$$T_1^{-1}|_{LD} \propto C(<S_z^2>-<S_z>^2) \frac{\tau_1}{1+\omega_n^2\tau_1^2} \qquad (2)$$

$$T_1^{-1}|_{BGS} \propto T_1^{-1}|_{Korr} C \frac{<S_z>}{H} \frac{\tau_2}{1+\omega_m^2\tau_2^2} \qquad (3)$$

where the τ's and the $<S_z>$ refer to the electronic life times and spins, and C is the impurity concentration. The GH, BGS, and LD are the three most likely dominant La^{139} relaxation mechanisms as discussed in ref. 1. The temperature and field dependences of $<S_z>$ and of $(<S_z^2>-<S_z>^2)$ are calculable from the crystal field levels and splittings deduced from the magnetization data (ref. 2) determined for the same alloys. It is known that $T_1^{-1}|Korr \propto T$ and is independent of the applied magnetic field. Lastly the dependences of τ_1 and τ_2 on C, T, and H are not known a priori, although only a limited number of possibilities for these seem likely. In this analysis, certain assumptions are made about τ_1 and τ_2. These are then subjected to various self-consistency tests.

DISCUSSION

The least complicated dependence to examine is that on impurity concnetration. From Eq. 1, the GH mechanism is seen to predict a simple linear concentration dependence in the enhancement for the spin lattice relaxation rate. The BGS and LD mechanisms also predict a linear dependence if τ_2 and τ_1 respectively are independent of the impurity level. Measurements are made of the concentration dependence at 4.2°K for various impurities, applied fields, and crystal structures. In each case the impurity enhancements of the relaxation rate, given by

$$T_1^{-1}|_{imp} = T_1^{-1}|_{meas.} - T_1^{-1}|_{Korr} \quad (4)$$

is plotted against the impurity concentration in atomic per cent. The most obvious feature of the concentration graphs is their initial linearity. Especially striking are the cases of Ce and Pr where linearity persists to at least 10 at %. This is significant because it indicates that τ_1 or τ_2 must be concentration independent if the LD or BGS mechanism respectively is dominant. Also of interest here, are the slopes of the graphs, which are given in Table I.

Impurity	H (koe)	Structure	$\frac{d}{dc}(T_1^{-1}\|_{imp})$ (sec at%)$^{-1}$	
Ce	15.6	dhcp	136	±5%
Ce	15.6	fcc	308	5%
Ce	50	fcc	140	5%
Pr	15.6	fcc	19.5	5%
Nd	15.6	fcc	1320	5%
Nd	50	fcc	600	5%
Gd	15.6	dhcp	2580	5%

Table I Normalized Relaxation Rate Enhancements

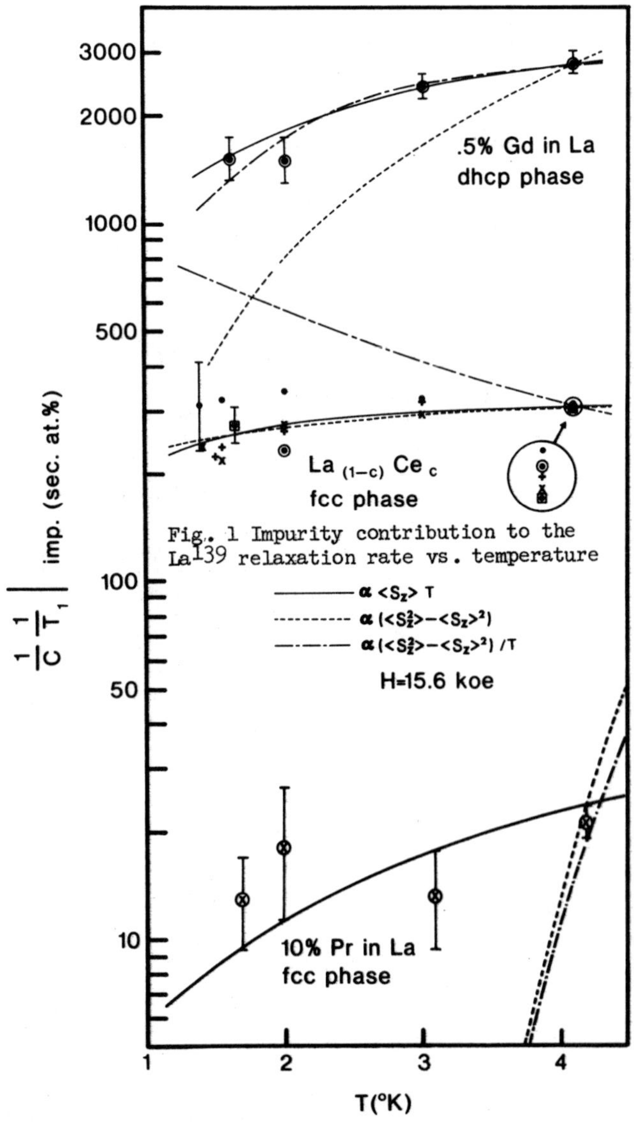

Fig. 1 Impurity contribution to the La^{139} relaxation rate vs. temperature

The temperature dependence data for the La:Ln alloys is summarized in Fig. 1. The solid curve, which is the functional form of $\langle S_z \rangle T$ gives the temperature dependence as predicted by the GH mechanism in Eq. 1. It also represents the BGS mechanism if $\tau_2 \neq f(T)$. This can be seen in Eq. 2. In both of these cases use has been made of the fact that $T_1^{-1}|_{Kor} \propto T$. The dashed curve gives the temperature dependence predicted by the LD mechanism if $\tau_1 \neq f(T)$. An examination of Eq. 3 also reveals that if $\omega_n \tau_1 << 1$ and if τ_1 is Korringa-like, then the LD mechanism predicts that $T_1^{-1}|_{imp} \propto (\langle S_z^2 \rangle - \langle S_z \rangle^2)/T$. This functional form is given by the broken line. The assumptions made here about τ_1 and τ_2 are the simplest possible and are intended to serve as a starting point. In Fig. 1, the relaxation rate enhancement due to 10% praseodymium in lanthanum is shown. The startling form of the temperature dependence predicted by the (LD) dipolar mechanism is a result of the singlet ground state of Pr. Because these temperatures are low compared to the splitting between the ground and first excited crystal field levels, only the ground state is significantly populated. As a result, $\langle S_z \rangle$ is temperature independent in this range, being equal to S_{zo} the induced ground state value. For the same reason, the average of the square becomes identical to the square of the average of S_z, hence, the rapid drop in their difference. It is most interesting to note that the indirect mechanisms (GH and BGS) correctly predict the temperature dependence in all cases. It has been presumed throughout that $\tau_2 \neq f(T)$ in the BGS equation. Supporting evidence for the validity of this claim, over the temperature range of interest here, will be given in the following section. The dipolar mechanism has not demonstrated any consistency in predicting the correct temperature dependences.

The method to be employed to test the predicted field dependence of the longitudinal dipolar mechanism is quite similar to the temperature test. Eq. 3 can be rewritten as

$$C(\langle S_z^2 \rangle - \langle S_z \rangle^2) T_1|_{imp} \propto \frac{1}{\tau_1} + \gamma_n^2 \tau_1 H^2 \qquad (5)$$

Here, if the LD mechanism is dominant a graph of $c(\langle S_z^2 \rangle - \langle S_z \rangle^2)T_1|_{imp}$ vs. H^2 should be linear, assuming τ_1 is field independent. Furthermore, if the graph is linear, τ_1 can be determined: $\tau_1 = (\text{Slope/Intercept})^{1/2}/\gamma_n$. As an example, when we make a plot for La-5%Ce, we find the best straight line falls just outside each of the error bars. In addition, this line has a negative slope which would predict an imaginary τ_1. In light of the many difficulties in fitting the temperature and field dependences with the LD mechanism does not dominate the relaxation enhancement.

Turning to the indirect mechanisms, it is first noted from Eq. 1, that the GH rate is proportional to $\langle S_z \rangle/H$. For each kind of impurity, the relaxation rate has been plotted vs. applied field. As an example we show this for La-5%Ce in Fig. 2. Also shown is the functional form of $\langle S_z \rangle/H$. These have been normalized to the 15.6 koe experimental points. Clearly the theoretical field dependence is too weak.

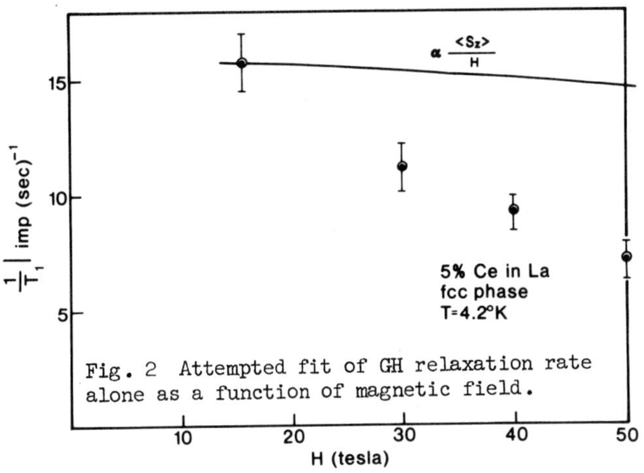

Fig. 2 Attempted fit of GH relaxation rate alone as a function of magnetic field.

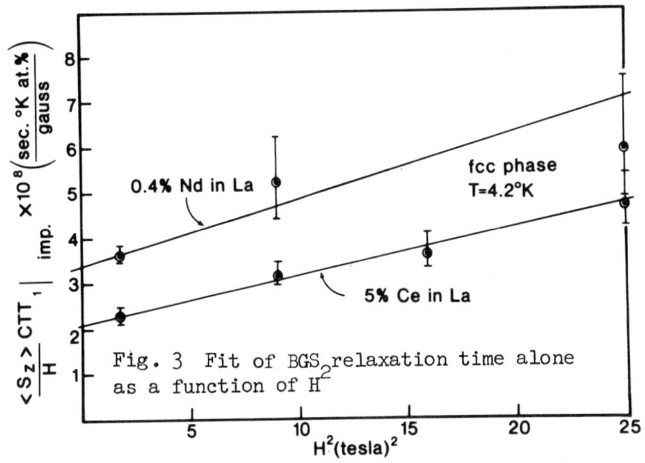

Fig. 3 Fit of BGS_2relaxation time alone as a function of H^2

To verify the BGS field dependence, a procedure similar to that used for LD is employed. Eq. 2 is rewritten as

$$\frac{C\langle S_z \rangle T}{H} T_1|_{imp} \propto \frac{1}{\tau_2} + \frac{\omega_m}{H}^2 \tau_2 H^2 \qquad (6)$$

again recalling that $T_1^{-1}|_{Korr} \propto T$. It is important to note that ω_m/H is field independent. Here ω_m is the impurity spin Larmour frequency. For BGS dominated relaxation enhancement, a graph of $[(C\langle S_z \rangle T)/H]T_1|_{imp}$ vs H^2 should be linear. From the best straight line:

$$\tau_2 = (H/\omega_m)(\text{Slope/Intercept})^{1/2} \qquad (7)$$

The assumption here is that $\tau_2 \neq f(H)$, which is reasonable. Graphs of this nature have been constructed, and for 5%Ce in La, Fig. 3 yields at 4.2°K $\tau_2 = (1.8 \pm .3) \times 10^{-12}$ sec. At 2°K a similar graph yields $\tau_2 = (1.7 \pm .2) \times 10^{-12}$ sec. This supports the assumption made in fitting the temperature dependence that $\tau_2 \neq f(T)$ over this range of temperature. Lastly, a 0.4% Nd in La alloy at 4.2°K is represented in Fig. 3. Again we can fit the graph with a straight line. In this case, $\tau_2 = (0.9 \pm .2) \times 10^{-12}$ sec. For both Ce and Nd, ω_m was taken to be the frequency characteristic of an electronic transition between levels of the magnetically split ground crystal field doublet.

The τ's quoted here fall within the typical range of theoretical and experimental electronic relaxation times (10^{-13} to 10^{-7} sec). The magnitudes indicate electronic linewidths on the order of 10 koe. Because of this great breadth, it is not surprising that reports of EPR studies in these alloy systems have not been found. This type of NMR experiment may provide the only measure of τ_2 for such materials.

REFERENCES

1. B. Giovannini, P. Pincus, G. Gladstone, and A. Heeger, J. de Physique 32, CL-163 1971 and references cited therein.
2. R. F. Wade, Ph.D. Thesis, Univ. of Ill. Sept. 1974 and R. F. Wade and D. S. Schreiber, to be published.

MAGNETIZATION OF Au-Gd AND Au-Yb AT VERY LOW TEMPERATURES

E. Jaehne and O. G. Symko
Dept. of Physics, Univ. of Utah, Salt Lake City, Utah 84112

ABSTRACT

The behavior of a rare-earth impurity in a metallic matrix is dominated by a well-localized 4f shell which interacts with its environment and in particular experiences the crystalline electric field of the host lattice. To investigate the interactions of such impurities with their environment, the magnetization of single crystals of very dilute Au-Gd and Au-Yb was measured from approximately 1K down to 0.01K in fields of 1 Oe and 10 Oe. The measurements were made in a ^3He-^4He dilution refrigerator using a SQUID magnetometer. For the Au-Gd crystal, the magnetization starts to deviate from a Curie law below 0.05K. Such behaviour is attributed to fine structure splitting of the localized moment. In Au-Yb, where because of the low temperatures only the splitting of the Γ_7 ground state doublet of Yb^{+3} is observed, the magnetization departs slightly from a Curie-law below 0.05K. The hyperfine interaction in the isotopes with nuclear spins 1/2 and 5/2 can cause such deviation. For both systems the results are compared with published EPR data.

ELECTRICAL RESISTIVITY OF $(Pd_{1-x}Ag_x)_{.99}Fe_{.01}$

K. V. Rao,[+] O. Rapp and Ch. Johannsson
Royal Institute of Technology, Stockholm, Sweden

J. I. Budnick and T. J. Burch
University of Connecticut,* Storrs, Conn. 06268

V. Cannella
Wayne State University, Detroit, Michigan 48202

ABSTRACT

Measurements of the temperature dependence of the resistivity, $\rho(T)$, have been made for the alloy system $Pd_{1-x}Ag_x$ containing 1 at.% Fe with x = 0.025, 0.10, 0.25, 0.33, 0.40 and 0.50. Temperature dependences of the low field susceptibility, $\chi(T)$, have shown that alloys with $0 \leq x \leq 0.25$ are ferromagnetic and that alloys with $x \geq 0.33$ have sharp susceptibility peaks thought to be characteristic of spin glass type ordering.[1] The $Pd_{1-x}Ag_x$ matrix resistivity increases with silver concentration up to about x = 0.40 and then decreases.[2,3] In the composition range $0 \leq x \leq 0.25$ addition of 1 at.% Fe increases the resistivity between 3 and 5 μΩcm. This is 3 to 4 times the resistivity increase per cent Ag in the same composition range. $\rho(T)$ for $Pd_{1-x}Ag_x$ has been found to approach $\rho(0)$ with positive or negative T^2 dependences depending on concentration. The negative slopes occur for alloys with x = 0.19, 0.33 and 0.50.[3] No negative slopes were observed in $\rho(T)$ for $(Pd_{1-x}Ag_x)_{.99}Fe_{.01}$ alloys. In $(Pd_{1-x}Ag_x)_{.99}Fe_{.01}$ with $x \geq 0.33$, alloys for which $\chi(T)$ shows sharp peaks and greatly reduced values, the resistivity data do not show any marked reduction in spin disorder scattering but do show possible weak effects at the temperatures of the sharp susceptibility peaks. Further experiments are in progress to carefully separate the various contributions to $\rho(T)$ in this region.

+Presently at Clarkson College of Technology.

*Work supported in part by the University of Connecticut Research Foundation.

1. J. I. Budnick, V. Cannella and T. J. Burch, AIP Conf. Proc. 18, 307 (1974).
2. L. R. Edwards, C. W. Chen and S. Legvold, Solid State Comm. 8, 1403 (1970).
3. A. P. Murani, Phys. Rev. Lett. 33, 91 (1974).

MAGNETIC SUSCEPTIBILITY STUDY OF THE CONTINUOUS DEMAGNETIZATION OF Ce IMPURITIES IN THE (La, Th)Ce SYSTEM*

J. G. Huber, J. Brooks, D. Wohlleben and M. B. Maple
Institute for Pure and Applied Physical Sciences
University of California, San Diego, La Jolla, California 92037

ABSTRACT

Measurements of the magnetic susceptibility as a function of temperature are reported for the (La,Th) Ce system which document directly the continuous demagnetization of Ce impurities which proceeds with increasing Th composition.

Ce impurities can undergo a continuous magnetic-nonmagnetic transition when the matrix into which they have been dissolved is subjected to an external pressure or alloyed with another element.[1] This effect was first observed during an experimental investigation of the pressure dependence of the superconducting transition temperature T_c of LaCe and related systems.[2] Subsequently, the effect of pressure on the low temperature resistivity anomaly $\rho_m(T)$ due to the Kondo effect in the LaCe system was independently studied by two groups.[3, 4] Unfortunately, experimental difficulties attendant to measurements at high pressure have so far prevented the acquisition of other data on this interesting system.

It was later found that the continuous demagnetization of Ce impurities could also be induced by alloying the La matrix with increasing amounts of Th.[5] This afforded the opportunity to measure various important physical parameters related to the demagnetization of Ce impurities in a matrix at zero applied pressure. Since then, extensive data have been accumulated for the depression of T_c with Ce concentration n,[6] the depression of the specific heat jump ΔC at T_c with T_c,[7] and $\rho_m(T)$[8] as a function of La, Th host composition. In this paper we present magnetic susceptibility data for the (La, Th)Ce system which document the continuous demagnetization of the Ce ions directly.

(La, Th)Ce alloys with La, Th host compositions of 10, 45, 65, 80, 90 and 100 at.% Th were fabricated from high purity elements by arc-melting in an argon atmosphere. Wrapped successively in foils of Ta, Zr, and Ta, the arc-melted ingots were annealed in evacuated quartz tubes for 10 days at 800-1000°C, depending on the composition of the La, Th host, and then "slow-cooled" (12-24 hrs) to room temperature. As demonstrated in a recent heat capacity study,[7] this procedure is effective in insuring fcc phase purity in the La, Th composition range from 10 to 100 at.% Th. Magnetization measurements were made in a Faraday magnetometer between 0.5 and 300°K in fields up to 11 kOe.

The magnetic susceptibility χ contributed by the Ce impurities was determined by correcting the susceptibility of each (La, Th)Ce alloy for the background susceptibility of the corresponding La, Th host. Plots of χ^{-1} vs T are shown in Fig. 1 for Ce in the six La, Th hosts. Alloys with two different Ce impurity concentrations were measured to determine whether χ scaled with Ce concentration for all but the $La_{0.90}Th_{0.10}$ and Th based systems where only one Ce impurity concentration was examined. Deviations of χ from linearity in Ce concentration were negligible for the $La_{0.55}Th_{0.45}$ and $La_{0.35}Th_{0.65}$ based alloys, and small, but noticeable, for the $La_{0.20}Th_{0.80}$ and $La_{0.10}Th_{0.90}$ based alloys. This can be seen in the χ vs T plots for the $La_{0.20}Th_{0.80}$, $La_{0.10}Th_{0.90}$ and Th based systems in Fig. 2. Since the overall temperature dependence of the susceptibility does not change significantly with Ce concentration, we believe that the general features are representative of single impurity behavior.

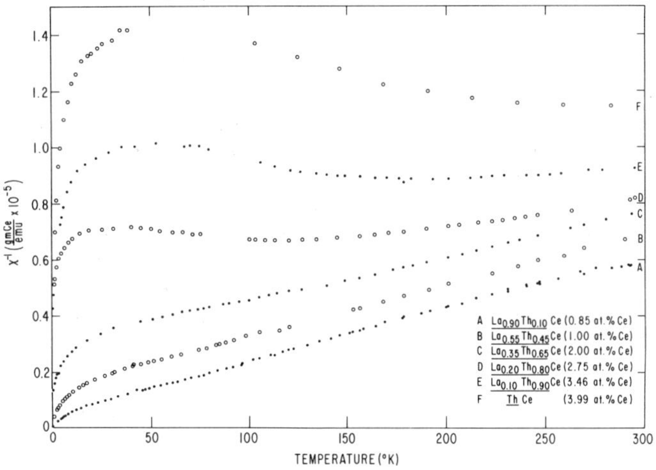

Fig. 1. Plots of χ^{-1} vs T for (La, Th)Ce alloys with La, Th matrix compositions of 10, 45, 65, 80, 90 and 100 at.% Th.

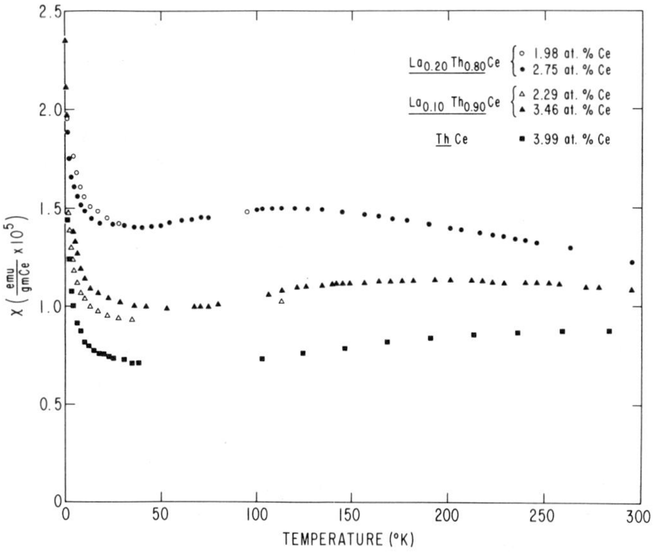

Fig. 2. Plots of χ vs T for (La, Th)Ce alloys with La, Th matrix compositions of 80, 90 and 100 at.% Th.

The data of Fig. 1 show that the Ce solute susceptibility evolves smoothly from strongly temperature dependent behavior to weakly temperature dependent behavior as the La, Th host composition is varied toward 100 at.% Th. This documents directly the continuous demagnetization of Ce ions with increasing Th concentration which was previously inferred from measurements of T_c vs n[5,6] and $\Delta C/\Delta C_o$ vs T_c/T_{c_o}[7] (ΔC_o and T_{c_o} refer to the matrix) in the superconducting state and $\rho_m(T)$[8] in the normal state. We note that the χ^{-1} vs T curves in Fig. 1 for the $La_{0.90}Th_{0.10}$, $La_{0.55}Th_{0.45}$ and $La_{0.35}Th_{0.65}$ based systems can be

Table I

Effective magnetic moments μ_{eff}, Curie-Weiss temperatures θ_p, and Kondo temperatures T_K (from ref. 7) of (La, Th)Ce alloys.

	$\mu_{eff}(\mu_B)$	θ_p(°K)	T_K(°K)
$La_{0.90}Th_{0.10}Ce$	2.42	-24	38
$La_{0.55}Th_{0.45}Ce$	2.48	-76	130
$La_{0.35}Th_{0.65}Ce$	2.56	-152	780

described phenomenologically by a Curie-Weiss law over an appreciable temperature range. Table I shows that the effective moment per Ce ion remains roughly constant, while the magnitude of the negative Curie-Weiss temperature θ_p increases rapidly with Th concentration. If $|\theta_p|$ is interpreted as a measure of the Kondo temperature T_K (i.e. $|\theta_p| \sim 3\ T_K$), this implies that T_K increases rapidly with increasing Th concentration; a trend which also emerged from an analysis of T_c vs n and $\Delta C/\Delta C_o$ vs T_c/T_{c_o} measurements on the same alloys.[6,7] However, comparison of the T_K values determined from the superconductivity studies[7] with the Curie-Weiss temperatures given in Table I shows large quantitative discrepancies. Apart from the Kondo effect, one needs to take into account the crystal field splitting of the Ce^{3+} J = 5/2 multiplet into a ground state doublet and an excited state quartet. Including both Kondo and crystal field effects, DeGennaro and Borchi[9] and Harris and Zuckermann[10] have calculated the susceptibility of LaCe[11] as a function of temperature, achieving a good description of the experimental results for reasonable values of the relevant parameters ($\Delta \sim 100$°K and $|\mathcal{J}| \sim 0.1$ eV, where $\mathcal{J}$ is the conduction electron-impurity spin exchange interaction parameter). Comparing our susceptibility measurements with the χ(T) function derived by Harris and Zuckermann, we find that the evolution of the χ^{-1} vs T curves of Fig. 1 with La, Th host composition up to 65 at.% Th is also consistent with an increase of T_K with Th concentration.

In the La, Th composition range above 65 at.% Th, the χ vs T curves take on a different aspect, exhibiting a maximum and then decreasing as the temperature is lowered before showing a low temperature upturn (Fig. 2). These low temperature tails are probably due to Ce ions which retain long-lived moments in a few statistically allowed clusters of La near neighbor atoms. Disregarding the low temperature tails, the overall weakly temperature dependent behavior of χ for these Th-rich (La, Th)Ce alloys is reminiscent of a recently recognized class of rare earth metallic alloys and compounds whose nonintegral valence and weakly magnetic behavior have been attributed to temporal fluctuations of the rare earth shells between two integral valence states.[12] The resultant magnetic moment fluctuations give rise to nonmagnetic behavior (i.e., saturation of χ to a constant value) below a characteristic temperature $T_o = \hbar/k_B\tau$ where τ is the mean magnetic moment lifetime. The decrease of the susceptibility with increasing Th concentration implies a corresponding increase of T_o, while the decrease of the susceptibility with decreasing temperature suggests a continuous depopulation of the 4f shell as the temperature is lowered. The existence of such valence fluctuations in Th-rich (La, Th)Ce alloys is further supported by the nonintegral valence (about 3.25) of Ce impurities in pure Th.[13] Again, these features are consistent with the superconducting properties of (La, Th)Ce alloys with La, Th host compositions above ~70 at.% Th which cannot be accounted for in terms of models based on exchange scattering of conduction electrons by well-defined impurity moments.[6,7] On the other hand, the superconducting properties of the ThCe system itself can be well represented by theories which consider the effect on superconductivity of nonmagnetic resonant impurity states.[14]

In conclusion, the results presented here document the behavior of the magnetic susceptibility during a magnetic-nonmagnetic transition of an impurity in a matrix. Thus (La, Th)Ce would appear to be a model system for testing theories of local moment formation in metals. It is apparent that a proper theory which spans the entire range between magnetic and nonmagnetic solute behavior must incorporate a variable solute magnetic moment lifetime.

REFERENCES

1. See, for example, M. B. Maple, in "MAGNETISM" A Treatise on Modern Theory and Materials," (edited by H. Suhl), Chapt. 10, Academic Press (1973).
2. M. B. Maple, J. Wittig and K. S. Kim, Phys. Rev. Letters 23, 1375 (1969); M. B. Maple and K. S. Kim, Phys. Rev. Letters 23, 118 (1969).
3. K. S. Kim and M. B. Maple, Phys. Rev. B2, 4696 (1969).
4. W. Gey and E. Umlauf, Z. Phys. 242, 241 (1971).
5. S. Ortega, M. Roth, C. Rizzuto and M. B. Maple, Solid State Commun. 13, 5 (1973).
6. J. G. Huber, W. A. Fertig and M. B. Maple, Solid State Commun. 15, 453 (1974).
7. C. A. Luengo, J. G. Huber, M. B. Maple and M. Roth, Phys. Rev. Letters 32, 54 (1974).
8. O. Peña and F. Meunier, Solid State Commun. 14, 1087 (1974).
9. S. DeGennaro and E. Borchi, Phys. Rev. Letters 30, 377 (1973).
10. R. Harris and M. J. Zuckermann, to be published.
11. A. S. Edelstein, Phys. Rev. Letters 20, 1348 (1968).
12. See, for example, M. B. Maple and D. Wohlleben, AIP Conference Proceedings (No. 18) on "Magnetism and Magnetic Materials" (edited by C. D. Graham, Jr. and J. J. Rhyne), pp. 447-462 (1974).
13. I. R. Harris and G. J. Raynor, J. Less-Common Metals 6, 70 (1964).
14. J. G. Huber and M. B. Maple, J. Low Temp. Phys. 3, 537 (1970).

*Supported by Air Force Grant No. AF-AFOSR-71-2073.

MOSSBAUER STUDIES OF DILUTE Ag:Fe AND Ag:Co ALLOYS

J. R. Thompson and J. O. Thomson
Department of Physics, The University of Tennessee, Knoxville 37916
and Oak Ridge National Laboratory, Oak Ridge, TN 37830†

ABSTRACT

We have studied the dilute Kondo alloys Ag:^{57}Fe and Ag:^{57}Co via the Mössbauer effect at temperatures between 0.022K and 20K and in applied magnetic fields from 0 to 60 kG. This allows study of both the thermal and magnetic breakup of the non-magnetic ground state. Our higher temperature results are consistent with those of Kitchens, Steyart and Taylor. As the temperature is lowered, a strong reduction from free spin behavior of the local moment is observed. At our lowest temperatures the dependence of the Fe hyperfine field, H_{hf}^{Fe}, on applied field is compared with several theoretical and semiempirical expressions, where each describes the data moderately well, none with marked superiority. We also report the first measurements of the hyperfine field of ^{57}Co in Ag, obtained via the intensity asymmetry of the spectra. The field can be described in terms of a large positive Knight shift K = (28±8)%. The small magnitudes of the hyperfine fields result from the negative core polarization fields being offset by positive orbital fields and illustrate the importance of orbital effects.

INTRODUCTION

Considerable theoretical and experimental effort in recent years has been devoted to the problem of an isolated magnetic impurity in a conducting system.[1] Presently it appears that in many such systems a low temperature "non-magnetic" state characterized by a temperature independent paramagnetism gives way gradually to a Curie-Weiss type of magnetic behavior as either the temperature or the applied field is increased above certain characteristic values T_c and $H_c \sim k_B T_c/\mu$. Here μ is the impurity moment derived from the high temperature susceptibility and k_B is Boltzmann's constant.

The measurement of the magnetic hyperfine structure coupling to the impurity nucleus has proven to be a valuable technique for the study of these systems. Such measurements can be obtained through Mössbauer effect or nuclear orientation studies of the impurity atom in samples of extremely low impurity concentration, so as to preclude impurity interaction effects. Also, the applicability of the Mössbauer effect is particularly evident in this very low concentration region where many other techniques lose sensitivity. Measurement of the hyperfine coupling to the nucleus gives direct information on the local impurity magnetization, which is perhaps the most significant parameter for study of the dilute impurity problem.

We have studied the magnetization of dilute ^{57}Fe and ^{57}Co impurities dissolved in Ag as a function of applied field from 0 to 60 kG and temperature from 0.022 to 20K using the Mössbauer effect for ^{57}Fe. These are the first measurements to be reported for this system in the very low temperature regions. Kitchens et al.[2] investigated the Ag:Fe dilute alloy in the higher temperature region above 1°K. Since the solubility of Fe and Co in Ag is only 10-13 ppm,[3] Mössbauer effect studies are particularly applicable, and furthermore, the likelyhood of impurity interaction effects is reduced.

EXPERIMENTAL

It is convenient to describe our results in terms of the hyperfine field H_{hf}, at the impurity nucleus. Such a description is valid when the electronic relaxation time is short compared with the nuclear precession time. This condition appears to prevail for ^{57}Co and ^{57}Fe in Ag. The dominant contributions to H_{hf} arise from core and conduction electron polarization and from an orbital field arising from spin-orbit coupling. When the atom remains in its spin-orbit ground state, then the local magnetization, M, is directly proportional to H_{hf}. One then has the relation $M = M_{sat} H_{hf}/H_{sat}$, where M_{sat} and H_{sat} are the saturation values of the moment and hyperfine field, respectively. In these studies H_{hf}^{Fe} was determined from the energy splittings obtained in the Mössbauer spectra. H_{hf}^{Co} was obtained from the intensity asymmetry of the spectra caused by the nuclear polarization of the ground state of the parent ^{57}Co.

Since the experiments were performed in a magnetic field whose axis was parallel to the propagation direction of the detected γ rays, the observed spectra consisted of the 8-line convolution of four circularly polarized source lines with four circularly polarized absorber lines split by the fringing magnetic field which was 7% of the central field. These spectra were computer fit using a code specifically modified for this purpose. An important feature of this code was that one of the fitting parameters was H_{hf}^{Co}/T. Therefore, it was unnecessary to estimate the intensity asymmetry of the lines in these unresolved or partially resolved spectra.

The spectra were obtained with a cold, stationary Ag:^{57}Co source and a room temperature absorber of $Na_2Fe(CN)_6 \cdot 3H_2O$. The source was a 2.5 x 10^{-3}cm Ag foil containing ∿ 1 ppm of ^{57}Co activity. The Ag starting material was quoted to be of 6N purity, while the total incidental magnetic impurities were found after the experiment to be < 60 ppm. The measurements were performed in a dilution refrigerator manufactured by SHE Corp. Low temperature thermometry was obtained via nuclear orientation measurements of the anisotropy of γ rays emitted from an hcp Co:^{60}Co single crystal. We monitored the γ ray count rate emitted parallel to both the crystalline C axis and the applied magnetic field. At low temperatures our estimated error in T^{-1} is ∿ $2K^{-1}$, e.g. 5% at 25 mK.

RESULTS AND DISCUSSION

In our spectra the intensity asymmetry is evident only for the lowest temperatures and the outermost groups of lines are resolved only for higher applied fields. In the computer analysis of the unresolved spectra a line width of 0.34 mm/sec was used, which was obtained from zero field spectra at 4.2 and 300K. This linewidth is in good agreement with that obtained for spectra taken at 60 kG at various temperatures where a width determination could be made.

In figure 1 the triangles show the absolute value of the measured field $|H_m^{Fe}|$ determined for those runs taken between 20 and 25 mK, plotted as a function of the applied field H_a. It appears that H_m passes through zero near H_a = 32 kG, with $H_m^{Fe}>0$ above 32 kG. Therefore, we have $H_{hf}^{Fe} = |H_m^{Fe}| - H_a$. The values of $-H_{hf}^{Fe}$ are shown as circles in figure 1. The size of the plotted symbols is somewhat larger than the statistical errors returned from the computer fits, but do not incorporate any systematic errors. It is evident that H_{hf}^{Fe} shows the non-magnetic behavior characteristic of a number of dilute impurity systems, since a free spin would be fully magnetized at values of H_a/T considerably smaller than those used here. The solid curve is a least squares fit to the data using the expression of Ishii,[4] derived for S = 1/2 and T = 0. This gives the saturation value H_{hf}^{Fe} = - 37.0 kG and taking g = 2.0 yields T_c = 2.4 K. The semiempirical expressions suggested by Gallop[5] and by Kitchens and Taylor[6] have also been fit to the data and give fits of similar quality. We have in addition, fit all our data at all temperatures and applied fields to Kitchens and Taylor's function:

$$H^{Fe}_{hf} = H_{sat}\, B_J\left(\frac{\mu H_a}{k_B(T + T_c)}\right),$$

where B_J is the Brillouin function. Using the statistical errors of 0.3 to 0.5 kG obtained in computer analysis of the spectra, we obtain $\chi^2/NDF \sim 4$. The resulting parameters are: $H_{sat} = -36.1$ kG, $\mu = gJ = 4.1\mu_B$ and $T_c = 2.4$K. To within errors, these results agree with their values, which were based on analysis of measurements of Kitchens, et al,[3] above 1K. This particular feature is not surprising since where they overlap the data are in essential agreement.

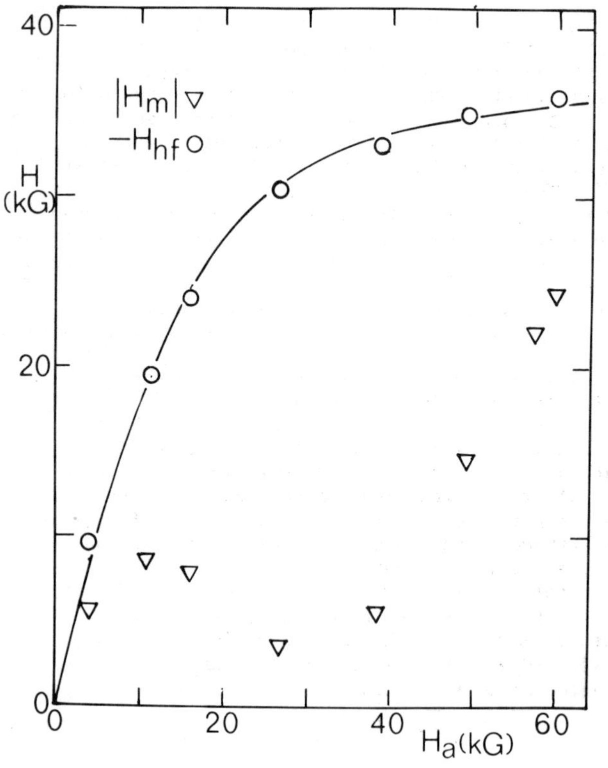

FIGURE 1. A PLOT OF THE NUCLEAR MAGNETIC FIELD VERSUS APPLIED FIELD H_A FOR $T \cong 0.022$K. THE TRIANGLES REPRESENT THE MEASURED FIELD $|H_M|$; THE CIRCLES ARE THE NEGATIVE OF THE DEDUCED HYPERFINE FIELD FOR ^{57}FE IN AG, $-H^{FE}_{HF}$. THE AVERAGE ERROR IS SOMEWHAT SMALLER THAN THE PLOTTED SYMBOLS.

In figure 2 are shown the results for the Co hyperfine field H_{hf}^{Co} when H_m^{Co}/T is plotted as a function of H_a/T. Over the temperature range from 20 to 80 mK and for $H_a > 40$ kG where there is a measurable asymmetry, our H_{hf}^{Co} can be described in terms of a large positive Knight shift, $H_m^{Co} = H_a(1+K)$ with $K = (28\pm8)\%$. This value is close to that found for Co in Au[7] of $K = 29.2\%$.

The above results for H_{hf}^{Fe} show that Fe in Ag is non-magnetic at low temperature and low field, similar to many other Kondo alloys, and that the magnetism is restored only for $T \sim T_c = 2.4$K or $H \sim H_c \sim 20$ kG. The low value obtained for H_{sat} probably arises from the incomplete quenching of the orbital magnetism which results in a positive orbital contribution to H_{hf}.

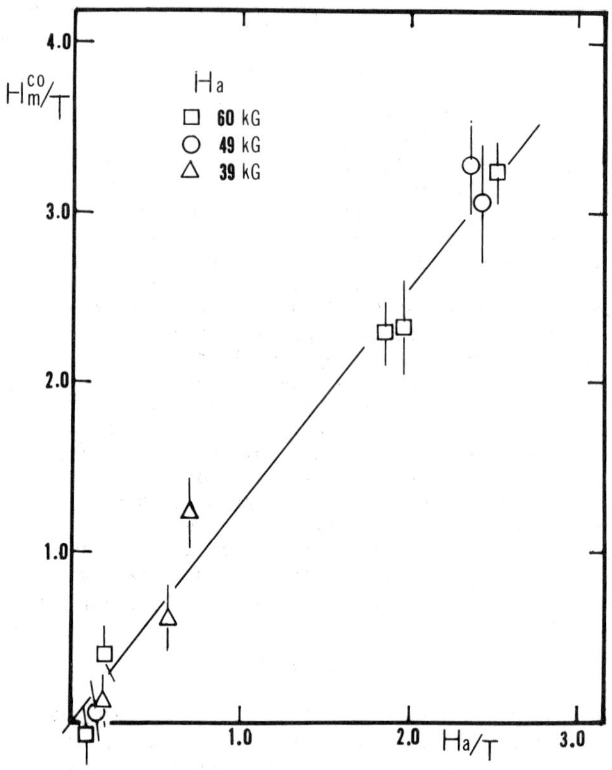

FIGURE 2. A PLOT OF THE MEASURED COBALT FIELD DIVIDED BY TEMPERATURE VERSUS THE APPLIED FIELD DIVIDED BY TEMPERATURE. UNITS ARE KG/MK. THE DATA IS DESCRIBED BY KNIGHT SHIFT $K = (28\pm8)\%$.

Although our results for H_{hf}^{Co} extend over only a small range of temperature, our data is consistant with the cobalt also being non-magnetic with T_c well above our measurement temperature. The fact that $H_{hf}^{Co} > 0$ implies that the orbital field for Co in Ag is even greater than for Fe in Ag.

ACKNOWLEDGEMENTS

One of us, J. R. Thompson, expresses his appreciation to the University of Tennessee Faculty Research Fund for support during a portion of this work. We also express our appreciation to P. G. Huray and F. E. Obenshain for many valuable discussions.

REFERENCES

1. Magnetism, vol. V, Ed. by H. Suhl, (Academic Press, New York, 1973), p. 1-416.
2. T. A. Kitchens, W. A. Steyert, and R. D. Taylor, Phys. Rev. 138, A467-A483 (1965).
3. M. Hanson and K. Anderko, Constitution of Binary Alloys, 2nd Ed. (McGraw-Hill, New York, 1958).
4. H. Ishii, Prog. of Theoretical Phys., 40, 201-209 (1968).
5. As quoted by J. Flouquet, Ann. Phys. 8, 5-53 (1973-1974).
6. T. A. Kitchens and R. D. Taylor, Phys. Rev. B9, 344-346 (1974).
7. A. Narath and D. C. Barham, Phys. Rev. B7, 2195-2199 (1973).

†Operated by Union Carbide for the USAEC.

MÖSSBAUER EXPERIMENTS ON DILUTE ^{57}Fe IN Cu,Ag AND Au

P. STEINER, D. GUMPRECHT, W.v. ZDROJEWSKI, AND S.HÜFNER
Institut für Atom- und Festkörperphysik, Freie Universität Berlin, Berlin (Germany)

ABSTRACT

The local susceptibility in CuFe, AgFe and AuFe is determined from Mössbauer experiments. CuFe and AgFe show nearly the same temperature dependence - a Curie-Weiss law at high temperatures and a T^2-behaviour at low temperatures. In contrast in AuFe we observe two Curie-Weiss regimes and again the T^2-behaviour at low temperatures.

INTRODUCTION

Mössbauer experiments on ^{57}Fe in nonmagnetic alloys have contributed considerably to the investigation of dilute magnetic alloys. The classical experiments e.g. of Frankel et al.[1] on ^{57}Fe in Cu demonstrated for the first time the vanishing of the local Fe moment at low temperatures and low external fields due to the Kondo effect[2]. In brief, this may be explained as follows: In the case of an antiferromagnetic coupling of the local magnetic moment of the conduction electrons below a characteristic temperature anomalies occur for various properties e.g. the well known minimum in the electrical resistivity, a maximum in the specific heat and a Curie-Weiss law for the magnetic susceptibility. These anomalies are caused by the strong correlations between the local electrons and the conduction electrons.

A quantity of primary interest from the experimental as well as from the theoretical point of view is the magnetic susceptibility of a local moment over a large temperature range. The theoretical treatment of the inherent complicated many-body problem gave only approximate solutions up to this point[3]. Experimentally the single impurity behaviour is often obscured by interactions between the impurities, which poses considerable experimental difficulties for bulk measurements, as e.g. macroscopic magnetization measurements. In this case hyperfine methods as e.g. the Mössbauer effect are of great advantage as they can often easily be performed in the very low impurity concentration limit of a few ppm, where interactions may be neglected. On the other hand the hyperfine fields contain a contribution of the orbital part of the moment which even in case of a quenched orbital moment in the ground state gives a temperature independent contribution to the local magnetization[4] in the sense of a Van Vleck paramagnetism. This part has to be subtracted from the data to obtain the d-spin contribution only, which one is mainly interested in. This can be a severe problem because of the large hyperfine coupling constant of the orbital moment compared to the core polarization term of the spin moment. Experiments over a large temperature range are needed to disentangle the different contributions.

In the following we will discuss the initial local susceptibility and summarize our results of Mössbauer experiments on the systems CuFe, AgFe and AuFe. In all cases the impurity concentration was less than 30ppm Co. Experimental details will be given elsewhere[5,6,7].

RESULTS AND DISCUSSION

For the systems CuFe and AgFe we found no linebroadening in the zero field Mössbauer spectra down to temperatures of 30 mK. The observed experimental width of about 0.40 mm/s is mainly due to the thick single line absorber of potassium ferrocyanide enriched in ^{57}Fe. This indicates that in these samples interactions between impurities play no significant role. For AuFe the line width increased slowly from 0.40 mm/s at 4.2K up to 0.58 mm/s at 50 mK. This could be an indication that in this AuFe sample at low temperatures indeed interactions between impurities are observed. In an earlier experiment with a different AuFe source the zero field spectra showed indeed considerable structure which we could interpret as arising from a local magnetic order[8] at the impurity sites. On the other hand the small linebroadening observed with the present AuFe source could also be due to relaxation effects of the local moment.

The local susceptibility ($\chi_{loc} = H_{hf}/H_{ext})_{H_{ext}\to 0}$) of the systems CuFe and AgFe shows a very similar temperature dependence over a large temperature range. At high temperatures we see a Curie-Weiss law and an additional temperature independent term.

$$(1) \quad \chi_{loc} = \frac{A}{T+\Theta} + B$$

The parameters are listed in Table I. The Curie-Weiss temperature Θ is due to the s-d exchange interaction between the local moment and the conduction electrons and may therefore stand as a measure of the characteristic Kondo temperature of the system. The temperature independent term B is the field induced Van Vleck paramagnetism by mixing higher states into the ground state.

At low temperatures we observe a levelling off of the susceptibility from the Curie-Weiss law as predicted by recent theories[3]. Especially the spin fluctuation theories predict a T^2-behaviour for $T \ll \Theta$.

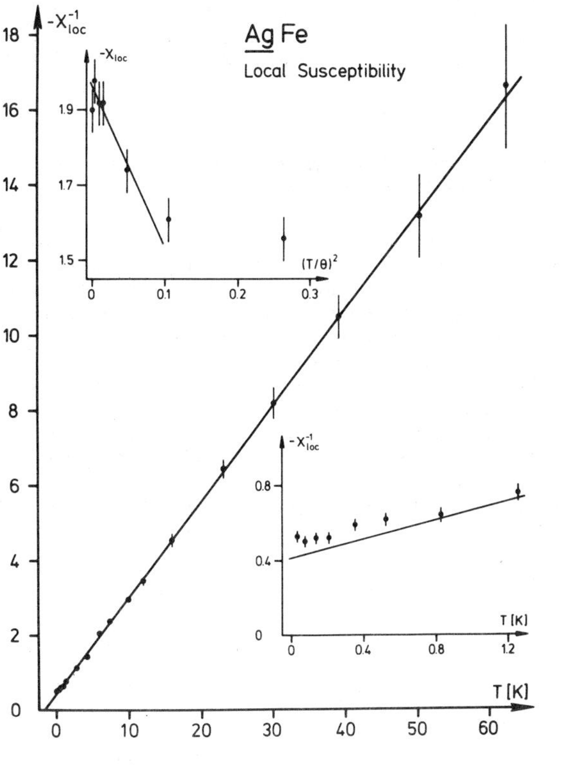

Fig.1 Initial local d-spin susceptibility for AgFe from Mössbauer experiments on ^{57}Fe

We get a reasonable good fit to our low temperature data with a T^2-law of the form

$$(2) \quad \chi_{loc} - B = \chi(0)\cdot(1-\alpha(T/\Theta)^2).$$

The constants $\chi(0)$ and α obtained from the data are given in table 1. Fig.1 shows as an example the temperature dependence of the susceptibility for AgFe. In the case of CuFe comparison with macroscopic magnetization results[9] shows that the total susceptibility has the same temperature dependence as the local susceptiblity obtained from the present experiments[5]. This shows clearly that no long ranged conduction electron polarization is present in the condensed Kondo stated, as first postulated by Heeger[12]. This finding is also in good agreement with recent NMR experiments of Boyce and Slichter on Cu nuclei being neighbours of Fe[11]. The Cu Knight shifts again show the same temperature dependence as the local and total susceptibility. For AgFe there exist as far as we know no macroscopic susceptibilities so that a detailed comparison as for CuFe is not possible.

The behaviour of AuFe deviates considerably from CuFe and AgFe. In this case (fig.2) we observe two Curie-Weiss regimes with different Curie-Weiss constants A (Table 1). From macroscopic experiments[13)

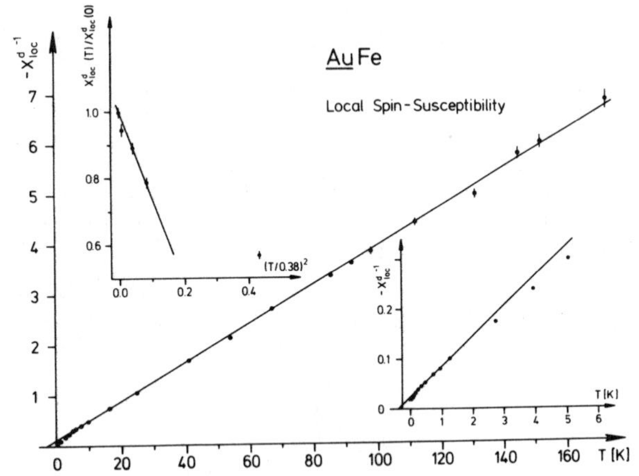

Fig.2 Initial local d-spin susceptibility for AuFe from Mössbauer experiments on ^{57}Fe

AuFe is believed to have a Kondo temperature of about 0.5 K. If we take the low temperature Curie-Weiss regime with Θ=0.38 K as characteristic for the Kondo behaviour of the system we get a similar T^2-behaviour at very low temperatures as in the case of CuFe and AgFe. The change in slope between 2K and 5K is not seen in the other two systems. This could indicate an effect of the crystalline electric field on the spin-orbit ground state of Fe as predicted by the ionic model given by Hirst[14]. In addition we will also mention without further discussion that the hyperfine coupling constants A (Table 1) are quite different for the three systems.

Table 1

	A[K]	Θ [K]	B	χ(0)	α	T[K]
FeCu	-15(1)	27.5(1.0)	0.056(9)			8<T<100
				-0.528(6)	2.9(4)	0<T<5
FeAg	-3.8(2)	1.6(3)	0.000(5)			1<T<60
				-1.96(5)	4.3(8)	0<T<0.4
FeAu	-26.0(5)	3.0(2)	0.038(6)			5<T<180
	-16.4(3)	0.38(2)				0.1<T<2
				-46.2(8)	2.5(5)	0<T<0.1

Summarizing the experimental results on the local susceptibility we can say: the low temperature behaviour for all three systems seems to be nearly the same - a T^2-law in the form of equ. (2) with a coefficient α roughly between 2 and 4. A similar behaviour is also found for AuV[15]. The high temperature behaviour for CuFe and AgFe is a single Curie-Weiss law while AuFe shows two different Curie-Weiss regimes.

This work was supported by the Bundesministerium für Wissenschaft und Technologie. The AgFe source was kindly supplied by W. Koch, Technische Universität München, and the AuFe source was obtained from Amersham Company, England.

REFERENCES

1) R.B. Frankel, N.A.Blum, B.B.Schwartz, and D.J.Kim, Phys. Rev. Lett. 18, 1050 (1967).
2) J. Kondo, Solid State Physics, ed. by N.Ehrenreich, F.Seitz, and D.Turnbull (Academic Press, New York, 1969), Vol.23, p.183.
3) For recent theoretical developments see e.g. W. Götze and P.Schlottmann, Solid State Comm. 13, 17 (1973); ibd. 13, 511 (1973); ibd. 13, 861 (1973) M.T.Béal-Monod and D.L.Mills, Solid State Comm. 13 1157 (1974).
K.D.Schotte, and K.Schotte, Phys. Rev. B4, 2228 (1971).
4) A.Narath, C.R.C.Critical Reviews in Solid State Sciences 3, 1 (1972).
5) P.Steiner, S.Hüfner, and W.v.Zdrojewski, to be published.
6) P.Steiner, W.v.Zdrojewski, D.Gumprecht, and S.Hüfner, Phys.Rev.Lett. 31, 355 (1973).
7) P.Steiner and S.Hüfner, Phys.Rev., to be published
8) P.Steiner, G.H.Beloserskij, D.Gumprecht, W.v.Zdrojewski, and S.Hüfner, Solid State Comm. 13, 1507 (1973); ibd. 14, 157 (1974).
9) J.L.Tholence and R.Tournier, Phys.Rev.Lett. 25, 867 (1970)
C.M.Hurd, J.Phys.Chem.Solids 28, 1345 (1967)
M.G.Hoeve and D.O.Van Ostenberg, Solid State Comm. 9, 941 (1971).
H.E.Ekström and H.P.Myers, Phys. Kondens.Materie 14, 265 (1972)
10) T.A.Kitchens and R.D.Taylor, Phys.Rev. B9, 344 (1974).
11) J.B.Boyce and C.P.Slichter, Phys.Rev.Lett. 32 61 (1974).
12) A.J.Heeger, Solid State Physics, ed. by N.Ehrenreich, F.Seitz, and D.Turnbull (Academic Press, New York, 1969), Vol.23, p. 283.
13) J.L.Tholence and R.Y.Tournier, J.Phys. Paris 326 201 (1971).
14) L.L.Hirst, Z.Phys. 244, 230 (1971); ibd. 241, 9 (1971).
15) J.E.Van Dam and P.C.M. Gubbems, Physics Letters 34A, 185 (1971)
S.Karzama, K.Kume, and K.Mitzano, Physics Letters 38A, 483 (1972)

LOW TEMPERATURE RESISTIVITY MEASUREMENTS OF DILUTE IRON IN EXCHANGE-ENHANCED PALLADIUM AND PALLADIUM-SILVER ALLOYS

R. P. Bittner, R. A. Levy, and J. A. Rayne*
Carnegie-Mellon University, Pittsburgh, Pennsylvania 15213

ABSTRACT

The effect of varying the host matrix susceptibility of the $Pd_{0.99}Fe_{0.01}$ alloy system is investigated by low temperature resistivity measurements on several $(Pd_{1-x}Ag_x)_{0.99}Fe_{0.01}$ alloys with silver concentrations given by x=0,0.005, 0.025, 0.10, 0.25, and 0.50. From the temperature dependence of the electrical resistivity, the magnetic ordering temperature is determined as a function of silver concentration and compared to reported data. For the x=0.50 sample, the incremental resistivity shows in the range 1.2-10K an anomalous temperature dependence that parallels the reported susceptibility behavior.

INTRODUCTION

The temperature dependent behavior of the electrical resistivity in dilute Pd-Fe alloys has been reported by several investigators.[1-4] Interest has generally focussed on the determination of the magnetic ordering temperature at various solute concentrations, on the examination of the critical behavior in the vicinity of the ferromagnetic phase transition, and on the comparison of the low temperature data with proposed theoretical models.[5] Since the enhanced susceptibility of the palladium host matrix is known to decrease upon alloying with silver[6], an extension of resistivity measurements from the Pd-Fe system to ternary alloys of dilute iron (1 at .%) in palladium-silver could reveal information about the possible changes in the ferromagnetic interactions caused by an altered host matrix susceptibility.

Recent Mossbauer[7] and low field magnetic susceptibility[8] measurements of the $(Pd_{1-x}Ag_x)_{0.99}Fe_{0.01}$ alloy system with silver concentrations up to 50 at % have indicated that the magnetic ordering temperatures in these alloys is essentially determined by changes in the spin paramagnetic susceptibility of the host matrix, in agreement with theoretical predictions. For values of $x \leq 0.25$ a linear dependence between those two variables is observed which, for a constant value of the iron impurity spin, would suggest a concentration-independent exchange interaction parameter with an approximate value of 0.10 eV. In this paper we present preliminary results for the temperature dependence of the electrical resistivity in alloys of the type $(Pd_{1-x}Ag_x)_{0.99}Fe_{0.01}$ with silver concentrations given by x=0, 0.005, 0.025, 0.10, 0.25, and 0.50.

EXPERIMENTAL RESULTS

Alloy ingots of the above compositions were prepared by induction melting 99.999%-pure palladium sponge, 99.999%-pure silver powder and 99.9% enriched Fe^{57} powder in zirconia crucibles under an argon atmosphere. The ingots were rolled and hand polished into foils approximately 50 microns thick. After being outgassed under vacuum, the foils were sealed into quartz tubes containing helium. They were then homogenized for one week at 1000°C and finally cooled rapidly in air followed by water. The samples were used for both the Mossbauer measurements previously reported[7] and the present work.

The resistivity measurements were carried out using a conventional four-probe technique over the temperature range 1.4-300K. Fig. 1 shows the experimental data as plots of $\Delta\rho=\Delta\rho_{alloy}-\Delta\rho_{host}$ versus temperature, where $\Delta\rho_{alloy}=\rho_{alloy}(T)-\rho_{alloy}(1.4)$ and $\Delta\rho_{host}=\rho_{host}(T)-\rho_{host}(1.4)$. Values of $\Delta\rho_{host}$ were obtained from the measurements of Edwards et al[9,10] and Murani.[11] Owing to their non-uniform dimensions, it was necessary to compute the geometrical form

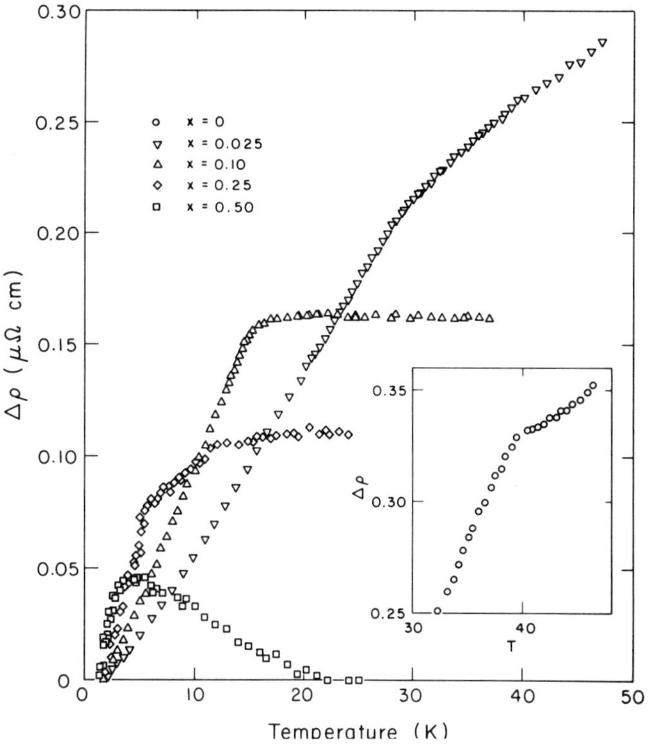

Fig. 1. Incremental resistivity $\Delta\rho=\Delta\rho_{alloy}-\Delta\rho_{host}$ as a function of temperature for alloys of the type $(Pd_{1-x}Ag_x)_{0.99}Fe_{0.01}$. The inset shows the behavior of the binary alloy $Pd_{0.99}Fe_{0.01}$.

factor and $\Delta\rho_{alloy}(T)$ for each specimen by an indirect method. The results shown in Fig. 1 are based on the assumption that the ideal resistivities of both the alloy and host are identical. Therefore, the change in alloy resistivity between 100 K and 300 K is the same as the corresponding change in the host ideal resistivity, which is known from the measurements of Kemp et al.[12] Although this procedure implies the validity of Matthiessen's rule, it appears to give consistent results. The value of $\Delta\rho(T)$ derived in this way gives the temperature dependence of the magnetic contribution to the resistivity arising from spin-disorder scattering.

The results for alloys with silver concentrations up to x=0.25 reveal the pronounced break generally associated with the onset of ferromagnetic ordering. The transition temperature is usually derived by either identifying it as the temperature at which $d\Delta\rho/dT$ reaches a maximum value as a function of temperature,[13] or as the temperature at which a knee appears in the $\Delta\rho$ versus T plot.[1] Evidently the two criteria coincide in cases where the knee is infinitely sharp. In the case of the ternary $(Pd_{1-x}Ag_x)_{0.99}Fe_{0.01}$ alloys the observed transitions are rather broad, making it difficult to evaluate absolute values for T_c. The inset in Fig. 1 shows the behavior of the binary alloy near T_c for comparison. Arbitrarily, we have chosen T_c as the temperature corresponding to the intersection of the linear sections just above and below the transition. This method yields values for T_c close to those derived from the maximum of $d\Delta\rho/dT$, and in reasonable agreement with those obtained through Mossbauer[7] and low field magnetic susceptibility[8] measurements. A compilation of all existing values of T_c for the $(Pd_{1-x}Ag_x)_{0.99}Fe_{0.01}$ alloy system, as obtained by various experimental techniques, is listed in Table I.

Another distinguishing feature of the data is the observed decrease in the values of the incremental resistivity with silver addition. The result reflects the substantial reduction in s-electron scattering by polarized electrons in the altered 4 d-band.

Atom % Silver x	Ordering Temperature T_c (K)			
	Mossbauer	Resistivity	Suscepti-bility	Magnet-ization
0	37.3[a]	40.0*(37[b])	-	39[e]
0.005	34.6[a]	36.0*	-	-
0.025	29.9[a]	29.0*	25.3[d]	-
0.10	20.1[a]	17.0*	17.6[d]	-
0.25	7.5[a]	6.0*(6[c])	6.5[d]	11[e]
0.33	-	-	4.0[d]	-
0.50	-	4.0*	3.3[d]	-

[a]See Ref. 7; [b]See Ref. 14; [c]See Ref. 15; [d]See Ref. 8; [e]See Ref. 16

Table I. Ordering temperature for alloys of the type $(Pd_{1-x}Ag_x)_{0.99}Fe_{0.01}$. Asterisks denote values obtained from present work.

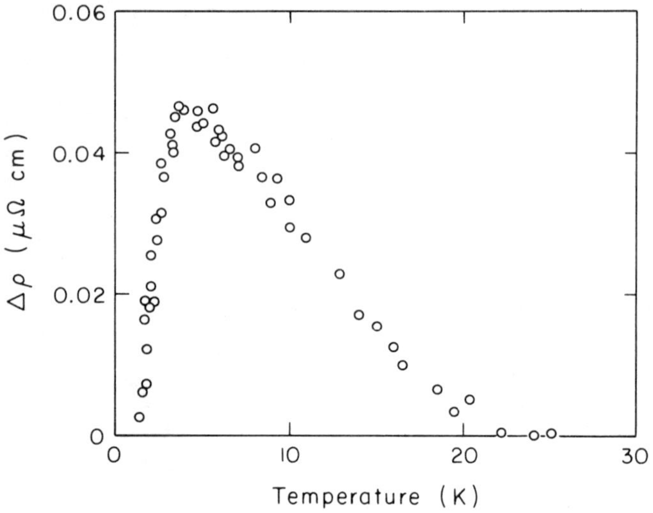

Fig. 2. Incremental resistivity $\Delta\rho=\Delta\rho_{alloy}-\Delta\rho_{host}$ as a function of temperature for ternary alloy $(Pd_{1-x}Ag_x)_{0.99}Fe_{0.01}$ with x=0.5.

As shown in Fig. 2, the sample with x=0.50 reveals a different resistive behavior from that observed in the alloys of lower silver concentrations. The incremental resistivity shows a relatively sharp peak at approximately 4.0 K, followed by a gradual decrease which tails off at approximately 20 K. This anomalous behavior parallels the reported temperature dependent variation of the susceptibility for this alloy in the same temperature range. Since the number of 4d-holes in the partially filled palladium band is substantially reduced at the silver concentration corresponding to x=0.50, the observed maximum in Δρ suggests magnetic ordering between the impurity iron spins via the indirect RKKY interaction[17]. Magnetoresistance measurement at this silver concentration is presently underway to verify this assumption.

Work is also progressing toward correlating the low temperature data with existing theoretical models and toward an examination of the critical behavior in the vicinity of the ordering temperature.

ACKNOWLEDGMENTS

The authors are very grateful to Dr. J.I. Budnick for his valuable support of this work and to Drs. L. Berger and R. Weiner for many helpful discussions.
*This work was supported by a grant from the National Science Foundation.

REFERENCES

1. G. Williams and J. W. Loram, J. Phys. Chem. Solids 30, 1827 (1969).
2. M. P. Kawatra, J. I. Budnick, and J. A. Mydosh, Phys. Rev. B 2, 1587 (1970).
3. J. A. Mydosh, J. I. Budnick, M. P. Kawatra, and S. Skalski, Phys. Rev. Letters 21, 1346 (1968).
4. M. P. Kawatra, S. Skalski, J. A. Mydosh, and J. I. Budnick, J. Appl. Phys. 40, 1202 (1969).
5. P. Rhodes and E. P. Wohlfarth, Proc. Roy. Soc. (London) A 273, 247 (1963); H. S. D. Coles and R. E. Turner, J. Phys. C2, 124 (1969); M. W. Stringfellow, J. Phys. C2, 1699 (1968).
6. F. E. Hoare, J. C. Matthews, and J. C. Walling, Proc. Roy. Soc. of London 216, 502 (1963); R.Daclo S. Foner, and A. Narath, J. Appl. Phys. 40, 1206 (1969).
7. R. A. Levy, J. J. Burton, D. I. Paul, and J. I. Budnick, Phys. Rev. B9, 1085 (1974).
8. J. I. Budnick, V. Cannella, and T. J. Burch, Proceedings of the 19th Conference on Magnetism and Magnetic Materials (1973).
9. C. W. Chen, L. R. Edwards, and S. Legvold, Phys. Stat. Sol. 26, 611 (1968).
10. L. R. Edwards, C. W. Chen, and S. Legvold, Solid State Commun. 8, 1403 (1970).
11. A. P. Murani, Phys. Rev. Letters 33, 91 (1974).
12. W. R. G. Kemp, P. G. Klemens, A. K. Sreedhar and G. K. White, Proc. Roy. Soc. A233, 480 (1956).
13. M. P. Kawatra and J. I. Budnick, Intern. J. Magnetism 1, 61 (1970).
14. G. Longworth and C. C. Tsuei, Phys. Letters A27, 258 (1968).
15. M. P. Sarachik, J. Appl. Phys. 39, 699 (1968).
16. A. M. Clogston, B. T. Matthias, M. Peter, M. J. Williams, E. Corenzwit, and R. C. Sherwood, Phys. Rev. 125, 541 (1962).
17. D. I. Paul, private communication.

MICROWAVE LOSSES OF Sn-SUBSTITUTED POLYCRYSTALLINE Ca-V-GARNETS WITH LARGE GRAINS

T. Inui and H. Takamizawa
Central Research Laboratories, Nippon Electric Company Ltd., Kawasaki, Japan

and

N. Ogasawara and T. Fuse
Faculty of Technology, Tokyo Metropolitan University, Tokyo, Japan.

ABSTRACT

Sn-substituted calcium vanadium garnets with narrow resonance linewidth ($\Delta H \leq 3$ Oe at X-band) composed of homogeneous large grains have been newly developed. Measurements of resonance linewidth as a function of frequency showed that the peak at the spinwave manifold ($\omega = 2|\gamma|4\pi Ms/3$) vanished for materials of 460 µm average grain size. A ΔH of 0.8 Oe, the smallest figure ever reported in this garnet series, was obtained at a frequency a little less than $2|\gamma|4\pi Ms/3$. As the grain size was varied from 8 to 30 µm, ΔH was found to be inversely proportional to the average grain size for $\omega < 2|\gamma|4\pi Ms/3$ and nearly independent of grain size for $\omega > 2|\gamma|4\pi Ms/3$.

Values of $\mu''_{\pm}$ were measured by using a cylindrical TE_{111} mode resonator filled with a sample measuring 10 mm diameter and 10 mm in length. A material with grain size of 460 µm has been found to exhibit values of $\mu''_{\pm} = 2 \times 10^{-4}$ off the resonance region implying that this material is highly promising as a potential candidate for low-noise circulators.

INTRODUCTION

Microwave ferrite losses inside the spinwave manifold are fairly well interpreted by the spin-wave theory[1], if such magnetic inhomogenities, such as grain-boundary dependent fluctuations of anisotropy field and localized demagnetizing fields due to pores and inclusions are low. Based on this theory, narrow resonance linewidth materials with the composition of

$$\{Ca_{3-y}Y_y\}[Fe_{2-x}Sn_x](Fe_{1.5+0.5x+0.5y}V_{1.5-0.5x-0.5y})O_{12}$$

have been developed[2], with values of ΔH at 9 GHz less than 3 Oe for saturation magnetizations ranging from 470 to 1950 Gauss and Curie temperature above 140°C.

Turning to the off-resonance loss, namely the loss for uniform precession outside the manifold, no conclusive elucidation has as yet been established except for the experimental finding that the loss is sensitive to the microstructure of the ferrite[3].

Inspired by the well-known facts that the magnetic resonance can occur outside the manifold as the exciting frequency is lowered[4] and that the micro-structure that narrows ΔH also reduces the magnetic loss off resonance, a series of experiments has been carried out for materials having low off-resonance losses.

This report deals with the microwave losses both around and off the magnetic resonance for a material $\{Ca_{1.2}Y_{1.8}\}[Fe_{1.5}Sn_{0.5}](Fe_{2.55}V_{0.45})O_{12}$, with grain sizes varied systematically.

EXPERIMENTAL

Polycrystalline specimens composed of homogeneous grains with the average size adjusted to 8, 14, 20, 33 or 460 µm were prepared by standard ceramic techniques. The material with the grain size of 460 µm was obtained by presintering the raw materials at 950°C and subsequent final sintering at 1410°C for 20 hours. The low presintering temperature combined with the prolonged sintering time was a key step for homogeneous large grain materials, in addition to the traditional close control of the stoichiometric composition. The quality of the materials is demonstrated in Fig. 1.

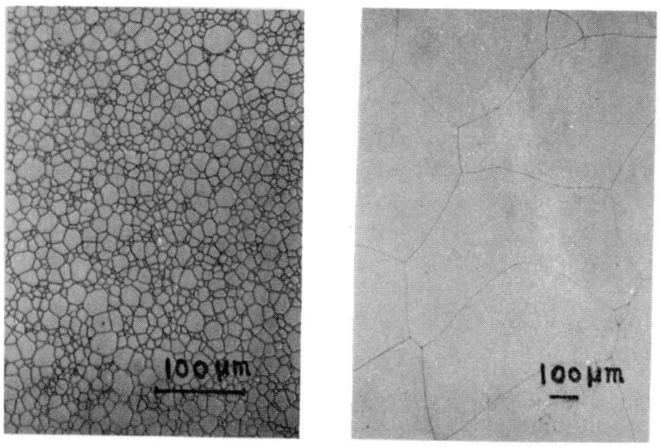

(a) (b)

Fig. 1. Microstructures of Sn-Ca-V-garnets with 8 µm and 460 µm average grain size.
Note that both structures are free of sizeable inclusions and pores.

The frequency dependence of ΔH gives information about the relaxation at the ceiling of the spinwave manifold. Values of ΔH of the Ca-V-garnets were measured as a function of frequency by the non-resonant method[5], using polished spherical samples measuring about 1 mm in diameter.

The grain size dependence of loss outside the manifold has also been measured in terms of $\mu''_{\pm}$ far from resonance. The quantity $\mu''_{\pm}$ is measured[6] using a cylindrical TE_{111} mode resonator filled with a sample measuring 10 mm in diameter and 10 mm in length. The resolution was 3×10^{-5} in $\mu''_{\pm}$.

RESULTS AND DISCUSSION

The frequency dependent linewidth for 8 µm grain size material is shown in Fig. 2. By contrast, the 460 µm grain size material does not have a peak at the frequency $2|\gamma|4\pi Ms/3$, as seen in Fig. 3. Fig. 3 also demonstrates that the value of ΔH reaches a value of 0.8 Oe, the smallest ever reported, at a frequency a little less than $2|\gamma|4\pi Ms/3$. This value is due to intrinisic polycrystalline loss in the sense that there is no two-magnon scattering from the anisotropy and porosity fields.

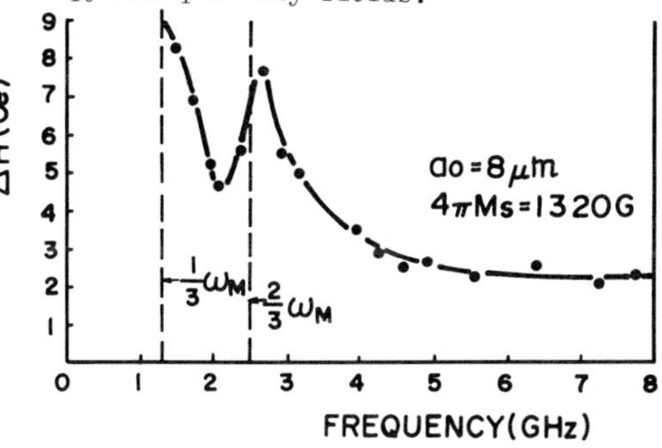

Fig. 2. Frequency dependence of ΔH for a grain size of 8 µm in Sn-Ca-V-garnet, with ω_M denoting $|\gamma|4\pi Ms$.

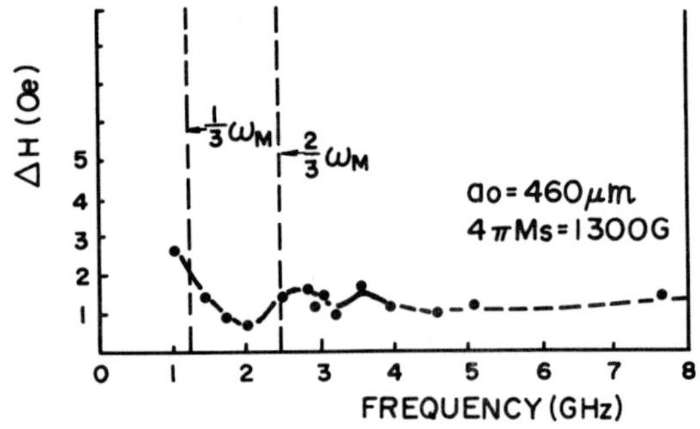

Fig. 3. Frequency dependence of ΔH for a grain size of 460 μm in Sn-Ca-V-garnet, with ω_M denoting $|\gamma| 4\pi Ms$.

Turning to the grain-size dependence of ΔH, a series of data are compiled in Fig. 4, for 8 GHz, for the critical frequencies, $2|\gamma|4\pi Ms/3$ and $|\gamma|4\pi Ms/3$, and for that frequency just below $2|\gamma|4\pi Ms/3$ that minimizing ΔH (resorting to interpolation if required). As this figure shows, the linewidths for the uniform precession frequencies outside the manifold are inversly proportional to grain-size, while the linewidths for the uniform precession frequency inside the manifold is almost constant. The physical origin of the grain-size-dependent ΔH still remains to be interpreted.

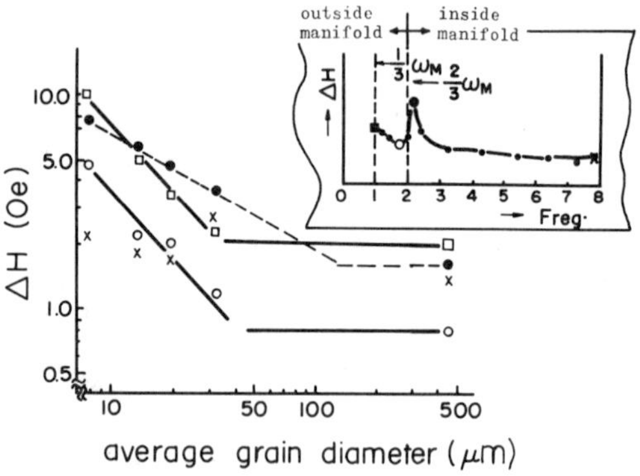

Fig. 4. ΔH versus grain-size.

x ; ΔH for 8 GHz
• ; ΔH for $2|\gamma|4\pi Ms/3$
□ ; ΔH for $|\gamma|4\pi Ms/3$
o ; minimum ΔH (for definition refer to the text.)

The preceding findings are concerned with the magnetic resonance loss both inside and outside the manifold. No definite correlation, however, has been identified between the grain-size and the loss off resonance, as far as the presently available experimental data are concerned, except that the value of μ''_{+} for 460 μm grain size Sn-Ca-V-garnet has been found to be less than those for 10 and 14 μm grain size material, as shown in Fig. 5 and Fig. 6.

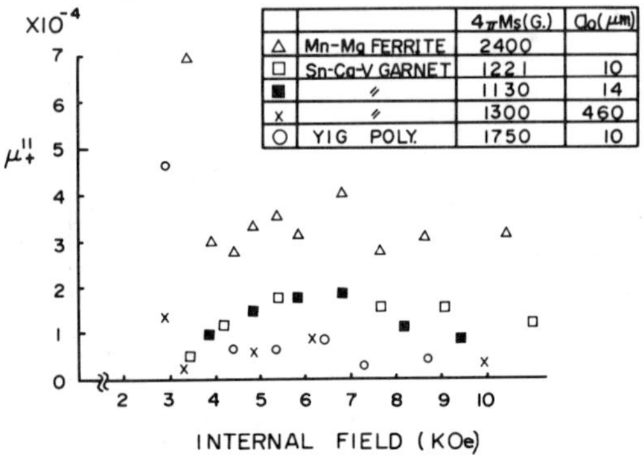

Fig. 5. μ''_{+} as a function of internal field.

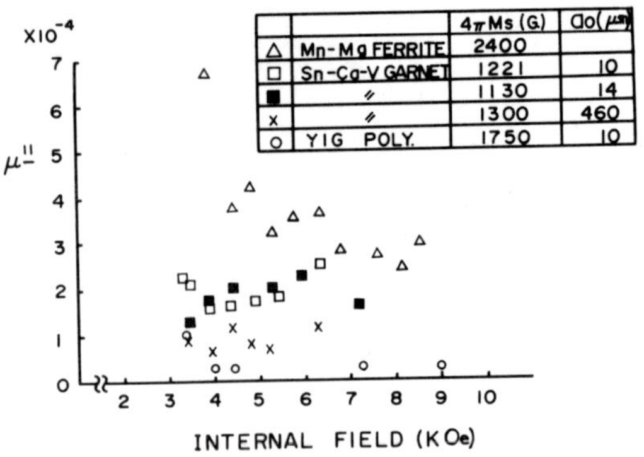

Fig. 6. μ''_{-} as a function of internal field.

Comparison is made between Sn-Ca-V-garnet and traditional garnet and spinel ferrites in Fig. 5 and Fig. 6 in terms of the off-resonance loss, $\mu''_{\pm}$, for samples of Sn-Ca-V-garnet of 10, 14 and 460 μm average grain size. The loss far from resonance is appreciably less than that of Mn-Mg-ferrite and almost the same as that of polycrystalline YIG, 2×10^{-4}, implying that this class of material has potential for practical use for low-noise circulators. An experimental circulator made of Sn-Ca-V-garnet with 460 μm grain size has proved to have a noise temperature 1.5°K less than that made of Sn-Ca-V-garnet with 20 μm grain size[7].

CONCLUSION

Summarising the results obtained, we conclude that:
(1) Physical origin of the grain-size dependent loss in materials with average grain size less than 30 μm still remains to be interpreted.
(2) Since it is free of the familiar loss steming from polycrystalline structure, Sn-Ca-V-garnet with average grain size of 460 μm has an extremely small microwave loss, both at and off magnetic resonance.
(3) A value of $\Delta H = 0.8$ Oe was obtained at 2 GHz, apparently an intrinsic loss of polycrystalline ferrite.
(4) The value of $\Delta H = 0.8$ Oe at 2 GHz makes the 460 μm Sn-Ca-V-garnet sample a promising candidate for applications to a family of low-noise circulators for S through L bands.

ACKNOWLEDGEMENTS

The authors wish to thank Mr. I. Saito for $\mu''_{\pm}$ measurements and Mr. N. Koyama for sample preparation. T. Inui and H. Takamizawa also wish to thank Dr. T. Tsuji and Dr. J. Kawai for continuing guidance and encouragement during this study.

REFERENCES

(1) E. Schlömann, J. Phys. Chem. Solids, Vol. 6, 242 (1958).
(2) H. Takamizawa, K. Yotsuyanagi and T. Inui, IEEE Trans., MAG-8, No.3, 446 (1972).
(3) C.E. Patton, IEEE Trans., MAG-8, No. 3, 433 (1972).
(4) C.R. Buffler, J. Appl. Phys., Vol. 30, No. 4, 172S (1959)
(5) N. Ogasawara, IEEE Trans., MAG-9, No. 3, 538 (1973).
(6) N. Ogasawara and J. Tanaka, Tech. Note, Special Tech. Committee on Microwaves, I.E.C.E., Japan, MW73-92 (1973).
(7) M. Kajikawa and W. Akinaga, private communication.

NOLINEAR PROPERTIES OF ALUMINUM - SUBSTITUTED LITHIUM FERRITE

Samuel Dixon Jr.
United States Army Electronics Command, Fort Monmouth, N.J.

ABSTRACT

Measurements have been taken on the ferromagnetic resonance properties of narrow linewidth single crystal aluminum-substituted lithium ferrite at room temperature and in the frequency range from 4.0 to 9.0 GHz. The composition of the starting mixture was as follows: Li_2CO_3 29.35%; Fe_2O_3 55.33%; and Al_2O_3 15.32%. The saturation magnetization of the samples was 2640 gauss at room temperature. The resonant linewidth measured 2.5 oersteds at 9.0 GHz and decreased to 1.0 oersteds at 4.0 GHz. The high power data obtained show the change in the microwave magnetic susceptibility as a function of incident power. From the room temperature threshold field (h_c) an intrinsic spin wave linewidth (ΔH_k) of 0.17 oersteds was calculated. This narrow linewidth, coupled with the relatively high saturation magnetization, indicates that the material can be utilized successfully in filters and low level power limiters.

INTRODUCTION

The ferromagnetic resonance properties have been investigated in narrow-linewidth single crystal aluminum-substituted lithium ferrite at room temperature and in the frequency range from 4.0 to 9.0 GHz. (1,2) Previous investigations have shown that the addition of aluminum into lithium ferrite has the initial effect of lowering the saturation magnetization and the curie temperature. The samples used in this investigation were grown in a $PbO-B_2O_3$ flux. The composition of the melt in mole percent was: 29.35% Li_2CO_3, 55.33% Fe_2O_3, 15.32% Al_2O_3. The actual single-crystal composition was as follows: $Li_{0.5}Fe_{1.75}Al_{.75}O_4$. This composition was determined by a linear interpolation of the lattice constants. The saturation magnetization of these samples was 2640 gauss at room temperature. The low resonance linewidth, of 2.5 oersteds at 9.0 GHz, observed in these samples was of considerable interest and therefore further characterization of the magnetic properties was initiated.

The investigation concentrated on determining the resonance linewidth and the variation of the microwave magnetic susceptibility as a function of incident power. Other investigators (3) have reported on the crystallographic and magnetic properties of polycrystalline lithium ferrite aluminate. The experimental data reported here represent the first measurements on the resonance properties of single crystal aluminum-substituted lithium ferrite.

EXPERIMENTAL DETAILS

Ferromagnetic resonance linewidth and high power rf magnetic susceptibility measurements were made on optically polished spheres with diameters in the order of 0.025 inches. The same sample was used to study the resonance linewidth (ΔH) variation over the frequency range from 4.0 to 9.0 GHz. The cavities used were of the transmission type operating in the TE_{104} mode. The samples were placed at the center of the cavity at a point of maximum rf magnetic field intensity. Data were taken on a point by point basis using calibrated attenuators and a bridge indication method.

EXPERIMENTAL RESULTS

Experimental measurements were made initially to evaluate the ferromagnetic resonance linewidth as a function of frequency at room temperature. The width of the line was determined from the resonance curve at the half-power points. These data are shown in Fig. 1. The measured linewidth is 1.0 oersteds at 4.0 GHz and increases to 2.5 oersteds at 9.0 GHz.

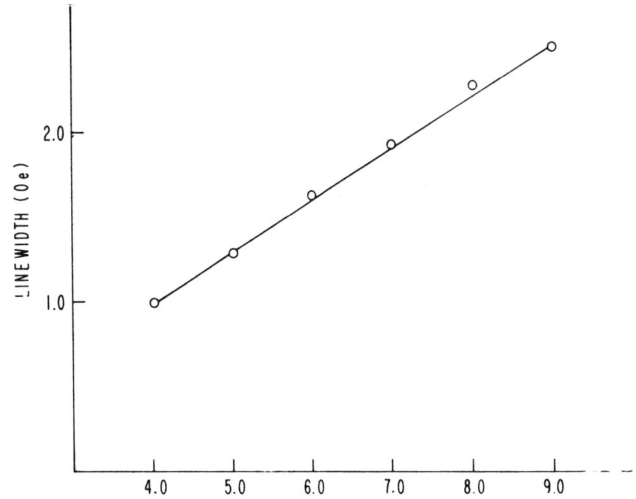

FIG. 1 SINGLE CRYSTAL ALUMINUM SUBSTITUTED LITHIUM FERRITE ($Li_{0.5}Fe_{1.75}Al_{0.75}O_4$) LINEWIDTH AS A FUNCTION OF FREQUENCY

To determine the threshold field for the onset of spin-wave instability, the susceptibility as a function of the microwave magnetic field intensity was measured at both 4.7 and 9.0 GHz. The low-power magnetic susceptability is used as a reference for normalizing the susceptibility at high power levels. As Fig. 2 and 3 indicates, there is a substantial decline in the main resonance absorption as the microwave magnetic field intensity is increased. At 4.7 GHz (Fig 2), it should be noted that the threshold field is two orders of magnitude lower than that at 9.0 GHz (Fig 3). This much lower threshold field is due to the fact that at 4.7 GHz the ferrite sample is operating in a coincidence mode. Spencer et al.(4) have shown both theoretically and experimentally that over a particular range of frequencies, the threshold field will occur at extremely low-power levels because the dc field required for the minimum threshold of the subsidiary resonance is coincident with the field required for the uniform precession mode. This range of frequencies is given by the following equation:

$$\gamma 4\pi M_s/3 \geqq \omega \leqq 2/3 (\gamma 4\pi M_s) \qquad (1)$$

where $4\pi M_s$ = saturation magnetization, and γ = gyromagnetic ratio. For these samples with a $4\pi M_s$ of 2640 gauss, the coincident range is from 2464 to 4928 MHz.

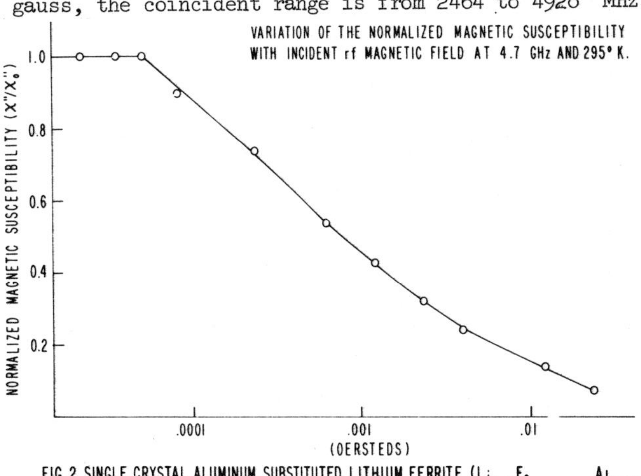

FIG. 2 SINGLE CRYSTAL ALUMINUM SUBSTITUTED LITHIUM FERRITE ($Li_{0.5}Fe_{1.75}Al_{0.75}O_4$)

However, at 9.0 GHz utilizing samples with a saturation magnetization of 2640 gauss, the second order process will be responsible for the observed nonlinear effects(5). That is, spin waves with a frequency equal to that of the test frequency will be the only ones excited. The decline in the resonance curves of Fig. 2 and 3 may be used to determine the critical

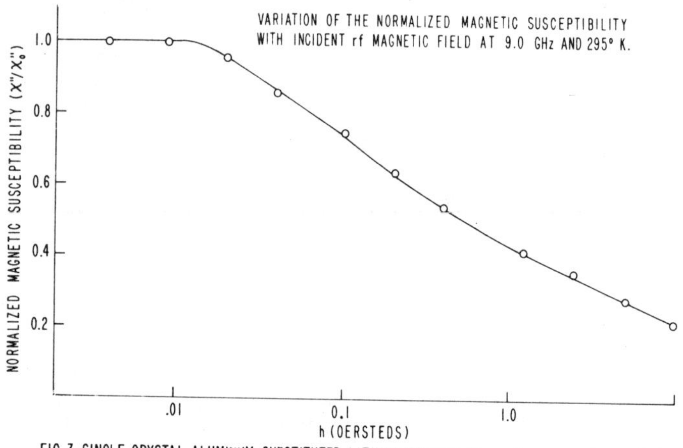

FIG. 3 SINGLE CRYSTAL ALUMINUM SUBSTITUTED LITHIUM FERRITE ($Li_{0.5}Fe_{1.75}Al_{0.75}O_4$)

field (h_c) for spin-wave instability by extrapolating back to the low-power magnetic susceptibility. From the critical field value, the spin-wave linewidth (ΔH_k) may be calculated using the following formula at 9.0 GHz:

$$h_c = \Delta H(\Delta H_k/4\pi M_s)^{\frac{1}{2}}, \qquad (2)$$

where ΔH is the measured linewidth, $4\pi M_s$ is the saturation magnetization, ΔH_k is the intrinsic linewidth characteristic of the unstable spin-wave mode, and h_c is the critical field. At 4.7 GHz, the experimentally measured threshold field was in the order of 0.1mOe. This instability threshold increased to 20 mOe at 9.0 GHz. From the room temperature threshold at both 4.7 and 9.0 GHz an intrinsic spin-wave linewidth of 170 mOe was calculated.

CONCLUSIONS

Information has been obtained concerning the ferromagnetic resonance and nonlinear properties of single crystal aluminum-substituted lithium ferrite. The resonance linewidth (ΔH) varies from 1.0 oersteds at 4.0 GHz to 2.5 oersteds at 9.0 GHz. The test frequency of 4.7 GHz and the saturation magnetization of 2640 gauss causes the uniform precession to be coincident with the subsidiary resonance mode. When the coincidence conditions exist the critical field (h_c) for the onset of spin-wave instability is extremely low. The experimentally measured threshold field at 4.7 GHz, which is in the coincidence frequency range, shows a reduction of approximately two orders of magnitude from that measured at 9.0 GHz. Future experimental device efforts will concentrate on developing low level frequency selective power limiters, in the 4.0 to 5.0 GHz range, for application in communication receiver systems.

ACKNOWLEDGEMENT

The author wishes to thank Thomas Swackhammer of Airtron, a division of Litton Industries, for his assistance in providing technical information on the growth and saturation magnetization of this material.

REFERENCES

1. Von Aulock W. H., Handbook of Microwave Ferrite Materials (Academic Press, New York, N.Y.), P. 423.
2. Vassiliev A., Thesis, University of Paris, 1962
3. Schulkes J.A., Blass G., J. Phys. Chem. Sol., Vol 24, 1963, P. 1651
4. Sansalone F.J., Spencer E.G., IRE Trans. Microwave Theory and Tech., MMT-9, 272 (1961)
5. Schloman E., Green J.J., and Milano U., J. Appl. Phys., Vol. 31, 387 (1960)

PROPERTIES OF SPIN WAVE EXCITATIONS IN LIQUID PHASE EPITAXY YIG FILMS

C. Vittoria and J. J. Krebs

Naval Research Laboratory, Washington, D.C. 20375

We have observed standing spin wave modes in LPE YIG films (~1μ thick) with a free surface. The spin wave mode intensities are very weak when compared with the corresponding intensities in CVD YIG films. For example, the intensity of the n=3 line is a factor of 300,000 below the main line. For most angles of the applied field the much stronger magnetostatic modes tend to mask the spin wave modes. The spin wave modes are well resolved for in-plane resonance when they are below and the magnetostatic modes are above the main line. Also the magnetostatic modes can be quenched by coating the YIG surface with a metal film. The field positions obey an n^2 law and the correct intensities are obtained if we assume negligible surface anisotropy on one surface and a uniaxial anisotropy of $K_s = 5\times10^{-3}$ ergs/cm^2 on the other surface. This represents the smallest surface anisotropy measured for either metal or insulating films. Thus, there is reason to believe that K_s may represent an intrinsic surface quantity. The much smaller surface anisotropy in LPE vs CVD films is consistent with the much thinner diffusion zone expected in the LPE films.[1]

[1] C.S. Guenzer, H. Lessoff and C. Vittoria, AIP Conf. Proc. 18, 1292 (1973).

HYSTERESIS LOOP PROPERTIES OF Li-FERRITE DOPED WITH Mn AND Co*

H. J. Van Hook
Raytheon Research Division, Waltham, MA 02154

G. F. Dionne
MIT Lincoln Laboratory, Lexington, MA 02173

ABSTRACT

Measurements of coercive force H_c and remanent magnetization $4\pi M_r$ have been made as a function of temperature ($-195 \leq T \leq 150°C$) on dense polycrystalline Li ferrite substituted with Mn ($0 \leq y \leq .20$) and Co ($0 \leq z \leq .02$). We report an abrupt degradation in loop squareness (decreased $4\pi M_r$, increased H_c) at a characteristic temperature T_θ, which increases with z but is independent of y. Single-crystal data on anisotropy K_1 show a correlation between T_θ and the temperature at which sign reversal in K_1 is found with Co substitution. The absence of a dependence on Mn content is in accord with the fact that K_1 is relatively insensitive to Mn substitutions, which have been shown to control magnetostriction effects. The deterioration in the hysteresis loop is discussed in terms of the nucleation of reverse domains as K_1 passes through zero.

INTRODUCTION

The use of Li-ferrite compositions in microwave ferrite devices has followed quite rapidly on the discovery of the effectiveness of Bi_2O_3 additions as a sintering aid,[1] and upon the observation that Mn suppresses dielectric loss and Co increases high power performance as effectively as in other spinel ferrites. Li-ferrite compositions have a hysteresis loop squareness and low magnetostriction which make them particularly attractive for latching-type phase shifter applications. Substitutions of Mn and Co, which are needed to control dielectric loss and high power properties in these materials, also affect loop properties by altering anisotropy and magnetostriction constants.[2] The problem of separating these intrinsic contributions of Co and Mn from microstructural influences has been circumvented in this study by measuring the temperature dependence of the hysteresis properties. In the range studied, $-195 \leq T \leq 200°C$, the microstructure is unchanged, whereas anisotropy and magnetostriction vary considerably. Single crystal literature data on Li-Mn ferrites and Li-Co ferrites[2,3] have been used in the interpretation of hysteresis properties.

*This work was supported in part by the Air Force Materials Laboratory, Air Force Systems Command, Wright-Patterson Air Force Base, under Contract No. F33615-72-C-1524.

SAMPLE PREPARATION AND MEASUREMENTS

Polycrystalline ceramics in the series

$$Li_{.5-z/2}Co_zMn_yFe_{2.5-z/2-y}O_4$$

with Bi_2O_3 additions in the range of 0.1 to 0.2 wt. percent were prepared by wet ball milling, and calcined at 850-900°C in flowing oxygen. The iron stoichiometry was adjusted in each case such that a trace of the iron deficient $LiFeO_2$ phase was observed microscopically. These conditions of Fe content combined with oxidizing firing conditions, i.e. 1000°C in flowing oxygen, were chosen to produce dense ceramics with a minimum Fe^{2+} content. Density measured by water displacement was > 99.5% of the x-ray value except for $z \geq .02$ where $\rho \geq 99\%$. Toroidal samples were measured on a standard 100 Hz loop tracer at temperatures between -195 and 200°C. Stress sensitivity was determined on bar-shaped samples using the method reported previously by one of the authors.[4] Single crystal results were obtained on spheres ground from PbO-flux grown crystals using standard microwave resonance techniques.

EXPERIMENTAL RESULTS

Mn-Doped Li-Ferrite

Even under oxidizing conditions used in this study, the dielectric loss (10 GHz) and dc conductivity data indicate Fe^{2+} ions are present in the polycrystalline ferrites. Conductivity is progressively reduced with Mn substitution to a saturation level at $y \approx 0.10$, (Table 1), presumably by an oxidation of Fe^{2+} by Mn or by the inhibition of electron mobility by some process. The table also indicates that these compositions are relatively insensitive to stress. The single crystal data in Fig. 1 show that $\langle\lambda\rangle = (2/5)\lambda_{100} + (3/5)\lambda_{111}$ is sensitive to Mn up to $y \simeq 0.1$. The unexpected absence of stress sensitivity in the polycrystals[5] below $y = 0.1$ may be explained by the Fe^{2+} at low Mn concentrations. The Fe^{2+} ions would give a $+\lambda_{111}$ contribution[6] to $\langle\lambda\rangle$, compensating the large $-\lambda_{100}$ in Li-ferrite. With Mn additions to Li-ferrite, the reduction in Fe^{2+} and the attendant $+\lambda_{111}$ increase are balanced by a similar decrease in magnitude of λ_{100} with Mn content and $\langle\lambda\rangle$ remains reasonably constant.

Table 1
Stress sensitivity and conductivity of $Li_{.5}Mn_yFe_{2.5-y}O_4$

y	Percent Change in $4\pi M_r$ at 800 psi	dc Conductivity at 20°C $(ohm\text{-}cm)^{-1}$
0	+1.1	4×10^{-4}
0.0125	+0.25	3.3×10^{-6}
0.025	+0.65	2×10^{-8}
0.050	+0.65	3.5×10^{-8}
0.10	+0.95	6×10^{-8}

On the other hand, the magnetic anisotropy constant K_1 remains large and negative with both Mn^{3+} and Fe^{2+} present[6] and no compensation in K_1 is expected.

The hysteresis properties, $4\pi M_r$ and H_c, versus temperature are plotted for a series of Li-Mn-ferrites in Fig. 2. The results show very little tem-

PBN-74-823

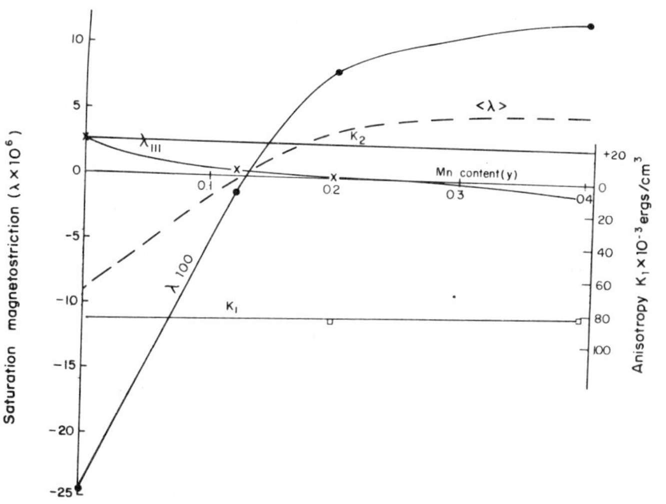

Fig. 1 Magnetostriction and Magnetic Anisotropy in $Li_{.5}Mn_yFe_{2.5-y}O_4$ Single Crystals

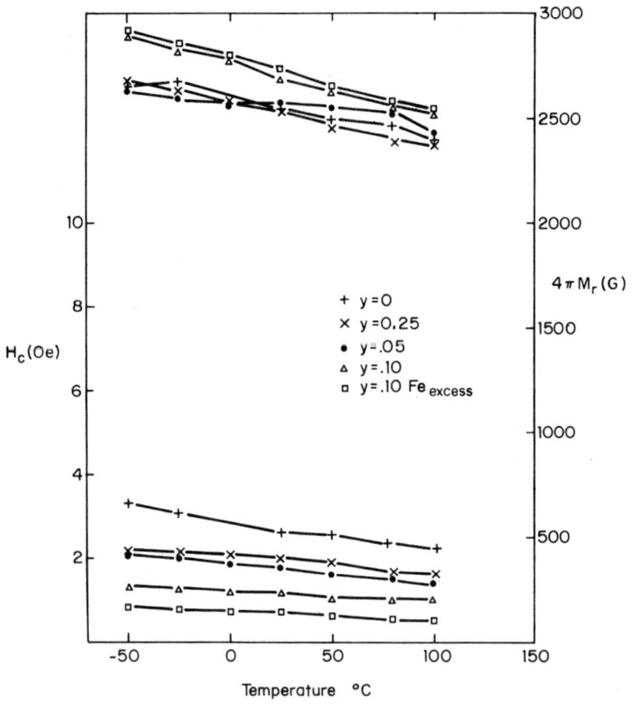

Fig. 2 H_c and $4\pi M_r$ vs. T. for various Mn concentrations

perature dependence. Differences in H_c between samples are believed to be predominantly grain size effects despite the efforts to prepare ceramics with identical microstructures. In the two samples with y = .10, the material with excess Fe (~4%) has an average grain size twice that of the larger H_c material. The constancy of the hysteresis properties in the Li-Mn ferrites is consistent with the dominance of magnetocrystalline anisotropy (K_1) over stress-induced anisotropy for the temperature interval shown.

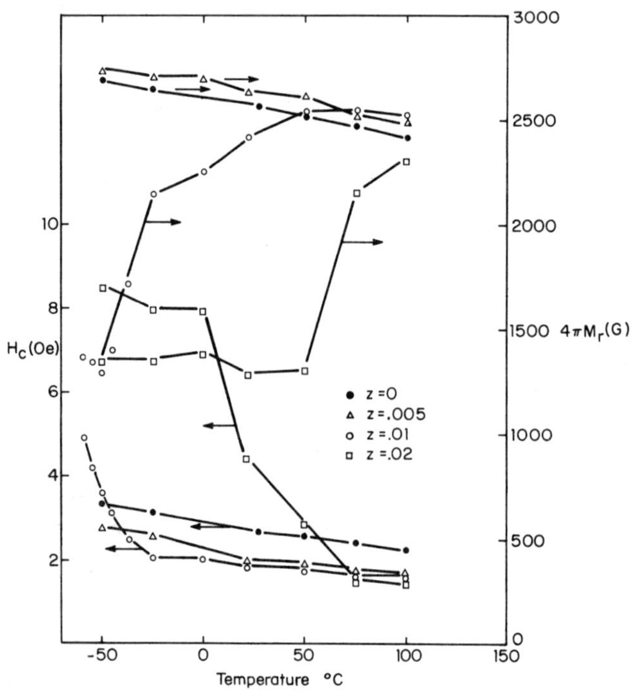

Fig. 3 H_c and $4\pi M_r$ vs. T for various Co concentrations

Co-Doped Li-Ferrite

Measurements of dielectric loss versus Co content in polycrystals Li-Co-ferrites indicate that Fe^{2+} is present in all compositions and unaffected by Co (z) content. From the above discussion, one should therefore expect a constant positive contribution of Fe^{2+} to λ_{111} and a negative contribution to K_1 relative to the single crystal values.

Figure 3 shows H_c and $4\pi M_r$ as functions of Co concentration (z) and temperature. Abrupt degradations in loop squareness take place with decreasing temperature on samples with z = 0.01 and .02, suggesting a change in intrinsic properties with either increasing z or decreasing T. The changes in H_c and 4π Mr both occur at the same temperature (T_θ) in each case.

The deterioration of the hysteresis loop, represented by the dramatic decrease in $4\pi M_r$ and increase in H_c shown in Fig. 3, is well correlated with the sign reversal in K_1. The temperatures at which the reversals occur agree closely with the T_θ values in this work. To explain the abruptness of the change in loop characteristics, the variation in K_1 with temperature must be considered. For pure Li-ferrite, $K_1 < 0$ and its magnitude increases at a moderate rate as T decreases. With Co-additions, this effect is offset by strong positive single-ion contributions to K_1 from Co^{2+} ions, which drive $|K_1|$ through zero as T decreases. The sharpness of the changes in H_c are caused by the very large increase in K_1 with decreasing temperature, once the reversal in sign takes place. According to Wijn et al.,[7] the deterioration of $4\pi M_r$ should be expected when $|K_1|$ is reduced. It is also probable that the passage of $|K_1|$ through zero produces the early onset of reverse domains. According to Goodenough,[8] the squareness of a hysteresis loop requires that the nucleation field $H_n > H_c$ and certainly greater than zero. From the physical situation involved here, it is possible for this inequality to break down, because a zero anistropy situation may easily create a $H_n < 0$ (see Goodenough's Eq. 6). Under conditions where $H_n < H_c$, the material will begin to switch its magnetic state before the coercive field is reached, rounding the knee of the loop. In situations where $H_n < 0$, the early onset of reverse domains will cause the switching to begin even before the driving field changes sign, resulting in a reduced remanent magnetization.

SUMMARY

The hysteresis loop effects of Mn and Co substitutions into Li-ferrite were examined for several concentrations. Manganese is believed to act in conjunction with ferrous iron to render the Li-Mn-compositions stress-insensitive and provide good hysteresis properties. Cobalt was found to cause a sharp deterioration in loop squareness, an effect attributed to a reversal in the sign of the magnetic anisotropy constant.

REFERENCES

1. P. D. Baba, G. M. Argentina, W. E. Courtney, G. F. Dionne, and D. H. Temme, IEEE Trans on Magnetics. MAG 8, 83 (1972).
2. G. F. Dionne, J. Appl. Phys. 40, 4486 (1969).
3. M. A. Stelmashenko and N. V. Seleznev, Sov. Phys. J. 2, 7 (1966).
4. G. F. Dionne, IEEE Trans on Magnetics MAG 5, 596 (1969).
5. G. F. Dionne, IEEE Trans on Magnetics MAG 10, 947 (1974).
6. A. P. Greifer, IEEE Trans on Magnetics MAG 5, 774 (1969).
7. H. P. J. Wijn, E. W. Gorter, C. J. Esveldt, and P. Geldermans, Philips Tech Rev 16, 49 (1954).
8. J. B. Goodenough, Phys, Rev. 95, 917 (1954).

SINGLE-PHASE INHOMOGENEITY IN CERAMIC GARNETS

Lionel M. Levinson, I. S. Jacobs, C. Greskovich and G. H. Glover
General Electric Corporate Research and Development
Schenectady, New York USA 12301

ABSTRACT

Sintered ceramic garnets of composition $(Y,Gd)_3(Fe,Al)_5O_{12}$ are of technological importance for the construction of microwave phase shifters. Mössbauer spectroscopy near T_c provides a unique monitor of the effect of ceramic processing variables, as well as sintering time at temperature, on chemical homogeneity. Near T_c we observe the coexistence of ferrimagnetic and paramagnetic phases, with the width of the coexistence region, ΔT_c, varying sharply between samples. The ΔT_c values are a quantitative measure of local chemical fluctuations, presumably in the Fe/Al ratio, which remain after the particular preparation. Roughly similar fluctuations are also inferred from x-ray line broadening, but with much less precision. Concurrent with the decrease in ΔT_c from 9°C to 3°C, the X-band resonance line width decreases from 67 to 45 Oe. This change is more than twice the maximum estimated for the observed slight decrease in porosity, and illustrates the influence of single-phase inhomogeneity on the microwave properties of substituted garnets.

Sintered ceramic garnets of composition $(Y,Gd)_3(Fe,Al)_5O_{12}$ are of technological importance[1] for the construction of microwave phase shifters. Careful control of the ceramic processing of these materials is essential[1-3] to obtain practical devices with low insertion loss. A variety of magnetic (e.g. ferrimagnetic resonance line width) and microstructural (e.g. porosity, grain size) parameters of the garnets are typically monitored to ensure adequate quality control. We report here the use of Mössbauer spectroscopy near the Curie temperature T_c as a unique monitor of the garnet chemical homogeneity.

Garnets of composition $(Y_{2.66}Gd_{0.34})(Fe_{4.19}Al_{0.67}Mn_{0.09})O_{12}$ were prepared by mixing high purity Y_2O_3, Gd_2O_3, Al_2O_3, Fe_2O_3, and $MnCO_3$ in a stainless steel ball mill with hardened steel balls. Fe pick-up from the ball mills was determined to be about 0.04 wt %, and was compensated for by starting with a slightly Fe deficient mix. The milled powder was then calcined in air at 1300°C for 4 hours to convert the mixed oxides to the garnet phase. The garnet powders were screened through a 60 mesh sieve to obtain better chemical homogeneity, and then either directly cold-pressed at 700 kg/cm² into disc shaped specimens (samples 1-4, Table I), or milled in a Trost-type[4] fluid energy mill followed by blending and cold compaction (sample 5, Table I).

Sintering experiments were performed on the powder compacts by firing in a Pt-40% Rh resistance wound furnace at 1450° in flowing O_2. The sintering times are listed in Table I. Temperature was stabilized to ± 2°C.

Careful metallographic and x-ray diffraction studies indicated all samples to be single phase, well-crystallized garnet solid solutions. In Table I we list the garnet lattice parameters, grain size, density and porosity (measured by Archimedes method using CCl_4 as immersion liquid). The room temperature magnetization $4\pi M_o$ (extrapolated to H=0 from high fields) and Curie temperature T_c of the garnets were measured with a vibrating sample magnetometer. No differences were detectable between the various samples by this method.

In Table I we give the ferrimagnetic resonance linewidth ΔH of the garnets as measured using a microwave (X-band, 9.468 GHz) spectrometer. We note that ΔH decreases with sintering time. Sample 5 (energy milled) has a low resonance linewidth even though it was sintered for only 3 hours (cf. sample 1).

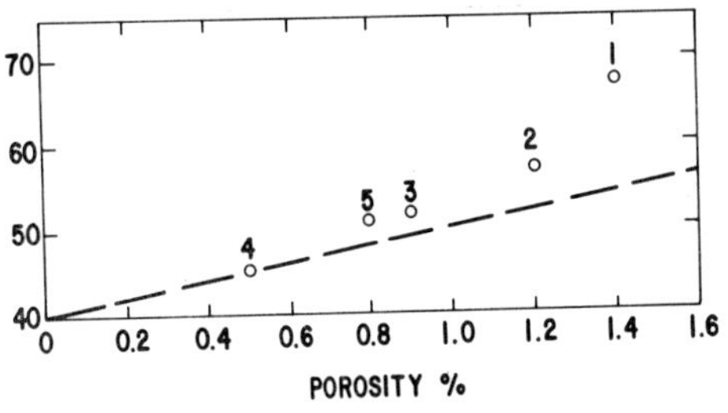

Fig. 1: Ferrimagnetic resonance linewidth ΔH and porosity P for samples 1-5 of Table I. The straight line Eq. (1) gives the estimated effect of pores upon ΔH.

In Fig. 1 we plot ΔH and porosity P for samples 1-5. The observed ferrimagnetic resonance line width ΔH increases with porosity, but at a rate greater than that predicted by the linear relation[5]

$$\Delta H = \Delta H_o + 1.5(4\pi M_o)P \quad (1)$$

In (1) ΔH_o is the broadening of the polycrystalline ferrimagnetic resonance at zero porosity, and M_o is the saturation magnetization. The excess variation in ΔH between samples 1-5 is thus unexpected in view of the otherwise closely similar microstructural[6] and macroscopic magnetic characteristics of these samples.

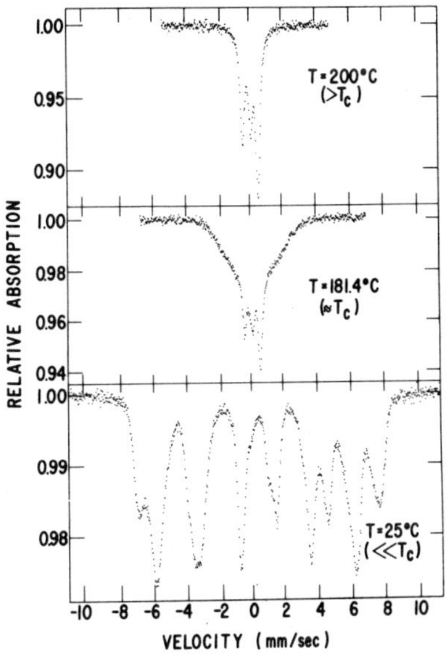

Fig. 2: Mössbauer spectra of $(Y_{2.66}Gd_{0.34})(Fe_{4.19}Al_{0.67}Mn_{0.09})O_{12}$ for $T \ll T_c$, $T \approx T_c$, $T > T_c$ (Sample 1). In the vicinity of T_c the spectra exhibit the coexistence of ordered and paramagnetic phases. Zero velocity is with respect to Fe-metal.

We have attempted to characterize the quality of samples 1-5 by Mössbauer spectroscopy in the vicinity of T_c. In Fig. 2 we give typical spectra of the garnets (sample 1) for $T \ll T_c$, $T \approx T_c$ and $T > T_c$. In the vicinity of T_c the Mössbauer spectra exhibit the coexistence of ferrimagnetic and para-

TABLE I

No.	Sint. Time (hrs)	L.P.[3] (Å)	Grain Size (μ)	Density (gm/cc)	P^4 (%)	$4\pi M_o$ (gauss)	T_c (°C) (mag.)	T_c (°C) (Möss.)	ΔT_c (°C) (Möss.)	ΔH (Oe)
1	3[1]	12.344	6.5	5.164	1.4	671±10	186±2	184	9	67
2	6	12.344	6.9	5.175	1.2	-	-	-	-	57
3	18	12.344	7.2	5.185	0.9	-	-	-	-	52
4	66	12.344	7.6	5.205	0.5	-	-	184	3.2	45
5	3[2]	12.344	7.1	5.180	0.8	668±10	184±1	185	2.3	51

1. Samples 1-4 mixed, calcined at 1300°C.
2. Sample 5 mixed, calcined at 1300°C, fluid energy milled.
3. L.P. = Lattice Parameter
4. P = Porosity.

magnetic phases. A similar coexistence of ferrimagnetic and paramagnetic phases has been observed[7] in $Y_3(Fe_{4.5}Sc_{0.5})O_{12}$ and $Y_3(Fe_{4.25}Ga_{0.75})O_{12}$, and ascribed to the presence of critical superparamagnetism.[8] This should be compared with the results of Van der Kraan et al.[9] who have prepared the latter garnet by a ceramic method differing from that used by Coey.[7] These authors find an almost negligible coexistence region and resolve the conflict between their results[9] and those of Coey[7] by contending that the ceramic processing of the latter produced inhomogeneous garnet material.

The width of the temperature region in which paramagnetic and ferrimagnetic phases coexist can be simply determined from the Mössbauer spectra.

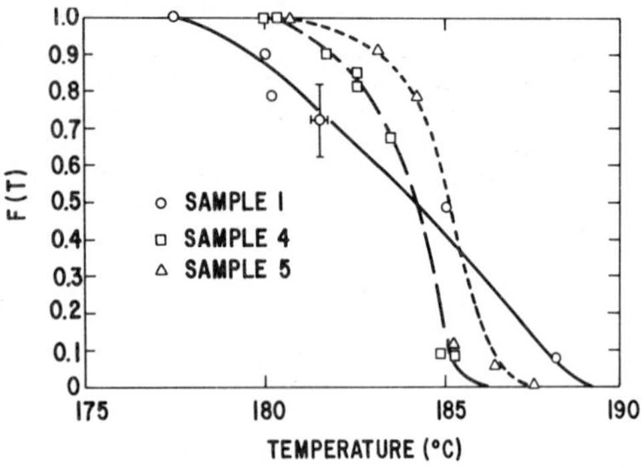

Fig. 3: Fraction of sample ferrimagnetic (F) versus temperature for samples 1, 4 and 5.

In Fig. 3 we plot F(T), the fraction of sample which is ferrimagnetic (area of spectrum which shows hyperfine splitting divided by total area) as a function of temperature for samples 1, 4 and 5. Clearly sample (1) has a much larger coexistence range than either samples 4 or 5. Defining ΔT_c as the temperature range $\Delta T_c = T_{c2} - T_{c1}$ where $F(T_{c2}) = 0.1$, $F(T_{c1}) = 0.9$, we obtain the values of ΔT_c listed in Table I. As expected, increased sintering time reduces sample inhomogeneity and is reflected in the reduction of ΔT_c. It is of practical interest that passage of the calcined sample through the fluid energy mill (sample 5) is at least as effective as about 60 hours of additional sintering time in reducing the width of the coexistence region. This illustrates the extreme importance of ceramic processing on the chemical homogeneity of the sintered product.

Although there is no direct evidence, it appears plausible to argue that the coexistence region in these samples is a result of local fluctuations in the Fe/Al ratio. Supporting evidence for this point of view can be derived from x-ray line broadening. Measurement of the $(8,8,0)K_\alpha$ line of Fe radiation gives a width of $\Delta_1(2\theta) = 0.28°$ (sample 1, 3 hours sintering) and $\Delta_4(2\theta) = 0.25°$ (sample 4, 66 hours sintering).

Thus $\Delta_1 - \Delta_4 = 0.03 \pm 0.01°$, and this can be associated with an excess internal lattice parameter distribution $\Delta a = 0.0022$ Å in sample 1. Using tabulated[10] data on the $Y_3(Fe,Al)_5O_{12}$ system we have $da/dx = 0.073$ Å, where x is the amount of Al^{3+} in the formula $Y_3Fe_{5-x}Al_xO_{12}$. From Ref. (10) we also have $dT_c/dx = 1.33 \times 10^2$ °C. Hence the x-ray line broadening implies ΔT_c(sample 1) - ΔT_c (sample 4) = 4 ± 2°C provided we ascribe the x-ray broadening to local variations in the Al/Fe ratio. This order of magnitude value is in reasonable agreement with the Mössbauer results (see Table I).

The variation in Al/Fe ratio may also be calculated from the coexistence region, ΔT_c, and the dT_c/dx data cited. A 7 to 10% spread in the Al concentration (sample 1) is inferred and a similar spread could be anticipated in the Gd concentration, although the latter would escape detection by Mössbauer or x-ray spectroscopy. By examining the variation of g_{eff} with composition in closely related series[11] we can estimate that these composition fluctuations could contribute some 5 to 8 Oe to ΔH. This is a significant part of the apparent excess in ΔH above that predicted by relation (1) (see Fig. 1).

The qualitative results by which homogeneity increases with sintering time or as a result of post-calcine milling are expected from general ceramic practice. The present work quantifies that part of the improvement obtained in the microwave parameter ΔH (often used for quality control), resulting from chemical homogenization. The application of this method and additional thermomagnetic measurements to other stages of garnet ceramic processing will be reported elsewhere.

REFERENCES

1. G. P. Rodrigue, J. Appl. Phys. 40, 929-937 (1969).
2. A. E. Paladino and E. A. Maguire, J. Amer. Cer. Soc. 53, 98-102 (1970).
3. G. H. Jonker and A. L. Stuijts, Philips Tech. Rev. 32, 79-95 (1971).
4. Helme Products, Inc., Helmetta, N. J.
5. E. Schlomann, Conf. on Magnetism and Magnetic Materials, AIEE Special Pub. T-91, 600-609 (1956). Equation (1) probably overestimates the effect of porosity P for P < 0.05. See Fig. 10 of this ref.
6. The very slight variation in grain size will not materially affect ΔH. See S. L. Blum, J. E. Zneimer and H. Zlotnick, J. Amer. Cer. Soc. 40, 143-149 (1957).
7. J. M. D. Coey, Phys. Rev. B6, 3240-3253 (1972).
8. L. M. Levinson, M. Luban and S. Shtrikman, Phys. Rev. 177, 864-871 (1969).
9. A. M. van der Kraan, J. J. van Loef and W. Tolksdorf, Phys. Stat. Solidi 17a, K79-K80 (1973).
10. "Handbook of Microwave Ferrite Materials", W. H. von Aulock, ed., Academic Press, N.Y., 1965.
11. G. R. Harrison and L. R. Hodges, in "Physics of Electronic Ceramics", L. L. Hench and D. B. Dove, eds., Marcel Dekker N.Y., 1972, Part B, p. 858-914.

MAGNETIC PROPERTIES OF SINGLE CRYSTAL CuNi-18H AND MgZn-18H SOLID SOLUTIONS

R. O. Savage, A. Tauber, J. R. Shappirio
US Army Electronics Technology and Devices Laboratory (ECOM), Fort Monmouth, N.J. 07703

ABSTRACT

The magnetization and anisotropy properties of single crystal CuNi-18H, $(Ba_{5.1}(Ni_{1.1}Cu_{0.4})Ti_{2.7}Fe_{12.3}Mn_{0.4}O_{31})$ and MgZn-18H, $(Ba_{5.4}(Mg_{1.3}Zn_{0.7})Ti_{2.9}Fe_{11.7}O_{31})$ solid solution hexagonal ferrites have been investigated in fields up to 16 kOe and temperatures from 4.2 to T_C. Both compositions exhibit an easy plane of magnetization at all temperatures except for CuNi-18H which has a complex behavior below 135 K. MgZn-18H exhibits a magnetization, σ_S = 33±1G cm³/g at 300 K. The magnetic moment $n_\beta(H = \infty, T = 0) = 12.5\mu_\beta$ and T_c=357±3 K. CuNi-18H could not be magnetically saturated below 135 K either ∥ to (0001) plane or [0001] axis. σ in 16 kOe field and 4.2 K equals 20±2G cm³/g, at 295 K the saturation value is equal to 18.5±1G cm³/g; T_C=564±3 K. The cation distribution is deduced from the magnetic moment and structure.

INTRODUCTION

Hexagonal ferrites have been under investigation at this laboratory for several years as both potential narrow and broad ferrimagnetic resonance linewidth materials for several device applications at microwave and millimeter-wave frequencies. Recently a new 18-layer structure $Ba_5Me_2Ti_3Fe_{12}O_{31}$ where Me=Zn, Mg, Co, and Cu was discovered and several compositions have been investigated.[1,2] In this paper the magnetic properties of two solid solutions MgZn-18H and CuNi-18H have been investigated as a function of temperature. The CuNi system is of interest for two reasons: Attempts to obtain end member Ni containing crystals have been unsuccessful and Ni containing hexagonal ferrites exhibit small temperature coefficients of magnetization. The latter is an important property for device applications. The second reason relates to the influence of Cu^{2+} ions in creating uniaxial anisotropy.[2]

EXPERIMENTAL PROCEDURE

Crystals were grown by the molten salt (flux) method. The flux was barium borate. Weighed amounts of oxides and/or carbonates were placed in platinum crucibles then raised to a maximum temperature of 1250 C, cooled at rates between 1.0 and 3.5 C/hour to 1050 C. The crucibles were air quenched and leached in dilute HNO_3 to free the crystals. Single crystal x-ray diffractometry was employed to determine that the crystals were single of the correct structure and lattice parameter. Metallography was used to determine the soundness (absences of inclusions) of the crystals. Electron-microprobe analysis of the crystals was employed to determine the composition. Empirical formulas of $Ba_{5.1}(Ni_{1.1}Cu_{0.4})Ti_{2.7}Fe_{12.3}Mn_{0.4}O_{31}$ for CuNi-18H and $Ba_{5.4}(Mg_{1.3}Zn_{0.7})Ti_{2.9}Fe_{11.7}O_{31}$ for MgZn-18H were obtained from the wt percents of Table I by normalizing for 22 cations and assuming 31 oxygen atoms.

TABLE 1

CHEMICAL COMPOSITION

Mg Zn - 18H		Cu Ni - 18H	
Element	wt percent	Element	wt percent
Ba	26.21 ± 0.60	Ba	33.37 ± 0.24
Ti	4.94 ± 0.03	Ti	6.16 ± 0.08
Fe	23.20 ± 0.12	Fe	32.85 ± 0.21
Zn	1.65 ± 0.09	Ni	3.08 ± 0.03
Mg	1.10 ± 0.02	Cu	1.23 ± 0.02
O	42.89 ± 0.77	Mn	0.92 ± 0.00
		O	22.39 ± 0.43

The crystals are typically black, thin hexagonal plates 0.1 to 0.5 mm thick, 1-3 mm in the longest dimension and weighed 5-30 mg. Magnetization isotherms were obtained with a P.A.R. vibrating sample magnetometer with H applied ∥ and ⊥ to c_o. Other details of measurement techniques have been described elsewhere.[1,2]

RESULTS AND DISCUSSION

MgZn-18H and CuNi-18H exhibit an easy plane of magnetization at all temperatures $(E_a = K_1 \sin^2\theta + K_2 \sin^4\theta)$ except for CuNi-18H which exhibits complex behavior below 135 K. The magnetization was invariant with rotation about [0001] for all temperatures at which the basal plane is the easy plane of magnetization. Crystals of CuNi-18H could not be saturated ∥ or ⊥ to c_o axis below 135 K. The magnetization curves H ∥ and ⊥ to c_o cross at 4.2 K and with increasing temperature cross at successively lower fields until 135 K. The composition of CuNi-18H in Table I indicates that Ni is predominant. This accounts for the absence of uniaxial anisotropy[2] associated with Cu^{2+}. It also accounts for the low temperature coefficient of magnetization below room temperature. NiY has a very small temperature coefficient of magnetization.[3] 18H, an 18-layer hexagonal structure, contains 12 layers of Y. Each Y block is separated from the next Y block by an intervening 3-layer hexagonal $BaTiO_3$ structure type block. Thus Y blocks comprising 2/3 of the structure impart a major character to 18H. T_c = 564±4 K for CuNi-18H and T_c = 357±3 K for MgZn-18H.

A magnetic moment of $n_\beta(H = \infty, T = 0) = 12.5\mu_\beta$ was found for MgZn-18H. The following cation distribution was computed from the previously deduced colinear spin model (9 oct →← 4 tet ← 4 oct)[1],

$$(7.1Fe^{3+}, 1.9Ti^{4+}) \rightarrow\leftarrow (0.7Zn^{2+}, 3.3Fe^{3+}) \leftarrow$$

$$(1.3Fe^{3+}, 1.3Mg^{2+}, 0.4Ba^{2+}, 1.0\ Ti^{4+})$$

with a moment distribution of

$$(35.5\mu_\beta)\rightarrow\leftarrow(16.5\mu_\beta) = \vec{12.5}\mu_\beta$$

The magnetic moment of $5\mu_\beta$ was assumed for $Fe^{3+}(d^5)$.

The indicated cation distribution is not unique. The Fe^{3+} distributed between the summed opposing sublattices is unique. This is forced by the agreement with the experimental moment and crystal chemical arguments. While small amounts of zinc may be found on octahedral sites, most experimental data supports zinc occupancy on tetrahedral sites. The distribution of the non-magnetic cations other than zinc over the remaining octahedral sites is arbitrary.

The anisotropy constants K_1 and K_2 were extracted from $H_i/I = (-2K_1+4K_2)/I_s^2 + (4K_2 I^2/I_s^4)$ where H_i is the internal field, I_s is the spontaneous magnetization, and I is the magnetization at some field H.[4] The fit to the magnetization isotherms with H ‖ to the hard c_o axis was obtained on a computer by regression analysis. The anisotropy field was computed from $H_a = -2(K_1+2K_2)/I_s$. The magnetization and anisotropy fields are plotted as a function of temperature in Figs. 1 and 2.

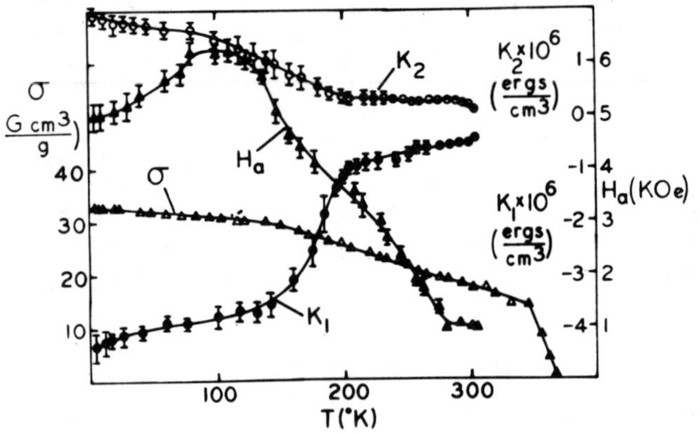

Fig. 1. MgZn-18H: Saturation magnetization σ, anisotropy constants K_1 and K_2, and anisotropy field H_a as a function of temperature. Error bars indicate the spread for four crystals.

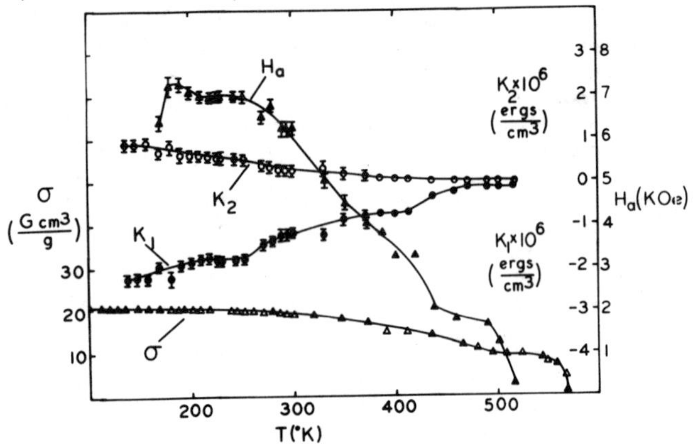

Fig. 2. CuNi-18H: Saturation magnetization σ, anisotropy constant K_1 and K_2, and anisotropy field H_a as a function of temperature. Error bars indicate the spread in data for three crystals.

ACKNOWLEDGMENT

The authors are grateful to Joseph Megill for assisting in many measurements.

REFERENCES

1. A. Tauber, R.O. Savage, Jr., and M.D. Grebenau, J. Appl. Phys. 42, 1738 (1971).
2. A. Tauber, R.O. Savage, Jr., and M.D. Grebenau, AIP Conference Proceedings No. 5, 1547 (1972).
3. J. Smit and H.P. Wijn, Ferrites, (Wiley, New York, 1959) Chapter IX, p. 197.
4. H. Zijlstra, Experimental Methods in Magnetism II, (Wiley, New York, 1967) p. 181.

MICROWAVE RESONANCE STUDY OF $Ba_3Zn_2Fe_{24}O_{41}$

R. W. Grant, M. D. Lind, G. P. Espinosa, and I. B. Goldberg
Science Center, Rockwell International, Thousand Oaks, CA 91360
and
F. E. Reisch
Collins Radio Group, Rockwell International, Dallas, TX 75207

ABSTRACT

Microwave resonance data at 35 GHz were obtained on single-crystal spheres of $Ba_3Zn_2Fe_{24}O_{41}$ (Zn_2Z) at room temperature. Values derived for the anisotropy field, saturation magnetization and first order anisotropy constants are $H_a = 4.81\pm0.04$ kOe, $M_o = 316\pm20$ G, and $K_1 = +0.76\pm0.05\times10^6$ erg/cm^3, respectively. The lowest resonance linewidth observed is $\Delta H = 38$ Oe. A phenomenological model is developed for predicting values of K_1 for zinc-containing hexagonal ferrites.

INTRODUCTION

Magnetically-tuned filters and oscillators employing single-crystal YIG spheres are widely used in microwave applications up to 18 GHz. The design of devices operating in the 18-26.5 and especially in the 26.5-40 GHz bands is severely hampered by the difficulty of obtaining electromagnets (with sufficient linearity and low hysteresis) capable of providing the large applied fields, H_o, necessary to achieve ferrimagnetic resonance. Hexagonal ferrites with large H_a offer an attractive solution for this application if materials with low ΔH can be obtained.

The majority of hexagonal ferrites belong to the system $BaO-Fe_2O_3-MeO$ where Me is one of several possible divalent cations. It is frequently observed that Zn^{2+}-containing members of a particular crystallographic structure exhibit exceptionally narrow ΔH. Most hexagonal ferrites have either uniaxial or planar magnetic anisotropies. For tunable resonator applications, uniaxial anisotropy is preferable because of the linear dependence of resonance frequency, f_o, on H_o. H_a for most unsubstituted Zn^{2+} containing uniaxial hexagonal ferrites is too large for use in the 18-40 GHz range. An exception is $Ba_3Zn_2Fe_{24}O_{41}$ (Zn_2Z) which, based on literature data for M_o (310 G)[1] and K_1 ($+0.66\times10^6$ erg/cm^3)[2] should have $H_a = 2K_1/M_o \approx 4.3$ kOe at room temperature. Ferrimagnetic resonance data reported for polycrystalline samples of this compound yielded $H_a = +7.3$ kOe and $\Delta H = 1115$ Oe.[3] The purpose of the present work was to obtain resonance data on single-crystal specimens of Zn_2Z and to assess its suitability for applications in the 18-40 GHz range.

EXPERIMENTAL

Single crystals of Zn_2Z were grown from $BaO-B_2O_3$ melts by methods similar to those used by Kerecman, et al.,[4] to prepare $Ba_4Zn_2Fe_{36}O_{60}$ (Zn_2U). The procedure used to obtain single-phase Zn_2Z will be discussed elsewhere.[5] Buerger precession X-ray diffraction photography was used to identify Zn_2Z. The identification was based both on measured lattice constants (a = 5.88 and c = 52.30Å) and the diffraction symmetry (space group $P6_3/mmc$).[6] X-ray fluorescence measurements in a scanning electron microscope qualitatively confirmed the chemical composition.

Single-crystal spheres of Zn_2Z were used for the microwave measurements. Cubes (cut from single crystals with a wire saw) were air-tumbled into rough spheres and subsequently ground into a spherical shape with a "race-track" type of grinding device (see, *e.g.*, Ref. 7). With care, spheres of 0.28-0.43 mm diameters were obtained. Because ΔH often depends on surface finish (see, *e.g.*, Ref. 8), final polishing was carried out with 0.05μm Al_2O_3 to obtain a highly-polished surface. The spheres were oriented in a magnetic field and attached with ambroid to the end of a wooden rod.

Ferrimagnetic resonance absorption measurements at 35 GHz were carried out by a transmission line technique. Zn_2Z was inserted into a reduced height (TE_{10}) waveguide section to provide good coupling to the sphere. The E-plane width at the sphere was 0.48 mm.

The microwave circuit consisted of a klystron (70 mW output), isolator and a 10 db directional coupler to split the power. The low-power arm was fed into a calibrated wavemeter and Si detector. This was used to lock the klystron frequency via an AFC circuit. The high-power arm was led to a rotary vane attenuator (±0.10 db), a slide screw tuner, and then to the stepped waveguide and detector. Most measurements were carried out between 9-15 db attenuation. Coils mounted around the stepped waveguide provided 15 kHz field modulation for phase-sensitive detection. A 15-inch magnet was used to sweep H_o.

RESULTS

Microwave resonance spectra at 35.155 GHz on visually-perfect Zn_2Z spheres of 0.28 and 0.38 mm diameter are shown in Fig. 1 (c was $\parallel H_o$). Both the absorption and first derivative curves are shown along

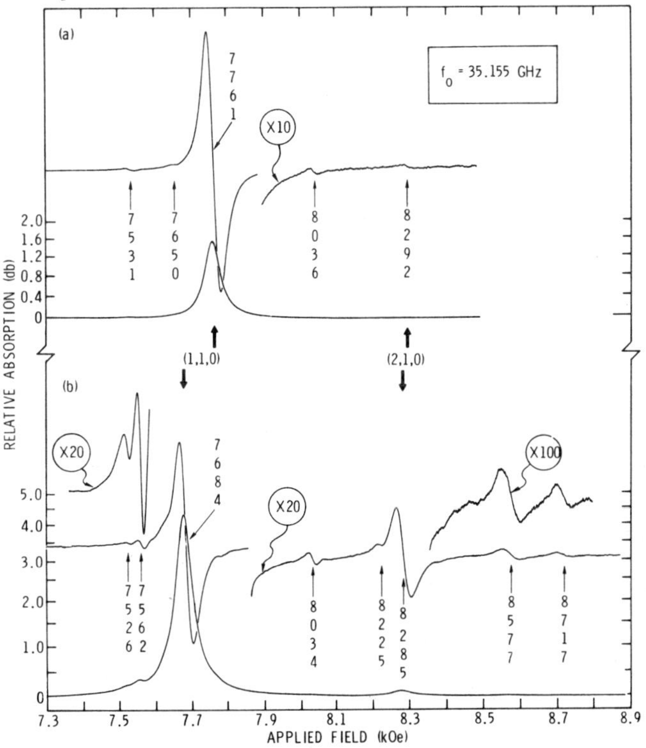

Fig. 1 Ferrimagnetic resonance absorption spectra for spheres of Zn_2Z at room temperature. Sphere diameters were (a) 0.28 mm, and (b) 0.38 mm. Resonance fields observed for several modes are indicated in the figure.

with some expanded regions of the derivative spectra. In addition to the fundamental resonance (1,1,0), several higher-order magnetostatic modes are observable, especially in the derivative spectra. The resonance at ≈ 8.29 kOe was observed in almost all measurements. This mode is more strongly excited in Fig. 1b than in 1a presumably because of the large sphere diameter and correspondingly larger nonuniformities in the RF field; we assume this corresponds to the (2,1,0) mode.[9] The value for H_o of this mode is $H_{(2,1,0)} = 8279\pm15$ Oe.

$H_{(1,1,0)}$ was found to be sensitive to the sample

diameter presumably because of induced currents in the waveguide walls[10]; a down-field shift of 77 Oe is observed for $H_{(1,1,0)}$ between Fig. 1b and 1a. Consequently, the value of $\Delta = H_{(2,1,0)} - H_{(1,1,0)}$ was obtained from the data of Fig. 1a where these effects should be minimized [a crude estimate indicated that possible residual wall effects expected for the 0.28 mm sphere are small and are within the uncertainty quoted for $H_{(1,1,0)}$]; the values are $\Delta = 530\pm30$ Oe and $H_{(1,1,0)} = 7749\pm34$ Oe. H_a calculated from $f_o = \gamma(H_o + H_a)$ is 4.81 ± 0.04 kOe; γ, the gyromagnetic ratio, was assumed to be 2.80 MHz/Oe. Δ is $8\pi M_o/15$; thus $M_o = 316\pm20$ G in good agreement with Ref. 1 (310 G). K_1 calculated from $K_1 = H_a M_o/2$ is $+0.76\pm0.05\times10^6$ erg/cm^3. This is somewhat higher than the previously-reported[2] value of $+0.66\times10^6$ erg/cm^3 but the measurement error limits overlap. The previously-reported[3] H_a value appears to be incorrect.

The indexing of other magnetostatic modes, especially below $H_{(1,1,0)}$ is considerably less certain. Above $H_{(1,1,0)}$ it is tempting to associate the resonances observed at 8.577, 8.225, and 8.035 kOe with the (3,1,1), (4,2,1) and (5,3,1) modes. By using Walker's calculations[9] and the derived material parameters, these modes are expected to occur at 8.55, 8.25 and 8.05 kOe, respectively; without additional experimental information, this assignment is very tenuous.

ΔH was evaluated by assuming that the absorption curve was Lorentzian. With this assumption, ΔH (full width of the absorption curve at 1/2 maximum absorbed power) is

$$\Delta H = \Delta H_j \left[\left(1 - 10^{0.1A_j}\right)\Big/\left(10^{0.1A_j} - 10^{0.1A_{max}}\right)\right]^{1/2} \qquad (1)$$

where A_j is the absorption at an arbitrary point j (with corresponding linewidth ΔH_j) and A_{max} is the maximum absorption; both A_j and A_{max} are expressed in db.

For spherical samples fabricated as outlined above, typical values of ΔH were $\approx$ 60 Oe. It has previously been noted[8] that substantial improvement in ΔH for some hexagonal ferrites can be obtained by annealing samples in O_2. We observed similar effects for Zn_2Z. For example, a specimen with a nominal diameter of 0.33 mm (sample was slightly aspherical) exhibited a ΔH of 62 Oe. After annealing in O_2 at 850°C for 1, 2.5 and 19.5 hrs, the measured ΔH were 38, 40 and 39 Oe, respectively. Thus, after 1 hr of annealing, no further ΔH reduction was obtained. In another experiment on a different specimen, it was observed that essentially all the ΔH improvement is obtained after only 15 min. of O_2 annealing at 850°C. The 38 Oe ΔH mentioned above is the narrowest ΔH achieved so far for Zn_2Z. For comparison, ΔH derived from the data of Fig. 1a and b are 47 and 40 Oe, respectively (both spheres were annealed in O_2 for 45 min. at 850°C).

Several determinations of K_1 have been made for zinc-containing members of the hexagonal ferrites. It is observed that K_1 monotonically decreases from $BaFe_{12}O_{19}$(M) to $Ba_2Zn_2Fe_{12}O_{22}$(Y), and from M to $ZnFe_2O_4$ in the phase diagram. There are over 60 structures in the BaO-Fe_2O_3-ZnO system.[11] These structures are formed from three basic crystallographic units which are conventionally referred to as R($BaFe_6O_{11}$), T($Ba_2Fe_8O_{14}$) and S(Fe_6O_8) blocks; the stacking arrangements of these blocks leads to the large number of structures. Within these structures Zn is expected to replace Fe in the tetrahedral sites of the S and T blocks to varying degrees. The relative amounts of the three structural blocks varies systematically along the M-Y and M-S joins of the phase diagram, which suggests the possibility of associating values of K_1 with R, T and S in an effort to roughly estimate the variation of K_1 in the system. For this purpose we used the expression

$$K_1 = \sum_i V_i K_i / V \qquad (2)$$

V_i is the volume of the ith block, K_i is the value of K_1 associated with the ith block, and V is the total unit cell volume of a particular structure; the sum is over all blocks within V. The values used for V_i were $V_S = 143.8$, $V_R = 203.6$ and $V_T = 290.4$ Å^3. The K_i were arbitrarily calculated from K_1 data for M(R_2S_2),[2] $Zn_2U(R_2S_3T)$,[12] and $Zn_2X(R_2S_3)$,[13] and are $K_R = +6.59\times10^6$, $K_S = -1.37\times10^6$ and $K_T = -1.78\times10^6$ erg/cm^3. With these parameters K_1 can be estimated for the more than 60 known members of the system. In particular, K_1 can be estimated for Zn_2Z and Zn_2W where experimental data exist; the estimated K_1 are $+0.56\times10^6$ and $+1.93\times10^6$ erg/cm^3, respectively. These estimates compare reasonably well with the K_1 for Zn_2Z (as measured herein) and the K_1 for Zn_2W[14] of $+1.8\times10^6$ erg/cm^3. The model indicates that only the R block (which contains the unusual five-fold coordinated Fe site) contributes a large positive contribution to K_1 for a given structure. It is expected that $Zn_2Y(S_3T_3)$ should have planar anisotropy (as observed) since it does not contain this block. Because estimates of M_o can be made for all of the structures,[1,15] it is also possible to estimate H_a throughout the system.

CONCLUSIONS

The magnetic properties of Zn_2Z suggest that it will be a useful resonator material in the 18-40 GHz range if additional ΔH reduction is achieved. The lowest observed value of ΔH (38 Oe) is exceptionally narrow for a uniaxial unsubstituted hexagonal ferrite at room temperature. Mn-substituted members of both the U and M structures are found to have narrower ΔH than we observe in Zn_2Z.[4,16] Also, ΔH in Zn_2Y was improved by Mn substitution.[17] Zn_2Y and Zn_2U bracket Zn_2Z in the phase diagram which suggests that ΔH in Zn_2Z may be improved by Mn substitution.

ACKNOWLEDGEMENTS

We thank Dr. B. R. Tittmann for discussions, and E. H. Cirlin for performing the SEM measurements.

REFERENCES

1. J. Smit and H. P. J. Wijn, Ferrites (John Wiley & Sons, Inc., New York, 1959).
2. H. B. G. Casimir, J. Smit, U. Enz, J. F. Fast, H. P. J. Wijn, E. W. Gorter, A. J. W. Duyvesteyn, J. D. Fast, and J. J. deJong, J. Phys. Radium 20, 360-73 (1959).
3. R. A. Braden, I. Gordon, and R. L. Harvey, IEEE Trans. Mag. MAG-2, 43-7 (1966).
4. A. J. Kerecman, T. R. AuCoin, and W. P. Dattilo, J. Appl. Phys. 40, 1416-7 (1969).
5. G. P. Espinosa, M. D. Lind, R. W. Grant, and F. E. Reisch, to be published.
6. P. B. Braun, Philips Res. Rep. 12, 491-548 (1957).
7. S. P. A. Marriott, Marconi Rev. 33, 79-110 (1970).
8. A. Tauber, S. Dixon, Jr., and R. O. Savage, Jr., J. Appl. Phys. 35, 1008-9 (1964).
9. L. R. Walker, Phys. Rev. 105, 390-9 (1957).
10. B. Lax and K. J. Button, Microwave Ferrites and Ferrimagnetics (McGraw-Hill, Inc., New York, 1962), Chapter 10.
11. J. A. Kohn, D. W. Eckart, and C. F. Cook, Jr., Science 172, 519-25 (1971).
12. A. J. Kerecman, A. Tauber, T. R. AuCoin, and R. O. Savage, J. Appl. Phys. 39, 726-7 (1968).
13. L. M. Silber and E. Tsantes, IEEE Trans. Mag. MAG-5, 600-4 (1969).
14. Calculated from M_o data of L. G. Van Uitert, M. H. Read, and F. J. Schnettler, J. Appl. Phys. 28, 280-1 (1957), and H_a data of L. R. Hodges, Jr., and G. R. Harrison, J. Am. Ceram. Soc. 47, 601-5 (1964).
15. E. W. Gorter, Proc. IEE (London) B104, 255-60 (1957)
16. S. Dixon, Jr., T. R. AuCoin, and R. O. Savage, J. Appl. Phys. 42, 1732-3 (1971).
17. R. O. Savage, Jr., S. Dixon, Jr., and A. Tauber, J. Appl. Phys. 36, 873-4 (1965).

INTERACTION OF SURFACE MAGNETIC WAVES IN ANISOTROPIC MAGNETIC SLABS

A. K. Ganguly, C. Vittoria, and D. Webb
Naval Research Laboratory
Washington, D. C. 20375

ABSTRACT

Group delay of magnetostatic surface wave propagation in parallel layers of magnetic slabs have been considered. The dispersive characteristics depend strongly on the separation between the slabs and their crystal plane orientations with respect to the applied field $\vec{H}_0$ where propagation is normal to $\vec{H}_0$ and in the plane of the slab. As an example we consider $\vec{H}_0$ along <100> direction in slab (Slab A) in contact with a metal surface and <110> direction in the other slab (Slab B) exposed to air. Both slabs are (100) planes. The variation of delay time and bandwidth with relative orientation of the two slabs and their thicknesses are shown. Much larger bandwidth is obtained with two magnetic slabs as compared to that of a single magnetic slab configuration.

INTRODUCTION

Magnetostatic surface wave devices offer the possibility of carrying out signal processing directly at microwave frequencies. However, the bandwidth of these devices is currently inadequate for many applications. Bongianni[1] has devised a scheme for increasing the bandwidth by using multiple magnetic films, each biased with a separate magnetic field. In a recent paper[2] we discussed a somewhat simpler configuration for accomplishing the same objective. We proposed applying a single external magnetic bias field to a multiple film structure, and using orientational variation of the magnetic anisotropy field to achieve the desired internal magnetic field values.

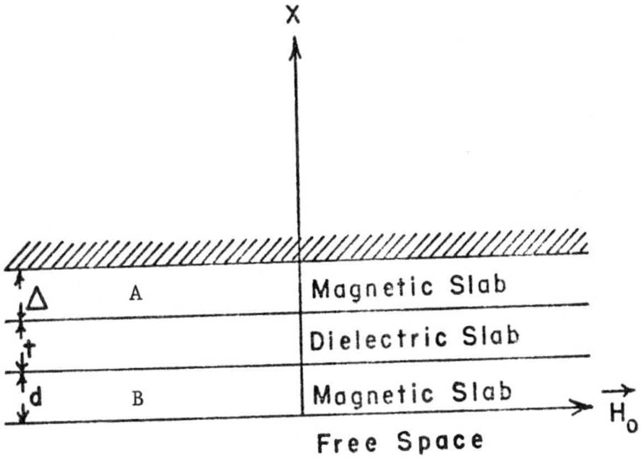

Fig. 1 A cross-view of the geometrical configuration of the three slabs. The slabs are in yz-plane. The x-axis is normal to the slab plane and the z-axis is parallel to $\vec{H}_0$. The magnetostatic waves propagate in y-direction.

As would be expected, we found[2] that there are two branches to the dispersive characteristics for a given value of applied magnetic field, H_0 (see Eq:(3) of Ref. 2). The shape of the dispersive characteristics can be significantly altered by 1) varying the separation between parallel layers of magnetic slabs and 2) varying the orientation of the slabs with respect to the applied field. This is due to the magnetostatic interaction between the magnetic moments in the two slabs. In this paper we calculate the dependence of some practical device quantities, delay time and bandwidths, upon physical and geometrical parameters. As a practical example we discuss characteristics of a geometry employing two single crystal yttrium iron garnet (YIG) slabs as shown in Fig. 1. However, we will demonstrate that it is possible to enhance the bandwidth by using materials with larger values of saturation magnetization and magnetic anisotropy, such as lithium ferrite.

RESULTS

The delay time (τ) is calculated for the case when the (100) slab next to the conductor (slab A) is aligned so that H_0 is in the slab plane and along the <100> axis. The other (100) slab (B) is free to

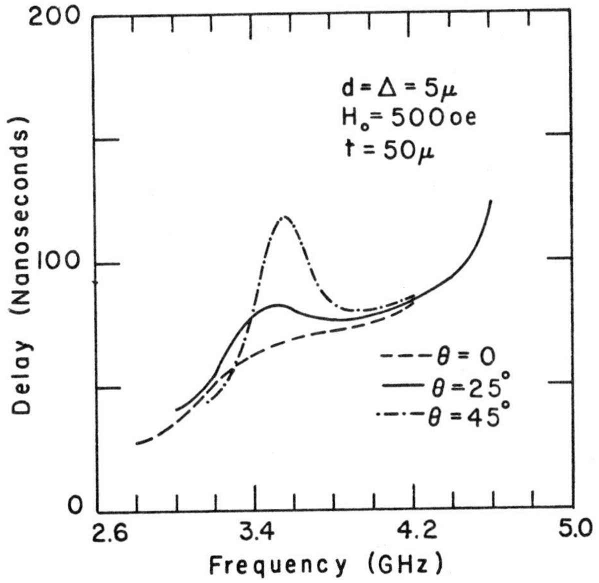

Fig. 2 Delay as a function of frequency for different values of θ.

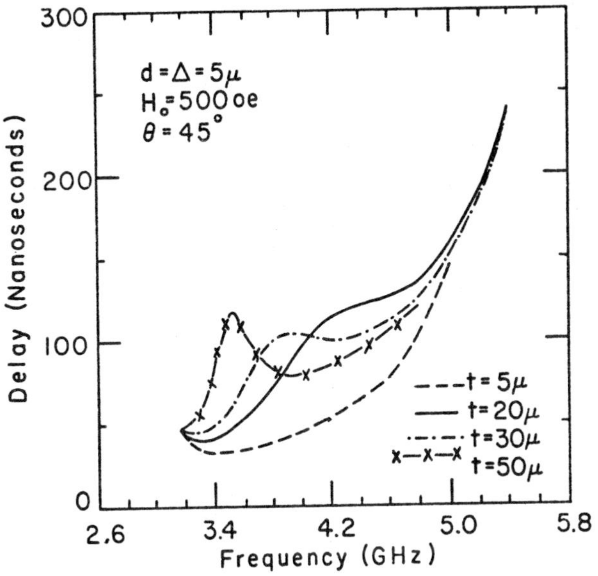

Fig. 3. Delay as a function of frequency for $\theta = 45^o$. Here $d = \Delta = 5\mu$. The values of t are 5μ, 20μ, 30μ and 50μ respectively for the curves as indicated in the figure.

rotate. In Fig. 2 we have plotted the delay time as a function of frequency for various values of the angle θ, between two respective <100> axis in slabs (A) and (B). The magnitude and direction of H_o is fixed. In the calculation of delay time the propagation length is taken as 1 cm. As seen from the figure the delay time remains reasonably constant (varying by $\pm$ 5%) over a wide frequency range. We will define this frequency range as the bandwidth (B). In this frequency range τ shows marked variation with θ. In Fig. 3 the delay time is plotted as a function of frequency for fixed values of H_o and different values of the dielectric slab thickness when $\theta = 45^o$. At very small values of t ($t \leq d$) another region of nearly constant delay appears near the lower end of the allowed frequency range with much smaller values of τ. Since the dispersion relation[2] is of the form F (kd, kΔ, kt, ω) = 0, it is easy to see that the group velocity $v_g = d\omega/dk$ is scaled to rv_g when d, Δ and t are simultaneously changed to rd, rΔ and rt respectively. Hence τ is scaled to τ/r. Thus the curves of Fig. 3 can be applied to other values of d, Δ and t by proper scaling. The bandwidth does not change under this scaling.

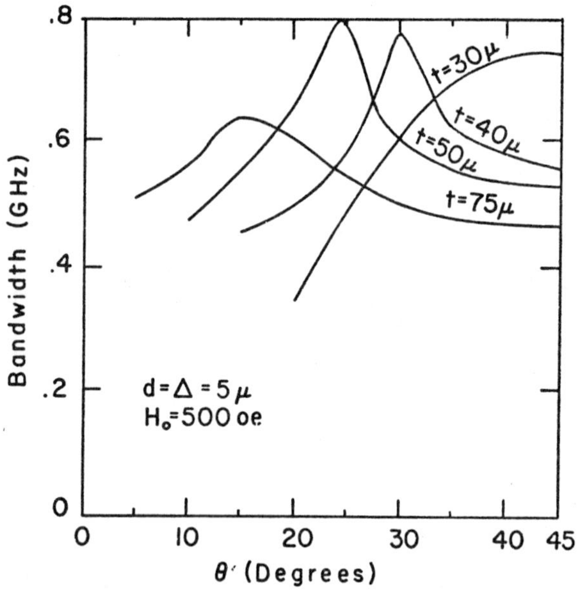

Fig. 4. Bandwidth vs θ for $d = \Delta = 5\mu$ and four different values of t.

In Fig. 4 we plot the bandwidth (B) as a function of θ for different values of dielectric slab thickness (t) with YIG films being 5μ in thickness. Maximum value of B first increases with t and then decreases. The angle θ where B is maximum depends on the parameters d, Δ and t. The bandwidth also decreases slowly with an increase in the applied d.c. magnetic field. The following table gives some typical values of B in YIG films with H_o = 500 Oe and $d = \Delta = 5\mu$.

B (GHz)	f_o (GHz)	τ (ns)	t (μ)	θ
.74	4.12	99.7	30	45°
.81	3.78	80.2	50	25°
.65	3.50	63.4	75	15°

Here f_o is the center frequency. When H_o = 250 Oe we get B = .84 GHz with τ = 72.5 ns and f_o = 2.96 GHz for t = 50μ . These values of B are much higher than the values obtained with a single magnetic film. Bongianni[3] reported a bandwidth of .25 GHz for the single film case.

We have so far shown the results for $d/\Delta = 1$ and various ratios of t/Δ . Calculations show that for given Δ and t/Δ, the bandwidth can be increased by selecting $d \neq \Delta$. As for example, when $\Delta = 5\mu$ and t = 50μ we obtain a bandwidth of .94 GHz when d = .5Δ for H_o = 375 Oe and $\theta = 25^o$.

Bandwidths can be increased by using magnetic materials with higher magnetization (M_o) and magnetic anisotropy ($-2K_1/M_o$) than those of YIG. As for example in lithium ferrite $4\pi M_o$ = 3640 Oe and $-2K_1/M_o$ = -535 Oe whereas in YIG, $4\pi M_o$ = 1750 Oe and $-2K_1/M_o$ = -82 Oe. For H_o = 500 Oe and $\theta = 45^o$ we obtain a bandwidth of 1.32 GHz in lithium ferrite slabs with $\Delta = d = 5\mu$ and t = 50μ .

CONCLUSIONS AND DISCUSSTIONS

It appears that some device potential can be realized using the double layer configuration for magnetostatic propagation. More importantly, it appears that greater flexibility can be exercised in designing a device using double layers. For, example the thickness of the magnetic slabs, magnetic anisotropy value, the dielectric separating the two magnetic slabs, and the slab orientations can be varied or adjusted to meet the need of a particular specification. In this paper we have discussed the properties of two surface waves traveling in the same direction. It should be interesting to study the properties of the device for both waves traveling in opposite directions.

REFERENCES

1. W.L. Bongianni, Microwave Journal, 17, 49 (1974).
2. A.K. Ganguly and C. Vittoria, J. Appl. Physics, 45, 4665 (1974).
3. W.L. Bongianni, J. Appl. Physics, 43, 2541 (1972).

PLANAR MICROWAVE MULTIPOLE FILTERS USING LPE YIG

J M Owens†, J H Collins* and J D Adam*
* Department of Electrical Engineering, University of Edinburgh, Edinburgh, Scotland
† Now at University of Texas, Arlington, Texas

ABSTRACT

YIG discs fabricated by LPE on GGG have been utilised in tunable microwave bandpass multipole filters. However, filter realisation in MIC format is complicated by conductor steps in excess of 10 μm. These result from the necessary microwave interconnections of the discrete YIG discs formed by etching the LPE YIG film. This paper describes a planar fabrication procedure which overcomes the conductor step problem without degrading the YIG properties.

A photolithographically defined SiO_2 mask is used to etch wells into the GGG substrate. LPE YIG is grown over the entire substrate to a thickness equal to the depth of the wells. The YIG outside the wells is removed by Syton polishing, leaving a planar surface with discrete YIG islands. YIG discs as thick as 50 μm, and YIG regions as narrow as 10 μm with 5 μm depth have been fabricated. Planar bandstop filters have exhibited linewidths below 1 Oe. Planar bandpass filters have shown external Q's down to 60 at 2.8 GHz.

INTRODUCTION

A wide range of microwave devices have been fabricated from bulk yttrium iron garnet (YIG), notably filters and limiters. These devices depend on shaping bulk flux grown crystals into spheres. The spheres are then placed and oriented relative to shaped conductors.

YIG produced by LPE on gadolinium gallium garnet (GGG) has resulted in bandstop[1] and bandpass[2] filters being fabricated with excellent characteristics. However, to date, filters produced from LPE-YIG have used cut discs with overlayed conductors. This process is little better than the sphere process in terms of ease of fabrication. Even YIG structures etched on a substrate are not totally acceptable since the resulting steps (10-100 μm) are difficult to bridge with conducting structures. The real appeal of LPE-YIG is the promise of planar fabrication of complex microwave circuits on a single substrate using photolithography - that is microwave integrated circuit (MIC) compatible.

This paper describes a technique for fabricating truly planar structures of LPE-YIG while retaining its high microwave quality. The technique involves masking and etching of wells in the substrate to the shape and depth required for the YIG elements. Single crystal YIG is then grown over the entire GGG substrate so filling the wells to the level of the original substrate polished surface. The LPE-YIG grown on the regions outside the wells is polished off to leave the required YIG elements.

The resolution limits of the technique have been studied and the quality of the resulting YIG elements evaluated in bandstop and bandpass filters.

FABRICATION

The fabrication procedure for planar YIG elements is shown in Figure 1. First, 0.5 μm of SiO_2 was sputtered on a polished GGG substrate. The shapes of the required YIG elements were etched completely through the SiO_2 layer using standard photoresist techniques. Next, the GGG was etched through the SiO_2 mask using fresh orthophosphoric acid at 270°C. This etch produces a polished surface in GGG with virtually zero undercutting (<10% of the depth) and square sides on <111> oriented GGG. 0.5 μm SiO_2 layers have been used to etch wells in excess of 100 μm deep with no noticeable degradation. The etching rate is typically 0.4 μm/sec. Aspect ratios of 2:1 (width:depth) are readily obtained. The technique has also been tried on <100> oriented substrates but the etch produces a rough surface resulting in a rough YIG surface after growth and poor pattern definition.

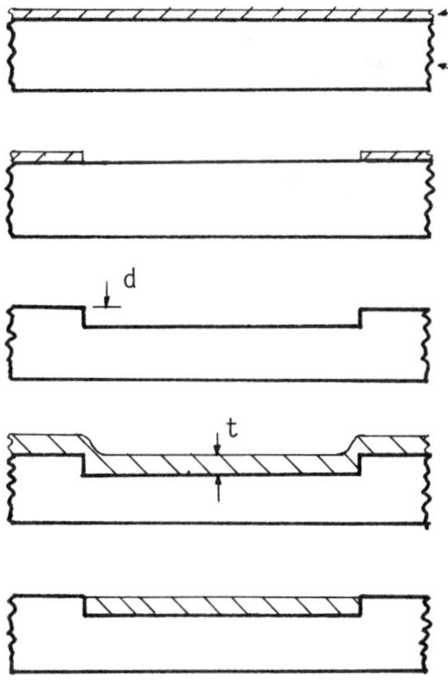

Fig 1. Processing procedure for planar YIG.

After etching the <111> GGG substrate in orthophosphoric acid the SiO_2 layer is stripped and an LPE-YIG layer grown of thickness equal to the well depth. The growth technique is described in reference 3 and utilised horizontal dipping with axial rotation at 100 rpm on the melt surface. The growth rate was typically 1 μm/min. Following growth excess flux was removed by a fast spin at 2000 rpm.

The growth process produces a single crystal YIG film with its <111> axis normal to the substrate <111> and with a defect density largely determined by the substrate defect density which can be as low as <5/cm^2. Wells in the range 5 μm to 50 μm deep have been filled. In general the coverage is uniform over the entire GGG surface except for a slight thinning at the steps at the edges of the wells. Next, the unwanted YIG outside the wells must be removed without damaging the YIG in the wells. This was performed using optical polishing jigs with ½ μm alumina abrasive for removal of the majority of the excess YIG. Syton flooded laps were used to remove the last 5 μm. Since the YIG in the wells has been grown to the level of the

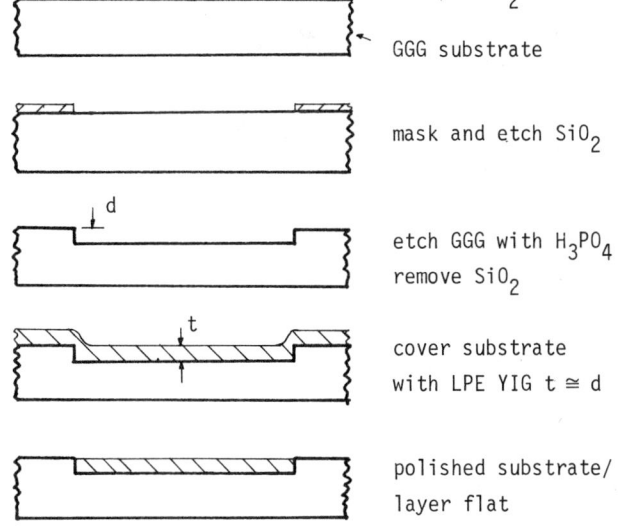

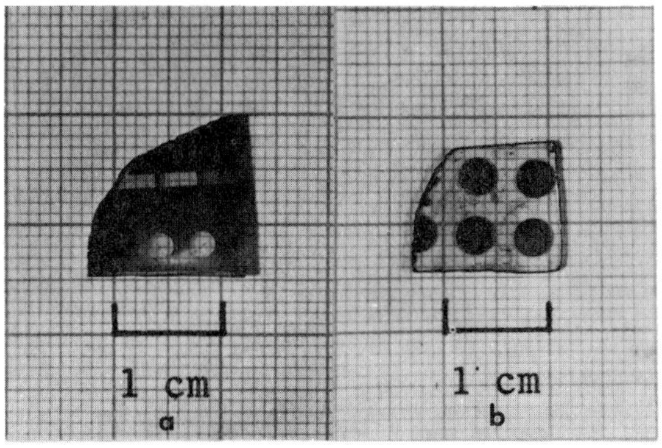

Fig 2. Planar YIG test samples
(a) Resolution test sample 10, 70, 1200 μm Lines, 2 mm discs.
(b) 3.5 mm discs, sample PL 4.

substrate surface, this ensures a minimum of damage to the active areas.

Figure 2 shows two practical examples of planar YIG elements. The design shown in Figure 2(a) was used to test the pattern resolution of the technique. It includes two extremes of 10 μm lines and 2.0 mm diameter discs. Figure 2(b) shows planar YIG discs of 3.5 mm diameter used in microwave filter evaluation studies. Figure 3(a) is a microphotograph showing the top view of the intersection of a 10 μm wide line with a 70 μm wide line in 5 μm thick YIG. The microphotograph was taken using phase contrast at 100X magnification. The channels are 20 μm apart and 25 μm deep.

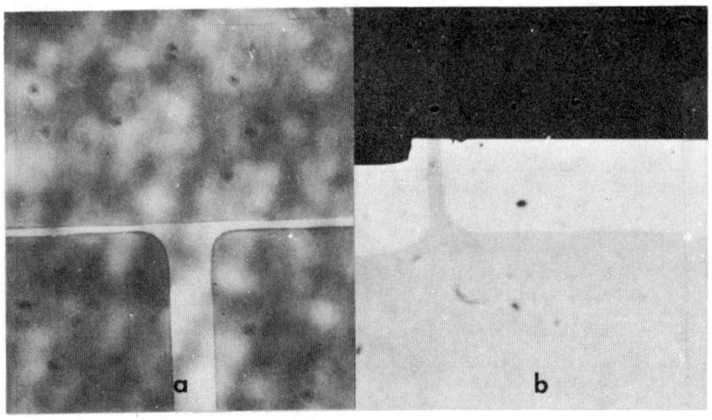

Fig 3. Microphotographs of planar YIG samples

(a) Intersection of 70 and 10 μm wide, 5 μm deep YIG channels.

(b) 10^0 angle lap of 25 μm deep adjacent YIG channels 20 μm channel spacing.

MICROWAVE EVALUATION

A range of planar LPE-YIG samples of different shapes and thicknesses were produced for microwave evaluation. The sample characteristics are given in Table I. The ferromagnetic resonance linewidth of these elements was measured using the bandstop filter technique[4] in a 50 Ω microstrip holder between 1 GHz and 4 GHz. The elements were biased in parallel resonance. Swept frequency transmission attenuation measurements were performed using an HP 8410A Network Analyser.

LPE-YIG SAMPLE	ELEMENT SHAPE (mm)	ELEMENT DEPTH (μm)	LINEWIDTH IN PARALLEL RESONANCE (Oe)
PL 1	3 x 3 square	45	0.70
PL 2	3 x 3 square	16	1.9
PL 4	3.5 dia circle	20	1.1
PL 5	3.5 dia circle	13	1.6

TABLE I

Results presented in Table I are average linewidth values over 1 GHz - 4 GHz. The thinner 13 μm and 16 μm samples show a somewhat larger linewidth than the 45μm thick sample, probably as a result of polishing damage which is more difficult to control in these samples. In addition the effects of resonances degenerate with the main resonance are also critical in thin films.

Samples have also been tested in the bandpass filter mode using the technique of Simpson et al[2]. Insulated crossed 50 Ω microstrip was used. Figure 4 shows the single pole bandpass characteristics of sample PL 4 at selected frequencies between 1.1 GHz and 1.9 GHz. This filter had a Q_L of 38 and a Qe of 160 with a midband insertion loss of 3 dB. Further work is required to reduce this loss to the typical Ga:YIG sphere filter value of 1 dB. Planar bandpass filters have been tested with Qe values as low as 60 at 2.8 GHz, and multipole devices have also been evaluated.

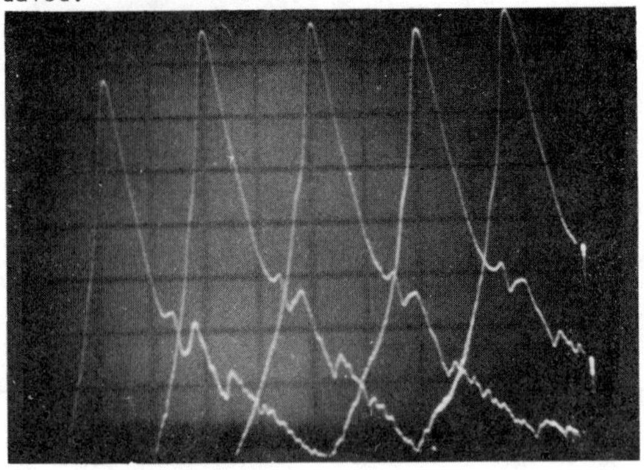

Fig 4. Bandpass response of sample PL 4 - vertical. 2.5 dB/Div, horizontal 100 MHz/Div, centre frequency 1.5 GHz.

CONCLUSIONS

A technique has been developed and evaluated for the fabrication of high quality LPE-YIG elements in a planar form on GGG substrates. The technique has high resolution capability, element shape versatility, and should be adaptable to mass production initially of MIC's. The technique overcomes the conductor step problems which have arisen in cut and etched LPE-YIG disc filters. Further, the technique promises applicability to any LPE garnet and substrate combination. Channel elements have applications in the areas of waveguiding and integrated waveguide components for signal processing. These include acoustic[5], magnetostatic[6] and optical[7] systems.

ACKNOWLEDGEMENTS

The authors wish to acknowledge contract support from the British Ministry of Defence Procurement Executive.

REFERENCES

1. J D Adam et al, Electronics Letters, 9, 325 (26th July 1973).
2. I T Simpson et al, Conf Proc 4th EMC/Microwave 74, pp 590-594, (1974).
3. W Tolksdorf et al, J Crys Growth, 17, pp 322-328, (1972).
4. J L Archer et al, J Appl Phys, 41, pp 1359-1360, (1970).
5. E A Ash et al, IEE Trans MTT, MTT-17, pp 882-892, (1969).
6. R E De Wames et al, Appl Phys Letters, 16, pp 305-308, (1970).
7. P K Tien, IEEE Trans Mag, MAG-10, No 3, (1974).

ERRATUM

STUDIES OF FMR LINEWIDTH IN THICK YIG FILMS GROWN BY LIQUID PHASE EPITAXY [AIP Conf Proc No 18, 1279 (1974)]

The observed high Q resonances in the presence of the stub were interpreted incorrectly as leading to a linewidth of 0.06 Oe. Besides mode segregation, these resonances can also arise from destructive interference between the output of the bandstop filter, containing a thick LPE YIG sample, and electromagnetic leakage across the microstrip holder. In open microstrip experiments this leakage is typically 30 dB below the input signal. These comments in no way invalidate the remaining results of Table I.

TAPPED MICROWAVE NONDISPERSIVE MAGNETOSTATIC DELAY LINES

J D Adam*, Z M Bardai*, J H Collins* and J M Owens†
* Department of Electrical Engineering, University of Edinburgh, Edinburgh, Scotland
† Now at University of Texas, Arlington, Texas

ABSTRACT

Nondispersive magnetostatic surface wave delay lines when spatially tapped form the basis for real time signal processing at microwave frequencies. Devices have been fabricated reproduceably from narrow linewidth LPE-YIG, spaced from a ground plane by a dielectric layer. Transduction and tapping of the magnetostatic waves was efficiently performed with fine line conductors at S-band. Experimental performance and design limitations are discussed with particular emphasis on delay, transduction loss, propagation loss and power saturation.

INTRODUCTION

Magnetostatic surface wave propagation in a structure composed of a liquid phase epitaxy (LPE) yttrium iron garnet (YIG) film and dielectric layer, placed on a ground plane has been investigated by Bongianni[1]. This geometry supports a wave which propagates with approximately nondispersive characteristics over a bandwidth of 200 MHz at centre frequencies up to at least 5 GHz. Bongianni[1] demonstrated nondispersive delays in the 80-150 nsec per cm range. Here we describe the conditions necessary to achieve long delays in the range 0.5-1 µsec per cm at S-band.

Investigation of the microwave signal processing applications of nondispersive tapped magnetostatic delay lines requires a knowledge of useable delay ranges and bandwidths, as well as insertion loss and power saturation characteristics which determine the device dynamic range. Detailed operation of a 250 nsec delay line tapped at its midpoint is described here.

LONG DELAY LINE OPERATION

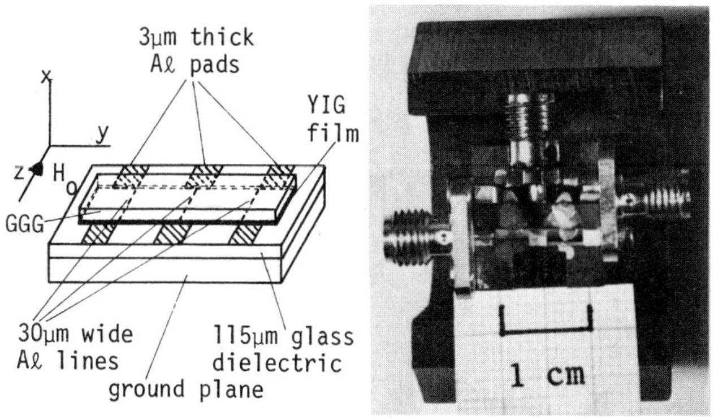

a b

Fig 1. Construction of non-dispersive YIG delay line

Fig 1 (a) and (b) show the construction of a typical nondispersive magnetostatic wave delay line in schematic and photographic forms respectively. The device illustrated consists of a 115 µm thick glass microscope slide sputtered on both sides with a 3 µm thick aluminium Aℓ layer, the lower side of which forms the ground plane. 3 mm long X 30 µm wide Aℓ conductors were used for magnetostatic wave transduction. These were photolithographically defined on the upper aluminised surface. The LPE-YIG film is cut to fit the length and spacing of the transducers, and laid down on the structure with the gadolinium gallium garnet (GGG) substrate uppermost. The magnetic bias field is parallel to the transducer conductors.

The delay characteristics of a 2 µm YIG film on a 115 µm thick glass dielectric (εr = 2.4) are shown in Fig 2. The measurements were performed directly using pulsed RF. Insertion loss was 46 dB at 3.2 GHz for a delay of 600 nsec. Above 3.3 GHz the insertion loss increased rapidly due to the magnetostatic half wavelength being less than the transducer width (30 µm). Operation with longer delays thus necessitates narrower transducer widths leading to increased RF conductor losses.

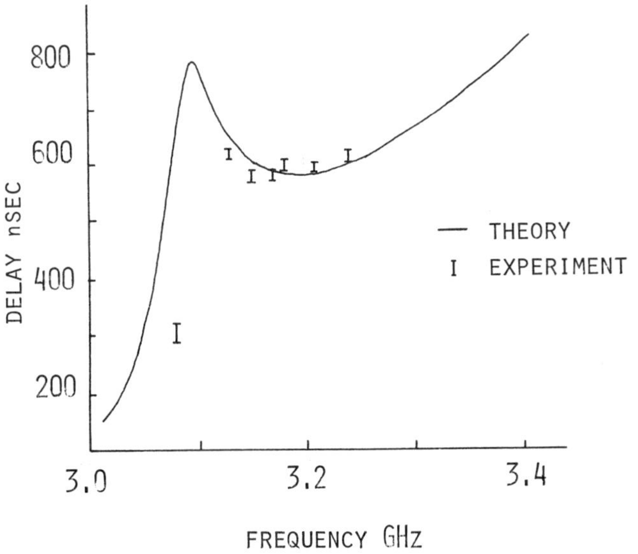

Fig 2. Plot of delay versus frequency for a 2 µm YIG film with a 115 µm glass dielectric-bias field, 500 Oe.

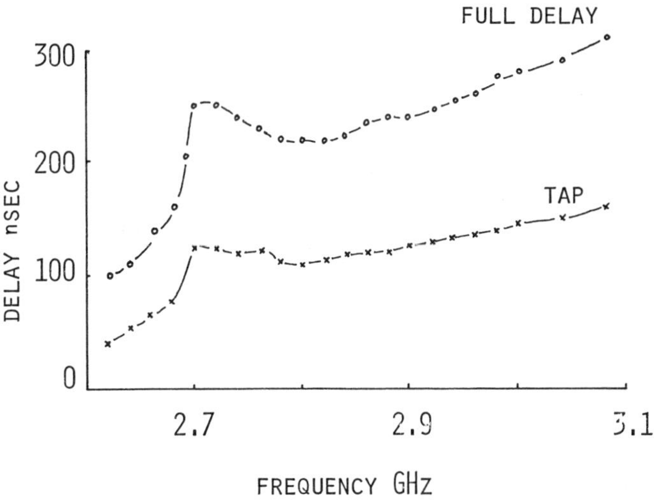

Fig 3. Experimental plot of delay versus frequency for tapped delay line - YIG film thickness = 4.5 µm, glass dielectric thickness = 115 µm, bias field = 500 Oe.

TAPPED DELAY LINE

Operation of a tapped delay line was achieved using the central coupler in Fig 1, as the tap, spaced 5 mm from both input and full delay couplers. The delay line to be described, was designed to achieve 250 nsec delay, tapped at 125 nsec. The LPE-YIG film thickness was 4.5 µm and the dielectric thickness was 115 µm. The input impedance was nominally 50 Ω, at a bias field of 500 Oe. Fig 3 shows the time delay for both tap and full delay in the frequency range, 2.6-3.1 GHz. Both taps and full delay are nondispersive within a ±10% tolerance over a bandwidth of 225 MHz.

Measurements of insertion loss are shown in Fig 4 over the same frequency range as Fig 3. The full delay results show a larger scatter of values due to interference by signals reflected from the end of the YIG film. The scatter is too large to allow a precise determination of propagation loss. However, an estimate derived from the extremes of delay and attenuation gives a value of 60 dB/μsec. This corresponds to a linewidth value of 0.8 Oe^2. Transduction loss is similarly estimated as 6 dB. 3 dB of this loss is attributable to the high shunt capacity of the bonding pads.

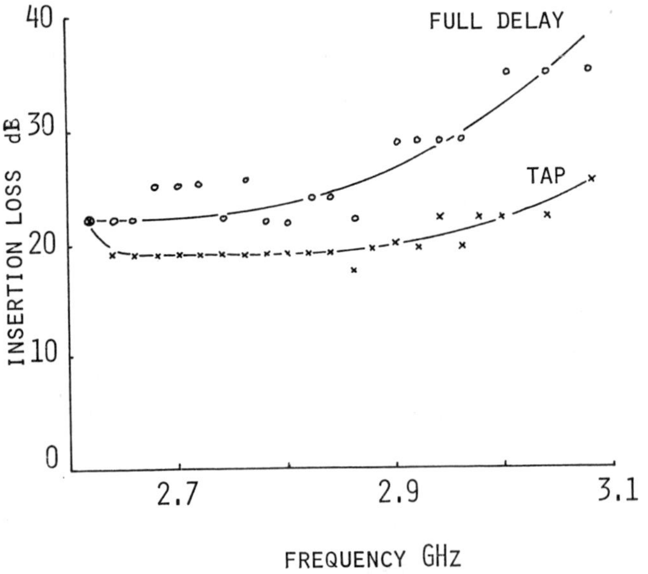

Fig 4. Experimental plot of insertion loss versus frequency for tapped delay line, shown in Fig 3.

POWER SATURATION

Power saturation phenomena in conventional YIG devices is well documented[3]. The instability causing the non-linearity takes a finite time to build up allowing transmission of a short pulse at higher power levels. This is the leading edge spike observed in YIG limiters. Theory[3] predicts for the low threshold first-order non-linear process that the unsaturated output pulse length, namely the leading edge spike width, is proportional to the inverse square root of input power.

Measurements of power saturation characteristics were performed on an LPE-YIG structure propagating magnetostatic surface waves. The leading edge spike width at the output of the delay line was determined as a function of input power with frequency as a parameter. Data is plotted in Fig 5. Note the dramatic increase in input power allowed (>10 dB) when the frequency is increased from 3.5 GHz to 3.6 GHz. This result should be compared with conventional pure YIG devices where the low threshold first order process ceases at 3.25GHz, for wave number zero, and the significantly higher threshold second order process takes over. Thus, it is necessary below 3.6 GHz to trade-off input power with pulse width to avoid distortion in delay lines.

CONCLUSIONS

Non-dispersive delays of 600 nsec/cm have been observed in 2 μm thick LPE-YIG/115 μm dielectric magnetostatic surface wave structures at S-band, in accord with theory. Further, a 250 nsec delay line tapped at its mid-point using a 4.5 μm thick LPE-YIG has been fabricated and tested. Measurements on this tapped line have yielded an estimated propagation loss of 60 dB/μsec. Transducer losses down to 6 dB have been demonstrated. Power saturation characteristics

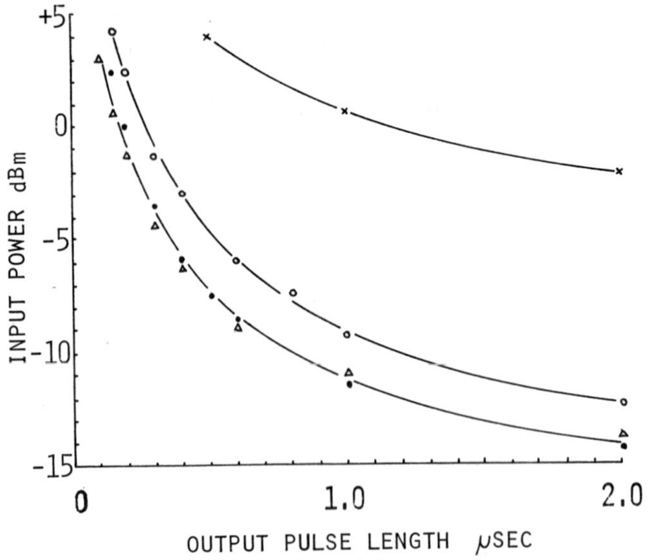

Fig 5. Experimental plot of input power versus unsaturated output pulse length (leading edge spike width) - YIG film thickness = 15 μm, glass dielectric = 100 μm, 50 μm gold wire couplers 1 cm apart : bias fields adjusted for 100 nsec delay at each frequency.

x - x	3.6 GHz	Insertion	Loss	22 dB
o - o	3.5 GHz	"	"	"	20 dB
• - •	3.0 GHz	"	"	"	22 dB
Δ - Δ	2.5 GHz	"	"	"	20 dB

of these delay lines have also been measured. Below 3.6 GHz it is necessary to trade-off input power with pulse width. However, for pulse lengths below 200 nsec input powers of several dBm can be used.

Thus, magnetostatic surface waves in LPE-YIG/dielectric structures can provide non-dispersive tapped delay lines at S-band frequencies, within well-defined parameter constraints. In consequence, real time signal processing functions at microwave frequencies are available similar to those already well-developed with surface acoustic wave technology[4] at VHF and UHF.

Two topics require further research. Low loss gallium substituted LPE-YIG should permit operation below 3.6 GHz at higher power levels. Improved transducer design is necessary to obtain efficient coupling in the thin LPE-YIG films required for delays in the 1 μsec region.

ACKNOWLEDGEMENTS

The authors wish to acknowledge contract support from the British Ministry of Defence Procurement Executive.

REFERENCES

1. W L Bongianni, J Appl Phys, 43, 6, 2541-48, 1972.
2. J B Merry and J C Sethares, IEEE Trans, MAG-9, 3, 527-9, 1973.
3. B Lax and K J Button, "Microwave Ferrites and Ferrimagnetics", McGraw Hill, 1962, p 673-702.
4. J H Collins, Invited Lectures of 8th ICA London, pp 107-122, 1974.

IMPROVEMENT IN BROAD BAND FERRITE ISOLATORS

L. Courtois, CNRS, 92190 Bellevue-France.
G. Forterre, B. Chiron, LTT, 78700 Conflans-Ste-Honorine-France.

ABSTRACT

A two and a half octave bandwidth edge mode isolator is described. The insertion loss is analysed in dielectric and magnetic ferrite loss, in conduction loss and in loss due to the dissipative load. This analysis leads to an optimization of the performance characteristics of edge mode devices.

INTRODUCTION

Research on surface wave in ferrite has been very active during these last years. Hines [1] was the first to explain the operation of strip line broad band devices. Afterwards it was shown that an edge mode isolator working below the gyromagnetic resonance has suitable characteristics in the VHF band [2] ; it was also shown that a surface mode propagates along a slot-line and a slot-line isolator was built [3] ; in this paper an isolator working from 2 to 12 GHz is presented. Such a bandwidth is not out of proportion with some market requirements but perhaps the best prospect for this device is to be sought in an improvement of insertion loss even at the expence of the bandwidth. To this purpose the influence of the different parameters on the insertion loss is analyzed.

The edge-mode device operation has been well described previously by Hines and other authors [1-4]. Cut-off frequencies of the surface mode TE_{oo} and the parasitic volume modes TE_{om} are given by graph and formula. Nevertheless this discussion is not complete : for instance, the experimental bandwidth is twice the theoretical one defined by the frequency band where the surface mode is the only one to propagate ; moreover, losses were never taken into account in previous papers

THEORY

The strip line section is divided into three parts by Hines. The first one is composed of the ferrite between the strip and the ground plate, and is double ; the two others being composed of the ferrite outside this first zone. Hines puts forward the hypothesis that boundary conditions between these three longitudinal zones are of a magnetic type (along yoz planes). Therefore the first zone can be considered as a rectangular flat guide of width S and of thickness h,

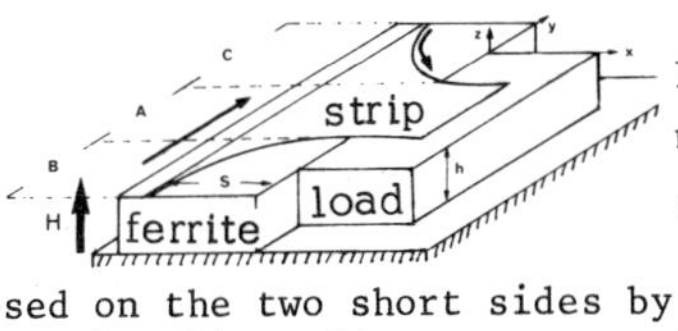

Fig. 1 - The edge mode isolator configuration.

closed on the two short sides by a magnetic wall (Fig. 1). This guide can propagate an unidirectional surface wave

$$e^{-k''_{xf}\cdot x + j(\omega t - k_y \cdot y)}$$

By using thermic coloration of liquid crystal a strong field displacement has been observed [5]. Energy is localized in the first zone, and is confined to the strip edge on the left or on the right according to the propagation direction. But the best confirmation of Hines' assumption is the realization of broad band isolators.

We limit this study to the rectangular guide of the first section zone, and we propose to divide this line into 3 lengths. In the length A the guide is limited on the right by the ferrite-dissipative load interface. The use of this load is to consume the reverse wave. The two other lengths B and C are transition zones, where the strip width is variable.

The two transition lengths B and C : The purpose of these two tapered zones is impedance matching and mode conversion. The understanding of this phenomenon is not complete so far. De Santis [6] showed that with a positive permeability the surface mode is a leaky mode, if the strip curvature is concave. It is an important but difficult problem which is at the origin of the failure in making circulators.

We shall consider this section of line as a rectangular guide of slowly varying width, this assumption being made valid by the fact that width S does not appear in the characteristic equation of the surface mode.

$$k_y = \omega\sqrt{\varepsilon_f \mu} \quad \text{and} \quad k''_{xf} = k_y\, K/\mu \qquad (1)$$

where :

$$\frac{\mu}{\mu_o} = \frac{HB-(\omega/\gamma)^2}{H^2-(\omega/\gamma)^2} \quad \text{and} \quad \frac{K}{\mu_o} = \frac{\omega}{\gamma}\,\frac{M}{H^2-(\omega/\gamma)^2} \qquad (2)$$

are the diagonal and off diagonal elements of the permeability tensor of the ferrite.

For the volume mode : $k_{xf} = m\pi/S$ (3)

is the characteristic equation.

The central length A : The zone A is treated as a rectangular guide closed on the left by a magnetic wall and limited on the right by the interface ferrite-dissipative load. The characteristic equation of the modes, the energy of which is confined in the slab of ferrite is :

$$\frac{\mu}{\mu_d} k_{xf} + \left\{ \frac{k_y^2-\omega^2\varepsilon_f\mu}{\sqrt{k_y^2-\omega^2\varepsilon_d\mu_d}} + \frac{K}{\mu_d} k_y \right\} \mathrm{tg}\,(k_{xf}\cdot S)=0 \qquad (4)$$

This equation is similar to formulas (1) and (3) making $\mu \to \infty$ and reducing k_{xf} to $-jk''_{xf}$ for the surface mode.

The dielectric losses : The ferrite dielectric losses, which are characterized by the dielectric loss tangent tg δ, involve a damped wave. The wave vector k_y is complex, its imaginary part is formally obtained from formula (1) (or from formula (4) if the field displacement is large enough). Its reduced form is

$$\frac{k''_{ye}}{\sqrt{\varepsilon'_f}\ \mathrm{tg}\delta} = \frac{\omega}{2C}\sqrt{\frac{HB-(\omega/\gamma)^2}{H^2-(\omega/\gamma)^2}} \quad \text{(in Np/cm)}. \qquad (5)$$

k''_{ye} is plotted in figure 2 for numerical data given in appendix.

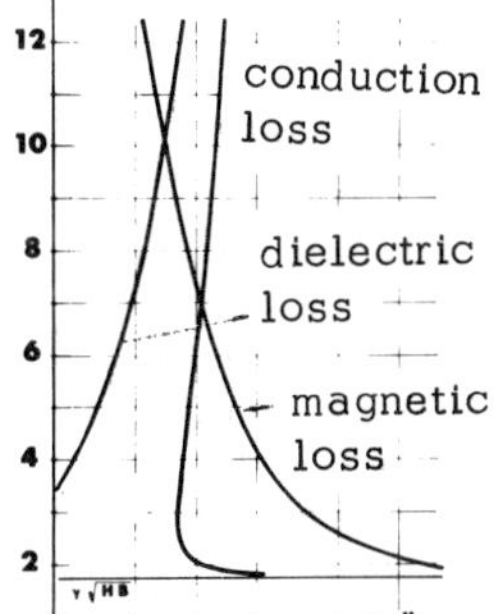

Fig. 2 - Theoretical analysis of the insertion loss.

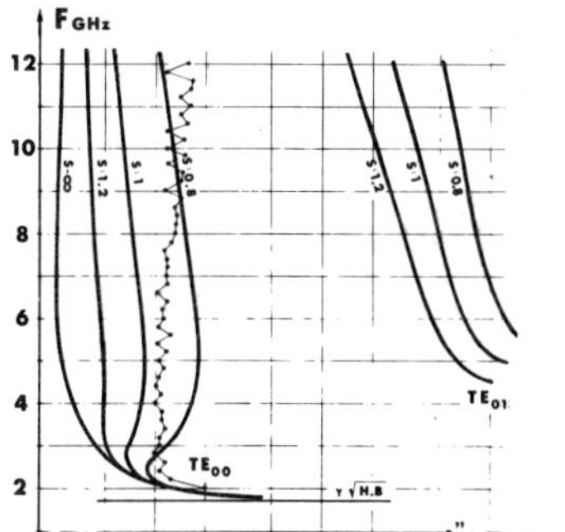

Fig. 3 - Influence of the width S on the insertion loss and on the bandwidth. Experimental points are compared.

The magnetic losses : in the same manner, magnetic losses are characterized by an effective linewidth ΔH_{eff} which is introduced in the calculation by means of a complex DC field $\bar{H} = H + j\Delta H_{eff}/2$. Then the imaginary part of the wave vector is formally obtained from 1. Its reduced form is

$$\frac{k''_{ym}}{\sqrt{\varepsilon'_f}\ \Delta H_{eff}} = \frac{\omega}{4C} M \frac{H^2+(\omega/\gamma)^2}{\sqrt{|H^2-(\omega/\gamma)^2|^3 |HB-(\omega/\gamma)^2|}} \qquad (6)$$

k''_{ym} is plotted in figure 2.

Conduction losses : The computation of Joule losses occuring in the strip and in the ground plate of resistivity ρ is not direct. But the method is well known in electromagnetic engineering. This loss involves an imaginary part of the wave vector, the reduced form of which is

$$\frac{k''_{yj}}{\sqrt{\varepsilon_f}\ \sqrt{\rho}} = \frac{1}{h}\frac{1}{\sqrt{\rho_o}}\sqrt{\frac{\omega}{2C}\ \frac{H^2-(\omega/\gamma)^2}{HB-(\omega/\gamma)^2}} \qquad (7)$$

where ρ_o is the vaccum resistance equal to 377 Ω. k''_{yj} is plotted on figure 2 with h taken as to 0.2 cm.

The losses due to the dissipative load : In the length A of the line, the presence of the dissipative load is a 4th cause of loss. The wave damping is calculated by the complex resolution of equation (4). The dissipative medium is characterized by the susceptibility : $\varepsilon_d=\varepsilon'_d-j\varepsilon''_d$ or $\varepsilon'_d-j\sigma/\omega$ according to whether the medium presents dielectric or conduction loss. It is also characterized by the permeability $\mu_d=\mu'_d-j\mu''_d$ the magnetic loss being obtained by the introduction of iron powder. If the displacement is large enough, the surface mode is not greatly damped ; in a more precise manner

$$k''_{yd}(S) = k''_{ydo}\ e^{-2k_{xf}\cdot S} \qquad (8)$$

where k''_{ydo}, which is in direct connection with the device isolation, must be large. If the displacement is not large enough the other parasitic modes are considerably damped and the three other origins of losses are negligible.

DISCUSSION OF THE RESULTS

For each mode the total damping is :

$$k''_y = k''_{ye} + k''_{ym} + k''_{yj} + \ell_A/\ell\ k''_{yd} \qquad (9)$$

where the length ratio ℓ_A/ℓ makes up the weighting function of the losses due to the load and which only occur along the line A. By utilising the experimental isolator data indicated in the appendix, it can be deduced from figure 2 that the dielectric losses are very slight except at high frequencies : the magnetic losses, which are very important at low frequencies, diminish rapidly and are negligible at the top of the band.

Lastly the conduction losses increase with the frequency ; they are preponderant at the top of the band.

The total value of k''_y is plotted on Fig. 3. The figure shows the importance of the width S. The main loss is the loss due to the dissipative load. For S=1.2 cm it is equal to the 3 others all together. The damping of the parasitic TE_{01} mode is plotted on Fig. 3. The cut off frequency which is given by equation (3) shifts all the damping curves. The higher the damping the weaker is the coupling between the main TE_{00} mode and the parasitic TE_{01} mode. Thus the line seems monomode. When the damping of the TE_{01} mode is not sufficient, coupling and interference phenomena occur. This can be observed in the experimental results at the top of the band in figures 3 and 4. These coupling phenomena explain the paradoxical observation that the isolator insertion loss increases on taking away the dissipative load. The choice of S is fundamental. S must be large enough to avoid high insertion loss of the main mode, but small enough to damp the TE_{01} mode.

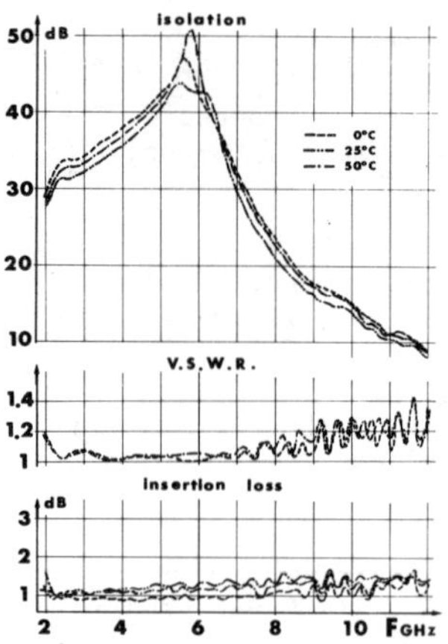

Fig. 4 - The characteristics of the broad band isolator measured on a H.P. network analyser.

It can be shown that the drop of the isolation at the top of the band can be explained by the propagation of the reverse TE_{01} mode. Here is another reason to choose S sufficiently small.

On Fig. 4 performances of a broad band isolator designed for operation from 2 to 12 GHz is presented. Ferrite material is chosen with the highest magnetization compatible with low field losses at 2 GHz. As expected by theory, this device presents a larger relative bandwidth than any other device optimized at a higher frequency.

CONCLUSION

The influence of the dissipative load on the bandwidth and on the insertion loss of an edge mode isolator is analyzed, making evident the difficulty of building broad band circulators.

To obtain more accurate predictions about on insertion loss, a knowledge of the effective linewidth and more particularly a knowledge of the dissipative load characteristics through all the frequency band are needed.

AKNOWLEDGEMENTS

This work was sponsored by the "Service Technique des Télécommunications de l'Air" in France.

Significant technical contributions were made by C. Rannou and J.F. Mayault.

APPENDIX : numerical data

Polycristalline : YIG+Al slab

Magnetization	$4\pi M$ = 1 000 Gauss
Internal field	H = 300 Oersteds
Gyromagnetic ratio	$\gamma = -2\pi$ 2.8 MHz/Oe
Permittivity	$\varepsilon_f = 14.5\varepsilon_o$
Electric loss tangent	$tg\delta = 2.10^{-4}$
Effective linewidth	ΔH_{eff} = 5 Oersteds

Dissipative medium : iron powder

Complex permittivity	$\varepsilon_d = (14-j)\ \varepsilon_o$
Complex susceptibility	$\mu_d = (1-j)\ \mu_o$

REFERENCES

1 Hines, M.E., IEEE Trans. MTT, 19, 482, (1971).
2 Courtois L., Bernard N., Chiron B., Forterre G., A new edge mode isolator in the VHF range. IEEE S-MTT Intern. Microwave Symposium, (1974).
3 Courtois L., de Vecchis M. A new class of non reciprocal composants using slot line. To be published at the IEEE-MTT.
4 Courtois L., Chiron B., Forterre G. Câbles et Transmission, 4, 416, (1973).
5 Puyhaubert J. L'Onde Electrique, 52, 213, (1972).
6 de Santis P., Appl. Phys., 4, 167, (1974).

THE SYNTHESIS OF CYLINDRICALLY SYMMETRIC STATIC MAGNETIC FIELDS IN A LOCALLY SATURATED FERROMAGNET

F.R. Morgenthaler

Department of Electrical Engineering and Center for Materials Science and Engineering, Massachusetts Institute of Technology, Cambridge, Mass. 02139

ABSTRACT

A procedure is given that permits an essentially exact synthesis of a cylindrically symmetric static magnetic field in a ferromagnet when the axial field,

$$H_z(r=0,z)$$

is specified over a **finite length.** The material is assumed to have a uniform magnetization (M) that is everywhere locally saturated. A variety of material shapes and corresponding magnetic pole piece designs is available, any of which will meet the synthesis requirement.

A simple modification of the procedure allows the incorporation of magnetic anisotropy that is uniaxial (or at least effectively so).

AN OVERVIEW OF THE SYNTHESIS PROCEDURE

The synthesis procedure can be divided into the following four steps:

1. Find the H-field that meets the on-axis field requirement for $z_1 < z < z_2$ inside of a cylindrically symmetric but as yet unbounded ferromagnet of specified magnetization M.
2. As shown schematically in Fig.1, choose a convenient boundary radius R(z) for the magnetic material subject to

$$R(z_1) = R(z_2) = 0$$

3. Choose a convenient series expansion for the Laplacian scalar potential outside of the ferromagnet and match boundary conditions over the entire R(z). Solve for the coefficients of the outer potential.
4. Plot the equipotentials of the outer field and choose two (or more) that are appropriate to serve as surface contours of high permeability magnetic pole pieces.

If the final design is not satisfactory, alter the boundary radius R(z) and/or the series expansion for the outer potential and carry out the alternate design.

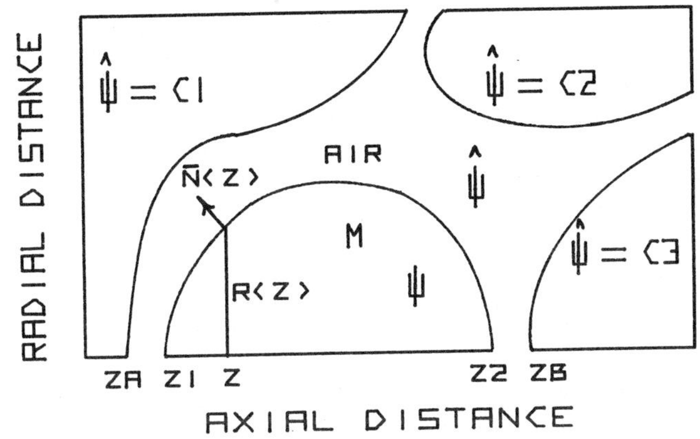

Fig.1 Synthesis Geometry

THE INNER FIELD DETERMINATION

In the absence of anisotropy, the well known equations governing the dc $\bar{H}$ field in a ferromagnet that is everywhere locally saturated (no domain structure) are

$$\bar{M} \times \bar{H} = 0 \quad (\bar{M} \cdot \bar{H} > 0) \qquad (1)$$

$$|\bar{M}| = M \qquad (2)$$

$$\nabla \times \bar{H} = 0 \qquad (3)$$

and

$$\nabla \cdot (\bar{H} + \bar{M}) = 0 \qquad (4)$$

In terms of the magnetic scalar potential ψ,

$$\bar{H} = -\nabla\psi \text{ and } \bar{M} = \frac{-\nabla\psi}{|\nabla\psi|}M$$

satisfy Eqs.(1),(2),(3). Therefore, Eq.(4) may be expressed as

$$\nabla \cdot [(1 + M/|\nabla\psi|)\nabla\psi] = 0 \qquad (5)$$

This nonlinear equation reduces to Laplace's Equation in the event that M = 0. Because the axis of the cylindrically symmetric ferromagnet is assumed non-singular, an appropriate expansion[1] of ψ is

$$-\psi = \sum_{n=0}^{\infty} a_{2n}(z) \; r^{2n} \qquad (6)$$

One could substitute Eq.(6) into Eq.(5) and expand the result so as to find a_2 in terms of a_o and its derivatives and so on. However, it is more convenient to separately expand $\bar{M}$ as

$$\bar{M} = M\,[\bar{i}_z \sum_{n=0}^{\infty} b_{2n}(z)\; r^{2n} + \bar{i}_r \sum_{n=0}^{\infty} b_{2n+1}(z) r^{2n+1}] \quad (7)$$

and use Eqs.(1),(2) and (4). The result is three sets of constraints, respectively

$$\sum_{n=0}^{s} [a'_{2n}\, b_{2(s-n)+1} - 2(s+1-n) a_{2(s+1-n)} b_{2n}] = 0 \quad (8a)$$

$$\sum_{n=0}^{s} [b_{2n+1} b_{2(s-n)+1} + b_{2n} b_{2(s+1-n)}] + b_o b_{2(s+1)} = 0 \quad (8b)$$

$$4(s+1)^2 a_{2(s+1)} + 2(s+1)Mb_{2s+1} + a''_{2s} + Mb'_{2s} = 0 \quad (8c)$$

where the primes denote differentiation with respect to z and s=0,1,2,3,... In addition,

$$b_o^2 = 1 \text{ and } b_o a'_o > 0$$

Because, by assumption, the field to be synthesized is known on the z-axis, a'_o (z) is specified over $z_1 < z < z_2$ and, in order to insure saturation, is either positive or negative definite over the same interval. (If a'_o were to reverse sign, the region where it went to zero would demagnetize and domain formation could take place). Without loss of generality, we take $a'_o > 0$ and $b_o = +1$.

The various functions a_{2k} and Eq.(6) constitute the desired solution of Eq.(5) and together generate the entire $\bar{H}$ field that is consistent with the on axis requirement. The number of terms required to satisfactorily approximate the field depends upon the extent of r, the value of M and the particular function a'_o; the lowest order terms (s=0) are

$$b_1 = -\frac{1}{2}\frac{a''_o}{a'_o + M} \qquad (9a)$$

$$b_2 = -\frac{1}{2} b_1^2 \qquad (9b)$$

and

$$a_2 = \frac{1}{2} a'_o b_1 \qquad (9c)$$

In the given example, terms up to and including b_6 and a_6 are sufficient.

If the material has uniaxial magnetic anisotropy oriented along the z-axis, the procedure can be generalized by replacing in Eq.(8b),

$$a'_{2n} \rightarrow a'_{2n} - \frac{2K_o}{\mu_o M} b_{2n}$$

where K_o is the uniaxial anisotropy constant. ($K_o < 0$ easy axis; $K_o > 0$ easy plane). If the anisotropy is not uniaxial, but effectively so when $\bar{M}$ is not greatly misaligned with the z-axis,(as for example a cubic material with [100] or [111] orientation) the formulation may still be used by replacing K_o with the appropriate effective value.

THE BOUNDARY RADIUS R(z)

The familiar boundary conditions that must be satisfied at the surface of the ferromagnet are continuity of ψ and the component of $\bar{B}$ normal to the surface. Natural choices for at least some portions of R(z) include

1. ψ_R = constant (equipotentials)
2. $(\frac{\partial\psi}{\partial r} - \frac{dR}{dz}\frac{\partial\psi}{\partial z})_R = 0$ ($\bar{n} \times \bar{M} = 0$)
3. $\frac{dR}{dz}$ = constant (cones, cylinders)

THE OUTER FREE SPACE POTENTIAL r>R(z)

If the z-axis passes through at least some portion of the outer field region, ($z_a < z_1$ and $z_b > z_2$),the Laplacian potential Φ may be taken non-singular over all z and expanded in the form

$$\Phi = \hat{a}_o(z) - \frac{\hat{a}''_o(z)}{4^1(1!)^2} r^2 + \frac{\hat{a}''''_o(z)}{4^2(2!)^2} r^4 - \dots \qquad (10)$$

Naturally if $\hat{a}_o$ is taken to be $\{{\sin \atop \cos}\}(kz)$ or $\{{\sinh \atop \cosh}\}(kz)$ Φ factors into the product of $\hat{a}_o$ and either $I_o(kr)$ or $J_o(kr)$. However, the usual cylinder functions are not especially convenient because our boundary specification does not lead to identification of eigenvalues of k. On the other hand, if $\hat{a}$ is taken to be a finite polynomial of order N, Φ will contain a sufficient finite number of terms to well approximate a suitable potential.

It is also permissible and often advantageous to utilize in the expansion axial multipoles of the form

$$\frac{d^n}{dz^n}\{(z - z_o)/[(z - z_o)^2 + r^2]^{3/2}\} \quad n = 0,1,2,\dots$$

as long as their locations are anywhere within the material boundary. Finally, if the outer region does not contain the z-axis, ($z_a > z_1$ and $z_b < z_2$),second solutions with logarithmic singularities over all r = 0, equal or equivalent to $K_o(kr)$ and $N_o(kr)$, are often convenient.

The matching of $\psi = \Phi$ and $(1 + \frac{M}{|\nabla\psi|})\frac{\partial\psi}{\partial n} = \frac{\partial\Phi}{\partial n}$ over the entire boundary is carried out in a least squares sense. Errors can be reduced to acceptably small values by choosing N, the total number of terms in the expansion of Φ,sufficiently large. If good judgement is shown in choosing the expansion, N need not be excessive; in the example that follows N = 10.

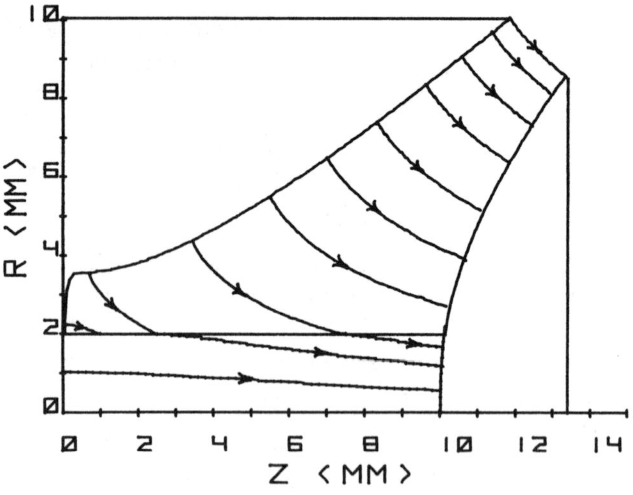

Fig.2 Parabolic Field Synthesis

EXAMPLE

Suppose the axial magnetic field is to be parabolic in form over a 10 mm. length. Specifically, we desire

$$H_z\ (0,\ z) = 200 + 50z^2\ (\text{oe.}) \qquad 0 < z < 10\text{mm}.$$

Further, we wish to create this field inside a ferromagnetic cylinder of radius 2 mm., with $4\pi M$ = 2000 Gauss. Fig.2 shows the field lines and boundary equipotentials for both the inner and outer regions. Note that due to surface magnetic poles the field lines bend as they pass through the boundary and that the ends of the rod,constrained by this particular design to be equipotentials,are nearly planes perpendicular to the z-axis. Therefore, the sample can be well approximated by a right circular cylinder.

ACKNOWLEDGEMENTS

The initial phase of this research was carried out at Chu Associates, Inc., Littleton,Mass. The author has benefitted from many helpful discussions with Dr. Aryeh Platzker.

REFERENCES

1. The expansion is a generalization of that used by the present author in connection with asymptotic spin wave focussing:
F.R. Morgenthaler, IEEE Trans. Magnetics, Mag.8, No.3, p. 550, Sept. 1972.

A more complete treatment of this work is contained in M.I.T. Microwave and Quantum Magnetics Group Technical Report No. 34, December 1974.

2. Proof of uniqueness is based upon the identity

$$\oint_S (\psi_1-\psi_2)(\lambda_1\frac{\partial\psi_1}{\partial n} - \lambda_2\frac{\partial\psi_2}{\partial n})da = \int|\nabla(\psi_1-\psi_2)|^2 dv$$
$$+ \int M(|\nabla\psi_1| + |\nabla\psi_2|)(1 - \frac{\nabla\psi_1 \cdot \nabla\psi_2}{|\nabla\psi_1||\nabla\psi_2|})dv$$

where $\lambda_i = 1 + M/|\nabla\psi_i|$ and ψ_1 and ψ_2 are assumed to be distinct solutions of Eq.(5) that satisfy the boundary conditions on S. If ψ is specified at every point on S for which M ≠ 0 and either ψ or $\partial\psi/\partial n$ is specified everywhere else, both sides of the identity are forced to vanish. Since M>0, each of the two volume integrals must separately vanish. Hence $\nabla(\psi_1 - \psi_2) = 0$, and for such boundary conditions the field is unique.

THEORY OF TRANSMISSION RESONANCE IN FERROMAGNETIC METALS*

Y.J. Liu and R.C. Barker
Department of Engineering and Applied Science
Yale University, New Haven, Connecticut 06520

ABSTRACT

The dispersion relations for a ferromagnetic metal applied to a plate configuration give for the ferromagnetic antiresonance, or transmission resonance, the characterizing frequency or field, the microwave penetration depth, and the transmitted line width. In addition, numerical calculation of the transmission spectrum yields the resonance line shape. The calculation is in satisfactory agreement with the experimental results reported by Heinrich and Cochran, using either spin-free or spin-pinned boundary conditions. The theoretical area density of transmitted power is very sensitive to the values of conductivity and magnetic damping. Thus, transmission resonance should be observable through 100 micron films of supermalloy but not ordinary permalloy or nickel. It is pointed out that although increased magnetic damping tends to broaden the transmitted line width, increased conductivity tends to narrow it.

DISPERSION RELATIONS

The dispersion relations $k(\omega)$ for a plane wave $\exp i(kx_3-\omega t)$ in a ferromagnetic metal occur in two sets of four branches; one set for each direction of energy propagation. The four values of k for the positive propagating waves all have positive imaginary parts. This set of four branches splits into two sets of two each. One pair corresponds to the positively polarized spin-wave branch (S) and its associated negatively polarized nonresonant branch (N). The other pair corresponds to the positively polarized "skin-depth" or "nearly uniform precession" branch (U), and the negatively polarized, nonresonant electromagnetic branch (E). As might be expected, the positively polarized branches account for the usual resonance behavior. Their admixture depends on the spin boundary conditions The negatively polarized solutions are required in general to satisfy the electromagnetic boundary conditions

As shown by Vittoria for nickel[1], the U branch expands out to large values of k as the S branch passes close to the ω-axis, eventually rejoining the E branch at very high frequencies[2]. This is often described as an exchange-conductivity effect. The projection of the U branch on the complex k-plane is nearly circular, passing through the imaginary k-axis (k_i) with a maximum k_i near the resonance frequency, and again with a minimum k_i near the frequency $\omega=\gamma B$. Perspective views of

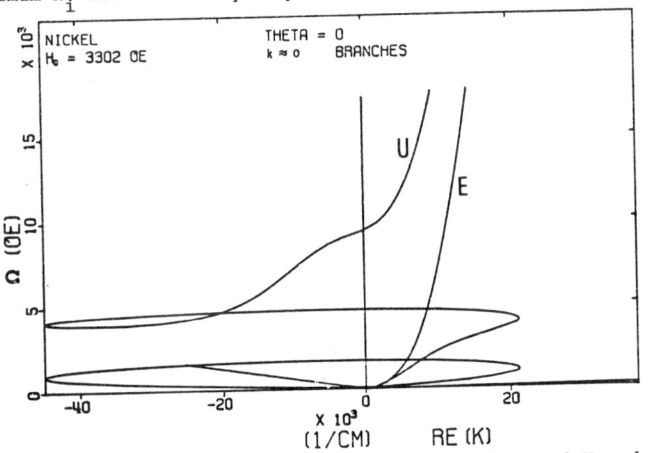

Fig. 1 Perspective view of positively polarized U and negatively polarized E branches of the dispersion relations for nickel, and their k-plane projections, including resonance and antiresonance frequency ranges.

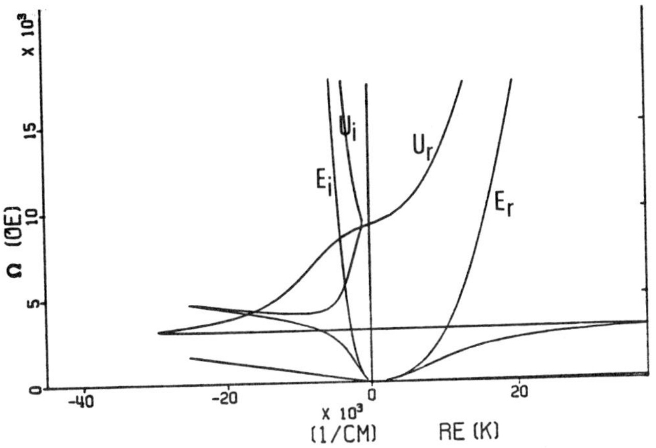

Fig. 2 Perspective view of Ω-k_r and Ω-k_i plane projections of the U and E branches of Fig. 1, to the same scales. Resonance and antiresonance correspond to maximum and minimum of k_{Ui} respectively.

the U and E branches and their projections for the perpendicular configuration are shown in Figs. 1 and 2, which were generated by a computer plotting routine. Permalloy exhibits similar crossings of the U branch, but behaves differently in detail and is discussed elsewhere[2].

The first crossing corresponds to the usual resonance, with maximum absorption and correspondingly minimum penetration depth. The second crossing is associated with an absorption minimum and a maximum penetration depth, and therefore corresponds to the antiresonance, first observed by Bloembergen[3] and observed and discussed by many others since that time[4-9].

SECULAR EQUATION

The secular equation, which has been derived under various degrees of generality elsewhere[2], is a quartic in k^2:

$$A_8 k^8 + A_6 k^6 + A_4 k^4 + A_2 k^2 + A_0 k^0 = 0 \quad (1)$$

where for zero anisotropy

$$A_8 = \xi D^2 \Delta^2$$

$$A_6 = 2[\xi(H_o + H') - i\eta\Omega]D\Delta^2$$

$$A_4 = [\xi H_o(H_o + 2H') - \Omega^2 - 2i\eta\Omega(H_o + H')]\Delta^2 \quad (2)$$

$$A_2 = 2[\xi(H_o+4\pi M)(H_o+H') - \Omega^2 - i\eta\Omega(2H_o+H'+4\pi M)]\Delta$$

$$A_0 = \xi(H_o + 4\pi M)^2 - \Omega^2 - 2i\eta\Omega(H_o + 4\pi M)$$

$$\eta = \frac{\lambda}{\gamma M} \qquad \xi = 1 + \eta^2 \qquad \Omega = \frac{\omega}{\gamma}$$

$$\Delta = \frac{i}{2}\delta^2 \qquad \delta^2 = \frac{c^2}{2\pi\sigma\omega} \qquad D = \frac{2A}{M}$$

$$H_o = |\vec{H}_a - 4\pi M\cos\theta\,\hat{e}_3| \qquad H' = 2\pi M\sin^2\theta$$

θ = angle of $\vec{M}$ with respect to the film normal, $\hat{e}_3$ and H_a, λ, γ, M, c, σ, and ω are as usual the applied field, Landau-Lifshitz damping parameter, gyromagnetic ratio (>0), saturation magnetization, velocity of light, conductivity, and angular frequency, respectively.

We note that A_4, A_2, and A_0 have zeroes near the following frequencies:

$$A_4 : \Omega_{A4}^2 = \xi H_o(H_o + 2H') \triangleq \Omega_R^2$$

$$A_2 : \Omega_{A2}^2 = \xi(H_o + 4\pi M)(H_o + H') \qquad (3)$$

$$A_0 : \Omega_{A0}^2 = \xi(H_o + 4\pi M)^2 \triangleq \Omega_T^2$$

Near Ω_R, A_4 is very small, resulting in the expected resonance condition. Near Ω_T, A_0 is very small, resulting in the antiresonance condition. Away from Ω_R, the magnitudes of k_U and k_E are much smaller than those of k_S and k_N. Therefore, one may make use of a small-root approximation to solve for k_U and k_E. The approximation is:

$$A_4 k^4 + A_2 k^2 + A_0 = 0 \qquad (4)$$

In the neighborhood of antiresonance ($A_0 \simeq 0$), we obtain for the U mode:

$$k_U^2 \equiv (k_{Ur} + ik_{Ui})^2 = -\frac{A_0}{A_2} \qquad (5)$$

which at $\Omega = \Omega_T$ yields, for small η,

$$k_{Ui} = \frac{1}{\delta}\left(\frac{\eta\Omega}{(2-\sin^2\theta)\pi M}\right)^{1/2} = \left(\frac{2\lambda\sigma}{2-\sin^2\theta}\right)^{1/2} \frac{\omega}{c\gamma M} \qquad (6)$$

A similar approximation may be obtained for the large roots in the same frequency range by setting the first three terms of (1) equal to zero:

$$A_8 k^4 + A_6 k^2 + A_4 = 0 \qquad (7)$$

which at $\Omega = \Omega_T$ yields for the spin-wave root:

$$k_{Sr} = \left(\frac{4\pi M - p}{D}\right)^{1/2} \qquad (8)$$

$$k_{Si} = \frac{\lambda\omega}{2\gamma^2 M[D(4\pi M - p)]^{1/2}} \qquad (9)$$

where $p = \Omega + H' - (\Omega^2 + H'^2)^{1/2}$. It should be noted that in the above expressions (6), (8), and (9), factors of $\xi^{1/2}$, ξ, ξ^2, all of which are close to unity, have been set equal to unity for simplicity.

The half-power width of the transmitted line is found to be

$$\Delta H_{1/2} = 4\pi M(2 - \sin^2\theta)\delta^2 k_r k_i \qquad (10)$$

where k_i satisfies

$$\exp[2(k_i - k_{Ui})L] - 4(k_i/k_{Ui})^2 + 2 = 0 \qquad (11)$$

L = film thickness

and

$$k_r = (k_i^2 - k_{Ui}^2)^{1/2} \qquad (12)$$

It can be seen that although the penetration depth ($=1/k_{Ui}$) decreases with both σ and λ, the linewidth decreases with σ but increases with λ. This is to be contrasted with the usual dependence of FMR linewidth on these two damping parameters.

APPLICATION TO SUPERMALLOY

The decay lengths and wavelength are given by

$$\delta_S = k_{Si}^{-1} \qquad \delta_T = k_{Ui}^{-1} \qquad \lambda_S = 2\pi k_{Sr}^{-1} \qquad (13)$$

For the supermalloy parameters of reference 9 at 23.7 GHz and $\theta=0$ we obtain, noting that $\sigma = 1.45\times10^{16}$ sec^{-1} ($\rho = 62$ μΩ-cm),

$$\delta = 2.57\ \mu \qquad \delta_T = 13.6\ \mu$$
$$\lambda_S = 453\ \text{Å} \qquad \delta_S = 0.808\ \mu \qquad (14)$$

Thus, the S mode decays much more rapidly than the U mode, and the E and N modes decay more rapidly still. For a film as thick as 107 μ[9] only the U mode has a significant amplitude at the far side of the film. In the 1 to 2 μ range (or 5 to 10 μ for 80-20 permalloy) the two modes are of comparable magnitude and the interferences reported by Phillips[7] can be expected. The half-power width of the transmitted line of the above example is obtained from (10) to be 181 Oe. The spectrum obtained from actual numerical calculation is shown in Fig. 3 for comparison.

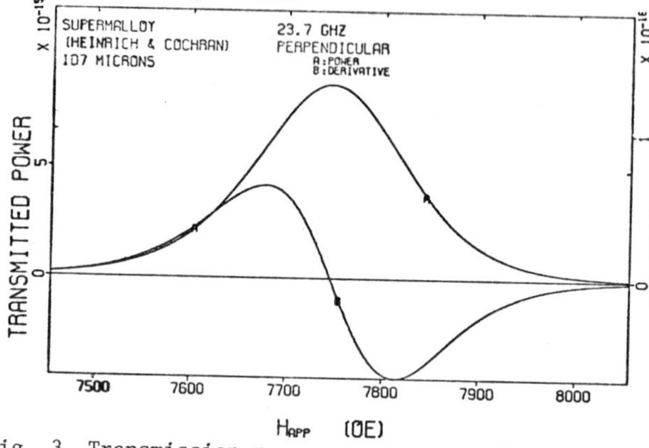

Fig. 3 Transmission resonance spectrum for supermalloy at the perpendicular configuration, with parameters of Heinrich and Cochran (1972).

The spin boundary conditions can be shown to determine the magnitude of the spin wave excited, but leave the U mode essentially undisturbed. Thus, as confirmed by detailed calculation, the power transmitted for thick samples of order 100 μ is independent of the spin boundary conditions. The transmitted U-mode field strength h_U can be related to the incident field strength by the value of k_{Ui} as given by (6). For the case of the 107 μ supermalloy film, the amplitude of h_U is reduced by a factor of $\exp(-7.87) = 3.82\times10^{-4}$ due to attenuation in traversing the film, and further reduced by a factor of 2.42×10^{-4} on emergence due to the surface impedance mismatch. The power transmission ratio is thus approximately 8.55×10^{-15}. The power transmission ratio depends strongly on both σ and λ. For example, an increase of σ by a factor of 2 would diminish the ratio by three orders of magnitude. The sensitivity of the power transmission ratio to σ, λ, and M makes it clear why transmission resonance behavior has not been observed in 100 μ films of nickel, iron, or 80-20 permalloy.

REFERENCES

1. C. Vittoria, Ph.D. Thesis, Yale University, 1970.
2. Y.J. Liu, Ph.D. Thesis, Yale University, 1974.
3. N. Bloembergen, Phys. Rev., 78, 572-580 (1950).
4. R.L. Cooper and E.A. Uehling, Phys. Rev., 164, 662-668 (1967).
5. M.I. Kaganov, JETP Letters, 10, 214-217 (1969).
6. O. Horan, G.C. Alexandrakis, and C.N. Manicopoulos, Phys. Rev. Lett, 25, 246-249 (1970).
7. T.G. Phillips, J. Appl. Phys., 41, 1109-1110 (1970).
8. B. Heinrich and V.F. Meshcheryakov, Sov. Phys. JETP, 32, 232-237 (1971).
9. B. Heinrich and J.F. Cochran, Phys. Rev. Lett, 29, 1175-1177 (1972).

* Supported by the National Science Foundation.

REMARKS ON ANTIRESONANCE IN FERROMAGNETIC METALS

P. Lubitz and C. Vittoria, Naval Research Laboratory
Washington, D. C. 20375

ABSTRACT

We have generalized earlier classical calculations[1], of the interaction of electromagnetic radiation with ferromagnetic metals to calculate the transmission efficiencies of ferromagnetic films as a function of the angle between the sample plane and an external DC magnetic field, including consideration of general surface pinning conditions. Using the parameters obtained by Heinrich and Cochran[2] we are able to duplicate the measured lineshapes and positions for DC applied fields parallel and perpendicular to the sample plane. At oblique angles, we find that the mechanical rotation of the magnetization broadens the line, but we are unable to duplicate the double transmission maxima seen in some earlier experiments.[2,3]

INTRODUCTION

The phenomenon[4] of ferromagnetic antiresonance (FMAR) occurs for frequencies somewhat above ferromagnetic resonance (FMR) at which the rf magnetic moment, $\vec{m}$, is out of phase with the driving field, $\vec{h}$, so that $\vec{b}=\vec{h}+4\pi\vec{m}=0$. For this condition, the dynamic permeability, μ, is very small and the effective skin depth is large, being limited only by the magnetic damping, so that the metal becomes electromagnetically transparent. The condition $\vec{b}=0$ combined with the magnetic equation of motion, $\dot{\vec{M}}=\gamma\vec{M}\times\vec{H}$, where $\vec{H}=\vec{H}_0+\vec{h}=\vec{H}_0-4\pi\vec{m}$, readily leads to the condition for FMAR:

$$\omega/\gamma=B_0=H_0+4\pi M_0 \quad (1)$$

where H_0 is the static internal field and M_0 is the saturation magnetization. Since the total field $\vec{H}$ at FMAR can always be defined as above for all angles, α, of the applied field with respect to the sample plane, Eq. 1 holds for all angles of the applied field.

Relevant experiments have been done for $\alpha=0$ and 90° in Ni-Fe films[2,3] and in pure Ni and Gd[5]. For these two geometries, the experimental lineshapes have been explained with varying success using the magnetic equation of motion, Maxwell's equations, and electromagnetic and magnetic boundary conditions. Solution of the secular equations and boundary condition equations is generally done numerically and no useful expression for the linewidth in terms of the magnetic and electronic parameters has been given.

Measurements of FMAR by Heinrich and Cochran[2] (107μ thick Ni-Fe with $4\pi M_0$=6950G) included observations at oblique angles of the fields; they observed that the transmission maximum splits into two peaks for $87\leq\alpha\leq 89.5°$. More recent experiments by Cochran et al.[6] indicate that this structure is an artifact caused by a slightly bent sample, and that for ideally planar thick film samples, only one transmission maximum is seen at any angle of the applied field. (Phillips[3] also reported a split line under similar conditions but with α nominally at 90°.) Heinrich and Cochran adequately described the lineshape of the transmission resonance of $\alpha=0$ and 90° in terms of the parameters appropriate to their material, but no lineshape calculation was done for oblique angles. We have thus undertaken a detailed calculation using a formalism described before[1] except that now the transmission efficiencies (defined as the ratio of the transmitted to the incident fields) are obtained and we have allowed for general pinning conditions at the surfaces and for the mechanical rotation of the magnetization as the applied field is swept in magnitude at a fixed angle. We have found that for thick films there is only one maximum of the transmitted amplitude of $\vec{h}$ in agreement with the most recent experiments.[6]

In addition to the general calculations, we have found, in an approximation appropriate to thick films, an analytical expression for the linewidth of sufficient accuracy to allow quantitative analysis of experimental data.

CALCULATIONS

In describing the dynamics of the magnetization in a conductor, we follow the approach of Ref. 1 and include the generalizations allowing for arbitrary angles of H_a and general spin pinning conditions. Within the framework of such calculations, we have neglected magnetocrystalline anisotropy and magnetoacoustic effects. Both are negligible for the materials and frequencies of interest, although the effects of anisotropy are easily included.

The outline of our calculation is as follows: (1) The boundary conditions are written in terms of the internal and external magnetic field strengths. Since we allow for internal reflections, there are 8 internal field variables. For each pair of field strengths, there corresponds a propagation constant, k, which is a solution of the secular equation of Ref. 1. (2) There are in addition 4 external rf field variables corresponding to the transverse components of the reflected and transmitted fields. The incident field components are considered as known. Thus there are 12 unknown field strengths at the two surfaces. (3) Eight of the 12 equations are obtained by requiring that the tangential electric and magnetic fields are continuous at the two surfaces. The other 4 equations follow from the pinning conditions (assuming uniaxial surface anisotropy with energy of the form $E_s=K_s\cos^2\theta$):

$$\partial m_z/\partial y+m_zK_s(\sin\theta)^2=0$$
$$\partial m_x/\partial y-m_xK_s\cos(2\theta)=0 \quad (2)$$

The surface anisotropy constant, K_s, is allowed to be different at the two surfaces. In writing Eq. 2 we have assumed that the orientation of the static magnetization is the same throughout the sample; this assumption may not be valid if K_s is large. Obviously, the pinning condition is angularly dependent, and the angle, θ, of the magnetization with respect to the sample plane is obtained from the equilibrium condition,

$$H_a\sin(\alpha-\theta)=2\pi M_o\sin(2\theta) \quad (3)$$

The internal field, H_o, is defined in terms of the external field as

$$H_o^2=H_a^2\cos^2\alpha+(H_a\sin\alpha-4\pi M_o\sin\theta)^2 \quad (4)$$

For oblique angles, the angle θ varies continuously as H_a is swept. This presents a unique situation in that the pinning conditions change as H_a is swept. We have taken this into account in our solutions. This problem does not arise for in plane or perpendicular cases for which the static magnetization remains fixed in orientation. (4) Finally, the transmitted rf magnetic fields are calculated using the twelve equations, as a function of H_a at a fixed frequency. For oblique angles, we have taken into account the mechanical rotation of the magnetically , which broadens the resonance line.

We find our numerical solutions of this

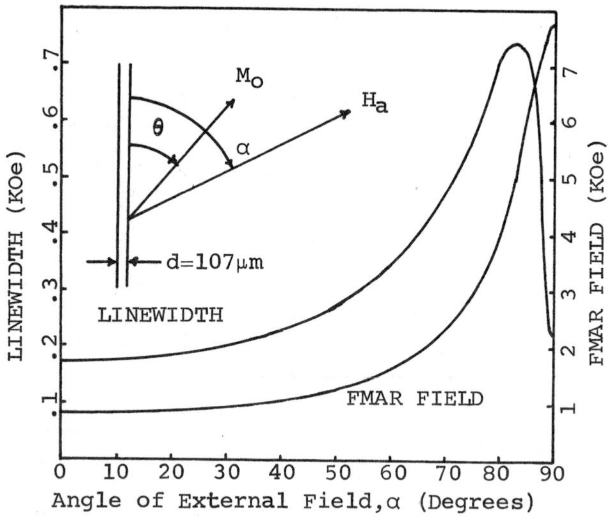

Fig. 1. Linewidth and Field for maximum transmission calculated from parameters of Heinrich and Cochran: $\omega/\gamma=7770$ Oe, $4\pi M_o=6950$G, $\lambda=1.7\times 10^8$/s, $d/\delta_o=60$, $A=10^{-6}$ erg/cm, $\omega=23.7$ GHz.

problem for $\alpha=0$ and 90° to be in essential agreement with those discussed in Ref. 2. In Fig. 1 we show the results of our detailed calculations for the position and width (at 1/2 of the maximum of h_{trans}) using the parameters reported by Heinrich and Cochran.[2] In our calculation, the width is seen to increase rapidly as $(90°-\alpha)$ increases as a result of the "mechanical" rotation of $\vec{M}$. This is a consequence of having H_{AR} only slightly above $4\pi M_0$ and makes the angular alinement of $\vec{H}_{ext}$ very important. Use of higher frequencies would considerably reduce such broadening. In our calculations, all of the exchange modes are included and the pinning conditions are found to be unimportant for thick films, for the parameters quoted in Fig. 1, variation of K_s from 0 to ∞ (including 1, 10, and 1000 erg/cm^2) produces only about .001% changes in transmitted intensity. However, in thin films on the order of 10μ, we find a number of transmission maxima. This contrasts to the results of Fig. 1 of Ref. 3 where only two peaks are shown. In fact, using Eq. 7 of Ref. 3 for the transmittance, we still find a number of transmission peaks.

APPROXIMATE CALCULATIONS

As pointed out previously, when the classical skin depth, $\delta_o^2=c^2/(2\pi\sigma\omega)$ is on the order of the film thickness, the transmission becomes very flat near FMAR except that the usual dimensional spin wave resonance can be seen in transmission. In this regime the results are quite sensitive to surface pinning, but the FMAR is poorly defined. The regime studied in most transmission experiments is $\delta_o<<d$. Under these conditions, the FMAR becomes progressively sharper and the effects of surface pinning are negligible. The waves with large wave vectors (k) are weighted correspondingly less so that only the long wavelength solution of the secular equation is important for FMAR in thick films. The approximate value of k near FMAR is given by $k\cong 1/\delta_o(\Delta H/2\pi M_o)^{1/2}$ where λ is the Landau-Lifshitz damping parameter, $\Delta H=2(\lambda/\gamma)(\omega/\gamma M_o)$ is the FMR half power linewidth neglecting exchange effects. This thickness regime can be further divided into a regime $\delta_o<d<\delta_o(2\pi M/\Delta H)^{1/2}$ and $(2\pi M/\Delta H)^{1/2}<d$. In the former case the film is still largely transparent at the antiresonance, and in the latter, the waves at antiresonance are strongly attenuated, but may still be easily observable. In the latter case, the transmitted wave amplitude is given essentially by e^{-kd} where k is the wave vector evaluated at antiresonance.

We have retained the exact low order terms in k from the secular equation, and have analyzed the field dependence of k near its minimum in order to determine the transmission lineshape. We make the approximation of neglecting terms of order k^4 compared to k^2 and neglecting reflected waves, and we consider the lineshape to be determined by $e^{-k(H_o)d}$ in the hope of generating a useful approximation.

We find that for oblique angles, the propagation constant at the transmission maximum is given by:

$$k\cong (1/\delta_o)(\lambda\omega/\pi\gamma^2 M_o^2)^{1/2}(1+\sin^2\theta)^{1/2} \qquad (5)$$

Thereby, the transmission efficiencies at FMAR are angularly dependent, reaching a minimum for $\alpha=0$. The FMAR linewidth can be determined from the field dependence of k near the transmission maximum. (Kaganov[4] has adapted a similar approach. However his approximations omit the damping terms and are not appropriate for most of the experiments.) The linewidth, which we define as the separation of the applied DC fields required to reduce the transmitted field amplitudes by 1/2 from the maximum, is found to be approximately

$$\Delta H_{FMAR}\cong .3(4\pi M_o)\sqrt{(\delta/d)\sqrt{(\Delta H/M_o)}}^{\,3} \qquad (6)$$

and the transmission maximum is shifted downward $H_o=\omega/\gamma-4\pi M_o-2(\Delta H)^2/(4\pi M_o(1+\sin^2\theta))$ where $\delta^2=\delta_o^2(1+\sin^2\theta)$. In this form, Eq. 6 is independent of the form of damping assumed. The field width is given in terms of the internal field of Eq. 4; it is equal to the external field width only for $\alpha=0$ or 90°. However, using Eqs. 3 and 4, the width can be approximately corrected for the effects of rotation of the magnetization.

CONCLUSIONS

We have undertaken a general classical calculation of the transmission of electromagnetic radiation through ferromagnetic conducting films including general surface pinning conditions and general angle of H_{ext} with respect to the sample plane. No qualitatively new behavior is found to occur at oblique angles of the applied DC fields, although the transmission maxima in thick films may be broadened considerably by the rotation of the magnetization as the external field is swept. However, it is apparent that the choice of antiresonant fields only slightly above magnetic saturation, which has often been used experimentally, can produce considerable broadening if the angular misalinement is even a fraction of a degree. Nevertheless the observation of FMAR must be considered a useful tool in observing the intrinsic relaxation rates.

1. C. Vittoria et al., J.Appl. Phys. 40, 1561 (1969).
2. B. Heinrich and J.F. Cochran, Phys. Rev. Letters 29, 1175 (1972).
3. T.G. Phillips, J. Appl. Phys. 41, 1109 (1970).
4. M.I. Kaganov, JETP Lett. 9, 378 (1969).
5. O. Horan, et al., Phys. Rev. Lett. 25, 246 (1970).
6. J.F. Cochran, et al. (this conference Proceedings).
7. In Phillips calculation, K_s, σ, and the magnon lifetime were all given unphysically large values. In addition he omits a factor of $\frac{4\pi\sigma}{c}$ which should multiply the free space wave vector (see also Horan et al.)

MICROWAVE TRANSMISSION AND ABSORPTION MEASUREMENTS ON SUPERMALLOY

J.F. Cochran, B. Heinrich, and G. Dewar
Simon Fraser University, Burnaby, British Columbia, Canada, V5A 1S6

ABSTRACT

We have studied the ferromagnetic resonance and anti-resonance microwave properties, at 24 GHz, of supermalloy at room temperature. The role of the exchange interaction and surface pinning was investigated by analyzing the asymmetry of the ferromagnetic resonance absorption line using intrinsic magnetic damping parameters obtained from transmission experiments. For the perpendicular configuration the FMR lineshapes are described very well by Landau-Lifshiftz damping and the exchange-conductivity mechanism with partly pinned surface spins.

INTRODUCTION

We have studied the magnetic properties of the soft ferromagnetic metal Supermalloy at room temperature using the transmission and absorption of 24 GHz microwaves. The transmission of microwave radiation through thick specimens ($d/\delta_0 > 10$, where d is the thickness and $\delta_0^2 = c^2\rho/4\pi\omega$ is the classical skin depth) is extremely small except in a small external magnetic field interval corresponding to ferromagnetic anti-resonance (FMAR)[2-5]. At the anti-resonance field the high frequency permeability becomes very small and would vanish in the absence of intrinsic magnetic damping. It can be shown[6] that the variation of the amplitude of the transmitted microwave signal with applied magnetic field depends upon magnetic damping, the saturation magnetization, $4\pi M_s$, the g-factor, and upon reduced thickness, d/δ_0; it is, however, effectively independent of exchange, diffusion, and the surface pinning mechanism for the high frequency component of the magnetization. On the other hand, the field dependence of the absorption at ferromagnetic resonance (FMR) depends upon all of the above parameters[7-9] (except thickness). Specifically, the g-factor is determined from the magnetic field at maximum transmission amplitude with the external magnetic field applied perpendicular to the plane of the specimen (perpendicular configuration, $H_{perp}=\omega/\gamma$), the saturation magnetization is determined from the magnetic field at maximum transmission amplitude with the external field in the plane of the specimen (parallel configuration, $H_{par}=\omega/\gamma-4\pi M_s$), and the damping parameter for either configuration is obtained from the variation of the transmitted microwave amplitude with magnetic field (the lineshape). These parameters, which were determined uniquely from the transmission experiments, were then used to analyze the magnetic field dependence of the ferromagnetic absorption signal in order to obtain information about the exchange parameter, A, and the dynamic surface pinning parameter, K_{surf}.

TRANSMISSION

Transmission studies at FMAR were carried out using paired transmitter and receiver cavities. This apparatus will be described in detail in a later publication. The specimens were in the form of 16mm diameter discs and were annealed (except as noted) in vacuum for 4 hours at 1000°C on molybdenum substrates. The resistivity, ρ, of this material was found to be 60.13×10^{-6} ohm cm at 25°C. These discs were electropolished to their final thicknesses. Average thickness of the central 3.5mm diameter central portions of the specimens, the portions which were exposed to the microwave radiation, were computed from their weights using a density which we determined to be 8.61±.01 g/cc. Demagnetizing factors were computed from the shift in EPR resonance field observed in a small piece of DPPH placed upon the surface of the specimen[10,11]. Typical results of transmission experiments are shown in Figs. 1 and 2. These lines exhibit none of the

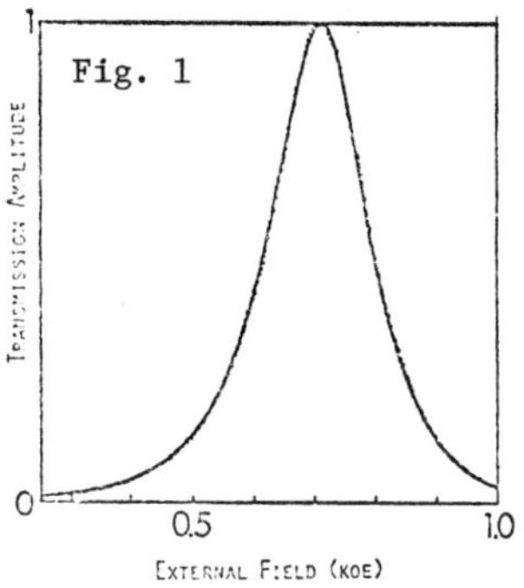

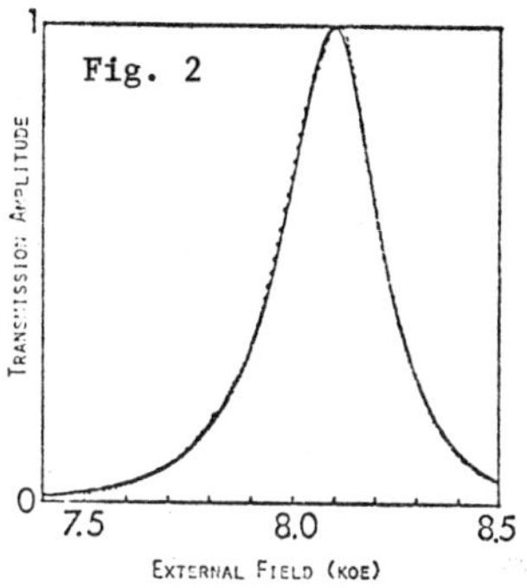

Figs. 1 & 2 Amplitude of 24 GHz transmission signal as a function of external magnetic field for specimen SPYJ27 in the parallel (Fig. 1) and perpendicular (Fig. 2) configurations. The solid line has been calculated using parameters listed in Table I.

splitting and anomalous broadening for angles near the perpendicular configuration which were reported previously[12-14]. We know now that these effects were due to a very slight curvature in the plane of the specimens, probably less than 1/10°. In this connection it should be noted that the transmission amplitude in the perpendicular configuration is extremely sensitive to the angle between applied field and specimen normal. A deviation of 1/10° from exact perpendicularity is readily observable. The theoretical curves shown in Figs. 1-4 were calculated using Maxwell's equations, the Landau-Lifshitz form for the equation of motion for the magnetization

$$\frac{d\vec{M}}{dt} = \gamma\, \vec{M}\times\vec{H}_{eff} - \frac{\lambda}{M_s^2}\,(\vec{M}\times\vec{M}\times\vec{H}_{eff}) \qquad (1)$$

with $$\vec{H}_{eff} = \vec{H} + \frac{2A}{M_s^2}\nabla^2\vec{M}, \qquad (2)$$

and the boundary condition for the high frequency magnetization

$$2A\frac{\partial\vec{m}}{\partial n} - K_{surf}\vec{m} = 0 \qquad (3)$$

$\vec{H}$ includes the high frequency magnetic field and demagnetizing fields. The results of fitting this theory to transmission data for four specimens of supermalloy are shown in Table I. The sensitivity of the fitting was such that $4\pi M_s$ and ω/γ could be readily determined to within ±2 gauss (although absolute

TABLE I

A description of the specimens used in this investigation along with the parameters which were adjusted to obtain the best agreement between theory and experimentally observed magnetic field dependence of transmission amplitudes at 24 GHz. The skin-depth $\delta_0=1.78\mu$.

Specimen	d [μm] (±.1)	g	$4\pi M_s$ (kg)	λ_{par} [10^{8}sec^{-1}]	λ_{per} [10^{8}sec^{-1}]
SPYJ19 (annealed)	67.6	2.104	7.444	0.95	1.05
SPYJ27 (annealed)	81.8	2.106	7.444	1.00	1.05
SPYU01 (not annealed)	89.1	2.109	7.341	1.05	1.10
SPY2 (annealed)	93.0	2.109	7.422	0.85	0.95

values are not known to better than ±5 gauss), and λ could be determined within $\pm 5\times 10^6$ sec^{-1}.

FERROMAGNETIC RESONANCE

Asymmetry in the FMR lineshape depends both upon exchange and the pinning parameter[8,9], K_{surf}. For fully pinned spins the magnetic field derivative of the absorption is essentially symmetric; for free spins ($K_{surf}=0$) the asymmetry in the absorption derivative is approximately 12% for our case, and the low field peak is stronger than the high field peak. In order to be able to use this asymmetry to investigate exchange and pinning we wished to use a technique which would enable us to precisely locate the zero of the absorption derivative. We therefore decided to measure power absorption in the specimen using a differential bolometric technique which exploited the fact that any increased absorption of microwave power by the specimen must of necessity be accompanied by a proportionate small increase in temperature. The bolometer of a matched pair was pressed against the center of the backside of the specimen, the other bolometer was attached to the water-cooled cavity. The specimen was clamped firmly to the outside of the cavity be a copper yoke, and was exposed to the radiation through a 3½mm coupling hole. A signal proportional to the temperature difference between the bolometers was digitalized, smoothed (corresponding to a 10 second time constant), and finally differentiated.

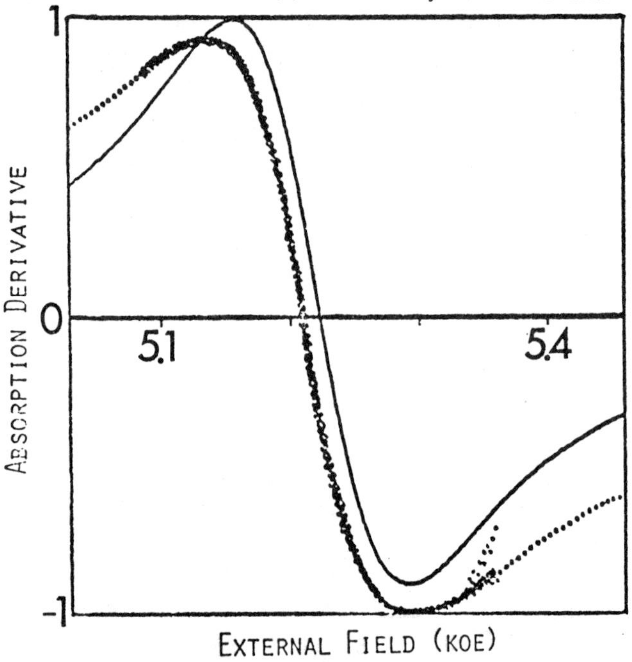

Fig. 3 Derivative of the FMR absorption signal at 24 GHz measured for specimen SPYJ19, parallel configuration. The solid line has been calculated using the parameters of Table I, and $A=1.5\times 10^{-6}$ ergs/cm, $K_{surf}=0.6$ ergs/cm^2. Data points for 8 independent runs are shown superimposed. No significance should be attributed to the relative positions of theoretical and experimental peaks because the absolute values of the field were uncertain within ±20 oe.

The results of measurements on SPY J19 are shown in Figs. 3 and 4. Electrical noise in these data is completely negligible, but the signal is very sensitive to fluctuations in the temperature of the water used to cool the cavity. The cooling water temperature was regulated to ±2 mdeg, and the noise visible in Figs. 3,4 is due to sudden jumps of a few mdeg in the water temperature. Nevertheless, it is clearly evident from the figures that these FMR absorption lines are not symmetric. The line shape for the perpendicular configuration is very well described by the theory using $A=1.5\times 10^{-6}$ ergs/cm, $K_{surf}=0.6$ ergs/cm^2.

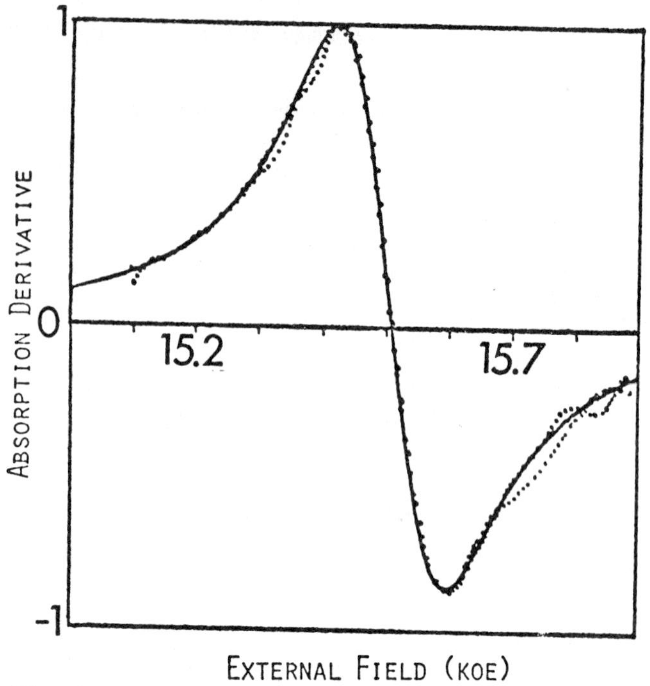

Fig. 4 Derivative of the FMR absorption signal at 24 GHz measured for specimen SPYJ19, perpendicular configuration. The solid line has been calculated using the parameters of Table I and the values $A=1.5\times 10^{-6}$ ergs/cm, $K_{surf}=0.6$ ergs/cm^2 chosen to obtain the best fit for the lineshape. Although most of the points are obscured by the theoretical line, data from 4 independent runs are shown superimposed. The visible noise is due to uncontrolled fluctuations of ~.002°C in the temperature of the water-cooled cavity. No significance should be attached to the relative positions of the theoretical and experimental peaks because the absolute values of the field were uncertain within ±20 oe.

The paired values $A=1.0\times 10^{-6}$ ergs/cm, $K_{surf}=0.5$ ergs/cm^2, or the pair $A=2.0\times 10^{-6}$ ergs/cm, $K_{surf}=0.6$ ergs/cm^2, gave noticeably poorer fits for the lineshape. The FMR lineshape for the parallel configuration is not well described by the theory because the absorption derivative has an asymmetry which is opposite to that calculated. We have no explanation for this observation, but we speculate that it may be due to relaxation via two-magnon scattering as discussed by Paton[15,16,17]. We plan to pursue this investigation using an improved temperature stabilizing system for our cavity cooling system. We also plan to improve the absolute precision of our magnetic field measurements - in the present instance we do not know our fields better than ±20 oe. For the perpendicular configuration, at least, it does appear that the FMR lineshape can be described very well using the intrinsic damping parameter obtained from transmission experiments and using reasonable values for exchange and surface pinning parameters. This conclusion is in disagreement with that of Cooper and Uehling[17]. On the basis of their FMR studies at 9.17 GHz these authors found it necessary to invoke spin-diffusion damping in order to explain their observed lineshapes.

REFERENCES

1. Supermalloy was obtained, in the form of rolled sheet 10^{-4} cm thick, from the Perfection Mica Co., 740 Thomas Drive, Bensenville, Ill., under the tradename "Conetic AA".
2. B. Heinrich and V.F. Meshcharyakov. Pis'ma Zh. Eksp. Teor. Fiz. 9, 618 (1969) [JETP Lett. 9, 378 (1969)].
3. B. Heinrich and V.F. Meshcharyakov. Zh. Eksp. Teor. Fiz. 59, 424 (1970) [Sov. Phys. JETP 32, 232 (1971)].

4. O. Horan, G.C. Alexandrakis, and C.N. Manicopolous Phys. Rev. Lett. 25, 246 (1970).
5. G.C. Alexandrakis, T.R. Carver, and O. Horan, Phys. Rev. B5, 3472 (1972).
6. J.F. Cochran and B. Heinrich. To be published.
7. G.T. Rado and J.R. Weertman. J. Phys. Chem. Solids 11, 315 (1959).
8. D.C. Rodbell. Physics 1, 279 (1965).
9. Z. Frait and H. Macfadden. Phys. Rev. 139 A1173 (1965).
10. Z. Frait and B. Heinrich. Archives des Sciences 14, 138 (1961).
11. L. Kraus and Z. Frait. Czech. J. Phys. B23, 188 (1973).
12. T.G. Phillips. J. Appl. Phys. 41, 1109 (1970).
13. B. Heinrich and J.F. Cochran, Phys. Rev. Lett. 29, 1175 (1972).
14. B. Heinrich, J.F. Cochran, and G. Dewar. Proc. Int. Conf. of Magnetism, ICM-73. NAUKA, Moscow 1974.
15. C.E. Patton. Phys. Rev. 179, 352 (1969).
16. C.E. Patton, F. Ono, and M. Takahashi. IEEE, Vol. MAG, 7760 (1971).
17. It has been suggested by the referees, C.H. Wilts and G. Ramer of California Institute of Technology, that the reversed asymmetry which we observe in the parallel configuration can be explained by a uniaxial surface anisotropy with the easy axis perpendicular to the sample surface.[19] We would like to thank our referees for bringing this point to our attention.
18. R.L. Cooper and E.A. Uehling. Phys. Rev. 164, 662 (1967).
19. G.C. Bailey, C. Vittoria. Phys. Rev. B, 8, 3247 (1973).

FERROMAGNETIC RESONANCE IN METALLIC MULTILAYERS*

G. Spronken, A. Friedmann, and A. Yelon
Département de Génie Physique, Ecole Polytechnique, Montréal, Québec, H3C 3A7 Canada.

ABSTRACT

We have developed a method for calculating the power absorbed by metallic multilayers in ferromagnetic resonance, for arbitrary directions of the applied magnetic field. This method also permits the calculation of the power absorbed in each layer of a symmetric structure. The calculation is based on the use of Hoffmann's equations for the continuity of the magnetization at the interfaces between films[1] within the approximate formalism for FMR in metals.[2] This formalism permits the reduction of the equation of motion for each medium to a pair of quadratic equations in the square of the wave vector, one for the resonant sense of polarization, and one for the non-resonant sense. The continuity equations are easily written in terms of these polarizations. Different levels of approximation are applied to the boundary value problem for two different cases. When the films are very similar, or the applied field is very strong (high resonant frequencies), one can neglect the non-resonant polarization and satisfy boundary conditions with the resonant polarization only. For dissimilar media at low fields, the non-resonant polarization must be included. From the simultaneous solution of boundary conditions (including those at the exterior) the absorbed power may be calculated. This is simpler for symmetric (e.g. 3-layer) structures than for asymmetric (e.g. 2-layer) ones. In the former case, application of Poynting's theorem yields relatively simple expressions for the total power absorbed, and for the power absorbed in each layer. The limitations of the model have been studied, and are discussed, as are illustrative numerical examples.

*Supported by the National Research Council of Canada.

1. F. Hoffmann, Phys. Stat. Sol. 41, 807 (1970); Thesis, Orsay, 1971.
2. A. Yelon, G. Spronken, T. Buithieu, R.C. Barker, Y.-J. Liu, and T. Kobayashi, Phys. Rev. B 10, 1070 (1974)

NONLINEAR EFFECTS IN FERROMAGNETIC METALS*

O.L.S. Lieu and G.C. Alexandrakis
Physics Department, University of Miami
Coral Gables, Florida 33124

ABSTRACT

A nonlinear theory of ferromagnetic transmission resonance in the case in which the static and microwave fields are mutually parallel and parallel to the sample surface is outlined. Its predictions are compared to experimental results obtained for iron and nickel. The theory predicts strong subharmonic generation of possible practical use.

THEORY

The effects described here are strictly ferromagnetic. Experimentally they are not observed in paramagnetic metals.[1] Assume the sample is in foil form of thickness d and is made to form the common wall of two identical microwave cavities. One cavity is used to store the incident microwaves of frequency ω_0 and the other to collect the power transmitted through the sample. Take the sample to lie on the xz-plane. The microwaves propagate along the y-axis. Both the applied static and microwave fields are along the z-axis.[1,2]

The Bloch-Bloembergen and Maxwell equations for this problem are written as:

$$\left.\begin{aligned}
\frac{\partial M_x}{\partial t} &= \gamma(M_y H_a - M_s H_y) - \frac{2\gamma A}{M_s}\nabla^2 M_y - \frac{M_x}{\tau}\\
\frac{\partial M_y}{\partial t} &= \gamma(M_s H_x - M_x H_a) + \frac{2\gamma A}{M_s}\nabla^2 M_x - \frac{M_y}{\tau}\\
\frac{\partial M_z}{\partial t} &= \gamma(M_x H_y - M_y H_x) + \frac{2\gamma A}{M_s^2}(M_x\nabla^2 M_y - M_y\nabla^2 M_x) - \frac{M_z}{T_1}\\
\vec{\nabla}\cdot(\vec{H}+4\pi\vec{M}) &= 0\\
\vec{\nabla}\times\vec{H} &= \frac{4\pi\sigma}{c}\vec{E}\\
\vec{\nabla}\times\vec{E} &= -\frac{1}{c}\frac{\partial}{\partial t}(\vec{H}+4\pi\vec{M})
\end{aligned}\right\} \qquad (1)$$

Here $\vec{H} = (H_x, H_y, H_a+H_z)$, $\vec{M} = (M_x, M_y, M_s+M_z)$ where H_a is the applied static field, M_s is the saturation magnetization, $H_{x,y,z}$ and $M_{x,y,z}$ are the high frequency components of the fields. Other symbols have their standard meanings.

Assume the plane wave solutions $H_{x,y} = h_\lambda^{x,y}\, e^{-(\lambda y + i\omega t)}$, $M_{x,y} = m^{x,y}\, e^{-(\lambda y+i\omega t)}$ for the transverse components of Eqs. (1). We obtain the secular equation for the propagation constant λ by using the transverse component equations from Eqs. (1), to be

$$\lambda^6 + a_2\lambda^4 + a_1\lambda^2 + a_0 = 0 \qquad (2)$$

where

$$a_2 = \frac{2i}{\delta^2} - \frac{(\omega_B+\omega_H)}{\Lambda M_s},\quad a_1 = \frac{\nu^2+\omega_B\omega_H}{\Lambda^2 M_s^2} - \frac{4i\,\omega_B}{\delta^2\Lambda M_s},\quad a_0 = \frac{2i}{\delta^2}\frac{(\nu^2+\omega_B^2)}{\Lambda^2 M_s^2}$$

$$\Lambda = \frac{2\gamma A}{M_s^2},\quad \nu = \frac{1}{\tau} - i\omega,\quad \delta = \frac{c}{\sqrt{2\pi\omega\sigma}}\ (\text{skin depth})$$

$$\omega_H = \gamma H_a,\quad \omega_M = 4\pi\gamma M_s,\quad \omega_B = \omega_H + \omega_M$$

Equation (2) is cubic in λ^2, the roots are $\pm\lambda_1$, $\pm\lambda_2$, $\pm\lambda_3$ which can be expressed in terms of a_0, a_1, a_2 analytically. They give the wave propagation along $+y$ and $-y$.

We now write

$$H_x = \sum_j (A_j e^{-\lambda_j y} + B_j e^{\lambda_j y})\, e^{-i\omega t} \quad (j=1,2,3) \qquad (3)$$

and similar expressions for M_x, M_y, H_y where A_j and B_j are the amplitudes to be determined by boundary conditions. Using these expressions of the transverse fields in the longitudinal components of (1) we solve for H_z. The solution is

$$H_z = \sum_{j,l}\left[\alpha_{jl}(A_jA_l e^{-a_{jl}y} + B_jB_l e^{a_{jl}y}) + \beta_{jl}(A_jB_l e^{-b_{jl}y} + B_jA_l e^{b_{jl}y})\right]e^{-i2\omega t} \qquad (4)$$

where

$$a_{jl} = \lambda_j+\lambda_l,\quad b_{jl} = \lambda_j - \lambda_l,\quad \alpha_{jl} = \frac{(-i8\pi\omega c_1)\epsilon_{jl}}{(\alpha_0^2 a_{jl}^2 + i2\omega)},\quad \beta_{jl} = \frac{(-i8\pi\omega c_1)\epsilon_{jl}}{(\alpha_0^2 b_{jl}^2 + i2\omega)}$$

$$\alpha_0^2 = \frac{c^2}{4\pi\sigma},\quad c_1 = \frac{T_1(1+i2\omega T_1)}{(1+4\omega^2T_1^2)},\quad \epsilon_{jl} = \frac{-1}{2}\left[\frac{\gamma}{4\pi}(\xi_j\eta_l + \eta_l) + \frac{\Lambda}{(4\pi)^2}\xi_j\eta_l(\lambda_j^2 - \lambda_l^2)\right]$$

$$\xi_j = -(\lambda^2 + 2i/\delta^2)/(2i/\delta^2),\quad \eta_j = \frac{\nu\xi_j}{(\omega_B - \Lambda M_s\lambda_j^2)} \quad (j,l = 1,2,3)$$

Notice that the field transmitted along the z-axis is at the applied frequency $\omega_0 = 2\omega$ whereas the one transmitted along the x-axis has the frequency $\omega = \omega_0/2$.

We now begin to solve the boundary value problem for the transmitted fields. The boundary conditions are (a) the tangential components of $\vec{H}$ and $\vec{E}$ are continuous across the sample surfaces at y=0 and y=d, (b) Kittel's boundary condition for the transverse magnetization, i.e. $M_x=0$ at y=o,d. We arrange the boundary conditions in two groups. The first group which contains the boundary conditions on H_x, E_z and M_x, is written in matrix form as

$$\left.\begin{aligned}
M_1\begin{pmatrix}A_1\\B_1\end{pmatrix} + M_2\begin{pmatrix}A_2\\B_2\end{pmatrix} + M_3\begin{pmatrix}A_3\\B_3\end{pmatrix} &= \begin{pmatrix}r\\t\end{pmatrix}\\
\lambda_1N_1\begin{pmatrix}A_1\\B_1\end{pmatrix} + \lambda_2N_2\begin{pmatrix}A_2\\B_2\end{pmatrix} + \lambda_3N_3\begin{pmatrix}A_3\\B_3\end{pmatrix} &= \frac{4\pi\sigma}{c}\begin{pmatrix}-r\\t\end{pmatrix}\\
\xi_1M_1\begin{pmatrix}A_1\\B_1\end{pmatrix} + \xi_2M_2\begin{pmatrix}A_2\\B_2\end{pmatrix} + \xi_3M_3\begin{pmatrix}A_3\\B_3\end{pmatrix} &= \begin{pmatrix}0\\0\end{pmatrix}
\end{aligned}\right\} \qquad (5)$$

where

$$M_j = \begin{pmatrix}1 & 1\\ e^{-\lambda_j d} & e^{\lambda_j d}\end{pmatrix},\quad N_j = \begin{pmatrix}1 & -1\\ e^{-\lambda_j d} & -e^{\lambda_j d}\end{pmatrix},\quad r = H_{rx},\quad t = H_{tx}e^{ik_0'd},\quad k_0' = \frac{\omega}{c}$$

H_{rx} and H_{tx} are the reflected and transmitted amplitudes along the x-axis.

From (5) we have

$$\left.\begin{aligned}
\begin{pmatrix}A_1\\B_1\end{pmatrix} &= -G^{-1}(\xi_3\lambda_2P_2 - \xi_2\lambda_3P_3)\begin{pmatrix}r\\t\end{pmatrix} + \frac{4\pi\sigma}{c}(\xi_3-\xi_2)G^{-1}\begin{pmatrix}-r\\t\end{pmatrix}\\
\begin{pmatrix}A_2\\B_2\end{pmatrix} &= \left[\frac{\xi_3M_2^{-1}}{(\xi_3-\xi_2)} + \frac{(\xi_3-\xi_1)}{(\xi_3-\xi_2)}M_2^{-1}M_1G^{-1}(\xi_3\lambda_2P_2 - \xi_2\lambda_3P_3)\right]\begin{pmatrix}r\\t\end{pmatrix}\\
&\quad - \frac{4\pi\sigma}{c}(\xi_3-\xi_1)M_2^{-1}M_1G^{-1}\begin{pmatrix}-r\\t\end{pmatrix}\\
\begin{pmatrix}A_3\\B_3\end{pmatrix} &= \left[\frac{-\xi_2M_3^{-1}}{(\xi_3-\xi_2)} + \frac{(\xi_1-\xi_2)}{(\xi_3-\xi_2)}M_3^{-1}M_1G^{-1}(\xi_3\lambda_2P_2 - \xi_2\lambda_3P_3)\right]\begin{pmatrix}r\\t\end{pmatrix}\\
&\quad + \frac{4\pi\sigma}{c}(\xi_2-\xi_1)M_3^{-1}M_1G^{-1}\begin{pmatrix}-r\\t\end{pmatrix}
\end{aligned}\right\} \qquad (6)$$

where

$$G = (\xi_3-\xi_2)\lambda_1N_1 + [(\xi_2-\xi_1)\lambda_3P_3 + (\xi_1-\xi_3)\lambda_2P_2]M_1,\quad P_j = N_jM_j^{-1}$$

We can write (6) as

$$\begin{pmatrix}A_j\\B_j\end{pmatrix} = \begin{pmatrix}\Delta_j & \theta_j\\ \Lambda_j & \varphi_j\end{pmatrix}\begin{pmatrix}r\\t\end{pmatrix} \quad (j=1,2,3) \qquad (7)$$

The second group of boundary conditions which contains the ones on H_z, E_x gives the following equations

$$\left.\begin{aligned}
r^2\psi_1(0) + t^2\psi_2(0) + rt\,\psi_3(0) &= H_{rf} + H_{rz}\\
r^2\psi_1(d) + t^2\psi_2(d) + rt\,\psi_3(d) &= H_{tz}e^{ik_0d}\\
r^2\phi_1(0) + t^2\phi_2(0) + rt\,\phi_3(0) &= -H_{rf} + H_{rz}\\
r^2\phi_1(d) + t^2\phi_2(d) + rt\,\phi_3(d) &= -H_{tz}e^{ik_0d}
\end{aligned}\right\} \qquad (8)$$

where

$$\psi_1(y) = \sum_{j,l}\{\alpha_{jl}(\Delta_j\Delta_l e^{-a_{jl}y} + \Lambda_j\Lambda_l e^{a_{jl}y}) + \beta_{jl}(\Delta_j\Lambda_l e^{-b_{jl}y} + \Lambda_j\Delta_l e^{b_{jl}y})\}$$

$$\psi_2(y) = \sum_{j,l}\{\alpha_{jl}(\theta_j\theta_l e^{-a_{jl}y} + \varphi_j\varphi_l e^{a_{jl}y}) + \beta_{jl}(\theta_j\varphi_l e^{-b_{jl}y} + \varphi_j\theta_l e^{b_{jl}y})\}$$

$$\psi_3(y) = \sum_{j,l}\{\alpha_{jl}[(\Delta_j\theta_l + \theta_j\Delta_l)e^{-a_{jl}y} + (\Lambda_j\varphi_l + \varphi_j\Lambda_l)e^{a_{jl}y}] + \beta_{jl}[(\Delta_j\varphi_l + \theta_j\Lambda_l)e^{-b_{jl}y} + (\Lambda_j\theta_l + \varphi_j\Delta_l)e^{b_{jl}y}]\}$$

$\phi_j(y) = \frac{c}{4\pi\sigma}\frac{d\psi_j}{dy}$, $k_0 = \frac{\omega_0}{c}$, H_{rf} is the incident microwave amplitude, H_{rz} and H_{tz} are the reflected and transmitted amplitudes along the z-axis.

The H_{tz} and H_{tx} are obtained from (8) to be

$$H_{tz} = H_{rf}e^{-ik_0d}Z_{(\pm)} \qquad (9)$$

$$H_{tx} = \sqrt{2H_{rf}}\,e^{-ik_0'd}X_{(\pm)} \qquad (10)$$

where

$$Z_{(\pm)} = \frac{\Pi_1 \Delta^2_{(\pm)} + \Pi_2 + \Pi_3 \Delta_{(\pm)}}{\Gamma_1 \Delta^2_{(\pm)} + \Gamma_2 + \Gamma_3 \Delta_{(\pm)}}, \quad X_{(\pm)} = \frac{1}{\sqrt{\Gamma_1 \Delta^2_{(\pm)} + \Gamma_2 + \Gamma_3 \Delta_{(\pm)}}}$$

$$\Delta_{(\pm)} = (-\Omega_3 \pm \sqrt{\Omega_3^2 - 4\Omega_1\Omega_2})/2\Omega_1$$

$$\Gamma_j = \psi_j(0) - \phi_j(0), \quad \Omega_j = \psi_j(d) + \phi_j(d), \quad \Pi_j = \psi_j(d) - \phi_j(d)$$

The theoretical signal is

$$H_T = Re\{A e^{-i(K_0 d + \phi)} Z_{(\pm)}\}$$

where A is the amplification factor and ϕ is an arbitrary phase introduced by the apparatus.

The power transmitted along the z-axis is then proportional to $|Z_{(\pm)}|^2$ while the one transmitted along the x-axis is proportional to $|X_{(\pm)}|^2$.

DISSCUSSION

In Fig. 1(a) and 1(d) we have plotted the experimental transmitted amplitudes along the z-direction for a 18μm thick Fe foil and a 10 μm thick Ni sample, at room temperature. The frequency of H_{zt} which is the same as the excitation frequency is close to ν_0 = 9.2 GHz. The theoretical results for $|H_{zt}|$ for Fe and Ni correspondingly are plotted in Fig. 1(b) and 1(e). The calculated $|H_{xt}|$ values for Fe and Ni are shown in Fig. 1(c) and 1(f). As was indicated above the H_{xt} field has half the applied frequency.

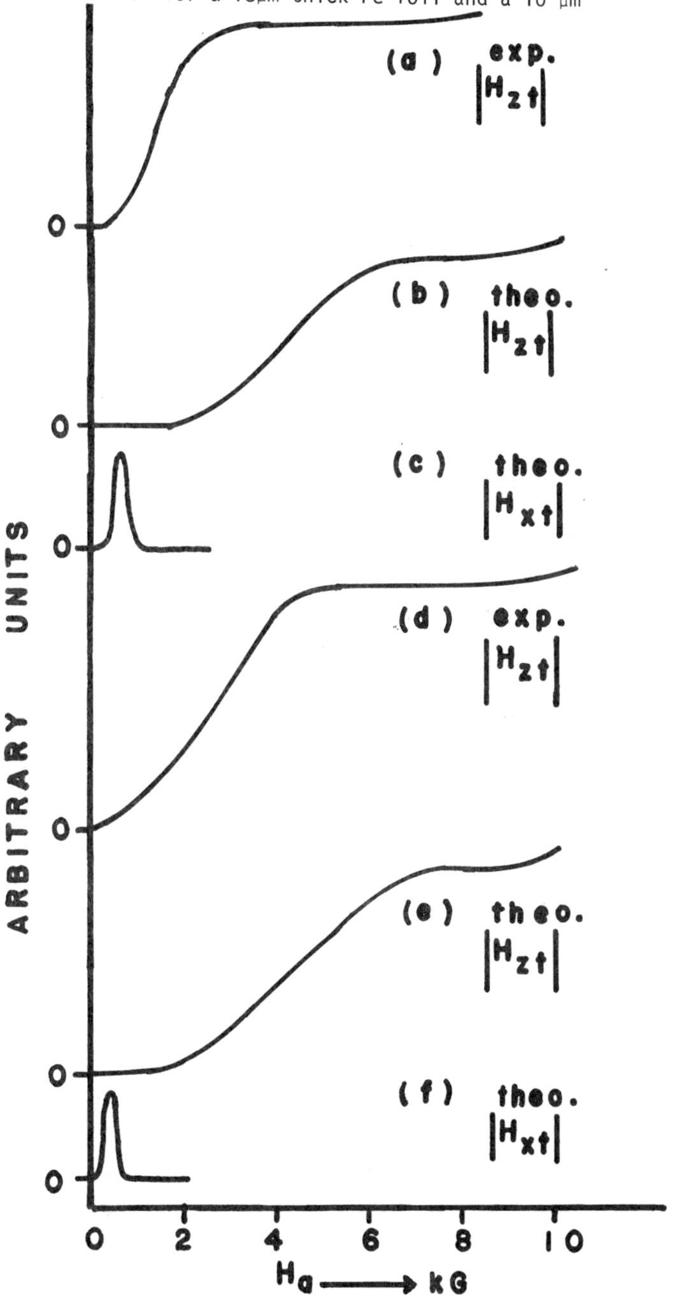

FIG. 1 (a) experimental, (b), (c) theoretical results for Fe. (d) experimental, (e), (f) theoretical results for Ni. Parameter values are given in text.

The parameter values used for the Fe plots are: M_s = 1707 G, g = 2.05, τ = 1.0 x 10^{-10} sec, T_1 = 1 x 10^{-10} sec, δ = 5 μm and A = 2.5 x 10^{-6} ergs/cm. For Ni we used: M_s = 485, g = 2.2, τ = 1.0 x 10^{-10}, T_1 = 1.0 x 10^{-10}, δ = 3.2μm and A = .9 x 10^{-6} ergs/cm. The values we used for δ which in our theory enters for the frequency $\nu_0/2$ are 50% larger than their suggested values from static resistivity data. This is in accordance with our past experience in metals. The A value for Ni used here is slightly larger than the published value[3] of A = .75 x 10^{-6} ergs/cm. The A value for Fe is the same as the published value.[3]

The calculated values for the $|H_{zt}|$'s are about one order of magnitude smaller than their experimental values. The $|H_{xt}|$'s are about two order of magnitude stronger than the $|H_{zt}|$'s. This is generally consistent with the fact that the H_{xt}'s have a larger skin depth than the H_{zt}'s.

In summary, what appears to be the case in the phenomena described here is that at low static field values most of the incident power along the z-axis, is pumped into the $H_{x,y}$ components through excitation of transverse spin waves, reminiscent of parallel pumping in insulators.[4] However at high static field values most of the power is retained by the H_{zt} component after allowing for normal attenuation in the metal.

We have observed the H_{zt} transmission resonance phenomenon in practically all ferromagnetic metals. Besides δ,on which all microwave propagations in metals have a strong dependence, the calculated H_{zt}'s are very strongly dependent on A contrary to other linear bulk resonance effects.[5] Thus the nonlinearity is predominantly dependant on A. We plan to fit the experimental results with the theory to try to establish the temperature dependence of the exchange constant. The predicted subharmonic generation might have practical use in special cases. Thin films will have to be used, although even then this method could not compete with the present harmonic generating diodes in most cases. We are presently preparing to detect the $|H_{xt}|$ and assess its usefulness.

We wish to thank Dr. M.A. Huerta for helpful suggestions regarding the calculations.

REFERENCES

*Work supported by the National Science Foundation.

1. G.C. Alexandrakis, O. Horan, T.R. Carver, and C.N. Manicopoulos, Phys. Rev. Lett. 25, 1758 (1970).
2. O.L.S. Lieu and G.C. Alexandrakis and M.A. Huerta, to be published.
3. B.E. Argyle, S. Charap, and E.W. Pugh, Phys. Rev. 132, 2051 (1963); F. Keffer, Handbuck der Physik XVIII/2, 30 (Springer-Verlag, Berlin, 1966); C. Herring, Rado and Suhl, Magnetism IV, Academic Press, New York, 1966, p. 357.
4. F.R. Morgenthaler, J. Appl. Phys. 31, 95S (1960); E. Schlomann, J.J. Green, and U. Milano, J. Appl. Phys. 31, 386S (1960); C. Kittel, Quantum Theory of Solids, (Wiley, New York, 1964),p. 69; M. Sparks, Ferromagnetic-Relaxation Theory, (McGraw-Hill, New York, 1964) contains an extensive bibliography on nonlinear effects in insulators.
5. O. Horan, G.C. Alexandrakis, and C.N. Manicopoulos, Phys. Rev. Lett. 25, 246 (1970).

FERROMAGNETIC RESONANCE RELAXATION PROCESSES IN Zn_2Y

M.MITA and H.SHIMIZU, Department of Physics, Faculty of Science and Technology, Science University of Tokyo, Noda, Chiba, Japan, 278

ABSTRACT

Experimentally obtained linewidth in F.M.R. of Zn_2Y is analyzed numerically on the basis of two-magnon, three-magnon and four-magnon relaxation processes. In the analyzing procedure of three-magnon linewidth, the effective exchange constants are determined to be $D_{\perp}=0.15\times10^{-9}$ Oe cm^2 and $D_{\parallel}=9.3\times10^{-9}$ Oe cm^2 within and between the crystallographic c planes. The two-magnon linewidth induced by surface imperfections is discussed in consideration of scattering due to multipole demagnetizations of the imperfections. The four-magnon linewidth is observed for the first time and analyzed successfully.

INTRODUCTION

Relaxation of the uniform mode precession of F.M.R. is studied on a hexagonal ferrite Zn_2Y. The material has an $(ST)_3$ crystal structure and has a uniaxial anisotropy field H_A=9900 Oe, $4\pi M$=2850 Gauss and gyromagnetic ratio γ=2.62 Mc/sec Oe at room temperature. In the experiment, a rf excitation field is in the crystallographic c plane. An external dc field is in the direction θ_H and the equilibrium magnetization is in the direction θ with respect to the crystallographic c axis. The rf field is always perpendicular to the external dc field. Polar angle θ_k of the spin wave propagation vector k is measured from the equilibrium magnetization and azimuthal angle ϕ_k is measured from the direction perpendicular to the plane containing the equilibrium magnetization and the external dc field.

EXPERIMENTAL RESULT

The absorption linewidth ΔH is measured on a spherically shaped single crystal of Zn_2Y. The relaxation frequency $1/T_2$, which is $\Delta H/(dH/d\omega_0)$ exhibits types of maxima A, B, C and D with increasing angle θ_H. The typical profile of $1/T_2$ against θ_H is shown in Fig.1. The angle positions θ_H of the maxima A, B and C and the starting angle position θ_H of maximum D are plotted against the driving frequency ω_0 in Fig.2.

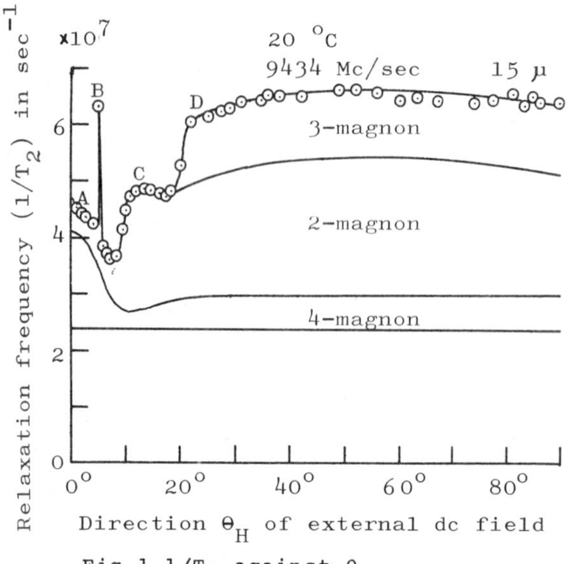

Fig.1 $1/T_2$ against θ_H

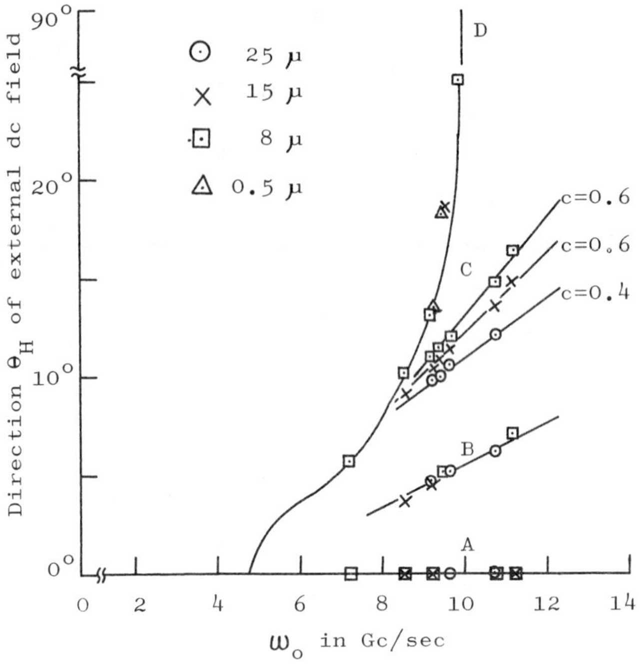

Fig.2 Position θ_H against ω_0

ANALYSIS

a) THREE-MAGNON LINEWIDTH: The large planar anisotropy field H_A enlarges the width of the magnon manifold. The calculated manifold contains half the uniform mode frequency, $\omega_0/2$, for angle θ_H larger than that shown by the curve D in Fig.2. For such angles, the three-magnon splitting can occur. The angle θ_H in the experiment, at which the maximum D starts, is just on the curve D. The relaxation frequency is calculated with the use of a transition probability method[1]:

$$(1/T_2)_{1-2}=(k_BTV/2\pi^2\hbar^3\omega_0)\int A_{-kko}\,\delta(\omega_0-2\omega_k)\,dk \qquad (1),$$

where $A_{-kko}=4|\tilde{\zeta}_{-kko}|^2$, $\tilde{\zeta}_{kk'k''}=(\zeta_{kk'k''}u_k-\zeta^*_{kk'k''}v^*_k)(u_{k'}u_{k''}+v^*_{k'}v_{k''})+(\zeta_{k''k'k}v_{k''}-\zeta^*_{k''k'k}u_{k''})v^*_ku_{k''}$, $\zeta_{kk'k''}=\mu_0(4\mu_0M/V)^{1/2}\{(-iH_A/2M)\sin2\theta-2\pi\sin2\theta_k e^{-i\phi_k}\}\delta_{k-k'-k''}$, $u_k=|B_k|/\{2\hbar\omega_k(A_k-\hbar\omega_k)\}^{1/2}$, $v_k=B^*_k(A_k-\hbar\omega_k)^{1/2}/|B_k|(2\hbar\omega_k)^{1/2}$, $A_k=2\mu_0[H\cos(\theta-\theta_H)-4\pi M/3-H_A\{1-(3/2)\sin^2\theta\}+2\pi M\sin^2\theta_k+Dk^2]$, $B_k=2\mu_0\{2\pi M\sin^2\theta_k e^{-i2\phi_k}-(H_A/2)\sin^2\theta\}$, $\hbar\omega_k=(A_k^2-|B_k|^2)^{1/2}$, $D=D_{\perp}(\cos^2\theta\sin^2\theta_k\sin^2\phi_k+\sin2\theta\sin\theta_k\cos\theta_k\sin\phi_k+\sin^2\theta\cos^2\theta_k+\sin^2\theta_k\cos^2\phi_k)+D_{\parallel}(\sin^2\theta\sin^2\theta_k\sin^2\phi_k-\sin2\theta\sin\theta_k\cos\theta_k\sin\phi_k+\cos^2\theta_k\cos^2\theta)$,

$2\mu_0=\hbar\gamma$. An asterisk denotes complex conjugate, k_B Boltzmann's constant, $\hbar$ Plank's constant divided by 2π, T absolute temperature and V is sample volume. By comparison of the calculated $(1/T_2)_{1-2}$ with the experimental maximum D, it is decided that the area named 3-magnon in Fig.1 corresponds to the three-magnon relaxation frequency. In the process of analysis, the effective exchange constants are determined to be $D_{\perp}=0.15\times10^{-9}$ Oe cm^2 and $D_{\parallel}=9.3\times10^{-9}$ Oe cm^2 within and between the c planes.

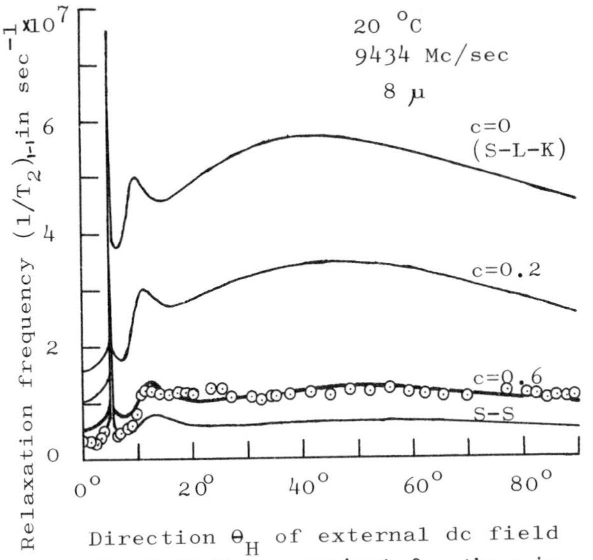

Fig.3 $(1/T_2)_{1-1}$ against θ_H, the c is the parameter in Eq.2. The curve named S-L-K denotes the calculated result by Sparks-Loudon-Kittel method and curve named S-S denotes that by Seiden-Sparks method.

b) TWO-MAGNON LINEWIDTH INDUCED BY SURFACE IMPERFECTIONS: The linewidths are measured on a sample polished with grits of various sizes. The linewidth for a fixed θ_H increases with grit size $G>0.5$ μ and is nearly constant for $G\lesssim 0.5$ μ. Consequently, the linewidth ΔH induced by surface imperfections would be the part obtained by direct subtraction of the linewidth for $G\lesssim 0.5$ μ from that for $G>0.5$μ. The corresponding relaxation frequencies $(1/T_2)_{1-1}$ are shown in Fig.1 with the area named two-magnon and in Fig.3. The $(1/T_2)_{1-1}$ is calculated[2] in consideration of the dipole as well as the quadrupole demagnetizations of the imperfections:

$$(1/T_2)_{1-1}=\{NV/(2\pi\hbar)^2\}\int|S\{F_d+cF_q\}|^2\delta(\omega_0-\omega_k)\ dk \qquad (2)$$

and $F_d=\{j_1(kR)/kR\}(3\cos^2\theta_k-1)$, $F_q=\{j_3(kR)/kR\}(35\cos^4\theta_k-30\cos^2\theta_k+3)$, $S=2\hbar\gamma(2\pi)^2R^3M(1/4)(u_0u_k+v_0v_k^*)/\sqrt{9c^2+(13/6)c+1}\}^{1/2}$, where N is the total number of pits and is equal to $4r^2/R^2$ for sample radius r. Pit radius R is decided to be G/3. The c value is 0.6, 0.4 or 0.3 corresponding to 8~15, 25 or 45 μ grit size, respectively. The result is shown in Fig.3 with a broad line. It is understood that the maxima B and C originate in the two-magnon process. The relaxation frequencies, which are calculated by methods developed by Sparks, Loudon and Kittel[3] and by Seiden and Sparks,[4] are also shown in Fig.3: the result by Sparks-Loudon-Kittel method always exceeds and that by Seiden-Sparks method is always below the experimental $(1/T_2)_{1-1}$. The present calculated result well reproduces the experimental $(1/T_2)_{1-1}$ for the large angles while the result by Seiden-Sparks method well reproduces the experimental $(1/T_2)_{1-1}$ near $\theta_H=0$. The value of c indicating the ratio of quadrupole contribution to dipole increases with decreases of G. This result reflects the fact the quadrupole field is a short range field in comparison with the dipole field. The extended calculation for the higher order multipole demagnetizations reproduces more correctly the experimental $(1/T_2)_{1-1}$.

c) FOUR-MAGNON LINEWIDTH: The linewidth by four-magnon processes is very small in ordinary ferrites. The linewidth in Zn_2Y, however, is expected to be fairly large owing to the following characters: 1) The four-magnon transition matrix element increases linearly with magnetocrystalline anisotropy. The large anisotropy induces a large four-magnon linewidth. 2) The four-magnon linewidth has a D^{-3} dependence for the isotropic effective exchange constant D. The effective exchange constants of Zn_2Y have large anisotropies for the magnon propagation direction. To exclude this difficulty, a cell, which is made of new 3-times large a-axis of the original and new c-axis of one (ST) length, is assumed to be a magnetic cell having nearly isotropic effective exchange constant. The corresponding D^3 is equal to $9D_{\parallel}D_{\perp}^2=2.1\times10^{-27}$ and is small compared with 1.45×10^{-25} for YIG.

2 IN 2 OUT SCATTERING: Two magnons involving one k=0 magnon are destroyed creating two magnons. The corresponding relaxation frequency is

$$(1/T_2)_{2-2}=\{\gamma(k_BT)^2/8(2\pi)^3D^3\}[(H_A/2M)(2\cos^2\theta-\sin^2\theta)+\pi]^2 \qquad (3).$$

3 IN 1 OUT CONFLUENCE: Three magnons involving one k=0 magnon are destroyed creating one magnon

$$(1/T_2)_{3-1}=\{\gamma(k_BT)^2/8(2\pi)^3D^3\}[(H_A/2M)\sin^2\theta-\pi]^2 \qquad (4).$$

A 1 in 3 out process, by which a k=0 magnon splits, never occurs because the bottom of the magnon manifold is always above the frequency $\omega_0/3$. The calculated relaxation frequency is shown in Fig.1 and 4. with lines.

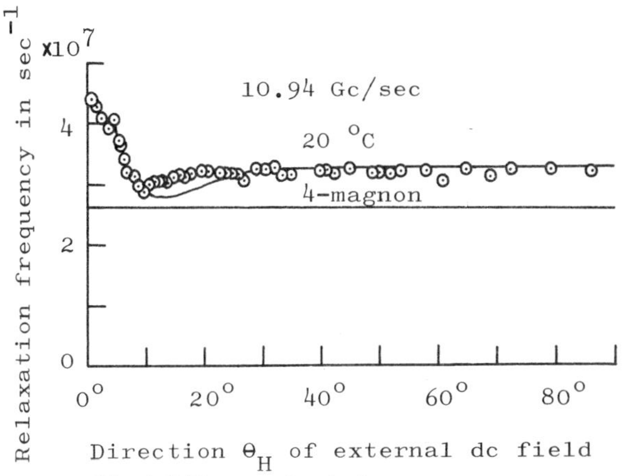

Fig.4 $1/T_2$ against θ_H

Fig.4 shows the total $(1/T_2)$ for 11 Gc/sec on the sample polished with grit less than 0.5 μ. The profile of $(1/T_2)$ against θ_H in the experiment well coincides with that of the calculated four-magnon $\{(1/T_2)_{2-2}+(1/T_2)_{3-1}\}$. It is understood that the maximum A originates in the four-magnon process.

d) REMAINDER OF THE LINEWIDTH: The remaining part of $(1/T_2)$ after subtracting the two-, three- and four-magnon linewidths is nearly constant against θ_H.

REFERENCES

1) M.MITA and H.SHIMIZU: J.Phys.Soc.Japan 35 (1973) 414.
2) M.MITA and H.SHIMIZU: to be published.
3) M.Sparks, R.Loudon and C.Kittel: Phys.Rev. 122 (1961) 791.
4) P.E.Seiden and M.Sparks: Phys.Rev. 137A (1965) 1278.

SPIN-WAVE RELAXATION AND PHENOMENOLOGICAL DAMPING IN FERROMAGNETIC RESONANCE

V. KAMBERSKY, Institute of Physics, Czechoslovak Academy of Sciences, Prague, Czechoslovakia, C. E. PATTON,* Department of Physics, Colorado State Univ., Fort Collins, Colorado, 80523, U.S.A.

ABSTRACT

Relaxation rates for the uniform precession mode in ferromagnetic resonance, with general elliptical polarization, have been calculated for several microscopic scattering processes using the spin-wave formalism. These results are compared with the widely used phenomenological formulations for FMR. The results demonstrate in relatively general terms the specific features of the Landau-Lifshitz and Gilbert phenomenological formulations on the one hand, and of what may be called "intrinsic" confluence processes in the microscopic formulation.

INTRODUCTION

The damping of ferromagnetic resonance is usually treated on the basis of the Landau-Lifshitz equation of motion[1] for the uniform precession mode (U.P.). For the frequency swept linewidth of ellipsoidal samples, one obtains:

$$(\Delta\omega/2)^{LL} = \alpha\omega_o P_A, \tag{1}$$

where ω_o is the Kittel resonance frequency,

$$\omega_o = (\omega_S^2 - \omega_A^2)^{1/2}, \tag{2}$$

with

$$\omega_{S,A} = (1/2)\ [(\omega_H + N_x\omega_M) \pm (\omega_H + N_y\omega_M)], \tag{3}$$

$$\omega_H = |\gamma|(H_o - 4\pi N_z M_o). \tag{4}$$

In the above, γ is the electron gyromagnetic ratio, M_o is the saturation magnetization, $N_{x,y,z}$ are the demagnetizing factors with the static applied field in the z-direction, and $\omega_M = |\gamma|4\pi M_o$. The quantity P_A is an ellipticity factor,

$$P_A = \partial\omega_o/\partial\omega_H = [1 - (\omega_A/\omega_S)^2]^{-1/2} \tag{5}$$

The demagnetizing factors satisfy $N_x + N_y + N_z = 1$. Equation (1) defines one half of the usual half-power linewidth in angular frequency units; this convention will make possible a direct comparison with relaxation rate expressions.

The importance of the ellipticity factor P_A has been pointed out previously in connection with linewidth data for Ni-Fe films.[2] On the other hand, Callen[3] has proposed that the phenomenological damping parameters themselves should depend on sample geometry. However, recent calculations of the damping in metals[4-6] indicate that the relaxation is consistent with the assumption of an intrinsic damping parameter, α, which is constant except for a possible temperature dependence.

In the present study, the U.P. decay rate has been examined as a function of the ellipticity of the U.P. polarization for several simple scattering models, using standard second order perturbation theory methods in calculating transition probabilities. The results show that: (1) In general, both the U.P. decay rate and the corresponding phenomenological parameters depend on the sample geometry, the U.P. polarization, and the frequency; (2) In the case of the general "confluence" process, the U.P. decay rate depends on an intrinsic (material) parameter α and on the field and frequency in exact accord with Eq. (1), provided that the participant excitations are unaffected by the magnetic field and that ω_o is small compared with their characteristic frequencies. This simple result reflects the qualitative features of the known damping mechanisms involving electron scattering, such as those studied in metals,[4-6] the fast-relaxing impurity mechanism in ferrites,[7] and in part, the multiparticle magnon-phonon processes.[8] It is not directly applicable to three-magnon confluence relaxation. With other scattering models, e.g., the two-magnon process, the decay rates are not consistent with any of the phenomenological formulations mentioned above or with the assumption of a constant (intrinsic) damping parameter.

MICROSCOPIC RELAXATION CALCULATION

The interaction of the U.P. with other excitations is assumed to be of the form,

$$\mathcal{H}_I = D^\dagger b_o + D b_o^\dagger \tag{6}$$

where the $b_o^\dagger$ and b_o are creation and destruction operators of the circularly polarized magnons, and the $(D^\dagger, D)$ are composed of operators of other excitations. The actual uniform precession eigenmodes in an ellipsoidal sample are represented by the operators $c_o^\dagger$ and c_o obtained from diagonalization of the U.P. Hamiltonian (in frequency units),

$$\mathcal{H}_o = \omega_S b_o^\dagger b_o + (\omega_A/2)(b_o^\dagger b_o^\dagger + b_o b_o), \tag{7}$$

by the transformation[9]

$$c_o = \rho\, b_o + \sigma\, b_o^\dagger, \tag{8}$$

where ρ and σ are real coefficients satisfying $\rho^2 - \sigma^2 = 1$ and $\rho^2 + \sigma^2 = P_A$. The parameters $\omega_{S,A}$ and the ellipticity factor P_A are the same as given above. The U.P. decay rate is calculated from the transition rate equation,

$$dn_o/dt = p^+ - p^- \tag{9}$$

where the "up" and "down" (p^+ and p^-, respectively) rates are;

$$p^\pm = 2\pi \sum_{i,f} |\langle f|\mathcal{H}_I|i\rangle|^2\ \delta(\omega_f - \omega_i \pm \omega_o). \tag{10}$$

Here (i,f) denote the initial and final states, $|n_o, i\rangle$ and $|n_o \pm 1, f\rangle$, respectively, of the system, n_o is the occupation number of the uniform precession eigenmode, and $\mathcal{H}_I$ is written in terms of the $(c_o^\dagger, c_o)$.

For a general confluence process, the interaction D is given by

$$D = \sum_{k,k'} f_{kk'}\, a_{k'}^\dagger\, a_k. \tag{11}$$

The $a_k^\dagger$ and a_k are either fermion or boson operators for the participant excitations with wave number k and frequency ω_k. The matrix elements $f_{kk'}$ are assumed known, even though their actual calculation is nontrivial and generally involves indirect processes,[4,6] so that the values are temperature dependent. With the usual assumptions that the occupation numbers $(n_k, n_{k'})$ are equal to their thermal values $(\bar{n}_k, \bar{n}_{k'})$, and that $dn_o/dt = 0$ when n_o is equal to its thermal value, one obtains,

$$dn_o/dt = -(n_o - \bar{n}_o)(1/\tau_o), \tag{12}$$

where $(1/\tau_o)$ generally depends on ρ and σ. We assume that the scattering has at least rotational symmetry, i.e., dn_o/dt is invariant under rotational transformations about the field axis. This allows retention of only terms proportional to ρ^2 and σ^2; those involving $\rho\cdot\sigma$ crossterms are not invariant.

$$1/\tau_o = 2\pi \sum_{k,k'} (\rho^2|f_{k'k}|^2 + \sigma^2|f_{kk'}|^2) \times (\bar{n}_k - \bar{n}_{k'})\, \delta(\omega_{k'} - \omega_k - \omega_o). \quad (13)$$

The factor $(\bar{n}_k - \bar{n}_{k'})$ may be expanded in the difference $(\omega_k - \omega_{k'})$ for $\omega_o \ll kT$; it is then equal to $-\omega_o d\bar{n}_k/d\omega_k$. If ω_o is small compared to the ω_k and $\omega_{k'}$ over which the matrix elements and density-of-states factors implied by the summation in Eq. (14) vary appreciably, the sum may be evaluated using $\delta(\omega_{k'} - \omega_k)$. With these assumptions, the result is

$$1/\tau_o = \alpha_c\, \omega_o\, P_A, \quad (14)$$

$$\alpha_c = 2\pi \sum_k (-\partial\bar{n}_k/\partial k)_k \sum_{k'} |f_{kk'}|^2\, \delta(\omega_{k'} - \omega_k), \quad (15)$$

which is in exact accordance with the phenomenological expression (1) based on the Landau-Lifshitz equation; $1/\tau_o$ is proportional to the frequency ω_o, the ellipticity factor P_A. and a scattering summation α_c, which is an intrinsic parameter under the above assumptions and the additional provision that the excitations (k,ω_k) and the $f_{kk'}$ are not appreciably affected by the magnetic field.

As an example of processes where such consistency is not found (even with other forms of the phenomenological damping), consider the magnon-boson process where, generally,

$$D = \sum_k (F_k b_k + G_k b_k^{\dagger}). \quad (16)$$

Proceeding in the same way as in the preceding case, one obtains the U.P. decay rate in the form of Eq. (12) with,

$$1/\tau_o = 2\pi \sum_k (\rho^2|F_k|^2 + \sigma^2|G_k|^2)\, \delta(\omega_k - \omega_o) \quad (17)$$

Obviously, there is no general "ellipticity factor" to be factored out, and no frequency factor. Even in concrete simple models, such as the isotropic approximation to the pseudo-dipolar two-magnon process[10] where it is possible to do the sums over $|F_k|^2$ and $|G_k|^2$ explicitly, the result is not proportional to $P_A = \rho^2 + \sigma^2$ since the two sums are not equal. Moreover, a significant and rather complicated variation of $1/\tau_o$ with frequency and sample geometry is known to be caused by the variation in the density of degenerate magnon states participating in the k sums as these parameters are changed. Equations (16) and (17) apply as well, at least formally, to a magnon-phonon process. As long as the scattering amplitudes are considered proportional to the deformation, $|F_k| = |G_k|$ is satisfied[11] so that $\rho^2 + \sigma^2 = P_A$ does factor out of Eq. (17).

The calculated relaxation rates may also be compared with the phenomenological linewidths obtained with other forms of the damping. The Bloch-Bloembergen damping[12] is characterized by a transverse relaxation parameter in an equation of the form,

$$d\underset{\sim}{M}/dt = -|\gamma|(\underset{\sim}{M} \times \underset{\sim}{H}) - (\nu/H_e^2)\, \underset{\sim}{H}_e \times (\underset{\sim}{M} \times \underset{\sim}{H}_e). \quad (18)$$

In the original formulation[12,3] $\underset{\sim}{H}_e$ was taken as the static internal field, in which case the frequency linewidth is simply

$$(\Delta\omega/2)^{BB} = \nu. \quad (19)$$

It is seen that none of the ellipticity corrections which appear in the microscopic decay rates of Eq. (13) and Eq. (14) are reflected. Moreover, the frequency dependences implicit in the microscopic results are totally inconsistent with an intrinsic (constant) value for ν. A modified form of the Bloch-Bloembergen analysis replaces $\underset{\sim}{H}_e$ in the damping term by the instantaneous field $\underset{\sim}{H}$ including both static and microwave fields.[13] This modification [originally proposed to eliminate the unphysical negative absorption for the antilarmor sense of circular precession in the original treatment of Eq. (18)] yields a frequency linewidth,

$$(\Delta\omega/2)^{MBB} = (\nu/\omega_H)\, \omega_o\, P_A. \quad (20)$$

This approach thus yields results formally similar to the Landau-Lifshitz result, with α replaced by ν/ω_H. The frequency and polarization dependence is, however, substantially different from that of Eq. (1) if ν instead of α is considered to be an intrinsic parameter. Neither of these phenomenological formulations appears to be consistent with any of the above physical relaxation processes considered from the microscopic point of view.

It is important also to mention the Gilbert formulation,[14]

$$d\underset{\sim}{M}/dt = -|\gamma|(\underset{\sim}{M} \times \underset{\sim}{H}) + (\alpha/M_o)\, \underset{\sim}{M} \times d\underset{\sim}{M}/dt, \quad (21)$$

which yields practically the same results as the Landau-Lifshitz equation, for small α, so that the above conclusions apply. For physical reasons the Gilbert formulation appears preferrable.[15]

CONCLUSION

The above results demonstrate in relatively general terms the specific features of the Landau-Lifshitz and Gilbert phenomenological formulations on the one hand, and of what may be called "intrinsic" confluence processes in the microscopic formulation. These formulations are consistent with the assumption of an intrinsic damping parameter describing the motion of the magnetization vector under sufficiently general conditions. The two-magnon process and the Bloch-Bloembergen phenomenological description of damping in ferromagnetic resonance are not consistent with such an assumption.

ACKNOWLEDGEMENTS

The authors are indebted to Drs. D. Fraitova and D. A. Krueger for helpful discussions in the course of this work.

REFERENCES

*Partially supported by a grant from the National Academy of Sciences (USA) and the Czechoslovak Academy of Sciences, and by a CSU Faculty Research Grant.

1. L. Landau and E. Lifshitz, Physik Z. Sowjetunion 8, 153 (1935).
2. C. E. Patton, J. Appl. Phys. 29, 3060 (1968).
3. H. B. Callen, J. Phys. Chem. Solids 4, 256 (1958).
4. B. Heinrich, D. Fraitova, V. Kambersky, Phys. Stat. Sol. 23, 501 (1966). D. Fraitova, V. Dvorak, Czech. J. Phys. B22, 413 (1972).
5. V. Kambersky, Int. Conf. on Magnetism, Moscow 1973; to be published.
6. V. Korenman, R. E. Prange, Phys. Rev. B6, 2769 (1972).
7. P. G. DeGennes, C. Kittel, A. M. Portis, Phys. Rev. 116, 323 (1959).
8. T. Kasuya, R. C. LeCraw, Phys. Rev. Lett. 6, 223 (1961).
9. T. Holstein and H. Primakoff, Phys. Rev. 58, 1098 (1940).
10. A. M. Clogston, H. Suhl, L. R. Walker, P. W. Anderson, J. Phys. Chem. Solids 1, 129 (1956).
11. J. M. Ziman, Electrons and Phonons, Oxford (1960).
12. N. Bloembergen, Phys. Rev. 78, 572 (1950).
13. See R. W. Davies, F. A. Blum, Phys. Rev. B3, 3321 (1971) and references therein.
14. T. L. Gilbert, Phys. Rev. 100, 1243 (1955).
15. S. Iida, J. Phys. Chem. Solids 24, 625 (1963).

FERROMAGNETIC RESONANCE OF TWO-DOMAIN SPHERICAL METALLIC IRON

D.L. Griscom, E.J. Friebele, and C.L. Marquardt
Naval Research Laboratory, Washington, D.C. 20375

ABSTRACT

Ferromagnetic resonance studies were carried out at 9, 16, and 35 GHz on spherical particles of metallic iron ~0.01-0.5 μm precipitated in an Fe^{2+} - containing silicate glass by hydrogen reduction. The 35 GHz spectrum was successfully computer simulated under the assumption that magnetocrystalline anisotropy was predominantly responsible for the powder line shape, leading to the conclusions that $2K_1/M_s \approx +600$ Oe for the precipitated particles and that the mean deviation from sphericity was $\lesssim 4\%$. In contrast, at the lower frequencies the spectra were characterized by (1) a continuous absorption extending from zero field to a mean cut-off field of ~ 7 kOe and (2) a hysteretic component appearing ~ 2.5 kOe below the free electron resonance field as a result of isothermal cycling to $\gtrsim$ 4 kOe *in situ*. The position of the latter feature is in good agreement with a theoretical calculation of the resonance fields of a two-domain sphere of iron with a thin domain wall. The results support the predictions of Amar[1] that magnetization of such fine particles by wall motion is a hard process.

INTRODUCTION

The magnetic behavior of a two-domain cube of iron has been considered theoretically by Amar[1], who concluded that the two-domain configuration should exist within a certain critical size range and that the wall characteristics are size dependent and quite different from those of bulk material. In particular, it was shown that magnetization of such small particles by wall motion is expected to be a hard process.

The present paper describes a ferromagnetic resonance (FMR) investigation of dispersions of metallic iron particles, some of which fall in the predicted size range for two-domain behavior. This experimental system differs from the structures considered by Amar mainly in that the particles are spheres, rather than rectangular prisms and that the magnetocrystalline anisotropy is cubic rather than uniaxial.

EXPERIMENT AND GENERAL RESULTS

Particles of metallic iron were precipitated in an Fe^{2+} -containing silicate glass[2] by reduction in an H_2- rich hydrogen-oxygen flame followed by quenching in liquid N_2. Transmission electron microscopy revealed generally well dispersed spherical particles in the size range ~ 0.01 - 0.05 μm. FMR measurements (below) indicated that particles $\lesssim$ 0.02 μm were rather rare, while static magnetization studies[2] showed that grains $\gtrsim$ 0.1 μm predominated volumetrically. For comparison, another glass[3] containing iron octahedra ~ 5 μm was also examined. Spectra were obtained on electron spin resonance spectrometers operating at 9,16, and 35 GHz.

Figure 1 shows the 9 GHz FMR absorption curves of dispersed iron particles in three general size ranges: multidomain (MD) (> 1 μm), small-MD (~ 0.02 - 0.5 μm), and single domain (SD) ($\lesssim$ 0.02 μm). The SD spectrum is hypothetical, as samples exhibiting this line shape at 9 GHz could not be produced. The hatched region in Fig. 1b is an isothermal remanent intensity which appears only in the case of small-MD particles after the sample is cycled to $\gtrsim$ 7 kOe *in situ*. The absorption due to MD particles is seen to be spread out between 0 and ~ 10 kOe, with an inflection point near 7 kOe.

THEORY

All of the line shapes of Fig. 1 are explainable on the basis of the ferromagnetic resonance condition derived by Kittel[4]. Resonance fields H_{res} were calculated by inserting the "effective" demagnetizing factors for a two-domain sphere with a centrally-located thin domain wall, following an analogous treatment by Smit and Wijn[5] (see Fig. 2). Assuming the applied DC field to be in the plane of the wall, the following resonance conditions were derived:

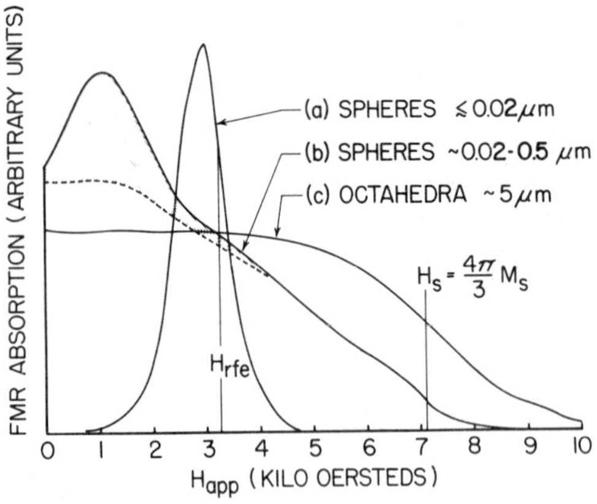

Fig. 1 FMR absorption spectra of fine-grained, equidimensional iron particles. ν= 9GHz, T≈ 300°K.

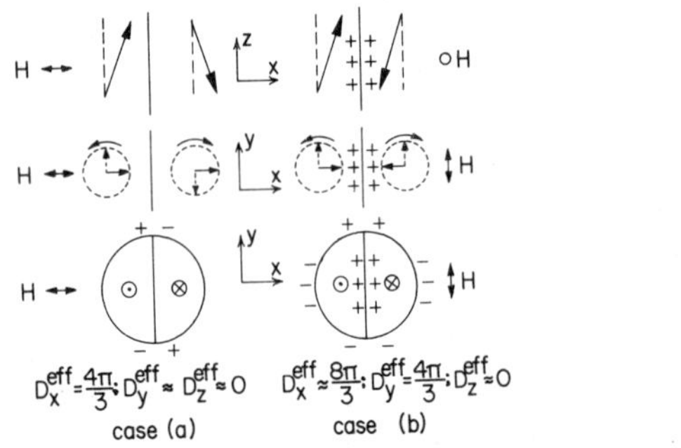

Fig. 2 Resonance modes of a two-domain sphere; (a) with AC field ⊥ to wall; (b) with AC field ∥ to wall (in the manner of Ref. 5).

$$H_{res}^{case\ a} = -(2\pi/3)M_s \pm \sqrt{(2\pi/3)^2 M_s^2 + H_{rfe}^2}\ , \qquad (1a)$$

$$H_{res}^{case\ b} = -2\pi M_s \pm \sqrt{(2\pi/3)^2 M_s^2 + H_{rfe}^2}\ , \qquad (1b)$$

where $H_{rfe} = h\nu/2\beta$ is the resonance field of the free electron.

DISCUSSION

In Fig. 3, attention is called to the relative derivative minimum labeled H_s near 7 kOe in the 9.5 and 16 GHz spectra. The frequency independence of this quantity, coupled with the fact that its magnitude is close to the theoretical maximum saturation field for spherical iron $(4\pi/3)\ M_s$, is suggestive that the "cut off" of FMR absorption above ~7 kOe in Fig. 1 may be due to the onset of magnetic saturation. Moreover, the temperature dependence of H_s as defined in Fig. 3a, was found to be similar to that of M_s for metallic iron from ~10 to 570°K. Thus, H_s can be formally identified as the saturation field of the experimental particles.

When ν = 35.5 GHz, H_{rfe} substantially exceeds H_s so that SD resonance behavior is expected and observed. As a confirmation of this fact, the spectrum of Fig. 3c was computer simulated[6-8] assuming angular averaging of magnetocrystalline anisotropy to be the principal determinant of the line shape. The success

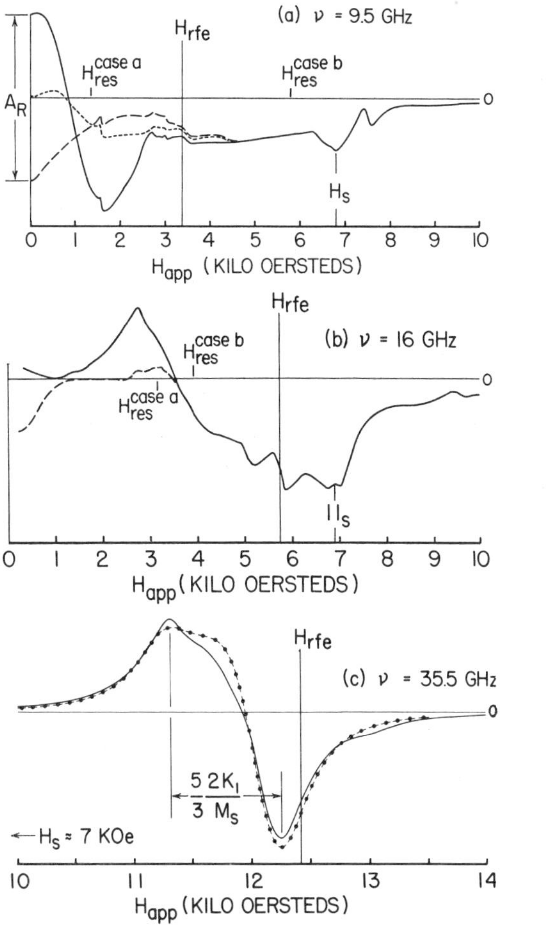

Fig. 3 FMR derivative spectra for iron spheres ~ 0.02-0.5 μm. Short-dashed curves represent first field scan; unbroken curves are second scan; long-dashed curves are third scan after 180° rotation at zero field. T ≈ 300°K. Dash-dot curve is a computer simulation[6-8].

of the simulation (dash-dot curve in Fig. 3c) implies that the particles are spherical within ~4%[7,9]. The peak-to-peak linewidth then provides a measure of the anisotropy field $2K_1/M_s$[6-8], and the value obtained (+600 Oe) is in good agreement with the literature for pure iron (e.g., +560 Oe[10], + 620 Oe[11]).

In contrast to the results at 35GHz, the 9 and 16 GHz spectra show absorption peaks (derivative zero crossings) ~ 2.5 kOe below H_{rfe} (Figs. 1b, 3a,b). Since this is not expected[4] for SD spheres—and the particles were shown to be spherical—it is concluded that these low-field peaks are due to MD effects apparent only in the lower-frequency experiments. As shown in Figs. 3a,b, there is good general agreement between the experimental zero-crossings and those values of $|H_{res}|$ calculated from Eqs. (1) which were smaller than H_s (≈ 7 kOe). The only exception to this rule (case b at 9 GHz) is reasonably ascribed to a displacement of the domain wall from its assumed central location when $H_{app} \approx 6$ kOe.[1] Thus, the spectra of Fig. 3 can be accounted for by assuming the presence of spherical particles which have two-domains below ~ 7 kOe and become SD in higher applied fields.

Based on the foregoing analysis, the following sequence of events is postulated to explain the magnetic hysteresis effects of Figs. 1b and 3a and b:

(i) The virgin sample contains single-crystal, two-domain spheres wherein the domain walls are "randomly" oriented in planes containing the cube edges. (ii) The first time the FMR experiment is performed, this randomness manifests itself as an absence of absorption peaks in the predicted locations. (iii) As the field is scanned above H_s (≈ 7 kOe at 300°K) all particles become saturated. (iv) When the field is lowered again, the re-nucleated domains are preferentially oriented near the direction of the applied field, i.e., the conditions assumed in deriving Eqs. (1) are fulfilled. (v) When the FMR experiment is next carried out, the absorption peaks predicted by Eqs. (1) are observed. (vi) Rotation of the sample by 90° at zero applied field would then nullify the effects of the domain alignment and the sample would behave for the next FMR experiment as though it had not been previously magnetized.

One additional effect is also observed: (vii) When the magnetized sample is rotated 180° at zero-applied-field the spectrum of Fig. 3a undergoes a further change which, by projection, appears tantamount to the remanent resonance peak having shifted from ~ + 1 kOe to ~ - 1 kOe. This is not predicted by Eqs. (1), since they do not depend on the algebraic sign of the DC field. This last effect is not understood in detail.

The acquisition of isothermal remanent magnetization by two-domain iron has been studied by FMR. A remanence amplitude A_R, defined in Fig. 3a as the difference in zero-field slope between the 0° and 180° orientations of the specimen, is plotted in Fig.4 versus the maximum applied field H_m. It is seen that the median magnetizing field for two-domain spherical iron is ≥ 4 kOe.

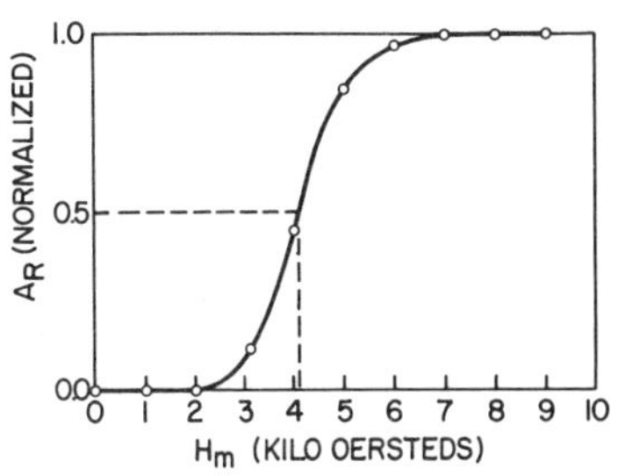

Fig. 4. FMR "remanence amplitude" A_R (Fig. 3a) versus maximum applied field H_m.

CONCLUSIONS

It has been shown that FMR can be used to elucidate domain effects in fine grained iron. In particular, it was found that isothermal magnetization of two-domain spheres is a hard process, in agreement with prediction[1] for idealized cubes in the same general size range.

REFERENCES

1. H. Amar, Phys. Rev. 111, 149-153 (1958).
2. D.L. Griscom, C.L. Marquardt, E.J. Friebele, and D.J. Dunlop, Earth Planet. Sci.Lett. (in press).
3. This sample and its characterization were provided by R.M. Housley.
4. C. Kittel, Phys.Rev. 73, 155-161 (1948).
5. J. Smit and H.P.J. Wijn, Ferrites, Wiley, New York (1959) pp. 82-84.
6. F.D. Tsay, S.I. Chan, and S.L. Manatt, Geochim. Cosmochim. Acta 35, 865-875 (1971).
7. D.L. Griscom, E.J. Friebele, and C.L. Marquardt, ibid. Suppl. 4, Vol. 3, 2709-2727, (1973)
8. D.L. Griscom, ibid. 38, 1509-1519 (1974).
9. E.J. Friebele, D.L. Griscom, C.L. Marquardt, R.A. Weeks, and D. Prestel, ibid., Suppl. 5, (in press).
10. R.W. Deblois and C.P. Bean, J. Appl. Phys. 30 225S (1959).
11. L.P. Tarasov, Phys. Rev. 56, 1231-1240 (1939).

ABSORPTION LINE SHAPE OF ELECTRON SPIN RESONANCE IN METALS

H. S. Jarrett, J. E. Gulley, & P. C. Hoell
E. I. duPont deNemours and Company, Wilmington, Delaware 19898

ABSTRACT

Calculation of the conduction electron spin resonance for a cylinder with the microwave H field parallel to the axis is given for normal skin effect conditions. The behavior of the line shape and area under the absorption curve is given specifically in the limit of no spin diffusion.

INTRODUCTION

The observation of electron spin resonance (ESR) in highly conducting organic compounds has revived interest in the details of resonance line shapes and the skin depth effect in metallic conductors. The original theoretical work of Bloembergen[1] on NMR in metals, and Dyson[2] on ESR in metals are basic to virtually all investigation in this area. Recently, the availability of rapid electronic computation has permitted extension of Dyson's work[3] and complete analysis of systems with arbitrary thickness and diffusion time.[4] All of these calculations pertain to an infinite flat plate sample geometry. This shape seldom occurs experimentally, however. Some experiments have been reported in which the sample is specifically stated to be cylindrical in shape[4,5], but the Dyson flat plate theory is applied. Recently, attempts have been made to study the spin susceptibility of a conducting organic crystal from CESR data[6]. If the experimental situation is such that skin depth effects are relevant, rather definite knowledge of geometrical effects is required for more than qualitative analysis.

The original treatment of the problem by Dyson was carried forward in general terms, hence it is relatively simple to extend his work to cylindrical geometry. In this paper we investigate the implications of geometric shape by performing calculations of CESR on a cylindrical rod with the H_1 field parallel to the long axis assuming normal skin effect conditions where $\omega_c\tau \ll 1$.

CESR in Cylindrical Rods

Expressions for resonance absorption in a metal can be obtained by solving for the impedance of the sample in a microwave field using Maxwell's equations, where the magnetization is due to the spin coupled by the Zeeman energy to the r.f. field, $\overline{H}_1(r)$ and is given by Dyson's Eqs (D54) and (D55). In this paper, we follow Dyson's procedure explicitly. We shall designate the equations taken from Dyson's paper by a "D". The equation expressing the magnetization in terms of the microwave $\overline{H}_1$ field and the surface electric field $\overline{E}_s$ is solved simultaneously with the Dyson equation for the r.f. magnetization Eq (D54) for $\overline{E}_s$ in terms of $\overline{H}_1$ only, from which is obtained the surface impedance

$$Z = \frac{1}{4\sigma r_o}\left[F + 4\pi\ \omega\chi T_2\lambda^2F^2G\right], \qquad (1)$$

to order χ^2, where $F^{-1} = \sum_0^\infty \frac{1}{\beta_n^2-u^2} = \frac{J_2(u)}{2uJ_1(u)} - \frac{1}{u^2}$,

and $G = 2a^2\sum_0^\infty \frac{1}{(\beta_n^2-u^2)^2}\,\frac{1}{\beta_n^2 + 2a^2(1-ix)}$. (2)

Here, $\lambda = \frac{r_o}{\delta}$, the ratio of the rod radius to skin depth, $u^2 = 2i\lambda^2$, $J_n(u)$ is the complex Bessel function,

$a = \lambda\ (\frac{T_D}{T_2})^{1/2}$ where T_D and T_2 are the diffusion and spin relaxation times, respectively,

$$x = T_2(\omega-\omega_o),$$

$J_1(\beta_n)=0$, and other terms have the usual definitions. Eq. (1) is essentially the same as Dyson's expression Eq (D71) for a flat plate except that $\beta_n = \pi n$, and the normalization factors are different.

Resonance Absorption

The first term of Eq 1 is the nonresonant impedance of a cylinder in a parallel $\overline{H}_1$ field carrying on induced current in the angular direction around the cylinder. For the usual ESR experimental configuration, the sample cannot be assumed to be an infinite plate, and the presence of this current alters significantly the spin resonance line shape for a given set of parameters λ and a from that presented by Dyson.

The spin resonance absorption is given by the real part of the second term of Eq. 1. For convenience, we introduce the normalized dimensionless quantity, Z_R, where

$$Z = \frac{4\pi^2 r_o}{c^2}\ \chi\omega\omega_o T_2 Z_R.$$

In the limit of zero diffusion $a\to\infty$, and Eq. 2 sums to

$$G = (\frac{1}{1-ix})\left\{\frac{1}{4u^2}\left[(\frac{J_2(u)}{J_1(u)})^2 - \frac{J_3(u)}{J_1(u)}\right] + \frac{1}{u^4}\right\}.$$

Letting,

$$H = \left[1 - \frac{J_3(u)J_1(u)}{J_2{}^2(u)}\right] \quad \text{and}$$

$K = uJ_1(u)/J_2(u)$, we obtain

$$Z_R = \frac{1}{1-ix}\left[\frac{H-(K/\lambda^2)^2}{(1 + iK/\lambda^2)^2}\right]. \qquad (3)$$

Taking the same limit in Eqs. (D72) and (D73) yields for the Dyson treatment

$$Z_R = \frac{1}{2}\ \frac{1}{1-ix}\left[\frac{\tan(u)}{u} + \frac{1}{\cos^2(u)}\right]. \qquad (4)$$

Although Eqs 3 and 4 have an entirely different dependence on skin depth, the shape of the resonance curve is identically the same. The coefficient in the bracket of either Eq 3 or Eq 4 is simply a complex number, r+is independent of applied field.

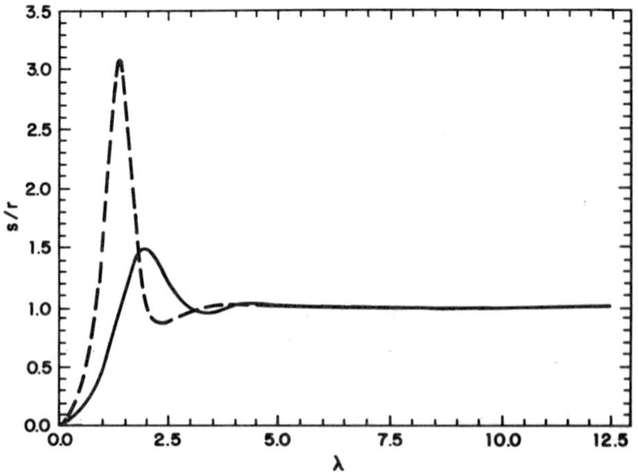

Figure 1 - Ratio of the dispersive to absorptive component, s/r of Eq. 5, of the CESR resonance line in the diffusionless limit for the cylinder (solid line) and the flat plate (dashed line).

Therefore, the absorption component is

$$ReZ_R = \frac{r-sx}{1+x^2}, \qquad (5)$$

and the line is asymmetric as expected.

Explicit dependence of r and s on λ can best be demonstrated by plotting in Figure 1 the ratio s/r as a function of λ for both the cylinder and the flat plate. We see that for the cylinder the contribution to the absorption line shape by the dispersive component s is less dominant as conductivity increases. Dispersion arises in ReZ_R because the phase of the H_1 field penetrating the skin depth oscillates relative to the phase of the external H_1 field. It is clear that in the diffusionless limit, a given s/r ratio may arise from widely different λ depending entirely on sample shape. The significance of sample geometry is greatest in the region where skin depth is approximately the sample dimension. Here, the greatest error would be incurred by analyzing line profile without regard for geometry.

Spin Susceptibility

For a Lorentz line, the integral of the absorption gives a number proportional to the spin susceptibility. However, we realize that in general the area under the absorption is proportional not only to the susceptibility, but also to a filling factor that depends on skin depth.

Integrating the series G given by Eq. 2 termwise, we obtain the area

$$A = \frac{4\pi^2}{c^2} r_o \omega\omega_o \chi T_2 \pi Re \left[\frac{H-(K/\lambda^2)^2}{[1+iK/\lambda^2]^2}\right]. \qquad (6)$$

Eq. 6 is independent of diffusion time. Thus, no matter how large the contribution of the dispersive component to the line shape, the area remains independent of T_D. The filling factor given by the function inside the brackets in Eq. 6 is plotted in Figure 2 and compared with the flat plate. When the skin depth diminishes, the absorption intensity decreases more rapidly for the flat plate than for the cylinder. Actual sample shapes generally lie between these two limits.

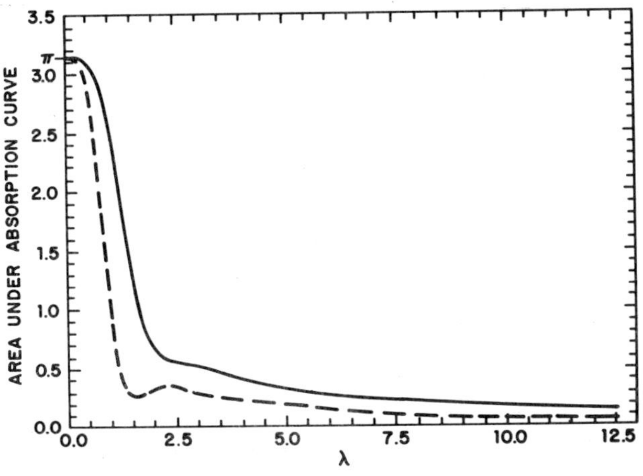

Figure 2 - Filling factor in the microwave H field as a function of the ratio of thickness to skin depth for the cylinder (solid line) and flat plate (dashed line).

Discussion

These results have immediate consequences to both ESR and NMR in metals. Not only does the line asymmetry itself present a problem in analysis of experimental data, but the filling factor is strongly sample shape dependent. Even in the limit of no diffusion, differences of as much as a factor of three occur in the s/r ratio, and the areas could differ by as much as a factor of six between cylindrical and blade-like sample shapes in the region near λ = 1.2.

REFERENCES

1. N. Bloembergen, J. Appl. Phys. 23, 1383 (1952).
2. F. Dyson Phys. Rev., 98, 349 (1955).
3. M. Lampe and P. M. Platzman, Phys. Rev. 150, 340 (1966).
4. H. Kodera, J. Phys. Society of Japan, 28, 89 (1970).
5. J. H. Pifer & R. Magno, Phys. Rev. B3, 663 (1971).
6. J. E. Gulley (To be published).

VAN VLECK SUSCEPTIBILITY AND CONDUCTION BAND EXCHANGE IN THE NOBLE METALS

R. E. Walstedt and Y. Yafet
Bell Laboratories
Murray Hill, New Jersey 07974

ABSTRACT

Residual Van Vleck orbital susceptibility χ_{VV} due to s-d hybridization in the noble metals and its associated NMR shift are discussed. The estimated orbital shift is large enough to seriously affect the value of the exchange enhancement of the spin susceptibility as deduced from the experimental Korringa product K^2T_1T and has the effect, in Cu metal, of reconciling this value with other measures of exchange enhancement.

INTRODUCTION

Conflicting evidence has come to our attention regarding the strength of band susceptibility enhancement effects in the noble metals. Experimental evaluations of the Korringa product K^2T_1T reported by Narath and Weaver[1] appear, on the reasonable assumption of a finite-range exchange potential, to give evidence for band enhancement factors of two or more. On the other hand, optical studies of Cu and Ag metals by Ehrenreich and Philipp[2] reveal an important dielectric screening effect in these metals due to the filled d-bands. This effect would presumably reduce the expected free-electron gas exchange interaction (as found, e.g., in the alkali metals) by a factor ~ 5 for Cu and somewhat less for Ag. Recent evidence tends to support this view: Transmission conduction electron spin resonance studies of pure copper[3] suggest that band enhancement is small. More quantitatively, the de Haas-van Alphen g factor measurements by Randles,[4] discussed more fully below, lead to $\alpha = 0.12$ for Cu metal, where $(1-\alpha)^{-1}$ is the susceptibility enhancement factor.

With the latter evidence in mind, we have reexamined the NMR shift and relaxation processes of nuclear spins in noble metal hosts. We are thereby led to suggest that the orbital shift mechanism, arising from residual Van Vleck paramagnetism[5] in these systems and not previously taken account of, allows one to reconcile the experimental Korringa products with greatly reduced band enhancement factors.

NMR SHIFT AND RELAXATION IN NOBLE METALS

It is well-known that the Korringa relation,[6] modified to take account of band exchange as in Moriya's RPA model,[7] must be obeyed regardless of Fermi surface topology if both shift and spin-lattice relaxation arise exclusively from the s-contact hyperfine interaction. Since the Fermi surface states of (e.g.) Cu metal appear to contain a considerable amount of 3d character due to hybridization,[8] there result contributions to both shift and relaxation due to d-spin and d-orbital hyperfine couplings. The observed shift is therefore made up of three terms.

$$K = K_s + K_d^{spin} + K_d^{orb}, \tag{1}$$

where K_d^{orb} is positive[5] but K_d^{spin} is very likely negative corresponding to a typically negative core-polarization hyperfine field. The spin-lattice relaxation rate consists of purely s-s and d-d terms, since s-d cross terms are prohibited by symmetry.[1] Estimates of the d-spin and d-orbital relaxation terms[9] yield numbers $\sim 1\%$ of the measured rate or smaller and are therefore considered negligible.

The magnitude of α has traditionally been estimated for simple metals using experimental evaluations of the quantity[7]

$$\mathcal{K}(\alpha) = \mathcal{S}/K^2T_1T \tag{2}$$

with $\mathcal{S} \equiv (\gamma_e/\gamma_n)^2(\hbar/4\pi k_B)$, where $\mathcal{K}(\alpha)$ goes monotonically from $1 \to 0$ as α goes from $0 \to 1$. Equation (2) is correct only for $K = K_s$. Thus we see from Eq. (1) that using the experimental shift in Eq. (2) will either over-or under-estimate $\mathcal{K}(\alpha)$ depending on whether the (negative) K_d^{spin} or (positive) K_d^{orb} predominates, respectively. It is our contention that K_d^{orb}, which has not hitherto been considered in this context, is very likely the predominating effect of d-admixture here, leading to underestimates of $\mathcal{K}(\alpha)$.

THE CASE OF CU METAL

To document this point we make estimates of K_d^{spin} and K_d^{orb} for Cu metal, the most clear-cut and well understood case. Considering K_d^{spin}, it is noted that $K_d^{spin}/K_s = H_{hf}^d f_d/H_{hf}^s f_s$, where H_{hf}^s and H_{hf}^d are the hyperfine fields per Bohr magneton of spin magnetization for s and d-electrons and f_s and f_d the fractional amount of s and d-character at the Cu Fermi surface, respectively. Estimating $|H_{hf}^d|/H_{hf}^s \sim 0.1$[10] and $f_d = 0.3$,[8] we see that K_d^{spin} is at most a few percent of K_s.

An expression for the Van Vleck orbital susceptibility in the tight-binding approximation has been given by Kubo and Obata.[5] For our present purposes this expression may be approximated as[11]

$$\chi_{VV} \sim 2A\mu_B^2 n_o n_u/\Delta, \tag{3}$$

where n_o and n_u are the number of occupied and unoccupied d-states, respectively, Δ is a typical energy splitting between such states, and A is Avogadro's number. The matrix elements of orbital angular momentum have been roughly estimated as unity to obtain (3). For Cu the numbers n_o and n_u have been calculated to be 9.73 and 0.27, respectively.[12] Since the d-band is centered at ~ 4 eV below the Fermi surface,[13] we take $\Delta \sim 8$ eV and find $\chi_{VV} \sim 2.1 \times 10^{-5}$ emu/mole.

To calculate K_d^{orb} we estimate the orbital hyperfine coefficient $\beta = \langle r^{-3}\rangle_{metal}/A$ (where $K_d^{orb} = \beta\chi_{VV}$)[11] by taking[14] $\langle r^{-3}\rangle_{atomic} = 7.25$ a.u. for Cu^+ and allowing a 25% reduction for the metallic environment.[11] This leads to $\beta \sim 120$ and $K_d^{orb} \sim 0.25\%$, which is as large as the observed copper shift. The foregoing procedure therefore clearly demonstrates the predominance of K_d^{orb} over K_d^{spin} and in fact leads to an overestimate of K_d^{orb}, as we shall now see.

One obtains a more realistic estimate of K_d^{orb} by deriving K_s from the experimental T_1 using Eq. (2) with the value of α obtained by Randles[4] and subtracting the result from the measured shift. Uncertainty in the theoretical value of $K(\alpha)$ is $\sim 1\%$ because of the smallness of α. First we note that the value $\alpha = 0.12$ quoted above is obtained by comparing the paramagnetic spin susceptibility $\chi_{spin} = 1.45\ \chi_{free\ electron}$ given by Randles[4] and the band structure density of states ($m^*/m = 1.27$) for Cu metal calculated by O'Sullivan et al.[15] Using the RPA model[7] or improved calculations[16] of $\mathcal{K}(\alpha)$, we find that $\alpha = 0.12$ corresponds to $\mathcal{K}(\alpha)_{theory} \sim 0.92 \to 0.94$ in contrast with $\mathcal{K}(\alpha)_{expt} \sim 0.52$.[8] Combining Eq. (1) and (2) with the experimental shift $K = 0.24\%$[8] and neglecting K_d^{spin}, $\mathcal{K}(\alpha)_{theory}$ is found to correspond to $K_d^{orb} = 0.06\%$ and $K_s = 0.18\%$. K_d^{orb} as determined in this fashion is seen to be only one-quarter of the estimate given in the previous paragraph. This may reflect, for example, an overestimate of the orbital matrix elements in χ_{vv}[5] by a factor ~2. The corresponding estimate $\chi_{vv} \sim \frac{1}{2}\chi_{spin}$ is not incompatible with the measured diamagnetic susceptibility of Cu metal.

χ_{vv} IN Ag AND Au METALS

For the case of Ag metal the orbital shift effect is probably somewhat diminished because of the more deeply buried d-band.[17] The orbital hyperfine field for Ag is somewhat larger[14] than for Cu, however, yielding an effect which may be significant. On the other hand, the dielectric screening effect is smaller,[2] so that α for silver may be considerably larger than for copper.

For Au metal both the dielectric shielding and orbital shift effects should again be large, owing to the proximity of the d-bands to the Fermi surface.[18] It is important to note, however, that the influence of orbital shift on the experimental Korringa product is diminished by the greatly increased s-contact hyperfine interaction (and shift) in gold[1] as compared with copper. The situation is further complicated by the possibility that higher-order quadrupole corrections[19] to the Korringa rate are important here because of the large ratio of electric quadrupole to magnetic dipole moments for ^{197}Au. The quadrupolar relaxation effect tends to increase $\mathcal{K}(\alpha)_{expt}$ (Eq. (2)). in opposition to the effect of the orbital shift.

An interesting piece of evidence concerning the cases of Au and Ag metals is found in the work of Narath[20] on Ag-Au alloys. $\mathcal{K}(\alpha)_{expt}$ for ^{109}Ag is found to be essentially constant across this alloy series and considerably smaller than that of ^{197}Au in pure gold. We consider this to be evidence for an orbital shift for ^{109}Ag and/or quadrupolar relaxation of the ^{197}Au.

In summary, Van Vleck orbital magnetism is concluded to play a greater role in the noble metals than previously realized and to clear up an apparent contradiction between NMR and other measurements as regards the strength of band exchange effects in these metals. In the case of Cu metal, the orbital effects are thought to have a considerable impact on a detailed interpretation of NMR measurements such as given by El Hanany and Zamir,[18] though a thorough discussion of these matters goes beyond the scope of the present paper. Resolution of the situation for Ag and Au awaits further data such as those provided by Randles' de Haas-van Alphen measurements.[4]

ACKNOWLEDGMENTS

The authors wish to thank T. M. Rice for pointing out the implication of the d-band screening effects here and U. El Hanany and W. W. Warren, Jr., for a useful discussion of this material.

REFERENCES

1. A. Narath and H. T. Weaver, Phys. Rev. 175, 373 (1968).
2. H. Ehrenreich and A. R. Philipp, Phys. Rev. 128, 1622 (1962).
3. D. Lubzens, M. R. Shanabarger and S. Schultz, Phys. Rev. Lett. 29, 1387; 1768 (1972).
4. D. L. Randles, Proc. R. Soc. Lond. A. 331, 85 (1972).
5. R. Kubo and Y. Obata, J. Phys. Soc. Japan 11, 547 (1956).
6. J. Korringa, Physica 16, 601 (1950).
7. T. Moriya, Prog. Theor. Phys. 18, 516 (1963).
8. U. El Hanany and D. Zamir, Phys. Rev. B5, 30 (1972).
9. Y. Yafet and V. Jaccarino, Phys. Rev. 133, A1630 (1964).
10. R. E. Walstedt, J. H. Wernick and V. Jaccarino, Phys. Rev. 162, 301 (1967).
11. A. M. Clogston, V. Jaccarino and Y. Yafet, Phys. Rev. 134, A650 (1964).
12. L. Hodges, H. Ehrenreich and N. D. Lang, Phys. Rev. 152, 505 (1966).
13. G. Burdick, Phys. Rev. 129, 138 (1963).
14. R. E. Watson and A. Freeman, in Magnetism, Vol. IIA (edited by G. T. Rado and H. Suhl) Academic Press, New York.
15. W. J. O'Sullivan, A. C. Switendick and J. E. Schirber, in Electronic Density of States (edited by L. H. Bennett), Nat. Bur. Stand. (U.S.), Special Publication 323, 1971.
16. R. W. Shaw and W. W. Warren, Phys. Rev. B3, 1562 (1971); P. Bhattacharyya, K. N. Pathak and K. S. Singwi, Phys. Rev. B3, 1568 (1971).
17. N. E. Christensen, Phys. Status Solidi 54, 551 (1972).
18. N. E. Christensen and B. O. Seraphin, Phys. Rev. B4, 3321 (1971).
19. E. Haga and S. Maeda, J. Phys. Soc. Japan 32, 324 (1972).
20. A. Narath, Phys. Rev. 163, 232 (1967); 175, 696 (1968).

THEORY OF SPIN-LATTICE RELAXATION FOR PARAMAGNETIC IONS*

R.S. Wilson, A.J. Fedro and D.C. Knauss
Northern Illinois University, DeKalb, Illinois 60115
and Argonne National Laboratory, Argonne, Illinois 60439

ABSTRACT

We consider the relaxation of a two spin-state paramagnetic ion embedded in an ionic lattice. The decay mechanism is assumed to be due to the vibrational modulation of the crystalline electric field which acts on the orbital motion of the paramagnetic electrons. We first express the relaxation rate in terms of a time-correlation function involving an effective interaction which is the matrix element of the interaction between the two spin states. The calculation of this correlation function is based on a projection operator method similar in spirit to our recent treatment of the Hubbard model. Our essential result is an exact expression for the relaxation time in terms of the Fourier-transform of another correlation function involving only the lattice variables, which do the coupling, thermally averaged over the uncoupled lattice in equilibrium.

DISCUSSION

Van Vleck in his pioneering work of 1940,[1] identified two basic mechanisms for paramagnetic spin relaxation in crystals. The first, the so-called direct process, represents stimulated emission and absorption of resonant phonons. The other is the inelastic scattering of more energetic, i.e., non-resonant phonons, the so-called Raman process. Recently, spin-lattice relaxation has served as an effective tool for the study of impurity-induced vibrational properties of crystals.[2] In addition, very recently, the very slow electron spin-lattice relaxation of Kramers ions dilutely dispersed in crystals in zero external magnetic field have been used to study the hyperfine levels.[3] With these considerations in mind, we see a general theory of the spin-lattice relaxation process is needed. The purpose of this work is to present such a method.

THEORY

We consider in this section the theory of a prototype model for the spin-lattice relaxation of a paramagnetic ion embedded in a lattice. The spin is assumed to have only two states. The decay mechanism is taken to be the vibrational modulation of the crystalline electric field which acts on the orbital motion of the paramagnetic electrons. Thus we consider the following Hamiltonian:[4]

$$H = H_L + H_S + H_I \ ; \qquad (1)$$

$$H_L = \sum_k \hbar\omega_k \, (a_k^\dagger a_k + \tfrac{1}{2}) + \lambda \sum_k \Gamma_k Q_k \ ; \qquad (2)$$

$$H_S = \varepsilon_\uparrow |\uparrow\rangle\langle\uparrow| + \varepsilon_\downarrow |\downarrow\rangle\langle\downarrow| \ ; \qquad (3)$$

$$H_I = \sum_f g_f O_f B_f \ ; \ B_f = \sum_{k,n} C_{kn}^f Q_k^n \ . \qquad (4)$$

Here the ω_k are the lattice normal mode frequencies, the $a_k^\dagger$ (a_k) the phonon creation (annihilation) operates for the kth phonon. The quantity Γ_k is assumed to be of the proper form to make an anharmonic contribution to the lattice Hamiltonian, λ is a coupling constant, usually assumed small and $Q_k = a_k + a_k^\dagger$. The term $\varepsilon\uparrow(\varepsilon\downarrow)$ is the energy of the spin up (down) state, and O_f is a complicated function depending on the spin and orbital angular momentum operators of the paramagnetic ion. Explicit expressions for them can be obtained from Ref. 4. The numbers g_f and C_{kn}^f are arbitrary coupling constants. The transition probability for the spin to flip from up to down can be written as follows:

$$W = \left(\frac{1}{\hbar^2}\right) \int_{-\infty}^{+\infty} dt \ e^{-i\omega_z t} \langle VV(t)\rangle \ , \qquad (5)$$

where $\hbar\omega_z \equiv \varepsilon_\uparrow - \varepsilon_\downarrow$, and $V \equiv \langle\uparrow|H_I|\downarrow\rangle$, or

$$V = \sum_f D_f \, B_f \ ; \qquad (6)$$

$$D_f \equiv g_f \langle\uparrow|O_f|\downarrow\rangle \ ; \qquad (7)$$

$$V(t) \equiv e^{i(t/\hbar)H_L} \, V(o) \, e^{-i(t/\hbar)H_L} \ . \qquad (8)$$

The brackets <...> denote a canonical ensemble average over the full decoupled lattice in equilibrium. We see that our basic task is to calculate the various lattice time correlation functions,

$$B_{ff'}(t) \equiv \langle B_f B_{f'}(t)\rangle \ , \qquad (9)$$

where $B_{f'}(t)$ is the Heisenberg operator. In fact, from Eq. 4 we see then that our task further boils down to finding

$$\underset{mn}{G}(t) \equiv \langle(\prod_{i=1}^{m} Q_i)\cdot(\prod_{j=1}^{n} Q_j(t))\rangle$$

$$= \langle(\prod_{i=1}^{m} Q_i)\cdot[\prod_{j=1}^{n} Q_j](t)\rangle = \langle q_m q_n(t)\rangle \ . \qquad (10)$$

For a general anharmonic lattice the calculation of the full-time dependence of G(t) is an enormous task. However, we are naturally interested for the relaxation process in the so-called long-time limit, and it is here where much progress can be made.

Calculation of G(t)

To calculate G(t) we introduce the following set of projection operators which project out of $q_n(t)$ the part which contributes to $G_{mn}(t)$, i.e., we choose

$$P_{mn}X(t) \equiv q_n \frac{\langle q_m X(t)\rangle}{\langle q_m q_n\rangle} \ . \qquad (11)$$

By the usual procedures[5] we immediately write down the exact equation of motion for G(t) in generalized Langevin form:

$$\dot{G}(t) = i\Omega G(t) - \int_0^t \gamma(s) G(t-s) ds \ , \qquad (12)$$

where the natural frequency term Ω and the damping or self-energy term $\gamma(s)$ are defined as follows:

$$\Omega \equiv \langle q_m L q_n\rangle / \langle q_m q_n\rangle \ , \qquad (13)$$

and

$$\gamma(s) \equiv \langle f^\dagger f(s)\rangle \ . \qquad (14)$$

where the "random-force" f(s) is taken as

$$f(s) = -e^{i(1-P_{mn})Ls}(1-P_{mn})Lq_n \qquad (15)$$

We have suppressed the subscripts m,n for convenience.

For a harmonic crystal only the first term in Eq. (12) survives, i.e., there is no damping. This approximation suffices for the direct and Raman processes if we neglect the localized or sharp resonance mode contributions. However, for the latter, the full anharmonic correlation functions are necessary.

Our primary result in solving Eq. (12) in the necessary long-time limit with the anharmonic coupling weak, is that Wick's theorem holds, i.e., all multi-time correlation functions break up into products of single-time correlation functions. The validity of this theorem in the long-time limit has tremendous calculational utility: we need calculate in fact only one correlation, e.g., $\langle Q_k Q_\ell(t)\rangle$ in the long-time weak coupling limit. We write down one simple example which arises naturally in the Raman spin-lattice relaxation process:

$$\langle Q_k Q_\ell Q_m(t) Q_n(t)\rangle \underset{t\to\infty}{\cong} \langle Q_k Q_\ell\rangle\langle Q_m Q_n\rangle$$
$$+\langle Q_k Q_m(t)\rangle\langle Q_\ell Q_n(t)\rangle+\langle Q_k Q_n(t)\rangle\langle Q_\ell Q_m(t)\rangle \ . \quad (16)$$

From previous work[6] we know that

$$\langle Q_k Q_m(t)\rangle =$$
$$\delta_{km}\ \langle a_k a_k^\dagger\rangle_o e^{+i\nu_k t} e^{-\tilde{\Gamma}_k(t)} + \langle a_k^\dagger a_k\rangle_o e^{-i\nu_k t} e^{-\tilde{\Gamma}_k(t)} \quad (17)$$

where
$$\nu_k = \omega_k + \gamma_2(\omega_k) \ ;$$
$$\tilde{\Gamma}_k = \gamma_1(\omega_k) > o \ , \quad (18)$$

where
$$\gamma(\omega) \equiv \gamma_1(\omega) - i\gamma_2(\omega)$$
$$= \left(\frac{\lambda^2}{\hbar^2}\right)\int_o^\infty \langle[\Gamma_k,\Gamma_k(t)]\rangle_o\ e^{-i\omega_k t}dt \ . \quad (19)$$

The zero subscript on the averages, $\langle\ldots\rangle_o$ denotes harmonic average over the lattice is to be taken.

If we restrict on calculation to the special case where:

$$B_f \cong \sum_k C_k^f Q_k^2 \ , \quad (20)$$

we find the well-known form for the spin-lattice relaxation time for a localized mode (Raman process):

$$\tau_R^{-1} = [1+e^{-\beta h\omega_z}]W \quad (21)$$

$$= \frac{16K}{h^2}\ [1+e^{-\beta h\omega_z}]n(\omega_o)[n(\omega_o)+1]\ \frac{\tilde{\Gamma}(\omega_o)}{\omega_z{}^2+4\tilde{\Gamma}^2(\omega_o)} \ , \quad (22)$$

where K is a constant which depends on the C_k^f's and

$$\tilde{\Gamma}(\omega_o) \equiv \frac{1}{2\hbar^2 n(\omega_o)}\int_{-\infty}^{+\infty}\langle\Gamma_o\Gamma_o(t)\rangle_o\ e^{i\omega_o t}dt \ . \quad (23)$$

Here $n(\omega_o) \equiv (e^{\beta h\omega_o}-1)^{-1}$, ω_o is the harmonic localized mode frequency, and Γ_o is the term in the anharmonic potential which couples the localized mode linearly to the rest of the lattice. Explicit expressions for Eq. (22) for various lattice models can be found in Ref. 6. The generalized form for the case of direct and Raman processes in anharmonic crystals for in-band modes and resonance modes will appear elsewhere. Eq. (22) represents the most general result to date for the localized mode Raman process; previous treatments have been restricted to specialized lattice models.[2]

*Supported in part through the auspices of the U.S. Atomic Energy Commission.

REFERENCES

1. J.H. Van Vleck, Phys. Rev. 47 426 (1940).
2. D.L. Mills, Phys. Rev. 146, 336 (1966).
3. C.A. Hutchinson, et al., Phys. Rev. Lett. 33 937 (1974).
4. R.D. Mattuck, M.W. Strandberg, Phys. Rev. 119, 1204 (1960).
5. A.J. Fedro and R.S. Wilson, Phys. Rev. (in press).
6. S.K. Kim and R.S. Wilson, Phys. Rev. A7, 1396 (1973).

MEASUREMENT OF INTERCHAIN DIFFUSION IN A QUASI-ONE DIMENSIONAL HEISENBERG MAGNET*

Peter M. Richards and R. C. Hughes
Sandia Laboratories, Albuquerque, New Mexico 87115

ABSTRACT

The characteristic time t_c for spin polarization to transfer between magnetically inequivalent chains has been measured in the quasi one-dimensional salt dichlorobis(pyridine)copper(II) [$CuCl_2 \cdot 2NC_5H_5$, referred to as CPC]. The method utilizes the large difference in ESR linewidths between directions of applied field where the chains appear to be inequivalent and symmetry directions where they are equivalent. Data at 23 GHz are presented between 300°K and 4°K and interpreted in terms of temperature variation of the intrachain diffusion coefficient for $T \geqslant 10$°K and the effect of antiferromagnetic correlations at lower temperatures. The time t_c does not appear to be influenced by spin lattice relaxation, whereas the linewidth along symmetry directions does appear to be so affected.

INTRODUCTION

Real, as opposed to ideal, one-dimensional (1d) systems possess finite coupling between the nearly isolated linear chains. This interchain coupling, J', is responsible for the ultimate 3d ordering of the chains at low temperature and also has been shown[1-3] to have profound effects on the spin dynamics as sampled by magnetic resonance. Although the ordering temperature T_c and the electron spin resonance (ESR) linewidth and lineshape can be related to J', the connection is generally through an approximate theoretical relation of questionable validity. T_c, for example, is usually related to J' by an RPA Green function analysis.[1,4] It is therefore desirable to have other measurements which can be used to infer J', hopefully in a more direct and less complicated way, and therefore serve as a check on the consistency of various theories as well as providing new techniques.

A quantity rather directly related to J' is the rate t_c^{-1} at which spin polarization transfers from one chain to another. We have shown in a recent paper,[5] referred to as I, that t_c can be measured in a system with magnetically inequivalent chains. In the absence of interchain coupling, the ESR spectrum would consist of two lines separated by an amount $\Delta\omega = \omega_A - \omega_B$ where ω_A and ω_B are the resonance frequencies of the inequivalent chains of type A and type B, respectively. The presence of J' produces the transfer rate t_c^{-1}, and if $\Delta\omega t_c \leqslant 1$, the two lines merge into a single one whose half-width is

$$\gamma\Delta H = \tfrac{1}{4}\,\Delta\omega^2 t_c + \gamma\Delta H_o \qquad (1)$$

where γ is the average gyromagnetic ratio of the two chains and where ΔH_o is the contribution of other sources, such as intrachain dipolar coupling, to the linewidth. Since $\Delta\omega$ is proportional to the microwave frequency and has an angular dependence which can readily be determined from the crystal symmetry (i.e. $\Delta\omega = 0$ for field directions along which the chains are magnetically equivalent), one can eliminate ΔH_o from (1) by studying the angular and frequency dependence of ΔH. Furthermore, $\Delta\omega$ can be determined from the measured g-tensor plus symmetry considerations,[5,6] so that a quantitative measurement of t_c is possible. This method was applied in I to the salt dichlorobis(pyridine)copper(II) [$CuCl_2 \cdot 2NC_5H_5$, referred to as CPC] which has inequivalent chains, and is one of the most extensively studied spin $\frac{1}{2}$ linear chain salts.[5-8]

The data of I, which yielded $t_c = 3 \times 10^{-11}$ sec, were for room temperature and the accompanying theory was for infinite temperature. Here we report on the variation of t_c in CPC between 300°K and 4°K and make appropriate modifications of the theory to account for finite temperature. Theory can be used to relate t_c to the strong intrachain diffusion coefficient D and to antiferromagnetic fluctuations and thereby provide insight into temperature dependence of the basic spin dynamics. We point out that, since ΔH_o is likely influenced by spin-lattice relaxation whereas t_c is not, the measurements here can possibly give a better handle on pure spin dynamics than can studies of ΔH_o alone.

EXPERIMENT

Data were taken at 23.0 GHz on a small needle-like single crystal of CPC orientated so that the applied field H could be rotated in the g_3-b plane. Here g_3 and b are principal axes of the crystal g-tensor as described in I and Ref. 6; b coincides with the crystal b-axis while g_3 is in the crystal a-c plane at an angle of -28° from c. The g_3-b plane is chosen because it contains the two inequivalent site $g_{\parallel}$ axes, and thus $\Delta\omega$ has its maximum value in this plane. We have, from I, $\Delta\omega/\gamma = 475 \sin 2\theta$, in Oersteds, where θ is the angle H makes with respect to the b axis. The b and g_3 axes are symmetry directions for which $\Delta\omega = 0$ so that measurements along these axes yield ΔH_o. The observed width at $\theta = 45°$, $\Delta H(45°)$, is much greater than ΔH_o.

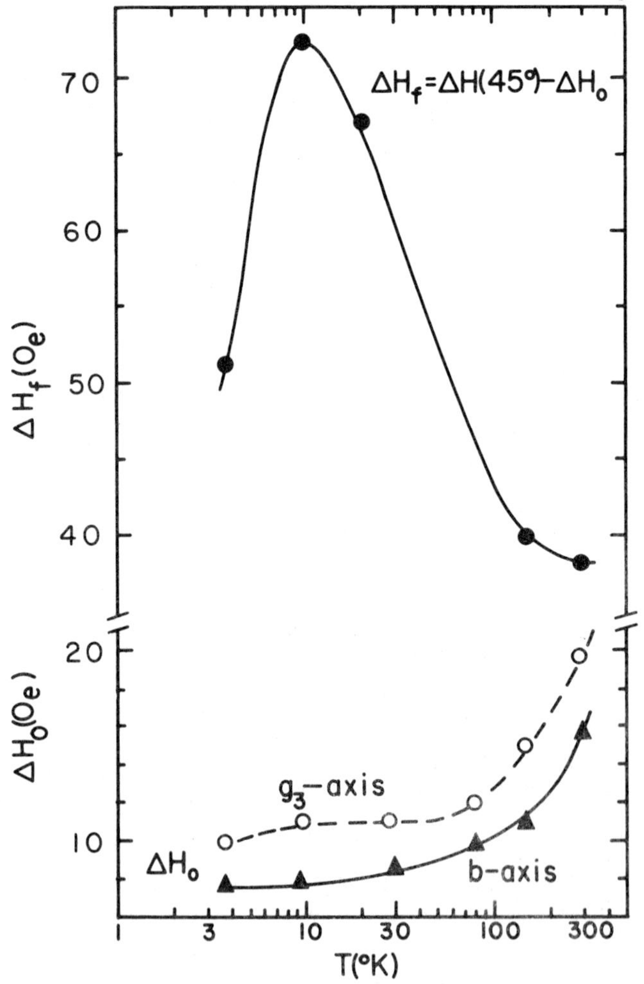

Fig. 1 Peak-to-peak linewidths for field in g_3-b plane. ΔH_f is the difference between the width at $\theta = 45°$ and the average ΔH_o. Note break in scales.

This, together with the data from I and here which show that ΔH_o has negligible angular dependence allow us to estimate $\frac{1}{4}\,\Delta\omega^2(45°)t_c \equiv \gamma\Delta H_f$ from the relation

$$\Delta H_f \approx \Delta H(45°) - \tfrac{1}{2}[\Delta H(0°) + \Delta H(90°)] \qquad (2)$$

Figure 1 shows ΔH_f, $\Delta H(0°)$ (b-axis), and $\Delta H(90°)$ (g_3-axis) as functions of temperature. Values quoted are for the peak-to-peak width on a derivative absorption curve.

THEORY

The general theory of interchain diffusion has been given elsewhere,[1,3,9] and it was specifically applied to the problem at hand in I. In essence, the rate t_c^{-1} is given by time-dependent perturbation theory in which the perturbation $2J'\vec{S}_{iA}\cdot\vec{S}_{jB}$ is modulated by intrachain motion which makes the spins $\vec{S}_{iA}$ and $\vec{S}_{jB}$ on A-type and B-type chains rapidly varying functions of time. A straightforward extension of the theory to finite temperature shows that

$$t_c^{-1} \propto (\chi_o T)^{2/3}\, D^{-1/3}\, |J'|^{4/3} + \Delta \qquad (3)$$

where χ_o is the uniform static susceptibility and D is the intrachain diffusion coefficient. The first term on the right-hand side of (3) represents the contribution from slowly decaying modes with wave vector $q \to 0$ to the modulation of the interchain interaction. The quantity Δ is the contribution from antiferromagnetic modes near $q^* = 0$ ($q^* = \pi/c - q$ where c is the intrachain spacing). Δ is unimportant at high temperatures, but it becomes dominant at temperatures $T \leqslant |J|/k_B$ in an antiferromagnetic chain such as CPC when the fluctuations become large near $q^* = 0$.

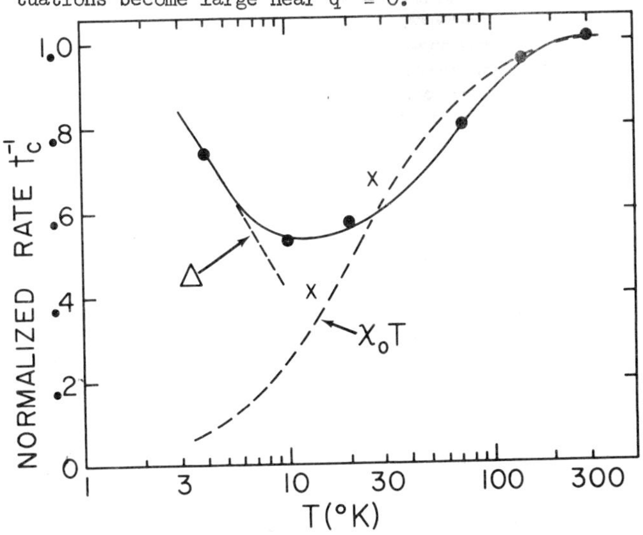

Fig. 2 Interchain diffusion rate $t_c^{-1} \propto 1/\Delta H_f$ normalized to its room temperature value. Crosses are theory of Ref 10 for $(\chi_o T)^{2/3}D^{-1/3}$. $\chi_o T$ curve is the expected high temperature behavior if $D \propto (\chi_o T)^{-1}$. Dashed curve labeled Δ is schematic of the contribution of antiferromagnetic fluctuations ($q^* \approx 0$) to t_c^{-1}.

DISCUSSION

Figure 2 gives a plot of $t_c^{-1} \propto \Delta H_f^{-1}$, normalized to 300°K. According to Eq. (3) this should be proportional to $(\chi_o T)^{2/3}D^{-1/3}$ at sufficiently high temperatures where Δ is negligible. If, as expected at high temperatures on the basis of simple moments arguments, $D \propto (\chi_o T)^{-1}$, we would have $t_c^{-1} \propto \chi_o T$, the curve for which[6,7] is shown. Recent "computer experiments" on a classical antiferromagnetic chain[10] predict a temperature dependence of D which, when combined with $\chi_o T$, gives a temperature dependent $(\chi_o T)^{2/3}D^{-1/3}$ as shown.

For $T \leqslant 10°K$ ($|J|/k_B = 13.4°K$ for CPC) antiferromagnetic fluctuations dominate and t_c^{-1} is controlled by Δ whose dependence is qualitatively as shown ($\Delta \to 0$ at 300°K).

We feel that the following conclusions can be drawn. First, t_c^{-1} and the inferred D follows $(\chi_o T)^{-1}$ between 300°K and 150°K. The small variations (4%) of D and ΔH_f in this region are to be contrasted with the 30-50% decreases[11] in ΔH_o. This may well show that ΔH_o is influenced by spin-lattice relaxation[12] in this relatively high temperature range. At 150°K, $|J|/k_B T < 0.1$ so that we would not expect the pure spin dynamics to produce a strong temperature dependence above 150°K, and this is borne out by ΔH_f. Spin-lattice relaxation should not directly affect ΔH_f since it is sensitive only to relaxation processes which transfer spin polarization from one chain to another. For this reason, studies of ΔH_f can be perhaps more meaningful than ΔH_o in terms of spin dynamics for non-S state systems such as Cu^{++} which have appreciable spin-lattice relaxation. The near constancy of ΔH_f between 150°K and 300°K also shows that there is negligible temperature dependence of the intra- and interchain exchange constants.

Secondly, the measurements clearly show that the transfer rate between chains t_c^{-1} initially decreases with lowering temperature. This further indicates that the intrachain motion, measured by D, is speeding up with decreasing temperature in qualitative agreement with recent calculations. (By the usual arguments of time-dependent perturbation theory, a decrease in an observed relaxation rate, t_c^{-1}, is caused by an increase in the rate at which the perturbation, J′, is modulated. In this case the modulation comes from intrachain motion.) The increase in t_c^{-1} below 10°K is indicative of the effects of antiferromagnetic correlations in slowing down the intrachain motion.

ACKNOWLEDGMENTS

Dr. B. Morosin aligned the crystal by x-ray techniques. Dr. R. Birgeneau informed us of the unpublished results in Ref. 11. Mr. D. Cooper provided valuable technical assistance.

REFERENCES

* Work supported by U. S. Atomic Energy Commission.
1. M. J. Hennessy, C. D. McElwee, and P. M. Richards, Phys. Rev. B7, 930 (1973).
2. F. Borsa and M. Mali, Phys. Rev. B9, 4981 (1974).
3. J. P. Boucher, F. Ferrieu, and M. Nechtschein, Phys. Rev. B9, 3871 (1974).
4. T. Oguchi, Phys. Rev. 133, A1098 (1964).
5. R. C. Hughes, B. Morosin, P. M. Richards, and W. Duffy, Jr., "ESR and Structure of Magnetically Inequivalent Chains in $CuCl_2\cdot 2NC_5H_5$," to be published.
6. W. Duffy, Jr., J. E. Venneman, D. L. Strandburg and P. M. Richards, Phys. Rev. B9, 2220 (1974).
7. K. Takeda, S. Matsukawa and T. Haseda, J. Phys. Soc. Jap. 30, 1330 (1971).
8. Y. Endoh, G. Shirane, R. J. Birgeneau, P. M. Richards and S. L. Holt, Phys. Rev. Lett. 32, 170 (1974).
9. G. F. Reiter, Phys. Rev. B8, 5311 (1973).
10. N. A. Lurie, D. L. Huber and M. Blume, Phys. Rev. B9, 2171 (1974).
11. Unpublished measurements by L. R. Rupp, Bell Telephone Laboratories, show a similar temperature dependence of ΔH_o.
12. T. G. Castner, Jr. and M. S. Seehra, Phys. Rev. B4, 38 (1971); M. S. Seehra, Bull. Am. Phys. Soc. 19, 368 (1974).

MAGNETORESISTIVE TRANSDUCERS IN HIGH-DENSITY MAGNETIC RECORDING

D. A. Thompson
IBM, Thomas J. Watson Research Center, Yorktown Heights, New York 10598

ABSTRACT

Magnetic recording heads fabricated using magnetic thin films have a number of well-known advantages over conventional heads. The magnetoresistive effect in thin permalloy films is especially attractive for the read function, because the signal amplitude available (about 40 mV per mm, of track width) is speed independent and is larger at all practical speeds than the output of comparably complex inductive film heads. Previously reported magnetoresistive heads have not been successfully used for high density magnetic recording because they lacked one or more of three essential ingredients: A reliable magnetoresistive stripe, a shielding magnetic yoke structure to provide linear resolution, and a magnetic bias method compatible with the shields. This paper will present data and analysis for the first such structure, and will discuss the present level of understanding of such transducers.

INTRODUCTION

Magnetic recording heads fabricated from magnetic thin films have a number of well-known advantages over conventional heads.[1,2] These include batch fabrication, improved dimensional reproducibility, increased linear and track density, and improved frequency response. Most of these devices are functionally equivalent to conventional inductive recording heads. However, there is a class of film heads, first proposed by Hunt,[3] which use the magnetoresistive effect in permalloy to effect the readback process. Because the signal voltage does not result from Faraday's Law, these devices have characteristics and limitations which are fundamentally different from those of inductive recording heads. This paper will review the present level of understanding of magnetoresistive heads, and will report the first head structures to simultaneously exhibit both high linear density and correct magnetic bias.

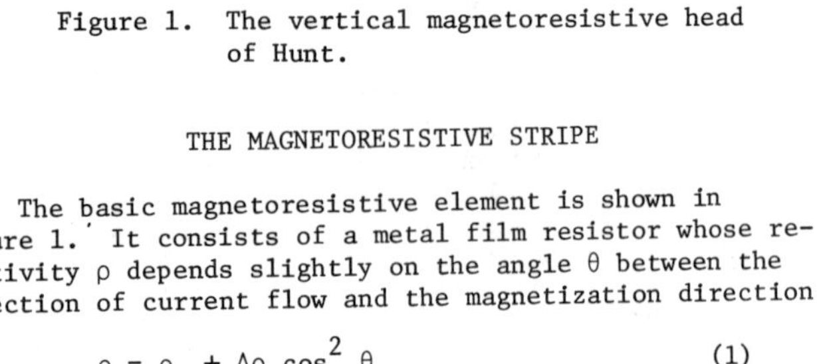

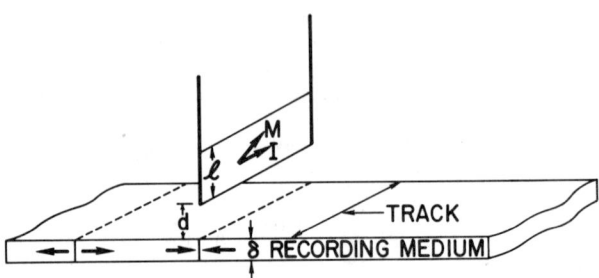

Figure 1. The vertical magnetoresistive head of Hunt.

THE MAGNETORESISTIVE STRIPE

The basic magnetoresistive element is shown in Figure 1. It consists of a metal film resistor whose resistivity ρ depends slightly on the angle θ between the direction of current flow and the magnetization direction:

$$\rho = \rho_o + \Delta\rho \cos^2 \theta \qquad (1)$$

In non-magnetostrictive permalloy (83-17 NiFe) the effect $\Delta\rho/\rho$ is 2 to 3%, depending on deposition conditions.[4,5,6] In 90-10 NiFe and 80-20 NiCo, the magnitude is closer to 5% at room temperature; however, strain sensitivity (and for the latter, high uniaxial anisotropy) prevents their use in most transducer applications.

When the current flow and magnetization directions are uniform, as they are in Figure 1, then the resistance dependence due to equation 1 can be quite simple. This occurs when the easy axis of the permalloy is perpendicular to the applied field, and is due to the well known magnetic properties[7] of uniaxial permalloy. Figure 2 shows the dependence of the resistance on magnetic field for a sheet film and a finite stripe. The basic resistance variation is parabolic until saturation. Demagnetizing effects at the edges of actual stripes cause a smoothing of saturation[8]. Also for demagnetizing reasons, the magnetoresistive stripe is essentially insensitive to applied field components perpendicular to the plane of the film.

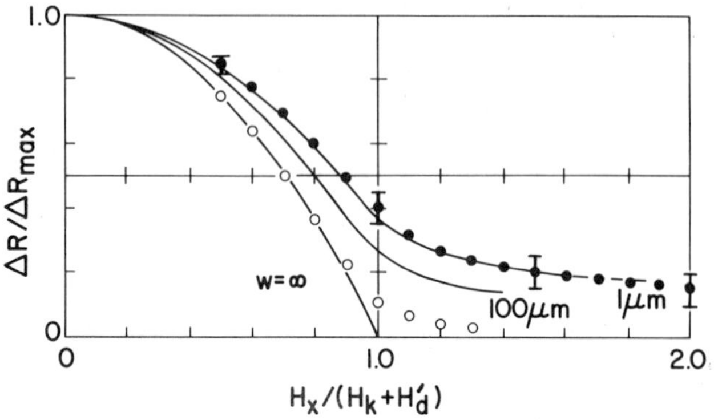

Figure 2. Magnetoresistance versus applied field in the hard direction. Open dots are data for a sheet film. Solid dots are data from stripes of width less than 100 microns, where the applied field is normalized to the anisotropy field H_k and the saturation demagnetizing field $H_d' = tM/w$. From reference 8.

In order to optimize the efficiency of the magnetoresistive stripe as a transducer, it is necessary to consider a number of factors. Since the magnetic field only modulates power provided by an external electrical source, the magnetoresistor does not have a fundamental maximum signal power as an inductive transducer does. It is necessary for the designer to determine whether current (in the interconnection and power supplies), current density in the head (which can cause electromigration[9]), or power dissipation in the head is the factor which will limit the applied power. The output variable will be either signal power, voltage, or current, depending upon whether the device impedance can be matched to existing preamplifiers and interconnection impedances. Magnetic design must reflect whether the magnetic signal is primarily a constant flux or a constant H source, and the effects of demagnetization fields.

In spite of these generalities, magnetic and impedance consideration normally favor the thinnest possible sensing film. The energy conversion efficiency $\Delta\rho/\rho$ is essentially independent of thickness down to about 200 Å[5,6], and is approximately inversely dependent on thickness below 200 Å. Magnetic and electrical properties are satisfactory above about 100 Å if the substrate is suitable, so most magnetoresistive designs are optimum at a permalloy thickness of about 200 Å. Film reliability and electromigration properties are exceptionally good at this thickness if the film is passivated against oxidation.[9] The sheet resistance is about 10 ohms per square, which leads to devices in the 10-500 ohm range, a good match to semiconductor amplifiers.

LINEAR RESOLUTION

Unfortunately, the vertical head of Hunt has a rather low linear resolving power. The output pulse width at half amplitude for an arctangent transition in a thin medium is

$$P_{50\%} = \ell + 2d + \delta \qquad (2)$$

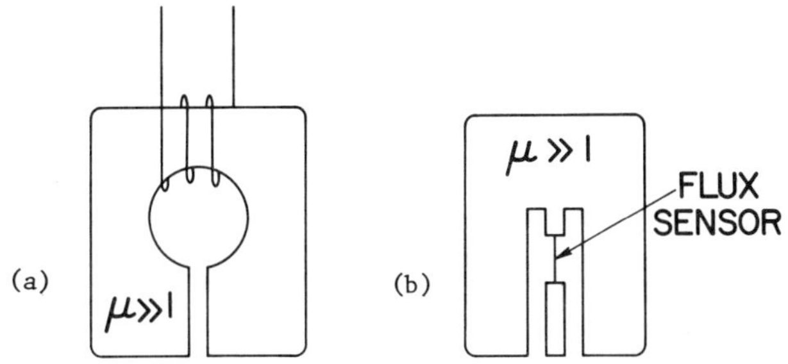

Figure 3. Idealized recording head geometries. (a) Inductive head. (b) Three legged flux sensing head.

For definition of symbols, see Figure 1. Mechanical grinding and wear tolerances dictate a lower limit for ℓ of a few microns (.1 mil), which limits the resolving power of the simple magnetoresistive head to less than that of state-of-the-art commercial recording systems. The horizontal head of Hunt[3] has no such dimensional limitations, but is impossible to protect against a catastrophic scratch, and produces a bipolar output pulse pair for each transition (which complicates detection).

The best method of overcoming these difficulties is to add a magnetic yoke structure which protects and encapsulates the magnetoresistive element, while eliminating the dependence of resolution on the stripe width. In order to understand the operation of such a device, consider the two cross-sectional schematic structures of Figure 3. Assume that the magnetic yokes have infinite permeability, that the magnetic gaps g are small compared to the flying height plus transition width, and that the magnetoresistive stripe is biased into a linear region of operation. The divergence of magnetization in the medium gives rise to magnetic flux into the face of the head which can be computed by a number of means.[10] If we define x = 0 to be at the center of the head, $B_\perp$ to be the perpendicular component at the face of the head, then

$$\phi(x) \cong \int_{-\infty}^{0} B_\perp (x)\, dx \qquad (3)$$

is the magnetic flux which produces an inductive readback voltage as the medium passes the head:

$$V_1 = -N \frac{d\phi}{dt} = -Nv \frac{d\phi}{dx} \qquad (4)$$

where v is the medium velocity.

The two gap head of Figure 3b is arranged so that the magnetoresistive element senses the fraction of the magnetic flux $\Delta\phi$ entering the middle leg

$$\Delta\phi \cong \int_{-(\frac{g+t}{2})}^{\frac{g+t}{2}} B\, (x)\, dx \cong (g+t) \frac{d\phi}{dx} \qquad (5)$$

The magnetoresistor is assumed to be operating in a linear, small signal mode, so

$$V_2 = K\, (g+t) \frac{d\phi}{dx} \qquad (6)$$

At constant velocity, an inductive head and a flux sensing head (with small magnetic gaps) will have the same output waveshape. One is a time differentiator (d/dt) and the other a space differentiator (d/dx). Their signals will be quite different, however, if the speed varies. A more formal and rigorous analysis based on the Karlquist approximation can be found in Reference 11.

Any flux sensing device, such as a Hall effect device, a semiconductor magnetoresistor, or a flux-gate sensor would give rise to a similar result. However, the permalloy magnetoresistor has a permeability in the film at least a thousand times greater than that of the magnetic gap material. This allows flux concentration in the sensor, and makes possible a sensing element, whose planar dimensions are substantially larger than the gap thickness.

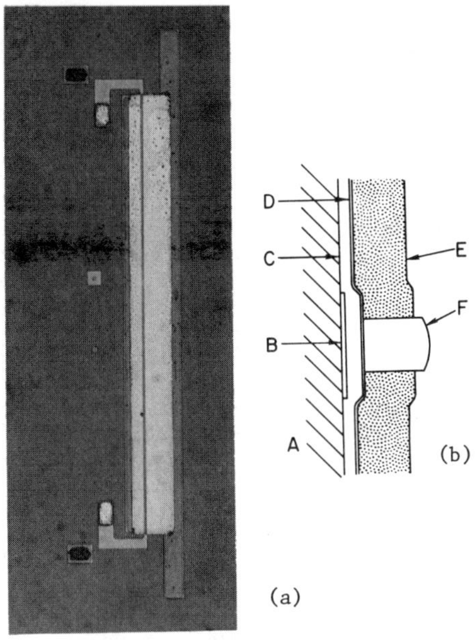

Figure 4. Middle leg of a three legged head. (a) Photograph of test structure fabricated on a silicon wafer (plan view). (b) Schematic cross-section of the central portion of the element. A is the substrate. B is the 200 Å thick by 10 micron high permalloy sensor. C is 1000 Å SiO_2 insulation. D is a non-magnetic 600 Å Ti-Cu evaporated metallization. E is the 2 micron thick plated permalloy leg. F is a photoresist plug applied prior to plating the permalloy leg. It prevents plating in that area, thus forming the internal gap bridged by the magnetoresistive sensor. Thicknesses not sketched to scale.

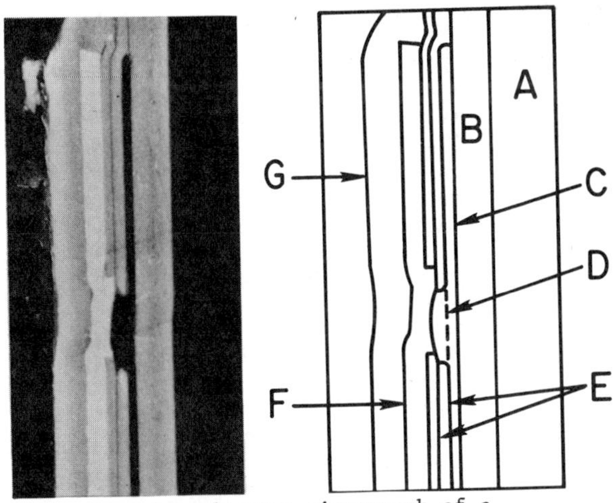

Figure 5. Scanning electron micrograph of a lapped cross section of a trident head structure (original 2400 x, reproduced here at about 1900 x). A is the oxidized silicon substrate. B is the first plated permalloy leg, 2 microns thick. C is the first gap, 7000 Å of SiO_2. D is the region containing the sensor (see Figure 4b). E is the middle leg, two 1/2 micron layers of permalloy with a 1000 Å SiO_2 spacer. F is the 7000 Å gold second gap and write conductor. G is the final 2 micron permalloy shield.

Figure 4 shows the middle leg of such a structure, fabricated separately for testing. Figure 5 shows the complete structure. Critical features of the design include a permalloy middle leg plated with a photoresist-filled internal gap and a magnetoresistive stripe bridging that gap, but insulated from it by an 800 Å sputtered SiO_2 film. The middle leg is laminated in two layers to reduce coercivity and improve frequency response.[12] The outer yoke members are also plated permalloy. Sputtered SiO_2 is used for the insulating, non-magnetic layer under the magnetoresistive stripe. Copper or gold is used for the other magnetic gap layer, which serves as the conductor for writing and for magnetically biasing the magnetoresistive element. A thick layer of sputtered SiO_2 protects the head element during wafer dicing, mounting, and head grinding. The fabrication methods are similar to those already described,[13,14] and will be reported elsewhere. For convenience, this three-legged structure will be referred as the trident.

An important feature of this design is the encapsulation of the sensing element away from the mechanical damage, corrosion, wear, and thermal transients[15,16] that occur at the face of the head. A principal problem with the head has been coercivity and remanence in the permalloy of the middle leg, which causes erratic shifting of the magnetic bias point.

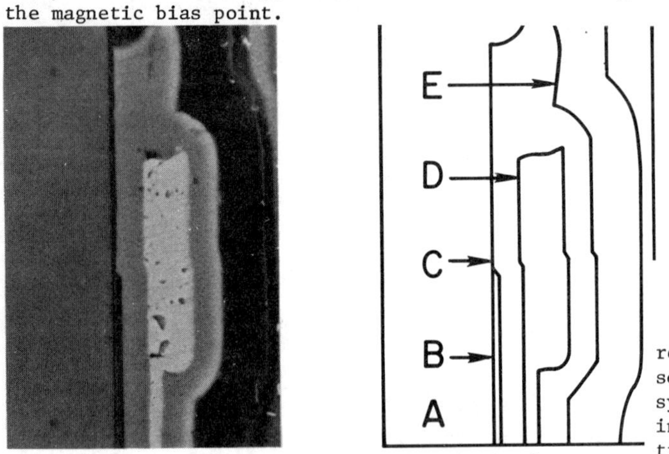

Figure 6. Scanning electron micrograph of a lapped cross section of a shield stripe head (2000 x, reproduced here 1600 x). A is the ferrite substrate and first shield. B is the magnetoresistive sensor region (see Figure 7). C is the permalloy shield and write head yoke. D is the gold write conductor. E is a sputtered SiO_2 protective layer. The "swiss cheese" appearance of the gold is due to embedded diamond flakes from the cross-sectioning process.

A simpler test element is shown in Figures 6 and 7. This device is suitable for applications where controllec environment and head flying eliminate the need for encaps lating the magnetoresistive element. A more detailed description appears elsewhere.[17] For convenience, this structure will be referred to as the shielded stripe.

The output waveforms of these devices are the same as those of an infinite pole tip (Karlquist) head. Waveshape and resolution data for the head of Figure 6 is presented in Reference 11. An interesting difference between inductive film heads and magnetoresistive heads is that finite pole tip lengths[18] have little effect on the resolution and isolated pulse shape for the magnetoresistive head. A naive explanation for this fact can b obtained from the integration limits in equations 3 and : Finite pole tips change the flux contribution in the region away from the gap. This alters the flux in the outer magnetic yoke members or "shields", but has essent ially no effect on the flux through the magnetoresistor, if the permeance of the outer legs is high enough. Thus the thin pole tip inductive head with gap g would seem to have some resolution advantage over the shielded magnetoresistive head with the same effective gap (g + t). This is illusionary, however, since in either case the gap can be fabricated small enough that flying height and medium properties determine the system resolution.

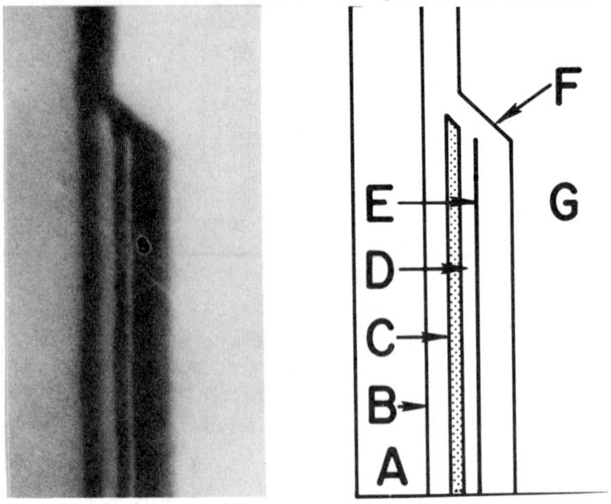

Figure 7. Close-up of the magnetoresistive sensor region of Figure 6 (20,000 x, reproduced 16,000 x). A is the ferrite substrate. B is the initial SiO_2 gap (2000 Å). C is the permanent magnet bias layer containing 75 Å Ti, 150 Å Fe (then oxidized), 250 Å NiFe and 75 Å Ti. D is a spacer layer of SiO_2 (1000 Å). E is the magnetoresistive film (200 Å). F is a SiO_2 gap (3500 Å). G is the permalloy shield, part of the write head.

MAGNETIC BIAS

There are useful recording systems, using magnetoresistive transducers, in which linearity and small-signal sensitivity are unimportant.[19] In high linear density systems, however, non-linearity accentuates inter-symbol interference and prevents the successful use of equalization. It is therefore necessary to apply a magnetic bias to the element to shift its operating point to the quasi-linear region near $\theta = 45°$. Hunt achieved bias by means of an external permanent magnet. This method cannot be used in the presence of the shielding structures necessary for high linear resolution. (Gorter, et al,[15] apparently used shielding pole tips so widely spaced that resolution was not affected). It is therefore necessary to use a source of magnetic bias within the head gap. Since the sensor thickness is several orders of magnitude less than the exchange length[20], the current which flows within the sensing permalloy produces an internal magnetic field which cannot bias the element. Symmetric placement of the other yoke members also produces no bias. However, the asymmetric placement of the stripe in the trident design results in a net biasing due to the magnetic circuit comprising the middle and the adjacent outer leg. Correct bias is observed with about 15 mA through the sense element. A current of comparable magnitude in the write conductor can be used to adjust the magnetic bias if other values of sense current are used. These current levels are insufficient to disturb data stored in the recording medium.

In the shielded stripe design, bias can be obtained either from additional current flowing in the non-magnetic portion of the gap[21] or by a magnetized film adjacent to the magnetoresistive film.[17] The great virtue of either method is that the biasing layer and the magnetoresistive layer can be etched into a single stripe in one sputter etching step, thus eliminating the precise alignment necessary in the trident design.

DESIGN CONSIDERATIONS

The most general and accurate method of analysis

of a magnetic structure is a self-consistent computer field mapping, taking into account all the demagnetizing fields and non-linearities of the magnetic material.[22] In the initial stages of design, however, this method is too cumbersome to be of much use. The heads described in this paper were all designed using an analysis based on transmission line methods.[23] This method allows an invaluable insight into the functioning of the head and allows simple, closed-form approximations for the relevant magnetic variables. Both read and write functions may be analysed, including the effects of write head saturation.

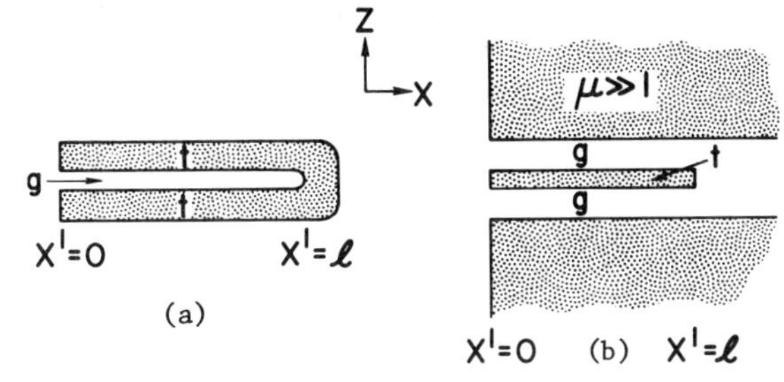

Figure 8. Two structures easy to analyse. (a) Single turn inductive film head. (b) Shielded magnetoresistive stripe.

The analysis is a two dimensional one based on the assumption that the relative permeability in the permalloy is much greater than one. The two simple geometries of Figure 8 will illustrate use of the method for the readback process. The extension to multiple layers of various thicknesses and to writing can be easily performed. The head is broken into several regions for analysis. The region to the left of x = 0 contains the magnetic medium and the write field. See Figure 8. The latter can be obtained for a given magnetic potential by conformal mapping or other analysis techniques which assume the permalloy pole pieces to be equipotential surfaces at the front. At any value of x in the region to the right of x = 0, the magnetic reluctance can be defined as

$$R(x) = \frac{U(x)}{\phi(x)} \qquad (7)$$

where the magnetic potential U across the gap is defined as

$$U = \int_{\text{gap } g} H_z \, dz. \qquad (8)$$

and the magnetic flux ϕ in the permalloy is defined as

$$\phi = \int_{\text{thickness } t} B_x \, dz \qquad (9)$$

In the central portions of the head, one notes that the magnetic drop through the thickness of a permalloy layer is negligible. It is thus possible to define the reluctance per unit length of the magnetic layers and the permeance per unit length of the internal non-magnetic gap, in a manner similar to that used in transmission line analysis,[24] to define the characteristic reluctance

$$R_o = \sqrt{\frac{2g}{(\mu_o)^2 \mu_r t}} \quad \text{for Figure 8a} \qquad (10)$$

$$= \sqrt{\frac{g}{2(\mu_o)^2 \mu_r t}} \quad \text{for Figure 8b}$$

where μ_o is the permeability of free space and μ_r the relative permeability of the permalloy, and to define the characteristic length.

$$\lambda = \sqrt{\frac{\mu_r t g}{2}} \qquad (11)$$

and the reflection coefficient

$$\rho(x) = \frac{R - R_o}{R + R_o} \qquad (12)$$

At higher frequencies the permeance $p = \mu_r t$ of the permalloy layer must be replaced with the correct complex value.[25] This leads to complex numbers for all the quantities (implying phase lag as well as attenuation) but does not require a change in the formalism.

The natural transmission line solutions for $\phi(x)$ and U(x) are exponentials for the form $e^{x/\lambda}$ or the equivalent hyperbolic functions. One familiar with transmission lines will recognize the concepts. At the back closure, front gap, or at a step in the yoke the reluctance differs from R_0, but as one moves away from the discontinuity, the reluctance decays toward that value at the rate $e^{-2x/\lambda}$. As an example, the reflection coefficient at the back closure is - 1. Therefore, the reluctance looking into the yoke of Figure 8 at the plane x= 0 is

$$R(0) = R_0 \frac{(1 + \rho(0)}{(1 - \rho(0)} = R_0 \frac{1 - e^{-2\ell/\lambda}}{1 + e^{-2\ell/\lambda}}$$

$$= R_0 \tanh(\ell/\lambda). \qquad (13)$$

This formalism is particularly convenient for calculating the field distributions in a head with sections of different thickness. The reluctance looking into a section is simply expressed in terms of the reluctance of the following section, taking into account the different R_0 of each section, by the relation

$$\rho(x_1) = \rho(x_2) \; e^{-2Wx/\lambda} \qquad (14)$$

The flux and magnetic potential are transmitted by similar relations:

$$\phi(h) = \phi(0) \, e^{-h/\ell} \left[\frac{1 - \rho(\ell)}{1 - \rho(0)}\right] \qquad (15)$$

$$U(h) = U(0) \, e^{-h/\ell} \left[\frac{1 + \rho(\ell)}{1 + \rho(0)}\right] \qquad (16)$$

Since equation (15) shows the fraction of flux transmitted fron one yoke section to another, one can express the total head efficiency of a stepped head (such as the write head of Figure 6) as the product of two efficiencies

$$\eta = \eta_A \eta_B \qquad (17)$$

where $\eta_B \approx \frac{\lambda_B}{\ell_B} \tanh\left(\frac{\ell_B}{\lambda_B}\right)$

is the efficiency of the back yoke and

$$\eta_A \approx e^{-\ell_A/\lambda_A} \left[\frac{1 - \rho_A(\ell_A)}{1 - \rho_A(0)}\right]$$

is the efficiency of the pole tips.

Here it has been assumed that the sense conductor is uniformly distributed in section B only.[2]

The efficiency for reading (fraction of signal flux linking the conductor) and for linear writing (fraction of the magnetic potential delivered to the pole tips) is the same, as might be expected from reciprocity. Above a certain current level, a portion of the head will saturate. In Figure 8a, this will occur at the back of the head first. As the current increases, the saturated portion will grow. Since anisotropic permalloy has a constant permeability upto an abrupt saturation, it is reasonable to postulate a moving boundary with permeability μ_r on one side and about 1 on the other. The portion of the write current flowing in the saturated portion of the head does not contribute to the write potential. It is as if the magnetic back closure were moving towards the front of the head. The result is that the write field in a head short compared to the characteristic length is linear until saturation occurs, then rises as the square root of the current after that. Saturation begins at a current

$$I \approx (2gtM_{SAT})/\ell \qquad (18)$$

here M_{SAT} is the saturation moment in the permalloy (MKS units).

The magnetic efficiency of a magnetoresistive head includes an additional factor having to do with the magnetic bias of the element. Consider the simple shielded stripe of Figure 8b. If a magnetic signal flux ϕ_o is injected into the stripe at X = 0, it will decay as

$$\phi(x) \approx \left(\frac{\phi_o}{\sinh \ell/\lambda}\right) \sinh \left(\frac{\ell - x}{\lambda}\right). \qquad (19)$$

For $\ell/\lambda < 1/2$, this is approximately a linear decay, $\phi = \phi_o(1 - x/\lambda)$. For $\ell/\lambda > 2$, this is approximately an exponential decay $\phi = \phi_o e^{-x/\lambda}$. In the example of Reference 17, g = 3700 Å, t = 200 Å, μ_r = 4000. Then the characteristic length is λ = 4 microns. As originally fabricated, ℓ = 7 microns, and the element is inefficiently excited. However, both read and write elements will allow further lapping, down to about ℓ = 2 microns. Thus the head can be tested at various throat heights as both read and write efficiencies improve.

The readback signal is not simply proportional to the average signal flux across the stripe. At each point in the stripe, the magnetic bias is different. Figure 9 shows the resulting magnetic bias levels and magnetic signal levels across a 7 micron and a 2 micron stripe for uniform H bias[21] and for permanent magnet bias.[17] It is possible to define a small-signal transducer efficiency by integrating the product of signal flux and the magnetoresistive sensitivity ($dR/d\phi$) determined by the bias level. However, the primary

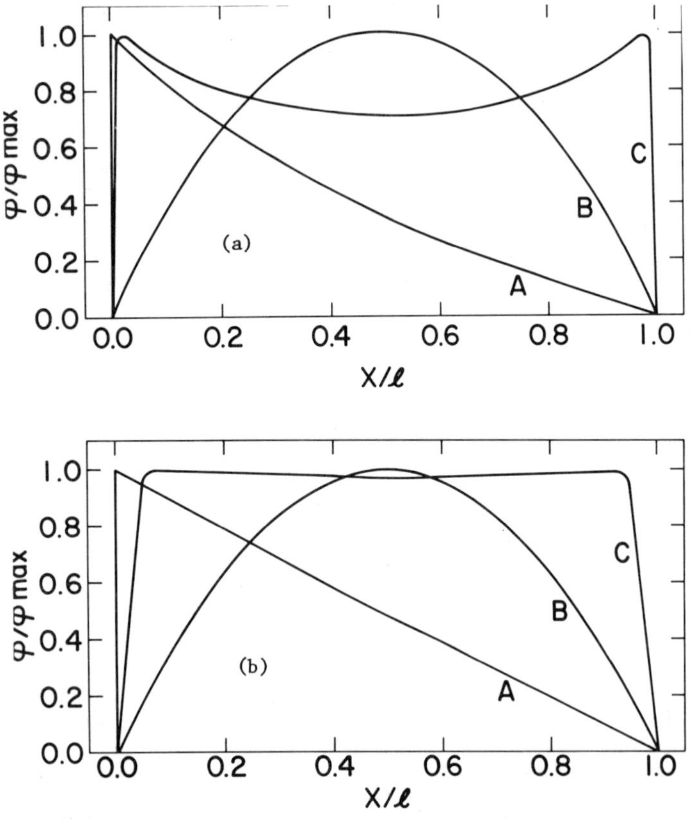

Figure 9. Flux distribution across the height of a shielded stripe. The signal flux A is injected at the edge (x = 0) by the recording medium. A uniform internally applied magnetic field produces a bias flux of shape B. A permanent magnet film produces an edge-injected bias flux of shape C. (a) Curves for ℓ/λ = 1.75, corresponding to he shielded stripe head of ℓ = 7 microns. (b) Curves for ℓ/λ = 0.5, corresponding to ℓ = 2 microns.

interest is at large signal levels, where the harmonic distortion arising from the incorrectly biased portion may be intolerable. Hence the permanent magnet bias is strongly preferred. The trident bias is intermediate between these two in bias uniformity, and the increased uniformity of signal flux is offset by an efficiency loss associated with reluctance of the front of the middle leg.

An important design consideration when using magnetoresistive transducers is saturation of the magnetoresistive element. In an inductive head, the output signal is proportional to the flux from the medium. At higher densities, where demagnetizing effects determine the transition width, one can obtain a proportional increase in readback signal by increasing H_c and the moment-thickness product $M_s\delta$ of the medium in the same ratio. This has lead to ever-increasing coercive forces in the medium, much to the consternation of the film write-head designers. In a magnetoresistive head, maximum output is obtained for a flux excursion of about ± 25% of saturation from the quiescent bias point (depending on tolerable distortion levels). For a 200 Å permalloy sensor, this is far less than the flux $2\mu_o M_R \delta$ emanating from a transition in any conventional medium. By decreasing the effective gap (g + t), it is possible to reduce the signal level in the magnetoresistive film to the desired level. However, an optimum system design would probably result in a lowering of H_c and $M_s\delta$, with resulting benefits for write head design. These considerations will have an important effect on media development, where higher H_c is presently a major goal.

OUTLOOK

Magnetoresistive read transducers have many advantages for high density magnetic recording. The output voltage depends on the power dissipation capabilities of the substrate, but we have had little trouble obtaining 40 mV peak-peak per millimeter of track width. This is more than a single turn inductive head can produce at any practical speed, and is especially advantageous at the speeds normally associated with flexible media (.01 to 3 meters per second). An important feature is that the read track width is determined only by the stripe dimensions. Thus, if the shielding outer legs are part of a write head, it can have a wider track width than the read element. The magnetoresistive element is basically simple to fabricate, and its reliability is well known from bubble sensing applications.

A significant advantage of shielded magnetoresistive heads is shielding from off-track flux sources. Thus the side resolution of a magnetoresistive head having shields wider than the track width is comparable to its linear resolution. Conventional inductive heads can have a significant coupling of long wavelength flux patterns from adjacent tracks.

Many problems remain to be solved. Some, like corrosion resistance and wear resistance, are part of the unglamorous materials problems associated with any new technology. In most cases, the same problems are shared by inductive film heads. The thermal transient problem[15,16] is peculiar to magnetoresistive heads. However, when flying over a smooth disk these noise spikes are not observed. We are confident that the highest areal densities of the future will be achieved using magnetoresistive transducers.

ACKNOWLEDGEMENTS

In a review article of this kind, it is impossible to acknowledge everyone who has contributed. All of the devices described were fabricated by a thin film memory fabrication group under L. T. Romankiw, including S. Krongelb, A. T. Pfeiffer, P. M. McCaffrey, B. J. Stoeber, and E. E. Castellani. Magnetic design and testing were performed by the author and C. H. Bajorek with the technical assistance of N. J. Mazzeo and A. J. Herrmann. E. W. Harden performed the sectioning for photography. Shielded stripe samples were evaluated on a disk test stand by R. I. Potter and colleagues. Their analysis[11,22] of the shielded magnetoresistive stripe has also strengthened the conclusions of earlier, less rigorous analyses. Many others have taken a lively interest in the magnetoresistive head, and I especially want to acknowledge the contributions of F. Shelledy, P. Simon, E. P. Valstyn, C. D. Cullum, A. F. Mayadas, I. M. Croll and S. Smith.

REFERENCES

1. E. P. Valstyn, Annals N.Y. Acad. Sci, 189, 21-51 (1972).
2. W. Chynoweth and W. Kayser, this conference.
3. R. P. Hunt, IEEE Trans. Mag., 7, 150-4 (1971). U.S. Patent 3,493,694.
4. S. Krongelb, J. Electron. Mat., 2, 227-38 (1973)
5. F. C. Williams, Jr., and E. N. Mitchell, Jap. JAP 7, 739-742 (1968).
6. K. Kuwahara, Trans. Jap. Inst. Met., 6, 192-3 (1965)
7. M. Prutton, Thin Ferromagnetic Films, Washington, Butterworths (1964) section 3.6.
8. R. L. Anderson, et al., AIP Conf. Proc., 10, 1445-9 (1972)
9. C. H. Bajorek and A. F. Mayadas, AIP Conf. Proc., 10, 212-6 (1972)
10. R. I. Potter, JAP 41, 1647-51 (1970)
11. R. I. Potter, IEEE Trans. Mag. , 10, 502-8 (1974)
12. D. A. Thompson, 792,917 Belgian patent
13. L. T. Romankiw, et al., 10, 828-31 (1974)
14. L. T. Romankiw and P. Simon, IEEE Trans. Mag., Jan-Feb. (1975)
15. F. W. Gorter, et al., IEEE Trans. Mag., 10, 899-902 (1974)
16. R. D. Hempstead, IBM J. Res. Dev., 18, 547-50 (1974)
17. C. H. Bajorek, et al, this Conference, U.S. Patent 3,840,898
18. R. I. Potter, et al., IEEE Trans. Mag., 7, 689-95 (1971)
19. C. H. Bajorek, et al., IBM J. Res. Dev., 18, 541-6, (1974).
20. S. Methfessel, et al., J. Phys. Soc. Jap. 17, supplement B-I, 607 (1962)
21. G. W. Brock and F. B. Shelledy, to be published. U.S. Patent 3,813,692.
22. R. W. Cole, et al., IBM J. Res. & Dev., 18, 551-5, (1974).
23. A. Paton and D. A. Thompson, to be published. Some of the concepts can be found in: A. Paton, JAP, 42, 5868-70 (1971).
24. E. M. Williams and J. B. Woodford, Jr., Transmission Circuits , New York, the MacMillan Co. (1957)
25. D. A. Thompson, IEEE Trans. Mag., 7, 674-5 (1971)

ELIMINATION OF DOMAIN WALL VELOCITY SATURATION AND LOCAL COERCIVITY VARIATIONS BY MULTI-LAYER GARNET MATERIALS

R.D.Henry, E.C.Whitcomb, and M.T.Elliott
Electronics Research Division, Rockwell International Corporation
Anaheim, California 92803

Experiments involving the growth of hard-bubble-suppressing 90° capping layers below, on top of, and surrounding the bubble film have been conducted. The investigation of the wall domain dynamics of the triple layer structure (suppression layer - bubble film - suppression layer) has shown that the velocity saturation associated with dynamic conversion is removed. Domain wall velocities of 12,000 cm/sec have been observed, presumably as a result of the large shape and stress anisotropy at the bubble surfaces preventing the formation of horizontal Bloch lines.

Further insights as to the origin of dynamic conversion are offerred by the data taken on structures involving the bubble film and a single 90° suppression layer. In these cases, the horizontal Bloch line model predicts dynamic behavior not dependent on the direction of the applied biasing field. However, experiments show that for both types of double-layer structures, velocity saturation is removed only when the biasing field is applied with the correct polarity. Application of the biasing field with the opposite polarity gives velocity saturation results identical to those observed with ion-implanted films. These results suggest either the horizontal Bloch line model is incorrect or that other complex wall structures are important in these layered films.

Multi-layered structures also offer improved static properties over ion-implanted films. A large variation in the coercivity has been seen over the surface of ion-implanted films. The origin of this local coercivity is unknown but may be related to the post-growth lead-rich film observed on the surface of garnet bubble films. The effects of this local coercivity become increasingly important as the device operation frequency tends toward the dynamic conversion point. No such local coercivity variations were observed in the triple-layer structures. The suppression layer also has been shown to improve the temperature stability properties of the stripwidth and the mobility, as well as to increase the wall energy.

THIN-FILM INDUCTIVE RECORDING HEADS

W. Chynoweth* and W. Kayser**
Honeywell Information Systems Inc.
*Oklahoma City, Okla. 73112;**Phoenix, Ariz. 85005

ABSTRACT

Thin-film magnetic recording head technologies have been investigated by a number of companies since about 1966 and it appears that now they may make the first team, because conventional magnetic heads, in spite of recent great advances, just cannot produce satisfactory data recovery at the desired high bit and track densities. In this paper,batch-fabricated, thin-film inductive transducers for magnetic disk recording are the major topic of discussion. The various proposed thin-film transducer structures are briefly reviewed and recent developments in the area of thin-film inductive transducers are discussed. Particular emphasis is placed on head-per-track applications and expected development trends are discussed briefly. It is concluded that thin-film heads in one form or another will become essential components in future magnetic recording systems.

INTRODUCTION

Magnetic recording systems which are in production today make use of recording heads which are made of ferrite or laminations of permalloy. These magnetic heads (or transducers) are difficult and costly to fabricate, involving many manual and precision, high-cost operations, and they are too bulky for many applications, for example, for application in high density head-per-track disk files.

Thin-film heads offer advantages over conventional heads because their magnetic structure can switch fast and because they can be batch fabricated by thin-film deposition and photolithographic techniques which in turn permits them to be made small with precise dimensions, tight spacings,and close tolerances, limited only by the above mentioned processing techniques. The fabrication process resembles thus the deposition and etching processes of integrated circuits and permits the fabrication of a large number of heads side-by-side at high density. An example for a wafer carrying many rows of tightly spaced transducers is shown in Figure 1.

Fig. 1. Transducer Wafer

Although conceptually known for more than a decade, these batch-fabricated thin-film heads (also called integrated heads) which promise considerable improvements in performance and cost over conventional transducers, have only in recent years received considerable attention in the literature [1,--,35]. However, they have still not surfaced in one of today's magnetic recording products. The reasons for the recently increasing interest in thin-film heads are that, 1) thin-film technology has made overwhelming progress and is today much more mature, and 2) in spite of recent great advances in conventional head technology, it is becoming increasingly difficult to fulfill the high-density recording systems needs with conventional transducer technology.

This paper will briefly discuss some of the proposed transducer structures. There have been several excellent reviews in the past [2, 3, 4], and another paper [5] will be published shortly. The details of design, fabrication and measurement of thin-film inductive transducers with emphasis on single turn devices, as applied to head-per-track recording systems will be discussed. Other developments, for example, magnetoresistive read-transducers will only be mentioned briefly, since a review of these will be given by D. A. Thompson.[6]

THIN-FILM RECORDING TRANSDUCERS

Magnetic recording transducers, batch-fabricated by vacuum deposition, electroplating and photofabrication techniques have been called thin-film or integrated transducers. Although thin-film counterparts of essentially all types of inductive and flux-sensing magnetic recording transducers could be made, only inductive write-read transducers and magnetoresistive read-only transducers have been reported on recently.

A multiturn thin-film inductive transducer was first proposed by Gregg in 1961, in a patent which was issued in 1967[7]. In 1961, production thin-film technology was in many respects in its infancy and, consequently, not much was done to implement such devices. Furthermore, with recording densities comparatively low at that time, conventional heads were adequate and there was no major reason or incentive to depart from the established, rather satisfactory, evolutionary path. Both factors have changed drastically today, which has led to the present activity in the field. Multiturn heads have been pursued by Lazzari and co-workers[8,9,10,11,12]. A number of papers[13,14,15,16,17,18,19] addressed inductive single-turn write/read heads which simplify the fabrication requirements by eliminating the multiturn winding. This was done at the expense of increased drive current requirements and reduced sense signals, the two main disadvantages of single-turn heads. Two basic structures have become known, the "Horizontal Head" [13,14,17,18,20] and the "Vertical Head" [9,15,19,21,22].

Because of the low signal level obtained from single-turn heads, attempts beyond the multiturn heads were made to improve the available output signal. Kaske et al [18] developed a horizontal head using anisotropic magnetic films with their Easy Axis parallel to the gap; an additional drive conductor is used to drive the head magnetization.

In the read process, the flux emerging from the transitions on the storage medium merely guides the magnetization as it relaxes back to the respective Easy Axis direction by which process the output signal is produced. The output signal is independent of medium surface speed; this head can thus be classified as a flux-sensitive head.

Another approach to solve the low signal level problem was suggested by Hunt [23] in 1971. He proposed the use of the magnetoresistive effect in an anisotropic thin-film of permalloy to implement a read transducer. He derived various parameters by means of a simple theory and gave some experimental results for tape operation of a simple magnetoresistive read head. Anderson et al [24], Bajorek and Mayadas[25] and Gorter et al [26] expanded on this work. Achievable signal level and signal-to-noise ratios are very good and better than on conventional or thin-film inductive heads, while fabrication is much simpler than for any of the previous heads. This is a most promising approach for read-head implementation. A disadvantage of magnetoresistive heads is that by definition they are just read-heads and must, therefore, be combined with another write structure to form a read/write head. Since a current through the device is needed to produce the sense signal, these devices are not passive read-heads, but are active devices. Magnetoresistive transducers are flux-sensitive, not flux-change sensitive devices.

All of the above work was aimed at an increase in signal level and a reduction in drive current level. As flying heights are reduced, and with better media available, even the single-turn drive current levels become more manageable and can be handled more easily with integrated circuits, which in turn leads to the required cost reduction per channel.

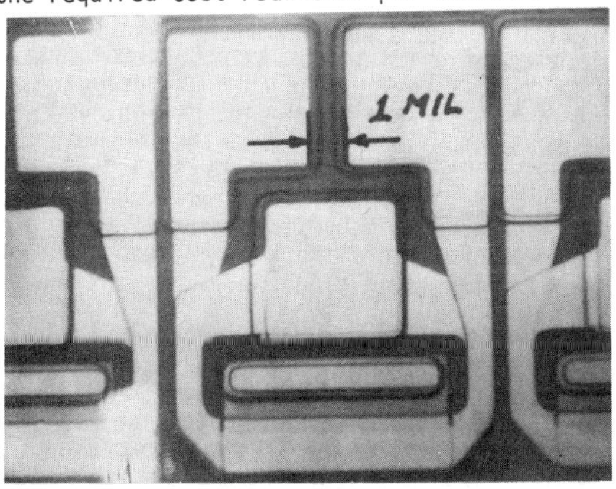

Fig. 2. Two-turn Transducer

INDUCTIVE TRANSDUCERS

Most of the proposed thin-film heads have been inductive read/write heads. Inductive heads can basically be grouped into single-turn and multi-turn devices of either the "Horizontal" or "Vertical" type. One of the early two-turn transducers made by General Electric/Honeywell is shown in Figure 2. This two-turn transducer provides a center tap, if desired, but the two turns can be fully utilized for "read" and "write". In this head, the turns are side-by-side in the gap, and as a consequence both turns are equally effective. A disadvantage is that the space in the gap is fixed by read requirements, and as a consequence the current density for the same ampere-turns is greater when

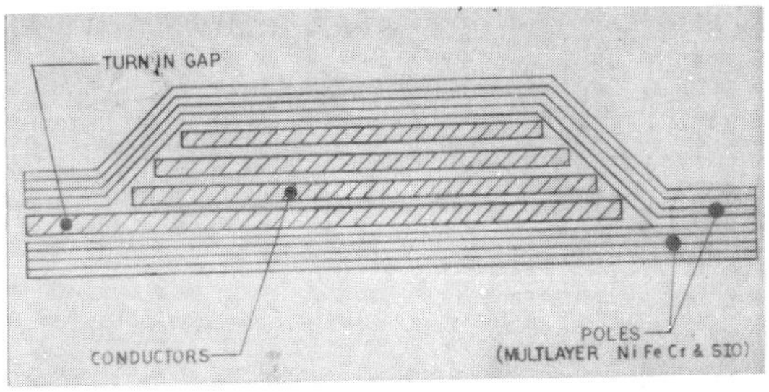

Fig. 3. Multi-turn Transducer (Lazzari)

two turns in the gap are used. The available cross-section of the metal conductor is reduced and the second turn is thermally isolated from the substrate through a second insulation layer; furthermore, the turn next to the substrate will already be at a higher temperature, when driven, thus representing a less-effective heat-transfer layer. An additional disadvantage of such a structure is the possibility of shorting of the two turns across the center insulation layer in the lapping process, as well as in actual operation. Letting only one conductor determine the gap and moving the other conductor away from the gap, as in Lazzari's [9, 10] multiturn heads (Figure 3) will eliminate the shorting but will reduce transducer efficiency. These particular heads in addition to being multi-turn devices have also multilayer magnetic poles, also formed by a series of vacuum deposition steps. These devices consist of a total of more than 30 layers deposited through appropriate masks. This particular approach to processing and the multi-turn design limits the realizable transducer dimensions and center-to-center spacings. In the limit, photolithographic techniques and single-turn devices can lead to smaller transducers on tighter center-to-center spacings. However, a remarkable density has been reached in Lazzari's devices, which according to Lazzari, have been successfully applied to magnetic disk and tape equipment. For the latter, the transducers were tailored to directly replace conventional tape heads. Objections to this fabrication approach include complex processing techniques which may result in lower yield, as well as geometrical restrictions imposed by the "deposition through a mask" process.

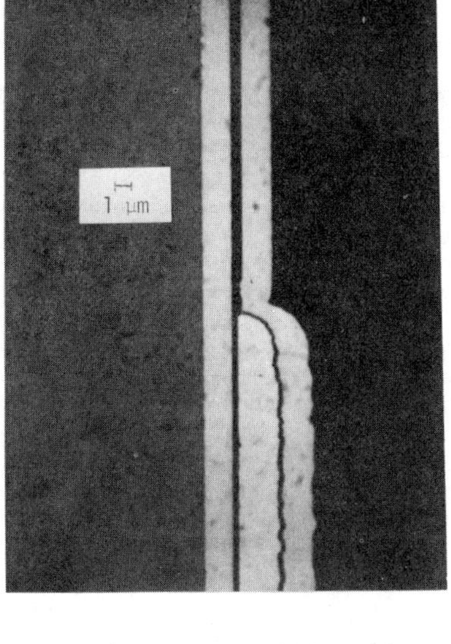

Fig. 4. Cross-section of Thin Film Head

The horizontal type of single turn inductive transducer requires the formation of a short gap in the order of one micron, probably by electron beam exposure or machining. This complication, the sensitivity of this transducer to abrasion, and the possibility of shorting the gap by magnetic debris have been factors in limiting the attention to the horizontal inductive head.

Because of the previously mentioned disadvantages of "Horizontal" heads and the complication of multiturn heads, most work has concentrated on "Vertical" single-turn heads. In the following, as an example, the particular approach followed by Honeywell in this area will be discussed. A cross-section through a Honeywell insulated single-turn head is shown in Figure 4. Insulation of the conductor from the magnetic poles is in principle not needed for the head to operate,however, the head efficiency would be reduced because current can now also flow through the magnetic poles. Since the resistivity of the magnetic material is an order of magnitude higher than that of the gold conductor and the thickness of the permalloy is less than that of the gold, only a small portion of the drive current would flow through the magnetic poles, and little degradation results.

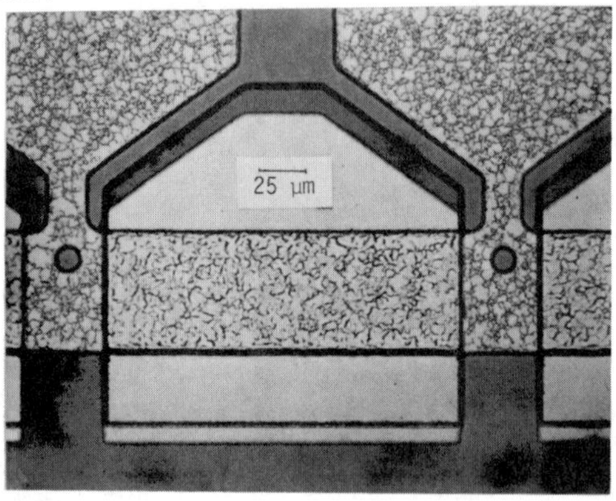

Fig. 5. Plan View of Thin Film Head

Figure 5 shows a plan view of the insulated single-turn inductive transducer. In this particular structure, the terminals of adjacent transducers have been merged into one (see also Figure 8). For head-per-track devices, this "shared terminal" construction permits the highest possible linear device density and provides for better area utilization of and better heat transfer in the intertransducer areas. It also results in a minimum number of connections to the chip; for n transducers, (n+1) connections are needed. For this structure, the associated drive and sense circuit requirements are somewhat more complex. In other configurations, the transducer leads may be carried out individually to their respective circuits (2n connections are needed) or one of each of the transducer leads may be tied to a common bus (n+1 connections). The point is, that once a structure has been decided on from a circuit point of view, part of the interconnection structure can be implemented on the transducer chip. Because of the high transducer density and the low transducer impedance, short, high density, low impedance strip leads are needed to connect the transducers to the circuits. These strip leads must be flexible since they mount on one end to the flying shoe and on the other end to a rigid circuit board. Figure 7 shows a picture of a flying head with the chip (carrying 32 transducers) and the strip leads attached to the chip. Figure 6 shows a single transducer as it appears when looking at the lapped surface of the flying head.

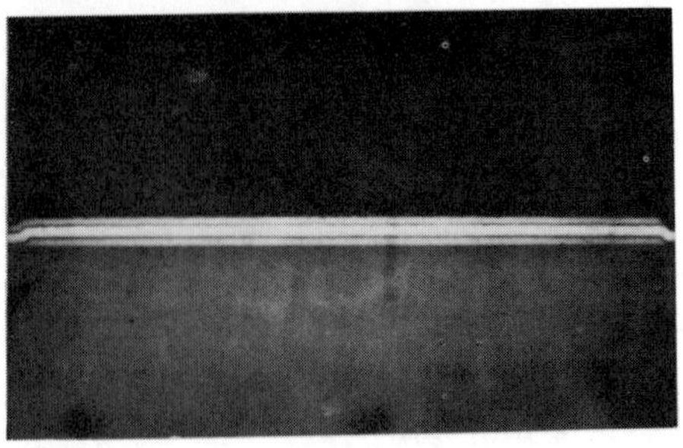

Fig. 6. Lapped Surface

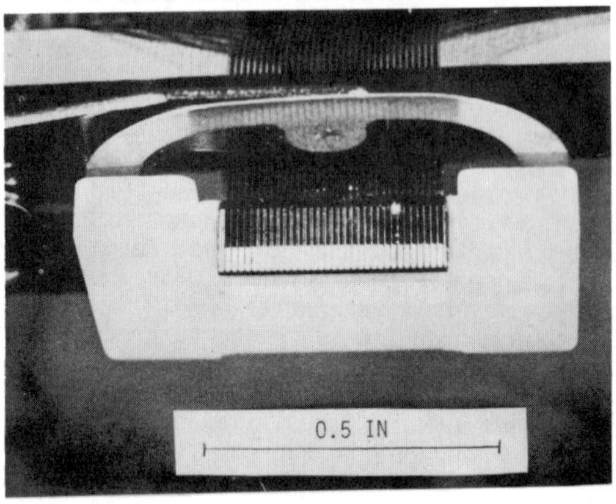

Fig. 7. 32-Track Thin Film Head

Flying heights in today's disk file products are in the order of 0.7 μm to 1.2 μm, depending on the specific design. For disk file application, the transducer chip is mounted to the flying head in such a way that the transducer poles will fly at the lowest point of the slider, closest to the disk. Having visual access to the chip is an advantage for precise mounting, lapping and inspection, and for making connections to the chip after the chip is mounted and lapped. Such a chip mounted on a slider of about 15 X 15 mm may carry up to 80 thin-film transducers side-by-side, whereby the transducers can cover every track of a group of tracks located under the flying head.

To increase the track density beyond 8 tracks per millimeter (200 TPI) and still populate every adjacent track becomes difficult, since the cross-section of the head connections on the chip, available to pass the required write current at some point becomes unacceptably small. Here again, the shared terminal transducer structure[21] is best. Interleaving for the head-per-track application would be required at densities higher than about ten tracks per millimeter (250 TPI). With write current requirements coming down, as flying height is reduced, minor track density extensions are possible. By spacing the transducers on the chip at one-half the track density, every other track can be served by a transducer from the same chip without need to move the flying head. In this way space is gained on the chip between the transducers to accommodate sufficient conductor cross-section to pass the required drive currents. With oxide media of the IBM 3336 type and flying heights of 1 μm a track density of 20 tracks per millimeter (500TPI) can be reached using a transducer density of 10 tracks per millimeter. Figure 9 shows an example for a transducer chip which carries 75 micron track width transducers on 167 micron centers. These transducers will perform adequately at 12 tracks per millimeter. To realize this density, transducer chips must be interleaved, either on the same flying head or on different flying heads or the heads must be made moveable into two positions. System requirements will dictate this choice. Interleaving is a problem because of the well-known static and dynamic mechanical alignment problems; batch-fabricated thin film heads significantly reduce some of these problems and permit the implementation of higher track densities.

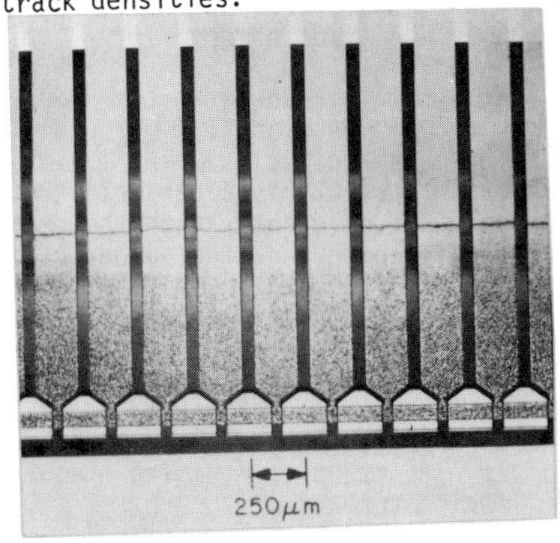

Fig. 8. 100 TPI Transducers

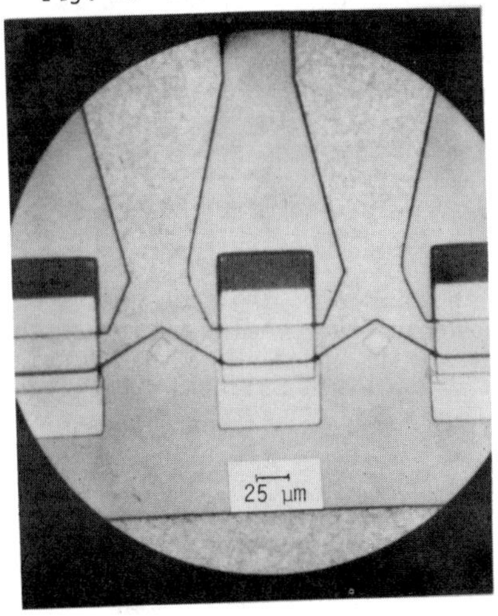

Fig. 9. 300 TPI Transducers

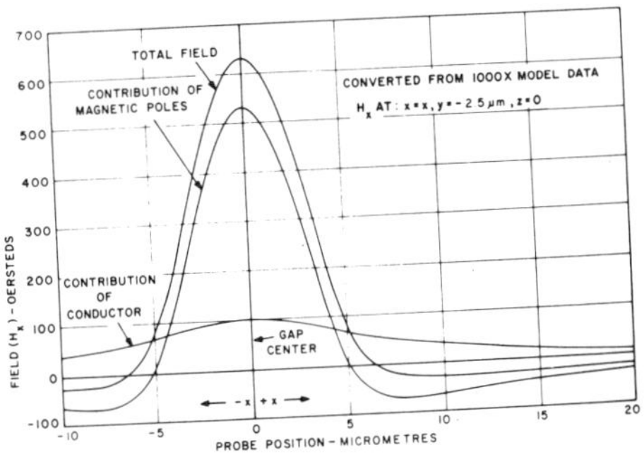

Fig. 10. Horizontal Field--Large Scale Analog

The small dimensions and the proportions of a typical vertical inductive thin-film head result in improved frequency response and isolated pulse performance. The small broadside dimension of the transducer results in low side-fringing and hence track definition is very good; this is important for high track density applications. The characteristic function [16,27,28], (longitudinal fringe field) of a vertical thin-film head has been reported in the literature. The magnitude of this function may go negative close to the main lobe; we have also verified this with large scale analog measurements (Figure 10). This characteristic function results in an improved resolution over conventional heads for the same gap length. In the frequency domain, it can be said that the head contour anomalies fall in the same region that the gap function is effective and this can result in an improved frequency response. This is fine for the usual digital recording applications, but results in a poor low frequency (long wavelength) response. Because of the turn-in-gap construction, there is also a direct contribution to both write and read by the gap conductor. The efficiency of the single turn head is high; however, high write currents are needed and the read voltage is low.

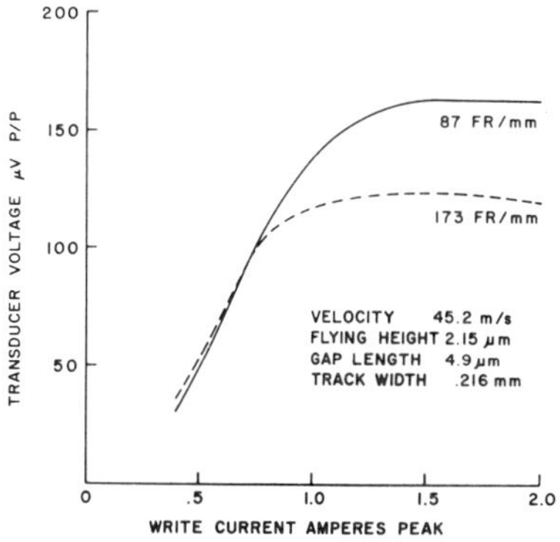

Fig. 11. Saturation Curves

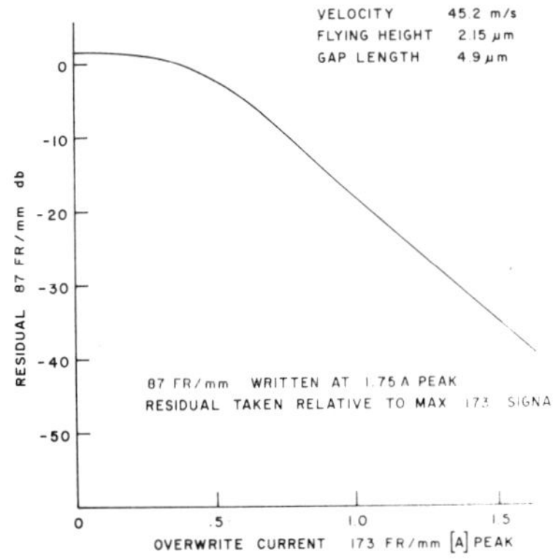

Fig. 12. Overwrite Performance

The read/write performance of a head/media combination is pretty well summed up in saturation curves, and a frequency response curve or isolated pulse. In Figure 11 is shown a pair of saturation curves. These curves do not drop off significantly above saturation, and as a consequence the write current can be chosen for good overwrite performance without much concern about resolution. For the head/media combination shown, an overwrite of -35 db can be achieved with an overwriting current of 1.5 amperes peak without degrading resolution (Figure 12). The read/write frequency response is shown in Figure 13. The resolution expressed as the ratio of the 174 flux reversals per millimeter (4400 FRI) to 87 flux reversals per millimeter (2200 FRI) outputs is 78% which compares to 50% for a 2314 ferrite head. This resolution superiority of the thin-film head is particularly striking when it is noted that the thin-film gap length is 5 microns compared to 2.5 microns for the 2314 head. The track definition of the thin-film transudcer is also superior. Lateral scan data has shown that the total side-fringing at the -30 db level on the lateral scan curve is less than 2.5 microns.

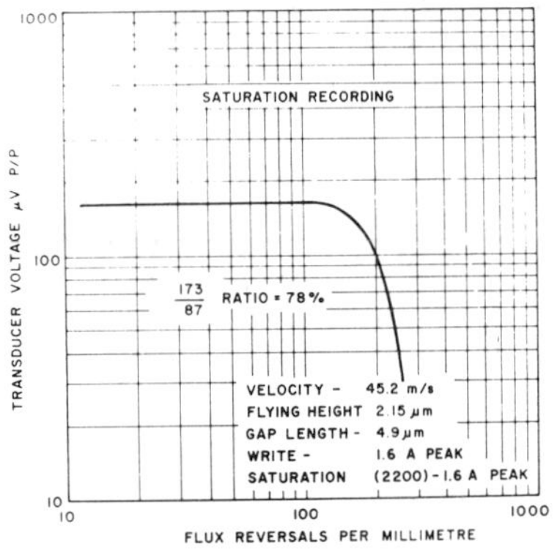

Fig. 13. Frequency Response

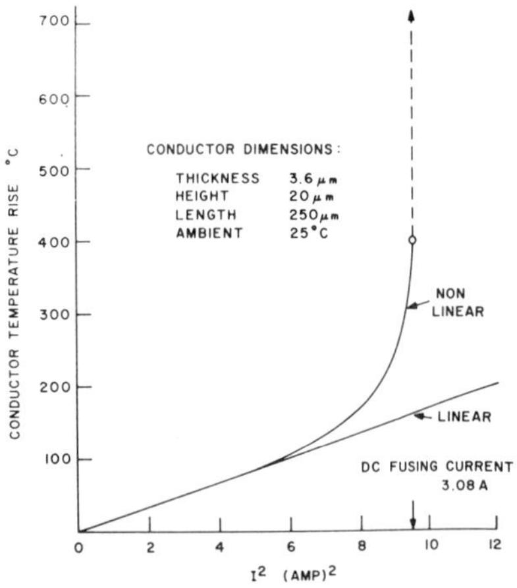

Fig. 14. Thermal Curve

Film transducers have a more serious thermal problem than conventional transducers. The current density in the conductor is around 3×10^6 A/cm^2 and the corresponding power density is 20 megawatts/cm^3. When the temperature dependence of the electrical and thermal conductivity is taken into account the curve shown in Figure 14 results. Direct measurements using the gold turn of the transducer as a resistance thermometer have shown that a temperature rise of 25-30°C above ambient is typical and that the thermal rise time is less than the 60 microseconds of the rise time of test current pulse. Fusing current for a 3.9 tracks per millimeter transducer having a 2.5 micron thick x 15 micron height gold conductor is 2.5 amperes. Calculated fusing currents have been verified experimentally.

TRANSDUCER AS A COMPONENT OF THE RECORDING SYSTEM

Single-turn thin-film transducers present a low impedance high-drive current, low-signal level component to the associated drive and sense circuits. For example, the Honeywell single-turn head including associated strip leads has an impedance of about 1 ohm at 2.5 MHz. At a flying height of 2 μm, the transducer requires more than 1 ampere of drive current to produce the required magnetic fields for saturation recording on conventional oxide media (coating thickness about 2.5 μm). From such a medium, the 197 FR/mm output signal obtained per millimeter of track width is about 380 μV P/P at a nominal flying height of 2 μm. Even for 216 micron wide tracks, such a transducer presents a challenge for the design of the associated sense circuits to obtain an adequate signal-to-noise ratio. Using integrated circuits, a special low-noise transistor input stage must be designed. Of primary concern is a sufficient reduction of the base resistance, R_b', of the input transistors, which is the major contributor to preamplifier noise.

As flying height is reduced[29], the required drive currents become less, and can more easily be handled with integrated circuits. The maximum frequency at which the IC driver can deliver the required current must be greater than the product of the media surface velocity and the number of flux reversals per inch. Reducing the flying height will permit the use of smaller transistors on the IC chip which reduces chip size and therefore cost and improves the high frequency response. Proper partitioning of the required drive and sense circuits and a reduction of the number of IC chips required is very important to minimize the overall cost per read/write track.

An alternate solution to direct IC drive and sense was used in an early Honeywell design[21]; a transformer was used with each transducer to reduce the drive current requirement and to increase the available signal level at the input stage of the preamplifier. This transformer, in addition to providing a "match" between transducer and circuits, was used as part of the head-selection system. For this particular case, the transformer reduced the drive current requirement from 1.5 ampere to 200 ma, and stepped up the signal level to more than 1 mV without introducing significant additional noise. Although a cost penalty must be paid using the transformer, it represents a viable solution to the problem of applying thin film inductive transducers. The use of a transformer, however, imposes constraints on the use of recording codes. Any code which requires processing of DC components, for example, the simple NRZ code, cannot be easily used. In other words, any code which permits an unbalanced write current sequence cannot be used with this approach. However, many of the commonly used codes, such as the "double frequency code" and the "phase modulation code" do not have this problem and are therefore usable with a transformer. The use of direct IC drive and sense circuits eliminates the transformer and the above-mentioned problems.

In a head-per-track disk file, a large number of tracks per surface must be covered. Partitioning results in flying heads carrying a large number of transducers, in order to reduce the cost per channel and optimize the use of the available space over the disk surface. Although many papers on thin-film heads have been published, Chynoweth, Jordan and Kayser [21] described the first working head-per track disk file, which incorporated 116 single-turn thin-film read/write transducers. Each of four flying heads per surface carried a chip with 29 transducers. A transformer per transducer was used to reduce the drive current level to 200 mA and to increase the sense signal to more than 1 mV. This same transformer was used as part of the selection system. The selection system is critical, since it must be capable of passing the required high- drive current levels and the very low level sense signals without noticeable degradation and its impact on system cost must be minimized in order to retain the cost advantage gained by the

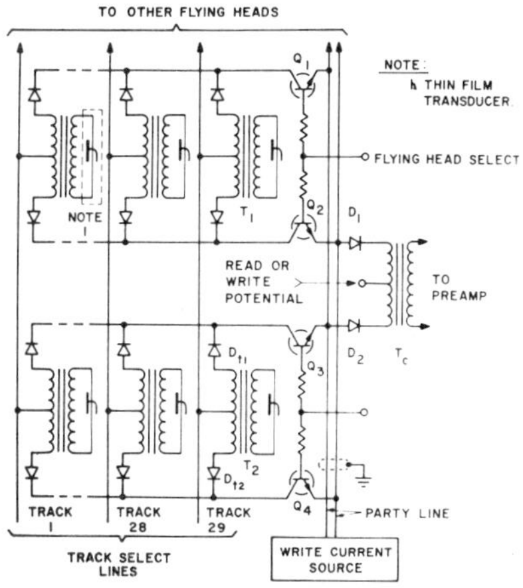

Fig. 15. Selection Circuit

Fig. 16. Feasibility Model

batch-fabricated thin-film heads. The selection system used was previously described[21] and is here shown in Figure 15. It permits the use of a single preamplifier with a transformer input for several flying heads in the read mode, and disconnects the preamplifier in the write mode. The "Track Select" (transducer selection) lines form the columns of the selection matrix, the "Flying Head" select lines form the rows of the matrix. The pair of write drivers serve as saturated switches during the read operation of a flying head. The transducer within a selected flying head is determined by the track select lines which are common to all heads. The selected track line is connected to a voltage source for write and a current source for read. Double frequency encoding was used in this system. Figure 16 shows a photo of the feasibility model which was operational with an exerciser in June 1970.[21]

The lower flying heights and improved media now in use make possible operation of single-turn heads without the transformer, using integrated circuit write/read chips.

SYSTEM PERFORMANCE

In Honeywell, an Engineering Model of a transducer per track 6-megabyte disk file was built using IBM-3336 type oxide disks. Each of the flying heads in this unit carried 32 transducers. The 32 transducers on a head were in four groups of eight for selection purposes. The track density was 3.9 tracks per millimeter (100 TPI) and the linear density was 196 flux reversals per millimeter (5100 FRI). Data recovery testing of this machine at digital level with an exerciser showed a data recovery performance of one error in 10^9 bits without error correction.

SUMMARY AND CONCLUSION

Thin-film transducer technology has been attacked on a broad front in the past several years with the goal to produce higher performance, lower cost digital recording systems for computer applications. For head-per-track devices, non-interleaved track densities of 7.8 tracks per millimeter (200 TPI) have been realized. Bit densities of up to 630 bits per millimeter (16,000 FRI) have been experimentally verified on individual transducers.[22] The first complete recording system with thin-film heads was described in 1973.[21] An Engineering Model, operating at 3.9 tracks per millimeter (100 TPI) and 196 flux reversals per millimeter (5100 FRI) demonstrated an error rate performance of 1 error in 10^9 bits without error correction.[19] A combination of inductive write transducers and magnetoresistive read-transducers, possibly in one physical device, shows promise for future advanced recording systems with densities approaching 790 bits per millimeter (20,000 BPI) and 39.4 tracks per millimeter (1000 TPI).

REFERENCES

1. S. Mitsuhashi, M. Ito, "New Multitrack Head for High Track Density", IEEE Trans. Magn. Vol MAG-10, No. 3, pp. 888-891, Sep 1974.
2. E. P. Valstyn, "Integrated Head Developments", Annals, N. Y. Acad. Sci, 18, 191, 1972.
3. A. J. Collins and Ivor Preece, "Factors Effecting the Design and Construction of Batch-Fabricated Recording Heads", Conf. Video and Data Recording, University of Birmingham, England, July 10-12, 1973.
4. J. P. Lazzari, "Integrated Magnetic Recording Heads: Review and Outlook", 1973 Conf. on Magnetism and Magnetic Material, Boston, Mass.
5. L. T. Romankiw, P. Simon, "Batch Fabrication of Thin Film Magnetic Recording Heads", A Literature Review and Process Description for Vertical Single-Turn Heads, to be published in IEEE Transactions on Magnetics, Jan-Feb Issue, 1975.
6. D. A. Thompson, "Magnetoresistive Recording Heads", 20th Conference on Magnetism and Magnetic Materials, San Francisco, December 3-6, 1974.
7. D. P. Gregg, "Deposited Film Transducing Apparatus and Method of Producing the Apparatus", US Patent 3,344,237 (1967 Sep 26), (filed April 19, 1961).
8. J. P. Lazzari, I. Melnik, "Integrated Magnetic Recording Heads", IEEE Trans. Magn., MAG-7, No. 1 146-150, Mar 1971.
9. J. P. Lazzari, I. Melnik, "Recording Integrated Magnetic Heads", IEEE Trans. Magn., MAG-6, 601-602, Sep 1970.
10. D. Augier, J. P. Lazzari, "Write Process Study on Integrated Magnetic Heads", IEEE Trans. Magn., MAG-7, 679-683, Sep 1971.
11. J. P. Lazzari, French Patent No. EN 69 36864.
12. J. P. Dejouhanet, J. P. Lazzari, "Theoretical Study of Integrated Magnetic Heads for High Recording Densities", IEEE Trans. Magn., Vol. MAG 10, No. 3, pp. 776-779, Sep 1974.

13. Y. Watanabe, S. Matsumoto, "Fabrication of Grouped Magnetic Heads", IEEE Trans. Magn., MAG-5, No. 3, p. 451, Sep 1969.
14. Y. Watanabe, S. Matsumoto, N. Yajima, "Fabrication of Grouped Magnetic Heads", IEEE Trans. Magn., Vol. MAG-5, No. 4, pp. 918-920, Dec 1969.
15. W. E. Tchon, D. S. Rodbell, "A New Magnetic Recording Head Fabricated from Coaxial Wire", IEEE Trans. Magn., MAG-6, pp. 593-597, Sep 1970.
16. R. I. Potter, R. J. Schmulian and K. Hartman, "Fringe Field and Readback Voltage Computations for Finite Pole Tip Length Recording Heads", IEEE Trans. Magn., MAG-7, 689-695, Sep 1971.
17. L. T. Romankiw, I. M. Croll and M. Hatzakis, "Batch-Fabricated Thin-Film Magnetic Recording Heads", IEEE Trans. Magn., MAG-6, 597-601, Sep 1970.
18. A. D. Kaske, P. E. Oberg, M. C. Paul and G. F. Sauter, "Vapor Deposited Thin-Film Recording Heads", IEEE Trans. Magn., MAG-7, 675-679, Sep 1971.
19. W. Chynoweth, J. Jordan, W. Kayser, "Small Thin-Film Transducers Point to Fast Dense Storage Systems", Electronics, Vol. 47, pp 122-127, July 25, 1974.
20. E. P. Valstyn, D. W. Kosy, "The Write Field of a Magnetic-Film Recording Head", IEEE Trans. Magn., MAG-5, 442-445, Sep 1969.
21. W. Chynoweth, J. Jordan, W. Kayser,"A Transducer-per-Track Recording System with Batch-Fabricated Magnetic Film Read/Write Transducers", Honeywell Computer Journal, Vol. 7, No. 2, pp. 103-117, 1973.
22. E. P. Valstyn and L. F. Shew, "Performance of Single-Turn Film Heads", IEEE Trans. Magn. Vol. MAG-9, pp. 317-322, Sep 1973.
23. R. P. Hunt,"A Magnetoresistive Readout Transducer", IEEE Trans. Magn., Vol. MAG-7, No. 1, pp. 150-154, March 1971.
24. R. L. Anderson, C. H. Bajorek, D. A. Thompson, "Numerical Analysis of a Magnetoresistive Transducer for Magnetic Recording Applications", AIP Conf. Proc. Magnetism and Magn. Materials 10, pp. 1445-1449, 1972.
25. C. H. Bajorek and A. F. Mayadas, "Reliability of Magnetoresistive Buble Sensors", AIP Conf. Proc. Magnetism and Magn. Mat. 10, p. 212, 1972.
26. F. W. Gorter, J. A. L. Potgiesser, D. L. A. Tjaden, "Magnetoresistive Reading of Information", IEEE Trans. Magn., MAG-10, No. 3, pp. 899-902, Sep 1974.
27. L. W. Brownlow, C. C. King, "Write Field Analysis for Integrated Heads of the Finite Pole-Tip Configuration", IEEE Trans. Magn., MAG-8, No. 3, 539-541, Sep 1972.
28. J. C. Mallison, "On Recording Head Field Theory", IEEE Trans. Magn., Vol. MAG-10, No. 3, pp. 773-775, Sep 1974.
29. F. E. Talke, R. C. Tseng, "An Experimental Investigation of the Effect of Submicron Transducer Spacings on the read-Back Signal", IBM Research Reprot RJ 1432, Aug. 16, 1974.
30. W. E. Proebster, "Magnetic Thin Film Transducer", US Patent 3,271,751 (1966 Sep), (filed Dec. 3, 1962).
31. National Cash Register Company, "Magnetic Transducer", UK Patent 1, 169, 869 (1969 Nov), (filed Sep 4, 1968).
32. R. I. Potter, "Digital Magnetic Recording Theory", IEEE Trans. Magn., Vol. MAG-10, No. 3, pp. 502-508, Sep 1974.
33. D. A. Thompson, "Exchange Eddy Current Effects in Thick Magnetic Films", IEEE Trans. Magn., MAG-7, 674-675, Sep 1971.
34. A. Paton, "Analysis of the Efficiency of Thin-Film Magnetic Recording Heads", JAP, Vol. 42, No. 13, pp. 5868-5870, Dec. 1971.
35. J. P. Lazzari, "Integrated Magnetic Recording Heads Applications", IEEE Trans. Magn., Vol. MAG-9, pp. 322-326, Sep 1973.

ADVANCES IN RECORDING TAPE MEDIA

B. Gustard
MPM Division, Pfizer Inc., Easton, Pa. 18042

ABSTRACT

This paper reviews new magnetic materials that are now in use in recording tape. The new high coercive force gamma iron oxides did not give any substantial improvement in audio recording characteristics down to 0.5 mil wavelength. New gamma iron oxides exhibiting much improved particle orientation continue to be developed in the U.S.A. and Europe. These materials offer the greatest improvement in recording properties.

Some success has been achieved in stabilizing the magnetic properties in magnetite and cobalt-doped magnetites by the addition of small percentages of zinc. Another attractive method has been to coat iron oxide particles with a metallic layer such as cobalt, which produces a high coercive force particle with temperature stability.

Our own empirical measurements continue to point to the importance of a high remanent magnetization and a narrow switching field distribution on a magnetic tape in order to achieve improved recording performance.

LORENTZ ELECTRON MICROSCOPY STUDIES OF BUBBLE DOMAINS

P.J. Grundy, G.A. Jones and R.S. Tebble
Dept.of Physics, University of Salford,
Salford M5 4WT, England

ABSTRACT

Lorentz electron microscopy is a technique which permits the imaging of magnetic structures in thin films. Its principal advantages are (i) the high resolution attainable which enables the study of bubbles well into the sub-micron region and (ii) the information it provides about the spin distribution in the domain walls. The latter point means that images can be obtained of discontinuites in magnetization i.e. vertical Bloch lines.

This paper describes the basic principles and practice of the Lorentz method and reviews some of its applications in bubble domain work e.g. the effects of temperature, pulsed and direct fields, and ion implantation. The conclusion contains a critical evaluation of Lorentz microscopy with regard to its usefulness and limitations in bubble studies.

1. INTRODUCTION

Lorentz microscopy is a special branch of electron microscopy in which images of magnetic domains and domain walls are obtained. The theory and practice of Lorentz microscopy is similar to the perhaps more familiar phase contrast microscopy of optics and this aspect has been discussed at length, e.g. see Wade[1]. At its most accessible level therefore the technique provides a means of directly observing domain walls in a wide range of specimen materials.

The first detailed reports of Lorentz microscopy were those of Fuller and Hale[2] and Boersch et al[3] who successfully imaged domain walls in thin films with a conventional transmission microscope in an out-of-focus mode. Since that time the scope of the technique has widened considerably to cover many new and sophisticated applications. The advantages of Lorentz microscopy which are responsible for this spread of interest may be summarised as follows (i) the very high resolution obtainable, (ii) the ability to distinguish the relative orientations of electron spins within domains or walls (iii) the ability to correlate physical and crystallographic detail with the magnetic structure (iv) the ease with which certain dynamic experiments may be carried out. Unfortunately there are concomitant disadvantages the most serious being that of the severe restriction placed upon specimens suitable for examination: they must be thin - of the order of 1000 A - in order to allow transmission of the electron beam. This limitation may not be too troublesome in the case of evaporated films of Permalloy or electropolished foils but is far more so in the case of garnets and other oxide material. Here special thinning procedures must be resorted to which are often lengthy and frequently unsuccessful and it is for this reason that so much study has been made of the hcp phase of cobalt. Despite preparation difficulties the advantages enumerated above make the investigation of bubble domain materials with the electron microscope very rewarding.

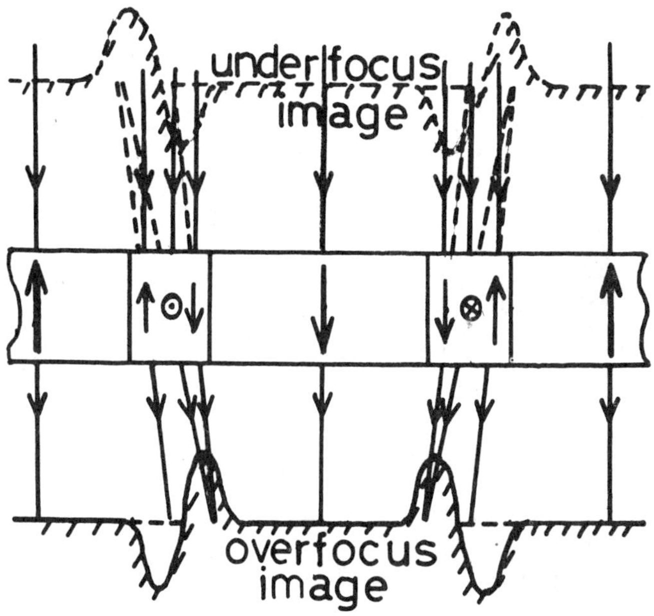

Fig 1. A schematic representation of the contrast expected in the images of strip and bubble domain walls in a thin uniaxial film.

2 PRINCIPLES OF LORENTZ MICROSCOPY

The method of imaging magnetic structure by transmission electron microscopy depends on the scattering of the incident electron radiation by the magnetic induction in the specimen. This is described classically as a deflection due to the Lorentz force and wave mechanically as a scattering or diffraction of the electron wave by the magnetic flux.

Fig 1 represents a uniaxial thin film specimen subdivided into domains which are magnetized perpendicular to the specimen surface. The electron beam is incident normally on the specimen and parallel or antiparallel to the domain magnetization. In this orientation the electrons interact with the magnetization only in the domain walls separating the domains (here two winding Bloch walls are chosen) where there is a component of magnetization perpendicular to the beam and are deflected through an angle ψ given by

$$\psi = \frac{4\pi e t}{mv} \cdot M_s \cos\theta = \psi_o \cos\theta. \quad (1)$$

ψ is derived from the Lorentz force under the approximation of small deflections and is a maximum ψ_o at the centre of the Bloch wall. With M_s=1430 Gauss for cobalt, t=10^{-5} cm and an accelerating voltage of 100 kV, v=1.63×10^{10} cm sec^{-1} giving ψ_o = 1.6×10^{-4} rad.

In general it is not possible to observe scattering of this strength, which is orders of magnitude less than the aperture of the objective lens of the electron microscope (typically 10^{-2} rad), in a conventional mode of operation. The most convenient method of observation is to defocus the objective lens, in analogy to the equivalent phase contrast mode of optical microscopy, so that the object plane for the lens is some distance z above or below the specimen. It can be seen in Fig 1 that the intensity distributions obtained in these planes should contain asymmetric "images" of the domain walls which change contrast from over to underfocus. Fig 2 shows images of domain walls separating strip domains in a thin magnetoplumbite ($PbFe_{12}O_{19}$) crystal. The micrograph was taken in overfocus and it can be seen that adjacent walls can be in a winding or unwinding relationship. The typical "black-white" (or vice-versa) nature of the Fresnel contrast at the domain walls is clear. It must be realised that the overlap or "underlap" between the scattered wavefront or beam at the walls and the undeviated fraction through the domains is $\sim \psi z$ and does not necessarily bear any real relationship to the real width of the wall.

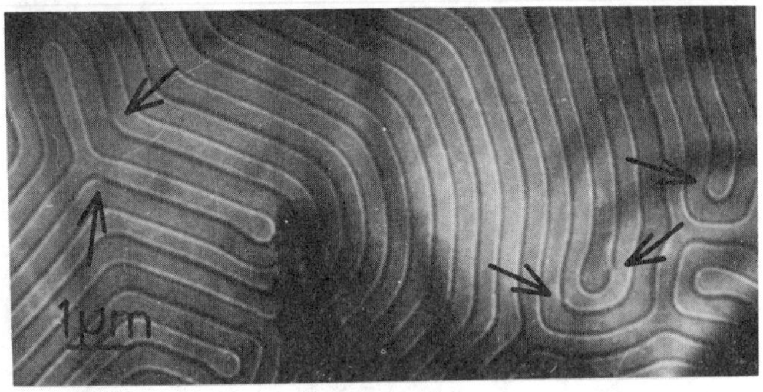

Fig 2. Out-of-focus Fresnel images of strip domain walls in a thin $PbFe_{12}O_{19}$ crystal. Bloch lines are marked by arrows . (Courtesy P. Dunk)

The details of the intensity distribution in the wall image can be related to the magnetization distribution in the domain wall by calculating the intensity expected in the Fresnel interference pattern from a trial function f(x), say,(here in one dimension) describing the rotation of the magnetization through the domain wall. A "fit" of the Fresnel integral is then made to the experimental intensity distribution. The function f(x) determines the phase difference ϕ between the interfering wavefronts as

$$\phi = \frac{e}{h}\,\Delta\Phi = \frac{4\pi Met}{h} \cdot \int_{wall} f(x)\,dx \,. \qquad (2)$$

$\Delta\Phi$ is the change in flux across the domain wall or magnetic feature explored. An example of the resulting interference fringes obtained at a domain wall in iron is shown in Fig 3. Investigations of domain wall structures have been carried out in this way with some success and notable references are the work of Hothersall[4] and Wohlleben.[5]

Strictly speaking the details of the intensity distribution should always be analysed in terms of the wave theory, as outlined above. However in some situations, e.g. near to focus and with convergent illumination, the geometrical description, as given in Fig 1, is adequate. These situations occur when the magnetic scattering is relatively strong and $\phi \gtrsim \pi$ or $\Delta\Phi \gtrsim h/2e$. For weak scattering at small flux changes the wave theory is essential and any geometrical interpretation of an image will probably be in error. The flux change $\Delta\Phi = h/2e$ has been termed the "fluxon" as it represents a limiting accuracy to which the magnetization distribution in any specimen can be determined[2] This is because each Fresnel fringe represents one "fluxon"[5] and provides a concept equivalent to that of resolution in the microscopy of spatial detail. The interpretation of Bloch line images is considered in this context in a later section.

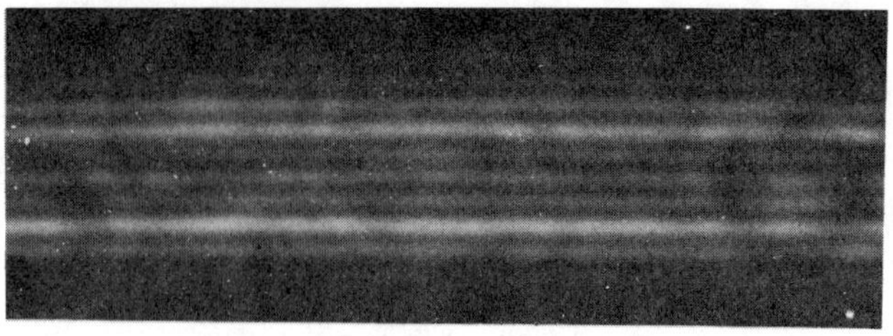

Fig 3. Fresnel interference fringes at a domain wall image in a thin crystal of iron (courtesy D.C. Hothersall)

An alternative method of observation is the displaced aperture method in which part of the objective aperture is obscured. This is an analogue of the Foucault method in optical microscopy and effectively cuts out that part of the magnetic Fraunhofer diffraction pattern produced by areas of the specimen magnetized in a specific direction leaving those arising from areas magnetized in different directions. Although useful in certain applications the displaced aperture method has not been of significant use in quantitative work.

3 EXPERIMENTAL RESULTS

In view of space restrictions the authors have made a choice of techniques and experiments which they consider to illustrate best the advantages of Lorentz electron microscopy in relation to bubble domain studies.

3.1 DOMAIN STRUCTURES AT REMANENCE

The virgin state of all uniaxial materials of suitable thickness and orientation consists of strip domains: an example has already been given, Fig 2 , of such a structure in $PbFe_{12}O_{19}$. There is an essential relationship between strip domains and the formation of bubble domains. The zero field strip width w is itself of some interest because it is easy to measure and goes some way to providing a value of the characteristic material parameter.[6] For high magnetization materials such as cobalt or even the hexaferrites, simple strip domains, as opposed to more complex surface domain structures, are only observed in very thin films which therefore make ideal specimens for electron microscopy. Moreover since the strip width is proportional to $(t/M_s)^{\frac{1}{2}}$ where t is the film thickness, the values of w obtained are small thus demanding a high resolution measurement technique. For example in basal cobalt crystals of thickness 1000 Å the zero field strip width is $\sim$ 1000 Å.

The principles of imaging have already been discussed in relation to the intensity distribution at domain walls. At certain well defined localities the sense of spin rotation across the wall reverses, an occurrence which is identified by an abrupt change in image contrast (see Fig 2). These reversals have been termed vertical Bloch lines[7] by analogy with the type of short range magnetization reversal seen in cross-tie walls. The incidence of Bloch lines may be increased by the use of chopped or pulsed fields.[8] The term Bloch line is somewhat of a misnomer since it infers that the magnetization reverses in a plane normal to the film. In practice the distribution is likely to be two-dimensional with a fairly complicated structure.[9]

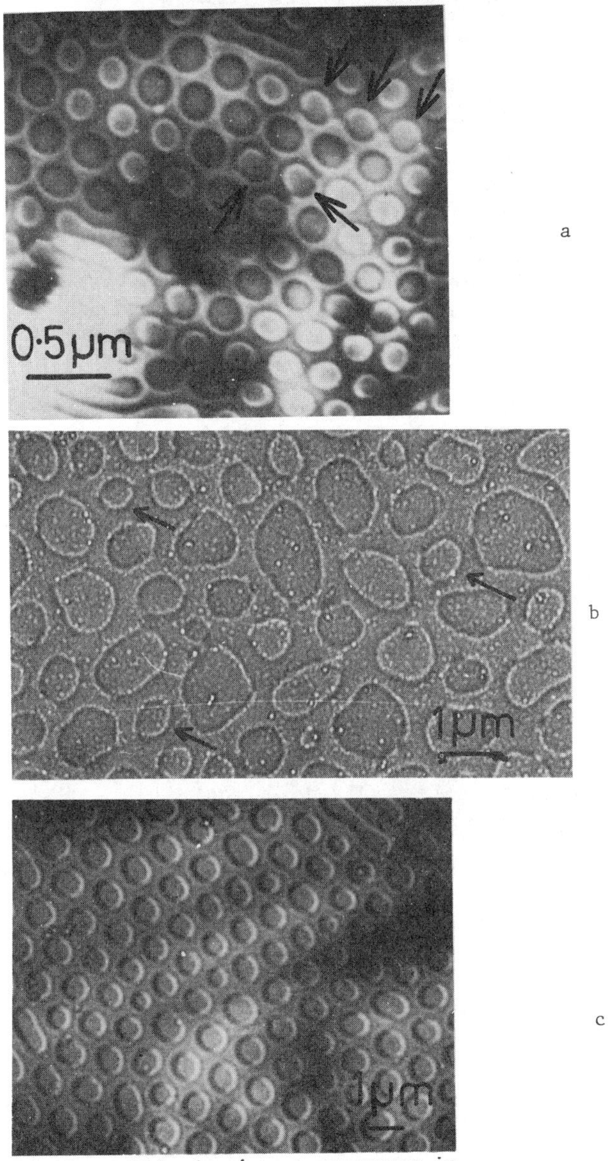

Fig 4. Bubble domains at remanence in (a) cobalt (b) GdCoAu (courtesy S.R. Herd) (c) $PbFe_{12}O_{19}$ (courtesy P. Dunk). Type 2 bubbles are marked by arrows.

Despite what might appear to be unfavourable energy considerations bubble domains may be retained at remanence in all uniaxial materials regardless of Q value $(K/2\pi M_s^2)$ although the conditions required to produce them do depend upon Q. Thus for $Q < 1$, e.g. cobalt (0.3) and cobalt-chromium alloys, a mixture of bubble and strip domains can be produced by first applying and then removing a saturating field normal to the specimen plane. For materials with $Q > 1$ e.g. garnets and hexaferrites, rafts of bubble domains can usually be produced if the saturating field is applied and removed in the plane of the film. Remanent domain structures containing bubbles are shown for cobalt , sputtered GdCoAu and magnetoplumbite in Fig 4. These micrographs reveal features which first stimulated the use of Lorentz microscopy in bubble studies. From Fig 4a it is evident that the bubbles can be classified into two types [9] depending upon the surrounding wall: (1) those in which the sense of rotation of the spins across the wall is conserved: (2) those containing two Bloch lines. As far as type (1) bubbles are concerned the magnetization may still rotate in either of two senses; thus bubbles of this type have a black circle surrounding an inner white one or vice versa. Notwithstanding the distorted shape of some of the bubbles in Fig 4b close examination shows that they also may be classified entirely into the same two categories. Fig 4c contains a raft consisting of type (2) bubbles only. It is clear that Bloch lines can exist only in pairs within a bubble wall although up to the present the maximum number of lines observed in an undistorted bubble of any electron microscope specimen - of whatever material - is two. A great deal of interest has been aroused in Bloch lines as a possible source of hardness in dynamic bubble behaviour. Various models for the Bloch line structure, both vertical[9] and horizontal,[10] have been proposed and the question as to how far electron microscopy can help elucidate these structures is discussed below.

3.2 CHARACTERIZATION OF BLOCH LINES

As mentioned previously, Lorentz microscopy has been used in several instances to investigate the magnetization distribution in domain walls in thin films. An obvious extension to these exercises is to attempt the determination of the spin distribution in vertical Bloch lines and the "handedness" of the line.[11] This characterization would,in principle, enable direct information on the interaction and disposition of vertical Bloch lines in bubbles to be obtained. It is probably true to say that a detailed analysis of the magnetization distribution is not required and all that is needed is the ability to distinguish between the four well known postulated alternatives of Fig 5 (in fact two of the lines are just the result of rotating the other two by π). This is fortunate because Bloch lines in all bubble materials of interest will be "weak" magnetic phase objects as the flux change across the distribution, $\Delta\Phi \simeq 4\pi M_s A$ (where A is the cross section of the B.L.)., is of the same order as the "fluxon limit" of 4×10^{-7} G cm^2. This means in practice that a detailed description of the spin distribution in the line by comparing the electron image with that calculated for a model distribution will be difficult. It would appear from a study of the images of Bloch lines that there are differences; whether these differences can be predicted is a question yet to be answered. Experimentally the task is difficult because a sufficient "amount" of scattering from the "weak"

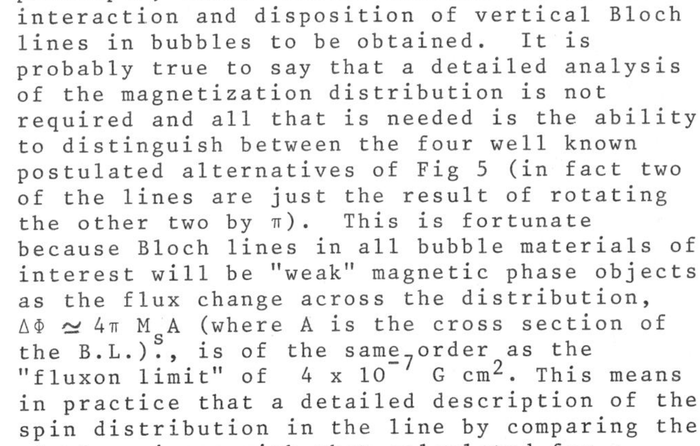

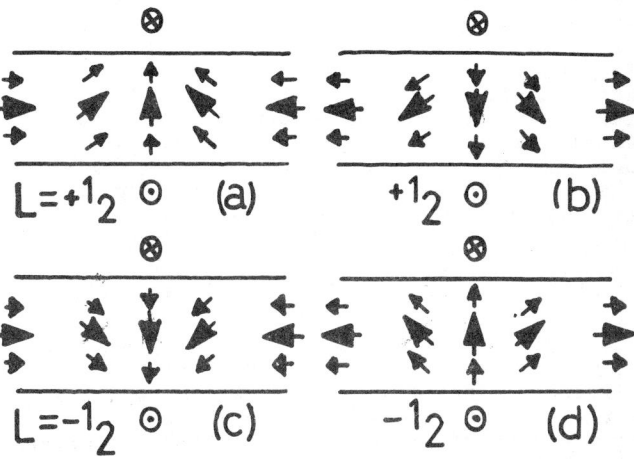

Fig 5. Showing the possible spin distributions in some Bloch lines.

structure must be recorded in a reasonable exposure time and theoretically a reasonable model must be chosen from which to compute the image (direct inversion of the experimental fringe pattern to give the spin distribution would be unreasonable in this case as the structure is a complex two dimensional distribution).

Attempts are in being to calculate the geometrical and wave-optical images expected from the magnetization distribution proposed for the Bloch line by Grundy et al[9]. This applies essentially to a material with large anisotropy and no "twisting" [12] in the domain wall. The method is essentially that developed by e.g. Wohlleben[5] and Hothersall[4] with a rather complex set of phase functions ϕ (equation 2) in the domain wall , Bloch line and surrounding domains.

3.3 BUBBLES IN STATIC FIELDS

One important feature of the electron microscope is that the vertical field associated with the magnetic objective lens can be used directly as a bias field. For a 100 kV instrument the maximum field available in this way is about 5-6 KOe whereas a value of perhaps 20 KOe might be achieved for a voltage of 1 MV. These very large fields are especially useful for high magnetization materials particularly cobalt for which $4\pi M_s = 18$ KOe. Other incidental advantages of high voltage microscopy are (i) thicker specimens may be examined and (ii) inelastic scattering is relatively reduced thus giving better images. Variable bias fields can be obtained by moving the specimen into or out of the lens field.[13] An alternative procedure possible on some microscopes e.g. the Philips 301 is to keep the specimen fixed, vary the objective lens current and focus with the 'diffraction' lens. The latter technique was used by Herd and Chaudhari[14] to form bubbles in thin films of sputtered GdCo.

Obviously the facility of investigating specimens under bias field conditions extends the scope of the Lorentz method because magnetization processes, including the creation and destruction of bubbles, can be studied. It is thus possible to demonstrate the pinching off of successive bubbles from a parent strip domain. Moreover if the microscope is suitably calibrated collapse fields can be measured and data of this nature has been collected from hcp CoCr alloys.[15] In another experiment of this genre the collapse fields of type (1) and type(2) bubbles in $PbFe_{12}O_{19}$ were shown to be identical.[16] This result is hardly surprising as the hardness of a bubble is expected to become significant only at a much higher packing density of lines. Also shown by applied field experiments is the fact that if a saturating bias field is reduced from a material with Q $<$ 1 then bubbles are renucleated whereas for Q $>$ 1 strips are nucleated. In cobalt therefore it is possible to modulate the bias field about saturation and continually destroy and renucleate bubbles[17]. There is some evidence that nucleation sites are associated with particular defect features on the specimen surface.

It has been known for several years that films of Permalloy evaporated at oblique incidence exhibit a species of 'strong stripe' structure in the remanent state. The stripe domains appear best in films exceeding 2000 Å in thickness and the system is therefore highly suited to high voltage microscope study.

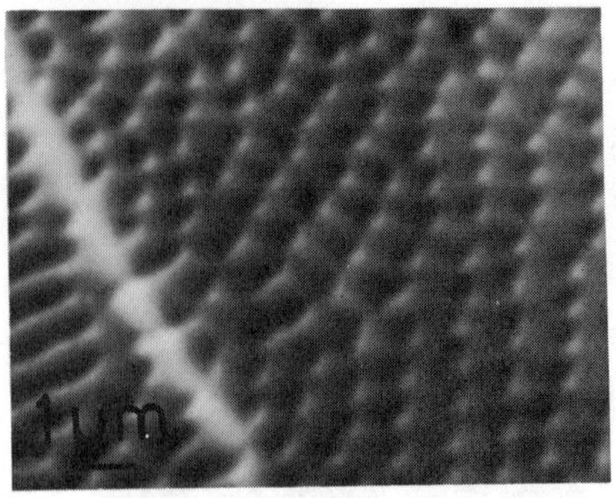

Fig 6. Bubble array in a Permalloy film of thickness 0.6 μm in a bias field of 1.8 kOe (Courtesy I.B. Puchalska)

The fact that bubble type domain structures could be obtained in such films was demonstrated by Puchalska and Jones[18] and Puchalska [19] using Lorentz microscopy, see Fig. 6. The application of planar fields also produces bubble domains. There is a difference between the bubble-like domains observed in Permalloy and those seen in more orthodox materials; they are less distinct, more elongated and tend to form in square arrays. These features may be attributed to the complex anisotropic nature of the films which contain a planar easy axis normal to the stripes and an out-of-plane anisotropy associated with a columnar growth structure. As a result the magnetization in the bubble domains is not normal to the film plane, an effect likely to complicate their dynamical behaviour. On the other hand the accumulated knowledge of Permalloy makes its investigation as a potential device material worthwhile.

3.4 PULSED FIELD EXPERIMENTS

So far in this paper we have considered in-situ experiments on bubble domains in the electron microscope using slowly varying magnetic fields. There have been several detailed investigations reported in the literature of the effect of pulsed fields in the "chopping" and propagation of bubbles in garnet epi-layers, e.g.[8]. The rapid expansion of bubbles followed by a rapid contraction and possibly "chopping" of the resulting, distorted domain has been proposed as a method of creating hard bubbles. These bubbles exhibit anomalous behaviour in collapse and propagation experiments. Preliminary attempts to observe the response of bubble domains to pulsed fields directly in the electron microscope have given some interesting results. Unfortunately the materials so far studied have relatively large values of $4\pi M_s$ ($\sim$ 4000 G) and require pulse fields of several hundreds of Oersteds superimposed on bias fields of 2500-3000 Oe. An extension of the observations to thinned garnet layers and substituted ferrites, and deposited Gd-Go films should prove more rewarding. The experiment to be described here can be considered as a first approximation to the admirable optical equivalent reported by Zimmer, Morris and Humphrey,[20] with the advantage that vertical Bloch lines can be seen and not just inferred.

Fig. 7 shows two consecutive micrographs in a sequence taken of a magnetoplumbite crystal from the work of Grundy and Herd.[21] The experiments were carried out in a Philips EM200, the objective lens provided the bias field and a small 50 turn coil (internal diameter 0.5mm) placed directly over the thin area of the

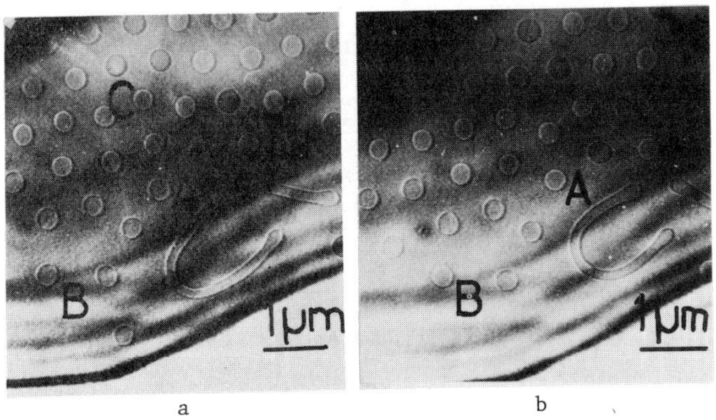

a b

Fig 7 Bubble domains in $PbFe_{12}O_{19}$ in a bias field of 2.5 kOe. The structure in (b) is the result of a 500 Oe 0.1 µs pulse on that in (a) . Bubbles A and B are converted from type 2 to type 1 and bubble C has collapsed.

specimen provided the pulsed component. The original paper details the difficulties of this kind of experiment on sub-micron bubbles in tapered, thin foils . The micrographs show, among other effects, small displacements of the bubbles within their lattice after the pulse, the collapse of one bubble and the possible change of state of two bubbles from the loss or gain of vertical Bloch lines. In other results changes of chirality between pulses was also observed. This state, S, has been defined as $S = 1 + \sum_i L_i$ where $L = \pm\frac{1}{2}$ is the 'handedness" or sense of the line.[11] The gain or loss of two Bloch lines changes S by 1,0 or -1 depending on the sense of the lines. These direct observations show that bubbles, at least in the materials studied and also, presumbaly, in the more 'conventional' bubble materials, can lose or gain vertical Bloch line when subjected to pulsed fields or field gradients. This phenomenon could be a hazard in high speed bubble domain devices where bubbles are displaced rapidly in straight lines or round corners, as in a register. The loss or gain of Bloch lines would also change the direction of propagation in devices utilising the fact that bubbles containing different numbers of lines propagate at different directions to a field gradient. Further experiments of the kind described here are called for with more suitable specimen materials, as mentioned previously, and attempts at a stroboscopic modification of the microscope wouldbe of interest.

3.5 HEATING EXPERIMENTS

This section, apart from presenting some unusual visual results, is meant to exemplify further the versatility of the electron microscope with regard to the performing of in-situ experiments.[22] Fig 8a shows a particular remanent state of a cobalt crystal at room temperature. On heating the specimen to 180°C (Fig 8b), a remarkable change of contrast occurs in that the bubbles develop a spot at their centre, either black or white. This behaviour is explicable when it is recalled that at a temperature just above 20°C the K_1 anisotropy constant of cobalt diminishes with the result that for a basally oriented foil the magnetization starts to leave the c-axis and tip towards the foil plane.

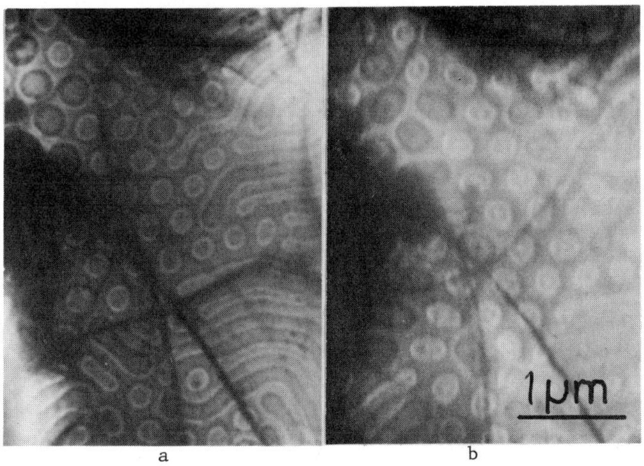

a b

Fig 8 Remanent array of bubbles in cobalt at (a) room temperature (b) 180°C

Consequent upon this spin reorganization, the Lorentz forces are modified with the result shown in Fig 8b. The central region of the bubble marked by the spot is similar to a circle Blochline as seen in a cross-tie wall. This region is some 100 Å across, which again testifies to the power of the technique. Although illustrated here for the case of cobalt similar behaviour might occur in other uniaxial materials.

3.6 ION IMPLANTATION

Ion implantation of garnet films has been used successfully as a means of reducing the incidence of hard bubbles. The suppression of these bubbles is due to a magnetostrictive mechanism which involves the formation of a planar layer of magnetization on the surface of the specimen. Domain structures have been revealed in the planar layer using Ferrofluid [23] but not as yet studied with the electron microscope. However Lorentz microscopy has been used[24] to observe directly the effect of implanting the bubble domain structure of cobalt with 40 KeV H^+ ions at doses between 10^{11} and 10^{16} cm^{-2}. The results are interesting, although not easy to explain as the simple magnetostrictive theory valid in garnets must for various reasons be discounted. At even quite low doses (10^{11} ions cm^{-2}) the Bloch lines may be stripped from type (2) bubbles thus converting them to type (1). The conversion process continues with increasing dosage until after 10^{16} ions cm^{-2} virtually all the bubbles are now Bloch line-free. It is important to note that the implanation does not affect the domain structure in any other way, in particular bubbles are not converted to strip domains. One final interesting fact is that if, subsequent to any dose level $> 10^{14}$, the specimen is saturated along the c-axis and returned to remanence, the bubbles produced are always type (2). Moreover their Bloch lines are immune to further implantation procedures.

4. DISCUSSION AND CONCLUSION

Many of the results described above represent extensions to optical measurements obtained by using the electron microscope as a 'super optical' microscope. Whereas previously domains were observed with a resolution scarcely better than 0.5 μm, Lorentz microscopy has detected bubbles as small as 500 Å. This extension of resolution should not be underestimated because with time there may be an increasing interest in smaller bubble sizes.

The combination of high resolution and the charged nature of the electron probe gives characteristic details about the spin distribution in domain walls, the most obvious example being that of the Bloch line. The existence of such lines in a domain structure is easily detected as well as the observation of their behaviour and motion under the influence of direct and pulsed fields. Unfortunately the data so far accumulated does not yet provide a definite link between this type of Bloch line and bubble hardness, primarily because materials with a sufficiently high value of Q to support close packed multi-line bubbles have not been examined due to preparation difficulties. It should also be remembered that under normal viewing conditions horizontal Bloch lines which also have been postulated as contributing to hardness will give zero contrast. Application of wave optical and the more simple geometric theories of magnetic phase contrast will certainly yield in the near future more detail about the spin distribution of vertical Bloch lines, in particular their handedness. Inspection of Bloch line images reveals asymmetry in contrast which is probably attributable to handedness. This type of spin structure study has been successfully applied to domain walls in the past.

In conclusion one might summarise by stating that although Lorentz microscopy has produced much interesting and perhaps unique information relative to the bubble problem, it has not realized its full potential. This failure is almost certainly due to the difficulties imposed by making thin specimens of garnet films suitable for transmission microscopy However this probably constitutes only a temporary stumbling block and could change the situation in the near future. With the provision of cine techniques and strip lines the link between hardness and bubble structure as well as other information should be directly open to experimental verification.

ACKNOWLEDGEMENTS

The authors wish to thank the Science Research Council for the provision of financial support, the Cobalt Information Centre, Brussels, and the Office of Scientific Research (AFSC) USAF under grant AFOSR 71-2102. They also wish to express their gratitude to those who have lent micrographs for this paper.

REFERENCES

1. R.H. Wade Adv in optical and electron microscopy 5, 239-296, 1973
2. H.W. Fuller and M.E. Hale J.Appl. Phys. 31, 238-248, 1960
3. H. Boersch, H. Hamisch, D. Wohlleben and K. Grohmann Z. für Phys. 159 397-404 1960
4. D.C. Hothersall, Phil Mag 24 241-258, 1971
5. D. Wohlleben, J. Appl. Phys. 38 3341-3352, 1967
6. J.A. Cape and G.W. Lehman J. Appl. Phys. 42, 5732-5747, 1971
7. P.J. Grundy, R.S. Tebble and D.C. Hothersall J. Phys. D 174-177, 1971
8. J.C. Slonczewski, A.P. Malozemoff and O. Voegeli AIP Conf Proc 10, 458-477, 1973
9. P.J. Grundy, D.C. Hothersall, G.A. Jones, B.K. Middleton and R.S. Tebble, Phys. Stat. Sol. (a) 9, 79-88, 1972
10. E. Schlömann AIP Conf Proc 10, 478-482, 1973
11. O. Voegeli and B.A. Calhoun IEEE. Trans. Magn. VOL MAG 9, 617-621, 1973
12. E. Schlömann, Appl. Phys. Lett., 21, 227-229, 1972
13. G.A. Jones, P.J. Grundy and M.J. Goringe J. Microscopy 97 147-152, 1973
14. S.R. Herd and P. Chaudhari, Phys. Stat. Sol. (a) 18, 603- 1973
15. D.C. Finbow and G.A. Jones, Phys. Stat. Sol. (a) 20, K91-94, 1973
16. P. Dunk and G.A. Jones, AIP Conf. Proc., 18, 162-166, 1974
17. P.J. Grundy and G.A. Jones, AIP Conf. Proc. 10, 369-368, 1973
18. I.B. Puchalska and G.A. Jones, J. Phys. D, L52-54, 1973
19. I.B. Puchalska, Phys. Stat. Sol. (a) 22, 647-658, 1979
20. G.J. Zimmer, T.M. Morris and F.B. Humphrey, IEEE Trans. Magn. (to be published)
21. P.J. Grundy and S.R. Herd, Phys. Stat. Sol. (a) 20, 295-307, 1973
22. P.J. Grundy, D.C. Hothersall, G.A. Jones, B.K. Middleton and R.S. Tebble, AIP Conf. Proc., 5, 155-159, 1972
23. R. Wolfe and J.C. North, Appl. Phys. Lett., 25, 122-124, 1974
24. F.A.J. Brown, G.A. Jones and W.A. Grant, Int. J. Mag. 5, 371-373, 1974

SOME ASPECTS OF PERMALLOY CHEVRON STRIP DETECTOR*

S. Yoshizawa, M. Kasai, M. Hiroshima and N. Saito
Central Research Laboratory, Hitachi, Ltd., Kokubunji, Tokyo, 185, Japan.

ABSTRACT

The response of a thick-permalloy chevron-strip detector (287 columns) on YSm garnet (6.5μ stripe width) has been examined for the cases of dc and pulse current drives in high detector current region.

Relations between a detector window Vw (defined as a voltage difference between the ONE and ZERO states) and a detector current Id are shown in Fig. 1. These data were obtained from a 100kHz module operation of 16k-bit major-minor organized chips by changing the ratio between a pulse width tw and a pulse period tp.

In all cases, saturation of Vw was observed. The maximum Vw for tw/tp=0.5/20 μs became 5 times higher than that for dc. Also, it was observed that a replicator acted as a nucleation type generator when the dc current of 8mA was applied. Present experiments have shown that these observed phenomena can be explained by considering the temperature rise of the garnet film heated by the detector.

* This work is contracted with the agency of Industrial Science and Technology, Ministry of International Trade and Industry, as a part of the National Research and Development Program "Pattern Information Processing System".

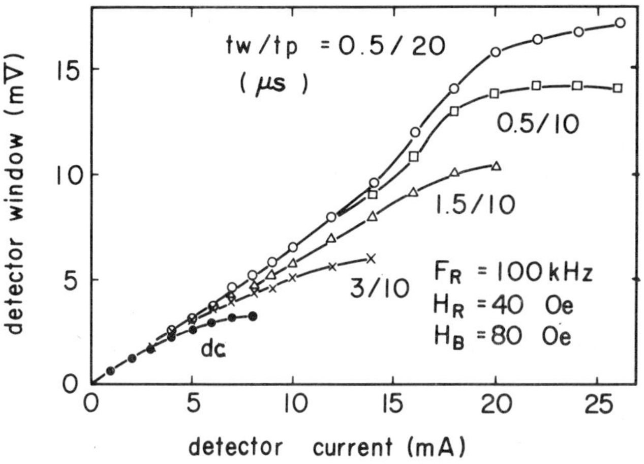

Fig. 1. Pulse drive response of detector.

ON MAGNETIC PATH SATURATION IN THIN-FILM HEADS

Kanji Kawakami, Eiji Kaneko, and Kunio Ono
Hitachi Research Laboratory of Hitachi Ltd.
Hitachi-shi, Ibaragi-ken 319-12, Japan

ABSTRACT

The distribution of magnetic field in the head gap and of flux density in the magnetic path of thin-film recording heads has been studied. It is shown that magnetic path saturation occurs first at the back end of the heads, which reduces efficiency. The relations which calculate the minimum current amplitude at magnetic path saturation are obtained. Results of the study are applied to thin-film heads having two-turn coils.

AN INTEGRATED MAGNETORESISTIVE READ, INDUCTIVE WRITE HIGH DENSITY RECORDING HEAD

C. H. Bajorek, S. Krongelb, L. T. Romankiw and D. A. Thompson
IBM, Thomas J. Watson Research Center, Yorktown Heights, New York 10598

ABSTRACT

The design, fabrication and performance of an experimental thin film recording head is described which has 25 μm and 125 μm track width magnetoresistive (MR) read elements combined with a single turn thin film inductive write head. The MR read elements are shielded to provide high linear resolution with an equivalent read gap of 1/2 μm and are internally biased by means o a permanent magnet film in the gap. The best head performance with a 0.5 μm air bearing on a high performanc disk was: write current of 400 mA peak, isolated pulse readback signal of 40 mV peak per mm of track width at 10 mA sense current, with a 6 db density of 440 flux changes per mm (11 kfci).

INTRODUCTION

A substantial increase in storage density in magnetic recording systems is likely to require transducers with dimensions smaller than can be easily achieved with conventional recording heads. One alternative has been the thin film inductive head.[1]

A readback transducer using a magnetoresistive (MR) element is a thin film structure, is planar and can be integrated with a thin film inductive write head. The MR transducer has a significant track width advantage over inductive heads, since it can produce signals far larger than those of equivalent heads.[2,3]

We have therefore developed an experimental recording head consisting of a one turn inductive write head built directly on top of a shielded MR head. Novel features in the integrated transducer are the internal biasing of the magnetoresistor about its linear operating point by means of a permanent magnet film in the gap, the use of one magnetic layer for both a write pole tip and magnetoresistive shielding member, and a stepped conductor to improve the write head efficiency.

DESIGN CRITERIA

A schematic crossection of the integrated head is shown in Figure 1. It consists of a one turn vertical thin film inductive head directly on top of a vertical shielded MR head. The center shield is shared by both transducers. Several design requirements for both transducers have been described before.[1-8] We discuss only the design features novel to this structure: Biasing the magnetoresistive element and provision of a stepped write conductor. The latter allows for a narrow write gap g_w with lower write current density and power dissipation.

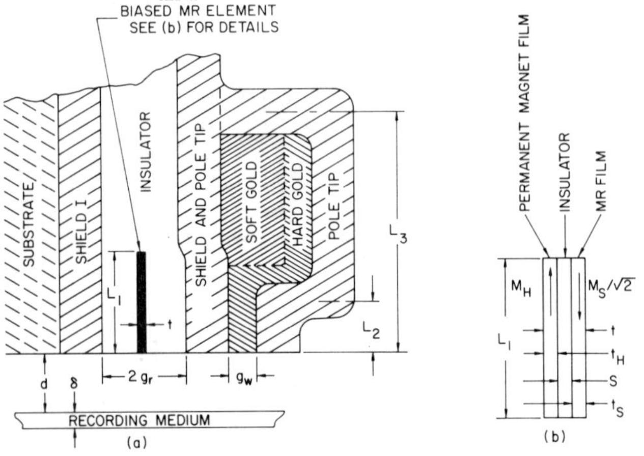

Fig. 1. Schematic Crossections of: (a) Integrated Head and (b) Biased MR film sandwich ($L_{1,2,3} >> g_{w,r}$).

The desirability of biasing an MR element, to maximize its sensitivity and linearize its response, has been described by Hunt[4] and others[5,6,8] for unshielded MR heads. For any high resolution application magnetic shields are required on both sides or the MR element; however these will also shield any fields originating from sources external to the heads. Thus a bias field source is required inside the shield structure.

We chose to implement internal biasing by using a permanent magnetic film.[3] Such a film in close flux coupling relationship ($t << L_1$, g_r) with a highly permeable (Permalloy MR) film will transfer most of its flux into the latter. In the simplest case where the track width (not shown in Fig. 1) is far larger than L_1 and the permanent magnet film is saturated transverse to the stripe, proper biasing of the magnetoresistor requires that

$$t_H M_H \simeq t_s M_s/\sqrt{2}, \quad (1)$$

where t_H and t_s and M_H and M_s represent the permanent magnet film and Permalloy film thickness and saturation magnetization respectively. Satisfying Eq. 1 insures that the Permalloy film is magnetized at 45° to the stripe direction, corresponding to the inflection point of its magnetoresistance vs. vertical flux response.

The composite sensor and bias film stripe can be defined by one photolithographic step, a processing advantage. The bias field is independent of changes in L_1, making the biasing condition rather insensitive to wear. The bias may be adjusted by providing excess hard film thickness and magnetizing it at an angle off the vertical. In the case of devices with small width to height aspect ratio (track width $\simeq L_1$) magnetizing the hard film at 45° to the vertical head direction can provide both vertical and horizontal bias fields and prevent the closure domains responsible for Barkhausen noise.[2,6]

The success of this biasing scheme depends on several other considerations such as electrical insulation, exchange coupling and preventing degradation of the coercivity of the Permalloy sense film as detailed in Refs. 3 and 9. Consideration of the field strengths expected from typical media as well as the stray field from the adjacent write head requires that the coercivity of the permanent magnet film be well in excess of 4×10^4 A/m (500 Oe) to avoid its demagnetization. The MR element could also be directly integrated into the gap of the write head. However, the coercivity of the permanent film would have to be well in excess of 2×10^5 A/m (2500 Oe) to prevent its demagnetization during writing. A feature of this read structure is that the bias-sensor film sandwich is considerably thinner (1,400 Å) than the other film thicknesses, resulting in a minimal step in the interhead region, and providing an adequate surface for subsequent fabrication of the inductive write yoke.

Write head design is dominated, not by the linear efficiency arguments of Paton,[7] but by more stringent requirements due to head saturation during writing. The design is suitable for fabrication over thickness ratios of at least a factor of two thicker or thinner, and for various media and flying heights, but is centered on 2 μm pole tips, a 1 μm gap, and a write field requirement of about 10^5 A/m (1900 Oe) in the medium. This is because twice H_c is required at the highest density to overcome the demagnetizing field and switching threshold of a perfectly square loop medium. For particulate media in the 30,000 to 50,000 A/m (400 to 600 Oe) range, with ordinary squareness ratios, this leads to the stated write field corresponding to a field in the internal gap of about $2.4\ 10^5$ A/m (3000 Oe).

Copper and pure gold are good conductors, but too soft for satisfactory lapping and wear. The hard gold used for the gap has about the same resistivity as Permalloy. Hence, nearly all the current in a simple head of uniform crossection would flow in the Permalloy, where its average efficiency is less than 50%. The thick soft gold conductor in the head provides a more magnetically efficient current flow. Saturation near the back of the yoke then limits the allowable pole tip length L_2 to 5μm for the required mmf at the pole tips of 240 mA-turns at a current of about 400 mA.

FABRICATION

The fabrication process uses sputter etching to define the MR stripe, electroplating through photoresist masks to form the conductors, and electroplating followed by chemical etching to make the Permalloy shields. Details and critical aspects of the individual process steps have already been given[10,11].

Shield 1 is a 1 to 2 μm layer of electroplated Permalloy. The plating base consists of 100 Å of Ti followed by 1000 Å of Permalloy evaporated on a thermally oxidized Si substrate. As an alternative, a polished ferrite slab may serve as both substrate and shield. The biasing and MR layers are deposited on a sputtered 2000 Å SiO_2 film which forms the first half of the gap. The permanent magnet film is a composite of 250 Å of α-Fe_2O_3 exchange coupled to a layer of 150 to 200 Å of Permalloy resulting in a combined coercivity greater than 180 Oe. Single layer films with even higher coercivities will be reported elsewhere.[9] The α-Fe_2O_3 film is formed first either by oxidizing an evaporated 150 Å Fe film in air (at 300°C for 15 minutes) or by reactively sputtering the Fe (at 200°C) in the presence of (10 μ) O_2. The exchange coupled Permalloy film, a 1000 Å separating layer of Schott glass and a 200 Å Permalloy MR film are then sequentially evaporated at 250°C.

The blanket MR film serves as a base for electroplating 2500 Å Au leads through a photoresist mask[10]. Sputter etching is then used to remove the unwanted deposits down to the SiO_2 as described in Ref. 10 with a photoresist mask designed to protect the gold leads as well as the sensor. The chemically non-selective sputter etch process is particularly advantageous for etching the multilayer structure.

A second layer of 3500 Å of sputtered SiO_2 completes the read gap resulting in $2\ g_r$ = 7500 Å. The second SiO_2 layer is etched to expose the MR leads for their full length to within about 0.1 mm of the sensor stripe. Thus subsequent metallizations used in fabricating the write head will deposit on the leads to increase their conductance. The openings in the SiO_2 are narrower than the leads so that some SiO_2 overlaps their edges as protection from subsequent etching steps.

The 2 μm shared shield is deposited over the read structure in the same manner as the first shield. The gold write conductor is now electroplated through a photoresist mask in a two stage process. The resist is first opened only in the throat and 2 μm of high conductivity gold (Selrex BDT-510) are plated. The gap and throat are both open during the second stage so that this plating defines the gap thickness of 1 μm and provides an additional 1 μm of conductor in the throat. A hard gold (Aurall 214, Knoop hardness 250 to 300) is used in the gap; the stepped structure compensates for the low conductivity inherent in this hard gold. If a positive resist is used for the first stage and plating done under safe-light conditions, the second plate-thru mask can be obtained by simply reexposing the first resist through a second photo mask. The outer shield is completed by electroplating 2 μm of Permalloy over the entire structure, followed by a 2 hour, 200°C stabilization treatment.

The shields are completed by spray etching in ferric chloride[11]. The resist mask protects both the shield and the leads so that the plated Permalloy will remain as part of the leads. The gold conductors must not be exposed during etching, since electrochemical couples would cause severe undercut and uncontrollable etching. An oversize etch mask ensures that the Permalloy in the final structure overlaps the gold on either side and encapsulates it. Etching is not critical since the front of the head will be defined by lapping after the head is mounted. The completed structure is shown in Figs. 2 and 3.

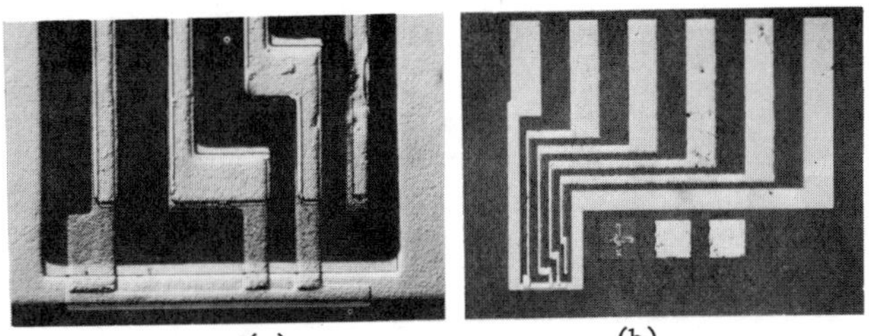

(a) (b)

Fig. 2. (a) Transducer area with 25 and 175 μm read track widths under a 375 μm wide write head and (b) overall planar view with contact pads.

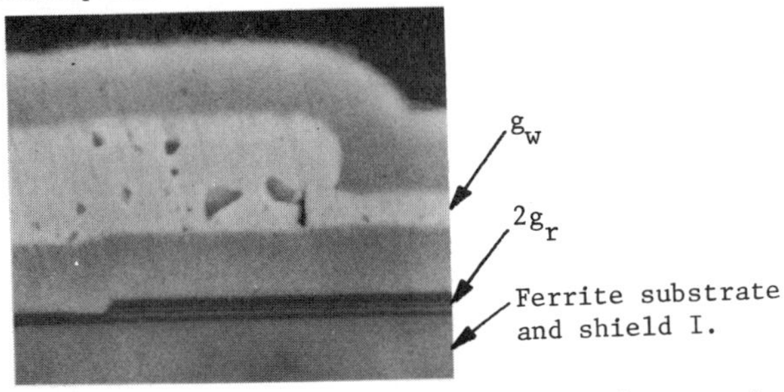

Fig. 3. SEM micrograph of lapped read/write yoke crossection (3000 X).

After dicing the processed substrate into individual elements, a layer of glass was epoxy bonded over the active portion of the head and the chip was then mounted on the end of an air bearing slider. The lower surface of the chip-slider assembly was lapped to a distance L_1 of 7 μm below the top edge of the MR stripe, with corresponding L_2 of about 5 μm and L_3 of 40 μm.

EXPERIMENTAL RESULTS

The transducer was tested with a conventional high performance particulate disk with a coercive force of approximately 480 Oe. The write current required was 400 mA peak and an isolated pulse readback signal of approximately 40 mV peak per mm of track width with a 10 mA sense current was achieved. The 6db resolution density was 440 flux changes per mm (11 kfci) at a 0.5 μm air bearing. Additional resolution data and a representative isolated transition pulse are found in Ref.2.

ACKNOWLEDGEMENTS

We gratefully acknowledge the technical assistance of N. Mazzeo, R. Anderson, D. Johnson, P. McCaffrey, E. Castellani, E. Harden, B. Stoeber and A. Pfeiffer.

REFERENCES

1. E. P. Valstyn, Advances in Magnetic Recording, Annals N.Y. Acad. Sci. 189, 21-51 (1972).
2. R. I. Potter, IEEE Trans. Mag., MAG-10, 502 (1974).
3. C. H. Bajorek, et al., U.S. Patent 3,840,898.
4. R. P. Hunt, IEEE Trans. Mag., MAG-7, 150 (1971); U.S. Patent 3,493,694.
5. R. L. Anderson,et al., AIP Conf. Proc.,10,1445(1973).
6. C. H. Bajorek,et al., IBM J. Res. Dev.,18,541,(1974).
7. A. Paton, J. Appl. Phys., 42, 5868 (1971).
8. F. W. Gorter,et. al.,IEEE Trans.Mag.,MAG-10,899(1974).
9. C. H. Bajorek and R. Hempstead, to be published.
10. L. T. Romankiw, et al., IEEE Trans. Mag., MAG-10, 828 (1974).
11. L. T. Romankiw and P. Simon, to be published in IEEE Trans. Mag.

NOVEL BUBBLE DRIVE

T. J. Walsh and S. H. Charap
Department of Electrical Engineering,
Carnegie-Mellon University,
Pittsburgh, Pennsylvania 15213

ABSTRACT

The interaction between a bubble and a distributed current flowing in a neighboring non-uniform conductor is proposed as the basis of a novel bubble drive. The effect of a conducting sheet perforated in a periodic pattern and excited by a rotating or oscillating current sequence is investigated by computer simulation. Results indicate that a bubble manipulation capability comparable to field-access drives may be accomplished in this way without coils.

INTRODUCTION

Consider a conducting layer perforated in a pattern resembling Permalloy T-I bars (Fig. 1). The bubble interaction is due to deviations from uniform current density which may be represented by current whorls whose positions and intensities depend upon the direction and magnitude of the applied current. Examination of the sequence shown in Fig. (1) indicates an analogy with the operation of the T-I bar circuit, with the in-plane field replaced by a rotating current differing in phase by 90°.

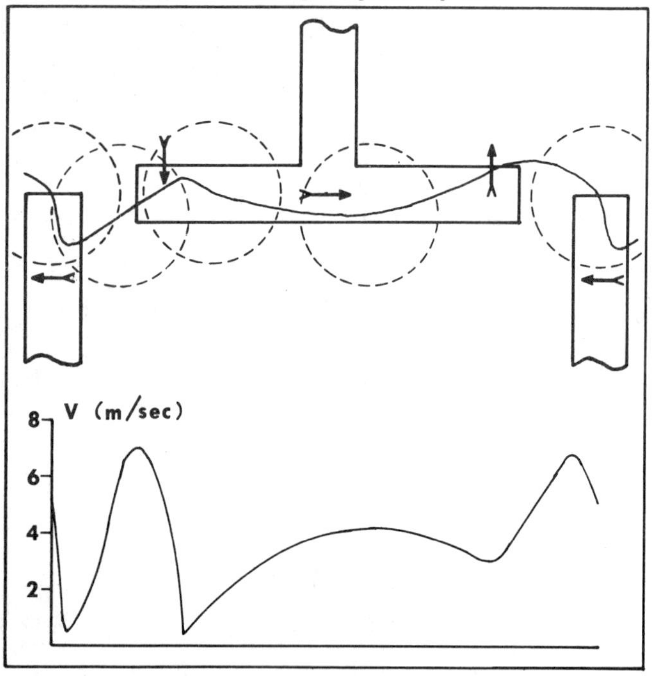

FIG. 1 Trajectory and velocity of 6 μm bubble with mobility 700 cm/sec/Oe and coercivity .05 Oe driven at 100 kHz by T-I circuit. Arrows locate bubble energy minima when impressed current is in directions indicated, while circles locate lagging bubble positions for the successive 90° increments (left-to-right).

Alternatively a current sheet perforated by a square array of round holes may be used. It produces bubble energy maxima and minima on opposite sides of the holes. As the current is rotated, these energy extrema also rotate about the holes driving the bubbles. Since the energy minima never hop from one hole to another, propagation using this overlay requires an asymmetric excitation and depends upon the bubbles lagging the current. However the direction of bubble propagation may be along any of the four directions.

ANALYSIS

A bubble, which produces a magnetic induction $\vec{B}(\vec{r})$, experiences a force exerted by a current density $\vec{J}(\vec{r})$ flowing in a neighboring conductor. The force is found by integrating the body force $-\vec{J} \times \vec{B}$ over the conductor volume. The sheet conductivity varies periodically in its plane so we may express this force as a sum of integrals over conductor unit cells (i,j). The force on a bubble located under cell (0,0) exerted by the current flowing in cell (i,j) is $-\int \vec{J} \times \vec{B}\, dv$, integrated over the volume of cell (i,j). If the sheet is of uniform thickness and infinite extent and if we neglect the influence of the bubble stray field on current density, then J takes the same form in each unit cell--only the magnetic induction B varies with cell indices.

It is convenient to compute each force $\vec{f}(i,j)$ by first translating the bubble to its equivalent position in cell (-i,-j) and integrating over cell (0,0). The sum of all forces computed in this manner is the total force and is equivalent to the force exerted by current flowing in cell (0,0) on an array of bubbles. This bubble array, introduced as a mathematical convenience, has the same periodicity as the sheet conductivity. The magnetic induction $\vec{B}'$ produced by the array[1] varies exponentially with distance from the bubble platelet and harmonically in the plane. Because $\vec{J}$ varies in the plane only, integration of $-\vec{J} \times \vec{B}'$ through the sheet thickness leaves integrals proportional to the Fourier coefficients of $\vec{J}$. The expressions for the forces are finally double series in harmonic terms which depend upon the lateral displacement of the bubble lattice from the conductor lattice. They may be evaluated without further reference to the conducting sheet. Conductor symmetry and the solenoidal nature of $\vec{J}$ reduces the number of coefficients to be calculated.

The current density components were found by solving Laplace's equation numerically over a square net. A polynomial, fitted to the potential along each net line, was differentiated to obtain the electric field along that net line. The Fourier coefficients of the current density components were evaluated by closed form integration along the net lines and numerical integration in the transverse direction.

SIMULATIONS

The operation of the T-I overlay was simulated by computing the force (and equivalent ΔH) and corresponding bubble velocity for given bubble position and applied current intensity and direction, translating the bubble a small distance in the force direction, then repeating the process with the appropriate incremental change in current direction.

An overlay period of four bubble diameters horizontally and eight bubble diameters vertically gives the maximum packing density allowed by mutual bubble interactions[2]. Bubble height was chosen as half the nominal bubble diameter, d. The width of the holes, sheet thickness, and the length of the horizontal T segments were varied to determine the conductor geometry which gives most efficient operation in the sense of requiring the least power input to sustain a given bit rate. With holes of width 0.8d the domain failed to transfer from the I to the T, being attracted instead by a local energy minimum at the upper left corner of the I. This local minimum appears when the applied current direction is between 180° and 270° and competes with the minimum at the lower left edge of the T. Successful operation was indicated by simulations using narrower holes, with power input per bit for a given bit rate decreasing with hole width. T-holes with horizontal segments 2.8d long in a sheet 0.55d thick were found to give the most efficient operation. With these dimensions and a hole width of 0.4d, a silver overlay consumes 0.8 μwatts per bit while driving the bubbles at 100 kHz. Further simulations keeping the same overlay and bubble platelet dimensions while varying bubble diameter indicate that larger diameter bubbles are readily propagated by this overlay but smaller diameter bubbles tend to stick to the upper left corner of the I and require higher currents for propagation. It is estimated that the bubble diameter will vary by ± 12% in this trajectory if $4\pi M$ = 150 G. The calculated bubble trajectory and velocity are shown in Fig. (1). The ratio of peak to average bubble velocity is 2.5:1. Comparison with the trajectory calculated by Hsin, et al,[3] and the velocity measured by Rossol[4] indicates that a bubble moves more smoothly when driven by this overlay than by its Permalloy counterpart.

The operation of an overlay containing a square array of round holes was also simulated. The conductor excitation could not be specified beforehand as for the T-I overlay. The applied current direction was found at each instant by requiring that it provide the maximum force in the propagation direction. Bubble motion parallel to a cell axis and also inclined at 45° to that direction was examined. The results of these simulations indicate that for propagation parallel to an axis the applied current must be rotated at an accelerating rate, then held in one direction for a time, and suddenly switched to a new direction before being rotated again. The accelerating rotation of current places an energy maximum directly behind the bubble in order to kick the domain to an adjacent cell. Small changes in the relative positions of the bubble and this energy maximum at the instant transfer begins would result in large changes in subsequent bubble motion. For propagation inclined 45° to the axis the current must oscillate at a nearly uniform rate between directions separated by about 270°. No accelerated rotation or abrupt changes in current direction are required to transfer the bubble to an adjacent hole, so propagation along these directions is preferred from the standpoint of reliability.

Hole radius and spacing and conductor height were varied for this overlay to determine the most efficient geometry. In these simulations the current direction was again chosen to maximize the force at 45° to the axis so that differences in operating efficiency indicated by the simulations could be attributed to geometry only. For most efficient operation the holes should be spaced 4d apart and be 2.8d across. Conductor height should be 0.6d.

The excitation sequence given by the simulation is not suitable for device operation since any variations in bubble size, mobility, or coercivity throw the bubble out of synchronism with the drive field. Such failure may be avoided by holding the current direction fixed for a time after the bubble begins to transfer between cells. Slow bubbles may then catch up with the drive field while fast bubbles can't race ahead.

The trajectory and velocity of a bubble driven by such a current sequence are shown in Fig. (2). The drive current has magnitude 12 mA per cell width, frequency 100 kHz, and is held parallel to each of the axes for 2.5 μ sec. between oscillations. An overlay made of silver would consume 1.7 μ watts per bit when excited by this drive current. Tolerances on bubble mobility and diameter are ± 20%, while the estimated change in bubble diameter is less than ± 7%. The ratio of peak to average velocity is 4.6:1. When the current is held along (a) the bubble comes to rest at (1) and when the current is held along (b) the bubble comes to rest at (2). If the direction of current rotation is reversed after the bubble comes to rest at (1) or (2), the direction of subsequent motion is changed by 90°. If the components of applied current are interchanged after bringing the bubble to rest at (3), bubble motion will be reversed and the domain will retrace its original path. By choosing an appropriate combination of 90° corners and/or reversals we may move the bubble from any one cell to any other cell in the lattice.

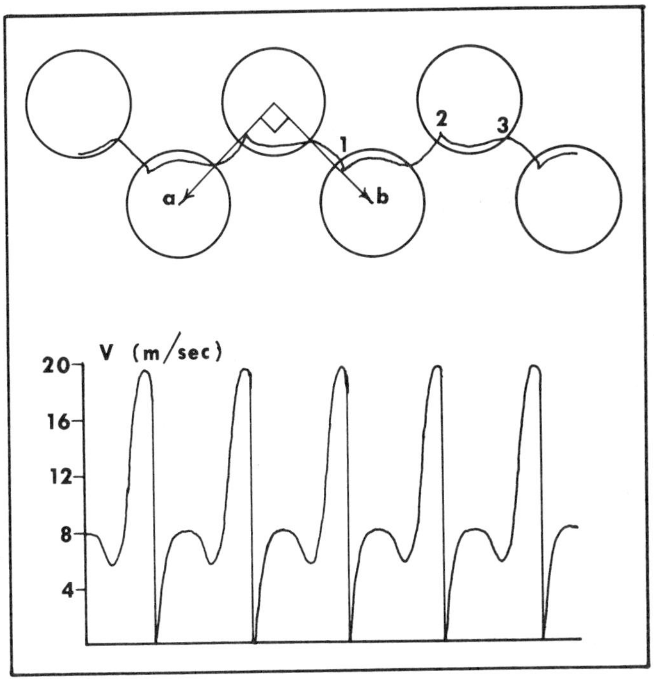

FIG. 2 Trajectory and velocity of 6 μm bubble with mobility 700 cm/sec/Oe and coercivity .05 Oe driven at 100 kHz by overlay containing round holes. Unit cell axes (a) and (b) are shown, as are bubble holding positions 1,2,3 (see text).

CONCLUSION

The computer simulations we have performed suggest that by using non-uniform current sheets one can reproduce all operations that are achieved in field-access drives. Power requirements, determined here for a bubble material with moderate mobility, compare favorably with the power requirements of field access drives as cited by Almasi.[2] This bubble propagation technique should provide an alternative to field access techniques which requires no coils, being more compatible with the planar geometry of present integrated circuit technology.

REFERENCES

1. W. F. Druyvesteyn, D. L. A. Tjaden and J. W. F. Dorleijn, Phil. Res. Rpts., 27, pp. 7-27 (1972).
2. G. S. Almasi, IEEE Proc., 61, pp. 438-444 (1973).
3. C. H. Hsin, T. J. Matcovich and R. L. Coren, AIP Conf. Proc., 5, pp. 244-248 (1971).
4. F. C. Rossol, IEEE Trans. Mag. MAG-7, pp. 142-145, (1971).

MAGNETIC BUBBLE PROPAGATION BY PULSED rf CURRENTS

W. C. Hubbell, Texas Instruments Incorporated,
P.O. Box 5936, M. S. 145, Dallas, Texas 75222

ABSTRACT

Propagation of bubbles along the edges of deposited conductors has been observed. In this mode of translation, bubbles move smoothly along the perimeter of the conductor. Bubbles of 5 μm diameter have been propagated in epitaxial films of $(Y,Sm)_3(FeGa)_5O_{12}$ with rf frequencies between 10 and 75 MHz. For pulses of 4 μsec duration, the minimum peak-to-peak rf current, I_o, required for propagation with a gold conductor 76.0 μm wide by 1 μm thick was found to be approximately 2 amp. Propagation has been observed in both ion-implanted and non-implanted samples. Collapse field measurements indicate that bubbles capable of being propagated by this technique are not hard bubbles.

EXPERIMENTAL

Propagation of magnetic bubbles in epitaxial garnet films along the edges of flat conductors carrying pulses of rf current has been observed. Bubbles are found to move with about one-half of their area under the current sheet (Fig. 1). They move parallel to the conductor's edge and in general follow its contour; for instance, propagation around 90° corners has been observed. For a given material, the direction of motion along the conductor's edge depends primarily on the magnetic polarity of the bubbles. Once the magnetic polarity is established, all of the bubbles on one edge of a long flat conductor move in the same direction (Fig. 1) while bubbles on the other side propagate in the opposite direction. If the magnetic polarity is then reversed, the subsequent bubbles are found to propagate in the reverse directions.

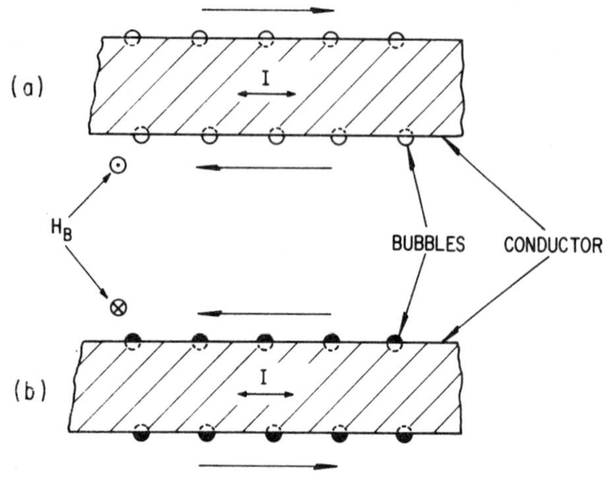

Fig. 1. Bubble Motion along the Edges of a Conductor Carrying Pulses of rf Current, I.
a) The bubbles on the top edge move to the right while at the same time the bubbles on the bottom edge move to the left.
b) When the bias field orientation is reversed and bubbles of the opposite magnetic polarity appear, those bubbles on the top edge of the conductor now move to the left while those on the bottom edge move to the right.

The above behavior has been observed for conductor patterns of various geometries including simple straight lines as well as meanderline patterns. Widths as narrow as 10 μm have been used successfully; however, the majority of the work reported here involved the use of 76 μm wide x 1 μm thick conductors. The conductor patterns can be deposited directly on the epitaxial garnet film or on an adjacent glass substrate. Propagation by this technique has been observed in the compositions $(Y,Gd,Tm)_3(FeGa)_5O_{12}$ and $(Y,Sm)_3(FeGa)_5O_{12}$ (SmYIGG). Similar results have been achieved in both ion-implanted and non-implanted films. Bubbles which propagate in this manner have been observed to collapse at the lower end of the spectrum of collapse fields and therefore are not hard bubbles.

The individual pulses responsible for this behavior contain many cycles of rf current and each pulse effects the translation of all the bubbles localized on the edge of the conductor. Bubble displacements due to individual pulses may range anywhere from less than one bubble diameter up to several diameters, depending on the pulse parameters.

The pulse parameters which affect propagation are
i) I = pulse amplitude (Amps, peak-to-peak)
ii) τ = pulse width (μsec)
iii) r = pulse repetition rate (Hz)
iv) f = rf frequency (MHz)

In addition to these parameters, the translation is influenced by the bias field, H_B, with propagation generally occurring over nearly the entire range of bubble stability from strip-out, H_{SO}, to collapse, H_o.

The propagation effects observed here have been most effectively achieved by repeated application of such pulses with repitition rates of up to 500 Hz.

For Sm YIGG, pulse widths of less than 1 μsec to more than 5 μsec have been used to propagate bubbles in the frequency range 10 - 75 MHz. At a given frequency, there is a lower value of pulse amplitude, I_o, below which bubble motion is virtually imperceptible. There is also an upper value of pulse amplitude, I_u, above which translation begins to degenerate and the bubbles chaotically reverse direction and spill off the conductor edge. These limiting values of peak-to-peak current are plotted as a function of frequency (Fig. 2) for the above ranges of the parameters, r, τ and H_B. For values of pulse amplitude

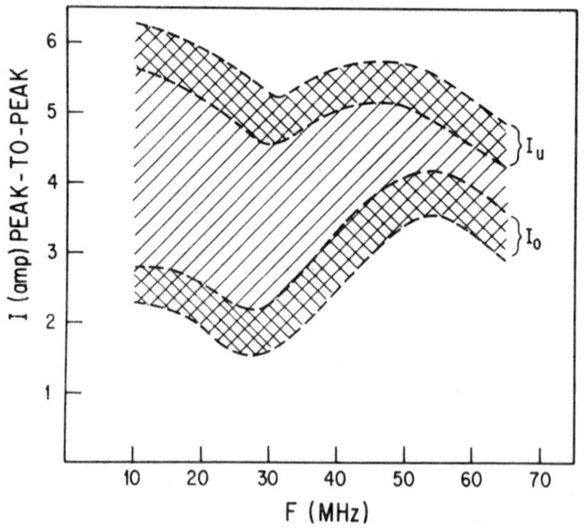

Fig. 2. The Limiting Values of Peak-to-peak Current, I, as a Function of rf Frequency, f, for the Range of Parameters $0 < r(H_z) < 500$, $0 < \tau(\mu sec) \leq 5$ and $H_{SO} < H_B^z\ (Oe) < H_o$. I_o and I_u are the minimum and maximum current amplitudes respectively, consistent with propagation.

in the range $I_o < I < I_u$, bubbles propagate uniformly with velocities parallel to the conductor which increase nonlinearly as I increases.

Experimental measurements of the velocities of bubbles propagated by this technique in Sm YIGG were made as a function of I, τ and H_B. These velocities were determined from measurements of the number of pulses of fixed width necessary to move bubbles a given distance. Although velocities in the neighborhood of 600 cm sec^{-1} have been measured, they correspond to pulse amplitudes near I_u where the translational behavior becomes erratic. For this reason, most of the measurements were restricted to lower values.

For constant pulse width and amplitude, the bubble velocity was found to increase linearly with bubble diameter. From measurements of the accumulated displacements after many pulses as a function of pulse width, it is possible to plot the bubble displacement per rf pulse, δ, vs. the number of cycles of rf per pulse, η. Fig. 3 contains plots of δ vs. η for representative rf frequencies. It is seen from these curves that δ is predominantly a linear function of η at all frequencies with, however, some departures for large η. This linear dependence, coupled with the fact that the curves pass through the origin, indicates that each cycle of rf in a pulse envelope causes an elementary displacement, δ_o, and that the total displacement of a bubble by a pulse of rf current is due to the cumulative effect of the many cycles in the envelope. From the slopes of the curves in Fig. 3 it is found that the corresponding values of δ_o lie in the range $8 \times 10^{-3}\ \mu m \leq \delta_o \leq 16 \times 10^{-3}\ \mu m$.

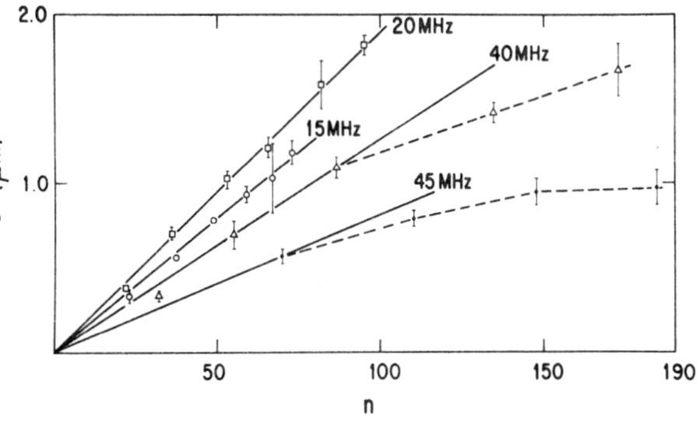

Fig. 3. The Bubble Displacement per rf Pulse, δ, vs. the Number of Cycles of rf per Pulse, η.

DISCUSSION

The description of the translational behavior presented in this work suggests that the mechanism responsible for the displacement of bubbles parallel to the conductor edges is associated with the effect of skewing, i.e. the deflection of bubbles away from the axis of a driving field gradient. Several authors[1-5] have observed these skewing effects which are attributed to the presence of vertical Bloch lines within the bubble wall. Recently Boxall[4] has described a method by which bubbles have been translated along the axis of a two dimensional potential well by unipolar current pulses. The translational motion was attributed to the skew component of bubble motion accompanying the pulsed field gradient. It seems that basically the same mechanism is responsible for the present behavior with two important differences. First, no potential well is necessary; and second, the current pulses are bipolar in nature.

Confinement of bubbles along the conductor edge apparently has to do with the fact that the oscillating field gradient, associated with the manifold cycles in an rf pulse, causes the bubbles to execute infinitesimally small, equal and opposite displacements normal to the conductor axis. Thus, the net motion normal to the conductor is zero. The net displacement parallel to the conductor axis appears to be the result of unidirectional skewing associated with the small displacements away from, and back towards, the conductor edge. This requires that the skew component of the bubble motion not reverse its direction when the component normal to the conductor reverses its direction. This type of skewing behavior has been observed by this author near the edge of conductors carrying rectangular current pulses. However, for these rectangular pulses, this type of skewing behavior is the exception rather than the rule. In most cases it has been observed that the skew component also reverses direction along with the normal component. On the other hand, for the rf translation effect described here, the condition of the high frequency gradient reversals and/or the associated reversing in-plane field component seems to have the effect of stabilizing this unidirectional skewing condition. It has been observed, however, that the application of an in-plane magnetic field of ∿ 40 Oe is sufficient to eliminate this effect and thus prevent propagation along the conductor edge.

The direction of propagation along the edge of a conductor is given by the vector $\hat{v}_\perp \times \hat{H}_z$ where $\hat{v}_\perp$ is a unit vector in the direction of the instantaneous bubble motion normal to the conductor edge and $\hat{H}_z$ is the direction of the instantaneous rf magnetic field as seen by the bubble.

Thus it appears that the effects of the combination of small alternating bubble displacements normal to the conductor edge along with the condition of unidirectional skewing are responsible for bubble propagation by rf currents.

ACKNOWLEDGMENTS

The author has benefited greatly from discussion with F. G. West, D. C. Bullock and C. T. M. Chang. Thanks are also due to G. G. Sumner for supplying the garnet films for this work.

REFERENCES

1. J. C. Slonczewski, A. P. Malozemoff and A. Voegeli, AIP Conference Proceedings, No. 10, Part 1, Magnetism and Magnetic Materials, New York, American Institute of Physics, pp. 458-477, 1973.

2. A. P. Malozemoff, J. Appl. Phys. 44, pp. 5080-5089, 1973.

3. W. J. Tabor, A. H. Bobeck, G. P. Vella-Colleiro and A. Rosencwaig, AIP Conference Proceedings, No. 10, Part 1, Magnetism and Magnetic Materials, New York, American Institute of Physics, pp. 442-457, 1973.

4. B. A. Boxall, IEEE Trans. Mag., MAG-10, pp. 648-650, 1974.

5. D. C. Bullock, AIP Conference Proceedings, No. 18, Part 1, Magnetism and Magnetic Materials, Boston, American Institute of Physics, pp. 232-236, 1974.

PERMALLOY FILM FOR SINGLE-LEVEL DETECTOR WITH HIGH SENSITIVITY

K. KOMENOU, H. NAKAJIMA and K. ASAMA
FUJITSU LABORATORIES LTD., KAWASAKI, JAPAN

ABSTRACT

The use of single-level circuit has simplified the fabrication process. A thick film detector, however, has weak points; it has a low detection efficiency due to the large demagnetizing field and unfavorable noise is inherent in it. The authors study the magnetoresistive characteristics of permalloy films and find out the detectors with high sensitivity, selecting the appropriate evaporation conditions. The significant results are as follows. (1) The large detection voltage is obtained from those elements whose magnetic easy axis deviates from the shape anisotropy direction, which serves to reduce the effective anisotropy field. (2) This angle of deviation α is increased with the raising of the substrate temperature and with the lowering of the evaporation rate. For example, α = 60° at 310°C, 3Å/sec, and α = 0° at 250°C, 10Å/sec. (3) Detection performances are investigated on the 30-column serpentine detector using 3 μm stripe domain width of $(YGdYb)_3(GaFe)_5O_{12}$ with a 4π Ms of 260 Gauss. The signal voltage of 1 mV/3 mA and S/N ratio of 5 are obtained at 100 kHz.

INTRODUCTION

In order to simplify the fabrication process and improve yield, the so-called single mask level circuit which uses the same layer of permalloy film for propagation and detection has been proposed.[1] As viewed from the detector, however, the detection efficiency becomes lower because of the large demagnetizing field and many chevrons in stretcher are required in comparison with thin film detectors. Calculating the demagnetizing field of a rectangular element whose width, length and thickness are 3.6 μm, 150 μm and 3000Å, respectively, it is approximately 1100 Oe crosswise (Y-axis) and approximately 60 Oe lengthwise (X-axis). Therefore, the magnetic easy axis is generally along the X-axis due to the shape anisotropy. The magnetic field Hs was defined as the field necessary to saturate the magnetoresistive effect and was equal to the effective anisotropy field[2] (anisotropy field + demagnetizing field). The Hs along the Y-axis is approximately 1100 Oe and 65 Oe along the X-axis. The change in resistivity, when magnetic field is applied lengthwise, is large if the easy axis of a element is crosswise. From these facts, detectors the easy axis of which deviate from the shape anisotropy direction are thought to be preferable. The authors studied the magnetoresistive characteristics of permalloy films and found out detectors with high sensitivity, the magnetic easy axis of which was deviated from the shape anisotropy direction, selecting the appropriate evaporation conditions.

EXPERIMENT AND RESULTS

All the films were evaporated on a pyrex glass and an LPE garnet with SiO_2 spacer in a vacuum of between 2 and 3 x 10^{-6} Torr. by the electron beam evaporation method. No external field was applied. The substrate was heated by a Ta-heater located behind the substrate holder. The film composition obtained was 79.6% Ni, 19.7% Fe and 0.5% Cu. The coercive force was about 5 Oe. The magnetoresistive elements were delineated by the usual etching technique.

Figure 1 shows the significant magnetoresistive characteristics of a typical element with high sensitivity. The change in resistivity was 2.8% when magnetic field was applied lengthwise (X-axis) and -1.0% when it was applied crosswise (Y-axis). Hs in both cases was 70 Oe and 1100 Oe, respectively. Here the angle α was defined as follows. In the direction with an angle of α to the X-axis, the resistivity under the application of magnetic field sufficient to saturate

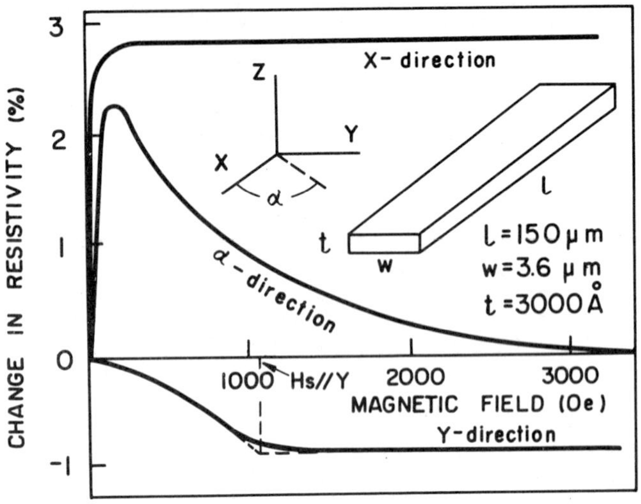

Fig. 1 Magnetoresistive characteristics of an element with high sensitivity.

the magnetoresistive effect was equal to the resistivity measured in zero magnetic field. This direction is considered to be the direction of the magnetic easy axis from the curve of B-H characteristics of magnetic films with uniaxial anisotropy. Because the remanent magnetization in the direction of the easy axis is equal to the saturation value. Except when α is 0° the characteristics of the element are symmetrical with respect to the X-axis, thus indicating its biaxial magnetic anisotropy.

Figure 2 shows the relationship between α and the film thickness when the substrate temperature and deposition rate were changed. The deviation of the easy axis from the magnetic shape anisotropy direction occured on those elements whose thickness exceeded 1300Å. The smaller the deposition rate and the higher the substrate temperature was, the greater α was obtained. For example, α was 60° when the film thickness was 3000Å and element width 3.6 μm, at the substrate temperature of 310°C and the deposition rate of 3Å/sec. α was 35° when the deposition rate was changed to 10Å/sec. While α was 8° at 280°C and 3Å/sec., respectively. α was 0° in all cases when the substrate temperature was 250°C or below. In other words, the easy axis coincided with that of the shape anisotropy direction. Figure 3 shows the dependency of α on element width. The magnetic film as evaporated has

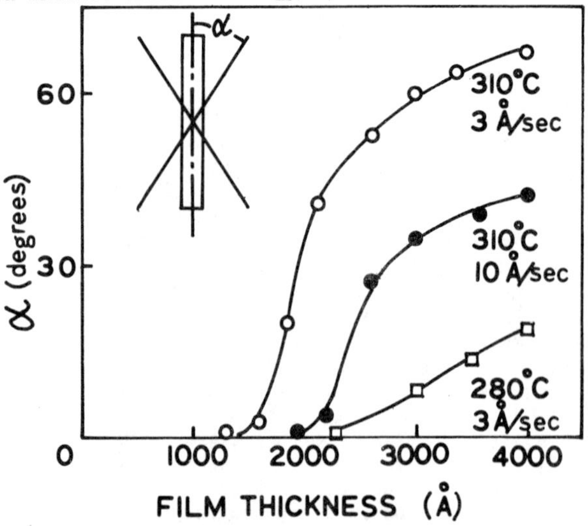

Fig. 2 Angle of deviation α versus film thickness in various evaporation conditions.

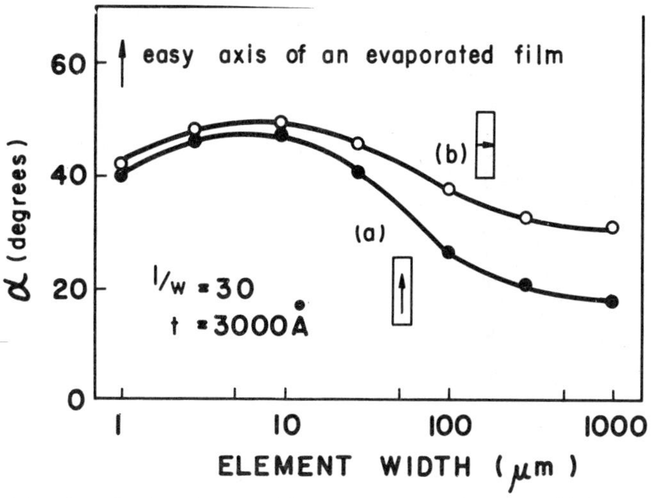

Fig. 3. Angle of deviation α versus element width.

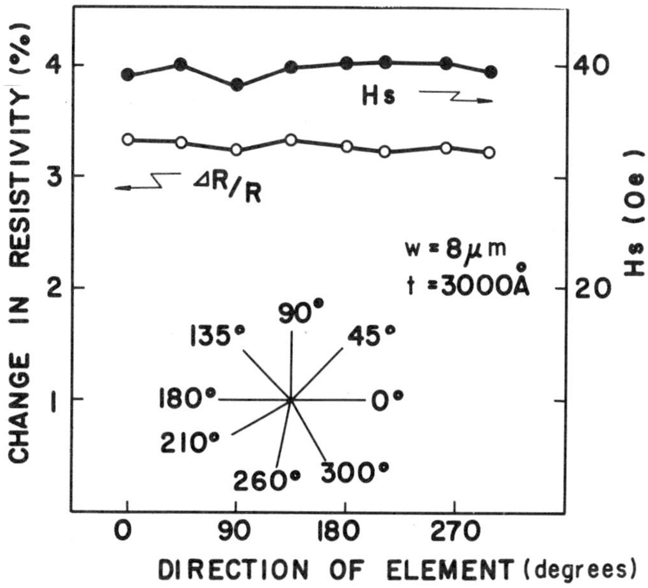

Fig. 4. Magnetoresistive characteristics of elements which are arranged radially.

slightly uniaxial anisotropy due to a considerable self-shadowing effect.[3] In the case of the specimen plotted as (a), the direction of the shape anisotropy axis coincided with the easy axis of the evaporated film. Plots (b) were the result on elements transverse to the easy axis of the evaporated film. The difference of α in wider element was thought to be caused by the above effect. The narrower the element, the smaller the difference. The magnetoresistive characteristics of elements, which were radially arranged on a substrate, are shown in Fig. 4. The change in resistivity and Hs were measured in the lengthwise direction of each element. It was evident that these characteristics were independent of the radial direction in the evaporated film. From Fig. 3 and Fig. 4, the characteristics of element with high sensitivity were thought to be caused by the microscopic characteristics of the film such as domain structure.

The serpentine detector was used for bubble detection. Film thickness was 4000Å for 6 μm-bubble domains and 2800Å for 3 μm-bubble domains, respectively. The bubble materials used were $(YSm)_3(GaFe)_5O_{12}$ ($4\pi Ms$: 250 Gauss, stripe-width: 6 μm) and $(YGdYb)_3(GaFe)_5O_{12}$ (260 Gauss, 3 μm), respectively. The signal voltage is shown as a function of the rotating field in Fig. 5, which shows the gradual increase of signal with the rotating field. A typical waveform for 3 μm-bubble domains is shown in Fig. 6. A signal voltage of 1 mV/3 mA, and S/N ratio of 5 were obtained at 100 kHz for 30-column serpentine detector whose resistance was about 100 Ω. The

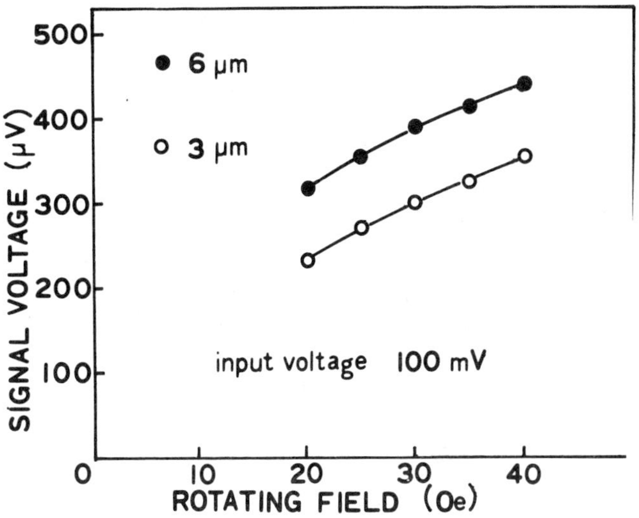

Fig. 5. Signal voltage as a function of rotating field for 6 μm, 3 μm-bubble domains.

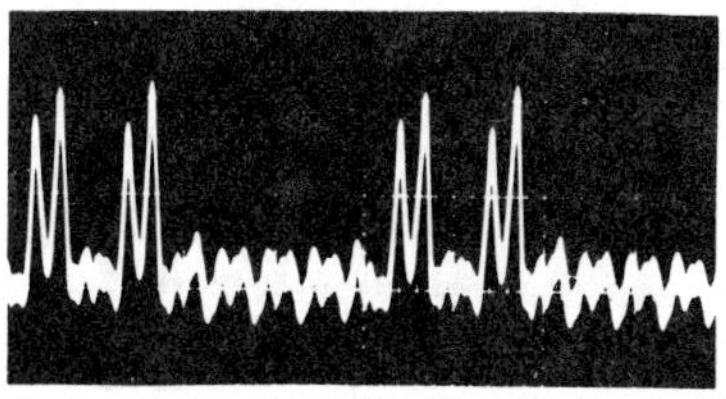

Fig. 6. A typical signal of a 3 μm-bubble domain.

vertical scale : 0.5 mV/div.
horizontal scale: 10 μsec/div.
detector current: 3 mA
rotating field : 35 Oe

change in resistivity at detection corresponds to a bubble-caused ΔR/R of 0.3%. Here α of the 1.8 μm width rectangular element was 50°. No undetectable region of the rotating field was observed. These detection performances were superior to those published previously.[4] The signal voltage was approximately proportional to α and was very small at the detector with α = 0°. In order to investigate the influence of these films on propagation performance, the start-stop test was performed on the 31 bits T-bar endless circuits. There was no obvious distinction in the operation margins between the elements with the different value of α = 0° and α = 50°.

CONCLUSION

These characteristics were also observed in 70 Ni-Co evaporated films and thought to be related to crystal magnetic anisotropy. The detection performance of this detector was satisfactory and no unexpected results in operation were observed. This detector is thought to be very suitable for the sub-micron bubble device.

ACKNOWLEDGMENT

The authors thank T. Miyashita for specimen preparation, T. Obokata and T. Mori for supplying LPE garnets and the device group for various suggestions.

REFERENCES

1. W. Strauss, A.H. Bobeck and F.J. Ciak, AIP Conf. Proc., 10, 202 (1972).
2. K. Asama, K. Takahashi and M. Hirano, AIP Conf. Proc., 18, 110 (1973).
3. D.O. Smith, M.S. Cohen and G.P. Weiss, J. Appl. Phys. 31, 1755 (1960)
4. A.H. Bobeck, I. Danylchuk, F.C. Rossol and W. Strauss, IEEE Trans. Magnetics, MAG-9, 474 (1973).

IMPROVED CHEVRON GAP DETECTOR

Arnulf Lill
Siemens AG, Unternehmensbereich Bauelemente, Balanstraße 73
D-8 München 80, Germany

ABSTRACT

For the chevron stretcher detectors usually employed, the fractional change in resistivity decreases compared to a detector bar without stretcher. By positioning the sensor between the gap of two successive chevron columns so that no metallic contact exists between bars and sensor, this decrease can be avoided. Characteristics of this improved gap detector are presented. Using an 11-fold chevron stretcher the bridge signal is 0.9 mV/mA for 9 µm bubble domains. The overall "ONE" to "ZERO" ratio is about 5:1 at 100 kHz drive field frequency.

INTRODUCTION

Magnetoresistive permalloy detectors[1] are usually employed in magnetic bubble devices. To increase the signal of small bubbles with diameters below 10 µm, stretching techniques are used. These bubble detectors consist of a rectangular permalloy sensor bar and a bubble stretcher similar to the Chinese character stretcher[2,3] or the chevron stretcher[4,5].

The purpose of this paper is to describe the characteristics of an improved thin film chevron stretcher detector.

SAMPLE FABRICATION

Both sandwich-like and planar samples were used for the experiments. For the latter ones the overlay was spaced from the garnet film by a 1 µm thick sputtered SiO_2 layer. The overlay fabrication was the same as described in ref.6 (see also ref.7): Onto a 0.03 µm-thick evaporated permalloy film a 0.4 µm-thick permalloy propagation structure and 0.5 µm-thick gold conductor lines were electroplated through photoresist masks. Finally the sensor with easy axis along its long edge was defined by sputter etching.

The permalloy film was evaporated from a Cermotherm-boat (40% ZrO_2, 60% Mo) in a vacuum of about 10^{-6}Torr under a magnetic field of 80 A/cm by resistance heating. The evaporation rate was about 4 nm/s. For the zero magnetostriction composition, $\Delta R/R$ was 3%. Anisotropy field H_k and coercivity H_c were 2 and 2.5 A/cm, respectively.

The bubble material was a$(YSm)_3(FeGa)_5O_{12}$ garnet film. Its saturation magnetization was 175×10^{-4}Vs/m^2, the width of zero field stripe domains 9.5 µm.

EXPERIMENTAL RESULTS AND DISCUSSION

For conventional chevron stretcher detectors as described first by Archer et al.[4], the sensor is positioned between two chevron columns, the bars of which overlap the sensor element (device B in Fig.2). For this device the saturation value of $\Delta R/R$ is decreased compared to an isolated sensor bar. The magnetoresistive properties of the detector material are only partly used for bubble sensing. By positioning the sensor between the gap of two successive chevron columns so that no metallic contact exists between bars and detector, this disadvantage can be avoided (device A in Fig.2). Fig.1 shows a portion of the gap detector test pattern. The propagation channel was designed

Fig.1: Sensor test pattern

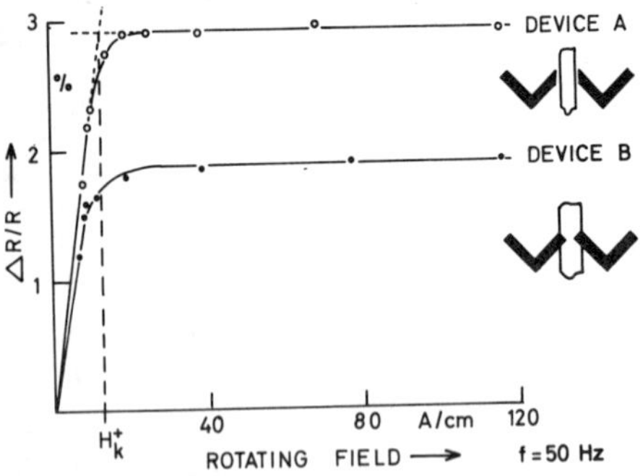

Fig.2: Sensor response to rotating field amplitude

to propagate and expand cylindrical bubble domains to an 11-fold chevron column. The bar width is 6 µm, the sensor is 7 µmx208 µm. Its resistance was 320 Ω including leads. The reference detector which cancels magnetoresistive and inductive noise is not shown in Fig.1.

The sensor response to a quasistatic rotating field H_{rot} is shown in Fig.2 for both devices mentioned above. The saturation value $\Delta R/R$ of the gap detector A is increased by about 50 % as compared with device B. Although the sensor width of the gap detector is reduced as compared with device B, the effective anisotropy field H_k^+, defined as the field to saturate the sensor element (see Fig.2), is comparable for both devices. It is about 16 A/cm. The decrease of H_k^+ with respect to an isolated bar is about 7 A/cm even for the gap detector.

Corresponding to the higher saturation value of $\Delta R/R$, the bubble signal of the gap detector A should be higher as compared with device B, which is shown in Fig.3. The bubble output signal of the detector bridge with

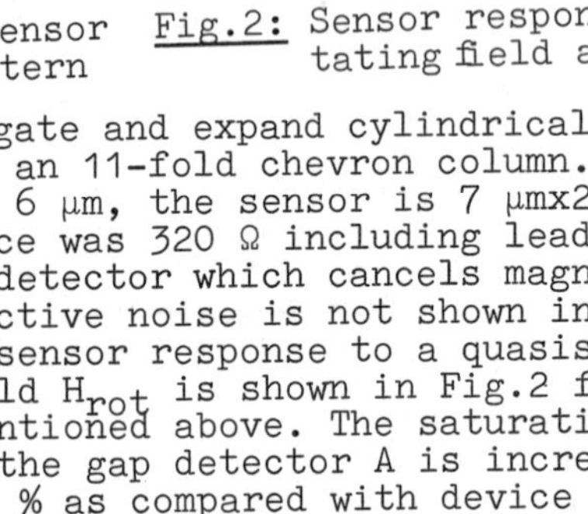

Fig.3: Bubble output vs. current density

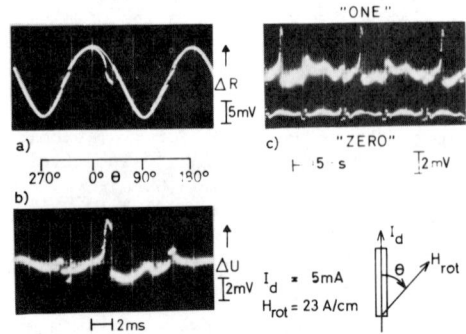

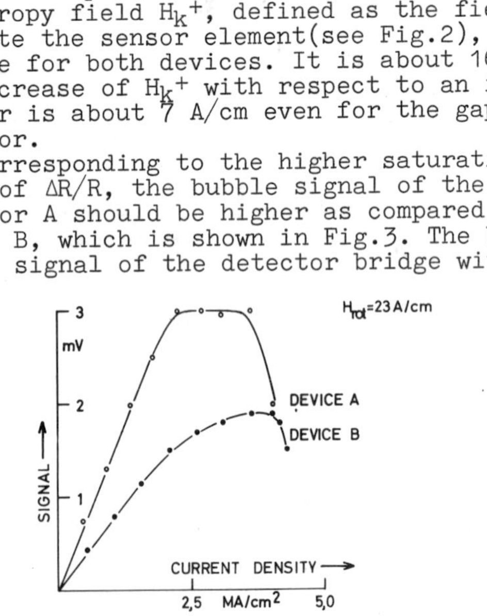

Fig.4: Oscilloscope traces of sensor outputs at 50 Hz rotating field

respect to ground is drawn as a function of current density for both devices.

Fig.4a shows the signals of the bubble sensor and the reference detector as a function of rotating field direction for a planar sample. For $\Theta=0^{o}$ and 180° the sensor magnetization lies parallel to the input current. When a domain approaches the sensor, its stray field switches the magnetization partly perpendicular to the current. The sensor resistance decreases with respect to the reference detector, resulting in a positive output signal of the detector bridge (Fig.4b). When the domain jumps over the chevron gap, the sign of the bubble stray field component which acts on the sensor magnetization is changed suddenly with respect to the drive field, resulting in the falling edge of the bridge signal for $\Theta \sim 30^{o}$. The discontinuous magnetoresistive "noise" for $\Theta \sim 310^{o}$ and $\Theta \sim 140^{o}$, which also occurs in the "ZERO" signal (see Fig.4c), is probably due to domain switching in the sensor pair.

Fig.4c shows the "ONE" (upper trace) and "ZERO" (lower trace) signals for 5 mA input current in each sensor and 23 A/cm drive field amplitude. The input current characteristic for the "ONE" and "ZERO" signals is shown in Fig.5. The "ONE" signal is linear with current up to 5 mA, that is, the sensor can be operated with current densities up to $2.5 \times 10^{6} A/cm^{2}$ without heating effects. The "ONE"/"ZERO" ratio is about 6:1 for this value. With further increase in current, the bubble signal peaks and then decreases because of heating effects in the sensor and garnet film, respectively.

Fig.6 shows the quasistatic operation margins of the gap detector test pattern for 5 mA sensor current. The data were taken from domain signals. The lower limit is defined by nucleation of bubbles at the reference detector, the upper limit by generator failing due to domain collapse. For 24 A/cm drive field amplitude the bias operation range is about 15%. The dashed line in Fig.6 shows the domain signal as a function of drive field. The signal decreases linearly with drive field amplitude due to sensor saturation (see Fig.2).

The gap detector was also operated at drive field frequency of 100 kHz. The "ONE" and "ZERO" signals are shown in Fig.7 for 5 mA input current and 27 A/cm rotating field amplitude. The domain signal of the detector bridge with respect to ground is about 5 mV. The dØ/dt noise in the conductor loops was about 2.5 mV p-p. By using a sensor pair in a differential mode this noise can be cancelled to about o.4 mV with respect to ground. The overall "ONE"/"ZERO" ratio is 5:1 which is more than adequate for electronic detection. The lower trace in Fig.7a and 7b shows the output signals at logical level. To trigger an AND-gate (SN 7525 N) the signals were amplified.

Chevron gap detectors designed for 5 µm bubble propagation (bar width=3 µm, sensor width=3.6 µm) were also tested. The maximum output signal of the detector bridge was about 3.5 mV for 105 µm sensor length. The effective anisotropy field H_k^{+} was about 35 A/cm. This value is quite low compared to about 60 A/cm for a sensor without bars. Although the stretcher bars do not overlap the sensor they also act as a yoke to reduce demagnetizing effects.

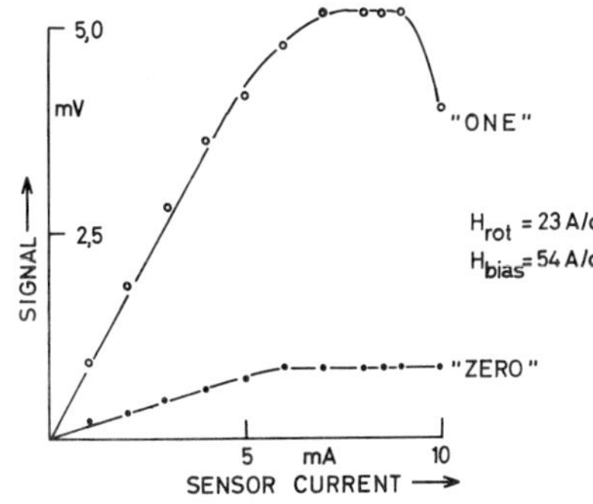

Fig.5: "ONE" and "ZERO" output vs. current

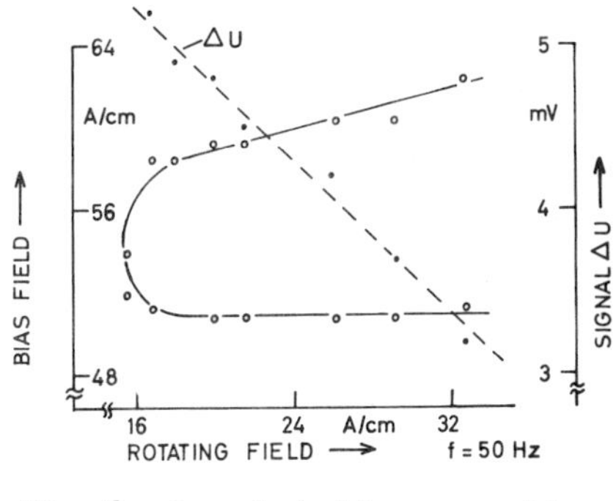

Fig.6: Quasistatic operation margins and "ONE" output vs. drive field (sensor current = 5 mA)

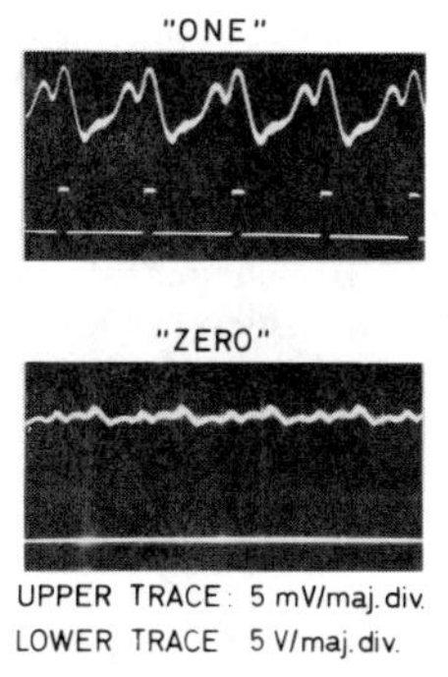

Fig.7: 100 kHz "ONE" and "ZERO" oscilloscope traces; lower traces = logical output level

CONCLUSIONS

An improved chevron stretcher detector has been described, the saturation value ΔR/R of which is increased about 50% as compared with conventional chevron detectors. Because of low effective anisotropy field, operation below 5 µm bubble diameter should be possible. The application of the gap detector down to 1 µm bubble diameter may be limited by alignment problems of photomasks.

ACKNOWLEDGEMENT

The author wishes to thank W.Metzdorf for encouragement and helpful discussions, B.Littwin and Miss G.Schuster for preparing the test patterns, F.Parzefall and J.Jary for supplying the garnet films.

REFERENCES

1. G.S.Almasi, G.E.Keefe, Y.S.Lin and D.A.Thompson, J.Appl.Phys.,42,1286(1971)
2. W.Strauss, P.W.Shumate, Jr. and F.J.Ciak, AIP Conf.Proc. 5,235(1972)
3. G.S.Almasi, G.E.Keefe and K.D.Terlep, AIP Conf.Proc. 10,207(1973)
4. I.L.Archer, L.R.Tocci, P.K.George and T.T.Chen, IEEE Trans.MAG-8,695(1972)
5. L.R.Tocci, P.K.George and I.L.Archer, AIP Conf.Proc.10,197(1973)
6. B.Littwin, to be published
7. R.E.Horstmann and I.V.Powers, "Overlay Fabrication for Bubble Domain Devices", 1973 INTERMAG invited paper, Abstract No.21.1

This work has been sponsored by the Data Processing Program (Sign DV 3301) of the Federal Department of Research and Technology of the FRG. The author alone is responsible for the contents.

MAGNETOSTRICTIVE-PIEZOELECTRIC BUBBLE DETECTORS*

W. Ishak, W. Kinsner and E. Della Torre
Group on Simulation, Optimization and Control,
Department of Electrical Engineering,
McMaster University, Hamilton, Ontario, L8S 4L7, Canada

ABSTRACT

A magnetic bubble detector using the principles of magneto-striction and piezoelectricity has been constructed and tested. It has higher signal to noise ratio than conventional detectors since it does not require any current to operate. This feature also avoids many other problems associated with currents flowing in thin conductors. A simplified analysis shows that this device is capable of outputs of tens of millivolts with presently developed materials. An experimental device produced an output 1.2 millivolts from a simulated bubble in orthoferrite material. The principle of operation of the device is that of electromechanical transformer. The field of a bubble causes a strain in a magnetostrictive overlay. This strain is transferred to a piezoelectric crystalline film which produces an electrical output at high electrical impedance.

INTRODUCTION

Four types of bubble-domain detection techniques based upon inductive[1], magnetooptic[2], galvanomagnetic[3], and magnetoresistive[4] sensors have been reported. The best of these techniques can give an output of many millivolts from a single stretched bubble. The galvanomagnetic and the magnetoresistive detectors require an external current source; hence this reduces their signal to noise ratio. Furthermore, these detectors are sensitive to transverse fields used to propagate bubbles.

The proposed magnetostrictive-piezoelectric sensor[5] does not require a current source to operate and can be made essentially insensitive to transverse magnetic fields. This device is principally an electromechanical transformer. The magnetization in a bubble is antiparallel to the magnetization in the host material. The field of a bubble is coupled to a magnetostrictive material and produces an elastic strain coupled to a piezoelectric material. The piezoelectric material produces the desired electric signal, indicating the presence of a bubble in the vicinity of the sensor. Calculations of field distributions[4] from a bubble indicate that at distance greater than two bubble diameters fields are negligible compared to the field directly over a bubble. This permits independent detection of bubbles in devices where bit locations are as close as three bubble diameters.

It is possible to design the detector so that it is essentially sensitive only to the bubble axial field, thereby minimizing the sensitivity to transverse propagating fields. Two techniques can be used to accomplish this: highly oriented materials can be used in the sensor and the geometry of the sensor can be appropriately chosen. Since both the magneto-strictive and the piezoelectric materials can have anisotropic crystal structures, they can be oriented during fabrication so that their transverse sensitivity is minimized. By choosing the proper symmetry in the geometry the sensor can be made to be essentially insensitive to the transverse field. Since the devices fabricated and reported in this paper were all manufactured with a single mask technique, it was not possible to achieve the optimum geometry.

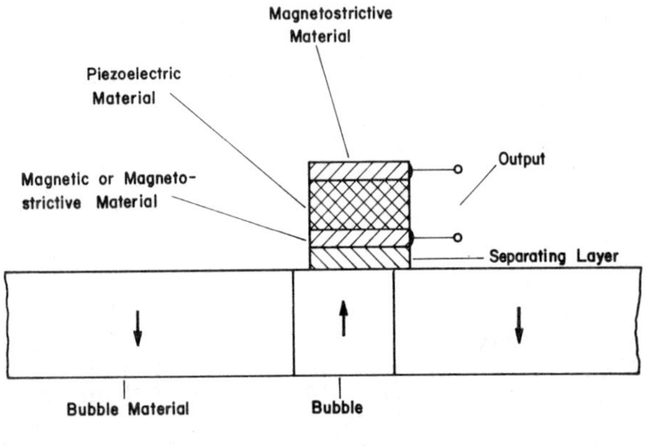

FIG. 1. Schematic representation of the sensor

Figure 1 is a schematic representation of the sensor. The piezoelectric material is sandwiched between two layers of magnetic material of which at least one layer is magnetostrictive. One of the magnetic material layers could be part of the bubble propagating circuit. Since the magnetic material layers are conductive, these layers themselves can serve as the output electrodes.

DEPOSITION OF FILMS

Detectors were fabricated on 2.5 cm x 2.5 cm substrates and were then placed in contact with orthoferrite plates for testing. Both 0.5 mm microscope cover glass and 2 mm alumina (glazed) were used for substrates The cross-section of a magnetostrictive-piezoelectric detector that has been fabricated is shown schematically in Fig. 2. The detector was fabricated by vacuum deposition in a single run using an Edwards 19E-2 high vacuum coating unit.

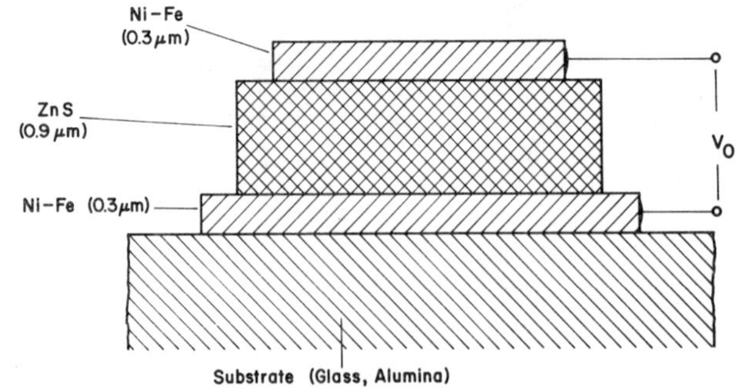

FIG. 2. Cross section of the detector

Two materials have been tried for the first magnetostrictive layer, i.e., 60% Ni - 40% Fe and 55% Ni - 30% Fe - 15% Co alloys. The percentages were chosen to obtain the maximum magnetostriction coefficient for each family of alloys[6]. Vacuum depositions were carried out in a dc magnetic field of 40 Oe at 2 x 10^{-6} Torr and substrate temperature was maintained at 260°C. The evaporation rate was 10 to 15 Å/sec. This low deposition rate was used to obtain low coercivity magnetostrictive-piezoelectric films 3000 Å thick.

The piezoelectric layers were vacuum deposited to a thickness of 1 - 2 μm. Only CdS and ZnS were used. It was observed that the ZnS displayed a better adhesion with the magnetostrictive layer. The substrate was held at 200°C for this evaporation and the rate was roughly 10 Å/sec during the first half of the process and 6 Å/sec during the second half. The graded evaporation rate was used because it has been reported[7] that it increases the alignment of the easy axis in an orientation perpendicular to the plane of the films.

Before evaporating the third layer the device was annealed at 200°C for 12 hours. The third layer was again magnetostrictive and 3000 Å thick. Its deposition was carried out in an identical manner to the first layer.

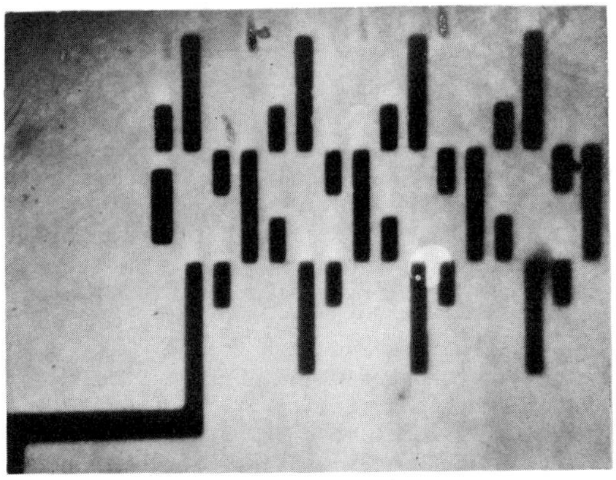

FIG. 3. Top view of the detector

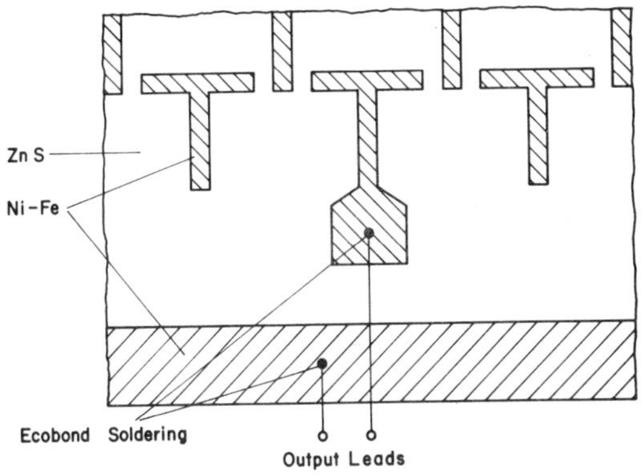

FIG. 4. Access to the detector

PATTERN PRODUCTION

Masks for the propagate and detection circuits were prepared by photolithography using two reduction steps of 20 times each with a Microkon 1700 high resolution camera. Positive photoresist was used as the mask and sulfuric acid was used to etch the pattern.

Figure 3 is a photograph of the circuit fabricated. The end of the long bar is the actual detector and the bar is used as the first output connection. The first layer is not etched and it serves as the other connection to the device. The "parallel-bar" circuit[8] is typical of the propagate circuits tested. The bars are 1 μm wide, the short bars are 2.5 μm long and the long bars are 6 μm long. Figure 4 indicates how the access is made to the sensor.

DEVICE CHARACTERISTICS

Of the detectors fabricated two were tested using both quasi-static and dynamic measurement techniques.

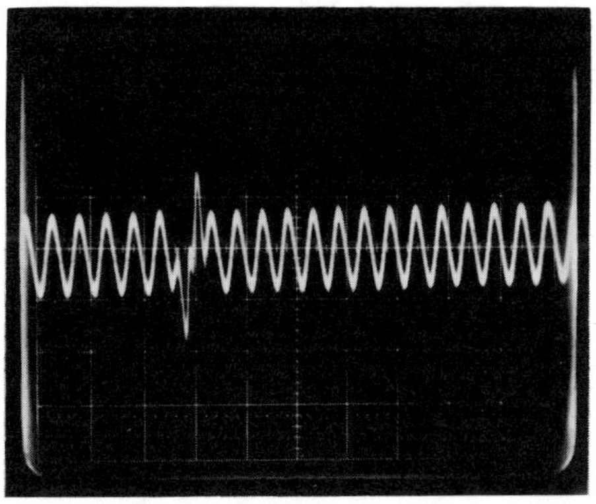

FIG. 5. Oscillogram of the output signal of simulated bubble

The response of the detectors to both the axial and transverse fields was investigated to determine the applicability of the device for use as a bubble detector. For 0.9 μm and 0.6 μm ZnS layers at 100 Hz the average sensitivities were: axial 39 μV/Oe and 26 μV/Oe respectively, and transverse 32 μV/Oe and 26 μV/Oe respectively.

Figure 5 shows the output of the detector in the presence of a transverse field and a simulated bubble field. A hairpin conductor whose width is of the order of a bubble diameter produced a 35 Oe field synchronized to the 100 kHz transverse field. Simulation was used because it is possible to vary parameters in a controllable way. Parallel-bar and T-bar circuits were constructed to be able to propagate the bubble past the detector; however, due to the length of the detector the circuit fabricated operated erraticaly with the bubble hanging up on the detector bar. New designs are being implemented to try to overcome this.

Assuming the bubble field (32.8 Oe) saturates the magnetostrictive layer, the resulting strain is given by

$$\lambda_{sat} = \alpha\lambda_{100}+(1-\alpha)\lambda_{111} \qquad (1)$$

where

$$\alpha = 0.4-0.125\ \ell n\ \frac{2C_{44}}{C_{11}-C_{12}} , \qquad (2)$$

λ_{100} and λ_{111} are the magnetostrictive coefficients, and the C's are the elastic constants[9]. For 60-40 Ni-Fe, λ_{sat} is approximately 8×10^{-6} cm/cm. The reception sensitivity of ZnS is 2×10^{10} V/m per unit strain. Assuming unity coupling a 0.9 μm layer should produce an output of 144 mV. The observed value of 1.3 mV indicates that the coupling is of the order of 1%.

CONCLUSIONS

Tests indicate that it is possible to produce detectors which have an output of tens of millivolts from garnet bubbles by using better device fabrication, geometry and better materials. It is also possible to make the detector relatively insensitive to transverse fields. It is noted that this detector is essentially field sensitive rather than flux sensitive, as in the case of magneto-resistive detectors, and will work better with smaller bubbles if their moments are higher. Tests at 0.5 MHz indicate no decrease in performance; this indicates that the device is essentially a high frequency transducer.

REFERENCES

1. A.H. Bobeck, R.F. Fisher, A.J. Perneski, J.P. Remeika, and L.G. Van Vitert, IEEE Trans. Magn., MAG-5 544-53, Sept. 1969.
2. G.S. Almasi, IEEE Trans. Magn., MAG-7, 370-3, Sept. 1971.
3. W. Strauss and G.E. Smith, J. Appl. Phys., 41, 1169-70, Mar. 1970.
4. G.S. Almasi, G.E. Keefe, Y.S. Lin, and D.A. Thompson, J. Appl. Phys., 42, 1268-9, Mar. 1971.
5. W. Kinsner and E. Della Torre, Faculty of Engineering, McMaster Univ., Hamilton, Ont., Canada, SOC-53, Sept. 1974 (Pat. Appl. May 3, 1973).
6. J.P. Reekstin, J. Appl. Phys., 38, 1449-51, Mar. 1967. H. Weinstein, L. Onyshkevych, K. Karstad, and R. Shahbender, RCA Rev., 28, 317-43, 1967.
7. N.F. Foster, G.A. Coquin, G.A. Rozgonyi, and F.A. Vannatton, IEEE Trans. Sonic Ultrasonic, SU-15, 28-41, Jan. 1968.
8. E. Della Torre and W. Kinsner, IEEE Trans. Magn., MAG-9, 298-303, Sept. 1973. G.Ng, W. Kinsner, and E. Della Torre, AIP Conf. Proc., 18, 127-31, 1974.
9. R.M. Horneich, H. Robinstein, and R.J. Spain, IEEE Trans. Magn., MAG-7, 29-48, Mar. 1971.

* This work was supported by the Defence Research Board, Ottawa, Canada.

CONCENTRATION EFFECTS IN FERROFLUIDS*

E. A. Peterson** and D. A. Krueger, Physics Dept., M. P. Perry and T. B. Jones, Dept.
of Electrical Engineering, Colorado State University, Fort Collins, Colorado 80523

ABSTRACT

The spatial and temporal development of concentration profiles in a 11 cm column of Ferrofluid under various gravitational and magnetic field conditions has been studied using a simple, but sensitive, Colpitts oscillator circuit. Waterbase Ferrofluid exhibits a 100% concentration effect within 1 hour even in a uniform 230 Gauss magnetic field. This effect may be explained qualitatively through formation of large aggregates which then settle gravitationally. An estimate of the aggregate size is made using the measured settling times.

INTRODUCTION

This research was originally intended as an investigation of the extent to which the small (20-200Å) magnetic particles comprising a Ferrofluid[1] concentrate when exposed to a magnetic field gradient. Ferrofluids are colloidal suspensions of single domain magnetic particles in a liquid carrier, such as water or kerosene. The technique used is a simple Colpitts oscillator circuit which has a resonant frequency of 1 MHz. When a column of Ferrofluid is inserted into the inductor coil of this circuit the resonant frequency of the circuit is changed. This frequency change is linearly related to the saturation magnetization of the Ferrofluid inserted and thus to the volume of magnetic material sensed by the coil. Thus by stepping the Ferrofluid column through the inductor coil, a concentration profile of the Ferrofluid column can be obtained. Fig. 1 is a schematic representation of the experimental arrangment and shows a typical profile after exposure to a magnetic field. This technique is sensitive enough to detect the concentration difference due solely to gravitational effects (less than 5% top-to-bottom).

RESULTS

After observing the expected concentration peak with a magnetic field which was strongest below a

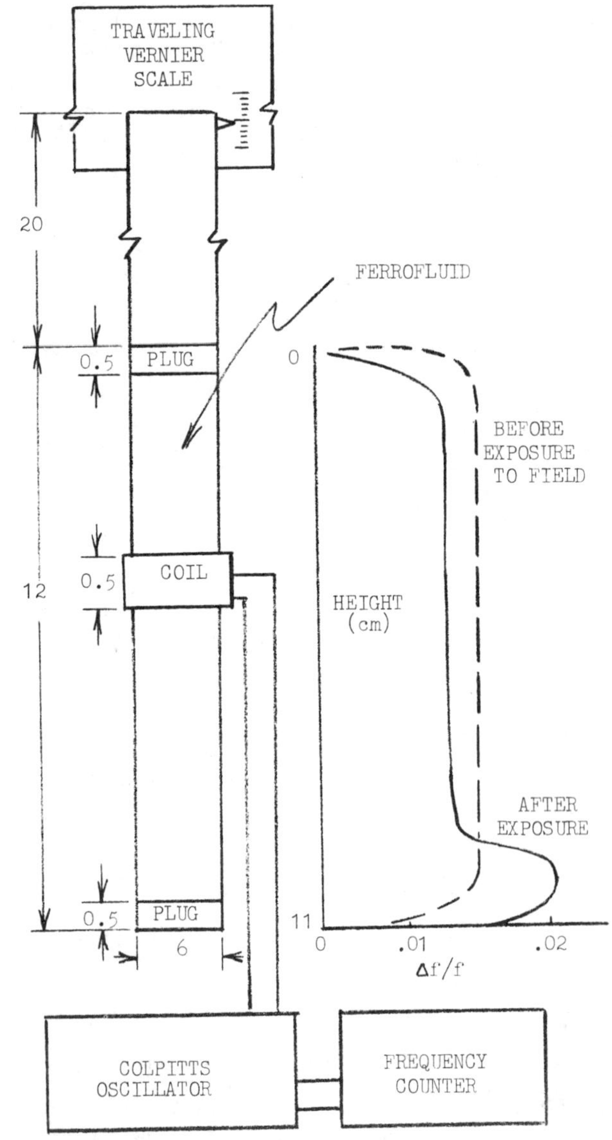

Fig. 1. Schematic diagram of apparatus (dimensions in centimeters)

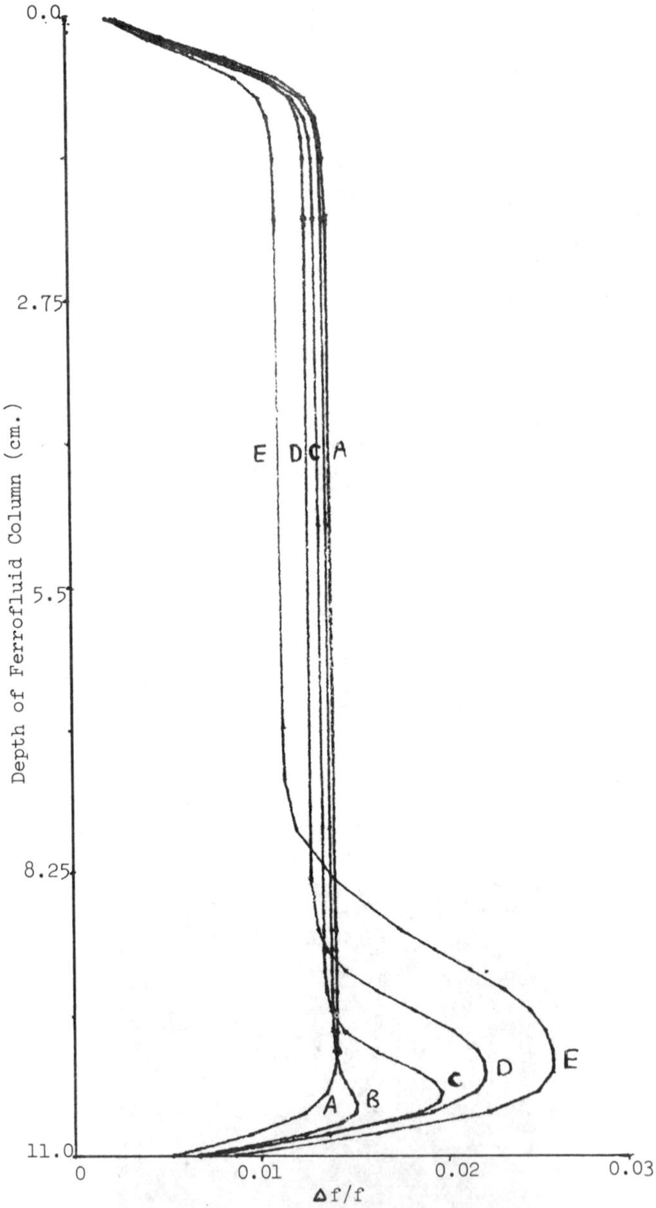

Figure 2 Results of Exposure of a 200 G. Waterbase Ferrofluid Column to a Uniform Magnetic Field Parallel to the Column Axis: A. No Field, B. 2.5 Gauss, C. 11.5 Gauss, D. 46 Gauss, E. 230 Gauss

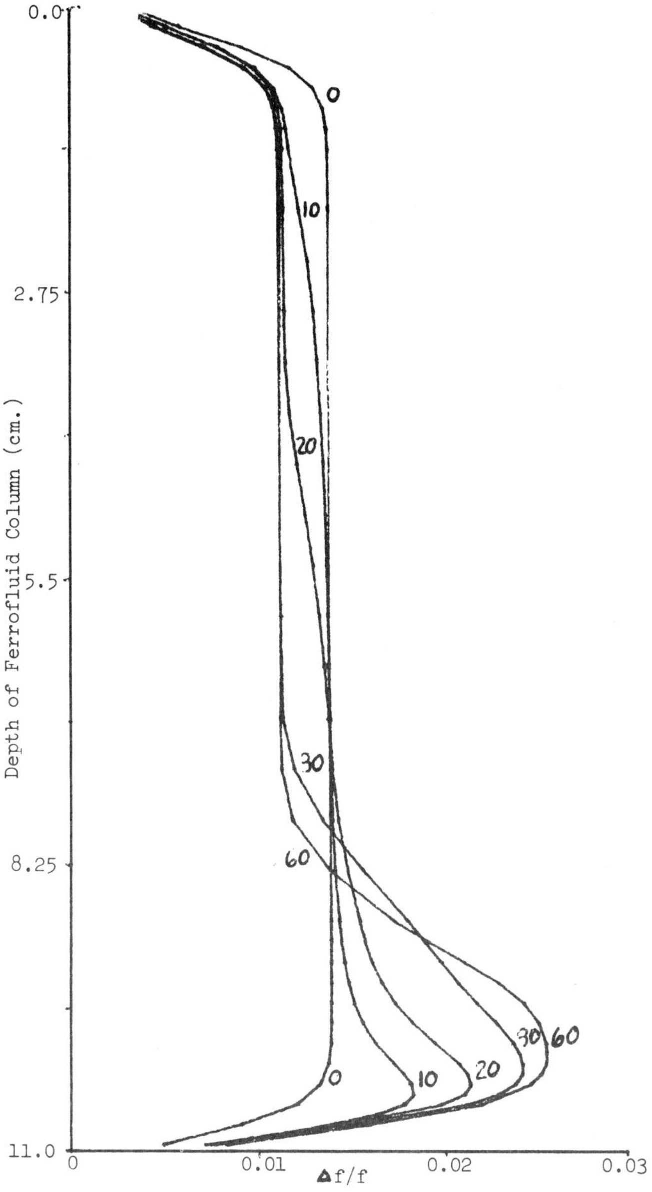

Figure 3 Results of Exposure of a 200 G. Waterbase Ferrofluid Column to a Uniform 230 Gauss Magnetic Field Parallel to the Column Axis. (Numbers indicate the exposure time in minutes.)

vertical Ferrofluid column, the strong field was applied above. In addition to a slight enhancement of the concentration near the top of the column, there appeared a large concentration peak at the bottom. This led to the discovery that even in a uniform magnetic field, either ac or dc, parallel or perpendicular to the column axis, a concentration peak occurs at the bottom. Fig. 2 shows the extent of this for a 200 G waterbase Ferrofluid exposed to various uniform field intensities. It should be noted that even low fields (2.5 G) alter the concentration profile of the Ferrofluid column in a 10 hour exposure. Fig. 3 shows the concentration effect as a function of time for a given uniform magnetic field intensity. During the first five minutes of exposure to a field of 230 Gauss, there are only slight changes in the Ferrofluid column. During the next 25 minutes there is a region of depletion of magnetic particles near the top of the column which progressively extends downward and there is a corresponding concentration enhancement at the bottom of the column. During the next few hours only slight changes in the profile occur indicating that the concentration effect is essentially complete after 1 hr. This time for completion is up to 10 hrs for the lowest field investigated (2.5 G).

The concentration peak shows no tendency to disperse if the column is undisturbed for 40 hrs, but a gentle inversion of the column several times or a vigorous shaking will redisperse the Ferrofluid. After 50 hrs it is still uniformly distributed.

The Quincke method is used to measure the magnetization in dc and ac (60 Hz) applied fields up to 700 G. Thoroughly mixed samples give results consistent with our magnetometer measurements. Samples with a prior exposure to a magnetic field in the Quincke apparatus show a decrease of magnetization which depends upon the cummulative exposure time.

CONCLUSIONS

This concentration effect in a uniform magnetic field is a somewhat unexpected result since in this case there are no magnetic forces present to induce the particles to move as they apparently do. The results can be explained qualitatively by hypothesizing the formation of aggregates under the influence of the magnetic field. Being more massive, the aggregates settle under the influence of gravity. The Quincke results are also consistent with particle aggregation and settling. Bibik[2] has presented optical evidence for aggregation in dilute magnetic liquids but does not indicate observation of any gravitational settling as a consequence of aggregation. Bibik's optical techniques are not amenable to the study of the concentrated Ferrofluids used in practical devices.

The settling time can be used as a rough estimate of the overall aggregate size. Assuming a spherical aggregate, a simple terminal velocity model yields a radius of 5 microns and approximately 10^8 particles per aggregate. DeGennes and Pincus[3] have discussed a theory of chaining in Ferrofluids. For a chain with an aspect ratio of 200 to 1, approximately 10^9 particles are required to give the observed settling time. The settling times indicate that larger fields produce larger aggregates.

We expect electron micrographs to show that the concentration peak has a particle size distribution skewed toward the larger particles while the upper regions will show a depletion of the larger particles. Inferences drawn by comparing the concentration peaks of Ferrofluids with different size distributions indicate that the larger size particles may serve as "condensation nuclei" about which the aggregates form. Unfortunately the dilution required to make a sample for the electron microscope breaks up the aggregates and thus precludes a direct measurement of aggregate size.

The views expressed herein are those of the authors and do not necessarily reflect the views of the United States Air Force or the Department of Defense.

*Sponsored in part by ONR Grant No. N00014-67-A-0299-0020.

**Supported under AFIT-CID program. Present address: Dept. of Physics, U.S. Air Force Academy, CO 80840.

1. Ferrofluidics Corp., Burlington, Mass. 200G Waterbase Ferrofluid Lot #F-417D.
2. E. E. Bibik, I. S. Lavrov and O. M. Merkushev Koll. Zh., 28. 631-634 (1966)
3. P. G. de Gennes and P. A. Pincus, Phys. kondens Materie 11. 189-198 (1970)

TERNARY AMORPHOUS ALLOYS FOR BUBBLE DOMAIN APPLICATIONS

P. Chaudhari, J. J. Cuomo, R. J. Gambino, S. Kirkpatrick, and L. J. Tao
IBM Research Center, Yorktown Heights, N.Y. 10598.

ABSTRACT

The drive field for bubble domain propagation in T-bar devices is proportional to $4\pi M_s$. From the definitions of the characteristic length, ℓ, and of Q it can be shown that $4\pi M_s$ is proportional to AQ, where A is the exchange stiffness.

We have found that the magnetic properties of ternary amorphous systems of the type Gd-Co-X, where X is a nonmagnetic element, eg. Au, Cu, Mo can be fine tuned for bubble device applications. It is possible to lower A by magnetic dilution as well as adjust $4\pi M_s$ by varying the Co/Gd ratio as in the binary system. Because the system is amorphous, we are not restricted by equilibrium solubility limits.

Kryder et al.[1] have shown that in T-I bar type bubble devices the drive field for propagation is directly proportional to the saturation magnetization of the bubble material. George, et al.[2] and Almasi, et al.[3] have shown this relationship follows from models which consider the potential well associated with the tip of a permalloy circuit element. From these considerations, it is clear that a low $4\pi M_s$ bubble material is advantageous. Furthermore, the bias field requirements for a high $4\pi M_s$ material impact the cost and complexity of device package.

In this paper, we discuss the design of amorphous magnetic alloy compositions with low magnetization for a given domain size. As an example, we will describe the design of a material suitable for the device discussed by Kryder[1] i.e., a T-I bar device with 2μm diameter (d_o) domains. The required ℓ parameter is given by[4] 1/8 d_o or 0.25μm and the thickness, h, is 4ℓ or 1μm.

Using the definations of the characteristic length, ℓ, and quality factor Q we can write ℓ as:

$$\ell = \frac{\sqrt{32\pi}\sqrt{AQ}}{4\pi M_s} \qquad (1)$$

The only material parameter in equation 1 not specified explicitely by device requirements is the exchange stiffness, A. However, once ℓ, $4\pi M_s$ and Q are specified, A is defined.

The exchange stiffness at absolute zero, A_o, of a magnetic system depends on all the exhange constants e.g. in Gd-Co on λCoCo, λGdCo and λGdGd. The A value is also temperature dependent going to zero at the Curie point. For a fixed ℓ and Q, a low $4\pi M_s$ implies a low A according to equation 1.

We have studied systems of the type Gd-Co-X where X is a non magnetic element such as Mo, Au, Cu and Cr. In these systems, A can be adjusted by several mechanisms. The exchange constants depend on the number of magnetic nearest neighbors, z, which is lowered by magnetic dilution. Further, there is evidence that the cobalt moment is lowered by d band filling when cobalt is alloyed with Gd[5], Mo or Cr[6]. Since the exchange constants depend on the individual atomic moments, A_o will decrease. In addition, lowering the Curie temperature will decrease A at room temperature.

Films of selected ternary compositions were prepared by bias sputtering. Arc melted button ingots of the ternary compositions were made from the elements. Several of these button ingots were then arc melted onto a molybdenum plate to form a thin layer. The molybdenum plate also serves as a means for mounting the target in the sputtering system. The film composition can be adjusted from the target composition by applying a negative bias voltage to the substrate during growth. Gadolinium resputters preferentially so the films are richer in cobalt than the target. It has been shown that the Co/Gd ratio in the film depends on the bias voltage according to the equation:[7]

$$Co/Gd_{film} = \frac{Co/Gd_{target}}{1 - K\frac{V_b}{V_t}} \qquad (2)$$

Where V_b is the effective bias voltage, V_t is the target voltage and K is a constant which depends on the system geometry.

The ℓ parameters of the films were determined from the thickness, h, and zero field stripwidth, w_s, in the usual manner. Magnetization as a function of temperature was measured on selected samples by means of a VSM or force balance.

A portion of the ternary system Gd-Co-Mo is shown in Fig. 1. The film compositions produced from several

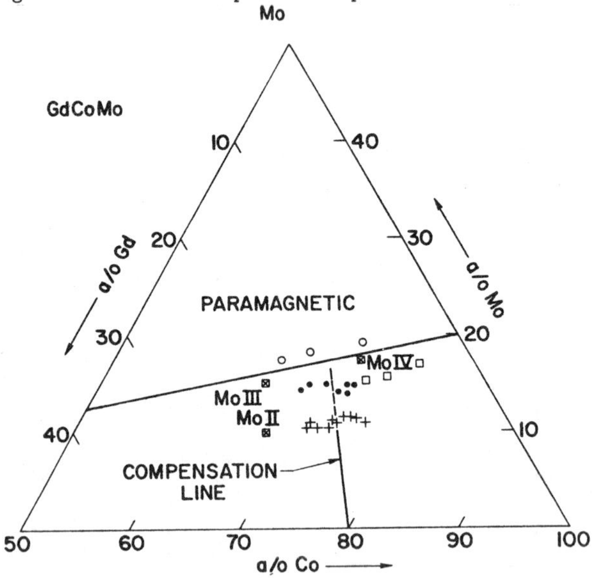

Figure 1 A portion of the Gd-Co-Mo system. Target compositions are indicated by ⊠. The film compositions shown were produced from targets containing 10 % Mo (+), 15% Mo(•) and 17.5% Mo(□)

different target compositions are shown. With increasing bias voltage, the film composition moves toward the Co-Mo binary axis, along a line of approximately constant Mo concentration. In the Co-Mo binary system, the Curie temperature extrapolates to room temperature at about 20 atomic percent Mo, i.e., at a Mo/Mo+Co ratio of 0.20. Compositions with Mo/Co+Mo greater than 0.20 are found to be paramagnetic at room temperature in the ternary system as well (open circles in figure 1). As the Co/Gd ratio decreases at a fixed Mo concentration, the magnetization at room temperature decreases to a compensated composition then increases at still lower Co concentrations. The line in figure 1 labeled compensation line defines compositions which are compensated at room temperature.

All the Gd-Co-X ternary systems studied can be similarly described in terms of a paramagnetic boundry, defined by compositions with Curie points at room temperature and a compensation line defined by compositions which are compensated at room temperature. The paramagnetic boundry terminates on the Gd-Co axis at about 50 atomic % Gd and on the Co-X axis at the composition in that binary system where the Curie point is at room temperature. The terminus on the Co-X axis can be estimated from the crystalline binary alloys.

The compensation line which terminates at the Gd-Co binary axis at about 80 atomic percent Cobalt is determined by the Co atomic moment which may be lowered by d band filling from the X element. One might expect the atomic fraction of Gd required to compensate the Cobalt to be lowered in this case. However, the magnetization of the Gd sublattice at room temperature depends on the effective field of the Cobalt sublattice and on the d-f exchange. We find that the compensation lie can shift either to lower or higher Gd fractions depending on which X element is used. This is illustrated in Table I. Apparently, two competing effects are at

TABLE I

System	Atomic fraction of X	Co/Gd ratio	Co/Co+Gd Atomic
Gd-Co	0	4.0	0.80
Gd-Co-Cu	0.11	3.5	0.785
Gd-Co-Au	0.11	3.2	0.760
Gd-Co-Mo	0.11	4.7	0.825
Gd-Co-Cr	0.11	4.6	0.820

work. In the Mo and Cr systems, the lowering of the Co moment dominates and less Gd is required to compensate. In the Au and Cu systems, the Gd-Co exchange is weakened by dilution faster than the Co moment is reduced, and more Gd is required to obtain a compensated composition at room temperature.

A graphical presentation[8] of equation 1 is a convienient method for assessing the effectiveness of an alloy addition for reducing the exchange stiffness. Figure 2

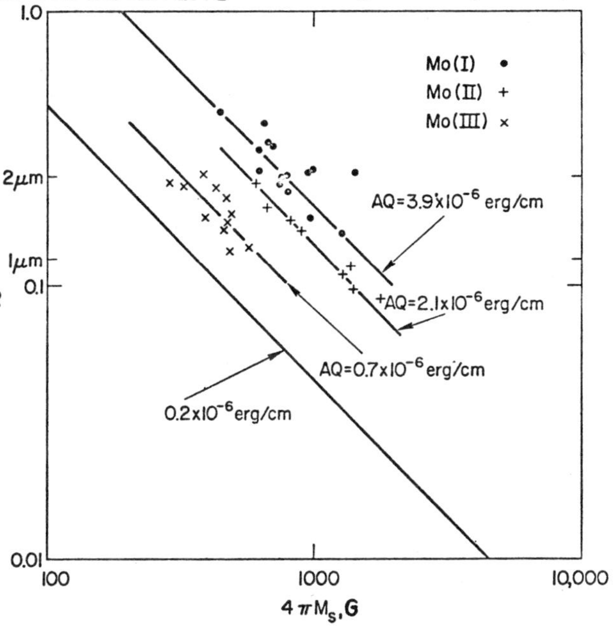

Figure 2 The ℓ parameter vs $4\pi M_s$ values are shown plotted for films prepared from targets containing various concentrations of Mo. The Mo(I) target (•) contained 7% Mo, Mo (II), (+), 10% Mo and Mo (III),(X), 15% Mo.

is a log - log plot of ℓ parameter vs $4\pi M_s$ for films in the Gd-Co-Mo system. The ℓ parameters appropriate for 1μm and 2μm domains are indicated. The ℓ parameters and $4\pi M_s$ values for various samples are plotted. Note that as the Mo concentration is increased, the AQ product of the films decrease. The Q depends somewhat on the conditions of deposition which accounts for the scatter of the data.

A number of films with ℓ parameters suitable for 2μm domains are shown in Table II. Note that the magnetiza-

TABLE II

System Gd-Co-X	Atomic Fraction of X	ℓμm	$4\pi M_s$ Units G	AQ Units erg/cm
Gd-Co	0	0.28	1260	12.5×10^{-6}
Gd-Co-Au	0.11	0.29	1138	10.9×10^{-6}
Gd-Co-Mo	0.09	0.26	970	6.4×10^{-6}
Gd-Co-Mo	0.14	0.26	580	2.2×10^{-6}
Ge-Co-Mo	0.16	0.21	480	1.0×10^{-6}

tion of a Gd-Co-Au sample with 11 atomic % is only slightly lower than that of a Gd-Co-film with approximately the same ℓ parameter. The addition of Mo is very effective in lowering the magnetization for films even with slightly smaller ℓ parameters. The film with 16 atomic %, Mo has an AQ product an order of magnitude lower than that of the binary alloy.

The temperature dependence of the magnetic properties of a bubble material is an important consideration. In particular, it is necessary that the magnetization be as nearly constant as possible near the operating point of the device. The required behavior of magnetization as a function of temperature can only be obtained if both the Curie temperature and the compensation temperature are far from room temperature. In the Gd-Co-Mo system, the best temperature behavior is obtained on the Co rich side of the compensation line in films containing 12 to 15 atomic percent Mo. Work is still in progress on the temperature dependence of the magnetization in the other systems.

It is clear from this study that ternary amorphous alloys offer considerable flexibility in materials design for bubble applications. The magnetization for 2μm domain materials can be lowered by a factor of three by Mo addition. The drive field for propagation is correspondingly reduced.[1]

ACKNOWLEDGEMENTS

The authors thank S. M. Kane for sample preparation, H. Lilienthal for magnetic measurements and F. Cardone, S. O. Ellmann and R. Schad for microprobe analysis.

REFERENCES

1. M. H. Kryder, K. Y. Ahn, G. S. Almasi, G. E. Keefe and J. V. Powers. IEEE Trans. on Mag. 10 825-827, (1974)
2. P. K. George, A. J. Hughes and J. L. Archer IEEE Trans. on Mag. 10 821-824 (1974)
3. G. S. Almasi, Y. S. Lin, E. Munro and S. Slusarczuk (These Proceedings.)
4. A. A. Thiele, Bell System Tech. J. 48 3287 (1969)
5. L. J. Tao, R. J. Gambino, S. Kirkpatrick, J. J. Cuomo and H. Lilienthal A.I.P. Conf. Proceedings 18 641-645 (1974)
6. R. M. Bozorth, Ferromagnetism, 2nd Printing D. Van Nostrand Co. Inc., New York, 1953 Chapter 8.
7. J. J. Cuomo and R. J. Gambino J. Vac. Science and Tech. To be published, Proceeding Issue. 21st Am. Vac. Soc. Symposium.
8. H. L. Hu, private communication April 1973.

POLAR KERR ROTATION AND SUBLATTICE MAGNETIZATION IN GdCoMo BUBBLE FILMS

B. E. Argyle, R. J. Gambino, and K. Y. Ahn
IBM Thomas J. Watson Research Center, Yorktown Heights, New York 10598

ABSTRACT

Kerr rotation in Gd-Co-Mo alloys is due mostly to the cobalt atoms. We investigate Polar Kerr loops vs composition and temperature, and determine total magnetization (from initial loop susceptibility and domain size) and the cobalt sublattice magnetization (from the saturation Kerr rotation angle). Deviations from the characteristic Kooy and Enz loop behavior due to thin ~100Å surface layer anomalies are first identified and removed in order to analyse the bulk film behavior.

INTRODUCTION

The discovery that amorphous GdCo films support mobile magnetic bubbles[1], naturally leads to questions about optimum compositions for temperature stability and operating range. However, in contrast to the history of the development of orthoferrite and garnet bubble materials where there was considerable background data and theory for sublattice moment behavior available for example from NMR studies on bulk crystals, the amorphous GdCo alloys are seemingly new "uncharted water". This paper discribes a technique to extract 'sublattice' behavior. We utilize the polar Kerr rotation to measure cobalt sublattice moment and the Kérr loop's shape to determine total magnetization. We apply the method here to the GdCoMo ternary which is currently of interest because, while contributing no moment of its own, the Mo modifies the Co moment via d-band filling. Chaudhari et al.[2] concurrently discuss properties and fabrication of GdCoMo while Hasegawa et al.[3] describe magnetic data in terms of a modified Néel two-sublattice model. After identifying and then avoiding anomalous loop shapes caused by surface inhomogeneities, we extract sublattice behavior and compare results with theory.

EXPERIMENTAL METHODS

The samples studied are amorphous films, 1-2 μm thick, deposited on glass by rf sputtering[1] from several arc-melted targets. The films ranged in composition over $11 < X_{Gd} < 18$, $66 < X_{Co} < 83$ and $8 < X_{Mo} < 17$ atomic percent as measured by electron microprobe analysis. Demagnetized strip widths W_s were measured using a technique[4] of diffracting laser light from a deposited ferrofluid film which decorates domain walls. Kerr rotation vs field perpendicular to the film was measured in a system equipped with a polarized He-Ne laser and a Wollaston polarizing beam splitter with two photodetectors followed by differential amplifier and X-Y recorder for plotting signals proportional to the Kerr rotation. A Bell 705 Hall probe plus Astrodata 121 RZ dc amplifier supplied signals proportional to field. Sample temperature was controlled between - 100 and +100°C using flowing nitrogen gas heated with an in-line nichrome resistor powered by a Eurotherm controller and a thermocouple located in the copper sample block.

RESULTS AND DISCUSSION

In the polar Kerr effect polarized light penetrating a skin depth undergoes rotation proportional to the magnetization component parallel to the light. In a bubble film measured with perpendicular applied fields the Kerr loop, θ_K vs $H^{\perp}$, has the characteristic shape first observed by Kooy and Enz[5] in Faraday loops taken on uniaxial hexaferrite platelets. Measurements on a GdCoMo amorphous alloy are presented in Fig. 1. With increasing $H^{\perp}$ an initial linear region is followed by curvature pointing always away from the field axis and ending abruptly at a plateau where saturation occurs due to collapse of stripes and/or bubbles. On decreasing $H^{\perp}$, θ_K shows a discontinuity of nucleation below saturation. The initial slope

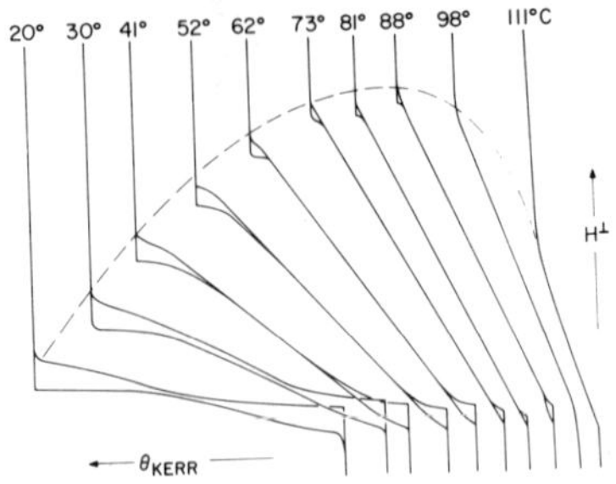

FIG. 1 Polar Kerr loops in $(Gd_{.15}Co_{.85})_{.85}Mo_{.15}$. Temperature behavior of total magnetization is reflected in collapse fields (dashed curve) or slopes $\partial\theta_K/\partial H^{\perp}$ near loop centers. Cobalt sublattice magnetization is reflected in saturation Kerr rotation.

$\partial\theta_K/\partial H^{\perp}$ is related to susceptibility $\chi_o = \partial M/\partial H^{\perp}$ which is uniquely defined in terms of strip width and height for a material with infinite uniaxial anisotropy H_K. Considering the Kerr loop as M/M_s plotted vs $H^{\perp}$ we determine total magnetization M_s from comparing the intial slope χ_o/M_s with χ_o theoretically calculated[6-8]. Shaw, et al.[8] show χ_o also depends weakly on the strength of the anisotropy field, H_K. This correction being small ($\lesssim 5\%$) for our films, we apply the approximate calculation of Craik[7]. While this procedure is an indirect method of determining total M_s, we have tested it satisfactorily for both garnets and amorphous bubble materials using a vibrating sample magnetometer.

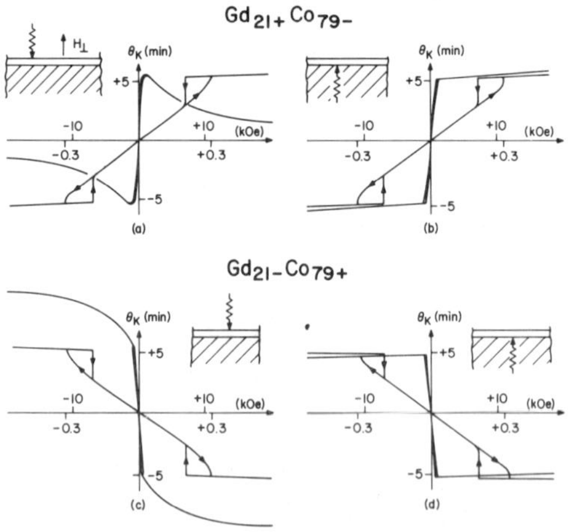

FIG. 2 Anomalous Kerr loops typical of sputtered amorphous films exposed to air. Sign of inner loop depends on composition relative to compensation $(Gd_{21}Co_{79})$. Loop outer tail is due to surface layer inhomogeneity (a, c) or glass substrate (b, d).

Anomalous deviations from the Kooy and Enz loop shape occur at both surfaces under certain conditions. The high field anomaly at the film-air interface exhibited in Figures 2a and 2c is associated with depletion of Gd metal by preferential oxidation. Earlier samples exhibited a film-glass interface anomaly due to an initial transient in the sputtering process. This anomaly is absent in Fig. 2b and 2d and subsequent materials because we presputtered the target into equilibrium for 1/2 hour before opening a shutter to deposit the film. The air-exposed surface anomaly can be eliminated by a post-sputter deposition of protective SiO_2 before the specimen is removed from the Argon atmosphere. The extra rotation of the oxidized surface after subtracting the inner loop always has (I) negative contribution to θ_K, (II) curvature pointing toward the H-axis, and (III) assymptotic approach to saturation at high kilo-Oersted fields (Fig. 2a and 2c). We note (II) and (III) are characteristics of a thin layer magnetized in-plane and having large M_s; (I) signifies this layer is dominated magnetically by the cobalt constituent. The thickness must be on the order of the optical skin depth or less (~ 100Å) in order to account for the presence of the inner loop. Auger spectroscopy[9] has detected both non-magnetic Gd_2O_3 and Co-enriched GdCoMo in the top 50 - 100 Å.

After removing surface anomalies, we investigated the compositional dependence of saturation θ_K^s. We observe that pure Gd exhibits weak response $\partial\theta_K/\partial H^{\perp} = 4 \times 10^{-4}$ min./Oe at 23°C, and $\theta_K^s \simeq 1$ min. at 0°C and $\simeq 4$ min. at −55°C. Values of −17 min. for sputtered cobalt[10] and −21 min. for pure crystalline cobalt[12] have been reported. We adopt the value −20 min. in this present analysis. For $(Gd_{1-x}Co_x)_{1-y}Mo_y$ we equate Kerr rotation and magnetic moment per cobalt atom each reduced by their respective values for pure cobalt, giving

$$\theta_K(x,y)/-20 = \mu_{Co}(x,y)/1.72 \qquad (1)$$

where $\mu_{Co}(x,y)$ can be estimated from the behavior of the binaries, GdCo and CoMo. In GdCo in the composition range of interest the Co-moment determined from high temperature susceptibility of several well ordered alloys[13] is $\sim 1.65\mu_B$. In the CoMo binary[14], μ_{Co} falls off rapidly with increasing atomic fraction of Mo. A linear combination suggests for the ternary

$$\mu_{Co}(x,y) = 1.65 - 5y/[y+(1-y)x] \quad (\mu_B/\text{atom}) \qquad (2)$$

This expression has also been found suitable by Hasegawa, et al.[3] in fitting magnetization data for the ternary to a modified Néel two-sublattice theory. Combining (1) and (2) gives for θ_K^s at T = 0°K,

$$\theta_K{}^{\circ}(x,y) = \frac{-20}{1.72}\,(1.65 - 5y/[y+(1-y)x] \text{ min.} \qquad (3)$$

In Fig. 3 this expression plotted as a function of composition exhibits rapid decreases in θ_K^s with Mo concentration. Experimental data at 300°K on several compositions are also presented along with points calculated from Eq. 3 and temperature shifts expected for 300°K from the fitted Néel theory[3].

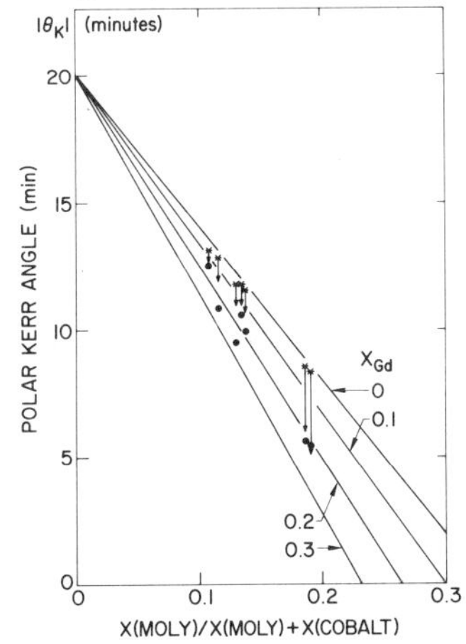

FIG. 3 $|\theta_K^s|$ vs. composition in GdCoMo. Solid lines from Eq. 3 for 0° K; (*) calculations for samples at 0° K (Eq. 3); (↓) 300° K temperature shifts predicted by Néel two-sublattice model (Ref. 3).

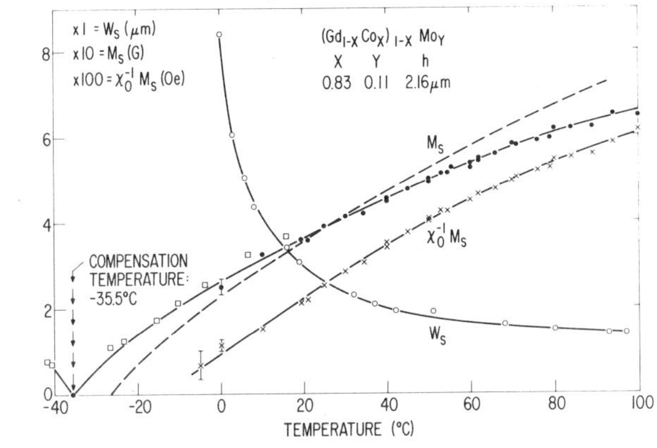

FIG. 4 Magnetization (·) derived from stripe widths and Kerr loop susceptibility (X). VSM data (□) taken near T_{comp}. Dashed line is theory based on Eq. 2 and Néel two-sublattice model (Ref. 3).

Encouraged by reasonable predictions for $\theta_K^s(x,y)$ according to this simple model (Eqs. 1-3), we proceed with a complete magnetic analysis. Results in Fig. 4 and 5 are for a 2.16 μm thick film having x=0.83 and y=0.11. Total M_s values (filled circles) were determined by combining $\chi_0^{-1}M_s$ (crosses) with theoretical χ_0 calculated[7] for strip domains of width W_s (open circles) and height h. Also shown in Fig. 4 are M_s data (squares) obtained directly using a vibrating sample magnetometer. The compensation point is obtained accurately by the temperature of maximum coercivity and/or where θ_K^s changes sign. The dashed line in Fig. 4 and solid lines in

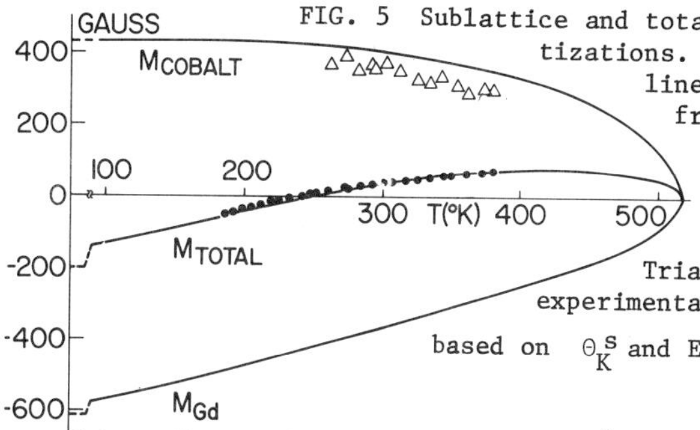

FIG. 5 Sublattice and total magnetizations. Solid line: theory from Ref. 3. Points: M_s data from Fig. 4. Triangles: experimental M_{Co} based on θ_K^s and Eq. 4.

Fig. 5 are theoretical results of the Néel two sublattice model in which Hasegawa[3] has used (1) coupled Brillouin functions for M_{Co} and M_{Gd} sublattices, (2) $\mu_{Gd} = 7\mu_B$ and $\mu_{Co} = 1.0035\mu_B$ from Eq. 2, and (3) fitted coupling constants: $J_{CoCo} = 1.5 \times 10^{-14}$ ergs, $J_{GdCo} = -2.3 \times 10^{-15}$ and $J_{GdGd} = 2.0 \times 10^{-16}$. The quality of fit in Fig. 4 is quite suitable when viewed in the broader perspective of Fig. 5 showing both theoretical sublattice magnetizations $\bar{M}_{Co}$ and $\bar{M}_{Gd}$ along with theoretical and experimental $\bar{M}_{TOTAL} = \bar{M}_{Co} - \bar{M}_{Gd}$.

Experimental cobalt sublattice data $M_{Co}(T)$ in Fig. 5 were determined from observed rotations $\theta_K^s(T)$ using

$$M_{Co}(T) = M_{Co}(0)\ \theta_K{}^s(T)/\theta_K{}^s(0) = 36.8\theta_K(T). \qquad (4)$$

The coefficient was determined from $\theta_K^s(0) = -11.8$ min. (Eq. 3) and from converting moment per cobalt atom, 1.0035 (Eq. 2), to sublattice magnetization, $M_{Co}(0) = \mu_{Co}\mu_B N(1-y)x = 434$ G where $N = 6.3 \times 10^{22}$ Cobalt atoms/cm^3 [15]. While the trend of $M_{Co}(T)$ in Fig. 5 agrees with theory, the ~15% discrepancy in magnitude may be related to the 20% separation in θ_K^s- values for sputtered vs crystalline cobalt. An estimated, smaller contribution from Gd, neglected in this analysis, has the wrong sign to account for this discrepancy. While further work is necessary to reduce this uncertainty we tentatively conclude that sublattice behavior in amorphous GdCoMo is measurable from the saturation Kerr rotation. The determination of total magnetization in the same localized area from initial slopes and strip widths suggests this new form of magnetic analysis may be useful in tailoring compositions for temperature stability in bubble devices.

We gratefully acknowledge L. Buszko, R. Ruf and S. Ellman for valuable assistance and R. Hasegawa for helpful discussions and computer calculated theory.

REFERENCES

1. P. Chandhari, J.J. Cuomo, and R.J. Gambino, Appl. Phys. Lett. 23, p. 337 (1973), IBM Jl. Res. Dev. 17, 66 (1973).
2. P. Chaudhari, J.J. Cuomo, R.J. Gambino, S. Kirkpatrick, and L.J. Tao (this Conference).
3. R. Hasegawa, B.E. Argyle and L.J. Tao, (this Conference).
4. R. Ruf and R.J. Gambino, to be published.
5. C. Kooy and U. Enz, Philips Res. Repts 15, 7 (1960).
6. J.A. Cape and G.W. Lehman, J. Appl. Phys. 42, 5732

(1971).
7. D.J. Craik, Physics Letters 39A, 45 (1972); D.J. Craik, P.V. Cogser and G. Myre, J. Phys. D. 6, 872 (1973).
8. R.W. Shaw, D.E. Hill, R.M. Landfort, and J.W. Moody, J. Appl. Phys. 44, 2346 (1973).
9. W. Molzen, private communication.
10. J.L. Erskine and E.A. Stern, Phys. Rev. B 8, 1239 (1973).
11. L.R. Ingersoll, J. Opt. Soc. Am. 8, 493 (1924).
12. S.G. Barker, Proc. Phys. Soc. (London) 29, 1 (1917).
13. E. Burzo, Phys. Rev. B 6, 2882 (1972).
14. R.M. Bozorth, Ferromagnetism (D. Van Nostrand Co., N.Y. 1951) P. 292.
15. R. Hasegawa, J. Appl. Phys. 45, 3109 (1974).

MAGNETIC PROPERTIES OF GdFe AMORPHOUS ALLOY FILMS PREPARED BY SPUTTERING

T. Kobayashi, N. Imamura, and Y. Mimura
KDD Research and Development Laboratory, Nakameguro, Tokyo, Japan

ABSTRACT

We studied Gd_xFe_{1-x} amorphous films ($0.22<x<0.34$) prepared by sputtering onto mirror polished surfaces of Si single crystal plates. The saturation magnetic moment M_s, coercive force H_c, perpendicular anisotropy field H_k, and nucleation field H_n were determined as a function of composition from M-H loops measured with a vibrating sample magnetometer. The average perpendicular anisotropy constant K_u was $4x10^5$ erg/cm^3 in the composition range $0.22<x<0.34$. Magnetic domains corresponding to various types of M-H loop were also observed. It was found that in the composition region having minimum M_s, the shape of the M-H loop was rectangular, and the magnetic domain structure in the remanent state was a spark-like pattern, that is, strip domains diverged radially from one point. In other composition regions stripe domains could be observed in the remanent state.

ANNEALING BEHAVIOR OF AMORPHOUS Gd-Co-Mo THIN FILMS

R. J. Kobliska and A. Gangulee
IBM Thomas J. Watson Research Center, Yorktown Heights, New York 10598

ABSTRACT

Long term thermal stability is a prerequisite for bubble film materials. Some results of an investigation of the isothermal annealing behavior of amorphous Gd-Co-Mo thin films, in the temperature range 150°-350°C, are being reported. All functional magnetic properties were retained by these films even after annealing at 350°C. The room temperature magnetization and the stripe collapse field decreased upon annealing and then stabilized, the magnitude of the decrement depending on the annealing temperature. The activation energy associated with the annealing process is about 0.4 eV. The results are interpreted in terms of corresponding changes in the compensation temperature, and they indicate that stabilization of the magnetic properties prior to bubble device fabrication may be a desirable process step.

INTRODUCTION

Long term thermal stability of the magnetic properties of amorphous thin films is a fundamental prerequisite to their eventual use in bubble domain memories. Specifically, it is crucial to understand the annealing behavior of the anisotropy K_u and the room temperature magnetization M, so as to be able to evaluate the probability of a device failure either during fabrication or in use.

Hasegawa and co-workers[1,2] have shown that it is possible to alter the magnetic properties of amorphous Gd-Co thin films by annealing them. In particular, they observed a large decrease in the anisotropy K_u and small changes in the room temperature magnetization M with annealing at temperatures above 180°C.

In order to improve device operation, it is necessary to reduce the room temperature magnetization of the amorphous Gd-Co alloys to values comparable to those of garnet films. One way of achieving this is by ternary alloy additions, such as Au, Cu or Mo.[3] In this paper, we report some results of an investigation of the effects of annealing on amorphous Gd-Co-Mo thin films.

EXPERIMENTAL PROCEDURES

The amorphous Gd-Co-Mo thin films were RF bias sputtered, details of the sputtering system and procedure have been discussed elsewhere[4]. The films were 2.5 μm thick with a nominal composition at 70 at% Co, 15 at% Gd and 15 at% Mo. The initial stripe width was about 2 μm with a compensation temperature of about -40°C; the room temperature magnetization (4πM) after deposition was about 375 Gauss and the corresponding stripe collapse field was about 250 Oersteds.

The collapse field was measured in a Kerr effect hysteresigraph with a lock-in detection scheme so as to improve the signal to noise ratio. The Kerr signal, synchronous with a small 300 Hz magnetic field applied perpendicular to the sample, was recorded against the applied field. The extraordinary Hall effect of Gd-Co alloys[5] was also used to obtain the perpendicular hysteresis loops of these films.

The room temperature magnetization M and the in-plane nucleation field H''_k were measured in an in-plane inductive hysteresigraph. Demagnetizing fields from the individual domains and the shape anisotropy of a thin platelet are responsible for the difference between the field H''_k and the anisotropy field H_k. This difference is significant when a single domain exists, in which case $H''_k = H_k - 4\pi M$. The perpendicular component of the in-plane field in the inductive hysteresigraph is small enough, so that a multi-domain situation exists and homogeneous nucleation should occur; details of such homogeneous nucleation have been discussed elsewhere.[6] The compensation temperature of these films was measured in a force balance magnetometer.

It should be noted here, that the Kerr method measured properties of the film only within a 2 mm diameter spot. In contrast, the Hall method measured averaged properties over much larger volumes, and the inductive method measured properties averaged over the entire film volume. Any non-uniformity within these films would make the correlation between these results quite complicated.[7] However, the specimens employed in this investigation were reasonably uniform.

The amorphous Gd-Co-Mo films were isothermally annealed at various temperatures in the range 150° - 350°C in a vacuum of 5 x 10^{-6} torr without any external magnetic field. The magnetic properties were measured at room temperature before and after each annealing step. The Gd, Co, Mo, Ar and O concentration in these films were measured before and after annealing in an electron probe microanalyzer; the distribution of these elements normal to the film plane was measured before and after annealing by He^+ back-scattering techniques.

RESULTS AND DISCUSSION

The room temperature magnetization M, and the stripe collapse field H_{col} decreased with annealing, whereas there were corresponding increases in the in-plane nucleation field H''_k and the stripe width. The change in the stripe collapse field at various temperatures (normalized to the initial or unannealed value) are shown in Figure 1. It appears that at each temperature, the stripe collapse field reaches an equilibrium on saturation value which is characteristic of that temperature. This saturation value of H_{col} is linear with temperature in the range 150° - 300°C, but at higher temperatures tends to approach a stable final value independent of the annealing temperature.

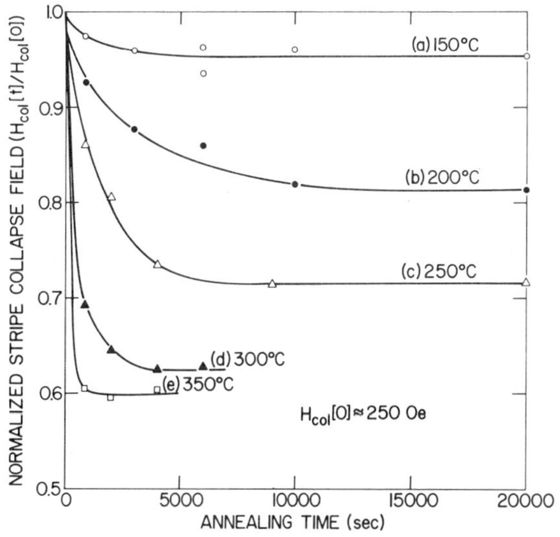

Figure 1

Changes in stripe collapse field as a result of annealing at various temperatures.

Additional experiments were carried out in order to investigate the dependence of the saturation value of H_{col} on the annealing temperature. Specimen b, previously stabilized at 200°C, was annealed for 10^3 seconds at 300°C, upon which its H_{col} decreased to a value close to what is characteristic for 300°C. On the other hand, specimen d which was previously stabilized at 300°C, did not show any change in H_{col} after annealing for 10^4 seconds at 200°C. These results indicate that the changes in the magnetic properties are apparently caused by irreversible changes in the amorphous Gd-Co-Mo thin films.

The compensation temperature of these films, as

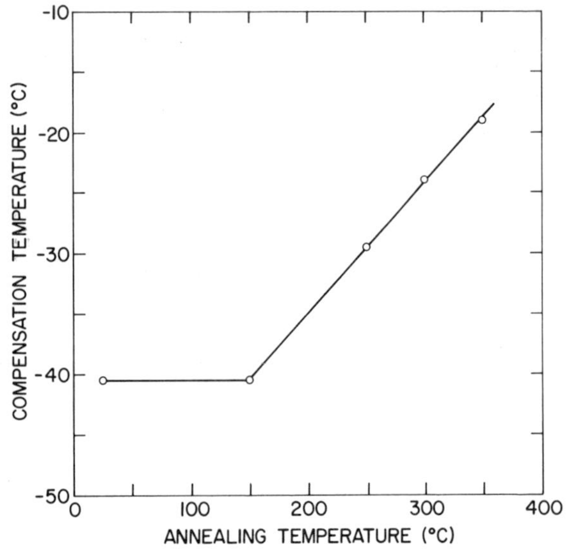

Figure 2

Changes in the compensation temperature after stabilizing at various annealing temperatures.

measured in a force balance magnetometer, showed definite changes as a result of annealing. The compensation temperatures of specimens stabilized at various annealing temperature are shown in Figure 2. The observed changes in the room temperature values of M and H''_k can immediately be interpreted as arising from shifts in the compensation temperature. In particular, if a simple functional dependence of the form (valid only when dM/dT at room temperature is positive, as in these films)

$$M = M(0).[295° - T_{comp}]/[295° - T_{comp}(0)] \qquad (1)$$

is assumed for the value of the room temperature magnetization, then the calculated changes are in good agreement with those observed experimentally. Similarly the observed increase in H''_k is also to be expected if K_u is not changed by annealing. In this case, since $H''_k \sim H_k = 2K_u/M$, the product $4\pi M \cdot H''_k$ would be expected to remain constant; this was indeed found to be true for all specimens within the limits of experimental error. This stability of K_u, however, is not surprising since the annealing temperatures were at or above the Curie temperature (∿ 150°C). It has been shown elsewhere[7] the stripe collapse field H_{col} is dramatically affected by changes in magnetization. The changes in H_{col} observed in this investigation are consistent with the concurrent changes in room temperature magnetization.

The kinetics of the annealing process were analyzed from the data shown in Fig. 1. It was assumed that this process is controlled by a thermally activated atomic mobility, and in view of the observed saturation of H_{col}, an appropriate rate equation may be written as

$$\frac{dH_{col}}{dt} = [H_{col}(t) - H_{col}(\infty)]^{\gamma} \cdot A \exp(-Q/kT) \qquad (2)$$

where $H_{col}(\infty)$ is the saturation value at temperature T, γ is an exponent defining the order of reaction, A is a constant, Q is the thermal activation energy and k is Boltzmann's constant. If t_f is defined as the time required to obtain a value of $H_{col} = f \cdot (H_{col}(0) - H_{col}(\infty))$, where $0 \leq f \leq 1$, then it may be shown[8] that $\ell n(t_f \cdot (1 - H_{col}(\infty)/H_{col}(0))^{\gamma-1})$ should be linearly proportional to the inverse absolute temperature 1/T. Such a linear fit was tried for f = 0.5, and various assumed integral values of γ; the best linear fit was found for $\gamma = 1$, which is shown in Figure 3, and from the slope of this curve the activation energy Q was obtained as 0.4 eV.

It should be emphasized here, that while the term $(H_{col}(t) - H_{col}(\infty))^{\gamma}$ in Eq. (2) represents the driving force, the origin of the driving force for the annealing process is not magnetic, since the annealing temperatures

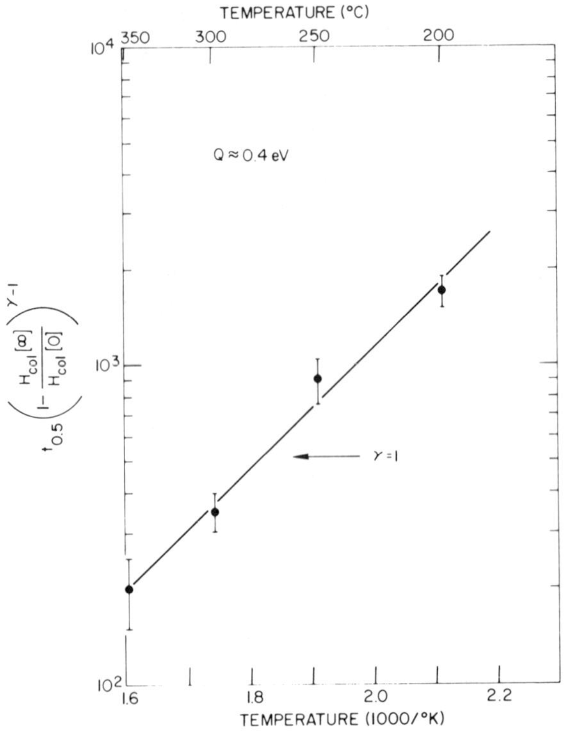

Figure 3

The function $\ell n\,[t_{0.5}\,(1 - H_{col}(\infty)/H_{col}(0))^{\gamma-1}]$ plotted against the inverse absolute temperature.

were mostly above the Curie temperature. Therefore, variations in the stripe collapse field merely reflects the changes in the state of the film upon annealing.

The phenomenology and the precise nature of the driving force for the annealing process, resulting in a shift in the compensation temperature, are not clear at this moment. A shift in the compensation temperature may be caused by (i) changes in composition and the composition gradient, (ii) changes in local and long-range atomic structure, and (iii) changes in stresses and stress distribution. The electron microprobe and the He^+ back scattering data indicate that there have been no detectable changes in the composition and the composition gradients in these films as a result of annealing. However, it still remains to be seen whether the changes in the magnetic properties observed in this investigation were caused by structural changes or changes in the stress system.

CONCLUSIONS

The amorphous Gd-Co-Mo thin films retain the magnetic characteristics of bubble films even after annealing at 350°C. These properties tend to stabilize at the annealing temperature, and if it is necessary to use a film composition which has a compensation temperature near room temperature, then a brief high temperature anneal as a processing step for stabilization may be beneficial for device fabrication.

The observed changes in some magnetic properties of these films are accounted for by changes in the compensation temperature. These changes in the compensation temperature are not caused by compositional fluctuations but may be caused by structural and/or stress changes.

The phenomenology and the precise nature of the driving force of the annealing process are not clear at this time, but the annealing kinetics may be interpreted as being controlled by a thermally activated process with an activation energy of about 0.4 eV.

ACKNOWLEDGEMENTS

The authors wish to thank C. H. Bajorek, R. J. Gambino and A. F. Mayadas for many stimulating discussions. They also wish to thank S. M. Kane and V. C. Richardson for specimen preparation, H. R. Lilienthal for the force balance magnetometry, F. Cardone for electron probe microanalysis, and A. Lurio and J. F. Ziegler for He^+ back scattering analysis.

REFERENCES

1. R. Hasagawa, J. Appl. Phys., 45 (1974), 3109-3112.
2. R. Hasegawa, R. J. Gambino, J. J. Cuomo and J. F. Ziegler, J. Appl. Phys., 45 (1974), 4036-4040.
3. P. Chaudhari, J. J. Cuomo, R. J. Gambino, S. Kirkpatrick and L. J. Tao, Paper 4D-1, this conference.
4. J. J. Cuomo and R. J. Gambino, J. Vac. Sci, Technol., 12 (1975), in press.
5. K. Okamoto, T. Shirakawa, S. Matsushita and Y. Sakurai, IEEE Trans., MAG-10 (1974), 799-802.
6. A. Hubert, A. P. Malozemoff and J. C. DeLuca, J. Appl. Phys., 45 (1974), 3562-3571.
7. R. J. Kobliska, R. Ruf and J. J. Cuomo, Paper 4D-4, this conference.
8. A. Gangulee, J. Appl. Phys., 43 (1972), 3943-3948.

UNIFORMITY OF AMORPHOUS BUBBLE FILMS

R. J. Kobliska, R. Ruf and J. Cuomo
IBM, Thomas J. Watson Research Center, Yorktown Heights, N.Y. 10598

ABSTRACT

Uniformity of the magnetic properties of a bubble film is requisite for proper device operation. We have evaluated the spatial uniformity of Gd-Co-Mo films which were rf sputtered from 12 and 20 cm dia. arc-melted targets. Measurements were made of variations in the collapse field (H_{col}), the coercive force field (H_c), the anisotropy field (H_k), the thickness h, as well as the stripwidth (w_s).

Films with H_{col} uniform to ± 3% over a 56 mm dia. wafer have been obtained with a 20 cm. dia. target and aspect ratio = 10 (AR = target diameter/electrode spacing). Modifications necessary to obtain better uniformity over large dia. wafers will also be discussed.

INTRODUCTION

The discovery of uniaxial anisotropy in amorphous metal alloy films[1] has created interest regarding their use as bubble film materials. Such films have been successfully used in T-I bar type structures.[2] As is the case in other bubble domain materials, it is desirable to maximize the uniformity of material and magnetic properties over planar dimensions as large as 5 cm.

In general, the sputtering of films from multi-component targets with a small aspect ratio produces films with a composition different from that of the target due to the dissimilar sputtering yields and losses of the target components.[3,4] In addition, inhomogeneities in the target and dark space can introduce compositional and thickness inhomogeneities in the film. In magnetic films these variations can cause substantial inhomogenéities in the film's magnetic properties.

The spatial uniformity of the material and magnetic properties of amorphous bubble domain films sputtered from Gd-Co-Mo targets has been evaluated as a function of film deposition conditions and sputtering geometry. These films exhibit compositional and magnetic nonuniformities which are primarily dependent on the aspect ratio (AR = target diameter/electrode spacing). It is observed that the thickness, compositional, and magnetic uniformities are improved with increasing aspect ratio.

EXPERIMENTAL DETAILS

The details of the r.f. bias sputtering system and the preparation of the arc-melted targets have been reported before.[4] Two modifications relevant to this experiment are: a rotating substrate holder which assures circumferential compositional uniformity of the deposited film and a special target holder designed to hold 12 cm and 20 cm dia. targets at variable distance from the substrate plane. The latter combination allowed for an AR in the range of 4 to 10.

Film thicknesses were measured with either a Taylor Hobsen Tallysurf or a beta backscattering apparatus calibrated with films of known composition and thickness. The stripwidths for a few anti-reflection coated films were measured with an optical microscope, whereas typically a diffraction technique[5] was used in which the domain walls are delineated with ferrofluid inks. The anisotropy field H_k, the coercivity H_c, and the stripe collapse field H_{col} were measured with a Kerr effect hysteresigraph. The latter apparatus has a spatial resolution of 2 mm which is adequate to establish the profiles of these film parameters over a diameter of 56 mm.

The Kerr apparatus has been modified to allow the implementation of a small perpendicular modulation field and an associated lock-in detection scheme. It can be shown, in the limit of very small modulation fields, that the Kerr signal synchronous with the modulation field is proportional to the reversible permeability of the film.[6] It is this sychronous Kerr signal which is recorded and used to determine the stripe collapse field. The coercive force and anisotropy field were measured using magneto-optic techniques described by Shumate et al.[7]

EXPERIMENTAL RESULTS

The thickness profiles of 2 Gd-Co-Mo films deposited under aspect ratio conditions of 4 and 10 are plotted in Fig. 1. Both films were nominally 1 μm thick and the

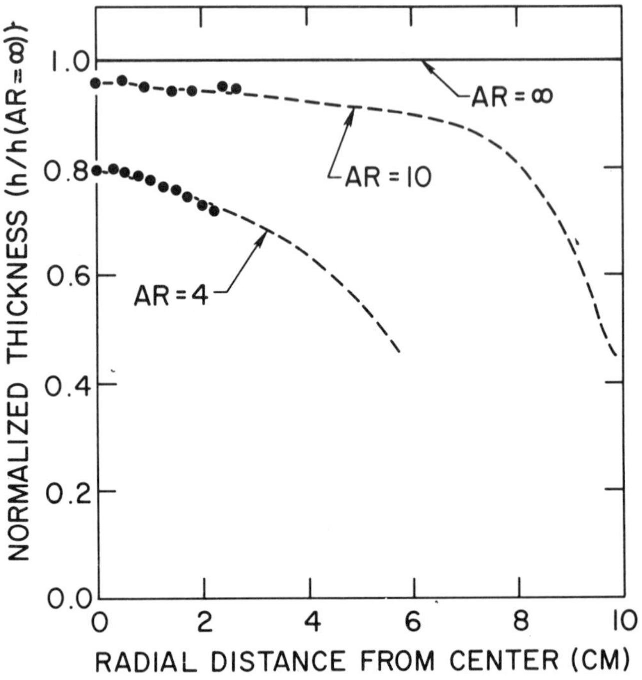

Figure 1

Thickness profile for Gd-Co-Mo films deposited under AR = 4 and 10 conditions.

stripwidth at the film centers for the AR = 4 and 10 films were 2 and 1 μm, respectively. The dotted lines represent the theoretical calculations of Maissel for these aspect ratios.[3] This calculation of the thickness profile is expected to be a worst case approximation since neither resputtering nor collisions of the sputtered material with the gas atoms has been considered.

The stripe collapse field, stripwidth, and coercivity as a function of radial position on the film are plotted in Fig. 2 for a sample prepared under AR = 4 conditions. Note the strong radial dependence of all of these film parameters. Over this same sample, the anisotropy field H_k (not shown) monotonically dropped by ∿ 30% to its lowest value at the film's edge.

In order to deduce the compositional variation of this film, it is necessary to relate the observed variation in H_{col} to the room temperature magnetization. The functional dependence of the bubble collapse field H_{bc} on the material length parameter $\ell = (AK_u)^{1/2}/\pi M_s^2$, the film thickness h, and $4\pi M_s$ is:[8]

$$H_{bc} = 4\pi M_s(1 + 3/4\ \ell/h - (3\ell/h)^{1/2}), \quad (1)$$

which when differentiated with respect to $4\pi M_s$ yields:

$$\left(\frac{\partial H_{bc}}{\partial 4\pi M_s}\right)_{A,K_u} = 1 - \frac{3}{4}(AK_u)^{1/2}/\pi M_s^2 = 1 - \frac{3}{4}\ell/h, \quad (2)$$

assuming constant exchange stiffness A and anisotropy K_u. Since for a typical bubble film, $\ell/h \sim .1$, the bubble collapse field closely tracks changes in M_s. Also note that H_{bc} is much less strongly dependent on A, K_u, and h than on M_s. Similarly,[9] it is expected that the stripe collapse field H_{col} will also exhibit this same strong dependence on M_s.

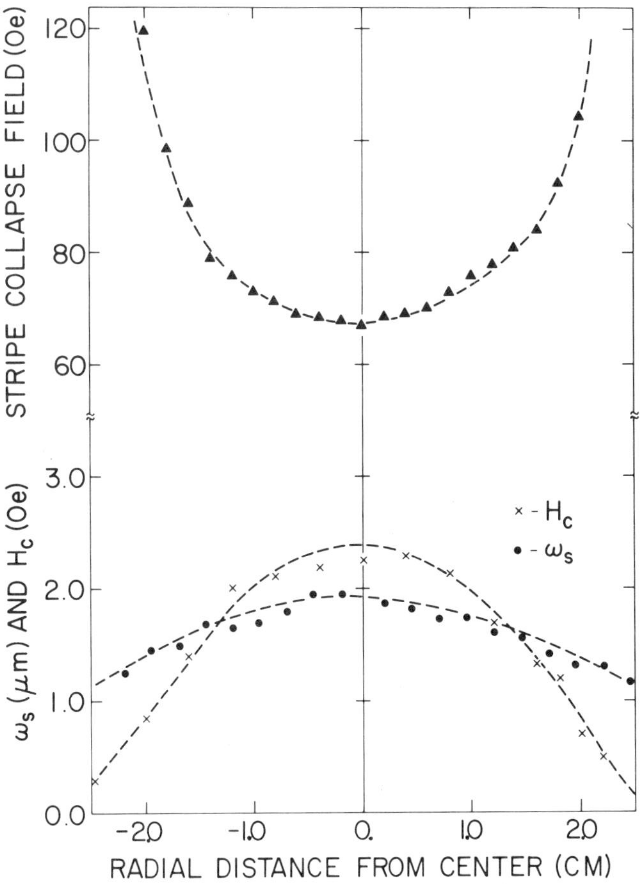

Figure 2

Stripe collapse field, coercivity, and stripwidth as a function of radial position on an AR = 4 film.

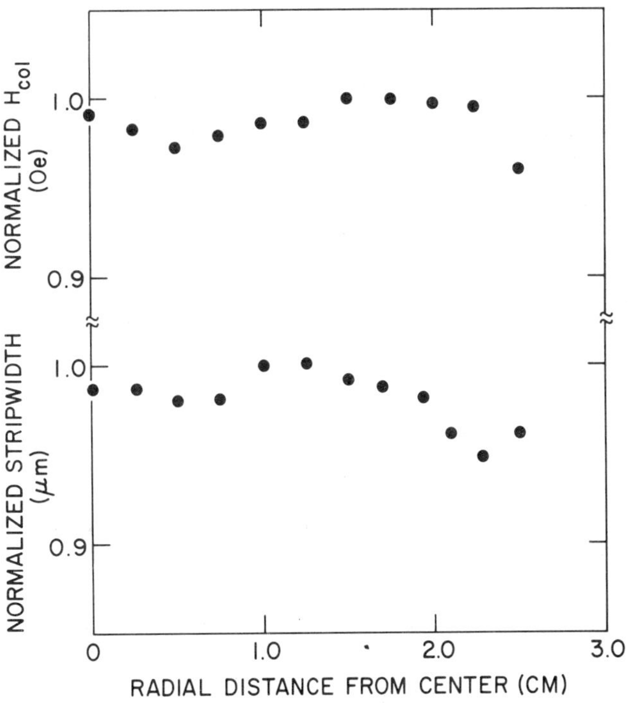

Figure 3

Stripe collapse and stripwidth profile for an AR = 10 deposition.

The value of the room temperature magnetization was calculated from the values of H_{col} and $2w_s/h$. The deduced M_s monotonically increased to a value ∿ 50% larger at the edge of the wafer than at the film's center. This non-uniformity in M_s is a result of a compositional variation which was verified by an electron microprobe analysis of the film. The film composition at the center is 12. at. % Gd, 70. at. % Co, and 18. at. % Mo, whereas at the edge the composition is 11. at. % Gd, 70. at. % Co, and 19. at. % Mo. Furthermore, the $(4\pi M_s)(H_k)$ product is essentially constant over the diameter suggesting the K_u is nearly constant for this compositional variation.

Considerable improvement in uniformity of all these properties has been achieved in films made with an AR = 10. In Fig. 3 are plotted w_s and H_{col} for a 56 mm dia. wafer. The variations in H_c and H_k are not shown since these are below the detection limit of our instrumentation (± 10%). In contrast to the 50% increase in M_s observed in the AR = 4 film, the M_s of this AR = 10 film varies by only ± 2 % over the entire 56 mm wafer.

DISCUSSION

The compositional non-uniformity which is responsible for the variation in magnetic properties plotted in Fig. 2 can be explained with straight-forward sputtering arguments. The sputtering yield of Gd is much larger than that of either Co or Mo.[4] The expected ratio of Co (or Mo) to Gd for any point on the film is:

$$(Co/Gd)_{film} = \frac{A_{Co}}{A_{Gd}} = \frac{S_{Co}\,F_{Co} - R_{Co}}{S_{Gd}\,F_{Gd} - R_{Gd}} \qquad (3)$$

where A_{Co} is the accumulation rate of Co. S_{Co} is the "sticking coefficient" which includes all loss mechanisms other than by resputtering R_{Co}, and F_{Co} is the flux of Co from the target. These quantities are similarly defined for Gd.

For sake of this argument, it will be assumed that the anode is an equipotential surface and that perfect target compositional and dark space uniformity exist. The thickness profiles plotted in Fig. 1 for different aspect ratio depositons can be explained by straightforward considerations of the system's finite aspect ratio and a cosine distribution for the emission of material from the cathode.[3] Therefore the flux of material incident at the film's edge is less than that at the center. Substituting $(R_{Gd}/R_{Co}) > (F_{Gd}/F_{Co})$ in Eq. 3, it is easily shown that the film's center is expected to be Gd-rich with respect to the edge. Samples with compensation temperatures below room temperature will be magnetically dominated by the Co moment component of this ferrimagnet and therefore the Gd-rich center of the wafer will have a lower room temperature magnetization. This is the case for the film discussed in Fig. 2. Conversely, a sample with compensation temperature above room temperature will have larger M_s at the film center. Both trends have been observed for films deposited with an AR = 4 geometry and for sputtering conditions appropriate for Gd and Co magnetically dominated compositions.

The uniformity of several film parameters has been assessed for AR = 4 and 10 depositions. Recognizing that it is desirable to share one common bias field for many devices, it is apparent that the most stringent requirements should be set for the collapse field uniformity. A complete assessment of uniformity should strive to establish the sputtering parameters governing compositional uniformity and therefore collapse field uniformity. This is beyond the scope of this paper, but such an analysis will enable exact definition of the sputtering system geometry necessary to achieve adequate uniformity over a prescribed area.

CONCLUSIONS

This is the first assessment of the film uniformity of rf bias sputtered amorphous bubble films. The nonuniformity observed for AR = 4 is consistent with a compositional gradient which is Gd-rich in the center relative to the edge of the substrate plane. This compositional gradient is ascribed to the higher sputtering yield of Gd relative to that of Co or Mo.

A comparison of the AR = 4 and AR = 10 cases clearly indicates that for adequate film uniformity over a substrate area 56 mm in diameter will require sputtering geometries with AR $\geq$ 10. A quantitative model which explains the magnitude of the compositional inhomogeneity does not exist. Recognizing the complexity and cost of such an empirical assessment suggests that such an analysis is desirable.

ACKNOWLEDGEMENTS

The authors wish to thank C. Bajorek, P. Chaudhari and R. J. Gambino for helpful discussions and useful information used in this work. The electron probe microanalysis was performed by J. Kuptsis. The film samples studied were prepared by S. Kane and P. Silano.

REFERENCES

1. P. Chaudhari, J. J. Cuomo, and R. J. Gambino, IBM J. Res. Dev. 17, No. 1, 66-68 (1973).
2. M. H. Kryder, et al., IEEE Trans. on Magnetics, MAG-10, 825-827, (1974). and H. L. Hu, et al., IBM Research Report RC-4277, March 22, 1973.
3. L. Maissel, Chapt. 4, Handbook of Thin Film Technology, edited by L. Maissel and R. Glang, McGraw-Hill, (1970), N.Y.
4. J. J. Cuomo and R. J. Gambino, J. Vac. Sci., Technol., 12, (1975), in press.
5. R. Ruf, to be published.
6. R. M. Bozorth, Ferromagnetism, Chapt. 11, Van Nostrand, N. J., (1951).
7. P. W. Shumate, D. H. Smith, F. B. Hagedorn, J. Appl. Phys. 44, 449-454, (1973).
8. R. M. Josephs, AIP Conference Proceeding 10, 286-303, (1972).
9. J. A. Cape and G. W. Lehman, J. Appl. Phys. 42, 5732-5756 (1971).

UNIAXIAL ANISOTROPY IN RARE EARTH (Gd, Ho, Tb) - TRANSITION METAL (Fe, Co) AMORPHOUS FILMS

Neil Heiman, A. Onton, D. F. Kyser, Kenneth Lee, and C. R. Guarnieri
IBM Research Laboratory
San Jose, California 95193

ABSTRACT

We present evidence that magnetic uniaxial anisotropy in amorphous RE-TM films is a more common feature than had been presumed. The incident atomic beam direction appears to define the anisotropy axis in thermally evaporated films while the electric field direction at the surface is the dominant influence in bias sputter deposition. The nature of the coupling of the magnetization to the structural anisotropy is also found to vary.

INTRODUCTION

The existence of amorphous magnetic films with uniaxial magnetic anisotropy ($K_u > 2\pi M_s^2$) was first reported for GdCo films prepared by bias sputter deposition.[1] It has since been reported[2] that evaporated amorphous Ho-Co and Ho-Fe films also possess $K_u > 2\pi M_s^2$. The studies reported here are directed toward a fundamental understanding of the magnetic uniaxial anisotropy in thin films of amorphous RE-TM alloys.

Gambino et al[3] performed a number of experiments to determine the origin of K_u in bias sputter deposited GdCo. Their results ruled out stress and shape anisotropy, leading them to suggest pair ordering as the likely mechanism. In the experiments reported here, stress and shape anisotropy have similarly been ruled out.

We present experimental evidence that magnetic uniaxial anisotropy in amorphous RE-TM films is a more common feature than had been presumed and that in general K_u does not arise simply from transition metal atom pair ordering. We further provide the basis for a structural model which could be responsible for the origin of K_u and present data that indicate the coupling of the magnetic properties to the structural anisotropy of the film varies according to film constituents and preparation methods.

K_u IN EVAPORATED AMORPHOUS RE-TM FILMS

We have found that evaporated amorphous films of Tb-Co, Tb-Fe, Ho-Ni, and Gd-Fe (in addition to the previously reported Ho-Co and Ho-Fe) possess a K_u sufficient to cause the magnetization to be normal to the film plane. A large number of amorphous GdCo samples were also prepared by thermal evaporation with greatly varied substrate temperatures and spanning a wide range in composition. All of these latter samples showed in-plane magnetization. Thus far amorphous GdCo is unusual in that bias sputter deposition is required to produce a $K_u > 2\pi M_s^2$ perpendicular to the film plane.

STRUCTURAL BASIS OF MAGNETIC ANISOTROPY

Structural uniaxial anisotropy in the amorphous film must have its origin in sources which also exhibit uniaxial symmetry. The anisotropic geometrical influences which exist during the growth of the film are the substrate (film) plane, the incident atomic beam axis, and, in the case of bias sputtered GdCo, the electric bias field direction. Any or all of these may act potentially on the film growth dynamics.[4]

In order to sort out the aspects of film growth which result in the uniaxial structural anisotropy of RE-TM alloys, we have prepared 900 Å thick $HoCo_2$ films with the substrate canted so that the atomic beam makes an angle, ϕ_o, of 30° or 60° with respect to the substrate normal.

Figure 1 shows polar Faraday rotation hysteresis loops at λ = 6328 Å on a canted anisotropy film with

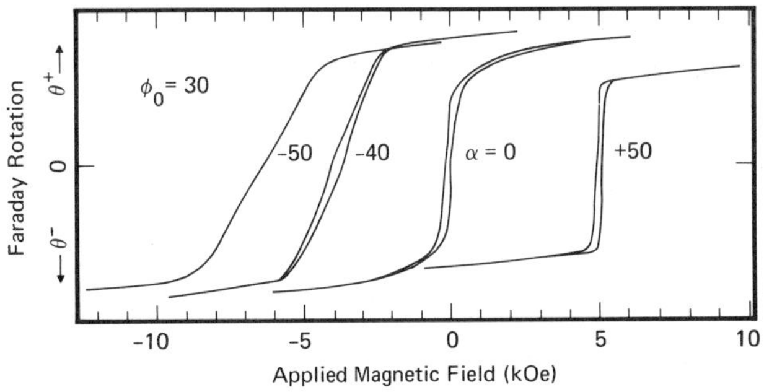

FIG. 1. Polar Faraday rotation hysteresis loops for amorphous canted anisotropy $HoCo_2$. The curves for $\alpha \neq 0$ have been displaced horizontally.

ϕ_o = 30° for various values of the angle of the applied field, α. All angles are measured from the normal to the substrate in the plane defined by the substrate normal and the growth beam direction. These data indicate that the magnetization is aligned at ψ= 45° $\pm$ 5°. The angle that K_u makes with the film normal, ϕ, can be obtained from the relationship:

$$2\tan 2\psi = Q\sin 2\phi/(Q\cos 2\phi - 1), \qquad (1)$$

which is obtained by minimizing the magnetic energy equation for the film, including both anisotropy and magnetostatic effects. Measurements on a magnetometer showed the film had a $Q = K_u/2\pi M^2$ of $\sim$ 1.4. This Q and Eq. (1) yields ϕ = 24° $\pm$ 10°. Data for a sample with ϕ_o = 60° gave ψ = 70° and Q = 1.9, giving ϕ = 60° $\pm$ 10°. These results are consistent with $\phi = \phi_o$, indicating that K_u is along the incident beam direction.

Similar experiments were performed with GdCo sputter deposited on canted substrates. However, magnetometer measurements showed no evidence of anisotropy canting. The films had either perpendicular or in-plane magnetization depending on the presence or absence of bias, respectively. We surmise that the bias electric field direction is the dominant influence on the anisotropy of GdCo sputtered films in that it remains normal to the film at the surface because of the high conductivity of these films.

We conclude that the predominant influences on the uniaxial anisotropy in these films of RE-TM alloys is the unidirectional incidence of the atomic beam in thermal evaporation and the bias field directed resputtering in bias-sputter deposition. The detailed nature of the anisotropy remains to be resolved. We expect that radial distribution function studies will be most useful from this point of view. Preliminary results indicate that easily observed changes are produced on the x-ray diffraction spectra of amorphous GdCo by the bias field during sputtering.[5] This would seem to rule out pair order and indicate a more uniform local anisotropic coordination. For example, a pseudo-crystalline coordination may occur possessing perhaps a preferential orientation or a longer range of coherence along the growth axis than in directions perpendicular to it.

K_u IN BIAS SPUTTER DEPOSITED GdCo

In order to determine how the coupling between magnetization and the structural anisotropy depends on the film composition and preparation conditions, we measured K_u in bias sputter deposited GdCo and thermally evaporated HoCo as a function of composition and temperature. One difficulty in the case of GdCo is the variation of composition with bias voltage, V_b. In an attempt to produce films of the same composition using different V_b, amorphous films of GdCo were sputter deposited from three targets ($Gd_{33}Co_{67}$, $Gd_{29}Co_{71}$, and $Gd_{25}Co_{75}$) using various bias voltages. The compositions of the resulting films were electron microprobe analyzed. The attempt was successful to some extent. For example, films sputter deposited with $V_b = 0$ V from the $Gd_{25}Co_{75}$ target had the same Co/Gd ratio (3.0) as films sputter deposited with $V_b \sim -100$ V from the $Gd_{29}Co_{71}$ target. However, the films contained considerably different amounts of Ar. Furthermore, the magnetic properties of the two films were noticeably different. The unbiased film had a compensation point (T_{comp}) of < 200 K whereas the bias sputtered film had T_{comp} > 400 K. A number of such compositional overlaps were obtained but in all cases the magnetic properties were somewhat different. Thus, although increasingly negative V_b produced films with increasing Co/Gd ratio and increasing Ar content (for a fixed target), the magnetic properties did not change nearly as much as would be expected from the resultant Co/Gd ratio and were in fact closer to the properties of the target composition.

By magnetization measurements we obtained a measure of K_u uncorrected for domain structure magnetostatics. This measured K_u' probably differs from the true K_u by at most 10-20%. K_u' was measured as a function of temperature from 4.2 K to 300 K for a large number of compositions prepared with various V_b. Figure 2 shows the main features of the results. First

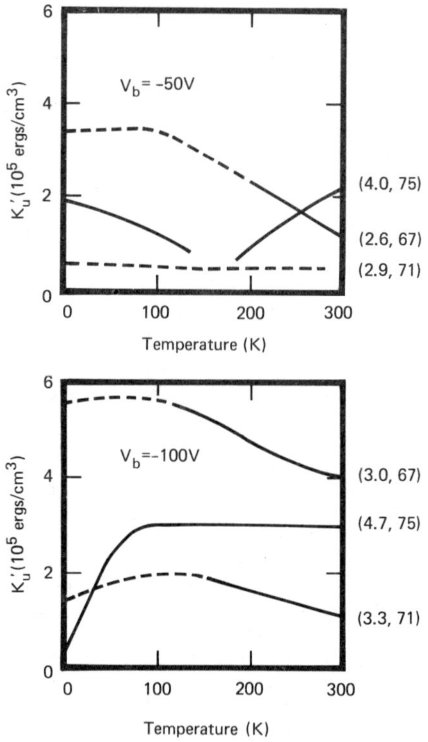

FIG. 2. K_u' vs. T for a number of samples. The numbers in ( ) indicate (Co/Gd ratio sample, at % Co in target). The dashed curve indicates $K_u' < 2\pi M_s^2$.

of all K_u' is seen to decrease sharply near T_{comp}, which is consistent with the observation of Cronemeyer.[6] Secondly K_u' increases monotonically with increasingly negative V_b for a fixed target. As a function of Co/Gd ratio, K_u' shows no specific trend; however, for a fixed V_b, K_u' was always a minimum for samples sputtered from the $Gd_{29}Co_{71}$ target. This result, coupled with T_{comp} behavior would seem to imply that the target composition plays a more important role than the final sample composition. K_u' has a relatively flat T dependence (excluding the region of T_{comp}) in these bias sputter deposited GdCo films, implying that K_u is most likely associated with the Co sublattice magnetization, as previously suggested; however, in light of the previous section, the K_u is more likely due to a preferred pseudo-crystalline coordination than to pair ordering.

T DEPENDENCE OF K_u' IN Ho-Co

To determine whether or not K_u in the evaporated amorphous RE-TM films results from the same mechanism as in bias sputter deposited GdCo, we examined the T dependence of K_u' in evaporated HoCo. The results are displayed in Fig. 3. Compared to GdCo K_u' in

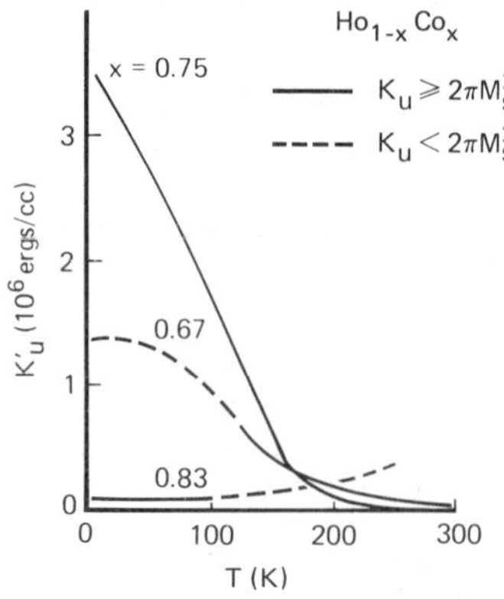

FIG. 3. K_u' of amorphous HoCo films for various compositions and temperatures.

HoCo is an order of magnitude larger at low temperature and exhibits strong T dependence. This suggests that K_u in HoCo is dependent at least in part on the orbital moment of the Ho atom.

CONCLUSIONS

We have presented results which demonstrate the existence of preferential anisotropic structure in amorphous RE-TM films. The influence producing the anisotropy appears to be different in bias sputter deposition and thermal evaporation. Thus the nature of the structural anisotropy produced may have significant differences. Preliminary x-ray diffraction results seem to rule out pair order and suggest a more prevalent anisotropic local coordination. Further, because of differences in electronic structure of Ho and Gd, the coupling mechanism of the magnetic system to the structural anisotropy may also be varying. What is clear from these results is that anisotropy in thin amorphous films, rather than being a curiosity, is a general feature. Thus its understanding and control must be a key consideration in amorphous materials.

ACKNOWLEDGMENTS

We would like to thank P. Chaudhari, J. Smit, R. F. Soohoo, and J. C. Suits for helpful discussions. The technical assistance of C. J. Breen, S. L. Lawrence, and J. A. Sandate is gratefully acknowledged.

REFERENCES

1. P. Chaudhari, J. J. Cuomo, and R. J. Gambino, IBM J. Res. Develop. 17, 66-68 (1973).
2. N. Heiman and K. Lee, Phys. Rev. Letters 33, 778-781 (1974).
3. R. J. Gambino, J. Ziegler, and J. J. Cuomo, Appl. Phys. Letters 24, 99-101 (1974).
4. N. G. Nahodkin and A. I. Shaldervan, Thin Solid Films 10, 109-122 (1972).
5. A. Onton, Neil Heiman, and J. C. Suits, Proc. of the International Conference on the Electronic and Magnetic Properties of Liquid Metals, Mexico City, 1975, to be published.
6. D. C. Cronemeyer, in Magnetism and Magnetic Materials 1973, AIP Conference Proceedings No. 18, edited by C. D. Graham, Jr. and J. J. Rhyne, pp. 85-89 (1974).

REVERSAL OF HALL EFFECT AND KERR ROTATION IN FERRIMAGNETIC RARE EARTH-COBALT SYSTEMS

A.Ogawa*, T.Katayama, M.Hirano, and T.Tsushima
Electrotechnical Laboratory, Tanashi, Tokyo 188, Japan
*Institute For Solid State Phys., Tokyo Univ., Tokyo 106

ABSTRACT

Hall effect in the ferrimagnetic systems of rare earth(R) and cobalt(Co) such as Gd_2Co_7 and its temperature dependence, especially near compensation temperature, is investigated both theoretically and experimentally. The polarity of the Hall voltage in these materials is reversed at the compensation temperature, and it is similar to the reversal in optical Kerr effect. The anomalous part of the Hall voltage is separated theoretically into two independent parts ; one is of R and the other of Co. The contribution of R to the observed Hall voltage is considered to be several hundreds times larger than that of Co in the ferrimagnetic temperature region by taking the quite large spin-orbit interaction of R into account, and the reversal of the Hall voltage at the temperature is explained essentially in terms of the anomalous Hall effect originated dominantly from R. On the other hand, the variation of Kerr rotation with temperature comes mainly from Co in the visible region.

INTRODUCTION

Hall resistivity in a magnetic material is expressed as

$$\rho_H = Vd/I = R_0H + R_1M \qquad (1)$$

where H,M,V,I and d are magnetic field, magnetization, Hall voltage, current, and the thickness of the sample, respectively. The first term is the normal Hall voltage and R_0 is the normal Hall coefficient. The second term is an anomalous one characteristic of a magnetic material, and R_1 is the anomalous Hall coefficient.

In this paper we treat the ferrimagnetic rare earth-cobalt systems. The magnetic moment of a more-than-half filled rare earth, Gd, for instance, is coupled antiferromagnetically with that of Co, and both moments are ordered as a ferrimagnet. In a certain composition region such as Gd_2Co_7, there exists a compensation temperature $T_{comp.}$ in this ferrimagnetic system. Sakurai et al.[1] measured Hall voltage and Kerr rotation on sputtered amorphous Gd-Co films as a magnetic bubble material. The Kerr angle reversed its polarity at $T_{comp.}$ and the Hall voltage was also found to reverse the sign at the same temperature. This observation suggests to us that the Hall voltage and the Kerr rotation are originated from each sublattice component separately.

THEORETICAL MODEL

There have been proposed two models for the anomalous Hall effect ; one is based on band electron model and the other on localized electron model. In both models, spin-orbit interaction of magnetic electrons plays an essential role in the anomalous Hall effect.

In transition metals such as Co taken as described by a band electron model[2,3], d-electrons are polarized magnetically and the electric current is carried mainly by s-electrons with a small fraction by the d-electrons. When d-electrons are scattered by impurities or phonons, spin-orbit interaction of d-electrons leads to the term proportional to $\lambda_1(k\times k')\sigma$ in the second Born term of transition probability. (Where k and k' are incoming and outgoing wave numbers of the electrons, σ is their spin and λ_1 is the spin orbit coupling constant.)

The spin-orbit interaction alters the density matrix of d-electrons and the off-diagonal parts of the density matrix also leads to $\lambda_1(k\times k')\sigma$ term[4,5]. The latter effect is called side-jump mechanism. The spin-orbit interaction is enhanced by band effect[6], and the theoretical order estimation for the magnitude of the anomalous Hall voltage agrees with the experiments. Its temperature dependence, however, has not yet been made clear sufficiently in this model. In rare earth metals such as Gd, which is thought to be described by a localized electron model[7,8], f-electrons are polarized magnetically, but the electric current is carried by the conduction electrons of the rare earth. The conduction electrons are scattered by the magnetic moment of the rare earth through the usual s-f exchange interaction. The exchange integral is modified by the spin-orbit interaction of f-electrons, and the second Born term of transition probability which includes inelastic scattering contains the term proportional to $\lambda_2(k\times k')r$. (Where $r=\langle(s_z-\langle s_z\rangle)^3\rangle$, s is the angular momentum of the rare earth, λ_2 is spin-orbit coupling constant of the 4f electron, and the magnetic field is parallel to z-direction.)

Magnetic dipole-dipole interaction is taken into account[9] between the orbital moment of the conduction electron and the magnetic moment of rare earth, and it gives the term $\lambda_1(k\times k')r$. The usual exchange interaction is considered together with the above dipole-dipole interaction.

By using these terms, the temperature dependence of the anomalous Hall voltage is derived as a function of magnetization ;

$$R_1M \propto \langle (M-\langle M\rangle)^3\rangle \qquad (2)$$

This dependence agrees well with the experiment on a single crystal of Gd, and it means that the spin fluctuation causes the anomalous Hall effect. Its order of magnitude, however, does not agree with the experimental data very well.

Thus, the theories on the anomalous Hall effect in a magnetic material are thought to be precise enough in explaining the effect qualitatively, but they do not always give good quantitative explanation for such properties as the temperature dependence, the order of magnitude, and the sign of the Hall voltage.

EXPERIMENTAL RESULTS

We prepared sputtered amorphous films of Gd-Co system, and measurements of Hall voltage and magnetization were done on several specimens between 77°K and room temperature. The composition expressed as Gd_xCo_{1-x} was varied, and there are shown three examples of the present results in Fig.1. As for the Hall voltage, the normal part is negligibly small and the temperature variation of the voltage is plotted in the figure. Vari-

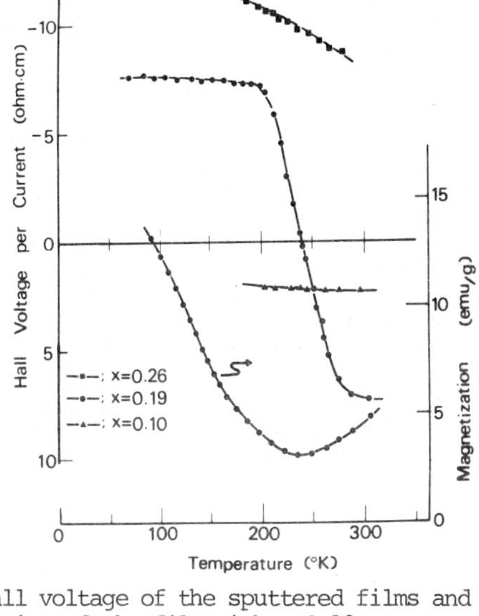

Fig.1 Hall voltage of the sputtered films and magnetization of the film with x=0.19 vs. temperature.

ation of magnetization with temperature is also shown for the film with x=0.19. This specimen has a compensation temperature at 240°K in its magnetization curve, and shows the reversal of sign in the Hall voltage at the same temperature. The observed gradual reversal is thought to be due to a microscopic compositional inhomogeneity in the sputtered film.

A relationship is shown in Fig.2 between the mag-

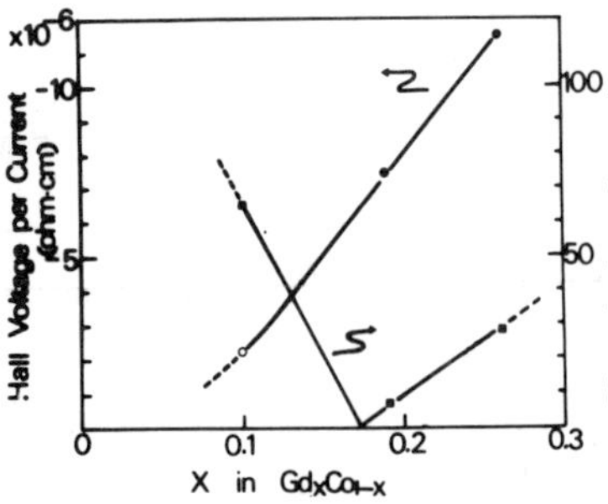

Fig.2 Hall voltage and the net magnetization at 200°K vs. Gd content. The observed Hall voltage is almost attributed to the anomalous part of Gd. For x=0.10 composition the magnetic moment of Gd is antiparallel to applied magnetic field so that the anomalous Hall voltage of Gd has negative sign.

nitudes of the Hall voltage and of the magnetization at 200°K. The magnitude of the Hall voltage does not depend on the magnetization but does depend on the amount of Gd x. As x increases, the Hall voltage increases linearly. The sign of the Hall voltage is always negative when the moment of Gd is parallel to the applied magnetic field for the measurements of the voltage.

DISCUSSION

Hall effect of a Gd single crystal of high purity is known[10]. In the ferromagnetic region, $|R_0^{Gd}| < 10^{-11}\Omega$ cm/emu. R_1^{Gd} has slight anisotropy, but obeys the formula $R_1^{Gd} = k[1-(M(T)/M(0))^2]$. When the magnetic field H is applied parallel to c-axis, $k=-1.0\times10^{-8}\Omega$cm/emu except near the Curie temperature at 292°K. When H is applied perpendicular to c-axis, k is slightly smaller than the previous case, but is the same order value. This temperature dependence of R_1^{Gd} varied through magnetization change agrees with that of the localized electron model theory. As temperature is lowered, spin fluctuation caused by temperature diminishes, and the Hall voltage is decreased.

Hall effect of a single crystal of hcp Co is also known[11]. Both R_0^{Co} and R_1^{Co} are less than $10^{-11}\Omega$cm/emu in the order of magnitude in the ferromagnetic region. The quite large value of R_1^{Gd} reflects the strong spin-orbit coupling in 4f electrons of the rare earth. From these facts, the anomalous Hall voltage from Gd component is supposed to give the dominant contribution to the total Hall voltage in Gd-Co system, and the other terms are thought to be negligible. Hall effect of such materials should be described by the localized electron model.

By the experiment on Gd-Co amorphous films, it is certain that the normal Hall voltage is negligibly small and that the anomalous part is contributed almost from Gd. It seems to agree well with the data of pure metals. For comparison, we take a Gd_2Co_7 single crystal which has Néel temperature at 770°K and $T_{comp.}$ at 470°K, and calculate the temperature dependence of the magnetic moment of each component within the molecular field approximation.

We also obtain the temperature dependence of the total Hall voltage using the value and the temperature dependence of the data of pure metals, where it is given as the sum of the additive contributions of each component. The computed results are shown in Fig.3. We discuss the present experimental results on some points comparing with the theoretical model.

In the Gd-Co system the Hall voltage is negative when the moment of Gd is parallel to the applied magnetic field. The Hall coefficient of the Gd component may

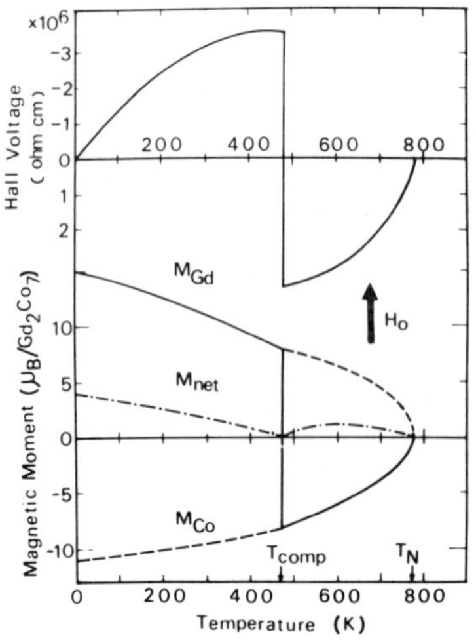

Fig.3 Calculated values of magnetization and Hall voltage of Gd_2Co_7 vs. temperature.

be negative, and the electric current is thought to be carried by the electron as in pure Gd metal.

The density of states of Co at the Fermi level is several times larger than that of Gd, and the electric current may be carried mainly by the electrons of Co. This fact explains the slightly larger value of the observed Hall resistivity than the calculated. The maximum value of the calculation is about $-3\times10^{-6}\Omega$cm, but the observed value is $-7\times10^{-6}\Omega$cm of the specimen with x =0.19 which has a little less Gd content than Gd_2Co_7.

CONCLUSION

Hall effect is measured on Gd-Co amorphous films, and a theoretical explanation is given for the temperature variation. The present treatment gives a fairly good agreement with the observation. The Hall voltage is explained as due to each component separately, and it is due almost to the anomalous part of Gd reflecting the large value of the spin-orbit coupling constant.

It shows the reversal of polarity at compensation temperature, which comes from the reversal of the magnetic moment of Gd at this temperature in the external field. As for the temperature dependence of Kerr rotation, especially the reversal of the rotation angle at $T_{comp.}$, it is explained as due to Co mainly.

It is concluded that the Hall effect in the ferrimagnetic rare earth-cobalt systems is explained mainly in terms of the quite strong spin-orbit coupling of the rare earths. We would like to thank Drs.J.Kondo and S. Ogawa for their helpful discussions.

REFERENCES

1) Y.Sakurai and others, to be published in Proceeding of 1974 Intermag.
2) R.Karplus and J.M.Luttinger, Phys.Rev.95, 1154 (1954)
3) J.M.Luttinger, Phys.Rev. 112, 739 (1958)
4) L.Berger, Phys.Rev. B2, 4559 (1970)
5) S.K.Lyo and T.Holstein, Phys.Rev.Lett. 29, 423 (1972)
6) J.Smit, Physica(Utr.) 24, 39 (1958)
7) J.Kondo, Prog.Theor.Phys. 27, 772 (1962)
8) Yu.P.Irkhin and Sh.Sh.Abel'skii, Sov.Phys.Solid State 6, 1283 (1964)
9) F.E.Maranzana, Phys.Rev. 160, 421 (1967)
10) N.V.Volkenstein, I.K.Grigonova and G.V.Fedrov, Sov. Phys. J.E.T.P. 50, 1003 (1966)
11) N.V.Volkenstein and G.V.Fedrov, Sov.Phys. J.E.T.P. 11, 48 (1960)

MAGNETIC BUBBLES IN $Ni_{1-x}Au_x$ ALLOYS

R. H. Geiss, Kenneth Lee and J. C. Suits
IBM Research Laboratory
San Jose, California 95193

ABSTRACT

Lorentz electron microscopy and small angle electron diffraction techniques as well as Kerr effect and magnetization measurements have been carried out on a series of $Ni_{1-x}Au_x$ thin films with $0 \leq x \leq 0.20$. These films consist of a Ni-Au fcc solid solution. The films showed a preferred orientation with a <111> axis normal to the film plane. Films with x = 0 and 0.20 showed typical thin film Bloch wall ripple domain structures. Films with x = 0.05, 0.10 and 0.15 showed "weak" stripe domain patterns in zero field ranging from parallel aligned domain walls at x = 0.05 to an almost random or "wickerwork" type structure at x = 0.15. An applied field caused the weak stripe contrast to increase until at ∿ 2 kOe, bubble domains could be nucleated. The bubbles existed predominantly in the 0 Bloch line state with both chiralities present. A few bubbles with 2 Bloch lines were also noted. In all cases the diameter of the bubbles was approximately 800 Å.

INTRODUCTION

The work reported herein concerns the observation of magnetic bubble domains in a series of evaporated fine grained polycrystalline nickel gold alloys. The observation of magnetic bubbles is not new as there are a magnitude of materials known to support bubble domains including such materials as single crystal cobalt metal,[1] large grained cobalt-chromium alloys,[2] orthoferrites, hexaferrites, ferrites, garnets and[3] amorphous gadolinium alloys.[4] Puchalska and Jones reported the occurrence of bubble domains in permalloy films evaporated at an oblique angle of incidence, however the direction of magnetization within the bubbles was thought to be skewed with respect to the film plane. What is new here is the observation of bubbles in a polycrystalline material evaporated at normal incidence where the grain size of the films is much less than the apparent bubble diameter. Also surprising is the mere existence of bubble domains in these Ni-Au alloys as $Q = K_u/2\pi M_s^2 < 1$ is expected from nominal magnetic parameters of pure Ni, where K_u is a uniaxial anisotropy energy. It is postulated that there exists a large K_u oriented normal to the film plane which cannot be accounted for by magnetocrystalline or magnetostrictive anisotropies.

EXPERIMENTAL DETAILS

The films were prepared by coevaporation of the two constituents in a UHV system at ∿ 1×10^{-7} Torr onto polished fused quartz and carbon coated electron microscope grids maintained at 0°. Although the equilibrium phase diagram shows immiscibility of the Ni and Au below 600°C in the composition range investigated, X-ray diffraction showed the films to consist of a Ni-Au fcc solid solution where the lattice parameter increases continuously as the gold content increases (see Fig. 1). The films were found to have a preferred orientation with a <111> axis normal to the film plane. Room temperature magnetization measurements by VSM showed that M_s for the pure Ni film agreed within 10% with bulk data. Table I lists some of the various physical and magnetic properties as determined by VSM of the individual alloy films. Film thickness was measured by the Tolansky technique and should be accurate to ± 50 Å. The decrease in moment with Au addition (see Fig. 1) is consistent with the filling of the Ni d-shell.

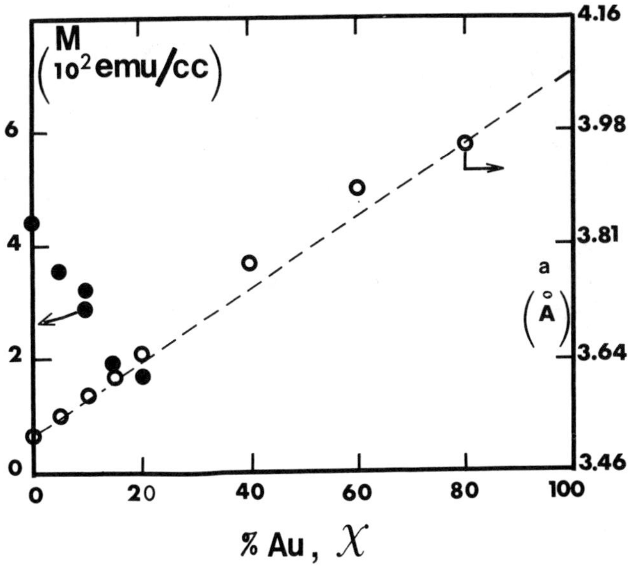

FIGURE 1

Lattice parameter, a(o), and magnetization, M(•), at room temperature for $Ni_{1-x}Au_x$ films. The dotted line connects the values of the lattice constants of bulk pure nickel and gold.

Table I

Magnetic characteristics and film thickness of the various Ni-Au alloys used in this study.

Ni Content of the Alloy (at %)	Film Thickness Å	$4\pi M_s$ (Gauss)	H_c (Oe)
100	900	5600	220
95	650	4500	145
90	850	3675	130
85	775	2350	100
80	1150	2250	55

Hysteresis loops with shapes characteristic of bubble materials have been observed by polar Kerr effect for alloys with 5, 10 and 15% Au. High resolution TEM of the films showed they were fine grained, with a typical grain size of 100-200 Å. A micrograph of $Ni_{.95}Au_{.05}$ in Figure 2 is typical of the microstructure. Note the evidence of porosity in the approximately 800 Å thick films. Electron diffraction determined that there was no preferred orientation direction lying in the plane of the films.

The magnetic structure of the films was studied by means of defocussed Lorentz electron microscopy. The films were positioned near the center of the objective lens of the microscope so that dc fields up to 6 kOe could be applied either normal or canted up to 60° to the film plane. Under zero applied field the films were all found to have a "weak" stripe domain structure, with a periodicity in the stripe of approximately 800 Å. Figure 3a shows the stripe structure obtained in the 5% Au sample with the corresponding small angle diffraction (SAD) pattern inserted. The SAD pattern spacing also confirms the observed stripe periodicity of 800 Å.

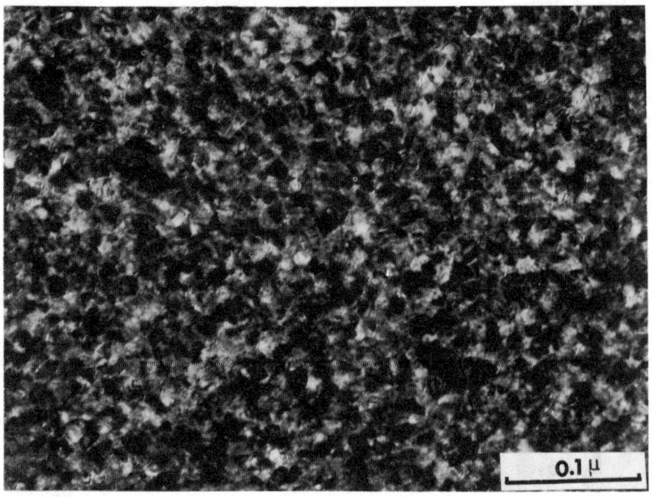

FIGURE 2

Microstructure of $Ni_{.95}Au_{.05}$ film showing 100-200 Å grain size. X-ray diffraction determined a <111> preferred orientation normal to the film.

By means of appropriate contrast experiments in the TEM it was determined that the net magnetization direction lies in the plane of the film and parallel to the domain walls. The origin of the magnetic contrast is due to lateral oscillations of the magnetization about the net direction. The magnetic contrast of these films is very weak, in fact, the films had to be tilted approximately 20° before the stripe structure could be recognized on the viewing screen of the TEM.

Applying a slowly increasing dc field normal to the film plane allowed visualization of the domain behavior during traversal of a hysteresis loop. The stripes come into improved contrast at 0° tilt as the field was increased indicating spin alignment parallel to the applied field. At an applied field of approximately 2000 Oe, bubble domains with M_s normal to the film plane were nucleated as shown in Figure 3b. These bubbles were found to coexist with the stripes until at approximately 2500 Oe the stripes disappeared. This is in agreement with the polar Kerr hysteresis loop observations. The bubbles then persisted until approximately 3000 Oe. As seen in Figure 3b the bubbles are predominantly of zero Bloch line state (S = 1) with both chiralities present. The black or white contrast of the bubbles is due to the opposite sense of the chirality in the walls (either clockwise or anti-clockwise) resulting in either a converging (white bubble) or diverging (black bubble) action on the incident electron beam. A few with two Bloch lines are also seen. The diameters of the bubbles are approximately 800 Å, in agreement with the stripe width.

By reducing the bias field from saturation, bubbles again were nucleated at approximately 3000 Oe and stripes at 2000 Oe. At remanence both were found to coexist although the apparent density of bubbles had decreased.

Similar dynamical domain behavior was observed in the 10% and 15% Au alloys as the bias field was increased from zero. However, the nucleation field for bubbles decreased to approximately 1600 Oe and 1200 Oe for the 10 and 15% alloys, respectively. In the 15% Au alloy the domain wall visibility was very weak due to the decrease in Lorentz deflection which is proportional to the magnetization. The stripe structure at remanence became increasingly disordered as the gold content increased, with that of the 15% Au alloy resembling a type of "wickerwork" structure and the SAD pattern forming a continuous circle. Such observations are also consistent with a lowering of K_u as the gold alloying was increased. The diameter of the bubbles was not measurably different in the three alloys. Pure Ni and $Ni_{.80}Au_{.20}$ showed the typical "fir tree" type of domain wall structure in zero field, which disappeared as the field was increased. No evidence of bubbles was found for these two compositions. If the films supporting bubbles were tilted from the normal while under an applied field, Bloch walls were nucleated which for the most part were parallel to the stripes, again a confirmation of "weak" stripes.

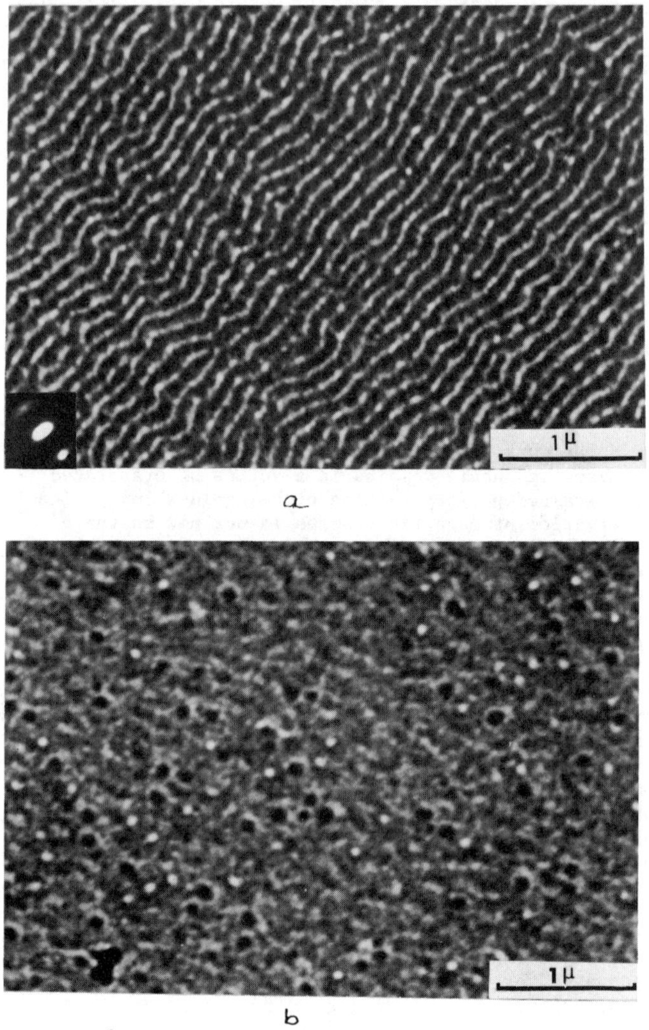

FIGURE 3

Defocussed Lorentz Electron Micrographs of $Ni_{.95}Au_{.05}$ Film

(a) Stripe structure obtained at zero applied field and 20° tilt. The net magnetization direction is parallel to the domain walls and oscillates laterally. The insert shows an SAD pattern resulting from the magnetic structure.

(b) Bubble domains of predominantly zero Bloch line state (S = 1) with both chiralities present. The bubble diameter is typically about 800 Å.

CONCLUSIONS

Magnetic bubble domains have been observed in a series of evaporated Ni-Au alloys. The compositions found to support bubbles were $Ni_{.95}Au_{.05}$, $Ni_{.90}Au_{.10}$ and $Ni_{.85}Au_{.15}$. Unique parameters of this alloy system are that the films are fine grained and polycrystalline with a fcc structure showing a preferred orientation normal to the film plane. The existence of bubble domains has been confirmed by VSM, polar Kerr and Lorentz transmission electron microscopy techniques. Magnetization in the bubbles lies normal to the film plane and the domain walls are found to be predominantly single chirality or in S = 1 states with both chiralities present. Bubble dimensions are considerably larger than the average grain size of the films and are found to coexist with stripe domains of the same dimensions at remanence.

ACKNOWLEDGMENTS

We would like to thank H. Arnal, G. Guthmiller, O. Navarro, and S. Lawrence for their able technical assistance.

REFERENCES

1. P. J. Grundy, D. C. Hothersall, G. A. Jones, B. K. Middleton and R. S. Tebble, Phys. Stat. Sol. (a) 9, 79 (1972).
2. D. C. Finbow and G. A. Jones, Phys. Stat. Sol. (a) 20, K91 (1973).
3. S. R. Herd and P. Chaudhari, Phys. Stat. Sol. (a) 18, 603 (1973).
4. I. B. Puchalska and G. A. Jones, J. Phys. D. 6, L52 (1973).

THE GROWTH AND MAGNETIC ANISOTROPY OF BULK AND EPITAXIAL TETRAGONAL $CuFe_2O_4$+

Dean A. Herman, Jr., and Robert L. White
Stanford Electronics Laboratories
Robert S. Feigelson, Brenton L. Mattes and Howard W. Swarts
Center for Materials Research
Stanford University, Stanford, California 94305

ABSTRACT

For copper ferrite to assume the tetragonal phase it is necessary that more than a certain fraction of the spinel octahedral sites be occupied by Cu^{2+} ions. To exceed this critical occupation, copper ferrite must be nearly stoichiometric. We have established that there is a systematic relationship between copper deficiency in (Bi_2O_3/B_2O_3) flux-grown copper ferrite and the growth temperature. In particular it is necessary to grow at low temperatures (<800°C) to obtain stoichiometric $CuFe_2O_4$. We have grown bulk single crystals and epitaxial films (on spinel substrates) from a stable melt at about 795°C. The bulk crystals evidence a microtwinning habit. Annealed epitaxial layers of $CuFe_2O_4$ display a uniaxial magnetic anisotropy with easy axis perpendicular to the films. The anisotropy is sufficiently strong ($K_u \cong 6.0 \times 10^5$ ergs/cc) to overcome the demagnetizing effects of the thin films, suggesting that $CuFe_2O_4$ may have application as a bubble domain material.

INTRODUCTION

The study of copper ferrite was undertaken in the anticipation that it might prove to be a suitable material for bubble domain applications. Copper ferrite is unique among the ferrospinels in that, although cubic at high temperatures, its stable room temperature phase is tetragonal with c/a>1. With proper annealing, the cubic-tetragonal phase transition occurs at about 360°C[1]. The Neel temperature of $CuFe_2O_4$ is about 455°C, above the crystallographic phase transition and well above room temperature.

TETRAGONAL PHASE TRANSITION

The cubic to tetragonal phase transition is believed to result through cooperation of local (prolate) tetragonal distortions of those octahedrally coordinated spinel "B" cation sites occupied by Cu^{2+} ions[2]. The local distortions occur because the electronic ground state of a Cu^{2+} ion (d^9) on an (undistorted) B site is doubly degenerate, satisfying the requirements for a Jahn-Teller type spontaneous distortion[3]. The local Jahn-Teller distortion for Cu^{2+} in an octahedral ligand field is usually prolate and this has been observed to be the case in copper ferrite[4]. The spinel B sites can be occupied by either the Cu^{2+} or the Fe^{3+} ions. The Cu^{2+} ion has a slight site preference energy advantage relative to a Fe^{3+} ion for locating at a B site[5]. This preference may be exploited by using high temperature annealing (circ. 800 to 750°C) to cause ion migration, resulting in more than the statistically disordered 1/3 of the B sites (in stoichiometric material) being occupied by Cu^{2+} ions[6]. Using sintered oxide copper ferrite-chromite ($CuFe_{2-x}Cr_xO_4$) Ohnishi, et al.[1] established experimentally that more than 37.5% of the B sites (the maximum possible is 50%) must be filled by Cu^{2+} for any overall cooperative distortion to occur in $CuFe_2O_4$.

CRYSTAL GROWTH

The growth of single crystal copper ferrite has been previously reported by a few groups, e.g., Okamura and Kojima[7], Miyadai, et al.[8], Evans and Hafner[4] and Petrakovskii, et al.[9]. All these groups grew their material at high temperatures (>900°C) from borax or lead oxide fluxes. Since 1971 a continuing program in the growth of copper ferrite, involving over seventy flux growth experiments, has been pursued at Stanford's Center for Materials Research. After exploring previously reported flux systems, a Bi_2O_3/B_2O_3 flux, with a liquidus between 970 on 1010°C, was developed yielding good quality copper ferrite octahedra, 4 to 6 mm on an edge[10]. However, careful chemical analysis (by atomic absorption) revealed that the Cu/Fe ratio on these crystals was near to 1/2.4 rather than the stoichiometric 1/2.0. The material being grown was a solid solution of $CuFe_2O_4$ and Fe_3O_4 which was approximately 20% copper deficient relative to stoichiometric $CuFe_2O_4$. Attempts were made to improve the stoichiometry by lowering the growth temperature in analogy with the phase diagram for the Fe_2O_3-CuO (solute only) system. This approach was successful. By reducing solute concentration and adjusting the solute (CuO/Fe_2O_3) ratio the liquidus temperature was reduced to approximately 795°C with $CuFe_2O_4$ as a stable phase (Table I). Samples grown at different growth temperatures were analyzed and the results are shown in Figure 1. The strong correlation between growth temperature and copper deficiency is apparent. X-ray powder diffraction of ground single crystals grown at various temperatures has verified this relationship and confirmed the importance of copper stoichiometry to the tetragonal transition.

TABLE I. Mixture for 100 gram crucible charge yielding copper ferrite at approximately 795°C.

Component	Wt.	Moles	Mole % relative to solute or solvent	Mole % total
Bi_2O_3	81.54	0.1750	80.00	47.55
B_2O_3	3.04	0.0437	20.00	11.88
Fe_2O_3	7.06	0.0442	29.60	12.01
CuO	8.36	0.1051	70.40	28.56
$CuFe_2O_4$	10.55	0.0442		13.65

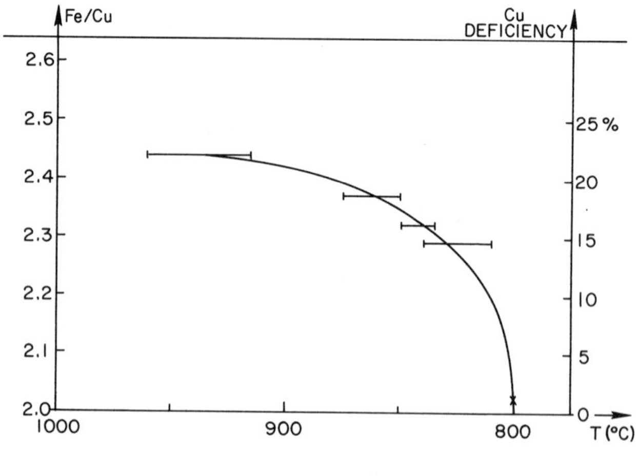

Figure 1. Iron to copper ratio and copper deficiency versus growth temperature for copper ferrite grown from Bi_2O_3/B_2O_3 flux. Note that the growth temperature decreases to the right. Analysis was done by atomic absorption with the samples initially dissolved in potassium pyrosulphate.

The bulk crystals grown near 795°C have yet to reach the quality of their copper deficient predecessors. Reasonably good octahedra (no flux inclusions, etc.) have been produced with edge dimensions on the order of 1 mm. When cut and polished, the bulk crystals (without annealing other than cooling with the furnace) exhibit a striped pattern which we interpret as representative of a microtwinning habit. The crystals are assumed to relax into "distortion domains" upon cooling to lower the strain component of free energy. Preliminary attempts at magnetic and non-magnetic annealing have only been marginally successful in removing these domains. A typical annealing procedure, here and for the epitaxial films described below, occurs in two stages; a "soak" or very slow cool (~1/2°C/hr) for perhaps 48 hours near 800°C to move the Cu^{2+} ions preferentially onto the octahedral sites, followed by a slow cool (~1/2°C/hr) through a temperature range of approximately 390°C - 350°C which embraces the cubic-to-tetragonal transition temperature. If a magnetic field was applied during the lower temperature anneal its magnitude was about 3.5 Kilogauss. In order to achieve untwinned tetragonal material (with thickness appropriate for bubble domain devices) liquid phase epitaxial (LPE) growth experiments were begun. Spinel gem stone was chosen for substrate material because of its availability in polished (100) oriented wafers and because its lattice constant (~8.09Å) was reasonably compatible with the smaller lattice constant of tetragonal copper ferrite (a~8.21Å). Layers were grown by dipping 0.5 cm diameter substrates into melts of the composition given in Table 1 at a temperature of about 795°C. The structure of the epitaxial layers has been difficult to verify. The intensity of the Laue pattern X-ray back reflections of the thin layers (circ. 20μm) tends to be small compared with that of the substrate. Nevertheless, based on the X-ray data we believe the layers to be single crystal, oriented with the substrate with c-axis perpendicular to the film. The ferromagnetic resonance data on the thin films also indicates the oriented nature of the films[11].

MAGNETIC CHARACTERIZATION

Ferromagnetic resonance experiments have been performed on bulk crystals and epitaxial layers before and after annealing. The data on the bulk crystals is difficult to interpret because of the microtwinned nature of the tetragonal phase. The epitaxial film data is easier to analyze. Figure 2 shows the reaonance data for a layer before and after annealing. From the downward displacement of the resonance field with the applied H along the [001] direction it is apparent that the effective anisotropy field exceeds

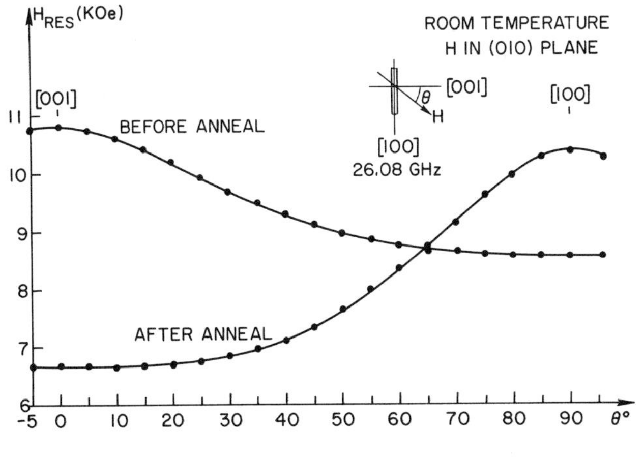

Figure 2. Resonance field (H_{RES}) versus orientation of an epitaxial layer with the H_{RES} in the (010) plane.

4πM by some 2,500 gauss. Though 4πM has not been measured on the epitaxial films, we believe from data on bulk samples and from ion site occupancy considerations that 4πM ≈2200 gauss. Fitting the response data to a magnetocrystalline energy of the form

$$f_K = K_c(\alpha_1^2\alpha_2^2 + \alpha_2^2\alpha_c^2 + \alpha_c^c\alpha_1^2) + K_u \sin^2\theta$$

where the α's are the direction cosines and θ is the angle of the magnetization with the c-axis, we then obtain for the annealed epitaxial film $K_c = 1.7 \times 10^4$ and $K_u = 6.0 \times 10^5$ ergs/cc, The epitaxial film therefore has an easy magnetic axis perpendicular to the film, and a q slightly greater than 2, indicating that tetragonal $CuFe_2O_4$ films should support bubble domains.

REFERENCES

+ This work suppored by the Army Research Office, Durham, through contract No. DAHCO-4-72-C-0014.

1. H. Ohnishi, and T. Teranishi, J. Phys. Soc. Japan 16, 35 (1961).
2. P. J. Wojtowicz, Phys. Rev. 116, 32 (1959).
3. J. Kanamori, J. Appl. Phys. 31, 14S (1960).
4. B. J. Evans and S. S. Hafner, J. Phys, Chem. Solids 29, 1979 (1968).
5. J. D. Dunitz and L. E. Orgel, J. Phys. Chem. Solids 3, 20 (1957).
6. E. F. Bertaut, J. Phys. Radium 12, 252 (1951).
7. T. Okamura and Y. Kojima, Phys. Rev. 86, 1040 (1952).
8. T. Miyadai, Y. Matsuo and S. Miyahara, J. Phys. Soc. Japan 19, 1747 (1964).
9. G. A. Petrakovskii, K. A. Sablina and E. M. Smokotin, Soviet Phys. Solid State 10, No. 8, 2005 (1969).
10. A review of high temperature flux growth of copper ferrite is to be published by R. S. Feigelson and H. W. Swarts of the Center for Materials Research, Stanford University.
11. Descriptions of the low temperature growth of copper ferrite and the growth and characterization of copper ferrite epitaxial layers are being prepared by the authors for separate publication.

$(EuTm)_3(FeGa)_5O_{12}$ GARNET FILMS WITH ONE-MICRON DIAMETER MAGNETIC BUBBLES

E. A. Giess, J. E. Davies, C. F. Guerci, H. L. Hu*, J. D. Kuptsis and J. M. Shaw
IBM T. J. Watson Research Center, Yorktown Heights, New York 10598
*** IBM Research Lab., San Jose, California 95193**

ABSTRACT

The $(EuTm)_3(FeGa)_5O_{12}$ garnet system is particularly useful in studies of small magnetic bubbles. Physical properties change relatively slowly with the Eu/Tm ratio, and the film composition changes little with isothermal LPE growth conditions. Uniform $Eu_{0.85}Tm_{2.15}Fe_{4.45}Ga_{0.55}O_{12}$ films with 1 μm diameter bubbles were grown on 38mm diameter substrates of GGG oriented (111). Using dilute, large volume (100 cc) melts undercooled 25°C and substrates rotated slowly (36 rpm) in a horizontal plane, the collapse field (H_o) and thickness are uniform to within 2 percent laterally across films. The temperature stability of H_o and of domain periodicity are good up to 100°C presumably because the films have a high Curie temperature (~215°C).

INTRODUCTION

Recently, the results of reproducibility and uniformity studies were reported[1,2] for the growth by isothermal LPE of magnetic garnet films with ~5 μm diameter bubbles using large area GGG discs as substrates. We present here the results of similar studies of a nominally $Eu_{0.85}$ $Tm_{2.15}$ $Fe_{4.45}$ $Ga_{0.55}$ O_{12} garnet with ~1 μm bubbles grown on 38 mm diameter GGG wafers. The present films show promise as high bit density materials for magnetic bubble devices.

EXPERIMENTAL

Good control of the film growth process was required so that both film thickness (h) and composition were reproducible from one run to the next and were uniform laterally across each film. Most of the equipment and procedures used were described earlier[3-5]. Since very thin (h < 1 μm) films were desired, results[5,6] for earlier submicron garnet film fabrication were especially helpful here.

The 876 gm. melt composition was dilute. It contained: 0.133 mole percent Eu_2O_3; 0.367 Tm_2O_3; 11.467 Fe_2O_3; 0.533 Ga_2O_3; 82.4 PbO and 5.1 B_2O_3. This 100 cc melt has a liquidus of ~955°C, so that undercooling (ΔT) was ~25°C for the growth temperature (T_g)=930°C most often employed. The Czochralski-grown substrates of 38 mm diameter GGG were supplied by Crystal Technology Inc., Mountain View, Ca. During growth, the wafers were held in a horizontal plane and were rotated axially continuously, except at the conclusion of the growth period when rotation was stopped while the wafer was being withdrawn from the melt.

Initially films were grown over the range T_g = 940 to 900°C (ΔT = 15 to 55°C), t_g = 20 to 80 sec., and wafer rotation rate, r = 9 to 196 rpm. The chemical composition of these films was evaluated by electron microprobe and the Ga in particular by Curie temperature (T_c) measurement. Blank *et al*[7] previously used magnetization data to compute Ga contents. Growth conditions of T_g = 930°C, r = 36 rpm and t_g = 50 sec. were selected for film reproducibility and uniformity studies. Both h and magnetic bubble collapse field (H_o) were measured both manually and automatically point-by-point. Properties were measured across each film diameter. Thickness measurements were interferometric[8] while H_o was determined visually in the manual mode and by a Kerr technique[9] in the automated mode.

FILM COMPOSITION

For growth conditions of T_g = 930°C and r = 36 rpm, the film composition is $Eu_{0.85}$ $Tm_{2.15}$ $Fe_{4.45}$ $Ga_{0.55}$ O_{12}. The present melts have Ga:Fe = 0.222: 4.778. Because of their lower Ga content, segregation in these 1 μm bubble melts is greater (α_{Ga} ~2.5) than for comparable 5 μm bubble melts (α_{Ga} < 2). Changing T_g from 900 to 950°C increased the film Ga content by about only 10 percent. The films grown at 930°C (r = 36 rpm) have < 1/2 wt. percent Pb, i.e. about 0.02 Pb ions per garnet formula unit. Changing T_g from 900 to 950°C (r = 100 rpm) decreased Pb from around 1.2 to 0.3 (± 0.05) wt. percent, while a change in r from 9 to 100 rpm (T_g = 930°C) only increased Pb from about 0.3 to 0.6 wt. percent; this was the maximum range because Pb peaks at r ~100 rpm and then falls off for r > 100 rpm. The films have slightly higher Eu:Tm ratios than the melt; α_{Eu} ~ 1.05. A similar rare earth segregation behavior was observed[6] for submicron bubble films with EuYb and EuLu rare earth pairs.

FILM MAGNETIC PROPERTIES

The prototype $Eu_{0.85}Tm_{2.15}Fe_{4.45}Ga_{0.55}O_{12}$ garnet has $4\pi M$ ~700 gauss at room temperature; characteristic length (l) ~0.13 μm; uniaxial anisotropy (K_u) ~53,500 erg/cm³ for an estimated magnetic exchange constant (A) ~3.0 x 10^{-7} erg/cm and a magnetic stability factor (q) ~2.8. It is believed that the favorable temperature stability of magnetic properties of this prototype garnet partly accrue from its high T_c (~215°C). Magnetic domain periodicity has a temperature coefficient of only -4 x 10^{-4}/°C and H_o decreases by ~20 percent, both over the 20 to 130°C range.

This EuTmFeGa garnet system is in many respects similar to the EuErFeGa garnets grown by Shick *et al*[10] with the main exception that the present films exhibit higher domain wall mobility[11]. Also, analogous EuYbFeGa garnet films were found to be nearly as good as the present films in all regards except possibly mobility, so that for some applications Tm could be replaced by Yb with only minor changes in processing conditions.

FILM REPRODUCIBILITY AND UNIFORMITY

A standard deviation of only 50 Angstroms was obtained for a linear least squares fit of thickness data versus t_g in a growth time series of 5 films made with t_g = 21.3 to 81.4 sec. at 930°C and for r = 100 rpm. In another series of 6 films grown under the "same" conditions, a growth strategy of slightly lowering T_g for each successive film run yielded the following film data:

#	(°C) T_g	(sec) t_g	(μm/min) f	(μm) h		(μm) P_o	(Oe) H_o
1	929.1	50.5	0.855	0.720		2.21	223
2	930.1	50.7	0.801	0.677	0.671 ± 0.005 (films 2–6)		210
3	928.1	52.6	0.767	0.672		2.22	211
4	929.0	51.9	0.766	0.663		2.18	205
5	927.8	50.2	0.808	0.676		2.14	209.5
6	927.2	49.7	0.806	0.668		2.14	211
(aver)		50.9	0.801	0.679		2.18	211.6
(σ)		1.0	0.030	0.019		0.03	5.5

Table I. Reproducibility series (r=36rpm) of 38mm diameter films.

The first film in the series of 6 is different from the 5 following it, as is seen in the growth rate (f), h (manual measurements), and H_o (automated measurements) data. The domain periodicity (P_o) of the first film, however, is not so different, presumably because its larger h compensated for the effect of a larger f. All 6 films have P_o and H_o with a standard deviation < 3 percent, and excluding the first film, agreement is even better.

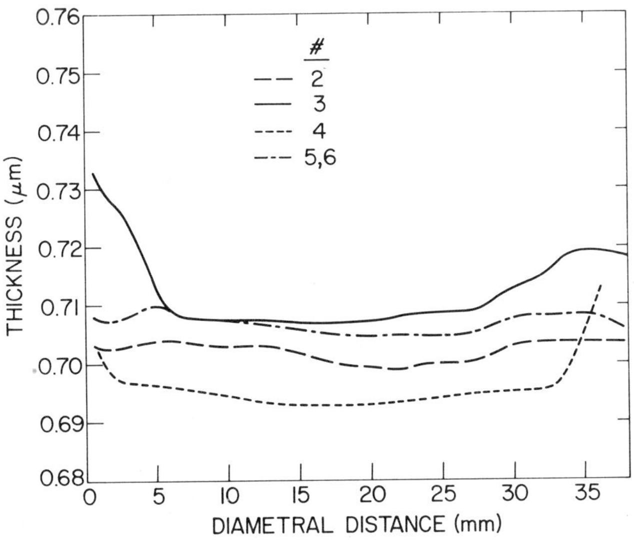

Fig. 1. Thickness versus diameter for 38mm diameter films.

Preliminary uniformity experiments lead to the choice of slow (r = 36 rpm) rotation in preference to fast (r = 121 rpm) rotation for the growth of 38 mm wafers in our 76 mm diameter crucibles. In a manual scan of h at 7 positions across the diameter of film No. 2, it was determined that h uniformity was of the same order as the 1 percent precision of this measurement up to within 2 mm of either edge of the film. Figure 1 shows the lateral uniformity of films No. 2 to 5 as measured automatically. The central portions of the films are quite uniform, but the outer portions are sometimes significantly thicker. Film No. 1 (not shown in Fig. 1) resembled film No. 3 and 4 in that it was thicker near the edges. The automated measurements were systematically larger than the manual measurements by a few percent. Finally, uniformity was assessed by manually measuring H_o at 7 positions across the diameter of a 38 mm film with h = 1.17 μm. All values of H_o were within a span of 2 percent.

CONCLUSIONS

It is possible to grow 1 μm bubble garnet films reproducibly and uniformly with h ~0.7 μm on 38 mm diameter substrates using the LPE dipping technique. The EuTmFeGa garnet system is a good medium for studying small magnetic bubbles because it has good temperature stability.

ACKNOWLEDGMENTS

F. Cardone and S. Ellmann helped with electron microprobe analyses. R. Kobliska measured the H_o values shown in Table I. J. Karasinski checked the lattice mismatch of these films. T. L. Hsu, M. Kryder, Y. Lin and W. Reynolds helped us evaluate films, and many other colleagues contributed advice as well as useful discussions.

REFERENCES

1. J. W. Nielsen, S. J. Licht and C. D. Brandle, IEEE Trans. Magn. *MAG-10*, 474 (1974).
2. P. J. Besser and R. G. Warren, Intermag Conf. 1974, Toronto, Paper No. 1-8.
3. E. A. Giess, J. D. Kuptsis and E. A. D. White, J. Cryst. Growth *16*, 36 (1972).
4. E. A. Giess and R. Ghez, "Epitaxy", J. W. Matthews, editor, Academic Press, New York (1974).
5. R. Ghez and E. A. Giess, J. Cryst. Growth *27*, 221 (1974).
6. E. A. Giess, C. F. Guerci, J. D. Kuptsis and H. L. Hu, Mater. Res. Bull. *8*, 1061 (1973).
7. S. L. Blank, B. S. Hewitt, L. K. Shick and J. W. Nielsen, A.I.P Conf. Proc. No. *10*, 256 (1973).
8. K. L. Konnerth and F. H. Dill, Solid-State Electronics *15*, 371 (1972).
9. R. J. Kobliska, R. R. Ruff, and J. J. Cuomo, this Conf.
10. L. K. Shick, J. W. Nielsen, A. H. Bobeck, A. J. Kurtzig, P. C. Michaelis, and J. P. Reekstin, Appl. Phys. Lett. *18*, 89 (1971).
11. H. L. Hu and E. A. Giess, this Conf.

LPE GROWTH OF GARNET FILMS UNDER ISOTHERMAL CONDITIONS FROM $PbO:B_2O_3$ BASED SOLUTIONS CONTAINING ORTHOFERRITE CRYSTALS

T. S. Plaskett and R. Ghez
IBM Thomas J. Watson Research Center, Yorktown Heights, New York 10598

ABSTRACT

The kinetics of isothermal growth from supersaturated solutions of composition in which the primary phase is orthoferrite is described. The film thickness increased as the square root of the growth time; the supersaturation decreased (aged) according to the Lifshitz-Slyozov theory of Ostwald ripening. The orthoferrite crystals were completely separated from the supersaturated solution at the growth temperature by a 45 mesh Pt screen for both the "dipping" and "tipping" methods. The orthoferrite was used as a source to replenish the solution in garnet constituents. The experimental data were taken on (100) $(EuY)_3Fe_5O_{12}$ films deposited on $Sm_3Ga_5O_{12}$ substrates.

INTRODUCTION

Garnet films are commonly grown isothermally by liquid phase epitaxy (LPE) from supersaturated $PbO:B_2O_3$ solutions by a technique developed by Levenstein et al[1]. In their method the composition of the solution is chosen so that the primary phase is garnet.

In this paper we describe the kinetics of isothermal growth of garnet films from solutions where the composition is such that the primary phase is orthoferrite. The solution at the growth temperature is therefore simultaneously saturated with respect to the orthoferrite and supersaturated with respect to the garnet phase. This unusual property of $PbO:B_2O_3$ solutions was initially reported by Blank and Nielsen[2]. Here, techniques are described to separate the orthoferrite crystals from the supersaturated solution for both the "dipping" and "tipping" methods. It is also shown that the orthoferrite crystals can be used to replenish the solution in garnet constituents after their depletion by film growth.

EXPERIMENTAL

Garnets of two compositions, $Eu_2Y_1Fe_5O_{12}$ and $Eu_1Y_2Fe_5O_{12}$, were grown. The growth kinetics for both compositions were found to be similar. The films were deposited on Syton[3]-polished (100) $Sm_3Ga_5O_{12}$ substrates, 12 mm. in diameter. The composition of the solution is shown in Table I. The molar ratio R_1, defined by Blank and Nielsen as the mole ratio of Fe_2O_3 to the sum of $Y_2O_3 + Eu_2O_3$, is here 4.0. For a number of garnet compositions, those authors reported that the orthoferrite-garnet phase boundary is between $R_1 = 12$ and 17; for smaller values of R_1 the primary phase is orthoferrite.

The films were grown by both a modified "dipping"[1] and by the "tipping"[4] methods. The presence of orthoferrite crystals at the growth temperature required that some means be provided to separate them from the supersaturated solution so as not to interfere with the garnet film growth. In the "tipping" method a 45 mesh Pt screen was placed across the mid-section of the Pt boat thus preventing the orthoferrite crystals from flowing onto the substrate. In the "dipping" method an inner Pt chamber, which had a 45 mesh screen covering an opening on its sides, was submerged into solution at the growth temperature. The inner chamber therefore remained free of orthoferrite crystals. In the "tipping" method the substrate was placed in a covered boat at the start of a run and brought up to temperature along with the charge. In the "dipping" method the substrate was preheated to the growth temperature and then dipped into the inner chamber.

The kinetic data reported in this paper were taken on films grown by this "tipping" method. For each growth the solution was heated to 300°C above the growth temperature and held at this temperature for

TABLE I. Typical charge composition for the growth of $(EuY)_3Fe_5O_{12}$ films.

$Eu_2O_3 + Y_2O_3$	2.6 mole %
Fe_2O_3	10.4
PbO	81.8
B_2O_3	5.2

three hours. The solution was then slowly cooled over a constant period of three hours to the growth temperature. The substrate was 2-5°C colder than the growth temperature to prevent its dissolution in the melt.

The stability of the supersaturated solution with time was studied. The solution was held at the growth temperature for times ranging from 0 to 36 hours before a film was grown isothermally. Time zero was taken at the end of the 3 hour cooling cycle. The film thicknesses were measured by optical interference techniques.

RESULTS

In previous studies[5,6,7], when films were grown using solutions shown in Table I, the solution was continually cooled during growth. It was assumed that it was not possible to form a supersaturated solution with respect to the garnet phase in the presence of orthoferrite crystals of sufficient degree to grow films isothermally. However, this assumption is invalid. Even for solutions of this composition, the solution was supersaturated with respect to the garnet phase at the growth temperature. This is evident in Fig. 1 where curves of film thickness (h) versus growth time (t_g) are shown for a solution cooled at 1.6°C/min and for isothermal growth. The initial growth temperature was the same in both cases. For thicknesses less than about 1 μm both curves are coincident. This clearly indicates that the solution at the growth temperature is supersaturated with respect to the garnet phase; cooling the solution may only affect the growth process after about 1 μm of film growth. The film thickness is directly proportional to the supersaturation (σ) and to $t_g^{1/2}$ according to[8],

$$h = 2\sigma(Dt_g/\pi)^{1/2}. \quad (1)$$

Here D is the diffusion coefficient and the supersaturation is defined as,

$$\sigma = (C_L - C_E)/\rho \quad (2)$$

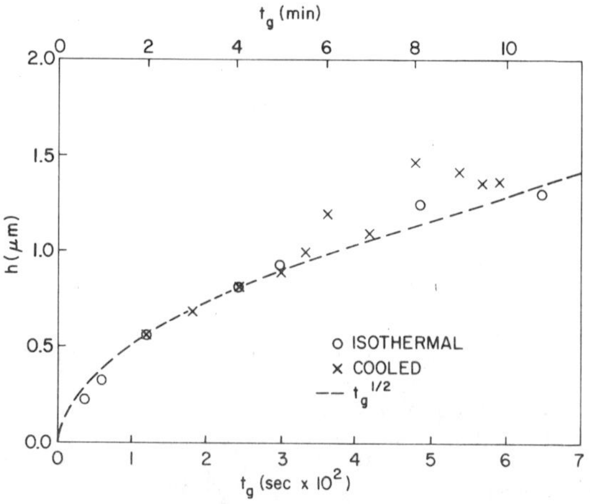

FIG. 1 Film thickness (h) as a function of growth time (t_g) for films grown isothermally and for films grown from a solution cooled at the rate of 1.6°C/min.

where C_L is the garnet concentration in solution, C_E its equilibrium value at T_g and ρ the density of solid garnet.

In a first set of experiments, films were grown immediately at the end of the cooling period. The film thicknesses versus $t_g^{1/2}$ is shown in Fig. 2 for three growth temperatures (T_g). For each temperature the data can be described by Eq. (1). The slope of these curves which is proportional to the supersaturation is steepest for data taken at 815°C. Thus the supersaturation passes through a maximum as T_g is lowered. The supersaturation does not continue to increase as the temperature is decreased below the orthoferrite-garnet phase boundary because, eventually, supersaturation is reached causing nucleation and growth of garnet crystallites in the solution. After nucleation, the rate of change of σ is a complex and largely unknown function of time and temperature.

To examine the effect of aging of the melt, a second set of films was grown at various times (t) after the end of the cooling period. The growth time for all films was 4 min. Lifshitz and Slyozov[9] developed a theory of grain growth from a supersaturated solid solution. As equilibrium is approached they predict a rate equation for supersaturation,

$$Z^3 = kt + B \tag{3}$$

where $Z = \sigma(0)/\sigma(t)$ (inverse fractional supersaturation) and k and B are constants. Since the film thickness is proportional to the supersaturation, Z is proportional to h^{-1}. In Fig. 3, h^{-3} is plotted against t. There is good agreement between the experimental data and the Lifshitz-Slyozov theory. Since their theory only applies after nucleation, i.e., during coalescence where smaller particles redissolve and larger particles grow, we can conclude that in our experiment nucleation of the garnet has occurred for all growth temperatures. In the "dipping" apparatus, it was possible to view the surface of the inner chamber at the growth temperature. No particles were observed on the surface of the inner chamber when it was first submerged in the solution after the three hour cooling period to the growth temperature. Therefore, the initial particle size must be extremely small. However, after holding

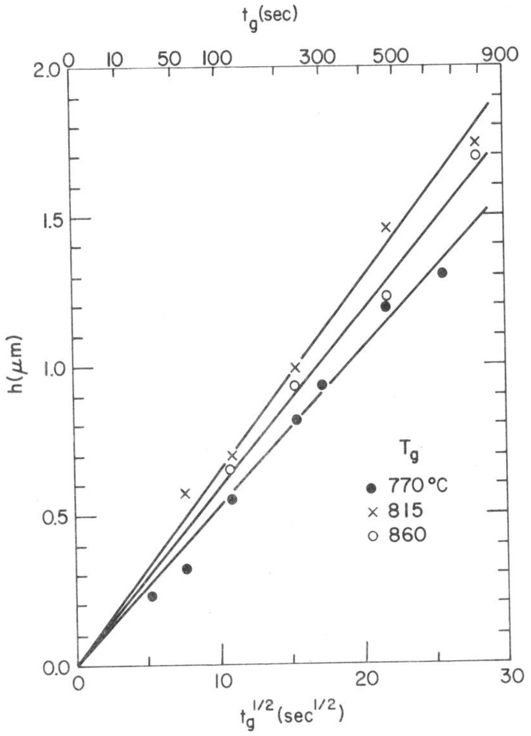

FIG. 2 Film thickness (h) as a function of growth time (t_g) under isothermal growth condition for three growth temperatures (T_g).

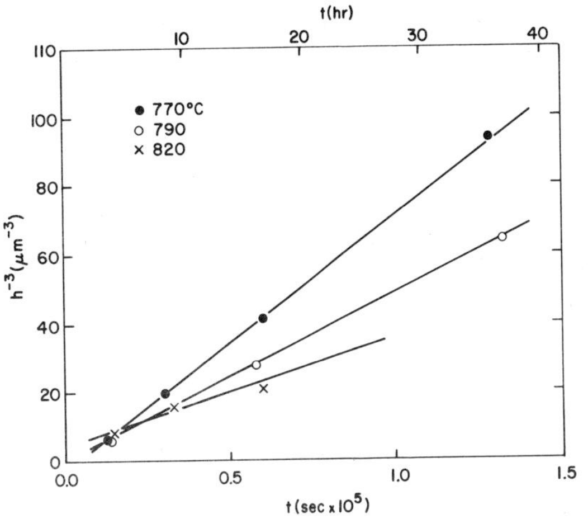

FIG. 3 Film thickness (h) as a function of aging time (t) of the solution at three temperatures (T_g).

the solution for about 12 hours at the growth temperature, a scum was evident on the surface. A detailed analysis of the model that applies to this case is in progress.

The presence of the orthoferrite crystals provides an in situ source to replenish the solution in garnet constituents. The solution is simply heated above the garnet-orthoferrite boundary and then lowered again to the growth temperature. About 40 films were grown in this manner by the "tipping" method from the same 50 gm charge. No detectable variation in the film composition was observed between films.

The film thickness was also duplicated after a series of about 40 film growths ("tipping") provided the film was grown under the same reheating, aging, and growth procedures. This means that the initial supersaturation is reproduced after the reheating step and is independent of the number of films grown. The supersaturation is given by Eq. (2). For the same degree of supersaturation we may conclude that C_L, determined initially by the orthoferrite-garnet phase boundary, is independent of the number of films grown. After each film growth the solution is slightly depleted in garnet constituents which alters the starting composition of the solution for the next growth. This variation, however, does not affect film thickness and film composition.

ACKNOWLEDGMENT

The authors are grateful to A. H. Parsons for growing and characterizing the films, and to E. A. Giess for useful discussions.

REFERENCES

1. H. J. Levinstein, S. Licht, R. W. Landorf and S. L. Blank, Appl. Phys. Letters 19, 486 (1971).
2. S. L. Blank and J. W. Nielsen, J. Crystal Growth 17, 302 (1972).
3. Syton is registered trademark of Monsanto, St. Louis, Mo.
4. L. K. Shick, J. W. Nielsen, A. H. Bobeck, A. J. Kurtzig, P. C. Michaelis and J. P. Reekstin, Appl. Phys. Letters 18, 89 (1971).
5. T. S. Plaskett, E. Klokholm, H. L. Hu and D. F. O'Kane, AIP Conf. Proc. 10, 319 (1974).
6. T. S. Plaskett, E. Klokholm and D. C. Cronemeyer, AIP Conf. Proc. 18, 75 (1974).
7. T. S. Plaskett, E. Klokholm, D. C. Cronemeyer, P. C. Yin and S. E. Blum, Appl. Phys. Letters 25, 357 (1974).
8. R. Ghez and E. A. Giess, Mater. Res. Bull. 8, 31 (1973).
9. I. M. Lifshitz and V. V. Slyozov, J. Phys. Chem. Solids 19, 35 (1961).

MAGNETOCRYSTALLINE ANISOTROPY OF Eu:YIG LPE FILMS GROWN ON (111) SUBSTRATES

D. C. Cronemeyer, T. S. Plaskett, & E. Klokholm
IBM Research Center, Yorktown Heights, New York 10598

ABSTRACT

As previously reported in part by us[1], epitaxial films of Eu:YIG exhibit a strong dependence of the uniaxial anisotropy K_u upon Pb-content. In the present work, magnetocrystalline and uniaxial anisotropies have been derived simultaneously from an analysis of the FMR rotation spectra of (111) films about an in-plane ⟨110⟩ axis. The (111) EuIG films are grown on NdGaG, Eu_2Y_1IG films on SmGaG, and Eu_1Y_2IG and YIG films on GdGaG substrates by the LPE tipping process. The Pb contents range from 0.5 to 6 wt. % for growth temperatures ranging from 895 to 740°C. Bulk values are assumed or extrapolated for the magnetization. For EuIG, the room temperature value of K_1 is -34,000 ergs/cm^3 for zero Pb; this is decreased in magnitude by about 30 % by a 6 wt. % Pb addition. Similar dramatic effects occur in the other members of the Eu:YIG system. In the case of YIG, the magnitude of K_1 is almost doubled by a similar addition of Pb. These changes in K_1 cannot be explained on the basis of a simple dilution model in which Pb displaces Eu or Y on the dodecahedral sites of the garnet.

INTRODUCTION

Eu:Ga:YIG and Eu:YIG epitaxial films hold some interest as bubble materials. The uniaxial anisotropy of such materials is of great importance in determining these bubble properties; this quantity has been measured directly by FMR or indirectly by bubble statics analysis. In the FMR determination, it has been necessary to assume values of the magnetocrystalline anisotropy K_1; such values are usually obtained by the FMR study of (110) disks of bulk material. Since epitaxial films are grown at much faster rates and lower temperatures than bulk materials, a great deal more Pb is incorporated in the films than in bulk materials; therefore bulk and film samples of exactly the same composition are almost impossible to make. Large effects of Pb have been seen upon the uniaxial anisotropy K_u, and at least somewhat analogous effects are to be expected for K_1 by Pb incorporation in epitaxial garnet films. These considerations point to the need for a direct measurement of K_1 on films, and it is the purpose of this paper to describe the results of the application of such a method to the measurement of K_1 (and K_u) in the Eu:YIG system. A separate paper gives the theory and derivation of the equations for the method[2]. Certain important references regarding K_1 may be noted[3, 4].

EQUATIONS AND EXPERIMENTAL METHOD

If one considers a (111) film crystallographically, it is readily apparent that the rotation of such a film about an in-plane ⟨110⟩ axis in FMR is most suitable for observing magnetocrystalline effects, since the dc magnetic field traverses easy, intermediate, and hard directions in this plane, whereas in other planes a less er variation is seen. Therefore the method utilized here employs such a rotation of a crystallographically oriented film. If one considers rotation by an angle in this (110) plane with the ⟨112⟩ direction as the reference axis, then it can be shown that the resonance equation to cover this case may be written as[2]:

$$\left[\frac{\omega}{\gamma}\right]^2 = \left(H_r - \sum_{i=0}^{2} H_i f_i\right)\left(H_r - \sum_{i=0}^{2} H_i g_i\right) \qquad (1)$$

where ω is the angular frequency, γ is proportional to the gyromagnetic ratio g, H_r is the resonance field, H_i represent $2K_i/M_s$, where K_i are $K_u - 2\pi M_s^2$, K_1, and K_2, the modified uniaxial, and first two magnetocrystalline anisotropy coefficients, respectively, and f_i and g_i are trigonometric polynomials: $f_o = -\sin^2\Phi$; $g_o = \cos 2\Phi$; $16f_1 = 3 - 16b_2 - 3b_4$; $-4g_1 = b_2 + 3b_4$; $128f_2 = 14 - 3b_2 - 14b_4 - 3b_6$; $-64g_2 = b_2 + 8b_4 + 9b_6$; and where $3b_2 = 2\sqrt{2}\sin 2\Phi + \cos 2\Phi$; $9b_4 = 4\sqrt{2}\sin 4\Phi - 7\cos 4\Phi$; $27b_6 = 10\sqrt{2}\sin 6\Phi + 23\cos 6\Phi$.

This quadratic equation for the resonance field may be solved explicitly for H_r as a function of the angle Φ; K_u, K_1, K_2, $4\pi M_s$, g, and ω all contribute to determining the shape of the curve. Incomplete saturation is also accounted for by a special procedure[2].

The samples are (111) films grown epitaxially on various (111) substrates, and are X-ray oriented.

They are either cut into small rectangles having (110) and (211) faces, or the ⟨110⟩ axis is marked so that the sample may be mounted on a quartz rod for rotation about this in-plane ⟨110⟩ axis in the microwave cavity. The dc magnetic field is thus confined to the (110) plane The data for the resonance field is recorded at intervals of 5, 10, or 15° over 360°. Values separated by 180° are averaged to minimize errors due to misorientation, setting, wobble, and temperature drift. A standard ESR equipment is utilized at about 9.1 GHz. The value of the resonance field is ascertained by an nmr probe. The cavity resonance field is calculated from observation of a standard DPPH powder sample resonance. An rms deviation of the theoretical equation prediction from the actual data set is obtained in five-space, since the parameters K_u, K_1, K_2, g, and δ are to be determined. δ is an epoch angle which corrects for the imprecisely known geometrical origin of Φ. Rms deviations of a few Oe are regularly obtained in the fashion outlined above.

RESULTS

The results of the measurement of K_1 as a function of composition and Pb content are shown in Fig. 1. For the sake of completeness, the behavior of g and K_u as functions of these parameters are also shown (Figs. 2 and 3). The results are very reproducible; for instance, a completely separate run on one sample yielded a K_1 of -30,990 ergs/cm^3 as compared with -31,030 previously obtained. K_2 data are not reported here, since it seems to be a rule, that at least one more anisotropy constant must be determined than may be believed. As shown in Fig. 1, the K_1 values of EuIG and Eu_2Y_1IG can be considerably reduced by Pb addition. It may also be noted that there is a region of the $Eu_yY_{3-y}IG$ system where both K_u and K_1 can be essentially independent of Pb content ($y \simeq 1.2$). The fit of eq. 1 to the experimental data is very good for Eu:YIG, but becomes poor for YIG itself with high Pb. The appearance of magnetostatic modes, etc. in YIG makes the identification of the uniform mode more difficult. The fact that the g-value of EuIG films was generally higher than the known room temperature bulk value[5] of

1.33 was puzzling until the Pb effect upon g was established. For zero Pb, or for a specific value of Pb, the variation of K_1 with change of composition from YIG to EuIG does not seem to be exactly linear[6].

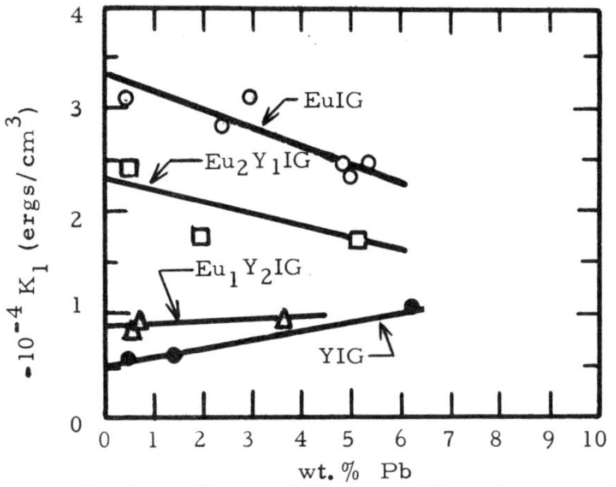

Fig. 1. The dependence of K_1 upon Pb and composition at room temperature.

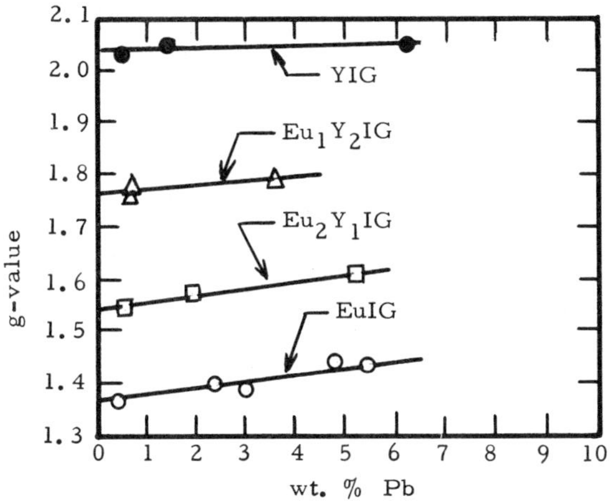

Fig. 2. The g-value as a function of Pb and composition at room temperature.

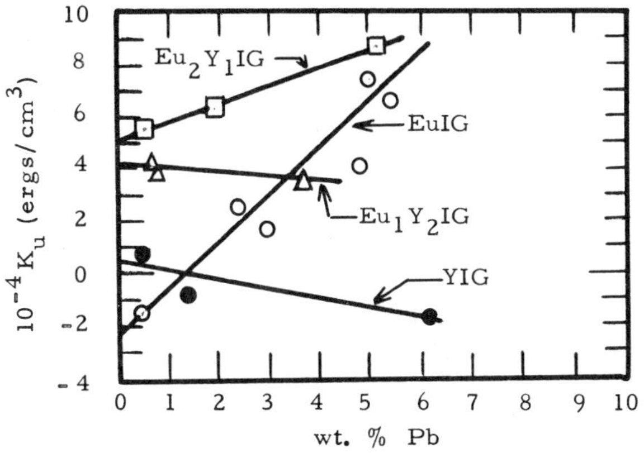

Fig. 3. The dependence of K_u upon Pb and composition at room temperature.

CONCLUSIONS

It is expected that Pb would have an appreciable effect upon the magnetocrystalline energy K_1, since by simple dilution, 6 wt. % Pb going into the Eu sites of EuIG would amount to 0.27 formula units out of 3, a 9 % change. The actual changes in K_1 induced by Pb at this level are as large as 30 %. K_1 itself is thought of mainly as a one ion contribution from Eu in EuIG, since the Fe lattice has only a small contribution of -5700 ergs/cm^3 (pure YIG, since Y is non-magnetic). However, the fact that YIG's K_1 is so greatly increased in magnitude by Pb rather than being slightly diminished as expected, indicates that Pb does not play a passive role in YIG either. Conjectures as to departure from cubic symmetry due to Pb incorporation do not seem to be supported by X-ray studies to date. These new effects of Pb in Eu:YIG were larger than expected from previous considerations of bulk values of K_1.

ACKNOWLEDGMENTS

We are greatly indebted to A. H. Parsons for growth and characterization of the films, as well as to J. Kuptsis for the electron microprobe determination of Pb content. J. C. Slonczewski was a constant help and inspiration in working out the method. A. Hubert supplied the numerical curve fitting computer program which was essential to utilizing the method.

REFERENCES

1. Plaskett, T. S., Klokholm, E., Cronemeyer, D. C., Yin, P. C., & Blum, S. E., Appl. Phys. Lett. 25, 357-9 (1974)
2. Cronemeyer, D. C., to be submitted to IEEE Trans. Magnetics
3. Heinz, D. M., Besser, P. J., Owens, J. M., Mee, J. E., & Pulliam, J. R., Jr., J. Appl. Phys. 42, 1243-51 (1971)
4. LeCraw, R. C., & Pierce, R. D., AIP Conf. Proc. 5, 200-4 (1971)
5. Miyadai, T., J. Phys. Soc. Japan 15, 2205-10 (1960)
6. Heilner, E. J. & Grodkiewicz, W. H., J. Appl. Phys. 44, 4218-19 (1973)

ANISOTROPY CONSTANTS OF GARNET FILMS FROM STRIPE DOMAIN MEASUREMENTS

M. W. Muller and S. K. Chung
Department of Electrical Engineering and Laboratory for Applied Electronic Sciences, Washington University, Saint Louis, Missouri 63130

ABSTRACT

We have measured the quasistatic behavior of stripe domains in (111) epitaxial garnet films placed in in-plane ($\bar{1}$10) and (11$\bar{2}$) fields. In films that possess appreciable cubic anisotropy, the domain walls line up perpendicular to applied fields along or near ($\bar{1}$10), and parallel to fields near (11$\bar{2}$). In (11$\bar{2}$) fields the domain magnetization in adjacent stripes is not symmetrical about the film plane, and adjacent stripes are of unequal width. The domain period first shrinks and then expands with increasing field. These results are interpreted by a theory that gives good agreement with the observed behavior, and yields accurate values of the cubic and uniaxial anisotropy constants.

INTRODUCTION

Four years ago at this conference it was reported that stripe domains in magnetic films could function as optical diffraction gratings, and that the period of such gratings could be controlled by an in-plane magnetic field[1]. Since adjustable gratings are potentially very useful in thin-film optics[2], we undertook a detailed study of the behavior of stripe domains in applied fields. Stripe domain patterns of great perfection and regularity are exhibited by the epitaxial garnet films that have been developed for application in bubble domain devices. These magnetic films are grown on (111) oriented substrates, and in addition to the dominant uniaxial anisotropy nominally normal to the film plane which gives rise to the gross features of the domain structure, they may exhibit cubic and orthorhombic magnetic anisotropies as well as tilting of the easy axis[3]. These additional anisotropies, especially the cubic magnetocrystalline component, can have striking effects on the field dependence of the domain structure. Conversely, observation of the domains in applied fields provides information about the anisotropy. In this note we deal with the determination of anisotropy constants from such observations.

THEORY

When a steady uniform in-plane field is applied parallel to the Bloch walls of a stripe domain structure in a purely uniaxial film the magnetization $\bar{M}$ in each domain "leans" in the direction of the field, and the magnetic surface pole density on the film faces and the wall rotation angle are both reduced. The resulting reduction of the wall energy is greater than that of the dipolar energy, and hence the lowest energy domain structure in the presence of a field has more closely spaced walls than the field-free structure. A simple calculation of the equilibrium wall spacing based on this reasoning[1] is confirmed, for films with large $Q = K_u/2\pi M^2$ (K_u = uniaxial anisotropy energy per unit volume), by a more accurate calculation[4] that includes a more careful evaluation of the stray field energy in the walls and of the exchange coupling of the domains to the walls.

This description of domain behavior in an applied in-plane field $\bar{H}$ is valid in the presence of cubic anisotropy only if the field is applied along (11$\bar{2}$). This direction is the bisectrix axis of the effectively trigonal crystal structure. Thus the anisotropy is mirror symmetric about the plane containing the field and the easy (111) axis, and $\bar{M}$ remains in this plane. Even so the cubic anisotropy affects the behavior of the domains because it breaks the mirror symmetry about the film plane. This results in an asymmetry between the volumes of the "upward" and "downward" sets of domains. In the lowest energy configuration the domains are of unequal width and the entire structure develops a net magnetic moment normal to the film. This transverse magnetic moment and the average dipolar field associated with it make it possible for the energy density in adjacent domains to be equal, and thus for the walls to be in stable equilibrium.

The cubic anisotropy affects the behavior of the domains not only by producing a width asymmetry, but also by modifying the approach to saturation. To understand qualitatively why this should be so, consider the path taken by $\bar{M}$ as it rotates from the easy axis to (11$\bar{2}$). The rotation from (111) to (11$\bar{2}$) passes through (11$\bar{1}$); the rotation from ($\bar{1}\bar{1}\bar{1}$) to (11$\bar{2}$) passes through (00$\bar{1}$). The former is an easy direction of the cubic anisotropy when K_1, the first order cubic anisotropy constant, is negative, the latter when K_1 is positive. This means not only that the set of domains lying originally in (111) will be favored when $K_1 > 0$, the opposed (111) domains when $K_1 < 0$, but also that saturation will take place by the favored set of domains expanding to fill the film rather than homogeneously, as one would expect when $K_1 = 0$. Detailed calculation confirms this expectation. We find[5] that the total domain period (width of two adjacent domains), after originally decreasing in the applied field, expands again in higher fields through the growth of the favored domains, and that saturation is accelerated. This phenomenon limits the range of grating spacing control, but it also provides a quantitative measure of the cubic anisotropy.

When the in-plane field $\bar{H}$ lies along any other direction than (11$\bar{2}$), the energy is no longer symmetric about the plane containing $\bar{H}$ and (111), and $\bar{M}$ in each domain develops a component normal to that plane. These transverse components are not equal in adjacent domains and they result in a finite magnetic pole density div $\bar{M}$ in any wall parallel to the field. If this "magnetic charge" is large enough, it gives rise to a dipolar energy that may be greater than the excess energy of a Néel wall oriented normal to $\bar{H}$ over the Bloch wall energy. Futhermore, as has been pointed out by Hubert[6], a stripe domain structure of perpendicular walls is magnetoelastically favored over parallel walls. Epitaxial films are partly constrained from deforming, so that magnetoelastic energy terms do not carry their full weight, but in combination with the dipolar energy they are probably capable of accounting for the observed preference[3,5] for perpendicular domain orientation in applied fields along or near ($\bar{1}$10).

The quantitative theory that corresponds with this description is being reported elsewhere[5]. The results may be represented in the form of curves of the two quantities $x = D/(d_1 + d_2)$ and $y = d_1/(d_1 + d_2)$ as functions of the reduced field $h = MH/2K_u$, with parameters Q, D/ℓ, and $k = K_1/K_u$. Here D is the film thickness, d_1 and d_2 are the domain widths, and ℓ is the "characteristic length" of Thiele[7]. In practice it is only necessary to calculate the curves for $\bar{H}$ along (11$\bar{2}$), since most of the useful information can be extracted from the domain behavior in this field orientation. This reduces the amount of computer time required for the analysis of experimental results to very reasonable proportions.

MEASUREMENTS

The orientation of the crystal axes in the films on which we have made extensive measurements was determined by X-ray diffraction. The film is then mounted between the pole pieces of an electromagnet in a plastic jig that allows it to be rotated. A small coil permits the application of both steady and alternating fields perpendicular to the film plane. The domain structure is observed and photographed in a polarizing microscope.

In films that possess appreciable cubic anisotropy, the asymmetry between the two sets of domains in even moderate applied fields is readily apparent for all orientations except the immediate vicinity of ($\bar{1}10$). Indeed, the vanishing of the asymmetry allows this direction to be determined to within approximately 1-2°, and effectively eliminates the need for prior X-ray orientation. With a small alternating normal field highly reproducible results are obtained, especially for field orientations at or near ($11\bar{2}$). With a ($\bar{1}10$) field most films exhibit domain patterns of stripes perpendicular to $\bar{H}$, but we have observed a tendency for the structures to break up into severed stripes or bubbles upon application of an alternating field, or to exhibit hysteresis.

All our results are derived from photographs of the domain structure in ($11\bar{2}$) and ($\bar{1}10$) fields taken at fixed (known) magnification. Domain widths and periods are measured using a reticle under a low-power magnifier. We have found that single domain widths of a few microns, characteristic of films in the bubble device range, cannot be measured with high accuracy. Even when the domains appear subjectively to show good optical contrast, it is difficult to locate the nominal wall position with confidence in the shaded region between the light and dark bands, and the contrast changes as the magnetization inclines in the applied field. It is much easier to measure precisely the total width of a group of domains, since the absolute error of such a measurement is no larger than that incurred in measuring a single domain.

Accordingly we base the quantitative determination of material parameters entirely upon fitting calculated curves to the experimentally determined values of x as a function of $\bar{H}$ along ($11\bar{2}$). The measurements of y, and of both x and y in a ($\bar{1}10$) field, are useful for determining other contributions to the anisotropy, notably tilting of the easy axis. Such tilting produces a finite slope of the curve of y vs h near zero field. Tilting of the axis and orthorhombic anisotropy can also be measured by comparing observations made with $\bar{H}$ lying along the three equivalent ($11\bar{2}$) directions.

We show in Fig. 1 our results on a film of nominal composition $Gd_{0.8}Er_{2.2}Fe_{4.5}Ga_{0.5}O_{12}$. Comparison of the

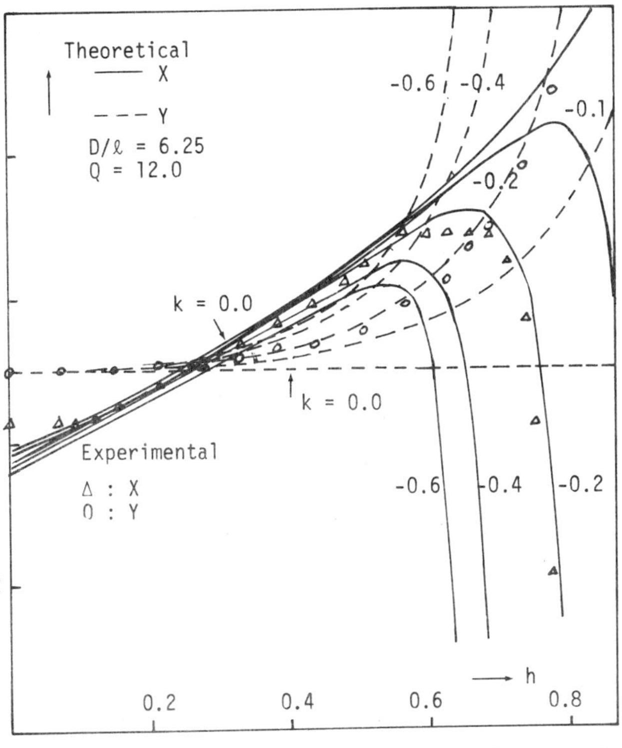

Figure 1. Domain width and asymmetry parameters x and y of a $Gd_{0.8}Er_{2.2}Fe_{4.5}Ga_{0.5}O_{12}$ film.

experimental data with calculated curves yields $K_u = 9.2x10^3$ erg cm^{-3}, a value quite close to other measurements reported on this material[8], and $K_1 = -1.85x10^3$ erg cm^{-3}, in qualitative agreement with estimates based on the single ion model[9].

The anisotropy measurement of Hubert, Malozemoff, and DeLuca, based on determining the field for homogeneous domain nucleation along ($11\bar{2}$) and ($\bar{1}10$), yields comparable results for films with negative cubic anisotropy, where it predicts a difference between the nucleation fields in the two orientations that is large and rapidly varying. For positive K_1 the difference is small and insensitive to the magnitude of the anisotropy. The method described here should be useful for such films; our calculations predict a sensitive dependence of the curve of x vs h on the magnitude of K_1 for both positive and negative values. We have not attempted a comparison of our measurements with Shumate's optical magnetometry[10]. The analysis of that technique as applied to cubic films neglects the transverse magnetic moment of the film, an approximation that may not be justifiable for appreciable K_1 values, especially for K_1 positive.

We have followed the usual practice of neglecting the second-order cubic anisotropy K_2 in the interpretation of these room-temperature measurements. Work is currently under way to assess the possible effect of this approximation on the results.

ACKNOWLEDGEMENTS

We are indebted to J. W. Moody and R. W. Shaw for the growth and characterization of the garnet films used in this study. This work was supported by the National Science Foundation, Grant GH32000, and by the Office of Naval Research, Contract N00014-67-A-0445-0005.

1. T. R. Johansen, D. I. Norman, and E. J. Torok, J. Appl. Phys. 42, 1715 (1971)
2. M. W. Muller, M. J. Sun, and S. K. Chung, MRI Symp. Proc. XXIII, 1974 (to be pub.)
3. A. Hubert, A. P. Malozemoff, and J. C. DeLuca, J. Appl. Phys. 45, 3562 (1974)
4. W. F. Druyvesteyn, J. W. F. Dorleijn, and P. J. Rijnierse, J. Appl. Phys. 44, 2397 (1973)
5. S. K. Chung and M. W. Muller, Int. J. Magn., to be pub.
6. A. Hubert, Phys. Stat. Sol. 6, 839 (1964)
7. A. A. Thiele, Bell Syst. Tech. J. 48, 3287 (1969)
8. J. W. Moody, R. M. Sandfort, and R. W. Shaw, Tech. Rept. "Magnetic Bubble Materials", ARPA Contract DAAH01-72-C-1098, February 11, 1973
9. R. D. Henry and D. M. Heinz, AIP Conf. Proc. 18, 194 (1974)
10. P. W. Shumate, Jr., J. Appl. Phys. 44, 3323 (1973)

TRANSVERSE MAGNETIZATION CURVES IN BUBBLE FILMS WITH PARAMETER DISPERSIONS

R. F. Soohoo* and Kenneth Lee
IBM Research Laboratory
San Jose, California 95193

ABSTRACT

A new method has been devised to relate the shape of the transverse or hard-axis magnetization curve to the dispersion of uniaxial anisotropy (in magnitude and direction) and magnetization in a bubble film. Using our theoretical results, the average value of H_K and $4\pi M$ as well as their dispersions can be determined by comparison with experiment. In our method, a field parallel to the easy axis, $H_{\parallel}$, sufficient to saturate the film as well as a varying field transverse to the easy axis (or in the film plane), $H_{\perp}$, are applied. In this case, the transverse loop has no hysteresis and its shape is dependent on the anisotropy and magnetization dispersion in the sample. Assuming normal distributions for the deviation from the mean for both the anisotropy and the magnetization, the standard deviation $\sigma_{H_\perp}$ of $H_\perp$ required for magnetic saturation, a quantity which is closely related to experiment, can be calculated. Experimental results on an amorphous gadolinium colbalt film is presented and compared with our theoretical predictions. We believe that our method will complement existing techniques of measurement.

INTRODUCTION

For a magnetic bubble film to be useful, its coercivity H_c must be relatively small. Since H_c may be dependent on the spatial dispersion in the anisotropy and magnetization of a sample, relating such dispersions to the shape of the hysteresis loop is of considerable interest to bubble physics and technology. In this paper, some theoretical and experimental results in this regard are presented.

Magnetization and intrinsic length of thin bubble films can be determined by measuring the thickness of the epitaxial layer (in the case of garnet bubble films), the demagnetizing strip width, and the bubble collapse field.[1] The magnetization and wall energy can also be measured using a magneto-optic technique.[2] These data can be used in the determination of the anisotropy field using an optical magnetometer.[3,4] A nucleation technique has also been used to measure the anisotropy in garnet films.[5,6] Most of these techniques can measure anisotropy from point to point in the film, but are rather tedious if data at a large number of points have to be taken. In our method, the average value of anisotropy and the range of its variation over the sample can be determined by a single measurement. In this respect, our method complements the other techniques in the measurement of anisotropy dispersion.

In the absence of applied fields, a bubble film with $q(=H_k/4\pi M)>1$ is not uniformly magnetized, but is divided up into a large number of oppositely magnetized vertical domains.[7] The exact configuration of these domains and remanence is dependent upon the magnetic history of the sample. Thus, when a field $H_\perp$ is applied transverse to the anisotropy easy axis (or parallel to the film plane), magnetization changes occur by a mixture of domain wall motion and rotation. This gives rise to an open hysteresis loop rendering its interpretation difficult, particularly when anisotropy dispersion is involved.

If a magnetic field $H_{\parallel}$ of the order of $4\pi M$ is simultaneously applied parallel to the easy axis (or perpendicular to the film plane) to saturate the film, the $M\cos\phi$-$H_\perp$ characteristic will be devoid of hysteresis. In what follows, we shall show that under this new experimental arrangement, the anisotropy and magnetization dispersion is intimately related to the curvature in the $M\cos\phi$ vs. $H_\perp$ characteristic.

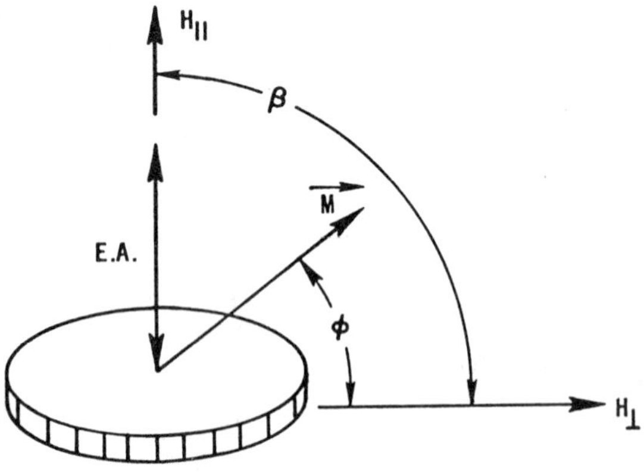

Figure 1. Bubble Film and Applied Fields. Note That $H_{\parallel} \sim 4\pi M$ is Fixed While $H_\perp$ is Varied.

THEORETICAL CONSIDERATIONS

Referring to the bubble film of Fig. 1, the free energy E of the system is given by:

$$E = K_1 \sin^2(\beta - \phi) + 2\pi M^2 \cos^2(\beta - \phi) - MH_\perp \cos\phi - MH_{\parallel} \sin\phi \qquad (1)$$

where K_1 is the uniaxial anisotropy constant and $H_{\parallel}$ and $H_\perp$ are applied perpendicular and parallel to the film plane respectively. To find the equation for the $M\cos\phi$ vs. $H_\perp$ or transverse magnetization characteristic, we set $\delta E/\delta\phi = 0$ to obtain:

$$\frac{H_k - 4\pi M}{2} \sin 2(\beta - \phi) = H_\perp \sin\phi - H_{\parallel} \cos\phi \qquad (2)$$

where $H_k = 2K_1/M$ is the anisotropy field. If there is no dispersion in $\vec{H}_k$ or $4\pi M$, $\beta = \pi/2$ and Eq. (2) reduces to:

$$(q - 1)\cos\phi + C\cot\phi = H_{\perp n}. \qquad (3)$$

where $q = H_k/4\pi M$, $C = H_{\parallel}/4\pi M$ and $H_{\perp n} = H_\perp/4\pi M$. Eq. (3) can easily be solved graphically for various values of q and C.[8] The results are shown in Fig. 2. It is seen from this figure that the application of a saturating field $H_{\parallel}$ parallel to the easy axis produces a transverse magnetization curve without hysteresis. Due to the presence of $H_{\parallel}$, however, the curve is no longer linear up to magnetic saturation in the film plant, as $H_\perp$ is increased. If dispersion is present, additional curvature will occur in $M\cos\phi$ vs. $H_\perp$ characteristic. Thus, to relate the experimentally observed characteristics to these dispersions, it is first necessary to correct the $M\cos\phi$ vs. $H_\perp$ characteristic to account for the presence of $H_{\parallel}$. To do this, we need only to add on to the experimentally determined $\cos\phi$ at any given $H_{\perp n}$ the corresponding Δ (see Fig. 2) representing the difference between $\cos\phi$ for $H_{\parallel} = 0$ and $H_{\parallel} \neq 0$.

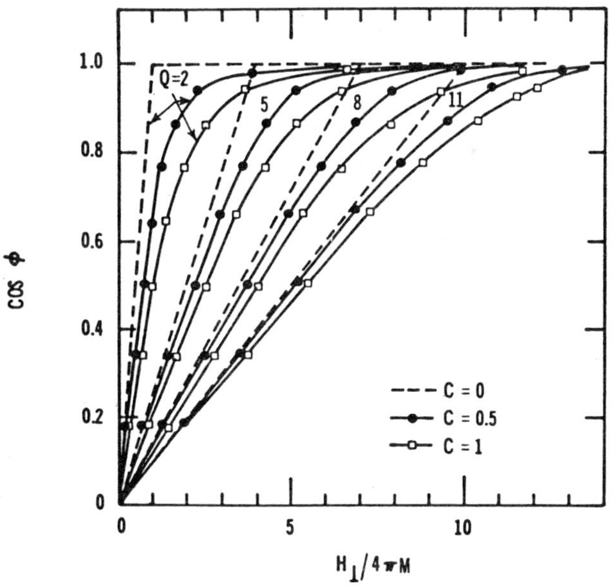

Figure 2. Theoretical M cos φ – H⊥ Curve

If anisotropy or magnetization dispersion is present, a straight forward calculation of the transverse magnetization curve is difficult, as Eq. (3) is quartic in cos ϕ. From an experimental standpoint, an alternative approach is more fruitful. In this approach, we relate the field separation between the boundaries of the curved portion of the M cos ϕ - $H_\perp$ characteristic to the dispersion in anisotropy or magnetization. Before we proceed with our general calculation, a simplified example of the effect of anisotropy dispersion will be instructive. Consider a film with two regions of volumes V_1 and V_2 having anisotropy fields H_{K1} and H_{K2} respectively with $H_{K1} < H_{K2}$. Assuming, for simplicity, a pure rotation process even in the absence of $H_\parallel$, we have from Eq. (2),

$$\cos\phi = \frac{H_\perp}{H_{K1} - 4\pi M}\left(\frac{V_1}{V}\right) + \frac{H_\perp}{H_{K2} - 4\pi M}\left(\frac{V_2}{V}\right) \qquad H_\perp < H_{K1}$$

$$= \frac{V_1}{V} + \frac{H_\perp}{H_{K2} - 4\pi M}\left(\frac{V_2}{V}\right) \qquad H_{K1} < H_\perp < H_{K2}$$

$$= 1 \qquad H_\perp > H_{K2} \qquad (4)$$

where $V = V_1 + V_2$ is the total volume of the film. Thus, we see that the M cos ϕ vs. $H_\perp$ characteristic first rises linearly until $H_\perp = H_{K1}$ whereby volume V_1 is saturated, then rises linearly with a smaller slope until $H_\perp = H_{K2}$ whereby the entire film is saturated. If the sample has a distribution of H_K varying between H_{K1} and H_{K2} instead, we can expect the M cos ϕ vs. $H_\perp$ characteristic in the $H_{K1} < H_\perp < H_{K2}$ region to be curved rather than straight. In other words, the field separation between the linear and flat portions of the transverse hysteresis curve is a measure of anisotropy dispersion.

Assuming normal distributions for the anisotropy (in magnitude and direction) and magnetization dispersion we have:

$$\frac{dv_h}{dh} = \frac{1}{\sqrt{2\pi}\,a} e^{-h^2/2a^2}$$

$$\frac{dv_\delta}{d\delta} = \frac{1}{\sqrt{2\pi}\,b} e^{-\delta^2/2b^2}$$

$$\frac{dv_m}{dm} = \frac{1}{\sqrt{2\pi}\,c} e^{-m^2/2c^2} \qquad (5)$$

where

$$H_k = \langle H_k\rangle + h$$

$$\beta = \langle\beta\rangle + \delta$$

$$M = \langle M\rangle + m \qquad (6)$$

and $\langle H_k\rangle$, $\langle\beta\rangle$, $\langle M\rangle$ are the average values of H_k, β, and M. Thus, dv_h/dh represents the differential volume dv with an anisotropy field range dH_k centered about H_k, etc. For $\delta \ll \langle\beta\rangle = \pi/2$, Eq. (2) simplifies to:

$$H_\perp = (H_k - 4\pi M)\left(\cos\phi - \frac{\cos 2\phi}{\sin\phi}\,\delta\right) + H_\parallel \cot\phi \qquad (7)$$

Multiplying the left hand side of Eq. (7) by dv_h, dv_δ and dv_m and the right hand side by their equivalent from Eq. (5) and integrating from $-\infty$ to $+\infty$, we find the standard deviation $\sigma_{H_\perp}$ in $H_\perp$ as:[9]

$$\sigma_{H_\perp} = \sqrt{(a - 4\pi c)^2 \cos^2\phi + (\langle H_k\rangle - 4\pi\langle M\rangle)^2 \left(\frac{\cos 2\phi}{\sin\phi}\right)^2 b^2} \qquad (8)$$

where

$$(\langle H_k\rangle - 4\pi\langle M\rangle)\cos\phi + H_\parallel \cot\phi = H_\perp \qquad (9)$$

A word about the physical meaning of $\sigma_{H_\perp}$ is in order. Due to dispersion in the film, a distribution in the applied field $H_\perp$ is required to obtain a uniform magnetization orientation (or ϕ) in the film. Thus, $\sigma_{H_\perp}$ is a measure of the distribution in anisotropy and magnetization and is therefore intimately related to the field separation between boundaries of the curved portion of the M cos ϕ vs. $H_\perp$ characteristic discussed above.

EXPERIMENTAL RESULTS

The transverse magnetization curve was obtained by using a vibrating sample magnetometer. The field required to saturate the film in the easy direction was provided by a water-cooled Helmoltz pair. Because of magnet gap space restrictions, the I.D. and O.D. of the coils were only 0.400" and 0.970" respectively with 660 turns/coil wound around aluminum bobbins 0.390" in length. These bobbins were in turn inserted into a water-cooled aluminum block. In this way, a field of 575 oe. was obtained with 1A through each coil giving a temperature at the sample of only 31° C.

Fig. 3 shows the Mcos ϕ vs. $H_\perp$ curves for an amorphous GdCo sample (19.6% Gd, 78.4% Co, 2% A) with a $4\pi M$ of about 1,000, with and without $H_\parallel$. There was considerable noise in our VSM signal; the curves of Fig. 3 were obtained after signal averaging. Much of the noise was traced to the current fluctuations in the electromagnets supplying the fields. It is seen that whereas the hysteresis loop at $H_\parallel = 0$ is open,[10] that for the $H_\parallel = 575$ oe. case is closed or devoid of hysteresis, consistent with theoretical prediction above providing easy axis dispersion is small. It then follows from Eq. (8) that $\sigma_{H_\perp} \simeq a$, or the field separation between the curved portion of the Mcos ϕ vs. $H_\perp$ characteristic is approximately 6a. From Fig. 2. $\sigma_{H_\perp}$ is found to be about 350 oe. or the standard deviation a is about 7% of the anisotropy value H_k of about 5,000 oe.

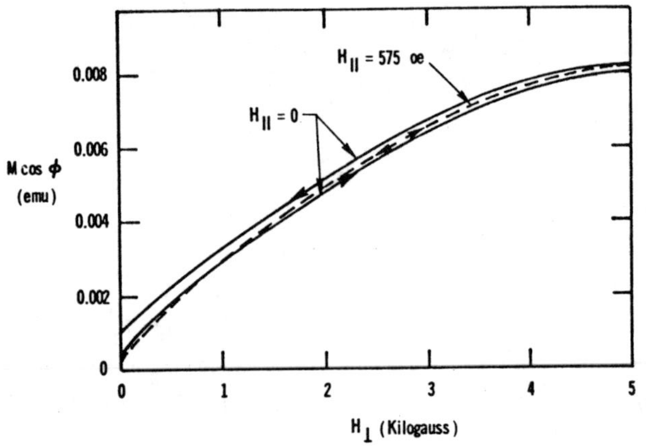

Figure 3. Experimental M cos ϕ-$H_\perp$ Characteristic With and Without $H_{\parallel}$

ACKNOWLEDGEMENT

We wish to thank N. Heiman for providing the GdCo sample and B. A. Reynolds, Jr. and G. E. Guthmiller for technical assistance.

REFERENCES

*NSF Faculty Research Participation Fellow; Permanent address: Univ. of Calif., Davis, California 95616.

1. D. C. Fowlis and J. A. Copeland, AIP Conf. Proc. No. 5, Magnetism and Magnetic Materials, 1971 (17th Annual Conference), p. 240.

2. D. J. Craik, P. V. Cooper and G. Myers, Jour. Physics D: Applied Physics, 6, 872 (1973).

3. P. W. Shumate, Jr., IEEE Trans. Magn. 7, 586 (1971).

4. P. W. Shumate, Jr. D. H. Smith, and F. B. Hagendorn, Jour. Appl. Phys. 44, 449 (1973).

5. A. Hubert, A. P. Malozemoff, and J.C. DeLuca, Journ. Appl. Phys. 45, 3562 (1974).

6. A. P. Malozemoff and J. C. DeLuca, Journ. Appl. Phys. 45, 4586 (1974).

7. See, e.g., Z. Malek, V. Kambersky, Czech. J. Phys. 8 , 416 (1956).

8. Since energy of the remanent state is less than $2\pi M^2$, it is only necessary to apply a field $H_{\parallel}$ a fraction of $4\pi M$ to saturate the film along the easy axis, providing anisotropy axis dispersion is not too large.

9. In this calculation, the exchange field due to $\nabla^2\vec{M}$ has been implicitly neglected. This is justified provided that the variation in $\vec{M}$ is reasonably gradual (see, e.g., S. Middlehoek, Ferromagnetic Domains in Thin Ni-Fe Films, Drukkerij Wed. G. Van Soest, N. V., Amsterdam, 1961).

10. In contrast to the case of a permalloy film whose transverse loop is linear and closed at low drive fields (see R. F. Soohoo, Magnetic Thin Films, Harper and Row Publishers, New York, N. Y., 1965, pps. 138-140), our GdCo sample exhibits hysteresis even at low $H_\perp$ in the absence of $H_{\parallel}$. This is consistent with Kerr and Faraday rotation results which indicate a complex mixture of rotation and wall motion even at low $H_\perp$ for $H_{\parallel} = 0$.

HIGH-SPEED PHOTOGRAPHY OF TOPOLOGICAL SWITCHING AND ANISOTROPIC SATURATION VELOCITIES IN A MISORIENTED GARNET FILM

A. P. Malozemoff, IBM Research Center, Yorktown Heights, N. Y. 10598
K. Papworth, Imperial College, London, England

ABSTRACT

The topological switching of a bubble array into one of the opposite polarity by a bias field pulse is studied in a garnet film by high-speed photography. The bubbles of the static array are elliptically distorted. When the pulse is applied, the walls move with highly anisotropic velocities, largest along the original minor axis of the bubbles in the array and smallest along the original major axis. The domain walls collide with four out of six of the nearest neighbor bubbles. At the moment the pulse terminates, the reformed array consists of dogbone-shaped domains, which later expand into an array of bubbles elliptically distorted along the original minor axis. Finally the major and minor axes interchange. The key requirement for regular topological switching is seen to be the anisotropic wall velocity. This velocity anisotropy is also observed by high-speed photography in isolated bubbles, pulsed to expand. The static ellipticity of the bubbles is related to an in-plane magnetic energy anisotropy arising from a crystallographic misorientation, and a qualitative argument shows that this same magnetic anisotropy may cause the observed velocity anisotropy.

INTRODUCTION

In this paper we study the phenomenon of topological switching in a garnet bubble film by high speed photography. It has been found[1,2] that if one starts with a regular demagnetized array of bubbles, a single bias field pulse of appropriate strength, length and polarity can switch the array into a regular array of the opposite polarity. That is, if the initial array, viewed in a polarizing microscope with a small polarizer-analyzer angle, is made up of black bubbles on a white background as in Fig. 1a, then the switched array is made up of white bubbles on a black background, as in Fig. 1o. In this paper we present high-speed photographs which show the switching mechanism; these photographs reveal an anisotropy in the saturation velocity which may be related to a crystallographic misorientation of the [111]-garnet film.

EXPERIMENTAL TECHNIQUES AND RESULTS

In this study we have used the same sample described earlier by one of us,[2] namely a $Eu_{0.6}Y_{2.4}Ga_{1.1}Fe_{3.9}O_{12}$ garnet, 17μ thick, with a room temperature magnetization $4\pi M_s$=197 Oe and uniaxial anisotropy K_u=12900 erg/cm^3. The material parameters are somewhat different from those reported earlier because of a difference in the area of the sample used and possibly because of an ambient temperature difference. The bubbles were pulsed with a 200 Oe, 420 nsec bias field pulse with 10 nsec rise and fall times (in contrast to the 145 Oe, 380 nsec pulses with somewhat longer rise and fall times used in the earlier work[2]). The starting array has an even more accentuated eccentricity than the arrays described in earlier work for reasons not fully understood but possibly due to the differences in material parameters and pulse conditions mentioned above.

To observe the actual dynamic switching process, we have used a high-speed photographic technique described earlier:[3] a single 10 nsec light pulse from a nitrogen-laser-pumped dye laser was synchronized to flash through the polarizing microscope at a given time after the onset of the bias field pulse. The light was sufficient to view an area 100μ across and expose high-speed Polaroid film in a single shot. Typical photographs are shown in Fig. 1. Although each of these photographs represents an individual experiment switching from a black to white array, the switching process is sufficiently consistent that this series of photographs represents the actual sequence of bubble shapes that occurs in a single pulse experiment.

Figs. 1a-1e, taken at 100 nsec intervals, show that the bubbles expand during the application of the pulse. The bubbles initially start out elliptically distorted and they tend to grow more rapidly along the direction which was the initial minor axis. Thus at

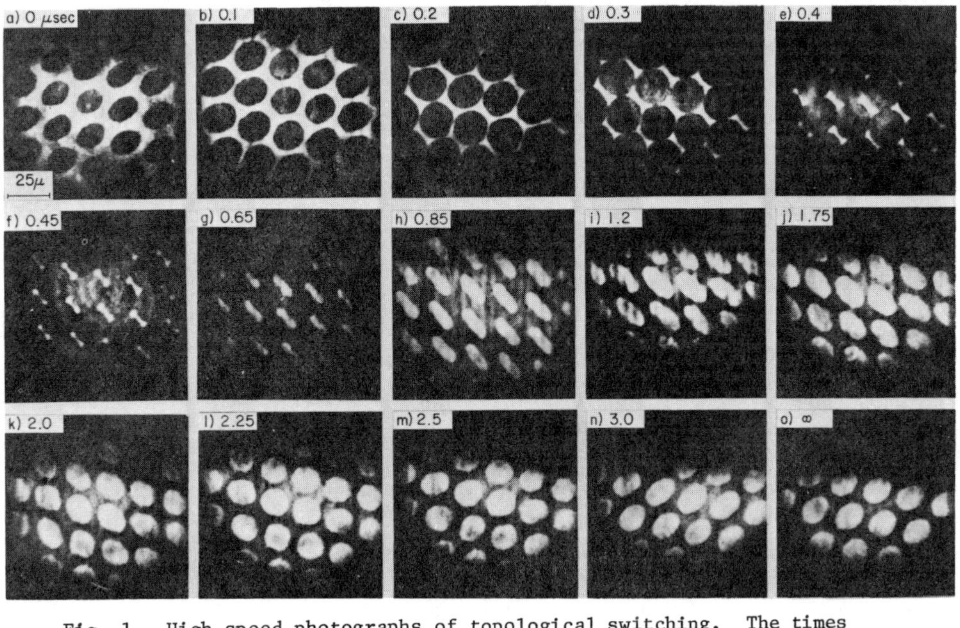

Fig. 1. High speed photographs of topological switching. The times indicate delay relative to the onset of the 200 Oe field pulse which lasts 420 nsec.

300 nsec they make contact with only four out of six of the nearest neighbor bubbles, failing to do so with the bubbles along the initial major axis. After Fig. 1e, at 420 nsec, the pulse turns off, leaving an array of "dog-bone" shaped domains which then re-expand to form a regular array of bubbles in the following way: In Fig. 1f the dogbone domains relax into a smoother dumbbell shape. In Fig. 1g-j these domains grow into ellipses whose major axes are perpendicular to the major axes of the original array. Finally in Figs. 1k-1o, from 2 to 3μsec after the onset of the original field pulse, these domains undergo an interchange of major and minor axes, leaving an array of the same proportions as the original array but of the opposite polarity.

DISCUSSION

The anisotropic nature of the topological switching process was previously deduced by one of us[2] from the fact that the static array was elliptically distorted and that the switched bubbles appeared half-way between the original bubbles along their elliptical major axes. However the high speed photographs suggest that the anisotropic nature of the topological switching process cannot be ascribed merely to the static ellipticity of the bubble array, since the distances from any bubble to all of its six neighbors are not very different to start with. Had the walls expanded uniformly in all directions, contact of the bubble walls along the initial distortion axis would also very likely have been achieved, giving rise to twice as many bubbles as present in the initial array and hence to an irregular switching. In fact, however, the wall velocities are highly anisotropic, 750$\pm$100 cm/sec along the original major axis of the bubbles and 1400$\pm$100, roughly twice as large along the minor. Thus the dominant anisotropy of the topological switching process seems to be in the wall velocity.

Experiments have also been performed on isolated bubbles pulsed to expand;[4] they will be described in more detail elsewhere.[5] The results show that the walls of isolated bubbles move initially at the same velocities and in the same anisotropic way as those of bubbles in the array. This confirms the fact that the anisotropy in wall velocity is an intrinsic property of this sample and not an effect of the initial starting eccentricity of an array. The results also indicate that the wall velocity in both directions is saturated with a saturation threshold of 30 Oe at most. This means that wall motion will proceed at a steady velocity as long as the effective fields of neighboring domains are insufficient to reduce the drive field below the saturation threshold. This concept explains an initially puzzling feature of Fig. 1 that as long as the topological switching pulse is on, the walls move as if oblivious of the proximity of other walls, thus generating the curious sharp-cornered "dogbone" domains of Fig. 1e.

The reasons for the velocity anisotropy, which seems to be primarily responsible for the phenomenon of topological switching in the sample, will also be discussed in detail elsewhere.[5] In brief, the velocity anisotropy is presumed to be related to the fact that bubble arrays as well as isolated bubbles are elliptical, for the principal axes of these two effects are the same. The cause of these two effects is an in-plane magnetic anisotropy K_p=2350$\pm$150 ergs/cm^3, arising from a 2° tilt of the crystallographic [111]-axis away from the surface normal.[6] For the in-plane anisotropy causes an anisotropy in wall energy[7] which in turn causes static bubbles to be elliptical.[8] The in-plane anisotropy can also cause the anisotropy in saturation velocity, as illustrated by Walker's limiting velocity v_w, derived for an infinite wall in the high Q limit:[9,10]

$$v_w = 2\pi\gamma M_s \sqrt{A/K_u}\ \left|1 \pm K_p/2\pi M_s^2\right|$$

Here, $\pm$ refers to walls lying parallel or perpendicular, respectively, to the in-plane easy axis; γ is the gyromagnetic ratio and A is the exchange stiffness constant. If $K_p=2\pi M_s^2$, the limiting velocity can actually go to zero along one direction. A qualitatively similar anisotropy should occur in the actual saturation velocity, for which there is unfortunately no simple quantitative theory including K_p.

The authors thank E. Hochberg for expert assistance in the bubble expansion and ellipticity experiments, J. C. DeLuca for measurement of the in-plane anisotropy, and T. H. O'Dell and J. C. Slonczewski for helpful conversations. One of us (K.P.) acknowledges the support of the Science Research Council, National Research and Development Corporation and the Plessey Co. of the U.K.

REFERENCES

1. T. H. O'Dell, Phil. Mag. 27, 595 (1973).
2. K. Papworth, IEEE Trans. on Magnetics, MAG-10, 638 (1974).
3. J. C. Slonczewski, A. P. Malozemoff and O. Voegeli, AIP Conf. Proc. 10, 458 (1972).
4. G. J. Zimmer, T. M. Morris and F. B. Humphrey, IEEE Trans. on Magnetics, MAG-10, 651 (1974).
5. A. P. Malozemoff and K. Papworth, to be published.
6. A. P. Malozemoff and J. C. DeLuca, J. Appl. Phys. 45, 4586 (1974).
7. A. A. Thiele, Phys. Rev. B7, 391 (1973).
8. E. Della Torre and M. Y. Dimyan, IEEE Trans. on Magnetics MAG-6, 487 (1970).
9. W. J. Tabor, G. P. Vella-Coleiro, F. B. Hagedorn and L. G. Van Uitert, J. Appl. Phys. 45, 3617 (1974).
10. J. C. Slonczewski, Int. J. of Magnetism 2, 85 (1972); J. Appl. Phys. 44, 1759 (1973).

DOMAIN WALL VELOCITY DURING MAGNETIC BUBBLE COLLAPSE

G. P. Vella-Coleiro
Bell Laboratories, Murray Hill, New Jersey 07974

ABSTRACT

Measurements of magnetic bubble collapse in a high mobility Lu Gd epitaxial garnet film have exhibited very high velocities when the bias field is made to approach the collapse field. A prominent peak in the velocity rising to ~10,000 cm/sec is interpreted in terms of Walker breakdown. The data also imply the existence of a characteristic time for dynamic conversion to occur. This time is ~10 nsec for our sample.

1. INTRODUCTION

Previous measurements[1] of domain wall velocities by bubble translation in a field gradient and by dynamic bubble collapse showed that much higher velocities could be obtained by the former technique. It was suggested that the different behavior was a result of the fact that the bubble collapse measurements necessitated the use of considerably greater driving fields than bubble translation, and that at these high fields Walker breakdown occurred during the initial stages of the wall motion, thus preventing large average velocities from being observed. Attempts to observe the initial fast motion by high speed photography with a 12 nsec exposure time yielded negative results, and this led us to speculate that the high velocity persisted for only a very brief period of time and was thus unobservable with a 12 nsec time resolution. The present experiments were designed in an attempt to observe this brief but fast wall motion. It was reasoned that if a dynamic collapse experiment was performed with the bias field set very close to the collapse field, the distance the wall would have to move for collapse to occur would be relatively small. Thus, provided the wall mobility was sufficiently large, the entire collapse process would take place in a time interval too short for dynamic conversion to occur, which would allow the initial high velocities to be observed.

For these experiments we selected an epitaxial garnet film having the highest wall mobility we have measured in any garnet composition, i.e., $Lu_1\ Gd_2\ Al_{0.6}\ Fe_{4.4}O_{12}$. The bubble mobility and collapse measurements in this film are described in Section 2. In Section 3, the results are discussed in terms of Walker breakdown and dynamic conversion. It is shown that bubble collapse measurements are greatly influenced by Walker breakdown, and that there exists a characteristic time for dynamic conversion of approximately 10 nsec in our sample.

2. EXPERIMENTAL RESULTS

The film of Lu Gd garnet used here was grown[2] on a (111) GGG substrate. The values of the relevant parameters are: thickness $h = 9.4\ \mu m$, saturation magnetisation $4\pi M = 189$ G, wall energy $\sigma = 0.15$ erg/cm^2, anisotropy field $H_k = 620$ Oe, gyromagnetic ratio $\gamma = 1.4 \times 10^7\ sec^{-1}\ Oe^{-1}$, microwave damping $\alpha = 0.023$. The value of H_k was obtained from optical magnetometer[3] measurements, while the values of γ and α were derived from ferromagnetic resonance measurements[4] at microwave frequencies. Combining H_k, $4\pi M$ and σ, we obtain a value for the exchange constant $A = 3.0 \times 10^{-7}$ erg/cm. The Callen and Josephs[5] approximation of the bubble stability theory was used to calculate the distance Δr the wall has to move in a dynamic collapse experiment for bubble collapse to occur. For future reference,

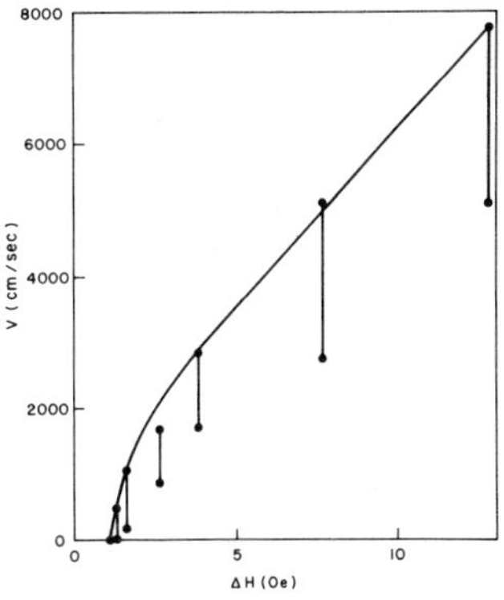

Fig. 1-Bubble translational velocity V versus the field difference across the bubble diameter, ΔH.

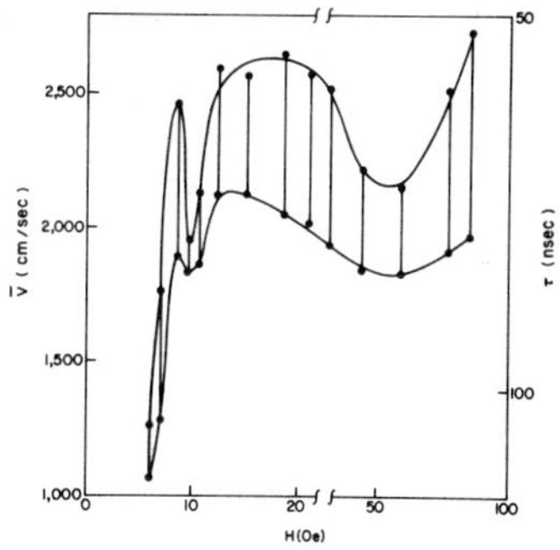

Fig. 2-Average wall velocity during bubble collapse $\bar{V}$ versus pulse field amplitude H for H_o-H_b= 4 Oe. τ is the pulse width required for collapse.

the calculated values of Δr are 1.40, 1.21 and 0.97 μm for $H_o - H_b = 4.0$, 3.0, and 2.0 Oe, respectively, where H_o is the static collapse field and H_b is the bias field. Although the Callen and Josephs expressions are correct in the limit $H_k >> 4\pi M$, DeBonte[6] has shown that the error in the bubble diameter is very small for $H_k > 1.5 \times 4\pi M$. Since in the present case $H_k \approx 3 \times 4\pi M$, we do not expect our Δr values to be significantly in error.

Measurements of bubble velocity in a field gradient were made using a technique described previously, and they are shown in Fig. 1. The bars parallel to the velocity axis represent the scatter observed among 10 successive measurements. These data were obtained using a compensation technique[7] which completely eliminates the possibility of strip-out occurring as the bubble moves in the field gradient.[8,9] Thus the velocity scatter is an intrinsic property of the bubble motion and its origin has been discussed previously.[10,11] From the upper envelope of the data in Fig. 1 we obtain a value for the wall mobility at low velocities $\mu_o = 4700$ cm/sec Oe., from which we derive $\alpha = 0.024$, in good agreement with the microwave value. On the other hand, the differential mobility for $V > 2000$ cm/sec is $\mu = 1100$ cm/sec Oe, with a corresponding value of $\alpha = 0.10$.

Bubble collapse measurements were made in the same location in the film in which the data of Fig. 1

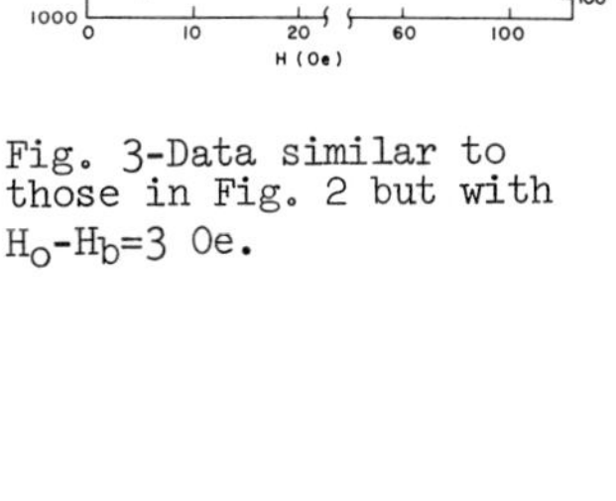

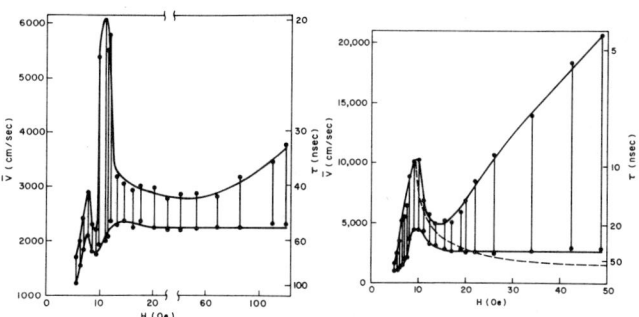

Fig. 3-Data similar to those in Fig. 2 but with H_o-H_b=3 Oe.

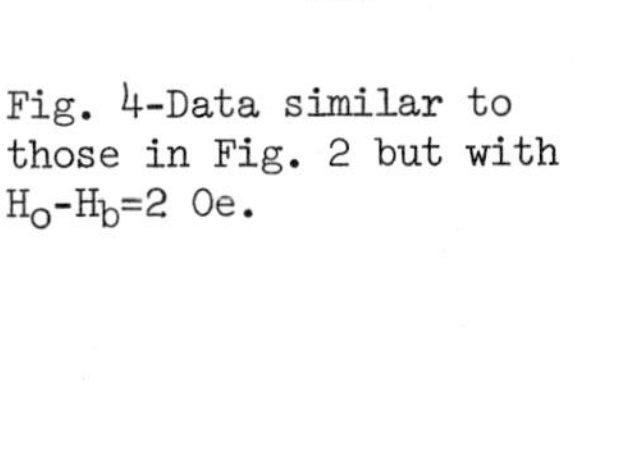

Fig. 4-Data similar to those in Fig. 2 but with H_o-H_b=2 Oe.

were taken. For $H_o - H_b$ = 12 Oe the curve of average velocity $\bar{V}$ vs. pulse field amplitude H is similiar to that obtained by several workers in other garnet compositions (see, for example, Ref. 1) with a saturation velocity of 2200 cm/sec. At $H_o - H_b$ = 4 Oe the velocity curve changes somewhat and shows one sharp and one broad peak, see Fig. 2. A more dramatic change occurs for $H_o - H_b$ = 3 Oe, as shown in Fig. 3, and further change occurs for $H_o - H_b$ = 2 Oe (Fig. 4). The data in Figs. 2-4 were taken by gradually increasing the pulse width at a fixed pulse amplitude until bubble collapse occurred. $\bar{V}$ is the average velocity obtained by dividing the appropriate value of Δr given above by the pulse width required for collapse. The bars parallel to the $\bar{V}$-axis are the scatter observed among a field of 10 widely spaced bubbles. Note that the magnitude of the scatter is a strong function of the pulse field amplitude H. This systematic dependence on H, which is very reproducible, precludes the possibility that the velocity scatter might be due to the presence of bubbles with different numbers of Bloch lines.[12,13] Rather, analogous to the scatter noted in connection with Fig. 1, it is an intrinsic property of bubble collapse and it is discussed further in Section 3.

As noted above, knowledge of the pulse width is required in order to obtain the value of $\bar{V}$. When wide pulses are used whose shape is very nearly rectangular, little uncertainty exists in defining the pulse width and pulse amplitude. However, the data reported here necessitated the use of pulses as narrow as 4.7 nsec. The shape of such narrow pulses is not rectangular but approximates a Gaussian curve. In this case the usual definition of pulse width is the full width at half peak amplitude (f.w.h.m.). However, some wall motion undoubtedly occurs while the field is rising to and falling from the half amplitude points, and so it is more appropriate to define the average velocity in terms of the total pulse width. Since the beginning and end of a Gaussian-like pulse are not well defined, we have chosen, somewhat arbitrarily, to define the pulse width as the width between the points at which the amplitude is 10% of the peak value. The pulse amplitude is then defined as the height of a hypothetical rectangular pulse whose area is equal to the area of the pulse used and whose width is equal to the "10% width" as defined above. These definitions lead to a value of $\bar{V}$ which is conservative; almost certainly, wall velocities somewhat greater than the $\bar{V}$ values given here occur during a portion of the wall motion. Although the definition of the 10% pulse width is arbitrary, little qualitative change in the data occurs if other reasonable definitions, such as f.w.h.m., are used. Based on the f.w.h.m., the value of the peak velocity in Fig. 4 is 14,200 cm/sec instead of 10,200 cm/sec, at H = 14.2 Oe instead of 9.5 Oe. The overall shape of the upper and lower envelopes, however, remains unchanged.

We have also obtained data similar to those in Fig. 4 with one prominent peak for $H_o - H_b$ = 1 Oe. In this case, however, we have $\Delta r = 0.68\ \mu$m, and with the shortest pulse width we have available, 4.7 nsec, the highest velocity we can observe is only 14,500 cm/sec, which is observed at H = 19.3 Oe.

3. DISCUSSION

The data in Fig. 2 and 3 show two peaks in the average velocity. The peak occurring at the lower value of H is similar to that reported by Malozemoff[13] and it probably has the same origin, i.e., motion of Bloch lines between the film surfaces, as discussed by Slonczewski[14]. We shall not discuss it further here other than to comment on its absence in Fig. 4, where $\bar{V}$ rises very steeply as H increases up to ~10 Oe. This rapid rise probably masks the lower peak and accounts for its absence.

The more prominent peak in the data, rising to ~10,000 cm/sec in Fig. 4, can be associated with Walker breakdown. From Walker's[15] expressions we calculate the value for the peak velocity V_m = 9800 cm/sec, while the driving field at which this velocity occurs is calculated to be H_m = 9.4 Oe, using α = 0.10 obtained from the high field data in Fig. 1. Both V_m and H_m are in good agreement with the data in Fig. 4. The closeness of the peak velocity in Fig. 4 and V_m may, of course, be fortuitous, since there is some uncertainty in the definition of $\bar{V}$, as explained in Section 2. Nonetheless, the fact that no significant discrepancy is observed between the measured and calculated values of V_m and H_m lends support to the identification of the velocity peak with Walker breakdown.

A further comparison with the Walker theory can be made in connection with the behavior of the velocity for $H > H_m$. The calculated velocities are shown by the dashed curve in Fig. 4 for α = 0.1, from which it can be seen that a large discrepancy exists between the upper envelope of the data and the calculated values. Since the Walker analysis is based on the assumption that the spin precession is uniform throughout the wall, this result implies that under the conditions of the present experiments a non-uniform spin precession occurs. Such a non-uniform spin precession might perhaps be associated with the spin-wave type of wall motion discussed by Schlömann,[16] or the corrugated-wall type of motion discussed by Slonczewski.[17]

The scatter in the data in Fig. 2-4 is analogous to that reported in Ref. 1. As was done there, we interpret it to be the result of a dynamic conversion of the spin configuration of the moving wall into a less mobile one. This process, which is presumed to arise from an interaction of the moving wall with imperfections,[11] occurs to a variable extent during successive pulses and for different bubbles, and thus it leads to the velocity scatter. An important consequence of the data in Fig. 2-4 and the interpretation above is that there exists a characteristic time associated with the occurrence of dynamic conversion, so that its effects can partially be suppressed if the measurement involves a short enough period of time. Thus when $H_o - H_b$, and consequently Δr, are large, the bubble collapse process necessarily takes a relatively long time, dynamic conversion takes place, and high average velocities are not observed. From a comparison of the data in Fig. 2-4 we make a crude estimate of 10 nsec for the characteristic time in our sample. A similar characteristic time for conversion of a moving wall from a high mobility into a low mobility state has been observed recently by de Leeuw[18] in straight wall dynamics experiments in the presence of a large in-plane magnetic field.

A remarkable feature of the data in Figs. 3 and 4 is the flatness of the lower envelope of the velocity for H > 20 Oe, with the value $\bar{V} \approx 2500$ cm/sec. Comparing this value with the saturation velocities derived by Slonczewski[14] (V_o) and by Hagedorn[11] (V_{av}) we find V_o = 540 cm/sec and V_{av} = 1000 cm/sec (a correction appropriate for a bubble diameter to film thickness ratio d/h = 0.5 has been applied to these values using the results of Ref. 14). The agreement between the observed and calculated velocities is not entirely satisfactory.

The data reported above may be compared qualitatively with the results of Josephs and Stein,[19] who reported a slight increase in the average wall velocity of the last bubble to collapse as the bias field approached the collapse field. The present data show a similar effect in that the lower limit of the average velocity increases from 2200 to 2500 cm/sec as $H_o - H_b$ decreases from 12 to 2 Oe.

It is a pleasure to thank F. B. Hagedorn for helpful discussions, S. L. Blank for the film growth, R. C. LeCraw for the microwave measurements, and R. J. Peirce for the optical magnetometer measurements.

REFERENCES

1. G. P. Vella-Coleiro, F. B. Hagedorn, Y. S. Chen and S. L. Blank, Appl. Phys. Lett. 22, 324 (1973).
2. J. W. Nielsen et al., J. Elec. Mat. 3, 693 (1974).
3. P. W. Shumate, Jr., D. H. Smith and F. B. Hagedorn, J. Appl. Phys. 44, 449 (1973).
4. R. C. LeCraw, unpublished work.
5. H. Callen and R. M. Josephs, J. Appl. Phys. 42, 1977 (1971).
6. W. J. DeBonte, J. Appl. Phys. 44, 1793 (1973).
7. G. P. Vella-Coleiro and W. J. Tabor, Appl. Phys. Lett. 21, 7 (1972).
8. R. M. Josephs, Appl. Phys. Lett. 25, 244 (1974).
9. G. P. Vella-Coleiro, F. B. Hagedorn and S. L. Blank, Appl. Phys. Lett., to be published.
10. G. P. Vella-Coleiro, AIP Conf. Proc. 18, 217 (1973).
11. F. B. Hagedorn, J. Appl. Phys. 45, 3129 (1974).
12. O. Voegeli and B. A. Calhoun, IEEE Trans. Magnetics MAG-9, 617 (1973).
13. A. P. Malozemoff, J. Appl. Phys. 44, 5080 (1973).
14. J. C. Slonczewski, J. Appl. Phys. 44, 1759 (1973).
15. N. L. Schryer and L. R. Walker, J. Appl. Phys. 45, 5406 (1974).
16. E. Schlömann, Appl. Phys. Lett. 19, 274 (1971).
17. J. C. Slonczewski, Intern. J. Magnetism 2, 85 (1972).
18. F. H. de Leeuw, J. Appl. Phys. 45, 3106 (1974).
19. R. M. Josephs and B. F. Stein, AIP Conf. Proc. 18, 227 (1973).

THEORY OF DOMAIN-WALL MOTION INDUCED BY MICROWAVE MAGNETIC FIELDS*

Ernst Schlömann
Raytheon Research Division, Waltham, MA 02154

ABSTRACT

A general theory is developed that applies to arbitrary polarization and takes account of damping and of the dipolar interaction between domains. The effect of the microwave field on the domain structure can be characterized by a pressure on the domain walls and by an alignment energy, both of which are proportional to the square of the rf magnetic field and become large in the vicinity of a resonance. For circular polarization the pressure tends to decrease the Larmor-domains (domains in which the imposed sense of polarization coincides with the sense of the natural spin precession) for frequencies outside of the resonance region. Inside the resonance region, however, the pressure tends to increase the Larmor-domains. A linearly polarized field also exerts a pressure on the domain walls, with the polarity dependent upon the orientation of the field to the wall normal. In a linearly polarized magnetic field the domain walls tend to become aligned parallel to the rf field at frequencies ω near the low-frequency resonance ($\omega = \gamma H_a$, γ = gyromagnetic ratio, H_a = anisotropy field) and perpendicular to the rf field at frequencies near the high frequency resonance ($\omega = \gamma[H_a(H_a + 4\pi M_o)]^{1/2}$, M_o = saturation magnetization).

*The research reported in this paper was accomplished in part with the support of the Office of Naval Research.

VELOCITY SCATTER IN BUBBLE DOMAIN TRANSPORT MEASUREMENTS

R. M. Josephs and B. F. Stein
Sperry Univac, P.O. Box 500,
Blue Bell, Pa. 19422

ABSTRACT

Bubble transport measurements, in which the effect of the radial motion on the effective drive field was incorporated theoretically, have previously been described. These earlier data revealed that excessive scatter in the velocity occurred when the bubble became elliptically unstable. The more complex mechanism of "dynamic conversion" was not required to explain the scatter. In the present work, the technique has been applied to ion-implanted films of $(YLaTm)_3(FeGa)_5O_{12}$ and $(YEuTm)_3(FeGa)_5O_{12}$ which have higher and lower mobilities, respectively, than the samples used previously. In each instance, the velocity data became anomalous when the bubble was in the run-out region, in agreement with the previous results. The origin of the spread in velocities is shown to arise from the mechanics of the stripping-out motion.

Bubble domain translational velocity data obtained with a single pair of drive conductors were recently described.[1] The bubble motion was not restricted to the region midway between the drive conductors, and the effect of the radial motion on the effective drive field was incorporated theoretically. These earlier measurements on ion implanted films of $Y_{2.4}Eu_{0.6}Ga_{1.2}Fe_{3.8}O_{12}$ (subsequently referred to as YEu) revealed that excessive scatter in the velocity data occurred when the effective bias field was ≃ the run-out field. It was concluded that this scatter could be attributed to the elliptic instability and did not require the phenomenon of "dynamic conversion"[2] invoked by Vella-Coleiro.[3] In the present work, the technique has been applied to ion-implanted films of $Y_{1.65}La_{0.35}Tm_{1.0}Fe_{4.0}Ga_{1.0}O_{12}$ (referred to as YLaTm) and $Y_{1.60}Eu_{0.75}Tm_{0.65}Fe_{4.2}Ga_{0.8}O_{12}$ (referred to as YEuTm) having higher and lower mobilities, respectively, than the samples used previously. In each instance, the velocity data became anomalous when the bubble was in the run-out region. The origin of this behavior was revealed by the measurements on the YLaTm film and will be described in detail.

The apparatus and the method have been described previously.[1] All the data refer to ion-implanted ($1.5 \times 10^{14} Ne^{+}_{20}/cm^2$, 100 Kev) epitaxial garnet films and were taken on bubbles generated by cutting worms with the field of the drive conductors. Although the general direction of propagation was always at an angle (~15° in the YEuTm film and ~50° in the YLaTm film) with respect to the direction of the field gradient,[4,5] only v_x, the average velocity in the direction of the gradient, was measured. The mobility μ_x, for this component of velocity, was 12m/sec-Oe for the YLaTm film and 4m/sec Oe for the YEuTm film.

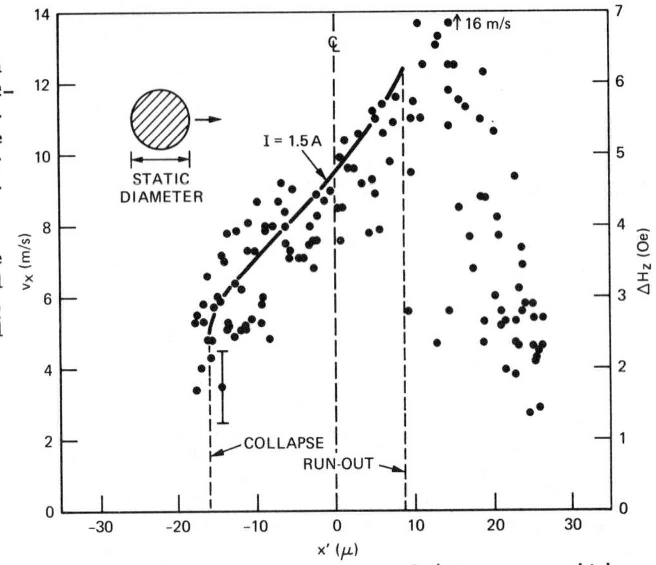

Fig. 1 Dependence of v_x and ΔH_z on position for the YEuTm film at I=1.5A. The ordinate on the left describes the data points for 10 traversals (taken with a pulse width of 0.43μ sec). The ordinate on the right refers to the solid curve.

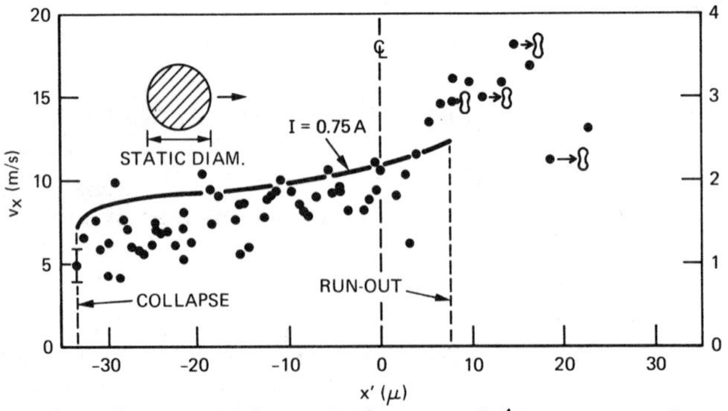

Fig. 2 Dependence of v_x and ΔH_z on position for the YLaTm film at I=0.75A. The ordinate on the left describes the data for 5 traversals (taken with a pulse width of 0.41μ sec.) The bubble to dumbbell conversion is indicated at the data point (x', v_x) for the pulse preceding the one in which the conversion occurred. The ordinate on the right refers to the solid curve.

In Fig. 1, the spatial variation of the v_x data is shown for the YEuTm sample having h=5.6μ, $4\pi M_s$=177G, q=13, and collapse field H_o=95.4 Oe. The solid curve represents the calculated value of the drive field ΔH_z between the limits of collapse and run-out for the drive current I=1.5A. It is seen that excessive scatter in the data occurred when the bubble was in the run-out region, in agreement with the previous results on YEu films. Data at lower drive currents also exhibited this behavior.

Measurements on the YLaTm film provided direct confirmation that the observed scatter was due to the elliptic instability. Whereas the data on the YEu and YEuTm films were usually taken with the bias field H_1 ≃ mid-range bias field, the data on the YLaTm sample were obtained at a value of H_1 slightly greater (by 3 Oe) than the run-out field. This value of H_1 was within the lower 20% of the static field margin where, as Copeland[6] has shown, both bubbles and dumbbells (short stripes) are stable configurations. If during the application of the drive field pulse, the bubble were to run out into a stripe, it need not collapse back into a bubble at the end of the pulse but could remain as a stripe. This is substantiated by the data in Fig. 2 which contains the spatial dependence of v_x and ΔH_z at a drive current of 0.75 A for the YLaTm film having $h=6.1\mu$, $4\pi M_s=164G$, $q=4$, and $H_O=103.8$ Oe. As indicated schematically in the figure, the bubble would frequently convert, in the run-out region, to a stable dumbbell whose length was ~2.5 times the static bubble diameter. When H_1 was increased by a few tenths of an oersted, the dumbbell would snap back to a bubble. These observations confirm that in the region where the velocity scatter is observed, the bubble runs out during the application of the pulse and then contracts when the pulse is turned off. If the length at the end of the pulse is greater than some minimum length, the resulting domain will be a dumbbell. If the length is shorter, a bubble results. Exactly how this contraction to a bubble occurs will depend delicately on local coercivity variations, stray fields, etc. It is this process which leads to the scatter in the velocity in the run-out region. It was noted above that a slight increase in H_1 caused the dumbbell to snap back into a bubble with a considerable variability in the final bubble position. In transport measurements, similar uncertainties in final bubble position would yield velocities ranging from ~3m/sec to ~17m/sec.

In every instance in which the bubble converted to a dumbbell, the apparent value of v_x (as determined from the position of the dumbbell relative to the initial bubble position) of the leading wall segment of the dumbbell was very much greater than that of the trailing wall segment and was typically 30m/sec. The presence of the dumbbells can explain another transport phenomenon - the jog in the path which can appear when a bubble is moved back and forth between the drive conductors. Such a trajectory is shown in Fig. 3 (c) and happens in the following manner. The stripping-out motion occurred along the same direction for both the forward and backward paths as indicated schematically in (a) and (b) of Fig. 3. In actuality, a portion of the dumbbell overlapped the preceding bubble position. When the motion was observed with a very low repetition rate pulse whose amplitude was inverted at the endpoints of the motion, the stripping-out process was blurred and the bubble locus was frequently in the form of a parallelepiped. The directional aspect of the stripping-out motion is presently under consideration.

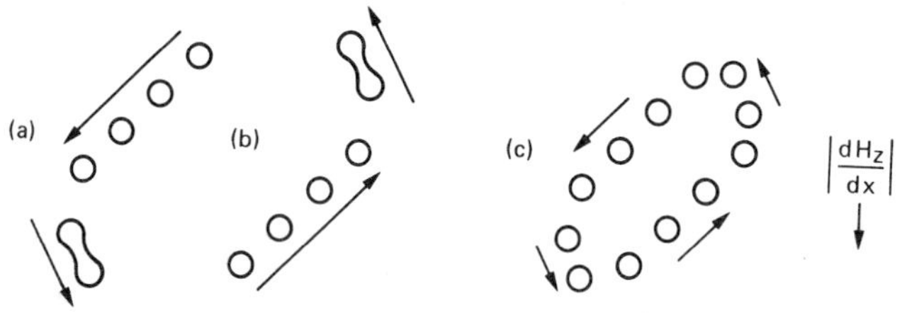

Fig. 3 Bubble motion in the region between the drive conductors.

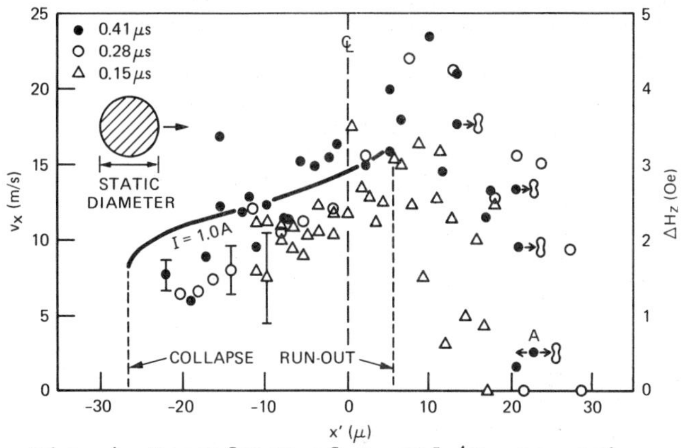

Fig. 4 Dependence of v_x and ΔH_z on position for the YLaTm film at I=1.0A. The ordinate on the left describes the data for 7 traversals (taken at several different pulse widths). The bubble to dumbbell conversion occurred only for the longest pulse width. The ordinate on the right refers to the solid curve.

In Fig. 4, the results at a higher drive current (I=1.0A) and at three different pulse widths (0.41, 0.28, and 0.15 μ sec) are presented. In the region between collapse and run-out, the velocity was independent of pulse width within the experimental uncertainty. The most interesting feature of the data occurred in the run-out region. Only for the longest pulse did the bubble convert to a stable dumbbell. For the shorter pulses, the bubble did not have sufficient time to strip out to the minimum length required for the conversion into a stable dumbbell and simply exhibited considerable uncertainty in its velocity. At several points, in the vicinity of $x'=20\,\mu$, there was no apparent motion. At point A, the bubble actually appeared to jump backwards by ~1μ during the pulse preceding the one in which the conversion to a dumbbell occurred.

Additional data on the YLaTm film were taken with a drive current of 1.25 A and a pulse width of 0.27μsec. At the highest drive field of 4 Oe, v_x was 20m/sec. No dumbbells were observed due to the narrow pulse width employed. Again excessive scatter occurred in the run-out region. Attempts at obtaining data at a higher drive current (I=1.5A) were frustrated by the appearance of extra bubbles under the edge of the forward drive conductor, i.e., the conductor towards which the bubble was heading. These extra bubbles were probably being nucleated at the drive conductors. The interaction with these additional bubbles, which stripped out along the drive conductor during the pulse and frequently converted to dumbbells, produced erratic behavior in the bubble under observation. It was found that even though the test bubble and an extra bubble were statically ~10 diameters apart, they would interact strongly during the application of the pulse if the extra bubble was near the edge of the forward drive conductor.

In conclusion, transport measurements have been made on a number of samples from several different compositions. Outside the run-out region, the scatter in the velocity data was consistent with the experimental uncertainty. In every instance in which the bubble was in the run-out region, excessive scatter in the data occurred. The origin of this scatter was shown to involve the mechanics of the stripping-out motion, without the need to introduce the complex phenomenon of "dynamic conversion".

The authors acknowledge the work of M. Kestigian, A. B. Smith, and W. Bekebrede who developed the YLaTm composition at the Sperry Research Center.

1. Richard M. Josephs, Appl. Phys. Lett. 25, 244 (1974).
2. F. B. Hagedorn, J. Appl. Phys. 45, 3129 (1974).
3. G. P. Vella-Coleiro, AIP Conf. Proc. 18, 217 (1974).
4. A. B. Smith, private communication.
5. D. C. Bullock, AIP Conf. Proc. 18, 232 (1974).
6. J. A. Copeland, IEEE Trans. Mag. MAG-9, 660 (1973).

BUBBLE VELOCITIES IN $(YLa)_3(FeGa)_5O_{12}$ FILMS

F.H. de LEEUW and J.M. ROBERTSON
Philips Research Laboratories, Eindhoven, The Netherlands

ABSTRACT

Bubble velocities in $(YLa)_3(FeGa)_5O_{12}$ films are measured by the two wire field gradient technique as a function of the field across the bubble (ΔH), an in-plane field (H_1) and an initial change of strength of the bias field. At zero in-plane field a skewed motion of the bubbles at an angle of 60° to 75° with the direction of negative field gradient was observed. The deflection was suppressed at in-plane fields of 85 Oe and above. In these in-plane fields we found that the bubble velocity depends on the change of strength of the bias field at the start of the bubble motion ($\overline{\Delta H}_{bias}$). We found that above a critical value of $\overline{\Delta H}_{bias}$ the velocity dropped considerably. This critical value increased with increasing in-plane field. We observed a bubble velocity of 671 m/s at ΔH=2.3 Oe, H_1= 285 Oe and $\overline{\Delta H}_{bias}$=0. The wall mobility was 4.6×10^4 cm $s^{-1}Oe^{-1}$ and was in agreement with the value from straight wall oscillation experiments on similar material.

INTRODUCTION

Recently it has been found that straight wall mobilities in Ga:YIG films are high[1-4]. In the earliest theory on the translation of cylindrical magnetic domains (bubbles) Thiele et al. showed[5] that the bubble velocity is proportional to the straight wall mobility. From recent experiments by Malozemoff[6] and the theory of Slonczewski[7] it is known that bubbles with a small number of Bloch lines (BL's) deflect from the direction of the field gradient that drives the bubble. This deflection is suppressed if an in-plane field of sufficient strength is applied as shown experimentally by Bullock[8].

We have measured the bubble velocity as a function of the field across the bubble and the in-plane field with the goal to investigate whether bubble velocities as high as straight wall velocities are possible or not. The deflection effects are also studied. We have found that in bubble translation experiments a critical initial bias field change can be defined. Above this critical field change the bubble velocity is considerably reduced. This is probably related to dynamic conversion during bubble motion as reported by Vella-Coleiro et al.[9], Vella-Coleiro[10], and Hagedorn[11].

EXPERIMENT

The bubble translation experiments are performed on a Ga:YIG film ($Y_{2.9}La_{0.1}Fe_{3.8}Ga_{1.2}O_{12}$). The film was grown on a (111)-substrate ($Gd_3Ga_5O_{12}$) by the usual technique of liquid phase epitaxy[12]. The melt composition and the film preparation are described in Ref. 4. The growth conditions were: growth temperature Tg, 894°C; saturation temperature T_s, 922°C and deposition time t_d, 60 minutes. The properties of the film were: thickness h, 4.5±0.1 μm; magnetization M, 6.2±0.6 G; uniaxial anisotropy K_u, 2600±260 erg cm^{-3}; cubic anisotropy K_1, -470±80 erg cm^{-3} and gyromagnetic ratio γ, $(18.5\pm0.1)\times10^6 s^{-1}Oe^{-1}$.

The velocity measurements are performed

ΔH [Oe]	T [ns]	Δs [μm]	sinβ	v [m/s]	n_r
1.14	880	19	0.96	22	0.91
1.14	880	22.4	0.92	25	0.74
2.23	440	23.1	0.84	55	0.62
2.23	440	19	0.98	43	0.98

in a two wire field gradient as described by Vella-Coleiro and Tabor[13]. The distance between the conducting wires and the film plane was about 10 μm. The technique of compensation of the bias field during bubble motion as described in Ref. 13 is not used in this work. The bubbles, with diameter of 7.0 μm, were translated less than about 20 μm. The rise time of the current (10%-90% time) through the conducting wires was 5 ns.

In the experiments at zero in-plane field we observed large deflection angles of the bubbles. Typical results are given in the table. Each data line represents a single measurement. Here ΔH is the field decrease across the bubble, T is the pulse duration, Δs is the distance travelled by the bubble, $v \equiv \Delta s/T$ and β is the angle between the propagation direction of the bubble and the direction of negative field gradient. The calculated quantity[6,7]

$$n_r = 1 + n_B/2 = \gamma r^2 (\nabla H_z) \sin\beta \,/ 2v, \qquad (1)$$

is the revolution number of the bubble, n_B is the number of Bloch lines (Bl's) and ∇H_z is the field gradient acting on the bubble. In Refs. (6,7) the quantity n_r is defined in a way that it can only be an integer. The results for n_r in the table are close to one. Their inaccuracy is difficult to assess, but the measurements suggest that in all four cases the bubble was a normal one (no BL's, n_r=1).

By applying an in-plane field (H_1) of 85 Oe, parallel to the field gradient, the deflection of the bubbles was suppressed completely[8] and the velocity increased considerably. Bubble velocity measurements are

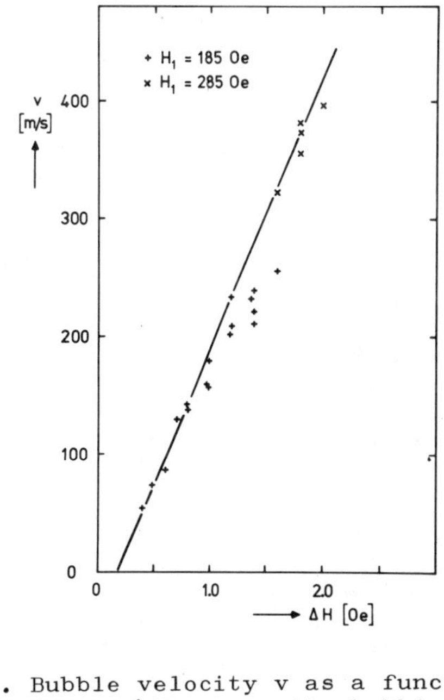

Fig. 1. Bubble velocity v as a function of field ΔH across the bubble.

performed at H_1=185 and 285 Oe, respectively. The bubble velocity as a function of field across the bubble is given in Fig. 1. Each point is the averaged result of ten bubble translations. When the field gradient is switched on there are positions where the change in bias field is equal to zero. These positions are located on the line exactly between the current conductors. The measurements in Fig. 1 are for bubbles which started $-\frac{1}{2}\Delta s$ before the latter positions. For some measurements it was checked that v was unchanged when T was halved. In Ref. 5 it was shown that

$$v = \frac{1}{2}\,\mu\,\left(\Delta H - \frac{8}{\pi}\,H_c\right), \qquad (2)$$

where μ is the straight wall mobility and H_c is the coercive force of the material. From Fig. 1 and Eq. (2) it follows $\mu = 4.6 \times 10^4$ cm $s^{-1}Oe^{-1}$ and H_c=0.08 Oe. From wall oscillation experiments on similar material[4] we found $\mu = (3.3 \pm 0.8) \times 10^4$ cm $s^{-1}Oe^{-1}$. The latter experiment was performed on a film having a somewhat larger uniaxial anisotropy constant (3100 ± 300 erg cm^{-3}). We therefore conclude that both mobility results are in good agreement with each other.

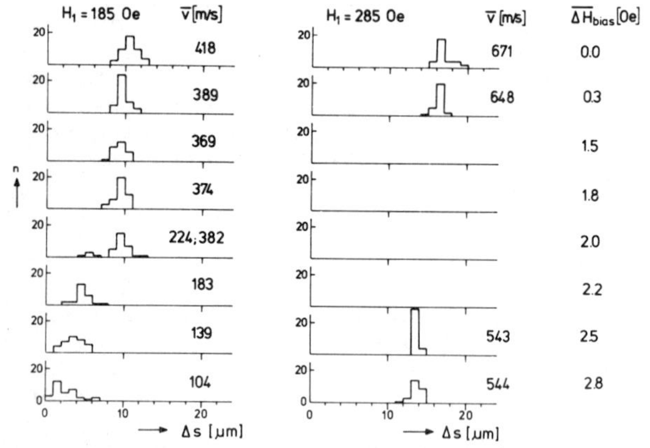

Fig. 2. Distributions of distances travelled by a bubble (Δs) and averaged bubble velocities ($\bar{v}$) as a function of the initial change in bias field ($\overline{\Delta H}_{bias}$) and for two in-plane fields H_1. The field across the bubble (ΔH) was 2.3 Oe.

We have measured distances travelled by a bubble at constant field gradient and as a function of the initial change in bias field ($\overline{\Delta H}_{bias}$) and the in-plane field H_1. We have not corrected for the change in diameter during motion as reported by Josephs[14]. The results of the measurements are given in Fig. 2. At H_1=185 Oe and $\overline{\Delta H}_{bias} <$ 1.8 Oe the bubble velocity was 390 ± 20 m/s. At $\overline{\Delta H}_{bias}$=2.0 Oe we observed that some bubbles had a lower velocity. When $\overline{\Delta H}_{bias}$ was increased to 2.2 Oe the velocity dropped considerably. At negative values of $\overline{\Delta H}_{bias}$ similar critical effects were observed. The fascinating point was that when the in-plane field was increased to 285 Oe the critical effect at $\overline{\Delta H}_{bias}$=2.0 Oe did not occur.

We do not believe that the effect is related to bubble collapse or bubble run-out instabilities. We feel, however, that the effect is related to the critical effects observed in straight wall motion experiments[1-4]. For this material the critical drive field[1,4]

$$H_{crit} = \frac{\pi}{2}\,\alpha\,H_1 \qquad (4)$$

is H_{crit}=1.5 Oe if α=0.005 and H_1=185 Oe. The value of the quantity H_{crit} is close to the value of $\overline{\Delta H}_{bias}$ obtained in this work at which the reduction in bubble velocities occurs. We therefore postulate that when a field across a bubble is applied, and if an initial change in bias field results which is higher than the critical straight wall drive field, a (complex) wall structure is created which will impede fast bubble motion.

DISCUSSION

The present study has shown that in Ga:YIG films bubble velocities up to 500 m/s are feasible. The important point is that these velocities can only be realized if the change in bias field at the start of bubble motion is lower than some critical value. This complicates the construction of bubble circuits. However, if a proper construction is chosen circuits can operate in the megahertz range. For the present material only small drive fields are then necessary.

Bubble motion above the critical field shows dynamic conversion effects as reported by Vella-Coleiro et al.[9] and Vella-Coleiro[10]. In his experiments Vella-Coleiro investigated[10] the influence of an in-plane field on the dynamic conversion effects. He found no reduction of the scatter in the observed velocities if an in-plane field was applied. The two peaks in Fig. 2 of Ref. 10 can be possibly explained as originating from bubbles having $|\overline{\Delta H}_{bias}| < H_{crit}$ and $|\Delta H_{bias}| > H_{crit}$.

Acknowledgements: the authors thank U. Enz for stimulating discussions, A.M.J. van de Heijden for his help in obtaining experimental data and B. Hoekstra for the anisotropy constants and γ determination.

REFERENCES

1. F.H. de Leeuw, IEEE Trans. Magn. Mag-9 614 (1973).
2. P.J. Rijnierse and F.H. de Leeuw, AIP Conf. Proc. 18, 199 (1974).
3. F.H. de Leeuw, J. Appl. Phys. 45, 3106 (1974).
4. F.H. de Leeuw and J.M. Robertson, to be published.
5. A.A. Thiele, A.H. Bobeck, E. Della Torre; and U.F. Gianola, Bell. Syst. Techn. J. 50, 711 (1971).
6. A.P. Malozemoff, J. Appl. Phys. 44, 5080 (1973).
7. J.C. Slonczewski, J. Appl. Phys. 45, 2705 (1974).
8. D.C. Bullock, AIP Conf. Proc. 18, 232 (1974).
9. G.P. Vella-Coleiro, F.B. Hagedorn, Y.S. Chen, and S.L. Blank, Appl. Phys. L. 22, 324 (1973).
10. G.P. Vella-Coleiro, AIP Conf. Proc. 18, 217 (1974).
11. F.B. Hagedorn, J. Appl. Phys. 45, 3129 (1974).
12. H.J. Levinstein, S. Licht, R.W. Landorf and S.L. Blank, Appl. Phys. L. 19, 486 (1971).
13. G.P. Vella-Coleiro and W.J. Tabor, Appl. Phys. L. 21, 7 (1972).
14. R.M. Josephs, Appl. Phys. L. 25, 244 (1974).

IN-PLANE PULSE RESPONSE OF BUBBLE DOMAINS

A. P. Malozemoff and J. C. Slonczewski, IBM Research Center, Yorktown Heights, N. Y. 10598

ABSTRACT

It is known that a pulsed field applied normally to a domain wall produces wall motion whose sign depends on the sense of twist. We predict that the presence of coercivity in combination with appropriate pulse shape produces a net displacement per pulse of at most $\pi\Delta/2\alpha$, where Δ is the wall thickness parameter and α is the damping. The effect should be observable in isolated bubble domains and its sign measures their chirality. Preliminary experiments in a garnet film show different states but with a more complex behavior than predicted by theory.

INTRODUCTION

In his original theory of domain-wall motion, Döring[1] noted that velocity is imparted to the wall by the precession of spins about a field-component H_n normal to the wall plane. He remarked that in the usual geometry the easy-axis drive field H_z is parallel rather than normal to the wall plane. In this case H_n, and hence wall motion, arises indirectly from the demagnetizing field which appears only after spins begin to precess about H_z and out of the wall plane.

A conceptually simpler way to cause a Bloch wall to move is to apply H_n directly as an external field H_p as indicated in Fig. 1. Stein and Feldtkeller[2] first proposed this mechanism to explain their experiments on "wall streaming" in permalloy films. Here we apply the same concept to bubble domain materials, referring to the effect as the "in-plane pulse response".

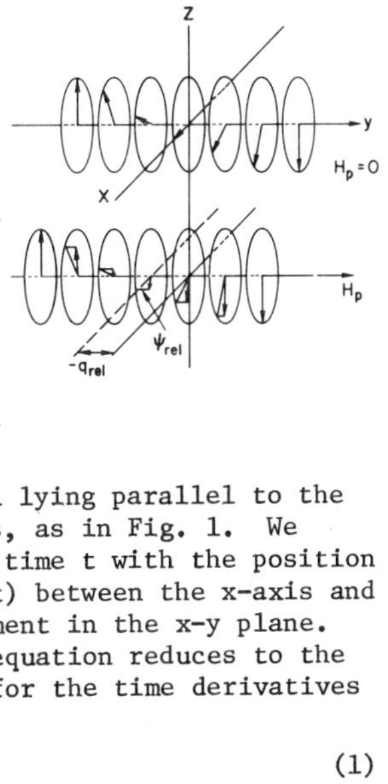

Fig. 1. Schematic Bloch wall lying in XZ plane, before and after application of a step field H_p normal to wall plane. Precession of spins around H_p causes an equilibrium wall displacement q_{rel}.

THEORY

Consider an infinite wall lying parallel to the xz plane, with z the easy axis, as in Fig. 1. We describe its configuration at time t with the position coordinate q(t), and angle $\psi(t)$ between the x-axis and the projection of the wall moment in the x-y plane. The Landau-Lifschitz-Gilbert equation reduces to the following coupled equations[3] for the time derivatives $\dot{q}$ and $\dot{\psi}$:

$$\dot{\psi} = \gamma H_z - \alpha\dot{q}/\Delta \quad , \tag{1}$$

$$\dot{q}/\Delta = 2\pi\gamma M \sin 2\psi - \tfrac{1}{2}\pi\gamma H_p \cos\psi + \alpha\dot{\psi} \quad . \tag{2}$$

Here γ is the gyromagnetic ratio, α the Gilbert damping constant, M the spontaneous magnetization, H_p and H_z the applied field components in the y and z directions. We assume a constant wall-width parameter $\Delta=\sqrt{A/K_u}$, where A is the exchange stiffness constant and K_u is the uniaxial anisotropy. Equation (1) states that the precession rate $\dot{\psi}$ of the wall moment is proportional to the pressure on the wall from an applied z-field and from viscous drag. Equation (2) states that velocity is proportional to the torque on the wall moment due to wall demagnetization, applied in-plane field, and viscous damping.

These equations, which neglect coercivity, have a curious consequence in the absence of an easy axis drive field ($H_z=0$). For then Eq. (1) integrates to

$$q(t) = -(\Delta/\alpha)\psi(t) + \text{constant} \ . \tag{3}$$

Whenever the wall is given sufficient time to relax to equilibrium in the presence of H_p, we have $\dot{q}=\dot{\psi}=0$, and Eq. (2) gives us

$$\psi_{rel} = \arcsin(H_p/8M) \text{ or } \pi-\arcsin(H_p/8M) \tag{4}$$

for the relaxed ψ, assuming $|H_p|\leq 8M$. This equation simply balances demagnetizing and in-plane field torques, thus implying motion has stopped. Equations (3) and (4) combine to give for the relaxed q,

$$q_{rel} = \mp (\Delta/\alpha)\arcsin(H_p/8M) \ , \tag{5}$$

where the sign depends on whether the wall twist is initially right-handed ($\psi=0$) or left-handed ($\psi=\pi$), and where the initial q is zero. For small ψ, the characteristic relaxation time for this behavior is

$$\tau_{rel} = (4\pi\alpha\gamma M)^{-1}(1+\alpha^2) \ . \tag{6}$$

Since the maximum value of the arcsin in Eq. (5) is $\pi/2$, application of a sufficiently large in-plane field ($H_p>8M$) can displace the wall a maximum of $\pi\Delta/2\alpha$. For typical low-loss garnets with $\alpha=0.03$ and $\Delta=0.05\mu m$, this gives a $2.6\mu m$ displacement. The inverse dependence of q_{rel} on α is plausible considering that in the absence of any loss, the spin precession about H_p, and therefore the wall motion, would continue as long as $H_p\neq 0$.

EFFECT OF COERCIVITY

Equation (5) shows that q_{rel} is a function of the applied field H_p; therefore, if H_p is applied as a pulse which begins and ends at the same value, the net equilibrium wall displacement per pulse must be zero, no matter what the pulse shape. However, a net displacement can be achieved in the presence of a coercive field H_c, which had been neglected above. To show this, we model coercivity conventionally with the expression

$$H_z = -H_c \operatorname{sgn} \dot{q} \tag{7}$$

for H_z in Eq. (1). Here sgn $\dot{q}$ is ± 1 if $\dot{q} \gtrless 0$, and if $\dot{q}=0$ it takes on some value between -1 and +1 determined by solving Eq. (2). According to this model, coercivity produces a constant effective field $H_z = \pm H_c$ tending to oppose any finite velocity.

Now, we ask, what is the critical form of $H_p(t) = H_{p,crit}$ required to make the wall barely move? Let $\dot{q}$ be a small negative quantity tending to zero. Then a solution of Eq. (1) is

$$\psi = \gamma H_c t \tag{8}$$

if we consider only a right-hand wall at t=0. For small ψ, Eqs. (2) and (8) reduce to

$$H_{p,crit} = 8\gamma M H_c t + (2\alpha/\pi)H_c \ . \tag{9}$$

Usually the term $(2\alpha/\pi)H_c$ may be neglected; thus we conclude $\dot{H}_p$ must exceed the critical value $8\gamma M H_c$ to produce wall motion. In particular, consider a triangular pulse with a peak field $H_{p,max}$; if its rise time is shorter than, and fall time longer than, $\tau_{crit} = H_{p,max}/8\gamma M H_c$, an initial motion can be induced but the reverse motion during the fall time will be suppressed. If in addition the fall time is greater than τ_{rel}, and if $H_{p,max}>H_{po}\equiv 2H_c/\pi\alpha$, it can be shown that the net displacement per pulse will approach

$(\Delta/\alpha)\cdot\arcsin(H_{p,max}/8M)$, as in Eq. (5). For the typical parameters mentioned earlier, and taking M=13 gauss, $\gamma=1.3\times10^{7}\,sec^{-1}Oe^{-1}$ and $H_c=0.5$ Oe, we have $\tau_{rel}=16$ nsec and $H_{po}=10.6$ Oe. If in addition we take $H_{p,max}=30$ Oe, we find $\tau_{crit}=45$ nsec and a net displacement of 0.5μ per pulse.

In fact any pulse which causes the forward and reverse motions to be different in nature can lead to some net displacement in the presence of coercivity. For example, one may also apply a square pulse of length T less than τ_{rel} so that the spins do not have time to reach the equilibrium of Eq. (5). Immediately following such a pulse, the demagnetizing torque $2\pi\gamma M\cdot\sin 2\psi$ acting on the wall moment is weaker than the in-plane field torques at the beginning of the motion; so the coercivity can cut in sooner to stop the motion. The net displacement, found by solving equations (1), (2) and (7) in the limit $T<<\tau_{rel}$, is $\frac{1}{2}\pi\Delta\gamma H_p T$.

DETECTION OF BUBBLE CHIRALITY

The free-wall condition assumed above is effectively met in isolated magnetic bubbles. Consider the two chiral states of the bubble, corresponding to the two possible senses of twist of the Bloch-wall structure, as illustrated in Fig. 2a. In these cases an in-plane field pulse translates diametrically opposite portions of bubble wall in the same direction. Thus

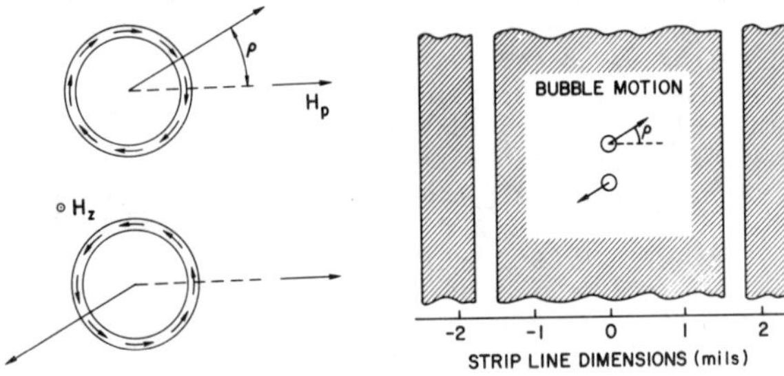

Fig. 2. a) Schematic spin structures for chiral bubble states and direction of in-plane pulse response. b) Strip-line dimensions for in-plane pulse response experiment.

bubbles of right-handed chirality will move in the opposite direction from bubbles of left-handed chirality. This result is of considerable interest because there has previously been no practical method for detecting bubble chirality in garnets: Gradient deflection experiments do not distinguish between these two states,[4] and transmission Lorentz microscopy, which was first used to demonstrate the existence of these states,[5] has so far been impractical for general employment because of the thickness of typical garnet films.

Just as chiral bubbles deflect in a gradient field,[4] they should also deflect from the in-plane field direction, as indicated in Fig. 2a. Indeed one finds the same deflection angle ρ in the absence of coercivity:

$$\rho=\arctan(2\Delta/\alpha r) \qquad (10)$$

where r is the radius of the bubble.

With the above theoretical motivation, we have attempted the experiment. Figure 2a shows the strip-line configuration used. The bubble is positioned under the central area of the central stripe. A pulse of current I through the central stripe of width W creates an in-plane field of magnitude I/2W in mks units, but it also creates a gradient $2I\pi/W^2$. The gradient is canceled out by the use of the adjacent pair of strip lines.

A garnet film was used which had a composition $Eu_{0.65}Y_{2.35}Ga_{1.08}Fe_{3.92}O_{12}$ and thickness 5.85 microns; other material parameters are approximately those given in the computations above. Dynamic properties of this sample have previously been described in Ref. 6. Bubbles were randomly generated and propagated by a pure

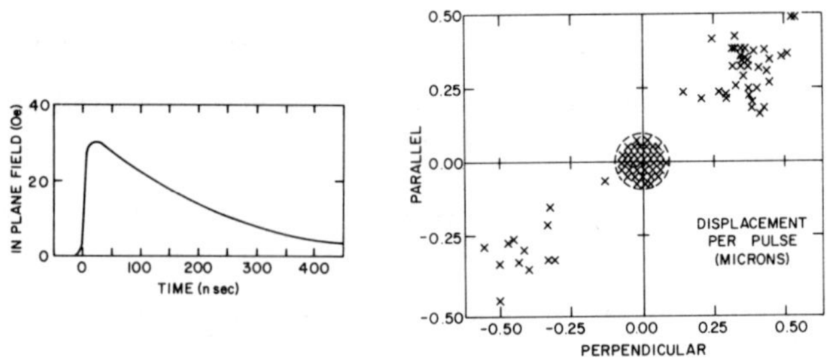

Fig. 3. a) Applied in-plane field pulse as measured from current pulse with Tektronix CT-1 current probe and sampling scope. b) Displacement vectors relative to strip lines for 100 bubbles, using the pulse shape of Fig. 3a. Each cross represents a different propagation experiment for a "randomly" generated $n_r=1$ bubble. Bubbles which did not move significantly after several hundred pulses are indicated within the dashed circle around the origin.

field gradient from the outer strip-lines until an $n_r=1$, zero-Bloch-line, bubble was found.[4,6] Then all of the conductors were simultaneously pulsed with a fast-rise slow-fall pulse of the shape and strength shown in Fig. 3a. The pulse was created by an SKL 503A pulse generator using a charged capacitor rather than a cable.

Bubble displacement following a single pulse was too small to measure, but when a train of pulses (e.g. 20 pulses per second) was applied, the bubbles were observed to behave in one of three ways - to jiggle about while remaining basically stationary, or to march in one of two directions, as shown by the average displacements per pulse plotted in Fig. 3b. All these bubbles behaved like $n_r=1$ bubbles in a gradient both before and after; thus it is clear that different states are subsumed under the $n_r=1$ category. If we attempt to interpret the marching bubbles as the chiral states according to the theory, we find a net displacement per pulse of the correct order of magnitude; however the deflection angle has the wrong sign. Furthermore when the polarity of in-plane field is reversed, the bubbles, instead of reversing direction, usually become stationary and remain so even when the original polarity is then restored. The existence of the stationary bubbles and the unexpected behavior of the marching bubbles suggests that the $n_r=1$ bubbles have more complex wall structures than previously assumed. A more complete report of this work will be published elsewhere.

The authors thank E. Giess for the garnet sample, W. Bogholtz, R. Anderson, and R. McGouey for the preparation of the striplines, E. Hochberg for experimental assistance, and O. Voegeli and L. Rosier for helpful conversations.

REFERENCES

1. W. Döring, Z. Naturforsch. 3a, 373 (1948).
2. K. U. Stein and E. Feldtkeller, J. Appl. Phys. 38, 4401 (1967).
3. J. C. Slonczewski, Intern. J. Magnetism 2, 85 (1972); J. C. Slonczewski, J. Appl. Phys. 45, 2705 (1974).
4. J. C. Slonczewski, A. P. Malozemoff, and O. Voegeli, AIP Conf. Proc. 10, 458 (1973); A. P. Malozemoff, J. Appl. Phys. 44, 5080 (1973).
5. P. J. Grundy, D. C. Hothersall, G. A. Jones, B. K. Middleton and R. S. Tebble, Phys. Status Solidi(a) 9, 19 (1972).
6. O. Voegeli and B. A. Calhoun, IEEE Trans. Mag. MAG-9, 617 (1973).

DYNAMICS OF MICRON SIZE BUBBLES

H. L. Hu
IBM Research Laboratory, San Jose, California 95123
E. A. Giess
IBM Thomas J. Watson Research Center, Yorktown Heights, New York 10598

ABSTRACT

Dynamic properties of micron diameter bubbles in garnet compositions $(EuYb)_3(FeGa)_5O_{12}$, $(EuTm)_3(FeGa)_5O_{12}$, and $(Eu)_3(FeGa)_5O_{12}$, are reported here. The coercivities and mobilities of these bubbles compare vary favorably with those of 5-6 micron bubbles. A mobility of 1000 cm/sec/Oe was found for the composition $Eu_{0.8}Tm_{2.2}Fe_{4.5}Ga_{0.5}O_{12}$. Despite the fact that Q (the ratio of anisotropy field to the saturation magnetization) is less than 5, these films contain all kinds of soft and hard bubbles, in contrast to what Smith et al. have reported in the case of 5 micron bubbles. However, it was found that when the Q is less than or equal to 2, there exists only one kind of bubble propagating at an angle with respect to the field gradient, in accord with their observation. The micron bubbles with no Bloch lines propagate at a larger angle with respect to the direction of the field gradient than 5 micron bubbles, in qualitative agreement with Slonczewski's theory.

I. INTRODUCTION

An extensive amount of work has been recently reported in the literature on the dynamics of magnetic bubbles in various kinds of materials. Mobilities ranging from a few cm/sec/Oe to several thousand cm/sec/Oe have been reported in various garnet compositions as well as amorphous GdCo.[1] The variety of bubble states, soft and hard, have also been under intensive investigation.[2] Many different means to suppress the generation of hard bubbles have been proposed and implemented in T-I bar devices. In particular, A. B. Smith[3] et al. have reported the suppression of hard bubbles by lowering the Q (the ratio of the uniaxial anisotropy field to the saturation magnetization) in the material. They indicated a Q of 5 or less is sufficient to suppress hard bubbles.

Except for the work in amorphous GdCo films, most work on bubble dynamics has been limited to materials having a bubble diameter of 4 microns or larger. As the bubble technology progresses along, it is apparent that further increases of bubble packing density by decreasing the bubble size are desirable. Growth of garnet films with micron size bubbles has been reported,[4] as well as the T-I bar device operation with these bubbles.[5] It has been reported that a large in-plane rotating field amplitude was required to propagate the bubbles. It is the intent of this paper to report the dynamics of these micron size bubbles in some garnet compositions. Specifically, coercivities, mobilities, and domain wall states in garnets $(EuYb)_3(FeGa)_5O_{12}$, $(EuTm)_3(FeGa)_5O_{12}$, and $(EuLu)_3(FeGa)_5O_{12}$ are discussed.

II. EXPERIMENTAL TECHNIQUES AND RESULTS

The bubble transport method, first proposed by Vella-Coleiro and Tabor[6] was used to study these micron size bubbles. The conductor lines used to produce the field gradients are shown, together with 5 micron and micron bubbles, in Fig. 1. They are fabricated on the garnet films with a spacer of 0.5 micron Schott glass. Thanks to the large saturation magnetization in these films, no inner conductor loop was used to compensate for the bias field change, which in this case caused very little change of bubble size during the measurements. Typically a 0.2 microseconds or longer, and up to 1 ampere current pulse was used so as to move the bubble at least one diameter in each measurement. Furthermore, the bubbles were always initiated, i.e. measurements were made only after the bubble had reversed its propagation direction and had also moved at least one bubble diameter. Due to the small dimension and low contrast of these bubble domains, viewing through the microscope using a low light level television camera equipped with an accurate video-micrometer was found both convenient and reliable.

TABLE 1

Composition	Film Thickness h (μm)	Demagnetizing Stripewidth W_s (μm)	Characteristic Length ℓ (μm)	Collapse Field H_o (Oe)	Saturation Magnetization $4\pi M_s$ (G)	Q	Mobility μ_w (cm/sec/Oe)
$Eu_{1.1}Yb_{1.9}Ga_{0.9}Fe_{4.1}O_{12}$	3.48	9	1.15	70	311	16	126
$Eu_1Yb_2Ga_{0.4}Fe_{4.6}O_{12}$	2.01	1.48	0.14	407	663	4	428
$Eu_{1.7}Yb_{1.3}Fe_5O_{12}$	2.8	1.14	0.062	1102	1450	2.1	404
$Eu_1Tm_2Ga_{.7}Fe_{4.3}O_{12}$	1.66	2.58	0.202	168	442	10	184
$Eu_{0.8}Tm_{2.2}Ga_{0.5}Fe_{4.5}O_{12}$	1.0	1.15	0.138	248	708	3	1000
$Eu_{1.2}Lu_{1.8}Ga_{.5}Fe_{4.5}O_{12}$	1.09	1.05	0.12	407	769	4	704

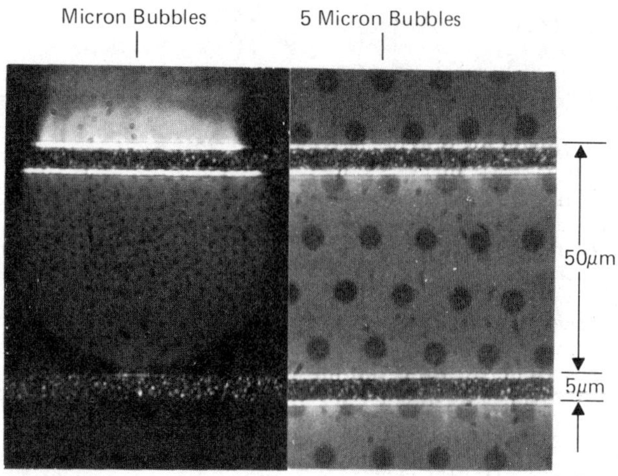

Figure 1. A photomicrograph of micron size and 5 micron bubbles, together with conductor lines. Notice the effect of the dust particles and light scattering from the edges of the conductor lines on the visibility of the bubbles in both cases. In the case of micron bubbles, the field aperture is stopped down so that only the upper conductor line is illuminated. The light scattering from it has completely obscured the bubbles located above it.

The results of the measurements of bubbles propagating in the direction of the field gradient are indicated in Fig. 2 for all three compositions. In Table 1, we list the properties of these micron bubble films. For comparison, we also list those compositions containing more gallium and which therefore support a larger bubble size. Notice in those compositions that support micron bubbles (the second, the fifth, and sixth rows in Table 1) the Q. measured with a technique similar to that Smith et al. have described,[7] is about 3 or 4 and certainly not larger than 5. But, in contrast to what Smith et al. have found in 5 micron bubble materials where hard bubbles do not exist with Q less than 5, we found bubbles which are not only hard to collapse but which also propagate slowly and at large angles (up to 80°) with respect to the direction of the field gradient. However, when the Q is reduced to 2 or so as is the case in $(EuYb)_3Fe_5O_{12}$, we found only one kind of bubble which propagates (the third row in Table 1) at about 60° with respect to the direction of the field gradient, in agreement with what Smith et al. has found in $Y_{0.9}Gd_{1.2}Tm_{0.9}Fe_{4.6}Al_{0.1}Ga_{0.3}O_{12}$.

The domain wall mobility in $(EuTm)_3Ga_{0.5}Fe_{4.5}O_{12}$ is quite large as compared with that of 5 micron bubble films. Even in the case of $(EuYb)_3(FeGa)_5O_{12}$ and $(EuLu)_3(FeGa)_5O_{12}$, the domain wall mobilities are acceptable, considering the fact that the bubble needs to move 1/5 the distance in each period in a device as compared with the 5 micron bubbles. As seen from Eq. 1,[1] when the bubble size is decreased from 5 microns to 1 micron in size, the decrease of Q and increase of exchange constant A are in the direction to increase the mobility. While this may quantitatively explain the (EuYb) system as listed in Table 1, a factor of 5 increase of mobility in the (EuTm) system from 3 μm bubble to 1 μm bubble (the fourth and fifth row of Table 1) is surprising if one assumes the damping parameter (λ/γ^2) remains unchanged.

$$\mu_w = \frac{1}{(\lambda/\gamma^2)} \left(\frac{A}{2\pi Q}\right)^{\frac{1}{2}} \qquad (1)$$

The coercivities (proportional to the intercepts of Fig. 2) are not so large either although the minimum field gradient required to move a bubble in a well behaved manner, is somewhat high, typically a ΔH of 2 Oe/bubble is needed. The velocity saturation in Fig. 2b may be due to the dynamic conversion effect that Vella-Coleiro[8] has reported and is less likely due to strip-out that Josephs[9] has reported. This is due to the fact that all the measurements were done in the central 8 micron region where a maximum of 25 Oe decrease in bias field occurred with a field gradient of 6.4 Oe/μm. The bias fields during the measurements were set at about mid-point between bubble collapse and elliptical instability, the range of which is typically 90 Oe. Therefore only a decrease of bias field of 45 Oe or larger may cause elliptical instability of the bubbles. With the exception of $(EuYb)_3Fe_5O_{12}$ as mentioned previously, soft bubbles with different states were observed in these films. In the case of $(EuLu)_3(FeGa)_5O_{12}$, the bubble motion is very erratic and reliable consistent bubble motion is very difficult to obtain. The data, shown in Fig. 1c, have actually contained a large amount of scattering. In the other two compositions, there is no state change during the bubble motion as long as one keeps the

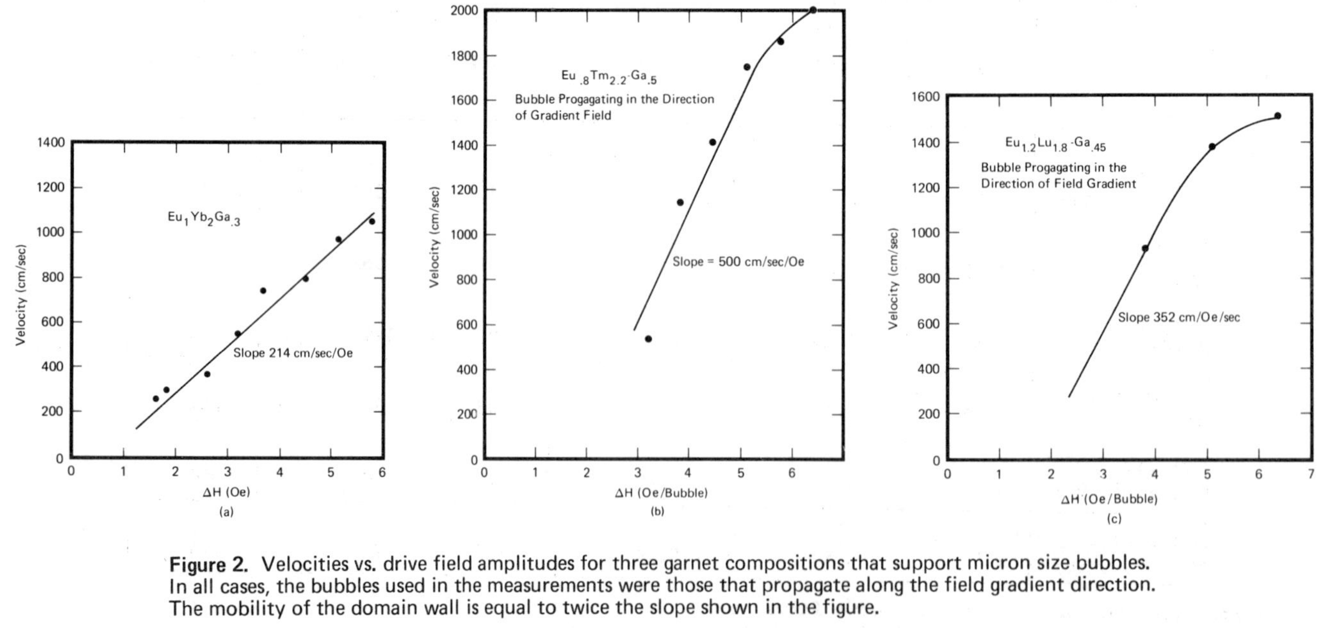

Figure 2. Velocities vs. drive field amplitudes for three garnet compositions that support micron size bubbles. In all cases, the bubbles used in the measurements were those that propagate along the field gradient direction. The mobility of the domain wall is equal to twice the slope shown in the figure.

TABLE 2

$$\text{Revolution number } n_r = \gamma \left(\frac{d\nabla H_z}{8}\right)\left(\frac{d\sin\rho}{v}\right) *$$

	Bubble Diameter (μm)	ρ angle	$(d\nabla H_z)$Oe	V(cm/sec)	n_r
$Eu_1Yb_2Ga_{0.4}Fe_{4.6}O_{12}$	1	30°	2.54	500	0.53
$Eu_{1.7}Yb_{1.3}Fe_5O_{12}$	0.8	60°	3.58	756	0.68
$Eu_{0.8}Tm_{2.2}Ga_{0.5}Fe_{4.5}O_{12}$	1.0	45°	4	1000	0.57
$Eu_{1.2}Lu_{1.8}Ga_{0.5}Fe_{4.5}O_{12}$	1.0	40°	3.84	900	0.60

*A γ of 1.64×10^7 Oe^{-1} sec^{-1} is assumed.

bubble propagation trajectory within the central 8 micron region where no more than 25 Oe of bias field change has occurred. The bubble with no Bloch lines showed a larger deflection angle than 5-6 micron bubbles, in qualitative agreement with Slonczewski's theory.[2] These data are listed in Table 2. The non-integer of revolution n_r was not understood.

ACKNOWLEDGEMENTS

The authors wish to express their appreciation for the fabrication of overlay done by R. L. Anderson, R. Beeck, D. Johnson, T. Montelbano and D. Saiki. Thanks are also due to W. A. Reynolds and C. Guerci for their valuable technical assistance.

REFERENCES

1. See for examples, G. P. Vella-Coleiro, AIP Conf. Proc. 10, 424 (1973), and M. H. Kryder and H. L. Hu, AIP Conf. Proc. 18 213 (1973).
2. See for examples, J. C. Slonczewski, A. P. Malozemoff and O. Voegeli, AIP Conf. Proc. 10, 442 (1972).
3. A. B. Smith et al., AIP Conf. Proc. 18, 167 (1973).
4. E. A. Giess et al., Mat. Res. Bull. 8, 1061 (1973) and this conference.
5. H. L. Hu et al., Intermag. Conf. 26-5, (1973).
6. G. P. Vella-Coleiro and W. J. Tabor, Appl. Phys. Letters 21, 7 (1972).
7. A. B. Smith et al. AIP Conf. 10, 308 (1972).
8. G. P. Vella-Coleiro, AIP Conf. 18, 217 (1973).
9. R. M. Josephs, Appl. Phys. Letters 25, 244 (1974).

ANNIHILATION OF BLOCH LINES IN HARD BUBBLES

R. W. Patterson
Westinghouse Research Laboratories
Pittsburgh, Pennsylvania 15235

ABSTRACT

When hard bubbles are translated at sufficiently high velocities, the Bloch lines in the bubbles' walls are annihilated, resulting in soft bubbles. Investigations of this annihilation phenomenon as a function of Bloch line density were performed in YEuIG and YGdTmIG films. It was found that the velocity required to cause annihilation is inversely proportional to the Bloch line density, while the drive field required to translate the bubbles at this velocity is directly proportional to the Bloch line density. The angle θ_c between the critical velocity and the field gradient is independent of Bloch line density, and virtually invariant from sample to sample, with a value of $\theta_c \approx 80°$. Evidence has been found which indicates the annihilation process involves sequential destruction of Bloch lines.

The presence of Bloch lines in bubble walls has been shown to account for the behavior of hard magnetic bubbles.[1-4] In general, Bloch lines will survive a bubble's being expanded, contracted or translated. However, since Bloch lines increase the total energy of a bubble, one would expect them to be only metastable. Tabor et al.[1] state that sufficiently rapid motion of a bubble wall, via either expansion or translation, produces Bloch line annihilation which results in a bubble with fewer, or no, Bloch lines.

This paper reports on further investigations into the annihilation phenomenon associated with bubble translation. The conditions leading to annihilation were studied as functions of Bloch line density in three films of $Y_{2.4}Eu_{0.6}Ga_{1.2}Fe_{3.8}O_{12}$ and one film of $Y_{1.3}Gd_1Tm_{0.7}Ga_{1.2}Fe_{3.8}O_{12}$.

The Bloch line density n/d is the ratio of the number of pairs of Bloch lines to the diameter of the bubble, and is obtained by fitting translational data to the equation[3,4]

$$\frac{V^2}{V_\perp} = \frac{\gamma d}{8n}(H_a - H) , \qquad (1)$$

where $V_\perp$ is the component of the velocity V perpendicular to the bias field gradient, and γ is the gyromagnetic ratio. The effective drive field, H_a, is the product of the gradient and the bubble diameter, and H is a phenomenological coercivity.[5] The angle θ between the velocity and the gradient increases as the velocity increases according to the relation[3,4]

$$\mathrm{ctn}\,\theta = f\alpha + \frac{\gamma d\, H_{cd}}{8n\, V} , \qquad (2)$$

where α is the Gilbert damping parameter, f is approximately equal to one when n/d is large,[4] and H_{cd} is another phenomenological coercivity.

Translational data were obtained by propagating bubbles having diameters of approximately 5 μm between a pair of stripe conductors carrying parallel currents. The conductors were 52 μm wide, 14 μm thick, and separated by 100 μm. The bubbles were propagated symmetrically about the centerline with total displacements kept to less than 10 μm. Displacements parallel and perpendicular to the gradient were measured with a pair of filar eyepieces. These measured displacements and the known pulse widths yielded the two components of the velocity.

The velocities were determined as functions of drive field amplitude, which was increased by incrementing the current pulse amplitude until Bloch line annihilation occurred, or until the pulser limit was reached. Annihilation was signified either by the disappearance of the bubble, or by a propagation anomaly —a marked increase in the bubble's displacement (velocity) coupled with an arbitrary change in the propagation angle θ. The anomalous velocity increments were as large as a factor of two, and correlated neither to the changes in θ, nor to the Bloch line densities. Disappearance generally occurred only when the static bias field was above the collapse field for soft bubbles, H_o.

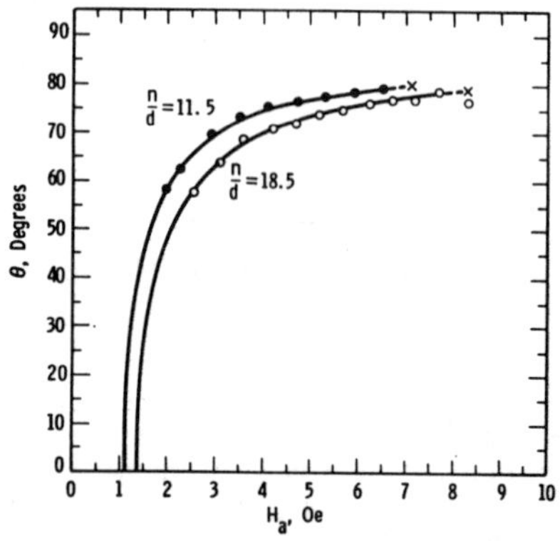

Figure 1 Angle vs Drive Field for two bubbles in a YEuIG film. Bloch line annihilation conditions are indicated by an X.

Figure 1 is a plot of θ vs drive field for two bubbles in one of the YEuIG films. The curves are obtained from the relationship

$$\sin\theta = (1 + f^2\alpha^2)^{-1}\left\{- f\alpha R + [1 - R^2(1 - f^2\alpha^2)]^{\frac{1}{2}}\right\} \qquad (3)$$

which results from eliminating V from (1) and (2), and defining $R = H_{cd}/(H_a - H)$. The values of $f\alpha$, H_{cd} and H, as well as n/d, are obtained from least-squares fits of the data to (1) and (2). The values so obtained are: n/d = 11.5, 18.5 μm^{-1}; $f\alpha$ = 0.05, 0.06; H_{cd} = 0.85, 1.10 Oe; H = 0.26, 0.28 Oe, respectively. Figure 1 also shows the extrapolated θ values the bubbles would have had, had annihilation not occurred, as well as the change in θ associated with the velocity anomaly of the bubble with n/d = 18.5. The bubble with n/d = 11.5 disappeared when Bloch line annihilation occurred. Although (3) does not show any explicit dependence on Bloch line density, there is an implicit dependence which enters through f (Ref. 4) and H_{cd}.[5]

Subsequent to a velocity anomaly, a bubble would usually appear "soft," i.e., it would propagate with a much-increased mobility, either parallel or nearly parallel to the drive field, and would collapse at a bias value of H_o. These bubbles were presumed to have either one or no Bloch line pairs. Occasionally, however, the resulting bubble had a reduced Bloch line density, but still displayed hard properties. This partial annihilation occurred in five of the forty-five annihilations observed in the YEuIG films, as well as in four of the sixteen annihilations observed in the YGdTmIG film. Complete annihilation was invariably, but not exclusively, associated with the largest anomalies. There was no observed instance of a propagation anomaly which resulted in an unchanged or increased Bloch line density.

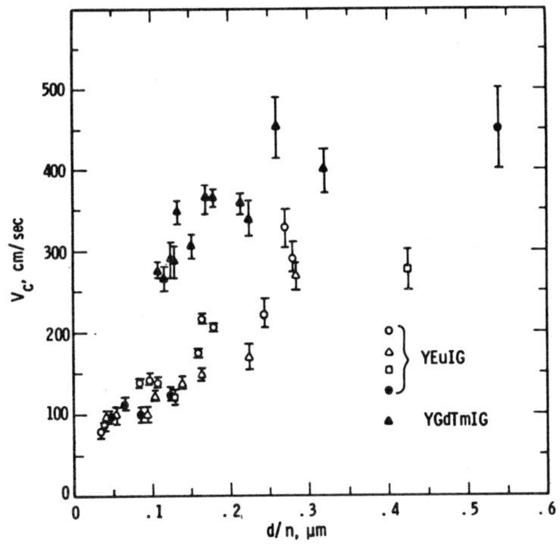

Figure 2 Critical velocity vs d/n for several films. Open circles and triangles represent data from different locations on the same film.

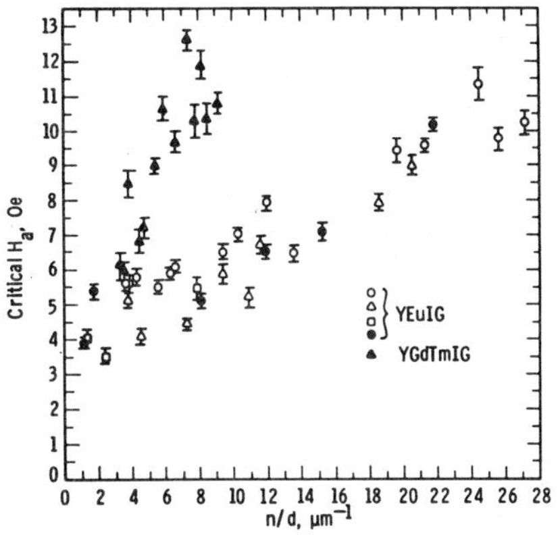

Figure 3 Critical drive field vs Bloch line density. The notation is the same as in Fig. 2.

The extent of Bloch line annihilation was unrelated to the conditions producing the annihilation. These conditions were, however, strongly dependent on the Bloch line density prior to annihilation. The critical velocity V_c shows an effectively linear dependence on the inverse of the Bloch line density, as demonstrated in Fig. 2. One straight line could be used to fit all the YEuIG data, and another, with a roughly equal slope, could fit the YGdTmIG data. The uncertainty bars in Fig. 2 represent the difference between the highest measured velocity and the extrapolated value the bubble would have had, had annihilation not occurred. Not all the observed annihilation phenomena are shown in Fig. 2 or subsequent figures because some annihilations were observed without obtaining sufficient data to determine n/d, critical H_a, and/or V_c.

The critical drive field showed a direct dependence on Bloch line density, as shown in Fig. 3. Here also two straight lines could fit the data—one for all the YEuIG films, another for the YGdTmIG film. The bars in Fig. 3 represent the amplitude of the drive field increments employed.

The angle the critical velocity made with the gradient shows no such dependence. The lack of variability of θ_c is demonstrated in Fig. 4, where the only significant variation is at very low Bloch line densities. A maximum in θ_c may occur around n/d = 5, but the rate of decrease with higher densities is insufficient for certainty. There are no discernible differences from sample to sample.

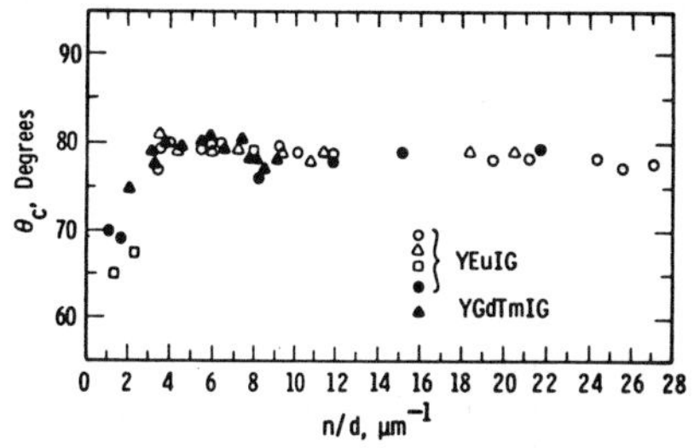

Figure 4 Critical angle vs Bloch line density. The notation is the same as in Fig. 2.

Although no suitable mechanism for Bloch line annihilation has been found, one can infer that the process is sequential. We can explain the fact that a significant fraction of bubbles underwent partial annihilation by postulating that the drive pulse terminated prior to the completion of a sequential annihilation process. That such a process is self-sustaining under a constant-amplitude drive field can be seen from the trend of the data in Fig. 3. A drive field amplitude which is sufficient to produce annihilation at a given Bloch line density would be more than sufficient at any reduced density. Thus, once initiated, a sequential annihilation process will run either to completion, or until the drive pulse is terminated.

The dependences of the critical velocities and the critical drive fields were related through the effective mobility. For large drive field amplitudes $V \approx (\gamma d/8n)H_a$. Thus, if the critical H_a is a linear function of n/d, V_c will be a linear function of d/n.

The relative independence of θ_c from n/d can be understood by reference to (3). As H_a becomes large, the terms containing H_{cd} become negligible; hence, there is little dependence on n/d at large drive fields. The sample-to-sample uniformity of θ_c is probably a result of the similarity among the samples' α values. At large velocities, θ approaches the limiting value of ctn^{-1} fα, according to (2). The α values for these samples all fall in the 0.05-0.1 range, so that the limiting angles are between 84 and 87°. Attempts to investigate the effects of higher damping in several films of $Er_3Ga_{0.7}Fe_{4.3}O_{12}$ and $Eu_1Er_2Ga_{0.7}Fe_{4.3}O_{12}$ were unsuccessful. Due to the high coercivities and low mobilities of these films, we were unable to propagate bubbles at velocities sufficient to produce annihilation.

ACKNOWLEDGMENTS

The author wishes to express his appreciation to A. I. Braginski, S. H. Charap and P. R. Emtage for many useful and informative discussions. The garnet films were grown by G. W. Roland.

REFERENCES

1. W. J. Tabor, A. H. Bobeck, G. P. Vella-Coleiro, and A. Rosencwaig, Bell Syst. Tech. J. 51, 1427-1430 (1972).
2. A. P. Malozemoff, Appl. Phys. Lett. 21, 149-150 (1972).
3. J. C. Slonczewski, Phys. Rev. Lett. 29, 1679-1682 (1972).
4. A. A. Thiele, J. Appl. Phys. 45, 377-393 (1974).
5. R. W. Patterson, J. Appl. Phys. (in press).

ELECTRON DIFFRACTION AT VERTICAL BLOCH LINES IN DOMAIN WALLS

M.I. Darby and P.J. Grundy
Department of Pure and Applied Physics
University of Salford, Salford M5 4WT,England

ABSTRACT

The dynamic properties of bubble domains are often determined to a considerable degree by the presence of Bloch lines in the domain wall. Current opinion is that the magnetization in the bubble wall can, in certain circumstances, rotate in Néel segments or vertical Bloch lines (BL's) of differing senses and that certain dynamic behaviour can be explained in terms of the sense or "state" of BL's. The object of this paper is to discuss theoretical calculations of the form of the Fresnel electron diffraction pattern or Lorentz microscopy image at vertical BL's. The calculations are based on a uniform rotation distribution in the BL and similar results are expected for other models.

INTRODUCTION

The possibility of determining experimentally the "handedness" or state(1) of vertical BL's in bubble or strip domain walls will be touched upon in another paper presented at this conference(2). This is an attractive proposition as it would help to prove or disprove the existence of different BL states. Many dynamic properties of bubble domains have been interpreted in terms of the reaction and interaction of BL's incorporated in their walls(3). Vertical BL's, or Néel segments, in bubbles were first observed in Lorentz electron micrographs(4) of thin, uniaxial crystals in which the BL is shown up by a reversal of contrast in the image or Fresnel electron diffraction pattern of the domain wall. The image detail in the region of this reversal is, in principle, subject to the way and sense in which the magnetization rotates between the Bloch wall sections.

The aim of this paper is to calculate the images expected from vertical BL's and to point out any differences between these images. In the following sections the theory of the image formation and the formulation of the wave from the BL are outlined and some results of the calculation discussed. Some suggestions for the rather exacting experiments that are required are also given and the materials of interest are discussed.

THEORY AND CALCULATION

The requirements for imaging magnetic phase objects, such as domain walls or BL's, in a transmission electron microscope are similar to those needed for detecting other sources of phase contrast. The magnetic object is observed out-of-focus and as long as a sufficiently small source size or coherent electron beam is used interference fringes will be formed by the waves "diffracted" at sufficiently large changes of magnetic flux in the object.

Consider a plane electron wave ψ_o of unit amplitude incident normally on a magnetic thin film in the (ξ,η) object plane. The transmitted wave in the (u,v) image plane a distance z below the film is given by

$$\psi_i(u,v,z) = A\,T\,\psi_o\,(\xi,\eta)\,.$$

T is the transmission function of the object and A the aperture function of the microscope. For constant amplitude $T = \exp(i\phi(\xi,\eta))$ where ϕ is the phase change produced by any change in magnetic flux. A contains phase contributions from the instrument; here it is essentially that introduced by defocussing the object.

Within the Fresnel approximation the amplitude of the diffracted waves in the image plane (u,v) is (5), for a point source,

$$\psi_i = K\int_{-\infty}^{\infty} d\xi \int_{-\infty}^{\infty} d\eta\, \exp i\{A(\xi^2+\eta^2) + C(u\xi+v\eta) + \phi(\xi,\eta)\},\ldots(1)$$

where $A = -\frac{1}{2}k(x^{-1}+z^{-1})$, $C = k/z$, K = constant, z is the object/image distance, x = source/object distance, and

$$\phi(\xi,\eta) = \frac{-e}{\hbar}\, D \int_0^R B_\theta(r)\,dr \qquad (2)$$

is the phase difference produced in the object plane (ξ,η). Here $R^2 = \xi^2 + \eta^2$; $\theta = \tan^{-1}(\eta/\xi)$, and D = film thickness.

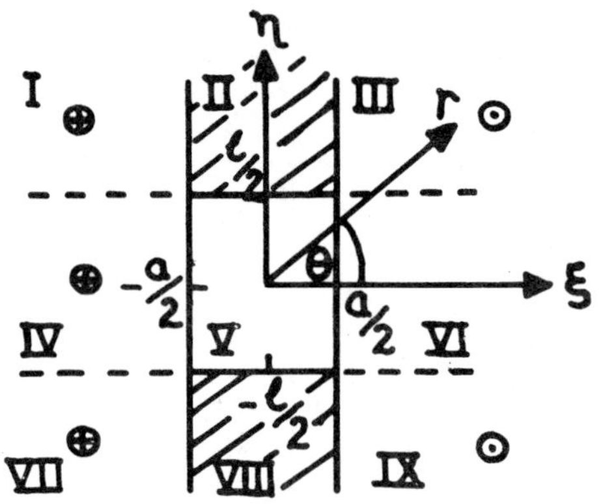

Fig.1. The division of the object plane (ξ,η) into regions for the purposes of evaluating the wave integral Eqn.(1). The phases in the various regions are (a) II and IV, $\phi = \phi(\xi)$ as for a single Bloch wall, (b) I, III, VII and IX, ϕ = constant, (c) V, in the Bloch line $\phi = \phi(\xi,\eta)$, and (d) IV and VI, $\phi = \phi(\eta)$.

It is convenient to divide the object plane into a number of regions as shown in Fig. 1. In the Bloch line, region V, the components of the magnetic induction have been taken to be

$$B_\xi(\xi,\eta) = \mp B_s \cos(\pi\xi/a)\cos(\pi\eta/\ell)$$
$$B_\eta(\xi,\eta) = \mp B_s \cos(\pi\xi/a)\sin(\pi\eta/\ell) \qquad (3)$$

At large values of $|\eta|$ the phase $\phi(\xi,\eta)$ is independent of η, as for a single Bloch wall, and this has been assumed to be the case for $|\eta|>\ell/2$. The magnetic induction in the Bloch wall regions has been taken as(6)

$$B_\eta(\xi) = \pm B_s \cos(\pi\xi/a) \qquad (4)$$

In regions IV and VI the phase has been approximated by a function independent of ξ, namely $\phi(\eta) = \phi(\mp\frac{a}{2},\eta)$.

Many of the integrals involved in the evaluation of the amplitude can be expressed in terms of Fresnel integrals and the remainder have been evaluated numerically. Constant K was chosen so that the intensity $(\psi_i{}^*\psi_i)$ has unity at large distances from the Bloch line. The effect of a finite source size, taken to be 160Å squared, has been included by suitably averaging the intensity distribution for a point source.

RESULTS AND DISCUSSION

The variable material parameter in the phase function ϕ of equation (2) is B_sD. In the present calculations this was fixed at 10^7 G Å and hence they are relevant to bubble films 500 Å thick for Co, 0.25 μm for magnetoplumbite, 0.8 μm for $EuYb_2Fe_5O_{12}$ garnet and ∿ 0.4 μm for $Gd_{20}Co_{80}$. These thicknesses compare to 4ℓ values (ℓ, the material length parameter familiar in bubble technology) of 120 Å, 0.12 and 0.2 μm for the first three materials and, therefore, correspond to films sufficiently thick to support bubble domains. The image profiles have been calculated numerically for arbitrary defocussing distances, z, of 50,200 and 600 μm. As the BL has a two dimensional distribution with respect to the incident electrons its image has been represented as a two dimensional plot of intensity using suitable computer line-printer symbols to form an intensity scale. This type of presentation is useful as it simulates a pictorial representation of the image which could be compared directly with the experimental image. For the present purposes it is also convenient to represent the one-dimensional Bloch wall on each side of the BL in the same way.

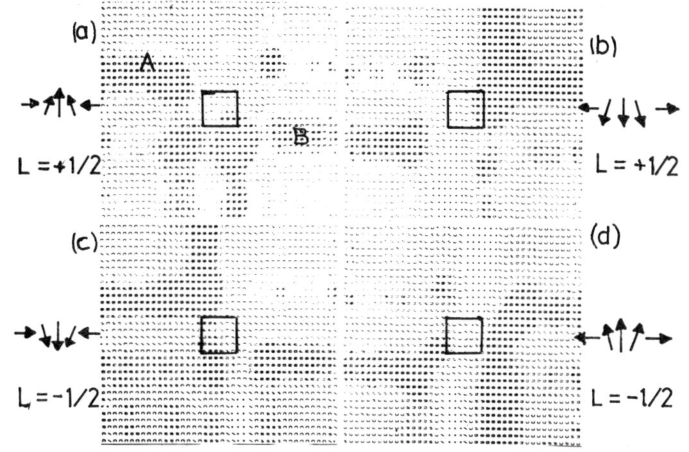

Fig.2. Computed electron micrographs for vertical BL's of different state and configuration (z = 200μm).

Figure 2 shows the results for BL's of state L = ±½. It should be noted that (a) and (b) have the same state (or sense of rotation) but different configuration in their magnetization; the same is true for (c) and (d). This configuration is marked on each micrograph as is the size of the BL in the (ξ,η) object plane. The four sign alternatives of equation (3) define these configurations; that of (b) was used previously[4] as a model for the calculation of the energy and dimensions of vertical BL's in cobalt. The symbols in the micrographs represent the intensity at points separated by 40Å in the (u,v) image plane and are presented slightly out of focus to enhance the contrast effect. Denser symbols represent greater electron wave intensity. Obvious differences exist between the four examples, principally the cross-over of contrast at the ends of the BL (e.g. regions A and B in (a)) and the appearance of features in the ±v direction as expected in diffraction from a two-dimensional structure. If all the necessary experimental requirements necessary to detect these changes were satisfied then BL's of this type could be distinguished one from the other.

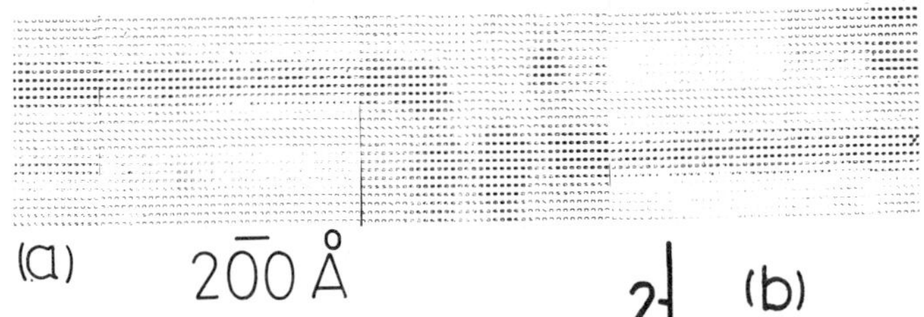

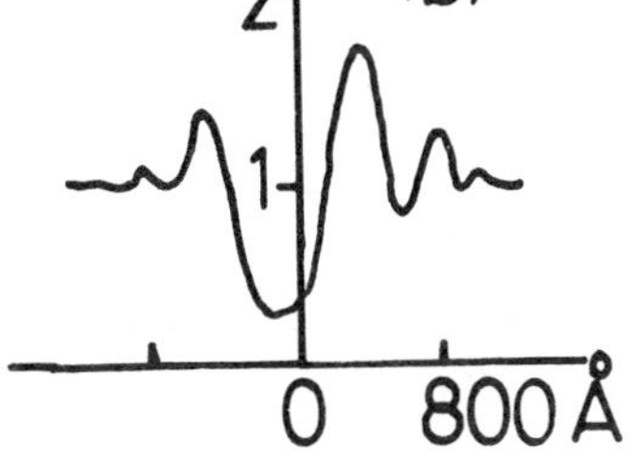

Fig.3.

Showing (a) a computed electron micrograph of a Bloch wall containing a vertical BL (as in 2(a)), (b) the intensity profile across the Bloch wall (z = 200μm).

The computed micrograph of a Bloch wall containing a vertical BL is shown in figure 3(a). It is seen that the character of the wall is regained at about 160 Å from the centre of the BL at this defocussing distance. This image can be compared with the BL alone in figure 2(a). A graphical intensity plot across the one dimensional wall sections is shown in figure 3(b). It is in essential agreement with similar plots obtained previously.[7] It is seen from this figure, and figure 2, that the "cross-over" in intensity from one section of the wall to the other occurs to one side of the BL; this result is also obtained from geometrical optical calculations of the intensity in the BL image. It is observed experimentally that BL's can be imaged with different intensity distributions at the same defocussing - pointing to the existence of BL's of different state and configuration - and also that there is some difference as a function of defocussing. This result is obtained in the calculations and the computed images for a particular BL at various values of z show definite differences. It is clear that in any comparison between theory and experiment all the microscope parameters determining the image should be well known.

The change in magnetic flux across BL's in the materials considered is between about 5 and 25 x 10^{-7} G $Å^2$. A convenient measure of resolution in wave-optical images of magnetic phase objects can be defined in terms of the fluxon[5]. The fluxon represents a limiting, detectable change of flux of h/2e ∿ 4 x 10^{-7}G $Å^2$ which contributes one interference fringe to the image. The characterization of detail in a magnetic flux change less than the fluxon in magnitude is therefore difficult. In experiment, conditions which approach this limit are ameliorated by an electron optical system with high coherency, i.e. practically a near point source, and high beam brightness. The microscope should also have the necessary stability for operation over lengthy exposure times at high magnifications where the image detail is, of necessity, greater in size than the grain size of the photographic emulsion. Any recording system which can reduce the "noise" level to below that in the photographic method would be an advantage.

CONCLUSION

The results presented in this paper show that it is possible, in principle, to distinguish stationary vertical BL's of different state and magnetization configuration (as defined by the examples in figure 2) one from the other by observation of their electron interference images. We assume, a priori of course, that these different magnetization distributions exist and it should be said that, in general, it will not be possible to do the reverse, i.e. determine the distribution from the image. We have used a convenient magnetization distribution, as given by equation (3), and it is well known from previous work[6] that electron images are not very sensitive to the fine details of any chosen distribution.

ACKNOWLEDGEMENTS

The authors would like to thank Dr D C Hothersall for helpful discussions. P.J.G. thanks IBM for a travel grant.

REFERENCES

1. O.Voegeli and B.A. Calhoun, IEEE Trans. Magn., MAG-9, 617-621, 1973.
2. P.J. Grundy, G.A. Jones and R.S. Tebble, paper submitted to this Conference.

3. e.g. A. Rosencwaig, W.J. Tabor and T.J. Nelson, Phys.Rev.Lett., 29, 946-948 1972. J.C. Sloncewski, A.P. Malozemoff and O.Voegeli, AIP Conf.Proc.10, 458-477, 1972.
4. P.J. Grundy, D.C. Hothersall, G.A. Jones, B.K. Middleton and R.S. Tebble, phys.stat.sol.(a) 9, 79-88, 1972.
5. D.K. Wohlleben, J.Appl.Phys., 38, 3341-3352 1967.
6. D.C. Hothersall, Phil.Mag., 20, 89-112 1969.
7. D.C. Hothersall, Phil.Mag., 24, 241-258, 1971.

INSTANTANEOUS RADIAL WALL VELOCITIES IN MAGNETIC GARNET BUBBLE DOMAINS*

G. J. Zimmer+, L. Gal+, K. Vural, and F. B. Humphrey
California Institute of Technology
Pasadena, California 91125

Using a stroboscopic microscope with a sample time of 10 nsec, the transient bubble domain configuration resulting from a uniform pulse field applied parallel to the bias field has been recorded photographically. The pulsed laser illumination is sufficient to allow a picture to be taken in a single 10 nsec exposure at a known time with respect to the applied pulse field. In this way the actual dynamic domain configuration of the bubble can be observed. Diameter measurements as a function of time have been made on both implanted and non-implanted $Y_{1.57}Eu_{0.78}Tm_{0.65}Ga_{1.05}Fe_{3.95}O_{12}$ garnet material. The material parameters for the non-implanted (implanted) sample are as follows: $4\pi M$ = 200 (201) Gauss, thickness = 6.2 (7.2) microns, Q = 15.4 (16.9), α = 0.026 (0.029), $\gamma = 1.1 \times 10^7$ rad/Oe-sec, $A = 2 \times 10^{-7}$ erg/cm. The dynamic behavior of the two samples were essentially the same except that the non-implanted one showed a considerably greater range of characteristics between bubbles. The radius of the bubble was plotted as a function of the time delay between the leading edge of the expanding field pulse and the exposure. The experimental curves seemed to be straight lines for a l but the very low pulse fields (<20 Oe) where the new equilibrium radius of the bubble was observed within the time scan of the experiment. The experimental data were corrected for the changing effective drive as the bubble expands. The corrected velocity increases linearly with drive from an initial velocity of about 1 m/sec to a velocity saturation at 18 Oe of 5.84 m/sec. It remains at this saturation value for drives up to about 180 Oe. For higher drives, the velocity increases again, linearly, with a slope of 0.023 m/sec to the end of the range of the experiment which was 280 Oe. The existing theories predict a peak velocity for these samples of about 5 m/sec at 0.5 Oe drive and V_o = 1.5 m/sec. If the lower drive mobility is associated with Malozemoff's predicted μ_{nl} = .1 m/sec-Oe, the value of the observed mobility in this region is μ = 0.27. It is concluded that the actual instantaneous expansion radial velocity cannot be explained by any existing theories.

*Work supported in part by the National Science Foundation.

+Permanent Address: Central Research Institute for Physics, Hungarian Academy of Sciences, Budapest, Hungary.

DYNAMICS OF HARD WALLS IN BUBBLE GARNET STRIPE DOMAINS*

T. M. Morris, G. J. Zimmer,+ and F. B. Humphrey
California Institute of Technology
Pasadena, California 91125

The motion of hard magnetic domain wall sections, intermixed with soft walls has been observed in stripe domains in bubble garnets. Using a faraday microscope with pulsed laser illumination, 10 nsec single exposure photographs of the stripe domains were taken while the stripes were in motion. When a short, low amplitude demagnetizing pulse is applied, well defined constrictions in the stripe become apparent before the critical width at which magnetostatic instabilities begin to develop. The constrictions appear either as a symmetric inward dent in both walls of the stripe or as a depression in one wall with the other remaining straight. The former is more prevalent. This effect has been observed in a wide variety of non-implanted materials but seems to be completely suppressed in implanted samples. The sample measured had a composition of $Y_{1.57}Eu_{0.78}Tm_{0.65}Ga_{1.05}Fe_{3.95}O_{12}$ with a thickness of 6.2 mm, $4\pi M$ = 200 Gauss, Q = 15.4 and α = 0.026. By photographing the hard stripe at a steadily incremented delay on successive field pulses wall velocity was obtained from the width vs. time data. Under the same pulse conditions the hard wall moves roughly an order of magnitude more slowly than a normal wall. The experimental velocity exhibits a strict linear dependence on drive field with a mobility of 1.3 cm/sec-Oe. The predicted mobility based on the continuous winding of vertical Bloch lines for this material is 0.9 cm/sec-Oe.

Observation of the constrictions on successive field pulses at a fixed delay shows their displacement along the length of the stripe. If the polarity of the pulse is reversed, the hard sections appear as bulges in the narrowing stripe and move in the opposite direction. Both left and right hand deflections are observed. As a function of pulse length, displacement per pulse of a bulge increases rapidly to a broad maximum and then falls slowly to a low constant value for pulses of sufficient duration to drive the stripe to a new equilibrium width.

If the bulges are photographed such that the laser delay alternates between two fixed values relative to each successive field pulse, the mobility of lateral displacement of the bulge during a single pulse can be determined to be about 75 cm/sec-Oe. The tight winding model predicts a lateral mobility of 170 cm/sec-Oe for an infinite string of tightly wound Bloch lines.

*Work supported in part by the National Science Foundation.

+Permanent Address: Central Research Institute for Physics, Hungarian Academy of Sciences, Budapest, Hungary

PROPERTIES OF BLOCH POINTS IN BUBBLE DOMAINS

J. C. Slonczewski, IBM Research Center, Yorktown Heights, N. Y. 10598

ABSTRACT

We summarize a derivation of the Bloch-point (BP) structure and energy appropriate to bubble domains. Stray and applied fields determine positions of BPs.. BPs can give rise to half-integer and continuous effective revolution numbers in bubble dynamics. Bullock's and Hasegawa's dynamic data provide evidence for the presence of BPs in bubbles. Possible roles of BPs in bubble-state transformation are proposed.

INTRODUCTION

In micromagnetism, one usually assumes that the magnetization vector $\vec{M}(\vec{x})$ varies continuously with position $\vec{x}$. However, Feldtkeller proved that some magnetic configurations demand topologically the presence of singularities he named Bloch points (BPs).[1] A defining property of a BP is that all possible directions of $\vec{M}$ are found on a sphere of sufficiently small radius centered on it. This paper describes properties of BPs in bubbles, cites experimental evidence of their presence, and sketches their essential role in wall-state transformations.

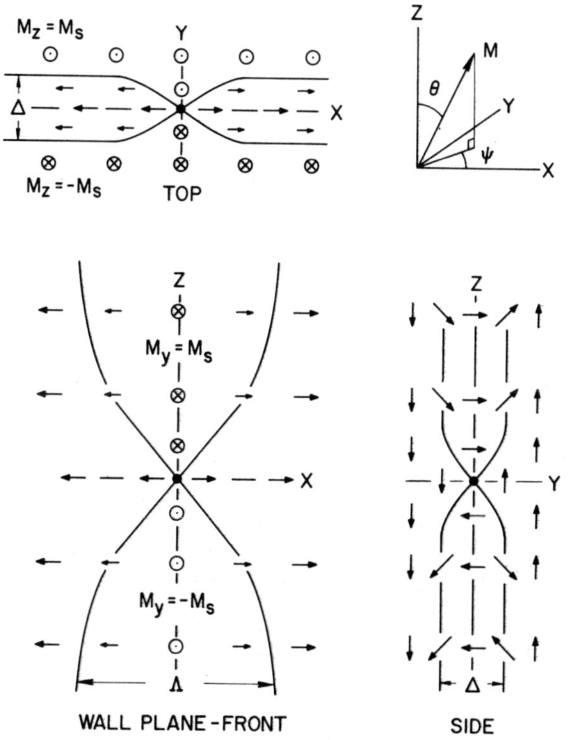

Fig. 1. Principal orthogonal sections of a micromagnetic configuration including one Bloch wall, one Bloch line, and one Bloch point.

STRUCTURE AND ENERGY OF A BLOCH POINT

We summarize here results of work to be published elsewhere. Assume a 180° wall (thickness parameter Δ) lying parallel to the x-z plane, and separating domains with $M_z = \pm M_s$, as shown in Fig. 1. Assume also one Bloch line (width parameter Λ),[2] lying parallel to the z-axis and separating wall regions in whose midplane $M_x = \pm M_s$. Finally, assume a BP region separating BL regions with $M_y = \pm M_s$. The singular point itself lies at the origin of the coordinate system. By inspection of the $\vec{M}$ arrows close to the origin one sees by continuity that all directions occur, thus verifying the BP definition given above. Note that the BP "pinches" the wall and BL thicknesses to a point, as indicated by the constant-angle contours shown in Fig. 1.

The vector field $\vec{M}(x,y,z)$ of this configuration is determined by minimizing the energy

$$W = \iiint dxdydz\{A[(\nabla\theta)^2+(\nabla\psi)^2\sin^2\theta]+K\sin^2\theta+(H_d^2/8\pi)\} \quad (1)$$

where θ and ψ are the orientation angles of $\vec{M}$, with z the polar axis (inset, Fig. 1). Here A is the exchange stiffness coefficient, K the uniaxial anisotorpy, and $\vec{H}_d(x,y,z)$ the stray-field due to divergence of $\vec{M}$. The quantity of interest is the Bloch-point insertion energy W_{bp}, obtained from the minimum $W=W_{min}$ by subtracting the normal Bloch-wall (density σ) and the Bloch-line (density ρ) contributions. Thus

$$W_{bp} = W_{min} - \sigma S - \rho L , \quad (2)$$

where S is the area of the wall and L the length of the BL.

This variational problem is difficult because it involves two fields $\vec{M}$ and $\vec{H}$ in three dimensions. However, the assumption $Q \equiv K/2\pi M_s^2 >> 0$, often made in bubble materials, makes the problem tractable. In this limit, the ratio $\Lambda/\Delta = Q^{1/2}$ also becomes large. One can then assume that θ depends only on y and $u = (x^2+z^2)^{1/2}$, and ψ depends only on x and z. One can also take the local stray-field approximation

$$H_{dy} = -4\pi M_y = -4\pi M\sin\theta\sin\psi , \; H_{dx}=H_{dz}=0 \quad (3)$$

as in simple discussions of Bloch lines.[2]

One can show that under these assumptions W_{bp} tends asymptotically to the limit

$$W_{bp} = 2\pi A^{3/2}K^{-1/2}(\ell nQ + C_1 + C_2) , \quad (4)$$

where C_1 and C_2 are pure numbers. They are obtainable by solving separate variational problems for θ(y,u) and ψ(x,z). Work on these problems is in progress.

STATIC POSITION OF A BLOCH POINT

In what uniform external field $\vec{H}$ can a BP remain at rest? For the ideal geometry of Fig. 1, $\vec{H}$ must vanish. To prove this, suppose $H_z \neq 0$; then the system would gain domain energy by wall displacement. Or, if $H_x \neq 0$, wall energy would be gained by BL displacement in the ±x direction. Finally, if $H_y \neq 0$, BL energy could be gained by BP displacement in the ±z direction. Thus we have $\vec{H}=0$.

Consider instead an isolated, straight, vertical BL in a circular bubble, with a uniform applied in-plane field H_p. The above argument is changed a little by the wall curvature. However one can see that the BL must lie on one end of the bubble diameter parallel to H_p (See Fig. 2). Moreover, if the BL has a BP, then the BP lies at the point z=Z, satisfying $H_p+H_r(Z)=0$, where H_r is the component of stray field normal to the wall (See Fig. 2). Computed plots of H_r are available,[3] permitting easy determination of Z.

In particular, note that for $H_p=0$ the BP lies on the midplane of the film (Z=0). Indeed this is the absolutely stable condition of a vertical BL. A simple BL without a BP has more energy because the moment of half of it opposes the stray field. As H_p increases from zero, each of two possible BPs displaces toward opposing film surfaces (Z = ± h/2, Fig. 2). For sufficiently large H_p, BPs should be driven out of the film, but estimating the critical H_p is difficult.

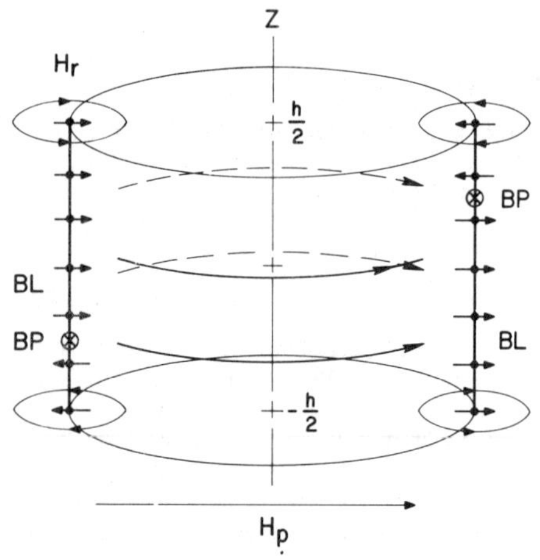

Fig. 2. A bubble domain in the presence of a field H_p.

BLOCH-POINT DYNAMICS

Simple considerations indicate that even in a uniformly moving bubble the position Z_i of the i-th BP is well approximated by the condition $H_p + H_r(Z_i) = 0$ if H_p is orthogonal to velocity $\vec{V}$. The main dynamic effect of BPs is their influence on the gyrotropic force

$$\vec{F} = \vec{G} \times \vec{V} , \quad (5)$$

where $\vec{G}$ is a constant vector. By a trivial extension of Thiele's theory[4] one has

$$G_z = 4\pi M h \, n_r / \gamma \quad (6)$$

where

$$n_r = h^{-1} \int_{-h/2}^{h/2} n(z) dz \quad (7)$$

is the mean revolution (or winding) number. Here h is the film thickness and n(z) is the number of revolutions made by the wall moment in a complete circuit of the domain perimeter lying in the plane given by z. Of course n(z) is an integer but n_r is generally not, if BPs are present, because n(z) changes by ± 1 at $z = Z_i$. According to Eq. (7) each BL contributes to n_r an amount $\pm Z_i/h$ which lies between -1/2 and 1/2.

During bubble translation, $\vec{F}$ deflects bubbles from the direction of the drive and the theory of this effect permits experimental determination of n_r.[2] Recently, Bullock observed deflection in an ion-implanted garnet.[5] He conducted each measurement after a preparatory application of a 40 Oe in-plane field, which was immediately turned off. From his data he calculated the spectrum of values n_r=0.50, 1.02, 1.42, 1.86, and his text indicates a value near 0 also occurred. The fact that these numbers are close to integers and half-integers, rather than only integers, suggested the theory was wrong by a factor 2.

We can more naturally explain these results by postulating that the prepared bubble states each have one pair of BLs, with or without BPs as in Fig. 2, held in place by some small H_p due to misalignment. For then the revolution number is clearly

$$n_r = 1 + (0 \text{ or } \pm 1/2) + (0 \text{ or } \pm 1/2), \quad (8)$$

where 1 comes from the winding of the moment in the wall segments between BLs, and 0 or $\pm 1/2$ comes from each BL depending on whether or not it has a BP. Allowing for all possible combinations, one has n_r=0, 1/2, 1, 3/2, 2, which are consistent with the data. Further dynamical evidence for BPs is given by Hasegawa.[6]

BUBBLE-STATE TRANSFORMATION

The dynamic intermediate states in the state transformation[2] represented by a change of n_r clearly involve singularities of $\vec{M}(x,t)$. To describe a plausible mechanism resembling Hagedorn's "vortex",[7] begin with a simple unichiral (no BL) bubble. Assume with Hagedorn that a large translational velocity causes horizontal BLs to nucleate and then deform into pairs of vertical BLs of opposing sense, which then migrate to and cluster on opposite sides of the bubble.[7] Assume also that the horizontal segments "tunnel" to the surface. Indeed, recent computations by Hubert[8] indicate that tunneling (rather than pile-up) occurs naturally, in some range of material parameters, even in the absence of any material defects.

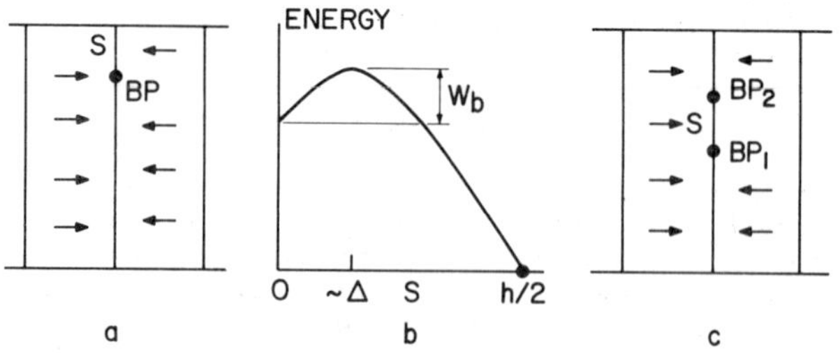

Fig. 3. Two hypothetical routes for bubble conversion. (a) One BP nucleates at surface. (b) Schematic potential energy. (c) Two BPs nucleate at center.

Such a BL cluster is illustrated in Fig. 3a. As indicated in Fig. 3b, a barrier energy $W = W_b$ must be supplied to nucleate a BP at the film surface and displace it a distance $s \sim \Delta$ into the film. Beyond this maximum the BP falls spontaneously to the film midplane s=h/2 because the energy gained in switching one-half of the BL into the stray-field direction (See Fig. 2) is far greater than W_b. The transformation $\delta n_r = \pm 1$ would then be completed whenever static or dynamic forces caused this partitioned BL to collide with and annihilate a neighboring BL by spontaneous unwinding. Unfortunately we cannot estimate W_b in this case because our Eq. (4) does not take into account the very substantial effect of surface fields. Consider, however, the possibility that 2 BPs of opposite sign nucleate near the midplane, and one of them, BP_2 in Fig. 3c, moves spontaneously to the surface. In this case $W_b = 2W_{bp}$, where W_{bp} is correctly estimated by Eq. (4).

For either mechanism, W_b is quite small, so that one may imagine it provided by the agitation of high velocity motion, perturbation of material imperfections, or interaction with additional magnetic layers. Indeed, W_{bp} is on the order of 1 eV for a bubble diamter of 0.1μ (neglecting C_1 and C_2) so that transformation by thermal activation of BP pairs at room temperature is likely. Since W_{bp} scales in proportion to bubble diameter, this should however be at virtual standstill for micron bubbles. Experiments of Hasegawa[8] and Hsu[9] support the role of BPs in state transformation.

The author warmly thanks A. Hubert, R. Hasegawa, and T.-L Hsu for helpful discussions and access to unpublished results.

REFERENCES

1. E. Feldtkeller, Z. Angew. Phys. 19, 530 (1965).
2. J. C. Slonczewski, A. P. Malozemoff and O. Voegeli, AIP Conf. Proc. 10, 458 (1973).
3. J. C. Slonczewski, J. Appl. Phys. 44, 1759 (1973).
4. A. A. Thiele, Phys. Rev. Lett 30, 230 (1973).
5. D. C. Bullock, AIP Conf. Proc. 18, 232 (1974).
6. R. Hasegawa, following paper.
7. F. B. Hagedorn, AIP Conf. Proc. 18, 222 (1974); J. Appl. Phys. 45, 3129 (1974).
8. A. Hubert, to be published.
9. T.-L. Hsu, these Proceedings.

EFFECT OF BLOCH POINTS ON THE DYNAMIC PROPERTIES OF BUBBLE DOMAIN

Ryusuke Hasegawa
IBM Thomas J. Watson Research Center, Yorktown Heights, New York 10598

ABSTRACT

In addition to the normal dynamic conversion of the bubble domain wall state, we have recently found in garnet films grown by an LPE technique a new anomaly which is best characterized by the observation of continuously varying, in-plane field dependent effective revolution number, n_r. Our results can be explained by introducing Bloch points within the vertical Bloch lines in the domain wall. In light of this interpretation (Slonczewski, this conference), a half-integer value for n_r corresponds to the ground state for the Bloch points. The value n_r is modified by an in-plane field, H_x, of the order of 1 to 10 times the saturation magnetization, M_s. Thus we expect the value n_r to be such as 0, 0.5, 1.5, 2.0 etc. when H_x=0, being consistent with the present observation. Increasing H_x reduces the value of n_r. For example, we observed a change of the average value of n_r from 1.0 to 0.5 when H_x/M_s=2.8 for a typical film, as predicted by theory.

INTRODUCTION

In order to understand the dynamic properties of bubble domains, considerable theoretical development has been advanced by works of Slonczewski[1,2], for example. Subsequent works[3-5] taking into consideration vertical Bloch lines in the domain wall seem to explain various experimental observations[6]. One of the most widely used formulas to characterize the motion of bubbles subject to a gradient field across the bubble diameter may be written as[5]

$$n_r = r^2 \gamma \nabla H_z \sin\phi / 2V \qquad (1)$$

where V is the velocity of a bubble of radius r moving with an angle ϕ with respect to the gradient of the applied field, ∇H_z, and γ is the gyromagnetic ratio. The quantity n_r can be regarded as a quantized value for the angle ϕ. Physically n_r represents the number of revolutions of the spin direction when the azimuthal angle of the spin position at the center of the cylindrical wall is varied from 0 to 2π. Thus we may write

$$n_r = (2\pi h)^{-1} \int_0^h \int_0^{2\pi} \frac{\partial\Phi(r_c,\theta,z)}{\partial\theta} dz d\theta \qquad (2)$$

where z and θ are the normal cylindrical coordinate variables, h is the thickness of the film and $\Phi(r_c,\theta,z)$ is the azimuthal angle of the spin at the center of the wall ($r=r_c$) with respect to a reference direction. Until recently it has been believed that the quantity n_r takes only integral valus[6], although some non-integral values for n_r have been obtained for ion-implanted garnet films[7]. Recently we have found half-integer values for n_r in the relatively high bubble drive field region. These new observations seem to be explained as we will show in this article by introducing Bloch points in the vertical Bloch lines as proposed by Slonczewski[8]. Since the present dynamic anomalies are observed in the drive field region in which normal dynamic conversion takes place most frequently, present study may shed light on the mechanism for the wall state change.

EXPERIMENTAL TECHNIQUES

We show here the results obtained for a garnet film grown by an LPE technique of composition $(Y_{2.3}Eu_{.66}Tb_{.04})(Fe_{3.85}Ga_{1.15})O_{12}$, for which detailed low field dynamic properties have already been reported[6]. The present film has a thickness of $h=5.25\,\mu m$, a material parameter $\ell=0.556\,\mu m$, a saturation magnetization M_s=14.3G, an anisotropy energy $K=1.02\times10^4$ erg/cm^3 and an exchange stiffness constant $A=2.5\times10^{-7}$ erg/cm. A pair of parallel strip lines with 70 μm wide and separated at a distance of 200 μm were overlayed on the sample film and supplied with pulsed currents to create gradient fields across the two strip-lines. A bubble of 5.9 μm in diameter was nucleated by a pulse field superimposed on a dc bias field and was subjected to the gradient field of up to about 1.7×10^4 Oe/cm. The pulse current with a width of 0.1 μ sec was used to translate a bubble over a distance of about 20 μm at the center between the two strip-lines. The total number of bubble propagations for each bubble was between 20 and 30. A set of Helmholtz coils was used to apply an inplane dc field H_x perpendicular to the direction of the pulse field gradient in order to further supply evidence for the Bloch points discussed here. The bubble translation behavior was detected by a polar Kerr contrast microscope whose optical output was displayed over a calibrated TV screen with a magnification of about 1000. With this technique, one can measure the bubble translation distance within an accuracy of ±0.2 μm.

RESULTS AND DISCUSSION

As shown in Fig. 1, the bubble velocity increases linearly with the gradient field up to about 1.1×10^3 cm/sec at $H_z=9\times10^3$ Oe/cm and undergoes a maximum followed by a minimum. It is noted that the maximum velocity is somewhat smaller than the peak velocity v_p = 1480 cm/sec predicted by Slonczewski's theory[2]. A recent result by Hubert[9], on the other hand, seems to predict the critical velocity v_{cl} closer to the experimental value. The predicted value of v_{cl} for the present film is 1140 cm/sec which is in excellent agreement with the experimental value shown in Fig. 1. In the low drive

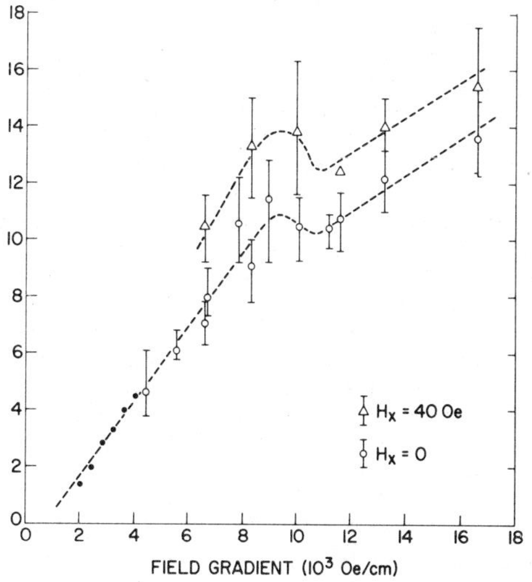

Fig. 1. Velocity versus gradient field for a 5.9 μm-diam bubble with n_r=0. Closed circles are taken from Ref. 6. Open circles and triangles represent the average velocities for the case with H_x=0 and 40 Oe respectively.

field region ($\nabla H_z \le 4\times10^3$ Oe/cm), previous data[6] show that the revolution number n_r obtained through Eq.(1) generally falls on some integral value. The situation appears to be more complex when the drive field is larger as seen in Fig. 2 in which histograms of revolution number n_r are shown for various values of ∇H_z. This figure indicates especially in the high drive field region that there exist some new domain wall states corresponding to the revolution numbers such as n_r=0.5, 1.5 etc. Based on calculation of the energy associated with a Bloch point, Slonczewski[8] has recently argued that creation of a Bloch point in a Bloch wall does not require much excess energy. Thus in a bubble domain wall with a pair of vertical Bloch lines, a Bloch point may be easily nucleated at one end of a Bloch line where the direction of magnetization changes by nearly 360° in a small region near the film surface, and then immediately moves its position to the mid-point of the Bloch line due to the stray fields emanating from the both surfaces of the film (See Fig. 8 of Ref. 2). Then from Eq. (2), one can see that the contributions of the upper and lower half of the Bloch line to the revolution number sum to zero. For example, take a bubble initially having a

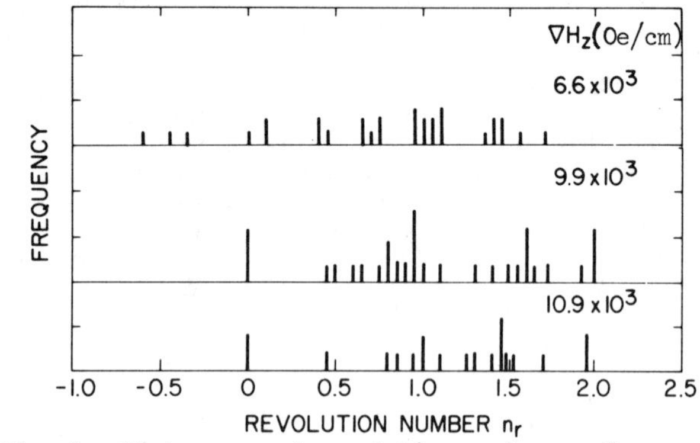

Fig. 2. Histograms of revolution number n_r for various values of ∇H_z. One division in the vertical scale corresponds to a frequency of 10% of finding n_r at a given value.

pair of Bloch lines with negative revolution, i.e. n_r=0. One Bloch point inserted into one of the two Bloch lines results in the case n_r=0.5. An interesting consequence follows immediately when each Bloch line contains one Bloch point resulting in n_r=1.0. This wall state seems to be indistinguishable from the wall state in which there are no Bloch lines at least in the present experimental procedure (with H_x=0). This may provide an explanation for the dynamic conversion between the state n_r=0 and 1, which is frequently observed in the present drive field region. As we will show in the following, this type of wall state degeneracy can be not only removed but also further evidence for the existence of the Bloch points will be given. The Bloch point situated at the mid-point (z=h/2) of a Bloch line shifts to a new position at $z=z_{BP}$ as an in-plane dc field is applied due to the shift of the stray field (H_s) potential minimum at which $H_x+H_s(z_{BP})=0$. Thus from Eq. (2) it is easily seen that the revolution number n_r for the wall state with Bloch points decreases with increasing in-plane field H_x. Since H_x does not affect the deflection angle for a bubble with no Bloch lines (n_r=1)[10], the presence of H_x removes the degeneracy mentioned above. The effect of the in-plane field on the quantity n_r is summarized in Fig. 3. The dotted lines represent n_r as a function of H_x/M_s predicted from Eq. (2) by using Fig. 8 of Ref. 2. It seems that the histograms at various values of H_x shown by the horizontal bars follow the prediction reasonably well. As an additional illustration, we show in Fig. 4 an example of the wall state change attributed to the existence of the Bloch points. The initial wall state is such that there is a pair of Bloch lines with no Bloch points (n_r=0). A Bloch point is nucleated in one of the Bloch lines at the second bubble propagation and because of the effect of the in-plane field (H_x=20 Oe) n_r takes a value 0.45 close to the predicted one, i.e. n_r=0.35 (point A). At a subsequent propagation, another Bloch point is introduced in the remaining Bloch line resulting in an average value of n_r of 0.74 in good

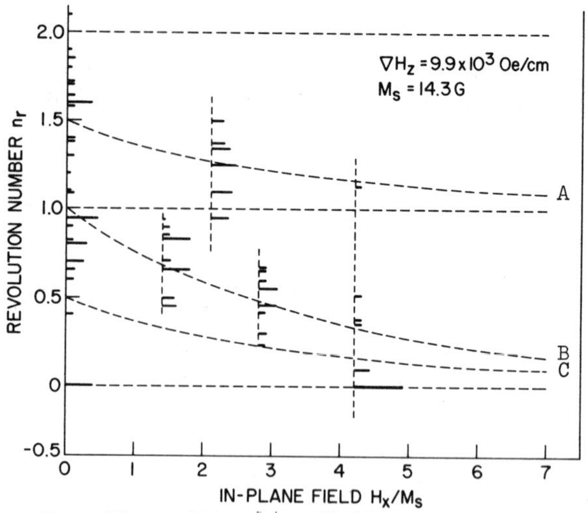

Fig. 3. Effect of in-plane field on n_r. Line A and C correspond to the case with one Bloch point in one of the two Bloch lines of negative and positive revolution respectively. Line B represents the case with one Bloch point in each of the two Bloch lines.

agreement with the predicted n_r=0.7 (point B in Fig. 4). It appears that the final wall state is relatively stable.

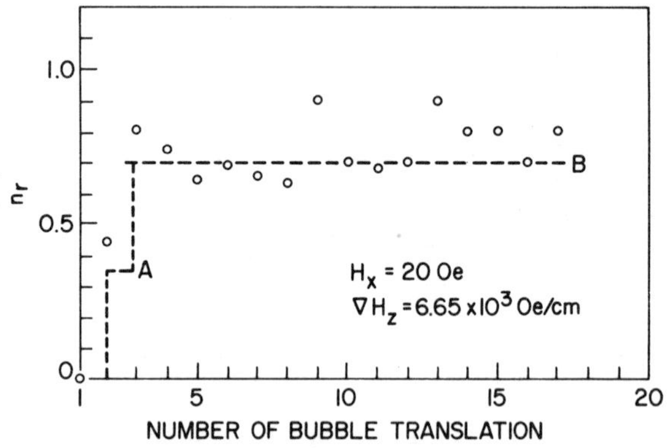

Fig. 4. Revolution number versus number of bubble translation for a given set of gradient and in-plane fields.

In summary, we conclude that the observed half-integer values for the revolution number could be attributed to the existence of Bloch points in the vertical Bloch lines in the domain wall. An in-plane magnetic field varies the new wall state resulting in a reduction of the revolution number.

ACKNOWLEDGEMENT

It is a pleasure to thank John Slonczewski and Alex Malozemoff for their helpful discussion and advice.

REFERENCES

1. J. C. Slonczewski, Int. J. Mag. 2, 85 (1972).
2. J. C. Slonczewski, J. Appl. Phys. 44, 1759 (1973).
3. G. P. Vella-Coleiro, A. Rosencwaig and W. J. Tabor, Phys. Rev. Lett. 29, 949 (1972).
4. A. P. Malozemoff and J. C. Slonczewski, Phys. Rev. Lett. 29, 952 (1972).
5. J. C. Slonczewski, A. P. Malozemoff and O. Vogelli, AIP Conf. Proc. 5, 458 (1972).
6. A. P. Malozemoff, J. Appl. Phys. 44, 5080 (1973).
7. D. C. Bullock, AIP Conf. Proc. 18, 232 (1973).
8. J. C. Slonczewski, (this conference).
9. A. Hubert, (unpublished).
10. Observations of $n_r \sim 1$ are shown in Fig. 3 for H_x/M_s=2.1.

THE USE OF BUBBLE LATTICES FOR INFORMATION STORAGE

O. Voegeli, B.A. Calhoun and L.L. Rosier
IBM Corporation, San Jose, California 95193

J.C. Slonczewski
IBM T.J. Watson Research Center
Yorktown Heights, New York 10598

ABSTRACT

A new approach to bubble memories, called a bubble lattice file (BLF), is described. This approach employs a periodic bubble lattice to define bit positions while the information is contained in the wall structure of the bubbles. The principal advantage of the BLF is the increased storage density, sixteenfold when the resolution capability of the fabrication process is the limiting factor. The device functions required for the BLF, including accessing and write/read operations, are described. Significant differences between BLF and the T-I bar bubble memories are discussed.

CONCEPT

Conventional bubble memories use a device structure, for example, permalloy T and I bars, to define bit positions and the presence or absence of bubbles to represent information. The storage density of such memories is limited by two factors. The finite resolution of lithographic fabrication processes places a lower limit on bubble size, and the aperiodic domain interactions necessitate a bit separation of at least four bubble diameters. These limitations are relaxed by a new approach: the bubble lattice file (BLF). A close-packed bubble lattice[1] is used to define bit positions, and information is stored by having the lattice composed of two kinds of bubbles.

First to be discussed are some properties of bubble lattices which are particularly pertinent to BLF operation, then the selection of the kinds of bubbles to be used for coding and, finally, the device functions utilized in a BLF.

Bubble domain lattices are equilibrium configurations and, because of stabilizing interactions among the domains, are stable over a wide range of external conditions. For example, a bubble lattice with a fixed period is stable with bias fields ranging from a small negative value to a value slightly lower than the collapse field for an isolated bubble.

Several criteria determine the choice of the kinds of bubbles to be used for information storage. They require similar magnetostatic properties; otherwise there will be local perturbations of the lattice, depending on the stored information content. The dynamic properties of the two types of bubbles must be compatible with the requirement of translating the bubbles for accessing informations. Finally, coding and storage of information requires convenient means of generating and discriminating between the different kinds of bubbles and stability of both kinds with compatible operating margins.

Bubbles which differ only in their wall states[2,3] are well suited to represent data in BLF devices. In view of the above criteria, our present choice from among the many possible wall states is the use of states S = 0 and S = 1 (S denotes the number of revolutions of the wall magnetization[3]). These states are magnetostatically identical, exhibit the highest mobilities among all states and have similar operating margins.[4] The two states can be reliably discriminated by their different deflection angles in a field gradient, and a practical write scheme is described by Hsu.[5]

DEVICE FUNCTIONS

The following device functions are required to implement the BLF concept: lattice isolation, accessing, initialization, and write/read functions.

Lattice Isolation -- To achieve maximum storage density, BLF devices employ as dense a lattice as possible. This condition as well as good lattice stability is obtained when the bias field is about 10% of $4\pi M_s$. At such a low field, the sample is nearly demagnetized and a magnetic domain pattern covers the whole area. Since data accessing requires translation, the storage lattice must be isolated from the stationary surrounding domains. This separation is done by the lattice isolation structure along two lines parallel to the translation direction. Typically, a separation of no more than one lattice constant is sufficient to render magnetostatic interactions across the boundaries ineffective.

Possible methods for lattice isolation include a local increase of bias field or a reduction of magnetization. Our present choice is a step change in thickness along the boundaries.[6] It possible to generate more densely packed domain patterns than would correspond to the minimum energy state. As long as the net magnetization is roughly equal inside and outside the storage area, there is negligible domain pressure on the isolation structure. We also note that lattices are stable only if their boundaries coincide with one of the three lattice symmetry axes. The resulting parallelogram-shaped isolation structure with 60 and 120° boundary angles is typical of all BLF designs.

Accessing -- Because BLF devices utilize rigid bubble lattices, domains need not be manipulated individually but retain their relative positions through magnetostatic interactions within the lattice. To translate lattices for data accessing, it suffices to apply driving forces at only a few select locations, as can be done with superposed current conductors. Present designs employ two periodic sets of conductor lines parallel to the lattice columns. Each conductor in this two-phase translation system, analyzed by Eggenberger,[7] is capable of shifting several bubble columns. The required driving force is proportional to the number of columns shifted. Coercive effects can be minimized

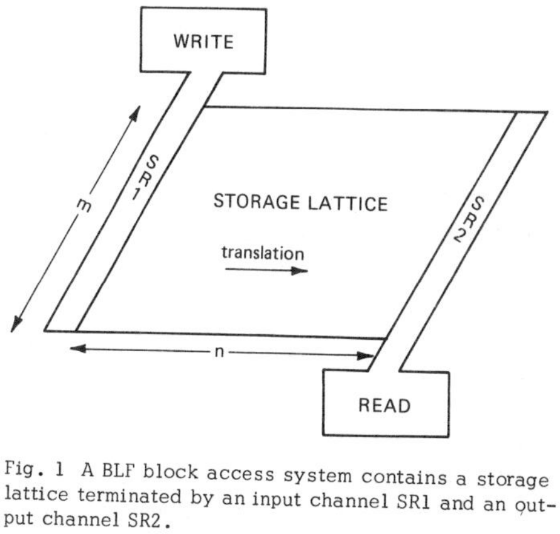

Fig. 1 A BLF block access system contains a storage lattice terminated by an input channel SR1 and an output channel SR2.

through a small modulation of the bias field. The resulting minor radial variation of domain size allows for almost anhysteretic lattice translation.

BLF devices have been designed for block access or for column access. A block access system is schematically shown in Fig. 1. It consists of a storage lattice having m rows and n columns of bubbles, which can be unidirectionally translated. During each translation step, a bubble column is transferred from shift register SR1 into the storage area while at the opposite end a column is transferred from the storage area into shift register SR2. Between translation steps, bubbles are shifted from the write station to fill SR1 while bubbles from SR2 are shifted into the read station. Average access time is roughly $m \cdot n/2f_s$, where f_s is the frequency of shift register operation.

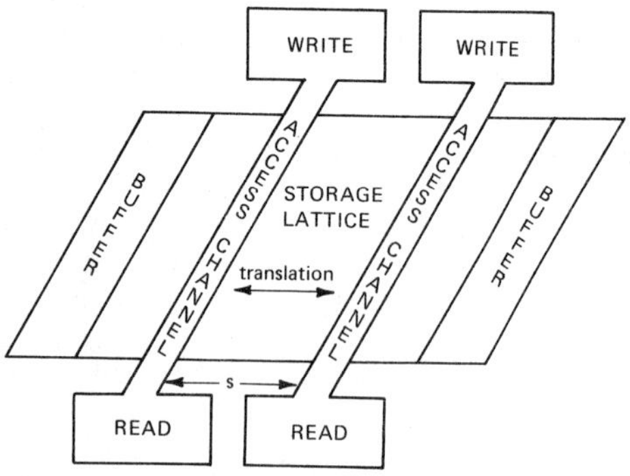

Fig. 2 A BLF column access system contains a storage lattice terminated by two buffer regions. Several access channels may cross the storage area.

A column access system is shown in Fig. 2. It avoids the long access times of the shift-register-like block-access scheme by having several access channels crossing the storage lattice. Each of these channels contains a shift register for propagating a single bubble column, and is terminated at opposite ends by read and write stations. During the operation of a shift register, two isolation conductors are activated to separate the column in the access channel from the rest of the storage lattice.

Data is accessed by translating the lattice either to the left or to the right, until the addressed column is located in the nearest access channel. With all channels separated by s bubble columns, the average access time is $s/4f_l$, where f_l is the lattice translation frequency.

The column access system requires a bidirectional translation distance over s bubble columns. The domain pattern remains invariant with translation as buffer columns are inserted and extracted at opposite ends of the storage area, corresponding to the direction of translation. There are s such buffer columns, which do not contain any information, in both indicated buffer regions. With this approach, it is necessary to generate and annihilate whole bubble columns at the translation frequency f_l.

In an alternative scheme, whose experimental operation is described in a separate paper,[8] the indicated buffer regions consist of an array of m stripe domains parallel to the translation direction. As the lattice is translated, one array expands and the other contracts while the energy of the system remains constant. Stability of both stripe and bubble arrays relies on the energy barrier associated with bubble runout.

Initialization -- In contrast to conventional bubble memories, BLF devices require the initial formation of a rather complex nonvolatile domain configuration. Our present approach for the initialization of the storage lattice and stripe arrays in the buffer regions involves the formation of an array of parallel stripes and the subsequent cutting of these stripes by means of deposited current conductors.[8] The density of domains depends on the initialization process and need not be a minimum energy configuration.

Write/Read Functions -- In BLF devices with wall state coding, these functions to date have been carried out on isolated domains which require bias conditions different from the lattice area. This difference is obtained by using current conductors to locally increase the bias field, or by a reduction of film thickness within the read/write areas.

The write scheme currently being explored[5] for the wall states $S = 0$ and $S = 1$ involves controlling the in-plane magnetization in a thin exchange-coupled layer by means of local fields. The read function requires, first, a discrimination between bubbles with different wall states and, subsequently, their sensing as in conventional bubble memories. In our current approach, the discrimination is based on the difference of deflection angles in a field gradient. After propagating over several bubble diameters, a bubble enters the propagation channel determined by its state. Their presence or absence in respective channels is sensed differentially by means of two magnetoresistive sensors.

SUMMARY

While conventional bubble memories require structural elements such as Permalloy T and I bars, whose width can be no more than about one-eight of the bit separation, the BLF approach does not necessitate any structural elements with a width less than about half the bit separation. For any given limiting resolution in the fabrication process, the attainable BLF bit separation is thus four times smaller, yielding a sixteenfold increase in storage density.

Several differences exist between our current approach to BLF design and conventional bubble memories. The use of current conductors instead of a rotating field for accessing the bubbles will simplify the packaging and substantially reduce the power consumption. However, it requires more connections to the bubble chip and the "on chip" power dissipation will increase. The confined lattices used in the BLF are much more stable than individual bubbles against variations in bias field and, presumably, temperature. The BLF, however, requires stability of both the wall states and the bubble positions.

At present, not enough information is available to say whether, in toto, the operating conditions of a BLF will be more or less restricted than those of T-I bar memories. Current BLF masks have very simple device configurations, largely linear patterns. This is obviously advantageous for device fabrication, but the multiple mask levels required with our present design is a serious disadvantage, especially compared with "single mask level" T-I bar memories.

REFERENCES

1. W.F. Druyvesteyn and J.W.F. Dorleyn, Philips Res. Repts. 26, 11 (1971).
2. J.C. Slonczewski, et al., AIP Conf. Proc., no. 10, 458 (1972).
3. O. Voegeli and B.A. Calhoun, IEEE Trans. Magn. MAG-9, 617 (1973).
4. T.J. Beaulieu and O. Voegeli, "Dynamic Penetration of Potential Barriers in Garnet Films," paper submitted to this conference.
5. Ta-lin Hsu, "Control of Domain Wall States for Bubble Lattice Devices," paper submitted to this conference.
6. R.M. Goldstein and J.A. Copeland, AIP Conf. Proc., no. 10, 383 (1972).
7. J.S. Eggenberger, "Bubble Lattice Translation-Analysis," paper submitted to this conference.
8. L.L. Rosier, et al., "Bubble Lattice Translation -- Experimental Results," Paper submitted to this conference.

ERROR GENERATION IN BUBBLE PROPAGATION

H. B. Callen*, Univ. of Penna., Phila. Pa.
W. D. Doyle & J. Seitchik, Sperry Univac
Box 500, Blue Bell, Pa. 19422

ABSTRACT

The occurence of errors during bubble propagation is described empirically by the relation [1,2]

$$f = A(H_o - H_B) - \epsilon \log n \qquad (1)$$

where f is the fraction of bubbles surviving after n cycles of the rotating field, and H_B is the applied bias field. This linearity in log n holds for a range of circuit parameters which brackets conventional design. The logarithmic dependence on n, at constant H_B, is in remarkable contrast to the exponential decay expected of independent bubbles. We have eliminated a class of statistical models for the n-dependence by measurements of variance, and history effects by direct observations. We infer that the logarithmic dependence is a consequence of bubble interaction. The interaction is of such a form that it can be represented by an effective field. The range is longer than the 17-bubble tracks which we have studied. A random phase approximation treatment of the interaction field leads to a result very similar to the observed logarithmic dependence. We have also corroborated the model by computer simulation. The effective interaction field is ≃ .1 or .2 oersteds. Such an interaction might be mediated by remanent effects in the permalloy overlay.

*Supported in part by the Office of Naval Research.

1. W. D. Doyle, W. E. Flannery and J. A. Coleman, AIP Conference Proceedings #18, p. 152 (1974).
2. P. W. Shumate, P. C. Michaelis and R. J. Peirce, AIP Conference Proceedings #18, p. 140 (1974).

BUBBLE LATTICE TRANSLATION - EXPERIMENTAL RESULTS

L. L. Rosier, D. M. Hannon, H. L. Hu, L. F. Shew, and O. Voegeli
IBM Corporation, San Jose, California 95193

ABSTRACT

One of the primary functions required for a bubble lattice device is the translation of the lattice for accessing information. This paper describes the results of a test device designed to study the translation of a bubble lattice. The design utilizes buffer regions at each end of the lattice. The buffer regions contain parallel stripe domains aligned along the direction in which the lattice is translated. The lattice initialization procedure involves first applying an a.c. in-plane field to form an array of parallel stripe domains. The stripe domains are then cut into bubble domains by conductor lines fabricated on the bubble material. The domains within the active area of the device are isolated from the domains in the surrounding area by selectively ion milling the garnet material. A 12 column X 28 row lattice was quasistatically translated by applying bipolar current pulses to pairs of conductor lines placed so as to provide a driving force on every fourth column of bubbles in the lattice. Using a 2.5 Oe peak, 1 MHz bias modulating field to minimize coercivity effects the minimum current required to translate the lattice was 3 mA which corresponds to a power dissipation in the conductor lines of 165 nW/bubble. The bias field operating margin at 3.5 mA was 10.5 Oe ± 25%.

INTRODUCTION

In order to access information stored in a bubble lattice device[1] the lattice must be translated. In the column access configuration the lattice is translated left or right so as to place the desired column of bubbles within one of the input/output channels. One approach for translating a lattice is the use of a set of conductor drive lines fabricated on the bubble material. In this paper we will report on the results of a test device designed to study the feasibility of this approach.

DEVICE DESIGN

The device design is shown in Fig. 1. A key feature of this design is the use of parallel stripe domains which serve as buffer regions at each end of the lattice. The total energy of the domain configuration, bubble lattice and parallel stripe buffer regions, is invariant with respect to translation of the lattice.

The bubble material used was a $Eu_{0.65}Y_{2.35}Fe_{3.8}Ga_{1.2}O_{12}$ film grown by the LPE process. The pertinent properties of this film are as follows: thickness of 3.3 μm, demagnetized stripe width of 5.1 μm, and a saturation magnetization of 143 gauss. The minimum gradient field required to move a 5.0 μm bubble in this film was measured to be 0.26 Oe/μm.

The lattice and stripe domains within the active area of the device are isolated from the domains in the surrounding area by the structure shown as the speckled region in Fig. 1. This speckled region represents an area of the garnet film which is 0.5 μm thicker than the surrounding area. This topology is produced by ion milling using photoresist to mask the desired area. This process produces a structure in which domains are free to move in the active area of the device but encounter a strong repulsive barrier if they attempt to penetrate the raised region surrounding the active area. This strong repulsive barrier is due to the increase in wall energy as the domain wall penetrates the thicker material.

The conductor drive line design shown in Fig. 1 consists of pairs of conductor lines spaced so as to provide force on every fourth column of bubbles in the lattice. As the bubbles under the conductor lines move, bubble-bubble interaction causes the bubbles located between the pairs of conductor lines to move. The conductor geometry was designed for a lattice having a lattice parameter of 11.6 μm. The spacing between pairs of conductor lines was chosen so as to provide a proper phasing of the lattice and the edges of the conductors when the conductors are connected in series as shown in Fig. 1. Application of a

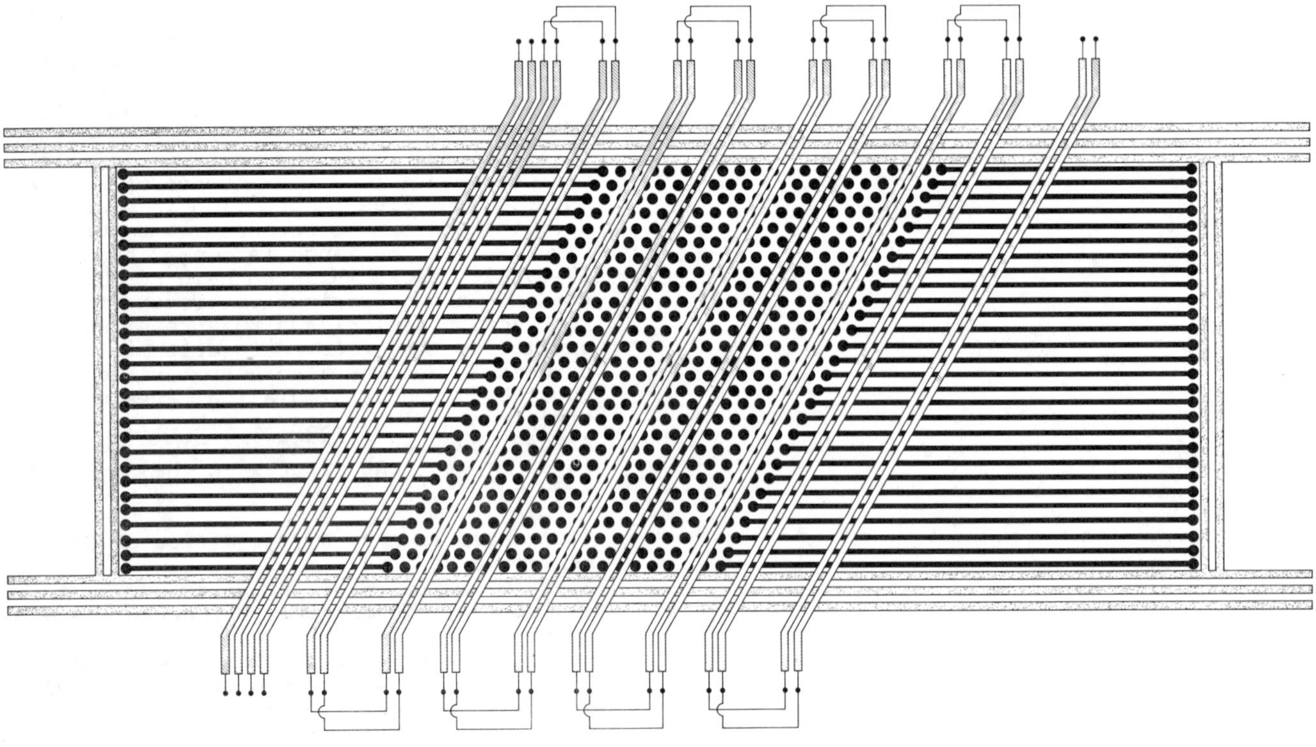

Figure 1. Device design used to study lattice translation. The speckled region is 0.5 μm thicker than the surrounding area. The conductor lines 4.3 μm wide, and the spacing between pairs of conductor lines is 3.3 μm.

bipolar sequence of current pulses causes the lattice to translate a distance of one lattice parameter. The direction of lattice translation may be reversed by reversing the order of the current pulses. The two conductors at the left of the array of conductors in Fig. 1 (the pair not connected in series) are used during lattice initialization. A 1.0 μm Schott glass spacer was used between the garnet film and the 1.0 μm thick gold conductor lines.

LATTICE INITIALIZATION

The domain configuration shown in Fig. 1 may be formed by the following procedure. A 1 MHz, 2.5 Oe peak amplitude bias modulating field is applied normal to the plane of the device. The function of this field is to produce anhysterietic motion which minimizes the effects of coercivity[2]. A 60 Hz, 300 Oe peak amplitude field is then applied in the plane of the device in the direction of the parallel stripe domains shown in the buffer region of Fig. 1. In the presence of an in-plane field the energy of the domain wall is lowest when the domain wall is aligned parallel to the direction of the in-plane field. Thus, the application of the in-plane field produces an array of parallel stripe domains. Irregularities in the stripe domains are forced to the ends of the lattice leaving a defect free array of parallel stripe domains in the active area of the device. The equilibrium spacing between parallel stripe domains is a function of bias field.[3] The density of stripe domains may therefore be varied by applying a bias field during the formation of the parallel stripe domains by the a.c. in-plane field. The density of stripe domains determines the spacing between bubbles in the lattice. Application of a bias field of 15 Oe during the formation of the parallel stripe domains resulted in a domain configuration having 28 parallel stripe domain within the active area of the device. This corresponds to a center-to-center distance between stripe domains of 9.9 μm.

After the parallel stripe domains are formed the a.c. in-plane field and the bias modulating field are removed, and the bias field is increased to 20 Oe. A 100 mA, 100 nS current pulse is then applied to the conductor line at the left of the array of conductors shown in Fig. 1, and each parallel stripe domain is cut into two segments. A 100 mA, 100 nS current pulse is then applied to the adjacent conductor line to cut the tips off of the ends of the parallel stripe domains thus forming a column of bubbles between the stripe domains. The bias field is then reduced to 12 Oe, and the column of bubbles is translated to the right by applying a bipolar sequence of current pulses. A 1 MHz, 2.5 Oe peak bias modulating field is applied during translation to minimize coercivity effects. As the column of bubbles moves to the right the tips of the stripe domains to the left of the bubbles advance under the conductor line. The bias field is then raised to 20 Oe, and the tips of the stripe domains are cut by applying a 100 mA, 100 nS current pulse to the conductor line thus forming a second column of bubbles. The bias field is then lowered to 12 Oe, and this process is repeated until a lattice of the desired size is formed. This technique has been used to form lattices with a capacity of up to 28 columns X 28 rows.

OPERATING MARGINS

In Fig. 2 quasistatic operating margins are shown for a 12 column X 28 row lattice which is translated left and right by a distance equal to ten lattice parameters. A 1 MHz, 2.5 Oe peak amplitude bias modulating field was used during translation. The minimum current required to translate the lattice was 3 mA. This 3 mA current level results in a power dissipation in the conductor lines of 165 nW/bubble. With a conductor current of 3.5 mA the bias field margin as shown in Fig. 2 is 10.5 Oe ± 25%.

The dominant failure mode at high drive currents is associated with the stripe-bubble interface as this interface moves past the conductor lines. When the stripe-bubble interface reaches the second line, the force exerted on the bubble by the conductor line exceeds that of the stripe domain, and the bubble domain is

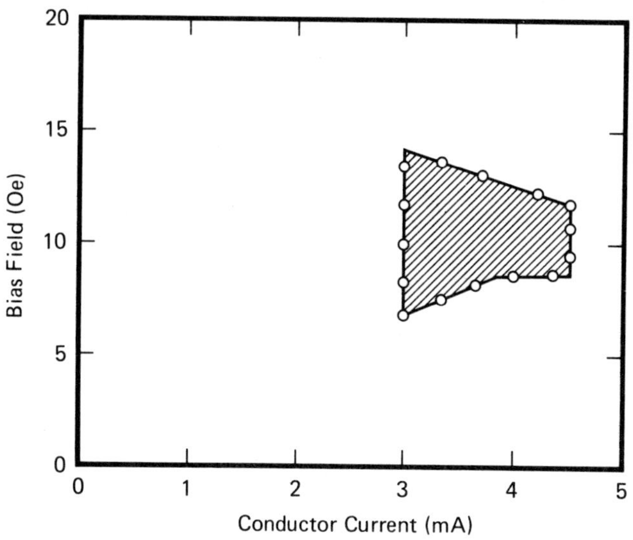

Figure 2. Quasistatic operating margins for the device shown in Fig. 1.

pulled backward as the remainder of the lattice moves forward. Stripe out of the bubble then follows. One means of avoiding this failure mode is to deactivate successive conductor lines as the stripe-bubble interface passes each conductor line. Under these conditions the operating margins are improved considerably. Reliable lattice translation was achieved over a bias field range of 10 Oe ± 50% and a current level of 6 mA ± 50%. This shows that the bubble lattice is inherently a very stable domain configuration. Unfortunately this technique of deactivating successive pairs of conductor lines is not a practical approach since it would require an excessive number of connections to the device. An alternative approach would be to use a third conductor line to provide a force on the bubble during this critical phase of the operation.

CONCLUSIONS

A technique has been developed for initializing a defect free lattice with parallel stripe domain buffer regions at each end of the lattice. This technique has been used for initializing lattices of a capacity of up to 28 columns X 28 rows and should be applicable for much larger lattices. A drive scheme consisting of pairs of conductor lines fabricated on the garnet film has been used to quasistatically translate a 12 column X 28 row lattice. The power required to translate the lattice was 165 nW/bubble.

ACKNOWLEDGEMENTS

The devices used in this work were fabricated by R. L. Anderson, R. Beeck, D. Johnson, T. Montelbano, and D. Saiki. Electronic test equipment was provided by J. S. Eggenberger, M. D. Montgomery, W. A. Reynolds, and P. Swartzle. The authors wish to thank B. A. Calhoun and J. S. Eggenberger for many helpful discussions.

REFERENCES

1. O. Voegeli, B. A. Calhoun, L. L. Rosier, and J. C. Slonczewski, "The Use of Bubble Lattices for Information Storage," paper submitted to this conference.
2. F. A. de Jonge and W. F. Druyvesteyn , Festkörperprobleme 12, 531 (1972).
3. C. Kooy and U. Enz, Phillips Research Report 15, 7 (1960).

BUBBLE LATTICE TRANSLATION-ANALYSIS

John S. Eggenberger
IBM Corporation, San Jose, California 95193

ABSTRACT

To access information stored in a bubble lattice effectively, it is necessary to translate the lattice to move the appropriate bubbles to the detector. One method of accomplishing this is by currents through appropriately spaced conductors on the surface of the bubble material. This paper presents a design for such a lattice translation system, including an evaluation of drive current requirements and power dissipation as a function of lattice translation rate.

INTRODUCTION

Companion papers describe the bubble lattice file concept[1] and experimental results[2] on a lattice translation system for such a file. The translation system is essentially a two-phase current conductor system consisting of two sets of conductors; Fig. 1 shows a diagram of one set. The conductors, of width approximately one-half the lattice spacing, all carry equal current and are spaced by an integral multiple of bubble lattice spacings such that the positioning of each conductor with respect to the lattice is identical. The second conductor set is similar but is spaced from the first set by an odd multiple of one-quarter of a lattice spacing.

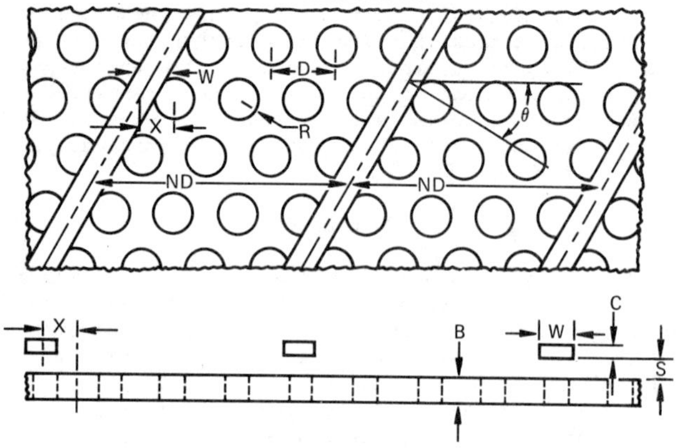

Fig. 1 Plan and cross section views of one set of lattice translation conductors.

Energizing one set of conductors with a current will shift the lattice to align columns of bubbles with the vertical field minima near one edge of the conductors. A reversal of the current causes this stable eqilibrium to become unstable. However, the spacing of the two conductor sets allows the polarity of a current in the second set to determine the direction of lattice motion when the current in the first set is reversed. Thus, application of alternating bipolar currents to each set of conductors causes a translation of the lattice, the direction of translation being determined by the time phasing between the currents in the conductor sets.

In this paper we analyze such a lattice translation system and calculate the expected performance and power dissipation.

ANALYSIS *

To analyze the effectiveness of the lattice translation system for translating a bubble lattice, we make the simplifying assumption that any distortions of the lattice from regular spacing are small enough that the applied field on a bubble can be approximated as the field at its undistorted position.** The force on each bubble is the sum of the applied force due to currents in the conductors and the interactive forces with the other bubbles. If we average the force over all the bubbles in the lattices, the interactive forces cancel. Hence, the average force on bubbles in the lattice is the average of the applied force.

The applied force on a bubble in the ith column due to the current in the jth conductor, $F_{ij}(x)$, can be written as:

$$F_{ij}(x) = H_e(x + iD - jND) \ , \tag{1}$$

where $H_e(\zeta)$ is the effective field produced by a current in a conductor as in Fig. 1 on a bubble whose center is at a distance ζ from the conductor center measured in the direction of translation, and X, R, D, and N are as shown in Fig. 1. For the case of a bubble with constant radius R, H_e can be written as:[3]

$$H_e(x) = \frac{1}{\pi}\int_0^{2\pi} H_z(x + R\cos\theta)\cos\theta \, d\theta \ , \tag{2}$$

where $H_z(\zeta)$ is the field perpendicular to the bubble material produced by a current in the conductor at distance ζ from the conductor center. We determine the average force on the bubbles in the lattice, F(x), by summing Eq. 1 over all the conductors (index j) and averaging over the N bubble colums (index i) in one period of the conductors, giving

$$F(x) = \frac{1}{N}\sum_{i=0}^{N-1}\sum_{j=-\infty}^{\infty} F_{ij}(x) \ . \tag{3}$$

The double summation in Eq. 3 reduces to a single infinite sum when Eq. 1 is substituted, giving:

$$F(x) = \frac{1}{N}\sum_{i=-\infty}^{\infty} H_e(x + iD) \ . \tag{4}$$

* All units are SI unless otherwise indicated.

** Lattice distortion effects result in an alternate translational mode involving a travelling wave of lattice density.

The force, $F(x)$, is periodic in x with period D, and symmetry implies that $H_e(\zeta)$ is an even function. We can then write:

$$F(x) = \frac{f_0}{2} + \sum_{n=1}^{\infty} f_n \cos(nkx) , \qquad (5)$$

where $k = 2\pi/D$ and

$$f_n = \frac{2}{D} \int_0^D F(x) \cos(nkx) dx . \qquad (6)$$

In evaluating the f_n, we substitute Eq. 4 into Eq. 6, reverse the order of summation and integration and recognize that the infinite sum of finite integrals is equivalent to an infinite integral, giving:

$$f_n = \frac{2}{ND} \int_{-\infty}^{\infty} H_e(x) \cos(nkx) dx . \qquad (7)$$

We evaluate the f_n first order for $H'(x)$ being due to a uniform current I in a zero-thickness conductor of width W in the direction of translation and at distance y from a zero-thickness bubble film. We further simplify by approximating $H_e(x)$ as $R dH_z(x)/dx$, giving:

$$f_n = \frac{2IR}{NWD\cos\theta} \left(\sin\frac{nkW}{2}\right) \exp\left(-\frac{nky}{\cos\theta}\right) , \qquad (8)$$

where θ is the angle between the normal to the conductors and the direction of translation. If the more exact form for $H_e(x)$ given by Eq. 2 is used, the f_n are as in Eq. 8 but multiplied by a factor $(2/nkR) J_1(nkR)$, where J_1 is the Bessel function of the first kind, order 1. To account for the nonzero conductor and bubble film thickness, we average this result from $y=y'$ to $y=y'+C$, and average this result from $y'=S$ to $y'=S+B$, where C, S, and B are as in Fig. 1, giving:

$$f_n = \frac{2I}{\pi nNW\cos\theta} \left(\sin\frac{nkW}{2}\right) \left(J_1(nkR)\right) L_n \qquad (9)$$

where L_n, the vertical geometry loss factors, are given by:

$$L_n = \left[\frac{1-\exp(-nkC')}{nkC'}\right]\left[\frac{1-\exp(-nkB')}{nkB'}\right]\exp(-nkS') , \qquad (10)$$

and $C' = C\sec\theta$, $B' = B\sec\theta$, and $S' = S\sec\theta$. The power dissipation in each set of conductors per unit area of lattice due to resistive losses is given by:

$$\frac{P}{A} = \frac{I^2\rho}{CWND\cos^2\theta} , \qquad (11)$$

where ρ is the conductor resistivity.

CALCULATIONS

Using the dimensions reported in the companion paper[2] (e.g., $R=3.5\ \mu m$, $N=4$, $W=5\ \mu m$, $D=11.6\ \mu m$, $C=1\ \mu m$, $B=2.8\ \mu m$, $S=1\ \mu m$, and $\theta=30^\circ$) we calculate from Eqs. 9 and 10:

$$f_1 = 0.049 \text{ Oe mA}^{-1}$$

and higher harmonic terms negligible. For high-frequency operation, each set of conductors translates the lattice from $x=-D/8$ to $x=D/8$ and, for the reverse current, from $x=3D/8$ to $x=5D/8$. Averaging Eq. 4 over these limits gives $\bar{F}=0.044$ Oe mA^{-1}.

We consider a 5 mA drive current and assume that the observed 3 mA threshold is due to a portion of the coercive force not eliminated by the oscillating bias field.[2] Hence, the average excess field applied to the lattice is that corresponding to 2 mA, or 0.088 Oe. For a mobility of 10^3 cm s^{-1} Oe^{-1}, this implies an average translational velocity of 88 cm s^{-1}. Since the lattice must translate by 11.6 μm each cycle, the implication is a maximum translation frequency of 75 kHz. Assuming that the drive currents have a 75% duty factor and that the conductors have a resistivity of 2.2 $\mu\Omega$ cm, the areal power dissipation density during lattice translation is 0.5 W cm^{-2}, consistent with relatively simple cooling.

Scaling all linear dimensions to extrapolate to smaller bubble sizes, we note that the f_n are proportional to $I R^{-1}$ and the areal power dissipation density is proportional to $I^2 R^{-3}$. Thus, the areal power dissipation density is proportional to $\bar{F}^2 R^{-1}$. For constant $\bar{F}$, the power density increases in inverse proportion to decreasing bubble size. However, for constant mobility and coercive force, a constant $\bar{F}$ implies a frequency limit which also increases in inverse proportion to decreasing bubble size. The tradeoff between translation frequency and power density involves the coercive force and mobility, we note that in the limit of low coercive force, the areal power dissipation density is directly proportional to the square of the frequency and directly proportional to bubble size, implying that smaller bubbles can give a lower areal power dissipation for the same performance.

CONCLUSIONS

A two-phase current conductor system for translating bubble lattices has been described and analyzed. Calculations have been presented which indicate that, for 7 μm diameter bubbles, such a system is capable of reasonable operating frequencies (~ 75 kHz) and power dissipations (~0.5 watts/cm^2). Extrapolating to smaller bubble diameters indicates that, if mobility and coercive force are constant, the translation frequency can be inversely proportional to bubble diameter at the expense of a power dissipation inversely proportional to bubble diameter. The ability to trade off performance and power dissipation improves with reduced coercive force.

ACKNOWLEDGEMENTS

I am indebted to T.J. Beaulieu, B.A. Calhoun, D.M. Hannon, and L.L. Rosier for their many helpful comments on this analysis and its presentation.

REFERENCES

1. O. Voegeli, B.A. Calhoun, L.L. Rosier, and J.C. Slonczewski, "The Use of Bubble Lattices for Information Storage," paper submitted to this conference.
2. L.L. Rosier, D.M. Hannon, H.L. Hu, L.F. Shew, and O.Voegeli, "Bubble Lattice Translation -- Experimental Results," paper submitted to this conference.
3. K. Kempter, "Cylindrical-Domain Propagation by Permalloy Bar Stray Fields," IEEE Trans. Magnetics, vol. Mag.-8, Dec. 1972, pp. 746-753.

CONTROL OF DOMAIN WALL STATES FOR BUBBLE LATTICE DEVICES

Ta-lin Hsu
IBM Research Laboratory, San Jose, California 95193

ABSTRACT

One of the approaches to represent a "1" or "0" in a bubble lattice device is by the presence or absence of a Bloch line (BL) pair inside the domain wall. In garnet films with a properly prepared exchange coupled layer, a method has been found by which a bubble with no BL's can be controllably switched to a bubble with one pair of BL's and vice versa. The method involves using a local in-plane magnetic field and a critical domain wall velocity. A model to explain the mechanism is proposed. Both types of bubbles are stable in the film as long as the wall velocity is kept below a certain critical value. Beyond the critical velocity, only bubbles with no BL's are stable. The latter mechanism can be used to switch a bubble with one pair of BL's back to one with no BL's.

INTRODUCTION

The concept of bubble lattice devices[1] outlined in a separate paper proposes a device structure employing closely packed bubbles. Information in the device can be represented by two different types of bubbles.

Bubbles with different wall states are known to exist in as grown garnet films. The wall state of a bubble can be defined[2] by an S number which is the number of revolutions of the magnetization vector along the center of the domain wall, i.e.,

$$S = \frac{1}{2\pi}\oint d\phi$$

where ϕ is the angle between the magnetization vector and some arbitrary direction in the film plane. Bubbles with different but small S numbers all have the same bubble diameters and static properties. However, they propagate at measurably different skew angles in a gradient field. If the magnetization of the bubble is pointing upward from the film the angle is clockwise for positive S numbers and counterclockwise for negative S numbers. The magnitude of the angle is given[2] by the relation

$$\sin\theta = \frac{8SV}{\gamma d \Delta H_z}$$

where V is the velocity of the bubble, γ is the gyromagnetic ratio of the material, d is the diameter of the bubble, and ΔH_z is the gradient field across the bubble.

The existence and distinguishability of bubbles with low S states suggests the possibility of using two of them for representing information in a bubble lattice device. In this paper, we report a method by which a bubble with no BL's can be controllably switched to a bubble with one pair of BL's and vice versa. The method involves using a magnetic layer exchange coupled to the bubble film as commonly used for hard bubble suppression and a local in-plane magnetic field.

EFFECT OF AN EXCHANGE COUPLED LAYER ON BUBBLES

Recent works on suppression of hard bubbles in magnetic garnet films have demonstrated that a magnetic layer exchange coupled to a garnet film suppresses the generation of hard bubbles. However, there have been different reports[3,4] on the type of bubbles that can be generated in a garnet film with an exchange coupled layer and the work of Henry et al.[5] shows that hard bubbles, once generated, actually can exist in such films. Furthermore, it seems difficult to understand why hard bubbles cannot be generated in low Q materials.[6,7]

In our study of the effect of an exchange coupled layer on bubbles in a garnet film we have used three different methods to prepare the exchange coupled layer. They are 1) ion-implantation, 2) a second magnetic garnet layer, and 3) thin permalloy film. The garnet films most often used in our experiments are of nominal composition $Y_{2.35}Eu_{0.65}Ga_{1.2}Fe_{3.8}O_{12}$ grown by the standard LPE technique. The films support 5μ diameter bubbles and have a bubble mobility in the linear region about 1000 cm/sec/Oe. If properly prepared, all three methods of preparing the exchange coupled layer produce essentially the same results. Bubbles generated by a pulse modulated bias field (PMBF) in the absence of an in-plane magnetic field all propagate at an angle to the direction of the gradient field as shown in Fig. 1a. From the sign and the magnitude of the angle, these bubbles are interpreted to be S = 1 bubbles, i.e., bubbles having no BL's in their walls. We have also extended the investigation to other garnet compositions such as SmYIG and $(YSmCa)_3(GeFe)_5O_{12}$ and found consistent results. In the case of SmYIG, the skew angle is small (~5°) due to the low mobility of the material. Based on these observations, we believe that bubbles generated by PMBF in garnet films having an exchange coupled layer all belong to the S = 1 state.

Rosencwaig[3] has proposed a model to explain the effect of suppression of hard bubbles by a second magnetic layer. The model assumes that the magnetic spins in the planar layer on top of a bubble are all aligned in one direction. Due to the exchange interaction, there is a perturbing molecular field acting on the spins in the domain wall to twist them into conforming with the direction of magnetization in the planar layer. Consequently, the presence of a second magnetic layer will result in only one type of bubble in the bubble medium. This bubble contains one pair of BL's in its domain wall and is an S = 0 bubble.

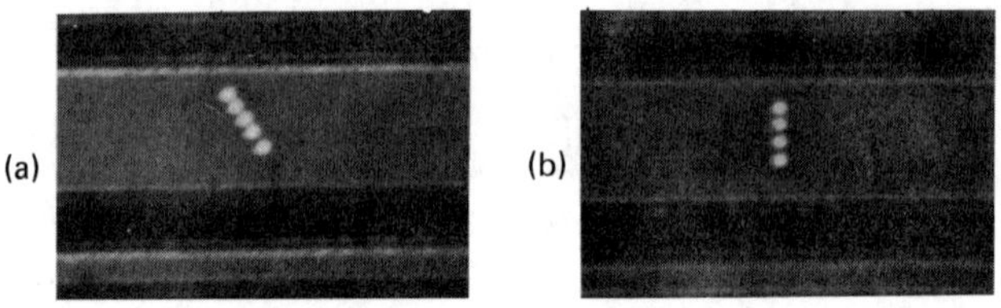

Figure 1. Multiple exposure of an (a) S=1 and (b) S=0 bubble propagating in a gradient drive field.

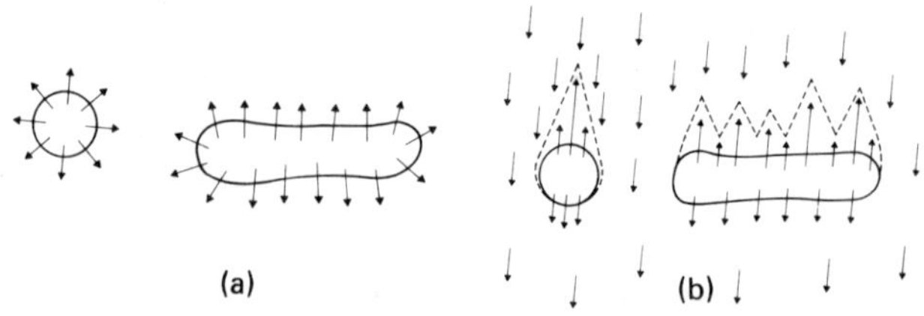

Figure 2. (a) The magnetization distribution on top of a bubble ("umbrella") and a stripe domain ("sea cucumber").
(b) Planar domain wall formed by placing the previous magnetization distribution in a magnetic planar layer.

Rosencwaig's model apparently fails to explain many of the reported results and our observations. Our understanding of the problem at this moment leads us to believe that the effect of the stray field on the exchange coupled layer cannot be ignored as in Rosencwaig's model. We speculate that the magnetization in the exchange coupled layer must be twisted by the stray field into an "umbrella" pattern on top of a bubble and a "sea cucumber" pattern on top of a stripe as shown in Fig. 2a. The magnetization in the rest of the planar layer not affected by the stray field should still follow some uniaxial direction. Combining the

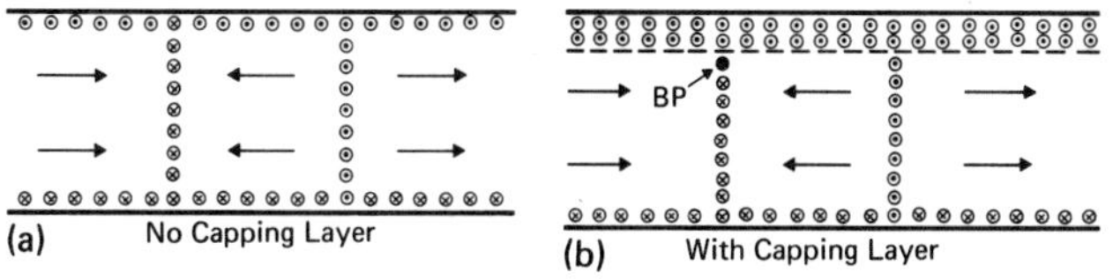

Figure 3. Schematic representation of magnetization in the center of a bubble domain wall that contains one pair of BLs for (a) a film without an exchange coupled layer and (b) a film with an exchange coupled layer.

two kinds of magnetization distribution in the planar layer necessarily creates planar domain walls in the layer. Minimization of the surface pole charges favors a saw-tooth geometry for the planar domain wall as shown in Fig. 2b. This model for the magnetization structure in the planar layer is consistent with the ferrofluid decoration work of Wolfe and North.[8]

A relevant question to ask at this point is why, with an "umbrella" pattern of magnetization distribution in the planar layer of the film, bubbles generated by PMBF contain no BL's. The dynamics of BL's in domain walls is a complicated matter and not as yet well understood. Nevertheless, a possible mechanism for the annihilation of a pair of BL's may involve the nucleation of a Bloch point[9] (BP) on one of the BL's and the passage of the BP from one surface of the film to the other surface. It is possible that a BP may first move part way down the BL and subsequently interact with another BL leading to annihilation of the BL pair. We speculate that in garnet films without an exchange coupled layer, there is normally no BP's on the BL's as shown in Fig. 3a. The twisted wall magnetization[10] near the film surfaces is also shown in Fig. 3a. The "umbrella" pattern of magnetization distribution in the exchange coupled layer on top of a bubble must have the effect of either spontaneously nucleating a BP just beneath the layer on one of the BL's as shown in Fig. 3b or making the nucleation of the BP there easier. Rapid wall motion as generated by PMBF is needed to complete the process of unwinding the BL pair. We therefore postulate that the mechanism of suppression of hard bubbles or the annihilation of BL's in a garnet film involves rapid wall motion of a bubble domain having an "umbrella" pattern of magnetization distribution in the exchange coupled layer on top of the bubble.

The results of Henry et al. concerning the existence of hard bubbles in ion-implanted garnet films can be explained by the present model. Suppression of hard bubbles requires not only an exchange coupled layer but also domain wall motion. Application of PMBF should annihilate the BL's in those hard bubbles. In low Q materials, both surface layers can be twisted by the stray field to resemble the exchange coupled layers. The "umbrella" pattern of magnetization distribution is naturally provided and rapid wall motion annihilates all the BL's.

CONTROL OF DOMAIN WALL STATES

The model proposed in the previous section seems to suggest two things: (1) if the "umbrella" pattern of magnetization distribution on top of a bubble is changed to a different pattern, a bubble with a different kind of wall structure may be generated; (2) if bubbles with a different kind of wall structure are generated, they can exist in films with exchange coupled layer as long as the wall velocity is not large. The "umbrella" pattern of magnetization distribution on top of a bubble can be destroyed by saturating the exchange coupled layer with an in-plane magnetic field. Underneath the saturated layer, there can exist a pair of BL's with their magnetic spins aligned in the same direction as that of the saturated layer. Generation of BP's on this pair of BL's is energetically unfavored. According to the proposed model, rapid wall motion will not be able to annihilate this pair of BL's. This seems to be indeed the case. It has been found that, in an ion-implanted garnet film, bubbles generated by PMBF in the presence of as little as 70 Oe in-plane magnetic field all belong to S = 0 state and propagate in the direction of the gradient field as shown in Fig. 1b.

It has also been observed that an S = 1 bubble can be switched to an S = 0 bubble by simply moving the bubble in the presence of some in-plane magnetic field. The velocity required to make the switch depends on the magnitude of the in-plane magnetic field. At H_{1p} = 70 Oe, a gradient field of 12 Oe across a bubble is needed while at H_{1p} = 130 Oe, hardly any wall motion is needed. When the in-plane field is removed, the bubbles thus generated still propagate in the direction of the gradient field as long as the velocity of the bubble is not large. The controlled generation and switching of bubble states have also been demonstrated in garnet films with an exchange coupled layer prepared by the other two methods. Previously, Smith[6] has reported the propagation of S = 0 bubbles generated by a large DC in-plane magnetic field in a low Q material. The suppression of deflection angles of bubbles by a constant in-plane magnetic field greater than 110 Oe in two ion-implanted garnet compositions has been discussed by Bullock.[4]

The stability of S = 0 bubbles in garnet films with exchange coupled layer varies with films but seems to depend on a critical velocity of the domain wall. In one sample, the S = 0 bubbles are stable as long as the gradient drive field is kept below 8 Oe across the bubble which is equivalent to a bubble velocity of about 1500 cm/sec. Beyond this value, the S = 0 bubble can lose its BL pair and convert to an S = 1 bubble. In our laboratory, the switching of the bubble state from S = 1 to S = 0 has been performed by using a stripe conductor overlay about 25μ wide. A single 100 nsec long and 500 mA amplitude current pulse can provide both the in-plane magnetic field and gradient drive field needed for switching the state of a bubble placed under the center of the conductor stripe. Switching an S = 0 bubble back to an S = 1 bubble requires only a large wall velocity which can be produced by sending a current pulse into a pulsing coil.

CONCLUSION

In garnet films with a properly prepared exchange coupled layer, S = 1 bubbles are generated by PMBF in the absence of an in-plane magnetic field. The S = 1 bubble can be switched to an S = 0 bubble by a combination of an in-plane magnetic field and domain wall motion. Both states are stable in the film as long as the wall velocity of the bubble is kept below a critical value. Beyond the critical velocity only S = 1 bubbles are stable.

ACKNOWLEDGEMENTS

The author wishes to thank L. L. Rosier, B. A. Calhoun, G. Henry, H. L. Hu, O. Voegeli, and J. Slonczewski for many valuable discussions. Thanks are also due to E. Giess for providing the double layered garnet films, to N. Penebre for ion-implantation of garnet films, and to D. Johnson for evaporating permalloy on garnets. The excellent technical assistance of M. Montgomery, W. Reynolds, and D. Saiki is deeply appreciated.

REFERENCES

1. O. Voegeli, B. A. Calhoun, L. L. Rosier, and J. C. Slonczewski, "The Use of Bubble Lattices for Information Storage," paper submitted to this conference.
2. J. C. Slonczewski, A. P. Malozemoff, and O. Voegeli, AIP Conf. Proc. 10, 458 (1972).
3. A. Rosencwaig, Bell. Syst. Tech. J. 51, 1440 (1972).
4. D. C. Bullock, AIP Conf. Proc. 18, 232 (1973).
5. R. D. Henry, P. L. Besser, R. G. Warren, and E. C. Whitcomb, IEEE Trans. Magn. MAG-9, 514 (1973).
6. A. B. Smith, M. Kestigian, and W. R. Bekebrede, AIP Conf. Proc. 18, 167 (1973).
7. M. Takahashi, H. Nishida, T. Kobayashi, and Y. Sugita, AIP Conf. Proc. 18, 172 (1973).
8. R. Wolfe and J. C. North, Appl. Phys. Letters 25, 122 (1974).
9. J. C. Slonczewski, "Properties of Bloch Points in Bubble Domains," paper submitted to this conference.
10. E. Schlömann, Appl. Phys. Letters 21, 227 (1972).

DYNAMIC PENETRATION OF POTENTIAL BARRIERS IN GARNET FILMS

T.J. Beaulieu and O. Voegeli
IBM Corporation, San Jose, California 95193

ABSTRACT

Propagation of bubble domains by means of sequentially pulsed pairs of current conductors provides convenient wall-state discrimination by means of deflection angle. Under certain drive and bias conditions, the final bubble position is found to overshoot the potential minimum located near the pulling conductor. This unexpected result is found to be associated with localized bubble stripe-out along the edge of the pulling conductor Drive-pulse fall time is found to be critical, with no overshoot occurring for fall times in excess of 10 μs. Experiments were done on an LPE film of nominal composition $Eu_{0.65}Y_{2.35}Fe_{3.8}Ga_{1.2}O_{12}$.

INTRODUCTION

Bubbles of differing wall structure can be conveniently identified by means of their deflection angle.[1,2] The S-state of a bubble has previously been defined as the number of 2π rotations of the wall magnetization.[3] This paper is restricted to bubbles having S = +1, 0, -1 for which the deflection angles are unambiguous.

This paper will discuss a series of measurements made on a device consisting of an array of parallel current conductors. The device is used for investigation of deflection angles and for establishing operating margins for single bubble propagation. While observing sequential bubble translation of one conductor spacing per pulse, we observed an unexpected result which we call overshoot. We found that under certain drive and bias conditions the bubble will stripe-out along the edge of the pulling conductor. Upon turn-off of the drive pulse, this collapsing stripe receives a forward kick which results in a final bubble position considerably beyond the pulling conductor.

SAMPLE CHARACTERIZATION

The reported data were all taken on an LPE film of $Eu_{0.65}Y_{2.35}Fe_{3.8}Ga_{1.2}O_{12}$ having thickness 3.54 μm, $4\pi M_s$ = 168 gauss, and characteristic length ℓ = 0.77 μm. Collapse and runout occurred at 59.7 and 42 Oe, respectively, with the bubble elliptically distorted below 44 Oe. One-micron-thick gold conductors of 6.6 μm width were deposited on 15.2 μm centers. A spacer layer of 0.7 μm photoresist was used to reduce stress from the conductors. Nucleation was by means of a pulsed current loop. The overall test area was 250 μm square as defined by a sputter-etched confinement groove.

Figure 1 shows three consecutive conductor lines and the associated vertical magnetic fields for current flow out of the page. Calculated fields[4] are shown for the garnet midplane. Two-conductor propagation consists of current pulses simultaneously applied to C1 and C3, with the bubble initially located at the left edge of C2.

EXPERIMENTAL RESULTS AND DISCUSSION

Operating margins for 4 μs square pulses are shown in Fig. 2 for S = +1, 0, -1 bubbles. Except for the fact that S = -1 bubbles failed first in the region of "insufficient displacement," little dependence upon S-state was observed. Reliable translation means that single pulses simultaneously applied to C1 and C3 caused the

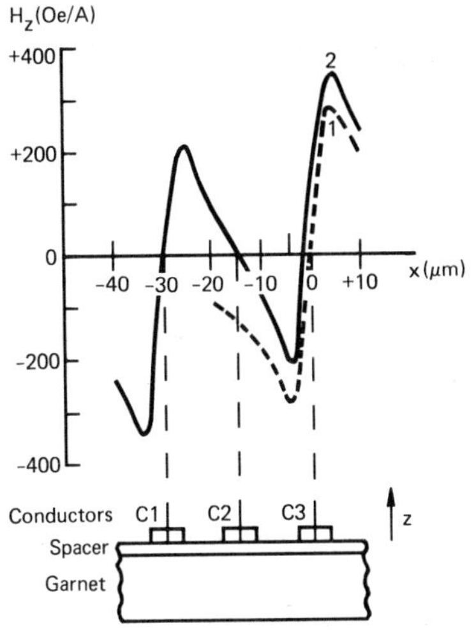

Fig. 1 Normalized vertical field existing at the mid-plane of the garnet. Curve 1: C3 carrying current out of the page. Curve 2: C1 and C3 carrying equal currents.

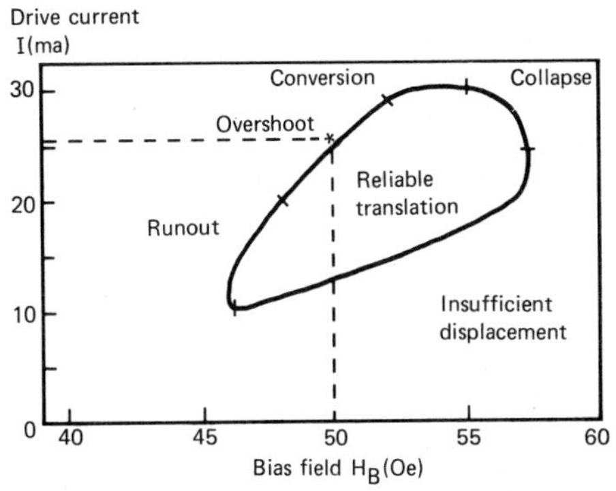

Fig. 2 Operating margins for reliable bubble translation of one conductor spacing per pulse using 4 μs duration square pulses.

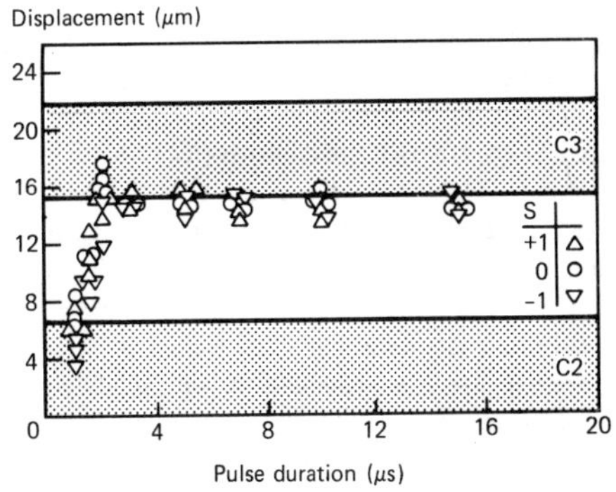

Fig. 3 Two-conductor propagation within the region of reliable translation. H_B = 52.0 Oe, I = 26 mA.

bubble to move from C2 to C3. Collapse occurs at high bias and square pulses because the bubble is initially located near x = -20 μm in Fig. 1. The resulting step change in bias field is consistent with the static collapse field of 59.7 Oe. State conversion is easily identified because it results in a change of deflection angle. For very low bias fields, the bubble experiences runout during translation and remains a stripe even after conclusion of the drive pulse.

Bubble displacement as a function of square pulse duration is given in Fig. 3 for an operating point within the region of reliable translation. The average deflection angles are +27°, 0°, and -22° for S = +1, 0, -1, respectively, with positive angles as defined by Malozemoff.[2] Except for some presumably inertial overshoot for S = 0 bubbles when pulse duration is just sufficient for the bubble to reach the pulling conductor, the final rest positions are as expected. On lowering the bias field from 52 to 50 Oe, the final rest position exhibits considerable overshoot as shown in Fig. 4. While the bias field difference between Fig. 3 and 4 results in different initial bubble diameters (5.7 μm and 6.2 μm, respectively), the following argument shows that this is not the cause of overshoot. Overshoot was observed for H_B = 52 Oe, D = 5.7 μm, I = 29 mA and for H_B = 48 Oe, D = 6.8 μm, I = 20 mA, whereas overshoot was not observed for H_B = 50 Oe, D = 6.2 μm, I = 24 mA. Normalized drives (DxI) for these three cases are 165, 136, and 148 μm-mA, respectively. The data for the latter case has the intermediate value of drive, and yet does not show overshoot. Calculations of the minimum field experienced by the bubble in the vicinity of the pulling conductor show that fields below runout (42 Oe) are attained for operating points along the overshoot boundary of Fig. 2. Josephs has recently reported on velocity scatter in bubble translation measurements whenever any portion of the bubble experiences a field below runout.[5] Assuming localized runout, the dependence of overshoot on pulse duration shown in Fig. 4 indicates a dependence upon stripe-out length. The maximum stripe length is limited to 250 μm by device geometry.

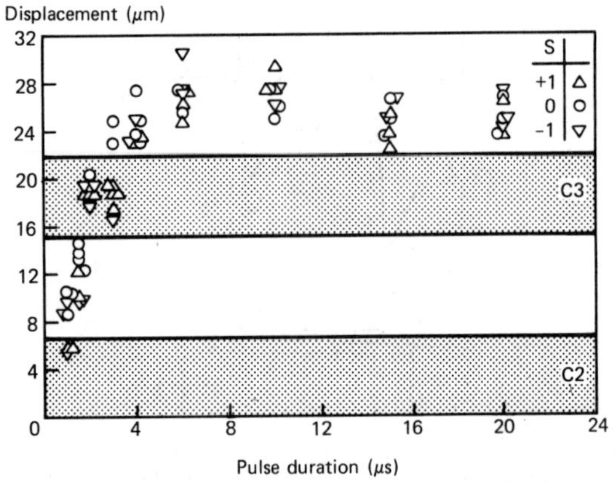

Fig. 4 Two-conductor propagation at the threshold of overshoot (asterisk in Fig. 2). H_B = 50.0 Oe, I = 26 mA. No state conversion.

The data of Fig. 4 was taken at the threshold of overshoot, where the bubble runs out under low drive conditions since the total field is only slightly below runout. Upon pulse turn-off, however, the stripe runs in under high drive conditions because the nominal bias field is 8 Oe above the runout field. To clarify the connection between overshoot and localized stripe-out, the pulse duration was extended to 50 ms. With the bubble initially located at C2 and with the same drive amplitude and bias as used in obtaining the data of Fig. 4, we were able to visually observe stripe-out at C3 and a final bubble position considerably beyond C3. This suggests that the overshoot kick occurs at the trailing edge of the drive pulse; i.e., as the stripe collapses. Identical results were obtained with the bubble initially at rest in the potential minimum near C3, so that inertial mechanisms seem to be ruled out.

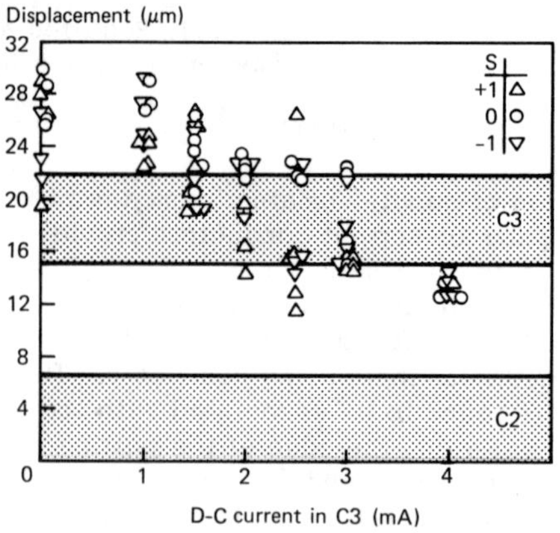

Fig. 5 Two conductor propagation with d-c current superimposed on the pulling conductor, C3. Pulse duration 20 μs, H_B = 50.0 Oe, I = 26 mA.

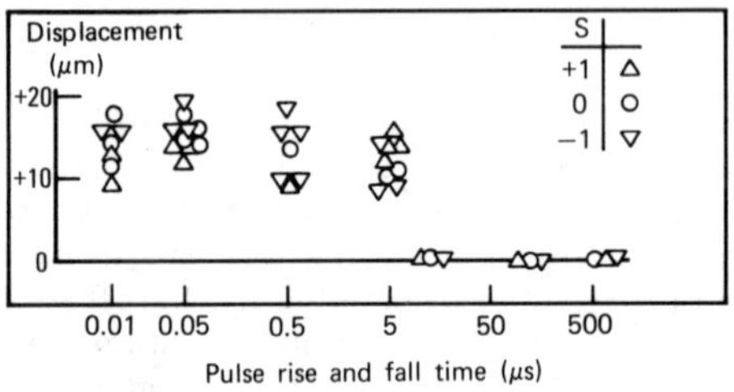

Fig. 6 Dependence of overshoot on pulse rise and fall time. Conductor C3 energized with a single 50 ms duration, 26 mA current pulse. H_B = 51.5 Oe.

A measure of the energy released during the overshoot kick can be obtained by superimposing a d-c current on the pulling conductor as shown in Fig. 5. Overshoot is entirely suppressed for a current of 4 mA.

The dependence of the overshoot kick on pulse rise and fall times (both equal) for 50 ms pulse duration and fixed bias is shown in Fig. 6. Only conductor C3 was pulsed. The bubble was initially located at the position of the potential minimum near C3, with displacement measured from this initial position. The critical rise and fall time is seen to be about 10 μs, with little other dependence. To separate the effects of rise and fall time, several different pulse shapes were tried. A pulse having 600 μs rise time and 5 μs fall time gave the same results as when both rise and fall times were 5 μs. A pulse having 0.5 μs rise time and 100 μs fall time resulted in no overshoot. Suddenly switching off 26 mA of slowly established d-c current gave overshoot similar to that obtained for 5 μs or less transition time. These observations establish the fall time of the drive pulse as the determining factor in overshoot once localized stripe-out is achieved.

CONCLUSIONS

Under the proper drive and bias conditions, a bubble will undergo stripe-out along the attractive edge of the pulling conductor. Upon turn-off of the drive pulse, this collapsing stripe receives a forward kick which results in a final bubble position far beyond the pulling conductor. This overshoot effect is critically dependent upon the fall time of the drive pulse, and appears to be independent of the initial forward velocity of the bubble.

ACKNOWLEDGEMENTS

We wish to thank Ms. J.A. Brown for making the measurements. The authors were aided by numerous discussions with B.A. Calhoun, A.P. Malozemoff, J.C. Slonczewski, and J.S. Eggenberger of IBM, and R.L. White of Stanford University.

REFERENCES

1. J.C. Slonczewski, A.P. Malozemoff, and O. Voegeli, AIP Conference Proceedings No. 10, Magnetism and Magnetic Materials (American Institute of Physics, New York, 1973), p. 458.
2. A.P. Malozemoff, J. Appl. Phys. 44, 5080 (1973).
3. O. Voegeli and B.A. Calhoun, IEEE Trans. Magn., MAG-9, 617 (Dec. 1973).
4. E. Weber, Electromagnetic Fields, Vol. I, (Wiley, New York, 1954), pp. 157-161.
5. R.M. Josephs, Appl. Phys. Lett. 25, 244 (1974).

BUBBLE DOMAIN PROPAGATION MECHANISMS IN ION-IMPLANTED STRUCTURES

G. S. Almasi, E. A. Giess, R. J. Hendel, G. E. Keefe, Y. S. Lin and M. Slusarczuk
IBM, Thomas J. Watson Research Center, Yorktown Heights, New York 10598

ABSTRACT

Experiments on disk and hole patterns in permalloy and in ion-implanted garnet layers[1,2] lead to the conclusion that the bubble domain propagation mechanism in the two structures is different. A "charged-wall" model explains the ion-implanted structure, and a single-domain model relates the minimum propagation field to the garnet's cubic anisotropy energy, in good agreement with experiment.

COMPARISON OF ION-IMPLANTED vs. PERMALLOY DISKS AND HOLES

Figure 1 shows a typical Bitter pattern and its interpretation around a 25 μm diameter hole in an 0.1 μm permalloy film on glass. Similar patterns were observed by Middlehoek[3]. Typically, two or more curved domain walls are tangent to the hole. The magnetization curls around the hole to avoid forming magnetic poles at the edge (Fig. 1b). An in-plane field parallel to the average magnetization causes the walls to bend and shrink, and finally disappear for fields over 25 Oe. When the field rotates, the domain patterns rearrange and become more complex and irregular.

In Fig. 2, the Ferrofluid technique[4] is used to show a typical domain pattern around a 25 μm hole in a layer formed by ion-implanting a 3.58 μm thick $Sm_{.1}Y_{1.92}(Ca,Ge)_{.98}$ iron garnet ($4\pi M_s$ = 160 G) with 4×10^{16} H^+/cm^2 at 50 KeV. Note that the domain walls are not tangent to the round hole, but radiate out from it like propeller blades. The polarity of these two walls (indicated by being lighter or darker than the background) is always opposite. Bubbles always attach to the wall providing favorable flux-matching, as shown in Fig. 2b. In response to an in-plane field, these two walls shorten and line up with the field direction. As the field rotates, the walls rotate around the disk, carrying the bubble along.

Our model for this behavior is different from the "reverse closure domain" model mentioned briefly in ref. (1). We assume the in-plane magnetization flows around the hole much like a slow stream of water flowing around a boulder, forming a diverging wall upstream and a converging wall downstream. Unlike conventional Bloch and Neel walls, the magnetization divergence of these two walls is nonzero, with opposite polarity, and so they are called "charged" walls. This type of wall can exist in

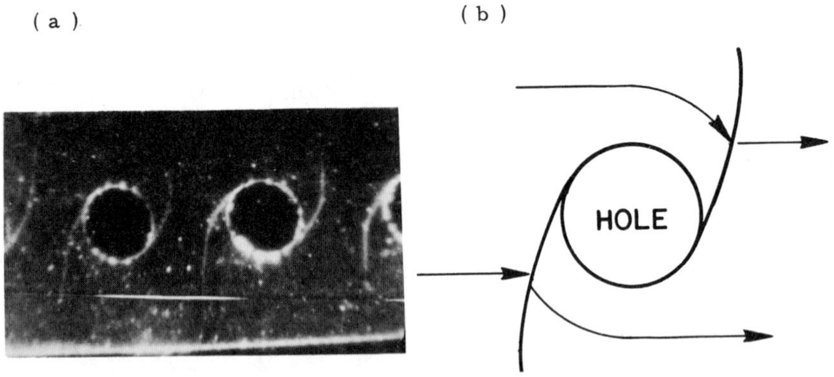

Fig. 1. Bitter pattern (a) and its interpretation (b) around a permalloy hole 25 μm in diameter in the presence of an 5 Oe in-plane field parallel to the direction of magnetization. Permalloy thickness is 1,000 Å.

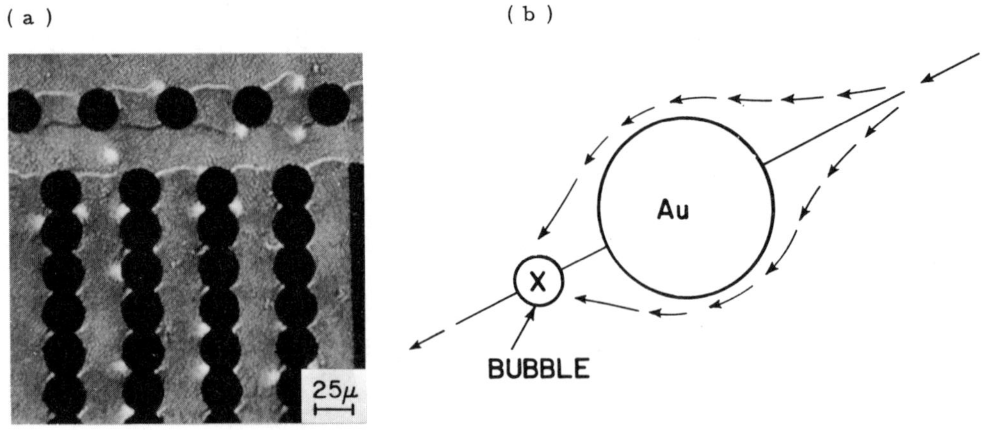

Fig. 2(a) Ferrofluid patterns of domains in an implanted $Sm_{.1}Y_{1.92}(Ca,Ge)_{.98}IG$ garnet film (3.58 μm thick, $4\pi M_s$ = 160 G). Dark areas are masked from the implantation (4×10^{16} H^+/cm^2, 50 KeV). (b) Interpretation of domain pattern around an isolated implanted hole. A 10 Oe in-plane field was applied parallel to the direction of magnetization.

the implanted layer because its low magnetization (160 G) makes the energy of a charged wall comparable to that of conventional Bloch walls. The magnetization of permalloy layers, however, is too large (10,000 G) to support charged walls. If the material around the hole of Fig. 2b were permalloy, the magnetization divergence at the right edge of the hole would create a positive charge and attract the bubble (whose top is negatively charged) to the right side. Because of the charged walls, however, the positive charge is created on the left side, attracting the bubble there. However, for in-plane fields large enough ($0.6 \times 4\pi M_s$) to saturate the in-plane layer, the charged walls are eliminated. Bubbles then jump to the adjacent hole.

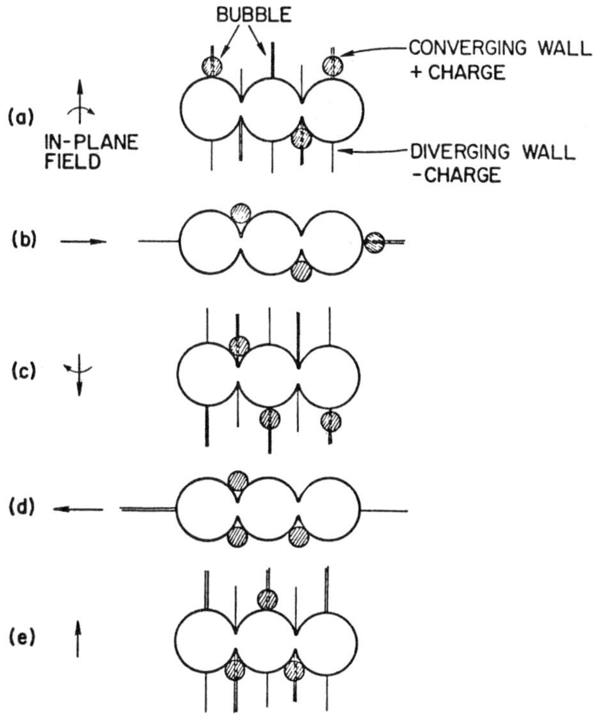

Fig. 3. Sequence of bubble propagation along a contiguous disk structure, non-implanted on the inside. Bubbles are attracted to + charged walls, which propagate along the perimeter of disk following the sequence of in-plane field (20 Oe).

Fig. 3 shows the mechanism of bubble propagation in the contiguous-hole structure of Fig. 2a. For an in-plane field perpendicular to the propagation track, a diverging wall (- charge) is formed where the magnetization meets a convex surface head-on, and a converging wall (+ charge) is formed where the magnetization meets a concave surface head-on (Fig. 3a). The situation is reversed on the other side of the propagation track. As the in-plane field rotates, the charged walls at the wide parts of the pattern move to the narrow parts (the "cusps") and disappear (Fig. 3b), only to reappear on the other side of the track 180° later (Fig. 3c). A 360° rotation completes one propagation step, although the bubble spends most of that time in the cusp. Note that this model explains why the bubble is not repelled from the cusp after it arrives there.

THRESHOLD NUCLEAR ENERGY LOSS DENSITY

In creating an implanted layer, ion species, energy and dosage are the three major parameters. For a given garnet composition and ion species, the dependence of the ion projected range (i.e., depth of penetration) and the nuclear stopping powers on the ion energy can be predicted from the LSS theory.[6] The predicted damage profile created in garnet films agrees reasonaly well with the profile deduced from X-ray measurements.[2] We thus concentrated our experiments

COMPOSITION	h (um)	$4\pi M_s$ (G)	ION SPECIES; ENERGY (KeV); DOSE $(cm)^{-2}$
○ $Sm_{.1}Y_{1.92}(Ca, Ge)IG$	3.58	160	H^+; 50; 2, 3, & 4×10^{16}
△ $Gd_{.6}Yb_{.7}Y_{1.7}Ga_1IG$	4	150	H^+; 100; 2, 3, & 4×10^{16}
X $Eu_{.8}Yb_{.2}Y_2Ga_1IG$	4.3	190	B^+; 100, 150, 200 & 300; 5×10^{14}
	4.1	210	B^+; 200; 2.5, & 10×10^{14}
			150; 10^{14}
□ $Sm_{.95}Yb_{2.05}Ga_1IG$	3.2	320	H^+; 100; 2×10^{16}
	4.5	340	H^+; 100; 3×10^{16}
	4.3	370	H^+; 100; 4×10^{16}

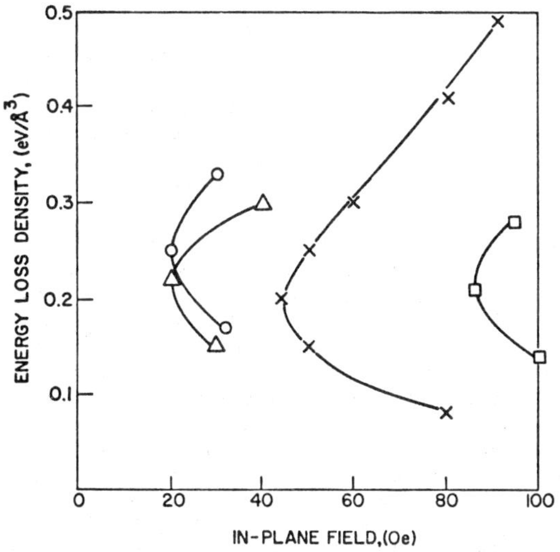

Fig. 4. Minimum in-plane field required to propagate bubbles quasistatically as function of nuclear energy loss density. Propagation circuits are indicated in Fig. 3a.

on varying the ion dosage. Figure 4 depicts the relationship between the nuclear energy loss density [= (ion dosage) x (linear nuclear energy loss)] and the minimum in-plane rotating field required to propagate bubbles. The garnet compositions, material parameters, and implantation parameters are indicated. The propagation circuit is an implanted contiguous disk 25 μm in diameter as shown in Fig. 3a. The ion energy is selected to provide a projected range (i.e., implanted layer thickness) around 0.1 h, where h is garnet film thickness. Note that a minimum in-plane propagation field is always obtained for conditions corresponding to a nuclear energy loss density of about 0.2 eV/Å.[3]

IN-PLANE FIELD vs. K_1

Figure 4 also reveals a relationship between the minimum propagation field and the garnet film composition. Conceptually, to propagate bubbles, the in-plane rotating field should first be strong enough to switch the magnetization of the implanted layer. Unlike the permalloy layer, in which the anisotropy energy is negligible during magnetization reversal, the cubic anisotropy energy is dominant in the implanted layer. To rotate the magnetization in the implanted layer, the in-plane field should overcome the effective cubic anisotropy field.

The magnetization reversal in an ion-implanted layer can be calculated by a single domain model similar to the Stoner-Wohlfarth model used in permalloy films, using energy minimization principles. Using the coordinates of Fig. 5a, in which the x, y, and z axes are the $[1 1 \bar{2}]$, $[\bar{1} 1 0]$, and $[1 1 1]$ crystallographic directions, the total energy "density" in the implanted

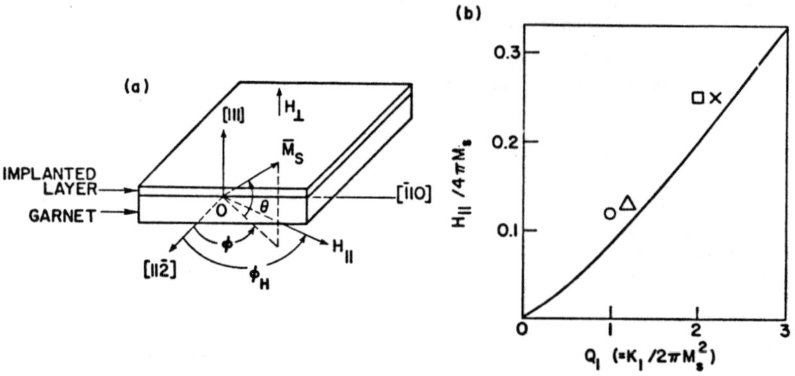

Fig. 5. Single domain model for computing the magnetization reversal in an implanted layer. (a) Coordinate systems and definition of symbols. (b) Threshold in-plane field vs. cubic anisotropy field. Solid line is calculated result (assuming bias field = $2\pi M_s$). Minimum in-plane field obtained from Fig. 4 is indicated for comparison. ($Q_1 = Q/4$ is assumed).

layer consists of (1) in-plane field energy = $- M_s H_{||} \cos\theta \cos(\phi_M - \phi)$, (2) bias field energy = $- M_s H^{\perp} \sin\theta$, and (3) the cubic anisotropy energy = $K_1(1/4 \cos^4\theta + 1/3 \sin^4\theta + \sqrt{2}/3 \cos^3\theta \sin\theta \cos 3\phi)$, where M_s, $H_{||}$, $H_{\perp}$ and K_1 are respectively saturation magnetization, in-plane rotating field, bias field and cubic anisotropy constant. Note that the uniaxial anisotropy (in the implanted layer) is assumed to be negligible (i.e., Q = 0).

The threshold in-plane field value required to rotate the magnetization, $\underline{M}_s$, by overcoming the cubic anisotropy field associated with 3-fold symmetry is plotted in Fig. 5b. (assuming $H = 1/2 \cdot 4\pi M_s$) Note that the calculated values are in reasonable agreement with the experimental results of Fig. 4. The discrepancy is due to the omission of coercive force in the calculation.

ACKNOWLEDGEMENTS

The authors are grateful to R. P. McGouey for the overlay fabrication used in these studies, to W. E. Bogholtz for the mask generation, to Dr. F. F. Morehead and N. A. Penebre for doing the ion-implantation. One of us (Y. S. L.) thanks Drs. A. Hubert and E. Klokholm for helpful discussions.

REFERENCES

1. R. Wolfe et al. AIP Conf. Proc. 10, 339 (1973).
2. J. C. North and R. Wolfe in "Ion Implantation in Semiconductors and other Materials" edited by B. L. Crowder, (Plenum Pub. New York, 1973), p. 505.
3. S. Middelhoek, "Ferromagnetic Domains in Thin Ni-Fe Films", (Ph.D. Thesis, Drukkeriz Wed. G. Van Soest N.V. Amsterdam. 1961). p. 40.
4. R. Wolfe and J. C. North, Appl. Phys. Letters, 25, 122 (1974).
5. H. E. Schiott, Mat. Fys. Medd. Dan. Vid. Selsk., 35, no. 9 (1966).

IMPROVED PROPAGATION MARGIN IN YIG COATED LPE GARNET FILMS FOR BUBBLE DEVICES

Y. Hidaka, K. Yoshimi, T. Hibiya and M. Mikami
Central Research Laboratories, Nippon Electric Co., Ltd.
Kawasaki Japan

ABSTRACT

YIG thin layers grown on $(Y,Eu,Yb)_3(Fe,Ga)_5O_{12}$ LPE films were found to be very effective for improvement of bubble propagation margin as well as for hard bubble suppression. In the ion implanted rare earth substituted Ga:YIG on (111) GGG with 8 micron bubble, T-bar propagation margin was diminished, because of stretching or oscillating of bubble along the patterns. A 600 Å YIG thin layer with in-plane magnetization, grown by CVD at 1065 °C, obviated these destructive shortcomings and guaranteed the minimum driving field for stable bubble propagation down to 10 Oe. This improvement can be attributed to the magnetostatic interaction between the YIG layer and the bubble supporting layer.

INTRODUCTION

Various kinds of garnets for a bubble film have been investigated from the view points of improvement of bubble velocity and of obtaining a higher bit density. One way to obtain garnet with high wall velocity is to diminish the rare earth ion contents. $Y_{2.02}Eu_{0.63}Yb_{0.35}Fe_{3.9}Ga_{1.1}O_{12}$ in the present experiment is favorable to such a purpose. However, in this material, bubbles were not propagated well, but caused propagation errors, because of stretching along the propagation patterns or oscillating in a pattern. These shortcomings might be due to the magnetic properties of a bubble layer, such as q-factor, wall energy or inhomogeneity of constituents along the thickness, but appropriate origins have not been found yet.

In this experiment, the above shortcomings were found to be removed by coating the bubble layer with a thin YIG layer with high in-plane magnetization.

EXPERIMENTAL

The bubble supporting layers were grown on (111) plane $Gd_3Ga_5O_{12}$ substrates by a usual dipping method. From 300 to 2,000 Å thick YIG layers were grown on the bubble layer by chemical vapor deposition at 1065 °C.

This YIG layer was verified to have an easy magnetization axis parallel to film plane due to the large shaped anisotropy by the torque measurement.

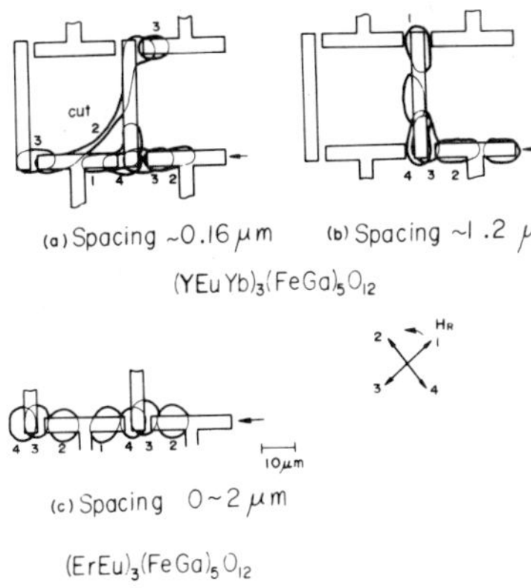

Fig. 1. Propagating bubbles in as-grown garnet films; (a),(b) $(Y,Eu,Yb)_3(Fe,Ga)_5O_{12}$, (c) $(Eu,Er)_3(Fe,Ga)_5O_{12}$. In-plane field H_r = 30 Oe.

Figure 1 shows successive sketches of a bubble in propagation along the patterns on the bubble layer without YIG coating. Depending on the spacing between the pattern and the bubble layer, propagation errors were classified into two categories: i) In the case of small spacing, the bubble did not transfer from position 4 to a successive T-pattern completely, due to strong I-pattern attractive force applied to the bubble, but stretched along both patterns and eventually split into two bubbles, as shown in Fig. 1(a) under a rotating in-plane field. ii) The larger spacing almost prevented bubbles from propagating through the I-pattern to the T-pattern and caused bubble oscillation in the I-pattern, as shown in Fig. 1(b). This suggests that there is inhomogeneity in magnetostatic coupling between the pattern and the bubble which shifts the stable bias field region. On the other hand, in $(Eu,Er)_3(Fe,Ga)_5O_{12}$ film, the bubble smoothly propagated as shown in Fig. 1(c) and ensured a larger propagation margin.

A 600 Å YIG layer grown on $Y_{2.02}Eu_{0.63}Yb_{0.35}Fe_{3.9}Ga_{1.1}O_{12}$ LPE film removed the above propagation errors. Figure 2 shows the improved bias field margin of the YIG coated film, compared to that of H^+ ion implanted film, in which hard bubbles were suppressed as well. Coating with YIG gives a stable bubble propagation region down to 10 Oe in driving field. A similar effect was also observed in films with a YIG layer at the bottom of the bubble layer. In Fig. 2, the other difference in a stable bubble region bias field between on H^+ ion implanted film and on YIG coated film was due to the annealing effect of bubble layer during the growth of YIG.

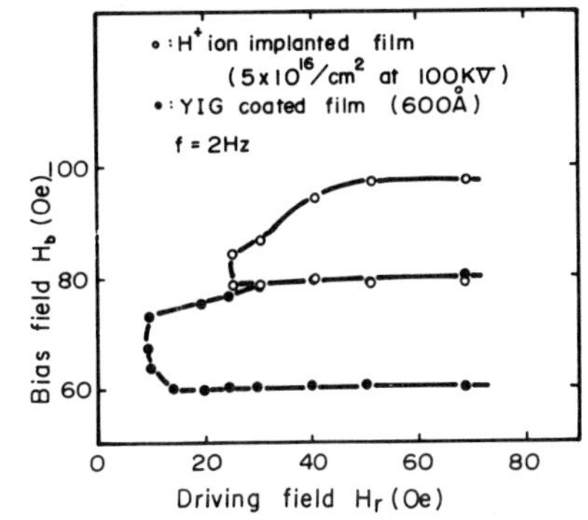

Fig. 2. Bubble propagation margin of YIG coated $(Y,Eu,Yb)_3(Fe,Ga)_5O_{12}$ films.

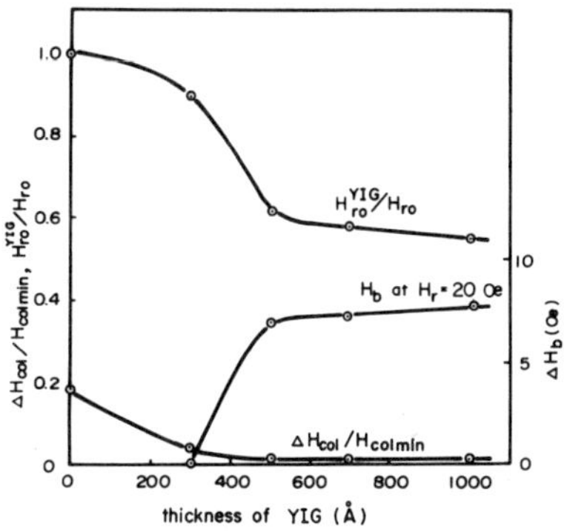

Fig. 3. YIG layer thickness dependence of H_{ro}, H_b and ΔH_{col}.

Figure 3 shows the minimum driving field dependence on the YIG layer thickness, in which H_{ro} and H_{ro}^{YIG} show minimum driving fields in T-bar patterns of the as-grown film and of the YIG coated film, respectively. H_{ro}^{YIG} was drastically lowered in films with a YIG layer thicker than 500 Å, compared to those with a YIG layer thinner than 500 Å. A YIG layer thicker than 2,000 Å, however, disturbed the proper interaction between the permalloy patterns and bubbles, and caused the bubbles to recede quickly, especially at the start or stop of bubble movement. Stable bubble propagation region ΔH_b at 20 Oe driving field was distinctly improved by a YIG coating thicker than 500 Å. $\Delta H_{col}/H_{col}^{min}$ shows the spread of the bubble collapse field reduced by H_{col}^{min}, the minimum collapse field in a given film. The spread of the collapse fields in YIG coating film decreased for a YIG layer thicker than 500 Å. This showed the hard bubble suppression effect of the YIG layer. A hard bubble suppression effect was reported on YGdIG coated films by Henry et al[1]. From these three measurements, the proper thickness of YIG layer was found to be between 500 and 2,000 Å.

In order to clarify the origins for the improvement, composite films comprised of permalloy and bubble layers, separated by SiO, were investigated, so that the interactions between the permalloy layer and the bubble layer were divided into magnetostatic coupling and exchange coupling. The propagation margin was improved in films with a SiO layer up to 1.16 micron thick. For example, 345 Å thick permalloy on 2,500 Å thick SiO diminished the minimum driving field from 32 Oe to 17 Oe and ΔH_b was almost independent of a driving field higher than 17 Oe. However, the improvement effect on propagation margin is more prominent in YIG coated film than in permalloy coated film.

DISCUSSION

Based on these experiments, magnetostatic coupling between the coated layer (YIG or permalloy) with in-plane magnetization and a bubble layer was found to play a very important role in the bubble propagation margin. This interaction is discussed in the light of the radial magnetization model[2]. The frame of this model is due to the magnetic charge distribution with cylindrical symmetry in the coated layer induced by the bubble field. These charges cause the excess magnetic field, which increases the demagnetizing field of the bubble layer, on the bubble domain wall. The average field $\overline{H}_z$, over the bubble layer thickness is analytically expressed, by more simplified calculation than that reported by Lin[3], as

$$\overline{H}_z = -2\pi M_{sp}\frac{t}{h}\ln\left[\sqrt{\left(\frac{2h}{d}\right)^2+1}+\left(\frac{2h}{d}\right)\right], \quad (1)$$

where M_{sp}, t, h and d are the magnetization of a coating layer, its thickness, bubble layer thickness and bubble diameter, respectively. Using $\overline{H}_z$, expressed by Eq. (1), the perturbation, due to the coated layer, to the usual force function, F, derived by Thiele[4], ΔF, is denoted by

$$\Delta F = -\frac{d}{4\pi M_s}\overline{H}z. \quad (2)$$

Here, M_s is the magnetization of the bubble layer. According to Eq. (1), a thin layer with in-plane magnetization, coating the bubble layer, increases the bubble collapse field by $-\overline{H}_z$. This trend, combined with flux matching between the two layers, explains collapse fields dependence on coated layer thickness. The appropriateness of this model is also reflected on the stable bubble region dependence on in-plane field H_r, which deforms the charge distribution of the coated layer. Figure 4 shows the dependence of H_{col} and H_{b-s}, which are the collapse field and the runout field of a free bubble, respectively, on H_r in a permalloy coated film. In lower H_r, both H_{col} and H_{b-s} decrease with increasing H_r. The dependence of the stable region, especially in a lower H_r region, was a contrast to that in a single layer film, shown by open circles in Fig. 4. The decrease of H_{col} in permalloy coated film in lower H_r is explained by the decrease of ΔF, considering the Hr dependence of the bubble diameter. In order words, application of in-plane field reduces the magnetostatic interaction between the coated layer and the bubble layer. This suggests the magnetization saturation of the coated layer to the H_r direction. The frame of the improvement of the bubble propagation due to this additional magnetostatic coupling is under investigation.

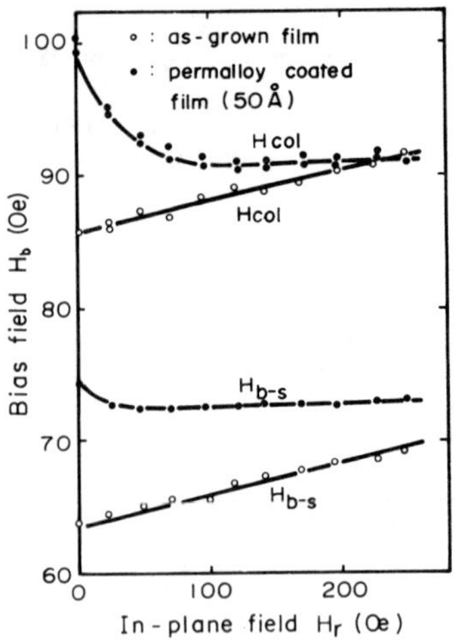

Fig. 4. Stable bubble region in-plane field dependence.

According to above discussions, the magnetostatic coupling between the coated layer and the bubble layer seems to be very effective not in extension of a stable bubble region bias field but in extension of the stable bubble region to the lower driving field.

The wall mobility measured by a bubble transport method[5] increased from 470 to 680 cm/sec·Oe by coating with the YIG layer. The linear portion in the dependence of bubble velocity on field gradient across the bubble was also improved from 3 Oe to 5 Oe by the thin YIG layer. Such effects were also observed in permalloy coated film[6]. This improvement suggests that the coating YIG layer has a profitable effect on the dynamic conversion suppression.

$(Y,Eu,Yb)_3(Fe,Ga)_5O_{12}$ film, coated with a YIG epitaxial layer 600 Å thick, has been used successfully for the 1,232 bit bubble memory chip driving at 100 kHz.

ACKNOWLEDGMENTS

The authors wish to thank T. Okada and T. Furuoya for their encouragement and S. Fujiwara for his valuable suggestions. Thanks also go to K. Suzuki for supplying many LPE garnet films and other colleagues for their encouragement.

This work was, in part, supported by the Agency of Industrial Sience and Technology of Japan.

REFERENCES

1. R.D. Henry, P.J. Besser, R.G. Warren and E.C. Whitcomb, IEEE Trans. Magn., MAG -9, 514 (1973).
2. N. Saito, Japanese Pat. Appl. Public Disclosure No.49-3535.
3. Y.S. Lin, Appl. Phys. Lett. 22, 29 (1973).
4. A.A. Thiele, Bell Syst. Tech. J. 48, 3287 (1969).
5. G.P. Vella-Coleiro and W.J. Tabor, Appl. Phys. Lett. 21, 7 (1972).
6. M. Takahashi, H. Nishida, T. Kobayashi and Y. Sugita, 3M Conf. 1973, 4B-3.

ANALYTICAL MODEL FOR BUBBLE PROPAGATION

G. S. Almasi, Y. S. Lin, E. Munro, and M. Slusarczuk
IBM T.J. Watson Research Center, Yorktown Hghts, NY 10598

ABSTRACT

A simple model for bubble domain propagation in a field-access device is obtained from a new analytical approach which, in contrast to previous treatments, concentrates on the energy of the bubble rather than of the overlay, and emphasizes flux flow rather than field distributions. This leads to a simple magnetic-circuit model for the bubble interacting with an overlay, and to analytical expressions for the trapping potential and effective drive field of permeable overlay features. These expressions are then applied to predicting the quasistatic operating margin (region of operation in the bias field-drive field plane) of a T + I bar device, the inputs being the device geometry and the bubble material parameters. Comparison with experimental results for a garnet device with 5 μm dia. bubbles shows remarkably good agreement.

INTRODUCTION

The design of magnetic bubble domain devices is still largely a cut-and-try process, relying on intuition and experience. This is especially true for field-access devices, in which the bubble is controlled by a magnetic overlay pattern under the influence of an external rotating field. Questions such as "what determines the minimum propagation field?" or "What is the optimum geometry?" are difficult to answer.

True, the stability of an isolated bubble has been treated, and its magnetic field distribution is available in the literature. There are also publications on the field distribution obtained from T-and-I bar permalloy overlay patterns, but most of these are correct only in the absence of bubble domains. The only correct analysis of the interaction between a bubble and a discrete overlay pattern is contained in the series of papers by P. George and co-workers[1..5]. Their analysis allows them to predict the energy well profile presented to the bubble by a permalloy bar, with and without an applied field, and thus allows them to make some useful comments relating device geometry and performance. But their analysis is numerical, with no closed-form expressions; correct answers are obtained for specific situations, but the trends and insights needed for a general, comprehensive design theory are difficult to obtain from these papers. In short, there is presently no simple model which allows the operating bias field and drive field region of a bubble device to be predicted from analytical expressions containing the device geometry and bubble magnetic properties.

Our approach is to consider a field-access bubble device as a magnetic circuit. Rather than focusing on energy and fields in the overlay, as in most previous work, we concentrate on the energy and flux in the bubble. This leads to a simple equivalent circuit for a bubble and eventually to two simple equations which, in conjunction with three graphs, allows the operating margin of a device to be predicted with about 10% accuracy. Because of space limitations, the derivation of the model will be given elsewhere. The present paper is written as a "user's guide" which presents the results of the model, describes their use, and compares predictions with experiment.

POTENTIAL WELLS AND DRIVE FIELDS

When a permalloy bar is brought near a bubble domain, the bubble's magnetostatic self-energy is decreased by an amount ΔE, and the bubble becomes larger. If the bias field is increased to restore the bubble to its original diameter, the bias-field increment H_w is related to ΔE by

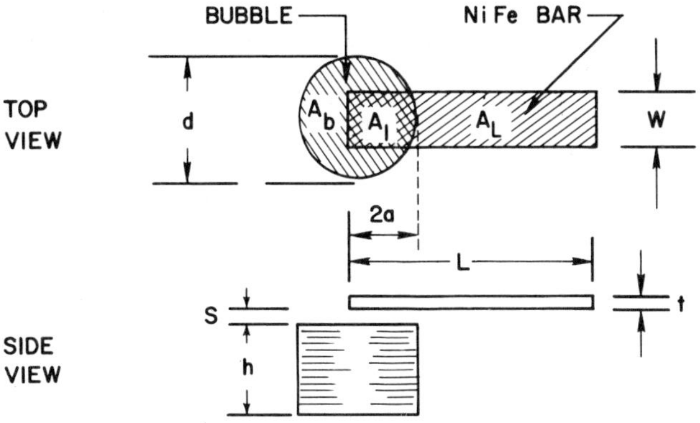

Fig. 1 Bubble, I-bar, and overlap areas.

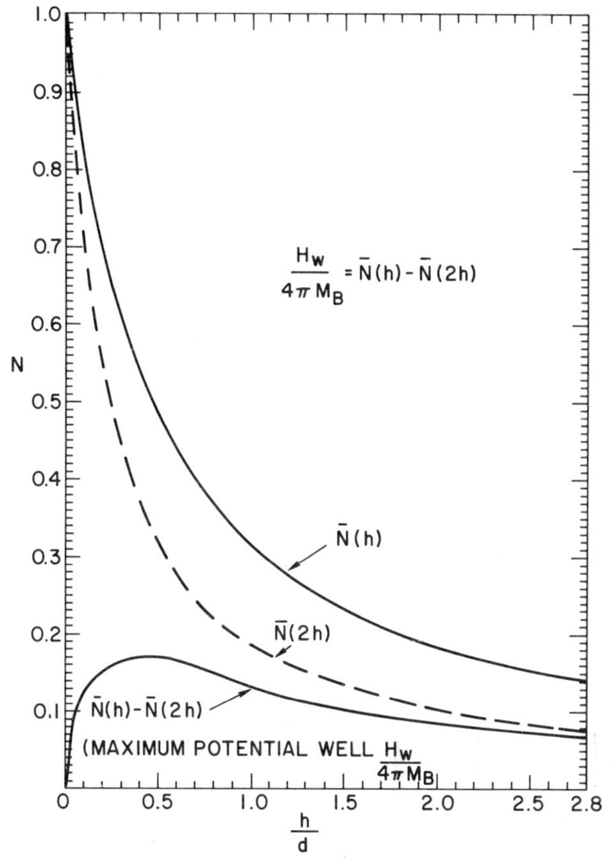

Fig. 2 Volume-averaged bubble demagnetizing factors

$$H_w = \frac{\Delta E}{2\mu_o M_B (\pi/4) d^2 h} \quad \text{(MKS units)} \tag{1}$$

where μ_o is the permeability of free space ($4\pi \times 10^{-7}$ henry/m), M_B is the bubble magnetization, and d and h are its diameter and height, respectively. H_w can be considered as the magnetostatic potential well depth of the permalloy bar,[1] and is a useful way to describe the bar's influence on the bubble. For the geometry depicted in Fig. 1, H_w for the bubble at any spot along the unsaturated bar is

$$H_w(A_1) = \frac{M_B[\bar{N}(h/d) - \bar{N}(2h/d)]}{\frac{A_b}{A_1}\frac{1}{1-A_1/A_L} + \frac{sd}{(R_B\mu_o d)A_1(1-s/d)}} \quad \text{(MKS)} \tag{2}$$

where $\bar{N}(h/d)$ is the volume-averaged demagnetizing factor of a bubble with aspect ratio h/d, and $\bar{N}(2h/d)$ is the same quantity for a bubble twice as tall. Both are plotted in Fig. 2. (For cgs units, M_B in Eq. 2 is replaced by $4\pi M_B$.) The numerator of Eq. 2 is the maximum well depth possible for a single-sided overlay and a bubble with a given h/d, obtained when the bubble is in intimate contact (s=0) with an infinite permeable sheet (overlap area A_1 = bubble area $A_b = \pi d^2/4$, overlay area $A_L = \infty$). $R_B \mu_o d$ is the normalized bubble reluctance plotted in Fig. 3, and varies from 0.13 to 0.26 as h/d increases.

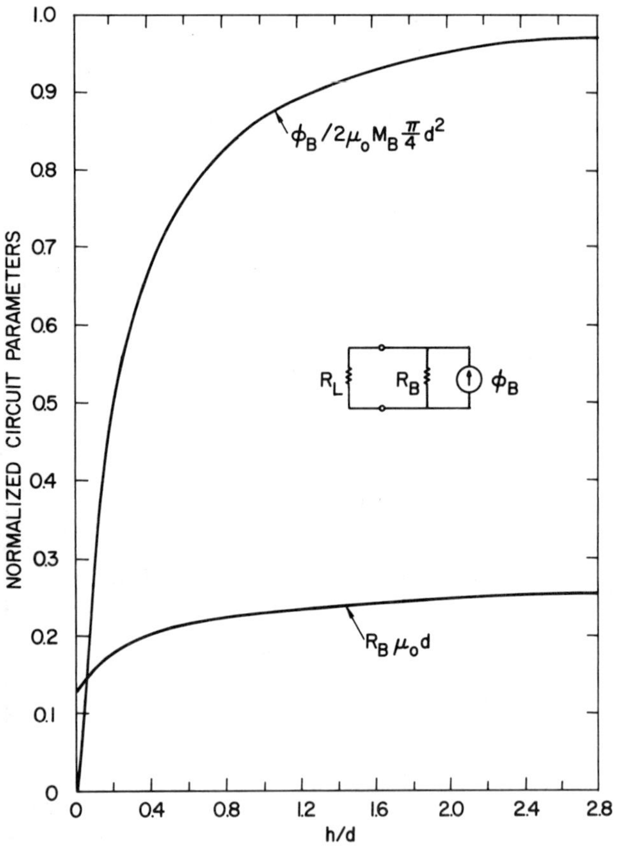

Fig. 3 Parameters of the bubble circuit model

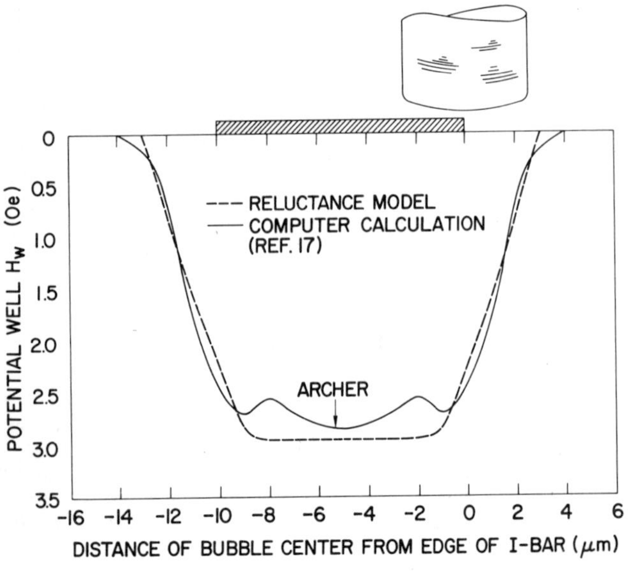

Fig. 4 Potential well profile of a 5 μm dia., 8 μm high bubble ($4\pi M_B$ = 150 G) 0.466 μm below a 10 μm x 2 μm x 0.4 μm permalloy bar.

Fig. 4 compares the profile predicted by Eq. 2 with that calculated in ref. 3 for a bubble with d=5 μm, h=8 μm, as it traverses a 10 μm x 2 μm x 0.4 μm I-bar at a separation of s = 0.466 μm. Agreement is better than 10%. For s/d < 0.5 and h/d > 0.5, eq. 2 also describes the interaction of a bubble with a permalloy disk of arbitrary diameter to about 10% accuracy, which improves for smaller s/d and larger h/d.

If an in-plane field H_{xy} is applied parallel to the bar of Fig. 1, the bubble size again changes, and the bias field must be increased by an additional amount ΔH_B to restore the bubble to its original size. The equivalent circuit model of Fig. 3 relates the flux flowing through the overlay to the bubble's energy change. Making an ellipsoidal approximation of the I-bar and performing flux matching, it can be shown that, for s/L < < 1,

$$\frac{\Delta H_B}{H_{xy}} = \sqrt{\frac{(2a + t)}{L/2 + t} - \left(\frac{a + t}{L/2 + t}\right)^2}\ \frac{wL}{\pi d^2 N'_p}\ \frac{\bar{N}(h/d)}{(L/w)} \left(1 - \frac{Cs}{L}\right) \quad (3)$$

where $C = 0.5 \ln(8L^2/wt)$ is a constant usually about 4 for practical I-bars, and $N'_p(L/w)$ is the permalloy bar's demagnetizing factor, plotted in Fig. 5.

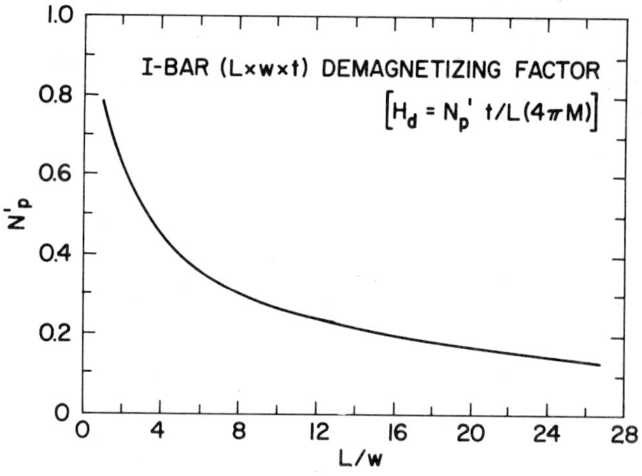

Fig. 5 I-Bar demagnetizing factor

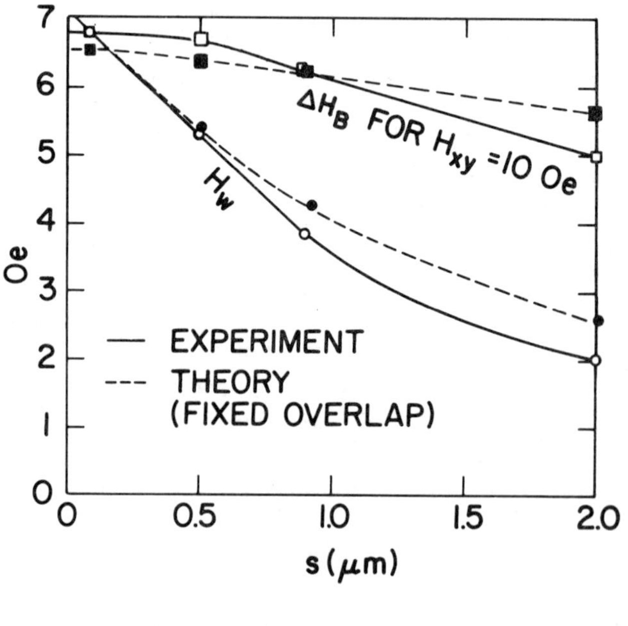

Fig. 6 Comparison of the bubble model's prediction with the experiments of ref. 6 on the spacing dependence of well depth and drive field. Bubble: $4\pi M$ = 115 G, d = 7.5 μm, h = 7 μm. Permalloy bar: 40 μm x 4 μm x 0.9 μm. In-plane field: 10 Oe. Overlap assumptions: $A_1 = 24\ (\mu m)^2$, a = 2.8 μm.

A comparison between the experiments of ref. 6 and the prediction of Eq. 3 (as well as Eq. 2) is shown in Fig. 6. It can be seen that both the potential well and drive field of the bar are predicted quite well.

PREDICTING THE OPERATING MARGIN

So far we have described the potential well and drive field experienced by a bubble interacting with a single permalloy bar. An assembly of bars, such as a T-I propagation track, can be described by superimposing the results for individual bars. This is sketched qualitatively in Fig. 7, based on the single-bar results of ref. 1 and 3 and Fig. 4.

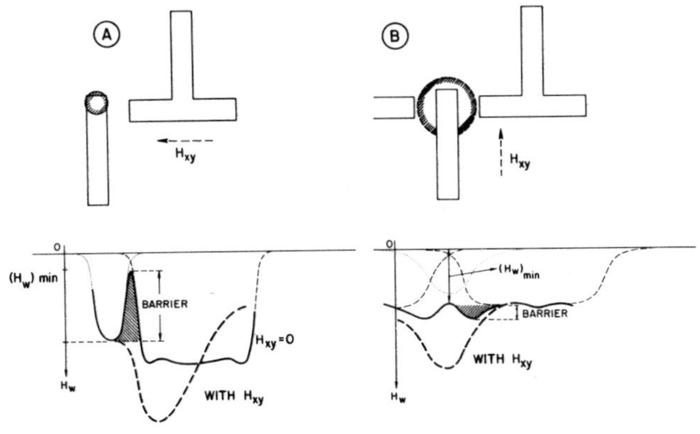

Fig. 7 Potential well for minimum bubble diameter (just before collapse) and maximum bubble diameter (just before runout) for a T-I bar propagation track with and without in-plane field.

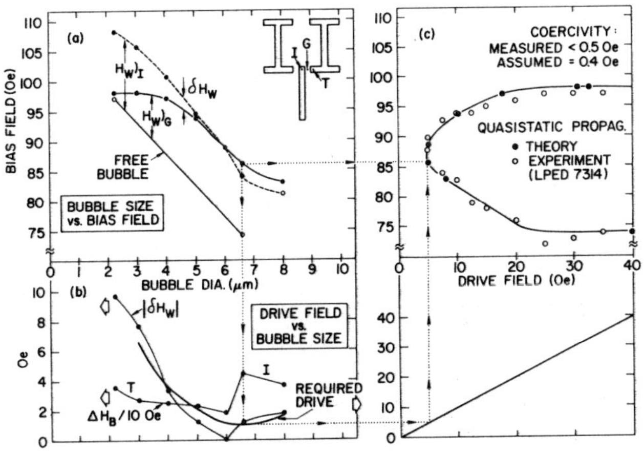

Fig. 8 Derivation of a T-I bar device margin and comparison with experiment. Garnet: $Eu_{.6}Y_{2.4}Fe_{3.9}Ga_{1.1}O_{12}$, h = 3.99 μm, stripwidth = 4.19 μm, collapse field = 96.9 Oe, runout field = 74.0* Oe, 4πM = 199* G, collapse diam. = 2.2* μm, runout diam. = 6.6* μm, coercivity H_d = 0.5 Oe. (* indicates calculquantities.) Permalloy: t = 0.2 μm, w = 3.0 μm, gapwidth G = 2.0 μm, I-bar length = 23.5 μm, T crossbar length = 13.5 μm, s = 0.2 μm.

At high bias, when the bubble is small, there is little overlap between the potential wells of the individual bars, and in the absence of an in-plane field, the gap between bars presents a large potential barrier (or hill) and the bubble prefers to sit on a bar (Fig. 7a). At low bias, the bubble is large, there is a large overlap between the individual wells, and the bubble prefers to sit in the gap (Fig. 7b). In both cases, a relatively large in-plane field is needed to depress the maximum or raise the minimum of the potential profile by an amount sufficient to move the bubble to the next position. But it seems that there should be some intermediate bias at which the undulations in the potential profile become minimum (perhaps zero), resulting in the minimum in-plane field for propagation.

Equations 2 and 3 are valid for a bubble interacting with multiple as well as single bars, so they can be used to put the above argument to a quantitative test. This is done by deriving the quasistatic operating margin of a real T-I bar device in the sequence of Fig. 8. The track is inset in Fig. 8a and the device parameters are given in the caption. At high bias, the margin-limiting step is assumed to be going from the I-bar (I) to the T-bar (T) across the gap (G). At low bias, the margin-limiting step is assumed to be going from G to I. The crossover occurs when the in-plane field required for both processes is the same.

The first step in Fig. 8a is to draw the free-bubble diameter-vs-bias-field (d vs H_B) curve. The straight-line approximation used here between collapse and runout is a reasonable approximation to the true curve.[7] Next, Eq. 2 is used to calculate H_W at I and G and to thus construct the d-vs-H_B curve for I and G. For a given d, the difference between these curves gives the potential hill height or valley depth δH_W which must be eliminated by H_{xy}, the in-plane field. δH_W is plotted vs. d in Fig. 8b.

Next, Eq. 3 is used to calculate the potential change ΔH_B caused at I and G by H_{xy} = 10 Oe (a value chosen purely for convenience). This value of ΔH_B is also shown in Fig. 8b. Quasistatic propagation occurs when ΔH_B exceeds δH_W and the coercive force term $(8/\pi)H_c$, which in this case is 1 Oe. The value of minimum H_{xy} determined this way is the third curve of Fig. 8b; it is read from the <u>right</u> vertical scale, whereas the two previous curves are read from the <u>left</u> vertical scale.

We now have H_{xy} vs. d, and it remains to relate d to H_B. Since failures usually occur in the gap, the $H_W)_G$-vs-d curve of Fig. 8a is used. The margin can now be derived point-by-point: for example, at d = 6.6 μm, the $H_W)_G$ curve of Fig. 8a gives a value of H_B = 86 Oe, while the "required drive" curve of Fig. 8b gives a corresponding value of H_{xy} = 5 Oe. Other points are derived in the same way, except at low bias, where the circular bubble assumption no longer holds. In this region (d > 8 μm), the curve derived for smaller d's is extrapolated in a straight line for H_B < 83 Oe, H_{xy} > 8 Oe until this line crosses the free-bubble runout line H_B = 74 Oe, after which the margin is assumed level. Fig. 8c compares this predicted margin curve with the experimental values obtained for this device, and the agreement is surprisingly good.

In conclusion, we have described a simple procedure which can be used to predict as well as to design and optimize the operation of field-access bubble domain devices.

REFERENCES

1. J. L. Archer et al., IEEE Trans. Mag. <u>8</u>, 695-700, Sept. 1972.
2. P. K. George and T. T. Chen, Appl. Phys. Lett. <u>21</u>, 263-264, Sept. 1972.
3. P. K. George and J. L. Archer, J. Appl. Phys. <u>44</u>, 444-448, Jan. 1973.
4. P. K. George and J. L. Archer, AIP Conf. Proc. <u>18</u>, 116-120, 1974.
5. P. K. George, A. J. Hughes, and J. L. Archer, IEEE Trans. Mag. <u>10</u>, 821-824, Sept. 1974.
6. M. E. Jones and R. D. Enoch, IEEE Trans. Mag. <u>10</u>, 832-835, Sept. 1974.
7. A. A. Thiele, Bell Sys. Tech. J. <u>50</u>, 725-773, Mar. 1971, Fig. 8.

AN EXPERIMENTAL TECHNIQUE FOR CHARACTERIZING LOCAL VARIATIONS OF THE VERTICAL MAGNETIC FIELDS IN BUBBLE CIRCUITS

S. K. Singh and W. C. Hubbell, Texas Instruments Incorporated, P.O. Box 5936, MS 145, Dallas, Texas 75222

ABSTRACT

A new experimental technique for characterizing the local fields associated with bubble propagation and control function circuit elements is presented. In this method, measurements of the local variations of bubble collapse field due to the circuit elements are made both statically and during high speed propagation. Short duration (∿ 1 μsec) increments of the bias field are applied synchronously with the drive field so that the effective bias field is temporarily increased as the bubble traverses a given location in the propagation circuit. By observing the minimum pulse amplitude necessary to produce collapse, the effective bias field can be determined. Such measurements are indicative of the strength of the local fields produced by the Permalloy elements.

INTRODUCTION

The motion of magnetic bubbles in a field access propagation circuit is produced by the space and time periodic magnetic fields resulting from the interaction of a rotating in-plane field with Permalloy circuit elements. The magnitude and spatial variation of these fields are dependent on element geometry and the strength of the rotating drive field. Progress in circuit element design has resulted largely from a combination of intuition and empiricism rather than analytical understanding.

Experimental techniques for characterizing circuit elements have been described. One method reported by George and Chen[1] involves the direct measurement of the vertical field component due to the Permalloy. In their technique the effective vertical field component is obtained from the difference between two measurements of the applied bias field - one for a bubble positioned under a Permalloy element, the other for a bubble of the same diameter in a Permalloy free region. For rectangular bar elements, the measurements are in reasonably good agreement with calculations based on a two-dimensional model.[2] In addition to the rectangular bar geometry, cheveron, Y-Bar and T-Bar elements have also been examined by this technique.[1,3] These measurements have been used to study the effects of drive field strength [1,2,3] as well as overlay-to-garnet spacing.[3]

A limitation of this technique has been its restriction to static situations where the bubbles are stabilized in fixed positions. Knowledge of dynamic behavior could, however, be useful in understanding the effects of high speed propagation on the bias margins. Another problem with the technique is the difficulty experienced in accurately measuring the diameter necessary for a determination of the vertical component of field. Further difficulties are encountered due to elliptical distortions of bubbles at low bias fields and the difficulty of observing bubbles at high bias fields where the bubbles are small.

An alternative technique for characterizing the vertical field component is presented here. In this method the difference between the applied bias field, H_o', necessary to collapse a bubble at a specific position under a Permalloy element and the free bubble collapse field, H_o, are used to deduce the maximum value of the vertical field (averaged across the bubble), ΔH_z, produced by the overlay element at that position. Thus, ΔH_z is effectively the difference in bias fields necessary to stabilize a free bubble and a trapped bubble at the collapse diameter. Since the vertical field components of the overlay elements have been shown to increase as the bubble diameter decreases,[1,2] the values measured by the collapse technique correspond to the maximum potential well depths. Using this collapse technique enables one to measure ΔH_z at quasistatic as well as high frequencies. Such measurements are useful in comparing different element designs, the effects of different drive fields and of different overlay-to-garnet spacings.

EXPERIMENTAL RESULTS

Static Measurements. Measurements of the applied bias field H_o' necessary for collapse were made on bubbles at various positions within a single period of a T-Bar propagating circuit. The positions were determined by the orientation of the in-plane drive field, H_D, which was changed by increments of either 18° or 36° for consecutive measurements. Data are presented in Fig. 1 for a bubble moving from left-to-right and from right to left. The effective vertical field due to the Permalloy elements, $\Delta H_z = H_o' - H_o$, is also indicated. The measured values corresponding

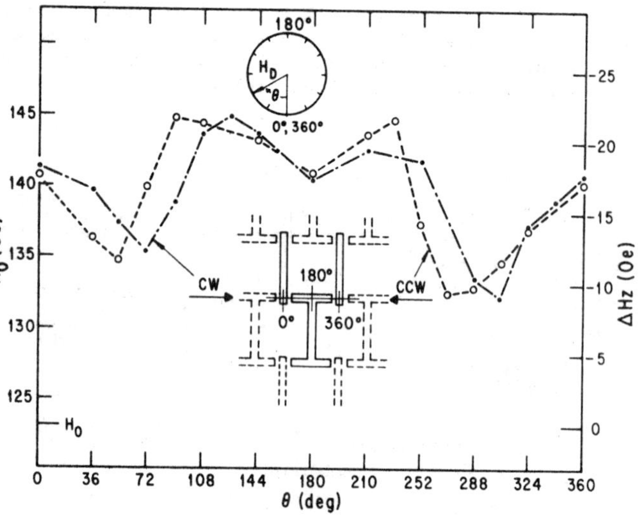

Fig. 1. The static collapse fields for bubbles under Permalloy elements, H_o', as a function of drive field orientation, Θ, for one period (20 μm) of a T-Bar circuit. The clockwise rotation (cw) corresponds to bubbles propagating from left-to-right while the counterclockwise rotation (ccw) is associated with the reverse motion. $\Delta H_z = H_o' - H_o$, H_D = 35 Oe, H_o = 123.2 Oe.

to bubble positions at the tips of the bar elements, Θ = 0°, 360°, are ∿ 17 - 18 Oe and are comparable with similar values observed by other authors.[2] The results reported here were obtained for circuits on $(YSm)_3(FeGa)_5O_{12}$ films with 4πM ∿ 213 G and H_o = 123.2 Oe. The sample was ion-implanted with Ne^+ at 100 keV to a dosage of ∿ 2 x 10^{14} ions cm^{-2}. Permalloy elements were sputtered (4,000 - 5,000 Å) over a 1 μm silicon oxide spacer and had a coercivity of less than 2 Oe. The nominal dimensions of the bar elements in Fig. 1 (insert) are 2.5 x 25 μm. The gaps are 1.25 μm.

An important feature of the curves in Fig. 1 is the pair of minima observed near $\Theta = 63°, 288°$. These valleys, corresponding to bubble positions in the gaps between elements, determine the upper bias margin for propagation, since the maximum applied bias field compatible with propagation cannot exceed the lowest value for which bubbles collapse within a circuit. Also noteworthy is the fact that angular orientations of the drive field for which these minima occur are not symmetric about $\Theta = 180°$. This indicates that the presence of a bubble perturbs the symmetry of the magnetic pole distribution in such a way that the motion of a bubble across the gap from the tip of the bar to the "T" is not equivalent to the reverse motion. This lack of symmetry was observed for both directions of propagation.

Dynamic Measurements. Dynamic collapse field measurements were made on the circuit elements used for static measurements. A bubble is propagated at 100 kHz through the circuit containing these elements and one complete cycle (10 μsec) of propagation is examined in detail. After the dc bias field is set to a given level within the operating margins, an auxiliary coil is pulsed to provide a temporary (1 μsec) increase in the net vertical field. The pulse rise time is < .2 μsec. The timing of the pulse is selected to correspond to various orientations of the 100 kHz drive field. If the instantaneous value of the net field during the pulse is less than H_o, then the continued existence of the bubble is readily observed after controlled shutdown of the drive field. By repeating this procedure with increasing net vertical field, the minimum value H'_o necessary to collapse a bubble at a given position can be determined to within ± 0.5 Oe. The contribution due to Permalloy $\Delta H_z = \overline{H}'_o - H_o$ is then determined for the given orientation of the drive field. The temporary increase in the total field can be achieved by adjusting the pulse height and/or raising the dc bias field. It has been verified that the free bubble collapse field as determined by a 1 μsec field pulse is the same as that obtained from static measurements and that measured values for H'_o are independent of pulse width for pulse widths ≥ 1 μsec.

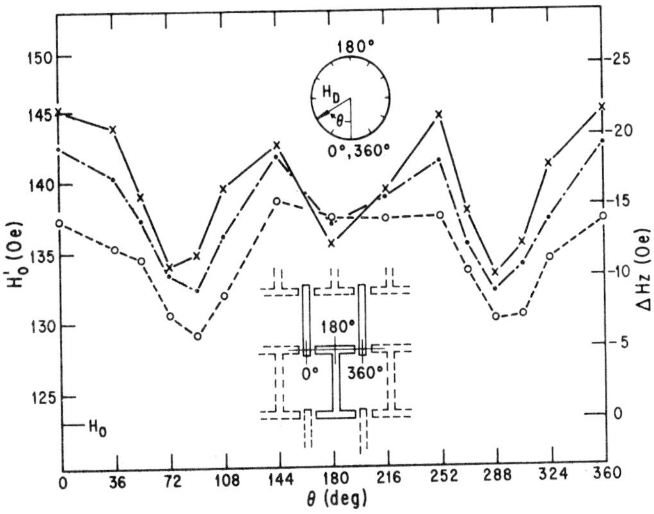

Fig. 2. The collapse fields for bubbles under Permalloy elements, H'_o, at 100 kHz as a function of drive field orientation, Θ, for one period (20 μm) of a T-Bar circuit. $\Delta H_z = H'_o - H_o$, $H_o = 123.2$, $H_D = 45$ Oe x, $H_D = 35$ Oe •, and $H_D = 25$ Oe o.

The data are plotted in Fig. 2 for three values of drive field. The main features of these curves are in good agreement with the previous static measurements (Fig. 1). It is seen that the upper propagation margin, corresponding to the minimum applied collapse field, increases with increasing drive field which is in general agreement with the results of others.[4] A degree of positional assymetry of the gap minima is also evident in the 100 kHz data, although it appears that the larger drive fields tend to reduce this effect. Another interesting feature of this data is that the three curves cross near $\Theta = 180°$, i.e., ΔH_z decreases as H_D increases. Such behavior at high drive fields has been previously observed for static measurements on bubbles trapped at the junction of the vertical and horizontal members of "T" elements.[5] On comparing the static measurements with those for 100 kHz, it is seen that the differences are mainly in the fine details.

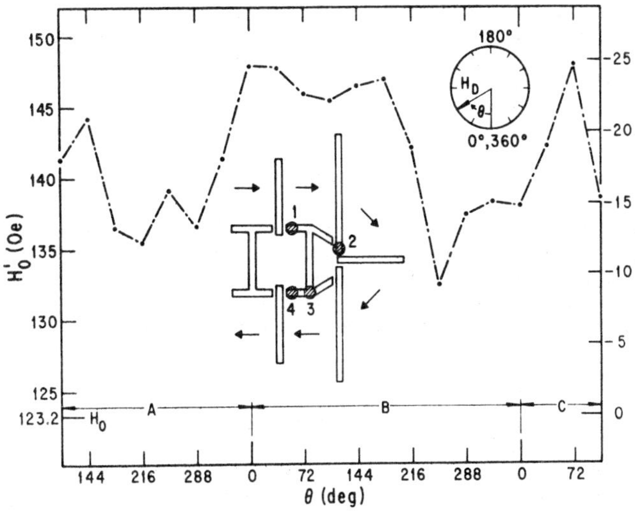

Fig. 3. The collapse fields for bubbles under Permalloy elements, H'_o, at 100 kHz as a function of drive field orientation, Θ, for the corner of a T-Bar circuit. Sections A, B and C correspond to the approximate bubble movement from positions 1 to 2, 2 to 3, and 3 to 4, respectively. $\Delta H_z = H'_o - H_o$, $H_o = 123.2$ Oe and $H_D = 35$ Oe.

Dynamic measurements of ΔH_z were also made on a corner of the same propagating circuit. Starting at $\Theta = 108°$, position 1 (Fig. 3), measurements were made at 36° intervals until position 4 was reached. The corresponding field orientations comprise portions of three cycles, viz:

cycle A, $108° \leq \Theta \leq 360°$, bubble position from 1 to 2;

cycle B, $0° \leq \Theta \leq 360°$, bubble position from 2 to 3;

cycle C, $0° \leq \Theta \leq 108°$, bubble position from 3 to 4.

The general features of the curve in Fig. 3 include the large ΔH_z values in the first half of cycle B, the following sharp minimum at 252° and the strong vertical field at 72° in cycle C. The large ΔH_z values noted in cycle B are associated with the three rectangular bar elements at the end of the loop. The exceptionally long bars seem to compensate for the effects of the gaps in their vicinity. Only one gap exhibits a strong vertical field minimum. It occurs near $\Theta = 252°$ in cycle B and corresponds to a bubble moving into the lower slanted element as it leaves the lower long vertical bar. In order for the corner element not to have a deleterious effect on the upper propagation margin of the circuit, this minimum value of ΔH_z must be at least as large as the minima found for the straight T-Bar elements. The other prominent behavior occurs between positions 3 and 4 and is comparable to the equivalent positions of a T-Bar circuit.

CONCLUSION

The technique described herein is capable of providing useful information to aid in the design of Permalloy circuit elements. It is able to pinpoint the positions which determine the upper bias margin for propagation and thereby detect potential design weaknesses. It can also be used to examine the behavior of a bubble circuit at high frequency and thereby point out any unusual effects that may not be apparent at quasistatic tests. Besides propagation, this technique can be used to probe control function circuit operation (e.g. transfer, replicate, etc). The data presented here for the particular T-Bar and corner elements show that the upper bias margins are quite sensitive to the gaps. Propagation at 100 kHz in Y,Sm- garnet (mobility $\sim$ 200 cm $sec^{-1}Oe^{-1}$) shows no unusual effects when compared with the static measurements.

ACKNOWLEDGMENTS

The authors are pleased to thank G. G. Sumner for providing the garnet material, D. C. Bullock for designing the circuits, and M. S. Shaikh for fabrication. We would also like to thank C. T. M. Chang and F. G. West for helpful discussions.

REFERENCES

1. P. K. George and T. T. Chen, Appl. Phys. Letts. 21, 263 (1972).
2. P. K. George and J. L. Archer, AIP Conference Proceedings No. 18, part 1, Magnetism and Magnetic Materials Boston, 1973, pp. 116-120.
3. M. E. Jones and R. D. Enoch, IEEE Trans. MAG-10, 832 (1974).
4. R. F. Fischer, IEEE Trans. MAG-7, 741 (1971).
5. C. T. M. Chang, Private Communication.

AN EXPERIMENTAL MAGNETIC BUBBLE TIME-SLOT INTERCHANGER

Y. S. Chen, P. I. Bonyhard and J. L. Smith
Bell Laboratories, Murray Hill, New Jersey 07974

ABSTRACT

One of the basic building blocks of a time-division switching network is a time-slot interchanger (TSI). This paper reports on the design and characterization of a small experimental magnetic bubble time slot interchanger (TSI) which handles 4 time slots per time-frame with 2 bits per time slot.

Current "state-of-the-art" technology for bubble memories is utilized. The bit period is 28.8 μm and the propagation rate is 100 kHz. The overall operating bias field margin range obtained with 25 Oe rotating field is 6 Oe. The limiting function is a passive AND/OR logic gate. A unidirectional transfer gate is used in the TSI to transfer routing bubbles into the storage loops. The operating bias margin for the transfer gate is nearly equal to that of loop propagation.

DESIGN CONSIDERATIONS

In this realization the TSI can be thought of as essentially a variable delay line, with locally stored switching information governing the number of steps of delay for the data in any given time slot. The design of the TSI is shown in Fig. 1. The input time-frame data are introduced into the bubble chip using bubble nucleate generator[1] marked G. From here the data propagate downwards in the vertical delay channel. Depending on the presence or absence of control bubbles merging into the delay channel from the left at bubble interaction gate positions G1, G2, G3, G4, these data will move to the right into the "chevron" output channel after some integer multiple of 2 steps of delay. If the 8 bits of the input time-frame are introduced at G on cycles 1 to 8, then at the end of cycle 16 the 8 bits of the output time-frame will appear in the 4 output time slot positions marked 4, 3, etc., with arrows in the figure. Table 1 should clarify this operation.

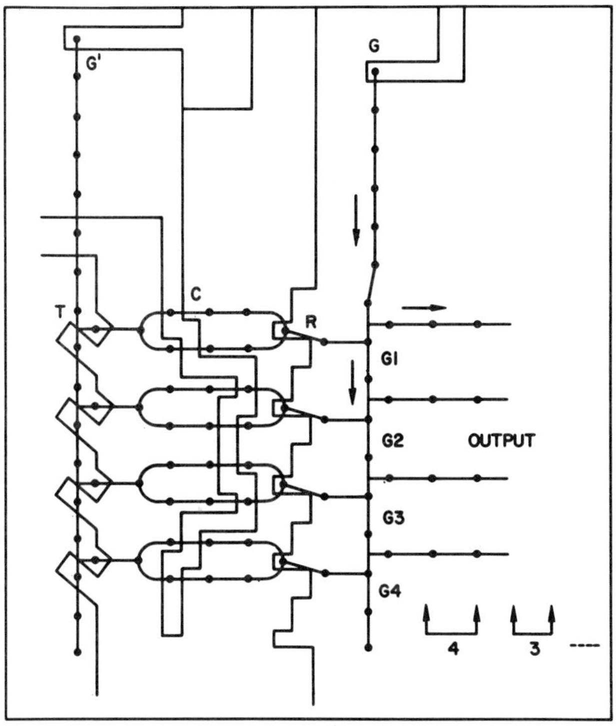

Fig. 1-Layout of a magnetic bubble time slot interchanger.

TABLE I

Routing Path for the 4 Time-Slot Magnetic Bubble TSI

Input Time Slot	Output Time Slot	Pass Via Gate
1	1	1
1	2	2
1	3	3
1	4	4
2	1	4
2	2	1
2	3	2
2	4	3
3	1	3
3	2	4
3	3	1
3	4	2
4	1	2
4	2	3
4	3	4
4	4	1

The logic gate is of a design already described.[2] The control bubbles are produced by the replication (at the four replicators marked R) of the switching information bubbles that circulate in the four closed storage loops. The replicator design has also been reported before.[3] Figure 2 shows a part of the gate and replicator region.

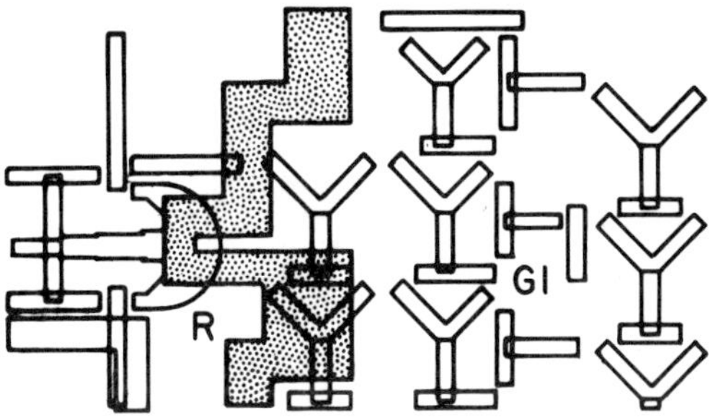

Fig. 2-Magnetic bubble logic gate (G1) and replicator (R) used in the TSI chip.

The remainder of the circuits is associated with the introduction and elimination of the switching information. Generator G' produces the bubbles and one of the four transfer gates marked T directs them into the appropriate storage loop. Bubbles must be generated, transferred and annihilated in pairs, as there are 2 bits per time slot. The circuit is designed for simplicity of control so that if one desires to establish a switching path from input time slot $1 \leq i \leq 4$ to output time slot $1 \leq j \leq 4$, then G' must be energized on the two cycles of time slot i in any time-frame whereas T must be energized on time slot j of the next time-frame. A set of bubble collapsers, C, in the storage loops are connected in series with the control conductor of G'. If the generate pulse is properly phased, then these collapsers will automatically eliminate any switching bubbles that may be left in the storage loops representing a previously used path from input time slot i to any output time slot. Thus, old paths are automatically "pulled down".

EXPERIMENTAL RESULTS

The chip has been fabricated using the same bubble material as the mass memory chips described before.[4] The operation of the TSI was tested at 100 kHz rotating field frequency and 25 Oe rotating field amplitude. Functions were tested by using a polarizing microscope test station with start-stop program cycles of less than 100 steps. A total of 6 Oe overall bias field margin was obtained at room temperature. The limiting factor for this chip design at high bias field is the passive logic gate.

CONCLUSION

Magnetic bubble TSI's utilizing bubble AND/OR gates have been shown to be feasible. A test unit having four time-slots was fabricated using current state-of-the-art bubble technology. At 100 kHz rotating field frequency and 25 Oe rotating field amplitude the bias margin range is 6 Oe.

REFERENCES

1. T. J. Nelson, et al., IEEE Trans. Magnetics MAG-9, 289-393, Sept. 1973.
2. P. I. Bonyhard, et al., IEEE Trans. Magnetics MAG-9, 708-709, Dec. 1973
3. P. I. Bonyhard, et al., AIP Conf. Proc. 18, 100-104, 1974.
4. P. I. Bonyhard, et al., IEEE Trans. Magnetics MAG-9, 433-436, Sept. 1973.

PASSIVE REPLICATOR FOR MAGNETIC BUBBLES

T. J. Nelson
Bell Laboratories, Murray Hill, New Jersey 07974

ABSTRACT

The design and operation of a passive magnetic bubble replicator are described. Operation has been achieved with 6 micron bubble materials in a 25 Oe field rotating at 100 kHz. The bias margins are comparable to those for propagation, although there is some loss at the high end, where the limit is analogous to that for detection. The number of interconnections to a magnetic bubble memory device, or the complexity of its drive circuitry, could be reduced by incorporation of the passive replicator. The design also represents an important step towards the development of devices that can be fabricated with a single level of fine-feature photolithography.

INTRODUCTION

The fundamental processes required in field access magnetic bubble devices are generation, propagation, detection and annihilation. To this basic list, in a recent description[1] of a magnetic bubble mass memory chip, are added: transfer from one propagation path to another and replication from a reentrant path to the detector. The need for a replicator is dictated by the detector design[2] which rapidly expands the domains to be detected into long strips. It is advantageous, in terms of chip area, to emit these strips through the guard rail surrounding the active area of the chip rather than to shrink them back to bubbles to be reused.

Propagation is achieved through field access, which means that the bubbles are transported along shift registers in response to the changing stray fields of the permalloy overlay features as an externally applied in-plane magnetic field rotates. The other functions are achieved by current access, through interconnections to the circuit and conducting paths, either in the permalloy level or in a nonmagnetic conductor level which underlies it. Information is transmitted between the chip and the outside world in the generator and detector connections, of course, but also as timing signals in the transfer and annihilator leads. The replicator, on the other hand, does not require any information exchange to perform its function. Rather, the fields of the current pulses in the replicator conductor were needed to accomplish the physical function of replication.

The number of interconnections to a magnetic bubble mass memory chip, or the complexity of its drive circuitry, could be reduced by the use of a passive replicator. Such a structure would be defined in the permalloy level only, would require no connections, and would produce domains on each of two outgoing propagate tracks for every bubble on its input track. A previous report[3] of such a structure, which was called a "splitter", operated within a small bias field range relative to the propagate margins. The purpose of this paper is to describe a passive replicator which has been successfully operated at 100 kHz with 6 micron bubbles in a 25 Oe rotating field. The bias margins are comparable to those for propagation, although there is some loss at the high end, where the limit is analogous to that for detection, and at the low end, where slightly increased rotating field strength still improves the margins.

DESIGN OF PASSIVE REPLICATOR

The design reported here is shown in Fig. 1. The chevron pattern has 28 micron period in the direction of propagation and 7 micron period in the orthogonal direction. The bar widths are 3.5 microns and the gaps are 2 microns. Domains enter from the right, elongated into strips, in a counter-clockwise rotating field.

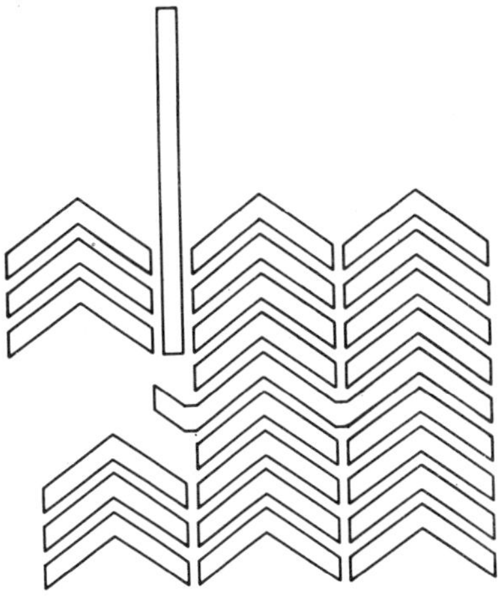

Fig. 1-Passive replicator design. Bubbles enter from the right elongated into strips, and exit on the two chevron propagate tracks on the left.

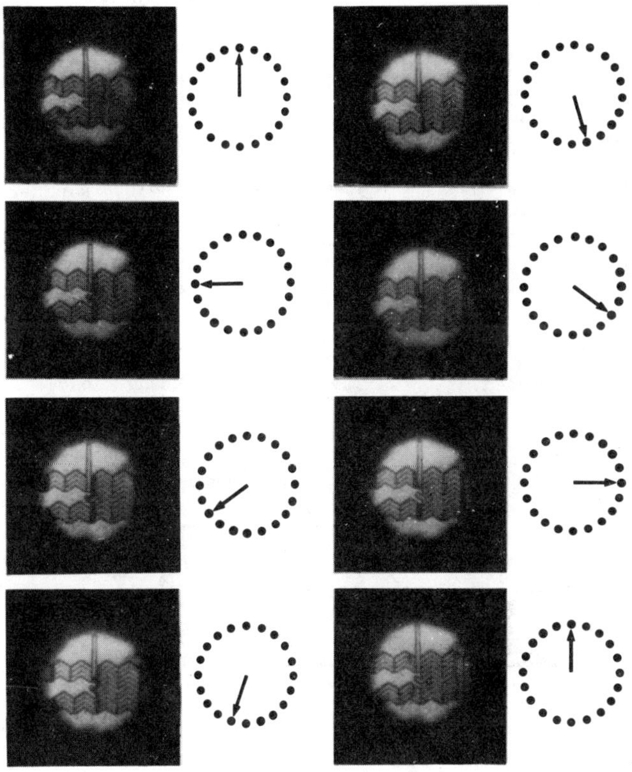

Fig. 2-Strobe photographs of the passive replicator in action. The analyzer was set for dark domains which enter the structure in every second cycle. The strobe pulse width was set at 0.5 microsec., giving good time resolution for the 100 kHz rotating field. The action is followed through a complete cycle, as indicated by the arrows to the right of the photographs.

The step-by-step operation can be followed in Fig. 2, which shows a series of stop action photographs taken at 100 kHz with the aid of an image intensifier strobe apparatus.[4] The basic principle of the design is best shown starting with the field oriented to the left. As the top and bottom ends move through the gaps to the adjacent chevrons, the middle of the strip undergoes a retrograde motion. That is, the middle of the strip retreats from the end of the extended "cutter" chevron, and becomes centered between the end of the cutter on its left and the ends of the "guard" chevrons, above and below the cutter, on its right, when the field points down. Thus as the field rotates to the right the middle of the strip is caught between the repulsive poles of the cutter and the guards and severs. The long vertical bar in the upper track helps to keep the upper end of the strip from being pulled downwards while this happens.

OPERATION OF PASSIVE REPLICATOR

The bias margins of the passive replicator were essentially unchanged from quasistatic (about 1 Hz) operation to 100 kHz. A slight difference in the high bias limit was observed in start-stop testing for a bubble which comes to rest one period in front of the replicator. Presumably this change, about 1 Oe, results from the limited time the bubble has to strip out to the full height of the column of chevrons at the replicator. Accordingly, the bias margins were taken with the field rotating continuously, using the strobe apparatus for observation. The strobe was pulsed once for each precession around the test loop containing the replicator, which was 48 cycles. Adequate intensity for this purpose was achieved by lengthening the strobe pulse to about 2 microsec. Bias margins vs rotating field intensity taken in this way are shown in Fig. 3. There are two relevant high bias limits, depending on whether bubbles are present in every other cycle, at most, as is the case for the chip described in Ref. 1, or may be present in every cycle. Mutual biasing effects evidently affect the performance when bubbles are one cycle apart. The difference found is about 2.5 Oe. The high bias failure mode appears to result from the strip pulling out of the lower track. One usually cannot follow a failure mode with the strobe, of course, but this tendency is present as the margin is approached. At low bias and low rotating field

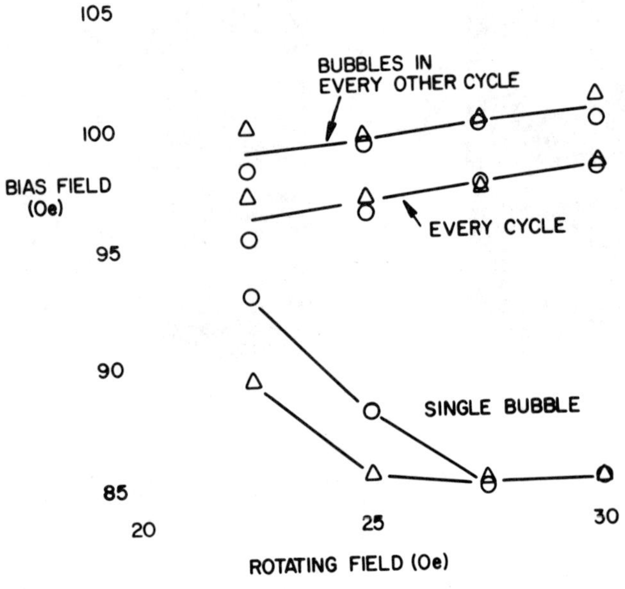

Fig. 3-Bias margins for passive replicator as a function of rotating field intensity for (△) 0.7 micron spacing and (O) 0.8 micron spacing with SmYIG bubble films.

intensity the strip appears not to cut. This deduction is supported also by the single bubble data given in Fig. 3 where the higher separation of 0.8 micron (circular data points) shows more loss of margins than does 0.7 micron (triangular data points) at low rotating field.

The film thickness h, material length ℓ and saturation induction $4\pi M$ for the two samples were △: h=6.6 μm, ℓ=0.58 μm, $4\pi M$=180G; O: h=6.5 μm, ℓ=0.61 μm, $4\pi M$=182G. The upper bias limit for propagation only in a 3-element chevron track is higher yet, at about 104-105 Oe. The lower bias limit, 85-86 Oe at 30 Oe drive, is considered good for any kind of propagate structure. The (one sided) rate of growth of strips as a function of bias field was observed in the chevron guard rail which isolates the test circuit containing the passive replicator. At 30 Oe drive, this rate drops below 10 chevrons per cycle (the vertical spacing of chevrons in the guard rail is 6.8 microns) at 101 Oe bias. At 25 Oe drive the observed drop off is at 99 Oe. The growth rate decreases rapidly with increasing bias above the points cited. An efficient detector design would contain just enough expander periods so that the strip would reach the desired length up to this limit of reasonably rapid growth. Thus the bias margins for replication are expected to be about the same as for detection. This may be contrasted with the chip described in Ref. 1, where the detector output drops rapidly in the top 2-3 Oe of the bias range due to insufficient expansion. Thus although the high bias limit of the passive replicator is inferior to that of the active replicator described in the literature, operation in the extra bias field range is not feasible.

APPLICATIONS OF PASSIVE REPLICATION

The active replicator described in Ref. 1 also performs annihilation, by means of an alternate pulsing scheme. Although the number of interconnections to the chip would not be reduced by incorporation of the passive replicator, the drive circuitry dedicated to the replicate functions would be eliminated. System noise during detection would be reduced to a minimum, because replication is the only function required during detection. Moreover thermal and electromigration problems associated with the relatively high current and duty cycle used in active replicators would be avoided, as would potential errors due to failures of the replicator drive circuitry.

The design of a passive replicator represents an important step towards development of magnetic bubble circuits that can be fabricated with a single level of fine-feature photolithography.[5] Such circuits, when developed, can be expected to give higher processing yields at the present density, or to eliminate alignment considerations from the design of circuits at higher density.

ACKNOWLEDGMENTS

It is a pleasure to acknowledge helpful discussions with P. I. Bonyhard, Y. S. Chen and J. L. Smith during the several iterations which resulted in the present design, and to thank F. B. Hagedorn for helpful comments on the manuscript.

REFERENCES

1. P. I. Bonyhard, J. E. Geusic, A. H. Bobeck, Y. S. Chen, P. C. Michaelis and J. L. Smith, IEEE Trans. Magnetics MAG-9, 433 (1973).
2. W. Strauss, A. H. Bobeck and F. J. Ciak, AIP Conf. Proc 10, 202 (1973).
3. H. Chang, J. Fox, D. Lu and L. L. Rosier, IEEE Trans. Magnetics MAG-8, 214 (1972).
4. G. P. Vella-Coleiro and T. J. Nelson, Appl. Phys. Letters 24, 397 (1974).
5. T. J. Nelson, AIP Conf. Proc. 18, 95 (1974).

TWO LAYER PERMALLOY CIRCUITS FOR 1.5-μm DIAMETER BUBBLE PROPAGATION

S. Matsuyama, R. Kinoshita, and M. Segawa
Fujitsu Laboratories Ltd., Kawasaki, Japan

ABSTRACT

It normally believed that the limitation of the practical bubble diameter will be around 3μm by means of conventional photolithography. In this paper, a new design of the overlay for bubble propagation is introduced for smaller diameter bubble propagation. T-bars are divided into two groups of x and y directed bars, each of which is arranged in different layers separated by SiO_2. The spacing of the I and the top bar of the T can be decreased down to the thickness of SiO_2 interlayer, because they are deposited in different layers.

By forming 1-μm wide bars and 1-μm spacings between adjacent bars, which is the limit of present photolithographic technology, it is possible to propagate 1.5-μm bubbles.

From the 1.5-μm bubble propagation experiment, bias field margin was about 18 Oe at 25 Oe, 2-Hz rotational field. An all-permalloy transfer gate was demonstrated to have 12 Oe bias field margin of 5-mA control current.

INTRODUCTION

Recently, extensive investigation has been made to obtain high density bubble memory devices. In order to increase packing density, needless to say the size of the bubble, as well as the linewidth of the permalloy pattern and the gap between adjacent patterns have to be reduced. The minimum pattern gap replicated by conventional photolithography is considered to be 1-μm, resulting in propagating around 3-μm diameter bubble. This paper describes a new arrangement of permalloy patterns introduced for smaller diameter bubble propagation. The advantages of this design are discussed, then the fabrication and the design of the overlay and circuit design without a nonmagnetic conductor layer are described. Also a 1.0-μm linewidth shiftregister is described which was fabricated using conventional photolithography.

PATTERN CONSTRUCTION AND CHARACTERIZATION

A typical layout of the two layer overlay memory chip is shown in Fig. 1. An SiO_2 layer is first applied to the garnet film and the first permalloy film, which acts both as a propagation element and as a current carrying conductor, is then deposited onto the SiO_2 layer by vacuum evaporation. A photoresist pattern is exposed on the permalloy layer and then the permalloy pattern is delineated using ion-milling

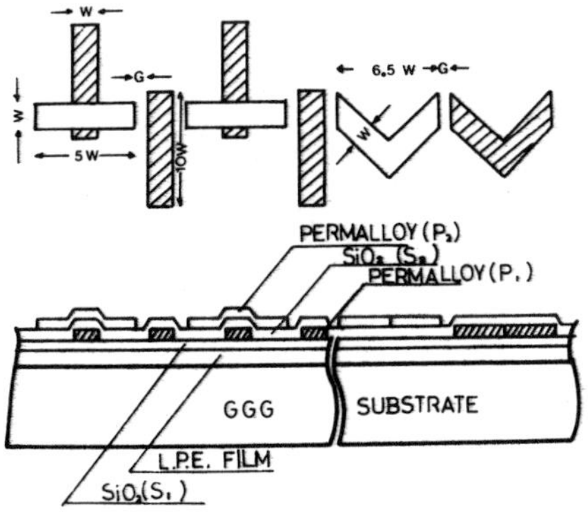

Fig. 1 Typical two layer pattern memory chip layout

techniques. The second SiO_2 layer is applied to the permalloy pattern and then the second permalloy film is deposited onto it. The upper permalloy pattern is obtained through the process of a photoregist alignment and ion-milling step.

The distance between adjacent patterns is determined by the thickness of SiO_2 interlayer because they are deposited in different layers, as shown in Fig. 1. The optimum spacing between garnet film and the first permalloy pattern should be decided by the magnitude of the anisotoropy field, the magnetization and the bubble diameter of the film. The two layer pattern method has the following advantages.

1. Zero gap pattern (means G = 0, see Fig. 1) for smaller diameter bubble propagation can be obtained easily. This will reduce the 4π Ms dependence of minimum drive field 1, 2.

2. Using the thick film expander sensor 3, twice the output voltage can be obtained and the stability of a propagating stretched bubble will be improved by the continuous potential well of zero gap stretcher chevrons.

3. Because the two bars of a T are deposited in different layers, the cross point of the T is a stable position, unlike the crosspoint of a conventional single layer T which is not a stable position.

TABLE 1

LPE film	CHARACTERISTICS			CHIP DESIGN (μm)					
	$4\pi M_s$ (gauss)	H_k (Oe)	Bubble diameter (μm)	W	G	P_1	P_2	S_1	S_2
a	348	700	1.5	1.0 1.2	-0.5	0.1	0.1	0.5	0.1
b	267	800	3.0	2.0	0	0.2	0.2	1.0	0.2
c	200	2000	6.0	4.0	0	0.4	0.4	1.1	0.4

a: $(YGdYb)_3(FeGa)_5O_{12}$
b: $(YGdYb)_3(FeGa)_5O_{12}$
c: $(YSm)_3(FeGa)_5O_{12}$

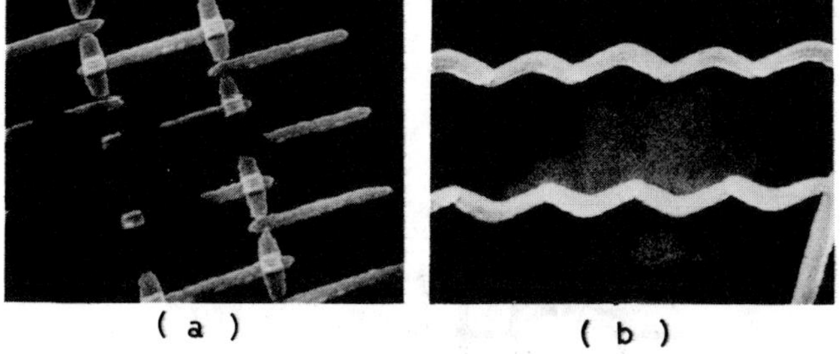

Fig. 2 Scanning electron micrograph of zero gap 1-μm linewidth T-bars (a) and chevrons (b) patterns fabricated by conventional photolithography

EXPERIMENTAL RESULTS

Typical magnetic properties of the garnet films and the overlay structure used in the experiment are summarized in Table 1. A scanning electron micrograph of a zero gap 1-μm linewidth T-bars and chevron patterns fabricated by conventional photolithographic technology are shown in Fig. 2. These patterns were fabricated on the film described in Table 1-a. The periodicity of the T-bars and the chevron are 6.0μm and 6.5μm, respectively. Quasistatic operating margin

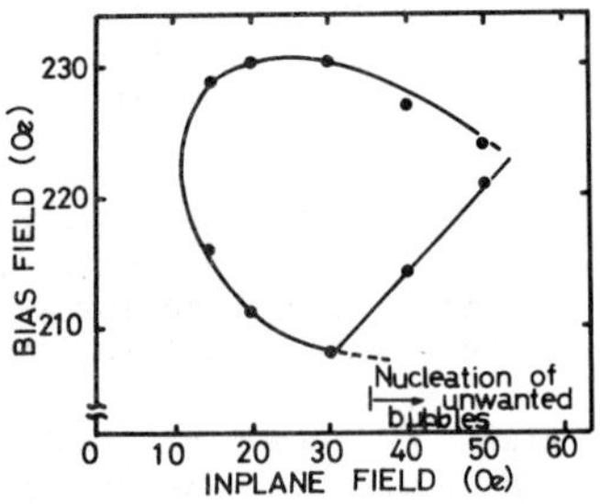

Fig. 3 Quasistatic operating margin curve of 1-μm linewidth chevron propagation patterns

curves of the chevron are shown in Fig. 3. T-bar margins could not be observed because the garnet film thickness used in the T-bars experiment was thinner than that used in the chevron experiment. At the rotating field of below 30 Oe, upper and lower thresholds of the bias margins are determined by collapse of the bubble under the pattern and by stretching out along the propagation path, respectively, as seen in a conventional pattern loop. Above 30 Oe of rotating field, they are determined by the bubbles being nucleated at the end of the corner bars because the anisotropy field of the garnet film is too small. This problem will be solved by increasing anisotropy.

The difficulty of aligning the two masks for the permalloy layers is canceled by the elimination of the conductor pattern by a permalloy conductor in either layer. The bubble control functions, such as sensor, splitter and generator, could have been designed without using a nonmagnetic conductor layer. The two layer method is especially suitable for the thick film expander type sensor. Here, only a transfer gate is described because the design without reducing the merits of the two layer structure is rather difficult. Typical design and operation of a transfer gate are shown in Fig. 4. White pattern segments are in the lower layer and the black ones are in the upper layer. The experimental circuit, consisting of 2μm width patterns, was constructed on the film shown in Table 1-b. The bias margins for propagation along the closed minor and the major loop were from 122 Oe to 140 Oe at 30-Oe rotating field (2Hz). The high end of transfer margins from the major loop to the minor loop and the minor loop to the major loop are at 140 Oe, and their low end is at 128 Oe. The loss in margins at low bias field is mainly caused by the transfered bubble streching out along the conductor. To improve the lower threshold margins of a transfer gate, conductor shape should be slightly changed.

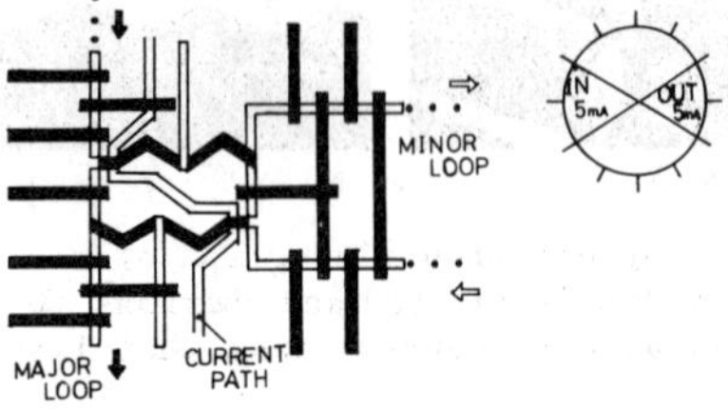

Fig. 4 Design and operation of two layer permalloy transfer gate

DISCUSSIONS

The 1-μm linewidth zero gap T-bar and chevron patterns were fabricated by conventional photolithographic technology. The bias field margins obtained for the propagation circuit are about 19 Oe for rotating field of 20 Oe. Also it was confirmed that bubble control functions necessary in major-minor loop memory chip could be designed without a nonmagnetic conductor layer. If this method is applied with electron beam lithography, further improvement for fabrication of small patterns is expected. The basic difficulty in the fabrication of the two layer pattern is only mask alignment of the small pattern. Figure 5 shows that the loss in margin caused by misalignment of the upper pattern and the lower one is about 10% at a quarter of a linewidth. Operating margins at 100kHz of conventional T-bars and two layer T-bars shifted about a quarter of a linewidth are shown in Figure 6. Although there is no obvious distinction between them, a pattern having more tolerance of misalignment has to be designed. However, mask alignment of small patterns remains a problem to be solved not only for this structure but also for the heretofore reported structures except one-mask structure 4.

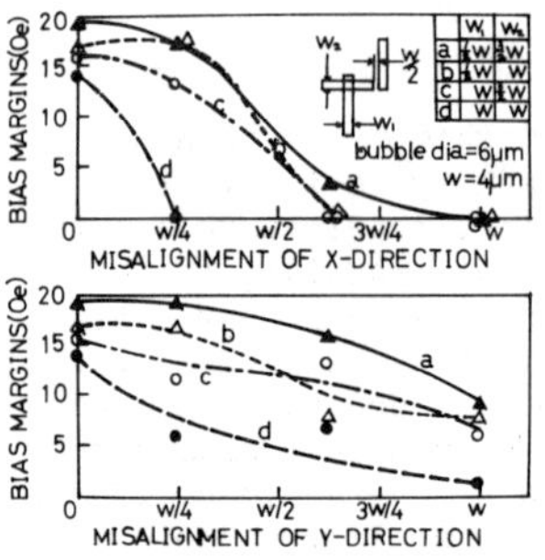

Fig. 5 Loss in margin by misalignment

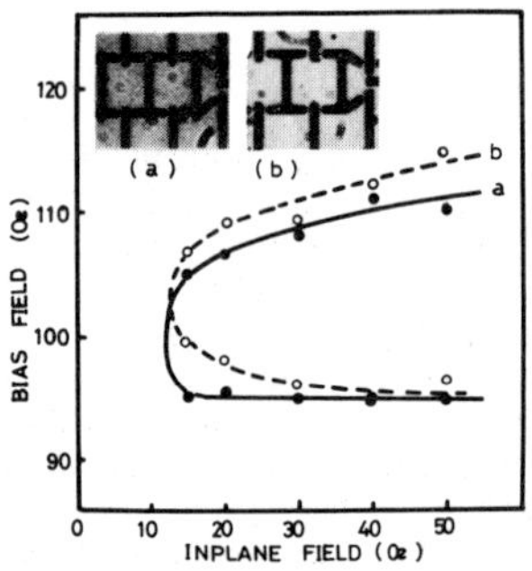

Fig. 6 Operating margin of typical two layer pattern(a) and conventional T-bars(b) at 100kHz

ACKNOWLEDGEMENT

The authors wish to thank Dr. S. Kojima, Dr. H. Sasaki, Mr. K. Yamagishi, Mr. A. Ihaya and Mr. S. Sakai for their advice and encouragement, and Mr. K. Asama and Mr. H. Tominaga for supplying various garnet films.

REFERENCES

(1) P. K. George, A. J. Hughs and J. L. Archer, "Submicron Bubble Domain Device Design", Paper No. 26-1, Intermag Conf., Toronto, Canada, 1974.
(2) M. H. Kryder, K. Y. Ahn, G. S. Almasi, G. E. Keefe and J. V. Powers, "Bubble to T-1 Bar Coupling in Amorphous Film Small Bubble Devices", Paper No. 26-2, Intermag Conf., Toronto, Canada, 1974.
(3) J. L. Archer, L. Tocci, P. K. George and T. T. Chen, "Magnetic Bubble Domain Devices", IEEE Trans. Magnetics, Vol. MAG-8 pp. 695 ~ 700, Sept. 1972.
(4) A. H. Bobeck, I. Danylchuk, F. C. Rossol and W. Strauss, "Evolution of Bubble Circuits Processed by a Single Level", IEEE Trans. Magnetics, Vol. MAG-9 pp. 474 ~ 480, Sept. 1973.

MATERIAL AND SUBMICRON BUBBLE DEVICE PROPERTIES OF $(LuSm)_3Fe_5O_{12}$*

D. C. Bullock, J. T. Carlo, D. W. Mueller and T. L. Brewer
Texas Instruments Incorporated, P.O. Box 5936, M. S. 145, Dallas, Texas 75222

ABSTRACT

The garnet, $Lu_xSm_{3-x}Fe_5O_{12}$, with x ranging from 1.5 to 1.9, has been epitaxially grown on gadolinium gallium garnet substrates by conventional LPE dipping techniques. This range of compositions has $q > 1$ and supports submicron diameter magnetic bubbles. For a x = 1.8 composition the coercivity, H_c, is 0.05 Oe and the anisotropy field, H_K, is 4360 Oe. The wall mobility, μ, was measured by bubble translation and is 192 cm/sec-Oe for a x = 1.5 composition. Uniform skewing but no hard bubble behavior was observed in the bubble translational studies. Magnetic bubble T-bar permalloy circuits have been fabricated by E-Beam patterning and electroplating on these films. Quasistatic bubble propagation has been observed for 0.8 μm diameter bubbles at a drive field of 100 Oe.

INTRODUCTION

The crystalline garnets[1] and the amorphous gadolinium cobalt alloys[2] are candidate materials for devices using submicron diameter magnetic bubbles. In this work the garnet, $Lu_xSm_{3-x}Fe_5O_{12}$, $(1.5 \leq x \leq 1.9)$ has been examined from both a material and a device viewpoint. Bulk crystals of $Lu_2Sm_1Fe_5O_{12}$ measured by Gyorgy[3] et al revealed this system to have high growth anisotropy. First the material growth and magnetic properties will be elucidated and then the device processing and behavior will be described.

MATERIAL GROWTH

The epitaxial films were grown between 840°C and 920°C by the horizontal LPE dipping method[4] in a super-saturated melt of lead oxide, boron oxide, vanadium oxide, and the necessary garnet nutrients. The rotation rate was 25 RPM and the growth rate ranged from 0.2 μm/min to 0.5 μm/min depending on the degree of supersaturation. Good quality films have been grown with a_f, the unstrained film lattice constant, between 12.369 Å and 12.402 Å. Once a_f is known, since we have only a two-component iron garnet, the amount of lutecium and samarium in the film can be calculated from the lattice constants[5] of SmIG and LuIG and Vegard's law. From these calculated compositions and the melt composition the segregation coefficient of samarium is 1.24 and lutecium is 0.84. The end-member technique has also been used to calculate the room temperature $4\pi M_s$ which ranges from 1745 G to 1765 G for the above films.

MATERIAL PROPERTIES

The coercivity, H_c, and the uniaxial anisotropy, H_K, have been measured for a x = 1.8 film with thickness, h = 4.2 μm. The coercivity was measured by the oscillating stripe domain technique.[6] In Fig. 1 the peak wall displacement is plotted versus the peak drive magnetic field which was applied perpendicular to the sample plane. The intercept with the horizontal axis gives the coercivity which is 0.05 Oe for this film.

The uniaxial anisotropy, H_K, was measured by the zero field susceptibility method.[7] The knee of the susceptibility, χ, curve versus $H_\perp$ yields $H_K' = H_K - 4\pi M_s = 2600$ Oe as shown in Fig. 2. (Cubic anisotropy has been neglected). Since $4\pi M_s$ is known and $H_K = 2 K_u/M_s = 4360$ Oe we find $K_u = 304{,}000$ ergs/cm³ and $q = H_K/4\pi M_s = 2.5$.

For the x = 1.8 film with h = 4.2 μm, the zero field stripe period is 2.5 μm. From this data and the

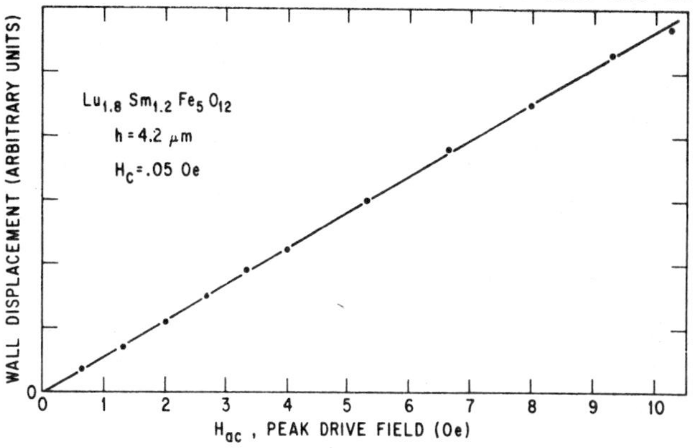

Fig. 1. Peak Wall Displacement Versus Peak Drive Field for $Lu_{1.8}Sm_{1.2}Fe_5O_{12}$ Yielding H_c = 0.05 Oe.

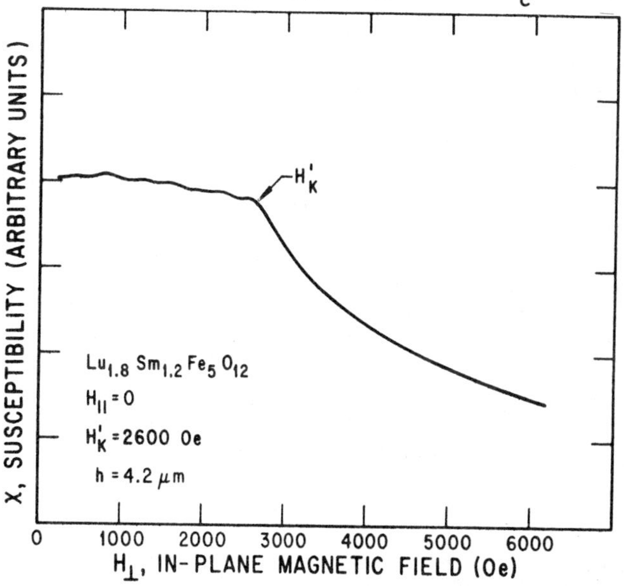

Fig. 2. Zero Field Susceptibility Versus In-plane Magnetic Field along the ⟨110⟩ Direction for $Lu_{1.8}Sm_{1.2}Fe_5O_{12}$.

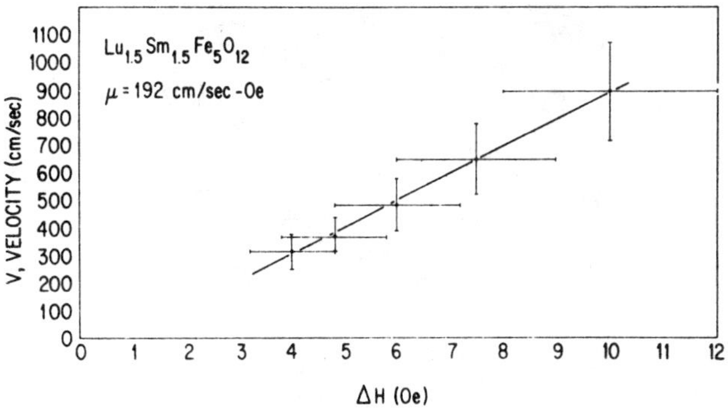

Fig. 3. Bubble Translational Velocity Versus ΔH for $Lu_{1.5}Sm_{1.5}Fe_5O_{12}$.

theory of Fowlis and Copeland[8] the characteristic length, ℓ, is 0.05 µm. The wall energy, $\sigma_w = 4\pi \ell M_s^2$ is 1.25 ergs/cm^2. From the additional wall energy relation, $\sigma_w = 4\sqrt{AK}$, the value of the exchange stiffness constant, A, is found to be 3.2×10^{-7} ergs/cm. From Thiele's theory[9] the bubble diameter stability range lies between 0.6 µm and 1.8 µm and the collapse field is 1450 Oe. For thinner films used in this work the bubble diameter spans 0.4 µm to 1.2 µm and the collapse field is lower.

In order to investigate data rate limitations the bubble translation experiment[10] was performed on a the sample with x = 1.5, h = 2.3 µm, q = 1.5 and a_f = 12.402 Å. The velocity, v, versus field difference, The ΔH, across the bubble is shown in Fig. 3. From the slope of this curve and the relationship

$$v = \frac{\mu}{2}\left(\Delta H - \frac{8}{\pi} H_c\right) \qquad 1)$$

whe

Landau-Lifshitz damping constant. If we assume the loss is dominated by samarium (since lutecium is a closed shell ion) λ/γ^2 can be calculated to be 12.5×10^{-7} Oe^2-sec from FMR data[11] which gives μ = 146 cm/sec-Oe. The agreement between the two data reductions is within experimental error.

During bubble translation the bubbles skew away from the maximum field gradient with a left-hand sense for the bias magnetic field pointing up similar to that reported in $(YGdTm)_3(FeGa)_5O_{12}$.[12] The skewing angle is 30° ± 5° for a 1 µm diameter bubble. This illustrates that the two-Bloch line bubble is not the ground state spin configuration. Although no effort has been made to suppress hard bubbles in this film by capping techniques or to suppress dynamic conversion no hard bubbles or dynamic conversion have been observed. This is probably due to the fact that q is low and that the velocities measured are less than the peak velocity[13,14]

$$v_p = 24|\gamma|A/K^{1/2}h \qquad 3)$$

which is 1390 cm/sec for this film assuming $|\gamma| = 1.76 \times 10^7$ Oe^{-1}-sec^{-1}.[5]

DEVICE PROCESSING AND BEHAVIOR

T-bar minor loop circuits with a linewidth of 0.4 µm and a period of 3.6 µm have been patterned in

Fig. 4. Scanning Electron Micrograph of a T-bar Permalloy Circuit; Bar Width is 0.4 µm.

PMMA resist by E-Beam slice writing. Minor loop patterns over a 27 x 27 mil^2 field have been defined. For the above bit density of 0.5×10^8 bits/in^2 this is a 36 k-bit storage register. To process the permalloy the scheme of Powers and Horstmann[15] was used. First a SiO_2 spacer of 5000 Å is deposited and then a 200 Å thick permalloy layer is sputtered for a plating base. After the resist application, E-Beam patterning and development, the permalloy propagation elements are electroplated in a galvanostatic mode using the Girard bath.[16] A SEM of part of a T-bar circuit is shown in Fig. 4. The film used for the quasistatic bubble propagation was 1.5 µm thick, the bubble diameter was 0.8 µm and the bias field was 1200 Oe. Straight line T-bar propagation was achieved for a drive field of 100 Oe at quasistatic frequencies. Kryder et al[17] have suggested that the high drive field observed in submicron devices is due to the high $4\pi M_s$ of the bubble which creates a large fringing field that couples strongly to the permalloy elements and hence preventing the bubbles from crossing the gaps between the T's and bars. Further work is in progress to reduce the drive field magnitude.

ACKNOWLEDGMENTS

The authors would like to express thanks to G. Rogers for help in growing the films, to L. Swink for the lattice constant measurements and to S. Shaikh for assistance in the device processing.

REFERENCES

1. E. A. Giess, C. F. Guerci, J. D. Kuptsis, and H. L. Hu, Mat. Res. Bull. 8, 1061 (1973).
2. P. Chaudhari, J. J. Cuomo, and R. J. Gambino, IBM J. Res. Dev. 17, 66 (1973).
3. E. M. Gyorgy, M. D. Sturge, L. G. Van Uitert, E. J. Heilner, and W. H. Grodkiewicz, J. Appl. Phys. 44, 438 (1973).
4. H. J. Levinstein, S. J. Licht, R. W. Landorf, and S. L. Blank, Appl. Phys. Letts. 19, 486 (1971).
5. W. H. von Aulock, "Handbook of Microwave Ferrite Materials", 1st edition, Academic Press, New York, 1965, Chap. 3.
6. J. A. Seitchik, W. D. Doyle, and G. K. Goldberg, J. Appl. Phys. 42, 1272 (1971).
7. P. W. Shumate, D. H. Smith, and F. B. Hagedorn, J. Appl. Phys. 44, 449 (1973).
8. D. C. Fowlis and J. A. Copeland, AIP Conf. Proc. No. 5, Magnetism and Magnetic Materials, Ed. C. D. Graham, Jr. and J. J. Rhyne, p. 240 (1971).
9. A. A. Thiele, B.S.T.J. 48, 3287 (1969).
10. G. P. Vella-Coleiro and W. J. Tabor, Appl. Phys. Letts. 21, 7 (1972).
11. G. P. Vella-Coleiro, AIP Conf. Proc. No. 10, Magnetism and Magnetic Materials, Ed. C. D. Graham, Jr. and J. J. Rhyne, p. 424 (1972).
12. D. C. Bullock, AIP Conf. Proc. No. 18, Magnetism and Magnetic Materials, Ed. C. D. Graham, Jr. and J. J. Rhyne, p. 232 (1973).
13. J. C. Slonczewski, J. Appl. Phys. 44, 1759 (1973).
14. F. B. Hagedorn, J. Appl. Phys. 45, 3129 (1974).
15. J. V. Powers and R. E. Horstmann, INTERMAG Conf. (1973), Washington, D. C., paper 21.1, unpublished.
16. R. Girard, J. Appl. Phys. 38, 1423 (1967).
17. M. H. Kryder, K. Y. Ahn, G. S. Almasi, G. E. Keefe, and J. V. Powers, IEEE Trans. Magnetics Vol. MAG-10, 825 (1974).

*This work was partially supported by Air Force Avionics Laboratory Contract No. F 33615-73-C-1029.

MAGNETIZATION REVERSAL STUDY IN PERMALLOY BUBBLE-PROPAGATION CIRCUITS WITH MAGNETO-OPTICAL EQUIPMENT OF MICRON RESOLUTION

G.S. Krinchik, E.E. Chepurova and U.N. Shamatov, Moscow State University, Lenin Hills, Moscow, 117234, USSR.

V.K. Raev and A.K. Andreev, Institute of Electronic Control Computers, Vavilova 24, Moscow, 117812, USSR.

ABSTRACT

Direct measurements of local magnetization in real-size Permalloy elements of bubble-domain propagation circuits with magneto-optical equipment of micron resolution were made. Magnetization reversal in T-bars with linear dimensions of 4 to 40 microns was studied in various transversal and lengthwise fields as well as in fields applied at an angle to the bar axis. Dependence of magnetization distribution in T-bars on the adjacent element magnetic state was found.

INTRODUCTION

Theoretical study of magnetization reversal in thin Permalloy film elements of bubble propagation circuits presents a problem of exceptional complexity even for the case of a uniform external field applied at an arbitrary angle to the Permalloy overlay. This problem becomes even more complex if one takes into account highly nonuniform time variable bubble fields and stray fields of adjacent elements in a propagation array.

The experimental approach to the estimation of magnetization distribution in T-bars looks more realistic. However, the task is not simple if one takes into account the small size of Permalloy elements (from 2 to 20 microns). In paper [1], large-scale simulation of propagating elements was applied to obtain a demagnetizing field pattern. One of the serious limitations of this approach is that it completely ignores the domain structure [2,8] and consequently, the real charge distribution in Permalloy. Hysteresis properties of Permalloy overlays obtained with VSM and Kerr-effect hysteresigraphs were considered in papers [2,3].

The purpose of this paper is to present the results of direct measurements of magnetization distribution in small size Permalloy elements of real bubble propagation pattern. In principle, the high resolution of the utilized magneto-optic equipment [4] allows an experimental solution of the Permalloy-bubble interaction problem [5] for submicron bubble propagation circuits.

MEASUREMENTS

Measurements were performed for typical T-bar propagation circuits with a periodicity of 30 to 40 microns and evaporated permalloy thickness of 1500 to 5000 Å on a glass substrate. The measured equatorial Kerr effect corresponds to the increment of the horizontal component of the magnetization of the illuminated region of the surface under sinusoidal external fields. A specially designed microtranslation system allows one to shift the photomultiplier with its aperture in the image plane of the microscope. The limiting resolution of the equipment is 0.2 to 0.3 microns and is determined by the microscope optical resolution. The results presented in this paper were obtained with a light spot diameter of 3 microns. The total error of the measurements is less than 0.05 to 0.1. A detailed description of the experimental equipment is contained in paper [4].

RESULTS AND DISCUSSION

The data presented in Fig. 1a illustrates the measured magnetization distribution along the x-axis of the Permalloy bar. The center of the bar reaches saturation at a field of 9 Oersteds. Y. Lin's ellipsoidal model [8] predicts saturation in the center of the bar of Fig. 1a at 8.75 Oe, in excellent agreement with the 9 Oe value found experimentally [9].

Fig. 1b shows the magnetization change along the x-axis in the horizontal part of the T-element. Due to the additional demagnetizing field from the vertical part of the "T", the center of it shows a noticeable decrease in the value of M, at the fields below saturation.

Magnetization curves for the T-element in the lengthwise (H) and transversal (H') fields are shown in Fig. 1c. As it is seen, one fails to achieve strong poles at the ends (point 2) in the fields used in practice. The same is true for the center of the "T" at a transversal field.

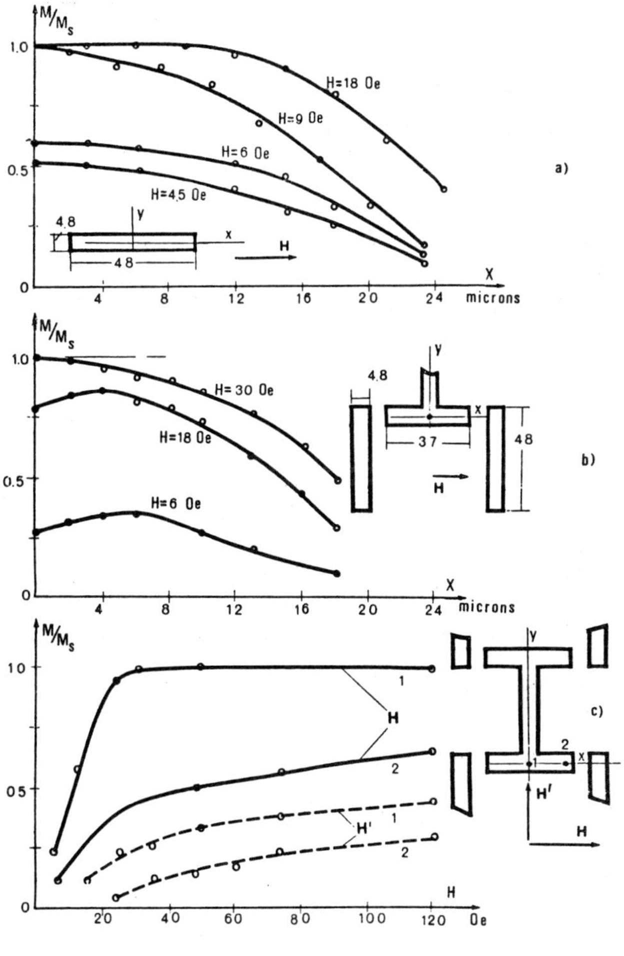

Fig. 1. Magnetization distribution in T-bar (a,b) and magnetization dependence on a bidirectional field in a horizontal part of the T-element (c).

Fig. 2 presents a magnetization dependence on the H and H' fields at three points of the bar. The end of the bar in the H' field (Fig. 2a) reaches saturation at H'>500, according to our data. Note that for the H field, the magnetization process at point 1 goes considerably easier than at points 2 and 3. We attribute such surprisingly high differences to the influence of additional fields from the ends of the neighboring T-elements. Approximately 30% of the bars in the tested patterns showed an interesting anomalous behaivor, presented in Fig. 2b. The anomalous bar is magnetized almost uniformly along the entire length in fields less than 30 Oe and reaches saturation at the value of 0.6M of a normal bar. This effect possibly is connected to the existence of an unusual domain structure [7] in such elements.

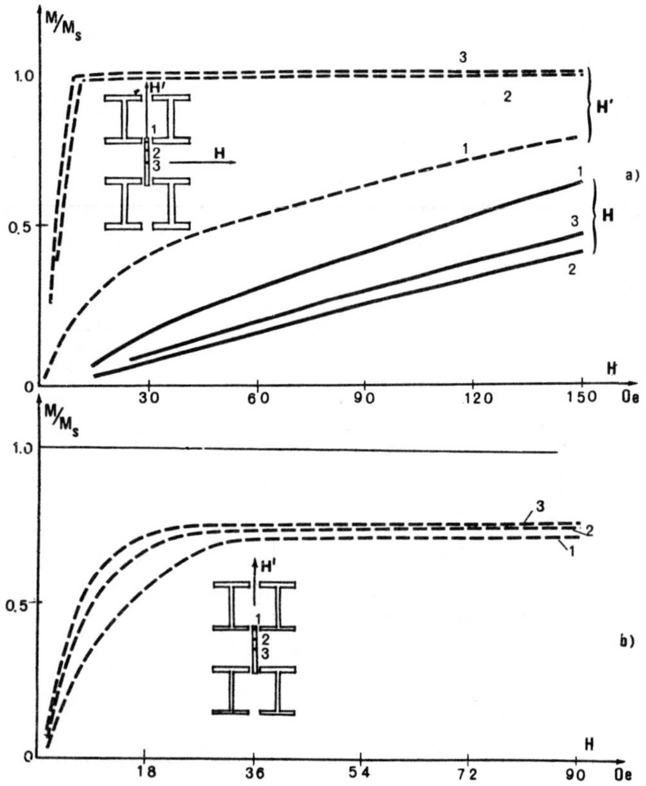

Fig. 2. Magnetization dependence on the external field for three points of the bar. Curves in Fig. 1a illustrate anamalous behavior of the magnetization. The bar is magnetized surprisingly uniformly and reaches a saturation on the entire length at comparatively low fields (above 30 Oe); however, saturation magnetization is only 70% of the normal bar in Fig. 2a.

The results of the magnetization of the T-element with the field H=18 Oe applied at different angles shown in Fig. 3 are of definite interest to a circuit designer. At α = 0 the magnetization distribution has the same character as in Fig. 1b. As the field is rotated to α = 45° the magnetization decreases and the curve becomes asymmetric. At this point the right part of the T-element is magnetized higher due to re-distribution of the demagnetizing field in the horizontal part of the "T". At further field rotation, up to an angle of 90°, the x-component of H becomes zero and along the x-axis, one can expect small magnetization in the direction of the field (y-axis). However, as shown in Fig. 3, the magnetization in both the right and left side of the horizonatal part of the "T" is perpendicular to the applied field and changes its sign at the center of the "T". This effect should be expected qualitatively for isolated T-bars, but should be quantitatively influenced by the presence of the adjacent bars [10]. A resultant flux closure for this angle is shown with solid lines in the upper right corner in Fig. 3. Thus in some T-bar propagation circuitries at angles of 45°<α<90°, some part of the horizontal bar of the "T" can be magnetized against the x-component of the field.

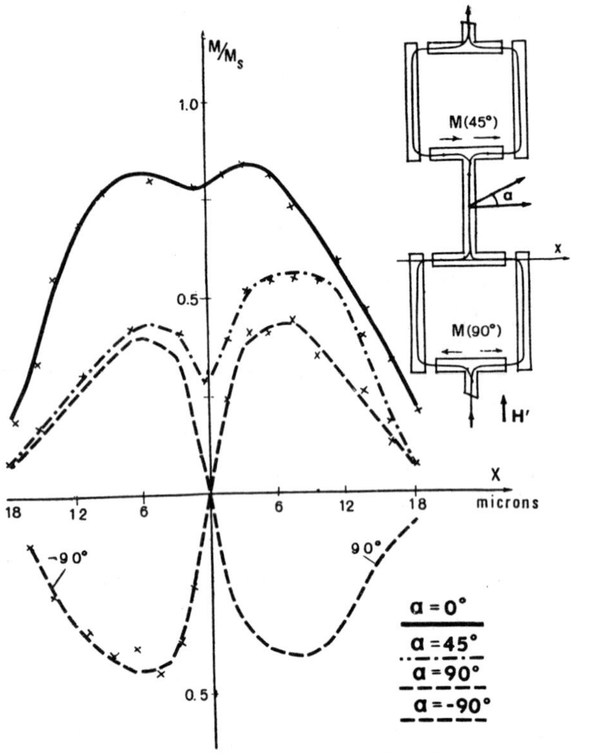

Fig. 3. Magnetization distribution in the horizontal part of the T-element for four directions of the external field H=18 Oe.

CONCLUSION

As it was shown above, a strong interaction between adjacent elements does exist in T-bar circuits, and in some cases this interaction determines the real magnetization distribution in Permalloy T-bars. In a previous paper in which a matrix of simple bars was considered [2], this interaction was not found. A noticeable percentage of the propagation elements obtained by the same technological process showed an anomalous magnetic behavior that can be considered as a limiting factor in the circuit performance. We hope to report on the magnetization measurements in the bubble-Permalloy system in our next paper.

ACKNOWLEDGEMENTS

The authors appreciate the interesting discussions with R.C. Barker and his valuable remarks to the results of this work. They also gratefully acknowledge the assistance of H. Omori and Y.J. Liu in preparing the galley.

REFERENCES

1. C.H. Hsin, T.J. Matcovich and R.L. Coren, AIP Conf. Proc. No. 5, p. 244 (1972).
2. W.D. Doyle and M. Casey, AIP Conf. Proc. No. 10 (1973).
3. K. Kempter, IEEE Trans. Magn., vol. MAG-8, p. 746 (1972).
4. G.S. Krinchik and O.I. Benidze, JETP 12, 2180 (1974).
5. P.K. George and J.L. Archer, JAP 44, 444 (1973).
6. J.A. Copeland, JAP 43, 1905 (1972).
7. A.K. Andreev, Proc. of Inst. Elec. Contr. Comp. 32, 63 (1973)(in Russian).
8. Y.S. Lin, IEEE Trans. Magn., vol. MAG-8, p. 375 (1972).
9. The authors are indebted to the reviewer for pointing out this agreement.
10. P.K. George, private communication.

MAGNETOELASTIC COUPLING IN PARAMAGNETIC DYSPROSIUM

B. M. Kale*, P. L. Donoho*, D. G. Pinatti, and O. Ferreira
Universidade Estadual de Campinas, Campinas, S. P., Brasil

ABSTRACT

The dependence of the elastic constants c_{11} and c_{33} on temperature and applied magnetic field has been measured for single-crystal dysprosium in its paramagnetic phase, in an effort to determine certain magnetoelastic coupling constants. This work was motivated by a theoretical treatment of magnetoelastic phenomena of Southern and Goodings. Similar measurements for the case of terbium have been reported by Salama et al, but the present work includes magnetostrictive corrections neglected by these authors. The values obtained for the magnetoelastic coupling constants of dysprosium, following the theory of Southern and Goodings, are considerably larger than expected; those estimated from the static magnetostriction results of Clark et al are smaller by a factor of about 20. A possible reason for this large difference between the statically and dynamically determined values for the magnetoelastic coupling constants may lie in the assumption by Southern and Goodings that the principal magnetoelastic contribution to the elastic constants arises from the inclusion of the nonlinear terms in the conventional definition of the finite strain tensor in the lowest order terms of the magnetoelastic Hamiltonian.

INTRODUCTION

Southern and Goodings[1] have carried out a theoretical treatment of dynamic magnetoelastic effects in rare-earth materials which extends the well known treatment of Callen and Callen[2] to include in the magnetoelastic Hamiltonian terms required to insure rotational invariance and terms arising from the nonlinear part of the conventionally defined finite lattice strain tensor. For temperatures in the paramagnetic range, the theory of Southern and Goodings is particularly easy to apply to the problem of calculating the magnetoelastic contributions to the elastic constants, and these authors show that careful measurements of the elastic constants c_{11} and c_{33} will permit the determination of certain one-ion and two-ion magnetoelastic coupling constants. Such measurements have previously been carried out by Salama, Melcher, and Donoho[3] for the case of single-crystal terbium, although they neglected to include a correction to account for static magnetostriction which may be appreciable. This paper describes similar work carried out on single-crystal dysprosium in the paramagnetic phase at temperatures from 180 K to 300 K in magnetic fields up to 18 kG. General quantitative agreement with the theory of Southern and Goodings was found, but the values obtained in this work for some of the magnetoelastic coupling constants were much larger than the values which may be estimated from the static magnetostriction data of Clark, DeSavage, and Bozorth[4].

THEORY

The theoretical treatment of dynamic magnetoelastic coupling in rare-earth materials of Southern and Goodings[1] includes the effects of the nonlinear part of the finite elastic strain tensor usually neglected in most small-strain approximations. In particular, these authors calculate, for the paramagnetic phase, the magnetoelastic contributions to the elastic constants c_{11} and c_{33}, which are determined through measurements of longitudinal elastic-wave velocities, under the assumption that the principal contribution to the magnetoelastic part of the internal energy quadratic in the components of the lattice-displacement gradient is due to these nonlinear terms in the definition of finite lattice strain. This assumption, however, neglects such contributions as the strain dependence of the magnetization, which increases as the magnetic ordering temperature is approached and higher order terms in the strain dependence of the crystal-field and exchange energies. Nevertheless, the experimental data reported here have been analyzed in terms of the theory of Southern and Goodings without attempting at this time any modification of their results to account for such additional contributions. The procedure of Southern and Goodings is reviewed briefly in this section in order to illustrate the method and arrive at their important qualitative results. Only the single-ion interaction is, however, considered in this section, since inclusion of the two-ion interaction would require excessive space and would be qualitatively similar. The data analysis of the following section, however, does include the two-ion interaction as presented in the work of Southern and Goodings.

In the presence of only longitudinal strain, the single-ion magnetoelastic Hamiltonian may be written to lowest order

$$H^{I}_{me} = - N[M^{\alpha 1}_{20} E^{\alpha 1} + M^{\alpha 2}_{20} E^{\alpha 2}][3J^2_z - J(J+1)]/2 - N M^{\gamma}_{22} \sqrt{\tfrac{3}{2}}\, E^{\gamma}_{1}[J^2_x - J^2_y]/2 , \qquad (1)$$

where the M-coefficients are magnetoelastic coupling constants, N is the number of ions per unit volume, and the E-quantities are the components of irreducible representations of the finite strain tensor whose Cartesian components are defined in the following way:

$$E_{\mu\nu} = \frac{1}{2}\left[\frac{\partial u_\mu}{\partial x_\nu} + \frac{\partial u_\nu}{\partial x_\mu}\right] + \frac{1}{2}\frac{\partial u_\lambda}{\partial x_\mu}\frac{\partial u_\lambda}{\partial x_\nu} . \qquad (2)$$

For the calculation of the magnetoelastic contribution to the elastic constants, it is necessary to calculate the second derivatives of the internal energy with respect to the components of the lattice strain. Southern and Goodings evaluated these derivatives in terms of the derivatives of the energy with respect to the "small-signal" strain, the derivatives of the lattice displacement, obtaining results which are particularly simple for the paramagnetic phase of the material of interest. They used the well-known high-temperature approximation[2] for the equilibrium values of the angular-momentum operators appearing in eq. (1), and they found expressions, not repeated here but given in detail in their paper[1], for the magnetoelastic contributions to c_{11} and c_{33} which are proportional to the square of the specimen magnetization and which depend upon the direction of propagation of the elastic waves used in the determination of the elastic constants and upon the direction of the applied magnetic field. These expressions involve linear combinations of certain one-ion and two-ion magnetoelastic coupling constants whose definition is given by Southern and Goodings[1].

It is not clear that the calculation of Southern and Goodings, as outlined briefly above, takes account of all contributions to the internal energy quadratic in the lattice-displacement gradient. For example, a simple molecular-field calculation reveals such a contribution arising from the strain dependence of the expectation values of the angular-momentum operators of eq. (1). It is also not completely clear that the elastic constants are properly given in terms of the second derivatives of the internal energy with respect to the infinitesimal strain components when the energy itself is expressed in terms of finite strain components. The term of the next higher order in the finite strain contains terms which are quadratic in the components of the infinitesimal strain.

EXPERIMENTAL RESULTS AND ANALYSIS

Measurements were made of the elastic constants c_{11} and c_{33} of single-crystal dysprosium for temperatures from 180 K to 300 K in magnetic fields up to 18 kG directed along the crystallographic a-, b-, and c-axes. The pulse-superposition method was employed to measure the velocity of 20-MHz longitudinal elastic waves propagating along the a- and c-axes, taking into account the dependence of the specimen dimensions on both temperature and field, using the data of Clark et al[4]. The susceptibility data of Behrendt et al[5] were used. The dependence of the elastic constants on the applied field at a constant temperature was found to follow quite well the prediction of Southern and Goodings[1], namely a variation with field proportional to the square of the field. Small deviations from this behavior were noted at high fields near the Néel temperature, although they were not as large as those observed by Salama et al[3] for the case of terbium.

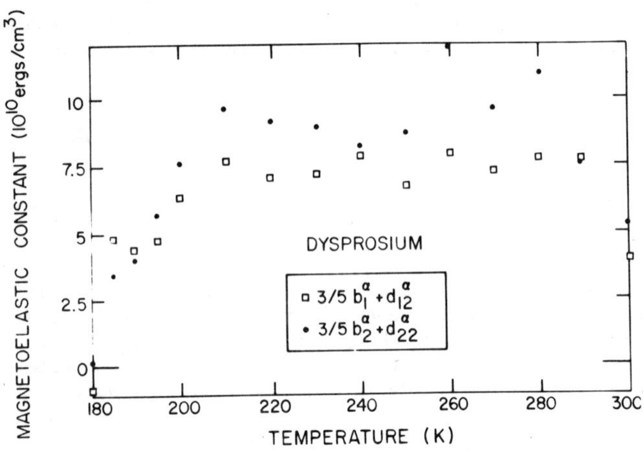

Fig. 1 Temperature dependence of $(3/5)b_\perp^\alpha + d_{12}^\alpha$ and $(3/5)b_2^\alpha + d_{22}^\alpha$ for dysprosium.

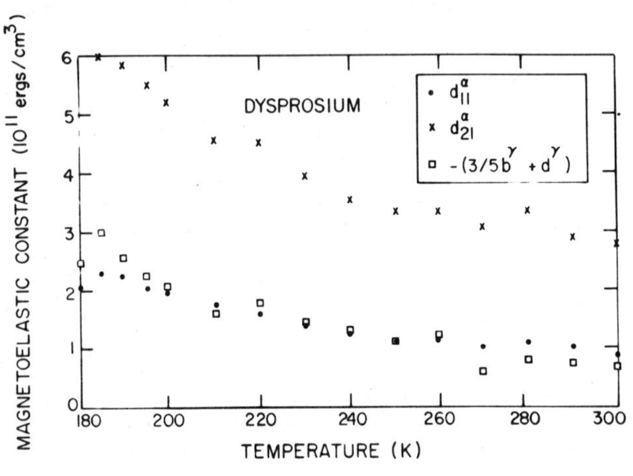

Fig. 2 Temperature dependence of d_{11}^α, d_{21}^α, and $(3/5)b^\gamma + d^\gamma$ for dysprosium.

The results of the measurements were analyzed in terms of the theory of Southern and Goodings[1], following the procedure outlined by Salama et al[3] to obtain values for the magnetoelastic coupling constants. These results are summarized in Figs. 1 and 2. It can be seen in these figures that the magnetoelastic constants exhibit considerable temperature dependence near the Néel temperature and that they become essentially temperature-independent at higher temperatures, as expected. The variation near the ordering temperature is undoubtedly due to the simplifying assumptions of the theory, since the static magnetostriction results of Clark et al[4] showed that these constants were essentially temperature independent over a much wider temperature range.

The magnitude of the magnetoelastic coupling constants obtained by following the theory of Southern and Goodings is somewhat larger than that which would be expected from a consideration of the static magnetostriction results of Clark et al. In particular, consider constants determined from a consideration of the velocity of waves propagating in the basal plane, namely the combination $(3/5)b^\gamma + d^\gamma$ shown in Fig. 2. If the value obtained at the higher temperatures is compared to the corresponding quantity obtained from the results of Clark et al, it is found to be approximately a factor of 18 times larger. It might reasonably be expected that better agreement between the static and dynamic results would be obtained since the static results fit the theory of Callen and Callen[2] so well and since the theory of Southern and Goodings[1] represents a rather simple extension in the present case of the theory of Callen and Callen. It is possible, therefore, that additional contributions to the magnetoelastic internal energy beyond that considered by Southern and Goodings are indeed important in the determination of the magnetoelastic contributions to the elastic constants of rare-earth materials.

Further work is in progress in an attempt to clarify the reasons for the above difference, including work at lower and higher temperatures and in much higher fields, together with simultaneous measurements of the specimen magnetization in order to avoid reliance on published susceptibility data.

ACKNOWLEDGMENTS

This work was begun with support from N. A. S. A. at Rice University, Houston, Texas, U. S. A. The authors wish to thank Mr. Carlos Pinelli for his valuable contributions in the construction of experimental apparatus.

REFERENCES

* Permanent address Rice University, Houston, Texas.

1. B. W. Southern and D. A. Goodings, Phys. Rev. B 7, 534 (1973).
2. Earl Callen and Herbert B. Callen, Phys. Rev. 139, A455 (1965).
3. K. Salama, C. L. Melcher, and P. L. Donoho, Proceedings of the 1973 Ultrasonics Symposium, p. 309.
4. A. E. Clark, B. F. DeSavage, and R. Bozorth, Phys. Rev. 138, A216 (1965).
5. D. R. Behrendt, S. Legvold, and F. H. Spedding, Phys. Rev. 109, 1544 (1958).

MAGNETOELASTIC ULTRASONIC GENERATION IN $TbFe_2$ THIN FILMS*

L. B. McLane
Collins Radio Group, Rockwell International, Inc., Dallas, Texas 75207

P. L. Donoho and W. L. Wimbush
Rice University, Houston, Texas 77001

ABSTRACT

Ultrasonic elastic waves have been magnetostrictively generated at room temperature in polycrystalline films of $TbFe_2$ deposited on quartz and sapphire substrates at a frequency of 690 MHz. The maximum generation efficiency was approximately 50 times that of nickel films deposited under similar conditions. In the case of deposition at substrate temperatures above 320°C, the films exhibited a strong magnetic anisotropy, with the easy axis invariably perpendicular to the plane of the film. This anisotropy led to coercive forces as high as 6 kOe and remanent magnetizations as high as 80% of the saturation magnetization. As a result, efficient generation of elastic waves could be achieved in films with no applied field if they had previously been magnetized. The field dependence of the ultrasonic generation and the magnetic anisotropy were found to be strongly dependent on the film thickness and on certain deposition parameters, particularly the substrate temperature, as indicated above. Annealing of the films subsequent to deposition also affected strongly the anisotropy and the ultrasonic generation efficiency. In all cases, however, the magnetic easy axis was perpendicular to the plane of the film (it should be pointed out that deposition was not carried out in the presence of a field).

INTRODUCTION

This paper reports the progress of an investigation into the magnetoelastic ultrasonic generation properties of $TbFe_2$ polycrystalline thin films. This investigation began with the study of several of the pure rare earths in single crystal as well as thin film form, [1,2] advanced into an exploration of the intermetallic compounds and mixtures of varying proportions of iron and selected rare earths, [2,3] and, finally, concentrated on the cubic Laves-phase compound $TbFe_2$. The earlier research with the pure rare earths was successful in terms of demonstrating the acoustic wave generation capability of these metals, though not with any appreciable degree of improvement over other, more well known materials such as nickel. It was evident, furthermore, that any serious practical application of the pure rare earths as ultrasonic transducers would be severely restricted by the cryogenic temperatures requisite to the magnetic ordering of these metals upon which the magnetoelastic generation mechanism is dependent, as well as by the substantial dc magnetic fields which are also necessary. The expansion of that study to include compounds and alloys of certain of the rare earths with iron was prompted by the discovery by Clark and Belson[4] of static magnetostrictions in these materials in bulk, polycrystalline form at room temperature of, for these conditions, unprecedented magnitudes. Investigation in this direction was fruitful and ultimately centered upon films of $TbFe_2$ which had proven to be the most effective as an ultrasonic transducer of all the materials examined (a result consistent with the bulk form, static data reported by Clark and Belson) and which, quite unexpectedly, exhibited a large magnetic anisotropy whose easy axis of magnetization was perpendicular to the plane of the film. Further study of this particular compound was then pursued to gain greater insight into the factors affecting these characteristics.

EXPERIMENTAL METHOD

All the films used in this study were prepared by vapor deposition methods. The sample materials were evaporated directly from resistance-heated tungsten boats. Deposition rates varied from an approximate 0.3 grams of vaporized material per minute to 3.0 grams per minute. Deposition chamber starting pressures were on the order of 2×10^{-7} Torr, while substrate temperatures were controlled by a time-proportional heating system and ranged from 220°C to 400°C. Film thickness was monitored during deposition by a commercial instrument employing a quartz-crystal sensor. Films made for thickness dependence study were deposited on substrates in a holder having an electrically operated, latching sequential shutter which progressively exposed an additional substrate upon each actuation. To avoid the inconsistencies which arise when films are made from different batches of $TbFe_2$ such as the smaller lots produced in this laboratory, all the data presented here were obtained with films prepared from a single large batch of commercially produced material. The films were deposited on single crystal substrates which also served as acoustic delay lines. Again, for consistency's sake, all the data in this paper were obtained with X-cut quartz rod substrates although observations were made with sapphire and ruby as well. Substrate preparation prior to deposition consisted of a multiple bath, acid/rinse/drying agent/degreaser process. Protective films of silicon monoxide of 0.5 μm thickness were immediately deposited over each sample film to inhibit chemical and physical deterioration.

The UHF pulse spectometer consisted of a commercial 100 watt transmitter-receiver with an ancillary gated boxcar integrator providing detection and measurement of selectable echos. A modulating rf field coplanar with the sample film was achieved with a resonant-length stripline "cavity". The biasing dc magnetic field was provided by a conventional iron electromagnet having a field capability of 18 kOe in a two inch gap. Due to the very small masses of the films involved, magnetization measurements required an instrument of high sensitivity and were made with a Foner-type vibrating sample magnetometer constructed locally.

EXPERIMENTAL RESULTS

The magnetostrictive ultrasonic generation data included here were obtained with transverse mode acoustic waves at 670 MHz. Generation in this mode requires the rf magnetic field to be parallel to the plane of the film and the dc bias field to be perpendicular to the film. Weak longitudinal mode signals have been observed with some films, however, and were obtained with the dc field being rotated to lie parallel with the rf field. Figure 1 shows the dc magnetic

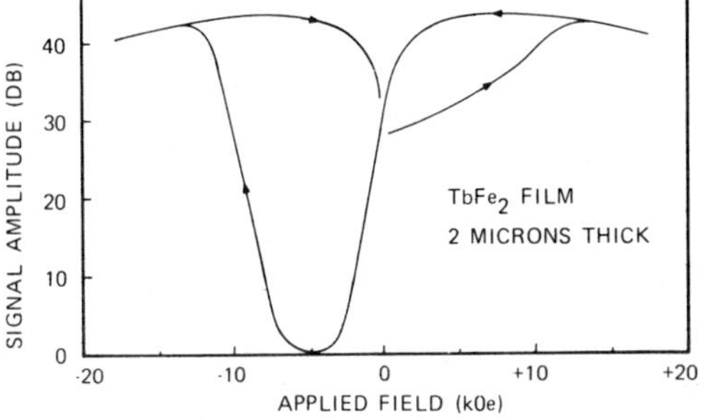

FIGURE 1. ULTRASONIC SIGNAL AMPLITUDE VERSUS APPLIED FIELD

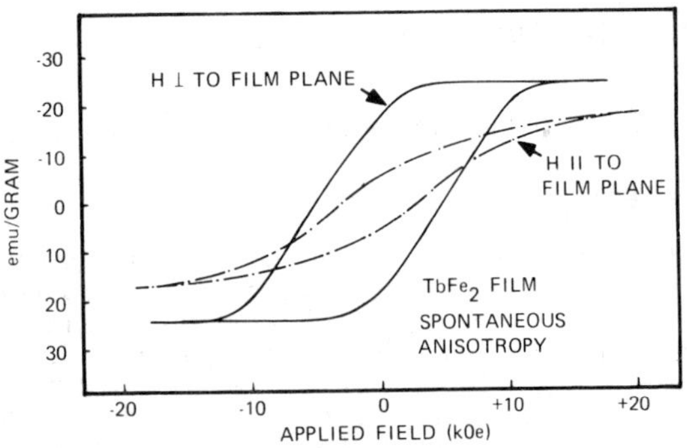

FIGURE 2. MAGNETIZATION VERSUS APPLIED FIELD

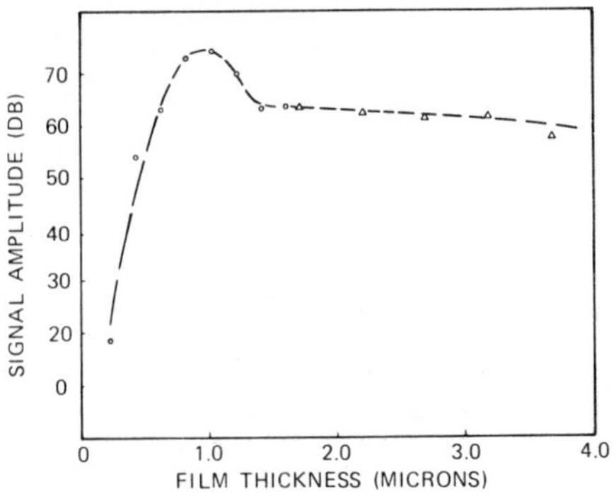

FIGURE 3. PEAK SIGNAL AMPLITUDE VERSUS FILM THICKNESS

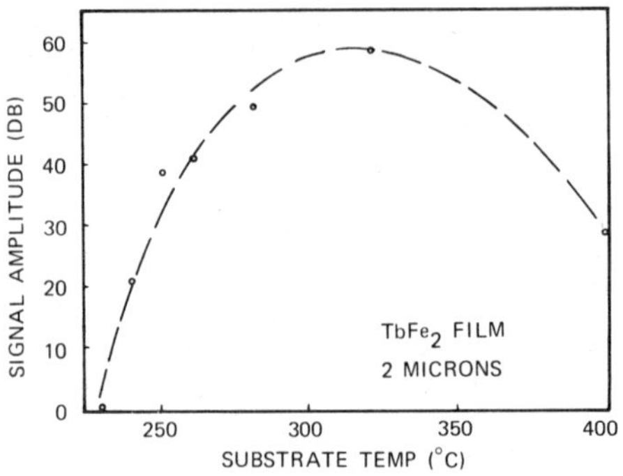

FIGURE 4. PEAK SIGNAL AMPLITUDE VERSUS SUBSTRATE TEMPERATURE

field dependence of the elastic wave generation capability of a typical $TbFe_2$ film and demonstrates those characteristics which make this material noteworthy. The peak generation efficiency is some 50 times greater than that of nickel (or more than 100 times more effective than the piezoelectric generation of longitudinal waves in the same substrate). One notes, moreover, that at zero applied field the acoustic signal level has dropped below the peak level by only 10 dB rather than falling to zero. This remanent signal implies the existence of a remanent magnetization which is confirmed by magnetization measurements. The arrows in the figure indicate the sense of the magnetic field sweep which was started in the same direction as the field that the film had just previously experienced. The remanent magnetization and large coercive force attendant to the aforementioned anisotropy as well as the perpendicularity of its easy axis relative to the plane of the film is evident in Figure 2. This figure shows the magnetization of a $TbFe_2$ film as a function of the applied external field with the applied fields perpendicular and parallel to the plane of the film.

Not all the films tested exhibited this anisotropy. Nor, it was discovered, did all those films having it possess it spontaneously without appropriate annealing, though many did. The reason for these differing conditions apparently lies in the deposition process, but numerous attempts to isolate the parameter or combination of parameters behind the cause were without success.

Films exhibiting a spontaneous anisotropy typically had coercive forces of 3.5 to 4.0 kOe with remanent signals 8 to 12 dB below peak, and film thickness did not noticeably affect these figures. The coercive forces and remanent magnetizations of the films having an induced anisotropy, on the other hand, were influenced by the annealing parameters. All the annealing was performed with a dc magnetic field of about 2 kOe perpendicular to the film. Annealing temperatures ranged between 325°C and 400°C, and annealing duration between one and twenty-four hours. Films thinner than 0.8 μm lost their ultrasonic generation properties during the annealing process, a 0.8 μm film had its efficiency significantly reduced, while thicker films of various thickness all appeared to respond alike as a group. The most interesting effect was associated with the annealing duration. Four hours at 350°C produced no apparent change in the characteristics of a film lacking spontaneous anisotropy; after 8 hours a film of the same thickness had a remanent signal 12 dB below peak with a coercive force of 4.5 kOe; 12 hours of annealing produced a coercive force of 6 kOe and a remanent signal only 4 dB down, but 24 hours at 400°C did not increase the coercive force beyond 6 kOe while the remanent signal dropped to 8 dB below maximum.

The annealing process, though effective in inducing a magnetic anisotropy where a spontaneous one was lacking, did not appear to contribute to the peak generation efficiencies of these films. A thickness study and a substrate temperature analysis were conducted to determine how these parameters might influence film efficiency. Figure 3 shows peak signal amplitude as a function of film thickness for two series of films. Films from the two, independent deposition runs are represented by different symbols, and the level of run A has been shifted to be contiuous with run B (discrepancies of several dB from one deposition run to the next were common, but all the films within a single run were consistent). A definite peak in efficiency is evident at 1 μm thickness and is presumably associated with a mechanical resonance of the films. The optimum substrate temperature is seen in Figure 4 to be 320°C.

CONCLUSIONS

$TbFe_2$ has applications potential in the areas of UHF and SHF ultrasonic transducers, thin film permanent magnets, and possibly in polycrystalline bubble domain devices. Additional study is indicated in the areas of annealing parameterization and deposition process. A careful study of sample material compositions of nominal $TbFe_2$ might also lead to interesting results. A simple theory has been developed which qualitatively accounts for the magnetoelastic acoustic wave generation facility of $TbFe_2$[3], but space limitations preclude its inclusion here. The large magnitude of the anisotropy and the perpendicularity of its easy axis to the plane of the film is, however, as yet unexplained. Further study should increase insight into that aspect of this material and, thereby, enhance utilization of its full potential.

REFERENCES

1. M. P. Maley, P. L. Donoho, and H. A. Blackstead, J. Appl. Phys. 37, 1006 (1966).
2. P. L. Donoho, L. V. Benningfield, P. K. Wunsch, L. B. McLane, and H. A. Blackstead, AIP Conf. Proc. 10, 762, 1973.
3. L. B. McLane, B.M. Kale, and P. L. Donoho, Proc. 1973 IEEE Ultrasonics Symposium, 313.
4. A. E. Clark and H. S. Belson, Phys. Rev. B 5, 3642 (1972).

* Supported in part by N.A.S.A., A.R.P.A., and N.S.F.

SECOND HARMONIC GENERATION OF ELASTIC WAVES IN $RbMnF_3$ THROUGH SPIN/ELASTIC INTERACTIONS

Jen King Jao and F.R. Morgenthaler

Department of Electrical Engineering and Center for Materials Science and Engineering, Massachusetts Institute of Technology, Cambridge, Mass. 02139

ABSTRACT

We report efficient second harmonic generation of elastic waves at a fundamental frequency of 524.2 MHz propagating in antiferromagnetic single crystal $RbMnF_3$ under spin flop conditions at a temperature of 4.6°K.

The energy conversion occurred whenever the magnon-phonon crossover was established by applying a suitable DC magnetic field. Under such conditions, elastic waves with frequencies below the hyperfine frequency couple strongly to spin waves, which in our case belong to one of the nuclear branches of spin-wave spectra. According to our model, the spin waves so excited then generate second harmonic elastic waves via the harmonic frequency component of the magnetoelastic force density. Based on this process a simple theory of nonlinear coupling among three waves is invoked to explain well the existing data. Careful analysis of the data yields an estimate of the effective nuclear spin relaxation frequency of 0.1 MHz for a wave number of 6.4×10^3 1/cm.

INTRODUCTION

Acoustic magnetic resonance has been observed by Merry and Bolef[1] in the cubic, low-anisotropy, antiferromagnetic $RbMnF_3$ under spin flop conditions. In their experiments, propagating longitudinal elastic waves with frequencies below the hyperfine frequency attenuated strongly at certain magnetic fields which were functions of the frequencies. They explained their observations as the excitation of nuclear spin resonances by elastic waves through the magnetoelastic fields. A more appropriate picture should, however, take proper consideration of the reactions of the nuclear spin waves on the elastic waves. These two waves couple together to form a mixture near the phonon-magnon crossover on their dispersion diagram.

Here we report additional observation of the generation of the second harmonic elastic waves under the same conditions mentioned above.

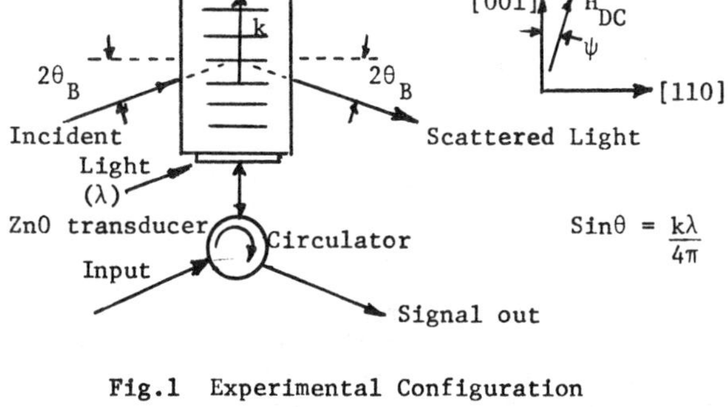

Fig.1 Experimental Configuration

EXPERIMENTAL

Fig. 1 illustrates the simplified experimental configurations. Our sample was an antiferromagnetic [001] oriented, single crystal $RbMnF_3$ bar of dimension 0.425" x 0.178" x 0.178" end faces optically polished.

A magnetic field larger than the spin flop field, which is of the order of $\sqrt{H_{ex}H_a}$, where $H_{ex} = 8.16 \times 10^5$ Oe is the exchange field and $H_a = 4.59$ Oe is the anisotropy field, is applied along the longer crystal axis. When the crystal is spin flopped, the sublattice magnetizations are virtually aligned along the easy direction in the plane perpendicular to the field. Because of the instrumentation problems, the absolute measurements of the fields were accurate to within only about 50 oe; the angle (ψ) only to within several degrees.

In the experiments, the whole sample was cooled in a helium exchange gas dewar to about 4.6°K. Square pulses of elastic energy at a fundamental frequency of 524.2 MHz, were injected into the sample through a ZnO thin film, longitudinal acoustic transducer deposited on one of the end faces. The harmonic elastic pulses generated were monitored by a microwave heterodyne receiver. Optical measurements were made of light scattering at a wavelength 6328 A at twice the Bragg angle of the fundamental.

We found that when the magnetic field was swept through spin-elastic crossover, a signal twice the input acoustic frequency appeared that coincided with the maximum attenuation of the fundamental elastic waves, as shown in Fig.2.

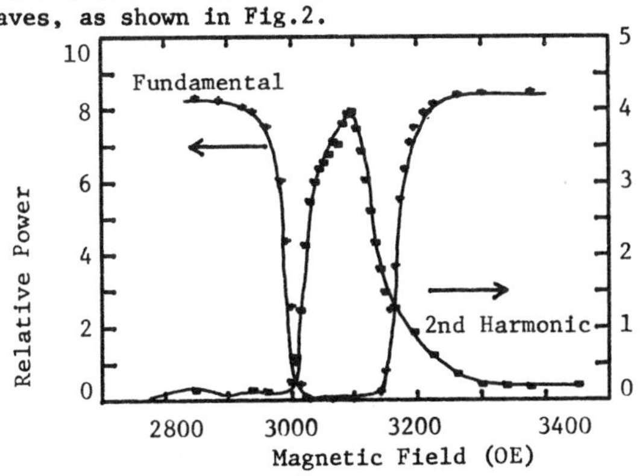

Fig.2 Acoustic power vs. field

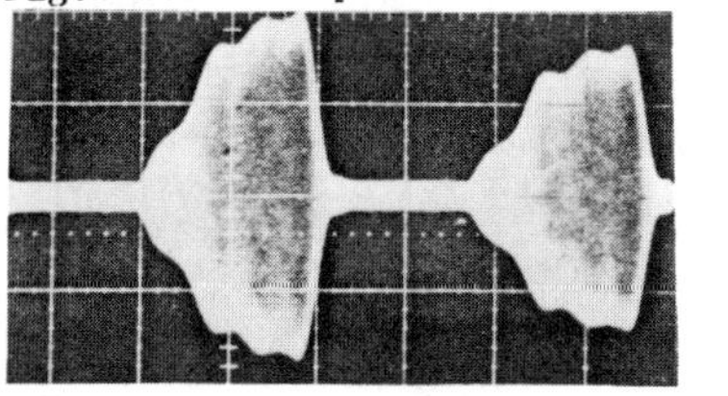

Fig.3 The Scope picture of the harmonic signal. Horizontal scale is 0.5 μ sec./cm.

The particularly interesting feature of this harmonic generation was the transient growth of the harmonic signal as shown in Fig.3. The signal build up is in periodic stages each followed by a plateau.

THE MODEL

Under the above experimental conditions, there exists a cascade process, namely, the elastic waves at the fundamental frequency of ω_p couple with the nuclear spin waves (referred to later only as spin waves unless otherwise specified); these spin waves then act as a source that pumps the harmonic elastic waves.

The links between the spin waves and the elastic waves are the fundamental and the harmonic frequency components of the magnetoelastic force density. The feedback is the effective magnetoelastic torque exerted by the elastic waves on the spin waves.

Despite the fact that the complete description

of the spin waves requires four coupled torque equations for the four sublattices[2], two electronic and two nuclear, a single equation for the electronic moment m_a along the major axis of the spin precession is sufficient to represent approximately its collective motion. There are several reasons that this simplification is permissible. First in the nuclear spin wave, the electronic magnetization is much larger than its nuclear counterpart and only the electronic moments couple directly with the elastic displacements. Also the spin precession ellipse has a large ellipticity[3] typically of the order of thousands. Therefore the magnetization along the minor axis is negligibly small compared with that along the major axis.

Define u, v, and p such that their absolute values are square roots of the average energies per unit volume of the fundamental elastic, the harmonic elastic and the spin waves respectively. Then

$$u = -\sqrt{\frac{D}{2}}\,\omega\rho^{\omega}\ ;\quad v = -\sqrt{2D}\,\omega\rho^{2\omega}\ ;\quad p = \frac{kb_1}{\sqrt{2D}}\cdot\frac{m_a}{KM} \qquad (1)$$

$$K = \left[\frac{\gamma_e^2\mu_o H_{ex}}{DM\,\omega\omega_n}\;\frac{\omega_{HF}^2-\omega_n^2}{\omega^2-\omega_n^2}\right]^{1/2} kb_1 \qquad (2)$$

Where γ_e is the gyromagnetic ratio, ρ^{ω}, $\rho^{2\omega}$ elastic displacements, k the wave number and M= 3820 Oe the sublattice magnetization; b_1 is the magnetoelastic constant, D = 4.356 gm/cm^3 the mass density $\omega_{HF}/2\pi$ = 686.2 MHz the hyperfine frequency, and ω_ρ, ω_n, ω_e are the fundamental elastic, the nuclear and the corresponding electronic spin wave frequencies.

Assume further that the medium is uniform; that variables have an e^{ikr} dependence; that all amplitudes are slowly varying; and that we are close to the crossover region $\omega \simeq \omega_n \simeq \omega_\rho$. A harmonic analysis yields

$$\dot{u} = j(\omega_\rho - \omega)u - 1/2bp \qquad (3a)$$

$$\dot{p} = [j(\omega_n-\omega)-\eta]p + 1/2\ bu - dp^{*}v \qquad (3b)$$

$$\dot{v} = 2j(\omega_\rho-\omega)v + dp^2 \qquad (3c)$$

$$1/2b = K\ f(\theta,\phi)\sin 2\theta_m \qquad (4a)$$

$$d = 2\sqrt{2D}\,\frac{K^2}{kb_1}\ f(\theta,\phi)\cos 2\theta_m \qquad (4b)$$

$$f(\theta,\phi) = \cos^2\theta_e^2 - \Phi\sin^2\theta_e \qquad (5a)$$

and

$$\Phi = \cos^2\phi_e\cos^2\phi_m + \sin^2\phi_e\sin^2\phi_m \qquad (5b)$$

Here η is the spin wave relaxation frequency. (θ_m,ϕ_m) and (θ_e,ϕ_e) are polar and azimuthal angles defining the direction of the magnetization and the elastic wave propagation.

The complete system has three additional equations conjugate to Eq.(3). Together, these equations imply that conservation of energy is satisfied. The paramter b is the magnetoelastic splitting frequency. Like any linearly coupled system, energies of the fundamental elastic waves and the spin waves exchange periodically with a period $2\pi/b$. The growth rate of the harmonic elastic signal, dependent upon d, is maximum when the energy stored in the spin waves is maximum. It not only determines the final conversion efficiency, but also affects the growth curve of the harmonic elastic waves.

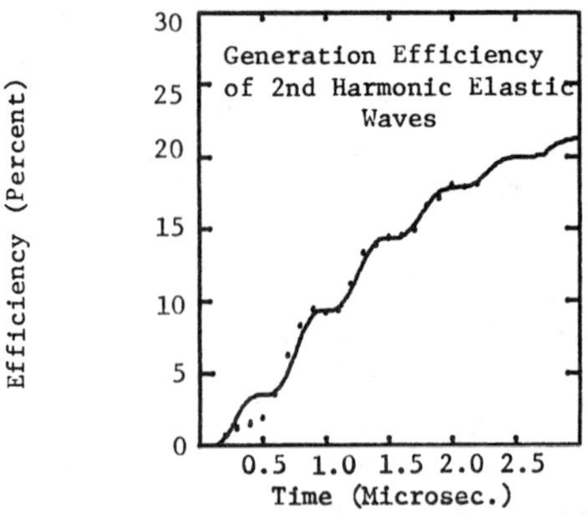

Fig.4 Generation efficiency vs. interaction time

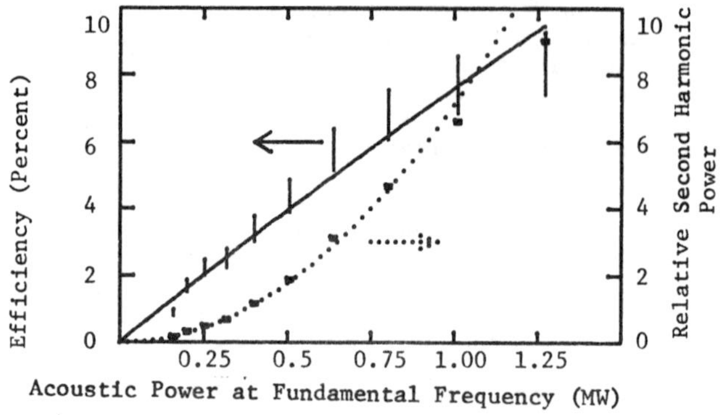

Fig.5 The harmonic power and generation efficiency vs. the acoustic power inside the sample

RESULTS

Numerical solutions of Eq.(3) are plotted in Fig.4 and Fig.5 and are chosen to best fit the experimental data. The acoustic power flux inside the sample was calculated by assuming a transducer area of 0.25 mm^2, a coupling loss of 21 db and by neglecting the elastic propagation loss. The angle ψ was taken to be 7 degrees. The estimated spin wave relaxation frequency was $\eta/2\pi \simeq$ 0.11 MHz.

Data points in Fig.4 are the measured conversion efficiencies for an input power of 160 mw that produces an acoustic power flux 0.5 watt/cm^2 as a function of the input acoustic pulse duration, or equivalently, the allowed interaction time. Essentially, this figure gives the transient build up of the harmonic power; the overall conversion efficiency, proportional to the input power, exceeded by 20%.

Fig. 5 shows the dependence of the conversion efficiencies and the harmonic signal power, on the input acoustic power in the sample for elastic pulses 1 μ sec. in duration.

The estimate of the nuclear spin relaxation frequency at the wave number of 6.4 x 10^{3}1/cm. appears reasonable as compared with the previously obtained parallel pumped relaxation frequency at a much larger wave number[4].

CONCLUSION

Second harmonic elastic waves were generated efficiently in antiferromagnetic $RbMnF_3$ through spin/elastic interactions at a moderate power level. Our model of coupled waves explains well the experimental data.

Although our work involved the particular field dependent nuclear branch of the spin wave spectra, our theoretical model is general in nature. Similar interactions should occur in other antiferromagnets and ferromagnets, especially those with large magnetoelastic constants or small spin wave relaxation frequencies. Therefore, this harmonic elastic generation process can provide an alternative probe of the relaxation time of spin waves with a preselected wave number and propagation direction.

Finally, in a system with a high spin-wave energy density, such as the surface spin waves which have attracted much interest recently, harmonic elastic generation may become an important process that should not be excluded from consideration.

ACKNOWLEDGEMENT

We would like to thank the Crystal Physics Laboratories at M.I.T. for supplying the $RbMnF_3$ crystal and Dr. A. Platzker for supplying the ZnO trans ducer.

REFERENCES

1. J.B. Merry and D.I. Bolef, Phys. Rev. Lett., 23, p. 126, 1969.
2. W.J. Ince, Phys. Rev. 184, p. 574, 1969.
3. L.W. Hinderks and P.M. Richards, Phys. Rev. 183, p. 575, 1969.
4. A. Platzker and F.R. Morgenthaler, Phys. Rev.Lett. 22, p. 1051, 1969.

MAGNETOELASTICALLY DRIVEN DOMAIN WALL RESONANCE AT MAGNETIC TRANSITIONS

D. T. Vigren
Institut für Theoretische Physik, Freie Universität Berlin,
1 Berlin 33, W. Germany

ABSTRACT

It is proposed that magnetoelastically driven domain wall resonance is the mechanism by which ultrasonic waves are absorbed near the ferromagnetic-spiral transition of Dy and Tb. Good agreement with the measured attenuation coefficient is obtained on the basis of a simple model. In Ho shear attenuation peaks observed at 24 K as well as some low temperature structure in the longitudinal attenuation appear to be due to this same mechanism.

INTRODUCTION

In heavy rare earth metals magnetic transitions are common where the ordered moment/ion is preserved with only a change in the spacial configuration of the moments. These transitions are usually of first order, and although the latent heat is extremely small[1], there are magnetostrictive volume changes and formation of domains[2]. The transitions are accomplished through the movement of Bloch walls, whereby domains of one spin structure grow at the expense of the other. In Ho a host of such transitions occurs in moderate applied fields, and neutron diffraction patterns give direct evidence of rather narrow regions of mixed magnetic phase[3]. In the mixed phase regions numerous Bloch walls are pinned to inclusions and other imperfections, which present local potential valleys to the walls. In this paper we show that sound waves are capable of driving the Bloch walls harmonically. Their motion creates local fluctuations of the magnetization, giving rise to eddy currents which provide the dissipative process for resonant absorption of the sound. That such domain wall resonance is possible at ultrasonic frequencies is supported by observation of wall resonance in Ni ferrite at 22 MHz with an oscillating magnetic field[4]. Here, however, the driving force is of magnetoelastic origin.

ATTENUATION IN TERBIUM AND DYSPROSIUM

Let us apply these ideas to Tb and Dy metals, which show dramatic longitudinal sound absorption at the spiral-ferromagnetic transition temperature, T_C[5,6]. We define T_C to be the temperature at which the magnetic phases are most highly mixed, and the number of Bloch walls most numerous. Significantly far from T_C only a small number of very strongly pinned walls remain, the others having merged to produce a predominance of the stable phase. For T somewhat below T_C the moments are ferromagnetically aligned and confined to the basal plane by large axial anisotropy. For T somewhat above T_C moments form the planar spiral structure[2]. Very near T_C the crystalis in a mixed phase of ferromagnetic and spiral domains separated by thin Bloch walls, lying in planes perpendicular to the c-axis. We assume that the spacing between walls is much larger than the range of the exchange coupling between moments, but much smaller than the wavelength of the ultrasound which propagates along the c-axis. In this way the strains induced may be taken as uniform in domains where a well-defined magnetic structure exists.

Consider a particular Bloch wall at the center of its potential well in the x-y plane. For $z < 0$, the region is ferromagnetic with magnetization $\vec{M} = M_0\hat{x}$; for $z > 0$, the region is spiral antiferromagnetic with $\vec{M} = 0$. An arbitrary magnetic field $\vec{H}$ is applied along $\hat{x}$. For a longitudinal sound wave propagating in the z direction we may write the strain in any domain as $\epsilon_{zz} = \epsilon_{zz}^{o}\sin\omega_{ph}t$, where ϵ_{zz}^{o} is the amplitude of the wave and ω_{ph} is its frequency. In the presence of this strain, a difference in the free-energy densities of the ferromagnetic and spiral regimes is created[7], giving rise to a strain-dependent pressure on the Bloch wall, $\vec{p}$:

$$\vec{P} = (1-\cos\psi)(c\lambda)_{\alpha}\,\epsilon_{zz}^{o}\sin\omega_{ph}t\;\hat{z} \qquad (1)$$

Here ψ is the spiral-turn angle and $(c\lambda)_{\alpha} \equiv c_{11}^{\alpha}\lambda_1 + c_{22}^{\alpha}\lambda_2/\sqrt{3}$. (Without altering the qualitative results we assume the elastic constants c_{ii}^{α} and the magnetoelastic constants λ_i to be independent of magnetic phase.)

Defining a damping constant β, associated with the eddy-current losses, we may write down the equation of forced hamronic oscillations of the Bloch wall:

$$m_o\ddot{z} + \beta\dot{z} + \kappa_o z = (1-\cos\psi)(c\lambda)_{\alpha}\,\epsilon_{zz}^{o}\sin\omega_{ph}t. \qquad (2)$$

Here m_0 is the effective mass/area of the wall, pinned elastically to a local imperfection with a stiffness constant, κ_o. The quantities m_0 and κ_o depend on the wall energy variation, which arises primarily from local demagnetizing effects of inclusions and nonuniform local stresses[8]. Therefore, we expect m_0 and κ_o to be quite sensitive to the application of a magnetic field. Eqn. (2) suggests that a strong resonant absorption of sound should occur when $\omega_{ph} \simeq \sqrt{\kappa_o(H)/m_o(H)}$. Thus, the application of a field may in some instances bring about an approach to resonance, and in others a reduction of the absorption.

In a very simple model of the situation we take all walls and pinning sites to be identical and assume that the number of walls pinned at temperature T is given by $n(\kappa_o, T) = n_0(\kappa_o)\,\mathcal{L}(T-T_C;\gamma_0)$, where $\mathcal{L}$ is a Lorentzian, centered at T_C, of width γ_0 that characterizes the temperature width of the phase transition.

The ultrasonic attenuation coefficient α_L is given by $\alpha_L(\vec{H},T) = (nP_W / v_{ph}U_{ph})$, where v_{ph} is the velocity of sound, $U_{ph} = \frac{1}{2}(c_{11}^{\alpha} + \frac{1}{3}c_{22}^{\alpha})(\epsilon_{zz}^{o})^2$ is the elastic-energy density of the longitudinal wave, and P_W is the forced harmonic power dissipation of a single wall. Substitution of the various quantities gives

$$\alpha_L(\vec{H},T) = \frac{n_o(1-\cos\psi)^2(c\lambda)_{\alpha}^2\, f(\omega_{ph},\vec{H})\,\mathcal{L}(T-T_c;\gamma_o)}{\beta v_{ph}(c_{11}^{\alpha} + \frac{1}{3}c_{22}^{\alpha})} \qquad (3)$$

where $f(\omega_{ph},\vec{H})$ characterizes the resonant nature of the absorption:

$$f(\omega_{ph},\vec{H}) = \left[1 + \left(\frac{m_o(\vec{H})}{\beta(\vec{H})\omega_{ph}}\right)^2\left(\omega_{ph}^2 - \frac{\kappa_o(\vec{H})}{m_o(\vec{H})}\right)^2\right]^{-1} \qquad (4)$$

This formula for α_L is compared with experiment[5,6] in Fig. 1. The values of n_0f/β needed to fit the experimental peak heights indicate an approach to resonance with increasing $\vec{H}$ for the Dy sample and nearly on-resonance absorption for the Tb sample in zero field[9]. For Dy, the value $\gamma_0 = 18$ K was used to fit all three curves shown in Fig. 1(a), whereas $\gamma_0 = 15$ K was used to fit the Tb data. These values compare well with the experimental widths of the magnetic transition.[10,11]

It should be noted that shear strains do not give rise to a pressure on Bloch walls that separate planar spiral from planar ferromagnetic domains. The strains ϵ_{xy}, ϵ_{yz}, and ϵ_{xz} couple magnetoelastically to the spin functions S^XS^Y, S^YS^Z, S^XS^Z, respectively[12]. For the magnetic structures found in Dy and Tb, the thermal average of these spin functions is zero, so that externally applied shear strains give no contribution to the free-energy densities in the spiral or ferromagnetic regions. Consequently, no absorption of shear sound is expected. This prediction has been confirmed in Tb[6].

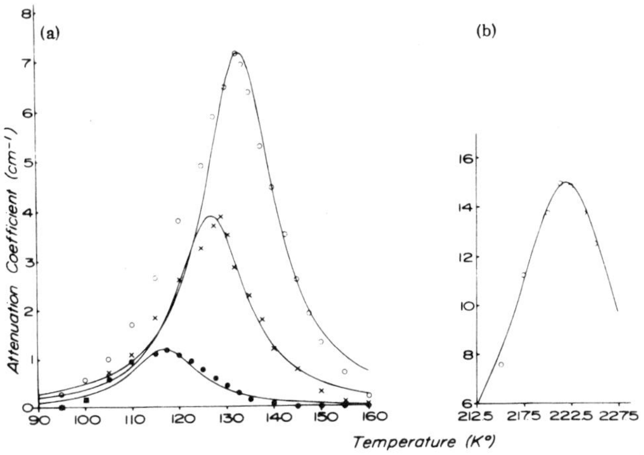

FIG. 1. Ultrasonic attenuation coefficient versus temperature. The solid curves are those predicted by the wall-resonance model. (a) Experimental points in Dy for three magnetic field strengths applied along the (easy) a-axis using the scheme 5.5 kOe (closed circles), 7.7 kOe (crosses), 9.2 kOe (open circles). (b) Experimental points in Tb for zero magnetic field.

ATTENUATION IN HOLMIUM

Longitudinal sound, propogating along the c-axis of Ho, is strongly absorbed[13,14] at 20 K, the temperature at which a flat cone structure stabilizes[3]. This peak is quite insensitive to an applied field in the basal plane and is most probably not due to domain wall resonance. However, at 28 K in a field of 5.3 kOe, applied in the a-direction, a second peak appears[13,14] and grows dramatically with increasing field, shifting to 38 K at 9.2 kOe. The magnetic phase diagram of Ho indicates that these peaks occur in regions of mixed magnetic phase[3], and magnetoelastically driven wall resonance is most likely responsible for them. All mixed phase structures above 20 K are such that the moments lie in the basal plane. We expect, therefore, that shear waves which create strains belonging to the ϵ -representation and propogating along the c-direction should not give rise to a pressure on the Bloch walls and no corresponding shear attenuation peaks should be seen. An investigation of the shear absorption using fields between 5-10 kOe above 20 K would be of interest in this respect.

Very recently[15], peaks in the attenuation of transverse sound were observed at 24 K in zero field, whereas longitudinal waves, propogated in a like manner, were not absorbed. Although neutron diffraction studies show no magnetic transition at 24 K in zero field, it has been predicted that a 'tilted-spiral' phase must stabilize before the cone does, if the anisotropic exchange is not too large[16,17]. The tilted-spiral (TS) may be envisioned by tilting the ferromagnetic planes of a normal spiral (NS) about an axis in the basal plane. Originally it was thought that the transition NS→TS was second order, occuring at a temperature where the spiral magnon modes of wavevector $2\vec{Q}$ soften, $\vec{Q}$ being the wavevector of the spiral. Recently[17], however, it was shown that before this mode completely softens, the TS is stabilized by a favorable balance of exchange, magnetoelastic and crystal anisotropy energies. Thus, the transition is probably first order and accomplished through a process of domain growth. A preliminary calculation based on the wall resonance model shows that the shear absorption is proportional to $\sin^2\theta$, whereas the longitudinal absorption is proportional to $\sin^4\theta$, θ being the tilt angle. Since θ is expected to be quite small, the model predicts a rather small shear peak, with the longitudinal absorption being negligible by comparison. The observed shear peaks are quite sensitive to an applied field, decreasing with increasing $\vec{H}$, implying that the domain walls are brought out of resonance by the field. The fact that the magnitude of these peaks is very sample dependent[15], should add credence to the belief that domain wall resonance is the responsible mechanism.

REFERENCES

1. R.J. Elliott, Phys. Rev. 124, 346 (1961)
2. B.R. Cooper, Solid State Phys. 21, 393 (1968)
3. W.C. Koehler et al., Phys. Rev. 158, 450 (1967)
4. A. Globus and P. Duplex, in Proceedings of the International Conference on Magnetism, Nottingham, England, 1964 (The Physical Society, London, England, 1965), pp 635-638
5. M.C. Lee and M. Levy, Phys. Lett. 32A, 88 (1970)
6. R.J. Pollina and B. Lüthi, Phys. Rev. 177, 841 (1969)
7. W.E. Evenson and S.H. Liu, Phys. Rev. 178, 783 (1969)
8. R.S. Tebble and D.J. Craik, in Magnetic Materials (Wiley, New York, 1969)
9. D.T. Vigren, Phys. Rev. Lett. 32, 1254 (1974)
10. Behrendt et al., Phys. Rev. 109, 1544 (1958)
11. Hegland et al., Phys. Rev. 131, 158 (1963)
12. E. Callen and H.B. Callen, Phys. Rev. 139, A455 (1965)
13. M. Levy and M.C. Lee, Phys. Lett. 32A, 294 (1970)
14. M.C. Lee and Moisés Levy, J. Phys. Chem. Solids 34, 995 (1973)
15. T. Tachiki et al., Solid State Comm. 15, 1o71 (1974)
16. David Sherrington, J. Phys. C: Solid State Phys. 6, 1037 (1973)
17. B.W. Southern and D. Sherrington, J. Phys. C: Solid State Phys. (to be published)

MOSSBAUER AND X-RAY STUDY OF THE SELF-INDUCED MAGNETOSTATIC DISTORTION OF $Tb_xY_{1-x}Fe_2$ LAVES PHASES†

R.S. Preston, A.E. Dwight, A.J. Fedro and C.W. Kimball,
Northern Illinois University, DeKalb, Illinois 60115

ABSTRACT

A comparison has been made between x-ray diffraction data and Mössbauer absorption data for a series of ferrimagnetic compounds $Tb_xY_{1-x}Fe_2$. Previous work has shown that the volume and the rhombohedral angle of the distorted "cubic" unit cell both vary with the Tb concentration x, but that the distortion disappears for $x \lesssim 0.25$. Having noted that this is close to the concentration where the rare-earth and the iron sublattice magnetizations are equal and opposite, we made a detailed study of the ^{57}Fe Mössbauer spectra for these compounds in the vicinity of x = 0.25. Our results confirm the previous finding that the magnetic hyperfine field varies continuously with the Tb concentration. Also, they reveal that the Fe sites are not all equivalent chemically and crystallographically, and that this is true whether or not the rhombohedral distortion is present. This suggests that in addition to the rhombohedral distortion there may be an independent distortion or shifting of the iron tetrahedra within the unit cube.

INTRODUCTION

As is being reported elsewhere,[1] x-ray diffraction measurements on a series of compounds $Tb_xY_{1-x}Fe_2$ show that these ostensibly cubic Laves phases experience a measurable rhombohedral distortion for values of x > 0.25. Buschow and van Stapele[2] studied spontaneous magnetization in the same system. They found that these compounds are ferrimagnetic at room temperature with the iron and rare-earth sublattice magnetizations aligned oppositely. For $x \simeq 0.35$ the sublattice magnetizations are equal and opposite (magnetic compensation) as shown in curve a of Figure 1. At compositions with larger Y concentrations the Fe sublattice magnetization dominates, reflecting the fact that the Y atoms have no moments. On the Tb-rich side of the compensation point the rare-earth sublattice dominates the magnetization.

The x-ray measurements show a rhombohedral stretching of the unit cell along a [111] direction. This is evident in the decrease of the rhombohedral angle α below its nominal value of 60°, as shown in curve b of Figure 1. The spontaneous magnetization is also along a [111] direction. This is shown by the Mössbauer data of Dariel et al.[3], as well as by our own. Thus the direction of the distortion appears to coincide with the direction of the magnetization, in agreement with expectations based on simple magnetoelastic theory. The surprising feature of this result is that this distortion disappears on the Y-rich side of the compensation point where the Fe magnetization dominates. It is also noteworthy that the variation of the volume with x, although small, appears to be greatest near the compensation point (curve c, Figure 1).

In an effort to learn more about the nature and causes of the rhombohedral distortion we have carried out detailed measurement and analysis of the ^{57}Fe Mössbauer absorption in samples with x values in the neighborhood of 0.25. Since a Y atom has essentially the same volume as a Tb atom, there should be little change in the volume of the mixed compounds over the entire range, as is confirmed by the data of curve c, Figure 1. Moreover, the Y atoms are in 3^+ valence states as are the Tb atoms. Consequently, variations of the crystal field, which might affect the magnetic behavior of atomic electrons as well as the nuclear quadrupole splittings, are minimized. However, each Tb atom has a magnetic moment of ∿ 8.3 Bohr magnetons, while the moments of the Y atoms are zero.[2] Thus the most immediate consequence of a change in x should be a change

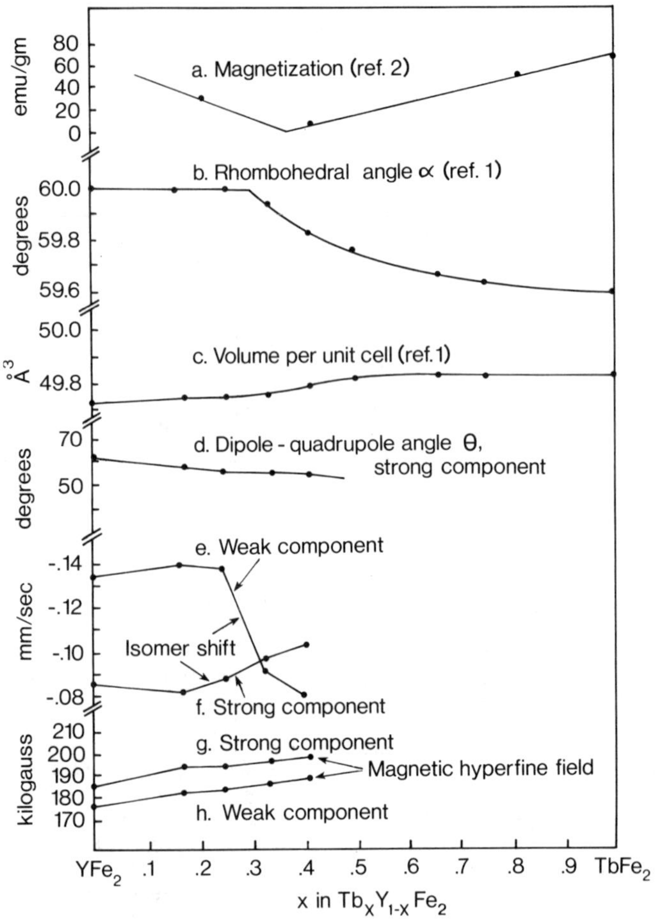

Fig. 1 Summary of magnetization, x-ray and Mössbauer parameters for the $Tb_xY_{1-x}Fe_2$ system. The isomer shifts are given with respect to a Co^{57} in Cr source at 295K.

in the average moment per atom on the rare-earth sublattice, and this should be reflected, to some extent, in the behavior of the Mössbauer spectrum. The previous Mössbauer results of Dariel et al. for this system showed that the magnetic hyperfine field associated with the more intense component of the two-component spectrum varies smoothly with x.

DISCUSSION

Our Mössbauer results indicate the possible presence of a second kind of distortion,--namely a compression, a stretching or a displacement of each tetrahedron of Fe atoms along the direction of magnetization. The Mössbauer spectrum is composed of two six-line components with an intensity ratio of 3:1. In the intense component the angle θ between the magnetic hyperfine field and the principal quadrupole field axis would be 70° 32' if the Fe atoms were in tetrahedra at the usual locations within the unit cell. This angle θ is usually close to 70° in other cubic Laves phases which magnetize along the [111] direction, but it is far from that value in this system, as may be seen in curve d of Figure 1. For the weak component the value of θ remains at or near 0°, as expected, so that it is unlikely that the magnetization has turned away from the [111] direction.

In addition, the values of the quadrupole interaction, $\frac{1}{2}e^2qQ$ (not shown) are different for the two components, as are the values of the isomer shift (curves e and f, Figure 1). As shown in curves g and h of Figure 1, the magnetic hyperfine fields for the two components also differ, but this would be expected even for an undistorted lattice. On the other hand, differences in isomer shift are not expected for an undistorted lattice. Nonetheless, differences do exist, and they are large in the concentration region where x-ray diffraction reveals no distortion.

All of these results imply that the iron sites corresponding to the two components of the spectrum are not equivalent crystallographically, which means that the iron atoms are displaced from their conventional positions. The fact that θ departs from 70° for one site, and remains at 0° for the other site points to a displacement of the iron atoms which leaves the quadrupole field axially symmetric about the [111] direction of magnetization. In the absence of detailed understanding of the sources of the quadrupole field it is impossible to use the values of θ, $\frac{1}{2}e^2qQ$ and isomer shift to determine the magnitudes or even the signs of the Fe displacements.

Calculations of expected x-ray diffraction intensities based on simple assumptions about the magnitudes of the Fe displacement show that such displacements would produce a small change in intensity of an already prominent Debye-Scherrer line in the diffraction pattern for $TbFe_2$ (x=1) and would therefore be difficult to detect. This line is too weak to be seen in the pattern for YFe_2 (x=0) and similar calculations show that it would remain too weak, even with displacement of the iron atoms. Thus we have as yet no independent corroboration of the existence of this internal distortion in the unit cell. A possible alternative explanation of our Mössbauer data is being pursued, namely that Tb and Y are not similar. This could lead to a distribution of the values of the parameters of the Hamiltonian for the hyperfine interaction. If so, our present computerized fitting procedure, which assumes that only two spectra are present, would give average values, at best, for the parameters, and might give spurious averages for some of them as a result of fitting the data to an oversimplified model.

†Based on work performed under the auspices of the National Science Foundation.

REFERENCES

1. Dwight, A.E. and Kimball, C.W., Acta. Cryst. (to be published).
2. Buschow, K.H.J. and Van Stapele, R.P., J. Appl. Phys. 41, 4066-4069 (1970).
3. Dariel, H.P., Atzmony, V. and Lebenbaum, D., Proceedings of the Tenth Rare Earth Conference, U.S. Atomic Energy Commission, Oak Ridge, 1973, pp. 439-447.

ELECTRIC FIELD DEPENDENCE OF THE MAGNETIC ANISOTROPY ENERGY IN MAGNETITE

G. T. Rado and J. M. Ferrari
Naval Research Laboratory, Washington, D.C. 20375

ABSTRACT

We show that in synthetic magnetite (Fe_3O_4) the macroscopic anisotropy coefficients depend on an external electric field $\vec{E}$. No such dependence has been reported previously for any material. Using $\vec{E}=0$ throughout, we observed that at 4.2°K an applied magnetic field $\vec{H}$ induces an electric polarization $\vec{P}$ whose components are nonlinear functions of the components of $\vec{H}$. For the case of a magnetically annealed circular disk whose faces are parallel to the orthorhombic bc plane (c=easy axis; b= intermediate axis), we replace the usual anisotropy coefficients K_b and K_{bb} by $K_b'=K_b+(\partial K_b'/\partial \vec{E})\cdot\vec{E}$ and $K_{bb}'=K_{bb}+(\partial K_{bb}'/\partial \vec{E})\cdot\vec{E}$, respectively, and express the free energy density by $\tilde{F}=K_b'\sin^2\theta+K_{bb}'\sin^4\theta-MH\cos(\psi-\theta)$. Here $\vec{M}$ is the magnetization while θ and ψ denote, respectively, the orientation of $\vec{M}$ and $\vec{H}$ with respect to the c axis. For $P_a=-(\partial\tilde{F}/\partial E_a)_{\vec{H}}$, we obtain $A_b\sin^2\theta+A_{bb}\sin^4\theta$ after relating H and ψ to θ via the equilibrium condition $(\partial\tilde{F}/\partial\theta)_{\vec{E},\vec{H}}=0$ and labeling the coefficients $\partial K_b'/\partial E_a$ and $\partial K_{bb}'/\partial E_a$ by $-A_b$ and $-A_{bb}$, respectively. If H is sufficiently large (i.e., $\theta\approx\psi$), then P_a becomes $A_b\sin^2\psi+A_{bb}\sin^4\psi$. Our measurements of P_a vs. ψ at constant H do verify this prediction and yield $A_{bb}/A_b=0.08\pm0.04$ with $|A_b|$ (which depends on the magnetic anneal) being approximately 4 statcoul/cm^2. Also in good agreement with the present model (without the use of adjustable parameters) are our measurements of P_a vs. H at constant ψ. Possible atomic mechanisms of A_b are briefly considered, and it is shown unambiguously that the crystallographic symmetry of magnetite at 4.2°K is triclinic with point group 1. The full text of this paper will be published elsewhere.

SPIN REORIENTATION STUDIES IN CUBIC LAVES RARE-EARTH IRON COMPOUNDS

U. Atzmony and M.P. Dariel
Atomic Energy Commission, Nuclear Research Centre-Negev
P.O.B.9001,Beer-Sheva 84190, Israel

A B S T R A C T

Several types of spin reorientation occur in $Ho_xEr_{1-x}Fe_2$, particularly that from the [110] toward the [100] direction. In all cases the general behavior of the hyperfine fields at the inequivalent iron sites is well accounted for by magnetic dipolar contributions. Analysis of the Mössbauer spectra taken in the spin rotation temperature interval indicates that the easy axis of magnetization deviates from the axes of symmetry of the cubic system. In the [110] to [111] rotation the spin is in the $(1\bar{1}0)$ plane, while in the [110] to [100] transition it rotates continuously within the (001) plane.

I N T R O D U C T I O N

Cubic rare-earth (R) - iron binary RFe_2 and ternary $(R^1_x R^2_{1-x} Fe_2)$ compounds were studied in recent years. The spin orientation diagrams (S.O.D) of the ternary systems were determined[1] by Mössbauer effect,(ME), studies on ^{57}Fe. Measurements of the elastic constants confirmed the ME findings in the $Ho_xTb_{1-x}Fe_2$ system[2]. The single ion model reproduced the main features of the experimentally determined S.O.D. This model predicts a first order transition, while ME measurements indicate that the spin rotation spreads over a wide temperature interval[3,4]. The behavior of the magnetic hyperfine fields was explained by magnetic dipolar contributions. The direction of the easy axis of magnetization, $\vec{n}$, during the spin rotation was then followed[3,4]. In the present study the S.O.D of the $Ho_xEr_{1-x}Fe_2$ system is revised and the mechanism of the spin rotation, is further examined.

RESULTS AND DISCUSSION

(a) Spin orientation diagram of the $Ho_xEr_{1-x}Fe_2$ system:

In the composition range $0.8 \leqslant x \leqslant 1.0$ sound velocity (S.V.) measurements revealed the existence of spin rotations, not detected in previous M.E. measurements. New M.E. measurements in samples with x = 0.8, 0.9 and 1 showed that spin rotations indeed occur in these compounds at low temperatures, i.e. for every compound there is a temperature below which the spectra ceases to be of the [100] type. For $Ho_{0.8}Er_{0.2}Fe_2$, spectra of the [110] type were obtained at 10 K, whereas for $Ho_{0.9}Er_{0.1}Fe_2$ and $HoFe_2$ this type was not obtained even at 4.2 K. A revised S.O.D is shown in Fig.1. The dotted-dashed lines are the theoretical boundaries[1]. Filled circles, filled triangles and filled squares correspond to experimentally determined spectra characteristic of the [111],[110] and [100] axes of magnetization, respectively. Open triangles correspond to intermediate spectra, i.e. spectra that could not be fitted to any of the three main types.

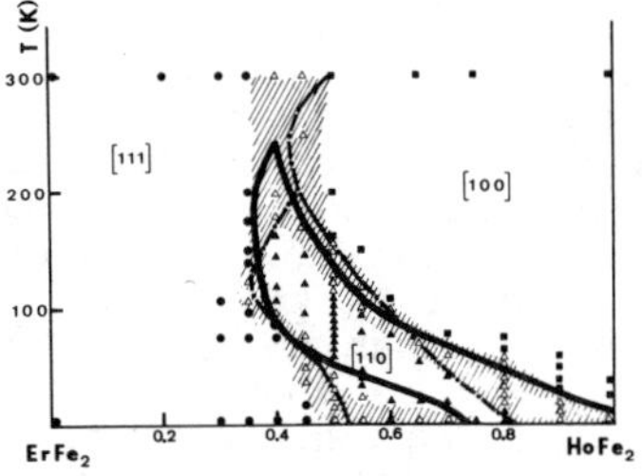

Fig.1: Spin orientation diagram of the $Ho_xEr_{1-x}Fe_2$ systems

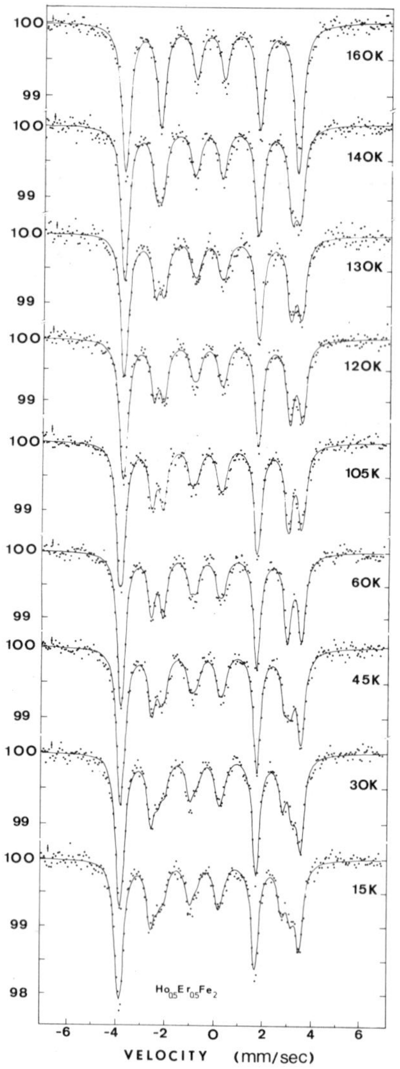

Fig.2: Mösbauer spectra of $Ho_{0.5}Er_{0.5}Fe_2$ at various temperatures. The solid lines are the best theoretical fits

The solid line in the S.O.D.is deduced from S.V. measurements. As can be seen,the transition region is relatively wide (shaded area in Fig.1). The theoretical boundaries are approximately in the middle of the transition region, whereas those obtained from S.V. measurements tend to be on it's upper temperature limit. Similar behavior was obtained in S.V. measurements in Gd[5]. There too the elastic constants have a minimum at the upper limit of the spin reorientation range.

(b) Spin reorientation mechanism: $Ho_{0.5}Er_{0.5}Fe_2$: Several spectra of this compound are shown in Fig.2. The solid lines are the best-fitted theoretical spectra. Four temperature ranges exist.

(1) T = 60 K: The spectra from 60 K down,resemble those of $SmFe_2$ in its transition range[3]. As the transition here is also from the [110] direction at 60 K towards the [111] direction at lower temperatures the same fitting procedure was applied. Assuming that the spin rotates continuously in the $(1\bar{1}0)$ plane, three inequivalent iron sites with population ratio 2:1:1 are obtained. The temperature behavior of the magnetic hyperfine fields in these sites is plotted in Fig.3.

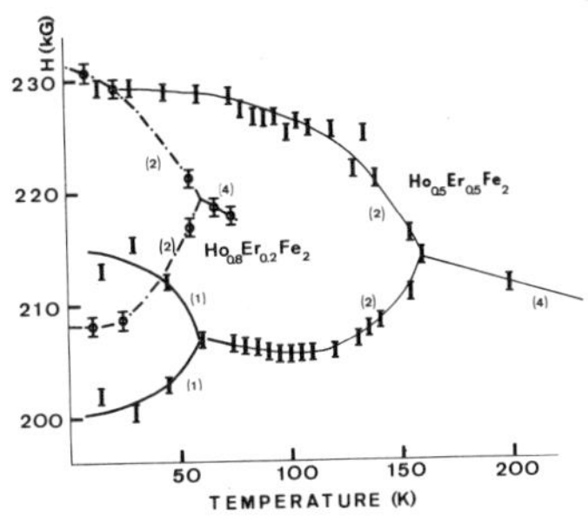

Fig.3: The hyperfine magnetic field acting on the various iron sites in $Ho_{0.5}Er_{0.5}Fe_2$ and $Ho_{0.8}Er_{0.2}Fe_2$ as function of temperature

The angle between the [110] direction and $\vec{n}$ changes almost linearly with temperature from 0° at 60 K to about 20° at 15 K. The behavior of the magnetic hyperfine fields is accounted bor by magnetic dipolar contributions.

(2) $60 < T < 100$ K : The spectra in this range indicate the presence of two equally populated inequivalent iron sites, which is typical of the [110] direction of magnetization. This is shown in Fig.2 by the equality of the intensities of the twin absorption peaks on the far right. The splitting between them, Δ, measures the difference between the hyperfine fields acting on the two iron sites. This difference is approximately constant in this range and it is therefore assumed that $\vec{n}$ is parallel to the [110] direction. The spectra were fitted accordingly. The obtained hyperfine fields are plotted in Fig.3. The fields decrease slightly with temperature as expected.

(3) $100 \leqslant T \leqslant 160$ K: In this range Δ decreases monotonically with T. This continuous change results from the decreasing dipolar field contributions as a result of a continuous spin rotation. If the direction of the spin is in the (001) plane (and is thus parallel to a [u v o] direction) then there are two equally populated inequivalent iron sites. In Fig.4 the calculated dipolar field contributions to the hyperfine magnetic fields, acting at the two sites (neglecting the temperature dependence of the dipole moments) are plotted in arbitrary units as a function of the angle $\phi = tg^{-1}u/v$. The splitting, Δ, behaves similarly. It can therefore be concluded that the spin rotates continuously from the [110] to the [100] axis in the (001) plane. Unfortunately the ϕ dependence of the theoretical spectra in the fitting procedure is weak and therefore the exact temperature dependence of ϕ could not be deduced. Nevertheless, allowing ϕ to change linearly with temperature the spectra obtained in this temperature range were fitted to two inequivalent iron sites. The obtained theoretical spectra and hyperfine fields are plotted in Figs.2 and 3, respectively. The resemblance between Fig.4 and the relevant portion of Fig.3 is remarkable. The slight differences are attributed to the temperature dependence of the fields.

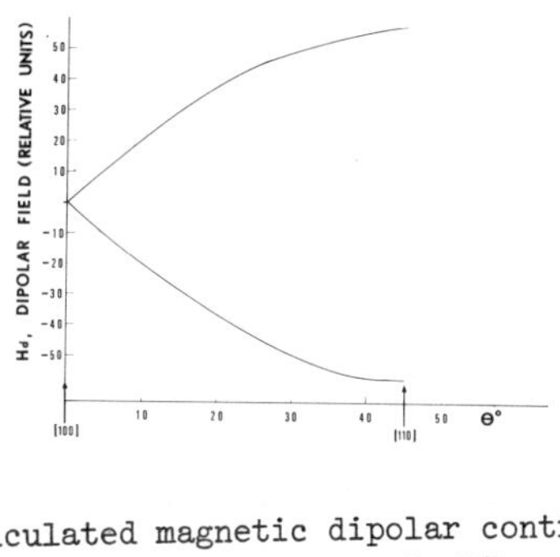

Fig.4: Calculated magnetic dipolar contributions to inequivalent iron sites in RFe_2 as function of ϕ; $\vec{n}$ is assumed to be in the (001) plane

(4) $T \geqslant 160$: Above T = 160 K the two peaks coincide. Though the spectrum obtained at 160 K was reasonably well fitted to one iron site with $\vec{n}$ in the [100] direction, a thorough inspection of the theoretical and experimental spectra reveals small discrepancies (mostly in the right hand peak). Similar discrepancies exist in spectra taken at 200 and 300 K. From the S.O.D. it is seen that these three points are near the transition region boundaries.

$Ho_{0.8}Er_{0.2}Fe_2$: Spectra for this compound at $10 \leqslant T \leqslant 65$ resemble those obtained for $Ho_{0.5}Er_{0.5}Fe_2$ at $60 \leqslant T \leqslant 160$. Therefore the same analysis was applied. The temperature dependence of the hyperfine fields is plotted in Fig.3. It is similar to that of $Ho_{0.5}Er_{0.5}Fe_2$ in the corresponding temperature range. The derived hyperfine fields at 10 K are of the order of those in $Ho_{0.5}Er_{0.5}Fe_2$ for $60 \leqslant T \leqslant 100$. It is therefore concluded that for $T \leqslant 10$ K, $\vec{n}$ is in the [110] direction. The spectrum at 65 K is of the [100] type and was fitted accordingly. The agreement between the theoretical [100] and the experimental spectrum is very good. It is therefore concluded that here too the direction of $\vec{n}$ changes continuously from the [110] to the [100] direction.

$Ho_{.9}Er_{.1}Fe_2$: The spectra for this compound at low temperatures are similar to those of $Ho_{.8}Er_{.2}Fe$. The largest Δ, is less than half of that for $Ho_{.8}Er_{.2}Fe_2$ at 10 K. On the other hand the spectra at $T \geqslant 25$ K undoubledly belong to the [100] type. It is concluded that in $Ho_{.9}Er_{.1}Fe_2$ the spin rotates continuously in the (001) plane from the [100] toward the [110] axis with decreasing temperature. However, $\vec{n}$ is not parallel to [110] even at 4.2 K. Comparison of Δ of these spectra with those of $Ho_{.8}Er_{.2}Fe_2$, yields ϕ at 4.2 K of about 20°.

$HoFe_2$: Spectra obtained at temperatures lower than 15 K are not of the [100] type. They resemble those obtained in the ternary compounds near the upper limit of the transition region. A theoretical fit to two iron sites with small ϕ is in much better agreement with the experimental result than a fit of the [100] type. It is thus concluded that in $HoFe_2$ a spin rotation also occurs at low temperatures, where $\vec{n}$ slightly deviates from the [100] direction, presumably in the (001) plane. The fact that the spin does not reach the [110] direction (or any other cubic axis) may explain the previously reported anomalous behavior of the elastic constants of $HoFe_2$ at temperatures below 20 K[2].

R E F E R E N C E S

1. U. ATZMONY, M.P. DARIEL, E.R. BAUMINGER, D. LEBENBAUM I. NOWIK and S. OFER, Phys. Rev.B,7, 4220 (1973).
2. H. KLIMKER, M. ROSEN, M.P. DARIEL and U. ATZMONY, to be published in Phys. Rev.B.
3. U. ATZMONY and M.P. DARIEL, Proc. 10th Rare-Earth, Conf. Carefree, Arizona, April 1973, p. 605 U S A E C
4. U. ATZMONY and M.P. DARIEL, Phys. Rev.B, 10, 2060 (1974).
5. H. KLIMKER and M. ROSEN, Phys. Rev.B, 7, 2054 (1973).

STRESS DEPENDENCE OF THE MAGNETIC SUSCEPTIBILITY IN PRASEODYMIUM BETWEEN 1.5 K AND 25 K.

H.R. Ott
Laboratorium für Festkörperphysik, ETH Zürich,
8049 Zürich, Switzerland.

ABSTRACT

The magnetostriction parallel and perpendicular to the hexagonal axis of single crystal Pr has been measured in magnetic fields up to 10 kOe. From these measurements the stress dependence of $\chi_\parallel$ and $\chi_\perp$ has been determined. The relative stress dependences $\partial \ln \chi_\parallel / \partial \sigma_c$ and $\partial \ln \chi_\perp / \partial \sigma_a$ are of the same order of magnitude. The data can be used to calculate the influence of changes of the axial ratio c/a on $\chi_\parallel$ and to estimate the same property of $\chi_\perp$. From these results we then conclude that the anisotropy of χ below 30 K can only partly be explained with the strong variation of ρ ($\rho = c/a$) below that temperature, and the increase of $\chi_\perp/\chi_\parallel$ with decreasing temperature is mainly due to crystal field effects as suggested by Rainford. It also seems very unlikely that the observed ordering effects in polycrystalline Pr are due to strain effects in these samples.

INTRODUCTION

Praseodymium is known for the pronounced anisotropy of its magnetic susceptibility at low temperatures [1,2]. At 25 K $\chi_\parallel$, the susceptibility parallel to the hexagonal axis is about 3 times smaller than $\chi_\perp$, the susceptibility in the basal plane of the double hexagonal close packed crystal lattice and it decreases slightly down to 1.5 K. $\chi_\perp$, however, increases with decreasing temperature and the ratio $\chi_\perp/\chi_\parallel$ is approximately 10 below 4 K. This behaviour could in general be explained with a crystal field scheme and with molecular field exchange constants due to Rainford [3].

Particular interest in the magnetic properties of praseodymium arises from the fact that polycrystalline Pr is reported to order antiferromagnetically below 25 K [4-8], whereas no ordering phenomena were observed in single crystal material above 1.5 K [9,10]. Such behaviour was tentatively ascribed to strain effects in polycrystalline samples due to e.g. anisotropic thermal expansion [9]. It is therefore of interest to study the stress dependence of the magnetic susceptibility $(\partial\chi/\partial\sigma)_T$ in order to obtain data for a possible explanation of the above mentioned discrepancy. Since magnetostriction measurements are a tool for investigating $(\partial\chi/\partial\sigma)_T$ we have studied this property of single crystal praseodymium.

EXPERIMENT AND RESULTS

The length change measurements were made on a single crystal of approximately 3mm length with a capacitance dilatometer. The crystal was cut from the same batch as the ones used for other experiments [2,11]. At fixed temperatures the external magnetic field was swept from 0 to 10.5 kOe and the induced strains were monitored on an x-y recorder as a function of the applied field. In fig. 1 we show the temperature dependence of the strains parallel and perpendicular to the hexagonal axis in a magnetic field of 10 kOe between 1.5 K and 25 K. The strains ε_i depend quadratically on H and are strictly reversible at all temperatures thus suggesting a paramagnetic behaviour. The anisotropy of the magnetic susceptibility is clearly reflected in these measurements. The strains parallel to the hexagonal axis are obviously much smaller than those observed perpendicular to the sixfold symmetry axis.

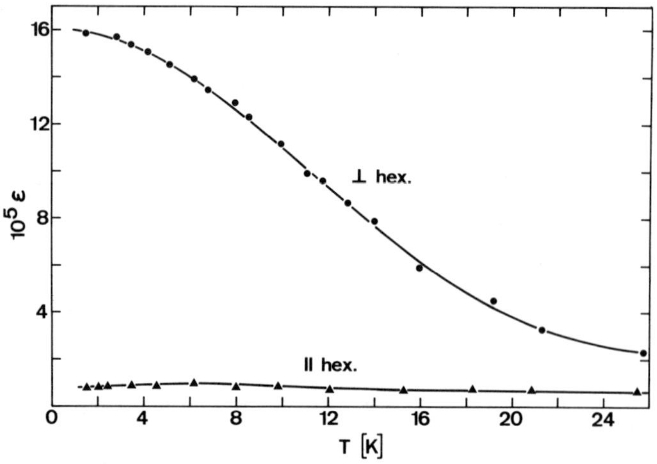

Fig. 1: Relative length change parallel and perpendicular to the hexagonal axis of a Pr single crystal in an external magnetic field of 10 kOe. Field and strain direction are parallel.

Integrating the Maxwell relation

$$\Omega \left(\frac{\partial \varepsilon_i}{\partial H}\right)_{\sigma,T} = \left(\frac{\partial M}{\partial \sigma_i}\right)_{H,T} \qquad (1)$$

where M is the molar magnetisation and Ω the molar volume, we obtain the uniaxial magnetostrictive strain which depends on the stress dependence of the magnetic susceptibility

$$\varepsilon_i = \frac{H_j^2}{2\Omega}\left(\frac{\partial \chi_j}{\partial \sigma_i}\right) \qquad (2)$$

In this relation σ_i is a stress in the direction in which the length change is observed. The index j denotes the direction of the external magnetic field H and is always parallel to the strain in our experiment. In the derivation of (2) we have assumed that χ_j is field independent and thus $M_j = \chi_j H_j$. This assumption is experimentally confirmed by the results of Lebech and Rainford [12]. In fig. 2 we give the temperature dependence of $\partial \ln \chi_i / \partial \sigma_i$ as calculated from (2) and using the susceptibility data of Andres and co-workers[2]. From this we

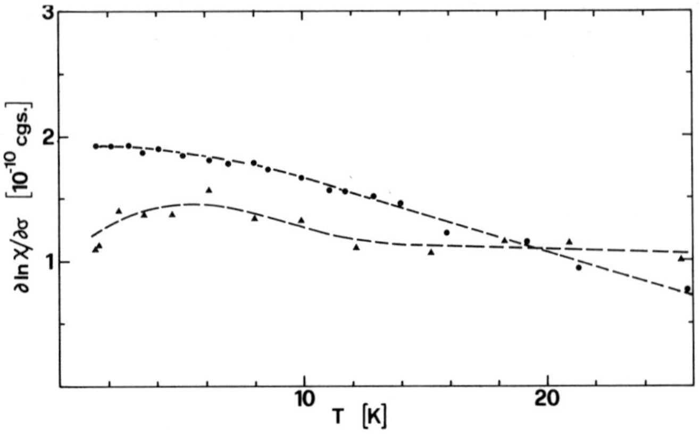

Fig. 2: Relative stress dependence of $\chi_\parallel$ and $\chi_\perp$ in Pr. Dots: $\partial \ln \chi_\perp / \partial \sigma_a$, triangles: $\partial \ln \chi_\parallel / \partial \sigma_c$.

learn that both $\partial \ln\chi_{\parallel}/\partial\sigma_c$ and $\partial \ln\chi_{\perp}/\partial\sigma_a$ are positive and that both relative stress dependences are of the same order of magnitude.

DISCUSSION

Several physical properties of Pr could be explaine by Rainford using a fitted crystal field scheme and two appropriate exchange constants. The model, initially developed by Bleaney [13], assumes two different crysta field schemes depending on the symmetry (cubic or hexagonal) of the surrounding of each ion. Since for both symmetry sites the ground state turns out to be a singlet, any ordering would have to be of the exchange induced type.

In Rainford's approximation the magnetisation for each sublattice is written in the form

$$M_c = \chi_c^\circ [H + \lambda_1(M_c + M_H) + \lambda_2 M_H]$$
$$M_H = \chi_H^\circ [H + \lambda_1(M_c + M_H) + \lambda_2 M_c] \qquad (3)$$

where λ_1 is the interaction of an ion with the nearest neighbors and λ_2 corresponds to the interaction with the next nearest neighbors. χ_C^o and χ_H^o are the susceptibilities of the cubic and hexagonal sites in the absence of any exchange. If both λ_1 and λ_2 are non zero the susceptibilitie take the form

$$2\chi_c = A[1 + \lambda_1 B] + B$$
$$2\chi_H = A[1 - \lambda_1 B] - B \qquad (4)$$

where

$$A = \frac{2\lambda_2\chi_c^\circ\chi_H^\circ + (\chi_c^\circ + \chi_H^\circ)}{1 - \lambda_1(\chi_c^\circ + \chi_H^\circ) - 2\lambda_1\lambda_2\chi_c^\circ\chi_H^\circ - \lambda_2^2\chi_c^\circ\chi_H^\circ} \qquad (5)$$

$$B = \frac{\chi_c^\circ - \chi_H^\circ}{1 - \lambda_2^2\chi_c^\circ\chi_H^\circ} \qquad (6)$$

These are rather complicated expressions and it is impossible to deduce the stress derivatives of the crystal field levels, which determine χ_C^o and χ_H^o, and of the exchange parameters λ_i.

From Rainford's work, however, we conclude that probably only the hexagonal sites would carry an ordered moment and that only the crystal field scheme for hexagonal symmetry is affected by the non-ideality of the c/a ratio. It is therefore interesting to check how the susceptibilities are influenced by a change of the axial ratio ρ, where ρ = c/a.

From our data of $\partial\chi_{\parallel}/\partial\sigma_C$ we can directly deduce $\partial\ln\chi_{\parallel}/\partial\ln\rho$, using the low temperature elastic constants reported by Greiner and co-workers [14]. This value is positive over the entire temperature range and is shown in fig. 3. $\partial\ln\chi_{\perp}/\partial\ln\rho$ cannot be calculated in this way, but the magnetostriction results suggest that $\partial\ln\chi_{\perp}/\partial\ln\rho$ is negative. Low temperature thermal expansion data [15] show that ρ increases with temperature and we would then expect the dependence of $\chi_{\parallel}$ and $\chi_{\perp}$ on ρ to make a contribution to $\chi_{\parallel}$ that increases and one to $\chi_{\perp}$ that decreases with temperature.

From our data we can also estimate the relative importance of a c/a-change to the low temperature behaviour of $\chi_{\parallel}$ and $\chi_{\perp}$. If we assume that

$$\chi_i = \chi_i[\rho(T), T] \qquad (7)$$

and

$$d\chi_i/dT = \left(\frac{\partial\chi_i}{\partial\rho}\right)_T \cdot \frac{\partial\rho}{\partial T} + \frac{\partial\chi_i}{\partial T} \qquad (8)$$

where $\partial\chi_i/\partial T$ and $\partial\rho/\partial T$ are experimentally determined

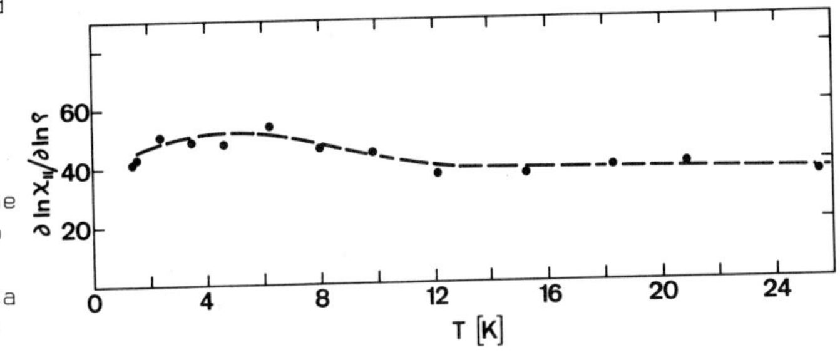

Fig. 3: Temperature variation of the axial ratio dependence of $\chi_{\parallel}$ in Pr . c/a = ρ .

quantities and $(\partial\chi_i/\partial\rho)_T$ can roughly be estimated, it then turns out that below 15 K the temperature dependence of χ_i is mainly governed by $\partial\chi_i/\partial T$, which we would then ascribe to crystal field effects.

In conclusion we may say that the magnetic susceptibilities of Pr depend strongly on the lattice parameters. In spite of this we consider it unlikely that internal strains could be responsible for the ordering phenomena in polycrystalline Pr. Even with a value of $\partial\ln\chi_i/\partial\ln\rho$ as high as 50 these strains would have to be considerable to account for such effects.

The author acknowledges several helpful discussions with Prof. J.L. Olsen, Prof. B. Lüthi and Dr. K. Andres.

REFERENCES

1.) T. Johansson, K.A. Mc Ewen and P. Touborg, J. Phys. 32, C1 372 (1971)
2.) K. Andres, E. Bucher, J.P. Maita, L.D. Longinotti and R. Flukiger, Phys. Rev. B6, 313 (1972)
3.) B.D. Rainford, AIP Conf. Proc. 5, 591 (1971)
4.) J.W. Cable, R.M. Moon, W.C. Koehler and E.D. Wollan, Phys. Rev. Letters 12, 553 (1964)
5.) B. Holmström, A.C. Anderson, and M. Krusius, Phys. Rev. 188, 888 (1969)
6.) W.H. Kapfhammer, W. Maurer, F.E. Wagner and P.Kienle, Z. Naturforsch. 26a, 357 (1971)
7.) A.S. Bulatov and N.S. Petrenko, Phys. Stat. sol. (b), 60, K91 (1973)
8.) G. Krithivas, G.T. Meaden and N.H. Sze, J. Phys. Soc. Jap. 33, 1584 (1972)
9.) T. Johansson, B. Lebech, M. Nielsen, H. Bjerrum Moller and A.R. Mackintosh, Phys.Rev. Letters 25, 524 (1970)
10.) B.D. Rainford and J. Gylden Houmann, Phys. Rev. Letters 26, 1254 (1971)
11.) B. Lüthi, M.E. Mullen and E. Bucher, Phys. Rev. Letters 31, 95 (1973)
12.) B. Lebech and B.D. Rainford, J. Phys. 32, C1 370 (1971)
13.) B. Bleaney, Proc. Roy. Soc. A276, 39 (1963)
14.) J.D. Greiner, R.J. Schiltz, Jr., J.J. Tonnies, F.H. Spedding and J.F. Smith, J. Appl. Phys. 44, 3862 (1973)
15.) H.R. Ott, to be published.

FIELD-DEPENDENT MAGNETIC ANISOTROPY OF SOME MIXED GARNETS WITH LIGHT RARE EARTHS

P. J. Flanders+ and T. Egami+*
Laboratory for Research on the Structure of Matter, University of Pennsylvania
Philadelphia, Pa., 19174
and
E. M. Gyorgy and L. G. Van Uitert
Bell Laboratories, Murray Hill, N. J. 07974

ABSTRACT

We have measured the magnetocrystalline anisotropy and the magnetization of mixed garnet single crystals with compositions $Y_{3-x}R_xM_yFe_{5-y}O_{12}$ (R is Nd or Sm, x from 0 to 2; M is Al or Sc, y from 0 to 1). It was found that the anisotropy and magnetization of the crystals containing Nd, Sm and/or Sc are significantly field dependent. The results are qualitatively explained in terms of the lattice effect on the Nd and Sm magnetic moment and the moment canting due to additions of Sc.

INTRODUCTION

Compared to Yttrium-iron-garnet (YIG) substituted with heavy rare earth elements, relatively little magnetic data has been accumulated for YIG with the light rare earth elements. The latter, however, should represent an interesting case to study, since the lattice effect is expected to be greater for the light rare earth elements(1). For instance, recent studies of the magnetocrystalline anisotropy of YIG substituted with Pr(2) and Sm(3) indicate that their anisotropy constants are extremely sensitive to the lattice parameter, and not necessarily linear with the concentration of light rare earth elements. In the present work we will extend these studies, particularly by examining the anisotropy and the magnetization in very high magnetic fields, and will attempt to understand the phenomena qualitatively in terms of the crystalline electric field (CEF) effect on the light rare earths and the Yaffet-Kittel effect due to Sc.

SAMPLE PREPARATION

Single crystals with eight compositions (Table I.) were grown by the standard flux method(4), and were annealed at 1200°C for 16 hrs in oxygen atmosphere to eliminate the growth induced anisotropy. The crystals other than those containing Sm were cut into cubes with edges parallel to the three principal crystallographic directions (110, 111, 112) weighing about 40 ~ 100 mg each. The crystals with Sm were prepared as (110) plane disks, weighing about 5 mg.

TABLE I.

Sample	Formula	$-K_1$	$-K_2$	H	$\frac{\partial\tau}{\partial H}$
1	$Y_3Fe_5O_{12}$	2.3	-	-	<0.1
2	$Nd_{1.8}Y_{1.2}Fe_5O_{1.2}$	261.	104.	2.2	5.7
3	$Y_3Al_{0.8}Fe_{4.2}O_{12}$	1.7	0.9	2.8	<0.1
4	$Nd_2YAlFe_4O_{12}$	199.	67.	6.4	5.5
5	$Y_3Sc_{0.9}Fe_{4.1}O_{12}$	~1.	-	2.2	4.
6	$Nd_2YScFe_4O_{12}$	55.	38.	2.2	50.
7	$Sm_{0.5}Y_{2.5}Fe_5O_{12}$	37.	14.	2.2	2.1
8	$Sm_{0.5}Y_{2.5}ScFe_4O_{12}$	16.	1.5	2.2	7.

Anisotropy constants K_1 and K_2 (10^4 erg/cm^3) are measured at 77 K in the magnetic field of H (Tesla). $\frac{\partial\tau}{\partial H}$ denotes the field dependence of the peak-to-peak torque, in % per Tesla.

MAGNETOCRYSTALLINE ANISOTROPY

The magnetocrystalline anisotropy of the crystals was measured at 77K in applied magnetic fields up to 6.8 Tesla (1 Tesla = 10 kOe) using the torque method. The description of the experimental set up is given elsewhere(5). The results are summarized in Table I. The principal findings are as follows.

1) Substitution of Nd or Sm for Y increases the anisotropy by 1 to 2 orders of magnitude.
2) The anisotropy is significantly field dependent when the crystal contains Sm, Nd and/or Sc.
3) The anistropies due to Sm or Nd are dependent not only on concentration but on the elements substituted for iron.

The points 2) and 3) make a distinction between the present case and YIG with the heavy rare earth elements. The torque and hence the anisotropy is always slightly field dependent, particularly near the Curie temperature(6). The magnitudes of the field dependence observed here are, however, much greater than usually found in most of the magnetic materials. Figure 1 shows the field dependence of the peak-to-peak torque of $YNd_2Fe_5O_{12}$ and $YNd_2ScFe_4O_{12}$ in the (110) plane at 77K. For the YIG with heavy rare earth elements it is well known that the anisotropy constant is proportional to the concentration of the magnetic rare earth. For the light rare earth YIG this does not appear to be the case.

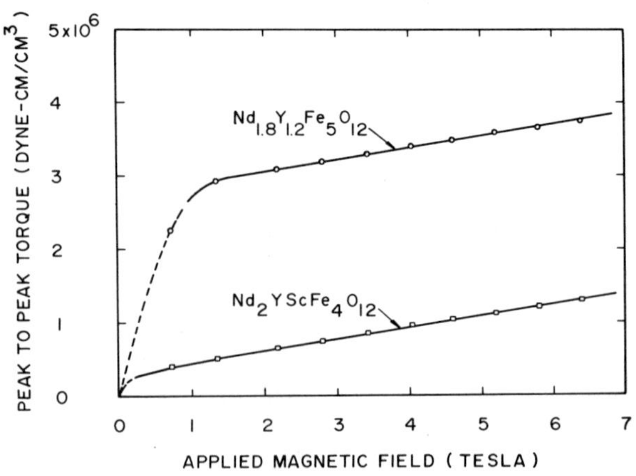

MAGNETIZATION

The most pronounced effect of the CEF on light rare earth elements is that the magnetic moment is often reduced below its saturation value, and in some cases become totally quenched to zero. For instance the magnetic moment of Nd in YIG is about 1.6 μ_B per atom at low concentrations[7], while its theoretical saturation is 3.6 μ_B. Quenching of the moment can also be detected by the appearance of the sizable temperature independent saturation susceptibility (Van Vleck susceptibility).

TABLE II.

Sample	Magnetic moment at 77K (emu/cm^3)	(μ_B/Formula)	Field dependence %/Tesla
1	184.2	4.70	-
2	255.2	6.74	0.46
3	68.9	1.74	-
4	105.9	2.75	0.52
5	209.6	5.45	~ 2
6	201.2	5.44	~ 2

The magnetization of sample 1 ~ 6 was measured at 77 K in fields up to 10 Tesla using a low frequency vibrating magnetometer. Table II summarizes the results. As expected, the crystals which contain Nd show an appreciable field dependence of the moment. The field dependence is essentially unchanged even at 4.2 K. The crystals with Sc show a signifi-cnat increase in the moment at higher field, strongly supporting the conjecture that the Yaffet-Kittel spin canting is taking place[2]. For these crystals the field dependence of the moment is non-linear.

DISCUSSION

The seemingly complicated magnetic behavior of YIG with light rare earth elements is to a greater extent elucidated in terms of CEF effect. Firstly it is obvious that the anisotropy of crystals containing Nd or Sm is due to these rare earth ions. It is also unlikely that the origin of the anisotropy in the rare earth garnet is different for heavy and light rare earths —— it is basically a single ion anisotropy due to the CEF, and the exchange interaction between rare earth and iron ions is molecular-field-like[8]. The crux of the matter here seems to be that the anisotropy of an induced moment system such as Nd-YIG depends upon the magnitude of the induced moment of Nd. It is well known that the variation in anisotropy constant (with temperature) is scaled by the magnetization[9]. A similar relation holds for the CEF spin defect[10], and possibly for the induced moment system, since the scaling relation depends only on the symmetry and not on the origin of the reduction of the moment. For instance, by comparing sample 1 with 2 and 3 with 4, it is found that the moment on Nd is about 1.0 μ_B per ion in YIG, and 0.5 μ_B in Al-YIG. The field dependence of the Nd moment, assuming that all the field dependence is due to Nd moment, is then about 1.5 %/Tesla. If we compare this with the field dependence of the anisotropy, which is about 6%/Tesla, the power law $(K/Ko)^n \sim (M/Mo)$; n = 4 appears to hold. This corresponds to the high temperature power law for cubic symmetry,[11] rather than the low temperature power law, n = 10.

Thus the following conclusions may be derived.

1) The magnetic moment of Nd is partially quenched by the CEF effect, and has a significant Van Vleck susceptibility. The anisotropy constant is also field dependent, presumably following the power law.

2) The anisotropy of Nd is dependent on the neighbouring atoms through the exchange field; the exchange field determines the magnitude of the induced moment, hence the anisotropy. When Sc is substituted for Fe, the molecular field on Nd appears to be very small, as is known from the small difference in the moment between sample 5 and 6. The anisotropy of Nd in Sc YIG is accordingly small.

3) The field dependent moment of Sc YIG and Nd Sc YIG is due to a change in the canting angle of Fe spins, rather than the susceptibility of Nd spin. The increase in Nd anisotropy is comparable to other Nd YIG in absolute terms, but percentage wise it is very large because of the small anisotropy in zero field.

4) The case of Sm YIG is probably similarly explained, except that J-mixing[12] further complicates the matter.

REFERENCES

* Also Department of Metallurgy and Materials Science.

\+ Supported by NSF through MRL Grant # 33633

1. B. R. Cooper, in Magnetic Properties of Rare Earth Metals, ed. R. J. Elliott (Plenum, New York, 1972).
2. E. M. Gyorgy, R. C. LeCraw, A. Rosencwaig, L. G. Van Uitert, R. D. Pierce, R. L. Barns, and E. Heilner, Phys. Rev. B8, 279 (1973).
3. E. J. Helner and W. H. Grodkiewicz, AIP Conf. Proc. 18, 1232 (1974).
4. L. G. Van Uitert, W. H. Grodkiewicz and E. F. Dearborn, J. Amer. Ceram. Soc., 48, 105 (1965).
5. C. D. Graham, Jr., P. J. Flanders, and T. Egami, AIP Conf. Proc. 10, 759 (1973).
6. H.-P. Klein and E. Kneller, Phys. Rev. 144, 372 (1966).
7. W. E. Henry, J. Phys. Soc. Japan, 17, S361 (1962).
8. A. B. Harris, Phys. Rev. 132, 2398 (1963).
9. F. Keffer in Handbuch für Physik XVIII/2 ed. S. Flügge (Springer, Berlin, 1966).
10. M. S. S. Brooks and T. Egami, J. Phys. C, 6, 513 (1973).
11. H. B. Callen and E. Callen, J. Phys. Chem. Solids, 27, 1271 (1966).
12. J. H. Van Vleck and A. Frank, Phys. Rev. 34, 1494 (1929).

MAGNETIC-ANNEAL-INDUCED UNIAXIAL ANISOTROPY IN $TbFe_2$

J.H. Schelleng, N.C. Koon, Naval Research Laboratory
Washington, D. C. 20375

ABSTRACT

Low temperature magnetic measurements were made on polycrystalline samples of $TbFe_2$. Hysteresis curves taken after the material was cooled to temperatures below 80K in the presence of a field were found to be oddly distorted when the measuring field direction and the annealing field direction differed. This behavior can qualitatively be explained in terms of a model based on the high cubic anisotropy and magnetostriction existing in this material, plus an added, anneal-induced uniaxial anisotropy. Activation energy studies at the annealing temperature indicate that dislocations are involved in the annealing process.

INTRODUCTION

The Laves phase intermetallic compound, $TbFe_2$, has been shown to have a very large cubic anisotropy[1] (as well as one of the highest magnetostrictive constants known[1,2]. The static magnetic properties of this compound have been studied by several workers[3,4], but the unusual low temperature hysteretic behavior discussed in this paper has not been described.

EXPERIMENT

A spherical sample of $TbFe_2$ was cut from a polycrystalline button prepared in an arc furnace. To insure uniformity, the button was turned and remelted several times and was annealed under argon at 1000K for 2 weeks. Measurements of moment versus applied field were first made using a vibrating sample magnetometer in a superconducting magnet capable of fields of 60 kOe. The spontaneous moments measured fell slightly below those given by Burzo[4], the value at 4.2°K being 95 emu/gm. As would be expected with a polycrystalline sample of highly anisotropic material, considerable susceptibility was evident above technical saturation. At 60 kOe and 4.2K the moment had increased to 99 emu/gm.

When the temperature of the sample was lowered with the magnet at full field, the hysteresis curves recorded were always narrow, with coercive forces of under 300 Oe. However, if the sample was reoriented and the temperature lowered without a field, the curves were broadened and distorted. Further experiments were made with a vibrating sample magnetometer in an iron core magnet, where the orientation of the sample with respect to the field could be varied at will. Fig. 1 shows some typical results of these experiments. As before, the sample was initially cooled in the presence of a field. Sharp narrow hysteresis curves were again recorded as long as the orientation of the field during the hysteresis curve measurements (H_m) was the same as that of the field applied during the cooldown (H_a). If, however, the orientation of H_m and H_a differed, the hysteresis curves were broadened and oddly deformed.

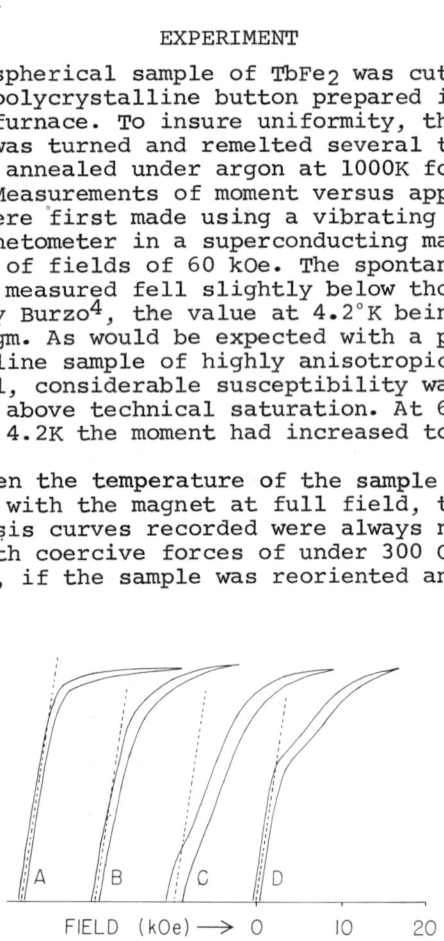

Fig. 1. Upper half of hysteresis curves of $TbFe_2$, taken at approximately 10K. Demagnetization lines are dotted. Curve A taken along annealing field; B, taken 45° from annealing field; C, taken 90° from annealing field, and D, after partial reannealing at 90°.

The observed behavior was apparently an effect of magnetic annealing during the cooldown. The original sharp curves could be obtained at any new orientation by heating the sample above 100K and again cooling it in a field at that orientation. However, heating to room temperature and recooling without any application of a field caused no appreciable reorientation of the hysteresis curves. At temperatures around 80K, the annealing process could be observed as a time dependent change in the presence of a field. It should be emphasized, however, that below 20K, where the curves of Fig. 1 were recorded, no time dependence of the moment was observed. The details of the shape of the hysteresis curves were found to depend not only on the relative angle between the annealing field (H_a) and the measuring field (H_m), but also on the magnitude of the annealing field and the manner in which it was applied. However, there was little evidence that the curve shape depended in any way upon the absolute direction of either H_a or H_m relative to the sample.

DISCUSSION

The character and magnitude of the anisotropy of $TbFe_2$ is such that, even at 60 kOe the domain moments, and therefore the magnetostrictive distortion will continue to lie close to one of the ⟨111⟩ easy directions. Due to the magnetostriction, considerable strain energy will be involved at all domain walls except those which involve a 180° change in moment direction. A polycrystal, particularly, will be highly strained because of the constraints imposed by the boundary between crystallites with varying ⟨111⟩ directions. When the temperature is sufficiently high, these strains can be reduced by diffusion of defects from the crystallite boundaries. If the temperature is then reduced, these defects will be trapped, lowering the energy of the prevailing moment orientation as opposed to orientation along any other ⟨111⟩ direction. As the magnetostriction distortion is uniaxial, moment orientation in either direction along this preferred ⟨111⟩ direction will have equal energy. The effect will be to add a small uniaxial anisotropy along this preferred direction, over and above the dominant cubic anisotropy.

To simplify the calculations, the cubic anisotropy will be assumed to be infinite with easy axis along the ⟨111⟩ directions. It will also be assumed that the moments will lie along that ⟨111⟩ direction which is energetically most favorable, neglecting the problem of how such a shift could take place. Writing the Zeeman and uniaxial energies out explicitly, the Hamiltonian for a single crystal can be written

$$\mathcal{H}=-HM_H(\alpha)-\frac{K_u(\alpha_1+\alpha_2+\alpha_3)^2}{3} \quad (1)$$

where H_M is the field applied in an arbitrary direction (θ,φ) K_u is the uniaxial anisotropy constant and $M_H(\alpha)$ is the component of the moment in the field direction, given by

$$M_H(\alpha)=M[\sin\varphi(\alpha_1\cos\theta+\alpha_2\sin\theta)+\alpha_3\cos\varphi]. \quad (2)$$

The direction cosine $(\alpha_1,\alpha_2,\alpha_3)$ of the moments will be permitted to take only the value $\pm 1/\sqrt{3}$, the signs determining which of the eight $\langle 111\rangle$ directions is being considered. The uniaxial energy term will take either of two values, K_u for the preferred $\langle 111\rangle$ axis, and $K_u/9$ for the other three axes. If the preferred $\langle 111\rangle$ axis is not the closest to the applied field, that field, as it increases, will reach a value at which it is favorable for the moment to be along the closest $\langle 111\rangle$ axis whose direction cosines will be labelled $(\alpha_1',\alpha_2',\alpha_3')$. This will happen at a field

$$H=\frac{8}{9}\frac{K_u}{[M_H(\alpha)-M_H(\alpha')]}. \quad (3)$$

The moment along the measuring field ($M_H(\alpha')$, will now be given by Eq. 2 with the primed cosines replacing the unprimed ones. An example of the steplike increase in moment seen

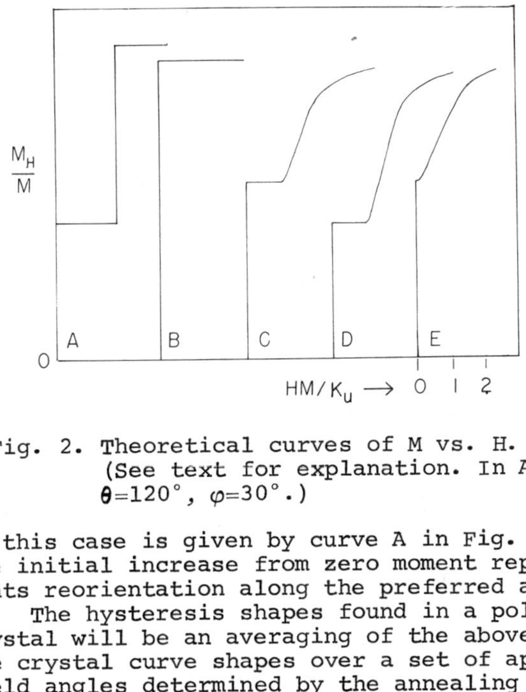

Fig. 2. Theoretical curves of M vs. H. (See text for explanation. In A, $\theta=120°$, $\varphi=30°$.)

in this case is given by curve A in Fig. 2. The initial increase from zero moment represents reorientation along the preferred axis.

The hysteresis shapes found in a polycrystal will be an averaging of the above single crystal curve shapes over a set of applied field angles determined by the annealing and measuring conditions being considered. If the preferred $\langle 111\rangle$ direction is the nearest one to the applied field, no moment shift will take place. This is precisely the situation for <u>all</u> crystallites if the measuring field and annealing field directions are the same. In this simplified model, the total moment (shown as line B) will be a constant with field, at $\sqrt{3}/2$ times the single crystal moment M. A second case that can be easily calculated is that of an unannealed sample, in which all directions for the preferred axis are equally probable. This result is shown in curve C. The solution for the case of a field applied at 90° to the annealing field is considerably more complicated as it will involve a double averaging, i.e., over all directions 90° to all directions in the octant about the preferred $\langle 111\rangle$ axis. Because of a matter of weighting, this is not quite the same as an average over all directions not in that octant. However, because of its simplicity this latter approach has been used, the result being shown in curve D.

Up until now, K_u has been assumed to be a constant. In reality, the value of K_u will vary considerably from crystallite to crystallite, depending on its size and shape, the angle of the annealing field to the $\langle 111\rangle$ axes and the manner in which the annealing is done. Curve E gives a theoretical result for an unannealed sample derived by permitting K_u to have any value between 0 and 1 with equal probability. While this exact distribution of anisotropies is unlikely, the qualitative similarities between the resulting curves and the observed curves is quite evident.

In an effort to determine the kinds of defects responsible for the annealing process, measurements were made of their activation energies. The sample was held at various temperatures between 74 and 84K and the field was held at 5 kOe, this being a field where the perpendicular and parallel moments were considerably different. Sufficient time was allowed to fully anneal the the sample along the applied field, as indicated by the fact that the moment no longer changed. The magnetometer and sample were then turned 90°. As the sample was reannealed to this new direction, the change in moment was followed as a function of time (t). Isothermal plots of this data as $\ln \partial m/\partial t=\beta-\alpha t$ proved not to be linear, indicating that several relaxation processes were important. The initial t=0 slopes (α) of the several curves were plotted using the relation $\ln \alpha=k-E/kT$. A similar plot was made of the slopes at high t, when reannealing was nearly complete. In both cases "activation energies", (E) of 0.08 eV were found.

While these results can only be suggestive, they do support the idea that dislocations are the important defects here, just as they are in other stress related phenomena, such as plastic flow, work hardening, and creep. In soft materials, dislocations move freely under stress. It is only under the effect of the high Peierls forces associated with brittle materials (and $TbFe_2$ is quite brittle) that dislocation motion becomes thermally activated. The activation energies found are consistent with dislocation motion. The activation energy of almost any other type of defect would be higher.

The preceding model describing the hysteresis curve shapes is, of course, oversimplified. However, the authors do believe that it qualitatively describes the observed phenomena.

The authors would like to acknowledge useful discussions with Drs. Jack Ayers and Conrad Williams.

1. A.E. Clark, H.S. Belson and N. Tamagawa, Phys. Letters <u>42A</u>, 160 (1972).
2. A.E. Clark, H.S. Belson, Phys. Rev. <u>B5</u>, 3642 (1972).
3. E. Burzo, Z. Angen. Physik <u>32</u>, 127 (1971).
4. K.H.J. Buschow and R.P. Van Stapele, J. Appl. Phys. <u>41</u>, 4066 (1970).

MAGNETOSTRICTION OF SINGLE CRYSTAL AND POLYCRYSTAL RARE EARTH-Fe_2 COMPOUNDS*

A. E. Clark and J. R. Cullen
Naval Ordnance Laboratory, White Oak, Maryland 20910 and
K. Sato
American University, Washington, D.C. 20016

ABSTRACT

Magnetostriction measurements on single crystal and polycrystal $TbFe_2$ and $ErFe_2$ are reported and compared. For $TbFe_2$ we find the largest known room temperature magnetostriction constant: $\lambda_{111} = 2400 \times 10^{-6}$. For $ErFe_2$: $\lambda_{111} = -300 \times 10^{-6}$. λ_{111} is the dominant magnetostriction constant and the major source of the magnetostriction of the polycrystalline rare earth-Fe_2 alloys. $TbFe_2$ possesses the largest known cubic anisotropy constant. At room temperature $K_1 = -7.6 \times 10^7$ erg/cm^3. The sign of the anisotropy plays an important role in determining the field dependence of the polycrystalline magnetostriction.

INTRODUCTION

Unlike the rare earth elements, the rare earth-Fe_2 compounds possess huge magnetostrictions[1] and magnetic anisotropies[2] which extend from low temperature to room temperature. These properties, in addition to a strong magnetic moment, make the rare earth-Fe_2 alloys attractive for many applications ranging from soft magnetic materials for magnetostrictive transducers[3] to hard magnetic materials for permanent magnets.[4] In most applications, polycrystalline materials will be used. However, it is important to know the basic magnetic parameters such as the saturation magnetic moment, σ_s, the magnetic anisotropy, K_1, and magnetostriction, λ_{100} and λ_{111}, which can best be obtained from single crystal measurements. The object of this paper is (a) to report our magnetic measurements on single crystal $TbFe_2$ and $ErFe_2$, and (b) to compare the magnetostriction of polycrystals to the single crystal magnetostriction.

MAGNETOSTRICTION

There is a striking difference between the field dependence of the magnetostriction of the RFe_2 compounds (R = Sm, Tb, Dy, Ho, Er, Tm) with [111] easy and with [100] easy. From this and the negative temperature dependence of the magnetostriction of $DyFe_2$, we inferred that $\lambda_{111} >> \lambda_{100}$.[3] In this section we illustrate the dominance of λ_{111} through a comparison of the single crystal magnetostriction constant, λ_{111}, to the magnetostriction of the polycrystal.

The saturation magnetostriction of isotropic polycrystals of cubic symmetry to lowest order is: $\lambda_s = (2\,\lambda_{100} + 3\,\lambda_{111})/5$. The magnetostriction of the same polycrystal at remanence is given by $0.637\,\lambda_{111}$ or $0.551\,\lambda_{100}$.[5,6] In Table I, we compare the remanent value of the magnetostriction with the remanent value of the magnetic moment for isotropic polycrystals having cubic symmetry and hexagonal symmetry ([0001] easy). Thus in theory it is possible to obtain the single crystal magnetostriction constants solely from polycrystal measurements of λ_s and λ_r.

TABLE I
Remanent Magnetostriction and Magnetic Moment

easy axis	[111]	[100]	[0001]
λ_r	$.637\,\lambda_{111}$	$.551\,\lambda_{100}$	0
σ_r	$.866\,\sigma_s$	$.832\,\sigma_s$	$.5\,\sigma_s$

In Fig. 1, we compare the single crystal[8] and polycrystalline magnetostriction for $ErFe_2$. Here [111] is easy. Note that the polycrystal magnetostriction is nearly field independent and remains close to the calculated remanent value of $0.637\,\lambda_{111}$ as the field is increased to 25 kOe. The magnetization over the same field range changes substantially by magnetization rotation.[2] Thus the magnetostriction is nearly independent of field rotation, affirming the smallness of λ_{100}. (The theoretical field dependence assuming $\lambda_{100} = 0$: $(\lambda_s - \lambda_r)/\lambda_s \simeq -6\%$).

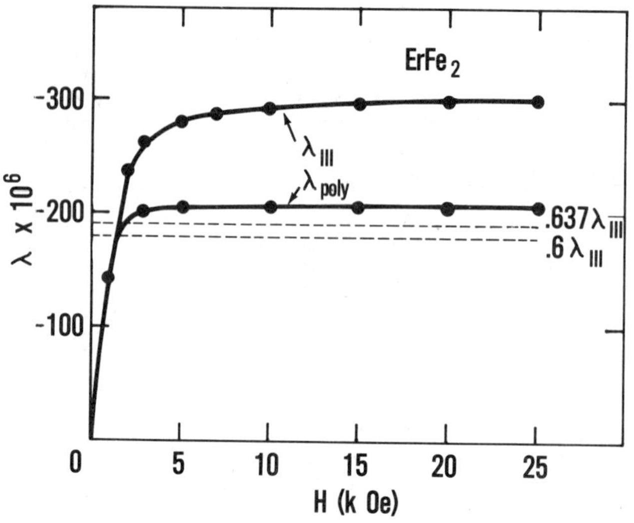

Fig. 1. Magnetostriction of $ErFe_2$ at room temperature. $\lambda_{poly} = 2(\lambda_{\parallel} - \lambda_{\perp})/3$.

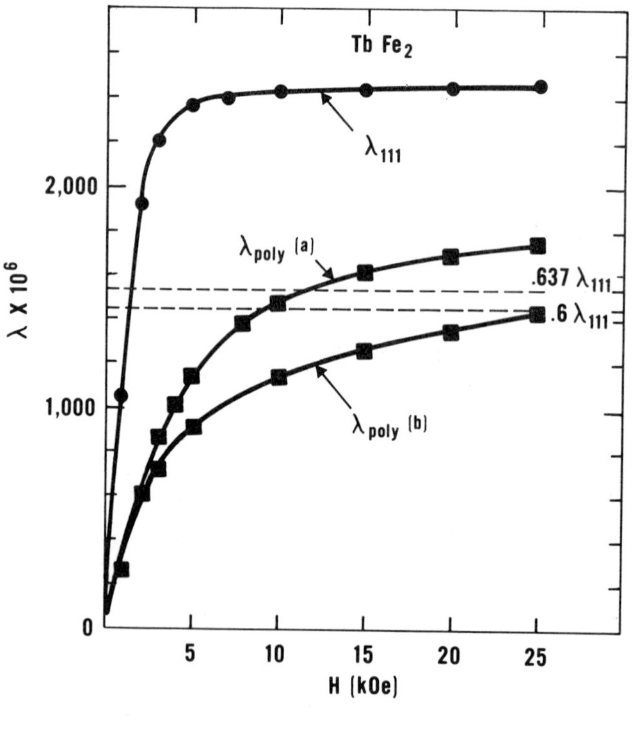

Fig. 2. Magnetostriction of $TbFe_2$ at room temperature. Curves (a) and (b) are two magnetostriction measurements of a polycrystal taken at different orientations.

In Fig. 2, we illustrate single crystal and polycrystal magnetostriction for $TbFe_2$. $TbFe_2$ has the largest known room temperature magnetostriction ($\lambda_{111} = 2400 \times 10^{-6}$). Two polycrystal curves are shown. Again we find the polycrystal magnetostrictions near $0.637\ \lambda_{111}$. The spread in the curves, which occurs from sample to sample, we attribute to preferential orientation of crystallites.

The room temperature values of λ_{111} for $DyFe_2$, $HoFe_2$, and $TmFe_2$, calculated from λ_{111} of $TbFe_2$ and $ErFe_2$ using Stevens Equivalent Operator Method and single-ion temperature dependences are shown in Table II.

TABLE II
Room Temperature Magnetostriction and Magnetic Anisotropy*

	$TbFe_2$	$DyFe_2$	$HoFe_2$	$ErFe_2$	$TmFe_2$
$\lambda_{111} \times 10^6$	2400	(1900)	(400)	-300	(-600)
$\Delta K_1 \times 10^{-7} erg/cm^3$	(-0.9)	----	----	(-0.02)	----
$K_1 \times 10^{-7} erg/cm^3$	-7.6	2	(1)	-0.3	(-0.3)

*Calculated values are shown in parenthesis.

Because the magnetostriction is large and anisotropic, there is a large magnetostrictive contribution to the anisotropy, ΔK_1. This contribution:

$$\Delta K_1 = 9[(C_{11}-C_{12})\ \lambda_{100}^2 - 2\ C_{44}\ \lambda_{111}^2]/4,$$

is negative, thereby increasing the magnitude of the negative anisotropies of $TbFe_2$, $ErFe_2$, and $TmFe_2$, and decreasing the positive anisotropies in $DyFe_2$ and $HoFe_2$. The calculated results, assuming λ_{100} small, are shown in Table II.

ANISOTROPY

$TbFe_2$ was predicted to have the largest magnetic anisotropy of all cubic compounds.[2] This huge anisotropy can be seen in the high field[9] moment measurements along the three principal directions in Fig. 3. $\sigma_{spon} \simeq 90$ emu/g. From the field dependence of σ_{100} and σ_{110} we calculate $K_1 = -7.6 \times 10^7$ erg/cm^3. This is the largest known cubic anisotropy at room temperature. We find $|K_2| < 2 \times 10^7$ ergs/cm^3. Our measured values for K_1 for $TbFe_2$, $DyFe_2$, and $ErFe_2$ are shown in Table II, along with calculated values for $HoFe_2$ and $TmFe_2$ based upon single-ion theory and Stevens Equivalent Operator calculation at T = 0 K.[3] Note the difference in sign of K_1 between $TbFe_2$ and $DyFe_2$ and between $HoFe_2$ and $ErFe_2$. Atzmony, Dariel, et. al.[10,11] have shown by Mossbauer measurements the easy direction changes in the ternary $Tb_xDy_{1-x}Fe_2$, $Tb_xHo_{1-x}Fe_2$, $Dy_xEr_{1-x}Fe_2$ and $Ho_xEr_{1-x}Fe_2$ systems. Assuming additive anisotropy constants for the rare earth ions in the ternary compounds, they calculated anisotropy constants which are close to our room temperature values.

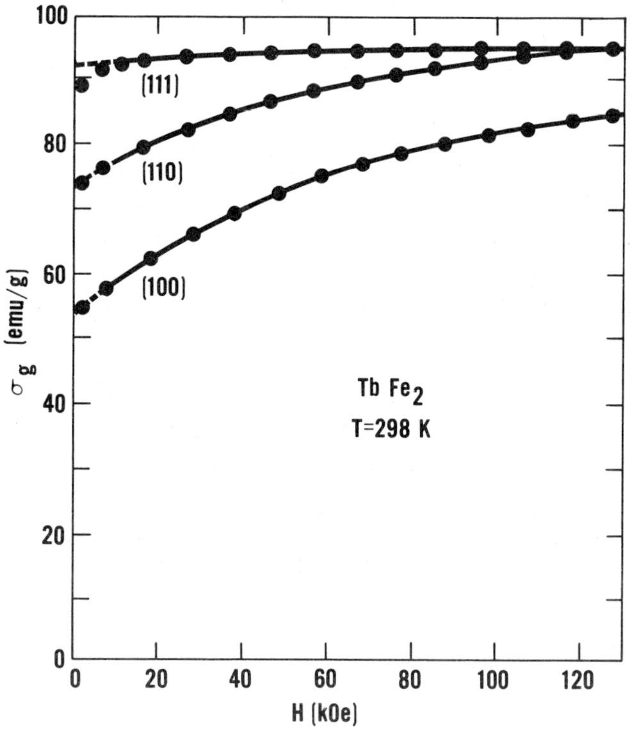

Figure 3 - Magnetic Moment of $TbFe_2$.

DISCUSSION

The RFe_2 compounds, in addition to a huge magnetostriction and magnetic anisotropy, possess a huge anisotropy in the magnetostriction constants themselves, i.e. $\lambda_{111} \gg \lambda_{100}$. Thus the magnetostriction of polycrystals arises principally from domain wall motion, rather than magnetization rotation.

The sign of the magnetic anisotropy plays an important role in determining the field dependence of the magnetostriction of RFe_2 polycrystals. When the anisotropy is large, as in the binary RFe_2 compounds, a rapid rise to a value close to saturation is found for [111] easy materials, since $\lambda_r \simeq \lambda_s$. However, for the [100] easy materials, a slow approach to saturation is found.[3] This is clearly evident in $DyFe_2$, where $\lambda_r \simeq 0$. In a material with strong magnetostriction anisotropy, such as the RFe_2 compounds, grain orientation is particularly important. Preferred orientation can lead to magnetostrictions greatly enhanced over those with isotropic crystallite distribution.

ACKNOWLEDGEMENT

*Supported by the Office of Naval Research, the Naval Ordnance Laboratory Internal Research Fund, and the Naval Sea Systems Command.

REFERENCES

1. A. E. Clark and H. S. Belson, 17th Conf. on Magnetism and Magnetic Materials, Nov. 1971, AIP Conf. Proc. 5, 1498 (1972); Phys. Rev. B5, 3642 (1972), N. C. Koon, A. I. Schindler, and F. L. Carter, Phys. Letters 37A, 413 (1971).
2. A. E. Clark, H. S. Belson and N. Tamagawa, Phys. Letters 42A, 160 (1972), 18th Conf. on Magnetism and Magnetic Materials, Nov. 1972, AIP Conf. Proc. 10, 749 (1973).
3. A. E. Clark, 19th Conf. on Magnetism and Magnetic Materials, Nov. 1973, AIP Conf. Proc. 18, 1015 (1974).
4. H. T. Savage, A. E. Clark, S. J. Pickart, J. J. Rhyne, and H. A. Alperin, IEEE Trans. on Magnetics, MAG-10 (1974).
5. R. Becker and W. Doring, Ferromagnetismus, Springer (1939) p. 287.
6. M. Yamamoto and T. Nakamichi, J. Phys. Soc. (Japan) 13, 228 (1958).
7. See S. Chikazumi and S. Charap, Physics of Magnetism, Wiley and Sons (1964) p. 250.
8. Single Crystals were grown by O. D. McMasters and K. Gschneidner, Iowa State Univ., Ames, Iowa.
9. Magnetic Fields were obtained at the NRL high field facility, Washington, D.C.
10. U. Atzmony, M. Dariel, E. Bauminger, D. Lebenbaum, I. Nowik, and S. Offer, Phys. Rev. B7, 4220 (1973)
11. M. P. Dariel and U. Atzmony, Int. J. Magnetism 4, 213 (1973).

MAGNETIZATION AND MAGNETIC ANISOTROPY IN $Lu_xTm_{2-x}Co_{17}$ INTERMETALLICS

A. E. Miller, K. Miura, H. Rodrigues, and T. D'Silva
Department of Metallurgical Engineering
and Materials Science
University of Notre Dame
Notre Dame, Indiana 46556

ABSTRACT

The temperature and field dependence of the magnetization of the hexagonal intermetallic compounds, $Lu_xTm_{2-x}Co_{17}$ (x = 0, 2/3, 4/3, 2) has been determined as a function of crystallographic direction. All specimens were single crystal spheres, and measurements were made in the temperature range 77.4K to 450K in applied fields up to 13 KOe. The anisotropy constants K_1 and K_2 were determined and the anisotropy behavior was found to obey the single ion model.

INTRODUCTION

For a hexagonal crystal, the anisotropy energy can be expressed as,

$$E = K_1 \sin^2\theta + K_2 \sin^4\theta \qquad (1)$$

where θ is the angle between the magnetization vector and the hexagonal c-axis, and K_1 and K_2 are the measured anisotropy constants. Previous investigations have indicated that hexagonal Lu_2Co_{17} has an easy basal plane (K_1 negative) while the easy magnetization direction in hexagonal Tm_2Co_{17} is along the c-axis (K_1 positive).[1] Thus the pseudobinary system $Lu_xTm_{2-x}Co_{17}$ affords an opportunity to study the nature of the change of the anisotropy from one component to the other and to ascertain if the Tm contribution to the anisotropy is of single ion nature.

EXPERIMENTAL

Magnetization measurements of $Lu_xTm_{2-x}Co_{17}$ (x = 0, 2/3, 4/3, 2) intermetallics were made in predetermined crystallographic directions for 3 mm diameter spherical single crystals with a vibrating sample magnetometer. Complete magnetization curves were obtained for all samples in both the a-axis and c-axis directions, in a temperature range of 77.4K to 450K at a maximum applied field of 13 KOe.

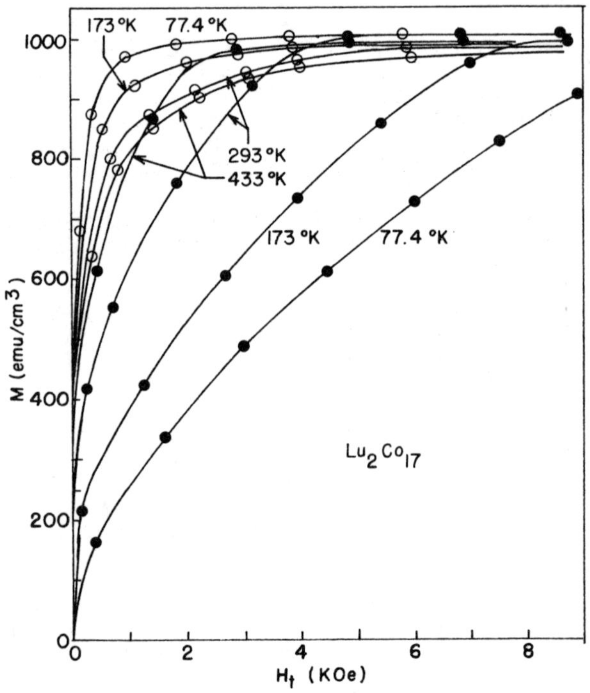

Figure 1. Magnetization vs. true field for Lu_2Co_{17} in the c-axis (●) and a-axis (○).

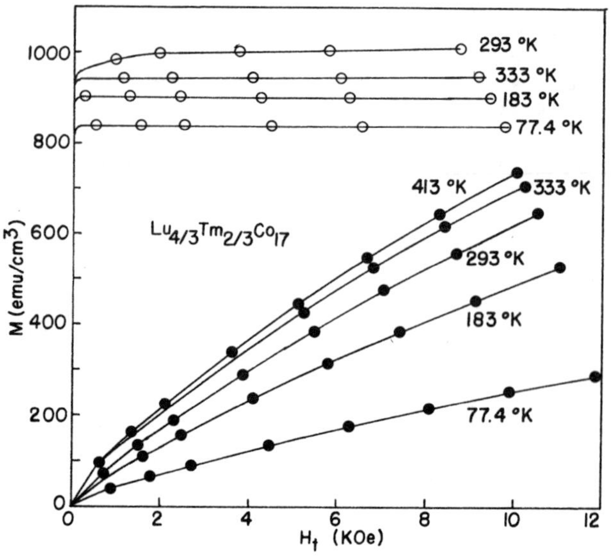

Figure 2. Magnetization vs. true field for $Lu_{4/3}Tm_{2/3}Co_{17}$ in the c-axis (○) and a-axis (●)

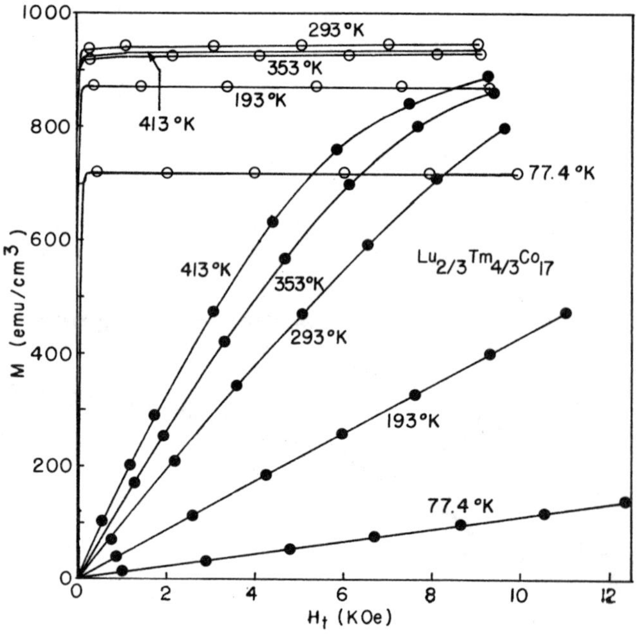

Figure 3. Magnetization vs. true field for $Lu_{2/3}Tm_{4/3}Co_{17}$ in the c-axis (○) and a-axis (●).

RESULTS AND DISCUSSION

Figures 1-4 show the magnetization behavior of the compounds studied. Figure 5 shows the temperature dependence of the saturation magnetization of these four compounds. The M_s behavior of the Tm_2Co_{17} and Lu_2Co_{17} is in good agreement with that reported for studies of aligned powders.[2] Figure 6 shows the temperature dependence of K_1 and K_2 obtained for the compounds using the method of Sucksmith and Thompson.[3]

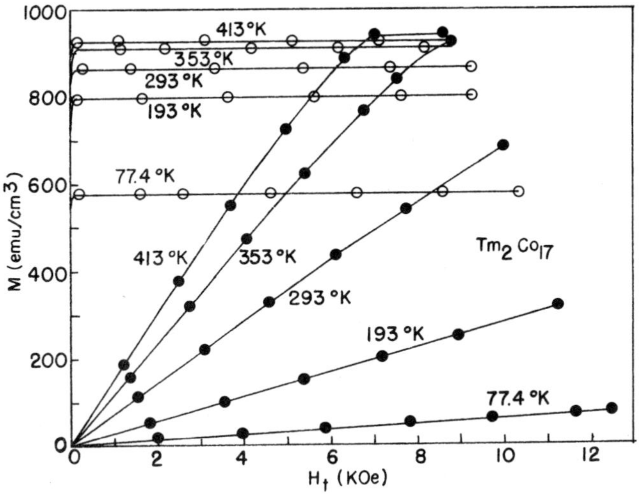

Figure 4. Magnetization vs. true field for Tm_2Co_{17} in the c-axis (o) and a-axis (●).

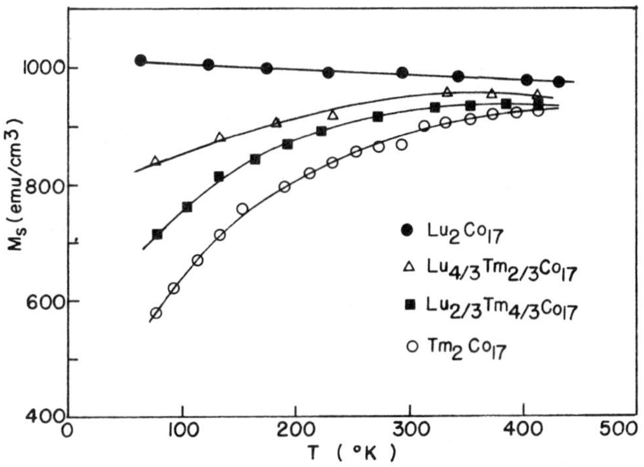

Figure 5. Saturation magnetization vs. temperature

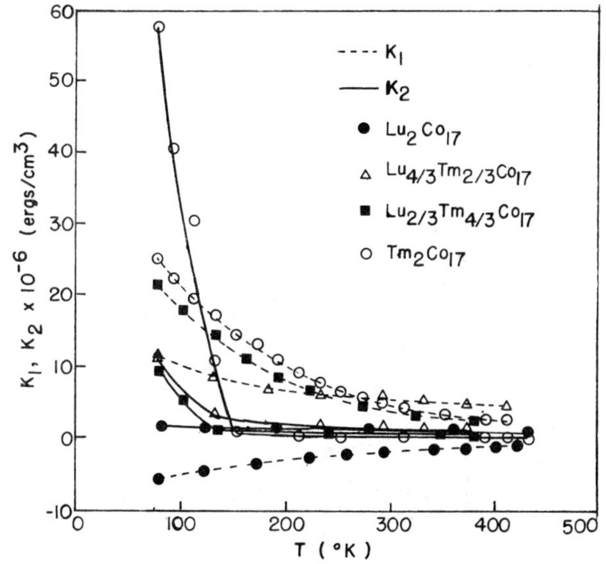

Figure 6. K_1 and K_2 vs. temperature

Analysis of the magnetization behavior of Lu_2Co_{17} shown in Figure 1 indicates that neither the a-axis nor the c-axis is preferred. Measurements of magnetization in the a-c plane at room temperature at an applied field of 3 KOe indicate that the magnetization is an easy cone about the c-axis with an apex angle of 120°. The temperature and field dependence of this cone configuration are currently under study.

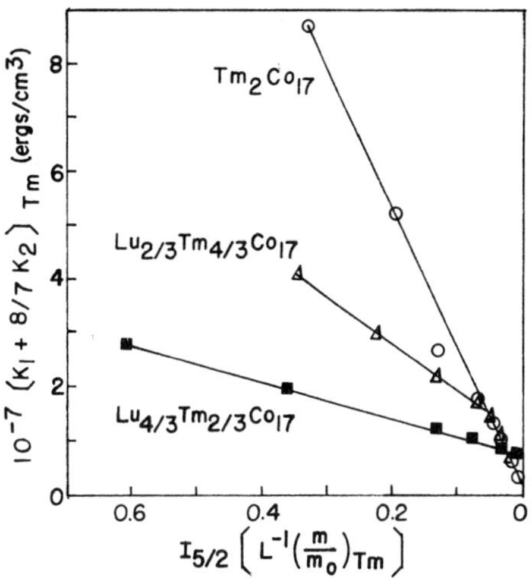

Figure 7. Test for single ion behavior of anisotropy constants.

According to the single ion model analyzed by Callen and Callen[4], the theoretical anisotropy coefficient to lowest order spin operator (ℓ = 2) has a temperature dependence given by

$$k_2(T) = k_2(0)\ I_{5/2}\ [L^{-1}\ (m)] \qquad (2)$$

where $I_{5/2}$ is a reduced hyperbolic Bessel Function whose argument is the inverse Langevin Function of the reduced magnetization. For a hexagonal crystal this anisotropy coefficient is related to the experimentally measured anisotropy constants by[5]

$$k_2 = K_1 + 8/7\ K_2 \qquad (3)$$

Figure 7 shows a plot of k_2 vs. the single ion dependence of the Tm in the rare-earth sublattice for the compound studied. The magnetization and the values of K_1 and K_2 for the Tm lattice were determined by subtracting the cobalt lattice contribution as given by $Lu_2\ Co_{17}$. The theoretical moment of pure Tm (253 emu/gm) was used to calculate the reduced magnetization. It is observed that below 350K the Tm in the rare-earth sublattice has single ion behavior.

ACKNOWLEDGMENT

The authors gratefully acknowledge the research support of the Office of Naval Research under contract number N00014-67-A-0242-0007.

REFERENCES

1. J. E. Greedan and V. U. S. Rao, J. Sol. State Chem., 6, 535 (1972).

2. K. Strnat, G. Hoffer, W. Ostertag and J. C. Olson, J. Appl. Phys., 37, 1252 (1966).

3. W. E. Sucksmith and J. E. Thompson, Proc. Roy, Soc., 225, 362 (1954).

4. E. Callen and H. Callen, Phys. Rev., 129, 578 (1963).

5. E. Kneller, Ferromagnetismus, Springer-Verlag, Berlin (1962).

MAGNETOELASTIC RELAXATION IN NICKEL-IRON FILMS

Jerzy TYMOWSKI
Institute of Computer Technology, Technical University,
ul. Nowowiejska 15/17, 00-665 Warszawa, Poland,

Henryk LACHOWICZ, Krystyna BURKIEWICZ, Maciej KWIECIEŃ
Institute of Physics, Polish Academy of Sciences,
Al. Lotników 32/46,02-668 Warszawa, Poland.

ABSTRACT

Cylindrical films of Ni-8%Fe alloy electrodeposited on a substrate wire, with circumferential easy axis of megnetization can be used as sensors of tension. The change of H_k is proportional to the strain produced in the film and can be observed as a change of the current in a measuring circuit. However, at a constant load (and strain), a considerable relaxation occurs, which can be represented by an exponential function of time, at least for periods shorter than 10 minutes. The rate of relaxation is temperature dependent with an apparent activation energy of 0.3 eV. It is suggested that the observed phenomena are due to the migration of defects and their interaction.

INTRODUCTION

Magnetic films, usually encountered in practice, exhibit uniaxial anisotropy. The mechanical stress applied induces stress anisotropy in the films. The direction of easy axis related to this anisotropy is determined by the kind of stresses applied, their direction and the sign of the magnetostriction coefficient λ_s. The resulting anisotropy consists of magnetic and stress anisotropy[1]. The magnetoelastic effect manifests itself, as a change of anisotropy field H_k of a film under the influence of the applied stress. For the cylindrical films considered here, the following result has been obtained[2]

$$\Delta H_k = \frac{2B}{M_s} \cdot (1+\nu)\epsilon_{22} \qquad (1)$$

where: $B = -3G\lambda_s$ - magnetoelastic coupling constant, G-shear modulus,

ν - Poissou's ratio,

$\epsilon_{22} = \Delta l/l$ - strain along the hard axis.

Other symbols have their usual meaning.

An attempt has been made to take advantage of the effect of linear changes of H_k as a function of strain applied along the hard axis of the film in the transducer, transforming force into an electrical signal.

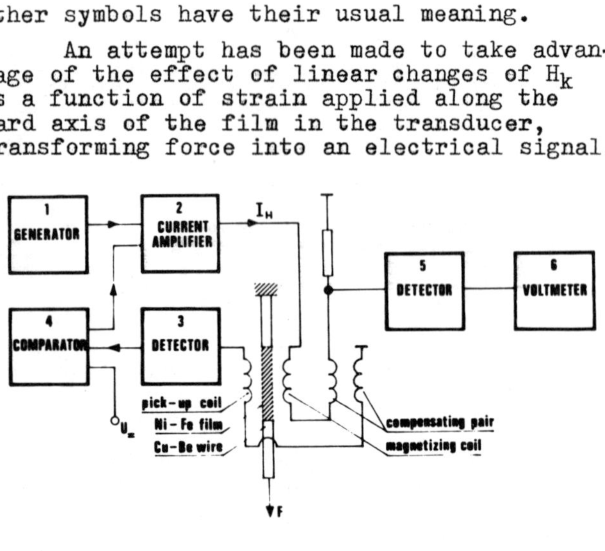

Fig 1 Schematic diagram of a force-electrical signal transducer with Ni-Fe film as a sensor.

EXPERIMENTAL

A schematic drawing of the transducer is shown in Fig.1. A cylindrical Ni-Fe film with circumferential easy axis of magnetization deposited on a beryllium bronze wire 0.6 mm in diameter is used as a sensor. Linear stress is produced by stretching the substrate wire along the hard axis. Variation of H_k is read as a change of the current I_H detected in circuit 5 and measured with the aid of a digital voltmeter 6. Details of the method and some additional results will be reported in a separate paper[3].

Nickel-iron films of 5 μm thickness containing 7-8% Fe were electrodeposited onto the chemically polished surface of the wire. The film composition was selected to obtain $\lambda_s < 0$, i.e., resulting in the increase of H_k with the elongation of the substrate wire. The axial load applied to the specimens was changed from 9.8 N (1 kG) to 39.2 N (4 kG) corresponding to a mean tensile stress value in the bronze wire ranging from 36.3 to 144 MN/m^2, and 67.5 to 268 MN/m^2 in the permalloy film. In the specimen, most of the load is carried by the mechanically stable substrate wire.

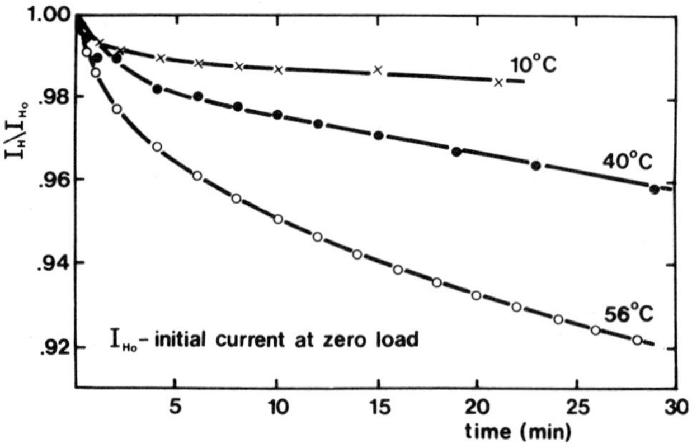

Fig 2 Plot of I_H/I_{H_o} for the samples loadec with 29.4 N (03 kG) at different temperature.
The plots were determined for the first period of loading.

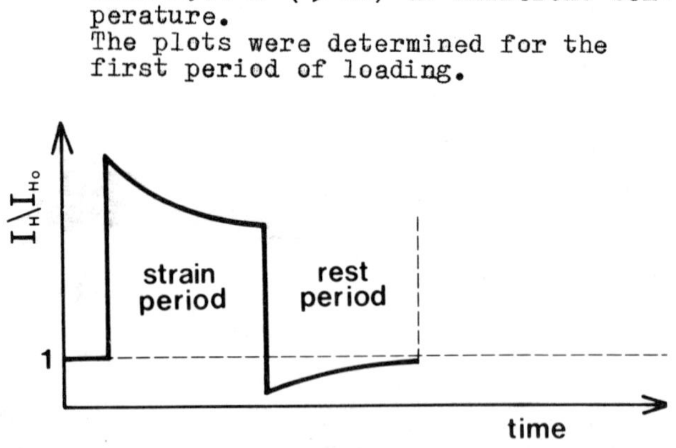

Fig. 3 Schematic plot of I_H/I_{H_o} for the succesive periods of loading and rest.

The electroplated film may be considered as being subjected to constant strain as long as the load remains constant. Two series of tests have been carried out:

a/ the specimens were held under constant load for periods of up to 430 hours at room temperature,

b/ 30 minute load periods were separated by 30 minute rest periods with the load removed.

The changes of H_k measured as (I_H/I_{H_o}) during the first 30 minutes under a load of 29.4 N (3 kG) and for different temperatures are shown in Fig.2.

Plotting the rate of the current decrease, taking place between the 2 nd and 3 rd minute of the test, in coordinates of $\ln(dI_H/dt)$ and $1/T$, a value of apparent activation energy of about 0.3 eV is obtained. These results were not consistently reproducible. After taking off the load at the end of the strain period, the readings indicate negative (compression) stresses in the film. They decrease slowly during the rest period as shown schematically in Fig.3.

DISCUSSION

It has been assumed that the change of H_k (or I_H) with time follows the equation

$$I_H = A + B \exp(-t/\tau), \qquad (2)$$

where, A,B and τ are constants, t is time.

The I_H values read at 1 minute intervals for all runs, were processed by computer in order to determine the constants in Eg.2. With regard to the relaxation time , it has been found, that for at least 10 minutes of every strain or rest period, remains fairly constant indicating good agreement with the assumed I_H vs. time function. The agreement between the shape of the experimental curve (in its initial part) and the time function given in Eg.2 suggests the effect of stress relaxation, according to the "standard linear solid" model[4]. One might expect this model to be valid at stress lovels sufficiently below the critical shear stress. In this case, the stress/time relationship for longer periods of strain should approach a steady-state value. However, it does not behave this way showing that the model is not wholly adequate.

The mechanism which is operative here, can be discussed only in hypothetical terms. Electrodeposited alloy films are characterized by their very fine-grained structure and high defect density, due to the inhibited growth and considerable inclusion of hydrogen, as well as non-metallic matter in the deposits. Diffusion of point defects and their interaction with dislocations may cause significant changes to occur, which would not be observed in bulk materials. One of the possible mechanisms is the Herring-Nabarro diffusion mechanism of creep. At a constant stress, the flow of vacancies from the tensile stressed to the compressed areas of individual grains produces the strain rate[5]

$$\dot{\epsilon} = \frac{2\, a^3 D}{d^2 kT}\,\sigma \qquad (3)$$

where all factors except σ are constant at constant temperature. The linear dependence of the strain rate on stress is consistent with the standard linear solid model. There is evidence of this mechanism at homological temperatures above 0.8, where D is sufficiently large. However, in the case of electrodeposited films, and films thermally deposited in vacuum, the decrease of D can be compensated by an extramely fine grain size and high density of defects. For the films described, we have found the average size of the structure blocks to be 30-50/um. Therefore, even at relatively low temperatures, a share of the vacancy mechanism may be expected, especially as the large number of structural faults and the influence of a free surface may facilitate the generation and anihilation of vacancies.

REFERENCES

1. D.O.Smith, J.Appl.Phys.,30, 264, /1959/.
2. N.Goldberg, J. Appl.Phys., 36, 966 /1965/.
3. J.Tymowski, K.Burkiewicz, M.Kwiecień, H.Lachowicz, to be published in Thin Solid Films.
4. C.Zener, Elasticity and Anelasticity of Metals, A.J.Kennedy, Process of Creep and Fatique in Metals, Oliver and Boyd, Edinburgh, 1962.
5. H.G. van Bueren, Imperfections in Crystals North-Holland Publishing Co., Amsterdam, 1960, 236.

PROPERTIES OF MICROSAMPLES OF SINTERED COBALT-SAMARIUM MAGNETS*

J. J. Becker
General Electric Research and Development Center, Schenectady, New York 12301

ABSTRACT

Hysteresis loops of small but still multi-grain samples taken from sintered Co_5Sm magnets show a wide range of behavior, samples from the same magnet bracketing the bulk properties. Magnetization discontinuities often appear and are associated with large grains. These jumps show a quantized dependence of jump field on magnetizing field and often a good $1/\cos\theta$ dependence, exactly as in particles prepared from cast material. A sample consisting of only one grain gave a completely rectangular loop. Measurements on sintered magnets indicate that large grains tend to reverse in small fields, detracting from the overall properties.

INTRODUCTION

It has become generally accepted that the magnetic reversal behavior of cobalt-rare-earth single particles is determined by defects through their influence on the nucleation of domain boundaries.[1,2] Strong support for this interpretation has come from direct observation of magnetization discontinuities and their dependence on magnetizing field,[3] angle,[1] and other variables. It is much less clear how the behavior of sintered magnets should be interpreted--whether they are like particles plus interactions, or whether some other model, involving perhaps grain-boundary phases[4] should be developed for them ab initio. This paper describes experiments on small samples taken from sintered magnets in which exactly the same behavior can be observed as in particles prepared from cast material.

EXPERIMENTAL PROCEDURE AND RESULTS

Samples were prepared by breaking from bulk sintered magnets. Large particles thus obtained were further broken into samples of the desired size, usually on the order of 50-100μ, under the microscope. Hysteresis loops were measured in a simple vibrating-sample magnetometer using a modification of the Mallinson coil configuration.[5] All hysteresis loops shown were traced twice, to illustrate the degree of reproducibility attained.

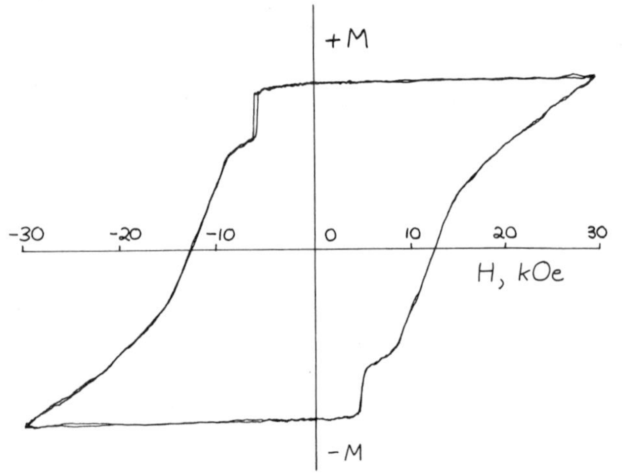

Fig. 1. Hysteresis loop of microsample from sintered magnet.

The first sample prepared in this way was relatively large, weighing about 10^{-4}g. Its reversal began with a magnetization discontinuity, as shown in Fig. 1. As the magnetizing field was decreased, the jumps suddenly appeared at lower fields. The angular dependence of the jump fields was an exact $1/\cos\theta$, but with angular offsets. Its behavior was very similar to that of a Co_5Gd particle reported on earlier.[1]

While there was nothing unusual in the appearance of this sample, the sintered magnet was observed to contain a number of grains that were conspicuously larger than the ~10μ average size. A sample was prepared so that one such grain comprised a substantial portion of it, the rest consisting of many small grains. Its hysteresis loop is shown in Fig. 2. The narrow loop within the major loop is the result of cycling in ±6 kOe just after the initial jump, showing the behavior of the free wall produced by the nucleating step. This behavior is exactly like that observed earlier in single particles of cast material. It thus appears that the large grain is reversing completely once nucleated, followed by the higher-H_{ci} multi-grain portion.

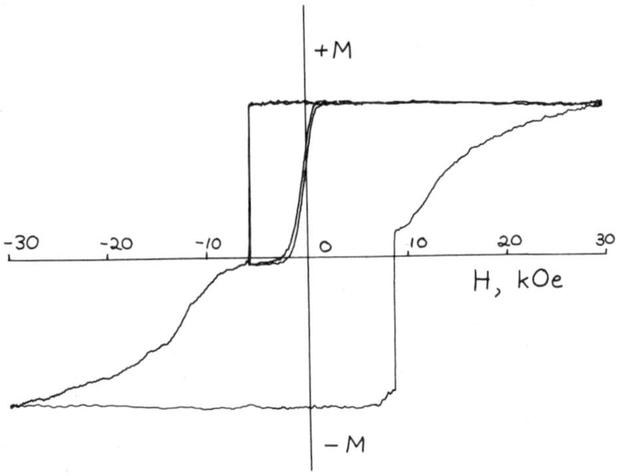

Fig. 2. Hysteresis loop of sample containing large grain and smaller grains.

Measurement of a number of particles from the same magnet showed a considerable spread in properties, especially in magnets showing many relatively large grains. This spread bracketed the properties of the bulk magnet. For example, eleven samples from a magnet whose bulk H_{ci} was only 2.5 kOe showed a spread in H_{ci} from 1.5 to 9.4 kOe, with an average of 2.8. Thus, the particles removed by fracturing appear to represent a reasonable sample of the overall properties.

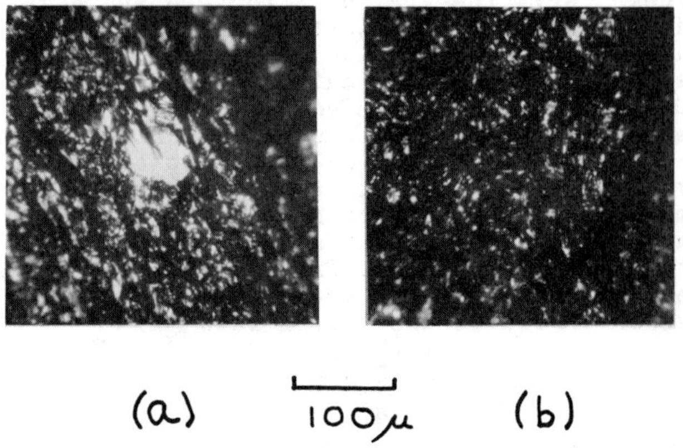

Fig. 3. Fracture surfaces of magnets sintered from (a) unsieved and (b) sieved powders.

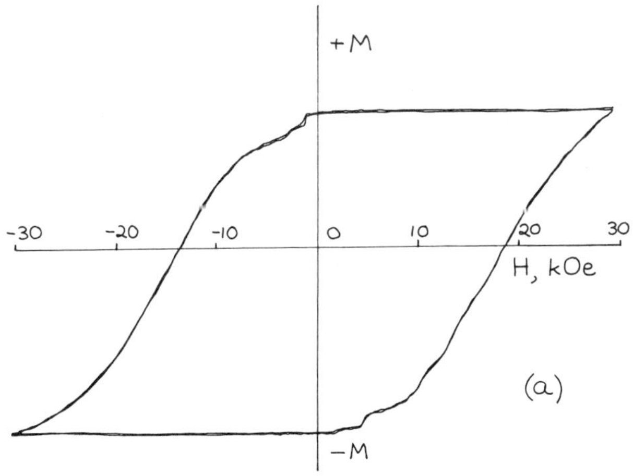

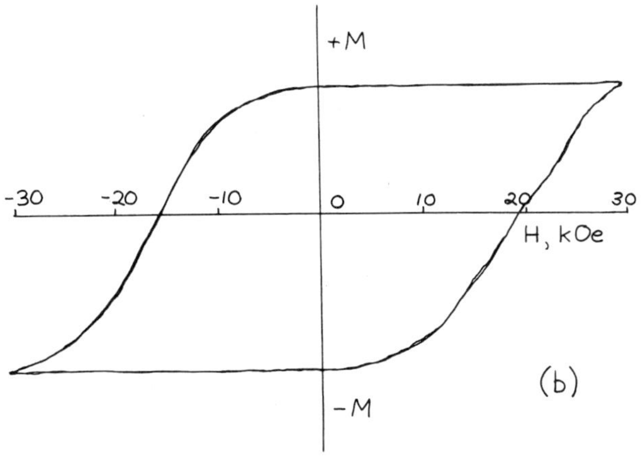

Fig. 4. Hysteresis loop of microsample from magnet sintered from (a) unsieved and (b) sieved powder.

The influence of large grains was studied more closely by investigating two magnets, one made from powder as received from the jet mill, the other made from powder passed through a 400 mesh (38μ) sieve after jet milling. Fracture surfaces of the magnet made from unsieved powder showed occasional large grains, as in Fig. 3a. They were not observable in the other magnet, Fig. 3b. The hysteresis loops of small samples of the first magnet characteristically showed a dropoff in magnetization early in the second quadrant, Fig. 4a. Samples from the second did not, Fig. 4b. In bulk properties, the second sample was somewhat superior to the first, having H_{ci} = 26.5 kOe and $(BH)_m$ = 22.2 MGOe, compared to 22.6 kOe and 19.8 MGOe. The complete loops for the small samples are drawn only to ±30 kOe magnetizing field and evidently are not completely saturated, as indicated by their asymmetry. Most loops showed this to some extent.

Then a sample consisting entirely of a thin portion of one large grain was excised from the first magnet. It is shown in Fig. 5. It weighed approximately 5 x 10^{-7}g. Its hysteresis loop is shown in Fig. 6. Again the complete reversal following nucleation is characteristic of individual grains. The fields at which this occurred are substantially below the bulk H_{ci}. However, they seem large considering that the surface of the particles as broken was not modified in any way.

Large discontinuities and occasional rectangular loops were observed in other samples, sometimes in very low fields, sometimes not. Of course, the nucleation situation may be different when the grain is surrounded by its neighbors in bulk.

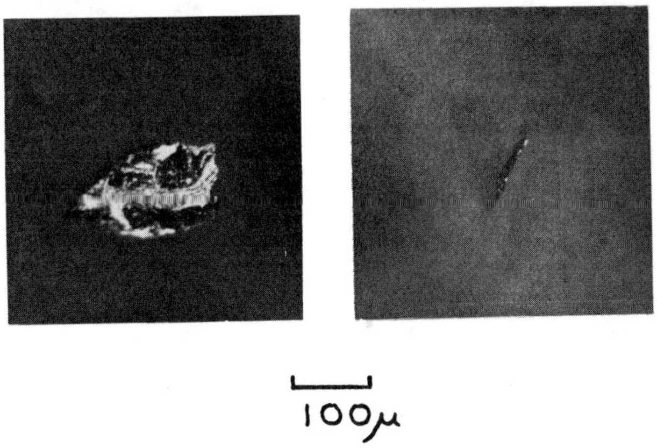

Fig. 5. Microsample consisting of portion of large grain. Dark-field illumination.

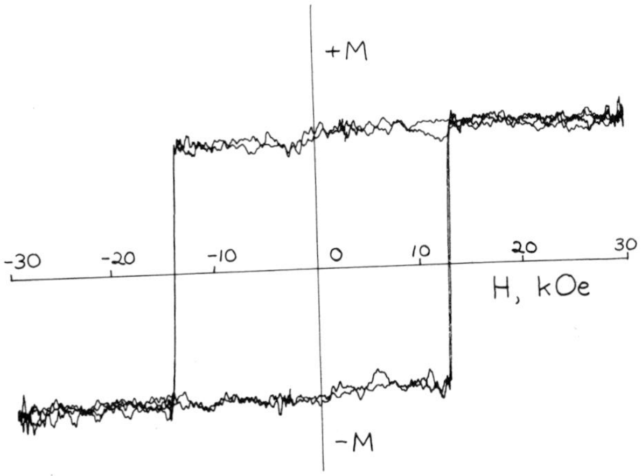

Fig. 6. Hysteresis loop of sample shown in Fig. 5.

CONCLUSIONS

Measurements on tiny samples of sintered Co_5Sm magnets indicate a behavior entirely similar to that of single particles made from cast material. Large magnetization jumps are associated with large grains in the sintered material, consistent with the observation that sieving the starting powder appears to eliminate low-field magnetization dropoff in sintered magnets.

REFERENCES

1. J. J. Becker, AIP Conf. Proc. 5, 1067-1071 (1972)
2. J. D. Livingston, AIP Conf. Proc. 10, 643-657 (1973)
3. J. J. Becker, IEEE Trans. Mag. MAG-5, 211-214 (1969)
4. K. J. Strnat, AIP Conf. Proc. 5, 1047-1066 (1972)
5. J. Mallinson, J. Appl. Phys. 37, 2514-2515 (1966)
6. J. J. Becker, IEEE Trans. Mag. MAG-7, 644-647 (1971)

*Supported in party by the Office of Naval Research

LOW TEMPERATURE MAGNETIC HARDNESS IN SUBSTITUTED $SmCo_5$*

H. OESTERREICHER
Department of Chemistry, University of California
San Diego-La Jolla, California 92037
and
D. McNEELY
Oregon Graduate Center, Beaverton, Oregon 97005

ABSTRACT

The mechanism of magnetization at liquid He temperatures is investigated in alloys based on SmCo5 with partial substitution for Co by Ni,Ag,Al and Si. Further materials Y0.167Co0.683Al0.15 and Gd0.167 Co0.683Al0.15 are included. Only homogeneous specimens of CaCu5 type are studied. Al and Si substituted materials when obtained by quenching from the melt exhibit stacking fault structures in the basal plane. These materials show, at 4.2K, outstanding coercive forces in bulk and powder form (with extrapolated values of about 60 kOe at the phase boundary Sm0.167 Co0.608Al0.255). When bulk alloys are annealed (700 C) demagnetization takes place spontaneously at critical fields with a slope comparable to the demagnetization factor rather than with the considerably smaller slope found in materials prepared by quenching from the melt. Ni and Ag substituted SmCo5, as well as Y0.167Co0.683Al0.15 and Gd0.167Co0.683Al0.15 show no appreciable coercive forces at 4.2K. Results are discussed in terms of a model of homogeneous pinning of domain walls of atomic dimensions on the crystal lattice. The role of substituent and rare earth in establishing highly energetic domain walls is discussed.

INTRODUCTION

Of the many individual properties of rare earth (R) and transition metals (T) that are enhanced on mutual compound formation the extraordinarly high anisotropies found in RT5 compounds are of special interest as they give rise to many unusual physical properties one of which is reflected in magnetic hardness.

Further modification of these properties can be engineered by the intermediary role of a third component. It has been shown previously that outstanding coercive forces can so be achieved at 4.2K in bulk homogeneous materials based on SmCo5 where Al substitutes for Co (1). In the present study we explore the chemical origin of this effect by variations either in substituent or in rare earth.

The role the rare earth plays is monitored by studies on analogous compounds with non magnetic Y and non anisotropic Gd. Substituents for Co are chosen such that electronic and spatial effects can be separated. Ni is closely related to Co both in electronic makeup and size. Al and Si contribute high electron concentration and large size. Ag is a low electron concentration component, however, with comparable size to Al.

RESULTS AND DISCUSSION

Materials are prepared by melting in MgO crucibles using induction techniques under inert atmosphere. Quenching from the melt is accelerated via a flow of purified argon. Deby-Scherrer diagrams are taken with filtered Cr radiation and show in all cases a homogeneous CaCu5 type structure. In materials based on Sm one faint additional line is observed which can be indexed as (111) on a doubled unit cell (2). A further observation pertains to the quality of reflections. In materials Sm0.167Co0.683Al0.15 and Sm0.167Co0.683Si0.15 reflections with low l are of considerably diffuse character. This is most notably so with prism planes such as (220) or (300) indicating stacking faults in the basal plane. No indication for stacking faults is found after annealing Al substituted compounds. Magnetization measurements are taken with a Foner magnetometer either on powders of <37μ or on bulk specimen. The individual single crystals composing the powder are oriented at room temperature with a conventional magnet either parallel ("parallel powder") or perpendicular ("perpendicular powder") to the direction of field in consecutive experiments. Coercive forces (Hc) correspond to the negative applied fields needed to decrease the magnetization to zero after prior exposure to fields of order 45kOe.

Table 1 - Coercive forces at 4.2K of various partially substituted compounds RT5.

Composition[1]	Structure Type	H_c (kOe) at 4.2K[2]
Sm0.167Co0.833	CaCu5	1.0 (10.0)
Sm0.167Co0.683Ni0.15	CaCu5	(6.58)
Sm0.167Co0.683Cu0.15	CaCu5	10.5 (12.5)
Sm0.167Co0.683Ag0.15	CaCu5	(2.55)
Sm0.167Co0.683Al0.15	CaCu5	22 (25)
Sm0.167Co0.683Si0.15	CaCu5	27
Gd0.167Co0.683Al0.15	CaCu5	4
Y0.167Co0.683Al0.15	CaCu5	3

[1] Materials were prepared by quenching from the melt.
[2] Coercive force measured on bulk materials. Bracketed values stand for powders (<37μ).

Coercive forces at 4.2K are summarized in Table I. Substitution of Al or Si for Co in SmCo5 results in a strong increase in coercive force compared to SmCo5 both in bulk and powder form. We have previously reported that values of Hc of about 60 kOe are extrapolated for bulk materials Sm0.167Co0.833-xAlx of CaCu5 type at the phase boundary with x = 0.225. Substitution of Co in SmCo5 by Ni and Ag has no significant effect on magnetic hardness. Values of Hc are of the order of a few kOe in Y0.167Co0.683Al0.15 and Gd0.167 Co0.683Al0.15.

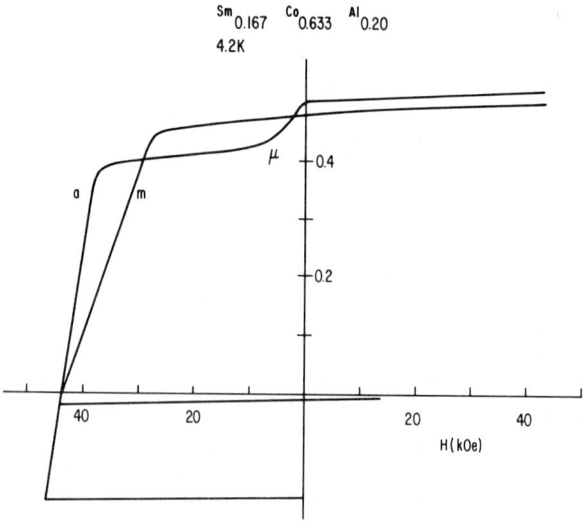

Fig. 1. Magnetic moment, μ (Bohr magnetons per normalized molecular weight) vs applied field H (kOe) at 4.2K for Sm0.167Co0.633Al0.20. "M" designates "as melted" material while "a" denotes preparation involving annealing (700 C).

The detailed mechanism of demagnetization is exemplified in Fig. 1 on "paralled powder" of Sm0.167 Co0.633Al0.20 of CaCu5 type. The origins of an initial drop in magnetization in the annealed material are not quite clear but could be connected with minor amounts of a second phase or with mechanical reorientation of part of the powder. Strong demagnetization in the second quadrant sets in at critical fields both in annealed and "as melted" materials. The slope of demagnetization of the annealed material corresponds approximately to the demagnetization factor while the "as melted" material shows a considerably smaller slope. Similar behavior is observed in "as melted" Sm0.167Co0.683 Si0.15. It is anticipated that after annealing this material the slope of demagnetization would also correspond to the demagnetization factor.

Only specimen SmCo5 with Al and Si substitution for Co show critical demagnetization fields. The absence of any noticeable dependence of Hc on particle size together with details in minor loop behavior indicate that the dominant mechanism of demagnetization involves domain wall pinning in Al and Si substituted materials rather than nucleation as in SmCo5. The extraordinary magnetic hardness in bulk materials points at highly energetic and thin domain walls. Moreover, the presence of critical fields suggests uniform barriers for domain wall motion and it is tempting to assign these barriers to the presence of stacking faults in "as melted" materials or to the periodicity of the lattice itself in the annealed specimens.

If the wall pinning is due to a decrease in wall thickness it is not entirely clear what the reasons are for the apparent variations in wall energy. In the straightforward Bloch model wall thickness is related to the competition between anisotropy and exchange. In view of the strong exchange due to Co it is clear that variations of wall energy cannot be accounted for by changes in anisotropy alone. In Zijlstra's model of a homogeneous intrinsic pinning (4) of domain walls of atomic dimensions we may rather have to consider a weak link in exchange for which Sm sites are a natural candidate. Weak exchange and high anisotropy due to crystal field effects on the rare earth appear to be prerequisites for highly energetic domain walls as is indicated by the absence of any dramatic effects in Gd and Y analog compounds. However, as these prerequisites are already met in SmCo5 the final clue for an understanding of the extraordinary magnetic hardness may be found in the state of partial magnetic disorder (ferromagnetic component magnetic glass) of the rare earth sublattice in pseudobinary regions (5). The disordered environment may greatly assist moment reversal within a domain wall. It is now not surprising to find strong effects only with substituents decisively different in electron concentration from Co, namely Al and Si. Size alone (Ag) does not make much difference.

We shall assume that the domain wall energy remains larger even at room temperature in "Al substituted SmCo5" compared to the pure binary. This is plausible particularily in view of generally higher anisotropies in these materials (1). Suitably substituted SmCo5 should therefore have potential for permanent magnet applications especially where low magnetizations are not an obstacle such as for coating of fine powders of SmCo5. Low temperature studies of magnetic hardness promise also to serve as a routine tool for identifying materials with highly energetic domain walls.

*This work was assisted by a grant from the Office of Naval Research.

REFERENCES

1. H. Oesterreicher, Solid State Comm. 14, 571 (1974).
2. Y. Khan and D. Feldmann, J. Less Comm. Metals 31, 221 (1973).
3. H. Oesterreicher, J. Less Comm. Metals 32, 385 (1973).
4. H. Zijlstra, Rare Earths and Actinides, Institute of Physics, Conference Digest pg. 158 (1971).
5. H. Oesterreicher, J. Appl. Physics 44, 2350 (1973).

Co_5Sm APPLIED

P. G. Frischmann
General Electric
Research and Development Center
Schenectady, New York 12301

While the physicists and metallurgists are developing an understanding of the material Co_5Sm, design engineers are rapidly evolving applications. Some of these applications have brought out new devices never before practical with Alnicos or ferrites, while other applications of Co_5Sm are more conventional magnetic designs, but still contribute significant device improvements.

It has rarely been found that a direct substitution of Co_5Sm for another permanent magnet material in a device leads to anything but higher cost. To take advantage of the properties of Co_5Sm, in spite of its intrinsically higher cost, it is, therefore, necessary to redesign a device. The benefits obtained from redesign must be explored broadly as they may contribute enough at the system level to offset the redesign and higher material costs.

Three designs each of which takes advantage of a different property of Co_5Sm were presented A tachometer generator utilizing the high energy product of $20x10^6$ gauss-oersted has been designed with a factor of 5 increase in mean time to failure and a weight reduction of 40%. A flux concentrator operating at zero B and using the characteristic 9000 oersted coercive force to reduce leakage flux of Alnico demonstrates an optimized usage of the two materials The third device, a switchable lifting magnet demonstrates repetitive use of the material along a unity recoil permeability line.

Device designs utilizing the unique Co_5Sm property of extremely high H_k have not been forthcoming, or at least not published. Potential areas for utilizing this property were discussed.

A plea to physicists and metallurgists working on Cobalt Rare Earth Material development incorporated the designers desire to trade some coercive force for higher inductions and the need for reducing magnet costs. The magnet users are challenged to find uses for the extremely high H_k and to design devices using these magnets in the repulsion mode.

EVIDENCE FOR NEW MAGNETIC RARE EARTH-COBALT PHASES*

K. J. Strnat and A. E. Ray
School of Engineering, University of Dayton in Dayton, Ohio 45469

ABSTRACT

An infinite series of hypothetical intermetallic compounds, with compositions converging toward RCo_5, can be derived by c-axis stacking of RCo_5 cells and selective substitution of R atoms for Co: RCo_2, RCo_3, R_2Co_7, R_5Co_{19}, RCo_4, R_7Co_{29}, R_4Co_{17}, R_3Co_{13}, R_5Co_{22}..... The first three compound types are well known. R_5Co_{19} phases were recently found for R=La, Ce, Pr and Nd. Dynamic permeability vs. temperature measurements (-190 to + 700°C) on several RCo_x alloys with R = La, Ce, Pr, Nd and Sm in the range $3.5<x<5$ yielded Curie points for the R_5Co_{19} and indicated the existence of several more phases. From the observed distinct permeability peaks Curie temperatures are determined and compared with expected values based on the compositions of the above phases. No discrete peaks were observed for Sm-Co, but a broad maximum between the Curie temperatures of Sm_2Co_7 and $SmCo_5$ indicates that certain heat treatments produce intermediate, disordered structures.

INTRODUCTION

Intermetallic phases between cobalt and some of the light rare-earth metals (R) are of practical importance as permanent magnet materials. For this reason, the phase relations in the binary R-Co alloy systems have received considerable attention in recent years.(1) However, certain details of the phase diagrams in the neighborhood of the technologically most important composition, RCo_5, are still incompletely known or subject to controversy.

Sintered magnets based on Sm-Co, Pr-Co, Ce-Co, etc., nearly always have overall compositions richer in the rare-earth component than RCo_5 and they must be rapidly cooled from temperatures between 800 and 1000 °C to obtain best coercive force and loop squareness.(2) One or more of the phases adjacent to RCo_5 on the R side are undoubtedly always present in these magnets, and we think that they play an important role in causing coercivity.(3) We have therefore recently re-investigated the binary phase diagrams of cobalt with La, Ce, Pr, Nd and Sm in the region R_2Co_7 to RCo_5.

The compounds with compositions between RCo_2 and RCo_5 all have closely related crystal structures that may be thought of as derived from the RCo_5 structure by an orderly substitution of R atoms at Co sites in some of the mixed layers and by layer shifts. This scheme was first proposed by Cromer and Larson(4) and discussed more recently for the case of R-Co by Buschow(5) and by Khan and Feldmann(6). Consequent development of the substitution scheme leads to an infinite series of compounds with compositions converging toward RCo_5 (Table I.) Stacking schemes for the initial members are shown in Figure 1. Hexagonal and rhombohedral modifications can be constructed for each phase, but only the rhomb. forms are shown beyond RCo_3

Table I: Possible R-Co phases derived from RCo_5 by substitution and shifting of layers. (Based on Ref. 4).

SCHEMATIC SUBSTITUTION	SIMPLE FORMULA	Co CONTENT ATOMIC %	NO. OF LAYERS HEX.	NO. OF LAYERS RHOMB.
RCo_5-Co+R=R_2Co_4	RCo_2	66.67	2	3 cubic
2 RCo_5-Co+R=R_3Co_9	RCo_3	75.00	4	6
3 RCo_5-Co+R=R_4Co_{14}	R_2Co_7	77.78	6	9
4 RCo_5-Co+R=R_5Co_{19}	R_5Co_{19}	79.17	8	12
5 RCo_5-Co+R=R_6Co_{24}	RCo_4	80.00	10	15
6 RCo_5-Co+R=R_7Co_{29}	R_7Co_{29}	80.56	12	18
7 RCo_5-Co+R=R_8Co_{34}	R_4Co_{17}	80.95	14	21
8 RCo_5-Co+R=R_9Co_{39}	R_3Co_{13}	81.25	16	24
9 RCo_5-Co+R=$R_{10}Co_{44}$	R_5Co_{22}	81.48	18	27
∞	RCo_5	83.33	1	

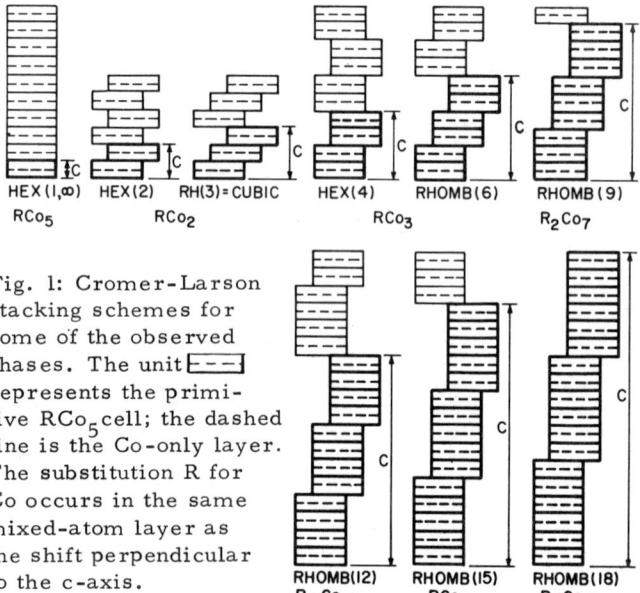

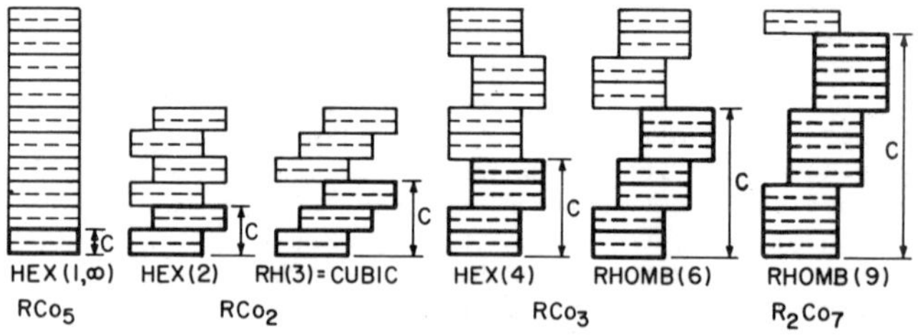

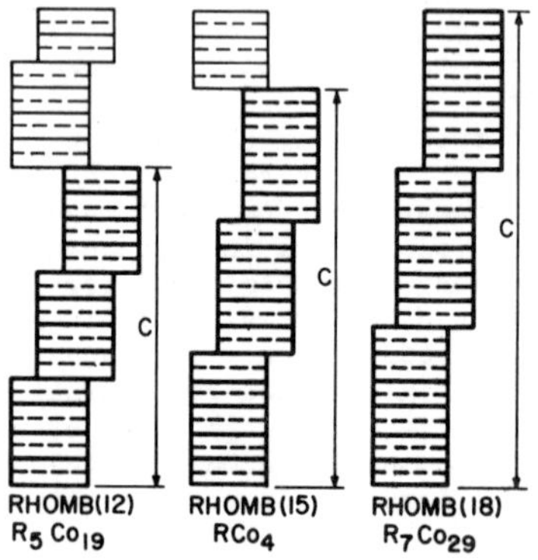

Fig. 1: Cromer-Larson stacking schemes for some of the observed phases. The unit ⊏---⊐ represents the primitive RCo_5 cell; the dashed line is the Co-only layer. The substitution R for Co occurs in the same mixed-atom layer as the shift perpendicular to the c-axis.

The RCo_2, RCo_3, R_2Co_7 and RCo_5 phases have long been known to exist in most R-Co alloy systems. Schweizer(7) found Pr_5Co_{19}, Ray(8) proved the existence of rhombohedral Ce_5Co_{19} and Nd_5Co_{19}; Khan and Feldmann(6) independently made the same report. Ray(1) later found La_5Co_{19} (rhomb.) as well. Modern versions of the R-Co phase diagrams show no other phases in this range, but Buschow(5) reported that $ErNi_4$, Er_4Ni_{17} and Er_5Ni_{22} in the Er-Ni system where the same stacking scheme is possible.

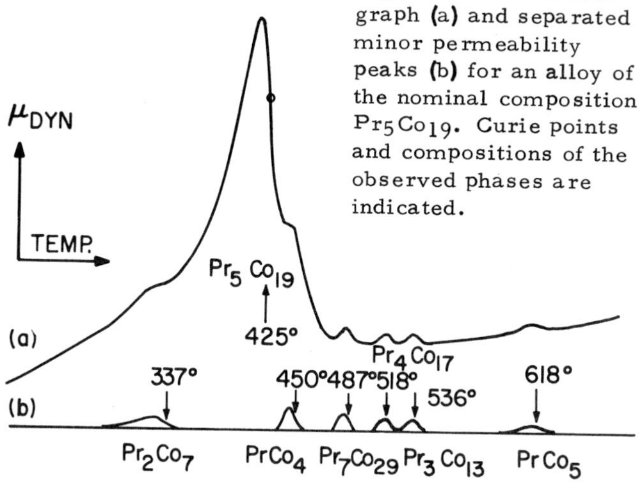

Fig. 2: TMA recorder graph (a) and separated minor permeability peaks (b) for an alloy of the nominal composition Pr_5Co_{19}. Curie points and compositions of the observed phases are indicated.

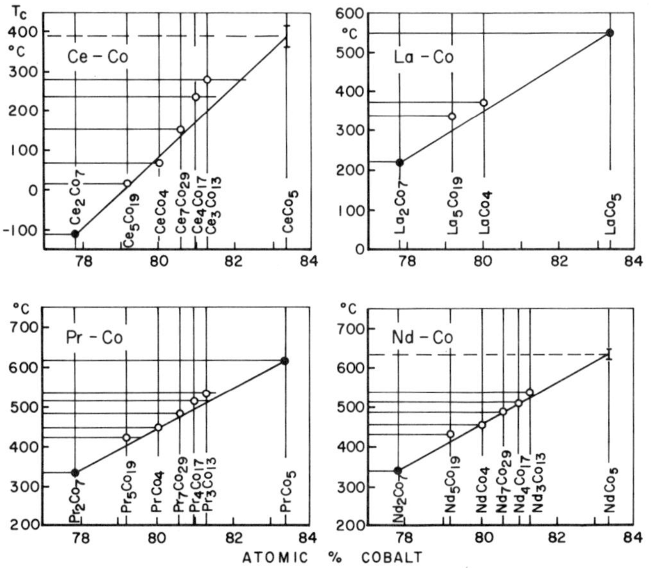

Fig. 3: Summary of observed Curie temperatures T_C values are sequentially assigned to the member of the series of theoretically possible phases.

EXPERIMENTAL WORK

We attempted to obtain several R_5Co_{19} compounds in single-phase form to better characterize their basic magnetic properties. The alloys were prepared by arc melting and vacuum annealing for 72 hours at 1100°C.

A thermomagnetic analysis (TMA) was performed or powders. We measured the low-field (1Oe), dynamic (5kHz) permeability as a function of temperature. This not only gave Curie-temperature values for the 5:19 phases with La, Ce, Pr and Nd, but it revealed the presence of several of the other theoretically possible compounds with higher Co content as minor, secondary phases.

The clearest example of such a "TMA spectrum", for an alloy of the nominal composition Pr_5Co_{19}, is reproduced in Figure 2. The highest permeability peak indicates that Pr_5Co_{19} is indeed the dominant phase. Taking the upper point of inflection as the Curie temperature, we can assign the lowest-temperature peak to Pr_2Co_7, the uppermost to $PrCo_5$. Both are present as minor phases. If the inflection temperatures of the other small peaks observed are considered Curie points of the hypothetical phases following R_5Co_{19} in Table I and plotted over the compositions of these, as in Figure 3, they fall very nearly on a straight line connecting the T_C of Pr_2Co_7 and $PrCo_5$, provided we take for Pr_2Co_7 the 337°C from the present measurement rather than the older literature value of 300°C.[9] Similar, but less well developed, permeability peaks were observed during TMA of La-Co, Ce-Co and Nd-Co alloys of the nominal 5:19 composition. They can be assigned in a similar manner to previously only hypothetical compounds. We feel therefore justified in saying that several compounds of the Cromer-and-Larson sequence tend to form in the systems where R=La, Ce, Pr and Nd. The distinct μ-peaks indicate good atomic order in small but finite volumes. It should be noted that the TMA method employed is capable of resolving smaller percentages and finer distributions of secondary phases than either differential thermal analysis or X-ray diffraction.

Several measurements on "Sm_5Co_{19}" alloys prepared by fusion techniques or by a reduction-diffusion method[10] failed to show the existence of either a 5:19 compound or any of the new phases in the Sm-Co system. Only the Curie temperatures of Sm_2Co_7 and $SmCo_5$ were observed in addition to a broad, low permeability peak between these that probably indicates the existence of ill-defined regions of intermediate composition whose crystal structure must be thought of as a modified hypostoichiometric $SmCo_5$ structure containing largely disordered substitutions of Sm for Co and irregular stacking faults. We later tried to prepare compounds of the 1:4, 7:29 and 4:17 phases of Ce, Pr and Nd in purer form by melting alloys of these compositions and annealing them for long periods just below their estimated peritectic temperatures. These attempts were generally unsuccessful and yielded phase mixtures dominated by RCo_5 or R_5Co_{19} which contained smaller amounts of the desired phase than the earlier alloys of the 5:19 nominal composition.

CONCLUSIONS

In view of the existence of the 5:19 and some of the other Cromer-Larson phases, the concept of a eutectoid decomposition of RCo_5 into R_2Co_7 and R_2Co_{17} will require re-thinking. The possibility of forming these new phases also has implications for the "epitaxial-shell model" of coercivity in sintered, RCo_5-based permanent magnets.[3] The infinite sequence of possible compositions converging at RCo_5 provides a mechanism for a quasi-continuous extension of the RCo_5 homogeneity region on the rare-earth side if one allows for stacking fault-like transitions between the theoretically discrete structures of the series. All Cromer-Larson structures and intermediate stacking-fault structures can grow epitaxially upon each other. Thus, during rapid cooling of a sintered magnet, a complex shell may grow on the surface of the RCo_5 grains in which, however, the c-axis is maintained as the common easy axis of magnetization and where anisotropy remains high. Stacking faults and R-for-Co substitutions would provide small local disturbances which may serve as pinning sites for magnetic domain walls. Also, if the energy of a domain wall decreases as it travels from the RCo_5 grain into this shell, the rare earth-rich layer could trap the wall even if such lattice imperfections are not effective pinning sites.[11]

ACKNOWLEDGMENT

The authors are grateful to the Messrs. R. Leasure, H. Mildrum and G. Plenge, all of the University of Dayton for assistance with the experimental work.

REFERENCES

1. A.E. Ray, Cobalt 1974-1, 13-20 (1974).
2. D.L. Martin, R.P. Laforce and M.G. Benz, General Electric R and D Center, Report No. 73 CRD140, Schenectady, New York, April 1973.
3. J. Schweizer, K.J. Strnat and J.B.Y. Tsui, IEEE Trans. Magnetics, MAG-7, 429-431 (1971).
4. D.T. Cromer and A.C. Larson, Acta Cryst. 12, 855-859 (1959).
5. K.H.J. Buschow, J. Less-Comm. Metals 16, 45-53 (1968).
6. Y. Khan and D. Feldmann, J. Less-Comm. Metals 33, 305-310 (1973).
7. J. Schweizer, USAF Materials Laboratory Tech. Report, AFML-TR-72-82 (May 1972), Chapter II.
8. A.E. Ray and K.J. Strnat, USAF Materials Laboratory Technical Report, AFML-TR-72-99(April 1972) Section III.
9. L.R. Salmans, K. Strnat and G.I. Hoffer, USAF Materials Laboratory Technical Report AFML-TR-68-159 (Sept. 1968).
10. R/D samples courtesy of Dr. F. Jones, Hitachi Magnetics, Inc., Edmore, Michigan.
11. Y. Tawara, Japan J. Appl. Phys. 11, 1578 (1972).

*) Work supported by the National Science Foundation

OXIDATION CONTROLLED AGING OF $SmCo_5$ MAGNETS

P. J. Jorgensen
Stanford Research Institute
Menlo Park, Ca., 94025

ABSTRACT

The minimum irreversible magnetic aging characteristics of samarium cobalt magnets can be calculated based on the kinetics of internal oxidation of $SmCo_5$. Experimentally measured open circuit remanent induction losses determined as a function of time show excellent correlation with the calculated curves for temperatures up to 200°C. Above 200°C, deviations occur from the calculated aging curves based on short term oxidation kinetics because of coalescence of the oxide fibers in the sub-scale. Measurement of internal oxidation kinetics at 300°C for times comparable to the magnetic aging times again allow accurate computation of the irreversible magnetic losses.

STRUCTURE TRANSFORMATIONS IN SPUTTER-DEPOSITED Sm_2Co_{17} ALLOYS

R. Wang and R. P. Allen
Battelle Memorial Institute
Pacific Northwest Laboratories
Richland, Washington 99352

ABSTRACT

A series of structure transformations were observed for thick, sputter-deposited amorphous Sm_2Co_{17} alloys heat treated between 500°C and 900°C for one hour. The amorphous phase crystallized at approximately 500°C to give a $CaCu_5$-type structure with a c/a ratio of 0.86, which is the highest value ever observed for these alloys. With increasing annealing temperature, the c parameter decreased and the a parameter increased resulting in a continuous reduction of the c/a ratio to a value of 0.84 at 800°C, which is the same as that of the Th_2Zn_{17}-type structure. These structure transformations and lattice parameter variations are attributed to a progressive change with annealing in the number and degree of order of Sm-atom/Co-atom pair substitutions. Based on a correlation found between the c/a ratio and composition for the $CaCu_5$-type Sm-Co alloys, the metastable phase crystallized from the amorphous state at 500°C has an approximate formula of Sm_2Co_{24} to Sm_2Co_{26}.

INTRODUCTION

The intermetallic compound Sm_2Co_{17} is reported[1] to have two crystal forms; a hexagonal Th_2Ni_{17}-type structure stable at high temperatures and a rhombohedral Th_2Zn_{17}-type structure at lower temperatures. Both of these structures are derived from the $CaCu_5$-type $SmCo_5$ structure by replacing every third Sm atom with a pair of Co atoms aligned parallel to the c-axis. The hexagonal and rhombohedral Sm_2Co_{17} modifications differ only in the arrangement of the ordered substitutions.

Recent studies of rare earth-transition metal phases (R_2M_{17}) with the Th_2Ni_{17}-type structure[2,3] have shown that Th_2Ni_{17} represents an ideal structural type and that some excess, disordered substitutions can also occur resulting in a shift in stoichiometry to higher transition metal contents. A high degree of substitutional disorder leads to the reestablishment of the basic $CaCu_5$-type structure, as observed for Gd_2Fe_{17}.[4] A reexamination of the hexagonal Sm_2Co_{17} compound formed by the rapid solidification of this alloy has established that its structure is also of the $CaCu_5$-type rather than the Th_2Ni_{17}-type structure.

In this paper, we shall present results of our investigation of Sm_2Co_{17} alloys made by high-rate sputter-deposition at room temperature. The structure transformations from the as-sputtered amorphous phase to the final equilbrium rhombohedral Sm_2Co_{17} structure through heat treatment provide further information regarding the characteristics of the structure ordering in Sm_2Co_{17} type phases.

EXPERIMENTAL

Thick deposits of amorphous Sm_2Co_{17} alloys were prepared on a 5-cm diameter water-cooled OFHC Cu substrate using high-rate sputter-deposition techniques.[6] The target for the Sm_2Co_{17} deposit was a 5-cm diameter disc of the cast alloy. The Sm_2Co_{17} deposit was 0.5-cm thick and had an oxygen content (by neutron activiation analysis) of 0.02 wt.%.

Specimens for the x-ray studies were electrical-discharge-machined from the deposits, wrapped in Ta foil, and heat-treated in an all-metal, high-vacuum furnace. An Enraf-Nonius Guinier camera calibrated with annealed Si powder was used for the lattice parameter measurements. The 4θ-Bragg angles were measured to an accuracy of ±0.1 , which is equivalent to an uncertainty in the lattice parameter of about 0.005Å. The accuracy of the measurements was limited by line broadening due to the very fine grain size of the crystallized materials.

RESULTS AND DISCUSSIONS

The as-deposited Sm_2Co_{17} alloys were amorphous having physical, mechanical, and magnetic properties significantly different from those of crystalline material of the same composition.[6,7] The structure of the amorphous alloys has not been investigated. Cargill,[8] however, conducted an x-ray scattering study of a thick, amorphous $GdFe_2$ alloy, which was prepared by high-rate sputtering in this laboratory. He proposed binary dense random packing of hard spheres as a structural model.

The amorphous alloy had a sharp crystallization temperature. The material was still amorphous after 72 hours at 475°C. X-ray patterns with broad, but well-defined diffraction lines were obtained after a one-hour heat treatment at 500°C. Figure 1 illustrates the diffraction patterns for Sm_2Co_{17} deposits heat treated for one hour at 500, 600, 650, 700 and 800°C. The crystallization of the alloy at 500°C results in a structure having a cubic-like x-ray diffraction pattern. Lattice parameter measurements of alloys annealed at 500, 600 and 700°C indicate that the cubic-like pattern has the hexagonal symmetry with a c/a ratio of 0.86. This leads to the overlapping of the hexagonal (200), (111) and (002) reflections and other high angle lines. The separation of the (200), (111) and (002) reflections increases continuously as the annealing temperature increases.

a
b
c
d
e

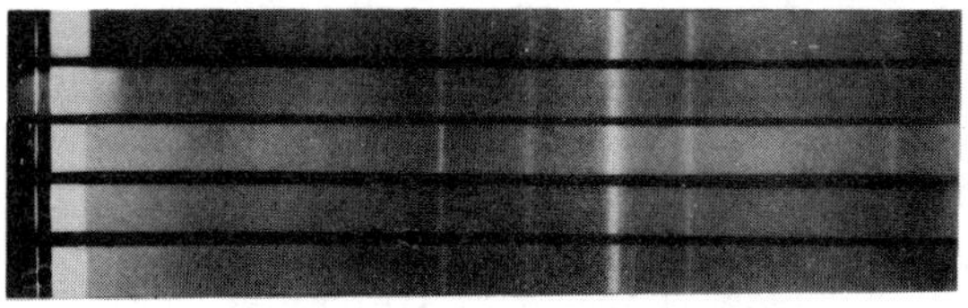

Figure 1 Guinier x-ray photographs of Sm_2Co_{17} sputter-deposited alloys heat-treated for one hour at (a) 500°C, (b) 600°C (c) 650°C (d) 700°C and (e) 800°C.

LATTICE PARAMETERS, c/a RATIOS AND UNIT CELL VOLUMES OF SPUTTERED Sm_2Co_{17} ALLOYS HEAT TREATED FOR ONE HOUR

Heat Treatment	a (Å)	c (Å)	c/a	V (Å)^3
500°C	4.842	4.160	0.859	84.46
600°C	4.862	4.162	0.856	85.20
650°C	4.861	4.161	0.856	85.15
700°C	4.844	4.132	0.853	83.96
750°C	4.853	4.135	0.852	84.34
800°C	4.870	4.081	0.838	83.82
850°C	4.872*	4.063*	0.834*	83.52*

*Reduce unit cell to match $CaCu_5$-type structure.

The lattice parameters, c/a ratio and unit cell volumes of the sputtered Sm_2Co_{17} alloys as a function of heat treatment temperatures are listed in Table I and shown in Figure 2. It is clear that increasing the heat treatment temperature caused an expansion of the a axis and a contraction of the c axis; a rapid decrease of the c/a ratio occurred. The c/a ratio drops from 0.86 at 500°C to 0.84 at 800°C. The rhombohedral Th_2Zn_{17}-type structure was formed above 800°C exhibited a sharp x-ray diffraction pattern. Thus the structural transformations observed in sputter-deposited Sm_2Co_{17} alloys followed the general sequence:

$$\text{amorphous} \xrightarrow{500^\circ C} \text{hexagonal-disordered-}$$
$$CaCu_5\text{-type} \xrightarrow{800^\circ C} \text{rhombohedral } Th_2Zn_{17}\text{-type}$$

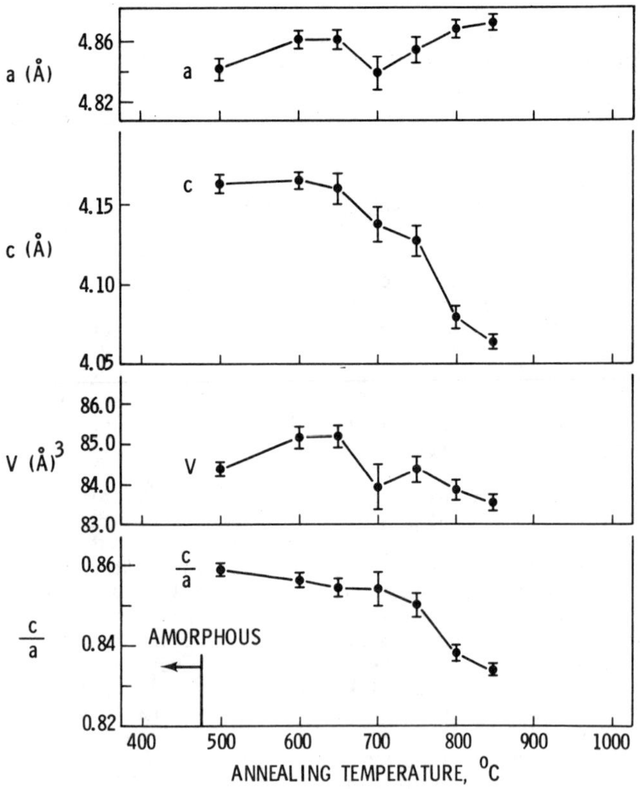

Figure 2. Lattice parameters, unit cell volume and c/a ratio of Sm_2Co_{17} sputtered alloys as function of annealing temperature.

The a parameter in RM_5 and R_2M_{17} structures increases with the number of R atoms alloyed, while structure ordering generally reduces the c parameter. The observed large c parameter and small a parameter for the as-crystallized phase at 500°C would indicate that a highly random substitution of Sm atoms by Co-atoms pairs occurs. This also implies a change of stoichiometry from 2:17 to a higher Co:Sm ratio. Since the structure is based on the $CaCu_5$-type structure, the hyper-stoichiometry $SmCo_{5+x}$ alloys can be expressed by a formula: $(Sm_{1-r}2Co_r)Co_5$ for which r is the measure of the substitution of Sm atoms by Co-atom pairs. Thus r = 0 represents $SmCo_5$. As r = 1, only the pure Co structure forms. A plot of lattice parameters and c/a ratios as a function of r is illustrated in Figure 3 for Sm-Co alloys having the $CaCu_5$- type structures. Extrapolation of these curves toward the highest c/a ratio (0.86) and lattice parameter values corresponds to a substitution factor r = 0.5 to 0.525. This suggests a composition range of Sm_2Co_{24} to Sm_2Co_{26} for as-crystallized phase. Heat treatment at high temperatures leads to the formation of partially ordered $(Sm_{2/3}2Co_{1/3})$ Co_5 structure with r = 0.333.[5]

Givord, et al[2,3] have found that for Th-Ni, Y-Ni, Er-Co, and Lu-Fe compounds, random substitution of Co,Ni or Fe atom pairs for rare-earth elements leads to a change of stoichiometry from 2:17 to 2:19. The c/a ratio for Lu-Fe compounds increase 0.51% from Lu_2Fe_{17} to Lu_2Fe_{19}. The magnitude of this c/a change is much less than the 3.3% change in sputtered Sm_2Co_{17} alloys resulting from heat treatment between 500 and 800°C. However, the result obtained by Givord, et al can be quantitively described by the extrapolated curves in Figure 3 as well.

The maximum number of Sm atoms that can be substituted by Co-atom pairs in the $CaCu_5$-type structure is restricted by the fact that no two Co-atom pairs can be positioned consecutively along the c axis. Thus, a maximum of one-half of the Sm atoms can be replaced by Co-atom pairs to form a complete random substitutional $CaCu_5$-type structure. This would suggest that the structural limit corresponds to a formula of $(Sm_{1/2}2Co_{1/2})$ Co_5 equivalent to a Sm_2Co_{24} composition. The linear correlation between the c parameters and the substitution factor r shown in Figure 3 indicates that the expansion of c axis is directly proportional to the substitution of Sm-atom by Co-atom pairs.

If the initial phase crystallized from the amorphous Sm_2Co_{17} alloy is truly hyper-stoichiometric, there should be a presence of a second Sm-rich phase. However, the grain size is too fine for metallographic examination. X-ray diffraction also fails to reveal the presence of the second phase. The absence of the diffraction lines of the second Sm-rich phase may be similar to the decomposition of RCo_5 into R_2Co_{17} and R_2Co_7 phases where the R_2Co_7 phase was not detectable by x-ray.[9]

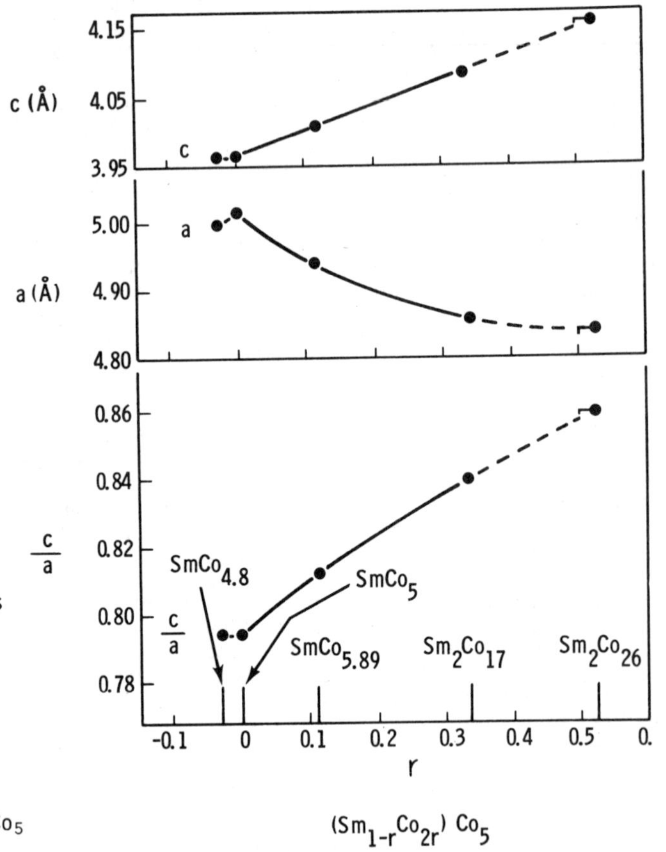

Figure 3. Correlations between c,a and c/a and the compositions of the Sm-Co alloys having $CaCu_5$-type structure in the range of $SmCo_5$ to Sm_2Co_{17}. Extrapolations of these curves to the c,a and c/a values for the metastable phase crystallized from the amorphous state at 500 C give a composition range of Sm_2Co_{24} to Sm_2Co_{26} for that phase.

REFERENCES

1. K. H. J. Buschow, J. Less-Common Metals, 11, 204-208 (1966)
2. D. Givord, R. Lemaire, J. M. Moreau and E. Rondant, J. Less Common Metals, 29, 361 (1972)
3. D. Givord, F. Givord and R. Lemaire, J. Less-Common Metals, 29, 389 (1972).
4. F. Givord and R. Leamaire, J. Less-Common Metals 21, 463 (1970).
5. Y. Khan and B. Mueller, J. Less-Common Metals, 32, 39-45 (1970).
6. R. P. Allen, S. D. Dahlgren, H. W. Arrowsmith and J. P. Heinrich, AFML-TR, 74-87, Wright-Patterson Air Force Base, Ohio, May 1974.
7. M. D. Merz, R. P. Allen and S. D. Dahlgren, J. of Appl Phys 45, 4126-4127 (1974)
8. C. S. Cargill III, AIP Conf. Proc. No. 18, Magnetism and Magnetic Materials - 1973, 631-635 (1974).
9. F. J. A. den Broeder and K. H. J. Bushchow, J. Less-Common Metals, 29, 65 (1972).

MECHANISM FOR COERCIVITY IN SPLAT-COOLED RCo_5 COMPOUNDS

E. Ziai and Y. B. Kim
University of Southern California
Los Angeles, California 90007

ABSTRACT

Splat-cooling technique is used to prepare fine powders of certain RCo_5 compounds. Powders produced in this way are free from most of the defects that are introduced in powders prepared by mechanical grinding. Furthermore, splat-cooled materials are shown to have large concentrations of grain boundaries. These powders have magnetic properties which are different from those of the powders produced by mechanical comminution. In particular, the intrinsic coercivities of $PrCo_5$ and $MMCo_5$ are increased significantly by splat cooling, while their maximum energy products are reduced. Heat treatments of splat-cooled powders, which reduces the density of grain boundaries, increases coercivity slightly but a large increase takes place in the maximum energy product. Mechanical grinding of splat-cooled materials produces similar effects, although this time it is due to the deliberate introduction of defects which act as pinning centers. By analyzing the hysteresis curves of as-ground, splat-cooled and heat-treated or ground splat-cooled $SmCo_5$ we have concluded that in the splat-cooled materials a nucleation mechanism is dominant in the initial stages of the magnetization reversal process. Nucleation takes place at the grain boundaries which are in abundance.

INTRODUCTION

The coercivities of rare earth cobalt compounds prepared by mechanical comminution are far below their theoretical values $2K_1/M_s$. The reduction in coercivities, particularly in the case of $PrCo_5$ and $MMCo_5$, are attributed to the metallurgical defects generated during grinding, mostly in the form of plastic deformation. In preparing fine RCo_5 particles without the "milling effect", we adopted splat-cooling technique[1] in which molten alloys are atomized into small liquid droplets and splat cooled on a copper substrate. The observed magnetic properties of splat-cooled materials are significantly different from that of as-ground materials as a result of changes in the metallurgical microstructures.

EXPERIMENTAL PROCEDURE AND RESULTS

The powders for both as-ground and splat-cooled materials were passed through a -325 mesh screen. The powders in both cases were thoroughly mixed with amber epoxy resin. The weight ratio of the powder to epoxy was 4 to 1 for as-ground and 3 to 1 for splat-cooled specimens. Each mixture was inserted into a capillary tube 10mm long and 1mm I.D. The sample was aligned and cured in an 18 kOe field for one hour at about 100°C. Magnetization at 18 kOe was 10.1 kG for as-ground $SmCo_5$ and 7.2 for splat-cooled $SmCo_5$. The low value for the splat-cooled sample is due to the multi-grain nature of its particles which makes perfect alignment impossible. Average grain size in splat-cooled particles is of the order of 1μ, which is an order of magnitude smaller than that observed in as-ground particles.

Hysteresis loops for as-ground and splat-cooled powder samples of $MMCo_5$ are shown in Fig. 1. The sample demagnetization factor is small (0.0172) and no correction has been applied. Hysteresis loops for splat-cooled $SmCo_5$ and $PrCo_5$ have similar features. For splat-cooled powders, the magnetization changes very slowly with the applied magnetic field and the saturation magnetization requires a very large field. Optical and electron microscope observations indicate that in the splat-cooled powder specimen the constituent particles are polycrystalline, unlike in as-ground powder. This characteristic, which is the result of rapid cooling, makes perfect alignment impossible; in attaining saturation a magnetizing field comparable to anisotropy field is required. Other common features of splat-cooled materials are: low remanent magnetization, a very rapid drop of magnetization at the beginning of demagnetization process, and a slow change at the latter part of the demagnetization curve. Unlike $SmCo_5$, the coercivities of $PrCo_5$ and $MMCo_5$ can be increased significantly by splat-cooling (see Table I). Maximum energy product generally decreases by splat-cooling. But it can be improved by either heat treatment or mechanical grinding of splat-cooled products without affecting the high coercivities of $PrCo_5$ and $MMCo_5$ (see Table II).

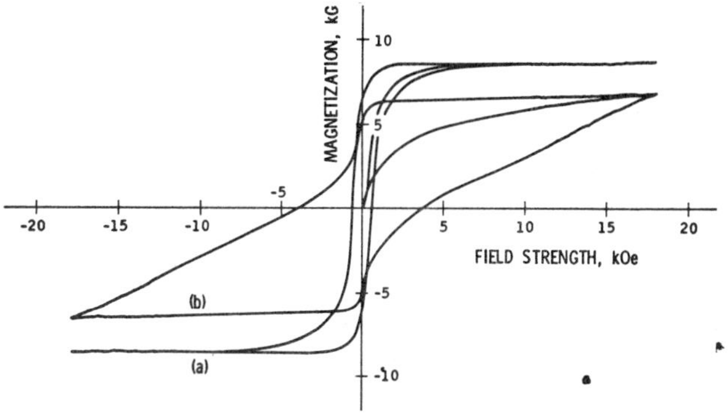

Fig. 1 Magnetization curves and major hysteresis loops of two $MMCo_5$ specimens: (a) as-ground; (b) splat-cooled

Table I

	$SmCo_5$		$PrCo_5$		$MMCo_5$	
	As-Ground	Splat-Cooled	As-Ground	Splat-Cooled	As-Ground	Splat-Cooled
$_MH_c$ (kOe)	8.40	5.50	1.10	3.60	0.60	3.90
$4\pi M_r$ (kG)	9.70	5.40	8.80	5.20	6.78	4.92
$(BH)_{max}$ (MG·Oe)	16.00	2.12	3.78	1.61	1.45	1.40

Table II

	Splat-Cooled $SmCo_5$		Splat-Cooled $PrCo_5$		Splat-Cooled $MMCo_5$	
	Heat Treated	Ground	Heat Treated	Ground	Heat Treated	Ground
$_MH_c$ (kOe)	6.23	7.50	4.63	4.90	–	4.40
$(BH)_{max}$ (MG·Oe)	4.80	4.57	3.00	3.80	–	4.07

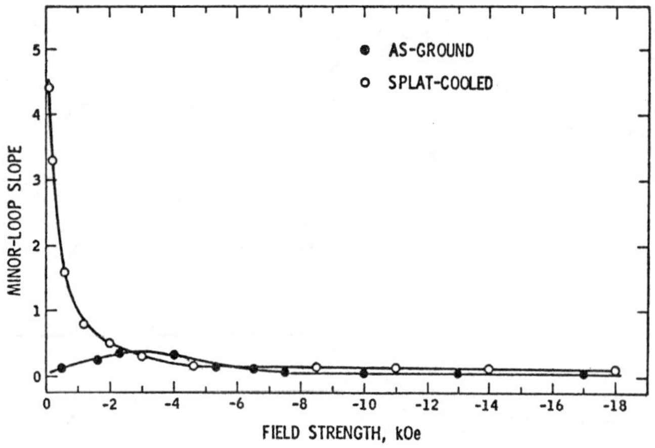

Fig. 2 Minor-loop slopes as functions of field strength of two types of $SmCo_5$

Metallurgically, splat-cooled powders are free from defects that are usually introduced during mechanical grinding. However, there exists a large concentration of grain boundaries in them. Traces of an unidentified second phase were observed in as-cast materials, and also in splat-cooled materials to a lesser degree. The large concentration of grain boundaries which act as nucleation sites and the absence of isolated metallurgical defects associated with mechanical grinding makes the magnetization reversal mechanism different from that of as-ground specimens. This is evident from the observed differences between magnetization reversal properties of as-ground and splat-cooled specimens of the same compound. Magnetization reversal is also studied by obtaining the minor loop slopes along the second and third quadrants of the hysteresis loop[2]. The change of the minor-loop slopes with the field for $SmCo_5$ is shown in Fig. 2. For the as-ground specimen the slope is small and nearly constant throughout the demagnetization process. However, in the case of splat-cooled specimen the slope is large at the beginning; and as the demagnetizing field increases, the slope decreases until it reaches a nearly constant value. The large and varying slopes at fields below the coercive force colaborate our interpretation that nucleation is the dominant mechanism for the magnetization reversal process in the splat-cooled rare-earth cobalt magnets.

DISCUSSION AND CONCLUSIONS

It is generally accepted that domain wall movement[3], rather than coherent rotation of domains, control the magnetization reversal process and thereby coercivity, in $SmCo_5$ powders. Zijlstra[2] has demonstrated that the coercivity of $SmCo_5$ powder is determined by pinning of the domain walls and their slow expansion with increasing applied field. Some of the observed dissimilarities in the magnetic properties of $SmCo_5$ powder specimens prepared by mechanical grinding and splat cooling may be accounted for by assuming that nucleation of the domain walls at the grain boundaries and their rapid movement through the interior of the grain is the predominant mechanism of the magnetization reversal in splat-cooled $SmCo_5$ specimens. In the splat-cooled specimen, the coercivity is reduced because of the reduction of pinning centers. In $PrCo_5$ and $MMCo_5$, like in $SmCo_5$, the ratio of nucleation sites to the pinning centers has increased by splat cooling and makes the nucleation mechanism dominant at the beginning of the demagnetization process. But, in this case, the concentration of defects in as-ground $PrCo_5$ and $MMCo_5$ that was damaging to coercivity has been reduced by splat cooling, making their coercivities larger by factors of three to five. Thus, splat-cooling technique may be used to obtain high coercivities which previously was not possible in these materials without going into elaborate and time-consuming powder treatments. Moreover, splat-cooled specimens may provide the means for the study of nucleation in a bulk specimen and therefore a better understanding of magnetization reversal process in rare-earth cobalt magnets.

REFERENCES

1. Duewez, P. and Williams, R. H., Trans. AIME 227, 362 (1969).
2. Zijlstra, H., J. Appl. Phys. 41, 4881 (1970).
3. Livingston, J. D., AIP Conference Proceedings No. 10, 643 (1972).

PREPARATION AND CHARACTERIZATION OF SPLAT-COOLED R-Co COMPOUNDS

D. V. Ratnam
Colt Ind., Crucible Materials Research Center, Pittsburgh, Pa. 15230

K. S. V. L. Narasimhan
Chemistry Dept., University of Pittsburgh, Pittsburgh, Pa. 15213

B. C. Giessen
Chemistry Dept., Northeastern University, Boston, Mass. 02115

ABSTRACT

Rapidly quenched alloys of Sm-Co, Pr-Co and MM-Co of the approximate composition RCo_5 were prepared by splat cooling droplets of the alloys in a modified arc furnace. X-ray diffraction analysis indicated a fine grained structure and also the possibility of some texture in the splat-cooled samples. Microscopic examination of the alloys in a light microscope revealed an equiaxed fine grained structure with a grain size of 2 to 5μm.

Magnetic measurements were performed on a vibrating sample magnetometer. Coercive force values were 5000 Oe for $SmCo_5$, 1500 Oe for $PrCo_5$ and 1250 Oe for $MMCo_5$. Thermomagnetic curves were obtained on the splat-cooled alloys by plotting magnetization versus temperature up to 900C. The data indicated that all the alloys were single phase with Curie points in fair agreement with published values for these compounds. Splat-cooled $PrCo_5$ was found to be unstable near its Curie temperature.

INTRODUCTION

Rare Earth (R) cobalt magnets based on the compound RCo_5 are currently produced using the powder metallurgical process. Simultaneously, research work is devoted to understanding the influence of grain size, oxygen content and RCo_5 phase stability on the permanent magnet properties.

Alternative preparation techniques such as sputtering[1] are being explored to develop fine grained, nearly oxygen free RCo_5 and R_2Co_{17} magnets. Some data was published[2] to indicate that high coercive forces can be developed by grain refinement in RCo_5 powders prepared by rapid quenching. In the present paper we report the results from a preliminary study on the influence of rapid quenching on the microstructure, magnetic properties and the phase stability of RCo_5 compounds.

EXPERIMENTAL

Sample Preparation - Four rare earth cobalt compounds were prepared in an arc furnace using the starting weight percentages: a) samarium 38, cobalt 62, b) samarium 35, cobalt 65, c) praseodymium 34, cobalt 66 and d) mischmetal 34, cobalt 66. Alloy a consisted of $SmCo_5$ and Sm_2Co_7 while alloys b, c and d were predominantly RCo_5 with small amounts of rare earth rich phase.

Using 10 to 20 milligram quantities of the above alloys, rapidly quenched samples were prepared in the form of thin flakes by splat cooling droplets of the alloys in a modified arc furnace[3]. The splat cooling technique employed consists of rapidly spreading the droplet of the alloy between two heat-conducting copper surfaces. This technique produces a foil of 20 to 50μm thickness with a quenching rate of the order of 10^4 C/sec. This quenching rate is slower than obtained in alternative quenching techniques[3] where cooling rates of 10^6 to 10^8 C/sec. were recorded. About twenty samples were obtained for each alloy.

Sample Characterization - X-ray diffraction data was obtained on the four alloy compositions before and after splat cooling using a diffractometer. A light microscope was used to examine the microstructural features of the samples.

Measurements of magnetization $4\pi M$ and coercive force H_{ci} at room temperature were obtained on a Princeton Applied Research vibrating sample magnetometer. A nickel sample was used to calibrate the $4\pi M$ measurements. The splat-cooled samples in the form of thin flakes were attached to the sample rod and measurements were made with the applied magnetic field at 0 and 90° to the surface of the flake. A maximum field of 19 kOe was used for the measurements.

Thermomagnetic analysis was performed on the samples before and after rapid quenching. Magnetic moment σ in emu/gm was measured from room temperature to 880C. The details of the procedure were described elsewhere[4]. The samples were in the form of coarse powder and were contained in a sealed quartz tube in a purified argon atmosphere.

RESULTS AND DISCUSSION

Microstructure - Examination of the various splat-cooled specimens using X-ray diffraction indicated that the samples are of single phase RCo_5. Diffraction peaks for R_2Co_7 or R_2Co_{17} phases were not present. The arc melted master alloys gave X-ray diffraction peaks for RCo_5 but also contained weak diffraction peaks corresponding to R_2Co_7 phase. The diffraction peaks for the splat-cooled samples exhibited line broadening indicating the possibility of very fine grain size and lattice strain.

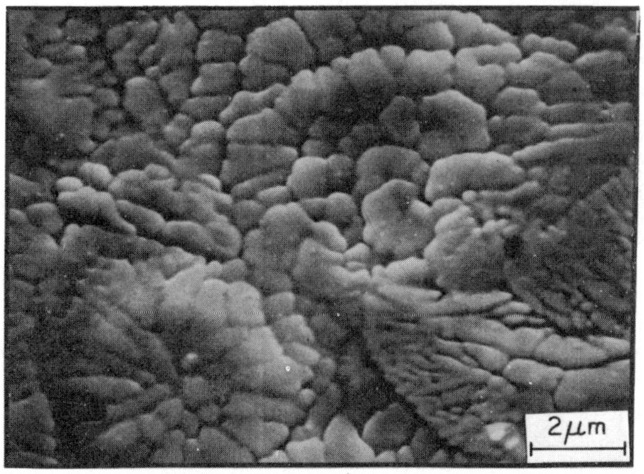

Fig. 1. Scanning Electron Micrograph of Rapidly Quenched $MMCo_5$ (8000X)

There was some difference in the relative intensity of the diffraction peaks for the arc melted and splat-cooled samples. But the reasons for the changes in relative intensity were not completely analyzed. Microscopic examination indicated that the splat-cooled samples have equiaxed grains in the size range of 1 to 5 μm. Magnetic domain structure could not be observed in the fine grains. However, small portions of the samples had grain sizes in the range of 5 to 20 μm and in these areas magnetic domain structure was observed.

The splat-cooled samples were also examined with AMR Model 1000 Scanning Electron Microscope and semi-quantitatively analyzed for cobalt and rare earth elements using PGT 1000 X-ray dispersive analyzer. In Fig. 1, a scanning electron micrograph revealing grain size in the range of 0.5 to 2 μm is shown for a $MMCo_5$ splat-cooled sample.

Magnetic Measurements - The properties measured on the vibrating sample magnetometer were $4\pi M$ in Gauss at an applied magnetic field of 19 kOe, $4\pi M_r$ in Gauss at zero field, and intrinsic coercive force H_{ci} in Oe.

Table I. Magnetic Properties of Rapidly Quenched R-Co Compounds

Sample	Cobalt Wt. %	$4\pi M$(G) at 19 kOe	$4\pi M_r$(G)	H_{ci}(Oe)
a) Sm-Co	62.2	5740	4550	2700
b) Sm-Co	65.5	6000	4720	5000
c) Pr-Co	67.1	7660	4920	1500
d) MM-Co	67.0	5550	3150	1250

Preferential crystalline orientation was not noticed in the rapidly quenched samples, as the angle between the applied field and the splat surface was varied between 0 and 90°.

In Table I, the magnetic properties of the four rapidly cooled compositions are given. Three splat-cooled flakes were used for the magnetic measurements of each composition. Results from chemical analysis as weight percent cobalt are also included in the table. The data indicates that the grain refinement achieved by rapid quenching may be responsible for the appreciable coercive force values of 1250 to 5000 Oe obtained. The coercive force of Sm-Co alloys was higher than for Pr-Co and MM-Co alloys. This observation is consistent with the much higher values of coercive force realized in Sm-Co alloy powders and sintered magnets. The coercive force values are in the range of 100 to 500 Oe for the conventionally cast RCo_5 alloys.

Thermomagnetic Analysis - Thermomagnetic curves were obtained for the arc melted alloys and also for the splat-cooled samples. Magnetic moment σ in emu/gm was plotted from room temperature to 900C. The thermomagnetic curves for single phase RCo_5 compounds are similar to those of iron and nickel. However, if the material being analyzed has more than one magnetic phase, deviations from the ideal curve occur near the Curie points for the secondary phases present. The observed thermomagnetic curves are interpreted based on the above assumption.

In the rapidly quenched state, the two Sm-Co alloys a and b exhibited very similar thermomagnetic behavior and the curve for sample a is shown in Fig. 2.

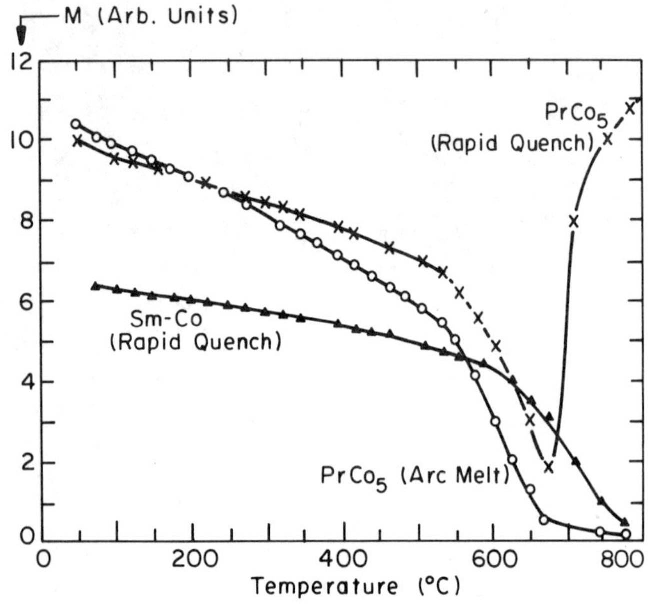

Fig. 2. Thermomagnetic Curves for Sm-Co and $PrCo_5$ Alloys

It appears that the rapid quenching substantially suppressed the formation of Sm_2Co_7 in alloy a, which is about 4 weight percent richer in samarium compared to stoichiometric $SmCo_5$ composition. In Fig. 2, the changes in the thermomagnetic behavior for $PrCo_5$ for 'slow' cooled and rapidly quenched samples is shown. Near the Curie point, the splat-cooled $PrCo_5$ transformed into a phase having higher magnetic moment and Curie point. This was observed in two measurements of splat-cooled $PrCo_5$. Two splat-cooled flakes were used for each of the two thermomagnetic measurements. After this transformation, thermomagnetic measurements were repeated from room temperature to 865C (the upper temperature limit of the equipment used). Curie point was not detected. X-ray diffraction data was obtained at room temperature after the samples were heated to 865C. Changes in diffraction peak profile were noticed but the location of the peaks was the same as in the splat-cooled $PrCo_5$ samples. Work is in progress to establish the instability of splat-cooled $PrCo_5$ compared to other alloys analyzed in this study.

In Fig. 3, the thermomagnetic behavior of a $MMCo_5$ alloy before and after rapid quenching is shown. The relatively rapid decrease in magnetization with temperature in the 200 to 400C region in the 'slow' cooled alloy indicates the presence of R_2Co_7 phases. The rapidly cooled alloy appears to be closer to single phase $MMCo_5$. This feature may also be noted for $PrCo_5$ in Fig. 2.

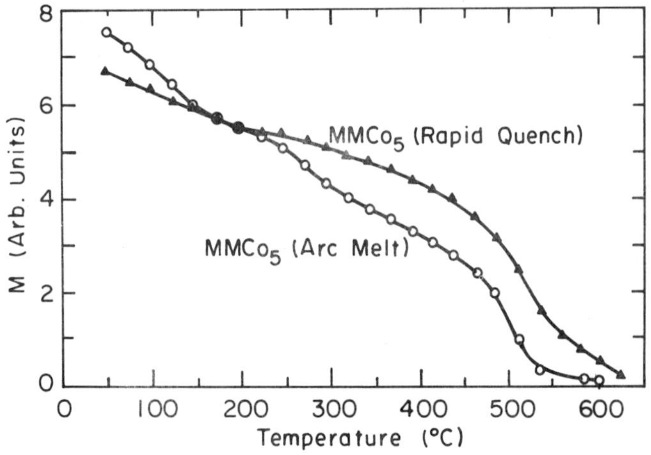

Fig. 3. Comparison of Thermomagnetic Curves for $MMCo_5$ After 'Slow' and Rapid Quench from the Melt

SUMMARY

Experimental results from the preliminary investigation indicate that by rapid quenching, grain size of 1 to 5 μm can be achieved in RCo_5 alloys to develop appreciable coercive forces. Crystalline texture is not readily obtained. Thermomagnetic data indicated the possible metastability of the rapidly quenched $PrCo_5$ phase.

ACKNOWLEDGMENTS

One of us (DVR) thanks Prof. L. N. Mulay, Materials Research Laboratory, Penn State University for the use of the Vibrating Sample Magnetometer and Dr. R. G. Wells, Crucible Materials Research Center, for valuable discussions. One of us (BCG) acknowledges support of this work by Office of Naval Research, under Contract N14-68-A-207-3.

REFERENCES

1. J. P. Heinrich, H. Garrett and R. P. Alien, J. Appl. Phys., Vol. 45, No. 4, April 1974, 1873.

2. R. Wang, K. S. Kim and Y. B. Kim, IEEE Trans. Magnetics, Mag.-8, 555 (1972) (Abstract).

3. B. C. Giessen and R. H. Willens, Phase Diagrams, Vol. 3, 1970, 103-141, Academic Press, New York.

4. R. A. Butera, R. S. Craig and L. V. Cherry, Rev. Sci. Instr. 32, 708 (1961).

PLASMA SPRAYED SAMARIUM-COBALT PERMANENT MAGNETS

M. C. Willson and R. J. Janowiecki
Monsanto Research Corporation, Dayton, Ohio 45407

ABSTRACT

Samarium-cobalt permanent magnets were fabricated by arc plasma spraying. This process involves the injection of relatively coarse powder particles into a high-temperature gas for melting and spraying onto a substrate. The technique is being investigated as an economical method for fabricating cobalt-rare earth magnets for advanced traveling wave tubes and cross-field amplifiers. Plasma spraying permits deposition of material at high rates over large areas with optional direct bonding to the substrate, and offers the ability to fabricate magnets in a variety of shapes and sizes. Isotropic magnets were produced with high coercivity and good reproducibility in magnetic properties. Post-spray thermal treatments were used to enhance the magnetic properties of sprayed deposits. Samarium-cobalt magnets, sprayed from samarium-rich powder and subjected to post-spray heat treatment, displayed energy products in excess of 9 million gauss-oersteds and coercive forces of approximately 6000 oersteds. Bar magnet arrays were constructed by depositing magnets on ceramic substrates.

INTRODUCTION

Two basic processes have been developed for preparing cobalt-rare earth magnets: (1) pressing and sintering including the optional use of a liquid phase; and (2) casting. Additional processes being investigated include sputtering[1] and plasma spraying.[2,3] Plasma spraying is a method for directly converting a powder into a solid magnet shape. Sprayed magnets, which are essentially isotropic magnetically, can be formed in various shapes and sizes as free-standing magnets or permanently bonded to substrates. Plasma spraying permits high deposition rates thus offering low cost potential. The process is being investigated as an economical method for fabricating samarium-cobalt magnets for advanced traveling wave tubes and cross-field amplifiers.

EXPERIMENTAL

In arc plasma spraying, an inert arc gas is continuously passed through an electric DC arc, located inside of a spray torch, for heating and ionization. The resulting ionized gas (plasma) leaving the torch nozzle resembles an open welding flame. Powder particles, transported in a carrier gas, are injected into the plasma stream for melting and spraying onto a substrate, as shown schematically in Figure 1. Process variables such as electrical energy to the torch, gas and powder flow rate, torch-to-substrate distance, and substrate temperature must be suitably controlled. For the processing of oxidation-sensitive rare earth-cobalt materials, plasma spray deposition should be conducted in a non-oxidizing atmosphere such as argon.

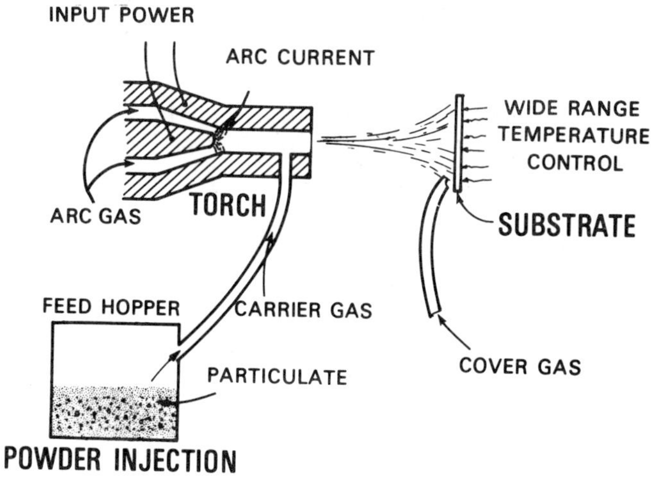

Fig. 1. Construction and operation of the plasma spray torch

Most spray powders were prepared by pulverizing prealloyed ingots to a nominal average particle size of 20 μm. Several alloy compositions, covering the range 33.8 to 39.1 wt.% Sm-Co, were systematically studied over a range of process conditions. A limited number of magnets was also produced using reduction-diffusion (R-D) alloy powders[4] containing 37.4 and 38 wt.% Sm in two particle sizes, 3.5 μm and 17 μm.

Most test magnets were produced by depositing Sm-Co alloy onto the ends of a cluster of 0.25-inch diameter graphite rods. Sprayed deposits were ground circumferentially, removed from the rod, and lapped. Several of the resulting free-standing disks were stacked to form a cylindrical magnet for testing using a hysteresigraph. Test magnets were pulse magnetized in a 60 KOe magnetic field after which demagnetization curves were plotted. Many magnets were subsequently heat treated and aged at elevated temperature in inert gas atmosphere in order to enhance magnetic properties and then retested. Special bonded bar magnets were also produced in this study by depositing Sm-Co through an aperture mask onto a flat ceramic substrate.

RESULTS AND DISCUSSION

Magnets displaying the highest energy products and best overall magnetic properties to date were produced from Sm-Co alloy ingot powder containing 38 to 39 wt.% Sm as determined by wet analyses. The optimum composition of reduction-diffusion alloy powder for use in this process has not yet been determined. However, magnets sprayed from unmilled

Table I

PROPERTIES OF SPRAY FABRICATED MAGNETS

Process Conditions	A	B	C
Powder type	Ingot	Ingot	R-D
%Sm;μm	37.9;-74	39.1;-74	38;17
Arc gas, type;cfh	A;190	A;150	A;190
Torch power, kW	25.2	16.2	15.2
Distance, in.	2.5	3	1.5
Heat treatment, hr;°C	1.5;1040	4.5;1080	1.5;1030
Aging, hr;°C	16;950	16;950	None
Properties			
As Sprayed			
B_r, kG	4.69	4.48	ND
H_c, kOe	2.49	2.08	ND
iH_c, kOe	6.2	5.07	ND
BH_{max}, MGOe	2.8	2.2	ND
Heat Treated and Aged			
ρ, g/cc	ND	7.96	7.82
B_r, kG	6.0	6.19	5.45
H_c, kOe	5.86	5.84	5.31
iH_c, kOe	∿30.0	∿30.5	∿47.8
BH_{max}, MGOe	9.0	9.3	7.1

ND - not determined

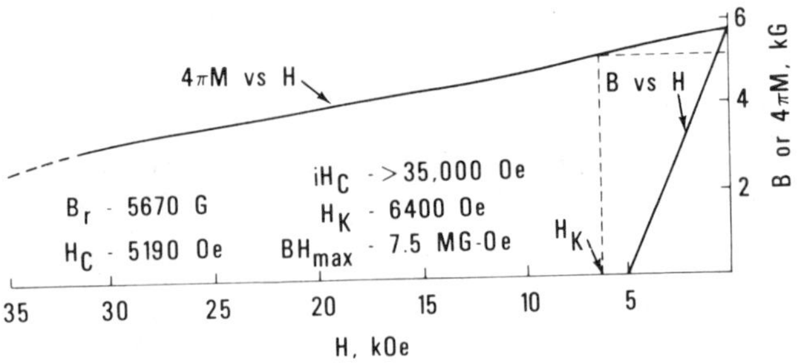

Fig. 2. Demagnetization curve for plasma-sprayed and heat-treated magnet

(~17μ) R-D powder were significantly better magnetically than those fabricated from milled (3.5μ) R-D powder. Properties of magnets recently produced from both ingot and R-D powders are presented in Table I together with the processing conditions employed for fabrication. The demagnetization curve for an earlier plasma-sprayed and heat treated magnet is shown in Figure 2.

The low B_r values of spray fabricated magnets are attributed to the isotropic nature of these materials. Attempts to enhance crystallographic orientation through the use of magnetic fields and temperature gradients have met with only limited success. However, the isotropic characteristic makes plasma-sprayed magnets suitable for use when extremely high magnetic fields are not required or when it is desired to magnetize in a particular direction without adversely affecting magnetic properties.

Oxygen, hydrogen and nitrogen content determined using the vacuum fusion procedure[5] for a magnet fabricated earlier from 38 wt.% Sm-Co ingot alloy powder showed 0.26 wt.% O, ~13 ppm hydrogen and ~175 ppm nitrogen. After heat treatment at 1050°C for 4 hr., the magnet displayed the following properties: B_r, 5.25 kG; H_c, 4.56 kOe; iH_c, 14.0 kOe; BH_{max}, 6.3 MGOe.

Fig. 3. Microstructure of as-sprayed Sm-Co magnet (bright field)

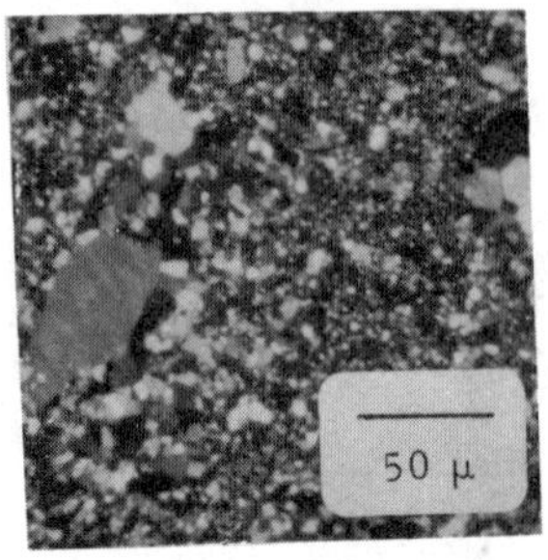

Fig. 4. Microstructure of sprayed Sm-Co magnet after heat treatment and aging (polarized light)

The microstructure of an "as-sprayed" Sm-Co magnet (Spec. A, Table I) is shown in Figure 3. It consists principally of elongated stringers, 2-3 μm thick and 50-200 μm long, aligned perpendicular to the spraying direction. At least four phases can be observed in the multiphase alloy: Sm_2Co_{17} (lightest), $SmCo_5$, Sm_2Co_7, and $SmCo_3$ (darkest). The microstructure of the same magnet after heat treatment and aging is shown in Figure 4. It consists principally of equiaxed grains of $SmCo_5$. Grain size range is 2-50 μm in diameter with most grains in the 2-10 μm range.

In this study, magnets have been fabricated in the shape of bars, tubes, rods, disks, plates, rings and arc segments as shown in Figure 5. Bar magnets were also directly bonded to ceramic substrates. Such bar arrays are being considered for use in advanced planar traveling wave tubes. The axial field profile of a magnet bar array containing 0.75x0.2x0.125 inch bar magnets on alumina substrate is shown in Figure 6.

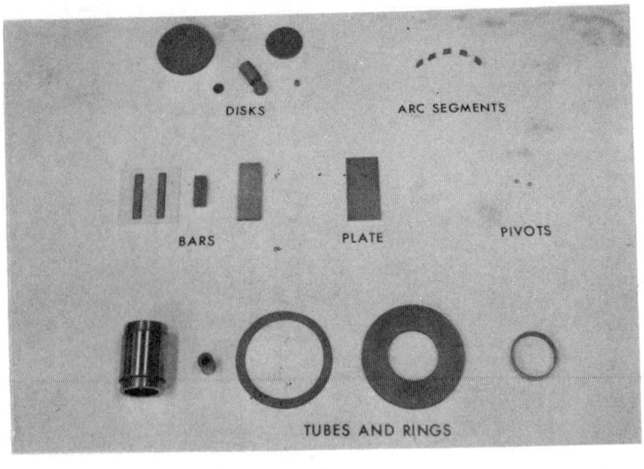

Fig. 5. Arc plasma-sprayed magnets

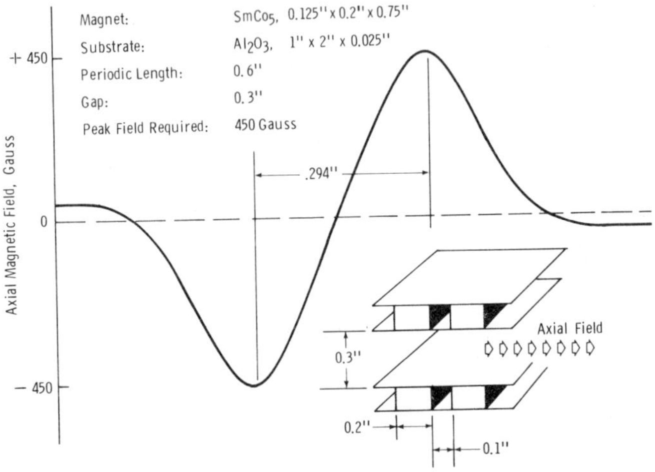

Fig. 6. Magnetic field profile of plasma-sprayed bar magnet array

ACKNOWLEDGMENT

This work was supported in part by the U.S. Army Electronics Command.

REFERENCES

1. Allen, R. P., Dahlgren, S. D. and Arrowsmith, H. W., "Research in the Production of Rare Earth-Cobalt Permanent Magnet Material by Sputter Deposition," AFML-TR-74-87, May 1974.

2. Janowiecki, R. J. and Willson, M. C., "Planar High Energy Permanent Magnets," Contract DAAB07-73-C-0112, Semi-Annual Report, TR-ECOM-0112-1, December 1973.

3. Willson, M. C. and Janowiecki R. J., "Planar High Energy Permanent Magnets," Contract DAAB07-73-C-0112, Semi-Annual Report, TR-ECOM-0112-2, May 1974.

4. McFarland, C. M., "Cobalt-Rare Earth Powders by the Reduction-Diffusion Process," General Electric Company Report No. 73CRD035, January 1973.

5. General Electric Company, Research and Development Center, Schenectady, New York.

KINETICS OF COERCIVE FORCE CHANGES DURING AGING OF Co-Fe-Cu-Ce PERMANENT MAGNETS

G. Y. Chin, M. L. Green*, and H. J. Leamy
Bell Laboratories, Murray Hill, N.J. 07974

ABSTRACT

The intrinsic coercive force $_iH_c$ of age-hardening Co-Fe-Cu-Ce permanent magnet alloys has been studied as a function of aging time at temperatures between 375 and 600°C following solution treatment at 1000°C. A new aging peak at ∿ 500°C has been observed in addition to that previously observed at ∿ 400°C. This new peak occurs at shorter times (20 mins. vs. 17 h) and the maximum $_iH_c$ obtainable is smaller than that obtained after aging at 400°C. It is absent in Ce-rich alloys, but persists and broadens to higher temperatures in Ce-poor alloys. The activation energy for the aging process in ∿30 kcal/mole for both peaks. The results are interpreted in terms of eutectoidal decomposition for the alloys.

INTRODUCTION

Previous studies of the age-hardening Co-Fe-Cu-Sm and Co-Fe-Cu-Ce permanent magnets indicated that high coercivities could be obtained by aging in the vicinity of 400°C.[1,2] Early work using optical metallography revealed the presence of fine precipitates along subgrain boundaries after this aging treatment.[2] Later structural studies showed that precipitation also occurs after aging at higher temperatures (e.g. 800°C) even though the coercive force remains rather low.[3] Hence, either the precipitates overage too rapidly at the higher temperatures, or the structures of the precipitates are different at the two temperature regimes. Since the early studies limited the aging times to two or four hours, we hoped to gain some additional insight regarding the coercive force - microstructure relationship by investigating the kinetics of aging. The microstructure of these alloys has been studied by transmission electron microscopy,[4] and will be discussed in terms of the observed magnetic behavior as reported here.

EXPERIMENTAL

Since a large number of samples were needed for this study, a modified Bridgman technique of directional solidification was used to grow aligned columnar crystals with good compositional homogeneity. The details have been reported elsewhere.[3] The as-grown crystals are 0.5 in. in diameter and 4 in. long. The middle 3 in. section was cut to 0.5 in. lengths for the aging study.

All samples were solution treated at 1000°C for 30 min. in an evacuated quartz ampoule, followed by water quenching with the ampoule unbroken. To further insure that all samples were structurally equivalent prior to aging, efforts were made to select only those which exhibit nearly the same value (± 10%) of coercive force. The intrinsic coercive force was measured by the torque method.

Specimens of three compositions were selected mainly to study the effect of Ce. They are nominally A-$Co_{3.5}Fe_{0.5}CuCe_{1.09}$, B-$Co_{3.5}Fe_{0.5}CuCe$, and C-$Co_{3.65}Fe_{0.5}Cu_{0.9}Ce_{0.96}$. It was thought that the $Co_{17}Ce_2$-type phase, which was identified as the high temperature precipitate phase,[3,4] might be suppressed as the Ce is increased.

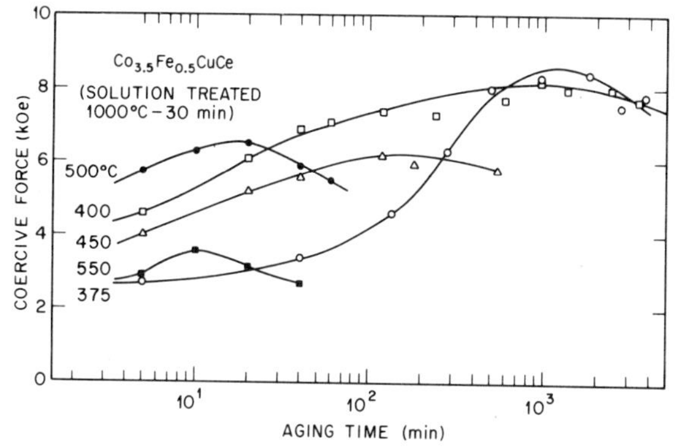

Fig. 1 Intrinsic coercive force vs. aging time for various aging temperatures, for the $Co_{3.5}Fe_{0.5}CuCe$ alloy.

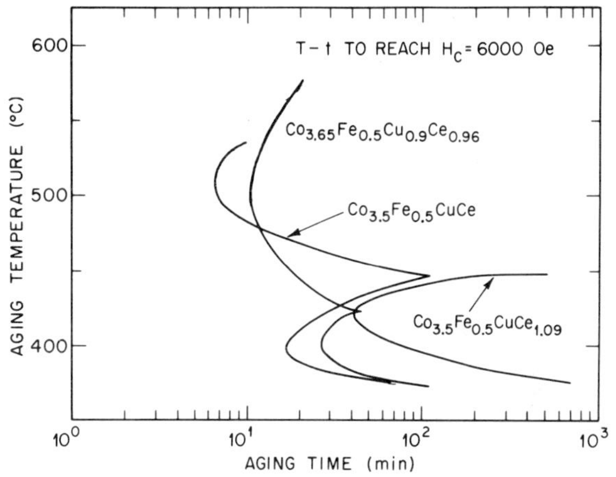

Fig. 2 Aging temperature vs. aging time required for $_iH_c$ to reach 6000 Oe.

RESULTS

Figure 1 shows a plot of the intrinsic coercive force of alloy B as a function of aging time at various temperatures. Results of aging at temperatures other than those plotted have been omitted for reason of clarity. It may be noted from these curves that as the aging temperature is increased, the kinetics of aging is first increased (375-400°C), then decreased (400-450°C), then increased (450-500°C) and finally decreased (500-550°C) again. Thus a double C-curve is expected in a T-t plot. This is shown in Fig. 2 along with results for the other two alloys. These curves were obtained by noting the aging time required to attain a value of $_iH_c$ = 6000 Oe at a given aging temperature. The double C-curve for alloys A and B is clear, while alloy C (high Ce) exhibits a single C-curve centered around 400°C only.

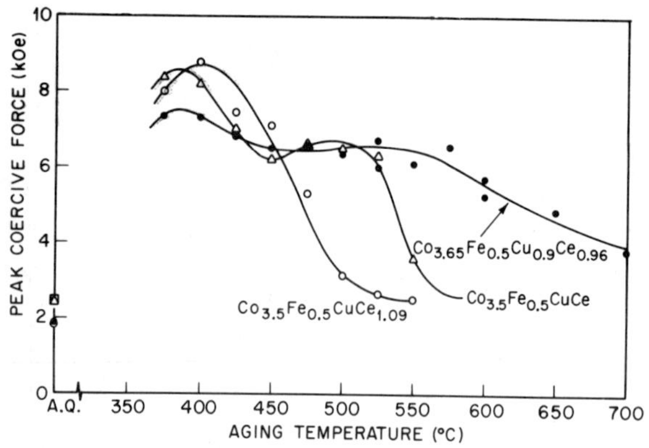

Fig. 3 Maximum value of intrinsic coercive force as a function of aging temperature.

Figure 3 is a plot of the peak value of coercive force vs. aging temperature for the alloys studied. It may be seen that the highest value of peak $_iH_c$ is obtained after aging at around 400°C and that it falls off at higher temperatures. The lower the Ce content, the higher is the temperature at which substantial coercivity can be attained.

An apparent activation energy for the aging process was obtained from an Arrhenius plot of the time required to achieve the peak value of $_iH_c$ at each aging temperature. The calculated activation energy is ∿30 kcal/mol for all three alloys.

DISCUSSION

The present investigation has revealed a new aging peak in the vicinity of 500°C. This peak was not detected previously because it occurs much more rapidly than the 400°C peak. Figure 1, for example, shows that for alloy B the 500°C peak occurs at 20 min., vs. 1000 min. for the 400°C peak. Therefore in the usual aging treatment of 2 to 4 h, the 500°C peak would have subsided already. The main reason for the early occurrence of the 500°C peak is faster diffusion, since the apparent activation energy for both temperature peaks is the same. The rather low value of activation energy, 30 kcal/mol, most likely reflects the diffusion of the smaller Co or Cu atoms in the relatively open $CaCu_5$-type lattice.

At the present time the most likely explanation for the two aging peaks is in terms of a eutectoidal transformation. Den Broeder and Buschow[5] have found a eutectoidal reaction for the binary Co_5R compounds, and additional experiments since then have given support to such a reaction.[6-8] For Co_5Ce, the eutectoidal transformation temperature is about 600°C, which may be depressed by the addition of Cu and increased with the Fe addition.[7] Based on the eutectoidal mechanism, the 400°C aging peak may be attributed to the eutectoidal decomposition of $(Co,Cu,Fe)_5Ce$ to the $(Co,Cu,Fe)_7Ce_2$ and $(Co,Cu,Fe)_{17}Ce_2$ phases, while the 500°C aging peak may be attributed to the proeutectoid precipitation of $(Co,Cu,Fe)_{17}Ce_2$ from the $(Co,Cu,Fe)_5Ce$ solid solution matrix. Our reasoning is as follows: the 400°C aging peak occurs over a wide-range of composition, on both Ce-rich and Ce-poor sides of the $(Co,Cu,Fe)_5Ce$ composition.[9] Transmission electron microscopy revealed essentially the same type of microstructure, i.e. dense arrays of fine coherent precipitates ∿150 A in diameter which on overaging coarsen to a 17:2 phase.[4] The emergence of the 7:2 phase is apparently extremely sluggish, as in the binary alloys. The alternative possibility for the 400°C peak is a spinodal decomposition as suggested by Hofer[10] for Co-Cu-Sm. Such a decomposition could conceivably occur where the eutectoidal reaction is suppressed.

The microstructure developed during 500°C aging is very similar to that produced by the 400°C treatment, except for the faster coarsening (and hence overaging) process. These are interpreted as proeutectoid precipitation of the 17:2 phase. With increasing Ce content, lesser amounts of proeutectoid precipitation is expected, accounting for the dropoff of peak $_iH_c$ as observed in Fig. 3. Furthermore, we have found no response to aging at 500°C and above in a Ce-rich alloy ($Co_{3.5}Fe_{.5}CuCe_{1.15}$), where the proeutectoid 7:2 phase is expected. The presence of the 7:2 phase was suggested in a previous optical metallographic study.[9] In this work transmission electron microscopy has revealed crystallographic shear features similarly to those attributed to the 7:2 phase in Co_5Sm.[6] Hence while the proeutectoid 17:2 phase results in an aging peak, such apparently is not the case for the proeutectoid 7:2 phase. One possible reason is that while the former appears as fine coherent precipitates, the latter appears as faulted patches which are less effective as pinning centers.

REFERENCES

1. E. A. Nesbitt, J. Appl. Phys. 40, 1259 (1969).
2. E. A. Nesbitt, G. Y. Chin, R. C. Sherwood and J. H. Wernick, J. Appl. Phys. 40, 4006 (1969).
3. G. Y. Chin, M. L. Green, E. A. Nesbitt, R. C. Sherwood and J. H. Wernick, IEEE Trans. Mag. MAG-8, 29 (1972).
4. H. J. Leamy and M. L. Green, IEEE Trans. Mag. MAG-9, 205 (1973).
5. F. J. A. den Broeder and K. H. J. Buschow, J. Less-Common Metals 29, 65 (1972).
6. J. G. Smeggil, P. Rao, J. D. Livingston and E. F. Koch, AIP Conf. Proc. No. 18, Part 2, 1144 (1974).
7. K. H. J. Buschow, J. Less-Common Metals 37, 91 (1974).
8. D. L. Martin, J. G. Smeggil, W. Hatfield, M. D. McCornel and A. Ritzer, 1974 Intermag Conf., paper 20.1.
9. E. A. Nesbitt, G. Y. Chin, R. C. Sherwood, M. L. Green and H. J. Leamy, IEEE Trans. Mag. MAG-9, 203 (1973).
10. F. Hofer, IEEE Trans. Mag. MAG-6, 221 (1970).

*Now at Department of Materials Science, Massachusetts Institute of Technology, Cambridge, Massachusetts

MAGNETIC PROPERTIES OF $R(Co_{1-y}Cu_y)_z$ COMPOUNDS

R.S.Perkins, A.J.Perry, H.Nagel and A.Menth
Brown Boveri Research Center, CH-5401 Baden, Switzerland

ABSTRACT

The hard magnetic characteristies of compounds having composition $Sm(Co_{1-y}Cu_y)_z$ and lying within the phase space $5{\cdot}0 \le z \le 8{\cdot}5$, $0 \le y \le 1{\cdot}0$, are strongly influenced by the nature of their metallurgical microstructure. However, by measurement of the anisotropy field, domain size, and angular dependence of the coercivity, and in conjunction with data obtained from transmission electron microscopy, as well as the implications of a theoretical model for the coercivity in these materials, and simple arguments concerning the diluting effect of copper upon the primary magnetic properties of the constituent phases, the principal features of this group of materials may be clarified.

The metallographic and hard magnetic properties of some $Sm(Co_{1-y}Cu_y)_z$ materials, which were presented in a previous paper[1], may be summarised as follows. Only within a relatively narrow composition range around the stoichiometries $SmCo_5$, $SmCu_5$ and Sm_2Co_{17} can single phase material be produced. In the intervening phase space, and depending upon the Cu concentration, the material is at least two phase with, however, the two principal phases distributed in a variety of ways e.g. macroscopic or microscopic precipitates of $Sm_2(Co_{1-y}Cu_y)_{17}$. In addition, for many of the compositions a long range segregation occurs with constant Sm but fluctuating Co:Cu content. Of the hard magnetic properties, the remanence, B_r, scales with the Cu content, as though the magnetisation were simply diluted. The coercivity $_iH_c$ changes in a much less predictable fashion, but crudely, passes through a maximum with Cu content, and decreases with increasing Co content. In many cases a double peak in the coercivity against heat treatment temperature is observed.

Figure 1 illustrates a hysteresis loop typical of samples from the mentioned phase space. The material represents approximately equal volume fractions of a primary phase, corresponding to $Sm_2(Co_{.9}Cu_{.1})_{17}$, situated in a matrix having composition $Sm(Co_{.72}Cu_{.23})_{6.4}$. The composition in the matrix phase fluctuates between the primary phase regions in a way typical of that observed in all compositions containing the former. In this case the variation is approximately 7 atomic % of the Co:Cu mixture. Figure 2 is a Kerr-effect picture of the domain structure in a bulk sample of this material.

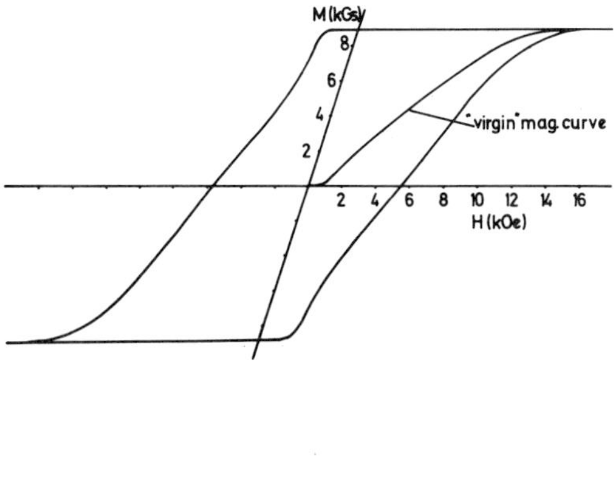

Fig.1 Hysteresis loop of a single crystal sample of $Sm(Co_{.84}Cu_{.16})_{6.8}$.

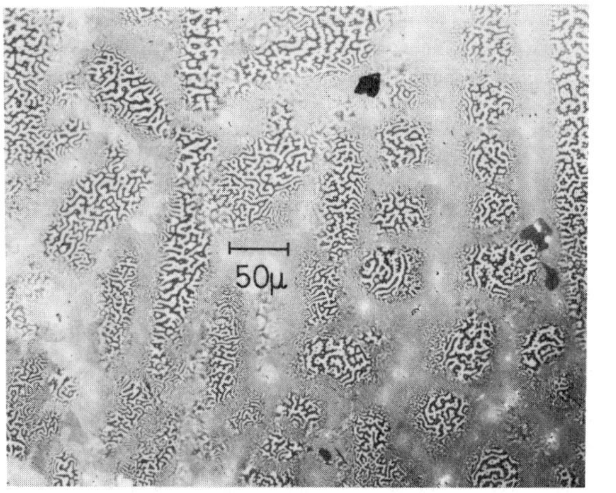

Fig.2 Domain patterns observed in a bulk sample of the same compound refered to in Fig.1 .

Although the above material is in fact multi-phase, it was shown in an electron microscope investigation[7] that the structures of the matrix and precipitates are coherent with one another. The specimen of Figure 1 was an approximately 2mm diameter sphere ground from a single grain of the as-cast material. The latter was prepared in an RF furnace and boron nitride crucible. The major differences between the two magnetic phases which comprise this material are visible in these two figures. The demagnetising curves in Figure 1 show both the presence of a magnetically softer phase, and a continuous variation of coercivity in a second phase. The change in the domain pattern of the thermally demagnetised sample of Figure 2 under an applied field, indicated that it was the precipitate phase which was magnetically softer and therefore responsible for the initial steeper drop in the demagnetising curve. This effect was also present in the "virgin" magnetisation curve. In addition, preparation of various $Sm_2(Co_{1-y}Cu_y)_{17}$ alloys showed, in addition to a phase boundary at $y \simeq 0{\cdot}1$, that none of these compositions has a coercivity in excess of about 0•5kOe. The coercivity of the precipitate phase is then intrinsically small.

If the substitution of Co in $SmCo_5$ by Cu is assumed to produce a simple magnetic dilution, then the composition fluctuation in the matrix phase will be responsible for the shape of the demagnetising curve, since both the anisotropy and magnetisation will also have a spactial variation. This is confirmed by the temperature dependence of the demagnetising slope which decreases with decreasing temperature - $K_1(T)$ varies faster in the Co rich regions. Additionally the "virgin" magnetisation curve shown in Figure 1 indicates a distribution of coercivities associated with a distribution of pinning fields. Nucleation mechanisms of coercivity require immediate saturation in a single crystal at quite low fields[2], whereas (homogeneous) pinning would do so at a(higher) pinning field[3].

The domain patterns of Figure 2 illustrate the variation of magnetic properties throughout the sample. The domains in the primary phase are of the same size as those observed in Sm_2Co_{17}, those in the matrix, however, are smaller and become even smaller with increasing Cu-content. The domain width in bulk samples is $W \propto \gamma/M_S^2$,[4] with domain wall energy, $\gamma = 4\sqrt{AK}$, in the usual notation. If M(Y), K(Y) and the exchange constant A(Y) are all assumed linearly decreasing functions of Y, vanishing at y=1•0, then W would increase with Y. However it has been observed[5] that K(Y) = 0 at Y~0•7 whereas M(·7)≠0. Secondly the Co-Co interaction must have quadratic dependence on Y, in comparison to the Sm-Co ones, so that in practice W can be a decreasing function of Y, in agreement with the observed variations. The reason for the small mean primary phase domain size lies in its own microstructure, which contains microscopic precipitates of the 2:17 phase,[6,7] Figure 3. The effect of these precipitates is to produce a true matrix composition even more Sm-rich than low resolution analysis would indicate, but also a more Cu-rich one due to the limited Cu solubility in Sm_2Co_{17}. This means that the matrix really has lower M_s, A, K values than would be imagined from the coarse composition. Calculation for the sample of Figure 3 indicates that K_{matrix} ~35% K_{SmCo_5} instead of the 65% derived on simpler assumptions.

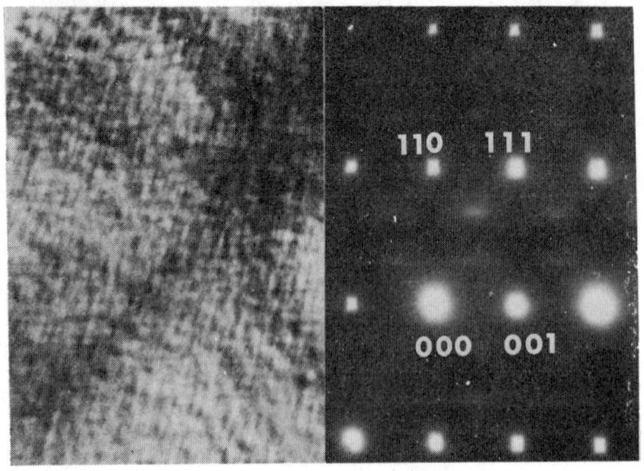

Fig.3

[1$\bar{1}$0] 1:5 Zone electron micrograph and diffraction pattern of a $Sm(Co_{.65}Cu_{.35})_{5.7}$ sample.

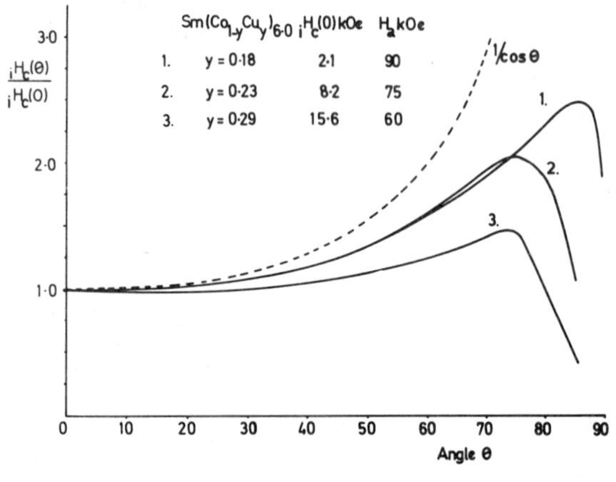

Fig.4 Normalised angular dependence of the coercivity observed in $Sm(Co_{1-y}Cu_y)_{6.0}$ samples.

Figure 4 shows the normalised angular dependence $_iH_c$ (θ) of the coercivity, between easy axis and basal plane, measured on single crystal samples of the shown compositions.These data may be compared with theoretical calculations[8], in an effort to understand their major features. The position of the maximum is determined by the anisotropy field, H_a of the sample. Values for these are included but expected to be only approximate. Nevertheless the respective maxima do occur at higher angles for higher H_a. The mechanism for the coercivity in these samples is expected to be pinning of the Bloch-walls at microscopic precipitates[3,7]. The theoretical model[8], however, considers the precipitate size as well as the shape and gradient of the anisotropy variation between precipitate and matrix, as competing mechanisms for the coercivity. The principitate volume fraction of the samples shown in Figure 4 is the same and determined by the small Sm:Co,Cu ratio, if it is assumed that the precipitates always have the composition $Sm_2(Co_{.9}Cu_{.1})_{17}$. Therefore, to first order, the inter-precipitate to precipitate radius is the same, as in the corresponding contribution to the coercivity. The mismatch between precipitate and matrix lattices will become larger and more anisotropic at higher Cu-concentrations, but the anisotropy difference will be smaller. The implications drawn from the theoretical model[8] show that the initial slope of the anisotropy at the precipitate effectively determines $_iH_c(0)$, whereas the subsequent slope essentially determines the angular dependence. These effects may be present in Figure 4. The greater lattice strain but smaller anisotropy difference of sample 3 may indeed fulfill the above mentioned criteria for high absolute coercivity and weak angular dependence.

On the basis of the above comments, some general remarks may be made with regard to the changes in coercivity and remanence throughout the mentioned phase space. The pinning mechanism varies in a complex manner with Cu and Sm concentration. The microscopic precipitate size will depend upon the particular Sm concentration according to phase splitting between the stable phases. However the effect of lattice mismatch is also concentration dependent. In progressing to higher transition metal content, the macroscopic precipitate volume fraction (Fig.2) increases, thereby increasing the contribution of the softer magnetic phase, and leading to lower coercivities[1]. However, as seen in Figure 1, the soft magnetic characteristics are to some extent reduced by the interaction with the matrix phase. Clearly these various contributions to the hard magnetic properties can be balanced in order to obtain material with optimal energy product.

REFERENCES

1. A.J. Perry, H. Nagel and A. Menth, 3rd European Conference on Hard Magnetic Materials, Amsterdam, Sept.1974.
2. E. Alff, D. Givord and J.P. Haberer, IEEE Trans.Mag., 9, 631 (1973).
3. E.A. Nesbitt, G.Y. Chin, G.W. Hull, R.C. Sherwood, M.L. Green, and J.H. Wernick, AIP Conf.Proc.10, 593 (1973).
4. J.D. Livingstone and M.D. McConnell, J.Appl.Phys., 43, 4756 (1972).
5. T. Katayama and T. Shibata, Japan J.Appl.Phys., 12, 762 (1973).
6. H.J. Leamy and M.L. Green, IEEE Trans.Mag.9, 205 (1973)
7. K.N. Melton and R.S. Perkins, to be published.
8. J. Bernasconi, S.Strässler and R.S. Perkins. Paper these proceedings and private communication.

PERMANENT MAGNETS ON THE BASIS OF MMCO5 AND THE RELATION BETWEEN THEIR PROPERTIES AND THE PRIMARY MAGNETIC PROPERTIES OF THESE COMPOUNDS

H. Nagel and H.P. Klein
Brown Boveri Research Centre CH-5401 Baden, Switzerland

ABSTRACT

The permanent magnetic properties of $MM_{1-x}Sm_xCo_5$ magnets, prepared by powder-metallurgical means, have been studied in the range $0<X<0.4$ and for various chemical compositions of MM. For both coercivity and energy product, a nearly linear dependence on the samarium content was found in the range investigated. Moreover, there is a very strong correlation between the permanent magnetic properties and the composition of the MM. By optimizing this composition, $MM_{0.85}Sm_{0.15}Co_5$ magnets can be produced with a coercivity $_IH_C > 16$ kOe and $H_K > 14$ kOe respectively, and a maximum energy product of up to 18 MGOe. These values will be discussed with respect to the primary magnetic properties of these compounds as obtained from single crystal measurements.

EXPERIMENTS

The substitution of samarium by the cheap cerium mischmetal (MM) is of great importance for a widespread application of $RECo_5$ magnets. However, compared with $SmCo_5$ two additional composition parameters must be considered in producing $RECo_5$ magnets on the basis of MM, namely the Sm-content[1,2,3] and the MM-composition[3]. We have investigated these two parameters with the aim of minimizing the samarium content and optimizing the mischmetal composition. The magnets were produced partly with commercial RE-Co alloys and partly from alloys prepared in our laboratory. These were cast from the components, the purity of which was 99.9 % or better, in boron nitride crucibles by r-f heating under a protective atmosphere of pure argon. All the alloys were analysed chemically and investigated metallographically. The magnets were made by powder metallurgical means appropriate for $SmCo_5$ magnets[4], including a post sintering heat treatment[3]. The densities of the magnets were always between 94 % and 97 % of the theoretical values. The hard magnetic properties of the samples were measured by a fluxmetric method in a superconducting solenoid in fields of up to 50 kOe. Only the results of magnets with rectangular shaped demagnetization curves are reported. To ensure constant production conditions magnets were made repeatedly from a standard powder mixture. The constant hard magnetic properties obtained on these test samples, proved that the production conditions remained unchanged during this investigation.

The single crystals of $MM_{1-x}Sm_xCo_5$ with $0\leq X\leq 0.4$ were obtained by fast cooling of the melts and annealing for 14 days at 1000 C. Single crystal grains, up to a few millimeters in size, were achieved. They were ground to spherical shape with diameters between 1 and 2 mm. The primary magnetic quantities e.g. the crystal anisotropy field H_A and the saturation magnetization M_s, of these samples were measured in a vibrating sample magnetometer as described elsewhere[5].

Both the primary and the secondary magnetic properties increase nearly linearly with the samarium content, as can be seen in Figure 1, where the ratios $_IH_C/H_A$ and B_R/M_s are plotted also. The results show that it is possible to produce $MMCo_5$ magnets without Sm-additives having coercive fields > 9 kOe and energy products > 14.5 MGOe. These values of $_IH_C$ and $(BxH)_{max}$ and those obtained with the $MM_{1-x}Sm_x$ Co_5 compounds are higher than corresponding ones published by other authors[2,6,7].

The addition of small amounts of samarium

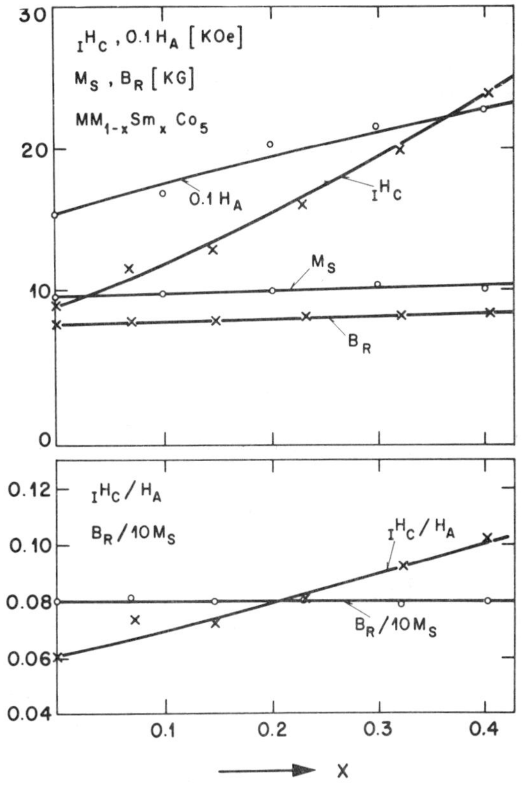

Fig. 1 The dependence on the samarium content of both the primary and secondary magnetic properties of $MM_{1-x}Sm_xCo_5$ compounds.

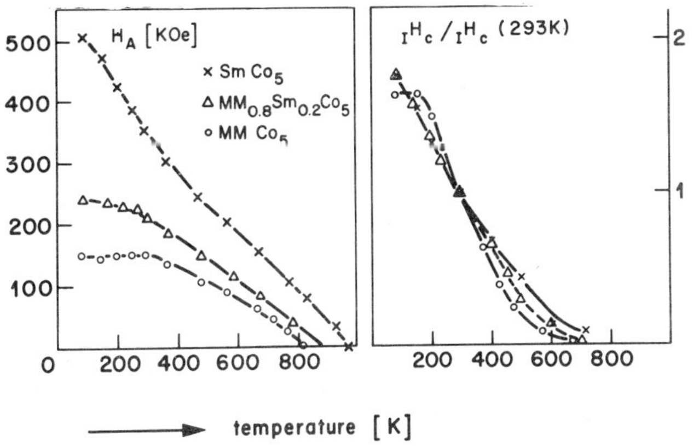

Fig. 2 Temperature dependence of the crystal anisotropy field H_A and the relative coercive field $_IH_C/_IH_C$ (293 K) of $ReCo_5$ compounds.

increases the hard magnetic properties to the same extent as the corresponding primary properties. In addition this relation becomes visible in the temperature dependence of the magnetic properties, which is shown in Figure 2 for three $RECo_5$-compounds. Whereas the temperature dependence of the relative coercive fields ${}_IH_C/{}_IH_C$ (293 K) of the $SmCo_5$ and the $MM_{0.8}Sm_{0.2}Co_5$ sample is very similar, it differs markedly for $MMCo_5$. As can be seen this difference is related to the temperature dependence of the crystal anisotropy field H_A.

Mischmetal consists mainly of the four rare earth metals Ce, La, Nd, Pr, its actual composition varying within wide limits. Since individual $RECo_5$ compounds possess different primary magnetic properties we investigated them as a function of the ratios of components in commercial MM by substituting the MM to some extent by the components separately. For the preparation of all these alloys the same MM ingot was used. Its composition was determined by means of x-ray fluoresence to be 53.7 at % Ce, 30.2 at % La, 12 at % Nd and 4 at % Pr with an accuracy of ± 1 %. With these alloys and a Sm 60 wt % Co 40 wt % alloy as a sinter additive we made magnets of the composition $(MM_{1-x}RE_x)_{0.85}Sm_{0.15}Co_5$ with RE equal to Ce, La, Nd, Pr and with $0<X<0.5$. The demagnetization curves of four magnets are shown in Figure 3. The dependence of the coercive field on the amount of the added rare earth is shown in Figure 4 from which the following conclusions can be drawn. 1.) An increase of both the cerium and the praseodymium content reduces ${}_IH_C$. 2.) Substitution of 30 % of the commercial MM by lanthanum increased ${}_IH_C$ from 8 kOe up to 13 kOe. 3.) The most remarkable change was observed by addition of neodymium. An increase of the Nd-concentration by 10 % doubles ${}_IH_C$. With further increasing Nd-content, however, ${}_IH_C$ is reduced to 5 kOe for a Nd-concentration of 50 %.

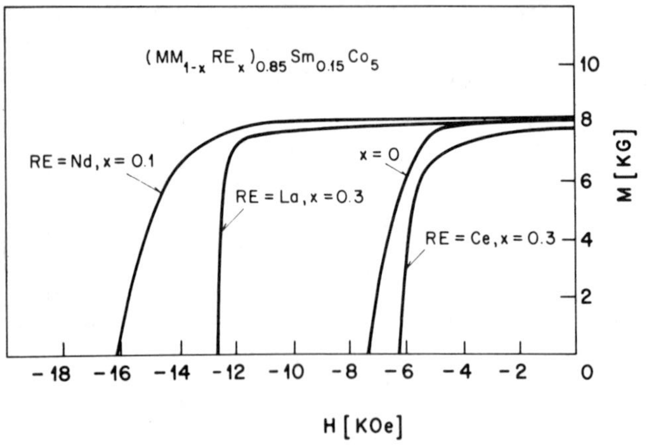

Fig. 3 Demagnetization curves of $(MM_{1-x}RE_x)_{0.85}Sm_{0.15}Co_5$ magnets

The magnetic remanence B_R depends on the MM-composition, too. With the initial $MMCo_5$ alloy magnets with a remanence B_R of 8 kG, and an energy product of 15.5 MGOe were achieved. This value was found to be reduced by the addition of cerium, unchanged by that of lanthanum and increased with both the neodymium and praseodymium concentration resulting in energy products up to 18 MGOe.

The composition dependence observed for the primary magnetic properties is less pro-

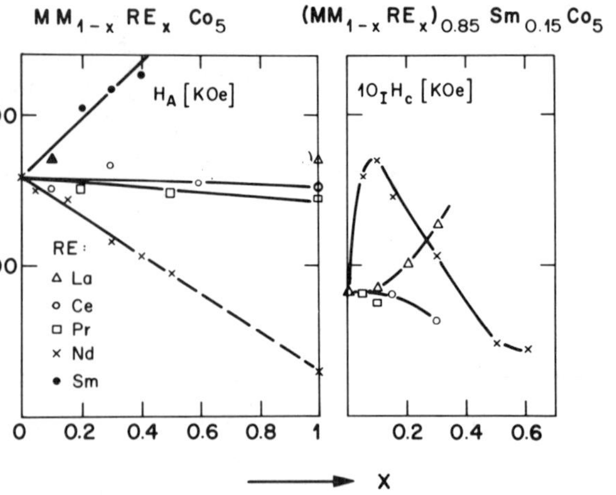

Fig. 4 Influence of the MM composition on the crystal anisotropy field H_A and the coercivity ${}_IH_C$ of MM-based materials. Point 1 is taken from Buschow et al[9].

nounced than for the secondary properties (Fig. 4). Indeed through an increase of cerium and praseodymium concentration the anisotropy field is linearly reduced, but the relative decrease is smaller than the measured coercivity reduction. Just the opposite behaviour is shown by H_A and ${}_IH_C$ as function of the neodymium concentration. For $MM_{1-x}Nd_xCo_5$ there is a steady decrease of H_A in the range $0<X<0.5$. This is in contrast to the ${}_IH_C$ values which have a maximum within this range of composition. The optimum value for the ratios ${}_IH_C/H_A$ we obtained sofar is 0.12 for the alloy with X = 0.1 compared to a value of 0.1 for $SmCo_5$. The saturation magnetization M_s is only very little influenced by both increasing the Ce and the La content. However, it increases with both the Nd and the Pr content.

M_s as well as B_R show an additive behaviour with respect to the corresponding values of the individual 1:5 compounds of the components of MM. The same is true for H_A, and this is accounted for by the single ion model[8]. However, for the composition dependence of the coercivity, especially with respect to Nd, no additive behaviour could be observed. We think that the reason for the anomaly of the Nd-dependence are structural defects, the nature of which must still be investigated.

REFERENCES

1. M.G. Benz, D.L. Martin; Jour. Appl. Physics 42, 1534 (1971).
2. D.K. Das; AIP Conf. Proc. 10, 628 (1973).
3. H. Nagel; Proceedings of the 3rd European Conf. on Hard Magnetic Materials, Amsterdam September 1974.
4. M.G. Benz et al; Report AFML TR-142, July 1971, Air Force Materials Laboratory, Wright Patterson Air Force Base Ohio.
5. H.P. Klein, A. Menth, R.S. Perkins; Int. Jour. Mag., to be published.
6. C.J. Fellows, R.E. Johnson; Cobalt, No. 56, 141, September 1972.
7. D.V. Ratnam and M.G.H. Wells; AIP Conf. Proc. 18, 1154 (1974).
8. J.E. Greedan, V.U.S. Rao; Jour. Sol. State Chem. 6, 387 (1973).
9. K.H.J. Buschow, W.A.J.J. Velge; Z. angew. Phys. 26, 157 (1969).

AMORPHOUS ALLOYS AS SOFT MAGNETIC MATERIALS

T. Egami*, P. J. Flanders, and C. D. Graham, Jr.*
Laboratory for Research on the Structure of Matter
and *Department of Metallurgy and Materials Science
University of Pennsylvania, Philadelphia, Pa. 19174

ABSTRACT

The preparation, structure, and mechanical properties of amorphous metallic alloys are briefly reviewed, and the magnetic properties, especially the low-field properties of engineering interest, are described. The materials reported are ribbons about 1.5 mm by 35 µm in cross-section, made under the trade name Metglas alloys by Allied Chemical Corp. These alloys have low coercive fields and very square hysteresis loops; the effective magnetization in low fields is greatly increased by the application of a tensile stress. When held under stress, amorphous ribbons have magnetic properties comparable to the best available crystalline materials. Some of the improvement due to stress can be made permanent by annealing under stress. Possible applications of these alloys are discussed.

INTRODUCTION

Many metallic alloys can be prepared in the form of amorphous solids--that is, solids that lack the regular repeating three-dimensional structure of crystals. Amorphous solids can be prepared in three ways: condensation from the vapor onto a cold substrate, electrodeposition from a liquid electrolyte, and very rapid freezing (splat cooling) from the melt.[1] Fairly recently, variations of the splat-cooling technique have been developed that allow the production of long lengths of amorphous alloy ribbons. The primary motivation for this development was the need for specimens suitable for measuring mechanical properties and electrical resistivity; however, such ribbons are also well-suited for the measurement of low-field magnetic properties.

Two somewhat different procedures for making amorphous ribbon have been reported. The Pond-Maddin process[2] squirts a jet of liquid metal against the inside of a rotating drum; the Chen-Miller process[3] squirts the jet into the gap between two rotating rolls. The Chen-Miller method gives more rapid cooling, since heat is removed from both sides of the strip; it also generally results in a more uniform cross-section. However, there may be plastic deformation after solidification, with undesirable effects on the magnetic properties.

The mechanical properties of a considerable range of amorphous alloys compositions are remarkably similar[4]: the yield stress is extremely high, frequently over 250,000 lb/in^2 (1725 MPa); the ductility is small but not zero, especially in rolling; and the tensile modulus is about 30% lower than corresponding crystalline materials. Densities are similar to those of liquid metals, and electrical resistivities are several times higher than equivalent crystalline alloys.

Only a limited range of alloy compositions can be made amorphous by the splat-cooling technique and its variations. The amorphous alloys are almost all near the composition $T_{80}M_{20}$, where T represents one or more of the transition metals Fe, Ni, Co, Cr, Mn, Pd, etc., and M represents one or more of the metalloid or glass-forming elements B, P, C, or Si. The ribbon-making process is relatively simple and inexpensive, since it requires a single operation on a small scale, yielding the finished product directly from the melt.

It is well-established that amorphous alloys can be ferromagnetic, and amorphous magnetic materials have been made by all the methods listed.[5] Generally it is found that an amorphous alloy is ferromagnetic if the corresponding crystalline alloy without the metalloids is ferromagnetic, although the saturation magnetization and Curie temperature of the amorphous materials are somewhat reduced. Most of the reported work on the magnetic properties of amorphous alloys has, not surprisingly, been on fundamental properties such as saturation magnetization, Curie temperature, neutron scattering behavior, etc. Notable exceptions are the investigation of sputtered GdCo and related alloys for bubble devices[6] and of $TbFe_2$ as a possible permanent magnet material.[7]

The possible usefulness of amorphous alloys as soft magnetic materials has already been recognized,[8,9] and efforts made to produce a material with zero magnetostriction.[9,10] The reasoning was that an amorphous material should have no macroscopic magnetic anisotropy, so that if zero magnetostriction could be acheived there would be no hindrance to magnetization in vanishingly small fields, and a material of infinite permeability would result. The experiments mentioned above, however, clearly show that amorphous alloys need not be isotropic,[6,7] in contrast to the obvious expectation. This raises the troublesome question of whether the alloys called amorphous are "really" amorphous, or whether they simply have a crystallite size smaller than the usual detection limit. There is no doubt that the word amorphous is sometimes loosely used; it is equally certain that many alloys can be prepared that appear amorphous in an ordinary x-ray diffraction experiment. This means that the crystallite size is no larger than about 20 A[11]. From a practical viewpoint the distinction between true amorphousness and microcrystallinity may be unimportant; however, in working with this class of materials, one must be alert to the possibility that the material preparation or subsequent treatment has led to partial crystallization with uncertain but possibly significant consequences.

We report here some of the first results of low-field measurements of amorphous magnetic alloys. We find that these materials have an anisotropy that detracts from their performance as soft magnetic materials; however, this anisotropy can be overcome by the application of an appropriate stress or (to some extent) by annealing under stress. Suitably prepared and treated, amorphous alloys can outperform conventional soft magnetic alloys such as nickel-irons.

MATERIALS; MAGNETIC MEASUREMENTS

The measurements reported here have all been made on amorphous ribbons produced by Allied Chemical Corp., Morristown, N.J., under the trade name Metglas alloys. The compositions of the ribbons are given in Table I, along with measured values of the saturation magnetization and magnetostriction. All the ribbons except composition 4 were rectangular in cross-section, about 1.5 mm by 35 µm; composition 4 was similar in size, but with one surface flat and the opposite surface convex. Details of the production method have not been published. Composition 3 had a Curie temperature of about 240K and was not extensively studied.

TABLE I

Alloy Number	Composition	Metglas* Number	Saturation Induction,G RT	Saturation Induction,G 77K	Saturation Magnetostr. RT
1	$Fe_{80}P_{16}B_1C_3$	2615	17100	18500	$29x10^{-6}$
2	$Fe_{40}Ni_{40}P_{14}B_6$	2826	8720	10250	11
3	$Fe_{32}Ni_{36}Cr_{14}P_{12}B_6$	2826A	--	4600	--
4	$Fe_{29}Ni_{49}P_{14}B_6Si_2$	2826B	4880	7650	5

*trade mark of Allied Chemical Corp.

Hysteresis loops were measured in two ways. Single lengths of ribbon, 10 to 20 cm long, were magnetized by a long solenoid, and the output from a small pickup coil around the center of the ribbon was electronically integrated. This procedure is quick and simple, especially for investigating the effect of stress (see below). However, the strip must be aligned perpendicular to the earth's field, and the demagnetizing field of the sample is not negligible.

Toroidal samples were wound from lengths of ribbon onto cylindrical forms 1 to 2 cm in diameter. Generally about 30 layers were wound, with primary and secondary electrical windings applied by hand. Such samples eliminate the problems of the earth's field and the demagnetizing field, and closely approximate practical devices. On the other hand, they are somewhat tedious to prepare, and do not lend themselves to the investigation of stress effects.

For both methods, drive fields were supplied from a bipolar operational amplifier which permitted manual control for dc loops, and which could be driven by a signal generator for ac loops.

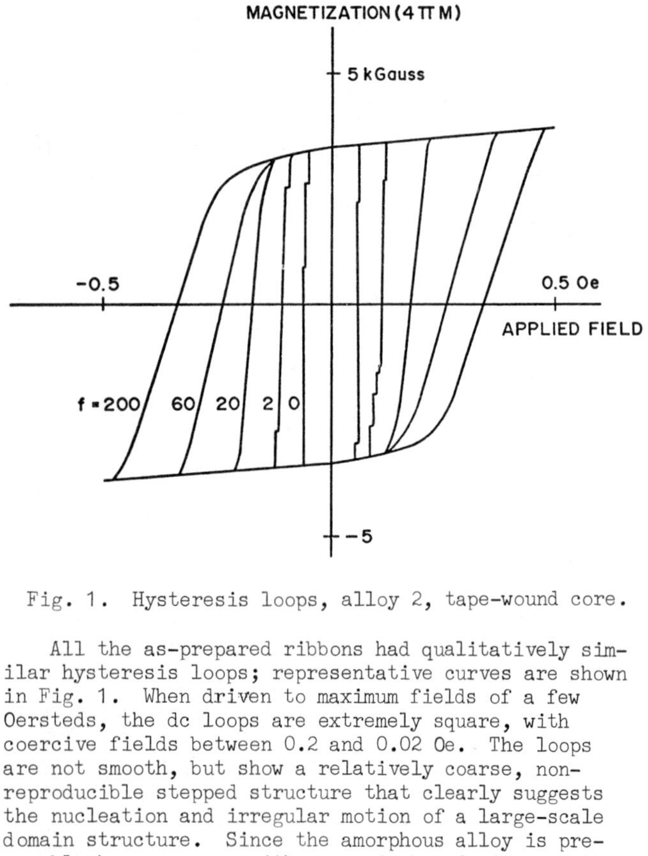

Fig. 1. Hysteresis loops, alloy 2, tape-wound core.

All the as-prepared ribbons had qualitatively similar hysteresis loops; representative curves are shown in Fig. 1. When driven to maximum fields of a few Oersteds, the dc loops are extremely square, with coercive fields between 0.2 and 0.02 Oe. The loops are not smooth, but show a relatively coarse, nonreproducible stepped structure that clearly suggests the nucleation and irregular motion of a large-scale domain structure. Since the amorphous alloy is presumably homogeneous, with no grain boundaries or other substructure, there is nothing to prevent a domain boundary from running the full length of the ribbon. Preliminary attempts to observe the domain structure have been inconclusive. Masumoto et al. have been able to observe domains only on some sections of a presumably uniform amorphous ribbon, and see stripe patterns suggesting magnetization normal to the ribbon surface.[8] Sherwood et al. have also observed domains in amorphous alloys.[10]

As the drive frequency increases, the steps in the hysteresis loop disappear and the coercive field increases (Fig. 1). The time rate of magnetization change, as determined from the unintegrated signal from the pickup coil, increases linearly with field according to

$$dM/dt=c(H-H_c),$$

suggesting that domain wall motion is controlled by eddy current losses. Total losses under ac excitation are substantially larger than predicted by the classical eddy-current calculation. Since the calculation assumes uniform magnetization, which is equivalent to vanishingly small domain size, this result tends to confirm the presence of a coarse domain structure. The amorphous ribbons have electrical resistivities of 150 to 200 μOhm-cm, about three times higher than crystalline Ni-Fe alloys, which tends to limit the eddy-current loss.

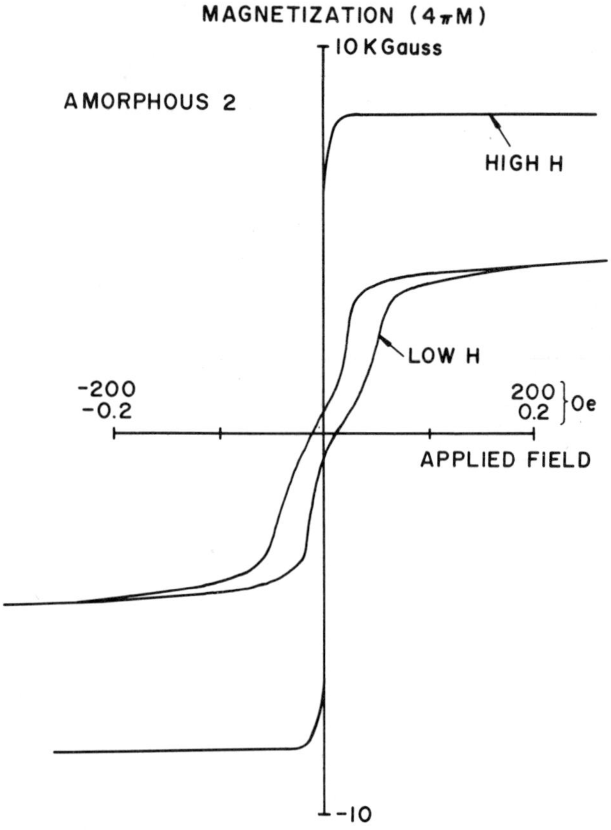

Fig. 2. Low-field and high-field loops, alloy 2, straight sample. Low-field curve shows effect of demagnetizing field.

The apparent saturation magnetizations measured in these low fields are lower by at least a factor of two than the values measured on similar alloys by other investigators.[5] The discrepancy is removed if maximum fields of a few hundred Oersteds are applied. (Fig. 2). This implies the presence of an anisotropy with a magnitude of order 100 Oe anisotropy field, opposing magnetization along the ribbon axis. At the same time, a very stable remanence, whose magnitude is 0.3 to 0.5 that of the high field saturation, is regularly observed. (On straight samples, a correction for the demagnetizing field must be made in order to measure this remanence.) A direct measurement of the anisotropy in the plane of the ribbon was made by recording magnetization curves parallel and perpendicular to the ribbon axis on small square samples cut from a ribbon. A vibrating-sample magnetometer was used. The demagnetizing field was about 275 Oe, and no difference between the parallel and perpendicular curves could be detected within an uncertainty of about 30 Oe caused by departure of the samples from perfect squareness. This result suggests that the magnetic anisotropy of the ribbon has an easy direction perpendicular to the plane of the ribbon. The analysis of certain magnetostriction measurements (see below) tends to confirm this result, as do the domain observations of Masumoto et al.[8]

MAGNETOSTRICTION AND STRESS EFFECTS

Magnetostriction measurements were made on short sections of ribbon, using a semiconductor strain gage and rotating the sample in a dc field. Saturation magnetostrictive strains along the ribbon axis are given in Table I. The values should be regarded as lower limits, since the adhesive and the gage itself may have constrained the very thin samples. Magnetostriction as a function of field both parallel and perpendicular to the ribbon axis has been measured; by comparing the strain in the sample in the demagnetized state and in the saturated state, we conclude that some portion of the sample has a component of its magnetization normal to the plane of the ribbon. This implies a (local) anisotropy with the easy axis normal to the plane of the ribbon.

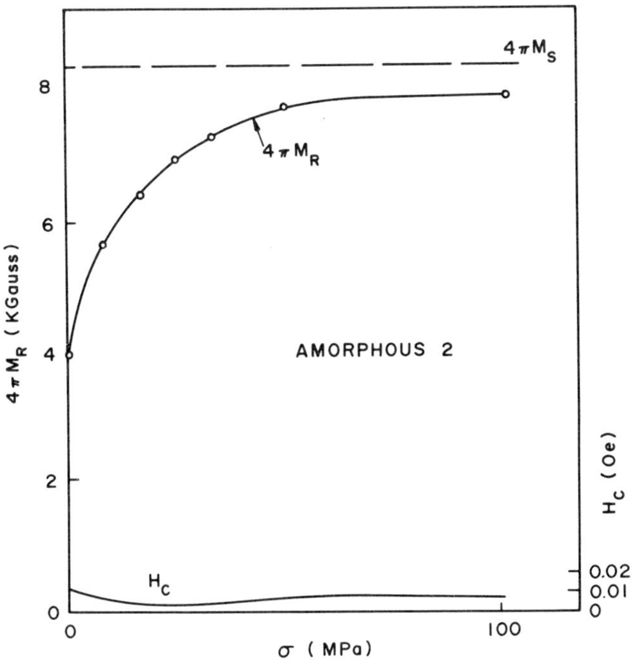

Fig. 3. Influence of stress on remanence and coercive field.

The existence of positive magnetostriction suggested that the application of a tensile stress to the ribbon should produce a magnetic anisotropy favoring magnetization parallel to the axis. This proved to be the case, as shown in Fig. 3. In all cases, the application of a tensile stress to the ribbon increased the low-field magnetization and the remanence. A sufficiently high stress gave a magnetization at 1 Oe essentially equal to the high-field saturation. The magnitude of the required stress varied with composition, and also between different lots of the same nominal composition, ranging from 40 to 200 MPa (6000 to 30,000 lb/in^2). In some cases, the stress caused a substantial drop in coercive field; in other cases there was little effect. Note that the stress levels are quite high, especially in comparison to the very low yield stresses of annealed crystalline Ni-Fe alloys. The stresses are, however, far below the yield stress of amorphous ribbons. The ribbons can be handled quite casually, and loaded many times, with no significant deterioration of the magnetic properties. We have, however, observed occasional sudden fracture of the ribbons, sometimes under quite moderate stress; this failure may be due to nicks or other surface imperfections.

Hysteresis loops with and without stress are shown in Fig. 4. This material needs a comparatively low stress in order to reach its optimum properties; it is accordingly quite sensitive to bending or twisting stresses, and must be carefully aligned to obtain zero-stress data. The effects of stress on the magnetic properties are reversible, at least for short applications of the stress. That is, removing the stress returns the magnetic properties to their original values.

TEMPERATURE EFFECTS

Amorphous alloys of the kind considered here are usually found by electrical resistivity and calorimetric measurements to undergo crystallization to the

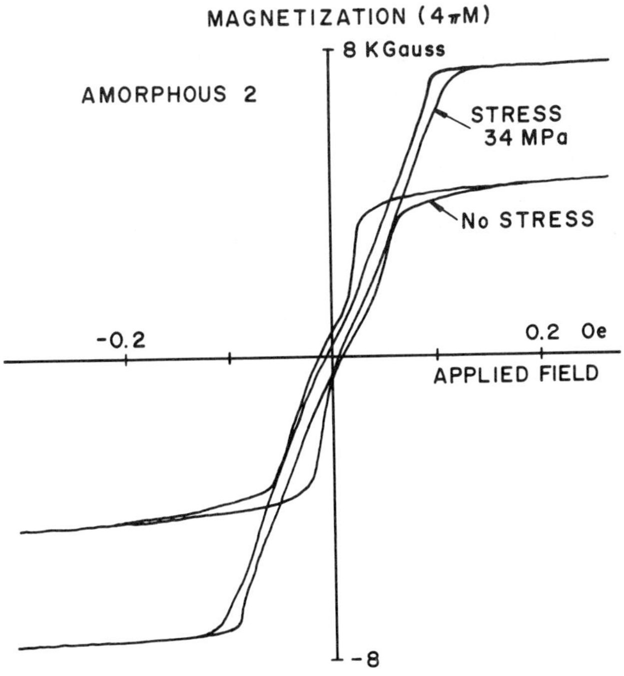

Fig. 4. Effect of stress on hysteresis loop. Alloy 2, straight sample. Slope and low remanence due to demagnetizing field.

equilibrium phases at temperatures near 400°C. Differential scanning calorimeter measurements on our ribbons confirms this behavior. Nevertheless, annealing for times of a few minutes to a few hours at much lower temperatures (200 to 300°C) leads to substantial degradation of the mechanical, and in many cases also the magnetic, properties. The remanence decreases, the coercive field increases, and the material becomes brittle--sometimes so brittle that it breaks even under careful handling. In a ribbon of composition 1, this transition occurs quickly at 100°C or lower, and x-ray diffraction shows the formation of the single metastable bcc crystalline phase first reported by Masumoto et al.[4] This phase appears to be α-iron supersaturated with metalloids; its lattice parameter is slightly less than that of pure Fe, suggesting that the metalloids are substitutionally located. The formation of this phase requires only very short-range rearrangement of atoms; it is not a precipitation reaction in the usual sense. Presumably for this reason there is only a small change in free energy (mostly change in configurational entropy), and ordinary calorimetric measurements detect no reaction.

MAGNETIC PROPERTIES AND DEGREE OF AMORPHOUSNESS

It is important to note that the low field magnetic properties (coercive field, remanence, and field required to reach saturation) vary from sample to sample, depending on the details of preparation even when chemical compositions are the same. The properties also vary somewhat in different sections of the same ribbon, indicating that the ribbon is not homogeneous in structure and/or composition. Since the diffraction patterns obtained in a conventional diffractometer are indistinguishable from sample to sample, all showing a broad maximum characteristic of an amorphous material, the ribbons are all amorphous in a broad sense. We suggest, however, that they have varying degrees of amorphousness.

If we assume that an ideally amorphous ferromagnet has a remanence-to-saturation ratio of unity, then a reduction in this ratio may indicate a departure from complete amorphousness. In ribbons of composition 1, a drop in remanence accompanies the appearance of crystallinity, as described above. In composition 2, no crystallinity is detected in a conventional diffraction analysis, but annealing at 250°C degrades the magnetic properties in a way completely analogous to changes in composition 1. We conjecture that composition 2 undergoes a transition to a metastable crystalline state with a grain size below the detection limit (microcrystalline state). If so, low-field magnetic measurements may be a highly sensitive and useful technique for the detection of crystallinity, with possible scientific and engineering applications. We are currently trying to substantiate this suggestion.

STRESS-ANNEALING

In an effort to make permanent the improved magnetic properties produced by elastic stress, we have annealed various ribbons under stress, at temperatures that do not lead to excessive brittleness. For the range of times, temperatures, and stresses investigated, the effect of such stress anneals is to increase the low-field magnetization, sometimes very substantially, but never to the full extent that is possible by stress alone. After stress-annealing, the ribbons are also more stress-sensitive; that is, it takes a lower elastic stress to acheive full saturation in low fields. Data are shown in Fig. 5, and in Table II.

APPLICATIONS

The ribbons measured here have not been optimized in composition or treatment for soft magnetic properties, but they are nevertheless competitive with available crystalline materials. The best properties measured to date on a ribbon taken directly from the production operation and wound into a core are: coercive field 0.01 Oe; B@1 Oe, 4600 G; B_r/B_1=0.9; μ_{max}=410,000; energy loss per cycle (dc), 15 erg/cm^3. A straight ribbon of the same material under an elastic stress of 56 MPa (8100 lb/in^2) shows H_c=0.007, B@1 Oe=7800 G; B_r/B_1=0.99; and μ_{max}=1,100,000. Magnetic materials with these properties, if available at low cost, should certainly have engineering applications. The currently available ribbon is near the minimum size normally used for tape-wound cores and should be directly applicable to such devices as inverter transformers, magnetic amplifiers, current transformers, pulse transformers, bi-stable switching devices, etc., replacing the crystalline alloys now used. Table II summarizes the properties of the amorphous alloys of compositions 1, 2 and 4 after various treatments. Generally composition 1 is similar to oriented 50Ni-50Fe alloys, composition 2 to molybdenum Permally, and composition 3 to Supermalloy.

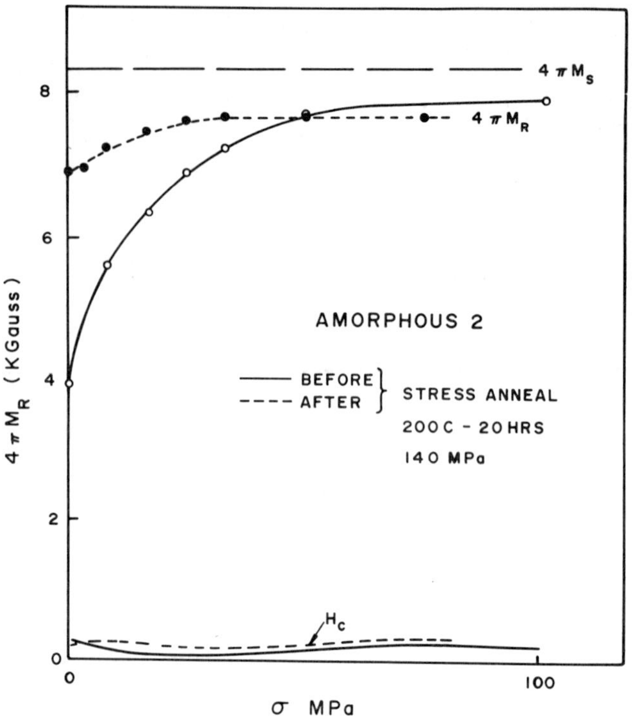

Fig. 5. Influence of stress on remanence and coercive field, before and after a stress anneal.

With the ribbons available at present, at least some applied stress is necessary to obtain the best magnetic properties. The obvious procedure of winding a toroid with the ribbon under tension does not work, because the inner layers of the core are under compression. Instead, the core must be wound with minimum tension, the free end secured against slipping,

TABLE II

Property	50Ni 50Fe	Amorphous 1 as recd.	Amorphous 1 stressed (d)	80Ni 15Fe 5Mo	Amorphous 2 as recd.	Amorphous 2 stressed (d)	Amorphous 2 (e)	Amorphous 2 (e,d)	80Ni 15Fe 5Mo(f)	Amorphous 4 as recd.	Amorphous 4 stressed (d)	Amorphous 4 (g)	Amorphous 4 (g,d)
$4\pi M_s$,kG	15.5	17.1	17.1	8	8.3	8.3	8.3	8.3	8	4.9	4.9	4.9	4.9
B_p,kG(a)	14-15.5	6.3	15.0	7.5	4.6	7.8	7.9	8.0	7-8	2.3	4.05	1.7	4.88
B_r,kG	12-15	4.0	14.6	4-6.5	4.1	7.7	7.0	7.6	5-7	1.85	3.8	1.18	4.86
B_r/B_p	.85-.95	.63	.97	.5-.9	.9	.99	.89	.95	.7-.9	.8	.94	.69	.995
H_c,Oe	.08	.065	.04	.03	.01	.007	.008	.008	.01	.01	.005	.015	.003
μ_{max}(b)	100	62	365	200	410	1100	875	950	500	185	760	790	1620
σ,MPa	--	0	75	--	0	56	0	37	--	0	140	0	19
σ,psi	--	0	11000	--	0	8100	0	5400	--	0	20000	0	2700

(a) Induction at H= 10xH_c.
(b) Taken as B_r/H_c for amorphous alloys. All values to be multiplied by 1000.
(c) Representative values from catalogs of Arnold Engineering, Carpenter Technology, and Magnetic Metals, Inc.
(d) Stressed to optimum level; value of stress given in last two rows of Table.
(e) Annealed 20 hr at 200C under 140 MPa=20000psi.
(f) Vacuum melted (Supermalloy).
(g) Annealed 1½ hr at 300C under 140 MPa=20000psi.

and the bobbin increased in diameter to stress the magnetic ribbon. This can be done mechanically, with an expanding bobbin. In production, it might be possible to use differential thermal expansion to give the same result. Alternatively, one can seek compositions or treatments for which stress is unneccesary.

Amorphous alloys with thicknesses greater than a few thousandths of an inch (about 100 µm) are unlikely, because of the very high freezing rates required. Amorphous sheets and strips several cm or more in width are entirely possible, and will greatly increase the range of applications. High-saturation amorphous alloys like composition 1 are candidates for power devices, including motors and generators, at least for relatively high frequency service where the small sheet thickness is desirable.

The high mechanical strength of amorphous alloys will make them easier to handle in manufacturing operations than conventional crystalline materials; the elimination of elaborate hydrogen annealing treatments is also clearly an advantage. On the other hand, the amorphous alloys are susceptible to magnetic and mechanical degradation on overheating to relatively low temperatures, so the use of some kind of temperature protective device may be required in service.

Other possible applications are in magnetostrictive delay lines, where the high strength and relatively high magnetostriction, plus low magnetizing fields, should be advantageous. The high permeability, good mechanical properties, and high electrical resistivity should make amorphous alloys useful for magnetic recording heads. The large stress-sensitivity of the as-prepared ribbons suggests their use as stress or strain sensors, in which the output from a pickup coil on a ribbon driven at constant ac field amplitude would vary with the stress or strain applied to the ribbon.

Finally, the low-field properties are seen to depend strongly on the details of preparation and heat treatment, and therefore on the degree of local order or amorphousness. Magnetic measurements could be a useful approach to the investigation of the scientific problems of amorphous materials, as well as a helpful quality-control procedure in manufacturing operations.

ACKNOWLEDGEMENTS

We wish to thank the Allied Chemical Corp., especially G. R. Bretts, J. J. Gilman, and Harry Knutson, for proving the amorphous ribbons. The work has been partly supported by the National Science Foundation through the Laboratory for Research on the Structure of Matter, University of Pennsylvania, and by the Department of Metallurgy and Materials Science. Valuable help with the experiments has been provided by Andrew Uri and Kenneth Yu.

REFERENCES

1. P. Duwez, Fizika-2, Suppl. 2 (1970) 1; H. Jones and C. Suryanarayana, J. Mat.Sci. 8 (1973) 705; H.Jones, Rep. Prog. Phys. 36 (1973) 1425.
2. R. Pond, Jr. and R. Maddin, Trans. Met. Soc. AIME, 245 (1969) 2474.
3. H. S. Chen and C. E. Miller, Rev. Sci. Instr. 41 (1970) 1237.
4. T. Masumoto and R. Maddin, Mat. Sci. Eng. 11 (1974) to be published.
5. T. Mizoguchi, in Amorphous Magnetism, ed. Hooper and deGraaf, Plenum Press, N.Y. (1973).
6. P. Chaudhari, J.J.Cuomo, and R.J.Gambino, Appl.Phys. Lett. 22 (1973) 337; IBM J. Res. and Dev. 17 (1973) 66; R.J.Gambino, P. Chaudhari, and J.J.Cuomo, AIP Conf.Proc. 18 (1974) 578.
7. H.T.Savage, A.E.Clark, S.J.Pickart, J.J.Rhyne, and H.A.Alperin, IEEE Trans. Mag. MAG-10 (1974) 807.
8. T. Masumoto et al., Jap. J. Appl. Phys. Nov. 1974 to be published
9. A.W.Simpson, German Patent 2126687, Dec. 7, 1972.
10. R.C.Sherwood, E.M.Gyorgy, H.S.Chen, S.D.Ferris, G. Norman, and H.J.Leary, this conference.
11. B.G.Bagley, H.S.Chen, and D. Turnbull, Mat.Res. Bull. 3 (1968) 159.

GRAIN-ORIENTED SILICON-IRON - ITS BASIC PROPERTIES, BEHAVIOUR AND APPLICATIONS

J.E. Thompson
Wolfson Centre, University of Wales, 30 The Parade, Cardiff, United Kingdom

ABSTRACT

Grain-oriented silicon-iron is used in large quantities in the electrical power industry, so that any improvements in its basic properties, its behaviour under operating conditions, and a critical examination of design considerations, are economically valuable: the commercial and technical importance of Hi-B material must be assessed on such grounds.

A significant improvement in the magnetostriction-stress sensitivity and a reduction in losses has been found (particularly at higher inductions), which are most advantageous in power transformer engineering. Rotational losses must be considered too in both power transformers and large machines, and recent experimental measurements in the laboratory will be discussed. In basic terms, losses are essentially governed by domain wall motion and recent work from various sources will be reported.

Both a critical examination and an understanding of design considerations is vital; taking the case of a large power transformer, the method of flux transfer between sheets, the dependence of localised losses on joint parameter, and the newly found importance of circulating fluxes will be compared for experimental cores built from 'conventional' and 'Hi-B' materials.

MAGNETIZATION PROCESSES IN IDEALLY SOFT MATERIALS

B. Heinrich and A.S. Arrott,
Simon Fraser University, Burnaby, British Columbia,
Canada V5A 1S6

ABSTRACT

Materials without magnetocrystalline anisotropy and internal imperfections are considered as ideally soft. Equilibrium magnetic configurations are determined from the competition between magnetostatic and exchange energies for given geometry and size. Approximate solutions have been found which show reversible nucleation processes, remanences, coercivities and critical fields at which irreversibility occurs. The configurations in the irreversible portion of the magnetization process are found both for eddy current and for local (Gilbert) damping. The approximation used should permit one to follow in detail in an irreversible process those effects usually referred to as hysteresis loss, macro-eddy-current loss and micro-eddy-current loss.

The ideally soft magnetic materials (ISM's) advertised here for future sale necessarily have exchange interactions to force alignment of atoms with magnetic moments. Without the magnetization the alignment would have little market value. The user of ideally soft materials will not have to grapple with the problems of magnetocrystalline anistropy and magnetostriction or with crystal structure and its defects. The material (ISM) to be offered does have losses. A certain amount of loss is necessary or the magnetic atoms would precess forever. Our ISM is patterned after iron, the most important real soft magnetic material (RSM). We have done away with crystal structure, magnetic anisotropy and magnetostriction, but we have kept the electrical conductivity. For those who would prefer not to think about eddy currents we also offer a model of the ISM with a local damping mechanism.

At present we plan to offer our ISM's in two shapes. One of them is in the form of a cylinder with a small hole down the axis. The other is in the form of a toroidal shell. The toroidal shell has its dimension ratios like those of a bicycle inner tube[1]. The hole down the middle is essential to produce materials which will reverse their magnetization while maintaining cylindrical symmetry of the magnetization pattern. The commercial importance of this is not evident, but it does help considerably in the manufacturing of these ISM's. For the manufacturing is primarily carried out by computer which finds the task much easier if cylindrical symmetry is maintained. At present we are experiencing production problems and our current prototypes have some serious defects. At best they only approximate the behaviour envisioned for the ultimate ISM's. Yet these prototypes seem interesting enough that we will discuss them at some length.

They have magnetization curves that show saturation, departure from saturation at a nucleation field, remanence, coercivity and critical fields at which the processes of magnetization becomes irreversible. A typical curve is shown in Fig. 1. By changing the dimensions of the ISM, their properties can be varied.

At present the magnetization and the exchange interaction of our ISM are the same as found in iron. We believe we have a way to add a bit of isotropic magnetostriction which will make our ISM's behave more like real soft materials (RSM's). This may facilitate the eventual adoption of ISM's by users of RSM's.

The process for manufacturing ISM's is to minimize an energy which has terms for the exchange interactions and the interaction of the magnetization with the applied field and its own stray fields generated at the surfaces by the normal component of the magnetization.

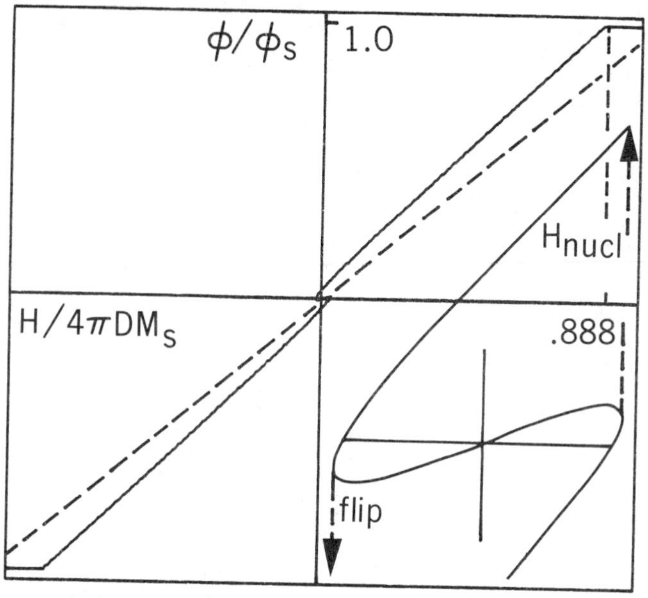

Fig. 1 The magnetic flux for an ideally soft material normalized to the magnetic flux at saturation as a function of applied field normalized to $4\pi DM_S$, the demagnetizing field at saturation for a tube with inside radius $r = 0.6\mu$, outside radius $R = 6\mu$ and length 1 cm. The inset shows the difference between the coercive field and the critical field for the onset of the irreversible flip of the magnetization. The dashed line is for $H = 4\pi DM_S\Phi/\Phi_S$.

The exchange energy is isotropic and looks like

$$E_{ex} = \frac{A}{M_s^2}\int\left\{(\mathrm{div}\,\vec{M})^2 + (\mathrm{curl}\,\vec{M})^2\right\}\,dv \qquad (1)$$

where A is 2×10^{-6} erg/cm and M_s is 1700 gauss. This is a piece of modern production machinery borrowed from the technology of liquid crystals[2]. It does everything the old fashion $\nabla^2\alpha + \nabla^2\beta + \nabla^2\gamma$ did, but with improved efficiency. It can be tested out for a magnetization pattern which just circulates about a cylindrical axis; $\vec{M} = M_s\hat{\Phi}$ where $\hat{\Phi}$ is a unit vector in the circumferential direction. Then div $\vec{M}=0$, curl $\vec{M}=M_s/\rho$ and

$$E_{ex} = A\int\frac{dv}{\rho^2} \qquad (2)$$

For a cylindrical tube of outside radius R and inside radius r and volume V this is

$$E_{ex} = \frac{2A}{R^2-r^2}\,\ell n\,\frac{R}{r}\,V \qquad (3)$$

It is to avoid the logarithimic singularity that the hole is put down the center of the cylinder. If the hole were not there, the magnetization would be forced to be along the axis at the axis and it would not be possible to completely reverse the magnetization without losing the cylindrical symmetry of the magnetization pattern. The curling pattern $M=M_s\hat{\Phi}$ produces no stray fields. Such a pattern is only an approximation to the pattern of an ISM cylinder in the absence of an applied field. An ISM cylinder will have a small remanence and it will be necessary to apply a reverse field to reach a demagnetized state. The remanence and the coercive field increase as the diameter decreases.

The remanence produces a stray field H_s which is

$$H_s(\vec{r}) = \int \frac{\operatorname{div}\vec{M}(r')(\vec{r}-\vec{r}')}{|r-r'|^3}\, dV' + \int \frac{\hat{n}\cdot\vec{M}(r')\ (\vec{r}-\vec{r}')}{|r-r'|^3}\, d\vec{S} \qquad (4)$$

where the first term arises from the volume magnetic charge density $\rho_M = \operatorname{div}\vec{M}$ and the second from the surface magnetic charge density $\sigma = \hat{n}\cdot\vec{M}$. We believe that our ISM's will have no volume charge. This conclusion is based upon our experience with iron whiskers[3]. Calculations show that the magnetic charge for an iron whisker in a magnetic field is almost all on the surface and is distributed in the same way as electrical charge is distributed on the surface of a conductor of the same shape in an electric field. The charge in the volume is all on the wall and is in the ratio $H_A/4\pi M_s$ to the charge on the surface. For iron at room temperature this is just over 1 to 50. As the anisotropy goes to zero in our ISM the charge should all be on the surface. Thus the magnetic charge behaves like real charge and goes to the surface. The surface charges produce stray fields which within the material are termed demagnetizing fields. Note that if $\operatorname{div}\vec{M}=0$ within the ISM's the expression for the exchange energy reduces to

$$E_{ex} = \frac{2\pi A}{M_s^2}\int (\operatorname{curl}\vec{M})^2\ \rho d\rho dz \qquad (5)$$

where we have built in cylindrical symmetry. The magnetostatic energy is

$$E_m = 2\pi M_s \int (\tfrac{1}{2}H_s - H_o)\cos\theta\ \rho d\rho dz \qquad (6)$$

The $\frac{1}{2}$ in front of the stray field appears because H_s arises from the magnetization and therefore this term is a self energy while H_o arises from external sources.

The H_o is along the cylinder axis. The magnetization is described in polar coordinates where θ is the angle between the magnetization and the cylinder axis

$$\vec{M} = M_s\left[\hat{z}\cos\theta + \hat{\Phi}\sin\theta\sin\varphi + \hat{\rho}\sin\theta\cos\varphi\right] \qquad (7)$$

The angle φ is measured from the radial vector $\hat{\rho}$ at each point in space. θ and φ are both functions of ρ and z. The vanishing of the volume charge requires $\operatorname{div}\vec{M}=0$ which yields

$$\frac{1}{\rho}\sin\theta\cos\varphi - \theta_z\sin\theta + \theta_\rho\cos\theta\cos\varphi - \varphi_\rho\sin\theta\sin\varphi = 0, \qquad (8)$$

where the subscripts denote partial derivatives. For an ISM in a magnetic field we expect the charge density on the surface of the cylinder to be very much like that on an equivalent conductor in a uniform electric field. Such a charge distribution is to a first approximation linear along the sides of the cylinder and constant across the ends, but more precisely it deviates gradually from this to produce integrable singularities on the side and ends where they meet at the corners. There are almost no charges on the inner surface. The solutions of Eq. (8) which match the boundary conditions that the charge be that of the conductor in a field are not known precisely. To a first approximation $\cos\theta$ varies quadratically in z from a maximum in the central cross section of the cylinder while $\cos\varphi$ starts at zero at the inner surface and increases linearly with ρ.

Minimization of the energy given by Eqs. 1, 4 and 6 produces the torque equations of micromagnetics[4]. Our process for producing ISM's minimizes the energy given by Eqs. 5 and 6 with the constraint of Eq. (8). Eq. (4) is used here also but without the term from the volume charge. This process still has a few bugs and seems quite expensive in computer time. For this reason we have developed a simplified process to produce our approximation to ISM's. We minimize the energy only for the central cross section assuming that

(1) the energy of the whole cylindrical tube scales in such a way that the total energy is the energy per unit length of the central cross section times an effective length of the cylinder,

(2) the stray field can be approximated by a demagnetizing field

$$H_D = D\Phi/\pi(R^2-r^2), \qquad (9)$$

where D is the ballistic demagnetizing factor of the cylinder and Φ is the flux arising from the magnetization in the central cross section,

$$\Phi = 4\pi M_s \cdot 2\pi \int \cos\theta\ \rho d\rho, \qquad (10)$$

(3) the condition

$$\hat{n}\cdot\vec{M} = 0 \qquad (11)$$

on the boundary of the central cross section follows from the symmetry, and

(4) a quadratic variation of $M_s\cos\theta$ with z away from the central cross section gives

$$M_s\cos\theta\ \theta_{zz} = -8M_s/\ell^2, \qquad (12)$$

where θ_{zz} is a second partial derivative.

The constraint $\operatorname{div}\vec{M}=0$ and the symmetry relation $\hat{n}\cdot\vec{M}=0$ are both satisfied if $\cos\varphi=0$, that is if

$$M = M_s\ (\hat{z}\cos\theta + \hat{\Phi}\sin\theta). \qquad (13)$$

The minimization of the energy with these assumptions produces a torque equation for θ

$$2A\left[\theta_{\rho\rho} + \frac{1}{\rho}\theta_\rho - \frac{\sin 2\theta}{2\rho^2} + \theta_{zz}\right] + M_s(H_D - H_o)\sin\theta = 0 \qquad (14)$$

From Eq. (12) it seems permissible to drop the term θ_{zz} for long cylinders. The solutions of this equation are found on the computer by the method of shooting using the condition that $\theta_\rho=0$ at both $\rho=r$ and $\rho=R$. One assumes a value of θ at the inside surface and varies the value of the dimensionless parameter

$$\lambda = R^2M_s(H_D-H_o)/2A = (R/R_{ex})^2(H_D-H_o)/M_s \qquad (15)$$

where $R_{ex}^2 \equiv 2A/M_s^2$ is the characteristic curling radius. $R_{ex} \approx 10^{-6}$ cm for iron. Newton's method is used to attain rapidly the value of λ for which $\theta_\rho=0$ at $\rho=R$.

For each such λ one finds $\theta(\rho)$ and calculates the flux from Eq. (10) and the demagnetizing field H_D from Eq. (9). Then from Eq. (15) one finds the applied field H_o that would give a self-consistent λ. The variation of θ with ρ depends upon the ratio of inner to outer radius r/R. The dependence of θ upon ρ for a set of λ's has been shown previously for r/R=0.1. Fig. 2 shows the behavior for r/R=0.1, 0.0368, 0.0135 and 0.0 for the condition of no net flux, $H_o=H_c$. In the limit as $r/R\to 1$ $\cos\theta\to 0$ giving the circumferential pattern. One can see from Fig. 2 that $r/R \approx .01$ is a good approximation to the behavior for r/R=0 at least near the coercive field. It is expected that a rod with r/R=0 would reverse by some process that does not preserve the cylindrical symmetry. It may be that the behavior for such a rod is not too different than a tube with some ratio $(r/R)_e$ which could be called the equivalent ratio.

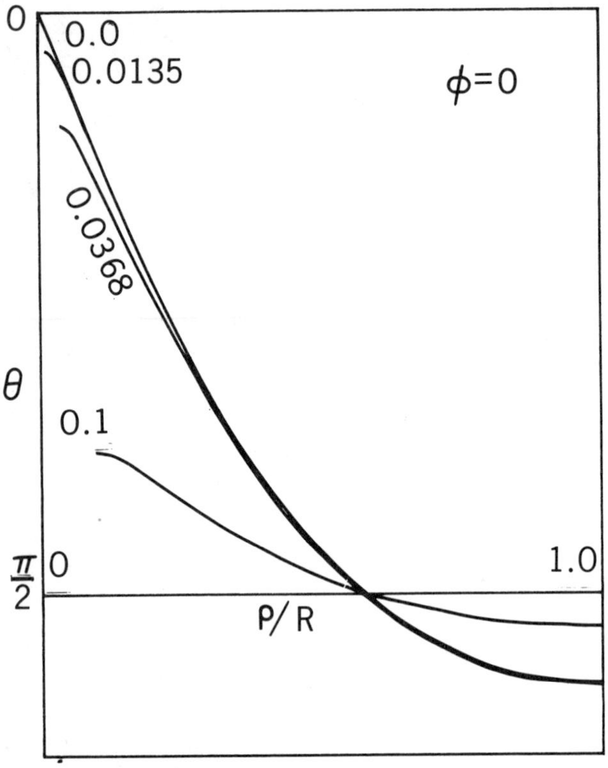

Fig. 2 The dependence of the polar angle θ upon distance from the center of the ISM tube for ratios of inside to outside radius of 0, 0.0135, 0.0368 and 0.1 for the condition that the net flux Φ = 0.

We are just guessing that $(r/R)_e = .1$. It is this ratio that we have chosen for our principle ISM.

The solutions of Eq. (14) for a series of values of λ give θ(ρ)'s from which we can calculate the magnetic fluxes. The dependence of magnetic flux upon λ is shown in Fig. 3 for several ratios r/R. The magnetic flux is normalized to the value at saturation. These curves can be used directly to calculate the magnetization dependence upon applied field once one specifies (R/R_{ex}) and D, the demagnetizing factor. D is approximately[3]

$$D = (2R/\ell)^2 \left\{\ln(2\ell/R) - 7/3\right\} \qquad (16)$$

where ℓ is the length of the tube. The procedure is to treat a particular curve in Fig. 3 as giving one equation in λ vs Φ/Φ_s to be solved simultaneously with Eq. (15) which is the straight line

$$\lambda = -(R/R_{ex})^2 H_o/M_s + D(R/R_{ex})^2 \Phi/\Phi_s \qquad (17)$$

A large demagnetizing factor gives a large slope. For a given value of H_o, the intersection of the straight line with the curve of Φ/Φ_s vs λ Fig. 3 gives the appropriate value of Φ/Φ_s.

The intercept on the λ axis is proportional to $-H_o$. As H_o is decreased from a positive value the line translates upward along the λ axis until it first crosses the Φ vs λ curve for positive values of Φ and λ. This determines the nucleation field for deviation from saturation. The remanent magnetization is determined by the line through the origin. The coercive field H_c is reached when

$$H_c = H_o = -\lambda_c (R_{ex}/R)^2 M_s \qquad (18)$$

where λ_c is the value of λ for which Φ=0. λ_c varies with r/R and approaches 2 as r/R→0. λ_c is 0, 0.744, 1.913 and 1.998 for r/R equal 1, 0.1, 0.0368 and 0.0135 respectively. The coercive field does not

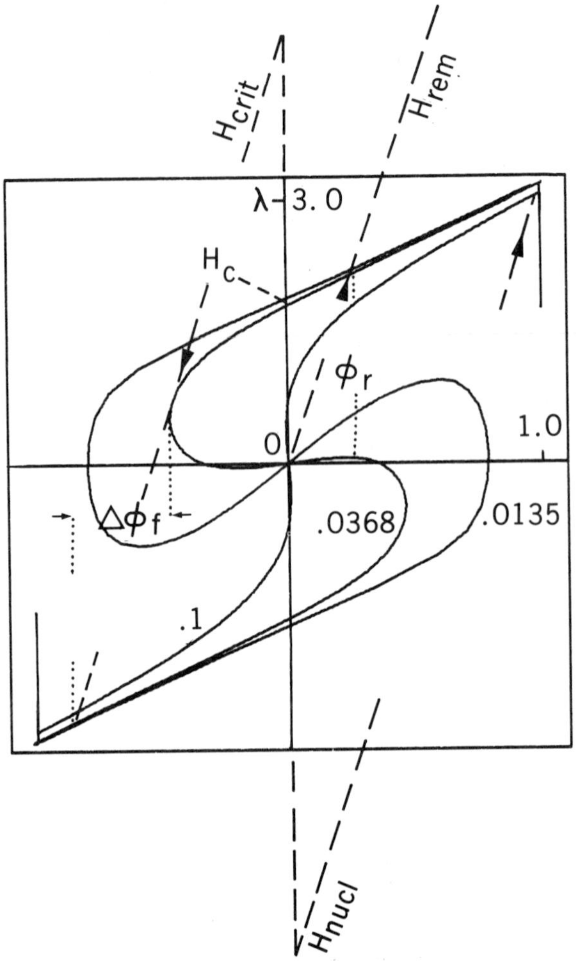

Fig. 3 The relation between the dimensionless parameter λ and the magnetic flux Φ (normalized) for three choices of r/R, 0.1, 0.0368 and 0.0135. The dashed lines represent Eq. (17) for the nucleation field, the critical field and H_o = 0. The coercive field is given by the intersections with λ axis. $\Delta\Phi_f$ gives the change in flux in the irreversible flip of the magnetization.

depend on the demagnetizing field for there is none at the coercive field in this approximation. Thus the coercive field varies as $1/R^2$ and is proportional to A. For a 25μ cylinder of iron Eq. (17) yields a coercive field of 10^{-3} Oersted. When the demagnetizing field line Eq. (17) is moved further up the λ axis by making H_o more negative a point is reached where the line is tangent to the λ vs Φ curve. This is H_{crit} the critical field. Beyond that field the only solution is the intersection with the negative portion of the λ vs Φ curve. This process of tracing out the intersections of the curves yields the magnetization curves of Figs. 1, 4 and 5. The dashed line in Fig. 1 shows the behavior for A=0 and corresponds to the demagnetizing field relation

$$H_o = H_D = 4\pi D M_s \Phi/\Phi_s \qquad (19)$$

For A=0 the solution has θ independent of ρ in the central cross section.

If the slope of the straight line in Fig. 3 is less than the slope of λ vs Φ at $\lambda=\lambda_c$, the critical field occurs before the coercive field. If the slope of the line is less than the slope of λ vs Φ at $\lambda=\lambda_{max}$ then the critical field coincides with the nucleation field and a completely square loop results.

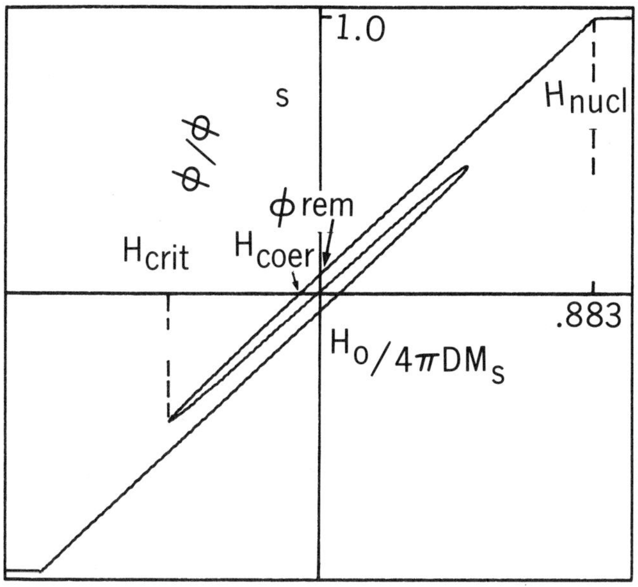

Fig. 4 The magnetization as in Fig. 1, but for $r/R = 0.0368$. $R = 6\mu$.

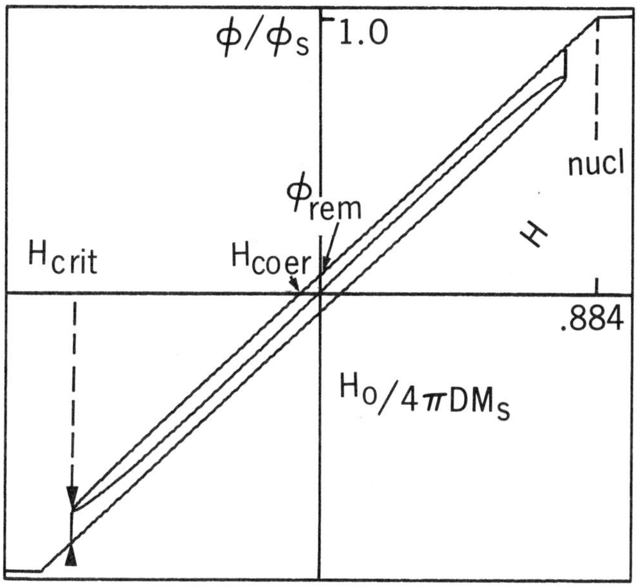

Fig. 5 The magnetization as in Fig. 1, but for $r/R = 0.0135$. $R = 6\mu$.

The slope $\Phi_s\, d\lambda/d\Phi$ approaches 1.445 as $r/R \to 0$ and it is approximately this even for $r/R = .1$ thus square loop behavior will occur whenever

$$(R/R_{ex})^2 < 1.445/D \qquad (20)$$

As D also decreases with R^2 this inequality leads to a critical radius R_c for a fixed length ℓ which is approximately given by

$$R_c^{\,2} = \tfrac{1}{4} R_{ex}\, \ell \qquad (21)$$

For ℓ=1 cm the critical radius is 5μ. The magnetization curves shown in Figs. 1, 4 and 5 were manufactured for ISM's with ℓ=1 cm and R=6μ which is just beyond the critical radius. The variation with R of the principle qualities of the magnetization curve are shown in Fig. 6 for the ISM's with r/R=0.1 and ℓ=1 cm. The remanent fraction Φ_r/Φ_s, the fractional change in flux at the critical field $\Delta\Phi_r/\Phi_s$ and the ratio $H_c/4\pi DM_s$ all go as R^{-4}. $H_{crit}/4\pi DM_s$ is essentially independent of R. The normalizing factor for the fields, $4\pi DM_s$, is just the demagnetizing field at saturation. As R increases the effects of R_{ex} becomes less and less important and the nucleation field H_n approaches $4\pi DM_s$.

The approximation to ISM behavior made here is particularly poor in the region of nucleation for it gives complete saturation at a finite field and does not take into account the difficulties of describing cylinder[5]. Procedures for producing better ISM's which are not yet perfected promise to remedy this defect.

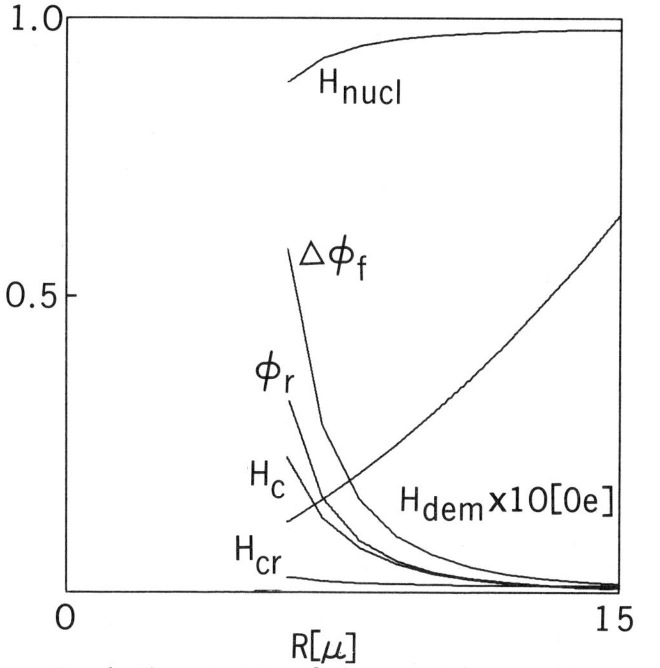

Fig. 6 The variation of the principle magnetic quantities with outside radius R for R > 6μ, r/R = .1, and ℓ = 1 cm. The nucleation field H_n, the critical field H_{cr}, and the coercive field H_c are normalized to the saturation demagnetizing field H_{dem} at each R. The remanent flux Φ_r and the change in flux in the irreversible process $\Delta\Phi_f$ are normalized to the saturation flux at each R.

The variation of θ with ρ can also be found for the irreversible portion of the magnetization process at each time during the switch from one stable configuration to another. This variation and its time evolution depend upon the loss mechanism. It is assumed that there is no precession about the applied field, that is cosφ remains zero. Such a precession would create large demagnetizing fields in the radial direction and force a rotation described in terms of variations in θ only. Thus to treat the case of local damping, one adds to the torque equation a term proportional to the rate of rotation of the magnetization. Thus Eq. (14) becomes

$$2A\left[\theta_{\rho\rho} + \frac{1}{\rho}\theta_\rho - \frac{\sin 2\theta}{2\rho^2}\right] + M_s(H_D - H_o)\sin\theta = \Gamma\sin\theta\cdot\dot{\theta} \qquad (22)$$

The damping parameter Γ is chosen to correspond to ferromagnetic resonance values of the intrinsic damping of iron platelets[6] for which $\Gamma = 3.5\times10^{-7}$ erg secs/cm^3.

The procedure for calculating $\theta(\rho,t)$ during an irreversible process starts from the solution for $H_o = H_{crit}$ which is denoted by $\theta(\rho,0)$. The left side of Eq. (19) at time $t=\Delta t$ is set equal to

$$\Gamma\sin\Big(\theta(\rho,\Delta t)\Big)\cdot\Big(\theta(\rho,\Delta t) - \theta(\rho,0)\Big)/\Delta t.$$

$\theta(\rho,\Delta t)$ is found by choosing values of $\theta(r,\Delta t)$ and $\lambda(\Delta t)$ and propagating the solution from the inner radius. One finds that $\theta_\rho(R,\Delta t)$ is not zero. $\lambda(\Delta t)$ is then varied until $\theta_\rho(R,\Delta t)=0$. But then the apparent applied field H_o' given by Eq. (15), that is

$$H_o' = -\lambda(\Delta t)\ (R_{ex}/R)^2 M_s + H_D \qquad (23)$$

is not the applied field, which is taken to be a field slightly more negative than H_{crit}. In this H_D is calculated as before from the flux (Eq. (10)) using Eq. (9). It is then necessary to choose a new value of $\theta(r,\Delta t)$ and again vary $\lambda(\Delta t)$ to fulfill $\theta_\rho(R,\Delta t)=0$.

Using Newton's method one soon finds a $\theta(r,\Delta t)$ which yields $H_o'=H_o$. This is then repeated for each time interval to follow the irreversible magnetization process. The results of a particular calculation have been shown previously[1] and are not repeated here in that they are similar to those shown in Figs. 7 and 8 but with some differences upon which comment is made.

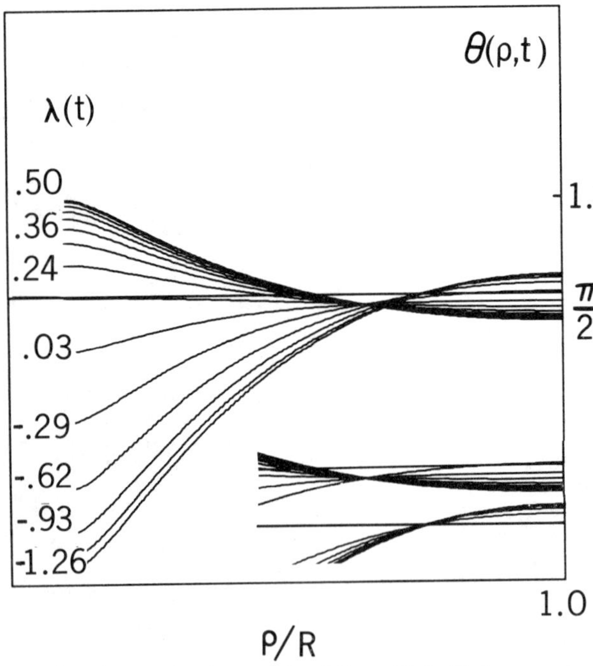

Fig. 7 The variation of θ with radial distance at a sequence of times during the irreversible flip of the magnetization pattern. The two insets show a change in the stationary point from one side to the other of the peak in the losses in time. This occurs for eddy current damping, but not for the local damping. In the latter case there is a single stationary point through out the entire flip.

The procedure for the eddy current case is more complicated. The torque equation

$$2A\left[\theta_{\rho\rho} + \frac{1}{\rho}\theta_\rho - \frac{\sin 2\theta}{2\rho^2}\right] + M_s\left[H_D - H_o - H_e\right]\sin\theta = 0 \qquad (24)$$

includes the eddy current field H_e which depends upon ρ and is itself a solution of a differential equation

$$(H_e)_{\rho\rho} + \frac{1}{\rho}(H_e)_\rho = -\frac{c^2}{(4\pi)^2\sigma M_s}\sin\theta\ \dot{\theta} \qquad (25)$$

with boundary conditions $(H_e)_\rho = 0$ at $\rho=r$ and $H_e = 0$ at $\rho=R$.

The time derivative is handled in the same way as for the local damping. The procedure involves a further nesting of the convergence processes. One chooses initial values for $\theta(r,\Delta t)$, $H(r,\Delta t)$ and $\lambda(\Delta t)$; $\lambda(\Delta t)$ is still given by Eq. (15). The two difference equations corresponding to Eq. (24) and Eq. (25) are then propagated simultaneously from $\rho=r$ using the boundary conditions $\theta_\rho(r,\Delta t)=0$ and $(H_e)_\rho(r,\Delta t)=0$. The values obtained for $\theta(\rho,\Delta t)$ and $H_e(\rho,\Delta t)$ do not at first satisfy the conditions $\theta_\rho(R,\Delta t)=0$ nor $H_e(R,\Delta t)=0$ nor the condition that H_o' (calculated from $\theta(\rho,\Delta t)$ by means of Eq. (23), Eq. (10) and Eq. (9)) equal H_o. First λ is varied to obtain $\theta_\rho(R,\Delta t)=0$ then new choices of $\theta(r,\Delta t)$ are made, λ again varied to obtain $\theta_\rho(R,\Delta t)=0$ and Newton's method used to obtain the $\theta(r,\Delta t)$ for which $H_o'=H_o$ just as for the case of local damping. But as $H_e(R,\Delta t)$ is not necessarily zero, $H_e(r,\Delta t)$ is changed and the whole procedure is repeated to get another value of $H_e(R,\Delta t)$. Again Newton's method is used to obtain the value of $H_e(r,\Delta t)$ that leads to $H_e(R,\Delta t)=0$. Each step in Δt requires the repetition of this process and these steps must be taken sufficiently small that the procedure converges.

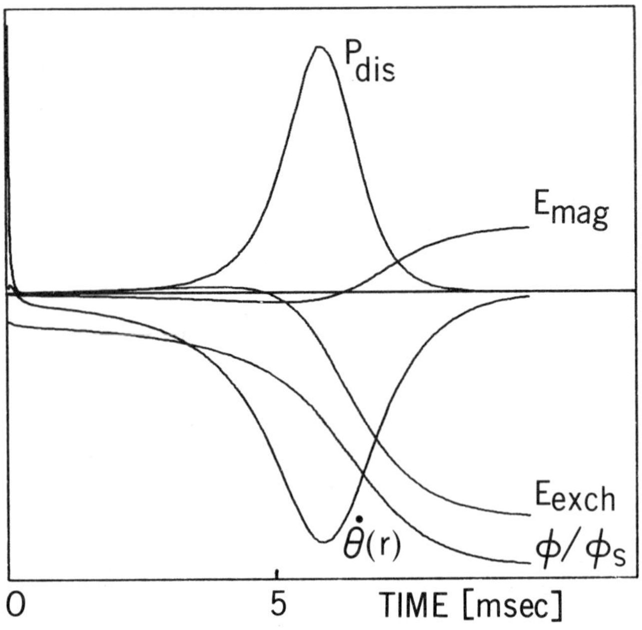

Fig. 8 The variation with time during the irreversible flip of the magnetization for the case of eddy current damping of the loss power P_{dis}, the exchange energy E_{exch}, the total magnetostatic energy E_{mag}; the flux Φ (normalized) and the angular velocity $\dot{\theta}$ of the magnetization at the inner radius. For eddy current damping $\dot{\theta}$ and P_{dis} peak together because the eddy current field is greatest at the inner radius. For local damping $\dot{\theta}$ peaks before P_{dis} peaks.

The difference equations for θ and H_e are set up on a grid of 129 points with closer spacing at the inner and outer surfaces to assure adequate treatment of the boundary conditions. The time intervals are in smaller steps at the start. A typical sequence has 10 steps of 27 μsecs, followed by 5 steps at 60 μsecs with the remaining 75 steps of 120 μsec each. Such a procedure requires 30 min. of computing time. At each step 258 simultaneous non-linear equations are being solved. The variations of θ with ρ for various times during the switching are shown in Fig. 7. The time evolution of the net flux is shown in Fig. 8 along with the changes in various energies.

At the critical field H_{crit} the exchange torque just compensates the magnetic torques. A slight change of H_o to a value more negative than H_{crit} upsets this balance. As the magnetization reversal starts it is driven by the magnetic torques and resisted by the exchange torque. As the flux crosses the demagnetizing line (dashed line of Fig. 1) the magnetic torque reverses, but before this happens the exchange torque changes sign and drives the process to completion. The losses peak after the exchange torques take over. The final decrease in exchange energy is greater than the increased magnetic energy with the difference going into the dissapative process. At present the calculation has been carried out for small step change in applied field. If carried out for a field changing in time it would yield the development of anomalous loss with increasing rate of change of field.

The approximation to an ISM is an interesting research material for studying the effects of eddy currents. The work to date is preliminary. It indicates that a complete study can be done with the techniques outlined above. The a.c. response of the ISM over the entire magnetization loop can be found. From this an explicit model for anomalous loss can be obtained. That is one can follow in detail in an irreversible process those effects usually referred to as hysteresis loss, macro eddy current loss and micro eddy current loss[7].

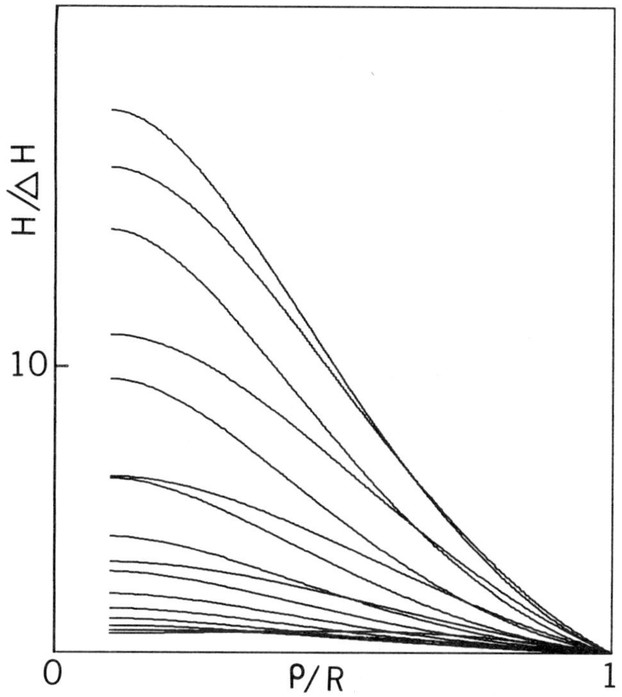

Fig. 9 The time evolution of the field from the eddy currents. The eddy current field peaks at the inner radius and vanishes at the outer radius. The field is normalized to the small step in applied field used to upset the equilibrium at the critical field. $\Delta H = .003 \cdot (4\pi D M_s)$.

In presenting our approximation to an ISM we have not given sufficient acknowledgement to the vast amount of work which preceeds and encompasses the problem of describing magnetic materials without anisotropy or magnetostriction. We have referred briefly to the subject of micromagnetics[4] for which the ISM is but a special case. The emphasis on patterns with $\text{div}\vec{M}=0$ also has a long history. We would in particular call attention to the work of La Bonte[8], Hubert[9] and Kronmüller[10] in regard the formulation of this problem in terms of finding solutions of $\text{div}\vec{M}=0$ with the constraint $\vec{M}\cdot\vec{M}=M_s^2$. The subject of eddy current losses has received even more attention and has been reviewed from time to time[7,11] Yet in all of the previous work we have not found another model wherein one can follow a non-uniform magnetization process through a complete hysteresis loop with irreversibility and the loss mechanisms explicity included[12]. From a study of our approximate ISM's we hope to be able to extract the essential behavior in a way that will allow us to construct simpler approximate ISM's which by their analytic nature have the virtue of lowering the production costs. For example if it becomes clear that a local damping mechanism gives essentially the same behavior as eddy current damping, it would be less expensive to use the local mechanism. The results of Figs. 2 and 3 look simple enough that analytic expressions may be found to characterize ISM's.

We have not gone into the behavior of the toroidal model of the ISM in that we have discussed this problem previously[1] up to the point where it becomes difficult and we have not made further progress in describing the Neél walls which we expect to occur.

We believe that the effect of isotropic magnetostriction can be included in the approximation that it is a local interaction. In that case it acts like an exchange term and would give rise to a larger value of A in all the expressions given here.

Finally we have confidence that there really are ISM's in nature. The particular example of most interest to us is the behavior of iron whiskers in a small range of temperatures just below the Curie temperature[13]. For very close to T_c, the magnetic anisotropy becomes vanishingly small. Another possible ISM would be an amorphous ferromagnetic material[14]. Each spin sees a local anisotropy field, but on the average these should balance out and only the fluctuations in the average would matter.

Our present program for the development of improved ISM's is to attempt to carry out the calculations of minimizing $\int |\text{div}\vec{M}| \cdot dV$ over an entire cylinder with the boundary condition that $\vec{n}\cdot\vec{M}$ on the surface gives surface charges as found for cylinders of infinite susceptibility. In striving for improved ISM's we have the collaboration of Dr. D.S. Bloomberg, Dr. M.J. Press and Dr. T.L. Templeton to whom we are grateful for contributions throughout the present work.

REFERENCES

[1] A.S. Arrott, B. Heinrich and D.S. Bloomberg, IEEE Transactions on Magnetics, Vol. Mag-10, No. 3, 950, 1974.

[2] P.G. de Gennes, The Physics of Liquid Crystals, Clarendon Press, Oxford 1974, p. 67.

[3] A.S. Arrott, B. Heinrich, D.S. Bloomberg, ASP Conf. Proc. 10, 941 (1973); D.S. Bloomberg thesis, Simon Fraser University (1973).

[4] W.F. Brown, Jr. Micromagnetics, Interscience publishers, 1963.
A. Aharoni, CRC Critical Reviews in Solid State Sciences, 121, August 1971.

[5] A. Holz, Phys. Stat. Sol. 26, 751, 1968.

[6] B. Heinrich, Z. Frait, Phys. Stat. Sol. 16, K11, 1966.

[7] J.W. Shilling and G.L. Mauze, Jr. IEEE Transition on Magnetics, Vol. Mag-10, 195 (1974).

[8] A.E. La Bonte, J. Appl. Phys. 40, 2450, 1969.

[9] A. Hubert, Phys. Stat. Solidi 32, 519, 1969.

[10] H. Kronmuller, Zeit. für Angew. Phys. 32, 49, 1971.

[11] J.J. Becker, J. App. Phys. 34, 1327 (1963).
J.E.L. Bishop, J. Phys. D:Appl. Phys. 6, 97, 1973.

[12] The work of M. Chang, Y.S. Lin, A. Priver, J. Appl. Phys. 38, 2294, 1967, on a two disc sandwich with different preferred axes has the features of the present calculation.

[13] T. Egami, P.J. Flanders and C.D. Graham, Jr., this conference.

[14] B. Heinrich and A.S. Arrott, this conference.

THE ORIGIN OF LOSSES IN GRAIN ORIENTED SILICON STEEL

G. L. Houze, Jr. and J. W. Shilling
Allegheny Ludlum Steel Corporation, Research Center
Brackenridge, Pa. 15014

ABSTRACT

The origin and minimization of losses in core materials used in power generation and distribution equipment have become increasingly important factors in the development, manufacturing, and application of such materials. The origin of losses in oriented Si-Fe is ultimately related to the changes in domain structure which occur during magnetization. In this paper, the domain structures which exist in oriented silicon steels (both conventional and high permeability steels) will be described and the factors which control these structures will be discussed (e.g., grain orientation, grain size, magnetizing frequency and induction, applied stress). The domain structure changes during 60 Hz magnetization will be illustrated using high-speed cinematography. The models of core loss based on domain wall motion will then be reviewed in the light of recent experiments which directly compare measured losses with those calculated from observed domain structures. It will be shown that the traditional separation of losses into hysteresis and eddy-current components is justified in some cases when the domain structure is known in detail. Present models frequently fail to predict the losses of commercial grain oriented silicon steels for several reasons which basically relate to the complexity of the actual domain structures existing in these materials compared to those assumed by the models.

SCIENCE-ENGINEERING INTERACTIONS IN DEVELOPING HIGH INDUCTION ELECTRICAL STEELS

H.W. Schadler and H.C. Fiedler
General Electric Company
Research and Development Center
Schenectady, New York

Recently there has been developed a new grade of grain-oriented soft magnetic steel commonly referred to as High B steel. This large grained, highly oriented and stress coated silicon iron contains only subtle differences from commercial grain-oriented silicon iron previously available which, nonetheless, can allow considerable improvement in the performance of transformers constructed from it. Most of the desirable engineering properties of this material and the metallurgical principles needed to achieve them were generally available at least a decade ago. This paper is an attempt to examine some of the problems in science-engineering interactions which may have delayed the development of such a material and to outline the possible driving forces that may have contributed to its ultimate successful introduction.

MAGNETIC PROPERTIES OF ORIENTED IRON-COBALT ALLOYS

K. Foster and D. R. Thornburg
Westinghouse Research Laboratories
Pittsburgh, Pennsylvania 15235

ABSTRACT

Magnetic characteristics of iron-cobalt sheets with strong (110)[001] and (100)[001] textures have been determined. Magnetization curves measured in the rolling direction for Co-Fe alloys with both textures were very similar to that of the commercial 49 Fe, 49 Co-2 V alloy. Both Cr and Si alloying additions were used to raise the resistivity of these alloys, with Cr being particularly effective in the 20 to 30% Co range. Core losses at 60 Hz were strongly texture dependent, while 400 Hz losses were more dependent on electrical resistivity and sheet thickness. The (110)[001] oriented sheets could be particularly useful in transformer applications, where the rolling direction properties can be utilized. Magnetic properties of punched ring samples were measured for (100)[001] or cube textured sheets, which would be more applicable to rotating apparatus.

INTRODUCTION

Iron-cobalt alloys in the range of 25 to 50% cobalt (all compositions are in wt %) have the highest room temperature magnetic saturation values of any known materials.[1,2] Thus, these alloys are becoming increasingly important as core materials in apparatus with minimum size and weight requirements. The commercial alloy Hiperco 50* (49% Co, 49% Fe, 2% V), which has a saturation induction of about 24 kG, is the most widely used iron-cobalt sheet material because its low magnetocrystalline anisotropy[3] results in high inductions for fields of 10 to 100 Oe. However, because of atomic ordering,[4] alloys containing more than 35% Co are quite brittle and require drastic quenching prior to cold rolling.[1,5] Considerable ductility is observed in alloys containing less than 30% Co,[5] and the commercial alloy Hiperco 27 (27% Co, 0.6% Cr), which has a saturation value similar to Hiperco 50, can be readily cold rolled without quenching; yet, Hiperco 27 sheet material is not widely used because its high positive anisotropy[3] prevents high inductions from being reached at low field strengths.

In the present work, strong (110)[001] and (100)[001] textures have been obtained in 10 to 27% Co-Fe sheets at thicknesses of 0.15 to 0.30 mm using primary recrystallization and normal grain growth processes. The magnetic properties of these materials are described and compared to commercial alloys. The substitution of textured alloys containing less than 30% Co for Hiperco 50 could lead to alloy cost reductions because of the decreased use of Co and relative ease of processing.

EXPERIMENTAL

All experimental alloys were prepared in a laboratory vacuum induction melting furnace using high purity starting materials. Most alloys contained additions of 0.05 to 0.15% Mn and 0.02 to 0.03% C. The alloys were cast as slab ingots, weighing approximately 7.5 kg, and hot rolled at about 1050°C (in the γ-phase region) to thicknesses of 2.5 and 4.5 mm. All subsequent annealing was done in the α-phase region. Samples were cold rolled in two stages to final thicknesses of 0.15 to 0.30 mm with intermediate anneals of 1 to 5 hr at 850 to 900°C. In some cases the initial cold rolling, to a thickness of about 1 mm, was carried out at about 250°C. Epstein samples, 3 cm by 30.5 cm, were sheared in the rolling direction from the cold rolled sheets; 25 mm diameter disc samples and 75 mm OD x 57 mm ID ring samples were also punched. Samples were given a final anneal of 24 to 48 hr at 850 or 900°C in dry H_2 (dew point <60°C) for texture development and optimum magnetic properties; the final anneal reduced C to <.003%. Some ring samples were subsequently field annealed by heating 2 hr at 850°C and furnace cooling in a field of about 10 Oe.

Standard dc and 60 and 400 Hz magnetic tests were done in an Epstein frame and on ring samples. Magnetic torque measurements[6] were made on annealed disc samples as a means of evaluating texture,[7] and x-ray reflection pole figures were made on selected samples using the Schulz method.[8] Electrical resistivity measurements were made on single Epstein strips.

ELECTRICAL RESISTIVITY EFFECTS

The electrical resistivity (ρ) of a soft magnetic sheet material is important in determining the core losses of the material, particularly at frequencies of 400 Hz and above. The addition of up to 18% Co to Fe in solid solution steadily increases ρ from about 10 to 20 μΩ-cm. Further addition of Co above 18% unexpectedly reduces ρ as shown in Fig. 1, with the decrease in ρ with increasing Co becoming very pronounced above 22% Co, such that the binary 27% Co-Fe alloy has a ρ of less than 14 μΩ-cm. These low ρ values above 25% Co greatly limit the use of binary alloys for high frequency applications, and ternary alloying additions are generally used to obtain increased ρ. Cr additions are very effective in this regard, as shown in Fig. 1 for a 0.6% Cr addition. Not only does the Cr addition significantly increase the ρ of the 18% Co alloy, but the negative slope of the ρ vs Co curve between 20 and 27% Co is greatly decreased.

The beneficial effect of Cr in increasing ρ of Co-Fe alloys is further shown in Fig. 2 for additions of up to 1.5% Cr to 18 and 27% Co-Fe alloys. Again Cr increases ρ for both Co contents, until at about 1.5% Cr, both alloys have the same ρ (∿32 μΩ-cm). Cr additions also increase ductility of alloys in the 25 to 35% Co range and are thus widely used in commercial alloys.[9]

Other alloying additions, such as Si and V, can

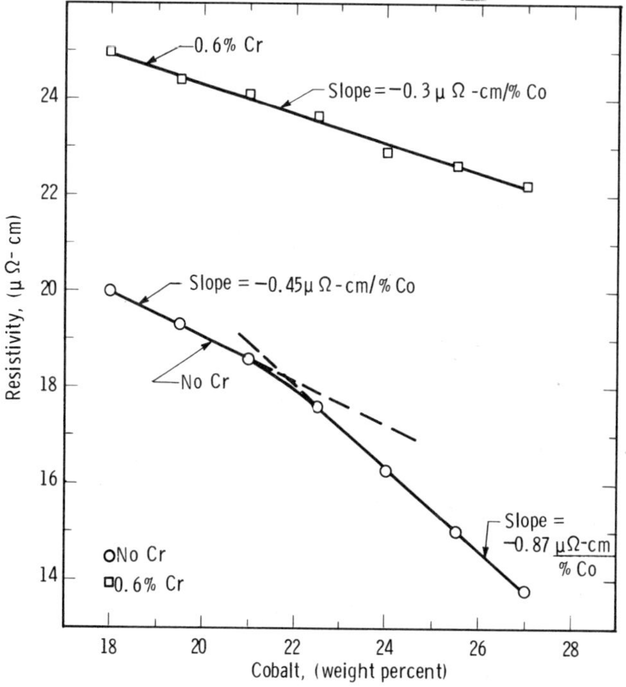

Fig. 1 Resistivity of Fe-Co alloys.

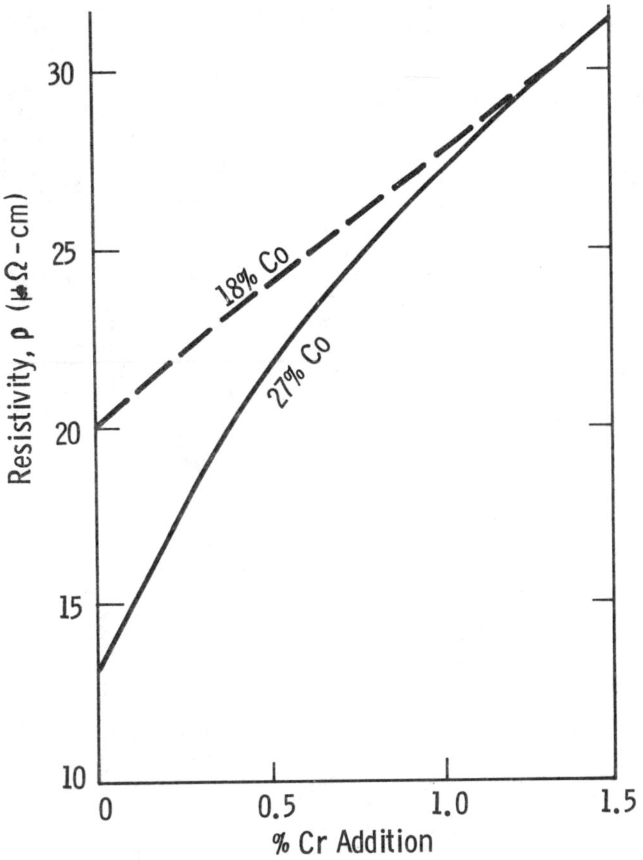

Fig. 2 Effect of Cr addition on the resistivity of 18 and 27% Co-Fe alloys.

TABLE I

Effect of Alloying Additions on Resistivity of 25% Co-Fe Alloys

% Cr	% Si	% V	ρ (μΩ-cm)
--	--	--	14.9
0.6	--	--	22.6
--	0.5	--	21.4
--	1.0	--	26.0
0.25	0.5	--	22.6
0.25	0.5	0.25	25.0

also be used to raise ρ, as shown in Table I for alloys containing 25% Co. Si additions increase ρ at almost the same rate as Cr, but Si additions above 0.5% tend to cause brittleness. Combinations of alloying additions tend to increase ρ to the same level as an equal amount of a single Cr addition.

(110)[001] TEXTURE

Strong (110)[001] or "cube-on-edge" textures were obtained in Co-Fe sheets with and without Cr additions using two stage cold rolling processes. Grain sizes after final annealing ranged between 0.15 and 0.25 mm. Figure 3 is a (110) pole figure of an 18% Co, 0.6% Cr alloy, showing a strong (110)[001] texture. The dc magnetization curve of a highly textured 0.30 mm thick 27% Co-Fe sample (containing 0.6% Cr) is shown in Fig. 4 together with curves for the commercial Hiperco 27 and Hiperco 50 alloys. At fields of 5 to 50 Oe, the magnetization curve for the oriented alloy is very similar to that of the Hiperco 50 alloy, and the inductions are much higher than those of the Hiperco 27 alloy.

Torque values and dc magnetic data for Epstein samples measured in the rolling direction are shown in Table II for several (110)[001] textured alloys containing 0.6% Cr; data for commercial alloys are

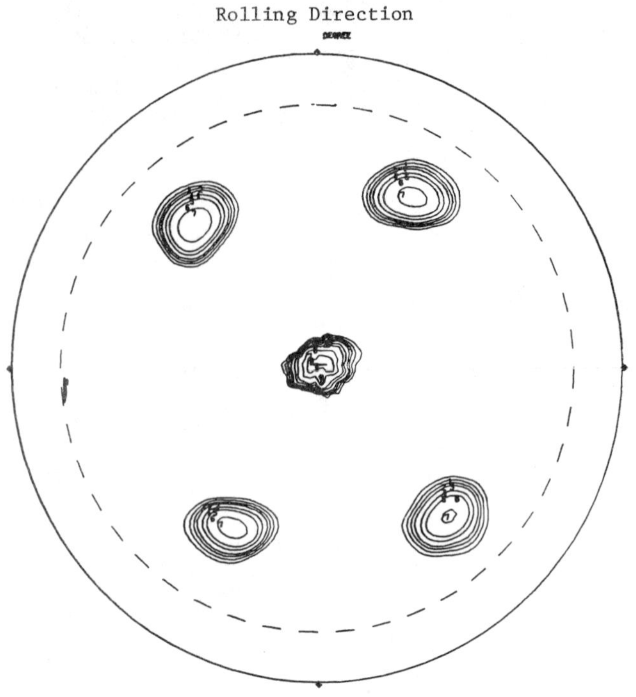

Fig. 3 (110) pole figure of an 18% Co, 0.6% Cr alloy, showing a strong (110)[001] texture. Contours are 1.5 to 12.2X random intensity.

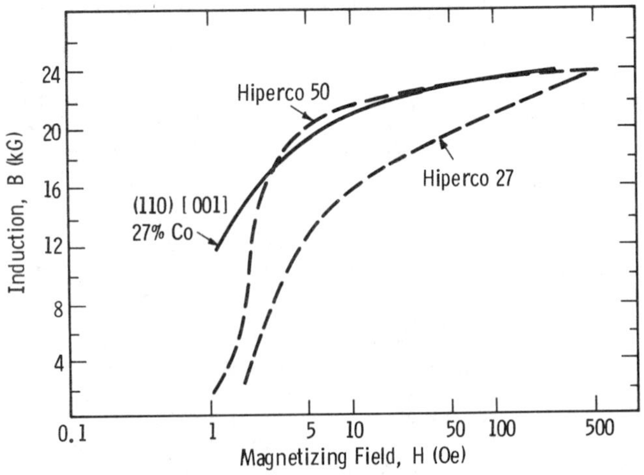

Fig. 4 DC magnetization curve of a (110)[001] oriented 27% Co, 0.6% Cr alloy (0.30 mm sheet thickness).

included for comparison. The high peak torque values and ratios of 0.34 to 0.40 indicate a high degree of (110)[001] texture in the experimental alloys.[6] The 18% Co alloy has a slightly higher ρ than either the 10 or 27% Co alloy, as indicated above. The commercial Hiperco 27 alloy has a somewhat higher ρ than the experimental 27% Co alloy because of a higher impurity content, while the Hiperco 50 alloy has a considerably higher ρ than any of the lower Co alloys. The coercive force (H_c) values of the textured alloys are considerably lower than those of the commercial alloys because of the texture, increased grain sizes, and decreased impurity contents. The induction values for a field of 100 Oe (B_{100}) of the oriented samples are very close to their saturation values, with B_{100} increasing with increasing Co content. The B_{10} values of the oriented samples are similarly dependent on the saturation values. Note the effect of texture on B_{10} for the 27% Co alloy, going from 15.9 kG for the nonoriented material to 21.2 kG for a highly oriented sample.

Ac magnetic data for the above samples are given

TABLE II

DC Magnetic Properties of (110)[001] Oriented Fe-Co Alloys Containing 0.6% Cr (Nominal Thickness 0.30 mm)

% Co	Peak Torque (erg/cm^3)	Peak Ratio	ρ (μΩ-cm)	DC H_c (Oe)	B_{10} (kG)	B_{100} (kG)
10	189,000	0.40	22.9	0.29	19.6	21.6
18	227,900	0.34	24.6	0.32	20.4	22.7
27	180,800	0.38	22.0	0.35	21.2	23.2
Hiperco 27*	--	--	26.5	1.69	15.9	20.9
Hiperco 50†	--	--	43.4	1.63	21.5	23.2

*Annealed 2 h-875°C.

†Annealed 2 h-760°C.

TABLE III

AC Magnetic Properties of (110)[001] Oriented Fe-Co Alloys Containing 0.6% Cr (Nominal Thickness 0.30 mm)

% Co	60 Hz P_{c15} (W/lb)	P_{c20} (W/lb)	P_{z15} (VA/lb)	P_{z20} (VA/lb)
10	1.0	1.9	1.2	24.2
18	1.0	1.8	1.5	7.6
27	1.0	1.8	1.8	7.1
Hiperco 27*	2.0	3.3	5.6	>30
Hiperco 50†	1.2	1.8	1.6	4.1

% Co	400 Hz P_{c15} (W/lb)	P_{c20} (W/lb)	P_{z15} (VA/lb)	P_{z20} (VA/lb
10	18.8	40.4	20.6	170.8
18	17.2	33.8	19.7	62.9
27	17.8	33.2	21.4	55.5
Hiperco 27*	27.8	48.3	44.6	--
Hiperco 50†	14.2	24.7	16.3	36.4

*Annealed 2 h-875°C.

†Annealed 2 h-760°C.

TABLE IV

Magnetic Properties of 0.15 mm Thick (110)[001] Oriented 25% Co Alloys

Alloy % Co	% Cr	ρ μΩ-cm	DC H_c (Oe)	B_{10} (kG)	B_{100} (kG)	400 Hz P_{c15} (W/lb)	P_{c20} (W/lb)
25	0.6	22.6	0.54	20.0	23.0	16.4	28.7
25	1.25	28.8	0.50	20.0	22.9	11.7	20.7
Hiperco	27	26.5	1.67	15.9	20.9	20.3	32.4
Hiperco	50	43.4	1.76	21.5	23.2	14.6	22.8

in Table III. The 60 Hz core losses (P_c) of the textured samples and the Hiperco 50 alloy were all very similar, both at 15 and 20 kG. The Hiperco 27 losses at 60 Hz were significantly higher than those of the textured materials. At 15 kG the exciting VA/lb (P_z) tended to increase with increasing Co content for the textured alloys, but P_z decreased with increasing Co at 20 kG. P_z was lower for Hiperco 50 than the textured alloys at 20 kG, 60 Hz, but was considerably higher at both inductions for Hiperco 27. At 20 kG, the 27% Co alloy had the best 400 Hz properties of the oriented samples. The 400 Hz properties of the oriented samples were considerably better than those of Hiperco 27, but somewhat inferior to Hiperco 50.

The high frequency losses of textured alloys can be further decreased by decreasing the sheet thickness or increasing ρ by increasing alloying additions. Test results are given in Table IV for 0.15 mm thick (110)[001] oriented 25% Co alloys with two Cr levels measured in the rolling direction; properties of 0.15 mm thick commercial alloys given the same annealing treatments as the 0.30 mm samples are also included. The two textured alloys have very similar dc properties, the B_{10} values of 20.0 kG indicating a high degree of texture but somewhat poorer than the 0.30 mm samples. The 400 Hz losses of both .15 mm samples are considerably lower than those of the 0.30 mm 27% Co sample. The 1.25% Cr sample has a higher ρ as expected and lower 400 Hz losses than either the 0.6% Cr alloy or the Hiperco 50 sample, while both textured samples have lower losses than the 0.15 mm Hiperco 27 material.

The above results show that (110)[001] oriented iron-cobalt alloys can have rolling direction properties approaching those of the Hiperco 50 alloy and significantly better than those of Hiperco 27. Alloys with this orientation should be of particular value in transformers and other static devices using wound cores, which take advantage of the rolling direction properties of the material.

CUBE TEXTURE

A high degree of cube texture, or (100)[001] orientation, has been obtained in binary iron-cobalt alloys or alloys with Si as a ternary addition using a two stage cold rolling process with a final reduction of over 80%. A (200) pole figure of a 27% Co-Fe sample is shown in Fig. 5, indicating a strong cube texture.

Rolling Direction

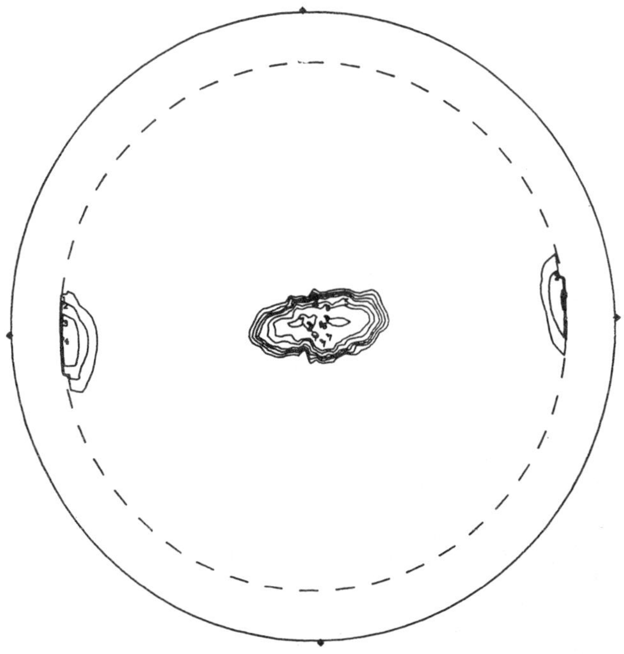

Fig. 5 (200) pole figure of a 27% Co-Fe alloy, showing a strong (100)[001] or cube texture. Contours are 1.5 to 13.7X random intensity.

Unfortunately, cube texture has not been obtained in alloys in which Cr has been added as a ternary addition for increased resistivity. Rather, these alloys tend to form strong (110)[001] textures using processes that produce strong cube texture in binary iron-cobalt alloys. Thus, in order to obtain cube texture in alloys with reasonable ρ values, compositions based on fairly large Si additions with smaller Cr additions were developed (see Table I). An alloy containing 25% Co, 0.5% Si, 0.25% Cr and 0.25% V developed excellent cube texture, a 0.30 mm thick annealed sample having a peak torque value of 157,000 erg/cm^3 and a peak ratio of 0.91. Figure 6 shows the microstructure of an annealed 0.30 mm sample of this alloy, indicating an average grain diameter of about 0.2 mm.

Magnetic properties of Epstein and ring samples of the above alloy are shown in Figs. 7-12, in which data for Hiperco 27 and Hiperco 50 alloys have again

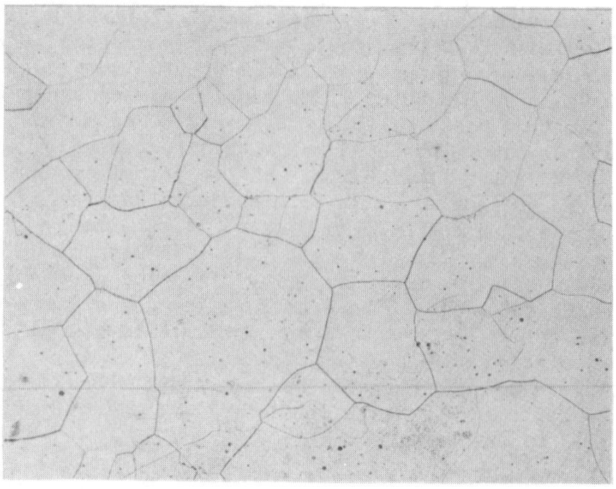

Fig. 6 Photomicrograph of the surface of a cube textured 25% Co-Fe alloy.
Etchant 2% Nital 100X

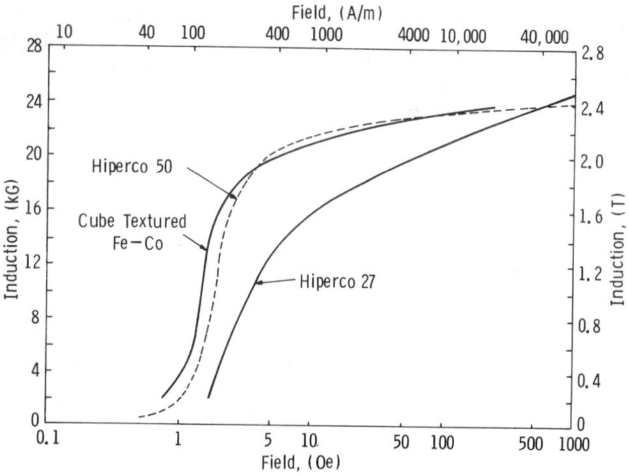

Fig. 7 DC magnetization curve of a cube textured 25% Co-Fe sample measured in the rolling direction.

been included for comparison. The dc magnetization curve for the cube textured alloy measured on an Epstein sample cut in the rolling direction is shown in Fig. 7. Between 5 and 50 Oe, the curve for the oriented alloy is very similar to that of Hiperco 50 and much superior to Hiperco 27. Core losses and exciting characteristics at 60 and 400 Hz are shown for Epstein samples in Figs. 8 and 9. The core losses for the cube textured material are lower than either Hiperco sample at 60 Hz, and fall between the two at 400 Hz. The exciting VA/lb for the cube textured sample are similar to the Hiperco 50 at both frequencies and are considerably lower than those of Hiperco 27. Dc magnetization curves for ring samples are shown in Fig. 10; here, the curve for the cube textured sample falls between the Hiperco curves. In Fig. 11, 60 and 400 Hz core loss curves for cube textured ring samples are compared with Hiperco Epstein samples; at 60 Hz, the losses of the textured sample are still lower than both Hiperco alloys, and at 400 Hz the cube texture loss curve falls between the two Hiperco curves. The exciting VA/lb curves for cube textured ring samples fall between the Hiperco 50 and Hiperco 27 at both 60 and 400 Hz, as shown in Fig. 12.

As shown previously for 27 to 50% Co-Fe alloys,[10] the magnetic properties of cube textured iron-cobalt alloys can be significantly improved by field annealing. The effect of field annealing on 0.30 mm ring samples of a cube textured alloy containing 25% Co, 0.5% Si and 0.25% Cr is shown in Table V (Epstein test values are included for comparison). Field annealing

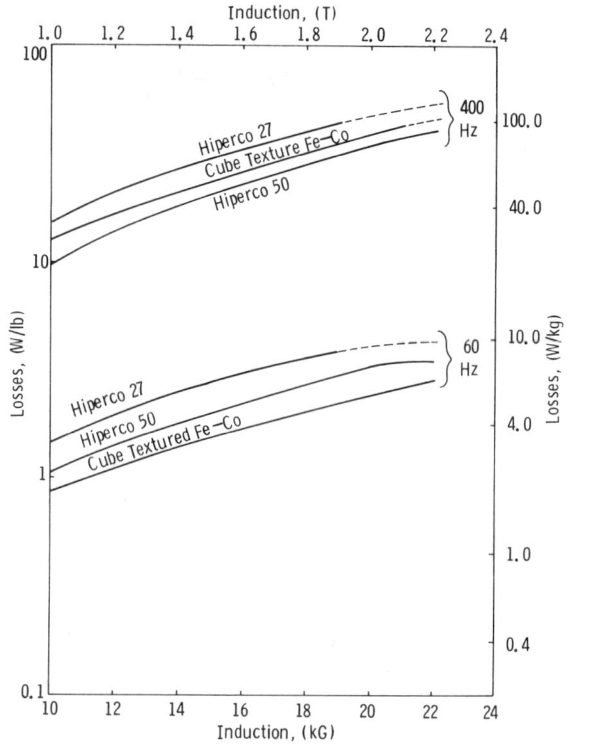

Fig. 8 Core loss curves of a cube textured 25% Co-Fe alloy measured in the rolling direction.

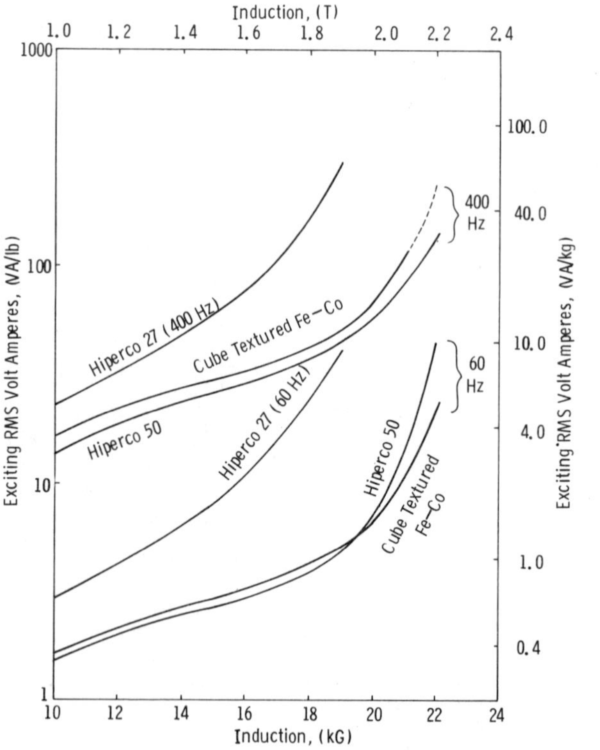

Fig. 9 Exciting VA/lb curves of a cube textured 25% Co-Fe alloy measured in the rolling direction.

of the ring sample resulted a very large decrease in H_c and increase in B_r. The B_{10} and B_{100} values also increased significantly. The 400 Hz properties were also substantially improved by field annealing.

Although in several cases the magnetization curves of cube textured samples in the rolling direction are as good as (110)[001] textured samples, it is unlikely that cube textured materials would be used in applications utilizing only the rolling direction,

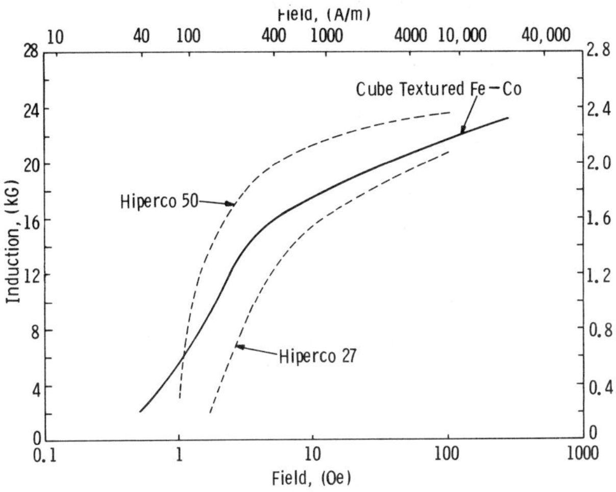

Fig. 10 DC magnetization curves of Fe-Co ring samples.

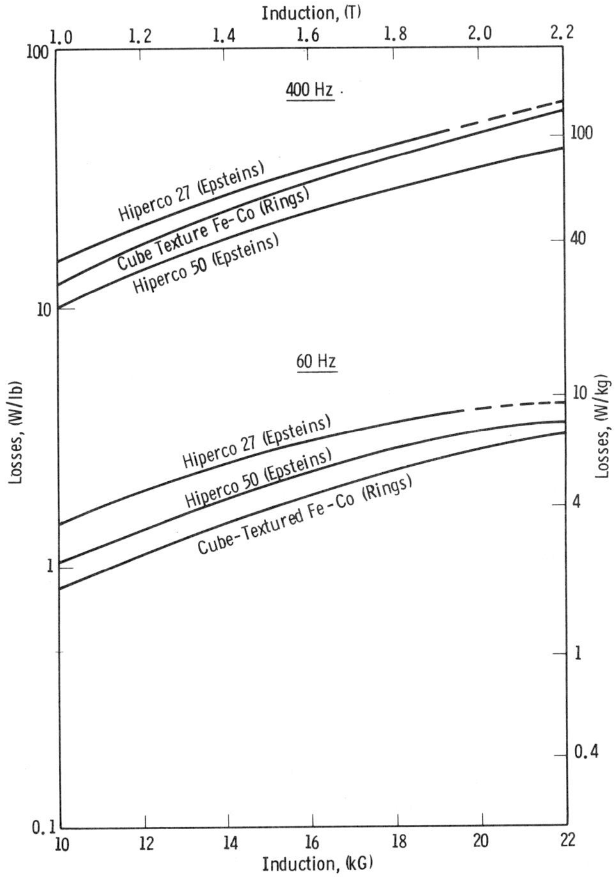

Fig. 11 Core loss curves of a cube textured 25% Co-Fe ring sample.

TABLE V

Effect Of Field Annealing On Magnetic Properties Of 0.30 mm Thick Cube Textured Sample (Composition-25% Co, 0.5% Si, 0.25% Cr)

	DC				400 Hz		
Sample	H_c (Oe)	B_r (kG)	B_{10} (kG)	B_{100} (kG)	P_{c15} (W/lb)	P_{c20} (W/lb)	P_{z15} (VA/lb)
Epstein	0.52	6.0	19.7	22.8	23.2	39.9	29.6
Ring	0.52	5.3	17.1	21.5	30.1	52.1	46.8
Ring-F.A.	0.23	16.0	18.2	21.9	22.9	46.9	24.6

F.A. - Field Annealed

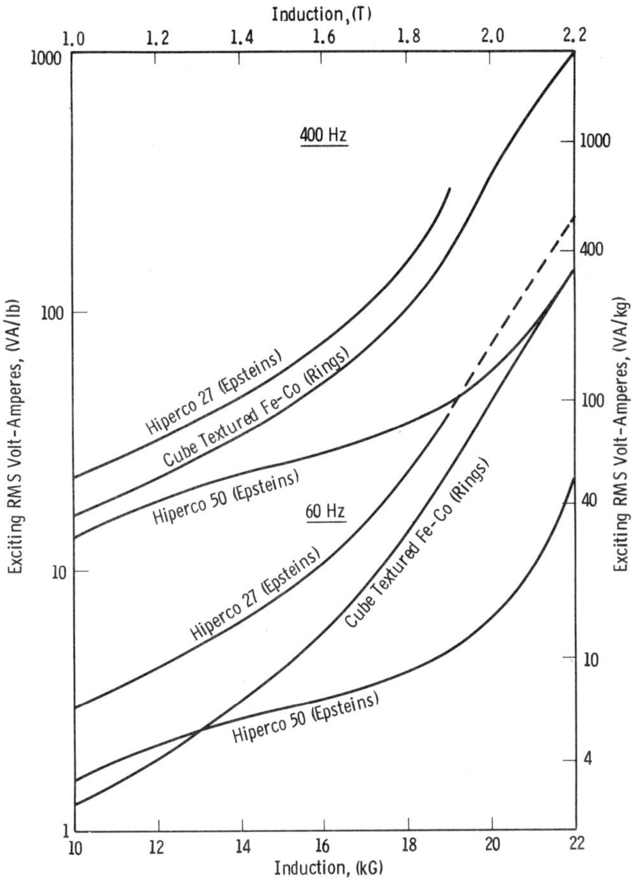

Fig. 12 Exciting VA/lb curves of a cube textured 25% Co-Fe ring sample.

such as wound transformer cores. Rather, the cube textured alloys would be used in E-I or D-U lamination configurations, or, more importantly, in laminations for rotating apparatus such as motors or generators. In these latter applications the cube textured iron-cobalt alloys would be somewhat inferior magnetically to the Hiperco 50 alloy, but could result in a considerable improvement over the Hiperco 27 alloy.

REFERENCES

1. C. W. Chen, Cobalt, No. 22, 1964, p. 1.
2. B. D. Cullity, "Introduction to Magnetic Materials," Addison-Wesley, Reading, Mass., 1972, p. 147.
3. R. C. Hall, J. Appl. Phys., Vol. 31, 1960, p. 157S.
4. T. L. Johnston, R. G. Davies and N. S. Stoloff, Phil. Mag., Vol. 12, 1965, p. 305.
5. J. K. Stanley, "Electrical and Magnetic Properties of Metals," ASM, Metals Park, Ohio, 1963, p. 300.
6. W. S. Byrnes and R. G. Crawford, J. Appl. Phys., Vol. 29, 1958, p. 493.
7. W. M. Swift, IEEE Trans. Mag., Vol. M9, 1973, p. 46.
8. C. G. Schulz, J. Appl. Phys., Vol. 20, 1949, p. 1030.
9. J. K. Stanley and T. D. Yensen, Trans. AIEE, Vol. 66, 1947, p. 714.
10. D. S. Shull, Jr., J. Appl. Phys., Vol. 32, 1961, p. 356S.

---

*Hiperco is a registered trademark of Carpenter Technology Corp.

RECENT DEVELOPMENTS IN MAGNETIC PROPERTIES OF GRAIN-ORIENTED SILICON STEEL WITH HIGH PERMEABILITY

A. Sakakura, T. Wada, F. Matsumoto, K. Ueno, K. Takashima, and M. Kawashima
Tech. Research Inst., Nippon Steel Corp., Kitakyushu, Japan

ABSTRACT

New developments of grain-oriented silicon steel with high permeability using AlN as a grain growth inhibitor are reported.

The most important factor in the improvement of core loss is crystal orientation, which is reflected directly in high permeability. The secondary grain size, purity, silicon content, the thickness and the tensile effect of surface coating are also important factors.

Recent developments in processing have made it possible to get higher permeability material with smaller grain size. Furthermore, success in new melting technique has decreased oxide inclusions which are remarkably detrimental to magnetic properties.

As a result of these developments, the following typical superior total loss values are obtained:

11 mil 0.57 W/lb (at 17kG, 60Hz)
12 mil 0.59 "
14 mil 0.68 "

INTRODUCTION

Demands for better grain-oriented silicon steel, with lower total loss and lower magnetostriction, have become quite strong from the standpoint of decrease in electrical energy loss and public nuisance (transformer noise).

Total iron loss, as is well known, consists mainly of two factors; hysteresis loss and eddy current loss. Hysteresis loss is related to crystal orientation, purity, internal stress, and surface condition. Eddy current loss depends on electrical resistivity of the steel, grain size, sheet thickness, and tensile effect of the coating.

Trying to obtain lower total loss, thinner gages of grain-oriented silicon steel, such as 11 mil or 9 mil, have been produced. However, such thinner gage has not improved transformer losses as expected, but has brought extra handling problems.

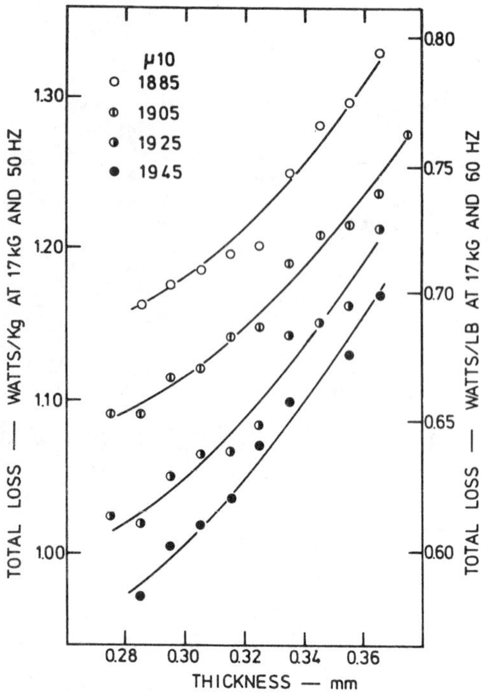

Fig. 1 Effect of thickness on total loss at 17 kG

The present study deals with the analysis of data on grain-oriented silicon steel collected over the last 20 years, and on recent development. About 50,000 Epstein samples were used in this study.

EXPERIMENTAL PROCEDURE

Epstein samples were made by the author's method[1] both in laboratory and mill. Magnetic properties measured were permeability μ_{10}, total iron loss W17/60 and magnetostriction λ_{19}. Magnetic tests were done by the standard Epstein method.

RESULTS

Fig. 1 shows the relationship between total loss and sheet thickness. In conventional grain-oriented silicon steel the total loss improvement by reducing thickness below 11 mil is very small, whereas in the material with high permeability, such as 1925 or 1945 at μ_{10}, total loss reduces linearly with thickness.

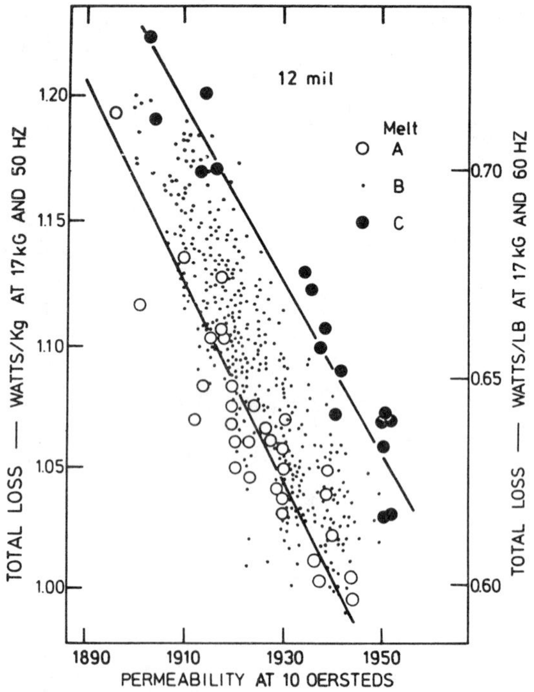

Fig. 2 Effect of melting method on total loss at 17 kG

Fig. 2 shows the effect of purity of the steel on the total loss. MnS or AlN which induces secondary recrystallization is removed during the final box anneal. However, inclusion such as SiO_2 or Al_2O_3 formed during melting practice can not be removed and stay in the final product as they were in the ingot. In Fig. 2, melts A, B and C represent the temperature of molten steel when alloying materials are added. Each temperature of A, B and C is 1660°C, 1630°C and 1600°C, respectively. In melting practice A an effort was made to decrease fine oxide inclusions.

Fig. 3 shows the typical chemical analysis of oxide inclusions for melts A and C. Even though the significant defference in the amount of oxide inclusions between A and C was not obtained, in the observation by electron micro scope, the dramatic difference on the dispersion of fine oxides was distinguished. Melt C, which has more fine inclusions, exhibited higher total loss than melt A even though μ_{10} is nearly equal. This is caused by the increase of hysteresis loss due to fine inclusions.

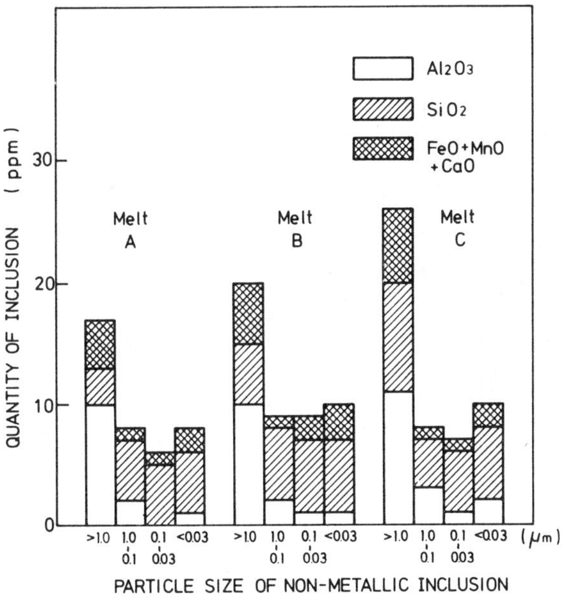

Fig. 3 Effect of melting method on quantity and particle size of non-metallic inclusions

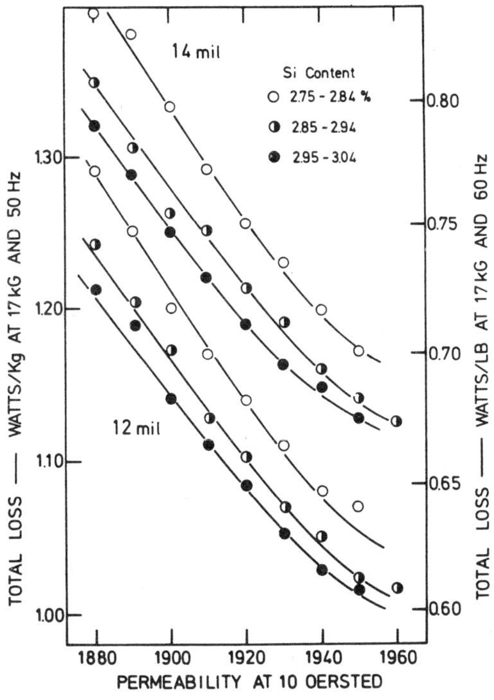

Fig. 4 Effect of [Si] content on total loss at 17 kG

Fig. 4 shows the effect of silicon content on the total loss. The effect of silicon content is largest on 14 mil, but on thinner strip the effect decreases.

Fig. 5 shows the effect of grain size on the total loss. As a general phenomenon, when permeability becomes higher the grain size becomes larger. Earlier conclusions that finer grain size is desirable to reduce loss[2] are also true for high permeability grain-oriented silicon steel. The key to the ideal grain-oriented silicon steel is to obtain high permeability steel with finer grain size. A recently developed technique in cold rolling[3] has enabled us to get a clue in this direction in high permeability steel. Tensile stress by surface coating[4] is also effective for the newly developed material.

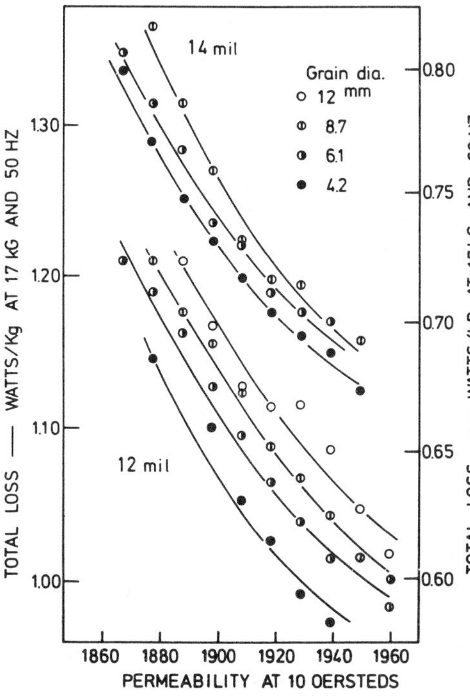

Fig. 5 Effect of permeability on total loss at 17 kG

CONCLUSION

The factors affecting the total loss and some new developments in high permeability grain-oriented were discussed, and are summarized as follows:

1. The effect of each factor on the total loss in high-permeability grain-oriented silicon steel has been evaluated and are:

Orientation	1 grade/250 Gauss in B at 10H
Si-content	1 grade/0.2% Si
Thickness	1 grade/30μm
Grain size	1 grade/3 No. in ASTM

where 1 grade = 0.06 W/kg at 17 kG, 50Hz
or 1 grade = 0.036 W/lb at 17 kG, 60Hz

2. Due to new developments in the manufacturing process, the magnetic properties of high-permeability grain-oriented silicon steel has been remarkably improved, and typical total loss values are as follows:

11 mil	0.57 W/lb (at 17kG, 60Hz)
12 mil	0.59 "
14 mil	0.68 "

ACKNOWLEDGMENT

The authors are grateful to Dr. S. Taguchi and Dr. T. Yamamoto for their valuable suggestions regarding preparation of the present study.

REFERENCES

1. S. Taguchi, A. Sakakura, and H. Takashima, U.S. Patent 3287183 (1966)
2. M.F. Littman, J. Appl. Phys. 38, 1104 (1967)
3. F. Matsumoto, K. Kuroki, K. Takashima, T. Takada, K. Honda, and H. Yasumoto, Belgian Patent. 816476 (1974)
4. T. Yamamoto, S. Taguchi, A. Sakakura, and T. Nozawa, IEEE Trans. Mag. 3, 677 (1972)

EFFECTS OF AlN ON THE SECONDARY RECRYSTALLIZATION BEHAVIOR OF 3% Si-Fe (110)〔001〕 SINGLE CRYSTALS

Fumio Matsumoto, Katsuro Kuroki, and Akira Sakakura
Tech. Research Inst., Nippon Steel Corp., Kitakyushu, Japan

ABSTRACT

Cold rolled (110)〔001〕 single crystals of 3% Si-Fe containing AlN controlled adequately in size, distribution and quantity develop the original (110)〔001〕 or cube-on-edge orientation by secondary recrystallization after an anneal at 550°C for 40h followed by an anneal at 1200°C. During the course of secondary recrystallization, (110)〔001〕 primaries grow at the expense of {111}<110> and {120}<001> main primaries.

INTRODUCTION

Many studies have been made of the secondary recrystallization phenomena in cold rolled (110)〔001〕 single crystals of 3% Si-Fe.

The secondary recrystallization texture of relatively pure (110)〔001〕 single crystals cold rolled and annealed consists of essentially three orientations, namely, {120}<001> , {121}<012> , and {111}<110> (called A, B and C by Dunn)[1-4].

In studies on the effects of precipitates such as AlN and MnS, it was reported that cold rolled (110)〔001〕 single crystals containing AlN in solution develop a doublet {111}<110> primary recrystallization texture after an anneal at 550°C for 40h followed by an anneal at 1200°C[5], and that the presence of fine MnS in cold rolled (110)〔001〕 single crystals inhibits primary grain growth and promotes secondary recrystallization but the texture is the same as that in pure crystals without the MnS[6]. To find the influence of AlN on secondary recrystallization behavior, studies on (100)〔001〕[7,8] and (100)〔011〕[8] single crystals of 3% Si-Fe have been done.

The present study was initiated to determine the effects of AlN precipitates on secondary recrystallization behavior of cold rolled (110)〔001〕 single crystals of 3% Si-Fe.

EXPERIMENTAL PROCEDURE

The (110)〔001〕 single crystals used were obtained from 0.95mm thick sheets prepared by secondary recrystallization method[9] from commercial grade hot rolled 3% silicon steel sheets containing a small amount of Al.

The crystals about 100mm diam with a (110) plane within 2° of the plane of the sheet were selected. These crystals were subjected to pre-treatment (a solution and precipitation treatment) to control AlN precipitates in quantity, size and distribution.

After cold rolling 80% to 0.19mm, these crystals were given a low temperature anneal to control primary recrystallization and AlN precipitation, and then given a high temperature anneal for primary grain growth control and secondary recrystallization.

Compositions, conditions of pretreatment and annealing, and quantity of AlN precipitated in each process are shown in Table I.

Fig. 1. Macrostructure of high temperature annealed crystals. (a) Q, (b) G, and (c) N, respectively.

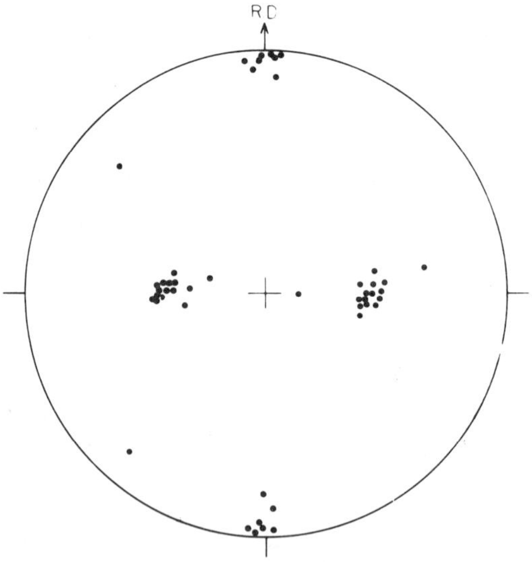

Fig. 2. The orientations of 16 secondaries in crystal G. Stereogram of (100) poles.

RESULTS AND DISCUSSION

Macrostructures of crystals Q, G and N after high temperature anneal are shown in Fig. 1. Secondary growth took place in crystals G and N but only normal growth of primaries in crystal Q.

The primary texture of crystal Q consists of a doublet {111}<110> orientation the same as reported by Sakakura[5], and the secondary texture of crystal N comprises B and C orientations as named by Dunn[2,3].

Almost all of the secondaries of crystal G consist of well oriented (110)〔001〕 grains. Fig. 2 gives the orientations of 16 secondaries in crystal G.

Such differences in recrystallization behavior of the crystals can be attributed to AlN behavior. As shown in Fig. 3 and Table I, after pretreatment crystal G contains cubic shaped AlN 100-150Å in size which precipitated in a cluster form with 2-3 precipitates

Table I. Conditions of pre-treatment and annealing.

Compositions of initial crystals. (%)	Crystals	Pre – treatment			Low temp. anneal		High temp. anneal
		Solution	Precipitation	N as AlN (%)	Anneal	N as AlN (%)	Anneal ✱
C 0.002 Si 2.95	Q	1300°C, 20min. (1) Water quench	——	0.0008	550°C, 40hrs ←(2)→	0.0089	50°C/hr, 900°C, 10hrs, 50°C/hr, 1200°C, 5hrs ←(2)→
Mn 0.11 S 0.005	G	〃	600°C, 30min Air cool (1)	0.0070		0.0085	
※Al 0.041 N 0.0189	N	——		0.0167		0.0170	

※ acid sol. Al (1) N_2 50% + H_2 50% (2) N_2 25% + H_2 75%
✱ Separator : MgO

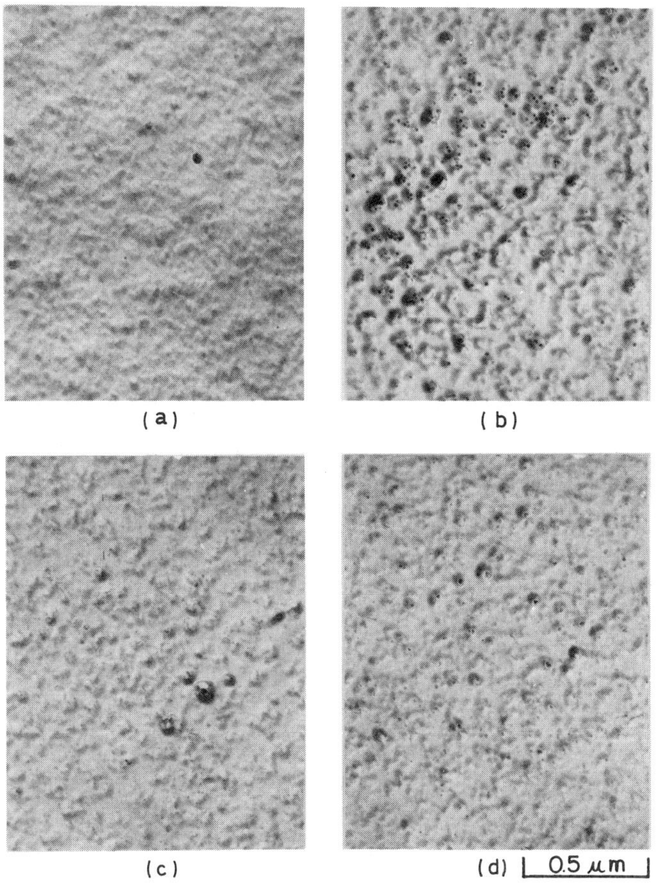

Fig. 3. AlN precipitates in crystals. (a) crystal Q and (b) crystal G pre-treated, (c) crystal Q and (d) crystal G cold rolled and annealed at 550°C for 40h. Carbon extraction replica.

per cluster, and contains 0.0070% N as AlN. These fine AlN were identified by electron diffraction patterns after coarsening them by an additional anneal. Crystal Q contains no AlN. Crystal N contains coarse needle AlN 1-2μm in length and 0.0167% N as AlN and almost all of the total N is as AlN.

After cold rolling and low temperature annealing at 550°C for 40h, some newly precipitated AlN 100Å or less in size, and pre-precipitated AlN are observed in crystal G containing 0.0085% N as AlN. Crystal Q contains AlN 100Å or less in size, in great numbers, and 0.0089% N as AlN. No change was observed in crystal N.

Furthermore, the effects of AlN on the recovery and recrystallization texture are very important.

The cold rolled textures of the crystals Q, G, and N consist mainly of the well known doublet {111}<112> orientation.

After the 550°C anneal, the main textures of crystals Q and N are sharp and of {111}<110> and {120}<001> type, respectively. Crystal G consists of {111}<110> and {120}<001> primaries as main components.

Crystal Q consists mainly of pancake-like coarse primaries (grain size 40 x 20μm; at longitudinal cross section), crystal G fine primaries (20 x 10μm), and crystal N very fine primaries (10 x 10μm).

It seems to be that AlN is apt to precipitate in the defect parts in the 550°C anneal and then influences the movement of dislocations to form and grow subgrains. However, it is not clear why the stronger the AlN inhibition the more {111}<110> subgrains develop.

During the subsequent high temperature anneal, the change in the textures of each crystal formed during low temperature anneal does not occur up to 900°C, as shown in Fig. 4. At 900°C, secondary growth took place in crystal N.

Holding for 10h at 900°C, in crystal Q main

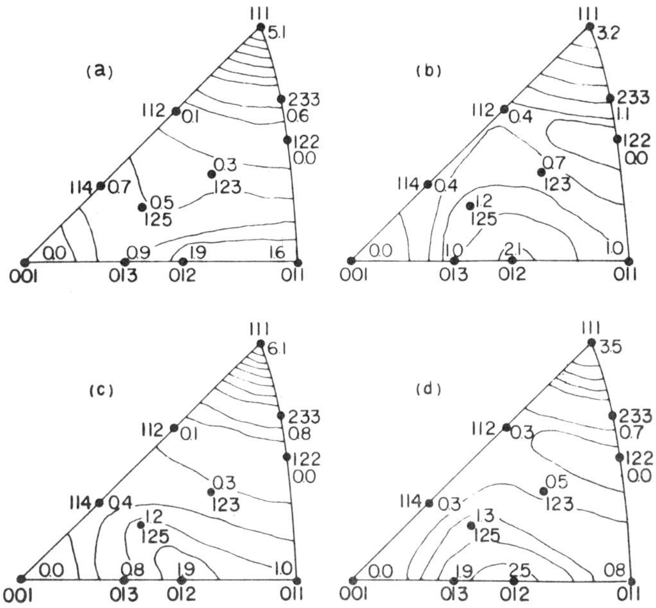

Fig. 4. Primary recrystallization textures during high temperature anneal. (a) crystal Q and (b) crystal G heated to 900°C, (c) crystal Q and (d) crystal G held for 10h at 900°C.

{111}<110> primaries further develop but (110)[001] primaries are largely consumed. On the other hand, in crystal G, a little change in the intensity of components {111}<110>, {120}<001>, and (110)[001] is observed. When the temperature is increased, the minor (110)[001] primaries grow to secondaries at the expense of other primaries in crystal G.

The result on crystal N shows that the minor {121}<012> and {111}<110> primary grains are more favored to grow for secondary recrystallization than the minor (110)[001] primary grain which exists in the matrix.

The difference in secondary recrystallization behavior between crystals G and N may be due to that of the primary recrystallization texture and primary grain growth by the presence of fine AlN.

In the case of crystal Q containing very fine AlN in great numbers, no secondary growth occurs because of the presence of coarse grained {111}<110> matrix primaries. Then the doublet {111}<110> primaries grow to about 0.5-1mm diameter at the expense of (110)[001] and {120}<001> primaries.

It is important to note that (110)[001] single crystals containing AlN controlled adequately in size (100-150Å), distribution, and quantity reproduce the original (110)[001] orientation by secondary recrystallization. This is because the (110)[001] primaries, which are the minor component of the primary recrystallization texture, grow into secondaries at the expense of a sharp textured, fine grained matrix which is stabilized by the presence of fine AlN as an inhibitor.

ACKNOWLEDGMENT

The authors are grateful to Dr. S. Taguchi and Mr. K. Takashima for their valuable suggestions regarding preparation of the present study.

REFERENCES

1. M.F. Littmann. U.S. Patent 2473156 (1949)
2. C.G. Dunn, Acta Met. 1, 163 (1953)
3. C.G. Dunn, Acta Met. 2, 173 (1954)
4. M.F. Littmann, J. Appl. Phys. 36, 1225 (1965)
5. A. Sakakura, J. Appl. Phys. 40, 1534 (1969)
6. M.F. Littmann, Trans. TMS-AIME 245, 2217 (1969)
7. S. Taguchi and A. Sakakura, Acta Met. 14, 405 (1966)
8. A. Sakakura and S. Taguchi, Met. Trans. 2, 205 (1971)
9. S. Taguchi, A. Sakakura, and H. Takashima, U.S. Patent 3287183 (1966)

THE MAGNETIC PROPERTIES OF JAPANESE HI-B (110) [001] GRAIN-ORIENTED SILICON-IRON

S.M. Pegler
Wolfson Centre, University of Wales, 30 The Parade, Cardiff, United Kingdom

ABSTRACT

A sophisticated apparatus has been built with which loss and magnetostriction can be measured in silicon-iron over the frequency range 20 to 400 Hz, at inductions up to 1.9T; and under either compressive or tensile stresses up to 7 x 10^6 N/m^2. Such measurements have been made on both Japanese HI-B grain-oriented silicon-iron and on other commercial-grade materials. A significant improvement in the magnetostriction stress-sensitivity has been observed in the newer material, and a reduction in the losses has been found (significant at 1.5T, but particularly marked at 1.7T and higher inductions). With the raising of the levels of magnetization in transformer cores, it is predicted that the new material should be considered seriously where the amortization of losses is important - as in large distribution transformers. A reduction in noise output would be expected from transformers built using this material.

INTRODUCTION

Over the past quarter-century or so, many efforts have been directed towards finding a substitute for the 3.25% (110) [001] oriented silicon-iron used in power transformer cores. The possibility of using (100) [001] oriented sheet has been discussed, as has the possibility of increasing the silicon content of the (110) [001] oriented material, but difficulties in economic production of such materials with core losses similar to or better than the conventional material have so far precluded them as serious competitors. The conclusion has been that efforts should be directed towards even further improvement of 3.25% (110) [001] oriented silicon-iron.

The recent appearance on the market of Japanese Orient-Core HI-B (110) [001] silicon-iron prompted comparison of this steel with existing grades produced in the Western world. The material is claimed to be particularly suitable for use at high inductions of the order of 1.8T, an interesting possibility in view of the high inductions currently being employed in large transformers.

Two important criteria in the performance of a finished transformer core are the overall loss and the noise output, the latter being closely associated with the magnetostriction of the steel used. Both loss and magnetostriction of oriented silicon-iron are severely affected by factors such as applied stresses[1,2], being most sensitive to compressive stress in the rolling-direction in Goss-oriented silicon-iron. Such stresses can be generated by techniques of construction of the transformer core and by use of non-flat laminations.

The apparatus used in this investigation was designed to measure the stress-sensitivity of both loss and magnetostriction in single Epstein-size (3 cm x 30.5 cm) samples. For high grade silicon-iron samples of 330μm thickness the available stress range is 0-7 x 10^6 N/m^2 compressive and tensile; the third harmonic distortion in dB/dt at 1.8T is 6% and a frequency range of 20-400 Hz can be covered. The stressing and magnetizing apparatus was developed from that designed by G.H. Simmons and J.E. Thompson[3]; a full description is to be published.

MEASUREMENTS AND DISCUSSION OF RESULTS

For the tests, two groups of samples were selected for comparison with a group of Orient-Core HI-B samples. One group, from a British manufacturer, had a nominal 50 Hz loss at 1.5T of 1.0 W/Kg and a thickness of 300μm, corresponding to type 30M6 or 46 grade. The second group, from another manufacturer,

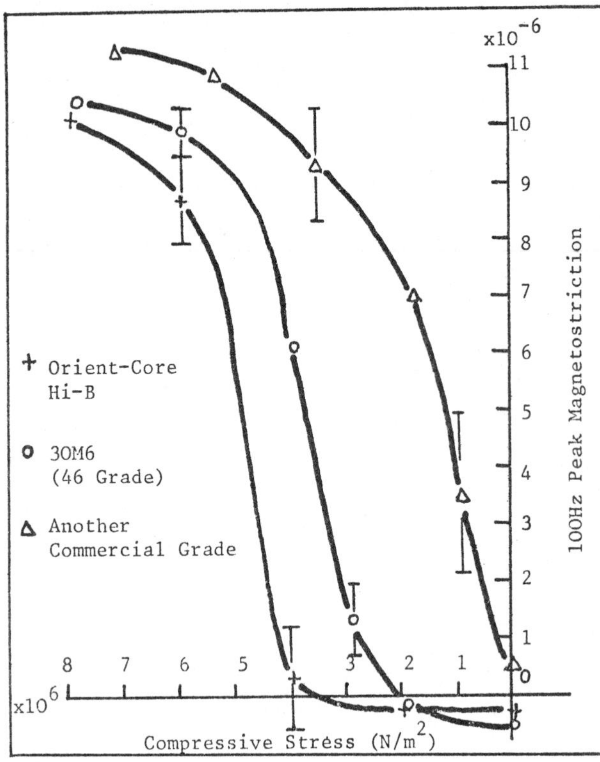

Fig. 1 - Magnetostriction curves at 1.5T, 50Hz.

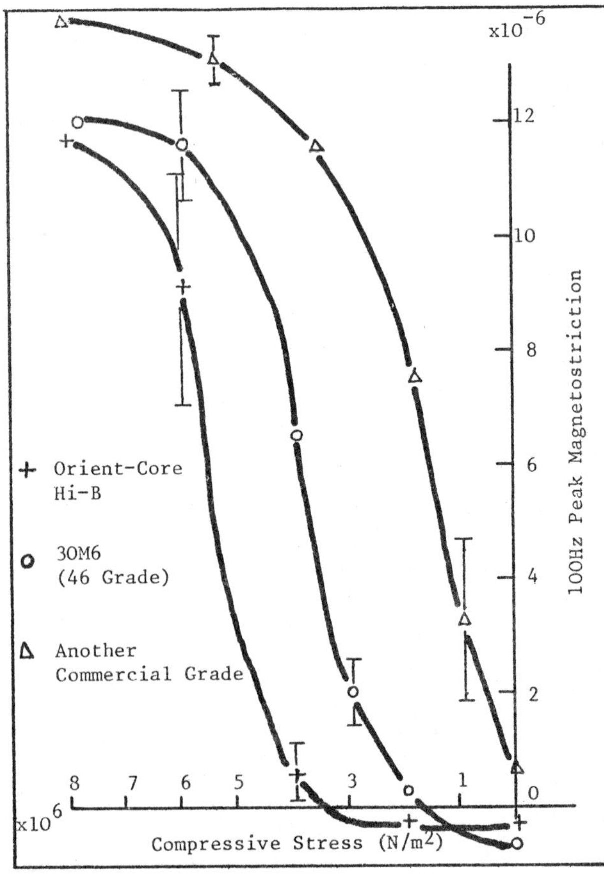

Fig. 2 - Magnetostriction curves at 1.8T, 50Hz.

had a nominal loss of 1.10 W/Kg and a thickness of 337μm, corresponding to 51 grade. The Orient-Core HI-B samples had a nominal loss of 0.85 W/Kg and a thickness of 300μm. All samples were given a stress-relief anneal at 800°C prior to being measured.

Magnetostriction and loss measurements were performed on the samples at two inductions, 1.5T and 1.8T, corresponding to the older and newer levels used in transformer cores, and the results (Figs. 1-4) are shown for the full range of applied compressive stress. In all cases the diagrams show the mean values obtained for each group of samples, together with the standard deviation from the mean (σ) near the ends of the curves. In the cases where the σ shown is not near the beginning of the curve, it is too small to be shown easily; to avoid a confused diagram no σ's are shown on the zero-stress axis.

The magnetostriction curves at the two flux densities are remarkably similar, the major difference being under high compression where a significant increase occurs at 1.8T. The curves for the 30M6 material may be taken as typical of a good modern material, the change from negative to positive magnetostriction occuring well into the compressive-stress region, at about 2×10^6 N/m^2. The tensile stress induced by the coating must be overcome by the applied compressive stress before rises in loss and magnetostriction can occur, and in this case the induced stress is evidently about 2×10^6 N/m^2. The curves for the HI-B material are even better than this, the corresponding stress value here being almost 4×10^6 N/m^2; this suggests good control of the magnetostriction by the coating. The 51 grade, 337μm material, by contrast, has positive initial magnetostriction, suggesting little control by the coating; such characteristics represent probably the worst material, magnetostrictively speaking, produced by Western manufacturers, and experience shows that nowadays only a relatively small proportion of high-grade silicon-iron is in this form.

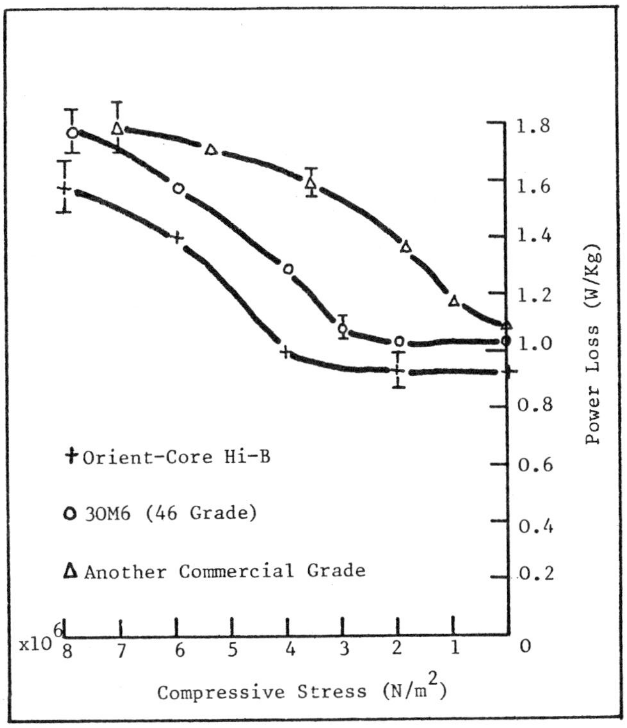

Fig. 3 - Loss curves at 1.5T, 50Hz.

The loss characteristics for the three materials bear similar relationships to each other as do the magnetostriction characteristics. The 30M6 and the HI-B results can be compared directly, since both materials are of the same thickness. The 51 grade material, however, is more than 10% thicker than the other two, and since losses increase with thickness even when the basic quality of the material remains unchanged, its curves should be read as rather lower for comparison purposes. This will make the extreme ends of the curves very similar to those of the 30M6 steel. Between the extremes, it is evident that the losses of the 30M6 steel rise less quickly than those of the 51 grade steel and, as with the magnetostriction, this can be attributed to a higher level of coating-induced tensile stress in the 30M6, inhibiting the formation of stress-induced domains magnetized at 90° to the rolling direction until a fairly high level of compressive stress has been applied.

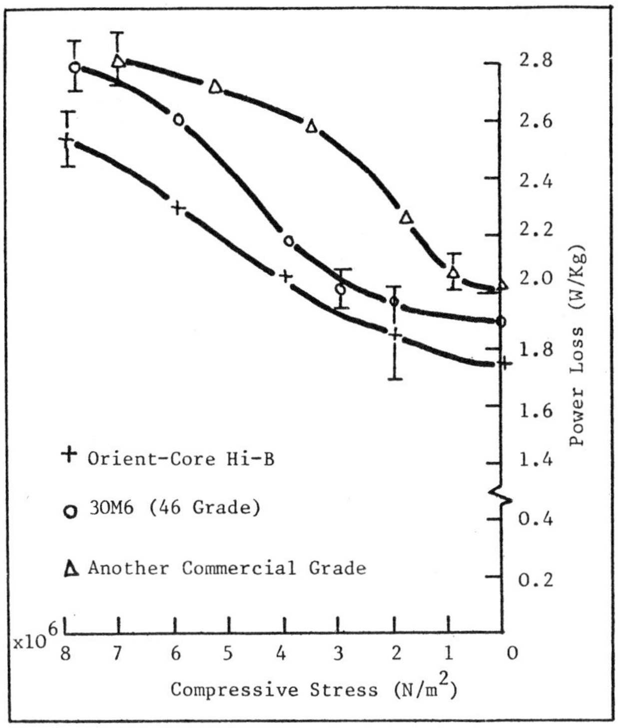

Fig. 4 - Loss curves at 1.8T, 50Hz.

At 1.5T, the same effect occurs in the HI-B steel, but to a greater extent, since the loss curve remains flat longer than that of the 30M6 steel. HI-B material has been shown to possess a better degree of orientation than the standard silicon-iron grades[4], and this is consistent with the lower levels of loss in these tests. The HI-B steel shows losses about 10% lower than the 30M6 steel throughout the loss curves at both inductions. It is interesting that the spread of results is considerably greater at 1.8T than at 1.5T, suggesting perhaps a more complex magnetization process than at 1.5T, owing to the tendency for alignment of badly-oriented domains to occur in high applied fields.

CONCLUSIONS

The magnetostriction and loss characteristics of Japanese Orient-Core HI-B silicon-iron are seen to be significantly better than those of a Western steel of grade 30M6 at an induction of 1.5T. This difference is still maintained at 1.8T, and transformers built using this material should be capable of increased efficiency. The noise output of such transformers should also show an improvement provided that the constructional techniques give rise to only intermediate stresses up to about 4×10^6 N/m^2. If high stresses are generated by such techniques, the magnetostriction would not show any improvement, since all the magnetostriction curves tend towards a saturation value at high stresses. The better orientation of the HI-B material, however, maintains a loss differential of about 10%, regardless of stress, in these single-strip measurements, and while scaling factors and effects of joints cannot be ignored, it is very likely that a complete transformer core would also show an improved performance.

It is interesting that, for any particular material, the only major difference between the magnetostriction curves at 1.5T and 1.8T occurs at high stresses. This suggests that different noise outputs observed between similar transformers operating at these two inductions are caused by the presence of high stresses in the cores.

REFERENCES

1. Mrs. D. Brown, C. Holt and J.E. Thompson, Proc. IEE, Vol. 112, pp183-188, 1965.
2. P.J. Banks and E. Rawlinson, Proc. IEE, Vol.114, pp1537-1546, 1967.
3. G.H. Simmons and J.E. Thompson, Proc. IEE, Vol.115, pp1835-1839, 1968.
4. T. Yamamoto, S. Taguchi, A. Sakakura and T. Nozawa, IEEE Trans. Mag., Vol. MAG-8, No. 3, pp677-681, 1972.

CORE LOSS OF GRAIN-ORIENTED 3% SILICON-IRON AT HIGH INDUCTIONS

M. F. Littmann
Research and Technology, Armco Steel Corp., Middletown, Ohio 45043

ABSTRACT

Development of commercial 2.9% silicon-iron with superior (110) [001] texture has permitted attainment of lower core loss than for conventional grain-oriented materials at 17 kG induction.

Core losses of regular 3.15% silicon-iron and high-permeability 2.9% silicon-iron specimens are compared at inductions ranging from 15 to 19 kG. Effects of texture and grain size increase when the induction exceeds 17 kG. Thickness tends to have a constant effect in the range of 9 to 14 mils.

Effects of tension-inducing coatings are described. Removal of coatings produces a large decrease in loss at high inductions but increases loss at low inductions.

INTRODUCTION

Development of techniques to employ AlN in addition to MnS as grain growth inhibitors has permitted attainment of a superior degree of (110)[001] texture than usually found in conventional grain-oriented silicon-iron (1). Originally named HI-B by Nippon Steel Corp., the same type of material is also produced by Armco Steel Corp. with the name "TRAN-COR H." The very high degree of orientation of HI-B and TRAN-COR H assure very low total core loss at B=17kG (1.7T) in spite of the unfavorably large grain size.

The purpose of this paper is to relate core loss values of conventional grain-oriented (G.O.) and TRAN-COR H at inductions from 15 to 19 kG as they are affected by thickness, grain size, texture, and coatings. Epstein specimens were prepared from commercial heats. The atmosphere during part of the final anneal was varied to broaden the range of textures and grain sizes available for study. All the samples with glass coating (formed by reaction with MgO during the final anneal) were reannealed at 1200°C in dry hydrogen to minimize effects of residual impurities such as nitrogen and sulfur. Glass coatings were removed with hot HCl and HF acids. Pickling removed about 0.25 mil (0.006mm) thickness. The pickled samples were stress-relief annealed at 790°C. Tests for core loss and permeability were made at 60Hz according to ASTM method A 343.

EFFECT OF GRAIN SIZE

Specimens 10.4 thick by weight (0.264mm) of relatively good G.O. and relatively poor TRAN-COR H were selected to evaluate the effect of grain size with glass coating and with permeability at H=10 Oe (800A/m) within a common range from 1850 to 1870. All samples had melt-silicon contents in the range of 2.89 to 3.08%. The average permeability of almost 1860 represents nearly the same degree of crystal texture.

Fig. 1 shows that the deleterious effect of grain size on total core loss increases with the test induction for glass coated samples. The slope at 17kG is in good agreement with Taguchi and Sakakura's data(2) for 14 mil (0.335mm by weight) HI-B with 1900 permeability at 10H. There seems to be no discontinuity between regular G.O. and TRAN-COR H when the parameters of thickness, grain size, permeability, and silicon content are normalized. Methods are not yet known to produce highly oriented TRAN-COR H with grain diameters as small as 2.8mm. However, data for TRAN-COR H with 1920 permeability at 10H (not plotted) show less effect of grain size, especially at 19kG, in the range of about 17 to 5mm dia.

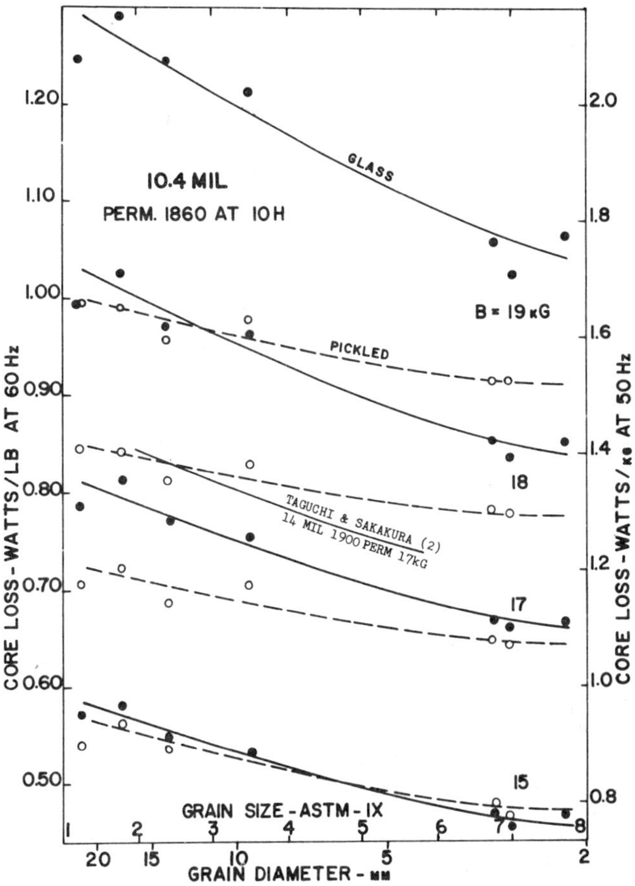

Fig. I Effect of grain size on core loss. Permeability is constant and average silicon content is 2.95%.

When the glass coatings were removed by pickling, core loss increased at 15kG but decreased markedly above 17kG. Also, the effect of grain size on core loss decreased at 19kG for the pickled samples. This suggests an interaction of grain size and surface effects with test induction.

EFFECT OF THICKNESS AND PERMEABILITY

The relations of total core loss to permeability for constant representative grain sizes of G.O. with 3.15% Si and TRAN-COR H with 2.9% Si are shown in Fig. 2 for 13.6 mil (0.346mm) specimens with glass coatings. The curves show the increasing effect of permeability, i.e. degree of texture, on core loss as the test induction is increased. There is little net advantage in core loss with TRAN-COR H compared to G.O. at 15kG. The maximum advantage of TRAN-COR H is near 18kG and then, surprisingly, the advantage decreases again at 19kG. The TRAN-COR H curves at 15 and 17kG are in good agreement with those of Taguchi and Sakakura (2) for HI-B with the same grain size. The slightly lower loss of the TRAN-COR H is attributed to the extra 1200°C anneal given the glass-coated specimens in this study.

Fig. 3 shows the same type of data for 10.4 mil specimens. Thickness decrease produces similar loss improvements for both material types. At 10.4 mils the core loss advantage of TRAN-COR H derived from the superior crystal texture is again greatest at 18kG. At both higher and lower inductions the large grain size of TRAN-COR H diminishes its advantage.

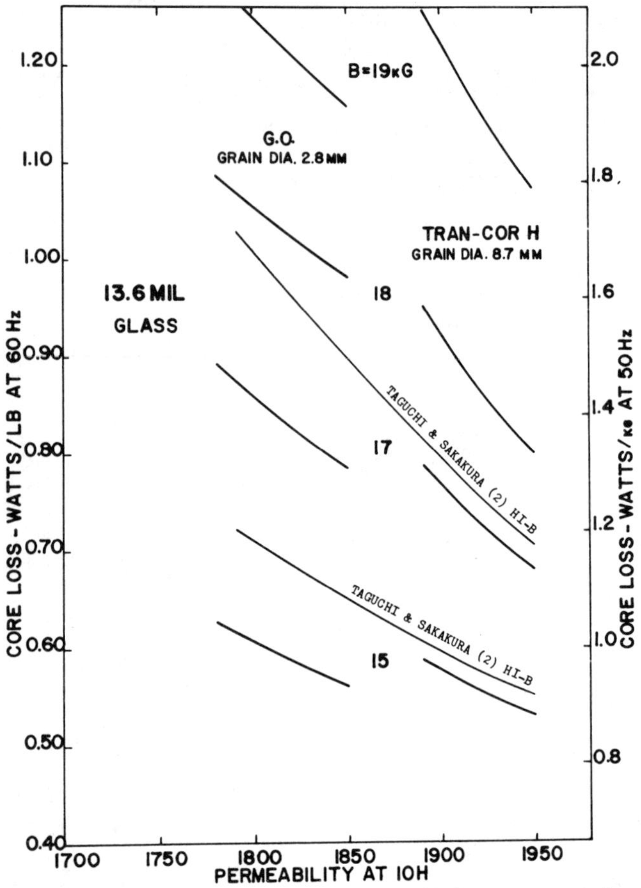

Fig. 2 Effect of permeability at H=10 Oe on core loss at B=15 to 19kG of 13.6 mil G.O. and TRAN-COR H with glass coating. Grain size and silicon content are constant for each type.

COATING EFFECTS

Removal of coatings involves several changes. Tension from the glass is eliminated; magnetically inert material is removed; and the surface roughness is changed. We concur with previous workers (2,3); removal of glass coatings increases core loss at 15kG, especially for samples with very high permeability. However, above 17kG remarkable improvements in core loss were obtained by acid removal of the glass to obtain a bright surface. From a theoretical view, very smooth surface permits easy domain wall movements and, therefore, low core loss - especially at high inductions.

Of course, insulative coatings are necessary for transformer use. The best types of commercial coatings known at this time are applied over the glass from the final high temperature anneal (4). Fig. 4 compares core loss for 10.4 mil specimens with glass, S2 over glass and no glass (pickled). The S2 coating improves core loss at all inductions, the benefit probably resulting from tension induced in the metal.

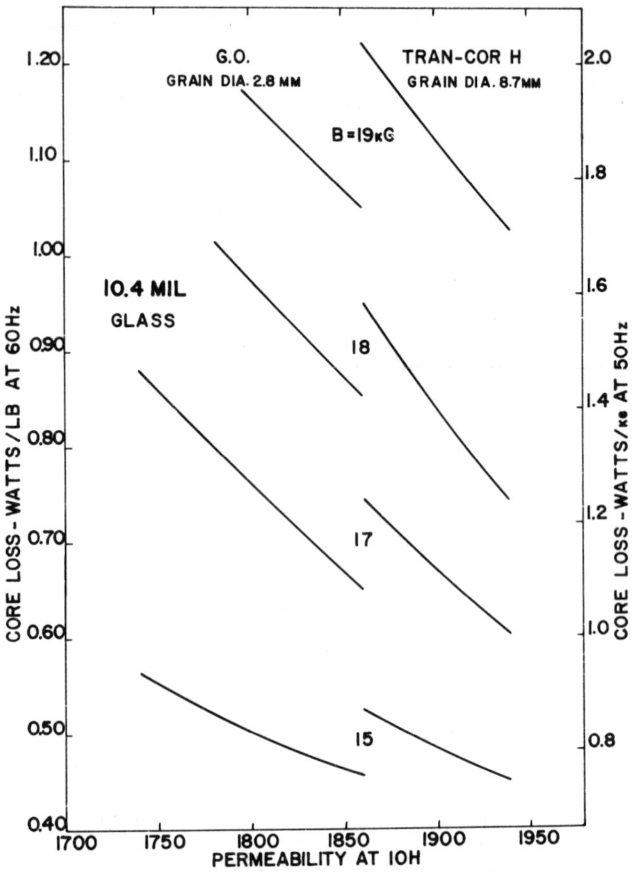

Fig. 3 Effect of permeability at H=10 Oe on core loss at B=15 to 19kG of 10.4 mil G.O. and TRAN-COR H with glass coating. Grain size and silicon content are constant for each type.

The S2 data represent stress relief annealed samples from commercial coils coated with S2 over the base glass coating resulting from the final coil anneal. Since the samples did not receive a second 1200°C anneal, the improvement in core loss from the S2 coating itself may be about 0.01 w/lb. greater than indicated in Fig. 4. Tests made under tension suggest that the S2 adds a tension of about 0.2-0.3 kg mm^2.

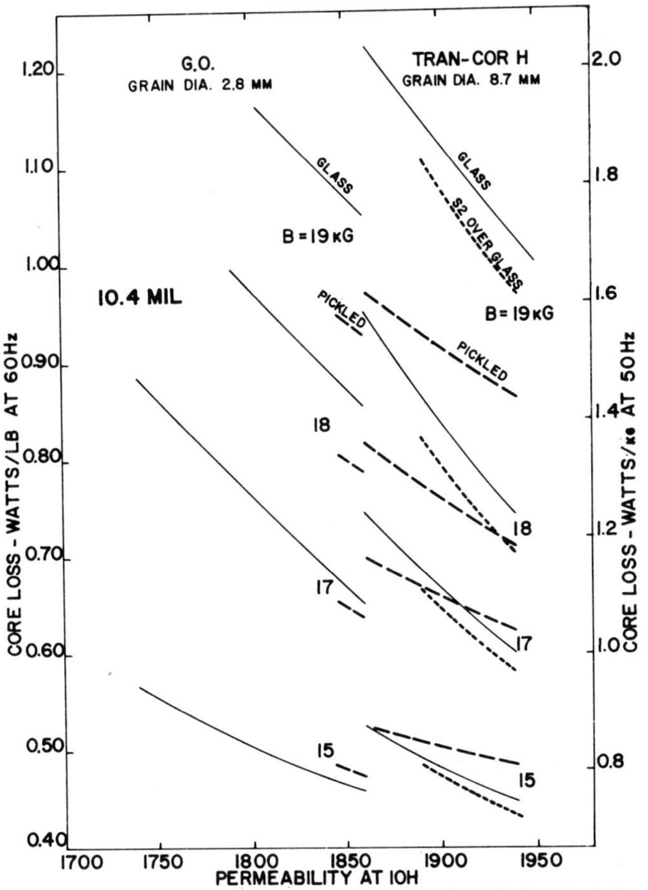

Fig. 4 Comparison of coating effects for 10.4 mil G.O. and TRAN-COR H on core loss at B=15 to 19kG.

CONCLUSIONS

This study shows that effects of grain size, permeability, tension and surface condition on core loss vary greatly with test induction, as the domain studies of Shilling (5) and others have suggested. The maximum core loss advantage of TRAN-COR H over G.O. appears to be near 18kG. The lowest core loss in commercial material will probably result from a combination of very high permeability with relatively small grain size and suitable coatings to preserve best surface conditions. The difficulty of accomplishing these goals increases with decreasing thickness.

REFERENCES

1) S. Taguchi, A. Sakakura and H. Takashima, U.S. Patent 3,287,183 (1966)
2) S. Taguchi and A. Sakakura, J. Appl. Phys. 40, 1539-1541 (1969)
3) T. Yamamoto and T. Nozawa, J. Appl. Phys. 41, 2981-2984 (1970)
4) S. Taguchi, T. Yamamoto and A. Sakakura, IEEE Trans. Magn. M10, 123-127 (1974)
5) J. W. Shilling IEEE Trans. Magn. M9, 351-356 (1973)

EFFECT OF ALUMINUM ON THE TEXTURE AND MAGNETIC PROPERTIES OF A COPPER BEARING NON-ORIENTED SILICON STEEL

P. K. Rastogi, Inland Steel Research Laboratories
3001 East Columbus Drive, East Chicago, Indiana 46312

ABSTRACT

It is generally known that aluminum additions in non-oriented silicon steels reduce the core loss, yet the detailed influence of aluminum on A. C. permeability and core loss at low and high inductions has not been documented. In the present work the influence of Al levels up to 0.65 Wt% on the texture and magnetic properties of a 1.50 Wt% silicon steel containing 0.17 Wt% Cu has been determined. The results indicate that the core loss at 10 and 15 kG decreases by about 20% with the addition of 0.43 Wt% Al and is essentially unaffected by further aluminum additions: the 10 kG permeability increases linearly with increasing aluminum content. The improvement in 10 and 15 kG core losses and 10 kG permeability is primarily associated with increasing grain size, decreasing eddy current loss and changes in texture. The 15 and 17 kG permeabilities are improved by 30 and 20%, respectively with the addition of 0.15 Wt% Al and above that level are essentially independent of composition. Since a slight reduction in saturation magnetization is observed with increasing aluminum content, the improvement in the 15 and 17 kG permeabilities is explained in terms of the increase in the {200} pole density and a decrease in the {222} pole density.

INTRODUCTION

Aluminum additions are generally made in non-oriented silicon steels to increase the electrical resistivity which reduces the classical eddy current loss and thereby improves the total core loss.[1] However, there is a lack of information in the literature regarding the influence of aluminum in dilute concentrations on the magnetic properties; namely, A. C. permeability and core loss at low and high induction levels in these steels. The purpose of this investigation is to examine the effect of aluminum levels up to 0.65 Wt% on the texture and magnetic properties of a 1.50 Wt% silicon steel containing 0.17 Wt% Cu.

MATERIAL PREPARATION AND EXPERIMENTAL PROCEDURE

Six 40 lb. vacuum melted alloys of 1.50 Wt% silicon steel containing 0.05 Wt% C, 0.38 Wt% Mn, 0.17 Wt% Cu, 0.006 Wt% S, and Al additions in the range of 0.015 to 0.65 Wt% were used in this investigation. The alloys were generally clean having fairly low inclusion content. These were hot rolled to .09" thick plates, cold rolled to 0.027" thick sheets and decarburized in the form of ring samples (3" O.D. x 2.5" I.D.) to a carbon level <0.006 Wt%. These ring samples were subsequently used for magnetic measurements. The A.C. permeability was measured at 10, 15 and 17 kG, while the core loss was measured only at 10 and 15 kG. The estimated 2σ limits for permeability and core loss measurements are ±7 and ±3%, respectively. In addition, measurements were made to determine the saturation magnetization and resistivity.

The grain size was characterized in terms of the average number of intercepts/mm which is inversely proportional to the average grain size.[2] Texture was monitored by the inverse pole figure technique in terms of the pole densities $(I/I_R)_{hkl}$ of eight planes at the mid-plane of decarburized samples.

RESULTS AND DISCUSSION

It is seen from Figure 1 that the 10 kG permeability increases linearly with increasing aluminum content. In contrast, the 15 and 17 kG permeabilities are improved by 30 and 20%, respectively with the addition of 0.15 Wt% Al and above that level are essentially independent of composition.

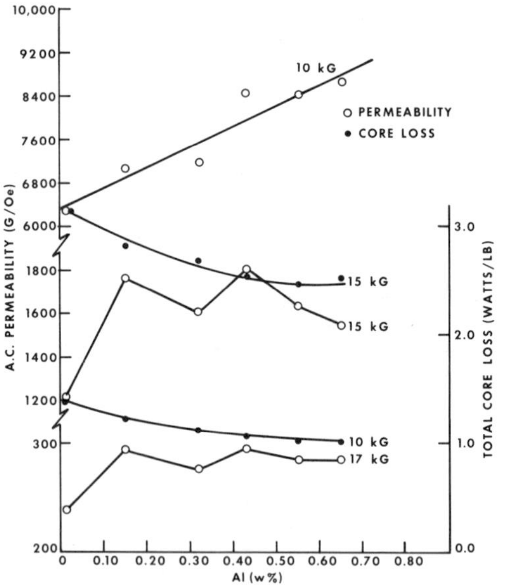

FIGURE 1

EFFECT OF Al CONTENT ON A.C. PERMEABILITY AND CORE LOSS AT DIFFERENT INDUCTIONS

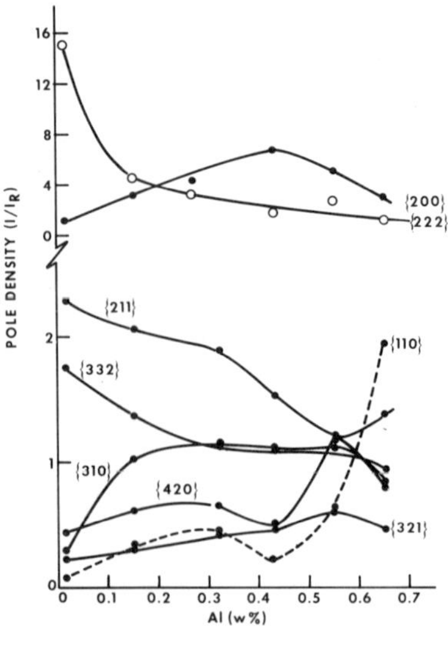

FIGURE 2

EFFECT OF Al CONTENT ON THE POLE DENSITIES OF VARIOUS ORIENTATIONS

TABLE I

Effect of Composition on Magnetic Properties and Grain Size of Copper Bearing 1.50 Wt% Silicon Steel

Al (Wt%)	Saturation Magnetization (kG)	Classical Eddy Current Loss at 10 kG (watts/lb)	Classical Eddy Current Loss at 15 kG (watts/lb)	Average Grain Boundary Intercepts/mm	Resistivity ($\mu\Omega$-cm)
0.015	20.71	0.47	1.08	18.6	34.00
0.15	20.65	0.45	1.02	14.0	36.00
0.32	20.57	0.43	0.99	12.4	37.35
0.43	20.46	0.42	0.96	11.2	38.34
0.55	20.40	0.40	0.91	10.8	40.38
0.65	20.36	0.39	0.88	9.5	41.29

The improvement in the 10 kG permeability is primarily attributed to the increase in grain size, decrease in eddy current loss and favorable changes in texture as shown in Table I and Fig. 2, which is consistent with the results of earlier investigations.[3-5]

The permeabilities of this steel at inductions of 15 and 17 kG, which lie above the knee of the magnetization curve, are largely dependent upon texture and, to some extent, on saturation magnetization. Since a slight reduction in saturation magnetization is observed with increasing aluminum content (Table I), the improvement in permeability is primarily associated with the changes in texture shown in Fig. 2.

The sharp drop in {222} together with an increase in {200} caused by the addition of 0.15 Wt% Al is believed to be the prime reason for the rapid increase in 15 and 17 kG permeabilities. The fact that little change in permeability occurs at higher levels of aluminum is somewhat surprising since the {200} continues to increase up to 0.43 Wt% Al while both the {332} and {211} decrease: a compilation would be expected to produce some improvement in permeability. The mechanism governing these textural changes as a function of aluminum addition is the subject of further investigation.

It is clear from Fig. 1 that the core loss at 10 and 15 kG decreases with the addition of aluminum up to 0.43 Wt% after which it remains essentially constant. In addition, it is important to note that at 0.43 Wt% Al level, the core loss at both inductions is improved by about 20%. It is seen from the data in Table I and Fig. 2 that the improvement in the core loss is primarily due to favorable changes in texture, a decrease in the classical eddy current loss and an increase in grain size, which is qualitatively consistent with the results of previous investigations.[5-6] In this study, the reduction in the classical eddy current loss[7] at a particular induction level is achieved through the increase in electrical resistivity (Table I).

CONCLUSION

The following conclusions are drawn from this investigation:

1. The 10 kG permeability increases linearly with increasing aluminum concentration in the range of 0.015 to 0.65 Wt%. The permeability at 15 and 17 kG is significantly improved with the addition of only 0.15 Wt% Al and above that remains essentially independent of aluminum content. Improvement in 15 and 17 kG permeabilities are associated, in part, with higher {200} and lower {222} pole densities.
2. At 10 and 15 kG, the total core loss decreases monotonically with the addition of aluminum up to 0.43 Wt%. Above this concentration, it is, to a first approximation, independent of composition. The improvement in this property is associated with increasing grain size, decreasing classical eddy current loss and changes in texture.

REFERENCES

1. L. J. Regitz, J. of App. Phys., 41, 1030 (1970).
2. R. T. Dehoff "Quantitative Microstructure Analysis", Fifty years of Progress in Metallographic Techniques, ASTM STP 430, 62 (1968).
3. P. L. Carpentier and J.H. Bucher, Mechanical Working and Steel Processing X, AIME, 61 (1972).
4. R. W. Easton, Written Discussion of Ref. (3), Mechanical Working and Steel Processing X, AIME, 89 (1972).
5. J. M. Shapiro, Written Discussion of Ref. (3), Mechanical Working and Steel Processing X, AIME, 91 (1972).
6. P. K. Rastogi and J. M. Shapiro, IEEE Trans. Magnetics, Vol. Mag-9, 122 (1973).
7. ASTM Special Technical Publication, No. 371, 31 (1964).

THE SHAPE OF INDIVIDUAL BARKHAUSEN PULSES

P. J. Coyne and J. J. Kramer
University of Delaware
Newark, Delaware 19711

ABSTRACT

Barkhausen pulses in single crystals of 3%Si-Fe, with well defined slab-like domains, were recorded on magnetic tape, digitally converted, and displayed by computer. Some of the pulses show a shape not previously reported, i.e., oscillations with a decreasing frequency and in general a decreasing amplitude each succeeding half cycle. The reproducibility of pulse shapes indicates that they are not pulse clusters.

The observation of a negative portion of the pulse, "a negative Barkhausen pulse", has been reported[1] but a detailed picture of pulse shape has not been shown. The negative portion of the pulse and the succeeding oscillations seem to indicate a single event and not the coupling of two or more events.

INTRODUCTION

While the Barkhausen effect has been known for fifty years, only recently has it been used to study material structure. Investigations of grain size,[2] strain,[3] crack nucleation and growth[4] and phase transformations[5] in ferromagnetic materials have been made. Correlations have involved measurements of the number and amplitude distribution of the pulses, their time constants or the autocorrelation function. Scant experimental data has been reported on the shape of the pulses.

Theoretical attempts to define pulse shape have involved simplifying assumptions to obtain mathematical solutions. Following Tebble, et al[6] or Polivanov, et al[7] one obtains

$$\nabla^2 \vec{B} = \mu\sigma \frac{\partial \vec{B}}{\partial t} \qquad (1)$$

as the defining equation for the evaluation of pulse shapes. Here B, σ, and μ are the magnetic induction, conductivity, and permeability respectively. For a thin sheet with slab-like domains and rigidity of the domain wall only in the magnetization direction, an instantaneous localized reversal in flux leads to a voltage, for $t>0$, in a multi-turn coil of the form

$$e(t) = \frac{-2NB_s A\pi^3}{\sigma\mu b^2} \sum_{n\ \mathrm{odd}} n \sin\frac{n\pi d}{b} \exp\frac{-\pi^2 n^2 t}{\sigma\mu b^2} \qquad (2)$$

Here N, B_s, A, b and d are the number of turns, the magnetic saturation, the effective cross-sectional area in the sample through which the flux changes, the sheet thickness, and the distance from the surface at which the flux change occurred, respectively. The expected pulse shape and its variation with d/b are shown in Fig. 1. The base time constant of the pulse should correlate with sheet thickness. More restricted conditions lead to pulses of the same general form. If the above analysis can be applied to a majority of the pulses it provides the potential for determining the spatial distribution of imperfections giving rise to the Barkhausen pulses.

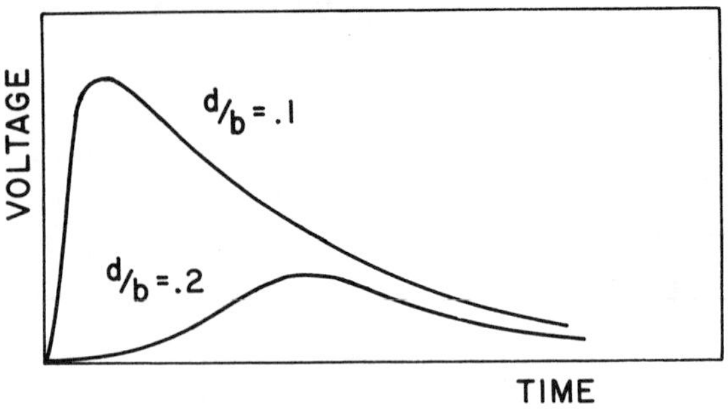

Fig. 1: Pulse shapes predicted by Eq. 2

The observation of negative pulses suggests that solutions of the form of Eq. 2 cannot describe all of the waveforms. Also defects other than nonmagnetic inclusions are expected to interact differently with domain walls.[8] In this investigation the shape of Barkhausen pulses found in single crystals of 3%Si-Fe are determined experimentally and compared with those predicted by Eq. 2.

EXPERIMENTAL

Rectangular single crystals (2.5cm X .5 cm X .01 or .02 cm) or window frames (2.5cm X 2.5 cm) (100) [001] orientation and a slab-like domain structure were used. They were annealed at 1250°C for 21 days in palladium-diffused hydrogen to reduce the inclusions and imperfections.

The rectangular samples were magnetized by a C-core electromagnet powered by a low noise operational amplifier used in the integrating mode. The magnetizing field was in the form of a .0033 Hertz triangular wave. The window frame samples were magnetized by applying power to a 10 turn winding on the sample.

The Barkhausen pulses were observed as a voltage induced in a search coil wound around the sample. The number of turns (~5000), gauge of wire, and the damping resistor were determined experimentally such that the time constant of the coil was 1-2μsec and oscillations were not observed. The voltage signal generated in the search coil was amplified using a low noise nanovolt amplifier (amplification 85 db) and recorded on a wideband instrumentation recorder (250KC) for either a time dilation or real time analysis. The analog data were converted to digital form and then displayed under computer control. A threshhold test allowed differentiation between noise and a Barkhausen pulse. The use of a computer as the control agent in the analysis allows flexibility in the sample interval and the length of time to be observed on the display equipment. The accompanying photos consist of 1024 samples at a sampling interval of 6.375μsec. The computer was programmed to compare the measured pulse with that expected from Eq. 2.

RESULTS AND DISCUSSION

The classes of pulses found are shown in Figs. 2-6 as plotted by the computer. (The pulse polarity is determined by the direction of magnetization. Therefore its initial half cycle, the conventional positive pulse, may

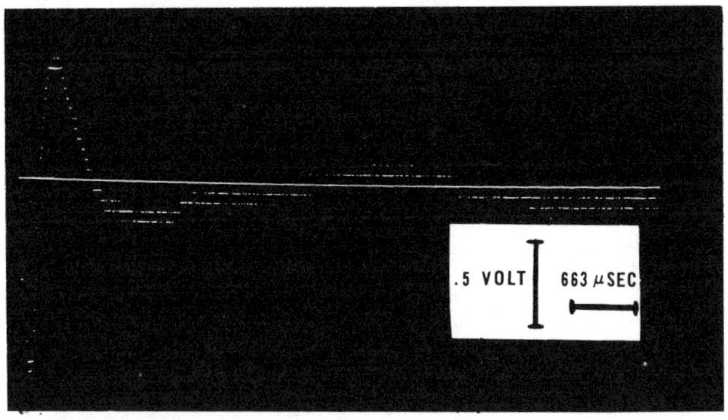

Fig. 2: Typical pulse - .012 cm. thick sample

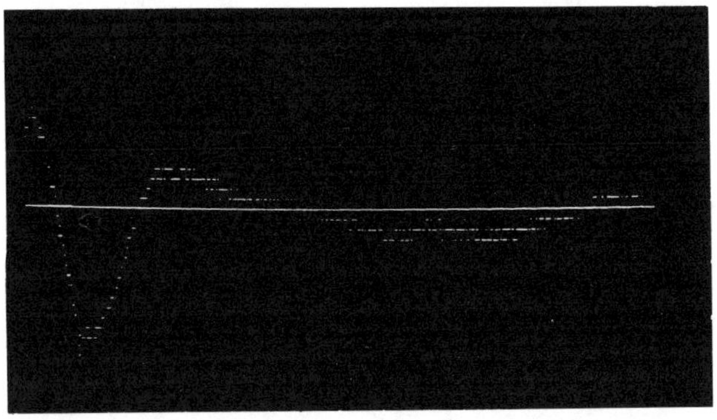

Fig. 3: Another typical pulse - .012 cm.

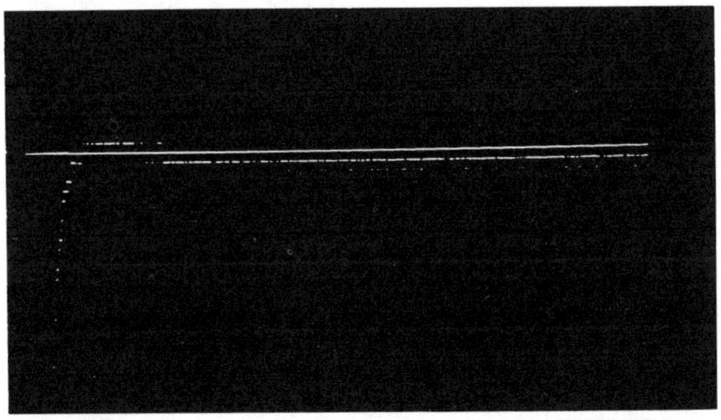

Fig. 4: Quick damped pulse - .015 cm.

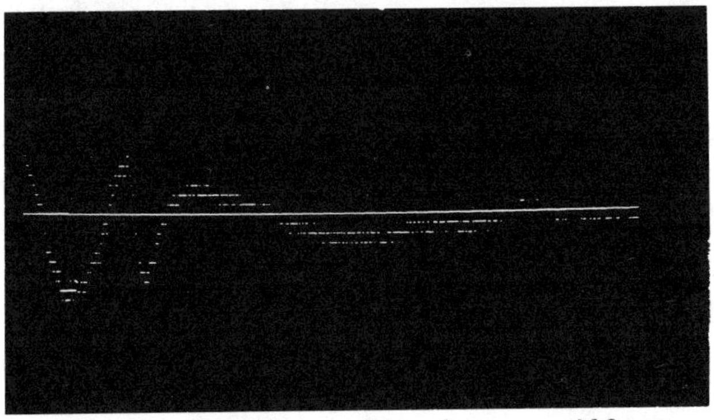

Fig. 5: Peculiar pulse shape - .012 cm.

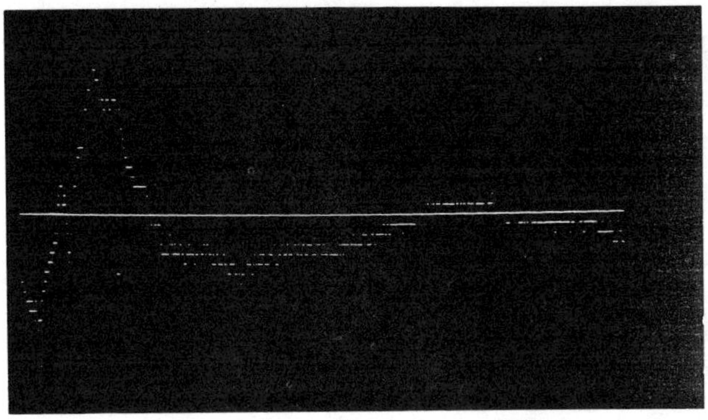

Fig. 6: Pulse cluster - .02 cm.

appear above or below the base line.) Most pulses with sufficient time duration to be detected (50μsec) show a damped oscillation of increasing time period (Figs. 2,3). Succeeding oscillations can extend to 2000μsec. A smaller number of pulses damped quickly (Fig. 4); a few show a peculiar shape (Fig. 5).

The form of a single pulse was not always well defined but shapes of a single oscillation were consistent with the form shown in Fig. 1. (See Fig. 2) Overlap of pulses, a pulse cluster, could also be observed (Fig. 6). Few pulses were coupled, an indication of the expected low material defect concentration. In a number of samples more pulses were observed as the magnetization was decreased from saturation than as it approached saturation.

The average volume region associated with the observed pulses was estimated to be 10^{-9} to $10^{-7}cm^3$. This compares favorably with previously reported values.[6] The threshold value used restricted pulse observations to those pulses with a high signal to noise ratio (i.e. large pulses) which permits better shape reproduction. Thus the total number of pulses found is at least an order of magnitude below that previously reported.[9] Also Eq. 2 predicts that the sheet thickness should play a role in the observed pulse lengths, but no correlation was readily evident.

The time constants of some of the first oscillations appear to be as short as 50-100 μsec. This compares with time constants reported by Tebble, et al[6] on iron samples, but not those given by Hajko, et al.[10]

For the pulses shown in Fig. 2-6 the relative magnitude of the negative (i.e. second pulse) to positive pulse is higher than that seen by Hajko, et al[11] for room temperature observations but compare more favorably with their reported values for lower temperatures.

The origin of the pulse shapes shown in Fig. 2-6 cannot be readily understood. While negative pulses have been reported earlier, their coupling with an initial positive pulse has only been implied.[1] Furthermore, a continual damped oscillation of increased period has not been anticipated. Also negative pulses have been explained by the influence of an eddy current induced field from a snapping wall on neighboring domain walls.[1] While this approach might be extended to deal with successive oscillations it would be difficult to explain the increased value of the observed time constants. Also preliminary observations on window frame samples, where a single wall can exist per leg, indicate that an oscillating pulse also occurs, but the amplitude of the negative portion seems to be a function of the low frequency cut-off of the amplifier.

REFERENCES

1. A. Zentkova', A. Zentko, V. Hajko, Czech. J. Phys., B19, 650-656, 1969.
2. M. Otala, S. Säynäjäkangas, J. Phys. E., Sci. Instru., 5, 669-672, 1972.
3. C. G. Gardner, G.A. Matzkanin, AIP-Conference Proc., 10, 1509-1513, 1971.
4. J. McClure, S. Bhattacharya, K. Schroëder, IEEE Trans. Mag., 10, 913-915, 1974.
5. K. Schroëder, T. Kunio, V. Weiss, IEEE Trans. Mag. 10, 916-918, 1974.
6. R.S. Tebble, I.C. Skidmore, W.D. Corner Proc. Phys. Soc., A63, 739-761, 1950.
7. K.M. Polivanov, et al, Fiz. Metal. Metalloved, 9, 130-140, 1960.
8. A. Seeger, H. Kronmüller, H. Rieger, H. Träuble, J. Appl. Phys., 35, 740-748, 1964.
9. V. Hajko, A. Zentko, S. Filka, Czech. J. Phys., B19, 547-548, 1969.
10. V. Hajko, et al, IEEE Trans. Mag., 10, 128-132, 1974.
11. V. Hajko, A. Zentko, Fyz. CAS., 20, 92-95, 1970.

This work was supported by NSF Grant GK10522.

BARKHAUSEN SPECTRA OF IRON AT 300°K AND 77°K

P. Deimel and H. Daniel
Technische Universität München, Munich, Germany

ABSTRACT

The Barkhausen spectra have been measured for two Fe-specimens. Particular emphasis was given to a low instrumental cut-off ($3.5\cdot10^{-8}$ magnetic cgs units). Although the spectra were measured with sensitivities never reached before, no maximum was found. This means that for the iron specimen used no preferred change in the magnetic moment existed. Furthermore, Barkhausen spectra were measured at 300°K and 77°K. Lowering the temperature to 77°K, a decrease in the number of Barkhausen jumps was registered, not only for large jumps but also for small ones. Also the dependence of the Barkhausen spectra on the actual slope of the hysteresis curve was measured. Theoretical ΔM-distributions were calculated and agree well with the measured distributions.

INTRODUCTION

Nearly all Barkhausen spectra measured up to now show a monotonic decrease with increasing jump size.[1] Only Ivlev and Rudjak[2] and Rudjak[3] found a pronounced maximum in the pulse size distribution in the region of approximately $1.5\cdot10^{-7}$ magnetic cgs units (mcgs). In order to clarify the situation, we had to make sure that the measured pulse size distribution was not marred by electronic noise produced in the search-coil circuit. Therefore we tried to diminish the noise by cooling the search coil and the specimen. With this apparatus the temperature dependence of the pulse size distribution of iron and the dependence on the magnetization were measured with more detection sensitivity ($8\cdot10^{-8}$mcgs) than was done by Lambeck[4] (10^{-5} mcgs).

EXPERIMENTAL SET-UP

Figure 1 shows the experimental set-up. The search coil S, surrounding the wire specimen Pr, was wound with 9694 turns of teflon insulated copper wire (0.063 mm diameter) and was 5 cm long. The magnetic field was applied to the specimen by a field coil F, surrounding both search coil and specimen, and could be slowly changed. The specimen was a hard drawn iron wire (0.1 mm diameter, 3 cm long, coercitive force 8.4 Oe at 77°K) of the Vakuumschmelze Hanau.
The Barkhausen time constant could be deduced from measurements made by Lambeck[4] to be ≤ 2.5 μsec at 77°K. The search coil circuit had a time constant of 30 μsec to fulfill the ballistic condition. The search coil pulses were amplified by means of a preamplifier with a transmission region from 4.15 to 23 kHz and a main amplifier. The output pulses were analyzed in a pulse height analyzer. The sensitivity reached with this experimental set-up was $3.5\cdot10^{-8}$ mcgs.

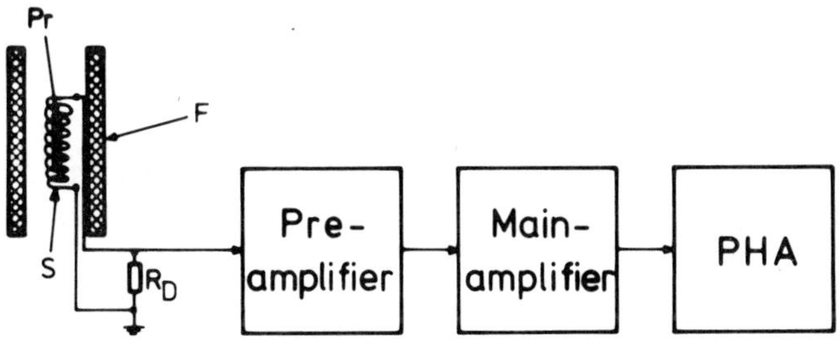

Fig. 1. Experimental set-up

EXPERIMENTAL RESULTS

Figure 2 shows the Barkhausen spectrum measured along the ascending part of the hysteresis curve. In the steep part the external magnetic field changed at 0.64 Oe/min.

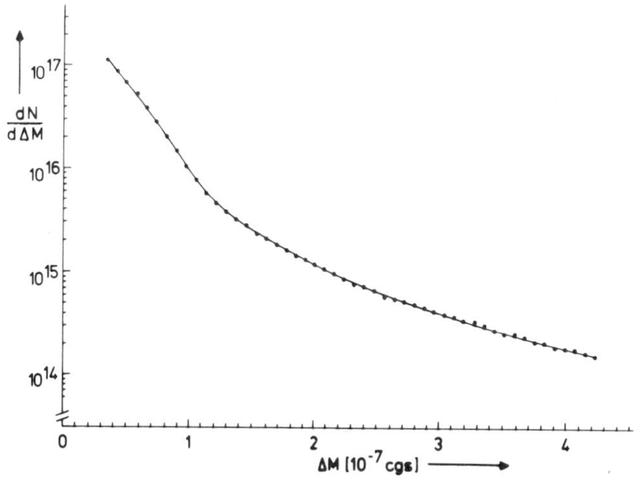

Fig. 2. Barkhausen spectrum (average of three measurements) at 77°K, noise spectrum subtracted. dN/dΔM, the number of Barkhausen jumps per cm³ and ΔM-unit in dependence on ΔM are shown. Fe-wire 0.1 mm diameter.

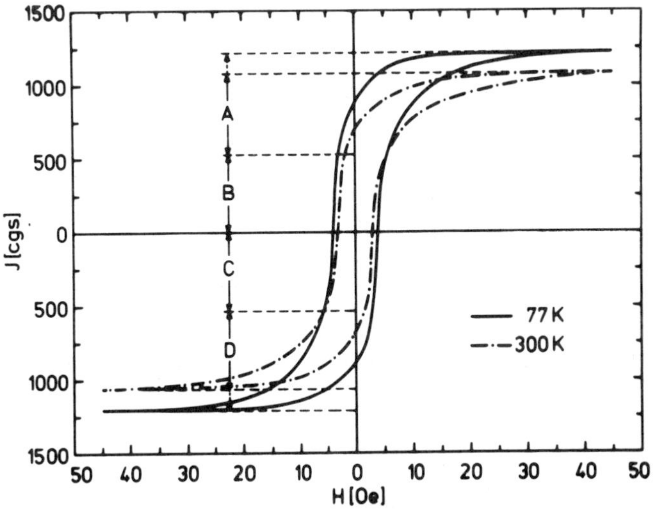

Fig. 3. Hysteresis loop of thermo-iron

To get Figure 2 it is necessary to measure the Barkhausen spectrum including the noise spectrum of the apparatus and then to measure the noise spectrum of the apparatus without the Fe-specimen. Both spectra have to be measured with the same "livetime", so that the influence of the noise can be practically eliminated if after measuring the Barkhausen spectrum the noise spectrum is subtracted. The result is shown in Figure 2.

The temperature dependence of the Barkhausen effect was measured with the same experimental set-up (Fig. 1). The specimen was a 3.5 cm long wire of annealed thermo-iron (0.1 mm diameter). Figure 3 shows the hysteresis loop at 300°K and 77°K. A, B, C and D are the magnetization intervals used in the measurement of the dependence of the pulse size distribution on the magnetization.
The pulse size distribution measured along one half of the hysteresis loop for 300°K and 77°K is shown in Figure 4.
The constants C_o, C_1 and C_2 are fitted parameters of a theoretical ΔM-distribution which is drawn as a solid curve and discussed in the following section. Lowering the temperature to 77°K, a decrease in the number of Barkhausen jumps was registered, not only for large but

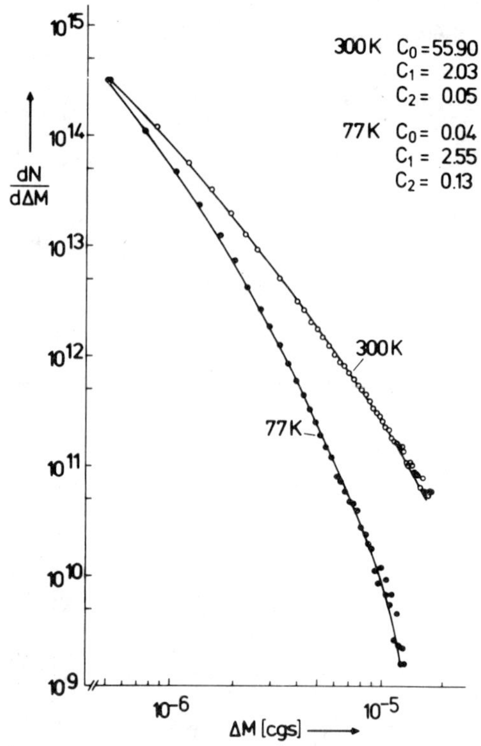

Fig. 4. Barkhausen spectra of annealed thermo-iron (average of 15 measurements) with temperature T as parameter, otherwise like Fig. 2.

also for small jumps. The large jumps disappeared almost completely. In a high sensitivity measurement ($8\cdot10^{-8}$ mcgs) no crossover of the distribution curves for 300°K and 77°K was found.

Figure 5 shows the dependence of the Barkhausen spectra on temperature and magnetization. A, B, C and D are the magnetization intervals ΔJ drawn in Fig. 3. C_o, C_1 and C_2 are the parameters of the theoretical ΔM-distribution (solid curve) for 300°K and the indicated ΔJ-intervals. Comparing the Barkhausen spectra measured at 77°K with those at 300°K one recognizes that almost all large jumps have disappeared at 77°K, independent of the magnetization interval. On the other hand, the dependence of the pulse size distribution on the magnetization interval is similar for both temperatures, almost all of the large jumps occurring in the steep intervals B, C.

DISCUSSION

The Barkhausen spectrum of iron (Fig. 2), measured with a sensitivity ($3.5\cdot10^{-8}$ mcgs) never reached before, is monotonically decreasing with increase in jump size and is in agreement with other measurements[1]. In contradiction to the results achieved by Ivlev and Rudjak[2,3], a maximum in the pulse size distribution in the sensitivity interval used (3.5 - $43\cdot10^{-8}$ mcgs) was not found for our specimen of wire.

The temperature dependence of the Barkhausen spectra, meaning the loss of large Barkhausen jumps at 77°K, becomes plausible if we consider the fact that lowering the temperature increases the number of Bloch walls[5]. In order to get more information from the measured spectra at both temperatures, we tried to approximate the ΔM-distribution in analogy to work done by Lieneweg and Grosse-Nobis[6], using a type of law like the following:

$$\frac{dN}{d\Delta M} = C_o\cdot(\Delta M)^{-C_1}\cdot\exp[-C_2(\Delta M/\Delta M_o)] \qquad (1)$$

where C_o, C_1 and C_2 are fitted parameters, $\Delta M_o = 4\cdot10^{-7}$ mcgs in our case and ΔM and $dN/d\Delta M$ are explained in the caption of Fig. 2. The ΔM-distributions calculated through Eq. (1) agree well with the majority of the measured spectra. There are also deviations which mean that Eq. (1) is only a first

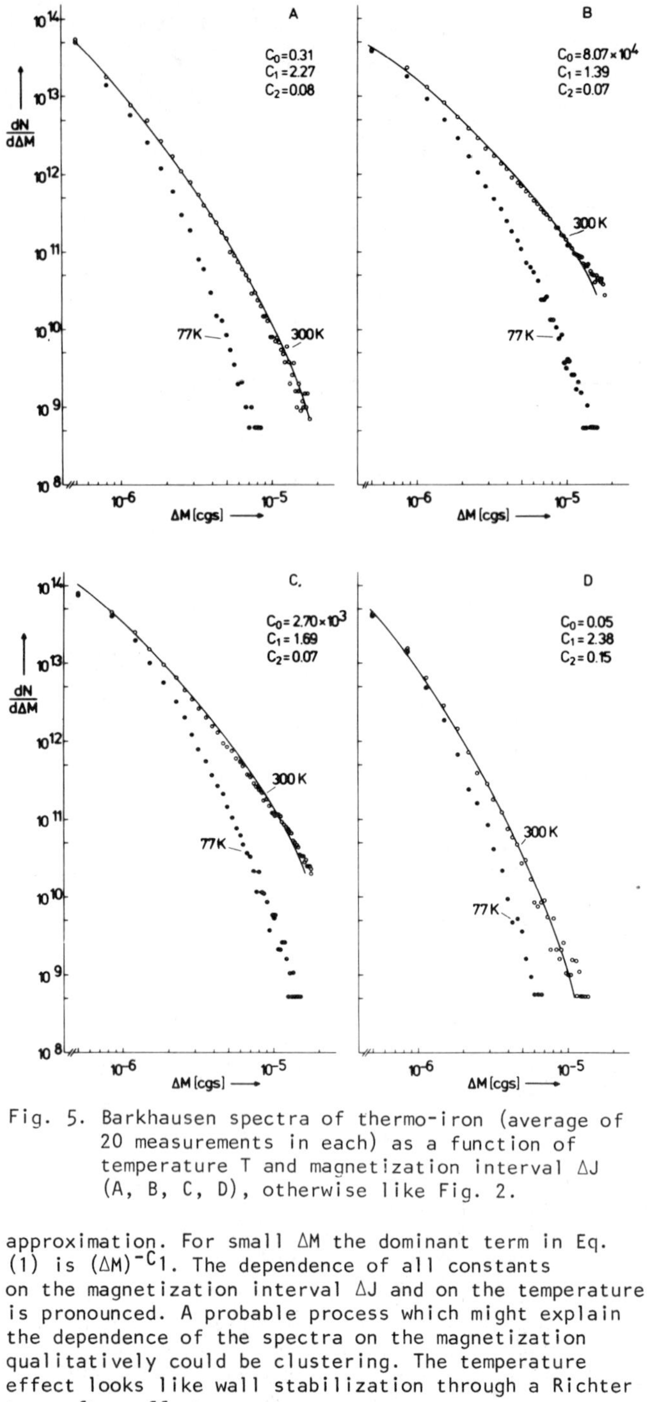

Fig. 5. Barkhausen spectra of thermo-iron (average of 20 measurements in each) as a function of temperature T and magnetization interval ΔJ (A, B, C, D), otherwise like Fig. 2.

approximation. For small ΔM the dominant term in Eq. (1) is $(\Delta M)^{-C_1}$. The dependence of all constants on the magnetization interval ΔJ and on the temperature is pronounced. A probable process which might explain the dependence of the spectra on the magnetization qualitatively could be clustering. The temperature effect looks like wall stabilization through a Richter type after effect.

We wish to thank Dr. B. Röde for his assistance during our computations.

REFERENCES

1. K. Stierstadt, Springer Tracts in Modern Physics 40, 29 (1966) and literature quoted therein.
2. V.F. Ivlev and V.M. Rudjak, Sov.Phys.Dokl.3, 571 (1958).
3. V.M. Rudjak, Sov.Phys.Usp.13, 461 (1971).
4. M. Lambeck, Barkhauseneffekt und Nachwirkung in Ferromagnetika, (Walter de Gruyter & Co., Berlin 1971).
5. W. Hampe and H. Bilger, Z.Angew.Physik 15, 391 (1963).
6. U. Lieneweg and W. Grosse-Nobis, Int.J.Mag.3, 11 (1972).

MAGNETIC BEHAVIOR OF HEUSLER CRYSTALS DISORDERED BY PLASTIC DEFORMATION

G. Y. Chin, M. L. Green*, R. C. Sherwood and E. M. Gyorgy
Bell Laboratories. Murray Hill, N.J. 07974

ABSTRACT

Single crystals of the Heusler alloy Cu_2MnAl were disordered by plastic deformation at room temperature. The saturation magnetization decreases with increasing strain, the magnitude of the decrease depending on orientation during deformation. Severe deformation imposed on drill turnings reduces the saturation moment to 10% of the initial value. Measurement of this sample with temperature shows a peak at ∿ 30°K indicative of antiferromagnetic behavior. The magnetocrystalline anisotropy constant K_1 decreases from the initial value of -1100 ergs/cm^3 to -4500 ergs/cm^3 after deformation. In addition, a uniaxial anisotropy energy ∿ 10^4 ergs/cm^3 was induced by the deformation and interpreted in terms of slip-induced directional order theory. The analysis indicates that directions of slip-induced nearest and next-nearest neighbor Mn-Mn atom pairs are hard and easy axes of magnetization, respectively.

INTRODUCTION

The Heusler alloy Cu_2MnAl is an ordered ferromagnetic compound with values of saturation magnetization and Curie temperature very nearly those of nickel. Its high temperature disordered b.c.c. phase orders to a B2 structure at ∿ 780°C and to the $L2_1$ structure at ∿ 620°C.(1) Not much is known about the magnetic behavior of disordered Cu_2MnAl since the magnetic moment is reduced by less than 10% with fast quenching from the disordered phase field.(2) Even with drastic quenching from the liquid state ("splat cooling") where the cooling rate reaches ∿ 10^6°C/s, we succeeded in reducing the magnetic moment by only 24%. In order to study the effect of controlled amount of disorder, we have decided to plastically deform the stoichiometric compound. In addition to changes in the technically interesting properties such as coercive force, we were interested in changes in basic parameters such as saturation magnetization and magnetocrystalline anisotropy. The disorder is expected to result in Mn atoms nearest and next-nearest neighbors to one another, whereas in the ordered state they are third neighbors only. Furthermore, as the disorder brought by plastic deformation is discrete across slip planes, one can expect an asymmetrical distribution of these Mn-Mn atom pairs and hence a slip-induced uniaxial magnetic anisotropy.(3,4)

EXPERIMENTAL

The constituents were first melted in an arc furnace. The cast buttons were crushed and placed in a high-purity alumina crucible for single crystal growth using a modified Bridgman technique. The analyzed composition for the as-grown crystal is very close to stoichiometry ($Cu_{1.998}Mn_{1.001}Al_{1.001}$). Oriented disks ∿ 4 mm dia × 0.4 mm thick were prepared by spark cutting for torque measurement of magnetocrystalline anisotropy. Rectangular specimens ∿ 5 mm wide × 8 mm long × 2.5 mm thick were also prepared for deformation by compressing in a channel (plane strain compression), where thinning in one dimension is accompanied by elongation in a second direction without widening in the third direction. Disks were prepared from samples deformed to various thickness strains for magnetic measurements. Crystals of several orientations were studied, although most crystals were oriented such that the compression plane is (110) and the elongation direction is [001].

Magnetic moment was measured using a pendulum magnetometer at an applied field of 15.3 kOe. Magnetic anisotropy was measured in a recording torque magnetometer and the torque curves Fourier analyzed using a computer program.

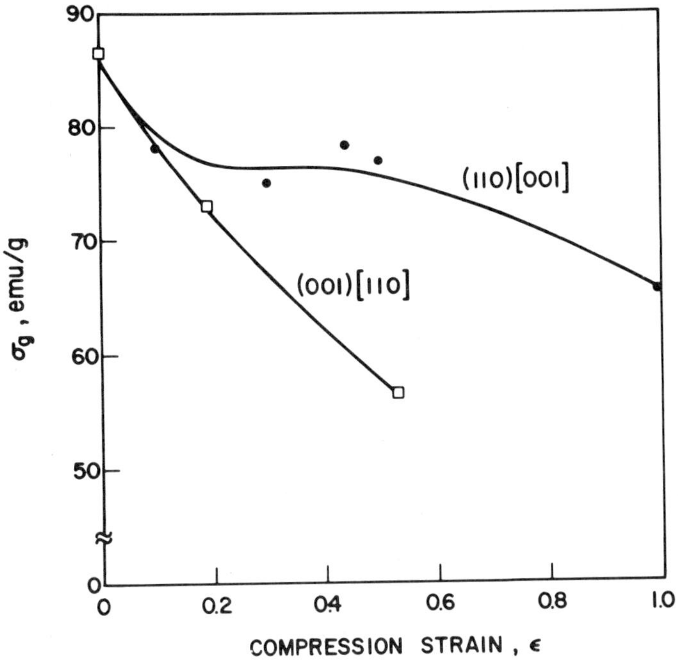

Fig. 1 Specific magnetic moment as function of strain.

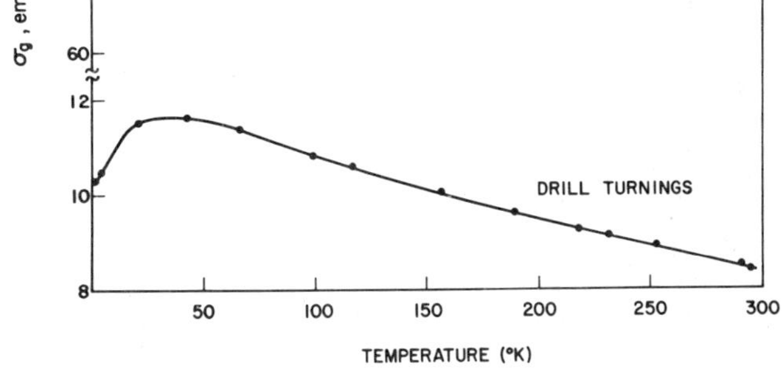

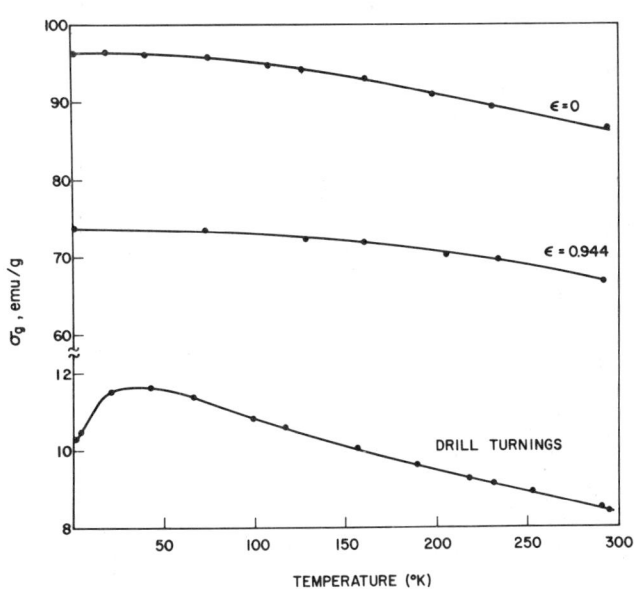

Fig. 2 Specific magnetic moment vs. temperature for three levels of strain.

*Now at Department of Materials Science, Massachusetts Institute of Technology Cambridge, Massachusetts

RESULTS

Saturation Magnetization - As expected, the saturation magnetization decreases with increasing deformation, Fig. 1. The magnitude of the decrease, however, depends on the orientation during deformation. The value of specific magnetic moment σ_g of crystals oriented for (001)[110] plane strain compression decreases faster than that of crystals oriented for (110)[001]. Figure 2 shows the temperature dependence of σ_g for an (110)[001] oriented crystal before and after compression to a strain of $\varepsilon = 0.944$ (upper two curves). No unusual behavior is noted. In order to reduce the value of σ_g drastically, we resorted to drilling the crystal with a ceramic drill and measuring the turnings. The results are shown as the bottom curve of Fig. 2. The room temperature value is down to less than 10% of the initial value. In addition, σ_g reaches a peak value near 30°K, suggestive of antiferromagnetic behavior.

Magnetocrystalline Anisotropy - In the unstrained condition, the value of the magnetocrystalline anisotropy constant K_1 was measured to be -1100 ergs/cm^3, in good agreement with the value of -1500 ergs/cm^3 previously reported by Aoyagi et al.[5] With plane strain compression in the (110)[001] orientation, K_1 decreases to $\sim$ 4500 ergs/cm^3 for a strain of 0.1 and remains constant thereafter as indicated in Fig. 3. This trend of decreasing K_1 with disorder is opposite to that observed by Aoyagi et al.[5] for thermally disordered Cu_2MnSn crystals.

Slip-Induced Uniaxial Anisotropy - In addition to changes in K_1, deformation also induces a uniaxial anisotropy. The value of induced anisotropy constant Ku is also shown in Fig. 3 as a function of compression strain, for the (110)[001] orientation. It rises to a maximum of about 12,000 ergs/cm^3 after a strain of about 0.35 and declines thereafter. The induced easy axis lies in the [001] direction in this case. The specimen geometry introduces a shape anisotropy energy of about 1000 ergs/cm^3 as measured on several annealed disks.

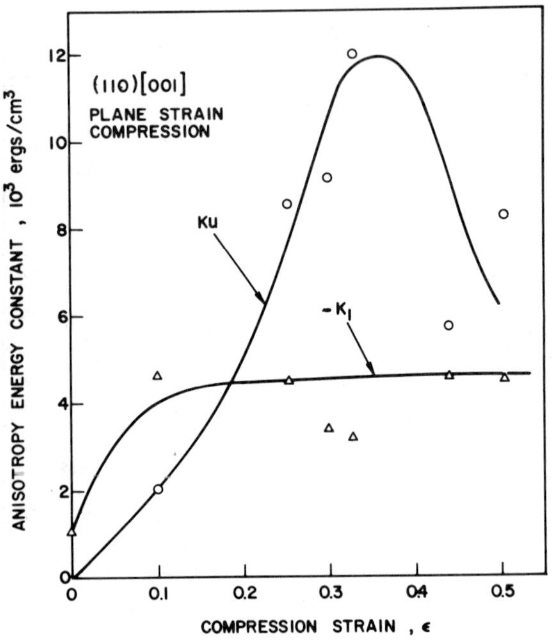

Fig. 3 Magnetocrystalline anisotropy K_1 and slip-induced anisotropy Ku as a function of strain.

DISCUSSION

Detailed studies of slip traces and direct observation of dislocations using transmission electron microscopy have established that Cu_2MnAl slips on {110} planes in <111> directions.[6] The dislocations tend to travel in groups of four, the number required to recover the $L2_1$ ordered lattice. Between dislocations 1 and 2, and 3 and 4, nearest neighbor (NN) Mn-Mn pairs are created in <111> directions, and between dislocations 2 and 3 next-nearest neighbor (NNN) Mn-Mn pairs are created in <100> directions, all across the slip plane. Undoubtedly these Mn-Mn pairs are responsible for the decrease in magnetic moment by deformation as indicated in Figs. 1 and 2. Since the activity of slip systems is different depending on orientation, a different dependence of magnetic moment on strain as shown in Fig. 1 is not unexpected. The low temperature peak in magnetic moment for the drill turnings (Fig. 2) suggests strongly that the NN and/or NNN Mn-Mn pairs are coupled antiferromagnetically. This is probably the first demonstration of antiferromagnetic behavior in disordered Cu_2MnAl. Lapworth and Jakubovics[7] had previously suggested that such antiferromagnetic coupling may be responsible for pinning domain walls.

The induced uniaxial anisotropy, shown in Fig. 3, most likely originates from the asymmetrical distribution of the slip-induced Mn-Mn atoms pairs. The theory of slip-induced directional order, as first proposed by Chikazumi, Suzuki and Iwata,[3] is now well established.[4] For plane strain compression in the (110)[001] orientation, calculation based on established procedures[3,4] shows that for Cu_2MnAl only NNN Mn-Mn pairs contribute to the induced anisotropy. The expression has the form $E_u = 1/2\ C_2Nl_2\ \varepsilon\cos^2\theta$, where C_2 contains terms involving the geometrical distribution of dislocations and an order parameter, N is the number of atoms per unit volume, l_2 is the so-called pseudodipolar coupling energy of NNN Mn-Mn pairs that give rise to the uniaxial anisotropy, ε is the strain, and θ is the angle on the (110) plane measured from the [001] direction. Since [001] is the observed induced easy axis, the theory predicts that $l_2 < 0$ and hence the NNN Mn-Mn direction is an easy axis. In order to obtain information on NN Mn-Mn interaction, we measured the induced anisotropy of crystals deformed in (111)[$\bar{1}\bar{1}$2] and (112)[$\bar{1}\bar{1}$1] orientations. The energy expressions are calculated to be $\frac{\varepsilon}{2}[-C_1Nl_1 + 2C_2Nl_2]\cos^2\theta$ and $\frac{\varepsilon}{24}[-13C_1Nl_1 + 8C_2Nl_2]\cos^2\theta$ respectively, where C_1 and l_1 are the corresponding expressions of C_2 and l_2 for NN Mn-Mn pairs and θ is measured from [$\bar{1}$10] on the (111) or (112) plane. It was found that [$\bar{1}$10] was the induced easy axis for both orientations, and the solution shows that $l_1 > 0$ This means that the nearest neighbor Mn-Mn direction is a hard axis direction. Unfortunately there are no means of independently testing the validity of these conditions at the present time.

REFERENCES

1. D. Chevereari, J-M Gras and B. Dubois, C. R. Acad. Sc. Paris Série C 276, 643 (1973).
2. R. M. Bozorth, "Ferromagnetism", D. Van Nostrand Co., Princeton, N.J., 1951, p.331.
3. S. Chikazumi, K. Suzuki and H. Iwata, J. Phys. Soc., Japan 12, 1259 (1957).
4. G. Y. Chin, in Adv. Mater. Res. Vol. 5, H. Herman, ed., John Wiley and Sons, New York City, 1971, p.217.
5. K. Aoyagi, M. Sugihara and C. Kuroda, Rev. Elect. Comm. Lab. NTT 19, 1249 (1971).
6. M. L. Green and G. Y. Chin, to be published.
7. A. J. Lapworth and J. P. Jakubovics, Phil. Mag. 29, 253 (1974).

MAGNETIC ANISOTROPY IN Co AND Co-Ni SINGLE CRYSTALS DEFORMED BY COLD ROLLING

M. Takahashi, T. Wakiyama, T. Anayama, M. Takahashi*, and T. Suzuki*
Department of Electrical Engineering and *Department of Applied Physics
Tohoku University, Sendai 980, Japan

ABSTRACT

A study of the magnetic anisotropy induced by cold rolling was made for h.c.p. Co and Co-Ni single crystals containing 10 and 20 at.% Ni, using a torque magnetometer. The observed induced uniaxial anisotropy K_{ur} for Co was found to reach a maximum of -5×10^5 erg/cm^3 at about R=5% (R=rolling reduction), with the easy axis parallel to the rolling direction. For the Co-10%Ni alloy, a similar dependence of K_{ur} on R was obtained, although the sign of K_{ur} changed from negative to positive at about R=30%. For the Co-20%Ni alloy, the value of K_{ur} was found to increase with R and to be about 4×10^5 erg/cm^3 at R=60%. The results are discussed in terms of magnetoelastic coupling energy and slip-induced directional order.

INTRODUCTION

It is known that ferromagnetic alloys exhibit strong uniaxial magnetic anisotropy when plastically deformed by cold rolling or cold drawing[1]. The origin of this induced magnetic anisotropy for cubic structures has been interpreted reasonably well in terms of slip-induced directional order[2]. However, induced magnetic anisotropy has been observed in pure Ni plastically deformed[3]. The origin of the anisotropy in pure metals has been thought to lie in the change of magnetoelastic coupling energy. However, very few systematic studies of induced magnetic anisotropy for hexagonal crystals have been reported in the literature[4]. The present work has been carried out to investigate the magnetic anisotropy induced by cold rolling of h.c.p. Co and Co-Ni alloys.

SPECIMENS AND EXPERIMENTAL PROCEDURE

The specimens used in this study were Co and Co-Ni single crystals made by the Bridgman method using electrolytic Co and Ni (Co: 99.5%; Ni: 99.95%). X-ray analysis of the crystals shows the coexistence of h.c.p. and f.c.c. structures. However, the intensity of f.c.c. lines is much weaker than that of h.c.p. The single crystals were sliced into 2 mm thick plates with the crystallographic orientation shown in Fig. 1, taking into account the slip system[5] $\{0001\}-\langle 2\bar{1}\bar{1}0\rangle$. The magnetic anisotropy was measured by a torque magnetometer using paper-backed strain gages, designed so that vertical torque does not influence the measurement of horizontal torque. For the measurement of torque as a function of reduction in thickness R ($R=(r_0-r)/r_0$, where r_0 and r are the thickness before and after rolling), specimens were recut to circular shape after each rolling step. Specimens were reduced from 10 to 8.5 mm in diameter and from 0.72 to 0.27 mm in thickness as rolling progressed. Observation of slip bands was made by optical microscopy on rectangular specimens (size before rolling, 10 x 3 x 1 mm).

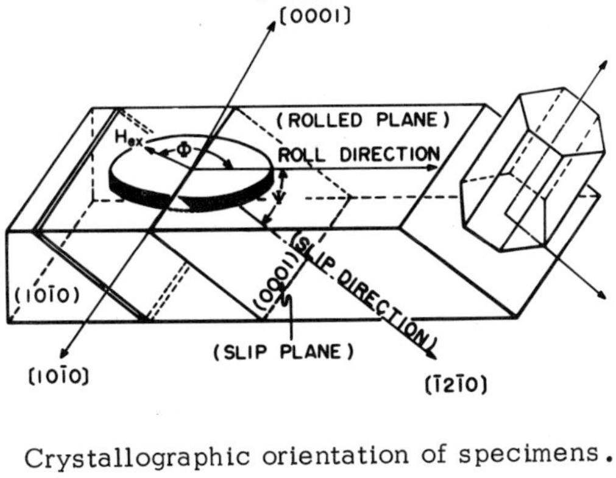

Fig. 1. Crystallographic orientation of specimens.

EXPERIMENTAL RESULTS

1. Magnetic anisotropy: Torque curves were measured in external fields H up to 12.4 kOe. To obtain the uniaxial and the biaxial anisotropy constants K_u and K_b, torque curves were Fourier analyzed. The values of K_u and K_b thus obtained were plotted vs. 1/H. By extrapolating these curves to 1/H=0, both $K_{u\infty}$ and $K_{b\infty}$ were obtained. As an example, $K_{u\infty}$ and $K_{b\infty}$ for the Co-20%Ni alloy are shown in Fig. 2 as a function of R.

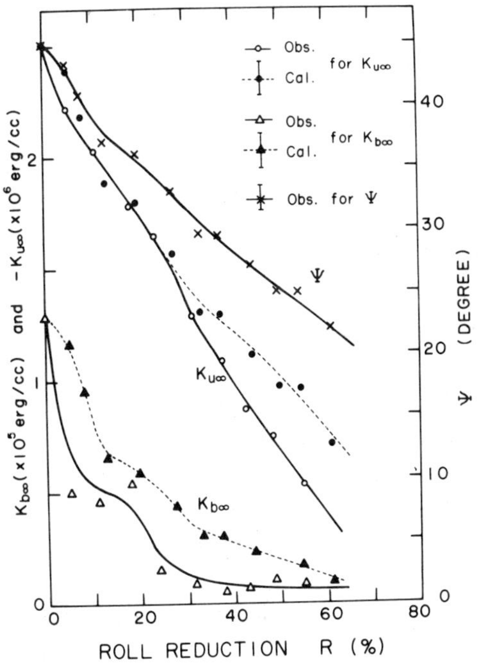

Fig. 2. $K_{u\infty}$, $K_{b\infty}$, and Ψ vs. reduction in thickness, for the Co-20at%Ni crystal.

2. Slip bands: It is noted first that only an elongation along the rolling direction took place. This means that basal-plane slip occurs, with single glide. With increasing R, the convex lenticular-shaped bands along the $\langle\bar{1}2\bar{1}0\rangle$ direction initially present became gradually straight and sharpened. The angle of inclination Ψ between the rolling plane and the slip plane (see Fig. 1) varied with R as shown in Fig. 2.

DISCUSSION

1. Dependence of $K_{u\infty}$, $K_{b\infty}$, and induced anisotropy K_{ur} on reduction: From Fig. 2, one sees that the decreas of $K_{u\infty}$ and $K_{b\infty}$ with R is similar to that for Ψ. Since the decrease in Ψ is connected with the rotation of the slip plane, the changes of $K_{u\infty}$ and $K_{b\infty}$ are thought to be due to the lattice rotation by rolling. Assuming that the h.c.p. structure is retained even after rolling, one readily derives the equation for torque L in the rolling plane as

$$L = K_{u\infty}\sin 2\phi + K_{b\infty}\sin 4\phi + \ldots \qquad (1)$$

where $K_u = -(K_{u1}+2K_{u2})\sin^2\Psi + K_{u2}\sin^4\Psi$, and

$$K_b = (K_{u2}/2)\sin^4\Psi. \qquad (2)$$

By using the observed values of Ψ, K_{u1}, and K_{u2}, $K_{u\infty}$ and $K_{b\infty}$ were calculated as a function of R, as shown in Fig. 2. (K_{u1}= 4.1, 3.5, and 1.7 x 10^6 erg/cm^3 and K_{u2}=1.4, 1.4, and 1.0 x 10^6 erg/cm^3 for Co, Co-10%Ni, and Co-20%Ni, respectively.) As shown, a disagreement exists for R above about 20%. This disagreement implies that the dependence of $K_{u\infty}$ on R can not be explained in terms of the lattice rotation only. On the other hand, in the case of Co, good agreement between the experimental and calculated results was obtained above about 20%, but not below 20%[6].

The uniaxial magnetic anisotropy induced by cold rolling, K_{ur}, not attributable to the effect of lattice rotation, is obtained by subtracting the calculated values of $K_{u\infty}$ from the experimental ones. K_{ur} thus obtained is shown vs. R on Fig. 3. Here, the negative value of K_{ur} is defined so that the easy axis is parallel to the rolling direction. For Co, negative K_{ur} was found below about R=30%, with a maximum value of about -5x10^5 erg/cm^3 at about R=5%. Similarly, for the Co-10%Ni alloy, negative K_{ur} was found in the region below about R=20%, and positive values above 20%. For the Co-20%Ni alloy, positive K_{ur} was found above about R=20%; below 20% no definite dependence of K_{ur} on R is found, although K_{ur} is not zero.

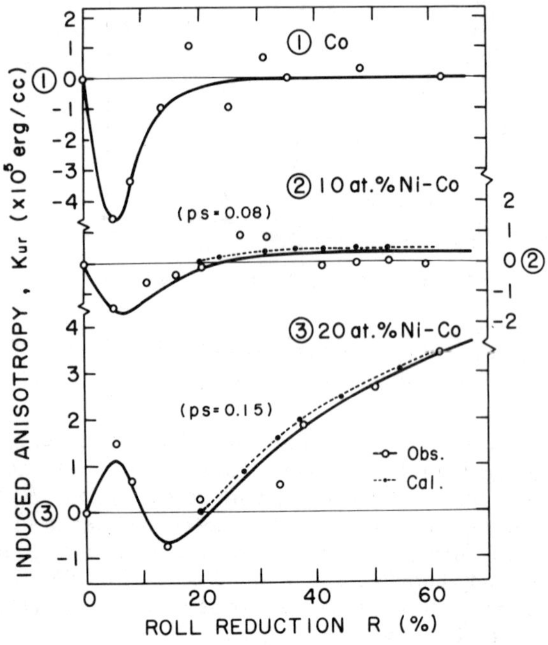

Fig.3. Induced magnetic anisotropy K_{ur} vs. reduction in thickness R for single crystals of three compositions.

2. Origin of the induced magnetic anisotropy: The macrostress caused by rolling may be of importance in discussing the induced anisotropy. However, the macrostress is considered to be a balanced force system, so we cannot expect induced anisotropy from this stress. One possible origin may be in a particular distribution of microstress resulting from cold rolling. In the present study the distribution of stress was conjectured based on the observation of slip bands. The micro-climb of slip bands was observed around the lenticular shaped bands, which means that the bands may act as barriers to glide dislocations. In this case, a strong shear stress might exist near the slip bands, and the component σ_{31} might be predominant. It should be noted that, considering the crystallographic orientation of the specimen, the induced anisotropy should be expected due not only to the normal stress but also to the shear stress. In order to explain the value of K_{ur} in Co, σ_{31} must be of order 10^9 dyne/ cm^2.[7]

The change in K_{ur} with R corresponds to the disappearing of the convex lenticular shaped bands, leading to straight slip bands. It is proposed that the convex lenticular shaped bands might be transformed to the h.c.p. structure by higher cold-rolling[8]. Such transformation might release the highly stressed state, which is responsible for the dependence of K_{ur} on R. On the other hand, the induced anisotropy was found at high R for the alloys but not for Co. This fact suggests that slip-induced directional order may be responsible. In Fig. 3, the calculated curves of K_{ur} obtained by assuming that directional order begins to occur beyond R=20% and that the values of ps, where p is the probability for one dislocation to start in a virgin atomic plane and s is the degree of short-range order, are 0.08 and 0.15 for the 10 and 20% Ni alloys respectively, are shown[7]. As seen, these values of ps give good agreement with experiment. However, it is difficult to make quantitative arguments because the state of ordering before the deformation is unknown.

REFERENCES

1. G.W.Rathenau and J.L.Snoek, Physica 8 555 (1941).
2. S.Chikazumi, K.Suzuki, and H.Iwata, J.Phys.Soc. Jap. 12 1259 (1957).
3. H.Gessinger, E.Köster, and H.Kronmüller, J.Appl. Phys. 39 986 (1968).
4. M.Takahashi and T.Kono, J.Phys.Soc.Jap. 15 936 (1960).
5. K.G.Davis and E.Teghtsoonian, Trans.AIME 227 762 (1963).
6. M.Takahashi, T.Wakiyama, T.Anayama et al., Proceedings Int.Conf.Mag, ICM-73, Moscow, III, 105 (1974).
7. To be submitted to J.Phys.Soc.Jap.
8. Y.Nakamura and T.Shinjo, Scripta Met. 2 647 (1968).

TEMPERATURE DEPENDENCE OF MAGNETIC ANISOTROPY INDUCED BY NEUTRON IRRADIATION IN Ni ALLOYS

P. Allia and G.P. Soardo
Istituto Elettrotecnico Nazionale Galileo Ferraris,
Gruppo Nazionale Struttura della Materia del C.N.R.
10125 Torino, Italy

ABSTRACT

It has been shown that only in some Fe alloys the temperature dependence of induced anisotropy energy K_u due to directional order of solute pairs fits the third power law of saturation magnetization $I_s(T)$ derived from theories of crystalline anisotropy energy. In order to further investigate the role of various solute atoms, comparison with the case of Ni alloys is of interest, since, differently from Fe case, the rate of decrease of I_s and of Curie temperature T_c with concentration is proportional to the solute valency. In this work the first K_u vs. T data are reported for NiCr5% and NiPd43%. Since T_c's are too low to permit atomic diffusion in the presence of a non zero I_s, directional order was obtained by fast neutron irradiation (1.3 10^{18} n/cm^2) under magnetic field (~5000 Oe) at about 60° C. K_u measurements were performed by means of a sensitive torque balance between about 100 and 320 K. From the present preliminary results it is found that both in NiCr and in NiPd the K_u curves decay with increasing temperature faster than $I_s^3(T)$, approximately according to a $I_s^6(T)$ law, independently of the solute atom valency, which is 0 for Pd and 5 for Cr.

INTRODUCTION

As is known, by annealing under magnetic field it is in general possible to induce in ferromagnetic alloys a uniaxial anisotropy energy K_u due to the formation of a directional order of solute pairs. When this order is frozen, K_u can in principle depend on the temperature T at which it is measured. From Zener's theory[1] which assumes that the decay of crystalline anisotropy energy with increasing T is only due to thermal fluctuations of the local spin directions with respect to the one of the average magnetization, one can predict[2] that K_u should vary, because of its uniaxial symmetry, proportionally to the cube of the saturation magnetization $I_s^3(T)$.

The first experimental results on the K_u dependence on T in various Fe alloys were reported by Ferro et al.[2], and recently confirmed by t'Hooft et al.[3]. These results show that while for some alloys (FeCr, FeNi, FeV) the experimental K_u vs. T curves are in very good agreement with the $I_s^3(T)$ law, strong deviations are observed for other alloys: in FeAl and FeSi, K_u varies rapidly with T approximately according to a I_s^6 law, while in FeCo it remains closely constant upon temperature variation.

Because of the observed discrepancies between experimental laws and with respect to theoretical predictions, it seems of interest to extend K_u vs. T measurements to other alloys, to possibly clarify the role of different solute atoms. The case of Ni alloys appears of special interest, since the behavior of Fe and Ni in solid solutions are rather different. As known in Fe alloys there is an approximately constant decrease in $I_s(0)$ per solute atom, while in the Ni ones the decrease of I_s and T_c is closely proportional to the number of valence electrons of the solute atom[4].

Since the Curie temperature in Ni alloys is generally low, in order to obtain directional ordering the diffusion below T_c must be enhanced either by neutron irradiation under magnetic field[5] or by quenching followed by convenient annealings[6]. The only data available in the literature on K_u vs. T for Ni alloys refer to NiCo[7] and NiFe[2,8], which are actually characterized by relatively high T_c.

In the present work the first results are reported on the temperature dependence of the induced anisotropy energy due to directional order obtained by the neutron irradiation technique in NiCr5% and NiPd43% whose T_c's are respectively found to be 190 and 270°C.

EXPERIMENTAL AND DISCUSSION

Polycrystalline disk specimens (Ø 12 mm thickness 0.5 mm), prepared with special care to obtain good compensation of crystalline anisotropy, were irradiated in the Grenoble swimming pool reactor with a fast neutron dose (E ≥1 MeV) of 1.3 10^{18} n/cm^2 under a magnetic field of about 5000 Oe applied by a convenient set-up[9]. The samples temperature during irradiation was maintained around 60°C, by direct cooling with the swimming pool water. The temperature dependence of K_u was followed by means of a sensitive magnetic torque balance between about 100 and 320 K. For each temperature, K_u was obtained from the difference between the torque curves (A) traced after irradiation under magnetic field and the ones (B) obtained after annealing in the torque balance at about 450°C for 2 hours, to destroy any formerly induced directional order.

Because of the uniaxial character of K_u the (A - B) curves should behave as pure second harmonics with respect to the torque angle, and have an initial phase strictly related to the direction of field during irradiation. For both examined alloys all (A - B) curves obey these criteria within the experimental accuracy which can be estimated to be of the order of ± 5%.

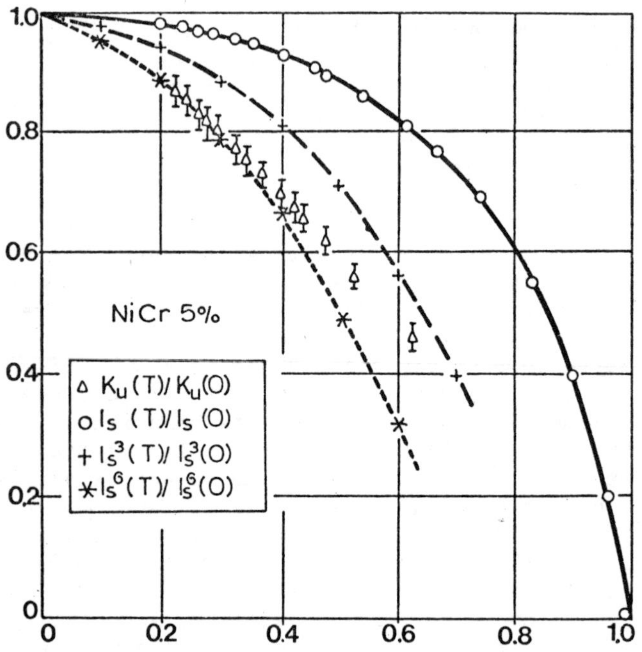

Fig. 1 - NiCr 5%: experimental dependence on temperature of induced anisotropy energy K_u (triangles), saturation magnetization I_s (dots, full line), I_s^3 (dashed line), and I_s^6 (dotted line). (Reduced coordinates are used).

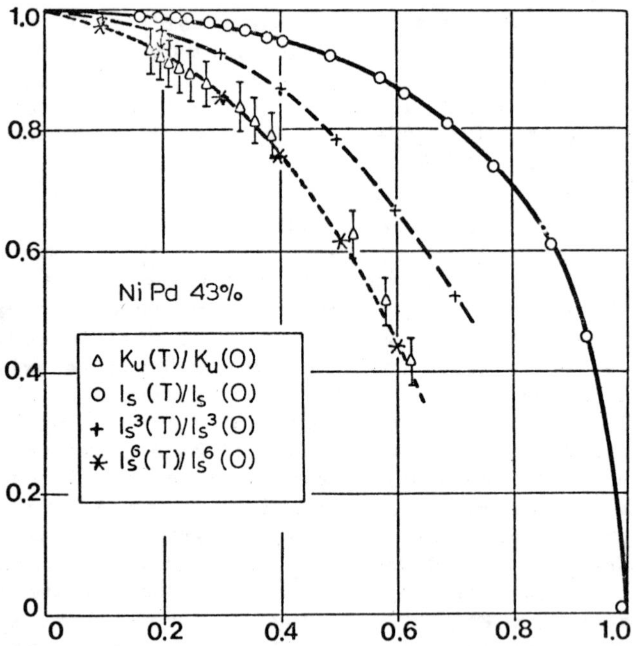

Fig. 2 - Same as Fig. 1 for NiPd 43%.

The experimental results are reported in form of reduced curves $K_u(T)/K_u(0)$ vs. T/T_c in Fig. 1 for NiCr5% and in Fig. 2 for Ni Pd 43%. From torque measurements on anisotropically shaped samples of the same alloys, the I_s vs. T curves were also determined. On Figs. 1 and 2 the reduced curves $I_s(T)/I_s(0)_3$ and the corresponding third power curves $I_s(T)/I_s^3(0)$ and the sixth power curves are also reported.

As is seen in both cases K_u decays with increasing T definitely faster than $I_s^3(T)$, approximately according to a I_s^6 law for NiPd and to a 5th or 6th power law for the low temperature range of NiCr. Furthermore from the present preliminary results the solute valency does not seem to play any particular role on directional order phenomena, or at least on the K_u dependence on T in Ni alloys.

It must be stressed that the observed effects should unequivocally be attributed to directional order of solute pairs and not to ordering of point defects caused by radiation damage. This is proved by the behavior of pure Ni control specimens, whose A and B curves after irradiation under magnetic field were closely coincident, but for minor effects having a clearly spurious character.

These results should, however, be confirmed by further measurements down to a temperature of about 20 K, in order to reduce the uncertainty on the K_u value extrapolated to 0 K which strongly affects the shape of the reduced curves. Moreover in order to check the present conclusions on the apparent independence of K_u vs. T behaviors from solute valency, similar measurements will be carried out on a Ni alloys with 20% Cu, whose valency has a value of 1, intermediate between the ones of Pd and Cr.

ACKNOWLEDGEMENTS

The authors are grateful to Prof. L. Néel and his collaborators for permitting the use of facilities for irradiation under magnetic field.

REFERENCES

1 Zener C., Phys. Rev. 96, 1335 - 1337 (1954).
2 Ferro A., Mazzetti P. and Montalenti G., J. Physique 32, 92 - 94 (1971).
3 't Hooft H.A., Andriessen T.C.M., Bustraan W. Fasten G. and Brommer P.E., Proc. Intern. Conf. on Magnetism (Moscow 1973), Vol. III, Publishing House Nauka, Moscow (1974), pp. 100-104.
4 See for instance: Morrish A.H., "The Physical Principles of Magnetism", J. Wiley and Sons, New York (1965), p. 307.
5 Ferro A. and Soardo G.P., J. Appl. Phys. 40, 3051 - 3053 (1969).
6 Ferro A., Griffa G. and Montalenti G., IEEE Trans. on Magnetics MAG-2, 764 - 768 (1966).
7 Chikazumi S. and Graham C.D.Jr., in Berkowitz A. and Kneller E., Eds., "Magnetism and Metallurgy", Vol. 2, Academic Press, New York (1969), p. 577.
8 Taniguchi S., Sci. Rept. Res. Inst. Tohoku Univ. A6, 269 - 281 (1955).
9 Néel L., Paulevé J., Pauthenet R., Laugier J. and Dautreppe D., J. Appl. Phys. 35, 873-876 (1964).

NEW CONCEPT OF MAGNETIC DOMAIN WALL STABILIZATION

William SIMONET
Groupe d'Etude et de Synthèse des Microstructures
C.N.R.S., 1 place A. Briand, 92190 BELLEVUE - FRANCE.

ABSTRACT

In the classical theory of magnetic domain wall stabilization, the induced anisotropy axis is assumed to lie parallel to the local direction of the magnetization inside the wall. This theory does not fit the experience since the theoretical field Ho, necessary to push a 180° domain wall out of its potential well, is too large compared with the experimental threshold field of the constricted hysteresis loops. We give here a new interpretation of domain wall stabilization based on the hypothesis that diffusion reduces only the magnetostrictive stresses existing in the 180° domain walls and that the induced anisotropy remains uniaxial in these walls. In our model the theoretical intensity and thermal variations of H_o fit in with the experimental values obtained on a Ni-Co-Fe ferrite. We show that the induced anisotropies measured under a constant strain and constant stress must be differentiated as in the case of magnetocrystalline anisotropies. Domain wall stabilization would be due only to that low part of induced anisotropy which proceeds from magnetostriction.

INTRODUCTION

When a cubic ferro or ferrimagnetic crystal of substitutional type is heated, an uniaxial anisotropy superposed to the magnetocrystalline anisotropy is generally induced. Thus the axis of the induced anisotropy becomes parallel to the magnetization vector. In the Neel's classical theory[1] of domain wall stabilization, this alignment occurs also inside the domain walls. Thus each wall creates an helicoïdal induced anisotropy which acts as a potential well.

In the temperature range of diffusion an with an induced anisotropy much smaller than the magnetocrystalline anisotropy, it is easy to calculate the H_o value of the magnetic field necessary to push a 180° wall out of its potential well[1,2]. Effectively the shape of a moving wall never changes in that case. H_o is found to be proportionnal to the induced anisotropy and to the inverse of the spontaneous magnetization. A material with stabilized walls exhibits a constricted hysteresis loop[3] and the threshold field H_s of the loop should be equal to H_o. In the diffusion temperature range, the agreement with experiment seems to be good for ferromagnetic materials having a weak induced anisotropy. But the induced anisotropy being weak, measurements of this energy are not accurate and the agreement is not so sure. Furthermore the comparison with the experiment cannot be made below the diffusion temperatures. After extension of this model to the general case of cubic materials with a strong induced anisotropy, we have shown[4] that the theoretical field H_o remains proportional to the induced anisotropy and to the inverse of the spontaneous magnetization, even for temperatures below the diffusion temperatures. But we have found that H_o is about 50 times larger than the H_s values. H_s measurements were made on Ni-Fe ferrites containing variable amount of cobalt[5]. A new interpretation of domain wall stabilization, based on the stress relaxation in the walls is given here. We suppose also that the induced anisotropy axis is not parallel to the spontaneous magnetization vector in each point of the crystal.

STRESS RELAXATION

Discussion will be restricted to a plane 180° Bloch wall separating two domains A and B. In the wall, the magnetization vector rotates from y = o in domain A to y = 180° in domain B. A consequence of magnetostriction is a change in dimensions when the direction of magnetization changes. However at the place of a 180° wall the crystal is usually assumed to be undistorted[6] because the domains A and B constrain the wall to be in the same deformation state as they. This leads to the existence of stresses inside the wall.

In contrast to the classical theory of domain wall stabilization, when ordering occurs inside the wall, we suppose that:
(i) the induced anisotropy axis remains parallel to y = o
(ii) each stress-tensor coefficient τ_{ij} is reduced by a factor t_{ij} independent of y.

In the actual state of knowledge, it is not possible to give a direct demonstration of the physical reality of this model. It is not sure that this is the only model which may occur. It may be that the classical model is valid for certain materials and the new model for others. Further, the two models may compete each other in the same material when temperature varies, but the following indirect information gives a good probability of existence for the stress relaxation model:
a) With this new model the magnetic field H_o necessary to push the wall out of its potential well is given[7] by:

$$H_o \doteq 0,6 \; (F_\lambda + G_\lambda) \; / \; M_s \qquad (1)$$

$$\text{with} \begin{cases} F_\lambda = 9 \; t_{22} \; (c_{11} - c_{12}) \; \lambda^2_{100} \; / \; 4 & (2) \\ G_\lambda = 9 \; t_{23} \; c_{44} \; \lambda^2_{111} & (3) \end{cases}$$

M_s, c_{ij}, λ_{ijk} are respectively the spontaneous magnetization, the moduli of elasticity and the coefficients of magnetostriction. Since the factors of stress reduction t_{ij} are unknown, we suppose t_{ij} = t and we give t such a value that H_o is equal to the field H_s measured at temperature T. In order to have $H_o = H_s$ at T = - 100°C, t must be chosen equal to 18,5 % for the ferrite of composition: $Ni_{0,65}\,Co_{0,001}\,Fe_{0,32}\,Fe_{2,03}\,O_{4,01}$. With this value of t, the slope of H_o = f (T) is nearly the same as the mean experimental slope of H_s = f (T), when temperature varies from - 100°C to + 400°C, figure 1.

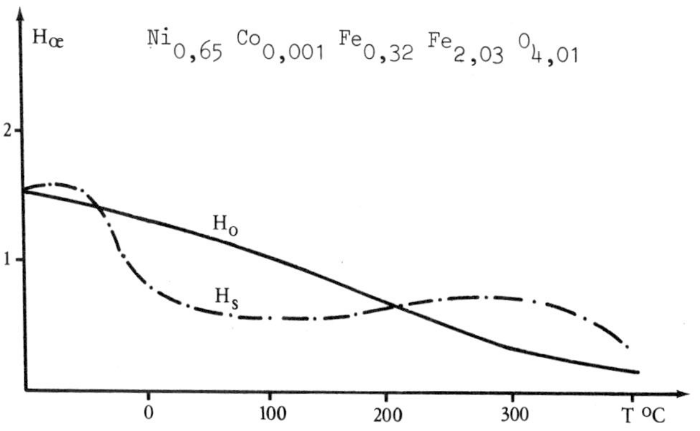

Figure 1 : Comparison between theory and experiment. H_s is the threshold field of the constricted hysteresis loops and H_o is the magnetic field necessary to push a 180° stabilized wall out of its potential well. With the classical model H_o is found to be 50 times larger than H_s.

Deviation from the mean slope is attributed to the fact that the easy magnetization directions of this ferrite deviate slighty from the <100> axes when temperature varies between 250°C and -50°C (maximum deviation is about 40° at room temperature). So the stabilized walls do not remain 180° walls and this fact has not been taken into account for the calculation of H_0. The potential wells being less efficient that explains the trough in the curve $H_S = f(T)$ and the large apparent misfit between H_S and H_0 in figure 1. In fact for other ferrites the easy magnetization directions remain always <100> and the variation of H_S is found to be more linear. In these cases the agreement between H_0 and H_S should be better but unfortunately the magnetostriction constants are unknown and we cannot estimate the F_λ and G_λ values for these ferrites.

b) We have shown,when the ferrite mentioned above is cooled from the Curie temperature, that wall stabilization occurs at first by stress relaxation. Later the classical model (rotation of the axis of induced anisotropy) becomes energetically more favorable but the previous equilibrium state cannot be altered because it would take too long time.

c) A phenomenological extension of the two hypothese (i) and (ii) from the particular case of the domain wall stabilization to the general case of a crystal exhibiting directional order, leads to the conclusion[7] that the energy of induced anisotropy (f_I) must be differentiated according to whether the anisotropy is measured on a crystal at constant stress (f_I^σ) or at constant strain (f_I^ϵ). These energies are linked together by the relation :

$$f_I^\sigma = f_I^\epsilon + f_I^\lambda \qquad (4)$$

where f_I^λ is the contribution of magnetostriction to induced anisotropy. Equation (4) is to be compared with the well known equation concerning cubic magnetocrystalline anisotropy :

$$f_K^\sigma = f_K^\epsilon + f_K^\lambda \qquad (5)$$

Generally f_K^λ is much smaller than f_K^ϵ and in the same way f_I^λ must be smaller than f_I^ϵ but its importance is that domain wall stabilization would only proceed from f_I^λ. In the temperature range of diffusion, when ordering occurs in the wall, the preceeding hypothese (i) and (ii)can be now interpreted as follows :

(j) induced anisotropy f_I remains uniaxial
(jj) the axis of f_I becomes parallel to the magnetization vector.

In the case of magnetocrystalline anisotropy, f_K^σ is measured and f_K^ϵ is obtained theoretically[8], after that further measurements have been made (magnetostriction, elasticity). The induced anisotropy is also measured at constant stress (f_I^σ) but the theory is still insufficient to separate f_I^ϵ and f_I^λ although equation (4) had been previously established in the special case of binary alloys[9].

CONCLUSION

A new interpretation of domain wall stabilization has been given. We assume that the stabilization occurs only by a process of stress relaxation inside the 180 walls. Application of this model leads to an interpretation which agrees with the thermal variations of the experimental threshold field H_S measured on a Ni-Fe-Co ferrite. Further experiments are necessary to confirm the experimental verification of the proposed model, but on other materials the knowledge of the thermal variation of all indispensable coefficients is unfortunately insufficient.

A phenomenological extension of our model leads to the conclusion that the induced anisotropy f_I^σ is split in two parts (f_I^σ is the induced anisotropy which is ususally obtained since the measurements are always made on a crystal at constant stress). The lowest part of f_I^σ proceeds from magnetostriction (f_I^λ) and we assume that f_I^λ is connected to the stress relaxation in the walls. The other part f_I^ϵ is the induced anisotropy at constant strain. In the classical model of wall stabilization, f_I^σ and f_I^ϵ were not differentiated and f_I^σ was assumed to be connected with wall stabilization. Until now the anisotropy energy measurements do not allow the experimental separation of f_I^ϵ and f_I^λ and we should to emphasize the high interest presented by experiments on the contribution of magnetostriction to the induced anisotropy. That is the only way to demonstrate theoretically either the proposed model is valid for a given material or not.

REFERENCES

1. L. NEEL : J. Phys. Radium 12, 339-351 (1951)
2. S. TANIGUGHI : Sc. Rep. Res. Inst. Tohoky Univ. A8 173-192 (1956)
3. W. SIMONET, M. PAULUS : Phys. Stat. Sol. Sol 22, 2, K87-90 (1967)
4. W. SIMONET : Phys. Stat. Sol. (a) 22, 231-238, (1974)
5. W. SIMONET, M. PAULUS : Inst. Conf. Magnetisme, Moscow (1973)
6. L. NEEL : Cahier de Phys. 25 1.20 (1944)
7. W. SIMONET : Thesis N° 1293 - C.N.R.S. France (1974)
8. C. KITTEL : Revue Mod. Phys. 21, 4, 541-583 (1949)
9. E.T. FERGUSSON : Thesis - C.N.R.S. - France (1962).

OPTIMIZING THE COMBINATION OF MAGNETIC INDUCTION AND YIELD STRENGTH OF VANADIUM PERMENDUR BY HEAT TREATMENT

H. C. Fiedler
Corporate Research and Development, General Electric Company
Schenectady, New York 12301

ABSTRACT

The high magnetic saturation of Vanadium Permendur, a soft magnetic alloy of 2% vanadium, 49% cobalt and 49% iron, makes this alloy particularly useful for laminations in generators and motors when weight is important. The normal heat treatment used to optimize the magnetic properties of this alloy produced a yield strength that was substantially below that required for the high speed rotor of a generator. Described are the results of a study of the relationship between microstructure and magnetic properties, and of the development of a heat treatment that provides a suitable combination of strength and high induction.

Vanadium Permendur, an alloy of 49% cobalt, 49% iron and 2% vanadium, has high magnetic saturation and zero magnetocrystalline anisotrophy[1]. These characteristics have made this alloy particularly useful for laminations in motors and generators when weight is important.

While physical strength is not normally an important requirement for soft magnetic materials, it may become so in high speed generators since the stress produced by centrifugal force may exceed the yield strength of the material. The study to be reported was prompted by the problem of building high-speed generators in which the laminations experience stresses substantially over the yield strength of Vanadium Permendur given a conventional anneal aimed solely at optimizing the magnetic properties. The design criteria led to the material requirements of a minimum yield strength of 70,000 psi with a minimum induction of 20,000 gauss in a field of 13 Oe and 22,400 gauss in a field of 85 Oe.

Mechanical and magnetic measurements were made on suitable specimens after heat treating in a hydrogen atmosphere. Load-elongation curves were made with samples having a gauge length of 0.7 inch and a width of about 0.2 inch. The strain rate was 0.03/min. In the fully recrystallized condition Vanadium Permendur shows an upper and lower yield point. The yield strength, defined as the stress at 0.2% strain, invariably equaled the lower yield point of recrystallized samples. Magnetic measurements were made on ¼inch high stacks of ring samples, usually 1-7/8 inches ID and 2½ inches OD. D-c hysteresis loops were traced for values of H between 5 and 85 Oe using an X-Y recorder.

The inability of a conventional batch-type heat treatment to provide the desired combination of strength and magnetic properties is illustrated by the properties in Table I of samples from a commercial heat (2.0% V, 49.1% Co, 48.7% Fe, 0.002% C, 0.0036% O and 0.0003% N) annealed for 2 hours at temperatures from 695°C to 750°C. The samples, sealed in a metal retort through which dry hydrogen was passed, were loaded into the hot zone of a hydrogen furnace which had been determined to be within 1° of the desired temperature. The furnace was equipped with a G.E. Reactrol Control, which maintained the temperature constant during the heat treatment. The yield strengths are from duplicate samples tested in the rolling direction, the yield strength generally being lower in this direction than in the transverse direction. It is seen that the only samples that meet the requirement of an induction of not less than 20,000 gauss with a field of 13 Oe were annealed for 2 hours at 710° or 750°C, but their yield strengths are too low to be useful.

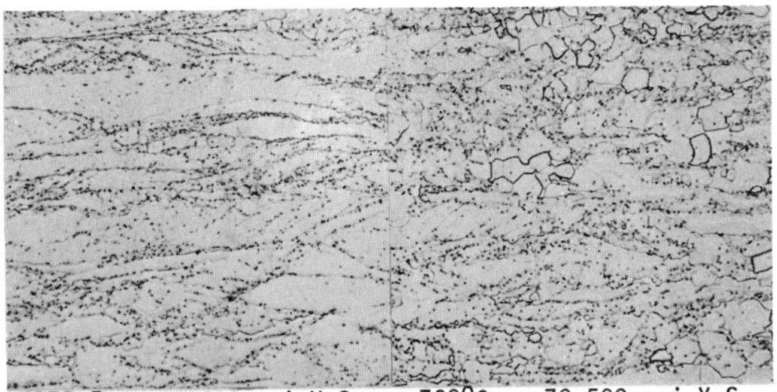

695° C -- 99,600 psi Y.S. 702°C -- 72,500 psi Y.S.

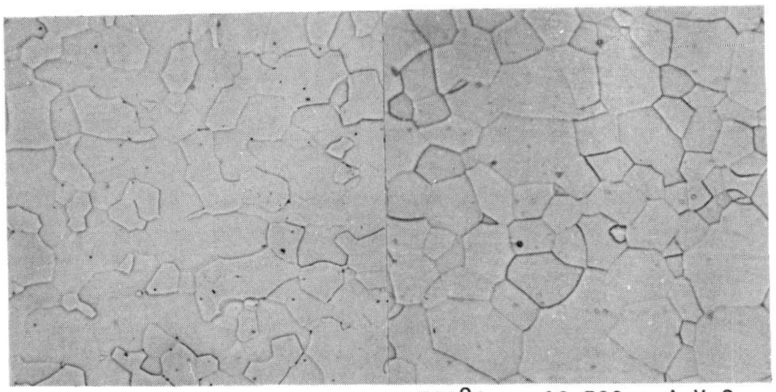

710°C -- 55,300 psi Y.S. 750°C -- 46,500 psi Y.S.

Fig. 1 Microstructures after heating for 2 hours at the temperatures indicated. The yield strengths associated with these widely differing grain structures are also indicated. The intercept grain size of the samples annealed at 710° and 750°C is 9 and 12 microns, respectively. 500X

The microstructures of some of these samples are shown in Fig.1. The partial recrystallization that occurred at 702°C is responsible for the wide variation in the yield strength of samples annealed at that temperature. Gamma, a non-magnetic phase comprised of approximately 22 pct V, 65 pct Co and 13 pct Fe and which readily forms in a cold worked structure[2], is prevalent in the samples annealed at 695° and 702°C, visible in the sample annealed at 710°C, and essentially absent from the sample annealed at 750°C. Although the degradation of the magnetic properties of the samples annealed at 702° is probably due primarily to the lack of complete recrystallization, the magnetic properties of the samples annealed at 704° and 710°C would presumably be improved by the absence of gamma.

Since in order to obtain the desired magnetic

TABLE I
TENSILE AND MAGNETIC PROPERTIES AFTER ANNEALING 2 HRS.

	Annealing Temperature °C				
	695	702	704	710	750
Y.S., psi	96,700	85,200	59,700	54,800	46,500
	99,600	72,500	60,000	55,300	46,700
B @ 10 Oe	-	10,700	17,400	19,400	21,900
B @ 85 Oe	-	22,000	22,600	22,900	23,100

properties the structure must be completely recrystallized and, preferably, free of gamma phase, the only strengthening mechanism is to reduce the grain size. McLean[3] cites many examples of this relationship, the yield strength increasing linearly with the reciprocal of the square root of the grain size.

Recrystallization and grain growth are intimately related processes, and it is well known that both are affected by the temperature and time of heat treatment. In general, a small recrystallized grain size is achieved by heat treating at a short time so as to minimize grain growth and at a relatively high temperature to promote the nucleation of many rather than a few recrystallized grains. Such conditions are readily achieved in a continuous heat treatment, which is characterized by rapid rates of heating and cooling as compared to a batch heat treatment.

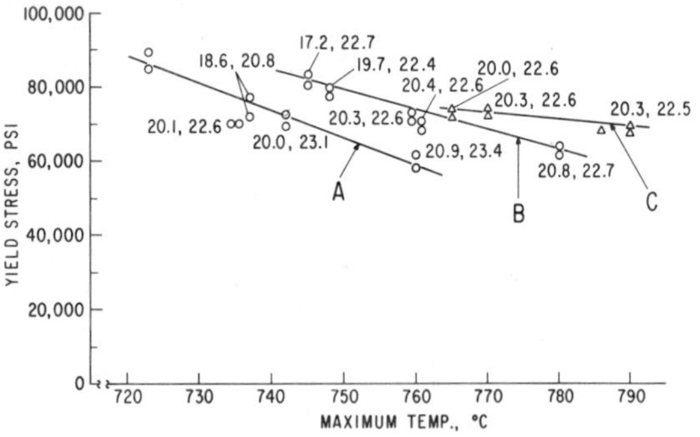

Fig. 2 The yield strength as affected by the maximum temperature reached with three different heat treating cycles. The times within 10^0 of the maximum temperatures were 3.5, 2 and 1 minutes for cycles A, B and C, respectively. Adjacent to the data points are the inductions (kilogauss) reached in fields of 13 and 85 Oe.

In Fig.2 are shown the longitudinal yield strengths and inductions of samples heat treated over a range of temperature with three heat treating cycles. Many samples with yield strengths in the 70,000 to 80,000 psi yield strength range meet the magnetic requirement of a minimum induction of 20 kG at 13 Oe and 22.4 kG at 85 Oe. In general, acceptable properties are obtained with a maximum temperature of between 736°and 742°when using cycle A, between 748° and 759°when using cycle B, and between 765°and 790°C when using cycle C, which allows the shortest time at temperature. The shorter the time at temperature the less is the opportunity for grain growth after recrystallization, and the less critical is the heat treating temperature.

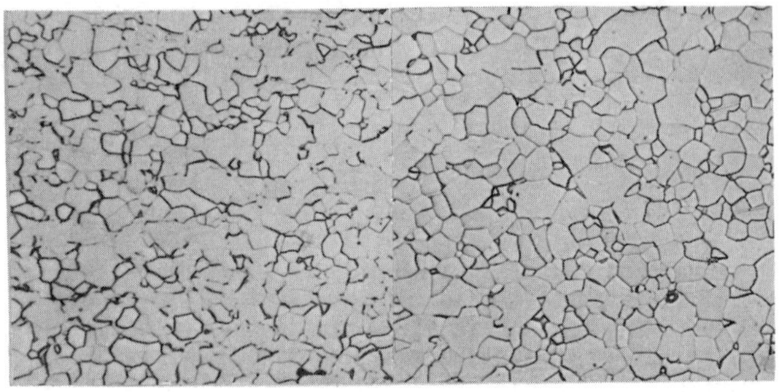

748°C -- 79,400 psi Y.S. 759°C -- 72,800 psi Y.S.

Fig. 3 Microstructures after continuous heat treatment cycle B. The maximum temperatures reached in the heat treatment and the yield strengths are indicated. 500X

Metallographic examination of many of these samples revealed a complete absence of gamma phase and, in those with an adequate strength, an intercept grain size of 8 microns or less. Photomicrographs of two samples heat treated using cycle B are shown in Fig.3.

SUMMARY

Batch type heat treatments were incapable of providing the desired combination of strength and high induction. High temperature heat treatments resulted in coarse grained, single phase microstructures with high inductions but low strengths. Lower temperature heat treatments resulted in high strengths but low inductions, a consequence of partial recrystallization and the occurrence of gamma phase. The desired combination of strength and high induction was achieved with high temperature, short time heat treatments, which resulted in a single phase microstructure of small grains.

REFERENCES

1. Hall, R.C., J. Appl.Phys., Vol. 31, 157S-158S, 1960.
2. Fiedler, H.C. and Davis, A.M., Met. Trans. Vol. 1, 1036-1037, 1970.
3. McLean, D., Mechanical Properties of Metals, John Wiley and Sons,Inc., New York, 1962, Chap. 4

ACKNOWLEDGEMENTS

The photomicrographs were made by C. R. Rodd and D. W. March; P.T. Hill made all of the mechanical tests. Helpful discussions were held with members of the Aircraft Equipment Div., General Electric Co.

KINETIC ENERGY OF A MOVING DOMAIN WALL IN ORTHOFERRITE

S. Konishi and T. Miyama
Faculty of Engineering, Hiroshima University, Hiroshima 730, Japan

ABSTRACT

The effective internal field responsible for spin precession and wall width are calculated by the molecular field approximation in a moving wall of canted antiferromagnet. The maximum effective field is limited by Dzyaloshinski-Moriya exchange field, but the wall width increases infinitely with increasing drive field. The calculated velocity-field relation is linear, and neither the velocity break-down nor saturation occurs.

INFLUENCE OF HEAT TREATMENT ON MAGNETIC PROPERTIES OF A SOFT IRON COBALT ALLOY

A.J. Moses
Wolfson Centre, University of Wales, 30 The Parade, Cardiff CF2 3AD, U.K.

ABSTRACT

Cobalt-iron-vanadium alloys are used in applications requiring high flux densities and good mechanical strength, such as in airborne power generating equipment. The alloy containing 49%Co, 49%Fe and 2%V, is extremely sensitive to the type of heat treatment carried out after fabrication. Changes in magnetic properties by varying the heat treatment temperature from 450°C to 800°C are shown. The variation of a.c. magnetostriction and power loss are discussed in terms of changes in anisotropy, magnetostriction constants, and structure.

The material is sensitive also to mechanical stresses of the type which could be expected in many devices. Tension or compression applied parallel to the magnetizing direction produce the largest changes. After final heat treatment at temperatures above 700°C compressive stress increases the power loss whereas tensile stress has little effect. After annealing between 450°C and 700°C, tensile stress can have a harmful effect and at some flux densities (up to about 1.0T) a "stress after effect" is present as indicated by a gradual change in loss after a stress is removed from a sample.

INTRODUCTION

Cobalt-iron-vanadium (nominally 49Co 49Fe 2V) is potentially an extremely useful soft magnetic alloy for applications in which operation under severe ambient conditions of temperature or mechanical stress is necessary. It also has a high saturation induction which allows a far better flux carrying capability than most materials. One application of the alloy where special requirements are necessary is as laminations in aircraft generators. The stator laminations carry a large a.c. flux and the rotor laminations must be mechanically very strong but also be able to carry a high d.c. flux for efficient operation. The rotor must be able to withstand high centrifugal stresses without degradation of magnetic properties or dimensional stability.

The cobalt-iron-vanadium alloy can satisfy most of the requirements to some extend by suitable heat treatment prior to fabrication. Both the magnetic and mechanical properties are extremely sensitive to heat treatment over the temperature range 350°C to 900°C. Depending on the requirement, either very good magnetic properties or good mechanical properties may be obtained by suitable heat treatment, but it is difficult to optimise the two together.

As part of a comprehensive study of various heat treatments on the material, measurements have shown the d.c. and a.c. magnetic properties to be very dependent on mechanical stress of the type expected to be present in a machine. The stress sensitivity is also dependent on the type of heat treatment carried out prior to measurement. In this paper, the effects of heat treatment temperature on the stress sensitivity of the a.c. properties is discussed.

EXPERIMENTAL RESULTS

Samples of cobalt-iron-vanadium alloy were obtained from two manufacturers in the cold worked state. The material, .014" thick, was cut into samples 7" long by 1" wide before heat treatment. The heat treatment was carried out in a vacuum of 10^{-5} torr in a furnace in which the temperature could be controlled to ± 0.5°C.

The magnetic properties of the samples were measured in a versatile test rig capable of applying mechanical stress of up to $\pm 50,000$ lbf/in^2 at ambient temperatures from -50°C to 350°C. The power loss could be measured under sinusoidal conditions up to 2.35T over a magnetizing frequency from 5 Hz to 800 Hz using an electronic wattmeter. Magnetostriction was measured using standard 350Ω strain gauges and a commercial strain gauge bridge. The mechanical design of the apparatus and measuring systems are described elsewhere[1].

Fig. 1 shows curves of power loss plotted against longitudinal stress (i.e. stress applied along the magnetizing direction) for a series of samples from one batch annealed over a temperature range from 450°C to 800°C. The material obtained from the second supplier behaved in an almost identical manner. The heat treatment cycle was to raise the samples to the appropriate temperature, hold for two hours, and then cool

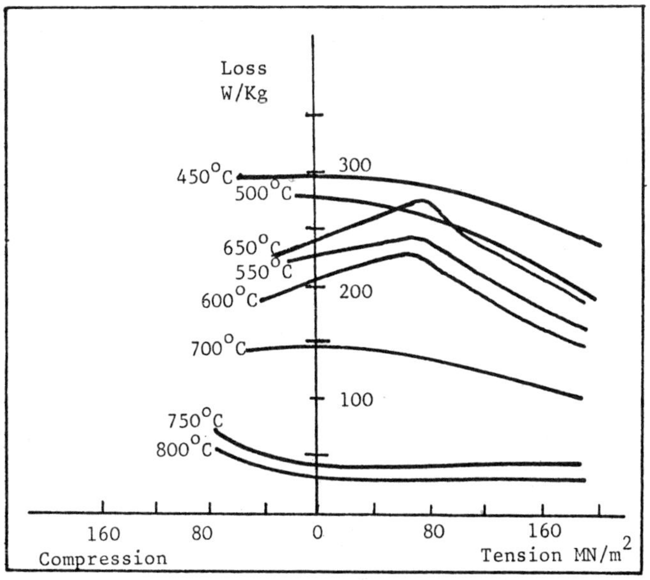

Fig. 1 - Variation of power loss measured under stress magnetized at 1.4T 400Hz, after annealing for 2 hours at various temperatures.

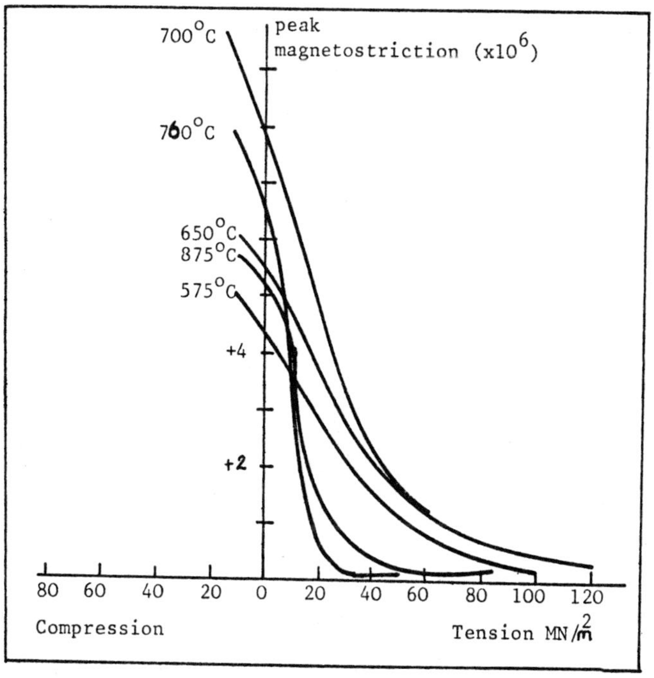

Fig. 2 - Variation of stress sensitivity of peak magnetostriction measured at 400Hz, 1.5T, after annealing for 2 hours at various temperatures.

at 100°C/hr to room temperature. The cooling rate and time at temperature did not affect the shapes of the curves shown in Fig. 1.

The magnetostriction of the samples was measured under the same conditions and Fig. 2 shows typical curves of the fundamental component (800 Hz) when magnetized at 400 Hz.

The stress sensitivity of the power loss is dependent on the flux density unlike most other metallic alloys. The curves are similar to those shown in Fig. 1 over the flux density range from 0.5T to 2.35T, below 0.5T the shapes of the curves after some heat treatments changed dramatically. Fig. 3 shows the variation of power loss with flux density in a sample annealed at 450°C. At low flux densities the loss increases under tension whereas at high values it rises when compression is applied.

After annealing in the temperature range 500°C to 700°C, a further change in stress sensitivity occurs as illustrated in Fig. 4, which is a typical stress sensitivity curve of a sample annealed at 700°C and magnetized at 0.35T. Before stress is applied, the power loss is the value at A. When tension is applied and then removed. it returns to the value at B.

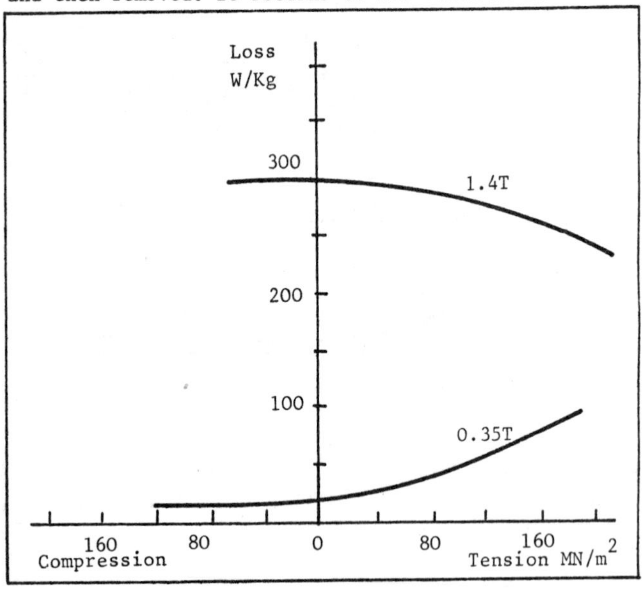

Fig. 3 - Variation of power loss under stress measured at two flux densities and 400Hz after annealing for 2 hours at 450°C.

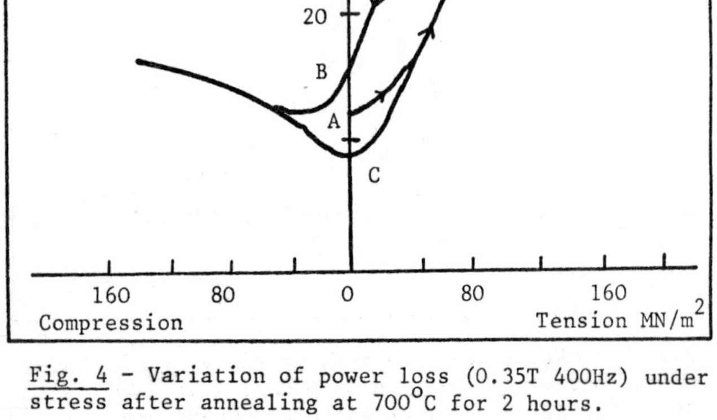

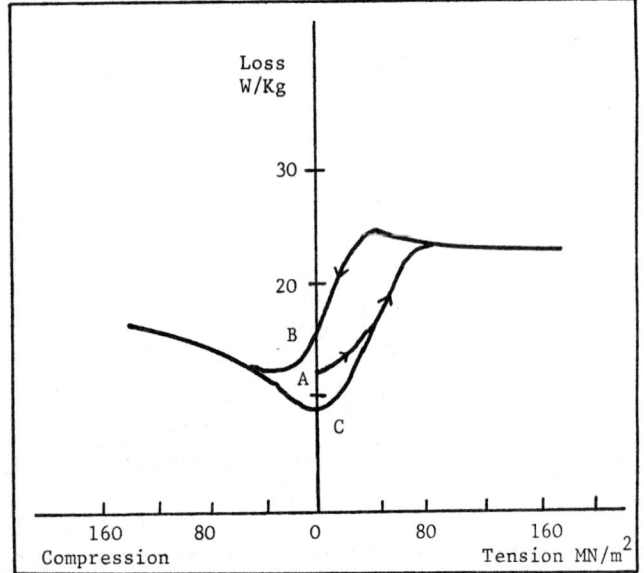

Fig. 4 - Variation of power loss (0.35T 400Hz) under stress after annealing at 700°C for 2 hours.

Similarly, after compression is applied, the loss returns to the value at C. If the sample is left in the condition at B or C, the loss slowly returns to the value at A after about 15 minutes.

DISCUSSION

The changes in a.c. loss and magnetostriction caused by the application of stress are dependent on the metallurgical structure of the material. It is known that the material undergoes several changes when heated to 900°C after cold rolling[2]. After annealing above 710°C, the material is in a recrystallized state and is totally ordered which causes a low iron loss. In this condition, the magnetostriction constants are positive (λ_{100} and λ_{111}) and the anisotropy constant (K_1) is also positive; because of this when a compressive force is applied parallel to the magnetizing direction, the domains tend to be aligned perpendicular to the stress and the loss and magnetostriction increase as shown.

After annealing at lower temperatures, the material is in a partially ordered state and high residual stresses might be present. The magnetostriction constant is still positive, but K_1 tends to become negative, thus causing compression to have a small effect, although the loss of the material is higher due to residual stress. The loss slowly drops with annealing temperature up to about 600°C and then it increases sharply before falling again. The increase is probably due to a large decrease in resistivity which occurs after annealing at 625°C. Precipitation of γ_2 phase has been suggested in the same temperature region[3], if this occurs it would cause an increase in loss.

The change in stress dependence of power loss at low flux density has not been reported in any material previously. The loss slowly reverts to its original value after removal of the stress, which suggests that it might be associated with a stress relaxation effect caused by interstituals which become more mobile after annealing in the range from 500°C to 700°C. The mechanism does not appear to operate at flux densities above about 1.5T, when magnetization by rotational processes begins to come into operation. It would be interesting to measure the stress sensitivity of magnetostriction at low flux density to see whether similar effects occur, but the apparatus used for this particular work was not sensitive enough to detect the small signals.

CONCLUSIONS

The measurements show that the cobalt-iron-vanadium alloy exhibits quite different properties under stress after annealing over the temperature range from 450°C to 800°C. The stress sensitivity curves can be explained qualitatively in terms of changes in anisotropy and structure. A stress relaxation effect occurring at low flux density is thought to be due to interstituals in the alloy.

ACKNOWLEDGEMENTS

This work was carried out in a programme sponsored by the UK Ministry of Defence (Procurement Executive). The author is grateful to Dr. B. Thomas, of the Wolfson Centre, who designed the apparatus and made many of the measurements.

REFERENCES

1. B. Thomas (to be published).
2. E. Josso, IEEE Trans. Mag-10, 2, 161-165, 1974.
3. C.W. Chen, Cobalt, 22, 3-21, 1964.

AN ELECTRON MICROSCOPIC STUDY OF THE ORIGIN OF COERCIVITY IN AN Fe-Co-V ALLOY

S. Mahajan and K. M. Olsen
Bell Laboratories, Murray Hill, N.J. 07974

ABSTRACT

To ascertain the origin of coercive force in an Fe-Co-V alloy (2.91 wt.% V, Fe=Co), we have studied by transmission electron microscopy the microstructures and domain wall configurations after the following treatments: A_1 - cold rolled to 97% reduction + strand annealed for 30 secs. at 950°C + aging at 600°C for 2 hrs; A_2-treatment A_1+ 5 secs at 1050°C; A_3-treatment A_1+ 5 secs at 950°C; A_4-treatment A_1+ 5 secs at 850°C. The values of coercive force are 25, 32, 38 and 13 Oe, respectively. The microstructure of A_1 samples shows precipitation within some grains. Lorentz microscopy reveals that domain walls are pinned by these precipitates. No precipitation is observed in A_2 samples, and instead they exhibit substructures characteristic of a structural transformation. The interaction between the transformation-induced substructure and domain walls appears to be weak. The structure of A_3 and A_4 samples consists of grains having relatively low and moderately high dislocation density. In A_3 samples, domain walls tend to lie along the interfaces of the latter grains and rarely penetrate them. It is envisioned that the overall coercivity of A_4 samples is lower because the volume fraction of the high dislocation density grains as well as the magnitude of internal strain are lower.

INTRODUCTION

Remendur, an alloy containing 2.5-3.0 wt.% V and Fe=Co, is a semihard magnetic material, and is presently being used in the fabrication of dry reed sealed contacts. The alloy undergoes a host of transformations, viz. a structural transformation (f.c.c.→b.c.c. on cooling from high temperature), order-disorder and precipitation reactions. The phase equilibria and transformations have previously been investigated by optical metallography, scanning electron microscopy-microprobe analysis and transmission electron microscopy.[1,2]

Recently, Pinnel and Bennett[3] have examined the influence of aging temperature on the structure and magnetic properties of reeds. It is observed that, depending upon the aging temperature, coercivity (Hc) may vary from 45 to 2 Oe, the minimum occurring in the range of 700-800°C. In the present study, the effect of a short time anneal at high temperatures on the magnetic and structural characteristics of strip samples, representative of those of the reeds, has been evaluated. An attempt has been made to rationalize the observed changes in Hc by examining the domain wall configurations by Lorentz electron microscopy.

EXPERIMENTAL

An alloy containing 2.91 wt.% V and Fe=Co was rolled into 0.007" thick strips. This material was strand annealed for 30 secs. at 950°C. The strand-annealed strip was subsequently aged at 600°C for 2 hrs. We will designate these samples as A_1, and they should, microstructurally as well as magnetically, be analogous to the reeds. These specimens were then subjected to a 5 sec anneal at 1050°, 950° and 850°C. The designations for these samples are A_2, A_3 and A_4. Magnetic properties were measured using the standard techniques and Hc's were evaluated from full hysteresis loop traces. To prepare thin sections, suitable for electron microscopy, samples were chemically thinned in a solution consisting of equal volumes of H_2O_2 and H_3PO_4. Chemically polished specimens were subsequently electropolished, using the window technique, in a 80% CH_3OH-$HClO_4$ solution maintained at -30°C. Suitable sections were cut from the thinned samples and examined in a Jem 200 microscope operating at 200 kV. Since the alloy is ferromagnetic and distorts the electron beam, very small sections were examined. To observe domain walls, the procedure of Hale, Fuller and Rubinstein[4] was followed.

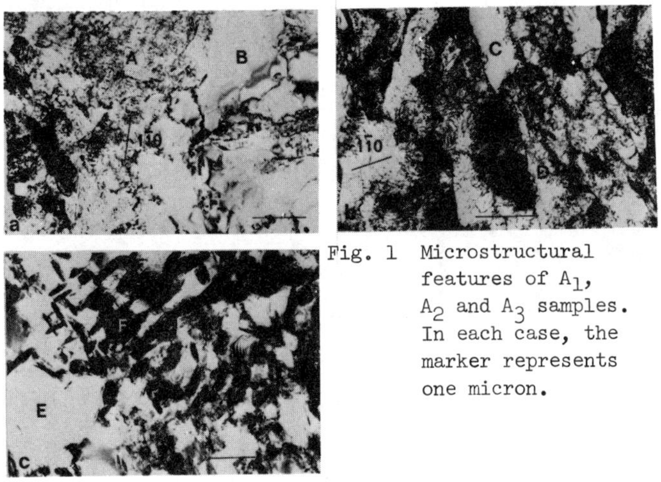

Fig. 1 Microstructural features of A_1, A_2 and A_3 samples. In each case, the marker represents one micron.

RESULTS AND DISCUSSION

Figures 1(a), (b) and (c) illustrate the microstructural characteristics of A_1, A_2 and A_3 samples, respectively. The unique feature of Fig. 1(a) is the presence of two types of grains, such as A and B. Extremely fine precipitates are observed only within A and dislocation density is relatively low in B. It has been shown previously[2] that these precipitates lie on the {110} planes, are semicoherent and may have an f.c.c. structure. Consequently, the material is likely to be internally stressed in the vicinity of those precipitates. When an A_1 specimen is annealed for 5 sec at 1050°C, the original microstructure is totally changed, compare Figs. 1(a) and (b). Precipitates are no longer observed and instead low and moderately high dislocation density regions, such as C and D, are seen. These observations are consistent with the assessment that: (i) V has been retained in a supersaturated solid solution; (ii) the sample has undergone a structural change, namely f.c.c.→ b.c.c. Both these factors result in internal stresses.

The microstructure observed following a 5 sec anneal at 950°C shows E and F type grains, Fig. 1(c), which differ markedly in their structural features. Dislocation density is fairly low in E, whereas it is quite high in F. It has been shown previously[1] that these microstructures develop during cooling from the two-phase field, and E and F grains represent the V-poor and V-rich regions existing at high temperature. The observed high dislocation density of F could be rationalized by assuming that on cooling the latter have undergone a structural transformation (f.c.c.→ b.c.c.). Since the lattice parameter of the V-rich regions may be different from that of the V-poor grains and since the former transform as well, internal stresses may exist at the interfaces between these two types of grains. The microstructural characteristics of A_4 samples is similar to that of A_3, except that the number of F type grains is less. As the compositions of the V-rich grains in A_3 and A_4 are likely to be different, the magnitude of internal stresses should not be the same in the two cases.

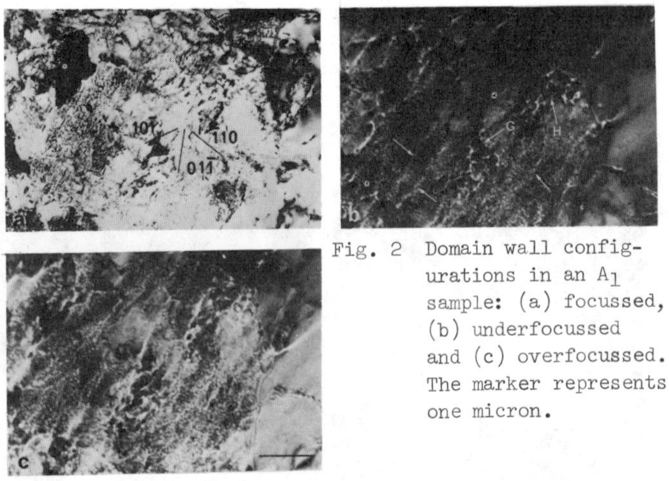

Fig. 2 Domain wall configurations in an A_1 sample: (a) focussed, (b) underfocussed and (c) overfocussed. The marker represents one micron.

Hc's of A_1, A_2, A_3 and A_4 samples are 25, 32, 38 and 13 Oe, respectively. Domain wall configurations in A_1, A_2 and A_3 specimens, observed by under and over-focussing the objective lens, are illustrated in Figs. 2, 3 and 4, respectively. In Fig. 2, the plane of the foil is close to {111}. Consequently, for the system to have a minimum energy, domain walls should lie along the <1$\bar{1}$0> directions lying in the {111} plane.[5] The domain topology at G and H is quite complex. Fissure-like features, some of which are arrowed in Fig. 2(b), could be the walls pinned by the precipitates. Except for some portions of the curved wall, the majority of the walls do not lie along the <1$\bar{1}$0> directions lying in the {111} plane. Thus the domain wall-precipitate interaction prevents them from attaining the minimum energy configuration. It is inferred that this interaction is likely to be responsible for the Hc in this material. Pinnel and Bennett[6] have advanced a similar suggestion.

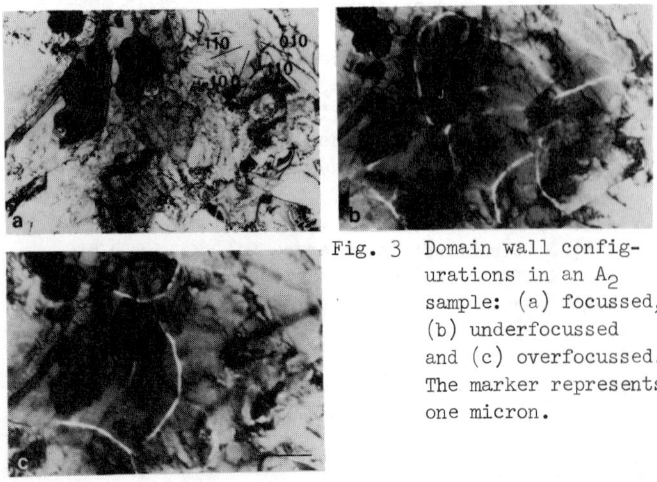

Fig. 3 Domain wall configurations in an A_2 sample: (a) focussed, (b) underfocussed and (c) overfocussed. The marker represents one micron.

In Fig. 3, the orientation of the foil is {001}, and the majority of the walls tend to lie along the <110> and <200> vectors lying in the {001} plane. Since the magnetocrystalline anisotropy in equiatomic Fe-Co alloy is quite small,[5] these results are not surprising. In addition, a wall is able to propagate across a highly dislocated area and its segments also lie along the <200> and <110> directions. This implies that the wall-substructure interaction is weak. This assessment is consistent with that of Silcox,[7] but is at variance with the results of Marcinkowski and Poliak.[5] A domain wall splits up into two during its interaction with grain J, but its components still lie along the <200> and <110> vectors. From the present observations, it is not clear which features are responsible for Hc in this material. It is conceivable that internal strains via the magnetostrictive interaction produce subtle changes in domain configurations which we have not resolved.

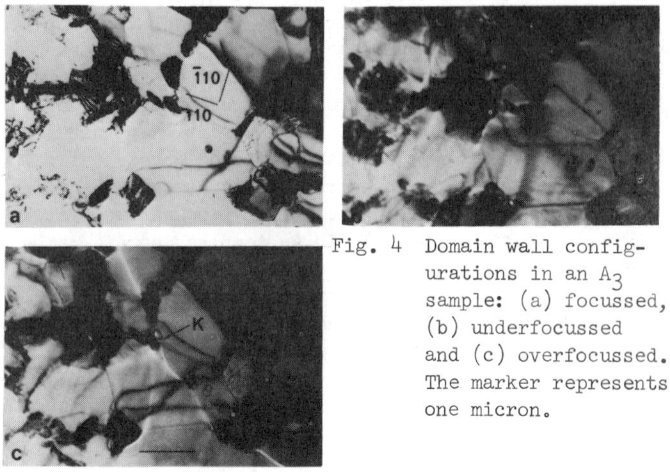

Fig. 4 Domain wall configurations in an A_3 sample: (a) focussed, (b) underfocussed and (c) overfocussed. The marker represents one micron.

In E type grains, Fig. 4, domains generally lie along the <110> and <200> vectors lying in the plane of the foil, i.e. {001}. The walls tend to lie along the interfaces between E and F type grains and rarely penetrate them. Occasionally, the interaction results in a complex domain topology, see K in Fig. 4(c). It is envisioned that in this material, Hc is caused by the internal stress-domain wall interaction. In A_4 samples, such an interaction is likely to be weak, and hence Hc is lower, for the following reasons: (i) the volume fraction of F type grains is smaller; (ii) the magnitude of internal strain may also be lower because of the compositional differences.

In conclusion, it has been shown that it is generally possible to rationalize the origin of Hc by correlating the domain wall configurations with microstructures.

REFERENCES

1. J. E. Bennett and M. R. Pinnel, J. Mater. Sci. 9 1083-1090, 1974.
2. S. Mahajan, M. R. Pinnel and J. E. Bennett, Met. Trans. 5, 1263-1272, 1974.
3. M. R. Pinnel and J. E. Bennett, Bell Laboratories, Columbus, Unpublished work, 1974.
4. M. E. Hale, H. W. Fuller and H. Rubinstein, J. Appl. Phys., 30, 789, 1959.
5. M. J. Marcinkowski and R. M. Poliak, Acta. Met. 12, 179-190, 1964.
6. M. R. Pinnel and J. E. Bennett, Met. Trans. 5, 1273-1283, 1974.
7. J. Silcox, Phil. Mag. 8, 7-28, 1963.

FERROMAGNETIC BEHAVIOR OF METALLIC GLASSES

R. C. Sherwood, E. M. Gyorgy, H. S. Chen, S. D. Ferris, G. Norman* and H. J. Leamy
Bell Laboratories, Murray Hill, New Jersey 07974
*Permanent Address: Southern University, Baton Rouge, La.

ABSTRACT

Magnetic properties of metallic glasses in the system (Fe-Co-Ni)$_{0.75}$ (P-B-Al)$_{0.25}$ have been investigated. Samples were made by the roller quenching technique in the form of ribbons 0.002 in. thick and about 0.125 in. wide. Room temperature saturation magnetization, $4\pi M_S$, ranged from a high value of 13,500 Gauss for $Fe_{0.75}P_{0.16}B_{0.06}Al_{0.03}$ to essentially zero for $Ni_{0.75}P_{0.16}B_{0.06}Al_{0.03}$ which exhibits only weak temperature independent paramagnetism. The highest Curie temperatures occurred in mixed Fe-Co alloys with a high value of 715°K for $(Fe_{0.4}Co_{0.6})_{0.75}P_{0.16}B_{0.06}Al_{0.03}$. The sign of the magnetostriction was found to change from positive for Fe rich alloys to negative for Co rich alloys so that low or zero magnetostriction compositions can be made near the Co rich side of the pseudo-ternary phase field. For example $(Fe_{0.04}Co_{0.96})_{0.75}P_{0.16}B_{0.06}Al_{0.03}$ has nearly zero magnetostriction. This alloy responds to heat treatment in a magnetic field, developing properties which compare favorably with commercial supermalloy.

INTRODUCTION

Glassy alloys containing transition metal elements have been formed as small samples by liquid state quenching with several different techniques.(1) recently H. S. Chen and C. E. Miller have produced long ribbon shaped samples of ferromagnetic glassy alloys by roller quenching.(2) The ribbons are formed from the melt by directing a fine stream of molten metal between two rapidly rotating rollers, held together with pressure, to obtain quenching rates of up to about 10^5 °C/sec. Ribbons produced by this technique are about 0.002 in thick and 0.125 in wide. Other details have been described elsewhere.(3) In the present investigation we have studied magnetic properties of roller quenched ribbons of metallic glasses in the system (Fe-Co-Ni)$_{0.75}$(P-B-Al)$_{0.25}$. The amorphous nature of the ribbons was confirmed by x-ray or electron diffraction.

MAGNETIZATION

Saturation magnetization, $4\pi M_S$, was measured with a pendulum magnetometer from room temperature to 1.5°K, with applied fields up to 15,300 Oe, which were sufficient to saturate all but the most nickel rich alloys. As shown in Fig. 1, the high value of 13,500 G obtained at room temperature for the Fe alloy decreased for solid solutions with Co or Ni to essentially zero for the Ni alloy which is a temperature independent paramagnet. The magnetic moment per transition metal atom measured at 1.5°K is about 2, 1, and 0 Bohr magnetons for Fe, Co, and Ni respectively, which can be compared with 2.2, 1.7, and 0.6 found for the crystalline elements. Similar results have been reported by Mizoguchi et al (4) for a series of alloys containing phosphorus and carbon. The moment of Fe is least affected by the addition of the glass forming elements (P-B-Al), while the moment of Co is considerably decreased; but Ni appears to act as a paramagnetic diluent in the mixed transition metal alloys.

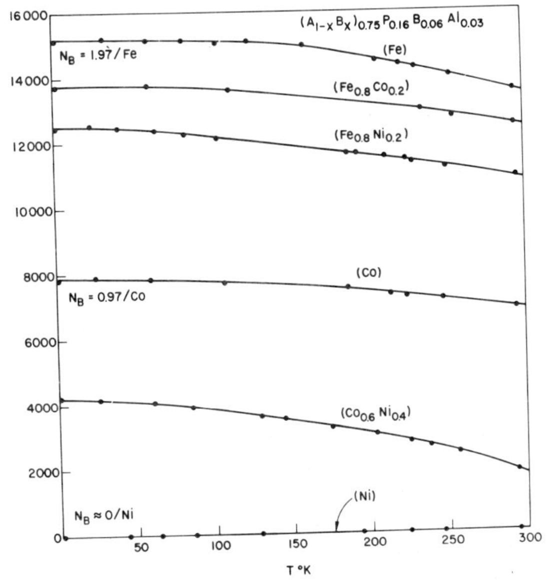

Fig. 1. Saturation magnetization vs temperature for some metallic glasses.

CURIE TEMPERATURES

Curie temperatures shown in Fig. 2 were measured above room temperature with a differential scanning calorimeter and below room temperature by the temperature dependence of the magnetization. A maximum was observed in the Fe-Co system with a high value of 715°K for $(Fe_{0.4}Co_{0.6})_{0.75}P_{0.16}B_{0.06}Al_{0.03}$. The Curie temperatures of many of the compositions are high enough for these materials to be considered for practical use.

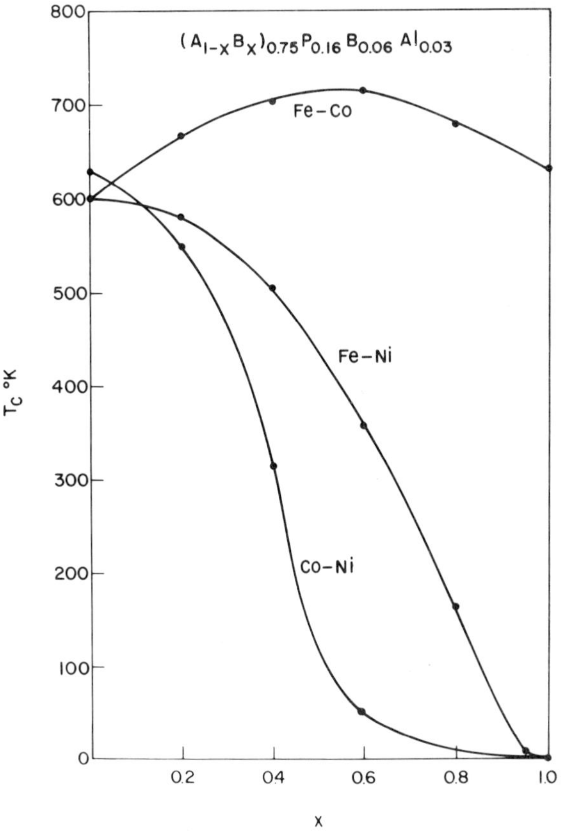

Fig. 2. Curie temperatures of some metallic glasses.

MAGNETOSTRICTION

Hysteresis loops measured on short samples with a vibrating sample magnetometer revealed that these materials were magnetically soft, as has been noted by other workers,(5,6,7) but were somewhat strain sensitive. Accordingly a search was made for compositions with zero magnetostriction, by observing changes in the loops caused by applying tension. The sign of the magnetostriction was found to change from positive for Fe rich alloys to negative for Co rich alloys so that low or zero magnetostrictive compositions can be made near the Co rich side of the pseudo-ternary phase field, as shown in Fig. 3. For example $(Fe_{0.04}Co_{0.96})_{0.75}P_{0.16}B_{0.06}Al_{0.03}$ has nearly zero magnetostriction and was therefore chosen for further study.

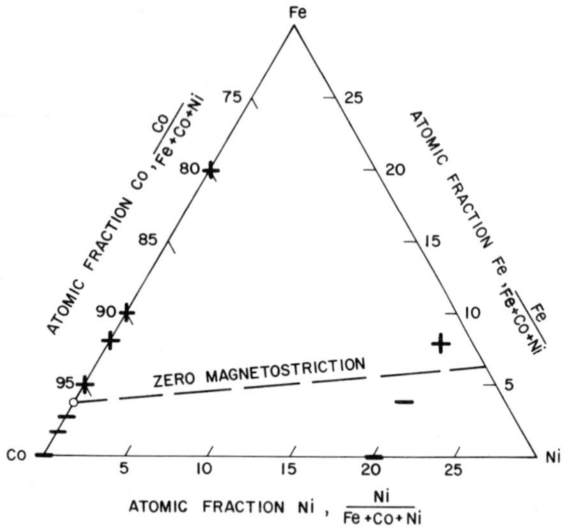

Fig. 3. Pseudo-ternary phase diagram of metallic glasses in the system $(Fe\text{-}Co\text{-}Ni)_{0.75}(P\text{-}B\text{-}Al)_{0.25}$, showing compositions with zero magnetostriction. Plus and minus signs indicate positive and negative magnetostriction.

MAGNETIC ANNEALING

To minimize demagnetizing fields, 150 cm lengths of ribbon were wound on ceramic bobbins. The low field hysteresis loop of an as quenched ribbon was quite square with a low coercive force of 0.023 Oe, but with a remanence value of 2840 G, which is only about half the saturation magnetization. The cause of the low remanence was made clear by the domain structure observed by scanning electron microscopy.[8] Domain regions with local uniaxial anisotropy were observed along various directions, with many oriented across the tape width. Annealing a sample only resulted in a slightly more skewed hysteresis loop. A sample was then annealed at 270°C for 45 min. with a magnetic field of 30 Oe applied along the tape length. The heat treatment in a magnetic field induced a dramatic improvement in properties.[9] The domain structure changed to relatively large domains with long straight boundaries aligned parallel to the tape length. The coercive force was reduced to 0.013 Oe and the remanence was increased to 4500 G, with a squareness ratio Br/Bm of 0.71. These properties compare favorably with commercial supermalloy.

Further work is in progress on other compositions with higher saturation magnetization. Roller quenching is essentially an inexpensive process and the promising magnetic properties together with ease of manufacture should lead to technologically useful soft magnetic materials.

REFERENCES

1. P. Duwez, Trans. ASM, 60, 607 (1967).
2. H. S. Chen and C. E. Miller, Rev. Sci, Instrum. 41, 1237 (1970).
3. H. S. Chen and D. E. Polk, J. Non-Crystalline Solids, 15, 165 (1974).
4. T. Mizoguchi, K. Yamauchi and H. Miyajima, Amorphous Magnetism, 325, ed. Hooper and deGraff, Plenum Press N.Y. (1973).
5. P. Duwez and S. C. H. Lin, J. Appl. Phys. 38, 4096 (1967)
6. G. S. Cargill III, and R. W. Cochrane, Amorphous Magnetism 313, (op. cit.).
7. T. Egami, P. J. Flanders, and C. D. Graham, Jr., to be published.
8. H. J. Leamy, S. D. Ferris, G. Norman, D. C. Joy, R. C. Sherwood, E. M. Gyorgy, and H. S. Chen to be published.
9. H. S. Chen, S. D. Ferris, E. M. Gyorgy, H. J. Leamy, and R. C. Sherwood, to be published.

TEMPERATURE AND FIELD DEPENDENCE OF BLOCH WALL THICKNESS MEASURED BY NEUTRON SMALL-ANGLE SCATTERING

O. Schärpf
Institut A für Physik der Technischen Universität
33 Braunschweig, Germany

ABSTRACT

Small-angle neutron scattering from regularly arranged Bloch walls in bulk single crystals of Fe-2.5% Si is observed. The scattering pattern is described and reasons are given to show that the Bloch walls cause the observed pattern. Three possible physical origins of the pattern are discussed. Temperature and field dependence give additional data for discussion.

EQUATION OF MOTION OF A FLEXIBLE DOMAIN WALL OF ARBITRARY SHAPE*

W. J. Carr, Jr.
Westinghouse Research Laboratories
Pittsburgh, Pennsylvania 15235

ABSTRACT

By means of a thermodynamic approach a general equation has been derived for the force density and equation of motion of each point on a flexible simple domain wall of arbitrary shape. A general expression is also given for the wall energy.

INTRODUCTION

In the real environment of a crystal, magnetic domain walls, whether static or in motion, seldom possess an ideal geometry, and the problem of calculating the force density acting at any point on the surface of the wall can be quite difficult. For irregularly curved surfaces and walls of arbitrary angle, magnetic terms in the energy, because of their long-range nature, must be treated with particular care. In addition, dissipation and wall damping are simply related only in non-metals. For a metal much of the dissipation occurs outside the wall. The purpose of the present calculation is to provide a reasonably rigorous derivation for a general equation of motion for each point on a flexible wall, in terms of all the force densities acting on the point. A thermodynamic treatment is used based upon consideration of entropy, energy conservation and certain symmetry arguments. The wall is assumed to be thin and for simplicity only a slowly moving "simple" wall is considered.

OUTLINE OF THE CALCULATION

If A is a fixed surface which instantaneously encloses the wall, the time rate of change of energy within the wall is

$$-\frac{c}{4\pi}\int_A \vec{E}\times\vec{H}\cdot d\underset{\sim}{A} + \int_A \vec{X}\cdot\frac{\partial\vec{u}}{\partial t}\,dA + \int_A \kappa\nabla T\cdot dA, \qquad (1)$$

where the first term gives the electromagnetic energy flowing into the wall, as given by the Poynting vector; the second term is the rate that mechanical work is done on the wall, where t is the time, $\vec{X}$ the surface traction and $\vec{u}$ the elastic displacement; and the third term gives the rate that heat flows into the wall, κ being the thermal conductivity and T the temperature. With the aid of Maxwell's equations and elasticity theory, these terms may be transformed to well-known volume integrals over the wall.[1] If an energy density $\bar{u}$ is defined, then

$$\frac{d}{dt}\int_V \bar{u}\,dV = \int_V \Big[\frac{1}{4\pi}\Big(\underset{\sim}{\vec{E}}\cdot\frac{\partial\vec{D}}{\partial t}+\underset{\sim}{\vec{H}}\cdot\frac{\partial\vec{B}}{\partial t}+4\pi\underset{\sim}{\vec{E}}\cdot\vec{J}\Big)+\vec{\sigma}:\frac{\partial\vec{e}}{\partial t}+\kappa\nabla^2 T\Big]dV \qquad (2)$$

where $\vec{\sigma}$ is the stress and $\vec{e}$ the strain tensor, and V the wall volume. The eddy current dissipation and the thermal term may be eliminated from (2) by use of the heat equation

$$\int_V T\,\frac{\partial s}{\partial t}\,dV = \int_V (q+\underset{\sim}{\vec{E}}\cdot\underset{\sim}{\vec{J}}+\kappa\nabla^2 T)\,dV \qquad (3)$$

where q is the rate that heat is generated by all processes except the eddy current dissipation $\underset{\sim}{\vec{E}}\cdot\underset{\sim}{\vec{J}}$, and s is the entropy density. For slowly moving walls, isothermal processes prevail and the free energy density $\bar{f} = \bar{u} - Ts$ is of interest. With neglect of the electric field term (2) becomes in this case

$$\frac{d}{dt}\int_V \bar{f}\,dV = \int_V \Big(\frac{\vec{H}}{4\pi}\cdot\frac{\partial\vec{B}}{\partial t} + \vec{\sigma}:\frac{\partial\vec{e}}{\partial t} - q\Big)dV. \qquad (4)$$

Finally if $\vec{B}$ is replaced by $\vec{H} + 4\pi\vec{M}$, and $\vec{e}$ by $\vec{s}:\vec{\sigma}+\vec{e}_m$ where $\vec{e}_m$ is the magnetostriction and $\vec{s}$ the elastic compliance tensor, (4) becomes

$$\frac{d}{dt}\int_V g_o\,dV = \int_V \Big(\vec{H}\cdot\frac{\partial\vec{M}}{\partial t} + \vec{\sigma}:\frac{\partial\vec{e}_m}{\partial t} - q\Big)dV, \qquad (5)$$

where g_o is $\bar{f} - H^2/8\pi - \frac{1}{2}\vec{\sigma}:\vec{s}:\vec{\sigma}$. In a uniformly magnetized material g_o, apart from a constant, is just the anisotropy of free energy for an unstressed crystal.[2] However, since non-uniform magnetization exists in a wall, the Landau-Lifshitz exchange energy must be included in g_o. Details of the evaluation of the integrals in (5) will be published elsewhere.[3] The following set of postulates and definitions are used: (1) q is an even function of velocity, (2) the stress is large compared with magnetostrictive effects so that $\vec{\sigma}$ is constant through the wall thickness, (3) the wall thickness is small compared with all other dimensions of interest, (4) in generalized coordinates dV = (1-Cn)dSdn where dS is an element of surface at the middle of the wall, n is distance measured normal to the wall and C is the cur- [illegible] be negligible outside the wall, and by requiring the normal component of $\vec{B}$ and tangential components of $\vec{H}$ to be constant through the wall, (6) $\partial/\partial t = D/dt - \vec{v}\cdot\nabla$ where $\vec{v}$ is the wall velocity and D/dt the time derivative following a point in the wall. D/dt can be non-vanishing due to a change in spin distribution with velocity, or due to the dependence of various crystal parameters on position in the crystal.

EQUATION OF MOTION

The integrands in (5) are functions of the normal component of velocity v_n. By rewriting (5) as a single vanishing integral, with integrand proportional to v_n, it follows that the factor multiplying the arbitrary function v_n must vanish. The resulting equation of motion contains all the expected terms, but with some modification. For a slowly moving wall each point on the surface must satisfy the equation

$$m\dot{v}_n = f_b - \frac{\partial\varepsilon}{\partial n} + C\varepsilon + g_o^+ - g_o^- - \vec{\sigma}:(\vec{e}_m^{\,+} - \vec{e}_m^{\,-}) - \tfrac{1}{2}(\vec{H}^+ + \vec{H}^-)\cdot(\vec{M}^+ - \vec{M}^-) - \beta v_n \qquad (6)$$

where $\dot{v}_n$ is the acceleration, m the (Döring) wall mass per unit area, ε the wall energy per unit area, β the damping factor due to the dissipation q, and ± indicates values just outside the wall, on the positive and negative sides. The meaning of the other surface force density terms is as follows: (1) f_b is a delta function force acting at the wall boundary, resulting from the increase in wall area due to surface roughness or non-parallel crystal surfaces.** (2) The term $\partial\varepsilon/\partial n$ gives the change in wall energy with displacement, resulting from variations within the crystal of the parameters determining ε. These variations may result from crystal imperfections or from gradients in thermodynamic variables such as stress and temperature. (3) $C\varepsilon$, where C is the wall curvature, is the force

*Work supported in part by AFOSR.

produced on a unit element of wall area by neighboring elements. In this respect the wall is analogous to a soap bubble. (4) $g_o^+ - g_o^-$ is the difference in anisotropy energy on either side of the wall. It can arise, for example, for a 90° wall in a crystal having uniaxial symmetry superimposed upon cubic symmetry. (5) $\vec{\sigma}:(\vec{e}_m^+ - \vec{e}_m^-)$ comes from the pressure produced by a stress, for example, on a 90° wall in a crystal of cubic symmetry. (6) The term $\frac{1}{2}(\vec{H}^+ + \vec{H}^-)\cdot(\vec{M}^+ - \vec{M}^-)$ is the magnetic pressure exerted on the wall, where $\vec{H}$ is the total magnetic field, being the sum of the applied field $\vec{H}_A$, the "demagnetizing" field $\vec{H}_D$ and the eddy current field $\vec{H}_e$, and written as $\frac{1}{2}(\vec{H}^+ + \vec{H}^-)$ because of possible discontinuities in $\vec{H}_D$ at the wall. The applied force per unit area is $-\vec{H}_A\cdot(\vec{M}^+ - \vec{M}^-)$, and that due to all the interactions among poles is $-\frac{1}{2}(\vec{H}_D^+ + \vec{H}_D^-)\cdot(\vec{M}^+ - \vec{M}^-)$. The latter tends to resist bending that leads to magnetic poles, and along with $C\varepsilon$, provides stiffness to the wall. The force per unit area $-\vec{H}_e\cdot(\vec{M}^+ - \vec{M}^-)$ is primarily a damping term which acts together with $-\beta v_n$. It can be shown,[4] however, that in metals the former also leads to inertia, adding to that explicitly given in $m\dot{v}_n$. This additional mass results from the energy in the surrounding eddy current field which must be increased to accelerate the wall.

Finally, for a static rather than a moving wall the wall configuration may be obtained by solving (6) with $v_n = 0$. However, stability questions have not been investigated.

THE WALL ENERGY

The expression obtained for the "wall energy" per unit area is

$$\varepsilon = \int\left\{\left[g_o - \tfrac{1}{2}(g_o^+ + g_o^-)\right] - \underset{\sim}{\vec{\sigma}}:\left[\vec{e}_m - \tfrac{1}{2}(\vec{e}_m^+ + \vec{e}_m^-)\right] - \tfrac{1}{2}(\vec{H}^+ + \vec{H}^-)\cdot\left[[\vec{M} - \tfrac{1}{2}(\vec{M}^+ + \vec{M}^-)]\right] + 2\pi\left[M_n - \tfrac{1}{2}(M_n^+ + M_n^-)\right]^2 - \frac{\pi}{2}(M_n^+ - M_n^-)^2\right\}dn \qquad (7)$$

where M_n is the normal component of magnetization and the integral is over the wall thickness. The energy described by this expression is not the energy within the wall, but rather the localized energy transported by the wall, which accounts for the appearance of correction terms that essentially subtract the energy density just outside of the wall.

REFERENCES

1. Magnetic ponderomotive forces and kinetic energy resulting from magnetostriction have been neglected in the elastic equation of motion.
2. W.J. Carr, Jr., Handbuch Der Physik, Vol. 18/2, Ed. by S. Flügge, Springer-Verlag (1966).
3. Submitted to Internat. J. Mag.
4. To be submitted to J. Appl. Phys.

** If ℓ is distance measured in the plane of the wall along an outward normal to the edge of the wall, the force normal to the wall is $f_b = -(v_\ell/v_n)\varepsilon\,\delta(\ell)$, where δ is a Dirac delta function, and v_ℓ the outward velocity of the wall edge, with the ratio v_ℓ/v_n determined by geometry.

OBSERVATIONS OF DOMAIN WALL VELOCITIES AND MOBILITIES IN $YFeO_3$[+]

Ching H. Tsang and Robert L. White
Stanford Electronics Laboratories
Stanford University, California 94305

ABSTRACT

Measurements have been made of the velocity and mobility of head-to-head domain walls in $YFeO_3$ single crystal bars using a Sixtus and Tonks technique. Measurements were made over a temperature range of 178°K to 600°K and at driving fields up to 80 Oe. Mobilities in excess of 150,000cm/sec-Oe and velocities in excess of 10^6 cm/sec were observed. The velocity at low fields is linear in applied field until a saturation velocity, $v_s = 4 \times 10^5$ cm/sec, equal to the velocity of transverse acoustic waves, is reached. The low field mobility is given by $\mu = (1.9 \times 10^6)\ e^{-T/80.5}$ where the mobility is in cm/sec-Oe and the temperature in degrees K.

INTRODUCTION

Yttrium orthoferrite is a canted antiferromagnet with a weak $4\pi M$ and a strong crystalline anisotropy along the easy c-axis. If two portions of a c-axis rod are magnetized oppositely in the ±c directions respectively, a head-to-head domain wall is created as the transition between the two sections. Using a Sixtus and Tonks transit time technique(1), we have measured velocities for such head-to-head walls in single-crystal $YFeO_3$, which was grown and zone refined, then cut into bars of dimensions 1 mm x 1 mm x 50 mm with the c axis oriented along the bar(2).

EXPERIMENT

The experimental set up is as follows: 2 single-turn copper coils acting as pick-up coils are mounted at a fixed distance (d) apart on the sample bar. At one end of the bar a small nucleation magnet is mounted. The sample is first magnetized completely along one c-direction by an applied bias field. An H field in the reversed direction follows, and the nucleation magnet is pulsed. A reverse domain is created by the nucleation device and propagates under the reverse H-field. Its passage through the pick-up coils induces electrical pulses which are displayed on a scope. Wall velocity is then computed from $v = d/\tau$, where τ is the time lapse between the two electrical pulses displayed.

RESULTS

Figure 1 shows a typical velocity-field plot for a c-axis $YFeO_3$ bar at room temperature (300°K). The distance between pick-up coils is 25.75 mm, wall passage signals are 1~10 mV in amplitude, and the transit times recorded vary from 1.5μs~150μs. The following prominent characteristics are observed:

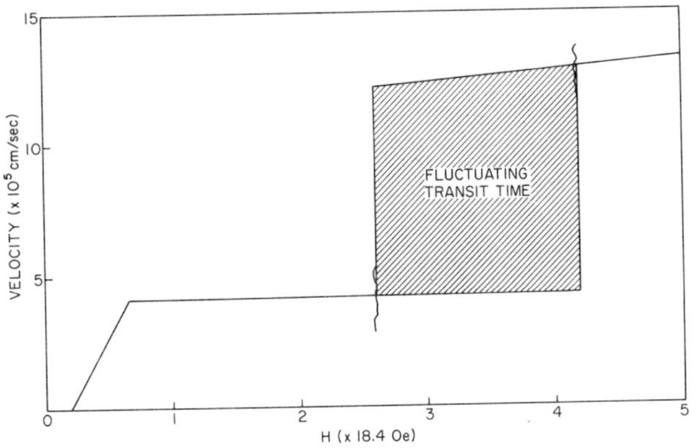

Figure 1. $YFeO_3$ Velocity-Field Characteristic at 300°K

(i) There is a small coercive field (Ho) of around 4 Oe. which must be overcome before the reverse domain starts propagating. Ho varies somewhat between samples depending upon the defects and impurities present.

(ii) At low fields (Ho<H<12 Oe.) the wall velocity v is linear with H; with a differential mobility (at room temperature) of $\mu_L = 4.2 \times 10^4$ cm/sec-Oe. For a set of 5 samples, $<\mu_L>$ is measured to be $(4.25 \pm 0.23) \times 10^4$ cm/sec-Oe, showing high consistency from sample to sample.

(iii) The linear behavior ends abruptly when wall velocity v reaches the saturation values $v_s \approx 4.1 \times 10^5$ cm/sec, at a reverse field of $H \cong 12$ Oe. Velocity v remains at v_s for values of H up to 48 Oe. Measurements on five different samples yield almost identical values of v_s (4.09×10^5 cm/sec ± 2.7%).

(iv) For $H \geq 48$ Oe., instabilities set in, and wall velocity v fluctuates between v_s and a very high value $v_H \cong 1.2 \times 10^6$ cm/sec.

(v) For H>65 Oe., v stabilizes at the higher velocity and increases roughly linearly with H giving high field mobility μ_H on the order of 10^4 cm/sec-Oe. Experimental accuracy is not good at the high velocities because of the low resolution associated with very short transit times. Spontaneous nucleation of reverse domains occurs when H>90 Oe., and controlled nucleation experiments cannot be done at drive fields higher than this.

To investigate temperature dependences of the above characteristics, the measurement process was repeated at 14 temperatures ranging from 178°K to 600°K. Figure 2

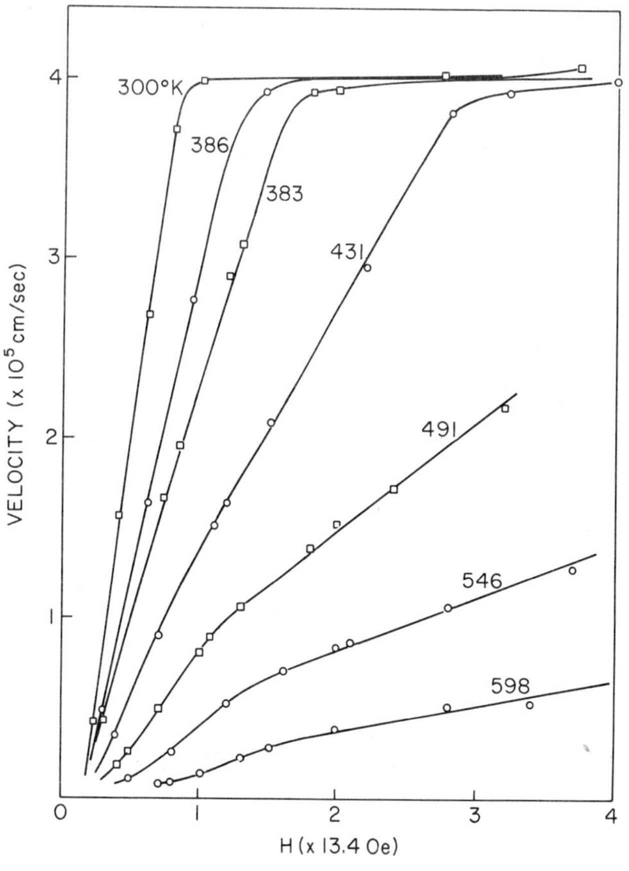

Figure 2. C-Axis $YFeO_3$ Mobility Measurements

is a plot of the high temperature data. Coercive field Ho does not vary much over these temperatures; neither does the value of v_s measured. Only one parameter, μ, varied drastically with T. μ_L increases to a high of 1.62×10^5 cm/sec-Oe at low temperature T = 178°K and

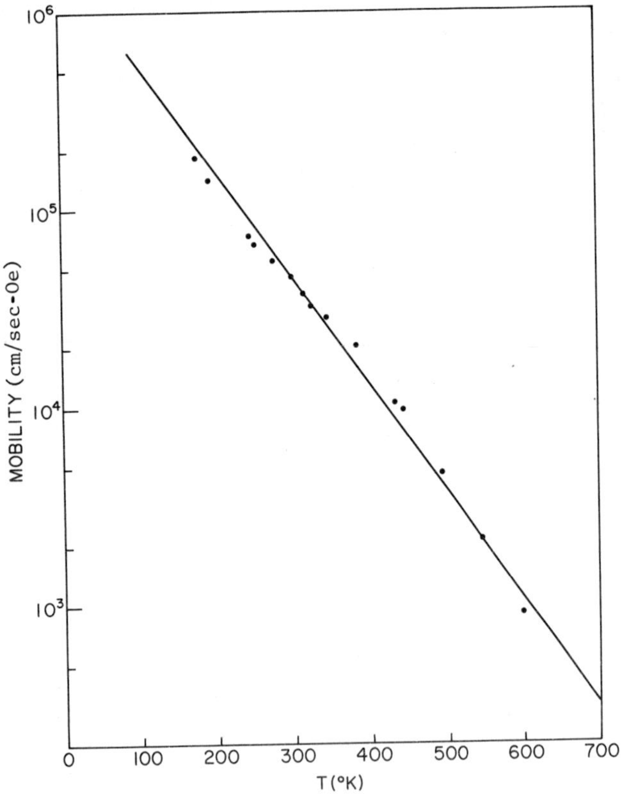

Figure 3. $YFeO_3$ Mobility-Temp.

$$\mu \approx (1.9 \times 10^6) e^{-T/80.5} \text{ cm/sec-Oe.}$$

diminishes to 9.7×10^2 cm/sec Oe at 600°K. Figure 3 is a semi-log plot of μ_L versus T; which shows that log μ_L varies remarkably linear with temperature, over four decades of mobility values, and can be fitted by

$$\mu_L(T) = (1.9 \times 10^6)\ e^{-T/80.5} \text{ cm/sec-Oe}$$

DISCUSSION

We believe that the abrupt limiting velocity v_s is the velocity of transverse acoustic waves in $YFeO_3$. We have measured the transverse acoustic velocity at room temperature to be 4.10×10^5 cm/sec, in good agreement with v_s. We have also measured the velocity of longitudinal sound waves to be 6.9×10^5 cm/sec, but have no evidence these waves play any role in the domain wall motion. The data suggests that the domain wall velocity rises until it hits Mach 1, then a variable but substantial overdrive is required before the velocity breaks through the "sound barrier" and resumes its proper course.

We have at present no explanation at a fundamental level for the magnitudes of the velocities and mobilities seen. The simple exponential (with temperature) dependence of the mobility suggests that a single process dominates and that the damping process involved may be collision with a thermally generated scatterer, perhaps a spin wave, whose activation energy is 80°K.

These experiments are being extended to the velocities of Bloch and Néel walls in single crystal $YFeO_3$. The results of these measurements will be reported subsequently.

REFERENCES

+ This work supported by the Air Force Office of Scientific Research through Grant AFOSR-72-2260C and by the National Science Foundation through the Center for Materials Research at Stanford University.

(1) K. J. Sixtus and L. Tonks, Phys. Rev. 37, 930 (1931); 42 419 (1932); 43, 70, 931 (1933).
(2) Crystals obtained from Tohoku Metal Ind., Ltd., Tokyo.

CALCULATION OF MICROMAGNETIC STRUCTURE BY A RELAXATION METHOD

G. R. Henry and B. R. Brown
IBM Research Laboratory
Monterey & Cottle Roads
San Jose, California 95193

ABSTRACT

We compute the detailed structure of domain walls by a relaxation method. Magnetic dipole moments are assigned to discrete lattice points in space, and the fundamental interactions (anisotropic, exchange, magnetic) are introduced. The dipoles are then iteratively "relaxed" in angular orientation to their lowest energy states until good convergence is obtained. The self field of the lattice is explicitly included in computing the magnetic energy of each dipole.

Calculations have been done on planar Bloch walls in uniaxial materials. The wall structure in bulk material (no demagnetizing fields) is first compared to the known analytic solution, as a indicator of the error introduced by the coarseness of the lattice. The detailed structure of a modified Bloch wall in a thin film is then examined.

INTRODUCTION

We compute planar Bloch wall structures in uniaxial thin films by introducing a lattice of magnetic dipoles with appropriate interactions (anisotropic, exchange, magnetic), and iteratively "relaxing" the dipoles to the minimum energy state. The only essential approximation made is related to the coarseness of the lattice spacing. Of course, if our calculational lattice spacing could be made equal to the magnetic lattice spacing of the crystal, this discrete approach would presumably be better than a continuum approximation. In practice, our lattice is very coarse on this scale.

Relaxation calculations tend to be inefficient, and there must be some specific advantages expected to justify such an approach. We believe there are indeed several appealing features:

1. Freedom from various simplifying approximations.
2. The explicit point-by-point calculation specifies quantities of interest (e.g., energy densities, magnetic field) over the height and width of the wall.
3. New physical situations are easily specified with such a basic starting point.
4. The very physical character of the calculation yields some insights which we have found helpful.

The case considered here is that of a thin film of uniaxial material, with the easy (z) axis orthogonal to the film plane. The axes x and y are chosen to complete a right-handed coordinate system, with y orthogonal to the planar Bloch wall. Such a wall is of course in unstable equilibrium in the absence of an external magnetic field gradient, and in this sense the planar wall may be regarded as an approximation to the cylindrical wall of a magnetic bubble. The direction of any given dipole is specified by the usual polar spherical angles, such that a vector from the origin to the point x,y,z lies along the θ,ϕ direction with $\cos\theta = z/\sqrt{(x^2+y^2+z^2)}$ and $\tan\phi = y/x$. Our ϕ is the Ψ of Slonczewski[1] and the χ of Hubert[2].

In the next Section the calculational method is briefly discussed, and in the following Section some of the results are tabulated.

CALCULATIONAL TECHNIQUE

As mentioned above, we impose a lattice of dipoles over the region of interest. There is no variation in the x direction, so the lattice is effectively two dimensional, with space divided into square prisms of infinite extent in the x direction, each containing a uniform dipole moment density; a given lattice point corresponds to the center of a square prism. We make use of the obvious symmetries of the static problem by computing quantities only in the y,z positive quadrant, with the mid-wall, mid-film point defined by y=z=0.

We specify material properties by choosing the saturation magnetic moment, M, the natural length, ℓ ($\ell=\sigma/4\pi M^2$, with σ the wall energy per unit area) and the dimensionless "quality factor", Q ($Q=K/2\pi M^2$, with K the uniaxial anisotropy constant). For convenience we set $\ell=M=1$ (equivalently we measure distances and fields in terms of ℓ and M, respectively). In the example below, Q=5. The only remaining specification needed for a completely defined physical structure is the film thickness, h, chosen here to be 4.

The calculational approach is to "freeze" all dipole moments except one, and to allow that moment to "relax" to the θ,ϕ of minimum energy. That moment then is frozen, another allowed to relax, and so on. Each moment must be relaxed a number of times to achieve good convergence. The magnetic field is computed from the "magnetic charges" arising from the divergence of the magnetic moment density (this charge resides in "sheets" between adjoining square prisms). The lattice extends over some three wall thicknesses, with the effect of surface charge outside this region computed analytically. The exchange energy is computed among the four nearest neighbors (the two nearest neighbors in the x direction of course make no contribution).

y	Bloch	Approx.
.585	.33	.00 (pinned)
.488	.87	.79
.390	2.31	2.35
.293	6.13	6.13
.195	16.18	15.70
.098	41.31	39.47
.000	90.00	90.00

Table I. Error Associated with Lattice Coarseness.

The key question in the calculation is the fineness of the lattice required to yield reasonably accurate results. An answer is suggested by "turning off" the magnetic interaction, which results in the classic Bloch wall structure for bulk material. The result from the relaxation calculation can then be compared to the analytic solution, as shown in Table I. Even with a lattice spacing so coarse that over half of the the 90 to 0 degree wall rotation occurs between the first two lattice points, the maximum error is less than two degrees. The results shown in Table I were computed twice, starting from a "thin" wall and from a "thick" wall, with computation continuing until both solutions agreed to four significant figures. The "pinned" moment was also moved further out, with no significant effect on the results. Once an acceptable lattice spacing has been found for the y (orthogonal to the wall) direction, the same spacing should be all right in the z direction, since for bubble materials the Bloch line width, Λ, is greater than the Bloch wall width, δ ($\Lambda=\delta\sqrt{Q}$): we expect slower variations in the z than in the y direction.

RESULTS

In this Section, results are presented in a tabular form such that the position in the Table corresponds to the y,z coordinates in the domain wall.

Z↑ \ Y→	0.0	.098	.195	.293	.390	.488
1.951	90.0	48.1	26.0	15.8	10.5	6.3
1.854	90.0	46.7	24.8	15.1	10.1	6.1
1.756	90.0	45.3	23.5	14.1	9.5	5.8
1.659	90.0	44.1	22.2	13.1	8.8	5.4
1.561	90.0	43.1	21.1	12.2	8.1	5.0
1.463	90.0	42.3	20.2	11.4	7.5	4.6
1.366	90.0	41.6	19.4	10.7	6.9	4.2
1.268	90.0	41.0	18.7	10.0	6.4	3.9
1.171	90.0	40.6	18.1	9.4	5.9	3.5
1.073	90.0	40.3	17.6	8.9	5.4	3.2
.976	90.0	40.0	17.2	8.3	4.9	2.9
.878	90.0	39.8	16.8	7.9	4.4	2.6
.780	90.0	39.7	16.5	7.5	4.0	2.3
.683	90.0	39.6	16.3	7.1	3.7	2.0
.585	90.0	39.5	16.1	6.8	3.3	1.7
.488	90.0	39.4	15.9	6.6	3.0	1.5
.390	90.0	39.4	15.8	6.4	2.8	1.3
.293	90.0	39.4	15.7	6.3	2.6	1.1
.195	90.0	39.4	15.7	6.2	2.5	.9
.098	90.0	39.4	15.6	6.1	2.4	.8
.000	90.0	39.4	15.6	6.1	2.3	.8

Table IIa. Magnetic Moment Direction; Theta in Degrees

Z↑ \ Y→	0.0	.098	.195	.293	.390	.488
1.951	-85.6	-85.9	-86.5	-87.5	-88.4	-89.0
1.854	-84.7	-85.0	-85.8	-87.1	-88.2	-88.9
1.756	-82.9	-83.3	-84.4	-86.1	-87.6	-88.6
1.659	-80.2	-80.7	-82.3	-84.6	-86.8	-88.1
1.561	-76.7	-77.4	-79.4	-82.6	-85.6	-87.5
1.463	-72.4	-73.2	-75.8	-80.0	-84.0	-86.6
1.366	-67.4	-68.3	-71.5	-76.8	-82.1	-85.5
1.268	-61.8	-62.9	-66.5	-72.9	-79.7	-84.2
1.171	-55.9	-57.1	-61.1	-68.6	-76.9	-82.6
1.073	-49.9	-51.1	-55.3	-63.7	-73.7	-80.8
.976	-43.9	-45.1	-49.4	-58.5	-70.0	-78.6
.878	-38.1	-39.3	-43.5	-53.0	-65.9	-76.1
.780	-32.7	-33.7	-37.7	-47.2	-61.2	-73.2
.683	-27.6	-28.5	-32.2	-41.3	-56.0	-69.7
.585	-22.9	-23.7	-26.9	-35.3	-50.2	-65.5
.488	-18.5	-19.2	-21.9	-29.4	-43.6	-60.2
.390	-14.4	-15.0	-17.2	-23.4	-36.3	-53.4
.293	-10.6	-11.0	-12.7	-17.5	-28.2	-44.5
.195	-6.9	-7.2	-8.4	-11.6	-19.3	-32.7
.098	-3.4	-3.6	-4.1	-5.8	-9.8	-17.6
.000	-.0	-.0	-.0	.0	.0	-.0

Table IIb. Magnetic Moment Direction; Phi in Degrees.

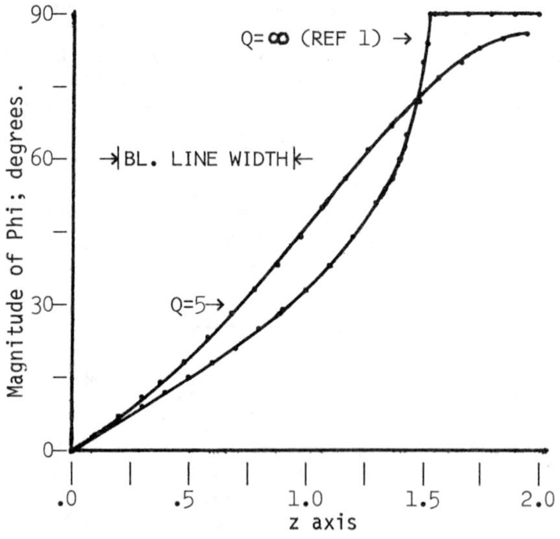

Figure 1. Magnitude, in degrees, of Phi averaged over y=0,±0.0976, in the Q=5 wall.

In all cases z is directed upward, y to the right, with the first row and column of the Table indicating y and z respectively. The last "pinned" column of moments is omitted from all Tables, and in each case only the upper right quadrant of the wall is shown.

The orientation of the magnetic moment throughout the domain wall is shown in Table II and Figure 1. Near the center of the film, the wall looks much like a Bloch wall in bulk material (see Table I). As the surface is approached, the wall "feathers out" and the moments rotate very nearly into the Néel direction (ϕ=-90). This is very roughly similar to the structure suggested by Schlömann[3], although several features are different. Qualitatively the ϕ behavior is similar to results obtained by Hubert (private communication), although we find somewhat greater rotation into the Néel direction at the surface.

In Table III the y component of the local magnetic field (H) is indicated. There are clearly regions near the center of the wall where the wall

Z↑ \ Y→	0.0	.098	.195	.293	.390	.488
1.951	-8.4	-9.8	-10.5	-10.0	-9.3	-8.7
1.854	-4.6	-7.0	-9.2	-9.4	-9.0	-8.5
1.756	-2.1	-5.1	-7.9	-8.6	-8.5	-8.2
1.659	-.3	-3.6	-6.8	-7.8	-7.9	-7.8
1.561	.9	-2.5	-5.9	-7.1	-7.3	-7.3
1.463	1.8	-1.7	-5.1	-6.4	-6.7	-6.8
1.366	2.5	-1.0	-4.5	-5.8	-6.1	-6.3
1.268	2.9	-.6	-3.9	-5.2	-5.6	-5.7
1.171	3.0	-.2	-3.4	-4.7	-5.1	-5.2
1.073	3.1	.0	-3.0	-4.2	-4.5	-4.7
.976	3.0	.2	-2.6	-3.7	-4.1	-4.2
.878	2.7	.2	-2.2	-3.2	-3.6	-3.8
.780	2.5	.3	-1.9	-2.8	-3.1	-3.3
.683	2.2	.3	-1.6	-2.4	-2.7	-2.8
.585	1.8	.3	-1.3	-2.0	-2.3	-2.4
.488	1.5	.2	-1.1	-1.6	-1.9	-2.0
.390	1.2	.2	-.8	-1.3	-1.5	-1.6
.293	.9	.1	-.6	-1.0	-1.1	-1.2
.195	.6	.1	-.4	-.6	-.7	-.8
.098	.3	.0	-.2	-.3	-.4	-.4
.000	.0	.0	.0	.0	-.0	-.0

Table III. y-Component of Local Magnetic Field, in Units of M.

demagnetizing fields overcome the fields produced by the surface charges.

Energy densities have also been computed point-by-point. Not unexpectedly, the exchange and anisotropic energy densities are no longer equal; the anisotropic term increases toward the surface, while the exchange energy decreases. The absolute magnetic energy density is typically substantially smaller than the other two in the central wall region.

More information than can be tabulated here is available from the authors. The calculations are being extended to other cases of interest.

We are indebted to A. Hubert and B. A. Calhoun for helpful discussions.

REFERENCES

1. Slonczewski, J. C., J. Appl. Phys. 44 1759-1770 (1973).
2. Hubert, A., to be published.
3. Schlömann, E., J. Appl. Phys. 44 1837-1849 (1973); 44 1850-1854 (1973); 45 369-373 (1974).

MAGNETIC CONFIGURATION AT A DOMAIN TIP AND ITS EFFECT ON THE ERASE THRESHOLD OF SMALL DOMAINS IN THIN FILMS

E. J. Torok
Sperry Univac
St. Paul, Mn. 55165

ABSTRACT

The tip of a domain in thin metallic film is not a simple intersection of two walls, but a dynamic configuration that determines the erase threshold, and the length and width of the smallest stable domain. The two walls at the domain tip do not intersect; one wall degenerates into a zero degree wall and fades out before it intersects the other. However, if that wall is extrapolated until it does intersect, the angle of intersection is given by $\beta = \pi/4 - \frac{1}{2} \arcsin (H_T/H_K)$, where H_K is the uniaxial anisotropy field and H_T is the hard axis component of applied field. The stray field on the wall is proportional to the curvature of the walls; the maximum allowed curvature and the boundary condition of the required value of β determine the minimum size domain. Experimental data taken via Lorentz microscopy agrees closely with the theory both for the minimum size domain and for the erase threshold as a function of domain length and H_T.

INTRODUCTION

This work was precipitated by an attempt to predict the erase threshold of domains in 100 Å permalloy films in an oligatomic film memory[1] as a function of an easy axis field from a digit line and a hard axis field from a word line. The key to resolving the paradoxes that arose and to matching the theoretical and experimental erase thresholds was finally found in the configuration of the magnetization at the tip of the domain.

THE DOMAIN TIP

An electron micrograph of a domain in an oligatomic film is shown in figure 1. Note that the walls at the domain tip do not meet; one stops short and the other curls around it. This type of domain tip is the lowest energy configuration, and is almost invariably found in permalloy and cobalt films of all thicknesses. The magnetization configuration is shown in figure 2a. There is also a higher energy configuration that is rarely found; this is illustrated in figure 2b. We will hereinafter consider only the low-energy configuration.

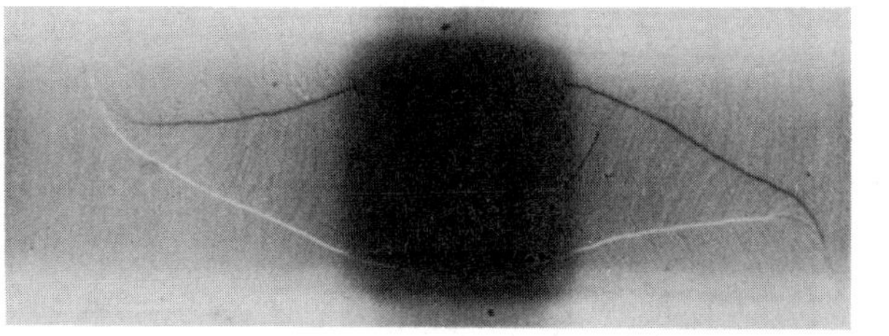

Figure 1. Domain in an oligatomic film; domain is 1.7 mils wide

Near the domain tip the two walls nearly intersect; in fact, they would if one of them did not fade out into a zero-degree wall. By extrapolating that wall in a straight line until it intersects the other wall, the angle of intersection, β, is formed as shown in figure 3. The problem then is to determine the value of β.

Note from figure 1 that the wall that fades out is parallel to the easy axis, and the magnetization directly off the end of that wall is 90° from the easy axis (the magnetization direction is perpendicular to the ripple). Note also that the magnetization outside the domain is straight and undeviated in two quadrants, namely the quadrants outside the overlapping walls (e.g., the lower left and upper right in figure 2a). The magnetization direction in those quadrants is given by the familiar static torque equation of the Stoner-Wohlfarth theory, i.e., the angle between the magnetization and easy axis is arc sin H_T/H_k where H_T is the applied hard axis field and H_k is the anisotropy field. Now domain theory tells us that walls are oriented parallel to the vector difference of the magnetization on either side to avoid poles on the wall. Since the

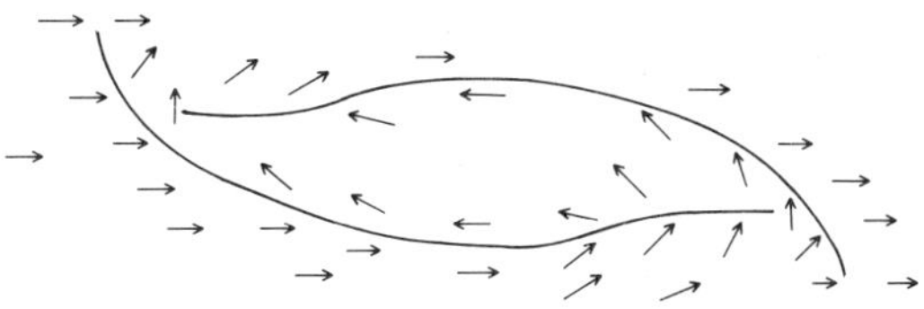

a. Lowest energy form, almost invariably found

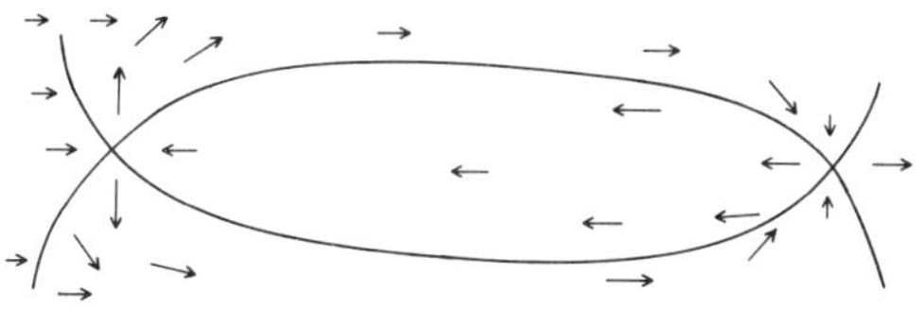

b. Higher energy form, rarely found

Figure 2. Tips of domains in oligatomic films

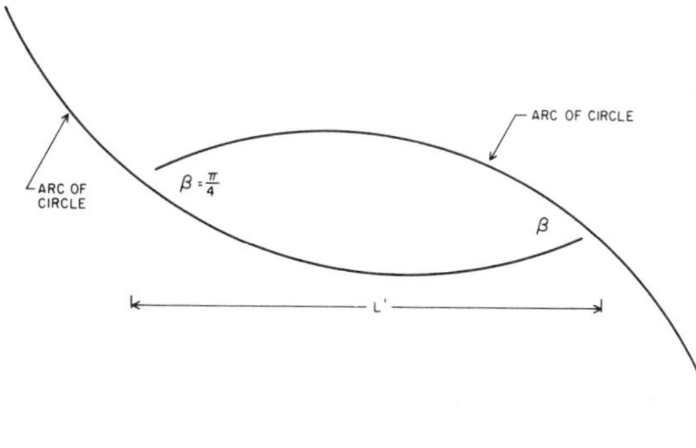

Figure 3. Minimum size domain; no applied field

magnetization on one side of the overlapping wall at the vertex is at 90° to the easy axis, and on the other side at $\sin^{-1} H_T/H_k$, the angle, β is given by

$$\beta = \pi/4 - \tfrac{1}{2} \sin^{-1} H_T/H_k \qquad (1)$$

Note that when the word current is zero, $\beta = 45°$; as the word field is increased, β becomes smaller, going to zero when $H_T = H_k$. The angle β provides an important boundary condition that will enable us to calculate the erase threshold.

STRAY FIELD FROM WALLS

In reference 2 is given an approximate theory for calculating the stray field of a slender domain as a function of distance along the domain. Let the distance along the domain be x, the width of the domain be W(x), and the x component of the stray field be H_x. The theory of reference 2, which is correct to third order, gives

$$H_x = -k \frac{d^2 W(x)}{dx^2} \qquad (2)$$

where k is the stray field coefficient. When domains are written into a 100Å permalloy film with a word line 2 mils wide, k is equal to 3 Oe mil.

THE MINIMUM SIZE DOMAIN

The angle, β, between walls at the domain tip is an important boundary condition that determines the minimum size of a stable domain. The free ends of the domain walls are very mobile, waving in the breeze of every passing transverse field. They wave outward if the β is too large, and inward if it is too small.

Consider a domain so small that it is on the verge of erasing itself with its own demagnetizing field. Let there be no external field applied. The stray field must then equal the coercivity:

$$H_x = H_c \tag{3}$$

$$-k \frac{d^2W}{dx^2} = H_c \tag{4}$$

The solution to this equation is going to be a domain whose walls are arcs of a circle, as in figure 3. The domain can't get any smaller without either increasing the curvature (not allowed because the demagnetizing field is already only infinitesimally less than H_c) or by decreasing β (also not allowed because the value of β given by equation 1 is the lowest energy configuration; if β is too small, the free ends of the walls bend inward to reduce the length and increase the curvature). We now integrate using as a boundry condition that when H_T is zero, $-dw/dx = 1$ at the two vertices of the domain. The length between vertices, L′, and the width, W, at the center are then for this minimum size domain

$$\frac{L}{2} = \frac{k}{H_c} \quad \text{and} \quad W = \frac{k}{2H_c} \tag{5}$$

Equations (5) state that a film having an H_c of 2 Oe and a k = 3 Oe mil, such as the one with which we did the electron microscope experiments would have a minimum width of 0.75 mils and a minimum length of 3 mils. These minimum values are indeed those reported in figure 9 of reference 1.

THE ERASE PARADOX

One of the experiments done in the electron microscope was to determine what happens when a domain is slowly erased. We found that as a reverse digit field is increased in coincidence with a fixed word amplitude, the domain becomes narrower and narrower but no shorter; the length remains the same until the domain suddenly disappears. The fields required to erase a domain depend on the length of the domain; the longer the domain, the larger the required fields. A domains longer than the minimum length can be erased with a coincidence of hard and easy axis fields each of which is too small to erase the domain by itself. Stated another way, a hard axis field helps. The paradox is that for domains of minimum length, the hard axis field doesn't help. A domain so small that its (easy axis) stray field is nearly strong enough to erase the domain by itself should be erased when a hard axis field is supplied. On the contrary, the domain remains stable for all values of $H_T < H_k$. However, the domain is erased with an infinitessimal easy axis field.

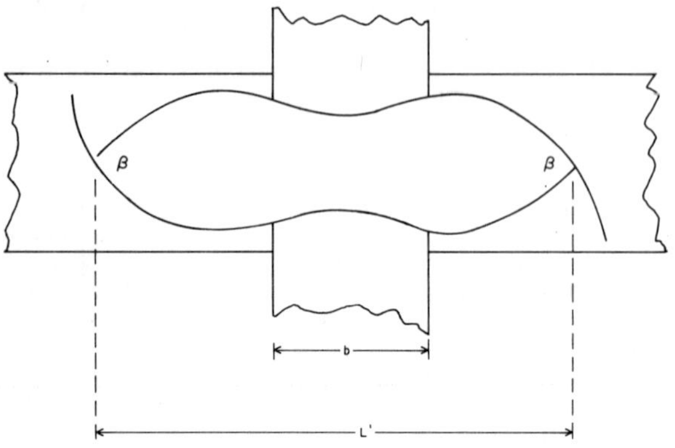

Figure 4. Domain just about to be erased with a combination word and digit field

Figure 4 shows a domain about to be erased with a combination of an oscillating field from the word line and a negative dc field from the digit line. Note that the curvature of that part of the domain outside the digit line is negative, indicating that the demagnetizing field in that region tends to erase the domain. The curvature inside the digit line may be either positive or negative depending on whether the domain is short or long.

Let H_L' be the field required to erase the domain, and let us adopt the sign convention that H_L' is chosen positive. The equation pertaining to the part of the domain under the digit line is obtained by equating the sum of applied and stray fields to the creep threshold, H_{cr}:

$$H_{cr} = H_L' - k\frac{d^2W}{dx^2} \tag{6}$$

The field from the digit line is zero for that part of the domain not covered by the digit line; therefore the differential equation in this region is

$$H_{cr} = -k\frac{d^2W}{dx^2} \tag{7}$$

When the solutions in these two regions are joined so that dw/dx is continuous, and the boundry condition is applied that $dw/dx = -\tan\beta$ at the vertices of the domain, the resulting solution is

$$\frac{H_L'}{H_{cr}} = \frac{L'}{b} - \frac{2k\tan\beta}{b\,H_{cr}} \tag{8}$$

which is the answer we were seeking.

Equation 8 contains the terms H_{cr} and $\tan\beta$. The creep threshold, H_{cr} will be taken for these very thin films as $H_c(1 - H_T/H_k)$ where H_T is the hard axis field amplitude. (For oligatomic films this assumption is pretty close to true, but for thicker films the reversible limit is considerably lower than H_k.) The small angle approximation to $\tan\beta$ is expressed by $1 - H_T/H_K$; this expression is exactly correct at $H_T = 0$ and at $H_T = H_K$ and is a good approximation in between. If we use these two expressions for H_{cr} and $\tan\beta$, equation 8 becomes:

$$\frac{H_L'}{H_c(1 - H_T/H_k)} = \frac{L'}{b} - \frac{2k}{b\,H_c} \tag{9}$$

This fits the experimental data of figure 9 of reference 1 almost exactly for the film and striplines on which the electron microscope data was taken (b = 2 mils, H_c = 2 Oe, k = 3 Oe mil, no keeper and no ground plane). Thus, the erase paradox is resolved by the theory; and as in the treatment of the minimum size domain, the essential part of the theory that does this is the boundary condition β, combined with the mobility of the tips of the walls.

REFERENCES

1. V.M. Benrud et al J. Appl. Phys. 42 1364 - 1373 (1971)
2. E.J. Torok et al A.I.P. Conf. Proc 5 722-726 (1971)

FERROMAGNETIC DOMAIN STRUCTURE OF d.h.c.p. Co-Fe ALLOYS

H. J. Leamy, T. Wakiyama* and G. Y. Chin
Bell Laboratories, Murray Hill, New Jersey 07974

ABSTRACT

Co-Fe alloys containing from ~1 to ~4 at. % Fe possess a double h.c.p. structure at room temperature.[1] These alloys exhibit a large negative uniaxial anisotropy.[3] We have studied the domain morphology in demagnetized d.h.c.p. alloys by scanning and transmission electron microscopy. A single crystal of Co-1.53 at. % Fe was revealed by SEM examination to consist of thin (~1 μm) lamellar domains parallel to the basal plane. Within each such domain, the magnetization is directed along a single $<100>_{d.h.c.p.}$ axis; consequently the lamella consist of a single set of large (100 μm) domains elongated in the direction of magnetization. These observations were confirmed by TEM study of partially transformed polycrystalline foils. The lamella were found to terminate at grain and transformation twin boundaries except those of the f.c.c./d.h.c.p. type.

INTRODUCTION

Alloys of Co-Fe have recently been shown to possess the d.h.c.p. (ABAC stacking) structure in the region of ~1 to ~4 at. % Fe at room temperature.[1,2] The d.h.c.p. phase was shown by Chikazumi et al.[3] to possess a large, negative, uniaxial anisotropy (K_{ul} = -10.2×10^6 erg/cm^3 at 4.2°K), which serves to hold the spins perpendicular to the hexagonal axis. Earlier studies[1] of the domain structure of the d.h.c.p. phase revealed large, elongated domains on the (0001) surface of a single crystal, while surfaces containing the [001] axis exhibited a very fine striation pattern parallel to the basal plane trace. In order to study the domain structure of this material at greater resolution, we have applied the scanning electron microscopy (SEM) technique of Fathers et al.[4-5] to a study of the same Co-1.53 at. % Fe d.h.c.p. single crystal described at this conference in 1972.[1] In addition, we have studied the domain structure in partially transformed, polycrystalline foils of Co-1.51 at. % Fe material by Lorentz[7] transmission electron microscopy (TEM). Electron transparent films were produced by ion milling the original 0.1 mm thick sheet, which had been subjected to thermal cycling between 200 and 500°C to promote transformation to the d.h.c.p. structure.

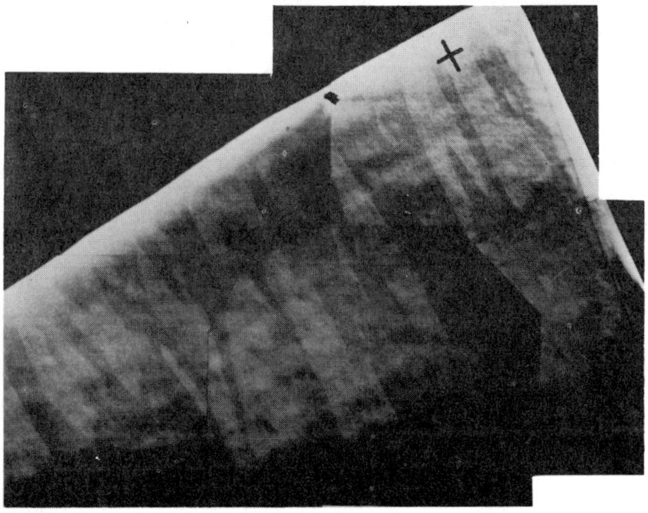

Fig. 1 SEM image of domains on a basal plane surface of a Co-1.53 at. % Fe single crystal. The marker is 100 μm.

*On leave from Tohoku University, Sendai, Japan.

In the domain observation technique of Fathers et al., contrast between domains of different magnetization direction arises via the effect the internal field, $\bar{B}$, upon the electron trajectories within this specimen. When a flat specimen is viewed in the SEM under conditions such that the tilt axis is parallel to $\bar{B}$, the Lorentz deflection of the incident beam within the sample is directed either towards or away from the sample surface, and the number of electrons which escape from the specimen is consequently increased or decreased respectively. Contrast is thus proportional to the component of $\bar{B}$ parallel to $\bar{n} \times \bar{e}$, where $\bar{n}$ is the specimen normal and $\bar{e}$ is the incident beam direction. Domains of oppositely directed magnetization thus appear as above and below background brightness regions when $\bar{B}$ is parallel to the tilt axis, and disappear when $\bar{B}$ is perpendicular to the tilt axis. The technique thus provides a method for the determination of $\bar{B}$ as well as for delineation of domain boundaries.

RESULTS

Figure 1 shows the remenant domain structure on the basal plane (0001) surface of the d.h.c.p. single crystal. The domains are elongated along [100], the direction of magnetization. Examination of this surface after rotation about its normal also revealed that $\bar{B}$ is directed along only one of the three possi-

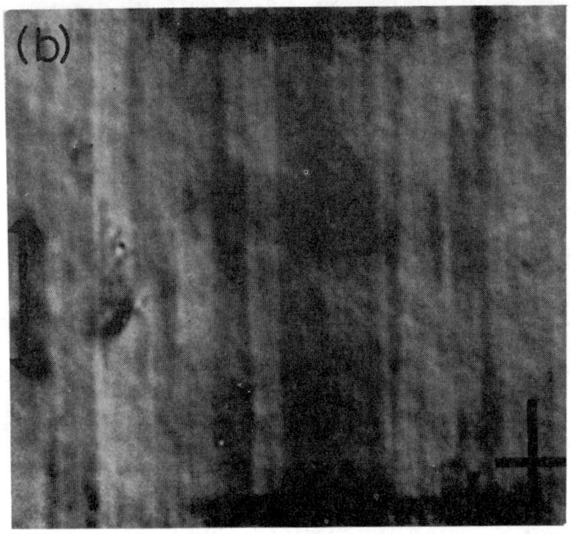

Fig. 2a SEM image of the domain structure on a (01$\bar{1}$0) surface. The domain lamella are parallel to [100]. The marker is 100 μm long.

Fig. 2b Higher magnification version of 2a. The arrow is parallel to [100], the basal plane trace, the marker is 25 μm long.

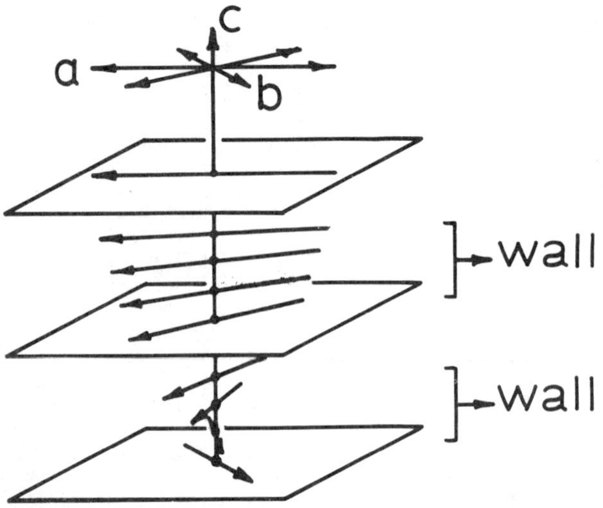

Fig. 2c Schematic illustration of interlamellar walls in d.h.c.p. Co-Fe alloys. Rectangular sheets indicate bulk domains and the spin configuration within the walls is shown in expanded scale.

ble directions of the <100> type. The boundaries are therefore 180° walls, and are presumably of the Neel type, in view of the large, negative K_{u1}. Further, the presence of only one magnetization direction indicates the existence of anisotropy within the easy basal plane.

Domain contrast on all surfaces perpendicular to the basal plane is similar. The contrast is strongest when the tilt axis lies within the basal plane and disappears when the sample is tilted about the hexagonal axis, indicating that $\bar{B}$ lies within the basal plane, as expected. The domain structure on a (01$\bar{1}$0) surface shown in Fig. 2a is seen to consist of fine lamella parallel to the basal plane. Higher magnification images (Fig. 2b) show that the structure contains lamellar features at the resolution limit of the technique; ∼0.5 μm. The domain contrast furthermore often consists of more than two distinct brightness levels, as shown in Fig. 2b. The darkest and lightest levels are produced in domains whose magnetization lies in the plane of the surface ($\bar{B}$//[100] in Fig. 2), while the mid grey levels are produced by domains whose component of $\bar{B}$ in the plane of the surface is smaller. Given the in plane anisotropy visible in Fig. 1 and the preponderance of light-grey, dark-grey boundaries in Fig. 2b, we are led to suggest that the lamella consist of domains magnetized along <100>, which are separated by Bloch type walls oriented parallel to the basal plane as shown schematically in Fig. 2c. Both the morphology and the high density of such walls suggest further that they are related to the defect substructure which accompanies the f.c.c.-d.h.c.p. transformation.

The transformation substructure in polycrystalline sheets was revealed by TEM observations to consist of: (a) retained, heavily faulted f.c.c. regions, (b) crystallographically related d.h.c.p. transformation twins, and (c) thin h.c.p. lamella in the d.h.c.p. phase. Domain boundaries were observed to terminate abruptly at all but d.h.c.p./f.c.c. boundaries, where continuity of the type observed in pure Co by Grundy and Tebble[8] was observed. Figure 3, for example, contains two twin related d.h.c.p. regions which formed in a (110) oriented f.c.c. parent grain. The hexagonal axis of each d.h.c.p. region lies in the plane of the foil as shown in Fig. 3b. The upper portion of the image contains a pair of domains of reverse magnetization, which are bounded by Bloch walls. The narrow transformation twin in the lower portion of the image also contains oppositely magnetized domains separated by Bloch walls. These close at the boundary as indicated in Fig. 3b. Note that the walls separating the twinned regions are 54.7° Bloch walls, while within the twin they possess 180° character.

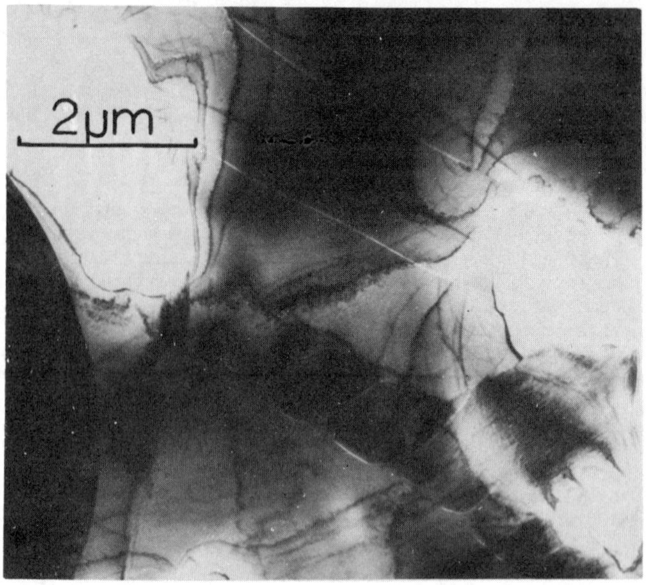

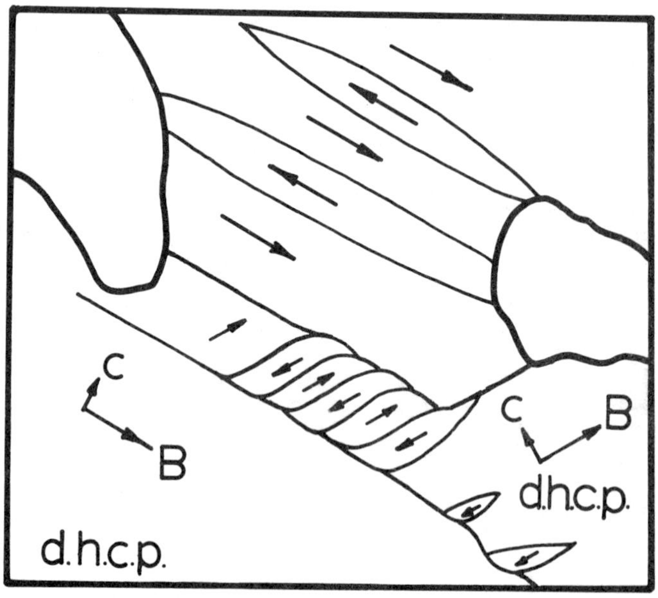

Fig. 3a Lorentz electron micrograph of a thin Co-1.51 at. % Fe foil.

Fig. 3b Schematic illustration of the domain configuration shown in Fig. 3a.

REFERENCES

1. T. Wakiyama, A.I.P. Conf. Proc. 10, 921 (1973).
2. T. Onozuka, S. Yamaguchi, M. Hirabayshi and T. Wakiyama, to be published in J. Phys. Soc. Japan.
3. S. Chikazumi, T. Wakiyama and K. Yosida, Prof. Int. Conf. Magnetism, Nottingham, 756 (1964).
4. D. J. Fathers, J. P. Jakubovics and D. C. Joy, Phil. Mag. 27, 765 (1973).
5. D. J. Fathers, J. P. Jakubovics, D. C. Joy, D. E. Newbury and H. Yakowitz, Phys. Stat. Sol. (a) 20, 535 (1973).
6. D. J. Fathers, J. P. Jakubovics, D. C. Joy, D. E. Newbury and H. Yakowitz, Phys. Stat. Sol. (a) 22, 609 (1974).
7. M. E. Hale, H. W. Fuller and H. Rubenstein, J. Appl. Phys. 30, 789 (1959).
8. P. J. Grundy and R. S. Tebble, Proc. Phys. Soc. 81, 971 (1963).

X-RAY TOPOGRAPHIC STUDY OF 71° and 109° MAGNETIC DOMAIN WALLS IN IRON GARNETS BETWEEN 300°K and 4.2°K

A. Mathiot
D.M.G. Centre d'Etudes Nucleaires
B.P. 85, 38041 Grenoble-Cedex, France

J.F. Petroff
Lab. Mineral. Cristallographie, Tour 16
Universite Paris VI, 75230 Paris Cedex 05

ABSTRACT

Domain wall observations have been carried out on several pure iron garnets (especially on TbIG) by means of x-ray transmission topography in the low-temperature region. The samples were (110) 100 μm slices cut from flux-grown crystals, in order to remove most of the stresses due to crystal defects (mostly growth bands). The room temperature structure of such slices consists of a rectangular array of 71° and 109° walls, separating domains where the magnetization lies in the plane of the plate. The domain contrast under I.R. polarised light is due to linear magnetic birefringence; the wall contrast on x-ray topographs is due to magnetostrictive stresses between adjacent domains. When the temperature is lowered, the number of walls increases with the increase in spontaneous magnetization, mostly between the compensation temperature and 100°K. One observes also a preferential occurence of the $(1\bar{1}0)$ (71°) walls, mostly on TbIG, but also on DyIG. These two points do not apply to YIG. The evolution of the wall contrast on x-ray topographs is also important. In order to explain it, one must first understand the intrinsic wall structure, which seems to be more complicated than the Bloch model.

INTRODUCTION

Most present applications of garnets use the 180° magnetic domain walls, for example in cylindrical form for bubble materials. This paper is a contribution to the study of 71° and 109° domain walls in bulk materials of good crystalline quality. In the case of perfect crystals the domain structure of (110) slices is quite simple and has been described in a previous article[1]. It consists of a rectangular network of plane walls along the $(1\bar{1}0)$ and (001) planes. This configuration corresponds to the equilibrium structure in perfect crystals and had previously been observed in Ni crystal surfaces[2]. The magnetization in both cases is along one of the two ⟨111⟩ directions which are contained in the (110) plane. Since no magnetic poles appear on the (110) surface of the plate, this pattern corresponds to a minimum of the magnetostatic energy. As we have only studied highly perfect samples (samples with few crystalline defects), this configuration has been found to be reproducible. Some 180° walls may exist, but they are in priciple not visible on x-ray topographs. The magnetic properties of rare earth garnets are known to change sharply with temperature[3], and we expected an influence on the domain configuration and on the wall images on x-ray topographs. We have studied three different garnets in order to check the influence of the rare earth on this evolution. The samples were (110) slices of TbIG, DyIG, and YIG, about 100 μm thick. They had been cut from flux-grown crystals (provided by the L.E.T.I. in Grenoble) and then carefully polished and fixed, in order to remove most of the non-magnetic stresses.

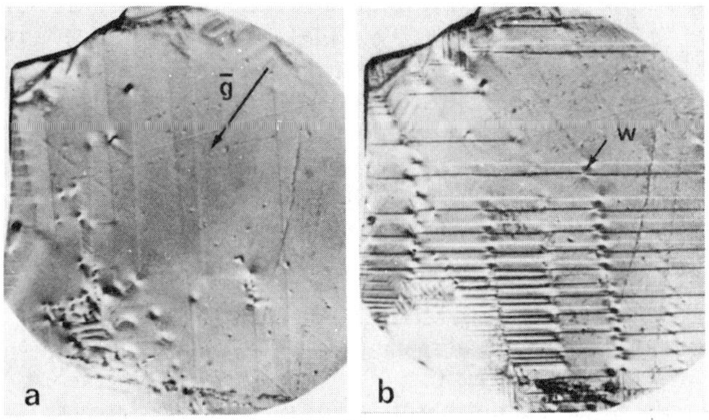

Fig.1-X-ray topographs of a (110) slice of DyIG(MoKα, 444 reflection, 11x). [001] direction is horizontal, $[1\bar{1}0]$ is vertical. a:T=295°K, b:T=116°K(w=180° wall im.)

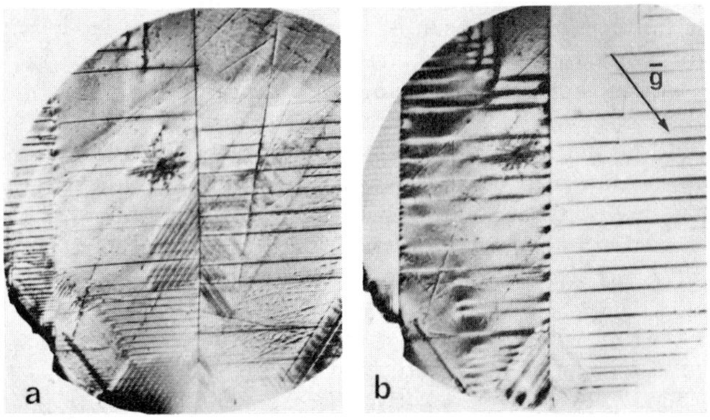

Fig.2-X-ray topographs of a (110) slice of TbIG(MoKα, 444 reflection, 11x). Same orientation as fig. 1. a:T=178°K, b:T=4.2°K.

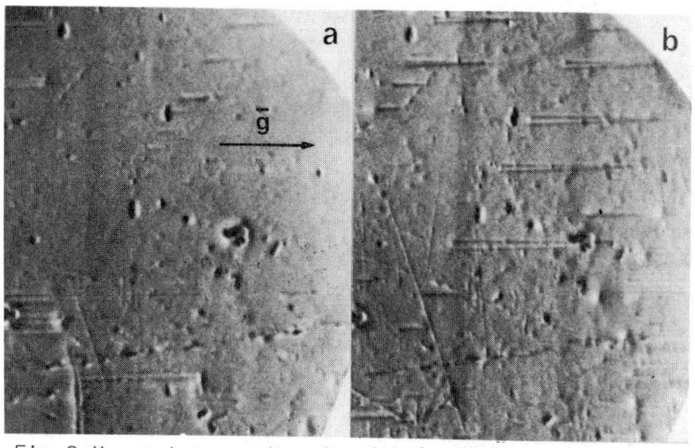

Fig.3-X-ray topographs of a (110) slice of YIG (MoKα, 008 reflection, 22x). Same orientation as fig. 1. a:T=294°K, b:T=4.5°K

RESULTS

The observations were made by transmission topography (Lang method). A variable temperature cryostat was mounted on an ordinary Lang camera, and x-ray topographs have been obtained down to 4.2°K. A complementary observation was performed with an I.R. microscope, using the Faraday and Cotton-Mouton effects. Optical observations have been made down to 80°K, and we obtained a good correlation with the x-ray topography results. We shall now try to compare the results obtained with the different garnets and to deduce some features concerning the influence of the rare earths on those results.

1. When the temperature is lowered, the number of domains increased sharply in the TbIG and DyIG samples, but no great variation was observed in YIG (see Figs. 1 and 3). This can be easily explained by the variations of the spontaneous magnetization of those garnets. For TbIG, $4\pi M_s$ increases from 0 at 246°K (which is the compensation temperature) to about 7500 G at 0°K. The corresponding values for DyIG are 0 at 221°K and 8000 G at 0°K. For YIG, $4\pi M$=1700 G at 300°K and 2400 G at 0°K. When the magnetization inside the domain increases, the domain width must decrease in order to minimize the free pole energy. A rough calculation of the domain width in TbIG, using the classical Kittel model[4], is in good agreement with our results (the width is proportional to $K_1^{.25}/M_s$, where M_s is the spontaneous magnetization and K_1 is the anisotropy constant).

2. The wall contrast on x-ray topographs is known to be related to magnetostrictive stresses[5]. This is why this method can in principle give information only for the domain walls which give rise to long-range stresses, and would be unable to show the 180° walls. However, some of the latter are visible on Fig. 1b, either because of a small residual contrast (which could be due to surface relaxation), or because of their interaction with other types of walls. Those junctions have been studied extensively by Miltat and Kleman[6]. When the temperature is lowered, the contrast at domain walls increases sharply in the DyIG and TbIG samples (Figs. 1 and 2), and this can be related to the increase of the λ_{111} coefficient of spontaneous magnetostriction. For TbIG, λ_{111} increases from 12×10^{-6} at 300°K to 2200×10^{-6} at 4.2°K. For DyIG, λ_{111} (300°K)=-6×10^{-6}, λ_{111} (4.2°K)= -875×10^{-6}.[7] In YIG, λ_{111} ($\approx 3\times10^{-6}$) remains quite constant and much lower than for TbIG or DyIG. At very low temperature, especially for TbIG, λ_{111} becomes so high that the misorientation between two adjacent domains becomes sufficient to be detected with the Lang camera (it produces a splitting of the Bragg peak). In Fig. 2b only one type of domain is in reflection. This difference in Bragg angle allowed us to calculate λ_{111} with good accuracy[8]. Our result of λ_{111}= $(2175\pm75)\times10^{-6}$ for TbIG at 4.2°K is in good agreement with that of Sokolov[7] (2300×10^{-6}) and of Sayetat[9] (2180×10^{-6}), obtained by other methods. The exact origin of the domain wall contrast is not quite clear now. We want to calculate the strains on both sides of the wall in order to clarify this point and to obtain a theoretical model of the wall for the x-ray diffraction contrast observed.

3. We have tried to calculate the surface energy density of the 71° and 109° walls as a function of temperature, assuming a classical Bloch model[10] of the wall structure. Our results seem to be in contradiction with the experimental observation that the 71° walls are more frequent in all samples at low temperatures than the 109° walls (except for YIG). Our calculation shows that the opposite should happen, because the 71° wall energy has been found to be higher than the 109° wall energy[11]. It could then be assumed that the 71° and 109° walls in garnets at low temperature might be somewhat different from classical Bloch walls. In any case, the microscopic structure of these walls has to be understood first. The presence of the rare earth ions is certainly very important, because this phenomenon has not been observed on the YIG samples. The topographs of YIG shown here have not been taken with the same reflecting planes as for the other garnets, and the wall contrast is quite different, but the optical observation shows good domain contrast. (In YIG the domains are more visible on optical photographs than on x-ray topographs.)

CONCLUSION

The aim of this study was to attempt to understand the 71° and 109° wall structure in garnets. At this point it seems that the nature of the rare-earth involved may be very important, and that the evolution of the domain texture is quite different for YIG and for the other garnets studied. In order to get a comprehensive view of the phenomena it seems necessary to extend our study to other rare earth garnets, such as ErIG and SmIG, in which the directions of easy magnetization are not $\langle 111\rangle$ at low temperature (they are $\langle 100\rangle$ and $\langle 110\rangle$, respectively). These studies could also help us in the understanding of the internal structure of the wall. The magnetic moments in this case must go from one to another equilibrium configuration. It is known that at low temperatures the local crystalline field and exchange anisotropies may become very high, so that it may be difficult for the magnetization vector to rotate simply from one position to another as given by the Bloch model. The computation of the structure could be done with the help of Tchéou's model[12]. This could give an explanation of the preferential occurence of 71° walls and allow a more rigorous calculation of the strains inside and outside the wall, giving a quantitative explanation of the wall contrast obtained on x-ray topographs.

REFERENCES

1. A. Mathiot, J.F. Petroff, Y. Bernard, Phys. stat. solidi (a) 20, K1 (1973).
2. M. Yamamoto, and T. Iwata, Sci. Rep. Res. Inst. Tohoku Univ. A5, 433 (1953).
3. W.H. Von Aulock, Handbook of Microwave Ferrite Materials (Academic Press, N.Y., 1965).
4. C. Kittel and J. K. Galt, Solid State Physics 3, 437 (1956).
5. M. Polcarová, IEEE Trans. Mag. MAG-5, 536 (1969); M. Schlenker, P. Brissonneau, and J.P. Perrier, Bull. Soc. Fr. Mineral. Cristall. 91, 653 (1968).
6. J. Miltat and M. Kleman, Phil. Mag. 28, 101 (1973).
7. S. Iida, J. Phys. Soc. Jap. 22, 1201 (1967) ; V.I. Sokolov and T. D. Hien, Sov. Phys.-JETP 25, 986 (1967).
8. J.F. Petroff and A. Mathiot, Mat. Res. Bull. 9, 319 (1974).
9. F. Sayetat, Thesis, Grenoble (1974).
10. L. Néel, Cahiers de Physique, Paris 25, 1 (1944); B.A. Lilley, Phil. Mag. 41, 792 (1950).
11. A. Mathiot and J.F. Petroff, to be published.
12. F. Tchéou, Thesis, Grenoble (1972), Ao 7632.

BEHAVIOR OF THE ACTION POTENTIAL OF NITELLA CLAVATA CELLS IN THE PRESENCE OF UNIFORM MAGNETIC FIELDS

Sigurds Arajs, Ling-Chia Lydia Yeh Lin, and T. E. Farrington
Department of Physics
Clarkson College of Technology
Potsdam, New York 13676

ABSTRACT

Single internodal cells of the alga Nitella clavata, of length 20-40 mm and of diameter of about 0.5 mm, have been studied at 25C in uniform magnetic fields up to 17 kOe. The action potentials were measured using cotton-padded Ag-AgCℓ electrodes. The cells were stimulated by square electric pulses of 5V and 2 msec duration with 5 minute rest periods. It has been found that the action potential is decreased when the cells are exposed to the magnetic fields, the effect being larger when the fields are parallel than perpendicular to the cell. In general, parallel fields above 10 kOe are particularly harmful to life of these algae cells, often resulting in death within 24 hours.

INTRODUCTION

It is well-known that large individual plant cells respond to electrical, chemical, or even mechanical stimulation in just the same way as a mammalian nerve cell does: with an electric impulse that sweeps along the surface of the cell. The detailed distribution of ions in and around a living cell accounts for its bio-electrical properties. One species of cells, whose electrical disturbances can be easily studied are those of the fresh-water algae Nitella. These calls can be up to 40 mm long with a diameter of about 0.5 mm. Recently there has been considerable interest in finding the effects of magnetic fields on biological systems.[1,2] One area which has been quite unexplored is the study of bioelectric signals in the presence of applied fields. In this paper we present new investigations on the behavior of the action potential in magnetic fields up to 17 kOe.

EXPERIMENTAL CONSIDERATIONS

The Nitella cells were kept in an aquarium filled with a culture solution having a composition given in Ref. 3 except for an addition of 1% $FeCℓ_3$ solution per liter. The algae were trimmed occasionally to promote the growth of larger cells.

The measuring of the bioelectric signals was done in a plexiglass moist chamber[4] with Ag-AgCℓ electrodes wrapped with cotton yarn. The stimulation was accomplished by a Grass SD9 stimulator and were recorded on a Bausch and Lomb 25 mV VOM6 recorder. The magnetic fields were produced by a 15-inch Magnion electromagnet, Model F-158, with a transistorized Magnion Model HS-10200 power supply.

The bioelectric measurements on the Nitella cells were done in the following manner. The cells were cleaned and cut several hours before usage. They were then placed across the dampened electrodes in the moist chamber. The cells were then stimulated with a dc pulse of 5 V and 2 msec duration and their responses (associated with the plasmalemma) observed. The intervals between successive stimulations were 5 min. After this, the moist chamber was placed in the zero field space between the magnet pole pieces. The intensity of light in this region was adjusted to that outside the magnet. After the bioelectric behavior of the cells was stabilized, the magnet was turned on and the responses were investigated as a function of the intensity of the applied field either parallel or perpendicular to the long axis of the cells. In this study the exposure time of the magnetic field was 5 min. Measurements were also made after the removal of the field in order to see whether the cells would return to their original state.

RESULTS

The average experimental results are shown in Fig. 1. The vertical axis represents the normalized action potential, i.e., the unperturbed potential (∿130 mV) is labelled as 1. The horizontal axis gives the time scale. All the measurements given in Fig. 1 are associated with a 5 min exposure to either parallel or perpendicular magnetic field. The solid part of the curve shows the action potential in the presence of a magnetic field. The broken curves represent the potential as a function of time after the removal of the field.

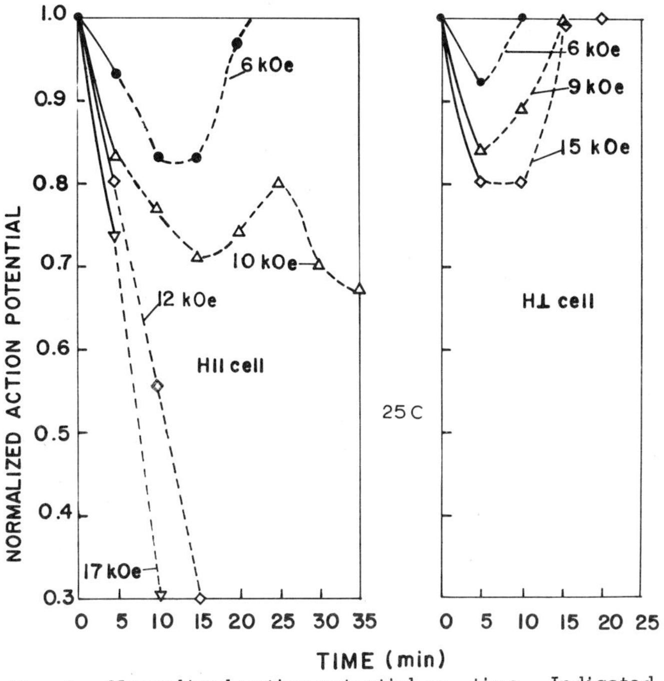

Fig. 1. Normalized action potential vs. time. Indicated field applied for first five minutes (solid lines).

Two observations can be made from Fig. 1. First, the effect is larger for parallel than for perpendicular fields. Second, magnetic fields can cause irreversible changes in the bioelectric behavior of the Nitella cells. When the perpendicular field of 5 min duration is removed, the cells eventually return to their original state. This is not true for the parallel fields larger than approximately 6 kOe. For instance, after the 10 kOe field has been shut off, the action potential still drops for about 10 min, then tries unsuccessfully to return to the original value. The exposure for 5 min to higher parallel fields, such as 12 and 17 kOe, causes a continuous decrease and eventual disappearance of the action potential. In general, we find that parallel fields above 10 kOe are particularly harmful to life of these algae cells, often resulting in death within 24 hours.

At the present time there are no satisfactory theories which could explain the observed effects. Our experiments strongly suggest that some parts of the membranes possess properties sensitive to magnetic fields. We propose that applied fields cause rearrangement of macromolecules which redistribute and neutralize some internal charges decreasing the potential difference across the membrane. Decrease in the action potential then follows directly from Hodgkin-Huxley equations.[5] Recently Gaffney and McConnell[6] have demonstrated that phospholipid molecules can, indeed, be oriented by a magnetic field whose direction lies in the plane of the biological membrane. This rotational effect is approximately proportional to the field strength. Further experimental and theoretical studies are needed to elucidate the detailed behavior of the action potential in the presence of magnetic fields.

REFERENCES

1. Becker, R. O., Med. Electron. Biol. Engrg. 1, 293 (1963).
2. Barnothy, M. F., Biological Effects of Magnetic Fields, Vol. 1 and 2 (Plenum Press, New York, 1964 and 1969).
3. Forsberg, C., Physiologia Plantarium 18, 275 (1965).
4. Goldstein, N. N., Project IMPS: Instrumentation Methods for Physiological Studies, Vol. 1 (Operations Service Department, University of California Extension, Berkeley, 1964).
5. Hodgkin, A. L. and Huxley, A. F., J. Physiol. 117, 500 (1952).
6. Gaffney, B. J. and McConnell, H. M., Chem. Phys. Letters 24, 310 (1974).

MOSSBAUER AND MAGNETIC MEASUREMENTS OF IRON PHASE DISTRIBUTIONS IN APOLLO LUNAR SAMPLES*

G. P. Huffman and F. C. Schwerer
U. S. Steel Research Laboratory, MS-98
Monroeville, Pennsylvania 15146

ABSTRACT

Knowledge of the distribution and valence of iron is required for meaningful interpretation of the basic physical properties of lunar surface samples (magnetic remanence, electrical conductivity, optical behavior, etc.) and also has important consequences for discussions of lunar regolith processes. The amounts of metallic Fe and ferrous Fe in various mineral phases are related to surface processes involving localized or wide-spread reheating with molten-state or subsolidus reduction, and the metallic-ferrous ratio (Fe^{o}/Fe^{2+}) appears to be a useful classification parameter for lunar samples. When combined with other data (e.g., chemical and petrographic analyses), the Fe phase distributions (the percentages of Fe contained in various phases) of lunar rocks may provide useful constraints concerning original source materials and thermal histories. Here we report Fe phase distributions as determined by Mossbauer spectroscopy and Fe^{o}/Fe^{2+} ratios as determined from both Mossbauer and magnetic measurements for a large suite of lunar samples. The (Fe^{o}/Fe^{2+}) values determined by the two techniques agree reasonably well with the exception of a number of samples which contain large amounts of Fe in olivine; for these samples, the magnetic analysis gave higher (Fe^{o}/Fe^{2+}) values than did the Mossbauer analysis. The most likely origins of the discrepancies are antiferromagnetic clustering in olivine and/or an abundance of large grain size metallic Fe ($\gtrsim 50\mu m$).

The percentage of Fe in olivine is found to decrease markedly with increasing wt.%FeO, which is believed to reflect higher crystallization temperatures for lunar highlands material. At $4^{o}K$, Mossbauer spectra of most high olivine samples exhibit a well-resolved hyperfine spectrum believed to arise from antiferromagnetic clusters of Fe^{2+} spins in olivine mirror sites. Some high olivine samples do not show this behavior and petrographic studies indicate that these samples underwent relatively rapid cooling. Presumably, this would allow less time for preferential Fe^{2+} occupation of mirror sites, which are slightly larger and more strongly coupled magnetically than the inversion sites in the olivine structure.

*Supported in part by NASA under Contract NAS-9-12271.
A detailed account of this work will appear in the Proceedings of the Fifth Lunar Science Conference, Geochim. Cosmochim. Acta, Suppl. 5, Vol. 3, 1974.

SIMPLE MODELS FOR THE COERCIVITY OF HARD MAGNETIC MATERIALS

J.Bernasconi, S.Strässler and R.S.Perkins
Brown Boveri Research Center, CH-5401 Baden, Switzerland

ABSTRACT

We investigate the possible effects of local regions with an appreciably lowered anisotropy constant K on the magnetization reversal in hard magnetic materials. Within the framework of very simple model-assumptions, and for different shapes of the defect-regions, we calculate the critical fields for the growing of a reversed nucleus (H_c^n) and for the pinning of an extended domain wall (H_c^p). In particular, we investigate the dependence of H_c^n on the form and size of the defect-regions, on the spatial gradient of K, and on the angle θ between magnetic field and easy axis. For the appropriate 1-dimensional geometry, the results are in reasonable agreement with Abraham and Aharoni's exact micromagnetic calculations (θ = 0). The variation of H_c^n with the angle θ resembles the Kondorsky - 1/cosθ behavior, and is markedly different from the predictions of the Stoner-Wohlfarth theory for coherent rotation. Pinning of extended domain walls turns out to be important only for very large K-gradients. The model-calculations are compatible with recent experiments on RE-Co systems.

The mechanisms that govern the magnetization reversal and the factors that determine the coercivity in hard magnetic materials are not yet fully understood[1]. The observed coercive forces H_c are much smaller than the theoretical upper limit given by the anisotropy field H_A[1,2], and the dependence of H_c on the angle θ between applied field and easy axis[2] is quite different from a Stoner-Wohlfarth behavior[3].

In the following, we shall investigate in some detail a simple theoretical model for the magnetization reversal in uniaxial hard magnetic systems. In particular, we shall determine the dependence of H_c/H_A on the model-parameters and the variation of H_c with the angle θ.

Based on earlier suggestions[1], we assume that the magnetization reversal is dominated by local regions with a low magnetocrystalline anisotropy (small anisotropy constant K). In RE-Co systems such regions could correspond, e.g., to inclusions with a different stochiometry or to defects consisting of a series of stacking faults. For simplicity, we shall assume that K = 0 inside the defect-region, so that only two critical fields may, in principle, become important for the determination of the coercivity. These are the critical field H_c^n for the growing of a reversed domain, especially against the spatial gradient of K at the boundaries of the defect-region, and the critical field H_c^p for the pinning of an extended domain wall to such low-anisotropy defects.

For a 1-dimensional geometry with a linear increase of K, and for θ = 0, Abraham and Aharoni[4] have determined H_c^n by exact micromagnetic calculations that are quite involved and require extended numerical computations. Here, we shall try to investigate the problem with much simpler model-assumptions that enable us to extend the calculations very easily to different geometries for the defect-region and, in particular, to obtain some information about the dependence of H_c^n on the angle θ.

For simplicity, we shall restrict most of our discussion to the 1-dimensional case. K(x) is assumed to be symmetrical with respect to the origin, to vanish for $x < R_o$, and to increase from zero to its bulk value $\bar{K}$ between R_o and R. The form of K(x) between R_o and R need not be linear, and we shall see that the behavior of the gradient dK/dx has a marked influence on the θ-dependence of H_c^n. The easy axis is perpendicular to the x-axis, and we assume the existence of a reversed domain inside the defect-region and denote the position of the Blochwall by x_o. The width δ of the Blochwall is assumed to be much smaller than the dimensions of the defect-region ($\delta \ll R_o$ and $R-R_o$). This allows us to include the energy of the Blochwall-region in a simple summary expression. The wall-energy γ per unit area is written as

$$\gamma(x_o) = 4\sqrt{AK(x_o)} \cdot F(\phi_i,\phi_a) \quad , \qquad (1)$$

where A is an exchange constant, and ϕ_i and ϕ_a are the directions of the magnetization (with respect to the easy axis) at the surfaces of the wall (F = 1 for a 180°-wall). Apart from the wall-energy, the total energy E includes the interaction with the applied field H and the anisotropy energy.

The growth of the reversed domain is discussed by investigating the variation of E with respect to a displacement δx_o of the Blochwall. The coercive field $-H_c^n$ is defined to be the smallest value of -H at which the Blochwall moves to infinity, i.e.

$$\frac{\partial E}{\partial x_o}(x_o,H,\theta,\bar{\phi}_i,\bar{\phi}_a) \le 0 \quad \text{for} \quad x_o \ge \bar{x}_o(H) \quad , \qquad (2)$$

where we have assumed that, for a given H, the reversed region ($x < x_o - \frac{1}{2}\delta$) extends at least to the point determined by the Stoner-Wohlfarth coherent rotation criterion[3],

$$-H = 2K(\bar{x}_o - \tfrac{1}{2}\delta) \cdot f(\theta)/I_s \quad . \qquad (3)$$

$\bar{\phi}_i(\theta)$ and $\bar{\phi}_a(\theta)$ are the corresponding metastable directions for the magnetization (evaluated for $\delta \to 0$, for simplicity), and I_s denotes the saturation magnetization.

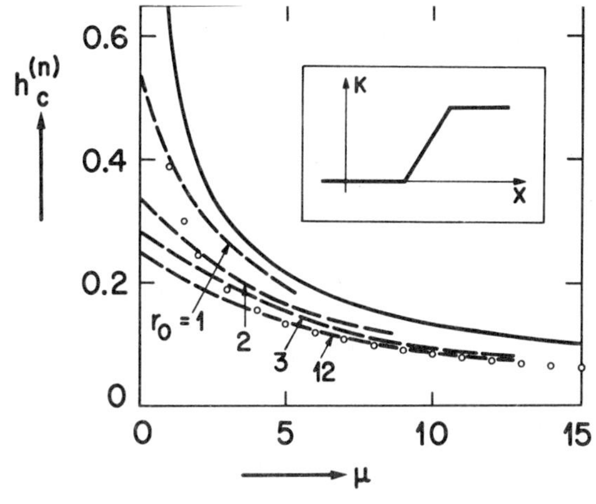

Fig.1 Reduced coercive field $h_c^n = -H_c^n I_s/2K$ as a function of $\mu = \pi(R-R_o)/\delta$ for a 1-dimensional geometry, where $R_o<x<R$ is the region of the linear increase of K(x) (see insert), and $\delta = \pi(A/\bar{K})^{1/2}$. $r_o = \pi R_o/\delta$ defines the size of the region with K(x) = 0. The circles and the full curve correspond to our simplified calculations, the broken curves to the exact micromagnetic calculations of Abraham and Aharoni.

The results for the 1-dimensional case with a linear increase of K(x) and with $\theta = 0$ (H parallel to the easy axis) are shown in Fig.1. The reduced coercive field $h_c^n = -H_c^n/H_A = -H_c^n I_s/2K$ is plotted versus μ, the inverse of the normalized K(x)-slope. Our results are represented by the circles (for a Blochwall-thickness of $\delta = \pi(A/\bar{K})^{1/2}$) and by the full curve (for $\delta = 0$). The broken curves correspond to the exact micromagnetic calculations of Abraham and Aharoni[4]. Contrary to our simplified calculations, these results depend on r_o (the reduced size of the region with K = 0), but converge very rapidly with increasing r_o. (We have investigated this convergence by extending the calculations of Abraham and Aharoni to larger values of r_o). For reasonably large values of r_o and μ, where our simplifying assumptions are satisfied, we have quite good agreement between our simple treatment and the exact results.

We have also calculated h_c^n for different geometries of the defect-region (disc, sphere). The results are quite close to those of the 1-dimensional case if the dimensions of the defect-region become reasonable large compared to the Blochwall-thickness.

The pinning of an extended domain-wall to such low-anisotropy defects has been investigated with a modified Kersten approach[5]. The corresponding critical field h_c^p turns out to be much smaller than the h_c^n for a defect with the same K(x)-behavior. Within our model, a pinning of extended domain-walls can thus only influence the coercivity if defects with both small K(x)-slopes and very steep K(x)-slopes are present. The nucleation of reversed domains could then start in defects with a flat K(x)-behavior, while the pinning could become important at the defects with a steep increase of K(x).

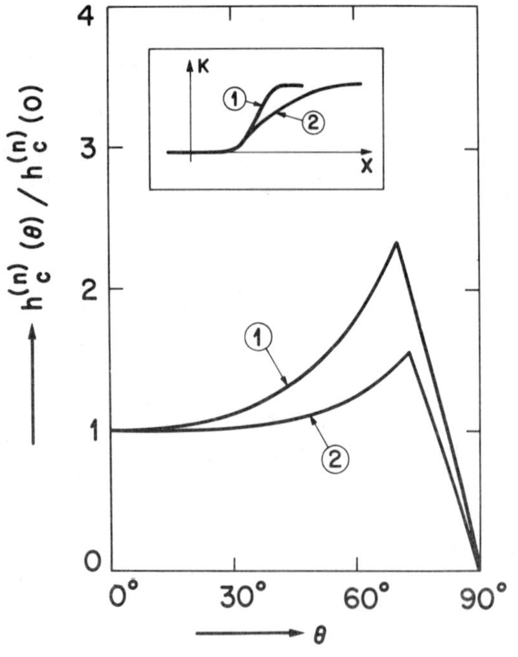

Fig.2 Variation of the reduced coercive field h_c^n with the angle θ between applied field and easy axis for two 1-dimensional examples with a K(x)-behavior as indicated in the insert.

The dependence of h_c^n on the angle θ between applied field and easy axis is illustrated by the two examples of Fig.2. This dependence is somewhat similar to a Kondorsky - $1/\cos\theta$ behavior[6], and markedly different from the predictions of the Stoner-Wohlfarth theory for coherent rotation[3]. (Note that for θ near 90° the coercive field is determined merely by a rotation of the saturated state). The specific form of the increase of K(x) has an evident influence on the θ-dependence of h_c^n. Roughly we can say that steep K-gradients lead to a steep increase of the h_c^n vs. θ curve and small gradients to a flat θ-dependence.

Recent measurements on $SmCo_{5-x}Cu_x$ single crystals and on $CeMMCo_5$ permanent magnets with different CeMM-compositions[7] show a sample-dependence of the H_c vs. θ behavior corresponding roughly to the situation of Fig.2. This suggests the possibility of explaining the different θ-dependences by a composition-dependence of the form, or even the type, of the defects.

While such conclusions, at the moment, are perhaps somewhat speculative, we can at least say that our simple model-calculations are not in contradiction with the known experimental results on RE-Co systems.

REFERENCES

1. For a review see e.g. J.D.Livingston, Proc. 1972 Conf. on Magnetism and Magnetic Materials, A.I.P. Conf. Proc. 10, 643 (1973), and A.Aharoni, Rev. Mod.Phys. 34, 227 (1962).
2. A.Menth, H.P.Klein, J.Bernasconi and S.Strässler, Proc. 1973 Conf. on Magnetism and Magnetic Materials, A.I.P. Conf.Proc. 18, 1182 (1974).
3. E.C.Stoner and E.P.Wohlfarth, Phil.Trans.Roy.Soc. London 240, 599 (1948).
4. C.Abraham and A.Aharoni, Phys.Rev.120, 1576 (1960).
5. M.Kersten, Physik.Z. 44, 63 (1943).
6. E.Kondorsky, J.Phys.USSR 2, 161 (1940).
7. H.Nagel, R.S.Perkins, A.Menth (private communication).

TEMPERATURE DEPENDENCE OF MAGNETIC HYSTERESIS IN MnBi THIN FILMS

Shigeo Honda, Masatoshi Koyama, and Tetsuzo Kusuda
Department of Electronics, Hiroshima University, Japan

ABSTRACT

MnBi polycrystalline films prepared on glass or mica substrates were studied with the aid of Faraday or Kerr effect. Major and minor magnetic hysteresis loops were measured as a function of temperature or film thickness. The nucleation field H_n is dependent on temperature and on the film thickness. These variations of H_n were discussed by introducing the effective field due to magnetostriction. The available information on the wall coercivity was furnished by the detailed measurements of the minor and the major loops.

INTRODUCTION

Thin films of MnBi with the c-axis perpendicular to the film plane can be prepared by vacuum evaporation.[1] MnBi films have been studied as a storage medium for laser beam addressable memories because of their strong uniaxial anisotropy perpendicular to the film plane.[2] For this application, the physical properties of the films have been studied by many workers above room temperature.[2] Studies, however, below room temperature are quite limited.[3] In this work, the magnetic hysteresis of MnBi films have been investigated in the temperature range from -180°C to +300°C.

EXPERIMENTAL PROCEDURE

MnBi films were prepared on glass or mica substrates by vacuum evaporation. Layers of Bi and Mn were evaporated successively on to a substrate held at room temperature, followed by a protective layer of SiO. After the evaporation, the films were annealed at about 270°C for 6 hours under a pressure of $0.5\text{-}1\times10^{-8}$ Torr.

The sample was loaded in a vacuum chamber and then it was set in an electromagnet having polepieces with a hole of 8mm diameter. The field was applied normal to the film. Hysteresis loops were recorded with the aid of Faraday or Kerr effect in the temperature range between -180 and +300°C.

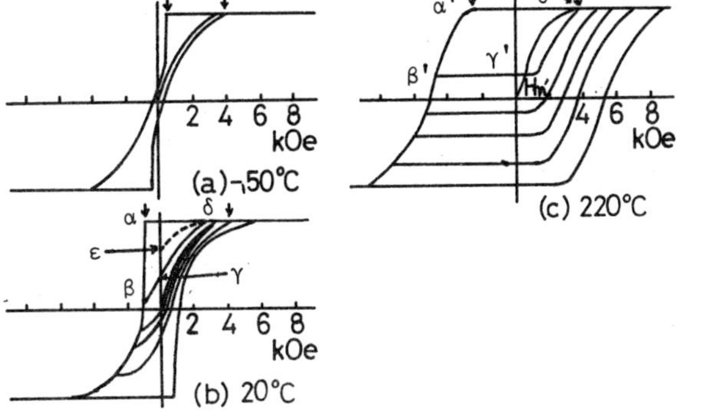

Fig. 1. Hysteresis loops for magnetization normal to a 800Å-thick MnBi film.

RESULTS AND DISCUSSION

Temperature Dependence of Hysteresis

Fig. 1 shows typical hysteresis loops of a MnBi film (800Å thickness) measured at various temperatures. The initial magnetization curve from the demagnetized state (origin) is only slightly dependent on temperature, and the film is nearly saturated at $H_s \simeq 3.3$kOe independent of temperature. The nucleation field H_n, where the flux reversal occurs, is dependent on temperature as shown in Fig. 2. At low temperature, H_n lies at the first or the third quadrant, in which H_n is defined as positive. H_n decreases with increasing of temperature and then lies at the second or the fourth quadrant (negative) above 0°C.

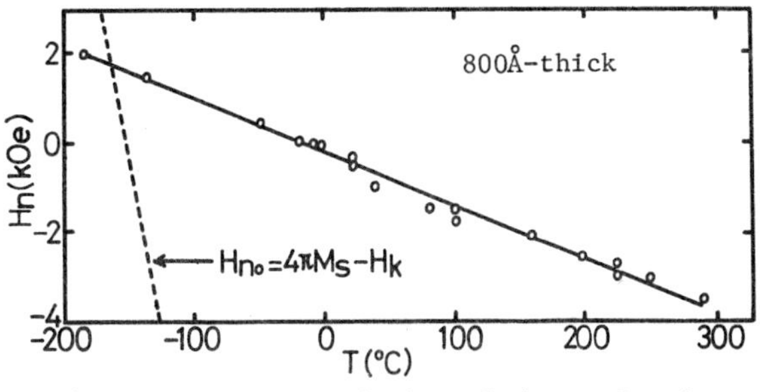

Fig. 2. Temperature variation of the nucleation field H_n. The broken line shows the calculated values $H_{no}=4\pi M_s-H_k$.

In an ideal film, the nucleation field H_{no} is given by $H_{no} = 4\pi M_s - H_k$, where $H_k = 2K_u/M_s$ is the effective field due to magnetocrystalline anisotropy. The broken line in Fig. 2 shows H_{no} obtained from Guillaud's data.[4] Both calculated values (H_{no}) and the experimantal values (H_n) decrease monotonically as temperature increases, but the temperature dependence of H_n is less than that of H_{no}. In the real films, a reverse domain nucleates at a local weak position, e.g., a lattice defect or a grain boundary, at the critical field H_n and then the flux switches by wall propagation.[5] Therefore, it is generally expected

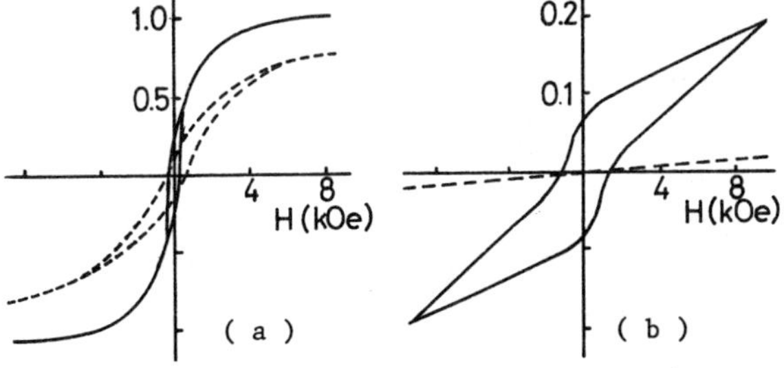

Fig. 3. Hysteresis in the film plane of 800Å-thick film (broken lines) and 8000Å-thick film (solid lines) measured by the sample vibrating magnetometer at LNT(a) and RT(b).

that $-H_n \lesssim -H_{no}$. But it should be noted that the value of $-H_n$ at -180°C (about LNT) is larger than $-H_{no}$ (at LNT). The value of H_n (at LNT) implies another effective field ($H_{\sigma 1}$) applied normal to the film.

H_n (at LNT) depends on the film thickness. It is nearly equal to H_{no} (at LNT) in films thicker than 5000Å, but it is negative as the film thickness decreases to less than 400Å. Since H_k is negligibly small and independent of the film thickness at LNT,[4] the thickness dependence of H_n (at LNT) presumably results from that of $H_{\sigma 1}$. The effect of $H_{\sigma 1}$ may be revealed from the hysteresis loops in the film plane as shown in Fig. 3(a). It is found that thinner films can hardly be magnetized in the film plane but in thicker films the magnetization in film plane is easy at LNT. The above facts show that $H_{\sigma 1}$ increases in thinner films. Therefore, $H_{\sigma 1}$ may be caused by the magnetostriction due to the stress between the film and the substrate.

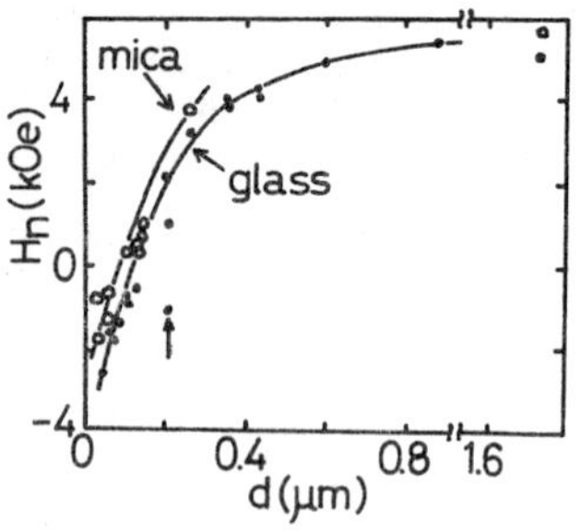

Fig. 4. Film thickness dependence of H_n at room temperature.

Effect of Film Thickness

Fig. 4 shows the values of H_n of polycrystalline films (grain size about 3~5μm) at room temperature (RT) as a function of the film thickness. H_n (at RT) increases as films become thicker and tends to about 6kOe. On the other hand, H_n of the single crystal film is only slightly dependent on the film thickness; $-H_n$ is larger than 7kOe even in films thicker than 1μm.[7] It should be emphasized that the thickness dependence of H_n is characteristic in the polycrystalline films. The internal stress, which acts between grains constituting the film, increases generally with increasing film thickness, as is well known.[7] If a new effective magnetostriction field $H_{\sigma 2}$ due to the stress between grains is applied in the film plane[5], H_n will become more positive as films become thicker.

$H_{\sigma 2}$ will also depend on the packing density of grains constituting a film, i.e. $H_{\sigma 2}$ will increase with increasing of the packing density. Therefore, if grains are isolated from each other, presumably the film will have negative H_n above 2000Å, as indicated by arrow in Fig. 4.

Hysteresis loops in the film plane were also measured at room temperature, as shown in Fig. 3(b). In thinner films, the hysteresis loop in the plane is hardly obtained because the anisotropy field perpendicular to the film plane is too large, but in thicker films it can be obtained. This fact suggests that the easy axis of the magnetization deviates slightly from the c-axis (the film normal) due to the effect of $H_{\sigma 2}$ in thicker films, even at room temperature.

In discussing the thickness dependence of H_n, we should take account of effects other than $H_{\sigma 2}$, e.g. the thickness dependence of the internal demagnetizing field applied to the residual domains at the lattice defects or the grain boundaries, as described in a previous paper.[9] Presumably, however, $H_{\sigma 2}$ is the dominant effect in this case as shown in Fig. 3(b). When we observe domains during the formation of MnBi films in situ, the effect of $H_{\sigma 2}$ can be clearly seen.[8] MnBi grains grow keeping the single domain state in the case of small packing density, whereas the grains grown in the single domain state are suddenly demagnetized by the internal stress between grains in the case of large packing density. When adjacent grains impinge upon each other, reverse domains nucleate at the grain boundaries and the grains are demagnetized by wall motion.[5,8]

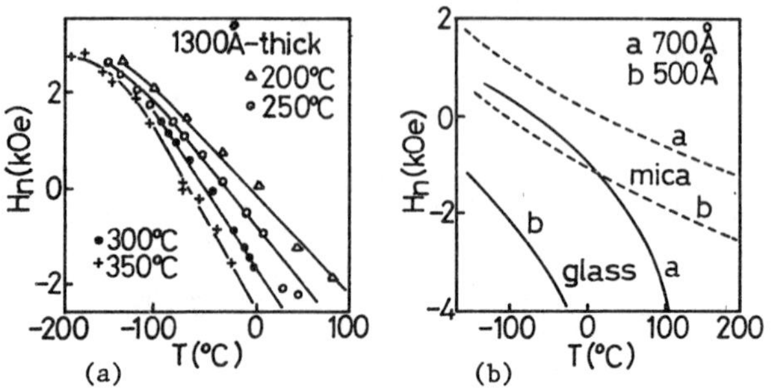

Fig. 5. Temperature dependence of H_n for various annealing temperatures in MnBi films (a), and for various substrates (b).

The temperature variations of $H_{\sigma 1}$ and $H_{\sigma 2}$ will affect the temperature dependence of H_n. For example, the temperature variation becomes steeper on increasing the annealing temperature in preparing MnBi films (Fig. 5(a)). This annealing effect may be concerned with the fact that grains grow to large size and then the internal stress between grains decreases, i.e., $H_{\sigma 2}$ decreases with increasing annealing temperature.[8] The temperature dependence of H_n is also affected by the substrate (Fig. 5(b)). This substrate effect may be caused by the fact that the thermal strain is different in each substrate[7], namely, the temperature variations of $H_{\sigma 1}$ or $H_{\sigma 2}$ are different in each substrate.

Minor Hysteresis Loop

The minor loops are shown in Fig. 1(b) and (c). The curve (β γ δ) in Fig. 1(b) follows the initial magnetization curve, but the loop (α'β'γ'δ') in Fig. 1 (c) is rectangular. The difference between minor loops can be explained by paying attention to the switched grains in the field range between point α(or α') and β(or β'). The domain structures at the points β, γ and ε in Fig. 1(b) are shown in Fig. 6(a), (b) and (c), respectively. The grains at the center portion were switched in the field range between α and β, but some grains remain their initial magnetized state (dark area) as shown in Fig. 6(a). Since the field of point β is less than the saturation field H_s, several narrow strip domains with initial magnetization (black) remain in the switched grains or at the grain boundaries. Their residual domains expand by the internal demagnetizing field when the external field is returned

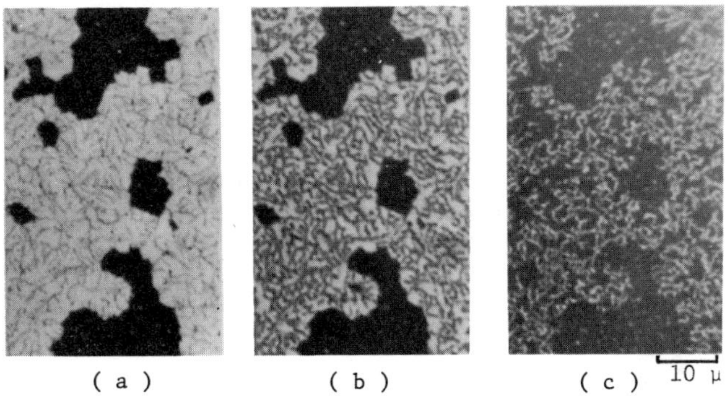

(a) (b) (c)

Fig. 6 Domain structures observed at room temperature with the aid of Kerr effect. (a), (b) and (c) show the domain configurations of the points β, γ and ε in Fig. 1(b), respectively.

to zero, and the switched grains produce almost a demagnetized state at the point γ, as shown in Fig. 6(b). Thus the magnetization curve from the point γ corresponds to the initial magnetization curve of the swithced grains. Some of the above switched grains are almost saturated positive at the point δ(about +2.5 kOe). Only the grains with the residual negative domains are demagnetized again by reducing the external field as shown in Fig. 6(c). Then the irreversible curve (δ ε) is produced.

On the other hand, the switched grains are almost saturated negative at the point β'(about 4.6 kOe) in Fig. 1(c). When the external field is returned to zero and then is applied in positive direction, the flux reversal of the switched grains occurs at H_n'. Thus, the rectangular minor loop appears at higher temperature.

The above results suggest the following: i) The wall coercivity in each grain is relatively small, and the grain boundaries act as a cause of high coercivity. ii) The major loop is determined by the distribution of the nucleation field H_n of each grain. iii) The minor loop is determined by the relation between H_n and H_s.

REFERENCES

1. D. Chen, J. Appl. Phys. 42, 3625 (1971).
2. see, for example, W. K. Unger and R. Räth, IEEE Trans. Magn. MAG-7, 885 (1971).
3. K. Egashira and S. Yoshii, J. Appl. Phys. 44, 3742 (1973).
4. P. C. Guillaud, J. Phys. Radium, 12, 492 (1951).
5. T. Kusuda and S. Honda, Appl. Phys. Let. 24, 516 (1974).
6. S. Honda, Y. Hosokawa, S. Konishi and T. Kusuda, Japan. J. Appl. Phys. 12, 1028 (1973).
7. K. L. Chopra, "Thin film Phenomena", McGraw-Hill Book Comp. New York (1969), pp. 266-327.
8. S. Honda and T. Kusuda, J. Appl. Phys. 45, 2689 (1974).
9. S. Honda, Y. Hosokawa and T. Kusuda, Trans. IECE of Japan, 56-C, 545 (1973).

SEMIHARD MAGNETIC ALLOY OF Fe-Mn-Ni SYSTEM

Toshio Takahashi and Kenichi Ono
Ibaraki Electrical Communication Laboratory,
N.T.T., Tokai, Ibaraki, Japan 319-11

ABSTRACT

To ascertain economical new semihard magnetic materials for latching relays or switches in an electronic switching system, alloys consisting of 4 ~ 16 wt % Mn, 2 ~ 18 wt % Ni and the balance Fe, to which up to 3 % Ti was added, were studied.

The magnetic properties of the alloy systems, a new working method to improve them and their structure are described.

As a result, new magnets were obtained whose magnetic properties are 40 ~ 283 Oe in coercive force and about 5.7 ~ 15 kG in residual induction.

INTRODUCTION

Magnetic latching relays and switches are used in an electronic switching system. For these cores, a semihard magnetic alloy having 20 ~ 100 Oe coercive force is used.

The Fe-Mn-Ni alloys dealt with in this report were studied for the purpose of developing economical alloys whose coercive force is within the above mentioned value range and whose residual induction Br is more than 10 kG.

The Fe-Mn permanent magnet was first reported by Jellinghaus in 1949.[1] Since then, this alloy system was studied as a semihard magnetic material for the electronic switching system and for the hysteresis motor. Alloys of Fe-Mn-Ti, Fe-Mn-Co, Fe-Mn-Co-Ti, Fe-Mn-Ti-Cu[2], and Fe-Mn-Cu-Cr[3] had been studied.

EXPERIMENTAL METHOD

The composition range of the test piece was considered to be $\alpha+\gamma$ phase, from the binary alloy phase diagrams of Fe-Mn and Fe-Ni. In addition, quarternary alloys were also studied, hoping to increase the coercive force by adding Ti.

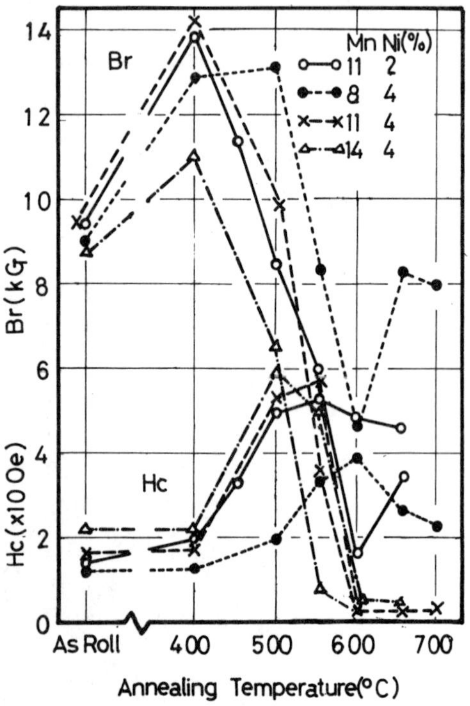

Fig. 1 Magnetic properties after annealing.

The ingot was made by melting electrolytic iron, nickel, manganese and titanium of a purity of more than 99 % in air or Argon atmosphere to bring their total weight to 1 kg.

The measured test pieces were cold rolled sheets 0.3 mm in thickness or cold drawn rods 5.5 mm in diameter.

Magnetic measurements were conducted in magnetic fields of 200 ~ 600 Oe.

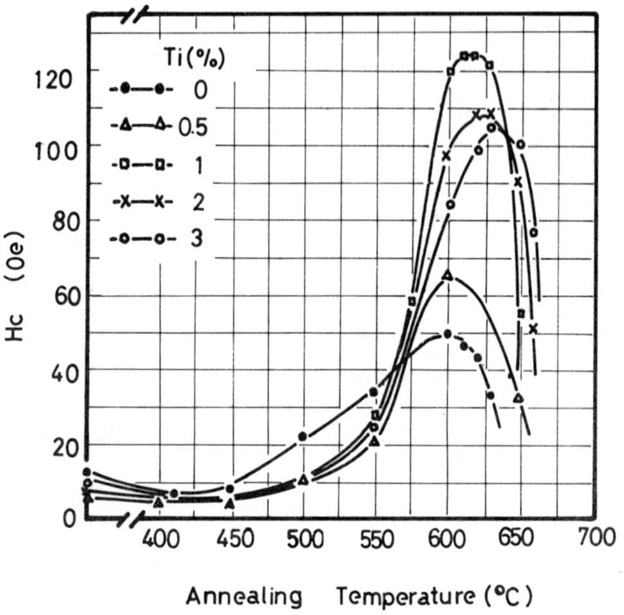

Fig. 2 Effects of Ti concentration on maximum coercive force.
(8%Mn-7%Ni-x%Ti-Fe alloys)

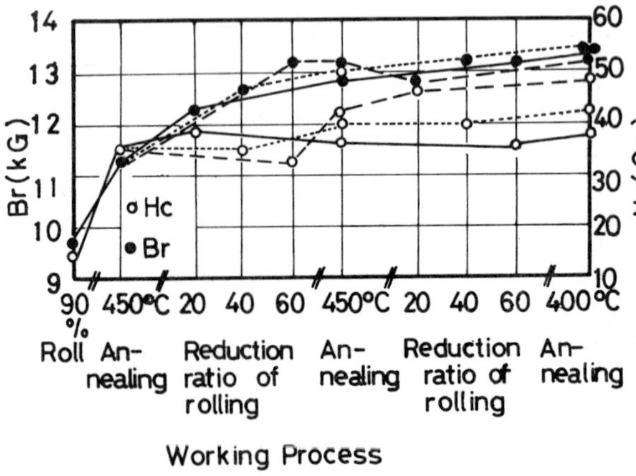

Fig. 3 Magnetic properties after several annealing and cold rolling cycles.
(11 % Mn-2 % Ni- Fe alloy)

EXPERIMENTAL RESULTS

The magnetic properties of Fe-Mn-Ni alloy are varied, as shown in Fig.1, by annealing after rolling. The annealing temperatures each of which give the maximum value for the coercive force (hereafter denoted as Hc) and for the residual induction (hereafter denoted as Br), are different from each other. In the composition range of these alloys, B_{200} becomes lower when Mn+Ni concentration is great, and it has the constant value of 18 kG for about 17 % Mn+Ni concentration, where Br is more than 10 kG. Hc has the constant values such as about 60 Oe for the 21 ~ 18 % Mn+Ni 79 ~ 84 % Fe and about 40 Oe for the 17 ~ 12 % Mn+Ni.

When Ti is added to these alloys, the coercive force becomes higher than in the original alloys. The effects of additional Ti are shown in Fig.2. Alloys which showed maximum Hc are those with 1 % Ti added.

The magnetic properties of these alloys can be improved by one more cycle of annealing and cold working. The alloy containing 11 % Mn-2 % Ni-Fe(as shown in Fig.1) reaches maximum Hc at an annealing temperature of 550 °C, where Hc value is 50 Oe and Br is 8.5 kG. After repeated cold rolling and annealing of the alloy, magnetic properties finally obtained are Hc 48 Oe and Br 13.5 kG, as shown in Fig.3. These values are equal to the maximum magnetic properties, as in Fig.1. Similarly, 13 % Mn-3 % Ni-1 % Ti-Fe alloy gives Hc 283 Oe and Br 5.7 kG through the annealing and working.

The magnetic properties of these alloys depend on the existence of the fine mixed phase of γ and elongated α', produced by the partial strain-induced transformation. These phases transformed as follows, at respective treatments:

as quenched:	mixture of cooling martensite α' and austenite γ,
first reduction:	γ -- α' transformation, induced by cold working,
annealing:	α' -- γ reverse transformation by annealing above 500 °C,
second reduction:	γ -- α' transformation induced by cold working the mixture phases of α' and γ.

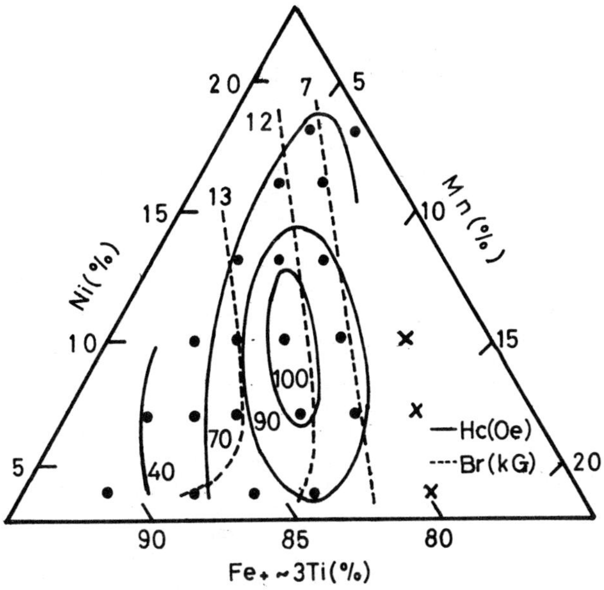

Fig. 4 Ti added (up to 3 %) Fe-Mn-Ni system magnetic properties diagram.
Dots show the experimental alloys (shown with additional concentration), x shows where the magnetic properties are extremely low, and shows where edge cracking occurred due to a working ratio of about 50 %.

In the composition range of these alloys the magnetic properties have been improved, in every case, by cyclic cold working and annealing. As a result, the relations between the alloy composition and magnetic properties are obtained as shown in Fig.4.

From the above results, a latching relay core (5.5 mm in diameter) of 9.5 % Mn-6.5 % Ni-0.2 % Ti-Fe alloy has been developed. The Hc of this alloy is 66 ± 4 Oe and Br is 14.4 ± 0.4 kG. The squareness ratio is 0.94 ~ 0.97.

CONCLUSION

Alloys consisting of Fe, Mn and Ni, with Ti added, were studied. Alloys having good workability and magnetic properties were obtained. The main experimental results are summarized as follows:

(1) The relations between magnetic properties and composition range of alloys were obtained. Ti added to the Fe-Mn-Ni alloy systems enables these alloys to have high Hc. The maximum Hc is obtained in the composition range of 7 ~ 11 % Mn, 7 ~ 12 % Ni and 80 ~ 83 % Fe, where Br is more than 12 kG.

(2) In this alloy system, excellent magnetic properties, due to the fine mixed structure consisting of γ and elongated α', can be obtained by a combination of higher than 20 % cold working ratio and suitable annealing temperature.

(1) D.Hadfield: Permanent Magnets, p.162,(1962), John Wiley & S..
(2) H.Kaneko,et al: Cold-worked Semihard Magnetic Alloys of the Fe-Mn System, Jour.Japan Inst. Metals, 34, p.441,(1970).
(3) T.Sada,et al: Semihard Magnetic alloys of the Fe-Mn System and its Applied Designs, Electronic Parts and Materials, 8, p.71,(1969).

CRYSTAL TRANSFORMATION AND ORIENTATION OF Mn-Al-C HARD MAGNETIC ALLOY

S. Kojima, T. Ohtani, N. Kato, K. Kojima, Y. Sakamoto,
I. Konno, M. Tsukahara and T. Kubo
Central Research Lab. Matsushita Electric Ind. Co., Ltd.
Kadoma, Osaka, Japan.

ABSTRACT

An anisotropic polycrystalline magnet of a Mn-Al-C alloy with (BH)max = 6 MG·Oe has been successfully produced by extrusion at about 700°C. During the study on this ternary alloy system, a ferromagnetic phase (τ-phase) single crystal was obtained by annealing a high-temperature phase (ε-phase) single crystal containing lamellae of the compound Mn_3AlC, under the action of a uniaxial compressive stress. The formation process of the τ-phase single crystal was analyzed. The process can be explained by the ε(h.c.p.) → ε'(orthorhomb.) → τ(f.c.t.) transformations with the following orientation relationships : $(0001)_\varepsilon \parallel (100)_{\varepsilon'} \parallel (111)_\tau$ and $[\bar{1}100]_\varepsilon \parallel [001]_{\varepsilon'} \parallel [\bar{1}\bar{1}2]_\tau$. The ε'-phase is new and has a B19-type structure. The ε → ε' transformation is a disorder-order transformation and the ε' → τ is martensitic. These single crystal results may be extended to the directionalization process for polycrystalline magnets.

INTRODUCTION

A ferromagnetic, metastable, tetragonal phase in the Mn-Al system was found by Nagasaki, Kono and Hirone[1] in 1955. Since then, basic and practical investigations of this alloy have been made by many authors[2-10]. Isotropic hard magnets of Mn-Al-C ternary alloy were found by Yamamoto[11] to have good magnetic properties. Recently, anisotropic hard magnets of this ternary alloy were produced by extrusion at about 700°C. They have good machinability and magnetic characteristics with Br = 6100G, bHc = 2400 Oe and (BH)max = 6 MG·Oe.

The metallugical and magnetic properties of the Mn-Al-C system were studied using single crystals. A single crystal of the ferromagnetic phase (τ-phase, f.c.t.) was obtained by annealing a single crystal of the high-temperature phase (ε-phase, h.c.p.) under the action of a uniaxial compressive stress in a specific crystallographic direction. In the present work, the formation process of the τ-phase single crystal was analyzed, and the transformation, orientation relationships and crystal orientations were investigated.

EXPERIMENTAL PROCEDURE

An ε-phase single crystal ingot (72.10 wt % Mn, 1.04 wt % C) containing lamellae of the compound Mn_3AlC was prepared. The ingot was prepared by the Tammann-Bridgman method. The compound Mn_3AlC was precipitated in lamellae parallel to the $(0001)_\varepsilon$ plane within the parent ε-phase by heat treatment at a temperature range of 830 to 900°C. Rectangular-shaped single crystals Nos. 1-11 which had the orientations shown in Fig. 1 were cut out from the ingot.

An experiment with the ε → τ transformation under a uniaxial compressive stress was performed. The ε-phase single crystals were annealed at a temperature of 560°C applying a pressure of 30 kg/mm^2 in the compression direction shown in Fig. 1. The orientation and the phase structure of the annealed specimens were determined by X-ray analysis and microscopic observation.

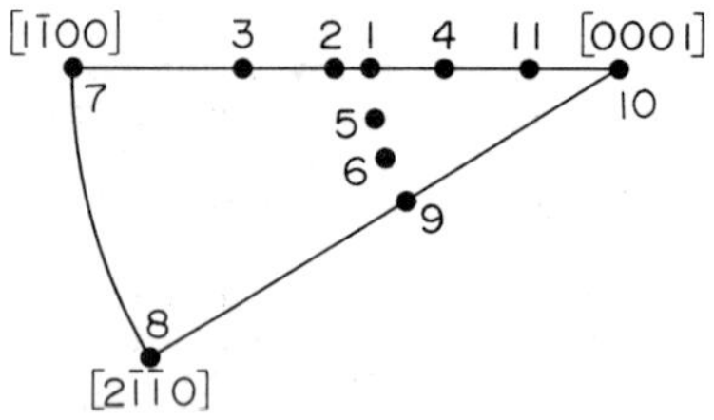

Fig.1 Orientations of the compression axes of the ε-phase single crystals Nos. 1-11.

τ-PHASE SINGLE CRYSTAL

A τ-phase single crystal nearly rectangular in shape (6.20 x 6.60 x 7.33 mm) was formed from the ε-phase single crystal 1 (7.19 x 5.79 x 7.18 mm). Rapid shrinkage in length was observed in the direction of the compression axis during the formation process. X-ray Laue diffraction patterns showed, when compared with those of the ε-phase single crystal, that $(\bar{1}10)_\tau$ was parallel to $(11\bar{2}0)_\varepsilon$ and $(111)_\tau$ was parallel to $(0001)_\varepsilon$ in which the compound Mn_3AlC was precipitated in lamellae. The $[001]_\tau$ axis, which was the direction of easy magnetization, made an angle of almost 84° to the compression axis. The magnetic hysteresis loop of the τ-phase single crystal had a wasp-waisted shape[12] in the $[001]_\tau$ direction, but the loop changed to a normal shape after being held at 600°C for 30 hours. The magnetic properties determined from the magnetization curve are as follows : saturation magnetization 4πIs = 6950G, and crystalline anisotropy constant $K = 1.03 \times 10^7$ $ergs/cm^3$.

In the case of the other ε-phase single crystals Nos. 2-11, τ-phase single crystals or nearly unidirectionally oriented τ-phase crystals were formed from the single crystals Nos. 2-6, and a bidirectionally oriented τ-phase crystal (twin) was formed from the single crystal No. 7. But highly oriented τ-phase specimens were not formed from the single crystals Nos. 8-11.

ORTHORHOMBIC PHASE ε'

A new ordered phase ε' was found in the intermediate step in the ε → τ transformation. This ε'-phase has an orthorhombic structure (B19-type) with lattice constants of a = 4.39A, b = 2.76A and c = 4.58A. The structure was determined from an X-ray analysis of the ε'-phase single crystal. The single crystal was obtained by annealing the ε-phase single crystal with the orientation 1 in Fig. 1 under the compressive stress for a time until just before rapid shrinkage began. The orientation relationships observed between the ε-phase and ε'-phase were that $(100)_{\varepsilon'}$ was parallel to $(0001)_\varepsilon$ and $(010)_{\varepsilon'}$ was parallel to $(11\bar{2}0)_\varepsilon$.

ε → ε' → τ TRANSFORMATIONS

From the experimental results above, the formation process of the τ-phase single crystal can be explained by a model of the ε → ε' → τ transformations

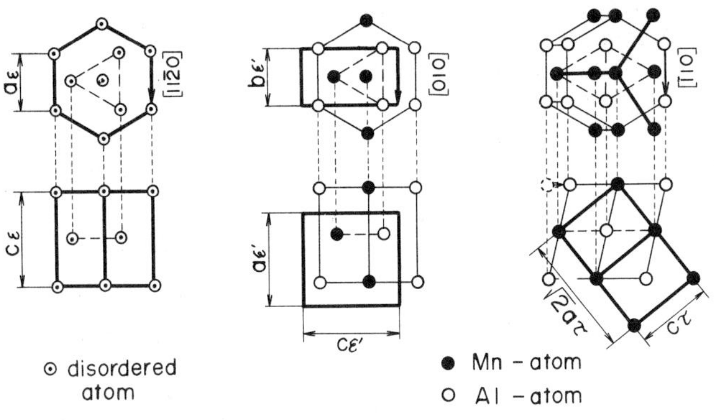

Fig.2 Schematic representation showing the process of the $\varepsilon \to \varepsilon' \to \tau$ transformations. (a) ε(h.c.p.), (b) ε'(orthorhomb.), (c) τ(f.c.t.).

(as shown in Fig. 2) with the following orientation relationships:

$(0001)_\varepsilon \parallel (100)_{\varepsilon'} \parallel (111)_\tau$,
$[\bar{1}100]_\varepsilon \parallel [001]_{\varepsilon'} \parallel [\bar{1}\bar{1}2]_\tau$,
$[11\bar{2}0]_\varepsilon \parallel [010]_{\varepsilon'} \parallel [\bar{1}10]_\tau$.

The $\varepsilon \to \varepsilon'$ is a disorder-order transformation with a shrinkage of 2 percent in the $[\bar{1}100]_\varepsilon$ direction. The $\varepsilon' \to \tau$ is a martensitic transformation where each atom in every other $(100)_{\varepsilon'}$ plane moves to the $[001]_{\varepsilon'}$ direction for a distance of about 1/3 $C_{\varepsilon'}$ as shown in Fig. 2-(c). Table 1 shows the calculated and observed values concerning the change in the shape and the orientations on the formation of the τ-phase single crystal from the rectangular-shaped ε-phase single crystal No. 1. Figure 3 shows the shrinkage rate in the compression axis in the cases of single crystals Nos. 1-4. The calculation was performed geometrically based on the transformation

Table 1 The change of the shape and the orientation on the formation of the τ-phase single crystal from the ε-phase single crystal 1.

	ratio of length (τ/ε)			angle between $[001]_\tau$ and compression axis
	ℓ'/ℓ*	m'/m**	n'/n***	
cal.	0.851	1.147	1.024	85.4°
obs.	0.862	1.140	1.021	84°

* in the compression direction.
** in the other rectangular direction.
*** in the $[11\bar{2}0]_\varepsilon$ direction.

model using the following lattice constants of each phase : a = 2.697A, c = 4.396A for the ε-phase, a = 4.39A, b = 2.76A, c = 4.58A for the ε'-phase, and a = 3.910A, c = 3.625A for the τ-phase. The observed values agree well with the calculated values.

The atom movement in the $(111)_\tau$ plane corresponding to the $\tau \to \varepsilon'$ martensitic transformation was examined by an experiment where a τ-phase single crystal was compressed in the $[001]_\tau$ direction at 620°C. As a result, rapid shrinkage in length in the compression direction, accompanied by the formation of a new $[001]_\tau$ axis, was observed. This was

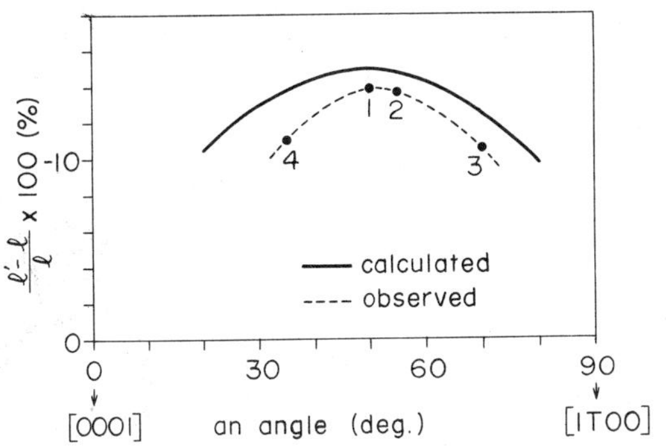

Fig.3 The relation between the shrinkage rate and the compression direction on the formation of τ-phase single crystals from ε-phase single crystals Nos.1-4.

interpreted as atom movement in the only $(111)_\tau$ plane in the $[11\bar{2}]_\tau$ direction (parallel to the $[00\bar{1}]_{\varepsilon'}$ direction). Atom movement in the other three $\{111\}_\tau$ planes, which were not parallel to the $(100)_{\varepsilon'}$ plane, was not observed.

The directionalization process induced by external stress has been clarified on single crystal, but not yet on extruded polycrystalline magnets. However, the process in polycrystalline magnets should be related to the atom movement in the special crystal plane observed in the single crystal.

The investigations of the Mn-Al-C alloy system are to be continued. Detailed results will be reported later.

ACKNOWLEDGEMENTS

The authors wish to thank Professor S. Chikazumi of Tokyo University, Dr. M. Fukuda, Dr. M. Asanuma, Mr. S. Nishimura and Mr. Y. Amano of this company for their guidance and encouragement during this work.

REFERENCES

1. S. Nagasaki, H. Kono and T. Hirone, Digest of the 10th Annual Conf. of the Physical Society of Japan 3 162 (1955).
2. H. Kono, J. Phys. Soc. Japan 13 1444 (1958).
3. A.J.J. Koch, P. Hokkeling, M.G.V.D. Steeg and K.J. De Vos, J. Appl. Phys. 31 75s (1960).
4. W. Köster and E. Wachtel, Z. Metallkde. 51 271 (1960).
5. M.A. Bohlmann, J. Appl. Phys. 33 1315 (1962).
6. T. Kawaguchi, M. Nagakura, K. Kamino and S. Yoshizawa, J. Japan Inst. Met. 28 384 (1964).
7. N. Makino, Y. Kimura and M. Suzuki, J. Japan Inst. Met. 28 483 (1964).
8. D. Ernst, J. Tydings and M. Pasnak, J. Appl. Phys. 36 1241 (1965).
9. H. Zijlstra and H.B. Haanstra, J. Appl. Phys. 37 2853 (1966).
10. L.M. Magat, Ya.S. Shur, G.S. Kandaurova, G.M. Makarova and N.I. Gusel'nikova, Phys. Met. Metallog. 23 32 (1967).
11. H. Yamamoto, U.S. Patent 3,661,567.
12. R.M. Bozorth, Ferromagnetism 173 (1951).

IMPROVEMENT OF TEMPERATURE DEPENDENCE OF REMANENCE IN FERRITE PERMANENT MAGNETS

K. Haneda, C. Miyakawa, and H. Kojima
Research Institute for Scientific Measurements
Tohoku University, Sanjo-machi, Sendai, Japan

ABSTRACT

A low temperature coefficient of remanence has been found for ferrite permanent magnets in which some of the Fe^{3+} ions are replaced by non-magnetic ions in the following ways: $2Fe^{3+} \rightarrow M^{2+} + M^{4+}$ or $3Fe^{3+} \rightarrow 2M^{2+} + M^{5+}$. By replacing Fe^{3+} by (Cu^{2+}, Ge^{4+}), (Cu^{2+}, Si^{4+}), (Cu^{2+}, V^{5+}), (Cu^{2+}, Nb^{5+}), or (Cu^{2+}, Ta^{5+}), the temperature coefficient of B_r between room temperature and 100°C was reduced from -0.2%/°C for $BaFe_{12}O_{19}$ to about -0.1%/°C. Substitution of (Cu, Nb) is especially effective, and gives practically no change in B_r between room temperature and 50°C. These results cannot be simply explained by a decrease in the temperature dependence of the saturation magnetization. Although further research is required to clarify the mechanism, the increase in BH_c with temperature may provide a phenomenological explanation.

INTRODUCTION

The temperature coefficient of the remanence of ferrite permanent magnets is about -0.2%/°C, and is an order of magnitude higher than that of Alnico magnets. Efforts have been made to improve this deficiency by replacing part of the Fe^{3+} ions with Ti^{4+}, Sb^{3+}, and As^{3+}.[1,2] In connection with this problem, investigations using Mössbauer spectroscopy[3,4] and calculation of the exchange parameters[5] for these ferrites have also been reported. But all these papers are concerned with the improvement of the temperature coefficient of the saturation magnetization.

In this report, the possibility of developing oxide magnets with more desirable temperature dependence of B_r will be shown for some substituted hexagonal ferrites which have temperature dependence of M_s similar to that of ordinary barium or strontium ferrites, but different temperature dependence of H_c.

EXPERIMENTAL RESULTS

Hexagonal ferrites with chemical formulae

$$AFe_{2n-x}(Cu_{.5x}B_{.5x})O_{3n+1}$$
or
$$AFe_{2n-x}(Cu_{.67x}C_{.33x})O_{3n+1}$$

were prepared by normal powder techniques. Here, A, B, and C denote Ba or Sr; Ge or Si; and V, Nb, or Ta, respectively. That is, some of the Fe^{3+} ions are replaced by two kinds of ions with different valency in the following ways: $2Fe^{3+} \rightarrow Cu^{2+} + M^{4+}$ or $3Fe^{3+} \rightarrow 2Cu^{2+} + M^{5+}$. The value of n varied from 5.3 to 6.0, and x from 0 to 1.0. The mixed powder was fired between 1100°C and 1250°C for 1 or 2 hours, milled, pressed at 3.5 tons/cm², and sintered between 1000° and 1250°C for 15 minutes in air. X-ray investigation showed that single-phase M-type hexagonal fer-

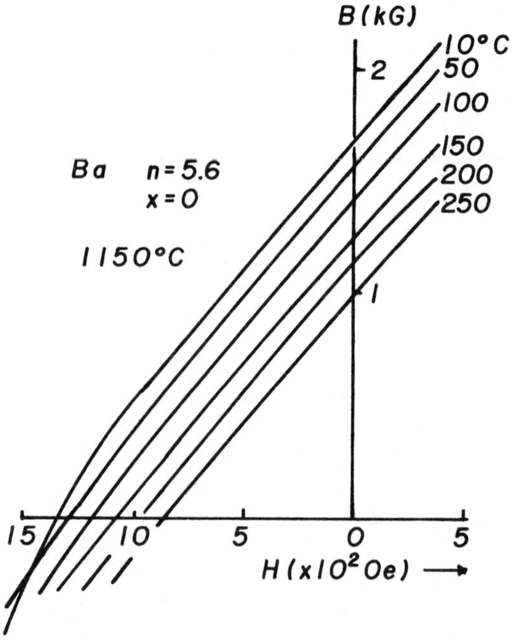

Fig. 1 Temperature dependence of B-H curve for $BaO \cdot 5.6Fe_2O_3$.

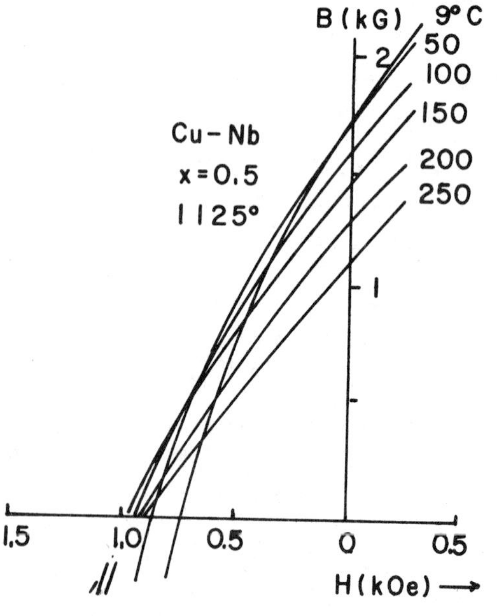

Fig. 2 Temperature dependence of B-H curve for $BaFe_{2n-x}(Cu_{2x/3}Nb_{x/3})O_{3n+1}$; n=5.6, x=0.5.

Table I Temperature coefficient of remanent magnetization between room temperature and 100° C and Curie point for several substituted ferrites.

Composition	Fe^{3+}	$Cu^{2+}-Si^{4+}$	$Cu^{2+}-Ge^{4+}$	$Cu^{2+}-V^{5+}$	$Cu^{2+}-Nb^{5+}$	$Cu^{2+}-Ta^{5+}$
$\Delta\sigma_r$ (%/°C)	-0.19	-0.13	-0.12	-0.14	-0.10	-0.13
Tc (°C)	450	439	421	435	414	416

Table II Temperature dependence of magnetic properties for some substituted ferrites

No.	Composition	n	x	React.T. (°C)	Sint.T. (°C)	ΔBr (%/°C) RT~50°	ΔBr (%/°C) RT~100°	ΔbHc (%/°C) RT~50°	ΔiHc (%/°C) RT~50°	$\Delta\sigma_s$ (%/°C) RT~50°
1	Ba	5.6	0	1100	1150	-0.19	-0.19	-0.09	+0.28	-0.16
2	Ba-Cu-Nb	5.6	0.5	1100	1200	-0.05	-0.11	+0.33	+0.55	-0.15
3	Ba-Cu-Nb	5.6	0.5	1100	1125	0	-0.09	+0.22	+0.42	-0.12
4	Ba-Cu-Ge	5.6	0.25	1100	1150	-0.12	-0.16	+0.09	+0.29	-0.16
5	Ba-Cu-Si	5.6	0.25	1100	1150	-0.13	-0.14	0	+0.22	-0.16
6	Sr-Cu-Nb	5.6	0.2	1200	1220	-0.09	-0.13	+0.11	+0.48	-0.21
7	Sr-Cu-Ge	5.6	0.2	1250	1180	-0.08	-0.16	0	+0.33	-0.16
8	Sr-Cu-Si	5.6	0.3	1250	1180	-0.09	-0.13	-0.05	+0.25	-0.11

Table III Magnetic properties of the same specimens as Table II.

No.	Composition	ρ (g/cm^3)	σ_s (emu/g)	Br (G)	bHc (Oe)	iHc (Oe)	(BH)max (MG·Oe)
1	Ba	3.86	60.1	1650	1400	3820	0.7
2	Ba-Cu-Nb	4.76	64.2	2170	1030	1300	0.7
3	Ba-Cu-Nb	4.53	63.9	1770	840	1110	0.6
4	Ba-Cu-Ge	4.51	63.3	2120	1450	2200	0.9
5	Ba-Cu-Si	4.47	61.4	2050	1500	2530	0.9
6	Sr-Cu-Nb	4.80	64.0	2650	1350	1570	1.0
7	Sr-Cu-Ge	4.76	60.9	2650	1650	2320	1.2
8	Sr-Cu-Si	4.66	58.9	2650	1675	2870	1.2

rite was formed and that the lattice constants changed linearly with the average ionic radius in all cases. B-H curves at room temperature were measured by the ordinary yoke method using a fluxmeter. The temperature change of magnetic properties was determined using a vibrating sample magnetometer from room temperature to 250°C.

Table I shows the temperature coefficient of σ_r, the remanent magnetization per unit mass, between room temperature and 100°C, which was roughly estimated from the demagnetizing curve by assuming that the demagnetizing field is $4\pi M_s/3 \doteq 1000$ Oe for each specimen. Substitution of non-magnetic ions reduces the Curie point, as also shown in Table I. The composition of these specimens is n = 6.0 and x = 0.5. They were fired at 1100°C for 2 hours and sintered at 1200°C for 15 minutes.

From these preliminary results, accurate B-H curves were measured for Cu-Si, Cu-Ge, and Cu-Nb substituted ferrites. Fig. 1 and Fig. 2 illustrate the changes in the curves with temperature for $BaO\cdot5.6Fe_2O_3$ and $BaFe_{2n-x}(Cu_{.67x}Nb_{.33x})O_{3n+1}$, n = 5.6 and x = 0.5. Demagnetizing curves are almost parallel for Ba ferrite, but they intersect in the second quadrant for Cu-Nb substituted ferrite. Some examples of the improved temperature coefficient of substituted hexagonal ferrites are given in Table II, and their magnetic properties at room temperature are given in Table III. Although the sintering conditions do not markedly affect the temperature coefficient of B_r, a rather low sintering temperature was chosen for most specimens to prevent the decrease of H_c.

From these results, it can be said that the temperature coefficient of B_r is remarkably reduced for substituted ferrites between room temperature and 100°C, and the permanent magnet properties are still satisfactory. Furthermore, it has been proved that the improvement by such substitution is also effective for anisotropic magnets.

DISCUSSION AND CONCLUSIONS

Heimke[1] and Esper and Kaiser[2] tried to improve the temperature coefficient of M_s by replacing Fe^{3+} in $4f_2$ sites of M-type ferrite, and obtained some successful results. But Kreber and Gonser[3] showed by Mössbauer studies that the substituted ions in Esper and Kaiser's specimens occupied 2b sites instead of $4f_2$ sites. Moreover, Grill and Harberey[5] reported that the replacement of Fe^{3+} ions in $4f_2$ sites by paramagnetic ions increases the temperature coefficient of M_s, and that the coefficient cannot be decreased without decreasing M_s. Therefore, it can be said that the mechanism of the improvement is still not clear.

However, the results obtained in this experiment are different from those described above in the following ways:

(1) The temperature coefficient of B_r can be reduced by replacement of a small fraction of the Fe^{3+} ions with nonmagnetic ions, even though the temperature dependence of M_s does not show a remarkable change.

(2) The magnetic saturation of the substituted ferrites is higher than that of the original ferrite under the same preparation conditions (see Table III).

Although further investigations are necessary to elucidate the mechanism of the improvement, it can be concluded from the phenomenological point of view that the increase of $_BH_c$ with temperature may offset the decrease of M_s, and this may be the main reason for the reduction of the temperature change of B_r.

REFERENCES

1. G. Heimke, J. Appl. Phys. 31, 271S (1960).
2. F. J. Esper and G. Kaiser, Inter. J. Magnetism 3, 189 (1972).
3. K. Kreber and U. Gonser, Appl. Phys. 1, 339 (1973).
4. G. Albanese, M. Carbucicchio, and A. Deriu, phys. stat. sol. (a) 23, 351 (1974).
5. A. Grill and F. Haberey, Appl. Phys. 3, 131 (1974).

NUCLEAR FERROMAGNETISM

A. Abragam
Service de Physique du Solide et de Résonance Magnétique
CEN-Saclay, BP n°2, 91190 Gif-sur-Yvette
France

ABSTRACT

The possibility of producing ordered states of nuclear spins by DNP followed by ADRF was first demonstrated in 1969. The spins of ^{19}F in a crystal of CaF_2 were cooled below one microdegree (with the applied field along the [100] axis) and their antiferromagnetic ordering was exhibited through the characteristic behaviour of their transverse and (later) longitudinal susceptibilities.

The "truncated" dipolar Hamiltonian responsible for the ordering, depends on the orientation of the applied field with respect to the crystal axes. Along the [111] axis and at a negative spin temperature we have observed a ferromagnetic domain structure, the domains being alternating planar layers perpendicular to the applied field. Direct evidence for it is provided by NMR of a magnetic probe, the rare (0.13%) isotope ^{43}Ca. Its resonance is split into two lines by opposite Weiss fields from both types of ferromagnetic domains of ^{19}F, in contradistinction to either paramagnetic or antiferromagnetic structures where a single line is observed. The behaviour of the spectrum of ^{43}Ca in the presence of a small effective field as well as the NMR spectrum of ^{19}F itself are consistent with this structure. An outstanding problem is the size of the domains.

INTRODUCTION

Five years ago, almost to a day, I was reporting at the Magnetism Conference in Philadelphia on the first observation of nuclear antiferromagnetism in a single crystal of CaF_2[1]. The mills of the gods grind slowly and today I am reporting on the production and observation, of nuclear ferromagnetism in the same crystal[2]. Some of you might say that it took us five years to rotate the crystal with respect to the applied magnetic field, through the magic angle from the [100] direction where antiferromagnetism was observed to the [111] direction where ferromagnetism is observed, surely one of the slowest rotations in Physics.

The theoretical and experimental background for our studies of nuclear ordered states can be found in a recent paper[3], and I will be content to list its most salient features.

a) The strength of the magnetic dipolar interaction between neighbouring spins of ^{19}F in CaF_2 is of the order of a microkelvin in temperature units and spin temperatures of that order are thus necessary to observe an ordering of the nuclear spins.

b) The cooling to such temperatures is limited to the nuclear spins, the lattice, temporarily isolated from the nuclear spin-system by the weakness of the relaxation mechanisms, remaining hot. The cooling is a two-step process. The first step is a dynamic nuclear polarization by the solid effect in a high field H_o (25 kOe), up to polarizations of the order of 90%, whereby the spin temperature T_S is lowered from the kelvin range into the millikelvin range ; the second step is an adiabatic demagnetization which lowers T_S down into the microkelvin range. Positive and negative temperatures which may lead to ordered states of quite different nature can be obtained with equal ease.

c) Although the demagnetization could in principle be performed in the laboratory frame by reducing H_o to zero, it is far easier experimentally and far more interesting theoretically to perform the adiabatic demagnetization in the rotating frame (ADRF). The effective spin-spin Hamiltonian is then the well-known truncated Hamiltonian of Van Vleck and the effective Zeeman field is $\Delta H_z = |H_o - (\omega/\gamma)|$ where ω and H_o are the values of the rf frequency and of the applied dc field H_o, at the end of the ADRF process and γ is the gyromagnetic ratio of the nuclei. The values of ΔH_z for which magnetic nuclear ordering can be observed range from zero to a few gauss, which is of the order of the local magnetic field.

d) The truncated spin-spin Hamiltonian, and therefore also the stable ordered structures depend on the orientation of H_o with respect to the crystal axes. This explains in particular the existence in the same crystal of antiferromagnetism along [100] and of ferromagnetism along [111]. Nuclear ordering in high external field offers thus a far wider field of study than ordinary dipolar ordering in zero external field. Furthermore the truncated Hamiltonian is simpler and easier to handle theoretically than the full dipolar Hamiltonian. Last but not least, ordered states in high field can be studied by NMR techniques and thus benefit from a large increase in sensitivity.

e) As for electronic magnetism two types of measurements are available for the study of ordered structures. Of the first type are large scale or macroscopic measurements which probe the sample with magnetic fields uniform over it such as the measurements of magnetic susceptibility, transverse and longitudinal. It is well known for instance that for antiferromagnetic structures the transverse susceptibility is approximately constant whilst the longitudinal susceptibility decreases with temperature below the Néel point. This is how the antiferromagnetic nature of the structure observed along [100] for $T_S<0$ was proven[1,3]. Of the second type are the microscopic or small scale measurements where the probing field is highly inhomogeneous over the sample. It can be the magnetic[4] (or nuclear, sometimes called pseudomagnetic)[5,6] field carried by a neutron spin in diffraction experiments. It can also be the field produced by a nuclear magnetic moment belonging to an atom present in the sample[7]. In the latter case the perturbation of the structure under study by the nuclear probe must be negligible. For electronic magnetism this condition is easily fulfilled even if the probe nuclei are of normal density[7] but for the study of nuclear magnetism such as that of ^{19}F in CaF_2 the concentration of the probe nuclei must be very small. We shall see that the rare isotope ^{43}Ca (0.13%) is adequate for that purpose.

NUCLEAR FERROMAGNETISM
THE FERROSANDWICH

The search for ferromagnetic structures is complicated by the long range of the magnetic dipolar forces. The truncated Hamiltonian is given by :

$$\mathcal{H} = \sum_{i<j} A_{ij}(\underset{\sim}{n}) \left\{ 3(\underset{\sim}{I}_i.\underset{\sim}{n})(\underset{\sim}{I}_j.\underset{\sim}{n}) - \underset{\sim}{I}_i.\underset{\sim}{I}_j \right\} \tag{1}$$

where n is the unit vector along the applied field

$$A_{ij}(\underset{\sim}{n}) = \gamma^2\hbar^2 r_{ij}^{-3}(1 - 3\cos^2\theta_{ij})/2 \tag{2}$$

and θ_{ij} is the angle $\widehat{\underset{\sim}{n}.\underset{\sim}{r}_{ij}}$.

As shown in refs.1,3 the search for ordered structures by the molecular field method is simpler in the reciprocal lattice where we define the Fourier transforms :

$$\underset{\sim}{I}(\underset{\sim}{k}) = N^{-1/2} \sum_j \exp(i\underset{\sim}{k}.\underset{\sim}{r}_i) \langle \underset{\sim}{I}_j \rangle \tag{3}$$

$$A(\underset{\sim}{k},\underset{\sim}{n}) = \sum_j A_{ij}(\underset{\sim}{n}) \exp\{i\underset{\sim}{k}(\underset{\sim}{r}_i - \underset{\sim}{r}_j)\} \quad . \tag{4}$$

One should note that because of the long range of the dipolar forces, $A(\underset{\sim}{k},\underset{\sim}{n})$ is uniquely defined by (4) only if the length 1/k is small compared to the size R of the sample. For a simple cubic lattice, $A(\underset{\sim}{k},\underset{\sim}{n})$ has been computed numerically for all vectors $\underset{\sim}{k}$ inside the first Brillouin zone for which it is well defined, and for all directions $\underset{\sim}{n}$ of the applied field. If $1/k$, whilst small compared to the sample dimension R, is large compared to the lattice spacing a, it turns out that $A(\underset{\sim}{k},\underset{\sim}{n})$ is a function of the angle $(\widehat{k,n})$ only and is given by the simple formula :

$$A(\underset{\sim}{k},\underset{\sim}{n}) = \frac{4\pi}{3}\left(3\cos^2(\widehat{k.n}) - 1\right)\frac{\gamma^2\hbar^2}{2a^3} \quad . \qquad (5)$$

Finally for $k=0$ $A(\underset{\sim}{k},\underset{\sim}{n})$ is uniquely defined only if the sample has the shape of an ellipsoid. In particular for a sphere (which was the shape of our sample) $A(0,n) = 0$.

It can be shown[3] that a necessary condition for the stability of a mode $\underset{\sim}{I}(\underset{\sim}{k})$ parallel to the applied field is that $A(\underset{\sim}{k},\underset{\sim}{n})$ be a minimum (for $T_S > 0$) or a maximum (for $T_S < 0$) in k space. It is clear that for an antiferromagnetic mode such a minimum (maximum) must occur for large values of $k \sim 1/a$. $A(\underset{\sim}{k},\underset{\sim}{n})$ is then well defined and structures of that type can sometimes be predicted with some confidence. Thus the theoretical prediction of an antiferromagnetic structure along [100] at a negative temperature has been checked experimentally in some detail through measurements of transverse[1] and longitudinal[3] susceptibility, critical field[3], etc...

We now look for a ferromagnetic structure. The uniform mode $k=0$ gives $A(0,n) = 0$. It turns out that for all directions $\underset{\sim}{n}$ of the field there are vectors $\underset{\sim}{k}$ and $\underset{\sim}{k}'$ such that $A(\underset{\sim}{k},\underset{\sim}{n}) > 0$ and $A(\underset{\sim}{k}',\underset{\sim}{n}) < 0$. A uniform ferromagnetic structure cannot be stable in a spherical sample whatever the sign of the temperature. On the other hand a ferromagnetic structure with domains cannot be excluded. If the size of the domains is small compared to that of the sample (whilst still large compared to the lattice spacing) we can argue that the structure is a superposition of modes I(k) with $1/a \gg k \gg 1/R$, for which $A(\underset{\sim}{k},\underset{\sim}{n})$ is given by (5). The largest value of $A(\underset{\sim}{k},\underset{\sim}{n})$ is then $A_o = (8\pi/3)(\gamma^2\hbar^2/2a^3)$, and according to (5) corresponds to domains which are slabs perpendicular to $\underset{\sim}{n}$. When the field has the direction $\underset{\sim}{n}_o$ parallel to the [111] axis, it turns out that for all k $A(\underset{\sim}{k},\underset{\sim}{n}_o) \leq A_o$.

This leads us to "guess" as a plausible stable structure (for a negative temperature T_S) a succession of alternating ferromagnetic slabs with polarizations p_A and p_B along the direction [111] of the applied field and relative thicknesses x and (1-x), with $0 < x < 1$. We call such a structure a "ferrosandwich". We shall not describe here a more detailed theoretical approach to the problem of this structure[8], aiming in particular at a prediction of the thickness of the domains and of the domain walls. Once the ferrosandwich structure is assumed, the following features are easily established (for a spherical sample).

a) The field "seen" by a spin placed in a site of cubic symmetry is given in a slab of either type by the classical formulae of magnetostatics.

$$H_A = \Delta H_z + \frac{8\pi}{3}\frac{\gamma\hbar}{2a^3}(1-x)(p_A - p_B)$$
$$H_B = \Delta H_z - \frac{8\pi}{3}\frac{\gamma\hbar}{2a^3}\, x(p_A - p_B) \qquad (6)$$

b) From the Weiss field approximation :

$$p_{A,B} = \tanh\left(\frac{\gamma\hbar H_{A,B}}{2kT_S}\right) \qquad (7)$$

it follows[8] that the free energy is minimum for :

$$p_A = -p_B = p > 0 \quad x = \frac{1}{2}\left\{1 + \Delta H_z \Big/ \frac{8\pi}{3}\left(\frac{\gamma\hbar p}{2a^3}\right)\right\} \quad x < 1 \qquad (8)$$

The polarizations of the two types of slabs are equal and opposite and as ΔH_z increases from zero to the critical value :

$$(\Delta H_z)_c = \frac{8\pi}{3}\left(\frac{\gamma\hbar p}{2a^3}\right) \quad , \qquad (8')$$

x grows from 1/2 to 1, the slabs A growing at the expense of the slabs B until the latter disappear and the whole sample goes paramagnetic.

EXPERIMENTAL PROOF OF THE VALIDITY OF THE FERROMAGNETIC STRUCTURE

a) The spectrum of ^{43}Ca in zero effective field.

The formulae (6) and (8) suggest that a nuclear probe will have different Larmor frequencies in the slabs A and B and thus should give rise to two separate resonance lines. Their separation in gauss is given by :

$$h = H_A - H_B = \frac{8\pi}{3}\frac{\gamma\hbar}{a^3}\, p = \frac{16\pi}{3} M \quad , \qquad (9)$$

where $\pm p$ and $\pm M$ are the nuclear polarization and magnetization of ^{19}F in either slab. We have used as a probe the spins of ^{43}Ca . Fig.1 shows a spectrum of ^{43}Ca when ^{19}F is demagnetized in zero effective field. Two resonance lines of equal intensity are observed as

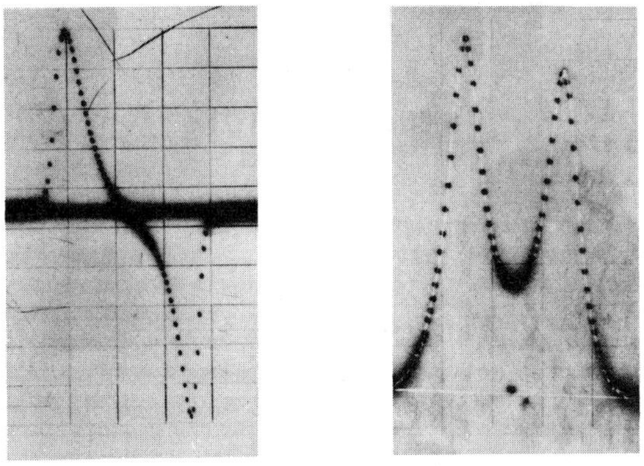

Fig.1 - Right : the ^{43}Ca doublet in the ferrosandwich structure along [111]. Left : the corresponding fluorine spectrum.

expected. Their distance h is 8.7 gauss which from (9) leads to $p(^{19}F) = 80\%$, in good agreement with the initial polarization of $^{19}F \sim 90\%$ prior to ADRF, considering the inevitable gain of entropy in the process. Such a splitting could not possibly be due to an antiferromagnetic structure since the Weiss fields produced by the two antiferromagnetic sublattices of ^{19}F cancel at a site of ^{43}Ca . Indeed a single line is observed for ^{43}Ca in the antiferromagnetic structure along [001], $T_S < 0$, as shown in Fig.2.

b) The spectrum of ^{43}Ca as a function of the effective field ΔH_z .

The formulae (6) and (8) predict that as long as ΔH_z is smaller than the critical value (8'), the intensities of the two lines of ^{43}Ca as well as their shifts vary linearly with ΔH_z whereas their separation remains unchanged. All these features have been observed within

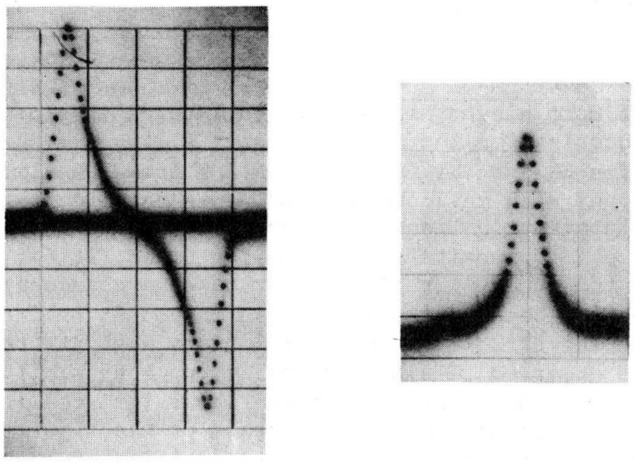

Fig.2 - Right : the ^{43}Ca single line in the antiferromagnetic structure along [100]. Left : the corresponding fluorine spectrum.

experimental error as shown in Figs.3 and 4. Both the continuous change in the relative intensities of the two lines with ΔH_z and the constance of their separation are again quite incompatible with an antiferromagnetic structure.

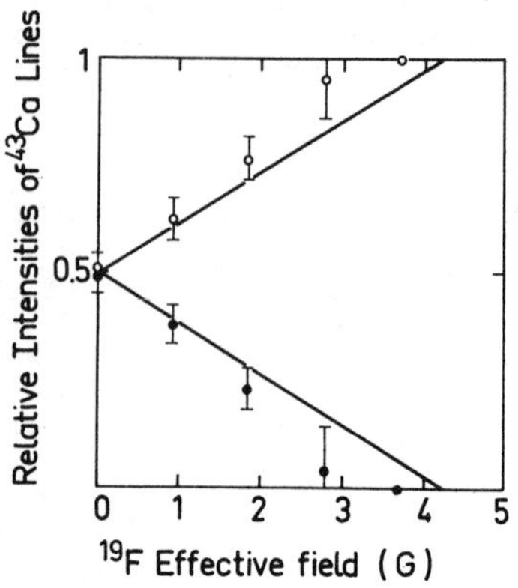

Fig.3 - Relative intensities of ^{43}Ca lines against the effective field.

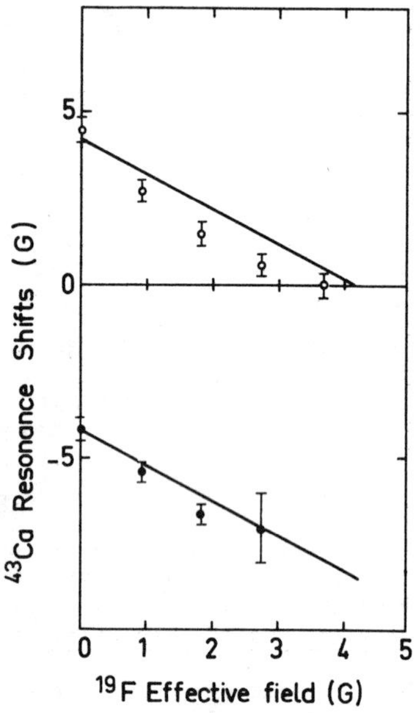

Fig.4 - Shifts of ^{43}Ca resonance lines against the effective field.

c) The spectrum of ^{19}F

A simple minded argument would lead us to expect also a doublet for the resonance of ^{19}F with the same separation h given by (9), originating from the slabs A and B but with one component being an emission rather than an absorption line since the fluorines in the slabs B point opposite to the applied field. This argument is too crude, for it does not take into account the coupling between the transverse component of the fluorine magnetizations in the two slabs when the resonance is observed. A simple calculation, very similar to that performed first by Kittel[9] shows that the actual separation between the two peaks of the fluorine spectrum in the ferrosandwich state is (for $\Delta H_z = 0$)

$$h(^{19}F) = \sqrt{\frac{3}{2}}\, h(^{43}Ca) = \sqrt{\frac{3}{2}}\, \frac{8\pi}{3}\, \frac{\gamma\hbar}{a^3}\, p = \sqrt{\frac{3}{2}}\, \frac{16\pi}{3}\, M \quad . \quad (10)$$

This relation is verified approximately, the better the larger the fluorine polarization. It should be noted however that in contrast to the ^{43}Ca spectrum, the ^{19}F spectrum does not discriminate unambiguously between a ferrosandwich and an antiferromagnetic structure for which a spectrum of ^{19}F of similar shape is observed.

It can be shown[8] that in either case the separation between the positive and the negative peak of the ^{19}F spectrum is given by :

$$h(^{19}F) = 2\sqrt{\frac{3}{2}}\, A(\underset{\sim}{k}_o, \underset{\sim}{n})p \quad . \quad (11)$$

Thus for the antiferromagnetic structure observed along [001] [3]

$$h(^{19}F)\,[001] = (3/2)^{1/2} \times 9.687 \left(\frac{\gamma\hbar}{a^3}\right) \quad (12)$$

greater by a factor 1.15 than that for the ferrosandwich for a polarization p of the two sublattices equal to that of the ferrosandwich slabs. This ambiguity of the fluorine spectrum explains why, although for a long time we have had cogent reasons to believe in a ferrosandwich structure we preferred to wait for the incontrovertible evidence of ^{43}Ca [10]. (Figs.1 and 2).

The fluorine spectrum provides yet another check of the model. For $\Delta H_z \neq 0$ it is possible to measure the average polarization of ^{19}F in the ferrosandwich :

$$P = xp + (-p)(1-x) = (2x-1)p$$

and to check its dependence on ΔH_z given by formula (8) that we rewrite as :

$$P = \frac{\Delta H_z}{4\pi\gamma\hbar/a^3} \quad \text{(Fig.5)} \quad (13)$$

d) Experimental method.

The sample of CaF_2 is a sphere with a diameter of

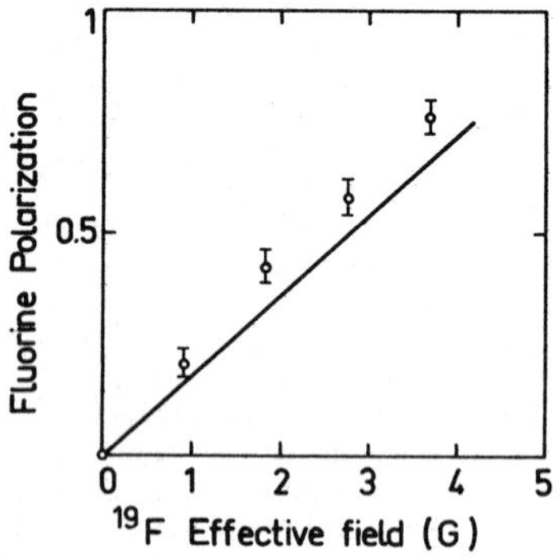

Fig.5 - Fluorine polarization against the effective field.

1.3 mm, doped with paramagnetic ions of Tm^{2+} [$N(Tm^{2+})/N(Ca^{2+}) = 3.10^{-5}$]. The dc field H_o is 27 kG, the ^{19}F frequency is 107 MHz, the microwave frequency for dynamic polarization of ^{19}F is 130 GHz, the lattice temperature after polarization 0.3 K , the fluorine polarization $p(^{19}F) \simeq 90\%$, the polarization time $\sim$3 hours.

The number of ^{43}Ca nuclei in the sample is $\sim 10^{16}$, their resonance frequency is $\sim$ 7 MHz, the polarization of ^{43}Ca $\sim$90% . The method for polarizing ^{43}Ca by thermal contact with the dipolar energy of ^{19}F in the ADRF state has been described elsewhere[10].

This method makes it possible to give opposite polarizations to ^{43}Ca nuclei in the domains A and B (Fig.6) thus materializing the structure of the domains. Some interesting results on the reversibility of domains upon remagnetization followed by a second demagnetization or upon rotation of the sample have been obtained but too preliminary to be reported here.

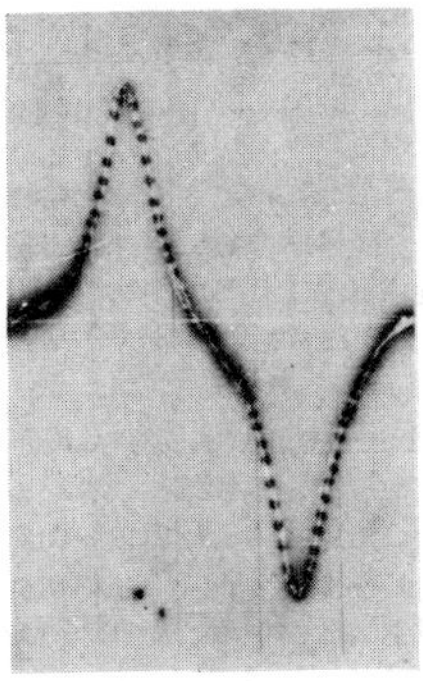

Fig.6 - The ^{43}Ca doublet with the polarization of one of the components reversed.

References

1. M. Chapellier, M. Goldman, Vu Hoang Chau and A. Abragam, J. Appl. Phys. 41, 849 (1970).
2. J.F. Jacquinot, W.Th. Wenckebach, M. Chapellier, M. Goldman and A. Abragam, C.R. Acad. Sci. Paris 278B, 96 (1974).
3. M. Goldman, M. Chapellier, Vu Hoang Chau and A. Abragam, Phys. Rev. B10, 226 (1974).
4. C.G. Shull, E.O. Wollan, W.C. Koehler, Phys. Rev. 84, 912 (1951).
5. A. Abragam, G.L. Bacchella, C.L. Long, P. Piesvaux, J. Pinot, Phys. Rev. Lett. 28, 805 (1972).
6. A. Abragam, in Trends in Physics (European Physical Society, Geneva, Switzerland, 1973), p.177.
7. M.J. Poulis, Physica 17, 392 (1951).
8. M. Goldman (unpublished).
9. Ch. Kittel, Introduction to Solid State Physics, Wiley (1971), p.597 and 603.
10. J.F. Jacquinot, W.Th. Wenckebach, M. Goldman and A. Abragam, Phys. Rev. Lett. 32, 1096 (1974).

THE MAGNETIC PROPERTIES OF LIQUID AND SOLID ^{3}HE

R. C. Richardson, Laboratory of Atomic and Solid State Physics and Materials Science Center
Cornell University, Ithaca, New York 14853

ABSTRACT

Recent experiments on the properties of ^{3}He at very low temperatures have revealed that both the liquid and the solid undergo phase transitions. The liquid becomes a superfluid at temperatures less than 2.75 mK. Three separate superfluid phases of the liquid have been identified in the presence of a magnetic field. The ^{3}He atoms apparently form Cooper pairs with an odd angular momentum and, hence, a triplet magnetic state. The three transitions correspond roughly with the onset of the separate paring of the three possible magnetic orientations. The solid modification of ^{3}He has a spin ordering transition at 1.1mK at a density of melting. On the basis of previous work an antiferromagnetic transition was anticipated in the solid at approximately twice the temperature of the observed transition.

INTRODUCTION

As a result of the refinements in the technology of cryogenics it has been possible in the past several years to study the properties of liquid and solid ^{3}He down to temperatures below 1 mK. ^{3}He is a material that is immensely rich in new physical phonomena in the temperature interval between 3 mK and 1 mK. Although the subject is one that is still under rapid development, several extensive review articles[1] have been written about it. In this paper we will give a very brief introduction to some of the things that are known about the magnetic properties of ^{3}He in this region.

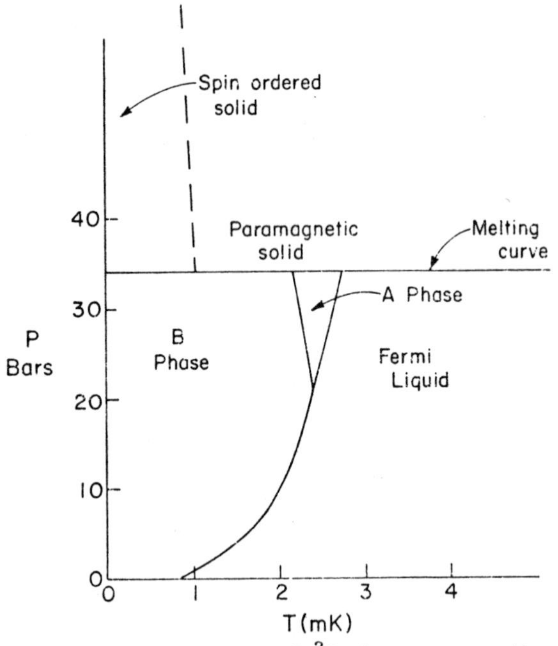

Fig. 1. The phase diagram of ^{3}He in zero magnetic field.

The boundaries of the region are illustrated in Fig. 1. A pressure of 34 bars is required to solidify ^{3}He at these temperatures. Both the liquid and solid undergo state changes affecting the magnetic properties. The state changes occur at a maximum temperature near the melting pressure. In zero magnetic field and at the melting pressure[2] the liquid enters the first of two superfluid phases, labeled the A phase, when cooled below the temperature of 2.75 mK. As the temperature is further lowered below 2.2 mK it enters a second superfluid phase, called the B phase, which has quite different magnetic properties from the A phase and the normal liquid. As the liquid pressure decreases from the melting pressure the temperature for the transition to the first superfluid phase decreases.[3] At pressures less than 21.5 bars the liquid goes directly from the normal phase into the superfluid B phase. The intersection of the three phases at 21.5 bars has been called a "polycritical point."[4] At zero applied pressure the transition occurs at 0.93 mK.[5] In the solid phase there is a transition to some state of spontaneous nuclear magnetic order which occurs at 1.1 mK at the melting pressure.[6] Observation of the nuclear spin ordered state has not been made at higher pressures of the solid but data about the magnetic properties obtained at higher temperatures indicate that the transition temperature should decrease rather rapidly as the solid density is increased.[7]

LIQUID ^{3}HE

The simplest picture of liquid ^{3}He at low temperatures emerges if we compare it with the electrons in a metal and think of it as a neutral electron gas. The Fermi temperature of liquid ^{3}He is of order 1 K so that at these low temperatures the normal liquid behaves much as the electrons in metals just above the superconducting temperature. The liquid is usually referred to as a "Fermi liquid" and the standard results from statistical mechanics for a gas of spin $\frac{1}{2}$ particles may be used to describe its properties.

As the liquid is cooled, it makes a transition to a superfluid state that is essentially a "neutral superconducting" phase. The electrons in metals form pairs to make the superconducting transition and here the ^{3}He atoms form similar pairs for the superfluid transition.

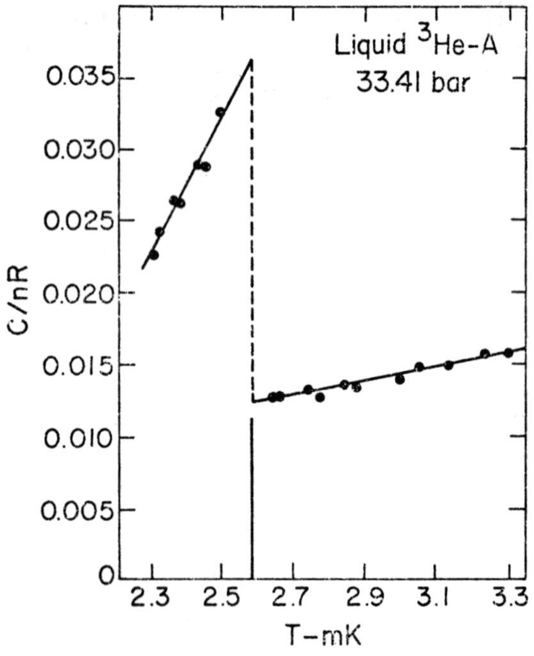

Fig. 2. The heat capacity of ^{3}He at the A transition (After Webb et al, Ref. 3)

The striking similarity of the two transitions was first shown in measurements of the heat capacity of the liquid.[3] One of the measurements by Webb et al[3] is illustrated in Fig. 2. The behavior of the heat capacity with temperature displays the classic features of a second order transition that is also seen in superconductors. There is virtually no thermal indication of the existence of a new phase until the actual point of the transition, at which there is a sudden discontinuity in the heat capacity. The "critical region," as in the case of bulk superconductors, is almost too narrow to be measured because of the large coherence length in the ordered phase. The magnitude of the ratio of the heat capacity discontinuity to the heat capacity of the normal liquid at the transition is approximately 30% larger than that of superconductors in the case of liquid measured at the melting pressure. As the liquid pressure is decreased to the polycritical pressure the transition heat capacity ratio decreases and approaches the value found in superconductors. These results suggest that some care must be exercised in applying a theory like the BCS theory of superconductivity to the ^{3}He transition but that the basic picture is quite similar.

Despite the strong analogy that exists between the superfluid transition in liquid ^{3}He and the superconducting transition in metals there is a rather fascinating difference between the two systems. In superconductivity, the electron pairs have antiparallel spin alignment (S = 0) no net magnetic moment or angular momentum. In ^{3}He the pairs form with parallel spin alignment (S = 1) and hence have both a magnetic moment and an orbital angular momentum. The orbital momentum arises because of the Pauli principle; the net wave function must be antisymmetric so that if the spins of the pair are in a symmetric state the spatial wave function must be antisymmetric and hence must have an odd angular momentum. The value of the orbital quantum number, ℓ, which appears to best describe the measured magnetic properties of superfluid ^{3}He is $\ell = 1$.

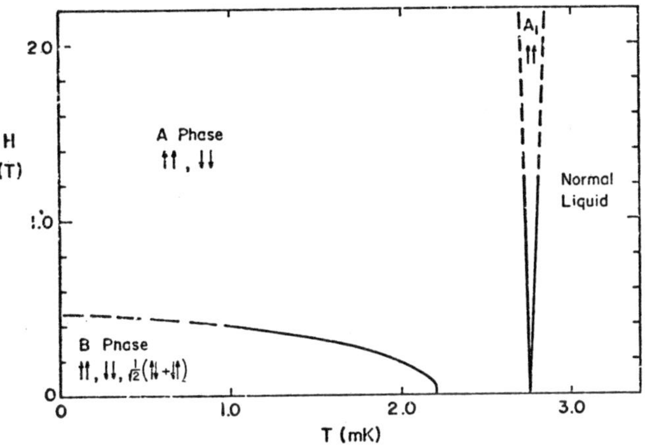

Fig. 3. The phase diagram of liquid ^{3}He at the melting pressure in applied magnetic fields.

The evidence for the parallel pairing of the spins in the liquid has emerged from various magnetic measurements, both of static and transient properties, and from the behavior of the phase diagram in magnetic fields.[8] The latter is illustrated in Fig. 3, where the magnetic field dependence of the phase boundaries is shown for liquid at the melting pressure. As a magnetic field is applied, the transition to the A phase splits into two phase transition, labeled A1 and A2, and the temperature of the B phase transition is depressed rapidly with increasing field. Thus, in finite magnetic fields less than 0.5 T (tesla), there are three distinctly different superfluid phases. The spin states of the various phases are indicated in the figure. The A1 phase, which occurs at the highest temperature, is formed by the pairing of spins with their moments aligned parallel to the applied field. At a slightly lower temperature it becomes energetically favorable to also form pairs with magnetic moments opposed to the applied field and a second transition, A2, occurs. Ambegaokar and Mermin[9] have used the linear dependence of the separation in temperature of the A1 and A2 phases as a function of field to make the identification of the phases with parallel spin pairing. The difference in the Fermi surfaces of two weakly interacting populations of spins aligned parallel and antiparallel to the field is linear in the strength of the field. Precise recent measurement[10] of the slopes of the A1 and A2 phase lines as a function of field support an argument by Brinkman and Anderson[11] that the A phase is a particular state, out of several otherwise equally likely possibilities, called the axial state,[12] that is energetically stabilized by the small ferromagnetic interaction that exists between ^{3}He atoms. In the absence of the ferromagnetic interaction and in zero magnetic field a state with all three spin components, $m_z = \pm 1, 0$, is expected to have the lowest free energy. Such a state was first discussed by Balian and Werthamer[13] and is essentially expected to have the magnetic properties exhibited by the B phase that we will discuss later. At pressures less than 21.5 bars, the polycritical pressure, the ferromagnetic interaction is not large enough to favor the formation of the A phase in zero applied field. The field dependence of the B phase[8] is similar to that of the superconductors. The rapid decrease in the transition temperature with increasing magnetic field results from the "pair breaking" effects of a field upon the pairs in the $m_z = 0$ state. Even at pressures below the polycritical pressure the A phase is found to be interposed between the normal liquid and the B phase in relatively small magnetic fields.[14]

Throughout this discussion, the term, superfluid, has been used in referring to the new low temperature phases of ^{3}He. The experimental evidence fo the superfluidity has been demonstrated beyond any reasonable doubt. ^{3}He is a superfluid in the same sense that liquid ^{4}He is a superfluid. Its apparent viscosity measured by the resistance that the liquid offers to the motion of a body dragged through it decreases somewhat at the A transition and falls rapidly at the B transition.[15] The liquid in the A and B phases has also been shown to flow through fine channels that do not permit the flow of the liquid in the normal phase.[16,17] Finally, the liquid thermal conductivity in the A and B phases[4] has a behavior that may readily be interpreted with a two fluid hydrodynamic model similar to that used to describe the heat flow in superlfuid ^{4}He. There are inconsistencies between the magnitude of the superfluid fraction ρ_s, determined in these various experiments[18] but the problems are probably more related to the character of the surfaces and to the details of the orientation of the liquid near walls and in magnetic fields, rather than with a more fundamental error in the applicability of a two fluid model to superfluid ^{3}He.

In ^{3}He the microscopic behavior can be probed in a way that is not possible in superconductors because of the screening effect of the electric changes in metals. ^{4}He, of course, has no magnetic moment that can be used to label the atoms. Thus, the studies of magnetism in ^{3}He present unique experimental opportunities for the detailed examination of such condensed phases. Quite a lot of work has already been completed in this direction.

The very first indication that the A and B transitions occured in liquid ^{3}He were found in nmr experiments.[19] The results of more recent variations of these measurements[20-25] are shown in Fig. 4 and 5. In the A phase of ^{3}He it is found that the usual "transverse" nmr absorption signal, measured by applying a small ac field perpendicular to the applied static field, H, becomes shifted when the liquid is cooled through the transition. The effect can be quite large; the shift is of the same order as the applied field for fields less than 3 mT at the lowest temperatures for which the A phase is stable.

In a fixed frequency spectrometer, with frequency ω_o, the field required to observe the resonance signal, H(T), decreases below that usually required by the gyromagnetic ratio, $H = \omega_o/\gamma$. The amount of the shift depends both upon the temperature and the spectrometer frequency, but a universal shift frequency, $\omega(T) = [\omega_o^2 - \gamma^2 H(T)^2]^{\frac{1}{2}}$ can be found to characterize the size of the shift at any temperature regardless of the spectrometer frequency (or initial field). This shift frequency is shown as a function of reduced temperature in Fig. 4. In accounting for this effect, Leggett[26]

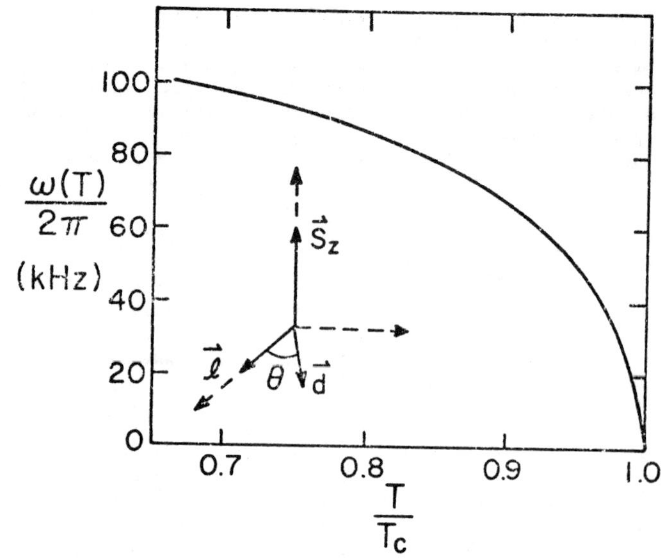

Fig. 4. The temperature dependence of the shift frequency, or longitudinal resonance, in the A phase at the melting pressure. The vector diagram illustrates the relationship between the vector indices of the A phase. In equilibrium, $\theta = 0$, and S_z is parallel to the applied field.

found that the scheme of pairing with $S_z \pm 1$ would produce a strong correlation in the magnetic dipole interaction between the spins so that the spins resonate at a higher frequency than under the usual conditions of nuclear magnetic resonance, $\omega = \gamma H$. Ordinarily, the magnetic dipole interaction only leads to the well known broadening of nmr lines without a significant shift in its position. Leggett also predicted that a "longitudinal" resonance could be observed at the shift frequency, $\omega(T)$, if the ac field were applied along the direction of the static field H_o. The longitudinal resonance has been observed at precisely the shift frequency[20,21] of the transverse resonance, both in the manner suggested by Leggett and by a novel ringing technique.[24] In the latter, oscillations in the magnetization parallel to the field were observed following the application of a small step change in the size of the static field. The magnetization eventually comes to equilibrium with the new field but a transient ringing is observed at the shift frequency.

These experiments provide an explicit demonstration of the anisotropic nature of the superfluid state of ^{3}He. The wave function of superfluid ^{3}He has tensor properties and there are two distinct vectors that characterize the description of the axial state for pairs in the A phase. There is an angular momentum vector $\underline{\ell}$ and a vector $\underline{d}$, along which the spin of the pair is zero ($\underline{S} \cdot \underline{d} = 0$). The BCS energy gap in this state is not spherically symmetric but instead goes to zero at the poles of the Fermi sphere defined by the $\underline{\ell}$ axis. Because of the weak dipole interaction it is energetically favorable for these two vectors to be parallel when in equilibrium. In the presence of an applied magnetic field the Zeeman interaction causes the magnetic dipoles to orient along the field and thus $\underline{d}$ and $\underline{\ell}$ become oriented perpendicular to the field in bulk liquid where interactions with the walls are not important. In the presence of a wall the $\underline{\ell}$ vector may be expected to align perpendicular to the wall and a rather long range order, similar to the nematic phases of liquid crystals,[27] may arise from the weak dipole interaction between the pairs.

The orientation between the $\underline{\ell}$ vector and the $\underline{d}$ vector, characterized by an angle θ, is coupled to S_z, the size of the difference between the number of spins in the population aligned with the field and the number opposed to the field, through the commutator relation $[S_z, \theta] = -2i\hbar$. Leggett[28] points out that this is analogous to the relation between the superconducting number density and the phase of the order parameter which is responsible for the Josephson effect in superconductors. In equilibrium, the angle θ is zero and S_z essentially takes on the value of the normal Fermi liquid. (It is in, in fact, slightly larger than the Fermi liquid[29] because of a shift in the Fermi energy that occurs as a result of the onset of pairing at the Al transition.[30]) If the value of S_z is suddenly perturbed from equilibrium, the angle between $\underline{\ell}$ and $\underline{d}$ becomes greater than zero and the weak dipole interaction acts to restore the orientation back to $\theta = 0$. The system responds in an underdamped manner, since the actual thermal equilibrium time, T_1, is quite long. Thus, there are oscillations in the angle θ with a characteristic frequency that is governed by the strength of the effective many body dipole interaction $g_d(T)$ that is temperature dependent. Because the magnetization is coupled to the angle θ, the rate of magnetization change $\dot{S}_z$ is proportional to both the size of the displacement between $\underline{\ell}$ and $\underline{d}$ and to the strength of the dipole interaction $\dot{S}_z \sim g_d(T) \sin\theta$. The oscillations in θ therefore produce oscillations in S_z. The differential equation for the motion in θ is $\ddot{\theta} = \omega^2(T) \sin\theta$, where $\omega(T)$ is the shift frequency in Fig. 4. The differential equation is the same as the equation for a simple pendulum and even the predicted[31] non-linearities associated with large initial displacements in θ have been verified in detail in the parallel ringing experiments.[32]

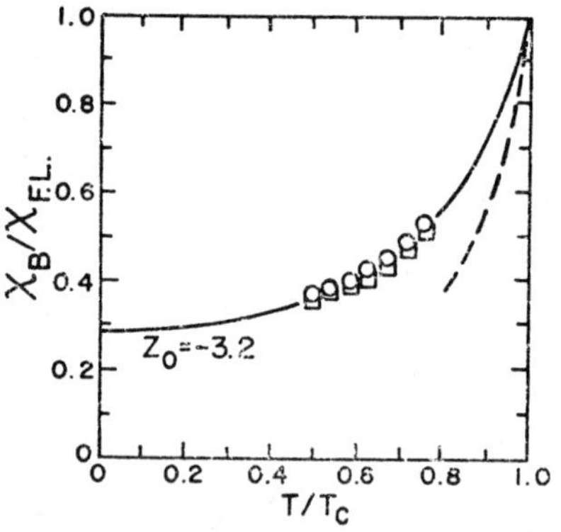

Fig. 5. The temperature dependence of the reduced magnteic susceptibility in the B phase of liquid ^{3}He. The solid line is the value calculated using the theory for the isotropic $\ell = 1$ state with the value $Z_o = -3.2$ for the Fermi liquid coefficient in Ref. 33. The circles and squares are the values derived from a theoretical extrapolation in the measurements of Ref. 25 and the dashed line is obtained from the magnetization measurements in Ref. 24.

In the B phase the magnetization decreases with temperature as a result of the pairing with $m_z = 0$. A theoretical curve[33] for the magnetization in the B phase for the "isotropic" state with $\ell = 1$ is shown in Fig. 5. There is currently some contention about whether experiments actually agree with the curve in Fig. 5. We will discuss these experiments later.

The pair wave function in the B phase (using the isotropic, or Balian and Werthamer, state) is also characterized by a vector, called $\underline{n}$, which is a symmetry axis for the spin and orbital coordinates. There is no real $\underline{\ell}$ vector that may be used to characterize the collective properties of this phase; and the "superconducting gap" for this phase has spherical symmetry with no nodes. The symmetry of the gap is the origin of the name "isotropic" for the phase. However, the $\underline{n}$ vector has a direction that can be influenced by interactions with walls and with magnetic fields so that the state is not truly isotropic. There is a longitudinal resonance in the B phase and there is a very slight shift in the transverse resonance.[25]

There have been several nmr measurements of the B phase susceptibility[23,25] and measurements of the bulk magnetization with a dc magnetometer.[24] The solid curve in Fig. 5 is the prediction for the ratio of the magnetization to the Pauli magnetization of the normal Fermi liquid. The prediction is based upon the assumption that the B phase is the state with $\ell = 1$ pairing that we have discussed. The dc measurements of the bulk magnetization measured in magnetic fields between 4 mT and 40 mT fall systematically below the solid curve. The measured decrease in the magnetization ratio is independent of the field strength and is 50% larger than predicted for the $\ell = 1$ isotropic state.

The susceptibility measured in nmr experiments is also less than the value indicated in the curve of Fig. 5; however, the ratio in nmr experiments was found to depend upon the strength of the applied magnetic field[25] and in fields of 0.15 T exceeds the value reported in the bulk magnetization measurements. The points shown on the curve are derived from the extrapolation of the measured ratios to very large magnetic fields. The justification for this analysis is given in a theory by Brinkman et al.[34] The model considers the effects of the wall interaction upon the orientation of the symmetry vector, n. Near surfaces it is energetically favorable for n to be perpendicular to the wall. In the bulk liquid n aligns parallel to the applied field H. Thus there is a competition between the surface effect and the bulk effect if the field is parallel to the surface. Liquid in a region where n is misoriented with respect to the applied field es expected to have a shift in the position of its transverse resonance signal and does not appear under the main resonance peak from the bulk liquid that is used to evaluate the susceptibility ratios. As the field is increased the shell of liquid influenced by the wall becomes thinner and a smaller fraction of the spins have their resonance shifted away from the peak.

Independent support of the model comes from the measured values of the longitudinal resonance frequency in the B phase.[25] The susceptibility ratio of the solid curve is required to account for the size of the longitudinal resonance frequency.[35] The difference between the nmr results and the dc magnetization measurements may be a result of an unexpected pressure dependence of the liquid properties. The puzzle must be resolved by future experiments.

SOLID ^{3}HE

Very much less is known about the details of the magnetically ordered state in solid ^{3}He than is known about the superfluid phases of ^{3}He. Solid ^{3}He is technically more difficult to study because it has a much poorer thermal conductivity than the liquid and because the maximum temperature of the transition occurs at lower temperatures than that of the liquid.

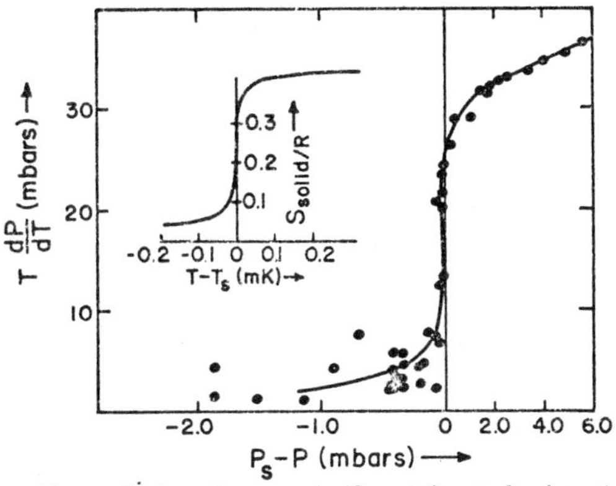

Fig. 6. The solid entropy at the spin ordering transition. The data points are from the direct measurement of the quantity $T\,(dP/dT)_m$ versus the melting pressure. The entropy, shown in the inset of the figure, is obtained from thermodynamic relations between the slope of the melting curve $(dP/dT)_m$ and the melting pressure. The quanitities P_s and T_s are the coordinates of the solid spin ordering transition on the melting curve. T_s = 1.1 mK. (After Halperin et al, Ref. 6.)

The temperature variation of the entropy of solid ^{3}He at the melting pressure near the transition[6] is shown in Fig. 6. The data shown in the figure were obtained from measurements of the latent heat of solid formation. The method takes advantage of the relatively rapid thermal equilibrium times in the liquid to obtain information about solid ^{3}He. The behavior shown in Fig. 6 is somewhat surprising. Most of the entropy decrease takes place in a very narrow temperature interval at the transition. It may well be that the transition is first order, with a latent heat. The experiments are not yet precise enough to determine whether or not there is a real discontinuity in the solid entropy at 1.1 mK.

A rather large body of information about solid ^{3}He has been gathered in magnetic and thermal measurements at higher temperatures.[7] In magnetic fields less than 1 T the high temperature data can be fitted to a Heisenberg model in which the ^{3}He spin interactions are described by a single exchange parameter J, for the nearest neighbor interactions. The exchange arises because of the relatively large zero point motion of the atoms in the ^{3}He solid. A rather precise high temperature expansion of the partition function for a bcc lattice has been calculated for this interaction.[36] Thermodynamic quantities derived from the expansion, such as the heat capacity, the isochoric expansion $dP/dT)_V$, and the magnetic susceptibility yield a consistent negative value of the exchange parameter when compared with the experiments. The model based upon these data predicts a lambda type phase transition to an antiferromagnetic state at a temperature near 2 mK. The actual phase transition is clearly different from that which was anticipated. In recent measurements of the heat capacity of the solid[37,38] a small and broad maximum is observed near 2 mK before the heat capacity diverges as it approaches the ordering transition at 1.1 mK. This behavior probably indicates that the actual magnetic interaction near the transition is much less isotropic than that of the simple Heisenberg model.

Even before the phase transition was observed it was recognized that the theoretical model for solid ^{3}He magnetism was probably too simple. Measurements of the magnetic susceptibility near the transition[2] yielded values far too large to fit the model, and measurements of the change in the solid pressure in large magnetic fields[39] yielded values too small to be consistent with the single exchange parameter high temperature expansion for a Heisenberg antiferromagnet. Attempts have been made to account for the magnetic behavior of the ^{3}He solid by postulating the existence of a relatively large antiferromagnetic exchange parameter in the second nearest neighbor.[40] The effect of such an interaction would be to lower the temperature of the phase transition but it does not force the field dependence of the solid pressure to be small enough to be consistent with the experimental values.

In view of the difficulties that the theory has in predicting any of the details of the transition, it remains an open question whether or not the actual state of the spin ordered phase is antiferromagnetic. Studies of this transition are really only in their infancy.

CONCLUSIONS

This work has contained a sketch of some of the results that have been obtained so far in studies of ^{3}He phases that were not even known to exist three years ago. As a result of rapidly developing cryogenic techniques, such as the use of the demagnetization of polarized nuclei to provide a cooling reservoir,[5,37] it will be possible to make much more detailed studies of the transitions in future work. Much of the emphasis in the liquid ^{3}He work will probably be placed upon studies of various textures that have been predicted to exist.[27,34] In solid ^{3}He the first order of business is to discover what the ordered spin state really looks like.

REFERENCES

1. O. V. Lounasmaa, Contemp. Physics 15, 353 (1974); D. M. Lee, Proceeding of European Low Temperature Conference (Haifa, 1974) to be published; J. C. Wheatley, Rev. Mod. Phys., to be published; A. J. Leggett, Rev. Mod. Phys., to be published.
2. D. D. Osheroff, R. C. Richardson, and D. M. Lee, Phys. Rev. Lett. 28, 885 (1972).
3. R. A. Webb, T. J. Greytak, R. T. Johnson and J. C. Wheatley, Phys. Rev. Lett. 30, 210 (1973).
4. T. J. Greytak, R. T. Johnson, D. N. Paulson, and J. C. Wheatley, Phys. Rev. Lett. 31, 452 (1973).
5. A. I. Ahonen, M. T. Haikala, M. Krusius and O. V. Lounasmaa, Phys. Rev. Lett. 33, 628 (1974).
6. W. P. Halperin, C. N. Archie, F. B. Rasmussen, R. A. Buhrman, and R. C. Richardson, Phys. Rev. Lett. 32, 927 (1974).
7. See, for instance, the review articles: R. A. Guyer, R. C. Richardson, L. I. Zone, Rev. Mod. Phys. 43, 532 (1971); and S. B. Trickey, W. P. Kirk and E. D. Adams, Rev. Mod. Phys. 44, 668 (1972).
8. D. D. Osheroff, Ph.D. thesis, Cornell University 1972 (unpublished); and W. J. Gully, D. D. Osheroff, D. T. Lawson, R. C. Richardson and D. M. Lee, Phys. Rev. A 8, 1633 (1973).
9. V. Ambegaokar and N. D. Mermin, Phys. Rev. Lett 30, 81 (1973).
10. D. D. Osheroff and P. W. Anderson, Phys. Rev. Lett. 33, 686 (1974).
11. W. F. Brinkman and P. W. Anderson, Phys. Rev. A 8, 2732 (1973); and P. W. Anderson and W. F. Brinkman, Phys. Rev. Lett. 30, 1108 (1973).
12. C. M. Varma and N. R. Werthamer, Phys. Rev. A 9, 1465 (1974).
13. R. Balian and N. R. Werthamer, Phys. Rev. 131, 1553 (1963).
14. D. N. Paulson, H. Kojima, and J. C. Wheatley, Phys. Rev. Lett. 32, 1098 (1974).
15. T. A. Alvesalo, Yu. P. Anufriyev, H. K. Collan, O. V. Lounasmaa, and P. Wennerström, Phys. Rev. Lett. 30, 962 (1973).
16. H. Kojima, D. N. Paulson, and J. C. Wheatley, Phys. Rev. Lett. 32, 141 (1974).
17. A. W. Yanof and J. D. Reppy, Phys. Rev. Lett. 33, 631 (1974).
18. T. A. Alvesalo, H. K. Collan, M. T. Loponen, and M. C. Veuro, Phys. Rev. Lett. 32, 981 (1974).
19. D. D. Osheroff, W. J. Gully, R. C. Richardson and D. M. Lee, Phys. Rev. Lett. 29, 920 (1972).
20. D. D. Osheroff and W. F. Brinkman, Phys. Rev. Lett. 32, 584 (1974).
21. H. M. Bozler, M. E. R. Bernier, W. J. Gully, R. C. Richardson and D. M. Lee, Phys. Rev. Lett. 32, 875 (1974).
22. R. A. Webb, R. L. Kleinberg, and J. C. Wheatley, Phys. Rev. Lett. 33, 145 (1974).
23. A. Il Ahonen, M. T. Haikala, M. Krusius, and O. V. Lounasmaa (to be published).
24. D. N. Paulson, R. T. Johnson and J. C. Wheatley, Phys. Rev. Lett. 31, 746 (1973).
25. D. D. Osheroff, Phys. Rev. Lett. 33, 1010 (1974).
26. A. J. Leggett, Phys. Rev. Lett. 31, 352 (1973).
27. P. G. deGennes, Phys. Lett. 44 A, 271 (1973); and V. Ambegaokar, P. G. deGennes, and D. Rainer, Phys. Rev. A 9, 2676 (1974).
28. A. J. Leggett, Proc. Phys. Soc., London 6, 3187 (1973);
29. D. N. Paulson, H. Kojima, and J. C. Wheatley, Phys. Lett. 47 A, 457 (1974).
30. S. Takagi, to be published.
31. K. Maki and T. Tsuneto, to be published.
32. D. N. Paulson, H. Kojima and J. C. Wheatley, Phys. Lett. A
33. A. J. Leggett, Phys. Rev. 140, A1869 (1965).
34. W. F. Brinkman, H. Smith, D. D. Osheroff and E. I. Blount, Phys. Rev. Lett. 33, 624 (1974).
35. S. Englesberg, W. F. Brinkman, and P. W. Anderson, Phys. Rev. A 9, 2592 (1974).
36. G. A. Baker, H. E. Gilbert, J. Eve and G. S. Rushbrooke, Phys. Rev. 164, 800 (1967).
37. J. M. Dundon and J. M. Goodkind, Phys. Rev. Lett. 32, 1342 (1974).
38. W. P. Halperin, C. A. Archie, F. B. Rasmussen and R. C. Richardson, to be published.
39. W. P. Kirk and E. D. Adams, Phys. Rev. Lett. 27, 392 (1971).
40. L. I. Zane and J. R. Sites, to be published.

Research supported by the National Science Foundation and by the Cornell Materials Science Center.

PLANETARY MAGNETISM

D. J. Stevenson
Physics Department, Cornell University
Ithaca, New York 14850

ABSTRACT

Magnetic fields are ubiquitous in the Universe because of the absence (or rarity) of free magnetic monopoles. However, the large scale concentration of magnetic flux in an essentially non-ferromagnetic body such as the Earth, requires a non-cosmogonical explanation. A regenerative process must exist because the magnetic diffusion time in the Earth is much less than the age of the Earth. The most popular explanation is the dynamo model in which convectional or precessional fluid motions in the core induce and sustain a magnetic field. Other mechanisms are possible in which, for example, thermoelectric effects are important. The advent of planetary probes has led to a dramatic increase in information on the magnetic fields of the five innermost planets. In particular, the recent fly-bys of Jupiter (by Pioneer 10) and Mercury (by Mariner 10) have yielded unexpected results. These observations are discussed and shown to be reconcilable with present theories of planetary magnetism and solar system chemistry. Some predictions are made for Saturn, Uranus and Neptune.

INTRODUCTION

One hundred years from now, the present decade will be looked upon as the beginning of an exciting new era in which Man first began to explore the environment of other planets in the solar system. Apart from the pure scientific knowledge that this exploration yields, there is hope that the understanding of our own environment will also be enhanced. One aspect of our environment that is as yet incompletely understood, is the geomagnetic field. In the past, theories of the geomagnetic field could be applied to only one body. Now, it is possible to test these theories on other bodies in the solar system.

The magnetic field of a planet is one of the few planetary properties that is externally measurable yet gives us information on the interior of the planet. If it is possible to deduce the interior composition of planets, then our knowledge of the chemistry of the solar system will be thereby enhanced. Since theories of planetary magnetism are strongly dependent on the assumed internal constitution, it should be possible to correlate models of planetary interiors with observed internal magnetic fields. Such a correlation is attempted here. We first discuss the existence of primordial fields and the need for regeneration. In subsequent sections, the various methods of regeneration are discussed, and their relevance for each planet considered.

PRIMORDIAL FIELDS

A turbulent era soon after the Big Bang could generate a magnetic field that reached energy equipartition with the kinetic energy of small scale motions.[1] This field persists because of the abundance of free electric charges and the apparent absence of free magnetic monopoles. It could account for almost all the magnetic field energy in the Universe. Subsequent gravitational contraction of a hot gas that can trap the field lines, leads to large localized concentrations of magnetic field. For example, the contraction from an interstellar medium in which the field is 10^{-5} gauss (the present field of our galaxy), would give fields as large as 10^{6} gauss in a body like Jupiter. Indeed, Hoyle[2] invoked a large magnetic field in the primordial solar nebula, to explain the distribution of angular momentum in the solar system.

The reason that such initial fields might not persist is to be found in Ohm's law and Maxwell's equations which together give

$$\nabla^2\vec{H} = \frac{4\pi\sigma}{c^2}\left[\frac{\partial\vec{H}}{\partial t} - \nabla \times (\vec{V} \times \vec{H})\right] \qquad (1)$$

where $\vec{H}$ is the magnetic field and $\vec{V}(\vec{r})$ the velocity of a medium of electrical conductivity σ. (Displacement currents are neglected). Consider a region with linear dimensions of order L, surrounded by an insulator (e.g. vacuum). If $\vec{V} = 0$ then equation (1) gives

$$\frac{H}{L^2} \sim \frac{4\pi\sigma}{c^2} \cdot \frac{H}{\tau} \qquad (2)$$

where the field is assumed to change on a time scale of order τ and a distance scale of order L. Thus,

$$\tau \sim \frac{4\pi\sigma L^2}{c^2}\,. \qquad (3)$$

Unless the term involving $\vec{V}$ in equation (1) is significant, τ is a measure of the decay time of an initial field. Fluid flow is important if

$$G_m \equiv \frac{V\tau}{L} \gtrsim 1 \qquad (4)$$

where V is a typical magnitude of $\vec{V}(\vec{r})$ and G_m is known as the magnetic Reynold's number.

The value of τ is typically only a fraction of a second for a laboratory sized object but it can by very large for an astronomical body. For the Sun,[3] τ is about 2×10^9 years. A similar value has been estimated for the metallic core of Jupiter.[4] It is possible that the present magnetic fields of the Sun and Jupiter are remnants of once large primordial fields. In contrast, the Earth has a decay time τ of about 10^3 or 10^4 years.[5] Similar decay times apply to the other terrestrial planets. The Earth's magnetic field is consequently not primordial, neither can it be explained by ferromagnetism.[6] A regenerative process must exist.

FIELD GENERATION

The most popular theory of field generation for planets is the dynamo mechanism. Before discussing that, we shall consider some of the other theories or explanations that have been proposed.

As we have already mentioned, ferromagnetism does not explain the Earth's dipole field. It could, however, be responsible for the much smaller fields of Mars and Mercury. Such planets might have had primordial dynamos when they were hotter, and then cooled off, leaving behind a permanent magnetic trace of their early history.[7]

In contrast, well known solid-state effects such as thermoelectricity and thermomagnetism offer ways of transforming thermal energy into magnetic field energy. As we have discussed elsewhere,[8] thermomagnetic effects are unimportant unless the electron mean free path greatly exceeds the inter-atomic spacing in the interior of a planet. None of the planets is cold enough for this criterion to be adequately satisfied at present. However there is no similar constraint on theories involving thermoelectric effects.

In 1939, Elsasser[9] proposed that the Earth's magnetic field could be maintained by electric currents driven by a thermoelectric e.m.f. In Elsasser's version, the thermocouple is obtained by assuming that the warm upward and cold downward streams of convecting fluid in the core have slightly different compositions. In a later version by Runcorn,[10] the e.m.f. is at the core-mantle boundary. In both cases, Inglis[11] found that to produce an electric current adequate to explain the Earth's magnetic moment, the convection currents required would transport more heat than the observed geothermal heat flux. For this reason, the theory is not favored.

Recently,[8] a more successful application of Elsasser's idea has been made to Jupiter. Jupiter satisfies two of the criteria that are necessary for large thermoelectric currents. One is a large internal heat flux, recently confirmed by Pioneer 10.[12] The other is rapid rotation. This is required to ensure a preferential alignment of the fluid flows and current loops, which in turn ensures a net magnetic dipole moment.

The thermoelectric current depends on the product of the electrical conductivity and the relative thermopower. The thermocouple in Jupiter could be between helium-rich and helium-poor metallic hydrogen fluid regions (in which case, the conductivity is large and the thermopower small) or between metallic and molecular hydrogen (small conductivity but large thermopower). We favor the former because of the very low electrical conductivity that molecular hydrogen is expected to have. (However, recent calculations by this author indicate that molecular hydrogen may even be a semi-metal at pressures just below the first order phase transition to metallic hydrogen. This might favor a thermoelectric mechanism.) With our present incomplete knowledge of the interior of Jupiter, a detailed model would be premature.

Amongst other ideas that have been proposed to explain planetary magnetism are: gyromagnetic effects, rotating electric charge, induction by magnetic storms, Hall currents, the compression effect (the differential motion of electrons and ions in a gravitational field), and new physical laws. These have all been discussed elsewhere.[8] The Hall effect is the only mechanism listed above that seems promising.

The mechanisms we have discussed so far share one feature: they all involve small perturbations on the dominant dynamic processes within a planet. Such mechanisms were worth considering only because the magnetic field energy in the Earth (for example) is much less than the thermal energy, or even the tidal energy that has been dissipated in a billion years. It seems that we should consider a more direct coupling between the magnetic field and the dynamic processes within a planet. This is the aim of the dynamo mechanism.

The existence of a homogeneous, planetary dynamo can be mathematically determined as follows: is there a velocity field $\vec{v}(\vec{r})$ that is a solution to the magnetohydrodynamic Navier-Stokes equation and to equation (1), such that the magnetic field does not decay with time? These are constraints on both the geometry of the velocity field and on the magnitude of a typical velocity.

We have already seen (equation (4)) that fluid flow is important if the magnetic Reynold's number exceeds unity. If the fluid flow is rapid enough, then the field does not have time to diffuse out of the fluid and is "frozen in" to the conductor. This is essential to all dynamo theories. This lower limit on the velocity is important since it limits dynamo action to a conducting fluid. Note, however, that the conductivity and fluid velocities required depend on the <u>size</u> of the planet. A very large planet does not require a large conductivity.

Cowling[13] considered the geometric constraints on the velocity field and showed that a dynamo can not exist if the fluid flow is axisymmetric. For this reason it is difficult to construct a dynamo in a non-rotating planet. Furthermore, dynamo models inevitably involve large toroidal, as well as poloidal, fields.

In addition to this qualitative constraint on the geometry of the velocity field, there is a quantitative constraint that expresses the requirement that the dissipation of electric currents be matched by production. In a laboratory dynamo, this requirement is met by attaining a certain angular velocity.[14] The corresponding criterion for a planetary dynamo is similar but not as simple. It should be emphasized that this requirement is a <u>kinematic</u> one, and not just a question of whether sufficient energy is available to run the dynamo.

The existence of many suitable velocity fields for dynamo action is now well established,[15] although no general existence proofs have been derived. We must now consider what mechanisms are available to sustain a suitable velocity field. Two are considered here: cyclonic convection and precession.

Thermal convection in a rotating planet leads to differential rotation of the fluid (cyclonic convection), and the resulting non-axisymmetric flow may sustain a dynamo. Parker[16] has analyzed this in detail and finds that dynamo modes do exist provided the quantity $\omega\tau$ exceeds some critical value. (Here, ω is the angular velocity of the planet and τ is defined by equation (3)). The critical value of $\omega\tau$ depends on the details of the flow pattern and the heat flux transported. If, as Levy has suggested,[17] the Earth only marginally satisfies this requirement, the critical value is $\omega\tau \sim 10^6$ or 10^7.

So far, we have discussed only the <u>kinematics</u> of the dynamo. This discussion has been independent of the <u>magnitude</u> of the generated field. The magnitude is determined by the available energy source. For thermal convection, the energy source is the heat flux emanating from the deep interior of the planet. This heat flux drives a heat engine that maintains the fluid flow. The inherent inefficiency of such an engine ensures that the Ohmic dissipation of the dynamo is much less than the heat flux.

We now consider the precessional model proposed by Malkus.[18] In the frame of reference in which a rotating, precessing planet is at rest, there is, in addition to the usual Coriolis force, a Poincare force per unit mass

$$\vec{F}_p = (\vec{\omega} \times \vec{\Omega}) \times \vec{r} \qquad (5)$$

where $\vec{\omega}$ is the angular velocity vector, $\vec{\Omega}$ the precession vector and $\vec{r}$ a radius vector. The precession rate Ω depends on the dynamic ellipticity of the body and consequently on the density of the body. If the planet has a density discontinuity in the interior, then the material on either side of the discontinuity tends to precess at different rates. The resulting fluid flow may be adequate to maintain a dynamo.

For fully magnetoturbulent flow in the core, Malkus found that the condition for field regeneration is

$$K \equiv (\Delta\Omega\tau \sin i)/2 \geq 1 \qquad (6)$$

where Δ is the fractional density discontinuity ($\Delta \sim \frac{1}{4}$ for the Earth) and i is the angle between $\vec{\Omega}$ and ω. This criterion is analogous to the condition on $\omega\tau$ that we discussed for the convectional dynamo. If this criterion is satisfied then an estimate for the poloidal field is

$$H_p \simeq (2K\omega\rho c^2/\sigma)^{\frac{1}{2}} \text{ gauss} \qquad (7)$$

where ρ is the density in g/cm^3, and σ the electrical conductivity in esu(sec^{-1}). Applications of these formulae are made in the next section.

THE PLANETS

In the following discussion, some of the numerical details are omitted since they have appeared elsewhere.[8]

<u>Mercury</u>: The high mean density of Mercury (5.4 g/cm^3) indicates a dense iron-rich core that occupies about 50% of the total volume. Even in the unlikely event that part of this core is still fluid, the smallness of the planet and its slow rotation do not favor a convectional dynamo. A precessional dynamo is even less likely because of the almost complete absence of precession.[19] Nevertheless, the recent fly-by of Mercury by Mariner 10 indicates a substantial magnetosphere.[20] The magnetic field could be intrinsic to the planet (in which case, it corresponds to a surface field of typically 0.02 gauss); or it could be induced by temporal or spatial variability in the local interplanetary field. If the field is intrinsic, then the most likely explanation is a ferromagnetic remnant of a primordial dynamo that existed when the planet was rotating faster and had a fluid core.

(A rotation rate comparable to the present Earth would be adequate.) Such a primordial state is not in conflict with our present understanding of the formation of the solar system. Indeed, the surface morphology indicates ancient lava flooding and volcanism.

Venus: The important difference between the Earth and its sister planet is that the angular velocity of Venus is over two orders of magnitude smaller. Since Venus does not undergo significant precession, the dynamo mechanisms that we discussed do not work. It is not surprising that Mariner 10 and other planetary probes have detected no magnetosphere. (An upper bound to the surface field of Venus is 10^{-4} gauss.) Since Venus is the nearest thing we have to a non-rotating planet in our solar system, it is appropriate to mention that there is but one theory of planetary magnetism regeneration that requires no rotation: Hall currents.

Earth: The geomagnetic field has been extensively studied and is most easily explained by the dynamo mechanism. Such features as the secular variation and prehistoric reversals have been qualitatively explained by the dynamo model, although the reversals are not fully understood. More controversal is the energy source for the dynamo. The Ohmic dissipation of the Earth's dynamo is about 4×10^{17} erg/sec.[8] If we suppose that cyclonic convection drives the dynamo then the geothermal heat flux of 3×10^{20} erg/sec would seem to be adequate. However, most of this heat is generated by radioactivity in the Earth's crust and is not available in the core. Furthermore, a convective engine is at most 4% efficient in transforming heat flux, through fluid motion, into Ohmic dissipation. It follows that unless more than about 5% of the Earth's heat flux originates from the core, a convective engine cannot run the Earth's dynamo. It has been suggested that the required heat flux could be supplied by radioactive ^{40}K in the core[21] or the release of latent heat as the inner solid core grows.[22]

The precession model has no similar problem. Malkus[18] finds that the critical parameter K exceeds unity and that the dipole field (from equation (7)) is about 7 gauss, roughly the value expected in the Earth's core. The precessional model seems to be very satisfactory for the Earth.

Mars: The rotation and precession of Mars are comparable to the Earth's. However, Mars is a much smaller planet and only a small fraction of it (if any) is fluid.[23] If the conditions on Earth are kinematically marginal for a convectional dynamo then they are certainly unfavorable for Mars. The precessional model is also in difficulty unless the conductivity in the deep interior of Mars is much greater than for the Earth. Our present understanding of solar system chemistry would not support that. It is not surprising that the observed magnetic field of Mars is only 10^{-4} gauss. This could be a ferromagnetic remnant of a primordial field or primordial dynamo.

Jupiter: The largest planet in our solar system also has the greatest range of possible explanations for its observed intrinsic field.[24] As we have already mentioned, the field could be primordial or a consequence of thermoelectric currents. It could also be easily driven by thermal convection because of the large heat flux[12] and rapid rotation. Indeed, there is enough energy within Jupiter to drive many smaller dynamos simultaneously. In this sense, Jupiter should be compared with the Sun rather than the Earth. However, there are not likely to be "Jovian spots" in analogy to Sun spots, since the outer layers of Jupiter are electrically insulating. The Jovian dynamo is expected to be confined to the highly conducting metallic hydrogen core.

The massive satellites of Jupiter lie almost exactly in the equatorial plane, and produce no precessional torques. However, the Sun causes precession that is sufficient to maintain a magnetic field.[8] Thus, Jupiter may have both precessional and convectional dynamos.

Saturn: Recent models for the interior of Saturn[25] indicate that Saturn is very similar to Jupiter. Not only is the deep interior expected to be metallic hydrogen, but the temperature is expected to exceed the melting point. Both precessional and convectional dynamos can be maintained without difficulty. Radio observations of Saturn indicate no synchrotron radiation, in contrast to Jupiter which is a strong radio source. However, the radiation belts may be at a very large radial distance from the planet, so as not to intersect with the rings. Luthey[26] has estimated that Saturn could have an intrinsic field of even twenty gauss and yet produce synchrotron radiation that is unobservable at Earth. When Pioneer 11 passes Saturn in 1979, it is likely to find a substantial intrinsic field.

Uranus: Unlike Jupiter and Saturn, Uranus does not seem to be endowed with a large internal heat flux.[27] The satellites of Uranus lie in the equatorial plane and the precessional torque of the Sun is too small to drive a precessional dynamo. It seems unlikely that Uranus has a large magnetic field.

Neptune: In contrast to Uranus, Neptune has a massive satellite Triton for which the angle between $\vec{\omega}$ and $\vec{\Omega}$ is about 160°. If conducting fluid is present, Neptune is likely to have a precessional dynamo.

Pluto: The outermost planet is almost certainly too small and too cold to have a dynamo.

CONCLUSION

We have found no conflict between the predictions of the dynamo theory and the observations of the planets. It is hoped that more detailed information on internal planetary fields in the future will lead to independent tests of interior models of planets.

I would like to thank Drs. N. W. Ashcroft, C. Sagan, E. E. Salpeter and S. Soter for their suggestions and comments. Support by NSF Grant GP-36425X and NASA Grant NGR-33-010-188 is gratefully acknowledged.

REFERENCES

1. E. R. Harrison, Monthly Notices Royal Astron. Soc. 165, 185 (1973).
2. F. Hoyle, Quart. J. Royal Astron. Soc. 1, 28 (1960).
3. S. M. Chitre, D. Ezer and R. Stothers, Astrophysics Letters 14, 37 (1973).
4. D. J. Stevenson, N. W. Ashcroft, Phys. Rev. A9, 782 (1974).
5. F. D. Stacey, "Physics of the Earth", J. Wiley and Sons, New York (1969), p. 153.
6. T. Rikitake, "Electromagnetism and the Earth's Interior", Elsevier, New York (1966), p. 14.
7. R. Smoluchowski, The Moon 7, 127 (1973).
8. D. J. Stevenson, Icarus, 22, 403 (1974).
9. W. M. Elsasser, Phys. Rev. 55, 489 (1939).
10. S. K. Runcorn, Trans. American Geophys. Union 35, 49 (1954).
11. D. R. Inglis, Rev. Mod. Phys. 27, 212 (1955).
12. S. C. Chase, R. D. Ruiz, G. Munch, G. Neugebauer, M. Schroeder and L. M. Trafton, Science 183, 315 (1974).
13. T. G. Cowling, Monthly Notices Royal Astron. Soc. 94, 39 (1934).
14. F. J. Lowes, I. Wilkinson, Nature 219, 717 (1968).
15. N. D. Weiss, Quart. J. Royal Astron. Soc. 12, 432 (1972).
16. E. N. Parker, Astrophys. J. 164, 491 (1970).
17. E. H. Levy, Astrophys. J. 171, 621 (1972).
18. W. V. R. Malkus, Science 160, 259 (1968).
19. S. J. Peale, Icarus 17, 168 (1972).
20. J. A. Dunne, Science 185, 141 (1974).
21. K. A. Goettel, Physics of Earth and Planetary Interiors 6, 161 (1972).
22. J. Verhoogen, Geophys. J. 4, 276 (1973).
23. A. E. Ringwood and S. P. Clark, Nature 234, 89 (1971).
24. E. J. Smith, L. Davis, Jr., D. E. Jones, D. S. Colburn, P. J. Coleman, Jr., P. Dyal and C. P. Sonett, Science 183, 305 (1974).
25. W. B. Hubbard and R. Smoluchowski, Space Science Reviews 14, 599 (1973).
26. J. L. Luthey, Icarus 20, 125 (1973).
27. R. L. Newburn, S. Gulkis, Space Science Reviews 14, 179 (1973).

DIRECT MEASUREMENTS OF THE SPIN CONTACT DENSITY IN MAGNETIC MATERIALS*

N. BENCZER-KOLLER, Rutgers University, New Brunswick, New Jersey 08903
J. M. TROOSTER, Katholieke Universität, Nijmegen, Netherlands and
C. J. SONG, Purdue University, Lafayette, Indiana

ABSTRACT

The existence of large hyperfine fields in transition metals has been successfully explained as due in part to the polarization of core s-electrons by the net magnetization of the d-electrons. The core contribution H_c to the hyperfine field can be written as a superposition of contributions from all s-shells:

$$H_c \sim \sum_n \{|\Psi^{\uparrow}_{ns}(o)|^2 - |\Psi^{\downarrow}_{ns}(o)|^2\} .$$

A variety of techniques have been used to measure the total hyperfine field at the nucleus, but the individual contribution to the hyperfine field from each s-shell are not known. We have developed a new technique which combines Mössbauer spectroscopy with electron spectroscopy, and which allows the ratio

$$|\Psi^{\uparrow}_{ns}(o)|^2/|\Psi^{\downarrow}_{ns}(o)|^2$$

to be measured for each of the s-electrons separately. Measurements have been carried out on ^{57}Fe in iron metal and Fe_2O_3. The results confirm the theoretically predicted sign and order of magnitude of the 2s and 3s electron spin polarizations. Details of the limitations and scope of the experimental technique, and comparison of the data with theoretical models and calculations are discussed.

INTRODUCTION

The Zeeman effect was observed for the first time in nuclei in 1960 when the high resolution available in Mössbauer resonance spectroscopy allowed observation of the 6 components of the γ-ray transition between the $3/2^-$, 14.4 keV, first excited state and the $1/2^-$ ground state of the 57-Fe nucleus embedded in an iron ferromagnetic matrix. The excited state magnetic moment was determined as well as the total hyperfine field at the nuclear site: H = -339 kGauss. Many terms contribute to H. However dipolar contributions vanish for S state ions and for ions in cubic symmetry. The main remaining contribution arises from the Fermi contact interaction which can be expressed in terms of an effective field at the nucleus

$$H_e = -\frac{8\pi}{3} g\mu_o \vec{s}|\Psi(o)|^2 ,$$

where $|\Psi(o)|^2$ is the s-electron density at the nucleus and $\vec{s}$ is the total electronic spin in the s shell in question. In closed shell systems with paired s-electrons H_e would vanish were it not for the polarization of core electrons (1s, 2s and 3s) by the net spin of the 3d electrons. Watson and Freeman[1] used unrestricted Hartree-Fock approximations to investigate the effects of the exchange interaction between 3d and core s-electrons. This interaction causes a repulsion of electrons with anti-parallel spin and an effective attraction between parallel spin electrons, resulting in different radial wave functions for electrons of opposite sign and a net spin density at the nucleus. The hyperfine field at the nucleus will be proportional to this spin density:

$$\sum_n \left\{ |\Psi^{\uparrow}_{ns}(o)|^2 - |\Psi^{\downarrow}_{ns}(o)|^2 \right\} ;$$

where n runs over all s shells. The term in brackets may be positive or negative depending on the relative charge distributions of the s and d electrons. In particular, calculations based on atomic wave functions[2,3] for iron predict a very small negative contribution to H_c from the 1s electrons, a large negative contribution from the 2s electrons but a positive and relatively large contribution from the 3s electrons.

The behaviour of the 4s conduction electrons is much more difficult to predict. Wakoh and Yamashita,[4] Callaway, Tawill and Wang,[5] Duff and Das,[6] and Ohtsubo, Yokoo and Morita[3] have estimated the spin density of 4s electrons at the nucleus from a variety of energy band and atomic model calculations and have obtained results varying widely in both sign and order of magnitude.

Experimentally, the total hyperfine field can be measured accurately in a variety of very different experiments. Conduction electron polarization data was obtained indirectly by M. B. Stearns[7] from NMR data on iron alloys. We have developed[8] a new technique which will allow the microscopic determination of each individual s-shell contribution to the hyperfine field at the nucleus.

THE MEASUREMENT OF $|\Psi^{\uparrow}_{ns}(o)|^2/|\Psi^{\downarrow}_{ns}(o)|^2$

The proposed technique relies on the fact that the probability of decay of an excited nuclear state by internal conversion, α, is proportional to the electronic charge density, and in particular for M1 transitions, $\alpha \sim |\Psi_{ns}(o)|^2$. Since only s electrons have an appreciable charge density at the nucleus, M1 conversion proceeds mainly by emission of s electrons. Thus M1 internal conversion is a particularly selective process and allows an unambiguous determination of the value of electronic wave functions at the origin. Consider the decay of the 14.4 keV excited state of ^{57}Fe, and examine in particular the spin selection rules for the transitions

$$m_{Ii} = -3/2 \rightarrow m_{If} = -1/2$$

and

$$m_{Ii} = +3/2 \rightarrow m_{If} = +1/2$$

where m_{Ii} and m_{If} are the initial and final nuclear magnetic quantum numbers respectively. (Fig. 1.) The first transition can occur only by emission of an electron with initial spin up, while the latter corresponds to emission of an electron with initial spin down. The direction of the 3d electron magnetization is chosen as the quantization axis. The main feature of the technique involves the selective excitation of either the $m_I = +3/2$ or the $m_I = -3/2$ substates of ^{57}Fe nuclei by resonant Mössbauer absorption of 14.4 keV γ rays emitted from an unsplit $^{57}Co(Pd)$ source, and the subsequent observation of the decay electrons with an electron spectrometer capable of resolving 1s (6.3 keV), 2s (13.6

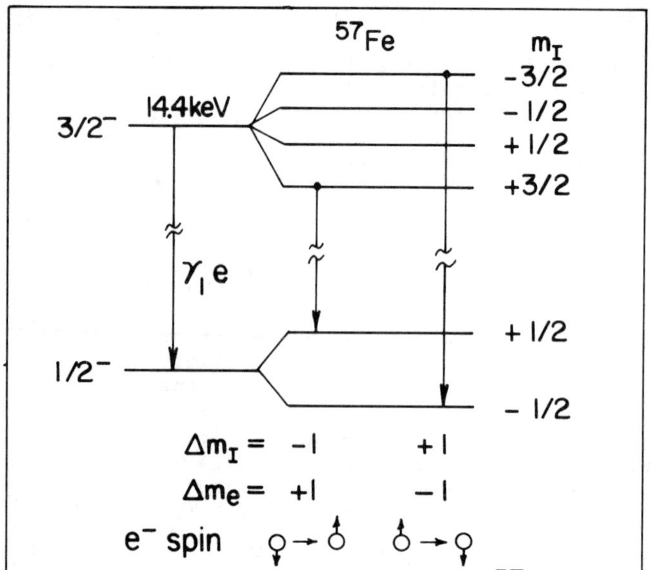

FIG.1. Energy level diagram of ^{57}Fe indicating the changes in spin angular momentum of selected internally converted electrons.

keV) or 3s (14.3 keV) internal conversion electrons (Fig. 2). An iron free magnetic solenoid spectrometer with a transmission of 1% at a momentum resolution of 2% was used. Absorber foils were 35 μg/cm² of 92% enriched ^{57}Fe or $^{57}Fe_2O_3$. The resonant absorption condition was achieved by moving a 50 mc $^{57}Co(Pd)$ source in a constant velocity mode that alternated between the two on-resonance velocities and off-resonance velocities with a period of either 60 msec or 120 msec. The experiments were carried out also at different source-absorber distances.

TABLE I. Spin density $\delta_{ns} = |\psi^{\uparrow}_{ns}(0)|^2/|\psi^{\downarrow}_{ns}(0)|^2 - 1$

Spin density x10²		δ_{2s}	δ_{3s}	δ_{4s}
Fe_2O_3	Expt.	-0.63±0.24	+2.17±0.88	0
Fe metal	Expt.	-0.63±0.15	+1.45±0.68 if	0
			or	
			-0.60±0.68 if	+ 86
Band Theory	Wakoh + Yamashita	− 0.28	+ 1.1	− 3.8
	Callaway et. al.			− 3.2
	Duff + Das			+2.6
Atomic Wave Functions	Morita et.al. Bagus+Liu	− 0.51	+ 2.1	+25.6

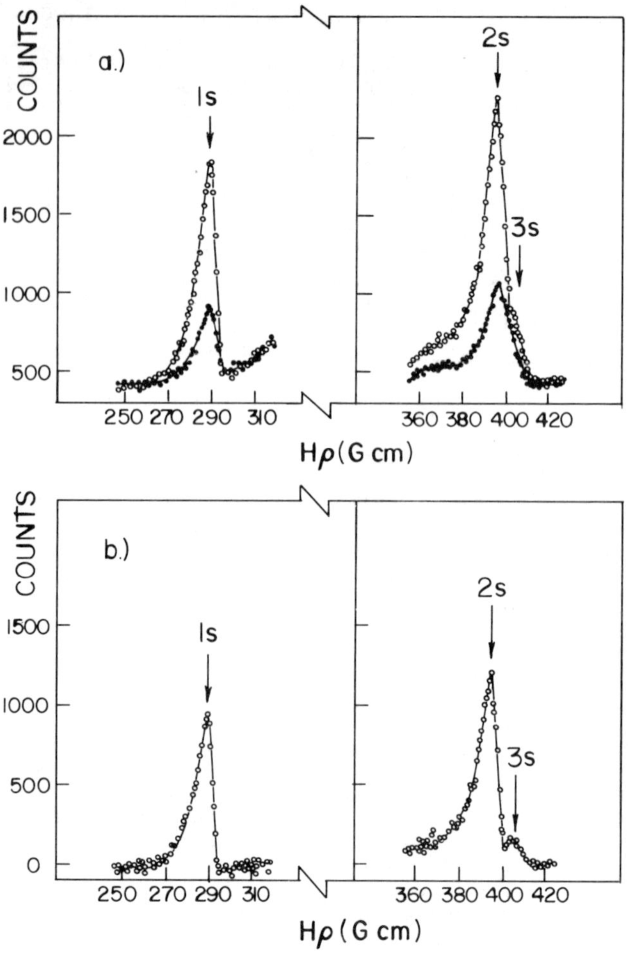

a.) Electron spectra from an $^{57}Fe_2O_3$ absorber in the region of the ls, 2s and 3s electron lines. The open circles correspond to data taken at resonant velocity while the solid circles correspond to data off resonant velocity.

b.) Net yield of internal conversion electrons after subtraction of nonresonant background.

Fig. 2.

The ratio of intensities of the two electron groups is directly proportional to

$$\alpha^{\uparrow}/\alpha^{\downarrow} = |\Psi^{\uparrow}_{ns}(o)|^2/|\Psi^{\downarrow}_{ns}(o)|^2$$

where the arrow refers to the initial spin direction of the emitted electron. The experiment is repeated for each n. Because of the very small effect expected for 1s electrons, $(\alpha^{\uparrow}/\alpha^{\downarrow})_{1s}$ was used for normalization. Appropriate corrections were applied to the data to account for the conversion of 2p, 3p, 3d and 4d electrons which could not be resolved from the 2s and 3s lines. It was assumed in the analysis that only the 4s conduction electrons might be polarized.

Data were obtained both for Fe metal and Fe_2O_3. The measured spin density of the "core" electrons defined here as:

$$\delta_{ns} = |\Psi^{\uparrow}_{ns}(o)|^2/|\Psi^{\downarrow}_{ns}(o)|^2 - 1 ,$$

is displayed in Table I. As expected theoretically, the 2s electrons which are distributed "within" the 3d charge distribution exhibit a negative spin polarization at the nucleus and of the right magnitude. Furthermore, at least for Fe_2O_3, the 3s electrons which presumably lie "outside" the 3d charge distribution yield a positive contribution to the hyperfine field.

In the case of Fe metal the interpretation is not as clear because the contribution of polarized 4s conduction electrons must be taken into account. Since theoretical predictions are not unique, the data were analyzed in terms of two possible values for the 4s polarization, namely zero or +0.86, a value suggested by M. B. Stearns[7] experimental data and analysis. Under these circumstances the determination of the 3s electron polarization in iron metal is as yet inconclusive.

However, we are building a new spectrometer with an order of magnitude better transmission and resolution which will allow completely independent accurate measurements of 4s spin and charge densities in addition to improved accuracy on the measurements of the core electrons.

*Work supported by the National Science Foundation.

REFERENCES

[1]R. E. Watson and A. J. Freeman, Phys. Rev. 123, 2027-2047 (1961).
[2]P. S. Bagus and B. Liu, Phys. Rev. 148, 79-82 (1966).
[3]H. Ohtsubo, Y. Yokoo and M. Morita, Prog. Theor. Phys. (Japan) 49, 701-702 (1973).
[4]S. Wakoh and J. Yamashita, J. Phys. Soc. Japan 25, 1272-1281 (1968).
[5]J. Callaway, R. A. Tawil and C. S. Wang, Phys. Lett. 46A, 161-162 (1973).
[6]K. S. Duff and T. P. Das, Phys. Rev. B3, 2294-2306 (1971).
[7]M. B. Stearns, Phys. Rev. B4, 4069-4080 (1971); Phys. Rev. B4, 4081-4091 (1971); Phys. Rev. B8, 4383-4398 (1973); Phys. Rev. B9, 2311-2327 (1974).
[8]C. J. Song, J. Trooster and N. Benczer-Koller, Phys. Rev. B9, 3854-3863 (1974).

$\mu^{\pm}$ PRECESSION STUDIES OF MAGNETIC MATERIALS

W. J. KOSSLER, College of William and Mary, Williamsburg, Virginia 23185.

ABSTRACT

The muon has recently begun to be used to study solid state. A review of this work to date and a description of current and proposed experiments will be made. The use of this proble is analogous to perturbed angular correlation with several special advantages: the muon is produced ≈ 100% polarized, the decay $\mu \rightarrow e + \nu + \nu$ produces highly anisotropic $(1 + 1/3 \cos \theta)$ energetic charged leptons (up to 50 MeV), the lifetime of the muon (2.2 μ sec) is sufficiently long for easy timing and short enough for high count rates. The μ^+ behaves as a light hydrogen ion and is thus the simplest impurity problem. Consequently the use of muons should not only yield fundamental solid state information, but also contribute to the understanding of radioactive ion implantation, particularly in the case of ferromagnetic samples.

INTRODUCTION

The precession of positive muons (μ^+) has recently been used to study magnetic material.[1,2,3] Related experiments have been carried out on superconductors.[4] The mass of the μ^+ is approximately 1/10 that of the proton but 200 times that of the positron. Thus the μ^+ behaves more like a proton than a positron, and could be considered just a special, light species of hydrogen. The muon's properties and fundamental interactions for their use as probes to solid state physics have been well determined and a selection of

TABLE I. PROPERTIES OF MUONS

Mass	m_μ = 105.6596(3) MeV 206.7684(6) $M_{electron}$ 0.1126123(6) M_{proton}.
Lifetime	$\tau_\mu = 2.1994(6) \times 10^{-6}$ seconds
Spin	1/2
Magnetic Moment	$\mu = \left\|g_\mu \delta_z\right\| \frac{e\hbar}{2M_\mu C} = 3.183347(9)\ \mu_p$
Precession Frequency	$\omega = 85.14 \times 10^3$ rad sec^{-1}gauss^{-1} ν = 13.55 MHz/kG
Hyperfine splitting	$\Delta\nu_0$ = 4463.248(31) MHz

$B_{HF} = 1/2\ \Delta\nu/\mu$ = 160 kG.

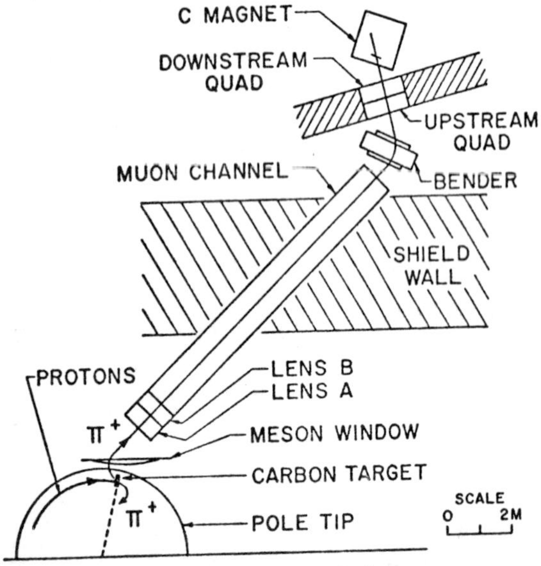

Fig. 1. Experimental Area

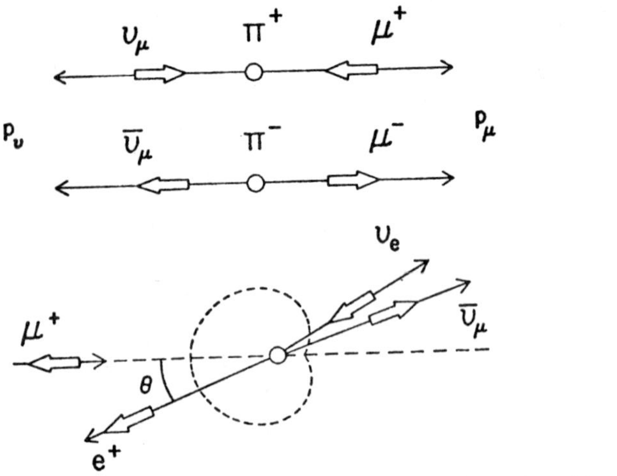

Fig. 2. Production of polarized muons from pion decay and positron decay distribution of the muon.

these properties are given in Table I. The μ^+ result from the decay of π^+ which in turn are produced, at least in the experiments of the author, by energetic protons (600 MeV) striking a carbon target inside the cyclotron (SREL synchrocyclotron) see Fig. 1. These experiments benefit from parity nonconservation twice. First, in the decay of the π^+ meson which produces in principle 100% polarized μ^+ if one can select those which decay only along the flight path of the π^+. See Fig. 2. In practice 70% polarization is not uncommon. The second time that parity nonconservation enters is in the μ^+ decay into an e^+ and two neutrinos. The angular correlation with respect to the μ^+ spin is:[5]

$$\frac{d^2W}{d\varepsilon\ d\cos\theta} = [(3-2\varepsilon) + (2\varepsilon-1)\cos\theta]\varepsilon^2 \qquad (1)$$

where $\varepsilon = 2Ee/M_{\mu^+}$ the ratio of the positron energy to its maximum value $M_{\mu^+}/2$. If one averages this expression over ε one obtains:

$$\frac{dW}{d\cos\theta} = \tfrac{1}{2}(1+1/3\cos\theta) \qquad (2)$$

If on the other hand one only chooses maximal energy μ^+ then:

$$\frac{dW}{d\cos\theta} = 1+\cos\theta \qquad (3)$$

A major difference from angular correlations of γ rays is that $\cos\theta$ enters and not $\cos^2\theta$. This is a consequence of parity nonconservation. Another difference is that the asymmetry is enormous.

If a magnetic field (b_μ) is at the site of the muon, the μ^+ spin and hence the e^+ angular correlation will precess. The μ^+ magnetic moment is approximately 10 nm, and hence $\omega = 85.1 \times 10^3$ rad sec^{-1}gauss^{-1}. This precession frequency is quite high, for example in 3000 gauss $\omega \approx 250 \times 10^6$ rad/sec and $\nu \approx 40$ MHz. The muon lifetime of 2.2 μsec thus allows hundreds of rotations to be observed.

In order to study internal fields the μ^+ are stopped in a sample with appropriate temperature control and in some applied magnetic field. See Fig. 3. This is very similar to PAC but some differing features need be pointed out. The scale of the apparatus is large. The pole gap is one foot and the sample has been, for instance, an ellipse 3" in major diameter and ½" thick. Positrons are detected by scintillators S3, S4 and S5. The stop of the μ^+ is indicated by the observation of essentially simultaneous signals in S1 and S2 and no signal in S3. Similarly to time differential PAC the time between the μ^+ stop and the e^+ detection is digitized and analyzed.

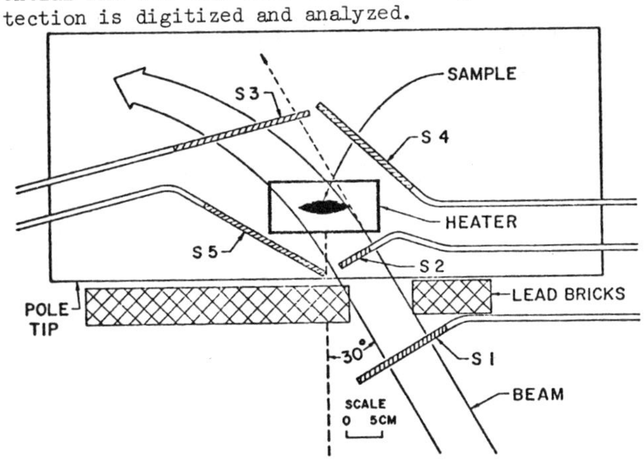

Fig. 3. Experimental Arrangement

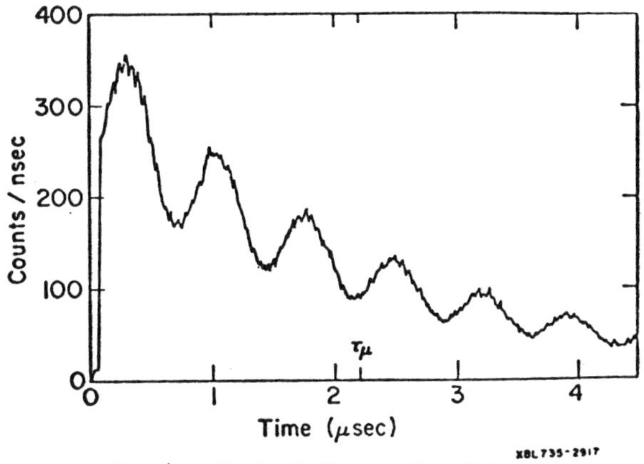

Fig. 4. Typical Precession Data

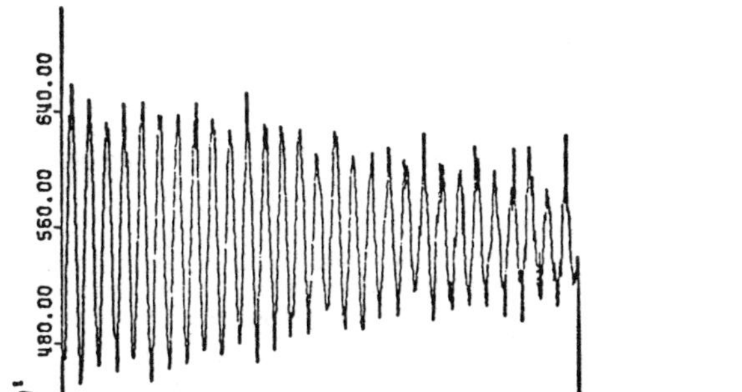

Fig. 5. Precession data for Ni with background subtracted and muon exponential decay removed

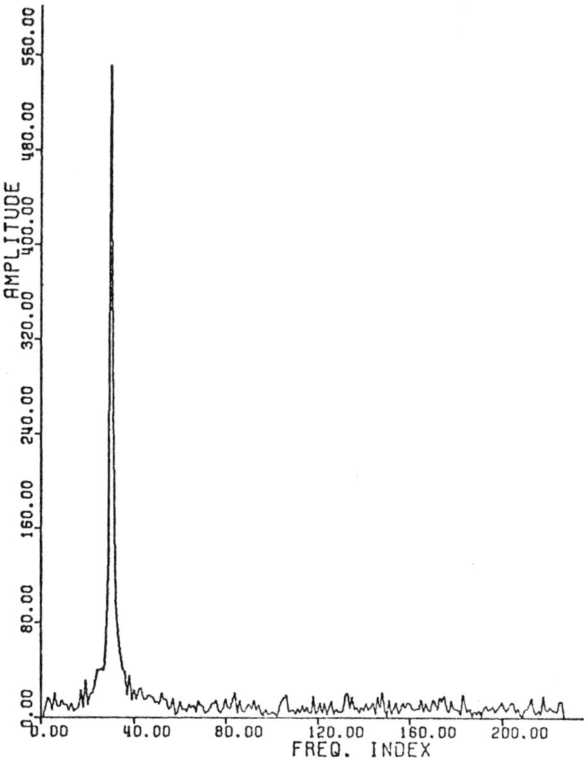

Fig. 6. Fourier transform of the data of Fig. 5.

Some precession data are presented in Fig. 4. Here one can see typical characteristics of the data: a large asymmetry, the rapid precession, the fall off with the μ^+ lifetime and some depolarization. If one subtracts background, multiplies by $\exp(+ t/\tau_\mu)$ one has e.g. data as in Fig. 5. There the depolarization is more clearly separated from the muon lifetime associated overall drop in count rate. One can take the fourier transform of the data in Fig. 5 to obtain the frequency spectrum shown in Fig. 6. This frequency distribution can in some cases be interpreted as the distribution of internal fields at the sites of the μ^+. We have obtained 20×10^3 μ^+ stops/sec in a sample such as that mentioned above, and an associated e^+ rate of approximately 2×10^3 e^+/sec in e.g. S5. This allows an excellent precession spectrum to be obtained in an hour. For smaller samples, longer times are necessary. For most of our data we have been limited by temperature stabilization time which is also nearly an hour. Further, for many of our samples, sample change over is also on the order of an hour. These rates are not exceeded at any accelerator which is in operation for experiments. It is expected in the relatively near future that the CERN SC, SIN, Nevis and TRIUMF meson facilities will have considerable higher rates, or what is the same thing, the ability to use very small samples.[6] Stopping rates much higher than 20×10^3/sec produce confusion concerning which e^+ is associated with a given μ^+.

REVIEW OF RESULTS: FERROMAGNETIC MATERIALS

Precession has been studied now for a variety of elemental ferromagnets. The first such case was Ni. Crowe and collaborators have studied single crystal Ni.[2] Gurevich and collaborators[3] have also studied Ni, Fe and to some extent Co. We have some preliminary results on Co which is not in complete agreement with that of Gurevich et al., although our Ni and Fe data are. The data were analyzed using a fitting routine to functional form:

$$N(T) = N_o e^{-T/\tau_\mu}(1+Pae^{-T/\tau}\cos(\omega T+\phi)) + \text{Background} \quad (4)$$

Where P is the polarization of the μ^+ and a is the asymmetry as measured for a copper sample with the same applied field. A phase angle ϕ is necessary in part due to uncertainties of the absolute stopping time. From the precession frequency the magnetic field at the site of the muon can be obtained. See Figs. 7, 8. The curve through the points below Tc is a Brillouin function with only the field at 0^oK fitted. The magnetic field as measured by the muon may be written:[7]

$$b_\mu = H_{applied} - DM + \frac{4\pi M}{3} + \Sigma H_{dipoles}(\text{over a sphere}) + H_{Hyperfine} \quad (5)$$

H applied is the magnetic field produced by the external magnet. DM, the demagnetization arises from the macroscopic magnetization of the actual sample geometry. For our first experiment this was zero since the sample was part of a magnetic circuit. For a sphere the result would be $- DM = \frac{4\pi M}{3}$. For an oblate ellipoid 3" Diameter 1/2" long $DM \approx \frac{4\pi M}{10}$.

Σ H dipole is a sum over the contribution[10] from the dipoles on the transition metal ions. For Ni this is zero, presuming the μ^+ is in the body centered site. For Fe the natural μ^+ sites would be the face center. The field would be +18.8kG or -9.4kG when the nearest neighbors are along or perpendicular to the magnetization respectively.

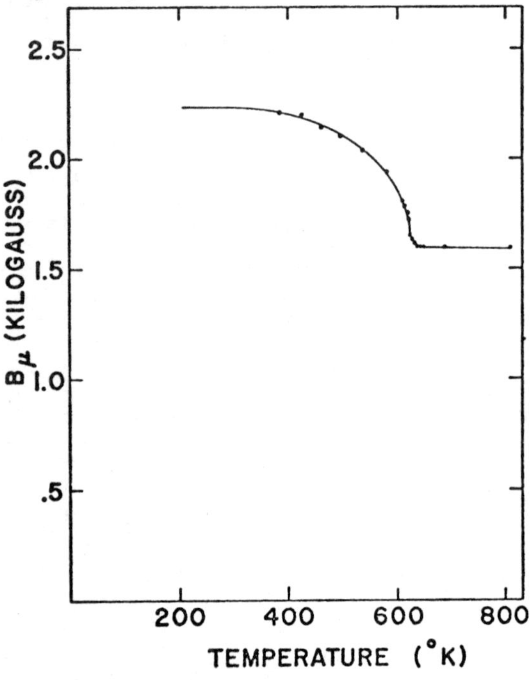

Fig. 7. Magnetic field b_μ as measured by the muon for Ni as a function of temperature

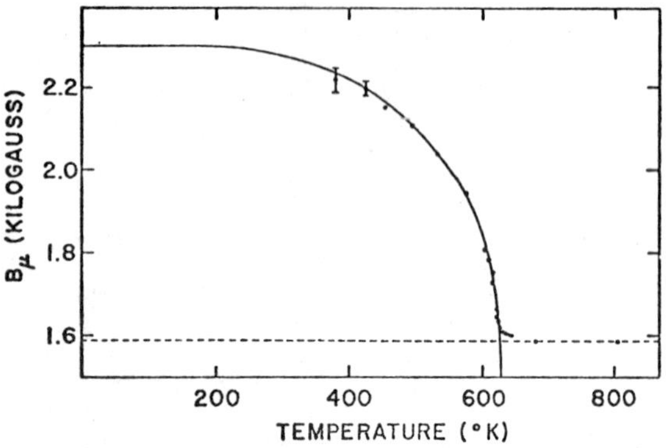

Fig. 8. The same data as in Fig. 7 but with emphasis on the ferromagnetic contribution. The solid curve is a Brillouin curve.

In iron the hopping time is about $\leq 10^{-9}$ seconds at room temperature much less than the precession time 10^{-7} sec hence one should average the fields weighted by the relative number of sites. This average: 4 x -9.4kG + 2 x 18.8 = 0.

$H_{hyperfine}$ the field produced by electrons on the muon, may be written[7] $H_{hyperfine} = H_{cep} + H_{cp} + H_v$. In this expression H_{cp} is due to the polarization of the inner core electron on the μ^+ by the magnetic electrons. A μ^+ ion has no core hence $H_{cp} = 0$. H_{cep} is due to the polarization of the conduction electrons by the moment of the transition metal atoms. H_v is a positive field due to the valence ns-like electrons which remain near the solute muon shielding the excess charge. These are polarized by an amount proportional to the volume misfit of the solute muon. This term is thought to be small for the muon.[8]

Keeping only those contributions which should be strong we have:

$$b_\mu = H_{applied} - DM + 4\pi M/3 + H_{cep}. \quad (6)$$

TABLE II. SUMMARY OF MAGNETIC FIELD RESULTS ALL EXTRAPOLATED TO 0 K

	b_μ with H_{ext}=1590G	H_{HF}	M_S	Magnetic Moment per atom
Ni	2300G	- 810	510G	0.616 μ_β
Co	?	?	1446G	1.715 μ_β
Fe	3400G	-3200	1752G	2.219 μ_β

We have measured b_μ for Ni and Fe and have begun on Co. These results are summarized Table II. We have extrapolated the Brillouin functions back to T = 0 K to obtain these values. It is interesting that in these cases H_{cep} is negative. For Ni we can compare against the neutron scattering conduction electron polarization measurements of Mook.[9] He obtains in the body center site a magnetization density of M = - .0085 μ_B/A^3. This corresponds to -660 gauss taking $H_{cep} = \frac{8\pi M}{3}$ in comparison to our -810 gauss. Crowe et al[2] have found a value even closer to -660 gauss for a single crystal at liquid nitrogen temperatures. Because the μ^+ measured magnetic field is so close to that from neutron scattering in this case where the muon site is relatively unambiguous it is tempting to infer from this that the μ^+ measures the magnetic field at its position, little disturbing the host. With this assumption one can also make comparisons for Fe. Natural positions for the μ^+ are the tetrahedral and octohedral sites on the faces of iron bcc lattice. Neutron scattering results of Shull et al.[10] indicate a spin density of -.05 μ/A^3 at the tetrahedral site, corresponding to about -4kG compared to our -3.2 kG. The octahedral site as measured by the neutrons has low magnetization. Thus it may be that the μ^+ averages the tetrahedral and octahedral sites. The saturation magnetizations M_s(0 K) are in relationship as 1:2.7:3.4 the magnetic moment per atom as 1:2.8:3.6. The b_μ as measured by the μ^+ as 1:?:3.9 and for further comparison the PAC result for ^{19}F in these materials yields hyperfine fields as 1:1.8:5 however with the opposite signs to the field. A preliminary result of ours on Co seems to place it out of this sequence; however, Gurevich et al.[2] have observed a rapidly depolarizing signal which is in accord with this monotonic behavior. In addition to the magnetic fields b_μ one also obtains τ the depolarization time. τ as a function of temperature for a fixed external field of 1590 gauss is shown in Fig. 9. The low temperature values are undoubtedly associated with macroscopic field inhomogeneities since as has been pointed out by Patterson et al.[11] the corresponding ΔB ($\approx$ 100 gauss) agrees with that of ^{61}Ni NMR but for a single crystal ΔB is less than 10 gauss even at 77K. In the vicinity of the Curie point τ drops again. This may be due to critical functuations or to a nonuniform temperature distribution. A 4 K temperature nonuniformity would explain our data. This magnitude of nonuniformity is not ruled out by the oven configuration used for these experiments.

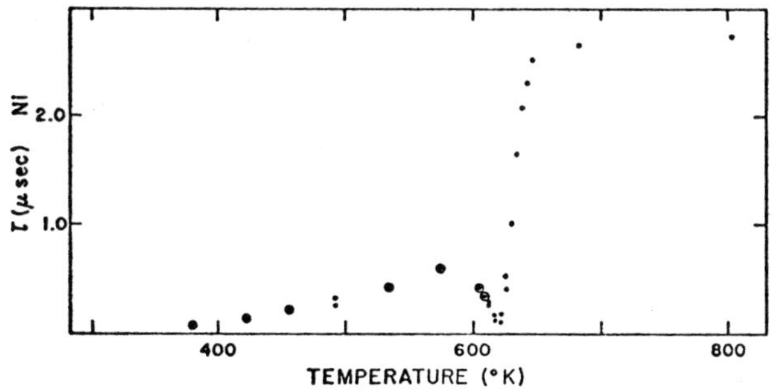

Fig. 9. Depolarization time τ as a function of temperature

When can critical fluctuations be seen? One would expect the depolarization time to be proportional to $(1/g\ H_{HF})^2 \times 1/\tau_z$ where g is the Landé g factor, and H_{HF} is the hyperfine field for the muon or other nucleus in a particular host; and τ_z is the spin correlation time.

Critical fluctuations for Rh and Ni have been measured by perturbed angular correlations (PAC)[12] above the Curie point. These can be compared to our measurements so far, at 3 K above T_c.

τ_μ = 1.66 μS g = 17.8 nm $B_{HF}(\mu)$ = - .81 kG

τ_{Rh} = .02 μS g = 2.1 nm $B_{HF}(R_H)$ = 225 kG

From the Rh data one predicts

$$\tau_\mu = \tau_{Rh}\ (g_{Rh}H_{HF}(Rh)/g_\mu H_{HF.}(\mu))^2 = 21_\mu \text{ Sec} \qquad (7)$$

From neutron scattering measurements[13] τ_z has been obtained which at 3K above the Curie point is 10 times larger than that from PAC. This would imply τ = 2.1 μsec near our observed result

From these results it is clear that we are close at present to being able to see critical fluctuations and in the next series of experiments with an improved oven we have expectations of being able to observe them unambiguously. An important difference between critical fluctuations observed with PAC and with μ^+ is that for μ^+ diffusion should play a role. In particular muons can probe the fourier transform of the spin correlation function.[14]

REVIEW OF RESULTS: DIFFUSION AND THE MEASUREMENT OF MAGNETIC FIELD DISTRIBUTIONS

Type II superconductors in the temperature and field region where flux penetration occurs provide an interesting testing ground for the use of μ^+ to determine magnetic field distributions. From e.g. neutron scattering experiments[15] vortex structures of the magnetic field distributions in Nb or PbIn can be considered known. The prominent features of the mixed state local field distribution expected in the muon-precession-frequency spectrum are a peak at the local field corresponding to the saddle-point location in the vortex lattice, B_s, and maximum and minimum cut-off fields.[16] The periodic inhomogeneity created is expected to lead to both line broadening and a shift, since $B_s < B$, the average of the internal field. In collaborative experiments with A. T. Fiory, D. E. Murnick and M. Leventhal[4] we obtained the fourier amplitudes shown in Figs. 10 and 11 for lead-indium and niobium respectively. Though it is not as pronounced as one might hope for, the saddle point

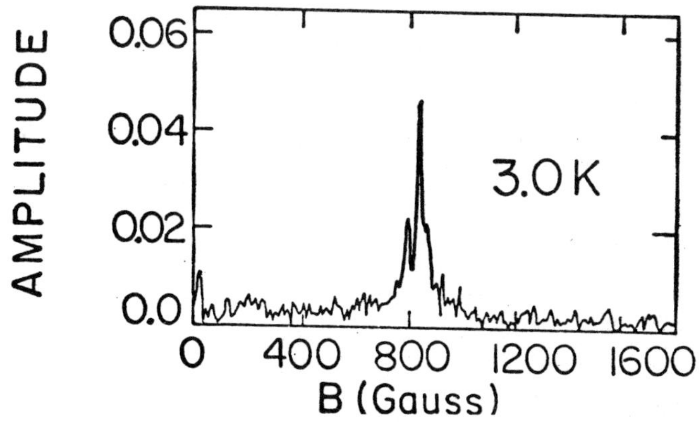

Fig. 10. Fourier amplitudes for Pb In

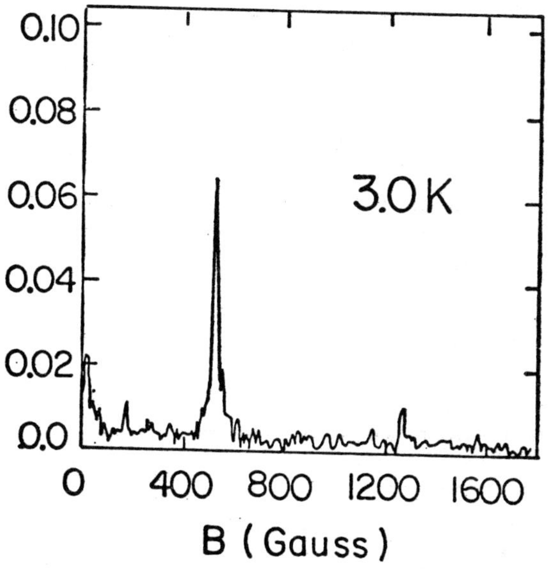

Fig. 11 Fourier amplitudes for Nb

field is quite in evidence and at the correct position in the lead indium, however no such field is observed for niobium. The only field present in niobium is the average field which coincides in the geometry used to the applied field. This can be understood if the muon is capable of rapid diffusion at 3.0 K.

That such a diffusion rate is at least plausible can be seen in various ways. One can start by noting that in this bcc lattice there are many tetrahedral sites available, see Fig. 12. To explain the temperature dependence of the elastic coefficients of Ta load-

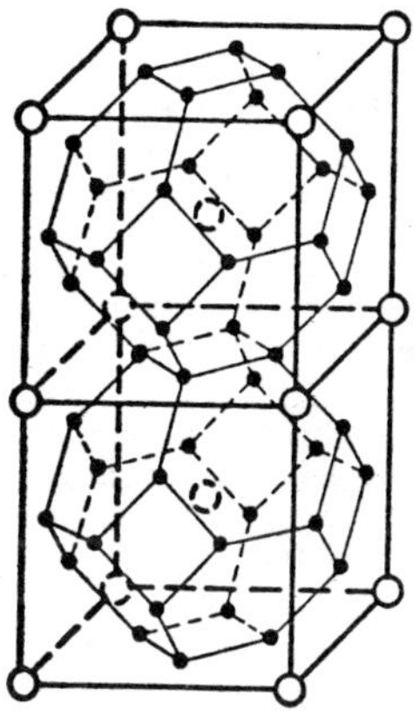

Fig. 12. Body centered cubic lattice. Small solid points represent the tetrahedral sites for the muon or hydrogen.

ed with H[17] as well as anomolously high exponents in the Debye-Waller factor[18] it has been suggested[17] that the position of the H is smeared out.

At room temperature H in N_B has a hop rate on the order of 10^{12} sec^{-1}, further from neutron scattering[19] the diffusion activation energy corresponds to the excitation to the first excited state of about .13 eV. The isotope shift in an harmonic oscillator going from H → μ puts the ground state of the μ just at this energy. This is so since $E_1 = 3E_o$ for an oscillator and $E_o(\mu) = \sqrt{\frac{M_\mu}{M_\mu}}\, E_o(H) \approx 3E_o(H)$. Thus the muon even in its ground state should diffuse rapidly. Also if one uses the 0.13 eV energy to characterize the potential well one can calculate the tunneling time and diffusion constant. One obtains $D \approx \frac{1}{2} \times 10^{-2}$ cm^2/sec. So it is at least plausible that we are seeing quantum diffusion of the μ^+ in Nb.

FUTURE EXPERIMENTS

A long list of possibilities were mentioned in a talk almost 2 years ago.[20] We are committed at present to study 1: Elemental ferromagnets through the rare earths; 2: Make measurements near the Curie point for various ferromagnets with a few millidegree temperature stability and uniformity; 3: Investigate the magnetic field at the site of the muon special intermetallic compounds, e.g., Fe_3Si; 4: Study the nature of the μ^+ wave function in material through motional narrowing of the μ^+ precession signal; and 5: Make a search for muonium systems, e.g. in TiO_2 and H_FO_2. For here I will restrict to a discussion of two.

The first involves a study of the possible quantum diffusion mentioned above. For the fcc copper lattice μ^+ depolarization has been observed[21] when the temperature of the sample is lowered implying a freezing out of motional narrowing. The depolarization mechanism is the coupling to the ^{63,65}Cu nuclear magnetic moments. Above the superconducting transition temperature but still cold on the scale of the Cu diffusion activation energy, both Nb and V should have rapid μ^+ depolarization times if no diffusion takes place. If rapid diffusion takes place at all there should be no depolarization. If some diffusion takes place, and some muons are trapped, two components of depolarization would be observed. In general it should be possible to follow the diffusion and trapping over times on the order of the μ^+ lifetime. In comparison to Nb and V, the alkali metals also form bcc lattices. These however are much larger and hence should have smaller couplings to the μ^+. A series of comparative measurements is planned.

The second is an investigation of the fields and field distributions in the elemental ferromagnets. The case of Co is interesting in that there are two phases. Above 500°C Co becomes fcc as is Ni; thus one might expect that for higher temperatures the μ^+ precession data would be similar. A preliminary run has observed precession for high temperature but not near room temperature. This would imply that the appearance of the hcp phase induces larger inhomogeneities. Gurevich et al.[3] have reported observation of precession even at room temperature, but with an extremely rapid depolarization time, less than one cycle. Following the depolarization as temperature is lowered will allow the development internal field inhomogeneities.

The study of Cr will be attempted at high temperatures well above the Néel point and as the Néel point is approached. Cr being bcc may also have rapid tunneling; thus the small scale rapid field variation should average out and it can be hoped to observe precession even in an antiferromagnet. Here we should not have a direct connection between a temperature nonuniformity and a range of ferromagnetic fields. Depolarization should be related to both spin correlation and diffusion.

The rare earths present several interesting questions as to the connection between their structure and field distributions as measured by muons. Above the Curie point precession has already been observed for Gd and Dy.

SUMMARY

I have attempted to illustrate the technique of μ^+ precession as applied to magnetic materials, summarize some of the results by no means exhaustively, and consider briefly future experiments that we intend. Since the μ^+ can be considered a member of the simplest atomic species one can say that μ^+ precession is related to the hydrogen atom problem of solid state physics.

REFERENCES

1. M. L. G. Foy, Neil Heiman, W. J. Kossler, C. E. Stronach, Phys. Rev. Letters 30, 1064 (1973).
N. Heiman, M. L. G. Foy, W. J. Kossler, C. E. Stronach , Magnetism and Magnetic Materials - 1973 19th Annual Conference AIP Conference Proceedings No. 18 p. 525.
2. J. H. Brewer, K. M. Crowe, F. N. Gygax, and A. Schenck in Muon Physics V. W. Hughes and C. S. Wu, Eds (Academic Press, New York, to be published). B. D. Patterson, K. M. Crowe, F. N. Gygax, R. F. Johnson, A. M. Portis and J. H. Brewer submitted to Physics Letters.
3. I. I. Gurevich, A. I. Klimov, V. N. Maiorov, E. A. Meleshko, B. A. Nikolsky, V. S. Roganov, V. I. Selivanov, and V. A. Suyetin JETP Letters 18, 564 (1973). I. I. Gurevich, A. J. Klimov, V. N. Maiorov, E. A. Meleshko, I. A. Muratova, B. A. Nikolsky, V. S. Roganov, V. I. Selivinov, and V. A. Suyetin JETP 66, 374 (1974).
4. A. T. Fiory, D. E. Murnick, M. Leventhal, and W. J. Kossler, Phys. Rev. Letters 33, 969 (1974).
5. F. Scheck in Meeting on Muons in Solid, Schweizerisches Institut fur Nuklearforschung (SIN Zurich (1971)).
6. μ^+ stopping rates much higher than 20 x 10^3/sec produce confusion concerning which e^+ is associated with a given μ^+.
7. C. P. Slichter Principles of Magnetic Reanance (Harper and Row, Publishers, Inc. New York, 1963) p. 115-160.
8. Mary Beth Stearns, Phys. Rev. B 8, 4383 (1973).
9. H. A. Mook, Phys. Rev. 148, 495 (1966).
10. C. Shull and T. Yamada, J. Phys. Soc. Japan, 17 (B-III): 1 (1962).
11. B. D. Patterson, K. M. Crowe, F. N. Gygax, R. F. Johnson, A. M. Porms and J. H. Brewer, Phys. Letters Submitted to Phys. Letters.
12. Albert M. Gottkieb and Cristoph Hohenemser, Phys. Rev. Lett. 31, 1222 (1973).
13. V. J. Minkiewicz, M. F. Collins, R. Nathans, and G. Shirane, Phys. Rev. 182, 624 (1969).
14. K. G. Petzinger. Private communication.
15. J. Schelton, M. Ullmaier, and W. Schmatz, Phys. Stat. Sol. (b) 48, 619 (1973).
16. P. Pincus, A. C. Gossard, V. Jaccarino, and J. H. Wernick Phys. Letters 13, 21 (1964); J. M. Delreiu and J. M. Winter, Solid State Commun 4, 545 (1966); A. G. Redfield, Phys. Rev. 162, 367 (1967).
17. J. Buchholz, J. Völkl, and G. Alefeld Phys. Rev. Lett. 30, 318 (1973).
18. K. Skold and G. Nelin J. Phys. Chem. Solids 28 2369 (1967).
19. J. H. L. Birchall and D. K. Ross International meeting on Hydrogen in Metals, Julich 1972.
20. W. J. Kossler Topical Meeting on Intermediate Energy Physics, Zuoz, Switzerland, April 4-13, 1973, CERN Report CERN 74-8 April 10, 1974.

AIP Conference Proceedings

		L.C. Number	ISBN
No. 1	Feedback and Dynamic Control of Plasmas Princeton 1970)	70-141596	0-88318-100-2
No. 2	Particles and Fields - 1971 (Rochester)	71-184662	0-88318-101-0
No. 3	Thermal Expansion - 1971 (Corning)	72-76970	0-88318-102-9
No. 4	Superconductivity in d- and f-Band Metals (Rochester 1971)	74-18879	0-88318-103-7
No. 5	Magnetism and Magnetic Materials-1971 (2 parts) (Chicago)	59-2468	0-88318-104-5
No. 6	Particle Physics (Irvine 1971)	72-81239	0-88318-105-3
No. 7	Exploring the History of Nuclear Physics (Brookline 1967, 1969)	72-81883	0-88318-106-1
No. 8	Experimental Meson Spectroscopy-1972 (Philadelphia)	72-88226	0-88318-107-X
No. 9	Cyclotrons - 1972 (Vancouver)	72-92798	0-88318-108-8
No.10	Magnetism and Magnetic Materials - 1972 (2 parts) (Denver)	72-623469	0-88318-109-6
No.11	Transport Phenomena - 1973 (Brown University Conference)	73-80682	0-88318-110-X
No.12	Experiments on High Energy Particle Collisions - 1973 (Vanderbilt Conference)	73-81705	0-88318-111-8
No.13	π-π Scattering - 1973 (Tallahassee Conference)	73-81704	0-88318-112-6
No.14	Particles and Fields - 1973 (APS/DPF Berkeley)	73-91923	0-88318-113-4
No.15	High Energy Collisions - 1973 (Stony Brook)	73-92324	0-88318-114-2
No.16	Causality and Physical Theories (Wayne State University, 1973)	73-93420	0-88318-115-0
No.17	Thermal Expansion - 1973 (Lake of the Ozarks)	73-94415	0-88318-116-9
No.18	Magnetism and Magnetic Materials - 1973 (2 parts) (Boston)	59-2468	0-88318-117-7
No.19	Physics and the Energy Problem - 1974 (APS Chicago)	73-94416	0-88318-118-5
No.20	Tetrahedrally Bonded Amorphous Semiconductors (Yorktown Heights, 1974)	74-80145	0-88318-119-3
No.21	Experimental Meson Spectroscopy - 1974 (Boston)	74-82628	0-88318-120-7
No.22	Neutrinos - 1974 (Philadelphia)	74-82413	0-88318-121-5
No.23	Particles and Fields - 1974 (APS/DPF Williamsburg)	74-27575	0-88318-122-3
No.24	Magnetism and Magnetic Materials - 1974 (20th Annual Conference San Francisco)	75-2647	0-88318-123-1